www.devbio.com

the website for

DEVELOPMENTAL BIOLOGY

Seventh Edition

This website is intended to supplement and enrich courses in developmental biology. It provides more information for advanced students as well as historical, philosophical, and ethical perspectives on issues in developmental biology. Included are movies, interviews, opinions (labeled as such), Web links, updates, and, of course, the *Developmental Biology* humor page, with its songs and abstracts. Supplementing the new Chapter 21, the site has five essays on bioethics and developmental biology. The topics are: (1) Cloning, (2) Stem Cells, (3) When Does Human Life Begin? (4) Human Sex Selection, and (5) What is "Normal"?

devbio.com is organized by textbook chapter. When you arrive at the home page, simply choose the chapter you are interested in from the chapter menu. You will then see the list of articles available for that chapter. Article numbers and names are identical to those listed below and throughout the book. Click an article name to view that article.

The following directory summarizes the numbered references to the website, along with references to vade mecum[2], that you will find throughout each of the book's chapters.

1.1 The reception of von Baer's principles
1.2 Conklin's art and science
1.3 Dr. Nicole Le Douarin and chick-quail chimeras
1.4 The mathematical background of pattern formation
1.5 How do zebras (and angelfish) get their stripes?
VM[2] The compound microscope
VM[2] Histotechniques

2.1 Immortal animals
2.2 The human life cycle
2.3 Protist differentiation
2.4 *Volvox* cell differentiation
2.5 Slime mold life cycle
VM[2] The amphibian life cycle
VM[2] Slime mold life cycle
VM[2] The amniote egg

3.1 Establishing experimental embryology
3.2 The hazards of environmental sex determination
3.3 Receptor gradients
3.4 "Rediscovery" of the morphogenetic field
3.5 How morphogenetic behaviors work
3.6 Demonstrating the thermodynamic model
3.7 Cadherins: Functional anatomy

3.8 Other cell adhesion molecules
VM[2] Sea urchins and UV radiation
VM[2] Sea urchin development
VM[2] Flatworm regeneration

4.1 The embryological origin of the gene theory
4.2 Creating developmental genetics
4.3 Amphibian cloning: Potency and deformity
4.4 Metaplasia
4.5 Antibody formation
4.6 DNA isolation techniques
4.7 Microarray technology
4.8 Knocking out specific genes at specific times and places

5.1 Displacing nucleosomes
5.2 Structure of the 5′ cap
5.3 Promoter structure and the mechanisms of transcription complex assembly
5.4 Families of transcription factors
5.5 Histone acetylation
5.6 Other recombinase methods
5.7 Enhancers and cancers
5.8 Further mechanisms of transcriptional regulation
5.9 Imprinting in humans and mice
5.10 Chromosome elimination and diminution

DEVELOPMENTAL BIOLOGY *Seventh Edition*

DEVELOPMENTAL BIOLOGY *Seventh Edition*

SCOTT F. GILBERT *Swarthmore College*

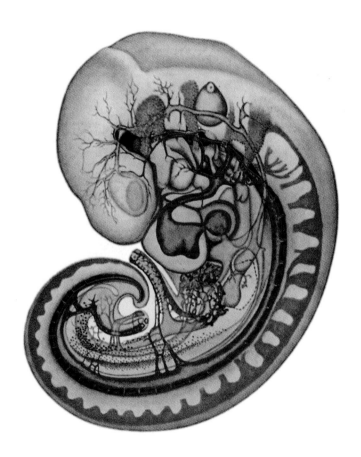

with a chapter on Plant Development by
SUSAN R. SINGER *Carleton College*

SINAUER ASSOCIATES, INC., PUBLISHERS
Sunderland, Massachusetts

About the cover and title page

The cover and frontispiece show human embryos between weeks 5 and 6 of gestation.

The human embryo has dominated the biological news during the past three years. Concerns over cloning, stem cells, genetic engineering, chemicals that disrupt development, and the possibilities of regenerating damaged organs have focused attention on human development. This edition of *Developmental Biology* includes a new chapter devoted to these topics.

The 44-day human embryo shown on the cover was "sectioned" by the technique of magnetic resonance microscopy (MRM). This technique, similar to magnetic resonance imaging (MRI), allows extremely fine resolution of anatomical details while not disturbing the specimen. Pictures such as these could not be obtained without computer reconstruction, and the resolution is all the more remarkable since the embryo is about 1.2 cm (0.5 inch) long. In this section, the vertebrae and dorsal root ganglia (of the peripheral nervous system) are beginning to develop, the forebrain is prominent, and the forelimbs and hindlimbs are forming digits. The embryo is in its amnion, and the umbilical cord attaches it to the placenta. More technical details of MRM analysis of stage 18 embryos can be found at http://embryo.soad.umich.edu/carnStages/stage18/stage18html.

The drawing on the title page depicts a human embryo at a slightly earlier stage, dissected and drawn by Dr. Emil Witschi (1890–1971), a professor of embryology at Iowa State University. Dr. Witschi had trained at the Art Institute of Munich before he became a biologist, and he was largely responsible for showing that the mammalian gonads are bipotential before becoming either an ovary or a testis. Before color photography, art training was important for embryologists. This drawing from Witschi's research was used as the frontispiece of his 1956 textbook, *Development of Vertebrates*.

Some landmarks on these embryos are labeled below.

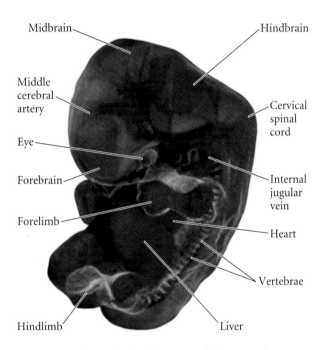

Image courtesy of Anatomical Travelogues, Inc. All rights reserved.
www.anatomicaltravel.com

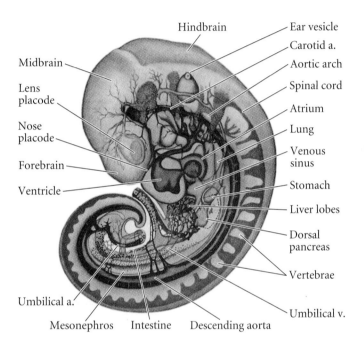

Reprinted from E. Witschi, *Development of Vertebrates* (1956), with permission from Elsevier.

Sinauer Associates Inc., 23 Plumtree Road, Sunderland, MA 01375 USA

   FAX: 413-549-1118

   email: publish@sinauer.com, orders@sinauer.com

   www.sinauer.com

Library of Congress Cataloging-in-Publication Data

Gilbert, Scott F., 1949-
    Developmental biology / Scott F. Gilbert; with a chapter on plant development
        by Susan R. Singer.-- 7th ed.
        p.   cm.
    Includes bibliographical references.
    ISBN 0-87893-258-5 (hardcover)
    1. Embryology. 2. Developmental biology. I. Title.
    QL955.G48 2003
    571.8--dc21   2003001670

Printed in U.S.A.

5 4 3 2

To Daniel, Sarah, and David

Brief Contents

part 1 Principles of developmental biology

Chapter 1   Developmental biology: The anatomical tradition 3
Chapter 2   Life cycles and the evolution of developmental patterns 25
Chapter 3   Principles of experimental embryology 51
Chapter 4   The genetic core of development 81
Chapter 5   The paradigm of differential gene expression 107
Chapter 6   Cell-cell communication in development 143

part 2 Early embryonic development

Chapter 7   Fertilization: Beginning a new organism 181
Chapter 8   Early development in selected invertebrates 221
Chapter 9   The genetics of axis specification in Drosophila 263
Chapter 10   Early development and axis formation in amphibians 305
Chapter 11   The early development of vertebrates: Fish, birds, and mammals 345

part 3 Later embryonic development

Chapter 12   The emergence of the ectoderm: The central nervous system and the epidermis 389
Chapter 13   Neural crest cells and axonal specificity 427
Chapter 14   Paraxial and intermediate mesoderm 465
Chapter 15   Lateral plate mesoderm and endoderm 491
Chapter 16   Development of the tetrapod limb 523
Chapter 17   Sex determination 547
Chapter 18   Metamorphosis, regeneration, and aging 575
Chapter 19   The saga of the germ line 613

part 4 Ramifications of developmental biology

Chapter 20   An overview of plant development 647
Chapter 21   Medical implications of developmental biology 683
Chapter 22   Environmental regulation of animal development 721
Chapter 23   Developmental mechanisms of evolutionary change 751

Contents

part 1 Principles of developmental biology

CHAPTER 1 Developmental biology: The anatomical tradition 3

The Questions of Developmental Biology 4

Approaches to Developmental Biology 5

The Anatomical Approach 5

Comparative embryology 5

Epigenesis and preformation 6

Naming the parts: The primary germ layers and early organs 8

The four principles of Karl Ernst von Baer 8

Fate mapping the embryo 10

Cell migration 13

Evolutionary Embryology 14

Embryonic homologies 15

Medical Embryology and Teratology 16

Mathematical Modeling of Development 18

The mathematics of organismal growth 18

The mathematics of patterning 19

CHAPTER 2 Life cycles and the evolution of developmental patterns 25

The Circle of Life: The Stages of Animal Development 25

The Frog Life Cycle 26

The Evolution of Developmental Patterns in Unicellular Protists 30

Control of developmental morphogenesis: The role of the nucleus 30

Unicellular protists and the origins of sexual reproduction 32

Multicellularity: The Evolution of Differentiation 33

The Volvocaceans 34

■ *Sidelights & Speculations: Sex and Individuality in Volvox 36*

Differentiation and morphogenesis in Dictyostelium: Cell adhesion 39

■ *Sidelights & Speculations: Rules of Evidence I 42*

Differentiation in Dictyostelium 43

Developmental Patterns among the Metazoa 44

The diploblasts 46

Protostomes and deuterostomes 46

CHAPTER 3 Principles of experimental embryology 51

Environmental Developmental Biology 51

Environmental sex determination 52

Adaptation of embryos and larvae to their environments 52

The Developmental Dynamics of Cell Specification 56

Autonomous specification 56

Conditional specification 58

■ *Sidelights & Speculations: Rules of Evidence II 67*

Stem cells and commitment 68

Syncytial specification 69

Morphogenesis and Cell Adhesion 69

Differential cell affinity 70

The thermodynamic model of cell interactions 71

Cadherins and cell adhesion 74

CHAPTER 4 *The genetic core of development 81*

The Embryological Origins of the Gene Theory 81
  Nucleus or cytoplasm: Which controls heredity? 81
  The split between embryology and genetics 83
  Early attempts at developmental genetics 84
Evidence for Genomic Equivalence 85
  Amphibian cloning: The restriction of nuclear potency 85
  Amphibian cloning: The totipotency of somatic cells 86
  Cloning mammals 87
■ *Sidelights & Speculations:* Why Clone Mammals? 89
■ *Sidelights & Speculations:* The Exception to the Rule: Immunoglobulin Genes 90
Differential Gene Expression 92
RNA Localization Techniques 93
  Northern blotting 93
  Reverse transcriptase-polymerase chain reaction 93
  Microarrays and macroarrays 96
  In situ hybridization 97
Determining the Function of Genes During Development 98
  Transgenic cells and organisms 98
  Determining the function of a message: Antisense RNA 101
Coda 104

CHAPTER 5 *The paradigm of differential gene expression 107*

Differential Gene Transcription 107
  Anatomy of the gene: Exons and introns 107
  Anatomy of the gene: Promoters and enhancers 111
  Transcription factors 114
  Transcription factor cascades 117
■ *Sidelights & Speculations:* Studying DNA Regulatory Elements 118
  Silencers 120
Methylation Pattern and the Control of Transcription 121
  DNA methylation and gene activity 121
  Chromatin modification 122
  Insulators 123
■ *Sidelights & Speculations:* Genomic Imprinting 123
  Transcriptional regulation of an entire chromosome: Dosage compensation 124
■ *Sidelights & Speculations:* The Mechanisms of X Chromosome Inactivation 126

Differential RNA Processing 127
  Control of early development by nuclear RNA selection 128
  Creating families of proteins through differential nRNA splicing 129
■ *Sidelights & Speculations:* Differential nRNA Processing and Drosophila *Sex Determination* 131
Control of Gene Expression at the Level of Translation 132
  Differential mRNA longevity 132
  Selective inhibition of mRNA translation 132
  Control of RNA expression by cytoplasmic localization 134
  Posttranslational regulation of gene expression 135

CHAPTER 6 *Cell-cell communication in development 143*

Induction and Competence 143
  Cascades of induction: Reciprocal and sequential inductive events 145
  Instructive and permissive interactions 145
  Epithelial-mesenchymal interactions 147
Paracrine Factors 149
  The fibroblast growth factors 150
  The Hedgehog family 151
  The Wnt family 152
  The TGF-β superfamily 153
  Other paracrine factors 153
Cell Surface Receptors and Their Signal Transduction Pathways 154
  The receptor tyrosine kinase (RTK) pathway 154
■ *Sidelights & Speculations:* The RTK Pathway and Cell-to-Cell Induction 155
  The Smad pathway 159
  The JAK-STAT pathway 159
  The Wnt pathway 161
  The Hedgehog pathway 161
Cell Death Pathways 164
Juxtacrine Signaling 166
  The Notch pathway: Juxtaposed ligands and receptors 166
■ *Sidelights & Speculations:* Cell-Cell Interactions and Chance in the Determination of Cell Types 167
  The extracellular matrix as a source of critical developmental signals 168
  Direct transmission of signals through gap junctions 171
Cross-Talk between Pathways 172
Maintenance of the Differentiated State 173
Coda 174

part 2 *Early embryonic development*

CHAPTER 7 *Fertilization: Beginning a new organism 183*

Structure of the Gametes 183
  Sperm 183
  The egg 186
Recognition of Egg and Sperm 189
  Sperm attraction: Action at a distance 189
  The acrosome reaction in sea urchins 191
■ *Sidelights & Speculations: Action at a Distance: Mammalian Gametes 192*
  Species-specific recognition in sea urchins 194
  Gamete binding and recognition in mammals 195
Gamete Fusion and the Prevention of Polyspermy 197
  Fusion of the egg and sperm cell membranes 197
  The prevention of polyspermy 198
The Activation of Egg Metabolism 203
  Early responses 204
■ *Sidelights & Speculations: The Activation of Gamete Metabolism 206*
  Late responses 208
Fusion of the Genetic Material 209
  Fusion of genetic material in sea urchins 209
  Fusion of genetic material in mammals 209
Rearrangement of the Egg Cytoplasm 210
■ *Sidelights & Speculations: The Nonequivalence of Mammalian Pronuclei 212*

CHAPTER 8 *Early development in selected invertebrates 221*

AN INTRODUCTION TO EARLY DEVELOPMENTAL PROCESSES 221

Cleavage 221
  From fertilization to cleavage 222
  The cytoskeletal mechanisms of mitosis 223
  Patterns of embryonic cleavage 224
  Specification of cell fates during cleavage 226
Gastrulation 226

Cell Specification and Axis Formation 226

THE EARLY DEVELOPMENT OF SEA URCHINS 227

Cleavage in Sea Urchins 227
  Blastula formation 227
  Fate maps and the determination of sea urchin blastomeres 228
Sea Urchin Gastrulation 233
  Ingression of primary mesenchyme 233
  First stage of archenteron invagination 236
  Second and third stages of archenteron invagination 237

THE EARLY DEVELOPMENT OF SNAILS 239

Cleavage in Snail Embryos 239
  Fate map of Ilyanassa obsoleta 241
  The polar lobe: Cell determination and axis formation 242
■ *Sidelights & Speculations: Adaptation by Modifying Embryonic Cleavage 242*
Gastrulation in Snails 245

EARLY DEVELOPMENT IN TUNICATES 246

Tunicate Cleavage 246
  The tunicate fate map 246
  Autonomous and conditional specification of tunicate blastomeres 247
  Specification of the embryonic axes 249
Gastrulation in Tunicates 250

EARLY DEVELOPMENT OF THE NEMATODE CAENORHABDITIS ELEGANS 251

Why *C. elegans*? 251
Cleavage and Axis Formation in *C. elegans* 251
  Rotational cleavage of the C. elegans egg 251
  Anterior-posterior axis formation 253
  Formation of the dorsal-ventral and right-left axes 253
  Control of blastomere identity 254
  Integration of autonomous and conditional specification: The differentiation of the C. elegans pharynx 257
Gastrulation in *C. elegans* 257
Coda 258

CHAPTER 9 *The genetics of axis specification in Drosophila 263*

EARLY DROSOPHILA DEVELOPMENT 263

Cleavage 264

The mid-blastula transition 265

Gastrulation 266

THE ORIGINS OF ANTERIOR-POSTERIOR POLARITY 268

The Maternal Effect Genes 270

Embryological evidence of polarity regulation by oocyte cytoplasm 270

The molecular model: Protein gradients in the early embryo 270

The anterior organizing center: The Bicoid gradient 273

The posterior organizing center: Localizing and activating nanos 276

The terminal gene group 277

The Segmentation Genes 278

The gap genes 279

The pair-rule genes 281

The segment polarity genes 283

The Homeotic Selector Genes 285

Patterns of homeotic gene expression 285

Initiating the patterns of homeotic gene expression 286

Maintaining the patterns of homeotic gene expression 288

Realisator genes 288

■ *Sidelights & Speculations:* The Homeodomain Proteins 289

THE GENERATION OF DORSAL-VENTRAL POLARITY 290

Dorsal: The Morphogenetic Agent for Dorsal-Ventral Polarity 291

Translocation of Dorsal to the nucleus 291

The signal cascade 293

Establishing the dorsal patterning gradient 294

Axes and Organ Primordia: The Cartesian Coordinate Model 297

Coda 297

CHAPTER 10 *Early development and axis formation in amphibians 305*

EARLY AMPHIBIAN DEVELOPMENT 305

Cleavage in Amphibians 305

Amphibian Gastrulation 307

The Xenopus fate map 308

Cell movements during amphibian gastrulation 309

The mid-blastula transition: Preparing for gastrulation 310

Positioning the blastopore 311

Invagination and involution 312

The convergent extension of the dorsal mesoderm 313

Migration of the involuting mesoderm 315

Epiboly of the ectoderm 315

■ *Sidelights & Speculations:* Fibronectin and the Pathways for Mesodermal Migration 316

AXIS FORMATION IN AMPHIBIANS: THE PHENOMENON OF THE ORGANIZER 317

The Progressive Determination of the Amphibian Axes 317

Hans Spemann and Hilde Mangold: Primary Embryonic Induction 320

Mechanisms of Axis Determination in Amphibians 321

The origin of the Nieuwkoop center 321

The molecular biology of the Nieuwkoop center 322

The Functions of the Organizer 325

The diffusible proteins of the organizer I: The BMP inhibitors 327

The diffusible proteins of the organizer II: The Wnt inhibitors 330

■ *Sidelights & Speculations:* BMP4 and Geoffroy's Lobster 330

Conversion of the ectoderm into neural plate cells 332

■ *Sidelights & Speculations:* Competence, Bias, and Neurulation 333

The Regional Specificity of Induction 334

The determination of regional differences 334

The posterior transforming proteins: Wnt signals and retinoic acid 334

The anterior transforming proteins: Insulin-like growth factors 336

Specifying the Left-Right Axis 337

CHAPTER 11 *The early development of vertebrates: Fish, birds, and mammals 345*

EARLY DEVELOPMENT IN FISH 345

Cleavage in Fish Eggs 347

Gastrulation in Fish Embryos 349

The formation of germ layers 349

Axis Formation in Fish Embryos 351

Dorsal-ventral axis formation: The embryonic shield 351

The fish Nieuwkoop center 352

Anterior-posterior patterning 353

EARLY DEVELOPMENT IN BIRDS 354

Cleavage in Bird Eggs 354

Gastrulation of the Avian Embryo 356

The hypoblast 356

The primitive streak 356

Epiboly of the ectoderm 358

Axis Formation in the Chick Embryo 360

The role of pH in forming the dorsal-ventral axis 360

The role of gravity in forming the anterior-posterior axis 360

Left-Right Axis Formation 363

EARLY MAMMALIAN DEVELOPMENT 363

Cleavage in Mammals 363

The unique nature of mammalian cleavage 364

Compaction 366

Escape from the zona pellucida 368

Gastrulation in Mammals 368

Modifications for development within another organism 368

Formation of extraembryonic membranes 370

■ *Sidelights & Speculations: Twins and Embryonic Stem Cells 373*

Mammalian Anterior-Posterior Axis Formation 375

Two signaling centers 375

Patterning the anterior-posterior axis: The Hox code hypothesis 377

Expression of Hox genes along the dorsal axis 377

Experimental analysis of the Hox code 377

The Dorsal-Ventral and Right-Left Axes in Mice 380

The dorsal-ventral axis 380

The left-right axis 381

part 3 *Later embryonic development*

CHAPTER 12 *The emergence of the ectoderm: The central nervous system and the epidermis 389*

Establishing the Neural Cells 391

Formation of the Neural Tube 393

Primary neurulation 393

Secondary neurulation 398

Differentiation of the Neural Tube 398

The anterior-posterior axis 398

The dorsal-ventral axis 401

Tissue Architecture of the Central Nervous System 402

Spinal cord and medulla organization 403

Cerebellar organization 404

Cerebral organization 405

■ *Sidelights & Speculations: The Unique Development of the Human Brain 408*

Adult neural stem cells 409

Differentiation of Neurons 410

Development of the Vertebrate Eye 413

The dynamics of optic development 413

Neural retina differentiation 414

Lens and cornea differentiation 414

The Epidermis and the Origin of Cutaneous Structures 416

The origin of epidermal cells 416

Cutaneous appendages 418

Developmental genetics of hair formation 419

CHAPTER 13 *Neural crest cells and axonal specificity 427*

THE NEURAL CREST 427

Specification and Regionalization of the Neural Crest 427

The Trunk Neural Crest 429

Migration pathways of trunk neural crest cells 429

The mechanisms of trunk neural crest migration 431

Trunk neural crest cell differentiation 432

The Cranial Neural Crest 434

Cranial neural crest cell migration and specification: The first wave 436

Intramembranous ossification 437

Innervation of the placodes: The second wave of cranial neural crest migration 438

■ *Sidelights & Speculations: Tooth Development 439*

The Cardiac Neural Crest 441

NEURONAL SPECIFICATION AND AXONAL SPECIFICITY 442

The Generation of Neuronal Diversity 442

Pattern Generation in the Nervous System 444

Cell adhesion and contact guidance by attractive and permissive molecules 445

Guidance by specific growth cone repulsion 445

Guidance by diffusible molecules 446

Target selection 450

Forming the synapse: Activity-dependent development 451

Differential survival after innervation: Neurotrophic factors 451

Paths to glory: Migration of the retinal ganglion axons 453

The Development of Behaviors: Constancy and Plasticity 457

CHAPTER 14 *Paraxial and intermediate mesoderm 465*

PARAXIAL MESODERM: THE SOMITES AND THEIR DERIVATIVES 466

The Formation of Somites 467

The periodicity of somite formation 467

The separation of somites from the unsegmented mesoderm 468

The epithelialization of the somite 469

Specification of the somite along the anterior-posterior axis 471

The derivatives of the somite 471

Determination of the sclerotome and dermatome 472

Determination of the myotome 473

Myogenesis: The Development of Muscle 473

Specification and differentiation by the myogenic bHLH proteins 473

Muscle cell fusion 474

Osteogenesis: The Development of Bones 474

Endochondral ossification 475

Osteoclasts 477

INTERMEDIATE MESODERM: THE UROGENITAL SYSTEM 477

The Specification of the Intermediate Mesoderm 478

Progression of Kidney Types 479

Reciprocal Interactions of Developing Kidney Tissues 480

The mechanisms of reciprocal induction 481

CHAPTER 15 *Lateral plate mesoderm and endoderm 491*

LATERAL PLATE MESODERM 491

The Heart 492

Specification of heart tissue and fusion of heart rudiments 492

■ *Sidelights & Speculations: Redirecting Blood Flow in the Newborn Mammal 498*

Formation of Blood Vessels 500

Constraints on the construction of blood vessels 500

Vasculogenesis: The initial formation of blood vessels 501

Angiogenesis: Sprouting of blood vessels and remodeling of vascular beds 503

Mechanisms of Arterial and Venous Differentiation 504

The Development of Blood Cells 505

The stem cell concept 505

Sites of hematopoiesis 506

Committed stem cells and their fates 509

Hematopoietic inductive microenvironments 509

ENDODERM 510

The Pharynx 511

The Digestive Tube and Its Derivatives 511

Specification of the gut tissue 512

Liver, pancreas, and gallbladder 512

■ *Sidelights & Speculations: Blood and Guts: The Specification of Liver and Pancreas 513*

The Respiratory Tube 515

The Extraembryonic Membranes 517

The amnion and chorion 517

The allantois and yolk sac 517

CHAPTER 16 *Development of the tetrapod limb 523*

Formation of the Limb Bud 524

Specification of the limb fields: Hox genes and retinoic acid 524

Induction of the early limb bud: Wnt proteins and fibroblast growth factors 525

Specification of forelimb or hindlimb: Tbx4 and Tbx5 526

Generating the Proximal-Distal Axis of the Limb 529

The apical ectodermal ridge 529

FGFs in the induction and maintenance of the AER 530

Specifying the limb mesoderm: Determining the proximal-distal polarity of the limb 531

Hox genes, meis genes, and the specification of the proximal-distal axis 532

Specification of the Anterior-Posterior Limb Axis 534
The zone of polarizing activity 534
Sonic hedgehog defines the ZPA 535
Specification of the ZPA 535
Specifying digit identity through cell interactions initiated by Sonic hedgehog 536
Generation of the Dorsal-Ventral Axis 537
Coordinating the Three Axes 537
Cell Death and the Formation of Digits and Joints 538
Sculpting the autopod 538
Forming the joints 539
Continued Limb Growth: Epiphyseal Plates 540
■ *Sidelights & Speculations:* Control of Cartilage Maturation at the Growth Plate 541

CHAPTER 17 *Sex determination 547*

CHROMOSOMAL SEX DETERMINATION IN MAMMALS 548

Primary and Secondary Sex Determination in Mammals 548
The Developing Gonads 549
The Mechanisms of Mammalian Primary Sex Determination 551
SRY: The Y chromosome sex determinant 551
SOX9: An autosomal testis-determining gene 552
Fibroblast growth factor 9 553
SF1: The link between SRY and the male developmental pathways 554
DAX1: A potential testis-suppressing gene on the X chromosome 554
WNT4: A potential ovary-determining gene on an autosome 556
Secondary Sex Determination: Hormonal Regulation of the Sexual Phenotype 556
Testosterone and dihydrotestosterone 557
Anti-Müllerian duct hormone 558
Estrogen 559
■ *Sidelights & Speculations:* Sex Determination and Behaviors 559

CHROMOSOMAL SEX DETERMINATION IN DROSOPHILA 561

The Sexual Development Pathway 561
Mechanisms of Sex Determination 562
The sex-lethal gene as the pivot for sex determination 562
The transformer genes 563
Doublesex: The switch gene of sex determination 564

ENVIRONMENTAL SEX DETERMINATION 567

Temperature-Dependent Sex Determination in Reptiles 567
Aromatase and the production of estrogens 567
Sex reversal, aromatase, and conservation biology 568
Location-Dependent Sex Determination in *Bonellia* and *Crepidula* 568

CHAPTER 18 *Metamorphosis, regeneration, and aging 575*

Metamorphosis: The Hormonal Reactivation of Development 575
Amphibian Metamorphosis 576
Morphological changes associated with metamorphosis 576
Biochemical changes associated with metamorphosis 577
Hormonal control of amphibian metamorphosis 578
Regionally specific developmental programs 580
■ *Sidelights & Speculations:* Variations on the Theme of Amphibian Metamorphosis 581
Metamorphosis in Insects 583
Types of insect metamorphosis 583
Imaginal discs 584
Determination of the wing imaginal discs 587
Hormonal control of insect metamorphosis 588
The molecular biology of 20-hydroxyecdysone activity 590
Regeneration 592
Epimorphic Regeneration of Salamander Limbs 592
Formation of the apical ectodermal cap and regeneration blastema 593
Proliferation of the blastema cells: The requirement for nerves 594
Proliferation of the blastema cells: The requirement for FGF10 595
Pattern formation in the regeneration blastema 595
Morphallactic Regeneration in Hydra 597
The head activation gradient 597
The head inhibition gradient 598
The hypostome as an "organizer" 598
The basal disc activation and inhibition gradients 600
Compensatory Regeneration in the Mammalian Liver 601
Aging: The Biology of Senescence 601
Maximum life span and life expectancy 601
Causes of aging 602
Genetically programmed aging 603

CHAPTER 19 *The saga of the germ line* 613

Germ Plasm and the Determination of the Primordial Germ Cells 613
Germ cell determination in nematodes 614
Germ cell determination in insects 615
Germ cell determination in amphibians 616
The inert genome hypothesis 617
Germ Cell Migration 617
Germ cell migration in amphibians 617
Germ cell formation and migration in mammals 618
■ *Sidelights & Speculations:* EG Cells, ES Cells, and Teratocarcinomas 621
Germ cell migration in birds and reptiles 622

Germ cell migration in Drosophila 622
Meiosis 624
■ *Sidelights & Speculations:* Big Decisions: Mitosis or Meiosis? Sperm or Egg? 627
Spermatogenesis 628
Forming the haploid spermatid 628
Spermiogenesis: The differentiation of the sperm 630
Oogenesis 631
Oogenic meiosis 631
Maturation of the oocyte in amphibians 632
Completion of amphibian meiosis: Progesterone and fertilization 633
Gene transcription in oocytes 634
Meroistic oogenesis in insects 637
Maturation of the mammalian oocyte 638

part 4 *Ramifications of developmental biology*

CHAPTER 20 *An overview of plant development* 649

Plant Life Cycles 651
Gamete Production in Angiosperms 654
Pollen 654
The ovary 655
Pollination 655
Fertilization 658
Embryonic Development 659
Experimental studies 659
Embryogenesis 659
Dormancy 664
Germination 664
Vegetative Growth 665
Meristems 665
Root development 667
Shoot development 667
Leaf development 668
The Vegetative-to-Reproductive Transition 671
Control of the reproductive transition 672
Inflorescence meristems 674
Floral meristem identity genes 675
Organ identity genes: The ABC model 676
Senescence 678

CHAPTER 21 *Medical implications of developmental biology* 683

INFERTILITY 683

Diagnosing Infertility 683
In Vitro Fertilization (IVF) 684
The IVF procedure 684
Variations on IVF 685
Success rates and complications of IVF 685
■ *Sidelights & Speculations:* Ethical Issues and Assisted Reproductive Technology 686

GENETIC ERRORS OF HUMAN DEVELOPMENT 686

Identifying the Genes for Human Developmental Anomalies 687
The Nature of Human Syndromes 689
Pleiotropy 689
Genetic heterogeneity 690
Phenotypic variability 690
Mechanisms of dominance 691
Gene Expression and Human Disease 691
Inborn errors in transcriptional regulation 691
Inborn errors of nuclear RNA processing 692
Inborn errors of translation 692

Preimplantation Genetics 693
  Amniocentesis, chorionic villus sampling, and preimplantation genetics 693
  Sperm separation and sex selection 694

TERATOGENESIS: ENVIRONMENTAL ASSAULTS ON HUMAN DEVELOPMENT 694

Teratogenic Agents 695
  Retinoic acid as a teratogen 696
  Alcohol as a teratogen 697
  Endocrine disruptors 698
  Other teratogenic agents 700
■ *Sidelights & Speculations:* The DES Story: Is It Happening Again? 701

DEVELOPMENTAL BIOLOGY AND THE FUTURE OF MEDICINE 703

Developmental Cancer Therapies 703
  Cancer as a disease of altered development 703
  Differentiation therapy 703
  Angiogenesis inhibition 704
Gene Therapy 705
  Somatic cell gene therapy 707
  Germ line gene therapy 707
Stem Cells and Therapeutic Cloning 708
  Cloning 708
  Embryonic stem cells and therapeutic cloning 708
  Adult stem cells 709
  Transgenic stem cells 711
  Regeneration therapy 712

CHAPTER 22 Environmental regulation of animal development 721

The Environment as Part of Normal Development 721
  Gravity and pressure 722
  Developmental symbiosis 722
  Larval settlement 725
  Sex in its season 726
  Diapause: Suspended development 727
Phenotypic Plasticity: Control of Development by Environmental Conditions 727
  Polyphenism and reaction norms 727
  Seasonal polyphenism in butterflies 728
  Nutritional polyphenism 730
  Environment-dependent sex determination 731
  Polyphenisms for alternative environmental conditions 732
■ *Sidelights & Speculations:* Genetic Assimilation 733
  Predator-induced polyphenisms 734

■ *Sidelights & Speculations:* Mammalian Immunity as a Predator-Induced Response 736
Learning: An Environmentally Adaptive Nervous System 737
  The formation of new neurons 737
  Experiential changes in mammalian visual pathways 737
Endocrine Disruptors 739
  Environmental estrogens 740
  Environmental thyroid hormone disruptors 742
  Chains of causation 742
Developmental Biology Meets the Real World 743
■ *Sidelights & Speculations:* Deformed Frogs 744

CHAPTER 23 Developmental mechanisms of evolutionary change 751

"Unity of Type" and "Conditions of Existence" 751
  Charles Darwin's synthesis 751
  "Life's splendid drama" 752
  The search for the Urbilaterian ancestor 753
Hox Genes: Descent with Modification 754
■ *Sidelights & Speculations:* How the Chordates Got a Head 761
Homologous Pathways of Development 763
  Instructions for forming the central nervous system 763
  Instructions for appendage formation 764
Modularity: The Prerequisite for Evolution through Development 766
  Dissociation: Heterochrony and allometry 766
  Duplication and divergence 768
  Co-option 768
Generating Evolutionary Novelty 769
Developmental Constraints 773
  Physical constraints 773
  Morphogenetic constraints 774
  Phyletic constraints 774
■ *Sidelights & Speculations:* Canalization and the Release of Developmental Constraints 776
A New Evolutionary Synthesis 777

Sources for Chapter-Opening Quotations 785

Author Index 787

Subject Index 805

Preface

Jean Rostand, the French embryologist and essayist, wrote, "Today my book is done. Will it be done tomorrow?" The answer, of course, is "No way in developmental biology!" Developmental biology has not stood still for a minute, and its new ideas and explanations have pushed it to the frontiers of our knowledge. Indeed, if you seek the intellectual challenges of building new concepts, developmental biology is the place to be. There is a twofold revolution going on in developmental biology. This present text is a document of that revolution.

The first phase of the revolution began in the 1970s. At that time, developmental biology began to make use of the new recombinant DNA technologies to explain how the genetic instructions specified phenotypes composed of different cell types and organs. This remains one of the great projects of developmental biology. The first edition of this book (1985) was written in response to the start of this revolution. Although hardly a gene was mentioned in that edition, the techniques of DNA cloning were discussed, and the first work from the Nüsslein-Volhard laboratory was mentioned. By the second edition (1988), the book discussed homeobox genes, enhancers, and promoters, and in situ hybridization was on its cover. Transcription factors made their appearance in the third edition (1991, and the fourth edition (1994) detailed paracrine factors. Subsequent editions went into the signal transduction pathways linking paracrine factors with transcription factors. This newly empowered science identified molecular mechanisms for the two most central processes of developmental biology: differentiation and induction.

During that time frame, the second phase of the revolution began. Developmental biology looked outward to other fields, applying recombinant DNA techniques and even newer technologies (bioinformatics and genomics) to bring developmental biology into areas that it had abandoned during its history. There are four fields of expansion and reconciliation: evolution, embryology, ecology, and medicine.

We have returned to evolutionary biology (just as Wilhelm Roux predicted we would) with a new paradigm for the causal mechanisms of evolution, seeing evolution as changes in gene expression as well as changes in allele frequency. The third edition of the textbook inaugurated the chapter on evolutionary developmental biology. Developmental biology had also abandoned the questions of morphogenesis in embryology, thinking it was largely a subset of differential gene expression. We are now returning to this area with new molecular and computational tools to analyze the cellular mechanisms that enable organogenesis and the mathematical constraints that permit only certain phenotypes to occur. The fourth edition (1994) highlighted these concerns. In addition, as the genetic mechanisms underlying such morphogenetic questions as axis specification became illuminated, the ubiquity of similar mechanisms became awe-inspiringly apparent. Even plant morphogenesis, long seen as so distinct from animal development as to constitute virtually a separate field, became part of this new incarnation of developmental biology. The sixth edition of this textbook (2000) saw the addition of a chapter on plant development.

Some of the first experimental embryologists were interested developmental plasticity and how the development was modified by the environment. Today we are once again bringing developmental biology to bear on issues of ecology. This ecological developmental biology seeks to find the proximate causes for the life history strategies well known to ecologists. The fifth edition of this book (1997) inaugurated our chapter on ecological developmental biology.

Many of the early investigations of experimental embryology were performed in order to discover the mechanisms of congenital birth defects. Today, after a long absence, developmental biology, with its new tools and insights, is reclaiming its medical heritage. We are identifying the causes for inherited malformations of development, seeking the ways by which exogenous chemicals disrupt normal development, and looking at developmental cures for such anomalies. This present edition inaugurates our chapter on the medical implications of developmental biology.

"To explore strange new worlds"

The new chapter on the medical implications of developmental biology had to be written. The stories are in every newspaper: cloning, stem cells, genetic engineering, in vitro fertilization, cancer therapies, organ regeneration, and protocols for prolonging our lifespan. In the past five years, developmental biology has usurped a place formerly occupied by science fiction. Our knowledge of development has enabled us to transform livers into pancreases and convert quail beaks into duck bills. It has allowed us to clone cats and place new genes into human embryonic stem cells. This ability to understand and even transform nature is revolutionary. A decade ago it was not even considered. The biology behind the headlines is hardly ever reported in the media, and this biology is every bit as fascinating as the technologies derived from it. Moreover, these technologies are bringing developmental biology into the social sphere as it never has been before. Students taking developmental biology classes should be able to explain to their classmates (and parents) the science behind the news stories, and I hope that this chapter will facilitate that transmission. I also believe that developmental biologists (both current and emergent) need to think about the implications of our research.

Along with the revolution in developmental biology has come the revolution in information technology. The information revolution has become integrated with the revolution in developmental biology and has greatly facilitated the merging of developmental biology with ecology, evolution, and medicine. This influence is also seen in this book. *Developmental Biology* may have been the first science text to have a website (in 1994), and in this edition, **devbio.com** has expanded to include an entire volume on the ethical issues of developmental biology. These may be useful for starting discussions in the laboratory while we wait for the gels to run or the embryos to cleave.

The ***Vade Mecum*** CD has undergone its own metamorphosis and has become even more integrally connected with the book. Not only has its coverage expanded, but it has incorporated an entire laboratory manual as well. Just as this is the first edition of the book to integrate all the diverse areas of developmental biology, it is the first edition to include a laboratory manual and a bioethics pamphlet in its electronic augmentations.

This is the first edition of the book that has brought together all these strands, and I hope it does justice to developmental biology and to the revolution that is changing our professional and personal lives. Throughout all this revolution, however, one always returns to the embryos. They are the source of both our curiosity and our awe. Some of us are lucky enough to be paid to think about them and to even get into the laboratory to see if our ideas are possibly correct. As Jean Rostand proclaimed, "What a profession this is—this daily inhalation of wonder."

Acknowledgments

This edition, like its earlier versions, has benefited enormously from the students of my embryology and developmental genetics classes. Much of the website material on bioethics were compiled by students in my history of biology course. The book was helped by the remarkably supportive staff and faculty of Swarthmore College. This includes the librarians, administrative assistants, mailroom personnel, and computer specialists as well as the academic staff and faculty. I also wish to thank those scientists who gave us permission to use their wonderful figures and photographs, and also those who sent in corrections and suggestions for this edition. There is no better compliment than to have somebody spend the time to go through the textbook and to point out things that can be made better. Charles Bieberich and Andrew Whipple provided especially useful feedback of this kind.

I am once again indebted to the many colleagues who vetted early versions of these chapters for errors and who steered me to new and important research papers being done in their respective specialties. Kathryn Anderson, Adam Antebi, Bruce Baker, Hans Bode, Donald Brown, Ann Burke, Blanche Capel, Judy Cebra-Thomas, Eddie De Robertis, Charles Emerson, John Fallon, David Gardiner, Laurinda Jaffe, Ray Keller, Marta Laskowski, Kirsti Linask, David McClay, Anthony-Samuel Lamantia, Philip Meneely, Mary Montgomery, Lisa Nagy, Fred Nijhout, John Opitz, Rudy Raff, Kirsi Saino, Hannu Sariola, Gary Schoenwolf, Billie Swalla, Carl Thummel, Kathy Tosney, Rocky Tuan, Mary Tyler, Adam Wilkins, Christopher Wright, and Christopher Wylie all provided valuable input, often including reprints and sketches of improved figures—the most concrete kind of help an author can ask for! Persisting errors are mine alone.

My editors Andy Sinauer and Carol Wigg shepherded the project to another successful completion while somehow managing (at least as far as I know) to maintain their sanity. The Sinauer production staff, including Chris Small, Jefferson Johnson, Janice Holabird, and Joanne Delphia worked their magic using modern digital publishing software to design and produce the pages, making changes and updating right up until the book "hit the presses." Fraser Tan provided back-up proofreading for a harried author. And as always, my thanks to Anne Raunio for her help with the medical chapter and especially for putting up with me during the past year.

Scott Gilbert
February 2003

PART **I** *Principles of Developmental Biology*

1 Developmental biology: The anatomical tradition

2 Life cycles and the evolution of developmental patterns

3 Principles of experimental embryology

4 The genetic core of development

5 The paradigm of differential gene expression

6 Cell-cell communication in development

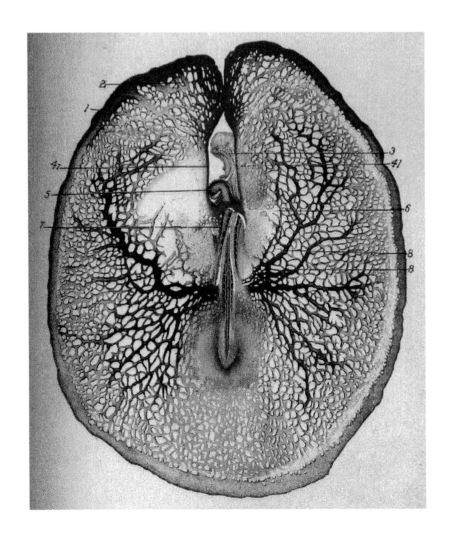

A two-day chick embryo seen from the ventral surface, allowing one to view the circulation of the blood to the yolk and back. The blood leaves the embryo through the two vitelline arteries, and it returns through the vitelline veins near the head of the embryo. (Figure 1.2B from F. R. Lillie, 1908).

chapter 1

Developmental biology:
The anatomical tradition

BETWEEN FERTILIZATION AND BIRTH, the developing organism is known as an embryo. The concept of an embryo is a staggering one, and forming an embryo is the hardest thing you will ever do. To become an embryo, you had to build yourself from a single cell. You had to respire before you had lungs, digest before you had a gut, build bones when you were pulpy, and form orderly arrays of neurons before you knew how to think. One of the critical differences between you and a machine is that a machine is never required to function until after it is built. Every animal has to function as it builds itself.

For animals, fungi, and plants, the sole way of getting from egg to adult is by developing an embryo. The embryo mediates between genotype and phenotype, between the inherited genes and the adult organism. Whereas most of biology studies adult structure and function, developmental biology finds the study of the transient stages leading up to the adult to be more interesting. Developmental biology studies the initiation and construction of organisms rather than their maintenance. It is a science of becoming, a science of process. To say that a mayfly lives but one day is profoundly inaccurate to a developmental biologist. A mayfly may be a winged adult for only a day, but it spends the other 364 days of its life as an aquatic juvenile under the waters of a pond or stream.

The questions asked by developmental biologists are often questions about becoming rather than about being. To say that XX mammals are usually females and XY mammals are usually males does not explain sex determination to a developmental biologist, who wants to know *how* the XX genotype produces a female and *how* the XY genotype produces a male. Similarly, a geneticist might ask how globin genes are transmitted from one generation to the next, and a physiologist might ask about the function of globin proteins in the body. But the developmental biologist asks how it is that the globin genes become expressed only in red blood cells and how they become active only at specific times in development. (We don't know the answers yet.)

Developmental biology is a great field for scientists who want to integrate different levels of biology. We can take a problem and study it on the molecular and chemical levels (e.g., How are globin genes transcribed, and how do the factors activating their transcription interact with one another on the DNA?); on the cellular and tissue levels (Which cells are able to make globin, and how does globin mRNA leave the nucleus?); on the organ and organ-system levels (How do the capillaries form in each tissue, and how are they instructed to branch and connect?); and even at the ecological and evolutionary levels (How do differences in globin gene activation enable oxygen to flow from mother to fetus, and how do environmental factors trigger the differentiation of more red blood cells?).

Developmental biology is one of the fastest growing and most exciting fields in biology, creating a framework that integrates molecular biology, physiology, cell biology, genetics, anatomy, cancer research, neurobiology, immunology, ecology, and evolutionary biology. The study of development has become essential for understanding any other area of biology.

The Questions of Developmental Biology

According to Aristotle, the first embryologist known to history, science begins with wonder: "It is owing to wonder that people began to philosophize, and wonder remains the beginning of knowledge" (Aristotle, *Metaphysics,* ca. 350 B.C.E.). The development of an animal from an egg has been a source of wonder throughout history. The simple procedure of cracking open a chick egg on each successive day of its 3-week incubation provides a remarkable experience as a thin band of cells is seen to give rise to an entire bird. Aristotle performed this procedure and noted the formation of the major organs. Anyone can wonder at this remarkable—yet commonplace—phenomenon, but the scientist seeks to discover how development actually occurs. And rather than dissipating wonder, new understanding increases it.

Multicellular organisms do not spring forth fully formed. Rather, they arise by a relatively slow process of progressive change that we call **development**. In nearly all cases, the development of a multicellular organism begins with a single cell—the fertilized egg, or **zygote**, which divides mitotically to produce all the cells of the body. The study of animal development has traditionally been called **embryology**, from that phase of an organism that exists between fertilization and birth. But development does not stop at birth, or even at adulthood. Most organisms never stop developing. Each day we replace more than a gram of skin cells (the older cells being sloughed off as we move), and our bone marrow sustains the development of millions of new red blood cells every minute of our lives. In addition, some animals can regenerate severed parts, and many species undergo metamorphosis (such as the transformation of a tadpole into a frog, or a caterpillar into a butterfly). Therefore, in recent years it has become customary to speak of **developmental biology** as the discipline that studies embryonic and other developmental processes.

Development accomplishes two major objectives: it generates cellular diversity and order within each generation, and it ensures the continuity of life from one generation to the next. Thus, there are two fundamental questions in developmental biology: How does the fertilized egg give rise to the adult body, and how does that adult body produce yet another body? These two huge questions have been subdivided into six general questions scrutinized by developmental biologists:

- **The question of differentiation.** A single cell, the fertilized egg, gives rise to hundreds of different cell types—muscle cells, epidermal cells, neurons, lens cells, lymphocytes, blood cells, fat cells, and so on (Figure 1.1). This generation of cellular diversity is called **differentiation**. Since each cell of the body (with very few exceptions) contains the same set of genes, how can this same set of genetic instructions produce different types of cells? How can the fertilized egg generate so many different cell types?
- **The question of morphogenesis.** Our differentiated cells are not randomly distributed. Rather, they are organized into intricate tissues and organs. During development, cells divide, migrate, and die; tissues fold and separate. The organs so formed are arranged in a particular way: our fingers are always at the tips of our hands, never in the middle; our eyes are always in our heads, not in our toes or gut. This creation of ordered form is called **morphogenesis**. How can the cells form such ordered structures?
- **The question of growth.** If each cell in our face were to undergo just one more cell division, we would be considered horribly malformed. If each cell in our arms underwent just one more round of cell division, we could tie our shoelaces without bending over. How do our cells know when to stop dividing? Our arms are generally the same size on both sides of the body. How is cell division so tightly regulated?
- **The question of reproduction.** The sperm and egg are very specialized cells. Only they can transmit the instructions for making an organism from one generation to the next. How are these cells set apart to form the next generation, and what are the instructions in the nucleus and cytoplasm that allow them to function this way?
- **The question of evolution.** Evolution involves inherited changes in development. When we say that today's one-toed horse had a five-toed ancestor, we are saying that changes in the development of cartilage and muscles occurred over many generations in the embryos of the horse's ancestors. How do changes in development create new body forms? Which heritable changes are possible, given the constraints imposed by the necessity of the organism to survive as it develops?
- **The question of environmental integration.** The development of many (and perhaps all) organisms is influenced by cues from the environment that surrounds the embryo or larvae. Certain butterflies, for instance, inherit the ability to produce different wing colors based on the temperature or the amount of daylight experienced by the caterpillar before it undergoes metamorphosis. Moreover, certain chemicals in the environment can disrupt normal development, causing malformations in the adult. How is the development of an organism integrated into the larger context of its habitat, and what properties enables certain chemicals to alter development?

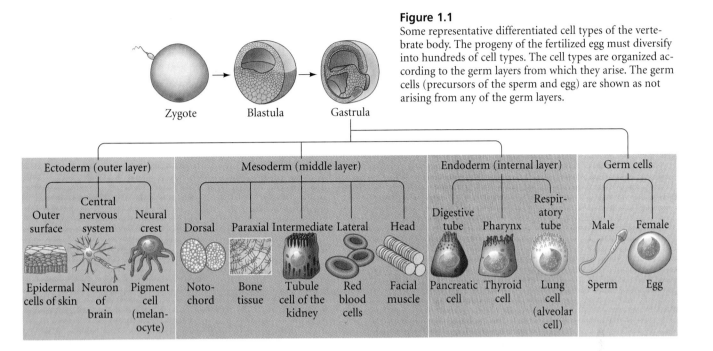

Figure 1.1
Some representative differentiated cell types of the vertebrate body. The progeny of the fertilized egg must diversify into hundreds of cell types. The cell types are organized according to the germ layers from which they arise. The germ cells (precursors of the sperm and egg) are shown as not arising from any of the germ layers.

Approaches to Developmental Biology

A field of science is defined by the questions it seeks to answer, and most of the questions in developmental biology have been bequeathed to it through its embryological heritage. There are numerous strands of embryology, each of which has predominated during a different era. Sometimes these traditions are distinct, and sometimes they blend. We can identify three major approaches to studying embryology:

• Anatomical approaches
• Experimental approaches
• Genetic approaches

While it is true that anatomical approaches gave rise to experimental approaches, and that genetic approaches built on the foundations of the earlier two approaches, all three traditions persist to this day and continue to play a major role in developmental biology. Chapter 3 of this text discusses experimental approaches, and Chapters 4 and 5 examine the genetic approaches in greater depth. In recent years, each of these traditions has become joined with molecular genetics to produce a vigorous and multifaceted science of developmental biology. But the basis of all research in developmental biology is the changing anatomy of the organism.

The Anatomical Approach

What parts of the embryo form the heart? How do the cells that form the retina position themselves the proper distance from the cells that form the lens? How do the tissues that form the bird wing relate to the tissues that form the fish fin or the human hand? What organs are affected by mutations in particular genes? Are there mathematical equations that relate the growth of different organs within the body? These are the questions asked by developmental anatomists.

Several strands weave together to form the anatomical approaches to development. The first strand is **comparative embryology**, the study of how anatomy changes during the development of different organisms. The second strand, based on the first, is **evolutionary embryology**, the study of how changes in development may cause evolutionary changes and of how an organism's ancestry may constrain the types of changes that are possible. The third strand of the anatomical approach to developmental biology is **teratology**, the study of birth defects. A fourth method used in the anatomical approach is **mathematical modeling**, which seeks to describe developmental phenomena in terms of equations.

Comparative embryology

The first known study of comparative developmental anatomy was undertaken by Aristotle in the fourth century B.C.E. In *The Generation of Animals* (ca. 350 B.C.E.), he noted the different ways that animals are born: from eggs (**oviparity**, as in birds, frogs, and most invertebrates), by live birth (**viviparity**, as in placental mammals), or by producing an egg that hatches inside the body (**ovoviviparity**, as in certain reptiles and sharks). Aristotle also identified the two major cell division patterns by which embryos are formed: the **holoblastic** pattern of cleavage (in which the entire egg is divided into smaller cells, as it is in frogs and mammals) and the **meroblastic** pattern of cleavage (as in chicks, wherein only part of the egg is destined to become the embryo, while the other portion—the yolk—serves as nutrition). And should anyone want to know who first figured out the functions of the placenta and the umbilical cord, it was Aristotle.

After Aristotle, there was remarkably little progress in embryology for the next two thousand years. It was only in 1651 that William Harvey concluded that all animals—even mammals—originate from eggs. *Ex ovo omnia* ("All from the egg") was the motto on the frontispiece of his *On the Generation of Living Creatures*, and this precluded the spontaneous generation of animals from mud or excrement. This statement was not made lightly, for Harvey knew that it went against the views of Aristotle, whom Harvey still venerated. (Aristotle thought that menstrual fluid formed the material of the embryo, while the semen acted to give it form and animation.) Harvey also was the first to see the **blastoderm** of the chick embryo (the small region of the egg containing the yolk-free cytoplasm that gives rise to the embryo) and he was the first to notice that "islands" of blood cells form before the heart does. Harvey also suggested that the amnionic fluid might function as a "shock absorber" for the embryo.

As might be expected, embryology remained little but speculation until the invention of the microscope allowed detailed observations. In 1672, Marcello Malpighi published the first microscopic account of chick development. Here, for the first time, the neural groove (precursor of the neural tube), the muscle-forming somites, and the first circulation of the arteries and veins—to and from the yolk—were identified (Figure 1.2).

Epigenesis and preformation

With Malpighi begins one of the great debates in embryology: the controversy over whether the organs of the embryo are formed *de novo* ("from scratch") at each generation, or whether the organs are already present, in miniature form, within the egg (or sperm). The first view is called **epigenesis**, and it was supported by Aristotle and Harvey. The second view, called **preformation**, was reinvigorated with support from Malpighi. Malpighi showed that the unincubated* chick egg already had a great deal of structure. This observation provided him with reasons to question epigenesis. According to the preformationist view, all the organs of the adult were prefigured in miniature within the sperm or (more usually) the egg. Organisms were not seen to be "developed," but rather "unrolled."

The preformationist hypothesis had the backing of eighteenth-century science, religion, and philosophy (Gould 1977; Roe 1981; Pinto-Correia 1997). First, because all organs were prefigured, embryonic development merely required the growth of existing structures, not the formation of new ones. No extra mysterious force was needed for embryonic develop-

ment. Second, just as the adult organism was prefigured in the germ cells, another generation already existed in a prefigured state within the germ cells of the first prefigured generation. This corollary, called *embôitment* (encapsulation), ensured that the species would always remain constant. Although certain microscopists claimed to see fully formed human miniatures within the sperm or egg, the major proponents of this hypothesis—Albrecht von Haller and Charles Bonnet—knew that organ systems develop at different rates, and that structures need not be in the same place in the embryo as they are in the newborn.

The preformationists had no cell theory to provide a lower limit to the size of their preformed organisms (the cell theory arose in the mid-1800s), nor did they view humankind's tenure on Earth as potentially infinite. Rather, said Bonnet (1764), "Nature works as small as it wishes," and the human species existed in that finite time between Creation and Resurrection. This view was in accord with the best science of its time, conforming to the French mathematician-philosopher René Descartes's principle of the infinite divisibility of a mechanical nature initiated, but not interfered with, by God. It also conformed to Enlightenment views of the Deity. The scientist-priest Nicolas Malebranche saw in preformationism the fusion of the rule-giving God of Christianity with Cartesian science (Churchill 1991; Pinto-Correia 1997).[†]

The embryological case for epigenesis was revived at the same time by Kaspar Friedrich Wolff, a German embryologist working in St. Petersburg. By carefully observing the development of chick embryos, Wolff demonstrated that the embryonic parts develop from tissues that have no counterpart in the adult organism. The heart and blood vessels (which, according to preformationism, had to be present from the beginning to ensure embryonic growth) could be seen to develop anew in each embryo. Similarly, the intestinal tube was seen to arise by the folding of an originally flat tissue. This latter observation was explicitly detailed by Wolff, who proclaimed (1767), "When the formation of the intestine in this manner has been duly weighed, almost no doubt can remain, I believe, of the truth of epigenesis." However, to explain how an organism is created anew each generation, Wolff had to postulate an unknown force, the *vis essentialis* ("essential force"), which, acting according to natural laws in the same way as gravity or magnetism, would organize embryonic development.

---

*As was pointed out by Maître-Jan in 1722, the egg examined by Malpighi may technically be called "unincubated," but as it was left sitting in the Bolognese sun in August, it was not unheated.

†Preformation was a conservative theory, emphasizing the lack of change between generations. Its principal failure was its inability to account for the variations known by the limited genetic evidence of the time. It was known, for instance, that matings between white and black parents produced children of intermediate skin color, an impossibility if inheritance and development were solely through either the sperm or the egg. In more controlled experiments, the German botanist Joseph Kölreuter (1766) produced hybrid tobacco plants having the characteristics of both species. Moreover, by mating the hybrid to either the male or female parent, Kölreuter was able to "revert" the hybrid back to one or the other parental type after several generations. Thus, inheritance seemed to arise from a mixture of parental components.

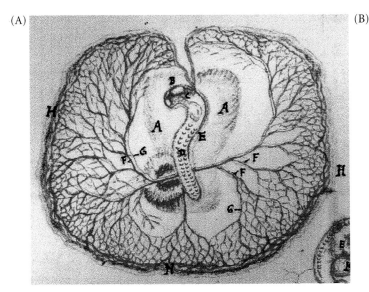

(A)

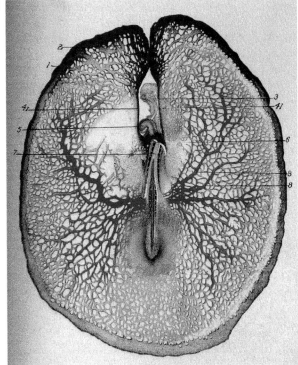

(B)

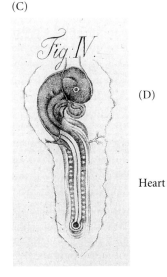

(C)

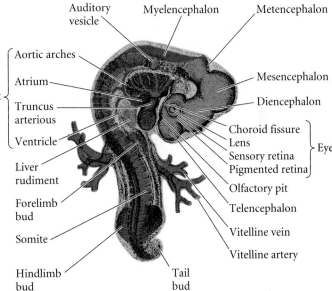

(D)

Figure 1.2
Depictions of chick developmental anatomy. (A) Dorsal view (looking "down" at what will become the back) of a 2-day chick embryo, as depicted by Marcello Malpighi in 1672. (B) Ventral view (looking "up" at the prospective belly) of a chick embryo at a similar stage, seen through a dissecting microscope and rendered by F. R. Lillie in 1908. (C) Eduard d'Alton's depiction of a later stage 2-day chick embryo in Pander (1817). (D) Modern rendering of a 3-day chick embryo. Details of the anatomy will be discussed in later chapters. (A from Malpighi 1672; B from Lillie 1908; C from Pander 1817, courtesy of Ernst Mayr Library of the Museum of Comparative Zoology, Harvard; D after Carlson 1981.)

A reconciliation of sorts was attempted by the German philosopher Immanuel Kant (1724–1804) and his colleague, biologist Johann Friedrich Blumenbach (1752–1840). Attempting to construct a scientific theory of racial descent, Blumenbach postulated a mechanical, goal-directed force called the *Bildungstrieb* ("developmental force"). Such a force, he said, was not theoretical, but could be shown to exist by experimentation. A hydra, when cut, regenerates its amputated parts by rearranging existing elements (see Chapter 18). Some purposive organizing force could be observed in operation, and this force was a property of the organism itself. This *Bildungstrieb* was thought to be inherited through the germ cells.

Thus, development could proceed through a predetermined force inherent in the matter of the embryo (Cassirer 1950; Lenoir 1980). Moreover, this force was believed to be susceptible to change, as demonstrated by the left-handed variant of snail coiling (where left-coiled snails can produce right-coiled progeny). In this hypothesis, wherein epigenetic development is directed by preformed instructions, we are not far from the view held by modern biologists that most of the instructions for forming the organism are already present in the egg.

Naming the parts: The primary germ layers and early organs

The end of preformationism did not come until the 1820s, when a combination of new staining techniques, improved microscopes, and institutional reforms in German universities created a revolution in descriptive embryology. The new techniques enabled microscopists to document the epigenesis of anatomical structures, and the institutional reforms provided audiences for these reports and students to carry on the work of their teachers. Among the most talented of this new group of microscopically inclined investigators were three friends, born within a year of each other, all of whom came from the Baltic region and studied in northern Germany. The work of Christian Pander, Karl Ernst von Baer, and Heinrich Rathke transformed embryology into a specialized branch of science.

Pander studied the chick embryo for less than two years (before becoming a paleontologist), but in those 15 months, he discovered the germ* layers, three distinct regions of the embryo that give rise to the specific organ systems (see Figure 1.1).

- The **ectoderm** generates the outer layer of the embryo. It produces the surface layer (epidermis) of the skin and forms the brain and nervous system.
- The **endoderm** becomes the innermost layer of the embryo and produces the epithelium of the digestive tube and its associated organs (including the lungs).
- The **mesoderm** becomes sandwiched between the ectoderm and endoderm. It generates the blood, heart, kidney, gonads, bones, muscles, and connective tissues.

These three layers are found in the embryos of all **triploblastic** (three-layer) organisms. Some phyla, such as the porifera (sponges), cnidarians (sea anemones, hydra, jellyfish), and ctenophores (comb jellies) lack a true mesoderm and are considered **diploblastic** animals.

Pander also made observations that weighted the balance in favor of epigenesis. The germ layers, he noted, did not form their organs independently (Pander 1817). Rather, each germ layer "is not yet independent enough to indicate what it truly is; it still needs the help of its sister travelers, and therefore, although already designated for different ends, all three influence each other collectively until each has reached an appropriate level." Pander had discovered the tissue interactions that we now call **induction**. No tissue is able to construct organs by itself; it must interact with other tissues. (We will discuss the principles of induction more thoroughly in Chapter 6.) Thus, Pander felt that preformation could not be true, since the organs come into being through interactions between simpler structures.

---

*From the same root as germination: the Latin *germen*, meaning "sprout" or "bud." The names of the three germ layers are from the Greek: ectoderm from *ektos* ("outside") plus *derma* (skin); mesoderm from *mesos* ("middle"), and endoderm from *endon* ("within").

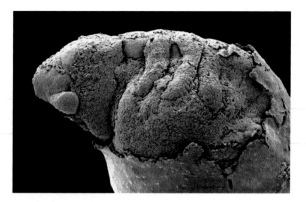

Figure 1.3
Pharyngeal arches (also called branchial arches and gill arches) in the embryo of the salamander *Ambystoma mexicanum*. The surface ectoderm has been removed to permit the easy visualization of these arches (highlighted) as they form. (Photograph courtesy of P. Falck and L. Olsson.)

Interestingly, the glory of Pander's book is its engravings; the artist, Eduard d'Alton, drew details for which the vocabulary had not yet been invented. Today we can look at these drawings and see the five regions of the embryonic chick brain, even though these regions had not yet been separately defined or given names (see Figure 1.2C; Churchill 1991). The ability to make precise observations has always been among the greatest skills of embryologists, and today modern developmental biologists looking at gene expression patterns are "discovering" regions of the embryo that were observed by embryologists a century ago.

Heinrich Rathke looked at the development of frogs, salamanders, fish, birds, and mammals, and emphasized the similarities in the development of all these vertebrate groups. During his 40 years of embryological research, he described for the first time the vertebrate **pharyngeal arches** (Figure 1.3), which become the gill apparatus of fish but in mammals become the jaws and ears, (among other things, as we will see in Figure 1.14). Rathke described the formation of the vertebrate skull, and the origin of the reproductive, excretory, and respiratory systems. He also studied the development of invertebrates, especially the crayfish. He is memorialized today in the name "Rathke's pouch," the embryonic rudiment of the glandular portion of the pituitary. That he could see such a structure using the techniques available at that time is testimony to his remarkable powers of observation and his steady hand.

The four principles of Karl Ernst von Baer

Karl Ernst von Baer extended Pander's studies of the chick embryo. He discovered the **notochord**, the rod of dorsalmost mesoderm that separates the embryo into right and left halves and which instructs the ectoderm above it to become the nervous system (Figure 1.4). He also discovered the mammalian

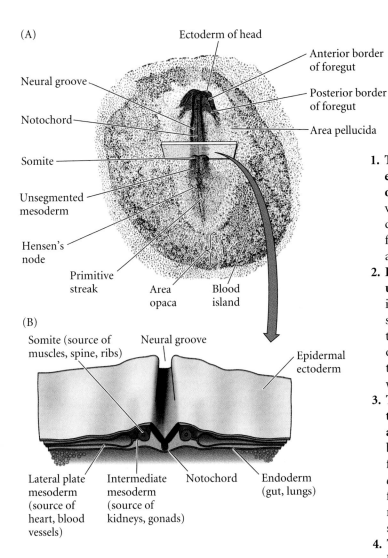

(A)

Ectoderm of head

Anterior border of foregut

Neural groove

Posterior border of foregut

Notochord

Area pellucida

Somite

Unsegmented mesoderm

Hensen's node

Primitive streak

Area opaca

Blood island

(B)

Somite (source of muscles, spine, ribs)

Neural groove

Epidermal ectoderm

Lateral plate mesoderm (source of heart, blood vessels)

Intermediate mesoderm (source of kidneys, gonads)

Notochord

Endoderm (gut, lungs)

Figure 1.4
The notochord in the chick embryo. (A) Dorsal view of the 24-hour chick embryo. (B) A cross-section through the trunk shows the notochord and developing neural tube. By comparing Figures 1.2 and 1.4, you should see the remarkable changes between days 1, 2, and 3 of chick egg incubation. (A after Patten 1951.)

1. **The general features of a large group of animals appear earlier in development than do the specialized features of a smaller group.** All developing vertebrates appear very similar shortly after gastrulation. It is only later in development that the special features of class, order, and finally species emerge. All vertebrate embryos have gill arches, notochords, spinal cords, and primitive kidneys.

2. **Less general characters develop from the more general, until finally the most specialized appear.** All vertebrates initially have the same type of skin. Only later does the skin develop fish scales, reptilian scales, bird feathers, or the hair, claws, and nails of mammals. Similarly, the early development of the limb is essentially the same in all vertebrates. Only later do the differences between legs, wings, and arms become apparent.

3. **The embryo of a given species, instead of passing through the adult stages of lower animals, departs more and more from them.**[†] The visceral clefts of embryonic birds and mammals do not resemble the gill slits of adult fish in detail. Rather, they resemble the visceral clefts of *embryonic* fish and other *embryonic* vertebrates. Whereas fish preserve and elaborate these clefts into true gill slits, mammals convert them into structures such as the eustachian tubes (between the ear and mouth).

4. **Therefore, the early embryo of a higher animal is never like a lower animal, but only like its early embryo.** Human embryos never pass through a stage equivalent to an adult fish or bird. Rather, human embryos initially share characteristics in common with fish and avian embryos. Later, the mammalian and other embryos diverge, none of them passing through the stages of the others.

Von Baer also recognized that there is a common pattern to all vertebrate development: the three germ layers give rise to different organs, and this derivation of the organs is constant whether the organism is a fish, a frog, or a chick.

WEBSITE 1.1 The reception of von Baer's principles.
The acceptance of von Baer's principles and their interpretation over the past hundred years has varied enormously. Recent evidence suggests that one important researcher in the 1800s even fabricated data when his own theory went against these postulates.

egg, that long-sought cell that everyone believed existed but no one had yet seen.[*]

In 1828, von Baer reported, "I have two small embryos preserved in alcohol, that I forgot to label. At present I am unable to determine the genus to which they belong. They may be lizards, small birds, or even mammals." Figure 1.5 allows us to appreciate his quandary. All vertebrate embryos (fish, reptiles, amphibians, birds, and mammals) begin with a basically similar structure. From his detailed study of chick development and his comparison of chick embryos with the embryos of other vertebrates, von Baer derived four generalizations. Now often referred to as "von Baer's laws," they are stated here with some vertebrate examples.

---

[*]von Baer could hardly believe that he had at last found it when so many others—Harvey, de Graaf, von Haller, Prevost, Dumas, and even Purkinje—had failed. "I recoiled as if struck by lightening … I had to try to relax a while before I could work up enough courage to look again, as I was afraid I had been deluded by a phantom. Is it not strange that a sight which is expected, and indeed hoped for, should be frightening when it eventually materializes?"

[†]von Baer formulated these generalizations prior to Darwin's theory of evolution. "Lower animals" would be those appearing earlier in life's history.

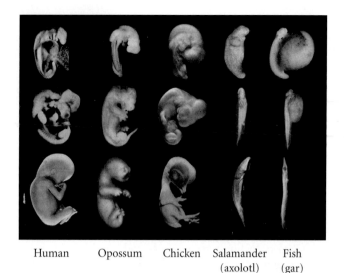

Human   Opossum   Chicken   Salamander   Fish
                              (axolotl)   (gar)

Figure 1.5
The similarities and differences among different vertebrate embryos as they proceed through development. They each begin with a basically similar structure, although they acquire this structure at different ages and sizes. As they develop, they become less like each other. (Adapted from Richardson et al. 1998; photograph courtesy of M. Richardson.)

Fate mapping the embryo

By the late 1800s, the cell had been conclusively demonstrated to be the basis for anatomy and physiology. Embryologists, too, began to base their field on the cell. One of the most important programs of descriptive embryology became the tracing of **cell lineages**: following individual cells to see what they become. In many organisms, this fine a resolution is not possible, but one can label groups of cells to see what that area of the embryo will become. By bringing such studies together, one can construct a **fate map**. These diagrams "map" the larval or adult structure onto the region of the embryo from which it arose. Fate maps are the bases for experimental embryology, since they provide researchers with information on which portions of the embryo normally become which larval or adult structures. Fate maps of some embryos at the early gastrula stage are shown in Figure 1.6. Fate maps have been generated in several ways.

OBSERVING LIVING EMBRYOS. In certain invertebrates, the embryos are transparent, have relatively few cells, and the daughter cells remain close to one another. In such cases, it is actually possible to look through the microscope and trace the descendants of a particular cell into the organs they generate. This type of study was performed about a century ago by Edwin G. Conklin. In one of these studies, he took eggs of the tunicate *Styela partita*, a sea squirt that resides in the waters off the coast of Massachusetts, and patiently followed the fates of every cell in the embryo until each differentiated into par-

ticular structures (Figure 1.7; Conklin 1905). He was helped in this endeavor by the peculiarity of the *Styela* egg, wherein the different cells contain different pigments. For example, the muscle-forming cells always had a yellow color. Conklin's fate map was confirmed by cell removal experiments. Removal of the B4.1 cell (which according to the map should produce all the tail musculature), for example, resulted in the absence of tail muscles (Reverberi and Minganti 1946).

> **WEBSITE 1.2 Conklin's art and science.** The plates from Conklin's remarkable 1905 paper are online. Looking at them, one can see the precision of his observations and how he constructed his fate map of the tunicate embryo.

> **VADE MECUM[2] The compound microscope.** The compound microscope has been the critical tool of developmental anatomists. Mastery of microscopic techniques allows one to enter an entire world of form and pattern. **[Click on Microscope]**

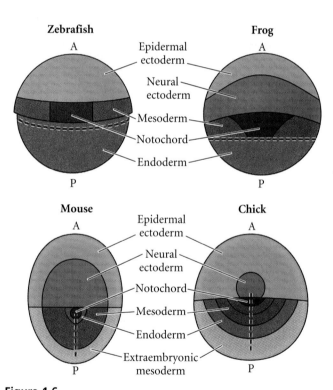

Figure 1.6
Fate maps of different vertebrate classes at the early gastrula stage. All are dorsal surface views (looking "down" on the embryo at what will become its back). Despite the different appearances of these adult animals, their fate maps show numerous similarities among the embryos. The cells that will form the notochord occupy a central dorsal position, while the precursors of the neural system lie immediately anterior to it. The neural ectoderm is surrounded by less dorsal ectoderm, which will form the epidermis of the skin. A indicates the anterior end of the embryo, P the posterior end. The dashed green lines indicate the site of ingression—the path cells will follow as they migrate from the exterior to the interior of the embryo.

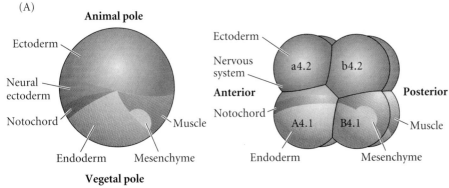

(A)

(B)

Figure 1.7
Fate map of the tunicate embryo. (A) The 1-cell embryo (left), shown shortly before the first cell division, with the fate of the cytoplasmic regions indicated. The 8-cell embryo on the right shows these regions after three cell divisions. (B) A linear version of the fate map, showing the fates of each cell of the embryo. Throughout this book, we will be using the color conventions of developmental anatomy: blue for ectoderm, red for mesoderm, and yellow for endoderm. (A after Nishida 1987 and Reverberi and Minganti 1946; B after Conklin 1905 and Nishida 1987.)

VITAL DYE MARKING. Most embryos are not so accommodating as to have cells of different colors. Nor do all embryos have as few cells as tunicates. In the early years of the twentieth century, Vogt (1929) traced the fates of different areas of amphibian eggs by applying vital dyes to the region of interest. Vital dyes will stain cells but not kill them. He mixed the dye with agar and spread the agar on a microscope slide to dry. The ends of the dyed agar would be very thin. He cut chips from these ends and placed them onto a frog embryo. After the dye stained the cells, the agar chip was removed, and cell movements within the embryo could be followed (Figure 1.8).

RADIOACTIVE LABELING AND FLUORESCENT DYES. A variant of the dye-marking technique is to make one area of the embryo radioactive. To do this, a donor embryo is usually grown in a solution containing radioactive thymidine. This base will be incorporated into the DNA of the dividing embryo. A second embryo (the host embryo) is grown under normal conditions. The region of interest is cut out from the host embryo and replaced by a radioactive graft from the donor embryo. After some time, the host embryo is sectioned for microscopy. The cells that are radioactive will be the descendants of the cells of the graft, and can be distinguished by **autoradiography**. Fixed microscope slides containing the sectioned tissues are dipped into photographic emulsion. The high-energy

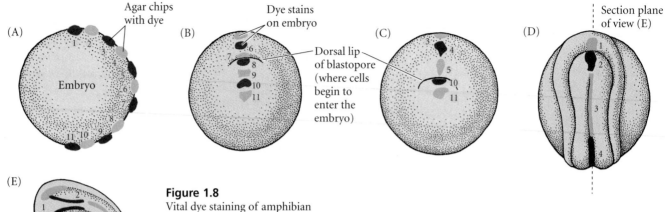

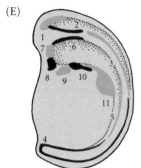

Figure 1.8
Vital dye staining of amphibian embryos. (A) Vogt's method for marking specific cells of the embryonic surface with vital dyes. (B–D) Dorsal surface views of stain on successively later embryos. (E) Newt embryo dissected in a medial sagittal section to show the stained cells in the interior. (After Vogt 1929.)

electrons from the radioactive thymidine will reduce the silver ions in the emulsion (just as light would). The result is a cluster of dark silver grains directly above the radioactive region. In this manner, the fates of different regions of the chick embryo have been determined (Rosenquist 1966).

One of the problems with vital dyes and radioactive labels is that as they become diluted at each cell division, they become difficult to detect. One way around this problem is the use of fluorescent dyes that are so intense that once injected into individual cells, they can still be detected in the progeny of these cells many divisions later. Fluorescein-conjugated dextran, for example, can be injected into a single cell of an early embryo, and the descendants of that cell can be seen by examining the embryo under ultraviolet light (Figure 1.9). More recently, *diI*, a powerfully fluorescent molecule that becomes incorporated into lipid membranes, has also been used to follow the fates of cells and their progeny.

GENETIC MARKING.   One permanent way of marking cells and following their fates is to create mosaic embryos having different genetic constitutions. One of the best examples of

this technique is the construction of **chimeric embryos**, consisting, for example, of a graft of quail cells inside a chick embryo. Chick and quail develop in a very similar manner (especially during early embryonic development), and a graft of quail cells will become integrated into a chick embryo and participate in the construction of the various organs. The substitution of quail cells for chick cells can be performed on an embryo while it is still inside the egg, and the chick that

Figure 1.9
Fate mapping using a fluorescent dye. (A) Specific cells of a zebrafish embryo were injected with a fluorescent dye that will not diffuse from the cells. The dye was then activated by laser in a small region (about 5 cells) of the late cleavage stage embryo. (B) After formation of the central nervous system had begun, cells that expressed the active dye were visualized by fluorescent light. The fluorescent dye is seen in particular cells that generate the forebrain and midbrain. (C) Fate map of the zebrafish central nervous system. Dye was injected into cells 6 hours after fertilization (left), and the results are color-coded onto the hatched fish (right). Overlapping colors indicate that cells from these regions of the 6-hour embryo contribute to two or more regions. (A, B from Kozlowski et al. 1998; photographs courtesy of E. Weinberg. C after Woo and Fraser 1995.)

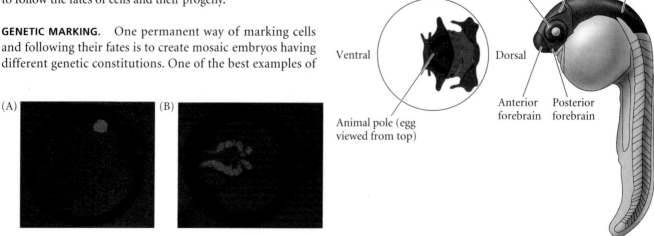

hatches will have quail cells in particular sites, depending upon where the graft was placed. The quail cells differ from the chick's in two important ways. First, the quail heterochromatin in the nucleus is concentrated around the nucleoli, making the quail nucleus easily distinguishable from chick nuclei. Second, there are cell-specific antigens that are quail-specific and can be used to find individual quail cells, even if they are in a large population of chick cells. In this way, fine-structure maps of the chick brain and skeletal system have been produced (Figure 1.10; Le Douarin 1969; Le Douarin and Teillet 1973).

WEBSITE 1.3 Dr. Nicole Le Douarin and chick-quail chimeras. We are fortunate to present here a movie made by Dr. Le Douarin of her chick-quail grafts. You will be able to see how these grafts are actually done.

VADE MECUM[2] Histotechniques. Most cells must be stained in order to see them; different dyes stain different types of molecules. Instructions on staining cells to observe particular structures (such as the nucleus) are given here. **[Click on Histotechniques]**

Cell migration

One of the most important contributions of fate maps has been their demonstration of extensive cell migration during development. Mary Rawles (1940) showed that the pigment cells (**melanocytes**) of the chick originate in the **neural crest**, a transient band of cells that joins the neural tube to the epidermis. When she transplanted small regions of neural crest-containing tissue from a pigmented strain of chickens into a similar position in an embryo from an unpigmented strain of chickens, the migrating pigment cells entered the epidermis and later entered the feathers (Figure 1.11A). Ris (1941) used similar techniques to show that while almost all of the external pigment of the chick embryo came from the migrating neural crest cells, the pigment of the retina formed in the retina itself and was not dependent on the migrating neural crest cells. By using radioactive marking techniques, Weston (1963) demonstrated that the migrating neural crest cells gave rise to the melanocytes, and also to the peripheral neurons and the epinephrine-secreting adrenal medulla (Figure 1.11B,C). This pattern was confirmed in chick-quail hybrids, in which the quail neural crest cells produced their own pigment and pattern in the chick feathers. More recently, fluorescent dye labeling has followed the movements of individual neural crest cells as they form their pigment, adrenal, and neuronal lineages (see Chapter 13).

In addition to the travels of pigment cells, other wide-scale migrations include those of the primordial germ cells (which migrate from the yolky cells to the gonads and form the sperm and eggs) and the blood cell precursors (which in vertebrates undergo several migrations to colonize the liver and bone marrow).

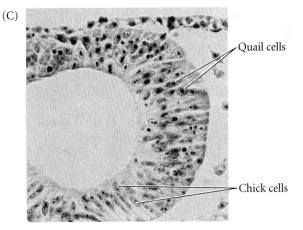

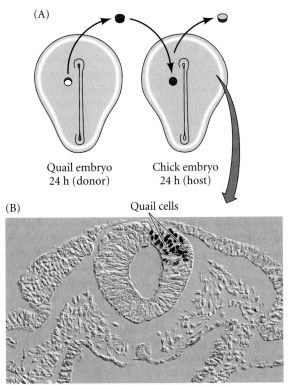

(A)

Quail embryo 24 h (donor)

Chick embryo 24 h (host)

(B)

Quail cells

Chick embryo with region of quail cells on the neural tube

(C)

Quail cells

Chick cells

Figure 1.10
Genetic markers as cell lineage tracers. (A) Grafting experiment wherein the cells from a particular region of a 1-day quail embryo have been placed into a similar region of a 1-day chick embryo. (B) After several days, the quail cells can be seen by using an antibody to quail-specific proteins. This region of the 3-day embryo produces cells that populate the neural tube. (C) Chick and quail cells can also be distinguished by the heterochromatin of their nuclei. Results of an experiment similar to (A) but using the heterochromatin of the nuclei as markers. The quail cells have a single large nucleolus (staining purple), distinguishing them from the diffuse nuclei of the chick. (After Darnell and Schoenwolf 1997; photographs courtesy of the authors.)

(A)

(B)

(C)

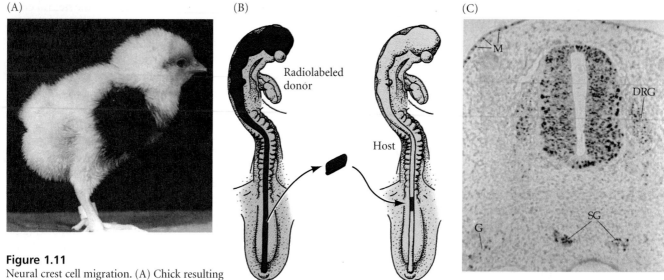

Radiolabeled donor

Host

M

DRG

G

SG

Figure 1.11
Neural crest cell migration. (A) Chick resulting from the transplantation of a trunk neural crest region from an embryo of a pigmented strain of chickens into the same region of an embryo of an unpigmented strain. The neural crest cells that gave rise to the pigment migrated into the wing epidermis and feathers. (B) Technique for following neural crest cells using radioactive tissue. (C) Autoradiograph showing locations of neural crest cells that have migrated from the radioactive donor cells. These cells form the pigment-forming melanocytes (M), sympathetic neural ganglia (SG), dorsal root ganglia (DRG), and glial cells (G). (A, original photograph from the archives of B. H. Willier; B after Weston 1963; C courtesy of J. Weston.)

Evolutionary Embryology

Charles Darwin's theory of evolution restructured comparative embryology and gave it a new focus. After reading Johannes Müller's summary of von Baer's laws in 1842, Darwin saw that embryonic resemblances would be a very strong argument in favor of the genetic connectedness of different animal groups. "Community of embryonic structure reveals community of descent," he would conclude in *On the Origin of Species* in 1859.

Even before Darwin, larval forms had been used for taxonomic classification. J. V. Thompson, for instance, had demonstrated that larval barnacles were almost identical to larval crabs, and he therefore counted barnacles as arthropods, not molluscs (Figure 1.12; Winsor 1969). Darwin, an expert on barnacle taxonomy, celebrated this finding: "Even the illustrious Cuvier did not perceive that a barnacle is a crustacean, but a glance at the larva shows this in an unmistakable manner." Darwin's evolutionary interpretation of von Baer's laws established a paradigm that was to be followed for many decades, namely, that relationships between groups can be discovered by finding common embryonic or larval forms. Kowalevsky (1871) would make a similar type of discovery (publicized in

Darwin's *The Descent of Man*, 1874) that tunicate larvae have notochords and pharyngeal pouches, and that they form their neural tubes and other organs in a manner very similar to that of the primitive chordate *Amphioxus*. The tunicates, another enigma of classification schemes (formerly placed, along with barnacles, among the molluscs), thereby found a home with the chordates.

Darwin also noted that embryonic organisms sometimes make structures that are inappropriate for their adult form but which show their relatedness to other animals. He pointed out the existence of eyes in embryonic moles, pelvic rudiments in embryonic snakes, and teeth in embryonic baleen whales.

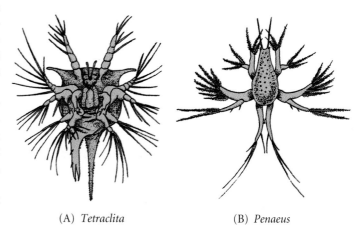

(A) *Tetraclita*

(B) *Penaeus*

Figure 1.12
Nauplius larvae of (A) a barnacle (*Tetraclita*, seen in ventral view) and (B) a shrimp (*Penaeus*, seen in dorsal view). The shrimp and barnacle share a similar larval stage despite their radical divergence in later development. (After Müller 1864.)

Darwin also argued that adaptations that depart from the "type" and allow an organism to survive in its particular environment develop late in the embryo.* He noted that the differences between species within genera become greater as development persists, as predicted by von Baer's laws. Thus, Darwin recognized two ways of looking at "descent with modification." One could emphasize the common descent by pointing out embryonic similarities between two or more groups of animals, or one could emphasize the modifications by showing how development was altered to produce structures that enabled animals to adapt to particular conditions.

Embryonic homologies

One of the most important distinctions made by the evolutionary embryologists was the difference between analogy and homology. Both terms refer to structures that appear to be similar. **Homologous** structures are those organs whose underlying similarity arises from their being derived from a common ancestral structure. For example, the wing of a bird and the forelimb of a human are homologous. Moreover, their respective parts are homologous (Figure 1.13). **Analogous** structures are those whose similarity comes from their performing a similar function, rather than their arising from a common ancestor. For example, the wing of a butterfly and the wing of a bird are analogous; the two types of wings share a common function (and therefore both are called wings), but the bird wing and insect wing did not arise from a common original ancestral structure that became modified through evolution into bird wings and butterfly wings.

Homologies must be made carefully and must always refer to the level of organization being compared. For instance, the bird wing and the bat wing are homologous as forelimbs, but not as wings. In other words, they share a common underlying structure of forelimb bones because birds and mammals share a common ancestry. However, the bird wing developed independently from the bat wing. Bats descended from a long line of nonwinged mammals, and the structure of the bat wing is markedly different from that of a bird's wing.

One of the most celebrated cases of embryonic homology is that of the fish gill cartilage, the reptilian jaw, and the mammalian middle ear (reviewed in Gould 1990). In all jawed vertebrates, including fish, the first pharyngeal arch generates the jaw apparatus. The neural crest cells of this arch migrate to form Meckel's cartilage, the precurser of the jaw (see Figure 1.3). In amphibians, reptiles, and birds, the posterior portion of this cartilage forms the quadrate bone of the upper jaw and the articular bone of the lower jaw. These

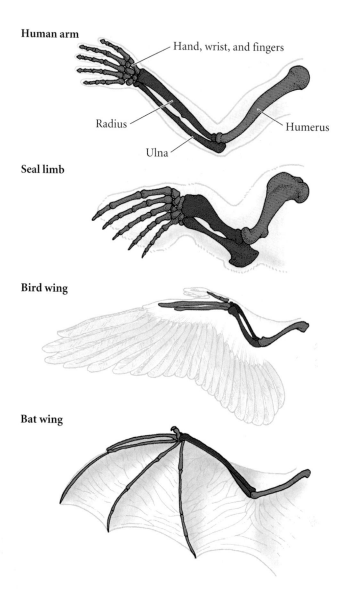

Human arm — Hand, wrist, and fingers; Radius; Ulna; Humerus

Seal limb

Bird wing

Bat wing

Figure 1.13
Homologies of structure among a human arm, a seal forelimb, a bird wing, and a bat wing; homologous supporting structures are shown in the same color. All four are homologous as forelimbs and were derived from a common tetrapod ancestor. The adaptations of bird and bat forelimbs to flight, however, evolved independently of each other, after the two lineages diverged from their common ancestor. Therefore, as wings they are not homologous, but analogous.

bones connect to each other and are responsible for articulating the upper and lower jaws. However, in mammals, this articulation occurs at another region (the dentary and squamosal bones), thereby "freeing" these bony elements to acquire new functions.

The quadrate bone of the reptilian upper jaw evolved into the mammalian incus bone of the middle ear, and the articular bone of the reptile's lower jaw has become our malleus

---

*Moreover, as first noted by Weismann (1875), larvae must have their own adaptations to help them survive. The adult viceroy butterfly mimics the monarch butterfly, but the viceroy caterpillar does not resemble the beautiful larva of the monarch. Rather, the viceroy larva escapes detection by resembling bird droppings (Begon et al. 1986).

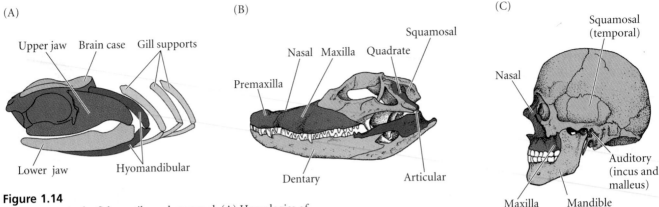

Figure 1.14
Jaw structure in the fish, reptile, and mammal. (A) Homologies of the jaws and gill arches as seen in the skull of the Paleozoic shark *Cobeledus aculentes*. (B) Lateral view of an alligator skull. The articular portion of the lower jaw articulates with the quadrate bone of the skull. (C) Lateral view of a human skull, showing the junction of the lower jaw with the squamosal (temporal) region of the skull. In mammals, the quadrate becomes internalized to form the incus of the middle ear. The articular bone retains its contact with the quadrate, becoming the malleus of the middle ear. (A after Zangerl and Williams 1975.)

(Goodrich 1930; Wang et al. 2001). This latter process was first described by Reichert in 1837, when he observed in the pig embryo that the mandible (jawbone) ossifies on the side of Meckel's cartilage, while the posterior region of Meckel's cartilage ossifies, detaches from the rest of the cartilage, and enters the region of the middle ear to become the malleus (Figure 1.14).

But the story does not end here. The upper portion of the second embryonic arch supporting the gill became the hyomandibular bone of jawed fishes. This element supports the skull and links the jaw to the cranium (Figure 1.14A). As vertebrates moved onto land, they had a new problem: how to hear in a medium as thin as air. The hyomandibular bone happens to be near the otic (ear) capsule, and bony material is excellent for transmitting sound. Thus, while still functioning as a cranial brace, the hyomandibular bone of the first amphibians also began functioning as a sound transducer (Clack 1989). As the terrestrial vertebrates altered their locomotion, jaw structure, and posture, the cranium became firmly attached to the rest of the skull and did not need the hyomandibular brace. The hyomandibular bone then seems to have become specialized into the stapes bone of the middle ear. What had been this bone's secondary function became its primary function.

Thus, the middle ear bones of the mammal are homologous to the posterior lower jaw of the reptile and to the gill arches of agnathan fishes. Chapter 22 will detail more recent information concerning the relationship of development to evolution.

Medical Embryology and Teratology

While embryologists could look at embryos to describe the evolution of life and how different animals form their organs, physicians became interested in embryos for more practical reasons. Between 2 and 5 percent of human infants are born with a readily observable anatomical abnormality (Thorogood 1997). These abnormalities may include missing limbs, missing or extra digits, cleft palate, eyes that lack certain parts, hearts that lack valves, and so forth. Physicians need know the causes of these birth defects in order to counsel parents as to the risk of having another malformed infant. In addition, the study of birth defects can tell us how the human body is normally formed. In the absence of experimental data on human embryos, we often must rely on nature's "experiments" to learn how the human body becomes organized.* Some birth defects are produced by mutant genes or chromosomes, and some are produced by environmental factors that impede development.

Abnormalities caused by genetic events (gene mutations, chromosomal aneuploidies and translocations) are called **malformations**. Malformations often appear as **syndromes** (from the Greek, "running together"), in which several abnormalities occur concurrently. For instance, a human malformation called piebaldism, shown in Figure 1.15A, is due to a dominant mutation in a gene (*KIT*) on the long arm of chromosome 4 (Spritz et al. 1992). The piebald syndrome includes anemia, sterility, unpigmented regions of the skin and hair, deafness, and the absence of the nerves that cause peristalsis in the gut. The common feature underlying these conditions is that the *KIT* gene encodes a protein that is expressed in the

---

*The word "monster," used frequently in textbooks prior to the mid-twentieth century to describe malformed infants, comes from the Latin *monstrare*, "to show or point out." This is also the root of the English word "demonstrate." It was realized by Meckel (of jaw cartilage fame) that syndromes of congenital anomalies demonstrated certain principles about normal development. Parts of the body that were affected together must have some common developmental origin or mechanism that was being affected.

(A)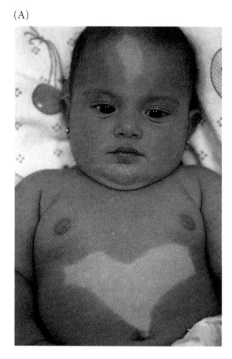

(B)

Figure 1.15
Developmental anomalies caused by genetic mutation. (A) Piebaldism in a human infant. This genetically produced condition results in sterility, anemia and underpigmented regions of the skin and hair, along with defective development of gut neurons and the ear. Piebaldism is caused by a mutation in the *KIT* gene. The Kit protein is essential for the proliferation and migration of neural crest cells, germ cell precursors, and blood cell precursors. (B) A piebald mouse with a mutation of the *Kit* gene. Mice provide important models for studying human developmental diseases. (Photographs courtesy of R. A. Fleischman.)

neural crest cells and in the precursors of blood cells and germ cells. The Kit protein enables these cells to proliferate. Without this protein, the neural crest cells—which generate the pigment cells, certain ear cells, and the gut neurons—do not multiply as extensively as they should (resulting in underpigmentation, deafness, and gut malformations), nor do the precursors of the blood cells (resulting in anemia) or the germ cells (resulting in sterility).

Developmental biologists and clinical geneticists often study human syndromes (and determine their causes) by studying animals that display the same syndrome. These are called **animal models** of the disease; the mouse model for piebaldism is shown in Figure 1.15B. It has a phenotype very similar to that of the human condition, and it is caused by a mutation in the *Kit* gene of the mouse.*

Abnormalities caused by exogenous agents (certain chemicals or viruses, radiation, or hyperthermia) are called **disruptions**. The agents responsible for these disruptions are called **teratogens** (Greek, "monster-formers"), and the study of how environmental agents disrupt normal development is called **teratology**. Teratogens were brought to the attention of the public in the early 1960s. In 1961, Lenz and McBride inde-

pendently accumulated evidence that the drug thalidomide, prescribed as a mild sedative to many pregnant women, caused an enormous increase in a previously rare syndrome of congenital anomalies. The most noticeable of these anomalies was phocomelia, a condition in which the long bones of the limbs are deficient or absent (Figure 1.16A). Over 7000 affected infants were born to women who took thalidomide, and a woman need only have taken one tablet to produce children with all four limbs deformed (Lenz 1962, 1966; Toms 1962). Other abnormalities induced by the ingestion of this drug included heart defects, absence of the external ears, and malformed intestines.

Nowack (1965) documented the period of susceptibility during which thalidomide caused these abnormalities. The drug was found to be teratogenic only during days 34–50 after the last menstruation (20–36 days postconception). The specificity of thalidomide action is shown in Figure 1.16B. From day 34 to day 38, no limb abnormalities are seen. During this period, thalidomide can cause the absence or deficiency of ear components. Malformations of upper limbs are seen before those of the lower limbs, since the arms form slightly before the legs during development. The only animal models for thalidomide, however, are primates, and we still do not know for certain the mechanisms by which this drug causes human developmental disruptions (although it seems to work by blocking certain molecules from the developing mesoderm). Thalidomide was withdrawn from the market in November of 1961, but it is beginning to be prescribed again (although not to pregnant women), as a potential anti-tumor and anti-autoimmunity drug (see Chapter 21; Raje and Anderson 1999).

---

*The mouse *Kit* and human *KIT* genes are considered homologous by their structural similarities and their presumed common ancestry. Human genes are usually italicized and written in all capitals. Mouse genes are italicized, but only the first letter is usually capitalized. Gene products—proteins—are not italicized. If the protein has no standard biochemical or physiological name, it is usually represented with the name of the gene in Roman type, with the first letter capitalized. These rules are frequently bent, however. One is reminded of Cohen's (1982) dictum that "Academicians are more likely to share each other's toothbrush than each other's nomenclature."

(A)

(B)

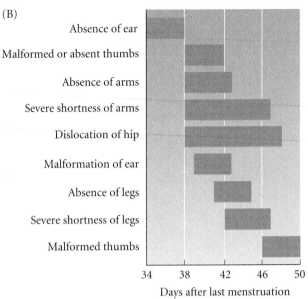

Figure 1.16
Developmental anomalies caused by an environmental agent. (A) Phocomelia, the lack of proper limb development, was the most visible of the birth defects that occurred in many children whose mothers took the drug thalidomide during pregnancy. (B) Thalidomide disrupts different structures at different times of human development. (Photograph © Deutsche Presse/Archive Photos; B after Nowack 1965.)

The integration of anatomical information about congenital malformations with our new knowledge of the genes responsible for development has had a revolutionary effect and is currently restructuring medicine. This integration is allowing us to discover the genes responsible for inherited malformations, and it permits us to identify the steps in development that are being disrupted by teratogens. We will see examples of this integration throughout this text, and Chapter 21 will detail some of the remarkable new discoveries in human teratology.

Mathematical Modeling of Development

Developmental biology has been described as the last refuge of the mathematically incompetent scientist. This phenomenon, however, is not going to last. While most embryologists have been content trying to analyze specific instances of development or even formulating general principles of embryology, some researchers are now seeking quantifiable laws of development. The goal of these investigators is to base embryology on formal mathematical or physical principles (see Held 1992; Webster and Goodwin 1996; Salazar-Ciudad et al. 2000, 2001). Pattern formation and growth are two areas in which such mathematical modeling has given biologists insights into some underlying laws of animal development.

The mathematics of organismal growth

Most animals grow by increasing their volume while retaining their proportions. Theoretically, an animal that increases its weight (volume) twofold will increase its length only 1.26 times (i.e., $1.26^3 = 2$). W. K. Brooks (1886) observed that this ratio was frequently seen in nature, and he noted that the deep-sea arthropods collected by the *Challenger* expedition increased about 1.25 times between molts. In 1904, Przibram and his colleagues performed a detailed study of mantises and found that the increase of size between molts was almost exactly 1.26 (see Przibram 1931). Even the hexagonal facets of the arthropod eye (which grow by cell expansion, not by cell division) increased by that ratio.

D'Arcy Thompson (1942) similarly showed that the spiral growth of shells (and fingernails) can be expressed mathematically as $r = a^\theta$, and that the ratio of the widths between two whorls of a shell can be calculated by the formula $r = e^{2\pi\cot\theta}$ (Figure 1.17; Table 1.1). Thus, if a whorl were 1 inch in breadth at one point on a radius and the angle of the spiral were 80°, the next whorl would have a width of 3 inches on the same radius. Most gastropod (snail) and nautiloid molluscs have an angle of curvature between 80° and 85°* Lower-angle curvatures are seen in some shells (mostly bivalves) and are common in teeth and claws.

Such growth, in which the shape is preserved because all components grow at the same rate, is called **isometric**

---

*If the angle were 90°, the shell would form a circle rather than a spiral, and growth would cease. If the angle were 60°, however, the next whorl would be 4 feet on that radius, and if the angle were 17°, the next whorl would occupy a distance of some 15,000 miles! Interestingly, the laboratory technician in Przibram's mantis studies was Alma Mahler, the academic muse of *fin de siecle* Vienna.

(B)

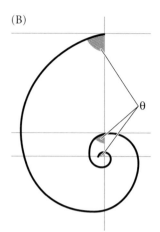

TABLE 1.1 Constant angle of an equiangular spiral and the ratio of widths between whorls

Constant angle	Ratio of widths[a]
90°	1.0
89°8′	1.1
86°18′	1.5
83°42′	2.0
80°5′	3.0
75°38′	5.0
69°53′	10.0
64°31′	20.0
58°5′	50.0
53°46′	10^2
42°17′	10^3
34°19′	10^4
28°37′	10^5
24°28′	10^6

Source: From Thompson 1942.

[a]The ratio of widths is calculated by dividing the width of one whorl by the width of the next larger whorl.

Figure 1.17
Equiangular spiral growth patterns. (A) A ram's horn and the shell of a chambered nautilus both show equiangular spiral growth. The nautilus shell is cut in cross section. (B) René Descartes' analysis of an equiangular spiral, showing that if the curve cuts each radius vector at a constant angle (symbolized θ), then the curve grows continuously without ever changing its shape. (A from the author's collection; B after Thompson 1942.)

growth. In many organisms, however, growth is not a uniform phenomenon. It is obvious that there are some periods in an organism's life during which growth is more rapid than in others. Physical growth during the first 10 years of a person's existence is much more dramatic than in the 10 years following one's graduation from college. Moreover, not all parts of the body grow at the same rate. This phenomenon of the different growth rates of parts within the same organism is called **allometric growth** (or **allometry**). Human allometry is depicted in Figure 1.18. Our arms and legs grow at a faster rate than our torso and head, such that adult proportions differ markedly from those of infants. Julian Huxley (1932) likened allometry to putting money in the bank at two different continuous interest rates.

The formula for allometric growth (or for comparing moneys invested at two different interest rates) is $y = bx^{a/c}$, where a and c are the growth rates of the two body parts, and b is the value of y when $x = 1$. If $a/c > 1$, then that part of the body represented by a is growing faster than that part of the body represented by c. In logarithmic terms (which are much easier to graph), $\log y = \log b + (a/c)\log x$.

One of the most vivid examples of allometric growth is seen in the male fiddler crab, *Uca pugnax*. In small males, the two claws are of equal weight, each constituting about 8% of

the crab's total weight. As the crab grows larger, its chela (the large crushing claw) grows even more rapidly, eventually constituting about 38% of the crab's weight (Figure 1.19). When these data are plotted on double logarithmic plots (with the body mass on the x axis and the chela mass on the y axis), one obtains a straight line whose slope is the a/c ratio. In the male *Uca pugnax* (whose name is derived from the huge claw), the a/c ratio is 6:1. This means that the mass of the chela increases 6 times faster than the mass of the rest of the body. In females of the species, the claw remains about 8% of the body weight throughout growth. It is only in the males (who use the claw for defense and display) that this allometry occurs.

Recent models of isometric and allometric growth have taken into account metabolic rates, life history changes, and cell death rates (see West et al. 2001). The relationship of growth to physical and genetic parameters and the coordination of growth rates throughout the organism remains a fascinating area that unifies development with physiology and medicine.

The mathematics of patterning

One of the most important mathematical models in developmental biology was formulated by Alan Turing (1952), one of the founders of computer science (and the mathematician who cracked the German "Enigma" code during World War II). He proposed a model wherein two homogeneously distributed solutions would interact to produce stable patterns during morphogenesis. These patterns would represent regional differences in the concentrations of the two substances.

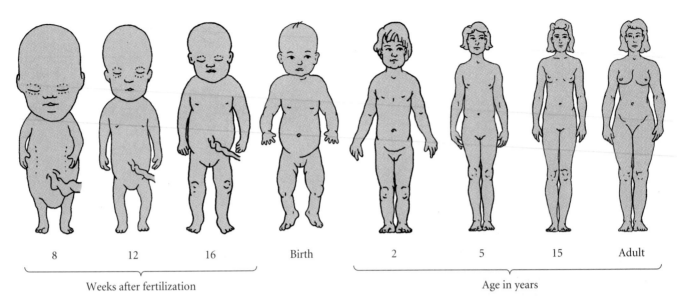

| 8 | 12 | 16 | Birth | 2 | 5 | 15 | Adult |

Weeks after fertilization

Age in years

Figure 1.18
Allometry in humans. The embryo's head is exceedingly large in proportion to the rest of the body. After the embryonic period, the head grows more slowly than the torso, hands, and legs. Human allometry has been represented in Western art only since the Renaissance. Before that, children resembled little adults. (After Moore 1983.)

Figure 1.19
Male specimens of the fiddler crab, *Uca pugnax*. Allometric growth occurs only in one of the male's claws. In females (not shown), both claws retain isometric growth. (Photograph courtesy of Swarthmore College Marine Biology laboratory.)

Their interactions would produce an ordered structure out of random chaos.

Turing's **reaction-diffusion model** involves two substances. Substance P promotes the production of more substance P as well as substance S. Substance S, however, inhibits the production of substance P. Turing's mathematics show that if S diffuses more readily than P, sharp waves of concentration differences will be generated for substance P (Figure 1.20). These waves have been observed in certain chemical reactions (Prigogine and Nicolis 1967; Winfree 1974).

The reaction-diffusion model predicts alternating areas of high and low concentrations of some substance. When the concentration of such a substance is above a certain threshold level, a cell (or group of cells) may be instructed to differentiate in a certain way. An important feature of Turing's model is that particular chemical wavelengths will be amplified while all others will be suppressed. As local concentrations of P increase, the values of S form a peak centering on the P peak, but becoming broader and shallower because of S's more rapid diffusion. These S peaks inhibit other P peaks from forming. But which of the many P peaks will survive? That depends on the size and shape of the tissues in which the oscillating reaction is occurring. (This pattern is analogous to the harmonics of vibrating strings, as in a guitar. Only certain resonance vibrations are permitted, based on the boundaries of the string.)

The mathematics describing which particular wavelengths are selected consist of complex polynomial equations. Such functions have been used to model the spiral patterning of slime molds, the polar organization of the limb, and the pigment patterns of mammals, fish, and snails (Figures 1.21 and 1.22; Kondo and Asai 1995; Meinhardt 1998). A computer simulation based on a Turing reaction-diffusion system can successfully predict such patterns, given the starting shapes and sizes of the elements involved.

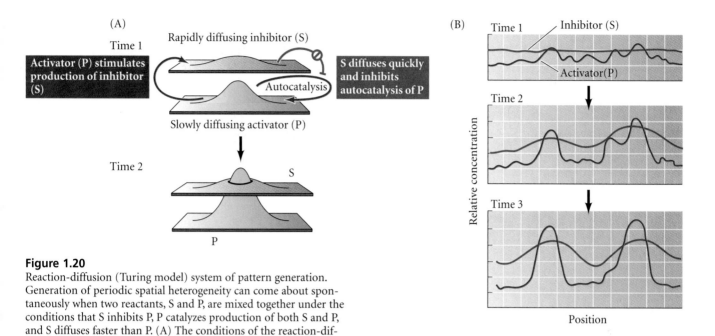

Figure 1.20
Reaction-diffusion (Turing model) system of pattern generation. Generation of periodic spatial heterogeneity can come about spontaneously when two reactants, S and P, are mixed together under the conditions that S inhibits P, P catalyzes production of both S and P, and S diffuses faster than P. (A) The conditions of the reaction-diffusion system yielding a peak of P and a lower peak of S at the same place. (B) The distribution of the reactants is initially random, and their concentrations fluctuate over a given average. As P increases locally, it produces more S, which diffuses to inhibit more peaks of P from forming in the vicinity of its production. The result is a series of P peaks ("standing waves") at regular intervals.

One way to search for the chemicals predicted by Turing's model is to find genetic mutations in which the ordered structure of a pattern has been altered. The wild-type alleles of these genes may be responsible for generating the normal pattern. Such a candidate is the *leopard* gene of zebrafish (Asai et al. 1999). Zebrafish usually have five parallel stripes along their flanks. However, in the different mutations, the stripes are broken into spots of different sizes and densities. Figure 1.22 shows fish homozygous for four different alleles of the *leopard* gene. If the *leopard* gene encodes an enzyme that catalyzes one of the reactions of the reaction-diffusion system, the different mutations of this gene may change the kinetics of synthesis or degradation. Indeed, all the mutant patterns (and those of their heterozygotes) can be computer-generated by changing a single parameter in the reaction-diffusion equation. The cloning of this gene should enable further cooperation between theoretical biology and developmental anatomy.

WEBSITE 1.4 The mathematical background of pattern formation. The equations modeling pattern formation are a series of partial derivatives depicting rates of synthesis, degradation, and diffusion of the activator and inhibitor molecules.

WEBSITE 1.5 How do zebras (and angelfish) get their stripes? No one knows for sure, but adding the Turing equations to what's known about equine embryology al-

lows one to model how each of the three known zebra species acquired its unique striping pattern. Similarly, changing some of the parameters of the Turing reaction enables one to predict the different pigmentation patterns of angelfish.

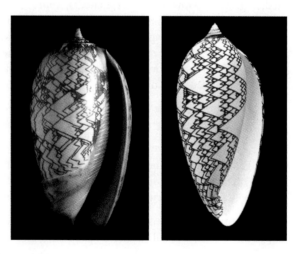

Figure 1.21
A photograph of the snail *Oliva porphyria* (left), and a computer model of the same snail (right) in which the growth parameters of the shell and its pigmentation pattern were both mathematically generated. (From Meinhardt 1998; computer image courtesy of D. Fowler, P. Prusinkiewicz, and H. Meinhardt.)

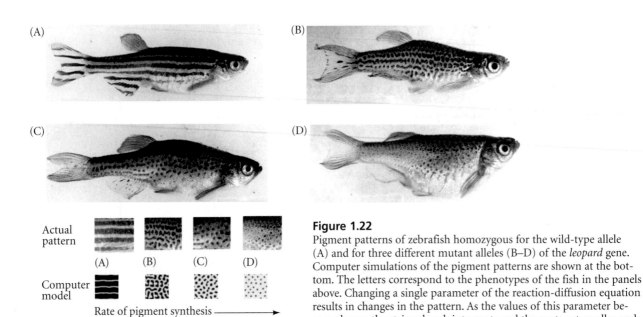

Figure 1.22
Pigment patterns of zebrafish homozygous for the wild-type allele (A) and for three different mutant alleles (B–D) of the *leopard* gene. Computer simulations of the pigment patterns are shown at the bottom. The letters correspond to the phenotypes of the fish in the panels above. Changing a single parameter of the reaction-diffusion equation results in changes in the pattern. As the values of this parameter become larger, the stripes break into spots and the spots get smaller and less dense. (From Asai et al. 1999; photographs courtesy of S. Kondo.)

Principles of Development: Developmental Anatomy

1. Organisms must function as they form their organs. They have to use one set of structures while constructing others.

2. The main question of development is, How does the egg becomes an adult? This question can be broken down into the component problems of differentiation (How do cells become different from one another and from their precursors?), morphogenesis (How is ordered form is generated?), growth (How is size regulated?), reproduction (How does one generation create another generation?), evolution (How do changes in developmental processes create new anatomical structures?), and environment (How is the developing organism affected by the physical and chemical conditions of the environment?).

3. Epigenesis happens. New organisms are created *de novo* each generation from the relatively disordered cytoplasm of the egg.

4. Preformation is not in the anatomical structures, but in the instructions to form them. The inheritance of the fertilized egg includes the genetic potentials of the organism.

5. The preformed nuclear instructions include the ability to respond to environmental stimuli in specific ways.

6. The ectoderm gives rise to the epidermis, nervous system, and pigment cells.

7. The mesoderm generates the kidneys, gonads, muscle, bones, heart, and blood cells.

8. The endoderm forms the lining of the digestive tube and the respiratory system.

9. Karl von Baer's principles state that the general features of a large group of animals appear earlier in the embryo than do the specialized features of a smaller group. As each embryo of a given species develops, it diverges from the adult forms of other species. The early embryo of a "higher" animal species is not like the adult of a "lower" animal.

10. Labeling cells with dyes shows that some cells differentiate where they form, while others migrate from their original sites and differentiate in their new locations. Migratory cells include neural crest cells and the precursors of germ cells and blood cells.

11. "Community of embryonic structure reveals community of descent" (Charles Darwin, *On the Origin of Species*).

12. Homologous structures in different species are those organs whose similarity is due to their sharing a common ancestral structure. Analogous structures are those organs whose similarity comes from their serving a similar function (but which are not derived from a common ancestral structure).

13. Congenital anomalies can be caused by genetic factors (mutations, aneuploidies, translocations) or by environmental agents (certain chemicals, certain viruses, radiation).

14. Syndromes consist of sets of developmental abnormalities that "run together."

15. Organs that are linked in developmental syndromes share either a common origin or a common mechanism of formation.

16. If growth is isometric, a twofold change in weight will cause a 1.26-fold expansion in length.

17. Allometric growth can create dramatic changes in the structure of organisms.

18. Complex patterns may be self-generated by reaction-diffusion events, wherein the activator of a local phenomenon stimulates the production of more of itself as well as the production of a more diffusible inhibitor.

Literature Cited

Aristotle. ca. 350 B.C.E. *Metaphysics*. Book 1, Part 2. D. Ross (trans.). Oxford University Press, New York, 1979.

Aristotle. ca. 350 B.C.E. *The Generation of Animals*. A. L. Peck (trans.); G. P. Goold (ed.). Harvard University Press, Cambridge, MA, 1990.

Asai, R., E. Taguchi, Y. Kume, M. Saito and S. Kondo. 1999. Zebrafish *Leopard* gene as a component of the putative reaction-diffusion system. *Mech. Dev.* 89: 87–92.

Begon, M., J. L. Harper and C. R. Townsend. 1986. *Ecology: Individuals, Populations, and Communities*. Blackwell Scientific, Oxford.

Bonnet, C. 1764. *Contemplation de la Nature*. Marc-Michel Ray, Amsterdam.

Brooks, W. K. 1886. Report in the Stomatopoda collected by *H.M.S. Challenger*. *Challenger Reports* 16: 1–114.

Carlson, B. M. 1981. *Patten's Foundations of Embryology*. McGraw-Hill, New York.

Cassirer, E. 1950. Developmental mechanics and the problem of cause in biology. In E. Cassirer (ed.), *The Problem of Knowledge*. Yale University Press, New Haven.

Churchill, A. 1991. The rise of classical descriptive embryology. In S. F. Gilbert (ed.), *A Conceptual History of Modern Embryology*, Plenum Press, New York, pp. 1–29.

Clack, J. A. 1989. Discovery of the earliest known tetrapod stapes. *Nature* 342: 425–427.

Cohen, M. M. Jr. 1982. *The Child with Multiple Birth Defects*. Raven, New York.

Conklin, E. G. 1905. The organization and cell lineage of the ascidian egg. *J. Acad. Nat. Sci. Phila.* 13: 1–119.

Darnell, D. K. and G. C. Schoenwolf. 1997. Modern techniques for labeling in avian and murine embryos. In G. P. Daston (ed.), *Molecular and Cellular Methods in Developmental Toxicology*. CRC Press, Boca Raton, FL, pp. 231–272.

Darwin, C. 1859. *On the Origin of Species by Means of Natural Selection, or the Preservation of Favoured Races in the Struggle for Life*. John Murray, London.

Darwin, C. 1874. *The Descent of Man, and Selection in Relation to Sex*. 2nd Ed. John Murray, London.

Goodrich, E. S. 1930. *Studies on the Structure and Development of Vertebrates*. Macmillan, London.

Gould, S. J. 1977. *Ontogeny and Phylogeny*. Belknap Press, Cambridge, MA.

Gould, S. J. 1990. An earful of jaw. *Nat. Hist.* 1990(3): 12–23.

Harvey, W. 1651. *Exercitationes de generatione animalium: quibus accedunt quaedum de partu, de membranis ac humoribus uteri et de conceptione*. London.

Held, L. I., Jr. 1992. *Models for Embryonic Periodicity*. Karger, New York.

Huxley, J. S. 1932. *Problems of Relative Growth*. Dial Press, New York.

Kölreuter, J. G. 1766. Vorläufige Nachricht von einigen das feschlecht der Planzen betreffenden Versuchen und Beobachtung, nebst Fortsetzungen, 1, 2, 3.

Kondo, S. and R. Asai. 1995. A reaction-diffusion wave on the skin of the marine angelfish *Pomacanthus*. *Nature* 376: 765–768.

Kowalevsky, A. 1871. Weitere Studien II. Die Entwicklung der einfachen Ascidien. *Arch. Micr. Anat.* 7: 101–130.

Kozlowski, D. J., T. Muramaki, R. K. Ho and E. S. Weinberg. 1998. Regional cell movement and tissue patterning in the zebrafish embryo revealed by fate mapping with caged fluorescein. *Biochem. Cell Biol.* 75: 551–562.

Le Douarin, N. M. 1969. Particularités du noyau interphasique chez la Caille japonaise (Coturnix coturnix japonica). Utilisation de ces particularités comme "marquage biologique" dans les recherches sur les interactions tissulaires et les migrations cellulaires au cours de l'ontogenèse. *Bull. Biol. Fr. Belg.* 103: 435–452.

Le Douarin, N. M. and M.-A. Teillet. 1973. The migration of neural crest cells to the wall of the digestive tract in avian embryo. *J. Embryol. Exp. Morphol.* 30: 31–48.

Lenoir, T. 1980. Kant, Blumenbach, and vital materialism in German biology. *Isis* 71: 77–108.

Lenz, W. 1962. Thalidomide and congenital abnormalities. *Lancet* 1: 45 (reported in a symposium in 1961.)

Lenz, W. 1966. Malformations caused by drugs in pregnancy. *Am. J. Dis. Child.* 112: 99–l06.

Lillie, F. R. 1908. *The Embryology of the Chick*. Henry Holt, New York.

Maître-Jan, A. 1722. *Observations sur la formation du poluet*. L. d'Houdry, Paris.

Malpighi, M. 1672. *De Formatione Pulli in Ovo* (London). Reprinted in H. B. Adelmannm, *Marcello Malpighi and the Evolution of Embryology*. Cornell University Press, Ithaca, NY, 1966.

McBride, W. G. 1961. Thalidomide and congenital abnormalities. *Lancet* 2: 1358.

Meinhardt, M. 1998. *The Algorhythmic Beauty of Sea Shells*. Springer, Berlin.

Moore, K. L. 1983. *The Developing Human*. 3rd Ed. Saunders, Philadelphia.

Müller, F. 1864. *Für Darwin*. Engelmann, Leipzig.

Nishida, H. 1987. Cell lineage analysis in ascidian embryos by intracellular injection of a tracer enzyme. III. Up to the tissue-restricted stage. *Dev. Biol.* 121: 526–541.

Nowack, E. 1965. Die sensible Phase bei der Thalidomide-Embryopathie. *Humangenetik* 1: 516–536.

Pander, C. 1817. *Beiträge zur Entwickelungsgeschichte des Hünchens im Eye*. Brönner, Würzburg.

Patten, B. M. 1951. *The Early Embryo of the Chick*. 4th Ed. McGraw-Hill, New York.

Pinto-Correia, C. 1997. *The Ovary of Eve*. University of Chicago Press, Chicago.

Prigogine, I. and G. Nicolis. 1967. On symmetry-breaking instabilities in dissipative systems. *J. Chem. Phys.* 46: 3542–3550.

Przibram, H. 1931. *Connecting Laws in Animal Morphology*. University of London Press, London.

Raje, N. and K. Anderson. 1999. Thalidomide: A revival story. *New Engl. J. Med.* 341: 1606–1609.

Rawles, M. E. 1940. The pigment forming potency of early chick blastoderm. *Proc. Natl. Acad. Sci. USA* 26: 86–94.

Reichert, C. B. 1837. Entwicklungsgeschichte der Gehörknöchelchen der sogenannte Meckelsche Forsatz des Hammers. *Müller's Arch. Anat. Phys. Wissensch. Med.* 177–188.

Reverberi, G. and A. Minganti. 1946. Fenomeni di evocazione nello sviluppo dell'uovo di Ascidie. Risultati dell'indagine spermentale sull'ouvo di Ascidiella aspersa e di Ascidia malaca allo stadio di 8 blastomeri. *Pubbl. Staz. Zool. Napoli* 20: 199–252.

Richardson, M. K., J. Hanken, L. Selwood, G. M. Wright, R. J. Richards, C. Pieau and A. Raynaud. 1998. Haeckel, embryos, and evolution. *Science* 280: 983–984.

Ris, H. 1941. An experimental study of the origins of melanophores in birds. *Physiol. Zool.* 14: 48–66.

Roe, S. 1981. *Matter, Life, and Generation: Eighteenth-Century Embryology and the Haller-Wolff Debate.* Cambridge University Press, Cambridge.

Rosenquist, G. C. 1966. A radioautographic study of labeled grafts in the chick blastoderm. Development from primitive streak stages to stage 12. *Contrib. Embryol. Carnegie Inst.* 38: 71–110.

Salazar-Ciudad, I., J. Garcia-Fernandez and R. V. Sole. 2000. Gene networks capable of pattern formation: from induction to reaction-diffusion. *J. Theor. Biol.* 205: 587–603.

Salazar-Ciudad, I., S. A. Newman and R. V. Sole. 2001. Phenotypic and dynamical transitions in model genetic networks. I. Emergence of patterns and genotype-phenotype relationships. *Evol. Dev.* 3: 84–94.

Spritz, R.A., S. A. Holmes, R. Ramesar, J. Greenberg, D. Curtis and P. Beighton. 1992. Mutations of the KIT (mast/stem cell growth factor receptor) proto-oncogene account for a continuous range of phenotypes in human piebaldism. *Am. J. Hum. Genet.* 51: 1058–1065.

Thompson, D. W. 1942. *On Growth and Form.* Cambridge University Press, Cambridge.

Thorogood, P. 1997. The relationship between genotype and phenotype: Some basic concepts. In P. Thorogood (ed.), *Embryos, Genes, and Birth Defects.* Wiley, New York, pp. 1–16.

Toms, D. A. 1962. Thalidomide and congenital abnormalities. *Lancet* 2: 400.

Turing, A. M. 1952. The chemical basis of morphogenesis. *Philos. Trans. R. Soc. Lond.* [B] 237: 37–72.

Vogt, W. 1929. Gestaltungsanalyse am Amphibienkeim mit örtlicher Vitalfärbung. II. Teil Gastrulation und Mesodermbildung bei Urodelen und Anuren. *Wilhelm Roux Arch. Entwicklungsmech. Org.* 120: 384–706.

von Baer, K. E. 1828. *Entwicklungsgeschichte der Thiere: Beobachtung und Reflexion.* Bornträger, Konigsberg.

Wang, Y., Y. Hu, J. Meng and C. Li. 2001. An ossified Meckel's cartilage in two Cretaceous mammals and origin of the mammalian middle ear. *Science* 294: 357–361.

Webster, G. and B. Goodwin. 1996. *Form and Transformation: Generative and Relational Principles in Biology.* Cambridge University Press, Cambridge.

Weismann, A. 1875. *Über den Saison-Dimorphismus der Schmetterlinge. In Studien zur Descendenz-Theorie.* Engelmann, Leipzig.

West, G. B., J. H. Brown and B. J. Enquist. 2001. A general model for ontogenetic growth. *Nature* 413: 628–630.

Weston, J. 1963. A radiographic analysis of the migration and localization of trunk neural crest cells in the chick. *Dev. Biol.* 6: 274–310.

Winfree, A. T. 1974. Rotating chemical reactions. *Sci. Am.* 230(6): 82–95.

Winsor, M. P. 1969. Barnacle larvae in the nineteenth century: A case study in taxonomic theory. *J. Hist. Med. Allied Sci.* 24: 294–309.

Wolff, K. F. 1767. De formatione intestinorum praecipue. *Novi Commentarii Academine Scientarum Imperialis Petropolitanae* 12: 403–507.

Woo, K. and S. E. Fraser. 1995. Order and coherence in the fate map of the zebrafish embryo. *Development* 121: 2595–2609.

Zangerl, R. and M. E. Williams. 1975. New evidence on the nature of the jaw suspension in Paleozoic anacanthus sharks. *Paleontology* 18: 333–341.

chapter 2 Life cycles and the evolution of developmental patterns

The view taken here is that the life cycle is the central unit in biology. … Evolution then becomes the alteration of life cycles through time, genetics the inheritance mechanisms between cycles, and development all the changes in structure that take place during one life cycle.

J. T. BONNER (1965)

It's the circle of life And it moves us all.

TIM RICE (1994)

TRADITIONAL WAYS OF CLASSIFYING ANIMALS catalog them according to their adult structure. But, as J. T. Bonner (1965) pointed out, this is a very artificial method, because what we consider an individual is usually just a brief slice of its life cycle. When we consider a dog, for instance, we usually picture an adult. But the dog is a "dog" from the moment of fertilization of a dog egg by a dog sperm. It remains a dog even as a senescent dying hound. Therefore, the dog is actually the entire life cycle of the animal, from fertilization through death.

The life cycle has to be adapted to its environment, which is composed of non-living objects as well as other life cycles. Take, for example, the life cycle of *Clunio marinus,* a small fly that inhabits tidal waters along the coast of western Europe. Females of this species live only 2–3 hours as adults, and they must mate and lay their eggs within this short time. To make matters even more precarious, they must lay their eggs on red algal mats that are exposed only during the lowest ebbing of the spring tide. Such low tides occur on four successive days shortly after the new and full moons (i.e., at about 15-day intervals). Therefore, the life cycle of these insects must be coordinated with the lunar cycle as well as the daily tidal rhythms such that the insects emerge from their pupal cases during the few days of the spring tide and at the correct hour for its ebb (Beck 1980; Neumann and Spindler 1991).

The Circle of Life: The Stages of Animal Development

One of the major triumphs of descriptive embryology was the idea of a generalizable life cycle. Each animal, whether earthworm, eagle, or beagle, passes through similar stages of development. The life of a new individual is initiated by the fusion of genetic material from the two gametes—the sperm and the egg. This fusion, called **fertilization**, stimulates the egg to begin development. The stages of development between fertilization and hatching are collectively called **embryogenesis**. Throughout the animal kingdom, an incredible variety of embryonic types exist, but most patterns of embryogenesis are variations on five themes:

1. Immediately following fertilization, **cleavage** occurs. Cleavage is a series of extremely rapid mitotic divisions wherein the enormous volume of zygote cytoplasm is divided into numerous smaller cells. These cells are called **blastomeres**, and by the end of cleavage, they generally form a sphere known as a **blastula**.

2. After the rate of mitotic division has slowed down, the blastomeres undergo dramatic movements wherein they change their positions relative to one another. This series of extensive cell rearrangements is called **gastrulation**, and the embryo is said to be in the **gastrula** stage. As a result of gastrulation, the embryo

25

contains three **germ layers**: the ectoderm, the endoderm, and the mesoderm.

3. Once the three germ layers are established, the cells interact with one another and rearrange themselves to produce tissues and organs. This process is called **organogenesis**. Many organs contain cells from more than one germ layer, and it is not unusual for the outside of an organ to be derived from one layer and the inside from another. For example, the outer layer of skin (epidermis) comes from the ectoderm, while the inner layer (the dermis) comes from the mesoderm. Also during organogenesis, certain cells undergo long migrations from their place of origin to their final location. These migrating cells include the precursors of blood cells, lymph cells, pigment cells, and gametes. Most of the bones of our face are derived from cells that have migrated ventrally from the dorsal region of the head.

4. In many species, a specialized portion of egg cytoplasm gives rise to cells that are the precursors of the **gametes** (sperm and egg). The gametes and their precursor cells are collectively called **germ cells**, and they are set aside for reproductive function. All the other cells of the body are called **somatic cells**. This separation of somatic cells (which give rise to the individual body) and germ cells (which contribute to the formation of a new generation) is often one of the first differentiations to occur during animal development. The germ cells eventually migrate to the gonads, where they differentiate into gametes. The development of gametes, called **gametogenesis**, is usually not completed until the organism has become physically mature. At maturity, the gametes may be released and participate in fertilization to begin a new embryo. The adult organism eventually undergoes senescence and dies.

5. In many species, the organism that hatches from the egg or is born into the world is not sexually mature. Indeed, in most animals, the young organism is a **larva** that may look significantly different from the adult. Larvae often constitute the stage of life that is used for feeding or dispersal. In many species, the larval stage is the one that lasts the longest, and the adult is a brief stage solely for reproduction. In the silkworm moths, for instance, the adults do not have mouthparts and cannot feed. The larvae must eat enough for the adult to survive and mate. Indeed, most female moths mate as soon as they eclose from their pupa, and they fly only once—to lay their eggs. Then they die.

The Frog Life Cycle

Figure 2.1 uses the development of the leopard frog, *Rana pipiens*, to show a representative life cycle. Let us look at this life cycle in a bit more detail.

In most frogs, gametogenesis and fertilization are seasonal events, because its life depends on the plants and insects in the pond where it lives and on the temperature of the air and water. A combination of photoperiod (hours of daylight) and temperature tells the pituitary gland of the female frog that it is spring. If the female is mature, her pituitary gland secretes hormones that stimulate her ovary to make estrogen. Estrogen is a hormone that can instruct the liver to make and secrete yolk proteins such as vitellogenin, which are then transported through the blood into the enlarging eggs in the ovary.* The yolk is transported into the bottom portion of the egg (Figure 2.2A). The bottom half of the egg usually contains more yolk than the top half and is called the **vegetal hemisphere** of the egg. Conversely, the upper half of the egg usually has less yolk and is called the **animal hemisphere** of the egg.†

Another ovarian hormone, progesterone, signals the egg to resume its meiotic division. This is necessary because the egg had been "frozen" in the metaphase of its first meiosis. When it has completed this first meiotic division, the egg is released from the ovary and can be fertilized. In many species, the eggs are enclosed in a jelly coat that acts to enhance their size (so they won't be as easily eaten), to protect them against bacteria, and to attract and activate sperm.

Sperm also occur on a seasonal basis. Male leopard frogs make their sperm in the summer, and by the time they begin hibernation in autumn, they have produced all the sperm that are to be available for the following spring's breeding season. In most species of frogs, fertilization is external. The male frog grabs the female's back and fertilizes the eggs as the female frog releases them (Figure 2.2B). *Rana pipiens* usually lays about 2500 eggs, while the bullfrog, *Rana catesbiana*, can lay as many as 20,000. Some species lay their eggs in pond vegetation, and the jelly adheres to the plants and anchors the eggs (Figure 2.2C). Other species float their eggs into the center of the pond without any support.

Fertilization accomplishes several things. First, it allows the egg to complete its second meiotic division, which provides the egg with a haploid **pronucleus**. The egg pronucleus and the sperm pronucleus meet in the egg cytoplasm to form the diploid zygote nucleus. Second, fertilization causes the cytoplasm of the egg to move such that different parts of the cytoplasm find themselves in new locations (Figure 2.2D). Third, fertilization activates those molecules necessary to begin cell cleavage and development (Rugh 1950). The sperm and egg die quickly unless fertilization occurs.

---

*As we will see in later chapters, there are numerous ways by which the synthesis of a new protein can be induced. Estrogen stimulates the production of vitellogenin protein in two ways. First, it uses transcriptional regulation to make new vitellogenin mRNA. Before estrogen stimulation, no vitellogenin message can be seen in the liver cells. After stimulation, there are over 50,000 vitellogenin mRNA molecules in these cells. Estrogen also uses translational regulation to stabilize these particular messages, increasing their half-life from 16 hours to 3 weeks. In this way, more protein can be translated from each message.

†The terms *animal* and *vegetal* reflect the movements of cells seen in some embryos (such as those of frogs). The cells derived from the upper portion of the egg divide more rapidly and are actively mobile (hence, animated), while the yolk-filled cells of the vegetal half were seen as being immobile (hence, like plants).

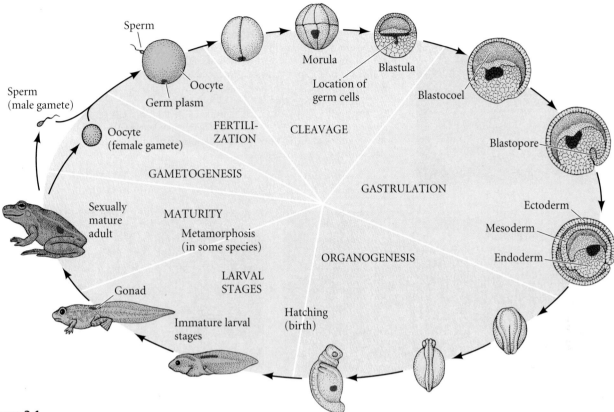

Figure 2.1
Developmental history of the leopard frog, *Rana pipiens*. The stages from fertilization through hatching (birth) are known collectively as embryogenesis. The region set aside for producing germ cells is shown in bright purple. Gametogenesis, which is completed in the sexually mature adult, begins at different times during development, depending on the species. (The sizes of the varicolored wedges shown here are arbitrary and do not correspond to the proportion of the life cycle spent in each stage.)

During cleavage, the volume of the frog egg stays the same, but it is divided into tens of thousands of cells (Figure 2.2E–H). The animal hemisphere of the egg divides faster than the vegetal hemisphere does, and the cells of the vegetal hemisphere become progressively larger the more vegetal the cytoplasm. A fluid-filled cavity, the **blastocoel**, forms in the animal hemisphere (Figure 2.2I). This cavity will be important for allowing cell movements to occur during gastrulation.

Gastrulation in the frog begins at a point on the embryo surface roughly 180 degrees opposite the point of sperm entry with the formation of a dimple, called the **blastopore**. At first, only a small slit is made. Cells migrating through this **dorsal blastopore lip** migrate toward the animal pole (Figure 2.3A,B). These cells become the dorsal mesoderm. The blastopore expands into a circle (Figure 2.3C), and cells migrating through the lateral and ventral lips of this circle become the lateral and ventral mesoderm. The cells remaining on the outside become the ectoderm, and this outer layer expands vegetally to enclose the entire embryo. The large yolky cells that re-

main at the vegetal hemisphere (until they are encircled by the ectoderm) become the endoderm. Thus, at the end of gastrulation, the ectoderm (the precursor of the epidermis and nerves) is on the outside of the embryo, the endoderm (the precursor of the gut lining) is on the inside of the embryo, and the mesoderm (the precursor of connective tissue, blood, skeleton, gonads, and kidneys) is between them.

Organogenesis begins when the notochord—a rod of mesodermal cells in the most dorsal portion of the embryo—signals the ectodermal cells above it that they are not going to become epidermis. Instead, these dorsal ectoderm cells form a tube and become the nervous system. At this stage, the embryo is called a **neurula**. The neural precursor cells elongate, stretch, and fold into the embryo (Figure 2.3D–F), forming the **neural tube**. The future back epidermal cells cover them. The cells that had connected the neural tube to the epidermis become the **neural crest cells**. The neural crest cells are almost like a fourth germ layer. They give rise to the pigment cells of the body (the melanocytes), the peripheral neurons, and the cartilage of the face.

Once the neural tube has formed, it induces changes in its neighbors, and organogenesis continues. The mesodermal tissue adjacent to the notochord becomes segmented into **somites**, the precursors of the frog's back muscles, spinal vertebrae, and dermis (the inner portion of the skin). These somites appear as blocks of mesodermal tissue (Figure 2.3F,G). The embryo develops a mouth and an anus, and it elongates into the typical tadpole structure (Figure 2.3H). The

(A)

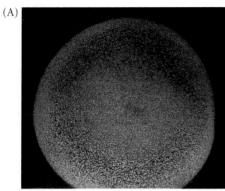

Figure 2.2

Early development of the frog *Xenopus laevis*. (A) As the egg matures, it accumulates yolk (here stained yellow and green) in the vegetal cytoplasm. (B) Frogs mate by amplexus, the male grasping the female around the belly and fertilizing the eggs as they are released. (C) A newly laid clutch of eggs. The brown area of each egg is the pigmented animal cap. The white spot in the middle of the pigment is where the egg's nucleus resides. (D) Cytoplasm rearrangement seen during first cleavage. Compare with the initial stage seen in (A). (E) A 2-cell embryo near the end of its first cleavage. (F) An 8-cell embryo. (G) Early blastula. Note that the cells get smaller, but the volume of the egg remains the same. (H) Late blastula. (I) Cross section of a late blastula, showing the blastocoel (cavity). (A–H courtesy of Michael Danilchik and Kimberly Ray; I courtesy of J. Heasman.)

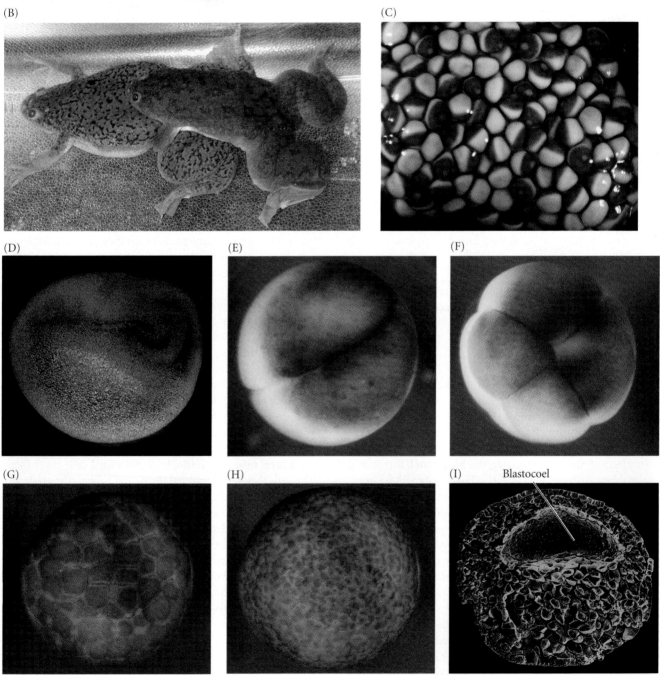

Figure 2.3

Continued development of *Xenopus laevis*. (A) Gastrulation begins with an invagination, or slit, in the future dorsal (top) side of the embryo. (B) This slit, the dorsal blastopore lip, as seen from the ventral surface (bottom) of the embryo. (C) The slit becomes a circle, the blastopore. Future mesoderm cells migrate into the interior of the embryo along the blastopore edges, and the ectoderm (future epidermis and nerves) migrates down the outside of the embryo. The remaining part, the yolk-filled endoderm, is eventually encircled. (D) Neural folds begin to form on the dorsal surface. (E) A groove can be seen where the bottom of the neural tube will be. (F) The neural folds come together at the dorsal midline, creating a neural tube. (G) Cross section of the *Xenopus* embryo at the neurula stage. (H) A pre-hatching tadpole, as the protrusions of the forebrain begin to induce eyes to form. (I) A mature tadpole, having swum away from the egg mass and feeding independently. (Photographs courtesy of Michael Danilchik and Kimberly Ray.)

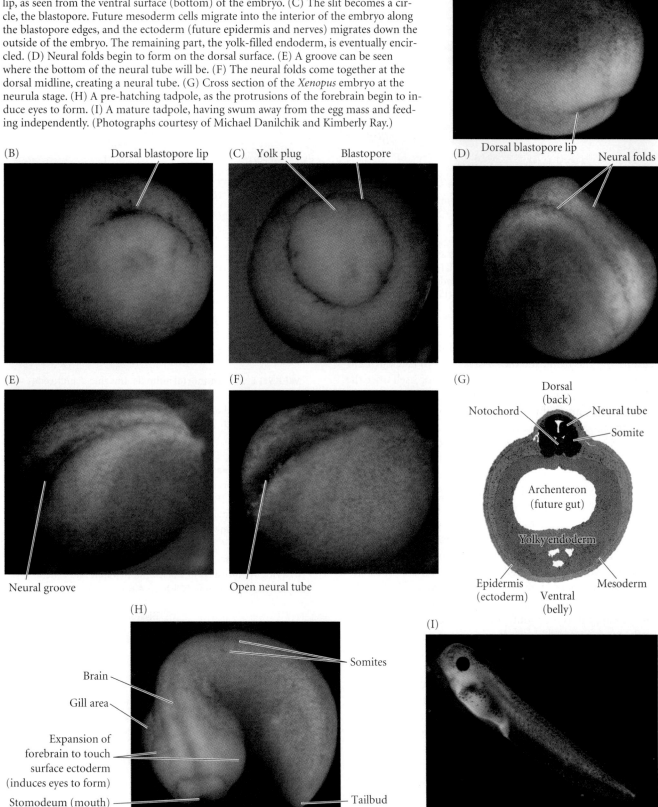

(A)

(B)

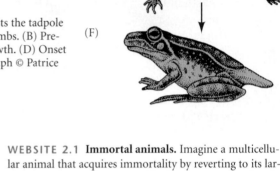

(C)

(D)

(E)

(F)

Figure 2.4
Metamorphosis of the frog. (A) Huge changes are obvious when one contrasts the tadpole and the adult bullfrog. Note especially the differences in jaw structure and limbs. (B) Pre-metamorphic tadpole. (C) Prometamorphic tadpole, showing hindlimb growth. (D) Onset of metamorphic climax as forelimbs emerge. (E, F) Climax stages. (Photograph © Patrice Ceisel/Visuals Unlimited.)

neurons make their connections to the muscles and to other neurons, the gills form, and the larva is ready to hatch from its egg jelly. The hatched tadpole will soon feed for itself once the yolk supply given it by its mother is exhausted (Figure 2.3I).

Metamorphosis of the tadpole larva into an adult frog is one of the most striking transformations in all of biology (Figure 2.4). These changes prepare an aquatic organism for a terrestrial existence. In amphibians, metamorphosis is initiated by hormones from the tadpole's thyroid gland. (The mechanisms by which thyroid hormones accomplish these changes will be discussed in Chapter 18.) In anurans (frogs and toads), almost every organ is subject to modification, and the resulting changes in form are striking and very obvious. The hindlimbs and forelimbs that the adult will use for locomotion differentiate and the tadpole's paddle tail recedes. The cartilaginous tadpole skull is replaced by the predominantly bony skull of the young frog. The horny teeth the tadpole uses to tear up pond plants disappear as the mouth and jaw take a new shape, and the fly-catching tongue muscle of the frog develops. Meanwhile, the large intestine characteristic of herbivores shortens to suit the more carnivorous diet of the adult frog. The gills regress and the lungs enlarge. As metamorphosis ends, the development of the first germ cells begins. In *Rana pipiens*, egg development lasts 3 years. At that time, the female frog is sexually mature and can produce offspring of her own.

The speed of metamorphosis is carefully keyed to environmental pressures. In temperate regions, for instance, *Rana* metamorphosis must occur before ponds freeze in winter. An adult leopard frog can burrow into the mud and survive the winter; its tadpole cannot.

WEBSITE 2.1 Immortal animals. Imagine a multicellular animal that acquires immortality by reverting to its larval form instead of growing old. That seems to be what the marine hydranth *Turritopsis* does.

WEBSITE 2.2 The human life cycle. The human animal provides a fascinating life cycle to study. Here are some websites that speculate about (A) when is an embryo or fetus "human"? (B) how might the strange way the human brain develops make childhood a necessity? and (C) do humans undergo metamorphosis?

VADE MECUM[2] The amphibian life cycle. The life cycle of a frog is illustrated in labeled photographs and time-lapse videomicroscopy. [**Click on Amphibian**]

The Evolution of Developmental Patterns in Unicellular Protists

Every living organism develops. Development can be seen even among the unicellular organisms. Moreover, by studying the development of unicellular protists, we can see the simplest forms of cell differentiation and sexual reproduction.

Control of developmental morphogenesis: The role of the nucleus

A century ago, it had not yet been proved that the nucleus of the cell contained hereditary or developmental information.

Some of the best evidence for this theory came from studies in which unicellular organisms were fragmented into nucleate and anucleate pieces (reviewed in Wilson 1896). When various protists were cut into fragments, nearly all the pieces died. However, the fragments containing nuclei were able to live and to regenerate entire complex cellular structures.

Nuclear control of cell morphogenesis and the interaction of nucleus and cytoplasm are beautifully demonstrated in studies of the protist *Acetabularia*. This enormous single cell (2–4 cm long) consists of three parts: a cap, a stalk, and a rhizoid (Figure 2.5A; Mandoli 1998). The rhizoid is located at the base of the cell and holds it to the substrate. The single nucleus of the cell resides within the rhizoid. The size of *Acetabularia* and the location of its nucleus allow investigators to remove the nucleus from one cell and replace it with a nucleus from another cell. In the 1930s, J. Hämmerling took advantage of these unique features and exchanged nuclei between two morphologically distinct species, *A. mediterranea** and *A. crenulata*. As Figure 2.5A shows, these two species have very different cap structures. Hämmerling found that when he transferred the nucleus from one species into the stalk of another species, the newly formed cap eventually assumed the form associated with the donor nucleus (Figure 2.5B). Thus, the nucleus was seen to control *Acetabularia* development.

The formation of a cap is a complex morphogenic event involving the synthesis of numerous proteins, which must be accumulated in a certain portion of the cell and then assembled into complex, species-specific structures. The transplanted nucleus does indeed direct the synthesis of its species-specific cap, but it takes several weeks to do so. Moreover, if the nucleus is removed from an *Acetabularia* cell early in its development, before it first forms a cap, a normal cap is formed weeks later, even though the organism will eventually die. These studies suggest that (1) the nucleus contains information specifying the type of cap to be produced (i.e., it contains the genetic information that specifies the proteins required for the production of a certain type of cap), and (2) material containing this information enters the cytoplasm long before cap production occurs. This information in the cytoplasm is not used for several weeks.

One current hypothesis proposed to explain these observations is that the nucleus synthesizes a stable mRNA that lies dormant in the cytoplasm until the time of cap formation (see Dumais et al. 2000). This hypothesis is supported by an observation that Hämmerling published in 1934. Hämmerling fractionated young *Acetabularia* into several parts (Figure 2.6). The portion with the nucleus eventually formed a new cap, as expected; so did the apical tip of the stalk. However, the intermediate portion of the stalk did not form a cap. Thus, Hämmerling postulated (nearly 30 years before the existence of mRNA was known) that the instructions for cap formation originated in the nucleus and were somehow stored in a dormant form near the tip of the stalk. Many years later, researchers established that nucleus-derived mRNA does accumulate in the tip of the

---

*After a recent formal name change, this species is now called *Acetabularia acetabulum*. For the sake of simplicity, however, we will use Hämmerling's designations here.

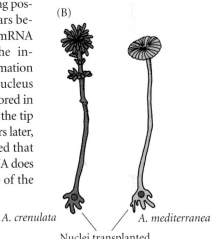

(B)

A. crenulata   *A. mediterranea*

Nuclei transplanted

Nucleus   Nucleus

Rhizoid

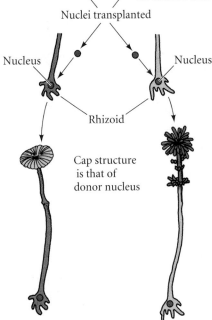

Cap structure is that of donor nucleus

(A)

Figure 2.5
(A) *Acetabularia crenulata* (left) and *A. mediterranea* (right). Each individual is a single cell. The rhizoid contains the nucleus. (B) Effect of exchanging nuclei between two species of *Acetabularia*. Nuclei were transplanted into enucleated rhizoid fragments. *A. crenulata* structures are darker, *A. mediterranea* structures lighter green. (Photographs courtesy of S. Berger.)

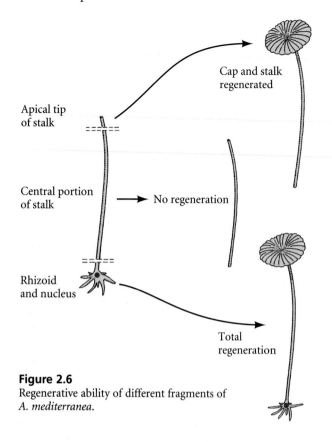

Apical tip
of stalk

Cap and stalk
regenerated

Central portion
of stalk

No regeneration

Rhizoid
and nucleus

Total
regeneration

Figure 2.6
Regenerative ability of different fragments of
A. mediterranea.

sential role in cap formation. The mRNAs are not translated for weeks, even though they are present in the cytoplasm. Something in the cytoplasm controls when the message is utilized. Hence, the expression of the cap is controlled not only by nuclear transcription, but also by the translation of the cytoplasmic RNA. In this unicellular organism, "development" is controlled at both the transcriptional and translational levels.

WEBSITE 2.3 **Protist differentiation.** Three of the most remarkable areas of protist development concern the control of sex type in fission yeast, the transformation of *Naegleria* amoebae into streamlined, flagellated cells, and the cortical inheritance of the cell surface in paramecia.

Unicellular protists and the origins of sexual reproduction

Sexual reproduction is another invention of the protists that has had a profound effect on more complex organisms. It should be noted that sex and reproduction are two distinct and separable processes. **Reproduction** involves the creation of new individuals. **Sex** involves the combining of genes from two different individuals into new arrangements. Reproduction in the absence of sex is characteristic of organisms that reproduce by fission (i.e., splitting into two); there is no sorting of genes when an amoeba divides or when a hydra buds off cells to form a new colony.

Sex without reproduction is also common among unicellular organisms. Bacteria are able to transmit genes from one individual to another by means of sex pili. This transmission is separate from reproduction. Protists are also able to reassort genes without reproduction. Paramecia, for instance, reproduce by fission, but sex is accomplished by **conjugation**. When two paramecia join together, they link their oral apparatuses and form a cytoplasmic connection through which they can

stalk, and that the destruction of this mRNA or the inhibition of protein synthesis in this region prevents cap formation (Kloppstech and Schweiger 1975; Garcia and Dazy 1986; Serikawa et al. 2001).

It is clear from the preceding discussion that nuclear transcription plays an important role in the formation of the *Acetabularia* cap. But note that the cytoplasm also plays an es-

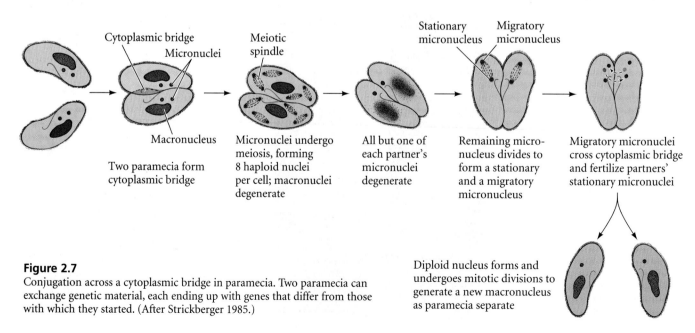

Cytoplasmic bridge

Micronuclei

Meiotic
spindle

Stationary
micronucleus

Migratory
micronucleus

Macronucleus

Two paramecia form
cytoplasmic bridge

Micronuclei undergo
meiosis, forming
8 haploid nuclei
per cell; macronuclei
degenerate

All but one of
each partner's
micronuclei
degenerate

Remaining micro-
nucleus divides to
form a stationary
and a migratory
micronucleus

Migratory micronuclei
cross cytoplasmic bridge
and fertilize partners'
stationary micronuclei

Diploid nucleus forms and
undergoes mitotic divisions to
generate a new macronucleus
as paramecia separate

Figure 2.7
Conjugation across a cytoplasmic bridge in paramecia. Two paramecia can exchange genetic material, each ending up with genes that differ from those with which they started. (After Strickberger 1985.)

exchange genetic material (Figure 2.7). The macronucleus of each individual (which controls the metabolism of the organism) degenerates, while each micronucleus undergoes meiosis to produce eight haploid micronuclei, of which all but one degenerate. The remaining micronucleus divides once more to form a stationary micronucleus and a migratory micronucleus. Each migratory micronucleus crosses the cytoplasmic bridge and fuses with ("fertilizes") the partner's stationary micronucleus, thereby creating a new diploid nucleus in each cell. This diploid nucleus then divides mitotically to give rise to a new micronucleus and a new macronucleus as the two partners disengage. Therefore, no reproduction has occurred, only sex.

The union of these two distinct processes, sex and reproduction, into **sexual reproduction** is seen in other unicellular eukaryotes. Figure 2.8 shows the life cycle of *Chlamydomonas*. This organism is usually haploid, having just one copy of each chromosome (like a mammalian gamete). The individuals of each species, however, are divided into two **mating types**: plus

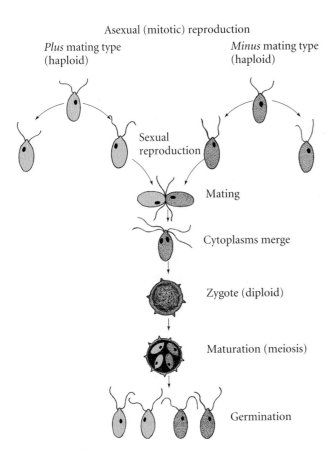

Asexual (mitotic) reproduction

Plus mating type (haploid)   *Minus* mating type (haploid)

Sexual reproduction

Mating

Cytoplasms merge

Zygote (diploid)

Maturation (meiosis)

Germination

Two *plus* and two *minus* mating types

Figure 2.8
Sexual reproduction in *Chlamydomonas*. Two mating types, both haploid, can reproduce asexually when separate. Under certain conditions, the two mating types can unite to produce a diploid cell that can undergo meiosis to form four new haploid organisms. (After Strickberger 1985.)

and minus. When a plus and a minus meet, they join their cytoplasms, and their nuclei fuse to form a diploid zygote. This zygote is the only diploid cell in the life cycle, and it eventually undergoes meiosis to form four new *Chlamydomonas* cells. This is true sexual reproduction, for chromosomes are reassorted during the meiotic divisions and more individuals are formed. Note that in this protist type of sexual reproduction, the gametes are morphologically identical; the distinction between sperm and egg has not yet been made.

In evolving sexual reproduction, two important advances had to be achieved. The first was the mechanism of meiosis (Figure 2.9), whereby the diploid complement of chromosomes is reduced to the haploid state (discussed in detail in Chapter 19). The second was a mechanism whereby the two different mating types could recognize each other. In *Chlamydomonas*, recognition occurs first on the flagellar membranes (Figure 2.10; Bergman et al. 1975; Wilson et al. 1997; Pan and Snell 2000). The flagella of two individuals twist around each other, enabling specific regions of the cell membranes to come together. These specialized regions contain mating type-specific components that enable the cytoplasms to fuse. Following flagellar agglutination, the plus individual initiates fusion by extending a fertilization tube. This tube contacts and fuses with a specific site on the minus individual. Interestingly, the mechanism used to extend this tube—the polymerization of the protein actin to form microfilaments—is also used to extend the processes of sea urchin eggs and sperm. In Chapter 7, we will see that the recognition and fusion of sperm and egg occur in an amazingly similar manner.

Unicellular eukaryotes appear to possess the basic elements of the developmental processes that characterize more complex organisms: protein synthesis is controlled such that certain proteins are made only at certain times and in certain places; the structures of individual genes and chromosomes are as they will be throughout eukaryotic evolution; mitosis and meiosis have been perfected; and sexual reproduction exists, involving cooperation between individual cells. Such intercellular cooperation becomes even more important with the evolution of multicellular organisms.

Multicellularity: The Evolution of Differentiation

One of evolution's most important experiments was the creation of multicellular organisms. There appear to be several paths by which single cells evolved multicellular arrangements; we will discuss only two of them here (see Chapter 22 for a fuller discussion). The first path involves the orderly division of the reproductive cell and the subsequent differentiation of its progeny into different cell types. This path to multicellularity can be seen in a remarkable series of multicellular organisms collectively referred to as the family Volvocaceae, or the volvocaceans (Kirk 1999, 2000).

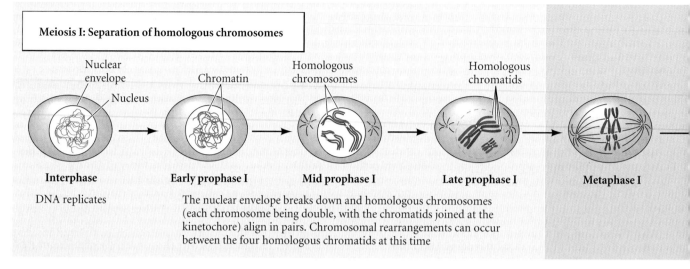

Meiosis I: Separation of homologous chromosomes

Nuclear envelope

Nucleus

Chromatin

Homologous chromosomes

Homologous chromatids

Interphase

DNA replicates

Early prophase I

Mid prophase I

The nuclear envelope breaks down and homologous chromosomes (each chromosome being double, with the chromatids joined at the kinetochore) align in pairs. Chromosomal rearrangements can occur between the four homologous chromatids at this time

Late prophase I

Metaphase I

Figure 2.9

Summary of meiosis. The DNA and its associated proteins replicate during interphase. During prophase, the nuclear envelope breaks down and homologous chromosomes (each chromosome is double, with the chromatids joined at the kinetochore) align in pairs. Chromosomal rearrangements between the four homologous chromatids can occur at this time. After the first metaphase, the two original homologous chromosomes are segregated into different cells. During the second meiotic division, the kinetochore splits and the sister chromatids separate, leaving each new cell with one copy of each chromosome.

(A)

(B)

Microfilaments

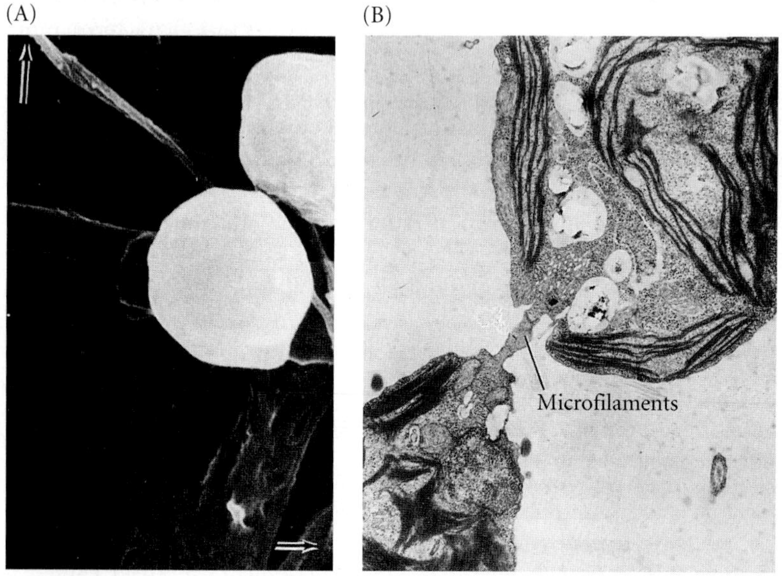

Figure 2.10

Two-step recognition in mating *Chlamydomonas*. (A) Scanning electron micrograph (7000×) of mating pair. The interacting flagella twist around each other, adhering at the tips (arrows). (B) Transmission electron micrograph (20,000×) of a cytoplasmic bridge connecting the two organisms. The actin microfilaments extend from the donor (lower) cell to the recipient (upper) cell. (From Goodenough and Weiss 1975 and Bergman et al. 1975; photographs courtesy of U. Goodenough.)

The Volvocaceans

The simpler organisms among the volvocaceans are ordered assemblies of numerous cells, each resembling the unicellular protist *Chlamydomonas*, to which they are related (Figure 2.11A). A single organism of the volvocacean genus *Gonium* (Figure 2.11B), for example, consists of a flat plate of 4 to 16 cells, each with its own flagellum. In a related genus, *Pandorina*, the 16 cells form a sphere (Figure 2.11C); and in *Eudorina*, the sphere contains 32 or 64 cells arranged in a regular pattern (Figure 2.11D). In these organisms, then, a very important developmental principle has been worked out: the ordered division of one cell to generate a number of cells that are organized in a predictable fashion. Like cleavage in most animal embryos, the cell divisions by which a single volvocacean cell produces an organism of 4 to 64 cells occur in very rapid sequence and in the absence of cell growth.

The next two genera of the volvocacean series exhibit another important principle of development: the differentiation of cell types within an individual organism. In these organisms, the reproductive cells become differentiated from the somatic cells. In all the genera mentioned earlier, every cell can, and normally does, produce a complete new organism by mitosis. In the genera *Pleodorina* and *Volvox*, however, relatively few cells can reproduce. In *Pleodorina californica* (Figure 2.11E), the cells in the anterior region are restricted to a somatic function; only those cells on the posterior side can reproduce. In *P. californica*, a colony usually has 128 or 64 cells, and the ratio of the number of somatic cells to the number of reproductive cells is usually 3:5. Thus, a 128-cell colony typically has 48 somatic cells, and a 64-cell colony has 24.

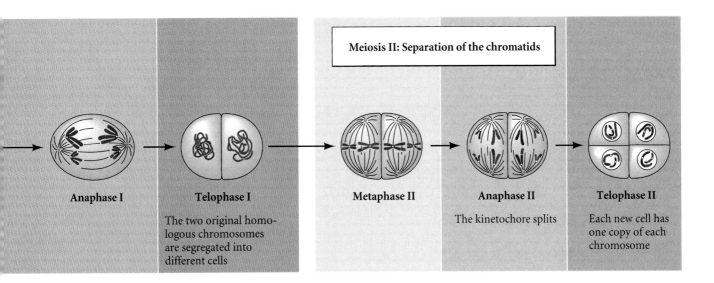

Meiosis II: Separation of the chromatids

Anaphase I	**Telophase I**	**Metaphase II**	**Anaphase II**	**Telophase II**
	The two original homologous chromosomes are segregated into different cells		The kinetochore splits	Each new cell has one copy of each chromosome

In *Volvox*, almost all the cells are somatic, and very few of the cells are able to produce new individuals. In some species of *Volvox*, reproductive cells, as in *Pleodorina*, are derived from cells that originally look and function like somatic cells before they enlarge and divide to form new progeny. However, in other members of the genus, such as *V. carteri*, there is a complete division of labor: the reproductive cells that will create the next generation are set aside during the division of the original cell that is forming a new individual. The reproductive cells never develop functional flagella and never contribute to motility or other somatic functions of the individual; they are entirely specialized for reproduction.

Thus, although the simpler volvocaceans may be thought of as colonial organisms (because each cell is capable of independent existence and of perpetuating the species), in *V. carteri* we have a truly multicellular organism with two distinct and interdependent cell types (somatic and reproductive), both of which are required for perpetuation of the species (Figure 2.11F). Although not all animals set aside the reproductive cells from the somatic cells (and plants hardly ever do), this separation of germ cells from somatic cells early in development is characteristic of many animal phyla and will be discussed in more detail in Chapter 19.

WEBSITE 2.4 *Volvox* cell differentiation. The pathways leading to germ cells or somatic cells are controlled by genes that cause cells to follow one or the other fate. Mutations can prevent the formation of one of these lineages.

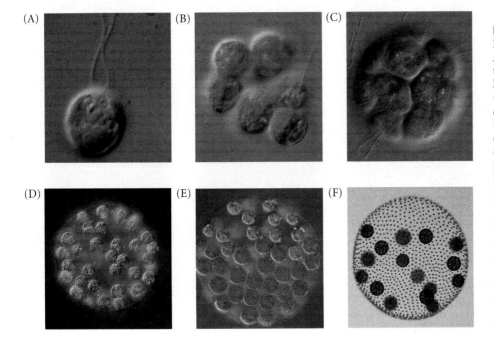

Figure 2.11
Representatives of the order Volvocales. All but *Chlamydomonas* are members of the family Volvocaceae. (A) The unicellular protist *Chlamydomonas reinhardtii*. (B) *Gonium pectorale*, with 8 *Chlamydomonas*-like cells in a convex disc. (C) *Pandorina morum*. (D) *Eudorina elegans*. (E) *Pleodorina californica*. Here, all 64 cells are originally similar, but the posterior ones dedifferentiate and redifferentiate as asexual reproductive cells called gonidia, while the anterior cells remain small and biflagellate, like *Chlamydomonas*. (F) *Volvox carteri*. Here, cells destined to become gonidia are set aside early in development and never have somatic characteristics. The smaller somatic cells resemble *Chlamydomonas*. Complexity increases from the single-celled *Chlamydomonas* to the multicellular *Volvox*. (Photographs courtesy of D. Kirk.)

Sex and Individuality in Volvox

Simple as it is, *Volvox* shares many features that characterize the life cycles and developmental histories of much more complex organisms, including ourselves. As already mentioned, *Volvox* is among the simplest organisms to exhibit a division of labor between two completely different cell types. As a consequence, it is among the simplest organisms to include death as a regular, genetically regulated part of its life history.

Death and Differentiation

Unicellular organisms that reproduce by simple cell division, such as amoebae, are potentially immortal. The amoeba you see today under the microscope has no dead ancestors. When an amoeba divides, neither of the two resulting cells can be considered either ancestor or offspring; they are siblings. Death comes to an amoeba only if it is eaten or meets with a fatal accident, and when it does, the dead cell leaves no offspring.

Death becomes an essential part of life, however, for any multicellular organism that establishes a division of labor between somatic (body) cells and germ (reproductive) cells. Consider the life history of *Volvox carteri* when it is reproducing asexually (Figure 2.12). Each asexual adult is a spheroid containing some 2000 small, bifla-gellated somatic cells along its periphery and about 16 large, asexual reproductive cells, called **gonidia**, toward one end of the interior. When mature, each gonidium divides rapidly 11 or 12 times. Certain of these divisions are asymmetrical and produce the 16 large cells that will become a new set of gonidia in the next generation. At the end of cleavage, all the cells that will be present in an adult have been produced from the gonidium. But the resulting embryo is "inside out": it is now a hollow sphere with its gonidia on the outside and the flagella of its somatic cells pointing toward the interior. This predicament is corrected by a process called **inversion**, in which the embryo turns itself right side out by a set of cell movements that resemble the gastrulation movements of animal embryos (Figure 2.13A–H). Clusters of bottle-shaped cells open a hole at one end of the embryo by producing tension on the interconnected cell sheet (Figure 2.13I). The embryo everts through this hole and then closes it up. About a day after this is done, the juvenile *Volvox* are enzymatically released from the parent and swim away.

What happens to the somatic cells of the "parent" *Volvox* now that its young have "left home"? Having produced offspring and being incapable of further reproduction, these somatic cells die. Actually, these cells commit suicide, synthesizing a set of proteins that cause the death and dissolution of the cells that make them (Pommerville and Kochert 1982). Moreover, in death, the cells release for the use of others, including their own offspring, all the nutrients that they had stored during life. "Thus emerges," notes David Kirk, "one of the great themes of life on planet Earth: 'Some die that others may live.'"

In *V. carteri*, a specific gene, *somatic regulator A*, or *regA*, plays a central role in regulating cell death (Kirk 1988, 2001a). This gene is expressed only in somatic cells,

Figure 2.12
Asexual reproduction in *V. carteri*. (A) When reproductive cells (gonidia) are mature, they enter a cleavage-like stage of embryonic development to produce juveniles within the adult. Through a series of cell movements resembling gastrulation, the embryonic *Volvox* invert and are eventually released from the parent. The somatic cells of the parent, lacking the gonidia, undergo senescence and undergo programmed cell death, while the juvenile *Volvox* mature. The entire asexual cycle takes 2 days. (B) Micrograph showing young adult spheres of *Volvox carteri* being released from parent to become free-swimming individuals. (A after Kirk 1988; B from Kirk 2001b.)

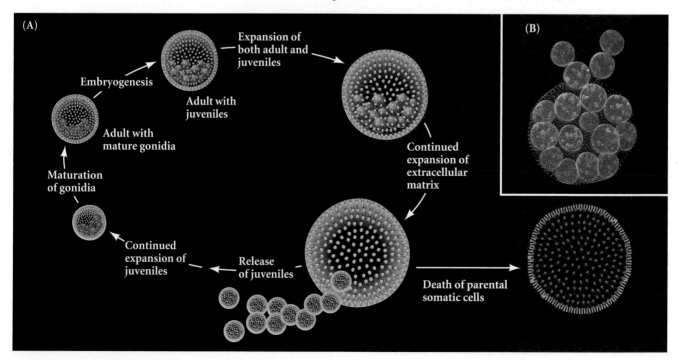

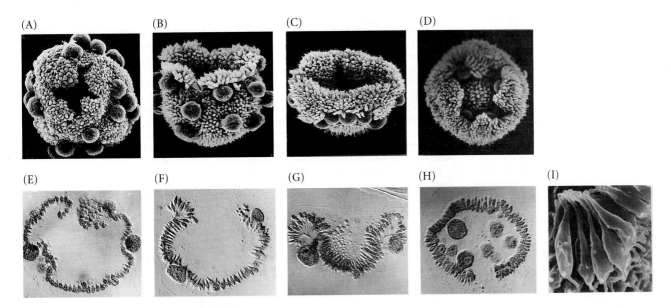

Figure 2.13

Inversion of embryos of *V. carteri*. A–D are scanning electron micrographs of whole embryos. E–H are sagittal sections through the center of the embryo, visualized by differential interference microscopy. Before inversion, the embryo is a hollow sphere of connected cells with the new gonidia on the outside. When the "bottle cells" change their shape, a hole (the phialopore) opens at the apex of the embryo (A, B, E, F). Cells then curl around and rejoin at the bottom (C, D, G, H). The new gonidia are now inside. (I) "Bottle cells" near the opening of the phialopore in a *V. carteri* embryo. These cells remain tightly interconnected through cytoplasmic bridges near their elongated apices, thereby creating the tension that causes the curvature of the interconnected cell sheet. (From Kirk et al. 1982; photographs courtesy of D. Kirk.)

and it prevents their expressing gonidial genes. In laboratory strains possessing regulatory mutations of this gene, somatic cells begin expressing *regA*, abandon their suicidal ways, gain the ability to reproduce asexually, and become potentially immortal (Figure 2.14). The fact that such mutants have never been found in nature indicates that cell death most likely plays an important role in the survival of *V. carteri* under natural conditions.

Enter Sex

Although *V. carteri* reproduces asexually much of the time, in nature it reproduces sexually once each year. When it does, one generation of individuals passes away and a new and genetically different generation is produced. The naturalist Joseph Wood Krutch (1956, pp. 28–29) put it more poetically:

> *The amoeba and the paramecium are potentially immortal. … But for Volvox, death seems to be as inevitable as it is in a mouse or in a man. Volvox must die as Leeuwenhoek saw it die because it had children and is no longer needed. When its time comes it drops quietly to the bottom and joins its ancestors. As Hegner, the Johns Hopkins zoologist, once wrote, 'This is the first advent of inevitable natural death in*

> *the animal kingdom and all for the sake of sex.' And he asked: 'Is it worth it?'*

For *Volvox carteri*, it most assuredly is worth it. *V. carteri* lives in shallow temporary ponds that fill with spring rains but dry out in the heat of late summer. Between those times, *V. carteri* swims about, reproducing asexually. These asexual volvoxes will die in minutes once the pond dries up. *V. carteri* is able to survive by turning sexual shortly before the pond disappears, producing dormant zygotes that survive the heat and drought of late summer and the cold of winter. When rain fills the pond in spring, the zygotes break their dormancy and hatch out a new generation of individuals that reproduce asexually until the pond is about to dry up once more.

How do these simple organisms predict the coming of adverse conditions so accurately that they can produce a sexual generation in the nick of time, year after year? The stimulus for switching from the asexual to the sexual mode of reproduction in *V.*

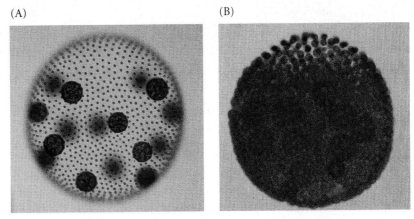

Figure 2.14

Mutation of a single gene (*somatic regenerator A*) abolishes programmed cell death in *V. carteri*. (A) A newly hatched *Volvox* carrying this mutation is indistinguishable from the wild-type spheroid. (B) Shortly before the time when the somatic cells of wild-type spheroids begin to die, the somatic cells of this mutant redifferentiate as gonidia (B). Eventually, every cell of the mutant will divide to regenerate a new spheroid that will repeat this potentially immortal developmental cycle. (Photographs courtesy of D. Kirk.)

carteri is known to be a 30-kDa sexual inducer protein. This protein is so powerful that concentrations as low as 6×10^{-17} M cause gonidia to undergo a modified pattern of embryonic development that results in the production of eggs or sperm, depending on the genetic sex of the individual (Sumper et al. 1993). The sperm are released and swim to a female, where they fertilize eggs to produce dormant zygotes (Figure 2.15). The sexual inducer protein is able to work at such remarkably low concentrations by causing slight modifications of the extracellular matrix. These modifications appear to signal the transcription of a whole battery of genes that form the gametes (Sumper et al. 1993; Hallmann et al. 2001).

What is the source of this sexual inducer protein? Kirk and Kirk (1986) discovered that the sexual cycle could be initiated by heating dishes of *V. carteri* to temperatures that might be expected in a shallow pond in late summer. When this was done, the somatic cells of the asexual volvoxes produced the sexual inducer protein. Since the amount of sexual inducer protein secreted by one individual is sufficient to initiate sexual development in over 500 million asexual volvoxes, a single inducing volvox can convert an entire pond to sexuality. This discovery explained an observation made over 90 years ago that "in the full blaze of Nebraska sunlight, *Volvox* is able to appear, multiply, and riot in sexual reproduction in pools of rainwater of scarcely a fortnight's duration" (Powers 1908). Thus, in temporary ponds formed by spring rains and dried up by summer's heat, *Volvox* has found a means of survival: it uses that heat to induce the formation of sexual individuals whose mating produces zygotes capable of surviving conditions that kill the adult organism. We see, too, that development is critically linked to the ecosystem in which the organism has adapted to survive.

Figure 2.15
Sexual reproduction in *V. carteri*. Males and females are indistinguishable in their asexual phase. When the sexual inducer protein is present, the gonidia of both mating types undergo a modified embryogenesis that leads to the formation of eggs in the females and sperm in the males. When the gametes are mature, sperm packets (containing 64 or 128 sperm each) are released and swim to the females. Upon reaching a female, the sperm packet breaks up into individual sperm, which can fertilize the eggs. The resulting dormant zygote has tough cell walls that can resist drying, heat, and cold. When spring rains cause the zygote to germinate, it undergoes meiosis to produce haploid males and females that reproduce asexually until heat induces the sexual cycle again.

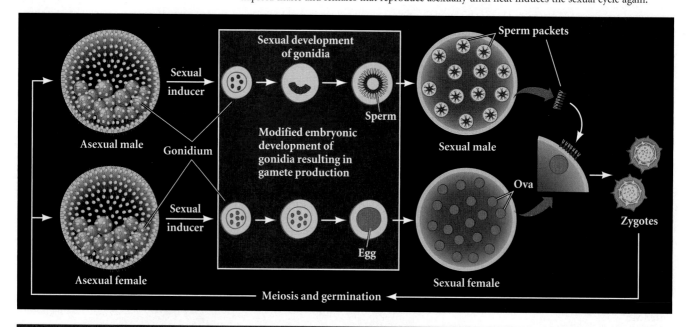

Although all the volvocaceans, like their unicellular relative *Chlamydomonas*, reproduce predominantly by asexual means, they are also capable of sexual reproduction, which involves the production and fusion of haploid gametes. In many species of *Chlamydomonas*, including the one illustrated in Figure 2.10, sexual reproduction is **isogamous** ("the same gametes"), since the haploid gametes that meet are similar in size, structure, and motility. However, in other species of *Chlamydomonas*—as well as many species of colonial volvocaceans—swimming gametes of very different sizes are produced by the different mating types. This pattern is called **heterogamy** ("different gametes"). But the larger volvocaceans have evolved a specialized form of heterogamy called **oogamy**, which involves the production of large, relatively immotile eggs by one mating type and small, motile sperm by the other (see Sidelights & Speculations). Here we see one type of gamete specialized for the retention of nutritional and developmental resources and the other type of gamete specialized for the transport of nuclei. Thus, the volvocaceans include the simplest organisms that have distinguishable male and female members of the species and that have distinct developmental pathways for the production of eggs or sperm.

In all volvocaceans, the fertilization reaction resembles that of *Chlamydomonas* in that it results in the production of a dormant diploid zygote that is capable of surviving harsh environmental conditions. When conditions allow the zygote to germinate, it first undergoes meiosis to produce haploid offspring of the two different mating types in equal numbers.

Differentiation and morphogenesis in Dictyostelium: *Cell adhesion*

THE LIFE CYCLE OF *DICTYOSTELIUM*.   Another type of multicellular organization derived from unicellular organisms is found in *Dictyostelium discoideum*.* The life cycle of this fascinating organism is illustrated in Figure 2.16. In its asexual cycle, solitary haploid amoebae (called myxamoebae or "social amoebae" to distinguish them from amoeba species that always remain solitary) live on decaying logs, eating bacteria and reproducing by binary fission. When they have exhausted their food supply,

---

*Though colloquially called a "cellular slime mold," *Dictyostelium* is not a mold, nor is it consistently slimy. It is perhaps best to think of *Dictyostelium* as a social amoeba.

Figure 2.16
Life cycle of *Dictyostelium discoideum*. Haploid spores give rise to myxamoebae, which can reproduce asexually to form more haploid myxamoebae. As the food supply diminishes, aggregation occurs and a migrating slug is formed. The slug culminates in a fruiting body that releases more spores. Times refer to hours since the onset of nutrient starvation. Prestalk cells are indicated in yellow. (Photographs courtesy of R. Blanton and M. Grimson.)

tens of thousands of these myxamoebae join together to form moving streams of cells that converge at a central point. Here they pile atop one another to produce a conical mound called a tight aggregate. Subsequently, a tip arises at the top of this mound, and the tight aggregate bends over to produce the migrating slug (with the tip at the front). The **slug** (often given the more dignified title of **pseudoplasmodium** or **grex**) is usually 2–4 mm long and is encased in a slimy sheath. The grex begins to migrate (if the environment is dark and moist) with its anterior tip slightly raised. When it reaches an illuminated area, migration ceases, and the culmination stages of the life cycle take place as the grex differentiates into a fruiting body composed of spore cells and a stalk. The anterior cells, representing 15–20% of the entire cellular population, form the tubed stalk. This process begins as some of the central anterior cells, the **prestalk** cells, begin secreting an extracellular cellulose coat and extending a tube through the grex. As the prestalk cells differentiate, they form vacuoles and enlarge, lifting up the mass of **prespore** cells that made up the posterior four-fifths of the grex (Jermyn and Williams 1991). The stalk cells die, but the prespore cells, elevated above the stalk, become spore cells. These spore cells disperse, each one becoming a new myxamoeba.

WEBSITE 2.5 Slime mold life cycle. Check out this website to see digitized videos of the *Dictyostelium* life cycle.

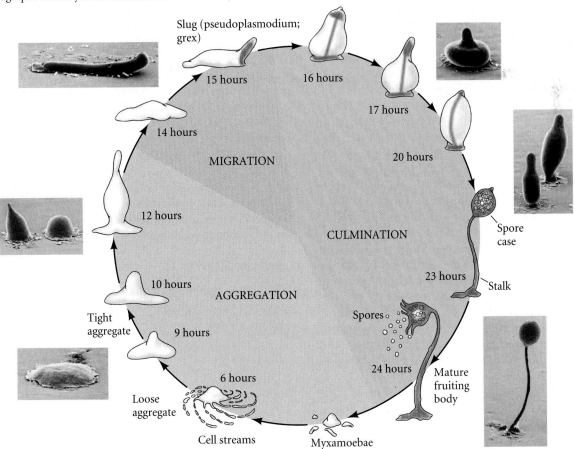

Slug (pseudoplasmodium; grex)

15 hours   16 hours

17 hours

14 hours

20 hours

MIGRATION

12 hours

Spore case

CULMINATION

10 hours

23 hours   Stalk

AGGREGATION

Tight aggregate   9 hours

Spores

6 hours

24 hours   Mature fruiting body

Loose aggregate

Cell streams   Myxamoebae

In addition to this asexual cycle, there is a possibility of sex for *Dictyostelium*. Two myxamoebae can fuse to create a giant cell, which digests all the other cells of the aggregate. When it has eaten all its neighbors, it encysts itself in a thick wall and undergoes meiotic and mitotic divisions; eventually, new myxamoebae are liberated.

Dictyostelium has been a wonderful experimental organism for developmental biologists because initially identical cells differentiate into two alternative cell types—spore and stalk. It is also an organism wherein individual cells come together to form a cohesive structure composed of differentiated cell types, a process akin to tissue formation in more complex organisms. The aggregation of thousands of myxamoebae into a single organism is an incredible feat of organization that invites experimentation to answer questions about the mechanisms involved.

VADE MECUM[2] **Slime mold life cycle.** The life cycle of *Dictyostelium*—the remarkable aggregation of myxamoebae, the migration of the slug, and the truly awesome culmination of the stalk and fruiting body—can best be viewed through movies. The Slime Mold segment in Vade Mecum[2] contains a remarkable series of videos.
[Click on Slime Mold]

AGGREGATION OF *DICTYOSTELIUM* CELLS. The first of these questions is, What causes the myxamoebae to aggregate? Time-lapse videomicroscopy has shown that no directed movement occurs during the first 4–5 hours following nutrient starvation. During the next 5 hours, however, the cells can be seen moving at about 20 mm/min for 100 seconds. This movement ceases for about 4 minutes, then resumes. Although the movement is directed toward a central point, it is not a simple radial movement. Rather, cells join with one another to form streams; the streams converge into larger streams, and eventually all streams merge at the center. Bonner (1947) and Shaffer (1953) showed that this movement is a result of chemotaxis: the cells are guided to aggregation centers by a soluble substance. This substance was later identified as **cyclic adenosine 3′5′-monophosphate (cAMP)** (Konijn et al. 1967; Bonner et al. 1969), the chemical structure of which is shown in Figure 2.17A.

Aggregation is initiated as each of the myxamoebae begins to synthesize cAMP. There are no dominant cells that begin the secretion or control the others. Rather, the sites of aggregation are determined by the distribution of the myxamoebae (Keller and Segal 1970; Tyson and Murray 1989). Neighboring cells respond to cAMP in two ways: they initiate a movement toward the cAMP pulse, and they release cAMP of their own (Robertson et al. 1972; Shaffer 1975). After this happens, the cell is unresponsive to further cAMP pulses for several minutes. The result is a rotating spiral wave of cAMP that is propagated throughout the population of cells (Figure 2.17B–D). As each wave arrives, the cells take another step toward the center.*

The differentiation of individual myxamoebae into either stalk (somatic) or spore (reproductive) cells is a complex matter. Raper (1940) and Bonner (1957) demonstrated that the anterior cells normally become stalk, while the remaining, posterior cells are usually destined to form spores. However, surgically removing the anterior part of a slug does not abolish its ability to form a stalk. Rather, the cells that now find themselves at the anterior end (and which originally had been destined to produce spores) now form the stalk (Raper 1940). Somehow a decision is made so that whichever cells are anterior become stalk cells and whichever are posterior become spores. This ability of cells to change their developmental fates according to their location within the whole organism and thereby compensate for missing parts is called **regulation**. We will see this phenomenon in many embryos, including those of mammals.

CELL ADHESION MOLECULES IN *DICTYOSTELIUM*. How do individual cells stick together to form a cohesive organism? This problem is the same one that embryonic cells face, and the solution that evolved in the protists is the same one used by embryos: developmentally regulated cell adhesion molecules.

While growing mitotically on bacteria, *Dictyostelium* cells do not adhere to one another. However, once cell division stops, the cells become increasingly adhesive, reaching a plateau of maximum adhesiveness about 8 hours after starvation. The initial cell-cell adhesion is mediated by a 24-kilodalton glycoprotein (gp24) that is absent in myxamoebae but appears shortly after mitotic division ceases (Figure 2.18A; Knecht et al. 1987; Wong et al. 1996). This protein is synthesized from newly transcribed mRNA and becomes localized in the cell membranes of the myxamoebae. If myxamoebae are treated with antibodies that bind to and mask this protein, the cells will not stick to one another, and all subsequent development ceases.

Once this initial aggregation has occurred, it is stabilized by a second cell adhesion molecule. This 80-kDa glycoprotein (gp80) is also synthesized during the aggregation phase. If it is defective or absent in the cells, small slugs will form, and their fruiting bodies will be only about one-third the normal size. Thus, the second cell adhesion system seems to be needed for

*The biochemistry of this reaction involves a receptor that binds cAMP. When binding occurs, specific gene transcription takes place, motility toward the source of the cAMP is initiated, and enzymes that synthesize cAMP from ATP are activated. cAMP activates the cell's own receptors as well as those of its neighbors. The cells in the area remain insensitive to new waves of cAMP until the bound cAMP is removed from the receptors by another cell surface enzyme, phosphodiesterase (Johnson et al. 1989). The mathematics of such oscillation reactions (similar to those shown in Figure 1.20) predict that the diffusion of cAMP should initially be circular. However, as cAMP interacts with the cells that receive and propagate the signal, the cells that receive the front part of the wave begin to migrate at a different rate than the cells behind them (see Nanjundiah 1997, 1998). The result is the rotating spiral of cAMP and migration seen in Figure 2.17. Interestingly, the same mathematical formulas predict the behavior of certain chemical reactions and the formation of new stars in rotating spiral galaxies (Tyson and Murray 1989).

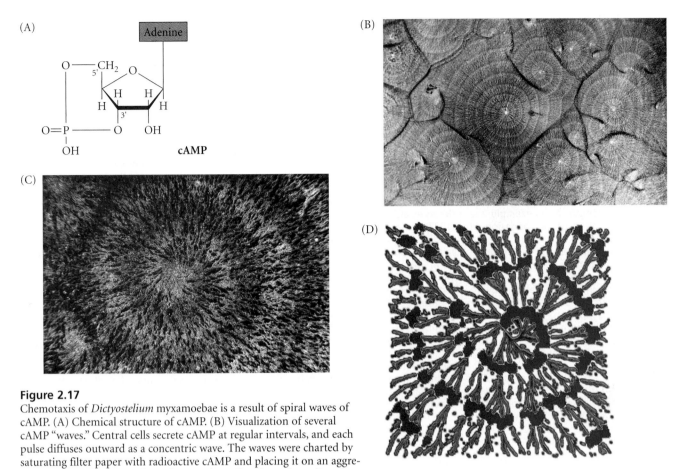

Figure 2.17

Chemotaxis of *Dictyostelium* myxamoebae is a result of spiral waves of cAMP. (A) Chemical structure of cAMP. (B) Visualization of several cAMP "waves." Central cells secrete cAMP at regular intervals, and each pulse diffuses outward as a concentric wave. The waves were charted by saturating filter paper with radioactive cAMP and placing it on an aggregating colony. The cAMP from the secreting cells dilutes the radioactive cAMP. When the radioactivity on the paper is recorded (by placing it over X-ray film), the regions of high cAMP concentration in the culture appear lighter than those of low cAMP concentration. (C) Spiral waves of myxamoebae moving toward the initial source of cAMP. Because moving and nonmoving cells scatter light differently, the photograph reflects cell movement. The bright bands are composed of elongated migrating cells; the dark bands are cells that have stopped moving and have rounded up. As cells form streams, the spiral of movement can still be seen moving toward the center. (D) Computer simulation of cAMP wave spreading across migrating *Dictyostelium* cells. The model takes into account the reception and release of cAMP, and changes in cell density due to the movement of the cells. The cAMP wave is plotted in dark blue. The population of amoebae goes from green (low) to red (high). Compare with the actual culture shown in (C). (B from Tomchick and Devreotes 1981; C from Siegert and Weijer 1989; D from Dallon and Othmer 1997.)

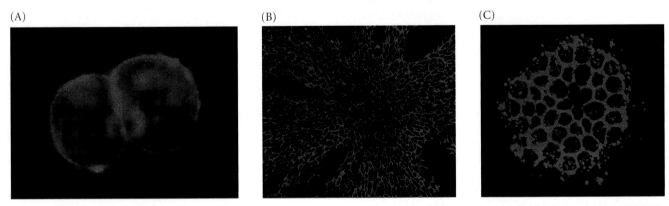

Figure 2.18

The three cell adhesion molecules of *Dictyostelium*. (A) *Dictyostelium* cells synthesize an adhesive 24-kDa glycoprotein (gp24) shortly after nutrient starvation. These *Dictyostelium* cells were stained with a fluorescently labeled (green) antibody that binds to gp24 and were then observed under ultraviolet light. This protein is not seen on myxamoebae that have just stopped dividing. However, as shown here—10 hours after cell division has ceased—individual myxamoebae have this protein in their cell membranes and are capable of adhering to one another. (B) The gp80 protein, stained by specific antibodies (green), is present at the cell membranes of streaming amoebae. (C) The gp150 protein (green) is present in the cells of the migrating grex (cross-sectioned). Photographs are not at the same magnification. (Photographs courtesy of W. Loomis.)

Rules of Evidence I

Biology, like any other science, does not deal with Facts, but with evidence. Several types of evidence will be presented in this book, and they are not equivalent in strength. As an example, we will use the analysis of cell adhesion in *Dictyostelium*.

The first, and weakest, type of evidence is **correlative evidence**. Here, correlations are observed between two or more events, and there is an inference that one event causes the other. Correlative evidence provides a starting point for investigations, but one cannot say with certainty that one event causes the other based solely on correlations. As we have seen, fluorescently labeled antibodies to a certain 24-kDa glycoprotein (gp24) do not bind to dividing myxamoebae, but they do find this protein in myxamoeba cell membranes soon after the cells stop dividing and become competent to aggregate (see Figure 2.18A). Thus, there is a correlation between the presence of this cell membrane glycoprotein and the ability to aggregate.

Although one might infer that the synthesis of gp24 causes the adhesion of the cells, it is also possible that cell adhesion causes the cells to synthesize this glycoprotein, or that cell adhesion and the synthesis of the glycoprotein are separate events initiated by the same underlying cause. The simultaneous occurrence of the two events could even be coincidental, the events having no relationship to each other.*

How, then, does one get beyond mere correlation? In the study of cell adhesion in

*In a tongue-in-cheek letter spoofing such correlative inferences, Sies (1988) demonstrated a remarkably good correlation between the number of storks seen in West Germany from 1965 to 1980 and the number of babies born during those same years.

Dictyostelium, the next step was to use the antibodies that bound to gp24 to block the adhesion of myxamoebae. Using a technique pioneered by Gerisch's laboratory (Beug et al. 1970), Knecht and co-workers (1987) isolated the antibodies' antigen-binding sites (the portions of the antibody molecule that actually recognize the antigen). This step was necessary because the whole antibody molecule contains two antigen-binding sites and would therefore artificially crosslink and agglutinate the myxamoebae. When these antigen-binding fragments (called Fab fragments) were added to aggregation-competent cells, the cells could not aggregate. The fragments inhibited the cells' adhesion, presumably by binding to gp24 and blocking its function. This type of evidence is called **loss-of-function evidence**.

While stronger than correlative evidence, loss-of-function evidence still does not make other inferences impossible. For instance, perhaps the antibody fragments kill the cells outright (as might have been the case if gp24 were a critical transport channel); this would also stop the cells from adhering. Or perhaps gp24 has nothing to do with adhesion itself, but is necessary for the real adhesive molecule to function (perhaps, for example, it stabilizes membrane proteins in general). In this case, blocking the glycoprotein would similarly cause the inhibition of cell aggregation. Thus, loss-of-function evidence must be bolstered by many controls demonstrating that the agents causing the loss of function specifically knock out the particular function and nothing else.

The strongest type of evidence is **gain-of-function evidence**. Here, the initiation of the first event causes the second event to happen even under circumstances where neither event usually occurs. For in-

stance, da Silva and Klein (1990) and Faix and co-workers (1990) obtained such evidence to show that the 80-kDa glycoprotein (gp80) is an adhesive molecule in *Dictyostelium*. They isolated the *gp80* gene and modified it in a way that would cause it to be expressed all the time. They then placed it back into well-fed, dividing myxamoebae, which do not usually express this protein and are not usually able to adhere to one another. The presence of this protein on the membranes of these dividing cells was confirmed by antibody labeling. Moreover, the treated cells now adhered to one another even in the proliferative stage (when they normally do not). Thus, they had gained adhesive function solely by expressing this particular glycoprotein on their cell surfaces. Similar experiments have recently been performed on mammalian cells to demonstrate the presence of particular cell adhesion molecules in the developing embryo. Such gain-of-function evidence is more convincing than other types of evidence.

This "find it; lose it; move it" progression of evidence is at the core of nearly all studies of developmental mechanisms (Adams 2003). Sometimes we find the entire progression in a single paper, but more often, as the case above illustrates, the evidence comes from many laboratories. Such evidence must be taken together. "Every scientist," writes Fleck (1979), "knows just how little a single experiment can prove or convince. To establish proof, an entire system of experiments and controls is needed." Science is a communal endeavor, and it is doubtful that any great discovery is the achievement of a single experiment, or of any individual. Correlative, loss-of-function, and gain-of-function evidence must consistently support each other to establish and solidify a conclusion.

retaining a large enough number of cells to form large fruiting bodies (Müller and Gerisch 1978; Loomis 1988). During late aggregation, the levels of gp80 decrease, and its role is taken over by a third cell adhesion protein, a 150-kDa protein (gp150) whose synthesis becomes apparent just prior to aggregation and which stays on the cell surface during grex migration (Wang et al 2000; Figure 2.18). If *Dictyostelium* cells lack functional genes for gp150, development is arrested at the loose aggregate stage, and the prespore and prestalk cells fail to sort out into their respective regions.* Thus, *Dictyostelium*

has evolved three developmentally regulated systems of cell-cell adhesion that are necessary for the morphogenesis of individual cells into a coherent organism. As we will see in sub-

*The gp150 cell adhesion protein of *Dictyostelium* may be critical for the sorting out of the prespore and prestalk cells in the grex. This protein is first expressed in prestalk cells and leads to their sorting out from prespore cells. A few hours later it is also expressed in prespore cells but at a lower level. If the protein is absent, no sorting out occurs. Thus, it appears that the temporal difference in expression and the levels of expression of this protein in the cell types allows them to sort out (W. Loomis, personal communication).

sequent chapters, metazoan cells also use cell adhesion molecules to form the tissues and organs of the embryo.

Dictyostelium is a "part-time multicellular organism" that does not form many cell types (Kay et al. 1989), and the more complex multicellular organisms do not form by the aggregation of formerly independent cells. Nevertheless, many of the principles of development demonstrated by this "simple" organism also appear in the embryos of more complex phyla (see Loomis and Insall 1999). The ability of individual cells to sense a chemical gradient (as in the myxamoeba's response to cAMP) is crucial for cell migration and morphogenesis during animal development. Moreover, the role of cell surface proteins in cell cohesion is seen throughout the animal kingdom, and differentiation-inducing molecules are now being isolated in metazoan organisms.

Differentiation in Dictyostelium

Differentiation into stalk cell or spore cell reflects another major phenomenon of embryogenesis: the cell's selection of a developmental pathway. Cells often select a particular developmental fate when alternatives are available. A particular cell in a vertebrate embryo, for instance, can become either an epidermal skin cell or a neuron. In *Dictyostelium*, we see a simple dichotomous decision, because only two cell types are possible. How is it that a given cell becomes a stalk cell or a spore cell? There appears to be a progressive commitment to

one of the two alternative pathways (Figure 2.19). At first there is a *bias* toward one path or another. Then, there is a *labile specification*, a time when the cell will normally become either a spore cell or a stalk cell, but when it can still change its fate if placed in a different position in the organism. The third and fourth stages are a *firm commitment* to a specific fate, followed by the cell's *differentiation* into a particular cell type, either a stalk cell or a spore cell.

BIAS. Although the details are not fully known, a cell's fate appears to be regulated by both internal and external agents. Pre-aggregation myxamoebae are not all the same; they can differ in several ways. The internal factors distinguishing individual myxamoebae include nutritional status, cell size, cell cycle phase at starvation, and intracellular calcium levels (Nanjundiah 1997; Azhar et al. 2001). Each of these factors can act to bias the cell toward a prespore or a prestalk pathway. For instance, cells starved in the S and early G2 phases of the cell cycle have relatively high levels of calcium and display a tendency to become stalk cells, while those starved in mid- or late G2 have lower calcium levels and tend to become spore cells.

LABILE SPECIFICATION. Several external factors are also important in specifying cells as stalk or spore. Ammonia, a product of protein degradation, stimulates prespore gene expression and suppresses the expression of those genes that would lead

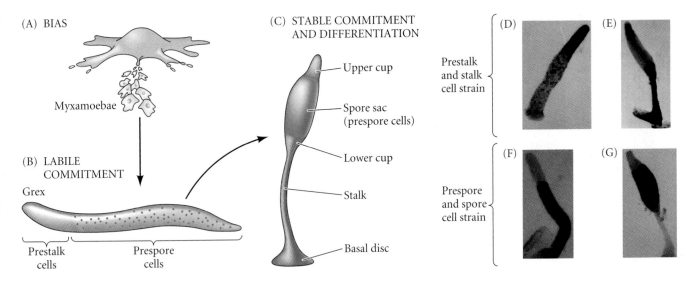

Figure 2.19
Alternative cell fates in *Dictyostelium discoideum*. (A–C) Progressive commitment of cells to become either spore or stalk cells. (A) Myxamoebae may have biases toward stalk or spore formation due to the stage of the cell cycle they were in when starved. (B) As the grex migrates, most prestalk cells are in the anterior third of the grex, while most of the posterior two-thirds are prespore cells. Some prestalk cells are also seen in the posterior, and these cells will contribute to the cups of the spore sac and to the basal disc at the bottom of the stalk. The cell fates are not yet fixed, however, and if the stalk-forming anterior is cut off, the anteriormost cells remaining will convert from stem to stalk. (C) At culmination, the spore-forming cells are massed together in the spore sac. The stalk cells form the cups of the spore sac, as well as the stalk and basal disc. (D, E) Grex and culminant stained with dye that recognizes the extracellular matrix of the prestalk and stalk cells. (F, G) Grex and culminant stained with a dye that recognizes the extracellular matrix of prespore and spore cells. (After Escalante and Vicente 2000. Photographs courtesy of R. Escalante.)

the cell to become a stalk (Oyama and Blumberg 1986). Prestalk cells are able to form only when ammonia is depleted, and this appears to occur by the diffusion of ammonia at the raised anterior tip of the grex. Cyclic AMP may also function to induce prespore cell formation (Ginsburg and Kimmel 1997). High concentrations of cAMP initiate the expression of spore-specific mRNAs in aggregated myxamoebae. Moreover, when slugs are placed in a medium containing an enzyme that destroys extracellular cAMP, the prespore cells lose their differentiated characteristics (Figure 2.20; Schaap and van Driel 1985; Wang et al. 1988a,b). Thus, cAMP works both as an extracellular signal (for chemotaxis) and an intracellular signal (to activate those genes responsible for spore formation).

In the pathway to stalk cells, calcium appears to play a critical role. High calcium levels appear to push cells into the prestalk pathway, and the percentage of stalk cells can be in-

creased by manipulating the slug to have higher calcium levels (Cubitt et al. 1995; Jaffe 1997). A secreted chlorinated lipid, DIF-1, also plays some role in making prestalk cells, and it may induce those genes that make the stalk-specific extracellular matrix. These two factors may act synergistically to push cells with high calcium levels into the prestalk pathway.

COMMITMENT AND DIFFERENTIATION. Two secreted proteins, spore differentiation factors SDF1 and SDF2, appear to be important in the final differentiation of the prespore cells into encapsulated spores (Anjard et al. 1998a,b). SDF1 is important in initiating culmination, while SDF2 seems to cause the prespore cells (but not prestalk cells) to become spores. The prespore cells appear to have a receptor that enables them to respond to SDF2, while the prestalk cells lack this receptor (Wang et al. 1999). The formation of stalk cells from prestalk cells is similarly complicated and may involve several factors working synergistically (Early 1999). The differentiation of stalk cells appears to need a signal from the intracellular enzyme PKA, and at least one type of stalk cell is induced by the DIF-1 lipid (Thompson and Kay 2000; Fukuzawa et al. 2001).

Developmental Patterns among the Metazoa

Since most of the remainder of this book concerns the development of **metazoans**—multicellular animals* that pass through embryonic stages of development—we will present an overview of their developmental patterns here. Figure 2.21 illustrates the major evolutionary trends of metazoan development. The most striking pattern is that life has not evolved in a straight line; rather, there are several branching evolutionary paths. We can see that metazoans belong to one of three major branches: diploblasts, protostomes, and deuterostomes.

The sponges (Porifera) develop in a manner so different from that of any other animal group that some taxonomists do not consider them metazoans at all, and call them "parazoans." A sponge has three major types of somatic cells, but one of these, the **archeocyte**, can differentiate into all the other cell types in the body. Individual cells of a sponge passed through a sieve can reaggregate to form new sponges. Moreover, in some instances, such reaggregation is species-specific: if individual sponge cells from two different species are mixed together, each of the sponges that re-forms contains cells from only one species (Wilson 1907). In these cases, it is thought that the motile archeocytes collect cells from their own species and not from others (Turner 1978). Sponges contain no mesoderm, so the Porifera have no true organ systems, nor do they have a digestive tube, circulatory system,

(A)

(B)

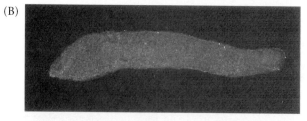

(C)

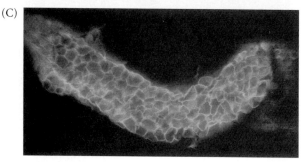

Figure 2.20
Chemicals controlling differentiation in *Dictyostelium*. (A, B) Effects of placing a *Dictyostelium* grex into a medium containing enzymes that destroy extracellular cAMP. (A) Control grex stained for the presence of a prespore-specific protein (white regions). (B) Similar grex stained after treatment with cAMP-degrading enzymes. No prespore-specific product is seen. (C) Higher magnification of a slug treated with DIF (in the absence of ammonia). The stain used here binds to the cellulose wall of the stalk cells. (A, B from Wang et al., 1988a; C from Wang and Schaap, 1989; courtesy of the authors.)

---

*Plants undergo equally complex and fascinating patterns of embryonic and postembryonic development. However, plant development differs significantly from that of animals, and the decision was made to focus this text on the development of animals. Readers who wish to discover some of the differences are referred to Chapter 20, which provides an overview of plant life cycles and the patterns of angiosperm (seed plant) development.

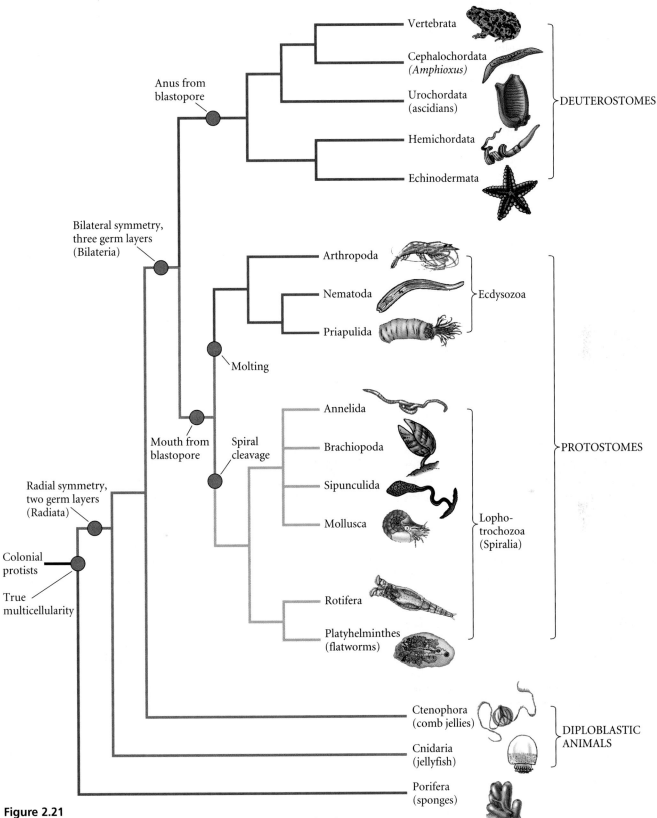

Figure 2.21
Major evolutionary divergences in extant animals. Other models of evolutionary relationships among the phyla are possible. This grouping of the Metazoa is based on embryonic, morphological, and molecular criteria. (Based on J. R. Garey, personal communication.)

nerves, or muscles. Thus, even though they pass through an embryonic and a larval stage, sponges are very unlike most metazoans (Fell 1997). However, sponges do share many features of development (including gene regulatory proteins and signaling cascades) with all the other animal phyla, suggesting that they share a common origin (Coutinho et al. 1998).

The diploblasts

Diploblastic animals are those that have ectoderm and endoderm, but no true mesoderm. The diploblasts include the cnidarians (jellyfish and hydras) and the ctenophores (comb jellies).

Cnidarians and ctenophores constitute the Radiata, so called because they have radial symmetry, like that of a tube or a wheel. In these animals, the mesoderm is rudimentary, consisting of sparsely scattered cells in a gelatinous matrix.

Protostomes and deuterostomes

Most metazoans have bilateral symmetry and three germ layers. The evolution of the mesoderm enabled greater mobility and larger bodies because it became the animal's musculature and circulatory system. The animals of these phyla are known collectively as the Bilateria. All Bilateria are thought to have descended from a primitive type of flatworm. These flatworms were the first to have a true mesoderm (although it was not hollowed out to form a body cavity), and they may have resembled the larvae of certain contemporary coelenterates.

Animals of the Bilataria are further classified as either protostomes or deuterostomes. **Protostomes** (Greek, "mouth first"), which include the mollusc, arthropod, and worm phyla, are so called because the mouth is formed first, at or near the opening to the gut, which is produced during gastrulation. The anus forms later at another location. The **coelom**, or body cavity, of these animals forms from the hollowing out of a previously solid cord of mesodermal cells. There are two major branches of the protostomes. The **Ecdysozoa** includes the animals that molt their exterior skeletons.* The major constituent of this group is Arthropoda, a phylum containing the insects, arachnids, mites, crustaceans, and millipedes. The second major group of protostomes is the **Lophotrochozoa**. These animals are characterized by a common type of cleavage (spiral), a common larval form (the trochophore), and a distinctive feeding apparatus (the lophophore) found in some species. Lophotrochozoan phyla include the flatworms, bryozoans, annelids, and molluscs.

*The name Ecdysozoa is derived from the Greek *ecdysis,* "to shed" or "to get clear of." Asked to provide a more dignified job description for a "stripper," editor H. L. Mencken suggested the term "ecdysiast."

Phyla in the **deuterostome** lineage include the chordates and echinoderms. Although it may seem strange to classify humans, fish, and frogs in the same group as starfish and sea urchins, certain embryological features stress this kinship. First, in deuterostomes ("mouth second"), the oral opening is formed after the anal opening. Also, whereas protostomes generally form their body cavities by hollowing out a solid block of mesoderm (**schizocoelous** formation of the body cavity), most deuterostomes form their body cavities from mesodermal pouches extending from the gut (**enterocoelous** formation of the body cavity). It should be mentioned that there are many exceptions to these generalizations.

The evolution of organisms depends on inherited changes in their development. One of the greatest evolutionary advances—the **amniote egg**—occurred among the deuterostomes. This type of egg, exemplified by that of a chicken (Figure 2.22), is thought to have originated in the amphibian ancestors of reptiles about 255 million years ago. The amniote egg allowed vertebrates to roam on land, far from existing ponds. Whereas most amphibians must return to water to lay their eggs, the amniote egg carries its own water and food supplies. It is fertilized internally and contains yolk to nourish the developing embryo. Moreover, the amniote egg contains four sacs: the **yolk sac**, which stores nutritive pro-

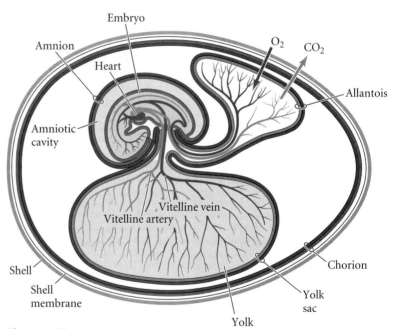

Figure 2.22
Diagram of the amniote egg of the chick, showing the membranes enfolding the 7-day chick embryo. The yolk is eventually surrounded by the yolk sac, which allows the entry of nutrients into the blood vessels. The chorion is derived in part from the ectoderm and extends from the embryo to the shell (where it will fuse with the blood vessel-rich allantois. This chorioallantoic membrane will exchange oxygen and carbon dioxide and absorb calcium from the shell). The amnion provides the fluid medium in which the embryo grows, and the allantois collects nitrogenous wastes that would be dangerous to the embryo. Eventually the endoderm becomes the gut tube and encircles the yolk.

teins; the **amnion**, which contains the fluid bathing the embryo; the **allantois**, in which waste materials from embryonic metabolism collect; and the **chorion**, which interacts with the outside environment, selectively allowing materials to reach the embryo.* The entire structure is encased in a shell that allows the diffusion of oxygen but is hard enough to protect the embryo from environmental assaults and dehydration. A similar development of egg casings enabled arthropods to be the first terrestrial invertebrates. Thus, the final crossing of the boundary between water and land occurred with the modification of the earliest stage in development: the egg.

> VADE MECUM[2] **The amniote egg.** The egg of the chick, as detailed in this sequence, is a beautiful and readily accessible example of the amniote egg—a remarkable adaptation to terrestrial life. **[Click on Chick-early]**

---

*In mammals, the chorion is modified to form the embryonic portion of the placenta—another example of the modification of development to produce evolutionary change.

Embryology provides an endless assortment of fascinating animals and problems to study. In this text, we will use but a small sample of them to illustrate the major principles of animal development. This sample is an incredibly small collection. We are merely observing a small tide pool within our reach, while the whole ocean of developmental phenomena lies before us.

After a brief outline of the experimental and genetic approaches to developmental biology, we will investigate the early stages of animal embryogenesis: fertilization, cleavage, gastrulation, and the establishment of the body axes. Later chapters will concentrate on the genetic and cellular mechanisms by which animal bodies are constructed. Although an attempt has been made to survey the important variations throughout the animal kingdom, a certain deuterostome chauvinism may be apparent. (For a more comprehensive survey of the diversity of animal development across the phyla, see Gilbert and Raunio 1997.)

Principles of Development: Life Cycles and Developmental Patterns

1. The life cycle can be considered a central unit in biology. The adult form need not be paramount. In a sense, the life cycle is the organism.

2. The basic life cycle consists of fertilization, cleavage, gastrulation, germ layer formation, organogenesis, metamorphosis, adulthood, and senescence.

3. Reproduction and sex are two separate processes that may but do not necessarily occur together. Some organisms, such as *Volvox* and *Dictyostelium*, exhibit both asexual reproduction and sexual reproduction.

4. Cleavage divides the zygote into numerous cells called blastomeres.

5. In animal development, gastrulation rearranges the blastomeres and forms the three germ layers.

6. Organogenesis often involves interactions between germ layers to produce distinct organs.

7. Germ cells are the precursors of the gametes. Gametogenesis forms the sperm and the eggs.

8. There are three main ways to provide nutrition to the developing embryo: (1) supply the embryo with yolk; (2) form a larval feeding stage between the embryo and the adult; or (3) create a placenta between the mother and the embryo.

9. Life cycles must be adapted to the nonliving environment and interwoven with other life cycles.

10. Don't regress your tail until you've formed your hindlimbs.

11. There are several types of evidence. Correlation between phenomenon A and phenomenon B does not imply that A causes B or that B causes A. Loss-of-function data (if A is experimentally removed, B does not occur) suggests that A causes B, but other explanations are possible. Gain-of-function data (if A happens where or when it does not usually occur, then B also happens in this new time or place) is most convincing.

12. Protostomes and deuterostomes represent two different sets of variations on development. Protostomes form the mouth first, while deuterostomes form their mouths later, usually forming the anus first.

Literature Cited

Adams, D. 2003. Teaching critical thinking in a developmental biology course at an American liberal arts college. *Int. J. Dev. Biol.* 47: 145–151.

Anjard, C., W. T. Chang, J. Gross and W. Nellen. 1998a. Production and activity of spore differ-

entiation factors (SDFs) in *Dictyostelium*. *Development* 125: 4067–4075.

Anjard, C., C. Zeng, W. F. Loomis and W. Nellen. 1998b. Signal transduction pathways leading to

spore differentiation in *Dictyostelium discoideum*. *Dev. Biol.* 193: 146–155.

Azhar, M., P. K. Kennady, G. Pande, M. Espiritu, W. Holloman, D. Brazill, R. H. Gomer and V. Nanjundiah. 2001. Cell cycle phase, cellular Ca^{2+}

and development in *Dictyostelium discoideum*. *Int. J. Dev. Biol.* 45: 405–414.

Beck, S. D. 1980. *Insect Photoperiodism*, 2nd Ed. Academic Press, New York.

Bergman, K., U. W. Goodenough, D. A. Goodenough, J. Jawitz and H. Martin. 1975. Gametic differentiation in *Chlamydomonas reinhardtii*. II. Flagellar membranes and the agglutination reaction. *J. Cell Biol.* 67: 606–622.

Beug, H., G. Gerisch, S. Kempff, V. Riedel and G. Cremer. 1970. Specific inhibition of cell contact formation in *Dictyostelium* by univalent antibodies. *Exp. Cell Res.* 63: 147–158.

Bonner, J. T. 1947. Evidence for the formation of cell aggregates by chemotaxis in the development of the slime mold *Dictyostelium discoideum*. *J. Exp. Zool.* 106: 1–26.

Bonner, J. T. 1957. A theory of the control of differentiation in the cellular slime molds. *Q. Rev. Biol.* 32: 232–246.

Bonner, J. T. 1965. *Size and Cycle*. Princeton University Press, Princeton, NJ.

Bonner, J. T., D. S. Berkley, E. M. Hall, T. M. Konijn, J. W. Mason, G. O'Keefe and P. B. Wolfe. 1969. Acrasin, acrasinase, and the sensitivity to acrasin in *Dictyostelium discoideum*. *Dev. Biol.* 20: 72–87.

Coutinho, C., J. Seack, G. de Vyler, R. Borojevic and W. E. G. Müller. 1998. Origin of the metazoan body plan: Characterization and functional testing of the promoter of the homeobox gene *EmH-3* from the freshwater sponge *Ephydatia muelleri* in mouse 3T3 cells. *Biol. Chem.* 379: 1243–1251.

Cubitt, A. B., R. A. Firtel, G. Fischer, L. F. Jaffe and A. L. Miller. 1995. Patterns of free calcium in multicellular stages of *Dictyostelium* expressing jellyfish apoaequorin. *Development*. 121: 2291–2301.

Dallon, J. C. and H. G. Othmer. 1997. A discrete cell model with adaptive signalling for aggregation of *Dictyostelium discoideum*. *Phil. Trans. Roy. Soc. Lond.* [B] 352: 391–417.

da Silva, A. M. and C. Klein. 1990. Cell adhesion transformed *D. discoideum* cells: Expression of *gp80* and its biochemical characterization. *Dev. Biol.* 140: 139–148.

Dumais, J., K. Serikawa and D. F. Mandoli. 2000. *Acetabularia*: A unicellular model for understanding subcellular localization and morphogenesis during development. *J. Plant Growth Reg.* 19: 253–264.

Early, A. 1999. Signalling pathways that direct prestalk and stalk cell differentiation in *Dictyostelium*. *Sem. Cell Dev. Biol.* 10: 587–595.

Escalante, R. and J. J. Vicente. 2000. *Dictyostelium discoideum*: A model system for differentiation and patterning. *Int. J. Dev. Biol.* 44: 819–835.

Faix, J., G. Gerisch and A. A. Noegel. 1990. Constitutive overexpression of the contact A glycoprotein enables growth-phase cells of *Dictyostelium discoideum* to aggregate. *EMBO J.* 9: 2709–2716.

Fell, P. E. 1997. Porifera: The sponges. *In* S. F. Gilbert and A. M. Raunio (eds.), *Embryology: Constructing the Organism*. Sinauer Associates, Sunderland, MA, pp. 39–54.

Fleck, L. 1979. *Genesis and Development of a Scientific Fact*. Translated by F. Bradley and T. J. Trenn. University of Chicago Press, Chicago.

Fukuzawa, M., T. Araki, I. Adrian and J. G. Williams. 2001. Tyrosine phosphorylation-independent nuclear translocation of a *Dictyostelium* STAT in response to DIF signaling. *Mol. Cell* 7: 779–788.

Garcia, E. and A.-C. Dazy. 1986. Spatial distribution of poly(A)$^+$ RNA and protein synthesis in *Acetabularia mediterranea*. *Biol. Cell* 58: 23–29.

Gilbert, S. F. and A. M. Raunio (eds.). 1997. *Embryology: Constructing the Organism*. Sinauer Associates, Sunderland, MA.

Ginsburg, G. T. and A. R. Kimmel. 1997. Autonomous and nonautonomous regulation of axis formation by antagonistic signaling via 7-span cAMP receptors and GSK3 in *Dictyostelium*. *Genes Dev.* 11: 2112–2123.

Goodenough, U. W. and R. L. Weiss. 1975. Gametic differentiation in *Chlamydomonas reinhardtii*. III. Cell wall lysis and microfilament associated mating structure activation in wild-type and mutant strains. *J. Cell Biol.* 67: 623–637.

Hallmann, A., P. Amon, K. Godl, M. Heitzer and M. Sumper. 2001. Transcriptional activation by the sexual pheromone and wounding: A new gene family from *Volvox* encoding modular proteins with (hydroxy)proline-rich and metalloproteinase homology domains. *Plant J.* 26: 583–593.

Hämmerling, J. 1934. Über formbildendend Substanzen bei *Acetabularia mediterranea*, ihre räumliche und zeitliche Verteilung und ihre Herkunft. *Wilhelm Roux Arch. Entwicklungsmech. Org.* 131: 1–82.

Jaffe, L. F. 1997. The role of calcium in pattern formation. *In* Y. Maeda, K. Inouye and I. Takeuchi (eds.), Dictyostelium: *A Model System for Cell and Developmental Biology*. Universal Academy Press, Tokyo, pp. 267–277.

Jermyn, K. A. and J. Williams. 1991. An analysis of culmination in *Dictyostelium* using prestalk and stalk-specific cell autonomous markers. *Development* 111: 779–787.

Johnson, R. L. and 7 others. 1989. G-protein-linked signal transduction systems control development in *Dictyostelium*. *Development* [Suppl.]: 75–81.

Kay, R. R., M. Berks and D. Traynor. 1989. Morphogen hunting in *Dictyostelium*. *Development* [Suppl.]: 81–90.

Keller, E. F. and L. A. Segal. 1970. Initiation of slime mold aggregation viewed as an instability. *J. Theor. Biol.* 26: 399–415.

Kirk, D. L. 1988. The ontogeny and phylogeny of cellular differentiation in *Volvox*. *Trends Genet.* 4: 32–36.

Kirk, D. L. 1999. Evolution of multicellularity in the Volvocine lineage. *Curr. Opin. Plant Biol.* 2: 496–501.

Kirk, D. L. 2000. *Volvox* as a model system for studying the ontogeny and phylogeny of multicellularity and cell differentiation. *J. Plant Growth Reg.* 19: 265–274.

Kirk, D. L. 2001a. Germ-soma differentiation in *Volvox*. *Dev. Biol.* 238: 213–223.

Kirk, D. L. 2001b. Cover photo, November 1, 2001. *Dev. Biol.* 239(1).

Kirk, D. L. and M. M. Kirk. 1986. Heat shock elicits production of sexual inducer in *Volvox*. *Science* 231: 51–54.

Kirk, D. L., G. I. Viamontes, K. J. Green and J. L. Bryant, Jr. 1982. Integrated morphogenetic behavior of cell sheets: *Volvox* as a model. *In* S. Subtelny and P. B. Green (eds.), *Developmental Order: Its Origin and Regulation*. Alan R. Liss, New York, pp. 247–274.

Kloppstech, K. and H. G. Schweiger. 1975. Polyadenylated RNA from *Acetabularia*. *Differentiation* 4: 115–123.

Knecht, D. A., D. Fuller and W. F. Loomis. 1987. Surface glycoprotein gp24 involved in early adhesion of *Dictyostelium discoideum*. *Dev. Biol.* 121: 277–283.

Konijn, T. M., J. G. Van De Meene, J. T. Bonner and D. S. Barkley. 1967. The acrasin activity of adenosine-3',5'-cyclic phosphate. *Proc. Natl. Acad. Sci. USA* 58: 1152–1154.

Krutch, J. W. 1956. *The Great Chain of Life*. Houghton Mifflin, Boston.

Loomis, W. F. 1988. Cell-cell adhesion in *Dictyostelium discoideum*. *Dev. Genet.* 9: 549–559.

Loomis, W. F. and R. H. Insall. 1999. A cell for all seasons. *Nature* 401: 440–441.

Mandoli, D. F. 1998. What ever happened to *Acetabularia*? Bringing a once-classic model system into the age of molecular genetics. *Int. Rev. Cytol.* 182: 1–67.

Müller, K. and G. Gerisch. 1978. A specific glycoprotein as the target of adhesion blocking Fab in aggregating *Dictyostelium* cells. *Nature* 274: 445–447.

Nanjundiah, V. 1997. Models for pattern formation in the dictyostelid slime molds. *In* Y. Maeda, K. Inouye and I. Takeuchi (eds.), *Dictyostelium: A Model System for Cell and Developmental Biology*. Universal Academy Press, Tokyo, pp. 305–322.

Nanjundiah, V. 1998. Cyclic AMP oscillations in *Dictyostelium discoideum*: Models and observations. *Biophys. Chem.* 72: 1–8.

Neumann, D. and K.-D. Spindler. 1991. Circasemilunar control of imaginal disc development in *Clunio marinus*: Temporal switching point, temperature-compensated developmental time, and ecdysteroid profile. *J. Insect Physiol.* 37: 101–109.

Oyama, M. and D. D. Blumberg. 1986. Cyclic AMP and NH_3/NH_4^+ both regulate cell-type-specific mRNA accumulation in the cellular slime mold, *Dictyostelium discoideum. Dev. Biol.* 117: 557–566.

Pan, J. and W. J. Snell. 2000. Signal transduction during fertilization in the unicellular green alga *Chlamydomonas. Curr. Opin. Microbiol.* 3: 596–602.

Pommerville, J. and G. Kochert. 1982. Effects of senescence on somatic cell physiology in the green alga *Volvox carteri. Exp. Cell Res.* 14: 39–45.

Powers, J. H. 1908. Further studies on *Volvox*, with description of three new species. *Trans. Am. Microsc. Soc.* 28: 141–175.

Raper, K. B. 1940. Pseudoplasmodium formation and organization in *Dictyostelium discoideum. J. Elisha Mitchell Sci. Soc.* 56: 241–282.

Robertson, A., D. J. Drage and M. H. Cohen. 1972. Control of aggregation in *Dictyostelium discoideum* by an external periodic pulse of cyclic adenosine monophosphate. *Science* 175: 333–335.

Rugh, R. 1950. *The Frog: Its Reproduction and Development.* McGraw-Hill, New York.

Schaap, P. and R. van Driel. 1985. The induction of post-aggregative differentiation in *Dictyostelium discoideum* by cAMP. Evidence for involvement of the cell surface cAMP receptor. *Exp. Cell Res.* 159: 388–398.

Serikawa, K. A., D. M. Porterfield, and D. F. Mandoli. 2001. Asymmetric subcellular mRNA distribution correlates with carbonic anhydrase activity in *Acetabularia acetabulum. Plant Physiol.* 125: 900–911.

Shaffer, B. M. 1953. Aggregation in cellular slime molds: In vitro isolation of acrasin. *Nature* 171: 975.

Shaffer, B. M. 1975. Secretion of cyclic AMP induced by cyclic AMP in the cellular slime mold *Dictyostelium discoideum. Nature* 255: 549–552.

Siegert, F. and C. J. Weijer. 1989. Digital image processing of optical density wave propagation in *Dictyostelium discoideum* and analysis of the effects of caffeine and ammonia. *J. Cell Sci.* 93: 325–335.

Sies, H. 1988. A new parameter for sex education. *Nature* 332: 495.

Strickberger, M. W. 1985. *Genetics*, 3rd Ed. Macmillan, New York.

Sumper, M., E. Berg, S. Wenzl and K. Godl. 1993. How a sex pheromone might act at a concentration below 10^{-16} *M. EMBO J.* 12: 831–836.

Thompson, C. R. L. and R. R. Kay. 2000. The role of DIF-1 signaling in *Dictyostelium* development. *Mol. Cell* 6: 1509–1514.

Tomchick, K. J. and P. N. Devreotes. 1981. Adenosine $3',5'$ monophosphate waves in *Dictyostelium discoideum. Science* 212: 443–446.

Turner, R. S., Jr. 1978. Sponge cell adhesions. *In* D. R. Garrod (ed.), *Specificity of Embryological Interactions.* Chapman and Hall, London, pp. 199–232.

Tyson, J. J. and J. D. Murray. 1989. Cyclic AMP waves during aggregation of *Dictyostelium* amoebae. *Development* 106: 421–426.

Wang, J., L. Hou, D. Awrey, W. F. Loomis, R. A. Firtel and C.-H. Siu. 2000. The membrane glycoprotein gp150 is encoded by the *lagC* gene and mediates cell-cell adhesion by heterophilic binding during *Dictyostelium* development. *Dev. Biol.* 227: 734–745.

Wang, M. R. and P. Schaap. 1989. Ammonia depletion and DIF trigger stalk cell differentiation in intact *Dictyostelium discoideum. Development* 105: 569–574.

Wang, M. R., R. J. Aerts, W. Spek and P. Schaap. 1988a. Cell cycle phase in *Dictyostelium discoideum* is correlated with the expression of cyclic AMP production, detection and degradation: Involvement of cAMP signaling in cell sorting. *Dev. Biol.* 125: 410–416.

Wang, M. R., R. van Driel and P. Schaap. 1988b. Cyclic AMP-phosphodiesterase induces dedifferentiation of prespore cells in *Dictyostelium discoideum* slugs: Evidence that cyclic AMP is the morphogenetic signal for prespore differentiation. *Development* 103: 611–618.

Wang, N., F. Soderbom, C. Anjard, G. Shaulsky and W. F. Loomis. 1999. SDF-2 induction of terminal differentiation in *Dictyostelium discoideum* is mediated by the membrane-spanning sensor kinase DhkA. *Mol. Cell. Biol.* 19: 4750–4756.

Wilson, E. B. 1896. *The Cell in Development and Inheritance.* Macmillan, New York.

Wilson, H. V. 1907. On some phenomena of coalescence and regeneration in sponges. *J. Exp. Zool.* 5: 245–258.

Wilson, N. F., M. J. Foglesong and W. J. Snell. 1997. The *Chlamydomonas* mating type plus fertilization tubule, a prototypic cell fusion organelle: Isolation, characterization, and in vitro adhesion to mating type minus gametes. *J. Cell Biol.* 137: 1537–1553.

Wong, E. F. S., S. K. Brar, H. Sesaki, C. Yang and C.-H. Siu. 1996. Molecular cloning and characterization of DdCAD-1, a calcium-dependent cell-cell adhesion molecule, in *Dictyostelium discoideum. J. Biol. Chem.* 2271: 16399–16408.

3 *Principles of experimental embryology*

DESCRIPTIVE EMBRYOLOGY AND EVOLUTIONARY EMBRYOLOGY both had their roots in anatomy. At the end of the nineteenth century, however, the new biological science of physiology made inroads into embryological research. The questions of "what?" became questions of "how?" A new generation of embryologists felt that embryology should not merely be a guide to the study of anatomy and evolution, but should answer the question, "How does an egg become an adult?" Embryologists were urged to study the mechanisms of organ formation (morphogenesis) and differentiation. This new program was called *Entwicklungsmechanik*, often translated as "causal embryology," "physiological embryology," or "developmental mechanics." Its goals were to find the molecules and processes that caused the visible changes in embryos. Experimentation was to supplement observation in the study of embryos, and embryologists were expected to discover the properties of the embryo by seeing how the embryonic cells responded to perturbations and disruptions. Wilhelm Roux (1894), one of the founders of this branch of embryology, saw it as a grand undertaking:

> We must not hide from ourselves the fact that the causal investigation of organisms is one of the most difficult, if not the most difficult, problem which the human intellect has attempted to solve … since every new cause ascertained only gives rise to fresh questions regarding the cause of this cause.

In this chapter, we will discuss three of the major research programs in experimental embryology. The first concerns how forces outside the embryo influence its development. The second concerns how forces within the embryo cause the differentiation of its cells. The third looks at how the cells order themselves into tissues and organs.

WEBSITE 3.1 **Establishing experimental embryology.** The foundations of *Entwicklungsmechanik* were laid by a group of young investigators who desired a more physiological approach to embryology. These scientists disagreed with one another concerning the mechanisms of development, but they cooperated to secure places to perform and publish their research.

Environmental Developmental Biology

The developing embryo is not isolated from its environment. In numerous instances, environmental cues are a fundamental part of the organism's life cycle. Moreover, removing or altering these environmental parameters can alter development.

Environmental sex determination

SEX DETERMINATION IN AN ECHIUROID WORM: BONELLIA.

When the field of developmental mechanics was first formulated, some of the obvious variables to manipulate were the temperature and media in which embryos were developing. These early studies initiated several experimental programs on the effects of the environment on development. For instance, Baltzer (1914) showed that the sex of the echiuroid worm *Bonellia viridis* depended on where the larva settled. The female *Bonellia* worm is a marine, rock-dwelling animal, with a body about 10 cm long (Figure 3.1). She has a proboscis that can extend over a meter in length. The male *Bonellia*, however, is only 1–3 mm long and resides within the uterus of the female, fertilizing her eggs. Baltzer showed that if a *Bonellia* larva settles on the seafloor, it becomes a female. However, should a larva land on a female's proboscis (which apparently emits chemical signals that attract larvae), it enters the female's mouth, migrates into her uterus, and differentiates into a male. Thus, if a larva lands on the seafloor, it becomes female; if it settles on a proboscis, it becomes male. Baltzer (1914) and Leutert (1974) were able to duplicate this phenomenon in the laboratory, incubating larvae in either the absence or presence of adult females (Figure 3.2).

SEX DETERMINATION IN A VERTEBRATE: ALLIGATOR.

The effects of the environment on development can have important consequences. Recent research has shown that the sex of alligators, crocodiles, and many other reptiles depends not on chromosomes, but on temperature. After studying the sex determination of the Mississippi alligator both in the laboratory

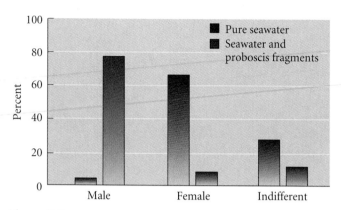

Figure 3.2

In vitro analysis of *Bonellia* sex determination. Larval *Bonellia* were placed either in normal seawater or in seawater containing fragments of the female proboscis. A majority of the animals cultured in the presence of the proboscis fragments became males, whereas in their absence, most became females. (After Leutert 1974.)

and in the field, Ferguson and Joanen (1982) concluded that sex is determined by the temperature of the egg during the second and third weeks of incubation. Eggs incubated at 30°C or below during this time period produce female alligators, whereas those eggs incubated at 34°C or above produce males. (At 32°C, 87% of the hatchlings were female.) Moreover, whereas nests built in wet marshes (close to 30°C) produce females, nests constructed on levees (close to 34°C) give rise to males. These findings are obviously important to wildlife managers and farmers who wish to breed this species. They also raise questions of environmental policy, since the shade of buildings or the heat of thermal effluents can have dramatic effects on sex ratios among reptiles. We will discuss the mechanisms of temperature-dependent sex determination further in Chapter 17.

> **WEBSITE 3.2 The hazards of environmental sex determination.** Ferguson and Joanen (1982) speculate that temperature-dependent sex determination may have been responsible for the extinction of the dinosaurs. The dependence on temperature for sex determination may also be dangerous for reptilian species in our present era of climate change.

Adaptation of embryos and larvae to their environments

PHENOTYPIC PLASTICITY.

Another program of environmental developmental biology concerns how the embryo adapts to its particular environment. August Weismann (1875) pioneered the study of larval adaptations, and recent research in this area has provided some fascinating insights into how an organism's development is keyed to its environment. Weismann noted that butterflies that hatched during different seasons were colored differently, and that this season-depen-

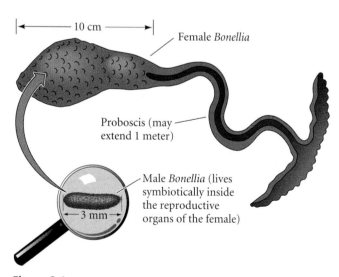

Figure 3.1

Sexual dimorphism in *Bonellia viridis*. The body of the mature female is about 10 cm in length, but the proboscis can extend up to a meter. The body of the symbiotic male is a minute 1–3 mm in length. While the body of the adult female is buried in the ocean sediments, her proboscis extends out of the sediments, where it can be used for feeding or attracting larvae.

progeny that hatch in the summer (Figure 3.4). The caterpillars that hatch in the spring and eat oak catkins (flowers) are yellow-brown, rugose, and beaded, resembling nothing else but an oak catkin. They are magnificently camouflaged against predation. But what of the caterpillars that hatch in the summer, after all the catkins are gone? They, too, are well camouflaged, resembling year-old oak twigs. What controls this difference? By doing reciprocal feeding experiments, Greene (1989) was able to convert spring caterpillars into summer morphs by feeding them oak leaves. The reciprocal experiment did not turn the summer morphs into catkin-like caterpillars. Thus, it appears that the catkin form is the "default state" and that something induces the twiglike morphology. That something is probably a tannin that is concentrated in oak leaves as they mature.

Embryologists have emphasized that what gets inherited is not a deterministic genotype, but rather a genotype that encodes a potential range of phenotypes. The environment is often able to select the phenotype that is adaptive for that sea-

Figure 3.3
Two morphs of *Araschnia levana*, the European map butterfly. The summer morph is represented at the top, the spring morph at the bottom. In this species, the phenotypic differences are elicited by differences in day length and temperature during the larval period. (Photographs courtesy of H. F. Nijhout.)

dent coloration could be mimicked by incubating larvae at different temperatures. Phenotypic variants that result from environmental differences are often called **morphs**. One example of such seasonal variation is the European map butterfly, *Araschnia levana*, which has two seasonal phenotypes so different that Linnaeus classified them as two different species (van der Weele 1995). The spring morph is bright orange with black spots, while the summer morph is mostly black with a white band (Figure 3.3). The shift from spring to summer morph is controlled by changes in both day length and temperature during the larval period. When researchers experimentally mimic spring conditions, summer caterpillars can give rise to "spring" butterflies (see Figure 22.10; Koch and Buchmann 1987; Nijhout 1991).

Another dramatic example of seasonal change in development occurs in the moth *Nemoria arizonaria*. This moth has a fairly typical insect life cycle. Eggs hatch in the spring, and the caterpillars feed on young oak flowers (catkins). These larvae metamorphose in the late spring, mate in the summer, and lay eggs on the oak trees, producing another brood of caterpillars. These caterpillars eat the oak leaves, metamorphose, and mate. Their eggs overwinter to start the cycle over again next spring. What is remarkable is that the caterpillars that hatch in the spring look nothing like their

(A)

(B)

Figure 3.4
Two morphs of *Nemoria arizonaria*. (A) Caterpillars that hatch in the spring eat oak catkins and develop a cuticle that resembles these flowers. (B) Caterpillars that hatch in the summer (after the catkins are gone) eat oak leaves. These caterpillars develop a cuticle that resembles young oak twigs. (Photographs courtesy of E. Greene.)

son or habitat. The continuous range of phenotypes expressed by a single genotype across a range of environmental conditions is called the **reaction norm** (Woltereck 1909; Schmalhausen 1949; Stearns et al. 1991; Schlichting and Pigliucci 1998).

PROTECTING THE EGG FROM UV RADIATION.   Survival in their environments poses daunting challenges for embryos. Indeed, as Darwin clearly noted, most eggs and embryos fail to survive. A sea urchin may broadcast tens of thousands of eggs into the seawater, but only one or two of the resulting embryos will become adult urchins. Most become food for other organisms. Moreover, if the environment changes, embryonic survival may increase or decrease dramatically. For instance, many eggs and early embryos lie in direct sunlight for long periods. If we lie in the sun for hours without sunscreen, we get a burn from the ultraviolet rays; this UV radiation is harmful to our DNA. So how do eggs survive all those hours of constant exposure to the sun (often on the same beaches where we sun ourselves)?

First, it seems that many eggs have evolved natural sunscreens. The eggs of many marine organisms contain high concentrations of mycosporine-like amino acid pigments, which absorb ultraviolet radiation (UV-B). Moreover, just like our melanin pigment, these pigments can be induced by exposure to UV-B radiation (Jokiel and York 1982; Siebeck 1988). The eggs of tunicates are very resistant to UV-B radiation, and much of this resistance comes from extracellular coats that are enriched with mycosporine compounds (Mead and Epel 1995). Adams and Shick (1996; 2001) experimentally manipulated the amount of mycosporine-like amino acids in sea urchin eggs and found that embryos from eggs with more of these compounds were better protected from UV damage than embryos with less. Moreover, when mycosporine-deficient eggs were exposed to ultraviolet radiation, significant developmental anomalies were seen (Figure 3.5). Thus, these mycosporine-like amino acids appear to play an important role in protecting the developing sea urchin embryo against ultraviolet radiation.

> **VADE MECUM[2]   Sea urchins and UV radiation.** This segment presents data documenting the protection of sea urchin embryos by mycosporine-like amino acids. This type of research is linking developmental biology with ecology and conservation biology. **[Click on Sea Urchin-UV]**

It is possible that increased UV-B exposure could be an important factor in the decline in amphibian populations seen throughout the world during the past two decades. Populations of amphibians in widely scattered locations have been drastically reduced in the past two decades, and some of these species (such as the golden toad of Costa Rica) have re-

	UV-filtered light	UV-filtered light + UV-A and UV-B
Early blastula	(A)	(D)
Gastrula	(B)	(E)
Pluteus larva	(C)	(F)

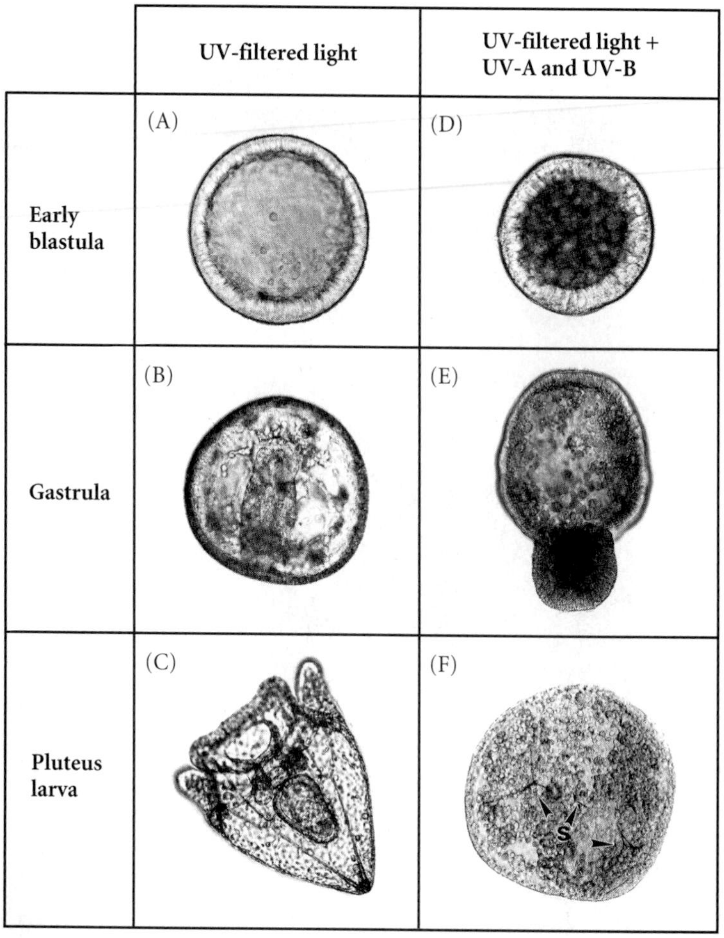

Figure 3.5
The effect of ultraviolet (UV) radiation on embryos of the sea urchin *Strongylocentrotus droebachiensis*. Eggs were fertilized and placed in seawater lacking sources of mycosporine-like amino acids. The first column (A–C) represents embryos grown in light lacking ultraviolet radiation. The second column (D–F) shows the same stage embryos grown in the presence of such filtered light, but with ultraviolet radiation added. The arrowheads at "s" show isolated skeletal spicules. (From Adams and Shick 2001; photographs courtesy of the authors.)

cently become extinct (Phillips 1994). Blaustein and his colleagues (1994) have looked at levels of photolyase, an enzyme that repairs UV damage to DNA by excising and replacing damaged thymidine residues, in amphibian eggs and oocytes. Levels of photolyase varied 80-fold among the tested species, and were correlated with the site of egg laying. Eggs more exposed to the sun had higher levels of photolyase (Table 3.1). These levels also correlated with whether or not the species was suffering population decline. The highest photolyase levels were found in those species (such as the Pacific tree frog, *Hyla regilla*) whose populations were not in decline. The lowest levels were seen in those species (such as the Western toad, *Bufo boreas*, and the Cascades frog, *Rana cascadae*) whose populations had declined dramatically.

TABLE 3.1 Photolyase activity correlated with exposure of eggs to ultraviolet radiation in 10 amphibian species

Species	Photolyase activity[a]	Mode of egg-laying	Exposure to sunlight
Plethodon dunni	<0.1	Eggs hidden	None
Xenopus laevis	0.1	Eggs laid in laboratory[b]	Limited
Triturus granulosa	0.2	Eggs hidden	Limited
Rana variegatus	0.3	Eggs hidden	None
Plethodon vehiculum	0.5	Eggs hidden	None
Ambystoma macrodactylum	0.8	Eggs often laid in open water	Some
Ambystoma gracile	1.0	Eggs often laid in open water	Some
Bufo boreas	1.3	Eggs laid in open, often in shallow water	High
Rana cascadae	2.4	Eggs laid in open shallow water	High
Hyla regilla	7.5	Eggs laid in open shallow water	High

Source: After Blaustein et al. 1994.

[a]Specific activity of photolyase, 1011 thymidine dimers separated per hour per mg. The values are averages of 6–8 assays.

[b]In nature, *Xenopus laevis* eggs are laid under vegetation with limited exposure to sunlight.

Blaustein and his colleagues tested whether or not UV-B could be a factor in lowering the hatching rate of amphibian eggs. At two field sites, they divided the eggs of each of three amphibian species into three groups (Figure 3.6). The first group developed without any sun filter. The second group developed under a filter that allowed UV-B to pass through. The third group developed under a filter that blocked UV-B from reaching the eggs. For *Hyla regilla*, the filters had no effect, and hatching success was excellent under all three conditions. For *Rana cascadea* and *Bufo boreas*, however, the UV-B blocking filter raised the percentage of eggs hatched from about 60% to close to 80%.

The effects of UV-B radiation in mediating amphibian population declines appears to be complex, involving climate change and fungal pathogens. Using long-term observational data and by manipulating the depth of pond water in which frogs laid their eggs, Kiesecker and colleagues (2001) showed that climate-induced reductions in pond water depth increase the exposure of eggs and embryos to UV-B radiation and, consequently, increase their vulnerability to fungal infection. In these studies, water depth and UV-B exposure in the Pacific Northwest of the United States was strongly linked to El Niño cycles. Moreover, elevated sea-surface temperatures in the southern Pacific since the mid-1970s, which have affected the climate over much of the world, could produce conditions for the pathogen-mediated amphibian declines in many regions. Therefore, UV-B radiation may be the factor that links changes in climate with changes in amphibian mortality.

The environmental programs of experimental embryology were a major part of the discipline when *Entwicklungsmechanik* was first established. However, it soon became obvious that experimental variables could be better controlled in the laboratory than in the field, and that a scientist could do many more experiments in the laboratory. Thus,

Figure 3.6
Hatching success rates in three amphibian species in the field. At each of two sites, eggs were placed in enclosures that were unshielded, shielded with an acetate screen that admitted UV-B radiation, or shielded with a Mylar screen that blocked UV-B radiation. Eggs of the tree frog *Hyla regilla* hatched successfully under all three conditions. Eggs of the frog *Rana cascadae* and the toad *Bufo boreas* hatched significantly better when protected from UV-B radiation. (After Blaustein et al. 1994.)

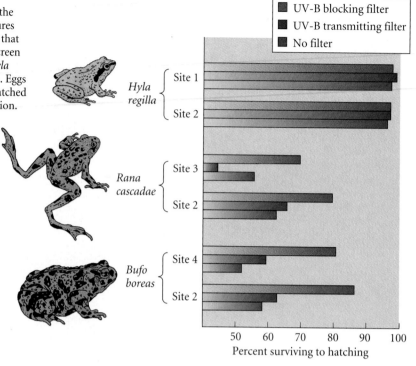

field experimentation in embryology dwindled in the first decades of the twentieth century (see Nyhart 1995). However, with our increasing concern about the environment, this area of developmental biology has become increasingly important. Other recent work in this field will be detailed in Chapter 22.

The Developmental Dynamics of Cell Specification

An embryo's environment may be a tide pool, a pond, or a uterus. As we saw above, the embryo interacts with its environment, and its developmental trajectory can be guided by information from its surroundings. On a smaller scale, the environment of an embryonic cell consists of the surrounding tissues within the embryo, and the fate of that cell (for instance, whether it becomes part of the skin or part of the lens) often depends upon its interactions with other components of its immediate "ecosystem."

Thus, a second research program of experimental embryology studies how interactions between embryonic cells generate the embryo. The development of specialized cell types is called **differentiation** (Table 3.2). These overt changes in cellular biochemistry and function are preceded by a process resulting in the **commitment** of the cell to a certain fate. At this point, even though the cell or tissue does not differ phenotypically from its uncommitted state, its developmental fate has become restricted.

The process of commitment can be divided into two stages (Harrison 1933; Slack 1991). The first stage is a labile phase called **specification**. The fate of a cell or a tissue is said to be specified when it is capable of differentiating autonomously when placed in a neutral environment, such as a petri dish or test tube. (The environment is neutral with respect to the developmental pathway.) At this stage, the commitment is still capable of being reversed. The second stage of commitment is **determination**. A cell or tissue is said to be determined when it is capable of differentiating autonomously even when placed into another region of the embryo. If it is able to differentiate according to its original fate even under these circumstances, it is assumed that the commitment is irreversible.*

Autonomous specification

Three basic modes of commitment have been described (Table 3.3; Davidson 1991). The first is called **autonomous specification**. In this case, if a particular blastomere is removed from an embryo early in its development, that isolated blastomere will produce the same types of cells that it would have made if it were still part of the embryo (Figure 3.7). Moreover, the embryo from which that blastomere is taken will lack those cells (and only those cells) that would have been produced by the missing blastomere. Autonomous specification gives rise to a pattern of embryogenesis referred to as **mosaic development**, since the embryo appears to be constructed like a tile mosaic of independent, self-differentiating parts. Invertebrate embryos, especially those of molluscs, annelids, and tunicates, often use autonomous specification to determine the fates of their cells. In these embryos, **morphogenetic determinants** (certain proteins or messenger RNAs) are placed in different regions of the egg cytoplasm and are apportioned to the different cells as the embryo divides. These morphogenetic determinants specify the cell type.

---

*Recall the discussion of specification of spore and stalk cells in the slime mold *Dictyostelium* in the previous chapter. This irreversibility of determination and differentiation is only with respect to normal development. As Dolly and other cloned animals have shown, the nucleus of a differentiated cell can be reprogrammed experimentally to give rise to any cell type in the body. We will discuss this phenomenon in detail in the next chapter.

TABLE 3.2   Some differentiated cell types and their major products

Cell type of cell	Differentiated cell product	Specialized function
Keratinocyte (epidermal cell)	Keratin	Protection against abrasion, desiccation
Erythrocyte (red blood cell)	Hemoglobin	Transport of oxygen
Lens cell	Crystallins	Transmission of light
B lymphocyte	Immunoglobulins	Antibody synthesis
T lymphocyte	Cytokines	Destruction of foreign cells; regulation of immune response
Melanocyte	Melanin	Pigment production
Pancreatic islet cell	Insulin	Regulation of carbohydrate metabolism
Leydig cell (♂)	Testosterone	Male sexual characteristics
Chondrocyte (cartilage cell)	Chondroitin sulfate; type II collagen	Tendons and ligaments
Osteoblast (bone-forming cell)	Bone matrix	Skeletal support
Myocyte (muscle cell)	Muscle actin and myosin	Contraction
Hepatocyte (liver cell)	Serum albumin; numerous enzymes	Production of serum proteins and numerous enzymatic functions
Neurons	Neurotransmitters (acetylcholine, epinephrine, etc.)	Transmission of electric impulses
Tubule cell (♀) of hen oviduct	Ovalbumin	Egg white proteins for nutrition and protection of embryo
Follicle cell (♀) of insect ovary	Chorion proteins	Eggshell proteins for protection of embryo

Autonomous specification was first demonstrated in 1887 by a French medical student, Laurent Chabry. Chabry desired to know the causes of birth defects, and he reasoned that such malformations might be caused by the lack of certain cells. He decided to perform experiments on tunicate embryos, since they have relatively large cells and were abundant in a nearby bay. This was a fortunate choice, because tunicate embryos develop rapidly into larvae with relatively few cells and cell types (Chabry 1887; Fischer 1991). Chabry set out to produce specific malformations by isolating or lancing specific blastomeres of the cleaving tunicate embryo. He discovered that each blastomere was responsible for producing a particular set of larval tissues (Figure 3.8). In the absence of particular blastomeres, the larva lacked just those structures normally formed by those cells. Moreover, he observed that when particular cells were isolated from the rest of the embryo, they formed their characteristic structures apart from the context of the other cells. Thus, each of the tunicate cells appeared to be developing autonomously.*

*This was not the answer Chabry expected, nor the one he had hoped to find. In nineteenth-century France, conservatives favored preformationist views, which were interpreted to support hereditary inequalities between members of a human community. What you were was determined by your lineage. Liberals, especially Socialists, favored epigenetic views, which were interpreted to indicate that everyone started off with an equal hereditary endowment, and that no one had a "right" to a higher position than any other person. Chabry, a Socialist who hated the inherited rights of the aristocrats, took pains not to extrapolate his data to anything beyond tunicate embryos (see Fischer 1991).

Figure 3.7

Autonomous specification (mosaic development). (A–C) Differentiation of trochoblast (ciliated) cells of the mollusc *Patella*. (A) 16-cell stage seen from the side; the presumptive trochoblast cells are shaded. (B) 48-cell stage. (C) Ciliated larval stage, seen from the animal pole. (D–G) Differentiation of a *Patella* trochoblast cell isolated from the 16-cell stage and cultured in vitro. (E, F) Results of the first and second divisions in culture. (G) Ciliated products of (F). Even in isolated culture, these cells divide and become ciliated at the correct time. (After Wilson 1904.)

Normal development of *Patella*

Presumptive trochoblast

Isolated trochoblast development

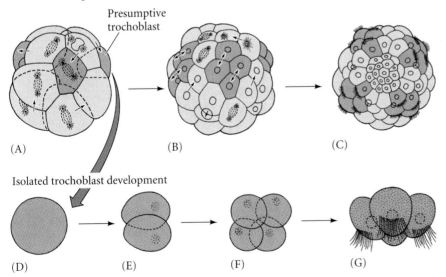

(A)   (B)   (C)

(D)   (E)   (F)   (G)

TABLE 3.3   Modes of cell type specification and their characteristics

I. *Autonomous specification*
   Characteristic of most invertebrates.
   Specification by differential acquisition of certain cytoplasmic molecules present in the egg.
   Invariant cleavages produce the same lineages in each embryo of the species. Blastomere fates are generally invariant.
   Cell type specification precedes any large-scale embryonic cell migration.
   Produces "mosaic" development: cells cannot change fate if a blastomere is lost.

II. *Conditional specification*
   Characteristic of all vertebrates and few invertebrates.
   Specification by interactions between cells. Relative positions are important.
   Variable cleavages produce no invariant fate assignments to cells.
   Massive cell rearrangements and migrations precede or accompany specification.
   Capacity for "regulative" development: allows cells to acquire different functions.

III. *Syncytial specification*
   Characteristic of most insect classes.
   Specification of body regions by interactions between cytoplasmic regions prior to cellularization of the blastoderm.
   Variable cleavage produces no rigid cell fates for particular nuclei.
   After cellularization, conditional specification is most often seen.

Source: After Davidson 1991.

ated muscle cells as well as its normal ectodermal progeny (Figure 3.10; Whittaker 1982). The mechanisms of autonomous specification will be detailed in Chapter 8.

Conditional specification

THE PHENOMENON OF CONDITIONAL SPECIFICATION. A second mode of commitment involves interactions among neighboring cells. In this type of specification, each cell originally has the ability to become any of many different cell types. However, interactions of the cell with other cells restrict the fate of one or more of the participants. This mode of commitment is sometimes called **conditional specification** because the fate of a cell depends upon the conditions in which the cell finds itself. If a blastomere is removed from an early embryo that uses conditional specification, the remaining embryonic cells alter their fates so that the roles of the missing cells can be taken

Recent studies have confirmed that when particular cells of the 8-cell tunicate embryo are removed, the embryo lacks those structures normally produced by the missing cells, and the isolated cells produce these structures away from the embryo. J. R. Whittaker provided dramatic biochemical confirmation of the cytoplasmic segregation of the morphogenetic determinants responsible for this pattern. Whittaker (1973) stained blastomeres for the presence of the enzyme acetylcholinesterase. This enzyme is found only in muscle tissue and is involved in enabling larval muscles to respond to repeated nerve impulses. From the cell lineage studies of Conklin and others (see Chapter 1), it was known that only one pair of blastomeres (the posterior vegetal pair, B4.1) in the 8-cell tunicate embryo is capable of producing tail muscle tissue. (As discussed in Chapter 1, the B4.1 blastomere pair contains the yellow crescent cytoplasm that correlates with muscle determination.) When Whittaker removed these two cells and placed them in isolation, they produced muscle tissue that stained positively for the presence of acetylcholinesterase (Figure 3.9). When he transferred some of the yellow crescent cytoplasm of the B4.1 (muscle-forming) blastomere into the b4.2 (ectoderm-forming) blastomere of an 8-cell tunicate embryo, the ectoderm-forming blastomere gener-

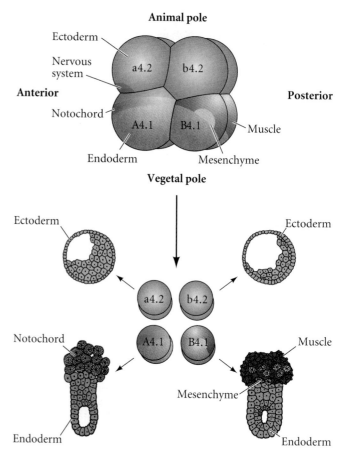

Figure 3.8
Autonomous specification in the early tunicate embryo. When the four blastomere pairs of the 8-cell embryo are dissociated, each forms the structures it would have formed had it remained in the embryo. (The fate map of the tunicate shows that the left and right sides produce identical cell lineages.) (After Reverberi and Minganti 1946.)

(A)

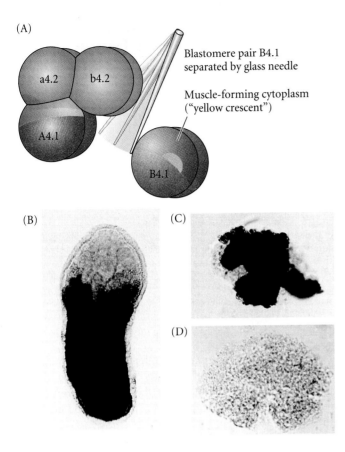

Blastomere pair B4.1
separated by glass needle

Muscle-forming cytoplasm
("yellow crescent")

over. This ability of embryonic cells to change their fates to compensate for the missing parts is called **regulation** (Figure 3.11). The isolated blastomere can also give rise to a wide variety of cell types (and sometimes generates cell types that the cell would normally not have made if it were still part of the embryo). Thus, conditional specification gives rise to a pattern of embryogenesis called **regulative development**.* Regulative development is seen in most vertebrate embryos, and it is obviously critical in the development of identical twins. In the formation of such twins, the cleavage-stage cells of a single embryo divide into two groups, and each group of cells produces a fully developed individual (Figure 3.12).

The research leading to the discovery of conditional specification began with the testing of a hypothesis claiming that there was no such thing. In 1893, August Weismann proposed the first testable model of cell specification, the **germ plasm theory**. Based on the scant knowledge of fertilization available at that time, Weismann boldly proposed that the sperm and egg provided equal chromosomal contributions, both quantitatively and qualitatively, to the new organism. Moreover, he postulated that the chromosomes carried the inherited potentials of this new organism.[†] However, not all the determinants on the chromosomes were thought to enter every cell of the embryo. Instead of dividing equally, the chromosomes were hypothesized to divide in such a way that different chromosomal determinants entered different cells. Whereas the fertil-

Figure 3.9
Acetylcholinesterase in the progeny of the muscle lineage blastomeres (B4.1) isolated from a tunicate embryo at the 8-cell stage. (A) Diagram of the isolation procedure. (B) Localization of acetylcholinesterase in the tail muscles of an intact tunicate larva. The presence of the enzyme is demonstrated by the dark staining. The same dark staining is seen in the progeny of the B4.1 blastomere pair (C), but not in the remaining 6/8 of the embryo (D) when incubated for the length of time it normally takes to form a larva. (From Whittaker 1977; photographs courtesy of J. R. Whittaker.)

*Sydney Brenner (quoted in Wilkins 1993) has remarked that animal development can proceed according to either the American or the European plan. Under the European plan (autonomous specification), you are what your progenitors were. Lineage is important. Under the American plan (conditional specification), the cells start off undetermined, but with certain biases. There is a great deal of mixing, lineages are not critical, and one tends to become what one's neighbors are.

[†]Embryologists were thinking in these terms some 15 years before the rediscovery of Mendel's work. Weismann (1892, 1893) also speculated that these nuclear determinants of inheritance functioned by elaborating substances that became active in the cytoplasm!

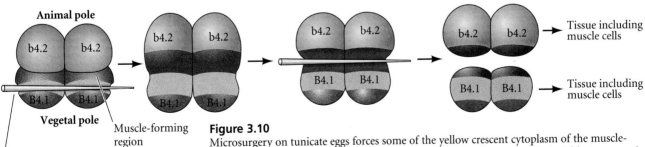

Needle pushes muscle-forming
cytoplasm into animal cells

Figure 3.10
Microsurgery on tunicate eggs forces some of the yellow crescent cytoplasm of the muscle-forming B4.1 blastomeres to enter the b4.2 (epidermis- and nerve-producing) blastomere pair. Pressing the B4.1 blastomeres with a glass needle causes the regression of the cleavage furrow. The furrow will re-form at a more vegetal position where the cells are cut with a needle. The new furrow will thereby separate the cells in such a way that the b4.2 blastomeres receive some of the muscle-forming ("yellow crescent") B4.1 cytoplasm. These modified b4.2 cells produce muscle cells as well as their normal ectodermal progeny. (After Whittaker 1982.)

(A)

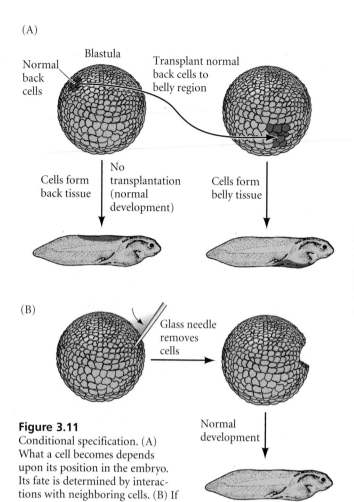

Figure 3.11
Conditional specification. (A) What a cell becomes depends upon its position in the embryo. Its fate is determined by interactions with neighboring cells. (B) If cells are removed from the embryo, the remaining cells can regulate and compensate for the missing part.

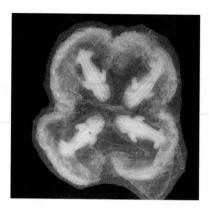

Figure 3.12
In the early developmental stages of many vertebrates, the separation of the embryonic cells into two parts can create twins. This phenomenon occurs sporadically in humans. However, in the nine-banded armadillo, *Dasypus novemcinctus*, the original embryo always splits into four separate groups of cells, each of which forms its own embryo. (Photograph courtesy of K. Benirschke.)

ized egg would carry the full complement of determinants, certain somatic cells would retain the "blood-forming" determinants while others would retain the "muscle-forming" determinants, and so forth (Figure 3.13). Only in the nuclei of those cells destined to become gametes (the germ cells) were all types of determinants thought to be retained. The nuclei of all other cells would have only a subset of the original determinant types.

Figure 3.13
Weismann's theory of inheritance. The germ cell gives rise to the differentiating somatic cells of the body (indicated in color), as well as to new germ cells (blue). Weismann hypothesized that only the germ cells contained all the inherited determinants. The somatic cells were each thought to contain a subset of the determinants. The types of determinants found in the nucleus would determine the cell type. (After Wilson 1896.)

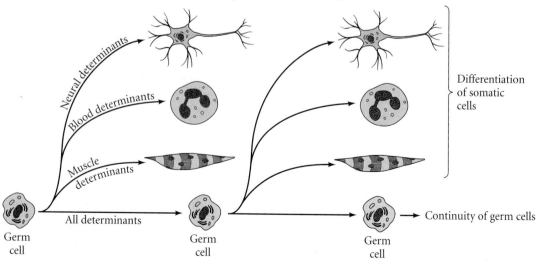

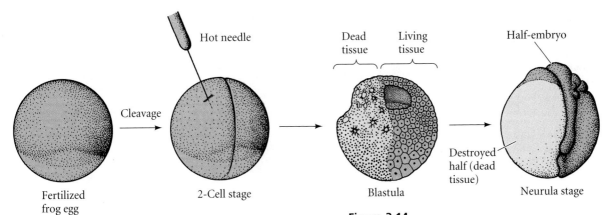

Figure 3.14
Roux's attempt to demonstrate mosaic development. Destroying (but not removing) one cell of a 2-cell frog embryo results in the development of only one-half of the embryo.

In postulating this model, Weismann had proposed a hypothesis of development that could be tested immediately. Based on the fate map of the frog embryo, Weismann claimed that when the first cleavage division separated the future right half of the embryo from the future left half, there would be a separation of "right" determinants from "left" determinants in the resulting blastomeres. The testing of this hypothesis pioneered three of the four major techniques involved in experimental embryology:

- The **defect experiment**, wherein one destroys a portion of the embryo and then observes the development of the impaired embryo.
- The **isolation experiment**, wherein one removes a portion of the embryo and then observes the development of the partial embryo and the isolated part.
- The **recombination experiment**, wherein one observes the development of the embryo after replacing an original part with a part from a different region of the embryo.
- The **transplantation experiment**, wherein one portion of the embryo is replaced by a portion from a different embryo. This fourth technique was used by some of the same scientists when they first constructed fate maps of early embryos (see Chapter 1).

One of the first scientists to test Weismann's hypothesis was Wilhelm Roux, a young German embryologist. In 1888, Roux published the results of a series of defect experiments in which he took 2- and 4-cell frog embryos and destroyed some of the cells of each embryo with a hot needle. Weismann's hypothesis predicted the formation of right or left half-embryos. Roux obtained half-blastulae, just as Weismann had predicted (Figure 3.14). These developed into half-neurulae having a complete right or left side, with one neural fold, one ear pit, and so on. He therefore concluded that the frog embryo was a mosaic of self-differentiating parts, and that each cell probably received a specific set of determinants and differentiated accordingly.

Nobody appreciated Roux's work and the experimental approach to embryology more than Hans Driesch. Driesch's goal was to explain development in terms of the laws of physics and mathematics. His initial investigations were similar to those of Roux. However, while Roux's studies were *defect* experiments that answered the question of how the remaining blastomeres of an embryo would develop when a subset of blastomeres was destroyed, Driesch (1892) sought to extend this research by performing *isolation* experiments. He separated sea urchin blastomeres from each other by vigorous shaking (or, later, by placing them in calcium-free seawater). To Driesch's surprise, each of the blastomeres from a 2-cell embryo developed into a complete larva. Similarly, when Driesch separated the blastomeres of 4- and 8-cell embryos, some of the isolated cells produced entire pluteus larvae (Figure 3.15). Here was a result drastically different from the predictions of Weismann or Roux. Rather than self-differentiating into its future embryonic part, each isolated blastomere regulated its development so as to produce a complete organism. Moreover, these experiments provided the first experimentally observable instance of regulative development.

Driesch confirmed regulative development in sea urchin embryos by performing an intricate recombination experiment. In sea urchin eggs, the first two cleavage planes are normally meridional, passing through both the animal and vegetal poles, whereas the third division is equatorial, dividing the embryo into four upper and four lower cells. Driesch (1893) changed the direction of the third cleavage by gently compressing early embryos between two glass plates, thus causing the third division to be meridional like the preceding two. After he released the pressure, the fourth division was equatorial. This procedure reshuffled the nuclei, causing a nucleus that normally would be in the region destined to form endoderm to now be in the presumptive ectoderm region. Some nuclei that would normally have produced dorsal structures were now found in the ventral cells (Figure 3.16). If segregation of nuclear determinants had occurred (as had been proposed by Weismann and Roux), the resulting embryo should

Figure 3.15

Driesch's demonstration of regulative development. (A) An intact 4-cell sea urchin embryo generates a normal pluteus larva. (B) When one removes the 4-cell embryo from its fertilization envelope and isolates each of the four cells, each cell can form a smaller, but normal, pluteus larva. (All larvae are drawn to the same scale.) Note that the four larvae derived in this way are not identical, despite their ability to generate all the necessary cell types. Such variations are also seen in adult sea urchins formed in this way (Marcus 1979). (Photograph courtesy of G. Watchmaker.)

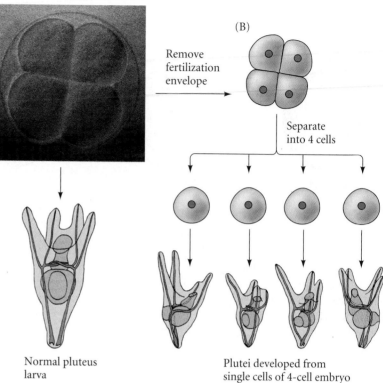

(A) Fertilization envelope

(B)

Remove fertilization envelope

Separate into 4 cells

Normal pluteus larva

Plutei developed from single cells of 4-cell embryo

have been strangely disordered. However, Driesch obtained normal larvae from these embryos. He concluded, "The relative position of a blastomere within the whole will probably in a general way determine what shall come from it."

The consequences of these experiments were momentous, both for embryology and for Driesch personally. First, Driesch had demonstrated that the prospective potency of an isolated blastomere (those cell types it was possible for it to form) is greater than its prospective fate (those cell types it would normally give rise to over the unaltered course of its development). According to Weismann and Roux, the prospective potency and the prospective fate of a blastomere should have been identical. Second, Driesch concluded that the sea urchin embryo is a "harmonious equipotential system" because all of its potentially independent parts functioned together to form a single organism. Third, he concluded that the fate of a nucleus depended solely on its location in the embryo. Driesch (1894) hypothesized a series of events wherein development proceeded by the interactions of the nucleus and cytoplasm:

> *Insofar as it contains a nucleus, every cell, during development, carries the totality of all primordia; insofar as it contains a specific cytoplasmic cell body, it is specifically enabled by this to respond to specific effects only. …When nuclear material is activated, then, under its guidance, the cytoplasm of its cell that had first influenced the nucleus is in turn changed, and thus the basis is established for a new elementary process, which itself is not only the result but also a cause.*

This strikingly modern concept of nuclear-cytoplasmic interaction and nuclear equivalence eventually caused Driesch to abandon science. Because the embryo could be subdivided into parts that were each capable of re-forming the entire organism, he could no longer envision it as a physical machine. In other words, Driesch had come to believe that development could not be explained by physical forces. Harking back to Aristotle, he invoked a vital force, entelechy ("internal goal-di-

rected force"), to explain how development proceeds. Essentially, he believed that the embryo was imbued with an internal psyche and the wisdom to accomplish its goals despite the obstacles embryologists placed in its path. Unable to explain his results in terms of the physics of his day, Driesch renounced the study of developmental physiology and became a philosophy professor, proclaiming vitalism (the doctrine that living things cannot be explained by physical forces alone) until his death in 1941. However, others, especially Oscar Hertwig (1894), were able to incorporate Driesch's experiments into a more sophisticated experimental embryology.*

---

*Driesch also became an outspoken opponent of the Nazis, and was one of the first non-Jewish professors to be forcibly retired when Hitler came to power (Harrington 1996). Hertwig used Driesch's experiments and some of his own to strengthen within embryology a type of materialist philosophy called **wholist organicism**. This philosophy embraces the views that (1) the properties of the whole cannot be predicted solely from the properties of the component parts, and (2) the properties of the parts are informed by their relationship to the whole. As an analogy, the meaning of a sentence obviously depends on the meanings of its component parts, words. However, the meaning of each word depends on the entire sentence. In the sentence, "The party leaders were split on the platform," the possible meanings of each noun and verb are limited by the meaning of the entire sentence and by their relationships to other words within the sentence. Similarly, the phenotype of a cell in the embryo depends on its interactions within the entire embryo. The opposite materialist view is **reductionism**, which maintains that the properties of the whole can be known if all the properties of the parts are known. Embryology has traditionally espoused wholist organicism as its ontology (model of reality) while maintaining a reductionist methodology (experimental procedures) (Needham 1943; Haraway 1976; Hamburger 1988; Gilbert and Sarkar 2000).

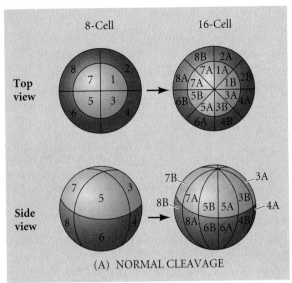 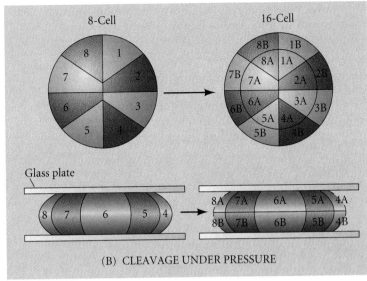

Figure 3.16
Driesch's pressure-plate experiment for altering the distribution of nuclei. (A) Normal cleavage in 8- to 16-cell sea urchin embryos, seen from the animal pole (upper sequence) and from the side (lower sequence). (B) Abnormal cleavage planes formed under pressure, as seen from the animal pole and from the side. (After Huxley and de Beer 1934.)

VADE MECUM[2] **Sea urchin development.** Roux's and Dreisch's experiments manipulated normal development. Normal sea urchin development is seen here in video and labeled photographs. [**Click on Sea Urchin**]

The differences between Roux's experiments and those of Driesch are summarized in Table 3.4. The difference between isolation and defect experiments and the importance of the interactions among blastomeres were highlighted in 1910, when J. F. McClendon showed that isolated frog blastomeres behave just like separated sea urchin cells. Therefore, the mosaic-like development of the first two frog blastomeres in Roux's study was an artifact of the defect experiment. Something in or on the dead blastomere still informed the live cells that it existed. Therefore, even though Weismann and Roux pioneered the study of developmental physiology, their proposition that differentiation is caused by the segregation of nuclear determinants was soon shown to be incorrect.

MORPHOGEN GRADIENTS. Cell fates may be specified by neighboring cells, but cell fates can also be specified by specific amounts of soluble molecules secreted at a distance from the target cells. Such a soluble molecule is called a **morphogen**, and a morphogen may specify more than one cell type by forming a **concentration gradient**.

The concept of morphogen gradients had been used to model another phenomenon of regulative development: **regeneration**. It had been known since the 1700s that when hydras and planarian flatworms were cut in half, the head half would regenerate a tail from the wound site, while the tail half would regenerate a head. Allman (1864) had called attention to the fact that this phenomenon indicated a polarity in the organization of the hydra. It was not until 1905, however, that Thomas Hunt Morgan (1905, 1906) realized that such polarity indicated an important principle in development. He

TABLE 3.4 Experimental procedures and results of Roux and Driesch

Investigator	Organism	Type of experiment	Conclusion	Interpretation concerning potency and fate
Roux (1888)	Frog (*Rana fusca*)	Defect	Mosaic development (autonomous specification)	Prospective potency equals prospective fate
Driesch (1892)	Sea urchin (*Echinus microtuberculatus*)	Isolation	Regulative development (conditional specification)	Prospective potency is greater than prospective fate
Driesch (1893)	Sea urchin (*Echinus* and *Paracentrotus*)	Recombination	Regulative development (conditional specification)	Prospective potency is greater than prospective fate

pointed out that if the head and tail were both cut off a flat-worm, leaving only the medial segment, that segment would regenerate a head from the former anterior end and a tail from the former posterior end— never the reverse (Figure 3.17A,B). Moreover, if the medial segment were sufficiently small, the regenerating portions would be abnormal (Figure 3.17C). Morgan postulated a gradient of anterior-producing materials concentrated in the head region. The middle segment would be told what to regenerate at both ends by the concentration gradient of these materials. If the piece were too small, however, the gradient would not be sensed within the segment. (It is possible that there are actually two gradients in the flatworm, one to direct the formation of a head and one to direct the production of a tail. Regeneration will be discussed in more detail in Chapter 18.)

VADE MECUM[2] **Flatworm regeneration.** You should see it for yourself. Flatworms are easy to obtain, and cutting the animal in half does nothing more than what the animal does to itself. Here are videos and easy instructions for experimenting with these fascinating animals. **[Click on Flatworm]**

In the 1930s through the 1950s, gradient models were used to explain conditional cell specification in sea urchin and amphibian embryos (Hörstadius and Wolsky 1936; Hörstadius 1939; Toivonen and Saxén 1955). In the 1960s, these gradient models were extended to explain how cells might be told their position along an embryonic axis (Lawrence 1966; Stumpf 1966; Wolpert 1968, 1969). According to such models, a soluble substance—the morphogen— diffuses from its site of synthesis (source) to its site of degradation (sink). Wolpert (1968) illustrated such a gradient of positional information using "the French flag analogy." Imagine a row of "flag cells," each of which is capable of differentiating into a red, white, or blue cell. Then imagine a morphogen whose source is on the left-hand edge of the blue stripe, and whose sink is at the other end of the flag, on the right-hand edge of the red stripe. A concentration gradient is thus formed, being highest at one end of the "flag tissue" and lowest at the other. The specification of the multipotential cells in this tissue is accomplished by the concentration of the morphogen. Cells sensing a high concentration of the morphogen become blue. Then there is a threshold of morphogen concentration below which cells become white. As the declining concentration of morphogen falls below another threshold, the cells become red (Figure 3.18A).

Different tissues may use the same gradient system, but respond to the gradient in a different way. If cells that would normally become the middle segment of a *Drosophila* leg are removed from the leg-forming area of the larva and placed in the region that will become the tip of the fly's antenna, they differentiate into claws. These cells retain their committed status as leg cells, but respond to the positional information of their environment. Thereby, they became leg tip cells—claws. This phenomenon, said Wolpert, is analogous to reciprocally transplanting portions of the American and French flags into each other. The segments will retain their identity (French or

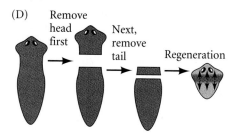

Figure 3.17

Flatworm regeneration and its limits. (A) If a flatworm is cut in two, the anterior portion of the lower half regenerates a head, while the posterior of the upper half regenerates a tail. The same tissue can generate a head (if it is at the anterior portion of the tail piece) or a tail (if it at the posterior portion of the head piece). (B) If a flatworm is cut into three pieces, the middle piece will regenerate a head from its anterior end and a tail from its posterior end. (C) However, if the middle piece is too thin, there is no morphogen gradient within it, and regeneration is abnormal. (D) If the second cut is delayed, however, an equally thin middle section forms a normal worm, since the time lag has allowed an anterior–posterior gradient to become established. (After Gosse 1969.)

(A)

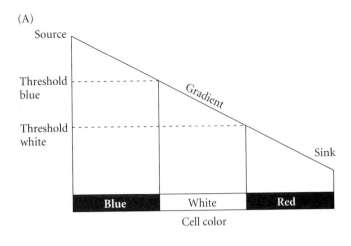

(B)

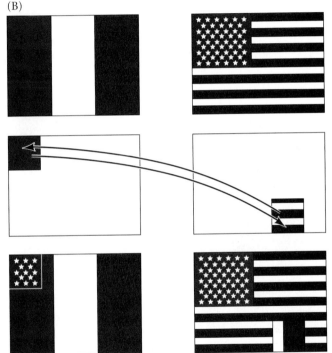

Reciprocal transplants develop
according to their final positions
in the "donor" flag

Figure 3.18
The French flag analogy for the operation of a gradient of positional information. (A) In this model, positional information is delivered by a gradient of a diffusible morphogen extending from a source to a sink. The thresholds indicated on the left are cellular properties that enable the gradient to be interpreted. For example, cells become blue at one concentration of the morphogen, but as the concentration declines below a certain threshold, cells become white. Where the concentration falls below another threshold, cells become red. The result is a pattern of three colors. (B) An important feature of this model is that a piece of tissue transplanted from one region of an embryo to another retains its identity (as to its origin), but differentiates according to its new positional instructions. This phenomenon is indicated schematically by reciprocal "grafts" between the flag of the United States of America and the French flag. (After Wolpert 1978.)

American), but will be positionally specified (develop colors) appropriate to their new positions (Figure 3.18B).

> **WEBSITE 3.3 Receptor gradients.** In addition to a gradient of morphogen, there can also be a gradient of those molecules that recognize the morphogen. The interplay of morphogen gradients and the gradients of molecules that interpret them can give rise to interesting developmental patterns.

The molecules involved in establishing morphogen gradients are beginning to be identified. For a diffusible molecule to be considered a morphogen, it must be demonstrated that cells respond directly to that molecule and that the differentiation of those cells depends upon the concentration of that molecule. One such system currently being analyzed concerns the ability of different concentrations of the protein activin to specify different fates in the frog *Xenopus*. In the *Xenopus* blastula, the cells in the middle of the embryo become mesodermal by responding to activin (or an activin-like compound) produced in the vegetal hemisphere. Makoto Asashima and his colleagues (Fukui and Asashima 1994; Ariizumi and Asashima 1994) have shown that the animal cap of the *Xenopus* embryo (which normally becomes ectoderm, but which can be induced to form mesoderm if transplanted into other regions within the embryo) responds differently to different concentrations of activin. If left untreated in saline solution, animal cap blastomeres form an epidermis-like mass of cells. However, if exposed to small amounts of activin, they form ventral mesodermal tissue—blood and connective tissue. Progressively higher concentrations of activin will cause the animal cap cells to develop into other types of mesodermal cells: muscles, notochord cells, and heart cells (Figure 3.19).

John Gurdon's laboratory has shown that these animal cap cells respond to activin by changing their expression of particular genes (Figure 3.20; Gurdon et al. 1994, 1995). Gurdon and his colleagues placed activin-releasing beads or control beads into "sandwiches" of *Xenopus* animal cap cells. They found that cells exposed to little or no activin failed to express any of the genes associated with mesodermal tissues; these cells differentiated into ectoderm. Higher concentrations of activin turned on genes such as *Brachyury*, which are responsible for instructing cells to become mesoderm. Still higher concentrations of activin caused the cells to express genes such as *goosecoid*, which are associated with the most dorsal mesodermal structure, the notochord. The expression of the *Brachyury* and *goosecoid* genes has been correlated with the number of activin receptors on each cell that are bound by activin. Each cell has about 500 activin receptors. If about 100 of them are bound, *Brachyury* expression is activated and the tissue becomes ventrolateral mesoderm, such as blood and connective tissue cells. If about 300 of these receptors are occupied, the cell turns on its *goosecoid* gene and differentiates into a more dorsal mesodermal cell type, such as notochord (Figure 3.20D; Dyson and Gurdon 1998; Shimizu and Gurdon 1999).

Figure 3.19

Activin (or a closely related protein, such as Nodal) is thought to be responsible for converting animal hemisphere cells into mesoderm. When animal cap cells were removed from *Xenopus* blastulae and placed in saline solutions containing activin, the activin conferred different fates on the cells at different concentrations. (After Fukui and Asashima 1994.)

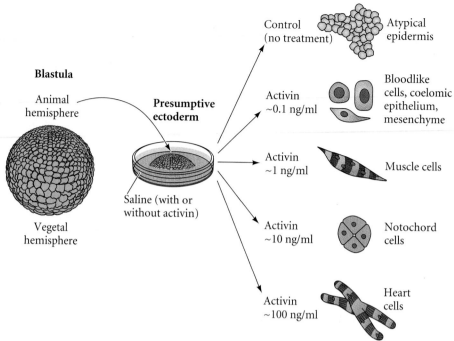

Blastula

Animal hemisphere

Vegetal hemisphere

Presumptive ectoderm

Saline (with or without activin)

Control (no treatment) — Atypical epidermis

Activin ~0.1 ng/ml — Bloodlike cells, coelomic epithelium, mesenchyme

Activin ~1 ng/ml — Muscle cells

Activin ~10 ng/ml — Notochord cells

Activin ~100 ng/ml — Heart cells

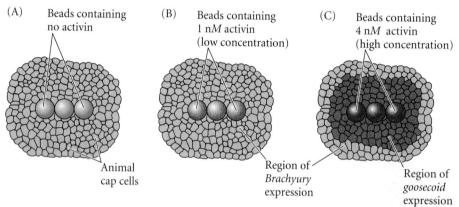

(A) Beads containing no activin

Animal cap cells

(B) Beads containing 1 nM activin (low concentration)

Region of *Brachyury* expression

(C) Beads containing 4 nM activin (high concentration)

Region of *goosecoid* expression

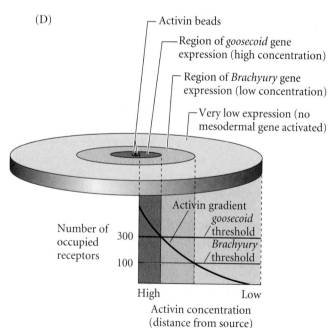

(D)

Activin beads

Region of *goosecoid* gene expression (high concentration)

Region of *Brachyury* gene expression (low concentration)

Very low expression (no mesodermal gene activated)

Number of occupied receptors

300

100

Activin gradient

goosecoid /threshold

Brachyury /threshold

High — Low

Activin concentration (distance from source)

Figure 3.20

A gradient of activin causes different gene expression in *Xenopus* animal cap cells. The mRNAs from the *Brachyury* and *goosecoid* genes can be monitored by hybridization techniques like those discussed in the next chapter. The cells containing these mRNAs appear darker than the cells not expressing them. Beads containing activin induce the expression (transcription of mRNA) of the *Brachyury* gene at distances removed from the beads. (A) Beads containing no activin did not elicit *Brachyury* gene expression. (B) Beads containing 1 nM activin elicited *Brachyury* expression near the beads. (C) Beads containing 4 nM activin elicited *Brachyury* expression several cell diameters away from the beads. However, *goosecoid* expression was seen where the concentration of activin was higher, near the source. Thus, it appears that *Brachyury* gene expression is induced at particular concentrations of activin, and that *goosecoid* is induced at higher concentrations of activin. (D) Interpretation of activin gradient by *Xenopus* animal cap cells. High concentrations of activin activate the *goosecoid* gene, while lower concentrations activate the *Brachyury* gene. This pattern correlates with the number of activin receptors occupied on the individual cells. A threshold value appears to exist that determines whether a cell will express *goosecoid*, *Brachyury*, or neither gene. (A–C after Gurdon et al. 1994, 1995; D after Gurdon et al. 1998.)

Rules of Evidence II

Science is both about concluding what something *is* and about concluding what something *is not*. A piece of science is only as good as the controls it uses to tell what something is not, and the difference between science and non-science is usually a matter of controls. Just because a particular compound causes cells in the same population to differentiate into particular cell types does not mean that it is a morphogen.

In *The Adventure of Black Peter*, Sherlock Holmes noted, "One should always look for a possible alternative and provide against it." What other possibilities could there be besides the morphogen hypothesis? A morphogen is expected to be a diffusible molecule that acts through a concentration gradient. However, there are at least two other possibilities besides gradients to explain the phenomenon whereby cells appear to differentiate with respect to their proximity to a source of activin. First, there could be a cascade of inductions. Activin might only induce nearby cells at high concentrations and do nothing at lower concentrations. Instead, the induced cells might put forth a second signal that induces the next set of cells, and these cells might put forth a third local signal, and so on. Such a sequence of events is sometimes called a relay amplification cascade. Second, the cells near the source of activin might respond not only by differentiating, but by migrating away to a distant position within the responding tissue (or by producing daughter cells that migrate). The stronger the signal, the farther away the induced cells or their daughters migrate.

Gurdon and his colleagues (1994) controlled for these alternative possibilities. First, to distinguish whether the effects of activin are due to the passive diffusion of activin or to a relay amplification cascade,

Gurdon and colleagues placed nonresponding cells (that is, cells that are not competent to respond to activin) between the source of activin and the responsive animal cap cells. If activin acted through a relay amplification cascade, the animal cap cells should not express *Brachyury*, since the cascade would have been broken by the intervening cells. However, if activin produced its effects through a concentration gradient, the animal cap cells should respond based on their distance from the source of activin, even if the cells between them and the source could not respond. In this experiment, the animal cap cells were found to express *Brachyury*, even when the cells closest to the source of activin could not respond to it.

Second, researchers controlled for cell movement by injecting animal cap cells (the responding cells) with green dye and placing these green cells between the vegetal cells (which secrete activin) and rhodamine (red dye)- injected animal cap cells. If there were migrations of the responding cells or their progeny, then green cells should be seen within the red cells. None were found. Moreover, cell division and movement were shown not to be critical factors. This was done by placing vegetal cell/animal cap conjugates into a medium containing cytochalasin, an inhibitor of cell movement and cell division. Expression of the *Brachyury* gene (the gene expressed at low concentrations of activin, but not at high concentrations) was seen several diameters away from the activin source, even though cell movement and cell division were suppressed.

These controls are **negative controls** because they tell the researcher what is *not* the right answer. Any scientific investigation has to have such controls; when they are not done, the investigation is suspect.

For instance, a group of investigators published an article claiming that a particular wormlike structure was found in the uteruses of women with a particular disease. However, no controls were mentioned. When other scientists read the original paper, they noted that the obvious control—to look for the wormlike structure in the uteruses of women *without* the disease—was not reported. It turned out that the structure is a common entity in most uteruses and was not associated with the disease.

Similarly, when scientists inject a specific antibody into an embryo to see if it inhibits a particular phenomenon, they will inject unrelated antibodies into other embryos as negative controls. This practice shows that the inhibition of the event is *not* a result of the solvent being injected along with the antibody, the presence of nonspecific antibody protein, or the fact of injection.

In addition to negative controls, experiments should also incorporate positive controls. **Positive controls** tell scientists that their techniques work and that they are measuring what they think they are measuring. Positive controls for the activin concentration gradient experiment would be to expose the responding tissue to known amounts of activin and see a change in gene expression as a result of increased activin concentrations; this would tell the researchers that their techniques work and that their cells can respond to activin. This means that negative results are not due to imperfections in the methods. These experiments have been done (Green and Smith 1990; McDowell et al. 1997), and their results support the conclusion that the cells are indeed responding to decreasing amounts of activin the farther away they are from its source.

MORPHOGENETIC FIELDS. One of the most interesting ideas to come from experimental embryology has been that of the **morphogenetic field.** A morphogenetic field can be described as a group of cells whose position and fate are specified with respect to the same set of boundaries (Weiss 1939; Wolpert 1977). The general fate of a morphogenetic field is determined; thus, a particular field of cells will give rise to its particular organ (forelimb, eye, heart, etc.) even when transplanted to a different part of the embryo. However, the individual cells *within* the field are not committed, and the

cells of the field can regulate their fates to make up for cells missing from the field (Huxley and de Beer 1934; Opitz 1985; De Robertis et al. 1991). Moreover, as described earlier (i.e., the case of presumptive *Drosophila* leg cells transposed to the tip region of the presumptive antennal field), if cells from one field are placed within another field, they can use the positional cues of their new location, even if they retain their organ-specific commitment.

The morphogenetic field has been referred to as a "field of organization" (Spemann 1921) and as a "cellular ecosystem" (Weiss 1923, 1939). The ecosystem metaphor is quite appropriate in that recent studies have shown that there are webs of interactions among the cells in different regions of a morphogenetic field.

WEBSITE 3.4 "Rediscovery" of the morphogenetic field. The morphogenetic field was one of the most important concepts of embryology during the early twentieth century. This concept was eclipsed by research on the roles of genes in development, but it is being "rediscovered" as a consequence of those developmental genetic studies.

Stem cells and commitment

One of the important principles derived from conditional specification is the notion of stem cells (National Institutes of Health 2001). **Stem cells** are cells that have the capacity to divide indefinitely and which can give rise to more specialized cells. When they divide, stem cells produce a more specialized type of cell *and also* generate more stem cells (Figure 3.21A).

Some single stem cells are capable of generating all the structures of the embryo. These cells, called **pluripotent stem cells,*** can generate ectoderm, endoderm, mesoderm, and germ cells. As well as giving rise to more pluripotent stem cells, these cells also generate **committed stem cells**. Committed stem cells can give rise to a smaller population of cells. For instance, one type of committed stem cell (the hemangioblast)

gives rise to all the blood vessels, blood cells, and lymphocytes. Another type of committed stem cell, the mesenchymal stem cell, can give rise to all the different connective tissues (cartilage, muscle, fat, etc). Committed stem cells can give rise to more specifically committed stem cells (such as a stem cell that generates only blood cells and lymphocytes, not blood vessels) or it can generate what are called progenitor cells (Figure 3.21B).

Progenitor cells, also called **precursor cells,** are no longer stem cells, since their divisions do not create another similar progenitor cell. Rather, the progenitor cell divides to form one or a few related types of cells, depending on the cellular environment it is in. For example, there is a blood cell precursor (the myeloid progenitor cell) that can generate all the different types of blood cells. Usually, the progenitor cell shows some evidence of differentiation, but the process is not complete until the differentiated cell has been formed.

The restriction on the potency of stem cells is gradual, and the potencies of these cells are determined by their surroundings. Once committed, though, they usually do not

---

*There are also stem cells called *totipotent* stem cells, and they will be discussed in Chapters 11 and 21. *Totipotent stem cells* are those very early mammalian cells that can form both the entire embryo and the fetal placenta (trophoblast) around it. The *pluripotent stem cells* of mammals can form the embryo, but not its surrounding tissues.

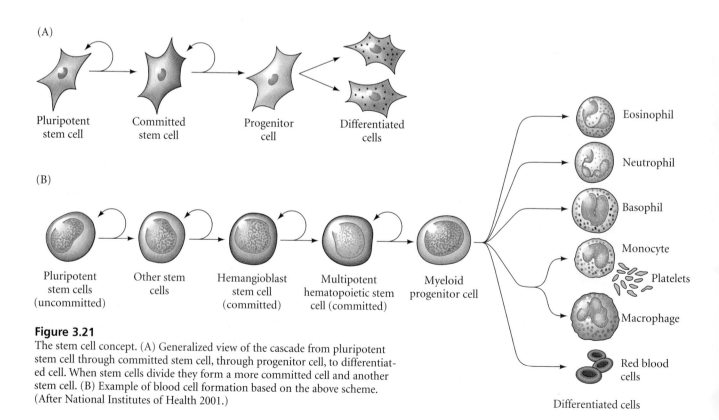

Figure 3.21
The stem cell concept. (A) Generalized view of the cascade from pluripotent stem cell through committed stem cell, through progenitor cell, to differentiated cell. When stem cells divide they form a more committed cell and another stem cell. (B) Example of blood cell formation based on the above scheme. (After National Institutes of Health 2001.)

switch commitment. When placed in a new environment, they will not change the type of cells they can generate. Stem cells are critical for maintaining cell populations that last for long periods of time and so must be renewed. Thus, stem cells are important for our continued production of blood, hair, epidermis, and intestinal epithelial cells.

Syncytial specification

Many insects also use a third means, known as **syncytial specification**, to commit cells to their fates. Here, interactions occur not between cells, but between parts of one cell. In early embryos of these insects, cell division is not complete. Rather, the nuclei divide within the egg cytoplasm, creating many nuclei within one large egg cell. A cytoplasm that contains many nuclei is called a **syncytium**. The egg cytoplasm, however, is not uniform. Rather, the anterior of the egg cytoplasm is markedly different from the posterior.

In the fruit fly, for instance, the anteriormost portion of the egg contains an mRNA that encodes a protein called Bicoid. The posteriormost portion of the egg contains an mRNA that encodes a protein called Nanos. When the egg is laid and fertilized, these two mRNAs are translated into their respective proteins. The concentration of Bicoid protein is highest in the anterior and declines toward the posterior; that of Nanos protein is highest in the posterior and declines as it diffuses anteriorly. Thus, the long axis of the *Drosophila* egg is spanned by two opposing morphogen gradients—one of Bicoid protein coming from the anterior, and one of Nanos protein coming from the posterior. The Bicoid and Nanos proteins form a coordinate system based on their ratios, such that each region of the embryo is distinguished by a different ratio of the two proteins. As the nuclei divide and enter different regions of the egg cytoplasm, they will be instructed by these ratios as to their position along the anterior-posterior axis. Those nuclei in regions containing high amounts of Bicoid and little Nanos will be instructed to activate those genes necessary for producing the head. Those nuclei in regions with slightly less Bicoid but with a small amount of Nanos will be instructed to activate those genes that generate the thorax. Those nuclei in regions that have little or no Bicoid and plenty of Nanos will be instructed to form the abdominal structures (Figure 3.22; Nüsslein-Volhard et al. 1987). The mechanisms of syncytial specification will be detailed in Chapter 9.

No embryo uses only autonomous, conditional, or syncytial mechanisms to specify its cells. One finds autonomous specification even in a "regulative embryo" such as the sea urchin, and the nervous system and some musculature of the "autonomously developing" tunicate have been shown to come from regulative interactions between its cells. Insects such as *Drosophila* use all three modes of specification to commit their cells to particular fates. Later chapters will detail the mechanisms by which cell fates are committed in these species.

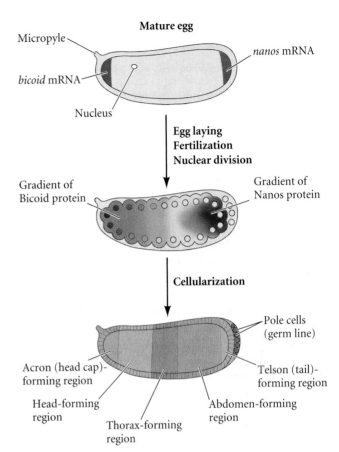

Figure 3.22

Syncytial specification in the fruit fly *Drosophila melanogaster*. Anterior-posterior specification originates from morphogen gradients within the egg cytoplasm. *Bicoid* mRNA is stabilized in the most anterior portion of the egg, while *Nanos* mRNA is tethered to the posterior end. (The anterior can be recognized by the micropyle on the shell; this structure permits sperm to enter.) When the egg is laid and fertilized, these two mRNAs are translated into proteins. The Bicoid protein forms a gradient that is highest at the anterior end, and the Nanos protein forms a gradient that is highest at the posterior end. These two proteins form a coordinate system based on their ratios. Each position along the axis is thus distinguished from any other position. When nuclear division occurs, each nucleus is given its positional information by the ratio of these proteins. The proteins forming these gradients activate the transcription of the genes specifying the segmental identities of the larva and the adult fly.

Morphogenesis and Cell Adhesion

A body is more than a collection of randomly distributed cell types. Development involves not only the differentiation of cells, but also their organization into multicellular arrangements such as tissues and organs. When we observe the detailed anatomy of a tissue such as the neural retina of the eye, we see an intricate and precise arrangement of many types of cells. How can matter organize itself so as to create a complex structure such as a limb or an eye?

There are five major questions for embryologists who study morphogenesis:

1. **How are tissues formed from populations of cells?** For example, how do neural retina cells stick to other neural retina cells rather than becoming integrated into the pigmented retina or iris cells next to them? How are the various cell types within the retina (the three distinct layers of photoreceptors, bipolar neurons, and ganglion cells) arranged so that the retina is functional?

2. **How are organs constructed from tissues?** The retina of the eye forms at a precise distance behind the cornea and the lens. The retina would be useless if it developed behind a bone or in the middle of the kidney. Moreover, neurons from the retina must enter the brain to innervate the regions of the brain cortex that analyze visual information. All these connections must be precisely ordered.

3. **How do organs form in particular locations, and how do migrating cells reach their destinations?** Eyes develop only in the head and nowhere else. What stops an eye from forming in some other area of the body? Some cells—for instance, the precursors of our pigment cells, germ cells, and blood cells—must travel long distances to reach their final destinations. How are cells instructed to travel along certain routes in our embryonic bodies, and how are they told to stop once they have reached their appropriate destinations?

4. **How do organs and their cells grow, and how is their growth coordinated throughout development?** The cells of all the tissues in the eye must grow in a coordinated fashion if one is to see. Some cells, including most neurons, do not divide after birth. In contrast, the intestine is constantly shedding cells, and new intestinal cells are regenerated each day. The mitotic rate of each tissue must be carefully regulated. If the intestine generated more cells than it sloughed off, it could produce tumorous outgrowths. If it produced fewer cells than it sloughed off, it would soon become nonfunctional. What controls the rate of mitosis in the intestine?

5. **How do organs achieve polarity?** If one were to look at a cross section of the fingers, one would see a certain organized collection of tissues—bone, cartilage, muscle, fat, dermis, epidermis, blood, and neurons. Looking at a cross section of the forearm, one would find the same collection of tissues. But they are arranged very differently in different parts of the arm. How is it that the same cell types can be arranged in different ways in different parts of the same structure?

All these questions concern aspects of cell behavior. There are two major types of cell arrangements in the embryo: **epithelial cells**, which are tightly connected to one another in sheets or tubes, and **mesenchymal cells**, which are unconnected to one another and operate as independent units. Morphogenesis is brought about through a limited repertoire of variations in cellular processes within these two types of arrangements: (1) the direction and number of cell divisions; (2) cell shape changes; (3) cell movement; (4) cell growth; (5) cell death; and (6) changes in the composition of the cell membrane or secreted products. We will discuss the last of these considerations here.

> **WEBSITE 3.5 How morphogenetic behaviors work.** Although the repertoire of morphogenetic behaviors is small, cells can do a great deal with this limited set of instructions. This website illustrates the epithelial and mesenchymal changes that effect development.

Differential cell affinity

Many of the answers to our questions about morphogenesis involve the properties of the cell surface. The cell surface looks pretty much the same in all cell types, and many early investigators thought that the cell surface was not even a living part of the cell. We now know that each type of cell has a different set of proteins at its surfaces, and that some of these differences are responsible for forming the structure of the tissues and organs during development. Observations of fertilization and early embryonic development made by E. E. Just (1939) suggested that the cell membrane differed among cell types, but the modern analysis of morphogenesis began with the experiments of Townes and Holtfreter in 1955. Taking advantage of the discovery that amphibian tissues become dissociated into single cells when placed in alkaline solutions, they prepared single-cell suspensions from each of the three germ layers of amphibian embryos soon after the neural tube had formed. Two or more of these single-cell suspensions could be combined in various ways. When the pH of the solution was normalized, the cells adhered to one another, forming aggregates on agar-coated petri dishes. By using embryos from species having cells of different sizes and colors, Townes and Holtfreter were able to follow the behavior of the recombined cells (Figure 3.23).

The results of their experiments were striking. First, they found that reaggregated cells become spatially segregated. That is, instead of the two cell types remaining mixed, each cell type sorts out into its own region. Thus, when epidermal (ectodermal) and mesodermal cells are brought together to form a mixed aggregate, the epidermal cells move to the periphery of the aggregate and the mesodermal cells move to the inside. In no case do the recombined cells remain randomly mixed, and in most cases, one tissue type completely envelops the other.

Second, the researchers found that the final positions of the reaggregated cells reflect their embryonic positions. The mesoderm migrates centrally with respect to the epidermis,

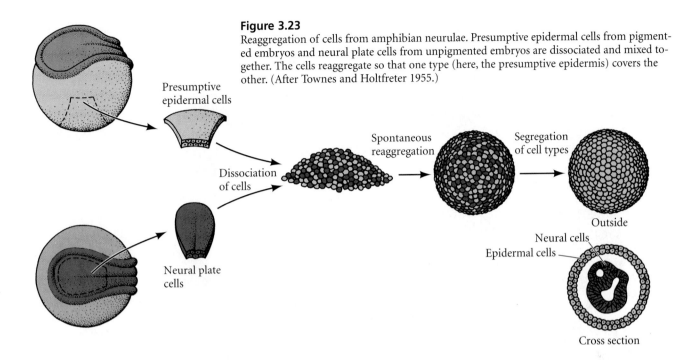

Figure 3.23
Reaggregation of cells from amphibian neurulae. Presumptive epidermal cells from pigmented embryos and neural plate cells from unpigmented embryos are dissociated and mixed together. The cells reaggregate so that one type (here, the presumptive epidermis) covers the other. (After Townes and Holtfreter 1955.)

Presumptive
epidermal cells

Dissociation
of cells

Neural plate
cells

Spontaneous
reaggregation

Segregation
of cell types

Outside

Neural cells
Epidermal cells

Cross section

adhering to the inner epidermal surface (Figure 3.24A). The mesoderm also migrates centrally with respect to the gut or endoderm (Figure 3.24B). However, when the three germ layers are mixed together, the endoderm separates from the ectoderm and mesoderm and is then enveloped by them (Figure 3.24C). In the final configuration, the ectoderm is on the periphery, the endoderm is internal, and the mesoderm lies in the region between them. Holtfreter interpreted this finding in terms of **selective affinity**. The inner surface of the ectoderm has a positive affinity for mesodermal cells and a negative affinity for the endoderm, while the mesoderm has positive affinities for both ectodermal and endodermal cells. Mimicry of normal embryonic structure by cell aggregates is also seen in the recombination of epidermis and neural plate cells (Figures 3.23 and 3.24D). The presumptive epidermal cells migrate to the periphery as before; the neural plate cells migrate inward, forming a structure reminiscent of the neural tube. When axial mesoderm (notochord) cells are added to a suspension of presumptive epidermal and presumptive neural cells, cell segregation results in an external epidermal layer, a centrally located neural tissue, and a layer of mesodermal tissue between them (Figure 3.24E). Somehow, the cells are able to sort out into their proper embryonic positions.

Such selective affinities were also noted by Boucaut (1974), who injected individual cells from specific germ layers into the body cavity of amphibian gastrulae. He found that these cells migrated back to their appropriate germ layer. Endodermal cells found positions in the host endoderm, whereas ectodermal cells were found only in host ectoderm. Thus, selective affinity appears to be important for imparting positional information to embryonic cells.

The third conclusion of Holtfreter and his colleagues was that selective affinities change during development. Such changes should be expected, because embryonic cells do not retain a single stable relationship with other cell types. For development to occur, cells must interact differently with other cell populations at specific times. Such changes in cell affinity are extremely important in the processes of morphogenesis.

The experimental reconstruction of aggregates from cells of later embryos of birds and mammals was accomplished by the use of the protease trypsin to dissociate the cells from one another (Moscona 1952). When the resulting single cells were mixed together in a flask and swirled so that the shear force would break any nonspecific adhesions, the cells sorted themselves out according to their cell type. In so doing, they reconstructed the organization of the original tissue (Moscona 1961; Giudice and Just 1962). Figure 3.25 shows the "reconstruction" of skin tissue from a 15-day embryonic mouse. The epidermal cells are separated by proteolytic enzymes and then aggregated in a rotary culture. The epidermal cells of each aggregate migrate to the periphery, and the dermal cells migrate toward the center. In 72 hours, the epidermis has been reconstituted, a keratin layer has formed, and interactions between these tissues form hair follicles in the dermal region. Such reconstruction of complex tissues from individual cells is called **histotypic aggregation**.

The thermodynamic model of cell interactions

Cells, then, do not sort randomly, but can actively move to create tissue organization. What forces direct cell movement during morphogenesis? In 1964, Malcolm Steinberg proposed the **differential adhesion hypothesis**, a model that sought to

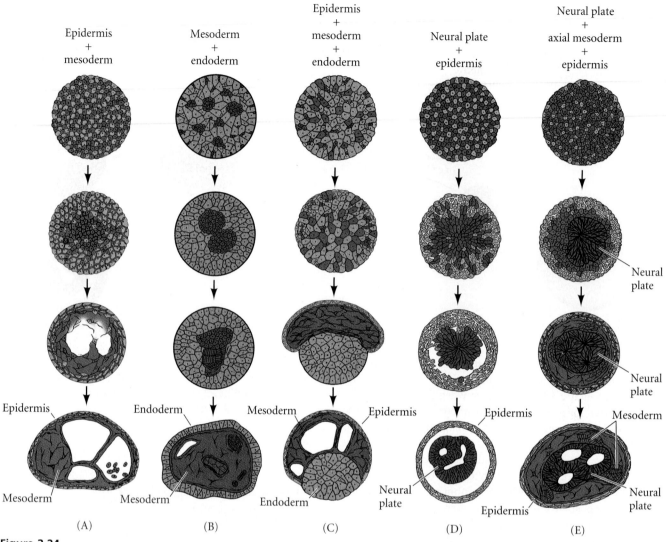

Epidermis
+
mesoderm

Mesoderm
+
endoderm

Epidermis
+
mesoderm
+
endoderm

Neural plate
+
epidermis

Neural plate
+
axial mesoderm
+
epidermis

Neural
plate

Epidermis

Neural
plate

Mesoderm

Epidermis

Mesoderm

Endoderm

Mesoderm

Mesoderm

Mesoderm

Epidermis

Endoderm

Epidermis

Neural
plate

Epidermis

Neural
plate

(A)   (B)   (C)   (D)   (E)

Figure 3.24
Sorting out and reconstruction of spatial relationships in aggregates of embryonic amphibian cells. (After Townes and Holtfreter 1955.)

explain patterns of cell sorting based on thermodynamic principles. Using cells derived from trypsinized embryonic tissues, Steinberg showed that certain cell types always migrate centrally when combined with some cell types, but migrate peripherally when combined with others. Figure 3.26 illustrates the interactions between pigmented retina cells and neural retina cells. When single-cell suspensions of these two cell types are mixed together, they form aggregates of randomly arranged cells. However, after several hours, no pigmented retina cells are seen on the periphery of the aggregates, and after 2 days, two distinct layers are seen, with the pigmented retina cells lying internal to the neural retina cells. Moreover, such interactions form a hierarchy (Steinberg 1970). If the final position of one cell type, A, is internal to a second cell type, B, and the final position of B is internal to a third cell type, C, then the final position of A will always be internal to C. For example, pigmented retina cells migrate internally to neural retina cells, and heart cells migrate internally to pigmented retina cells. Therefore, heart cells migrate internally to neural retina cells.

This observation led Steinberg to propose that cells interact so as to form an aggregate with the smallest interfacial free energy. In other words, the cells rearrange themselves into the most thermodynamically stable pattern. If cell types A and B have different strengths of adhesion, and if the strength of A-A connections is greater than the strength of A-B or B-B connections, sorting will occur, with the A cells becoming central. On the other hand, if the strength of A-A connections is less than or equal to the strength of A-B connections, then the aggregate will remain as a random mix of cells. Finally, if the strength of A-A connections is far greater than the strength of A-B connections—in other words, if A and B cells show essentially no adhesivity toward one another—then A cells and B cells will form separate aggregates. According to this hypothe-

(A) Epidermis   Dermis   Primary hair follicle

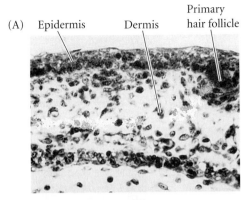

(B)

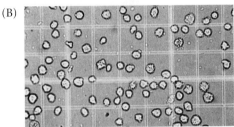

(C)

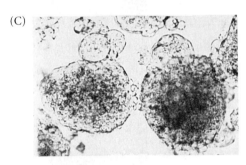

(D)

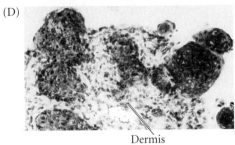

Dermis

(E)   Dermis   Epidermis   Keratinized layer

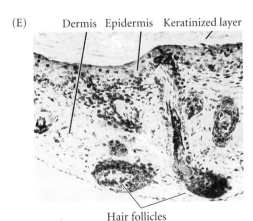

Hair follicles

Figure 3.25
Reconstruction of skin from a suspension of skin cells from a 15-day embryonic mouse. (A) Section through intact embryonic skin, showing epidermis, dermis, and primary hair follicle. (B) Suspension of single skin cells from both the dermis and the epidermis. (C) Aggregates after 24 hours. (D) Section through an aggregate, showing migration of epidermal cells to the periphery. (E) Further differentiation of aggregates after 72 hours, showing reconstituted epidermis and dermis, complete with hair follicles and keratinized layer. (From Monroy and Moscona 1979; photographs courtesy of A. Moscona.)

sis, the early embryo can be viewed as existing in an equilibrium state until some change in gene activity changes the cell surface molecules. The movements that result seek to restore the cells to a new equilibrium configuration. All that is needed for sorting to occur is that cell types differ in the strengths of their adhesion.

(A)

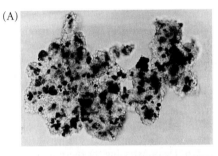

(B)

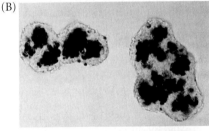

(C)

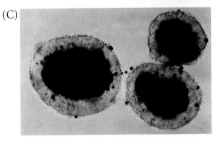

Figure 3.26
Aggregates formed by mixing 7-day chick embryo neural retina (unpigmented) cells with pigmented retina cells. (A) Five hours after the single-cell suspensions are mixed, aggregates of randomly distributed cells are seen. (B) At 19 hours, the pigmented retina cells are no longer seen on the periphery. (C) At 2 days, a great majority of the pigmented retina cells are located in a central internal mass, surrounded by the neural retina cells. (The scattered pigmented cells are probably dead cells.) (From Armstrong 1989; photographs courtesy of P. B. Armstrong.)

Figure 3.27
Hierarchy of cell sorting in order of decreasing surface tensions. The equilibrium configuration reflects the strength of cell cohesion, with the cell types having the greater cell cohesion segregating inside the cells with less cohesion. The images were obtained by sectioning the aggregates and assigning colors to the cell types by computer. The black areas represent cells whose signal was edited out in the program of image optimization. (From Foty et al. 1996; photograph courtesy of M. S. Steinberg and R. A. Foty.)

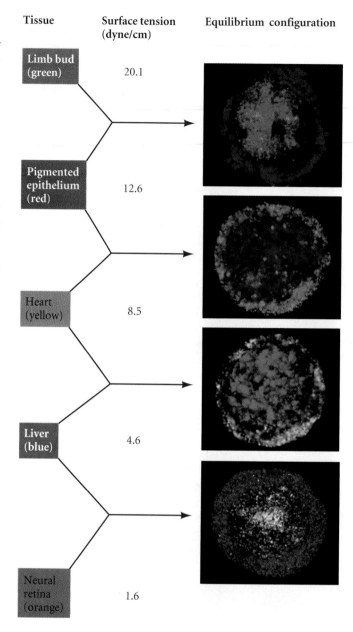

Tissue	Surface tension (dyne/cm)	Equilibrium configuration
Limb bud (green)	20.1	
Pigmented epithelium (red)	12.6	
Heart (yellow)	8.5	
Liver (blue)	4.6	
Neural retina (orange)	1.6	

In 1996, Foty and his colleagues in Steinberg's laboratory demonstrated that this was indeed the case: the cell types that had greater surface cohesion sorted within those cells that had less surface tension (Figure 3.27; Foty et al. 1996). In the simplest form of this model, all cells could have the same type of "glue" on the cell surface. The amount of this cell surface product, or the cellular architecture that allows the substance to be differentially distributed across the surface, could cause a difference in the number of stable contacts made between cell types. In a more specific version of this model, the thermodynamic differences could be caused by different types of adhesion molecules (see Moscona 1974). When Holtfreter's studies were revisited using modern techniques, Davis and colleagues (1997) found that the tissue surface tensions of the individual germ layers were precisely those required for the sorting patterns observed both in vitro and in vivo.

WEBSITE 3.6 Demonstrating the thermodynamic model. The original in vivo evidence for the thermodynamic model of cell adhesion came from studies of limb regeneration. This website goes into some of the details of these experiments and how they are interpreted.

Cadherins and cell adhesion

Recent evidence shows that such boundaries between tissues can indeed be created by different cell types having both different types and different amounts of cell adhesion molecules. Several classes of molecules can mediate cell adhesion, but he major cell adhesion molecules appear to be the **cadherins**.

As their name suggests, cadherins are calcium-dependent adhesion molecules. They are critical for establishing and maintaining intercellular connections, and they appear to be crucial to the spatial segregation of cell types and to the organization of animal form (Takeichi 1987). Cadherins interact with other cadherins on adjacent cells, and they are anchored into the cell by a complex of proteins called **catenins** (Figure 3.28). The cadherin–catenin complex forms the classic adherens junctions that help connect epithelial cells together. Moreover, since the catenins bind to the actin cytoskeleton of the cell, they integrate the epithelial cells into a mechanical unit. Univalent (Fab) antibody fragments (Fab; see Chapter 2) against cadherins will convert a three-dimensional, histotypic aggregate of cells into a single layer of cells (Takeichi et al. 1979).

In vertebrate embryos, several major cadherin classes have been identified:

- **E-cadherin** (epithelial cadherin, also called uvomorulin and L-CAM) is expressed on all early mammalian embryonic cells, even at the zygote stage. Later, this molecule is restricted to epithelial tissues of embryos and adults.

- **P-cadherin** (placental cadherin) appears to be expressed primarily on the trophoblast cells (those placental cells of the mammalian embryo that contact the uterine wall) and on the uterine wall epithelium (Nose and Takeichi 1986). It is possible that P-cadherin facilitates the connection of the embryo to the uterus, since P-cadherin on

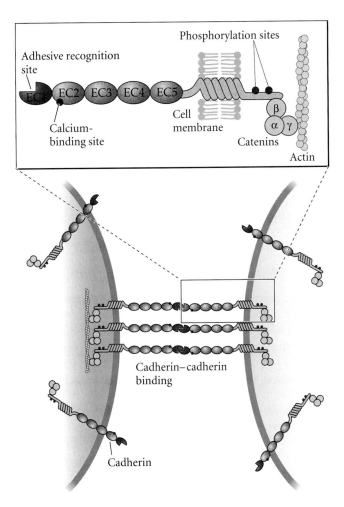

Figure 3.28
Cadherin-mediated cell adhesion. Cadherins are associated with three types of catenins. The catenins can become associated with the actin microfilament system within the cell. (After Takeichi 1991.)

Cadherins join cells together by binding to the same type of cadherin on another cell. Thus, cells with E-cadherin stick best to other cells with E-cadherin, and they will sort out from cells containing N-cadherin in their membranes. This adhesion pattern, which is called **homophilic binding**, may have important consequences in the embryo. In the gastrula of the frog *Xenopus*, for example, the neural tube expresses N-cadherin, while the epidermis expresses E-cadherin. Normally, these two tissues separate from each other such that the neural tube is inside the body and the epidermis covers the body. If the epidermis is experimentally manipulated to remove its E-cadherin, the epidermal epithelium cannot hold together. If the epidermis is made to express N-cadherin, or if the neural cells are made to lose it, the neural tube will not separate from the epidermis (Figure 3.30; Detrick et al. 1990; Fujimori et al. 1990). Thus, boundaries can be created between cell populations expressing different cadherins.

The amount of cadherin can also mediate the formation of embryonic structures. This was first shown to be a possibility when Steinberg and Takeichi (1994) collaborated on an experiment using two cell lines that were identical except that they synthesized different amounts of P-cadherin. When these two groups of cells, each expressing a different amount of cadherin, were mixed, the cells that expressed more cadherin had a higher surface cohesion and migrated internally to the lower-expressing group of cells. This process also appears to work in the embryo as well (Godt and Tepass 1998; González-Reyes and St. Johnston 1998).

the uterine cells is seen to contact P-cadherin on the trophoblast cells of mouse embryos (Kadokawa et al. 1989).

• **N-cadherin** (neural cadherin) is first seen on mesodermal cells in the gastrulating embryo as they lose their E-cadherin expression. It is also highly expressed on the cells of the developing central nervous system (Hatta and Takeichi 1986).

• **EP-cadherin** (C-cadherin) is critical for maintaining adhesion between the blastomeres of the *Xenopus* blastula and is required for the normal movements of gastrulation (Figure 3.29; Heasman et al. 1994; Lee and Gumbiner 1995).

• **Protocadherins** are calcium-dependent adhesion proteins that differ from the classic cadherins in that they lack connections to the cytoskeleton through catenins. Protocadherins are very important in separating the notochord from the other mesodermal tissues during *Xenopus* gastrulation (see Chapter 10).

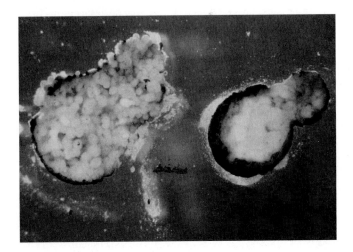

Figure 3.29
The importance of cadherins for maintaining cohesion between developing cells can be demonstrated by interfering with their production. When an oocyte is injected with an antisense oligonucleotide against a maternally inherited cadherin mRNA (thus preventing the synthesis of the cadherin), the inner cells of the resulting embryo disperse when the animal cap is removed (left). In control embryos (right), the inner cells remain together. (After Heasman et al. 1994; photograph courtesy of J. Heasman.)

Figure 3.30
The importance of N-cadherin in the separation of neural and epidermal ectoderm. At the 4-cell stage, the blastomeres that formed the left side of this *Xenopus* embryo were injected with an mRNA for N-cadherin that lacks the extracellular region of the cadherin. This mutation blocks N-cadherin function. During neurulation, the cells with the mutant protein did not form a coherent layer distinct from the epidermis. (From Kintner 1993; photograph courtesy of C. Kintner.)

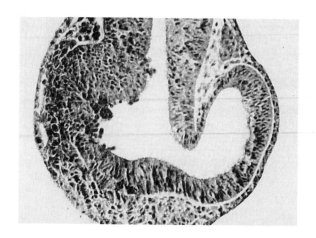

The timing of particular developmental events can depend on cadherin expression. For instance, N-cadherin appears in the mesenchyme of the developing chick leg just before these mesenchymal cells condense and form nodules of cartilage (which are the precursors to the bone tissue). It is not seen prior to condensation, nor is it seen afterwards. If the limbs are injected just prior to mesenchyme condensation with antibodies that block N-cadherin, the mesenchyme cells fail to condense and cartilage fails to form (Oberlander and Tuan 1994). Therefore, it appears that the signal to begin cartilage formation in the chick limb is the appearance of N-cadherin.

> **WEBSITE 3.7 Cadherins: Functional anatomy.** The cadherin molecule has several functional domains that mediate its activities, and the mechanisms of homophilic adhesion are currently being resolved.

> **WEBSITE 3.8 Other cell adhesion molecules.** There are more types of cell adhesion molecules than cadherins. This website looks at some of the other cell adhesion and substrate adhesion molecules that have been discovered.

During development, the cadherins often work with other adhesion systems. For instance, one of the most critical times in a mammal's life is when the embryo is passing through the uterus. If development is to continue, the embryo must adhere to the uterus and embed itself in the uterine wall. That is why the first differentiation event in mammalian development distinguishes the **trophoblast** cells (the outer cells that bind to the uterus) from the **inner cell mass** (those cells that will generate the adult organism). This process occurs as the embryo travels from the upper regions of the oviduct on its way to the uterus. The trophoblast cells are endowed with several adhesion molecules to anchor the embryo to the uterine wall. First, they contain both E-cadherins and P-cadherins (Kadokawa et al. 1989), and these cadherins recognize similar cadherins on the uterine cells. Second, they have receptors (integrin proteins) for the collagen and the heparan sulfate glycoproteins of the uterine wall (Farach et al. 1987; Carson et al. 1988, 1993; Cross et al. 1994). Third, the trophoblast cells also have a modified glycosyltransferase enzyme that extends out from the cell membrane and can bind to specific carbohydrate residues on uterine glycoproteins (Dutt et al. 1987). For something as important as the implantation of the mammalian embryo, it is not surprising that several cell adhesion systems appear to be working together.

The questions of morphogenesis remain some of the most fascinating of all developmental biology. Think, for example, of the thousands of specific connections made by the millions of cells within the human brain, or ponder the mechanisms by which the heart chambers form on the correct sides and become connected to the appropriate arteries and veins. These and other questions will be specifically addressed in later chapters.

Principles of Development: Experimental Embryology

1. Norms of reaction describe an embryo's inherited ability to develop a range of phenotypes. The environment can play a role in selecting which phenotype is expressed. (Examples include temperature-dependent sex determination and seasonal phenotypic changes in caterpillars and butterflies.)

2. Developing organisms are adapted to the ecological niches in which they develop. (Examples include the ability of frog eggs exposed to sunlight to repair DNA damage.)

3. Before cells overtly differentiate into the many cell types of the body, they undergo a "covert" commitment to a certain fate. This commitment is first labile (the specification step) but later becomes irreversible (the determination step).

4. In some embryos, removal of a blastomere from an embryo causes the absence in the embryo of those tissues formed by that blastomere. This autonomous specification produces a mosaic pattern of development. (Examples include early snail and tunicate embryos.)

5. In autonomous specification, morphogenetic determinants in the egg cytoplasm are apportioned to different blastomeres during cleavage. (An example is the yellow crescent cytoplasm that is found in the muscle-forming cells of tunicate embryos.)

6. In some embryos, the removal of a blastomere can be compensated for by the other cells' changing their fates. Each cell has the potential to give rise to more cell types than it normally does. This conditional specification produces a regulative pattern of development wherein cell fates are determined relatively late. (Examples include frog and mammalian embryos.)

7. In conditional specification, the fate of a cell often depends on interactions with its neighbors

8. In conditional specification, groups of cells can have their fates determined by a concentration gradient of a morphogen. The cells specified by such a morphogen can constitute a morphogenetic field.

9. Stem cells are able to generate another stem cell as well as a more committed cell type. By generating more stem cells, the population continues.

10. In syncytial specification, the fates of cells are determined by gradients of morphogens within the egg cytoplasm.

11. Different cell types can sort themselves into regions by means of cell surface molecules such as cadherins. These molecules can be critical in patterning cells into tissues and organs.

12. Negative controls tell investigators that a given agent does not cause an observed phenomenon. Such controls are critical in arguing that some other agent does cause the phenomenon.

13. Positive controls confirm that an investigator's protocols are working, so that if a negative result is observed, it is not merely due to faulty procedure.

Literature Cited

Adams, N. L. and J. M. Shick. 1996. Mycosporine-like amino acids provide protection against ultraviolet radiation in eggs of the green sea urchin *Strongylocentrotus droebachiensis*. *Photochem. Photobiol.* 64: 149–158.

Adams, N. L. and J. M. Shick. 2001. Mycosporine-like amino acids prevent UVB-induced abnormalities during early development of the green sea urchin *Strongylocentrotus droebachiensis*. *Mar. Biol.* 138: 267–280.

Allman, J. A. 1864. Reproductive systems in the Hydroidaea. Published in *Report of the Thirty-third Meeting of the British Association*, 1869.

Ariizumi, T. and M. Asashima. 1994. In vitro control of the embryonic form of *Xenopus laevis* by activin A: Time- and dose-dependent inducing properties of activin-treated ectoderm. *Dev. Growth Diff.* 36: 499–507.

Armstrong, P. B. 1989. Cell sorting out: The self-assembly of tissues in vitro. *CRC Crit. Rev. Biochem. Mol. Biol.* 24: 119–149.

Baltzer, F. 1914. Die Bestimmung und der Dimorphismus des Geschlechtes bei *Bonellia*. *Sber. Phys.-Med. Ges. Würzb.* 43: 1–4.

Blaustein, A. R., P. D. Hoffman, D. G. Hokit, J. M. Kiesecker, S. C. Walls and J. B. Hays. 1994. UV repair and resistance to solar UV-B in amphibian eggs: A link to population declines? *Proc. Natl. Acad. Sci. USA* 91: 1791–1795.

Boucaut, J. C. 1974. Étude autoradiographique de la distribution de cellules embryonnaires isolées, transplantées dans le blastocèle chez *Pleurodeles waltii* Michah (Amphibien, Urodele). *Ann. Embryol. Morphol.* 7: 7–50.

Carson, D. D., J.-P. Tang and S. Gay. 1988. Collagens support embryo attachment and outgrowth in vitro: Effects of the Arg-Gly-Asp sequence. *Dev. Biol.* 127: 368–375.

Carson, D. D., J.-P. Tang and J. Julian. 1993. Heparan sulfate proteoglycan (perlecan) expression by mouse embryos during acquisition of attachment competence. *Dev. Biol.* 155: 97–106.

Chabry, L. M. 1887. Contribution a l'embryologie normale tératologique des ascidies simples. *J. Anat. Physiol. Norm. Pathol.* 23: 167–321.

Cross, J. C., Z. Werb and S. J. Fisher. 1994. Implantation and the placenta: Key pieces of the development puzzle. *Science* 266: 1508–1518.

Davidson, E. H. 1991. Spatial mechanisms of gene regulation in metazoan embryos. *Development* 113: 1–26.

Davis, G. S., H. M. Phillips and M. S. Steinberg. 1997. Germ-layer surface tension and "tissue affinities" in *Rana pipiens*: Quantitative measurement. *Dev. Biol.* 192: 630–644.

De Robertis, E. A., E. M. Morita and K. W. Y. Cho. 1991. Gradient fields and homeobox genes. *Development* 112: 669–678.

Detrick, R. J., D. Dickey and C. R. Kintner. 1990. The effects of N-cadherin misexpression on morphogeneies in *Xenopus* embryos. *Neuron* 4: 493–506.

Driesch, H. 1892. The potency of the first two cleavage cells in echinoderm development: Experimental production of partial and double formations. In B. H. Willier and J. M. Oppenheimer (eds.), *Foundations of Experimental Embryology*. Hafner, New York, 1974.

Driesch, H. 1893. Zur Verlagerung der Blastomeren des Echinideneies. *Anat. Anz.* 8: 348–357.

Driesch, H. 1894. *Analytische Theorie de organischen Entwicklung*. W. Engelmann, Leipzig.

Dutt, A., J.-P. Tang and D. D. Carson. 1987. Lactosaminoglycans are involved in uterine epithelial cell adhesion in vitro. *Dev. Biol.* 119: 27–37.

Dyson, S. and J. B. Gurdon. 1998. The interpretation of position in a morphogen gradient as revealed by occupancy of activin receptors. *Cell* 93: 557–568.

Farach, M. C., J. P. Tang, G. L. Decker and D. D. Carson. 1987. Heparin/heparan sulfate is involved in attachment and spreading of mouse embryos in vitro. *Dev. Biol.* 123: 401–410.

Ferguson, M. W. J. and T. Joanen. 1982. Temperature of egg incubation determines sex in *Alligator mississippiensis. Nature* 296: 850–853.

Fischer, J.-L. 1991. Laurent Chabry and the beginnings of experimental embryology in France. In S. Gilbert (ed.), *A Conceptual History of Modern Embryology.* Plenum, New York, pp. 31–41.

Foty, R. A., C. M. Pfleger, G. Forgacs and M. S. Steinberg. 1996. Surface tensions of embryonic cells predict their mutual envelopment behavior. *Development* 122: 1611–1620.

Fujimori, T., S. Miyatani and M. Takeichi. 1990. Ectopic expression of N-cadherin perturbs histogenesis in *Xenopus* embryos. *Development* 110: 97–104.

Fukui, A. and M. Asashima. 1994. Control of cell differentiation and morphogenesis in amphibian development. *Int. J. Dev. Biol.* 38: 257–266.

Gilbert, S. F. and S. Sarkar. 2000. Embryo complexity: Organicism for the 21st century. *Dev. Dynam.* 219: 1–9.

Giudice, G. and R. Just. 1962. Restitution of whole larvae from disaggregated cells of sea urchin embryos. *Dev. Biol.* 5: 402–411.

Godt, D. and U. Tepass. 1998. *Drosophila* oocyte localization is mediated by differential cadherin-based adhesion. *Nature* 395: 387–391.

González-Reyes, A. and D. St. Johnston. 1998. The *Drosophila* AP axis is polarised by the cadherin-mediated positioning of the oocyte. *Development* 125: 3635–3644.

Gosse, R. J. 1969. *Principles of Regeneration.* Academic Press, New York.

Green, J. B. A. and J. C. Smith. 1990. Graded changes in dose of a *Xenopus* activin A homologue elicit stepwise transitions in embryonic cell fate. *Nature* 347: 391–394.

Greene, E. 1989. A diet-induced developmental polymorphism in a caterpillar. *Science* 243: 643–646.

Gurdon, J. B., P. Harger, A. Mitchell and P. Lemaire. 1994. Activin signalling and responses to a morphogen gradient. *Nature* 371: 487–493.

Gurdon, J. B., A. Mitchell and D. Mahony. 1995. Direct and continuous assessment by cells of their position in a morphogen gradient. *Nature* 376: 520–521.

Gurdon, J. B., S. Dyson and D. St. Johnston. 1998. Cells' perception of position in a concentration gradient. *Cell* 95: 159–163.

Hamburger, V. 1988. *The Heritage of Experimental Embryology: Hans Spemann and the Organizer.* Oxford University Press, New York, p. vii.

Haraway, D. J. 1976. *Crystals, Fabrics and Fields: Metaphors of Organicism in Twentieth-Century Biology.* Yale University Press, New Haven.

Harrington, A. 1996. *Reenchanted Science: Holism in German Culture from Wilhelm II to Hitler.* Princeton University Press, Princeton, NJ.

Harrison, R. G. 1933. Some difficulties of the determination problem. *Am. Nat.* 67: 306–321.

Hatta, K. and M. Takeichi. 1986. Expression of N-cadherin adhesion molecules associated with early morphogenetic events in chick development. *Nature* 320: 447–449.

Heasman, J., D. Ginsberg, K. Goldstone, T. Pratt, C. Yoshidanaro and C. Wylie. 1994. A functional test for maternally inherited cadherin in *Xenopus* shows its importance in cell adhesion at the blastula stage. *Development* 120: 49–57.

Hertwig, O. 1894. *Zeit- und Streitfragen der Biologie I. Präformation oder Epigenese? Grundzüge einer Entwicklungstheorie der Organismen.* Gustav Fischer, Jena.

Hörstadius, S. 1939. The mechanics of sea urchin development studied by operative methods. *Biol. Rev.* 14: 132–179.

Hörstadius, S. and A. Wolsky. 1936. Studien über die Determination der Bilateralsymmetrie des jungen Seeigelkeimes. *Wilhelm Roux Arch. Entwicklungsmech. Org.* 135: 69–113.

Huxley, J. and G. R. de Beer. 1934. *The Elements of Experimental Embryology.* Cambridge University Press, Cambridge.

Jokiel, P. L. and R. H. York, Jr. 1982. Solar ultraviolet photobiology of the reef coral *Pocillopora damicornis* and symbiotic zooxanthellae. *Bull. Mar. Sci.* 32: 301–315.

Just, E. E. 1939. *The Biology of the Cell Surface.* Blackiston, Philadelphia.

Kadokawa, Y., I. Fuketa, A. Nose, M. Takeichi and N. Nakatsuji. 1989. Expression of E- and P-cadherin in mouse embryos and uteri during the periimplantation period. *Dev. Growth Diff.* 31: 23–30.

Kiesecker, J. M., A. R. Blaustein and L. K. Belden. 2001. Complex causes of amphibian population declines. *Nature* 410: 681–684.

Kintner, C. 1993. Regulation of embryonic cell adhesion by the cadherin cytoplasm domain. *Cell* 69: 225–236.

Koch, P. B. and D. Buchmann. 1987. Hormonal control of seasonal morphs by the timing of ecdysteroid release in *Araschnia levana* (Nymphalidae: Lepidoptera). *J. Insect Physiol.* 36: 159–164.

Lawrence, P. A. 1966. Gradients in the insect segment: The orientation of hairs in the milkweed bug *Oncopeltus fasciatus. J. Exp. Zool.* 44: 602–620.

Lee, C.-H. and B. M. Gumbiner. 1995. Disruption of gastrulation movements in *Xenopus* by a dominant-negative mutant for C-cadherin. *Dev. Biol.* 171: 363–373.

Leutert, T. R. 1974. Zur Geschlechtsbestimmung und Gametogenese von *Bonellia viridis* Rolando. *J. Embryol. Exp. Morphol.* 32: 169–193.

Marcus, N. H. 1979. Developmental aberrations associated with twinning in laboratory-reared sea urchins. *Dev. Biol.* 70: 274–277.

McClendon, J. F. 1910. The development of isolated blastomeres of the frog's egg. *Am. J. Anat.* 10: 425–430.

McDowell, N., A. M. Zorn, D. J. Crease and J. B. Gurdon. 1997. Activin has direct long-range signalling activity and can form a concentration gradient by diffusion. *Curr. Biol.* 7: 671–681.

Mead, K. S. and D. Epel. 1995. Beakers versus breakers: How fertilization in the laboratory differs from fertilization in nature. *Zygote* 3: 95–99.

Monroy, A. and A. A. Moscona. 1979. *Introductory Concepts in Developmental Biology.* University of Chicago Press, Chicago.

Morgan, T. H. 1905 An attempt to analyze the phenomenon of polarity in *Tubularia. J. Exp. Zool.* 1: 589–591.

Morgan, T. H. 1906. "Polarity" considered as a phenomenon of gradation of materials. *J. Exp. Zool.* 2: 495–506.

Moscona, A. A. 1952. Cell suspension from organ rudiments of chick embryos. *Exp. Cell Res.* 3: 535–539.

Moscona, A. A. 1961. Rotation-mediated histogenetic aggregation of dissociated cells: A quantifiable approach to cell interaction in vitro. *Exp. Cell Res.* 22: 455–475.

Moscona, A. A. 1974. Surface specification of embryonic cells: Lectin receptors, cell recognition, and specific cell ligands. In A. Monroy (ed.), *The Cell Surface in Development.* Wiley, New York, pp. 67–99.

National Institutes of Health. 2001. *Stem Cells: Scientific Progress and Future Research Directions.* http://www.nih.gov/news/stemcell/fullrptstem.pdf.

Needham, J. 1943. *Time: The Refreshing River.* Allen and Unwin, London.

Nijhout, H. F. 1991. *The Development and Evolution of Butterfly Wing Patterns.* Smithsonian Institution Press, Washington, D.C.

Nose, A. and M. Takeichi. 1986. A novel cadherin adhesion molecule: Its expression patterns associated with implantation and organogenesis of mouse embryos. *J. Cell Biol.* 103: 2649–2658.

Nüsslein-Volhard, C., H. G. Fröhnhofer and R. Lehmann. 1987. Determination of anterioposterior polarity in *Drosophila. Science* 238: 1675–1681.

Nyhart, L. K. 1995. *Biology Takes Form: Animal Morphology and the German Universities, 1800–1900.* University of Chicago Press, Chicago.

Oberlender, S. A. and R. S. Tuan. 1994. Expression and functional involvement of N-cadherin in embryonic limb chondrogenesis. *Development* 120: 177–187.

Opitz, J. M. 1985. The developmental field concept. *Am. J. Med. Genet.* 21: 1–11.

Phillips, K. 1994. *Tracking the Vanishing Frogs: An Ecological Mystery.* St. Martin's Press, New York.

Reverberi, G. and A. Minganti. 1946. Fenomeni di evocazione nello sviluppo dell'uovo di Ascidie. Risultati dell'indagine spermentale sull'ouvo di *Ascidiella aspersa* e di *Ascidia malaca* allo stadio di 8 blastomeri. *Pubbl. Staz. Zool. Napoli* 20: 199–253. (Quoted in G. Reverberi,. *Experimental Embryology of Marine and Freshwater Invertebrates.* North-Holland, Amsterdam, 1971, p. 537.)

Roux, W. 1888. Contributions to the developmental mechanics of the embryo. On the artificial production of half-embryos by destruction of one of the first two blastomeres and the later development (postgeneration) of the missing half of the body. In B. H. Willier and J. M. Oppenheimer (eds.) 1974, *Foundations of Experimental Embryology.* Hafner, New York, pp. 2–37.

Roux, W. 1894. Einleitung zum Archiv für Entwicklungsmechanik. *Arch. Embryol.* 1: 1–142. English translation reprinted in J. M. Maienschein (ed.), 1986, *Defining Biology,* Harvard University Press, Cambridge, MA, pp. 107–148.

Schlichting, C. D. and M. Pigliucci. 1998. *Phenotypic Evolution: A Reaction Norm Perspective.* Sinauer Associates, Sunderland, MA.

Schmalhausen, I. I. 1949. *Factors of Evolution: The Theory of Stabilizing Selection.* University of Chicago Press, Chicago.

Shimizu, K. and J. B. Gurdon. 1999. A quantitative analysis of signal transduction from activin receptor to nucleus and its relevance to morphogen gradient interpretation. *Proc. Natl. Acad. Sci. USA* 96: 6791–6796.

Siebeck, O. 1988. Experimental investigation of UV tolerance in hermatypic corals. *Mar. Ecol. Prog. Ser.* 43: 95–103.

Slack, J. M. W. 1991. *From Egg to Embryo: Regional Specification in Early Development.* Cambridge University Press, New York.

Spemann, H. 1921. Die Erzeugung tierischer Chimaeren durch heteroplastische embryonale Transplantation zwischen *Triton cristatus* u. *taeniatus. Wilhelm Roux Arch. Entwicklungsmech. Org.* 48: 533–570.

Stearns, S. C., G. de Jong and R. A. Newman. 1991. The effects of phenotypic plasticity on genetic correlations. *Trends Ecol. Evol.* 6: 122–126.

Steinberg, M. S. 1964. The problem of adhesive selectivity in cellular interactions. In M. Locke (ed.), *Cellular Membranes in Development.* Academic Press, New York, pp. 321–434.

Steinberg, M. S. 1970. Does differential adhesion govern self-assembly processes in histogenesis? Equilibrium configurations and the emergence of a hierarchy among populations of embryonic cells. *J. Exp. Zool.* 173: 395–434.

Steinberg, M. S. and M. Takeichi. 1994. Experimental specification of cell sorting, tissue spreading and specific spatial patterning by quantitative differences in cadherin expression. *Proc. Natl. Acad. Sci. USA* 91: 206–209.

Stumpf, H. 1966. Mechanism by which cells estimate their location within the body. *Nature* 212: 430–431.

Takeichi, M. 1987. Cadherins: A molecular family essential for selective cell-cell adhesion and animal morphogenesis. *Trends Genet.* 3: 213–217.

Takeichi, M. 1991. Cadherin cell adhesion receptors as a morphogenetic regulator. *Science* 251: 1451–1455.

Takeichi, M., H. S. Ozaka, K. Tokunaga and T. S. Okada. 1979. Experimental manipulation of cell surface to affect cellular recognition mechanisms. *Dev. Biol.* 70: 195–205.

Toivonen, S. and L. Saxén. 1955. The simultaneous inducing action of liver and bone marrow of the guinea pig in implantation and explantation experiments of *Triturus.* *Exp. Cell Res.* [Suppl.] 3: 346–357.

Townes, P. L. and J. Holtfreter. 1955. Directed movements and selective adhesion of embryonic amphibian cells. *J. Exp. Zool.* 128: 53–120.

van der Weele, C. 1995. *Images of Development: Environmental Causes in Ontogeny.* Elinkwijk, Utrecht.

Weismann, A. 1875. Über den Saison-Dimorphismus der Schmetterlinge. In. *Studien zur Descendenz-Theorie.* Engelmann, Leipzig.

Weismann, A. 1892. *Essays on Heredity and Kindred Biological Problems.* Translated by E. B. Poulton, S. Schoenland and A. E. Shipley. Clarendon, Oxford.

Weismann, A. 1893. *The Germ-Plasm: A Theory of Heredity.* Translated by W. Newton Parker and H. Ronnfeld. Walter Scott Ltd., London.

Weiss, P. 1923. Die Regeneration der Urodelenextremität als Selbst differenzierung des Organrestes. *Naturwissenschaften* 11: 669–677.

Weiss, P. 1939. *Principles of Development.* Holt, New York.

Whittaker, J. R. 1973. Segregation during ascidian embryogenesis of egg cytoplasmic information for tissue-specific enzyme development. *Proc. Natl. Acad. Sci. USA* 70: 2096–2100.

Whittaker, J. R. 1977. Segregation during cleavage of a factor determining endodermal alkaline phosphatase development in ascidian embryos. *J. Exp. Zool.* 202: 139–153.

Whittaker, J. R. 1982. Muscle cell lineage cytoplasm can change the developmental expression in epidermal lineage cells of ascidian embryos. *Dev. Biol.* 93: 463–470.

Wilkins, A. S. 1993. *Genetic Analysis of Animal Development,* 2nd Ed. Wiley-Liss, New York.

Wilson, E. B. 1896. *The Cell in Development and Inheritance.* Macmillan, New York.

Wilson, E. B. 1904. Experimental studies on germinal location. *J. Exp. Zool.* 1: 1–72.

Wolpert, L. 1968. The French flag problem: A contribution to the discussion on pattern formation and regulation. In C. H. Waddington (ed.), *Towards a Theoretical Biology.* Edinburgh University Press, Edinburgh, pp. 125–133.

Wolpert, L. 1969. Positional information and the spatial pattern of cellular differentiation. *J. Theor. Biol.* 25: 1–47.

Wolpert, L. 1977. *The Development of Pattern and Form in Animals.* Carolina Biological, Burlington, NC.

Wolpert, L. 1978. Pattern formation in biological development. *Sci. Am.* 239(4): 154–164.

Woltereck, R. 1909. Weitere experimentelle Untersuchungen über Artveränderung, speziell über das Wesen quantitativer Artunderscheide bei Daphniden. *Versuch. Deutsch. Zool. Ges.* 1909: 110–173.

chapter 4 The genetic core of development

The secrets that engage me—that sweep me away—are generally secrets of inheritance: how the pear seed becomes a pear tree, for instance, rather than a polar bear.

CYNTHIA OZICK (1989)

Here, there is one central field. Development. How the egg turns into the organism. But development ultimately includes all of biology: and it will have to be put on a molecular basis.

SYDNEY BRENNER (1979)

"BETWEEN THE CHARACTERS THAT FURNISH THE DATA for the theory, and the postulated genes, to which the characters are referred, lies the whole field of embryonic development." Here Thomas Hunt Morgan noted in 1926 that the only way to get from genotype to phenotype is through developmental processes.

In the early twentieth century, embryology and genetics were not considered separate sciences. They diverged in the 1920s, when Morgan redefined genetics as the science studying the *transmission* of inherited traits, as opposed to embryology, the science studying the *expression* of those traits. In recent years, however, the techniques of molecular biology have effected a rapprochement of embryology and genetics. In fact, the two fields have become linked to a degree that makes it necessary to discuss molecular genetics early in this text. Problems in animal development that could not be addressed a decade ago are now being solved by a set of techniques involving nucleic acid synthesis and hybridization. This chapter seeks to place these new techniques within the context of the ongoing dialogues between genetics and embryology.

The Embryological Origins of the Gene Theory

Nucleus or cytoplasm: Which controls heredity?

Mendel called them *bildungsfähigen Elemente*, "form-building elements" (Mendel 1866, p. 42); we call them genes. It is in Mendel's term, however, that we see how closely intertwined were the concepts of inheritance and development in the nineteenth century. Mendel's observations, however, did not indicate where these hereditary elements existed in the cell, or how they came to be expressed. The gene theory that was to become the cornerstone of modern genetics originated from a controversy within the field of physiological embryology (see Chapter 3). In the late 1800s, a group of scientists began to study the mechanisms by which fertilized eggs give rise to adult organisms.

Two young American embryologists, Edmund Beecher Wilson and Thomas Hunt Morgan (Figure 4.1), became part of this group of "physiological embryologists," and each became a partisan in the controversy over which of the two compartments of the fertilized egg—the nucleus or the cytoplasm—controls inheritance. Morgan allied himself with those embryologists who thought the control of development lay within the cytoplasm, while Wilson allied himself with Theodor Boveri, one of the biologists who felt that the nucleus contained the instructions for

(A)

(B)

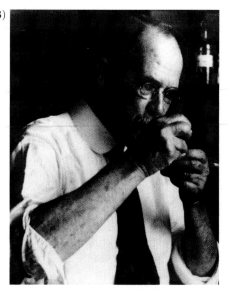

Figure 4.1
(A) E. B. Wilson (1856–1939; shown here around 1899), an embryologist whose work on early embryology and sex determination greatly advanced the chromosomal hypotheses of development. (Wilson was also acknowledged to be among the best amateur cellists in the United States.) (B) Thomas Hunt Morgan (1866–1945), who brought the gene theory out of embryology. This photo—taken in 1915, as the basic elements of the gene theory were coming together—shows Morgan using a hand lens to sort fruit flies. (A courtesy of W. N. Timmins; B courtesy of G. Allen.)

development. In fact, Wilson (1896, p. 262) declared that the processes of meiosis, mitosis, fertilization, and unicellular regeneration (only from the fragment containing the nucleus) "converge to the conclusion that the chromatin is the most essential element in development."* He did not shrink from the consequences of this belief. Years before the rediscovery of Mendel or the gene theory, Wilson (1895, p. 4) noted, "Now, chromatin is known to be closely similar, if not identical with, a substance known as nuclein … which analysis shows to be a tolerably definite chemical composed of a nucleic acid (a complex organic acid rich in phosphorus) and albumin. And thus we reach the remarkable conclusion that inheritance may, perhaps, be effected by the physical transmission of a particular chemical compound from parent to offspring."

Some of the major support for the chromosomal hypothesis of inheritance was coming from the embryological studies of Theodor Boveri (Figure 4.2A), a researcher at the Naples Zoological Station. Boveri fertilized sea urchin eggs with large concentrations of their sperm and obtained eggs that had been fertilized by two sperm. At first cleavage, these eggs formed four mitotic poles and divided into four cells instead of two (see Chapter 7). Boveri then separated the blastomeres and demonstrated that each cell developed abnormally, and in a different way, as a result of each of the cells having different types of chromosomes. Thus, Boveri claimed that each chromosome had an individual nature and controlled different vital processes.

Adding to Boveri's evidence, E. B. Wilson (1905) and Nettie Stevens (1905a,b; Figure 4.2B) demonstrated a critical correlation between nuclear chromosomes and organismal development: XO or XY embryos became male; XX embryos became female. Here was a nuclear property that correlated with development. Eventually, Morgan began to obtain mutations that correlated with sex and with the X chromosome, and he began to view the genes as being physically linked to

---

*Notice that Wilson was writing about form-building units in chromatin in 1896—before the rediscovery of Mendel's paper or the founding of the gene theory. For further analysis of the interactions between Morgan and Wilson that led to the gene theory, see Gilbert 1978, 1987; and Allen 1986.

Figure 4.2
Chromosomal uniqueness was shown by Boveri and Stevens. (A) Theodor Boveri (1862–1915), whose work, Wilson (1896) said, "accomplished the actual amalgamation between cytology, embryology, and genetics—a biological achievement which … is not second to any of our time." This photograph was taken in 1908, when Boveri's chromosomal and embryological studies were at their zenith. (B) Nettie M. Stevens (1861–1912), who trained with both Boveri and Morgan, seen here in 1904 when she was a postdoctoral student pursuing the research that correlated the number of X chromosomes with sexual development. (A from Baltzer 1967; B courtesy of the Carnegie Institute of Washington.)

one another on the chromosomes. The embryologist Morgan had shown that nuclear chromosomes are responsible for the development of inherited characters.*

WEBSITE 4.1 **The embryological origin of the gene theory.** The emergence of the gene theory from embryological research is a fascinating story and complements the history of genetics that begins with Mendel's experiments.

The split between embryology and genetics

Morgan's evidence provided a material basis for the concept of the gene. Originally, this type of genetics was seen as being part of embryology, but by the 1930s, genetics became its own discipline, developing its own vocabulary, journals, societies, favored research organisms, professorships, and rules of evidence. Hostility between embryology and genetics also emerged. Geneticists believed that the embryologists were old-fashioned and that development would be completely explained as the result of gene expression. Conversely, the embryologists regarded the geneticists as uninformed about how organisms actually develop and felt that genetics was irrelevant to embryological questions. Embryologists such as Frank Lillie (1927), Ross Granville Harrison (1937), Hans Spemann (1938), and Ernest E. Just (1939) (Figure 4.3) claimed that there could be no genetic theory of development until at least three major challenges had been met by the geneticists:

1. Geneticists had to explain how chromosomes—which were thought to be identical in every cell of the organism—produce different and changing types of cell cytoplasms.
2. Geneticists had to provide evidence that genes control the early stages of embryogenesis. Almost all the genes known at the time affected the final modeling steps in development (eye color, bristle shape, wing venation in *Drosophila*). As Just said (quoted in Harrison 1937), embryologists were interested in how a fly forms its back, not in the number of bristles on its back.
3. Geneticists had to explain phenomena such as sex determination in certain invertebrates (and vertebrates such as reptiles), in which the environment determines sexual phenotype.

The debate became quite vehement. In rhetoric reflecting the political anxieties of the late 1930s, Harrison (1937) warned:

> *Now that the necessity of relating the data of genetics to embryology is generally recognized and the Wanderlust of geneticists is beginning to urge them in our direction, it may not be inappropriate to point out a danger of this threatened invasion. The prestige of success enjoyed by the gene theory might easily become a hindrance to the understanding of development by directing our attention solely to the genom, whereas cell movements, differentiation, and in fact all of developmental processes are actually effected by cytoplasm. Already we have theories that refer the processes of development to gene action and regard the whole performance as no more than the realization of the potencies of genes. Such theories are altogether too one-sided.*

Until geneticists could demonstrate the existence of inherited variants during early development, and until geneticists had a

---

*Morgan's evidence for nuclear control of development went against his expectations; until 1910, he was the leading proponent of the cytoplasmic control of development. Wilson was one of Morgan's closest friends, and Morgan considered Stevens his best graduate student at that time. Both were against Morgan on this issue. Even though they disagreed, Morgan wholeheartedly supported Stevens's request for research funds, saying that her qualifications were the best possible. Wilson wrote an equally laudatory letter of support, even though she would be a rival researcher (see Brush 1978).

(A)

Figure 4.3
Embryologists attempted to keep genetics from "taking over" their field in the 1930s. (A) Frank Lillie headed the Marine Biological Laboratory at Woods Hole and was a leader in fertilization research and reproductive endocrinology. (B) Hans Spemann (left) and Ross Harrison (right) perfected the transplantation operations used to discover when the body and limb axes are determined. They argued that geneticists had no mechanism for explaining how the same nuclear genes could create different cell types during development. (C) Ernest E. Just made critical discoveries on fertilization. He spurned genetics and emphasized the role of the cell membrane in determining the fates of cells. (A courtesy of V. Hamburger; B courtesy of T. Horder; C courtesy of the Marine Biological Laboratory, Woods Hole.)

(B)

(C)

well-documented theory explaining how the same chromosomes could produce different cell types, embryologists generally felt no need to ground their science in gene action.

Early attempts at developmental genetics

Some scientists, however, felt that neither embryology nor genetics was complete without the other. There were several attempts to synthesize the two disciplines, but the first successful reintegration of genetics and embryology came in the late 1930s from Salome Glucksohn-Schoenheimer (now S. Gluecksohn Waelsch) and Conrad Hal Waddington (Figure 4.4). Both were trained in European embryology and had learned genetics in the United States from Morgan's students.

Glucksohn-Schoenheimer and Waddington attempted to find mutations that affected early development and to discover the processes that these genes affected. Glucksohn-Schoenheimer (1938, 1940) showed that mutations in the *Brachyury* gene of the mouse caused the aberrant development of the posterior portion of the embryo, and she traced the effects of this mutant gene to the notochord, which would normally have helped induce the dorsal-ventral axis.*

At the same time, Waddington (1939) isolated several genes that caused wing malformations in fruit flies (*Drosophila*). He, too, analyzed these mutations in terms of how the genes might affect the developmental primordia that give rise to these structures. The *Drosophila* wing, he correctly

---

*Glucksohn-Schoenheimer's observations took 50 years to be confirmed by DNA hybridization. However, when the *Brachyury* (*T*-locus) gene was cloned and its expression detected by the in situ hybridization technique (discussed later in this chapter), Wilkinson and co-workers (1990) found that "the expression of the *T* gene has a direct role in the early events of mesoderm formation and in the morphogenesis of the notochord." While a comprehensive history of early developmental genetics needs to be written, more information on its turbulent origins can be found in Oppenheimer 1981; Sander 1986; Gilbert 1988, 1991, 1996; Burian et al. 1991; Harwood 1993; Keller 1995; and Morange 1996.

(A)

(B)

Figure 4.4
Two of the founders of developmental genetics. (A) Salome Gluecksohn-Schoenheimer (now S. Gluecksohn-Waelsch; b. 1907) received her doctorate in Spemann's laboratory. Fleeing Hitler's Germany, she brought her embryological acumen to Leslie Dunn's genetics laboratory in the United States. (B) Conrad Hal Waddington (1905–1975) did not believe in the distinction between genetics and embryology and sought to find mutations that were active during development. (A courtesy of S. Gluecksohn-Waelsch; B from Waddington 1948.)

claimed, "appears favorable for investigations on the developmental action of genes." Thus, one of the main objections of embryologists to the genetic model of development—that genes appeared to be working only on the final modeling of the embryo and not on its major outlines—was countered.

WEBSITE 4.2 **Creating developmental genetics.** The resynthesis of embryology and genetics into developmental genetics did not come easily. These websites look at some of the ways researchers attempted to join the two fields together.

Evidence for Genomic Equivalence

The other major objection to a genetically based embryology remained: How could nuclear genes direct development when they were the same in every cell type? The existence of this **genomic equivalence** was not so much proved as assumed (because every cell is the mitotic descendant of the fertilized egg), so one of the first problems of developmental genetics was to determine whether every cell of an organism indeed had the same set of genes, or **genome**, as every other cell.

Amphibian cloning: The restriction of nuclear potency

The ultimate test of whether the nucleus of a differentiated cell has undergone any irreversible functional restriction is to have that nucleus generate every other type of differentiated cell in the body. If each cell's nucleus is identical to the zygote nucleus, then each cell's nucleus should be **totipotent** (capable of directing the entire development of the organism) when transplanted into an activated enucleated egg. Before

such an experiment could be done, however, three techniques for transplanting nuclei into eggs had to be perfected: (1) a method for enucleating host eggs without destroying them; (2) a method for isolating intact donor nuclei; and (3) a method for transferring such nuclei into the host egg without damaging either the nucleus or the oocyte.

All three techniques were developed in the 1950s by Robert Briggs and Thomas King. First, they combined the enucleation of the host egg with its activation. When an oocyte (a developing egg cell) from the leopard frog (*Rana pipiens*) is pricked with a clean glass needle, the egg undergoes all the cytological and biochemical changes associated with fertilization. The internal cytoplasmic rearrangements of fertilization occur, and the completion of meiosis takes place near the animal pole of the cell. The meiotic spindle can easily be located as it pushes away the pigment granules at the animal pole, and puncturing the oocyte at this site causes the spindle and its chromosomes to flow outside the egg (Figure 4.5). The host egg is now considered both activated (the fertilization reactions necessary to initiate development have been completed) and enucleated.

The transfer of a nucleus into the egg is accomplished by disrupting a donor cell and transferring the released nucleus into the oocyte through a micropipette. Some cytoplasm accompanies the nucleus to its new home, but the ratio of donor to recipient cytoplasm is only $1:10^5$, and the donor cytoplasm does not seem to affect the outcome of the experiments. In 1952, Briggs and King, using these techniques, demonstrated that blastula cell nuclei could direct the development of complete tadpoles when transferred into the cytoplasm of an activated enucleated frog egg. This procedure is called **somatic nuclear transfer**, or more commonly, **cloning**.

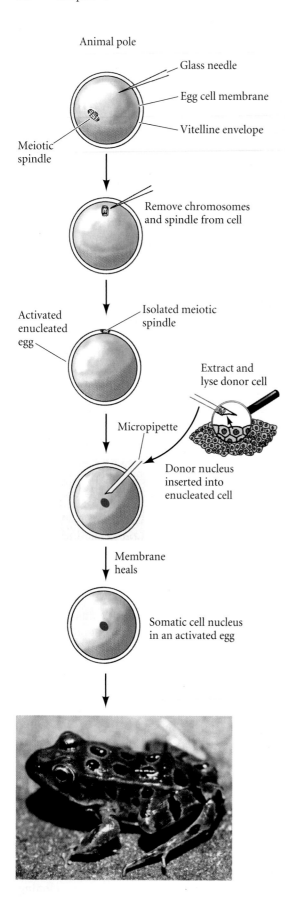

Animal pole

Glass needle

Egg cell membrane

Vitelline envelope

Meiotic spindle

Remove chromosomes and spindle from cell

Isolated meiotic spindle

Activated enucleated egg

Extract and lyse donor cell

Micropipette

Donor nucleus inserted into enucleated cell

Membrane heals

Somatic cell nucleus in an activated egg

Figure 4.5
Procedure for transplanting blastula nuclei into activated enucleated *Rana pipiens* eggs. The relative dimensions of the meiotic spindle have been exaggerated to show the technique. "Freddy," the handsome and mature *R. pipiens* in the photograph, was derived in this way by M. DiBerardino and N. Hoffner Orr. The vitelline envelope is the extracellular matrix surrounding the egg. (After King 1966; photograph courtesy of M. DiBerardino.)

What happens when nuclei from more advanced developmental stages are transferred into activated enucleated oocytes? King and Briggs (1956) found that whereas most blastula nuclei could produce entire tadpoles, there was a dramatic decrease in the ability of nuclei from later stages to direct development to the tadpole stage (Figure 4.6). When nuclei from the somatic cells of tailbud-stage tadpoles were used as donors, normal development did not occur. However, nuclei from the germ cells of tailbud-stage tadpoles (which could give rise to a complete organism after fertilization) were capable of directing normal development in 40% of the blastulae that developed (Smith 1956). Thus, most somatic cells appeared to lose their ability to direct development as they became determined and differentiated.

Amphibian cloning: The totipotency of somatic cells

Is it possible that some differentiated cell nuclei differ from others in their ability to direct development? John Gurdon and his colleagues, using slightly different methods of nuclear transplantation on the frog *Xenopus*, obtained results suggesting that the nuclei of some differentiated cells can remain

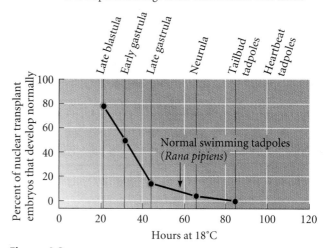

Developmental stage from which nuclei were taken

Late blastula · Early gastrula · Late gastrula · Neurula · Tailbud tadpoles · Heartbeat tadpoles

Percent of nuclear transplant embryos that develop normally

Normal swimming tadpoles (*Rana pipiens*)

Hours at 18°C

Figure 4.6
Percentage of successful nuclear transplants as a function of the developmental age of the donor nucleus. The abscissa represents the developmental stage at which a donor nucleus (from *R. pipiens*) was isolated and inserted into an activated enucleated oocyte. The ordinate shows the percentage of those transplants capable of producing blastulae that could then direct development to the swimming tadpole stage. (After McKinnell 1978.)

totipotent. Gurdon, too, found a progressive loss of potency with increasing developmental age, although *Xenopus* cells retained their potencies for a longer period than did the cells of *Rana* (Figure 4.7).

To clone amphibians from the nuclei of cells known to be differentiated, Gurdon and his colleagues cultured epithelial cells from adult frog foot webbing. These cells were shown to be differentiated: each of them contained a specific keratin, the characteristic protein of adult skin cells. When nuclei from these cells were transferred into activated enucleated *Xenopus* oocytes, none of the first-generation transfers progressed further than the formation of the neural tube shortly after gastrulation. By serial transplantation, however, numerous tadpoles were generated (Gurdon et al. 1975). Although these tadpoles all died prior to feeding, they showed that a single differentiated cell nucleus still retained incredible potencies. A nucleus of a skin cell could produce all the cells of the young tadpole.

> **WEBSITE 4.3 Amphibian cloning: Potency and deformity.** The ability of amphibian nuclei to produce normal frogs has been a controversial area. The definition of what is a differentiated cell nucleus and what constitutes a "normal" frog have both been questioned.

> **WEBSITE 4.4 Metaplasia.** Before cloning became possible, evidence for genomic equivalence came from a phenomenon called metaplasia, the regeneration of tissue from a different source.

Cloning mammals

In 1997, Ian Wilmut announced that a sheep had been cloned from a somatic cell nucleus from an adult female sheep. This was the first time that an adult vertebrate had been successfully cloned from another adult.* To do this, Wilmut and his colleagues (1997) took cells from the mammary gland of an adult (6-year-old) pregnant ewe and put them into culture. The culture medium was formulated to keep the nuclei in these cells at the intact diploid stage of the cell cycle (G_1). This cell-cycle stage turned out to be critical. They then obtained oocytes from a different strain of sheep and removed their nuclei. These oocytes had to be in the second meiotic metaphase (which is the stage at which they are usually fertilized). As with the somatic (mammary gland) cell, cell-cycle stage turned out to be a critical factor for the success of the procedure.

---

*The creation of Dolly was the result of a combination of scientific and social circumstances. These circumstances involved job security, people with different areas of expertise meeting each other, childrens' school holidays, international politics, and who sits near whom in a pub. The complex interconnections giving rise to Dolly are told in *The Second Creation* (Wilmut et al. 2000), a book that should be read by anyone who wants to know how contemporary science actually works. As Wilmut acknowledged (p. 36), "The story may seem a bit messy, but that's because life is messy, and science is a slice of life."

Wild-type donor　　　　Albino parents
of enucleated eggs　　　of nucleus donor

Figure 4.7
A clone of *Xenopus laevis* frogs. The nuclei for all the members of this clone came from a single individual—a female tailbud-stage tadpole whose parents (upper panel) were both marked by albino genes. The nuclei (containing these defective pigmentation genes) were transferred into activated enucleated eggs from a wild-type female (upper panel). The resulting frogs were all female and albino (lower panel). (Photographs courtesy of J. Gurdon.)

The fusion of the donor cell and the enucleated oocyte was accomplished by bringing the two cells together and sending electric pulses through them. The electric pulses destabilized the cell membranes, allowing the cells to fuse together. Moreover, the same pulses that fused the cells activated the egg to begin development. The resulting embryos were eventually transferred into the uteri of pregnant sheep.

Of the 434 sheep oocytes originally used in this experiment, only one survived: Dolly (Figure 4.8). DNA analysis confirmed that the nuclei of Dolly's cells were derived from the strain of sheep from which the donor nucleus was taken (Ashworth et al. 1998; Signer et al. 1998). Thus, it appears that the nuclei of vertebrate adult somatic cells can indeed be totipotent. No genes necessary for development have been lost or mutated in a way that would make them nonfunctional.

Cloning has been confirmed in cows (Kato et al. 1998), mice (Wakayama et al. 1998), cats (Shin et al. 2002), and other mammals. However, certain caveats must be applied. First, although it appears that all the organs were properly formed in the cloned animals, many of the clones developed debilitating diseases as they matured (Humphreys et al. 2001; Jaenisch and Wilmut 2001; Kolata 2001). Second, the phenotype of the cloned animal is sometimes not identical to that of the animal from which the nucleus was derived. There is variability due to random chromosomal events and the effects of environment. The pigmentation of calico cats, for instance, is due to the random inactivation of one or the other X chromosome in each somatic cell of the female cat embryo (to be discussed in the next chapter). Therefore, the markings of the first cloned cat, a calico named "CC," were different from those of "Rainbow," the calico cat whose cells provided the implanted nucleus that generated the clone (Figure 4.9). The same genotype gives rise to multiple phenotypes in cloned sheep as well. Wilmut noted that four sheep cloned from blastocyst nuclei from the same

(A)

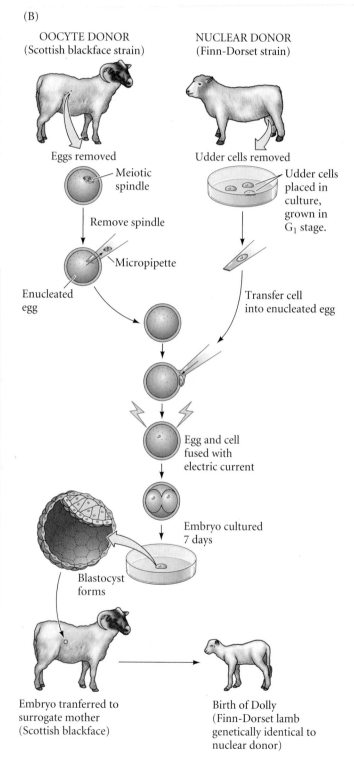

(B)

Figure 4.8
Cloned mammals, whose nuclei came from adult somatic cells. (A) Dolly, the adult sheep on the left, was derived by fusing a mammary gland cell nucleus with an enucleated oocyte, which was then implanted in a surrogate mother (of a different breed of sheep) who gave birth to Dolly. Dolly has since produced a lamb (Bonnie, at right) by normal reproduction. (B) Procedure used for cloning sheep. (A, photograph by Roddy Field, © Roslin Institute; B after Wilmut et al. 2000.)

Figure 4.9
The kitten "CC" (A) is a clone produced using somatic nuclear transfer from "Rainbow" (B). Their markings are not identical because the pigmentation pattern in calico cats is affected by the random inactivation of the second X chromosome (see Chapter 5). Their behaviors are also quite different. (Photographs courtesy of the College of Veterinary Medicine, Texas A&M University.)

embryo "are genetically identical to each other and yet are very different in size and temperament, showing emphatically that an animal's genes do not 'determine' every detail of its physique and personality" (Wilmut et al. 2000, p. 5). Wilmut concludes that for this and other reasons, the "resurrection" of lost loved ones by cloning is not feasible. (The possibility of cloning humans will be discussed in Chapter 21.)

Sidelights & Speculations

Why Clone Mammals?

Given that we already knew from amphibian studies in the 1960s that nuclei were totipotent, why clone mammals? Many of the reasons are medical and commercial, and there are good reasons why these techniques were first developed by pharmaceutical companies rather than at universities. Cloning is of interest to some developmental biologists who study the relationships between nucleus and cytoplasm during fertilization or who study aging (and the loss of totipotency that appears to accompany it), but cloned mammals are of special interest to the people and corporations concerned with the creation of protein pharmaceuticals.

Protein drugs such as human insulin, protease inhibitors, and clotting factors are difficult to manufacture. Because of immunological rejection problems, human proteins are usually much better tolerated by patients than proteins from other animals. Similarly, we reject the products of

human genes synthesized in bacteria. The problem thus becomes how to obtain large amounts of human proteins. One of the most efficient ways to produce these proteins is to insert the human genes encoding them into the oocyte DNA of sheep, goats, or cows. Animals containing a gene from another individual (often of a different species)—a **transgene**—are called **transgenic animals**. A transgenic female sheep or cow not only might contain the gene for the human protein, but might also be able to express the gene in her mammary tissue and thereby secrete the protein in her milk (Figure 4.10; Prather 1991).

Producing transgenic sheep, cows, or goats is not an efficient undertaking, however. Only 20% of the treated eggs survive the technique. Of these, only about 5% express the human gene. And of those transgenic animals expressing the human gene, only half are female, and only a small percentage of these actually secrete a high

level of the protein into their milk. (And it often takes years before they first first produce milk). Moreover, they die after several years of milk production, and their offspring are usually not as good at secreting the human protein as the originals. Cloning would enable pharmaceutical companies to make numerous copies of such an "elite transgenic animal," all of which should produce high yields of the human protein in their milk. The medical importance of such a technology would be great, since such proteins could become much cheaper for the patients who need them for survival. The economic incentives for cloning are therefore enormous (Meade 1997). Thus, shortly after the announcement of Dolly, the same laboratory announced the birth of Polly (Schnieke et al. 1997). Polly was cloned from transgenic fetal sheep fibroblasts that contained the gene for human clotting factor IX, a gene whose function is deficient in hereditary hemophilia.

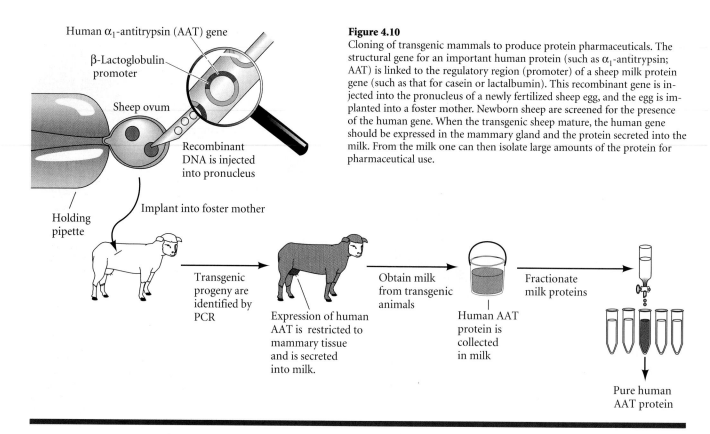

Human α₁-antitrypsin (AAT) gene

β-Lactoglobulin promoter

Sheep ovum

Recombinant DNA is injected into pronucleus

Holding pipette

Implant into foster mother

Transgenic progeny are identified by PCR

Expression of human AAT is restricted to mammary tissue and is secreted into milk.

Obtain milk from transgenic animals

Human AAT protein is collected in milk

Fractionate milk proteins

Pure human AAT protein

Figure 4.10
Cloning of transgenic mammals to produce protein pharmaceuticals. The structural gene for an important human protein (such as α₁-antitrypsin; AAT) is linked to the regulatory region (promoter) of a sheep milk protein gene (such as that for casein or lactalbumin). This recombinant gene is injected into the pronucleus of a newly fertilized sheep egg, and the egg is implanted into a foster mother. Newborn sheep are screened for the presence of the human gene. When the transgenic sheep mature, the human gene should be expressed in the mammary gland and the protein secreted into the milk. From the milk one can then isolate large amounts of the protein for pharmaceutical use.

Sidelights & Speculations

The Exception to the Rule: Immunoglobulin Genes

While the rule is that the genome is the same in every cell in the body, some white blood cells that function as part of the immune system provide exceptions to that rule. The plasma cell is able to synthesize proteins called **immunoglobulins** that can function as antibodies. For decades, immunologists puzzled over how the immune system could possibly generate so many different types of antibodies. Could all the 10 million different types of antibody proteins be encoded in the genome? This would take up an enormous amount of chromosomal space. Moreover, how could the immune system "know" how to make an antibody to some foreign molecule (**antigen**) that isn't even found outside the laboratory? Researchers eventually discovered that the genome of the B cell does not contain DNA encoding any of the antibody proteins. Rather, plasma cell DNA is rearranged during the cell's development from a B cell

to create the antibody-encoding genes. Moreover, while the mammalian organism has the ability to synthesize over 10 million different types of antibody proteins, each plasma cell can synthesize only one.

All immunoglobulin proteins have a similar structure. Each consists of two pairs of polypeptide subunits. There are two identical **heavy chains** and two identical **light chains**; the chains are linked together by disulfide bonds (Figure 4.11). The specificity of the immunoglobulin molecule (i.e., whether it will bind to a poliovirus, an *E. coli* cell, or some other antigen) is determined by the amino acid sequence of the **variable regions** at the amino-terminal ends of the heavy and light chains. The variable regions of the immunoglobulin molecule are attached to **constant regions** that give the antibody the effector properties needed for inactivating the antigen.

The genes that encode the immunoglobulin heavy and light chains are orga-

nized in segments. Mammalian light chain genes contain three segments (Figure 4.11). The first gene segment, V, encodes the first 97 amino acids of the light chain variable region. There are about 300 different V sequences linked tandemly on the mouse genome. The second segment, J, consists of 4 or 5 possible DNA sequences for the last 15–17 amino acid residues of the variable region. (These residues are often the most important contact regions for binding the antigen). The third gene segment, C, encodes the constant region of the light chain. During B cell differentiation to plasma cells, which occurs as the B cells are maturing in the bone marrow, one of the 300 V segments and one of the 5 J segments combine to form the variable region of the antibody gene. This is done by moving a V segment sequence next to a J segment sequence—a rearrangement that eliminates the intervening DNA (Hozumi and Tonegawa 1976).

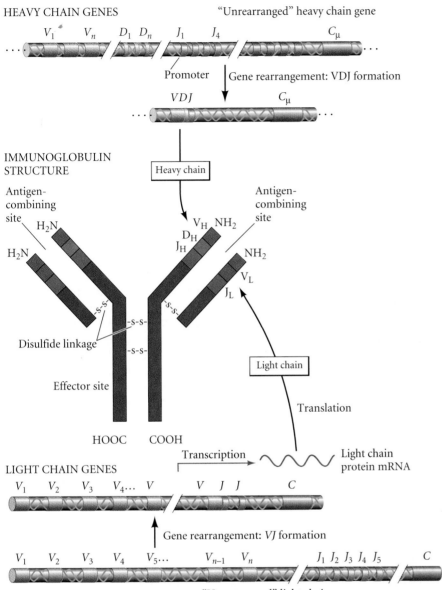

HEAVY CHAIN GENES

"Unrearranged" heavy chain gene

IMMUNOGLOBULIN
STRUCTURE

LIGHT CHAIN GENES

"Unrearranged" light chain gene

Figure 4.11
(Center) Structure of a typical immunoglobulin (antibody) protein. Two identical heavy chains and two identical light chains are connected by disulfide linkages. The antigen-binding site is composed of the variable regions of the heavy and light chains, whereas the effector site of the antibody is determined by the amino acid sequence of the heavy chain constant region. (Bottom) Rearrangement of the light chain genes during B lymphocyte differentiation. While the developing B cell is still maturing in the bone marrow, one of the 300 or more V gene segments combines with one of the 5 J gene segments and moves closer to the constant (C) gene segment. (Top) Rearrangement of the heavy chain genes. A heavy chain gene contains three segments (V, D, and J) that come together to form the variable region, as well as a constant region.

The B cell is not the only cell type that alters its genome during differentiation. The other major cell type of the immune system, the T cell, also recombines and deletes a portion of its genome in the construction of its antigen receptor (Fujimoto and Yamagishi 1987). The enzymes responsible for mediating these DNA recombination events appear to be the same in the B and T cell lineages. Called **recombinases** (Agrawal et al. 1998; Bassing et al. 2002), these two proteins recognize the signal regions of DNA immediately upstream from the recombinable DNA segments and form a complex there that initiates double-stranded breaks. Moreover, the genes for these recombinase enzymes are active only in pre-B cells and pre-T cells, in which the genes are being recombined.* Mutations that eliminate the function of either of the recombinases lead to severe immunodeficiency syndromes that are manifest at birth (Schwarz et al. 1996; Villa et al. 1998).

WEBSITE 4.5 **Antibody formation.** How the immunoglobulin genes eventually produce antibody proteins is a fascinating story, full of exceptions to the rules. More details are given here about DNA rearrangements.

*Recombinase proteins were once thought to be found solely in lymphocytes, but there is evidence that recombination events and recombinases exist in brain tissue as well (Chun et al. 1991; Matsuoka et al. 1991). It is not known what their function might be in neural cells, but it is fascinating to speculate that some of the receptors that bind a nerve cell axon to its specific target might be made by the recombination of several gene regions.

The heavy chain genes contain even more segments than the light chain genes. Heavy chain genes include a V segment (200 different sequences for the first 97 amino acids), a D segment (10–15 different sequences encoding 3–14 amino acids), and a J segment (4 sequences for the last 15–17 amino acids of the variable region). The next segment, C, codes for the constant region. The heavy chain variable region is formed by joining one V segment and one D segment to one J segment (Figure 4.11). This VDJ variable region sequence is now adjacent to the first constant region of the heavy chain genes—the C_μ region, which is specific for antibodies that can be inserted into the cell membrane.

Thus, an immunoglobulin molecule is formed from two genes created during the antigen-independent stage of B lymphocyte development. About 10^3 different light chain genes and about 10^4 different heavy chain genes can be formed. Since each is formed independently of the other, about 10^7 types of immunoglobulins can be created from the union of the light chain and the heavy chain within a cell. Each cell makes only one of these 10^7 antibody types.

Differential Gene Expression

If the genome is the same in all somatic cells within an organism (with the exception of the lymphocytes; see Sidelights & Speculations), how do the cells become different from one another? If every cell in the body contains the genes for hemoglobin and insulin proteins, how are the hemoglobin proteins made only in the red blood cells, the insulin proteins made only in certain pancreas cells, and neither made in the kidney or nervous system? Based on the embryological evidence for genomic equivalence (as well as on bacterial models of gene regulation), a consensus emerged in the 1960s that cells differentiate through **differential gene expression**. The three postulates of differential gene expression are as follows:

1. Every cell nucleus contains the complete genome established in the fertilized egg. In molecular terms, the DNAs of all differentiated cells are identical.
2. The unused genes in differentiated cells are not destroyed or mutated, and they retain the potential for being expressed.
3. Only a small percentage of the genome is expressed in each cell, and a portion of the RNA synthesized in each cell is specific for that cell type.

The first two postulates have already been discussed. The third postulate—that only a small portion of the genome is active in making tissue-specific products—was first tested in insect larvae. Fruit fly larvae have certain cells whose chromosomes become **polytene**. These chromosomes, beloved by *Drosophila* geneticists, undergo DNA replication in the absence of mitosis and therefore contain 512 (2^9), 1024 (2^{10}), or even more parallel DNA double helices instead of just one (Figure 4.12A). These cells do not undergo mitosis, and they grow by expanding to about 150 times their original volume. Beermann (1952) showed that the banding patterns of polytene chromosomes were identical throughout the larva, and that no loss or addition of any chromosomal region was seen when different cell types were compared. However, he and others showed that in different tissues, different regions of these chromosomes were making tissue-specific mRNA. In certain cell types, particular regions of the chromosomes would loosen up, "puff" out, and transcribe mRNA. In other cell types, these regions would be "silent," but other regions would puff out and synthesize mRNA.

The idea that the genes of chromosomes were differentially expressed in different cell types was confirmed using DNA-RNA hybridization (Figure 4.12B). This technique involves annealing single-stranded pieces of RNA and DNA to allow complementary strands to form double-stranded hybrids. If one of the nucleic acids is marked with a dye or radioactive tracer, it can be used to indicate the presence of its complement. While some mRNAs from one cell type were also found in other cell types (as expected for mRNAs encoding enzymes concerned with cell metabolism), many mRNAs were found to be specific for a particular type of cell and were not expressed in other cell types, even though the genes encoding them were present (Wetmur and Davidson 1968). Thus, differential gene expression was shown to be the way a single genome derived from the fertilized egg could generate the hundreds of different cell types in the body.

The question then became, How does this differential gene expression occur? The answers to that question will be the topic of the next chapter. To understand the results that

(A)

(B)

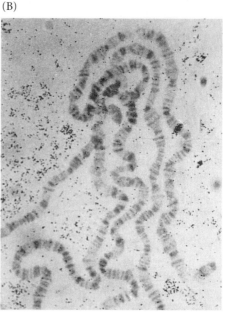

Figure 4.12
Polytene chromosomes. (A) Electron micrograph of a small region of a *Drosophila* polytene chromosome. The bands (dark) are highly condensed compared with the interband (lighter) regions. (B) Hybridization of a yolk protein mRNA with the polytene chromosome of a larval *Drosophila* salivary gland. The dark grains (arrow) show where the radioactive yolk protein message has bound to the chromosomes. Note that the gene for the yolk protein is present in the salivary gland chromosomes, even though yolk protein is not synthesized there. (A from Burkholder 1976, photograph courtesy of G. D. Burkholder; B from Barnett et al. 1980; photograph courtesy of P. C. Wensink.)

will be presented there, however, one must become familiar with some of the techniques of molecular biology that are being applied to the study of development. These include techniques to determine the spatial and temporal location of specific mRNAs, as well as techniques to determine the functions of these messages.

WEBSITE 4.6 **DNA isolation techniques.** The basic techniques of DNA analysis—gene cloning, sequencing, and Southern blotting—are discussed in most introductory biology books. This site gives a review of these procedures.

RNA Localization Techniques

Northern blotting

We can determine the temporal and spatial locations of RNA expression by "running" an RNA blot (often called a **northern blot**). First, an investigator extracts RNA from embryos at different stages of development, or from different organs of the same embryo. The investigator then places these RNA samples side by side at one end of a gel and runs an electric current through the gel. The smaller the RNA, the faster it moves through the gel. Thus, different RNAs are separated by their sizes. This technique is called **electrophoresis.***

The investigator then transfers the separated RNAs to a nitrocellulose paper or nylon membrane filter. The RNA-containing filter is incubated in a solution containing a radioactive single-stranded DNA fragment from a particular gene (Figure 4.13A). This radioactively labeled DNA probe binds only to regions of the filter where the target RNA (to which it is complementary) is located. If the mRNA for that gene is present in a sample, the labeled DNA will bind to it and can be detected by autoradiography. X-ray film is placed above the filter and incubated in the dark. The localized radioactivity in the probe reduces the silver in the X-ray film, and grains form. The resulting spots, which appear directly above the places where the radioactive DNA has bound, appear black when viewed directly. Autoradiographs of this type, in which RNAs from several stages or tissues are compared simultaneously, are called **developmental northern blots.**

Figure 4.13F shows a developmental northern blot used to investigate the expression of the Pax6 protein in the mammalian embryo. Pax6 is critical for normal eye development; mutations in the *Pax6* gene result in small eyes (in heterozygous mice) or no eyes or nose (in mice or humans homozygous for the loss-of-function mutation). The northern blot shows that this gene is expressed in the embryo in the brain, eyes, and pancreas, but in no other tissue.

Reverse transcriptase-polymerase chain reaction

The **polymerase chain reaction (PCR)** is a method of in vitro gene cloning[†] that can generate enormous quantities of a specific DNA fragment from a small amount of starting material (Saiki et al. 1985). It can be used for cloning a specific gene or for determining whether a specific gene is actively transcribing mRNA in a particular organ or cell type. Standard methods of gene cloning use living microorganisms to amplify recombinant DNA. PCR, however, can amplify a single segment of a DNA molecule several million times in a few hours, and can do it in a test tube. The techniques of PCR are extremely useful in cases where there is very little sample to study.

Preimplantation mouse embryos, for instance, have very little mRNA, and we cannot obtain millions of such embryos to study. If we wanted to know whether a single preimplantation mouse embryo contained the mRNA for a particular protein, it would be very difficult to find out using standard methods—we would have to lyse thousands of mouse embryos in order to obtain enough mRNA. However, by combining PCR techniques with the ability of the **reverse transcriptase (RT)** enzyme to make DNA out of mRNA, we can get around the problem of scarce messages. The **RT-PCR** technique allows us convert the mRNA into DNA and to copy the specific DNA sequence of interest (Rappolee et al. 1988).

The use of RT-PCR to find rare mRNAs is illustrated in Figure 4.14. First, the mRNAs from a sample are purified and converted into complementary DNA (cDNA) using the RT enzyme. Next, a specific cDNA is targeted for amplification. Two small oligonucleotide primers that are complementary to a portion of the message being looked for are added to the population of cDNA. Oligonucleotides are relatively short stretches of DNA (about 20 bases). If the oligonucleotides bind to sequences in the cDNA, this means that the mRNA being sought was present in the original sample. The oligonucleotide primers are made so that they hybridize to opposite strands at opposite ends of the targeted sequence. (If we are trying to isolate the gene or mRNA for a specific protein of known sequence, we can synthesize oligonucleotides that are complementary to the sequences encoding the amino end and the carboxyl end of the protein.) The 3′ ends of the primers face each other, so that replication will run through the target DNA.

Once the first primer has hybridized with the cDNA, DNA polymerase can be used to synthesize a new strand. The DNA polymerase used in this process is from thermophilic (heat-loving) bacteria such as *Thermus aquaticus* or *Thermococcus littoralis*. These bacteria live in hot springs (such as those in Yellowstone National Park) or in submarine thermal vents, where the temperature reaches nearly 90°C. These DNA polymerases can withstand temperatures near boiling, and

---

*Given the same charge-to-mass ratio, smaller RNA fragments obtain a faster velocity than larger ones when propelled by the same energy (i.e., larger fragments move more slowly than smaller fragments). This property is a function of the kinetic energy equation, $E = 1/2\ mv^2$. Solving for velocity, we find that velocity is inversely proportional to the square root of the mass.

[†]Gene cloning—making numerous copies of the same DNA sequence—should not be confused with organism cloning—making genetically identical copies of an organism.

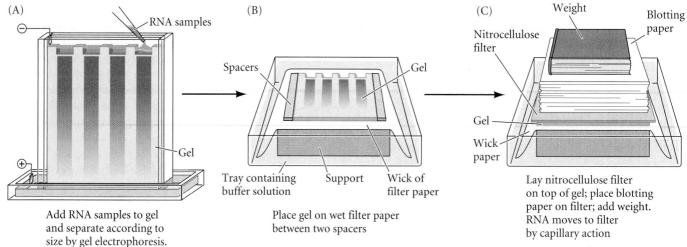

(A) RNA samples

Gel

Add RNA samples to gel and separate according to size by gel electrophoresis.

(B) Spacers

Gel

Tray containing buffer solution   Support   Wick of filter paper

Place gel on wet filter paper between two spacers

(C) Weight

Nitrocellulose filter

Blotting paper

Gel

Wick paper

Lay nitrocellulose filter on top of gel; place blotting paper on filter; add weight. RNA moves to filter by capillary action

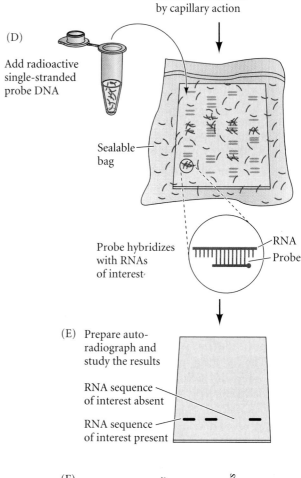

(D) Add radioactive single-stranded probe DNA

Sealable bag

Probe hybridizes with RNAs of interest

RNA
Probe

(E) Prepare auto-radiograph and study the results

RNA sequence of interest absent

RNA sequence of interest present

(F)

Retina
Total brain
Temporal lobe
Cerebellum
Choroid plexus
Liver
Spleen
Thymus
Lung
Cardiac muscles
Skin
Pancreas
Kidney

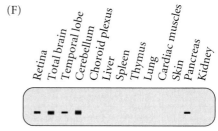

Figure 4.13
Developmental Northern blotting. (A–E) Procedure for Northern blotting. (A) RNA is isolated from various tissues and is separated by size using gel electrophoresis. (B) The gel is then placed on a paper wick, which absorbs an ionic solution from a trough. (C) A filter that traps RNA is placed above the gel, and blotting paper is placed above the filter. Capillary action draws the solution through the gel, trapping the RNA on the filter. (D) The filter is incubated with radioactive single-stranded DNA complementary to the mRNA of interest. (E) After any unbound DNA is washed off, autoradiography localizes the mRNA in the samples that contain it. (F) Drawing of a developmental Northern blot showing the presence of Pax6 mRNA in the eye, brain, and pancreas of a mammalian embryo. (F after Ton et al. 1991.)

RT-PCR takes advantage of this evolutionary adaptation. Once the second strand of DNA is made, it is heat-denatured from its complement. (The temperatures used would inactivate the more usual *E. coli* DNA polymerase, but the thermostable polymerases are not damaged.)

The second primer is added, and now both strands can synthesize new DNA. Repeated cycles of denaturation and synthesis amplify the DNA sequence exponentially. After 20 such rounds, that specific sequence has been amplified 2^{20} (a little more than a million) times. When the DNA is subjected to electrophoresis, the presence of such an amplified fragment is easily detected. Its presence shows that there was an mRNA with the sequence of interest present in the original sample.

Figure 4.14

The reverse transcriptase-polymerase chain reaction (RT-PCR) technique for determining whether a particular type of mRNA is present. First, the mRNA from a small sample is converted to double-stranded cDNA using the enzymes reverse transcriptase and RNase H. A primer is added to the cDNA and the second strand is completed using thermostable DNA polymerase from *T. aquaticus* (*Taq* polymerase). This "target" DNA is then denatured and two sets of primers are added; the primers hybridize to opposite ends of the target sequence if the sequence is present. (This will happen only if the mRNA of interest was originally present.) When *Taq* polymerase is added to the denatured DNA, each strand synthesizes its complement. These strands are then denatured, and the primers are hybridized to them, starting the cycle again. In this way, the number of new strands having the sequence of interest between the two primers increases exponentially.

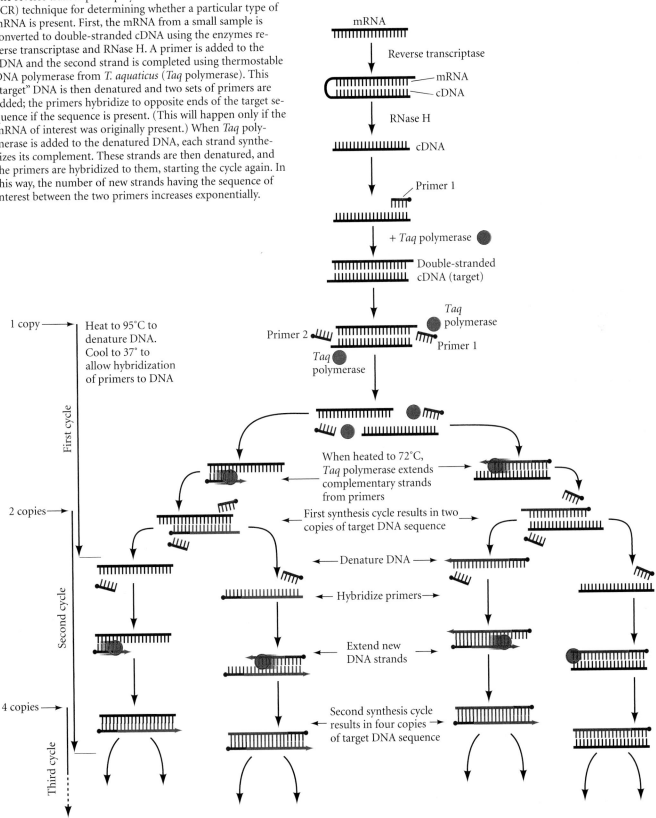

PREPARE cDNA
"TARGETS"

PREPARE MICROARRAY
"PROBES" (by robot)

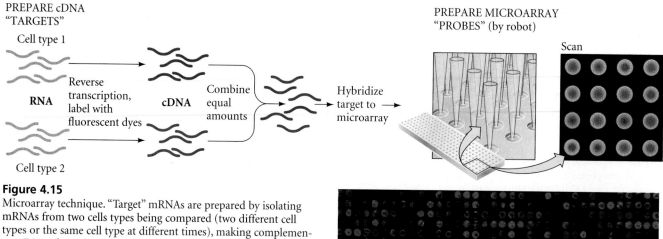

Figure 4.15
Microarray technique. "Target" mRNAs are prepared by isolating mRNAs from two cells types being compared (two different cell types or the same cell type at different times), making complementary DNAs from them, and adding two different types of fluorescent tags (one color for each cell type). In this case, cDNAs from cell type 1 are labeled with a green dye, while cDNAs from cell type 2 are tagged with a red dye. Microarray probes are made by taking a number of different cDNAs made from the mRNA of one cell type and adhering them to a glass microscope slide. The probes and the target are hybridized together. If the mRNA in a probe is abundant in cell type 1, the signal is green. If the mRNA is abundant in cell type 2, the signal is red. If the mRNA is present in both cell types, the signal is yellow. The bottom shows a portion of such a microarray. (After http://www.genetics.ucla.edu/microarray/instruction.html)

Microarrays and macroarrays

Northern blots generally work on a "one gene in one experiment" basis, which means that the "whole picture" of differential gene expression between cells or tissues is difficult to obtain. In the past several years, a new technology, called **DNA microarrays**, has allowed scientists to monitor changes in the transcription of thousands of genes simultaneously (Wan et al. 1996). Microarrays combine northern blot and PCR technologies with high-speed robotics. First, one takes mRNA from a tissue and converts the mRNAs into their complementary DNAs (using reverse transcriptase). Each individual cDNA can then be cloned, denatured, and amplified by PCR. The resulting cDNAs serve as probes in the microarrays. Each of these cDNA clone probes is robotically printed onto glass slides in a particular order, and the slides are subsequently hybridized to two "targets" with different fluorescent labels. The targets are pools of cDNAs that are generated after isolating mRNA from cells or tissues in two states that one wishes to compare. For instance, if the aim is to compare cell type A with cell type B, one would take mRNAs from these both cell types, convert the mRNAs into cDNAs, and label the cDNAs from cell type A with fluorescein (green) and the cDNAs from cell type B with rhodamine (red). The two DNA pools would then be equally mixed, and the mixture placed onto each spot of probe DNA. The resulting fluorescent intensities are produced using a laser confocal fluorescent microscope, and ratio information is obtained following image processing. By com-

paring the fluorescent intensities, one can tell if a particular cDNA (and hence, the mRNA of interest) is present in higher amounts in one cell type or another (Figure 4.15).

For instance, one can look at the genes active in the prospective dorsal part of a frog blastula and compare them with the genes active in the future ventral portion of the same blastula (Altmann et al. 2001). Alternatively, one can look at entire animals at different stages of their development to see which genes are active at each stage(White et al. 1999). This allows one to focus one's research on those genes whose expression differs between the two sets of cDNAs.

In addition to being able to screen thousands of genes at a time, microarrays have another important advantage over northern blots. Rare mRNAs are not outcompeted by prevalent mRNA species, and this allows those important messages that are present in small amounts to be found.

A less expensive modification of microarrays is the **macroarray**. The technology is similar, but the spots are larger (about 1 mm as compared to 250 μm). This means that macroarrays can be visually interpreted without a microscope, and that radioisotopes as well as fluorescent labels can be used. For instance, in Chapter 3, we saw that activin caused the expression of certain genes in the animal cap of *Xenopus*. One of the naturally occurring activin-like compounds that is found in early *Xenopus* embryos is called *Xenopus* Nodal related 1 (Xnr1). To see if Xnr1 is activated particular genes, Naoto Ueno's laboratory injected *Xnr1* mRNA into some early embryos but not others. They then harvested mRNAs from the animal caps of the Xnr1-secreting and control embryos. (The animal cap would not be exposed to Xnr1 under normal development). After converting the mRNAs into radioactively labeled cDNAs, Ueno and colleagues hybridized these probes

(A)

Xnr1 mRNA or H₂O
microinjection at 2-cell stage

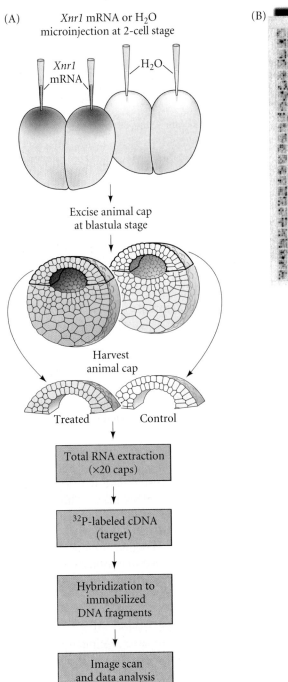

(B)

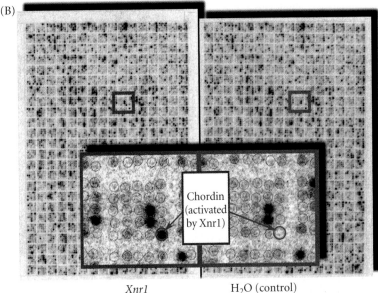

Chordin
(activated
by Xnr1)

Xnr1   H₂O (control)

Figure 4.16
Macroarray analysis of those genes whose expression in the early *Xenopus* embryo is caused by the activin-like protein Nodal-related 1 (Xnr1). (A) Creation of targets for macroarray analysis. (B) In the macroarrays, certain radioactive spots (representing hybridizations) were seen in samples from the Xnr1-stimulated cells, but not in the control cells. These spots represent the genes activated by Xnr1; the insert shows one of these, the *chordin* gene. Most of the thousands of genes observed were not activated. The DNA of the hybridized spots can be sequenced and identified. (Photographs courtesy of N. Ueno.)

to slides containing bound DNA. One macroarray was hybridized with radioactive cDNAs from the control animal caps; an identical macroarray was hybridized with radioactive cDNAs from the Xnr1-treated caps. One of the results is shown in Figure 4.16. Here, the gene for chordin is shown to be expressed in the treated animal cap, but not in the untreated cap.

> **WEBSITE 4.7 Microarray technology.** There are many variations on the scheme described here. Microarray technology may become essential for developmental biology if we are to understand the complex changes in gene expression that occur as cell types differentiate.

In situ hybridization

Northern blot analyses and microarrays can give only an approximate location and time for gene expression. A more detailed map of gene expression patterns can be obtained by using a process called **in situ hybridization**. Instead of using a DNA probe to seek mRNA on a filter, an antisense mRNA probe (which can be either a DNA or an RNA) is hybridized with the mRNA in the organ itself. **Antisense mRNA** is made from a cloned gene in which the gene is reversed with respect to a promoter within the vector (we will see how genes can be "reversed" in a later section). The mRNA transcribed from such a gene encodes a sequence complementary to the normal mRNA made by that gene. Such antisense mRNA can be used as a probe, since it will recognize the "sense" mRNA in the cell.

In in situ hybridization, the antisense RNA is labeled (either made radioactive or attached to a dye), which allows the probe to be visualized. Thus, one makes a sequence-specific stain that will label only those cells that have accumulated mRNAs of a particular sequence. When radioactive probes are

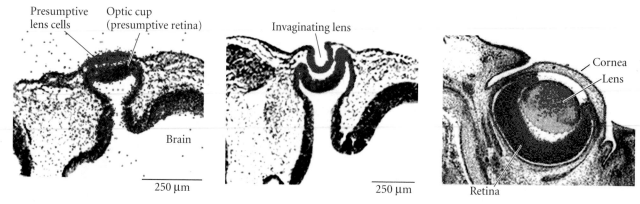

Presumptive lens cells   Optic cup (presumptive retina)   Invaginating lens   Cornea   Lens

Brain   250 μm   250 μm   Retina

Figure 4.17
In situ hybridization showing the expression of the *Pax6* gene in the developing mouse eye. Transverse microscopic sections were taken through the developing heads of 9-, 10-, and 15-day embryonic mice. At this time, the optic cup is touching the outer ectoderm and inducing it to form a lens. The radioactive *Pax6* antisense DNA probe binds only where *Pax6* mRNA is present, and can be visualized by developing the photographic emulsion. The locations where the probe has bound, depicted here as yellow dots (by computer imaging), show that the *Pax6* message is expressed in both the presumptive lens ectoderm and the optic stalk, which forms the retina and optic nerve. (From Grindley et al. 1995; photograph courtesy of R. E. Hill.)

used, embryos or organs are first fixed to preserve their structure and to prevent their mRNA from being degraded. These tissues are then sectioned for microscopy and placed on a slide. When the radioactive sequence is added, it binds only where the target mRNA (to which it is complementary) is present. After any unbound probe is washed off, the slide is covered with a transparent photographic emulsion for autoradiography. By using dark-field microscopy (or computer-mediated bright-field imaging), the reduced silver grains can be shown in a color that contrasts with the background stain. Thus, we can visualize those cells (or even regions within cells) that have accumulated a specific type of mRNA. Figure 4.17 shows an in situ hybridization for *Pax6* mRNA in mice. One can see that *Pax6* mRNA is found in the region where the presumptive retina meets the presumptive lens tissue. As development proceeds, it is seen in the developing retina, lens, and cornea of the eye.

In **whole-mount in situ hybridization,** the entire embryo (or a part thereof) can be stained for certain mRNAs. This technique, which uses dyes rather than radioactivity, allows researchers to look at entire embryos (or their organs) without sectioning them, thereby observing large regions of gene expression. Figure 4.18 shows an in situ hybridization performed on a whole chick embryo that has been fixed without being sectioned. The embryo has also been permeabilized by lipid and protein solvents so that the probe can get in and out of its cells. The probe used in this experiment recognizes the mRNA encoding *Pax6* in the chick embryo. This probe is la-

beled not with a radioactive isotope, but by modifying the nucleotide substrate uridine triphosphate (UTP). To create this probe, a region of the cloned *Pax6* gene was transcribed into mRNA, but with two important modifications. First, in addition to regular UTP, the nucleotide mix also contained UTP conjugated with *digoxigenin*. Digoxigenin—a compound made by particular groups of plants and not found in animal cells—does not interfere with the coding properties of the resulting mRNA, but does make it recognizably different from any other RNA in the cell.

The digoxigenin-labeled probe is incubated with the embryo. After several hours, numerous washes remove any probe that has not bound to the embryo. Then the embryo is incubated in a solution containing an antibody against digoxigenin. The only places where digoxigenin should exist is where the probe has bound (i.e., where it recognized its mRNA), so the antibody sticks in those places. This antibody, however, is not in its natural state. It has been conjugated covalently to an enzyme, such as alkaline phosphatase. After repeated washes to remove all the unbound enzyme-conjugated antibody, the embryo is incubated in another solution that will be converted into a dye by the enzyme. The enzyme should be present only where the digoxigenin is present, and the digoxygenin should be present only where the specific complementary mRNA is found.

Thus, in Figure 4.18, the dark blue precipitate formed by the enzyme indicates the presence of the target mRNA. The figure reveals mRNA for the Pax6 protein to be present in the roof of the brain region that will form the eyes, as well as in the head ectoderm that will form the lenses. It is also expressed in a more caudal region of the neural tube as well as in the pancreas.

Determining the Function of Genes During Development

Transgenic cells and organisms

While it is important to know the sequence of a gene and its temporal-spatial pattern of expression, what's really crucial is to

(A)

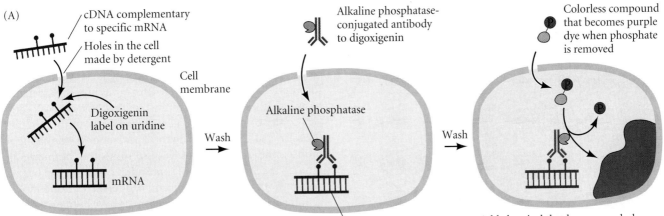

cDNA complementary to specific mRNA

Holes in the cell made by detergent

Cell membrane

Digoxigenin label on uridine

mRNA

1. Add digoxigenin-labeled probe complementary to RNA of interest

Alkaline phosphatase-conjugated antibody to digoxigenin

Alkaline phosphatase

Wash

2. Add alkaline phosphatase-conjugated antibody that binds to digoxigenin

Colorless compound that becomes purple dye when phosphate is removed

Wash

3. Add chemical that becomes a dark purple dye when phosphate is removed; dye colors the cell.

Figure 4.18
Whole-mount in situ hybridization localizing *Pax6* mRNA in early chick embryos. (A) Schematic of the procedure. A digoxigenin-labeled antisense probe hybridizes to a specific mRNA. Alkaline phosphatase-conjugated antibodies to digoxigenin recognize the digoxigenin-labeled probe. The enzyme is able to convert a colorless compound into a dark purple precipitate. (B) *Pax6* mRNA can be seen to accumulate in the roof of the brain region that will form the eyes, as well as in the ectoderm that will form lenses. More caudal expression of this gene in the nervous system is also seen. (After Li et al. 1994; photograph courtesy of O. Sundin.)

(B)

know the functions of that gene during development. Recently developed techniques have enabled us to study gene function by moving certain genes into and out of embryonic cells.

INSERTING NEW DNA INTO A CELL. Cloned pieces of DNA can be isolated, modified (if so desired), and inserted into cells by several means. One direct technique is **microinjection**, in which a solution containing the cloned gene is injected very carefully into the nucleus of a cell (Capecchi 1980). This technique is especially useful for injecting genes into newly fertilized eggs, since the haploid nuclei of the sperm and egg are relatively large (Figure 4.19). In **transfection**, DNA is incorporated directly into cells by incubating them in a solution that makes them "drink" it in. The chances of a DNA fragment being incorporated into the chromosomes in this way are relatively small, however, so the DNA of interest is usually mixed with another gene, such as a gene encoding resistance to a particular antibiotic, that enables those rare cells that incorporate the DNA to survive under culture conditions that will kill all the other cells (Perucho et al. 1980; Robins et al. 1981). Another similar technique is **electroporation**, in which a high-voltage pulse "pushes" the DNA into the cells.

A more "natural" way of getting genes into cells is to insert a cloned gene into a **transposable element** or **retroviral vector**. These naturally occurring mobile regions of DNA can integrate themselves into the genome of an organism. Retroviruses are RNA-containing viruses. They enter a host cell, where they make a DNA copy of themselves (using their own

virally encoded reverse transcriptase); the copy then becomes double-stranded and integrates itself into a host chromosome. The integration is accomplished by two identical sequences (long terminal repeats) at the ends of the retroviral DNA. Retroviral vectors can be made by removing the viral packaging genes (needed for the exit of viruses from the cell) from the center of a mouse retrovirus. This extraction creates a vacant site where other genes can be placed. By using the appropriate restriction enzymes researchers can excise a gene of interest (such as a gene isolated by PCR) and insert it into a retroviral vector. These retroviral vectors infect mouse cells with an efficiency approaching 100 percent.

Similarly, in *Drosophila*, new genes can be carried into a fly via **P elements**. These DNA sequences are naturally occurring transposable elements that can integrate like viruses into any region of the *Drosophila* genome. Moreover, they can be isolated, and cloned genes can be inserted into the center of the isolated P element. When the recombined P element is in-

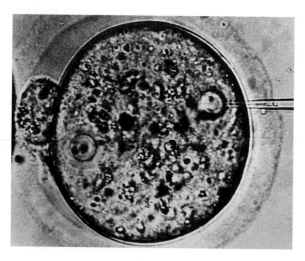

Figure 4.19
Insertion of new DNA into embryonic cells. Here, DNA (from cloned genes) is injected into the pronucleus of a mouse egg. (From Wagner et al. 1981; photograph courtesy of T. E. Wagner.)

jected into a *Drosophila* oocyte, it can integrate itself into the embryo's DNA, providing the organism with the new gene (Spradling and Rubin 1982).

CHIMERIC MICE. The techniques described above have been used to transfer genes into every cell of the mouse embryo (Figure 4.20). During early mouse development, there is a stage (the blastocyst) when only two cell types are present: the outer trophoblast cells, which will form the fetal portion of the placenta, and the inner cell mass, whose cells will give rise to the embryo itself. The inner cells are the cells whose separation can lead to twins (see Chapters 3 and 11), and if an inner cell mass blastomere of one mouse is transferred into the embryo of a second mouse, that donor cell can contribute to every organ of the host embryo.

Inner cell mass blastomeres can be isolated from an embryo and cultured in vitro; such cultured cells are called **embryonic stem cells (ES cells)**. ES cells are almost totipotent, since each of them can contribute to all tissues except the trophoblast if injected into a host embryo (Gardner 1968; Moustafa and Brinster 1972). Moreover, once in culture, these cells can be treated as described in the preceding section so that they will incorporate new DNA. This added gene (the transgene) can come from any eukaryotic source. A treated ES cell (the entire cell, not just the DNA) can then be injected into another early-stage mouse embryo, and will integrate into this host. The result is a **chimeric mouse.** * Some of the chimera's cells will be derived from the host's own embryonic stem cells,

*It is critical to note the difference between a chimera and a hybrid. A **hybrid** results from the union of two different genomes within the same cell: the offspring of an *AA* genotype parent and an *aa* genotype parent is an *Aa* hybrid. A **chimera** results when cells of different genetic constitutions appear in the same organism. The term is apt: it refers to a mythical beast with a lion's head, a goat's body, and a serpent's tail.

but some portion of its cells will be derived from the treated embryonic stem cell. If the treated cells become part of the germ line of the mouse, some of its gametes will be derived from the donor cell. If such a chimeric mouse is mated with a wild-type mouse, some of its progeny will carry one copy of the inserted gene. When these heterozygous progeny are mated to one another, about 25% of the resulting offspring will carry two copies of the inserted gene in every cell of their bodies (Gossler et al. 1986). Thus, in three generations—the chimeric mouse, the heterozygous mouse, and the homozygous mouse—a gene cloned from some other organism will be present in both copies of the chromosomes within the mouse genome. Strains of such transgenic mice have been particularly useful in determining how genes are regulated during development.

GENE TARGETING ("KNOCKOUT") EXPERIMENTS. The analysis of early mammalian embryos has long been hindered by our inability to breed and select animals with mutations that affect early embryonic development. This block has been circumvented by the techniques of gene targeting (or, as it is sometimes called, gene knockout). These techniques are similar to those that generate transgenic mice, but instead of adding genes, gene targeting *replaces* wild-type alleles with mutant ones. As an example, we will look at the gene knockout of bone morphogenetic protein 7 (BMP7). Bone morphogenetic proteins are involved in numerous developmental interactions whereby one set of cells interacts with other neighboring cells to alter their properties. BMP7 has been implicated as a protein that prevents cell death and promotes cell division in several developing organs.

Dudley and his colleagues (1995) used gene targeting to find the function of *Bmp7* in the development of the mouse. First, they isolated the *Bmp7* gene, cut it at one site with a restriction enzyme, and inserted a bacterial gene for neomycin resistance into that site (Figure 4.21). In other words, they mutated the *Bmp7* gene by inserting into it a large piece of foreign DNA, destroying the ability of the BMP7 protein to function. These mutant *Bmp7* genes were electroporated into ES cells that were sensitive to neomycin. Once inside the nucleus of an ES cell, the mutated *Bmp7* gene may replace a normal allele of *Bmp7* by a process called homologous recombination. In this process, the enzymes involved in DNA repair and replication incorporate the mutant gene in the place of the normal copy. It's a rare event, but such cells can be selected by growing the ES cells in neomycin. Most of the cells are killed by the drug, but the ones that have acquired resistance from the incorporated gene survive. The resulting cells have one normal *Bmp7* gene and one mutated *Bmp7* gene. These heterozygous ES cells were then microinjected into mouse blastocysts, where they were integrated into the cells of the embryo. The resulting mice were chimeras composed of wild-type cells from the host embryo and heterozygous *Bmp7*-containing cells from the donor ES cells. The chimeras

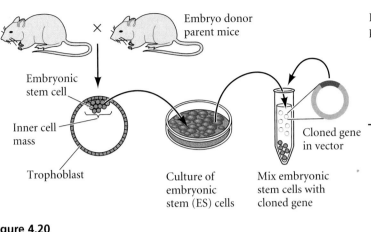

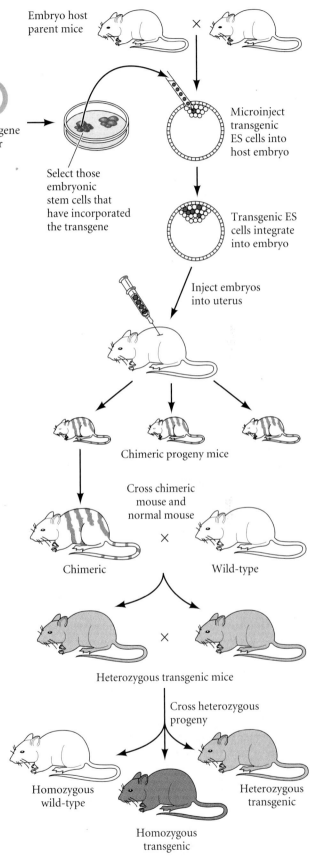

Figure 4.20
Production of transgenic mice. Embryonic stem cells from a mouse are cultured and their genome altered by the addition of a cloned gene. These transgenic cells are selected and then injected into the early stages of a host mouse embryo. Here, the transgenic embryonic stem cells integrate with the host's embryonic stem cells. The embryo is placed into the uterus of a pregnant mouse, where it develops into a chimeric mouse. The chimeric mouse is then crossed with a wild-type mouse. If the donor stem cells have contributed to the germ line, some of the progeny will be heterozygous for the added allele. By mating heterozygotes, a strain of transgenic mice can be generated that is homozygous for the added allele. The added gene (the transgene) can be from any eukaryotic source.

were mated to wild-type mice, producing progeny that were heterozygous for the *Bmp7* gene. These heterozygous mice were then bred with each other, and about 25% of their progeny carried two copies of the mutated BMP7 gene. These homozygous mutant mice lacked eyes and kidneys (Figure 4.22). In the absence of BMP7, it appears that many of the cells that normally form these two organs stop dividing and die. In this way, gene targeting can be used to analyze the roles of particular genes during mammalian development.

WEBSITE 4.8 **Knocking out specific genes at specific times and places.** Some genes are active early in development and also later in development. If their early function is critical to the life of the embryo, one cannot find out what the later function is by knocking out the gene. So techniques have been developed to knock out a gene only in particular cell types.

Determining the function of a message: Antisense RNA

Another method for determining the function of a gene during development is to use "antisense" copies of its message to block the function of that message. Antisense RNA allows developmental biologists to analyze the action of genes that would otherwise be inaccessible for genetic analysis.

Antisense messages can be generated by inserting cloned DNA into vectors that have promoters at both ends of the inserted gene. When the vector is incubated with nucleotide

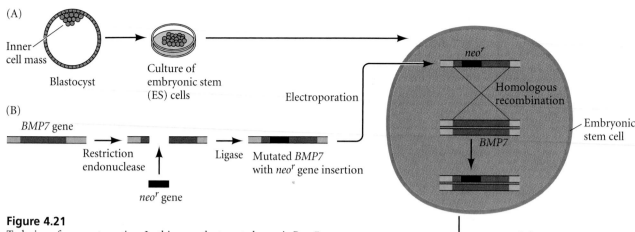

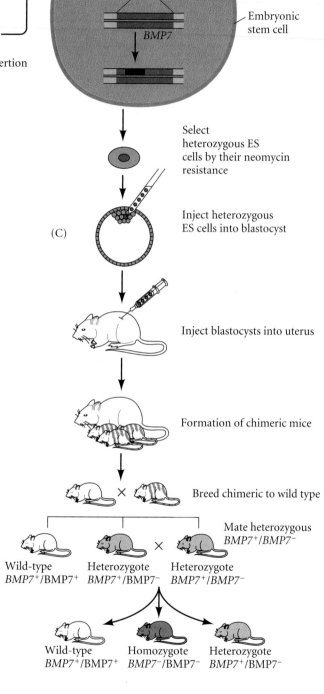

Figure 4.21
Technique for gene targeting. In this case, the targeted gene is *Bmp7*. (A) Embryonic stem (ES) cells from a mouse blastocyst are cultured. (B) Cloned *Bmp7* genes are cut with a restriction enzyme, and a neomycin resistance gene is inserted into the region that encodes the DNA-binding site of the protein. These mutant *Bmp7* genes are electroporated into ES cells. In some of the cells, homologous recombination exchanges a wild-type gene for the mutant copy. These cells are selected by their neomycin resistance. (C) The selected heterozygous ES cells are inserted into the inner cell mass of a wild-type embryo, and the blastocyst is returned to the uterus. The resulting mouse is a chimera composed of heterozygous *Bmp7* tissues and wild-type *Bmp7* tissues. Mating the chimeric mouse to a wild-type mouse produces heterozygous *Bmp7* offspring if the ES cells have contributed to the germ line. These heterozygous mice can be bred together, and about 25% of their progeny should be homozygous for the mutant *Bmp7*.

triphosphates (such as UTP) and a particular RNA polymerase, one of the promoters will initiate transcription of the message "in the wrong direction." In so doing, it synthesizes a transcript that is complementary to the natural one (Figure 4.23A). This complementary transcript is called antisense RNA because it is the complement of the original ("sense") message. When large amounts of antisense RNA are injected or transfected into cells containing the normal mRNA from the same gene, the antisense RNA binds to the normal message, and the resulting double-stranded nucleic acid is degraded by enzymes in the cell cytoplasm, causing a functional deletion of the message—just as if there were a deletion mutation for that gene.

The similarities between the phenotypes produced by a loss-of-function mutation and by antisense RNA treatment were demonstrate by making antisense RNA to the *Krüppel* gene of *Drosophila*. *Krüppel* is critical for forming the thorax and abdomen of the fly. If this gene is absent, fly larvae die because they lack thoracic and anterior abdominal segments (Figure 4.23B). A similar defect can be created by injecting large amounts of antisense RNA against the *Krüppel* message into early fly embryos (Rosenberg et al. 1985).

MORPHOLINO ANTISENSE OLIGOMERS. An important modification of antisense technology is the use of **morpholino anti-**

sense oligomers (Summerton and Weller 1997). These molecules differ from traditional antisense oligonucleotides in that they contain six-member morpholine rings instead of five-member ribose or deoxyribose sugars. This structure gives

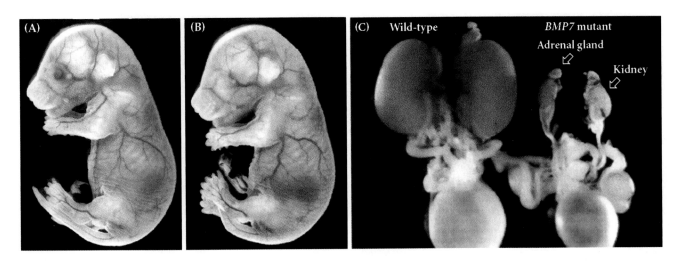

Figure 4.22
Morphological analysis of *Bmp7* knockout mice. (A) Wild-type and (B) homozygous *Bmp7*-deficient mouse at day 17 of their 21-day gestation. The *Bmp7*-deficient mouse lacks eyes. (C) The kidneys of these mice at day 19 of gestation. The kidney of the *Bmp7*-deficient mouse (right) is severely atrophied. Microscopic sections reveal the death of the cells that would otherwise have formed the nephrons. (From Dudley et al. 1995; photographs courtesy of E. Robertson.)

them complete resistance to nucleases, enabling them to stay intact and function longer. Moreover, they can hybridize with their target mRNAs independently of the salt concentration and over a large concentration range. Their stability allows them to initiate events many cell generations after they are first injected into the cell. (Heasman et al. 2000). Morpholino antisense oligomers work by inhibiting the initiation of translation, and they are therefore made against sequences very near the translation initiation site of the message.

RNA INTERFERENCE. Another type of sequence-specific targeting of mRNA that leads to inhibition of its expression is **RNA interference** (**RNAi**). Here, the introduction of homologous double-stranded RNA (dsRNA) results in phenotypes that appear the same as those we would expect if the mRNA had been degraded or absent. Rather than inhibiting translation, RNAi works by degrading the targeted message. Biologists have used RNAi to study development in nematodes, *Drosophila*, plants, fungi, and mice.

Figure 4.23
Use of antisense RNA to examine the roles of genes in development. (A) An antisense message (in this case, to the *Krüppel* gene of *Drosophila*) is produced by placing a cloned cDNA fragment encoding the *Krüppel* message between two strong promoters (where RNA polymerases bind to initiate transcription). The two promoters are in opposite orientation with respect to the *Krüppel* cDNA. In this case, the T3 promoter is in normal orientation and the T7 promoter is reversed. These promoters are recognized by different RNA polymerases (from the T3 and T7 bacteriophages, respectively). T3 polymerase enables the transcription of "sense" mRNA, whereas T7 polymerase transcribes antisense transcripts. (B) Result of injecting the *Krüppel* antisense message into an early embryo (syncytial blastoderm stage) of *Drosophila* before the normal *Krüppel* message is produced. The central figure is a wild-type embryo just prior to hatching. Above it is a mutant lacking the *Krüppel* gene. Below it is a wild-type embryo that was injected with *Krüppel* antisense message at the syncytial blastoderm stage. Both the mutant and the antisense-treated embryos lack thoracic and anterior abdominal segments. (B after Rosenberg et al. 1985.)

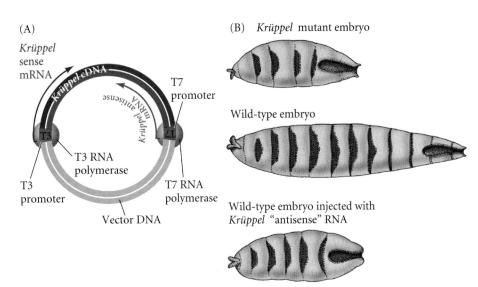

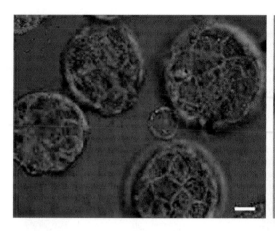
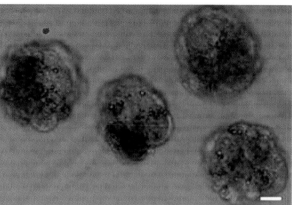

Figure 4.24
Injection of dsRNA for E-cadherin into the mouse zygote blocks E-cadherin expression. (A) Staining of E-cadherin in 4-day mouse embryos. These embryos were injected with a control dsRNA that does not recognize any message in the mouse embryo. The staining was done by a fluorescent (red) antibody that binds to E-cadherin. (B) 4-day mouse embryos injected at the zygote stage with E-cadherin dsRNA. Hardly any E-cadherin is recognized by the antibody stain, and the embryos are malformed because they lack this cell adhesion protein. The same type of abnormal embryos can be produced by knocking out the E-cadherin gene. Scale bars are 20 mm. (From Wianny and Zernicka-Goetz 2000.)

The first evidence that dsRNA could lead to gene silencing came from work on the nematode *Caenorhabditis elegans*. Guo and Kemphues (1995) were attempting to use antisense RNA to shut down the expression of the *par-1* gene in order to assess its function, and one of their controls didn't work. Or rather, it worked too well. As expected, injection of antisense RNA disrupted expression of the *par-1* gene. However, so did the injection of the sense-strand control. The mystery was solved by Fire and colleagues (1998), who injected dsRNA into *C. elegans*. Injection of dsRNA resulted in much more efficient silencing of gene expression than the injection of either the sense or the antisense strands alone. Indeed, just a few molecules of dsRNA per cell were sufficient to completely silence expression of the targeted gene. Furthermore, injecting dsRNA into the gut of the worm not only silenced the gene's expression, but also stopped its expression in the next generation. This result turned out to be a general phenomenon.

The dsRNAs apparently work by activating an enzyme ("Dicer") that breaks them down into small pieces. These small pieces then bind to another enzyme complex that destroys any RNAs bound by that small fragment (Figure 4.24; Hunter 2000; Hammond 2001). These enzymes probably evolved as a way of preventing viral infections, since viruses often have double-stranded RNA intermediates, and this technique is proving applicable in many organisms.

Coda

In the past decade, new techniques have enabled us to isolate individual genes, to localize the mRNAs of particular genes, and to delete or silence the expression of particular genes. This technology has enabled biologists to answer questions that were unanswerable only a few years ago. For the first time in human history, we are confronted with the ability to understand how the hereditary potentials of the nucleus become expressed in the formation of our organs.

Principles of Development: Genetic Approaches to Development

1. Development connects genotype and phenotype.

2. A given genotype can produce a limited range of phenotypes, depending on random processes and environmental interactions with the developing organism.

3. The ability of nuclei from differentiated cells to direct the development of complete adult organisms has recently confirmed the principle of genomic equivalence.

4. Nuclear genes are not lost or mutated during development. The genome of each cell is equivalent to that of every other cell.

5. The exceptions to the rule of genomic equivalence are the lymphocytes. During differentiation, these cells rearrange their DNA to create new immunoglobulin and antigen receptor genes.

6. Only a small percentage of the genome is expressed in any particular cell.

7. Polytene chromosomes, in which the DNA has replicated without separating (as in larval *Drosophila* salivary glands), show regions where DNA is being transcribed. Different cell types show different regions of DNA being transcribed.

8. Northern blots, polymerase chain reaction techniques, and in situ hybridization can show which cells are transcribing particular genes.

9. Microarrays and macroarrays allow thousands of genes in different types of cells to be compared simultaneously.

10. The functions of a gene often can be ascertained by techniques that manipulate the gene's expression. These techniques include methods for inhibiting a gene's expression, such as antisense RNA; methods for eliminating a gene, such as gene knockouts in mice; and the use of transgenes to overexpress or misexpress a gene.

11. Double-stranded RNAs can specifically delete messages from a particular gene.

Literature Cited

Agrawal, A., Q. M. Eastman and D. G. Schatz. 1998. Transposition mediated by RAG1 and RAG2 and its implications for the evolution of the immune system. *Nature* 394: 744–751.

Allen, G. E. 1986. T. H. Morgan and the split between embryology and genetics, 1910–1935. In T. J. Horder, J. A. Witkowski and C. C. Wylie (eds.), *A History of Embryology*. Cambridge University Press, New York, pp. 113–146.

Altmann, C. R., E. Bell, A. Sczyrba, J. Pun, S. Bekiranov, T. Gaasterland and A. H. Brivanlou. 2001. Microarray-based analysis of early development in *Xenopus laevis*. *Dev. Biol.* 236: 64–75.

Ashworth, D. and 10 others. 1998. DNA satellite analysis of Dolly. *Nature* 394: 329.

Baltzer, F. 1967. *Theodor Boveri: Life and Work of a Great Biologist*. (Trans. D. Rudnick.) University of California Press, Berkeley.

Barnett, T., C. Pachl, J. P. Gergen and P. C. Wensink. 1980. The isolation and characterization of *Drosophila* yolk protein genes. *Cell* 21: 729–738.

Bassing, C. H., W. Swat and F. W. Alt. 2002. The mechanism and regulation of chromosomal V(D)J recombination. *Cell* 109: S45–S55.

Beermann, W. 1952. Chromomerenkonstanz und spezifische Modifikationen der Chromosomenstruktur in der Entwicklung und Organdifferenzierung von *Chironomus tentans*. *Chromosoma* 5: 139–198.

Briggs, R. and T. J. King. 1952. Transplantation of living nuclei from blastula cells into enucleated frogs' eggs. *Proc. Natl. Acad. Sci. USA* 38: 455–464.

Brush, S. 1978. Nettie Stevens and the discovery of sex determination. *Isis* 69: 132–172.

Burian, R., J. Gayon and D. T. Zallen. 1991. Boris Ephrussi and the synthesis of genetics and embryology. In S. Gilbert (ed.), *A Conceptual History of Modern Embryology*. Plenum, New York, pp. 207–227.

Burkholder, G. D. 1976. Whole mount electron microscopy of polytene chromosome from *Drosophila melanogaster*. *Can. J. Genet. Cytol.* 18: 67–77.

Capecchi, M. R. 1980. High efficiency transformation by direct microinjection of DNA into cultured mammalian cells. *Cell* 22: 479–488.

Chun, J. J. M., D. G. Schatz, M. A. Oettinger, R. Jaenisch and D. Baltimore. 1991. The recombination activating gene 1 (*RAG-1*) is present in the murine central nervous system. *Cell* 64: 189–200.

Dudley, A. T., K. M. Lyons and E. J. Robertson. 1995. A requirement for bone morphogenetic protein 7 during development of the mammalian kidney and eye. *Genes Dev.* 9: 2795–2807.

Fire, A., S. Xu, M. K. Montgomery, S. A. Kostas, S. E. Driver and C. C. Mello. 1998. Potent and specific genetic interference by double-stranded RNA in *Caenorhabditis elegans*. *Nature* 391: 806–811.

Fujimoto, S. and H. Yamagishi. 1987. Isolation of an excision product of T cell receptor α-chain gene rearrangements. *Nature* 327: 242–244.

Gardner, R. L. 1968. Mouse chimeras obtained by the injection of cells into the blastocyst. *Nature* 220: 596–597.

Gilbert, S. F. 1978. The embryological origins of the gene theory. *J. Hist. Biol.* 11: 307–351.

Gilbert, S. F. 1987. In friendly disagreement: Wilson, Morgan, and the embryological origins of the gene theory. *Am. Zool.* 27: 797–806.

Gilbert, S. F. 1988. Cellular politics: Ernest Everett Just, Richard B. Goldschmidt, and the attempts to reconcile embryology and genetics. In R. Rainger, K. R. Benson and J. Maienschein (eds.), *The American Development of Biology*. University of Pennsylvania Press, Philadelphia, pp. 311–346.

Gilbert, S. F. 1991. Induction and the origins of developmental genetics. In S. F. Gilbert (ed.), *A Conceptual History of Modern Embryology*. Plenum, New York, pp. 181–206.

Gilbert, S. F. 1996. Enzyme adaptation and the entrance of molecular biology into embryology. In S. Sarkar (ed.), *The Molecular Philosophy and History of Molecular Biology: New Perspectives*.

Kluwer Academic Publishers, Dordrecht, pp. 101–124.

Glücksohn-Schoenheimer, S. 1938. The development of two tailless mutants in the house mouse. *Genetics* 23: 573–584.

Glücksohn-Schoenheimer, S. 1940. The effect of an early lethal (t^o) in the house mouse. *Genetics* 25: 391–400.

Gossler, A., T. Doetschman, R. Korn, E. Serfling and R. Kemler. 1986. Transgenesis by means of blastocyst-derived stem cell lines. *Proc. Natl. Acad. Sci. USA* 83: 9065–9069.

Grindley, J. C., D. R., Davidson and R. E. Hill. 1995. The role of *Pax-6* in eye and nasal development. *Development* 121: 1433–1442.

Guo, S. and K. J. Kemphues. 1995. *Par-1*, a gene required for establishing polarity in *C. elegans* embryos, encodes a putative Ser/Thr kinase that is asymmetrically distributed. *Cell* 81: 611–620.

Gurdon, J. B., R. A. Laskey and O. R. Reeves. 1975. The developmental capacity of nuclei transplanted from keratinized cells of adult frogs. *J. Embryol. Exp. Morphol.* 34: 93–112.

Hammond, S. M., A. A. Caudy and G. J. Hannon. 2001. Posttranscriptional gene silencing by double-stranded RNA. *Nature Rev. Gen.* 2: 110–119.

Harrison, R. G. 1937. Embryology and its relations. *Science* 85: 369–374.

Harwood, J. 1993. *Styles of Scientific Thought: The German Genetics Community 1900–1934*. University of Chicago Press, Chicago.

Heasman, J., M. Kofron and C. Wylie. 2000. β-Catenin signaling activity dissected in the early *Xenopus* embryo: A novel antisense approach. *Dev. Biol.* 222: 124–134.

Hozumi, N. and S. Tonegawa. 1976. Evidence for somatic rearrangement of immunoglobulin genes coding for variable and constant regions. *Proc. Natl. Acad. Sci. USA* 73: 3628–3632.

Humphreys, D. and 7 others. 2001. Epigenetic instability in ES cells and cloned mice. *Science* 293: 95–97.

Hunter, C. P. 2000. Shrinking the black box of RNAi. *Curr. Biol.* 10: R137–R140.

Jaenisch, R. and I. Wilmut. 2001. Don't clone humans! *Science* 291: 2662.

Just, E. E. 1939. *The Biology of the Cell Surface.* Blakiston, Philadelphia.

Kato, Y. and 7 others. 1998. Eight calves cloned from somatic cells of a single adult. *Science* 282: 2095–2098.

Keller, E. F. 1995. *Refiguring Life: Metaphors of Twentieth-Century Biology.* Columbia University Press, New York.

King, T. J. 1966. Nuclear transplantation in amphibia. *Methods Cell Physiol.* 2: 1–36.

King, T. J. and R. Briggs. 1956. Serial transplantation of embryonic nuclei. *Cold Spring Harb. Symp. Quant. Biol.* 21: 271–289.

Kolata, G. 2001. Researchers find big risk of defect in cloning animals. *New York Times*, 25 March.

Li, H.-S., J.-M. Yang, R. D. Jacobson, D. Pasko and O. Sundin. 1994. *Pax-6* is first expressed in a region of ectoderm anterior to the early neural plate: Implications for stepwise determination of the lens. *Dev. Biol.* 162: 181–194.

Lillie, F. R. 1927. The gene and the ontogenetic process. *Science* 64: 361–368.

Matsuoka, M. and 8 others. 1991. Detection of somatic DNA recombination in the transgenic mouse brain. *Science* 254: 81–86.

McKinnell, R. G. 1978. *Cloning—Nuclear Transplantation in Amphibia.* University of Minnesota Press, Minneapolis.

Meade, H. M. 1997. Dairy gene. *The Sciences* (Sept./Oct.), 20–25.

Mendel, G. 1866. *Versuche über Pflanzenhybriden. Verh. Naturf. Vereines* (Brünn) 4: 3–47.

Morange, M. 1996. Construction of the developmental gene concept. The crucial years: 1960–1980. *Biol. Zent.* bl. 115: 132–138.

Morgan, T. H. 1926. *The Theory of the Gene.* Yale University Press, New Haven.

Moustafa, L. A. and R. L. Brinster. 1972. Induced chimaerism by transplanting embryonic cells into mouse blastocysts. *J. Exp. Zool.* 181: 193–202.

Oppenheimer, J. M. 1981. Walter Landauer and developmental genetics. In S. Subtelny and U. K. Abbott (eds.), *Levels of Genetic Control in Development.* Alan R. Liss, New York, pp. 1–14.

Perucho, M., D. Hanahan and M. Wigler. 1980. Genetic and physical linkage of exogenous sequences in transformed cells. *Cell* 22: 309–317.

Prather, R. S. 1991. Nuclear transplantation and embryo cloning in mammals. *Int. Lab. Anim. Res. News* 33: 62–68.

Rappolee, D. A., C. A. Brenner, R. Schultz, D. Mark and Z. Werb. 1988. Developmental expression of *PDGF*, *TGF-α* and *TGF-β* genes in preimplantation mouse embryos. *Science* 241: 1823–1825.

Robins, D. M., S. Ripley, A. S. Henderson and R. Axel. 1981. Transforming DNA integrates into the host chromosome. *Cell* 23: 29–39.

Rosenberg, U. B., A. Preiss, E. Seifert, H. Jäckle and D. C. Knipple. 1985. Production of phenocopies by *Krüppel* antisense RNA injection into *Drosophila* embryos. *Nature* 313: 703–706.

Saiki, R. K., S. Scharf, F. Faloona, K. B. Mullis, G. T. Horn, H. A. Erlich and N. Arnheim. 1985. Enzymatic amplification of β-globin genomic sequences and restriction site analysis for diagnosis of sickle cell anemia. *Science* 230: 1350–1354.

Sander, K. 1986. The role of genes in ontogenesis: Evolving concepts from 1883 to 1983 as perceived by an insect embryologist. In T. J. Horder, J. A. Witkowski and C. C. Wylie (eds.), *A History of Embryology.* Cambridge University Press, New York, pp. 363–395.

Schnieke, A. and 8 others. 1997. Human factor IX transgenic sheep produced by transfer of nuclei from transfected fetal fibroblasts. *Science* 278: 2130–2134.

Schwarz, K. and 11 others. 1996. *Rag* mutations in human B cell-negative SCID. *Science* 274: 97–99.

Shin T, and 9 others. 2002. A cat cloned by nuclear transplantation. *Nature* 415: 859.

Signer, E. N., Y. E. Dubrova, A. J. Jeffreys, C. Wilde, L. M. B. Finch, M. Wells and M. Peaker. 1998. DNA fingerprinting Dolly. *Nature* 394: 329–330.

Smith, L. D. 1956. Transplantation of the nuclei of primordial germ cells into enucleated eggs of *Rana pipiens. Proc. Natl. Acad. Sci. USA* 54: 101–107.

Spemann, H. 1938. *Embryonic Development and Induction.* Yale University Press, New Haven.

Spradling, A. C. and G. M. Rubin. 1982. Transposition of cloned P elements into *Drosophila* germ line chromosomes. *Science* 218: 341–347.

Stevens, N. M. 1905a. *Studies in Spermatogenesis with Especial Reference to the "Accessory Chromosome."* Carnegie Institute of Washington, Washington, D.C.

Stevens, N. M. 1905b. A study of the germ cells of *Aphis rosae* and *Aphis oenotherae. J. Exp. Zool.* 2: 371–405, 507–545.

Summerton, J. and D. Weller. 1997. *Morpholino* antisense oligomers: Design, preparation, and properties. *Antisense Nucleic Acid Drug Dev.* 7: 187–195.

Ton, C. C. and 10 others. 1991. Positional cloning and characterization of a paired box- and homeobox-containing gene from the *aniridia* region. *Cell* 67: 1059–1074.

Villa, A. and 11 others. 1998. Partial V(D)J recombination activity leads to Omenn syndrome. *Cell* 93: 885–896.

Waddington, C. H. 1939. Preliminary notes on the development of wings in normal and mutant strains of *Drosophila. Proc. Natl. Acad. Sci. USA* 25: 299–307.

Waddington, C. H. 1948. *The Scientific Attitude.* Pelican Books, New York.

Wagner, T. E., P. Hoppe, J. D. Jollick, D. R. Scholl, R. L. Hodinka and J. B. Gault. 1981. Microinjection of rabbit β-globin gene into zygotes and its subsequent expression in adult mice and offspring. *Proc. Natl. Acad. Sci. USA* 78: 6376–6380.

Wan, J. S. and 11 others. 1996. Cloning differentially expressed mRNAs. *Nature Biotech.* 14: 1685–1691.

Wakayama, T., A. C. F. Perry, M. Zuccotti, K. R. Johnson and R. Yanagimachi. 1998. Full-term development of mice from enucleated oocytes injected with cumulus cell nuclei. *Nature* 394: 369–374.

Wetmur, J. G. and N. Davidson. 1968. Kinetics of renaturation of DNA. *J. Mol. Biol.* 31: 349–370.

White, K. P., S. A. Rifkin, P. Hurban and D. S. Hogness. 1999. Microarray analysis of *Drosophila* development during metamorphosis. *Science* 286: 2179–2184.

Wianny, F. and M. Zernicka-Goetz. 2000. Specific interference with gene function by double-stranded RNA in early mouse development. *Nature Cell Biol.* 2: 70–75.

Wilkinson, D. G., S. Bhatt and B. G. Herrmann. 1990. Expression pattern of the mouse *T* gene and its role in mesoderm formation. *Nature* 343: 657–659.

Wilmut, I., A. E. Schnieke, J. McWhir, A. J. Kind and K. H. S. Campbell. 1997. Viable offspring from fetal and adult mammalian cells. *Nature* 385: 810–814.

Wilmut, I., K. Campbell and C. Tudge. 2000. *The Second Creation: Dolly and the Age of Biological Control.* Harvard University Press, Cambridge, MA.

Wilson, E. B. 1895. *An Atlas of the Fertilization and Karyogenesis of the Ovum.* Macmillan, New York.

Wilson, E. B. 1896. *The Cell in Development and Inheritance.* Macmillan, New York.

Wilson, E. B. 1905. The chromosomes in relation to the determination of sex in insects. *Science* 22: 500–502.

chapter 5 The paradigm of differential gene expression

DIFFERENT CELL TYPES make different sets of proteins, even though their genomes are identical. Each human being has roughly 50,000 genes in each nucleus, but each cell uses only a small subset of those genes. Moreover, different cell types use different gene subsets. Red blood cells make globin proteins, lens cells make crystallins, melanocytes make melanin, and endocrine glands make their specific hormones. **Developmental genetics** is the discipline that examines how the genotype is transformed into the phenotype, and the major paradigm of developmental genetics is *differential gene expression from the same nuclear repertoire*. This chapter will discuss the mechanisms of cell differentiation: how the various genes present in the nucleus are activated or repressed at particular times and places to cause cells to become different from one another.

Gene expression can be regulated at several levels:

- **Differential gene transcription**, regulating which of the nuclear genes are transcribed into RNA
- **Selective nuclear RNA processing**, regulating which of the transcribed RNAs (or which parts of such a nuclear RNA) enter into the cytoplasm to become messenger RNAs
- **Selective messenger RNA translation**, regulating which of the mRNAs in the cytoplasm become translated into proteins
- **Differential protein modification**, regulating which proteins are allowed to remain or function in the cell

Some genes (such as those coding for the globin proteins of hemoglobin) are regulated at each of these levels.

Differential Gene Transcription

Anatomy of the gene: Exons and introns

Two fundamental differences distinguish most eukaryotic genes from most prokaryotic genes. First, eukaryotic genes are contained within a complex of DNA and protein called **chromatin**. The protein component constitutes about half the weight of chromatin and is composed largely of **histones**. The **nucleosome** is the basic unit of chromatin structure. It is composed of an octamer of histone proteins (two molecules each of histones H2A, H2B, H3, and H4) wrapped with two loops containing approximately 140 base pairs of DNA (Kornberg and Thomas 1974). Chromatin can be visualized as a string of nucleosome beads linked by ribbons of DNA.

While classical geneticists have likened genes to "beads on a string," molecular geneticists liken genes to "string on the beads." Most of the time, the nucleosomes are themselves wound into tight "solenoids" that are stabilized by histone H1. Histone H1 is bound to the 60 or so base pairs of "linker" DNA between the nucleosomes (Figure 5.1; Weintraub 1984). This H1-dependent conformation of nucleosomes inhibits the transcription of genes in somatic cells by packing adjacent nucleosomes together into tight arrays that prevent transcription factors and RNA polymerases from gaining access to the genes (Thoma et al. 1979; Schlissel and Brown 1984). It is generally thought, then, that the "default" condition of chromatin is a repressed state, and that tissue-specific genes become activated by local interruption of this repression (Weintraub 1985).

WEBSITE 5.1 **Displacing nucleosomes.** Transcription can occur even in a region of nucleosomes. If a gene's promoter is accessible to RNA polymerase and transcription factors, the presence of nucleosomes will not inhibit the elongation of the message.

The second difference is that eukaryotic genes are not co-linear with their peptide products. Rather, the single nucleic acid strand of eukaryotic mRNA comes from noncontiguous regions on the chromosome. Between **exons**—the regions of DNA that code for a protein—are intervening sequences called **introns** that have nothing whatsoever to do with the

amino acid sequence of the protein.* The structure of a typical eukaryotic gene can be illustrated by the human β-globin gene, shown in Figure 5.2. This gene, which encodes part of the hemoglobin protein of the red blood cells, consists of the following elements:

1. A **promoter** region, which is responsible for the binding of RNA polymerase and for the subsequent initiation of transcription. The promoter region of the human β-globin gene has three distinct units and extends from 95 to 26 base pairs before ("upstream from")[†] the transcription initiation site (i.e., from –95 to –26).

2. The **transcription initiation site**, which for human β-globin is ACATTTG. This site is often called the **cap sequence** because it represents the 5′ end of the RNA, which will receive a "cap" of modified nucleotides soon after it is transcribed. The specific cap sequence varies among genes.

3. The **translation initiation site**, ATG. This codon (which becomes AUG in the mRNA) is located 50 base pairs after the transcription initiation site in the human β-globin

---

*The term *exon* refers to a nucleotide sequence whose RNA "exits" the nucleus. It has taken on the functional definition of a protein-encoding nucleotide sequence. Leader sequences and 3′ untranslated sequences are also derived from exons, even though they are not translated into protein.

[†]By convention, upstream, downstream, 5′, and 3′ directions are specified in relation to the RNA. Thus, the promoter is upstream of the gene, near its 5′ end.

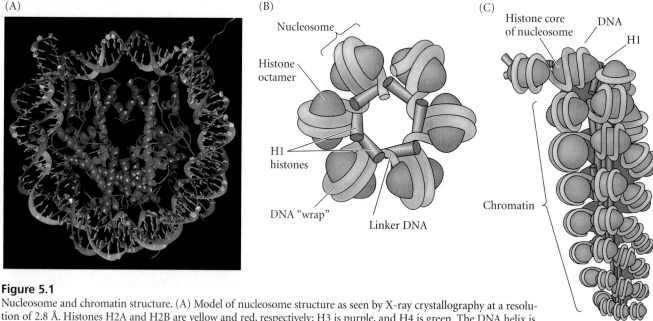

Figure 5.1
Nucleosome and chromatin structure. (A) Model of nucleosome structure as seen by X-ray crystallography at a resolution of 2.8 Å. Histones H2A and H2B are yellow and red, respectively; H3 is purple, and H4 is green. The DNA helix is wound around the protein core. The histone tails that extend from the core are the sites of acetylation and methylation, which may disrupt or stabilize, respectively, the formation of nucleosome assemblages. (B) Histone H1 can draw nucleosomes together into compact forms. About 140 base pairs of DNA encircle each histone octamer, and about 60 base pairs of DNA link the nucleosomes together. (C) Model for the arrangement of nucleosomes in the highly compacted solenoidal chromatin structure. (A from Luger et al. 1997, photograph courtesy of the authors; B, C after Wolfe 1993.)

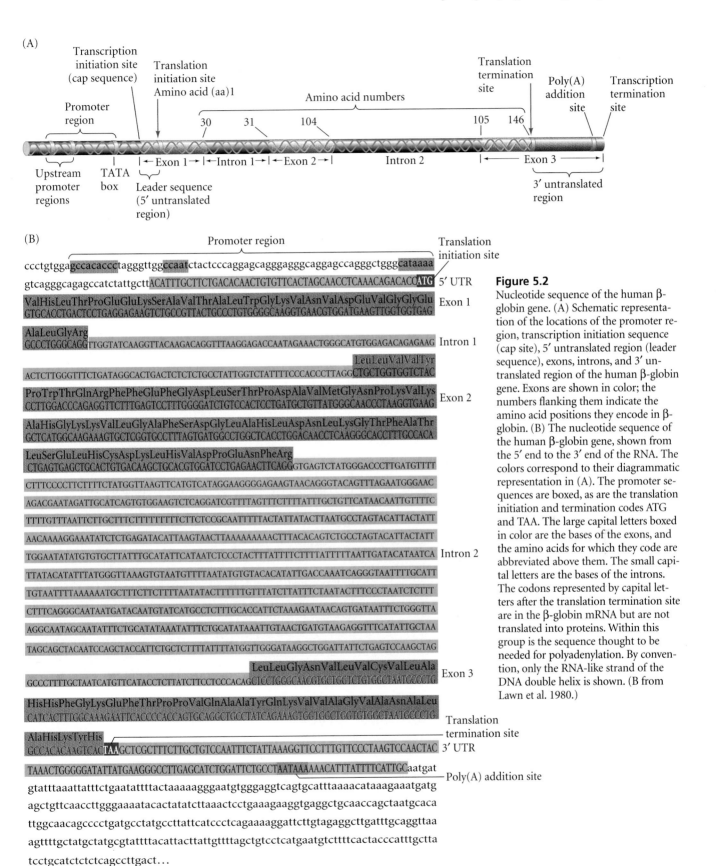

(A)

(B)

Figure 5.2

Nucleotide sequence of the human β-globin gene. (A) Schematic representation of the locations of the promoter region, transcription initiation sequence (cap site), 5′ untranslated region (leader sequence), exons, introns, and 3′ untranslated region of the human β-globin gene. Exons are shown in color; the numbers flanking them indicate the amino acid positions they encode in β-globin. (B) The nucleotide sequence of the human β-globin gene, shown from the 5′ end to the 3′ end of the RNA. The colors correspond to their diagrammatic representation in (A). The promoter sequences are boxed, as are the translation initiation and termination codes ATG and TAA. The large capital letters boxed in color are the bases of the exons, and the amino acids for which they code are abbreviated above them. The small capital letters are the bases of the introns. The codons represented by capital letters after the translation termination site are in the β-globin mRNA but are not translated into proteins. Within this group is the sequence thought to be needed for polyadenylation. By convention, only the RNA-like strand of the DNA double helix is shown. (B from Lawn et al. 1980.)

gene (although this distance differs greatly among different genes). The sequence of 50 base pairs intervening between the initiation points of transcription and translation is the **5′ untranslated region**, often called the **5′ UTR** or **leader sequence**. The 5′ UTR can determine the rate at which translation is initiated.

4. The first **exon**, which contains 90 base pairs coding for amino acids 1–30 of human β-globin.

5. An **intron** containing 130 base pairs with no coding sequences for the β-globin protein. The structure of this intron is important in enabling the RNA to be processed into messenger RNA and exit from the nucleus.

6. An exon containing 222 base pairs coding for amino acids 31–104.

7. A large intron—850 base pairs—having nothing to do with globin protein structure.

8. An exon containing 126 base pairs coding for amino acids 105–146.

9. A **translation termination codon**, TAA. This codon becomes UAA in the mRNA. The ribosome dissociates at this codon, and the protein is released.

10. A **3′ untranslated region** (3′ UTR) that, although transcribed, is not translated into protein. This region includes the sequence AATAAA, which is needed for **polyadenylation**: the placement of a "tail" of some 200 to 300 adenylate residues on the RNA transcript. This **poly(A) tail** (1) confers stability on the mRNA, (2) allows the mRNA to exit the nucleus, and (3) permits the mRNA to be translated into protein. The poly(A) tail is inserted into the RNA about 20 bases downstream of the AAUAAA sequence. Transcription continues beyond the AATAAA site for about 1000 nucleotides before being terminated.

The original nuclear RNA transcripts for genes such as β-globin contain the cap sequence, the 5′ untranslated region, exons, introns, and the 3′ untranslated region (Figure 5.3).

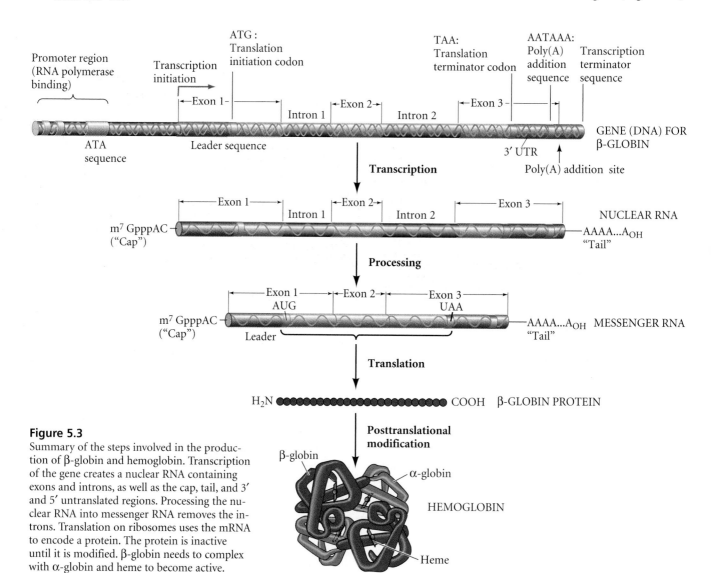

Figure 5.3
Summary of the steps involved in the production of β-globin and hemoglobin. Transcription of the gene creates a nuclear RNA containing exons and introns, as well as the cap, tail, and 3′ and 5′ untranslated regions. Processing the nuclear RNA into messenger RNA removes the introns. Translation on ribosomes uses the mRNA to encode a protein. The protein is inactive until it is modified. β-globin needs to complex with α-globin and heme to become active.

Both ends of these transcripts are modified before these RNAs leave the nucleus. A cap consisting of methylated guanosine is placed on the 5′ end of the RNA in opposite polarity to the RNA itself. This means that there is no free 5′ phosphate group on the nuclear RNA. The 5′ cap is necessary for the binding of mRNA to the ribosome and for subsequent translation (Shatkin 1976). The 3′ terminus is usually modified in the nucleus by the addition of a poly(A) tail. The adenylate residues in this tail are put together enzymatically and are added to the transcript; they are not part of the gene sequence. Both the 5′ and 3′ modifications may protect the mRNA from exonucleases that would otherwise digest it (Sheiness and Darnell 1973; Gedamu and Dixon 1978). The modifications thus stabilize the message and its precursor.

> **WEBSITE 5.2 Structure of the 5′ cap.** The capping and methylation of the 5′ end are critical steps in the synthesis of mRNA. If the cap is missing or unmethylated, translation may fail to occur.

Anatomy of the gene: Promoters and enhancers

In addition to the protein-encoding region of the gene, there are regulatory sequences that can be located on either end of the gene (or even within it). These sequences—the promoters and enhancers—are necessary for controlling where and when a particular gene is transcribed.

Promoters are the sites where RNA polymerase binds to the DNA to initiate transcription. Promoters of genes that synthesize messenger RNAs (i.e., genes that encode proteins*) are typically located immediately upstream from the site where the RNA polymerase initiates transcription begins. Most of these promoters contain the sequence TATA, to which RNA polymerase will be bound (Figure 5.4). This site, known as the **TATA box**, is usually about 30 base pairs upstream from the site where the first base is transcribed. Eukaryotic RNA polymerases, however, will not bind to this naked DNA sequence. Rather, they require the presence of additional proteins, called **transcription factors**, to bind efficiently to the promoter. Protein-encoding genes are transcribed by RNA polymerase II, and at least six nuclear proteins have been shown to be necessary for the proper initiation of transcription by this polymerase (Buratowski et al. 1989; Sopta et al. 1989). These nuclear proteins are called **basal transcription factors**.

The first basal transcription factor, **TFIID**,[†] recognizes the TATA box through one of its subunits, **TATA-binding protein (TBP)**. TFIID serves as the foundation of the transcrip-

---

*There are several types of RNA that do not encode proteins. These include the ribosomal RNAs and transfer RNAs (which are used in protein synthesis) and the small nuclear RNAs (which are used in RNA processing). In addition, there are regulatory RNAs (such as *Xist* and *lin-4*, which we will discuss later in this chapter) that are involved in regulating gene expression (and which are not translated into peptides).

[†]TF stands for transcription factor; II indicates that the factor was first found to be needed for RNA polymerase II; and the letter designations refer to the active fractions from the phosphocellulose columns used to purify these proteins.

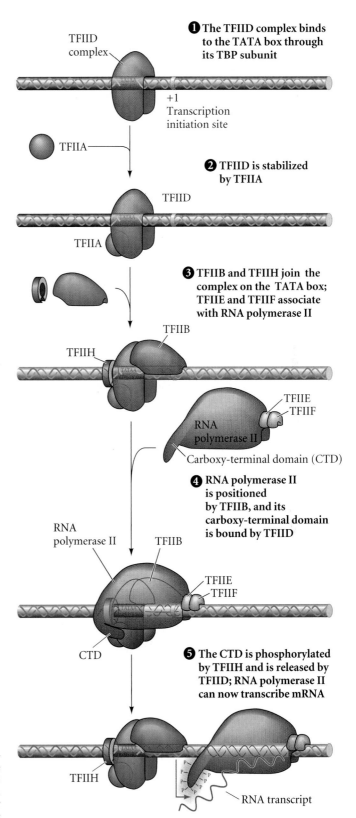

❶ The TFIID complex binds to the TATA box through its TBP subunit

❷ TFIID is stabilized by TFIIA

❸ TFIIB and TFIIH join the complex on the TATA box; TFIIE and TFIIF associate with RNA polymerase II

❹ RNA polymerase II is positioned by TFIIB, and its carboxy-terminal domain is bound by TFIID

❺ The CTD is phosphorylated by TFIIH and is released by TFIID; RNA polymerase II can now transcribe mRNA

Figure 5.4
Formation of the active eukaryotic transcription initiation complex. The diagrams represent the complex formed on the TATA box by the basal transcription factors and RNA polymerase II.

tion initiation complex, and it also prevents nucleosomes from forming in this region. Once TFIID is stabilized by the basal transcription factor, **TFIIA**, it becomes able to bind **TFIIB**. Once TFIIB is in place, RNA polymerase can bind to the complex. Other transcription factors (TFIIE, F, and H) are then used to release RNA polymerase from the complex so that it can transcribe the gene, and to unwind the DNA helix so that the RNA polymerase will have a free template from which to transcribe.

The ability of the basal transcription factors to interact with RNA polymerase is regulated by two sets of proteins, **TBP-associated factors** (**TAFs**) and the **mediator complex** (Myers and Kornberg 2000; Baek et al. 2002). The mediator complex contains about twenty-five proteins that can modulate the activity of RNA polymerase II and TFIIH, a basal transcription factor that is critical in allowing RNA polymerase II to start transcription (Woychik and Hampsey 2002). The TAFs stabilize the TBP (Figure 5.5; Buratowski 1997; Lee and Young 1998). This function is critical for gene transcription, for if the TBP is not stabilized, it can fall off the small TATA sequence. The TAFs are bound by **upstream promoter elements** on the DNA. These DNA sequences are near the TATA box (usually upstream from it). Every TAF need not be present in every cell of the body, however. **Cell-specific transcription factors** (such as the Pax6 protein mentioned in Chapter 4) can also activate the gene by stabilizing the transcription initiation complex. They can do so by binding to the TAFs, or by binding directly to other factors such as TFIIB. They can also facilitate transcription by destabilizing nucleosomes (as we will see soon).

WEBSITE 5.3 Promoter structure and the mechanisms of transcription complex assembly. Getting RNA polymerase to a promoter is not an easy task. The transcription initiation complex is a major protein complex that must be created at each round of transcription. The delineation of the different parts of promoters is worked out via mutations and transgenes.

An **enhancer** is a DNA sequence that activates the utilization of a promoter, controlling the efficiency and rate of transcription from that promoter. Enhancers can activate only *cis*-linked promoters (i.e., promoters on the same chromosome*), but they can do so at great distances (some as great as 50 kilobases away from the promoter). Moreover, enhancers do not need to be on the 5′ (upstream) side of the gene; they can also be at the 3′ end, in the introns, or even on the complementary DNA strand (Maniatis et al. 1987). The human β-globin gene has an enhancer in its 3′ UTR, roughly 700 base pairs downstream from the AATAAA site. This enhancer sequence is necessary for the temporal- and tissue-specific expression of the β-globin gene in adult red blood cell precursors (Trudel and Constantini 1987). Like promoters, enhancers function by binding specific regulatory proteins called transcription factors.

Enhancers can regulate the temporal and tissue-specific expression of any differentially regulated gene, but different types of genes normally have different enhancers. For instance, genes encoding the *exocrine* proteins of the pancreas (the digestive proteins chymotrypsin, amylase, and trypsin) have enhancers different from that of the genes encoding the pancreatic *endocrine* proteins (such as insulin). These enhancers both lie in the 5′ flanking sequences of their genes (Walker et al 1983).

One of the principal methods of identifying enhancer sequences is to clone DNA sequences flanking the gene of interest and fuse them to **reporter genes** whose products are not usually made in the cells and that can be readily identified. Constructs of possible enhancers and reporter genes can be inserted into embryos, and the expression of the reporter gene monitored. If the sequence contains an enhancer, the reporter gene should become active at particular times and places. For instance, the *E. coli* **β-galactosidase gene** (the *lacZ* gene) can be used as a reporter gene and fused to an enhancer that normally directs the expression of a particular mouse gene in muscles. If the resulting transgene is injected into a newly fer-

---

Cis- and *trans*-regulatory elements are so named by analogy with *E. coli* genetics and organic chemistry. There, *cis*-elements are regulatory elements that reside on the same strand of DNA (*cis-*, "on the same side as"), while *trans*-elements are those that could be supplied from another chromosome (*trans-*, "on the other side of"). The term *cis*-regulatory elements now refers to those DNA sequences that regulate a gene on the same stretch of DNA (i.e., the promoters and enhancers). *Trans*-regulatory factors are soluble molecules whose genes are located elsewhere in the genome and which bind to the *cis*-regulatory elements. They are usually transcription factors.

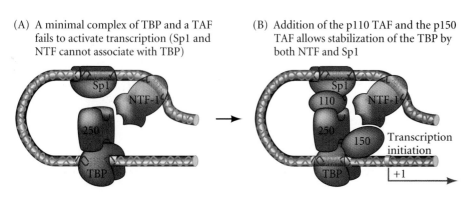

(A) A minimal complex of TBP and a TAF fails to activate transcription (Sp1 and NTF cannot associate with TBP)

(B) Addition of the p110 TAF and the p150 TAF allows stabilization of the TBP by both NTF and Sp1

Figure 5.5
Model of TAF stabilization of TBP. (A) A minimal complex near the promoter, containing TBP on the TATA box of the promoter and two upstream promoter elements occupied by two transcription factors, Sp1 and NTF-1. One TAF, TAF$_{II}$250, is bound to the TBP, but this complex is not stable enough to activate transcription. (B) When other TAFs bind to these proteins, however, they form bridges that stabilize the TBP on the promoter. (After Chen et al. 1995.)

Figure 5.6

The genetic elements regulating tissue-specific transcription can be identified by fusing reporter genes to suspected enhancer regions of the genes expressed in particular cell types. (A) The enhancer region of the gene encoding muscle-specific protein Myf-5 is fused to a β-galactosidase reporter gene and incorporated into a mouse embryo. When stained for β-galactosidase activity (darkly staining region), the 13.5-day mouse embryo shows that the reporter gene is expressed in the eye muscles, facial muscles, forelimb muscles, neck muscles, and segmented myotomes (which give rise to the back musculature). (B) The gene for green fluorescent protein (GFP) is fused to a lens crystallin enhancer in *Xenopus tropicalis*. The result is the expression of GFP in the tadpole lens. (A, photograph courtesy of A. Patapoutian and B. Wold; B from Offield et al. 2000, photograph courtesy of R. Grainger.)

(A)

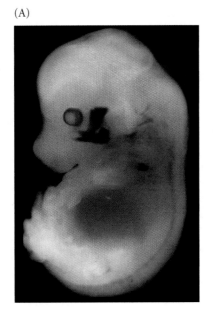

(B)

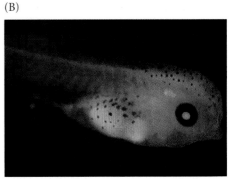

tilized mouse egg and becomes incorporated into its DNA, β-galactosidase will be expressed in the mouse muscle cells. By staining for the presence of β-galactosidase, the expression pattern of that muscle-specific gene can be seen (Figure 5.6A). Similarly, a sequence flanking a lens crystallin protein in *Xenopus* was shown to be an enhancer. When this sequence was fused to a reporter gene for **green fluorescent protein** (**GFP**, a protein usually found only in jellyfish), GFP expression was seen solely in the lens (Figure 5.6B; Offield et al. 2000). GFP reporter genes are very useful because they can be monitored in live embryos and because the changes in gene expression can be seen in single organisms.

Enhancers are modular. As we will see shortly, if a protein is synthesized in several tissues, there can be separate enhancers directing the gene to be expressed at these different sites. The mouse *Pax6* gene, for instance, is expressed in the lens, retina, neural tube and pancreas. The regulatory regions of the mouse *Pax6* gene were discovered by taking regions from its 5′ flanking sequence and introns and fusing them to a *lacZ* reporter gene. Each of these transgenes was then microinjected into newly fertilized mouse pronuclei, and the resulting embryos were stained for β-galactosidase (Figure 5.7A; Kammandel et al. 1998; Williams et al. 1998). The results are summarized in Figure 5.7B. The enhancer farthest upstream from the promoter contains the regions necessary for *Pax6* expression in the pancreas, while a second enhancer activates *Pax6* expression in surface ectoderm (lens, cornea, and conjunctiva). A third enhancer resides in the leader sequence; it contains the sequences that direct *Pax6* expression in the neural tube. A fourth enhancer sequence, located in an intron shortly after

(A)

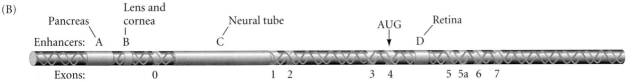

β-glactosidase
(reporter gene)

Figure 5.7

Regulatory regions of the mouse *Pax6* gene. (A) A sequence in the upstream enhancer of the *Pax6* gene directs the expression of a *lacZ* reporter transgene in the surface ectoderm overlying the optic cup, as shown by the dark staining in this area. (B) Map of the enhancer sites of the *Pax6* gene, based on reporter gene studies. The gene has two major promoters (in exons 0 and 1, respectively) either of which can initiate transcription of the same RNA. The translation initiation codon is in exon 4. (Exon 5a is an alternatively spliced exon, discussed in the text.) Regions A–D are enhancers that activate the *Pax6* gene in the pancreas, lens, neural tube, and retina, respectively. The B region expression pattern is shown in (A). (A from Williams et al. 1998; B after Kammandel et al. 1998.)

(B)

Pancreas Lens and cornea Neural tube AUG Retina

Enhancers: A B C D

Exons: 0 1 2 3 4 5 5a 6 7

the translation initiation site, determines the expression of *Pax6* in the retina.

Enhancers are critical in the regulation of normal development. Over the past decade, eight generalizations have emerged that emphasize their importance for differential gene expression:

1. Most genes require enhancers for their transcription.
2. Enhancers are the major determinant of differential transcription in space (cell type) and time.
3. The ability of an enhancer to function while far from the promoter means that there can be multiple signals to determine whether a given gene is transcribed. A given gene can have several enhancer sites linked to it, and each enhancer can be bound by more than one transcription factor.
4. An interaction between the transcription factors bound to the enhancer sites and the transcription initiation complex assembled at the promoter is thought to regulate transcription. The mechanism of this association is not fully known, nor do we comprehend how the promoter integrates all these signals.
5. Enhancers are combinatorial. Various DNA sequences regulate temporal and spatial gene expression, and these can be mixed and matched. For example, the enhancers for endocrine hormones such as insulin and for lens-specific proteins such as crystallins both have sites that bind Pax6 protein. But Pax6 alone doesn't tell the lens to make insulin or the pancreas to make crystallins; there are other transcription factor proteins that also must bind. It is the *combination* of transcription factors that causes particular genes to be transcribed.
6. Enhancers are modular. A gene can have several enhancer elements, each turning it on in a different set of cells.
7. Enhancers generally activate transcription by one of two means: they either remodel chromatin to expose the pro-

moter, or they facilitate the binding of RNA polymerase to the promoter by stabilizing TAFs.

8. Enhancers can also *inhibit* transcription. In some cases, the same transcription factors that activate the transcription of one gene can repress the transcription of other genes. These "negative enhancers" are sometimes called **silencers**.

Transcription factors

As we have seen, transcription factors are proteins that bind to enhancer or promoter regions and interact to activate or repress the transcription of a particular gene. Most transcription factors can bind to specific DNA sequences. These proteins can be grouped together in families based on similarities in structure (Table 5.1). The transcription factors within such a family share a common framework structure in their DNA-binding sites, and slight differences in the amino acids at the binding site can cause the binding site to recognize different DNA sequences.

WEBSITE 5.4 **Families of transcription factors.** There are several families of transcription factors grouped together by their structural similarities and their mechanisms of action. Homeodomain transcription factors are important in specifying anterior-posterior axes, and hormone receptors mediate the effects of hormones on genes.

Transcription factors have three major domains. The first is a **DNA-binding domain** that recognizes a particular DNA sequence. The second is a *trans*-activating domain that activates or suppresses the transcription of the gene whose promoter or enhancer it has bound. Usually, this *trans*-activating domain enables the transcription factor to interact with the proteins involved in binding RNA polymerase (such as TFIIB or TFIIE; see Sauer et al. 1995). In addition, there may be a **protein-protein interaction domain** that allows the transcription factor's activity to be modulated by TAFs or other

TABLE 5.1 Some major transcription factor families and subfamilies

Family	Representative transcription factors	Some functions
Homeodomain:		
Hox	Hoxa-1, Hoxb-2, etc.	Axis formation
POU	Pit-1, Unc-86, Oct-2	Pituitary development; neural fate
LIM	Lim-1, Forkhead	Head development
Pax	Pax1, 2, 3, 6, etc.	Neural specification; eye development
Basic helix-loop-helix (bHLH)	MyoD, MITF, daughterless	Muscle and nerve specification; *Drosophila* sex determination; pigmenttion
Basic leucine zipper (bZip)	C/EBP, AP1	Liver differentiation; fat cell specification
Zinc finger:		
Standard	WT1, Krüppel, Engrailed	Kidney, gonad, and macrophage development; *Drosophila* segmentation
Nuclear hormone receptors	Glucocorticoid receptor, estrogen receptor, testosterone receptor, retinoic acid receptors	Secondary sex determination; craniofacial development; limb development
Sry-Sox	Sry, SoxD, Sox2	Bend DNA; mammalian primary sex determination; ectoderm differentiation

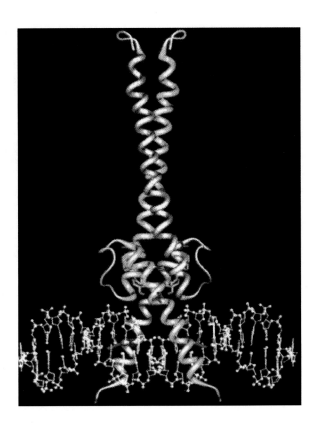

Figure 5.8
Three-dimensional model of the MITF homodimer (one protein in red, the other in blue) binding to a promoter element in DNA (white). The amino termini are located at the bottom of the figure and form the DNA-binding domain. The protein-protein interaction domain is located immediately above. The carboxyl end of the molecule is thought to be the *trans*-activating domain that binds the p300/CBP co-activator protein. (After Steingrímsson et al. 1994; photograph courtesy of N. Jenkins.)

transcription factors. We will discuss the functioning of transcription factors by analyzing two examples: MITF and *Pax6*.

MITF. The microphthalmia (MITF) protein is a transcription factor that is active in the ear and in the pigment-forming cells of the eye and skin. Humans heterozygous for a mutation of the gene that encodes MITF are deaf, have multicolored irises, and have a white forelock in their hair (see Chapter 21). The MITF protein is a basic helix-loop-helix transcription factor (see Table 5.1) that has three functionally important domains. First, MITF has a protein-protein interaction domain that enables it to dimerize with another MITF protein (Ferré-D'Amaré et al. 1993). The resulting homodimer (two MITF proteins bound together) forms the functional protein that can bind to DNA and activate the transcription of certain genes (Figure 5.8). The second region, the DNA-binding domain, is close to the amino-terminal end of the protein and contains numerous basic amino acids that make contact with

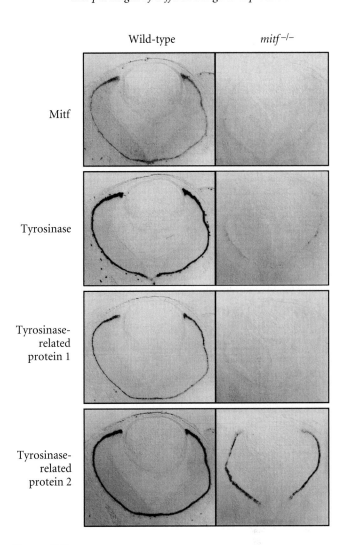

Wild-type   *mitf*⁻/⁻

Mitf

Tyrosinase

Tyrosinase-related protein 1

Tyrosinase-related protein 2

Figure 5.9
MITF is required for the transcription of pigmention genes. Serial sections of the eye in 15.5-day mouse embryos are shown. In the wild-type mouse embryo, in situ hybridization reveals the presence of *Mitf* mRNA (dark staining) in the retinal pigment epithelial layer. In the mi mutant embryo (*mitf*⁻/⁻), no MITF is present. MITF recognizes promoter sites in three genes that encode enzymes involved in melanin production. In the *mitf*⁻/⁻ mutant, the genes encoding these enzymes are not transcribed well, compared to the wild-type mice. Transcription of the tyrosinase-related protein 2 gene is significantly reduced in the mutant. (From Nakayama et al. 1998; photographs courtesy of H. Arnheiter.)

the DNA (Hemesath et al. 1994; Steingrímsson et al. 1994). This assignment was confirmed by the discovery of various human and mouse mutations that map within the DNA-binding site for MITF and which prevent the attachment of the MITF protein to the DNA. Sequences for MITF binding have been found in the promoter regions of genes encloding three pigment-cell-specific proteins of the tyrosinase family (Bentley et al. 1994; Yasumoto et al. 1997). Without MITF, these proteins are not synthesized properly (Figure 5.9) and

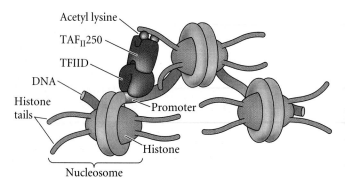

Figure 5.10
TAF$_{II}$250, a TAF that binds TBP, can function as a histone acetyl-transferase. Thus, it can acetylate histones and thereby disrupt nucleosomes. It can also bind to acetylated lysine residues on the histones and enable the positioning of TBP on the promoter. These abilities relate to its function as proposed in Figure 5.5, where the TBP is part of the TFIID complex. (After Pennisi 2000.)

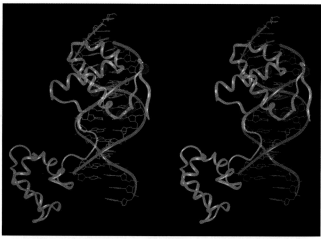

Figure 5.11
Stereoscopic model of Pax6 protein binding to its enhancer element in DNA. The DNA-binding region of Pax6 is shown in yellow; the DNA double helix is blue. Red dots indicate the sites of loss-of-function mutations in the *Pax6* gene that give rise to nonfunctional Pax6 proteins. It is worth trying to cross your eyes to get the central three-dimensional figure. (From Xu et al. 1995; photograph courtesy of S. O. Pääbo.)

the melanin pigment is not made. These promoters all contain the same 11-base-pair sequence, including the core sequence CATGTG, recognized by MITF.

The third functional region of MITF is its *trans*-activating domain. This domain includes a long stretch of amino acids in the center of the protein. When the MITF dimer is bound to its target sequence in a promoter or enhancer, the *trans*-activating region is able to bind a TAF called p300/CBP. The p300/CPB protein is a **histone acetyltransferase** enzyme that can transfer acetyl groups to each histone in the nucleosomes (Ogryzko et al. 1996; Price et al. 1998). Acetylation of the nucleosomes destabilizes them and allows the genes for pigment-forming enzymes to be expressed.

Recent discoveries have shown that numerous transcription factors operate by recruiting histone acetyltransferases. When histone proteins are acetylated, the nucleosome disperses, allowing other transcription factors and RNA polymerase access to the site. Some TAFs, such as TAF$_{II}$250, are themselves histone acetyltransferases, so they can activate transcription both by destabilizing the nucleosome and by enabling the TBP protein to bind to the DNA to establish a site for RNA polymerase binding (Figure 5.10; Mizzen et al. 1996).

WEBSITE 5.5 **Histone acetylation.** Histone acetylation is a critical step in clearing the way for the transcription initiation complex. Derepressed chromatin is characterized by acetylated histones.

PAX6. The Pax6 transcription factor, which is needed for mammalian eye, nervous system, and pancreas development, contains two potential DNA-binding domains. Its major DNA-binding domain resides at its amino-terminal end; the

amino acids of this domain interact with a specific 20–26-base-pair sequence of DNA (Figure 5.11; Xu et al. 1995). Such Pax6-binding sequences have been found in the enhancers of vertebrate lens crystallin genes and in the genes expressed in the endocrine cells of the pancreas (those that release insulin, glucagon, and somatostatin) (Cvekl and Piatigorsky 1996; Andersen 1999). When Pax6 binds to a gene's enhancer or promoter, it can either activate or repress the gene. The *trans*-activating domain of Pax6 is rich in proline, threonine, and serine. Mutations in this region cause severe nervous system, pancreatic, and optic abnormalities in humans (Glaser et al. 1994).

The use of Pax6 by different organs demonstrates the combinatorial manner by which transcripts factors work. Figure 5.12 shows two gene regulatory regions that bind Pax6. The first is that of the chick δ1 lens crystallin gene (Cvekl and Piati-gorsky, 1996; Muta et al. 2002). This gene has a promoter containing a site for TBP binding and an upstream promoter element that binds Sp1 (a general transcriptional activator found in all cells). The gene also has an enhancer in its third intron that controls the time and place of crystallin gene expression. This enhancer has two Pax6-binding sites. The crystallin gene will not be expressed unless Pax6 is present in the nucleus and bound to these enhancer sites. As mentioned in Chapter 4, Pax6 is present during early development in the central nervous system and head surface ectoderm of the chick. Moreover, this enhancer has a binding site for another transcription factor, the Sox2 protein. Sox2 is not usually found in the outer ectoderm, but it appears in those outer ectodermal cells that will become lens by virtue of their being induced by the optic vesicle evaginating from the brain (Kamachi et al. 1998). Thus, only those cells that contain both Sox2 and Pax6 can express the lens crystallin gene. In addition, a *third* site on the enhancer can bind either an activator (the δEF3 protein) or a repressor (the δEF1 protein) of transcription. It is thought that the repressor may be critical in preventing crystallin expression in the nervous system. Thus, enhancers function in a combinatorial manner, wherein several transcription factors work together to promote or inhibit transcription.

Another set of regulatory regions that use Pax6 are the enhancers regulating the transcription of the insulin, glucagon, and somatostatin genes in the pancreas (Figure 5.12B). Here, Pax6 is essential for gene expression, and it works in cooperation with other transcription factors such as Pdx1 (specific for the pancreatic region of the endoderm) and Pbx1 (Andersen et al. 1999; Hussain and Habener 1999). In the absence of Pax6 (as in the homozygous small eye mutation in mice and rats), the endocrine cells of the pancreas do not develop properly, and the production of hormones by those cells is deficient (Sander et al. 1997). One can see that the genes for specific proteins use numerous transcription factors in various combinations. Thus, enhancers are modular (such that the *Pax6* gene is expressed in the eye, pancreas, and nervous system, as shown in Figure 5.7); but within these modules, transcription factors work in a combinatorial fashion (such that in some cells Pax6 helps activate the crystalline genes but in other cells Pax6 helps transcribe to insulin genes). In this way, transcription factors can regulate the timing and place of gene expression.

There are other genes that are activated by Pax6 binding, and one of them is the *Pax6* gene itself. Pax6 protein can bind to the *Pax6* gene promoter (Plaza et al. 1993). This means that once the *Pax6* gene is turned on, it will be continue to be expressed, even if the signal that originally activated it is no longer given.

Within the past decade, our knowledge of transcription factors has progressed enormously, giving us a new, dynamic view of gene expression. The gene itself is no longer seen as an independent entity controlling the synthesis of proteins. Rather, the gene both directs and is directed by protein synthesis. Natalie Angier (1992) has written, "A series of new discoveries suggests that DNA is more like a certain type of politician, surrounded by a flock of protein handlers and advisers that must vigorously massage it, twist it and, on occasion, reinvent it before the grand blueprint of the body can make any sense at all."

Transcription factor cascades

We now know that the tissue-specific expression of a particular gene is the result of the presence of a particular constellation of transcription factors in the cell nucleus. It is important to see that their combinatorial mode of operation means that none of these transcription factors must be tissue-specific.

But how do the transcription factors themselves come to be expressed in a tissue-specific manner? In many cases, the genes for transcription factors are activated by other transcription factors. For instance, the gene for Pax6, which is a transcription factor, is regulated (in the anterior head) by the Mbx transcription factor that becomes expressed at the end of gastrulation (Kawahara et al. 2002). In other cases (as we will see in the case of MITF), the transcription factor is present, but must be activated by signals provided by induction. This will be discussed more fully in the next chapter.

Wilhelm Roux (1894) described this situation eloquently in his manifesto for experimental embryology when he stated

Figure 5.12
Modular transcriptional regulatory regions using Pax6 as an activator. (A) Promoter and enhancer of the chick δ1 lens crystallin gene. Pax6 interacts with two other transcription factors, Sox2 and Maf, to activate this gene. (B) Promoter and enhancer of the rat somatostatin gene. Pax6 activates this gene by cooperating with the Pdx1 transcription factor. (A after Cvekl and Piatigorsky 1996; B after Andersen et al. 1999.)

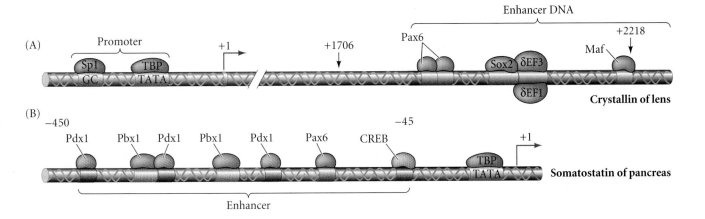

Studying DNA Regulatory Elements

Identifying DNA regulatory elements.
How do we know that a particular DNA fragment binds a particular transcription factor? One of the simplest ways is to perform a **gel mobility shift assay**. The basis for this assay is gel electrophoresis. Recall from Chapter 4 that fragments of DNA can be placed in a depression at one end of a gel and an electric current run through the gel. The fragments will move toward the negative pole, and the distance each fragment travels in a given time will depend upon its mass and conformation. Larger fragments will run more slowly than smaller fragments. If the fragments of interest are incubated in a solution of Pax6 protein before being placed in the gel, one of two things will happen. If the Pax6 does not recognize any sequence in a DNA fragment, the DNA will not be bound and the fragment will migrate through the gel as it normally would. Alternatively, if the Pax6 protein does recognize a sequence in the DNA, it will bind to it, increasing the mass of the fragment and cause it to run more slowly through the gel. Figure 5.13A shows the use of such a gel to locate the binding site of Pax6 protein.

The results of this procedure are often confirmed by a **DNase protection assay**. If a DNA-binding protein such as Pax6 finds its target sequence in a DNA fragment, it will bind to it. If the fragment is then placed in a solution of DNase I (an enzyme that cleaves DNA randomly), the bound transcription factor will protect that specific region of DNA from being cleaved. Figure 5.13B shows the results of one such assay for Pax6 binding.

Reporter genes such as the *lacZ* gene mentioned above are also used to determine the functions of various DNA fragments. If a sequence of DNA is thought to contain an enhancer, it can be fused to a reporter gene and injected into an egg by various means. If the tested sequence is able to direct the expression of the reporter gene in the appropriate tissues, it is assumed to contain an enhancer.

One of the interesting concepts emerging from this work is that the transcription factors involved in the mechanisms of cell differentiation in a given region are often used to specify that region as a certain type of tissue. Thus, the Pax6 transcription factor is used in the specification of the eye-forming region as well as in the differentiation of the retina, and the Pdx transcription factor is used both in the specification of the pancreatic rudiment and in the subsequent activation of the genes for insulin and somatostatin derived from that tissue (see Hui and Perfetti 2002).

Conditional knockouts: Floxed mice

One critically important experimental use of enhancers has been the conditionally elimination of gene expression in certain cell types. For example, the transcription factor Hnf4α is expressed in liver cells. It looks like a hormone-binding protein, and it might be important in the transcription of certain liver proteins. But if this gene is deleted from mouse embryos, the embryos die before they even form a liver. This is because Hnf4α is critical in forming the visceral endoderm of the yolk sac; if this tissue fails to form properly, the animal dies very early in development. So one needs to create a mutation that will be *conditional*—that is, a mutation that will appear only in the liver and nowhere else. How can this be done?

The **Cre-Lox** technique uses homologous recombination (see Chapter 4) to place two Cre-recombinase recognition sites (*loxP* sequences) within the gene of interest, usually flanking important exons (see Kwan 2002). Such a gene is said to be "floxed" ("*loxP*-flanked"). For example, using cultured mouse ES cells, Parvis and colleagues (2002) placed two *loxP* sequences around the second exon of the mouse *Hnf4α* gene (Figure 5.14). These ES cells were then used to generate mice that had this floxed allele. A second strain of mice was generated that had a gene encoding bacteriophage Cre-recombinase (the enzyme that recognizes the *loxP* sequence) attached to the promoter of an albumin gene that is expressed very early in liver development. Thus, during mouse development, Cre-recombinase would be made only in the liver cells. When the two strains of mice were crossed, some of their offspring carried both additions. In these double-marked mice, Cre-recombinase bound to its recognition sites—the *loxP* sequences—flanking the second exon of the *Hnf4α* genes. It then acted as a recombinase and deleted this second exon. The resulting DNA would encode a nonfunctional protein, since the second exon has a critical function in *HNF4α*. Thus, the *HNF4α* gene was "knocked out" only in liver cells.

WEBSITE 5.6 Other recombinase methods. The Cre-Lox system is patented, and each researcher using it is expected to pay the owner of the patent. This has caused much legal concern over whether such procedures should be private property. It has also sparked the search for other ways of making conditional mutants.

Figure 5.13
Procedures for determining the DNA-binding sites of transcription factors. (A) Gel mobility shift assay. A DNA fragment containing the Pax6-binding site changes its mobility in the gel when Pax6 protein binds to it. Lanes 1 and 3 show the positions of a DNA fragment containing a Pax6-binding site. When Pax6 is added to the fragment, it moves more slowly. Lanes 2 and 4 show the positions of a similar-sized fragment that does not bind Pax6. (B) A DNase protection assay of the intron between the third and fourth exons of the chick δ1 lens crystallin gene. DNA with and without Pax6 is labeled at one end and is subjected to varying concentrations of DNase I, which randomly cleaves the DNA. The resulting fragments are run on an electrophoretic gel and autoradiographed. No cleavage is seen in the region where a bound protein (Pax6) has prevented DNase from binding. There are two sites (heavy bars) where Pax6 binding is able to protect the DNA from digestion with DNase; these sites are the enhancers. (A after Beimesche et al. 1999; B from Cvekl et al. 1995.)

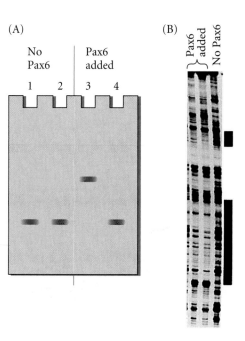

In most cells: No recombination

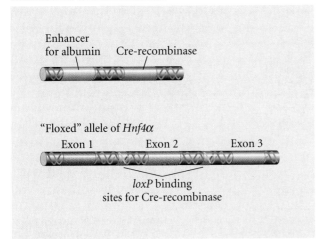

In liver cells only (expressing albumin)

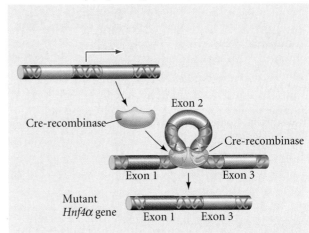

Figure 5.14
The Cre-lox technique for conditional mutagenesis, by which gene mutations can be generated in specific cells only. Mice are made wherein wild-type alleles (in this case, the genes encoding the Hnf4α transcription factor) have been replaced by alleles in which the second exon is flanked by *loxP* sequences. These mice are mated with mice having the gene for Cre-recombinase transferred onto a promoter that is active only in particular cells. In this case, the promoter is that of an albumin gene that functions early in liver development. In mice with both these altered alleles, Cre-recombinase is made only in the cells where that promoter was activated (i.e., in these cells synthesizing albumin). The Cre-recombinase binds to the *loxP* sequences flanking the exons and removes those exons. Thus, in the case depicted here, only the developing liver cells lack a functional *Hnf4α* gene.

Enhancer traps: Being at the right place at the right time

The ability of an enhancer from one gene to activate other genes has been used by scientists to find new enhancers and the genes regulated by them. To do this, one makes an **enhancer trap**, consisting of a reporter gene (such as the *E. coli lacZ* gene jellyfish GFP gene) fused to a relatively weak promoter. The weak promoter will not initiate the transcription of the reporter gene without the help of an enhancer. This recombinant enhancer trap is then introduced into an egg or oocyte, where it integrates randomly into the genome. If the reporter gene is expressed, it means that the reporter has come within the domain of an active enhancer (Figure 5.15). By isolating this activated region of the genome in wild-type flies or mice, the normal gene activated by the enhancer can be discovered (O'Kane and Gehring 1987).

Activating genes in all the wrong places: GAL4 activation

One of the most powerful uses of this genetic technology has been to *activate regulatory genes such as Pax6 in new places.* Using *Drosophila* embryos, Halder and his colleagues (1995) in Walter Gehring's laboratory placed a gene encoding the yeast GAL4 transcriptional activator protein downstream from an enhancer that was known to function in the labial imaginal discs (those parts of the *Drosophila* larva that become the adult mouth parts). In other words, the gene for the GAL4 tran-

(A)

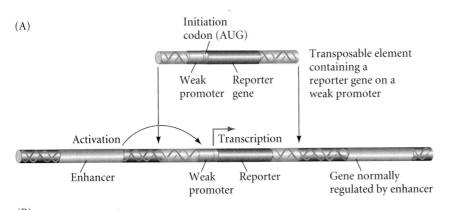

(B)

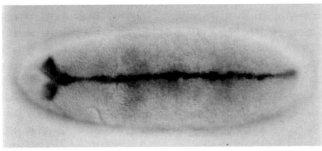

Figure 5.15
The enhancer trap technique. (A) A reporter gene is fused to a weak promoter that cannot direct transcription on its own. This recombinant gene is injected into the nucleus of an egg and integrates randomly into the genome. If it integrates near an enhancer, the reporter gene will be expressed when that enhancer is activated, showing the normal expression pattern of a gene normally associated with that enhancer. (B) Reporter gene expression (dark region) in a *Drosophila* embryo injected with an enhancer trap. This expression pattern demonstrated the presence of an enhancer that is active in the development of the insect nervous system and which was unrecognized before the procedure. (Photograph courtesy of Y. Hiromi.)

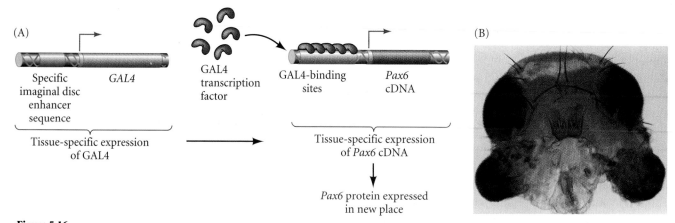

Figure 5.16
Targeted expression of the *Pax6* gene in a *Drosophila* non-eye imaginal disc. (A) A strain of *Drosophila* was constructed wherein the gene for the yeast GAL4 transcription factor was placed downstream from an enhancer sequence that normally stimulates gene expression in the imaginal discs for mouthparts. If the embryo also contains a transgene that places GAL4-binding sites upstream of the *Pax6* DNA gene, the *Pax6* gene will be expressed in whichever imaginal disc the GAL4 protein is made. (B) *Drosophila* ommatidia (compound eye) emerging from the mouthparts of a fruit fly in which the *Pax6* gene was expressed in the labial (jaw) discs. (Photograph courtesy of W. Gehring and G. Halder.)

scription factor was placed next to an enhancer for genes normally expressed in the developing jaw. Therefore, GAL4 should also be expressed in jaw tissue. Halder and his colleagues then constructed a second transgenic fly, placing the cDNA for the *Drosophila Pax6* gene downstream from a sequence composed of five GAL4-binding sites. The GAL4 protein should be made only in a particular group of cells destined to become the jaw, and when that protein was made, it should cause the transcription of the *Pax6* gene in those particular cells (Figure 5.16A). In flies in which the *Pax6* gene was expressed in the incipient jaw cells, part of the jaw gave rise to eyes (Figure 5.16B). Pax6 in *Drosophila* and frogs (but not in mice) is able to turn several developing tissue types into eyes (Chou et al. 1999). It appears that in *Drosophila*, Pax6 not only activates those genes that are necessary for the construction of eyes, but also represses those genes that are used to construct other organs.

WEBSITE 5.7 Enhancers and cancers. Enhancer trapping may occur accidentally during development and place a new gene next to a particular regulatory element. When growth factor genes are placed next to the genes involved in making antibodies, the results are leukemias.

that the causal analysis of development may be the greatest problem the human intellect has attempted to solve, "since every new cause ascertained only gives rise to fresh questions concerning the cause of this cause."

Silencers

Silencers are DNA regulatory elements that actively repress the transcription of a particular gene. They can be viewed as "negative enhancers." For instance, in the mouse, there is a sequence that prevents a promoter's activation in any tissue except neurons. This sequence, given the name **neural restrictive silencer element** (**NRSE**), has been found in several mouse genes whose expression is limited to the nervous system: those encoding synapsin I, sodium channel type II, brain-derived neurotrophic factor, Ng-CAM, and L1. The protein that binds to the NRSE is a zinc finger transcription factor called neural restrictive silencer factor (NRSF). NRSF appears to be expressed in every cell that is *not* a mature neuron (Chong et al. 1995; Schoenherr and Anderson 1995).

To test the hypothesis that NRSE is necessary in the normal repression of neural genes in non-neural cells, transgenes were made by fusing a *lacZ* gene with part of the *L1* neural cell adhesion gene. (L1 is a protein whose function is critical for brain development, as we will see in later chapters.) In one case, the *L1* gene, from its promoter through the fourth exon, was fused to the *lacZ* sequence. A second transgene was made just like the first, except that the NRSE had been deleted from the L1 promoter. The two transgenes were separately inserted into the pronuclei of fertilized oocytes, and the resulting transgenic mice were analyzed for the expression of β-galactosidase (Kallunki et al. 1995, 1997). In the embryos receiving the complete transgene (which included the NSRE), expression was seen only in the nervous system (Figure 5.17A). However, in those mice whose transgene lacked the NRSE, expression was seen in the heart, limb mesenchyme and limb ectoderm, kidney mesoderm, ventral body wall, and cephalic mesenchyme (Figure 5.17B).

The repressive activity of NRSF works by the binding of DNA-bound NRSF to a **histone deacetylase**. Just as histone acetyltransferases release nucleosomes, histone deacetylases stabilize them. This prevents the promoter from being recognized (Roopra et al. 2000; Jepson et al. 2000).

(A)

L1
promoter · · · · · NRSE

lacZ

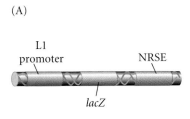

(B)

L1
promoter · · · · · No NRSE

lacZ

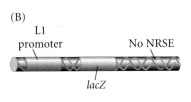

Figure 5.17
Silencers. Analysis of β-galactosidase staining patterns in 11.5-day embryonic mice. (A) Embryo containing a transgene composed of the *L1* promoter, a portion of the *L1* gene, and a *lacZ* gene fused to the second exon (which contains the NRSE region). (B) Embryo containing a similar transgene, but lacking the NRSE sequence. The dark areas reveal the presence of β-galactosidase (the *lacZ* product). (From Kallunki et al. 1997.)

WEBSITE 5.8 Further mechanisms of transcriptional regulation. These three websites cover (1) DNase I hypersensitivity, (2) locus control regions and the mechanisms by which LCRs may regulate the temporal expression of linked genes, and (3) the association of active genes with the nuclear matrix.

Figure 5.18
Methylation of globin genes in human embryonic blood cells. (A) Structure of 5-methylcytosine (B) The activity of the human β-globin genes correlates inversely with the methylation of their promoters. (After Mavilio et al. 1983.)

*Methylation P[...]
the Control of [...]*

DNA methylation[...]

How does a pattern [...] can a lens cell conti[...] muscle-specific gene[...] sis and still maintai[...] answer appears to b[...] active genes becom[...] and the resulting m[...] prevents transcripti[...]

It is often assu[...] nucleotides whethe[...] gene in a red blood [...] a β-globin gene in a fibroblast or retinal cell of the same animal. There is, however, a subtle difference in the DNA. In 1948, R. D. Hotchkiss discovered a "fifth base" in DNA, 5-methylcytosine. In vertebrates, this base is made enzymatically after DNA is replicated. At this time, about 5% of the cytosines in mammalian DNA are converted to 5-methylcytosine (Figure 5.18A). This conversion can occur only when the cytosine residue is followed by a guanosine. Recent studies have shown that the degree to which the cytosines of a gene are methylated can control the level of the gene's transcription. Cytosine methylation appears to be a major mechanism of transcriptional regulation in vertebrates; however, *Drosophila*, nematodes, and perhaps most invertebrates do not methylate their DNA.

In vertebrates, the presence of methylated cytosines in the promoter of a gene correlates with the repression of transcription from that gene. In developing human and chick red blood cells, the DNA of the globin gene promoters is almost completely unmethylated, whereas the same promoters are highly methylated in cells that do not produce globins. Moreover, the methylation pattern changes during development (Figure 5.18B). The cells that produce hemoglobin in the

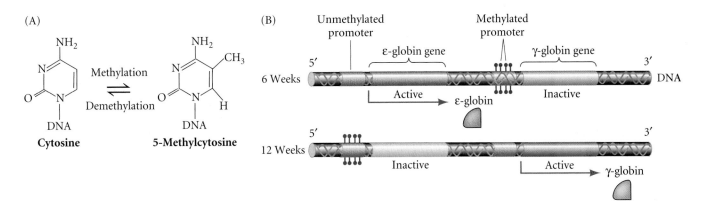

human embryo have unmethylated promoters in the genes encoding the ε-globins ("embryonic globin chains") of embryonic hemoglobin. These promoters become methylated in the fetal tissue, as the genes for fetal-specific γ-globin (rather than the embryonic chains) become activated (van der Ploeg and Flavell 1980; Groudine and Weintraub 1981; Mavilio et al. 1983). Similarly, when fetal globin gives way to adult (β) globin, the γ-globin gene promoters of the fetal globin genes become methylated.

The correlation between methylated cytosines and transcriptional repression has been confirmed experimentally. By adding transgenes with different methylation patterns to cells, Busslinger and co-workers (1983) showed that methylation in the promoter or enhancer of a gene correlates extremely well with the repression of gene transcription. In vertebrate development, the absence of DNA methylation correlates well with the tissue-specific expression of many genes.

Chromatin modification

How are some genes methylated, and how does methylation prevent gene expression? One proposal is that there is a reciprocal interaction between the histones in chromatin and the DNA that surrounds them. There seems to be competition for modification at the lysine residue on the ninth position of histone H3. If this site is acetylated, the histone is destabilized and can cause the nucleosome to disperse. If this site is not acetylated, it can be methylated. Methylation of this histone increases the nucleosome's stability, preventing its dissociation or movement. So the modification of the histone H3 tail might act as a "switch" between activated (dispersed nucleosomes) and inactivated (stable nucleosomes) states of a gene. The methylated histone may be able to recruit to its vicinity the enzymes that methylate the DNA (Rea et al. 2000; Tamaru and Selker 2001). Once the DNA is methylated, it may stabilize the nucleosomes even further.

A protein called MeCP2 selectively binds to methylated regions of DNA; it also binds to histone deacetylases. Thus, when MeCP2 binds to methylated DNA it can stabilize the nucleosomes in that particular region of chromatin (Keshet et al. 1986; Jones et al. 1998; Nan et al. 1998). Methylated DNA may also be preferentially bound by histone H1, the histone that associates nucleosomes into higher-order folded complexes (McArthur and Thomas 1996). In this way, an inactive state can be propagated along the DNA. Transcription factors may act to prevent higher-order folding or to release the nucleosomes by demethylating and acetylating the histones, thereby allowing the DNA to become unmethylated and the promoter to become accessible.

The methyl groups that distinguish active from the inactive genes are placed on the DNA during the differentiation of the cell types. The totipotent cells of the inner cell mass, for instance, do not have methyl groups on cell type-specific genes such as those for the globins. Rather, the DNA of these early cells lacks the tissue-specific methylation patterns that characterize the differentiated cells. (There are, however, other, rarer methylation patterns that they do carry; see Sidelights & Speculations.) When mammals are cloned (see Chapter 4), we are actually asking the egg cytoplasm to erase or ignore the methylation patterns that have been established in a differentiated

Figure 5.19

Insulators. (A) Insulators flanking the chick β-globin genes prevent the β-globin enhancers from activating the folate receptor gene or the odorant receptor gene. (B) In developing red blood cells, the folate receptor gene is found in highly condensed chromatin, wherein the histone H3 tails are methylated. The β-globin insulator is thought to block the extension of histone methylation by recruiting histone acetyltransferases, which prevents histone methylation by the neighboring methyltransferases. (C) The BEAF32 protein binds to hundreds of insulator sites in the *Drosophila* genome. The DNA is stained red with propidium iodide. The antibody to BEAF32 stains green, and the overlap of signals makes the bound BEAF32 antibody appear yellow. (B after Litt et al. 2001; C, photograph courtesy of U. K. Laemmli.)

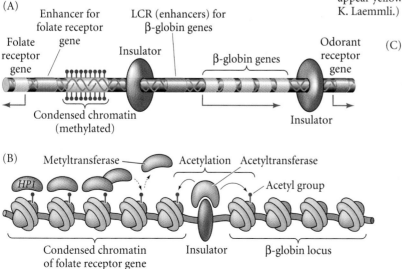

(A)

Enhancer for folate receptor gene
LCR (enhancers) for β-globin genes
Folate receptor gene
Insulator
β-globin genes
Odorant receptor gene
Condensed chromatin (methylated)
Insulator

(B)

Metyltransferase
Acetylation   Acetyltransferase
HP1
Acetyl group
Condensed chromatin of folate receptor gene
Insulator
β-globin locus

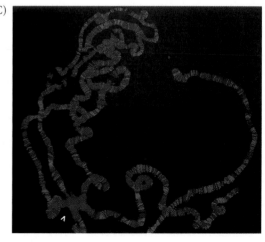

(C)

cell's nucleus. This is probably why cloning works less than 5% of the time, even though the genes themselves should be normal, and why those cloned animals that do survive often have developmental anomalies (Rideout et al. 2001).

The normal DNA methylation patterns of differentiated cells are maintained throughout cell division by the enzyme DNA (cytosine-5)-methyltransferase. During DNA replication, one strand (the template strand) retains the methylation pattern, while the newly synthesized strand does not. The enzyme DNA (cytosine-5)-methyltransferase has a strong preference for DNA that has one methylated strand, and when it sees a methyl-CpG on one side of the DNA, it methylates the new C on the other side (Gruenbaum et al. 1982; Bestor and Ingram 1983).

Insulators

The effects of enhancers must be told where to stop. Since enhancers can work at relatively long distances, it is possible for them to activate several nearby promoters. In order to stop the spreading of the enhancer's power, there are **insulator** se-

quences in the DNA (Figure 5.19; Zhao et al. 1995; Bell et al. 2001). Insulators bind proteins that prevent the enhancer from activating an adjacent promoter; they are often located between the enhancer and the promoter. For instance, in the chick, the genes of the β-globin family are flanked by insulators on both sides. On one side is an insulator that prevents the globin gene enhancers from activating odorant receptor genes (which are active in the nasal neurons), and on the other side is an insulator preventing the globin gene enhancers from activating the promoter of the folate receptor gene.

These insulators also prevent the condensed chromatin of the neighboring loci from repressing the actively transcribing globin genes. In the precursors of the red blood cells, the globin genes are active, but the adjacent folate receptor gene is in a highly condensed state. Litt and colleagues (2001) have shown that the globin genes are in an active configuration, and that the nucleosomes around the enhancer are highly acetylated. The folate receptor genes, however, are tightly packaged in condensed nucleosomes, with methyl groups on their histone H3 tails. These nucleosomes are thought to re-

Sidelights & Speculations

Genomic Imprinting

It is usually assumed that the genes one inherits from one's father and the genes one inherits from one's mother are equivalent. In fact, the basis for Mendelian ratios (and the Punnett square analyses used to teach them) is that it does not matter whether the genes came from the sperm or from the egg. But in mammals, there are at least 30 known genes for which it *does* matter. In these cases, only the sperm-derived or only the egg-derived allele of the gene is expressed. This means that a severe or lethal condition arises if a mutant allele is derived from one parent, but the same mutant allele will have no deleterious effects if inherited from the other parent. It also means (as will be shown in Chapter 7) that both maternal and paternal chromosomes are required for normal mammalian development.

For instance, in mice, the gene for insulin-like growth factor II (*Igf2*) on chromosome 7 is active in early embryos only on the chromosome transmitted by the father. Conversely, the gene for the protein that binds this growth factor, *Igf2r*, is located on chromosome 17 and is active only in the chromosome transmitted by the mother (Barlow et al. 1991; DeChiara et al. 1991; Bartolomei and Tilghman 1997). The Igf2r protein binds and degrades excess Igf2. A mouse pup that inherits a deletion of the *Igf2r* gene from its father is normal, but if

the same deletion is inherited from the mother, the fetus experiences a 30% increase in growth and dies late in gestation.*

In humans, the loss of a particular segment of the long arm of chromosome 15 results in different phenotypes, depending on whether the loss is in the male- or the female-derived chromosome (Figure 5.20). If the chromosome with the defective or missing segment comes from the father, the child is born with Prader-Willi syndrome, a disease associated with mild mental retardation, obesity, small gonads, and short stature. If the defective or missing segment comes from the mother, the child has Angelman syndrome, characterized by severe mental retardation, seizures, lack of speech, and inappropriate laughter (Knoll et al. 1989; Nicholls et al. 1998).

In such cases, the differences between the expressing and the nonexpressing alleles usually involve the methylation of cytosine residues (Arney et al. 2001). In the primordial germ cells that give rise to the sperm and egg, all methylation differences are wiped out; the DNA is almost entirely

---

*The increase in growth is caused by an excess of Igf2. The lethality is probably due to lysosomal defects, since the Igf2 receptor also targets lysosomal enzymes into that organelle. In humans, misregulation of Igf2 methylation causes Beckwith-Wiedemann growth syndrome.

Figure 5.20
Inheritance patterns for Prader-Willi and Angelman syndrome. A region in the long arm of chromosome 15 contains the genes whose absence causes these syndromes. However, they are imprinted in reverse fashion. In Prader-Willi syndrome, the paternal genes are active, while in Angelman syndrome, the maternal genes are active.

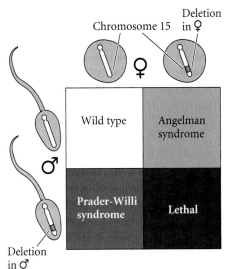

unmethylated (Monk et al. 1987; Driscoll and Migeon 1990). However, as the germ cells develop into sperm or eggs, their genes undergo extensive methylation. Moreover, the pattern of methylation on a given gene can differ between egg and sperm. These gene-specific methylation differences can be seen in the chromosomes of embryonic cells (Sanford et al. 1987; Chaillet et al. 1991; Kafri et al. 1992). The methylation of an enhancer region near the mouse *Igf2* gene determines whether or not that gene will be expressed, and this enhancer is methylated differently in the sperm and the egg. The expression of the maternal or paternal loci on human chromosome 15 also depends on the methylation differences at specific regions in this chromosome (Zeschingk et al. 1997; Ferguson-Smith and Surani 2001). This marking of a gene as coming either from the father or the mother is called **genomic imprinting**.

Genomic imprinting adds information to the inherited genome, information that may regulate spatial and temporal patterns of gene activity. It also provides a reminder that the organism cannot be explained solely by its genes. One needs knowledge of developmental parameters as well as genetic ones.

WEBSITE 5.9 **Imprinting in humans and mice.** The mechanisms of imprinting may be very different. In humans, methylation patterns affect the transmission of certain diseases, and unmethylating β-globin genes may be a cure for certain blood diseases.

---

cruit methyltransferases that would extend the methylation to more nucleosomes. However, the insulator appears to recruit acetyltransferases that acetylate the histone tails and prevent the inactivation of the chromatin (Figure 5.19B).

Transcriptional regulation of an entire chromosome: Dosage compensation

In *Drosophila* and mammals, females are characterized as having two X chromosomes per cell, while males are characterized as having a single X chromosome per cell. Unlike the Y chromosome, the X chromosome contains thousands of genes that are essential for cell activity. Yet, despite the female's cells having double the number of X chromosomes the male's have, male and female cells contain approximately equal amounts of X chromosome-encoded gene products. This equalization is called **dosage compensation**. In *Drosophila*, the transcription rates of the X chromosomes are altered so that male and female cells transcribe the same amount of RNAs from their X chromosomes. Both X chromosomes in the female are active, but there is increased transcription from the male's X chromosome, so that the single X chromosome of male cells produces as much RNA as the two X chromosomes in female cells (Lucchesi and Manning 1987). This is accomplished by acetylation of the nucleosomes throughout the male's X chromosomes, which gives RNA polymerase more efficient access to that chromosome's promoters (Akhtar et al. 2000; Smith et al. 2001).

In mammals, dosage compensation occurs through the inactivation of one X chromosome in each female cell. Thus, each mammalian somatic cell, whether male or female, has only one functioning X chromosome. This phenomenon is called **X chromosome inactivation**. The chromatin of the inactive X chromosome is converted into **heterochromatin**—chromatin that remains condensed throughout most of the cell cycle and replicates later than most of the other chromatin (the **euchromatin**) of the nucleus. The heterochromatic (inactive) X chromosome, which can often be seen on the nuclear envelope of female cells, is referred to as a **Barr body** (Figure 5.21A,B; Barr and Bertram 1949). By monitoring the expression of X-linked genes whose products can be detected in the early embryo, it was shown that X chromosome inactivation occurs early in development (Figure 5.21C,D). This inactivation appears to be critical. Using a mutant X chromosome that could not be inactivated, Tagaki and Abe (1990) showed that the expression of two X chromosomes per cell in mouse embryos leads to ectodermal cell death and the absence of mesoderm formation, eventually causing embryonic death at day 10 of gestation.

The early inactivation of one X chromosome per cell has important phenotypic consequences. One of the earliest analyses of X chromosome inactivation was performed by Mary Lyon (1961), who observed coat color patterns in mice. If a mouse is heterozygous for an autosomal gene controlling hair pigmentation, then it resembles one of its two parents, or has a color intermediate between the two. In either case, the mouse is a single color. But if a female mouse is heterozygous for a pigmentation gene on the X chromosome, a different result is seen: patches of one parental color alternate with patches of the other parental color (Figure 5.22). Lyon proposed the following hypothesis to account for these results:

1. Very early in the development of female mammals, both X chromosomes are active.

2. As development proceeds, one X chromosome is inactivated in each cell.

3. This inactivation is random. In some cells, the paternally derived X chromosome is inactivated; in other cells, the maternally derived X chromosome is shut down.

4. This process is irreversible. Once an X chromosome has been inactivated in a cell, the same X chromosome is inactivated in all that cell's progeny. Since X inactivation happens relatively early in development, an entire region of cells derived from a single cell may all have the same X chromosome inactivated. Thus, all tissues in female mammals are mosaics of two cell types.

The Lyon hypothesis of X chromosome inactivation provides an excellent account of differential gene inactivation at the level of transcription. Some interesting exceptions to the general rules further show its importance. First, X chromosome inactivation holds true only for somatic cells, not germ cells. In female germ cells, the inactive X chromosome is reactivated

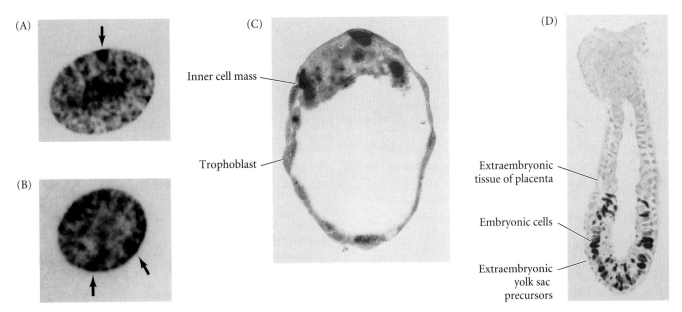

Figure 5.21
Inactivation of a single X chromosome in mammalian XX cells. (A, B) Barr bodies in the nuclei of human oral epithelial cells stained with Cresyl violet. (A) Cell from a normal XX female, showing a single Barr body (arrow). (B) Cell from a female with three X chromosomes. Two Barr bodies can be seen, and only one X chromosome per cell is active. (C–D) X-chromosome inactivation in the early mouse embryo. The paternally derived X chromosome contained a *lacZ* transgene that is active in the early embryo. Those cells in which the chromosome is active make β-galactosidease and can be stained blue. The other cells are counterstained and appear pink. (C) In the early blastocyst stage at day 4, both X chromosomes are active in all cells; thus all cells appear blue. (D) At day 6, random inactivation of one of the chromosomes occurs. Thus embryonic cells where the maternal X is active appear pink, while those where the paternal X is active stain blue. In the mouse (but not the human) trophoblast, the paternally derived X chromosome in preferentially inactivated. Thus, the trophoblast cells are uniformly pink. (A, B, photographs courtesy of M. L. Barr; C, D after Sugimoto et al. 2000, photographs courtesy of N. Takagi.)

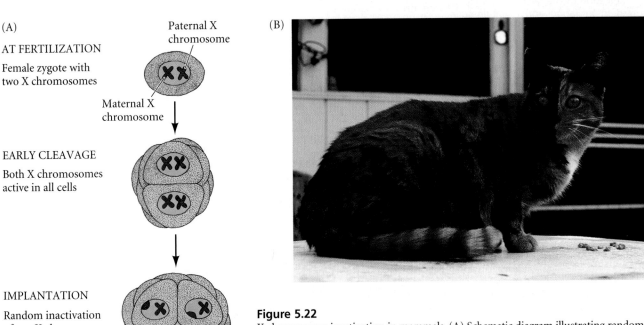

Figure 5.22
X chromosome inactivation in mammals. (A) Schematic diagram illustrating random X chromosome inactivation. The inactivation is believed to occur at about the time of implantation. (B) A calico cat, whose cells have one X chromosome containing an orange allele for a pigmentation gene, and the other X chromosome bearing a black allele for that pigmentation gene. The regions of different color correspond to the inactivation of one or the other X chromosome. (Photograph courtesy of R. Loredo and G. Loredo.)

shortly before the cells enter meiosis (Gartler et al. 1973; Migeon and Jelalian 1977). Thus, in early oocytes, both X chromosomes are unmethylated (and active). In each generation, X chromosome inactivation has to be established anew.

Second, there are some exceptions to the rule of randomness in the inactivation pattern. For instance, the first X chromosome inactivation in the mouse is seen in the fetal portion of the placenta, where the paternally derived X chromosome is specifically inactivated (Tagaki 1974; see Figure 5.21C,D). Third, X chromosome inactivation does not extend to every gene on the human X chromosome. There are a few genes (such as that encoding steroid sulfatase) on both arms of the X chromosome that "escape" X inactivation (Brown et al. 1997).

The fourth exception really ends up proving the rule. There are a few male mammals with coat color patterns we would not expect to find unless the animals exhibited X chromosome inactivation. Male calico and tortoiseshell cats are among these examples. These orange and black coat patterns are normally seen in females and are thought to result from random X chromosome inactivation* (see Figure 5.22B). But rare males exhibit these coat patterns as well. How can this be? It turns out that these cats are XXY. The Y chromosome makes them male, but one X chromosome undergoes inactivation, just as in females, so there is only one active X per cell (Centerwall and Benirschke 1973). Thus, these cats undergo random X chromosome inactivation and their cells have a Barr body.

It is clear, then, that one mechanism for transcriptional control of gene expression is to make a large number of genes heterochromatic and thus transcriptionally inert.

WEBSITE 5.10 **Chromosome elimination and diminution.** The inactivation or the elimination of entire chromosomes is not uncommon among invertebrates and is sometimes used as a mechanism of sex determination.

---

*Although the terms *calico* and *tortoiseshell* are sometimes used synonymously, calico cats usually have white patches (i.e., patches with no pigment) as well.

Sidelights & Speculations

The Mechanisms of X Chromosome Inactivation

The mechanisms of X chromosome inactivation are still poorly understood, but new research is giving us some indication of the factors that may be involved in initiating and maintaining a heterochromatic X chromosome.

Initiation of X Chromosome Inactivation: Xist RNA

In 1991, Brown and her colleagues found an RNA transcript that was made solely from the *inactive* X chromosome of humans (Brown 1991a,b). This transcript, *XIST*, does not encode a protein. Rather, it stays within the nucleus and interacts with the inactive X chromatin, forming an *XIST*-Barr body complex (Brown et al. 1992). A similar situation exists in the mouse, in which the transcript of the *Xist* gene is seen to coat the inactive X chromosome (Figure 5.23; Borsani et al. 1991; Brockdorrf et al. 1992).

The *Xist* gene is an excellent candidate for the initiator of X inactivation. First, transcripts from the *Xist* gene are seen in mouse embryos prior to X chromosome inactivation, which would be expected if this gene plays a role in initiating inactivation

(A)

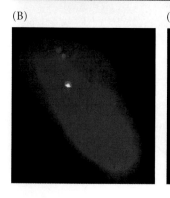

(B) (C) 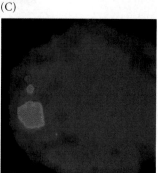 (D)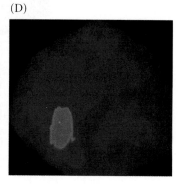

Figure 5.23
Xist RNA associates with the inactive X chromosome that made it. (A) Mouse *Xist* RNA (red) on an inactive X chromosome (blue) in metaphase. (In humans, *XIST* RNA is not seen to coat metaphase X chromosomes.) (B–D) XX murine embryonic stem cells undergoing differentiation and X chromosome inactivation. *Xist* RNA is stained light blue; mRNA from the X-linked phosphoglycerate kinase (*Pgk*) gene is stained red; the background DNA is stained blue. (B) XX cell before differentiating shows both X chromosomes transcribing *Pgk* and *Xist*. (C) As X chromosome inactivation begins, the *Xist* RNA is stabilized on one of the chromosomes. This chromosome no longer transcribes the *Pgk* gene. (D) As the cell finishes differentiating, X chromosome inactivation is complete. One X chromosome continues to make Xist protein. The other chromosome no longer makes Xist, but continues to transcribe *Pgk*. (A courtesy of R. Jaenisch; B–D from Sheardown et al. 1997, photograph courtesy of N. Brockdorff.)

(Kay et al. 1993). Second, knocking out one *Xist* locus in an XX cell prevents X inactivation from occurring on the targeted chromosome (Penny et al. 1996). Third, the transfer of a 450-kilobase segment containing the mouse *Xist* gene into an autosome of a male embryonic stem cell causes the random inactivation of either that autosome or the endogenous X chromosome (Lee et al. 1996). The autosome is thus "counted" as an X chromosome. Fourth, *Xist* appears to be involved in "choosing" which X chromosome is inactivated. Female mice heterozygous for a deletion of a particular region of the *Xist* gene preferentially inactivate the wild-type chromosome (Marahrens et al. 1998). *Xist* expression is needed only for the initiation of X chromosome inactivation; once inactivation occurs, *Xist* transcription is dispensable (Brown and Willard 1994).

The *Xist* RNA works only in *cis*; that is, on the chromosome that made it. When the cells begin to differentiate, the *Xist* RNA is stabilized on one of the two X chromosomes (Figure 5.23B–D; Sheardown et al. 1997; Panning et al. 1997). However, the mechanism by which *Xist* is preferentially stabilized on one particular chromosome appears to differ between humans and mice. In the preimplantation *mouse* embryo, both X chromosomes synthesize *Xist* RNA, but this RNA is quickly degraded on one of the X chromosomes. This differential stabilization appears to be affected by an X-linked gene called *Tsix* ("Xist" spelled backwards). *Tsix* encodes a small, nuclear, non-protein-coding RNA that contains the *antisense* sequence to *Xist* (Lee and Lu 1999; Chao et al. 2002). *Tsix* RNA is associated with the future *active* X chromosome, where it appears to degrade *Xist*. The *human TSIX* gene, however, lacks the region that is antisense to *XIST*, and human *TSIX* does not appear to play a role in X chromosome inactivation (Migeon et al. 2002). While female mice with *Tsix* mutations die due to faulty X chromosomes inactivation (Lee 2000; Sado et al. 2001), human female cells (with their naturally "mutant" *TSIX*) must have some other mechanism to insure differential *XIST* stabilization.

Maintaining X Chromosome Inactivation

Once *Xist* initiates the inactivation of an X chromosome, the silencing of that chromosome is maintained in at least two ways. The first way involves DNA methylation. The *Xist* locus on the *active* X chromosome becomes methylated, while the active *Xist* gene (on the otherwise "inactive" X chromosome) remains unmethylated (Norris et al. 1994). Conversely, the promoter regions of numerous genes are methylated on the inactive X chromosome and unmethylated on the active X chromosome (Wolf et al. 1984; Keith et al. 1986; Migeon et al. 1991). The second method of maintaining X-chromosome inactivation appears to involve histone modification. The methylation of a lysine residue on histone H3 occurs almost immediately after *Xist* RNA is seen to coat what will be the inactive X chromosome. This histone methylation occurs before transcriptional silencing is observed, and it appears to begin at the nucleosomes immediately upstream form the *Xist* gene. Soon afterward, transcription stops, and other histone modifications occur on the inactive X chromosome, including the removal of acetyl groups from histone H4 (Figure 5.24; Jeppesen and Turner 1993; Heard et al. 2001; Mermoud et al. 2002). These nucleosome changes may create the heterochromatin that is characteristic of the Barr body.

We are still ignorant of the mechanisms by which the *Xist* transcript regulates the state of the chromatin and by which the spreading of inactivation occurs. We still do not understand the ways in which *Xist* transcription is linked to DNA methylation. We do not yet know how the choice between the two X chromosomes is originally made, nor how the *Xist* RNA is transcribed from a region surrounded by inactivated genes. There is still much to learn about dosage compensation in mammals, as well as in other animals that have chromosomal sex determination.

WEBSITE 5.11 The medical importance of X chromosome inactivation. The mechanisms responsible for human X chromosome inactivation may differ significantly from those that inactivate the mouse X chromosome. Moreover, random X chromosome inactivation in women heterozygous for a mutant X-linked gene (e.g., hemophilia A) may cause the predominant expression of the mutant allele, causing such women to have the disease, as would a male.

(A)

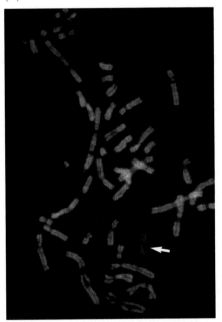

(B)

Attributes of the inactive X chromosome
XIST RNA production
XIST/Barr body complex
Histone H4 hypoacetylation
Histone H3 methylation
Histone macroH2A1 concentration
Nuclear envelope association
Late replication
Heterochromatin
Methylated promoters

Figure 5.24
The inactive X chromosome of human female cells contains underacetylated histone H4. (A) Chromosomes from a human female fibroblast cell stained green with fluorescent antibody to acetylated histone H4. While all the other chromosomes are stained green, the inactive X is not and thus appears red (arrow). (B) List of attributes characterizing the inactive X chromosome. (From Jeppesen and Turner 1993; photograph courtesy of the authors.)

Differential RNA Processing

The regulation of gene expression is not confined to the differential transcription of DNA. Even if a particular RNA transcript is synthesized, there is no guarantee that it will create a functional protein in the cell. To become an active protein, the RNA must be (1) processed into a messenger RNA by the removal of introns, (2) translocated from the nucleus to the cytoplasm, and (3) translated by the protein-synthesizing apparatus. In some cases, the synthesized protein is not in its mature form and must be (4) posttranslationally modified to become active. Regulation during development can occur at any of these steps.

The essence of differentiation is the production of different sets of proteins in different types of cells. In bacteria, differential gene expression can be effected at the levels of transcription, translation, and protein modification. In eukaryotes, however, another possible level of regulation exists—namely, control at the level of RNA processing and transport. There are two major ways in which differential RNA processing can regulate development. The first involves "censorship"—selecting which nuclear transcripts are processed into cytoplasmic messages. Different cells select different nuclear transcripts to be processed and sent to the cytoplasm as messenger RNA. Thus, the same pool of nuclear transcripts can give rise to different populations of cytoplasmic mRNAs in different cell types (Figure 5.25A).

The second mode of differential RNA processing is the splicing of mRNA precursors into messages for different proteins by using different combinations of potential exons. If an mRNA precursor had five potential exons, one cell might use exons 1, 2, 4, and 5; a different cell might utilize exons 1, 2, and 3; and yet another cell type might use yet another combination (Figure 5.25B). Thus, a single gene can produce an entire family of related proteins.

Control of early development by nuclear RNA selection

In the late 1970s, numerous investigators found that mRNA was not the primary transcript from the genes. Rather, the initial transcript is **nuclear RNA (nRNA)**, sometimes called heterogeneous nuclear RNA (hnRNA) or pre-messenger RNA (pre-mRNA). This nRNA is usually many times longer than the messenger RNA because the nuclear RNA contains introns that get spliced out during the passage from nucleus to cytoplasm. Originally, investigators thought that whatever RNA was transcribed in the nucleus was processed into cytoplasmic mRNA. But studies of sea urchins showed that different cell types could be *transcribing* the same type of nuclear RNA, but *processing* different subsets of this population into mRNA in different types of cells (Kleene and Humphreys 1977, 1985). Wold and her colleagues (1978) showed that sequences present in sea urchin blastula *messenger* RNA, but absent in gastrula and adult tissue mRNA, were nonetheless present in the *nuclear* RNA of the gastrula and adult tissues.

More genes are transcribed in the nucleus than are allowed to become mRNAs in the cytoplasm. This "censoring" of RNA transcripts has been confirmed by probing for the introns and exons of specific genes. Gagnon and his colleagues (1992) performed such an analysis on the transcripts from the *SpecII* and *CyIIIa* genes of the sea urchin *Strongylocentrotus purpuratus*. These genes encode calcium-binding and actin proteins, respectively, which are expressed only in a particular part of the ectoderm of the sea urchin larva. Using probes that bound to an exon (which is included in the mRNA) and to an intron (which is not included in the mRNA), they found that these genes were being transcribed not only in the ectodermal cells, but also in the mesoderm and endoderm. The analysis of the *CyIIIa* gene showed that the concentration of introns was the same in both the gastrula ectoderm and the mesoderm/endoderm samples, suggesting that this gene was being transcribed at the same rate in the nuclei of all cell types, but was made into cytoplasmic mRNA only in ectodermal cells (Figure 5.26). The unprocessed nRNA for *CyIIIa* is degraded while still in the nuclei of the endodermal and mesodermal cells.

WEBSITE 5.12 Differential nRNA censoring. Studies of differential nRNA censoring overturned the paradigm that differential gene transcription was the ultimate means of regulating embryonic differentiation. It "freed" embryology from microbiological ways of thinking about gene expression.

WEBSITE 5.13 An inside-out gene. Some RNAs stay in the nucleus to function. In one interesting case, the exons go outside the nucleus to be degraded, while the introns stay and help construct the nucleolus.

(A) RNA selection

(B) Differential splicing

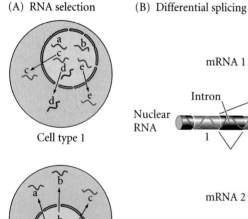

Cell type 1

Cell type 2

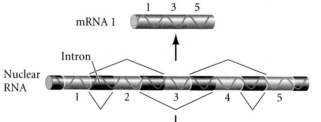

mRNA 1

Intron

Nuclear RNA

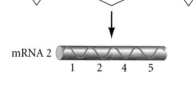

mRNA 2

Figure 5.25
Roles of differential RNA processing during development. By convention, splicing paths are shown by fine V-shaped lines. (A) RNA selection, whereby the same nuclear RNA transcripts are made in two cell types, but the set that becomes cytoplasmic messenger RNAs is different. (B) Differential splicing, whereby the same nuclear RNA is spliced into different proteins by selectively removing possible exons.

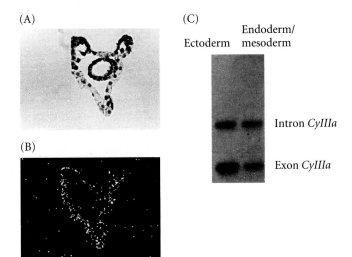

(A)

(B)

(C)

Ectoderm

Endoderm/
mesoderm

Intron *CyIIIa*

Exon *CyIIIa*

Figure 5.26
Regulation of ectoderm-specific gene expression by RNA process-ing. (A, B) *CyIIIa* mRNA is seen by autoradiography to be present only in the ectoderm. (A) Phase contrast micrograph. (B) In situ hy-bridization using a probe that binds to a *CyIIIa* exon. (C) The *CyIIIa* nuclear transcript, however, is found in both ectoderm and endoderm/mesoderm. The left lane of the gel represents RNA iso-lated from the gastrula ectodermal tissue; the right lane represents RNA isolated from endodermal and mesodermal tissues. The upper band is the RNA bound by a probe that binds to an intron sequence (which should be found only in the nucleus) of *CyIIIa*. The lower band represents the RNA bound by a probe complementary to an exon sequence. The presence of the intron indicates that the *CyIIIa* nuclear RNA is being made in both groups of cells, even if the mRNA is seen only in the ectoderm. (From Gagnon et al. 1992; photographs courtesy of R. and L. Angerer.)

Creating families of proteins through differential nRNA splicing

Alternative nRNA splicing is a means of producing a wide variety of proteins from the same gene. The average vertebrate nRNA consists of several relatively short exons (averaging about 140 bases) separated by introns that are usually much longer. Most mammalian nRNAs contain numerous exons. By splicing together different sets of exons, different cells can make different types of mRNAs, and hence, different proteins. Recognizing a sequence of nRNA as either an exon or an in-tron is a crucial step in gene regulation. What is an intron in one cell's nucleus may be an exon in another cell's nucleus.

Alternative nRNA splicing is based on the determination of which sequences will be spliced out as introns. This can occur in several ways (Figure 5.27). Most genes contain "con-sensus sequences" at the 5′ and 3′ ends of the introns. These sequences are also called the "splice sites" of the intron. The splicing of nRNA is mediated through complexes known as **spliceosomes** that bind to these splice sites. Spliceosomes are made up of small nuclear RNAs (snRNAa) and proteins called splicing factors that bind to splice sites or to the areas adjacent to them. By their production of specific splicing factors, cells

can differ in their ability to recognize the 5′ splice site (at the beginning of the intron) or the 3′ splice site (at the end of the intron). Some cells may fail to recognize a sequence as an in-tron at all, thus retaining it within the message.

The 5′ splice site is normally recognized by small nuclear RNA U1 (U1 snRNA) and splicing factor 2 (SF2; also known as alternative splicing factor). The choice of alternative 3′ splice sites is often controlled by which splice site can best bind a protein called U2AF. The spliceosome forms when the proteins that accumulate at the 5′ splice site contact those proteins bound to the 3′ splice site. Once the 5′ and 3′ ends are brought together, the intervening intron is excised and the two exons are ligated together.

WEBSITE 5.14 The mechanism of differential nRNA splicing. Differential nRNA splicing depends on the assem-bly of the nucleosome and upon the ratio of certain pro-teins in the nucleus of the cell.

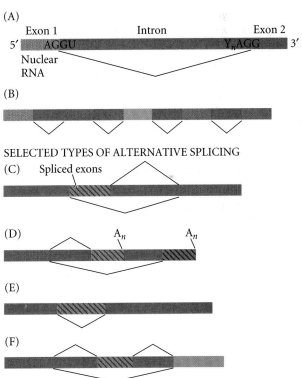

CONSTITUTIVE SPLICING

(A)
Exon 1   Intron   Exon 2
5′ AGGU   Y$_n$AGG 3′
Nuclear
RNA

(B)

SELECTED TYPES OF ALTERNATIVE SPLICING
(C) Spliced exons

(D)   A$_n$ A$_n$

(E)

(F)

Figure 5.27
Schematic diagram of alternative nRNA splicing. Colored portions of the bars represent exons; alternatively spliced exons are indicated by hatching. The gray portions of the bars represent introns. (A) Exon-intron borders, showing the consensus sequences at the 5′ and 3′ ends of the intron. Y represents pyrimidine. (B) The splicing of an nRNA that has five exons. (C–F) Alternative splicing by (C) alterna-tive 5′ splice sites; (D) alternative 3′ splice sites (in some cases, this would provide different termini to the mRNA, and both sites would need a polyadenylation sequence, shown here as A$_n$); (E) a splice/no splice decision; and (F) exon inclusion/exon skipping. (After Horowitz and Krainer 1995.)

Differential RNA processing has been found to control the alternative forms of expression of the genes encoding over 100 proteins. The deletion of certain potential exons in some cells but not in others enables one gene to create a family of closely related proteins. Instead of one gene-one polypeptide, one can have one gene-one family of proteins. For instance, alternative RNA splicing enables the α-tropomyosin gene to encode brain, liver, skeletal muscle, smooth muscle, and fibroblast forms of this protein (Figure 5.28; Breitbart et al. 1987). The nuclear RNA for α-tropomyosin contains 11 potential exons, but different sets of exons are used in different cells. Such different proteins encoded by the same gene are called **splicing isoforms** of the protein.

In some instances, alternatively spliced RNAs yield proteins that play similar yet distinguishable roles in the same cell. Different isoforms of the WT1 protein perform different functions in the development of the gonads and kidneys. The isoform without the extra exon functions as a transcription factor during kidney development, while the isoform containing the extra exon appears to be involved in splicing different nRNAs and may be critical in testis development (Hastie 2001; Hammes et al. 2001). In *Drosophila*, the *Held-out wings* gene transcript can be spliced in two ways. One of the resulting mRNAs encodes a protein that inhibits tendon differentiation, while the other encodes a protein that stimulates tendon differentiation. The relative amounts of these two transcripts during development regulate the state of differentiation in the tendon cells so that these cells mature at the appropriate times (Nabel-Rosen et al. 2002).

If you get the impression from this discussion that a gene with dozens of introns could create literally thousands of different, related proteins through differential splicing, you are probably correct. For example, proteins derived from the neurexin genes are found on the cell surfaces of developing neurons and may be important in specifying the connections that these neurons make.* The RNAs transcribed from neurexin genes can be alternatively spliced at several different sites, creating hundreds of different proteins from the same gene (Ullrich et al. 1995; Ichtchenko et al. 1995).

The current champion at making multiple proteins from the same gene is the *Drosophila Dscam* gene, which is involved in guiding axons to their targets during the insect's development. The *Dscam* gene contains 24 exons. However, more than a dozen different adjacent DNA sequences can be selected to be exon 4. Similarly, at least a dozen mutually exclusive adjacent DNA sequences can become exons 6 and 9 (Figure 5.29; Schmucker et al. 2000). The nRNA of *Dscam* has been found to be alternatively spliced in different axons and may control the specificity of axon attachments (Celotto and Graveley 2001). If all possible combinations of exons are used, this one gene can produce 38,016 different proteins, and random searches for these combinations indicate that a large fraction of them are in fact made. The *Drosophila* genome is thought to contain only 14,000 genes, but here is a single gene that encodes three times that number of proteins! It is estimated that at least 35% of human genes produce alternatively spliced RNAs (Croft et al. 2000). Therefore, even though the human genome may contain only 35,000–80,000 genes, its **proteome**—the number and type of proteins encoded by the genome—is probably far more complex.

---

*The neurexins are cell recognition molecules and appear to be involved in neuron-neuron adhesion and recognition. The venom of the black widow spider works by binding to neurexins, causing massive neurotransmitter release (Rosenthal and Meldolesi 1989).

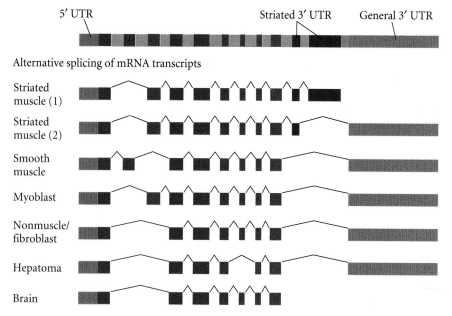

Figure 5.28
Alternative RNA splicing to form a family of rat α-tropomyosin proteins. The α-tropomyosin gene is represented on top. The thin lines represent the sequences that become introns and are spliced out to form the mature mRNAs. Constitutive exons (found in all tropomyosins) are shown in green. Those expressed only in smooth muscle are red; those expressed only in striated muscle are purple. Those that are variously expressed are yellow. Note that in addition to the many possible combinations of exons, two different, alternative 3′ ends are possible. (After Breitbart et al. 1987.)

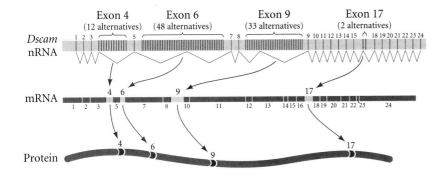

Figure 5.29
The *Dscam* gene of *Drosophila* can produce 38,016 different types of proteins by alternative nRNA splicing. The gene contains 24 exons. Exons 4, 6, 9, and 17 are encoded by sets of mutually exclusive possible sequences. Each messenger RNA will contain one of the 12 possible exon 4 sequences, one of the 48 possible exon 6 alternatives, one of the 33 possible exon 9 alternatives, and one of the 2 possible exon 17 sequences. The *Drosophila Dscam* gene is homologous to a human DNA sequence on chromosome 21 that is expressed in the human nervous system. Disturbances of this gene in humans may lead to the neurological defects of Down syndrome (Yamakawa et al. 1998; Saito 2000).

Sidelights & Speculations

Differential nRNA Processing and Drosophila Sex Determination

Sex determination in *Drosophila* is regulated by a cascade of RNA processing events (Baker et al. 1987; MacDougall et al. 1995). As we will see in Chapter 17, the development of the sexual phenotype in *Drosophila* is mediated by the ratio of X chromosomes to autosomes (non-sex chromosomes). The X chromosomes produce transcription factors that activate the *Sex-lethal* (*Sxl*) gene,* while the autosomes produce transcription factors that repress the *Sxl* gene. Thus, these two sets of transcription factors—activators from the X and repressors from the autosomes—compete for the enhancer sites of the *Sxl* gene. When the X-to-autosome ratio is 1 (i.e., when there are two X chromosomes per diploid cell), the activators dominate, the *Sxl* gene is active, and the embryo develops into a female. When the ratio is 0.5 (i.e., when the fly is XY with only one X chromosome per diploid cell), the repressors dominate, the *Sxl* gene is not active, and the embryo develops into a male (Figure 5.30).

*The name of this gene, *Sex-lethal*, comes from the deadly decoupling of the dosage compensation mechanism that arises when this gene is mutated. When this happens, the fly will become male even if it is XX. However, its two X chromosomes will be instructed to transcribe their genes at the higher (male) rate, creating regulatory defects that kill the embryo (Cline 1986).

But what is the Sxl protein doing to determine sex? It is acting as a differential splicing factor on the nRNA transcribed from the *transformer* (*tra*) gene. Throughout the larval period, the *tra* gene actively synthesizes a nuclear transcript that is processed into either a general mRNA (found in both females and males) or a female-specific mRNA. The female-specific message is made when Sxl protein binds to the nRNA and inhibits spliceosome formation on the general 3′ splice site of the first intron (Sosnowski et al. 1989; Valcárcel et al. 1993). Instead, the spliceosome forms on another, less efficient 3′ site and allows splicing to occur there. As a result, the female form of the *transformer* mRNA lacks an exon found in the general form. And that difference is crucial, because this exon contains a translational stop codon (UGA) that causes the message to make a small, nonfunctional protein (Belote et al. 1989). However, in the female-specific message, the UGA codon is spliced out during mRNA formation and does not interfere with the translation of the message. In other words, the female-specific *tra* transcript is the only functional transcript of this gene.

The Tra protein is itself an alternative splicing factor that regulates the splicing of the nuclear transcript of the *doublesex* (*dsx*) gene. This gene is needed for the production of either sexual phenotype, and mutations of *dsx* can reverse the expected sexual phenotype, causing XX embryos to become males or XY embryos to become females. During pupation, the *dsx* gene makes a nuclear transcript that can be processed in two alternative ways. It can generate a female-specific or a male-specific mRNA (Nagoshi et al. 1988). The

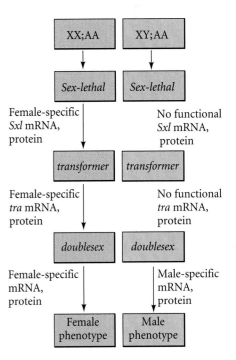

Figure 5.30
Sex determination in *Drosophila*. This simplified scheme shows that the X-to-autosome ratio is monitored by the *Sex-lethal* gene. If this gene is active, it processes *transformer* nRNA into a functional female-specific message. In the presence of the female-specific Transformer protein, the *doublesex* gene transcript is processed in a female-specific fashion. The female-specific Doublesex protein is a transcription factor that leads to the production of the female phenotype. If the *transformer* gene does not make a female-specific product (i.e., if the *Sex-lethal* gene is not activated), the *doublesex* transcript is spliced in the male-specific manner, leading to the formation of a male-specific Doublesex protein. This is a transcription factor that generates the male phenotype.

first three exons of *dsx* mRNA are the same in females and males (Tian and Maniatis 1992, 1993). But if Tra protein is present, it converts a weak 3′ splice site into a strong site (e.g., a more efficient binder of U2AF), and exon 4 is retained, resulting in female-specific *dsx* mRNA. If Tra protein is not present, U2AF will not bind to this 3′ site, and exon 4 will not be included, resulting in the male-specific *dsx* message. Doublesex proteins made by the male and female mRNAs are both transcription factors, and they recognize the same sequence of DNA. However, while female Dsx protein activates female-specific enhancers (such as those on the genes encoding yolk proteins), male Dsx protein inhibits transcription from those same enhancers (Coschigano and Wensink 1993; Jursnich and Burtis 1993). Conversely, the female protein can inhibit transcription from genes that are otherwise activated by the male protein. Research into *Drosophila* sex determination shows that differential RNA processing plays enormously important roles throughout development.

Control of Gene Expression at the Level of Translation

The splicing of nuclear RNA is intimately connected with its export through the nuclear pores and into the cytoplasm. As the introns are being removed, specific proteins bind to the spliceosome and attach the spliceosome-RNA complex to the nuclear pores (Luo et al. 2001; Sträßer and Hurt 2001). But once the RNA has reached the cytoplasm, there is still no guarantee that it will be translated. The control of gene expression at the level of translation can occur by many means; some of the most important of these are described below.

Differential mRNA longevity

The longer an mRNA persists, the more protein can be translated from it. If a message with a relatively short half-life were selectively stabilized in certain cells at certain times, it would make large amounts of its particular protein only at those times and places. The stability of a message is often dependent upon the length of its poly(A) tail. This, in turn, appears to depend upon sequences in the 3′ untranslated region. Certain 3′ UTR sequences allow longer poly(A) tails than others. If these 3′ UTR regions are experimentally traded, the half-lives of the resulting mRNAs are altered: long-lived messages will decay rapidly, while normally short-lived mRNAs will remain around longer (Shaw and Kamen 1986; Wilson and Treisman 1988; Decker and Parker 1994).

In some instances, messenger RNAs are selectively stabilized at specific times in specific cells. The mRNA for casein, the major protein of milk, has a half-life of 1.1 hours in rat mammary gland tissue. However, during periods of lactation, the presence of the hormone prolactin increases this half-life to 28.5 hours (Figure 5.31; Guyette et al. 1979). In *Drosophila*, the Stripe protein is a transcription factor that positively regulates *Drosophila* tendon differentiation. It is regulated at the translational level by the proteins encoded by the *Held-out-wings* gene mentioned above. The longer Held-out-wings isoform leads to the degradation of the *Stripe* mRNA, while the shorter isoform of the Held-out-wings protein stabilizes that same message (Nabel-Rosen et al. 2002).

WEBSITE 5.15 **Mechanisms of mRNA translation and degradation.** Translation is a complex process involving the initiation, elongation, and termination of protein synthesis. It has numerous points at which regulation can occur. Similarly, the degradation of mRNA is a tightly regulated event.

Selective inhibition of mRNA translation

Some of the most remarkable cases of translational regulation of gene expression occur in the oocyte. The oocyte often makes and stores mRNAs that will be used only after fertilization occurs. These messages stay in a dormant state until they are activated by ionic signals (discussed in Chapter 7) that spread through the egg during ovulation or sperm binding. Table 5.2 gives a partial list of mRNAs that are stored in the oocyte cytoplasm. Some of these stored mRNAs encode proteins that will be needed during cleavage, when the embryo makes enormous amounts of chromatin, cell membranes, and cytoskeletal components. Some of them encode cyclin pro-

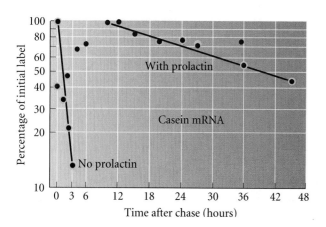

Figure 5.31
Degradation of casein mRNA in the presence and absence of prolactin. Cultured rat mammary cells were given radioactive RNA precursors (pulse) and, after a given time, were washed and given nonradioactive precursors (chase). This procedure labeled the casein mRNA synthesized during the pulse time. Casein mRNA was then isolated at different times following the chase and its radioactive label measured. In the absence of prolactin, the labeled (i.e., newly synthesized) casein mRNA decayed rapidly, with a half-life of 1.1 hours. When the same experiment was done in a medium containing prolactin, the half-life was extended to 28.5 hours. (After Guyette et al. 1979.)

TABLE 5.2 Some mRNAs stored in oocyte cytoplasm and translated at or near fertilization

mRNAs encoding	Function(s)	Organism(s)
Cyclins	Cell division regulation	Sea urchin, clam, starfish, frog
Actin	Cell movement and contraction	Mouse, starfish
Tubulin	Formation mitotic spindles, cilia, flagella	Clam, mouse
Small subunit of ribo-nucleotide reductase	DNA synthesis	Sea urchin, clam, starfish
Hypoxanthine phospho-ribosyl-transferase	Purine synthesis	Mouse
Vg1	Mesodermal determination(?)	Frog
Histones	Chromatin formation	Sea urchin, frog, clam
Cadherins	Blastomere adhesion	Frog
Metalloproteinases	Implantation in uterus	Mouse
Growth factors	Cell growth; uterine cell growth(?)	Mouse
Sex determination factor FEM-3	Sperm formation	*C. elegans*
PAR gene products	Segregate morphogenetic determinants	*C. elegans*
SKN-1 morphogen	Blastomere fate determination	*C. elegans*
Hunchback morphogen	Anterior fate determination	*Drosophila*
Caudal morphogen	Posterior fate determination	*Drosophila*
Bicoid morphogen	Anterior fate determination	*Drosophila*
Nanos morphogen	Posterior fate determination	*Drosophila*
GLP-1 morphogen	Anterior fate determination	*C. elegans*
Germ cell-less protein	Germ cell determination	*Drosophila*
Oskar protein	Germ cell localization	*Drosophila*
Ornithine transcarbamylase	Urea cycle	Frog
Elongation factor 1a	Protein synthesis	Frog
Ribosomal proteins	Protein synthesis	Frog, *Drosophila*

Sources: Compiled from numerous sources.

teins that regulate the timing of early cell division (Rosenthal et al. 1980; Standart et al. 1986). Indeed, in many species (including sea urchins and *Drosophila*), maintenance of the normal rate and pattern of early cell divisions does not require a nucleus; rather, it requires continued protein synthesis from stored maternal mRNAs (Wagenaar and Mazia 1978; Edgar et al. 1989). Other stored messages encode proteins that determine the fates of cells. These include the *bicoid* and *nanos* messages that provide positional information in the *Drosophila* embryo, as we saw in Chapter 3, and the *glp-1* mRNA of the nematode *C. elegans*.

WEBSITE 5.16 **The discovery of stored mRNAs.** The existence of maternally transcribed mRNAs stored in the oocyte was one of the first discoveries of molecular embryology. Even before gene cloning became available, the identity of several of these mRNAs was known.

The 5′ cap and the 3′ untranslated region (UTR) seem especially important in regulating the accessibility of mRNA to ribosomes. If the 5′ cap is not made or if the 3′ UTR lacks a polyadenylate tail, the message probably will not be translated. The oocytes of many species have "used these ends as means" to regulate the translation of mRNAs. For instance, the oocyte of the tobacco hornworm moth makes some of its mRNAs without their methylated 5′ caps. In this state, they cannot be efficiently translated. However, at fertilization, a methyltransferase completes the formation of the caps, and these mRNAs can be translated (Kastern et al. 1982).

In oocytes, having a short poly(A) tail does not lead to the degradation of a message; however, such messages are not translated. In the *Drosophila* oocyte, the *bicoid* message remains untranslated until signals at fertilization allow the Cortex and Grauzone proteins to add poly(A) residues to the *bicoid* mRNA (Sallés et al. 1994; Lieberfarb et al. 1996). At that point, the *bicoid* message becomes translatable (and its product determines which part of the embryo becomes the head and thorax).

In amphibian oocytes, the 5′ and 3′ ends of many mRNAs are tethered together by a protein called **maskin** (Stebbins-Boaz et al. 1999; Mendez and Richter 2001). Maskin links the 5′ and 3′ ends into a circle by binding to two other proteins, each at opposite ends of the message. First, it binds to the CPEB protein attached to the UUUUAU sequence in the 3′ UTR; second, maskin binds to translation initiation factor 4E (eIF4E) that is attached to the cap sequence (Figure 5.32A). In this configuration, the mRNA cannot be translated. The binding of eIF4E to maskin is thought to prevent the binding of eIF4E to initiation factor 4G (eIF4G), a critically important translation initiation factor that brings the small ribosomal subunit to the mRNA.

Mendez and Richter (2001) have proposed a intricate scenario to explain how mRNAs bound together by maskin become translated about the time of fertilization (Figure 5.32B). At ovulation (when the hormone progesterone stimulates the last meiotic divisions of the oocyte and the oocyte is released for fertilization), a kinase activated by progesterone phosphorylates the CPEB protein. The phosphorylated CPEB can now bind to CPSF, the cleavage and polyadenylation specificity factor (Mendez et al. 2000; Hodgman et al. 2001).

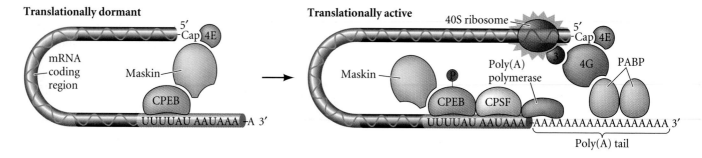

Figure 5.32

Translational regulation in the *Xenopus* oocyte. In the oocyte, the 3′ and 5′ ends (gray) of the mRNA are brought together by maskin, a protein that binds to CPEB on the 3′ end and translation initiation factor 4E (eIF4E) on the 5′ end. Maskin blocks the initiation of translation by preventing eIF4E from binding eIF4G. Upon stimulation by progesterone during ovulation. CPEB is phosphorylated and can bind CPSF. CPSF can bind poly(A) polymerase and initiate the growth of the poly(A) tail. Poly(A) binding protein (PABP) can bind to this tail and then bind eIF4G in a stable manner. This initiation factor can then bind eIF4E and, through its association with eIF3, position a 40S ribosomal subunit on the mRNA. (After Mendez and Richter 2001.)

The bound CPSF protein sits on a particular sequence of the 3′ UTR that has been shown to be critical for polyadenylation, and it complexes with the poly(A) polymerase that elongates the poly(A) tail of the mRNA. Once the poly(A) tail is extended, molecules of the poly(A) binding protein (PABP) can attach to the growing tail. The PABP proteins stabilize eIF4G, allowing it to outcompete maskin for the binding site on the eIF4E protein at the 5′ end of the mRNA. The eIF4G protein can then bind translation initiator protein 3 (eIF3), which can position the small ribosomal subunit onto the mRNA. The small (40S) ribosomal subunit will then find the initiator tRNA, complex with the large ribosomal subunit, and initiate translation.

In some instances, messages in the oocyte are prevented from being translated by the binding of some specific inhibitory protein. For instance, the Smaug protein binds to the 3′ UTR of the *Drosophila nanos* message, and this binding prevents the translation of *nanos* mRNA in the oocyte until fertilization (Smibert et al. 1996). (At that time, the Nanos protein will become critical for determining which part of the fly will be its abdomen.)

One of the most remarkable ways of regulating the translation of a specific message is seen in *Caenorhabditis elegans*. This nematode lives up to its name, having evolved a particularly elegant solution to the problem of controlling larval gene expression (Lee et al. 1993; Wightman et al. 1993). It makes a naturally occuring antisense mRNA to one of its own messages. High levels of the LIN-14 transcription factor are important in the development of early larval organs. Thereafter, the LIN-14 protein is no longer seen, although *lin-14* mes-

sages can be detected throughout development. *C. elegans* is able to inhibit the synthesis of LIN-14 from these messages by activating the *lin-4* gene. The *lin-4* gene does not encode a protein. Rather, it encodes two small RNAs (the most abundant being 25 nucleotides long, the other continuing for 40 more nucleotides) that are complementary to an imperfectly repeated site in the *lin-14* 3′ UTR. Figure 5.33 shows a hypothetical sketch of what might be happening. It appears that the binding of the *lin-4* transcripts to the *lin-14* mRNA 3′ UTR does not signal the destruction of the message, but rather prevents the message from being translated.

The *lin-4* RNA is now thought to be a member of a very large group of **small regulatory RNAs** (sometimes called **"micro-RNAs"**). These RNAs are made from longer precursors and are processed by the enzyme Dicer (the same enzyme used in the RNA interference technique; see Chapter 4). Such small regulatory RNAs can bind to the 3′UTR of messages and inhibit their translation. In many instances, these messages are stabilized, not degraded by this binding (Grosshans and Slack 2002). The abundance of such micro-RNAs and their apparent conservation among flies, nematodes, and vertebrates suggest that such RNA regulation is a previously unrecognized but potentially very important means of regulating gene expression (Lagos-Quintana et al. 2001; Lau et al. 2001; Lee and Ambros 2001).

Control of RNA expression by cytoplasmic localization

Not only is the time of mRNA translation regulated, but so is the place of RNA expression. Just like the selective repression of mRNA translation, the selective localization of messages is often accomplished through their 3′ UTRs, and it is often performed in oocytes. Rebagliati and colleagues (1985) showed that there are certain mRNAs in *Xenopus* oocytes that are selectively transported to the vegetal pole (Figure 5.34). After fertilization, these messages make proteins that are found only in the vegetal blastomeres. In *Drosophila*, the *bicoid* and *nanos* messages are each localized to different ends of the oocyte. The 3′ UTR of the *bicoid* mRNA allows this message to bind to the microtubules through its association with two other proteins (swallow and staufen). If the *bicoid* 3′ UTR is attached to some other message, that mRNA will also be bound

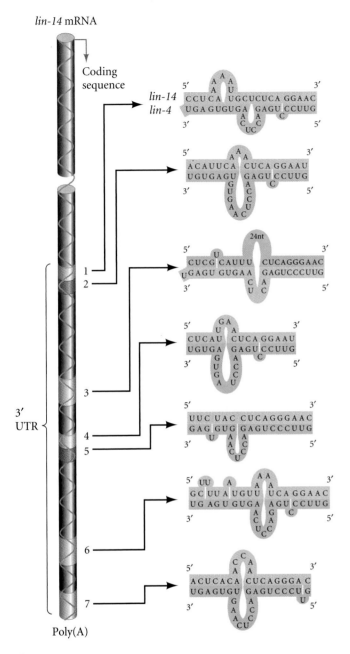

Figure 5.33
Hypothetical model of the regulation of *lin-14* mRNA translation by *lin-4* RNAs. The *lin-4* gene does not produce an mRNA. Rather, it produces small RNAs that are complementary to a repeated sequence in the 3′ UTR of the *lin-14* mRNA, which bind to it and prevent its translation. (After Wickens and Takayama 1995.)

while the Nanos protein forms a gradient with its peak at the posterior pole (see Figure 3.25). The ratio of these two proteins will eventually determine the anterior-posterior axis of the *Drosophila* embryo and adult. (The ability of cells to be specified by gradients of proteins is a critical phenomenon and is discussed in more detail in Chapters 3 and 9.)

> **WEBSITE 5.17 Other examples of translational regulation of gene expression.** There are numerous other fascinating examples wherein mRNA is selectively translated under different conditions. The 1:1 ratio of α- and β-globins in adult blood comes from the differential translation of the respective globin messages. The production of heme for the hemoglobin is also regulated at the level of translation.

Posttranslational regulation of gene expression

When a protein is synthesized, the story is still not over (Figure 5.35). Once a protein is made, it becomes part of a larger level of organization. For instance, it may become part of the structural framework of the cell, or it may become involved in one of the myriad enzymatic pathways for the synthesis or breakdown of cellular metabolites. In any case, the individual protein is now part of a complex "ecosystem" that integrates it

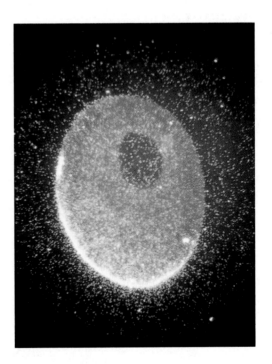

Figure 5.34
Localization of *Vg1* mRNA to the vegetal portion of the *Xenopus* oocyte. The white crescent at the bottom of the egg represents the tethered *Vg1* message. The black area is the haploid nucleus of the oocyte. At fertilization, the *Vg1* message is translated into an inactive protein. If that protein is processed to its active form, it can become an important signaling protein in the embryo. (Photograph courtesy of D. Melton.)

to the anterior pole of the oocyte (Driever and Nüsslein-Volhard 1988a,b; Ferrandon et al. 1994). The 3′ UTR of the *nanos* message similarly allows it to accumulate at the posterior pole of the egg (Gavis and Lehmann 1994). As we saw in Chapter 3, this localization allows the Bicoid protein to form a gradient wherein the highest amount of it is at the anterior pole,

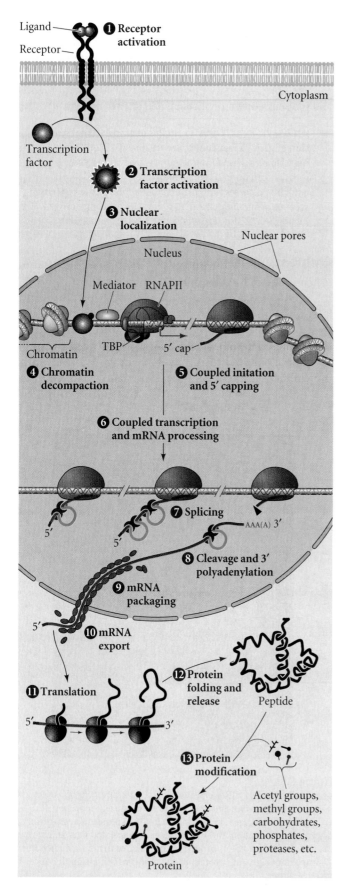

Ligand
Receptor
❶ **Receptor activation**

Cytoplasm

Transcription factor

❷ **Transcription factor activation**

❸ **Nuclear localization**

Nuclear pores

Nucleus

Mediator   RNAPII

Chromatin

TBP   5′ cap

❹ **Chromatin decompaction**

❺ **Coupled initation and 5′ capping**

❻ **Coupled transcription and mRNA processing**

5′

❼ **Splicing**

5′

AAA(A) 3′

❽ **Cleavage and 3′ polyadenylation**

❾ **mRNA packaging**

5′

❿ **mRNA export**

⓫ **Translation**

⓬ **Protein folding and release**

Peptide

5′   3′

⓭ **Protein modification**

Acetyl groups, methyl groups, carbohydrates, phosphates, proteases, etc.

Protein

into a relationship with numerous other proteins. Thus, several changes can still take place that determine whether or not the protein will be active.

Some newly synthesized proteins are inactive without the cleaving away of certain inhibitory sections. This is what happens when insulin is made from its larger protein precursor. Some proteins must be "addressed" to their specific intracellular destinations in order to function. Proteins are often sequestered in certain regions of the cell, such as membranes, lysosomes, nuclei, or mitochondria. Some proteins need to assemble with other proteins in order to form a functional unit. The hemoglobin protein, the microtubule, and the ribosome are all examples of numerous proteins joining together to form a functional unit. And some proteins are not active unless they bind an ion (such as Ca^{2+}), or are modified by the covalent addition of a phosphate or acetate group. This last type of protein modification will become very important in the next chapter, since many important proteins in embryonic cells are just sitting there until some signal activates them. We turn next to how the embryo develops by activating certain proteins in specific cells.

Figure 5.35
An integrated account of gene expression. Each step regulating gene expression is a subdivision of a continuous process from transcription through protein modification. The activation of transcription factors by membrane receptors will be the subject of the next chapter. (After Orphanides and Reinberg 2002.)

Principles of Development: Differential Gene Expression

1. Differential gene expression from genetically identical nuclei creates different cell types. Differential gene expression can occur at the levels of gene transcription, nuclear RNA processing, mRNA translation, and protein modification. Notice that RNA processing and export occur while the RNA is still being transcribed from the gene.

2. Genes are usually repressed, and activating a gene often means inhibiting its repressor. This fact leads to thinking in double and triple negatives: Activation is often the inhibition of the inhibitor; repression is the inhibition of the inhibitor of the inhibitor.

3. Eukaryotic genes contain promoter sequences to which RNA polymerase can bind to initiate transcription. To accomplish this, the eukaryotic RNA polymerases are bound by a series of proteins called basal transcription factors.

4. Eukaryotic genes expressed in specific cell types contain enhancer sequences that regulate their transcription in time and space.

5. Specific transcription factors can recognize specific sequences of DNA in the promoter and enhancer regions. These proteins activate or repress transcription from the genes to which they have bound.

6. Enhancers work in a combinatorial fashion. The binding of several transcription factors can act to promote or inhibit transcription from a certain promoter. In some cases transcription is activated only if both factor A and factor B are present, while in other cases, transcription is activated if either factor A or factor B is present.

7. A gene encoding a transcription factor can keep itself activated if the transcription factor it encodes also activates its own promoter. Thus, a transcription factor gene can have one set of enhancer sequences to initiate its activation and a second set of enhancer sequences (which bind the encoded transcription factor) to maintain its activation.

8. Often, the same transcription factors that are used during the differentiation of a particular cell type are also used to activate the genes for that cell type's specific products.

9. Enhancers can act as silencers to suppress the transcription of a gene in inappropriate cell types.

10. Transcription factors act in different ways to regulate RNA synthesis. Some transcription factors stabilize RNA polymerase binding to the DNA, some disrupt nucleosomes, increasing the efficiency of transcription.

11. Transcription correlates with a lack of methylation on the promoter and enhancer regions of genes. Methylation differences can account for examples of genomic imprinting, wherein a gene transmitted through the sperm is expressed differently from the same gene transmitted through the egg.

12. Dosage compensation enables the X chromosome-derived products of males (which have one X chromosome per cell in fruit flies and mammals) to equal the X chromosome-derived products of females (which have two X chromosomes per cell). This compensation is accomplished at the level of transcription, either by accelerating transcription from the lone X chromosome in males (*Drosophila*) or by inactivating a large portion of one of the two X chromosomes in females (mammals).

13. X chromosome inactivation in mammals is generally random and involves the activation of the Xist gene on the chromosome that will be inactivated.

14. Differential nuclear RNA selection can allow certain transcripts to enter the cytoplasm and be translated while preventing other transcripts from leaving the nucleus.

15. Differential RNA splicing can create a family of related proteins by causing different regions of the nRNA to be read as exons and introns. What is an exon in one set of circumstances may be an intron in another.

16. Some messages are translated only at certain times. The oocyte, in particular, uses translational regulation to set aside certain messages that are transcribed during egg development but used only after the egg is fertilized. This activation is often accomplished either by the removal of inhibitory proteins or by the polyadenylation of the message.

17. Many messenger RNAs are localized to particular regions of the oocyte or other cells. This localization appears to be regulated by the 3′ untranslated region of the mRNA.

Literature Cited

Akhtar, A., D. Zink and P. B. Becker. 2000. Chromodomains are protein-RNA interaction modules. *Nature* 407: 405–409.

Andersen, F. G., J. Jensen, R. S. Heller, H. V. Petersen, L.-I. Larsson, O. D. Madsen and P. Serup. 1999. *Pax6* and *Pdx1* form a functional complex on the rat somatostatin gene upstream enhancer. *FEBS Lett.* 445: 315–320.

Angier, N. 1992. A first step in putting genes into action: Bend the DNA. *New York Times*, Aug. 4, C1, C7.

Arney, K. L., S. Erhardt, R. A. Drewell and M. A. Surani. 2001. Epigenetic reprogramming of the genome: From the germ line to the embryo and back again. *Int. J. Dev. Biol.* 45: 533–540.

Baek, H. J., S. Malik, J. Qin and R. G. Roeder. 2002. The requirement of TRAP/mediator for both activator-independent and activator-dependent transcription in conjunction with TFIID-associated TAF$_{II}$s. *Mol. Cell Biol.* 22: 2842–2852.

Baker, B., R. N. Nagoshi and K. C. Burtin. 1987. Molecular genetic aspects of sex determination in *Drosophila*. *BioEssays* 6: 66–70.

Barlow, D. P., R. Stoger, B. G. Herrmann, K. Saito and N. Schweifer. 1991. The mouse insulin-like growth factor type-2 receptor is imprinted and closely linked to the *Tme* locus. *Nature* 349: 84–87.

Barr, M. L. and E. G. Bertram. 1949. A morphological distinction between neurones of the male and female, and the behavior of the nucleolar satellite during accelerated nucleoprotein synthesis. *Nature* 163: 676.

Bartolomei, M. S. and S. M. Tilghman. 1997. Genomic imprinting in mammals. *Annu. Rev. Genet.* 31: 493–525.

Beimesche, S. and 8 others. 1999. Tissue-specific transcriptional activity of a pancreatic islet cell-specific enhancer sequence/Pax6 binding site determined in normal adult tissues in vivo using transgenic mice. *Mol. Endocrinol.* 13: 718–728.

Bell, A. C., A. G. West and G. Felsenfeld. 2001. Insulators and boundaries: Versatile regulatory elements in the eukaryotic genome. *Science* 291: 447–450.

Belote, J. M., M. McKeown, R. T. Boggs, R. Ohkawa and B. A. Sosnowski. 1989. Molecular genetics of *transformer*, a genetic switch controlling sexual differentiation in *Drosophila*. *Dev. Genet.* 10: 143–155.

Bentley, N. J., T. Eisen and C. R. Goding. 1994. Melanocyte-specific expression of the human tyrosinase promoter: Activation by the *microphthalmia* gene product and the role of the initiator. *Mol. Cell. Biol.* 14: 7996–8006.

Bestor, T. H. and V. M. Ingram. 1983. Two DNA methyltransferases from murine erythroleukemia cells: Purification, sequence speci-

ficity, and mode of interaction with DNA. *Proc. Natl. Acad. Sci. USA* 82: 2674–2678.

Borsani, G. and 13 others. 1991. Characterization of a murine gene expressed from the inactive X chromosome. *Nature* 351: 325–329.

Breitbart, R. A., A. Andreadis and B. Nadal-Ginard. 1987. Alternative splicing: A ubiquitous mechanism for the generation of multiple protein isoforms from single genes. *Annu. Rev. Biochem.* 56: 481–495.

Brockdorff, N. and 7 others. 1992. The product of the mouse *Xist* gene is a 15-kb inactive X-specific transcript containing no conserved ORF and located in the nucleus. *Cell* 71: 515–526.

Brown, C. J. and H. F. Willard. 1994. The human X-inactivation centre is not required for maintenance of X-chromosome inactivation. *Nature* 368: 154–156.

Brown, C. J., A. Ballabio, J. L. Rupert, R. G. Lafreniere, M. Grompe, R. Tonlorenzi and H. F. Willard. 1991a. A gene from the region of the human X inactivation center is expressed exclusively from the inactive X chromosome. *Nature* 349: 38–45.

Brown, C. J. and 9 others. 1991b. Localization of the X chromosome inactivation center on the human X chromosome. *Nature* 349: 82–85.

Brown, C. J., B. D. Hendrich, J. L. Rupert, R. G. Lafreniere, Y. Xing, J. Lawrence and H. F. Willard. 1992. The human *XIST* gene: Analysis of a 17-kb inactive X-specific RNA that contains conserved repeats and is highly localized within the nucleus. *Cell* 71: 527–542.

Brown, C. J., L. Carrel and H. F. Willard. 1997. Expression of genes from the human active and inactive X chromosomes. *Am. J. Hum. Genet.* 60: 1333–1343.

Buratowski, S. 1997. Multiple TATA-binding factors come back into style. *Cell* 91: 13–15.

Buratowski, S., S. Hahn, L. Guarente and P. A. Sharp. 1989. Five initiation complexes in transcription initiation by RNA polymerase II. *Cell* 56: 549–561.

Busslinger, M., J. Hurst and R. A. Flavell. 1983. DNA methylation and the regulation of globin gene expression. *Cell* 34: 197–206.

Celotto, A. M. and B. R. Graveley. 2001. Alternative splicing of the *Drosophila Dscam* pre-mRNA is both temporally and spatially regulated. *Genetics* 159: 599–608.

Centerwall, W. R. and K. Benirschke. 1973. Male tortoiseshell and calico (T-C) cats. *J. Hered.* 64: 272–278.

Chaillet, J. R., T. F. Vogt, D. R. Beier and P. Leder. 1991. Parental-specific methylation of an imprinted transgene is established during gametogenesis and progressively changes during embryogenesis. *Cell* 66: 77–83.

Chen, J.-L., L. D. Attardi, C. P. Verrijzer, K. Yokomori and R. Tjian. 1995. Assembly of recombinant TFIID reveals differential cofactor requirements for distinct transcriptional activators. *Cell* 79: 93–105.

Chong, J. A. and 9 others. 1995. REST: A mammalian silencer protein that restricts sodium channel gene expression to neurons. *Cell* 80: 949–957.

Chou, R. L., C. R. Altmann, R. A. Lang and A. Hemmati-Brivanlou. 1999. Pax6 induces ectopic eyes in a vertebrate. *Development* 126: 4213–4222.

Chou, W., K. D. Huynh, R. J. Spencer, L. S. Davidow and J. T. Lee. 2002. CTCF, a candidate Trans-acting factor for X-inactivation choice. *Science* 295: 345–347.

Cline, T. 1986. A female-specific lethal lesion in an X-linked positive regulator of the *Drosophila* sex determination gene, *Sex-lethal*. *Genetics* 113: 641–663.

Coschigano, K. T. and P. Wensink. 1993. Sex-specific transcriptional regulation by the male and female doublesex proteins of *Drosophila*. *Genes Dev.* 7: 42–55.

Croft, L., S. Schandorff, F. Clark, K. Burrage, P. Arctander and J. S. Mattick. 2000. ISIS, the intron information system, reveals the high frequency of alternative splicing in the human genome. *Nature Genet.* 24: 340–341.

Cvekl, A. and J. Piatigorsky. 1996. Lens development and crystallin gene expression: Many roles for *Pax-6*. *BioEssays* 18: 621–630.

Cvekl, A., C. M. Sax, X. Li, J. B. McDermott and J. Piatigorsky. 1995. Pax-6 and lens-specific transcription of the chicken δ1-crystallin gene. *Proc. Natl. Acad. Sci. USA* 92: 4681–4685.

DeChiara, T. M., E. J. Robertson and A. Efstratiadis. 1991. Parental imprinting of the mouse insulin-like growth factor II gene. *Cell* 64: 849–859.

Decker, C. J. and R. Parker. 1995. Mechanisms of mRNA degradation in eukaryotes. *Trends Biochem.* 19: 336–340.

Driever, W. and C. Nüsslein-Volhard. 1988a. The bicoid protein determines position in the *Drosophila* embryo in a concentration-dependent manner. *Cell* 54: 95–105.

Driever, W. and C. Nüsslein-Volhard. 1988b. A gradient of bicoid protein in *Drosophila*. *Cell* 54: 83–93.

Driscoll, D. J. and B. R. Migeon. 1990. Sex difference in methylation of single-copy genes in human meiotic germ cells: Implications for X chromosome inactivation, parental imprinting, and the origin of PGC mutations. *Somat. Cell Mol. Genet.* 16: 267–268.

Edgar, B., F. Sprenger, R. J. Duronio, P. Leopold and P. O'Farrell. 1994. MPF regulation during the embryonic cell cycles of *Drosophila*. *Genes Dev.* 8: 440–453.

Epstein, J., T. Cai, T. Glaser, L. Jepeal and R. Maas. 1994. Identification of a *Pax* paired domain recognition sequence and evidence for DNA-dependent conformational changes. *J. Biol. Chem.* 269: 8355–8361.

Ferguson-Smith, A. C. and A. M. Surani. 2001. Imprinting and the epigenetic asymmetry between parental genomes. *Science* 293: 1086–1089.

Ferrandon, D., L. Elphick, C. Nüsslein-Volhard and D. St. Johnston. 1994. Staufen protein associates with the 3′ UTR of *bicoid* to form particles that move in a microtubule-dependent manner. *Cell* 79: 1221–1232.

Ferré-D'Amaré, A. R., G. C. Predergast, E. B. Ziff and S. K. Burley. 1993. Recognition by Max of its cognate DNA through a dimeric bHLH/Z domain. *Nature* 363: 38–45.

Gagnon, M. L., L. M. Angerer and R. C. Angerer. 1992. Posttranscriptional regulation of ectoderm-specific gene expression in early sea urchin embryos. *Development* 114: 457–467.

Gartler, S. M., R. M. Liskay and N. Grant. 1973. Two functional X chromosomes in human fetal oocytes. *Exp. Cell Res.* 82: 464–466.

Gavis, E. R. and R. Lehmann. 1994. Translational regulation of *nanos* by RNA localization. *Nature* 369: 315–318.

Gedamu, L. and G. H. Dixon. 1978. Effect of enzymatic decapping on protamine messenger RNA translation in wheat-germ S-30. *Biochem. Biophys. Res. Comm.* 85: 114–125.

Glaser, T., L. Jepeal, J. G. Edwards, S. R. Young, J. Favor and R. L. Maas. 1994. *PAX6* gene dosage effects in a family with congenital cataracts, aniridia, anophthalmia, and central nervous system defects. *Nature Genet.* 8: 463–471.

Grosshans, H. and F. J. Slack. 2002. MicroRNAs: Small is plentiful. *J. Cell Biol.* 156: 17–21.

Groudine, M. and H. Weintraub. 1981. Activation of globin genes during chick development. *Cell* 24: 393–401.

Gruenbaum, Y., H. Ceder and A. Razin. 1982. Substrate and sequence specificity of a eukaryotic DNA methylase. *Nature* 295: 620–622.

Guyette, W. A., R. J. Matusik and J. M. Rosen. 1979. Prolactin-mediated transcriptional and post-transcriptional control of casein gene expression. *Cell* 17: 1013–1023.

Halder, G., P. Callaerts and W. J. Gehring. 1995. Induction of ectopic eyes by targeted expression of the *eyeless* gene in *Drosophila*. *Science* 267: 1788–1792.

Hammes, A. and 7 others. 2001. The splice variants of the Wilms' tumor 1 gene have distinct functions during sex determination and nephron formation. *Cell* 106: 319–329.

Hastie, N. D. 2001. Life, sex, and WT1 isoforms: Three amino acids can make all the difference. *Cell* 106: 391–394.

Heard, E., C. Rougeulle, D. Arnaud, P. Avner, C. D. Allis and D. L. Spector. 2001. Methylation of histone H3 at Lys-9 is an early mark on the X chromosome during X inactivation. *Cell* 107: 727–738.

Hemesath, T. J. and 9 others. 1994. Microphthalmia, a critical factor in melanocyte development, defines a discrete transcription factor family. *Genes Dev.* 8: 2770–2780.

Hodgman, R., J. Tay, R. Mendez and J. D. Richter. 2001. CPEB phosphorylation and cytoplasmic polyadenylation are catalyzed by the kinase IAK1/Eg2 in maturing mouse oocytes. *Development* 128: 2815–2822.

Horowitz, D. S. and A. R. Krainer. 1995. Mechanisms for selecting 5′ splice sites in mammalian pre-mRNA splicing. *Trends Genet.* 10: 100–106.

Hotchkiss, R. D. 1948. The quantitative separation of purines, pyrimidines, and nucleosides by paper chromatography. *J. Biol. Chem.* 175: 315–332.

Hui, H. and R. Perfetti 2002. Pancreas duodenum homeobox-1 regulates pancreas development and islet cell function in adulthood. *Eur. J. Endocrinol.* 146: 129–141.

Hussain, M. A. and J.-F. Habener. 1999. Glucagon gene transcription activation mediated by synergistic interactions of Pax6 and Cdx2 with the p300 co-activator. *J. Biol. Chem.* 274: 28950–28957.

Ichtchenko, K., Y. Hata, T. Nguyen, B. Ullrich, M. Missler, C. Moomaw and T. C. Südhof. 1995. Neuroglian 1: A splice-site specific ligand for β-neurexins. *Cell* 81: 435–443.

Jeppesen, P. and B. M. Turner. 1993. The inactive X chromosome in female mammals is distinguished by a lack of histone H4 acetylation, a cytogenetic marker for gene expression. *Cell* 74: 281–289.

Jepson, K. and 14 others. 2000. Combinatorial roles of the nuclear receptor corepressor in transcription and development. *Cell* 102: 753–763.

Jiang, Y.-H., T.-F. Tsai, J. Bressler and A. L. Beaudet. 1998. Imprinting in Angelman and Prader-Willi syndromes. *Curr. Opin. Genet. Dev.* 8: 334–342.

Jones, P. L. and 7 others. 1998. Methylated DNA and MeCP2 recruit histone deacetylase to repress transcription. *Nature Genet.* 19: 187–191.

Jursnich, V. A. and K. C. Burtis. 1993. A positive role in differentiation of the male doublesex protein of *Drosophila*. *Dev. Biol.* 155: 235–249.

Kafri, T. and 7 others. 1992. Developmental pattern of gene-specific DNA methylation in the mouse embryo and germ line. *Genes Dev.* 6: 705–715.

Kallunki, P., S. Jenkinson, G. M. Edelman and F. S. Jones. 1995. Silencer elements modulate the expression of the gene for neuron-glia cell adhesion molecule, Ng-CAM. *J. Biol. Chem.* 270: 21291–21298.

Kallunki, P., G. M. Edelman and F. S. Jones. 1997. Tissue-specific expression of the L1 cell adhesion molecule is modulated by the neural restrictive silencer element. *J. Cell Biol.* 138: 1343–1355.

Kamachi, Y., M. Uchikawa, J. Collignon, R. Lovell-Badge and H. Kondoh. 1998. Involvement of Sox1, 2, and 3 in the early and subsequent molecular events of lens induction. *Development* 125: 2521–2532.

Kammandel, B., K. Chowdhury, A. Stoykova, S. Aparicio, S. Brenner and P. Gruss. 1998. Distinct *cis*-essential modules direct the time-space pattern of *Pax6* gene activity. *Dev. Biol.* 205: 79–97.

Kastern, W. H., M. Swindlehurst, C. Aaron, J. Hooper and S. J. Berry. 1982. Control of mRNA translation in oocytes and developing embryos of giant moths. I. Functions of the 5′ terminal "cap" in the tobacco hornworm *Manduca sexta*. *Dev. Biol.* 89: 437–449.

Kawahara, A., C. B. Chien and I. B. Dawid. 2002. The homeobox gene *mbx* is involved in eye and tectum development. *Dev. Biol.* 248:107–117.

Kay, G. F., G. D. Penny, D. Patel, A. Ashworth, N. Brockdorrf and S. Rastan. 1993. Expression of *Xist* during mouse development suggests a role in the initiation of X chromosome inactivation. *Cell* 72: 171–182.

Keith, D. H., J. Singersam and A. D. Riggs. 1986. Active X-chromosome DNA is unmethylated at eight CCGG sites clustered in a guanine-plus-cytosine-rich island at the 5′ end of the gene for phosphoglycerate kinase. *Mol. Cell Biol.* 6: 4122–4125.

Keshet, I., J. Lieman-Hurwitz and H. Cedar. 1986. DNA methylation affects the formation of active chromatin. *Cell* 44: 535–543.

Kleene, K. C. and T. Humphreys. 1977. Similarity of hnRNA sequences in blastula and pluteus stage sea urchin embryos. *Cell* 12: 143–155.

Kleene, K. C. and T. Humphreys. 1985. Transcription of similar sets of rare maternal RNAs and rare nuclear RNAs in sea urchin blastulae and adult coelomocytes. *J. Embryol. Exp. Morphol.* 85: 131–149.

Knoll, J. H. M., R. D. Nicholls, R. E. Magenis, J. M. Graham, Jr., M. Lalande and S. A. Latt. 1989. Angelman and Prader-Willi syndromes share a common chromosome 15 deletion but differ in the parental origin of the deletion. *Am. J. Med. Genet.* 32: 285–290.

Kornberg, R. D. and J. D. Thomas. 1974. Chromatin structure: Oligomers of histones. *Science* 184: 865–868.

Kwan, K.-M. 2002. Conditional alleles in mice: Practical considerations for tissue-specific knockouts. *genesis* 32: 49–62.

Lagos-Quintana, M., R. Rauhut, W. Lendeckel and T. Tuschl. 2001. Coding for small expressed RNAs. *Science* 294: 853–858.

Lau, N. C., L. P. Lim, E. G. Weinstein and D. P. Bartel. 2001. An abundant class of tiny RNAs with probable regulatory roles in *Caenorhabditis elegans*. *Science* 294: 858–862.

Lawn, R. M., A. Efstratiadis, C. O'Connell and T. Maniatis. 1980. The nucleotide sequence of the human β-globin gene. *Cell* 21: 647–651.

Lee, J. T. 2000. Disruption of imprinted X inactivation by parent-of-origin effects of *Tsix*. *Cell* 103: 17–27.

Lee, J. T. and N. Lu. 1999. Targeted mutagenesis of *Tsix* leads to nonrandom X inactivation. *Cell* 99: 47–57.

Lee, J. T., W. M. Strauss, J. A. Dausman and R. Jaenisch. 1996. A 450-kb transgene displays properties of the mammalian X-inactivation center. *Cell* 86: 83–95.

Lee, R. C. and V. Ambros. 2001. An extensive class of small RNAs in *Caenorhabditis elegans*. *Science* 294: 862–864.

Lee, R. C., R. L. Feinbaum and V. Ambros. 1993. The *C. elegans* heterochromatic gene *lin-4* encodes small RNAs with antisense complementarity to *lin-14*. *Cell* 75: 843–855.

Lee, T. I. and R. A. Young. 1998. Regulation of gene expression by TBP-associated proteins. *Genes Dev.* 12: 1398–1408.

Lieberfarb, M. E., T. Chu, C. Wreden, W. Theurkauf, J. P. Gergen and S. Strickland. 1996. Mutations that perturb poly(A)-dependent maternal mRNA activation block the initiation of development. *Development* 122: 579–588.

Litt, M. D., M. Simpson, M. Gaszner, C. D. Allis and G. Felsenfeld. 2001. Correlation between histone lysine methylation and developmental changes at the chicken β-globin locus. *Science* 293: 2453–2455.

Lucchesi, J. C. and J. E. Manning. 1987. Gene dosage and compensation in *Drosophila melanogaster*. *Adv. Genet.* 24: 371–429.

Luger , K., A. W. Mäder, R. K. Richmond, D. F. Sargent and T. J. Richmond. 1997. Crystal structure of the nucleosome core particle at 2.8 Å resolution. *Nature* 389: 251–260.

Luo, M. L., Z. Zhou, K. Magni, C. Christoforides, J. Rappsilber, M. Mann and R. Reed. 2001. Pre-mRNA splicing and mRNA export linked by direct interactions between UAP56 and Aly. *Nature* 413: 644–647.

Lyon, M. F. 1961. Gene action in the X chromosome of the mouse (*Mus musculus* L.). *Nature* 190: 372–373.

MacDougall, C., D. Harbison and M. Bownes. 1995. The developmental consequences of alternative splicing in sex determination and differentiation in *Drosophila*. *Dev. Biol.* 172: 353–376.

Maniatis, T., S. Goodbourn and J. A. Fischer. 1987. Regulation of inducible and tissue-specific gene expression. *Science* 236: 1237–1245.

Marahrens, Y., J. Loring and R. Jaenisch. 1998. Role of the *Xist* gene in X chromosome choosing. *Cell* 92: 657–665.

Mavilio, F. and 9 others. 1983. Molecular mechanisms for human hemoglobin switching: Selective undermethylation and expression of globin genes in embryonic, fetal, and adult erythroblasts. *Proc. Natl. Acad. Sci. USA* 80: 6907–6911.

McArthur, M. and J. O. Thomas. 1996. A preference of histone H1 for methylated DNA. *EMBO J.* 15: 1705–1715.

Mendez, R. and J. D. Richter. 2001. Translational control by CPEB: A means to the end. *Nature Rev. Mol. Cell Biol.* 2: 521–529.

Mendez, R., K. G. Murthy, K. Ryan, J. L. Manley and J. D. Richter. 2000. Phosphorylation of CPEB by Eg2 mediates the recruitment of CPSF into an active cytoplasmic polyadenylation complex. *Mol. Cell* 6: 1253–1259.

Mermoud, J. E., B. Popova, A. H. Peters, T. Jenuwein and N. Brockdorff. 2002. Histone H3 lysine 9 methylation occurs rapidly at the onset of random X chromosome inactivation. *Curr. Biol.* 12: 247–251.

Migeon, B. R. and K. Jelalian. 1977. Evidence for two active X chromosomes in germ cells of female before meiotic entry. *Nature* 269: 242–243.

Migeon, B. R., M. M. Holland, D. J. Driscoll and J. C. Robinson. 1991. Programmed demethylation in CpG islands during human fetal development. *Somatic Cell Mol. Genet.* 17: 159–168.

Migeon, B. R., C. H. Lee, A. K. Chowdhury and H. Carpenter. 2002. Species differences in *TSIX/Tsix* reveal the roles of these genes in X-chromosome inactivation. *Am. J. Hum. Genet.* 71: 286–293.

Mizzen, C. A. and 11 others. 1996. The $TAF_{II}250$ subunit of TFIID has histone acetyltransferase activity. *Cell* 87: 1261–1270.

Monk, M., M. Boubelik and S. Lehnert. 1987. Temporal and regional changes in DNA methylation in the embryonic, extraembryonic, and germ cell lineages during mouse embryo development. *Development* 99: 371–382.

Muta, M., Y. Kamachi, A. Yoshimoto, Y. Higashi and H. Kondoh. 2002. Distinct roles of Sox2, Pax6, and Maf transcription factors in the regulation of lens-specific δ1 crystallin enhancer. *Genes Cells* 7: 791–805.

Myers, L. C. and R. D. Kornberg. 2000. Mediator of transcriptional regulation. *Annu. Rev. Biochem.* 69: 729–749.

Nabel-Rosen, H., G. Volohonsky, A. Reuveny, R. Zaidel-bar and T. Volk. 2002. Two isoforms of the *Drosophila* RNA binding protein How act in opposite directions to regulate tendon cell differentiation. *Dev. Cell* 2: 183–193.

Nagoshi, R. N., M. McKeown, K. C. Burtis, J. M. Belote and B. S. Baker. 1988. The control of alternative splicing at genes regulating sexual differentiation in *D. melanogaster*. *Cell* 53: 229–236.

Nakayama, A., M.-T. Nguyen, C. C. Chen, K. Opdecamp, C. A. Hodgkinson and H. Arnheiter. 1998. Mutations in *microphthalmia*, the mouse homolog of the human deafness gene *MITF*, affect neuroepithelial and neural crest-derived melanocytes differently. *Mech. Dev.* 70: 155–166.

Nan, X., H.-H. Ng, C. A. Johnson, C. D. Laherty, B. M. Turner, R. N. Eisenman and A. Bird. 1998. Transcriptional repression by the methyl-CpG-binding protein MeCP2 involves a histone deacetylase complex. *Nature* 393: 386–389.

Nicholls, R. D. 1998. Imprinting in Prader-Willi and Angelman syndromes. *Trends Genet.* 14: 194–200.

Norris, D. P., D. Patel, G. F. Kay, G. D. Penny, N. Brockdorff, S. A. Sheardown and S. Rastan. 1994. Evidence that random and imprinted *Xist* expression is controlled by preemptive methylation. *Cell* 77: 41–51.

Offield, M., F. N. Hirsch and R. M. Grainger. 2000. The use of *Xenopus tropicalis* transgenic lines for studying lens developmental timing in living embryos. *Development* 127: 1789–1797.

Ogryzko, V. V., R. L. Schlitz, V. Russanova, B. H. Howard and Y. Nakatani. 1996. The transcriptional coactivators p300 and CBP are histone acetyltransferases. *Cell* 87: 953–959.

O'Kane, C. J. and W. J. Gehring. 1987. Detection in situ of genomic regulatory elements in *Drosophila*. *Proc. Natl. Acad. Sci. USA* 84: 9123–9127.

Orphanides, G. and D. Reinberg. 2002. A unified theory of gene expression. *Cell* 108: 439–451.

Panning, B., J. Dausman and R. Jaenisch. 1997. X chromosome inactivation is mediated by *Xist* stabilization. *Cell* 90: 907–916.

Parvis, F., J. Li, K. H. Kaestner and S. A. Duncan. 2002. Generation of conditionally null allele of *hnf4α*. *genesis* 32: 130–133.

Pennisi, E. 2000. Matching the transcription machinery to the right DNA. *Science* 288: 1372–1373.

Penny, G. D., G. F. Kay, S. A. Sheardown, S. Rastan and N. Brockdorff. 1996. Requirement for *Xist* in X chromosome inactivation. *Nature* 379: 131–137.

Plaza, S., C. Dozier and S. Saule. 1993. Quail *PAX6* (*PAX-QNR*) encodes a transcription factor able to bind and transactivate its own promoter. *Cell Growth Diff.* 4: 1041–1050.

Price, E. R. and 7 others. 1998. Lineage-specific signaling in melanocytes: c-Kit stimulation recruits p300/CBP to microphthalmia. *J. Biol. Chem.* 273: 33042–33047.

Rea, S. and 10 others. 2000. Regulation of chromatin structure by site-specific histone H3 methyltransferases. *Nature* 406: 593–599.

Rebagliati, M. R., D. L. Weeks, R. P. Harvey and D. A. Melton. 1985. Identification and cloning of localized maternal RNAs from *Xenopus* eggs. *Cell* 42: 769–777.

Rideout, W. M. III, K. Eggan and R. Jaenisch. 2001. Nuclear cloning and epigenetic reprogramming of the genome. *Science* 293: 1093–1098.

Roopra, A., S. Sharling, I. C. Wood, T. Briggs, U. Bachfischer, A. J. Paquette and N. J. Buckley. 2000. Transcriptional repression by neuron-restrictive silencer factor is mediated via the Sin3-histone deacetylase complex. *Mol. Cell Biol.* 20: 2147–2157.

Rosenthal, E., T. Hunt and J. V. Ruderman. 1980. Selective translation of mRNA controls the pattern of protein synthesis during early development of the surf clam, *Spisula solidissima*. *Cell* 20: 487–495.

Rosenthal, L. and J. Meldolesi. 1989. α-Latrotoxin and related toxins. *Pharm. Ther.* 42: 115–134.

Roux, W. 1894. The problems, methods, and scope of developmental mechanics. *In* W. M. Wheeler (trans.), *Biological Lectures of the Marine Biology Laboratory, Woods Hole*. Ginn, Boston, pp. 149–190.|

Sado, T., Z. Wang, H. Sasaki and E. Li. 2001. Regulation of imprinted X-chromosome inactivation in mice by *Tsix*. *Development* 128: 1275–1286.

Saito, Y. and 8 others. 2000. The developmental and aging changes of Down's syndrome cell adhesion molecule expression in normal and Down's syndrome brains. *Acta Neuropathol.* 100: 654–664.

Sallés, F. J., M. E. Lieberfarb, C. Wreden, J. P. Gergen and S. Strickland. 1994. Coordinate initiation of *Drosophila* development by regulated polyadenylation of maternal messenger RNAs. *Science* 266: 1996–1999.

Sander, M., A. Neubüser, H. Ee, G. R. Martin and M. S. German. 1997. Genetic analysis reveals that *Pax6* is required for normal transcription of pancreatic hormone genes and islet development. *Genes Dev.* 11: 1662–1673.

Sanford, J. P., H. J. Clark, V. M. Chapman and J. Rossant. 1987. Differences in DNA methylation during oogenesis and spermatogenesis and their persistence during early embryogenesis in the mouse. *Genes Dev.* 1: 1039–1046.

Sauer, F., J. D. Fondell, Y. Ohkuma, R. G. Roeder and H. Jäckle. 1995. Control of transcription by Krüppel through interactions with TFIIB and TFIIE. *Nature* 375: 162–165.

Schlissel, M. S. and D. D. Brown. 1984. The transcriptional regulation of *Xenopus* 5S RNA genes in chromatin: The roles of active stable transcription complex and histone H1. *Cell* 37: 903–913.

Schmucker, D. and 7 others. 2000. *Drosophila* Dscam is an axon guidance receptor exhibiting extraordinary molecular diversity. *Cell* 101: 671–684.

Schoenherr, C. J. and D. J. Anderson. 1995. The neuron-restrictive silencer factor (NRSF): A coordinate repressor of multiple neuron-specific genes. *Science* 267: 1360–1363.

Shatkin, A. J. 1976. Capping of eucaryotic mRNAs. *Cell* 9: 645–653.

Shaw, G. and R. Kamen. 1986. A conserved AU sequence from the 3′ untranslated region of *GM-CSF* mRNA mediates selective mRNA degradation. *Cell* 46: 659–667.

Sheardown, S. A. and 9 others. 1997. Stabilization of *Xist* RNA mediates initiation of X chromosome inactivation. *Cell* 91: 99–107.

Sheiness, D. and J. E. Darnell. 1973. Polyadenylic segment in mRNA becomes shorter with age. *Nature New Biol.* 241: 265–268.

Smibert, C. A., J. E. Wilson, K. Kerr and P. M. Macdonald. 1996. Smaug protein represses translation of unlocalized *nanos* mRNA in the *Drosophila* embryo. *Genes Dev.* 10: 2600–2609.

Smith, E. R., C. D. Allis and J. C. Lucchesi. 2001. Linking global histone acetylation to the transcription enhancement of X-chromosomal genes in *Drosophila* males. *J. Biol. Chem.* 276: 31483–31486.

Sopta, M., Z. F. Burton and J. Greenblatt. 1989. Structure and associated DNA helicase activity of a general transcription factor that binds to RNA polymerase II. *Nature* 341: 410–415.

Sosnowski, B. A., J. M. Belote and M. McKeown. 1989. Sex-specific alternative splicing of RNA from the *transformer* gene results from sequence-specific splice site blockage. *Cell* 58: 449–459.

Standart, N., T. Hunt and J. V. Ruderman. 1986. Differential accumulation of ribonucleotide reductase subunits in clam oocytes: The large subunit is stored as a polypeptide, the small subunit as untranslated mRNA. *J. Cell Biol.* 103: 2129–2136.

Stebbins-Boaz, B., Q. Cao, C. H. de Moor, R. Mendez and J. D. Richter. 1999. Maskin is a CPEB-associated factor that transiently interacts with eIF-4E. *Mol. Cell* 4: 1017–1027.

Steingrímsson, E. and 10 others. 1994. Molecular basis of mouse *microphthalmia* (*mi*) mutations helps explain their developmental and phenotypic consequences. *Nature Genet.* 8: 256–263.

Sträßer, K. and E. Hurt. 2001. Splicing factor Sub2p is required for nuclear mRNA export through its interaction withYra1p. *Nature* 413: 648–652.

Sugimoto, M., S.-S. Tan and N. Takagi. 2000. X chromosome inactivation revealed by the X-linked *lacZ* transgene activity in periimplantation mouse embryos. *Int. J. Dev. Biol.* 44: 177–182.

Tagaki, N. 1974. Differentiation of X chromosomes in early female mouse embryos. *Exp. Cell Res.* 86: 127–135.

Tagaki, N. and K. Abe. 1990. Detrimental effects of two active X chromosomes on early mouse development. *Development* 109: 189–201.

Tamaru, H. and E. U. Selker. 2001. A histone H3 methyltransferase controls DNA methylation in *Neurospora crassa*. *Nature* 414: 277–283.

Thoma, F., T. Koller and A. Klug. 1979. Involvement of histone H1 in the organization of the nucleosome and of the salt-dependent superstructures of chromatin. *J. Cell Biol.* 83: 403–427.

Tian, M. and T. Maniatis. 1992. Positive control of pre-mRNA splicing in vitro. *Science* 256: 237–240.

Tian, M. and T. Maniatis. 1993. A splicing enhancer complex controls alternative splicing of *doublesex* pre-mRNA. *Cell* 74: 105–115.

Trudel, M. and F. Constantini. 1987. A 3′ enhancer contributes to the stage-specific expression of the human β-globin gene. *Genes Dev.* 1: 954–961.

Ullrich, B., Y. A. Uskaryov and T. C. Südhof. 1995. Cartography of neurexins: More than 1000 isoforms generated by alternative splicing and expressed in distinct subsets of neurons. *Neuron* 14: 497–507.

Valcárcel, J., R. Singh, P. D. Zamore and M. R. Greene. 1993. The protein Sex-lethal antagonizes the splicing factor U2AF to regulate alternative splicing of *transformer* pre-mRNA. *Nature* 362: 171–175.

van der Ploeg, L. H. T. and R. D. Flavell. 1980. DNA methylation in the human γ–δ–β-globin locus in erythroid and non-erythroid cells. *Cell* 19: 947–958.

Wagenaar, E. B. and D. Mazia. 1978. The effect of emetine on the first cleavage division of the sea urchin, *Strongylocentrotus purpuratus*. *In* E. R. Dirksen, D. M. Prescott and L. F. Fox (eds.), *Cell Reproduction: In Honor of Daniel Mazia*. Academic Press, New York, pp. 539–545.

Walker, M. D., T. Edlund, A. M. Boulet and W. J. Rutter. 1983. Cell-specific expression controlled by the 5′ flanking region of the insulin and chymotrypsin genes. *Nature* 306: 557–561.

Weintraub, H. 1984. Histone H1-dependent chromatin superstructures and the suppression of gene activity. *Cell* 38: 17–27.

Weintraub, H. 1985. Assembly and propagation of repressed and derepressed chromosomal states. *Cell* 42: 705–711.

Wickens, M. and K. Takayama. 1995. Deviants —or emissaries? *Nature* 367: 17–18.

Wightman, B., I. Ha and G. Ruvkun. 1993. Posttranslational regulation of the heterochronic gene *lin-14* by lin-4 mediates temporal pattern formation in *C. elegans*. *Cell* 75: 855–862.

Williams, S. C., C. R. Altmann, R. L. Chow, A. Hemmati-Brivanlou and R. A. Lang. 1998. A highly conserved lens transcriptional control element from the *Pax-6* gene. *Mech. Dev.* 73: 225–229.

Wilson, T. and R. Treisman. 1988. Removal of poly(A) and consequent degradation of *c-fos* mRNA facilitated by 3′ AU-rich sequences. *Nature* 336: 396–399.

Wold, B. J., W. H. Klein, B. R. Hough-Evans, R. J. Britten and E. H. Davidson. 1978. Sea urchin embryo mRNA sequences expressed in nuclear RNA of adult tissues. *Cell* 14: 941–950.

Wolf, S. F., S. Dintgis, D. Toniolo, G. Persico, K. D. Lunnen, J. Axelman and B. R. Migeon. 1984. Complete concordance between glucose-6-phosphate dehydrogenase activity and hypomethylation of 3′ CpG clusters: Implication for X chromosome dosage compensation. *Nucleic Acids Res.* 12: 9333–9348.

Wolfe, S. L. 1993. *Molecular and Cellular Biology*. Wadsworth, Belmont, CA.

Woychik, N. A. and M. Hampsey. 2002. The RNA polymerase II machinery: Structure illuminates function. *Cell* 108: 453–463.

Xu, W., M. A. S. Rould, S. Jun, C. Desplan and S. O. Pääbo. 1995. Crystal structure of a paired domain-DNA complex at 2.5 Å resolution reveals structural basis for *Pax* developmental mutations. *Cell* 80: 639–650.

Yamakawa, K., Y. K. Huot , M. A. Haendelt, R. Hubert, X.-N. Chen, G. E. Lyons and J. R. Korenberg. 1998. *DSCAM:* A novel member of the immunoglobulin superfamily maps in a Down syndrome region and is involved in the development of the nervous system. *Hum. Mol. Genet.* 7: 227–237.

Yasumoto, K.-I., K. Yokoyama, K. Takahashi, Y. Tomita and S. Shibihara. 1997. Functional analysis of *microphthalmia*-associated transcription factor in pigment cell-specific transcription of human tyrosinase family genes. *J. Biol. Chem.* 272: 503–509.

Zeschingk, M., B. Schmitz, B. Dittrich, K. Buiting, B. Horsthemke and W. Doerfler. 1997. Imprinted segments in the human genome: Different DNA methylation patterns in the Prader-Willi/Angelman syndrome region as determined by the genomic sequencing method. *Hum. Mol. Genet.* 6: 387–395.

Zhao, K., C. M. Hart and U. K. Laemmli. 1995. Visualization of chromosomal domains with boundary-element associated factor BEAF-32. *Cell* 81: 879–889.

chapter 6 *Cell-cell communication in development*

*All that you touch
You Change.
All that you Change
Changes you.
The only lasting truth
Is Change.*

OCTAVIA BUTLER (1998)

In dealing with such a complex system as the developing embryo, it is futile to inquire whether a certain organ rudiment is "determined" and whether some feature of its surroundings, to the exclusion of others, "determines" it. A score of different factors may be involved and their effects most intricately interwoven. In order to resolve this tangle we have to inquire into the manner in which the system under consideration reacts with other parts of the embryo at successive stages of development and under as great a variety of experimental conditions as it is possible to impose.

R. G. HARRISON (1933)

THE FORMATION OF ORGANIZED BODIES has been one of the great sources of wonder for humankind. Indeed, the "miracle of life" seems just that—matter has become organized in such a way that it lives.* While each organism starts off as a single cell, the progeny of that cell form complex structures—tissues and organs—that are themselves integrated into larger systems. Probably no one better recognizes how remarkable life actually is than the developmental biologists who get to study how all this complexity arises. In the past decade, developmental biologists have started to answer some of the most important questions of natural science: We have begun to understand how organs form.

Induction and Competence

Organs are complex structures composed of numerous types of tissues. In the vertebrate eye, for example, light is transmitted through the transparent corneal tissue and focused by the lens tissue (the diameter of which is controlled by muscle tissue), eventually impinging on the tissue of the neural retina. The precise arrangement of tissues in this organ cannot be disturbed without impairing its function. Such coordination in the construction of organs is accomplished by one group of cells changing the behavior of an adjacent set of cells, thereby causing them to change their shape, mitotic rate, or fate. This kind of interaction at close range between two or more cells or tissues of different histories and properties is called proximate interaction, or **induction**.[†] There are at least two components to every inductive interaction. The first component is the inducer: the tissue that produces a signal (or signals) that changes the cellular behavior of the other tissue. The second component, the tissue being induced, is the responder.

Not all tissues can respond to the signal being produced by an inducer. For instance, if the optic vesicle (presumptive retina) of *Xenopus laevis* is placed in an ec-

---

*The twelfth-century rabbi and physician Maimonides (1190) framed the question of morphogenesis beautifully when he noted that the pious persons of his day believed that an angel of God had to enter the womb to form the organs of the embryo. How much more powerful a miracle would life be, he asked, if the Deity had made matter such that it could generate such remarkable order without a matter-molding angel having to intervene in every pregnancy? The idea of an angel was still part of the embryology of the Renaissance. The problem addressed today is the secular version of Maimonides' question: How can matter alone construct the organized tissues of the embryo?

[†]These inductions are often called "secondary" inductions, while the tissue interactions that generate the neural tube are called "primary embryonic induction." However, there is no difference in the molecular nature of "primary" and "secondary" induction. Primary embryonic induction will be detailed in Chapters 10 and 11.

Figure 6.1
Ectodermal competence and the ability to respond to the optic vesicle inducer in *Xenopus*. The optic vesicle is able to induce lens formation in the anterior portion of the ectoderm (1), but not in the presumptive trunk and abdomen (2). If the optic vesicle is removed (3), the surface ectoderm forms either an abnormal lens or no lens at all. (4) Most other tissues are not able to substitute for the optic vesicle.

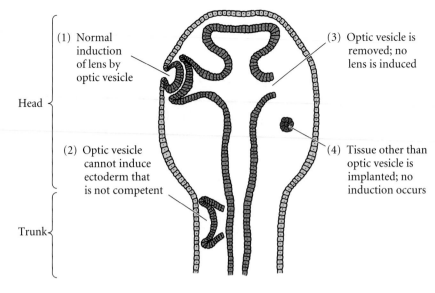

(1) Normal induction of lens by optic vesicle

(3) Optic vesicle is removed; no lens is induced

Head

(2) Optic vesicle cannot induce ectoderm that is not competent

(4) Tissue other than optic vesicle is implanted; no induction occurs

Trunk

topic location (i.e., in a different place from where it normally forms) underneath the *head* ectoderm, it will induce that ectoderm to form lens tissue. Only the optic vesicle appears to be able to do this; therefore, it is an inducer. However, if the optic vesicle is placed beneath ectoderm in the *flank* or *abdomen* of the same organism, that ectoderm will not be able to form lens tissue. Only the head ectoderm is competent to respond to the signals from the optic vesicle by producing a lens* (Figure 6.1; Saha et al. 1989; Grainger 1992).

This ability to respond to a specific inductive signal is called **competence** (Waddington 1940). Competence is not a passive state, but an actively acquired condition. For example, in the developing mammalian eye, the Pax6 protein appears to be important in making the ectoderm competent to respond to the inductive signal from the optic vesicle. *Pax6* gene expression is seen in the head ectoderm, which can respond to the optic vesicle by forming a lens, and it is not seen in other regions of the surface ectoderm (see Figure 4.17; Li et al. 1994). The importance of Pax6 as a **competence factor** was demonstrated by recombination experiments using embryonic rat eye tissue (Fujiwara et al. 1994). The homozygous *Pax6* mutant rat has a phenotype similar to the homozygous *Pax6* mutant mouse (see Chapter 4), lacking eyes and nose. It has been shown that part of this phenotype is due to the failure of lens induction (Figure 6.2). But which is the defective component—the optic vesicle or the surface ectoderm?

*When describing lens induction, one has to be careful to mention which species one is studying because there are numerous species-specific differences. In some species, lens induction will not occur at certain temperatures. In other species, the entire ectoderm can respond to the optic vesicle by forming lenses. These species-specific differences have made this area very difficult to study (Jacobson and Sater 1988; Saha et al. 1989; Saha 1991).

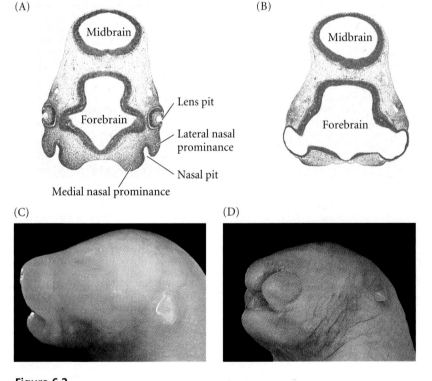

(A) Midbrain / Forebrain / Medial nasal prominance

(B) Midbrain / Forebrain

Lens pit
Lateral nasal prominance
Nasal pit

(C)

(D)

Figure 6.2
Induction of optic and nasal structures by Pax6 in the rat embryo. (A, B) Histology of wild-type (A) and homozygous *Pax6* mutant (B) embryos at day 12 of gestation shows induction of lenses and retinal development in the wild-type embryo, but not in the mutant. Similarly, neither the nasal pit nor the medial nasal prominence is induced in the mutant rats. (C) Newborn wild-type rats show prominent nose as well as (closed) eyes. (D) Newborn *Pax6* mutant rats show neither eyes nor nose. (From Fujiwara et al. 1994; photographs courtesy of M. Fujiwara.)

When head ectoderm from *Pax6*-mutant rat embryos was combined with a wild-type optic vesicle, no lenses were formed. However, when the head ectoderm from wild-type rat embryos was combined with a *Pax6*-mutant optic vesicle, lenses formed normally (Figure 6.3). Therefore, Pax6 is needed for the surface ectoderm to respond to the inductive signal from the optic vesicle; the inducing tissue does not need it. It is not known how Pax6 becomes expressed in the anterior ectoderm of the embryo, although it is thought that its expression is induced by the anterior regions of the neural plate. Competence to respond to the optic vesicle inducer (and *Pax6* expression) can be conferred on ectodermal tissue by incubating it next to anterior neural plate tissue (Henry and Grainger 1990; Li et al. 1994; Zygar et al. 1998).

There is no single inducer of the lens. Studies on amphibians suggest that the first inducers may be the pharyngeal endoderm and heart-forming mesoderm that underlie the lens-forming ectoderm during the early- and mid-gastrula stages (Jacobson 1963, 1966). The anterior neural plate may produce the next signals, including a signal that promotes the synthesis of Pax6 in the anterior ectoderm (Zygar et al. 1998; Figure 6.4). Thus, the optic vesicle appears to be *the* inducer, but the anterior ectoderm has already been induced by at least two other factors. (The situation is like that of the player who kicks the "winning goal" of a soccer match.) The optic vesicle appears to secrete two induction factors, one of which may be BMP4 (Furuta and Hogan 1998), a protein that induces the production of the Sox2 and Sox3 transcription factors. The other is thought to be FGF8, a signal that induces the appearance of the L-Maf transcription factor (Ogino and Yasuda 1998; Vogel-Höpker et al. 2000). The combination of Pax6, Sox2, Sox3, and L-Maf ensures the production of the lens.

Cascades of induction: Reciprocal and sequential inductive events

Another feature of induction is the reciprocal nature of many inductive interactions. Once the lens has formed, it can then induce other tissues. One of these responding tissues is the optic vesicle itself. Now the inducer becomes the induced. Under the influence of factors secreted by the lens, the optic vesicle becomes the optic cup, and the wall of the optic cup differentiates into two layers, the pigmented retina and the neural retina (Figure 6.5; Cvekl and Piatigorsky 1996). Such interactions are called reciprocal inductions.

At the same time, the lens is also inducing the ectoderm above it to become the cornea. Like the lens-forming ectoderm, the cornea-forming ecto-

derm has achieved a particular competence to respond to inductive signals, in this case the signals from the lens (Meier 1977; Thut et al. 2001). Under the influence of the lens, the corneal ectoderm cells become columnar and secrete multiple layers of collagen. Mesenchymal cells from the neural crest use this collagen matrix to enter the area and secrete a set of proteins (including the enzyme hyaluronidase) that further differentiate the cornea. A third signal, the hormone thyroxine, dehydrates the tissue and makes it transparent (Hay 1980; Bard 1990). Thus, there are sequential inductive events, and multiple causes for each induction.

Instructive and permissive interactions

Howard Holtzer (1968) distinguished two major modes of inductive interaction. In **instructive interaction**, a signal from

Optic vesicles	Surface ectoderm	Lens induction
Wild-type	Wild-type	Yes
Pax6⁻/Pax6⁻	Wild-type	Yes
Wild-type	*Pax6⁻/Pax6⁻*	No
Pax6⁻/Pax6⁻	*Pax6⁻/Pax6⁻*	No

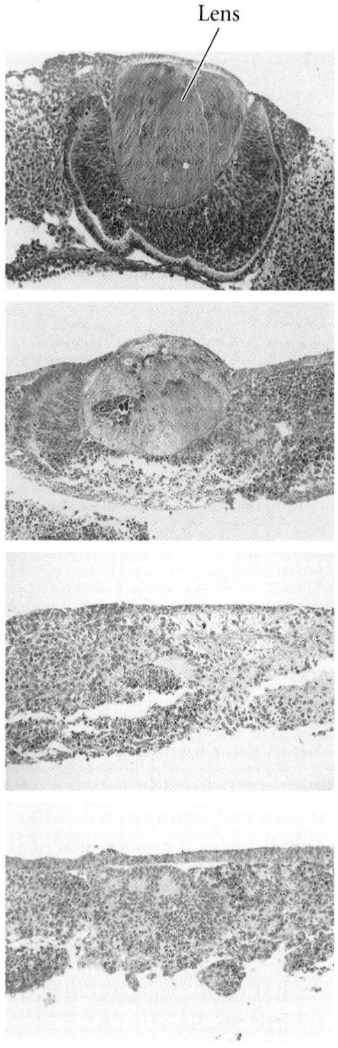

Lens

Figure 6.3
Recombination experiments show that the induction deficiency of *Pax6*-deficient rats is caused by the inability of the surface ectoderm to respond to the optic vesicle. (Photographs courtesy of M. Fujiwara.)

(A)

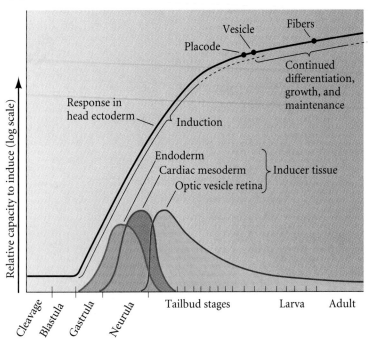

(B)

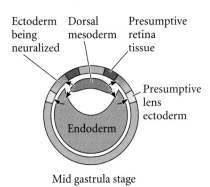 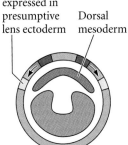 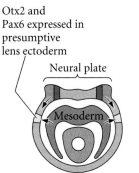 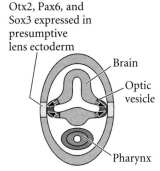

Mid gastrula stage Late gastrula stage Early neurula stage Late neurula stage

Figure 6.4
Lens induction in amphibians. (A) The additive effects of inducers, as shown by transplantation and extirpation (removal) experiments on the newt *Taricha torosa*. The ability to produce lens tissue is first induced by pharyngeal endoderm, then by cardiac mesoderm, and finally by the optic vesicle. The optic vesicle eventually acquires the ability to induce the lens and retain its differentiation. (B) Sequence of induction postulated by similar experiments performed on embryos of the frog *Xenopus laevis*. Unidentified inducers (possibly from the pharyngeal endoderm and heart-forming mesoderm) cause the synthesis of the Otx2 transcription factor in the head ectoderm during the late gastrula stage. As the neural folds rise, inducers from the anterior neural plate (including the region that will form the retina) induce *Pax6* expression in the anterior ectoderm that can form lens tissue. Expression of Pax6 protein may constitute the competence of the surface ectoderm to respond to the optic vesicle during the late neurula stage. The optic vesicle secretes factors (probably of the BMP family) that induce the synthesis of the Sox transcription factors and initiate observable lens formation. (A after Jacobson 1966; B after Grainger 1992.)

the inducing cell is necessary for initiating new gene expression in the responding cell. Without the inducing cell, the responding cell is not capable of differentiating in that particular way. For example, when the optic vesicle is experimentally placed under a new region of the head ectoderm and causes that region of the ectoderm to form a lens, that is an instructive interaction. Wessells (1977) has proposed three general principles characteristic of most instructive interactions:

1. In the presence of tissue A, responding tissue B develops in a certain way.
2. In the absence of tissue A, responding tissue B does not develop in that way.
3. In the absence of tissue A, but in the presence of tissue C, tissue B does not develop in that way.

The second type of inductive interaction is **permissive interaction**. Here, the responding tissue contains all the potentials that are to be expressed, and needs only an environment that allows the expression of these traits.* For instance, many tissues need a solid substrate containing fibronectin or laminin in order to develop. The fibronectin or laminin does not alter the type of cell that is produced, but only enables what has been determined to be expressed.

---

*It is easy to distinguish permissive and instructive interactions by an analogy with a more familiar situation. This textbook is made possible by both permissive and instructive interactions. The reviewers can convince me to change the material in the chapters. This is an instructive interaction, as the information expressed in the book is changed from what it would have been. However, the information in the book could not be expressed at all without permissive interactions with the publisher and printer.

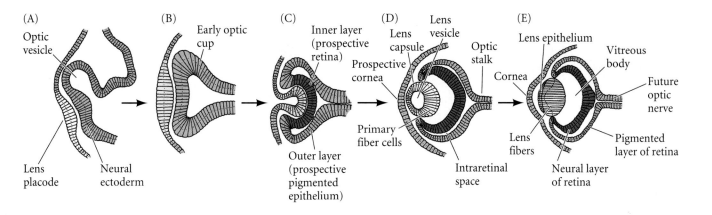

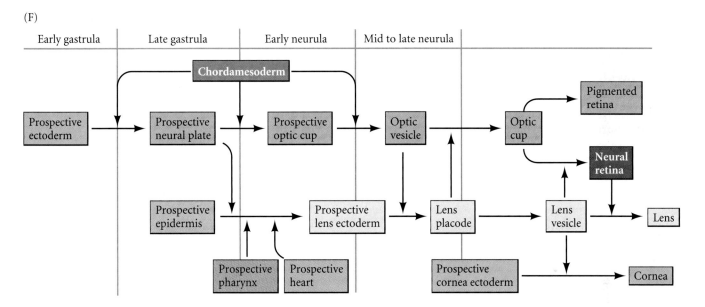

Figure 6.5

Schematic diagram of the induction of the mouse lens. (A) At embryonic day 9, the optic vesicle extends toward the surface ectoderm from the forebrain. The lens placode (the prospective lens) appears as a local thickening of the surface ectoderm near the optic vesicle. (B) By the middle of day 9, the lens placode has enlarged and the optic vesicle has formed an optic cup. (C) By the middle of day 10, the central portion of the lens-forming ectoderm invaginates, while the two layers of the retina become distinguished. (D) By the middle of day 11, the lens vesicle has formed. (E) By day 13, the lens consists of anterior cuboidal epithelial cells and elongating posterior fiber cells. The cornea develops in front of the lens. (F) Summary of some of the inductive interactions during eye development. (A–E after Cvekl and Piatigorsky 1996.)

Epithelial-mesenchymal interactions

Some of the best-studied cases of induction are those involving the interactions of sheets of epithelial cells with adjacent mesenchymal cells. These interactions are called **epithelial-mesenchymal** interactions. Epithelia are sheets or tubes of connected cells; they can originate from any germ layer. Mesenchyme refers to loosely packed, unconnected cells. Mesenchymal cells are derived from the mesoderm or neural crest. All organs consist of an epithelium and an associated mesenchyme, so epithelial-mesenchymal interactions are among the most important phenomena in nature. Some examples are listed in Table 6.1.

REGIONAL SPECIFICITY OF INDUCTION. Using the induction of cutaneous structures as our examples, we will look at the properties of epithelial-mesenchymal interactions. The first of these properties is the regional specificity of induction. Skin is composed of two main tissues: an outer epidermis (an epithelial tissue derived from ectoderm), and a dermis (a mesenchymal tissue derived from mesoderm). The chick epidermis signals the underlying dermal cells to form condensations (probably by secreting Sonic hedgehog and TGF-β2 proteins, which will be discussed below), and the condensed dermal mesenchyme responds by secreting factors that cause the epidermis to form regionally specific cutaneous structures (Figure 6.6; Nohno et al. 1995, Ting–Berreth and Chuong 1996).

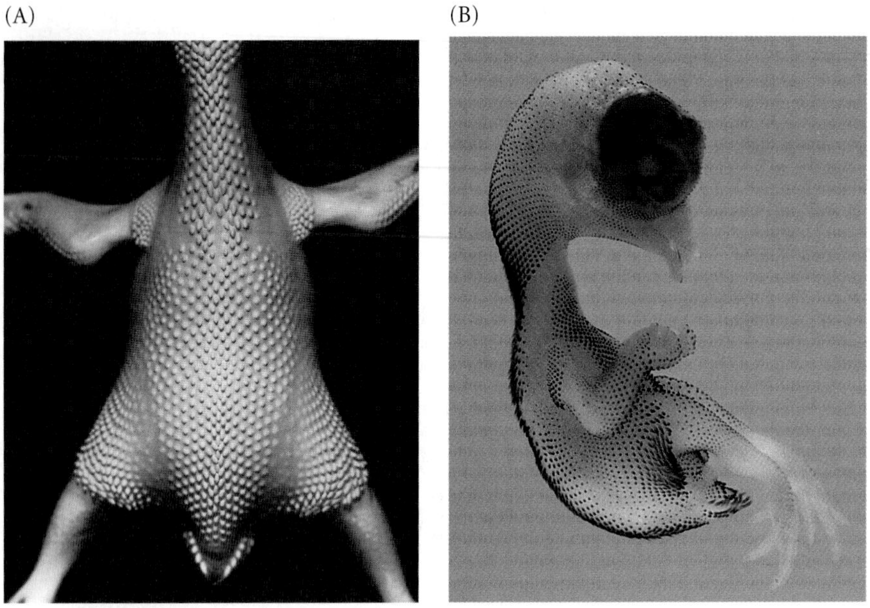

(A)

(B)

Figure 6.6
(A) Feather tracts on the dorsum of a day 9 chick embryo. Note that each feather primordium is located between the primordia of adjacent rows. (B) In situ hybridization of a day 10 chick embryo shows *Sonic hedgehog* expression (dark spots) in the ectoderm of the developing feathers and scales. (A courtesy of P. Sengal; B courtesy of W.-S. Kim and J. F. Fallon.)

These structures can be the broad feathers of the wing, the narrow feathers of the thigh, or the scales and claws of the feet. As Figure 6.7 demonstrates, the dermal mesenchyme is responsible for the regional specificity of induction in the competent epidermal epithelium. Researchers can separate the embryonic epithelium and mesenchyme from each other and recombine them in different ways (Saunders et al. 1957). The same epithelium develops cutaneous structures according to the region from which the mesenchyme was taken. Here, the

mesenchyme plays an instructive role, calling into play different sets of genes in the responding epithelial cells.

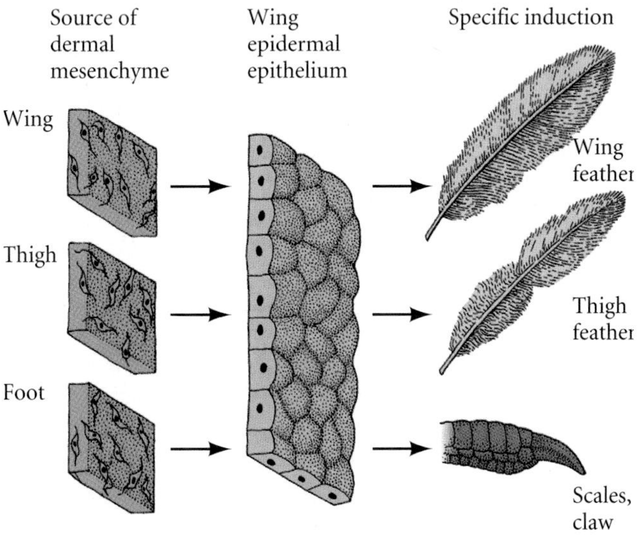

GENETIC SPECIFICITY OF INDUCTION. The second property of epithelial-mesenchymal interactions is the genetic specificity of induction. Whereas the mesenchyme may instruct the epithelium as to what sets of genes to activate, the responding epithelium can comply with these instructions only so far as its genome permits. This property was discovered through experiments involving the transplantation of tissues from one species to another. In one of the most dramatic examples of interspecific induction, Hans Spemann and Oscar Schotté (1932) transplanted flank ectoderm from an early *frog* gas-

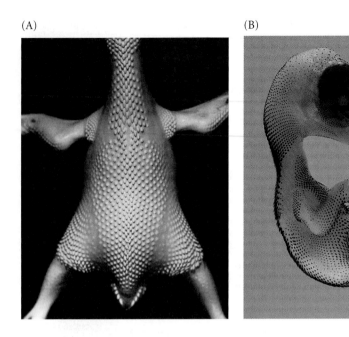

Figure 6.7
Regional specificity of induction in the chick. When cells from different regions of the dermis (mesenchyme) are recombined with the epidermis (epithelium), the type of cutaneous structure made by the epidermal epithelium is determined by the original source of the mesenchyme. (After Saunders 1980.)

TABLE 6.1 Some epithelial-mesenchymal interactions

Organ	Epithelial component	Mesenchymal component
Cutaneous structures (hair, feathers, sweat glands, mammary glands)	Epidermis (ectoderm)	Dermis (mesoderm)
Limb	Epidermis (ectoderm)	Mesenchyme (mesoderm)
Gut organs (liver, pancreas, salivary glands)	Epithelium (endoderm)	Mesenchyme (mesoderm)
Pharyngeal and respiratory associated organs (lungs, thymus, thyroid)	Epithelium (endoderm)	Mesenchyme (mesoderm)
Kidney	Ureteric bud epithelim (mesoderm)	Mesenchyme (mesoderm)
Tooth	Jaw epithelium (ectoderm)	Mesenchyme (neural crest)

trula to the region of a *newt* gastrula destined to become parts of the mouth. Similarly, they placed presumptive flank ectodermal tissue from a *newt* gastrula into the presumptive oral regions of *frog* embryos. The structures of the mouth region differ greatly between salamander and frog larvae. The salamander larva has club-shaped balancers beneath its mouth, whereas the frog tadpole produces mucus-secreting glands and suckers (Figure 6.8). The frog tadpole also has a horny jaw without teeth, whereas the salamander has a set of calcareous teeth in its jaw. The larvae resulting from the transplants were chimeras. The salamander larvae had froglike mouths, and the frog tadpoles had salamander teeth and balancers. In other words, the mesodermal cells instructed the ectoderm to make a mouth, but the ectoderm responded by making the only kind of mouth it "knew" how to make, no matter how inappropriate.*

Thus, the instructions sent by the mesenchymal tissue can cross species barriers. Salamanders respond to frog signals, and chick tissue responds to mammalian inducers. The

---

*Spemann is reported to have put it this way: "The ectoderm says to the inducer, 'you tell me to make a mouth; all right, I'll do so, but I can't make your kind of mouth; I can make my own and I'll do that'" (quoted in Harrison 1933).

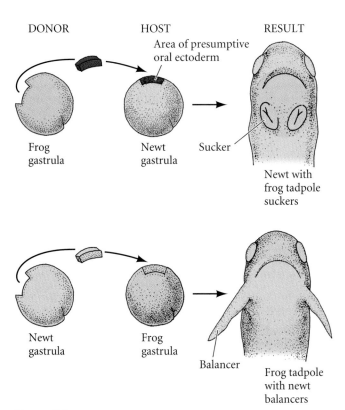

Figure 6.8
Genetic specificity of induction in amphibians. Reciprocal transplantation between the presumptive oral ectoderm regions of salamander and frog gastrulae leads to newts with tadpole suckers and tadpoles with newt balancers. (After Hamburgh 1970.)

response of the epithelium, however, is species-specific. So, whereas organ type specificity (e.g., feather or claw) is usually controlled by the mesenchyme within a species, species specificity is usually controlled by the responding epithelium. As we will see in Chapters 22 and 23, large evolutionary changes can be brought about by changing the response to a particular inducer.

> **WEBSITE 6.1 Hen's teeth.** Some inductive events between species can bring forth lost structures. Mouse molar mesenchyme may be able to induce teeth in the bird jaw.

Paracrine Factors

How are the signals between inducer and responder transmitted? While studying the mechanisms of induction that produce the kidney tubules and teeth, Grobstein (1956) and others (Saxén et al. 1976; Slavkin and Bringas 1976) found that some inductive events could occur despite a filter separating the epithelial and mesenchymal cells. Other inductions, however, were blocked by the filter. The researchers therefore concluded that some of the inductive molecules were soluble factors that could pass through the small pores of the filter, and that other inductive events required physical contact between the epithelial and mesenchymal cells (Figure 6.9). When cell membrane proteins on one cell surface interact with receptor proteins on adjacent cell surfaces, these events are called **juxtacrine interactions** (since the cell membranes are juxtaposed). When proteins synthesized by one cell can diffuse over small distances to induce changes in neighboring cells, the event is called a **paracrine interaction**, and the diffusible proteins are called **paracrine factors** or **growth and differentiation factors** (GDFs). We will consider paracrine interactions first and return to juxtacrine interactions later in the chapter.

Whereas **endocrine factors** (hormones) travel through the blood to exert their effects, paracrine factors are secreted into the immediate spaces around the cell producing them.* These proteins are the "inducing factors" of the classic experimental embryologists. During the past decade, developmental biologists have discovered that the induction of numerous organs is actually effected by a relatively small set of paracrine factors. The embryo inherits a rather compact "tool

---

*There is considerable debate as to the distances at which paracrine factors can operate. Activin, for instance, can diffuse over many cell diameters and can induce different sets of genes at different concentrations (Gurdon et al. 1994, 1995; see Chapter 3). The Vg1, BMP4, and Nodal proteins, however, probably work only on their adjacent neighbors (Jones et al. 1996; Reilly and Melton 1996). These factors may induce the expression of other short-range factors from these neighbors, and a cascade of paracrine inductions can be initiated.

In addition to endocrine, paracrine, and juxtacrine regulation, there is also autocrine regulation. Autocrine regulation occurs when the same cells that secrete paracrine factors also respond to them. In this case, the cell synthesizes a molecule for which it has its own receptor. Although autocrine regulation is not common, it is seen in placental cytotrophoblast cells; these cells synthesize and secrete platelet-derived growth factor, whose receptor is on the cytotrophoblast cell membrane (Goustin et al. 1985). The result is the explosive proliferation of that tissue.

(A)

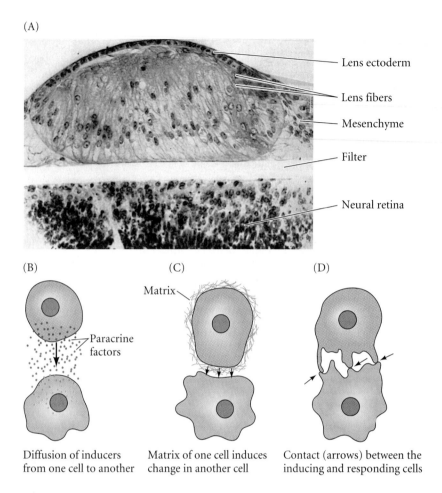

— Lens ectoderm

— Lens fibers

— Mesenchyme

— Filter

— Neural retina

(B)   (C)   (D)

Matrix

Paracrine factors

Diffusion of inducers from one cell to another

Matrix of one cell induces change in another cell

Contact (arrows) between the inducing and responding cells

Figure 6.9
Mechanisms of inductive interaction. (A) A paracrine interaction. Presumptive mouse lens ectoderm and mesenchyme were placed on a filter. Retinal tissue was placed beneath it. After 3 days, a lens had developed from the surface ectoderm. In the absence of a signal from the retinal tissue, the surface ectoderm would have become epidermal. (B–D) Paracrine and juxtracrine modes of signaling. (B) Paracrine modes of signaling involve the secretion of diffusible molecules from one cell and their reception by a nearby cell. (C) In some cases, the paracrine signal can come from an extracellular matrix protein secreted by a cell. (D) In juxtracrine interactions, contact is made between a signaling molecule on the surface of one cell and its receptor on another cell. (A from Muthukkarapan 1965, photograph courtesy of R. Auerbach; B–D after Grobstein 1956.)

kit" and uses many of the same proteins to construct the heart, the kidneys, the teeth, the eyes, and other organs. Moreover, the same proteins are utilized throughout the animal kingdom; the factors active in creating the *Drosophila* eye or heart are very similar to those used in generating mammalian organs. Many of these paracrine factors can be grouped into four major families on the basis of their structures. These families are the fibroblast growth factor (FGF) family, the Hedgehog family, the Wingless (Wnt) family, and the TGF-β superfamily.

The fibroblast growth factors

The **fibroblast growth factor** (**FGF**) family currently has nearly two dozen structurally related members. FGF1 is also known as acidic FGF; FGF2 is sometimes called basic FGF; and FGF7 sometimes goes by the name of keratinocyte growth factor. Nearly two dozen distinct FGF genes are known in vertebrates, and they can generate hundreds of protein isoforms by varying their RNA splicing or initiation codons in different tissues (Lappi 1995). FGFs can activate a set of receptor tyrosine kinases called the **fibroblast growth factor receptors** (**FGFRs**). As we will detail later in this chapter, receptor tyrosine kinases are proteins that extend through

the cell membrane (Figure 6.10A). On the extracellular side of the receptor is the ligand-binding of the protein, which binds the paracrine factor. On the intracellular side is a dormant tyrosine kinase (i.e., a peptide domain that can phosphorylate another protein by splitting ATP). When the FGF receptor binds an FGF (and only when it binds an FGF), the dormant kinase is activated and phosphorylates certain proteins within the responding cell. These proteins, once activated, can perform new functions. FGFs are associated with several developmental functions, including angiogenesis (blood vessel formation), mesoderm formation, and axon extension. While FGFs can often substitute for one another, their expression patterns give them separate functions. FGF2 is especially important in angiogenesis, and FGF8 is important for the development of the midbrain, eyes, and limbs (Figure 6.10B; Crossley et al. 1996).

FGF8 has been implicated in the induction of the lens used as an example earlier. FGF8 is usually expressed in the optic vesicle that contacts the outer ectoderm of the head (Figure 6.11; Vogel-Höpker et al. 2000). After contact with the outer ectoderm occurs, *fgf8* expression becomes concentrated in the region of the presumptive neural retina—the tissue directly apposed to the presumptive lens. Moreover, if FGF8-containing beads are placed adjacent to head ectoderm, the FGF8 will induce this ectoderm to produce ectopic lenses and to express the lens-associated transcription factor L-Maf.

WEBSITE 6.2 **FGF binding.** The binding of FGFs to their receptors is a complex acrobatic act involving an interesting cast of cell surface molecules. Glycoproteins play a major supporting role in this event.

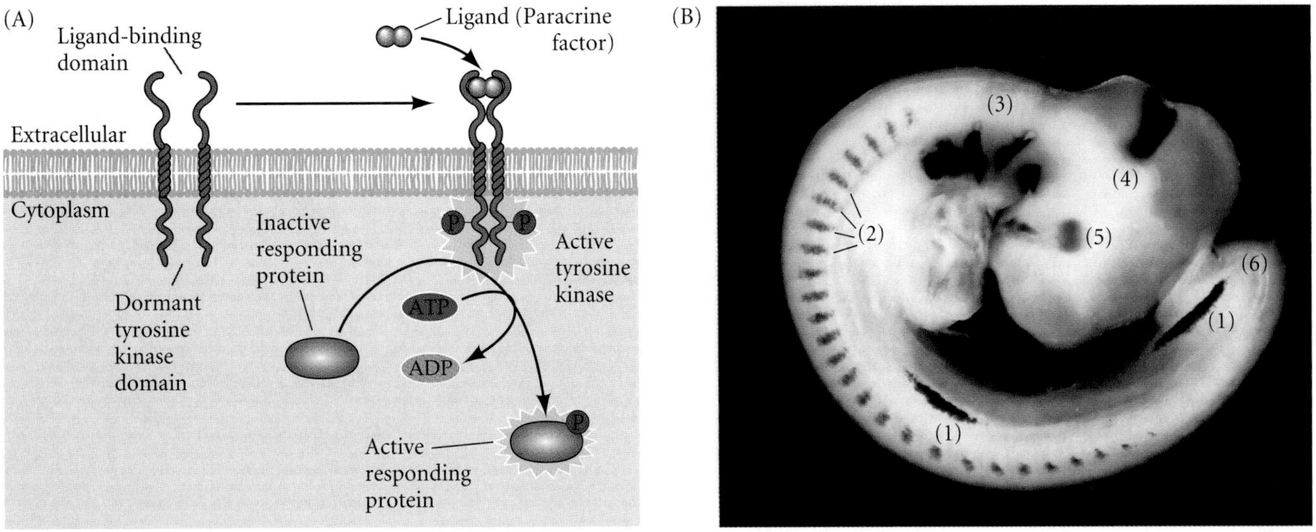

(A)

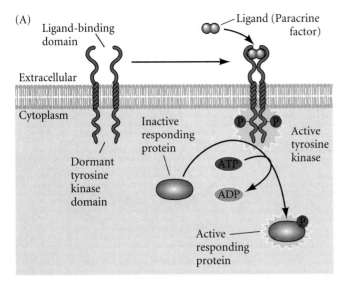

(B)

Figure 6.10

FGF expression and signaling. (A) Structure of a receptor tyrosine kinase. The binding of FGF by the extracellular portion of the receptor protein activates the dormant tyrosine kinase, whose enzyme activity phosphorylates specific tyrosine residues of certain proteins. (B) *Fgf8* expression pattern in the 3-day chick embryo, shown by in situ hybridization. FGF8 (dark areas) is seen in the distalmost limb bud ectoderm (1), in the somitic mesoderm (2; the segmented blocks of cells along the anterior-posterior axis), in the branchial arches of the neck (3), at the boundary between the midbrain and hindbrain (4), in the developing eye (5) and in the tail (6). (Photograph courtesy of E. Laufer, C.-Y. Yeo, and C. Tabin.)

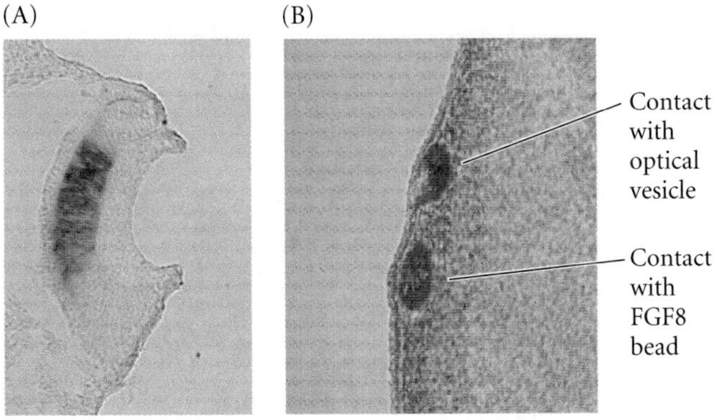

(A)   (B)

— Contact with optical vesicle

— Contact with FGF8 bead

Figure 6.11

FGF8 function in the developing chick eye. (A) In situ hybridization of *fgf8* in the optic vesicle. The *fgf8* mRNA (purple) is localized to the presumptive neural retina of the optic cup and is in direct contact with the outer ectoderm cells that will become the lens. (B) Ectopic expression of L-Maf in competent ectoderm can be induced by the optic vesicle (above) and by an FGF8-containing bead (below). (Photographs courtesy of A. Vogel-Höpker.)

The Hedgehog family

The Hedgehog proteins constitute a family of paracrine factors that are often used by the embryo to induce particular cell types and to create boundaries between tissues. Vertebrates have at least three homologues of the *Drosophila hedgehog* gene: *sonic hedgehog* (*shh*), *desert hedgehog* (*dhh*), and *indian hedgehog* (*ihh*). The Desert hedgehog protein is expressed in the Sertoli cells of the testes, and mice homozygous for a null allele of *dhh* exhibit defective spermatogenesis. Indian hedgehog protein is expressed in the gut and in cartilage and is important in postnatal bone growth (Bitgood and McMahon 1995; Bitgood et al. 1996).

Sonic hedgehog* has the greatest number of functions of the three vertebrate homologues. Made by the notochord, it is processed so that only the amino-terminal two-thirds of the molecule is secreted. This peptide is responsible for patterning the neural tube such that motor neurons are formed from the ventral neurons and sensory neurons are formed from the dorsal neurons (see Chapter 12; Yamada et al. 1993). Sonic hedgehog is also responsible for patterning the somites so that the portion of the somite closest to the notochord becomes the cartilage of the spine (Fan and Tessier-Lavigne 1994; Johnson et al. 1994). As we will see in later chapters, Sonic hedgehog has been shown to mediate the formation of the left-right axis in chicks, to initiate the anterior-posterior

---

*Yes, it is named after the Sega Genesis character. The original *hedgehog* gene was found in *Drosophila*, in which genes are named after their mutant phenotypes. The loss-of-function *hedgehog* mutation in *Drosophila* causes the fly embryo to be covered with pointy denticles on its cuticle; hence, it looks like a hedgehog. The vertebrate *hedgehog* genes were discovered by searching vertebrate gene libraries (chick, rat, zebrafish) with probes that would find sequences similar to that of the fruit fly *hedgehog* gene. Riddle and his colleagues in Cliff Tabin's laboratory (1993) discovered three genes homologous to *Drosophila hedgehog*. Two were named after species of hedgehogs, the third was named after the cartoon character.

Figure 6.12
The *sonic hedgehog* gene is shown by in situ hybridization to be expressed in the chick nervous system (red arrow), gut (blue arrow), and limb bud (black arrow) of a 3-day chick embryo. (Photograph courtesy of C. Tabin.)

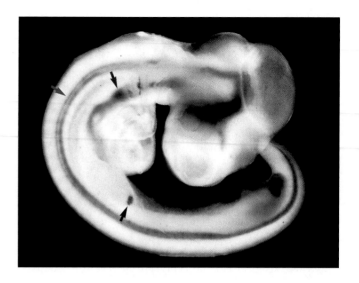

axis in limbs, to induce the regionally specific differentiation of the digestive tube, and to induce feather formation (Figure 6.12; see also Figure 6.6). Sonic hedgehog often works with other paracrine factors, such as Wnt and FGF proteins. In the developing tooth, Sonic hedgehog, FGF4, and other paracrine factors are concentrated in the region where cell interactions are creating the cusps of the teeth (see Figure 13.9; Vaahtokari et al. 1996a).

> **WEBSITE 6.3 Functions of the Hedgehog family.** While Sonic hedgehog protein induces and specifies numerous tissues in the embryo, Indian hedgehog and desert hedgehog function postnatally to regulate bone growth and sperm production.

The Wnt family

The Wnt's constitute a family of cysteine-rich glycoproteins. There are at least 15 members of this family in vertebrates. Their name is a fusion of the name of the *Drosophila* segment polarity gene *wingless* with the name of one of its vertebrate homologues, *integrated*. While Sonic hedgehog is important in patterning the ventral portion of the somites (causing the cells to become cartilage), Wnt1 appears to be active in inducing the dorsal cells of the somites to become muscle and is involved in the specification of the midbrain cells (McMahon and Bradley 1990; Stern et al. 1995). Wnt proteins also are critical in establishing the polarity of insect and vertebrate limbs, and they are used in several steps of urogenital system development (Figure 6.13).

> **WEBSITE 6.4 Wnts: An ancient family.** The biochemistry of the Wnt proteins and the mechanism of their actions is a fascinating tale of theme and variations. The Wnt family may be one of the oldest groups of signaling molecules in the animal kingdom.

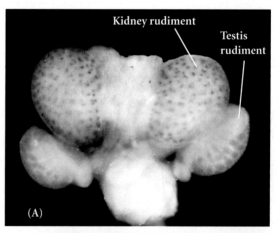

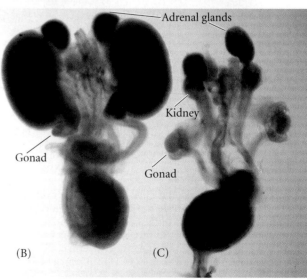

Figure 6.13
Wnt proteins play several roles in the development of the urogenital organs. Wnt4 is necessary for kidney development and for female sex determination. (A) Whole-mount in situ hybirdization of *Wnt4* expression in a 14-day mouse embryonic male urogenital rudiment. Expression (dark purple-blue staining) is seen in the mesenchyme that condenses to form the kidney's nephrons. (B) The urogenital rudiment of a wild-type newborn female mouse. (C) The urogenital rudiment of a newborn female mouse with targeted knockout of the Wnt4 gene shows that the kidney fails to develop. In addition, the ovary starts synthesizing testosterone and becomes surrounded by a modified male duct system. (Photographs courtesy of J. Perasaari and S. Vainio.)

The TGF-β superfamily

There are over 30 structurally related members of the TGF-β superfamily,* and they regulate some of the most important interactions in development (Figure 6.14). The proteins encoded by TGF-β superfamily genes are processed such that the carboxy-terminal region contains the mature peptide. These peptides are dimerized into homodimers (with themselves) or heterodimers (with other TGF-β peptides) and are secreted from the cell. The TGF-β superfamily includes the TGF-β family, the activin family, the bone morphogenetic proteins (BMPs), the Vg1 family, and other proteins, including glial-derived neurotrophic factor (necessary for kidney and enteric neuron differentiation) and Müllerian inhibitory factor (which is involved in mammalian sex determination).

TGF-β family members TGF-β1, 2, 3, and 5 are important in regulating the formation of the extracellular matrix between cells and for regulating cell division (both positively and negatively). TGF-β1 increases the amount of extracellular matrix epithelial cells make (both by stimulating collagen and fibronectin synthesis and by inhibiting matrix degradation). TGF-β proteins may be critical in controlling where and when epithelia branch to form the ducts of kidneys, lungs, and salivary glands (Daniel 1989; Hardman et al. 1994; Ritvos et al. 1995). The effects of the individual TGF-β family members are difficult to sort out, because members of the TGF-β family appear to function similarly and can compensate for losses of the others when expressed together. Moreover, targeted deletions of the *Tgf-β1* gene in mice are difficult to interpret, since the mother can supply this factor through the placenta and milk (Letterio et al. 1994).

The members of the BMP family were originally discovered by their ability to induce bone formation; hence, they are known as **bone morphogenetic proteins**. Bone formation, however, is only one of their many functions. They have been found to regulate cell division, apoptosis (programmed cell death), cell migration, and differentiation (Hogan 1996). BMPs can be distinguished from other members of the TGF-β superfamily by having seven (rather than nine) conserved cysteines in the mature polypeptide. The BMPs include proteins such as Nodal (responsible for left-right axis formation) and BMP7 (important in neural tube polarity, kidney development, and sperm formation; see Figure 4.22). As it turns out, BMP1 is not a member of the BMP family at all; it is a protease.

The *Drosophila* Decapentaplegic protein is homologous to the vertebrate BMP4, and human BMP4 can replace the *Drosophila* homologue, rescuing those flies deficient in Dpp (Padgett et al. 1993). BMPs are thought to work by diffusion from the cells producing them. Their range is determined by the amino acids in their N-terminal region, which determine whether the specific BMP will be bound by proteoglycans, thereby restricting its diffusion (Ohkawara et al. 2002).

---

* TGF stands for "transforming growth factor." The designation "superfamily" is often given when each of the different classes of molecules constitutes a "family." The members of a superfamily all have similar structures, but are not as close as the molecules within a family are to one another.

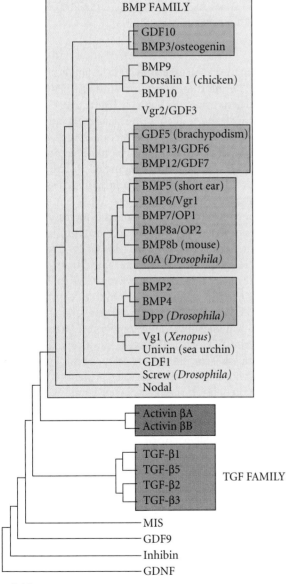

Figure 6.14
Relationships among members of the TGF-β superfamily. (After Hogan 1996.)

Other paracrine factors

Although most of the paracrine factors are members of the above-mentioned four families, some have few or no close relatives. Factors such as epidermal growth factor, hepatocyte growth factor, neurotrophins, and stem cell factor are not included among these families, but each plays important roles during development. In addition, there are numerous factors involved almost exclusively with developing blood cells: erythropoietin, the cytokines, and the interleukins. These factors will be discussed when we detail blood cell formation in Chapter 14.

Cell Surface Receptors and Their Signal Transduction Pathways

We now turn from the paracrine factors, which are inducer proteins, to the molecules involved in the response to induction. These molecules include receptors in the membrane of the responding cell, which bind the paracrine factor; and the cascade of interacting proteins that transmit a signal through a pathway from the bound receptor to the nucleus. These pathways between the cell membrane and the genome are called **signal transduction pathways**. Several types of signal transduction pathways have been discovered; we will outline some of the major ones here

As you will see, the major signal transduction pathways all appear to be variations on a common and rather elegant theme. Each receptor spans the cell membrane and has an extracellular region, a transmembrane region, and a cytoplasmic region. When a ligand (the paracrine factor) binds its receptor in the extracellular region, the ligand induces a conformational change in the receptor's structure. This shape change is transmitted through the membrane and changes the shape of the cytoplasmic domains. The conformational change in the cytoplasmic domains gives them enzymatic activity—usually a kinase activity that can use ATP to phosphorylate proteins, including the receptor molecule itself. The active receptor can now catalyze reactions that phosphorylate other proteins, and this phosphorylation activates their latent activities in turn. Eventually, the cascade of phosphorylation activates a dormant transcription factor, which activates (or represses) a particular set of genes.

Thus, transcription factors have been divided into three catagories (Table 6.2; Brivanlou and Darnell 2002). The first group includes those transcription factors that are constitutively present in all cells. These include the basal transcription factors. The second group includes those transcription factors that are active whenever a cell acquires them by cytoplasmic localization (e.g. Bicoid) or by induction (Pax6). The third group includes those transcription factors (such as MITF) whose functions are activated by cell signaling pathways.

The receptor tyrosine kinase (RTK) pathway

The RTK signal transduction pathway was one of the first pathways to unite various areas of developmental biology. Researchers studying *Drosophila* eyes, nematode vulvae, and human cancers found that they were all studying the same genes. The RTK pathway begins at the cell surface, where a **receptor tyrosine kinase** (RTK) binds its specific ligand. Ligands that bind to RTKs include the fibroblast growth factors, epidermal growth factors, platelet-derived growth factors, and stem cell factor. Each RTK can bind only one or a small set of these ligands. (Stem cell factor, for instance, will bind to only one RTK, the Kit protein.) The RTK spans the cell membrane, and when it binds its ligand, it undergoes a conformational change that enables it to dimerize with another RTK. This

TABLE 6.2 A classification of transcription factors based on their modes of activation

Class	Examples
CONSTITUTIVE	Sp1, NF1, other TBP-associated factors
REGULATORY (CONDITIONAL)	
Developmental (Active in cells that acquire them by cytoplasmic localization or induction)	GATA4 (blood) Pit1 (pituitary) MyoD, Myf (muscles) Hox proteins (body axes) Bicoid (anterior segments of *Drosophila*) Pax6 (eye, pancreas)
Signal-dependent (Not active unless signal is given to cell)	Steroid receptor family (estrogen, testosterone, retinoids, thyroid hormone) Cell surface receptor-dependent (STATs, SMADs, β-catenin, GLI, MITF, etc.)

Source: After Brivanlou and Darnell 2002.

conformational change activates the latent kinase activity of each RTK, and these receptors phosphorylate each other on particular tyrosine residues (see Figure 6.10). Thus, the binding of the ligand to the receptor causes the autophosphorylation of the cytoplasmic domain of the receptor.

The phosphorylated tyrosine on the receptor is then recognized by an adaptor protein (Figure 6.15). The adaptor protein serves as a bridge that links the phosphorylated RTK to a powerful intracellular signaling system. While binding to the phosphorylated RTK through one of its cytoplasmic domains, the adaptor protein also activates a **G protein**, such as **Ras**. Normally, the G protein is in an inactive, GDP-bound state. The activated receptor stimulates the adaptor protein to activate the **guanine nucleotide releasing factor** (GNRP). This protein exchanges a phosphate from a GTP to transform the bound GDP into GTP. The GTP-bound G protein is an active form that transmits the signal to the next molecule. After the signal is delivered, the GTP on the G protein is hydrolyzed back into GDP. This catalysis is greatly stimulated by the complexing of the Ras protein with the **GTPase-activating protein** (GAP). In this way, the G protein is returned to its inactive state, where it can await further signaling. Without the GAP protein, Ras protein cannot catalyze GTP well, and so remains in its active configuration (Cales et al. 1988; McCormick 1989). Mutations in the *RAS* gene account for a large proportion of cancerous human tumors (Shih and Weinberg 1982), and the mutations of *RAS* that make it oncogenic all inhibit the binding of the GAP protein

The active Ras G protein associates with a kinase called Raf. The G protein recruits the inactive Raf protein to the cell membrane, where it becomes active (Leevers et al. 1994; Stokoe et al. 1994). The Raf protein is a kinase that activates the MEK protein by phosphorylating it. MEK is itself a kinase, which activates the ERK protein by phosphorylation. In turn, ERK is a kinase that enters the nucleus and phosphorylates certain transcription factors.

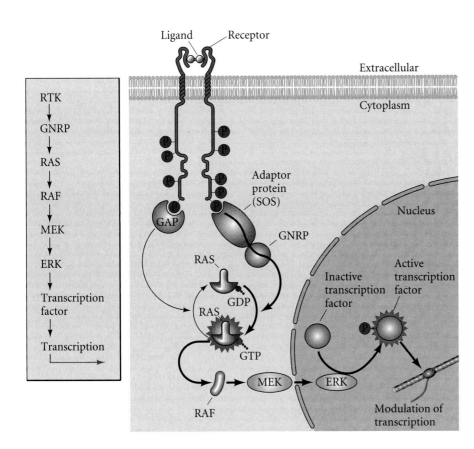

Figure 6.15
The widely used RTK signal transduction pathway. The receptor tyrosine kinase is dimerized by the ligand, which causes the autophosphorylation of the receptor. The adaptor protein recognizes the phosphorylated tyrosines on the RTK and activates an intermediate protein, GNRP, which activates the Ras G protein by allowing the phosphorylation of the GDP-bound Ras. At the same time, the GAP protein stimulates the hydrolysis of this phosphate bond, returning Ras to its inactive state. The active Ras activates the Raf protein kinase C (PKC), which in turn phosphorylates a series of kinases. Eventually, the activated ERK kinase alters gene expression in the nucleus of the responding cell by phosphorylating certain transcription factors (which can then enter the nucleus to change the types of genes transcribed) and certain translation factors (which alter the level of protein synthesis). In many cases, this pathway is reinforced by the release of calcium ions. A simplified version of the pathway is depicted on the left.

Sidelights & Speculations

The RTK Pathway and Cell-to-Cell Induction

Recent research into the development of *Drosophila* and *Caenorhabditis elegans* has shown that induction does indeed occur on the cell-to-cell level. Some of the best-studied examples involve the formation of the retinal photoreceptors in the *Drosophila* eye and the formation of the vulva in *C. elegans*. Remarkably, the signal transduction pathways involved turn out to be the same in both cases; only the targeted transcription factors are different. In both cases, an epidermal growth factor-like inducer activates the RTK pathway.

Photoreceptor induction in Drosophila

The *Drosophila* retina consists of about 800 units called **ommatidia** (Figure 6.16). Each ommatidium is composed of 20 cells arranged in a precise pattern. Eight of those cells are photoreceptors; the rest are lens cells. The eye develops in the flat epithelial layer of the eye imaginal disc of the larva.

There are no cells directly above or below this layer, so the interactions are confined to neighboring cells in the same plane. The differentiation of these randomly arranged epithelial cells into the retinal photoreceptors and their surrounding lens tissue occurs during the last (third) larval stage. An indentation forms at the posterior margin of the imaginal disc, and this **morphogenetic furrow** begins to travel forward toward the anterior of the epithelium (Figure 6.17). The movement of the furrow depends on interactions between two paracrine factors, Hedgehog and Decapentaplegic. Hedgehog is expressed by the cells immediately posterior to the furrow (i.e., those that have just differentiated), and it induces the expression of the Decapentaplegic protein within the furrow (Heberlein et al. 1993; Ma et al. 1993). Thus, as retinal cells begin to differentiate behind the furrow, they secrete the Hedgehog protein, which drives the furrow anteriorly (Brown et al. 1995).

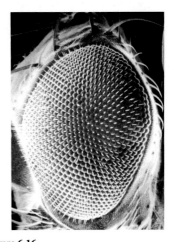

Figure 6.16
Scanning electron micrograph of a compound eye in *Drosophila*. Each facet is a single ommatidium. A sensory bristle projects from each ommatidium. (Photograph courtesy of T. Venkatesh.)

VADE MECUM[2] *Drosophila* imaginal **discs.** Imaginal discs are groups of larval cells from which adult structures form. You can easily dissect imaginal discs from the *Drosophila* larva to see the magnificent eye-antenna disc for yourself. This segment shows how it is done.

[Click on Fruit Fly]

As the morphogenetic furrow passes through a region of cells, those cells begin to differentiate in a specific order. The first cell to differentiate is the central (R8) photoreceptor (Chen and Chien 1999). Hedgehog proteins in the furrow region appear to induce R8 determination (Hsiung and Moses, 2002; Dokucu et al., 1996), and the R8 photoreceptor expresses the Atonal transcription factor (as do retinal neurons in vertebrates). The R8 cell is thought to induce the cell anterior to it and the cell posterior to it (with respect to the furrow) to become the R2 and R5 photoreceptors, respectively. The R2 and R5

Figure 6.17
Differentiation of photoreceptors in the *Drosophila* compound eye. The morphogenetic furrow (arrow) crosses the disc from posterior (left) to anterior (right). (A) Confocal micrograph of a triple-labeled late larval eye/antennal imaginal disc, showing *hairy* expression in green ahead of the morphogenetic furrow (arrow). Within the furrow, the Ci protein (red) is expressed as a consequence of the Hedgehog signal. (It will activate the decapentaplegic gene.) The neural specific protein, 22C10, is stained blue in the differentiating photoreceptors behind the morphogenetic furrow. (The blue horizontal line of staining is Bolweg's nerve.) (B) Behind the furrow, the photoreceptor cells differentiate in a defined sequence. The first photoreceptor cell to differentiate (shown in blue) is R8. R8 appears to induce the differentiation of R2 and R5, and a cascade of induction continues until the R7 photoreceptor is differentiated. (A, photograph courtesy of N. Brown, S. Paddock, and S. Carroll; B after Tomlinson 1988.)

photoreceptors are functionally equivalent, so the signal from R8 is probably the same to both cells (Tomlinson and Ready 1987). Signals from these cells induce four more adjacent cells to become the R3, R4, and then the R1 and R6 photoreceptors. Last, the R7 photoreceptor appears. The other cells around these photoreceptors become the lens cells. Lens determination is the "default" condition if the cells are not induced.

A series of mutations has been found that blocks some of the steps of this induction cascade. Mutations in the *sevenless* (*sev*) gene or in the *bride of sevenless* (*boss*) gene can each prevent the R7 cell from differentiating into a photoreceptor. (It becomes a lens cell instead.) Analysis of these mutations has shown that they affect the inductive process. The *sev* gene is required in the R7 cell itself. If mosaic embryos are made such that some of the cells of the eye imaginal disc are heterozygous (normal) and some are homozygous for the *sevenless* mutation, the R7 photoreceptor develops only if the R7 precursor cell has the wild-type *sev* allele (Basler and Hafen 1989; Bowtell et al. 1989). Antibodies to the Sevenless protein have found it in the cell membrane, and the sequence of the *sev* gene suggests that it encodes a trans-

membrane protein with a tyrosine kinase site in its cytoplasmic domain (Banerjee et al. 1987; Hafen et al. 1987). This finding is consistent with the protein's being a receptor for some signal—an RTK.

The signal that tells the R7 precursor to differentiate into an R7 photoreceptor comes from a protein encoded by the wild-type *boss* gene. Flies homozygous for the *boss* mutation also lack R7 photoreceptors. Genetic mosaic studies wherein some of the cells of the eye imaginal disc are normal and some are homozygous for the *boss* mutation show that the wild-type *boss* gene is not needed in the R7 precursor cell itself. Rather, the R7 photoreceptor differentiates only if the wild-type *boss* gene is expressed in the R8 cell. Thus, the *boss* gene encodes a protein whose existence in the R8 cell is necessary for the differentiation of the R7 cell.

In fact, the Boss protein is the ligand for the Sev RTK. Boss probably works in a juxtacrine fashion (it remains attached to the R8 cell), with its extracellular domain binding to the the extracellular domain of Sev (Reinke and Zipursky 1988; Hart et al. 1993). In *Drosophila* eyes, the RTK cascade initiated by the binding of Boss to Sev activates the Sevenless-in-Absentia (Sina) transcription factor, whose activity is necessary

(A)

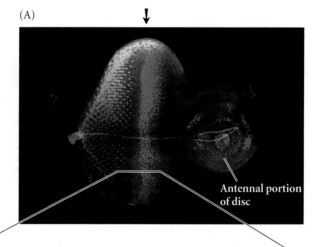

Antennal portion of disc

(B)

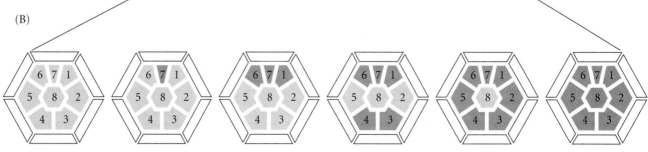

Later differentiation
(posterior to morphogenetic furrow)

Early differentiation
(entering morphogenetic furrow)

for the differentiation of photoreceptor R7 (Carthew and Rubin 1990; Dickson et al. 1992). Once R7 is induced, it reciprocally induces the expression of opsin proteins in the R8 cell (Chou et al. 1999). A summary of the some of the cell-to-cell inductions in the *Drosophila* retina (Figure 6.18) shows that individual cells are able to induce other individual cells to create the precise arrangement of cells in particular tissues.

WEBSITE 6.5 Eye formation: A conserved pathway. The proteins of the Sevenless pathway are seen not only in ommatidial development, but in vertebrate eye development as well. This appears to be a remarkably conserved pathway for photoreceptor differentiation. Moreover, there appear to be "safeguards" preventing the R8 cell from inducing R7 differentiation in other ommatidial cells.

Vulval induction in Caenorhabditis elegans

Most *Caenorhabditis elegans* individuals are hermaphrodites. In their early development, they are male, and the gonad produces sperm, which is stored for later use. As they grow older, they develop ovaries. The eggs "roll" through the region of sperm storage, are fertilized inside the ne-

Figure 6.18
Summary of the major genes known to be involved in the induction of *Drosophila* photoreceptors. For development to continue beyond the differentiation of the R8, R2, and R5 photoreceptors, the *rough* gene (*ro*) must be present in both the R2 and R5 cells. For the differentiation of the R7 photoreceptor, the *sevenless* gene (*sev*) has to be active in the R7 precursor cell, while the *bride of sevenless* gene (*boss*) must be active in the R8 photoreceptor. (After Rubin 1989.)

matode, and then pass out of the body through the vulva (see Figure 8.43).

The vulva of *C. elegans* represents a case in which one inductive signal generates a variety of cell types. This organ forms during the larval stage from six cells called the **vulval precursor cells** (**VPCs**). The cell connecting the overlying gonad to the vulval precursor

cells is called the **anchor cell**. The anchor cell secretes the LIN-3 protein, a relative of epidermal growth factor (EGF) and the Boss protein (Hill and Sternberg 1992). If the anchor cell is destroyed (or if the *lin-3* gene is mutated), the VPCs will not form a vulva; they will instead become part of the hypodermis (skin) (Kimble 1981).

The six VPCs influenced by the anchor cell form an **equivalence group**. Each member of this group is competent to become induced by the anchor cell and can assume any of three fates, depending on its proximity to the anchor cell (Figure 6.19). The cell directly beneath the anchor cell divides to form the central vulval cells. The two cells flanking that central cell divide to become the lateral vulval cells, while the three cells farther away from the anchor cell generate hypodermal cells. If the anchor cell is destroyed, all six cells of the equivalence group divide once and contribute to the hypodermal tissue. If the three central VPCs are destroyed, the three outer cells, which normally form hypodermal cells, generate vulval cells instead. The LIN-3 protein is received by the LET-23 receptor tyrosine kinase on the VPCs, and the

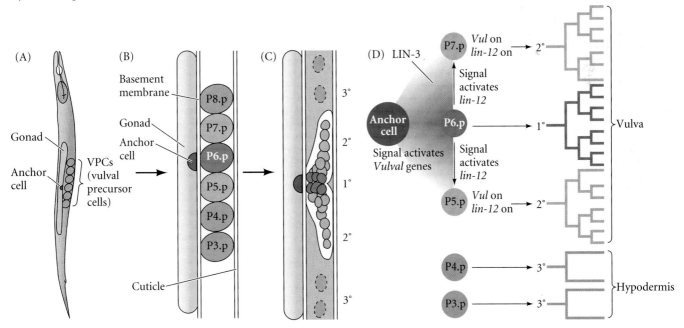

Figure 6.19
The *C. elegans* vulval precursor cells and their descendants. (A) Location of the gonad, anchor cell, and VPCs in the second instar larva. (B, C) Relationship of the anchor cell to the six VPCs and their subsequent lineages. 1° lineages result in the central vulval cells; 2° lineages constitute the lateral vulval cells; 3° lineages generate hypodermal cells. (C) Outline of the vulva in the fourth instar larva. The circles represent the positions of the nuclei. (D) Model for the determination of vulval cell lineages in *C. elegans*. The LIN-3 signal from the anchor cell causes the determination of the P6.p cell to generate the central vulval lineage (dark purple). Lower concentrations of LIN-3 cause the P5.p and P7.p cells to form the lateral vulval lineages. The P6.p (central lineage) cell also secretes a short-range juxtacrine signal that induces the neighboring cells to activate the LIN-12 (Notch) protein. This signal prevents the P5.p and P7.p cells from generating the primary, central vulval cell lineage. (After Katz and Sternberg 1996.)

signal is transferred to the nucleus through the RTK pathway. The target of the kinase cascade is the LIN-31 protein (Tan et al. 1998). When this protein is phosphorylated in the nucleus, it loses its inhibitory protein partner and is able to function as a transcription factor, promoting vulval cell fates.

Two mechanisms coordinate the formation of the vulva through this induction (Figure 6.20D; Katz and Sternberg 1996):

1. The LIN-3 protein forms a concentration gradient. Here, the VPC closest to the anchor cell (i.e., the P6.p cell) receives the highest concentration of LIN-3 protein and generates the central vulval cells. The two VPCs adjacent to it (P5.p and P7.p) receive a lower amount of LIN-3 and become the lateral vulval cells. The VPCs farther away from the anchor cell do not receive enough LIN-3 to have an effect, so they become hypodermis (Katz et al. 1995).

2. In addition to forming the central vulval lineage, the VPC closest to the anchor cell also signals laterally to the two adjacent cells and instructs them not to generate the central vulval lineages. These lateral cells do not instruct the peripheral VPCs to do anything, so they become hypodermis (Koga and Ohshima 1995; Simske and Kim 1995). This lateral inhibition of the "secondary" vulval precursor cells by the "primary" VPC is accomplished through the LIN-12 proteins (which are discussed more thoroughly below; Sternberg 1988).

Both of these mechanisms function during normal development. As Kenyon (1995) notes, "together they could produce the ever-perfect tiny vulvae that *C. elegans* is so famous for."

The RTK pathway is critical in numerous developmental processes. In the migrating neural crest cells of humans and mice, the pathway is important in activating the microphthalmia transcription factor (Mitf) to produce the pigment cells. Mitf, whose mechanism of action was described in Chapter 5, is transcribed in the pigment-forming melanoblast cells that migrate from the neural crest into the skin and in the melanin-forming cells of the pigmented retina. But we have not yet discussed what proteins signal this transcription

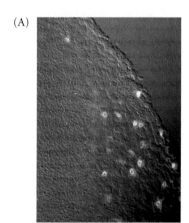

(A)

(B)

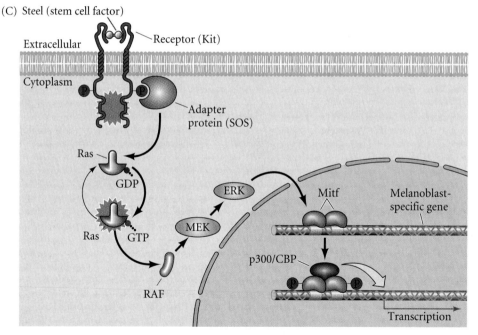

(C) Steel (stem cell factor)

Receptor (Kit)

Extracellular

Cytoplasm

Adapter protein (SOS)

Ras

GDP

Ras   GTP

MEK

ERK

RAF

Mitf

Melanoblast-specific gene

p300/CBP

Transcription

Figure 6.20
Activation of the Mitf transcription factor through the binding of stem cell factor by the Kit RTK protein. The information received at the cell membrane is sent to the nucleus by the RTK signal transduction pathway. (A, B) Demonstration that Kit protein and Mitf are present in the same cells. Antibodies to these proteins stain the Kit protein (red) and Mitf (green). The overlap is yellow or yellow-green. Both proteins are present in the migrating melanocyte precursor cells (melanoblasts). (A) Migrating melanoblasts can be seen in a wild-type mouse embryo at day 10.5. (B) No melanoblasts are visible in a *microphthalmia* mutant embryo of the same age. The lack of melanoblasts in the mutant is due to the relative absence of Mitf. (C) Signal transduction pathway leading from the cell membrane to the nucleus. When the receptor domain of the Kit RTK protein binds the Steel paracrine factor, Kit dimerizes and becomes phosphorylated. This phosphorylation is used to activate the Ras G protein, which activates the chain of kinases that will phosphorylate the Mitf protein. Once phosphorylated, Mitf can bind the cofactor p300/CBP, acetylate the nucleosome histones, and initiate transcription of the genes needed for melanocyte development. (A, B from Nakayama et al. 1998, photographs courtesy of H. Arnheiter; C after Price et al. 1998.)

factor to become active. The clue lay in two mouse mutants whose phenotypes resemble those of mice homozygous for *microphthalmia* mutations. Like those mice, homozygous *White* mice and homozygous *Steel* mice are white because their pigment cells have failed to migrate. Perhaps all three genes (*Mitf*, *Steel*, and *White*) are on the same developmental pathway. In 1990, several laboratories demonstrated that the *Steel* gene encodes a paracrine protein called **stem cell factor** (see Witte 1990). Stem cell factor binds to and activates the **Kit** receptor tyrosine kinase encoded by the *White* gene (Spritz et al. 1992; Wu et al. 2000). The binding of stem cell factor to the Kit RTK dimerizes the Kit protein, causing it to become phosphorylated. The phosphorylated Kit activates the pathway whereby phosphorylated ERK is able to phosphorylate the Mitf transcription factor (Hsu et al. 1997; Hemesath et al. 1998). Only the phosphorylated form of Mitf is able to bind the p300/CBP histone acetyltransferace protein that enables it to activate transcription of the genes encoding tyrosinase and other proteins of the melanin-formation pathway (Figure 6.20; Price et al. 1998).

The Smad pathway

Members of the TFG-β superfamily of paracrine factors activate members of the **Smad** family of transcription factors (Figure 6.21; Heldin et al. 1997). The TGF-β ligand binds to a type II TGF-β receptor, which allows that receptor to bind to a type I TGF-β receptor. Once the two receptors are in close contact, the type II receptor phosphorylates a serine or threonine on the type I receptor, thereby activating it. The activated

type I receptor can now phosphorylate the Smad proteins. (Researchers named the Smad proteins by merging the names of the first identified members of this family: the *C. elegans* Sma protein and the *Drosophila* Mad protein.) Smads 1 and 5 are activated by the BMP family of TGF-β factors, while the receptors binding activin and the TGF-β family phosphorylate Smads 2 and 3. These phosphorylated Smads bind to Smad 4 and form the transcription factor complex that will enter the nucleus.

The JAK-STAT pathway

Another important pathway for transducing information received at the cell's membrane to the nucleus is the JAK-STAT pathway. Here the set of transcription factors consists of the **STAT** (**s**ignal **t**ransducers and **a**ctivators of **t**ranscription) proteins (Ihle 1996; 2001). STATs are phosphorylated by certain tyrosine kinases, including fibroblast growth factor receptors and the JAK family of tyrosine kinases. The JAK-STAT path-

Figure 6.21
The Smad pathway activated by TGF-β superfamily ligands. (A) An activation complex is formed by the binding of the ligand by the type I and type II receptors. This allows the type II receptor to phosphorylate the type I receptor on particular serine or threonine residues (of the "GS box"). The phosphorylated type I receptor protein can now phosphorylate the Smad proteins. (B) Those receptors that bind TGF-β family proteins or members of the activin family phosphorylate Smads 2 and 3. Those receptors that bind to BMP family proteins phosphorylate Smads 1 and 5. These Smads can complex with Smad 4 to form active transcription factors. A simplified version of the pathway is shown at the left.

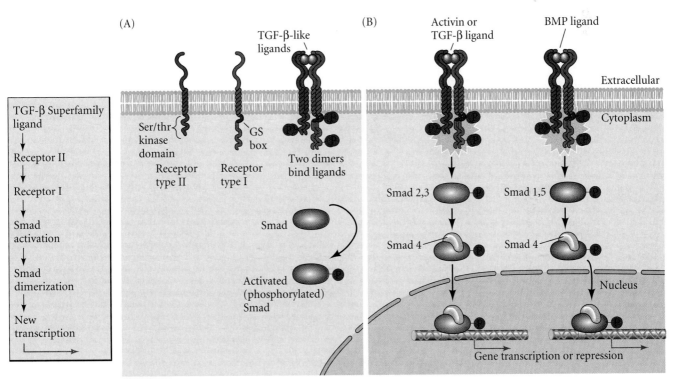

way is extremely important in the differentiation of blood cells and in the activation of the casein gene during milk production (Briscoe et al. 1994; Groner and Gouilleux 1995).

The STAT pathway's role in casein production is shown in Figure 6.22. Here, the hormone prolactin binds to the extracellular domains of prolactin receptors, causing them to dimerize. A JAK protein kinase is bound to each of the receptors (in their respective cytoplasmic regions), and these JAK proteins are now brought together, where they can phosphorylate the receptors at several sites. The receptors are now activated and have their own protein kinase activity. Therefore, the JAK proteins convert a receptor into a receptor tyrosine kinase. The activated receptors can now phosphorylate particular inactive STATs and cause them to dimerize. These dimers are the active form of the STAT transcription factors, and they are translocated to the nucleus, where they bind to specific regions of DNA. In this case, they bind to the upstream promoter elements of the casein gene, causing it to be transcribed.

The STAT pathway is very important in the regulation of human fetal bone growth. Mutations that prematurely activate the STAT pathway have been implicated in some severe forms of dwarfism, such as the lethal **thanatophoric dysplasia**, wherein the growth plates of the rib and limb bones fail to proliferate. The short-limbed newborn dies because its

ribs cannot support breathing. The genetic lesion responsible is in the gene encoding fibroblast growth factor receptor 3 (FGFR3) (Figure 6.23; Rousseau et al. 1994; Shiang et al. 1994). This protein is expressed in the cartilage precursor cells—known as **chondrocytes**—in the growth plates of the long bones. Normally, the FGFR3 protein (a receptor tyrosine kinase) is activated by a fibroblast growth factor, and it signals the chondrocytes to stop dividing and begin differentiating into cartilage. This signal is mediated by the StAT1 protein, which is phosphorylated by the activated FGFR3 and then translocated into the nucleus. Inside the nucleus, StAT1

Figure 6.22
A STAT pathway: the casein gene activation pathway activated by prolactin. The casein gene is activated during the last (lactogenic) phase of mammary gland development, and its signal is the secretion of the hormone prolactin from the anterior pituitary gland. Prolactin causes the dimerization of prolactin receptors in the mammary duct epithelial cells. A particular JAK protein (Jak2) is "hitched" to the cytoplasmic domain of these receptors. When the receptors bind prolactin and dimerize, the JAK proteins phosphorylate each other and the dimerized receptors, activating the dormant kinase activity of the receptors. The activated receptors add a phosphate group to a tyrosine residue (Y) of a particular STAT protein—in this case, Stat5. This allows Stat5 to dimerize, be translocated into the nucleus, and bind to particular regions of DNA. In combination with other transcription factors (which presumably have been waiting for its arrival), the Stat5 protein activates transcription of the casein gene. GR is the glucocorticoid receptor, OCT1 is a general transcription factor, and TBP is the TATA-binding protein (see Chapter 5) responsible for binding RNA polymerase. A simplified diagram is shown to the right. (For details, see Groner and Gouilleux 1995.)

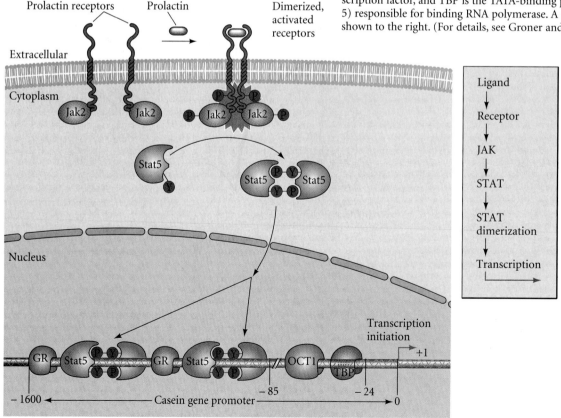

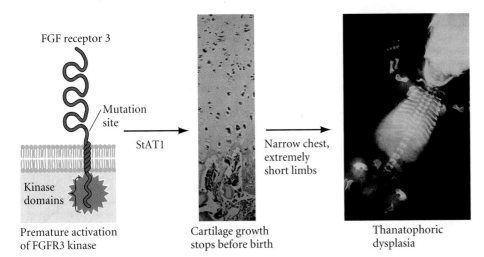

Premature activation
of FGFR3 kinase

Cartilage growth
stops before birth

Thanatophoric
dysplasia

Figure 6.23
A mutation in the gene for FGFR3 causes the premature constitutive activation of the STAT pathway and the production of phosphorylated Stat1 protein. This transcription factor activates genes that cause the premature termination of chondrocyte cell division. The result is a condition of failed bone growth that results in the death of the newborn infant, since the thoracic cage cannot expand to take breaths. (From Gilbert–Barness and Opitz 1996.)

activates the genes encoding a cell cycle inhibitor, the p21 protein (Su et al. 1997). The mutations causing thanatophoric dwarfism result in a gain-of-function phenotype, wherein the mutant FGFR3 is active constitutively—that is, without the need to be activated by an FGF (Deng et al. 1996; Webster and Donoghue 1996). This causes the chondrocytes to stop proliferating shortly after they are formed, and the bones fail to grow. Mutations that prematurely activate FGFR3 to a lesser degree produce achondroplasic (short-limbed) dwarfism, the most prevalent of the human dominant syndromes.

WEBSITE 6.6 **FGFR mutations.** Mutations of the human FGF receptors have been associated with several skeletal malformation syndromes, including syndromes wherein skull cartilage, rib cartilage, or limb cartilage fails to grow or differentiate.

The Wnt pathway

Members of the Wnt family of paracrine factors interact with transmembrane receptors of the **Frizzled** family of proteins. In most instances, the binding of Wnt by a Frizzled protein causes the Frizzled protein to activate the Disheveled protein. Once the Disheveled protein is activated, it inhibits the activity of the glycogen synthase kinase-3 (GSK-3) enzyme. GSK-3, if it were active, would prevent the dissociation of the β-catenin protein from the APC protein, which targets β-catenin for degradation. However, when the Wnt signal is given and GSK-3 is inhibited, β-catenin can dissociate from the APC protein and enter the nucleus. Once inside the nucleus, it can form a heterodimer with an LEF or TCF DNA-binding protein, becoming a transcription factor. This complex binds to and activates the Wnt-responsive genes (Figure 6.24A; Behrens et al. 1996; Cadigan and Nusse 1997).

This model is undoubtedly an oversimplification, because different cells use this pathway in different ways (see McEwen and Peifer 2001). Moreover, its components can have more than one function in the cell. In addition to being part of the Wnt signal transduction cascade, GSK-3 is also an enzyme that regulates glycogen metabolism. The β-catenin protein was recognized as being part of the cell adhesion complex on the cell surface before it was also found to be a transcription factor. The APC protein also functions as a tumor suppressor in adults. The transformation of normal colon epithelial cells into colon cancer is thought to occur when the *APC* gene is mutated and can no longer keep the β-catenin protein out of the nucleus (Figure 6.24B; Korinek et al. 1997; He et al. 1998). Once in the nucleus, β-catenin can bind with another transcription factor and activate genes for cell division. One principle readily seen in the Wnt pathway (and one that is also evident in the Hedgehog pathway) is that *activation is often accomplished by inhibiting an inhibitor*. Thus, the GSK-3 protein is an inhibitor that is itself repressed by the Wnt signal.

In addition to sending signals to the nucleus, Wnt can also affect the actin and microtubular cytoskeleton. Here, Wnt activates alternative, "non-canonical," pathways. For instance, when Wnt activates Disheveled, the Disheveled protein can interact with a Rho GTPase. This GTPase can activate the kinases that phosphorylate cytoskeletal proteins and thereby alter cell shape, cell polarity (where the upper and lower portions of the cell differ), and motility (Shulman et al. 1998; Winter et al. 2001). A third Wnt pathway diverges earlier than Disheveled. Here, the Frizzled receptor protein activates a phospholipase that synthesizes a compound that releases calcium ions from the endoplasmic reticulum.

The Hedgehog pathway

Members of the Hedgehog protein family function as paracrine factors by binding to a receptor called Patched. The Patched protein, however, is not a signal transducer. Rather, it is bound to a signal transducer, the Smoothened protein. The Patched protein prevents Smoothened from functioning. In the absence of Hedgehog binding to Patched, Smoothened is inactive, and the Cubitus interruptus (Ci) protein is tethered to the microtubules of the responding cell. While on the micro-

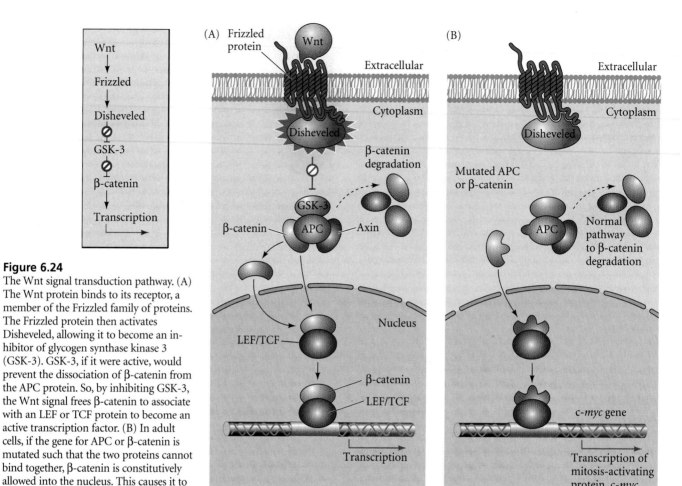

Figure 6.24
The Wnt signal transduction pathway. (A) The Wnt protein binds to its receptor, a member of the Frizzled family of proteins. The Frizzled protein then activates Disheveled, allowing it to become an in-hibitor of glycogen synthase kinase 3 (GSK-3). GSK-3, if it were active, would prevent the dissociation of β-catenin from the APC protein. So, by inhibiting GSK-3, the Wnt signal frees β-catenin to associate with an LEF or TCF protein to become an active transcription factor. (B) In adult cells, if the gene for APC or β-catenin is mutated such that the two proteins cannot bind together, β-catenin is constitutively allowed into the nucleus. This causes it to activate certain cell division genes and ini-tiate tumors. (B after Pennisi 1998.)

tubules, it is cleaved in such a way that a portion of it enters the nucleus and acts as a transcriptional repressor. This por-tion of the Ci protein binds to the promoters and enhancers of particular genes and acts as an inhibitor of transcription. When Hedgehog binds to Patched, the Patched protein's shape is altered such that it no longer inhibits Smoothened. Smoothened acts (probably by phosphorylation) to release the Ci protein from the microtubules and to prevent its being cleaved. The intact Ci protein can now enter the nucleus, where it acts as a transcriptional *activator* of the same genes it used to repress (Figure 6.25; Aza-Blanc et al. 1997).

The Hedgehog pathway is extremely important in verte-brate limb and neural differentiation. When mice were made homozygous for a mutant allele of Sonic hedgehog, they had major limb abnormalities as well as **cyclopia**—a single eye in the center of the forehead (Chiang et al. 1996). The vertebrate homologues of the Ci protein in *Drosophila* are the Gli pro-teins. Severe truncations of the human *GLI3* gene produce a nonfunctional protein that gives rise to Grieg cephalopolysyn-dactyly, a condition involving a high forehead and extra digits. A less severe truncation retains the DNA-binding domain of

the GLI3 protein but deletes the activator region. This mutant GLI3 protein can act only as a repressor. This protein is found in infants with Pallister-Hall syndrome, a much more severe syndrome (indeed, lethal soon after birth) involving not only extra digits, but also poor development of the pituitary gland, hypothalamus, anus, and kidneys (see Shin et al. 1999).

While mutations that inactivate the Hedgehog pathway can cause malformations, mutations that activate the pathway ectopically can cause cancers. If the Patched protein is mutated in somatic tissues such that it can no longer inhibit Smoothened, it can cause tumors of the basal cell layer of the epidermis (basal cell carcinomas). Heritable mutations of the *patched* gene cause basal cell nevus syndrome, a rare autosomal dominant condition characterized by both developmental anomalies (fused fingers, rib and facial abnormalities) and multiple malignant tumors such as basal cell carcinoma (Hahn et al. 1996; Johnson et al. 1996).

One remarkable feature of the Hedgehog signal transduc-tion pathway is the importance of cholesterol. First, cholesterol is critical for the catalytic cleavage of Sonic hedgehog protein. Only the amino-terminal portion of the protein is functional

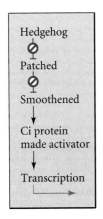

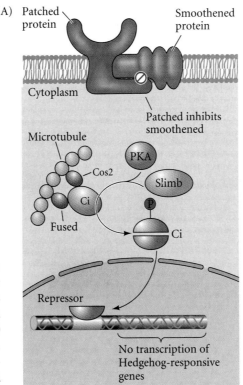

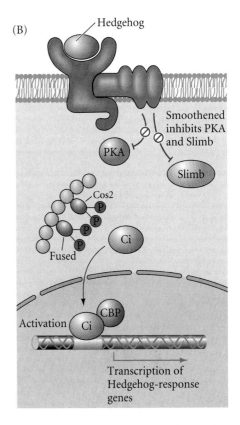

Figure 6.25
The Hedgehog signal transduction pathway. The Patched protein in the cell membrane is an inhibitor of the Smoothened protein. (A) In the absence of Hedgehog protein binding to Patched, the Ci protein is tethered to the microtubules (by the Cos2 and Fused proteins). This binding allows the PKA and Slimb proteins to cleave Ci into a transcriptional repressor that blocks the transcription of particular genes. (B) When Hedgehog binds to Patched, its conformation changes, releasing the inhibition of the Smoothened protein. Smoothened releases Ci from the microtubules (probably by adding more phosphates to the Cos2 and Fused proteins) and inactivates the cleavage proteins PKA and Slimb. The Ci protein enters the nucleus, binds a CBP protein and acts as a transcriptional activator of particular genes. (After Johnson and Scott 1998.)

and secreted. The cholesterol also binds to the active N-terminus of the Sonic hedgehog protein and allows this paracrine factor to diffuse over a range of a few hundred μm (about 30 cell diameters in the mouse limb). Without this cholesterol modification, diffusion is severely hampered (Lewis et al. 2001). Second, the Patched protein that binds Sonic hedgehog also needs cholesterol in order to function. Some human cyclopia syndromes are caused by mutations in genes that encode either Sonic hedgehog or the enzymes that synthesize cholesterol (Kelley et al. 1996; Roessler et al. 1996). Moreover, certain chemicals that induce cyclopia* do so by interfering with the cholesterol biosynthetic enzymes (Figure 6.26; Beachy et al. 1997; Cooper et al. 1998).

VADE MECUM[2] **Cyclopia induced in zebrafish.** Alcohol can act as a teratogen and can induce cyclopia in zebrafish embryos. **[Click on Zebrafish]**

---

*Environmental factors that cause developmental anomalies are called **teratogens**, and they will discussed in more detail in Chapters 21 and 22. Two teratogens known to cause cyclopia in vertebrates are jervine and cyclopamine. Both substances are found in the plant *Veratrum californicum* (Keeler and Binns 1968), and both block the synthesis of cholesterol.

Figure 6.26
Head of a cyclopic lamb born of a ewe who had eaten *Veratrum californicum* early in pregnancy. The cerebral hemispheres fused, forming only one central eye and no pituitary gland. The jervine alkaloid made by this plant inhibits cholesterol synthesis, which is needed for Hedgehog production and reception. (Photograph courtesy of L. James and the USDA Poisonous Plant Laboratory.)

Cell Death Pathways

"To be, or not to be: that is the question." While we all are poised at life-or-death decisions, this existential dichotomy is exceptionally stark for embryonic cells. **Programmed cell death**, or **apoptosis**,[*] is a normal part of development (see Baehrecke 2002). In the nematode *C. elegans*, in which we can count the number of cells, exactly 131 cells die according to the normal developmental pattern. All the cells of this nematode are "programmed" to die unless they are actively told not to undergo apoptosis. In humans, as many as 10^{11} cells die in each adult each day and are replaced by other cells. (Indeed, the mass of cells we lose each year through normal cell death is close to our entire body weight!) Within the uterus, we were constantly making and destroying cells, and we generated about three times as many neurons as we eventually ended up with when we were born. Lewis Thomas (1992) has aptly noted,

> By the time I was born, more of me had died than survived. It was no wonder I cannot remember; during that time I went through brain after brain for nine months, finally contriving the one model that could be human, equipped for language.

---

[*]The term *apoptosis* (both "p"s are pronounced) comes from the Greek word for the natural process of leaves falling from trees or petals falling from flowers. Apoptosis is an active process that can be subject to evolutionary selection. A second type of cell death, *necrosis*, is a pathological death caused by external factors such as inflammation or toxic injury.

Apoptosis is necessary not only for the proper spacing and orientation of neurons, but also for generating the middle ear space, the vaginal opening, and the spaces between our fingers and toes (Saunders and Fallon 1966; Roberts and Miller 1998; Rodriguez et al. 1997). Apoptosis prunes unneeded structures (frog tails, male mammary tissue), controls the number of cells in particular tissues (neurons in vertebrates and flies), and sculpts complex organs (palate, retina, digits, and heart).

Different tissues use different signals for apoptosis. One of the signals often used in vertebrates is bone morphogenetic protein 4 (BMP4). Some tissues, such as connective tissue, respond to BMP4 by differentiating into bone. Others, such as the frog gastrula ectoderm, respond to BMP4 by differentiating into skin. Still others, such as neural crest cells and tooth primordia, respond by degrading their DNA and dying. In the developing tooth, for instance, numerous growth and differentiation factors are secreted by the enamel knot. After the cusp has grown, the enamel knot synthesizes BMP4 and shuts itself down by apoptosis (see Chapter 13; Vaahtokari et al. 1996b).

In other tissues, the cells are "programmed" to die, and will remain alive only if some growth or differentiation factor is present to "rescue" them. This happens during the development of mammalian red blood cells. The red blood cell precursors in the mouse liver need the hormone erythropoietin in order to survive. If they do not receive it, they undergo apoptosis. The erythropoietin receptor works through the JAK-STAT pathway, activating the Stat5 transcription factor. In this way, the amount of erythropoietin present can determine how many red blood cells enter the circulation.

Figure 6.27
Apoptosis pathways in nematodes and mammals. (A) In *C. elegans*, the CED-4 protein is a protease activating factor that can activate the CED-3 protease. The CED-3 protease initiates the cell destruction events. CED-9 can inhibit CED-4 (and CED-9 can be inhibited upstream by EGL-1). (B) In mammals, a similar pathway exists, and appears to function in a similar manner. In this hypothetical scheme for the regulation of apoptosis in mammalian neurons, BclXL (a member of the Bcl2 family) binds Apaf1 and prevents it from activating the precursor of caspase-9. The signal for apoptosis allows another protein (here, Bik) to inhibit the binding of Apaf1 to BclXL. Now, Apaf-1 is able to bind to the caspase-9 precursor and cleave it. Caspase-9 now dimerizes and activates caspase-3, which initiates apoptosis. (C) There are other apoptosis pathways in mammals, such as the one initiated by the CD95 protein in the cell membranes of lymphocytes. The same colors are used to represent homologous proteins. (After Adams and Cory 1998.)

(A) *C. elegans*

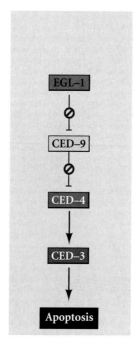

(B) Mammalian neurons

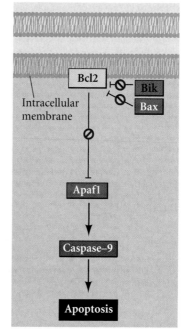

(C) Mammalian lymphocytes

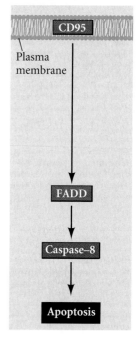

One of the pathways for apoptosis was largely delineated through genetic studies of *C. elegans*. Indeed, the importance of this pathway was recognized by the awarding of a Nobel Prize to Sydney Brenner, Bob Horvitz, and Jonathan Sulston in 2002. It was found that the proteins encoded by the *ced-3* and *ced-4* genes were essential for apoptosis, but that in the cells that did not undergo apoptosis, those genes were turned off by the product of the *ced-9* gene (Figure 6.27A; Hengartner et al. 1992). The CED-4 protein is a protease activating factor that activates CED-3, a protease that initiates the destruction of the cell. The CED-9 protein can bind to and inactivet CED-4. Mutations that inactivate the *ced-9* gene cause numerous cells that would normally survive to activate their *ced-3* and *ced-4* genes and die, leading to the death of the entire embryo. Conversely, gain-of-function mutations of *ced-9* cause CED-9 protein to be made in cells that would normally die, resulting in their survival. Thus, the *ced-9* gene appears to be a binary switch that regulates the choice between life and death on the cellular level. It is possible that every cell in the nematode embryo is poised to die, and those cells that survive are rescued by the activation of the *ced-9* gene.

The CED-3 and CED-4 proteins form the center of the apoptosis pathway that is common to all animals studied. The trigger for apoptosis can be a developmental cue such as a particular molecule (such as BMP4 or glucocorticoids) or the loss of adhesion to a matrix. Either type of cue can activate the CED-3 or CED-4 proteins or inactivate the CED-9 molecules. In mammals, the homologues of the CED-9 protein are members of the **Bcl-2 family** of genes. This family includes *Bcl-2*, *Bcl-X*, and similar genes. The functional similarities are so strong that if an active human Bcl-2 gene is placed in C. elegans embryos, it prevents normally occurring cell deaths in the nematode embryos (Vaux et al. 1992). One of the promoters recognized by the Mitf transcription in melanocytes is the promoter for *Bcl-2*. By activating the *Bcl-2* gene, Mitf prevents the apoptotic death of these pigment cells (McGill et al. 2002; see Figure 6.16B). In vertebrate red blood cell development, the Stat5 transcription factor activated by erythropoietin functions by binding to the promoter of the *Bcl-X* gene, where it activates the synthesis of that anti-apoptosis protein (Socolovsky et al. 1999).

The mammalian homologue of CED-4 is called **Apaf1** (**a**poptotic **p**rotease **a**ctivating **f**actor **1**)*, and it participates in the cytochrome *c*-dependent activation of the mammalian CED-3 homologues, the proteases **caspase-9** and **caspase-3**

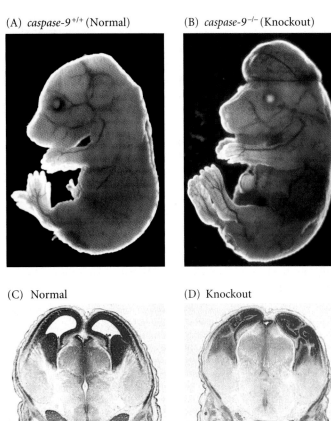

(A) *caspase-9⁺/⁺* (Normal)   (B) *caspase-9⁻/⁻* (Knockout)

(C) Normal   (D) Knockout

Figure 6.28
Disruption of normal brain development by blocking apoptosis. In mice in which the genes for caspase-9 have been knocked out, normal neural apoptosis fails to occur, and the overproliferation of brain neurons is obvious. (A) 16-day embryonic wild-type mouse. (B) A *caspase-9* knockout mouse of the same age. The enlarged brain protrudes above the face, and the limbs are still webbed. (C,D) This effect is confirmed by cross-sections through the forebrain at day 13.5. The knockout exhibits thickened ventricle walls and the near-obliteration of the ventricles. (From Kuida et al. 1998.)

(Shaham and Horvitz 1996; Cecconi et al. 1998; Yoshida et al. 1998). Activation of the caspases causes the autodigestion of the cell. Caspases are strong proteases that digest the cell from within, cleaving cellular proteins and fragmenting the DNA.

While apoptosis-deficient nematodes deficient for CED-4 are viable (despite their having 15% more cells than wild-type worms), mice with loss-of-function mutations for either *caspase-3* or *caspase-9* die around birth from massive cell overgrowth in the nervous system (Figure 6.28; Kuida et al. 1996, 1998; Jacobson et al. 1997). Mice homozygous for targeted deletions of *Apaf1* have similarly severe craniofacial abnormalities, brain overgrowth, and webbing between their toes.

---

*There is some evidence (see Barinaga 1998a,b; Saudou et al. 1998) that activation of the apoptosis pathway in adult neurons may be responsible for the pathology of Alzheimer disease and stroke. DNA fragmentation is one of the major ways in which apoptosis is recognized, and it is seen in regions of the brain affected by these diseases. The fragmentation of DNA into specific sized pieces (protected and held together by nucleosomes) may be caused by the digestion of poly(ADP-ribose) polymerase (PARP) by caspase-3 (Lazebnik et al. 1994). PARP recognizes and repairs DNA breaks. For more on other pathways leading to apoptosis, see Green 1998.

In mammals, there is more than one pathway to apoptosis. Apoptosis of the lymphocytes, for instance, is not affected by the deletion of *Apaf1* or *caspase-9*, and works by a separate pathway initiated by the CD95 protein (see Figure 6.27C). Different caspases may be functioning in different cell types to mediate the apoptotic signals (Hakem et al. 1998; Kuida et al. 1998).

WEBSITE 6.7 **The uses of apoptosis.** Apoptosis is used for numerous processes throughout development. This website explores the role of apoptosis in such phenomena as *Drosophila* germ cell development and the eyes of blind cave fish.

Juxtacrine Signaling

In juxtacrine interactions, proteins from the inducing cell interact with receptor proteins of adjacent responding cells without diffusing from the cell producing it. There are three types of juxtacrine interactions.

1. A protein on one cell membrane binds to its receptor on the adjacent cell membrane. We saw this type of juxtacrine interaction when we discussed the interaction between the Bride-of-sevenless protein and its receptor, Sevenless.
2. A receptor on one cell binds to its ligand on the extracellular matrix secreted by another cell.
3. A signal is transmitted directly from the cytoplasm of one cell through small conduits into the cytoplasm of an adjacent cell.

The Notch pathway: Juxtaposed ligands and receptors

While most known regulators of induction are diffusible proteins, some inducing proteins remain bound to the inducing cell surface. In one such pathway, cells expressing the **Delta**, **Jagged**, or **Serrate** proteins in their cell membranes activate neighboring cells that contain the **Notch** protein in their cell membranes. Notch extends through the cell membrane, and its external surface contacts Delta, Jagged, or Serrate proteins extending out from an adjacent cell. When complexed to one of these ligands, Notch undergoes a conformational change that enables a part of its cytoplasmic domain to be cut off by the presenilin-1 protease. The cleaved portion enters the nucleus and binds to a dormant transcription factor of the CSL family. When bound to the Notch protein, the CSL transcription factors activate their target genes (Figure 6.29; Lecourtois and

Schweisguth 1998; Schroeder et al. 1998; Struhl and Adachi 1998). This activation is thought to involve the recruitment of histone acetyltransferases (Wallberg et al. 2002).

Notch proteins are extremely important receptors in the nervous system. In both the vertebrate and *Drosophila* nervous system, the binding of Delta to Notch tells the receiving cell not to become neural (Chitnis et al. 1995; Wang et al. 1998). In the vertebrate eye, the interactions between Notch and its ligands seem to regulate which cells become optic neurons and which become glial cells (Dorsky et al. 1997; Wang et al. 1998). Notch proteins are also important in the patterning of the nematode vulva (see Figure 6.20). The primary VPC (closest to the anchor cell) is able to inhibit its neighbors from becoming central vulval cells by signaling to them through its Notch homologue, the LIN-12 receptor. The LIN-12 signal blocks the protein kinase pathway initiated by the LIN-3 signal (Berset et al. 2001).

WEBSITE 6.8 **Notch mutations.** Mutations in Notch proteins can cause nervous system abnormalities in humans. Humans have more than one Notch gene and more than one ligand. Their interactions may be critical in neural development. Moreover, the association of Notch with the presenilin protease suggests that disruption of Notch functioning might lead to Alzheimer disease.

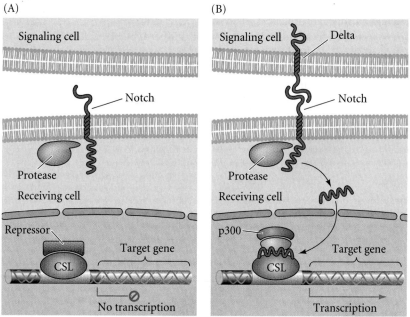

Figure 6.29
Mechanism of Notch activity. (A) Prior to Notch signaling, a CSL transcription factor (such as Suppressor of hairless or CBF1) is on the enhancer of Notch-regulated genes. The CSL binds repressors of transcription. (B) Model for the activation of Notch. A ligand (Delta, Jagged, or Serrate protein) on one cell binds to the extracellular domain of the Notch protein on an adjacent cell. This binding causes a shape change in the intracellular domain of Notch, which activates a protease. The protease cleaves Notch and allows the intracellular region of the Notch protein to enter the nucleus and bind the CSL transcription factor. This intercellular region of Notch displaces the repressor proteins and binds activators of transcription, including the histone acetyltransferase p300. The activated CSL can then transcribe its target genes (After Koziol-Dube, personal communication.)

Cell-Cell Interactions and Chance in the Determination of Cell Types

The development of the vulva in *C. elegans* offers several examples of induction on the cellular level. We have already discussed the reception of the EGF-like LIN-3 signal by the cells of the equivalence group that can form the vulva. But before this induction occurs, there is an earlier interaction that forms the anchor cell. The formation of the anchor cell is mediated by the *lin-12* gene, the *C. elegans* homologue of the *Notch* gene. In wild-type *C. elegans* hermaphrodites, two adjacent cells, Z1.ppp and Z4.aaa, have the potential to become the anchor cell. They interact in a manner that causes one of them to become the anchor cell while the other one becomes the precursor of the uterine tissue. In loss-of-function *lin-12* mutants, both cells become anchor cells, while in gain-of-function mutations, both cells become uterine precursors (Greenwald et al. 1983). Studies using genetic mosaics and cell ablations have shown that this decision is made in the second larval stage, and that the *lin-12* gene needs to function only in that cell destined to become the uterine precursor cell. The presumptive anchor cell

does not need it. Seydoux and Greenwald (1989) speculate that these two cells originally synthesize both the signal for uterine differentiation (the LAG-2 protein, homologous to Delta in *Drosophila*) and the receptor for this molecule (the LIN-12 protein, homologous to Notch) (Figure 6.30; Wilkinson et al. 1994). During a particular time in larval development, the cell that, by chance, is secreting more LAG-2 causes its neighbor to cease its production of this differentiation signal and to increase its production of LIN-12 protein. The cell secreting LAG-2 becomes the gonadal anchor cell, while the cell receiving the signal through its LIN-12 protein becomes the ventral uterine precursor cell. Thus, the two cells are thought to determine each other prior to their respective differentiation events. When the LIN-12 protein is used again during vulva formation, it is activated by the primary vulval lineage to stop the lateral vulval cells from forming the central vulval phenotype (see Figure 6.19).

The anchor cell/ventral uterine precursor decision illustrates two important aspects of determination in two originally equivalent cells. First, the initial difference between the two cells is created by chance. Second, this initial difference is reinforced by feedback. Such a mechanism is also seen in the determination of which of the originally equivalent epidermal cells of the in-

sect embryo will generate the neurons of the peripheral nervous system. Here, the choice is between becoming a skin (hypodermal) cell or a neural precursor cell (a **neuroblast**). The *Notch* gene of *Drosophila*, like its *C. elegans* homologue, *lin-12*, channels a bipotential cell into one of two alternative paths. Soon after gastrulation, a region of about 1800 ectodermal cells lies along the ventral midline of the *Drosophila* embryo. These cells, known as neurogenic ectodermal cells, all have the potential to form the ventral nerve cord of the insect. About one-fourth of these cells will become neuroblasts, while the rest will become the precursors of the hypodermis. The cells that give rise to neuroblasts are intermingled with those that become hypodermal precursors. Thus, each neurogenic ectoderm cell can give rise to either hypodermal or neural precursor cells (Hartenstein and Campos-Ortega 1984). In the absence of *Notch* gene transcription in the embryo, these cells develop exclusively into neuroblasts, rather than into a mixture of hypodermal and neural precursor cells (Artavanis-Tsakonis et al. 1983; Lehmann et al. 1983). These embryos die, having a gross excess of neural cells at the expense of the ventral and head hypodermis (Poulson 1937; Hoppe and Greenspan 1986).

Heitzler and Simpson (1991) proposed that the Notch protein, like LIN-12, serves as a receptor for intercellular signals involved in the differentiation of equivalent cells. Moreover, they provided evidence that Delta is the ligand for Notch. Genetic mosaics show that whereas Notch is needed in the cells that are to become hypodermis, Delta is needed in the cells that *induce* the hypodermal phenotype.

Greenwald and Rubin (1992) have proposed a model based on the LIN-12 hypothesis to explain the spacing of neuroblasts in these proneural clusters of epidermal and neural precursors (Figure 6.31). Initially, all the neurogenic ectodermal cells have equal potentials and produce the same signals. However, when one of the cells, by chance, produces more signal (say, Delta protein), it activates the receptors on adjacent cells and reduces their level of signaling. As their signaling levels are lowered, the neighbors of these low-signaling cells will tend to become high-level signalers. In this way, a spacing of neuroblasts is produced.

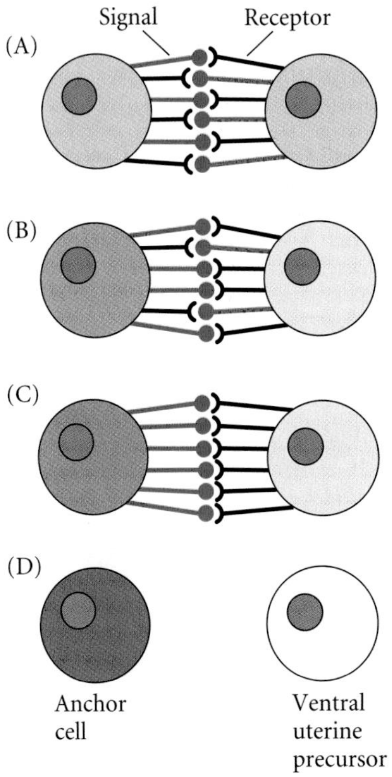

Figure 6.30
Model for the generation of two cell types (anchor cell and ventral uterine precursor) from two equivalent cells (Z1.ppp and Z4.aaa) in *C. elegans*. (A) The cells start off as equivalent, producing fluctuating amounts of signal and receptor (inverted arrow). The *lag-2* gene is thought to encode the signal; the *lin-12* gene is thought to encode the receptor. Reception of the signal turns down LAG-2 (Delta) production and up-regulates LIN-12 (Notch). (B) A stochastic (chance) event causes one cell to produce more LAG-2 than the other cell at some particular critical time. This stimulates more LIN-12 production in the neighboring cell. (C) This difference is amplified, since the cell producing more LIN-12 produces less LAG-2. Eventually, just one cell is delivering the LAG-2 signal, and the other cell is receiving it. (D) The signaling cell becomes the anchor cell; the receiving cell becomes the ventral uterine precursor. (After Greenwald and Rubin 1992.)

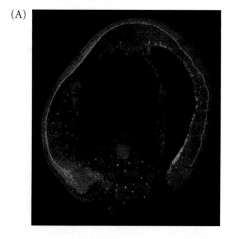

Figure 6.31
Model to explain the spacing pattern of neuroblasts among initially equivalent neurogenic ectodermal cells. (A) A field of equivalent cells, all of which signal and receive equally. (B) A chance event causes one of the cells (darker shading) to produce more signal. The cells surrounding it receive this higher amount of signal and reduce their own signaling level (lighter shading). (C) The rest of the pattern is now constrained. Those cells that have downregulated their own signaling (in response to the events in B) are less likely to express more signal than their neighboring cells. The cells surrounded by down-regulated signalers are more likely to become signalers. (D, E) The fates of the cells throughout the field become specified as the amplification of the signal creates populations of signalers surrounded by populations of receivers. In the case of the neurogenic cells, the signal is thought to be the Delta protein, and the receiver is the Notch protein. (After Greenwald and Rubin 1992.)

The extracellular matrix as a source of critical developmental signals

PROTEINS AND FUNCTIONS OF THE EXTRACELLULAR MATRIX. The **extracellular matrix** consists of macromolecules secreted by cells into their immediate environment. These macromolecules form a region of noncellular material in the interstices between the cells. The extracellular matrix is a critical region for much of animal development. Cell adhesion, cell migration, and the formation of epithelial sheets and tubes all depend on the ability of cells to form attachments to extracellular matrices. In some cases, as in the formation of epithelia, these attachments have to be extremely strong. In other instances, as when cells migrate, attachments have to be made, broken, and made again. In some cases, the extracellular matrix merely serves as a permissive substrate to which cells can adhere, or upon which they can migrate. In other cases, it provides the directions for cell movement or the signal for a developmental event.

Extracellular matrices are made up of collagen, proteoglycans, and a variety of specialized glycoprotein molecules, such as fibronectin and laminin. These large glycoproteins are responsible for organizing the matrix and the cells into an ordered structure. **Fibronectin** is a very large (460 kDa) glycoprotein dimer synthesized by numerous cell types. One function of fibronectin is to serve as a general adhesive molecule, linking cells to one another and to other substrates such as collagen and proteoglycans. Fibronectin has several distinct binding sites, and their interaction with the appropriate molecules results in the proper alignment of cells with their extracellular matrix (Figure 6.32). As we will see in later chapters, fibronectin also has an important role in cell migration. The "roads" over which certain migrating cells travel are paved

Figure 6.32
Fibronectin in the developing embryo. (A) Fluorescent antibodies to fibronectin show fibronectin deposition as a green band in the *Xenopus* embryo during gastrulation. The fibronectin will orient the movements of the mesoderm cells. (B) Structure and binding domains of fibronectin. The rectangles represent protease-resistant domains. The fibroblast-binding domain consists of two units, the RGD site and the high-affinity site, both of which are essential for cell binding. Another binding site for avian neural crest cells is necessary for these cells to migrate on a fibronectin substrate. Other regions of fibronectin enable it to bind to collagen, heparin, and other molecules of the extracellular matrix. (A, photograph courtesy of M. Marsden and D. W. DeSimone; B after Dufour et al. 1988.)

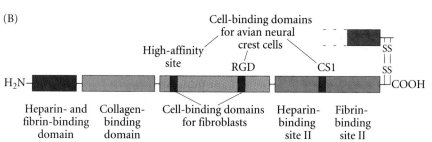

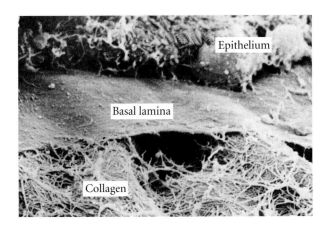

Figure 6.33
Location and formation of extracellular matrices in the chick embryo. This scanning electron micrograph shows the extracellular matrix at the junction of the epithelial cells (above) and mesenchymal cells (below). The epithelial cells synthesize a tight, laminin-based basal lamina, while the mesenchymal cells secrete a loose reticular lamina made primarily of collagen. (Photograph courtesy of R. L. Trelstad.)

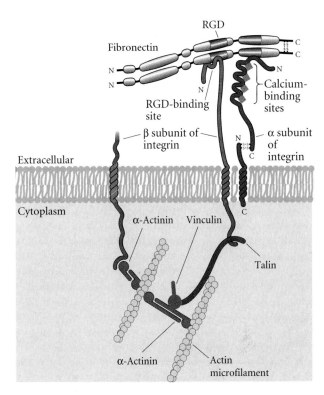

Figure 6.34
Speculative diagram of the binding of cytoskeleton to the extracellular matrix through the integrin molecule. (After Luna and Hitt 1992.)

with this protein. Fibronectin paths lead germ cells to the gonads and lead heart cells to the midline of the embryo. If chick embryos are injected with antibodies to fibronectin, the heart-forming cells fail to reach the midline, and two separate hearts develop (Heasman et al. 1981; Linask and Lash 1988).

Laminin and **type IV collagen** are major components of a type of extracellular matrix called the **basal lamina**. The basal lamina is characteristic of the closely knit sheets that surround epithelial tissue (Figure 6.33). The adhesion of epithelial cells to laminin (upon which they sit) is much greater than the affinity of mesenchymal cells for fibronectin (to which they must bind and release if they are to migrate). Like fibronectin, laminin plays a role in assembling the extracellular matrix, promoting cell adhesion and growth, changing cell shape, and permitting cell migration (Hakamori et al. 1984).

**INTEGRINS, THE RECEPTORS FOR EXTRACELLULAR MATRIX MOLE-
CULES.** The ability of a cell to bind to adhesive glycoproteins depends on its expressing membrane receptors for the cell-binding sites of these large molecules. The main fibronectin receptors were identified by using antibodies that block the attachment of cells to fibronectin (Chen et al. 1985; Knudsen et al. 1985). The fibronectin receptor complex was found not only to bind fibronectin on the outside of the cell, but also to bind cytoskeletal proteins on the inside of the cell. Thus, the fibronectin receptor complex appears to span the cell membrane and unite two types of matrices. On the outside of the cell, it binds to the fibronectin of the extracellular matrix; on the inside of the cell, it serves as an anchorage site for the actin microfilaments that move the cell (Figure 6.34). Horwitz and co-workers (1986; Tamkun et al. 1986) have called this

family of receptor proteins **integrins** because they integrate the extracellular and intracellular scaffolds, allowing them to work together. On the extracellular side, integrins bind to the sequence arginine-glycine-aspartate (RGD), found in several adhesive proteins in extracellular matrices, including fibronectin, vitronectin (found in the basal lamina of the eye), and laminin (Ruoslahti and Pierschbacher 1987). On the cytoplasmic side, integrins bind to talin and α-actinin, two proteins that connect to actin microfilaments. This dual binding enables the cell to move by contracting the actin microfilaments against the fixed extracellular matrix.

Bissell and her colleagues (1982; Martins-Green and Bissell 1995) have shown that the extracellular matrix is capable of inducing specific gene expression in developing tissues, especially those of the liver, testis, and mammary gland. In these tissues, the induction of specific transcription factors depends on cell-substrate binding (Figure 6.35; Liu et al. 1991; Streuli et al. 1991; Notenboom et al. 1996). Often, the presence of bound integrin prevents the activation of genes that specify apoptosis (Montgomery et al. 1994; Frisch and Ruoslahti 1997). The chondrocytes that produce the cartilage of our vertebrae and limbs can survive and differentiate only if they are surrounded by an extracellular matrix and are joined to that matrix through their integrins (Hirsch et al. 1997). If chondrocytes from the developing chick sternum are incubated with anti-

(A)

(B)

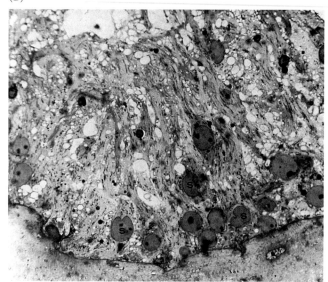

Figure 6.35
Role of the extracellular matrix in cell differentiation. Light micrographs of rat testis Sertoli cells grown for 2 weeks (A) on tissue culture plastic dishes and (B) on dishes coated with basal lamina. The two photographs were taken at the same magnification, 1200×. (From Hadley et al. 1985; photographs courtesy of M. Dym.)

pressed, and the cell division genes remain turned off. By this time, the mammary gland cells have enveloped themselves in a basal lamina, forming a secretory epithelium reminiscent of the mammary gland tissue. The binding of integrins to laminin is essential for the transcription of the casein gene,

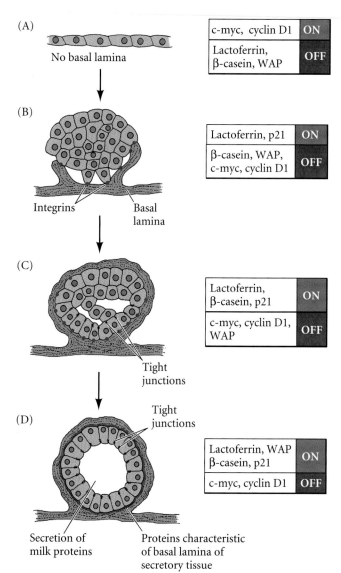

Figure 6.36
Basement membrane-directed gene expression in mammary gland tissue. (A) Mouse mammary gland tissue divides when placed on tissue culture plastic. The genes encoding cell division proteins are on, and the genes capable of synthesizing the differentiated products of the mammary gland lactoferrin, casein, whey acidic protein (WAP) are off. (B) When these cells are placed on a basement membrane that contains laminin (basal lamina), the genes for cell division proteins are turned off, while the gene inhibiting cell division (p21) and the gene for lactoferrin are turned on. (C, D) The mammary gland cells wrap the basal lamina around them, forming a secretory epithelium. The genes for casein and whey acidic protein are sequentially activated. (After Bissell, et al. 2003.)

bodies that block the binding of integrins to the extracellular matrix, they shrivel up and die. While the mechanisms by which bound integrins inhibit apoptosis remain controversial (see Howe et al. 1998), the extracellular matrix is obviously an important source of signals that can be transduced into the nucleus to produce specific gene expression.

Some of the genes induced by matrix attachment are being identified. When plated onto tissue culture plastic, mouse mammary gland cells will divide (Figure 6.36). Indeed, the genes for cell division (*c-myc*, *cyclinD1*) are expressed, while the genes for the differentiated products of the mammary gland (casein, lactoferrin, whey acidic protein) are not expressed. If the same cells are plated onto plastic coated with a laminin-containing basement membrane, the cells stop dividing and the differentiated genes of the mammary gland are expressed. This happens only after the integrins of the mammary gland cells bind to the laminin of the extracellular basement membrane. Then the gene for lactoferrin is expressed, as is the gene for p21, a cell division inhibitor. The *c-myc* and *cyclinD1* genes become silent. Eventually, all the genes for the developmental products of the mammary gland are ex-

(A)

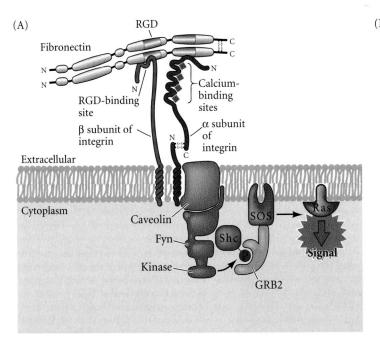

(B)

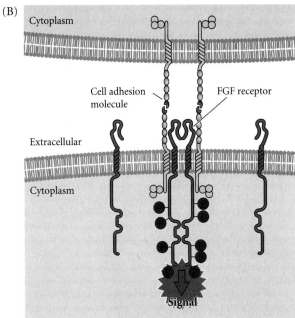

Figure 6.37
Two types of activation by cell adhesion molecules. (A) Cell-substrate adhesion molecules such as integrins may transmit a signal from the cytoplasmic portion of the integrin protein to the Ras G protein through a cascade involving the caveolin and Fyn proteins. (B) FGF receptors may be "hijacked" by cell adhesion molecules and dimerized. They may be brought together by the interaction of opposite cell adhesion molecules, or the "crosslinking" of FGF receptors by the apposing cell membrane may activate their kinase domains. (A after Wary et al. 1998.)

and the integrins act in concert with prolactin (see Figure 6.22) to activate that gene's expression (Roskelley et al. 1994; Muschler et al. 1999).

Recent studies have shown that the binding of integrins to an extracellular matrix can stimulate the RTK pathway. When an integrin on the cell membrane of one cell binds to the fibronectin or collagen secreted by a neighboring cell, the integrin can activate the RTK cascade through an adaptor protein-like complex that connects the integrin to the Ras G protein (Figure 6.37A; Wary et al. 1998). Cadherins and other cell adhesion molecules (see Chapter 3) can also transmit signals by "hijacking" the FGF receptors (Williams et al. 1994b; Clark and Brugge 1995). Cadherins, for example, are able to bind to the cytoplasmic region of FGF receptors and thereby dimerize these receptors, just as normal FGF ligands do (Figure 6.37B; Williams et al. 1994a; Doherty and Walsch 1996).

Direct transmission of signals through gap junctions

Throughout this chapter we have been discussing the transduction of signals by cell membrane receptors. In these cases, the receptor is altered in some manner by binding a ligand, so

that its cytoplasmic domain transmits the signal into the cell. Another signaling mechanism transmits small, soluble signaling molecules directly through the cell membrane, for the membrane is not continuous in all places; structures called **gap junctions** serve as communication channels between adjacent cells (Figure 6.38A,B). Cells linked by gap junctions are said to be "coupled," and small molecules (molecular weight <1500) and ions can freely pass from one cell to the other. In most embryos, at least some of the early blastomeres are connected by gap junctions. These junctions allow epithelial sheets (i.e., closely connected cells) to act together. Moreover, the ability of cells to form gap junctions with some cells and not with others creates physiological "compartments" within the developing embryo (Figure 6.38C).

Gap junction channels are made of **connexin** proteins. In each cell, six identical connexins group together in the membrane to form a transmembrane channel with a central pore. The channel complex of one cell connects to the channel complex of another cell, joining the cytoplasms of both cells (Figure 6.38D). Different types of connexin proteins have separate, but overlapping, roles in normal development. For example, connexin-43 is found in nearly every tissue of the developing mouse embryo,. However, the mouse embryo will still develop even if the connexin-43 genes are "knocked out" by gene targeting (see Chapter 4). It appears that most of the functions of connexin-43 can be taken over by other, related connexin proteins—but not all. Shortly after birth, connexin-43-deficient mice take gasping breaths, turn bluish, and die. Autopsies of these mice show that the right ventricle (the chamber that pumps blood through the pulmonary artery to the lungs) is filled with tissue that occludes the chamber and obstructs blood flow (Reaume et al. 1995; Huang et al. 1998).

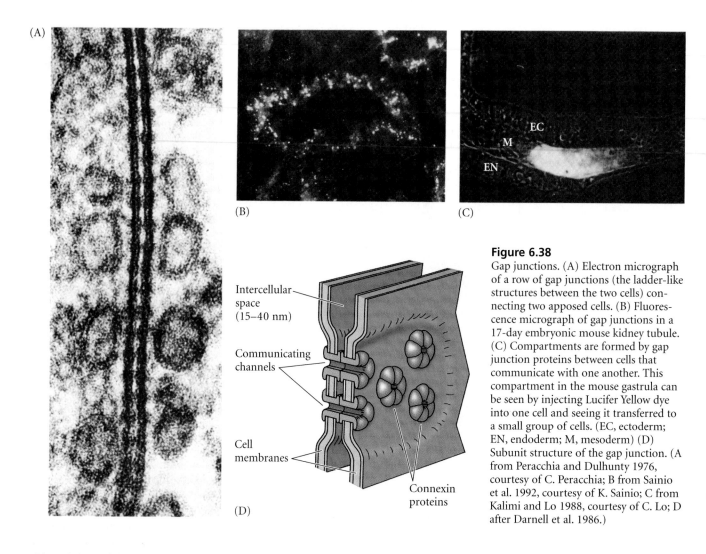

(A)

(B)

(C)

Intercellular space (15–40 nm)

Communicating channels

Cell membranes

Connexin proteins

(D)

Figure 6.38
Gap junctions. (A) Electron micrograph of a row of gap junctions (the ladder-like structures between the two cells) connecting two apposed cells. (B) Fluorescence micrograph of gap junctions in a 17-day embryonic mouse kidney tubule. (C) Compartments are formed by gap junction proteins between cells that communicate with one another. This compartment in the mouse gastrula can be seen by injecting Lucifer Yellow dye into one cell and seeing it transferred to a small group of cells. (EC, ectoderm; EN, endoderm; M, mesoderm) (D) Subunit structure of the gap junction. (A from Peracchia and Dulhunty 1976, courtesy of C. Peracchia; B from Sainio et al. 1992, courtesy of K. Sainio; C from Kalimi and Lo 1988, courtesy of C. Lo; D after Darnell et al. 1986.)

Although loss of the connexin-43 protein can be compensated for in many tissues, it appears to be critical for normal heart development.

WEBSITE 6.9 **Connexin mutations.** Mutations in human connexin proteins cause congenital malformations of the heart and ear. In many cases, one connexin can substitute for another, but when the connexins cannot compensate, a mutant phenotype results.

The importance of gap junctions in development has been demonstrated in amphibian and mammalian embryos (Warner et al. 1984). When antibodies to connexins were microinjected into one specific cell of an 8-cell *Xenopus* blastula, the progeny of that cell, which are usually coupled through gap junctions, could no longer pass ions or small molecules from cell to cell. The tadpoles that resulted from such treated blastulae showed defects specifically relating to the developmental fate of the injected cell (Figure 6.39). The progeny of the injected cell did not die, but they were unable to develop normally (Warner et al. 1984). In the mouse embryo, the first eight blastomeres are also connected to one another by gap

junctions. Although loosely associated with one another, these eight cells move together to form a compacted embryo. If compaction is inhibited by antibodies against connexins, the treated blastomeres continue to divide, but further development ceases (Lo and Gilula 1979; Lee et al. 1987). If antisense RNA to connexin messages is injected into one of the blastomeres of a normal mouse embryo, that cell will not form gap junctions and will not be included in the embryo (Bevilacqua et al. 1989).

Cross-Talk between Pathways

We have been representing the major signal transduction pathways as if they were linear chains through which information flows in a single conduit. However, these pathways are just the major highways of information flow. Between them, avenues and streets connect one pathway with another. (This may be why there are so many steps between the cell surface and the nucleus. Each step is a potential regulatory point as well as a potential intersection.) This **cross-talk** can be seen in numerous tissues, wherein two signaling pathways reinforce each

(A)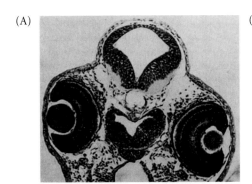

(B)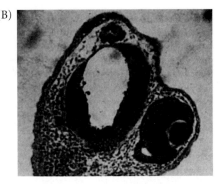

Figure 6.39
Developmental effects of gap junctions. Sections through *Xenopus* tadpole in which one of the blastomeres at the 8-cell stage was injected with (A) a control antibody or (B) an antibody against connexins. In the treated tadpole, the side formed by the injected blastomere lacks its eye and has abnormal brain morphology. (From Warner et al. 1984; photographs courtesy of A. E. Warner.)

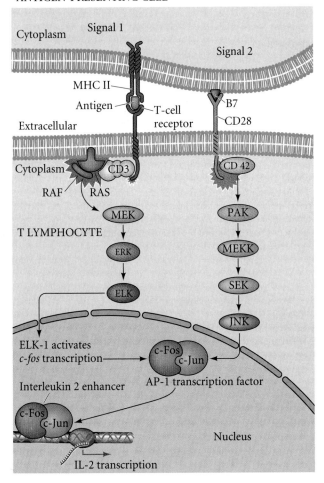

Figure 6.40
Two signals are needed to effect the differentiation of the T lymphocyte. The first signal comes from the receptors that bind the antigen. The second signal comes from the binding of the CD28 transmembrane protein of the T cell to the B7 protein on the surface of the antigen-presenting cell. The first signal directs the synthesis of one subunit of the AP-1 transcription factor, c-Fos. The second signal directs the synthesis of the other subunit, c-Jun. Together c-Fos and c-Jun form the AP-1 transcription factor, which can activate T cell-specific enhancers such as that regulating interleukin 2 production.

other. We must remember that a cell has numerous receptors and is constantly receiving many signals simultaneously.

In some cells, gene transcription requires two different signals. This pattern is seen during lymphocyte differentiation, for which two signals are needed, each producing one of the two peptides of a transcription factor needed for the production of interleukin 2 (IL-2, also known as T cell growth factor). One of these peptides, c-Fos, is induced by the binding of the T-cell receptor to an antigen (Figure 6.40). This signal activates the RTK pathway, creating a transcription factor, Elk-1, that activates the *c-fos* gene to synthesize c-Fos. The second signal comes from the B7 glycoprotein on the surface of the cell presenting the antigen. This signal activates a second cascade of kinases, eventually producing c-Jun. The two peptides—c-Fos and c-Jun—join to make the AP-1 protein, a transcription factor that binds to the IL-2 enhancer and activates its expression (Li et al. 1996).

We also have seen that one receptor can activate several different pathways. The fibroblast growth factors, for instance, can activate the RTK pathway, the STAT pathway, or even a third pathway that involves lipid turnover and increases the levels of calcium ions in the cell.

Maintenance of the Differentiated State

Development obviously means more than initiating gene expression. For a cell to become committed to a particular phenotype, gene expression must be maintained. There are four major ways that nature has evolved for maintaining differentiation once it has been initiated (Figure 6.41). First, the transcription factor whose gene is activated by a signal transduction cascade can bind to the enhancer of its own gene. In this way, once the transcription factor is made, its synthesis becomes independent of the signal that induced it originally. The MyoD transcription factor in muscle cells is produced in this manner. Second, a cell can stabilize its differentiation by synthesizing proteins that act on chromatin to keep the gene accessible. Such proteins include the Trithorax family. Third, a cell can maintain its differentiation in an autocrine fashion. If differentiation is dependent on a particular signaling molecule, the cell can make both that signaling molecule and that

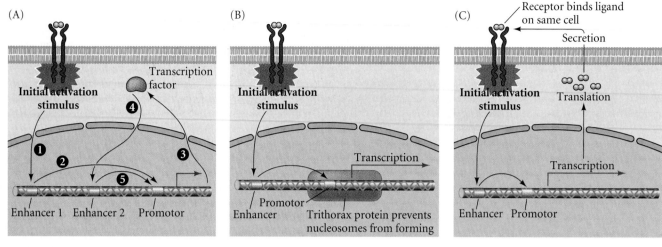

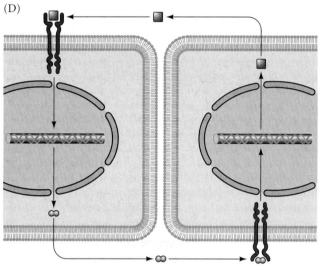

Figure 6.41
Four ways of maintaining differentiation after the initial signal has been given. (A) The initial stimulus (1) activates enhancer 1, which stimulates the promoter (2) to transcribe the gene. The gene product (3) is a transcription factor, and one of its targets is enhancer 2 of its own gene (4). This activated enhancer can now stimulate the promoter (5) to make more of this protein. (B) Proteins such as those of the Trithorax group prevent nucleosomes from forming so that the gene remains accessible. (C) Autocrine stimulation of the differentiated state. The cell is stimulated to activate those genes that enable it to synthesize and bind its own stimulatory proteins. (D) Paracrine loop between two cells such that a paracrine factor from one cell stimulates the differentiated state of the second cell, and that differentiated state includes the secretion of a paracrine factor that maintains the first cell's differentiated state.

molecule's receptor. The Sertoli cells of the mammalian testis may maintain their differentiation through such a self-stimulatory autocrine loop. Fourth, the cell may interact with its neighboring cells such that each one stimulates the differentiation of the other, and part of each neighbor's differentiated phenotype is the production of a paracrine factor that stimulates the other's phenotype. This type of I-scratch-your-back-you-scratch-mine strategy is found in the neighboring cells of the developing vertebrate limb and insect segments.

Coda

In 1782, the French essayist Denis Diderot posed the question of morphogenesis in the fevered dream of a noted physicist. This character could imagine that the body was formed from myriad "tiny sensitive bodies" that collected together to form an aggregate, but he could not envision how this aggregate could become an animal. A hundred years later, Charles Darwin wrote that the eye must have evolved by small useful variations, but he did not know the mechanisms by which the intricate coordination of eye development occurred.

Recent studies have shown that this coordination of cells to form organs and organisms depends on the molecules on the embryonic cell surfaces. Inducers and their competent responders interact with one another to instruct and coordinate the further development of the component parts. In subsequent chapters, we will look more closely at some of these morphogenetic interactions. As we continue our study of animal development, we will find that induction and competence provide the core of morphogenesis. Moreover, we will find the same signal transduction pathways wherever we look, both throughout the animal kingdom and within each developing embryo. We are now at the stage where we can begin our study of early embryogenesis and see the integration of the organismal, genetic, and cellular processes of animal development.

Principles of Development: Cell-Cell Communication

1. Inductive interactions involve inducing and responding tissues.

2. The ability to respond to inductive signals depends upon the competence of the responding cells.

3. Reciprocal induction occurs when the two interacting tissues are both inducers and are competent to respond to each other's signals.

4. Cascades of inductive events are responsible for organ formation.

5. Regionally specific inductions can generate different structures from the same responding tissue.

6. The specific response to an inducer is determined by the genome of the responding tissue.

7. Juxtacrine interactions are inductive interactions that take place between the cell membranes of adjacent cells or between a cell membrane and an extracellular matrix secreted by another cell.

8. Paracrine interactions occur when a cell or tissue secretes proteins that induce changes in neighboring cells.

9. Paracrine factors are proteins secreted by inducing cells. These factors bind to cell membrane receptors in competent responding cells.

10. Competent cells respond to paracrine factors through signal transduction pathways. Competence is the ability to bind and to respond to the inducers, and it is often the result of a prior induction.

11. Signal transduction pathways begin with a paracrine or juxtacrine factor causing a conformational change in its cell membrane receptor. The new shape results in enzymatic activity in the cytoplasmic domain of the receptor protein. This activity allows the receptor to phosphorylate other cytoplasmic proteins. Eventually, a cascade of such reactions activates a transcription factor (or set of factors) that activates or represses specific gene activity.

12. Programmed cell death is one possible response to inductive stimuli. Apoptosis is a critical part of life.

13. Gap junctions allow ions and small molecules to move between cells and facilitate coordinated action of coupled cells.

14. There is cross-talk between signal transduction pathways, which allows the cell to respond to multiple inputs simultaneously.

15. The maintenance of the differentiated state can be accomplished by positive feedback loops involving transcription factors, autocrine factors, or paracrine factors.

Literature Cited

Adams, J. M. and S. Cory. 1998. The Bcl-2 protein family: Arbiters of cell survival. *Science* 281: 1322–1326.

Artavanis-Tsakonis, S., M. A. T. Muskavitch and Y. Yedvobnick. 1983. Molecular cloning of *Notch*, a locus affecting neurogenesis in *Drosophila melanogaster*. *Proc. Natl. Acad. Sci. USA* 80: 1977–1981.

Aza-Blanc, P., F. A. Ramirez-Weber, M. P. Laget and T. B. Kornberg. 1997. Proteolysis that is inhibited by hedgehog targets Cubitus interruptus protein to the nucleus and converts it to a repressor. *Cell* 89: 1043–1053.

Baehrecke, E. H. 2002. How death shapes life during development. *Nature Rev. Mol. Cell Biol.* 3: 779–787.

Banerjee, U., P. J. Renfranz, J. A. Pollock and S. Benzer. 1987. Molecular characterization and expression of *sevenless*, a gene involved in neuronal pattern formation in the *Drosophila* eye. *Cell* 49: 281–291.

Bard, J. B. L. 1990. *Morphogenesis: The Cellular and Molecular Processes of Developmental Anatomy*. Cambridge University Press, Cambridge.

Barinaga, M. 1998a. Is apoptosis key in Alzheimer's disease? *Science* 281: 1303–1304.

Barinaga, M. 1998b. Stroke-damaged neurons may commit cellular suicide. *Science* 281: 1302–1303.

Basler, K. and E. Hafen. 1989. Ubiquitous expression of *sevenless*: Position-dependent specification of cell fate. *Science* 243: 931–934.

Beachy, P. A. and 7 others. 1997. Multiple roles of cholesterol in hedgehog protein biosynthesis and signaling. *Cold Spring Harb. Symp. Quant. Biol.* 62: 191–204.

Behrens, J., J. P. von Kries, M. Kühl, L. Bruhn, D. Wedlich, R. Grosschedl and W. Birchmeier. 1996. Functional interaction of β-catenin with the transcription factor LRF-1. *Nature* 382: 638–642.

Berset, T., E. Fröhl, G. Battu, S. Canevacini and A. Hajnal. 2001. Notch inhibition of RAS signaling through MAP kinase phosphatase LIP-1 during *C. elegans* vulval development. *Science* 291: 1055–1058.

Bevilacqua, A., R. Loch-Caruso and R. P. Erickson. 1989. Abnormal development and dye coupling produced by antisense RNA to gap junction protein in mouse preimplantation embryos. *Proc. Natl. Acad. Sci. USA* 86: 5444–5448.

Bissell, M. J., H. G. Hall and G. Parry. 1982. How does the extracellular matrix direct gene expression? *J. Theor. Biol.* 99: 31–68.

Bissell, M., I. S. Mian, D. Raditsky and E. Turley. 2003. Tissue-specificity: Structural cues allow diverse phenotypes form a constant genotype. In G. B. Müller and S. A. Newman (eds.), *Origin of Organismal Form*. MIT Press, Cambridge.

Bitgood, M. J. and A. P. McMahon. 1995. *Hedgehog* and *BMP* genes are coexpressed at many diverse sites of cell-cell interaction in the mouse embryo. *Dev. Biol.* 172:126–158.

Bitgood, M. J., L. Shen and A. P. McMahon. 1996. Sertoli cell signalling by desert hedgehog regulates the male germline. *Curr. Biol.* 6: 298–304.

Bowtell, D. D. L., M. A. Simon and G. M. Rubin. 1989. Ommatidia in the developing *Drosophila* eye require and can respond to *sevenless* for only a restricted period. *Cell* 56: 931–936.

Briscoe, J., D. Guschin and M. Müller. 1994. Just another signalling pathway. *Curr. Biol.* 4: 1033–1036.

Brivanlou, A. H. and J. E. Darnell. 2002. Signal transduction and the control of gene expression. *Science.* 295: 813–818.

Brown, N. L., C. A. Sattler, S. W. Paddock and S. B. Carroll. 1995. *Hairy* and *Emc* negatively regulate morphogenetic furrow progression in the *Drosophila* eye. *Cell* 80: 879–887.

Cadigan, K. M. and R. Nusse. 1997. Wnt signaling: A common theme in animal development. *Genes Dev.* 24: 3286–3306.

Cales, C., J. F. Hancock, C. J. Marshall and A. Hall. 1988. The cytoplasmic protein GAP is implicated as a target for regulation by the *ras* gene product. *Nature* 332: 548–551.

Carthew, R. W. and G. M. Rubin. 1990. *Seven-in-absentia*, a gene required for the specification of R7 cell fate in the *Drosophila* eye. *Cell* 63: 561–577.

Cecconi, F., G. Alvarez–Bolado, B. I. Meyer, K. A. Roth and P. Gruss. 1998. Apaf-1 (CED-4 homologue) regulates programmed cell death in mammalian development. *Cell* 94: 727–737.

Chen, K. C. and C. T. Chien. 1999. Negative regulation of *atonal* in proneural cluster formation in *Drosophila* R8 photoreceptors. *Proc. Natl. Acad. Sci. USA* 96: 5055–5060.

Chen, W. T., E. Hasegawa, T. Hasegawa, C. Weinstock and K. M. Yamada. 1985. Development of cell-surface linkage complexes in cultured fibroblasts. *J. Cell Biol.* 100: 1103–1114.

Chiang, C., L. T. Ying, E. Lee, K. E. Young, J. L. Corden, H. Westphal and P. A. Beachy. 1996. Cyclopia and defective axial patterning in mice lacking *Sonic hedgehog* gene function. *Nature* 383: 407–413.

Chitnis, A., D. Henrique, J. Lewis, D. Ish-Horowicz and C. Kintner. 1995. Primary neurogenesis in *Xenopus* embryos regulated by a homologue of the *Drosophila* gene *Delta*. *Nature* 375: 761–766.

Chou, W. H. and 7 others. 1999. Patterning of the R7 and R8 photoreceptor cells of *Drosophila*: Evidence for induced and default cell-fate specification. *Development* 126: 607–616.

Clark, E. A. and J. S. Brugge. 1995. Integrin and signal transduction pathways: The road taken. *Science* 268: 233–239.

Cooper, M. K., J. A. Porter, K. E. Young and P. A. Beachy. 1998. Teratogen-mediated inhibition of target tissue response to hedgehog signaling. *Science* 280: 1603–1607.

Crossley, P. H., G. Monowada, C. A. MacArthur and G. Martin. 1996. Roles for FGF8 in the induction, initiation, and maintenance of the chick limb development. *Cell* 84: 127–136.

Cvekl, A. and J. Piatigorsky. 1996. Lens development and crystallin gene expression: Many roles for Pax-6. *BioEssays* 18: 621–630.

Daniel, C. W. 1989. TGFβ1 induced inhibition of mouse mammary ductal growth: Developmental specificity and characterization. *Dev. Biol.* 134: 20–30.

Darnell, J., H. Lodish and D. Baltimore. 1986. *Molecular Cell Biology*. Scientific American Books, New York.

Deng, C., A. Wynshaw-Boris, F. Zhou, A. Kuo and P. Leder. 1996. Fibroblast growth factor receptor-3 is a negative regulator of bone growth. *Cell* 84: 911–921.

Dickson, B., F. Sprenger, D. Morrison and E. Hafen. 1992. Raf functions downstream of Ras1 in the Sevenless signal transduction pathway. *Nature* 360: 600–603.

Diderot, D. 1782. *D'Alembert's Dream*. Reprinted in J. Barzun and R. H. Bowen (eds.), *Rameau's Nephew and Other Works* (1956). Doubleday, Garden City, NY, p. 114.

Doherty, P. and F. S. Walsh. 1996. CAM-FGF receptor interactions: A model for axonal growth. *Mol. Cell Neurosci.* 8: 99–111.

Dokucu, M. E., S. L. Ziqursky and R. L. Cagan. 1996. Atonal, rough, and the resolution of proneural clusters in developing *Drosophila* retinal. *Development* 122: 139–147.

Dorsky, R. I., W. S. Chang, D. H. Rapaport and W. A. Harris. 1997. Regulation of neuronal diversity in the *Xenopus* retina by Delta signalling. *Nature* 385: 67–74.

Dufour, S., J.-L. Duband, M. J. Humphries, M. Obara, K. M. Yamada and J. P. Thiery. 1988. Attachment, spreading and locomotion of avian neural crest cells are mediated by multiple adhesion sites on fibronectin molecules. *EMBO J.* 7: 2661–2671.

Fan, C. M. and M. Tessier-Lavigne. 1994. Patterning of mammalian somites by surface ectoderm and notochord: Evidence for sclerotome induction by a hedgehog homolog. *Cell* 79: 1175–1186.

Frisch, S. M. and E. Ruoslahti. 1997. Integrins and anoikis. *Curr. Opin. Cell Biol.* 9: 701–706.

Fujiwara, M., T. Uchida, N. Osumi-Yamashita and K. Eto. 1994. Uchida rat (*rSey*): A new mutant rat with craniofacial abnormalities resembling those of the mouse *Sey* mutant. *Differentiation* 57: 31–38.

Furuta, Y. and B. L. M. Hogan. 1998. BMP4 is essential for lens induction in the mouse embryo. *Genes Dev.* 12: 3764–3775.

Gilbert-Barness, E. and J. M. Opitz. 1996. Abnormal bone development: Histopathology and skeletal dysplasias. *In* M. E. Martini-Neri, G. Neri and J. M. Opitz (eds.), *Gene Regulation and Fetal Development*. March of Dimes Birth Defects Foundation Original Article Series 30: (1). Wiley-Liss, New York, pp. 103–156.

Goustin, A. S. and 9 others. 1985. Coexpression of the *sis* and *myc* proto-oncogenes in developing human placenta suggests autocrine control of trophoblast growth. *Cell* 41: 301–312.

Graff, J. M., A. Bansal and D. A. Melton. 1996. *Xenopus* Mad proteins transduce distinct subsets of signals for the TGF-β superfamily. *Cell* 85: 479–487.

Grainger, R. M. 1992. Embryonic lens induction: Shedding light on vertebrate tissue determination. *Trends Genet.* 8: 349–356.

Green, D. R. 1998. Apoptotic pathways: Roads to ruin. *Cell* 94: 695–698.

Greenwald, I. and G. M. Rubin. 1992. Making a difference: The role of cell-cell interactions in establishing separate identities for equivalent cells. *Cell* 68: 271–281.

Greenwald, I., P. W. Sternberg and H. R. Horvitz. 1983. The *lin-12* locus specifies cell fates in *Caenorhabditis elegans*. *Cell* 34: 435–444.

Grobstein, C. 1956. Trans-filter induction of tubules in mouse metanephrogenic mesenchyme. *Exp. Cell Res.* 10: 424–440.

Groner, B. and F. Gouilleux. 1995. Prolactin-mediated gene activation in mammary epithelial cells. *Curr. Opin. Genet. Dev.* 5: 587–594.

Gurdon, J. B., P. Harger, A. Mitchell and P. Lemaire. 1994. Activin signalling and response to a morphogen gradient. *Nature* 371: 487–492.

Gurdon, J. B., A. Mitchell and D. Mahony. 1995. Direct and continuous assessment by cells of their position in a morphogen gradient. *Nature* 376: 520–521.

Hadley, M. A., S. W. Byers, C. A. Suárez-Quian, H. Kleinman and M. Dym. 1985. Extracellular matrix regulates Sertoli cell differentiation, testicular cord formation, and germ cell development in vitro. *J. Cell Biol.* 101: 1511–1512.

Hafen, E., K. Basler, J. E. Edstrom and G. M. Rubin. 1987. *sevenless*, a cell-specific homeotic gene of *Drosophila*, encodes a putative transmembrane receptor with a tyrosine kinase domain. *Science* 236: 55–63.

Hahn, H. and 20 others. 1996. Mutations of the human homolog of *Drosophila patched* in the nevoid basal cell carcinoma syndrome. *Cell* 85: 841–851.

Hakamori, S., M. Fukuda, K. Sekiguchi and W. G. Carter. 1984. Fibronectin, laminin, and other extracellular glycoproteins. *In* K. A. Picz and A. H. Reddi (eds.), *Extracellular Matrix Biochemistry*. Elsevier, New York, pp. 229–276.

Hakem, R. and 15 others. 1998. Differential requirement for caspase-9 in apoptotic pathways in vivo. *Cell* 94: 339–352.

Hamburgh, M. 1970. *Theories of Differentiation.* Elsevier, New York.

Hardman, P., E. Landels, A. S. Woolf and B. S. Spooner. 1994. TGF-β1 inhibits growth and branching morphogenesis in embryonic mouse submandibular and sublingual glands in vitro. *Dev. Growth Diff.* 36: 567–577.

Harrison, R. G. 1933. Some difficulties of the determination problem. *Am. Nat.* 67: 306–321.

Hart, A. C., H. Krämer and S. L. Zipursky. 1993. Extracellular domain of the boss transmembrane ligand acts as an antagonist of the *sev* receptor. *Nature* 361: 732–736.

Hartenstein, V. and J. A. Campos-Ortega. 1984. Early neurogenesis in wild-type *Drosophila melanogaster. Wilhelm Roux Arch. Dev. Biol.* 193: 308–326.

Hay, E. D. 1980. Development of the vertebrate cornea. *Int. Rev. Cytol.* 63: 263–322.

He, T.-C. and 8 others. 1998. Identification of c-MYC as a target of the APC pathway. *Science* 281: 1509–1512.

Heasman, J., R. D. Hines, A. P. Swan, V. Thomas and C. C. Wylie. 1981. Primordial germ cells of *Xenopus* embryos: The role of fibronectin in their adhesion during migration. *Cell* 27: 437–447.

Heberlein, U., T. Wolff and G. M. Rubin. 1993. The TGF-β homolog *dpp* and the segment polarity gene *hedgehog* are required for propagation of a morphogenetic wave in the *Drosophila* retina. *Cell* 75: 913–926.

Heitzler, P. and P. Simpson. 1991. The choice of cell fate in the epidermis of *Drosophila. Cell* 64: 1083–1092.

Heldin, C.-H., K. Miyazono and P. ten Dijke. 1997. TGF-β signaling from cell membrane to nucleus through SMAD proteins. *Nature* 390: 465–471.

Hemesath, T. J., E. R. Price, C. Takemoto, T. Badalian and D. E. Fisher. 1998. MAP kinase links the transcription factor Microphthalmia to c-Kit signalling in melanocytes. *Nature* 391: 298–301.

Hengartner, M. O., R. E. Ellis and H. R. Horvitz. 1992. *Caenorhabditis elegans* gene *ced-9* protects from programmed cell death. *Nature* 356: 494–499.

Henry, J. J. and R. M. Grainger. 1990. Early tissue interaction leading to embryonic lens formation in *Xenopus laevis. Dev. Biol.* 141: 149–163.

Hill, R. J. and P. W. Sternberg. 1992. The gene *lin-3* encodes an inductive signal for vulval development in *C. elegans. Nature* 358: 470–476.

Hirsch, M. S., L. E. Lunsford, V. Trinkaus-Randall and K. K. H. Svoboda. 1997. Chondrocyte survival and differentiation in situ are integrin medicated. *Dev. Dynam.* 210: 249–263.

Hogan, B. L. M. 1996. Bone morphogenesis proteins: Multifunctional regulators of vertebrate development. *Genes Dev.* 10: 1580–1594.

Holtzer, H. 1968. Induction of chondrogenesis: A concept in terms of mechanisms. *In* R. Gleischmajer and R. E. Billingham (eds.), *Epithelial-Mesenchymal Interactions.* Williams & Wilkins, Baltimore, pp. 152–164.

Hoppe, P. E. and R. J. Greenspan. 1986. Local function of the *Notch* gene for embryonic ectodermal pathway choice in *Drosophila. Cell* 46: 773–783.

Horwitz, A., K. Duggan, R. Greggs, C. Decker and C. Buck. 1986. The cell substrate attachment (CSAT) antigen has properties of a receptor for laminin and fibronectin. *J. Cell Biol.* 101: 2134–2144.

Howe, A., A. E. Aplin, S. K. Alahari and R. L. Juliano. 1998. Integrin signaling and cell growth control. *Curr. Opin. Cell Biol.* 10: 220–231.

Hsiung, F. and K. Moses. 2002. Retinal development in *Drosophila*: Specifiying the first neuron. *Hum. Mol. Genet.* 11: 1207–1214.

Hsu, Y.-R. and 10 others. 1997. The majority of stem cell factor exists as monomer under physiological conditions. *J. Biol. Chem.* 272: 6406–6416.

Huang, G. Y., A. Wessels, B. R. Smith, K. K. Linask, J. L. Ewart and C. W. Lo. 1998. Alteration in connexin 43 gap junction gene dosage impairs conotruncal heart development. *Dev. Biol.* 198: 32–44.

Ihle, J. N. 1996. STATs: Signal transducers and activators of transcription. *Cell* 84: 331–334.

Ihle, J. N. 2001. The Stat family in cytokine signaling. *Curr. Opin. Cell Biol.* 13: 211–217.

Jacobson, A. G. 1963. The determination and positioning of the nose, lens and ear. I. Interactions within the ectoderm, and between ectoderm and underlying tissues. *J. Exp. Zool.* 154: 273–283.

Jacobson, A. G. 1966. Inductive processes in embryonic development. *Science* 152: 25–34.

Jacobson, A. G. and A. K. Sater. 1988. Features of embryonic induction. *Development* 104: 341–359.

Jacobson, M. D., M. Weil and M. C. Raff. 1997. Programmed cell death in animal development. *Cell* 88: 347–354.

Johnson, R. L. and M. P. Scott. 1998. New players and puzzles in the hedgehog signaling pathway. *Curr. Opin. Genet. Dev.* 8: 450–456.

Johnson, R. L., E. Laufer, R. D. Riddle and C. Tabin. 1994. Ectopic expression of *Sonic hedgehog* alters dorsal-ventral patterning of somites. *Cell* 79: 1165–1173.

Johnson, R. L. and 10 others. 1996. Human homolog of *patched*, a candidate gene for the basal cell nevus syndrome. *Science* 272: 1668–1671.

Jones, C. M., N. Armes and J. C. Smith. 1996. Signalling by TGF-β family members: Short-range effects of Xnr-2 and BMP4 contrast with the long-range effects of activin. *Curr. Biol.* 6: 1468–1475.

Kalimi, G. H. and C. Lo. 1988. Communication compartments in the gastrulating mouse embryo. *J. Cell Biol.* 107: 241–256.

Katz, W. S. and P. W. Sternberg. 1996. Intercellular signalling in *Caenorhabditis elegans* vulval pattern formation. *Semin. Cell Dev. Biol.* 7: 175–183.

Katz, W. S., R. J. Hill, T. R. Clandenin and P. W. Sternberg. 1995. Different levels of the *C. elegans* growth factor LIN-3 promote distinct vulval precursor fates. *Cell* 82: 297–307.

Keeler, R. F. and W. Binns. 1968. Teratogenic compounds of *Veratrum californicum* (Durand). V. Comparison of cyclopian effects of steroidal alkaloids from the plant and structurally related compounds from other sources. *Teratology* 1: 5–10.

Kelley, R. I. and 7 others. 1996. Holoprosencephaly in RSH/Smith-Lemli-Opitz syndrome: Does abnormal cholesterol metabolism affect the function of *Sonic hedgehog? Am. J. Med. Genet.* 66: 478–484.

Kenyon, C. 1995. A perfect vulva every time: Gradients and signaling cascades in *C. elegans. Cell* 82: 171–174.

Kimble, J. 1981. Alterations in cell lineage following laser ablation of cells in the somatic gonad of *Caenorhabditis elegans. Dev. Biol.* 87: 286–300.

Knudsen, K., A. F. Horwitz and C. Buck. 1985. A monoclonal antibody identifies a glycoprotein complex involved in cell-substratum adhesion. *Exp. Cell Res.* 157: 218–226.

Koga, M. and Y. Ohshima. 1995. Mosaic analysis of the *let-23* gene function in vulval induction of *Caenorhabditis elegans. Development* 121: 2655–2666.

Korinek, V. and 7 others. 1997. Constitutive transcriptional activation by a β-catenin-Tcf complex in APC/colon carcinoma. *Science* 275: 1784–1786.

Kuida, K. and 7 others. 1996. Decreased apoptosis in the brain and premature lethality in CPP32-deficient mice. *Nature* 384: 368–372.

Kuida, K. and 8 others. 1998. Reduced apoptosis and cytochrome *c*-mediated caspase activation in mice lacking caspase 9. *Cell* 94: 325–337.

Lappi, D. A. 1995. Tumor targeting through fibroblast growth factor receptors. *Semin. Cancer Biol.* 6: 279–288.

Lazebnik, Y. A., S. H. Kaufmann, S. Desnoyers, G. G. Poirer and W. C. Earnshaw. 1994. Cleavage of poly(ADP-ribose) polymerase by a proteinase with properties like ICE. *Nature* 371: 346–347.

Lecourtois, M. and F. Schweisguth. 1998. Indirect evidence for Delta-dependent intercellular processing of Notch in *Drosophila* embryos. *Curr. Biol.* 8: 771–774.

Lee, S., N. B. Gilula and A. E. Warner. 1987. Gap junctional communication and compaction during preimplantation stages of mouse development. *Cell* 51: 851–860.

Leevers, S. J., H. F. Paterson and C. J. Marshall. 1994. Requirement for Ras in Raf activation is overcome by targeting Raf to the plasma membrane. *Nature* 369: 411–414.

Lehmann, R., F. Jimenez, U. Dietrich and J. A. Campos-Ortega. 1983. On the phenotype and development of mutants of early neurogenesis in *Drosophila melanogaster*. *Wilhelm Roux Arch. Dev. Biol.* 192: 62–74.

Letterio, J. J., A. G. Geiser, A. B. Kulkarni, A. B. Roche, N. S. Sporn and A. B. Roberts. 1994. Maternal rescue of *TGF-β1*-null mice. *Science* 264: 1936–1938.

Lewis, P. M., M. P. Dunn, J. A. McMahon, M. Logan, J. F. Martin, B. St.-Jacques and A. P. McMahon. 2001. Cholesterol modification of Sonic hedgehog is required for long-range signaling activity and effective modulation of signaling by Ptc1. *Cell* 105: 599–612.

Li, H.-S., J.-M. Yang, R. D. Jacobson, D. Pasko and O. Sundin. 1994. *Pax-6* is first expressed in a region of ectoderm anterior to the early neural plate: Implications for stepwise determination of the lens. *Dev. Biol.* 162: 181–194.

Li, W., C. D. Whaley, A. Mondino and D. L. Mueller. 1996. Blocked signal transduction to the ERK and JNK protein kinases in anergic CD4[+] T cells. *Science* 271: 1272–1274.

Linask, K. L. and J. W. Lash. 1988. A role for fibronectin in the migration of avian precardiac cells. I. Dose-dependent effects of fibronectin antibody. *Dev. Biol.* 129: 315–323.

Liu, J.-K., M. C. Di Persio and K. S. Zaret. 1991. Extracellular signals that regulate liver transcription factors during hepatic differentiation in vitro. *Mol. Cell Biol.* 11: 773–784.

Lo, C. and N. B. Gilula. 1979. Gap junctional communication in the preimplantation mouse embryo. *Cell* 18: 399–409.

Luna, E. J. and A. L. Hitt. 1992. Cytoskeleton-plasma membrane interactions. *Science* 258: 955–964.

Ma, C., Y. Zhou, P. A. Beachy and K. Moses. 1993. The segment polarity gene *hedgehog* is required for progression of the morphogenetic furrow in the developing *Drosophila* retina. *Cell* 75: 927–938.

Maimonides (Moshe ben Maimon). 1190. *A Guide for the Perplexed* (M. Friedländer, trans.). Dover, New York (1956).

Martins-Green, M. and M. J. Bissell. 1995. Cell-ECM interactions in development. *Semin. Dev. Biol.* 6: 149–159.

McCormick, F. 1989. Ras GTPase activating protein: Signal transmitter and signal terminator. *Cell* 56: 5–8.

McEwen, D. G. and M. Peifer. Wnt signaling: The naked truth. *Curr. Biol.* 11:R524–R526.

McGill, G. G. and 14 others. 2002. Bcl2 regulation by the melanocyte master regulator Mitf modulates lineage survival and melanoma cell viability. *Cell* 109: 707–718.

McMahon, A. P. and A. Bradley. 1990. The *Wnt-1* (*int 1*) proto-oncogene is required for the development of a large region of the mouse brain. *Cell* 62: 1073–1086.

Meier, S. 1977. Initiation of corneal differentiation prior to cornea-lens association. *Cell Tissue Res.* 184: 255–267.

Montgomery, A. M. P., R. A. Reisfeld and D. A. Cheresh. 1994. Integrin αvβ3 rescues melanoma cells from apoptosis in a three-dimensional dermal collagen. *Proc. Natl. Acad. Sci. USA* 91: 8856–8860.

Muschler, J., A. Lochter, C. D. Roskelley, P. Yurchenko and M. J. Bissell. 1999. Division of labor among the α6β4 integrins, β1 integrins, and an E3 laminin receptor to signal morphogenesis and β-casein expression in mammary epithelial cells. *Mol. Biol. Cell* 10: 2817–2828.

Muthukkarapan, V. R. 1965. Inductive tissue interaction in the development of the mouse lens in vitro. *J. Exp. Zool.* 159: 269–288.

Nakayama, A., M.-T. Nguyen, C. C. Chen, K. Opdecamp, C. A. Hodgkinson and H. Arnheiter. 1998. Mutations in *microphthalmia*, the mouse homolog of the human deafness gene *MITF*, affect neuroepithelial and neural crest-derived melanocytes differently. *Mech. Dev.* 70: 155–166.

Nohno, T. W., Y. Kawakami, H. Ohuchi, A. Fujiwara, H. Yoshioka and S. Noji. 1995. Involvement of the *sonic hedgehog* gene in chick feather formation. *Biochem. Biophys. Res. Comm.* 206: 33–39.

Nomura, M. and E. Li. 1998. Smad2 role in mesoderm formation, left-right patterning, and craniofacial development. *Nature* 393: 786–790.

Notenboom, R. G. E., P. A. J. de Poer, A. F. M. Moorman and W. H. Lamers. 1996. The establishment of the hepatic architecture is a prerequisite for the development of a lobular pattern of gene expression. *Development* 122: 321–332.

Ogino, H. and K. Yasuda. 1998. Induction of lens differentiation by activation of a bZIP transcription factor, L-maf. *Science* 280: 115–118.

Ohkawara, B., S.-I. Iemura, P. ten Dijke and N. Ueno. 2002. Action range of BMP is defined by its *N*-terminal basic amino acid core. *Curr. Biol.* 12: 205–209.

Padgett, R. W., J. M. Wozney and W. M. Gelbart. 1993. Human BMP sequences can confer normal dorsal-ventral patterning in the *Drosophila* embryo. *Proc. Natl. Acad. Sci. USA* 90: 2905–2809.

Pennisi, E. 1998. How a growth control path takes a wrong turn to cancer. *Science* 281: 1439–1441.

Peracchia, C. and A. F. Dulhunty. 1976. Low resistance junctions in crayfish: Structural changes with functional uncoupling. *J. Cell Biol.* 70: 419–439.

Poulson, D. F. 1937. Chromosomal deficiencies and the embryonic development of *Drosophila*

melanogaster. *Proc. Natl. Acad. Sci. USA* 23: 133–137.

Price, E. R. and 7 others. 1998. Lineage-specific signaling in melanocytes: c-Kit stimulation recruits p300/CBP to microphthalmia. *J. Biol. Chem.* 273: 17983–17986

Reaume, A. G. and 8 others. 1995. Cardiac malformation in neonatal mice lacking connexin-43. *Science* 267: 1831–1834.

Reilly, K. M. and D. A. Melton. 1996. Short-range signaling by candidate morphogens of the TGF-β family and evidence for a relay mechanism of induction. *Cell* 86: 743–754.

Reinke, R. and A. L. Zipursky. 1988. Cell-cell interaction in the *Drosophila* retina: The *bride of sevenless* gene is required in photoreceptor cell R8 for R7 cell development. *Cell* 55: 321–330.

Riddle, R. D., R. L. Johnson, E. Laufer and C. Tabin. 1993. Sonic hedgehog mediates the polarizing activity of the ZPA. *Cell* 75: 1401–1416.

Ritvos, O., T. Tuuri, M. Erämaa, K. Sainio, K. Hilden, L. Saxén and S. F. Gilbert. 1995. Activin disrupts epithelial branching morphogenesis in developing murine kidney, pancreas, and salivary gland. *Mech. Dev.* 50: 229–246.

Roberts, D. S. and S. A. Miller. 1998. Apoptosis in cavitation of middle ear space. *Anatomical Rec.* 251: 286–289.

Rodriguez, I., K. Araki, K. Khatib, J.-C. Martinou and P. Vassalli. 1997. Mouse vaginal opening is an apoptosis-dependent process which can be prevented by the overexpression of Bcl2. *Dev. Biol.* 184: 115–121.

Roessler, E. and 7 others. 1996. Mutations in the human *Sonic hedgehog* gene cause holoprosencephaly. *Nature Genet.* 14: 357–360.

Roskelley, C. D., P. Y. Desprez and M. J. Bissell. 1994. Extracellular matrix-dependent tissue-specific gene expression in mammary epithelial cells requires both physical and biochemical signal transduction. *Proc. Natl. Acad. Sci. USA* 91: 12378–12382.

Rousseau, F. and 7 others. 1994. Mutations in the gene encoding fibroblast growth factor receptor-3 in achondroplasia. *Nature* 371: 252–254.

Rubin, G. M. 1989. Development of the *Drosophila* retina: Inductive events studied at single cell resolution. *Cell* 57: 519–520.

Ruoslahti, E. and M. D. Pierschbacher. 1987. New perspectives in cell adhesion: RGD and integrins. *Science* 238: 491–497.

Saha, M. S. 1991. Spemann seen through a lens. *In* S. F. Gilbert (ed.), *A Conceptual History of Modern Embryology*. Plenum Press, New York, pp. 91–108.

Saha, M. S., C. L. Spann and R. M. Grainger. 1989. Embryonic lens induction: More than meets the optic vesicle. *Cell Diff. Dev.* 28: 153–172.

Sainio, K., S. F. Gilbert, E. Lehtonen, M. Nishi, N. M. Kumar, N. B. Gilula and L. Saxén. 1992. Differential expression of gap junction mRNAs and proteins in the developing murine kidney and in experimentally induced nephric mesenchymes. *Development* 115: 827–837.

Saudou, F., S. Finkbeiner, D. Devys and M. E. Greenberg. 1998. Huntingtin acts in the nucleus to induce apoptosis but death does not correlate with the formation of intranuclear inclusions. *Cell* 95: 55–66.

Saunders, J. W., Jr. 1980. *Developmental Biology.* Macmillan, New York.

Saunders, J. W., Jr. and J. F. Fallon. 1966. Cell death and morphogenesis. *In* M. Locke (ed.), *Major Problems of Developmental Biology.* Academic Press, New York, pp. 289–314.

Saunders, J. W., Jr., J. M. Cairns and M. T. Gasseling. 1957. The role of the apical ectodermal ridge of ectoderm in the differentiation of the morphological structure of and inductive specificity of limb parts of the chick. *J. Morphol.* 101: 57–88.

Saxén, L., E. Lehtonen, M. Karkinen-Jääskeläinen, S. Nordling and J. Wartiovaara. 1976. Are morphogenetic tissue interactions mediated by transmissable signal substances or through cell contacts. *Nature* 259: 662–663.

Schroeder, E. H., J. A. Kisslinger and R. Kopan. 1998. Notch-1 signalling requires ligand-induced proteolytic release of intracellular domain. *Nature* 393: 382–386.

Seydoux, G. and I. Greenwald. 1989. Cell autonomy of *lin-12* function in a cell fate decision in *C. elegans. Cell* 57: 1237–1246.

Shaham, S. and H. R. Horvitz. 1996. An alternatively spliced *C. elegans ced-4* RNA encodes a novel cell death inhibitor. *Cell* 86: 201–208.

Shiang, R. and 7 others. 1994. Mutations in the transmembrane domain of FGFR3 cause the most common genetic form of dwarfism, achondroplasia. *Cell* 78: 335–342.

Shih, C. and R. A. Weinberg. 1982. Isolation of a transforming sequence from a human bladder carcinoma cell line. *Cell* 29: 161–169.

Shin, S. H., P. Kogerman, E. Lindström, R. Toftgård and L. G. Biesecker. 1999. GLI3 mutations in human disorders mimic Cubitus interruptus protein functions and localization. *Proc. Natl. Acad. Sci. USA* 96: 2880–2884.

Shulman, J. M., N. Perrimon and J. D. Axelrod. 1998. *Frizzled* signaling and the developmental control of cell polarity. *Trends Genet.* 14: 452–458.

Simske, J. S. and S. K. Kim. 1995. Sequential signaling during *Caenorhabditis elegans* vulval induction. *Nature* 375: 142–146.

Slavkin, H. C. and P. Bringas, Jr. 1976. Epithelial mesenchymal interactions during odontogenesis. IV. Morphological evidence for direct heterotypic cell-cell contacts. *Dev. Biol.* 50: 428–442.

Socolovsky, M., A. E. J. Fllon, S. Wang, C. Brugnara and H. F. Lodish. 1999. Fetal anemia and apoptosis of red cell progenitors in *Stat5a^{-/-}5b^{-/-}* mice: A direct role for Stat5 in Bcl-X$_L$ induction. *Cell* 98: 181–191.

Spemann, H. and O. Schotté. 1932. Über xenoplatische Transplantation als Mittel zur Analyse der embryonalen Induction. *Naturwissenschaften* 20: 463–467.

Spritz, R. A., L. B. Gielbel and S. A. Holmes. 1992. Dominant negative and loss-of-function mutations of the c-kit (mast/Stem cell growth factor receptor) proto-oncogene in human piebaldism. *Am. J. Hum. Genet.* 50: 261–269.

Stern, H. M., A. M. C. Brown and S. D. Hauschka. 1995. Myogenesis in paraxial mesoderm: Preferential induction by dorsal neural tube and by cells expressing Wnt-1. *Development* 121: 3675–3686.

Sternberg, P. W. 1988. Lateral inhibition during vulval induction in *Caenorhabditis elegans. Nature* 335: 551–554.

Stokoe, D., S. G. Macdonald, K. Cadwallader, M. Symons and J. F. Hancock. 1994. Activation of raf as well as recruitment to the plasma membrane. *Science* 264: 1463–1467.

Streuli, C. H., N. Bailey and M. J. Bissell. 1991. Control of mammary epithelial differentiation: Basement membrane induces tissue specific gene expression in the absence of cell-cell interactions and morphological polarity. *J. Cell Biol.* 115: 1383–1396.

Struhl, G. and A. Adachi. 1998. Nuclear access and action of Notch in vivo. *Cell* 93: 382–386.

Su, W.-C. S. and 8 others. 1997. Activation of Stat1 by mutant fibroblast growth factor receptor in thanatophoric dysplasia type II dwarfism. *Nature* 386: 288–292.

Tamkun, J. W., D. W. DeSimone, D. Fonda, R. S. Patel, C. Buck, A. F. Horwitz and R. O. Hynes. 1986. Structure of integrin, a glycoprotein involved in transmembrane linkage between fibronectin and actin. *Cell* 46: 271–282.

Tan, P. B., M. R. Lackner and S. K. Kim. 1998. MAP kinase signaling specificity mediated by the LIN-1/Ets/LIN31 WH transcription factor complex during *C. elegans* vulval induction. *Cell* 93: 569–580.

Thomas, L. 1992. *The Fragile Species.* Macmillan, New York

Thut, C. J., R. B. Rountree, M. Hwa and D. M. Kingsley. 2001. A large-scale *in situ* screen provides molecular evidence for the induction of eye anterior structures by the developing lens. *Dev. Biol.* 231: 63–76.

Ting-Berreth, S. A. and C.-M. Chuong. 1996. Local delivery of TGFβ2 can substitute for placode epithelium to induce mesenchymal condensation during skin morphogenesis. *Dev. Biol.* 179: 347–359.

Tomlinson, A. 1988. Cellular interactions in the developing *Drosophila* eye. *Development* 104: 183–193.

Tomlinson, A. and D. F. Ready. 1987. Cell fate in the *Drosophila* ommatidium. *Dev. Biol.* 123: 264–276.

Vaahtokari, A., T. Aberg, J. Jernvall, S. Keränen and I. Thesleff. 1996a. The enamel knot as a signalling center in the developing mouse tooth. *Mech. Dev.* 54: 39–43.

Vaahtokari, A., T. Aberg and I. Thesleff. 1996b. Apoptosis in the developing tooth: Association with an embryonic signaling center and suppression by EGF and FGF-4. *Development* 122: 121–129.

Vaux, D. L., I. L. Weissman and S. K. Kim. 1992. Prevention of programmed cell death in *Caenorhabditis elegans* by human *bcl-2. Science* 258: 1955–1957.

Vogel-Höpker, A., T. Momose, H. Rohrer, K. Yasua, L. Ishihara and D. H. Rappaport. 2000. Multiple functions of fibroblast growth factor-8 (FGF-8) in chick eye development. *Mech. Dev.* 94: 25–36.

Waddington, C. H. 1940. *Organisers and Genes.* Cambridge University Press, Cambridge.

Wallberg, A. E., K. Pedersen, U. Lendahl and R. G. Roeder. 2002. p300 and PCAF act cooperatively to mediate transcriptional activation from chromatin templates by Notch intracellular domains in vitro. *Mol. Cell Biol.* 22: 7812–7819.

Wang, S. and 7 others. 1998. Notch receptor activation inhibits oligodendrocyte differentiation. *Neuron* 21: 63–76.

Warner, A. E., S. C. Guthrie and N. B. Gilula. 1984. Antibodies to gap junctional protein selectively disrupt junctional communication in the early amphibian embryo. *Nature* 311: 127–131.

Wary, K. K., A. Mariotti, C. Zurzolo and F. Giancotti. 1998. A requirement for caveolin-1 and associated kinase Fyn in integrin signaling and anchorage-dependent cell growth. *Cell* 94: 625–634.

Webster, M. K. and D. J. Donoghue. 1996. Constitutive activation of fibroblast growth factor receptor 3 by the transmembrane domain point mutation found in achondroplasia. *EMBO J.* 15: 520–527.

Wessells, N. K. 1977. *Tissue Interaction and Development.* Benjamin Cummings, Menlo Park, CA.

Wilkinson, H. A., K. Fitzgerald and I. Greenwald. 1994. Reciprocal changes in expression of the receptor lin-12 and its ligand lag-2 prior to commitment in a *C. elegans* cell fate decision. *Cell* 79: 1187–1198.

Williams, E. J., J. Furness, F. S. Walsh and P. Doherty. 1994a. Activation of the FGF receptor underlies neurite outgrowth stimulated by L1, N-CAM, and N-cadherin. *Neuron* 13: 583–594.

Williams, E. J., F. S. Walsch and P. Doherty. 1994b. Tyrosine kinase inhibitors can differentially inhibit integrin-dependent and CAM-dependent neurite outgrowth. *J. Cell Biol.* 124: 1029–1037.

Winter, C. G., B. Wang, A. Ballew, A. Royou, R. Karess, J. D. Axelrod and L. Luo. 2001. *Drosophila* Rho-associated kinase (Drok) links Frizzled-mediated planar cell polarity signaling to the actin cytoskeleton. *Cell* 105: 81–91.

Witte, O. N. 1990. *Steel* locus defines new multipotent growth factor. *Cell* 63: 5–6.

Wu, M. and 7 others. 2000. c-Kit triggers dual phosphorylation, which couples activation and degradation of the essential melanocyte factor Mi. *Genes Dev.* 14: 301–312.

Yamada, T., S. L. Pfaff, T. Edlund and T. M. Jessell. 1993. Control of cell pattern in the neural tube: Motor neuron induction by diffusible factors from notochord and floor plate. *Cell* 73: 673–686.

Yoshida, H. and 7 others. 1998. Apaf1 is required for mitochondrial pathways of apoptosis and brain development. *Cell* 94: 739–750.

Zygar, C. A., T. L. Cook, Jr. and R. M. Grainger. 1998. Gene activation during early stages of lens induction in *Xenopus*. *Development* 125: 3509–3519.

PART II Early Embryonic Development

7 Fertilization: Beginning a new organism

8 Early development in selected invertebrates

9 The genetics of axis specification in Drosophila

10 Early development and axis formation in amphibians

11 The early development of vertebrates:
    Fish, birds, and mammals

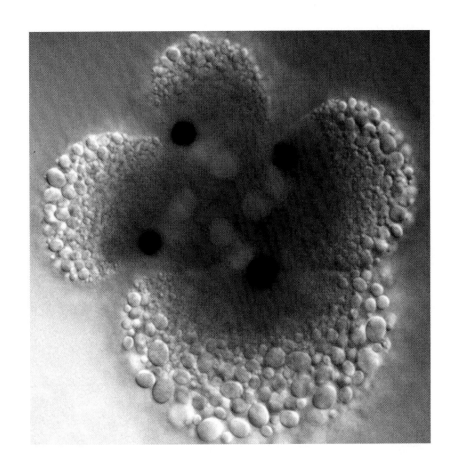

During early development proteins are often actively apportioned to particular cells. As the four cells of this snail embryo undergo division, the mRNA for the Decapentaplegic protein becomes associated with only one of the two centrosomes of each cell. This message follows the centrosomes and becomes localized in only four of the eight cells. Thus, this protein is synthesized only in a portion of the embryo. (Figure 8.31 courtesy of Dr. L. Nagy.)

7 *Fertilization: Beginning a new organism*

FERTILIZATION IS THE PROCESS whereby two sex cells (gametes) fuse together to create a new individual with a genome derived from both parents. Fertilization accomplishes two separate ends: sex (the combining of genes derived from two parents) and reproduction (the creation of a new organism). Thus, the first function of fertilization is to transmit genes from parent to offspring, and the second is to initiate in the egg cytoplasm those reactions that permit development to proceed.

Although the details of fertilization vary from species to species, conception generally consists of four major events:

1. Contact and recognition between sperm and egg. In most cases, this ensures that the sperm and egg are of the same species.
2. Regulation of sperm entry into the egg. Only one sperm can ultimately fertilize the egg. This is usually accomplished by allowing only one sperm to enter the egg and inhibiting any others from entering.
3. Fusion of the genetic material of sperm and egg.
4. Activation of egg metabolism to start development.

Structure of the Gametes

A complex dialogue exists between egg and sperm. The egg activates the sperm metabolism that is essential for fertilization, and the sperm reciprocates by activating the egg metabolism needed for the onset of development. But before we investigate these aspects of fertilization, we need to consider the structures of the sperm and egg—the two cell types specialized for fertilization.

Sperm

It is only within the past 125 years that the sperm's role in fertilization has been known. Anton van Leeuwenhoek, the Dutch microscopist who co-discovered sperm in 1678, first believed them to be parasitic animals living within the semen (hence the term *spermatozoa*, meaning "sperm animals"). He originally assumed that they had nothing at all to do with reproducing the organism in which they were found, but he later came to believe that each sperm contained a preformed embryo. Leeuwenhoek (1685) wrote that sperm were seeds (both *sperma* and *semen* mean "seed") and that the female merely provided the nutrient soil in which the seeds were planted. In this, he was returning to a notion of procreation promulgated by Aristotle 2000 years earlier. Try as he might, Leeuwenhoek was continually disappointed in his attempts to find preformed embryos within the spermatozoa. Nicolas Hartsoeker, the other co-discoverer of sperm, drew a picture of what he hoped to find: a preformed human ("homunculus") within the human sperm (Figure 7.1). This belief that the sperm contained the entire embryonic organism never gained much acceptance, as it implied an enormous waste of potential life. Most investigators regarded the sperm as unimportant (see Pinto-Correia 1997 for details of this remarkable story).

> **WEBSITE 7.1 Leeuwenhoek and images of homunculi.**
> Scholars in the 1600s thought that either the sperm or the egg carried the rudiments of the adult body. Moreover, these views became distorted by contemporary commentators and later historians.

The first evidence suggesting the importance of sperm in reproduction came from a series of experiments performed by Lazzaro Spallanzani in the late 1700s. Spallanzani demonstrated that filtered toad semen devoid of sperm would not fertilize eggs. He concluded, however, that the viscous fluid retained by the filter paper, and not the sperm, was the agent of fertilization. He, like many others, felt that the spermatic "animals" were parasites.

The combination of better microscopic lenses and the cell theory led to a new appreciation of spermatic function. In 1824, J. L. Prevost and J. B. Dumas claimed that sperm were not parasites, but rather the active agents of fertilization. They noted the universal existence of sperm in sexually mature males and their absence in immature and aged individuals. These observations, coupled with the known absence of spermatozoa in the sterile mule, convinced them that "there exists an intimate relation between their presence in the organs and the fecundating capacity of the animal." They proposed that the sperm entered the egg and contributed materially to the next generation.

These claims were largely disregarded until the 1840s, when A. von Kolliker described the formation of sperm from cells within the adult testes. He ridiculed the idea that the semen could be normal and yet support such an enormous number of parasites. Even so, von Kolliker denied that there was any physical contact between sperm and egg. He believed that the sperm excited the egg to develop, much as a magnet communicates its presence to iron.

It was not until 1876 that Oscar Hertwig and Herman Fol independently demonstrated sperm entry into the egg and the union of the two cells' nuclei. Hertwig had been seeking an organism suitable for detailed microscopic observations, and he found that the Mediterranean

Figure 7.1
The human infant preformed in the sperm, as depicted by ▶
Nicolas Hartsoeker (1694).

sea urchin, *Toxopneustes lividus*, was perfect. Not only was it common throughout the region and sexually mature throughout most of the year, but its eggs were available in large numbers and were transparent even at high magnifications. After mixing sperm and egg suspensions together, Hertwig repeatedly observed a sperm entering an egg and saw the two nuclei unite. He also noted that only one sperm was seen to enter each egg, and that all the nuclei of the embryo were mitotically derived from the nucleus created at fertilization. Fol made similar observations and detailed the mechanism of sperm entry. Fertilization was at last recognized as the union of sperm and egg, and the union of sea urchin gametes remains one of the best-studied examples of fertilization.

> **WEBSITE 7.2 The origins of fertilization research.**
> Studies by Hertwig, Fol, Boveri, and Auerbach investigated fertilization by integrating cytology with genetics. The debates over meiosis and nuclear structure were critical in these investigations of fertilization.

Each sperm consists of a haploid nucleus, a propulsion system to move the nucleus, and a sac of enzymes that enable the nucleus to enter the egg. Most of the sperm's cytoplasm is eliminated during maturation, leaving only certain organelles that are modified for spermatic function (Figure 7.2). During the course of sperm maturation, the haploid nucleus becomes very streamlined, and its DNA becomes tightly compressed. In front of this compressed haploid nucleus lies the **acrosomal vesicle**, or **acrosome**, which is derived from the Golgi apparatus and contains enzymes that digest proteins and complex sugars; thus, it can be considered a modified secretory vesicle. The enzymes stored in the acrosome are used to lyse the outer coverings of the egg. In many species, such as sea urchins, a region of globular actin molecules lies between the nucleus and the acrosomal vesicle. These proteins are used to extend a fingerlike **acrosomal process** from the sperm during the early stages of fertilization. In sea urchins and several other species, recognition between sperm and egg involves molecules on the acrosomal process. Together, the acrosome and nucleus constitute the **head** of the sperm.

The means by which sperm are propelled vary according to how the species has adapted to environmental conditions. In some species (such as the parasitic roundworm *Ascaris*), the sperm travel by the amoeboid motion of lamellipodial extensions of the cell membrane. In most species, however, each sperm is able to travel long distances by whipping its **flagellum**. Flagella are complex structures. The major motor portion of the flagellum is called the **axoneme**. It is formed by microtubules emanating from the centriole at the base of the sperm nucleus (Figures 7.2 and 7.3). The core of the axoneme consists of two central microtubules surrounded by a row of nine doublet micro-

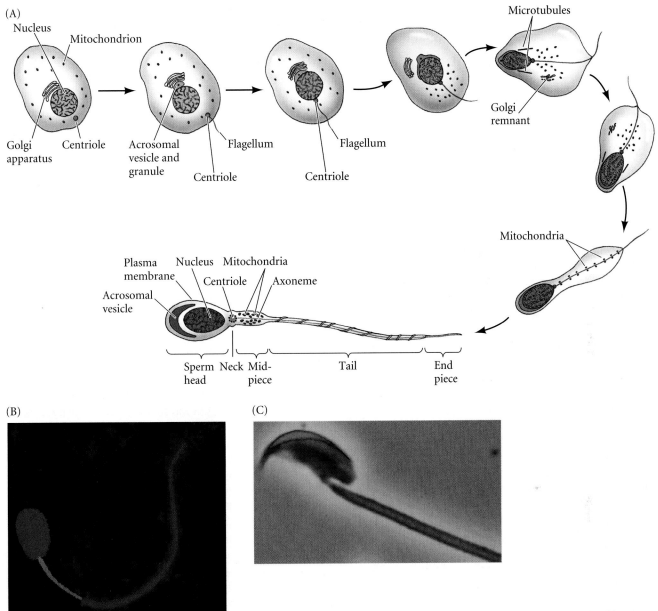

Figure 7.2
The modification of a germ cell to form a mammalian sperm. (A) The centriole produces a long flagellum at what will be the posterior end of the sperm, and the Golgi apparatus forms the acrosomal vesicle at the future anterior end. The mitochondria (dots) collect around the flagellum near the base of the haploid nucleus and become incorporated into the midpiece of the sperm. The remaining cytoplasm is jettisoned, and the nucleus condenses. The size of the mature sperm has been enlarged relative to the other stages. (B) Mature bull sperm. The DNA is stained blue with DAPI; the mitochondria are stained green, and the tubulin of the flagellum is stained red. (C) Acrosome of mouse sperm, stained green by GFP. A construct whereby the GFP gene was combined to the proacrosin promoter caused GFP to accumulate in the acrosome. (A after Clermont and Leblond 1955; B from Sutovsky et al. 1996, photograph courtesy of G. Schatten; C, photograph courtesy of K.-S. Kim and G. L. Gerton.)

tubules. Actually, only one microtubule of each doublet is complete, having 13 protofilaments; the other is C-shaped and has only 11 protofilaments (Figure 7.3B). A three-dimensional model of a complete microtubule is shown in Figure 7.3C. Here we can see the 13 interconnected protofilaments, which are made exclusively of the dimeric protein **tubulin**.

Although tubulin is the basis for the structure of the flagellum, other proteins are also critical for flagellar function. The force for sperm propulsion is provided by **dynein**, a protein attached to the microtubules (Figure 7.3B). Dynein hydrolyzes molecules of ATP and converts the released chemical energy into the mechanical energy that propels the sperm. This energy allows the active sliding of the outer doublet microtubules, causing the flagellum to bend (Ogawa et al. 1977; Shinyoji et al. 1998). The importance of dynein can be seen in individuals with the genetic syndrome called the Kartagener

(A) Sperm

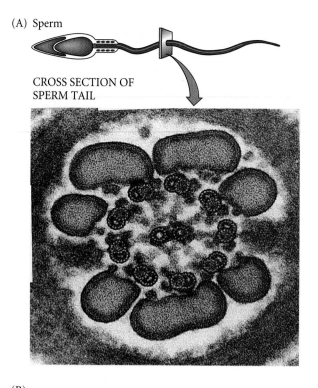

CROSS SECTION OF
SPERM TAIL

(B)

Plasma membrane
Radial spoke
Spoke head
Nexin
Subfiber A
Subfiber B
Central singlet
microtubule
Inner dynein arm
Outer dynein arm

AXONEME

MICROTUBULE
DOUBLET

(C)

"A"
MICROTUBULE

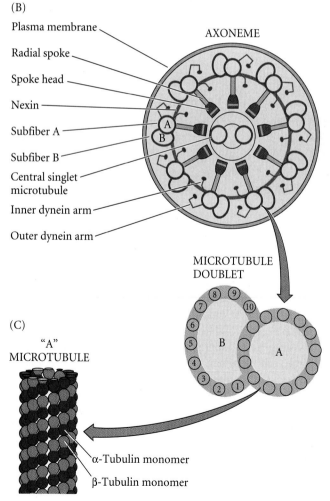

α-Tubulin monomer
β-Tubulin monomer

Figure 7.3
The motile apparatus of the sperm. (A) Cross section of the flagellum of a mammalian spermatozoon, showing the central axoneme and the external fibers. (B) Interpretive diagram of the axoneme, showing the "9 + 2" arrangement of the microtubules and other flagellar components. The schematic diagram shows the association of tubulin protofilaments into a microtubule doublet. The first ("A") portion of the doublet is a normal microtubule comprising 13 protofilaments. The second ("B") portion of the doublet contains only 11 (occasionally 10) protofilaments. The dynein arms contain the ATPases that provide the energy for flagellar movement. (C) A three-dimensional model of an "A" microtubule. The α- and β-tubulin subunits are similar but not identical. The microtubule can change size by polymerizing or depolymerizing tubulin subunits at either end. (A, photograph courtesy of D. M. Phillips; B after De Robertis et al. 1975 and Tilney et al. 1973.)

triad. These individuals lack dynein on all their ciliated and flagellated cells, rendering these structures immotile. Males with this disease are sterile (immotile sperm), are susceptible to bronchial infections (immotile respiratory cilia), and have a 50% chance of having the heart on the right side of the body (see Chapter 11; Afzelius 1976).

Another important flagellar protein appears to be histone H1. This protein is usually found inside the nucleus, where it folds the chromatin into tight clusters. However, Multigner and colleagues (1992) found that histone H1 also stabilizes the flagellar microtubules so that they do not disassemble.

The "9 + 2" microtubule arrangement with the dynein arms (Figure 7.2B) has been conserved in axonemes throughout the eukaryotic kingdoms, suggesting that this arrangement is extremely well suited for transmitting energy for movement. The ATP needed to whip the flagellum and propel the sperm comes from rings of mitochondria located in the midpiece of the sperm (see Figure 7.2). In many species (notably mammals), a layer of dense fibers has interposed itself between the mitochondrial sheath and the axoneme. This fiber layer stiffens the sperm tail. Because the thickness of this layer decreases toward the tip, the fibers probably prevent the sperm head from being whipped around too suddenly. Thus, the sperm has undergone extensive modification for the transport of its nucleus to the egg.

In mammals, the sperm released during ejaculation are able to move, but they do not yet have the capacity to bind to and fertilize an egg. These final stages of sperm maturation (called **capacitation**) do not occur until the sperm has been inside the female reproductive tract for a certain period of time.

The egg

All the material necessary for the beginning of growth and development must be stored in the mature egg (the **ovum**). Whereas the sperm has eliminated most of its cytoplasm, the developing egg (called the **oocyte** before it reaches the stage of meiosis at which it is fertilized) not only conserves the material it has, but is actively involved in accumulating more. The meiotic divisions that form the oocyte conserve its cytoplasm (rather than giving half of it away; see Figure 19.22), and the oocyte either synthesizes or absorbs proteins, such as yolk, that act as food reservoirs for the developing embryo. Thus, birds' eggs are enormous single cells, swollen with their accumulated yolk. Even eggs with relatively sparse yolk are large compared with sperm. The volume of a sea urchin egg (Figure 7.4) is about 200 picoliters (2×10^{-4} mm³), more than 10,000 times the volume of the sperm. So, while sperm and egg have equal haploid nuclear components, the egg also accumulates a remarkable cytoplasmic storehouse during its maturation. This cytoplasmic trove includes the following:*

- **Proteins.** It will be a long while before the embryo is able to feed itself or even obtain food from its mother. The early embryonic cells need a supply of energy and amino acids. In many species, this is accomplished by accumulating yolk proteins in the egg. Many of these yolk proteins are made in other organs (liver, fat bodies) and travel through the maternal blood to the egg.
- **Ribosomes and tRNA.** The early embryo needs to make many of its own proteins, and in some species, there is a burst of protein synthesis soon after fertilization. Protein synthesis is accomplished by ribosomes and tRNA, which exist in the egg. The developing egg has special mecha-

nisms for synthesizing ribosomes; certain amphibian oocytes produce as many as 10^{12} ribosomes during their meiotic prophase.
- **Messenger RNA.** In most organisms, the instructions for proteins to be made during early development are packaged in the oocyte. It is estimated that the eggs of sea urchins contain thousands of different types of mRNA. This mRNA, however, remains dormant until after fertilization (see Chapter 5).
- **Morphogenetic factors.** Molecules that direct the differentiation of cells into certain cell types are present in the egg. In many species, they are localized in different regions of the egg and become segregated into different cells during cleavage (see Chapter 8).
- **Protective chemicals.** The embryo cannot run away from predators or move to a safer environment, so it must come equipped to deal with threats. Many eggs contain ultraviolet filters and DNA repair enzymes that protect them from sunlight. Some eggs contain molecules that potential predators find distasteful, and the yolk of bird eggs even contains antibodies.

WEBSITE 7.3 **The egg and its environment.** The laboratory is not where most eggs are found. Eggs have evolved remarkable ways to protect themselves in particular environments.

---

*The contents of the egg vary greatly from species to species. The synthesis and placement of these materials will be addressed in Chapter 19, when we discuss the differentiation of germ cells.

Within this enormous volume of cytoplasm resides a large nucleus. In some species (e.g., sea urchins), the nucleus is already haploid at the time of fertilization. In other species (including many worms and most mammals), the egg nucleus is still diploid—the sperm enters before the egg's meiotic divisions are completed. The stage of the egg nucleus at the time of sperm entry in different species is illustrated in Figure 7.5.

Enclosing the cytoplasm is the egg **cell membrane.** This membrane must regulate the flow of certain ions during fertilization and must be capable of fusing with the sperm cell membrane. Outside the cell membrane is an extracellular envelope that forms a fibrous mat around the egg and is often involved in sperm-egg recognition (Correia and Carroll 1997). In invertebrates, this structure is usually called the **vitelline envelope** (Figure 7.6). The vitelline envelope contains several different glycoproteins. It is supplemented by extensions of membrane glycoproteins from the cell membrane and by proteinaceous vitelline "posts" that adhere the vitelline envelope to the membrane (Mozingo and Chandler 1991). The vitelline envelope is essential for the species-specific binding of sperm. In mammals, the vitelline envelope is a separate and thick extracellular matrix called the **zona pellucida.** The mammalian egg is also surrounded by a layer of cells called the **cumulus** (Figure 7.7), which is made up of the ovarian follicular cells that were nurturing the egg at the time of its release from the ovary. Mammalian sperm have to get past these cells to fertilize the egg. The innermost layer of cumulus cells,

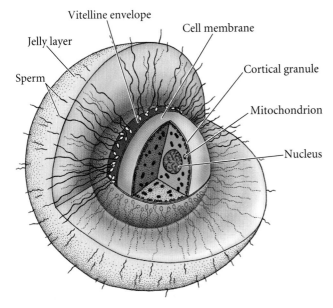

Figure 7.4
Structure of the sea urchin egg at fertilization. The drawing shows the relative sizes of egg and sperm. (After Epel 1977.)

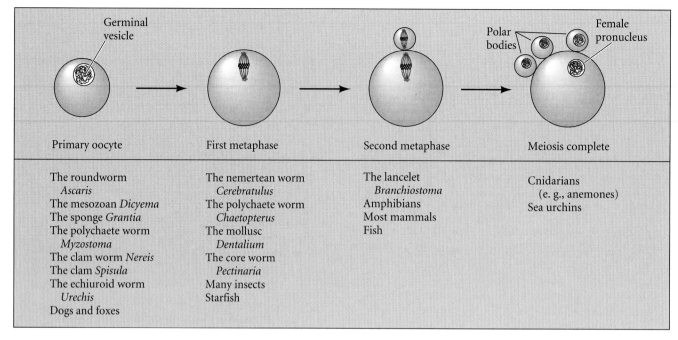

Figure 7.5
Stages of egg maturation at the time of sperm entry in different animal species. The germinal vesicle is the name given to the large diploid nucleus of the primary oocyte. The polar bodies are seen as smaller cells. (After Austin 1965.)

immediately adjacent to the zona pellucida, is called the **corona radiata**.

Lying immediately beneath the cell membrane of the egg is a thin shell (about 5 μm) of gel-like cytoplasm called the **cortex**. The cytoplasm in this region is stiffer than the internal cytoplasm and contains high concentrations of globular actin molecules. During fertilization, these actin molecules poly-merize to form long cables of actin known as **microfilaments**. Microfilaments are necessary for cell division, and they are also used to extend the egg surface into small projections called **microvilli**, which may aid sperm entry into the cell (see Figure 7.6B; also see Figure 7.19).

Figure 7.6
The sea urchin egg cell surface. (A) Scanning electron micrograph of an egg before fertilization. The plasma membrane is exposed where the vitelline envelope has been torn. (B) Transmission electron micrograph of an unfertilized egg, showing microvilli and plasma membrane, which are closely covered by the vitelline envelope. A cortical granule lies directly beneath the plasma membrane. (From Schroeder 1979; photographs courtesy of T. E. Schroeder.)

(A)

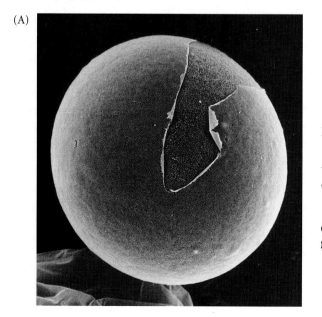

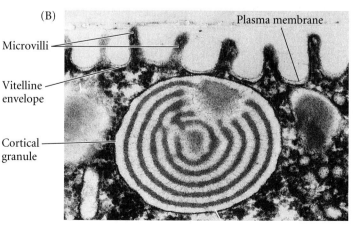

(A)

(B)

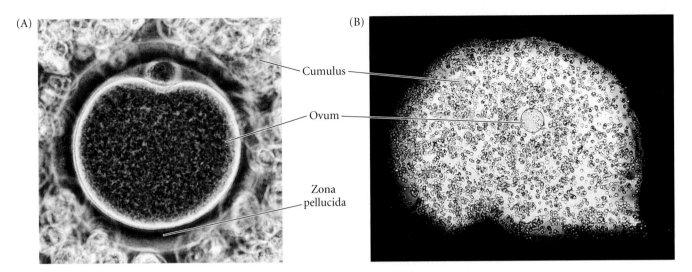

Cumulus

Ovum

Zona pellucida

Figure 7.7
Hamster eggs immediately before fertilization. (A) The hamster egg, or ovum, is encased in the zona pellucida. This, in turn, is surrounded by the cells of the cumulus. A polar body cell, produced during meiosis, is also visible within the zona pellucida. (B) At lower magnification, a mouse oocyte is shown surrounded by the cumulus. Colloidal carbon particles (India ink) are excluded by the hyaluronidate matrix. (Photographs courtesy of R. Yanagimachi.)

Also within the cortex are the **cortical granules** (see Figures 7.4 and 7.6B). These membrane-bound structures, which are homologous to the acrosomal vesicle of the sperm, are Golgi-derived organelles containing proteolytic enzymes. However, whereas each sperm contains just one acrosomal vesicle, each sea urchin egg contains approximately 15,000 cortical granules. Moreover, in addition to digestive enzymes, the cortical granules contain mucopolysaccharides, adhesive glycoproteins, and hyalin protein. The enzymes and mucopolysaccharides prevent any additional sperm from entering the egg after the first sperm has entered, and the hyalin and adhesive glycoproteins surround the early embryo and provide support for the cleavage-stage blastomeres.

Many types of eggs also have a layer of **egg jelly** outside the vitelline envelope (see Figure 7.4). This glycoprotein meshwork can have numerous functions, but most commonly is used either to attract or to activate sperm. The egg, then, is a cell specialized for receiving sperm and initiating development.

> **VADE MECUM**[2] **Gametogenesis.** Stained sections of testis and ovary illustrate the process of gametogenesis, the streamlining of developing sperm, and the remarkable growth of the egg as it stores nutrients for its long journey. You can see this in movies and labeled photographs that take you at each step deeper into the mammalian gonad.
> **[Click on Gametogenesis]**

Recognition of Egg and Sperm

The interaction of sperm and egg generally proceeds according to five basic steps (Figure 7.8; Vacquier 1998):

1. The chemoattraction of the sperm to the egg by soluble molecules secreted by the egg
2. The exocytosis of the acrosomal vesicle to release its enzymes
3. The binding of the sperm to the extracellular envelope (vitelline layer or zona pellucida) of the egg
4. The passage of the sperm through this extracellular envelope
5. Fusion of egg and sperm cell membranes

Sometimes steps 2 and 3 are reversed (as in mammalian fertilization), and the sperm binds to the egg before releasing the contents of the acrosome. After these five steps are accomplished, the haploid sperm and egg nuclei can meet and the reactions that initiate development can begin.

In many species, the meeting of sperm and egg is not a simple matter. Many marine organisms release their gametes into the environment. That environment may be as small as a tide pool or as large as an ocean. Moreover, it is shared with other species that may shed their sex cells at the same time. Such organisms are faced with two problems: How can sperm and eggs meet in such a dilute concentration, and how can sperm be prevented from trying to fertilize eggs of another species? Two major mechanisms have evolved to solve these problems: species-specific attraction of sperm and species-specific sperm activation.

Sperm attraction: Action at a distance

Species-specific sperm attraction has been documented in numerous species, including cnidarians, molluscs, echinoderms, and urochordates (Miller 1985; Yoshida et al. 1993). In many species, sperm are attracted toward eggs of their species by

(A) SEA URCHIN

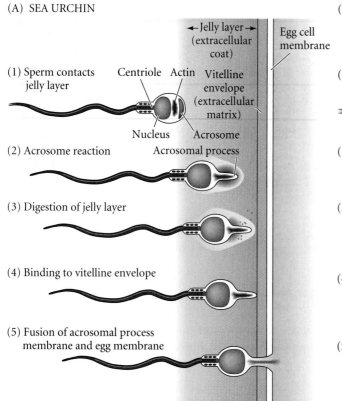

(1) Sperm contacts jelly layer

Centriole Actin

Nucleus Acrosome

Vitelline envelope (extracellular matrix)

← Jelly layer → (extracellular coat)

Egg cell membrane

(2) Acrosome reaction

Acrosomal process

(3) Digestion of jelly layer

(4) Binding to vitelline envelope

(5) Fusion of acrosomal process membrane and egg membrane

(B) MOUSE

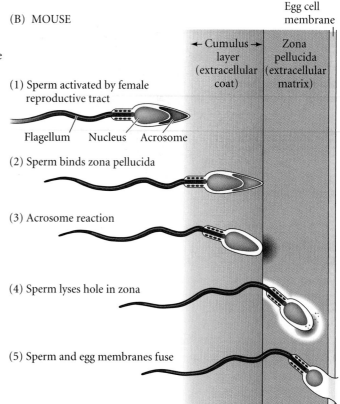

(1) Sperm activated by female reproductive tract

Flagellum Nucleus Acrosome

Egg cell membrane

← Cumulus → layer (extracellular coat)

Zona pellucida (extracellular matrix)

(2) Sperm binds zona pellucida

(3) Acrosome reaction

(4) Sperm lyses hole in zona

(5) Sperm and egg membranes fuse

Figure 7.8
Summary of events leading to the fusion of egg and sperm plasma membranes in (A) the sea urchin and (B) the mouse. (A) Sea urchin fertilization is external. (1) The sperm is chemotactically attracted to and activated by the egg. (2, 3) Contact with the egg jelly triggers the acrosome reaction, allowing the acrosomal process to form and release proteolytic enzymes. (4) The sperm adheres to the vitelline envelope and lyses a hole in it. (5) The sperm adheres to the egg plasma membrane and fuses with it. The sperm pronucleus can now enter the egg cytoplasm. (B) Mammalian fertilization is internal. (1) The contents of the female reproductive tract capacitate, attract, and activate the sperm. (2) The acrosome-intact sperm binds to the zona pellucida, which is thicker than the vitelline envelope of sea urchins. (3) The acrosome reaction occurs on the zona pellucida. (4) The sperm digests a hole in the zona pellucida. (5) The sperm adheres to the egg, and their plasma membranes fuse.

chemotaxis—that is, by following a gradient of a chemical secreted by the egg. In 1978, Miller demonstrated that the eggs of the cnidarian *Orthopyxis caliculata* not only secrete a chemotactic factor but also regulate the timing of its release. Developing oocytes at various stages in their maturation were fixed on microscope slides, and sperm were released at a certain distance from the eggs. Miller found that when sperm were added to oocytes that had not yet completed their second meiotic division, there was no attraction of sperm to eggs. However, after the second meiotic division was finished and the eggs were ready to be fertilized, the sperm migrated toward them. Thus, these oocytes control not only the type of sperm they attract, but also the time at which they attract them.

The mechanisms of chemotaxis differ among species (see Metz 1978; Ward and Kopf 1993). One chemotactic molecule, a 14-amino acid peptide called **resact**, has been isolated from the egg jelly of the sea urchin *Arbacia punctulata* (Ward et al. 1985). Resact diffuses readily in seawater and has a profound effect at very low concentrations when added to a suspension of *Arbacia* sperm (Figure 7.9). When a drop of seawater containing *Arbacia* sperm is placed on a microscope slide, the sperm generally swim in circles about 50 μm in diameter. Within seconds after a small amount of resact is injected into the drop, sperm migrate into the region of the injection and congregate there. As resact continues to diffuse from the area of injection, more sperm are recruited into the growing cluster. Resact is specific for *A. punctulata* and does not attract sperm of other species. *A. punctulata* sperm have receptors in their cell membranes that bind resact (Ramarao and Garbers 1985; Bentley et al. 1986) and can swim up a concentration gradient of this compound until they reach the egg.

Resact also acts as a **sperm-activating peptide**. Sperm-activating peptides cause dramatic and immediate increases in mitochondrial respiration and sperm motility (Tombes and Shapiro 1985; Hardy et al. 1994). The sperm receptor for resact is a transmembrane protein, and when it binds resact on the extracellular side, a conformational change on the cytoplasmic side activates the receptor's enzymatic activity. This activates the mitochondrial ATP-generating apparatus as well

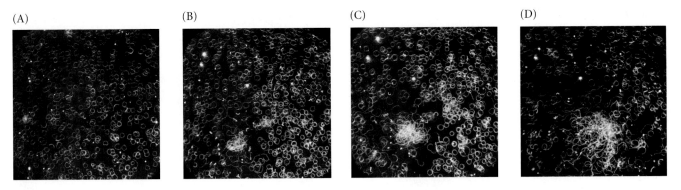

Figure 7.9

Sperm chemotaxis in the sea urchin, *Arbacia punctulata*. One nano-liter of a 10-n*M* solution of resact is injected into a 20-μl drop of sperm suspension. The position of the micropipette is indicated (white lines). (A) A 1-second photographic exposure showing sperm swimming in tight circles before the addition of resact. (B–D) Similar 1-second exposures showing migration of sperm to the center of the resact gradient 20, 40, and 90 seconds after injection. (From Ward et al. 1985; photographs courtesy of V. D. Vacquier.)

as the dynein ATPase that stimulates flagellar movement in the sperm (Shimomura et al. 1986; Cook and Babcock 1993).

The acrosome reaction in sea urchins

A second interaction between sperm and egg is the **acrosome reaction**. In most marine invertebrates, the acrosome reaction has two components: the fusion of the acrosomal vesicle with the sperm cell membrane (an exocytosis that results in the release of the contents of the acrosomal vesicle) and the extension of the acrosomal process (Colwin and Colwin 1963). The acrosome reaction in sea urchins is initiated by contact of the

sperm with the egg jelly. This contact causes the exocytosis of the sperm's acrosomal vesicle and the release of proteolytic enzymes that can digest a path through the jelly coat to the egg surface (Dan 1967; Franklin 1970; Levine et al. 1978). The sequence of these events is outlined in Figure 7.10.

In sea urchins, the acrosome reaction is initiated by the interactions of the sperm cell membrane with at least three compounds in the egg jelly. These compounds bind to specific receptors located on the sperm cell membrane directly above the acrosomal vesicle. This binding opens calcium ion chan-

Figure 7.10

Acrosome reaction in sea urchin sperm. (A–C) The portion of the acrosomal membrane lying directly beneath the sperm plasma membrane fuses with the plasma membrane to release the contents of the acrosomal vesicle. (D) The actin molecules assemble to produce microfilaments, extending the acrosomal process outward. Actual photographs of the acrosome reaction in sea urchin sperm are shown below the diagrams. (After Summers and Hylander 1974; photographs courtesy of G. L. Decker and W. J. Lennarz.)

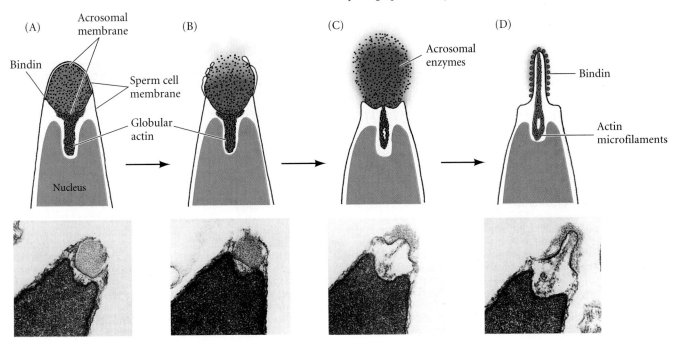

Figure 7.11
Acrosome reaction in hamster sperm. (A) Transmission electron micrograph of hamster sperm undergoing the acrosome reaction. The acrosomal membrane can be seen to form vesicles. (B) Interpretive diagram of electron micrographs showing the fusion of the acrosomal and cell membranes in the sperm head. (A from Meizel 1984, photograph courtesy of S. Meizel; B after Yanagimachi and Noda 1970.)

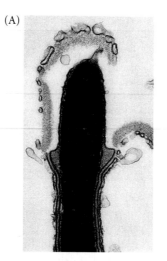

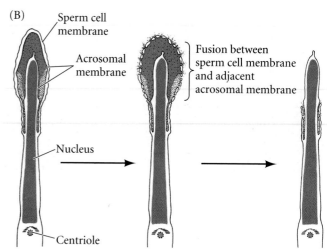

nels in the cell membrane, permitting calcium to enter the sperm head (Schackmann and Shapiro 1981; Alves et al. 1997; Hirohashi and Vacquier 2002a). The exocytosis of the acrosomal vesicle is caused by calcium-mediated fusion of the acrosomal membrane with the adjacent sperm cell membrane (Figures 7.10 and 7.11). The egg jelly factors that initiate the acrosome reaction in sea urchins are often highly specific to each species, and egg jelly carbohydrates from one species fail to activate the acrosome reaction even in closely related species (Hirohashi and Vacquier 2002b; Hirohashi et al. 2002).* Thus, the activation of the acrosome reaction constitutes a barrier to interspecies (and thus unviable) fertilizations (Summers and Hylander 1975; Vilela-Silva et al. 2002).

The second part of the acrosome reaction involves the extension of the acrosomal process (see Figure 7.10). This protrusion arises through the polymerization of globular actin molecules into actin filaments (Tilney et al. 1978). The influx

of calcium ions (Ca^{2+}) is thought to activate the RhoB protein, located in the acrosomal region and midpiece of the sea urchin sperm (Castellano et al. 1997). This GTP-binding protein helps organize the actin cytoskelton in many types of cells, and it is thought to be active in polymerizing the actin to make the acrosomal process.

---

*Such exocytotic reactions are seen in the release of insulin from pancreatic cells and in the release of neurotransmitters from synaptic terminals. In all cases, there is a calcium-mediated fusion between the secretory vesicle and the cell membrane. Indeed, the similarity of acrosomal vesicle exocytosis and synaptic vesicle exocytosis may actually be quite deep. Studies of acrosome reactions in sea urchins and mammals (Florman et al. 1992; González-Martínez et al. 1992) suggest that when the receptors for the sperm-activating ligands bind these molecules, they cause a depolarization of the membrane that would open voltage-dependent calcium ion channels in a manner reminiscent of synaptic transmission. The proteins that dock the cortical granules of the egg to the cell membrane also appear to be homologous to those used in the axon terminal (Bi et al. 1995).

Sidelights & Speculations

Action at a Distance: Mammalian Gametes

It is very difficult to study the interactions that might be occurring between mammalian gametes prior to sperm-egg contact. One obvious reason for this is that mammalian fertilization occurs inside the oviducts of the female. While it is relatively easy to mimic the conditions surrounding sea urchin fertilization (using either natural or artificial seawater), we do not yet know the components of the various natural environments that mammalian sperm encounter as they travel to the egg. A second reason for this difficulty is that the sperm population ejaculated into the female is probably very heterogeneous,

containing spermatozoa at different stages of maturation. Of the 280×10^6 human sperm normally ejaculated into the vagina, only about 200 reach the ampullary region of the oviduct, where fertilization takes place (Ralt et al. 1991). Since fewer than 1 in 10,000 sperm get close to the egg, it is difficult to assay those molecules that might enable the sperm to swim toward the egg and become activated. There is a great deal of controversy concerning the mechanisms underlying the translocation of mammalian sperm to the oviduct, the possibility that the egg may be attracting the sperm through chemotaxis, and the ca-

pacitation and hyperactivation reactions that appear necessary for some species' sperm to bind with the egg.

Translocation and Capacitation

The reproductive tract of female mammals plays a very active role in the mammalian fertilization process. While sperm motility is required for mouse sperm to encounter the egg once it is in the oviduct, sperm motility is probably a minor factor in getting the sperm into the oviduct in the first place. Sperm are found in the oviducts of mice, hamsters, guinea pigs, cows, and humans within 30 minutes of sperm deposition in the vagina, a time "too short to have been attained by even the most Olympian sperm relying on their own flagellar power" (Storey 1995). Rather, the sperm appear to be transported to the oviduct by the muscular activity of the uterus.

By whatever means, mammalian sperm pass through the uterus and oviduct, interacting with the cells and secretions of the female reproductive tract as they do so. Newly ejaculated mammalian sperm are unable to undergo the acrosome reaction without residing for some time in the female reproductive tract (Chang 1951; Austin 1952). The set of physiological changes by which the sperm becomes competent to fertilize the egg is called **capacitation**. The requirement for capacitation varies from species to species (Gwatkin 1976). Capacitation can be mimicked in vitro by incubating sperm in tissue culture media (containing calcium ions, bicarbonate, and serum albumin; see Chapter 21) or in fluid from the oviducts. Sperm that are not capacitated are "held up" in the cumulus and so do not reach the egg (Austin 1960; Corselli and Talbot 1987).

Contrary to the opening scenes of the *Look Who's Talking* movies, "the race is not always to the swiftest." Although some human sperm reach the ampullary region of the oviduct within a half-hour after intercourse, those sperm may have little chance of fertilizing the egg. Wilcox and colleagues (1995) found that nearly all human pregnancies result from sexual intercourse during a 6-day period ending on the day of ovulation. This means that the fertilizing sperm could have taken as long as 6 days to make the journey. Eisenbach (1995) has proposed a hypothesis wherein capacitation is a transient event, and sperm are given a relatively brief window of competence in which they can successfully fertilize the egg. As the sperm reach the ampulla, they acquire competence—but they lose it if they stay around too long. Sperm may also have different survival rates depending on their location within the reproductive tract, so that some late-arriving sperm may have a better chance of success than those that arrived days earlier.

The molecular changes that account for capacitation are still unknown, but there are four sets of molecular changes that may be important. First, the sperm cell membrane is altered by the removal of cholesterol by albumin proteins in the female reproductive tract (Davis 1978; Cross 1998). If serum albumin is experimentally pre-loaded with cholesterol, capacitation will not occur in vitro. The effect of the cholesterol is not known, but recent evidence suggests that the loss of cholesterol leads to a rise in pH, which in turn enables the sperm to undergo the acrosome reaction. Second, particular proteins or carbo-

hydrates on the sperm surface are lost during capacitation (Lopez et al. 1985; Wilson and Oliphant 1987). It is possible that these compounds block the recognition sites for the proteins that bind to the zona pellucida. It has been suggested (Benoff 1993) that the unmasking of these sites might be one of the effects of cholesterol depletion.

Third, the membrane potential of the sperm cell membrane becomes more negative as potassium ions leave the sperm. This change in membrane potential may allow calcium channels to be opened and permit calcium to enter the sperm. Calcium and bicarbonate ions may be critical in activating cAMP production and in facilitating the membrane fusion events of the acrosome reaction (Visconti et al. 1995; Arnoult et al. 1999). Fourth, protein phosphorylation occurs (Galantino-Homer et al. 1997). It is still uncertain whether these four events are independent of one another and to what extent each of them contributes to sperm capacitation (Figure 7.12).

There may be an important connection between sperm translocation and capacita-

tion. Timothy Smith (1998) and Susan Suarez (1998) have documented that before entering the ampulla of the oviduct (where mammalian fertilization occurs), the uncapacitated sperm bind actively to the membranes of the oviduct cells in the narrow passage (isthmus) preceding it (Figure 7.13). This binding is temporary and appears to be broken when the sperm become capacitated. Moreover, the life span of the sperm is significantly lengthened by this binding, and its capacitation is slowed down. This restriction of sperm entry into the ampulla, the slowing down of capacitation, and the expansion of sperm life span may have important consequences (Töpfer-Petersen et al. 2002). First, this binding may function as a block to polyspermy by preventing many sperm from reaching the egg at the same time. If the isthmus is excised in cows, a much higher rate of polyspermy results. Second, slowing the rate of sperm capacitation and extending the active life of sperm may maximize the probability of there being some sperm in the ampulla to meet the egg if ejacula-

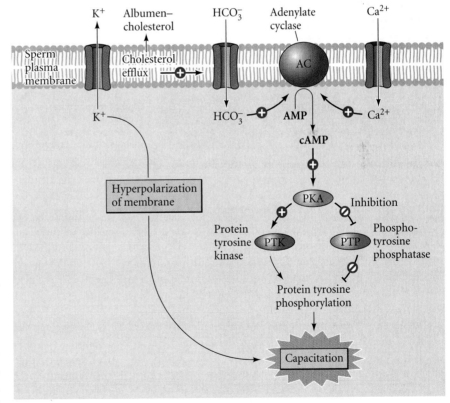

Figure 7.12
Hypothetical model for mammalian sperm capacitation. The efflux of potassium (whose cause we do not know) results in a change in the resting potential of the sperm cell membrane. The removal of cholesterol by albumin stimulates ion channels that enable calcium and bicarbonate ions to enter the sperm. These ions promote the activity of adenylate cyclase, which makes cAMP from AMP. The rise in cAMP activates protein kinase A, causing it to activate the protein tyrosine kinases (while inactivating the protein phosphatases). The kinases phosphorylate proteins that are essential for capacitation. (After Visconti and Kopf 1998.)

tion does not occur at the same time as ovulation.

Hyperactivation and Chemotaxis

Different regions of the female reproductive tract may secrete different, regionally specific molecules. These factors may influence sperm motility as well as capacitation. For instance, when sperm of certain mammals (especially hamsters, guinea pigs, and some strains of mice) pass from the uterus into the oviducts, they become **hyperactivated**, swimming at higher velocities and generating greater force than before. This hyperactivation appears to be mediated

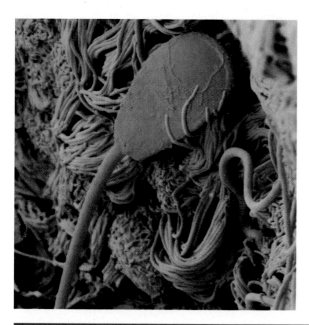

through the activation by cAMP of a sperm-specific calcium channel located in the sperm tail (Ren et al. 2001). Suarez and co-workers (1991) have shown that while this behavior is not conducive to traveling through low-viscosity fluids, it appears to be extremely well suited for linear sperm movement in the viscous fluid that sperm might encounter in the oviduct. Hyperactivation, plus a hyaluronidase enzyme on the outside of the sperm cell membrane, enable the sperm to digest a path through the extracellular matrix of the cumulus cells (Lin 1994; Primakoff and Myles 2002).

In addition to increasing the activity of sperm, soluble factors in the oviduct may also provide the directional component of sperm movement. There has been speculation that the ovum (or, more likely, the ovarian follicle in which it developed) may secrete chemotactic substances that attract the sperm toward the egg during the last stages of sperm migration (see Hunter 1989). Ralt and colleagues (1991) tested this hypothesis using follicular fluid from human follicles whose eggs were being used for in vitro fertiliza-

tion. Performing an experiment similar to the one described earlier with sea urchins, they microinjected a drop of follicular fluid into a larger drop of sperm suspension. When they did this, some of the sperm changed their direction to migrate toward the source of follicular fluid. Microinjection of other solutions did not have this effect. These studies did not rule out the possibility that the effect was due to a general stimulation of sperm movement or metabolism. However, these investigations uncovered a fascinating correlation: the fluid from only about half the follicles tested showed a chemotactic effect, and in nearly every case, the egg was fertilizable if, and only if, the fluid showed chemotactic ability ($P < 0.0001$). Further research has confirmed the ability of follicular fluid to attract human sperm in vitro, and has shown that only capacitated sperm will be attracted (Cohen-Dayag et al. 1995; Eisenbach and Tur-Kaspa 1999; Wang et al. 2001). It is possible, then, that like certain invertebrate eggs, the human egg secretes a chemotactic factor only when it is capable of being fertilized, and that sperm are attracted to such a compound only when they are capable of fertilizing the egg.

Thus the female reproductive tract is not a passive conduit through which sperm race, but a highly specialized set of tissues that regulate the timing of sperm capacitation and access to the egg.

Figure 7.13
Mammalian sperm in the female reproductive tract. Bull sperm adhere to the membranes of the epithelial cells in the oviduct of a cow prior to entering the ampulla. (From Lefebvre et al. 1995; photograph courtesy of S. Suarez.)

Species-specific recognition in sea urchins

While the contact of the sperm with the egg jelly can provide the first species-specific recognition event, another critical species-specific binding event must occur once the sea urchin sperm has penetrated the egg jelly and the acrosomal process of the sperm contacts the surface of the egg (Figure 7.14A). The acrosomal protein mediating this recognition in sea urchins is called **bindin**. In 1977, Vacquier and co-workers isolated this nonsoluble, 30,500-Da protein from the acrosome of *Strongylocentrotus purpuratus* and found it to be capable of binding to dejellied eggs of the same species (Figure 7.14B; Vacquier and Moy 1977). Further, its interaction with eggs is relatively species-specific (Glabe and Vacquier 1977; Glabe and Lennarz 1979): bindin isolated from the acrosomes of *S. purpuratus* binds to its own dejellied eggs, but not to those of *Arbacia punctulata*. Using immunological techniques, Moy and Vacquier (1979) demonstrated that bindin is located specifically on the acrosomal process—exactly where it should be for sperm-egg recognition (Figure 7.15).

Biochemical studies have shown that the bindins of closely related sea urchin species are indeed different.[*] This finding implies the existence of species-specific bindin receptors on the egg, vitelline envelope, or cell membrane. Such receptors were also suggested by the experiments of Vacquier and Payne (1973), who saturated sea urchin eggs with sperm. As seen in Figure 7.16A, sperm binding does not occur over the entire egg surface. Even at saturating numbers of sperm (approximately 1500), there appears to be room on the ovum for more sperm heads, implying a limiting number of sperm-binding sites. A 350-kDa glycoprotein that has been isolated from sea urchin eggs has some of the properties expected of a bindin receptor (Giusti et al. 1997; Stears and Lennarz 1997; Hirohashi and Lennarz 2001). The bindin receptors are

---

[*]Bindin and other gamete adhesion glycoproteins are probably the fastest evolving proteins known (Metz and Palumbi 1996; Vacquier 1998). Closely related species may have near-identity of every other protein, but their bindins may have diverged significantly. There appear to be two species-specific domains in the bindin protein.

Figure 7.14
Species-specific binding of acrosomal process to egg surface in sea urchins. (A) Actual contact of a sea urchin sperm acrosomal process with an egg microvillus. (B) In vitro model of species-specific binding. The agglutination of dejellied eggs by bindin was measured by adding bindin aggregates to a plastic well containing a suspension of eggs. After 2–5 minutes of gentle shaking, the wells were photographed. Each bindin bound to and agglutinated only eggs from its own species. (A from Epel 1977, photograph courtesy of F. D. Collins and D. Epel; B based on photographs of Glabe and Vacquier 1977.)

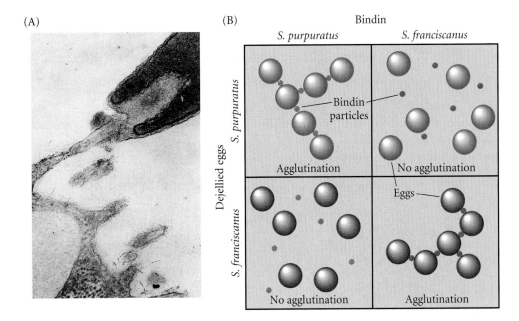

thought to be aggregated into complexes on the egg cell surface, and hundreds of these complexes may be needed to tether the sperm to the egg (Figure 7.16B). Thus, species-specific recognition of sea urchin gametes occurs at the levels of sperm attraction, sperm activation, and sperm adhesion to the egg surface.

WEBSITE 7.4 The Lillie-Loeb dispute over sperm-egg binding. In the early 1900s, fertilization research was framed by a dispute between F. R. Lillie and Jacques Loeb, who disagreed over whether the sperm recognized the egg through soluble factors or through cell-cell interactions.

Gamete binding and recognition in mammals

ZP3: THE SPERM-BINDING PROTEIN OF THE MOUSE ZONA PELLUCIDA. The zona pellucida in mammals plays a role analogous to that of the vitelline envelope in invertebrates. This glycoprotein matrix, which is synthesized and secreted by the growing oocyte, plays two major roles during fertilization: it binds the sperm, and it initiates the acrosome reaction after the sperm is bound (Saling et al. 1979; Florman and Storey 1982; Cherr et al. 1986). The binding of sperm to the zona is relatively, but not absolutely, species-specific. (Species-specific gamete recognition is not a major problem when fertilization occurs internally.)

The mouse zona pellucida is made of three major glycoproteins, ZP1, ZP2, and ZP3. There are several pieces of evi-

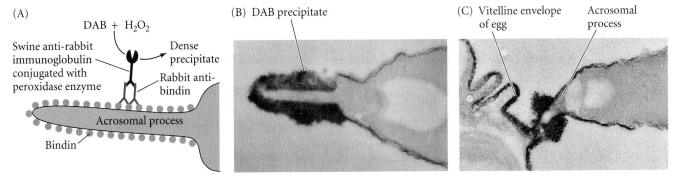

Figure 7.15
Localization of bindin on the acrosomal process. (A) Immunochemical technique used to localize bindin. Rabbit antibody was made to the bindin protein, and this antibody was incubated with sperm that had undergone the acrosome reaction. If bindin were present, the rabbit antibody would remain bound to the sperm. After any unbound antibody was washed off, the sperm were treated with swine antibody that had been covalently linked to peroxidase enzymes. The swine antibody bound to the rabbit antibody, placing peroxidase molecules wherever bindin was present. Peroxidase catalyzes the formation of a dark precipitate from diaminobenzidine (DAB) and hydrogen peroxide. Thus, this precipitate formed only where bindin was present. (B) Localization of bindin to the acrosomal process after the acrosome reaction (33,200×). (C) Localization of bindin to the acrosomal process at the junction of the sperm and the egg. (B and C from Moy and Vacquier 1979; photographs courtesy of V. D. Vacquier.)

(A)

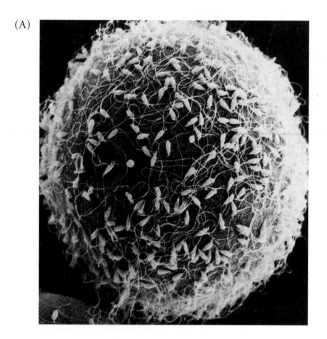

(B)

Figure 7.16
Bindin receptors on the egg. (A) Scanning electron micrograph of sea urchin sperm bound to the vitelline envelope of an egg. Although this egg is saturated with sperm, there appears to be room on the surface for more sperm, implying the existence of a limited number bindin receptors. (B) Binding of *S. purpuratus* sperm to polystyrene beads that have been coated with purified bindin receptor protein. (A, photograph courtesy of C. Glabe, L. Perez, and W. J. Lennarz; B from Foltz et al. 1993.)

dence demonstrating that **ZP3** (zona protein 3) is the glycoprotein that initially binds the sperm. The binding of mouse sperm to the mouse zona pellucida can be inhibited by first incubating the sperm with solubilized zona glycoproteins. Using this inhibition assay, Bleil and Wassarman (1980, 1986, 1988) found that ZP3 was the active competitor for sperm binding (Figure 7.17). In other words, it appeared from this assay that purified ZP3 (but not ZP1 or ZP2) can bind to the sperm and prevent the sperm from binding to the zona pellucida. This was confirmed by the finding that radiolabeled ZP3 (but not ZP1 or ZP2) bound to the heads of mouse sperm with intact acrosomes (Figure 7.18B).

The cell membrane overlying the sperm head can bind to thousands of ZP3 glycoproteins in the zona pellucida. Moreover, there appear to be several different proteins on sperm that are capable of binding ZP3 (Figure 7.18A; Wassarman et al. 2001). Many of these sperm proteins bind to the serine- and threonine-linked carbohydrate chains of ZP3, suggesting

Figure 7.17
Mouse ZP3, the zona protein that binds sperm. (A) Diagram of the fibrillar structure of the mouse zona pellucida. The major strands of the zona are composed of repeating dimers of proteins ZP2 and ZP3. These strands are occasionally crosslinked by ZP1, forming a mesh-like network. (B) Inhibition assay showing the specific decrease of mouse sperm binding to zonae pellucidae when sperm and zonae were first incubated with increasingly large amounts of the glycoprotein ZP3. The importance of the carbohydrate portion of ZP3 is also indicated by this graph. (A after Wassarman 1989; B after Bleil and Wassarman 1980 and Florman and Wassarman 1985.)

(A)

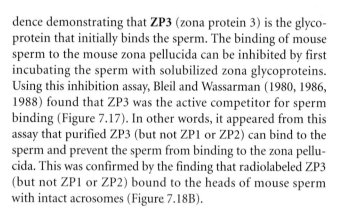

ZP1

ZP2

ZP3

Carbohydrate residues

(B)

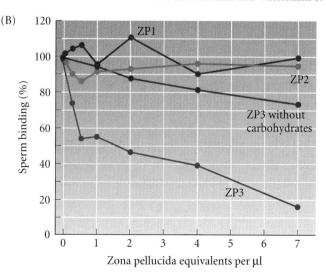

(A)

(B)

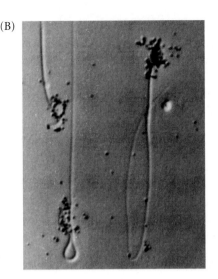

Figure 7.18
Sperm ZP3-binding proteins at the zona pellucida. (A) ZP3-binding proteins on the mouse sperm are located in the plasma membrane, overlying the acrosome. In this confocal image, a ZP3-binding protein is stained red by antibody immunofluorescence. (B) Radioactively labeled ZP3 binds to capacitated mouse sperm. (A photograph by F. Suzuki, M. Toyota, J. Maekawa, J. Bleil and J. Cheng, courtesy of J. Bleil; B from Bleil and Wassarman 1986, photograph courtesy of the authors.)

that the carbohydrate moieties of ZP3 are critical for sperm attachment to the zona. This conclusion has been confirmed by the finding that if these carbohydrate groups are removed from ZP3, it will not bind sperm as well as intact ZP3 (see Figure 7.17B; Florman and Wassarman 1985; Kopf 1998).

INDUCTION OF THE MAMMALIAN ACROSOME REACTION BY ZP3.
ZP3 is the specific glycoprotein in the mouse zona pellucida to which sperm bind. ZP3 also initiates the acrosome reaction after sperm have bound to it. The mouse zona pellucida, unlike the sea urchin vitelline envelope, is a thick structure. By undergoing the acrosome reaction on the zona pellucida, the mouse sperm can concentrate its proteolytic enzymes directly at the point of attachment and digest a hole through this extracellular layer (see Figure 7.8B). Indeed, mouse sperm that undergo the acrosome reaction before they reach the zona pellucida are unable to penetrate it (Florman et al. 1998).

The mouse sperm acrosome reaction is induced when ZP3 crosslinks the receptors on the sperm cell membrane (Endo et al. 1987; Leyton and Saling 1989). One of the sperm proteins that is crosslinked is **galactosyltransferase-I,** an intramembrane enzyme whose active site faces outward and binds to the carbohydrate residues of ZP3. This crosslinking activates specific G proteins in the sperm cell membrane, initiating a cascade that opens the membrane's calcium channels and causes the calcium-mediated exocytosis of the acrosomal vesicle (Leyton et al. 1992; Florman et al. 1998; Shi et al. 2001). The calcium appears to activate the same cytoskeletal fusion proteins (the SNARE complex) that are activated in the exocytoses of pancreatic cells (releasing digestive enzymes), neurons (releasing neurotransmitters), and mast cells (releasing histamine) (Tomes et al. 2002).

TRAVERSING THE ZONA PELLUCIDA. The exocytosis of the acrosomal vesicle releases a variety of proteases that lyse the zona pellucida (Yamagata et al. 1999). These enzymes create a

hole through which the sperm can travel toward the egg. However, during the acrosome reaction, the anterior portion of the sperm cell membrane (i.e., the region containing the ZP3 binding sites) is shed from the sperm (see Figure 7.11). But if sperm are going to penetrate the zona pellucida, they must somehow retain some adhesion to it. In mice, it appears that this secondary binding to the zona is accomplished by proteins in the inner acrosomal membrane that bind specifically to the **ZP2** glycoprotein (Bleil et al. 1988). Whereas acrosome-intact sperm will not bind to ZP2, acrosome-reacted sperm will. Moreover, antibodies against the ZP2 glycoprotein will not prevent the binding of acrosome-intact sperm to the zona, but will inhibit the attachment of acrosome-reacted sperm. The structure of the zona consists of repeating units of ZP3 and ZP2, occasionally crosslinked by ZP1 (see Figure 7.17A). It appears that the acrosome-reacted sperm transfer their binding from ZP3 to the adjacent ZP2 molecules.*

Gamete Fusion and the Prevention of Polyspermy

Fusion of the egg and sperm cell membranes

Once the sperm has undergone the acrosome reaction and has travelled to the egg, the fusion of the sperm cell membrane with the cell membrane of the egg can begin.

---

*In guinea pigs, secondary binding to the zona is thought to be mediated by the protein PH-20. Moreover, when this inner acrosomal membrane protein was injected into adult male or female guinea pigs, 100% of them became sterile for several months (Primakoff et al. 1988). The blood sera of these sterile guinea pigs had extremely high concentrations of antibodies to PH-20. The antiserum from guinea pigs sterilized in this manner not only bound specifically to PH-20, but also blocked sperm-zona adhesion in vitro. The contraceptive effect lasted several months, after which fertility was restored. These experiments show that the principle of immunological contraception is well founded.

The entry of a sperm into a sea urchin egg is illustrated in Figure 7.19. Sperm-egg fusion appears to cause the polymerization of actin in the egg to form a **fertilization cone** (Summers et al. 1975; Schatten and Schatten 1980; Terasaki 1996). Homology between the egg and the sperm is again demonstrated, since the acrosomal process also appears to be formed by the polymerization of actin. The actin from the gametes forms a connection that widens the cytoplasmic bridge between the egg and the sperm, and the sperm nucleus and tail pass through this bridge. A similar process occurs during the fusion of mammalian gametes (Yanagimachi and Noda 1970; Figure 7.20).

In the sea urchin, all regions of the egg cell membrane are capable of fusing with sperm. In several other species, certain regions of the membrane are specialized for sperm recognition and fusion (Vacquier 1979). Fusion is an active process, often mediated by specific "fusogenic" proteins. It has been suggested that sea urchin sperm bindin plays a second role as a fusogenic protein. In addition to recognizing the egg, bindin contains a long stretch of hydrophobic amino acids near its amino terminus, and this region is able to fuse phospholipid vesicles in vitro (Ulrich et al. 1998, 1999).

In mammals, the sperm contacts the egg not at its tip (as in the case of sea urchins), but on the side of the head, at a region called the equatorial domain of the sperm head (see Figure 7.20). The mechanism of mammalian gamete fusion is still controversial (see Primakoff and Myles 2002). Gene knockout experiments suggest that mammalian gamete fusion may depend on interaction between a sperm protein and integrin-associated CD9 protein on the egg (Le Naour et al. 2000; Miyado et al. 2000; Evans 2001). Female mice carrying gene knockouts for CD9 are infertile because their eggs fail to fuse with sperm. This infertility can be reversed by the microinjection of mRNA encoding either mouse or human CD9 (Kaji et al. 2002). It is not known exactly how these proteins facilitate membrane fusion, but CD9 is also known to be critical for the fusion of myocytes (the muscle cell precursors) to form the multinucleated myotube of striated muscle (Tachibana and Hemler 1999).

The prevention of polyspermy

As soon as one sperm has entered the egg, the fusibility of the egg membrane, which was so necessary to get the sperm inside the egg, becomes a dangerous liability. In sea urchins, as in most animals studied, any sperm that enters the egg can provide a haploid nucleus and a centriole to the egg. In normal **monospermy**, in which only one sperm enters the egg, a haploid sperm nucleus and a haploid egg nucleus combine to form the diploid nucleus of the fertilized egg (zygote), thus restoring the chromosome number appropriate for the species. The centriole, which is provided by the sperm, divides to form the two poles of the mitotic spindle during cleavage.

The entrance of multiple sperm—**polyspermy**—leads to disastrous consequences in most organisms. In the sea urchin, fertilization by two sperm results in a triploid nucleus, in which each chromosome is represented three times rather than twice. Worse, since each sperm's centriole divides to form the two poles of a mitotic apparatus, instead of a bipolar mitotic spindle separating the chromosomes into two cells, the triploid chromosomes may be divided into as many as four cells. Because there is no mechanism to ensure that each of the four cells receives the proper number and type of chromo-

Figure 7.19
Scanning electron micrographs of the entry of sperm into sea urchin eggs. (A) Contact of sperm head with egg microvillus through the acrosomal process. (B) Formation of fertilization cone. (C) Internalization of sperm within the egg. (D) Transmission electron micrograph of sperm internalization through the fertilization cone. (A–C from Schatten and Mazia 1976, photographs courtesy of G. Schatten; D, photograph courtesy of F. J. Longo.)

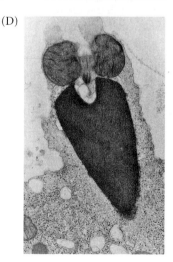

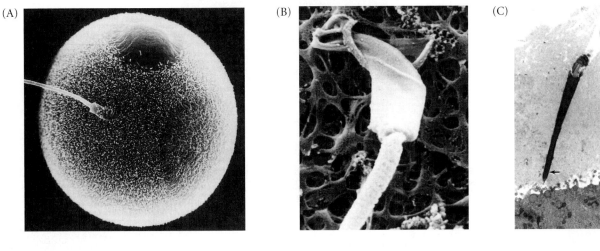

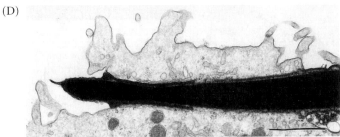

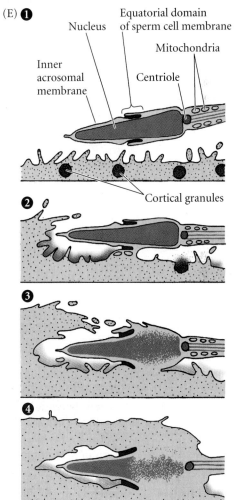

Figure 7.20
Entry of sperm into golden hamster egg. (A) Scanning electron micrograph of sperm fusing with egg. The "bald" spot (without microvilli) is where the polar body has budded off. (B) Close-up of sperm-zona binding. (C) Transmission electron micrograph showing the sperm head passing through the zona. (D) Transmission electron micrograph of a hamster sperm fusing parallel to the egg plasma membrane. (E) Diagram of the fusion of the sperm and egg plasma membranes. (A–D from Yanagimachi and Noda 1970 and Yanagimachi 1994; photographs courtesy of R. Yanagimachi.)

somes, the chromosomes are apportioned unequally. Some cells receive extra copies of certain chromosomes and other cells lack them. Theodor Boveri demonstrated in 1902 that such cells either die or develop abnormally (Figure 7.21).

Species have evolved ways to prevent the union of more than two haploid nuclei. The most common way is to prevent the entry of more than one sperm into the egg. The sea urchin egg has two mechanisms to avoid polyspermy: a fast reaction, accomplished by an electric change in the egg cell membrane, and a slower reaction, caused by the exocytosis of the cortical granules (Just 1919).

THE FAST BLOCK TO POLYSPERMY. The **fast block to polyspermy** is achieved by changing the electric potential of the egg cell membrane. This membrane provides a selective barrier between the egg cytoplasm and the outside environment, so that the ionic concentration of the egg differs greatly from that of its surroundings. This concentration difference is es-

pecially significant for sodium and potassium ions. Seawater has a particularly high sodium ion concentration, whereas the egg cytoplasm contains relatively little sodium. The reverse is the case with potassium ions. This condition is maintained by the cell membrane, which steadfastly inhibits the entry of

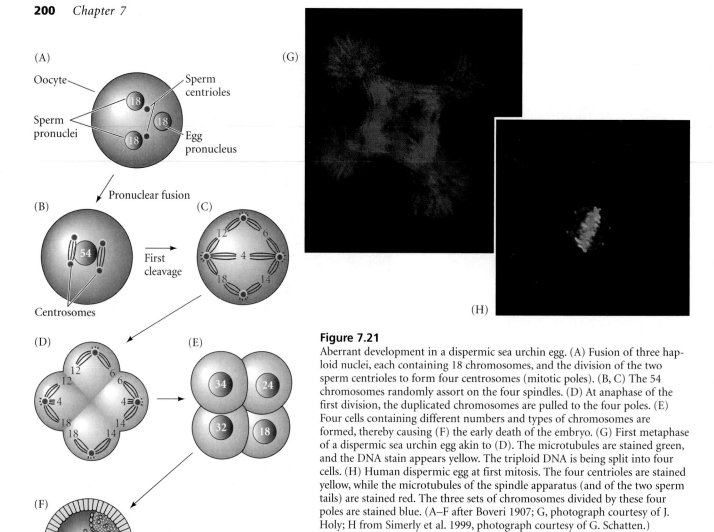

Figure 7.21
Aberrant development in a dispermic sea urchin egg. (A) Fusion of three haploid nuclei, each containing 18 chromosomes, and the division of the two sperm centrioles to form four centrosomes (mitotic poles). (B, C) The 54 chromosomes randomly assort on the four spindles. (D) At anaphase of the first division, the duplicated chromosomes are pulled to the four poles. (E) Four cells containing different numbers and types of chromosomes are formed, thereby causing (F) the early death of the embryo. (G) First metaphase of a dispermic sea urchin egg akin to (D). The microtubules are stained green, and the DNA stain appears yellow. The triploid DNA is being split into four cells. (H) Human dispermic egg at first mitosis. The four centrioles are stained yellow, while the microtubules of the spindle apparatus (and of the two sperm tails) are stained red. The three sets of chromosomes divided by these four poles are stained blue. (A–F after Boveri 1907; G, photograph courtesy of J. Holy; H from Simerly et al. 1999, photograph courtesy of G. Schatten.)

sodium ions into the oocyte and prevents potassium ions from leaking out into the environment. If we insert an electrode into an egg and place a second electrode outside it, we can measure the constant difference in charge across the egg cell membrane. This **resting membrane potential** is generally about 70 mV, usually expressed as –70 mV because the inside of the cell is negatively charged with respect to the exterior.

Within 1–3 seconds after the binding of the first sperm, the membrane potential shifts to a positive level, about +20 mV (Longo et al. 1986). This change is caused by a small influx of sodium ions into the egg (Figure 7.22A). Although sperm can fuse with membranes having a resting potential of –70 mV, they cannot fuse with membranes having a positive resting potential, so no more sperm can fuse to the egg. It is not known whether the increased sodium permeability of the egg is due to the binding of the first sperm or to the fusion of the first sperm with the egg (Gould and Stephano 1987, 1991; McCulloh and Chambers 1992).

The importance of sodium ions and the change in resting potential was demonstrated by Laurinda Jaffe and colleagues. They found that polyspermy can be induced if sea urchin eggs are artificially supplied with an electric current that keeps their membrane potential negative. Conversely, fertilization can be prevented entirely by artificially keeping the membrane potential of eggs positive (Jaffe 1976). The fast block to polyspermy can also be circumvented by lowering the concentration of sodium ions in the surrounding water (Figure 7.22B–D). If the supply of sodium ions is not sufficient to cause the positive shift in membrane potential, polyspermy occurs (Gould-Somero et al. 1979; Jaffe 1980).

It is not known how the change in membrane potential acts on the sperm to block secondary fertilization. Most likely, the sperm carry a voltage-sensitive component (possibly a positively charged fusogenic protein), and the insertion of this component into the egg cell membrane could be regulated by the electric charge across the membrane (Iwao and Jaffe 1989). An electric block to polyspermy also occurs in frogs (Cross and Elinson 1980), but probably not in most mammals (Jaffe and Cross 1983).

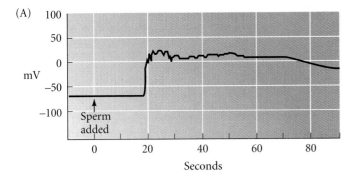

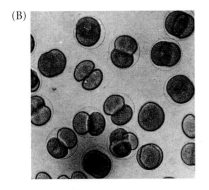 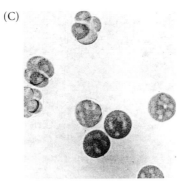

Figure 7.22
Membrane potential of sea urchin eggs before and after fertilization. (A) Before the addition of sperm, the potential difference across the egg plasma membrane is about −70 mV. Within 1–3 seconds after the fertilizing sperm contacts the egg, the potential shifts in a positive direction. (B, C) *Lytechinus* eggs photographed during first cleavage. (B) Control eggs developing in 490 mM Na$^+$. (C) Polyspermy in eggs fertilized in similarly high concentrations of sperm in 120 mM Na$^+$ (choline was substituted for sodium). (D) Table showing the rise of polyspermy with decreasing sodium ion concentration. (From Jaffe 1980; photographs courtesy of L. A. Jaffe.)

Na$^+$ (mM)	Percentage of polyspermic eggs
490	22
360	26
120	97
50	100

THE SLOW BLOCK TO POLYSPERMY. The eggs of sea urchins (and many other animals) have a second, **slow block to polyspermy** to ensure that multiple sperm do not enter the egg cytoplasm (Just 1919). The fast block to polyspermy is transient, since the membrane potential of the sea urchin egg remains positive for only about a minute. This brief potential shift is not sufficient to prevent polyspermy permanently, and polyspermy can still occur if the sperm bound to the vitelline envelope are not somehow removed (Carroll and Epel 1975). This removal is accomplished by the **cortical granule reaction**, a slower, mechanical block to polyspermy that becomes active about a minute after the first successful sperm-egg fusion.

Directly beneath the sea urchin egg cell membrane are about 15,000 cortical granules, each about 1 μm in diameter (see Figure 7.6B). Upon sperm entry, these cortical granules fuse with the egg cell membrane and release their contents into the space between the cell membrane and the fibrous mat of vitelline envelope proteins. Several proteins are released by this cortical granule exocytosis. The first is a trypsin-like protease called **cortical granule serine protease**. This enzyme dissolves the protein posts that connect the vitelline envelope proteins to the cell membrane, and it clips off the bindin receptors and any sperm attached to them (Vacquier et al. 1973; Glabe and Vacquier 1978; Haley and Wessel 1999). Second, mucopolysaccharides released by the cortical granules produce an osmotic gradient that causes water to rush into the space between the cell membrane and the vitelline envelope, causing the envelope to expand and become the **fertilization envelope** (Figures 7.23 and 7.24). A third protein released by the cortical granules, a peroxidase enzyme, hardens the fertilization envelope by crosslinking tyrosine residues on adjacent proteins (Foerder and Shapiro 1977; LaFleur et al. 1998). As shown in Figure 7.23, the fertilization envelope starts to form at the site of sperm entry and continues its expansion around the egg. As it forms, bound sperm are released from the envelope. This process starts about 20 seconds after sperm attachment and is complete by the end of the first minute of fertilization. Finally, a fourth set of cortical granule proteins, including **hyalin**, forms a coating around the egg (Hylander and Summers 1982). The egg extends elongated microvilli whose tips attach to this **hyaline layer**. This layer provides support for the blastomeres during cleavage.

WEBSITE 7.5 **Building the egg's extracellular matrix.** In sea urchins, the cortical granules secrete not only hyalin but a number of proteins that construct the extracellular matrix of the embryo. This highly coordinated process results in sequential layers.

VADE MECUM^2 **Sea urchin fertilization.** The remarkable reactions that prevent polyspermy in a fertilized sea urchin egg can be seen in the raising of the fertilization envelope. This segment contains movies of this event shown in real time. [Click on Sea Urchin]

In mammals, the cortical granule reaction does not create a fertilization envelope, but its ultimate effect is the same. Released enzymes modify the zona pellucida sperm receptors such that they can no longer bind sperm (Bleil and Wassarman 1980). Cortical granules of mouse eggs have been found to contain *N*-acetylglucosaminidase enzymes capable of cleaving *N*-acetylglucosamine from ZP3 carbohydrate chains. *N*-acetyl-

(A)

(B)

(C)

(D)

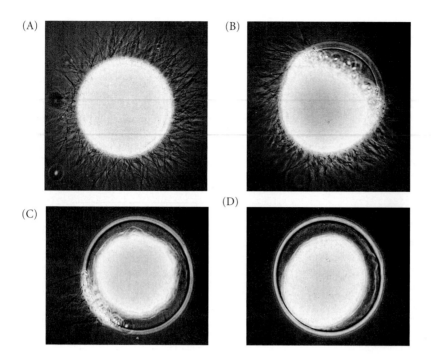

Figure 7.23
Formation of the fertilization envelope and removal of excess sperm. To create these photographs, sperm were added to sea urchin eggs, and the suspension was fixed in formaldehyde to prevent further reactions. (A) At 10 seconds after sperm addition, sperm are seen surrounding the egg. (B, C) At 25 and 35 seconds after insemination, a fertilization envelope is forming around the egg, starting at the point of sperm entry. (D) The fertilization envelope is complete, and excess sperm have been removed. (From Vacquier and Payne 1973; photographs courtesy of V. D. Vacquier.)

glucosamine is one of the carbohydrate groups that sperm can bind to. Miller and co-workers (1992, 1993) have demonstrated that when the *N*-acetylglucosamine residues are removed at fertilization, ZP3 will no longer serve as a substrate for the binding of other sperm. ZP2 is clipped by another cortical granule protease, and it loses its ability to bind sperm as well (Moller and Wassarman 1989). Thus, once a sperm has entered the egg, other sperm can no longer initiate or maintain their binding to the zona pellucida and are rapidly shed.

CALCIUM AS THE INITIATOR OF THE CORTICAL GRANULE REACTION. The mechanism of the cortical granule reaction is similar to that of the acrosome reaction, and it may involve the same molecules. Upon fertilization, the intracellular free Ca^{2+} concentration of the egg increases greatly. In this high-calcium environment, the cortical granule membranes fuse with the egg cell membrane, releasing their contents (see Figure 7.24). Once the fusion of the cortical granules begins near the point of sperm entry, a wave of cortical granule exocytosis propagates around the cortex to the opposite side of the egg.

In sea urchins and mammals, the rise in calcium concentration responsible for the cortical granule reaction is not due to an influx of calcium into the egg, but rather comes from within the egg itself. The release of calcium from intracellular storage can be monitored visually using calcium-activated luminescent dyes such as aequorin (isolated from luminescent jellyfish) or fluorescent dyes such as fura-2. These dyes emit light when they bind free Ca^{2+}. When a sea urchin egg is injected with dye and then fertilized, a striking wave of calcium release propagates across the egg (Figure 7.25). Starting at the point of sperm entry, a band of light traverses the cell (Stein-

hardt et al. 1977; Gilkey et al. 1978; Hafner et al. 1988). The calcium ions do not merely diffuse across the egg from the point of sperm entry. Rather, the release of Ca^{2+} starts at one end of the cell and proceeds actively to the other end. The entire release of Ca^{2+} is complete in roughly 30 seconds in sea urchin eggs, and free Ca^{2+} is resequestered shortly after being released. If two sperm enter the egg cytoplasm, Ca^{2+} release can be seen starting at the two separate points of entry on the cell surface (Hafner et al. 1988).

Several experiments have demonstrated that calcium ions are directly responsible for propagating the cortical granule reaction, and that these ions are stored within the egg itself. The drug A23187 is a calcium ionophore (a compound that transports free Ca^{2+} across lipid membranes, allowing these cations to traverse otherwise impermeable barriers). Placing unfertilized sea urchin eggs into seawater containing A23187 causes the cortical granule reaction and the elevation of the fertilization envelope. Moreover, this reaction occurs in the absence of any Ca^{2+} in the surrounding water. Therefore, the A23187 must be causing the release of Ca^{2+} already sequestered in organelles within the egg (Chambers et al. 1974; Steinhardt and Epel 1974). The cortical granules themselves are tethered to the cell membrane by a series of integral membrane proteins that facilitate calcium-mediated exocytosis (Conner et al. 1997; Conner and Wessel 1998).

In sea urchins and vertebrates (but not snails and worms), the calcium ions responsible for the cortical granule reaction are stored in the endoplasmic reticulum of the egg (Eisen and Reynolds 1985; Terasaki and Sardet 1991). In sea urchins and frogs, this reticulum is pronounced in the cortex and surrounds the cortical granules (Figure 7.26; Gardiner and Grey 1983; Luttmer and Longo 1985). In *Xenopus*, the cortical endoplasmic reticulum becomes ten times more abundant during the maturation of the egg and disappears locally within a minute after the wave of cortical granule exocytosis occurs in any region of the cortex. Once initiated, the release of calcium is self-propagating. Free calcium is able to release sequestered calcium from its storage sites, thus causing a wave of Ca^{2+} release and cortical granule exocytosis.

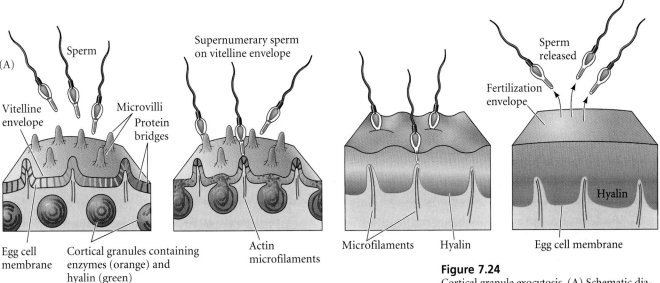

(A)

Sperm

Vitelline envelope

Microvilli

Protein bridges

Supernumerary sperm on vitelline envelope

Sperm released

Fertilization envelope

Hyalin

Egg cell membrane

Cortical granules containing enzymes (orange) and hyalin (green)

Actin microfilaments

Microfilaments

Hyalin

Egg cell membrane

(B)

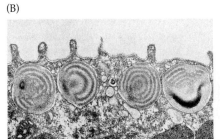

(C)

(D)

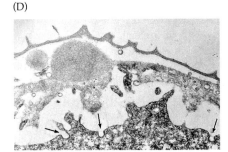

(E)

Figure 7.24
Cortical granule exocytosis. (A) Schematic diagram showing the events leading to the formation of the fertilization envelope and the hyaline layer. As cortical granules undergo exocytosis, they release proteases that cleave the proteins linking the vitelline envelope to the cell membrane. Mucopolysaccharides released by the cortical granules form an osmotic gradient, thereby causing water to enter and swell the space between the vitelline envelope and the plasma membrane. Other enzymes released from the cortical granules harden the vitelline envelope (now the fertilization envelope) and release sperm bound to it. (B, C) Transmission and scanning electron micrographs of the cortex of an unfertilized sea urchin egg. (D, E) Transmission and scanning electron micrographs of the same region of a recently fertilized egg, showing the raising of the fertilization envelope and the points at which the cortical granules have fused with the plasma membrane of the egg (arrows in D). (A after Austin 1965; B–E from Chandler and Heuser 1979, photographs courtesy of D. E. Chandler.)

WEBSITE 7.6 Blocks to polyspermy. Theodore Boveri's analysis of polyspermy is a classic of experimental and descriptive biology. E. E. Just's delineation of the fast and slow blocks was a critical paper in embryology. Both papers are reprinted here, along with commentaries.

VADE MECUM[2] **E. E. Just.** This segment contains videos of Just's work on sea urchin fertilization.
[Click on Sea Urchin]

The Activation of Egg Metabolism

Although fertilization is often depicted as merely the means to merge two haploid nuclei, it has an equally important role in initiating the processes that begin development. These events happen in the cytoplasm and occur without the involvement of the nuclei.*

The mature sea urchin egg is a metabolically sluggish cell that is activated by the sperm. This activation is merely a stimulus, however; it sets into action a preprogrammed set of metabolic events. The responses of the egg to the sperm can be divided into "early" responses, which occur within seconds of the cortical granule reaction, and "late" responses, which take place several minutes after fertilization begins (Table 7.1; Figure 7.27).

---

*In certain salamanders, this developmental function of fertilization has been totally divorced from the genetic function. The silver salamander (*Ambystoma platineum*) is a hybrid subspecies consisting solely of females. Each female produces an egg with an unreduced chromosome number. This egg, however, cannot develop on its own, so the silver salamander mates with a male Jefferson salamander (*A. jeffersonianum*). The sperm from the male Jefferson salamander only stimulates the egg's development; it does not contribute genetic material (Uzzell 1964). For details of this complex mechanism of procreation, see Bogart et al. 1989.

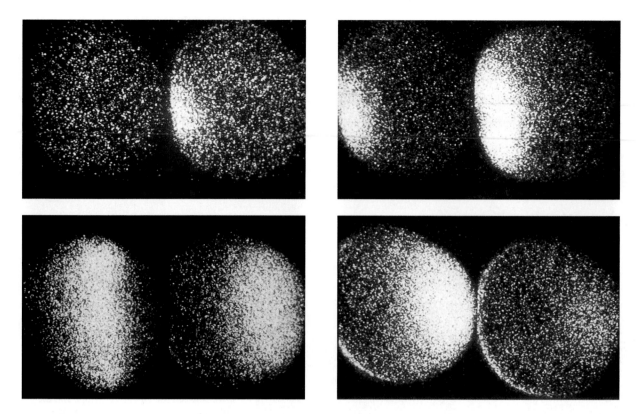

Figure 7.25
Wave of calcium release across sea urchin eggs during fertilization. A sea urchin egg is preloaded with a dye that fluoresces when it binds calcium. When a sperm fuses with the egg, a wave of calcium release can be seen, beginning at the site of sperm entry and propagating across the egg. The wave takes 30 seconds to traverse the egg. (Photograph courtesy of G. Schatten.)

Early responses

As we have seen, contact or fusion between sea urchin sperm and egg activates the two major blocks to polyspermy: the fast block, initiated by sodium influx into the cell; and the slow block, initiated by the intracellular release of Ca^{2+}. The activa-

tion of all eggs appears to depend on an increase in the concentration of free Ca^{2+} within the egg. The same release of Ca^{2+} responsible for the cortical granule reaction is also responsible for the re-entry of the egg into the cell cycle and the reactivation of egg protein synthesis. Ca^{2+} levels in the egg increase from 0.1 to 1 μM, and in almost all species, this occurs as a wave or succession of waves that sweep across the egg beginning at the site of sperm-egg fusion (Jaffe 1983; Terasaki and Sardet 1991; see Figure 7.25).

In mammals, several waves of calcium traverse the egg. Recent studies by Ducibella and colleagues (2002) have shown

Figure 7.26
Endoplasmic reticulum surrounding cortical granules in sea urchin eggs. (A) The endoplasmic reticulum has been stained with osmium-zinc iodide to allow visualization by transmission electron microscopy. The cortical granule is seen to be surrounded by the endoplasmic reticulum. (B) An entire egg stained with fluorescent antibodies to calcium-dependent calcium release channels. The antibodies show these channels in the cortical endoplasmic reticulum. (A from Luttmer and Longo 1985, photograph courtesy of S. Luttmer; B from McPherson et al. 1992, photograph courtesy of F. J. Longo.)

(A)
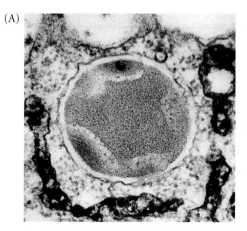

(B)
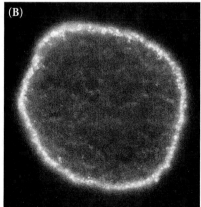

TABLE 7.1 Events of sea urchin fertilization

Event	Approximate time postinsemination[a]
EARLY RESPONSES	
Sperm-egg binding	0 seconds
Fertilization potential rise (fast block to polyspermy)	within 1 sec
Sperm-egg membrane fusion	within 1 sec
Calcium increase first detected	10 sec
Cortical granule exocytosis (slow block to polyspermy)	15–60 sec
LATE RESPONSES	
Activation of NAD kinase	starts at 1 min
Increase in NADP[+] and NADPH	starts at 1 min
Increase in O_2 consumption	starts at 1 min
Sperm entry	1–2 min
Acid efflux	1–5 min
Increase in pH (remains high)	1–5 min
Sperm chromatin decondensation	2–12 min
Sperm nucleus migration to egg center	2–12 min
Egg nucleus migration to sperm nucleus	5–10 min
Activation of protein synthesis	starts at 5–10 min
Activation of amino acid transport	starts at 5–10 min
Initiation of DNA synthesis	20–40 min
Mitosis	60–80 min
First cleavage	85–95 min

Main sources: Whitaker and Steinhardt 1985; Mohri et al. 1995.

[a]Approximate times based on data from *S. purpuratus* (15–17°C), *L. pictus* (16–18°C), *A. punctulata* (18–20°C), and *L. variegatus* (22–24°C). The timing of events within the first minute is best known for *Lytechinus variegatus*, so times are listed for that species.

that the different activities of egg activation are initiated at different numbers of calcium waves. Moreover, events that are

initiated by one calcium wave might not go to completion without additional calcium waves.

The presence of Ca^{2+} is essential for activating the development of the embryo. If the calcium-chelating chemical EGTA is injected into sea urchin eggs, there is no cortical granule reaction, no change in membrane resting potential, and no reinitiation of cell division upon fertilization (Runft et al. 2002). Conversely, eggs can be activated artificially in the absence of sperm by procedures that release free calcium into the oocyte. Steinhardt and Epel (1974) found that injection of micromolar amounts of the calcium ionophore A23187 into a sea urchin egg elicits most of the responses characteristic of a normally fertilized egg. The elevation of the fertilization envelope, a rise in intracellular pH, a burst of oxygen utilization, and increases in protein and DNA synthesis are all generated in their proper order. In most of these cases, development ceases before the first mitosis because the egg is still haploid and lacks the sperm centriole needed for division.

Calcium release activates a series of metabolic reactions (Figure 7.27). One of these is the activation of the enzyme NAD[+] kinase, which converts NAD[+] to NADP[+] (Epel et al. 1981). This change may have important consequences for lipid metabolism, since NADP[+] (but not NAD[+]) can be used as a coenzyme for lipid biosynthesis. Thus, the conversion of NAD[+] to NADP[+] may be important in the construction of the many new cell membranes required during cleavage. Calcium release also affects oxygen consumption. A burst of oxygen reduction (to hydrogen peroxide) is seen during fertilization, and much of this "respiratory burst" is used to crosslink the fertilization envelope. The enzyme responsible for this reduction of oxygen is also NADPH-dependent (Heinecke and Shapiro 1989). Lastly, NADPH helps regenerate glutathione and ovothiols, which may be crucial for scavenging free radicals that could otherwise damage the DNA of the egg and early embryo (Mead and Epel 1995).

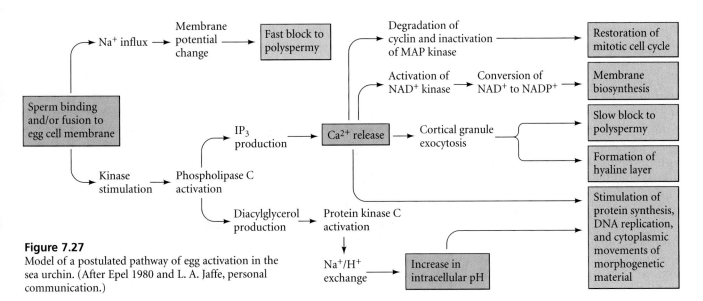

Figure 7.27
Model of a postulated pathway of egg activation in the sea urchin. (After Epel 1980 and L. A. Jaffe, personal communication.)

The Activation of Gamete Metabolism

In 1937, Yale embryologist Ross Granville Harrison lamented, "The liaison between genetics and embryology is now established, but can we say the same of embryology and physiology? Perhaps we are still under the spell of the doctrine that more than one liaison at a time is sin." Indeed, for all our knowledge of the genes that instruct the embryo, the physiological mechanisms by which the instructions are carried out have been largely ignored. This situation is changing as our knowledge of signaling pathways becomes integrated with our knowledge of the cytoskeleton and membrane ion channels. One of the most exciting places of integration involves the ionic mechanisms of fertilization.

IP_3: Releaser of Calcium Ions

A decade ago, Berridge (1993) pointed out that "Just how the sperm triggers the explosive release of calcium in the egg is still something of a mystery." We still do not know the mechanisms whereby the union of sperm and egg initiates a massive Ca^{2+}

release. However, one conclusion that has been reached is that the production of **inositol 1,4,5-trisphosphate (IP_3)** is the primary mechanism for releasing Ca^{2+} from intracellular storage. The IP_3 pathway is shown in Figure 7.28. The membrane **phospholipid phosphatidylinositol 4,5-bisphosphate (PIP_2)** is split by the enzyme **phospholipase C (PLC)** to yield two active compounds: IP_3 and **diacylglycerol (DAG)**. IP_3 is able to release Ca^{2+} into the cytoplasm by opening the Ca^{2+} channels of the endoplasmic reticulum. DAG activates **protein kinase C**, which in turn activates a protein that exchanges sodium ions for hydrogen ions, raising the pH of the egg (Swann and Whitaker 1986; Nishizuka 1986). This Na^+/H^+ exchange pump also needs Ca^{2+} for activity. The result of PLC activation is therefore the liberation of Ca^{2+} and the alkalinization of the egg, and both of the compounds it creates, IP_3 and DAG, are involved in the initiation of development.

IP_3 is formed at the site of sperm entry in sea urchin eggs and can be detected

within seconds of their being fertilized. The inhibition of IP_3 synthesis prevents calcium release (Lee and Shen 1998; Carroll et al. 2000), while injected IP_3 can release sequestered Ca^{2+} in sea urchin eggs, leading to the cortical granule reaction (Whitaker and Irvine 1984; Busa et al. 1985). Moreover, these IP_3-mediated effects can be thwarted by preinjecting the egg with calcium-chelating agents (Turner et al. 1986).

IP_3-responsive calcium channels have been found in the egg endoplasmic reticulum. The IP_3 formed at the site of sperm entry is thought to bind to the IP_3 receptors of these channels, effecting a local release of calcium (Ferris et al. 1989; Furuichi et al. 1989; Terasaki and Sardet 1991). Once released, Ca^{2+} can diffuse directly, or they can facilitate the release of more Ca^{2+} by binding to calcium-sensitive receptors located in the cortical endoplasmic reticulum (McPherson et al. 1992). The binding of Ca^{2+} to these receptors releases more calcium, and this released calcium binds to more receptors, and so on. The resulting wave of calcium release is propagated throughout the cell, starting at the point of sperm entry. The cortical granules,

Figure 7.28
The roles of inositol phosphates in releasing calcium from the endoplasmic reticulum and the initiation of development. Phospholipase C splits PIP_2 into IP_3 and DAG. The IP_3 releases calcium from the endoplasmic reticulum, and the DAG, with assistance from the released Ca^{2+}, activates the sodium-hydrogen exchange pump in the membrane.

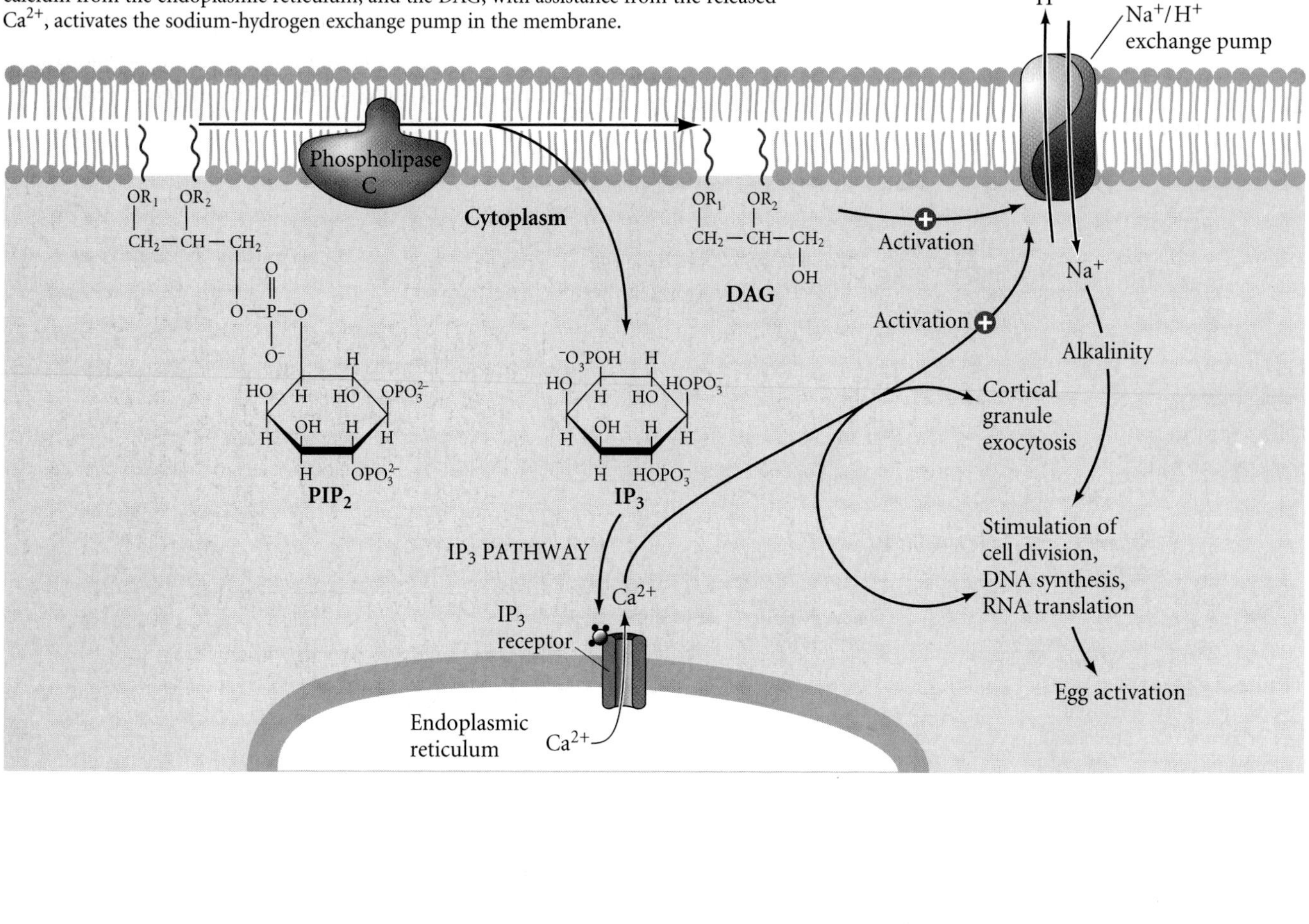

which fuse with the cell membrane in the presence of high calcium concentrations, respond with a wave of exocytosis that follows the Ca^{2+}. Mohri and colleagues (1995) have shown that IP_3-released Ca^{2+} is both necessary and sufficient for initiating the wave of calcium release.

IP_3 has been similarly found to release Ca^{2+} in vertebrate eggs. As in sea urchins, waves of IP_3 are thought to mediate calcium release from sites within the endoplasmic reticulum (Miyazaki et al. 1992; Ducibella et al. 2002). Blocking the IP_3 receptor in hamster and mouse eggs prevents the release of calcium at fertilization. Xu and colleagues (1994) found that blocking IP_3-mediated calcium release blocks every aspect of sperm-induced egg activation, including cortical granule exocytosis, mRNA recruitment, and cell cycle resumption.

Phospholipase C: Generator of IP_3

The question then becomes, What initiates the production of IP_3? In other words, what activates the phospholipase C enzymes? This question has not been easy to address, since (1) there are numerous types of PLC, (2) they can be activated through different pathways, and (3) different species can use different mechanisms to activate them. Results from recent studies of sea urchin eggs suggest that the active PLC in this group of animals is a member of the γ (gamma) family of PLCs (Carroll et al. 1997, 1999; Shearer et al. 1999). Inhibitors that specifically block this family of PLCs inhibit IP_3 production as well as Ca^{2+} release. Moreover, these inhibitors can be circumvented by microinjecting IP_3 into the egg. The mechanism by which IP_3 is generated in vertebrates is largely unknown, but may involve cytoplasmic activators coming from the sperm nucleus (Kuretake et al. 1996; Perry et al. 2000).

Kinases: Link between Sperm and PLC?

The finding that the γ class of PLC was responsible for generating IP_3 during echinoderm fertilization caused investigators to look at the proteins that activated this class of phospholipases. This work soon came to focus on the Src family of protein kinases. Src proteins are found in the cortical cytoplasm of sea urchin and starfish eggs, and such proteins can form a complex with PLCγ. Inhibition of Src protein kinases lowered and delayed the amount of calcium released (Giusti et al. 1999a,b; Kinsey and Shen 2000).

One possibility is that the PLCγ-Src complex is connected to and activated by a sperm receptor in the cell membrane of the egg (Figure 7.29B). A second possibility is that the activation of the IP_3 pathway is caused not by the binding of sperm and egg, but by the fusion of the sperm and egg cell membranes. McCulloh and Chambers (1992) have electrophysiological evidence that sea urchin egg activation does not occur until after sperm and egg cytoplasms are joined. They suggest that the egg-activating components are located on the sperm cell membrane or in the sperm cytoplasm. It is even possible that when the fusion of gamete membranes occurs, the sperm receptor kinases or PLCs (activated by the egg jelly to initiate the acrosome reaction) activate the IP_3 cascade resulting in calcium release in the egg (see Gilbert 1994). In this scenario, shown in Figure 7.29C, bindin serves for cell-cell adhesion and membrane fusion, but not for signal-

Figure 7.29
Possible mechanisms of egg activation. (A) A schematic outline of sea urchin egg activation. (B–D) Possible mechanisms by which this scheme might be accomplished. (B) The bindin receptor activates a cytoplasmic Src kinase. (C) An activated Src kinase or PLC in the sperm plasma membrane activates the egg pathways. (D) Calcium release and egg activation by activated PLC from the sperm or by a substance from the sperm that activates egg PLC.

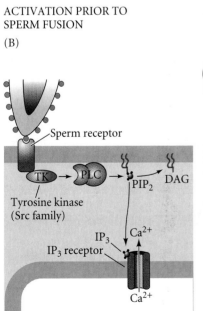

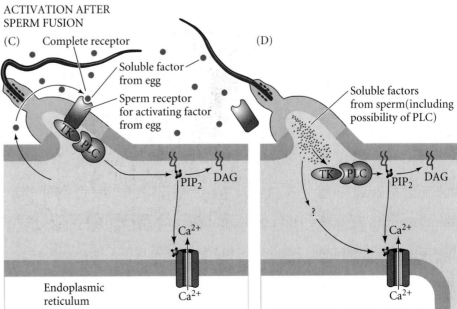

ing. Rather, the egg "activates itself" through the sperm. There is evidence in mammals that sperm soluble PLC may be sufficient to initiate calcium release in oocytes, but the data are difficult to interpret due to the concentrations of inhibitors and activators used (see Runft et al. 2002).

Small soluble activator hypothesis

Still another possibility is that the agent active in releasing the sequestered calcium is a small molecule from the sperm cytosol (Figure 7.29D). Such a molecule may be important in mammalian fertilization. Support for this hypothesis comes from the clinical procedure of intracytoplasmic sperm injection (ICSI), used to treat infertility when a man's sperm count is low. In this procedure, a single intact human sperm is injected directly into the cytoplasm of the egg (see Figure 21.2). The injection results in egg activation, the formation of a male pronucleus, and normal embryonic development (Van Steirtinghem 1994). Kimura and colleagues (1998) have shown that the isolated head of a mouse sperm is capable of activating a mouse oocyte, and that the active components of the sperm head appears to be the proteins surrounding the haploid nucleus. It is not known what role these perinuclear components may play in the normal physiology of egg activation, but a recent set of experiments by Saunders and colleagues (2002) suggests that the active factor may be a novel sperm-specific form of PLC. This protein, called PLC$_\zeta$ (PLC-zeta), triggers Ca^{2+} waves indistinguishable from those of normal fertilization, and removal of PLC$_\zeta$ from mouse sperm extracts abolishes Ca^{2+} release in eggs.

Different species have evolved different ways of obtaining Ca^{2+} during fertilization. It is also possible that any particular species may use more than one means of initiating Ca^{2+} release at fertilization.

Late responses

The late responses of fertilization include the activation of DNA synthesis and protein synthesis. The fusion of egg and sperm in sea urchins causes the intracellular pH to increase.* This rise in intracellular pH begins with a second influx of sodium ions, which causes a 1:1 exchange between sodium ions from the seawater and hydrogen ions from the egg. The loss of hydrogen ions causes the pH of the egg to rise (Shen and Steinhardt 1978). It is thought that the pH increase and the Ca^{2+} elevation act together to stimulate new protein synthesis and DNA synthesis (Winkler et al. 1980; Whitaker and Steinhardt 1982; Rees et al. 1995). If one experimentally elevates the pH of an unfertilized egg to a level similar to that of a fertilized egg, DNA synthesis and nuclear envelope breakdown ensue, just as if the egg were fertilized (Miller and Epel 1999).

Calcium ions are also critical to DNA synthesis. Ca^{2+} inactivates the enzyme MAP kinase, converting it from a phosphorylated (active) to an unphosphorylated (inactive) form. The inactivation of MAP kinase removes an inhibition on DNA synthesis (Carroll et al. 2000). Thus, the wave of free Ca^{2+} inactivates MAP kinase, and DNA synthesis can resume.

In sea urchins, a burst of protein synthesis usually occurs within several minutes after sperm entry. This protein synthesis does not depend on the synthesis of new messenger RNA; rather, it utilizes mRNAs already present in the oocyte cytoplasm (Figure 7.30; see Table 5.2). These mRNAs encode proteins such as histones, tubulins, actins, and morphogenetic factors that are utilized during early development. Such a burst of protein synthesis can be induced by artificially raising the pH of the cytoplasm using ammonium ions (Winkler et al. 1980). One mechanism for this global rise in the translation of stored messages appears to be the release of

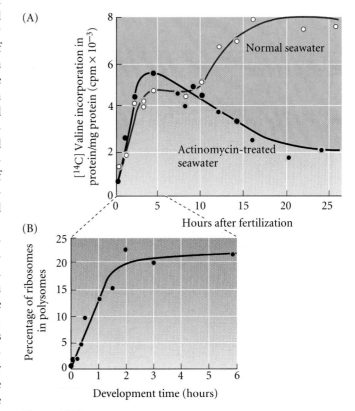

Figure 7.30

A burst of protein synthesis at fertilization uses mRNAs stored in the oocyte cytoplasm. (A) Protein synthesis in embryos of the sea urchin *Arbacia punctulata* fertilized in the presence or absence of actinomycin D, an inhibitor of transcription. For the first few hours, protein synthesis occurs without any new transcription from the zygote or embryo nuclei. A second burst of protein synthesis occurs during the mid-blastula stage. This burst represents translation of newly transcribed messages, and therefore is not seen in embryos growing in actinomycin. (B) Increase in the percentage of ribosomes recruited into polysomes during the first hours of sea urchin development, especially during the first cell cycle. (A after Gross et al. 1964; B after Humphreys 1971.)

---

*This is due to the production of diacylglycerol (as mentioned above). Again, variation among species may be prevalent. In the much smaller egg of the mouse, there is no elevation of pH after fertilization. In the mouse, there is no dramatic increase in protein synthesis immediately following fertilization (Ben-Yosef et al. 1996).

inhibitors from the mRNA. In Chapter 5, we discussed maskin, an inhibitor of translation in the unfertilized amphibian oocyte. In sea urchins, a similar inhibitor also binds translation initiation factor eIF4E and prevents translation from occurring. Upon fertilization, however, this protein, the 4E-binding protein, becomes phosphorylated and degraded, allowing translation of the stored sea urchin mRNAs (Cormier et al. 2001). It is thought that one of the kinases activated at fertilization is responsible for phosphorylating the 4E-binding protein.

Fusion of the Genetic Material

Fusion of genetic material in sea urchins

In sea urchins, the sperm nucleus enters the egg perpendicular to the egg surface. After fusion of the sperm and egg cell membranes, the sperm nucleus and its centriole separate from the mitochondria and the flagellum. The mitochondria and the flagellum disintegrate inside the egg, so very few, if any, sperm-derived mitochondria are found in developing or adult organisms. Thus, although each gamete contributes a haploid genome to the zygote, the mitochondrial genome is transmitted primarily by the maternal parent. Conversely, in almost all animals studied (the mouse being the major exception), the centrosome needed to produce the mitotic spindle of the subsequent divisions is derived from the sperm centriole (see Figure 7.21; Sluder et al. 1989, 1993).

In sea urchins, fertilization occurs after the second meiotic division, so there is a haploid **female pronucleus** in the cytoplasm of the egg when the sperm enters. Once inside the egg, the sperm nucleus undergoes a dramatic transformation as it decondenses to form the **male pronucleus**. First, the nuclear envelope vesiculates into small packets, exposing the compact sperm chromatin to the egg cytoplasm (Longo and Kunkle 1978; Poccia and Collas 1997). Then proteins holding the sperm chromatin in its condensed, inactive state are exchanged for other proteins derived from the egg cytoplasm. This exchange permits the decondensation of the sperm chromatin. In sea urchins, decondensation appears to be initiated by the phosphorylation of the nuclear envelope lamin protein and the phosphorylation of two sperm-specific histones that bind tightly to the DNA. This process begins when the sperm comes into contact with a glycoprotein in the egg jelly that elevates the level of cAMP-dependent protein kinase activity. These protein kinases phosphorylate several of the basic residues of the sperm-specific histones and thereby interfere with their binding to DNA (Garbers et al. 1980, 1983; Porter and Vacquier 1986; Stephens et al. 2002). This loosening is thought to facilitate the replacement of the sperm-specific histones with other histones that have been stored in the oocyte cytoplasm (Poccia et al. 1981; Green and Poccia 1985). Once decondensed, the DNA can begin transcription and replication.

After the sea urchin sperm enters the egg cytoplasm, the male pronucleus rotates 180° so that the sperm centriole is between the sperm pronucleus and the egg pronucleus. The sperm centriole then acts as a microtubule organizing center, extending its own microtubules and integrating them with egg microtubules to form an aster.* These microtubules extend throughout the egg and contact the female pronucleus, and the two pronuclei migrate toward each other (Hamaguchi and Hiramoto 1980; Bestor and Schatten 1981). Their fusion forms the diploid **zygote nucleus** (Figure 7.31). DNA synthesis can begin either in the pronuclear stage (during migration) or after the formation of the zygote nucleus, and depends on the calcium released earlier in fertilization (Jaffe et al. 2001).

Fusion of genetic material in mammals

In mammals, the process of pronuclear migration takes about 12 hours, compared with less than 1 hour in the sea urchin. The mammalian sperm enters almost tangentially to the surface of the egg rather than approaching it perpendicularly, and it fuses with numerous microvilli (see Figure 7.20). The DNA of the sperm nucleus is bound by basic proteins called protamines, which are tightly compacted through disulfide bonds. Glutathione in the egg cytoplasm reduces these disulfide bonds and allows the uncoiling of the sperm chromatin (Calvin and Bedford 1971; Kvist et al. 1980; Perreault et al. 1988).

The mammalian sperm enters the oocyte while its nucleus is "arrested" in metaphase of its second meiotic division (Figure 7.32A; see Figure 7.5). The calcium oscillations inactivate MAP kinase and allow DNA synthesis. But unlike the sea urchin egg, which is already in a haploid state, the mammalian oocyte still has chromosomes in the middle of meiotic metaphase. The calcium oscillations activate another kinase that leads to the proteolysis of cyclin (thus allowing the cell cycle to continue) and securin (the protein holding the metaphase chromosomes together) (Watanabe et al. 1991; Johnson et al. 1998).

DNA synthesis occurs separately in each pronucleus, and the centrosome (new centriole) accompanying the male pronucleus produces its asters (largely from proteins stored in the oocyte). The microtubules join the two pronuclei and enable them to migrate toward one another other. Upon meeting, the two nuclear envelopes break down (Figure 7.32B). However, instead of producing a common zygote nucleus (as happens in sea urchin fertilization), the chromatin condenses into chromosomes that orient themselves on a common mitotic spindle (Figure 7.32C). Thus, a true diploid nucleus in mammals is first seen not in the zygote, but at the two-cell stage.

---

*When Oscar Hertwig observed this radial array of sperm asters forming in his newly fertilized sea urchin eggs, he called it "the sun in the egg," and he thought it the happy indication of a successful fertilization (Hertwig 1877). More recently, Simerly and co-workers (1999) found that certain types of human male infertility are due to defects in the centriole's ability to form these microtubular asters. This deficiency causes the failure of pronuclear migration and the cessation of further development.

(A)

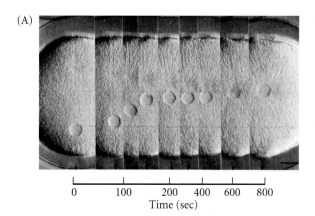

(C)

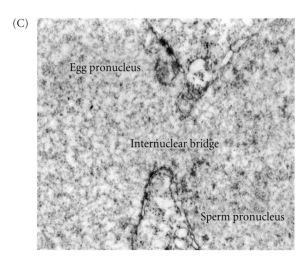

Egg pronucleus

Internuclear bridge

Sperm pronucleus

(B)

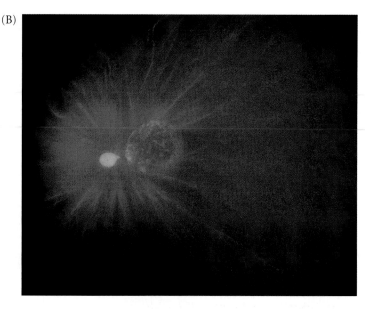

Figure 7.31
Nuclear events in the fertilization of the sea urchin. (A) Sequential photographs showing the migration of the egg pronucleus and the sperm pronucleus toward each other in an egg of *Clypeaster japonicus*. The sperm pronucleus is surrounded by its aster of microtubules. (B) The two pronuclei migrate toward each other on these microtubular processes. (The pronuclear DNA is stained blue by Hoechst dye.) The microtubules (stained green with fluorescent antibodies to tubulin) radiate from the centrosome associated with the (smaller) male pronucleus and reach toward the female pronucleus. (C) Fusion of pronuclei in the sea urchin egg. (A from Hamaguchi and Hiramoto 1980, courtesy of the authors; B from Holy and Schatten 1991, courtesy of J. Holy; C courtesy of F. J. Longo.)

The sperm mitochondria and their DNA are degraded in the egg cytoplasm, so that all of the body's mitochondria are derived from the mother. The egg and embryo appear to get rid of the paternal mitochondria both by dilution and by actively targeting the sperm mitochondria for destruction (Cummins et al. 1998; Shitara et al. 1998; Schwartz and Vissing 2002).

WEBSITE 7.7 Sperm decondensation. The formation of the male pronucleus involves changes in both the chromatin and the nuclear envelope. The nuclear annulus may be critical to uncoiling the DNA.

Rearrangement of the Egg Cytoplasm

Fertilization can initiate radical displacements of the egg's cytoplasmic materials. While these cytoplasmic movements are not obvious in mammalian or sea urchin eggs, there are several species in which these rearrangements of oocyte cytoplasm are crucial for cell differentiation later in development. In the eggs of tunicates, as we will see in Chapter 8, cytoplasmic rearrangements are particularly obvious because of the differing pigmentation of the different regions of the egg. Such cytoplasmic movements are also easy to see in amphibian eggs. In frogs, a single sperm can enter anywhere on the animal hemisphere of the egg; when it does, it changes the cytoplasmic pattern of the egg. Originally, the egg is radially symmetrical about the animal-vegetal axis. After sperm entry, however, the cortical (outer) cytoplasm shifts about 30° toward the point of sperm entry, relative to the inner cytoplasm (Manes and Elinson 1980; Vincent et al. 1986). In some frogs (such as *Rana*), a region of the egg that was formerly covered by the dark cortical cytoplasm of the animal hemisphere is exposed by this shift (Figure 7.33). This underlying cytoplasm, located near the equator on the side opposite the point of sperm entry, contains diffuse pigment granules and therefore appears gray. Thus, this region has been referred to as the **gray crescent** (Roux 1887; Ancel and Vintenberger 1948). As we will see in subsequent chapters, the gray crescent marks the region where gastrulation is initiated in amphibian embryos. In frogs such as *Xenopus*, in which no gray crescent appears, dye labeling confirms that the cortical cytoplasm rotates 30° relative to the internal, subcortical cytoplasm (Vincent et al. 1986).

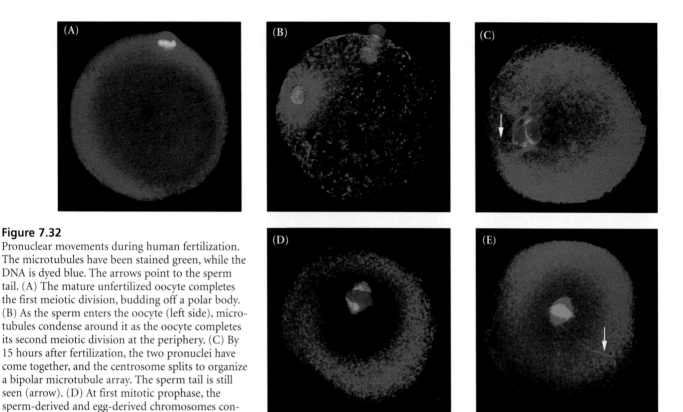

Figure 7.32

Pronuclear movements during human fertilization. The microtubules have been stained green, while the DNA is dyed blue. The arrows point to the sperm tail. (A) The mature unfertilized oocyte completes the first meiotic division, budding off a polar body. (B) As the sperm enters the oocyte (left side), microtubules condense around it as the oocyte completes its second meiotic division at the periphery. (C) By 15 hours after fertilization, the two pronuclei have come together, and the centrosome splits to organize a bipolar microtubule array. The sperm tail is still seen (arrow). (D) At first mitotic prophase, the sperm-derived and egg-derived chromosomes condense separately as the mitotic spindle poles form. (E) At prometaphase, chromosomes from the sperm and egg intermix on the metaphase equator and a mitotic spindle initiates the first mitotic division. The sperm tail can still be seen. (From Simerly et al. 1995; photographs courtesy of G. Schatten.)

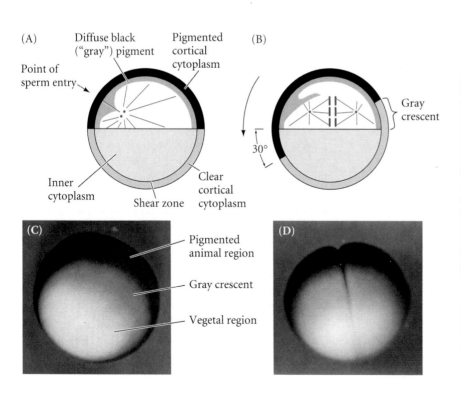

Figure 7.33

Reorganization of cytoplasm in the newly fertilized frog egg. (A) Schematic cross section of an egg midway through the first cleavage cycle. The egg has radial symmetry about its animal-vegetal axis. The sperm has entered at one side, and the sperm nucleus is migrating inward. The cortex represented is like that of *Rana*, with a heavily pigmented animal hemisphere and an unpigmented vegetal hemisphere. (B) About 80% of the way into first cleavage, the cortical cytoplasm rotates 30° relative to the internal cytoplasm. Gastrulation will begin in the gray crescent, the region opposite the point of sperm entry where the greatest displacement of cytoplasm occurs. (C) Gray crescent of *R. pipiens* immediately after cortical rotation exposes the different pigmentation beneath the heavily pigmented cortical cytoplasm. (D) The first cleavage furrow bisects the gray crescent.
(A, B after Gerhart et al. 1989; C, D photographs courtesy of R. P. Elinson.)

The Nonequivalence of Mammalian Pronuclei

It is generally assumed that males and females carry equivalent haploid genomes. Indeed, one of the fundamental tenets of Mendelian genetics is that genes derived from the sperm are functionally equivalent to those derived from the egg. However, as we saw in Chapter 5, genomic imprinting can occur in mammals such that the sperm-derived genome and the egg-derived genome may be functionally different and play complementary roles during certain stages of development.

The first evidence for nonequivalence came from studies of a human tumor called a **hydatidiform mole**, which resembles placental tissue. A majority of such moles have been shown to arise from a haploid sperm fertilizing an egg in which the female pronucleus is absent. After entering the egg, the sperm chromosomes duplicate themselves, thereby restoring the diploid chromosome number. Thus, the entire genome is derived from the sperm (Jacobs et al. 1980; Ohama et al. 1981). Here we see a situation in which the cells survive, divide, and have a normal chromosome number, but development is abnormal. Instead of forming an embryo, the egg becomes a mass of placenta-like cells. Normal development does not occur when the entire genome comes from the male parent.

Conversely, normal development does not occur when the genome is derived totally from the egg. The ability to develop an embryo without spermatic contribution is called **parthenogenesis** (Greek, "virgin birth"). The eggs of many invertebrates and some vertebrates are capable of developing normally in the absence of sperm

(see Chapters 2 and 19). Mammals, however, do not exhibit parthenogenesis. Placing mouse oocytes in a culture medium that artificially activates the oocyte while suppressing the formation of the second polar body produces diploid mouse eggs whose inheritance is derived from the oocyte alone (Kaufman et al. 1977). These cells divide to form embryos with spinal cords, muscles, skeletons, and organs, including beating hearts. However, development does not continue, and by day 10 or 11 (halfway through the mouse's gestation), these parthenogenetic embryos deteriorate. Neither human nor mouse development can be completed solely with egg-derived chromosomes.

The hypothesis that male and female pronuclei are both needed for normal development gains support from pronuclear

transplantation experiments (Surani and Barton 1983; Surani et al. 1986; McGrath and Solter 1984). Either male or female pronuclei can be removed from recently fertilized mouse eggs and added to other recently fertilized eggs. (The two pronuclei can be distinguished at this stage because the female pronucleus is the one beneath the polar bodies.) Thus, zygotes with two male or two female pronuclei can be constructed. Although embryonic cleavage occurs, neither of these types of eggs develops to birth, whereas some control eggs (containing one male pronucleus and one female pronucleus from different zygotes) undergoing such transplantation develop normally (Table 7.2). Thus, for mammalian development to occur, both the sperm-derived and the egg-derived pronuclei are critical.

TABLE 7.2 Pronuclear transplantation experiments

Class of reconstructed zygotes	Operation	Number of successful transplants	Number of progeny surviving
Bimaternal		339	0
Bipaternal		328	0
Control		348	18

Source: McGrath and Solter 1984.

The motor for cytoplasmic movements in amphibian eggs appears to be an array of parallel microtubules that form between the cortical and inner cytoplasm of the vegetal hemisphere parallel to the direction of cytoplasmic rotation. These microtubular tracks are first seen immediately before rotation commences, and they disappear when rotation ceases (Figure 7.34; Elinson and Rowning 1988). Treating the egg with colchicine or ultraviolet radiation at the beginning of rotation stops the formation of these microtubules, thereby inhibiting the cytoplasmic rotation. Using antibodies that bind to the microtubules, Houliston and Elinson (1991a) found that these tracks are formed from both sperm- and egg-derived microtubules, and that the sperm centriole directs the polymerization of the microtubules so that they grow into the vegetal region of the egg. Upon reaching the vegetal cortex, the microtubules angle away from the point of sperm entry, toward the vegetal pole. The off-center position of the sperm centriole as it initiates microtubule polymerization provides the directionality of the rotation. The microtubules appear to be tethered to the cortical cytoplasm by a kinesin-related ATPase (Marrari et al. 2000). As we will see in Chapter 10, these cytoplasmic movements initiate a cascade of events that determine the dorsal-ventral axis of the frog.

Toward the end of the first cell cycle, then, the cytoplasm has been rearranged, the pronuclei have met, DNA is replicating, and new proteins are being translated. The stage is set for the development of a multicellular organism.

Figure 7.34
Parallel arrays of microtubules extend along the vegetal hemisphere along the future dorsal-ventral axis. These arrays are seen here in the second half of the first cell cycle using fluorescent antibodies to tubulin. (A) View looking at the vegetal hemisphere. (B) Higher magnification of the parallel microtubule arrays. Prior to cytoplasmic rotation (about midway through the first cell cycle), no array of microtubules can be seen, and at the end of cytoplasmic rotation, the microtubules depolymerize. (Photographs courtesy of R. P. Elinson.)

VADE MECUM[2] **Amphibian fertilization.** The shifting of cytoplasm in response to fertilization in the amphibian egg is often illustrated in models. Here you will see 3-D models of this event compiled into a movie.
[Click on Amphibian]

Snapshot Summary: Fertilization

1. Fertilization accomplishes two separate activities: sex (the combining of genes derived from two parents) and reproduction (the creation of a new organism).

2. The events of fertilization usually include (1) contact and recognition between sperm and egg; (2) regulation of sperm entry into the egg; (3) fusion of genetic material from the two gametes; and (4) activation of egg metabolism to start development.

3. The sperm head consists of a haploid nucleus and an acrosome. The acrosome is derived from the Golgi apparatus and contains enzymes needed to digest extracellular coats surrounding the egg. The midpiece and neck of the sperm contain mitochondria and the centriole that generates the microtubules of the flagellum. Energy for flagellar motion comes from mitochondrial ATP and a dynein ATPase in the flagellum.

4. The female gamete can be an egg (with a haploid nucleus, as in sea urchins) or an oocyte (in an earlier stage of development, as in mammals). The egg or oocyte has a large mass of cytoplasm storing ribosomes, mRNAs, and nutritive proteins. Other mRNAs and proteins, used as morphogenetic factors, are also stored in the egg. Cortical granules lie beneath the egg's cell membrane. Many eggs also contain protective agents needed for survival in their particular environment.

5. Surrounding the egg cell membrane is an extracellular layer often used in sperm recognition. In most animals, this extracellular layer is the vitelline envelope. In mammals, it is the much thicker zona pellucida.

6. In many species, eggs secrete diffusible molecules that attract and activate the sperm.

7. In sea urchins, the acrosome reaction is initiated by compounds in the egg jelly. The acrosomal vesicle undergoes exocytosis to release its enzymes. Globular actin polymerizes to extend the acrosomal process. Bindin on the acrosomal process is recognized by a protein complex on the sea urchin egg surface.

8. In mammals, sperm must be capacitated in the female reproductive tract before they are capable of fertilizing the egg.

9. Mammalian sperm bind to the zona pellucida before undergoing the acrosome reaction. In the mouse, this binding is mediated by ZP3 (zona protein 3) and several sperm proteins that recognize it. The mammalian acrosome reaction is initiated on the zona pellucida, and the acrosomal enzymes are concentrated there.

10. Fusion between sperm and egg is probably mediated by protein molecules whose hydrophobic groups can merge the sperm and egg cell membranes. In sea urchins, bindin may mediate gamete fusion. In mammals, the mechanism of fusion remains controversial.

11. Polyspermy results when two or more sperm fertilize an egg. It is usually lethal, since it blastomeres with different numbers and types of chromosomes.

12. Many species have two blocks to polyspermy. The fast block is electric and is mediated by sodium ions: the egg membrane resting potential rises, and sperm can no longer fuse with the egg. The slow block is physical and is mediated by calcium ions. A wave of Ca^{2+} propagates from the point of sperm entry, causing the cortical granules to fuse with the egg cell membrane. The released contents of the granules cause the vitelline envelope to rise and to harden into the fertilization envelope.

13. In mammals, blocks to polyspermy include the modification of the zona proteins by the contents of the cortical granules so that sperm can no longer bind to the zona.

14. Inositol 1,4,5-trisphosphate (IP_3) is responsible for releasing Ca^{2+} from storage in the endoplasmic reticulum. DAG (diacylglycerol) is thought to initiate the rise in egg pH. Free Ca^{2+}, supported by the alkalization of the egg, activates egg metabolism, protein synthesis, and DNA synthesis.

15. IP_3 is generated by phospholipases. Different species may use different mechanisms to activate the phospholipases.

16. The rise in intracellular free Ca^{2+} at fertilization in amphibians and mammals causes the degradation of cyclin and the inactivation of MAP kinase, allowing the second meiotic metaphase to be completed and the formation of the haploid female pronucleus.

17. The male pronucleus and the female pronucleus migrate toward each other. In some species, such as mammals, DNA replicates as they move. In other species (such as sea urchins), DNA replication occurs after pronuclear fusion.

18. In sea urchins, the two pronuclei merge and a diploid zygote nucleus is formed. In mammals, the pronuclear membranes disintegrate as the pronuclei approach each other, and their chromosomes gather around a common metaphase plate.

19. Microtubular changes cause cytoplasmic movements. These rearrangements of cytoplasm can be critical in specifying which portions of the egg are going to develop into which organs.

20. The centriole, usually inherited from the sperm, organizes microtubules in the egg and can cause the movement of cytoplasm during the first cell cycle.

Literature Cited

Afzelius, B. A. 1976. A human syndrome caused by immotile cilia. *Science* 193: 317–319.

Alves, A.-P., B. Mulloy, J. A., Diniz and P. A. S. Mourao. 1997. Sulfated polysaccharides from the egg jelly layer are species-specific inducers of acrosome reactions in sperms of sea urchins. *J. Biol. Chem.* 272: 6965–6971.

Ancel, P. and P. Vintenberger. 1948. Recherches sur le determinisme de la symmetrie bilatérale dans l'oeuf des amphibiens. *Bull. Biol. Fr. Belg.* [Suppl.] 31: 1–182.

Arnoult, C., I. G. Kazam, P. E. Visconti, G. Kopf, M. Villaz and H. Florman. 1999. Control of the low-voltage-activated calcium channel of mouse sperm by egg ZP3 and by membrane hyperpolarization during capacitation. *Proc. Natl. Acad. Sci. USA* 96: 6757–6762.

Austin, C. R. 1952. The "capacitation" of mammalian sperm. *Nature* 170: 327.

Austin, C. R. 1960. Capacitation and the release of hyaluronidase. *J. Reprod. Fertil.* 1: 310–311.

Austin, C. R. 1965. *Fertilization.* Prentice-Hall, Englewood Cliffs, NJ.

Benoff, S. 1993. The role of cholesterol during capacitation of human spermatozoa, *Hum. Reprod.* 8: 2001–2008.

Bentley, J. K., H. Shimomura and D. L. Garbers. 1986. Retention of a functional resact receptor in isolated sperm plasma membranes. *Cell* 45: 281–288.

Ben-Yosef, D., Y. Oron and R. Shalgi. 1996. Intracellular pH in rat eggs is not affected by fertilization and the resulting calcium oscillations. *Biol. Reprod.* 55: 461–468.

Berridge, M. J. 1993. Inositol triphosphate and calcium signalling. *Nature* 361: 315–325.

Bestor, T. M. and G. Schatten. 1981. Anti-tubulin immunofluorescence microscopy of microtubules present during the pronuclear movements of sea urchin fertilization. *Dev. Biol.* 88: 80–91.

Bi, G.-Q., J. M. Alderton and R. A. Steinhardt. 1995. Calcium-mediated exocytosis is required for cell membrane resealing. *J. Cell Biol.* 131: 1747–1758.

Bleil, J. D. and P. M. Wassarman. 1980. Mammalian sperm and egg interaction: Identification of a glycoprotein in mouse-egg zonae pellucidae possessing receptor activity for sperm. *Cell* 20: 873–882.

Bleil, J. D. and P. M. Wassarman. 1986. Autoradiographic visualization of the mouse egg's sperm receptor bound to sperm. *J. Cell Biol.* 102: 1363–1371.

Bleil, J. D. and P. M. Wassarman. 1988. Galactose at the nonreducing terminus of O-linked oligosaccharides of mouse egg zona pellucida glyco-protein ZP3 is essential for the glycoprotein's sperm receptor activity. *Proc. Natl. Acad. Sci. USA* 85: 6778–6782.

Bleil, J. D., J. M. Greve and P. M. Wassarman. 1988. Identification of a secondary sperm receptor in the mouse egg zona pellucida: Role in maintenance of binding of acrosome-reacted sperm to eggs. *Dev. Biol.* 28: 376–385.

Bogart, J. P., R. P. Elinson and L. E. Licht. 1989. Temperature and sperm incorporation in polyploid salamanders. *Science* 246: 1032–1034.

Boveri, T. 1902. On multipolar mitosis as a means of analysis of the cell nucleus. [Translated by S. Glueckshon-Waelsch.] *In* B. H. Willier and J. M. Oppenheimer (eds.), *Foundations of Experimental Embryology.* Hafner, New York, 1974.

Boveri, T. 1907. Zellenstudien VI. Die Entwicklung dispermer Seeigeleier. Ein Beiträge zur Befruchtungslehre und zur Theorie des Kernes. *Jena Z. Naturwiss.* 43: 1–292.

Busa, W. B., J. E. Ferguson, S. K. Joseph, J. R. Williamson and R. Nuccitelli. 1985. Activation of frog (*Xenopus laevis*) eggs by inositol triphosphate. I. Characterization of Ca^{2+} release from intracellular stores. *J. Cell Biol.* 100: 677–682.

Calvin, H. I. and J. M. Bedford. 1971. Formation of disulfide bonds in the nucleus and accessory structures of mammalian spermatozoa during maturation in the epididymis. *J. Reprod. Fertil.* [Suppl.] 13: 65–75.

Carroll, D. J. and D. Epel. 1975. Isolation and biological activity of the proteases released by sea urchin eggs following fertilization. *Dev. Biol.* 44: 22–32.

Carroll, D. J., C. S. Ramarao, L. Mehlmann, S. Roche, M. Terasaki and L. A. Jaffe. 1997. Calcium release at fertilization in starfish eggs is mediated by phospholipase Cγ. *J. Cell Biol.* 138: 1303–1311.

Carroll, D. J., D. T. Albay, M. Terasaki, L. A. Jaffe and K. R. Foltz. 1999. Identification of PLCγ-dependent and independent events during fertilization of sea urchin eggs. *Dev. Biol.* 206: 232–247.

Carroll, D. J., D. T. Albay, K. M. Hoag, F. J. O'Neill, M. Kumano and K. R. Foltz. 2000. The relationship between calcium, MAP kinase, and DNA synthesis in the sea urchin egg at fertilization. *Dev. Biol.* 217: 179–191.

Castellano, L. E., G. Martinez-Cadena, J. Lopez-Godinez, A. Obregon and J. Garcia-Soto. 1997. Subcellular localization of the GTP-binding protein Rho in the sea urchin sperm. *Eur. J. Cell Biol.* 74: 329–335.

Chambers, E. L., B. C. Pressman and B. Rose. 1974. The activity of sea urchin eggs by the divalent ionophores A23187 and X-537A. *Biochem. Biophys. Res. Comm.* 60: 126–132.

Chandler, D. E. and J. Heuser. 1979. Membrane fusion during secretion: Cortical granule exocytosis in sea urchin eggs as studied by quick-freezing and freeze fracture. *J. Cell Biol.* 83: 91–108.

Chang, M. C. 1951. Fertilizing capacity of spermatozoa deposited into the fallopian tubes. *Nature* 168: 697–698.

Cherr, G. N., H. Lambert, S. Meizel and D. F. Katz. 1986. In vitro studies of the golden hamster sperm acrosome reaction: Completion on zona pellucida and induction by homologous zonae pellucidae. *Dev. Biol.* 114: 119–131.

Clermont, Y. and C. P. Leblond. 1955. Spermiogenesis of man, monkey, and other animals as shown by the "periodic acid-Schiff" technique. *Am. J. Anat.* 96: 229–253.

Cohen-Dayag, A., I. Tur-Kaspa, J. Dor, S. Mashiach and M. Eisenbach. 1995. Sperm capacitation in humans is transient and correlates with chemotactic responsiveness to follicular factors. *Proc. Natl. Acad. Sci. USA* 92: 11039–11043.

Colwin, A. L. and L. H. Colwin. 1963. Role of the gamete membranes in fertilization in *Saccoglossus kowalevskii* (Enteropneustra). I. The acrosome reaction and its changes in early stages of fertilization. *J. Cell Biol.* 19: 477–500.

Conner, S. and G. M. Wessel. 1998. rab3 mediates cortical granule exocytosis in the sea urchin egg. *Dev. Biol.* 203: 334–344.

Conner, S., D. Leaf and G. M. Wessel. 1997. Members of the SNARE hypothesis are associated with cortical granule exocytosis in the sea urchin egg. *Mol. Reprod. Dev.* 48: 106–118.

Cormier, P., S. Pyronnet, J. Morales, O. Mulner-Lorillon, N. Sonenberg and R. Bellé. 2001. eIF4E association with 4E-BP decreases rapidly following fertilization in sea urchin. *Dev. Biol.* 232: 275–283.

Cook, S. P. and D. F. Babcock. 1993. Selective modulation by cGMP of the K^+ channel activated by speract. *J. Biol. Chem.* 268: 22402–22407.

Correia, L. M. and E. J. Carroll, Jr. 1997. Characterization of the vitelline envelope of the sea urchin *Strongylocentratus purpuratus*. *Dev. Growth Diff.* 39: 69–85.

Corselli, J. and P. Talbot. 1987. In vivo penetration of hamster oocyte-cumulus complexes using physiological numbers of sperm. *Dev. Biol.* 122: 227–242.

Cross, N. L. 1998. Role of cholesterol in sperm capacitation. *Biol. Reprod.* 59: 7–11.

Cross, N. L. and R. P. Elinson. 1980. A fast block to polyspermy in frogs mediated by changes in the membrane potential. *Dev. Biol.* 75: 187–198.

Cummins, J. M., T. Wakayama and R. Yanagimachi. 1998. Fate of microinjected spermatid mitochondria in the mouse oocyte and embryo. *Zygote* 5: 301–308.

Dan, J. C. 1967. acrosome reaction and lysins. *In* C. B. Metz, Jr., and A. Monroy (eds.), *Fertilization*, vol. 1. Academic Press, New York, pp. 237–367.

Davis, B. K. 1978. Inhibition of fertilizing capacity in mammalian spermatozoa by natural and synthetic vesicles. *In* J. J. Kabara (ed.), *Symposium of the Pharmacological Effect of Lipids*. #Amer. Oil Chem Soc., Champaign, IL, pp. 145–157.

De Robertis, E. D. P., F. A. Saez and E. M. F. De Robertis. 1975. *Cell Biology*, 6th Ed. Saunders, Philadelphia.

Ducibella, T. and 8 others. 2002. Egg-to-embryo transition is driven by differential responses to Ca^{2+} oscillation number. *Dev. Biol.* 250: 280–291.

Eisen, A. and G. T. Reynolds. 1985. Sources and sinks for the calcium release during fertilization of single sea urchin eggs. *J. Cell Biol.* 100: 1522–1527.

Eisenbach, M. 1995. Sperm changes enabling fertilization in mammals. *Curr. Opin. Endocrinol. Diabetes* 2: 468–475.

Eisenbach, M. and I. Tur-Kaspa. 1999. Do human eggs attract spermatozoa? *Bioessays* 21: 203–210.

Elinson, R. P. and B. Rowning. 1988. A transient array of parallel microtubules in frog eggs: Potential tracks for a cytoplasmic rotation that specifies the dorso-ventral axis. *Dev. Biol.* 128: 185–197.

Endo, Y. G., G. S. Kopf and R. M. Schultz. 1987. Effects of phorbol ester on mouse eggs: Dissociation of sperm receptor activity from acrosome reaction-inducing activity of the mouse zona pellucida protein, ZP3. *Dev. Biol.* 123: 574–577.

Epel, D. 1977. The program of fertilization. *Sci. Am.* 237(5): 128–138.

Epel, D. 1980. Fertilization. *Endeavour* N.S. 4: 26–31.

Epel, D., C. Patton, R. W. Wallace and W. Y. Cheung. 1981. Calmodulin activates NAD kinase of sea urchin eggs: An early response. *Cell* 23: 543–549.

Evans, J. P. 2001. Fertilin beta and other ADAMs as integrin ligands: Insights into cell adhesion and fertilization. *BioEssays* 23: 628–639.

Ferris, C. D., R. L. Huganir, S. Supattapone and S. H. Snyder. 1989. Purified inositol 1,4,5-trisphosphate receptor mediates calcium flux in reconstituted lipid vesicles. *Nature* 342: 87–89.

Florman, H. M. and B. T. Storey. 1982. Mouse gamete interactions: The zona pellucida is the site of the acrosome reaction leading to fertilization in vitro. *Dev. Biol.* 91: 121–130.

Florman, H. M. and P. M. Wassarman. 1985. O-linked oligosaccharides of mouse egg ZP3 account for its sperm receptor activity. *Cell* 41: 313–324.

Florman, H. M., M. E. Corron, T. D.-H. Kim and D. F. Babcock. 1992. Activation of voltage-dependent calcium channels of mammalian sperm is required for zona pellucida-induced acrosomal exocytosis. *Dev. Biol.* 152: 304–314.

Florman, H. M., C. Arnoult, I. Kazam, C. Li and C. M. B. O'Toole. 1998. A perspective on the control of mammalian fertilization by egg-activated ion channels in sperm: A tale of two channels. *Biol. Reprod.* 59: 12–17.

Foerder, C. A. and B. M. Shapiro. 1977. Release of ovoperoxidase from sea urchin eggs hardens fertilization membrane with tyrosine crosslinks. *Proc. Natl. Acad. Sci. USA* 74: 4214–4218.

Fol, H. 1877. Sur le commencement de l'hémogénie chez divers animaux. *Arch. Zool. Exp. Gén.* 6: 145–169.

Foltz, K. R., J. S. Partin and W. J. Lennarz. 1993. Sea urchin egg receptor for sperm: Sequence similarity of binding domain and hsp 70. *Science* 259: 1421–1425.

Franklin, L. E. 1970. Fertilization and the role of the acrosome reaction in non-mammals. *Biol. Reprod.* [Suppl.] 2: 159–177.

Furuichi, T., S. Yoshikawa, A. Miyawaki, K. Wada, N. Maeda and K. Mikoshiba. 1989. Primary structure and functional expression of the inositol 1,4,5-trisphosphate-binding protein P400. *Nature* 342: 32–38.

Galantino-Homer, H. L., P. E. Visconti and G. S. Kopf. 1997. Regulation of protein tyrosine kinase phosphorylation during bovine capacitation by a cyclic adenosine 3',5'-monophosphate-dependent pathway. *Biol. Reprod.* 56: 707–719.

Garbers, D. L., D. J. Tubb and G. S. Kopf. 1980. Regulation of sea urchin sperm cAMP-dependent protein kinases by an egg associated factor. *Biol. Reprod.* 22: 526–532.

Garbers, D. L., G. S. Kopf, D. J. Tubb and G. Olson. 1983. Elevation of sperm adenosine 3':5'-

monophosphate concentrations by a fucose sulfate-rich complex associated with eggs. I. Structural characterization. *Biol. Reprod.* 29: 1211–1220.

Gardiner, D. M. and R. D. Grey. 1983. Membrane junctions in *Xenopus* eggs: Their distribution suggests a role in calcium regulation. *J. Cell Biol.* 96: 1159–1163.

Gerhart, J., M. Danilchik, T. Doniach, S. Roberts, B. Rowning and R. Stewart. 1989. Cortical rotation of the *Xenopus* egg: Consequences for the anteriorposterior pattern of embryonic dorsal development. *Development* 1989 [Suppl.]: 37–51.

Gilbert, S. F. 1994. *Developmental Biology*, 4th Ed. Sinauer Associates, Sunderland, MA.

Gilkey, J. C., L. F. Jaffe, E. G. Ridgway and G. T. Reynolds. 1978. A free calcium wave traverses the activating egg of the medaka, *Oryzias latipes*. *J. Cell Biol.* 76: 448–467.

Giusti, A. F., K. M. Hoang and K. R. Foltz. 1997. Surface localization of the sea urchin egg receptor for sperm. *Dev. Biol.* 184: 10–24.

Giusti, A. F., D. J. Carroll, Y. A. Abassi and K. R. Foltz. 1999a. Evidence that a starfish egg Src family tyrosine kinase associates with PLC-γl SH2 domains at fertilization. *Dev. Biol.* 208: 189–199.

Giusti, A. F., D. J. Carroll, Y. A. Abassi, M. Terasaki, K. R. Foltz and L. A. Jaffe. 1999b. Requirement of a Src family kinase for initiating calcium release at fertilization in starfish eggs. *J. Biol. Chem.* 274: 29318–29322.

Glabe, C. G. and W. J. Lennarz. 1979. Species-specific sperm adhesion in sea urchins: A quantitative investigation of bindin-mediated egg agglutination. *J. Cell Biol.* 83: 595–604.

Glabe, C. G. and V. D. Vacquier. 1977. Species-specific agglutination of eggs by bindin isolated from sea urchin sperm. *Nature* 267: 836–838.

Glabe, C. G. and V. D. Vacquier. 1978. Egg surface glycoprotein receptor for sea urchin sperm bindin. *Proc. Natl. Acad. Sci. USA* 75: 881–885.

González-Martínez, M. T., A. Guerrero, E. Morales, L. de la Torre and A. Darszon. 1992. A depolarization can trigger Ca²⁺ uptake and the acrosome reaction when preceded by a hyperpolarization in *L. pictus* sea urchin sperm. *Dev. Biol.* 150: 193–202.

Gould, M. and J. L. Stephano. 1987. Electrical response of eggs to acrosomal protein similar to those induced by sperm. *Science* 235: 1654–1657.

Gould, M. and J. L. Stephano. 1991. Peptides from sperm acrosomal protein that activate development. *Dev. Biol.* 146: 509–518.

Gould-Somero, M., L. A. Jaffe and L. Z. Holland. 1979. Electrically mediated fast polyspermy block in eggs of the marine worm, *Urechis caupo*. *J. Cell Biol.* 82: 426–440.

Green, G. R. and E. L. Poccia. 1985. Phosphorylation of sea urchin sperm H1 and H2B histones precedes chromatin decondensation

and H1 exchange during pronuclear formation. *Dev. Biol.* 108: 235–245.

Gross, P. R., L. I. Malkin and W. Moyer. 1964. Templates for the first proteins of embryonic development. *Proc. Natl. Acad. Sci. USA* 51: 407–414.

Gwatkin, R. B. L. 1976. Fertilization. *In* G. Poste and G. L. Nicolson (eds.), *The Cell Surface in Animal Embryogenesis and Development*. Elsevier North-Holland, New York, pp. 1–53.

Hafner, M., C. Petzelt, R. Nobiling, J. B. Pawley, D. Kramp and G. Schatten. 1988. Wave of free calcium at fertilization in the sea urchin egg visualized with Fura-2. *Cell Motil. Cytoskel.* 9: 271–277.

Haley, S. A. and G. M. Wessel. 1999. The cortical granule serine protease CGSP1 of the sea urchin, *Strongylocentrotus purpuratus*, is autocatalytic and contains a low-density lipoprotein receptor-like domain. *Dev. Biol.* 211: 1–10.

Hamaguchi, M. S. and Y. Hiramoto. 1980. Fertilization process in the heart-urchin, *Clypeaster japonicus*, observed with a differential interference microscope. *Dev. Growth Diff.* 22: 517–530.

Hardy, D. M., T. Harumi and D. L. Garbers. 1994. Sea urchin sperm receptors for egg peptides. *Semin. Dev. Biol.* 5: 217–224.

Harrison, R. G. 1937. Embryology and its relations. *Science* 85: 369–374.

Hartsoeker, N. 1694. *Essai de dioptrique*. Paris.

Heinecke, J. W. and B. M. Shapiro. 1989. Respiratory oxygen burst of fertilization. *Proc. Natl. Acad. Sci. USA* 86: 1259–1263.

Hertwig, O. 1877. Beiträge zur Kenntniss der Bildung, Befruchtung, und Theilung des theirischen Eies. *Morphol. Jahr.* 1: 347–452.

Hirohashi, N. and W. J. Lennarz. 2001. Role of a vitelline layer-associated 350 kDa glycoprotein in controlling species-specific gamete interaction in the sea urchin. *Dev. Growth Diff.* 43: 247–255.

Hirohashi, N. and V. D. Vacquier. 2002a. High molecular weight fucose sulfate polymer is required for opening both Ca²⁺ channels involved in triggering the sea urchin acrosome reaction. *J. Biol. Chem.* 277: 1182–1189.

Hirohashi, N. and V. D. Vacquier. 2002b. Egg fucose sulfate polymer, sialoglycan, and speract all trigger the sea urchin sperm acrosome reaction. *Biochem. Biophys. Res. Commun.* 296: 833–839.

Hirohashi, N., A. C. Vilela-Silva, P. A. Mourao and V. D. Vacquier. 2002. Structural requirements for species-specific induction of the sperm acrosome reaction by sea urchin egg sulfated fucan. *Biochem. Biophys. Res. Commun.* 298: 403–407.

Holy, J. and G. Schatten. 1991. Spindle pole centrosomes of sea urchin embryos are partially composed of material recruited from maternal stores. *Dev. Biol.* 147: 343–353.

Houliston, E. and R. P. Elinson. 1991a. Evidence for the involvement of microtubules, endo-

plasmic reticulum, and kinesin in cortical rotation of fertilized frog eggs. *J. Cell Biol.* 114: 1017–1028.

Humphreys, T. 1971. Measurements of messenger RNA entering polysomes upon fertilization in sea urchins. *Dev. Biol.* 26: 201–208.

Hunter, R. H. F. 1989. Ovarian programming of gamete progression and maturation in the female genital tract. *Zool. J. Linn. Soc.* 95: 117–124.

Hylander, B. L. and R. G. Summers. 1982. An ultrastructural and immunocytochemical localization of hyaline in the sea urchin egg. *Dev. Biol.* 93: 368–380.

Iwao, Y. and L. A. Jaffe. 1989. Evidence that the voltage-dependent component in the fertilization process is contributed by the sperm. *Dev. Biol.* 134: 446–451.

Jacobs, P. A., C. M. Wilson, J. A. Sprenkle, N. B. Rosenshein and B. R. Migeon. 1980. Mechanism of origin of complete hydatidiform moles. *Nature* 286: 714–717.

Jaffe, L. A. 1976. Fast block to polyspermy in sea urchins is electrically mediated. *Nature* 261: 68–71.

Jaffe, L. A. 1980. Electrical polyspermy block in sea urchins: Nicotine and low sodium experiments. *Dev. Growth Diff.* 22: 503–507.

Jaffe, L. A. and N. L. Cross. 1983. Electrical properties of vertebrate oocyte membranes. *Biol. Reprod.* 30: 50–54.

Jaffe, L. A., A. F. Giusti, D. J. Carroll and K. R. Foltz. 2001. Ca²⁺ signaling during fertilization of echinoderm eggs. *Semin. Cell Dev. Biol.* 12: 45–51.

Jaffe, L. F. 1983. Sources of calcium in egg activation: A review and hypothesis. *Dev. Biol.* 99: 265–277.

Johnson, J., B. M. Bierle, G. I. Gallicano and D. G. Capco. 1998. Calcium/calmodulin-dependent protein kinase II and calmodulin: Regulators of the meiotic spindle in mouse eggs. *Dev. Biol.* 204: 464–477.

Just, E. E. 1919. The fertilization reaction in *Echinarachnius parma*. *Biol. Bull.* 36: 1–10.

Kaji, K., S. Oda, S. Miyazaki and A. Kudo. 2002. Infertility of CD9-deficient mouse eggs is reversed by mouse CD9, human CD9, or mouse CD81: Polyadenylated mRNA injection developed for molecular analysis of sperm-egg fusion. *Dev. Biol.* 247: 327–334.

Kaufman, M. H., S. C. Barton and M. A. H. Surani. 1977. Normal postimplantation development of mouse parthenogenetic embryos to the forelimb bud stage. *Nature* 265: 53–55.

Kimura, Y., R. Yanagimachi, S. Kuretake, H. Bortiewicz, A. C. F. Perry and H. Yanagimachi. 1998. Analysis of mouse oocyte activation suggests the involvement of sperm perinuclear material. *Biol. Reprod.* 58: 1407–1415.

Kinsey, W. H. and S. S. Shen. 2000. Role of the Fyn kinase in calcium release during fertilization of the sea urchin egg. *Dev. Biol.* 225: 253–264.

Kopf, G. S. 1998. acrosome reaction. *In* E. Knobil and J. D. Neill (eds.), *Encyclopedia of Reproduction*, vol. 1. Academic Press, San Diego, pp. 17–27.

Kuretake, S., Y. Kimura, K. Hoshi and R. Yanagimachi. 1996. Fertilization and development of mouse oocytes injected with isolated sperm heads. *Biol. Reprod.* 55: 789–795.

Kvist, U., B. A. Afzelius and L. Nilsson. 1980. The intrinsic mechanism of chromatin decondensation and its activation in human spermatozoa. *Dev. Growth Diff.* 22: 543–554.

LaFleur, G. J., Jr., Y. Horiuchi and G. M. Wessel. 1998. Sea urchin ovoperoxidase: Oocyte-specific member of a heme-dependent peroxidase superfamily that functions in the block to polyspermy. *Mech. Dev.* 70: 77–89.

Lee, S.-J. and S. S. Shen. 1998. The calcium transient in sea urchin eggs during fertilization requires the production of inositol 1,4,5-trisphosphate. *Dev. Biol.* 193: 195–208.

Leeuwenhoek, A. van. 1685. Letter to the Royal Society of London. Quoted in E. G. Ruestow, 1983, Images and ideas: Leeuwenhoek's perception of the spermatozoa. *J. Hist. Biol.* 16: 185–224.

Lefebvre, R., P. J. Chenoweth, M. Drost, C. T. LeClear, M. MacCubbin, J. T. Dutton and S. S. Suarez. 1995. Characterization of the oviductal sperm receptor in cattle. *Biol. Reprod.* 53: 1066–1074.

Le Naour, F., E. Rubinstein, C. Jasmin, M. Prenant and C. Boucheix. 2000. Severely reduced female fertility in CD9-deficient mice. *Science* 287: 319–321.

Levine, A. E., K. A. Walsh and E. J. B. Fodor. 1978. Evidence of an acrosin-like enzyme in sea urchin sperm. *Dev. Biol.* 63: 299–307.

Leyton, L. and P. Saling. 1989. Evidence that aggregation of mouse sperm receptors by ZP3 triggers the acrosome reaction. *J. Cell Biol.* 108: 2163–2168.

Leyton, L., P. Leguen, D. Bunch and P. M. Saling. 1992. Regulation of mouse gametic interaction by a sperm tyrosine kinase. *Proc. Natl. Acad. Sci. USA* 93: 1164–1169.

Lin, Y., K. Mahan, W. F. Lathrop, D. G. Myles and P. Primakoff 1994. A hyaluronidase activity of the sperm plasma membrane protein PH-20 enables sperm to penetrate the cumulus cell layer surrounding the egg. *J. Cell Biol.* 125: 1157–1163.

Longo, F. J. and M. Kunkle. 1978. Transformation of sperm nuclei upon insemination. *In* A. A. Moscona and A. Monroy (eds.), *Current Topics in Developmental Biology*, vol. 12. Academic Press, New York, pp. 149–184.

Longo, F. J., J. W. Lynn, D. H. McCulloh and E. L. Chambers. 1986. Correlative ultrastructural and electrophysiological studies of sperm-egg interactions of the sea urchin *Lytechinus variegatus*. *Dev. Biol.* 118: 155–167.

Lopez, L. C., E. M. Bayna, D. Litoff, N. L. Shaper, J. H. Shaper and B. D. Shur. 1985. Receptor function of mouse sperm surface galactosyltransferase during fertilization. *J. Cell Biol.* 101: 1501–1510.

Luttmer, S. and F. J. Longo. 1985. Ultrastructural and morphometric observations of cortical endoplasmic reticulum in *Arbacia*, *Spisula*, and mouse eggs. *Dev. Growth Diff.* 27: 349–359.

Manes, M. E. and R. P. Elinson. 1980. Ultraviolet light inhibits gray crescent formation in the frog egg. *Wilhelm Roux Arch. Dev. Biol.* 189: 73–77.

Marrari, Y., M. Terasaki, V. Arrowsmith and E. Houliston. 2000. Local inhibition of cortical rotation in *Xenopus* eggs by an anti-KRP antibody. *Dev. Biol.* 224: 250–262.

McCulloh, D. H. and E. L. Chambers. 1992. Fusion of membranes during fertilization. *J. Gen. Physiol.* 99: 137–175.

McGrath, J. and D. Solter. 1984. Completion of mouse embryogenesis requires both the maternal and paternal genome. *Cell* 37: 179–183.

McPherson, S. M., P. S. McPherson, L. Mathews, K. P. Campbell and F. J. Longo. 1992. Cortical localization of a calcium release channel in sea urchin eggs. *J. Cell Biol.* 116: 1111–1121.

Mead, K. S. and D. Epel. 1995. Beakers and breakers: How fertilisation in the laboratory differs from fertisation in nature. *Zygote* 3: 95–99.

Meizel, S. 1984. The importance of hydrolytic enzymes to an exocytotic event, the mammalian sperm acrosome reaction. *Biol. Rev.* 59: 125–157.

Metz, C. B. 1978. Sperm and egg receptors involved in fertilization. *Curr. Top. Dev. Biol.* 12: 107–148.

Metz, E. C. and S. R. Palumbi 1996. Positive selection and sequence rearrangements generate extensive polymorphism in the gamete recognition protein bindin. *Mol. Biol. Evol.* 13: 397–406.

Miller, B. S. and D. Epel. 1999. The roles of changes in NADPH and pH during fertilization and artificial activation of the sea urchin egg. *Dev. Biol.* 216: 394–405.

Miller, D. J., M. B. Macek and B. D. Shur. 1992. Complementarity between sperm surface β-1,4-galactosyltransferase and egg-coat ZP3 mediates sperm-egg binding. *Nature* 357: 589–593.

Miller, D. J., X. Gong, G. Decker and B. D. Shur. 1993. Egg cortical granule N-acetylglucosaminidase is required for the mouse zona block to polyspermy. *J. Cell Biol.* 123: 1431–1440.

Miller, R. L. 1978. Site-specific agglutination and the timed release of a sperm chemoattractant by the egg of the leptomedusan, *Orthopyxis caliculata*. *J. Exp. Zool.* 205: 385–392.

Miller, R. L. 1985. Sperm chemo-orientation in the metazoa. *In* C. B. Metz, Jr., and A. Monroy (eds.), *Biology of Fertilization*, vol. 2. Academic Press, New York, pp. 275–337.

Miyado, K. and 11 others. 2000. Requirement of CD9 on the egg plasma membrane for fertilization. *Science* 287: 321–319.

Miyazaki, S.-I., M. Yuzaki, K. Nakada, H. Shirakawa, S. Nakanishi, S. Nakade and K. Mikoshiba. 1992. Block of Ca^{2+} wave and Ca^{2+} oscillation by antibody to the inositol 1,4,5-trisphosphate receptor in fertilized hamster eggs. *Science* 257: 251–255.

Mohri, T., P. I. Ivonnet and E. L. Chambers. 1995. Effect of sperm-induced activation current and increase of cytosolic Ca^{2+} by agents that modify the mobilization of $[Ca^{2+}]$. I. Heparin and pentosan polysulfate. *Dev. Biol.* 172: 139–157.

Moller, C. C. and P. M. Wassarman. 1989. Characterization of a proteinase that cleaves zona pellucida glycoprotein ZP2 following activation of mouse eggs. *Dev. Biol.* 132: 103–112.

Moy, G. W. and V. D. Vacquier. 1979. Immunoperoxidase localization of bindin during the adhesion of sperm to sea urchin eggs. *Curr. Top. Dev. Biol.* 13: 31–44.

Mozingo, N. M. and D. E. Chandler. 1991. Evidence for the existence of two assembly domains within the sea urchin fertilization envelope. *Dev. Biol.* 146: 148–157.

Multigner, L., J. Gagnon, A. van Dorsselaer and D. Job. 1992. Stabilization of sea urchin flagellar microtubules by histone H1. *Nature* 360: 33–39.

Nishizuka, Y. 1986. Studies and perspectives of protein kinase C. *Science* 233: 305–312.

Ogawa, K., T. Mohri and H. Mohri. 1977. Identification of dynein as the outer arms of sea urchin sperm axonemes. *Proc. Natl. Acad. Sci. USA* 74: 5006–5010.

Ohama, K. and 7 others. 1981. Dispermic origin of XY hydatidiform moles. *Nature* 292: 551–552.

Perreault, S. D., R. R. Barbee and V. L. Slott. 1988. Importance of glutathione in the acquisition and maintenance of sperm nuclear decondensing activity in maturing hamster oocytes. *Dev. Biol.* 125: 181–187.

Perry, A. C. F., T. Wakayama, I. M. Cook and R. Yaganimachi. 2000. Mammalian oocyte activation by the synergistic action of discrete sperm head components: Induction of calcium transients and involvement of proteolysis. *Dev. Biol.* 217: 386–393.

Pinto-Correia, C. 1997. *The Ovary of Eve: Eggs and Sperm and Preformation*. University of Chicago Press, Chicago.

Poccia, D. and P. Collas. 1997. Nuclear envelope dynamics during male pronuclear development. *Dev. Growth Diff.* 39: 541–550.

Poccia, D., J. Salik and G. Krystal. 1981. Transitions in histone variants of the male pronucleus following fertilization and evidence for a maternal store of cleavage-stage histones in the sea urchin egg. *Dev. Biol.* 82: 287–297.

Porter, D. C. and V. D. Vacquier. 1986. Phosphorylation of sperm histone H1 is induced by the egg jelly layer in the sea urchin *Strongylocentrotus purpuratus*. *Dev. Biol.* 116: 203–212.

Prevost, J. L. and J. B. Dumas. 1824. Deuxieme mémoire sur la génération. *Ann. Sci. Nat.* 2: 129–149.

Primakoff, P. and D. G. Myles. 2002. Penetration, adhesion, and fusion in mammalian sperm-egg interaction. *Science* 296: 2183–2185.

Primakoff, P., W. Lathrop, L. Woolman, A. Cowan and D. Myles. 1988. Fully effective contraception in male and female guinea pigs immunized with the sperm protein PH-20. *Nature* 335: 543–547.

Ralt, D. and 8 others. 1991. Sperm attraction to a follicular factor(s) correlates with human egg fertilizability. *Proc. Natl. Acad. Sci. USA* 88: 2840–2844.

Ramarao, C. S. and D. L. Garbers. 1985. Receptor-mediated regulation of guanylate cyclase activity in spermatozoa. *J. Biol. Chem.* 260: 8390–8397.

Rees, B. B., C. Patton, J. L Grainger and D. Epel. 1995. Protein synthesis increases after fertilization of sea urchin eggs in the absence of an increase in intracellular pH. *Dev. Biol.* 169: 683–698.

Ren, D. and 7 others. 2001. A sperm ion channel required for sperm motility and male fertility. *Nature* 413: 603–609.

Roux, W. 1887. Beiträge zur Entwicklungsmechanik des Embryo. *Arch. Mikrosk. Anat.* 29: 157–212.

Runft, L. L., L. A. Jaffe and L. M. Mehlmann. 2002. Egg activation at fertilization: Where it all begins. *Dev. Biol.* 245: 237–254.

Saling, P. M., J. Sowinski and B. T. Storey. 1979. An ultrastructural study of epididymal mouse spermatozoa binding to zonae pellucida in vitro: Sequential relationship to acrosome reaction. *J. Exp. Zool.* 209: 229–238.

Saunders, C. M. and 7 others. 2002. PLCζ: A sperm-specific trigger of Ca^{++} oscillations in eggs and embryo development. *Development* 129: 3533–3544.

Schackmann, R. W. and B. M. Shapiro. 1981. A partial sequence of ionic changes associated with the acrosome reaction of *Strongylocentrotus purpuratus*. *Dev. Biol.* 81: 145–154.

Schatten, G. and D. Mazia. 1976. The penetration of the spermatozoan through the sea urchin egg surface at fertilization: Observations from the outside on whole eggs and from the inside on isolated surfaces. *Exp. Cell Res.* 98: 325–337.

Schatten, H. and G. Schatten. 1980. Surface activity at the plasma membrane during sperm incorporation and its cytochalasin-B sensitivity: Scanning electron micrography and time-lapse video microscopy during fertilization of the sea urchin *Lytechinus variegatus*. *Dev. Biol.* 78: 435–449.

Schroeder, T. E. 1979. Surface area change at fertilization: Resorption of the mosaic membrane. *Dev. Biol.* 70: 306–327.

Schwartz, M. and J. Vissing. 2002. Paternal inheritance of mitochondrial DNA. *New Engl. J. Med.* 347: 576–580.

Shearer, J., C. D. Nadal, F. Emily-Fenouil, C. Gache, M. Whitaker and B. Ciapa. 1999. Role of phospholipase Cγ at fertilization and during meiosis in sea urchin eggs and embryos. *Development* 126: 2273–2284.

Shen, S. S. and R. A. Steinhardt. 1978. Direct measurement of intracellular pH during metabolic depression of the sea urchin egg. *Nature* 272: 253–254.

Shi, X., S. Amindari, K. Paruchuru, D. Skalla, H. Burkin, B. D. Shur and D. J. Miller. 2001. Cell surface β-1,4-galactosyltransferase-I activates G protein-dependent exocytotic signaling. *Development* 128: 645–654.

Shimomura, H., L. J. Dangott and D. L. Garbers. 1986. Covalent coupling of a resact analogue to guanylate cyclase. *J. Biol. Chem.* 261: 15778–15782.

Shinyoji, C., H. Higuchi, M. Yoshimura, E. Katayama and T. Yanagida. 1998. Dynein arms are oscillating force generators. *Nature* 393: 711–714.

Shitara, H., H. Kaneda, A. Sato and H. Honekawa. 1998. Selective and continuous elimination of mitochondria microinjected into mouse eggs from spermatids, but not from liver cells, occurs throughout embryogenesis. *Genetics* 156: 1277–1284.

Simerly, C. and 7 others. 1995. The paternal inheritance of the centrosome, the cell's microtubule-organizing center, in humans, and the implications for infertility. *Nature Med.* 1: 47–52.

Simerly, C. and 7 others. 1999. Biparental inheritance of γ-tubulin during human fertilization: Molecular reconstitution of functional zygote centrosomes in inseminated human oocytes and in cell-free extracts nucleated by human sperm. *Mol. Cell Biol.* 10: 2955–2969.

Sluder, G., F. J. Miller, K. K. Lewis, E. D. Davison and C. L. Reider. 1989. Centrosome inheritance in starfish zygotes: Selective loss of the maternal centrosome after fertilization. *Dev. Biol.* 131: 567–579.

Sluder, G., F. J. Miller and K. Lewis. 1993. Centrosome inheritance in starfish zygotes. II. Selective suppression of the maternal centrosome during meiosis. *Dev. Biol.* 155: 58–67.

Smith, T. T. 1998. The modulation of sperm function by the oviductal epithelium. *Biol. Reprod.* 58: 1102–1104.

Stears, R. L. and W. J. Lennarz. 1997. Mapping sperm binding domains on the sea urchin egg receptor for sperm. *Dev. Biol.* 187: 200–208.

Steinhardt, R. A. and D. Epel. 1974. Activation of sea urchin eggs by a calcium ionophore. *Proc. Natl. Acad. Sci. USA* 71: 1915–1919.

Steinhardt, R., R. Zucker and G. Schatten. 1977. Intracellular calcium release at fertilization in the sea urchin egg. *Dev. Biol.* 58: 185–197.

Stephens, S. and 7 others. 2002. Two kinase activities are sufficient for sea urchin sperm chromatin decondensation in vitro. *Mol. Reprod. Dev.* 62: 496–503.

Storey, B. T. 1995. Interactions between gametes leading to fertilization: The sperm's eye view. *Reprod. Fertil. Dev.* 7: 927–942.

Suarez, S. S. 1998. The oviductal sperm reservoir in mammals: Mechanisms of formation. *Biol. Reprod.* 58: 1105–1107.

Suarez, S. S., D. F. Katz, D. H. Owen, J. B. Andrew and R. L. Powell. 1991. Evidence for the function of hyperactivated motility in sperm. *Biol. Reprod.* 44: 375–381.

Summers, R. G. and B. L. Hylander. 1974. An ultrastructural analysis of early fertilization in the sand dollar, *Echinarachnius parma*. *Cell Tissue Res.* 150: 343–368.

Summers, R. G. and B. L. Hylander. 1975. Species-specificity of acrosome reaction and primary gamete binding in echinoids. *Exp. Cell Res.* 96: 63–68.

Summers, R. G., B. L. Hylander, L. H. Colwin and A. L. Colwin. 1975. The functional anatomy of the echinoderm spermatozoon and its interaction with the egg at fertilization. *Am. Zool.* 15: 523–551.

Surani, M. A. H. and S. C. Barton. 1983. Development of gynogenetic eggs in the mouse: Implications for parthenogenetic embryos. *Science* 222: 1034–1037.

Surani, M. A. H., S. C. Barton and M. L. Norris. 1986. Nuclear transplantation in the mouse: Heritable differences between parental genomes after activation of the embryonic genome. *Cell* 45: 127–137.

Sutovsky, P., C. S. Navara and G. Schatten. 1996. Fate of the sperm mitochondria and the incorporation, conversion, and disassembly of the sperm tail structures during bovine fertilization. *Biol. Reprod.* 55: 1195–1205.

Swann, K. and M. Whitaker. 1986. The part played by inositol trisphosphate and calcium in the propagation of the fertilization wave in sea urchin eggs. *J. Cell Biol.* 103: 2333–2342.

Tachibana, I. and M. E. Hemler. 1999. Role of transmembrane 4 superfamily (TM4SF) proteins CD9 and CD81 in muscle cell fusion and myotube maintenance. *J. Cell Biol.* 146: 893–904.

Terasaki, M. 1996. Actin filament translocations in sea urchin eggs. *Cell Motil. Cytoskel.* 34: 48–56.

Terasaki, M. and C. Sardet. 1991. Demonstration of calcium uptake and release by sea urchin egg cortical endoplasmic reticulum. *J. Cell Biol.* 115: 1031–1037.

Tilney, L. G., J. Bryan, D. J. Bush, K. Fujiwara, M. S. Mooseker, D. B. Murphy and D. H. Snyder. 1973. Microtubules: Evidence for 13 protofilaments. *J. Cell Biol.* 59: 267–275.

Tilney, L. G., D. P. Kiehart, C. Sardet and M. Tilney. 1978. Polymerization of actin. IV. Role of Ca^{2+} and H$^+$ in the assembly of actin and in membrane fusion in the acrosome reaction of echinoderm sperm. *J. Cell Biol.* 77: 536–560.

Tombes, R. M. and B. M. Shapiro. 1985. Metabolite channeling: A phosphocreatine shuttle to mediate high-energy phosphate transport between sperm mitochondrion and tail. *Cell* 41: 325–334.

Tomes, C. N., M. Michaut, G. D. Blas, P. Visconti, U. Matti and L. S. Mayoga. 2002. SNARE complex assembly is required for human sperm acrosome reaction. *Dev. Biol.* 243: 326–338.

Töpfer-Petersen, E., A. Wagner, J. Friedrich, A. Petrunkina, M. Ekhlasi-Hundrieser, D. Waberski and W. Drommer. 2002. Function of the mammalian oviductal sperm reservoir. *J. Exp. Zool.* 292: 210–215.

Turner, P. R., L. A. Jaffe and A. Fein. 1986. Regulation of cortical vesicle exocytosis in sea urchin eggs by inositol 1,4,5-trisphosphate and GTP-binding protein. *J. Cell Biol.* 102: 70–77.

Ulrich, A. S., M. Otter, C. G. Glabe and D. Hoekstra. 1998. Membrane fusion is induced by a distinct peptide sequence of the sea urchin fertilization protein bindin. *J. Biol. Chem.* 273: 16748–16755.

Ulrich, A. S., W. Tichelaar, G. Forster, O. Zschornig, S. Weinkauf and H. W. Meyer. 1999. Ultrastructural characterization of peptide-induced membrane fusion and peptide self-assembly in the lipid bilayer. *Biophys. J.* 77: 829–841.

Uzzell, T. M. 1964. Relations of the diploid and triploid species of the *Ambystoma jeffersonianum* complex. *Copeia* 1964: 257–300.

Vacquier, V. D. 1979. The interaction of sea urchin gametes during fertilization. *Am. Zool.* 19: 839–849.

Vacquier, V. D. 1998. Evolution of gamete recognition proteins. *Science* 281: 1995–1998.

Vacquier, V. D. and G. W. Moy. 1977. Isolation of bindin: The protein responsible for adhesion of sperm to sea urchin eggs. *Proc. Natl. Acad. Sci. USA* 74: 2456–2460.

Vacquier, V. D. and J. E. Payne. 1973. Methods for quantitating sea urchin sperm in egg binding. *Exp. Cell Res.* 82: 227–235.

Vacquier, V. D., M. J. Tegner and D. Epel. 1973. Protease release from sea urchin eggs at fertil-ization alters the vitelline layer and aids in preventing polyspermy. *Exp. Cell Res.* 80: 111–119.

Van Steirtinghem, A. 1994. IVF and micromanipulation techniques for male-factor fertility. *Curr. Opin. Obstet. Gynaecol.* 6: 173–177.

Vilela-Silva, A. C., M. O. Castro, A. P. Valente, C. H. Biermann and P. A. Mourao. 2002. Sulfated fucans from the egg jelly of sea urchins: Sulfated fucans from the egg jelly of the closely related sea urchins *Strongylocentrotus droebachiensis* and *S. pallidus* ensure species-specific fertilization. *J. Biol. Chem.* 277: 379–387.

Vincent, J. P., G. F. Oster and J. C. Gerhart. 1986. Kinematics of gray crescent formation in *Xenopus* eggs: The displacement of subcortical cytoplasm relative to the egg surface. *Dev. Biol.* 113: 484–500.

Visconti, P. E. and G. S. Kopf. 1998. Regulation of protein phosphorylation during sperm capacitation. *Biol. Reprod.* 59: 1–6.

Visconti, P. E. and 7 others. 1995. Capacitation of mouse spermatozoa. II. Protein tyrosine phosphorylation and capacitation are regulated by a cAMP-dependent pathway. *Development* 121: 1139–1150.

von Kolliker, A. 1841. *Beiträge zur Kenntnis der Geschlectverhältnisse und der Samenflüssigkeit wirbelloser Thiere, nebst einem Versuch Über Wesen und die Bedeutung der sogenannten Samenthiere.* Berlin.

Wang, Y., R. Storeng, P. O. Dale, T. Abyholm and T. Tanbo. 2001. Effects of follicular fluid and steroid hormones on chemotaxis and motility of human spermatozoa in vitro. *Gynecol. Endocrinol.* 15: 286–292.

Ward, C. R. and G. S. Kopf. 1993. Molecular events mediating sperm activation. *Dev. Biol.* 158: 9–34.

Ward, G. E., C. J. Brokaw, D. L. Garbers and V. D. Vacquier. 1985. Chemotaxis *of Arbacia punctulata* spermatozoa to resact, a peptide from the egg jelly layer. *J. Cell Biol.* 101: 2324–2329.

Wassarman, P. M. 1989. Fertilization in mammals. *Sci. Am.* 256(6): 78–84.

Wassarman, P. M., L. Jovine and E. S. Litscher. 2001. A profile of fertilization in mammals. *Nature Cell Biol.* 3: E59–E64.

Watanabe, N., T. Hunt, Y. Ikawa and N. Sagata. 1991. Independent inactivation of MPF and cy-tostatic factor (Mos) upon fertilization of *Xenopus* eggs. *Nature* 352: 247–249.

Whitaker, M. and R. F. Irvine. 1984. Inositol 1,4,5-triphosphate microinjection activates sea urchin eggs. *Nature* 312: 636–639.

Whitaker, M. and R. Steinhardt. 1982. Ionic regulation of egg activation. *Q. Rev. Biophys.* 15: 593–667.

Wilcox, A. J., C. R. Weinberg and D. D. Baird. 1995. Timing of sexual intercourse in relation to ovulation: Effects on the probability of conception, survival of pregnancy, and the sex of the baby. *N. Engl. J. Med.* 333: 1517–121.

Wilson, W. L. and G. Oliphant. 1987. Isolation and biochemical characterization of the subunits of the rabbit sperm acrosome stabilizing factor. *Biol. Reprod.* 37: 159–169.

Winkler, M. M., R. A. Steinhardt, J. L. Grainger and L. Minning. 1980. Dual ionic controls for the activation of protein synthesis at fertilization. *Nature* 287: 558–560.

Xu, Z., G. S. Kopf and R. M. Schultz. 1994. Involvement of inositol-1,4,5-trisphosphate-mediated Ca^{2+} release in early and late events of mouse egg activation. *Development* 120: 1851–1859.

Yamagata, K., A. Honda, S. I. Kashiwabara and T. Baba. 1999. Difference of acrosomal serine protease system between mouse and other rodent sperm. *Dev. Genet.* 25: 115–122.

Yanagimachi, R. 1994. Mammalian fertilization. *In* E. Knobil and J. D. Neill (eds.), *The Physiology of Reproduction*, 2nd Ed. Raven Press, New York.

Yanagimachi, R. and Y. D. Noda. 1970. Electron microscope studies of sperm incorporation into the golden hamster egg. *Am. J. Anat.* 128: 429–462.

Yoshida, M., K. Inabar and M. Morisawa. 1993. Sperm chemotaxis during the process of fertilization in the ascidians *Ciona savignyi* and *Ciona intestinalis. Dev. Biol.* 157: 497–507.

chapter **8** *Early development in selected invertebrates*

REMARKABLE AS IT IS, fertilization is but the initiating step in development. The zygote, with its new genetic potential and its new arrangement of cytoplasm, now begins the production of a multicellular organism. Between the events of fertilization and the events of organ formation are two critical stages: cleavage and gastrulation. During cleavage, rapid cell divisions divide the cytoplasm of the fertilized egg into numerous cells. These cells then undergo dramatic displacements during gastrulation, a process whereby they move to different parts of the embryo and acquire new neighbors (see Chapter 2). During cleavage and gastrulation, the major axes of the embryo are determined, and the cells begin to acquire their respective fates.

While cleavage always precedes gastrulation, axis formation can begin as early as oocyte formation. It can be completed during cleavage (as in *Drosophila*) or extend all the way through gastrulation (as it does in *Xenopus*). There are three axes that need to be specified: the anterior-posterior (head-anus) axis, the dorsal-ventral (back-belly) axis, and the left-right axis. Different species specify these axes at different times, using different mechanisms.

AN INTRODUCTION TO EARLY DEVELOPMENTAL PROCESSES

Cleavage

After fertilization, the development of a multicellular organism proceeds by a process called **cleavage**, a series of mitotic divisions whereby the enormous volume of egg cytoplasm is divided into numerous smaller, nucleated cells. These cleavage-stage cells are called **blastomeres**. In most species (mammals being the chief exception), the rate of cell division and the placement of the blastomeres with respect to one another is completely under the control of the proteins and mRNAs stored in the oocyte by the mother. During cleavage, cytoplasmic volume does not increase. Rather, the enormous volume of zygote cytoplasm is divided into increasingly smaller cells. First the zygote is divided in half, then quarters, then eighths, and so forth. This division of cytoplasm without increasing its volume is accomplished by abolishing the growth period between cell divisions (that is, the G_1 and G_2 phases of the cell cycle). Meanwhile, the cleavage of nuclei occurs at a rapid rate never seen again (not even in tumor cells). A frog egg, for example, can divide into 37,000 cells in just 43 hours. Mitosis in cleavage-stage *Drosophila* embryos occurs every 10 minutes for over 2 hours and in just 12 hours forms some 50,000 cells. This dramatic in-

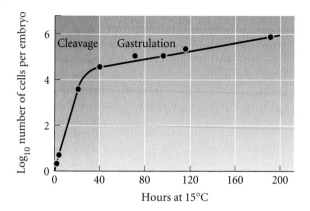

Figure 8.1
Rate of formation of new cells during the early development of the frog *Rana pipiens*. (After Sze 1953.)

crease in cell number can be appreciated by comparing cleavage with other stages of development. Figure 8.1 shows the logarithm of cell number in a frog embryo plotted against the time of development (Sze 1953). It illustrates a sharp discontinuity between cleavage and gastrulation.

From fertilization to cleavage

The transition from fertilization to cleavage is caused by the activation of **mitosis-promoting factor (MPF)**. MPF was first discovered as the major factor responsible for the resumption of meiotic cell divisions in the ovulated frog egg. It continues to play a role after fertilization, regulating the biphasic cell cycle of early blastomeres. Blastomeres generally progress through a cell cycle consisting of just two steps: M (mitosis) and S (DNA synthesis) (Figure 8.2). Gerhart and co-workers (1984) showed that MPF undergoes cyclical changes in its level of activity in mitotic cells. The MPF activity of early blastomeres is highest during M and undetectable during S.

Newport and Kirschner (1984) demonstrated that DNA replication (S) and mitosis (M) are driven solely by the gain and loss of MPF activity. Cleaving cells can be experimentally trapped in S phase by incubating them in an inhibitor of protein synthesis. When MPF is microinjected into these cells, they enter M. Their nuclear envelope breaks down and their chromatin condenses into chromosomes. After an hour, MPF is degraded and the chromosomes return to S phase.

What causes this cyclical activity of MPF? Mitosis-promoting factor contains two subunits. The large subunit is called **cyclin B**. It is this component that shows a periodic behavior, accumulating during S and then being degraded after the cells have reached M (Evans et al. 1983; Swenson et al. 1986). Cyclin B is often encoded by mRNAs stored in the oocyte cytoplasm, and if the translation of this message is specifically inhibited, the cell will not enter mitosis (Minshull et al. 1989). The cyclical appearance and degradation of cyclin B is a key factor in mitotic regulation. Cyclin B regulates the small subunit of MPF, the **cyclin-dependent kinase**. This kinase activates mitosis by phosphorylating several target proteins, including histones, the nuclear envelope lamin pro-

Figure 8.2
Cell cycles of somatic cells and early blastomeres. (A) The simple biphasic cell cycle of the early amphibian blastomeres has only two states, S and M. Cyclin B synthesis allows progression to M (mitosis), while degradation of cyclin B allows cells to pass into S (synthesis) phase. (B) Cell cycle of a typical somatic cell. Mitosis (M) is followed by an "interphase" stage. This latter period is subdivided into G_1, S (synthesis), and G_2 phases. Cells that are differentiating are usually taken "out" of the cell cycle and are in an extended G_1 phase called G_0. The cyclins responsible for the progression through the cell cycle and their respective kinases are shown at their point of cell cycle regulation. (B after Nigg 1995.)

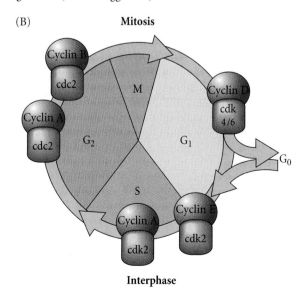

teins, and the regulatory subunit of cytoplasmic myosin. These actions bring about chromatin condensation, nuclear envelope depolymerization, and the organization of the mitotic spindle.

Without cyclin B, the cyclin-dependent kinase will not function. The presence of cyclin is controlled by several proteins that ensure its periodic synthesis and degradation. In most species studied, the regulators of cyclin B (and thus of MPF) are stored in the egg cytoplasm. Therefore, the cell cycle is independent of the nuclear genome for numerous cell divisions. These early divisions tend to be rapid and synchronous. However, as the cytoplasmic components are used up, the nucleus begins to synthesize them. The embryo now enters the **mid-blastula transition**, in which several new phenomena are added to the biphasic cell divisions of the embryo. First, the "gap" stages (G_1 and G_2) are added to the cell cycle, permitting the cells to grow (Figure 8.2B). (Recall that before this time, the cytoplasm was being divided among smaller and smaller cells, but the total volume of the organism remained unchanged.) *Xenopus* embryos add G_1 and G_2 phases to the cell cycle shortly after the twelfth cleavage. *Drosophila* adds G_2 during cycle 14 and G_1 during cycle 17 (Newport and Kirschner 1982a; Edgar et al. 1986). Second, the synchronicity of cell division is lost, because different cells synthesize different regulators of MPF. Third, new mRNAs are transcribed. Many of these messages encode proteins that will become necessary for gastrulation. In many species, if transcription is blocked cell division will still occur at normal rates and times, but the embryo will not be able to initiate gastrulation.

> **WEBSITE 8.1 Regulating the cell cycle.** Cyclins and the cyclin-dependent kinases are critical in regulating cell division and integrating it with the activities of the nuclear envelope and DNA synthesis.

The cytoskeletal mechanisms of mitosis

Cleavage is actually the result of two coordinated cyclical processes. The first of these is **karyokinesis**: the mitotic division of the nucleus. The mechanical agent of karyokinesis is the **mitotic spindle**, with its microtubules composed of tubulin (the same type of protein that makes up the sperm flagellum). The second process is **cytokinesis**: the division of the cell. The mechanical agent of cytokinesis is a **contractile ring** of microfilaments made of actin (the same type of protein that extends the egg microvilli and the sperm acrosomal process). Table 8.1 presents a comparison of these agents of cell division. The relationship and coordination between these two systems during cleavage is depicted in Figure 8.3A, in which a sea urchin egg is shown undergoing first cleavage. The mitotic spindle and contractile ring are perpendicular to each other, and the spindle is internal to the contractile ring. The contractile ring creates a **cleavage furrow**, which eventually bisects the plane of mitosis, thereby creating two genetically equivalent blastomeres.

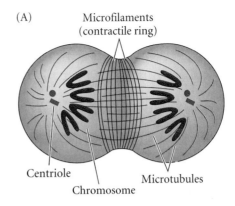

(A)

Microfilaments (contractile ring)

Centriole

Chromosome

Microtubules

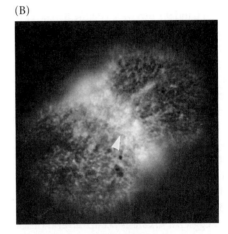

(B)

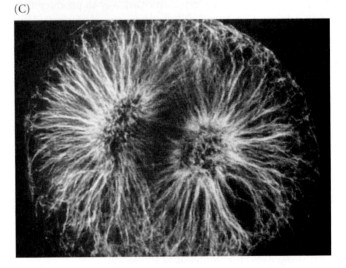

(C)

Figure 8.3
Role of microtubules and microfilaments in cell division. (A) Diagram of first-cleavage telophase. The chromosomes are being drawn to the centrioles by microtubules, while the cytoplasm is being pinched in by the contraction of microfilaments. (B) Fluorescent labeling of the actin microfilaments shows the contractile ring in the first-cleavage furrow (arrowhead) of a telophase sea urchin egg. (C) Fluorescent labeling of tubulin shows the microtubular asters of a sea urchin egg during telophase of first cleavage. (B from Bonder et al. 1988; C from White et al. 1988; photographs courtesy of the authors.)

TABLE 8.1 Karyokinesis and cytokinesis

Process	Mechanical agent	Major protein composition	Location	Major disruptive drug
Karyokinesis	Mitotic spindle	Tubulin microtubules	Central cytoplasm	Colchicine, nocodazole[a]
Cytokinesis	Contractile ring	Actin microfilaments	Cortical cytoplasm	Cytochalasin B

[a]Because colchicine has been found to independently inhibit several membrane functions, including osmoregulation and the transport of ions and nucleosides, nocodazole has become the major drug used to inhibit microtubule-mediated processes (see Hardin 1987).

The actin microfilaments are found in the cortex (outer cytoplasm) of the egg rather than in the central cytoplasm. Under the electron microscope, the ring of microfilaments can be seen forming a distinct cortical band 0.1 μm wide (Figure 8.3B). This contractile ring exists only during cleavage and extends 8–10 μm into the center of the egg. It is responsible for exerting the force that splits the zygote into blastomeres; if it is disrupted, cytokinesis stops. Schroeder (1973) likened the contractile ring to an "intercellular purse-string," tightening about the egg as cleavage continues. This tightening of the microfilamentous ring creates the cleavage furrow. Microtubules are also seen near the cleavage furrow (in addition to their role in creating the mitotic spindle), since they are needed to bring membrane material to the site of membrane addition (Danilchik et al. 1998).

Although karyokinesis and cytokinesis are usually coordinated, they are sometimes separated by natural or experimental conditions. In insect eggs, karyokinesis occurs several times before cytokinesis takes place. Another way to produce this state is to treat embryos with the drug cytochalasin B. This drug inhibits the formation and organization of microfilaments in the contractile ring, thereby stopping cleavage without stopping karyokinesis (Schroeder 1972).

Patterns of embryonic cleavage

In 1923, embryologist E. B. Wilson reflected on how little we knew about cleavage: "To our limited intelligence, it would seem a simple task to divide a nucleus into equal parts. The cell, manifestly, entertains a very different opinion." Indeed, different organisms undergo cleavage in distinctly different ways. The pattern of embryonic cleavage peculiar to a species is determined by two major parameters: (1) the amount and distribution of yolk protein within the cytoplasm, and (2) factors in the egg cytoplasm that influence the angle of the mitotic spindle and the timing of its formation.

The amount and distribution of yolk determines where cleavage can occur and the relative size of the blastomeres. When one pole of the egg is relatively yolk-free, cellular divisions occur there at a faster rate than at the opposite pole. The yolk-rich pole is referred to as the **vegetal pole**; the yolk concentration in the **animal pole** is relatively low. The zygote nucleus is frequently displaced toward the animal pole. In general, yolk inhibits cleavage. Figure 8.4 provides a classification of cleavage types and shows the influence of yolk on cleavage symmetry and pattern.

At one extreme are the eggs of sea urchins, mammals, and snails. These eggs have sparse, equally spaced yolk and are thus **isolecithal** (Greek, "equal yolk"). In these species, cleavage is **holoblastic** (Greek *holos*, "complete"), meaning that the cleavage furrow extends through the entire egg. With little yolk, these embryos must have some other way of obtaining food. Most will generate a voracious larval form, while mammals get their nutrition from the maternal placenta.

At the other extreme are the eggs of insects, fishes, reptiles, and birds. Most of their cell volumes are made up of yolk. The yolk must be sufficient to nourish these animals throughout embryonic development. Zygotes containing large accumulations of yolk undergo **meroblastic** cleavage (Greek *meros*, part), wherein only a portion of the cytoplasm is cleaved. The cleavage furrow does not penetrate into the yolky portion of the cytoplasm, since the yolk platelets impede membrane formation there. The eggs of insects have their yolk in the center (i.e., they are **centrolecithal**), and the divisions of the cytoplasm occur only in the rim of cytoplasm around the periphery of the cell (i.e., **superficial cleavage**). The eggs of birds and fishes have only one small area of the egg that is free of yolk (**telolecithal** eggs), and therefore, the cell divisions occur only in this small disc of cytoplasm, giving rise to the **discoidal** pattern of cleavage. These are general rules, however, and closely related species can evolve different patterns of cleavage in different environments.

Yolk, however, is just one factor influencing a species' pattern of cleavage. There are also inherited patterns of cell division that are superimposed upon the constraints of the yolk. This can readily be seen in isolecithal eggs, in which very little yolk is present. In the absence of a large concentration of yolk, four major cleavage types can be observed: radial holoblastic, spiral holoblastic, bilateral holoblastic, and rotational holoblastic cleavage. We will see examples of these cleavage patterns below when we take a more detailed look at the early development of four different invertebrate groups.

VADE MECUM[2] **Cleavage patterns.** The quantities of yolk within an egg influence the pattern of cleavage. Eggs from various organisms commonly studied in developmental biology illustrate this, and are seen in labeled photographs and time-lapse movies.
[Click on Sea Urchin; Fruit Fly; Chick-Early; and Amphibian]

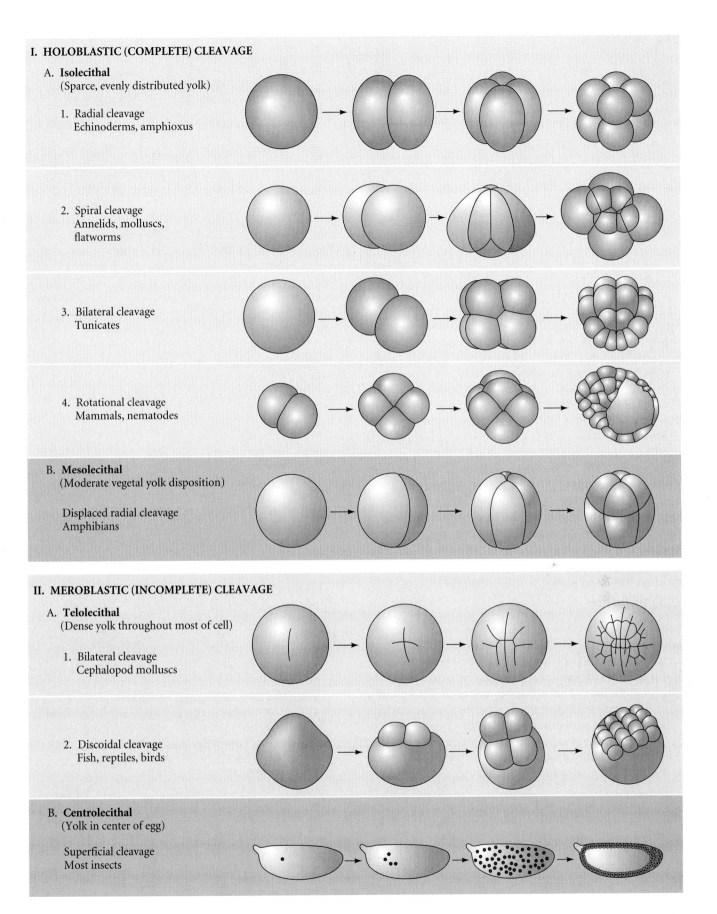

I. HOLOBLASTIC (COMPLETE) CLEAVAGE

  A. **Isolecithal**
    (Sparce, evenly distributed yolk)

      1. Radial cleavage
         Echinoderms, amphioxus

      2. Spiral cleavage
         Annelids, molluscs,
         flatworms

      3. Bilateral cleavage
         Tunicates

      4. Rotational cleavage
         Mammals, nematodes

  B. **Mesolecithal**
    (Moderate vegetal yolk disposition)

    Displaced radial cleavage
    Amphibians

II. MEROBLASTIC (INCOMPLETE) CLEAVAGE

  A. **Telolecithal**
    (Dense yolk throughout most of cell)

      1. Bilateral cleavage
         Cephalopod molluscs

      2. Discoidal cleavage
         Fish, reptiles, birds

  B. **Centrolecithal**
    (Yolk in center of egg)

    Superficial cleavage
    Most insects

Figure 8.4
Summary of the main patterns of cleavage.

Invagination:
Infolding of cell sheet into embryo

Example:
Sea urchin endoderm

Involution:
Inturning of cell sheet over the basal surface of an outer layer

Example:
Amphibian mesoderm

Ingression:
Migration of individual cells into the embryo

Example:
Sea urchin mesoderm, *Drosophila* neuroblasts

Delamination:
Splitting or migration of one sheet into two sheets

Example:
Mammalian and bird hypoblast formation

Epiboly:
The expansion of one cell sheet over other cells

Example:
Ectoderm formation in amphibians, sea urchins, and tunicates

Figure 8.5
Types of cell movements during gastrulation. The gastrulation of any particular organism is an ensemble of several of these movements.

Specification of cell fates during cleavage

Cell fates can be specified by cell-cell interactions or by the asymmetric distribution of patterning molecules into particular cells. These unevenly distributed molecules are usually transcription factors that activate or repress the transcription of specific genes in those cells that acquire them. Such asymmetric distributions of patterning molecules can happen during cleavage and generally follow one of three mechanisms: (1) the molecules are bound to the egg cytoskeleton and are passively acquired by the cells that obtain this cytoplasm; (2) the molecules are actively transported along the cytoskeleton to one particular cell; or (3) the molecules becomes associated with a specific centrosome and follow that centrosome into one of the two mitotic sister cells (Lambert and Nagy 2002). Once asymmetry has been established, one cell can specify a neighboring cell by paracrine or juxtacrine interactions at the cell surface (see Chapter 6).

Gastrulation

Gastrulation is the process of highly coordinated cell and tissue movements whereby the cells of the blastula are dramatically rearranged. The blastula consists of numerous cells, the positions of which were established during cleavage. During gastrulation, these cells are given new positions and new neighbors, and the multilayered body plan of the organism is established. The cells that will form the endodermal and mesodermal organs are brought to the inside of the embryo, while the cells that will form the skin and nervous system are spread over its outside surface (see Chapter 2). Thus, the three germ layers—outer ectoderm, inner endoderm, and interstitial mesoderm—are first produced during gastrulation. In addition, the stage is set for the interactions of these newly positioned tissues.

The movements of gastrulation involve the entire embryo, and cell migrations in one part of the gastrulating embryo must be intimately coordinated with other movements occurring simultaneously. Although patterns of gastrulation vary enormously throughout the animal kingdom, there are only a few basic types of cell movements. Gastrulation usually involves some combination of the following types of movements (Figure 8.5):

- **Invagination.** The infolding of a region of cells, much like the indenting of a soft rubber ball when it is poked.
- **Involution.** The inturning or inward movement of an expanding outer layer so that it spreads over the internal surface of the remaining external cells.
- **Ingression.** The migration of individual cells from the surface layer into the interior of the embryo. The cells become mesenchymal (separate) and migrate independently.
- **Delamination.** The splitting of one cellular sheet into two more or less parallel sheets. While on a cellular basis it resembles ingression, the result is the formation of a new sheet of cells.
- **Epiboly.** The movement of epithelial sheets (usually of ectodermal cells) that spread as a unit, rather than individually, to enclose the deeper layers of the embryo. Epiboly can occur by the cells dividing, by the cells changing their shape, or by several layers intercalating into fewer layers. Often, all three mechanisms are used.

VADE MECUM[2] **Gastrulation movements.** These segments show time-lapse and real-time movies of gastrulation in several organisms, in which color coding of germ layers has been superimposed on the living embryo as well as on 3-D models. [**Click on Sea Urchin; Fruit Fly; Chick-Mid; and Amphibian**]

Cell Specification and Axis Formation

Embryos must develop three crucial axes that are the foundation of the body: the anterior-posterior axis, the dorsal-ventral axis, and the right-left axis (Figure 8.6).

The **anterior-posterior** (or **anteroposterior**) **axis** is the line extending from head to tail (or mouth to anus in those organisms that lack heads and tails). The **dorsal-ventral** (**dorsoven-**

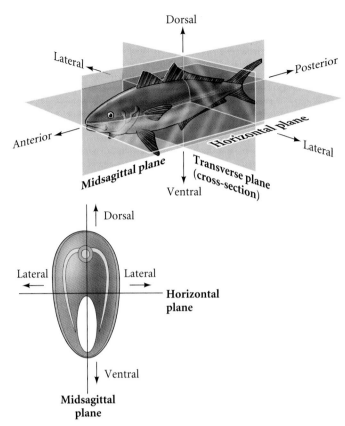

Figure 8.6
Axes of a bilaterally symmetrical animal. A single plane, the midsagittal plane, divides the animal into left and right halves. Cross sections are taken along the anterior-posterior axis.

tral) **axis** is the line extending from back (dorsum) to belly (ventrum). For instance, in vertebrates, the neural tube is a dorsal structure. In insects, the neural cord is a ventral structure. The **right-left axis** is a line between the two lateral sides of the body. Although humans (for example) may look symmetrical, recall that in most of us, the heart and liver are in the left half of the body only. Somehow, the embryo knows that some organs belong on one side and other organs go on the other.

In this chapter, we will look at how four invertebrates—the sea urchin, the tunicate, the snail, and a species of nematode—undergo cleavage, gastrulation, axis specification, and cell fate determination. These four invertebrates were chosen because they have been important **model systems** for developmental biologists. In other words, they can be studied easily in the laboratory, and they have special properties that allow their mechanisms of development to be readily observed. They also represent a wide variety of cleavage types, patterns of gastrulation, and ways of specifying axes and cell fates.*
Despite their differences, these invertebrate embryos are all

---

*However, model systems—by their very ability to develop in the laboratory—sometimes preclude our asking certain questions concerning the relationship of development to its habitat. These questions will be addressed in Chapter 22.

characterized by what Eric Davidson (2001) has called "Type I embryogenesis" (see Chapter 3), which includes:

- the immediate activation of the zygotic genes;
- the rapid specification of the blastomeres by the products of the zygotic genes and by maternally active genes;
- the relatively small number of cells (a few hundred or less) at gastrulation.

THE EARLY DEVELOPMENT OF SEA URCHINS

Cleavage in Sea Urchins

Sea urchins exhibit **radial holoblastic cleavage**. The first seven divisions are "stereotypic" in that the same pattern is followed in each individual within a species. The first and second cleavages are both meridional and are perpendicular to each other. That is to say, the cleavage furrows pass through the animal and vegetal poles. The third cleavage is equatorial, perpendicular to the first two cleavage planes, and separates the animal and vegetal hemispheres from each other (Figures 8.7 and 8.8). The fourth cleavage, however, is very different from the first three. The four cells of the animal tier divide meridionally into eight blastomeres, each with the same volume. These eight cells are called **mesomeres**. The vegetal tier, however, undergoes an unequal equatorial cleavage to produce four large cells, the **macromeres**, and four smaller **micromeres** at the vegetal pole (Figures 8.8, 8.9E; Summers et al. 1993). As the 16-cell embryo cleaves, the eight "animal" mesomeres divide equatorially to produce two tiers, an_1 and an_2, one staggered above the other. The macromeres divide meridionally, forming a tier of eight cells below an_2. The micromeres also divide, albeit somewhat later, producing a small cluster beneath the larger tier. At the sixth division, the animal hemisphere cells divide meridionally, while the vegetal cells divide equatorially; this pattern is reversed in the seventh division. At that time, the embryo is a 128-cell blastula, and the pattern of divisions becomes less regular.

Blastula formation

The blastula stage of sea urchin development begins at the 128-cell stage. Here the cells form a hollow sphere surrounding a central cavity, or **blastocoel** (Figure 8.8 D–F). By this time, all the cells are the same size, the micromeres having slowed down their cell divisions. Every cell is in contact with the proteinaceous fluid of the blastocoel on the inside and with the hyaline layer on the outside. Tight junctions unite the once loosely connected blastomeres into a seamless epithelial sheet that completely encircles the blastocoel (Dan-Sohkawa and Fujisawa 1980). As the cells continue to divide, the blastula remains one cell layer thick, thinning out as it expands. This is accomplished by the adhesion of the blastomeres to the hyaline layer and by an influx of water that expands the blastocoel (Dan 1960; Wolpert and Gustafson 1961; Ettensohn and Ingersoll 1992).

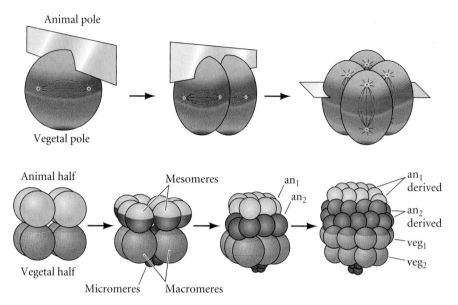

Animal pole

Vegetal pole

Animal half

Mesomeres

an_1
an_2

an_1 derived

an_2 derived

veg_1

veg_2

Vegetal half

Micromeres Macromeres

Figure 8.7
Cleavage in the sea urchin. Planes of cleavage in the first three divisions and the formation of tiers of cells in divisions 3–6.

Fate maps and the determination of sea urchin blastomeres

CELL FATE DETERMINATION. The first fate maps of the sea urchin embryo followed the descendants of each of the 16-cell-stage blastomeres. More recent investigations have refined these maps by following the fates of individual cells that have been injected with fluorescent dyes such as diI. These dyes "glow" in the injected cells' progeny for many cell divisions (see Chapter 1). Such studies have shown that by the 60-cell stage, most of the embryonic cell fates are specified, but the cells are not irreversibly committed. In other words, particular blastomeres consistently produce the same cell types in each embryo, but these cells remain pluripotent and can give rise to other cell types if experimentally placed in a different part of the embryo.

A fate map of the 60-cell sea urchin embryo is shown in Figure 8.9 (Logan and McClay 1999; Wray 1999). The animal half of the embryo consistently gives rise to the ectoderm—the larval skin and its neurons. The veg_1 layer produces cells that can enter into either the ectodermal or the endodermal organs. The veg_2 layer gives rise to cells that can populate three different structures—the endoderm, the coelom (the internal body wall), and secondary mesenchyme (pigment cells, immunocytes, and muscle cells). The first tier of micromeres produces the primary mesenchyme cells that form the larval skeleton,

These rapid and invariant cell cleavages last through the ninth or tenth division, depending on the species. By this time, the cells have become specified (discussed in the next section) and they end develop cilia. The ciliated blastula begins to rotate within the fertilization envelope. Soon afterward, differences are seen in the cells. The cells at the vegetal pole of the blastula begin to thicken, forming a **vegetal plate** (Figure 8.8F). The cells of the animal hemisphere synthesize and secrete a hatching enzyme that digests the fertilization envelope (Lepage et al. 1992). The embryo is now a free-swimming **hatched blastula**.

Figure 8.8
Photomicrographs of cleavage in live embryos of the sea urchin *Lytechinus variegatus*, seen from the side. (A) The 1-cell embryo (zygote). The site of sperm entry is marked with a black arrow, while a white arrow marks the vegetal pole. The fertilization envelope surrounding the embryo is clearly visible. (B) The 2-cell stage. (C) The 8-cell stage. (D) The 16-cell stage. (E) The 32-cell stage. Micromeres have formed at the vegetal pole. (F) The blastula has hatched from the fertilization envelope. The future vegetal plate (arrow) is beginning to thicken. (Photographs courtesy of J. Hardin.)

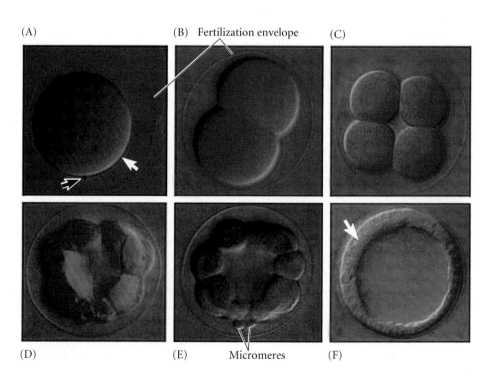

(A)   (B) Fertilization envelope   (C)

(D)   (E) Micromeres   (F)

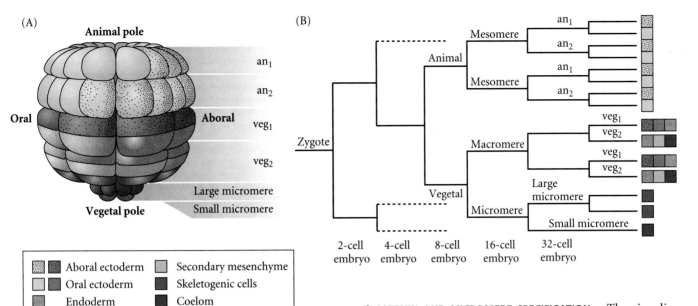

Figure 8.9

Fate map and cell lineage of the sea urchin *Strongylocentrotus purpuratus*. (A) The 60-cell embryo is shown, with the left side facing the viewer. Blastomere fates are segregated along the animal-vegetal axis of the egg. (B) Cell lineage map of the embryo. For simplicity, only one of the four embryonic cells is shown beyond second cleavage (solid lines). The veg$_1$ tier gives rise to both ectodermal and endodermal lineages, and the coelom comes from two sources: the second tier of micromeres and some veg$_2$ cells. (After Wray 1999.)

while the second tier of micromeres contributes cells to the coelom (Logan and McClay 1997 1999).

Although the early blastomeres have consistent fates in the larva, most of these fates are achieved by conditional specification. The only cells whose fates are determined autonomously are the skeletogenic micromeres. If these micromeres are isolated from the embryo and placed in test tubes, they will still form skeletal spicules. Moreover, if these micromeres are transplanted into the animal region of the blastula, not only will their descendants form skeletal spicules, but the transplanted micromeres will alter the fates of nearby cells by inducing a secondary site for gastrulation. Cells that would normally have produced ectodermal skin cells will be respecified as endoderm and will produce a secondary gut (Figure 8.10; Hörstadius 1973; Ransick and Davidson 1993). The micromeres appear to produce a signal that tells the cells adjacent to them to become endoderm and induces them to invaginate into the embryo. Their ability to reorganize the embryonic cells is so pronounced that if the isolated micromeres are recombined with an isolated animal cap (the top two animal tiers), the animal cap cells will generate endoderm, and a more or less normal larva will develop (Figure 8.11; Hörstadius 1939).

β-CATENIN AND MICROMERE SPECIFICATION. The signaling molecules involved in cell specification are just now being identified (Ettensohn and Sweet 2000). The molecule responsible for specifying the micromeres (and their ability to induce the neighboring cells) appears to be **β-catenin**. As we saw in Chapter 6, β-catenin is a transcription factor that is often activated by the Wnt pathway. Several pieces of evidence suggest it specifies the micromeres. First, during normal sea urchin development, β-catenin accumulates in the nuclei of those cells fated to become endoderm and mesoderm (Figure 8.12A). This accumulation is autonomous and can occur even if the micromere precursors are separated from the rest of the embryo. Second, this nuclear accumulation appears to be responsible for specifying the vegetal half of the embryo. It is possible that the levels of nuclear β-catenin accumulation help to determine the mesodermal and endodermal fates of the vegetal cells. Treating sea urchin embryos with lithium chloride causes the accumulation of β-catenin in every cell, and this treatment also transforms the presumptive ectoderm into endoderm. Conversely, experimental procedures that inhibit β-catenin accumulation in the vegetal cell nuclei prevent the formation of endoderm and mesoderm (Figure 8.12B,C; Logan et al. 1998; Wikramanayake et al. 1998).

Finally, β-catenin is essential for giving the micromeres their inductive ability.* The experiments described in the previous section demonstrated that the micromeres were able to induce a second embryonic axis when transplanted to the animal hemisphere. However, micromeres from embryos in which β-catenin was prevented from entering the nucleus were unable to induce the animal cells to form endoderm, and a second axis was not formed (Logan et al. 1998). The β-catenin binds to a TCF transcription factor in the nucleus, allowing it to function in gene activation (Vonica et al. 2000).

---

*β-catenin probably activates the genes necessary for producing the inducing signal, and a Notch ligand (such as Delta) may be one of the signal's components.

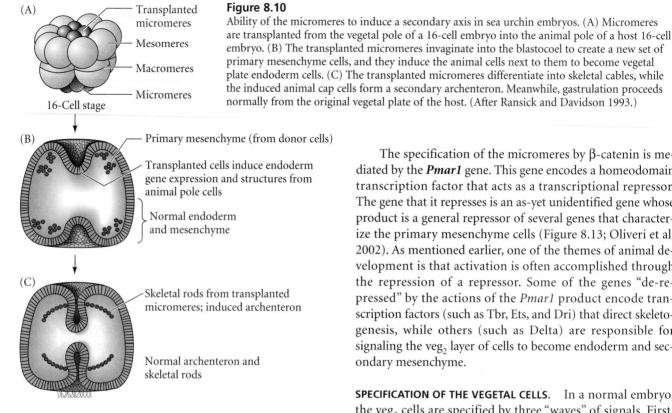

(A)
Transplanted micromeres
Mesomeres
Macromeres
Micromeres
16-Cell stage

(B)
Primary mesenchyme (from donor cells)
Transplanted cells induce endoderm gene expression and structures from animal pole cells
Normal endoderm and mesenchyme

(C)
Skeletal rods from transplanted micromeres; induced archenteron
Normal archenteron and skeletal rods

Figure 8.10

Ability of the micromeres to induce a secondary axis in sea urchin embryos. (A) Micromeres are transplanted from the vegetal pole of a 16-cell embryo into the animal pole of a host 16-cell embryo. (B) The transplanted micromeres invaginate into the blastocoel to create a new set of primary mesenchyme cells, and they induce the animal cells next to them to become vegetal plate endoderm cells. (C) The transplanted micromeres differentiate into skeletal cables, while the induced animal cap cells form a secondary archenteron. Meanwhile, gastrulation proceeds normally from the original vegetal plate of the host. (After Ransick and Davidson 1993.)

The specification of the micromeres by β-catenin is mediated by the ***Pmar1*** gene. This gene encodes a homeodomain transcription factor that acts as a transcriptional repressor. The gene that it represses is an as-yet unidentified gene whose product is a general repressor of several genes that characterize the primary mesenchyme cells (Figure 8.13; Oliveri et al. 2002). As mentioned earlier, one of the themes of animal development is that activation is often accomplished through the repression of a repressor. Some of the genes "de-repressed" by the actions of the *Pmar1* product encode transcription factors (such as Tbr, Ets, and Dri) that direct skeletogenesis, while others (such as Delta) are responsible for signaling the veg$_2$ layer of cells to become endoderm and secondary mesenchyme.

SPECIFICATION OF THE VEGETAL CELLS. In a normal embryo, the veg$_2$ cells are specified by three "waves" of signals. First, moderate levels of β-catenin specify the cells to be "endomesoderm." Then two sets of inductive signals issue from the micromeres. The first of these is an "early veg$_2$ signal" secreted by the micromeres immediately upon their formation at fourth cleavage. This signal amplifies the mesendoderm specification established by the β-catenin. Next, the Delta protein on the micromeres activates the Notch pathway in the adjacent veg$_2$ cells. The Notch pathway causes these cells to become secondary mesenchyme rather than endoderm. It appears that both these signals are needed in this particular order, and if either one is not present, the veg$_2$ cells fail to make secondary mesenchyme or endoderm (Ransick and Davidson 1995; Sherwood and McClay 1999; Sweet et al. 1999). The identity of the early signal is not yet known, but it may create competence in the veg$_2$ cells to respond to the second signal. Finally, Wnt8 appears to be made by the endoderm cells (i.e., those endomesoderm cells *not* receiving the Delta signal). Wnt8 appears to act in an autocrine manner to boost the specification of the endoderm (D. McClay, personal communication).

(A) Normal development
an$_1$
an$_2$ } Animal hemisphere
veg$_1$
veg$_2$
Micromeres
Pluteus larva

(B) Animal hemishere alone
Dauerblastula
Complete animalization

(C) Animal hemishere and micromeres
Recognizable larva; endoderm from animal layers

Figure 8.11

Ability of the micromeres to induce presumptive ectodermal cells to acquire other fates. (A) Normal development of the 64-cell sea urchin embryo, showing the fates of the different layers. (B) An isolated animal hemisphere becomes a ciliated ball of ectodermal cells. (C) When an isolated animal hemisphere is combined with isolated micromeres, a recognizable pluteus larva is formed, with all the endoderm derived from the animal hemisphere. (After Hörstadius 1939.)

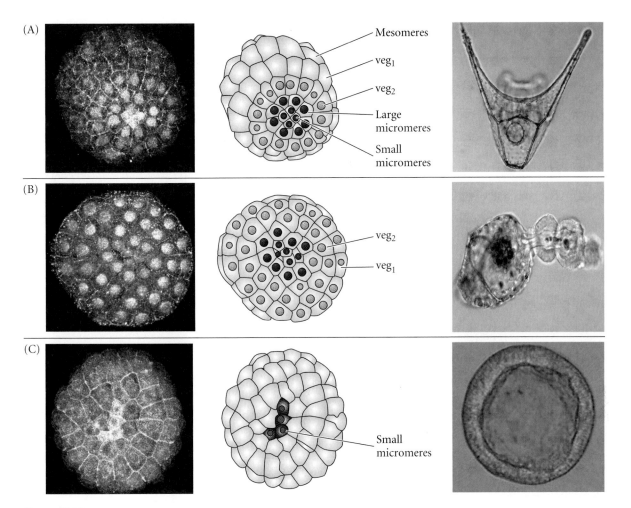

(A)

Mesomeres
veg₁
veg₂
Large micromeres
Small micromeres

(B)

veg₂
veg₁

(C)

Small micromeres

Figure 8.12
The role of β-catenin in specifying the vegetal cells of the sea urchin embryo. β-catenin is stained by a fluorescently labeled antibody. (A) During normal development, β-catenin accumulates predominantly in the micromeres and somewhat less in the veg₂ tier cells. (B) When lithium chloride treatment permits β-catenin to accumulate in the nuclei of all blastula cells (probably by blocking the GSK-3 enzyme of the Wnt pathway), the animal cells become specified as endoderm and mesoderm. (C) When β-catenin is prevented from entering the nuclei, (i.e., it remains in the cytoplasm), the vegetal cell fates are not specified, and the entire embryo develops as a ciliated ectodermal ball. (After Logan et al. 1998; photographs courtesy of D. McClay.)

DIFFERENTIATION: COMBINATIONS OF TRANSCRIPTION FACTORS. Once these transcription factors are present in their specific regions, they can activate the genes that characterize the different cell types in the embryo. One of the best-studied of these genes is that for the endodermal protein endo16. Endo16 appears to be a secreted product of the endodermal cells and has no known function. However, it serves as a marker protein for gut-specific tissues. Writing about the *endo16* research, Eric Davidson (2001, p. 54) notes, "What emerges is astounding: a network of logic interactions programmed into the DNA sequence that amounts essentially to a hardwired computational device." An outline of that computational logic is shown in Figure 8.14. Figure 8.14A outlines the specification of the endoderm, while Figure 8.14B shows a portion of the regulatory network that activates the *endo16* gene. The *endo16* upstream regulatory region contains seven modular elements to which at least thirteen different transcription factors bind. Some of these modular regions activate *endo16* transcription, while other modules act to inhibit *endo16* transcription. In Figure 8.14B, modules 1 and 2 activate *endo16* expression in the vegetal cells (prior to gastrulation) and in the midgut (during and after gastrulation). The inhibitory modules are activated in those cells other than endoderm. Thus, the different regions of the *endo16* regulatory region act synergistically to make certain that the *endo16* gene is expressed only in the midgut endoderm (Figure 8.14C) and in no other cell type (Yuh et al. 2001; Davidson et al. 2002).

AXIS SPECIFICATION. In the sea urchin blastula, the cell fates line up along the animal-vegetal axis established in the egg cytoplasm prior to fertilization. The animal-vegetal axis also appears to structure the future anterior-posterior axis, with the vegetal region sequestering those maternal components necessary for posterior development (Boveri 1901; Maruyama et al. 1985).

In most sea urchins, the dorsal-ventral and left-right axes are specified after fertilization, but the manner of their specification is not well understood. The oral-aboral (ventral-

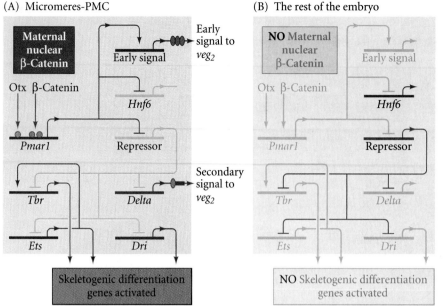

(A) Micromeres-PMC

Maternal nuclear β-Catenin

Early signal → Early signal to veg₂

Otx β-Catenin

Hnf6

Pmar1 Repressor

Tbr Delta → Secondary signal to veg₂

Ets Dri

Skeletogenic differentiation genes activated

(B) The rest of the embryo

NO Maternal nuclear β-Catenin

Early signal

Otx β-Catenin

Hnf6

Pmar1 Repressor

Tbr Delta

Ets Dri

NO Skeletogenic differentiation genes activated

Figure 8.13

The micromere regulatory network proposed by Davidson and colleagues (2002). (A) In the precursors of the primary mesenchyme cells in the 16-cell embryo, β-catenin and the ubiquitous maternal Otx transcription factor activate the *Pmar1* gene. The *Pmar1* product is a transcriptional repressor that specifically represses a gene encoding a "global repressor" of numerous genes that characterize the primary mesenchyme cells. Pmar1 therefore allows the de-repression of genes such as *Tbr*, *Ets*, and *Dri* (which turn on the skeleton-forming genes) and allows the de-repression of genes encoding Delta and the "early veg₂ signal" (which instruct the veg₂ layer to form endoderm and secondary mesenchyme cells). (B) In all the other cells of the embryo, *Pmar1* is not activated, since β-catenin is not expressed there and the maternal Otx protein is not translocated into those cells. Therefore, the repressor protein is transcribed and blocks activation of the micromere specification pathway. (After Oliveri et al. 2002.)

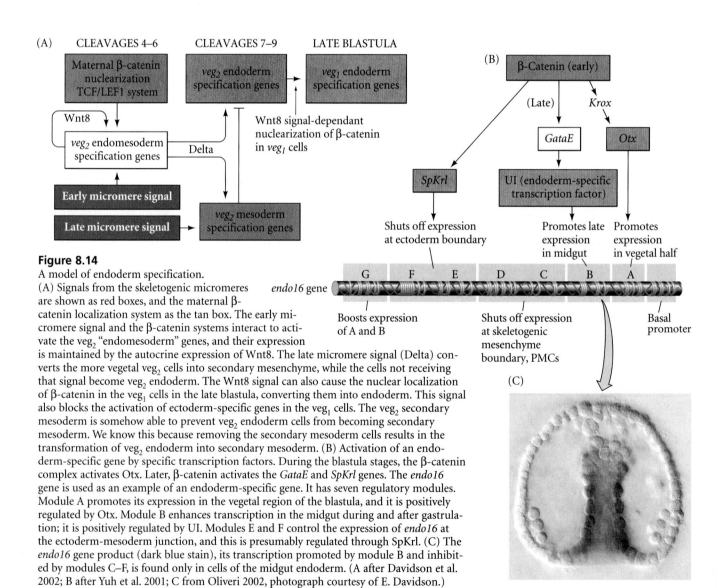

(A)

CLEAVAGES 4–6   CLEAVAGES 7–9   LATE BLASTULA

Maternal β-catenin nuclearization TCF/LEF1 system

veg₂ endoderm specification genes → veg₁ endoderm specification genes

Wnt8

veg₂ endomesoderm specification genes

Delta

Wnt8 signal-dependant nuclearization of β-catenin in veg₁ cells

Early micromere signal

Late micromere signal

veg₂ mesoderm specification genes

(B)

β-Catenin (early)

(Late)   Krox

GataE   Otx

SpKrl   UI (endoderm-specific transcription factor)

Shuts off expression at ectoderm boundary

Promotes late expression in midgut

Promotes expression in vegetal half

G F E D C B A

endo16 gene

Boosts expression of A and B

Shuts off expression at skeletogenic mesenchyme boundary, PMCs

Basal promoter

(C)

Figure 8.14

A model of endoderm specification.
(A) Signals from the skeletogenic micromeres are shown as red boxes, and the maternal β-catenin localization system as the tan box. The early micromere signal and the β-catenin systems interact to activate the veg₂ "endomesoderm" genes, and their expression is maintained by the autocrine expression of Wnt8. The late micromere signal (Delta) converts the more vegetal veg₂ cells into secondary mesenchyme, while the cells not receiving that signal become veg₂ endoderm. The Wnt8 signal can also cause the nuclear localization of β-catenin in the veg₁ cells in the late blastula, converting them from endoderm. This signal also blocks the activation of ectoderm-specific genes in the veg₁ cells. The veg₂ secondary mesoderm is somehow able to prevent veg₂ endoderm cells from becoming secondary mesoderm. We know this because removing the secondary mesoderm cells results in the transformation of veg₂ endoderm into secondary mesoderm. (B) Activation of an endoderm-specific gene by specific transcription factors. During the blastula stages, the β-catenin complex activates Otx. Later, β-catenin activates the *GataE* and *SpKrl* genes. The *endo16* gene is used as an example of an endoderm-specific gene. It has seven regulatory modules. Module A promotes its expression in the vegetal region of the blastula, and it is positively regulated by Otx. Module B enhances transcription in the midgut during and after gastrulation; it is positively regulated by UI. Modules E and F control the expression of *endo16* at the ectoderm-mesoderm junction, and this is presumably regulated through SpKrl. (C) The *endo16* gene product (dark blue stain), its transcription promoted by module B and inhibited by modules C–F, is found only in cells of the midgut endoderm. (A after Davidson et al. 2002; B after Yuh et al. 2001; C from Oliveri 2002, photograph courtesy of E. Davidson.)

dorsal) axis of the sea urchin embryo usually is delineated by the first cleavage plane. Lineage tracer dye injection into one blastomere at the 2-cell stage demonstrated that in nearly all cases, the oral pole of the future oral-aboral axis lay 45 degrees clockwise from the first cleavage plane as viewed from the animal pole (Cameron et al. 1989). Interestingly, in those sea urchins that bypass the larval stage to develop directly into juveniles, the dorsal-ventral axis is specified maternally in the egg cytoplasm (Henry and Raff 1990).

> **WEBSITE 8.2 Sea urchin cell specification.** The specification of sea urchin cells was one of the first major research projects in experimental embryology and remains a fascinating area of research. It appears that the initial signaling parses the blastula into domains characterized by the expression of specific transcription factors.

Sea Urchin Gastrulation

The late sea urchin blastula consists of a single layer of about 1000 cells that form a hollow ball, somewhat flattened at the vegetal end. The blastomeres, derived from different regions of the zygote, have different sizes and properties. Figures 8.15 and 8.16 show the fates of the various regions of the blastula as it develops through gastrulation to the **pluteus larva** stage characteristic of sea urchins. The fate of each cell layer can be seen through its movements during gastrulation.

Ingression of primary mesenchyme

FUNCTION OF PRIMARY MESENCHYME CELLS. Shortly after the blastula hatches from its fertilization envelope, the vegetal side of the spherical blastula begins to thicken and flatten (Figure 8.16, 9 hours). At the center of this flat vegetal plate, a cluster of small cells begins to change. These cells begin extending and contracting long, thin (30×5 μm) processes called

filopodia from their inner surfaces. The cells then dissociate from the epithelial monolayer and ingress into the blastocoel (Figure 8.16, 9–10 hours). These cells, derived from the micromeres, are called the **primary mesenchyme**. They will form the larval skeleton, so they are sometimes called the **skeletogenic mesenchyme**. At first the cells appear to move randomly along the inner blastocoel surface, actively making and breaking filopodial connections to the wall of the blastocoel. Eventually, however, they become localized within the prospective ventrolateral region of the blastocoel. Here they fuse into syncytial cables, which will form the axis of the calcium carbonate spicules of the larval skeleton (Figure 8.17).

IMPORTANCE OF EXTRACELLULAR LAMINA INSIDE THE BLASTOCOEL. The ingression of the micromere descendants into the blastocoel is a result of these cells losing their affinity for their neighbors and for the hyaline membrane; instead they acquire a strong affinity for a group of proteins that line the blastocoel. This model was first proposed by Gustafson and Wolpert (1967) and was confirmed in 1985, when Rachel Fink and David McClay measured the strengths of sea urchin blastomere adhesion to the hyaline layer, to the basal lamina lining the blastocoel, and to other blastomeres.

Originally, all the cells of the blastula are connected on their outer surface to the hyaline layer and on their inner surface to a basal lamina secreted by the cells. On their lateral surfaces, each cell has another cell for a neighbor. Fink and McClay found that the prospective ectoderm and endoderm cells (descendants of the mesomeres and macromeres, respectively) bind tightly to one another and to the hyaline layer, but adhere only loosely to the basal lamina (Table 8.2). The micromeres originally display a similar pattern of binding. However, the micromere pattern changes at gastrulation. Whereas the other cells retain their tight binding to the hyaline layer and to their neighbors, the primary mesenchyme precursors

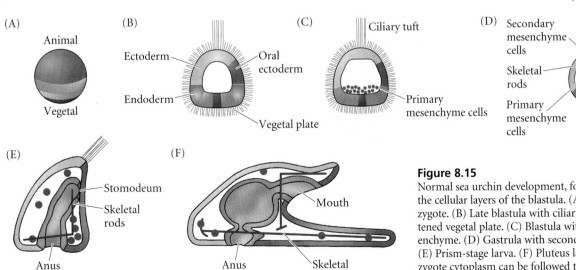

Figure 8.15
Normal sea urchin development, following the fate of the cellular layers of the blastula. (A) Fate map of the zygote. (B) Late blastula with ciliary tuft and flattened vegetal plate. (C) Blastula with primary mesenchyme. (D) Gastrula with secondary mesenchyme. (E) Prism-stage larva. (F) Pluteus larva. Fates of the zygote cytoplasm can be followed through the color pattern (see Figure 8.11). (Courtesy of D. McClay.)

lose their affinity for these structures (which drops to about 2% of its original value), while their affinity for components of the basal lamina and extracellular matrix (such as fibronectin) increases a hundredfold. This change in affinity causes the micromeres to release their attachments to the external hyaline layer and to their neighboring cells and, drawn in by the basal lamina, to migrate up into the blastocoel (Figures 8.17 and 8.18). These changes in affinity have been correlated with changes in cell surface molecules that occur during this time (Wessel and McClay 1985), and proteins such as fibronectin, integrin, laminin, L1, and cadherins have been shown to be involved in cellular ingression.

As shown in Figure 8.17, there is a heavy concentration of extracellular material around the ingressing primary mesenchyme cells (Galileo and Morrill 1985; Cherr et al. 1992).

TABLE 8.2 Affinities of mesenchymal and nonmesenchymal cells to cellular and extracellular components[a]

Cell type	Dislodgment force (in dynes)		
	Hyaline	Gastrula cell monolayers	Basal lamina
16-cell-stage micromeres	5.8×10^{-5}	6.8×10^{-5}	4.8×10^{-7}
Migratory-stage mesenchyme cells	1.2×10^{-7}	1.2×10^{-7}	1.5×10^{-5}
Gastrula ectoderm and endoderm	5.0×10^{-5}	5.0×10^{-5}	5.0×10^{-7}

Source: After Fink and McClay 1985.

[a] Tested cells were allowed to adhere to plates containing hyaline, extracellular basal lamina, or cell monolayers. The plates were inverted and centrifuged at various strengths to dislodge the cells. The dislodgement force is calculated from the centrifugal force needed to remove the test cells from the substrate.

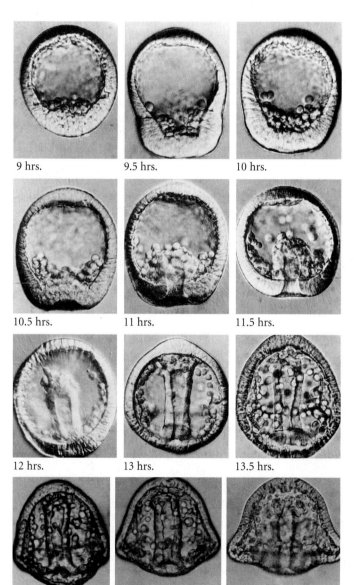

9 hrs. 9.5 hrs. 10 hrs.

10.5 hrs. 11 hrs. 11.5 hrs.

12 hrs. 13 hrs. 13.5 hrs.

15 hrs. 17 hrs. 18 hrs.

Once inside the blastocoel, the primary mesenchyme cells appear to migrate along the extracellular matrix of the blastocoel wall, extending their filopodia in front of them (Galileo and Morrill 1985; Karp and Solursh 1985). Several proteins (including fibronectin and a particular sulfated glycoprotein) are necessary to initiate and maintain this migration (Wessel et al. 1984; Sugiyama 1972; Lane and Solursh 1991; Berg et al. 1996).

But these guidance cues cannot be sufficient, since the migrating cells "know" when to stop their movement and form spicules near the equator of the blastocoel. The primary mesenchyme cells arrange themselves in a ring at a specific position along the animal-vegetal axis. At two sites near the future ventral side of the larva, many of these primary mesenchyme cells cluster together, fuse with one another, and initiate spicule formation (Figure 8.17A; Hodor

Figure 8.16
Entire sequence of gastrulation in *Lytechinus variegatus*. The times show the length of development at 25°C. (Photographs courtesy of J. Morrill; pluteus larva courtesy of G. Watchmaker.)

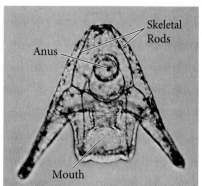

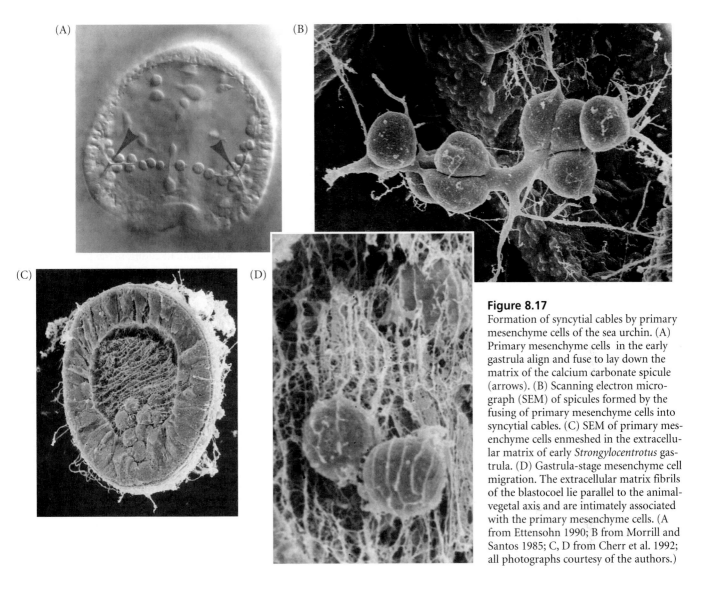

Figure 8.17
Formation of syncytial cables by primary mesenchyme cells of the sea urchin. (A) Primary mesenchyme cells in the early gastrula align and fuse to lay down the matrix of the calcium carbonate spicule (arrows). (B) Scanning electron micrograph (SEM) of spicules formed by the fusing of primary mesenchyme cells into syncytial cables. (C) SEM of primary mesenchyme cells enmeshed in the extracellular matrix of early *Strongylocentrotus* gastrula. (D) Gastrula-stage mesenchyme cell migration. The extracellular matrix fibrils of the blastocoel lie parallel to the animal-vegetal axis and are intimately associated with the primary mesenchyme cells. (A from Ettensohn 1990; B from Morrill and Santos 1985; C, D from Cherr et al. 1992; all photographs courtesy of the authors.)

Figure 8.18
Ingression of primary mesenchyme cells. (A–E) Interpretative diagrams depicting changes in the adhesive affinities of the presumptive primary mesenchyme cells (pink). These cells lose their affinities for hyalin and for their neighboring blastomeres while gaining an affinity for the proteins of the basal lamina. Non-mesenchymal blastomeres retain their original high affinities for the hyaline layer and neighboring cells. (F) SEM montage showing the ingression of the primary mesenchyme cells in *Lytechinus variegatus*. (F courtesy of J. B. Morrill and D. Flaherty.)

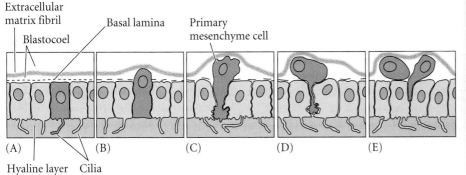

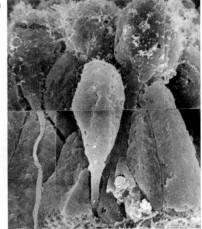

and Ettensohn 1998). If a labeled micromere from another embryo is injected into the blastocoel of a gastrulating sea urchin embryo, it migrates to the correct location and contributes to the formation of the embryonic spicules (Figure 8.19; Ettensohn 1990). It is thought that the necessary positional information is provided by the prospective ectodermal cells and their basal laminae (Harkey and Whiteley 1980; Armstrong et al. 1993; Malinda and Ettensohn 1994). Only the primary mesenchyme cells (and not other cell types or latex beads) are capable of responding to these patterning cues (Ettensohn and McClay 1986). Miller and colleagues (1995) have reported the existence of extremely fine (0.3 μm diameter) filopodia on the skeleton-forming mesenchyme cells. These filopodia are not thought to function in locomotion; rather, they appear to explore and sense the blastocoel wall and may be responsible for picking up dorsal-ventral and animal-vegetal patterning cues from the ectoderm (Figure 8.19; Malinda et al. 1995).

First stage of archenteron invagination

As the primary mesenchyme cells leave the vegetal region of the blastocoel, important changes are occurring in the cells that remain at the vegetal plate. These cells remain bound to

(A)

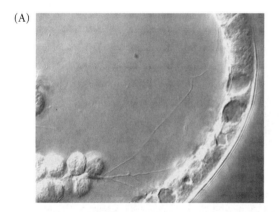

(B)

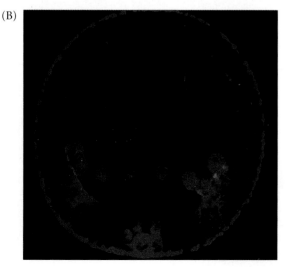

one another and to the hyaline layer of the egg, and they move to fill the gaps caused by the ingression of the primary mesenchyme. Moreover, the vegetal plate bends inward and invaginates about one-fourth to one-half the way into the blastocoel (Figure 8.20A; see also Figure 8.16, 10.5–11.5 hours). Then invagination suddenly ceases. The invaginated region is called the **archenteron** (primitive gut), and the opening of the archenteron at the vegetal pole is called the **blastopore**.

Invagination appears to be caused by shape changes in the vegetal plate cells and in the extracellular matrix underlying them. Kimberly and Hardin (1998) have shown that a group of vegetal plate cells surrounding the 2-8 cells at the vegetal pole become bottle-shaped, constricting their apical ends. This change causes the cells to pucker inward. Destroying these cells with lasers retards gastrulation. In addition, the hyaline layer at the vegetal plate buckles inward due to changes in its composition (Lane et al. 1993). The hyaline layer is actually made up of two layers, an outer lamina made primarily of hyalin protein and an inner lamina composed of fibropellin proteins (Hall and Vacquier 1982; Bisgrove et al. 1991). Fibropellins are stored in secretory granules within the oocyte and are secreted from those granules after cortical granule exocytosis releases the hyalin protein. By the blastula stage, the fibropellins have formed a meshlike network over the embryo surface. At the time of invagination, the vegetal plate cells (and only those cells) secrete a chondroitin sulfate proteoglycan into the inner lamina of the hyaline layer directly beneath them. This hygroscopic (water-absorbing) molecule swells the inner lamina, but not the outer lamina, which causes the vegetal region of the hyaline layer to buckle (Figure 8.20B,C). Slightly later, a second force arising from the movements of epithelial cells adjacent to the vegetal plate may facilitate invagination by drawing the buckled layer inward (Burke et al. 1991).

At the stage when the skeletogenic mesenchyme cells begin ingressing into the blastocoel, the fates of the vegetal plate cells have already been specified (Figure 8.21; Ruffins and Ettensohn 1996). The endodermal cells adjacent to the micromere-derived mesenchyme become foregut, migrating the farthest distance into the blastocoel. The next layer of endodermal cells becomes midgut, and the last circumferential row to invaginate forms the hindgut and anus.

Figure 8.19
Localization of the primary mesenchyme cells. (A) Nomarski videomicrograph showing a long, thin filopodium extending from a primary mesenchyme cell to the ectodermal wall of the gastrula, as well as a shorter filopodium extending inward from the ectoderm. The mesenchymal filopodia extend through the extracellular matrix and directly contact the cell membrane of the ectodermal cells. (B) The localization of the micromeres to form the calcium carbonate skeleton is determined by the ectodermal cells. The primary mesenchyme cells are stained green, while β-catenin is stained red. The primary mesenchyme cells appear to accumulate in those regions characterized by high β-catenin concentrations. (A from Miller et al. 1995; photographs courtesy of J. R. Miller and D. McClay.)

(A)

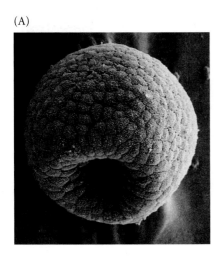

(B)

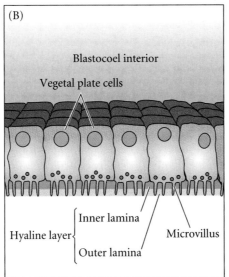

Blastocoel interior

Vegetal plate cells

Hyaline layer { Inner lamina / Outer lamina

Microvillus

(C)

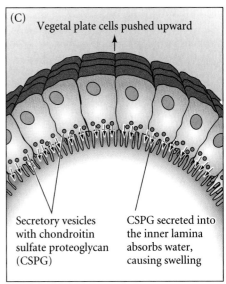

Vegetal plate cells pushed upward

Secretory vesicles with chondroitin sulfate proteoglycan (CSPG)

CSPG secreted into the inner lamina absorbs water, causing swelling

Figure 8.20
Invagination of the vegetal plate. (A) Vegetal plate invagination in *Lytechinus variegatus*, seen by scanning electron microscopy of the external surface of the early gastrula. The blastopore is clearly visible. (B) The hyaline layer consists of inner and outer laminae. Microvilli from the vegetal plate cells extend through the hyaline layer, and their cytoplasm contains secretory vesicles that store a chondroitin sulfate proteoglycan (CSPG). (C) The storage granules secrete CSPG into the inner lamina of the hyaline layer. The CSPG absorbs water and swells the inner lamina, while the outer lamina, to which it is attached, does not swell. This causes the hyaline layer and its attached epithelium to buckle inward. (A from Morrill and Santos 1985, courtesy of J. B. Morrill; B and C after Lane et al. 1993.)

Second and third stages of archenteron invagination

The invagination of the vegetal cells occurs in discrete stages. After a brief pause following the initial invagination, the second phase of archenteron formation begins. During this stage, the archenteron extends dramatically, sometimes tripling its length. In this process of extension, the wide, short gut rudiment is transformed into a long, thin tube (Figure 8.22; see also Figure 8.16, 12 hours). To accomplish this extension, the cells of the archenteron rearrange themselves by migrating over one an-

other and by flattening themselves (Ettensohn 1985; Hardin and Cheng 1986). This phenomenon, wherein cells intercalate to narrow the tissue and at the same time move it forward, is called **convergent extension**. Moreover, cell division continues, producing more endodermal and secondary mesenchyme cells as the archenteron extends (Figure 8.23; Martins et al. 1998).

In some species of sea urchins, a third stage of archenteron elongation occurs. This final phase is initiated by the tension provided by secondary mesenchyme cells, which form at the tip of the archenteron and remain there (Figure 8.24; see also Figure 8.16, 12 hours). These cells extend filopodia through the blastocoel fluid to contact the inner surface of the blastocoel wall (Dan and Okazaki 1956; Schroeder 1981). The filopodia attach to the wall at the junctions between the blastomeres and then shorten, pulling up the archenteron (see Figure 8.16, 13 hours). Hardin (1988) ablated the secondary mesenchyme cells with a laser, with the result that the archenteron could elongate to only about two-thirds of the normal length. If a few secondary mesenchyme cells were left, elongation continued, although at a slower rate. The secondary mes-

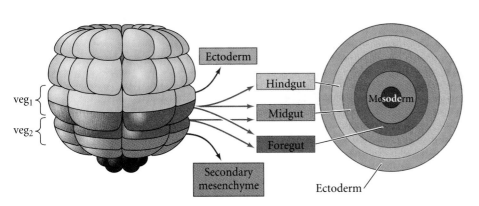

veg₁

veg₂

Ectoderm

Hindgut

Midgut

Foregut

Secondary mesenchyme

Mesoderm

Ectoderm

Figure 8.21
Fate map of the vegetal plate of the sea urchin embryo, looking "upward" at the vegetal surface. The central portion becomes the secondary mesenchyme cells, while the concentric layers around it become the foregut, midgut, and hindgut, respectively. The boundary where the endoderm meets the ectoderm marks the anus. The secondary mesenchyme and foregut come from the veg₂ layer, the midgut comes from veg₁ and veg₂ cells, and the hindgut (and the ectoderm in contact with it) comes from the veg₁ layer. (After Logan and McClay 1999.)

Figure 8.22
Cell rearrangement during the extension of the archenteron in sea urchin embryos. In this species, the early archenteron has 20 to 30 cells around its circumference. Later in gastrulation, the archenteron has a circumference made by only 6 to 8 cells. Fluorescently labeled clones can be seen to stretch apically. (After Hardin 1990.)

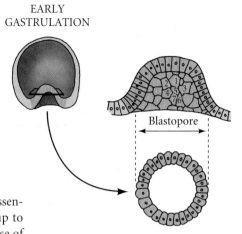

EARLY
GASTRULATION

Blastopore

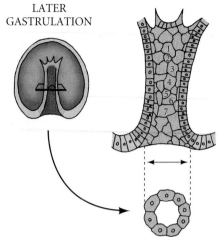

LATER
GASTRULATION

enchyme cells, in this species, play an essential role in pulling the archenteron up to the blastocoel wall during the last phase of invagination.

But can the secondary mesenchyme filopodia attach to any part of the blastocoel wall, or is there a specific target in the animal hemisphere that must be present for attachment to occur? Is there a region of the blastocoel wall that is already committed to becoming the ventral side of the larva? Studies by Hardin and McClay (1990) show that there is a specific "target" site for the filopodia that differs from other regions of the animal hemisphere. The filopodia extend, touch the blastocoel wall at random sites, and then retract. However, when the filopodia contact a particular region of the wall, they remain attached there, flatten out against this region, and pull the archenteron toward it. When Hardin and McClay poked in the other side of the blastocoel wall so that contacts were made most readily with that region, the filopodia continued to extend and retract after touching it. Only when the filopodia found their "target" did they cease these movements. If the gastrula was constricted so that the filopodia never reached the target area, the secondary mesenchyme cells continued to explore until they eventually moved off the archenteron and found the target tissue as

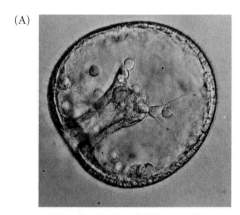

(A)

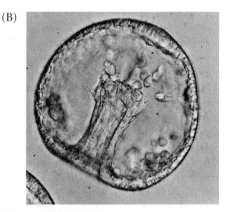

(B)

Figure 8.24
Mid-gastrula stage of *Lytechinus pictus*, showing filopodial extensions of secondary mesenchyme extending from the archenteron tip to the blastocoel wall. (A) Secondary mesenchyme cells extending filopodia from the tip of the archenteron. (B) Filopodial cables connecting the blastocoel wall to the archenteron tip. The tension of the cables can be seen as they pull on the blastocoel wall at the point of attachment. (Photographs courtesy of C. Ettensohn.)

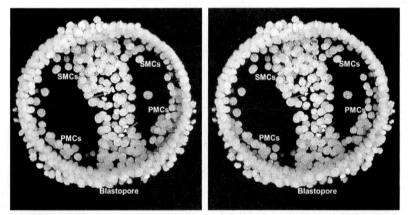

Figure 8.23
Stereo pair showing a *Lytechinus* gastrula during the dispersal of the secondary mesenchyme cells. Convergent extension has occurred, and a few mitotic spindles (small arrows) can be seen. (PMC, primary mesenchyme cell; SMC, secondary mesenchyme cell.) If you cross your eyes, a three-dimensional "central image" will form. (From Martins et al. 1998; photographs courtesy of the authors.)

freely migrating cells. There appears, then, to be a target region on what is to become the ventral side of the larva that is recognized by the secondary mesenchyme cells, and which positions the archenteron in the region where the mouth will form.

As the top of the archenteron meets the blastocoel wall in the target region, the secondary mesenchyme cells disperse into the blastocoel, where they proliferate to form the mesodermal organs (see Figure 8.16, 13.5 hours). Where the archenteron contacts the wall, a mouth is eventually formed. The mouth fuses with the archenteron to create a continuous digestive tube. Thus, as is characteristic of deuterostomes, the blastopore marks the position of the anus.

> VADE MECUM[2] **Sea urchin development.** The CD-ROM provides an excellent review of sea urchin development as well as questions on the fundamentals of echinoderm cleavage and gastrulation. **[Click on Sea Urchin]**

THE EARLY DEVELOPMENT OF SNAILS

Cleavage in Snail Embryos

Spiral holoblastic cleavage is characteristic of several animal groups, including annelid worms, some flatworms, and most molluscs. It differs from radial cleavage in numerous ways. First, the cleavage planes are not parallel or perpendicular to the animal-vegetal axis of the egg; rather, cleavage is at oblique angles, forming a "spiral" arrangement of daughter blas-

tomeres. Second, the cells touch one another at more places than do those of radially cleaving embryos. In fact, they assume the most thermodynamically stable packing orientation, much like that of adjacent soap bubbles. Third, spirally cleaving embryos usually undergo fewer divisions before they begin gastrulation, making it possible to follow the fate of each cell of the blastula. When the fates of the individual blastomeres from annelid, flatworm, and mollusc embryos were compared, many of the same cells were seen in the same places, and their general fates were identical (Wilson 1898). Blastulae produced by spiral cleavage have no blastocoel and are called **stereoblastulae**.

Figures 8.25 and 8.26 depict the cleavage pattern typical of many molluscan embryos. The first two cleavages are nearly meridional, producing four large macromeres (labeled A, B, C, and D). In many species, these four blastomeres are different sizes (D being the largest), a characteristic that allows them to be individually identified. In each successive cleavage, each macromere buds off a small micromere at its animal pole. Each successive quartet of micromeres is displaced to the right or to the left of its sister macromere, creating the characteristic spiral pattern. Looking down on the embryo from the animal pole, the upper ends of the mitotic spindles appear to alternate clockwise and counterclockwise. This arrangement causes alternate micromeres to form obliquely to the left and to the right of their macromeres.

At the third cleavage, the A macromere gives rise to two daughter cells, macromere 1A and micromere 1a. The B, C, and D cells behave similarly, producing the first quartet of micromeres. In most species, these micromeres are to the *right* of

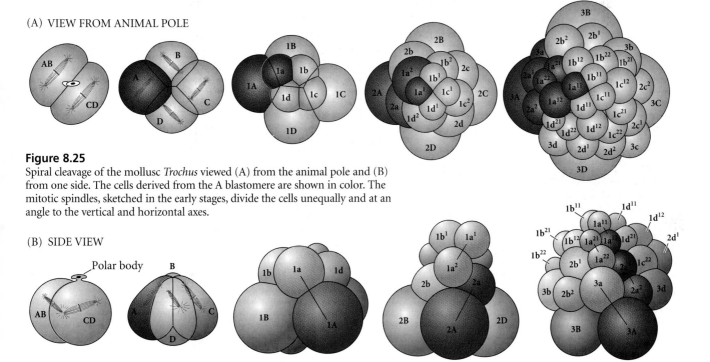

Figure 8.25
Spiral cleavage of the mollusc *Trochus* viewed (A) from the animal pole and (B) from one side. The cells derived from the A blastomere are shown in color. The mitotic spindles, sketched in the early stages, divide the cells unequally and at an angle to the vertical and horizontal axes.

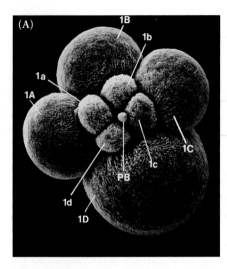
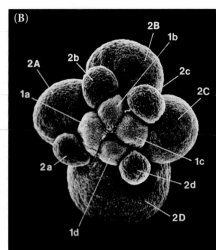
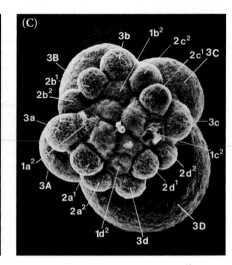

Figure 8.26
Spiral cleavage of the snail *Ilyanassa*. The D blastomere is larger than the others, allowing the identification of each cell. Cleavage is dextral. (A) 8-cell stage. PB, polar body. (B) Mid-fourth cleavage (12-cell embryo). The macromeres have already divided into large and small spirally oriented cells; 1a–d have not divided yet. (C) 32-cell embryo. (From Craig and Morrill 1986; photographs courtesy of the authors.)

their macromeres (looking down on the animal pole). At the fourth cleavage, macromere 1A divides to form macromere 2A and micromere 2a, and micromere 1a divides to form two more micromeres, $1a^1$ and $1a^2$. The micromeres of this second quartet are to the *left* of the macromeres. Further cleavage yields blastomeres 3A and 3a from macromere 2A, and micromere $1a^2$ divides to produce cells $1a^{21}$ and $1a^{22}$. In normal development, the first-quartet micromeres form the head structures, while the second-quartet micromeres form the statocyst (balance organ) and shell. These fates are specified both by cytoplasmic localization and by induction (Clement 1967; Cather 1967; Render 1991; Sweet 1998).

The orientation of the cleavage plane to the left or to the right is controlled by cytoplasmic factors within the oocyte. This was discovered by analyzing mutations of snail coiling. Some snails have their coils opening to the right of their shells (**dextral coiling**), whereas other snails have their coils opening to the left (**sinistral coiling**). Usually the direction of coiling is the same for all members of a given species, but occasionally mutants are found. For instance, in a population of snails in which the coils open on the right, some individuals will be found with coils that open on the left. Crampton (1894) analyzed the embryos of such aberrant snails and found that their early cleavage differed from the norm. The orientation of the cells after the second cleavage was different in the sinistrally coiling snails owing to a different orientation of the mitotic apparatus (Figure 8.27). All subsequent divisions in left-coiling embryos were mirror images of those in dextrally coiling embryos. In Figure 8.27, one can see that the position of the

4d blastomere (which is extremely important, as its progeny will form the mesodermal organs) is different in the two types of spiraling embryos.

The direction of snail shell coiling is controlled by a single pair of genes (Sturtevant 1923; Boycott et al. 1930). In the snail *Limnaea peregra*, most individuals are dextrally coiled. Rare mutants exhibiting sinistral coiling were found and mated with wild-type snails. These matings showed that the right-coiling allele, *D*, is dominant to the left-coiling allele, *d*. However, the direction of cleavage is determined not by the genotype of the developing snail, but by the genotype of the snail's mother. A *dd* female snail can produce only sinistrally coiling offspring, even if the offspring's genotype is *Dd*. A *Dd* individual will coil either left or right, depending on the genotype of its mother. Such matings produce a chart like this:

		Genotype	Phenotype
DD ♀ × *dd* ♂	→	*Dd*	All right-coiling
DD ♂ × *dd* ♀	→	*Dd*	All left-coiling
Dd × *Dd*	→	1*DD*:2*Dd*:1*dd*	All right-coiling

The genetic factors involved in snail coiling are brought to the embryo by the oocyte cytoplasm. It is the genotype of the ovary in which the oocyte develops that determines which orientation cleavage will take. When Freeman and Lundelius (1982) injected a small amount of cytoplasm from dextrally coiling snails into the eggs of *dd* mothers, the resulting embryos coiled to the right. Cytoplasm from sinistrally coiling snails did not affect right-coiling embryos. These findings confirmed that the wild-type mothers were placing a factor into their eggs that was absent or defective in the *dd* mothers.

WEBSITE 8.3 Alfred Sturtevant and the genetics of snail coiling. By a masterful thought experiment, Sturtevant demonstrated the power of applying genetics to embryology. To do this, he brought Mendelian genetics into the study of snail coiling.

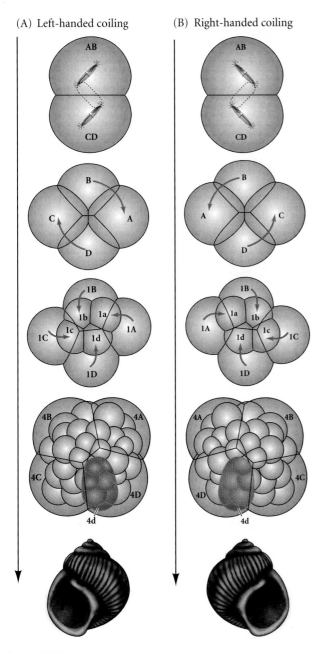

Figure 8.27
Looking down on the animal pole of (A) left-coiling and (B) right-coiling snails. The origin of sinistral and dextral coiling can be traced to the orientation of the mitotic spindle at the second cleavage. Left- and right-coiling snails develop as mirror images of each other. (After Morgan 1927.)

Fate map of *Ilyanassa obsoleta*

Joanne Render (1997) has constructed a detailed fate map of the snail *Ilyanassa obsoleta* by injecting specific micromeres with large polymers conjugated to the fluorescent dye Lucifer Yellow. The fluorescence is maintained over the period of embryogenesis and can be seen in the larval tissue derived from the injected cells. Render's fate map (Figure 8.28) shows that the second-quartet micromeres (2a–d) generally contribute to the shell-forming mantle, the velum, the mouth, and the heart. The third-quartet micromeres (3a–d) generate large regions of the foot, velum, esophagus, and heart. The 4d cell—the mesentoblast—contributes to the larval kidney, heart, retractor muscles, and intestine.

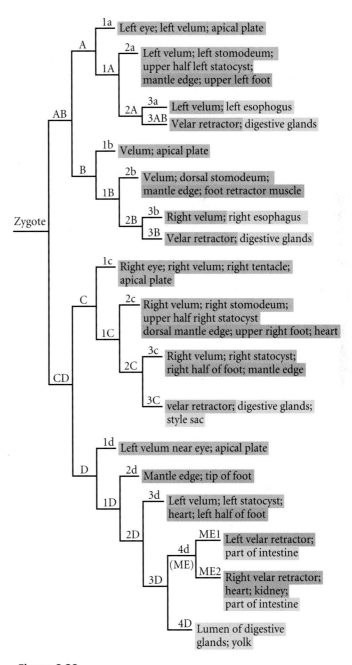

Figure 8.28
Fate map of the snail *Ilyanassa obsoleta*. Beads containing Lucifer Yellow were injected into individual blastomeres at the 32-cell stage. When the embryos developed into larvae, their descendants could be identified by their fluorescence. (After Render 1997.)

Adaptation by Modifying Embryonic Cleavage

Evolution is caused by the hereditary alteration of embryonic development. Sometimes we are able to identify a modification of embryogenesis that has enabled the organism to survive in an otherwise inhospitable environment. One such modification, discovered by Frank Lillie in 1898, is brought about by an alteration of the typical pattern of spiral cleavage in the unionid family of clams.

Unlike most clams, *Unio* and its relatives live in swift-flowing streams. Streams create a problem for the dispersal of larvae: because the adults are sedentary, free-swimming larvae would always be carried downstream by the current. These clams, however, have adapted to this environment via two modifications of their development. The first is an alteration in embryonic cleavage. In typical molluscan cleavage, either all the macromeres are equal in size or the 2D blastomere is the largest cell at that embryonic stage. However, cell division in *Unio* is such that the *2d* blastomere gets the largest amount of cytoplasm (Figure 8.29). This cell divides to produce most of the larval structures, including a gland capable of producing a large shell. The resulting larvae (called **glochidia**) resemble tiny bear traps; they have sensitive hairs that cause the valves of the shell to snap shut when they are touched by the gills or fins of a wandering fish. The larvae attach themselves to a fish

and "hitchhike" with it until they are ready to drop off and metamorphose into adult clams. In this manner, they can spread upstream.

In some unionid species, glochidia are released from the female's brood pouch and then wait passively for a fish to swim by. Some other species, such as *Lampsilis ventricosa*, have increased the chances of their larvae finding a fish by yet another developmental modification (Welsh 1969). Many clams develop a thin mantle that flaps around the shell and surrounds the

Figure 8.30
Phony fish atop the unionid clam *Lampsilis ventricosa*. The "fish" is actually the brood pouch and mantle of the clam. (Photograph courtesy of J. H. Welsh.)

brood pouch. In some unionids, the shape of the brood pouch (marsupium) and the undulations of the mantle mimic the shape and swimming behavior of a minnow. To make the deception even better, they develop a black "eyespot" on one end and a flaring "tail" on the other. The "fish" in Figure 8.30 is not a fish at all, but the brood pouch and mantle of the female clam beneath it. When a predatory fish is lured within range of this "prey," the clam discharges the glochidia from the brood pouch. Thus, the modification of existing developmental patterns has permitted unionid clams to survive in challenging environments.

WEBSITE 8.4 Modifications of cell fate in spiralian eggs. Within the gastropods, differences in the timing of cell fate result in significantly different body plans. Furthermore, in the leeches and nemerteans, the spiralian cleavage pattern has been modified to produce new types of body plans.

Figure 8.29
Formation of glochidium larva by the modification of spiral cleavage. After the 8-cell embryo is formed (A), the placement of the mitotic spindle causes most of the D cytoplasm to enter the 2d blastomere (B). This large 2d blastomere divides (C), eventually giving rise to the large "bear-trap" shell of the larva (D). (After Raff and Kaufman 1983.)

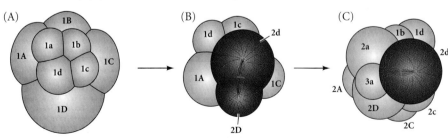
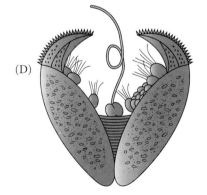

The polar lobe: Cell determination and axis formation

Molluscs provide some of the most impressive examples of mosaic development, in which the blastomeres are specified autonomously, and of cytoplasmic localization, wherein morphogenetic determinants are placed in a specific region of the oocyte (see Chapter 3). Mosaic development is widespread throughout the animal kingdom, especially in protostomes such as annelids, nematodes, and molluscs, all of which initiate gastrulation at the future anterior end after only a few cell divisions.

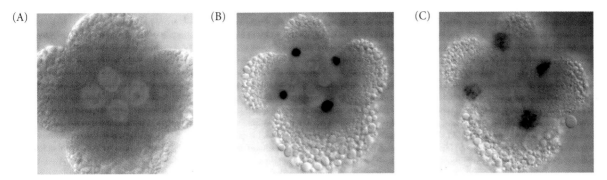

Figure 8.31
Association of *decapentaplegic* (*dpp*) mRNA with specific centrosomes of *Ilyanassa*. (A) In situ hybridization of the mRNA for the BMP-like paracrine factor Dpp in the 4-cell snail embryo shows no *Dpp* accumulation. (B) At prophase of the 4- to 8-cell stage, *dpp* mRNA (black) accumulates at one centrosome of the pair forming the mitotic spindle. (C) As mitosis continues, *dpp* mRNA is seen to attend the centrosome in the macromere rather than the centrosome in the micromere of each cell. The *dpp* message encodes a BMP-like paracrine factor critical to molluscan development. (From Lambert and Nagy 2002; photographs courtesy of L. Nagy.)

In molluscs, the mRNAs for some transcription factors and paracrine factors are placed in particular cells by associating with certain centrosomes (Figure 8.31; Lambert and Nagy 2002). In other cases, the patterning molecules appear to be bound to a certain region of the egg that will form the **polar lobe**.

E. B. Wilson and his student H. E. Crampton demonstrated that mosaic development characterizes the early development of mollusc embryos (see Figure 3.7). They observed that certain spirally cleaving embryos (mostly in the mollusc and annelid phyla) extrude a bulb of cytoplasm immediately before first cleavage (Figure 8.32). This protrusion is the **polar lobe**. In some species of snails, the region uniting the polar lobe to the rest of the egg becomes a fine tube. The first cleavage splits the zygote asymmetrically, so that the polar lobe is connected only to the CD blastomere. In several species, nearly one-third of the total cytoplasmic volume is contained in this anucleate lobe, giving it the appearance of another cell. The resulting three-

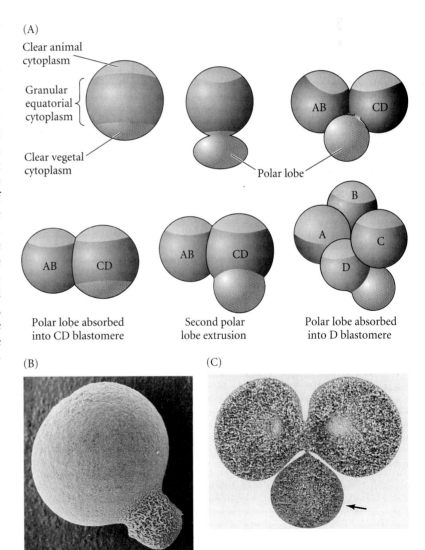

Figure 8.32
Polar lobe formation in certain mollusc embryos. (A) Cleavage. Extrusion and reincorporation of the polar lobe occur twice. (B) Scanning electron micrograph of the extending polar lobe in the uncleaved egg of *Buccinum undatum*. The surface ridges are confined to the polar lobe region. (C) Section through first-cleavage, or trefoil-stage, embryo of *Dentalium*. The arrow points to the large polar lobe. (A after Wilson 1904; photographs courtesy of M. R. Dohmen.)

lobed structure is often referred to as the **trefoil-stage** embryo (Figure 8.32C). The CD blastomere absorbs the polar lobe material, but extrudes it again prior to second cleavage. After this division, the polar lobe is attached only to the D blastomere, which absorbs its material. Thereafter, no polar lobe is formed.

Crampton (1896) showed that if one removes the polar lobe at the trefoil stage, the remaining cells divide normally. However, instead of producing a normal trochophore larva,* they produce an incomplete larva, wholly lacking its endoderm (intestine), mesodermal organs (such as the heart and retractor muscles), as well as some ectodermal organs (such as eyes) (see Figure 8.33). Moreover, Crampton demonstrated that the same type of abnormal larva can be produced by removing the D blastomere from the 4-cell embryo. Crampton concluded that the polar lobe cytoplasm contains the endodermal and mesodermal determinants, and that these determinants give the D blastomere its endomesoderm-forming capacity. Crampton also showed that the localization of the mesodermal determinants is established shortly after fertilization, thereby demonstrating that a specific cytoplasmic region of the egg, destined for inclusion in the D blastomere, contains whatever factors are necessary for the special cleavage rhythms of the D blastomere and for the differentiation of the mesoderm.

Centrifugation studies (Clement 1968) demonstrated that the morphogenetic determinants sequestered within the polar lobe are probably located in the cytoskeleton or cortex, not in the lobe's diffusible cytoplasm. Van den Biggelaar (1977) obtained similar results when he removed the cytoplasm from the polar lobe with a micropipette. Cytoplasm from other regions of the cell flowed into the polar lobe, replacing the portion that he had removed. The subsequent development of these embryos was normal. In addition, when he added the diffusible polar lobe cytoplasm to the B blastomere, no duplicated structures were seen (Verdonk and Cather 1983). Therefore, the diffusible part of the polar lobe cytoplasm does not contain the morphogenetic determinants; they probably reside in the nonfluid cortical cytoplasm or on the cytoskeleton.

---

*The **trochophore** (Greek, *trochos*, "wheel") is a planktonic (free-swimming) larval form found among the molluscs and several other protostome phyla with spiral cleavage, most notably the marine annelid worms.

Figure 8.33
Importance of the polar lobe in the development of *Ilyanassa*. (A) Normal trochophore larva. (B) Abnormal larva, typical of those produced when the polar lobe of the D blastomere is removed. (E, eye; F, foot; S, shell; ST, statocyst, a balancing organ; V, velum; VC, velar cilia; Y, residual yolk; ES, everted stomodeum; DV, disorganized velum.) (From Newrock and Raff 1975; photographs courtesy of K. Newrock.)

Clement (1962) also analyzed the further development of the D blastomere in order to observe the further appropriation of these determinants. The development of the D blastomere can be traced in Figure 8.26. This macromere, having received the contents of the polar lobe, is larger than the other three. When one removes the D blastomere or its first or second macromere derivatives (1D or 2D), one obtains an incomplete larva, lacking heart, intestine, velum (the ciliated border of the larva), shell gland, eyes, and foot. This is essentially the same phenotype one gets when one removes the polar lobe. Since the D blastomeres do not directly contribute cells to many of these structures, it appears that the D-quadrant macromeres are involved in inducing other cells to have these fates.

When one removes the 3D blastomere shortly after the division of the 2D cell to form the 3D and 3d blastomeres, the larva produced looks similar to those formed by the removal of the D, 1D, or 2D macromeres. However, ablation of the 3D blastomere at a later time produces an almost normal larva, having eyes, foot, velum, and some shell gland, but no heart or intestine (Figure 8.33). After the 4d cell is given off (by the division of the 3D blastomere), removal of the D derivative (the 4D cell) produces no qualitative difference in development. In fact, all the essential determinants for heart and intestine formation are now in the 4d blastomere, and removal of that cell results in a heartless and gutless larva (Clement 1986). The 4d blastomere is responsible for forming (at its next division) the two **mesentoblasts**, the cells that give rise to both the mesodermal (heart) and endodermal (intestine) organs.

The mesodermal and endodermal determinants of the 3D macromere, then, are transferred to the 4d blastomere, while the inductive ability of the 3D blastomere (to induce eyes and shell gland, for instance) is needed during the time the 3D cell is formed but is not required afterward. The 3D cell appears to activate the MAP kinase signaling pathway in the micromeres above it (Figure 8.34; Lambert and Nagy 2001; see also Chap-

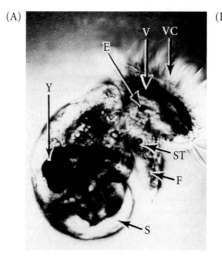

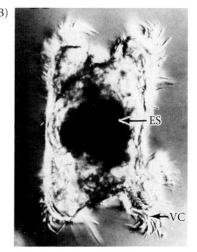

Figure 8.34
The 3D blastomere activates MAP kinase activity in adjacent micromeres. (A) Activated MAP kinase (blue stain) can be seen in the 3D macromere and in the micromeres above it ($1a$–d^1, $1d^2$, $2d^1$, $2d^2$, 3d). The nuclei are counterstained green, and the cell boundaries have been superimposed on the photographic image. Staining was done 30 minutes after the formation of the 3D macromere. (B) Control larva grown to veliger larval stage. (C) Same age larva treated with MAP kinase inhibitor 15 minutes after 3D blastomere formed. The shell, eye, statocyst, and operculum have not developed. (D) Same age larva treated with MAP kinase inhibitor at 150 minutes after 3D had formed (shortly before 4D formation). This larva had all the organs induced by the 3D macromere. (After Lambert and Nagy 2001.)

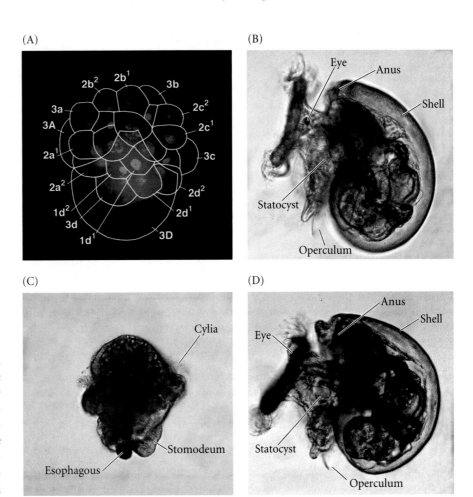

ter 6). If cells are stained for activated MAP kinase, the stain is seen in those cells that require the signal from the 3D macromere for their normal differentiation. Removal of 3D prevents MAP kinase signaling, and if the MAP kinase signaling is blocked by specific inhibitors, the resulting larvae look precisely like those formed by the deletion of the D blastomeres (see Figure 8.33). Thus, the 3D macromere appears to activate the MAP kinase cascade in the ectodermal (eye- and shell gland-forming) micromeres above it.

In addition to its role in cell differentiation, the material in the polar lobe is also responsible for specifying the dorsal-ventral (back-belly) polarity of the embryo. When polar lobe material is forced to pass into the AB blastomere as well as into the CD blastomere, twin larvae are formed that are joined at their ventral surfaces (Guerrier et al. 1978; Henry and Martindale 1987).

To summarize, experiments have demonstrated that the nondiffusible polar lobe cytoplasm is extremely important in normal molluscan development for a number of reasons:

- It contains the determinants for the proper cleavage rhythm and cleavage orientation of the D blastomere.
- It contains certain determinants (those entering the 4d blastomere and hence leading to the mesentoblasts) for autonomous mesodermal and intestinal differentiation.
- It is responsible for permitting the inductive interactions (through the material entering the 3D blastomere) leading to the formation of the shell gland and eye.
- It contains determinants needed for specifying the dorsal-ventral axis of the embryo.

Although the polar lobe is clearly important in normal snail development, we still do not know the mechanisms for most of its effects. One possible clue has been provided by Atkinson (1987), who has observed differentiated cells of the velum, digestive system, and shell gland within lobeless embryos. But even though lobeless embryos can produce these cells, they appear unable to organize them into functional tissues and organs. Tissues of the digestive tract can be found, but they are not connected; muscle cells are scattered around the lobeless larva, but are not organized into a functional muscle tissue. Thus, the developmental functions of the polar lobe are probably very complex and may be essential for axis formation.

Gastrulation in Snails

The snail stereoblastula is relatively small, and its cell fates have already been determined by the D series of macromeres. Gastrulation is accomplished primarily by epiboly, wherein the micromeres at the animal cap multiply and "overgrow" the vegetal macromeres. Eventually, the micromeres will cover the entire embryo, leaving a small slit at the vegetal pole (Figure 8.35; Collier 1997).

Figure 8.35
Gastrulation in *Crepidula*. The ectoderm undergoes epiboly from the animal pole and envelops the other cells of the embryo. (After Conklin 1897.)

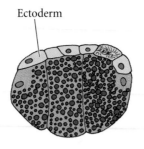
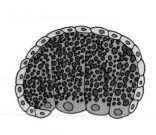
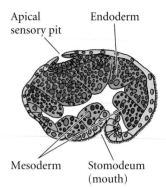

Ectoderm

Apical sensory pit   Endoderm

Mesoderm   Stomodeum (mouth)

EARLY DEVELOPMENT IN TUNICATES

Tunicate Cleavage

Ascidians, members of the tunicate subphylum, are fascinating animals for several reasons, but the foremost is that they are invertebrate chordates. They have a notochord as larvae (and therefore are chordates), but they lack vertebrae. As larvae, they are free-swimming tadpoles; but when the tadpole undergoes metamorphosis, it sticks to the sea floor, its nerve cord and notochord degenerate, and it secretes a cellulose tunic (which gave the name "tunicates" to these creatures).

These animals are characterized by **bilateral holoblastic cleavage**, a pattern found primarily in tunicates (Figure 8.36). The most striking feature of this type of cleavage is that the first cleavage plane establishes the earliest axis of symmetry in the embryo, separating the embryo into its future right and left sides. Each successive division orients itself to this plane of symmetry, and the half-embryo formed on one side of the first cleavage plane is the mirror image of the half-embryo on the other side. The second cleavage is meridional, like the first,

but unlike the first division, it does not pass through the center of the egg. Rather, it creates two large anterior cells (the A and a blastomeres) and two smaller posterior cells (blastomeres B and b). Each side now has a large and a small blastomere. During the next three divisions, differences in cell size and shape highlight the bilateral symmetry of these embryos. At the 64-cell stage, a small blastocoel is formed, and gastrulation begins from the vegetal pole. The cell lineages of the tunicate *Styela partita* are shown in Figure 1.7.

The tunicate fate map

As mentioned in Chapter 3, early tunicate embryos are specified autonomously, each cell acquiring a specific type of cytoplasm that will determine its fate. In tunicates such as *Styela*, the different regions of cytoplasm have distinct pigmentation, and the cell fates can easily be seen to correspond to the type of cytoplasm taken up by each cell. These cytoplasmic regions are apportioned to the egg during fertilization. In the unfertilized egg of *Styela partita*, a central gray cytoplasm is enveloped by a cortical layer containing yellow lipid inclusions (Figure 8.37A). During meiosis, the breakdown of the nucleus releases a clear substance that accumulates in the animal hemisphere of the egg. Within 5 minutes of sperm entry, the inner clear and cortical yellow cytoplasms contract into the vegetal (lower) hemisphere of the egg (see Figure 8.38). As the male pronucleus migrates from the vegetal pole to the equator of the cell along the future posterior side of the embryo, the yellow lipid inclusions migrate with it. This migration forms a

Figure 8.36
Bilateral symmetry in the egg of the tunicate *Styela partita*. (A) Uncleaved egg. The regions of cytoplasm destined to form particular organs are labeled here and coded by color throughout the diagrams. (B) 8-cell embryo, showing the blastomeres and the fates of various cells. The embryo can be viewed as two 4-cell halves; from here on, each division on the right side of the embryo has a mirror-image division on the left. (C, D) Views of later embryos from the vegetal pole. The dashed line shows the plane of bilateral symmetry. (A after Balinsky 1981.)

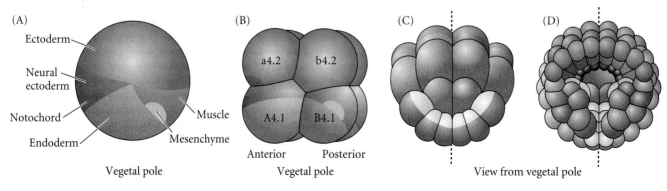

(A)
Ectoderm
Neural ectoderm
Notochord
Endoderm
Muscle
Mesenchyme
Vegetal pole

(B)
a4.2   b4.2
A4.1   B4.1
Anterior   Posterior
Vegetal pole

(C)

(D)
View from vegetal pole

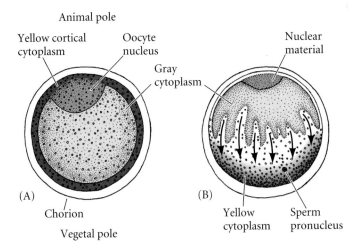

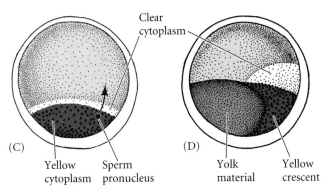

Figure 8.37
Cytoplasmic rearrangement in the fertilized egg of *Styela partita*. (A) Before fertilization, yellow cortical cytoplasm surrounds the gray yolky inner cytoplasm. (B) After sperm entry (in the vegetal hemisphere of the oocyte), the yellow cortical cytoplasm and the clear cytoplasm derived from the breakdown of the oocyte nucleus contract vegetally toward the sperm. (C) As the sperm pronucleus migrates animally toward the newly formed egg pronucleus, the yellow and clear cytoplasms move with it. (D) The final positions of the yellow cytoplasm marks the location where cells give rise to tail muscles. (After Conklin 1905.)

WEBSITE 8.5 The experimental analysis of tunicate cell specification. Researchers analyzing tunicate development are using biochemical and molecular probes to find the morphogenetic determinants that are segregated to different regions of the egg cytoplasm.

yellow crescent, extending from the vegetal pole to the equator (Figure 8.37B–D); this region will produce most of the tail muscles of the tunicate larva. The movement of these cytoplasmic regions depends on microtubules that are generated by the sperm centriole and on a wave of calcium ions that contracts the animal pole cytoplasm (Sawada and Schatten 1989; Speksnijder et al. 1990; Roegiers et al. 1995).

Edwin Conklin (1905) took advantage of the differing coloration of these regions of cytoplasm to follow each of the cells of the tunicate embryo to its fate in the larva (see Figure 1.7). He found that cells receiving clear cytoplasm become ectoderm; those containing yellow cytoplasm give rise to mesodermal cells; those that incorporate slate-gray inclusions become endoderm; and light gray cells become the neural tube and notochord. The cytoplasmic regions are localized bilaterally around the plane of symmetry, so they are bisected by the first cleavage furrow into the right and left halves of the embryo. The second cleavage causes the prospective mesoderm to lie in the two posterior cells, while the prospective neural ectoderm and chordamesoderm (notochord) will be formed from the two anterior cells (see Figure 8.36). The third division further partitions these cytoplasmic regions such that the mesoderm-forming cells are confined to the two vegetal posterior blastomeres, while the chordamesoderm cells are restricted to the two vegetal anterior cells.

Autonomous and conditional specification of tunicate blastomeres

As mentioned in Chapter 3, the autonomous specification of tunicate blastomeres was one of the first observations in the field of experimental embryology (Chabry 1888). Reverberi and Minganti (1946) extended this analysis in a series of isolation experiments, and they, too, observed the self-differentiation of each isolated blastomere and of the remaining embryo. The results of one of these experiments are shown in Figure 3.8. When the 8-cell embryo is separated into its four doublets (the right and left sides being equivalent), both mosaic and conditional specification are seen. The animal posterior pair of blastomeres gives rise to the ectoderm, and the vegetal posterior pair produces endoderm, mesenchyme, and muscle tissue, just as expected from the fate map. Autonomous specification is seen in the tunicate gut endoderm, muscle mesoderm, skin ectoderm, and neural cord. Conditional specification (by induction) is seen in the formation of the brain, notochord, and mesenchyme cells.

AUTONOMOUS SPECIFICATION OF THE MYOPLASM: THE YELLOW CRESCENT. From the cell lineage studies of Conklin and others, it was known that only one pair of blastomeres (posterior vegetal; B4.1) in the 8-cell embryo is capable of producing tail muscle tissue. These cells contain the yellow crescent cytoplasm (Figure 8.38). When yellow crescent cytoplasm is transferred from the B4.1 (muscle-forming) blastomere to the b4.2 (ectoderm-forming) blastomere of an 8-cell tunicate embryo, the ectoderm-forming blastomere generates muscle cells as well as its normal ectodermal progeny (Whittaker 1982; see Figure 3.10). Moreover, cytoplasm from the yellow crescent

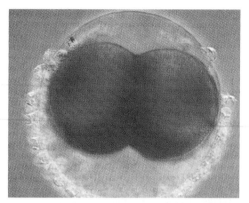

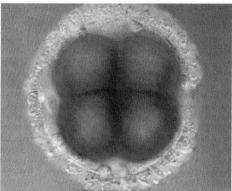

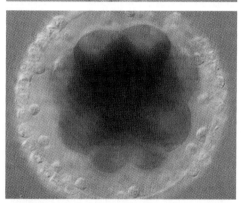

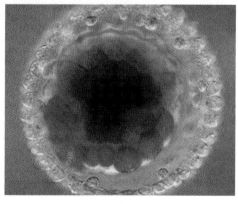

Figure 8.38
Cytoplasmic segregation in the egg of *Styela partita*. The yellow crescent, originally seen in the vegetal pole, becomes segregated into the B4.1 blastomere pair and thence into the muscle cells. (Photographs courtesy of J. R. Whittaker.)

area of the fertilized egg can cause the a4.2 blastomere to express muscle-specific proteins (Nishida 1992a). Conversely, Tung and colleagues (1977) showed that when larval cell nuclei are transplanted into enucleated tunicate egg fragments, the newly formed cells show the structures typical of the egg regions providing the cytoplasm, not of those cells providing the nuclei. We can conclude, then, that certain determinants that exist in the egg cytoplasm cause the formation of certain tissues. These morphogenetic determinants appear to work by selectively activating (or inactivating) specific genes. The determination of the blastomeres and the activation of certain genes are controlled by the spatial localization of the morphogenetic determinants within the egg cytoplasm.

Using RNA hybridization techniques, Nishida and Sawada (2001) found particular mRNAs to be highly enriched in the vegetal hemisphere of the tunicate *Halocynthia roretzi*. One of these messages encodes a zinc-finger transcription factor called **macho-1**. *Macho-1* mRNA was found to be concentrated in the vegetal hemisphere in the unfertilized egg and during early fertilization, and it migrated with the yellow crescent cytoplasm into the posterior vegetal region of the egg during the second half of the first cell cycle. By the 8-cell stage, *macho-1* mRNA was found only in the B4.1 blastomeres. At the 16- and 32-cell stages, it was seen only in those blastomeres that give rise to the muscle cells (Figure 8.39).

When antisense oligonucleotides were injected into unfertilized eggs to deplete *macho-1* mRNA from the embryo, the resulting tunicate larvae lacked all the muscles usually formed by the descendants of the B4.1 blastomeres. (They did have the secondary muscles that are generated through the interactions of A4.1 and b4.2 blastomeres.) The tails of these *macho-1*-depleted larvae were severely shortened, but the other regions of the tadpoles appeared normal, both structurally and biochemically. Moreover, B4.1 blastomeres isolated from *macho-1*-depleted embryos failed to produce muscle tissue. Nishida and Sawada then injected *macho-1* mRNA into cells that would *not* normally form muscle, and found that these ectoderm or endoderm precursors will generate muscle cells when they are given *macho-1* mRNA. Thus, the *macho-1* message is found at the right place and at the right time, and these experiments suggest that macho-1 protein is both necessary and sufficient to promote muscle differentiation in certain ascidian cells.

Erives and Levine (2000) found another mRNA localized to the yellow crescent. This message encodes the T-box transcription factor **CiVegTR**. The *CiVegTR* message migrates with the yellow cytoplasm, eventually becoming incorporated into the B4.1 blastomere, and CiVegTR protein binds to the enhancer of a muscle-specific gene *Snail*. Snail protein is important in preventing *Brachyury* (*T*) expression in presumptive muscle cells, and is therefore needed to prevent the muscle precursors from becoming notochord cells. It appears, then, that the *CiVegTR* and *macho-1* transcription factors are components of the tunicate yellow crescent muscle-forming

Figure 8.39
Autonomous specification by a morphogenetic factor. The *macho-1* mRNA message is localized to the muscle-forming tunicate cytoplasm. In situ hybridization shows the *macho-1* message found first in the vegetal pole cytoplasm (A), then migrating up the presumptive posterior surface of the egg (B) and becoming localized in the B4.1 blastomere (C). (From Nishida and Sawada 2001; photographs courtesy of H. Nishida and N. Satoh.)

(A)
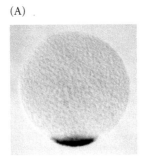

(B)
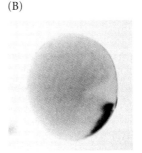

(C)

cytoplasm. It is possible that they function together, with the macho-1 protein instructing the cells to become muscle cells while the CiVegTR protein instructs the cells *not* to become notochord.

AUTONOMOUS SPECIFICATION OF THE ENDODERM: β-CATENIN.
Presumptive endoderm originates from the vegetal A4.1 and B4.1 blastomeres. The specification of these cells coincides with the localization of β-catenin, a transcription factor that we saw earlier in our discussion of sea urchin endoderm specification (see Figure 8.13). Inhibition of β-catenin results in the loss of endoderm and its replacement by ectoderm in the ascidian embryo (Figure 8.40; Imai et al. 2001). Conversely, increasing β-catenin synthesis causes an increase in the endoderm at the expense of the ectoderm (just as in sea urchins). The β-catenin transcription factor appears to function by activating the synthesis of the homeobox transcription factor Lhx-3. Inhibition of the *Lhx-3* message prohibits the differentiation of endoderm (Satou et al. 2001).

CONDITIONAL SPECIFICATION OF THE MESENCHYME AND NOTO-CHORD. While most of the muscles are specified autonomously from the yellow crescent cytoplasm, the most posterior muscle cells form through conditional specification by cell interactions with the descendants of the A4.1 and b4.2 blastomeres (Nishida 1987, 1992a,b). Moreover, the other two major types of mesoderm—the notochord and mesenchyme—also form through inductive interactions. In fact, the noto-

chord and the mesenchyme appear to be induced by the same fibroblast growth factor that is secreted by the endoderm cells (Nakatani et al. 1996; Kim et al. 2000; Imai et al. 2002). The anterior cells that will become notochord respond by becoming competent to receive the BMP signal from the endoderm (Darras and Nishida 2001). The BMP signal causes the notochord precursor cells to transcribe the *Brachyury* gene, which encodes a transcription factor that activates notochordal genes throughout the chordate phyla (Jeffery et al. 1998; Hotta et al. 1999; Di Gregorio et al. 2001). The posterior cells that will become mesenchyme respond differently, due to the presence of factors from the posterior vegetal cytoplasm. The FGF signal to the mesenchyme also represses muscle formation, another role that is conserved in vertebrates.

Specification of the embryonic axes

The axes of the tunicate larva are among its earliest commitments. Indeed, all of its embryonic axes are determined by the

Figure 8.40
Antibody staining of β-catenin protein shows its involvement with endoderm formation. (A) No β-catenin is seen in the animal pole nuclei of a 110-cell *Ciona* embryo. (B) In contrast, nuclear β-catenin is readily seen in the nuclei in the vegetal endoderm precursors at the 110-cell stage (C) When β-catenin is expressed in notochordal precursor cells, those cells will become endoderm and express endodermal markers such as alkaline phosphatase. The white arrows show normal endoderm; the black arrows show notochordal cells that are expressing endodermal enzymes. (Fom Imai et al. 2000; photographs courtesy of H. Nishida and N. Satoh.)

(A)
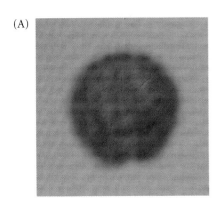

(B)
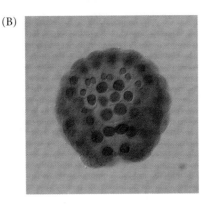

(C)
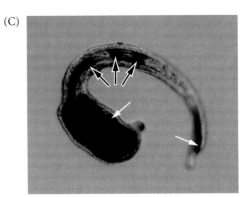

cytoplasm of the zygote prior to first cleavage. The first axis to be determined is the dorsal-ventral axis, which is defined by the cap of cytoplasm at the vegetal pole. This vegetal cap defines the future dorsal side of the larva and the site where gastrulation is initiated (Bates and Jeffery 1988). When small regions of vegetal pole cytoplasm were removed from zygotes (between the first and second waves of zygote cytoplasmic movement), the zygotes neither gastrulated nor formed a dorsal-ventral axis.

The anterior-posterior axis is the second axis to appear and is also determined during the migration of the oocyte cytoplasm. The yellow crescent forms in the region of the egg that will become the posterior side of the larva. When roughly 10% of the cytoplasm from this posterior vegetal region of the egg was removed after the second wave of cytoplasmic movement, most of the embryos failed to form an anterior-posterior axis. Rather, these embryos developed into radially symmetrical larvae with anterior fates (Figure 8.41). This posterior vegetal cytoplasm (PVC) is "dominant" to other cytoplasms in that when it was transplanted into the anterior vegetal region of zygotes that had had their own PVC removed, the anterior of the cell became the new posterior, and the axis was reversed (Nishida 1994).

Figure 8.41
Comparison of normal tunicate embryos and embryos from which posterior vegetal cytoplasm has been removed. (A) Wild-type larva. (B) Radially symmetrical larva from egg in which PVC was removed. The larva has no anterior-posterior axis. It consists of an outer epidermal layer, a central notochordal mass, and an intermediately placed endodermal layer. (C) Vegetal view of normal 76-cell embryo. (D) Vegetal view of radially symmetrical embryo whose PVC was removed. (From Nishida 1994; photographs courtesy of H. Nishida.)

The left-right axis is specified as a consequence of these first two axes, and the first cleavage divides the embryo into its future right and left sides.

Gastrulation in Tunicates

Tunicates, like sea urchins, follow the deuterostome pattern of gastrulation, in which the blastopore will eventually become the anus. Tunicate gastrulation is characterized by the invagination of the endoderm, the involution of the mesoderm, and the epiboly of the ectoderm. About 4–5 hours after fertilization, the vegetal (endoderm) cells assume a wedge shape, expanding their apical margins and contracting near their vegetal margins (Figure 8.42). The A8.1 and B8.1 pairs of blastomeres appear to lead this invagination into the center of the embryo. The invagination forms a blastopore whose lips will become the mesodermal cells. The presumptive notochord cells are now on the anterior portion of the blastopore lip, while the presumptive tail muscle cells (from the yellow crescent) are on the posterior lip of the blastopore. The lateral lips of the blastopore comprise those cells that will become mesenchyme.

The second step of gastrulation involves the involution of the mesoderm. The presumptive mesoderm cells involute over the lips of the blastopore and, by migrating upon the basal surfaces of the ectodermal cells, move inside the embryo. The ectodermal cells then flatten and epibolize over the mesoderm and endoderm, eventually covering the embryo. After gastrulation is completed, the embryo elongates along its anterior-posterior axis. The dorsal ectodermal cells that are the precur-

(A)

(B)   (C)   (D)

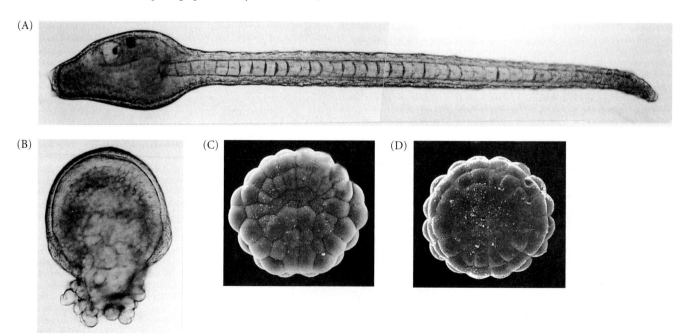

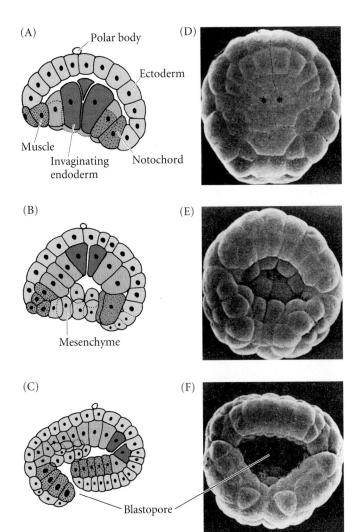

Figure 8.42
Gastrulation in the tunicate. Cross-sections (A–C) and scanning electron micrographs viewed from the vegetal pole (D–F) illustrate the invagination of the endoderm (A, D), the involution of the mesoderm (B, E), and the epiboly of the ectoderm (C, F). Cell fates are color-coded as in Figure 8.36. (From Satoh 1978 and Jeffery and Swalla 1997; photographs courtesy of N. Satoh.)

sors of the neural tube invaginate into the embryo and are enclosed by neural folds. This process forms the neural tube, which will form a brain anteriorly and a spinal chord posteriorly. Meanwhile, the presumptive notochord cells on the right and left sides of the embryo migrate to the midline and interdigitate to form the notochord, a single row of 40 cells. The muscle cells of the tail differentiate on either side of the neural tube and notochord (Jeffery and Swalla 1997). A tadpole is formed and will now search for a substrate on which to settle and metamorphose into an adult tunicate.

EARLY DEVELOPMENT OF THE NEMATODE *CAENORHABDITIS ELEGANS*

Why C. elegans?

Our ability to analyze development requires appropriate organisms. Sea urchins have long been a favorite organism of embryologists because their gametes are readily obtainable in large numbers, their eggs and embryos are transparent, and fertilization and development can occur under laboratory conditions. But sea urchins are difficult to rear in the laboratory for more than one generation, making their genetics difficult to study. Geneticists (at least those working with multicellular eukaryotes), have always favored *Drosophila*. Its rapid life cycle, its readiness to breed, and the polytene chromosomes of the larva (which allow gene localization) make the fruit fly superbly suited for hereditary analysis. But *Drosophila* development is complex and difficult to study.

A research program spearheaded by Sydney Brenner (1974) was set up to identify an organism wherein it might be possible to identify each gene involved in development as well as to trace the lineage of each and every cell. Nematodes seemed like a good place to start, since embryologists such as Goldschmidt and Boveri had shown that several species had a relatively small number of chromosomes and a small number of cells with invariant cell lineages. Eventually, Brenner and his colleagues settled upon *Caenorhabditis elegans*, a small (1 mm long), free-living soil nematode (Figure 8.43A). It has a rapid period of embryogenesis (about 16 hours), which it can accomplish in a petri dish, and relatively few cell types. Moreover, its predominant adult form is hermaphroditic, with each individual producing both eggs and sperm. These roundworms can reproduce either by self-fertilization or by cross-fertilization with the infrequently occurring males. The body of an adult *C. elegans* hermaphrodite contains exactly 959 somatic cells, whose entire lineage has been traced through its transparent cuticle (Figure 8.43B; Sulston and Horvitz 1977; Kimble and Hirsh 1979; Sulston et al. 1983). Furthermore, unlike vertebrate cell lineages, the cell lineage of *C. elegans* is almost entirely invariant from one individual to the next. There is little room for randomness (Sulston et al. 1983). *C. elegans* also has a small number of genes for a multicellular organism—about 17,500—and its genome has been entirely sequenced (the first complete sequence ever obtained for a multicellular organism; *C. elegans* Sequencing Consortium 1999).

Cleavage and Axis Formation in C. elegans

Rotational cleavage of the C. elegans *egg*

The *C. elegans* zygote exhibits **rotational holoblastic cleavage** (Figure 8.43C,D). During early cleavage, each asymmetrical

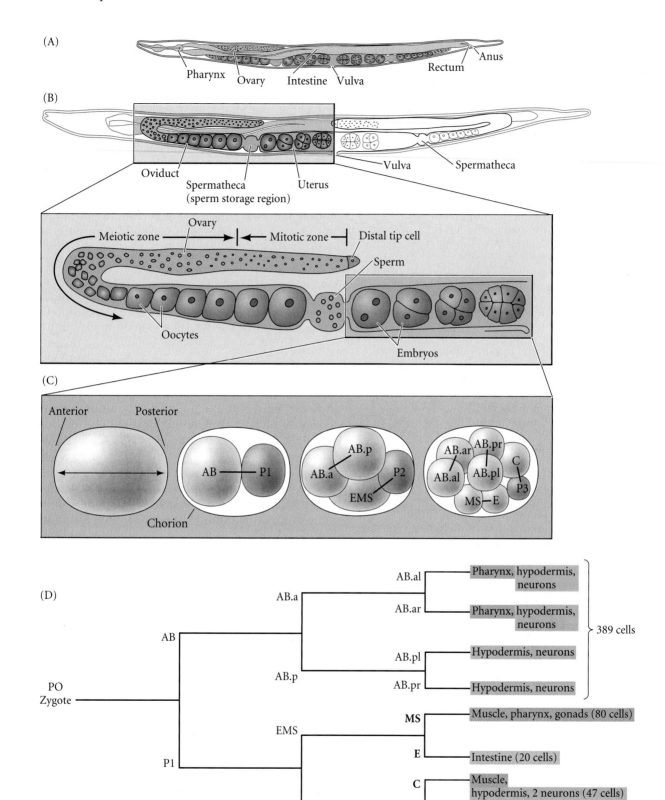

(A)

Pharynx Ovary Intestine Vulva Rectum Anus

(B)

Oviduct Spermatheca (sperm storage region) Uterus Vulva Spermatheca

Meiotic zone ⟶ Ovary ⟵ Mitotic zone Distal tip cell

Sperm

Oocytes

Embryos

(C)

Anterior Posterior

AB P1

AB.p AB.a P2 EMS

AB.ar AB.pr C AB.al AB.pl P3 MS E

Chorion

(D)

PO Zygote

AB

AB.a

AB.al — Pharynx, hypodermis, neurons
AB.ar — Pharynx, hypodermis, neurons

AB.p

AB.pl — Hypodermis, neurons
AB.pr — Hypodermis, neurons

389 cells

P1

EMS

MS — Muscle, pharynx, gonads (80 cells)
E — Intestine (20 cells)

P2

C — Muscle, hypodermis, 2 neurons (47 cells)

P3

D — Muscle (20 cells)

P4

Z2
Z3

Germ line

◀ **Figure 8.43**
Caenorhabditis elegans. (A) Side view of adult hermaphrodite. Sperm are stored such that a mature egg must pass through the sperm on its way to the vulva. (B) The gonads. Near the distal end, the germ cells undergo mitosis. As they move farther from the distal tip, they enter meiosis. Early meioses form sperm, which are stored in the spermatheca. Later meioses form eggs, which are fertilized as they roll through the spermatheca. (C) Early development, as the egg is fertilized and moves toward the vulva. The P-lineage consists of stem cells that will eventually form the germ cells. (D) Abbreviated cell lineage chart. The germ line segregates into the posterior portion of the most posterior (P) cell. The first three cell divisions produce the AB, C, MS, and E lineages. The number of derived cells (in parentheses) refers to the 558 cells present in the newly hatched larva. Some of these continue to divide to produce the 959 somatic cells of the adult. (After Pines 1992, based on Sulston and Horvitz 1977 and Sulston et al. 1983.)

division produces one **founder cell** (denoted AB, MS, E, C, and D), that produces differentiated descendants, and one **stem cell** (the P1–P4 lineage). In the first cell division, the cleavage furrow is located asymmetrically along the anterior-posterior axis of the egg, closer to what will be the posterior pole. It forms an anterior founder cell (AB) and a posterior stem cell (P1). During the second division, the founder cell (AB) divides equatorially (longitudinally; 90° to the anterior-posterior axis), while the P1 cell divides meridionally (transversely) to produce another founder cell (EMS) and a posterior stem cell (P2). The stem cell lineage always undergoes meridional division to produce (1) an anterior founder cell and (2) a posterior cell that will continue the stem cell lineage.

The descendants of each founder cell can be observed through the transparent cuticle and divide at specific times in ways that are nearly identical from individual to individual. In this way, the exactly 558 cells of the newly hatched larva are generated. Descendant cells are named according to their positions relative to their sister cells. For instance, ABal is the "left-hand" daughter cell of the Aba cell, and ABa is the "anterior" daughter cell of the AB cell.

Anterior-posterior axis formation

The elongated axis of the *C. elegans* egg defines the future anterior-posterior axis of the nematode's body. The decision as to which end will become the anterior and which the posterior seems to reside with the position of sperm pronucleus. When it enters the oocyte cytoplasm, the centriole associated with the sperm pronucleus initiates cytoplasmic movements that push the male pronucleus to the nearest end of the oblong oocyte. That end becomes the posterior pole (Goldstein and Hird 1996). The sperm centriole continues to organize microtubules in the egg, and these microtubules allow the positioning of several maternal proteins at the future anterior and future posterior ends of the egg. PAR-2 becomes localized in the cortical cytoplasm at the pole closest to the sperm nucleus. Conversely, PAR-3 localizes to the cortical cytoplasm at the opposite (the future anterior) end. The placement of these two proteins ensures that a third protein, PAR-1, will be localized to the future posterior pole. The PAR proteins were discovered by screening mutants that were defective in their cytoplasmic *partitioning*. They work together to align the mitotic spindles properly and to partition the maternal transcription factors to their appropriate cells (Kemphues et al. 1988; Kirby et al. 1990; Bowerman 1999; Wallenfang and Seydoux 2000).

Some of the most important entities localized by the PAR proteins are the **P-granules**, ribonucleoprotein complexes that specify the germ cells. The P-granules appear to be a collection of translation regulators. The proteins of these granules include RNA helicases, poly(A) polymerases, and translation initiation factors (Amiri et al. 2001; Smith et al. 2002; Wang et al. 2002).

Using fluorescent antibodies to a component of the P-granules, Strome and Wood (1983) discovered that shortly after fertilization, the randomly scattered P-granules move toward the posterior end of the zygote, so that they enter only the blastomere (P1) formed from the posterior cytoplasm (Figure 8.44). The P-granules of the P1 cell remain in the posterior of the P1 cell when it divides and are thereby passed to the P2 cell. During the division of P2 and P3, however, the P-granules become associated with the nucleus that enters the P3 cytoplasm. Eventually, the P-granules will reside in the P4 cell, whose progeny become the sperm and eggs of the adult. The localization of the P-granules requires microfilaments, but can occur in the absence of microtubules. Treating the zygote with cytochalasin D (a microfilament inhibitor) prevents the segregation of these granules to the posterior of the cell, whereas demecolcine (a colchicine-like microtubule inhibitor) fails to stop this movement.

WEBSITE 8.6 **P-granule migration.** Movies of P-granule migration under natural and experimental conditions were taken in the laboratory of Susan Strome. They show P-granule segregation to the P-lineage blastomeres except when perturbed by mutations or chemicals that inhibit microfilament function.

WEBSITE 8.7 **The PAR proteins.** The polarity of cell divisions and the distribution of the morphogenetic determinants are specified by the positions of the PAR proteins. These proteins are critical in coordinating cell division and cytoplasmic localization.

Formation of the dorsal-ventral and right-left axes

The dorsal-ventral axis of the nematode is seen in the division of the AB cell. As the AB cell divides, it becomes longer than the eggshell is wide. This causes the cells to slide, resulting in one AB daughter cell being anterior and one being posterior (hence their names, ABa and ABp, respectively; see Figure 8.46). This

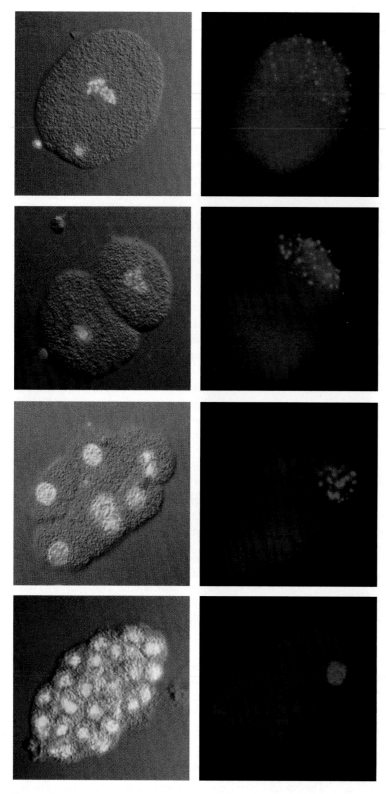

Figure 8.44
Segregation of the P-granules into the germ line lineage of the *C. elegans* embryo. The left column shows the cell nuclei (the DNA is stained blue by Hoescht dye), while the right column shows the same embryos stained for P-granules. At each successive division, the P-granules enter the P-lineage blastomere, the one that will form the germ cells. (Photographs courtesy of S. Strome.)

squeezing also causes the ABp cell to take a position above the EMS cell that results from the division of the P1 blastomere. The ABp cell defines the future dorsal side of the embryo, while the EMS cell, the precursor of the muscle and gut cells, marks the future ventral surface of the embryo. The left-right axis is specified later, at the 12-cell stage, when the MS blastomere (from the division of the EMS cell) contacts half the "granddaughters" of the ABa cell, distinguishing the right side of the body from the left side (Evans et al. 1994).

Control of blastomere identity

C. elegans demonstrates both the conditional and autonomous modes of cell specification. Both modes can be seen if the first two blastomeres are experimentally separated (Priess and Thomson 1987). The P1 cell develops autonomously without the presence of AB. It makes all the cells that it would normally make, and the result is the posterior half of an embryo. However, the AB cell, in isolation, makes only a fraction of the cell types that it would normally make. For instance, the resulting ABa blastomere fails to make the anterior pharyngeal muscles that it would have made in an intact embryo. Therefore, the specification of the AB blastomere is conditional, and it needs the descendants of the P1 cell to interact with it.

AUTONOMOUS SPECIFICATION. The determination of the P1 lineages appears to be autonomous, with the cell fates determined by internal cytoplasmic factors rather than by interactions with neighboring cells. The P-granules of *C. elegans* are localized in a way consistent with a role as a morphogenetic determinant, and they act through translational regulation (the exact mechanism is not well understood) to specify germ cells. Meanwhile, the SKN-1, PAL-1, and PIE-1 proteins are thought to encode transcription factors that act intrinsically to determine the fates of cells derived from the four P1-derived somatic founder cells, MS, E, C, and D.

The **SKN-1** protein is a maternally expressed polypeptide that may control the fate of the EMS blastomere, the cell that generates the posterior pharynx. After first cleavage, only the posterior blastomere, P1, has the ability to autonomously produce pharyngeal cells when isolated. After P1 divides, only EMS is able to generate pharyngeal muscle cells in isolation (Priess and Thomson 1987). Similarly, when the EMS cell divides, only one of its progeny, MS, has the intrinsic ability to generate pharyngeal tissue. These findings suggest that pharyngeal cell fate may be determined autonomously by maternal factors residing in the cytoplasm that are parceled out to these particular cells. Bowerman and his co-workers (1992a,b, 1993) found maternal effect mutants lacking pharyngeal cells and were able to isolate a mutation in the *skn-1* gene. Embryos from homozygous *skn-1*-deficient mothers lack both pharyngeal mesoderm and endoderm derivatives of EMS (Figure 8.45). Instead of making the normal intestinal and pharyngeal structures, these embryos seem to make extra hypodermal (skin) and body wall tissue where their intestine and pharynx should be. In other words, EMS appears to be respecified as C. Only those cells destined to form pharynx or intestine are affected by this mutation. Moreover, the protein encoded by the *skn-1* gene has a DNA-binding site motif similar to that seen in the bZip family of transcription factors (Blackwell et al. 1994).

SKN-1 is a maternal protein, and it activates the transcription at least two genes, *med-1* and *med-2*, whose products are also transcription factors. The MED transcription factors appear to specify the entire fate of the EMS cell, since expression of the *med* genes in other cells can cause non-EMS cells to become EMS even if SKN-1 is absent (Maduro et al. 2001).

A second putative transcription factor, **PAL-1**, is also required for the differentiation of the P1 lineage. PAL-1 activity is needed for the normal development of the somatic descendants of the P2 blastomere. Thus, embryos lacking PAL-1 have no somatic cell types derived from the C and D stem cells (Hunter and Kenyon 1996). PAL-1 is regulated by the MEX-3 protein, an RNA-binding protein that appears to inhibit the translation of *pal-1* mRNA. Wherever *mex-3* is expressed, PAL-1 is absent. Thus, in *mex-3*-deficient mutants, PAL-1 is seen in every blastomere. SKN-1 also inhibits PAL-1 (thereby preventing it from becoming active in the EMS cell).

A third putative transcription factor, **PIE-1,** is necessary for the germ line fate. PIE-1 is placed into the P blastomeres through the action of the PAR-1 protein, and it appears to inhibit both SKN-1 and PAL-1 function in the P2 and subsequent germ line cells (Hunter and Kenyon 1996). Mutations of the maternal *pie-1* gene result in germ line blastomeres adopting somatic fates, with the P2 cell behaving similarly to a wild-type EMS blastomere. The localization and the genetic properties of PIE-1 suggest that it represses the establishment of somatic cell fate and preserves the totipotency of the germ cell lineage (Mello et al. 1996; Seydoux et al. 1996).

WEBSITE 8.8 Mechanisms of cytoplasmic localization in *C. elegans.* Analyses of *C. elegans* mutations have isolated genes whose protein products are essential in specifying cell fate. In some instances, these proteins are involved in the placement of morphogenetic determinants.

CONDITIONAL SPECIFICATION. As we saw above, the *C. elegans* embryo uses both autonomous and conditional modes of specification. Conditional specification can be seen in the development of the endoderm cell lineage. At the 4-cell stage, the EMS cell requires a signal from its neighbor (and sister), the P2 blastomere. Usually, the EMS cell divides into an MS cell (which produces mesodermal muscles) and an E cell (which produces the intestinal endoderm). If the P2 cell is removed at the early 4-cell stage, the EMS cell will divide into two MS cells, and endoderm will not be produced. If the EMS cell is recombined with the P2 blastomere, however, it will form endoderm; it will not do so, however, when combined with ABa, ABp, or both AB derivatives (Figure 8.46; Goldstein 1992).

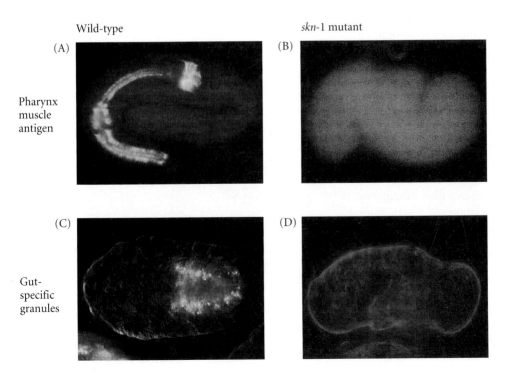

Wild-type

skn-1 mutant

(A)

(B)

Pharynx muscle antigen

(C)

(D)

Gut-specific granules

Figure 8.45
Deficiencies of intestine and pharynx in *skn-1* mutants of *C. elegans.* Embryos derived from wild-type females (A, C) and from females homozygous for mutant *skn-1* (B, D) were tested for the presence of pharyngeal muscles (A, B) and gut-specific granules (C, D). A pharyngeal muscle-specific antibody labels the pharynx musculature of those embryos derived from wild-type females (A), but does not bind to any structure in the embryos from *skn-1* mutant females (B). Similarly, the birefringent gut granules characteristic of embryonic intestines (C) are absent from embryos derived from the *skn-1* mutant females (D). (From Bowerman et al. 1992a; photographs courtesy of B. Bowerman.)

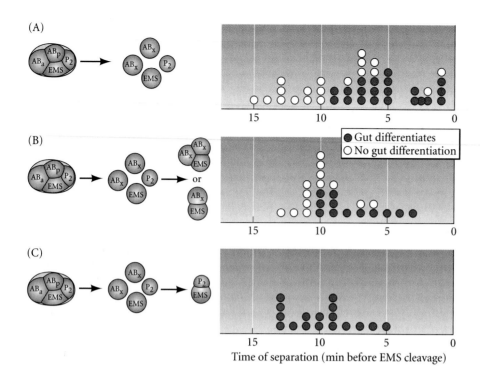

(A)

(B)

(C)

● Gut differentiates
○ No gut differentiation

Time of separation (min before EMS cleavage)

Figure 8.46
Isolation and recombination experiments show that cell-cell interactions are required for the EMS cell to form intestinal lineage determinants. (A) When isolated shortly after its formation, the EMS blastomere cannot produce gut-specific granules. If left in place for longer periods, it can. (B) If the EMS cell is recombined with either or both derivatives of the AB blastomere, it will not form gut-specific granules. (C) If recombined with the P2 blastomere, the EMS cell gives rise to gut-specific structures. (After Goldstein 1992.)

The P2 cell produces a signal that interacts with the EMS cell and instructs the EMS daughter that is next to it to become the E cell. This message is transmitted through the Wnt signaling cascade (Figure 8.47; Rocheleau et al. 1997; Thorpe et al. 1997). The P2 cell produces a *C. elegans* homologue of a Wnt protein, the MOM-2 peptide. The MOM-2 peptide is received in the EMS cell by the MOM-5 protein, a *C. elegans* version of the Wnt receptor protein Frizzled. The result of this signaling cascade is to down-regulate the expression of the *pop-1* gene in the EMS daughter destined to become the E cell. In *pop-1*-deficient embryos, both EMS daughter cells become E cells (Lin et al. 1995).

The P2 cell is also critical in giving the signal that distinguishes ABp from its sister, ABa (see Figure 8.47). ABa gives rise to neurons, hypodermis, and the anterior pharynx cells, while ABp makes only neurons and hypodermal cells. However, if one experimentally reverses the positions of these two cells, their fates are similarly reversed, and a normal embryo is formed. In other words, ABa and ABp are equivalent cells whose fate is determined by their positions within the embryo (Priess and Thomson 1987). Transplantation and genetic studies have shown that ABp becomes different from ABa through its interaction with the P2 cell. In an unperturbed embryo, both ABa and ABp contact the EMS blastomere, but only ABp contacts the P2 cell. If the P2 cell is killed at the early 4-cell stage, the ABp cell does not generate its normal complement of cells (Bowerman et al. 1992a,b). Contact between ABp and P2 is essential for the specification of ABp cell fates, and the ABa cell can be made into an ABp-type cell if it is forced into contact with P2 (Hutter and Schnabel 1994; Mello et al. 1994).

This interaction is mediated by the GLP-1 protein on the ABp cell and the APX-1 (anterior pharynx excess) protein on the P2 blastomere. In embryos whose mothers have mutant *glp-1*, ABp is transformed into an ABa cell (Hutter and Schnabel 1994; Mello et al. 1994). The GLP-1 protein is a member of a widely conserved family called the Notch proteins, which serve as cell membrane receptors in many cell-cell interactions, and it is seen on both the ABa and ABp cells (Evans et

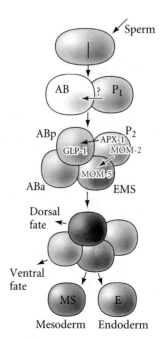

Figure 8.47
Cell-cell signaling in the 4-cell embryo of *C. elegans*. The P2 cell produces two signals: (1) the juxtacrine protein APX-1 (Delta), which is bound by GLP-1 (Notch) on the ABp cell, and (2) the paracrine protein MOM-2 (Wnt), which is bound by the MOM-5 (Frizzled) protein on the EMS cell. (After Han 1998.)

al. 1994).* As mentioned in Chapter 5, one of the most important ligands for Notch proteins such as GLP-1 is another cell surface protein called Delta. In *C. elegans*, the Delta-like protein is APX-1, and it is found on the P2 cell (Mango et al. 1994a; Mello et al. 1994). This APX-1 signal breaks the symmetry between ABa and ABp, since it stimulates the GLP-1 protein solely on the AB descendant that it touches—namely, the ABp blastomere. In doing this, the P2 cell initiates the dorsal-ventral axis of *C. elegans* and confers on the ABp blastomere a fate different from that of its sister cell.

Integration of autonomous and conditional specification: The differentiation of the C. elegans pharynx

In the above discussion, it should become apparent that the pharynx is generated by two sets of cells. One group of pharyngeal precursors comes from the EMS cell and is dependent on the maternal *skn-1* gene. The second group of pharyngeal precursors comes from the ABa blastomere and is dependent on GLP-1 signaling from the EMS cell. In both cases, the pharyngeal precursor cells (and only these cells) are instructed to activate the *pha-4* gene (Mango et al. 1994b). This gene encodes a transcription factor that resembles the mammalian HNF-3β protein. Using microarrays to determine the targets of this protein, Gaudet and Mango (2002) found that the PHA-4 transcription factor activates almost all of the pharynx-specific genes. It appears that the PHA-4 transcription factor may be the node that takes the maternal inputs and transforms them into a signal that transcribes those zygotic genes necessary for pharynx development.

Gastrulation in C. elegans

Gastrulation in *C. elegans* starts extremely early, just after the generation of the P4 cell in the 24-cell embryo (Skiba and Schierenberg 1992). At this time, the two daughters of the E cell (Ea and Ep) migrate from the ventral side into the center of the embryo. There they will divide to form a gut consisting of 20 cells. There is a very small and transient blastocoel prior to the movement of the Ea and Ep cells, and their inward migration creates a tiny blastopore. The next cell to migrate through this blastopore is the P4 cell, the precursor of the germ cells. It migrates to a position beneath the gut primordium. The mesodermal cells move in next: the descendants of the MS cell migrate inward from the anterior side of the blastopore, and the C- and D-derived muscle precursors enter from the posterior side. These cells flank the gut tube on the left and right sides (Figure

8.48; Schierenberg 1997). Finally, about 6 hours after fertilization, the AB-derived cells that contribute to the pharynx are brought inside, while the hypoblast (hypodermal precursor) cells move ventrally by epiboly, eventually closing the blastopore. The two sides of the hypodermis are sealed by E-cadherin on the tips of the leading cells that meet at the ventral midline (Raich et al. 1999). During the next 6 hours, the cells move and develop into organs, while the ball-shaped embryo stretches out to become a worm (see Priess and Hirsh 1986; Schierenberg 1997). This hermaphroditic worm will have 558 somatic cells. An additional 115 cells will have formed, but undergone apoptosis (see Chapter 6). After four molts, this worm will be a sexually mature adult, containing 959 somatic cells as well as numerous sperm and eggs.

The *C. elegans* research program was designed to integrate genetics and embryology to provide an understanding of the networks that govern cell differentiation and morphogenesis. In addition to giving us some remarkable insights into how gene expression can change during development, studies of *C. elegans* have also humbled us by demonstrating how complex these networks are. Even in an organism as "simple" as *C. elegans*, with only a few genes and cell types, the right side of the body is made in a different manner from the left! The identification of the genes mentioned above is just the beginning of our effort to understand the complex interacting systems of development.

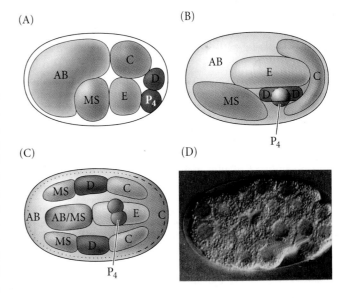

Figure 8.48

Gastrulation in *C. elegans*. (A) Positions of founder cells and their descendants at the 26-cell stage, just prior to gastrulation. (B) 102-cell stage, after the migration of the E, P4, and D descendants. (C) Positions of the cells near the end of gastrulation. The dotted and dashed lines represent regions of the hypodermis contributed by AB and C, respectively. (D) Early gastrulation, as the two E cells start moving inward. (After Schierenberg 1998; photograph courtesy of E. Schierenberg.)

---

*The GLP-1 protein is localized in the ABa and ABp blastomeres, but the maternally encoded *glp-1* mRNA is found throughout the embryo. Evans and colleagues (1994) have postulated that there might be some translational determinant in the AB blastomere that enables the glp-1 message to be translated in its descendants. The *glp-1* gene is also active in regulating postembryonic cell-cell interactions. It is used later by the distal tip cell of the gonad to control the number of germ cells entering meiosis; hence the name GLP, "germ line proliferation."

Coda

This chapter has described early embryonic development in four invertebrate species, each of which develops in a different pattern. The largest group of animals on this planet, however, is another invertebrate group—the insects. We probably know more about the development of one particular insect, *Drosophila melanogaster*, than any other organism. The next chapter details the early development of this particularly well-studied creature.

Snapshot Summary: Early Invertebrate Development

1. During cleavage, most cells do not grow. Rather, the volume of the oocyte is cleaved into numerous cells. The major exceptions to this rule are mammals.

2. "Blast" vocabulary: A *blastomere* is a cell derived from cleavage in an early embryo. A *blastula* is an embryonic structure composed of blastomeres. The cavity within the blastula is the *blastocoel*. If the blastula lacks a blastocoel, it is a *stereoblastula*. (A mammalian blastula is called a *blastocyst*; see Chapter 11.) The invagination where gastrulation begins is the *blastopore*.

3. The blastomere cell cycle is governed by the synthesis and degradation of cyclin B. Cyclin B synthesis promotes the formation of mitosis-promoting factor, and MPF promotes mitosis. Degradation of cyclin B brings the cell back to the S phase. The G phases are added at the mid-blastula transition.

4. The movements of gastrulation include invagination, involution, ingression, delamination, and epiboly.

5. Three axes form the foundations of the body: the anterior-posterior axis (head to tail or mouth to anus), the dorsal-ventral axis (back to belly), and the right-left axis (between the two lateral sides of the body).

6. Body axes are established in different ways in different species. In some, such as the sea urchin and tunicate, the axes are established at fertilization through determinants in the egg cytoplasm. In others, such as the nematode and snail, the axes are established by cell interactions later in development.

7. In all four invertebrates described in this chapter, cleavage is holoblastic. In the sea urchin, cleavage is radial; in the snail, spiral; in the tunicate, bilateral; and in the nematode, rotational.

8. In the sea urchin, gastrulation occurs only after thousands of cells have formed, and the blastopore becomes the anus. This is the deuterostome mode of gastrulation, and is characteristic of echinoderms, tunicates, and chordates (including vertebrates).

9. In the sea urchin, cell fates are determined by cell-cell signaling. The micromeres constitute a major signaling center. Nuclear β-catenin is important for the inducing capacity of the micromeres.

10. Differential cell adhesion is important in regulating sea urchin gastrulation. The micromeres detach first from the vegetal plate and move into the blastocoel. They form the primary mesenchyme, which becomes the skeletal rods of the pluteus larva. The vegetal plate invaginates to form the endodermal archenteron, with a tip of secondary mesenchyme cells. The archenteron elongates by convergent extension and is guided to the future mouth region by the secondary mesenchyme.

11. Snails exhibit spiral cleavage and form stereoblastulae, with no blastocoels. The direction of spiral cleavage is regulated by a factor encoded by the mother and placed in the oocyte. Spiral cleavage can be modified by evolution, and adaptations of spiral cleavage have allowed some molluscs to survive in otherwise harsh environments.

12. The polar lobe of certain molluscs contains the morphogenetic determinants for mesoderm and endoderm. These determinants enter the D blastomere.

13. In the snail and the nematode, gastrulation occurs when there are relatively few cells, and the blastopore becomes the mouth. This is the protostome mode of gastrulation. The tunicate, like the sea urchin, is a deuterostome; thus the blastopore is posterior while the mouth is formed elsewhere.

14. The tunicate fate map is identical on its right and left sides. The yellow cytoplasm contains muscle-forming determinants; these act autonomously. The nervous system of tunicates is formed conditionally, by interactions between blastomeres.

15. The soil nematode *Caenorhabditis elegans* was chosen as a model organism because it has a small number of cells, has a small genome, is easily bred and maintained, has a short lifespan, can be genetically manipulated, and has a cuticle through which one can see cell movements.

16. In the early divisions of the *C. elegans* zygote, one daughter cell becomes a founder cell (producing differentiated descendants) and the other becomes a stem cell (producing other founder cells and the germ line).

17. Blastomere identity in *C. elegans* is regulated by both autonomous and conditional specification.

Literature Cited

Amiri, A., B. D. Keiper, I. Kawasaki, Y. Fan, Y. Kohara, R. E. Rhoads and S. Strome. 2001. An isoform of eIF4E is a component of germ granules and is required for spermatogenesis in *C. elegans*. *Development* 128: 3899–3912.

Angerer, L. M., D. W. Oleksyn, A. M. Levine, X. Li, W. H. Klein and R. C. Angerer. 2001. Sea urchin goosecoid function links fate specification along the animal-vegetal and oral-aboral embryonic axes. *Development* 128: 4393–4404.

Armstrong, N., J. Hardin and D. R. McClay. 1993. Cell-cell interactions regulate skeleton formation in the sea urchin embryo. *Development* 119: 833–840.

Atkinson, J. W. 1987. An atlas of light micrographs of normal and lobeless larvae of the marine gastropod *Ilyanassa obsoleta*. *Int. J. Invert. Reprod. Dev.* 9: 169–178.

Balinsky, B. I. 1981. *Introduction to Embryology*, 5th Ed. Saunders, Philadelphia.

Bates, W. R. and W. R. Jeffery. 1988. Polarization of ooplasmic segregation and dorsal-ventral axis determination in ascidian embryos. *Dev. Biol.* 130: 98–107.

Berg, L. K., S. W. Chen and G. M. Wessel. 1996. An extracellular matrix molecule that is selectively expressed during development is important for gastrulation in the sea urchin embryo. *Development* 122: 703–713.

Bisgrove, B. W., M. E. Andrews and R. A. Raff. 1991. Fibropellins, products of an EGF repeat-containing gene, form a unique extracellular matrix structure that surrounds the sea urchin embryo. *Dev. Biol.* 146: 89–99.

Blackwell, T. K., B. Bowerman, J. R. Priess and H. Weintraub. 1994. Formation of a monomeric DNA binding domain by SKN bZIP and homeodomain elements. *Science* 266: 621–628.

Bonder, E. M., D. J. Fishkind, J. H. Henson, N. M. Cotran and D. A. Begg. 1988. Actin in cytokinesis: Formation of the contractile apparatus. *Zool. Sci.* 5: 699–701.

Boveri, T. 1901. Die Polaritat von Ovocyte, Ei, und Larve des *Strongylocentrotus lividus*. *Zool. Jahrb. Abt. Anat. Ontog. Tiere* 14: 630–653.

Bowerman, B. 1999. Maternal control of polarity and patterning during embryogenesis in the nematode *Caenorhabditis elegans*. *In* S. A. Moody (ed.), *Cell Lineage and Determination*. Academic Press, New York, pp. 97–118.

Bowerman, B., B. A. Eaton and J. R. Priess. 1992a. *skn-1*, a maternally expressed gene required to specify the fate of ventral blastomeres in the early *C. elegans* embryo. *Cell* 68: 1061–1075.

Bowerman, B., F. E. Tax, J. H. Thomas and J. R. Priess. 1992b. Cell interactions involved in the development of the bilaterally symmetrical intestinal valve cells during embryogenesis of *Caenorhabditis elegans*. *Development* 116: 1113–1122.

Bowerman, B., B. W. Draper, C. C. Mello and J. R. Priess. 1993. The maternal gene *skn-1* encodes a protein that is distributed unequally in early *C. elegans* embryos. *Cell* 74: 443–452.

Boycott, A. E., C. Diver, S. L. Garstang and F. M. Turner. 1930. The inheritance of sinistrality in *Limnaea peregra* (Mollusca: Pulmonata). *Phil. Trans. R. Soc. Lond.* [B] 219: 51–131.

Brenner, S. 1974. The genetics of *Caenorhabditis elegans*. *Genetics* 77: 71–94.

Burke, R. D., R. L. Myers, T. L. Sexton and C. Jackson. 1991. Cell movements during the initial phase of gastrulation in the sea urchin embryo. *Dev. Biol.* 146: 542–558.

Cameron, R. A., S. E. Fraser, R. J. Britten and E. H. Davidson. 1989. The oral-aboral axis of a sea urchin embryo is specified by the first cleavage. *Development* 106: 641–647.

Cather, J. N. 1967. Cellular interactions in the development of the shell gland of the gastropod *Ilyanassa*. *J. Exp. Zool.* 166: 205–224.

C. elegans Sequencing Consortium. 1999. Genome sequence of the nematode *C. elegans*: A platform for investigating biology. *Science* 282: 2012–2018. www.sanger.ac.uk; genome.wustl.edu/gsc/gschmpg.html

Chabry, L. M. 1888. Contribution a l'embryologie normale tératologique des ascidies simples. *J. Anat. Physiol. Norm. Pathol.* 23: 167–321.

Cherr, G. N., R. G. Summers, J. D. Baldwin and J. B. Morrill. 1992. Preservation and visualization of the sea urchin blastocoelic extracellular matrix. *Microsc. Res. Tech.* 22: 11–22.

Clement, A. C. 1962. Development of *Ilyanassa* following removal of the D micromere at successive cleavage stages. *J. Exp. Zool.* 149: 193–215.

Clement, A. C. 1967. The embryonic value of the micromeres in *Ilyanassa obsoleta*, as determined by deletion experiments. I. The first quartet cells. *J. Exp. Zool.* 166: 77–88.

Clement, A. C. 1968. Development of the vegetal half of the *Ilyanassa* egg after removal of most of the yolk by centrifugal force, compared with the development of animal halves of similar visible composition. *Dev. Biol.* 17: 165–186.

Clement, A. C. 1986. The embryonic value of the micromeres in *Ilyanassa obsoleta*, as determined by deletion experiments. III. The third quartet cells and the mesentoblast cell, 4d. *Int. J. Invert. Reprod. Dev.* 9: 155–168.

Collier, J. R. 1997. Gastropods, the snails. *In* S. F. Gilbert and A. Raunio (eds.), *Embryology: Constructing the Organism*. Sinauer Associates, Sunderland, MA, pp. 189–218.

Conklin, E. G. 1897. The embryology of *Crepidula*. *J. Morphol.* 13: 3–209.

Conklin, E. G. 1905. The orientation and cell-lineage of the ascidian egg. *J. Acad. Nat. Sci. Phila.* 13: 5–119.

Craig, M. M. and J. B. Morrill. 1986. Cellular arrangements and surface topography during early development in embryos of *Ilyanassa obsoleta*. *Int. J. Invert. Reprod. Dev.* 9: 209–228.

Crampton, H. E. 1894. Reversal of cleavage in a sinistral gastropod. *Ann. NY Acad. Sci.* 8: 167–170.

Crampton, H. E. 1896. Experimental studies on gastropod development. *Arch. Entwicklunsmech.* 3: 1–19.

Dan, K. 1960. Cytoembryology of echinoderms and amphibia. *Int. Rev. Cytol.* 9: 321–368.

Dan, K. and K. Okazaki. 1956. Cyto-embryological studies of sea urchins. III. Role of secondary mesenchyme cells in the formation of the primitive gut in sea urchin larvae. *Biol. Bull.* 110: 29–42.

Danilchik, M. V., W. C. Funk, E. E. Brown and K. Larkin. 1998. Requirement for microtubules in new membrane formation during cytokinesis of *Xenopus* embryos. *Dev. Biol.* 194: 47–60.

Dan-Sohkawa, M. and H. Fujisawa. 1980. Cell dynamics of the blastulation process in the starfish, *Asterina pectinifera*. *Dev. Biol.* 77: 328–339.

Darras, S. and H. Nishida. 2001. The BMP signaling pathway is required together with the FGF pathway for notochord induction in the ascidian embryo. *Development* 128(14): 2629–2638.

Davidson, E. H. 2001. *Genomic Regulatory Systems*. Academic Press, New York.

Davidson, E. H. and 24 others. 2002. A provisional regulatory gene network for specification of endomesoderm in the sea urchin embryo. *Dev. Biol.* 246: 162–190.

Di Gregorio, A., J. C. Corbo and M. Levine. 2001. The regulation of *forkhead/HNF-3β* expression in the *Ciona* embryo. *Dev. Biol.* 229: 31–43.

Edgar, B., C. P. Kiehle and G. Schubiger. 1986. Cell cycle control by the nucleo-cytoplasmic ratio in early *Drosophila* development. *Cell* 44: 365–372.

Erives, A. and M. Levine. 2000. Characterization of a maternal T-box gene in *Ciona intestinalis*. *Dev. Biol.* 225: 169–178.

Ettensohn, C. A. 1985. Gastrulation in the sea urchin embryo is accompanied by the rearrangement of invaginating epithelial cells. *Dev. Biol.* 112: 383–390.

Ettensohn, C. A. 1990. The regulation of primary mesenchyme cell patterning. *Dev. Biol.* 140: 261–271.

Ettensohn, C. A. and E. P. Ingersoll. 1992. Morphogenesis of the sea urchin embryo. *In* E. F. Rossomondo and S. Alexander (eds.), *Morphogenesis*. Marcel Dekker, New York, pp. 189–262.

Ettensohn, C. A. and D. R. McClay. 1986. The regulation of primary mesenchyme cell migration in the sea urchin embryo: Transplantations of cells and latex beads. *Dev. Biol.* 117: 380–391.

Ettensohn, C. A. and H. Sweet. 2000. Patterning of the early sea urchin embryo. *Curr. Top. Dev. Biol.* 50: 1–44.

Evans, T., E. Rosenthal, J. Youngblom, D. Distel and T. Hunt. 1983. Cyclin: A protein specified by maternal mRNA in sea urchin eggs that is destroyed at each cleavage division. *Cell* 33: 389–396.

Evans, T. C., S. L. Crittenden, V. Kodoyianni and J. Kimble. 1994. Translational control of maternal *glp-1* mRNA establishes an asymmetry in the *C. elegans* embryo. *Cell* 77: 183–194.

Fink, R. D. and D. R. McClay. 1985. Three cell recognition changes accompany the ingression of sea urchin primary mesenchyme cells. *Dev. Biol.* 107: 66–74.

Freeman, G. and J. W. Lundelius. 1982. The developmental genetics of dextrality and sinistrality in the gastropod *Limnea peregra*. *Wilhelm Roux Arch. Dev. Biol.* 191: 69–83.

Galileo, D. S. and J. B. Morrill. 1985. Patterns of cells and extracellular material of the sea urchin *Lytechinus variegatus* (Echinodermata; Echinoidea) embryo, from hatched blastula to late gastrula. *J. Morphol.* 185: 387–402.

Gaudet, J. and S. E. Mango. 2002. Regulation of organogenesis by the *Caenorhabditis elegans* FoxA protein PHA-4. *Science* 295: 821–825.

Gerhart, J. C., M. Wu and M. Kirschner. 1984. Cell dynamics of an M-phase-specific cytoplasmic factor in Xenopus laevis oocytes and eggs. *J. Cell Biol.* 98: 1247–1255.

Goldstein, B. 1992. Induction of gut in *Caenorhabditis elegans* embryos. *Nature* 357: 255–258.

Goldstein, B. and S. N. Hird. 1996. Specification of the anterioposterior axis in *Caenorhabditis elegans*. *Development* 122: 1467–1474.

Guerrier, P., J. A. M. van den Biggelaar, C. A. M. Dongen and N. H. Verdonk. 1978. Significance of the polar lobe for the determination of dorsoventral polarity in *Dentalium vulgare* (da Costa). *Dev. Biol.* 63: 233–242.

Gustafson, T. and L. Wolpert. 1967. Cellular movement and contact in sea urchin morphogenesis. *Biol. Rev.* 42: 442–498.

Hall, H. G. and V. D. Vacquier. 1982. The apical lamina of the sea urchin embryo: Major glycoprotein associated with the hyaline layer. *Dev. Biol.* 89: 168–178.

Han, M. 1998. Gut reaction to Wnt signaling in worms. *Cell* 90: 581–584.

Hardin, J. D. 1988. The role of secondary mesenchyme cells during sea urchin gastrulation studied by laser ablation. *Development* 103: 317–324.

Hardin, J. D. 1990. Context-dependent cell behaviors during gastrulation. *Semin. Dev. Biol.* 1: 335–345.

Hardin, J. D. and L. Y. Cheng. 1986. The mechanisms and mechanics of archenteron elongation during sea urchin gastrulation. *Dev. Biol.* 115: 490–501.

Hardin, J. D. and D. R. McClay. 1990. Target recognition by the archenteron during sea urchin gastrulation. *Dev. Biol.* 142: 87–105.

Harkey, M. A. and A. M. Whiteley. 1980. Isolation, culture and differentiation of echinoid primary mesenchyme cells. *Wilhelm Roux Arch. Dev. Biol.* 189: 111–122.

Henry, J. J. and M. Q. Martindale. 1987. The organizing role of the D quadrant as revealed through the phenomenon of twinning in the polychaete *Chaetopterus variopedatus*. *Wilhelm Roux Arch. Dev. Biol.* 196: 449–510.

Henry, J. J. and R. A. Raff. 1990. Evolutionary change in the process of dorsoventral axis determination in the direct developing sea urchin, *Heliocidaris erythrogramma*. *Dev. Biol.* 141: 55–69.

Hodor, P. G. and C. A. Ettensohn. 1998. The dynamics and regulation of mesenchymal cell fusion in the sea urchin. *Dev. Biol.* 199: 111–124.

Hörstadius, S. 1939. The mechanics of sea urchin development, studied by operative methods. *Biol. Rev.* 14: 132–179.

Hörstadius, S. 1973. *Experimental Embryology of Echinoderms.* Clarendon Press, Oxford.

Hotta, K., H. Takahashi, A. Erives, M. Levine and N. Satoh. 1999. Temporal expression patterns of 39 *Brachyury*-downstream genes associated with notochord formation in the *Ciona intestinalis* embryo. *Dev. Growth Diff.* 41: 657–664.

Howard, E. W., L. A. Newman, D. W. Oleksyn, R. C. Angerer and L. M. Angerer. 2001. SpKrl: A direct target of β-catenin regulation required for endoderm differentiation in sea urchin embryos. *Development* 128:365–375.

Hunter, C. P. and C. Kenyon. 1996. Spatial and temporal controls target *pal-1* blastomere-specification activity to a single blastomere in *C. elegans* embryos. *Cell* 87: 217–226.

Hutter, H. and R. Schnabel. 1994. *glp-1* and inductions establishing embryonic axes in *C. elegans*. *Development* 120: 2051–2064.

Imai, K. S., N. Takada, N. Satoh and Y. Satou. 2000. β-Catenin mediates the specification of endoderm cells in ascidian embryos. *Development* 127: 3009–3020.

Imai, K. S., N. Satoh and Y. Satou. 2002. Early embryonic expression of *FGF4/6/9* gene and its role in the induction of mesenchyme and notochord in *Ciona savignyi* embryos. *Development* 129: 1729–1738.

Jeffery, W. R. and B. J. Swalla. 1997. Tunicates. *In* S. F. Gilbert and A. M. Raunio (eds.), *Embryology: Constructing the Organism.* Sinauer Associates, Sunderland, MA, pp. 331–364.

Jeffery, W. R., N. Ewing, J. Machula, C. L. Olsen and B. J. Swalla. 1998. Cytoskeletal actin genes function downstream of HNF-3β in ascidian

notochord development. *Int. J. Dev. Biol.* 42:1085–1092.

Karp, G. C. and M. Solursh. 1985. Dynamic activity of the filopodia of sea urchin embryonic cells and their role in directed migration of the primary mesenchyme in vitro. *Dev. Biol.* 112: 276–283.

Kemphues, K. J., J. R. Priess, D. G. Morton and N. Cheng. 1988. Identification of genes required for cytoplasmic localization in early *C. elegans* embryos. *Cell* 52: 311–320.

Kenny, A. P., D. Kozlowski, D. W. Oleksyn, L. M. Angerer and R. C. Angerer. 1999. SpSoxB1, a maternally encoded transcription factor asymmetrically distributed among early sea urchin blastomeres. *Development* 126: 5473–5483.

Kim, G. J., A. Yamada and H. Nishida. 2000. An FGF signal from endoderm and localized factors in the posterior-vegetal egg cytoplasm pattern the mesodermal tissues in the ascidian embryo. *Development* 127: 2853–2862.

Kimberly, E. L. and J. Hardin.1998. Bottle cells are required for the initiation of primary invagination in the sea urchin embryo. *Dev. Biol.* 204: 235–250.

Kimble, J. and D. Hirsh. 1979. The postembryonic cell lineages of the hermaphrodite and male gonads in *Caenorhabditis elegans*. *Dev. Biol.* 70: 396–418.

Kirby, C. M., M. Kusch and K. Kemphues. 1990. Mutations in the par genes of *Caenorhabditis elegans* affect cytoplasmic reorganization during the first cell cycle. *Dev. Biol.* 142: 203–215.

Lambert, J. D. L. and L. M. Nagy. 2001. MAPK signaling by the D quadrant embryonic organizer of the mollusk *Ilyanassa obsoleta*. *Development* 128: 45–56.

Lambert, J. D. and L. M. Nagy. 2002. Asymmetric inheritance of centrosomally localized mRNAs during embryonic cleavage. *Nature* 420: 682–686.

Lane, M. C. and M. Solursh. 1991. Primary mesenchyme cell migration requires a chondroitin sulfate/dermatan sulfate proteoglycan. *Dev. Biol.* 143: 389–398.

Lane, M. C., M. A. R. Koehl, F. Wilt and R. Keller. 1993. A role for regulated secretion of apical matrix during epithelial invagination in the sea urchin. *Development* 117: 1049–1060.

Lepage, T., C. Sardet and C. Gache. 1992. Spatial expression of the hatching enzyme gene in the sea urchin embryo. *Dev. Biol.* 150: 23–32.

Lillie, F. R. 1898. Adaptation in cleavage. In *Biological Lectures of the Marine Biological Laboratory of Woods Hole.* Ginn, Boston, pp. 43–68.

Lillie, F. R. 1902. Differentiation without cleavage in the egg of the annelid *Chaetopterus pergamentaceus*. *Wilhelm Roux Arch. Entwicklungsmech. Org.* 14: 477–499.

Lin, R., S. Thompson and J. R. Priess. 1995. *pop-1* encodes an HMG box protein required for the specification of a mesoderm precursor in early *C. elegans* embryos. *Cell* 83: 599–609.

Logan, C. Y. and D. R. McClay. 1997. The allocation of early blastomeres to the ectoderm and endoderm is variable in the sea urchin embryo. *Development* 124: 2213–2223.

Logan, C. Y. and D. R. McClay. 1999. Lineages that give rise to endoderm and mesoderm in the sea urchin embryo. *In* S. A. Moody (ed.), *Cell Lineage and Determination.* Academic Press, New York, pp. 41–58.

Logan, C. Y., J. R. Miller, M. J. Ferkowicz and D. R. McClay. 1998. Nuclear β-catenin is required to specify vegetal cell fates in the sea urchin embryo. *Development* 126: 345–358.

Maduro, M. F., M. D. Meneghini, B. Bowerman, G. Broitman-Maduro and J. H. Rothman. 2001. Restriction of mesendoderm to a single blastomere by the combined action of SKN-1 and a GSK-3β homolog is mediated by MED-1 and -2 in *C. elegans. Mol. Cell* 7: 475–485.

Malinda, K. M. and C. A. Ettensohn. 1994. Primary mesenchyme cell migration in the sea urchin embryo: Distribution of directional cues. *Dev. Biol.* 164: 562–578.

Malinda, K. M., G. W. Fisher and C. A. Ettensohn. 1995. Four-dimensional microscopic analysis of the filopodial behavior of primary mesenchyme cells during gastrulation in the sea urchin embryo. *Dev. Biol.* 172: 552–566.

Mango, S. E., C. J. Thorpe, P. R. Martin, S. H. Chamberlain and B. Bowerman. 1994a. Two maternal genes, *apx-1* and *pie-1*, are required to distinguish the fates of equivalent blastomeres in early *C. elegans* embryos. *Development* 120: 2305–2315.

Mango, S. E., E. J. Lambie and J. Kimble. 1994b. The *pha-4* gene is required to generate the pharyngeal primordium of *Caenorhabditis elegans. Development* 120: 3019–3031.

Martins, G. G., R. G. Summers and J. B. Morrill. 1998. Cells are added to the archenteron during and following secondary invagination in the sea urchin *Lytechinus variegatus. Dev. Biol.* 198: 330–342.

Maruyama, Y. K., Y. Nakeseko and S. Yagi. 1985. Localization of cytoplasmic determinants responsible for primary mesenchyme formation and gastrulation in the unfertilized eggs of the sea urchin *Hemicentrotus pulcherrimus. J. Exp. Zool.* 236: 155–163.

Mello, C. C., B. W. Draper and J. R. Priess. 1994. The maternal genes *apx-1* and *glp-1* and establishment of dorsal-ventral polarity in the early *C. elegans* embryo. *Cell* 77: 95–106.

Mello, C. C., C. Schubert, B. Draper, W. Zhang, R. Lobel and J. R. Priess. 1996. The PIE-1 protein and germline specification in *C. elegans* embryos. *Nature* 382: 710–712.

Miller, J. R., S. E. Fraser and D. R. McClay. 1995. Dynamics of thin filopodia during sea urchin gastrulation. *Development* 121: 2505–2511.

Minshull, J., J. J. Blow and T. Hunt. 1989. Translation of cyclin mRNA is necessary for extracts of activated *Xenopus* eggs to enter mitosis. *Cell* 56:497–956.

Morgan, T. H. 1927. *Experimental Embryology.* Columbia University Press, New York.

Morrill, J. B. and L. L. Santos. 1985. A scanning electron micrographical overview of cellular and extracellular patterns during blastulation and gastrulation in the sea urchin, *Lytechinus variegatus. In* R. H. Sawyer and R. M. Showman (eds.), *The Cellular and Molecular Biology of Invertebrate Development.* University of South Carolina Press, Columbia, pp. 3–33.

Nakatani, Y., H. Yasuo, N. Satoh and H. Nishida. 1996. Basic fibroblast growth factor induces notochord formation and the expression of As-T, a *Brachyury* homolog, during ascidian embryogenesis. *Development* 122: 2023–2031.

Newport, J. W. and M. W. Kirschner. 1982a. A major developmental transition in early *Xenopus* embryos: I. Characterization and timing of cellular changes at midblastula stage. *Cell* 30: 675–686.

Newport, J. W. and M. W. Kirschner. 1984. Regulation of the cell cycle during *Xenopus laevis* development. *Cell* 37: 731–742.

Newrock, K. M. and R. A. Raff. 1975. Polar lobe specific regulation of translation in embryos of *Ilyanassa obsoleta. Dev. Biol.* 42: 242–261.

Nigg, E. A. 1995. Cyclin-dependent protein kinases: Key regulators of the eukaryotic cell cycle. *BioEssays* 17: 471–480.

Nishida, H. 1987. Cell lineage analysis in ascidian embryos by intracellular injection of a tracer enzyme. III. Up to the tissue restricted stage. *Dev. Biol.* 121: 526–541.

Nishida, H. 1992a. Determination of developmental fates of blastomeres in ascidian embryos. *Dev. Growth Diff.* 34: 253–262.

Nishida, H. 1992b. Regionality of egg cytoplasm that promotes muscle differentiation in embryo of the ascidian *Halocynthia roretzi. Development* 116: 521–529.

Nishida, H. 1994. Localization of determinants for formation of the anterior-posterior axis in eggs of the ascidian *Halocynthia roretzi. Development* 120: 3093–3104.

Nishida, H. and K. Sawada. 2001. *macho-1* encodes a localized mRNA in ascidian eggs that specifies muscle fate during embryogenesis. *Nature* 409: 724–729.

Oliveri, P., D. M. Carrick and E. H. Davidson. 2002. A regulatory gene network that directs micromere specification in the sea urchin embryo. *Dev. Biol.* 246: 209–228.

Pines, M. (ed.). 1992. *From Egg to Adult.* Howard Hughes Medical Institute, Bethesda, MD, pp. 30–38.

Priess, R. A. and D. I. Hirsh. 1986. *Caenorhabditis elegans* morphogenesis: The role of the cytoskeleton in elongating the embryo. *Dev. Biol.* 117: 156–173.

Priess, R. A. and J. N.Thomson. 1987. Cellular interactions in early *C. elegans* embryos. *Cell* 48: 241–250.

Raff, R. A. and T. C. Kaufman. 1983. *Embryos, Genes, and Evolution: The Developmental-Genetic Basis of Evolutionary Change.* Macmillan, New York.

Raich, W. B., C. Agbunag and J. Hardin 1999. Rapid epithelial-sheet sealing in the *Caenorhabditis elegans* embryo requires cadherin-dependent filopodial priming. *Curr. Biol.* 9: 1139–1146.

Ransick, A. and E. H. Davidson. 1993. A complete second gut induced by transplanted micromeres in the sea urchin embryo. *Science* 259: 1134–1138.

Ransick, A. and E. H. Davidson. 1995. Micromeres are required for normal vegetal plate specification in sea urchin embryos. *Development* 121: 3215–3222.

Render, J. 1991. Fate maps of the first quartet micromeres in the gastropod *Ilyanassa obsolete. Development* 113: 495–501.

Render, J. 1997. Cell fate maps in the *Ilyanassa obsoleta* embryo beyond the third division. *Dev. Biol.* 189: 301–310.

Reverberi, G. and A. Minganti. 1946. Fenomeni di evocazione nello sviluppo dell'uovo di Ascidie. Risultati dell'indagine spermentale sull'ouvo di Ascidiella aspersa e di Ascidia malaca allo stadio di 8 blastomeri. *Pubbl. Staz. Zool. Napoli* 20: 199–252.

Rocheleau, C. E. and 8 others. 1997. Wnt signaling and an APC-related gene specify endoderm in early *C. elegans* embryos. *Cell* 90: 707–716.

Roegiers, F., A. McDougall and C. Sardet. 1995. The sperm entry point defines the orientation of the calcium-induced contraction wave that directs the first phase of cytoplasmic reorganization in the ascidian egg. *Development* 121: 3457–3466.

Ruffins, S. W. and C. A. Ettensohn. 1996. A fate map of the vegetal plate of the sea urchin (*Lytechinus variegatus*) mesenchyme blastula. *Development* 122: 253–263.

Satoh, N. 1978. Cellular morphology and architecture during early morphogenesis of the ascidian egg: An SEM study. *Biol. Bull.* 155: 608–614.

Satou, Y., K. S. Imai and N. Satoh. 2001. Early embryonic expression of a LIM-homeobox gene *Cs-lhx3* is downstream of β-catenin and responsible for the endoderm differentiation in *Ciona savignyi* embryos. *Development* 128: 3559–3570.

Sawada, T. and G. Schatten. 1989. Effects of cytoskeletal inhibitors on ooplasmic segregation and microtubule organization during fertilization and early development in the ascidian *Molgula occidentalis. Dev. Biol.* 132: 331–342.

Schierenberg, E. 1997. Nematodes, the roundworms. *In* S. F. Gilbert and A. M. Raunio (eds.), *Embryology: Constructing the Organism.* Sinauer Associates, Sunderland, MA, pp. 131–148.

Schroeder, T. E. 1972. The contractile ring. II. Determining its brief existence, volumetric changes, and vital role in cleaving *Arbacia* eggs. *J. Cell Biol.* 53: 419–434.

Schroeder, T. E. 1973. Cell constriction: Contractile role of microfilaments in division and development. *Am. Zool.* 13: 687–696.

Schroeder, T. 1981. Development of a "primitive" sea urchin (*Eucidaris tribuloides*): Irregularities in the hyaline layer, micromeres, and primary mesenchyme. *Biol. Bull.* 161: 141–151.

Seydoux, G., C. C. Mello, J. Pettitt, W. B. Wood, J. R. Priess and A. Fire. 1996. Repression of gene expression in the embryonic germ lineage of *C. elegans*. *Nature* 382: 713–716.

Sherwood, D. R. and D. R. McClay 1999. LvNotch signaling mediates secondary mesenchyme specification in the sea urchin embryo. *Development* 126:1703–1713.

Skiba, F. and E. Schierenberg. 1992. Cell lineage, developmental timing, and spatial pattern formation in embryos of free-living soil nematodes. *Dev. Biol.* 151: 597–610.

Smith, P., W. M. Leung-Chiu, R. Montgomery, A. Orsborn, K. Kuznicki, E. Gressman-Coberly, L. Mutapcic and K. Bennet. 2002. The GLH proteins, *C. elegans* P-granule components, associate with CSN-5 and KGB-1, proteins necessary for fertility, and with ZYX-1, a predicted cytoskeletal protein. *Dev. Biol.* 251: 333–347.

Speksnijder, J. E., C. Sardet and L. F. Jaffe. 1990. The activation wave of calcium in the ascidian egg and its role in ooplasmic segregation. *J. Cell Biol.* 110: 1589–1598.

Strome, S. and W. B. Wood. 1983. Generation of asymmetry and segregation of germ-like granules in early *Caenorhabditis elegans* embryos. *Cell* 35: 15–25.

Sturtevant, M. H. 1923. Inheritance of direction of coiling in *Limnaea*. *Science* 58: 269–270.

Sugiyama, K. 1972. Occurrence of mucopolysaccharides in the early development of the sea urchin embryo and its role in gastrulation. *Dev. Growth Diff.* 14: 62–73.

Sulston, J. E. and H. R. Horvitz. 1977. Postembryonic cell lineages of the nematode *Caenorhabditis elegans*. *Dev. Biol.* 56: 110–156.

Sulston, J. E., J. Schierenberg, J. White and N. Thomson. 1983. The embryonic cell lineage of the nematode *Caenorhabditis elegans*. *Dev. Biol.* 100: 64–119.

Summers, R. G., J. B. Morrill, A. Leith, M. Marko, D. W. Piston and A. T. Stonebraker. 1993. A stereometric analysis of karyogenesis, cytokinesis, and cell arrangements during and following fourth cleavage period in the sea urchin, *Lytechinus variegatus*. *Dev. Growth Diff.* 35: 41–58.

Sweet, H. C. 1998. Specification of first quartet micromeres in *Ilyanassa* involves inherited factors and position with respect to the inducing D macromere. *Development* 125: 4033–4044.

Sweet, H. C., P. G. Hodor and C. A. Ettensohn. 1999. The role of micromere signaling in *Notch* activation and mesoderm specification during sea urchin embryogenesis. *Development* 126: 5255–5265.

Swenson, K. L., K. M. Farrell and J. V. Ruderman. 1986. The clam embryo protein cyclin A induces entry into M phase and the resumption of meiosis in *Xenopus* oocytes. *Cell* 47: 861–870.

Sze, L. C. 1953. Changes in the amount of deoxyribonucleic acid in the development of *Rana pipiens*. *J. Exp. Zool.* 122: 577–601.

Thorpe, C. J., A. Schlesinger, J. C. Carter and B. Bowerman. 1997. Wnt signaling polarizes an early *C. elegans* blastomere to distinguish endoderm from mesoderm. *Cell* 90: 695–705.

Trinkaus, J. P. 1984. *Cells into Organs: The Forces that Shape the Embryo*, 2nd Ed. Prentice-Hall, Englewood Cliffs, NJ.

Tung, T. C., S. C. Wu, Y. F. Yel, K. S. Li and M. C. Hsu. 1977. Cell differentiation in ascidians studied by nuclear transplantation. *Scientia Sinica* 20: 222–233.

van den Biggelaar, J. A. M. 1977. Development of dorsoventral polarity and mesentoblast determination in *Patella vulgata*. *J. Morphol.* 154: 157–186.

Verdonk, N. H. and J. N. Cather. 1983. Morphogenetic determination and differentiation. *In* N. H. Verdonk, J. A. M. van den Biggelaar and A. S. Tompa (eds.), *The Mollusca*. Academic Press, New York, pp. 215–252.

Vonica, A., W. Weng, B. M. Gumbiner and J. M. Venuti. 2000. TCF is the nuclear effector of the β-catenin signal that patterns the sea urchin animal-vegetal axis. *Dev. Biol.* 217: 230–243.

Wallenfang, M. R. and G. Seydoux. 2000. Polarization of the anterior-posterior axis of *C. elegans* is a microtubule-directed process. *Nature* 408: 89–92.

Wang, L., C. R. Eckmann, L. C. Kadyk, M. Wickens and J. Kimble. 2002. A regulatory cytoplasmic poly(A) polymerase in *Caenorhabditis elegans*. *Nature* 419: 312–316.

Welsh, J. H. 1969. Mussels on the move. *Nat. Hist.* 78: 56–59.

Wessel, G. M. and D. R. McClay. 1985. Sequential expression of germ layer specific molecules in the sea urchin embryo. *Dev. Biol.* 111: 451–463.

Wessel, G. M., R. B. Marchase and D. R. McClay. 1984. Ontogeny of the basal lamina in the sea urchin embryo. *Dev. Biol.* 103: 235–245.

White, J. C., W. B. Amos and M. Fordham. 1988. An evaluation of confocal versus conventional imaging of biological structures by fluorescence light microscopy. *J. Cell Biol.* 105: 41–48.

Whittaker, J. R. 1982. Muscle cell lineage can change the developmental expression in epidermal lineage cells of ascidian embryos. *Dev. Biol.* 93: 463–470.

Wikramanayake, A. H., L. Huang and W. H. Klein. 1998. β-catenin is essential for patterning the maternally specified animal-vegetal axis in the sea urchin embryo. *Proc. Nat. Acad. Sci. USA* 95: 9343–9348.

Wilson, E. B. 1898. Cell lineage and ancestral reminiscences. *In Biological Lectures of the Marine Biological Laboratory of Woods Hole.* Ginn, Boston, pp. 21–42.

Wilson, E. B. 1904. Experimental studies on germinal localization. I. The germ regions of the egg of *Dentalium*. II. Experiments on the cleavage-mosaic in *Patella* and *Dentalium*. *J. Exp. Zool.* 1: 1–72.

Wilson, E. B. 1923. *The Physical Basis of Life.* Yale University Press, New Haven. p. 10.

Wolpert, L. and T. Gustafson. 1961. Studies in the cellular basis of morphogenesis of the sea urchin embryo: The formation of the blastula. *Exp. Cell Res.* 25: 374–382.

Wray, G. A. 1999. Introduction to sea urchins. *In* S. A. Moody (ed.), *Cell Lineage and Determination.* Academic Press, New York, pp. 3–9.

Yuh, C.-H, H. Bolouri and E. H. Davidson. 2001. *Cis*-regulatory logic in the *endo16* gene: Switching from a specification to a differentiation mode of control. *Development* 128: 617–629.

chapter 9 *The genetics of axis specification in* Drosophila

THANKS LARGELY TO THE STUDIES by Thomas Hunt Morgan's laboratory during the first decade of the twentieth century, we know more about the genetics of *Drosophila* than about any other multicellular organism. The reasons for this have to do with both the flies and the people who first studied them. *Drosophila* is easy to breed, hardy, prolific, tolerant of diverse conditions, and the polytene chromosomes of its larvae are easy to identify (see Chapter 4). The techniques for breeding and identifying mutants are easy to learn. Moreover, the progress of *Drosophila* genetics was aided by the relatively free access of every scientist to the mutants and the techniques of every other researcher. Mutants were considered the property of the entire scientific community, and Morgan's laboratory established the database and exchange network whereby anyone could obtain them.

Undergraduates (starting with Calvin Bridges and Alfred Sturtevant) played important roles in *Drosophila* research, which achieved its original popularity as a source of undergraduate research projects. As historian Robert Kohler noted (1994), "Departments of biology were cash poor but rich in one resource: cheap, eager, renewable student labor." The *Drosophila* genetics program was "designed by young persons to be a young person's game," and the students set the rules for *Drosophila* research: "No trade secrets, no monopolies, no poaching, no ambushes."

But *Drosophila* was a difficult organism on which to study embryology. Although Jack Schultz and others attempted to relate the genetics of *Drosophila* to its development, the fly embryos proved too complex and intractable to study, being neither large enough to manipulate experimentally nor transparent enough to observe. It was not until the techniques of molecular biology allowed researchers to identify and manipulate the genes and RNAs of the insect that its genetics could be related to its development. And when that happened, a revolution occurred in the field of biology. The merging of our knowledge of the molecular aspects of *Drosophila* genetics with our knowledge of the fly's development built the foundations on which the current sciences of developmental genetics and evolutionary developmental biology are based.

EARLY DROSOPHILA DEVELOPMENT

In the last chapter, we discussed the specification of early embryonic cells by cytoplasmic determinants stored in the oocyte. The cell membranes that form during cleavage establish the region of cytoplasm incorporated into each new blastomere, and the incorporated morphogenetic determinants then direct differential gene expression in each blastomere. During *Drosophila* development, however, cellular membranes do not form until after the thirteenth nuclear division. Prior to this time, all the nuclei share a common cytoplasm, and material can diffuse throughout the embryo. In these embryos, the specification of cell types along the anterior-posterior and dorsal-ventral axes is accomplished by the interactions of cytoplasmic materials *within* the single multinucleated cell. Moreover, the initiation of the anterior-posterior and dorsal-ventral differences is controlled by the position of the egg within the mother's ovary. Whereas the sperm entry site may fix the axes in ascidians and nematodes, the fly's anterior-posterior and dorsal-ventral axes are specified by interactions between the egg and its surrounding follicle cells.

WEBSITE 9.1 *Drosophila* fertilization. Fertilization of *Drosophila* can only occur in the region of the oocyte that will become the anterior of the embryo. Moreover, the sperm tail appears to stay in this region.

Cleavage

Most insect eggs undergo **superficial cleavage**, wherein a large mass of centrally located yolk confines cleavage to the cytoplasmic rim of the egg. One of the fascinating features of this cleavage pattern is that cells do not form until after the nuclei have divided. Cleavage in a *Drosophila* egg is shown in Figure 9.1. The zygote nucleus undergoes several mitotic divisions within the central portion of the egg. In *Drosophila*, 256 nuclei are produced by a series of eight nuclear divisions averaging 8 minutes each. The nuclei then migrate to the periphery of the egg, where the mitoses continue, albeit at a progressively slower rate. During the ninth division cycle, about five

nuclei reach the surface of the posterior pole of the embryo. These nuclei become enclosed by cell membranes and generate the **pole cells** that give rise to the gametes of the adult. Most of the other nuclei arrive at the periphery of the embryo at cycle 10 and then undergo four more divisions at progressively slower rates. During these stages of nuclear division, the embryo is called a **syncytial blastoderm**, meaning that all the cleavage nuclei are contained within a common cytoplasm. No cell membranes exist other than that of the egg itself.

Although the nuclei divide within a common cytoplasm, this does not mean that the cytoplasm is itself uniform. Karr and Alberts (1986) have shown that each nucleus within the syncytial blastoderm is contained within its own little territory of cytoskeletal proteins. When the nuclei reach the periphery of the egg during the tenth cleavage cycle, each nucleus becomes surrounded by microtubules and microfilaments. The nuclei and their associated cytoplasmic islands are called **energids**. Figure 9.2 shows the nuclei and their essential microfilament and microtubule domains in prophase of the mitotic division 12.

Following division cycle 13, the oocyte plasma membrane folds inward between the nuclei, eventually partitioning off each somatic nucleus into a single cell (Figure 9.3). This process creates the **cellular blastoderm**, in which all the cells are arranged in a single-layered jacket around the yolky core of the egg (Turner and Mahowald 1977; Foe and Alberts 1983). Like any other cell formation, the formation of the cellular blastoderm involves a delicate interplay between microtubules and microfilaments. The first phase of blastoderm cellularization is characterized by the invagination of cell membranes and their underlying actin microfilament net-

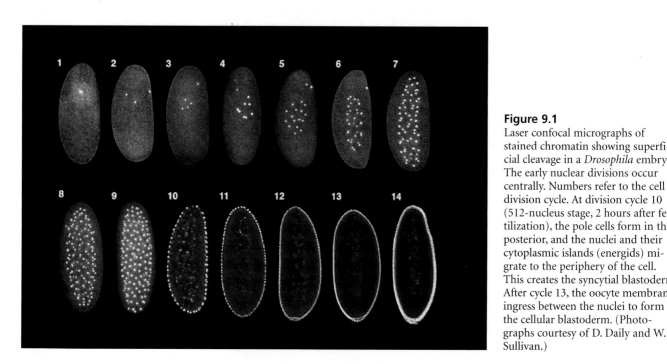

Figure 9.1
Laser confocal micrographs of stained chromatin showing superficial cleavage in a *Drosophila* embryo. The early nuclear divisions occur centrally. Numbers refer to the cell division cycle. At division cycle 10 (512-nucleus stage, 2 hours after fertilization), the pole cells form in the posterior, and the nuclei and their cytoplasmic islands (energids) migrate to the periphery of the cell. This creates the syncytial blastoderm. After cycle 13, the oocyte membranes ingress between the nuclei to form the cellular blastoderm. (Photographs courtesy of D. Daily and W. Sullivan.)

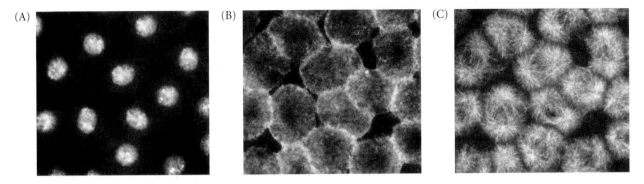

Figure 9.2
Localization of the cytoskeleton around nuclei in the syncytial blastoderm of *Drosophila*. A *Drosophila* embryo entering the mitotic prophase of its twelfth division was sectioned and triple-stained. (A) The nuclei were localized by a dye that binds to DNA. (B) Microfilaments were identified using a fluorescent antibody to actin. (C) Microtubules were recognized by a fluorescent antibody to tubulin. Cytoskeletal domains can be seen surrounding each nucleus. (From Karr and Alberts 1986; photographs courtesy of T. L. Karr.)

work into the regions between the nuclei to form furrow canals. This process can be inhibited by drugs that block microtubules. After the furrow canals have passed the level of the nuclei, the second phase of cellularization occurs. Here, the

rate of invagination increases, and the actin-membrane complex begins to constrict at what will be the basal end of the cell (Schejter and Wieschaus 1993; Foe et al. 1993; Mazumdar and Mazumdar 2002). In *Drosophila*, the cellular blastoderm consists of approximately 6000 cells and is formed within 4 hours of fertilization.

The mid-blastula transition

After the nuclei reach the periphery, the time required to complete each of the next four divisions becomes progressively longer. While cycles 1–10 are each 8 minutes long, cycle 13, the last cycle in the syncytial blastoderm, takes 25 minutes to complete. Cycle 14, in which the *Drosophila* embryo forms cells (i.e., after 13 divisions), is asynchronous. Some groups of cells complete this cycle in 75 minutes, whereas other groups of cells take 175 minutes (Figure 9.4; Foe 1989). Transcription from the nuclei (which begins around the eleventh cycle) is greatly enhanced at this stage. This slowdown of nuclear division and the concomitant increase in RNA transcription is

(A)

— Egg surface

— Mitotic spindle

— Cleavage furrow
— Aster
— Nucleus

— Furrow canal

— Microtubules

— Yolk membrane

Figure 9.3
Formation of the cellular blastoderm in *Drosophila*. (A) Developmental series showing progressive cellularization. (B) Confocal fluorescence photomicrographs of nuclei dividing during the cellularization of the blastoderm. While there are no cell boundaries, actin (green) can be seen forming regions within which each nucleus divides. The microtubules of the mitotic apparatus are stained red with antibodies to tubulin. (C) Cross section during cellularization. As the cells form, the domain of the actin expands into the egg. (A after Fullilove and Jacobson 1971; B and C from Sullivan et al. 1993; photographs courtesy of E. Theurkauf and W. Sullivan.)

(B)   (C)

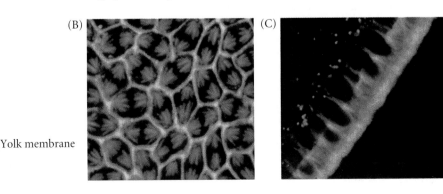

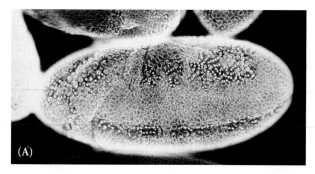

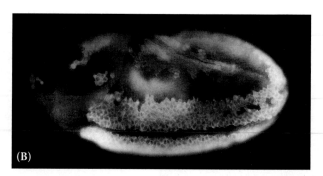

Figure 9.4
Differences in regional rates of cell division in *Drosophila* embryos. (A) Early gastrula embryo is stained with fluorescent antibodies to tubulin to show the microtubules of the mitotic spindles. (B) Antibodies to the cyclin A protein show that it is degraded after mitosis and is not seen in those regions undergoing cell division. (From Edgar and O'Farrell 1989; photographs courtesy of B. A. Edgar.)

often referred to as the mid-blastula transition (see Chapter 8). Such a transition is also seen in the embryos of numerous vertebrate and invertebrate phyla.

The control of this mitotic slowdown (most obvious in *Xenopus* and *Drosophila* embryos) appears to be effected by the ratio of chromatin to cytoplasm (Newport and Kirschner 1982; Edgar et al. 1986a). Edgar and his colleagues compared the early development of wild-type *Drosophila* embryos with that of a haploid mutant. These haploid *Drosophila* embryos have half the wild-type quantity of chromatin at each cell division. Hence a haploid embryo at cell division cycle 8 has the same amount of chromatin that a wild-type embryo has at cycle 7. The investigators found that whereas wild-type embryos formed a cellular blastoderm immediately after the thirteenth division, the haploid embryos underwent an extra, fourteenth, division before cellularization. Moreover, the lengths of cycles 11–14 in wild-type embryos corresponded to those of cycles 12–15 in the haploid embryos. Thus, the haploid embryos follow a pattern similar to that of the wild-type embryos—only they lag by one cell division.

WEBSITE 9.2 **The regulation of *Drosophila* cleavage.** The control of the cell cycle in *Drosophila* is a story of how the zygote nucleus gradually takes control from the mRNAs and proteins stored in the oocyte cytoplasm.

WEBSITE 9.3 **The early development of other insects.** *Drosophila* is a highly derived species. There are other insect species that develop in ways very different from the "standard" fruit fly.

Gastrulation

Gastrulation begins at the time of mid-blastula transition. The first movements of *Drosophila* gastrulation segregate the presumptive mesoderm, endoderm, and ectoderm. The prospective mesoderm—about 1000 cells constituting the ventral midline of the embryo—folds inward to produce the **ventral furrow** (Figure 9.5A). This furrow eventually pinches off from the surface to become a ventral tube within the embryo. It then flattens to form a layer of mesodermal tissue beneath the ventral ectoderm (see Figure 9.40). The prospective endoderm invaginates as two pockets at the anterior and posterior ends of the ventral furrow (Figure 9.5B,C). The pole cells are internalized along with the endoderm. At this time, the embryo bends to form the **cephalic furrow.**

The ectodermal cells on the surface and the mesoderm undergo convergence and extension, migrating toward the ventral midline to form the **germ band**, a collection of cells along the ventral midline that includes all the cells that will form the trunk of the embryo. The germ band extends posteriorly and, perhaps because of the egg case, wraps around the top (dorsal) surface of the embryo (Figure 9.5D). Thus, at the end of germ band formation, the cells destined to form the most posterior larval structures are located immediately behind the future head region. At this time, the body segments begin to appear, dividing the ectoderm and mesoderm. The germ band then retracts, placing the presumptive posterior segments at the posterior tip of the embryo (Figure 9.5E).

While the germ band is in its extended position, several key morphogenetic processes occur: organogenesis, segmentation, and the segregation of the imaginal discs* (Figure 9.5E). In addition, the nervous system forms from two regions of ventral ectoderm. As described in Chapter 6, neuroblasts differentiate from this neurogenic ectoderm within each segment (and also from the nonsegmented region of the head ectoderm). Therefore, in insects like *Drosophila*, the nervous system is located ventrally, rather than being derived from a dorsal neural tube as in vertebrates (Figure 9.6).

The general body plan of *Drosophila* is the same in the embryo, the larva, and the adult, each of which has a distinct head end and a distinct tail end, between which are repeating seg-

---

*Imaginal discs are those cells set aside to produce the adult structures. The details of imaginal disc differentiation will be discussed in Chapter 18. For more information on *Drosophila* developmental anatomy, see Bate and Martinez-Arias 1993; Tyler and Schetzer 1996; and Schwalm 1997.

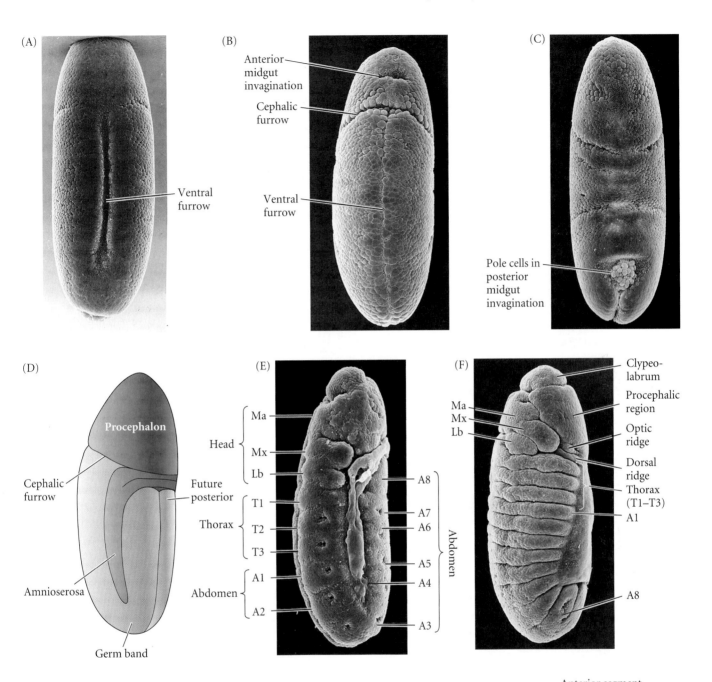

(A)

(B)
Anterior midgut invagination
Cephalic furrow
Ventral furrow

(C)
Pole cells in posterior midgut invagination

Ventral furrow

(D)
Procephalon
Cephalic furrow
Future posterior
Amnioserosa
Germ band

(E)
Head { Ma, Mx, Lb }
Thorax { T1, T2, T3 }
Abdomen { A1, A2 }
A8
A7
A6
A5
A4
A3 } Abdomen

(F)
Ma
Mx
Lb
Clypeo-labrum
Procephalic region
Optic ridge
Dorsal ridge
Thorax (T1–T3)
A1
A8

Figure 9.5
Gastrulation in *Drosophila*. (A) Ventral furrow beginning to form as cells flanking the ventral midline invaginate. (B) Closing of ventral furrow, with mesodermal cells placed internally and surface ectoderm flanking the ventral midline. (C) Dorsal view of a slightly older embryo, showing the pole cells and posterior endoderm sinking into the embryo. (D) Diagram of a dorsolateral view of *Drosophila* embryo at fullest germ band extension, just prior to segmentation. The cephalic furrow separates the future head region (procephalon) from the germ band, which will form the thorax and abdomen. (E) Lateral view, showing fullest extension of germ band and the beginnings of segmentation. Subtle indentations mark the incipient segments along the germ band: Ma, Mx, and Lb correspond to the mandibular, maxillary, and labial head segments; T1–T3, the thoracic segments; A1–A8, the abdominal segments. (F) Germ band reversing direction. The true segments are now visible, as well as the other territories of the dorsal head, such as the clypeolabrum, procephalic region, optic ridge, and dorsal ridge. (G) Newly hatched first-instar larva. (Photographs courtesy of F. R. Turner. D after Campos-Ortega and Hartenstein 1985.)

Anterior segment

(G)
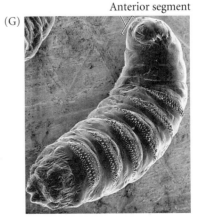

mental units (Figure 9.7). Three of these segments form the thorax, while another eight segments form the abdomen. Each segment of the adult fly has its own identity. The first thoracic segment, for example, has only legs; the second thoracic segment has legs and wings; and the third thoracic segment has legs and halteres (balancers). Thoracic and abdominal segments can also be distinguished from each other by differences in the cuticle. How does this pattern arise? During the past decade, the combined approaches of molecular biology, genetics, and embryology have led to a detailed model describing how a segmented pattern is generated along the anterior-posterior axis and how each segment is differentiated from the others.

The anterior-posterior and dorsal-ventral axes of *Drosophila* form at right angles to each other, and they are both determined by the position of the oocyte within the follicle cells of the ovary. The rest of this chapter is divided into three main parts. The first part concerns how the anterior-posterior axis is specified and how it determines the identity of each segment. The second part concerns how the dorsal-ventral axis is specified by the interactions between the oocyte and its surrounding follicle cells. The third part concerns how embryonic tissues are specified to become particular organs by their placement along these two axes.

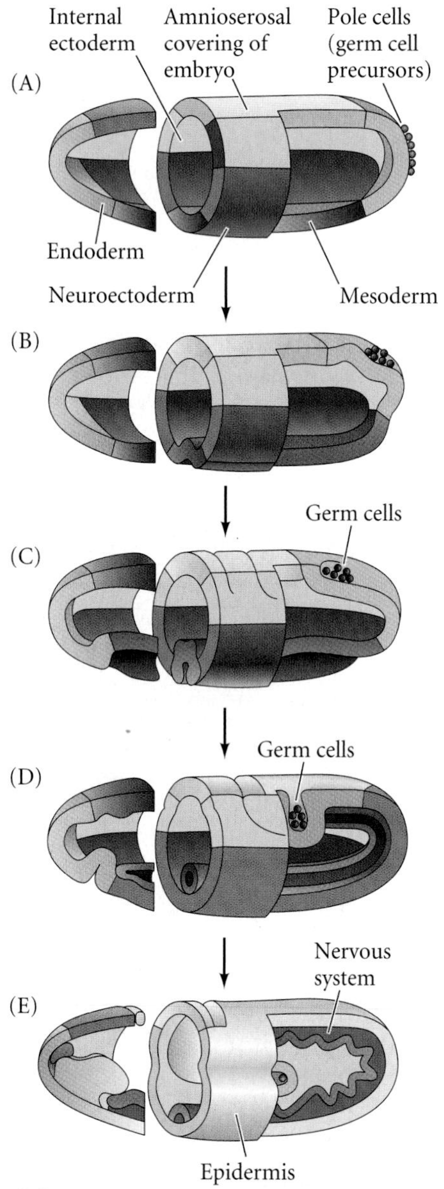

VADE MECUM[2] ***Drosophila* development.** The CD-ROM contains some remarkable time-lapse sequences of *Drosophila* development, including cleavage and gastrulation. This segment also provides access to the fly life cycle. The color coding superimposed on the germ layers allows you to readily understand tissue movements. **[Click on Fruit Fly]**

THE ORIGINS OF ANTERIOR-POSTERIOR POLARITY

The anterior-posterior polarity of the embryo, larva, and adult has its origin in the anterior-posterior polarity of the egg (Figure 9.8). The **maternal effect genes** expressed in the mother's ovaries produce messenger RNAs that are placed in different

Figure 9.6
Schematic representation of gastrulation in *Drosophila*. (A) and (B) are surface and cutaway views showing the fates of the tissues immediately prior to gastrulation. (C) shows the beginning of gastrulation as the ventral mesoderm invaginates into the embryo. (D) corresponds to Figure 9.5A, while (E) corresponds to 9.5B and C. In (E), the neuroectoderm is largely differentiated into the nervous system and the epidermis. (After Campos-Ortega and Hartenstein 1985.)

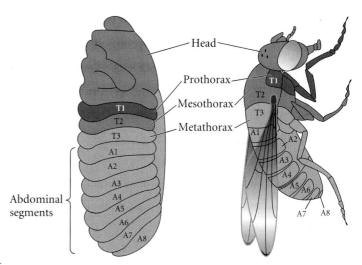

Figure 9.7
Comparison of larval and adult segmentation in *Drosophila*. The three thoracic segments can be distinguished by their appendages: T1 (prothorax) has legs only; T2 (mesothorax) has wings and legs; T3 (metathorax) has halteres and legs.

regions of the egg. These messages encode transcriptional and translational regulatory proteins that diffuse through the syncytial blastoderm and activate or repress the expression of certain zygotic genes. Two of these proteins, **Bicoid** and **Hunchback**, regulate the production of anterior structures, while another pair of maternally specified proteins, **Nanos** and **Caudal**, regulates the formation of the posterior parts of the em-

bryo. Next, the zygotic genes regulated by these maternal factors are expressed in certain broad (about three segments wide), partially overlapping domains. These genes are called **gap genes** (because mutations in them cause gaps in the segmentation pattern), and they are among the first genes transcribed in the embryo. Differing concentrations of the gap gene proteins cause the transcription of **pair-rule genes**, which divide the embryo into periodic units. The transcription of the different pair-rule genes results in a striped pattern of seven vertical bands perpendicular to the anterior-posterior axis. The pair-rule gene proteins activate the transcription of the **seg-**

Figure 9.8

Generalized model of *Drosophila* anterior-posterior pattern formation. (A) The pattern is established by maternal effect genes that form gradients and regions of morphogenetic proteins. These morphogenetic determinants create a gradient of Hunchback protein that differentially activates the gap genes, which define broad territories of the embryo. The gap genes enable the expression of the pair-rule genes, each of which divides the embryo into regions about two segments wide. The segment polarity genes then divide the embryo into segment-sized units along the anterior-posterior axis. Together, the actions of these genes define the spatial domains of the homeotic genes that define the identities of each of the segments. In this way, periodicity is generated from nonperiodicity, and each segment is given a unique identity. (B) Maternal effect genes. The anterior axis is specified by the gradient of Bicoid protein (yellow through red). (C) Gap gene protein expression and overlap. The domain of Hunchback protein (orange) and the domain of Krüppel protein (green) overlap to form a region containing both transcription factors (yellow). (D) Products of the *fushi tarazu* pair-rule gene form seven bands across the embryo. (E) Products of the segment polarity gene *engrailed*, seen here at the extended germ band stage. (B courtesy of C. Nüsslein-Volhard; C courtesy of C. Rushlow and M. Levine; D courtesy of T. Karr; E courtesy of S. Carroll and S. Paddock.)

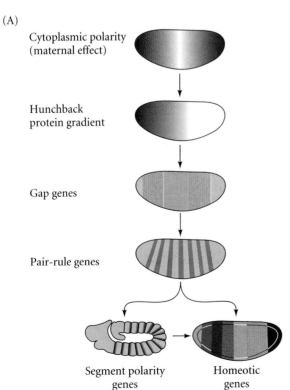

(A)

Cytoplasmic polarity (maternal effect)

Hunchback protein gradient

Gap genes

Pair-rule genes

Segment polarity genes → Homeotic genes

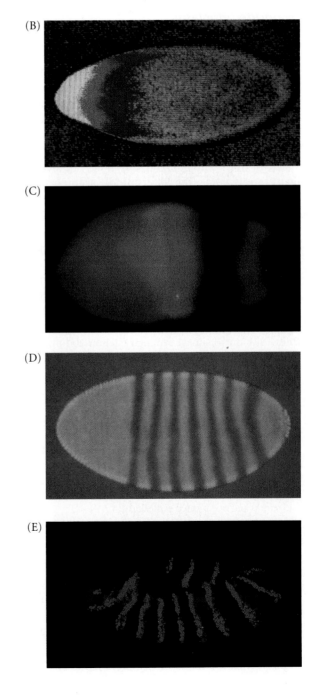

(B)

(C)

(D)

(E)

ment polarity genes, whose mRNA and protein products divide the embryo into 14 segment-wide units, establishing the periodicity of the embryo. At the same time, the protein products of the gap, pair-rule, and segment polarity genes interact to regulate another class of genes, the **homeotic selector genes**, whose transcription determines the developmental fate of each segment.

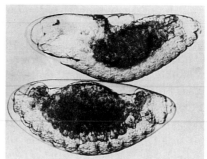

Figure 9.9
Normal and irradiated embryos of the midge *Smittia*. The normal embryo (top) shows a head on the left and abdominal segments on the right. The UV-irradiated embryo (bottom) has no head region, but has abdominal segments at both ends. (From Kalthoff 1969; photographs courtesy of K. Kalthoff.)

The Maternal Effect Genes

Embryological evidence of polarity regulation by oocyte cytoplasm

Classical embryological experiments demonstrated that there are at least two "organizing centers" in the insect egg, one in the anterior of the egg and one in the posterior. For instance, Klaus Sander (1975) found that if he ligated the egg early in development, separating the anterior half from the posterior half, one half developed into an anterior embryo and one half developed into a posterior embryo, but neither contained the middle segments of the embryo. The later in development the ligature was made, the fewer middle segments were missing. Thus, it appeared that there were indeed morphogenetic gradients emanating from the two poles during cleavage, and that these gradients interacted to produce the positional information determining the identity of each segment.

Moreover, when the RNA in the anterior of insect eggs was destroyed (by either ultraviolet light or RNase), the resulting embryos lacked a head and thorax. Instead, these embryos developed two abdomens and telsons (tails) with mirror-image symmetry: telson-abdomen-abdomen-telson (Figure 9.9; Kalthoff and Sander 1968; Kandler-Singer and Kalthoff 1976). Thus, Sander's laboratory postulated the existence of a gradient at both ends of the egg, and hypothesized that the egg sequesters an mRNA that generates a gradient of anterior-forming material.

WEBSITE 9.4 **Evidence for gradients in insect development.** The original evidence for gradients in insect development came from studies providing evidence for two "organization centers" in the egg, one located anteriorly and one located posteriorly.

The molecular model: Protein gradients in the early embryo

In the late 1980s, the gradient hypothesis was united with a genetic approach to the study of *Drosophila* embryogenesis. If there were gradients, what were the morphogens whose concentrations changed over space? What were the genes that shaped these gradients? And did these morphogens act by activating or inhibiting

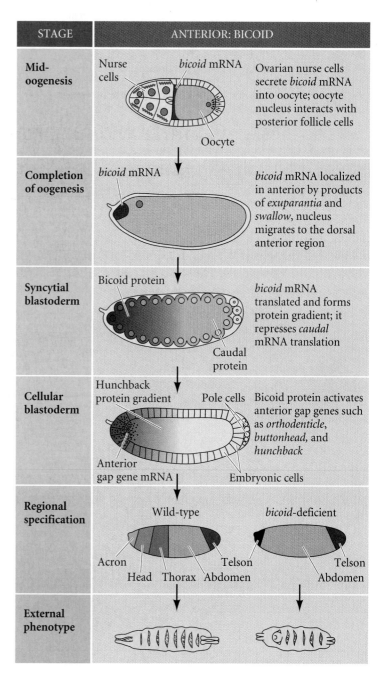

STAGE	ANTERIOR: BICOID	
Mid-oogenesis	Nurse cells · *bicoid* mRNA · Oocyte	Ovarian nurse cells secrete *bicoid* mRNA into oocyte; oocyte nucleus interacts with posterior follicle cells
Completion of oogenesis	*bicoid* mRNA	*bicoid* mRNA localized in anterior by products of *exuparantia* and *swallow*, nucleus migrates to the dorsal anterior region
Syncytial blastoderm	Bicoid protein · Caudal protein	*bicoid* mRNA translated and forms protein gradient; it represses *caudal* mRNA translation
Cellular blastoderm	Hunchback protein gradient · Pole cells · Anterior gap gene mRNA · Embryonic cells	Bicoid protein activates anterior gap genes such as *orthodenticle*, *buttonhead*, and *hunchback*
Regional specification	Wild-type: Acron · Head · Thorax · Abdomen · Telson	*bicoid*-deficient: Telson · Abdomen
External phenotype		

certain genes in the areas where they were concentrated? Christiane Nüsslein-Volhard led a research program that addressed these questions. The researchers found that one set of genes encoded morphogens for the anterior part of the embryo, another set of genes encoded morphogens responsible for organizing the posterior region of the embryo, and a third set of genes encoded proteins that produced the terminal regions at both ends of the embryo (Figure 9.10; Table 9.1). This work resulted in a Nobel Prize for Nüsslein-Volhard and her colleague, Eric Wieschaus, in 1995.

The anterior-posterior axis of the *Drosophila* embryo is patterned before the nuclei even begin to function. The nurse cells of the ovary deposit mRNAs in the developing oocyte, and these mRNAs are apportioned to different regions of the cell. Four maternal messenger RNAs are critical to the formation of the anterior-posterior axis:

- *bicoid* and *hunchback* mRNAs, whose protein products are critical for head and thorax formation
- *nanos* and *caudal* mRNAs, whose protein products are critical for the formation of the abdominal segments

The *bicoid* mRNAs are located in the anterior portion of the unfertilized egg, tethered to the anterior microtubules. The *nanos* messages are bound to the cytoskeleton in the posterior

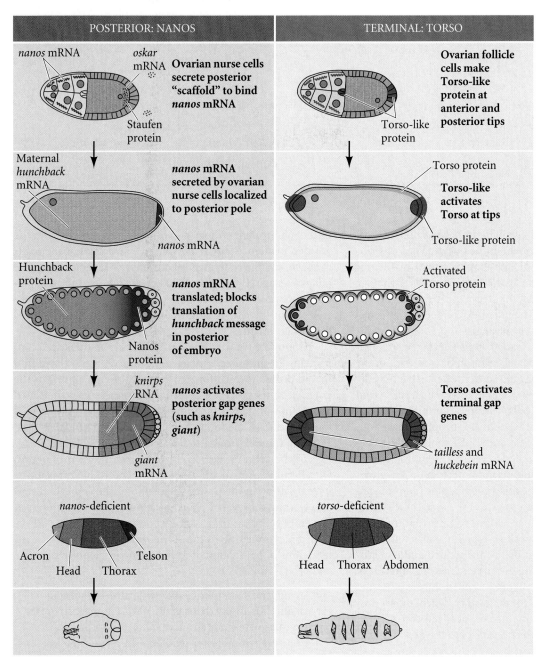

Figure 9.10
Three independent genetic pathways interact to form the anterior-posterior axis of the *Drosophila* embryo. In each case, the initial asymmetry is established during oogenesis, and the pattern is organized by maternal proteins soon after fertilization. The realization of the pattern comes about when the localized maternal proteins activate or repress specific zygotic genes in different regions of the embryo. (After St. Johnston and Nüsslein-Volhard 1992.)

TABLE 9.1 Maternal effect genes that effect the anterior-posterior polarity of the *Drosophila* embryo

Gene	Mutant phenotype	Proposed function and structure
ANTERIOR GROUP		
bicoid (bcd)	Head and thorax deleted, replaced by inverted telson	Graded anterior morphogen; contains homeodomain; represses caudal
exuperantia (exu)	Anterior head structures deleted	Anchors *bicoid* mRNA
swallow (swa)	Anterior head structures deleted	Anchors *bicoid* mRNA
POSTERIOR GROUP		
nanos (nos)	No abdomen	Posterior morphogen; represses *hunchback* mRNA
tudor (tud)	No abdomen, no pole cells	Localization of Nanos protein
oskar (osk)	No abdomen, no pole cells	Localization of Nanos protein
vasa (vas)	No abdomen, no pole cells; oogenesis defective	Localization of Nanos protein
valois (val)	No abdomen, no pole cells; cellularization defective	Stabilization of the Nanos localization complex
pumilio (pum)	No abdomen	Helps Nanos protein bind *hunchback* message
caudal (cad)	No abdomen	Activates posterior terminal genes
TERMINAL GROUP		
torso (tor)	No termini	Possible morphogen for termini
trunk (trk)	No termini	Transmits Torso-like signal to Torso
fs(1)Nasrat[fs(1)N]	No termini; collapsed eggs	Transmits Torso-like signal to Torso
fs(1)polehole[fs(1)ph]	No termini; collapsed eggs	Transmits Torso-like signal to Torso

Source: After Anderson 1989.

region of the unfertilized egg. The *hunchback* and *caudal* mRNAs are distributed throughout the oocyte. These distributions are accomplished in the developing oocyte by the microtubules. The nurse cells at the anterior end of the egg chamber synthesize mRNAs, that travel by means of the cytoskeleton into the oocyte. Here they are connected to the microtubules by a series of motor proteins.

The microtubules of the early *Drosophila* oocyte are organized by the fact that the oocyte itself is in the posterior of the egg chamber. Here, Gurken protein from the oocyte tells the terminal follicle cells that they will differentiate as *posterior*, rather than *anterior* follicle cells. The posterior follicle cells respond by sending a signal that activates protein kinase A in the oocyte cell membrane (Figure 9.11; Lane and Kalderon 1994; Roth et al. 1995; González-Reyes et al. 1995). This activation causes the microtubule orientation to shift. The growing ends ("plus ends") of the microtubules now point toward the *posterior* pole rather than towards the nurse cells. The *bicoid* mRNA contains a sequence in its 3′ UTR that interacts with the **Exuperantia** and **Swallow** proteins, which tether this message to a dynein protein that is maintained at the microtubule organizing center (the "minus end") at the anterior of the oocyte (Cha et al. 2001). While *bicoid* message is bound to the anchored end of the microtubules via a dynein motor protein, the posterior determinants are localized through the motor protein **Kinesin I**, which localizes to the growing ends

of the microtubules. Kinesin I will bind *oskar* mRNA and the Staufen protein. Staufen allows the translation of the *oskar* message, and the resulting Oskar protein is capable of binding *nanos* message (Brendza et al. 2000). Therefore, at the completion of oogenesis, the *bicoid* message is anchored at the anterior end of the oocyte, and the *nanos* message is tethered to the posterior end.

Upon fertilization, these mRNAs can be translated into proteins. At the anterior pole, *bicoid* mRNA is translated into Bicoid protein, which forms a gradient that is highest at the anterior. At the posterior pole, the *nanos* message is translated into Nanos protein, which forms a gradient that is highest at the posterior. Bicoid inhibits the translation of *caudal* mRNA, allowing Caudal protein to be synthesized only in the posterior of the cell. Conversely, Nanos protein, in conjunction with Pumilio protein, binds to *hunchback* mRNA, preventing its translation in the posterior portion of the embryo. Bicoid also elevates the level of Hunchback protein in the anterior of the embryo by binding to the enhancers of the *hunchback* gene and stimulating its transcription. The result of these interactions is the creation of four protein gradients in the early embryo (Figure 9.12):

- An anterior-to-posterior gradient of Bicoid protein
- An anterior-to-posterior gradient of Hunchback protein
- A posterior-to-anterior gradient of Nanos protein
- A posterior-to-anterior gradient of Caudal protein

(A)

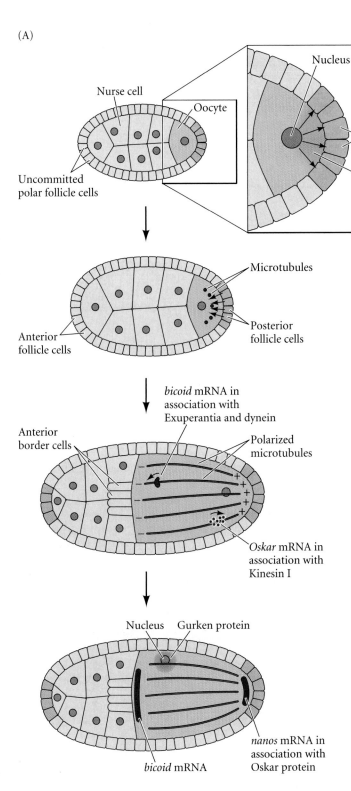

(B)

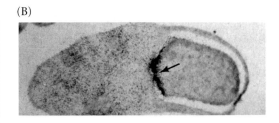

Figure 9.11

Specifying the anterior-posterior axis of the *Drosophila* embryo during oogenesis. (A) The oocyte moves into the posterior region of the egg chamber, while the nurse cells fill the anterior portion. The oocyte nucleus moves toward the terminal follicle cells and synthesizes the Gurken protein (green). The terminal follicle cells have receptors for the Gurken protein; when they receive it, these cells differentiate into posterior follicle cells. The posterior follicle cells reciprocate by synthesizing a molecule that activates protein kinase A in the egg. Protein kinase A orients the microtubules such that the growing end is at the posterior. The *bicoid* message binds to dynein, a motor protein associated with the non-growing end of microtubules. Dynein moves the *bicoid* message to the anterior end of the egg. The *oskar* message becomes complexed with kinesin I, a motor protein that moves it toward the growing end of the microtubules at the posterior region, where the Oskar protein can bind *nanos* message. The nucleus (and its Gurken protein) migrates along the microtubules, inducing the adjacent follicle cells to become the dorsal follicles. (B) Photograph showing *bicoid* mRNA (stained black) passing into the oocyte from the nurse cells (arrow) during oogenesis. (B from Stephanson et al. 1988; photograph courtesy of the authors.)

WEBSITE 9.5 Christiane Nüsslein-Volhard and the molecular approach to development. The research that revolutionized developmental biology had to wait for someone to synthesize molecular biology, embryology, and *Drosophila* genetics.

The Bicoid, Hunchback, and Caudal proteins are transcription factors whose relative concentrations can activate or repress particular zygotic genes. The stage is now set for the activation of zygotic genes in those nuclei that were busy dividing while this gradient was being established.

The anterior organizing center: The Bicoid gradient

In *Drosophila*, the phenotype of the *bicoid* mutant provides valuable information about the function of morphogenetic gradients. Instead of having anterior structures (acron, head, and thorax) followed by abdominal structures and a telson, the structure of the *bicoid* mutant is telson-abdomen-abdomen-telson (Figure 9.13). It would appear that these embryos lack whatever substances are needed for the formation of anterior structures. Moreover, one could hypothesize that the substance these mutants lack is the one postulated by

Figure 9.12
A model of anterior-posterior pattern generation by the *Drosophila* maternal effect genes. (A) The *bicoid*, *nanos*, *hunchback*, and *caudal* messenger RNAs are placed in the oocyte by the ovarian nurse cells. The *bicoid* message is sequestered anteriorly; the *nanos* message is sent to the posterior pole. (B) Upon translation, the Bicoid protein gradient extends from anterior to posterior, while the Nanos protein gradient extends from posterior to anterior. Nanos inhibits the translation of the *hunchback* message (in the posterior), while Bicoid prevents the translation of the *caudal* message (in the anterior). This inhibition results in opposing Caudal and Hunchback gradients. The Hunchback gradient is secondarily strengthened by the transcription of the *hunchback* gene in the anterior nuclei (since Bicoid acts as a transcription factor to activate *hunchback* transcription). (C) Parallel interactions whereby translational gene regulation establishes the anterior-posterior patterning of the *Drosophila* embryo. (C after Macdonald and Smibert 1996.)

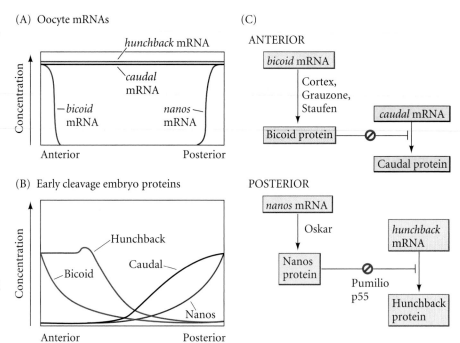

(A) Oocyte mRNAs

(B) Early cleavage embryo proteins

(C)

Sander and Kalthoff to turn on genes for the anterior structures and turn off genes for the telson structures (compare Figures 9.9 and 9.13).

Further studies have strengthened the view that the product of the wild-type *bicoid* gene is the morphogen that controls anterior development. First, *bicoid* is a maternal effect gene. Messenger RNA from the mother's *bicoid* genes is placed in the embryo by the mother's ovarian cells (see Figure 9.11B; Frigerio et al. 1986; Berleth et al. 1988). The *bicoid* mRNA is strictly localized in the anterior portion of the oocyte (Figure 9.14A), where the anterior cytoskeleton anchors it through the message's 3′ untranslated region (Ferrandon et al. 1997; Macdonald and Kerr 1998). This mRNA is dormant until fertilization, at which time it receives a longer polyadenylate tail and can be translated. Driever and Nüsslein-Volhard (1988b) have shown that when Bicoid protein is translated from this RNA during early cleavage, it forms a gradient, with the highest concentration in the anterior of the egg and the lowest in the posterior third of the egg. Moreover, this protein soon becomes concentrated in the embryonic nuclei in the anterior portion of the embryo (Figure 9.13C–E; see also Figure 5.35).

WEBSITE 9.6 Mechanism of *bicoid* mRNA localization.
One of the most critical steps in *Drosophila* pattern formation is the binding of the *bicoid* mRNA to the anterior microtubules. Several genes are involved in this process, wherein the *bicoid* message forms a complex with several proteins.

Further evidence that Bicoid protein is the anterior morphogen came from experiments that altered the steepness of the gradient. As we have seen, the *exuperantia* and *swallow*

genes are responsible for keeping the *bicoid* message at the anterior pole of the egg. Exuperantia and Swallow proteins appear to link the *bicoid* message to dynein ATPases on the microtubules, enabling the *bicoid* mRNA to travel on the microtubules to the anterior cortex of the egg (Schnorrer et al.

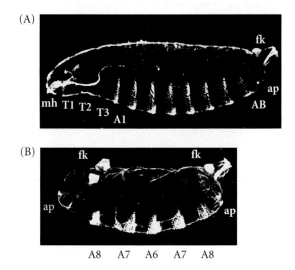

Figure 9.13
Phenotype of a strongly affected embryo from a female fly deficient in the *bicoid* gene. (A) Wild-type cuticle pattern. (B) *bicoid* mutant. The head and thorax have been replaced by a second set of posterior telson structures. Abbreviations: fk, filzkörper neurons; ap, anal plates (both telson structures); T1–T3, thoracic segments; A1, A8, the two terminal abdominal segments; mh, head structures. (From Driever et al. 1990; photographs courtesy of W. Driever.)

(A)

(B)

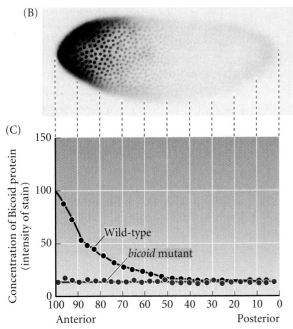

(C)

Figure 9.14
Gradient of Bicoid protein in the early *Drosophila* embryo. (A) Localization of *bicoid* mRNA to the anterior tip of the embryo. (B) Bicoid protein gradient shortly after fertilization. Note that the concentration is greatest anteriorly and trails off posteriorly. Notice also that Bicoid is concentrated in the nuclei. (C) Densitometric scan of the Bicoid protein gradient. The upper curve represents the Bicoid gradient in wild-type embryos. The lower curve represents Bicoid in embryos of *bicoid* mutant mothers. (A from Kaufman et al. 1990; B and C from Driever and Nüsslein-Volhard 1988b; photographs courtesy of the authors.)

mutants is similar to that of *bicoid*-deficient embryos, but less severe. These embryos lack their most anterior structures and have an extended mouth and thoracic region. Thus, by altering the gradient of Bicoid protein, one correspondingly alters the fate of the embryonic regions.

Confirmation that the Bicoid protein is crucial for initiating head and thorax formation came from experiments in which purified *bicoid* mRNA was injected into early-cleavage embryos (Figure 9.15; Driever et al. 1990). When injected into the anterior of *bicoid*-deficient embryos (whose mothers lacked *bicoid* genes), the *bicoid* mRNA rescued the embryos and caused them to have normal anterior-posterior polarity. Moreover, any location in an embryo where the *bicoid* mes-

2000; Cha et al. 2001). In their absence, the *bicoid* message diffuses farther into the posterior of the egg, and the gradient of Bicoid protein is less steep (Driever and Nüsslein-Volhard 1988a). The phenotype produced by *exuperantia* and *swallow*

Figure 9.15
Schematic representation of the experiments demonstrating that the *bicoid* gene encodes the morphogen responsible for head structures in *Drosophila*. The phenotypes of *bicoid*-deficient and wild-type embryos are shown at the left. When *bicoid*-deficient embryos are injected with *bicoid* mRNA, the point of injection forms the head structures. When the posterior pole of an early-cleavage wild-type embryo is injected with *bicoid* mRNA, head structures form at both poles. (After Driever et al. 1990.)

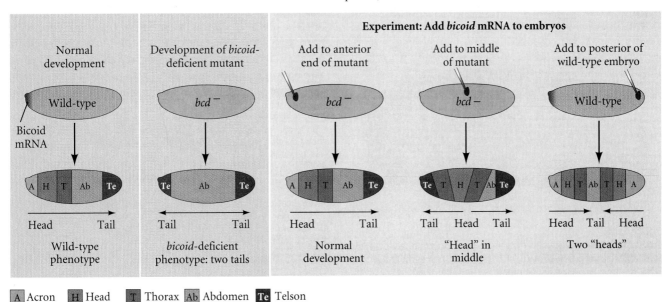

A Acron · H Head · T Thorax · Ab Abdomen · Te Telson

sage was injected became the head. If *bicoid* mRNA was injected into the center of an embryo, that middle region became the head, and the regions on either side of it became thorax structures. If a large amount of *bicoid* mRNA was injected into the posterior end of a wild-type embryo (with its own endogenous *bicoid* message in its anterior pole), two heads emerged, one at either end.

The next question then emerged: How might a gradient in Bicoid protein control the determination of the anterior-posterior axis? Recent evidence suggests that Bicoid acts in two ways to specify the anterior of the *Drosophila* embryo. First, it acts as a repressor of posterior formation. It does this by binding to and suppressing the translation of *caudal* mRNA, which is found throughout the egg and early embryo. The Caudal protein is critical in specifying the posterior domains of the embryo, and it activates the genes responsible for the invagination of the hindgut (Wu and Lengyel 1998). Bicoid binds to a specific region of the *caudal* message's 3′ untranslated region, thereby preventing translation of this message in the anterior section of the embryo (Figure 9.16; Chan and Struhl 1997; Rivera-Pomar et al. 1996; Niessing et al. 2000). This suppression is necessary, for if Caudal protein is made in the anterior, the head and thorax are not properly formed.

Second, Bicoid protein functions as a transcription factor. Bicoid enters the nuclei of early-cleavage embryos, where it activates the *hunchback* gene. Transcription of *hunchback* is seen only in the anterior half of the embryo—the region where Bicoid is found. Mutants deficient in maternal and zygotic Hunchback protein lack mouth parts and thorax structures. In the late 1980s, two laboratories independently demonstrated that Bicoid binds to and activates the *hunchback* gene (Driever and Nüsslein-Volhard 1989; Struhl et al. 1989). The *hunchback* mRNA is initially present throughout the embryo, although more can be made from zygotic nuclei if they are activated by Bicoid. Hunchback protein derived from the synthesis of new *hunchback* mRNA joins the Hunchback synthesized by the translation of maternal messages in the anterior of the embryo. Hunchback is also a transcription factor that represses abdominal-specific genes, thereby allowing the region of *hunchback* expression to form the head and thorax.

Hunchback also works with Bicoid to generate the anterior pattern of the embryo. Based on two pieces of evidence, Driever and co-workers (1989) predicted that Bicoid must activate at least one other anterior gene besides *hunchback*. First, deletions of *hunchback* produce only some of the defects seen in the *bicoid* mutant phenotype. Second, the *swallow* and *exuperantia* experiments showed that only moderate levels of Bicoid protein are needed to activate thorax formation (i.e., *hunchback* gene expression), but head formation requires higher Bicoid concentrations. Driever and co-workers (1989) predicted that the promoters of such a head-specific gap gene would have low-affinity binding sites for Bicoid, so that this gene would be activated only at extremely high concentrations of Bicoid—that is, near the anterior tip of the embryo. Since then, three gap genes of the head have been discovered that are dependent on very high concentrations of Bicoid for their expression (Cohen and Jürgens 1990; Finkelstein and Perrimon 1990; Grossniklaus et al. 1994). The *buttonhead*, *empty spiracles*, and *orthodenticle* genes specify progressively anterior regions of the head. In addition to needing high Bicoid levels for activation, these genes also require the presence of Hunchback protein to be transcribed (Simpson-Brose et al. 1994; Reinitz et al. 1995). The Bicoid and Hunchback proteins act synergistically at the enhancers of these "head genes" to promote their transcription.[*]

The posterior organizing center: Localizing and activating nanos

The posterior organizing center is defined by the activities of the *nanos* gene (Lehmann and Nüsslein-Volhard 1991; Wang and Lehmann 1991; Wharton and Struhl 1991). The *nanos* RNA is produced by the ovarian nurse cells and is transported into the posterior region of the egg (farthest away from the nurse cells). The *nanos* message is bound to the cytoskeleton in the posterior region of the egg through its 3′ UTR and its association with the products of several other genes (*oskar, valois, vasa, staufen,* and *tudor*).[†] If *nanos* or any other of these

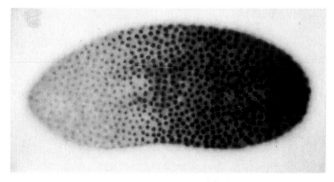

Figure 9.16
Gradient of Caudal protein in the syncitial blastoderm of a wild-type *Drosophila* embryo. The protein (stained darkly) enters the nuclei and helps specify posterior fates. Compare with the complementary gradient of Bicoid protein in Figure 9.14. (From Macdonald and Struhl 1986; photograph courtesy of the authors.)

---

[*]*Bicoid* appears to be a relatively "new" gene that evolved in flies but is not found in other insects. The anterior determinant of these other insect groups has not yet been found but appears to have *bicoid*-like properties (Wolff et al. 1998; Brown et al. 2001).

[†]Like the placement of the *bicoid* message, the location of the *nanos* message is determined by its 3′ untranslated region. If the *bicoid* 3′ UTR is experimentally placed on the protein-encoding region of *nanos* mRNA, the *nanos* message gets placed in the anterior of the egg. When the RNA is translated, the Nanos protein inhibits the translation of *hunchback* and *bicoid* mRNAs, and the embryo forms two abdomens—one in the anterior of the embryo and one in the posterior (Gavis and Lehmann 1992). Interestingly, much, if not most, of the *nanos* message normally remains in the nonpolar region of the egg. However, lacking the Oskar and Stufen proteins, it cannot be translated upon fertilization (Brendza et al. 2000). We will see these proteins again in Chapter 19, since they are also critical in forming the germ cells of *Drosophila*.

maternal effect genes are absent in the mother, no embryonic abdomen forms (Lehmann and Nüsslein-Volhard 1986; Schüpbach and Wieschaus 1986).

The *nanos* message is dormant in the unfertilized egg because it is repressed by the binding of the Smaug protein to its 3′ UTR (Smibert et al. 1996). At fertilization, this repression is removed in the posterior pole of the embryo, probably via the Oskar protein, and Nanos can be synthesized (Bergsten and Gavis 1999; Dahanukar et al. 1999). The Nanos protein forms a gradient that is highest at the posterior end. Nanos functions by inactivating *hunchback* mRNA translation (Figure 9.17; see also Figure 9.12; Tautz 1988). In the anterior of the cleavage-stage embryo, the *hunchback* message is bound by its 3′ UTR to the Pumilio protein, and the message can be translated into Hunchback protein. In the posterior of the early embryo, however, the bound Pumilio can be joined by Nanos. Nanos binds to Pumilio and deadenylates the *hunchback* mRNA, preventing its translation (Barker et al. 1992; Wreden et al. 1997). Thus, the combination of Bicoid and Nanos proteins causes a gradient of Hunchback across the egg. Bicoid activates hunchback gene transcription in the anterior part of the embryo, while Nanos inhibits the translation of *hunchback* mRNA in the posterior part of the embryo.

The terminal gene group

In addition to the anterior and posterior morphogens, there is third set of maternal genes whose proteins generate the unsegmented extremities of the anterior-posterior axis: the **acron** (the terminal portion of the head that includes the brain) and the **telson** (tail). Mutations in these terminal genes result in the loss of the acron and the most anterior head segments as well as the telson and the most posterior abdominal segments (Degelmann et al. 1986; Klingler et al. 1988). A critical gene here appears to be *torso*, a gene encoding a receptor tyrosine kinase. The embryos of mothers with mutations of

torso have neither acron nor telson, suggesting that the two termini of the embryo are formed through the same pathway. The *torso* mRNA is synthesized by the ovarian cells, deposited in the oocyte, and translated after fertilization. The transmembrane Torso protein is not spatially restricted to the ends of the egg, but is evenly distributed throughout the plasma membrane (Casanova and Struhl 1989). Indeed, a dominant mutation of *torso*, which imparts constitutive activity to the receptor, converts the entire anterior half of the embryo into an acron and the entire posterior half into a telson. Thus, Torso must normally be activated only at the ends of the egg.

Stevens and her colleagues (1990) have shown that this is the case. Torso protein is activated by the follicle cells only at the two poles of the oocyte. Two pieces of evidence suggest that the activator of Torso is probably the **Torso-like** protein: first, loss-of-function mutations in the *torso-like* gene create a phenotype almost identical to that produced by *torso* mutants; and second, ectopic expression of Torso-like protein activates Torso in the new location. The *torso-like* gene is usually expressed only in the anterior and posterior follicle cells, and secreted Torso-like protein can cross the perivitelline space to activate Torso in the egg membrane (Martin et al. 1994; Furriols et al. 1998). In this manner, Torso-like activates Torso in the anterior and posterior regions of the oocyte membrane.

The end products of the RTK-kinase cascade activated by Torso diffuse into the cytoplasm at both ends of the embryo (Figure 9.18; Gabay et al. 1997; see also Chapter 6). These kinases are thought to inactivate the Groucho protein, a transcriptional inhibitor of the *tailless* and *huckebein* gap genes (Paroush et al. 1997); it is these two gap genes that specify the termini of the embryo. The distinction between the anterior and posterior termini depends on the presence of Bicoid. If *tailless* and *huckebein* act alone, the terminal region differentiates into a telson. However, if Bicoid is also present, the terminal region forms an acron (Pignoni et al. 1992).

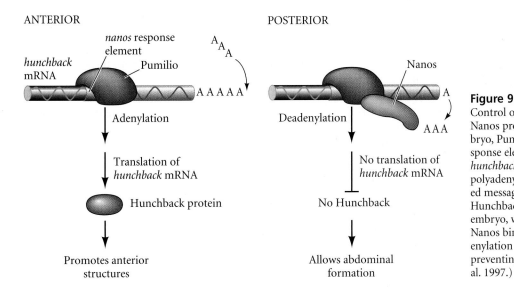

Figure 9.17

Control of *hunchback* mRNA translation by Nanos protein. In the anterior of the embryo, Pumilio protein binds to the Nanos response element (NRE) in the 3′ UTR of the *hunchback* message, and the message is polyadenylated normally. This polyadenylated message can be translated into Hunchback protein. In the posterior of the embryo, where Nanos protein is found, Nanos binds to Pumilio to cause the deadenylation of the *hunchback* message, thus preventing its translation. (After Wreden et al. 1997.)

(A)

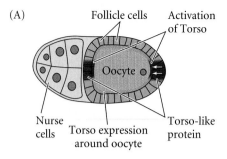

(B)

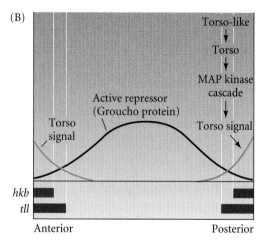

Figure 9.18
Formation of the unsegmented extremities by *torso* signaling. (A) Torso-like protein is expressed by the follicle cells at the poles of the oocyte. Torso protein is expressed around the entire oocyte. Torso-like activates Torso at the poles (see Casanova et al. 1995). (B) Inactivation of the transcriptional suppression of *huckebein* (*hkb*) and *tailless* (*tll*) genes. The Torso signal antagonizes the Groucho protein. Groucho acts as a repressor of *tailless* and *huckebein* expression. The gradient of Torso is thought to provide the information that allows *tailless* to be expressed farther into the embryo than *huckebein*. (A after Gabay et al. 1997; B after Paroush et al. 1997.)

The anterior-posterior axis of the embryo is therefore specified by three sets of genes: (1) those that define the anterior organizing center, (2) those that define the posterior organizing center, and (3) those that define the terminal boundary region. The anterior organizing center is located at the anterior end of the embryo and acts through a gradient of Bicoid protein that functions as a transcription factor to activate anterior-specific gap genes and as a translational repressor to suppresses posterior-specific gap genes. The posterior organizing center is located at the posterior pole and acts translationally through the Nanos protein to inhibit anterior formation and transcriptionally through the Caudal protein to activate those genes that form the abdomen. The activation of those genes responsible for constructing the posterior is performed by Caudal, a protein whose synthesis (as we have seen above) is inhibited in the anterior portion of the embryo. The boundaries of the acron and telson are defined by the product of the *torso* gene, which is activated at the tips of the embryo.

The next step in development will be to use these gradients of transcription factors to activate specific genes along the anterior-posterior axis.

The Segmentation Genes

The process of cell fate commitment in *Drosophila* appears to have two steps: specification and determination (Slack 1983). Early in development, the fate of a cell depends on environmental cues, such as those provided by the protein gradients mentioned above. This specification of cell fate is flexible and can still be altered in response to signals from other cells. Eventually, however, the cells undergo a transition from this loose type of commitment to an irreversible determination. At this point, the fate of a cell becomes cell-intrinsic.*

The transition from specification to determination in *Drosophila* is mediated by the **segmentation genes**. These genes divide the early embryo into a repeating series of segmental primordia along the anterior-posterior axis. Mutations in segmentation genes cause the embryo to lack certain segments or parts of segments. Often these mutations affect **parasegments**, regions of the embryo that are separated by mesodermal thickenings and ectodermal grooves. The segmentation genes divide the embryo into 14 parasegments (Martinez-Arias and Lawrence 1985). The parasegments of the embryo do not become the segments of the larva or adult; rather, they include the posterior part of an anterior segment and the anterior portion of the segment behind it (Figure 9.19). While the segments are the major anatomical divisions of the larval and adult body plan, they are built according to rules that use the parasegment as the basic unit of construction.

There are three classes of segmentation genes, and they are expressed sequentially (see Figure 9.8). The transition from an embryo characterized by gradients of morphogens to an embryo with distinct units is accomplished by the products of the **gap genes**. The gap genes are activated or repressed by the maternal effect genes, and they divide the embryo into broad regions, each containing several parasegment primordia. The *Krüppel* gene, for example, is expressed primarily in parasegments 4–6, in the center of the *Drosophila* embryo (Figures 9.20A and 9.8C); in the absence of the Krüppel protein, the embryo lacks these regions.

The protein products of the gap genes interact with neighboring gap gene proteins to activate the transcription of the **pair-rule genes**. The products of these genes subdivide the broad gap gene regions into parasegments. Mutations of pair-rule genes such as *fushi tarazu* usually delete portions of alter-

---

*Aficionados of information theory will recognize that the process by which the anterior-posterior information in morphogenetic gradients is transferred to discrete and different parasegments represents a transition from analog to digital specification. Specification is analog, determination digital. This process enables the transient information of the gradients in the syncytial blasto-derm to be stabilized so that it can be utilized much later in development (Baumgartner and Noll 1990).

Figure 9.19
Segments and parasegments. A and P represent the anterior and posterior compartments of the segments. The parasegments are shifted one compartment forward. Ma, Mx, and Lb represent three of the head segments (mandibular, maxillary, and labial), the T segments are thoracic, and the A segments are abdominal. The parasegments are numbered 1 through 14. The bars below the map show the boundaries of gene expression observed by the in situ hybridization of radioactive cDNA from the pair-rule gene *fushi tarazu* (*ftz*). (After Martinez-Arias and Lawrence 1985.)

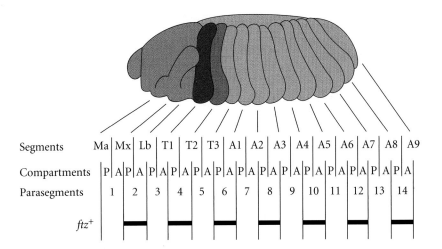

nate segments (Figures 9.21, 9.20B, 9.8C). Finally, the **segment polarity genes** are responsible for maintaining certain repeated structures within each segment. Mutations in segment polarity genes cause a portion of each segment to be deleted and replaced by a mirror-image structure of another portion of the segment. For instance, in *engrailed* mutants, portions of the posterior part of each segment are replaced by duplications of the anterior region of the subsequent segment (Figures 9.20C, 9.8E). Thus, the segmentation genes are transcription factors that use the gradients of the early-cleavage embryo to transform the embryo into a periodic, parasegmental structure.

After the parasegmental boundaries are set, the pair-rule and gap genes interact to regulate the homeotic selector genes, which determine the identity of each segment. By the end of the cellular blastoderm stage, each segment primordium has been given an individual identity by its unique constellation of gap, pair-rule, and homeotic gene products (Levine and Harding 1989).

The gap genes

The gap genes were originally discovered through a series of mutant embryos that lacked groups of consecutive segments (Figure 9.22; Nüsslein-Volhard and Wieschaus 1980). Deletions caused by mutations of the *hunchback*, *Krüppel*, and *knirps* genes span the entire segmented region of the *Drosophila* embryo. The *giant* gap gene overlaps with these three, and mutations of the gap genes *tailless* and *huckebein* genes delete portions of the unsegmented termini of the embryo.

The expression of the gap genes is dynamic. There is usually a low level of transcriptional activity across the entire em-

Figure 9.20
Three types of segmentation gene mutations. The left panel shows the early-cleavage embryo, with the region where the particular gene is normally transcribed in wild-type embryos shown in color. These areas are deleted as the mutants develop.

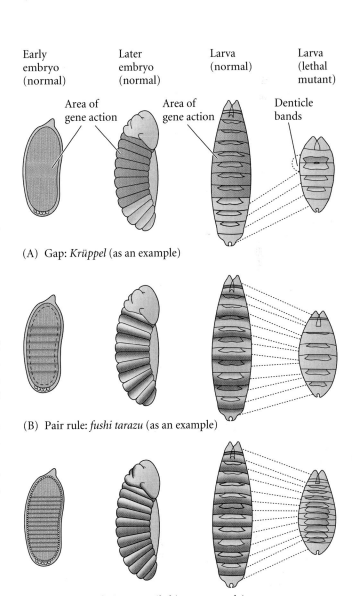

(A) Gap: *Krüppel* (as an example)

(B) Pair rule: *fushi tarazu* (as an example)

(C) Segment polarity: *engrailed* (as an example)

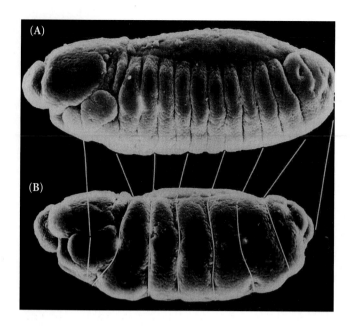

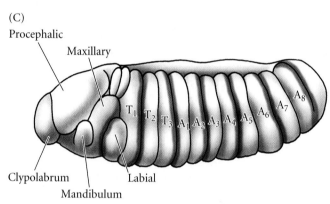

Figure 9.21
Defects seen in the *fushi tarazu* mutant. (A) Scanning electron micrograph of a wild-type embryo, seen in lateral view. (B) Same stage of a *fushi tarazu* mutant embryo. The white lines connect the homologous portions of the segmented germ band. (C) Diagram of wild-type embryonic segmentation. The shaded areas show the parasegments of the germ band that are missing in the mutant embryo. (After Kaufman et al. 1990; photographs courtesy of T. Kaufman.)

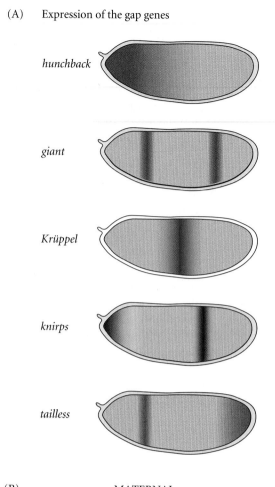

(A)   Expression of the gap genes

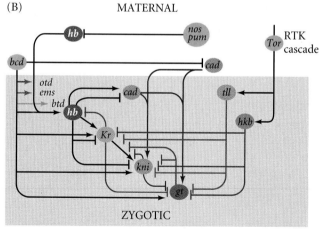

Figure 9.22
Conversion of maternal protein gradients into zygotic gap gene expression. (A) Gap gene expression patterns. (B) The gradients of maternal transcription factors Bicoid, Caudal, and Hunchback regulate the transcription of the gap genes. Hunchback and Caudal proteins come from both maternal messages and new zygotic transcription. These gap gene-encoded proteins diffuse, and the interactions between them are critical in activating the transcription of the pair-rule genes. At the two termini of the embryo, the interaction between Torso and Torso-like activates the *tailless* and *huckebein* gap genes. (B after Rivera-Pomar and Jäckle 1996.)

bryo that becomes consolidated into discrete regions of high activity as cleavage continues (Jäckle et al. 1986). The critical element appears to be the expression of the Hunchback protein, which by the end of nuclear division cycle 12 is found at high levels across the anterior part of the embryo, and then forms a steep gradient through about 15 nuclei (see Figure 9.12). The last third of the embryo has undetectable Hunchback levels. The transcription patterns of the anterior gap genes are initiated by the different concentrations of the Hunchback and Bicoid proteins. High levels of Hunchback induce the expression of *giant*, while the *Krüppel* transcript appears over the region where Hunchback begins to decline.

High levels of Hunchback also prevent the transcription of the posterior gap genes (such as *knirps*) in the anterior part of the embryo (Struhl et al. 1992). It is thought that a gradient of the Caudal protein, highest at the posterior pole, is responsible for activating the abdominal gap genes *knirps* and *giant* in the posterior part of the embryo. The *giant* gene thus has two methods for its activation, one for its anterior expression band and one for its posterior expression band (Rivera-Pomar 1995; Schulz and Tautz 1995).

After these patterns have been established by the maternal effect genes and Hunchback, the expression of each gap gene is stabilized and maintained by interactions between the different gap gene products themselves (Figure 9.22B).* For instance, *Krüppel* gene expression is negatively regulated on its anterior boundary by Hunchback and Giant, and on its posterior boundary by Knirps and Tailless (Jäckle et al. 1986; Harding and Levine 1988; Hoch et al. 1992). If Hunchback activity is lacking, the domain of *Krüppel* expression extends anteriorly. If Knirps activity is lacking, *Krüppel* gene expression extends posteriorly.

The boundaries between the regions of gap gene transcription are probably created by mutual repression. Just as the Giant and Hunchback proteins can control the anterior bound-

---

*The interactions between genes and gene products are facilitated by the fact that these reactions occur within a syncytium, in which the cell membranes have not yet formed.

ary of *Krüppel* transcription, so Krüppel protein can determine the posterior boundaries of *giant* and *hunchback* transcription. If an embryo lacks the *Krüppel* gene, *hunchback* transcription continues into the area usually allotted to Krüppel (Jäckle et al. 1986; Kraut and Levine 1991). These boundary-forming inhibitions are thought to be directly mediated by the gap gene products because all four major gap genes (*hunchback, giant, Krüppel,* and *knirps*) encode DNA-binding proteins that can activate or repress the transcription of other gap genes (Knipple et al. 1985; Gaul and Jäckle 1990; Capovilla et al. 1992).

The pair-rule genes

The first indication of segmentation in the fly embryo comes when the pair-rule genes are expressed during cell division cycle 13. The transcription patterns of these genes are striking in that they divide the embryo into areas that are the precursors of the segmental body plan. As can be seen in Figure 9.23B–E and Figure 9.8D, one vertical band of nuclei (the cells are just beginning to form) expresses a pair-rule gene, the next band of nuclei does not express it, and then the next band expresses it again. The result is a "zebra stripe" pattern along the anterior-posterior axis, dividing the embryo into 15 subunits (Hafen et al. 1984). Eight genes are currently known to be capable of dividing the early embryo in this fashion; they are listed in Table 9.2. It is important to note that not all nuclei express the same pair-

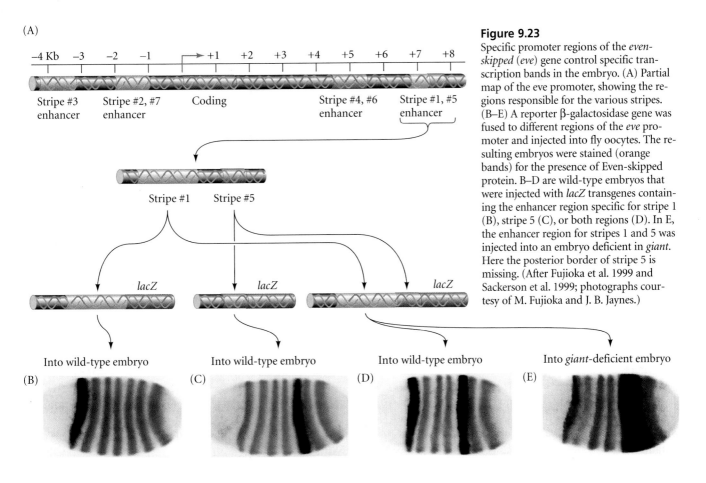

Figure 9.23
Specific promoter regions of the *even-skipped* (*eve*) gene control specific transcription bands in the embryo. (A) Partial map of the eve promoter, showing the regions responsible for the various stripes. (B–E) A reporter β-galactosidase gene was fused to different regions of the *eve* promoter and injected into fly oocytes. The resulting embryos were stained (orange bands) for the presence of Even-skipped protein. B–D are wild-type embryos that were injected with *lacZ* transgenes containing the enhancer region specific for stripe 1 (B), stripe 5 (C), or both regions (D). In E, the enhancer region for stripes 1 and 5 was injected into an embryo deficient in *giant*. Here the posterior border of stripe 5 is missing. (After Fujioka et al. 1999 and Sackerson et al. 1999; photographs courtesy of M. Fujioka and J. B. Jaynes.)

rule genes. In fact, within each parasegment, each row of nuclei has its own constellation of pair-rule gene expression that distinguishes it from any other row.

How are some nuclei of the *Drosophila* embryo told to transcribe a particular gene while their neighbors are told not to transcribe it? The answer appears to come from the distribution of the protein products of the gap genes. Whereas the mRNA of each of the gap genes has a very discrete distribution that defines abutting or slightly overlapping regions of expression, the protein products of these genes extend more broadly. In fact, they overlap by at least 8–10 nuclei (which at this stage accounts for about 2–3 segment primordia). This was demonstrated in a striking manner by Stanojević and co-workers (1989). They fixed cellularizing blastoderms (i.e., the stage when cells are beginning to form at the rim of the syncytial embryo), stained Hunchback protein with an antibody carrying a red dye, and simultaneously stained Krüppel protein with an antibody carrying a green dye. Cellularizing regions that contained both proteins bound both antibodies and were stained bright yellow (see Figure 9.8C). Krüppel overlaps with Knirps protein in a similar manner in the posterior region of the embryo (Pankratz et al. 1990).

Three genes are known to be the primary pair-rule genes. These genes—*hairy*, *even-skipped*, and *runt*—are essential for the formation of the periodic pattern, and they are directly controlled by the gap gene proteins. The enhancers of the primary pair-rule genes are recognized by gap gene proteins, and it is thought that the different concentrations of gap gene proteins determine whether a pair-rule gene is transcribed or not. The enhancers of the primary pair-rule genes are often modular: control over expression in each stripe is located in a discrete region of the DNA.

One of the best-studied enhancers is that for the *even-skipped* gene; its structure is shown in Figure 9.23A. It is composed of modular units arranged such that each unit regulates a separate stripe. For instance, the second *even-skipped* stripe is repressed by both Giant and Krüppel proteins and is activated by Hunchback and low concentrations of Bicoid (Figure 9.24; Small et al. 1991, 1992; Stanojević et al. 1991). DNase I footprinting (see Chapter 5) showed that the enhancer region for this stripe contains six binding sites for Krüppel protein, three for Hunchback protein, three for Giant protein, and five for Bicoid protein. Similarly, *even-skipped* stripe 5 is regulated negatively by Krüppel protein (on its anterior border) and by Giant protein (on its posterior border) (Small et al. 1996; Fujioka 1999).

The importance of these enhancer elements can be shown by both genetic and biochemical means. First, a mutation in a particular enhancer can delete its particular stripe and no other. Second, if a reporter gene such as *lacZ* (encoding β-galactosidase) is fused to one of the enhancers, the *lacZ* gene is expressed only in that particular stripe (see Figure 9.23; Fujioka et al. 1999). Third, the placement of the stripes can be altered by deleting the gap genes that regulate them.

TABLE 9.2 Major genes affecting segmentation pattern in *Drosophila*

Category	Gene name
Gap genes	*Krüppel (Kr)*
	knirps (kni)
	hunchback (hb)
	giant (gt)
	tailless (tll)
	huckebein (hkb)
	buttonhead (btd)
	empty spiracles (ems)
	orthodenticle (otd)
Pair-rule genes Primary	*hairy (h)*
	even-skipped (eve)
	runt (run)
Pair-rule genes Secondary	*fushi tarazu (ftz)*
	odd-paired (opa)
	odd-skipped (odd)
	sloppy-paired (slp)
	paired (prd)
Segment polarity genes	*engrailed (en)*
	wingless (wg)
	cubitus interruptusD (ciD)
	hedgehog (hh)
	fused (fu)
	armadillo (arm)
	patched (ptc)
	gooseberry (gsb)
	pangolin (pan)

Thus, the placement of the stripes of pair-rule gene expression is a result of (1) the modular *cis*-regulatory enhancer elements of the pair-rule genes, and (2) the *trans*-regulatory gap gene proteins that bind to these enhancer sites.

Once initiated by the gap gene proteins, the transcription pattern of the primary pair-rule genes becomes stabilized by interactions among their products (Levine and Harding 1989). The primary pair-rule genes also form the context that allows or inhibits the expression of the later-acting secondary pair-rule genes. One such secondary pair-rule gene is *fushi tarazu* (*ftz*; Japanese, "too few segments;" see Figures 9.8, 9.20, and 9.21). Early in division cycle 14, *ftz* mRNA and protein are seen throughout the segmented portion of the embryo. However, as the proteins from the primary pair-rule genes begin to interact with the *ftz* enhancer, the *ftz* gene is repressed in certain bands of nuclei to create interstripe regions. Meanwhile, the Ftz protein interacts with its own promoter to stimulate more transcription of *ftz* where it is already present (Figure 9.25; Edgar et al. 1986b; Karr and Kornberg 1989; Schier and Gehring 1992).

(A)

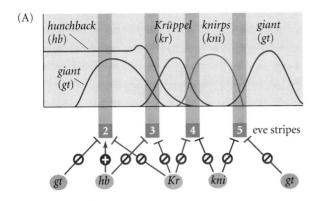

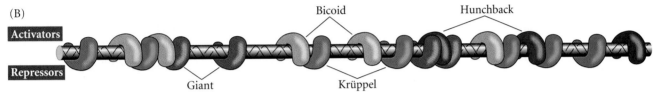

Figure 9.24
Hypothesis for the formation of the second stripe of transcription from the *even-skipped* gene. (A) "Stripes" 2–5 of even-skipped (*eve*) form at particular junctions of gap genes. Thus, the boundaries of *eve* transcription are determined by high concentrations of these proteins. Different enhancer elements contain binding sequences for different transcription factors. In the enhancer for the second *eve* transcription band, the binding of Hunchback protein stimulates transcription. In the enhancer for the third band, it inhibits transcription. (B) Enhancer element for stripe 2 regulation, containing binding sequences for Krüppel, Giant, Bicoid, and Hunchback proteins. Note that nearly every activator site is closely linked to a repressor site, suggesting competitive interactions at these positions. (A and B after Reinitz and Sharp 1995; C after Stanojevíc et al. 1991.)

The expression of each pair-rule gene in seven stripes divides the embryo into 14 parasegments, with each pair-rule gene being expressed in alternate parasegments. Moreover, each row of nuclei within each parasegment expresses a particular and unique combination of pair-rule products. These products will activate the next level of segmentation genes, the segment polarity genes.

The segment polarity genes

So far, our discussion has described interactions between molecules within the syncytial embryo. But once cells form, interactions take place between the cells. These intercellular interactions are mediated by the segment polarity genes, and they accomplish two important tasks. First, they reinforce the parasegmental periodicity established by the earlier transcription factors. Second, through this cell-to-cell signaling, cell fates are established within each parasegment.

The segment polarity genes encode proteins that are constituents of the Wingless and Hedgehog signal transduction pathways (see Chapter 6). Mutations in these genes lead to defects in segmentation and in gene expression pattern across each parasegment. The development of the normal pattern relies on the fact only one row of cells in each parasegment is permitted to express the Hedgehog protein, and only one row

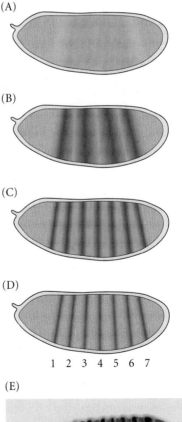

1 2 3 4 5 6 7

Figure 9.25
Transcription of the *fushi tarazu* gene in the *Drosophila* embryo. (A–D) At the beginning of division cycle 14, there is low-level transcription of *ftz* in each of the nuclei in the segmented region of the embryo. Within the next 30 minutes, the expression pattern alters as *ftz* transcription is enhanced in certain regions (which form stripes) and repressed in the interstripe regions. (E) Double labeling of the *even-skipped* (blue bands) and *fushi tarazu* (brown bands) transcripts, showing that *ftz* is expressed between the *eve* bands at this stage. (A–D after Karr and Kornberg 1989; E courtesy of J. B. Jaynes and M. Fujioka.)

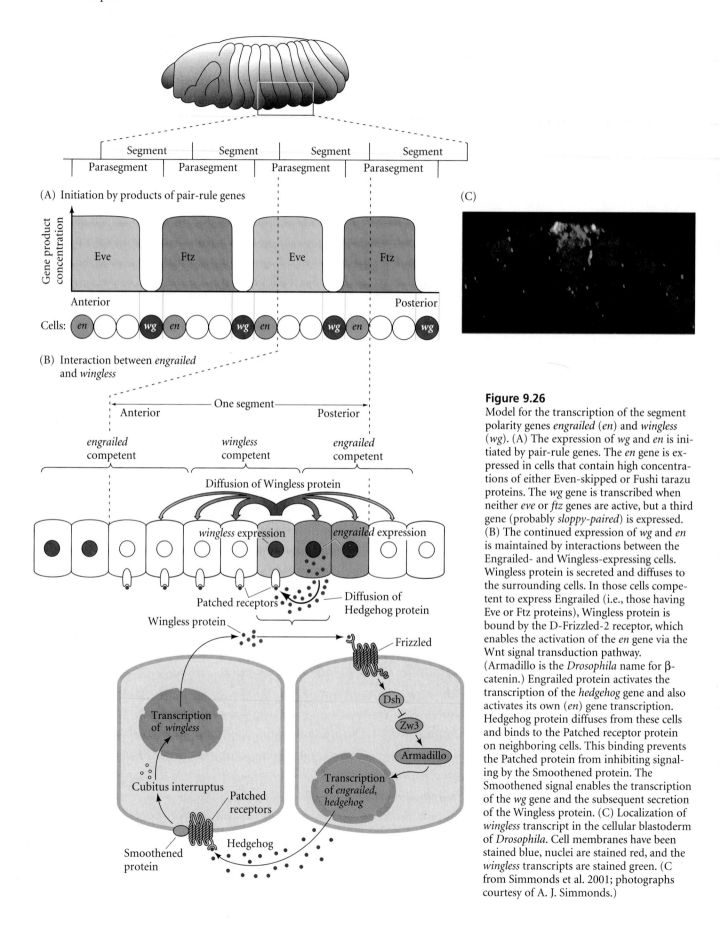

(A) Initiation by products of pair-rule genes

(B) Interaction between *engrailed* and *wingless*

(C)

Figure 9.26
Model for the transcription of the segment polarity genes *engrailed* (*en*) and *wingless* (*wg*). (A) The expression of *wg* and *en* is initiated by pair-rule genes. The *en* gene is expressed in cells that contain high concentrations of either Even-skipped or Fushi tarazu proteins. The *wg* gene is transcribed when neither *eve* or *ftz* genes are active, but a third gene (probably *sloppy-paired*) is expressed. (B) The continued expression of *wg* and *en* is maintained by interactions between the Engrailed- and Wingless-expressing cells. Wingless protein is secreted and diffuses to the surrounding cells. In those cells competent to express Engrailed (i.e., those having Eve or Ftz proteins), Wingless protein is bound by the D-Frizzled-2 receptor, which enables the activation of the *en* gene via the Wnt signal transduction pathway. (Armadillo is the *Drosophila* name for β-catenin.) Engrailed protein activates the transcription of the *hedgehog* gene and also activates its own (*en*) gene transcription. Hedgehog protein diffuses from these cells and binds to the Patched receptor protein on neighboring cells. This binding prevents the Patched protein from inhibiting signaling by the Smoothened protein. The Smoothened signal enables the transcription of the *wg* gene and the subsequent secretion of the Wingless protein. (C) Localization of *wingless* transcript in the cellular blastoderm of *Drosophila*. Cell membranes have been stained blue, nuclei are stained red, and the *wingless* transcripts are stained green. (C from Simmonds et al. 2001; photographs courtesy of A. J. Simmonds.)

of cells in each parasegment is permitted to express the Wingless protein. The key to this pattern is the activation of the *engrailed* gene in those cells that are going to express the Hedgehog protein. The *engrailed* gene is activated in cells that have high levels of the Even-skipped, Fushi tarazu, or Paired transcription factors. Moreover, it is repressed in those cells that receive high levels of Odd-skipped, Runt, or Sloppy-paired proteins. As a result, Engrailed is expressed in 14 stripes across the anterior-posterior axis of the embryo (see Figure 9.8E). (Indeed, in mutations that cause the embryo to be deficient in Fushi tarazu, only seven bands of Engrailed are expressed.)

These stripes of *engrailed* transcription mark the anterior boundary of each parasegment (and the posterior border of each segment). The *wingless* gene is activated in those bands of cells that receive little or no Even-skipped or Fushi tarazu protein, but which do contain Sloppy-paired. This pattern causes *wingless* to be transcribed solely in the row of cells directly anterior to the cells where *engrailed* is transcribed (Figure 9.26A).

Once *wingless* and *engrailed* expression are established in adjacent cells, this pattern must be maintained to retain the parasegmental periodicity of the body plan established by the pair-rule genes. It should be remembered that the mRNAs and proteins involved in initiating these patterns are short-lived, and that the patterns must be maintained after their initiators are no longer being synthesized. The maintenance of these patterns is regulated by reciprocal interaction between neighboring cells: Those cells secreting Hedgehog activate the expression of *wingless* in its neighbor; the wingless protein is received by the cell that secreted Hedgehog and maintains *hedgehog* expression.

In the cells transcribing the *wingless* gene, *wingless* mRNA is translocated by its 3′ UTR to the apex of the cell (see Figure 6.25B–D; Simmonds et al. 2001; Wilkie and Davis 2001). Like the *bicoid* message in the egg, *wingless* mRNA appears to be transported along microtubules by dynein proteins. At the apex, the *wingless* message is translated and secreted from the cell. The cells expressing *engrailed* can bind this protein because they contain the *Drosophila* membrane receptor protein for Wingless, D-Frizzled-2 (see Figure 6.23C; Bhanot et al. 1996). This receptor activates the Wnt signal transduction pathway, resulting in the continued expression of *engrailed* (Siegfried et al. 1994).

Moreover, this activation starts another portion of this reciprocal pathway. The Engrailed protein activates the transcription of the *hedgehog* gene in the *engrailed*-expressing cells. The Hedgehog protein can bind to the Hedgehog receptor (the Patched protein) on neighboring cells. When it binds to the adjacent posterior cells, it stimulates the expression of the *wingless* gene. The result is a reciprocal loop wherein the Engrailed-synthesizing cells secrete the Hedgehog protein, which maintains the expression of the *wingless* gene in the neighboring cells, while the Wingless-secreting cells maintain the expression of the *engrailed* and *hedgehog* genes in their neighbors

in turn (Heemskerk et al. 1991; Ingham et al. 1991; Mohler and Vani 1992). In this way, the transcription pattern of these two types of cells is stabilized. This interaction creates a stable boundary, as well as a signaling center from which Hedgehog and Wingless proteins diffuse across the parasegment.

The diffusion of these proteins is thought to provide the gradients by which the cells of the parasegment acquire their identities. This process can be seen in the dorsal epidermis, where the rows of larval cells produce different cuticular structures depending on their position within the segment. The 1° row of cells consists of large, pigmented spikes called denticles. Posterior to these cells, the 2° row produces a smooth epidermal cuticle. The next two cell rows have a 3° fate, making small, thick hairs, and these are followed by several rows of cells that adopt the 4° fate, producing fine hairs (Figure 9.27).

The *wingless*-expressing cells lie within the region producing the fine hairs, while the *hedgehog*-expressing cells are near the 1° row of cells. By using *hedgehog* genes fused to heat-shock promoters, the fates of the cells can be altered by experimentally increasing or decreasing the levels of Hedgehog or Wingless protein (Heemskerek and DiNardo 1994; Bokor and DiNardo 1996; Porter et al. 1996). Thus, Hedgehog and Wingless appear necessary for elaborating the entire pattern of cell types across the parasegment. The gradients of Hedgehog and wingless proteins are interpreted by a second series of protein gradients within the cells. This second set of gradients provides certain cells with the receptors for Hedgehog and often with the receptor for wingless (Lander et al. 2002; Casal et al. 2002). The resulting pattern of cell fates also changes the focus of patterning from parasegment to segment. There are now external markers, as the *engrailed*-expressing cells become the most posterior cells of each segment.

WEBSITE 9.7 **Asymmetrical spread of morphogens.** It is unlikely that morphogens such as Wingless spread by free diffusion. The asymmetry of Wingless diffusion suggests that neighboring cells play a crucial role in moving the protein.

WEBSITE 9.8 **Getting a head in the fly.** The segment polarity genes may act differently in the head than in the trunk. Indeed, the formation of the *Drosophila* head may differ significantly from the way the rest of the body is formed.

The Homeotic Selector Genes

Patterns of homeotic gene expression

After the segmental boundaries have been established, the characteristic structures of each segment are specified. This specification is accomplished by the **homeotic selector genes** (Lewis 1978). Two regions of *Drosophila* chromosome 3 contain most of these homeotic genes (Figure 9.28). One of these, the **Antennapedia complex**, contains the homeotic genes

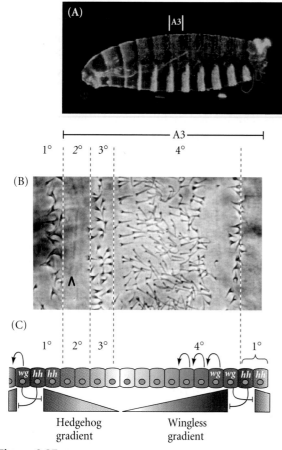

Figure 9.27
Cell specification by the Wingless/Hedgehog signaling center. (A) Bright-field photograph of wild-type *Drosophila* embryo, showing the position of the third abdominal segment. (B) Close-up of the dorsal area of the A3 segment, showing the different cuticular structures made by the 1°, 2°, 3°, and 4° rows of cells. The diagram below shows a model for the role of Wingless and Hedgehog. Each signal is responsible for about half the pattern. Either each signal acts in a graded manner (shown here as gradients decreasing with distance from their respective sources) to specify the fates of cells at a distance from these sources, or each signal acts locally on the neighboring cells to initiate a cascade of inductions (shown here as sequential arrows). (After Heemskerk and DiNardo 1994; photographs courtesy of the authors.)

labial (lab), Antennapedia (Antp), sex combs reduced (scr), Deformed (dfd), and *proboscipedia (pb). The labial* and *Deformed* genes specify the head segments, while *sex combs reduced* and *Antennapedia* contribute to giving the thoracic segments their identities. The *proboscipedia* gene appears to act only in adults, but in its absence, the labial palps of the mouth are transformed into legs (Wakimoto et al. 1984; Kaufman et al. 1990).

The second region of homeotic genes is the **bithorax complex** (Lewis 1978). Three protein-coding genes are found in this complex: *Ultrabithorax (Ubx)*, which is required for the identity of the third thoracic segment; and the *abdominal A*

(abdA) and *Abdominal B (AbdB)* genes, which are responsible for the segmental identities of the abdominal segments (Sánchez-Herrero et al. 1985). The lethal phenotype of the triple-point mutant *Ubx⁻, abdA⁻, AbdB⁻* is identical to that resulting from a deletion of the entire bithorax complex (Casanova et al. 1987). The chromosome region containing both the Antennapedia complex and the bithorax complex is often referred to as the **homeotic complex** (**Hom-C**).

Because the homeotic selector genes are responsible for the specification of fly body parts, mutations in them lead to bizarre phenotypes. In 1894, William Bateson called these organisms **homeotic mutants,** and they have fascinated developmental biologists for decades.* For example, the body of the normal adult fly contains three thoracic segments, each of which produces a pair of legs. The first thoracic segment does not produce any other appendages, but the second thoracic segment produces a pair of wings in addition to its legs. The third thoracic segment produces a pair of wings and a pair of balancers known as **halteres**. In homeotic mutants, these specific segmental identities can be changed. When the *Ultrabithorax* gene is deleted, the third thoracic segment (characterized by halteres) is transformed into another second thoracic segment. The result is a fly with four wings (Figure 9.29)—an embarrassing situation for a classic dipteran.†

Similarly, Antennapedia protein usually specifies the second thoracic segment of the fly. But when flies have a mutation wherein the *Antennapedia* gene is expressed in the head (as well as in the thorax), legs rather than antennae grow out of the head sockets (Figure 9.30). In the recessive mutant of *Antennapedia*, the gene fails to be expressed in the second thoracic segment, and antennae sprout in the leg positions (Struhl 1981; Frischer et al. 1986; Schneuwly et al. 1987).

The major homeotic selector genes have been cloned and their expression analyzed by in situ hybridization (Harding et al. 1985; Akam 1987). Transcripts from each gene can be detected in specific regions of the embryo and are especially prominent in the central nervous system (see Figure 9.28).

Initiating the patterns of homeotic gene expression

The initial domains of homeotic gene expression are influenced by the gap genes and pair-rule genes. For instance, ex-

---

*"Homeo" vocabulary: *Homeo* means "similar." *Homeotic mutants* are mutants in which one structure is replaced by another (as where an antenna is replaced by a leg). *Homeotic genes* are genes whose mutation can cause such transformations; thus, they are genes that specify the identity of a particular body segment. The *homeobox* is a conserved DNA sequence of about 180 base pairs that is shared by many homeotic genes. This sequence encodes the 60-amino-acid homeodomain, which recognizes specific DNA sequences. The *homeodomain* is an important region of the transcription factors encoded by homeotic genes (see Sidelights & Speculations). Not all genes containing homeoboxes are homeotic genes, however.

†Dipterans (two-winged insects such as flies) are thought to have evolved from four-winged insects; it is possible that this change arose via alterations in the bithorax complex. Chapter 23 includes more speculation on the relationship between the homeotic complex and evolution.

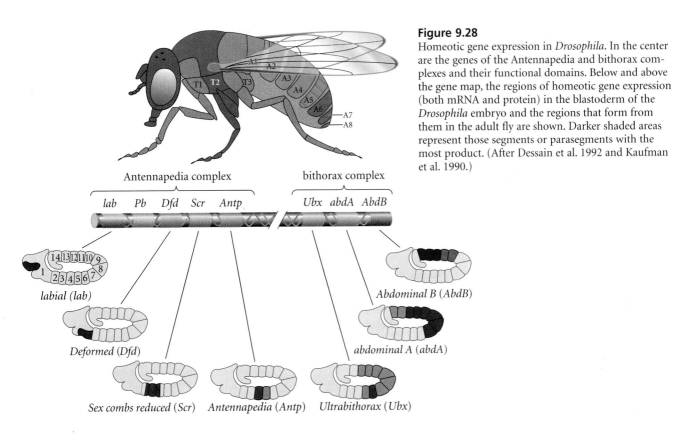

Figure 9.28
Homeotic gene expression in *Drosophila*. In the center are the genes of the Antennapedia and bithorax complexes and their functional domains. Below and above the gene map, the regions of homeotic gene expression (both mRNA and protein) in the blastoderm of the *Drosophila* embryo and the regions that form from them in the adult fly are shown. Darker shaded areas represent those segments or parasegments with the most product. (After Dessain et al. 1992 and Kaufman et al. 1990.)

pression of the *abdA* and *AbdB* genes is repressed by the gap gene proteins Hunchback and Krüppel. This inhibition prevents these abdomen-specifying genes from being expressed in the head and thorax (Casares and Sánchez-Herrero 1995). Conversely, the *Antennapedia* gene is activated by particular

Figure 9.29
A four-winged fruit fly constructed by putting together three mutations in *cis* regulators of the *Ultrabithorax* gene. These mutations effectively transform the third thoracic segment into another second thoracic segment (i.e., halteres into wings). (Photograph courtesy of E. B. Lewis.)

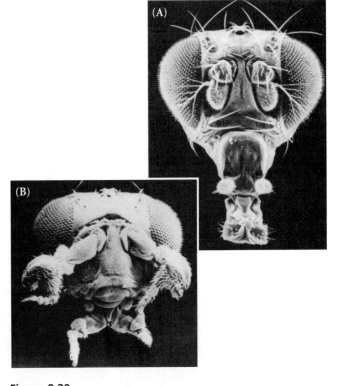

Figure 9.30
(A) Head of a wild-type fruit fly. (B) Head of a fly containing the *Antennapedia* mutation that converts antennae into legs. (From Kaufman et al. 1990; photographs courtesy of T. C. Kaufman.)

levels of Hunchback (needing both the maternal and the zygotically transcribed messages), so *Antennapedia* is originally transcribed in parasegment 4, specifying the mesothoracic (T2) segment (Wu et al. 2001). The boundaries of homeotic gene expression are soon confined to the parasegments defined by the Fushi tarazu and Even-skipped proteins (Ingham and Martinez-Arias 1986; Müller and Bienz 1992).

Maintaining the patterns of homeotic gene expression

The expression of homeotic genes is a dynamic process. The *Antennapedia* gene, for instance, although initially expressed in presumptive parasegment 4, soon appears in parasegment 5. As the germ band expands, *Antp* expression is seen in the presumptive neural tube as far posterior as parasegment 12. During further development, the domain of *Antp* expression contracts again, and *Antp* transcripts are localized strongly to parasegments 4 and 5. Like that of other homeotic genes, *Antp* expression is negatively regulated by all the homeotic gene products expressed posterior to it (Harding and Levine 1989; González-Reyes and Morata 1990). In other words, each of the bithorax complex genes represses the expression of Antennapedia. If the *Ultrabithorax* gene is deleted, *Antp* activity extends through the region that would normally have expressed *Ubx* and stops where the *Abd* region begins. (This allows the third thoracic segment to form wings like the second thoracic segment, as in Figure 9.29.) If the entire bithorax complex is deleted, *Antp* expression extends throughout the abdomen. (Such a larva does not survive, but the cuticle pattern throughout the abdomen is that of the second thoracic segment.)

As we have seen, gap gene and pair-rule gene proteins are transient, but the identities of the segments must be stabilized so that differentiation can occur. Thus, once the transcription patterns of the homeotic genes have become stabilized, they are "locked" into place by alteration of the chromatin conformation in these genes. The repression of homeotic genes appears to be maintained by the Polycomb family of proteins, while the active chromatin conformation appears to be maintained by the Trithorax proteins (Ingham and Whittle 1980; McKeon and Brock 1991; Simon et al. 1992).

Realisator genes

The homeotic genes work by activating or repressing a group of "realisator genes," those genes that are the targets of the homeotic gene proteins and which function to form the specified tissue or organ primordia. For example, *Antennapedia* is expressed in the formation of the second thoracic segment. The Antennapedia protein binds to and represses the enhancers of at least two genes, *homothorax* and *eyeless*, which encode transcription factors that are critical for antenna and eye formation, respectively[*] (Casares and Mann 1998; Plaza et al. 2001). Therefore, one of *Antennapedia's* functions is to suppress the genes that would trigger antenna and eye development.

The Ultrabithorax protein is able to repress the expression of the *wingless* gene in those cells that will become the halteres of the fly. One of the major differences between the dorsal imaginal disc cells of the second and the third thoracic segments is that *wingless* expression occurs in the imaginal discs cells of the second thoracic segment, but not in those of the third thoracic segment. Wingless acts as a growth promoter and morphogen in these tissues. Ubx protein is found in the cells of the third thoracic segment, and Ubx prevents expression of the *wingless* gene (Figure 9.31; Weatherbee et al. 1998). Thus, one of the ways in which Ubx protein specifies the third thoracic segment is by preventing the expression of those genes that would generate wing tissue.

Another target of the homeotic proteins, the *distal-less* gene (itself a homeobox-containing gene: see Sidelights & Speculations), is necessary for limb development and is active solely in the thorax. Distal-less expression is repressed in the abdomen by a combination of Ubx and AbdA proteins, which bind to its enhancer and block its transcription (Vachon et al. 1992; Castelli-Gair and Akam 1995). This expression pattern presents a paradox, since parasegment 5 (entirely thoracic and leg-producing) and parasegment 6 (which includes most of the legless first abdominal segment) both express Ubx. How can these two very different segments be specified by the same gene?

---

[*]The Homothorax protein appears to be a cofactor for the Distal-less protein. While mutations affecting either of these proteins cause antenna-to-leg transformations, both of these proteins are needed for the expression of antenna-specific genes (Dong et al. 2000).

(A)

(B)

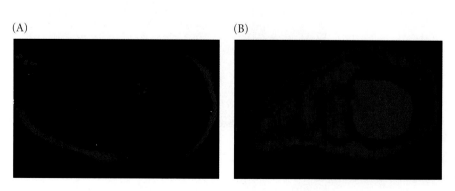

Figure 9.31
Antibody staining of the Ultrabithorax protein in (A) the wing disc and (B) the haltere disc of third instar *Drosophila* larvae. The cells of these discs will give rise to a wing and a haltere, respectively. In the wing disc, Ultrabithorax staining can be seen only on the cells that form the peripodial membrane, and not on those that form the wing itself. In the haltere disc, Ultrabithorax is found in those cells that will produce the major portion of the haltere. (From Weatherbee et al. 1998; photographs courtesy of S. D. Weatherbee and S. Carroll.)

Castelli-Gair and Akam (1995) have shown that the mere presence of Ubx protein in a group of cells is not sufficient for specification. Rather, the time and place of its expression within the parasegment can be critical. Before Ubx expression, parasegments 4–6 have similar potentials. At division cycle 10, Ubx expression in the anterior parts of parasegments 5 and 6 prevents those parasegments from forming structures (such as the anterior spiracle) characteristic of parasegment 4. Moreover, in the posterior compartment of parasegment 6 (but not parasegment 5), Ubx blocks the formation of the limb primordium by repressing the *Distal-less* gene. At division cycle 11, by which time Ubx has pervaded all of parasegment 6, the *Distal-less* gene has become self-regulatory and cannot be repressed by Ubx (Figure 9.32

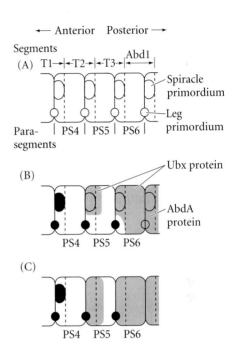

Figure 9.32
Schematic representation of the differences between *Ubx* expression in parasegments 5 and 6. (A) Before *Ubx* expression, each parasegment is competent to make both spiracles and legs. (B) At division cycle 10, early *Ubx* expression blocks the formation of the anterior spiracle in PS5 and PS6, and prevents limb formation in the posterior compartment of PS6. AbdA protein plays the same role in other abdominal segments. (C) At division cycle 11, the *Ubx* expression domain extends to the limb primordia of PS5 and PS6, but it is "too late" to repress *Distal-less* gene expression. (After Castelli-Gair and Akam 1995.)

Sidelights & Speculations

The Homeodomain Proteins

Homeodomain proteins are a family of transcription factors characterized by a 60-amino-acid domain (the **homeodomain**) that binds to certain regions of DNA. The homeodomain was first discovered in those proteins whose absence or misregulation caused homeotic transformations of *Drosophila* segments. It is thought that homeodomain proteins activate batteries of genes that specify the particular properties of each segment. The homeodomain proteins include the products of the eight genes of the homeotic complex (Hom-C), as well as other proteins such as Fushi tarazu, Caudal, Distal-less and Bicoid. Homeodomain proteins are important in determining the anterior-posterior axes of both invertebrates and vertebrates. In *Drosophila*, the presence of certain homeodomain proteins is also necessary for the determination of specific neurons. Without these transcription factors, the fates of these neuronal cells are altered (Doe et al. 1988).

Structure of the Homeodomain

The homeodomain is encoded by a 180-base-pair DNA sequence known as the **homeobox**. The homeodomains appear to be the DNA-binding sites of these proteins, and they are critical in specifying cell fates. For instance, if a chimeric protein is constructed mostly of Antennapedia but with the carboxyl terminus (including the homeodomain) of Ultrabithorax, it can substitute for Ultrabithorax and specify the appropriate cells as parasegment 6 (Mann and Hogness 1990). The isolated homeodomain of Antennapedia will bind to the same promoters as the entire Antennapedia protein, indicating that the binding of this protein is dependent on its homeodomain (Müller et al. 1988).

The homeodomain folds into three α helices, the latter two folding into a helix-turn-helix conformation that is characteristic of transcription factors that bind DNA in the major groove of the double helix

(Otting et al. 1990; Percival-Smith et al. 1990). The third helix is the recognition helix, and it is here that the amino acids make contact with the bases of the DNA. A four-base motif, TAAT, is conserved in nearly all sites recognized by homeodomains; it probably distinguishes those sites to which homeodomain proteins can bind. The 5′ terminal T appears to be critical in this recognition, as mutating it destroys all homeodomain binding. The base pairs following the TAAT motif are important in distinguishing between similar recognition sites. For instance, the next base pair is recognized by amino acid 9 of the recognition helix. Mutation studies have shown that the Bicoid and Antennapedia homeodomain proteins use lysine and glutamine, respectively, at position 9 to distinguish related recognition sites. The lysine of the Bicoid homeodomain recognizes the G of CG pairs, while the glutamine of the Antennapedia homeodomain recognizes the A of AT pairs (Figure 9.33; Hanes and

Brent 1991). If the lysine in Bicoid is replaced by glutamine, the resulting protein will recognize Antennapedia-binding sites (Hanes and Brent 1989, 1991). Other homeodomain proteins show a similar pattern, in which one portion of the homeodomain recognizes the common TAAT sequence, while another portion recognizes a specific structure adjacent to it.

Cofactors for the Hom-C Genes

The genes of the *Drosophila* homeotic complex specify segmental fates, but they may need some help in doing so. The DNA-binding sites recognized by the homeodomains of the Hom-C proteins are very similar, and there is some overlap in their binding specificity. In 1990, Peifer and Wieschaus discovered that the product of the *Extradenticle* (*Exd*) gene interacts with several Hom-C proteins and may help to specify segmental identities. For instance, the Ubx protein is responsible for specifying the identity of the first abdominal segment (A1). Without Extradenticle protein, it will transform this

segment into A3. Moreover, the Exd and Ubx proteins are both needed for the regulation of the *decapentaplegic* gene, and the structure of the *decapentaplegic* promoter suggests that the Exd protein may dimerize with the Ubx protein on the enhancer of this target gene (Raskolb and Wieschaus 1994; van Dyke and Murre 1994). The Extradenticle protein includes a homeodomain; the human protein PBX1, which resembles the Extradenticle protein, may play a similar role as a cofactor for human homeotic genes.

The product of the *teashirt* gene may also be an important cofactor. This zinc finger transcription factor is necessary for the functioning of the Sex combs reduced protein, which distinguishes between the labial and first thoracic segments. It is critical for the specification of anterior prothoracic (parasegment 3) identity, and it may be the gene that specifies the "ground state" condition of the homeotic complex. If the Bithorax complex and the *Antennapedia* gene are removed, all the segments become anterior prothorax. The product of the *teashirt* gene appears to work with the Scr protein to distinguish thorax from head and to work

throughout the trunk to prevent head structures from forming (Roder et al. 1992).

In addition to the homeodomain, other regions of the homeotic proteins are critical for normal development. Indeed, some of these regions also contact the DNA of their target genes, and the binding specificity of these non-homeodomain domains may explain how a single homeodomain protein can have different effects in cells of different types. For instance, in addition to specifying the ectodermal cuticle, the *abdA* expression domain also specifies the midgut chambers of the visceral mesoderm. By making chimeric proteins between AbdA and Ubx, Chauvet and colleagues (2000) showed that, whereas the epidermal specificity of the AbdA protein resides mostly in its homeodomain, the mesodermal specificity of AbdA resides primarily in the adjacent peptide sequence closer to the amino terminus.

WEBSITE 9.9 Homeotic genes and their protein products. The Hom-C genes have fascinating structures. The protein products of these genes bind to DNA in the presence of other proteins that may allow them to recognize specific sequences of DNA.

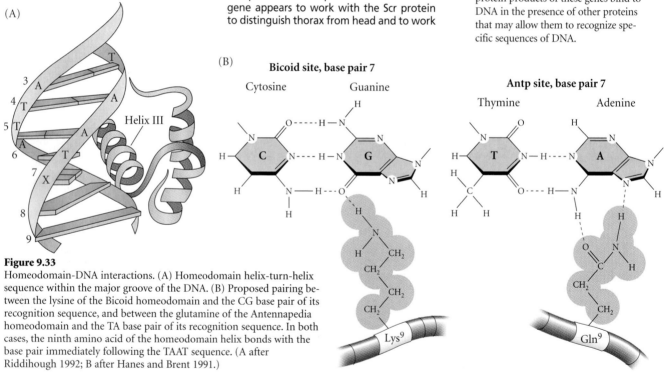

Figure 9.33
Homeodomain-DNA interactions. (A) Homeodomain helix-turn-helix sequence within the major groove of the DNA. (B) Proposed pairing between the lysine of the Bicoid homeodomain and the CG base pair of its recognition sequence, and between the glutamine of the Antennapedia homeodomain and the TA base pair of its recognition sequence. In both cases, the ninth amino acid of the homeodomain helix bonds with the base pair immediately following the TAAT sequence. (A after Riddihough 1992; B after Hanes and Brent 1991.)

THE GENERATION OF DORSAL-VENTRAL POLARITY

In 1936, the embryologist E. E. Just criticized those geneticists who sought to explain *Drosophila* development by looking at specific mutations affecting eye color, bristle number, and wing shape. He said that he wasn't interested in the development of

the bristles of a fly's back; rather, he wanted to know how the fly embryo makes the back itself. Fifty years later, embryologists and geneticists are finally answering that question.*

---

*In a manner that Just could not have predicted, it turns out that some of the genes (such as *decapentaplegic*) that are involved in regulating bristle number or wing shape also have earlier functions regulating dorsal-ventral polarity.

Dorsal: The Morphogenetic Agent for Dorsal-Ventral Polarity

Dorsal-ventral polarity is established by the gradient of a transcription factor called **Dorsal**. Unlike Bicoid protein, whose gradient is established within a syncytium, Dorsal forms a gradient over a field of cells that is established as a consequence of cell-to-cell signaling events.

The specification of the dorsal-ventral axis takes place in several steps. The critical step is the translocation of the Dorsal protein from the cytoplasm into the nuclei of the ventral cells during the fourteenth division cycle. Anderson and Nüsslein-Volhard (1984) isolated 11 maternal effect genes, each of whose absence is associated with a lack of ventral structures (Figure 9.34). The absence of another maternal effect gene, *cactus*, causes the ventralization of all cells. The proteins encoded by these maternal genes are critical for making certain that Dorsal protein gets into only those nuclei on the ventral surface of the embryo.[†] After its translocation, Dorsal acts on cell nuclei to specify the different regions of the embryo. Different concentrations of Dorsal protein in the nuclei appear to specify different fates for cells.

Translocation of Dorsal to the nucleus

The protein that actually distinguishes dorsum (back) from ventrum (belly) is the product of the *dorsal* gene. The mRNA transcript of the mother's *dorsal* gene is placed in the oocyte by her ovarian cells. However, Dorsal protein is not synthesized from this maternal message until about 90 minutes after fertilization. When Dorsal is translated, it is found throughout the embryo, not just on the ventral or dorsal side. How can this protein act as a morphogen if it is located everywhere in the embryo?

In 1989, the surprising answer to this question was found (Roth et al. 1989; Rushlow et al. 1989; Steward 1989). While Dorsal is found throughout the syncytial blastoderm of the early *Drosophila* embryo, it is translocated into nuclei only in the ventral part of the embryo (Figure 9.35A, B). In the nu-

---

[†]Remember that a gene in *Drosophila* is usually named after its mutant phenotype. Thus, the product of the *dorsal* gene is necessary for the differentiation of ventral cells. That is, in the absence of the *dorsal* gene, the ventral cells become dorsalized.

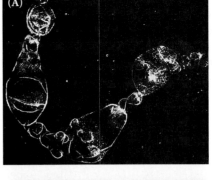

Figure 9.34
Effect of mutations affecting the distribution of the Dorsal protein. (A) Deformed larva consisting entirely of dorsal cells. Larvae like these developed from the eggs of a female homozygous for a mutation of the *snake* gene, one of the maternal effect genes involved in the signaling cascade that establishes a gradient of Dorsal in the embryo. (B) Larvae developed from *snake* mutant eggs that received injections of mRNA from wild-type eggs. These larvae have a wild-type appearance. (From Anderson and Nüsslein-Volhard 1984; photographs courtesy of C. Nüsslein-Volhard.)

Figure 9.35
Translocation of Dorsal protein into ventral, but not lateral or dorsal, nuclei. (A) Fate map of a lateral cross section through the *Drosophila* embryo. The most ventral part becomes the mesoderm; the next higher portion becomes the neurogenic (ventral) ectoderm. The lateral and dorsal ectoderm can be distinguished in the cuticle, and the dorsalmost region becomes the amnioserosa, the extraembryonic layer that surrounds the embryo. (B–D) Transverse sections of embryos stained with antibody to show the presence of Dorsal protein (dark-stained area). (B) A wild-type embryo, showing Dorsal protein in the ventralmost nuclei. (C) A dorsalized mutant, showing no localization of Dorsal protein in any nucleus. (D) A ventralized mutant, in which Dorsal protein has entered the nucleus of every cell. (A from Rushlow et al. 1989; B–D from Roth et al. 1989, photographs courtesy of the authors.)

(A)
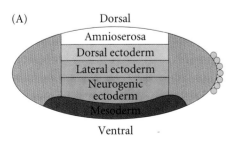

LATERAL VIEW   TRANSVERSE SECTION

(B) 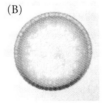   (C) 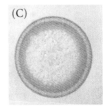   (D)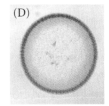

cleus, Dorsal binds to certain genes to activate or suppress their transcription. If Dorsal does not enter the nucleus, the genes responsible for specifying ventral cell types (*snail* and *twist*) are not transcribed, the genes responsible for specifying dorsal cell types (*decapentaplegic* and *zerknüllt*) are not repressed, and all the cells of the embryo become specified as dorsal cells.

This model of dorsal-ventral axis formation in *Drosophila* is supported by analyses of mutations that give rise to an entirely dorsalized or an entirely ventralized phenotype (see Figures 9.34A and 9.35). In those mutants in which all the cells are dorsalized (evident from their dorsal cuticle), Dorsal protein does not enter the nucleus in any cell. Conversely, in those

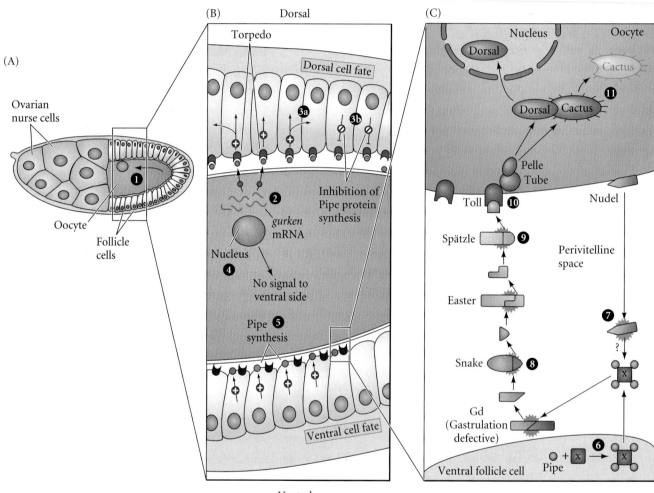

❶ Oocyte nucleus travels to anterior dorsal side of oocyte. It synthesizes *gurken* mRNA, which remains between the nucleus and the follicle cells.

❷ *gurken* messages are translated. The Gurken protein is received by Torpedo proteins during mid-oogenesis.

❸ₐ Torpedo signal causes follicle cells to differentiate to a dorsal morphology.

❸ᵦ Synthesis of Pipe protein is inhibited in dorsal follicle cells.

❹ Gurken protein does not diffuse to ventral side.

❺ Ventral follicle cells synthesize Pipe protein.

❻ In ventral follicle cells, Pipe completes the modification of an unknown factor (x).

❼ Nudel and factor (x) interact to split the Gastrulation-deficient (Gd) protein.

❽ The activated Gd protein splits the Snake protein, and the activated Snake protein cleaves the Easter protein.

❾ The activated Easter protein splits Spätzle; activated Spätzle binds to Toll receptor protein.

❿ Toll activation activates Tube and Pelle, which phosphorylate the Cactus protein. Cactus is degraded, releasing it from Dorsal.

⓫ Dorsal protein enters the nucleus and ventralizes the cell.

mutants in which all cells have a ventral phenotype, Dorsal is found in every cell nucleus.

The signal cascade

SIGNAL FROM THE OOCYTE NUCLEUS TO THE FOLLICLE CELLS.
If Dorsal is found throughout the embryo, but gets translocated into the nuclei of only ventral cells, then something else must be providing asymmetrical cues. It appears that this signal is mediated through a complex interaction between the oocyte and its surrounding follicle cells (Figure 9.36).

The follicular epithelium surrounding the developing oocyte is initially symmetrical, but this symmetry is broken by a signal from the oocyte nucleus. The oocyte nucleus is originally located at the posterior end of the oocyte, away from the nurse cells. It then moves to an anterior dorsal position and signals the overlying follicle cells to become the more columnar dorsal follicle cells (Montell et al. 1991; Schüpbach et al. 1991; see Figure 9.11). The dorsalizing signal from the oocyte nucleus is the product of the **gurken** gene, the only gene known to be transcribed from the haploid oocyte nucleus (Schüpbach 1987; Forlani et al. 1993). The *gurken* message becomes localized in a crescent between the oocyte nucleus and the oocyte cell membrane, and its protein product forms an anterior-posterior gradient along the dorsal surface of the oocyte (Figure 9.37; Neuman-Silberberg and Schüpbach 1993). Since it can diffuse only a short distance, the Gurken protein reaches only those follicle cells closest to the oocyte nucleus. Mutations of the *gurken* gene in the mother (and thus in the oocyte) cause the ventralization of both the embryo and its surrounding follicle cells. (If the mutation is in the follicle cells and not in the egg, the embryo is normal.)

The Gurken signal is received by the follicle cells through a receptor encoded by the **torpedo** gene. Molecular analysis has now established that *gurken* encodes a homologue of the vertebrate epidermal growth factor (EGF), while *torpedo* encodes a homologue of the vertebrate EGF receptor (Price et al. 1989; Neuman-Silberberg and Schüpbach 1993). Maternal deficiency of *torpedo* causes ventralization of the embryo. Moreover, the *torpedo* gene is active in the ovarian follicle cells, not in the embryo. This was discovered by experiments with germ line/so-

(A)

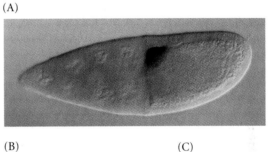

(B)   (C)

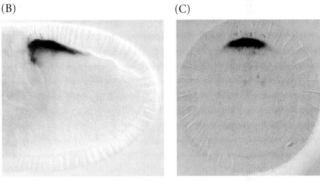

(D)

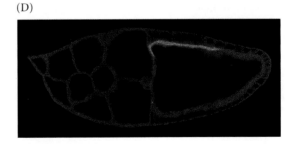

Figure 9.37
Expression of the *gurken* message and protein between the oocyte nucleus and the dorsal anterior cell membrane. (A) The *gurken* mRNA is localized between the oocyte nucleus and the dorsal follicle cells of the ovary. (B) The Gurken protein is similarly located (shown here is a younger stage than A). (C) Cross section of the egg through the region of Gurken protein expression. (D) A more mature oocyte, showing Gurken protein (yellow) across the dorsal region. The actin is stained red, showing cell boundaries. As the oocyte grows, follicle cells migrate across the top of the oocyte, becoming exposed to Gurken. (A from Ray et al. 1996, photograph courtesy of T. Schüpbach; B and C from Peri et al. 1999, photograph courtesy of S. Roth; D, photograph courtesy of C. van Buskirk and T. Schüpbach.)

◀ **Figure 9.36**
Schematic representation of the generation of dorsal-ventral polarity in *Drosophila*. (A) The oocyte develops in an ovarian follicle consisting of 15 nurse cells (which supply maternal proteins and messages to the developing egg) and numerous follicle cells. (B) The nucleus of the oocyte travels to what will become the dorsal side of the embryo. The *gurken* genes of the oocyte synthesize mRNA that becomes localized between the oocyte nucleus and the cell membrane, where it is translated into Gurken protein. The Gurken signal is received by the receptor protein made by the torpedo gene of the follicle cells. Given the short diffusibility of the signal, only the follicle cells closest to the oocyte nucleus (i.e., the dorsal follicle cells) receive this signal. The signal from the Torpedo receptor causes the follicle cells to take on a characteristic dorsal follicle morphology and (somehow) inhibit the synthesis of Pipe protein. Therefore, this protein is made only by the ventral follicle cells. (C) Pipe modifies an unknown protein (X) and allows it to be secreted from the ventral follicle cells. Nudel protein interacts with this modified factor to split the products of the gastrulation defective and snake genes to create an active enzyme that will split the zymogen form of the Easter protein into an active Easter protease. The Easter protease splits the Spätzle protein into a form that can bind to the Toll receptor (which is found throughout the embryonic cell membrane). Thus, only the ventral cells receive the Toll signal. This signal separates the Cactus protein from the Dorsal protein, allowing Dorsal to be translocated into the nuclei and ventralize the cells. (After van Eeden and St. Johnston 1999.)

matic chimeras. Schüpbach (1987) transplanted germ cell precursors from wild-type embryos into embryos whose mothers carried the *torpedo* mutation. Conversely, she transplanted the germ cells from *torpedo* mutants into wild-type embryos (Figure 9.38). The wild-type eggs produced mutant, ventralized embryos when they developed within the *torpedo* mutant mothers' follicles. The *torpedo* mutant eggs were able to produce normal embryos if they developed within a wild-type ovary. Thus, unlike the *gurken* gene product, the wild-type *torpedo* gene is needed in the follicle cells, not in the egg itself.

SIGNAL FROM THE FOLLICLE CELLS TO THE OOCYTE CYTOPLASM. The activated Torpedo receptor protein inhibits the expression of the **pipe** gene. As a result, the Pipe protein is made only in the ventral follicle cells (Sen et al. 1998; Amiri and Stein 2002). Pipe (in some as yet unknown way) activates the Nudel protein, which is secreted to the cell membrane of the ventral embryonic cells. A few hours later in development, the activated Nudel protein initiates the activation of three serine proteases that are secreted by the embryo into the perivitelline fluid (see Figure 9.36C; Hong and Hashimoto 1995). These three serine proteases are the products of the *gastrulation defective* (*gd*), *snake* (*snk*), and *easter* (*ea*) genes. Like most extracellular proteases, they are secreted in an inactive form and are activated by peptide cleavage. In a complex cascade of events, activated Nudel protein activates the Gastrulation defective protease. This protease cleaves the Snake protein, activating the Snake protease, which in turn cleaves the Easter protein. This cleavage activates the Easter protease, which cleaves the Spätzle protein (Chasan et al. 1992; Hong and Hashimoto 1995; LeMosy et al. 2001).

The cleaved Spätzle protein is now able to bind to its receptor in the oocyte cell membrane, the product of the *toll* gene. Toll protein is a maternal product that is evenly distributed throughout the cell membrane of the egg (Hashimoto et al. 1988, 1991), but it becomes activated only by binding the Spätzle protein, which is produced only on the ventral side of the egg. Therefore, the Toll receptors on the ventral side of the egg are transducing a signal into the egg, while the Toll receptors on the dorsal side of the egg are not.

Establishing the dorsal patterning gradient

SEPARATION OF THE DORSAL AND CACTUS PROTEINS. The crucial outcome of signaling through the Toll protein is the establishment of a gradient of Dorsal protein in the ventral cell nuclei. How is this gradient established? It appears that a protein called Cactus is blocking the portion of the Dorsal protein that enables the Dorsal protein to get into nuclei. As long as Cactus protein is bound to it, Dorsal protein remains in the cytoplasm. Thus, this entire complex signaling system is organized to split the Cactus protein from the Dorsal protein in the ventral region of the egg. When Spätzle binds to and activates the Toll protein, Toll can activate the Pelle protein kinase. (The Tube protein is probably necessary for bringing Pelle to the cell membrane, where it can be activated: Galindo et al. 1995.) The activated Pelle protein kinase can (probably through an intermediate) phosphorylate Cactus. Once phosphorylated, the Cactus protein is degraded and the Dorsal protein can enter the nucleus (Kidd 1992; Shelton and Wasserman 1993; Whalen and Steward 1993; Reach et al. 1996). Since the signal transduction cascade creates a gradient of Spätzle protein that is highest in the most ventral region, there is a gradient of Dorsal

Figure 9.38
Germ-line chimeras made by interchanging pole cells (germ cell precursors) between wild-type embryos and embryos from mothers homozygous for a mutation of the *torpedo* gene. These transplants produce wild-type females whose eggs come from the mutant mothers, and *torpedo*-deficient females that lay wild-type eggs. The *torpedo*-deficient eggs produced normal embryos when they developed in the wild-type ovary, while the wild-type eggs produced ventralized embryos when they developed in the mutant mother's ovary

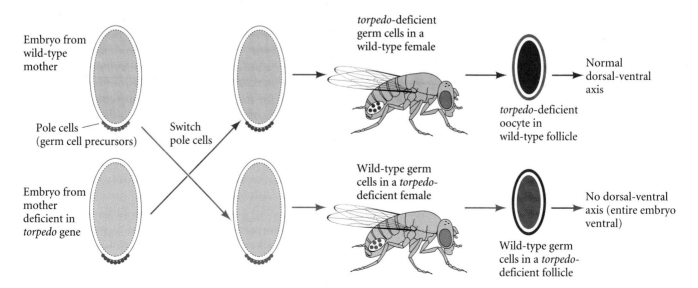

translocation in the ventral cells of the embryo, with the highest concentrations of Dorsal in the most ventral cell nuclei.

The process described for the translocation of Dorsal protein into the nucleus is very similar to the process for the translocation of the NF-κB transcription factor into the nucleus of mammalian lymphocytes. In fact, there is substantial homology between NF-κB and Dorsal, between I-B and Cactus, between the Toll protein and the interleukin 1 receptor, between Pelle protein and an IL-1-associated protein kinase, and between the DNA sequences recognized by Dorsal and by NF-κB* (González-Crespo and Levine 1994; Cao et al. 1996). Thus, the biochemical pathway used to specify dorsal-ventral polarity in *Drosophila* appears to be homologous to that used to differentiate lymphocytes in mammals (Figure 9.39).

EFFECTS OF THE DORSAL PROTEIN GRADIENT. What does the Dorsal protein do once it is located in the nuclei of the ventral cells? A look at the fate map of a cross section through the *Drosophila* embryo at the division cycle 14 (see Figure 9.35A) makes it obvious that the 16 cells with the highest concentration of Dorsal are those that generate the mesoderm. The next cell up from this region generates the specialized glial and neural cells of the midline. The next 2 cells are those that give rise to the ventral epidermis and ventral nerve cord, while the 9 cells above them produce the dorsal epidermis. The most dorsal group of 6 cells generates the amnioserosal covering of the embryo (Ferguson and Anderson 1991).

This fate map is generated by the gradient of Dorsal protein in the nuclei. Large amounts of Dorsal instruct the cells to become mesoderm, while lesser amounts instruct the cells to become glial or ectodermal tissue (Jiang and Levine 1993). The first morphogenetic event of *Drosophila* gastrulation is the invagination of the 16 ventralmost cells of the embryo (Figure 9.40). All of the body muscles, fat bodies, and gonads derive from these mesodermal cells (Foe 1989). The Dorsal protein specifies these cells to become mesoderm in two ways. First, Dorsal activates specific genes that create the mesodermal phenotype. Three of the target genes for Dorsal are *twist*, *snail*,

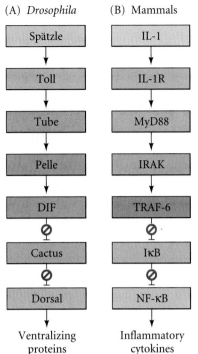

(A) *Drosophila*   (B) Mammals

Drosophila	Mammals
Spätzle	IL-1
↓	↓
Toll	IL-1R
↓	↓
Tube	MyD88
↓	↓
Pelle	IRAK
↓	↓
DIF	TRAF-6
⊘	⊘
Cactus	IκB
⊘	⊘
Dorsal	NF-κB

Ventralizing proteins   Inflammatory cytokines

Figure 9.39
Model of a conserved pathway for regulating nuclear transport of transcription factors in *Drosophila* and mammals. (A) In *Drosophila*, the Toll protein binds the signal from the Spätzle protein and activates the kinase region of the Pelle protein (probably with the help of the Tube protein). Pelle protein phosphorylates Cactus and Dorsal (through the dorsal-related immunity factor, DIF), causing the two proteins to separate from each other. The Dorsal protein can then enter the nucleus and regulate the transcription of ventrally specific genes. (B) In mammalian lymphocytes, the IL-1 receptor can cause the phosphorylation of IκB (through a cascade involving the IRAK kinase). This enables the NF-κB protein to enter the nucleus and effect the transcription of several lymphocyte-specific genes involved in the inflammatory response. The particular colors indicate homologous proteins in these homologous pathways. (After Qureshi et al. 1999.)

Figure 9.40
Gastrulation in *Drosophila*. In this cross section, the mesodermal cells at the ventral portion of the embryo buckle inward, forming a tube, which then flattens and generates the mesodermal organs. The nuclei are stained with antibody to the Twist protein. (From Leptin 1991a; photographs courtesy of M. Leptin.)

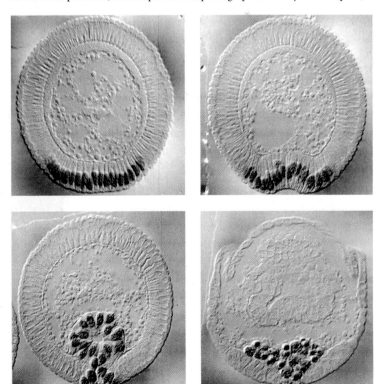

---

*Lemaitre and colleagues (1996) have shown that Toll and its ligand (Spätzle) are also involved in the *Drosophila* immune response against fungal infections. Moreover, several *toll*-related genes have been discovered in humans, which may be involved in both the immune response and early development (Rock et al. 1998; Qureshi et al. 1999).

Figure 9.41

Subdivision of the *Drosophila* dorsal-ventral axis by the gradient of Dorsal protein in the nuclei. (A) Dorsal protein activates the zygotic genes *rhomboid*, *twist*, and *snail*, depending on its nuclear concentration. The mesoderm forms where Twist and Snail are present, and the glial cells form where Twist and Rhomboid interact. Those cells with Rhomboid, but no Snail or Twist, form the neurogenic ectoderm. (B) Interactions in the specification of the ventral portion of the *Drosophila* embryo. Dorsal protein inhibits those genes that would give rise to dorsal structures (*tolloid*, *decapentaplegic*, and *zerknüllt*) while activating the three ventral genes. Snail protein, formed most ventrally, inhibits the transcription of *rhomboid* and prevents ectoderm formation. Twist activates *dMet2* and *bagpipe* (which activate muscle differentiation) as well as *tinman* (heart muscle development). (A after Steward and Govind 1993; B after Furlong et al. 2001.)

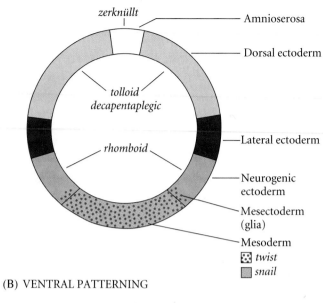

(A) DORSAL PATTERNING

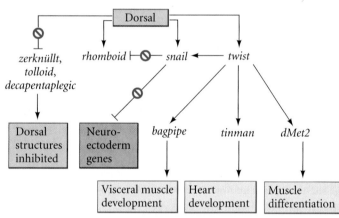

(B) VENTRAL PATTERNING

and *rhomboid* (Figure 9.41). These genes are transcribed only in nuclei that have received high concentrations of Dorsal protein, since their enhancers do not bind Dorsal with a very high affinity (Thisse et al. 1988, 1991; Jiang et al. 1991; Pan et al. 1991). The Twist protein activates mesodermal genes, while the Snail protein represses particular non-mesodermal genes that might otherwise be active. The *rhomboid* gene is interesting because it is activated by Dorsal but repressed by Snail. Thus, *rhomboid* is not expressed in the most ventral cells (i.e., the mesodermal precursors), but is expressed in the cells adjacent to the mesoderm that form the presumptive neural ectoderm (Figure 9.41; Jiang and Levine 1993). Both Snail and Twist are needed for the complete mesodermal phenotype and proper gastrulation (Leptin et al. 1991b). The sharp border between the mesodermal cells and those cells adjacent to them that generate glial cells (mesectoderm) is produced by the presence of Snail and Twist in the ventralmost cells, but of only Twist in the next cell up (Kosman et al. 1991). In *snail* mutants, the ventralmost cells still have the *twist* gene activated, and they resemble the more lateral cells (Nambu et al. 1990).

Meanwhile, intermediate levels of nuclear Dorsal activate the transcription of the *short gastrulation* (*sog*) gene in two lateral stripes that flank the ventral *twist* expression

Figure 9.42

The products of the *short gastrulation* gene. The mRNA for *sog* (yellow) is transcribed only in the ventrolateral region of the blastoderm, the region that is destined to become the neural ectoderm. The protein (red) synthesized from the *sog* message is found predominantly in the cells fated to become the neural ectoderm and those cells beneath this region that are fated to become the mesoderm. The Short gastrulation protein is degraded in those more dorsal cells that will become the epidermis of the larva. The figure to the side is a computer-reconstructed image of a late blastoderm embryo seen in cross-section (rotated 90° from the larger image). It shows the ventral-to-dorsal gradient of Sog protein. (After Srinivasan et al. 2002; photographs courtesy of E. Bier).

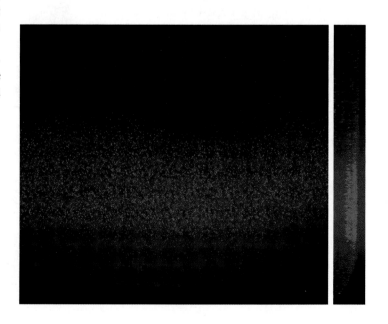

domain, each 12–14 cells wide (Figure 9.42; François et al. 1994; Srinivasan et al. 2002). The *sog* gene encodes a protein that prevents the ectoderm in this region from becoming hypodermis and begins the processes of neural differentiation.

The Dorsal protein also determines the mesoderm directly. In addition to activating the mesoderm-stimulating genes (*twist* and *snail*), it directly inhibits the dorsalizing genes *zerknüllt* (*zen*) and *decapentaplegic* (*dpp*). Thus, in the same cells, Dorsal can act as an activator of some genes and a repressor of others. Whether Dorsal protein activates or represses a given gene depends on the structure of the genes' enhancers. The *zen* enhancer contains a silencer region that contains a binding site for Dorsal and a second binding site for two other DNA-binding proteins. These two other proteins enable the Dorsal protein to bind a transcriptional repressor protein (Groucho) and bring it to the DNA (Valentine et al. 1998). Mutants of *dorsal* express *dpp* and *zen* genes throughout the embryo (Rushlow et al. 1987), and embryos deficient in *dpp* and *zen* fail to form dorsal structures (Irish and Gelbart 1987). Thus, in wild-type embryos, the mesodermal precursors express *twist* and *snail* (but not *zen* or *dpp*); precursors of the dorsal epidermis and amnioserosa express *zen* and *dpp* but not *twist* or *snail*. Glial (mesectoderm) precursors express *twist* and *rhomboid*, while the lateral neural ectodermal precursors do not express any of these four genes (Kosman et al. 1991; Ray and Schüpbach 1996). Thus, as a consequence of the responses to the Dorsal protein gradient, the axis becomes subdivided into mesoderm, mesectoderm, neurogenic ectoderm, epidermis, and amnioserosa.

Axes and Organ Primordia: The Cartesian Coordinate Model

The anterior-posterior and dorsal-ventral axes of *Drosophila* embryos form a coordinate system that can be used to specify positions within the embryo. Theoretically, cells that are initially equivalent in developmental potential can respond to their position by expressing different sets of genes. This type of specification has been demonstrated in the formation of the salivary gland rudiments (Panzer et al. 1992; Bradley et al. 2001; Zhou et al. 2001).

Drosophila salivary glands form only in the strip of cells defined by the activity of the *sex combs reduced* (*scr*) gene along the anterior-posterior axis (parasegment 2). No salivary glands form in *scr*-deficient mutants. Moreover, if *scr* is experimentally expressed throughout the embryo, salivary gland primordia form in a ventrolateral stripe along most of the length of the embryo. The formation of salivary glands along the dorsal-ventral axis is repressed by both Decapentaplegic and Dorsal. These proteins inhibit salivary gland formation both dorsally and ventrally. Thus, the salivary glands form at the intersection of the vertical *scr* expression band (parasegment 2) and the horizontal region in the middle of the em-

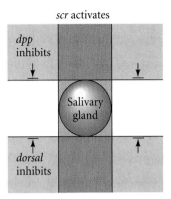

Figure 9.43
Cartesian coordinate system for the expression of genes giving rise to *Drosophila* salivary glands. These genes are activated by the protein product of the *sex combs reduced* (*scr*) homeotic gene in a narrow band along the anterior-posterior axis, and they are inhibited in the regions marked by *decapentaplegic* (*dpp*) and *dorsal* gene products along the dorsal-ventral axis. This pattern allows salivary glands to form in the midline of the embryo in the second parasegment. (After Panzer et al. 1992.)

bryo's circumference that has neither Decapentaplegic nor Dorsal gene products (Figure 9.43). The cells that form the salivary glands are directed to do so by the intersecting gene activities along the anterior-posterior and dorsal-ventral axes.

A similar situation is seen with tissues that are found in every segment of the fly. Neuroblasts arise from ten clusters of four to six cells each that form on each side in every segment in the strip of neural ectoderm at the midline of the embryo (Skeath and Carroll 1992). The potential to form neural cells is conferred on these cells by the expression of proneural genes from the achaete-scute gene complex: *achaete* (*ac*), *scute* (*sc*), and *lethal of scute* (*l'sc*). The cells in each cluster interact (via the Notch pathway discussed in Chapter 6) to generate a single neural cell from the cluster. Skeath and colleagues (1993) have shown that the pattern of *achaete* and *scute* transcription is imposed by a coordinate system. Their expression is repressed by the Decapentaplegic and Snail proteins along the dorsal-ventral axis, while positive enhancement by pair-rule genes along the anterior-posterior axis causes their repetition in each half-segment. The enhancer recognized by these axis-specifying proteins lies between the *achaete* and *scute* genes and appears to regulate both of them. It is very likely, then, that the positions of organ primordia are specified throughout the fly through a two-dimensional coordinate system based on the intersection of the anterior-posterior and dorsal-ventral axes.

Coda

Genetic studies on the *Drosophila* embryo have uncovered numerous genes that are responsible for the specification of the anterior-posterior and dorsal-ventral axes. We are far from a complete understanding of *Drosophila* pattern formation, but we are much more aware of its complexity than we were five years ago. The mutations of *Drosophila* genes have given us our first glimpses of the multiple levels of pattern regulation in a complex organism and have enabled us to isolate these genes and their products. Moreover, as we will see in forthcoming chapters, these genes provide clues to a general mechanism of pattern formation used throughout the animal kingdom.

We are beginning to learn how the genome influences the construction of the organism. The genes regulating pattern formation in *Drosophila* operate according to certain principles:

• There are *morphogens*—such as Bicoid and Dorsal— whose gradients determine the specification of different cell types. These morphogens can be transcription factors.

• There is a *temporal order* wherein different classes of genes are transcribed, and the products of one gene often regulate the expression of another gene.

• *Boundaries* of gene expression can be created by the interaction between transcription factors and their gene targets. Here, the transcription factors transcribed earlier regulate the expression of the next set of genes.

• *Translational control* is extremely important in the early embryo, and localized mRNAs are critical in patterning the embryo.

• Individual cell fates are not defined immediately. Rather, there is a stepwise specification wherein a given field is divided and subdivided, eventually regulating individual cell fates.

Snapshot Summary: Drosophila *Development and Axis Specification*

1. *Drosophila* cleavage is superficial. The nuclei divide 13 times before forming cells. Before cell formation, the nuclei reside in a syncytial blastoderm. Each nucleus is surrounded by actin-filled cytoplasm.

2. When the cells form, the *Drosophila* embryo undergoes a mid-blastula transition, wherein the cleavages become asynchronous and new mRNA is made. The amount of chromatin determines the timing of this transition.

3. Gastrulation begins with the invagination of the most ventral region (the presumptive mesoderm), which causes the formation of a ventral furrow. The germ band expands such that the future posterior segments curl just behind the presumptive head.

4. Maternal effect genes are responsible for the initiation of anterior-posterior polarity. *Bicoid* mRNA is bound by its 3′ UTR to the cytoskeleton in the future anterior pole; *nanos* mRNA is sequestered by its 3′ UTR in the future posterior pole. *Hunchback* and *caudal* messages are seen throughout the embryo.

5. At fertilization, *bicoid* and *nanos* messages are translated. A gradient of Bicoid protein activates more *hunchback* transcription in the anterior. Moreover, Bicoid inhibits the translation of *caudal* mRNA. A gradient of Nanos in the posterior inhibits the translation of *hunchback* mRNA. Caudal protein is made in the posterior.

6. The Bicoid and Hunchback proteins activate the genes responsible for the anterior portion of the fly; Caudal activates genes responsible for posterior development.

7. The unsegmented anterior and posterior extremities are regulated by the activation of the Torso protein at the anterior and posterior poles of the egg.

8. The gap genes respond to concentrations of the maternal effect gene proteins. Their protein products interact with each other such that each gap gene protein defines specific regions of the embryo.

9. The gap gene proteins activate and repress the pair-rule genes. The pair-rule genes have modular promoters such that they become activated in seven "stripes." Their boundaries of transcription are defined by the gap genes. The pair-rule genes form seven bands of transcription along the anterior-posterior axis, each one comprising two parasegments.

10. The pair-rule gene products activate *engrailed* and *wingless* expression in adjacent cells. The *engrailed*-expressing cells form the anterior boundary of each parasegment. These cells form a signaling center that organizes the cuticle formation and segmental structure of the embryo.

11. The homeotic selector genes are found in two complexes on chromosome 3 of *Drosophila*. Together, these regions are called Hom-C, the homeotic gene complex. The genes are arranged in the same order as their transcriptional expression. The Hom-C genes specify the individual segments, and mutations in these genes are capable of transforming one segment into another.

12. The expression of each homeotic selector gene is regulated by the gap and pair-rule genes. Their expression is refined and maintained by interactions whereby their protein products prevent the transcription of neighboring Hom-C genes.

13. In *Ultrabithorax* mutations, the third thoracic segment becomes specifed as the second thoracic segment. This converts the halteres into wings. When *Antennapedia* is ex-

pressed in the head as well as in the thorax, it represses antenna formation, allowing legs to form where the antenna should be.

14. The targets of the Hom-C proteins are the realisator genes. These genes include *distal-less* and *wingless* (in the thoracic segments).

15. Dorsal-ventral polarity is regulated by the entry of Dorsal protein into the nucleus. Dorsal-ventral polarity is initiated when the nucleus moves to the dorsal-anterior of the oocyte and transcribes the *gurken* message. The *gurken* message is then transported to the region above the nucleus and adjacent to the follicle cells.

16. The *gurken* mRNA is translated into Gurken protein, which is secreted from the oocyte and binds to its receptor, Torpedo, on the follicle cells. This dorsalizes the follicle cells, preventing them from synthesizing Pipe.

17. The Pipe protein in the ventral follicle cells modifies an as yet unknown factor that modifies the Nudel protein. This

modification allows Nudel to activate a cascade of proteolysis in the space between the ventral follicle cells and the ventral cells of the embryo.

18. As a result of this cascade, the Spätzle protein is activated and binds to the Toll protein on the ventral embryonic cells.

19. Activated Toll protein activates Pelle and Tube to phosphorylate the Cactus protein, which has been bound to the Dorsal protein. Phosphorylated Cactus is degraded, allowing Dorsal to enter the nucleus.

20. Once in the nucleus, Dorsal activates the genes responsible for the ventral cell fates and represses those genes whose proteins would specify dorsal cell fates. Since a gradient of Dorsal protein enters the various nuclei, those at the most ventral surface become mesoderm, and those more lateral become neurogenic ectoderm.

21. Organs form at the intersection of dorsal-ventral and anterior-posterior regions of gene expression.

Literature Cited

Akam, M. E. 1987. The molecular basis for metameric pattern in the *Drosophila* embryo. *Development* 101: 1–22.

Amiri, A. and D. Stein. 2002. Dorsoventral patterning: A direct route from ovary to embryo. *Curr. Biol.* 12: R532–R534.

Anderson, K. V. 1989. *Drosophila*: The maternal contributions. *In* D. M. Glover and B. D. Hames (eds.), *Genes and Embryos*. IRL, New York, pp. 1–37.

Anderson, K. V. and C. Nüsslein-Volhard. 1984. Information for the dorsal-ventral pattern of the *Drosophila* embryo is stored as maternal mRNA. *Nature* 311: 223–227.

Barker, D. D., C. Wang, J. Moore, L. K. Dickinson and R. Lehmann. 1992. Pumilio is essential for function but not for distribution of the *Drosophila* abdominal determinant, Nanos. *Genes Dev.* 6: 2312–2326.

Bate, M. and A. Martinez-Arias. 1993. *The Development of* Drosophila melanogaster. Cold Spring Harbor Laboratory Press, Cold Spring Harbor, NY.

Bateson, W. 1894. *Materials for the Study of Variation*. Macmillan, London.

Baumgartner, S. and M. Noll. 1990. Networks of interaction among pair-rule genes regulating paired expression during primordial segmentation of *Drosophila*. *Mech. Dev.* 1: 1–18.

Bergsten, S. E. and E. R. Gavis. 1999. Role for mRNA localization in translational activation but not spatial restriction of *nanos* RNA. *Development* 126: 659–669.

Berleth, T. and 7 others. 1988. The role of localization of *bicoid* RNA in organizing the anterior pattern of the *Drosophila* embryo. *EMBO J.* 7: 1749–1756.

Bhanot, P. and 8 others. 1996. A new member of the frizzled family from *Drosophila* functions as a Wingless receptor. *Nature* 382: 225–230.

Bokor, P. and S. DiNardo. 1996. The roles of Hedgehog and Wingless in patterning the dorsal epidermis in *Drosophila*. *Development* 122: 1083–1092.

Bradley, P.L., A. S. Haberman and D. J. Andrew. 2001.Organ formation in *Drosophila*: Specification and morphogenesis of the salivary gland. *BioEssays* 23: 901–911.

Brendza, R. P., L. R. Serbus, J. B. Duffy and W. M. Saxtons. 2000. A function for kinesin I in the posterior transport of Oskar mRNA and Staufen protein. *Science* 289: 2120–2122.

Brown, S., J. Fellers, T. Shippy, R. Denell, M. Stauber and U. Schmidt-Ott. 2001. A strategy for mapping *bicoid* on the phylogenetic tree. *Curr. Biol.* 11: R43–44.

Campos-Ortega, J. A. and V. Hartenstein. 1985. *The Embryonic Development of* Drosophila melanogaster. Springer-Verlag, New York.

Cao, Z., W. J. Henzel and X. Gao. 1996. IRAK: A kinase associated with the interleukin-1 receptor. *Science* 271: 1128–1131.

Capovilla, M., E. D. Eldon and V. Pirrotta. 1992. The *Giant* gene of *Drosophila* encodes a bZIP DNA-binding protein that regulates the expression of other segmentation gap genes. *Development* 114: 99–112.

Casal, J., G. Struhl, and P. A. Lawrence. 2002. Developmental compartments of planar polarity in *Drosophila*. *Curr. Biol.* 12: 1189–1198.

Casanova, J. and G. Struhl. 1989. Localized surface activity of torso, a receptor tyrosine kinase, specifies body pattern in *Drosophila*. *Genes Dev.* 3: 2025–2038.

Casanova, J., E. Sánchez-Herrero, A. Busturia and G. Morata. 1987. Double and triple mutant combination of the bithorax complex of *Drosophila*. *EMBO J.* 6: 3103–3109.

Casanova, J., M. Furrioles, C. A. McCormick and G. Struhl. 1995. Similarities between trunk and spätzle, putative extracellular ligands specifying body pattern in *Drosophila*. *Genes Dev.* 9: 2539–2544.

Casares, F. and R. S. Mann. 1998. Control of antennal versus leg development in *Drosophila*. *Nature* 392: 723–726.

Casares, F. and E. Sánchez-Herrero. 1995. Regulation of the infraabdominal regions of the bithorax complex of *Drosophila* by gap genes. *Development* 121: 1855–1866.

Castelli-Gair, J. and M. Akam. 1995. How the Hox gene *Ultrabithorax* specifies two different segments: The significance of spatial and temporal regulation within metameres. *Development* 121: 2973–2982.

Cha, B. J., B. S. Koppetsch and W. E. Theurkauf. 2001. In vivo analysis of *Drosophila bicoid* mRNA localization reveals a novel microtubule-dependent axis specification pathway. *Cell* 106: 35–46.

Chan, S. K. and G. Struhl. 1997. Sequence-specific RNA binding by Bicoid. *Nature* 388: 634.

Chasan, R., Y. Jin and K. V. Anderson. 1992. Activation of the easter zymogen is regulated by five other genes to define dorsal-ventral polarity in the *Drosophila* embryo. *Development* 115: 607–615.

Chauvet, S., S. Merabet, D. Bilder, M. P. Scott, J. Pradel and Y. Graba. 2000. Distinct hox protein sequences determine specificity in different tissues. *Proc. Natl. Acad. Sci. USA* 97: 4064–4069.

Cohen, S. M. and G. Jürgens. 1990. Mutations of *Drosophila* head development by gap-like segmentation genes. *Nature* 346: 482–485.

Dahanukar, A., J. A. Walker and R. P. Wharton. 1999. Smaug, a novel RNA-binding protein that operates a translational switch in *Drosophila*. *Mol. Cell* 4: 209–218.

Degelmann, A., P. A. Hardy, N. Perrimon and A. P. Mahowald. 1986. Developmental analysis of the *torsolike* phenotype in *Drosophila* produced by a maternal-effect locus. *Dev. Biol.* 115: 479–489.

Dessain, S., C. T. Gross, M. A. Kuziora and W. McGinnis. 1992. Antp-type homeodomains have distinct DNA-binding specificities that correlate with their different regulatory functions in embryos. *EMBO J.* 11: 991–1002.

Doe, C. Q., Y. Hiromi, W. J. Gehring and C. S. Goodman. 1988. Expression and function of the segmentation gene *fushi tarazu* during *Drosophila* neurogenesis. *Science* 239: 170–175.

Dong, P. D., J. Chu and G. Panganiban. 2000. Coexpression of the homeobox genes *distal-less* and *homothorax* determines *Drosophila* antennal identity. *Development* 127: 209–216.

Driever, W. and C. Nüsslein-Volhard. 1988a. The Bicoid protein determines position in the *Drosophila* embryo in a concentration-dependent manner. *Cell* 54: 95–104.

Driever, W. and C. Nüsslein-Volhard. 1988b. A gradient of Bicoid protein in *Drosophila* embryos. *Cell* 54: 83–93.

Driever, W. and C. Nüsslein-Volhard. 1989. The Bicoid protein is a positive regulator of *hunchback* transcription in the early *Drosophila* embryo. *Nature* 337: 138–143.

Driever, W., G. Thoma and C. Nüsslein-Volhard. 1989. Determination of spatial domains of zygotic gene expression in the *Drosophila* embryo by the affinity of binding sites for the Bicoid morphogen. *Nature* 340: 363–367.

Driever, W., V. Siegel and C. Nüsslein-Volhard. 1990. Autonomous determination of anterior structures in the early *Drosophila* embryo by the Bicoid morphogen. *Development* 109: 811–820.

Edgar, B. A. and P. H. O'Farrell. 1989. Genetic control of cell division patterns in the *Drosophila* embryo. *Cell* 57: 177–187.

Edgar, B. A., C. P. Kiehle and G. Schubiger. 1986a. Cell cycle control by the nucleo-cytoplasmic ratio in early *Drosophila* development. *Cell* 44: 365–372.

Edgar, B. A., M. P. Weir, G. Schubiger and T. Kornberg. 1986b. Repression and turnover pattern of *fushi tarazu* RNA in the early *Drosophila* embryo. *Cell* 47: 747–754.

Ferguson, E. L and K. V. Anderson. 1991. Dorsal-ventral pattern formation in the *Drosophila* embryo: The role of zygotically active genes. *Curr. Top. Dev. Biol.* 25: 17–43.

Ferrandon, D., I. Koch, E. Westhof and C. Nüsslein-Volhard. 1997. RNA-RNA interaction is required for the formation of specific *bicoid* mRNA 3′ UTR-Stufen ribonucleoprotein particles. *EMBO J.* 16: 1751–1758.

Finkelstein, R. and N. Perrimon. 1990. The *orthodenticle* gene is regulated by *bicoid* and *torso* and specifies *Drosophila* head development. *Nature* 346: 485–488.

Foe, V. E. 1989. Mitotic domains reveal early commitment of cells in *Drosophila* embryos. *Development* 107: 1–22.

Foe, V. E. and B. M. Alberts. 1983. Studies of nuclear and cytoplasmic behavior during the five mitotic cycles that precede gastrulation in *Drosophila* embryogenesis. *J. Cell Sci.* 61: 31–70.

Foe, V. E., G. M. Odell and B. A. Edgar. 1993. Mitosis and morphogenesis in the *Drosophila* embryo: Point and counterpoint. In B. M. Bate and A. Martinez-Arias (eds.), *The Development of Drosophila melanogaster*. Cold Spring Harbor Laboratory Press, Cold Spring Harbor, NY.

Forlani, S., D. Ferrandon, O. Saget and E. Mohier. 1993. A regulatory function for *K10* in the establishment of dorsoventral polarity in the *Drosophila* egg and embryo. *Mech. Dev.* 41: 109–120.

François, V., M. Solloway, J. W. O'Neill, J. Emery and E. Bier. 1994. Dorsal-ventral patterning of the *Drosophila* embryo depends on a putative negative growth factor encoded by the *short gastrulation* gene. *Genes Dev.* 8: 2602–2616.

Frigerio, G., M. Burri, D. Bopp, S. Baumgartner and M. Noll. 1986. Structure of the segmentation gene *paired* and the *Drosophila* PRD gene set as part of a gene network. *Cell* 47: 735–746.

Frischer, L. E., F. S. Hagen and R. L. Garber. 1986. An inversion that disrupts the *Antennapedia* gene causes abnormal structure and localization of RNAs. *Cell* 47: 1017–1023.

Fujioka, M., Y. Emi-Sarker, G. L. Yusibova, T. Goto and J. B. Jaynes. 1999. Analysis of an *even-skipped* rescue transgene reveals both composite and discrete neuronal and early blastoderm enhancers, and multi-stripe positioning by gap gene repressor gradients. *Development* 126: 2527–2538.

Fullilove, S. L. and A. G. Jacobson. 1971. Nuclear elongation and cytokinesis in *Drosophila montana*. *Dev. Biol.* 26: 560–578.

Furlong, E. E. M., E. C. Andersen, B. Null, K. P. White and M. P. Scott. 2001. Patterns of gene expression during *Drosophila* mesoderm development. *Science* 293: 1629–1633.

Furriols, M., A. Casali and J. Casanova. 1998. Dissecting the mechanism of torso receptor activation. *Mech. Dev.* 70: 111–118.

Gabay, L., R. Seger and B. Z. Shilo. 1997. MAP kinase in situ activation atlas during *Drosophila* embryogenesis. *Development* 124: 3535–3541.

Galindo, R. L., D. N. Edwards, S. K. H. Gillespie and S. A. Wasserman. 1995. Interaction of the *pelle* kinase with the membrane-associated protein tube is required for transduction of the dorsoventral signal in *Drosophila* embryos. *Development* 121: 2209–2218.

Gaul, U. and H. Jäckle. 1990. Role of gap genes in early *Drosophila* development. *Annu. Rev. Genet.* 27: 239–275.

Gavis, E. R. and R. Lehmann. 1992. Localization of *nanos* RNA controls embryonic polarity. *Cell* 71: 301–313.

González-Crespo, S. and M. Levine. 1994. Related target enhancers for dorsal and NF-κB signalling pathways. *Science* 264: 255–258.

González-Reyes, A. and G. Morata. 1990. The developmental effect of overexpressing a Ubx product in *Drosophila* embryos is dependent on its interactions with other homeotic products. *Cell* 61: 515–522.

Gonzáles-Reyes, A., H. Elliott and D. St. Johnson. 1995. Polarization of both major body axes in *Drosophila* by *gurken-torpedo* signalling. *Nature* 375: 654-658.

Grossniklaus, U., K. M. Cadigan and W. J. Gehring. 1994. Three maternal coordinate systems cooperate in the patterning of the *Drosophila* head. *Development* 120: 3155–3171.

Hafen, E., M. Levine and W. J. Gehring. 1984. Regulation of Antennapedia transcript distribution by the bithorax complex in *Drosophila*. *Nature* 307: 287–289.

Hanes, S. D. and R. Brent. 1989. DNA specificity of the *bicoid* activator protein is determined by homeodomain recognition helix residue 9. *Cell* 57: 1275–1283.

Hanes, S. D. and R. Brent. 1991. A genetic model for interaction of the homeodomain recognition helix with DNA. *Science* 251: 426–430.

Harding, K. and M. Levine. 1988. Gap genes define the limits of Antennapedia and Bithorax gene expression during early development in *Drosophila*. *EMBO J.* 7: 205–214.

Harding, K. and M. Levine. 1989. *Drosophila*: The zygotic contribution. In D. M. Glover and B. D. Hames (eds.), *Genes and Embryos*. IRL, New York, pp. 38–90.

Harding, K., C. Wedeen, W. McGinnis and M. Levine. 1985. Spatially regulated expression of homeotic genes in *Drosophila*. *Science* 229: 1236–1242.

Hashimoto, C., K. L. Hudson and K. V. Anderson. 1988. The *toll* gene of *Drosophila*, required for dorsal-ventral embryonic polarity, appears to encode a transmembrane protein. *Cell* 52: 269–279.

Hashimoto, C., S. Gerttula and K. V. Anderson. 1991. Plasma membrane localization of the Toll protein in the syncytial *Drosophila* embryo: Importance of transmembrane signalling for dorsal-ventral pattern formation. *Development* 11: 1021–1028.

Heemskerk, J. and S. DiNardo. 1994. *Drosophila* hedgehog acts as a morphogen in cellular patterning. *Cell* 76: 449–460.

Heemskerk, J., S. DiNardo, R. Kostriken and P. H. O'Farrell. 1991. Multiple modes of engrailed regulation in the progression towards cell fate determination. *Nature* 352: 404–410.

Hoch, M., N. Gerwin, H. Taubert and H. Jäckle. 1992. Competition for overlapping sites in the regulatory region of the *Drosophila* gene *Krüppel*. *Science* 256: 94–97.

Hong, C. C. and C. Hashimoto. 1995. An unusual mosaic protein with a protease domain, encoded by the *nudel* gene, is involved in defining embryonic dorsoventral polarity in *Drosophila*. *Cell* 82: 785–794.

Ingham, P. W. and A. Martinez-Arias. 1986. The correct activation of Antennapedia and bithorax complex genes requires the *fushi tarazu* gene. *Nature* 324: 592–597.

Ingham, P. W. and R. Whittle. 1980. Trithorax: A new homeotic mutation of *Drosophila* causing transformations of abdominal and thoracic imaginal segments. I. Putative role during embryogenesis. *Mol. Gen. Genet.*[1] 179: 607–614.

Ingham, P. W., A. M. Taylor and Y. Nakano. 1991. Role of *Drosophila patched* gene in positional signalling. *Nature* 353: 184–187.

Irish, V. F. and W. M. Gelbart. 1987. The *decapentaplegic* gene is required for dorsal-ventral patterning of the *Drosophila* embryo. *Genes Dev.* 1: 868–879.

Jäckle, H., D. Tautz, R. Schuh, E. Seifert and R. Lehmann. 1986. Cross-regulatory interactions among the gap genes of *Drosophila*. *Nature* 324: 668–670.

Jiang, J. and M. Levine. 1993. Binding affinities and cooperative interactions with bHLH activators delimit threshold responses to the dorsal gradient morphogen. *Cell* 72: 741–752.

Jiang, J., D. Kosman, Y. T. Ip and M. Levine. 1991. The dorsal morphogen gradient regulates the mesoderm determinant twist in early *Drosophila* embryos. *Genes Dev.* 5: 1881–1891.

Just, E. E. 1936. Quoted in R. G. Harrison, 1937. Embryology and its relations. *Science* 85: 369–374.

Kalthoff, K. 1969. Der Einfluss vershiedener Versuchparameter auf die Häufigkeit der Missbildung "Doppelabdomen" in UV-bestrahlten Eiern von *Smittia* sp. (Diptera, Chironomidae). *Zool. Anz. Suppl.* 33: 59–65.

Kalthoff, K. and K. Sander. 1968. Der Enwicklungsgang der Missbildung "Doppelabdomen" im partiell UV-bestrahlten Ei von *Smittia parthenogenetica* (Diptera, Chironomidae). *Wilhelm Roux Arch. Entwicklungsmech. Org.* 161: 129–146.

Kandler-Singer, I. and K. Kalthoff. 1976. RNase sensitivity of an anterior morphogenetic determinant in an insect egg (*Smittia* sp., Chironomidae, Diptera). *Proc. Natl. Acad. Sci. USA* 73: 3739–3743.

Karr, T. L. and B. M. Alberts. 1986. Organization of the cytoskeleton in early *Drosophila* embryos. *J. Cell Biol.* 102: 1494–1509.

Karr, T. L. and T. B. Kornberg. 1989. Fushi tarazu protein expression in the cellular blastoderm of *Drosophila* detected using a novel imaging technique. *Development* 105: 95–103.

Kaufman, T. C., M. A. Seeger and G. Olsen. 1990. Molecular and genetic organization of the Antennapedia gene complex of *Drosophila melanogaster*. *Adv. Genet.* 27: 309–362.

Kidd, S. 1992. Characterization of the *Drosophila cactus* locus and analysis of interactions between cactus and dorsal proteins. *Cell* 71: 623–635.

Klingler, M., M. Erdélyi, J. Szabad and C. Nüsslein-Volhard. 1988. Function of *torso* in determining the terminal anlagen of the *Drosophila* embryo. *Nature* 335: 275–277.

Knipple, D. C., E. Seifert, U. B. Rosenberg, A. Preiss and H. Jäckle. 1985. Spatial and temporal patterns of *Krüppel* gene expression in early *Drosophila* embryos. *Nature* 317: 40–44.

Kohler, R. 1994. *Lords of the Fly*. University of Chicago Press, Chicago.

Kosman, D., Y. T. Ip, M. Levine and K. Arora. 1991. Establishment of the mesoderm-neuroectoderm boundary in the *Drosophila* embryo. *Science* 254: 118–122.

Kraut, R. and M. Levine. 1991. Mutually repressive interactions between the gap genes *giant* and *Krüppel* define middle body regions of the *Drosophila* embryo. *Development* 111: 611–621.

Lane, M. E. and D. Kalderon. 1994. RNA localization along the anteroposterior axis of the *Drosophila* oocyte requires PKA-mediated signal transduction to direct normal microtubule organization. *Genes Dev.* 8: 2986–2995.

Lander, A. D., G. Nie and F. Y. Wan. 2002. Do morphogen gradients arise by diffusion? *Dev. Cell* 2: 785–796.

Lehmann, R. and C. Nüsslein-Volhard. 1986. Abdominal segmentation, pole cell formation, and embryonic polarity require the localized activity of *oskar*, a maternal gene in *Drosophila*. *Cell* 47: 141–152.

Lehmann, R. and C. Nüsslein-Volhard. 1991. The maternal gene *nanos* has a central role in posterior pattern formation of the *Drosophila* embryo. *Development* 112: 679–691.

Lemaitre, B., E. Nicolas, L. Michaut, J.-M. Reichhart and J. A. Hoffmann. 1996. The dorsoventral regulatory gene casette *spätzle/Toll/cactus* controls the potent antifungal response in *Drosophila* adults. *Cell* 86: 973–983.

LeMosy, E. K., Y. Q. Tan and C. Hashimoto. 2001. Activation of a protease cascade involved in patterning the *Drosophila* embryo. *Proc. Natl. Acad. Sci. USA.* 98: 5055–5060.

Leptin, M. 1991a. Mechanics and genetics of cell shape changes during *Drosophila* ventral furrow formation. In R. Keller et al. (eds.), *Gastrulation: Movements, Patterns, and Molecules.* Plenum, New York, pp. 199–212.

Leptin, M. 1991b. *twist* and *snail* as positive and negative regulators during *Drosophila* mesoderm development. *Genes Dev.* 5: 1568–1576.

Levine, M. S. and K. W. Harding. 1989. *Drosophila*: The zygotic contribution. In D. M. Glover and B. D. Hames (eds.), *Genes and Embryos.* IRL, New York, pp. 39–94.

Lewis, E. B. 1978. A gene complex controlling segmentation in *Drosophila*. *Nature* 276: 565–570.

Macdonald, P. M. and K. Kerr. 1998. Mutational analysis of an RNA recognition element that mediates localization of *bicoid* mRNA. *Mol. Cell Biol.* 18: 3788–3795.

Macdonald, P. M. and C. A. Smibert. 1996. Translational regulation of maternal mRNAs. *Curr. Opin. Genet. Dev.* 6: 403–407.

Macdonald, P. M. and G. Struhl. 1986. A molecular gradient in early *Drosophila* embryos and its role in specifying body pattern. *Nature* 324: 537–545.

Mann, R. S. and D. S. Hogness. 1990. Functional dissection of Ultrabithorax proteins in *D. melanogaster*. *Cell* 60: 597–610.

Martin, J. R., A. Railbaud and R. Ollo. 1994. Terminal elements in *Drosophila* embryo induced by torsolike protein. *Nature* 367: 741–745.

Martinez-Arias, A. and P. A. Lawrence. 1985. Parasegments and compartments in the *Drosophila* embryo. *Nature* 313: 639–642.

Mazumdar, A. and M. Mazumdar. 2002. How one becomes many: blastoderm cellularization in *D. malanogaster*. *BioEssay* 24: 1012–1022.

McKeon, J. and H. W. Brock. 1991. Interactions of the *Polycomb* group of genes with homeotic loci of *Drosophila*. *Wilhelm Roux Arch. Dev. Biol.* 199: 387–396.

Mohler, J. and K. Vani. 1992. Molecular organization and embryonic expression of the *hedgehog* gene involved in cell-cell communication in segmental patterning in *Drosophila*. *Development* 115: 957–971.

Montell, D. J., H. Keshishian and A. C. Spradling. 1991. Laser ablation studies of the role of the *Drosophila* oocyte nucleus in pattern formation. *Science* 254: 290–293.

Müller, J. and M. Bienz. 1992. Sharp anterior boundary of homeotic gene expression conferred by the fushi tarazu protein. *EMBO J.* 11: 3653–3661.

Müller, M., M. Affolter, W. Leupin, G. Otting, K. Wüthrich and W. J. Gehring. 1988. Isolation and sequence-specific DNA binding of the Antenapedia homeodomain. *EMBO J.* 7: 4299–4304.

Nambu, J. R., R. G. Franks, S. Hong and S. Crews. 1990. The *single-minded* gene of *Drosophila* is required for the expression of genes important for the development of CNS midline cells. *Cell* 63: 63–75.

Neuman-Silberg, F. S. and T. Schüpbach. 1993. The *Drosophila* dorsoventral patterning gene *gurken* produces a dorsally localized RNA and encodes a TGF-α-like protein. *Cell* 75: 165–174.

Newport, J. W. and M. W. Kirschner. 1982. A major developmental transition in early *Xenopus* embryos: I. Characterization and timing of cellular changes at mid-blastula stage. *Cell* 30: 687–696.

Niessing, D., W. Driever, F. Sprenger, H.Taubert, H. Jäckle and R. Rivera-Pomar. 2000. Homeodomain position 54 specifies transcriptional versus translational control by Bicoid. *Mol. Cell* 5: 395–401.

Nüsslein-Volhard, C. and E. Wieschaus. 1980. Mutations affecting segment number and polarity in *Drosophila*. *Nature* 287: 795–801.

Otting, G., Y. Q. Qian, M. Billeter, M. Müller, M. Affolter, W. J. Gehring and K. Wüthrich. 1990. Protein-DNA contacts in the structure of a homeodomain-DNA complex determined by nuclear magnetic resonance spectroscopy in solution. *EMBO J.* 9: 3085–3092.

Pan, D., J.-D. Huang and A. J. Courey. 1991. Functional analysis of the *Drosophila twist* promoter reveals a dorsal-binding ventral activator region. *Genes Dev.* 5: 1892–1901.

Pankratz, M. J., E. Seifert, N. Gerwin, B. Billi, U. Nauber and H. Jäckle. 1990. Gradients of *Krüppel* and *knirps* gene products direct pairrule gene stripe patterning in the posterior region of the *Drosophila* embryo. *Cell* 61: 309–317.

Panzer, S., D. Weigel and S. K. Beckendorf. 1992. Organogenesis in *Drosophila melanogaster*: Embryonic salivary gland determination is controlled by homeotic and dorsoventral patterning genes. *Development* 114: 49–57.

Paroush, Z., S. M. Wainwright and D. IshHorowitz. 1997. Torso signaling mediates terminal patterning in *Drosophila* by antagonizing Groucho-mediated repression. *Development* 124: 3827–3834.

Peifer, M. and E. Wieschaus. 1990. Mutations in the *Drosophila* gene *extradenticle* affect the way specific homeodomain proteins regulate segment identity. *Genes Dev.* 4: 1209–1223.

Percival-Smith, A., M. Müller, M. Affolter and W. J. Gehring. 1990. The interaction with DNA of wild-type and mutant *fushi tarazu* homeodomains. *EMBO J.* 9: 3967–3974.

Peri, F., C. Bökel and S. Roth. 1999. Local Gurken signaling and dynamic MAPK activation during *Drosophila* oogenesis. *Mech. Dev.* 81: 75–88.

Pignoni, F., E. Steingrímsson and J. A. Lengyel. 1992. *bicoid* and the terminal system activate *tailless* expression in the early *Drosophila* embryo. *Development* 115: 239–251.

Plaza, S., F. Prince, J. Jaeger, U. Kloter, S. Flister, C. Benassayag, D. Cribbs and W. J.Gehring. 2001. Molecular basis for the inhibition of *Drosophila* eye development by Antennapedia. *EMBO J.* 20: 802–811.

Porter, J. A. and 10 others. 1996. Hedgehog patterning activity: Role of a lipophilic modification by the carboxy-terminal autoprocessing domain. *Cell* 86: 21–34.

Price, J. V., R. J. Clifford and T. Schüpbach. 1989. The maternal ventralizing gene *torpedo* is allelic to *faint little ball*, an embryonic lethal, and encodes the *Drosophila* EGF receptor homolog. *Cell* 56: 1085–1092.

Qureshi, S. T., P. Gros and D. Malo. 1999. Host resistance to infection. *Trends Genet.* 15: 291–294.

Raskolb, C. and E. Wieschaus. 1994. Coordinate regulation of downstream genes by extradenticle and homeotic selector proteins. *EMBO J.* 15: 3561–3569.

Ray, R. P., and T. Schüpbach. 1996. Intercellular signaling and the polarization of body axes during *Drosophila* oogenesis. *Genes Dev.* 10: 1711–1723.

Ray, R. P., K. Arora, C. Nüsslein-Volhard and W. M. Gelbart. 1991. The control of cell fate along the dorsal-ventral axis of the *Drosophila* embryo. *Development* 113: 35–54.

Reach, M., R. L. Galindo, P. Towb, J. L. Allen, M. Karin and S. A. Wasserman. 1996. A gradient of Cactus protein degradation establishes dorsoventral polarity in the *Drosophila* embryo. *Dev. Biol.* 180: 353–364.

Reinitz, J. and D. H. Sharp. 1995. Mechanism of *eve* stripe formation. *Mech. Dev.* 49: 133–158.

Reinitz, J., E. Mjolsness and D. H. Sharp. 1995. Model for cooperative control of positional information in *Drosophila* by *bicoid* and maternal *hunchback*. *J. Exp. Zool.* 271: 47–56.

Riddihough, G. 1992. Homing in on the homeobox. *Nature* 357: 643–644.

Rivera-Pomar, R. and H. Jäckle. 1996. From gradients to stripes in *Drosophila* embryogenesis: Filling in the gaps. *Trends Genet.* 12: 478–483.

Rivera-Pomar, R., X. Lu, N. Perrimon, H. Taubert and H. Jackle. 1995. Activation of posterior gap gene expression in the *Drosophila* blastoderm. *Nature* 376: 253–256.

Rivera-Pomar, R., D. Niessling, U. Schmidt-Ott, W. J. Gehring and H. Jäckle. 1996. RNA binding and translational suppression by *bicoid*. *Nature* 379: 746–749.

Rock, F. L., G. Hardiman, J. C. Timans, R. A. Kasterlein and J. F. Bazan. 1998. A family of human receptors structurally related to *Drosophila* Toll. *Proc. Natl. Acad. Sci. USA* 95: 588–593.

Roder, L., C. Vola and S. Kerridge. 1992. The role of *teashirt* in trunk segmental identity in *Drosophila*. *Development* 115: 1017–1033.

Roth, S., D. Stein and C. Nüsslein-Volhard. 1989. A gradient of nuclear localization of the

dorsal protein determines dorsoventral pattern in the *Drosophila* embryo. *Cell* 59: 1189–1202.

Roth, S., F. S. Neuman-Silberg, G. Barcelo and T. Schupbach. 1995. Cornichon and the EGF receptor signaling process are necessary for both anterior-posterior and dorsal-ventral pattern formation in *Drosophila*. *Cell* 81: 967–978.

Rushlow, C., J. Frasch, H. Doyle and M. Levine. 1987. Maternal regulation of a homeobox gene controlling differentiation of dorsal tissues in *Drosophila*. *Nature* 330: 583–586.

Rushlow, C. A., K. Han, J. L. Manley and M. Levine. 1989. The graded distribution of the dorsal morphogen is initiated by selective nuclear transport in *Drosophila*. *Cell* 59: 1165–1177.

Sackerson, C. M. Fujioka and T. Goto. 1999. The *even-skipped* locus is contained in a 16-kb chromatin domain. *Dev. Biol.* 210: 39–52.

Sánchez-Herrero, E., I. Verños, R. Marco and G. Morata. 1985. Genetic organization of *Drosophila* bithorax complex. *Nature* 313: 108–113.

Sander, K. 1975. Pattern specification in the insect embryo. In *Cell Patterning*. CIBA Foundation Symposium, new series, 29. Associated Scientific Publishers, Amsterdam, NY, pp. 241–263.

Schejter, E. D. and E. Wieschaus. 1993. Bottleneck acts as a regulator of the microfilament network governing cellularization of the *Drosophila* embryo. *Cell* 75: 373–385.

Schier, A. F. and A. J. Gehring. 1992. Direct homeodomain-DNA interaction in the autoregulation of the *fushi tarazu* gene. *Nature* 356: 804–807.

Schneuwly, S., A. Kuroiwa and W. J. Gehring. 1987. Molecular analysis of the dominant homeotic *Antennapedia* phenotype. *EMBO J.* 6: 201–206.

Schnorrer, F., K. Bohmann and C. NüssleinVolhard. 2000. The molecular motor dynein is involved in targeting *swallow* and *bicoid* RNA to the anterior pole of *Drosophila* oocytes. *Nature Cell Biol.* 2: 185–190.

Schulz, C. and D. Tautz. 1995. Zygotic caudal regulation by hunchback and its role in abdominal segment formation of the *Drosophila* embryo. *Development* 121: 1023–1028.

Schüpbach, T. 1987. Germ line and soma cooperate during oogenesis to establish the dorsoventral pattern of egg shell and embryo in *Drosophila melanogaster*. *Cell* 49: 699–707.

Schüpbach, T. and E. Wieschaus. 1986. Maternal effect mutations altering the anterior-posterior pattern of the *Drosophila* embryo. *Wilhelm Roux Arch. Dev. Biol.* 195: 302–317.

Schüpbach, T., R. J. Clifford, L. J. Manseau and J. V. Price. 1991. Dorsoventral signaling processes in *Drosophila* oogenesis. In J. Gerhart (ed.), *Cell-Cell Interactions in Early Development*. Wiley-Liss, New York, pp. 163–174.

Schwalm, F. 1997. The insects. *In* S. F. Gilbert and A. M. Raunio (eds.), *Embryology: Construct-*

ing the Organism. Sinauer Associates, Sunderland, MA, pp. 259–278.

Sen, J., J. S. Goltz, L. Stevens and D. Stein. 1998. Spatially restricted expression of *pipe* in the *Drosophila* egg chamber defines embryonic dorsal-ventral polarity. *Cell* 95: 471–481.

Shelton, C. A. and S. A. Wasserman. 1993. *pelle* encodes a protein kinase required to establish dorsoventral polarity in the *Drosophila* embryo. *Cell* 72: 515–525.

Siegfried, E., E. L. Wilder and N. Perrimon. 1994. Components of *wingless* signalling in *Drosophila. Nature* 367: 76–80.

Simmonds, A. J., G. dosSantos, I. Livne-Bar and H. M. Krause. 2001.Apical localization of *wingless* transcripts is required for *wingless* signaling. *Cell* 105: 197–207.

Simon, J., A. Chiang and W. Bender. 1992. Ten different Polycomb genes are required for spatial control of the *abdA* and *AbdB* homeotic products. *Development* 114: 493–505.

Simpson-Brose, M., J. Treisman and C. Desplan. 1994. Synergy between the *hunchback* and *bicoid* morphogens is required for anterior patterning in *Drosophila. Cell* 78: 855–865.

Skeath, J. B. and S. B. Carroll. 1992. Regulation of proneural gene expression and cell fate during neuroblast segregation in the *Drosophila* embryo. *Development* 114: 939–946.

Skeath, J. B., G. Panganiban, J. Selegue and S. B. Carroll. 1993. Gene regulation in two dimensions: The proneural *achaete* and *scute* genes are controlled by combinations of axis-patterning genes through a common intergenic control region. *Genes Dev.* 6: 2606–2619.

Slack, J. M. W. 1983. *From Egg to Embryo: Determinative Events in Early Development.* Cambridge University Press, Cambridge.

Small, S., R. Kraut, T. Hoey, R. Warrior and M. Levine. 1991. Transcriptional regulation of a pair-rule stripe in *Drosophila. Genes Dev.* 5: 827–839.

Small, S., A. Blair and M. Levine. 1992. Regulation of *even-skipped* stripe 2 in the *Drosophila* embryo. *EMBO J.* 11: 4047–4057.

Small, S., A. Blair and M. Levine. 1996. Regulation of two pair-rule stripes by a single enhancer in the *Drosophila* embryo. *Dev. Biol.* 175: 314–324.

Smibert, C. A., J. E. Wilson, K. Kerr and P. M. Macdonald. 1996. Smaug protein represses translation of unlocalized *nanos* mRNA in the *Drosophila* embryo. *Genes Dev.* 10: 2600–2609.

Stanojevíc, D., T. Hoey and M. Levine. 1989. Sequence-specific DNA-binding activities of the gap proteins encoded by *hunchback* and *Krüppel* in *Drosophila. Nature* 341: 331–335.

Stanojevíc, D., S. Small and M. Levine. 1991. Regulators of a segmentation stripe by overlapping activators and repressors in the *Drosophila* embryo. *Science* 254: 1385–1387.

Stephanson, E. C., Y.-C. Chao and J. D. Frackenthal. 1988. Molecular analysis of the *swallow* gene of *Drosophila melanogaster. Genes Dev.* 2: 1655–1665.

Srinivasan, S., K. E. Rashka and E. Bier. 2002. Creation of a Sog morphogen gradient in the *Drosophila* embryo. *Dev. Cell* 2: 91–101.

Stevens, L. M., H. G. Frohnhöfer, M. Klingler and C. Nüsslein-Volhard. 1990. Localized requirement for *torsolike* expression in follicle cells for development of terminal anlagen of the *Drosophila* embryo. *Nature* 346: 660–662.

Steward, R. 1989. Relocalization of the dorsal protein from the cytoplasm to the nucleus correlates with its function. *Cell* 59: 1179–1188.

Steward, R. and S. Govind. 1993. Dorsal-ventral polarity in the *Drosophila* embryo. *Curr. Opin. Genet. Dev.* 3: 556–561.

St. Johnston, D. and C. Nüsslein-Volhard. 1992. The origin of pattern and polarity in the *Drosophila* embryo. *Cell* 68: 201–219.

Struhl, G. 1981. A homeotic mutation transforming leg to antenna in *Drosophila. Nature* 292: 635–638.

Struhl, G., K. Struhl and P. M. Macdonald. 1989. The gradient morphogen *bicoid* is a concentration-dependent transcriptional activator. *Cell* 57: 1259–1273.

Struhl, G., P. Johnson and P. Lawrence. 1992. Control of a *Drosophila* body pattern by the Hunchback morphogen gradient. *Cell* 69: 237–249.

Sullivan, W., P. Fogarty and W. E. Theurkauf. 1993. Mutations affecting the cytoskeletal organization of syncytial *Drosophila* embryos. *Development* 118: 1245–1254.

Tautz, D. 1988. Regulation of the *Drosophila* segmentation gene *hunchback* by two maternal morphogenetic centers. *Nature* 332: 281–284.

Thisse, B., C. Stoetzel, C. Gorostiza-Thisse and F. Perrin-Schmidt. 1988. Sequence of the *twist* gene and nuclear localization of its protein in endomesodermal cells of early *Drosophila* embryos. *EMBO J.* 7: 2175–2183.

Thisse, C., F. Perrin-Schmidt, C. Stoetzel and B. Thisse. 1991. Sequence-specific *trans* activation of the *Drosophila twist* gene by the *dorsal* gene product. *Cell* 65: 1191–1201.

Turner, F. R. and A. P. Mahowald. 1977. Scanning electron microscopy of *Drosophila melanogaster* embryogenesis. *Dev. Biol.* 57: 403–416.

Tyler, M. S. and J. W. Schetzer. 1996. *The Lives of a Fly.* (Videocassette.) ASAP Media Services, Orono, ME, and Sinauer Associates, Sunderland, MA.

Vachon, G., B. Cohen, C. Pfeifle, M. E. McGuffin, J. Botas and S. M. Cohen. 1992. Homeotic genes of the bithorax complex repress limb development in the abdomen of the *Drosophila* embryo through the target gene *Distal-less. Cell* 71: 437–450.

Valentine, S. A., G. Chen, T. Shandala, J. Fernandez, S. Mische, R. Saint and A. J. Courney. 1998. Dorsal-mediated repression requires the formation of a multiprotein repression complex at the ventral silencer. *Mol. Cell Biol.* 18: 6584–6594.

van Dyke, M. A. and C. Murre. 1994. Extradenticle raises the DNA binding specificity of homeotic selector gene products. *Cell* 78: 617–624.

Van Eeden, F. and D. St. Johnston. 1999. The polarisation of the anterior-posterior and dorsal-ventral axes during *Drosophila* oogenesis. *Curr. Opin. Genet. Dev.* 9: 396–404.

Wakimoto, B. T., F. R. Turner and T. C. Kaufman. 1984. Defects in embryogenesis in mutants associated with the Antennapedia gene complex of *Drosophila melanogaster. Dev. Biol.* 102: 147–172.

Wang, C. and R. Lehman. 1991. Nanos is the localized posterior determinate in *Drosophila. Cell* 66: 637–647.

Weatherbee, S. D., G. Halder, J. Kim, A. Hudson and S. Carroll. 1998. Ultrabithorax regulates genes at several levels of the wing-patterning hierarchy to shape the development of the *Drosophila* haltere. *Genes Dev.* 12: 1474–1482.

Whalen, A. M. and R. Steward. 1993. Dissociation of the dorsal-cactus complex and phosphorylation of the dorsal protein correlate with the nuclear localization of dorsal. *J. Cell. Biol.* 123: 523–534.

Wharton, R. P. and G. Struhl. 1991. RNA regulatory elements mediate control of *Drosophila* body pattern by the posterior morphogen Nanos. *Cell* 67: 955–967.

Wilkie, G. S. and I. Davis. 2001. *Drosophila wingless* and *pair-rule* transcripts localize apically by dynein-mediated transport of RNA particles. *Cell* 105: 209–219.

Wolff, C., R. Schroder, C. Schulz, D. Tautz and M. Klingler. 1998. Regulation of the *Tribolium* homologues of *caudal* and *hunchback* in *Drosophila:* Evidence for maternal gradient systems in a short germ embryo. *Development* 125: 3645–3654.

Wreden, C., A. C. Verrotti, J. A. Schisa, M. E. Lieberfarb and S. Strickland. 1997. Nanos and pumilio establish embryonic polarity in *Drosophila* by promoting posterior deadenylation of *hunchback* mRNA. *Development* 124: 3015–3023.

Wu, L. H. and J. A. Lengyel. 1998. Role of *caudal* in hindgut specification and gastrulation suggests homology between *Drosophila* amnioproctodeal invagination and vertebrate blastopore. *Development* 125: 2433–2442.

Wu, X., V. Vasisht, D. Kosman, J. Reinitz and S. Small. 2001. Thoracic patterning by the *Drosophila* gap gene hunchback. *Dev. Biol.* 237: 79–92.

Zhou, B., A. Bagri and S. K. Beckendorf. 2001. Salivary gland determination in *Drosophila:* A salivary-specific, *forkhead* enhancer integrates spatial pattern and allows *forkhead* autoregulation. *Dev. Biol.* 237: 54–67.

chapter 10 Early development and axis formation in amphibians

AMPHIBIAN EMBRYOS WERE THE ORGANISMS of choice for experimental embryology. With their large cells and their rapid development, salamander and frog embryos were excellently suited for transplantation experiments. However, amphibian embryos fell out of favor during the early days of developmental genetics, since frogs and salamanders undergo a long period of growth before they become fertile, and their chromosomes are often found in several copies, precluding easy mutagenesis.* However, new molecular techniques such as in situ hybridization, antisense oligonucleotides, and dominant negative proteins have allowed researchers to return to studying amphibian embryos and to integrate molecular analyses of development with earlier experimental findings. The results have been spectacular, and we are enjoying new vistas of how vertebrate bodies are patterned and structured.

EARLY AMPHIBIAN DEVELOPMENT

Cleavage in Amphibians

Cleavage in most frog and salamander embryos is radially symmetrical and holoblastic, just like echinoderm cleavage. The amphibian egg, however, contains much more yolk. This yolk, which is concentrated in the vegetal hemisphere, is an impediment to cleavage. Thus, the first division begins at the animal pole and slowly extends down into the vegetal region (Figure 10.1; see also Figures 2.2D and 8.4). In the axolotl salamander, the cleavage furrow extends through the animal hemisphere at a rate close to 1 mm per minute. The cleavage furrow bisects the gray crescent and then slows down to a mere 0.02–0.03 mm per minute as it approaches the vegetal pole (Hara 1977).

Figure 10.2A is a scanning electron micrograph showing the first cleavage in a frog egg. One can see the difference in the furrow between the animal and the vegetal hemispheres. Figure 10.2B shows that while the first cleavage furrow is still cleaving the yolky cytoplasm of the vegetal hemisphere, the second cleavage has already started near the animal pole. This cleavage is at right angles to the first one and is also meridional. The third cleavage, as expected, is equatorial. However, because of the vegetally placed yolk, this cleavage furrow in amphibian eggs is not actually at the

---

*In the 1960s, *Xenopus laevis* replaced the *Rana* frogs and the salamanders because it could be induced to mate throughout the year. Unfortunately, *Xenopus laevis* has four copies of each chromosome rather than the more usual two, and it takes 1–2 years to reach sexual maturity. These attributes make genetic studies difficult. Recently another *Xenopus* species, *X. tropicalis,* has begun to be used in the laboratory. It has all the advantages of *X. laevis,* plus it is diploid and reaches sexual maturity in a mere 6 months (Hirsch et al. 2002.).

Figure 10.1
Cleavage of a frog egg. Cleavage furrows, designated by Roman numerals, are numbered in order of appearance. (A, B) Because the vegetal yolk impedes cleavage, the second division begins in the animal region of the egg before the first division has divided the vegetal cytoplasm. (C) The third division is displaced toward the animal pole. (D–H) The vegetal hemisphere ultimately contains larger and fewer blastomeres than the animal half. H represents a cross section through a mid-blastula stage embryo. (After Carlson 1981.)

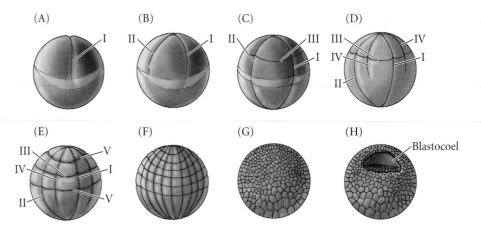

equator, but is displaced toward the animal pole (Figure 10.3A; Valles et al. 2002). It divides the frog embryo into four small animal blastomeres (micromeres) and four large blastomeres (macromeres) in the vegetal region. This unequal holoblastic cleavage establishes two major embryonic regions: a rapidly dividing region of micromeres near the animal pole and a more slowly dividing vegetal macromere area (Figure 10.2C; see also Figure 2.2E). As cleavage progresses, the animal region becomes packed with numerous small cells, while the vegetal region contains only a relatively small number of large, yolk-laden macromeres.

The cell cycles of the early *Xenopus* blastomeres are regulated by the levels of mitosis promoting factor (MPF) in the cytoplasm. As discussed in Chapter 8, MPF has two components, cyclin B and the cyclin-dependent kinase cdc2. In *Xenopus*, there are no G stages in the cell cycle for the first 12 divisions. Thus, there is no growth between cell divisions. The permissive events that allow the cell cycle to proceed (in which the DNA replication complexes begin to form on the chromatin) actually occur while the cells are in mitosis. Active MPF stimulates the cells to enter mitosis, and the cells enter S phase when that MPF is degraded. Cyclin B increases linearly after its degradation at the end of mitosis (Figure 10.3B). The cdc2 protein, which is inactived in mitosis, becomes activated (by dephosphorylation) during S phase. However, this increase in activity is not linear. Toward the end of S phase, cdc2 undergoes a dramatic increase in activity, suddenly creating a large increase in the amount of active MPF. This initiates the mitotic events, as well as preparing the chromatin for replicating in the next S phase (Jares and Blow 2000; Iwabuchi et al. 2002).

An amphibian embryo containing 16 to 64 cells is commonly called a **morula** (plural: **morulae**; Latin, "mulberry," whose shape it vaguely resembles). At the 128-cell stage, the blastocoel becomes apparent, and the embryo is considered a blastula. Actually, the formation of the blastocoel has been traced back to the very first cleavage furrow. Kalt (1971) demonstrated that in the frog *Xenopus laevis*, the first cleavage furrow widens in the animal hemisphere to create a small intercellular cavity that is sealed off from the outside by tight intercellular junctions. This cavity expands during subsequent cleavages to become the blastocoel (Figure 10.4A).

Figure 10.2
Scanning electron micrographs of the cleavage of a frog egg. (A) First cleavage. (B) Second cleavage (4 cells). (C) Fourth cleavage (16 cells), showing the size discrepancy between the animal and vegetal cells after the third division. (A from Beams and Kessel 1976, photograph courtesy of the authors; B and C, photographs courtesy of L. Biedler.)

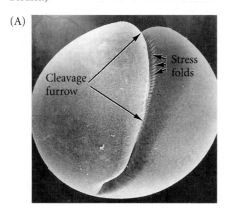
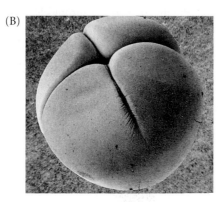
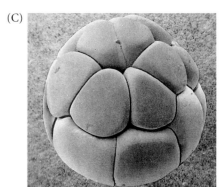

(A)

(B)

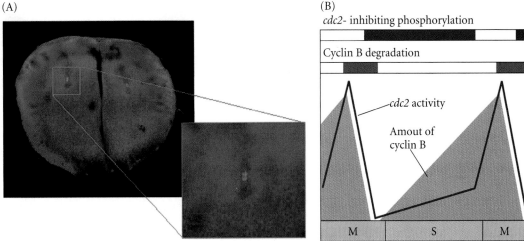

cdc2- inhibiting phosphorylation

Cyclin B degradation

cdc2 activity

Amout of
cyclin B

M   S   M

Figure 10.3
Mitosis and DNA replication in *Xenopus*. (A) Third cleavage of
Xenopus showing displacement of mitotic spindle towards the animal pole. Microtubules are red; chromatin is green; and the overlap
is yellow). (B) Regulation of the cell cycle by levels of cdc2 activity
in early *Xenopus* blasomeres. (A after Valles et al. 2002; photograph
courtesy of J. M. Valles, Jr.; B after Iwabuchi et al. 2002.)

The blastocoel serves two major functions in frog embryos: (1) it permits cell migration during gastrulation, and
(2) it prevents the cells beneath it from interacting prematurely with the cells above it. When Nieuwkoop (1973) took
embryonic newt cells from the roof of the blastocoel in the
animal hemisphere (a region often called the "animal cap"),
and placed them next to the yolky vegetal cells from the base
of the blastocoel, these animal cells differentiated into mesodermal tissue instead of ectoderm. Mesodermal tissue is normally formed from those animal cells that are adjacent to the
vegetal endoderm precursors, because vegetal cells induce adjacent cells to differentiate into mesodermal tissues. Thus, the
blastocoel prevents the contact of the vegetal cells destined to
become endoderm with those cells fated to give rise to the
skin and nerves.

While these cells are dividing, numerous cell adhesion
molecules keep the blastomeres together. One of the most important of these molecules is EP-cadherin. The mRNA for this
protein is supplied in the oocyte cytoplasm. If this message is
destroyed (by injecting antisense oligonucleotides complementary to this mRNA into the oocyte), the EP-cadherin is
not made, and the adhesion between the blastomeres is dramatically reduced (Heasman et al. 1994a,b), resulting in the
obliteration of the blastocoel (Figure 10.4).

Amphibian Gastrulation

The study of amphibian gastrulation is both one of the oldest
and one of the newest areas of experimental embryology. Even
though amphibian gastrulation has been extensively studied
for the past century, most of our theories concerning the
mechanisms of these developmental movements have been revised over the past decade. The study of amphibian gastrulation has been complicated by the fact that there is no single
way amphibians gastrulate. Different species employ different
means toward the same goal (Smith and Malacinski 1983;

(A)

(B)

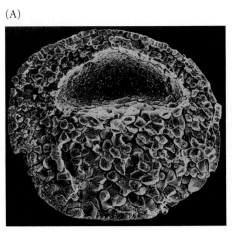
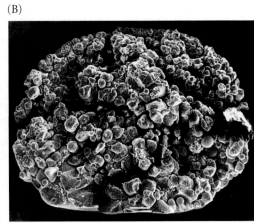

Figure 10.4
Depletion of EP-cadherin mRNA
in the *Xenopus* oocyte results in
the loss of adhesion between blastomeres and the obliteration of
the blastocoel. (A) Control embryo. (B) EP-cadherin-depleted
embryo. (From Heasman et al.
1994b; photographs courtesy of
J. Heasman.)

Minsuk and Keller 1996). In recent years, the most intensive investigations have focused on the frog *Xenopus laevis*, so we will concentrate on its mode of gastrulation.

The Xenopus *fate map*

Amphibian blastulae are faced with the same tasks as the invertebrate blastulae we followed in Chapters 8 and 9—namely, to bring inside the embryo those areas destined to form the endodermal organs, to surround the embryo with cells capable of forming the ectoderm, and to place the mesodermal cells in the proper positions between them. The movements whereby this is accomplished can be visualized by the technique of vital dye staining (see Chapter 1). Fate mapping by Løvtrup (1975; Landstrom and Løvtrup 1979) and by Keller (1975, 1976) has shown that cells of the *Xenopus* blastula have different fates depending on whether they are located in the deep or the superficial layers of the embryo (Figure 10.5). In *Xenopus*, the mesodermal precursors exist mostly in the deep layer of cells, while the ectoderm and endoderm arise from the superficial layer on the surface of the embryo. Most of the precursors for the notochord and other mesodermal tissues are located beneath the surface in the equatorial (marginal) region of the embryo. In urodeles (salamanders such as *Triturus* and *Ambystoma*) and in some frogs other than *Xenopus*, many more of the notochord and mesoderm precursor cells are among the surface cells* (Purcell and Keller 1993; Shook et al. 2002).

As we have seen, the unfertilized egg has a polarity along the animal-vegetal axis. Thus, the germ layers can be mapped onto the oocyte even before fertilization. The surface of the animal hemisphere will become the cells of the ectoderm (skin and nerves), the vegetal hemisphere surface will form the cells of the gut and associated organs (endoderm), and the mesodermal cells will form from the internal cytoplasm around the equator. This general fate map is thought to be imposed upon the egg by the transcription factor **VegT** and the TGF-β family paracrine factor **Vg1**. The mRNAs for these proteins are located in the cortex of the vegetal hemisphere of *Xenopus* oocytes, and they are apportioned to the vegetal cells during cleavage (see Figure 5.34). By using antisense oligonucleotides, Zhang and colleagues (1998) were able to deplete maternal VegT protein in early embryos. The resulting embryos lacked the normal fate map. The animal third of the embryo produced only ventral epidermis, while the

---

*In urodeles, the superficial mesoderm cells move into the deeper layers by forming a primitive streak similar to that of amniote embryos (see the next chapter). As in chick and mouse embryos, the cells leave the epithelial sheet to become mesenchymal cells (Shook et al. 2002).

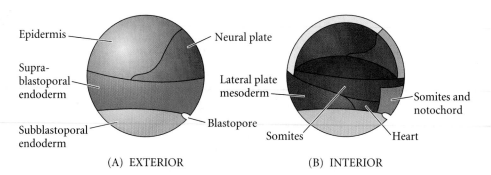

(A) EXTERIOR (B) INTERIOR

Figure 10.5

Fate maps of the blastula of the frog *Xenopus laevis*. (A) Exterior. (B) Interior. Most of the mesodermal derivatives are formed from the interior cells. The placement of ventral mesoderm remains controversial (see Kumano and Smith 2002; Lane and Sheets 2002). (After Lane and Smith 1999; Newman and Krieg 1999.)

marginal cells (which normally produced mesoderm) generated epidermal and neural tissue. The vegetal third (which usually produces endoderm) produced a mixture of ectoderm and mesoderm (Figure 10.6). Joseph and Melton (1998) demonstrated that embryos that lacked functional Vg1 lacked endoderm and dorsal mesoderm. Thus, the allocation of cells to the three germ layers depends on pre-localized cytoplasmic determinants laid down in the egg.

These findings tell us nothing, however, about which part of the egg will form the belly and which the back. The anterior-posterior, dorsal-ventral, and left-right axes are specified by events triggered at fertilization and realized during gastrulation.

Figure 10.6

The fates of the three regions of the *Xenopus* blastula are altered by the depletion of VegT. In normal embryos, the animal third forms epidermal and neural ectoderm, the equatorial third forms mesoderm, and the vegetal third contains the VegT protein and forms the endoderm. In VegT-depleted embryos, the animal cap forms only ventral epidermis, while the equatorial third produces epidermal and neural ectoderm. The vegetal third of these embryos produces ectoderm (both epidermal and neural) as well as mesoderm. No endoderm is produced. (After Zhang et al. 1998.)

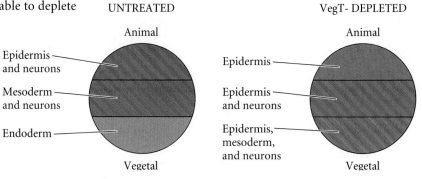

VADE MECUM[2] **Amphibian development.** The events of cleavage and gastrulation are difficult to envision without three-dimensional models. You can see movies of such 3-D models, as well as footage of a living *Xenopus* embryo, in the segments on amphibian development.
[Click on Amphibian]

Cell movements during amphibian gastrulation

Before we look at the process of gastrulation in detail, let us first trace the general movement patterns of the germ layers. Gastrulation in frog embryos is initiated on the future dorsal side of the embryo, just below the equator in the region of the gray crescent (Figure 10.7). Here, the cells invaginate to form a slit-like blastopore. These cells change their shape dramatically. The main body of each cell is displaced toward the inside of the embryo while the cell maintains contact with the outside surface by way of a slender neck (Figure 10.8). These **bottle cells** line the archenteron as it forms. Thus, as in the gastrulating sea urchin, an invagination of cells initiates archenteron formation. However, unlike gastrulation in sea urchins, gastrulation in the frog begins not in the most vegetal region, but in the **marginal zone**: the zone surrounding the equator of the blastula, where the animal and vegetal hemispheres meet. Here the endodermal cells are not as large or as yolky as the most vegetal blastomeres.

The next phase of gastrulation involves the involution of the marginal zone cells while the animal cells undergo epiboly

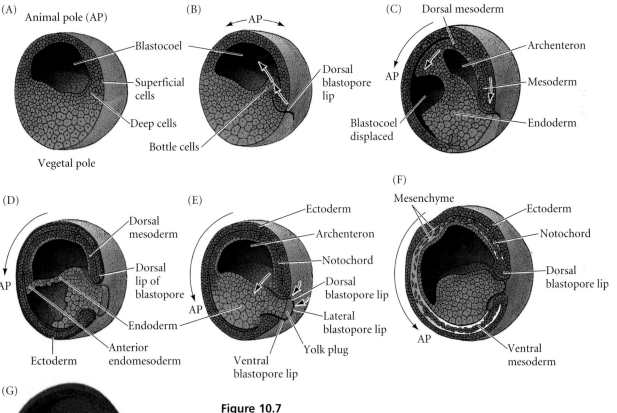

Figure 10.7

Cell movements during frog gastrulation. The meridional sections are cut through the middle of the embryo and positioned so that the vegetal pole is tilted toward the observer and slightly to the left. The major cell movements are indicated by arrows, and the superficial animal hemisphere cells are colored so that their movements can be followed. (A, B) Early gastrulation. The bottle cells of the margin move inward to form the dorsal lip of the blastopore, and the mesodermal precursors involute under the roof of the blastocoel. AP marks the position of the animal pole, which will change as gastrulation continues. (C, D) Mid-gastrulation. The archenteron forms and displaces the blastocoel, and cells migrate from the lateral and ventral lips of the blastopore into the embryo. The cells of the animal hemisphere migrate down toward the vegetal region, moving the blastopore to the region near the vegetal pole. (E, F) Toward the end of gastrulation, the blastocoel is obliterated, the embryo becomes surrounded by ectoderm, the endoderm has been internalized, and the mesodermal cells have been positioned between the ectoderm and endoderm. (G, H) Modern visualizations of the views shown in (A) and (D), respectively, imaged with a surface imaging microscope (Ewald et al. 2002) and rendered using ResView 3.1 software. (A–F after Keller 1986; G, H courtesy of Andrew Ewald and Scott Fraser.)

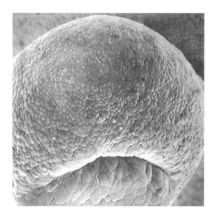

Figure 10.8
Surface view of an early dorsal blastopore lip of *Xenopus*. The size difference between the animal and vegetal blastomeres is readily apparent. (Micrograph courtesy of C. Phillips.)

and converge at the blastopore (Figure 10.7C,D). When the migrating marginal cells reach the **dorsal lip of the blastopore**, they turn inward and travel along the inner surface of the outer animal hemisphere cells. Thus, the cells constituting the lip of the blastopore are constantly changing. The first cells to compose the dorsal blastopore lip are the bottle cells that invaginated to form the leading edge of the archenteron. These cells later become the pharyngeal cells of the foregut. As these first cells pass into the interior of the embryo, the dorsal blastopore lip becomes composed of cells that involute into the embryo to become the **prechordal plate** (the precursor of the head mesoderm). The next cells involuting into the embryo through the dorsal blastopore lip are called the **chordamesoderm** cells. These cells will form the **notochord**, a transient mesodermal "backbone" that plays an important role in inducing and patterning the nervous system.

As the new cells enter the embryo, the blastocoel is displaced to the side opposite the dorsal lip of the blastopore. Meanwhile, the blastopore lip expands laterally and ventrally as the processes of bottle cell formation and involution continue around the blastopore (Figure 10.9). The widening blastopore "crescent" develops lateral lips and finally a ventral lip over which additional mesodermal and endodermal precursor cells pass. With the formation of the ventral lip, the blastopore has formed a ring around the large endodermal cells that remain exposed on the vegetal surface. This remaining patch of endoderm is called the **yolk plug**; it, too, is eventually internalized (Figure 10.9). At that point, all the endodermal precursors have been brought into the interior of the embryo, the ectoderm has encircled the surface, and the mesoderm has been brought between them.

The mid-blastula transition: Preparing for gastrulation

Now that we have seen an overview of amphibian gastrulation, we can look more deeply into its mechanisms. The first precondition for gastrulation is the activation of the genome. In *Xenopus*, only a few genes appear to be transcribed during early cleavage (Yang et al. 2002). For the most part, nuclear genes are not activated until late in the twelfth cell cycle (Newport and Kirschner 1982a,b). At that time, different genes begin to be transcribed in different cells, and the blastomeres acquire the capacity to become motile. This dramatic change is called the **mid-blastula transition (MBT)** (see

Figure 10.9
Epiboly of the ectoderm. (A) Changes in the region around the blastopore as the dorsal, lateral, and ventral lips are formed in succession. When the ventral lip completes the circle, the endoderm becomes progressively internalized. Numbers ii–v correspond to Figures 10.7 B–E, respectively. (B) Summary of epiboly of the ectoderm and involution of the mesodermal cells migrating into the blastopore and then under the surface. The endoderm beneath the blastopore lip (the yolk plug) is not mobile and is enclosed by these movements. (A from Balinsky 1975; photographs courtesy of B. I. Balinsky.)

(A)

i   Dorsal lip   ii   iii

Lateral lip

iv   Dorsal lip   v   Ventral lip

(B)

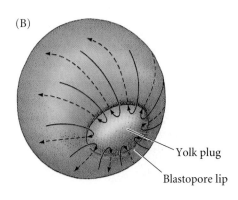

Yolk plug

Blastopore lip

Chapters 8 and 9). It is thought that some factor in the egg is being titrated by the newly made chromatin, because the time of this transition can be changed experimentally by altering the ratio of chromatin to cytoplasm in the cell (Newport and Kirschner 1982a,b). Thus, cleavage begins soon after fertilization and ends shortly after the stage at which the embryo achieves a new balance between nucleus and cytoplasm.

One of the events that triggers the mid-blastula transition is the demethylation of certain promoters. In *Xenopus* (unlike mammals), high levels of methylated DNA are seen in both the paternally and maternally derived chromosomes. However, during the late blastula stages, there is a loss of methylation on the promoters of genes that are activated at the MBT. This demethylation is not seen on promoters that are not activated at the MBT, nor is it observed in the actual coding regions of the MBT-activated genes. It appears, then, that demethylation of certain promoters may play a pivitol role in regulating the timing of gene expression at the MBT (Stancheva et al. 2002).

It is thought that once the chromatin at the promoters has been remodeled, various transcription factors (such as the VegT protein, mentioned above) bind to the promoters and initiate new transcription. For instance, the vegetal cells (probably under the direction of the maternal VegT protein) become the endoderm and begin secreting the factors that cause the cells above them to become the mesoderm (Wylie et al. 1996; Agius et al. 2000).

Positioning the blastopore

The dorsal-ventral axis can be traced to the sperm entry point. Microtubules nucleated by the sperm centriole direct cytoplasmic movements that will empower the vegetal cells opposite the point of sperm entry to induce the blastopore in the mesoderm above them. This region of cells opposite the point of sperm entry will form the blastopore and become the dorsal portion of the body.

In Chapter 7, we saw that the internal cytoplasm of the fertilized egg remains oriented with respect to gravity because of its dense yolk accumulation, while the cortical cytoplasm actively rotates 30° animally ("upward"), toward the point of sperm entry (see Figure 7.33). In this way, a new state of symmetry is acquired. Whereas the unfertilized egg was radially symmetrical about the animal-vegetal axis, the fertilized egg now has a dorsal-ventral axis. It has become bilaterally symmetrical (having right and left sides). The inner cytoplasm moves as well. These cytoplasmic movements activate the cytoplasm opposite the point of sperm entry, enabling it to initiate gastrulation. The side where the sperm enters marks the future ventral (belly) surface of the embryo; the opposite side, where gastrulation is initiated, marks the future dorsum (back) of the embryo (Gerhart et al. 1981, 1986; Vincent et al. 1986). If cortical rotation is blocked, there is no dorsal development, and the embryo dies as a "belly piece" in which the three germ layers are formed, but all dorsal structures (neural tube, notochord, and somites) are absent (Vincent and Gerhart 1987).

Although the sperm is not needed to induce these movements in the egg cytoplasm, it is important in determining the direction of the rotation. If an egg is artificially activated, the cortical rotation still takes place at the correct time. However, the direction of this movement is unpredictable. The directional bias provided by the point of sperm entry can be overridden by mechanically redirecting the spatial relationship between the cortical and internal cytoplasms. When a *Xenopus* egg is turned 90°, so that the point of sperm entry faces upward, the cytoplasm rotates such that the embryo initiates gastrulation on the same side as sperm entry (Gerhart et al. 1981; Cooke 1986). One can even produce two gastrulation initiation sites by combining the natural sperm-oriented rotation with an artificially induced rotation of the egg. Black and Gerhart (1985, 1986) let the initial sperm-directed rotation occur, but then immobilized eggs in gelatin and gently centrifuged them so that the internal cytoplasm would flow *toward* the point of sperm entry. When the centrifuged eggs were then allowed to develop in normal water, two dorsal blastopore lips formed, leading to the formation of conjoined twin larvae (Figure 10.10). A dorsal determinant is redistributed by this centrifugation.

Figure 10.10
Twin blastopores produced by rotating dejellied *Xenopus* eggs ventral side (sperm entry point) up at the time of first cleavage. (A) Two blastopores are instructed to form: the original one (opposite the point of sperm entry) and the new one created by the displacement of cytoplasmic material. (B) These eggs develop two complete axes, which form twin tadpoles, joined ventrally. (Photographs courtesy of J. Gerhart.)

(A)

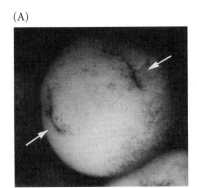

(B)

It appears that cortical rotation enables the vegetal blastomeres opposite the point of sperm entry to induce the cells above them to initiate gastrulation. Gimlich and Gerhart (1984), using transplantation experiments on 64-cell *Xenopus* embryos, showed that the three vegetal blastomeres opposite the point of sperm entry are able to induce the formation of the dorsal lip of the blastopore and of a complete dorsal axis when transplanted into UV-irradiated embryos (which otherwise would have failed to properly initiate gastrulation: Figure 10.11A). Moreover, these three blastomeres, which underlie the prospective dorsal lip region, can also induce a secondary blastopore and axis when transplanted into the ventral side of a normal, unirradiated embryo (Figure 10.11B). Holowacz and Elinson (1993) found that cortical cytoplasm from the dorsal vegetal cells of the 16-cell *Xenopus* embryo was able to induce the formation of secondary axes when injected into ventral vegetal cells. Neither cortical cytoplasm from animal cells nor the deep cytoplasm from ventral cells could induce such axes. Later in this chapter, we will provide evidence that this dorsal signal is the transcription factor β-catenin.

Invagination and involution

Amphibian gastrulation is first visible when a group of marginal endoderm cells on the dorsal surface of the blastula sinks into the embryo. The outer (**apical**) surfaces of these cells contract dramatically, while their inner (**basal**) ends expand. The apical-basal length of these cells greatly increases to yield the characteristic "bottle" shape. In salamanders, these bottle cells appear to have an active role in the early movements of gastrulation. Johannes Holtfreter (1943, 1944) found that bottle cells from early salamander gastrulae could attach to glass coverslips and lead the movement of those cells attached to them. Even more convincing were Holtfreter's recombination experiments. When dorsal marginal zone cells (which would normally give rise to the dorsal lip of the blastopore) were excised and placed on inner prospective endoderm tissue, they formed bottle cells and sank below the surface of the inner endoderm (Figure 10.12). Moreover, as they sank, they created a depression reminiscent of the early blastopore. Thus, Holtfreter claimed that the ability to invaginate into the deep endoderm is an innate property of the dorsal marginal zone cells.

Figure 10.11
Transplantation experiments on 64-cell amphibian embryos demonstrating that the vegetal cells underlying the prospective dorsal blastopore lip region are responsible for causing the initiation of gastrulation. (A) Rescue of irradiated embryos by transplanting the dorsal vegetal blastomeres of a normal embryo into a cavity made by the removal of a similar number of vegetal cells. An irradiated zygote without this transplant fails to undergo normal gastrulation. (B) Formation of a new gastrulation site and body axis by the transplantation of the most dorsal vegetal cells of one embryo into the ventralmost vegetal region of another embryo. (After Gimlich and Gerhart 1984.)

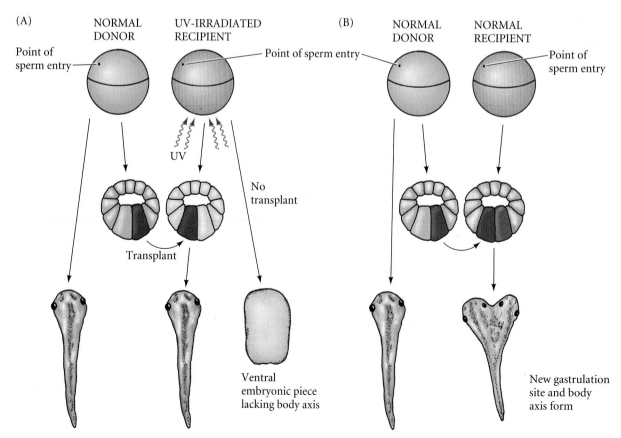

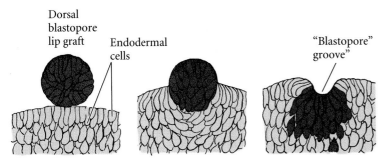

Dorsal blastopore lip graft   Endodermal cells   "Blastopore" groove"

Figure 10.12
A graft of cells from the dorsal marginal zone of a salamander embryo sinks into a layer of endodermal cells and forms a blastopore-like groove. (After Holtfreter 1944.)

WEBSITE 10.1 **Demonstrating tissue affinities.** The tissue affinities that Holtfreter predicted have been demonstrated quantitatively by new studies that measure the surface tensions of different cell layers.

The situation in the frog embryo is somewhat different. R. E. Keller and his students (Keller 1981; Hardin and Keller 1988) have shown that although the bottle cells of *Xenopus* may play a role in initiating the involution of the marginal zone as they become bottle-shaped, they are not essential for gastrulation to continue. The peculiar shape change of the bottle cells is needed to initiate gastrulation; it is the constriction of these cells that first forms the slit-like blastopore. However, after starting these movements, the *Xenopus* bottle cells are no longer needed for gastrulation. When bottle cells are removed after their formation, involution and blastopore formation and closure continue.

The major factor in the movement of cells into the embryo appears to be the involution of the subsurface marginal cells, rather than the superficial ones. Internalization of the endoderm and mesoderm is initated by a movement called **vegetal rotation.** At least 2 hours before the bottle cells form, internal cell movements propel the cells of the dorsal floor of the blastocoel toward the animal cap (Figure 10.13). This rotation places the prospective pharyngeal endoderm adjacent to the blastocoel and immediately above the involuting mesoderm. These cells then migrate along the basal surface of the blastocoel roof (Figure 10.13A–D; Nieuwkoop and Florschütz 1950; Winklbauer and Schürfeld 1999; Ibrahim and Winklbauer 2001). The superficial layer of marginal cells is pulled inward to form the endodermal lining of the archenteron, merely because it is attached to the actively migrating deep cells. While experimental removal of the bottle cells does not affect the involution of the deep or superficial marginal zone cells into the embryo, the removal of the deep **involuting marginal zone (IMZ)** cells stops archenteron formation.

The convergent extension of the dorsal mesoderm

Involution begins dorsally, led by the pharyngeal endomeso-derm* and the prechordal plate. These tissues will migrate most anteriorly beneath the surface ectoderm. The next tissues to enter the dorsal blastopore lip contain notochord and somite precursors. Meanwhile, as the lip of the blastopore expands to have dorsolateral, lateral, and ventral sides, the prospective heart mesoderm, kidney mesoderm, and ventral mesoderm enter into the embryo.

Figures 10.13D–F depict the behavior of the IMZ cells at successive stages of *Xenopus* gastrulation (Keller and Schoenwolf 1977; Keller 1980, 1981; Hardin and Keller 1988). The IMZ is originally several layers thick. Shortly before their involution through the blastopore lip, the several layers of deep IMZ cells intercalate radially to form one thin, broad layer. This intercalation further extends the IMZ vegetally. At the same time, the superficial cells spread out by dividing and flattening. When the deep cells reach the blastopore lip, they involute into the embryo and initiate a second type of intercalation. This intercalation causes a **convergent extension** along the mediolateral axis (Figure 10.13F) that integrates several mesodermal streams to form a long, narrow band. This movement is reminiscent of traffic on a highway when several lanes must merge to form a single lane. The anterior part of this band migrates toward the animal cap. Thus, the mesodermal stream continues to migrate toward the animal pole, and the overlying layer of superficial cells (including the bottle cells) is passively pulled toward the animal pole, thereby forming the endodermal roof of the archenteron (see Figures 10.7 and 10.13E). The radial and mediolateral intercalations of the deep layer of cells appear to be responsible for the continued movement of mesoderm into the embryo.

Two major forces appear to drive convergent extension. The first force is differential cell cohesion. During gastrulation, the genes encoding adhesion proteins **paraxial protocadherin** and **axial protocadherin**, become expressed specifically in the paraxial mesoderm (which gives rise to the somites) and the notochord, respectively (Figure 10.14). An experimental dominant negative form of paraxial protocadherin (which is secreted instead of being bound to the cell membrane) prevents convergent extension[†] (Kim et al. 1998). Moreover, the expression domain of paraxial protocadherin separates the trunk mesodermal cells, which undergo convergent extension, from the head mesodermal cells, which do not.

The second factor regulating convergent extension is calcium flux. Wallingford and colleagues (2001) found that dramatic waves of calcium ions surge across the dorsal tissues un-

---

*The pharyngeal endoderm and head mesoderm cannot be separated experimentally at this stage, so they are therefore sometimes referred to collectively as the pharyngeal endomesoderm. The notochord is the basic unit of the dorsal mesoderm, but it is thought that the dorsal portion of the somites may also have similar properties.

[†]Dominant negative proteins are mutated forms of the wild-type protein that interfere with the normal functioning of the wild-type protein. Thus, a dominant negative protein will have an effect similar to a loss-of-function mutation in the gene encoding the particular protein.

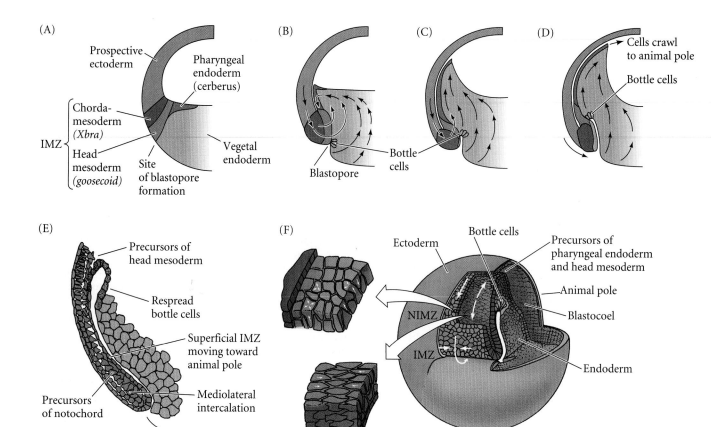

Figure 10.13

Early movements of *Xenopus* gastrulation. The yellow area is vegetal endoderm. Orange represents the prospective pharyngeal endoderm (as seen by Cerberus expression). Dark orange represents the prospective head mesoderm (as seen by Goosecoid expression), and the chordamesoderm (indicated by Xbra expression) is red. The prospective ectoderm is blue. (A) At the beginning of gastrulation, the inner marginal zone (IMZ) is formed. (B) The vegetal rotation (white arrows) pushes the prospective pharyngeal endoderm to the side of the blastocoel. (C, D) The vegetal endoderm movements push the pharyngeal endoderm forward, driving the mesoderm passively into the embryo and toward the animal pole. The ectoderm begins epiboly. (E) As gastrulation continues, the deep marginal cells flatten, and the formerly superficial cells form the wall of the archenteron. (F) Radial intercalation, looking down at the dorsal blastopore lip from the dorsal suface. In the noninvoluting marginal zone (NIMZ) and the upper portin of the IMZ, deep (mesodermal) cells are intercalating radially to make a thin band of flattened cells. This thinning of several layers into a few causes extension toward the blastopore lip. Just above the lip, mediolateral intercalation of the cells produces stresses that pull the IMZ over the lip. After involuting over the lip, mediolateral intercalation continues, elongating and narrowing the axial mesoderm. (After Wilson and Keller 1991 and Winklbauer and Schürfeld 1999.)

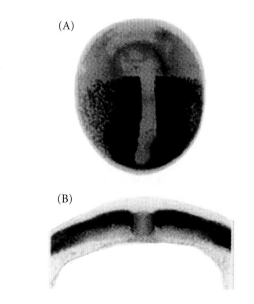

Figure 10.14

The expression of paraxial protocadherin. (A) Expression of paraxial protocadherin during late gastrulation (dark areas) shows the distinct downregulation in the notochord and the absence of expression in the head region. (B) Double-stained cross section through a late *Xenopus* gastrula shows the separation of notochord (brown staining for chordin) and the paraxial mesoderm (blue staining for paraxial protocadherin). (Photographs courtesy of E. M. De Robertis.)

dergoing convergent extension, causing waves of contraction within the tissue. The calcium ions are released from intracellular stores and are required for convergent extension. If the release of calcium ions is blocked, normal cell specification still occurs, but the dorsal mesoderm neither converges nor

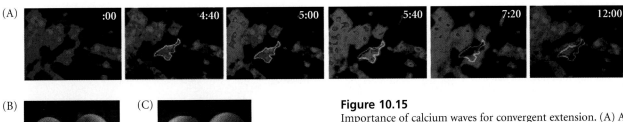

Figure 10.15

Importance of calcium waves for convergent extension. (A) A wave of calcium ions in the dorsal marginal zone of *Xenopus* visualized by calcium-activated dyes and confocal microscopy. The wave moves across the cells from left to right, and as the wave moves across and dissipates, the cells move dramatically. First they move in the direction of the wave, and then they move toward it (from right to left), moving farther than their original position. (B) Two *Xenopus* control embryos at the late gastrula stage, stained for the notochordal mRNA *Xnot*. Convergent extension has occurred, and the blastopore is almost completely closed. (C) Two embryos of the same age treated with a drug that prevented calcium release. Staining for *Xnot* reveals that the notochords have neither converged nor extended, and the blastopore remains open. (After Wallingford et al. 2001; photographs courtesy of S. F. Fraser.)

extends (Figure 10.15). This calcium is thought to regulate the contraction of the actin microfilaments.

Interestingly, the region specified to become the head remains outside the convergent extension. This phenomenon is due to the expression of the *Otx2* gene. This gene is the vertebrate homologue of the *otd* (*orthodenticle*) gene of *Drosophila*, and like the *otd* gene, *Otx2* encodes a transcription factor expressed in the most anterior region of the embryo. The Otx2 protein is critical in head formation, and it activates those genes involved in forebrain formation. In addition to specifying the anterior tissues of the embryo, Otx2 prevents the cells expressing it from undergoing convergence and extension. One of the genes Otx2 activates is the *Xenopus calponin* gene (Morgan et al. 1999). Calponin is a protein that binds to actin and myosin and prevents actin microfilaments from contracting.

Migration of the involuting mesoderm

As mesodermal movement progresses, convergent extension continues to narrow and lengthen the involuting marginal zone. The IMZ contains the prospective endodermal roof of the archenteron in its superficial layer (IMZ$_S$) and the prospective mesodermal cells, including those of the notochord, in its deep region (IMZ$_D$). During the middle third of gastrulation, the expanding sheet of mesoderm converges toward the midline of the embryo. This process is driven by the continued mediolateral intercalation of cells along the anterior-posterior axis, thereby further narrowing the band. Toward the end of gastrulation, the centrally located notochord separates from the somitic mesoderm on either side of it, and the notochord cells elongate separately (Wilson and Keller 1991). This may in part be a consequence of the different adhesion molecules in the axial and paraxial mesoderms (see Figure 10.14; Kim et al. 1998). This convergent extension of the mesoderm appears to be autonomous, because the movements of these cells occur even if this region of the embryo is experimentally isolated from the rest of the embryo (Keller 1986).

During gastrulation, the animal cap and **noninvoluting marginal zone** (**NIMZ**) cells expand by epiboly to cover the entire embryo. These cells will form the surface ectoderm. The dorsal portion of the NIMZ extends more rapidly toward the blastopore than the ventral portion, thus causing the blasto-

pore lips to move toward the ventral side. While those mesodermal cells entering through the dorsal lip of the blastopore give rise to the dorsal axial mesoderm (notochord and somites), the remainder of the body mesoderm (which forms the heart, kidneys, blood, bones, and parts of several other organs) enters through the ventral and lateral blastopore lips to create the **mesodermal mantle**. The endoderm is derived from the IMZ$_S$ cells that form the lining of the archenteron roof and from the subblastoporal vegetal cells that become the archenteron floor (Keller 1986).

Epiboly of the ectoderm

While involution is occurring at the blastopore lips, the ectodermal precursors are expanding over the entire embryo. The major mechanism of epiboly in *Xenopus* gastrulation appears to be an increase in cell number (through division) coupled with a concurrent integration of several deep layers into one (Figure 10.16; Keller and Schoenwolf 1977; Keller 1980). The result of these expansions is the epiboly of the superficial and deep cells of the animal cap and NIMZ over the surface of the embryo (Keller and Danilchik 1988; Saka and Smith 2001). Most of the marginal zone cells, as previously mentioned, involute to join the mesodermal cell stream within the embryo. As the ectoderm epibolizes over the entire embryo, it eventually internalizes all the endoderm within it. At this point, the ectoderm covers the embryo, the endoderm is located within the embryo, and the mesoderm is positioned between them.

WEBSITE 10.2 **Migration of the mesodermal mantle.** Different growth rates coupled with the intercalation of cell layers allows the mesoderm to expand in a tightly coordinated fashion.

Figure 10.16
Epiboly of the ectoderm is accomplished by cell division and intercalation. (A, B) Cell division in the presumptive ectoderm. Cell division is shown by staining for phosphorylated histone 3, a marker of mitosis. Stained nuclei appear black. In early gastrulae (A; stage 10.5), most cell division occurs in the animal hemisphere presumptive ectoderm. In late gastrulae (B; stage 12), cell division can be seen throughout the ectodermal layer. (Interestingly, the dorsal mesoderm shows no cell division). (C) Scanning electron micrographs of the *Xenopus* blastocoel roof, showing the changes in cell shape and arrangement. Stages 8 and 9 are blastulae; stages 10–11.5 represent progressively later gastrulae. (A, B after Saka and Smith 2001, photographs courtesy of the authors; C from Keller 1980; photographs courtesy of R. E. Keller.)

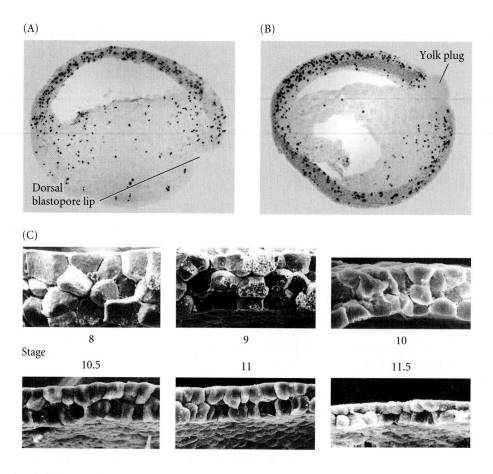

Sidelights & Speculations

Fibronectin and the Pathways for Mesodermal Migration

How are the involuting cells informed where to go once they enter the inside of the embryo? In many amphibians, it appears that the involuting mesodermal precursors migrate toward the animal pole on a fibronectin lattice secreted by the cells of the blastocoel roof (Figure 10.17A,B). Shortly before gastrulation, the presumptive ectoderm of the blastocoel roof secretes an extracellular matrix that contains fibrils of fibronectin (Boucaut et al. 1984; Nakatsuji et al. 1985). The involuting mesoderm appears to travel along these fibronectin fibrils. Confirmation of this hypothesis was obtained by chemically synthesizing a "phony" fibronectin that can compete with the genuine fibronectin of the extracellular matrix. Cells bind to a region of the fibronectin protein that contains a three-amino acid sequence (Arg-Gly-Asp; RGD). Boucaut and co-workers injected large amounts of a small peptide containing this sequence

into the blastocoels of salamander embryos shortly before gastrulation began. If fibronectin were essential for cell migration, then cells binding this soluble peptide fragment instead of the real extracellular fibronectin should stop migrating. Unable to find their "road," the mesodermal cells should cease their involution. That is precisely what happened (Figure 10.17C–F). No migrating cells were seen along the underside of the ectoderm in the experimental embryos. Instead, the mesodermal precursors remained outside the embryos, forming a convoluted cell mass. Other small synthetic peptides (including other fragments of the fibronectin molecule) did not impede migration. Thus the fibronectin-containing extracellular matrix appears to provide both a substrate for adhesion as well as cues for the direction of cell migration. Shi and colleagues (1989) showed that salamander IMZ cells would migrate in the wrong direction if extra

fibronectin lattices were placed in their path.

In *Xenopus*, fibronectin is similarly secreted by the cells lining the blastocoel roof. The result (Figure 10.17.A,B) is a band of fibronectin lining the roof, including Brachet's cleft, the part of the blastocoel roof extending vegetally on the dorsal side. The vegetal rotation places the pharyngeal endoderm and invouting mesoderm into contact with these fibronectin fibrils (Winklbauer and Schürfeld 1999). Convergent extension pushes the migrating cells upward toward the animal pole. The fibronectin fibrils are necessary for the head mesodermal cells to flatten and to extend broad (lamelliform) processes in the direction of migration (Winklbauer et al. 1991; Winklbauer and Keller 1996). Studies using inhibitors of fibronectin formation have shown that these fibronectin fibrils are necessary for the direction of mesoderm migration, the maintenance of intercalation of animal cap cells, and the initiation of radial intercalation in the marginal zone (Marsden and DeSimone 2001).

The mesodermal cells are thought to adhere to fibronectin through the $\alpha_V\beta_1$ integrin protein (Alfandari et al. 1995). Mesodermal migration can also be arrested by the microinjection of antibodies against either fibronectin or the α_V subunit of integrin, which serves as part of the fibronectin receptor (Darribère et al. 1988, 1990). Alfandari and colleagues (1995) have shown that the α_V subunit of integrin appears on the mesodermal cells just prior to gastrulation, persists on their surfaces throughout gastrulation, and disappears when gastrulation ends. The integrin coordinates the interaction of the blastocel roof with actin microfilaments in the migrating mesendodermal cells. This interaction allows for increased traction and spread of migration (Davidson et al. 2002). It seems, then, that the coordinated synthesis of fibronectin and its receptor may signal the times for the mesoderm to begin, continue, and stop migration.

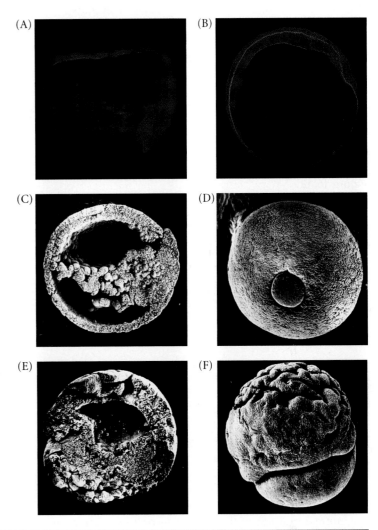

Figure 10.17
Fibronectin and amphibian gastrulation. (A, B) Sagittal section of *Xenopus* embryos at (A) early and (B) late gastrulation. The fibronectin lattice on the blastocoel roof is identified by fluorescent antibody labeling (yellow) while the embryonic cells are counterstained red. (C–F) Scanning electron micrographs of (C, D) a normal salamander embryo injected with a control solution at the blastula stage; and (E, F) an embryo of the same stage injected with the cell-binding fragment of fibronectin. (C) Section during mid-gastrulation. (D) The yolk plug toward the end of gastrulation. (E, F) The finishing stages of the arrested gastrulation, wherein the mesodermal precursors, having bound the synthetic fibronectin, cannot recognize the normal fibronectin-lined migration route. The archenteron fails to form, and the noninvoluted mesodermal precursors remain on the surface. (A, B from Marsden and DeSimone 2001, photographs courtesy of the authors; C–F from Boucaut et al. 1984, photographs courtesy of J.-C. Boucaut and J.-P. Thiery.)

AXIS FORMATION IN AMPHIBIANS: THE PHENOMENON OF THE ORGANIZER

The Progressive Determination of the Amphibian Axes

Vertebrate axes do not form from localized determinants, as in *Drosophila*. Rather, they arise progressively through a sequence of interactions between neighboring cells. Amphibian axis formation is an example of regulative development. In Chapter 3, we discussed the concept of regulative development, wherein (1) an isolated blastomere has a potency greater than its normal embryonic fate, and (2) a cell's fate is determined by interactions between neighboring cells. Such interactions are called inductions (see Chapter 6). That such inductive interactions were responsible for amphibian axis determination was demonstrated by the laboratory of Hans Spemann at the University of Freiburg (see De Robertis and Aréchaga 2001; Sander and Fässler 2001). The experiments of Spemann and his students framed the questions that experimental embryologists asked for most of the twentieth century, and they resulted in a Nobel Prize for Spemann in 1935. More recently, the discoveries of the molecules associated with these inductive processes have provided some of the most exciting moments in contemporary science.

The experiment that began this research program was performed in 1903, when Spemann demonstrated that early newt blastomeres have identical nuclei, each capable of producing an entire larva. His procedure was ingenious: Shortly after fertilizing a newt egg, Spemann used a baby's hair taken from his daughter to lasso the zygote in the plane of the first cleavage. He then partially constricted the egg, causing all the nuclear divisions to remain on one side of the constriction. Eventually, often as late as the 16-cell stage, a nucleus would escape across the constriction into the non-nucleated side. Cleavage then began on this side, too, whereupon Spemann tightened the lasso until the two halves were completely separated. Twin larvae developed, one slightly older than the other

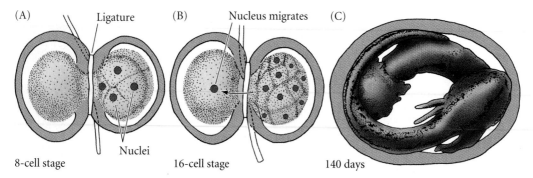

8-cell stage   16-cell stage   140 days

Figure 10.18
Spemann's demonstration of nuclear equivalence in newt cleavage. (A) When the fertilized egg of the newt *Triturus taeniatus* was constricted by a ligature, the nucleus was restricted to one half of the embryo. The cleavage on that side of the embryo reached the 8-cell stage, while the other side remained undivided. (B) At the 16-cell stage, a single nucleus entered the as yet undivided half, and the ligature was further constricted to complete the separation of the two halves. (C) After 140 days, each side had developed into a normal embryo. (After Spemann 1938.)

passes into only one of the two blastomeres. Spemann found that when these two blastomeres are separated, only the blastomere containing the gray crescent develops normally.

WEBSITE 10.3 **Embryology and individuality.** One egg usually makes only one adult; however, there are exceptions to this rule. Spemann was drawn into embryology through the paradoxes of creating more than one individual from a single egg.

(Figure 10.18). Spemann concluded from this experiment that early amphibian nuclei were genetically identical and that each cell was capable of giving rise to an entire organism.

However, when Spemann performed a similar experiment with the constriction still longitudinal, but perpendicular to the plane of the first cleavage (separating the future dorsal and ventral regions rather than the right and left sides), he obtained a different result altogether. The nuclei continued to divide on both sides of the constriction, but only one side—the future dorsal side of the embryo—gave rise to a normal larva. The other side produced an unorganized tissue mass of ventral cells, which Spemann called the *Bauchstück*—the belly piece. This tissue mass was a ball of epidermal cells (ectoderm) containing blood and mesenchyme (mesoderm) and gut cells (endoderm), but no dorsal structures such as nervous system, notochord, or somites (Figure 10.19).

Why should these two experiments give different results? One possibility was that when the egg was divided perpendicular to the first cleavage plane, some *cytoplasmic* substance was not equally distributed into the two halves. Fortunately, the salamander egg was a good place to test that hypothesis. As we have seen, there are dramatic movements in the cytoplasm following the fertilization of amphibian eggs, and in some amphibians these movements expose a gray, crescent-shaped area of cytoplasm in the region directly opposite the point of sperm entry. This area has been called the **gray crescent**. Moreover, the first cleavage plane normally splits the gray crescent equally into the two blastomeres. If these cells are then separated, two complete larvae develop. However, should this cleavage plane be aberrant (either in the rare natural event or in an experiment), the gray crescent material

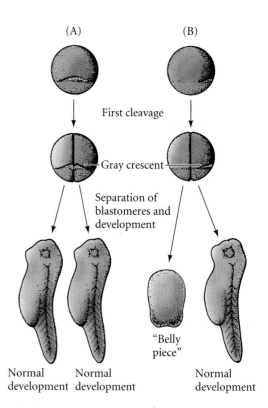

Figure 10.19
Asymmetry in the amphibian egg. (A) When the egg is divided along the plane of first cleavage into two blastomeres, each of which gets one-half of the gray crescent, each experimentally separated cell develops into a normal embryo. (B) When only one of the two blastomeres receives the entire gray crescent, it alone forms a normal embryo. The other blastomere produces a mass of unorganized tissue lacking dorsal structures. (After Spemann 1938.)

It appeared, then, that something in the gray crescent region was essential for proper embryonic development. But how did it function? What role did it play in normal development? The most important clue came from the fate map of this area of the egg, for it showed that the gray crescent region gives rise to the cells that initiate gastrulation. These cells form the dorsal lip of the blastopore. The cells of the dorsal lip are committed to invaginate into the blastula, thus initiating gastrulation and the formation of the notochord. Because all future amphibian development depends on the interaction of cells rearranged during gastrulation, Spemann speculated that the importance of the gray crescent material lies in its ability to initiate gastrulation, and that crucial developmental changes occur during gastrulation.

TABLE 10.1 Results of tissue transplantation during early- and late-gastrula stages in the newt

Donor region	Host region	Differentiation of donor tissue	Conclusion
EARLY GASTRULA			
Prospective neurons	Prospective epidermis	Epidermis	Dependent (conditional) development
Prospective epidermis	Prospective neurons	Neurons	Dependent (conditional) development
LATE GASTRULA			
Prospective neurons	Prospective epidermis	Neurons	Independent (autonomous) development (determined)
Prospective epidermis	Prospective neurons	Epidermis	Independent (autonomous) development (determined)

In 1918, Spemann demonstrated that enormous changes in cell potency do indeed take place during gastrulation. He found that the cells of the early gastrula were uncommitted, but that the fates of late gastrula cells were determined. Spemann demonstrated this by exchanging tissues between the gastrulae of two species of newts whose embryos were differently pigmented (Figure 10.20). When a region of prospective epidermal cells from an *early* gastrula was transplanted into an area in another early gastrula where the neural tissue normally formed, the transplanted cells gave rise to neural tissue. When prospective neural tissue from early gastrulae was transplanted to the region fated to become belly skin, the neural tissue became epidermal (Table 10.1). Thus, these early newt gastrula cells were not yet committed to a specific fate. Such cells are said to exhibit **regulative** (i.e., **conditional** or **dependent**) **development** because their ultimate fates depend on their location in the embryo. However, when the same interspecies transplantation experiments were performed on *late* gastrulae, Spemann obtained completely different results. Rather than differentiating in accordance with their new location, the transplanted cells exhibited **autonomous** (or **independent**, or **mosaic**) **development**. Their prospective fate was determined, and the cells developed independently of their new embryonic location. Specifically, prospective neural cells now developed into brain tissue even when placed in the region of prospective epidermis, and prospective epidermis formed skin even in the region of the prospective neural tube. Within the time separating early and late gastrulation, the potencies of these groups of cells had become restricted to their eventual paths of differentiation. Something was causing them to become committed to epidermal and neural fates. What was happening?

(A) TRANSPLANTATION IN EARLY GASTRULA

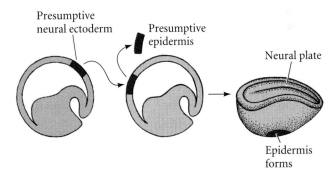

Presumptive neural ectoderm

Presumptive epidermis

Neural plate

Epidermis forms

(B) TRANSPLANTATION IN LATE GASTRULA

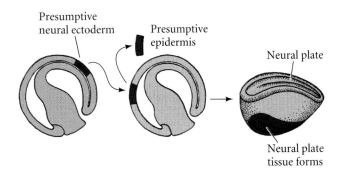

Presumptive neural ectoderm

Presumptive epidermis

Neural plate

Neural plate tissue forms

Figure 10.20
Determination of ectoderm during newt gastrulation. Presumptive neural ectoderm from one newt embryo is transplanted into a region in another embryo that normally becomes epidermis. (A) When the tissues are transferred between early gastrulae, the presumptive neural tissue develops into epidermis, and only one neural plate is seen. (B) When the same experiment is performed using late-gastrula tissues, the presumptive neural cells form neural tissue, thereby causing two neural plates to form on the host. (After Saxén and Toivonen 1962.)

Hans Spemann and Hilde Mangold: Primary Embryonic Induction

The most spectacular transplantation experiments were published by Hans Spemann and Hilde Mangold in 1924.* They showed that, of all the tissues in the early gastrula, only one has its fate determined. This self-differentiating tissue is the dorsal lip of the blastopore, the tissue derived from the gray crescent cytoplasm. When this dorsal lip tissue was transplanted into the presumptive belly skin region of another gastrula, it not only continued to be blastopore lip, but also initiated gastrulation and embryogenesis in the surrounding tissue (Figure 10.21). Two conjoined embryos were formed instead of one!

In these experiments, Spemann and Mangold used differently pigmented embryos from two newt species—the darkly pigmented *Triturus taeniatus* and the nonpigmented *Triturus cristatus*—so that when Spemann and Mangold prepared these transplants, they were able to identify host and donor tissues on the basis of color.[†] When the dorsal lip of an early *T. taeniatus* gastrula was removed and implanted into the region of an early *T. cristatus* gastrula fated to become ventral epidermis (belly skin), the dorsal lip tissue invaginated just as it would normally have done (showing self-determination) and disappeared beneath the vegetal cells. The pigmented donor tissue then continued to self-differentiate into the chordamesoderm (notochord) and other mesodermal structures that normally form from the dorsal lip. As the new donor-derived mesodermal cells moved forward, host cells began to participate in the production of a new embryo, becoming organs that normally they never would have formed. In this secondary embryo, a somite could be seen containing both pigmented (donor) and unpigmented (host) tissue. Even more spectacularly, the dorsal lip cells were able to interact with the host tissues to form a complete neural plate from host ectoderm. Eventually, a secondary embryo formed, face to face with its host. The results of these technically difficult experiments have been confirmed many times (Capuron 1968; Smith and Slack 1983; Recanzone and Harris 1985).

WEBSITE 10.4 Spemann, Mangold, and the Organizer. Spemann did not see the importance of this work the first time he and Mangold did it. This website provides a more detailed account of why Spemann and Mangold did this experiment.

*Hilde Proescholdt Mangold died in a tragic accident in 1924, when her kitchen's gasoline heater exploded. At the time she was 26 years old, and her paper was just being published. Hers is one of the very few doctoral theses in biology that have directly resulted in the awarding of a Nobel Prize. For more information about Hilde Mangold, her times and the experiments that identified the organizer, see Hamburger 1984, 1988, and Fässler and Sander 1996.

[†]*T. cristatus* is the northern crested newt, and *T. taeniatus* (also known as *T. vulgaris*) is the common (or smooth) newt. It was fortunate that Spemann's laboratory, and those of his students, usually used salamander embryos for their experiments. It turns out that frog ectoderm is much more difficult to induce than that of these urodeles.

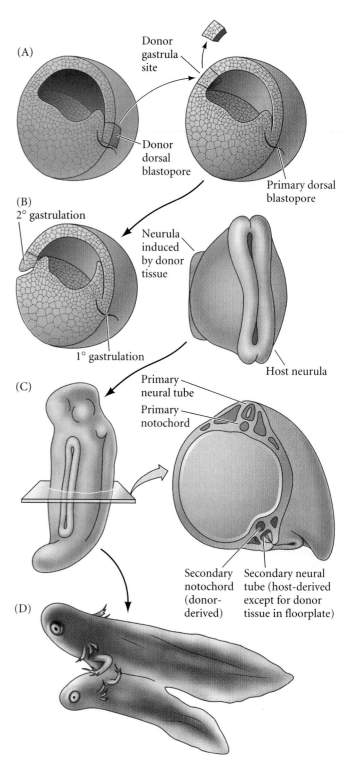

Figure 10.21
Organization of a secondary axis by dorsal blastopore lip tissue. (A) Dorsal lip tissue from an early gastrula is transplanted into another early gastrula in the region that normally becomes ventral epidermis. (B) The donor tissue invaginates and forms a second archenteron, and then a second embryonic axis. Both donor and host tissues are seen in the new neural tube, notochord, and somites. (C) Eventually, a second embryo forms that is joined to the host. (After Hamburger 1988.)

Spemann (1938) referred to the dorsal lip cells and their derivatives (notochord, prechordal mesoderm) as the **organizer** because (1) they induced the host's ventral tissues to change their fates to form a neural tube and dorsal mesodermal tissue (such as somites), and (2) they organized host and donor tissues into a secondary embryo with clear anterior-posterior and dorsal-ventral axes. He proposed that during normal development, these cells organize the dorsal ectoderm into a neural tube and transform the flanking mesoderm into the anterior-posterior body axis. It is now known (thanks largely to Spemann and his students) that the interaction of the chordamesoderm and ectoderm is not sufficient to "organize" the entire embryo. Rather, it initiates a series of sequential inductive events. As discussed in Chapter 6, the process by which one embryonic region interacts with a second region to influence that second region's differentiation or behavior is called induction. Because there are numerous inductions during embryonic development, this key induction wherein the progeny of dorsal lip cells induce the dorsal axis and the neural tube is traditionally called **primary embryonic induction.**[*]

Mechanisms of Axis Determination in Amphibians

The experiments of Spemann and Mangold showed that the dorsal lip of the blastopore, and the dorsal mesoderm and pharyngeal endoderm that form from it, constituted an "organizer" that could instruct the formation of new embryonic axes. But the mechanisms by which the organizer was constructed and through which it operated remained a mystery. Indeed, it is said that Spemann and Mangold's paper posed more questions than it answered. Among these questions were:

- How did the organizer get its properties? What caused the dorsal blastopore lip to differ from any other region of the embryo?
- What factors were being secreted from the organizer to cause the formation of the neural tube and to create the anterior-posterior, dorsal-ventral, and left-right axes?
- How did the different parts of the neural tube become established, with the most anterior becoming the sensory organs and forebrain, and the most posterior becoming spinal cord?

We will take up each of these questions in turn.

The origin of the Nieuwkoop center

The major clue in determining how the dorsal blastopore lip obtained its properties came from the experiments of Pieter Nieuwkoop (1969, 1973, 1977) and Osamu Nakamura. Naka-

mura and Takasaki (1970) showed that the mesoderm arises from the marginal (equatorial) cells at the border between the animal and vegetal poles. He and his colleagues in the Netherlands demonstrated that the properties of this newly formed mesoderm were induced by the vegetal (presumptive endoderm) cells underlying them. He removed the equatorial cells (i.e., presumptive mesoderm) from a blastula and showed that neither the animal cap (presumptive ectoderm) nor the vegetal cap (presumptive endoderm) produced any mesodermal tissue. However, when the two caps were recombined, the animal cap cells were induced to form mesodermal structures such as notochord, muscles, kidney cells, and blood cells (Figure 10.22).

(A) Dissected blastula fragments give rise to different tissue in culture:

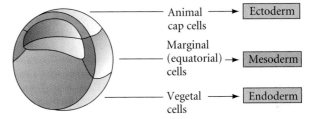

(B) Animal and vegetal fragments give mesoderm

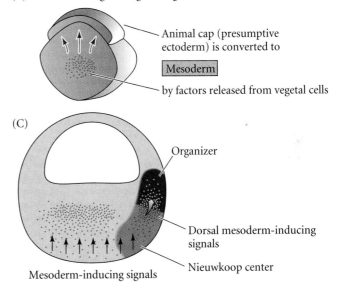

Figure 10.22
Summary of experiments by Nieuwkoop and by Nakamura and Takasaki, showing mesodermal induction by vegetal endoderm. (A) Isolated animal cap cells become a mass of ciliated epidermis, isolated vegetal cells generate gutlike tissue, and isolated equatorial (marginal zone) cells become mesoderm. (B) If animal cap cells are combined with vegetal cap cells, many of the animal cells generate mesodermal tissue. (C) Simplified model for mesoderm induction in *Xenopus*. A ventral signal (probably a complex set of signals from activin-like TGF-β factors and FGFs) is released throughout the vegetal region of the embryo. This signal induces the marginal cells to become mesoderm. On the dorsal side (away from the point of sperm entry), a signal is released by the vegetal cells of the Nieuwkoop center. This dorsal signal induces the formation of the Spemann organizer in the overlying marginal zone cells. The possible identity of this signal will be discussed later in this chapter. (C after De Robertis et al. 1992.)

---

[*]This classical term has been a source of confusion because the induction of the neural tube by the notochord is no longer considered the first inductive process in the embryo. We will soon discuss inductive events that precede this "primary" induction.

The polarity of this induction (whether the animal cells formed dorsal mesoderm or ventral mesoderm) depended on the dorsal-ventral polarity of the endodermal (vegetal) fragment. While the ventral and lateral vegetal cells (those closer to the side of sperm entry) induced ventral (mesenchyme, blood) and intermediate (muscle, kidney) mesoderm, the dorsalmost vegetal cells specified dorsal mesoderm components (somites, notochord), including those having the properties of the organizer. These dorsalmost vegetal cells of the blastula, which are capable of inducing the organizer, have been called the **Nieuwkoop center** (Gerhart et al. 1989).

The Nieuwkoop center was demonstrated in the 32-cell *Xenopus* embryo by transplantation and recombination experiments. First, Gimlich and Gerhart (Gimlich and Gerhart 1984; Gimlich 1985, 1986) performed an experiment analogous to the Spemann and Mangold studies, except that they used blastulae rather than gastrulae. When they transplanted the dorsalmost vegetal blastomere from one blastula into the ventral vegetal side of another blastula, two embryonic axes were formed (see Figure 10.11B). Second, Dale and Slack (1987) recombined single vegetal blastomeres from a 32-cell *Xenopus* embryo with the uppermost animal tier of a fluorescently labeled embryo of the same stage. The dorsalmost vegetal cell, as expected, induced the animal pole cells to become dorsal mesoderm. The remaining vegetal cells usually induced the animal cells to produce either intermediate or ventral mesodermal tissues (Figure 10.23). Thus, dorsal vegetal cells can induce animal cells to become dorsal mesodermal tissue.

The Nieuwkoop center is created by the cytoplasmic rotation that occurs during fertilization (see Chapter 7). When this rotation is inhibited by UV light, the resulting embryo will not form dorsal-anterior structures such as the head or neural tube (Vincent and Gerhart 1987). However, these UV-treated embryos can be rescued by transplantation of the dorsalmost vegetal blastomeres from a normal embryo at the 32-cell stage (Dale and Slack 1987; see Figure 10.11A). If eggs are rotated toward the end of the first cell cycle so that the future ventral side is upward, two Nieuwkoop centers are formed, leading to two dorsal blastopore lips and two embryonic axes (see Figure 10.10). Therefore, the specification of the dorsal-ventral axis begins at the moment of sperm entry.

The molecular biology of the Nieuwkoop center

In *Xenopus*, the endoderm is able to induce the formation of mesoderm by causing the presumptive mesodermal cells to express the *Xenopus Brachyury* (*Xbra*) gene. The mechanism of this induction involves the activation of the *Xbra* gene by one or more of the activin-like factors (such as the Vg1 and Nodal-related proteins) that are secreted by endodermal cells (Latinkic et al. 1997; Smith 2001). The Nodal-related protein **Derrière** is a particularly promising candidate for a mesoderm inducer, since it is induced by the VegT transcription factor that is found in the endoderm, and it can induce animal cap cells to become mesoderm over the long-range distances predicted by Nieuwkoop's experiments (see Figure 10.22; White et al. 2002). As we saw in Chapter 3, the *Xbra* gene is transcribed when a particular concentration of such proteins is present. Once expressed, the Xbra protein acts as a transcription factor to induce eFGF, a paracrine factor that maintains *Xbra* expression. This keeps the *Xbra* gene active long after the signals that initiated its transcription have ceased. The Xbra protein also activates the genes that produce mesoderm-specific proteins.

While all the vegetal cells appear to be able to induce the overlying marginal cells to become mesoderm (and express *Xbra*), only the dorsalmost vegetal cells can instruct the overlying marginal cells to become the organizer. The major candidate for the factor that forms the Nieuwkoop center in these dorsalmost

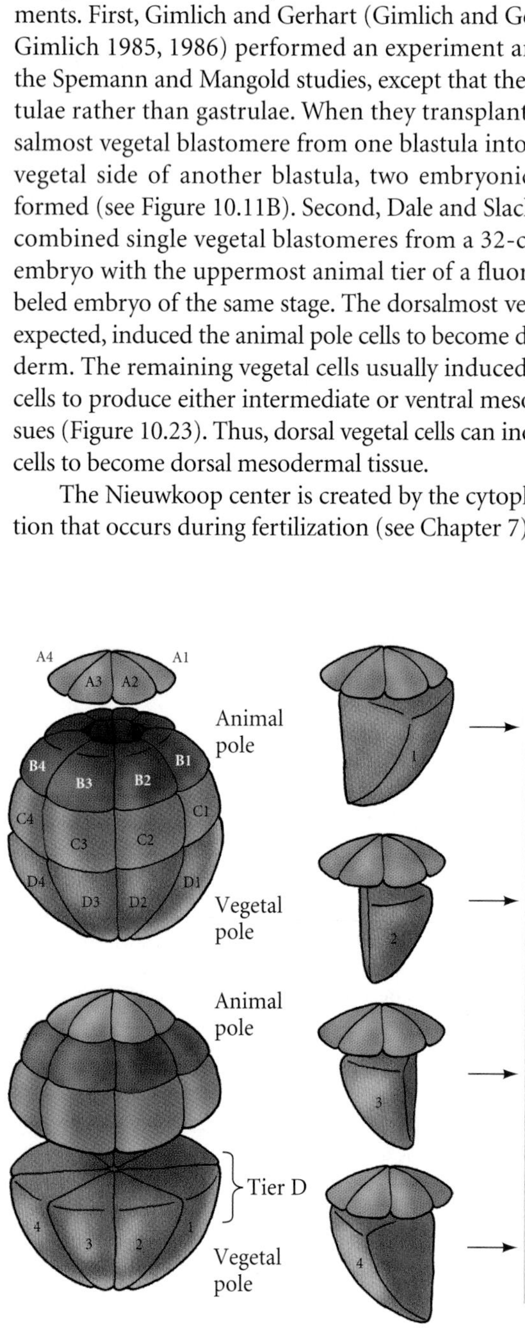

	Percentage of total inductions		
	Dorsal	Intermediate	Ventral
1	77	23	0
2	11	61	28
3	5	45	50
4	16	42	42

Figure 10.23
The regional specificity of mesoderm induction can be demonstrated by recombining blastomeres of 32-cell *Xenopus* embryos. Animal pole cells were labeled with fluorescent polymers so that their descendants could be identified, then combined with individual vegetal blastomeres. The inductions resulting from these recombinations are summarized at the right. D1, the dorsalmost vegetal blastomere, was the most likely to induce the animal pole cells to form dorsal mesoderm. (After Dale and Slack 1987.)

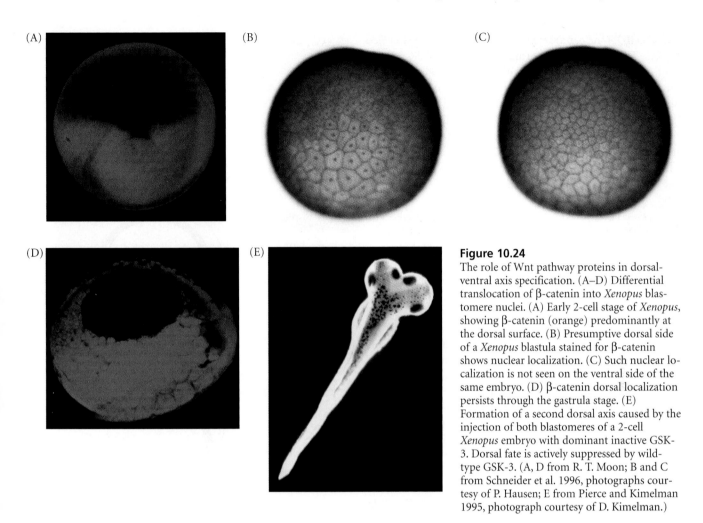

(A) (B) (C)

(D) (E)

Figure 10.24

The role of Wnt pathway proteins in dorsal-ventral axis specification. (A–D) Differential translocation of β-catenin into *Xenopus* blastomere nuclei. (A) Early 2-cell stage of *Xenopus*, showing β-catenin (orange) predominantly at the dorsal surface. (B) Presumptive dorsal side of a *Xenopus* blastula stained for β-catenin shows nuclear localization. (C) Such nuclear localization is not seen on the ventral side of the same embryo. (D) β-catenin dorsal localization persists through the gastrula stage. (E) Formation of a second dorsal axis caused by the injection of both blastomeres of a 2-cell *Xenopus* embryo with dominant inactive GSK-3. Dorsal fate is actively suppressed by wild-type GSK-3. (A, D from R. T. Moon; B and C from Schneider et al. 1996, photographs courtesy of P. Hausen; E from Pierce and Kimelman 1995, photograph courtesy of D. Kimelman.)

vegetal cells is **β-catenin.** β-catenin is a multifunctional protein that can act as an anchor for cell membrane cadherins (see Chapter 3) or as a nuclear transcription factor (see Chapter 6). In *Xenopus* embryos, β-catenin begins to accumulate in the dorsal region of the egg during the cytoplasmic movements of fertilization. β-catenin continues to accumulate preferentially at the dorsal side throughout early cleavage, and this accumulation is seen in the nuclei of the dorsal cells (Figure 10.24A–D; Schneider et al. 1996; Larabell et al. 1997). This region of β-catenin accumulation originally appears to cover both the Nieuwkoop center and organizer regions. During later cleavage, the cells containing high levels of β-catenin may reside specifically in the Nieuwkoop center (Heasman et al. 1994a; Guger and Gumbiner 1995).

WEBSITE 10.5 Mesoderm induction. There are numerous theories concerning how the generic mesoderm is induced by the endoderm. Evidence points to three molecules as possible mesoderm inducers: bFGF, Vg1, and an activin-like protein. These proteins can activate *Xbra* as well as other mesodermal proteins.

β-catenin is a key player in the formation of the dorsal axis, and experimental depletion of β-catenin transcripts with

antisense oligonucleotides results in the lack of dorsal structures (Heasman et al. 1994a). Moreover, the injection of exogenous β-catenin into the ventral side of an embryo produces a secondary axis (Funayama et al. 1995; Guger and Gumbiner 1995). β-catenin is part of the Wnt signal transduction pathway and is negatively regulated by glycogen synthase kinase 3 (GSK-3; see Chapter 6). GSK-3 also plays a critical role in axis formation by suppressing dorsal fates. Activated GSK-3 blocks axis formation when added to the egg (Pierce and Kimelman 1995; He et al. 1995; Yost et al. 1996). If endogenous GSK-3 is knocked out by a dominant negative protein in the ventral cells of the early embryo, a second axis forms (Figure 10.24E).

So how can β-catenin become localized to the future dorsal cells of the blastula? Labeling experiments (Yost et al. 1996; Larabell et al. 1997) suggest that β-catenin is initially synthesized (from maternal messages) throughout the embryo, but is degraded by GSK-3-mediated phosphorylation specifically in the ventral cells. The critical event for axis determination may be the movement of an inhibitor of GSK-3 to the cytoplasm opposite the point of sperm entry (i.e., to the future dorsal cells). One candidate for this agent is the Disheveled protein. This protein is the normal suppressor of GSK-3 in the Wnt pathway (see Figure 6.24), and it is originally found in the vegetal cortex of the unfertilized *Xenopus* egg. However, upon fertilization, Disheveled is translocated along the microtubular array to the dorsal side of the embryo (Figure 10.25; Miller et al. 1999). Thus, on the dorsal side of the embryo, β-catenin should be stable, since GSK-3 is not able to degrade it; while in the ventral portion of the embryo, GSK-3 should initiate the degradation of β-catenin.

> **WEBSITE 10.6 GBP.** In addition to Disheveled, a second inhibitor of GSK-3 has been identified in *Xenopus* eggs. This protein, GBP, can rescue axis formation in UV-treated eggs.

β-catenin is a transcription factor that can associate with other transcription factors to give them new properties. It is known that *Xenopus* β-catenin can combine with a ubiquitous transcription factor known as **Tcf3**, and that a mutant form of Tcf3 lacking the β-catenin binding domain results in embryos without dorsal axes (Molenaar et al. 1996). The β-catenin/Tcf3 complex appears to bind to the promoters of several genes whose activity is critical for axis formation. One of these genes is *siamois.* This gene encodes a homeodomain transcription factor, and it is expressed in the Nieuwkoop center immediately following the mid-blastula transition. If this gene is ectopically expressed in the ventral vegetal cells, a secondary axis emerges on the former ventral side of the embryo, and if cortical rotation is prevented, *siamois* expression is eliminated (Lemaire et al. 1995; Brannon and Kimelman 1996). The Tcf3 protein is thought to inhibit *siamois* transcription when it binds to that gene's promoters in the absence of β-catenin. However, when the Tcf3/β-catenin complex binds to its promoter, *siamois* is activated (Figure 10.26; Brannon et al. 1997).

The Siamois protein is critical for the expression of organizer-specific genes (Fan and Sokol 1997; Kessler 1997). Siamois protein binds to and activates the promoters of several genes involved in organizer function. These include genes encoding the transcription factors Goosecoid and Xlim1, and the paracrine factors Cerberus and Frzb (Laurent et al. 1997; Engleka and Kessler 2001). Goosecoid appears to be essential for activating numerous genes in the organizer. So one could

Figure 10.25
Model of the mechanism by which the Disheveled protein stabilizes β-catenin in the dorsal portion of the amphibian egg. (A) Disheveled (Dsh) associates with a particular set of proteins at the vegetal pole of the unfertilized egg. (B) Upon fertilization, these protein vesicles are translocated dorsally along subcortical microtubule tracks. (C) Disheveled is then released from its vesicles and is distributed in the future dorsal third of the 1-cell embryo. (D) Disheveled binds to and blocks the action of GSK-3, thereby preventing the degradation of β-catenin on the dorsal side of the embryo. (E) The nuclei of the blastomeres in the dorsal region of the embryo receive β-catenin, while the nuclei of those in the ventral region do not.

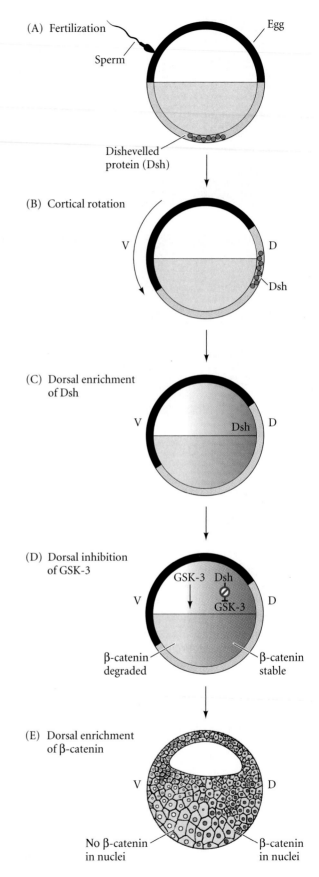

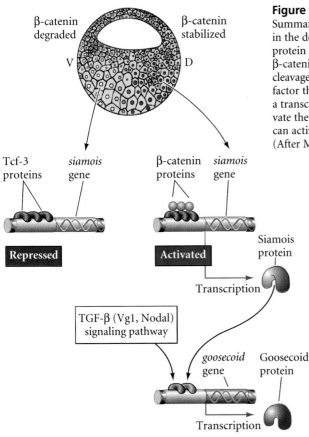

Figure 10.26
Summary of events hypothesized to bring about the induction of the organizer in the dorsal mesoderm. Cortical rotation causes the translocation of Disheveled protein to the dorsal side of the embryo. Dsh binds GSK-3, thereby allowing β-catenin to accumulate in the future dorsal portion of the embryo. During cleavage, β-catenin enters the nuclei and binds with Tcf3 to form a transcription factor that activates genes encoding proteins such as Siamois. Siamois and Xlim1, a transcription factor activated by the TGF-β pathway, function together to activate the *goosecoid* gene in the organizer. Goosecoid is a transcription factor that can activate genes whose proteins are responsible for the organizer's activities. (After Moon and Kimelman 1998).

the Nodal-related factors. When they repeated the Nieuwkoop animal-vegetal recombination experiments (see Figure 10.22) but included a specific inhibitor of Nodal-related proteins, the induction by the vegetal cells failed to occur. (The inhibitor did not inhibit Vg1, VegT, or activin.) Moreover, they found that during the late blastula stages, several Nodal-related proteins (including Xnr1, Xnr2, and Xnr4) are expressed in a dorsal-to-ventral gradient in the endoderm. This gradient is formed by the activation of *Xenopus* Nodal-related gene expression by the synergistic action of VegT and Vg1 with β-catenin. Agius and his colleagues present a model, shown in Figure 10.27, in which the dorsally located β-catenin and the vegetally located Vg1 signals interact to create a gradient of Nodal-related proteins (Xnr1, 2, 4) across the endoderm. These Nodal-related proteins specify the mesoderm such that those regions with little Nodal-related protein become ventral mesoderm, those regions with some Nodal-related protein become lateral mesoderm, and those regions with a great deal of Nodal-related protein become the organizer.

The Functions of the Organizer

While the Nieuwkoop center cells remain endodermal, the cells of the organizer become the dorsal mesoderm and migrate underneath the dorsal ectoderm. There, the dorsal mesoderm induces the central nervous system to form. The properties of the organizer tissue can be divided into four major functions:

1. The ability to self-differentiate dorsal mesoderm (prechordal plate, chordamesoderm, etc.)
2. The ability to dorsalize the surrounding mesoderm into paraxial (somite-forming) mesoderm (when it would otherwise form ventral mesoderm)
3. The ability to dorsalize the ectoderm, inducing the formation of the neural tube
4. The ability to initiate the movements of gastrulation

In *Xenopus* (and other vertebrates), the formation of the anterior-posterior axis follows the formation of the dorsal-ventral axis. Once the dorsal portion of the embryo is established, the movement of the involuting mesoderm establishes the anterior-posterior axis. The mesoderm that migrates first through the dorsal blastopore lip gives rise to the anterior

expect that if the dorsal side of the embryo contained β-catenin, that β-catenin would allow this region to express Siamois, and that Siamois would initiate the formation of the organizer. However, Siamois alone is not sufficient for generating the organizer; another protein also appears to be critical in the activation of *goosecoid* and the formation of the organizer. Recent studies suggest that maximum *goosecoid* expression occurs when there is synergism between the Siamois protein and a vegetally expressed TGF-β signal (see Chapter 6; Brannon and Kimelman 1996; Engleka and Kessler 2001). While the cortical rotation may activate the β-catenins and allow the expression of *siamois* in the dorsal region of the embryo, the translation of vegetally localized messages encoding paracrine factors of the TGF-β family may generate a protein that stimulates the greatest activation of *goosecoid* in the cells that will become the organizer. A TGF-β family protein in the Nieuwkoop center could induce the cells in the dorsal marginal zone above them to express some transcription factors that would also bind to the promoter of the *goosecoid* gene and cooperate with Siamois to activate it (see Figure 10.26).

The candidates for this TGF-β factor include the Vg1 and *Xenopus* Nodal-related (Xnr) proteins. Each of these proteins is made in the endoderm (see Figure 5.34). Agius and colleagues (2000) have provided evidence that all of these proteins may act in a pathway, and that the critical proteins are

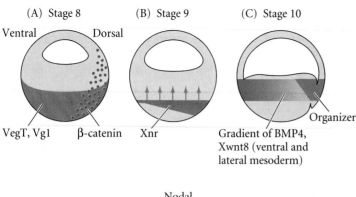

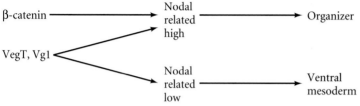

Figure 10.27

Model for mesoderm induction and organizer formation by the interaction of β-catenin and TGF-β proteins. (A) At late blastula stages, Vg1 and VegT are found in the vegetal hemisphere, while β-catenin is located in the dorsal region. (B) β-catenin acts synergistically with Veg1 and VegT to activate the *Xenopus* Nodal-related (*Xnr*) genes. This creates a gradient of Xnr proteins across the endoderm, highest in the dorsal region. (C) The mesoderm is specified by the gradient of Xnr proteins. Mesodermal regions with little or no Xnr proteins have high levels of BMP4 and Xwnt8; they become ventral mesoderm. Those having intermediate concentrations of Xnrs become lateral mesoderm. Where there is a high concentration of Xnrs, the *goosecoid* gene and other dorsal mesodermal genes are activated, and the mesodermal tissue becomes the organizer. (These results may explain the activity concentration experiments mentioned in Chapter 3.) (After Agius et al. 2000.)

structures; the mesoderm migrating through the lateral and ventral lips forms the posterior structures.

It is now thought that the cells of the organizer ultimately contribute to four cell types—pharyngeal endoderm, head mesoderm (prechordal plate), dorsal mesoderm (primarily the notochord), and the dorsal blastopore lip (Keller 1976; Gont et al. 1993). The pharyngeal endoderm and prechordal plate lead the migration of the organizer tissue and appear to induce the forebrain and midbrain. The dorsal mesoderm induces the hindbrain and trunk. The dorsal blastopore lip forms the dorsal mesoderm and eventually becomes the chordaneural hinge that induces the tip of the tail.

The first description of the organizer started one of the first truly international scientific research programs: the search for the organizer molecules. Researchers from Britain, Germany, France, the United States, Belgium, Finland, Japan, and the Soviet Union all tried to find these remarkable substances (see Gilbert and Saxén 1993). R. G. Harrison (quoted by Twitty 1966, p. 39) referred to the amphibian gastrula as the "new Yukon to which eager miners were now rushing to dig for gold around the blastopore." Unfortunately, their picks and shovels proved too blunt to uncover the molecules involved. The proteins responsible for induction were present in concentrations too small for biochemical analyses, and the large quantity of yolk and lipids in the amphibian egg further interfered with protein purification (Grunz 1997). The analysis of organizer molecules had to wait until recombinant DNA technologies enabled investigators to make cDNA clones from blastopore lip mRNA and to see which of these clones encoded factors that could dorsalize the embryo.

WEBSITE 10.7 **Early attempts to locate the organizer molecules.** While Spemann did not believe that molecules alone could organize the embryo, his students began a long quest for these factors.

The formation of the dorsal (organizer) mesoderm involves the activation of several genes. The secreted proteins of the Nieuwkoop center are thought to activate a set of transcription factors in the mesodermal cells above them. These transcription factors then activate the genes encoding the secreted products of the organizer. Several organizer-specific transcription factors have been found and are listed in Table 10.2.

As mentioned earlier, one of the important targets of the Nieuwkoop center appears to be the *goosecoid* gene. The area of expression of *goosecoid* mRNA correlates with the organizer domain in both normal and experimentally treated animals. When lithium chloride treatment is used to increase the organizer mesoderm throughout the marginal zone, the expression of *goosecoid* likewise is expanded. Conversely, when eggs are treated with UV light prior to first cleavage, both dorsal-anterior induction and *goosecoid* expression are significantly inhibited. Injection of the full-length *goosecoid* message into

TABLE 10.2 Proteins expressed solely or almost exclusively in the organizer (partial list)

Nuclear proteins	Secreted proteins
Xlim1	Chordin
Xnot	Dickkopf
Otx2	ADMP
XFD1	Frzb
XANF1	Noggin
Goosecoid	Follistatin
HNF3β	Sonic hedgehog
	Cerberus
	Nodal-related proteins (several)

(A)   (B)   (C)   (D)

Figure 10.28
Ability of *goosecoid* mRNA to induce a new axis. (A) At the gastrula stage, a control embryo (either uninjected or given an injection of *goosecoid*-like mRNA but lacking the homeobox) has one dorsal blastopore lip (arrow). (B) An embryo whose ventral vegetal blastomeres were injected at the 16-cell stage with *goosecoid* message. Note the secondary dorsal lip. (C) The top two embryos, which were injected with *goosecoid* mRNA, show two dorsal axes; the bottom two control embryos do not. (D) Twinned embryo produced by *goosecoid* injection. Two complete sets of head structures have been induced. (After Cho et al. 1991a; Niehrs et al. 1993; photographs courtesy of E. De Robertis.)

the two ventral blastomeres of a 4-cell *Xenopus* embryo causes the progeny of those blastomeres to involute, undergo convergent extension, and form the dorsal mesoderm and head endoderm of a secondary axis (Figure 10.28; Niehrs et al. 1993). Labeling experiments (Niehrs et al. 1993) have shown that such *goosecoid*-injected cells are able to recruit neighboring host cells into the dorsal axis as well. Thus, the Nieuwkoop center activates the *goosecoid* gene in the organizer tissues, and this gene encodes a DNA-binding protein that (1) activates the migration properties (involution and convergent extension) of the dorsal blastopore lip cells, (2) autonomously determines the dorsal mesodermal fates of those cells expressing it, and (3) enables the *goosecoid*-expressing cells to recruit neighboring cells into the dorsal axis. Goosecoid also has been found to activate *Otx2*, a gene that is critical for brain formation, in the anterior mesoderm and in the presumptive brain ectoderm (Blitz and Cho 1995).

The diffusible proteins of the organizer I: The BMP inhibitors

Goosecoid protein works in the nucleus. It must activate (either directly or indirectly) those genes encoding the soluble proteins that function to organize the dorsal-ventral and anterior-posterior axis. Early evidence for such diffusible signals from the notochord came from several sources. First, Hans Holtfreter (1933) showed that if amphibian embryos are placed in a high salt solution, the mesoderm will evaginate rather than invaginate, and will not underlie the ectoderm. In this case, the ectoderm is not underlain by the notochord, and

it does not form neural structures. Further evidence for soluble factors came from the transfilter studies of Finnish investigators (Saxén 1961; Toivonen et al. 1975; Toivonen and Wartiovaara 1976). Newt dorsal lip tissue was placed on one side of a filter fine enough so that no processes could fit through the pores, and competent gastrula ectoderm was placed on the other side. After several hours, neural structures were observed in the ectodermal tissue (Figure 10.29). The identities of the factors diffusing from the organizer, however, took another quarter of a century to identify.

Recent studies on induction have resulted in a remarkable and non-obvious conclusion: The ectoderm is actually induced to become epidermal. The agents of this induction are bone morphogenetic proteins (BMPs). The nervous system forms from that region of the ectoderm that is protected from this epidermal induction (Hemmati-Brivanlou and Melton 1997). In other words, (1) the "default fate" of the ectoderm is to become neural tissue; (2) certain parts of the embryo induce the ectoderm to become epidermal tissue, and (3) the organizer tissue acts by secreting molecules that block this induction, thereby allowing the ectoderm "protected" by these factors to become neural.

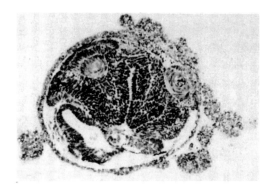

Figure 10.29
Neural structures induced in presumptive ectoderm by newt dorsal lip tissue, separated from the ectoderm by a Nucleopore filter with an average pore diameter of 0.05 mm. Anterior neural tissues are evident, including some induced eyes. (From Toivonen 1979; photograph courtesy of L. Saxén.)

Figure 10.30
Rescue of dorsal structures by Noggin protein. When *Xenopus* eggs are exposed to ultraviolet radiation, cortical rotation fails to occur, and the embryos lack dorsal structures (top). If such an embryo is injected with *noggin* mRNA, it develops dorsal structures in a dosage-related fashion (top to bottom). If too much *noggin* message is injected, the embryo produces dorsal anterior tissue at the expense of ventral and posterior tissue, becoming little more than a head (bottom). (Photograph courtesy of R. M. Harland.)

NOGGIN. In 1992, the first of the soluble organizer molecules was isolated. Smith and Harland (1992) constructed a cDNA plasmid library from dorsalized (lithium chloride-treated) gastrulae. Messenger RNAs synthesized from sets of these plasmids were injected into ventralized embryos (having no neural tube) produced by irradiating early embryos with ultraviolet light. Those sets of plasmids whose mRNAs rescued the dorsal axis in these embryos were split into smaller sets, and so on, until single-plasmid clones were isolated whose mRNAs were able to restore the dorsal axis in such embryos. One of these clones contained the ***noggin*** gene (Figure 10.30). Smith and Harland showed that newly transcribed *noggin* mRNA is first localized in the dorsal blastopore lip region and then becomes expressed in the notochord (Figure 10.31). Injection of *noggin* mRNA into 1-cell, UV-irradiated embryos completely rescued the dorsal axis and allowed the formation of a complete embryo. Noggin is a secreted protein, and in 1993, Smith and his colleagues found that Noggin could accomplish two of the major functions of the organizer: it induced dorsal ectoderm to form neural tissue, and it dorsalized mesoderm cells that would otherwise contribute to the ventral mesoderm. Noggin binds to BMP4 and BMP2 and inhibits their binding to receptors (Zimmerman et al. 1996).

CHORDIN AND NODAL-RELATED 3. The second organizer protein found was **chordin**. It was isolated from clones of cDNA whose mRNAs were present in dorsalized, but not in ventralized, embryos (Sasai et al. 1994). These clones were tested by injecting them into ventral blastomeres and seeing whether they induced secondary axes. One of the clones capable of inducing a secondary neural tube contained the *chordin* gene. The *chordin* mRNA was found to be localized in the dorsal blastopore lip and later in the dorsal mesoderm of the notochord (Figure 10.32), and morpholino antisense oligomers (see Chapter 4) directed

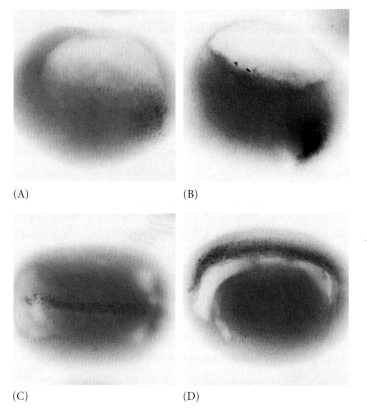

(A)

(B)

(C)

(D)

Figure 10.31
Localization of *noggin* mRNA in the organizer tissue, shown by in situ hybridization. (A) At gastrulation, *noggin* message (dark areas) accumulates in the dorsal marginal zone. (B) When cells involute, *noggin* mRNA is seen in the dorsal blastopore lip. (C) During convergent extension, *noggin* message is expressed in the precursors of the notochord, prechordal plate, and pharyngeal endoderm, which extend (D) beneath the ectoderm in the center of the embryo. (Photographs courtesy of R. M. Harland.)

(A) (B) (C)

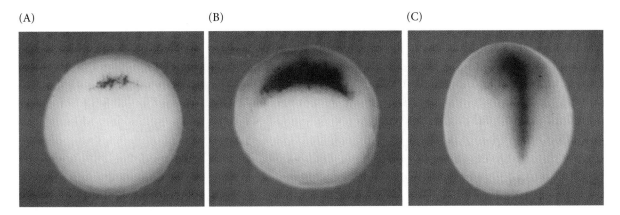

Figure 10.32
Localization of *chordin* mRNA. (A) Whole-mount in situ hybridization shows that just prior to gastrulation, *chordin* message (dark area) is expressed in the region that will become the dorsal blastopore lip. (B) As gastrulation begins, *chordin* is expressed at the dorsal blastopore lip. (C) In later stages of gastrulation, *chordin* message is seen in the organizer tissues. (From Sasai et al. 1994; photographs courtesy of E. De Robertis.)

against the *chordin* message completely blocked the induction of the *Xenopus* central nervous system (Oelgeschläger et al. 2003). Like Noggin, chordin binds directly to BMP4 and BMP2 and prevents their complexing with their receptors (Piccolo et al. 1996). In zebrafish, a loss-of-function mutation of *chordin* (the *chordino* mutant) has a greatly reduced neural plate and an enlarged region of ventral mesoderm (Hammerschmidt et al. 1996). **Nodal-related protein 3** (Xnr-3) is synthesized by the superficial cells of the organizer and is also able to block BMP4 (Smith et al. 1995; Hansen et al. 1997).

FOLLISTATIN AND ADMP. The fourth organizer-secreted protein, **follistatin**, was found in the organizer through an unexpected result of an experiment that was looking for something else. Ali Hemmati-Brivanlou and Douglas Melton (1992, 1994) wanted to see whether the protein activin was crucial for mesoderm induction, so they constructed a dominant negative activin receptor and injected it into *Xenopus* embryos. Remarkably, the ectoderm of these embryos began to express neural-specific proteins. It appeared that the activin receptor (which also binds other structurally similar molecules such as the bone morphogenetic proteins) normally functioned to bind an inhibitor of neurulation. When its function was blocked, all the ectoderm became neural. In 1994, Hemmati-Brivanlou and Melton proposed a "default model of neurulation" whereby the organizer functioned by producing inhibitors of whatever was inducing epidermis formation. That is to say, the "normal" fate of an ectodermal cell was to become a neuron; it had to be induced to become an epidermal skin cell. The organizer somehow prevented the ectodermal cells from being induced. This model was supported by, and explained, some cell dissociation experi-

ments that had also produced odd results. Three studies, by Grunz and Tacke (1989), Sato and Sargent (1989), and Godsave and Slack (1989) had shown that when whole embryos or their animal caps were dissociated, they formed neural tissue. This result would be explainable if the "default state" of the ectoderm was not epidermal, but neural, and the tissue had to be induced to have an epidermal phenotype. The organizer, then, would block this epidermalizing induction.

Since the naturally occurring protein follistatin binds to and inhibits activin (and other related proteins), it was hypothesized that it might be one of the factors secreted by the organizer. Using in situ hybridization, Hemmati-Brivanlou and Melton (1994) found the mRNA for follistatin in the dorsal blastopore lip and notochord. So it appeared that there might be a neural default state and an actively induced epidermal fate. This hypothesis was counter to the neural induction model that had preceded it for 70 years. But what proteins were inducing the epidermis, and were they really being blocked by the molecules secreted by the organizer?

In *Xenopus*, the epidermal inducer is **bone morphogenesis protein 4** (**BMP4**). It was known that there is an antagonistic relationship between BMP4 and the organizer. If the mRNA for BMP4 is injected into *Xenopus* eggs, all the mesoderm in the embryo becomes ventrolateral mesoderm, and no involution occurs at the blastopore lip (Dale et al. 1992; Jones et al. 1992). Conversely, overexpression of a dominant negative BMP4 receptor resulted in the formation of two dorsal axes (Graff et al. 1994; Suzuki et al. 1994). In 1995, Wilson and Hemmati-Brivanlou demonstrated that BMP4 induced ectodermal cells to become epidermal. By 1996, several laboratories had demonstrated that Noggin, chordin, and follistatin each was secreted by the organizer and that each prevented BMP from binding to the ectoderm and mesoderm near the organizer (Piccolo et al. 1996; Zimmerman et al. 1996; Iemura et al. 1998).

BMP4 is initially expressed throughout the ectodermal and mesodermal regions of the late blastula.* However, during gastrulation, *bmp4* transcripts are restricted to the ventro-

---

*The details of BMP4 production and degradation are discussed in Chapter 23.

lateral marginal zone. This is probably because the Xiro1 protein is made in the dorsal mesoderm (organizer) region starting at the beginning of gastrulation, and this transcription factor represses *bmp4* transcription (Hemmati-Brivanlou and Thomsen 1995; Northrop et al. 1995; Glavic et al. 2001).

When BMP4 binds to cells, it activates the expression of genes encoding transcription factors such as *Xvent1, Xmsx1, Vox,* and *Xom,* which induce the expression of epidermal and ventral mesoderm differentiation. At the same time, these transcription factors suppress those genes that would produce a neural phenotype (Suzuki et al. 1997b; Yamamoto et al. 2000). Low doses of BMP4 appear to activate muscle formation; intermediate levels instruct cells to become kidney, while high doses activate those genes that instruct the mesoderm to become blood cells (Hemmati-Brivanlou and Thomsen 1995; Gawantka et al. 1995; Dosch et al. 1997). The varying doses are created by the interaction of BMP4 (coming from the ventral and lateral mesoderm) with the BMP antagonists coming from the organizer* (Figure 10.33).

Thus, by 1996, there was a consensus that BMP4 was the active inducer of ventral ectoderm (epidermis) and the ventralizer of the mesoderm (blood cells and connective tissue), and that Noggin, chordin, and follistatin could prevent its function. The organizer worked by secreting inhibitors of BMP4, not by directly inducing neurons.

The diffusible proteins of the organizer II: The Wnt inhibitors

It had been thought that all the neural tissue induced by the organizer was induced to become forebrain, and that the notochord represented the most anterior portion of the organizer itself. However, the most anterior regions of the head and brain are underlain not by notochord, but by pharyngeal endoderm and head (prechordal) mesoderm. This "endome-

---

*It is possible that BMP4 is produced through the actions of a "noncanonical Wnt pathway" that activates the NF-AT transcription factor (Saneyoshi et al. 2002). In this way, the actions of different Wnt proteins have opposite functions in the *Xenopus* embryo.

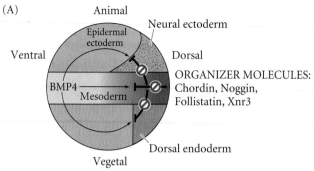

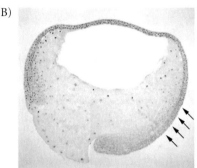

Figure 10.33
Model for the action of the organizer. (A) BMP4 (and certain other molecules) is a powerful ventralizing factor. Organizer proteins such as chordin, Noggin, and follistatin block the action of BMP4. The antagonistic effects of these proteins can be seen in all three germ layers. BMP4 may elicit the expression of different genes in a concentration-dependent fashion. Thus, in the regions of Noggin and chordin expression, BMP4 is totally prevented from binding, and these tissues become notochord (organizer) tissue. Slightly farther away from the organizer, *myf5*, a marker for the dorsolateral muscles, is activated. As more and more BMP4 molecules are allowed to bind to the cells, *Xvent2* (ventrolateral) and *Xvent1* (ventral) genes become expressed. (B) BMP pathway activation visualized in an early gastrulating *Xenopus* embryo. Antibodies specific to the phosphorylated form of Smad1 (i.e., to Smad1 that has been activated by BMP4) can be seen binding to the ectodermal cells in the ventral portion of the embryo. Close to the dorsal lip there is a gradient of BMP4 inactivation (arrows). (After Dosch et al. 1997; De Robertis et al. 2000; B from Kurata et al. 2000, photograph courtesy of N. Ueno.)

Sidelights & Speculations

BMP4 and Geoffroy's Lobster

The hypothesis that the organizer secretes proteins that block BMP4 received further credence from an unexpected source—the emerging field of evolutionary developmental biology. De Robertis, Kimelman, and others (Holley et al. 1995; Schmidt et al. 1995; De Robertis and Sasai 1996) found that the same chordin-BMP4 interaction that instructed the formation of the neural tube in vertebrates also formed neural tissue in fruit flies. The dorsal neural tube of the vertebrate and the ventral neural cord of the fly appear to be generated by the same set of instructions, conserved throughout evolution. This was the second paradigm shift occasioned by the newly acquired information on the molecular biology of induction.

The *Drosophila* homologue of the *bmp4* gene is *decapentaplegic* (*dpp*). As discussed in the previous chapter, the Dpp protein is responsible for the patterning of the dorsal-ventral axis of *Drosophila*, and it is present in the dorsal portion of the embryo and diffuses ventrally. It is opposed by a protein called Short-gastrulation (Sog), which is the *Drosophila* homologue of chordin. These homologues not only appear to be similar, they can actually substitute for each other. When *sog* mRNA is injected into ventral regions of *Xenopus* embryos, it induces the *Xenopus* notochord and neural tube. Injection of chordin mRNA into *Drosophila* embryos produces ventral nervous tissue. Although chordin usually dorsalizes the *Xenopus* embryo, it ventralizes the *Drosophila* embryo. In *Drosophila*, Dpp is made dorsally; in *Xenopus*, BMP4 is made ventrally. In both cases, Sog/chordin makes neural tissue by blocking the effects of Dpp/BMP4. In *Drosophila*, Sog interacts with Tolloid and several other proteins that create a gradient of Sog proteins. In *Xenopus*, the homologue of the same proteins act to create a gradient of chordin (see Figure 23.14; Hawley et al. 1995; Holley et al. 1995; De Robertis et al. 2000).

In 1822, the French anatomist Étienne Geoffroy Saint-Hilaire provoked one of the most heated and critical confrontations in biology when he proposed that the lobster was but the vertebrate upside down. He claimed that the ventral side of the lobster (with its nerve cord) was homologous to the dorsal side of the vertebrate (Appel 1987). It seems that he was correct on the molecular level, if not on the anatomical. De Robertis and Sasai (1996) have proposed that there was a common ancestor for all the bilateral phyla—a hypothetical creature (dubbed Urbilateria) of some 600 million years ago that was the ancestor of both the protostome and the deuterostome subkingdoms. The BMP4(Dpp)/chordin(Sog) interaction is an example of "homologous processes," suggesting a unity of developmental principles among all animals (Gilbert and Bolker 2001).

---

soderm" constitutes the leading edge of the dorsal blastopore lip. Recent studies have shown that these cells not only induce the most anterior head structures, but that they do it by blocking the Wnt pathway as well as by blocking BMP4.

CERBERUS. In 1993, Christian and Moon showed that Xwnt8, a member of the Wnt family of growth and differentiation factors, inhibited neural induction. Xwnt8 was found to be synthesized throughout the marginal mesoderm—except in the region forming the dorsal lip. Thus, in addition to BMP4, there was a second anti-neuralizing protein being secreted from the non-organizer mesoderm. Were there any proteins in the organizer that countered this activity?

In 1996, Bouwmeester and colleagues showed that the induction of the most anterior head structures could be accomplished by a secreted protein called **Cerberus.*** Unlike the other proteins secreted by the organizer, Cerberus promotes the formation of the cement gland (the most anterior region of tadpole ectoderm), eyes, and olfactory (nasal) placodes. When *cerberus* mRNA was injected into a vegetal ventral *Xenopus* blastomere at the 32-cell stage, ectopic head structures were formed (Figure 10.34). These head structures were made from the injected cell as well as from neighboring cells. The *cerberus* gene is expressed in the pharyngeal endomesoderm cells that arise from the deep cells of the early dorsal lip, and the Cerberus protein can bind BMPs, Nodal-related proteins, and Xwnt8 (Figure 10.35; Glinka et al. 1997; Piccolo et al. 1999).

FRZB AND DICKKOPF. Shortly after the attributes of Cerberus were demonstrated, two other proteins, Frzb and Dickkopf,

---

*"Cerberus" is another name out of Greek mythology; the protein is named after the three-headed dog that guarded the entrance to Hades.

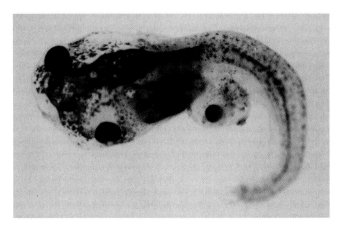

Figure 10.34
Cerberus mRNA injected into a single D4 (ventral vegetal) blastomere of a 32-cell *Xenopus* embryo induces head structures as well as a duplicated heart and liver. The secondary eye (a single cyclopic eye) and olfactory placode can be readily seen. (From Bouwmeester et al. 1996; photograph courtesy of E. M. De Robertis.)

were discovered to be synthesized in the involuting endomesoderm. **Frzb** (pronounced "frisbee") is a small, soluble form of Frizzled, the Wnt receptor, that is capable of binding Wnt proteins in solution (Figure 10.35, 10.36; Leyns et al. 1997; Wang et al. 1997). It is synthesized predominantly in the en-

Figure 10.35
Paracrine factors from the organizer are able to block certain other paracrine factors. The pharyngeal endoderm that underlies the head secretes Dickkopf and Cerberus. Dickkopf blocks Wnt proteins; Cerberus blocks Wnts, Nodal-related progeins, and BMPs. The prechordal plate secretes the Wnt blockers Dickkopf and Frzb, as well as BMP-blockers chordin and Noggin. The notochord contains BMP-blockers chordin, Noggin, and follistatin.

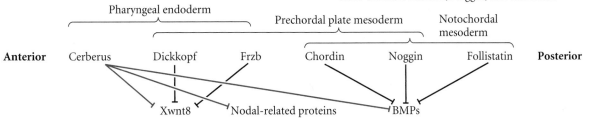

(A)

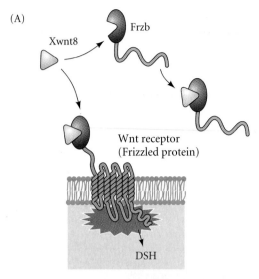

(B)

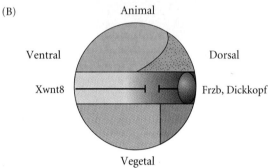

Figure 10.36
Xwnt8 is capable of ventralizing the mesoderm and preventing anterior head formation in the ectoderm. (A) Frzb is a protein secreted by the anterior region of the organizer. It binds to Xwnt8 before that inducer can bind to its receptor. Frzb resembles the Wnt-binding domain of the Wnt receptor (Frizzled protein), but is a soluble molecule. (B) Xwnt8 is made throughout the marginal zone.

domesoderm cells beneath the head (Figure 10.37A). If embryos are made to synthesize excess Frzb, Wnt signaling fails to occur, and the embryos lack ventral posterior structures, becoming solely head (Figure 10.37B). The **Dickkopf** (German; "big head," "stubborn") protein also appears to interact directly with the Wnt receptors, preventing Wnt signaling (Mao et al. 2001, 2002). Injection of antibodies against Dickkopf protein into the blastocoel causes the resulting embryos to have small, deformed heads with no forebrain (Glinka et al. 1998).

Glinka and colleagues (1997) have thus proposed a new model for embryonic induction. The induction of trunk structures may be caused by the blockade of BMP signaling from the notochord. However, to produce a head, both the BMP signal and the Wnt signal must be blocked. This blockade comes from the endomesoderm, now considered the most anterior portion of the organizer.

Conversion of the ectoderm into neural plate cells

So far, we have discussed the factors that prevent the dorsal ectoderm from becoming epidermis. Obviously, once that is accomplished, other genes must transform the ectoderm into neural tissue. The key protein involved in activating the neural phenotype in the ectoderm appears to be **neurogenin** (Ma et al. 1996). The transcription factors that appear in the ectoderm in the absence of BMP are able to induce the expression of neurogenin, and the transcription factors (such as Msx1) induced in the ectoderm by BMP signals are able to repress neurogenin expression (Figure 10.38; see Sasai 1998). Neurogenin is itself a transcription factor, and it activates a series of genes whose products are responsible for the neural phenotype. One of the genes activated by neurogenin is the gene for NeuroD, a transcription factor that activates the genes producing the structural neural-specific proteins (Lee et al. 1995). In addition, Noggin or Cerberus can induce another transcription factor in the ectoderm. This protein is called *Xenopus* brain factor 2 (Xbf2), and it appears to repress the epidermal genes (Mariani and Harland 1998). By these pushes and pulls, the dorsal ectoderm is converted into neural plate tissue.

(A)

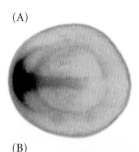

(B)

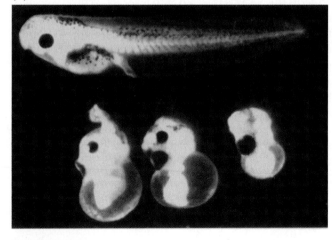

Figure 10.37
Frzb expression and function. (A) Double in situ hybridization localizing Frzb (dark blue) and chordin (brown) messages. The *frzb* mRNA is seen to be transcribed in the head endomesoderm of the organizer, but not in the notochord (where chordin is expressed). (B) Microinjection of *frzb* mRNA into the marginal zone leads to the inhibition of trunk formation. The control embryo is on the top; injected embryos are at the bottom. (From Leyns et al. 1997; photographs courtesy of E. M. De Robertis.)

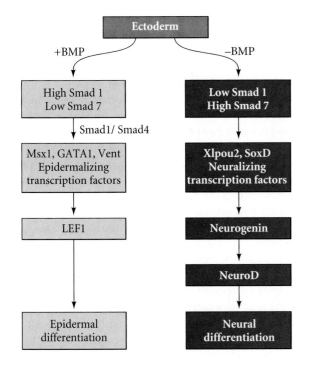

WEBSITE 10.8 **The specification of the endoderm.** While the mesoderm is induced, and the differences between neural and epidermal ectoderm are induced, the endoderm appears to be specified autonomously. Recent studies show that the maternally derived VegT transcription factor in the vegetal cells initiates a cascade of events leading to endoderm formation.

Figure 10.38
Hypothetical pathways differentiating ectoderm into epidermis or neural ectoderm. In the presence of BMP signaling, epidermalizing transcription factors are generated, leading to the activation of the pathway enabling the cell to become an epidermal keratinocyte. In the absence of BMP signaling, neuralizing transcription factors are produced. These factors activate the gene for neurogenin. Neurogenin acts as a transcription factor to activate the *NeuroD* gene, and NeuroD acts as a transcription factor to cause the differentiation of the cell into a neuron.

Sidelights & Speculations

Competence, Bias, and Neurulation

In addition to the signals coming from the underlying chordal plate and dorsal mesoderm, there may also be a bias in the cells of the dorsal part of the embryo toward becoming neural. Phillips and colleagues (London et al. 1988; Savage and Phillips 1989) have shown that the dorsal and ventral animal pole cells of the early-cleavage embryo differ in their expression of the Epi1 protein. Not only do the presumptive epidermal cells express this protein, which is not expressed in the presumptive neural cells, but the region of cells failing to express Epi1 increases during gastrulation. Other differences between dorsal and ventral ectoderm also become apparent at this time, prior to the notochord's movement beneath the ectoderm (Otte and Moon 1992; Gamse and Sive 2000). Moreover, in the ventral mesoderm, proteins encoded by *Xvex1* and *Xvex2* block the expression of dorsal genes (Shapira et al. 2000).

The cues for this "bias" toward neurulation may be provided by signals from the dorsal lip traveling in a planar (horizontal) fashion through the ectoderm (Figure 10.39; Sharpe et al. 1987; Dixon and Kintner 1989; Doniach 1993). The molecular basis of this planar "ectodermal" signaling system is not fully known, but Wessley and colleagues (2001) have shown that a second signaling center may be induced in the animal cap blastula ectoderm by β-catenin, and that these cells also activate chordin, independently of the mesoderm.

WEBSITE 10.9 **Planar induction.** Several studies suggest that the planar mode of signal transduction—from the dorsal lip through the ectoderm—may help pattern the ectoderm. Other studies argue against planar induction playing any role in patterning the ectoderm.

Figure 10.39
Ectodermal bias toward neurulation. (A) Neural gene expression occurs more readily in dorsal ectoderm than in ventral ectoderm. Ectoderm fragments from the dorsal and ventral thirds of a *Xenopus* embryo were wrapped around dorsal mesoderm (notochord) tissue. The fragments were separated and tested for the induction of neural mRNAs. The mesoderm had none, the ventral ectoderm had some, and the dorsal ectoderm expressed high concentrations of neural messages. (B) Model showing the two modes of signaling from the dorsal blastopore lip: a vertical mechanism through the dorsal mesoderm, and a horizontal (planar) mechanism through the ectoderm. (A after Sharpe et al. 1987; B after Doniach 1993.)

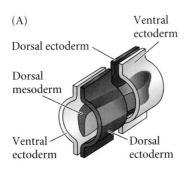

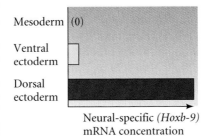

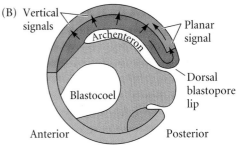

The Regional Specificity of Induction

The determination of regional differences

One of the most fascinating phenomena in neural induction is the regional specificity of the neural structures that are produced. Forebrain, hindbrain, and spinocaudal regions of the neural tube must all be properly organized in an anterior-to-posterior direction. The organizer tissue not only induces the neural tube, but also specifies the regions of the neural tube. This region-specific induction was demonstrated by Hilde Mangold's husband, Otto Mangold (1933). He transplanted four successive regions of the archenteron roof of late-gastrula newt embryos into the blastocoels of early-gastrula embryos (Figure 10.40). The most anterior portion of the archenteron roof induced balancers and portions of the oral apparatus (Figure 10.40A); the next most anterior section induced the formation of various head structures, including nose, eyes, balancers, and otic vesicles (Figure 10.40B); the third section induced the hindbrain structure (Figure 10.40C); and the most posterior section induced the formation of dorsal trunk and tail mesoderm* (Figure 10.40D). Moreover, when dorsal blastopore lips from early salamander gastrulae were transplanted into other early salamander gastrulae, they formed secondary heads. When dorsal lips from later gastrulas were transplanted into early salamander gastrulae, however, they induced the formation of secondary tails (Figure 10.41; Mangold 1933). These results show that the first cells of the organizer to enter the embryo induce the formation of brains and heads, while those cells that form the dorsal lip of later-stage embryos induce the cells above them to become spinal cords and tails.

---

*The induction of dorsal mesoderm—rather than the dorsal ectoderm of the nervous system—by the posterior end of the notochord was confirmed by Bijtel (1931) and Spofford (1945), who showed that the posterior fifth of the neural plate gives rise to tail somites and the posterior portions of the pronephric kidney duct.

The posterior transforming proteins: Wnt signals and retinoic acid

In 1938, Sulo Toivonen found that certain organs induced only part of the amphibian axis. Moreover, these organs did not have to come from an amphibian. Guinea pig *liver*, for instance, when placed in an early salamander gastrula, would induce only forebrain structures. Guinea pig *bone marrow*, however, would transform all of the neural tube into spinal cord cells. Putting them together in the blastocoel would generate all the structures of the neural tube, including hindbrain-like tissue that wasn't seen in either case alone. Toivonen and Saxén (1955, 1968; reviewed in Saxén 2001; see Website 10.10) postulated two gradients in amphibian embryos: a dorsal gradient of "neuralizing" activity and a caudal gradient of posteriorizing ("mesodermalizing") activity. The neuralizing activity came from the organizer and induced the ectoderm to be neural. The posteriorizing activity originated in the posterior of the embryo and weakened anteriorly. Recent studies have extended this model and have proposed candidates for the neuralizing and posteriorizing molecules. As predicted, the two signaling systems—neuralizing and posteriorizing—work independently (Kolm and Sive 1997).

The primary protein involved in posteriorizing the neural tube is thought to be a member of the Wnt family of paracrine factors, most likely Xwnt8 (Domingos et al. 2001; Kiecker and Niehrs 2001). In *Xenopus* embryos, an endogenous gradient of Wnt signaling and β-catenin is highest in the posterior and absent in the anterior (Figure 10.42A). Moreover, if Xwnt8 is added to developing embryos, spinal cord-

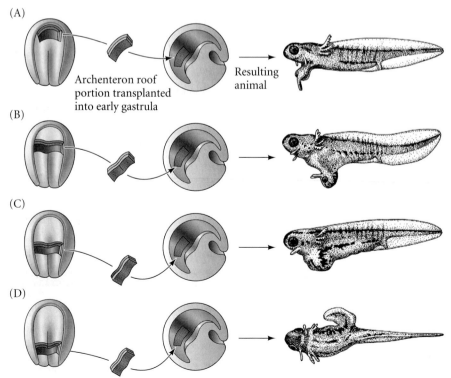

(A)

Archenteron roof portion transplanted into early gastrula

Resulting animal

(B)

(C)

(D)

Figure 10.40
Regional specificity of induction can be demonstrated by implanting different regions (color) of the archenteron roof into early *Triturus* gastrulae. The resulting embryos develop secondary dorsal structures. (A) Head with balancers. (B) Head with balancers, eyes, and forebrain. (C) Posterior part of head, diencephalon, and otic vesicles. (D) Trunk-tail segment. (After Mangold 1933.)

(A) Transplantation of young gastrula dorsal lip

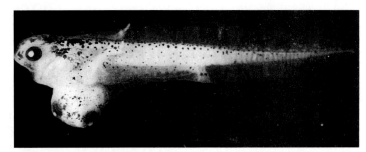

(B) Transplantation of advanced gastrula dorsal lip

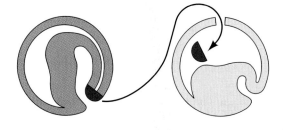

Figure 10.41
Regionally specific inducing action of the dorsal blastopore lip.
(A) Young dorsal lips (which will form the anterior portion of the organizer) induce anterior dorsal structures when transplanted into early newt gastrulae. (B) Older dorsal lips transplanted into early newt gastrulae produce more posterior dorsal structures. (From Saxén and Toivonen 1962; photographs courtesy of L. Saxén.)

WEBSITE 10.10 Regional specification. The research into regional specification has been a fascinating endeavor involving scientists from all over the world. Before molecular biology gave us the tools to uncover morphogenetic proteins, embryologists developed ingenious ways of finding out what those proteins were doing.

(A) **Anterior**

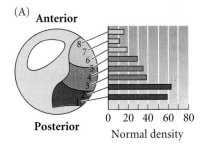

Posterior

Normal density

(E)

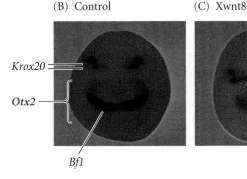

Anterior

Ventral — **Dorsal**

BMPs — Wnts

Posterior

(B) Control

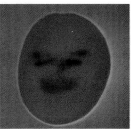

Krox20

Otx2

Bf1

(C) Xwnt8

(D) Xfrzb1

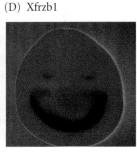

Figure 10.42
The Wnt signaling pathway and posteriorization of the neural tube. (A) The gradient of β-catenin in the presumptive neural plate during gastrulation. Gastrulating embryos were stained for β-catenin and the density of the stain compared between regions of the ectodermal cells. (B–D) Changing the level of Wnt signaling changes the expression of regionally specific markers. In the control neurula (B) *Bf1* marks the neural cells of the forebrain and is the most anterior marker. The *Otx2* gene is expressed in the forebrain and midbrain (red), and the *Krox20* gene is expressed in two regions of the hindbrain. (C) Microinjecting plasmids that produce more Xwnt8 causes posteriorization of the neural plate: the disappearance of the *Bf1* expression band and the reduction of *Otx2* expression to its midbrain region. (D) Microinjecting plasmids making Xfrzb1 (*Xenopus* Frzb, a Wnt antagonist) anteriorizes the neural plate, causing the expansion of the regions of *Bf1* and *Otx2* expression at the expense of the more posterior regions. (E) Model whereby a gradient in BMP expression specifies the dorsal-ventral axis of the frog, while a gradient of Wnt proteins specifies the anterior-posterior axis. (After Kiecker and Niehrs 2001.)

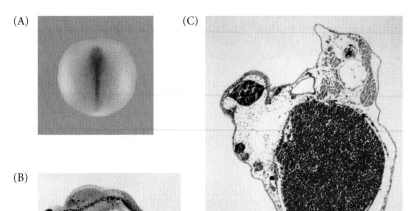

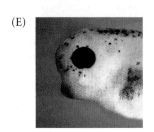

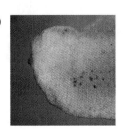

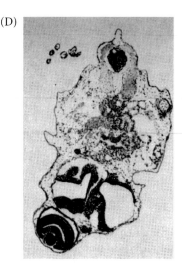

Figure 10.43
Insulin-like growth factors enhance anterior neural development. (A) Expression pattern of IGF3, showing its accumulation in the anterior neural tube. (B) Ectopic head-like structure (complete with eyes and cement gland) formed when *Igf2* mRNA was injected into ventral marginal zone blastomeres. (C) Section through the embryo of (B), showing secondary brain and nasal placodes. Compare with (D). (D) Section through a salamander larva with an extra forebrain, produced by implanting guinea pig bone marrow into the blastocoel. (E) Anterior of 3-day control tadpole. (F) Anterior of tadpole whose 4-cell embryonic blastomeres were injected with an inhibitor of IGF signaling. The cement gland and eyes are absent. (A–C, E, F from Pera et al. 2001; photographs courtesy of E. M. De Robertis. D from Toivonen 1938.)

like neurons are seen more anteriorly in the embryo, and the most anterior markers of the forebrain are absent. Conversely, suppressing Wnt signaling (by adding Frzb or Dickkopf to the developing embryo) leads to the expression of more neural cells in the anteriormost markers (Figure 10.42B–D). Therefore, while the BMP gradient specifies the dorsal-ventral axis of the frog embryo and induces cells to become neural, the Wnt gradient may similarly specify the anterior-posterior axis.

While the Wnt proteins probably play a major role in specifying the anterior-posterior axis, they are probably not the only agent involved. Fibroblast growth factors appear to be critical in allowing the cells to respond to the Wnt signal (Domingos et al. 2001). Retinoic acid also has been found in a gradient highest at the posterior end of the neural plate, and it can also posteriorize the neural tube in a concentration-dependent manner (Cho and De Robertis 1990; Sive and Cheng 1991; Chen et al. 1994). Retinoic acid signaling appears to be especially important in patterning the hindbrain, but not the forebrain (Blumberg et al. 1997; Kolm et al. 1997; Dupé and Lumsden 2001).

The anterior transforming proteins: Insulin-like growth factors

In addition to the proteins that block BMP and Wnt signaling in the head, there is a *positive* signal that promotes anterior head development. Recent studies by Pera and colleagues (2001) suggest that **insulin-like growth factors** (**IGFs**) are required to form the anterior neural tube with its brain and sensory placodes. IGFs accumulate in the dorsal midline and are especially prominent in the anterior neural tube (Figure 10.43A). Moreover, IGFs are able to convert chordin-induced trunk neural tissue into anterior tissue. When injected into ventral mesodermal blastomeres, mRNA from IGFs caused the formation of ectopic heads, while blocking the IGF receptors resulted in the lack of head formation (Figure 10.43B–E). When such inhibitors of IGF signaling were added to these embryos, the anterior neural cell markers (such as Otx2) were almost completely abolished.

It appears, then, that the inhibition of BMP4 signaling causes cells to become neural, but their fates along the anterior-posterior axis depend on their position relative to axes of Wnt. Cells in the head region are protected against Wnt signaling and are fated to become brain and placode-type neural cells. The completion of this determination seems to be due to insulin-like growth factors that act within the neural tube. The basic model of neural induction, then, looks like the diagram in Figure 10.44.

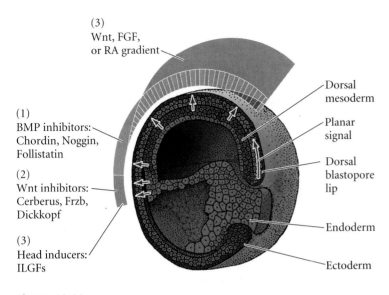

(3)
Wnt, FGF,
or RA gradient

(1)
BMP inhibitors:
Chordin, Noggin,
Follistatin

(2)
Wnt inhibitors:
Cerberus, Frzb,
Dickkopf

(3)
Head inducers:
ILGFs

Dorsal
mesoderm

Planar
signal

Dorsal
blastopore
lip

Endoderm

Ectoderm

Figure 10.44
Model of organizer function and axis specification in the *Xenopus* gastrula. (1) BMP inhibitors from organizer tissue (dorsal mesoderm and pharyngeal mesendoderm) block the formation of epidermis, ventrolateral mesoderm, and ventrolateral endoderm. (2) Wnt inhibitors in the anterior of the organizer (pharyngeal mesendoderm) allow the induction of head structures. (3) Head structures are induced through insulin-like growth factor signaling. (4) A gradient of caudalizing factors (Wnts, retinoic acid) causes the regional expression of Hox genes, specifying the regions of the neural tube.

pression of *Xnr1* in both the right and left lateral plates and to the randomization of heart and gut positions (Hyatt et al. 1996; Kramer and Yost 2002). Moreover, when active Vg1 protein is injected into a particular right-side vegetal blastomere (the third vegetal cell to the right of the dorsal midline), the entire left-right axis is inverted. No other TGF-β family member is able to accomplish this reversal (Hyatt and Yost 1998). It is not yet known how the events at fertilization lead to the expression of *Xnr1* in the left lateral plate mesoderm during gastrulation.

The pathway by which the Xnr1 protein instructs the heart and gut to fold properly is unknown, but one of the key genes activated by Xnr1 appears to be ***pitx2***. Since *pitx2* is activated by Xnr1, it is normally expressed only on the left side of the embryo. However, if the Pitx2 protein is injected into the right side, too, the placement of the heart and the coiling of the gut are randomized (Figure 10.45; Ryan et al. 1998). Pitx2 persists on the left side of the embryo as the heart and gut develop, controlling their respective positions. Pitx2 may be at the "heart of the heart" (Strauss 1998).

We are finally putting names to the "agents" and "soluble factors" of the experimental embryologists. We are also finally delineating the intercellular path-

Specifying the Left-Right Axis

Although the developing tadpole looks symmetrical from the outside, there are several internal organs, such as the heart and the gut tube, that are not evenly balanced on the right and left sides. In other words, in addition to its dorsal-ventral and anterior-posterior axes, the embryo has a left-right axis. Somehow, the body must be given clues as to which half is right and which half is left.

In all vertebrates studied so far, the crucial event in left-right axis formation is the expression of a *nodal* gene in the lateral plate mesoderm on the *left* side of the embryo. In *Xenopus*, this gene is *Xnr1* (*Xenopus nodal-related 1*). If the expression of this gene is also permitted to occur on the right-hand side, the position of the heart (which is normally on the left side) and the coiling of the gut are randomized.

But what causes the expression of *Xnr1* solely on the left-hand side? In *Xenopus*, it is possible that the first cue is given at fertilization. The microtubules involved in cytoplasmic rotation appear to be crucial, since if their formation is inhibited, no left-right axis appears (Yost 1998). Moreover, the Vg1 protein, which appears to be expressed throughout the vegetal hemisphere, seems to be processed into its active form predominantly on the left-hand side of the embryo. Injecting active Vg1 protein into the left vegetal blastomeres has no effect, but adding it to the right vegetal blastomeres leads to the ex-

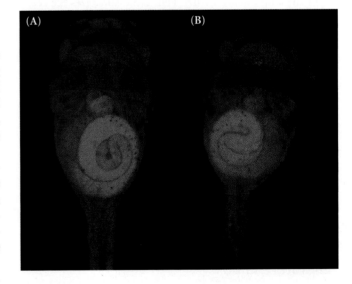

(A)　　　　　　　　　(B)

Figure 10.45
Pitx2 determines the direction of heart looping and gut coiling. (A) Wild-type *Xenopus* tadpole viewed from the ventral side, showing rightward heart looping and counterclockwise gut coiling. (B) If an embryo is injected with Pitx2 so that this protein is present in the mesoderm of both the right and left sides (instead of just the left side), heart looping and gut coiling are random with respect to each other. Sometimes this treatment results in complete reversals, as in this embryo, in which the heart loops to the left and the gut coils in a clockwise manner. (From Ryan et al. 1998; photograph courtesy of J. C. Izpisúa-Belmonte.)

ways of paracrine factors and transcription factors that constitute the first steps in the processes of organogenesis. The international research program initiated by Spemann's laboratory in the 1920s is reaching toward a conclusion. But this research has found levels of complexity far deeper than Spemann would have conceived, and just as Spemann's experiments told us how much we didn't know, so today, we are faced with a whole new set of problems generated by our solutions to older ones.

Surveying the field in 1927, Spemann remarked:

We still stand in the presence of riddles, but not without hope of solving them. And riddles with the hope of solution—what more can a scientist desire?

The challenge still remains.

Snapshot Summary: Early Development and Axis Formation in Amphibians

1. Amphibian cleavage is holoblastic, but it is unequal due to the presence of yolk in the vegetal hemisphere.

2. Amphibian gastrulation begins with the invagination of the bottle cells, followed by the coordinated involution of the mesoderm and the epiboly of the ectoderm. Vegetal rotation plays a significant role in directing the involution.

3. The driving forces for ectodermal epiboly and the convergent extension of the mesoderm are the intercalation events in which several tissue layers merge. Fibronectin plays a critical role in enabling the mesodermal cells to migrate into the embryo.

4. The dorsal lip of the blastopore forms the organizer tissue of the amphibian gastrula. This tissue dorsalizes the ectoderm, transforming it into neural tissue, and it transforms ventral mesoderm into lateral mesoderm.

5. The organizer consists of pharyngeal endoderm, head mesoderm, notochord, and dorsal blastopore lip tissues. The organizer functions by secreting proteins (Noggin, chordin, and follistatin) that block the BMP signal that would otherwise ventralize the mesoderm and activate the epidermal genes in the ectoderm.

6. The organizer is itself induced by the Nieuwkoop center, located in the dorsalmost vegetal cells. This center is formed by cortical rotation during fertilization, which translocates the Dishevelled protein to the dorsal side of the egg.

7. The Dishevelled protein stabilizes β-catenin in the dorsal cells of the embryo. Thus, the Nieuwkoop center is formed by the accumulation of β-catenin, which can complex with Tcf3 to form a transcription factor complex that can activate the transcription of the *siamois* gene.

8. The siamois protein and a TGF-β signal (perhaps from Vg1) can activate the *goosecoid* gene in the organizer. The *goosecoid* gene can activate other genes that cause the organizer to function.

9. In the head region, an addition set of proteins (Cerberus, Frzb, Dickkopf) block the Wnt signal from the ventral and lateral mesoderm.

10. Wnt signaling causes a gradient of β-catenin along the anterior-posterior axis of the neural plate. This graded signaling appears to specify the regionalization of the neural tube.

11. Insulin-like growth factors appear to transform the neural tube into anterior (forebrain) tissue.

12. The left-right axis appears to be initiated at fertilization through the Vg1 protein. In a still unknown fashion, this protein activates a Nodal protein solely on the left side of the body. In *Xenopus*, as in other vertebrates, the Nodal protein activates expression of Pitx2, which is critical in distinguishing left-sidedness from right-sidedness in the heart and gut tubes.

Literature Cited

Agius, E., M. Oelgeschläager, O. Wessely, C. Kemp and E. M. De Robertis. 2000. Endodermal Nodal-related signals and mesoderm induction in *Xenopus*. *Development* 127: 1151–1159.

Alfandari, D., C. A. Whittaker, D. W. DeSimone and T. Darribère. 1995. Integrin α₅ subunit is expressed on mesodermal cell surfaces during amphibian gastrulation. *Dev. Biol.* 170: 249–261.

Appel, T. A. 1987. *The Cuvier-Geoffroy Debate: French Biology in the Decades before Darwin.* Oxford University Press, New York.

Arendt, D. and K. Nübler-Jung. 1999. Rearranging gastrulation in the name of yolk: Evolution of gastrulation in yolk-rich amniote eggs. *Mech. Dev.* 81: 3–22.

Balinsky, B. I. 1975. *Introduction to Embryology,* 4th Ed. Saunders, Philadelphia.

Beams, H. W. and R. G. Kessel. 1976. Cytokinesis: A comparative study of cytoplasmic division in animal cells. *Am. Sci.* 64: 279–290.

Bijtel, J. H. 1931. Über die Entwicklung des Schwanzes bei Amphibien. *Wilhelm Roux Arch. Entwicklungsmech. Org.* 125: 448–486.

Black, S. D. and J. Gerhart. 1985. Experimental control of the site of embryonic axis formation in *Xenopus laevis* eggs centrifuged before first cleavage. *Dev. Biol.* 108: 310–324.

Black, S. D. and J. Gerhart. 1986. High frequency twinning of *Xenopus laevis* embryos from eggs centrifuged before first cleavage. *Dev. Biol.* 116: 228–240.

Blitz, I. L. and K. W. Y. Cho. 1995. Anterior neurectoderm is progressively induced during gastrulation: The role of the *Xenopus* homeobox gene *orthodenticle*. *Development* 121: 993–1004.

Blumberg, B., J. Bolado, T. Moreno, C. Kintner, R. Evans and N. Papalopulu. 1997. An essential role for signaling in anteroposterior neural patterning. *Development* 124: 373–379.

Boucaut, J.-C., T. D'Arribère, T. J. Poole, H. Aoyama, K. M. Yamada and J.-P. Thiery. 1984. Biologically active synthetic peptides as probes of embryonic development: A competitive peptide inhibition of fibronectin function inhibits gastrulation in amphibian embryos and neural crest cell migration in avian embryos. *J. Cell Biol.* 99: 1822–1830.

Boucaut, J.-C., T. D'Arribère, S. D. Li, H. Boulekbache, K. M. Yamada and J.-P. Thiery. 1985. Evidence for the role of fibronectin in amphibian gastrulation. *J. Embryol. Exp. Morphol.* 89 [Suppl.]: 211–217.

Bouwmeester, T., S.-H. Kim, Y. Sasai, B. Lu and E. M. De Robertis. 1996. Cerberus is a head-inducing secreted factor expressed in the anterior endoderm of Spemann's organizer. *Nature* 382: 595–601.

Brannon, M. and D. Kimelman. 1996. Activation of *siamois* by the Wnt pathway. *Dev. Biol.* 180: 344–347.

Brannon, M., M. Gomperts, L. Sumoy, R. T. Moon and D. Kimelman. 1997. β-catenin /XTcf-3 complex binds to the *siamois* promoter to regulate dorsal axis specification in *Xenopus*. *Genes Dev.* 11: 2359–2370.

Capuron, A. 1968. Marquage autoradiographique et conditions de l'organogenèse générale d'embryons induits par de la greffe de la lèvre dorsale du blastopore chez l'amphibien urodèle *Pleurodeles waltii* Michah. *Ann. Embryol. Morphol.* 1: 271–293.

Carlson, B. M. 1981. *Patten's Foundations of Embryology.* McGraw-Hill, New York.

Chen, Y. P., L. Huang and M. Solursh. 1994. A concentration gradient of retinoids in the early *Xenopus laevis* embryo. *Dev. Biol.* 161: 70–76.

Cho, K.W. and De Robertis, E. M. 1990. Differential activation of *Xenopus* homeobox genes by mesoderm-inducing growth factors and retinoic acid. *Genes Dev.* 4: 1910–1916.

Cho, K. W. Y., B. Blumberg, H. Steinbeisser and E. De Robertis. 1991a. Molecular nature of Spemann's organizer: The role of the *Xenopus* homeobox gene *goosecoid*. *Cell* 67: 1111–1120.

Cho, K. W. Y., A. A. Morita, C. V. E. Wright and E. M. De Robertis. 1991b. Overexpression of a homeodomain protein confers axis-forming activity to uncommitted *Xenopus* embryonic cells. *Cell* 65: 55–64.

Christian, J. L. and R. T. Moon. 1993. Interactions between Xwnt8 and Spemann organizer signaling pathways generate dorsoventral pattern in the embryonic mesoderm of *Xenopus*. *Genes Dev.* 7: 13–28.

Cooke, J. 1986. Permanent distortion of positional system of *Xenopus* embryo by brief early perturbation in gravity. *Nature* 319: 60–63.

Dale, L. and J. M. W. Slack. 1987. Regional specificity within the mesoderm of early embryos of *Xenopus laevis*. *Development* 100: 279–295.

Dale, L., G. Howes, B. M. J. Price and J. C. Smith. 1992. Bone morphogenetic protein 4: A ventralizing factor in early *Xenopus* development. *Development* 115: 573–585.

Darribère, T., K. M. Yamada, K. E. Johnson and J.-C. Boucaut. 1988. The 140-kD fibronectin receptor complex is required for mesodermal cell adhesion during gastrulation in the amphibian *Pleurodeles waltii*. *Dev. Biol.* 126: 182–194.

Darribère, T., K. Guida, H. Larjava, K. E. Johnson, K. M. Yamada, J.-P. Thiery and J.-C. Boucaut. 1990. In vivo analysis of integrin β1 subunit function in fibronectin matrix assembly. *J. Cell Biol.* 110: 1813–1823.

Davidson, L. A., B. G. Hoffstrom, R. Keller and D. W. DeSimone. 2002. Mesendoderm extension and mantle closure in *Xenopus laevis* gastrulation: Combined roles for integrin α5β1, fibronectin, and tissue geometry. *Dev. Biol.* 242: 109–129.

De Robertis, E. M. and J Aréchaga (eds.). 2001. The Spemann-Mangold Organizer: 75 Years On. *Int. J. Dev. Biol.* 45 (1) (Special Issue).

De Robertis, E. M. and Y. Sasai. 1996. A common plan for dorsoventral patterning in Bilateria. *Nature* 380: 37–40.

De Robertis, E. M., M. Blum, C. Niehrs and H. Steinbeisser. 1992. Goosecoid and the organizer. *Development* 1992 [Suppl.]: 167–171.

De Robertis, E. M., J. Larraín, M. Oelgeschländer and O. Wessley. 2000. The establishment of Spemann's organizer and patterning of the vertebrate embryo. *Nature Rev. Genet.* 1: 171–181.

Dixon, J. E. and C. R. Kintner. 1989. Cellular contacts required for neural induction in *Xenopus laevis* embryos: Evidence for two signals. *Development* 106: 749–757.

Domingos, P. M., N. Itasaki, C. M. Jones, S. Mercurio, M. G. Sargent, J. C. Smith and R. Krumlauf. 2001. The Wnt/β-catenin pathway posteriorizes neural tissue in *Xenopus* by an indirect mechanism requiring FGF signalling. *Dev. Biol.* 239: 148–160.

Doniach, T. 1993. Planar and vertical induction of anteroposterior pattern during the development of the amphibian central nervous system. *J. Neurobiol.* 24: 1256–1275.

Dosch, R., V. Gawantka, H. Delius, C. Blumenstock and C. Niehrs. 1997. BMP-4 acts as a morphogen in dorsolateral mesoderm patterning in *Xenopus*. *Development* 124: 2325–2334.

Dupé, V. and A. Lumsden. 2001. Hindbrain patterning involves graded responses to retinoic acid signalling. *Development* 128: 2199–2208.

Engleka, M. J. and D. S. Kessler. 2001. Siamois cooperates with TGFβ signals to induce the complete function of the Spemann-Mangold Organizer. *Int. J. Dev. Biol.* 45: 241–250.

Ewald, A.J., H. McBride, M. Reddington, S. E. Fraser and R. Kerschmann. 2002. Surface imaging microscopy: An automated method for visualizing whole embryo samples in three dimensions at high resolution. *Dev. Dynam.* 225: 369–375.

Fan, C.-M. and S. Y. Sokol. 1997. A role for Siamois in Spemann organizer formation. *Development* 124: 2581–2589.

Fässler, P. E. and K. Sander. 1996. Hilde Mangold (1898–1924) and Spemann's organizer: Achievement and tragedy. *Wilhelm Roux Arch. Dev. Biol.* 205: 323–332.

Funayama, N., F. Fagotto, P. McCrea and B. M. Grumbiner. 1995. Embryonic axis induction by the armadillo repeat domain of β-catenin: Evidence for intracellular signalling. *J. Cell Biol.* 128: 959–968.

Gamse, J. and H. Sive. 2000. Vertebrate anteroposterior patterning: The *Xenopus* neurectoderm as a paradigm. *BioEssays* 22: 976–986.

Gawantka, V., H. Delius, K. Hirschfeld, C. Blumenstock and C. Niehrs. 1995. Antagonizing the Spemann organizer: Role of the homeobox gene *Xvent-1*. *EMBO J.* 14: 6268–6279.

Geoffroy Saint-Hilaire, E. 1822. Considérations générales sur la vertèbre. *Mém. Mus. Hist. Nat.* 9: 89–1110.

Gerhart, J., G. Ubbels, S. Black, K. Hara and M. Kirschner. 1981. A reinvestigation of the role of the grey crescent in axis formation in *Xenopus laevis*. *Nature* 292: 511–516.

Gerhart, J. and 7 others. 1986. Amphibian early development. *BioScience* 36: 541–549.

Gerhart, J. C., M. Danilchik, T. Doniach, S. Roberts, B. Rowning, and R.Stewart. 1989. Cortical rotation of the *Xenopus* egg: Consequences for the anteroposterior pattern of embryonic dorsal development. *Development* [Suppl.] 107: 37–51.

Gilbert, S. F. and J. A. Bolker. 2001. Homologies of process and modular elements of embryonic construction. *J. Exp. Zool.* 291: 1–12.

Gilbert, S. F. and L. Saxén. 1993. Spemann's organizer: Models and molecules. *Mech. Dev.* 41: 73–89.

Gimlich, R. L. 1985. Cytoplasmic localization and chordamesoderm induction in the frog embryo. *J. Embryol. Exp. Morphol.* 89: 89–111.

Gimlich, R. L. 1986. Acquisition of developmental autonomy in the equatorial region of the *Xenopus* embryo. *Dev. Biol.* 116: 340–352.

Gimlich, R. L. and J. C. Gerhart. 1984. Early cellular interactions promote embryonic axis formation in *Xenopus laevis*. *Dev. Biol.* 104: 117–130.

Glavic, A., J. L. Gomez-Skarmeta and R. Mayor. 2001. *Xiro-1* controls mesoderm patterning by repressing *bmp-4* expression in the Spemann organizer. *Dev. Dyn.* 222: 368–376.

Glinka, A., W. Wu, D. Onichtchouk, C. Blumenstock and C. Niehrs. 1997. Head induction by simultaneous repression of BMP and Wnt signalling in *Xenopus. Nature* 389: 517–519.

Glinka, A., W. Wu, A. P. Monaghan, C. Blumenstock and C. Niehrs. 1998. Dickkopf-1 is a member of a new family of secreted proteins and functions in head induction. *Nature* 391: 357–362.

Godsave, S. F. and J. M. W. Slack. 1989. Clonal analysis of mesoderm induction in *Xenopus. Dev. Biol.* 134: 486–490.

Gont, L. K., H. Steinbeisser, B. Blumberg and E. M. De Robertis. 1993. Tail formation as a continuation of gastrulation: The multiple tail populations of the *Xenopus* tailbud derive from the late blastopore lip. *Development* 119: 991–1004.

Graff, J. M., R. S. Thies, J. J. Song, A. J. Celeste and D. A. Melton. 1994. Studies with a *Xenopus* BMP receptor suggest that ventral mesoderm-inducing signals override dorsal signals in vivo. *Cell* 79: 169–179.

Grunz, H. 1997. Neural induction in amphibians. *Curr. Topics Dev. Biol.* 35: 191–228.

Grunz, H. and L. Tacke. 1989. Neural differentiation of *Xenopus laevis* ectoderm takes place after disaggregation and delayed reaggregation without inducers. *Cell Diff. Dev.* 32: 117–124.

Guger, K. A. and B. M. Gumbiner. 1995. β-catenin has wnt-like activity and mimics the Nieuwkoop signaling center in *Xenopus* dorsal-ventral patterning. *Dev. Biol.* 172: 115–125.

Hamburger, V. 1984. Hilde Mangold, co-discoverer of the organizer. *J. Hist. Biol.* 17: 1–11.

Hamburger, V. 1988. *The Heritage of Experimental Embryology: Hans Spemann and the Organizer.* Oxford University Press, Oxford.

Hammerschmidt, M., G. N. Serbedzija and A. P. McMahon. 1996. Genetic analysis of dorsoventral pattern formation in the zebrafish: Requirement of a BMP-like ventralizing activity and its dorsal repressor. *Genes Dev.* 10: 2452–2461.

Hansen, C. S., C. D. Marion, K. Steele, S. George and W. C. Smith. 1997. Direct neural induction and selective inhibition of mesoderm formation and epidermis inducers by Xnr3. *Development* 124: 483–492.

Hara, K. 1977. The cleavage pattern of the axolotl egg studied by cinematography and cell counting. *Wilhelm Roux Arch. Entwicklungsmech. Org.* 181: 73–87.

Hardin, J. D. and R. Keller. 1988. The behaviour and function of bottle cells during gastrulation of *Xenopus laevis. Development* 103: 211–230.

Hawley, S. H. B., K. Wünnenberg-Stapleton, C. Hashimoto, M. N. Laurent, T. Watabe, B. W. Blumberg and K. W. Y. Cho. 1995. Disruption of BMP signals in embryonic *Xenopus* ectoderm

leads to direct neural induction. *Genes Dev.* 9: 2923–2935.

He, X., J.-P. Saint-Jeannet, J. R. Woodgett, H. E. Varmus and I. B. Dawid. 1995. Glycogen synthase kinase-3 and dorsoventral patterning in *Xenopus* embryos. *Nature* 374: 617–622.

Heasman, J. M. and 8 others. 1994a. Overexpression of cadherins and underexpression of β-catenin inhibit dorsal mesoderm induction in early *Xenopus* embryos. *Cell* 79: 791–803.

Heasman, J., D. Ginsberg, K. Goldstone, T. Pratt, C. Yoshidanaro and C. Wylie. 1994b. A functional test for maternally inherited cadherin in *Xenopus* shows its importance in cell adhesion at the blastula stage. *Development* 120: 49–57.

Hemmati-Brivanlou, A. and D. A. Melton. 1992. A truncated activin receptor inhibits mesoderm induction and formation of axial structures in *Xenopus* embryos. *Nature* 359: 609–614.

Hemmati-Brivanlou, A. and D. A. Melton. 1994. Inhibition of activin signalling promotes neuralization in *Xenopus. Cell* 77: 273–281.

Hemmati-Brivanlou A. and D. A. Melton. 1997. Vertebrate embryonic cells will become nerve cells unless told otherwise. *Cell* 88: 13–17.

Hemmati-Brivanlou, A. and G. H. Thomsen. 1995. Ventral mesodermal patterning in *Xenopus* embryos: Expression patterns and activities of BMP-2 and BMP-4. *Dev. Genet.* 17: 78–810.

Hirsch, N., L. B. Zimmerman and R. M. Grainger. 2002. *Xenopus*, the next generation: *X. tropicalis* genetics and genomics. *Dev. Dynam.* 225: 422–433.

Holley, S. A., P. D. Jackson, Y. Sasai, B. Lu, E. M. De Robertis, F. M. Hoffmann and E. L. Ferguson. 1995. A conserved system for dorsal-ventral patterning in insects and vertebrates involving sog and chordin. *Nature* 376: 249–253.

Holowacz, T. and R. P. Elinson. 1993. Cortical cytoplasm, which induces dorsal axis formation in *Xenopus*, is inactivated by UV irradiation of the oocyte. *Development* 119: 277–285.

Holtfreter, H. 1933. Die totale Exogastrulation, eine Selbststablösung des Ektoderms von Entomesoderm. Entwicklung und funktionelles Verhalten nervenloser Organe. *Arch. Entwick. Mech. Org.* 129: 669–793.

Holtfreter, J. 1943. A study of the mechanics of gastrulation, Part I. *J. Exp. Zool.* 94: 261–318.

Holtfreter, J. 1944. A study of the mechanics of gastrulation, Part II. *J. Exp. Zool.* 95: 171–212.

Hyatt, B. A. and H. J. Yost. 1998. The left/right coordinator: The role of Vg1 in organizing left-right axis formation. *Cell* 93: 37–46.

Hyatt, B. A., J. L. Lohr and H. J. Yost. 1996. Initiation of vertebrate left-right axis by maternal Vg1. *Nature* 384: 62–65.

Ibrahim, H. and R. Winklbauer. 2001. Mechanisms of mesendoderm internalization in *Xenopus* gastrula: Lessons from the vegetal side. *Dev. Biol.* 240: 108–122.

Iemura, S.-I. and 7 others. 1998. Direct binding of follistatin to a complex of bone morphogenetic protein and its receptor inhibits ventral and epidermal cell fates in early *Xenopus* embryo. *Proc. Natl. Acad. Sci. USA* 95: 9337–9342.

Iwabuchi, M., K. Ohsumi, T. M. Yamamoto and T. Kishimoto. 2002. Coordinated regulation of M phase exit and S phase entry by the Cdc2 activity level in the early embryonic cell cycle. *Dev. Biol.* 243: 34–43.

Jares, P. and J. J. Blow. 2000. *Xenopus* Cdc7 function is dependent on licensing but not on XORC, XCdc6, or CDK activity and is required for XCdc45 loading. *Genes Dev.* 14: 1528–1540.

Jones, C. M., K. M. Lyons, P. M. Lapan, C. V. E. Wright and B. L. M. Hogan. 1992. DVR-4 (bone morphogenetic protein-4) as a posterior ventralizing factor in *Xenopus* mesoderm induction. *Development* 115: 639–647.

Joseph, E. M. and D. A. Melton. 1998. Mutant Vg1 ligands disrupt endoderm and mesoderm formation in *Xenopus* embryos. *Development* 125: 2677–2685.

Kalt, M. R. 1971. The relationship between cleavage and blastocoel formation in *Xenopus laevis.* I. Light microscopic observations. *J. Embryol. Exp. Morphol.* 26: 37–410.

Keller, R. E. 1975. Vital dye mapping of the gastrula and neurula of *Xenopus laevis.* I. Prospective areas and morphogenetic movements of the superficial layer. *Dev. Biol.* 42: 222–241.

Keller, R. E. 1976. Vital dye mapping of the gastrula and neurula of *Xenopus laevis.* II. Prospective areas and morphogenetic movements of the deep layer. *Dev. Biol.* 51: 118–137.

Keller, R. E. 1980. The cellular basis of epiboly: An SEM study of deep cell rearrangement during gastrulation of *Xenopus laevis. J. Embryol. Exp. Morphol.* 60: 201–243.

Keller, R. E. 1981. An experimental analysis of the role of bottle cells and the deep marginal zone in the gastrulation of *Xenopus laevis. J. Exp. Zool.* 216: 81–101.

Keller, R. E. 1986. The cellular basis of amphibian gastrulation. *In* L. Browder (ed.), *Developmental Biology: A Comprehensive Synthesis,* Vol. 2. Plenum, New York, pp. 241–327.

Keller, R. and M. Danilchik. 1988. Regional expression, pattern and timing of convergence and extension during gastrulation of *Xenopus laevis. Development* 103: 193–209.

Keller, R. E. and G. C. Schoenwolf. 1977. An SEM study of cellular morphology, contact, and arrangement as related to gastrulation in *Xenopus laevis. Wilhelm Roux Arch. Dev. Biol.* 182: 165–186.

Kessler, D. S. 1997. Siamois is required for formation of Spemann's organizer. *Proc. Natl. Acad. Sci. USA* 94: 13017–13022.

Kiecker, C. and C. Niehrs. 2001. A morphogen gradient of Wnt/β-catenin signalling regulates anteroposterior neural patterning in *Xenopus. Development* 128: 4189–4201.

Kim, S.-H., A. Yamamoto, T. Bouwmeester, E. Agius and E. M. De Robertis. 1998. The role of paraxial protocadherin in selective adhesion and cell movements of the mesoderm during *Xenopus* gastrulation. *Development* 125: 4681–4691.

Kolm, P. J. and H. L. Sive. 1997. Retinoids and posterior neural induction: A reevaluation of Nieuwkoop's two-step hypothesis. *Cold Spring Harb. Symp. Quant. Biol.* 62: 511–521.

Kolm, P. J., V. Apekin and H. Sive. 1997. *Xenopus* hindbrain patterning requires retinoic acid signaling. *Dev. Biol.* 192: 1–16.

Kramer, K. L. and H. J. Yost. 2002. Ectodermal syndecan-2 mediates left-right axis formation in migrating mesoderm as a cell-nonautonomous Vg1 cofactor. *Dev. Cell* 2: 115–124.

Kumano, G. and W. C. Smith. 2002. Revisions to the *Xenopus* gastrula fate map: implications for mesoderm induction and patterning. *Dev. Dynam.* 225: 409–421.

Kurata, T., J. Nakabayashi, T. S. Yamamoto, M. Mochii and N. Ueno. 2000. Visualization of endogenous BMP signaling during *Xenopus* development. *Differentiation* 67: 33–40.

Landstrom, U. and S. Løvtrup. 1979. Fate maps and cell differentiation in the amphibian embryo: An experimental study. *J. Embryol. Exp. Morphol.* 54: 113–130.

Lane, M. C. and W. C. Smith. 1999. The origins of primitive blood in *Xenopus*: Implications for axial patterning. *Development* 126: 423–434.

Lane, M. C. and M. D. Sheets. 2002. Rethinking axial patterning in amphibians. *Dev. Dynam.* 225: 434–447.

Larabell, C. A. and 7 others. 1997. Establishment of the dorsal-ventral axis in *Xenopus* embryos is presaged by early asymmetries in β-catenin which are modulated by the Wnt signaling pathway. *J. Cell Biol.* 136: 1123–1136.

Latinkic, B. V., M. Umbhauer, K. A. Neal, W. Lerchner, J. C. Smith and V. Cunliffe. 1997. The *Xenopus Brachyury* promoter is activated by FGF and low concentrations of activin and suppressed by high concentrations of activin and by paired-type homeodomain proteins. *Genes Dev.* 11: 3265–3276.

Laurent, M. N., I. L. Blitz, C. Hashimoto, U. Rothbacher and K. W.-Y. Cho. 1997. The *Xenopus* homeobox gene *twin* mediates Wnt induction of *goosecoid* in establishment of Spemann's organizer. *Development* 124: 4905–4916.

Lee, J. E., S. M. Hollenberg, L. Snider, D. L. Turner, N. Lipnick and H. Weintraub. 1995. Conversion of *Xenopus* ectoderm into neurons by neuroD, a basic helix-loop-helix protein. *Science* 268: 836–844.

Lemaire, P., N. Garrett and J. B. Gurdon. 1995. Expression cloning of *Siamois*, a *Xenopus* homeobox gene expressed in dorsal-vegetal cells of blastulae and able to induce a complete secondary axis. *Cell* 81: 85–94

Leyns, L., T. Bouwmeester, S.-H. Kim, S. Piccolo and E. M. De Robertis. 1997. Frzb-1 is a secreted antagonist of Wnt signaling expressed in the Spemann organizer. *Cell* 88: 747–756.

London, C., R. Akers and C. Phillips. 1988. Expression of Epi-1, an epidermis-specific marker in *Xenopus laevis* embryos, is specified prior to gastrulation. *Dev. Biol.* 129: 380–389.

Løvtrup, S. 1975. Fate maps and gastrulation in amphibia: A critique of current views. *Can. J. Zool.* 53: 473–479.

Ma, Q. F., C. Kintner and D. J. Anderson. 1996. Identification of neurogenin, a vertebrate neuronal determination gene. *Cell* 87: 43–52.

Mangold, O. 1933. Über die Induktionsfahigkeit der verschiedenen Bezirke der Neurula von Urodelen. *Naturwissenschaften* 21: 761–766.

Mao, B., W. Wu, D. Hoope, P. Stannek, A. Glinka and C. Niehrs. 2001. LDL-receptor-related protein 6 is a receptor for Dickkopf proteins. *Nature* 411: 321–325.

Mao, B. W. and 11 others. 2002. Kremen proteins are Dickkopf receptors that regulate Wnt/β-catenin signalling. *Nature* 417: 664–667.

Mariani, F. and R. Harland. 1998. XBF-2 is a transcriptional repressor that converts ectoderm into neural tissue. *Development* 125: 5019–5031.

Marsden, M. and D. W. DeSimeone. 2001. Regulation of cell polarity, radial intercalation, and epiboly in *Xenopus*: Novel roles for integrin and fibronectin. *Development* 128: 3635–3647.

Miller, J. R., B. A. Rowning, C. A. Larabell, J. A. Yang-Snyder, R. L. Bates and R. T. Moon. 1999. Establishment of the dorsal-ventral axis in *Xenopus* embryos coincides with the dorsal enrichment of Disheveled that is dependent on cortical rotation. *J. Cell Biol.* 146: 427–437.

Minsuk, S. B. and R. E. Keller. 1996. Dorsal mesoderm has a dual origin and forms by a novel mechanism in *Hymenochirus*, a relative of *Xenopus*. *Dev. Biol.* 174: 92–103.

Molenaar, M. and 8 others. 1996. Xtcf-3 transcription factor mediates β-catenin-induced axis formation in *Xenopus* embryos. *Cell* 86: 391–399.

Moon, R. T. and D. Kimelman. 1998. From cortical rotation to organizer gene expression: Toward a molecular explanation of axis specification in *Xenopus*. *BioEssays* 20: 536–545.

Morgan, R. and 8 others. 1999. Calponin modulates the exclusion of *Otx*-expressing cells from convergence extension movements. *Nature Cell Biol.* 1: 404–408.

Nakamura, O. and H. Takasaki. 1970. Further studies on the differentiation capacity of the dorsal marginal zone in the morula of *Triturus pyrrhogaster*. *Proc. Jpn. Acad.* 46: 700–705.

Nakatsuji, N., M. A. Smolira and C. C. Wylie. 1985. Fibronectin visualized by scanning electron microscope immunocytochemistry on the

substratum for cell migration in *Xenopus laevis* gastrulae. *Dev. Biol.* 107: 264–268.

Newman, C. S. and P. A. Krieg. 1999. Specification and differentiation of the heart in amphibia. *In* S. A. Moody, *Cell Lineage and Fate Determination*. Academic Press, New York, pp. 341–351.

Newport, J. W. and M. W. Kirschner. 1982a. A major developmental transition in early *Xenopus* embryos: I. Characterization and timing of cellular changes at midblastula stage. *Cell* 30: 675–686.

Newport, J. W. and M. W. Kirschner. 1982b. A major developmental transition in early *Xenopus* embryos. II. Control of the onset of transcription. *Cell* 30: 687–696.

Niehrs, C., R. Keller, K. W. Y. Cho and E. M. De Robertis. 1993. The homeobox gene *goosecoid* controls cell migration in *Xenopus* embryos. *Cell* 72: 491–503.

Nieuwkoop, P. D. 1969. The formation of the mesoderm in urodele amphibians. I. Induction by the endoderm. *Wilhelm Roux Arch. Entwicklungsmech. Org.* 162: 341–373.

Nieuwkoop, P. D. 1973. The "organisation center" of the amphibian embryo: Its origin, spatial organisation and morphogenetic action. *Adv. Morphogenet.* 10: 1–310.

Nieuwkoop, P. D. 1977. Origin and establishment of embryonic polar axes in amphibian development. *Curr. Top. Dev. Biol.* 11: 115–132.

Nieuwkoop, P. D. and P. A. Florschütz. 1950. Quelques caractèrer spéciaux de le gastrulation et de la neurulation de l'oeuf de *Xenopus laevis*, Daud. et de quelques autres anoures. *Arch. Biol.* 61: 113–150.

Northrop, J., A. Woods, R. Seger, A. Suzuki, N. Ueno, E. Krebs and D. Kimelman. 1995. BMP-4 regulates the dorsal-ventral differences in FGF/MAPKK-mediated mesoderm induction in *Xenopus*. *Dev Biol.* 172: 242–252.

Oelgeschläger, M., H. Kuroda, B. Reversade and E. M. de Robertis. 2003. Chordin is required for the Spemann organizer transplantation phenomenon in *Xenopus* embryos. *Dev. Cell.* In press.

Otte, A. P. and R. T. Moon. 1992. Protein kinase C isozymes have distinct roles in neural induction and competence in *Xenopus*. *Cell* 68: 1021–1029.

Pera, E. M., O. Wessely, S.-S. Li and E. M. De Robertis. 2001. Neural and head induction by insulin-like growth factor signals. *Dev. Cell* 1: 655–665.

Piccolo, S., Y. Sasai, B. Lu and E. M. De Robertis. 1996. Dorsoventral patterning in *Xenopus*: Inhibition of ventral signals by direct binding of chordin to BMP-4. *Cell* 86: 589–598.

Piccolo, S., E. Agius, L. Leyns, S. Bhattacharyya, H. Grunz, T. Bouwmeester and E. M. DeRobertis. 1999. The head inducer Cerberus is a multifunctional antagonist of Nodal, BMP, and Wnt signals. *Nature* 397: 707–710.

Pierce, S. B. and D. Kimelman. 1995. Regulation of Spemann organizer formation by the intracellular kinase Xgsk-3. *Development* 121: 755–765.

Purcell, S. M. and R. Keller. 1993. A different type of amphibian mesoderm morphogenesis in *Ceratophrys ornata*. *Development* 117: 307–317.

Recanzone, G. and W. A. Harris. 1985. Demonstration of neural induction using nuclear markers in *Xenopus*. *Wilhelm Roux Arch. Dev. Biol.* 194: 344–354.

Ryan, A. and 14 others. 1998. Pitx2 determines left-right asymmetries in vertebrates. *Nature* 394: 54–55.

Saka, Y. and J. C. Smith. 2001. Spatial and temporal patterns of cell division during early *Xenopus* embryogenesis. *Dev. Biol.* 229: 307–318.

Sander, K. and P. Fässler. 2001. Introducing the Spemann-Mangold organizer: Experiments and insights that generated a key concept in developmental biology. *Int. J. Dev. Biol.* 45: 1–11.

Saneyoshi, T., S. Kume, Y. Amasaki and K. Mikoshiba. 2002. The Wnt/calcium pathway activates NF-AT and promotes ventral cell fate in *Xenopus* embryos. *Nature* 417: 295–299.

Sasai, Y. 1998. Identifying the missing links: Genes that connect neural induction and primary neurogenesis in vertebrates. *Neuron* 21: 455–458.

Sasai, Y., B. Lu, H. Steinbeisser, D. Geissert, L. K. Gont and E. M. De Robertis. 1994. *Xenopus* chordin: A novel dorsalizing factor activated by organizer-specific homeobox genes. *Cell* 79: 779–790.

Sato, S. M. and T. D. Sargent. 1989. Development of neural inducing capacity in dissociated *Xenopus* embryos. *Dev. Biol.* 134: 263–366.

Savage, R. and C. R. Phillips. 1989. Signals from the dorsal blastopore lip region during gastrulation bias the ectoderm toward a nonepidermal pathway of differentiation in *Xenopus laevis*. *Dev. Biol.* 133: 157–168.

Saxén, L. 1961. Transfilter neural induction of amphibian ectoderm. *Dev. Biol.* 3: 140–152.

Saxén, L. 2001. Spemann's heritage in Finnish developmental biology. *Int. J. Dev. Biol.* 45: 51–55.

Saxén, L. and S. Toivonen. 1962. *Embryonic Induction*. Prentice-Hall, Englewood Cliffs, NJ.

Schmidt, J., V. Francoise, E. Bier and D. Kimelman. 1995. *Drosophila short gastrulation* induces an ectopic axis in *Xenopus*: Evidence for conserved mechanisms of dorsoventral patterning. *Development* 121: 4319–4328.

Schneider, S., H. Steinbeisser, R. M. Warga and P. Hausen. 1996. β-catenin translocation into nuclei demarcates the dorsalizing centers in frog and fish embryos. *Mech. Dev.* 57: 191–198.

Shapira, E., K. Marom, V. Levy, R. Yelin and A. Fainsod. 2000. The Xvex-1 antimorph reveals the temporal competence for organizer forma-

tion and an early role for ventral homeobox genes. *Mech. Dev.* 90: 77–87.

Sharpe, C. R., A. Fritz, E. M. De Robertis and J. B. Gurdon. 1987. A homeobox-containing marker of posterior neural differentiation shows importance of predetermination in neural induction. *Cell* 50: 749–758.

Shi, D.-L., T. D'Arribère, K. E. Johnson and J.-C. Boucaut. 1989. Initiation of mesodermal cell migration and spreading relative to gastrulation in the urodele amphibian *Pleurodeles walti*. *Development* 105: 351–363.

Shook, D. R., C. Majer and R. Keller. 2002. Urodeles remove mesoderm from the superficial layer by subduction through a bilateral primitive streak. *Dev. Biol.* 248: 220–239.

Sive, H. L. and P. F. Cheng. 1991. Retinoic acid perturbs the expression of *Xhox-lab* genes and alters mesodermal determination in *Xenopus laevis*. *Genes Dev.* 5: 1321–1332.

Smith, J. C. 2001. Making mesoderm: Upstream and downstream of *Xbra*. *Int. J. Dev. Biol.* 45: 219–224.

Smith, J. C. and G. M. Malacinski. 1983. The origin of the mesoderm in an anuran, *Xenopus laevis*, and a urodele, *Ambystoma mexicanum*. *Dev. Biol.* 98: 250–254.

Smith, J. C. and J. M. W. Slack. 1983. Dorsalization and neural induction: Properties of the organizer in *Xenopus laevis*. *J. Embryol. Exp. Morphol.* 78: 299–317.

Smith, W. C. and R. M. Harland. 1992. Expression cloning of noggin, a new dorsalizing factor localized to the Spemann organizer in *Xenopus* embryos. *Cell* 70: 829–840.

Smith, W. C., A. K. Knecht, M. Wu and R. M. Harland. 1993. Secreted noggin mimics the Spemann organizer in dorsalizing *Xenopus* mesoderm. *Nature* 361: 547–5410.

Smith, W. C., R. McKendry, S. Ribisi and R. M. Harland. 1995. A *nodal*-related gene defines a physical and functional domain within the Spemann organizer. *Cell* 82: 37–46.

Spemann, H. 1903. Entwicklungsphysiologische Studien am Tritonei. III. *Arch. Entwicklungsmech.* 16: 551–631.

Spemann, H. 1918. Über die Determination der ersten Organanlagen des Amphibienembryo. *Wilhelm Roux Arch. Entwicklungsmech. Org.* 43: 448–555.

Spemann, H. 1927. Neue Arbieten über Organisatoren in der tierischen Entwicklung. *Naturwissenschaften* 15: 946–951.

Spemann, H. 1938. *Embryonic Development and Induction*. Yale University Press, New Haven.

Spemann, H. and H. Mangold. 1924. Induction of embryonic primordia by implantation of organizers from a different species. (Trans. V. Hamburger.) *In* B. H. Willier and J. M. Oppenheimer (eds.), *Foundations of Experimental Embryology*. Hafner, New York, pp. 144–184. Reprinted in *Int. J. Dev. Biol.* 45: 13–38.

Spofford, W. R. 1945. Observations on the posterior part of the neural plate in *Ambystoma*. *J. Exp. Zool.* 99: 35–52.

Stancheva, I., O. El-Maarri, J. Walter, A. Niveleau and R. R. Meehan. 2002. DNA methylation at promoter regions regulates the timing of gene activation in *Xenopus laevis* embryos. *Dev. Biol.* 243(1): 155–165.

Strauss, E. 1998. How embryos shape up. *Science* 281: 166–167.

Suzuki, A., R. S. Thies, N. Yamaji, J. J. Song, J. M. Wozney, K. Muramaki and N. Ueno. 1994. A truncated bone morphogenetic protein receptor affects dorsal-ventral patterning in early *Xenopus* embryo. *Proc. Natl. Acad. Sci. USA* 91: 10255–10259.

Suzuki, A., N. Ueno and A. Hemmati-Brivanlou. 1997b. *Xenopus* msx1 mediates epidermal induction and neural inhibition by BMP4. *Development* 124: 3037–3044.

Toivonen, S. 1938. Spezifische Induktionsleistungen von abnormen Induktoren im Implantatversuch. *Ann. Soc. Zool.-Bot. Fenn. Vanano* 6 (5):1–12.

Toivonen, S. 1979. Transmission problem in primary induction. *Differentiation* 15: 177–181.

Toivonen, S. and L. Saxén. 1955. The simultaneous inducing action of liver and bone marrow of the guinea pig in implantation and explantation experiments with embryos of *Triturus*. *Exp. Cell Res.* [Suppl.] 3: 346–357.

Toivonen, S. and L. Saxen 1968. Morphogenetic interaction of presumptive neural and mesodermal cells mixed in different ratios. *Science* 159: 539–540.

Toivonen, S. and J. Wartiovaara. 1976. Mechanism of cell interaction during primary induction studied in transfilter experiments. *Differentiation* 5: 61–66.

Toivonen, S., D. Tarin, L. Saxén, P. J. Tarin and J. Wartiovaara. 1975. Transfilter studies on neural induction in the newt. *Differentiation* 4: 1–7.

Twitty, V. C. 1966. *Of Scientists and Salamanders*. Freeman, San Francisco.

Valles, J. M., Jr., S. R. Wasserman, C. Schweidenback, J. Edwardson, J. M. Denegre and K. L. Mowry. 2002. Processes that occur before second cleavage determine third cleavage orientation in *Xenopus*. *Exp. Cell Res.* 274: 112–118.

Vincent, J. P. and J. C. Gerhart. 1987. Subcortical rotation in *Xenopus* eggs: An early step in embryonic axis specification. *Dev. Biol.* 123: 526–539.

Vincent, J. P., G. F. Oster and J. C. Gerhart. 1986. Kinematics of gray crescent formation in *Xenopus* eggs: Displacement of subcortical cytoplasm relative to the egg surface. *Dev. Biol.* 113: 484–500.

Wallingford, J. B., A. J. Ewald, R. M. Harland and S. F. Fraser. 2001. Calcium signaling during convergent extension in *Xenopus*. *Curr. Biol.* 11: 652–661.

Wang, S., M. Krinks, K. Lin, F. P. Luyten and M. Moos, Jr. 1997. Frzb, a secreted protein expressed in the Spemann organizer, binds and inhibits Wnt-8. *Cell* 88: 757–766.

Wessley, O., E. Agius, M. Oelgeschläger, E. M. Pera and E. M. De Robertis. 2001. Neural induction in the absence of mesoderm: β-catenin-dependent expression of secreted BMP antagonists at the blastula stage in *Xenopus*. *Dev. Biol.* 234: 161–173.

White, R. J., B. I. Sun, H. L. Sive and J. C. Smith. 2002. Direct and indirect regulation of *derrière*, a *Xenopus* mesoderm-inducing factor, by VegT. *Development* 129: 4867–4876.

Wilson, P. A. and A. Hemmati-Brivanlou. 1995. Induction of epidermis and inhibition of neural fate by BMP-4. *Nature* 376: 331–333.

Wilson, P. and R. Keller. 1991. Cell rearrangement during gastrulation of *Xenopus*: Direct observation of cultured explants. *Development* 112: 289–300.

Winklbauer, R. and R. E. Keller. 1996. Fibronectin, mesoderm migration, and gastrulation in *Xenopus. Dev. Biol.* 177: 413–42

Winklbauer, R. and M. Nagel. 1991. Directional mesoderm cell migration in the *Xenopus* gastrula. *Dev. Biol.* 148: 573–589.

Winklbauer, R. and M. Schürfeld. 1999. Vegetal rotation, a new gastrulation movement involved in the internalization of the mesoderm and endoderm in *Xenopus. Development* 126: 3703–3713.

Wylie, C., M. Kofron, C. Payne, R. Anderson, M. Hosobuchi, E. Joseph and J. Heasman. 1996. Maternal β-catenin establishes a "dorsal signal" in early *Xenopus* embryos. *Development* 122: 2987–2996.

Yamamoto, T. S., C. Takagi and N. Ueno. 2000. Requirement of Xmsx-1 in the BMP-triggered ventralization of *Xenopus* embryos. *Mech. Dev.* 91: 131–141.

Yang, J., C. Tan, R. S. Darken, P. A. Wilson and P. S. Klein. 2002. β-Catenin/Tcf-regulated transcription prior to the midblastula transition. *Development* 129: 5743–5752.

Yost, C., M. Torres, J. R. Miller, E. Huang, D. Kimelman and R. T. Moon. 1996. The axis-inducing ability, stability, and subcellular localization of β-catenin are regulated in *Xenopus* embryos by glycogen synthase kinase-3. *Genes Dev.* 10: 1443–1454.

Yost, H. J. 1998. Left-right development in *Xenopus* and zebrafish. *Semin. Cell Dev. Biol.* 9: 61–66.

Zhang, J., D. W. Houston, M. L. King, C. Payne, C. Wylie and J. Heasman. 1998. The role of maternal VegT in establishing the primary germ layers in *Xenopus* embryos. *Cell* 94: 515–524.

Zimmerman, L. B., J. M. de Jesús Escobar and R. M. Harland. 1996. The Spemann organizer signal noggin binds and inactivates bone morphogenesis protein 4. *Cell* 86: 599–606.

11 The early development of vertebrates: Fish, birds, and mammals

THIS FINAL CHAPTER ON THE PROCESSES of early development will extend our survey of vertebrate development to include fish, birds, and mammals. The amphibian embryos described in the previous chapter divide by means of radial holoblastic cleavage. Cleavage in bird, reptile, and fish eggs is meroblastic, with only a small portion of the egg cytoplasm being used to make cells. Mammals modify their holoblastic cleavage to make a placenta, which enables the embryo to develop inside another organism. Although methods of gastrulation also differ among the vertebrate classes, there are some underlying principles common to all vertebrates.

EARLY DEVELOPMENT IN FISH

In recent years, the teleost fish *Danio rerio*, commonly known as the zebrafish, has become a favorite model organism of those who study vertebrate development. Zebrafish have large broods, breed all year, are easily maintained, have transparent embryos that develop outside the mother (an important feature for microscopy), and can be raised so that mutants can be readily screened and propagated. In addition, they develop rapidly, so that at 24 hours after fertilization, the embryo has formed most of its tissues and organ primordia and displays the characteristic tadpole-like form (Figure 11.1; see Granato and Nüsslein-Volhard 1996; Langeland and Kimmel 1997). Therefore, much of the description of fish development below is based on studies of this species.

The zebrafish is also the first vertebrate for which intensive mutagenesis has been attempted. By treating parents with mutagens and selectively breeding their progeny, scientists have found thousands of mutations whose normally functioning genes are critical for zebrafish development. The traditional method of genetic screening (modeled after large-scale screens in *Drosophila*) begins when the male parental fish are treated with a chemical mutagen that will cause random mutations in their germ cells (Figure 11.2). Each mutagenized male is then mated with a wild-type female fish to generate F_1 lines. Individuals in the F_1 generation carry the mutations inherited from their father. If the mutation is dominant, it will become expressed in the F_1 generation. If these mutations are recessive, the F_1 fish will not show a mutant phenotype, since the wild-type allele will mask the mutation. The F_1 fish are then mated with wild-type fish to produce the F_2 generation. This F_2 generation will include both males and females that have the mutant allele. When two F_2 parents contain the same recessive mutation, there is a 25% chance that their offspring will show the mutant phenotype (Figure 11.2). Since zebrafish development occurs in the open (not in an

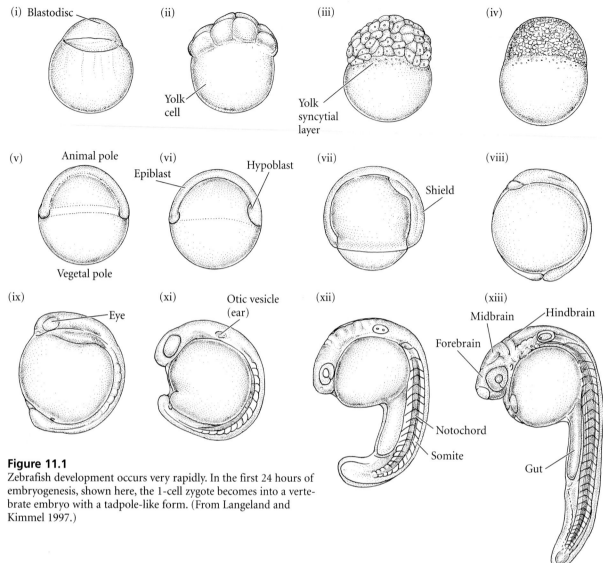

Figure 11.1
Zebrafish development occurs very rapidly. In the first 24 hours of embryogenesis, shown here, the 1-cell zygote becomes into a vertebrate embryo with a tadpole-like form. (From Langeland and Kimmel 1997.)

opaque shell or within the mother), abnormal developmental stages can be readily observed and the defects in development can often be traced to changes in a particular group of cells (Driever et al. 1996; Haffter et al. 1996).

The similarity of developmental programs among all vertebrates has given the zebrafish an important role in investigating the genes that operate during human development. The genetic screening method described above has enabled us to identify thousands of genes that may also be active in human development. For instance, the zebrafish *mariner* gene encodes the myosin VIIA protein found in the otic vesicle. Mutations of this gene in zebrafish impair hearing by preventing the sensory cells of the otic vesicle from forming properly. Humans with a disruption of the gene that encodes myosin VIIA have a syndrome of congenital deafness (Ernest et al. 2000).

Zebrafish genes are extremely amenable to study. Zebrafish embryos are susceptible to morpholino antisense molecules (Zhong et al. 2001; see Chapter 4), and researchers can use this method to test whether a particular gene is required

for a particular function. Furthermore, the green fluorescent protein (GFP) reporter gene can be fused with specific zebrafish promoters and enhancers and inserted into the fish embryos. The resulting transgenic fish will express GFP at the same times and places as the actual proteins controlled by these regulatory sequences. The amazing thing is that one can observe the reporter protein in living embryos (Figure 11.3).

And, as one more attribute, zebrafish embryos are permeable to small molecules placed in its water. This property allows us to test drugs that may be deleterious to vertebrate development. For instance, zebrafish development can be altered by the addition of ethanol or retinoic acid in manners that resemble human developmental syndromes caused by these molecules (Blader and Strähle 1998).

Given these advantages of research on this organism, many developmental biologists believe that zebrafish will allow us to acquire a detailed outline of vertebrate development.

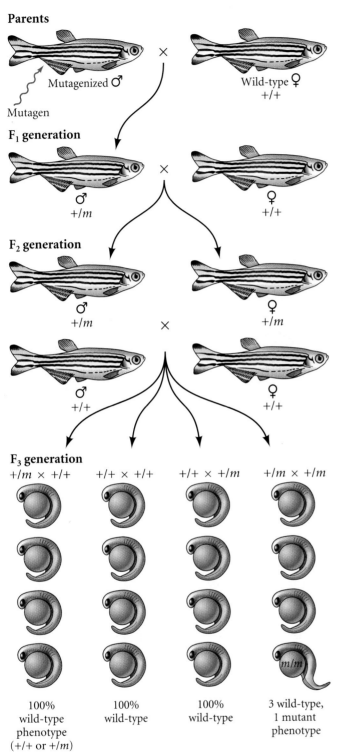

Parents

Mutagenized ♂

Mutagen

Wild-type ♀
+/+

F₁ generation

♂
+/m

♀
+/+

F₂ generation

♂
+/m

♀
+/m

♂
+/+

♀
+/+

F₃ generation

+/m × +/+   +/+ × +/+   +/+ × +/m   +/m × +/m

m/m

100%
wild-type
phenotype
(+/+ or +/m)

100%
wild-type

100%
wild-type

3 wild-type,
1 mutant
phenotype

Figure 11.2
Screening protocol for identifying mutations of zebrafish development. The male parent is mutagenized and mated with a wild-type (+/+) female. If some of the male's sperm carry a recessive mutant allele (m), then some of the F₁ progeny of the mating will inherit that allele. F₁ individuals (here shown as a male carrying the mutant allele m) are then mated with wild-type partners. This creates an F₂ generation wherein some males and some females carry the recessive mutant allele. When the F₂ fish are mated, some of their progeny will show the mutant phenotype. (After Haffter et al. 1996.)

WEBSITE 11.1   GFP zebrafish movies and photographs. The ability to photograph and film living embryos expressing the GFP reporter gene at specific promoter sites has opened a new dimension in developmental biology and has allowed us to link gene structure to developmental anatomy.

VADE MECUM² Zebrafish development. A full account of zebrafish development includes time-lapse movies of the beautiful and rapid development of this organism. **[Click on Zebrafish]**

Cleavage in Fish Eggs

Fish eggs are **telolecithal**, meaning that most of the egg cell is occupied by yolk. Cleavage can take place only in the **blastodisc**, a thin region of yolk-free cytoplasm at the animal pole of the egg. The cell divisions do not completely divide the egg, so this type of cleavage is called **meroblastic** (Greek *meros*,

(A)

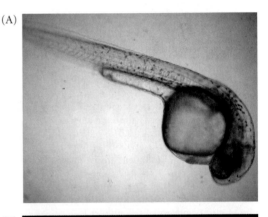

(B)

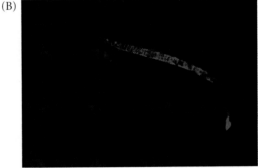

Figure 11.3
Regulated GFP in living zebrafish embryos. The gene for green fluorescent protein (GFP) was fused to the regulatory region of a zebrafish *hedgehog* gene. As a result, GFP was synthesized wherever the Hedgehog protein is normally expressed in the fish embryo. Here its expression is seen in the notochord of the living 24-hour zebrafish embryo. (A) Bright-field photograph of the embryo. (B) Photograph of the same embryo in fluorescent light. (From Du and Dienhart 2001; photographs courtesy of S. J. Du.)

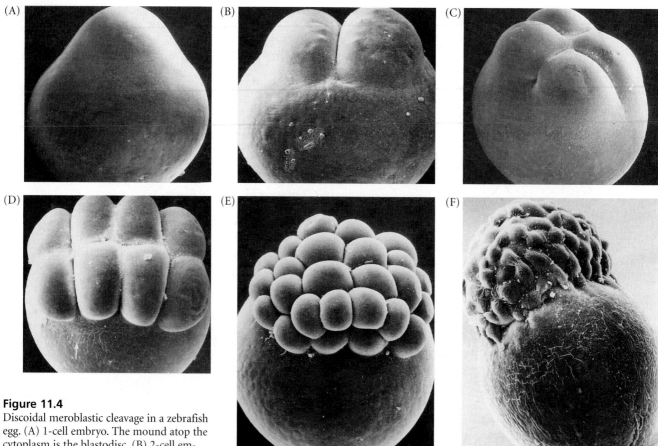

Figure 11.4
Discoidal meroblastic cleavage in a zebrafish egg. (A) 1-cell embryo. The mound atop the cytoplasm is the blastodisc. (B) 2-cell embryo. (C) 4-cell embryo. (D) 8-cell embryo, wherein two rows of four cells are formed. (E) 32-cell embryo. (F) 64-cell embryo, wherein the blastodisc can be seen atop the yolk cell. (From Beams and Kessel 1976; photographs courtesy of the authors.)

"part"). Since only the blastodisc becomes the embryo, this type of meroblastic cleavage is called **discoidal**.

Scanning electron micrographs show beautifully the incomplete nature of discoidal meroblastic cleavage in fish eggs (Figure 11.4). The calcium waves initiated at fertilization stimulate the contraction of the actin cytoskeleton to squeeze nonyolky cytoplasm into the animal pole of the egg. This process converts the spherical egg into a pear-shaped structure with an apical blastodisc (Leung et al. 1998, 2000). The first cell divisions follow a highly reproducible pattern of meridional and equatorial cleavages. These divisions are rapid, taking about 15 minutes each. The first 12 divisions occur synchronously, forming a mound of cells that sits at the animal pole of a large **yolk cell**. These cells constitute the **blastoderm**. Initially, all the cells maintain some open connection with one another and with the underlying yolk cell, so that moderately sized (17-kDa) molecules can pass freely from one blastomere to the next (Kimmel and Law 1985).

Beginning at about the tenth cell division, the onset of the mid-blastula transition can be detected: zygotic gene transcription begins, cell divisions slow, and cell movement becomes evident (Kane and Kimmel 1993). At this time, three distinct cell populations can be distinguished. The first of these is the **yolk syncytial layer** (**YSL**). The YSL is formed at the ninth or tenth cell cycle, when the cells at the vegetal edge of the blastoderm fuse with the underlying yolk cell. This fusion produces a ring of nuclei within the part of the yolk cell cytoplasm that sits just beneath the blastoderm. Later, as the blastoderm expands vegetally to surround the yolk cell, some of the yolk syncytial nuclei will move under the blastoderm to form the **internal YSL**, and some of them will move vegetally, staying ahead of the blastoderm margin, to form **external YSL** (Figure 11.5A,B). The YSL will be important for directing some of the cell movements of gastrulation. The second cell population distinguished at the mid-blastula transition is the **enveloping layer** (**EVL**; Figure 11.5A). It is made up of the most superficial cells from the blastoderm, which form an epithelial sheet a single cell layer thick. The EVL eventually becomes the **periderm**, an extraembryonic protective covering that is sloughed off during later development. Between the EVL and the YSL are the **deep cells**. These are the cells that give rise to the embryo proper.

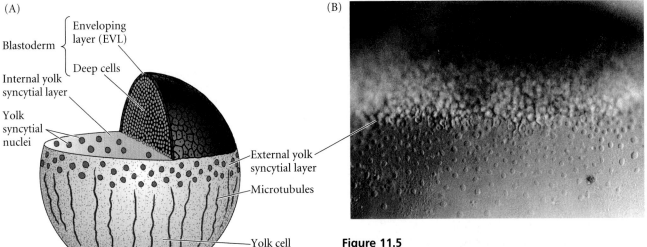

(A)

Blastoderm { Enveloping layer (EVL)
Deep cells

Internal yolk syncytial layer

Yolk syncytial nuclei

External yolk syncytial layer

Microtubules

Yolk cell

(B)

Figure 11.5

Fish blastula. (A) Prior to gastrulation, the deep cells are surrounded by the enveloping layer (EVL). The animal surface of the yolk cell is flat and contains the nuclei of the yolk syncitial layer (YSL). Microtubules extend through the yolky cytoplasm and through the external region of the YSL. (B) Late blastula-stage embryo of the minnow *Fundulus*, showing the external YSL. The nuclei of these cells were derived from cells at the margin of the blastoderm, which released their nuclei into the yolky cytoplasm. (C) Fate map of the deep cells after cell mixing has stopped. The lateral view is shown, and not all organ fates are labeled (for the sake of clarity). (A and C after Langeland and Kimmel 1997; B from Trinkaus 1993, photograph courtesy of J. P. Trinkaus.)

(C)

Animal pole

Nose, eye

Epidermis

Brain

Neural crest

Spinal cord

Ectoderm

Somite muscle

Head

Mesoderm

Ventral

Pronephros

Noto-chord

Dorsal

Blood Fins Heart Muscle

Intestine Liver Pharynx

Endoderm

Blastoderm margin

Yolk cell

Vegetal pole

The fates of the early blastoderm cells are not determined, and cell lineage studies (in which a nondiffusible fluorescent dye is injected into a cell so that its descendants can be followed) show that there is much cell mixing during cleavage. Moreover, any one of these early blastomeres can give rise to an unpredictable variety of tissue descendants (Kimmel and Warga 1987; Helde et al. 1994). The fate of the blastoderm cells appears to be fixed shortly before gastrulation begins. At this time, cells in specific regions of the embryo give rise to certain tissues in a highly predictable manner, allowing a fate map to be made (Figure 11.5C; see also Figure 1.6; Kimmel et al. 1990).

Gastrulation in Fish Embryos

The first cell movement of fish gastrulation is the epiboly of the blastoderm cells over the yolk. In the initial phase of this movement, the deep cells of the blastoderm move outward to intercalate with the more superficial cells (Warga and Kimmel 1990). Later, this concatenation of cells moves vegetally over the surface of the yolk and envelops it completely (Figure 11.6). This movement is not due to the active crawling of the blastomeres, however. Rather, the downward movement toward the vegetal pole is a result of the autonomous expansion of the YSL "within" the yolk cell cytoplasm. The EVL is tightly joined to the YSL and is dragged along with it. The deep cells of the blastoderm then fill in the space between the YSL and the EVL as epiboly proceeds. This mechanism can be demonstrated by severing the attachments between the YSL and the EVL. When this is done, the EVL and the deep cells spring back to the top of the yolk, while the YSL continues its expansion around the yolk cell (Trinkaus 1984, 1992). The expansion of the YSL depends on a network of microtubules within it, and radiation or drugs that block the polymerization of tubulin inhibit epiboly (Strahle and Jesuthasan 1993; Solnica-Krezel and Driever 1994).

During epiboly, one side of the blastoderm becomes noticeably thicker than the other. Cell labeling experiments indicate that the thicker side marks the site of the future dorsal surface of the embryo (Schmitz and Campos-Ortega 1994).

The formation of germ layers

After the blastoderm cells have covered about half the zebrafish yolk cell (and earlier in fish eggs with larger yolks), a thicken-

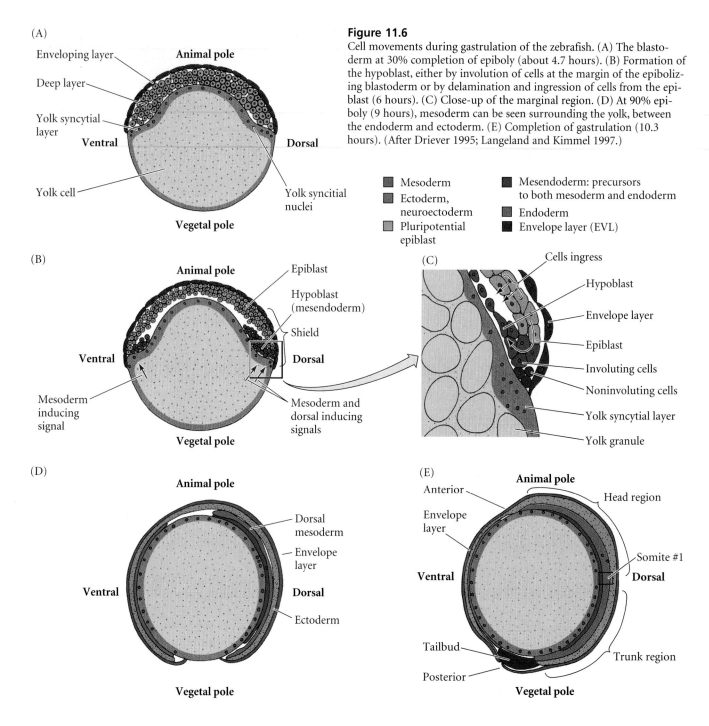

(A)
Enveloping layer
Animal pole
Deep layer
Yolk syncytial layer
Ventral
Dorsal
Yolk cell
Yolk syncytial nuclei
Vegetal pole

Figure 11.6
Cell movements during gastrulation of the zebrafish. (A) The blastoderm at 30% completion of epiboly (about 4.7 hours). (B) Formation of the hypoblast, either by involution of cells at the margin of the epibolizing blastoderm or by delamination and ingression of cells from the epiblast (6 hours). (C) Close-up of the marginal region. (D) At 90% epiboly (9 hours), mesoderm can be seen surrounding the yolk, between the endoderm and ectoderm. (E) Completion of gastrulation (10.3 hours). (After Driever 1995; Langeland and Kimmel 1997.)

Mesoderm
Ectoderm, neuroectoderm
Pluripotential epiblast
Mesendoderm: precursors to both mesoderm and endoderm
Endoderm
Envelope layer (EVL)

(B)
Animal pole
Epiblast
Hypoblast (mesendoderm)
Shield
Ventral
Dorsal
Mesoderm inducing signal
Mesoderm and dorsal inducing signals
Vegetal pole

(C)
Cells ingress
Hypoblast
Envelope layer
Epiblast
Involuting cells
Noninvoluting cells
Yolk syncytial layer
Yolk granule

(D)
Animal pole
Dorsal mesoderm
Envelope layer
Ventral
Dorsal
Ectoderm
Vegetal pole

(E)
Animal pole
Anterior
Head region
Envelope layer
Ventral
Dorsal
Somite #1
Tailbud
Posterior
Trunk region
Vegetal pole

ing occurs throughout the margin of the epibolizing blastoderm. This thickening is called the **germ ring**, and it is composed of a superficial layer, the **epiblast**, and an inner layer, the **hypoblast**. We do not understand how the hypoblast is made. Some research groups claim that the hypoblast is formed by the involution of superficial cells from the blastoderm under the margin followed by their migration toward the animal pole (see Figure 11.6C); in this scenario, involution begins at the future dorsal portion of the embryo, but occurs all around the margin. Other laboratories claim that superficial cells ingress

to form the hypoblast (see Trinkaus 1996). It is possible that both mechanisms are at work, with different modes of hypoblast formation predominating in different species.

Once the hypoblast has formed, cells of the epiblast and hypoblast intercalate on the future dorsal side of the embryo to form a localized thickening, the **embryonic shield** (Figure 11.7). As we will see, the embryonic shield is functionally equivalent to the dorsal blastopore lip of amphibians, since it can organize a secondary embryonic axis when transplanted to a host embryo (Oppenheimer 1936; Ho 1992). Thus, as the

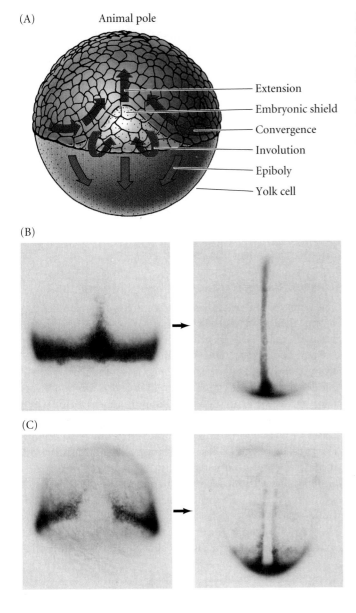

(A) Animal pole

— Extension
— Embryonic shield
— Convergence
— Involution
— Epiboly
— Yolk cell

(B)

(C)

Figure 11.7
Convergence and extension in the zebrafish gastrula. (A) Dorsal view of convergence and extension movements during zebrafish gastrulation. Epiboly spreads the blastoderm over the yolk; involution or ingression generates the hypoblast; convergence and extension bring the hypoblast and epiblast cells to the dorsal side to form the embryonic shield. Within the shield, intercalation extends the chordamesoderm toward the animal pole. (B) Convergent extension of the chordamesoderm is shown by those cells expressing the gene *no tail*, a gene that is expressed by notochord cells. (C) Convergent extension of paraxial mesodermal cells to flank the notochord. These cells are marked by their expression of the *snail* gene (dark areas). (From Langeland and Kimmel 1997; photographs courtesy of the authors.)

cells of the blastoderm undergo epiboly around the yolk, they are also involuting at the margins and converging anteriorly and dorsally toward the embryonic shield (Trinkaus 1992). The hypoblast cells of the embryonic shield itself converge and extend anteriorly, eventually narrowing along the dorsal midline of the hypoblast. This movement forms the **chordamesoderm**, the precursor of the **notochord** (Figure 11.7B). The cells adjacent to the chordamesoderm, the **paraxial mesoderm** cells, are the precursors of the mesodermal somites (Figure 11.7C). Concomitant convergence and extension in the epiblast brings presumptive neural cells from all over the epiblast into the dorsal midline, where they form the **neural keel**. Those cells remaining in the epiblast become the ectoderm. The zebrafish fate map, then, is not much different from that of the frog or other vertebrates (as we will soon

see). If one conceptually opens a *Xenopus* blastula at the vegetal pole and stretches the opening into a marginal ring, the resulting fate map closely resembles that of the zebrafish embryo at the stage when half of the yolk has been covered by the blastoderm (see Figure 1.9; Langeland and Kimmel 1997).

Meanwhile, the endoderm arises from the most marginal blastomeres of the late blastula- stage embryo. These blastomeres involute early in gastrulation and occupy the deep layers of the hypoblast, directly above the YSL.

Axis Formation in Fish Embryos

Dorsal-ventral axis formation: The embryonic shield

The embryonic shield is critical in establishing the dorsal-ventral axis in fish. It can convert lateral and ventral mesoderm (blood and connective tissue precursors) into dorsal mesoderm (notochord and somites), and it can cause the ectoderm to become neural rather than epidermal. This was shown by transplantation experiments in which the embryonic shield of one early-gastrula embryo was transplanted to the ventral side of another (Figure 11.8; Oppenheimer 1936; Koshida et al. 1998). Two axes formed, sharing a common yolk cell. Although the prechordal plate and notochord were derived from the donor embryonic shield, the other organs of the secondary axis came from host tissues that would have formed ventral structures. This new axis had been induced by the donor cells. In the embryo that had had its embryonic shield removed, no dorsal structures formed, and the embryo lacked a nervous system. These experiments are similar to those performed on amphibian gastrulae by Spemann and Mangold (1924; see Chapter 10), and they demonstrate that the embryonic shield is the homologue of the dorsal blastopore lip, the amphibian organizer.

Like the amphibian dorsal blastopore lip, the embryonic shield forms the prechordal plate and the notochord of the developing embryo. The precursors of these two regions are responsible for inducing the ectoderm to become neural ectoderm. Moreover, the presumptive notochord and prechordal plate appear to do this in a manner very much like the homol-

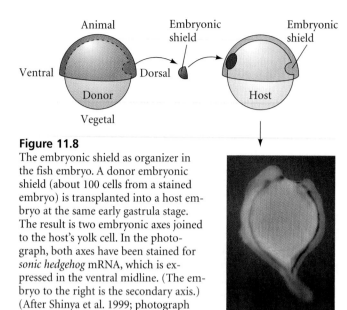

Figure 11.8
The embryonic shield as organizer in the fish embryo. A donor embryonic shield (about 100 cells from a stained embryo) is transplanted into a host embryo at the same early gastrula stage. The result is two embryonic axes joined to the host's yolk cell. In the photograph, both axes have been stained for *sonic hedgehog* mRNA, which is expressed in the ventral midline. (The embryo to the right is the secondary axis.) (After Shinya et al. 1999; photograph courtesy of the authors.)

ogous structures in amphibians.* In both fish and amphibians, BMPs and certain Wnt proteins made in the ventral and lateral regions of the embryo would normally induce the ectoderm to become epidermis. The notochord of both fish and amphibians secretes factors that block this induction and thereby allow the ectoderm to become neural. In fish, **BMP2B** induces embryonic cells to acquire ventral and lateral fates and **Wnt8** ventralizes, lateralizes, and posteriorizes the embryonic tissues (see Schier 2001). The protein secreted by the chordamesoderm that binds with and inactivates BMP2B is a

---

*Another similarity between the amphibian and fish organizers is that they can be duplicated by rotating the egg and changing the orientation of the microtubules (Fluck et al. 1998). One difference in the axial development of these groups is that in amphibians (see Chapter 10), the prechordal plate is necessary for inducing the anterior brain to form. In *Danio*, the prechordal plate appears to be necessary for forming ventral neural structures, but the anterior regions of the brain can form in its absence (Schier et al. 1997; Schier and Talbot 1998).

chordin-like paracrine factor called **Chordino** (Figure 11.9B; Kishimoto et al. 1997; Schulte-Merker et al. 1997). If the *chordino* gene is mutated, the neural tube fails to form. It is hypothesized (Nguyen et al. 1998) that different concentrations of BMP2B pattern the ventral and lateral regions of the zebrafish ectoderm and mesoderm, and that the ratio of Chordino to BMP2B may specify the position along the dorsal-ventral axis. In fish, however, the notochord may not be the only structure capable of producing the proteins that block BMP2B. If the notochord fails to form (as in the *floating head* or *no tail* mutations), the neural tube will still be produced. It is possible that the notochordal precursor cells (which are produced in these mutations) are still able to induce the neural tube, or that the dorsal portion of the somite precursors can compensate for the lack of a notochord (Halpern et al. 1993, 1995; Hammerschmidt et al. 1996).

The fish Nieuwkoop center

The embryonic shield appears to acquire its organizing ability in much the same way as its amphibian counterparts. In amphibians, the endoderm cells beneath the dorsal blastopore lip (i.e., the Nieuwkoop center) accumulate β-catenin synthesized from maternal messages. This protein is critical in amphibians for the ability of the endoderm to induce the cells above it to become the dorsal lip (organizer) cells. In zebrafish, the nuclei in that part of the yolk syncytial layer that lies beneath the cells that will become the embryonic shield similarly accumulate β-catenin. This protein distinguishes the dorsal YSL from the lateral and ventral YSL regions (Figure 11.10; Schneider et al. 1996). Inducing β-catenin accumulation on the ventral side of the egg causes dorsalization and a second embryonic axis (Kelly et al. 1995).

The β-catenin of the embryonic shield combines with the zebrafish homologue of Tcf3 to become an active transcrip-

Figure 11.9
Axis formation in the zebrafish embryo. (A) Prior to gastrulation, the zebrafish blastoderm is arranged with the presumptive ectoderm near the animal pole, the presumptive mesoderm beneath it, and the presumptive endoderm sitting atop the yolk cell. The yolk syncitial layer (and possibly the endoderm) sends two signals to the presumptive mesoderm. One signal (lighter arrows) induces the mesoderm, while a second signal (heavy arrow) specifically induces an area of mesoderm to become the dorsal mesoderm (embryonic shield). (B) Formation of the dorsal-ventral axis. During gastrulation, the ventral mesoderm secretes BMP2B (arrows) to induce the ventral and lateral mesodermal and epidermal differentiation. The dorsal mesoderm secretes factors (such as Chordino) that block BMP2B and dorsalize the mesoderm and ectoderm (converting the latter into neural tissue). (After Schier and Talbot 1998.)

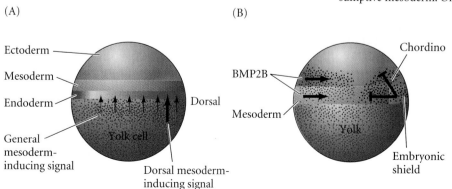

(A)

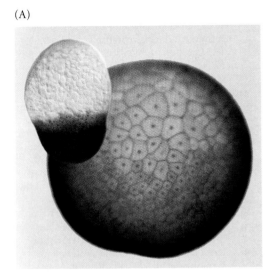

(B)

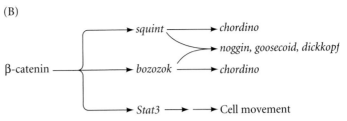

Figure 11.10
β-Catenin activates organizer genes in the zebrafish. (A) Nuclear localization of β-catenin marks the dorsal side of the *Xenopus* blastula (larger image) and helps form its Nieuwkoop center beneath the organizer. In the zebrafish late blastula (smaller image), nuclear localization of β-catenin is seen in the yolk syncytial layer nuclei beneath the future embryonic shield. (B) β-catenin activates *squint* and *bozozok*, whose proteins activate organizer-specific genes as well as the *Stat3* gene whose product is necessary for gastrulation movements. (A, photograph courtesy of S. Schneider; B after Schier and Talbot 2001.)

tion factor. It activates the genes encoding two proteins, **Squint** and **Bozozok** that are very similar to the proteins that pattern the amphibian mesoderm. Squint is a Nodal-like paracrine factor that acts to pattern the mesoderm, while Bozozok is a homeodomain protein similar to Siamois.

The Bozokok protein works in several ways.* First, acting alone, it can repress *bmp* and *wnt* genes that would promote ventral functions (Solnica-Krezel and Driever 2001). Second, Bozozok suppresses the *vega1* gene that is active in all other blastomeres and which acts to repress the *bozozok* gene (Kawahara et al. 2000). Third, Bozozok and Squint act individually to activate the *chordino* genes, and they act synergistically to activate the other organizer genes such as *goosecoid*, *noggin*, and *dickkopf* (Figure 11.10B; Sampath et al. 1998; Gritsman et al. 2000; Schier and Talbot 2001.) These genes encode the proteins that block BMPs and Wnts and allow the specification of the dorsal mesoderm and neural ectoderm. Thus, the embryonic shield is considered equivalent to the amphibian organizer, and the dorsal part of the yolk cell can be thought of as the **Nieuwkoop center** of the fish embryo.

In addition to its function in forming the fish organizer, Nodal-like signaling is also critical for the formation of the endoderm. Here, Squint and Cyclops (another Nodal-related protein) induce the Bon and GATA5 transcription factors. These two transcription factors are co-expressed in the marginal domain of the late blastula and regulate the expression of the downstream genes that form the endoderm (Reiter et al. 2001).

Maternal β-catenin also initiates a third pathway, one that coordinates the cell movements of the mesoderm and endoderm during gastrulation. Dorsal mesoderm cells derived from the embryonic shield generate the notochord posteriorly and the prechordal plate mesoderm anteriorly. The anterior cells are highly motile and lead the mesoderm into the embryo. The cells that form the notochord are not motile, but they undergo convergent extension similar to the processes seen in *Xenopus*. Both the activity of the anterior cells and the convergent extension of the notochordal cells are regulated by **Stat3**, a transcription factor subunit that is itself regulated positively by β-catenin (Figure 11.10B; Yamashita et al. 2002). Embryos injected with antisense morpholinos to Stat3 had a mispositioned head and a shortened anterior-posterior axis (Figure 11.11). Without Stat3, the anterior movements of the dorsal mesoderm and the dorsal convergence of the nonaxial mesoderm do not occur.

Anterior-posterior patterning

Just as in *Xenopus*, the patterning of the neural ectoderm and mesoderm in the zebrafish appears to be due to Wnt proteins. The Wnt8 protein is especially important for the patterning of these areas. Morpholinos against Wnt8 (preventing its translation) or deletions of the Wnt8 gene cause the expansion of the neural ectoderm and dorsal mesoderm at the expense of the ventral and posterior regions (Levken et al. 2001).

Zebrafish have meroblastic cleavage and a huge yolk cell with syncytial nuclei. Amphibians such as *Xenopus* have holoblastic cleavage and a blastocoel. Yet both types of embryos use similar molecular tools to undergo cleavage, gastrulation, and axis specification. Both use β-catenin and Nodal-related proteins to form the dorsal mesoderm and enable this mesoderm to express the organizer genes. Both use BMPs and Wnts to lateralize and vegetalize the embryo, and the organizer genes encode proteins such as chordin, Noggin, and Dickkopf that antagonize the BMPs and Wnts. Furthermore,

---

*Bozozok is Japanese slang for an arrogant youth on a motorcycle. The name comes from the severe *loss-of-function* phenotype wherein a single-lensed embryo curves ventrally over the yolk cell. However, this gene is also called *Dharma* (the name of a famous Buddhist priest) because embryos with *too much* of this protein (resulting from the injection of its mRNA into the embryo) develop huge eyes and head, but no trunk or tail. They thus resemble Japanese Dharma dolls (Yamanaka et al. 1998; Fekany et al. 1999). So perhaps this gene regulates moral consciousness, as well.

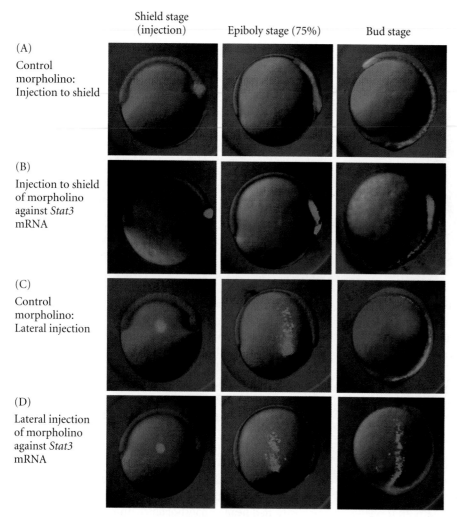

Shield stage
(injection)   Epiboly stage (75%)   Bud stage

(A)
Control
morpholino:
Injection to shield

(B)
Injection to shield
of morpholino
against *Stat3*
mRNA

(C)
Control
morpholino:
Lateral injection

(D)
Lateral injection
of morpholino
against *Stat3*
mRNA

Figure 11.11
Requirement of Stat3 for coordinated cell movement during zebrafish gastrulation. (A) Injection of fluoresceinated dextran (green) into the embryonic shield along with a control morpholino. The axial mesoderm migrates anteriorly. (B) Injection of fluorescienated dextran into the shield along with an antisense morpholino against *Stat3* mRNA. The migration of the axial mesoderm fails to occur. (C) Injection of fluoresceinated dextran and control morpholino into lateral mesoderm. The lateral mesoderm undergoes convergent extension to the paraxial region. (D) Injection of fluoresceinated dextran and *Stat3* antisense morpholino into lateral mesoderm. Here the mesoderm fails to undergo convergent extension and remains lateral. (After Yamashita et al. 2002; photograph courtesy of T. Hirano.)

both zebrafish and *Xenopus* use a particular Wnt protein later in development to posteriorize the ectoderm to form the trunk neural tube. In some instances, they use these proteins in different ways; but the result is a structure that is definitely recognizable as a vertebrate embryo.

EARLY DEVELOPMENT IN BIRDS

Cleavage in Bird Eggs

Ever since Aristotle first followed its 3-week development, the domestic chicken (*Gallus*) has been a favorite organism for embryological studies. It is accessible year-round and is easily raised. Moreover, at any particular temperature, its developmental stage can be accurately predicted. Thus, large numbers of embryos can be obtained at the same stage. The chick embryo can be surgically manipulated and, since chick organ formation is accomplished by genes and cell movements similar

to those of mammalian organ formation, the chick embryo has often served as an inexpensive (and more morally acceptable) surrogate for human embryos.

Fertilization of the chick egg occurs in the oviduct, before the albumen and the shell are secreted upon it. The egg is telolecithal (like that of the fish), with a small disc of cytoplasm sitting atop a large yolk. Like fish eggs, the yolky eggs of birds undergo discoidal meroblastic cleavage. Cleavage occurs only in the blastodisc, a small disc of cytoplasm 2–3 mm in diameter at the animal pole of the egg cell. The first cleavage furrow appears centrally in the blastodisc, and other cleavages follow to create a single-layered blastoderm (Figure 11.12). As in the fish embryo, these cleavages do not extend into the yolky cytoplasm, so the early-cleavage cells are continuous with one another and with the yolk at their bases (Figure 11.12E). Thereafter, equatorial and vertical cleavages divide the blastoderm into a tissue five to six cell layers thick. The cells become linked together by tight junctions (Bellairs et al. 1975; Eyal-Giladi 1991). Between the blastoderm and the yolk is a space called the **subgerminal cavity** (Figure 11.13A). This space is created when the blastoderm cells absorb water from the albumen ("egg white") and secrete the fluid between themselves and the yolk (New 1956). At this stage, the deep cells in the center of the blastoderm are shed and die, leaving behind a one-cell-thick **area pellucida**. This part of the blastoderm forms most of the actual embryo. The peripheral ring of blastoderm cells that have not shed their deep cells constitutes the **area opaca**. Between the area pellucida and the area opaca is a thin layer of

Figure 11.12
Discoidal meroblastic cleavage in a chick egg. (A–D) Four stages viewed from the animal pole (the future dorsal side of the embryo). (E) An early-cleavage embryo viewed from the side. (After Bellairs et al. 1978.)

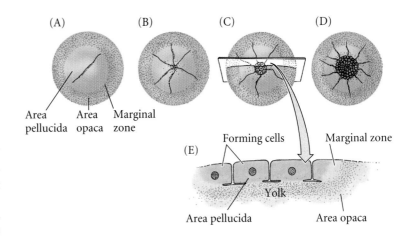

cells called the **marginal zone** (or **marginal belt**) (Eyal-Giladi 1997; Arendt and Nübler-Jung 1999).* Some of the marginal zone cells become very important in determining cell fate during early chick development.

---

*Arendt and Nübler-Jung (1999) have argued that the region should be called the marginal *belt* to distinguish it from the marginal zone of amphibians. Here we will continue to use the earlier nomenclature.

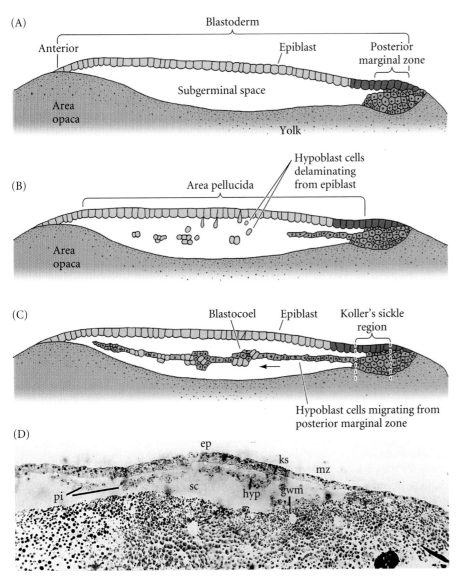

Figure 11.13
Formation of the two-layered blastoderm of the chick embryo. (A, B) Primary hypoblast cells delaminate individually to form islands of cells beneath the epiblast. (C) Secondary hypoblast cells from the posterior margin (Koller's sickle and the posterior marginal cells behind it) migrate beneath the epiblast (ep) and incorporate the polyinvagination islands. As the hypoblast moves anteriorly, epiblast cells collect at the region anterior to Koller's sickle to form the primitive streak. (D) This sagittal section of an embryo near the posterior margin shows an upper layer consisting of a central epiblast (ep) that trails into the cells of Koller's sickle (ks) and the posterior marginal zone (mz). Certain cells have delaminated from the epiblast to form polyinvagination islands (pi) of 5 to 20 cells each. These cells will be joined by those hypoblast cells (hyp) migrating anteriorly from Koller's sickle to form the lower (secondary hypoblastic) layer. (sc, subgerminal cavity; gwm, germ wall margin.) (From Eyal-Giladi et al. 1992; photograph courtesy of H. Eyal-Giladi.)

Gastrulation of the Avian Embryo

The hypoblast

By the time a hen has laid an egg, the blastoderm contains some 20,000 cells. At this time, most of the cells of the area pellucida remain at the surface, forming the epiblast, while other area pellucida cells have delaminated and migrated individually into the subgerminal cavity to form the **polyinvagination islands** (**primary hypoblast**), an archipelago of disconnected clusters containing 5–20 cells each (Figure 11.13B). Shortly thereafter, a sheet of cells from the posterior margin of the blastoderm (distinguished from the other regions of the margin by **Koller's sickle**, a local thickening) migrates anteriorly and pushes the primary hypoblast cells anteriorly, thereby forming the **secondary hypoblast** (Eyal-Giladi et al. 1992; Bertocchini and Stern 2002). The resulting two-layered blastoderm (epiblast and hypoblast) is joined together at the marginal zone of the area opaca, and the space between the layers forms a blastocoel. Thus, although the shape and formation of the avian blastodisc differ from those of the amphibian, fish, or echinoderm blastula, the overall spatial relationships are retained.

The avian embryo comes entirely from the epiblast. The hypoblast does not contribute any cells to the developing embryo (Rosenquist 1966, 1972). Rather, the hypoblast cells form portions of the external membranes, especially the yolk sac and the stalk that links the yolk mass to the endodermal digestive tube. All three germ layers of the embryo proper (plus a considerable amount of extraembryonic membrane) are formed from epiblast cells. Fate maps of the chick epiblast are shown in Figures 11.14 and 1.6 (Schoenwolf 1991).

The primitive streak

The major structural characteristic of avian, reptilian, and mammalian gastrulation is the **primitive streak***. Dye marking experiments and scanning electron micrographs indicate that the primitive streak cells arise locally in the posterior marginal region and that cells from other areas of the embryo are not involved in its formation (Lawson and Schoenwolf 2001a,b). The streak is first visible as a thickening of the epiblast at the posterior marginal zone, just anterior to Koller's sickle (Figure 11.14A). This thickening is initiated by an increase in the height (thickness) of the cells forming the center of the primitive streak. The presumptive streak cells around them become globular and motile, and they appear to digest away the extracellular matrix underlying them. This process allows their intercalation (mediolaterally) and convergent extension. Such convergent extension is responsible for the progression of the streak: a doubling in streak length is accompanied by a concomitant halving of its width. Those cells that

*Although, as we saw in the last chapter, a structure resembling the primitive streak has recently been discovered in certain salamander embryos (Shook et al. 2002).

initiated streak formation (i.e., that were in the midline of the epiblast overlying Koller's sickle) appear to migrate anteriorly and may constitute an unchanging cell population that directs the movement of epiblast cells into the streak.

As cells converge to form the primitive streak, a depression forms within the streak. This depression is called the **primitive groove**, and it serves as an opening through which migrating cells pass into the blastocoel. Thus, the primitive groove is analogous to the amphibian blastopore. At the anterior end of the primitive streak is a regional thickening of cells called the **primitive knot** or **Hensen's node**. The center of this node contains a funnel-shaped depression (sometimes called the **primitive pit**) through which cells can pass into the blastocoel. Hensen's node is the functional equivalent of the dorsal lip of the amphibian blastopore (i.e., the organizer) and the fish embryonic shield (Boettger et al. 2001).

The first cells that ingress through the primitive streak and into the blastocoel are endodermal precursors from the epiblast (Figure 11.14B; Vakaet 1984; Bellairs 1986; Eyal-Giladi et al. 1992). These cells undergo an epithelial-to-mesenchymal transformation, and the basal lamina beneath them breaks down. As these cells enter the primitive streak, the streak elongates toward the future head region. Cell division adds to the length produced by convergent extension, and some of the cells from the anterior portion of the epiblast contribute to the formation of Hensen's node (Streit et al. 2000; Lawson and Schoenwolf 2001b).

At the same time, the secondary hypoblast cells continue to migrate anteriorly from the posterior margin of the blastoderm. The elongation of the primitive streak appears to be coextensive with the anterior migration of these secondary hypoblast cells, and the hypoblast directs the movement of the primitive streak (Waddington 1933; Foley et al. 2000). The streak eventually extends to 60–75% of the length of the area pellucida.

The primitive streak defines the axes of the embryo. It extends from *posterior* to *anterior*; migrating cells enter through its *dorsal* side and move to its *ventral* side; and it separates the *left* portion of the embryo from the *right*. Those elements close to the streak will be the **medial** (central) structures, while those farther from it will be the **distal** (lateral) structures (Figure 11.14C–E).

As soon as the primitive streak has formed, epiblast cells begin to migrate through it and into the blastocoel (Figure 11.15). The primitive streak thus has a continually changing cell population. Cells migrating through the anterior end pass down into the blastocoel and migrate anteriorly, forming foregut, head mesoderm, and notochord; cells passing through the more posterior portions of the primitive streak give rise to the majority of endodermal and mesodermal tissues (Schoenwolf et al. 1992). Unlike the *Xenopus* mesoderm, which migrates as sheets of cells into the blastocoel, cells entering the inside of the avian embryo ingress as individuals after undergoing an epithelial-to-mesenchymal transforma-

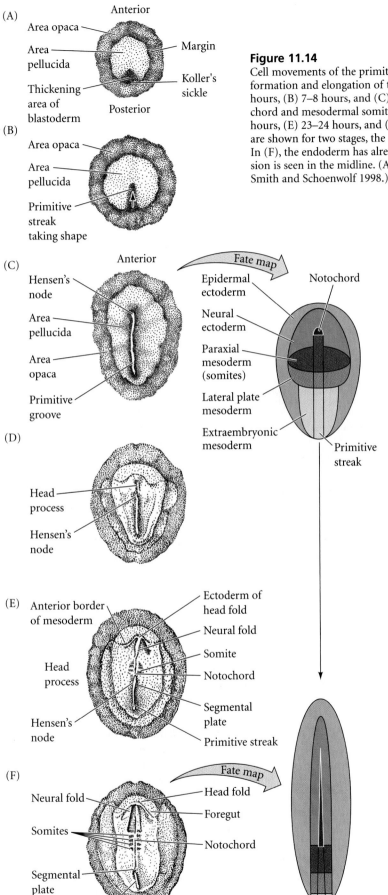

(A)

Anterior

Area opaca

Area pellucida

Margin

Thickening area of blastoderm

Koller's sickle

Posterior

(B)

Area opaca

Area pellucida

Primitive streak taking shape

(C)

Anterior

Hensen's node

Area pellucida

Area opaca

Primitive groove

Fate map

Epidermal ectoderm

Neural ectoderm

Paraxial mesoderm (somites)

Lateral plate mesoderm

Extraembryonic mesoderm

Notochord

Primitive streak

(D)

Head process

Hensen's node

(E)

Anterior border of mesoderm

Head process

Hensen's node

Ectoderm of head fold

Neural fold

Somite

Notochord

Segmental plate

Primitive streak

(F)

Neural fold

Somites

Segmental plate

Primitive streak

Fate map

Head fold

Foregut

Notochord

Figure 11.14
Cell movements of the primitive streak of the chick embryo. (A–C) Dorsal view of the formation and elongation of the primitive streak. The blastoderm is seen at (A) 3–4 hours, (B) 7–8 hours, and (C) 15–16 hours after fertilization. (D–F) Formation of notochord and mesodermal somites as the primitive streak regresses, shown at (D) 19–22 hours, (E) 23–24 hours, and (F) the four-somite stage. Fate maps of the chick epiblast are shown for two stages, the definitive primitive streak stage (C) and neurulation (F). In (F), the endoderm has already ingressed beneath the epiblast, and convergent extension is seen in the midline. (Adapted from several sources, especially Spratt 1946 and Smith and Schoenwolf 1998.)

tion. At Hensen's node and throughout the primitive streak, the breakdown of the basal lamina and the release of these cells into the embryo is thought to be accomplished by **scatter factor**, a 190-kDa protein secreted by the cells as they enter the streak (Stern et al. 1990; DeLuca et al. 1999). Scatter factor can convert epithelial sheets into mesenchymal cells in several ways, and it is probably involved both in downregulating E-cadherin expression and in preventing E-cadherin from functioning

MIGRATION THROUGH THE PRIMITIVE STREAK: FORMATION OF ENDODERM AND MESODERM. The first cells to migrate through Hensen's node are those destined to become the pharyngeal endoderm of the foregut. Once inside the blastocoel, these endodermal cells migrate anteriorly and eventually displace the hypoblast cells, causing the hypoblast cells to be confined to a region in the anterior portion of the area pellucida. This region, the **germinal crescent**, does not form any embryonic structures, but it does contain the precursors of the germ cells, which later migrate through the blood vessels to the gonads (see Chapter 19). The next cells entering the blastocoel through Hensen's node also move anteriorly, but they do not move as far ventrally as the presumptive foregut endodermal cells. Rather, they remain between the endoderm and the epiblast to form the **head mesenchyme** and the **prechordal plate mesoderm** (see Psychoyos and Stern 1996). These early-ingressing cells all move anteriorly, pushing up the anterior midline region of the epiblast to form the **head process** (Figure 11.16). Thus, the head of the avian embryo forms anterior (rostral) to Hensen's node. The next cells migrating through Hensen's node become **chordamesoderm** (notochord) cells. These cells extend up to the presumptive midbrain, where they meet the prechordal plate. The hindbrain and trunk form from the chordamesoderm at the level of Hensen's node and caudal to it.

Meanwhile, cells continue migrating inwardly through the lateral portions of the primitive streak. As they enter the blastocoel, these cells separate into two layers. The deep layer joins the hypoblast along

(A)

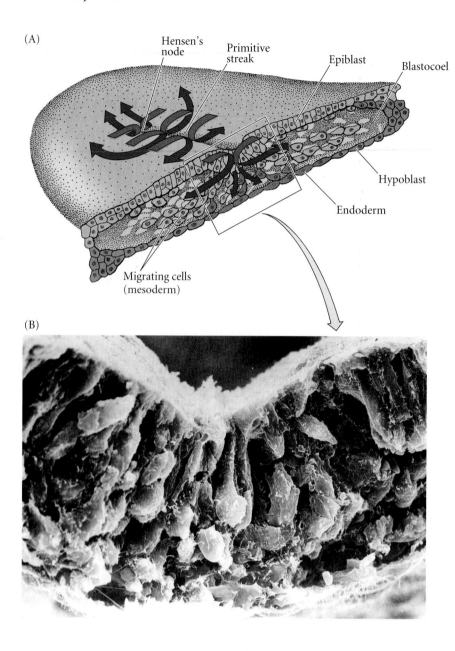

Hensen's node

Primitive streak

Epiblast

Blastocoel

Hypoblast

Endoderm

Migrating cells (mesoderm)

(B)

Figure 11.15
Migration of endodermal and mesodermal cells through the primitive streak. (A) Scanning electron micrograph shows epiblast cells passing into the blastocoel and extending their apical ends to become bottle cells. (B) Stereogram of a gastrulating chick embryo, showing the relationship of the primitive streak, the migrating cells, and the two original layers of the blastoderm. The lower layer becomes a mosaic of hypoblast and endodermal cells; the hypoblast cells eventually sort out to form a layer beneath the endoderm and contribute to the yolk sac. (A from Solursh and Revel 1978, photograph courtesy of M. Solursh; B after Balinsky 1975.)

from near the center of the area pellucida to a more posterior position (Figure 11.16). It leaves in its wake the dorsal axis of the embryo and the notochord. As the node moves posteriorly, the notochord is laid down, starting at the level of the future midbrain. While the anterior (head) portion of the notochord is formed by the ingression of cells through Hensen's node, the posterior (trunk) notochord (after somite 17 in the chick) forms from the condensation of mesodermal tissue that has ingressed through the primitive streak (i.e., not through Hensen's node). This portion of the notochord extends posteriorly to form the tail of the embryo (Le Douarin et al. 1996). Finally, Hensen's node regresses to its posterior position, forming the anal region. By this time, all the presumptive endodermal and mesodermal cells have entered the embryo, and the epiblast is composed entirely of presumptive ectodermal cells.

its midline and displaces the hypoblast cells to the sides. These deep-moving cells give rise to all the endodermal organs of the embryo as well as to most of the extraembryonic membranes (the hypoblast forms the rest). The second migrating layer spreads between this endoderm and the epiblast, forming a loose layer of cells. This middle layer of cells generates the mesodermal portions of the embryo and mesoderm lining the extraembryonic membranes. By 22 hours of incubation, most of the presumptive endodermal cells are in the interior of the embryo, although presumptive mesodermal cells continue to migrate inward for a longer time.

REGRESSION OF THE PRIMITIVE STREAK. Now a new phase of gastrulation begins. While mesodermal ingression continues, the primitive streak starts to regress, moving Hensen's node

As a consequence of the sequence in which the head mesoderm and notochord are established, avian (and mammalian) embryos exhibit a distinct anterior-to-posterior gradient of developmental maturity. While cells of the posterior portions of the embryo are undergoing gastrulation, cells at the anterior end are already starting to form organs (see Darnell et al. 1999). For the next several days, the anterior end of the embryo is more advanced in its development (having had a "head start," if you will) than the posterior end.

Epiboly of the ectoderm

While the presumptive mesodermal and endodermal cells are moving inward, the ectodermal precursors proliferate and migrate to surround the yolk by epiboly. The enclosure of the

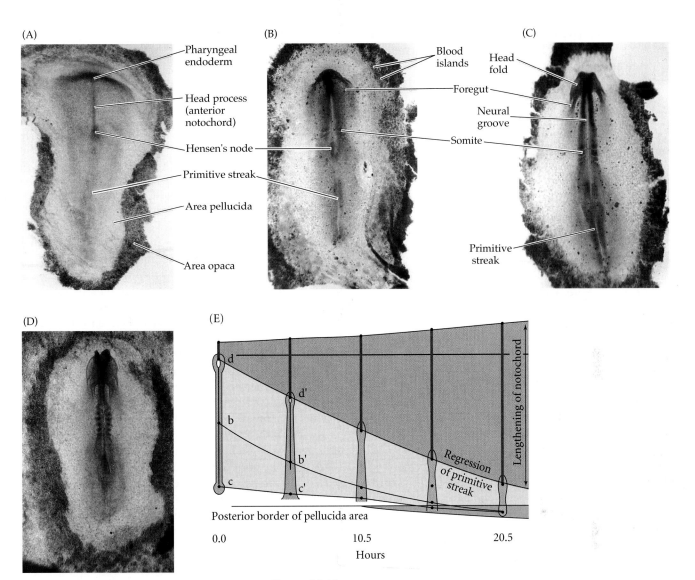

Figure 11.16
Chick gastrulation 24–28 hours after fertilization. (A) The primitive streak at full extension (24 hours). The head process (anterior notochord) can be seen extending from Hensen's node. (B) Two-somite stage (25 hours). Pharyngeal endoderm is seen anteriorly, while the anterior notochord pushes up the head process beneath it. The primitive streak is regressing. (C) Four-somite stage (27 hours). (D) At 28 hours, the primitive streak has regressed to the caudal portion of the embryo. (E) Regression of the primitive streak, leaving the notochord in its wake. Various points of the streak (represented by letters) were followed after it achieved its maximum length. The *x* axis (time) represents hours after achieving maximum length (the reference line is about 18 hours of incubation). (A–D, photographs courtesy of K. Linask; E after Spratt 1947.)

yolk by the ectoderm (again reminiscent of the epiboly of the amphibian ectoderm) is a Herculean task that takes the greater part of four days to complete. It involves the continuous production of new cellular material and the migration of the presumptive ectodermal cells along the underside of the vitelline envelope (New 1959; Spratt 1963). Interestingly, only the cells of the outer margin of the area opaca attach firmly to the vitelline envelope. These cells are inherently different from the other blastoderm cells, as they can extend enormous (500 μm) cytoplasmic processes onto the vitelline envelope. These elongated filopodia are believed to be the locomotor apparatus of these marginal cells, by which they pull the other ectodermal cells around the yolk (Schlesinger 1958). The filopodia appear to bind to fibronectin, a laminar protein that is a component of the chick vitelline envelope (see Chapter 6).

If the contact between the marginal cells and the fibronectin is experimentally broken (by adding a soluble polypeptide similar to fibronectin), the filopodia retract, and epidermal migration ceases (Lash et al. 1990).

Thus, as avian gastrulation draws to a close, the ectoderm has surrounded the yolk, the endoderm has replaced the hypoblast, and the mesoderm has positioned itself between these two regions. We have identified many of the processes involved in avian gastrulation, but we are now just beginning to understand the mechanisms by which many of these processes are carried out.

> WEBSITE 11.2 **Epiblast cell heterogeneity.** Although the early epiblast appears uniform, different cells have different molecules on their cell surfaces. This variability allows some of them to remain in the epiblast while others migrate into the embryo.

> VADE MECUM[2] **Chick development.** Viewing these movies of 3-D models of chick cleavage and gastrulation will help you understand these phenomena.
> **[Click on Chick-Early]**

Axis Formation in the Chick Embryo

While the formation of the body axes is accomplished during gastrulation, their specification occurs earlier, during the cleavage stage.

The role of pH in forming the dorsal-ventral axis

The dorsal-ventral (back-belly) axis is critical to the formation of the hypoblast and to the further development of the embryo. This axis is established when the cleaving cells of the blastoderm establish a barrier between the basic (pH 9.5) albumen above the blastodisc and the acidic (pH 6.5) subgerminal cavity below it. Water and sodium ions are transported from the albumen through the cells and into the subgerminal cavity, creating a membrane potential difference of 25 mV across the epiblast cell layer (positive at the ventral side of the cells). This process distinguishes two sides of the epiblast: a side facing the negative and basic albumen, which becomes the dorsal side, and a side facing the positive and acidic fluid of the subgerminal cavity, which becomes the ventral side. This axis can be reversed experimentally either by reversing the pH gradient or by inverting the potential difference across the cell layer (reviewed in Stern and Canning 1988).

The role of gravity in forming the anterior-posterior axis

The conversion of the radially symmetrical blastoderm into a bilaterally symmetrical structure is determined by gravity. As the ovum passes through the hen's reproductive tract, it is rotated for about 20 hours in the shell gland. This spinning, at a rate of 10 to 12 revolutions per hour, shifts the yolk such that its lighter components lie beneath one side of the blastoderm. This tips up that end of the blastoderm, and that end becomes the posterior portion of the embryo—the part where primitive streak formation begins (Figure 11.17; Kochav and Eyal-Giladi 1971; Eyal-Giladi and Fabian 1980).

It is not known what interactions cause this portion of the blastoderm to become the **posterior marginal zone (PMZ)** and to initiate gastrulation. Early on, the ability to initiate a primitive streak is found throughout the marginal zone, and if the blastoderm is separated into parts, each having its own marginal zone, each part will form its own primitive streak (Spratt and Haas 1960). However, once a PMZ has formed, it controls the other regions of the margin. Not only do these PMZ cells initiate gastrulation, but they also prevent other regions of the margin from forming their own primitive streaks* (Khaner and Eyal-Giladi 1989; Eyal-Giladi et al. 1992). Recent studies (Bertocchini and Stern 2002) suggest that the rotation causes the secondary hypoblast to form in the area and that the 2° hypoblast cells push away the 1° hypoblast cells that secrete Cerberus. As the 1° hypoblast cells are moved away from the posterior marginal zone, the absence of Cerberus allows Nodal protein to be expressed in the PMZ.

It now seems apparent that the PMZ contains cells that act as the equivalent of the amphibian Nieuwkoop center. When placed in the *anterior* region of the marginal zone, a graft of *posterior* marginal zone tissue (posterior to and not

---

*Earlier investigators (Waddington 1932) thought that the hypoblast induced the formation of the primitive streak and provided it with anterior-posterior polarity. However, Khaner (1995) rotated the epiblast with respect to the hypoblast at different stages of chick development and showed that the epiblast initiates primitive streak formation and maintains its polarity independently of the orientation of the hypoblast.

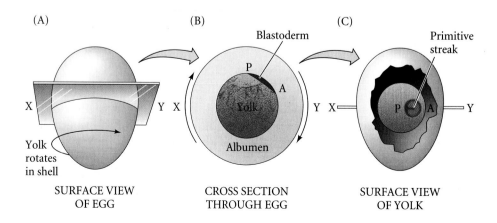

Figure 11.17
Specification of the chick anterior-posterior axis by gravity. (A) Rotation in the shell gland results in (B) the lighter components of the yolk pushing up one side of the blastoderm (C) This more elevated region becomes the posterior of the embryo. (After Wolpert et al. 1998.)

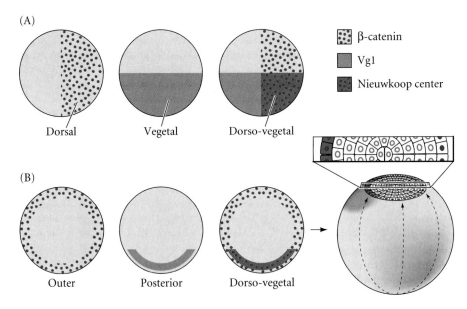

Figure 11.18
Formation of the Nieuwkoop center in frogs and chicks. (A) Formation of the Nieuwkoop center in *Xenopus*. Gravity and site of sperm entry determine the placement of β-catenin in the dorsal region. The vegetal cells secrete Vg1. The overlap becomes the Nieuwkoop center. (B) In the chick, β-catenin is found in the outer cells of the blastodisc (which, being farthest from the center and adjacent to the yolk, are equivalent to the ventral cells of the frog embryo). The Vg1-producing cells are restricted (probably by gravity) to the presumptive posterior cells of the blastoderm. The overlap becomes the Nieuwkoop center. (After Boettger et al. 2001.)

including Koller's sickle) is able to induce a primitive streak and Hensen's node without contributing cells to either structure (Bachvarova et al. 1998; Khaner 1998). Like the amphibian Nieuwkoop center, this region is thought to be the place where β-catenin localization in the nucleus and a TGF-β family signal coincide. β-catenin is found throughout the marginal zone (but not in the interior cells), while *Vg1* expression is seen only in the posterior cells of this region (Mitrani et al. 1990; Hume and Dodd 1993; Seleiro et al. 1996). The overlap of these two regions constitutes the posterior marginal zone Nieuwkoop center (Figure 11.18; Boettger et al. 2001).

The "organizer" of the chick embryo forms just anteriorly to the Nieuwkoop center. The epiblast and middle layer cells in the anterior portion of Koller's sickle become Hensen's node (Bachvarova et al. 1998). The lateral portions of Koller's sickle contribute to the posterior portion of the primitive

streak (Figure 11.19). Hensen's node has long been known to be the avian equivalent of the amphibian dorsal blastopore lip, since it is (1) the site where gastrulation begins, (2) the region whose cells become the chordamesoderm, and (3) the region whose cells can organize a second embryonic axis when transplanted into other locations of the gastrula (Figure 11.20; Waddington 1933; 1934; Dias and Schoenwolf 1990).

Gene expression in the chick organizer can be categorized into two sets of genes (Lawson et al. 2001). The first set contains those genes that are first expressed in the posterior portion of Koller's sickle and which probably help form the Nieuwkoop center. These genes, which include *Vg1* and *nodal*, then appear throughout the entire length of the primitive streak (Figure 11.21). Recent studies (Skromne and Stern 2002) have shown that Vg1 plays a crucial role in forming the primitive streak, and if Vg1 is ectopically expressed in the *anterior* marginal zone, a secondary axis will form there.

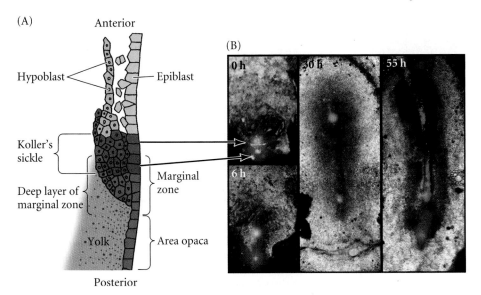

Figure 11.19
Formation of Hensen's node from Koller's sickle. (A) Diagram of the posterior end of an early (pre-streak) embryo, showing the cells labeled with fluorescent dyes in the photographs. (B) Just before gastrulation, cells in the anterior end of Koller's sickle (the epiblast and middle layer) were labeled with green dye. Cells of the posterior portion of Koller's sickle were labeled with red dye. As the cells migrated, the anterior cells formed Hensen's node and its notochord derivatives. The posterior cells formed the posterior region of the primitive streak. The time after dye injection is labeled on each photograph. (After Bachvarova et al. 1998; photographs courtesy of R. F. Bachvarova.)

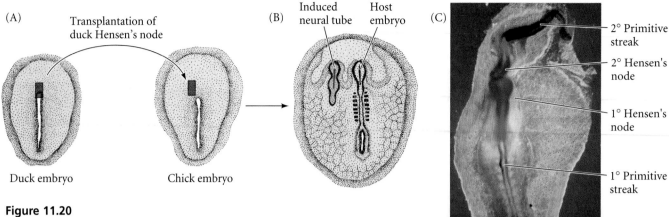

Figure 11.20
Induction of a new embryo by transplantation of Hensen's node. (A) A Hensen's node from a duck embryo is transplanted into the epiblast of a chick embryo. (B) A secondary embryo is induced (as is evident by the neural tube) from host tissues at the graft site. (C) Graft of Hensen's node from one embryo into the periphery of a host embryo. After further incubation, the host embryo has a neural tube whose regionalization can be seen by in situ hybridization. Probes to *otx2* (red) recognize the head region, while probes to *hoxb1* (blue) recognize the trunk neural tube. The donor node has induced the formation of a secondary axis, complete with head and trunk regions. (A, B after Waddington 1933; C from Boettger et al. 2001.)

The second set of genes comprises those whose expression is are confined to the anterior portion of the primitive streak, and finally to Hensen's node. These genes include *chordin* and *sonic hedgehog*. As is the case in all vertebrates, the dorsal mesoderm is able to induce the formation of the central nervous system in the ectoderm overlying it. The cells of Hensen's node and its derivatives act like the amphibian organizer, and they secrete the chordin, Noggin, and Nodal proteins. These proteins antagonize BMPs and dorsalize the ectoderm and mesoderm (Figure 11.22). However, repression of the BMP signal does not appear to be sufficient for neural induction (see Streit and Stern 1999). Fibroblast growth factors (FGFs) synthesized in Hensen's node precursor cells just prior to gastrulation appear to be critical for preparing the epiblast to generate neuronal phenotypes* (Alvarez et al. 1998; Storey et al. 1998; Streit et al. 2000).

*A precondition for the ability of the BMP-chordin system to neuralize ectoderm is that dissociated (uninduced) presumptive ectoderm cells will become neural. In chick embryos, however, the dissociated epiblast cells develop into muscles (George-Weinstein et al. 1996).

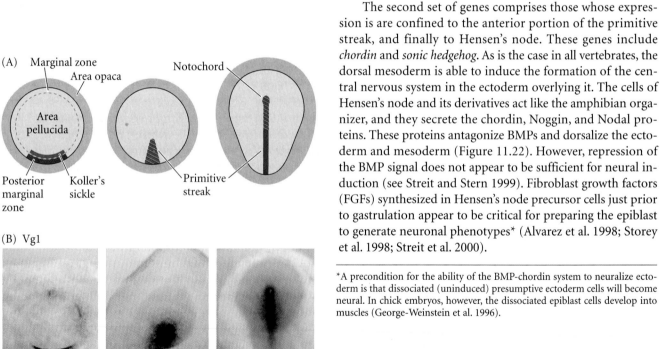

Figure 11.21
Gene expression in the primitive streak. (A) Schemata of the two general gene expression patterns. The early chick epiblast (left panel) shows the area opaca, area pellucida, marginal zone, Koller's sickle (red), and posterior marginal zone (blue). At a slightly later stage (middle panel), cells expressing both the Nieuwkoop center genes and organizer genes extend into the primitive streak. At later stages (right panel), the Nieuwkoop center genes are expressed throughout the streak, while the organizer genes are expressed in the most anterior region. (B) Expression of Vg1 protein as the primitive streak forms. (C) Expression of chordin as the primitive streak forms. (A after Boettger et al. 2001; B, C after Lawson et al. 2001, photographs courtesy of G. Schoenwolf.)

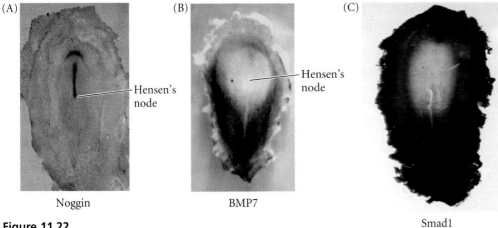

(A) Noggin

(B) BMP7

(C) Smad1

Figure 11.22
Possible contribution to chick neural induction by the inhibition of BMP signaling. (A) In a neurulating embryo, Noggin protein (dark purple) is expressed in Hensen's node, the notochord, and the pharyngeal endoderm. (B) *Bmp7* expression (dark purple), which had encompassed the entire epiblast, becomes restricted to the nonneural regions of the ectoderm. (C) Similarly, the product of BMP signaling, the phosphorylated form of Smad1 (recognized by antibodies to the phosphorylated form of the protein; dark brown) is not seen in the neural plate. (After Faure et al. 2002; photographs courtesy of the authors.)

Left-Right Axis Formation

As we have seen, the vertebrate body is not symmetrical. Rather, it has distinct right and left sides. The heart and spleen, for instance, are generally on the left side of the body, while the liver is usually on the right side. The distinction between the right and left sides of vertebrates is regulated by two major proteins: the paracrine factor **Nodal** and the transcription factor **Pitx2**. However, the mechanism by which *nodal* gene expression is activated in the left side of the body differs among vertebrate classes. The ease with which chick embryos can be manipulated has allowed scientists to elucidate the pathways of left-right determination in birds more readily than in other vertebrates.

As the primitive streak reaches its maximum length, transcription of the *sonic hedgehog* gene ceases on the right side of the embryo, due to the expression on this side of activin and its receptor (Figure 11.23). Activin signaling blocks the expression of *sonic hedgehog* and activates the expression of *fgf8*. FGF8 prevents the transcription of the *caronte* gene. In the absence of Caronte, bone morphogenetic proteins (BMPs) are able to block the expression of *nodal* and *lefty-2*. This allows the expression of the *snail* gene, which is characteristic of the right side of avian embryonic organs. On the left side of the body, the Lefty-1 protein blocks the expression of *fgf8*, while Sonic hedgehog activates *caronte* (Figure 11.24). **Caronte** is a paracrine factor that prevents BMPs from repressing the *nodal* and *lefty-2* genes, and also inhibits BMPs from blocking the expression of *lefty-1* on the ventral midline structures (Ro-driguez-Esteban et al. 1999; Yokouchi et al. 1999). Nodal and Lefty-2 activate *pitx2* and repress *snail*. Lefty-1 in the ventral midline prevents the Caronte signal from passing to the right side of the embryo. As in *Xenopus*, Pitx2 is crucial in directing the asymmetry of the embryonic structures. Experimentally induced expression of either *nodal* or *pitx2* on the right side of the chick is able to reverse the asymmetry or cause randomization of the asymmetry on the right or left sides* (Logan et al. 1998; Ryan et al. 1998).

The course to either left- or right-sidedness can be interfered with at any point along this pathway. If activin is blocked by the experimental addition of follistatin to the embryo, the asymmetry of Sonic hedgehog expression vanishes, and the heart has an equal chance of looping either way (Levin et al. 1995, 1997). Conversely, when activin-soaked beads are placed to the left side of Hensen's node, the *sonic hedgehog* gene (usually expressed only on the left side) is repressed. This, in turn, suppresses the transcription of *nodal*. In this situation, too, the heart tube forms randomly, having an equal probability of being left or right. A similar condition can be produced by implanting cells secreting Sonic hedgehog into the right side of the node. In these cases, nodal is induced symmetrically in the lateral plate mesoderm, and the heart has a 50 percent chance of having a left-sided tube (Figure 11.25).

EARLY MAMMALIAN DEVELOPMENT

Cleavage in Mammals

It is not surprising that mammalian cleavage has been the most difficult to study. Mammalian eggs are among the smallest in the animal kingdom, making them hard to manipulate experimentally. The human zygote, for instance, is only 100

---

*In humans, homozygous loss of *PITX2* causes Rieger syndrome, a condition characterized by asymmetry anomalies. A similar condition is caused by knocking out this gene in mice (Fu et al. 1999; Lin et al. 1999).

μm in diameter—barely visible to the eye and less than one-thousandth the volume of a *Xenopus* egg. Also, mammalian zygotes are not produced in numbers comparable to sea urchin or frog zygotes, so it is difficult to obtain enough material for biochemical studies. Usually, fewer than ten eggs are ovulated by a female at a given time. As a final hurdle, the development of mammalian embryos is accomplished within another organism, rather than in the external environment.

Only recently has it been possible to duplicate some of these internal conditions and observe mammalian development in vitro.

The unique nature of mammalian cleavage

With all these difficulties, knowledge of mammalian cleavage was worth waiting for, as mammalian cleavage has turned out to be strikingly different from most other patterns of embryonic cell division. The mammalian oocyte is released from the ovary and swept by the fimbriae into the oviduct (Figure 11.26). Fertilization occurs in the ampulla of the oviduct, a region close to the ovary. Meiosis is completed at this time, and first cleavage begins about a day later (see Figure 7.32). Cleavages in mammalian eggs are among the slowest in the animal kingdom—about 12–24 hours apart. Meanwhile, the cilia in the oviduct push the embryo toward the uterus; the first cleavages occur along this journey.

In addition to the slowness of cell division, several other features of mammalian cleavage distinguish it from other cleavage types. The second of these differences is the unique orientation of mammalian blastomeres with relation to one another. The first cleavage is a normal meridional division; however, in the second cleavage, one of the two blastomeres

(A)

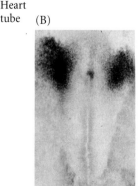

Figure 11.23
Pathway for left-right asymmetry in the chick embryo. (A) Left side: On the left side of Hensen's node, *sonic hedgehog* (*shh*) and *lefty-1* are expressed. Lefty-1 protein blocks *fgf8* expression, while Sonic hedgehog activates the expression of *caronte* (*car*). Meanwhile, retinoic acid permits the expression of *lefty-1* along the ventral midline. Caronte protein blocks expression of bone morphogenetic proteins (BMPs) on the left side, which would otherwise block the expression of *nodal* and *lefty-2*. In the presence of Nodal and Lefty-2, the *pitx2* gene is activated and the *snail* gene (*cSnR*) is repressed. Pitx2 is active in the various organ primordia and specifies the side to be left. In the right side of the embryo, activin is expressed, along with activin receptor IIa. This activates FGF8, a protein that blocks the expression of *caronte*. In the absence of Caronte protein, BMP represses the activation of *nodal* and *lefty-2*. This allows the *snail* gene to be active while the *pitx2* gene remains repressed. (B, C) Dorsal and close-up views of in situ hybridization of *sonic hedgehog* mRNA. (D, E) Dorsal and close-up views of the activin receptor IIa message. (A after Rodriguez-Esteban et al. 1999; B–E from Levin et al. 1995, photographs courtesy of C. Tabin and C. Stern.)

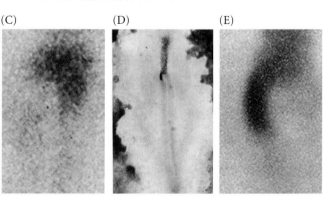

(B)　　(C)　　(D)　　(E)

(A)   (B)   (C)

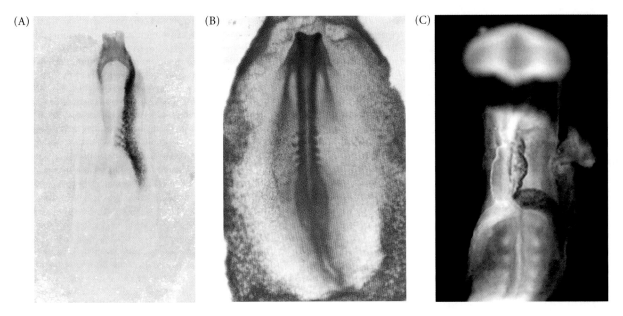

Figure 11.24
Transmission of the left-side signal from Hensen's node to the lateral plate mesoderm results in the asymmetrical expression of *caronte*, *nodal*, and *pitx2* genes in the left side of the chick embryo. (A) Whole mount in situ hybridization of *caronte* mRNA. This view is from the ventral surface, "from below," so the expression seems to be on the right. Dorsally, the expression pattern would be on the left. (B) Whole mount in situ hybridization using probes for the chick *nodal* message (stained purple) shows its expression in the lateral plate mesoderm only on the left side of the embryo. This view is from the dorsal side (looking "down" at the embryo). (C) A similar in situ hybridization, using the probe for *pitx2* at a later stage of development. (A)The embryo is seen from its ventral surface. At this stage, the heart is forming, and *pitx2* expression can be seen on the left side of the heart tube (as well as symmetrically in more anterior tissues). (A from Rodriguez-Esteban et al. 1999, photograph courtesy of J. Izpisúa-Belmonte; B, photograph courtesy of C. Stern; C from Logan et al. 1998, photograph courtesy of C. Tabin.)

(A)

Notochord ——

Hensen's —— node

(B)

Sonic hedgehog —— pellet

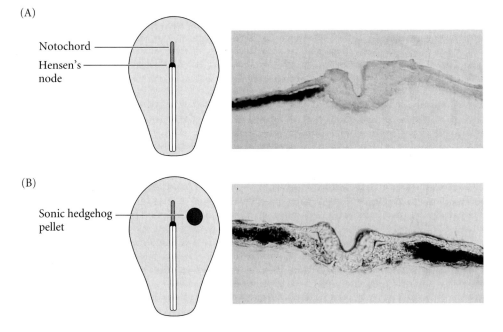

Figure 11.25
Ectopic expression of *sonic hedgehog* leads to symmetrical *nodal* expression and randomization of heart looping. (A) Wild-type expression of the chick *nodal* gene, showing expression on the left side (dark area). Nearly all hearts develop right-sided loops. This pattern is also seen when pellets containing control substances are implanted into the right side of Hensen's node, or when a Sonic hedgehog-containing pellet is implanted on the left side (where *sonic hedgehog* is usually expressed). (B) When a Sonic hedgehog pellet is implanted on the right side of the node, the expression of *nodal* becomes bilaterally symmetrical, and the heart can be on either side. (From Levin et al. 1995; photographs courtesy of the authors.)

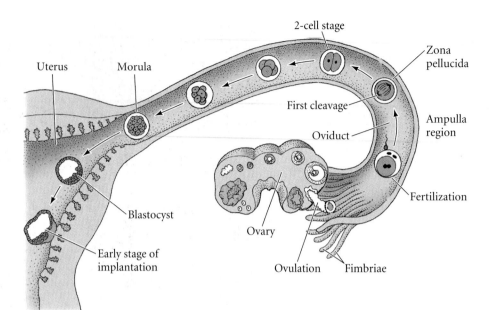

Figure 11.26
Development of a human embryo from fertilization to implantation. Compaction of the human embryo occurs on day 4, when it is at the 10-cell stage. The embryo "hatches" from the zona pellucida upon reaching the uterus. During its migration to the uterus, the zona prevents the embryo from prematurely adhering to the oviduct rather than traveling to the uterus. (After Tuchmann-Duplessis et al. 1972.)

divides meridionally and the other divides equatorially (Figure 11.27). This type of cleavage is called **rotational cleavage** (Gulyas 1975).

The third major difference between mammalian cleavage and that of most other embryos is the marked asynchrony of early cell division. Mammalian blastomeres do not all divide at the same time. Thus, mammalian embryos do not increase exponentially from 2- to 4- to 8-cell stages, but frequently contain odd numbers of cells. Fourth, unlike almost all other animal genomes, the mammalian genome is activated during early cleavage, and produces the proteins necessary for cleavage and development to occur. In the mouse and goat, the switch from maternal to zygotic control occurs at the 2-cell stage (Piko and Clegg 1982; Prather 1989).

Most research on mammalian development has focused on the mouse, since mice are relatively easy to breed throughout the year, have large litters, and can be housed easily. Thus, most of the studies discussed here will concern murine (mouse) development.

Compaction

The fifth, and perhaps the most crucial, difference between mammalian cleavage and all other types involves the phenomenon of **compaction**. As seen in Figure 11.28, mouse blastomeres through the 8-cell stage form a loose arrangement with plenty of space between them. Following the third cleav-

age, however, the blastomeres undergo a spectacular change in their behavior. Cell adhesion proteins such as E cadherin become expressed, and the blastomeres suddenly huddle together, forming a compact ball of cells (Figure 11.28C, D; Figure 11.29; Peyrieras et al. 1983; Fleming et al. 2001). This tightly packed arrangement is stabilized by tight junctions that form between the outside cells of the ball, sealing off the inside of the sphere. The cells within the sphere form gap junctions, thereby enabling small molecules and ions to pass between them.

The cells of the compacted 8-cell embryo divide to produce a 16-cell **morula** (Figure 11.28E). The morula consists of a small group of internal cells surrounded by a larger group of external cells (Barlow et al. 1972). Most of the descendants of the external cells become the **trophoblast** (**trophectoderm**) cells. This group of cells produces no embryonic structures. Rather, it forms the tissue of the **chorion**, the embryonic por-

Figure 11.27
Comparison of early cleavage in (A) echinoderms and amphibians (radial cleavage) and (B) mammals (rotational cleavage). Nematodes also have a rotational form of cleavage, but they do not form the blastocyst structure characteristic of mammals. (After Gulyas 1975.)

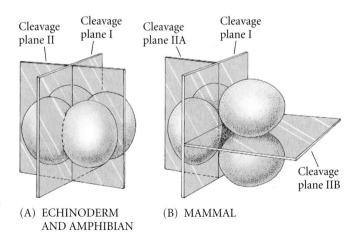

(A) ECHINODERM AND AMPHIBIAN

(B) MAMMAL

(A) (B) (C)

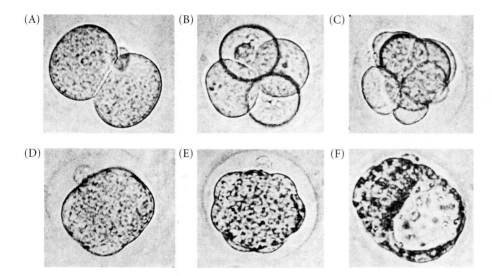

(D) (E) (F)

Figure 11.28
The cleavage of a single mouse embryo in vitro. (A) 2-cell stage. (B) 4-cell stage. (C) Early 8-cell stage. (D) Compacted 8-cell stage. (E) Morula. (F) Blastocyst. (From Mulnard 1967; photographs courtesy of J. G. Mulnard.)

tion of the **placenta**. The chorion enables the fetus to get oxygen and nourishment from the mother. It also secretes hormones that cause the mother's uterus to retain the fetus, and produces regulators of the immune response so that the mother will not reject the embryo as she would an organ graft.

The mouse embryo proper is derived from the descendants of the inner cells of the 16-cell stage, supplemented by cells dividing from the trophoblast during the transition to the 32-cell stage (Pedersen et al. 1986; Fleming 1987). These cells generate the **inner cell mass** (**ICM**), which will give rise to the embryo and its associated yolk sac, allantois, and amnion. By the 64-cell stage, the inner cell mass (approximately 13 cells) and the trophoblast cells have become separate cell layers, with neither contributing cells to the other group (Dyce et al. 1987; Fleming 1987). Thus, the distinction between trophoblast and inner cell mass blastomeres represents the first differentiation event in mammalian development. This differentiation is required for the early mammalian embryo to adhere to the uterus; the development of the embryo proper can wait until after that attachment occurs. The inner cell mass actively supports the trophoblast, secreting proteins (such as FGF4) that cause the trophoblast cells to divide (Tanaka et al. 1998). Once the decision to become trophoblast or inner cell mass is made, different genes are expressed by the cells of these two regions. The inner cell mass retains the expression of **Oct4 and Foxd3**, two transcription factors associated with **pluripotentiality**, the ability to form all the cell types of the body. The trophoblast cells synthesize the transcription factor **eomesodermin**, which activates those proteins characteristic of the trophoblast layer (Russ et al. 2000; Hanna et al. 2002).

Initially, the morula does not have an internal cavity. However, during a process called **cavitation**, the trophoblast cells secrete fluid into the morula to create a blastocoel. The membranes of trophoblast cells contain a sodium pump (an Na^+/K^+-ATPase) facing the blastocoel, and this transmem-

brane protein pumps sodium ions into the central cavity. The subsequent accumulation of sodium ions draws in water osmotically, thus enlarging the blastocoel (Borland 1977; Wiley 1984). The inner cell mass is positioned on one side of the ring of trophoblast cells (see Figures 11.28F). The resulting type of blastula, called the **blastocyst**, is another hallmark of mammalian cleavage.

(A)

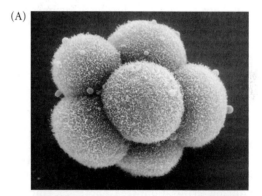

(B)

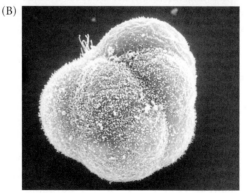

Figure 11.29
Scanning electron micrographs of (A) uncompacted and (B) compacted 8-cell mouse embryos. (Photographs courtesy of C. Ziomek.)

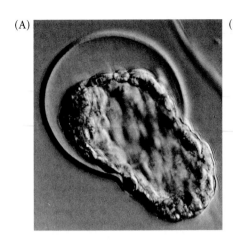
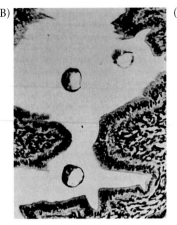
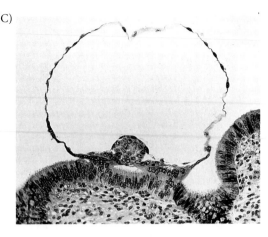

Figure 11.30
Hatching from the zona and implantation of the mammalian blastocyst in the uterus. (A) Mouse blastocyst hatching from the zona pellucida. (B) Mouse blastocysts entering the uterus. (C) Initial implantation of the blastocyst in a rhesus monkey. (A from Mark et al. 1985, photograph courtesy of E. Lacy; B from Rugh 1967; C, photograph courtesy of the Carnegie Institution of Washington, Chester Reather, photographer.)

WEBSITE 11.3 **Mechanisms of compaction and the formation of the inner cell mass.** What determines whether a cell is to become a trophoblast cell or a member of the inner cell mass? It may just be a matter of chance. However, once the decision is made, different genes are switched on.

WEBSITE 11.4 **Human cleavage and compaction.** XY blastomeres have a slight growth advantage that may have had profound effects on in vitro fertility operations.

Escape from the zona pellucida

While the embryo is moving through the oviduct en route to the uterus, the blastocyst expands within the **zona pellucida** (the extracellular matrix of the egg that was essential for sperm binding during fertilization; see Chapter 7). During this time, the zona pellucida prevents the blastocyst from adhering to the oviduct walls. When such adherence does take place in humans, it is called an **ectopic ("tubal") pregnancy**. This condition is dangerous because the implantation of the embryo into the oviduct can cause a life-threatening hemorrhage. When the embryo reaches the uterus, however, it must "hatch" from the zona so that it can adhere to the uterine wall.

The mouse blastocyst hatches from the zona by lysing a small hole in it and squeezing through that hole as the blastocyst expands (Figure 11.30A). **Strypsin**, a trypsin-like protease on the trophoblast cell membranes, lyses a hole in the fibrillar matrix of the zona (Perona and Wassarman 1986; Yamazaki and Kato 1989; O'Sullivan et al. 2001). Once out of the zona, the blastocyst can make direct contact with the uterus. The

uterine epithelium (**endometrium**) "catches" the blastocyst on an extracellular matrix containing collagen, laminin, fibronectin, hyaluronic acid, and heparan sulfate receptors. The trophoblast cells contain integrins that bind to the uterine collagen, fibronectin, and laminin, and they synthesize heparan sulfate proteoglycan precisely prior to implantation (see Carson et al. 1993). Once in contact with the endometrium, the trophoblast secretes another set of proteases, including collagenase, stromelysin, and plasminogen activator. These protein-digesting enzymes digest the extracellular matrix of the uterine tissue, enabling the blastocyst to bury itself within the uterine wall (Strickland et al. 1976; Brenner et al. 1989).

WEBSITE 11.5 **The mechanisms of implantation.** The molecular mechanisms by which the blastocyst adheres to and enters into the uterine wall constitute a fascinating story of cell adhesion and reciprocal interactions between two organisms, the mother and the embryo.

Gastrulation in Mammals

Birds and mammals are both descendants of reptilian species. Therefore, it is not surprising that mammalian development parallels that of reptiles and birds. What *is* surprising is that the gastrulation movements of reptilian and avian embryos, which evolved as an adaptation to yolky eggs, are retained even in the absence of large amounts of yolk in the mammalian embryo. The mammalian inner cell mass can be envisioned as sitting atop an imaginary ball of yolk, following instructions that seem more appropriate to its reptilian ancestors.

Modifications for development within another organism

The mammalian embryo obtains nutrients directly from its mother and does not rely on stored yolk. This adaptation has entailed a dramatic restructuring of the maternal anatomy (such as expansion of the oviduct to form the uterus) as well

Figure 11.31
Schematic diagram showing the derivation of tissues in human and rhesus monkey embryos. The dashed line indicates a possible dual origin of the extraembryonic mesoderm. (After Luckett 1978; Bianchi et al. 1993.)

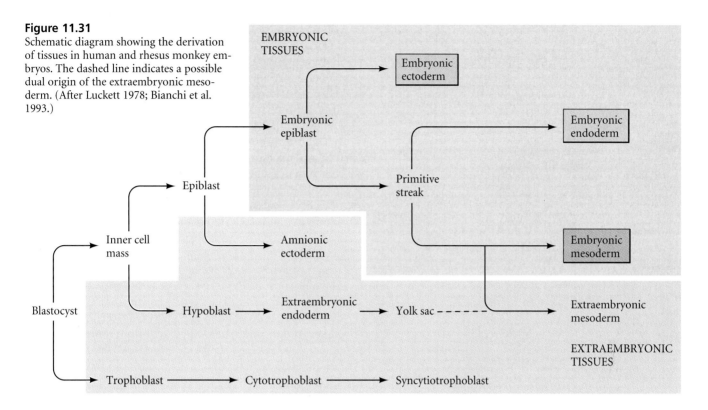

as the development of a fetal organ capable of absorbing maternal nutrients. This fetal organ—the chorion—is derived primarily from embryonic trophoblast cells, supplemented with mesodermal cells derived from the inner cell mass. The chorion forms the fetal portion of the placenta. It also induces the uterine cells to form the maternal portion of the placenta, the **decidua**. The decidua becomes rich in the blood vessels that will provide oxygen and nutrients to the embryo.

The origins of early mammalian tissues are summarized in Figure 11.31. The first segregation of cells within the inner cell mass forms two layers: the lower layer, the **hypoblast** (sometimes called the **primitive endoderm**), and the remaining inner cell mass tissue above it, the **epiblast** (Figure 11.32A). The epiblast and hypoblast form a structure called the **bilaminar germ disc**. The hypoblast cells delaminate from the inner cell mass to line the blastocoel cavity, where they give rise to the **extraembryonic endoderm**, which forms the yolk sac. As in avian embryos, these cells do not produce any part of the newborn organism. The epiblast cell layer is split by small clefts that eventually coalesce to separate the **embryonic epiblast** from the other epiblast cells that line the **amnionic cavity** (Figures 11.32 B, C). Once the lining of the amnion is completed, the amniotic cavity fills with a secretion called **amnionic (amniotic) fluid**, which serves as a shock absorber for the developing embryo while preventing its desiccation. The embryonic epiblast is thought to contain all the cells that will generate the actual embryo, and it is similar in many ways to the avian epiblast.

By labeling individual cells of the epiblast with horseradish peroxidase, Kirstie Lawson and her colleagues (1991) were able to construct a detailed fate map of the mouse epiblast (see Figure 1.6). Gastrulation begins at the posterior end of the embryo, and this is where the **node** forms[*] (Figure 11.33). Like the chick epiblast cells, the mammalian mesoderm and endoderm migrate through a primitive streak, and like their avian counterparts, the migrating cells of the mammalian epiblast lose E-cadherin, detach from their neighbors, and migrate through the streak as individual cells (Burdsal et al. 1993). Those cells migrating through the node give rise to the notochord. However, in contrast to notochord formation in the chick, the cells that form the mouse notochord are thought to become integrated into the endoderm of the primitive gut (Jurand 1974; Sulik et al. 1994). These cells can be seen as a band of small, ciliated cells extending rostrally from the node (Figure 11.34). They form the notochord by converging medially and folding off in a dorsal direction from the roof of the gut.

Cell migration and specification appear to be coordinated by FGFs. The cells of the primitive streak appear to be capable of both synthesizing and responding to FGFs (Sun et al. 1999; Ciruna and Rossant 2001). In embryos that are homozygous for the loss of the *fgf8* gene, cells fail to migrate through the

---

[*]In mammalian development, Hensen's node is usually just called "the node," despite the fact that Hensen discovered this structure in rabbit embryos.

primitive streak, and neither mesoderm nor endoderm are formed. FGF8 (and perhaps other FGFs) probably control cell movement into the primitive streak by downregulating the E-cadherin that holds the epiblast cells together, and it controls cell specification by regulating *snail*, *Brachyury* (*T*), and *Tbx6*, three genes that are essential for mesodermal specification and patterning.

The ectodermal precursors are located anterior to the fully extended primitive streak, as in the chick epiblast; in some instances, however, a single cell gives rise to descendants in more than one germ layer, or to both embryonic and extraembryonic derivatives. Thus, at the epiblast stage, these lineages have not become separate from one another. As in avian embryos, the cells migrating into the space between the hypoblast and epiblast layers become coated with hyaluronic acid, which they synthesize as they leave the primitive streak. This substance acts to keep them separate while they migrate (Solursh and Morriss 1977). It is thought (Larsen 1993) that the replacement of human hypoblast cells by endoderm precursors occurs on days 14–15 of gestation, while the migration of cells forming the mesoderm does not start until day 16 (Figure 11.33C).

Formation of extraembryonic membranes

While the embryonic epiblast is undergoing cell movements reminiscent of those seen in reptilian or avian gastrulation, the extraembryonic cells are making the distinctly mammalian tissues that enable the fetus to survive within the maternal uterus. Although the initial trophoblast cells of mice and humans divide like most other cells of the body, they give rise to a population of cells wherein nuclear division occurs in the absence of cytokinesis. The original trophoblast cells constitute a layer called the **cytotrophoblast**, whereas the multinucleated cell type forms the **syncytiotrophoblast**. The cytotrophoblast initially adheres to the endometrium through a series of adhesion molecules, as we saw above. Moreover, these cells contain proteolytic enzymes that enable them to enter the uterine wall and remodel the uterine blood vessels so that

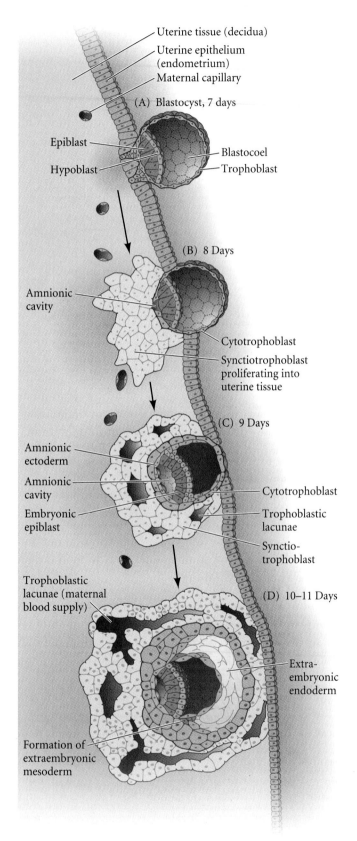

Figure 11.32

Tissue formation in the human embryo between days 7 and 11. (A, B) Human blastocyst immediately prior to gastrulation. The inner cell mass delaminates hypoblast cells that line the blastocoel, forming the extraembryonic endoderm of the primitive yolk sac and a two-layered (epiblast and hypoblast) blastodisc similar to that seen in avian embryos. The trophoblast divides into the cytotrophoblast, which will form the villi, and the syncytiotrophoblast, which will ingress into the uterine tissue. (C) Meanwhile, the epiblast splits into the amnionic ectoderm (which encircles the amnionic cavity) and the embryonic epiblast. The adult mammal forms from the cells of the embryonic epiblast. (D) The extraembryonic endoderm forms the yolk sac. The actual size of the embryo at this stage is about that of the period in "Figure 11.32."

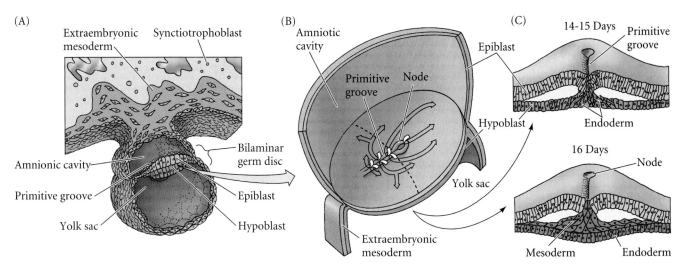

Figure 11.33
Amnion structure and cell movements during human gastrulation. (A, B) Human embryo and uterine connections at day 15 of gestation. (A) Sagittal section through the midline. (B) View looking down on the dorsal surface of the embryo. (C) The movements of the epiblast cells through the primitive streak and Hensen's node and underneath the epiblast are superimposed on the dorsal surface view. At days 14 and 15, the ingressing epiblast cells are thought to replace the hypoblast cells (which contribute to the yolk sac lining), while at day 16, the ingressing cells fan out to form the mesodermal layer. (After Larsen 1993.)

the maternal blood bathes fetal blood vessels. The syncytiotrophoblast tissue is thought to further the progression of the embryo into the uterine wall by digesting uterine tissue (Fisher et al. 1989). The uterus, in turn, sends blood vessels into this area, where they eventually contact the syncytiotrophoblast. Shortly thereafter, mesodermal tissue extends outward from the gastrulating embryo (see Figure 11.32D). Stud-

ies of human and rhesus monkey embryos have suggested that the yolk sac (and hence the hypoblast) as well as primitive streak-derived cells contribute this extraembryonic mesoderm (Bianchi et al. 1993).

The extraembryonic mesoderm joins the trophoblastic extensions and gives rise to the blood vessels that carry nutrients from the mother to the embryo. The narrow connecting stalk of extraembryonic mesoderm that links the embryo to the trophoblast eventually forms the vessels of the **umbilical cord**. The fully developed extraembryonic organ, consisting of trophoblast tissue and blood vessel-containing mesoderm, is called the chorion, and it fuses with the uterine wall to create the placenta. Thus, the placenta has both a maternal portion (the uterine endometrium, which is modified during pregnancy) and a fetal component (the chorion). The chorion may be very closely apposed to maternal tissues while still

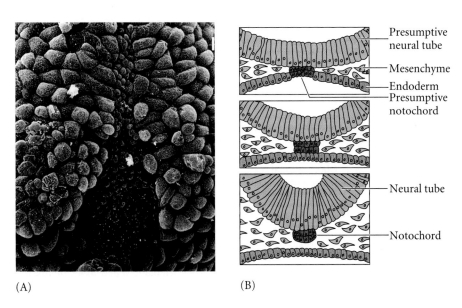

Figure 11.34
Formation of the notochord in the mouse. (A) The ventral surface of a 7.5-day mouse embryo, seen by scanning electron microscopy. The presumptive notochord cells are the small, ciliated cells in the midline that are flanked by the larger endodermal cells of the primitive gut. The node (with its ciliated cells) is seen at the bottom. (B) The formation of the notochord by the dorsal infolding of the small, ciliated cells. (From Sulik et al. 1994; photograph courtesy of K. Sulik and G. C. Schoenwolf.)

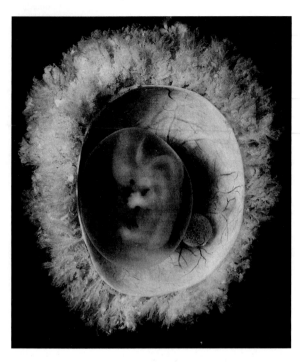

Figure 11.35
Human embryo and placenta after 50 days of gestation. The embryo lies within the amnion, and its blood vessels can be seen extending into the chorionic villi. The small sphere to the right of the embryo is the yolk sac. (The Carnegie Institution of Washington, courtesy of C. F. Reather.)

being readily separable from them (as in the contact placenta of the pig), or it may be so intimately integrated with maternal tissues that the two cannot be separated without damage to both the mother and the developing fetus (as in the deciduous placenta of most mammals, including humans).*

Figure 11.35 shows the relationships between the embryonic and extraembryonic tissues of a 6.5-week human embryo. The embryo is seen encased in the amnion and is further shielded by the chorion. The blood vessels extending to and from the chorion are readily observable, as are the villi that project from the outer surface of the chorion. These villi contain the blood vessels and allow the chorion to have a large area exposed to the maternal blood. Although fetal and maternal circulatory systems normally never merge, diffusion of soluble substances can occur through the villi (Figure 11.36). In this manner, the mother provides the fetus with nutrients and oxygen, and the fetus sends its waste products (mainly carbon dioxide and urea) into the maternal circulation. The maternal and fetal blood cells, however, usually do not mix.

WEBSITE 11.6 Placental functions. Placentas are nutritional, endocrine, and immunological organs. They provide hormones that enable the uterus to retain the pregnancy and also accelerate mammary gland development. Placentas also block the potential immune response of the mother against the developing fetus. Recent studies suggest that the placenta uses several mechanisms to block the mother's immune response.

---

*There are numerous types of placentas, and the extraembryonic membranes form differently in different orders of mammals (see Cruz and Pedersen 1991). Although mice and humans gastrulate and implant in a similar fashion, their extraembryonic structures are distinctive. It is very risky to extrapolate developmental phenomena from one group of mammals to another. Even Leonardo da Vinci got caught (Renfree 1982). His remarkable drawing of the human fetus inside the placenta is stunning art, but poor science: the placenta is that of a cow.

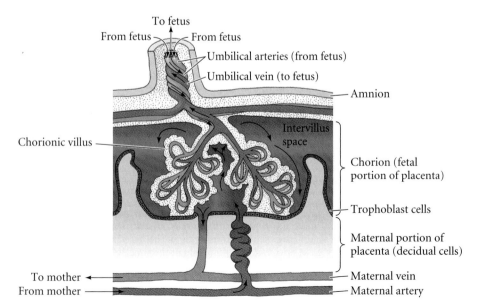

Figure 11.36
Relationship of the chorionic villi to the maternal blood supply in the uterus. In the umbilicus, there are two arteries and a single vein.

Twins and Embryonic Stem Cells

The early cells of the embryo can replace each other and compensate for a missing cell. This was first demonstrated in 1952, when Seidel destroyed one cell of a 2-cell rabbit embryo and the remaining cell produced an entire embryo. Once the inner cell mass (ICM) has become separate from the trophoblast, the ICM cells constitute an equivalence group. In other words, each ICM cell has the same potency (in this case, each cell can give rise to all the cell types of the embryo, but not to the trophoblast), and their fates will be determined by interactions among their descendants. Gardiner and Rossant (1976) also showed that if cells of the ICM (but not trophoblast cells) are injected into blastocysts, they contribute to the new embryo. Since the ICM blastomeres can generate any cell type in the body, the cells of the blastocyst are referred to as pluripotent (see Chapter 4).

This regulative capacity of the ICM blastomeres is also seen in humans. Human twins are classified into two major groups: monozygotic (one-egg, or identical) twins and dizygotic (two-egg, or fraternal) twins. Fraternal twins are the result of two separate fertilization events, whereas identical twins are formed from a single embryo whose cells somehow become dissociated from one another. Identical twins may be produced by the separation of early blastomeres, or even by the separation of the inner cell mass into two regions within the same blastocyst.

Identical twins occur in roughly 0.25% of human births. About 33% of identical twins have two complete and separate chorions, indicating that separation occurred before the formation of the trophoblast tissue at day 5 (Figure 11.37A). The remaining identical twins share a common chorion, suggesting that the split occurred within the inner cell mass after the trophoblast formed. By day 9, the human embryo has completed the construction of another extraembryonic layer, the lining of the amnion. This tissue forms the amnionic sac, which surrounds the embryo with amnionic fluid and protects it from desiccation and abrupt movement. If the separation of the embryo were to come after the formation of the chorion on day 5 but before the formation of the amnion on day 9, then the resulting embryos should have one chorion and two amnions (Figure 11.37B). This happens in about two-thirds

Figure 11.37
Diagram showing the timing of human monozygotic twinning with relation to extraembryonic membranes. (A) Splitting occurs before the formation of the trophoblast, so each twin has its own chorion and amnion. (B) Splitting occurs after trophoblast formation but before amnion formation, resulting in twins having individual amnionic sacs but sharing one chorion. (C) Splitting after amnion formation leads to twins in one amnionic sac and a single chorion. (After Langman 1981).

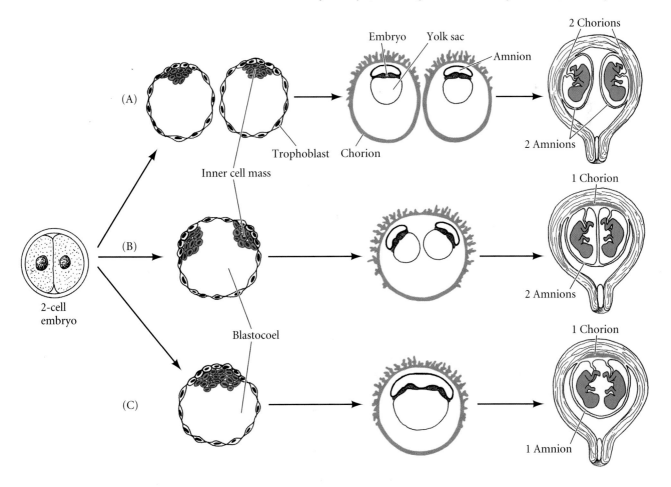

of human identical twins. A small percentage of identical twins are born within a single chorion and amnion (Figure 11.37C). This means that the division of the embryo came after day 9. Such newborns are at risk of being conjoined ("Siamese") twins.

The ability to produce an entire embryo from cells that normally would have contributed to only a portion of the embryo is called regulation, and is discussed in Chapter 3. Regulation is also seen in the ability of two or more early embryos to form one chimeric individual rather than twins, triplets, or a multiheaded individual. Chimeric mice can be produced by artificially aggregating two or more early-cleavage (usually 4- or 8-cell) embryos to form a composite embryo. As shown in Figure 11.38A, the zonae pellucidae of two genetically different embryos can be artificially removed

(A)

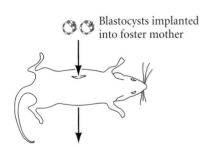

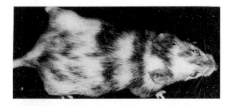

and the embryos brought together to form a common blastocyst. These blastocysts are then implanted into the uterus of a foster mother. When they are born, the chimeric offspring have some cells from each embryo. This is readily seen when the aggregated blastomeres come from mouse strains that differ in their coat colors. When blastomeres from white and black strains are aggregated, the result is commonly a mouse with black and white bands. There is even evidence (de la Chappelle et al. 1974; Mayr et al. 1979) that human embryos can form chimeras. Some individuals have two genetically different cell types (XX and XY) within the same body, each with its own set of genetically defined characteristics. The simplest explanation for such a phenomenon is that these individuals resulted from the aggregation of two embryos, one male and one female, that were developing at the same time. If this explanation is correct, then two fraternal twins have fused to create a single composite individual.

Markert and Petters (1978) have shown that three early 8-cell embryos can unite to form a common compacted morula and that the resulting mouse can have the coat colors of the three different strains (Figure 11.38B). Moreover, they showed that each of the three embryos gave rise to precursors of the gametes. When a chimeric (black/brown/white) female mouse was mated to a white-furred (recessive) male, offspring of each of the three colors were produced.

According to our observations of twin formation and chimeric mice, each blastomere of the inner cell mass should be able to produce any cell of the body. This hypothesis has been confirmed, and it has very important consequences for the study

(B)

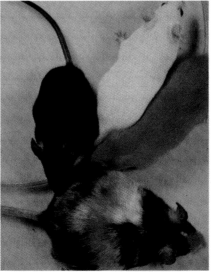

of mammalian development. When ICM cells are isolated and grown under certain conditions, they remain undifferentiated and continue to divide in culture (Evans and Kaufman 1981; Martin 1981). Such cells are called embryonic stem cells (ES cells). As shown in Chapter 4, cloned genes can be inserted into the nuclei of these cells, or the existing genes can be mutated. When these manipulated ES cells are injected into a mouse blastocyst, they can integrate into the host inner cell mass. The resulting embryo has cells coming from both the host and the donor tissue. This technique has become extremely important in determining the function of genes during mammalian development.

WEBSITE 11.7 **Nonidentical monozygotic twins.** Although monozygotic twins have the same genome, random developmental factors or the uterine environment may give them dramatically different phenotypes.

WEBSITE 11.8 **Conjoined twins.** There are rare events in which more than one set of axes is induced in the same embryo. These events can produce conjoined twins—twins that share some parts of their bodies. The medical and social issues raised by conjoined twins provide a fascinating look at what has constituted "individuality" throughout history.

Figure 11.38
Production of chimeric mice. (A) The experimental procedures used to produce chimeric mice. Early 8-cell embryos of genetically distinct mice (here, with coat color differences) are isolated from mouse oviducts and brought together after their zonae are removed by proteolytic enzymes. The cells form a composite blastocyst, which is implanted into the uterus of a foster mother. The photograph shows one of the actual chimeric mice produced in this manner. (B) An adult female chimeric mouse (bottom) produced from the fusion of three 4-cell embryos: one from two white-furred parents, one from two black-furred parents, and one from two brown-furred parents. The resulting mouse has coat colors from all three embryos. Moreover, each embryo contributed germ line cells, as is evidenced by the three colors of offspring (above) produced when this chimeric female was mated with recessive (white-furred) males. (A, photograph courtesy of B. Mintz; B from Markert and Petters 1978, photograph courtesy of C. Markert.)

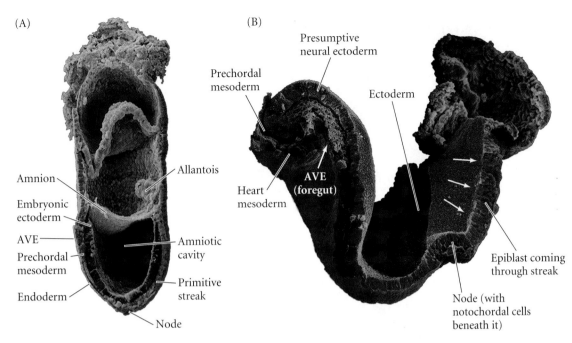

(A)

Amnion

Embryonic
ectoderm

AVE

Prechordal
mesoderm

Endoderm

Allantois

Amniotic
cavity

Primitive
streak

Node

(B)

Presumptive
neural ectoderm

Prechordal
mesoderm

Ectoderm

Heart
mesoderm

**AVE
(foregut)**

Epiblast coming
through streak

Node (with
notochordal cells
beneath it)

Figure 11.39
Axis formation in the mouse. (A) In the day 7 mouse embryo, the dorsal surface of the epiblast (embryonic ectoderm) is in contact with the amnionic cavity. The ventral surface of the epiblast contacts the newly formed mesoderm. In this cuplike arrangement, the endoderm covers the surface of the embryo. The node is at the bottom of the cup, and it has generated chordamesoderm. The two signaling centers, the node and the anterior visceral endoderm, are located on opposite sides of the cup. Eventually, the notochord will link them. The caudal side of the embryo is marked by the presence of the allantois. (B) By embryonic day 8, the anterior visceral endoderm lines the foregut, and the prechordal mesoderm is now in contact with the forebrain ectoderm. The node is now farther caudal, due largely to the rapid growth of the anterior portion of the embryo. The cells in the midline of the epiblast migrate through the primitive streak (white arrows). (Photographs courtesy of K. Sulik.)

Mammalian Anterior-Posterior Axis Formation

Two signaling centers

The formation of the mammalian anterior-posterior axis has been studied most extensively in the mouse. The structure of the mouse epiblast, however, differs from that of humans in that it is cup-shaped rather than disc-shaped. The dorsal surface of the epiblast (the embryonic ectoderm) contacts the amnionic cavity, while the ventral surface of the epiblast contacts the newly formed mesoderm. In this cuplike arrangement, the endoderm covers the surface of the embryo on the "ouside" of the cup (Figure 11.39A).

The mammalian embryo appears to have two signaling centers: one in the **node** ("the organizer") and one in the **anterior visceral endoderm (AVE)** (Figure 11.39B; Beddington 1994). The node (at the "bottom of the cup" in the mouse) appears to be responsible for the creation of all of the body, and the two signaling centers work together to form the anterior region of the embryo (Bachiller et al. 2000). The positions of the two signaling centers appears to be regulated by interactions between the epiblast and the extraembryonic membranes. The positioning of the AVE is probably initiated by signals from the extraembryonic ectoderm, which induce *nodal* gene expression in the epiblast beneath it. The Nodal proteins within the epiblast activate the expression of posterior genes that are required for mesoderm formation. They also induce the patterning of the visceral endoderm surrounding the mesoderm, promoting the formation of the AVE (Brennan et al. 2001). Formation of the node is dependent upon the trophoblast. The trophoblast cells make the Arkadia

protein, a nuclear protein that is critical in forming a signal that induces the node in the epiblast (Episkopou et al. 2001). Thus in mammals, the extraembryonic tissues are critical in polarizing the embryo and forming its two major signaling centers.

Both the mouse node and the anterior visceral endoderm express many of the genes known to be expressed in the chick and frog organizer tissues. The node produces chordin and Noggin (which the AVE does not), while the AVE expresses several genes that are necessary for head formation. These include the genes for transcription factors Hesx1, Lim-1, and Otx2, as well as the gene for the paracrine factor Cerberus (Perea-Gomez et al. 2001). The anterior visceral endoderm is established before the node, and the primitive streak (with its node) always forms on the side of the epiblast *opposite* the AVE. Homozygosity for mutant alleles of any of the above-mentioned head-specific genes of the AVE produces mice lacking forebrains (Thomas and Beddington 1996; Bedding-

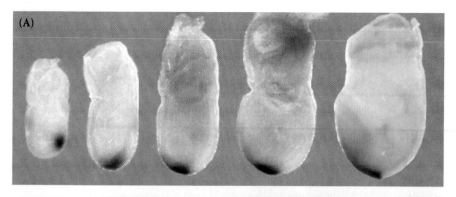

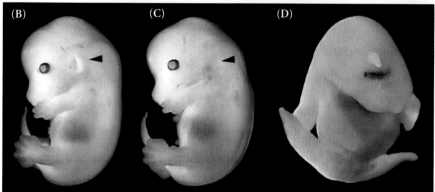

Figure 11.40
Expression of BMP antagonists in the mammalian node. (A) Expression of chordin during gastrulation is seen in the anterior primitive streak, the node, and axial mesoderm. It is not expressed in the anterior visceral endoderm. (B–D) Phenotypes of 12.5-day embryos. (B) Wild-type embryo. (C) Embryo with the *chordin* gene knocked out has a defective ear but an otherwise normal head. (D) Phenotype of a mouse deficient in both *chordin* and *noggin*. There is no jaw and a single centrally located eye, over which protrudes a large proboscis (nose). (After Bachiller et al. 2000; photographs courtesy of E. M. De Robertis.)

ton and Robertson 1999). While knockouts of either *chordin* or *noggin* do not affect development, mice missing *both* genes develop a body lacking forebrain, nose, and facial structures (Figure 11.40). It is probable that the AVE functions in the epiblast to restrict the Nodal signal, thereby allowing anterior genes to be expressed in the anterior portion of the epiblast. By the middle of day 6, the region around the node has been induced to express the Crypto transcription factor, while the area around the AVE expresses the Otx transcription factor. In this way, the embryo is delineated into anterior-posterior regions (Figure 11.41; Perea-Gomez et al. 2001; Kimura et al. 2001).

Figure 11.41
Schematic model of interactions between the visceral endoderm and epiblast in mice. (A) Nodal protein from the epiblast patterns the visceral endoderm surrounding it. All cells of the epiblast synthesize Oct4, Crypto, and Otx2 transcription factors. Head-forming genes *Cerberus* and *otx2* are expressed in the distal tip of the epiblast, and begin to pattern it. (B) At 5.5 days (1 day before primitive streak formation), the distal visceral endoderm (DVE) may prevent *crypto* expression in the distal portion of the epiblast. (C) As the rotation of the visceral endoderm causes the distal regions to become the AVE, the Crypto transcription factor becomes reduced in those cells that will be in the anterior of the embryo. (After Kimura et al. 2001.)

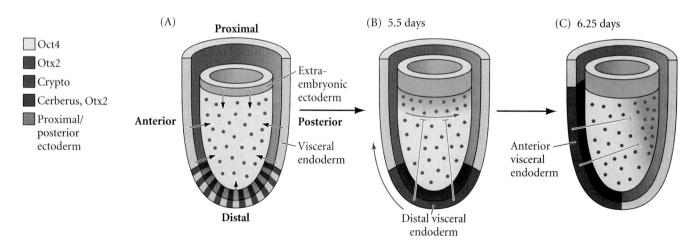

Patterning the anterior-posterior axis: The Hox code hypothesis

Once gastrulation begins, anterior-posterior polarity in all vertebrates becomes specified by the expression of **Hox genes.** Hox genes are homologous to the homeotic selector genes (Hom-C) of the fruit fly (see Chapter 9). The *Drosophila* homeotic gene complex on chromosome 3 contains the Antennapedia and bithorax clusters of homeotic genes (see Figure 9.27), and can be seen as a single functional unit. (Indeed, in other insects, such as the flour beetle *Tribolium*, it is physically a single unit.) The Hom-C genes are arranged in the same general order as their expression pattern along the anterior-posterior axis, the most 3′ gene (*labial*) being required for producing the most anterior structure, and the most 5′ gene (*AbdB*) specifying the development of the posterior abdomen. Mouse and human genomes contain four copies of the Hox complex per haploid set, located on four different chromosomes (*Hoxa* through *Hoxd* in the mouse, *HOXA* through *HOXD* in humans; see Boncinelli et al. 1988; McGinnis and Krumlauf 1992; Scott 1992). Not only are the same general types of homeotic genes found in both flies and mammals, but the order of these genes on their respective chromosomes is remarkably similar. In addition, the expression of these genes follows the same pattern: those mammalian genes homologous to the *Drosophila labial, proboscipedia,* and *deformed* genes are expressed anteriorly, while those genes that are homologous to the *Drosophila Abd-B* gene are expressed posteriorly. Another set of genes that controls the formation of the fly head (*orthodenticle* and *empty spiracles*) has homologues in the mouse that show expression in the midbrain and forebrain.

While Hox genes appear to specify the anterior-posterior axis throughout the vertebrates, we shall discuss mammals here, since the experimental evidence is particularly strong for this class. The mammalian Hox/HOX genes are numbered from 1 to 13, starting from that end of each complex that is expressed most anteriorly. Figure 11.42 shows the relationships between the *Drosophila* and mouse homeotic gene sets. The equivalent genes in each mouse complex (such as *Hoxa-1, Hoxb-1,* and *Hoxd-1*) are called a **paralogous group.** It is thought that the four mammalian Hox complexes were formed from chromosome duplications. Because there is not a one-to-one correspondence between the *Drosophila* Hom-C genes and the mouse Hox genes, it is likely that independent gene duplications have occurred since these two animal branches diverged (Hunt and Krumlauf 1992; see Chapter 23).

Expression of Hox genes along the dorsal axis

Hox gene expression can be seen along the dorsal axis (in the neural tube, neural crest, paraxial mesoderm, and surface ectoderm) from the anterior boundary of the hindbrain through the tail. The different regions of the body from the midbrain through the tail are characterized by different con-stellations of Hox gene expression, and the pattern of Hox gene expression is thought to specify the different regions. In general, the genes of paralogous group 1 are expressed from the tip of the tail to the most anterior border of the hindbrain. Paralogue 2 genes are expressed throughout the spinal cord, but the anterior limit of expression stops two segments more caudally than that of the paralogue 1 genes (see Figure 11.42; Wilkinson et al. 1989; Keynes and Lumsden 1990). The higher-numbered Hox paralogues are expressed solely in the posterior regions of the neural tube, where they also form a "nested" set.

Experimental analysis of the Hox code

The expression patterns of the murine Hox genes suggest a code whereby certain combinations of Hox genes specify a particular region of the anterior-posterior axis (Hunt and Krumlauf 1991). Particular sets of paralogous genes provide segmental identity along the anterior-posterior axis of the body. Evidence for such a code comes from three sources:

- Gene targeting or "knockout" experiments (see Chapter 4), in which mice are constructed that lack both copies of one or more Hox genes
- Retinoic acid teratogenesis, in which mouse embryos exposed to retinoic acid show an atypical pattern of Hox gene expression along the anterior-posterior axis and abnormal differentiation of their axial structures
- Comparative anatomy, in which the types of vertebrae in different vertebrate species are correlated with the constellation of Hox gene expression

GENE TARGETING. When Chisaka and Capecchi (1991) knocked out the *Hoxa-3* gene from inbred mice, the resulting homozygous mutants died soon after birth. Autopsies of these mice revealed that their neck cartilage was abnormally short and thick and that they had severely deficient or absent thymuses, thyroids, and parathyroid glands (Figure 11.43). The heart and major blood vessels were also malformed. Further analysis showed that the number and migration of the neural crest cells that normally form these structures were not affected. Rather, it appears that the *Hoxa-3* genes are responsible for specifying cranial neural crest cell fate and for enabling these cells to differentiate and proliferate into neck cartilage and the glands that form from the fourth and sixth pharyngeal pouches (Manley and Capecchi 1995).

Knockout of the *Hoxa-2* gene also produces mice whose neural crest cells have been respecified, but the defects in these mice are anterior to those in the *Hoxa-3* knockouts. Cranial elements normally formed by the neural crest cells of the second pharyngeal arch (stapes, styloid bones) are missing and are replaced by duplicates of the structures of the first pharyngeal arch (incus, malleus, etc.) (Gendron-Maguire et al. 1993; Rijli et al. 1993). Thus, without certain Hox genes, some regionally specific organs along the anterior-posterior axis fail to form, or become respecified as other regions. Similarly, when the *Hoxc-8* gene is knocked out (Le Mouellic et al. 1992), several axial skeletal segments resemble more anterior segments, much like what is seen in *Drosophila* loss-of-function homeotic mutations. As can be seen in Figure 11.44, the first lumbar vertebra of such a mouse has formed a rib—something characteristic of the thoracic vertebrae anterior to it.

One can obtain severe axial transformations by knocking out two or more genes of a paralogous group. Mice homozygous for the *Hoxd-3* deletion have mild abnormalities of the first cervical vertebra (the atlas), while mice homozygous for the *Hoxa-3* deletion have no abnormality of this bone, though they have other malformations (see above). When both sets of mutations are bred into the same mouse, both sets of problems become more severe. Mice with neither *Hoxa-3* nor *Hoxd-3* have no atlas bone at all, and the hyoid and thyroid cartilage is so reduced in size that there are holes in the skeleton (Condie and Capecchi 1994; Greer et al. 2000). It appears that there are interactions between the products of the Hox genes,

(A)

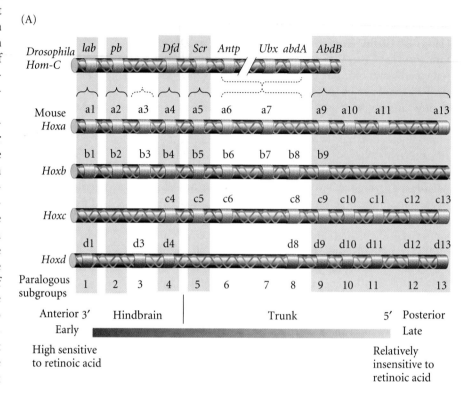

(B)

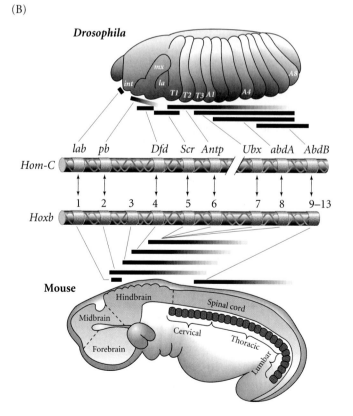

Figure 11.42
Evolutionary conservation of homeotic gene organization and transcriptional expression in fruit flies and mice. (A) Similarity between the Hom-C cluster on *Drosophila* chromosome 3 and the four Hox gene clusters in the mouse genome. The shaded regions show particularly strong structural similarities between species, and one can see that the order of the genes on the chromosomes has been conserved. Those genes at the 5′ end (since all mouse Hox genes are transcribed in the same direction) are those that are expressed more posteriorly, are expressed later, and can be induced only by high doses of retinoic acid. Genes having similar structures, the same relative positions on each of the four chromosomes, and similar expression patterns belong to the same paralogous group. (B) Comparison of the transcription patterns of the *Hom-C* and *Hoxb* genes of *Drosophila* (at 10 hours) and mice (at 12 days), respectively. The homologous human genes are called *HOX* (capitalized) genes. (A after Krumlauf 1993; B after McGinnis and Krumlauf 1992.)

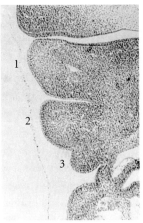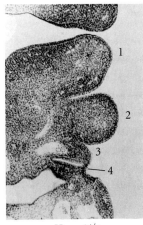

Hoxa-3⁻/⁻ *Hoxa-3⁺/⁻*

Figure 11.43
Deficient development of neural crest-derived pharyngeal arch and pouch structures in *Hoxa-3*-deficient mice. The arches are numbered. (Right) A 10.5-day embryo of a heterozygous *Hoxa-3* mouse (wild-type), showing normal development of pouch 3 (thymus), pouch 4 (parathyroid), and other structures. (Left) A homozygous mutant *Hoxa-3*-deficient mouse lacks the proper development of these structures. (From Chisaka and Capecchi 1991.)

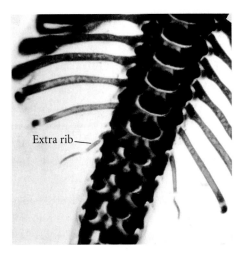

Figure 11.44
Partial transformation of the first lumbar vertebra into a thoracic vertebra by the knockout of the *Hoxc-8* gene. The first lumbar vertebra of this mouse has formed a rib—a structure normally formed only by the thoracic vertebrae anterior to it. (From Le Mouellic et al. 1992; photograph courtesy of the authors.)

and that in some functions, one of the paralogues can replace the other.

Thus, the evidence from gene knockouts supports the hypotheses (1) that different sets of Hox genes are necessary for the specification of any region of the anterior-posterior axis, (2) that the members of a paralogous group of Hox genes may be responsible for different subsets of organs within these regions, and (3) that the defects caused by knocking out particular Hox genes occur in the most anterior region of that gene's expression.

RETINOIC ACID TERATOGENESIS. Homeotic changes are also seen when mouse embryos are exposed to teratogenic doses of retinoic acid (RA), a derivative of vitamin A. By day 7 of development, a gradient of RA has been established that is high in the posterior regions and low in the anterior portions of the embryo (Sakai et al. 2001). This gradient appears to be controlled by the differential synthesis or degradation of RA in the different parts of the embryo. Several Hox genes have retinoic acid receptor sites in their enhancers, and exogenous retinoic acid given to mouse embryos in utero can cause certain Hox genes to become expressed in groups of cells that usually do not express them (Conlon and Rossant 1992; Kessel 1992). Moreover, craniofacial abnormalities of mouse embryos exposed to teratogenic doses of RA (Figure 11.45) can be mimicked by causing the expression of *Hoxa-7* throughout the embryo (Balling et al. 1989). Exogenous RA should mimic the RA concentrations normally encountered only by the posterior cells. If high doses of RA can activate Hox genes in inappropriate locations along the anterior-pos-

terior axis, and if the constellation of active Hox genes specifies the region of the anterior-posterior axis, then mice given RA in utero should show homeotic transformations manifested as rostralizing malformations occurring along that axis.

Kessel and Gruss (1991) found this to be the case. Wild-type mice have 7 cervical (neck) vertebrae, 13 thoracic (ribbed) vertebrae, and 6 lumbar (abdominal) vertebrae, in addition to the sacral and caudal (tail) vertebrae. In embryos exposed to RA on day 8 of gestation (during gastrulation), the first one or two lumbar vertebrae were transformed into thoracic (ribbed) vertebrae, while the first sacral vertebra often became a lumbar vertebra. In some cases, the entire posterior region of the embryo failed to form (Figure 11.45E). These changes in structure were correlated with changes in the constellation of Hox genes expressed in these tissues. For example, when RA was given to embryos on day 8, *Hoxa-10* expression was shifted posteriorly, and an additional set of ribs formed on what had been the first lumbar vertebra. When posterior Hox genes were not expressed at all, the caudal part of the embryo failed to form. The evidence points to a Hox code wherein different constellations of Hox genes, activated by different retinoic acid concentrations, specify the regional characteristics along the anterior-posterior axis. The effects of retinoic acid on human embryos will be discussed in Chapter 21.

VADE MECUM² **Retinoic acid as a teratogen.** See the teratogenic effects of retinoic acid on zebrafish development. **[Click to Zebrafish]**

(A)　(B)　(C)　(D)

Figure 11.45

The effect of retinoic acid on mouse embryos. Mouse embryos cultured (A) in control medium or (B) in medium containing retinoic acid, seen at day 10. The first pharyngeal arch of the treated embryos has a shortened and flattened appearance and has apparently fused with the second pharyngeal arch. The ossification of some of the skull has failed, and there are limb abnormalities as well. (C, D) Skeletal formation in (C) control mouse and (D) a mouse exposed to retinoic acid while in the uterus, seen at day 17. Craniofacial malformations can be seen in the neural crest-derived cartilage of the treated embryo. Meckel's cartilage has been completely displaced from the mandibular (lower jaw) to the maxillary (upper mouth) region, and the malleus and incus cartilages have not formed. (E) In some cases, RA exposure causes the loss of the lumbar, sacral, and caudal vertebrae. (A, B courtesy of G. Morriss-Kay; C, D from Morriss-Kay 1993, courtesy of G. Morriss-Kay; E from Kessel 1992, photographs courtesy of M. Kessel.)

(E)

COMPARATIVE ANATOMY. A new type of comparative embryology is emerging based on the comparison of gene expression patterns among species. Gaunt (1994) and Burke and her collaborators (1995) have compared the vertebrae of the mouse and the chick. Although the mouse and the chick have a similar number of vertebrae, they apportion them differently. Mice (like all mammals, be they giraffes or whales) have only 7 cervical vertebrae. These are followed by 13 thoracic vertebrae, 6 lumbar vertebrae, 4 sacral vertebrae, and a variable (20+) number of caudal vertebrae (Figure 11.46). The chick, on the other hand, has 14 cervical vertebrae, 7 thoracic vertebrae, 12 or 13 (depending on the strain) lumbosacral vertebrae, and 5 coccygeal (fused tail) vertebrae. The researchers asked, Does the constellation of Hox gene expression correlate with the type of vertebra formed (e.g., cervical or thoracic) or with the relative position of the vertebrae (e.g., number 8 or 9)?

The answer is that the constellation of Hox gene expression predicts the type of vertebra formed. In the mouse, the transition between cervical and thoracic vertebrae is between vertebrae 7 and 8; in the chick, it is between vertebrae 14 and 15. In both cases, the Hox-5 paralogues are expressed in the last cervical vertebra, while the anterior boundary of the Hox-6 paralogues extends to the first thoracic vertebra. Similarly, in both animals, the thoracic-lumbar transition is seen at the boundary between the Hox-9 and Hox-10 paralogous groups.

It appears there is a code of differing Hox gene expression along the anterior-posterior axis, and that code determines the type of vertebra formed.

> **WEBSITE 11.9 Why do mammals have only seven cervical vertebrae?** Recent speculation predicts that the mammalian Hox genes function simultaneously in several processes. To alter a Hox gene's expression so as to change vertebral type might lead to lethal changes in the other processes.

The Dorsal-Ventral and Right-Left Axes in Mice

The dorsal-ventral axis

Very little is known about the mechanisms of dorsal-ventral axis formation in mammals. In mice and humans, the hypoblast forms on the side of the inner cell mass that is exposed to the blastocyst fluid, while the dorsal axis forms from those ICM cells that are in contact with the trophoblast. Thus, the dorsal-ventral axis of the embryo is, in part, defined by the embryonic-abembryonic axis of the blastocyst. The **embryonic region** contains the ICM, while the **abembryonic region** is that part of the blastocyst opposite the ICM. The placement of the inner cell mass (and hence, the embryonic-abembry-

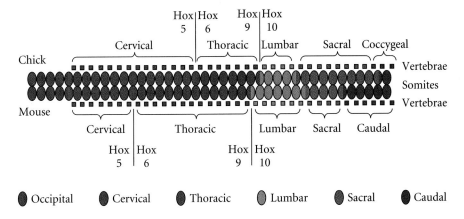

Figure 11.46
Schematic representation of the mouse and chick vertebral pattern along the anterior-posterior axis. The boundaries of expression of certain Hox gene paralogous groups have been mapped onto these domains. (After Burke et al. 1995.)

onic axis) appears to be specified by the point of sperm entry. Marking experiments have shown that the first cleavage plane defines the border between the embryonic and the abembryonic regions of the blastocyst, and that the position of the first cleavage plane correlates with the position of sperm entry (Figure 11.47; Gardner 2001; Piotrowska and Zernicka-Goetz 2001; Plusa et al. 2002). One cell of the first two blastomeres seems biased to become ICM, while the other appears to be biased to form the cells of the trophoblast. As development proceeds, the notochord maintains dorsal-ventral polarity by inducing specific dorsal-ventral patterns of gene expression in the overlying ectoderm (Goulding et al. 1993). Thus, "there is a memory of the first cleavage in our life" (M. Zernicka-Goetz, quoted in Pearson 2002).

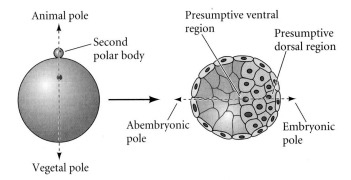

Figure 11.47
Relationship between the animal-vegetal axis of the egg and the embryonic-abembryonic axis of the blastocyst. The polar body marks the animal pole of the embryo. The dorsal-ventral axis of the embryo appears to form at right angles to the animal-vegetal axis of the egg.

The left-right axis

The mammalian body is not symmetrical. Although the human heart begins its formation at the midline of the embryo, it moves to the left side of the chest cavity and loops to the right (Figure 11.48). The spleen is found solely on the left side of the abdomen, the major lobe of the liver forms on the right side of the abdomen, the large intestine loops right to left as it traverses the abdominal cavity, and the right lung has one more lobe than the left lung.

Mutations in mice have shown that there are two levels of regulation of the left-right axis: a global level and an organ-specific level. Mutation of the gene *situs inversus viscerum (iv)* randomizes the left-right axis for each asymmetrical organ independently (Hummel and Chapman 1959; Layton 1976). This means that the heart may loop to the left in one homozygous animal, but to the right in another. Moreover, the direction of heart looping is not coordinated with the placement of the spleen or the stomach. This lack of coordination can cause serious problems, even death. A second gene, *inversion of embryonic turning (inv)*, causes a more global phenotype. Mice homozygous for an insertion mutation at this locus had all their asymmetrical organs on the wrong side of the body (Yokoyama et al. 1993).* Since all the organs were reversed, this asymmetry did not have dire consequences for the mice.

Several additional asymmetrically expressed genes have recently been discovered, and their influence on one another has enabled scientists to arrange them into a possible pathway.

---

*This gene was discovered accidentally when Yokoyama and colleagues (1993) made transgenic mice in which the transgene (for the tyrosinase enzyme) was inserted randomly into the genome. In one instance, this gene inserted itself into a region of chromosome 4, knocking out the existing *inv* gene. The resulting homozygous mice had laterality defects.

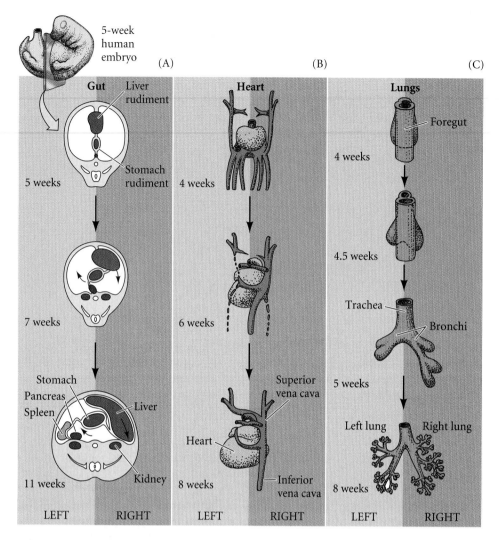

Figure 11.48
Left-right asymmetry in the developing human. (A) Abdominal cross sections show that the originally symmetrical organ rudiments acquire asymmetric positions by week 11. The liver moves to the right and the spleen moves to the left. (B) Not only does the heart move to the left side of the body, but the originally symmetrical veins of the heart regress differentially to form the superior and inferior venae cavae, which connect only to the right side of the heart. (C) The right lung branches into three lobes, while the left lung (near the heart) forms only two lobes. In human males, the scrotum also forms asymmetrically. (After Kosaki and Casey 1998.)

The end of this pathway—the activation of Nodal proteins and the Pitx2 transcription factor on the left side of the lateral plate mesoderm—appears to be the same as in frog and chick embryos, although the path leading to this point differs between the species (Figure 11.49; see Figure 11.23; Collignon et al. 1996; Lowe et al. 1996; Meno et al. 1996). In frogs, the pathway begins with the placement of Vg1; in chicks it begins with the suppression of *sonic hedgehog* expression.

In mammals, the distinction between left and right sides begins in the ciliary cells of the node (Figure 11.49B). The cilia cause fluid in the node to flow from right to left. When Nonaka and colleagues (1998) knocked out a mouse gene en-

coding the ciliary motor protein dynein (see Chapter 7), the nodal cilia did not move, and the situs (lateral position) of each asymmetrical organ was randomized. This finding correlated extremely well with other data. First, it had long been known that humans having a dynein deficiency had immotile cilia and a random chance of having their hearts on the left or right side of the body (Afzelius 1976). Second, when the *iv* gene described above was cloned, it was found to encode the ciliary dynein protein (Supp et al. 1997). Third, when Nonaka and colleagues (2002) cultured early mouse embryos under an artificial flow of medium from left to right, they obtained a reversal of the left-right axis. Moreover, the flow was able to

(A)

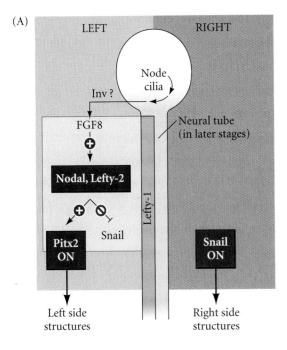

(B)

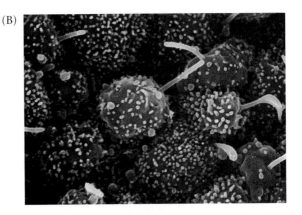

Figure 11.49

Situs formation in mammals. (A) Proposed pathway for left-right axis formation in the mouse. The leftward movement of cilia in the node activates some as yet unidentified factor (possibly the product of the *inv* gene). This product activates the *nodal* and *lefty-2* genes. The diffusion of Nodal and Lefty-2 proteins to the right-hand side is restricted by the product of the *lefty-1* gene, which coats the bottom of the neural tube on the left side. Nodal activates *pitx2*, the gene whose product activates left-sided properties in the various organs containing it. Either Nodal or Lefty-2 signaling (perhaps both) represses the *snail* gene, whose product is needed to instruct right-sidedness. (B) Ciliated cells of the mammalian node. This photograph is a close-up of the node seen in Figure 11.34A. (Photograph courtesy of K. Sulik and G. C. Schoenwolf.)

direct the polarity of the left-right axis in *iv* mutant mice, whose cilia are otherwise immotile.*

In some way (perhaps mediated through the product of the *rotatin* gene, which encodes a protein that is essential for the expression of *nodal*, *lefty2*, and *pitx2*), the leftward motion of the cilia activates the downstream pathway (Faisst et al. 2002). One possibility is that this leftward motion of the cilia activates the *fgf8* gene. FGF8 signaling could then activate the genes for two paracrine factors, Nodal and Lefty-2, in the lateral plate mesoderm on the left side of the embryo (Figure 11.49A). The proteins produced by these factors spread

throughout the left side of the embryo and appear to be restrained to that side by the Lefty-1 protein, which is secreted by the bottom left side of the neural tube (Meno et al. 1998). Lefty-2 appears to be able to block the Snail protein (which becomes specific for the right side of the body), while Nodal activates *pitx2* gene expression (Pedra et al. 1998). Thus, the genes for right-left axis formation appear to be similar throughout the vertebrates, but in some cases they are used differently. One of the big differences is that Fgf8 helps specify the right side of chicks and the left side of mammals (Meyers and Martin 1999).

Studies of mammalian development have enormous importance for understanding numerous human diseases. In the next chapters, we will discuss later aspects of vertebrate development and the relationship between genetics and development during organ formation.

---

*A recent report (Essner et al. 2002) suggests that nodal cilia may be responsible for left-right axis formation in all the vertebrate classes. When the gene for a dynein subunit was cloned, it was found to be expressed in the ventral portion of the node (or organizer) in mouse, chick, *Xenopus*, and zebrafish embryos.

Snapshot Summary: Early Vertebrate Development

1. Fish, reptiles, and birds undergo discoidal meroblastic cleavage, wherein the early cell divisions do not cut through the yolk of the egg. These early cells form a blastoderm.

2. In fish, the deep cells of the blastoderm form between the yolk syncytial layer and the enveloping layer. These cells

migrate over the top of the yolk, forming the hypoblast and epiblast layers. On the future dorsal side, these layers intercalate to form the embryonic shield, a structure homologous to the amphibian organizer. Transplantation of the embryonic shield into the ventral side of another embryo will cause the formation of a second embryonic axis.

3. In each class of vertebrates, neural ectoderm is permitted to form where the BMP-mediated induction of epidermal tissue is prevented

4. In chick embryos, early cleavage forms an area opaca and an area pellucida. The region between them is the marginal zone. Gastrulation begins at the posterior marginal zone, as the hypoblast and primitive streak both start there.

5. The primitive streak is derived from anterior epiblast cells and the central cells of the posterior marginal zone. As the primitive streak extends rostrally, Hensen's node is formed. Cells migrating through Hensen's node become chordamesoderm (notochord) cells. These cells extend up to the presumptive midbrain, where they meet the prechordal plate.

6. The prechordal plate induces the formation of the forebrain; the chordamesoderm induces the formation of the midbrain, hindbrain, and spinal cord. The first cells migrating laterally through the primitive streak become endoderm, displacing the hypoblast. The mesoderm cells then migrate through the primitive streak. Meanwhile, the surface ectoderm undergoes epiboly around the entire yolk.

7. In birds, gravity is critical in determining the anterior-posterior axis, while pH differences appear crucial for distinguishing dorsal from ventral. The left-right axis is formed by the expression of Nodal protein on the left side of the embryo, which signals Pitx2 expression on the left side of developing organs.

8. Mammals undergo holoblastic rotational cleavage, characterized by a slow rate of cell division, a unique cleavage orientation, lack of divisional synchrony, and the formation of a blastocyst.

9. The blastocyst forms after the blastomeres undergo compaction. It contains outer cells—the trophoblast cells—that become the chorion, and an inner cell mass that becomes the amnion and the embryo.

10. The chorion forms the fetal portion of the placenta, which functions to provide oxygen and nutrition to the embryo, to provide hormones for the maintenance of pregnancy, and to provide barriers to the mother's immune system.

11. Mammalian gastrulation is not unlike that of birds. There appear to be two signaling centers, one in the node and one in the anterior visceral endoderm. The latter center is critical for generating the forebrain, while the former is critical in inducing the axial structures caudally from the midbrain.

12. Hox genes pattern the anterior-posterior axis and help to specify positions along that axis. If Hox genes are knocked out, segment-specific malformations can arise. Similarly, causing the ectopic expression of Hox genes can alter the body axis.

13. The homology of gene structure and the similarity of expression patterns between *Drosophila* and mammalian Hox genes suggests that this patterning mechanism is extremely ancient.

14. The mammalian left-right axis is specified similarly to that of the chick, but with some significant differences in the roles of certain genes.

Literature Cited

Afzelius, B. A. 1976. A human syndrome caused by immotile cilia. *Science* 193: 317–319.

Alvarez, I. S., M. Araujo and M. A. Nieto. 1998. Neural induction in whole embryo cultures by FGF. *Dev. Biol.* 199: 42–54.

Arendt, D. and K. Nübler-Jung. 1999. Rearranging gastrulation in the name of yolk: Evolution of gastrulation in yolk-rich amniote eggs. *Mech. Dev.* 81: 3–22.

Bachiller, D. and 10 others. 2000. The organizer factors Chordin and Noggin are required for mouse forebrain development. *Nature* 403: 658–661.

Bachvarova, R. F., I. Skromne and C. D. Stern. 1998. Induction of primitive streak and Hensen's node by the posterior marginal zone in the early chick embryo. *Development* 125: 3521–3534.

Balinsky, B. I. 1975. *Introduction to Embryology*, 4th Ed. Saunders, Philadelphia.

Balling, R., G. Mutter, P. Gruss and M. Kessel. 1989. Craniofacial abnormalities induced by ectopic expression of the homeobox gene *Hox-1.1* in transgenic mice. *Cell* 58: 337–347.

Barlow, P., D. A. J. Owen and C. Graham. 1972. DNA synthesis in the preimplantation mouse embryo. *J. Embryol. Exp. Morphol.* 27: 432–445.

Beams, H. W. and R. G. Kessel. 1976. Cytokineses: A comparative study of cytoplasmic division in animal cells. *Am. Sci.* 64: 279–290.

Beddington, R. S. P. 1994. Induction of a second neural axis by the mouse node. *Development* 120: 613–620.

Beddington, R. S. P. and E. J. Robertson. 1999. Axis development and early asymmetry in mammals. *Cell* 96: 195–209.

Bellairs, R. 1986. The primitive streak. *Anat. Embryol.* 174: 1–14.

Bellairs, R., A. S. Breathnach and M. Gross. 1975. Freeze–fracture replication of junctional complexes in unincubated and incubated chick embryos. *Cell Tissue Res.* 162: 235–252.

Bellairs, R., F. W. Lorenz and T. Dunlap. 1978. Cleavage in the chick embryo. *J. Embryol. Exp. Morphol.* 43: 55–69.

Benson, G. V., H. Lim, B. C. Paria, I. Satokata, S. K. Dey and R. L. Maas. 1996. Mechanisms of reduced fertility in *Hoxa-10* mutant mice: Uterine homeosis and loss of maternal *Hoxa-10* expression. *Development* 122: 2687–2696.

Bertocchini, F. and C. D. Stern. 2002. The hypoblast of the chick embryo positions the primitive streak by antagonizing Nodal signaling. *Dev. Cell* 3: 735–744.

Bianchi, D. W., L. E. Wilkins-Haug, A. C. Enders and E. D. Hay . 1993. Origin of extraembryonic mesoderm in experimental animals: Relevance to chorionic mosaicism in humans. *Am. J. Med. Genet.* 46: 542–550.

Blader, P. and U. Strähle 1998. Ethanol impairs migration of the prechordal plate in the zebrafish embryo. *Dev. Biol.* 201: 185–201.

Boettger, T., H. Knoetgen, L. Wittler and M. Kessel. 2001. The avian organizer. *Int. J. Dev. Biol.* 45: 281–287.

Boncinelli, E., R. Somma, D. Acampora, M. Pannese, M. D'Esposito, A. Faiella and A. Simeone. 1988. Organization of human homeobox genes. *Hum. Reprod.* 3: 880–886.

Borland, R. M. 1977. Transport processes in the mammalian blastocyst. *Dev. Mammals* 1: 31–67.

Brennan, J., C. C. Lu, D. P. Norris, T. A. Rodriguez, R. S. Beddington and E. J. Robertson. 2001. Nodal signalling in the epiblast patterns the early mouse embryo. *Nature* 411: 965–969.

Brenner, C. A., R. R. Adler, D. A. Rappolee, R. A. Pedersen and Z. Werb. 1989. Genes for extracellular matrix-degrading metalloproteases and their inhibitor, TIMP, are expressed during early mammalian development. *Genes Dev.* 3: 848–859.

Burdsal, C. A., C. H. Damsky and R. A. Pedersen. 1993. The role of E-cadherin and integrins in mesoderm differentiation and migration at the mammalian primitive streak. *Development* 118: 829–844.

Burke, A. C., A. C. Nelson, B. A. Morgan and C. Tabin. 1995. Hox genes and the evolution of vertebrate axial morphology. *Development* 121: 333–346.

Carson, D. D., J.-P. Tang and J. Julian. 1993. Heparan sulfate proteoglycan (perlecan) expression by mouse embryos during acquisition of attachment competence. *Dev. Biol.* 155: 97–106.

Chisaka, O. and M. Capecchi. 1991. Regionally restricted developmental defects resulting from targeted disruption of the mouse homeobox gene *Hox-1.5*. *Nature* 350: 473–479.

Ciruna, B. and J. Rossant. 2001. FGF signaling regulates mesoderm cell fate specification and morphogenetic movement at the primitive streak. *Dev. Cell* 1: 37–49.

Collignon, J., I. Varlet and E. J. Robertson. 1996. Relationship between asymmetric nodal expression and the direction of embryonic turning. *Nature* 381: 155–158.

Condie, B. G. and M. R. Capecchi. 1994. Mice with targeted disruptions in the paralogous genes *hoxa-3* and *hoxd-3* reveal synergistic interactions. *Nature* 370: 304–307.

Conlon, R. A. and J. Rossant. 1992. Exogenous retinoic acid rapidly induces anterior ectopic expression of murine Hox-2 genes in vivo. *Development* 116: 357–368.

Cruz, Y. P. and R. A. Pedersen. 1991. Origin of embryonic and extraembryonic cell lineages in mammalian embryos. *In Animal Applications of Research in Mammalian Development.* Cold Spring Harbor Press, Cold Spring Harbor, NY, pp. 147–204.

Darnell, D. K., M. R. Stark and G. C. Schoenwolf. 1999. Timing and cell interactions underlying neural induction in the chick embryo. *Development* 126: 2505–2514.

de la Chappelle, A., J. Schroder, P. Rantanen, B. Thomasson, M. Niemi, A. Tilikainen, R. Sanger and E. E. Robson. 1974. Early fusion of two human embryos? *Ann. Hum. Genet.* 38: 63–75.

DeLuca, S. M. and 7 others. 1999. Hepatocyte growth factor/scatter factor promotes a switch from E- to N-cadherin in chick embryo epiblast cells. *Exp. Cell Res.* 251: 3–15.

Dias, M. S. and G. C. Schoenwolf. 1990. Formation of ectopic neuroepithelium in chick blastoderms: Age-related capacities for induction and self-differentiation following transplantation of quail Hensen's nodes. *Anat. Rec.* 229: 437–448.

Driever, W. 1995. Axis formation in zebrafish. *Curr. Opin. Genet. Dev.* 5: 610–618.

Driever, W. and 11 others. 1996. A genetic screen for mutations affecting development in zebrafish. *Development* 123: 37–46.

Du, S. J. and M. Dienhart. 2001. Zebrafish Tiggywinkle hedgehog promoter directs notochord and floorplate green fluorescence protein expression in transgenic zebrafish embryos. *Dev. Dynam.* 222: 655–666.

Dyce, J., M. George, H. Goodall and T. P. Fleming. 1987. Do trophectoderm and inner cell mass cells in the mouse blastocyst maintain discrete lineages? *Development* 100: 685–698.

Episkopou, V., R. Arkell, P. M. Timmons, J. J. Walsh, R. L. Andrew and D. Swan. 2001. Induction of the mammalian node requires Arkadia function in the extraembryonic lineages. *Nature* 410: 825–830.

Ernest, S., G. J. Rauch, P. Haffter, R. Geisler, C. Petit and T. Nicolson. 2000. *Mariner* is defective in myosin VIIA: A zebrafish model for human hereditary deafness. *Hum. Mol. Genet.* 9: 2189–2196.

Essner, J. J., K. J. Vogan, M. K. Wagner, C. J. Tabin, H. J. Yost and M. Brueckner. 2002. Conserved function for embryonic nodal cilia. *Nature* 418: 37–38.

Evans, M. J. and M. H. Kaufman. 1981. Establishment in culture of pluripotent cells from mouse embryos. *Nature* 292: 154–156.

Eyal-Giladi, H. 1991. The early embryonic development of the chick, an epigenetic process. *Crit. Rev. Poultry Biol.* 3: 143–166.

Eyal-Giladi, H. 1997. Establishment of the axis in chordates: Facts and speculations. *Development* 124: 2285–2296.

Eyal-Giladi, H. and B. Fabian. 1980. Axis determination in uterine chick blastocysts under changing spatial positions during the sensitive period of polarity. *Dev. Biol.* 77: 228–232.

Eyal-Giladi, H., A. Debby and N. Harel. 1992. The posterior section of the chick's area pellucida and its involvement in hypoblast and primitive streak formation. *Development* 116: 819–830.

Faisst, A. M., G. Alvarez-Bolado, D. Treichel and P. Gruss. 2002. *Rotatin* is a novel gene required for axial rotation and left-right specification in mouse embryos. *Mech. Dev.* 113: 15–28.

Faure, S., P. de Santa Barbara, D. J. Roberts and M. Whitman. 2002. Endogenous patterns of BMP signaling during early chick development. *Dev. Biol.* 244: 44–65.

Fekany, K. and 14 others. 1999. The zebrafish *bozozok* locus encodes Dharma, a homeodomain protein essential for induction of gastrula organizer and dorsoanterior embryonic structures. *Development* 126: 1427–1438.

Fisher, S. J., T.-Y. Cui, L. Zhang, K. Grahl, Z. Guo-Yang, J. Tarpey and C. H. Damsky. 1989. Adhesive and degradative properties of the human placental cytotrophoblast cells in vitro. *J. Cell Biol.* 109: 891–902.

Fleming, T. P. 1987. Quantitative analysis of cell allocation to trophectoderm and inner cell mass in the mouse embryo. *Dev. Biol.* 119: 520–531.

Fleming, T. P., B. Sheth and I Fesenko. 2001. Cell adhesion in the preimplantation mammalian embryo and its role in trophectoderm differentiation and blastocyst morphogenesis. *Front. Biosci.* 6: d1000–1007.

Fluck, R. A., K. L. Krok, B. A. Bast, S. E. Michaud and C. E. Kim. 1998. Gravity influences the position of the dorsoventral axis in medaka fish embryos (*Oryzias latipes*). *Dev. Growth Diff.* 40: 509–518.

Foley, A. C., I. Skromne and C. D. Stern. 2000. Reconciling different models of forebrain induction and patterning: A dual role for the hypoblast. *Development* 127: 3839–3854.

Fu, M.-F., C. Pressman, R. Dyer, R. L. Johnson and J. F. Martin. 1999. Function of Rieger syndrome gene in left-right asymmetry and craniofacial development. *Nature* 401: 276–278.

Gardiner, R. C. and J. Rossant. 1976. Determination during embryogenesis in mammals. *Ciba Found. Symp.* 40: 5–18.

Gardner, R. L. 2001. Specification of embryonic axes begins before cleavage in normal mouse development. *Development* 128: 839–847.

Gaunt, S. J. 1994. Conservation in the Hox code during morphological evolution. *Int. J. Dev. Biol.* 38: 549–552.

Gendron-Maguire, M., M. Mallo, M. Zhang and T. Gridley. 1993. *Hoxa-2* mutant mice exhibit homeotic transformation of skeletal elements derived from cranial neural crest. *Cell* 75: 1317–1331.

George-Weinstein, M. and 9 others. 1996. Skeletal myogenesis: The preferred pathway of chick embryo epiblast cells in vitro. *Dev. Biol.* 173: 279–291.

Gilbert, S. G. 1989. *Pictorial Human Embryology.* University of Washington Press, Seattle.

Goulding, M. D., A. Lumsden and P. Gruss. 1993. Signals from the notochord and floor plate regulate the region-specific expression of two Pax genes in the developing spinal cord. *Development* 117: 1001–1016.

Granato, M. and C. Nüsslein-Volhard. 1996. Fishing for genes controlling development. *Curr. Opin. Genet. Dev.* 6: 461–468.

Greer, J. M., J. Puetz, K. Thomas and M. R. Capecchi. 2000. Maintenance of functional equivalence during paralogous Hox gene evolution. *Nature* 403: 661–664.

Gritsman, K., W. S. Talbot and A. F. Schier. 2000. Nodal signaling patterns the organizer. *Development* 127: 921–932.

Gulyas, B. J. 1975. A reexamination of the cleavage patterns in eutherian mammalian eggs: Rotation of the blastomere pairs during second cleavage in the rabbit. *J. Exp. Zool.* 193: 235–248.

Haffter, P. and 16 others. 1996. The identification of genes with unique and essential functions in the development of the zebrafish, *Danio rerio*. *Development* 123: 1–36.

Halpern, M., R. Ho, C. Walker and C. Kimmel. 1993. Induction of somitic muscle pioneers and floor plate is distinguished by the zebrafish *no tail* mutation. *Cell* 75: 99–111.

Halpern, M. E., C. Thisse, R. K. Ho, B. Thisse, B. Riggeleman, E. S. Trevarrow and J. H. Weinberg. 1995. Cell-autonomous shift from axial to paraxial mesoderm development in zebrafish *floating head* mutants. *Development* 121: 4257–4264.

Hammerschmidt, M. and 14 others. 1996. *dino* and *mercedes*, two genes regulating dorsal development in the zebrafish embryo. *Development* 123: 95–102.

Hanna, L. A., R. K. Foreman, I. A. Tarasenko, D. S. Kessler, and P. A. Labosky. 2002. Requirement for Foxd3 in maintaining pluripotent cells of the early embryo. *Genes Dev.* 16: 2650–2661.

Helde, K. A., E. T. Wilson, C. J. Cretekos and D. J. Grunwald. 1994. Contribution of early cells to the fate map of the zebrafish gastrula. *Science* 265: 517–520.

Ho, R. K. 1992. Axis formation in the embryo of the zebrafish, *Brachydanio rerio*. *Sem. Dev. Biol.* 3: 53–64.

Hume, C. R. and J. Dodd. 1993. *Cwnt-8C*: A novel Wnt gene with a potential role in primitive streak formation and hindbrain organization. *Development* 119: 1147–1160.

Hummel, K. P. and D. B. Chapman. 1959. Visceral inversion and associated anomalies in the mouse. *J. Hered.* 50: 9–13.

Hunt, P. and R. Krumlauf. 1991. Deciphering the Hox code: Clues to patterning branchial regions of the head. *Cell* 66: 1075–1078.

Hunt, P. and R. Krumlauf. 1992. Hox codes and positional specification in vertebrate embryonic axes. *Annu. Rev. Cell Biol.* 8: 227–256.

Jurand, A. 1974. Some aspects of the development of the notochord in mouse embryos. *J. Embryol. Exp. Morphol.* 32: 1–33.

Kane, D. A. and C. B. Kimmel. 1993. The zebrafish midblastula transition. *Development* 119: 447–456.

Kawahara, A., T. Wilm, L. Solnica-Krezel and I. B. Dawid. 2000. Antagonistic role of *vega1* and *bozozok/dharma* homeobox genes in organizer formation. *Proc. Natl. Acad. Sci. USA* 97: 12121–12126.

Kelly, G. M., D. F. Erezyilmaz and R. T. Moon. 1995. Induction of a secondary embryonic axis in zebrafish occurs following the overexpression of β-catenin. *Mech. Dev.* 53: 261–273.

Kessel, M. 1992. Respecification of vertebral identities by retinoic acid. *Development* 115: 487–501.

Kessel, M. and P. Gruss. 1991. Homeotic transformations of murine vertebrae and concomitant alteration of Hox codes induced by retinoic acid. *Cell* 67: 89–104.

Keynes, R. and A. Lumsden. 1990. Segmentation and the origin of regional diversity in the vertebrate central nervous system. *Neuron* 2: 1–9.

Khaner, O. 1995. The rotated hypoblast of the chicken embryo does not initiate an ectopic axis in the epiblast. *Proc. Natl. Acad. Sci. USA* 92: 10733–10737.

Khaner, O. 1998. The ability to initiate an axis in the avian blastula is concentrated mainly at a posterior site. *Dev. Biol.* 194: 257–266.

Khaner, O. and H. Eyal-Giladi. 1989. The chick's marginal zone and primitive streak formation. I. Coordinative effect of induction and inhibition. *Dev. Biol.* 134: 206–214.

Kimmel, C. B. and R. D. Law. 1985. Cell lineage of zebrafish blastomeres. II. Formation of the yolk syncytial layer. *Dev. Biol.* 108: 86–93.

Kimmel, C. B. and R. M. Warga. 1987. Indeterminate cell lineage of the zebrafish embryo. *Dev. Biol.* 124: 269–280.

Kimmel, C. B., R. M. Warga and T. F. Schilling. 1990. Origin and organization of the zebrafish fate map. *Development* 108: 581–594.

Kimura, C., M. M. Shen, N. Takeda, S. Aizawa and I. Matsuo. 2001. Complementary functions of Otx2 and Cripto in initial patterning of mouse epiblast. *Dev. Biol.* 235: 12–32.

Kishimoto, Y., K. H. Lee, L. Zon, M. Hammerschmidt and S. Schulte-Merker. 1997. The molecular nature of zebrafish *swirl*: BMP2 function is essential during early dorsoventral patterning. *Development* 124: 4457–4466.

Kochav, S. M. and H. Eyal-Giladi. 1971. Bilateral symmetry in the chick embryo determination by gravity. *Science* 171: 1027–1029.

Kosaki, K. and B. Casey. 1998. Genetics of human left-right axis malformations. *Semin. Cell Dev.* 9: 89–99.

Koshida, S., M. Shinya, T. Mizuno, A. Kuroiwa and H. Takeda. 1998. Initial anteroposterior pattern of zebrafish central nervous system is determined by differential competence of the epiblast. *Development* 125: 1957–1966.

Krumlauf, R. 1993. Hox genes and pattern formation in the branchial region of the vertebrate head. *Trends Genet.* 9: 106–112.

Langeland, J. and C. B. Kimmel. 1997. The embryology of fish. *In* S. F. Gilbert and A. M. Raunio (eds.), *Embryology: Constructing the Organism*. Sinauer Associates, Sunderland, MA, pp. 383–407.

Langman, J. 1981. *Medical Embryology*, 4th Ed. Williams & Wilkins, Baltimore.

Larsen, W. J. 1993. *Human Embryology*. Churchill Livingston, New York.

Lash, J. W., E. Gosfield III, D. Ostrovsky and R. Bellairs. 1990. Migration of chick blastoderm under the vitelline membrane: The role of fibronectin. *Dev. Biol.* 139: 407–416.

Lawson, A. and G. C. Schoenwolf. 2001a. New insights into critical events of avian gastrulation. *Anat. Rec.* 262: 238–252.

Lawson, A. and G. C. Schoenwolf. 2001b. Cell populations and morphogenetic movements underlying formation of the avian primitive streak. *genesis* 29: 188–195.

Lawson, A., J.-F. Colas and G. C. Schoenwolf. 2001. Classification scheme for genes expressed during formation and progression of the anterior primitive streak. *Anat. Rec.* 262: 221–226.

Lawson, K. A., J. J. Meneses and R. A. Pedersen. 1991. Clonal analysis of epiblast fate during germ layer formation in the mouse embryo. *Development* 113: 891–911.

Layton, W. M., Jr. 1976. Random determination of a developmental process. *J. Hered.* 67: 336–338.

Le Douarin, N., A. Grapin-Botton and M. Catala. 1996. Patterning of the neural primordium in the avian embryo. *Semin. Dev. Biol.* 7: 157–167.

Le Mouellic, H., Y. Lallemand and P. Brûlet. 1992. Homeosis in the mouse induced by a null mutation in the *Hox-3.1* gene. *Cell* 69: 251–264.

Leung, C., S. E. Webb and A. Miller. 1998. Calcium transients accompany ooplasmic segregation in zebrafish embryos. *Dev. Growth Diff.* 40: 313–326.

Leung, C., S. E. Webb and A. Miller. 2000. On the mechanism of ooplasmic segregation in single-cell zebrafish embryos. *Dev. Growth Diff.* 42: 29–40.

Levin, M., R. L. Johnson, C. Stern, M. Kuehn and C. Tabin. 1995. A molecular pathway determining left-right asymmetry in chick embryogenesis. *Cell* 82: 803–814.

Levin, M., S. Pagan, D. Roberts, J. Cooke, M. R. Kuehn and C. J. Tabin. 1997. Left/Right patterning signals and the independent regulation of different aspects of situs in the chick embryo. *Dev. Biol.* 189: 57–67.

Levken, A. C., C. J. Thorpe, J. S. Waxman and R. T. Moon. 2001. Zebrafish *wnt8* encodes two Wnt8 proteins on a bicistronic transcript and is required for mesoderm and neurectoderm patterning. *Dev. Cell* 1: 103–114.

Lin, C. R. and 7 others. 1999. Pitx2 regulates lung asymmetry, cardiac positioning, and pituitary and tooth morphogenesis. *Nature* 401: 279–282.

Logan, M., S. M. Pagán-Westphal, D. M. Smith, L. Paganessi and C. J. Tabin. 1998. The transcription factor Pitx2 mediates situs-specific morphogenesis in response to left-right asymmetric signals. *Cell* 94: 307–317.

Lowe, L. A. and 8 others. 1996. Conserved left-right asymmetry of *nodal* expression and alterations in murine *situs inversus*. *Nature* 381: 158–161.

Luckett, W. P. 1978. Origin and differentiation of the yolk sac and extraembryonic mesoderm in presomite human and rhesus monkey embryos. *Am. J. Anat.* 152: 59–98.

Manley, N. R. and M. R. Capecchi. 1995. The role of *Hoxa-3* in mouse thymus and thyroid development. *Development* 121: 1989–2003.

Mark, W. H., K. Signorelli and E. Lacy. 1985. An inserted mutation in a transgenic mouse line results in developmental arrest at day 5 of gestation. *Cold Spring Harb. Symp. Quant. Biol.* 50: 453–463.

Markert, C. L. and R. M. Petters. 1978. Manufactured hexaparental mice show that adults are derived from three embryonic cells. *Science* 202: 56–58.

Martin, G. R. 1981. Isolation of a pluripotent cell line from early mouse embryos cultured in medium conditioned by teratocarcinoma stem cells. *Proc. Natl. Acad. Sci. USA* 78: 7634–7638.

Mayr, W. R., V. Pausch and W. Schnedl. 1979. Human chimaera detectable only by investigation of her progeny. *Nature* 277: 210–211.

McGinnis, W. and R. Krumlauf. 1992. Homeobox genes and axial patterning. *Cell* 68: 283–302.

Meno, C. and 7 others. 1996. Left-right asymmetric expression of the TGFβ-family member *lefty* in mouse embryos. *Nature* 381: 151–155.

Meno, C. and 8 others. 1998. *lefty-1* is required for left-right determination as a regulator of *lefty-2* and *nodal*. *Cell* 94: 287–297.

Meyers, E. N. and G. R. Martin. 1999. Differences in left-right axis pathways in mouse and chick: Functions of FGF8 and SHH. *Science* 285: 403–406.

Mitrani, E., T. Ziv, G. Thomsen, Y. Shimoni, D. A. Melton and A. Bril. 1990. Activin can induce the formation of axial structures and is expressed in the hypoblast of the chick. *Cell* 63: 495–501.

Morriss-Kay, G. 1993. Retinoic acid and craniofacial development: Molecules and morphogenesis. *BioEssays* 15: 9–15.

Mulnard, J. G. 1967. Analyse microcinematographique du developpement de l'oeuf de souris du stade II au blastocyste. *Arch. Biol.* (Liege) 78: 107–138.

New, D. A. T. 1956. The formation of subblastodermic fluid in hens' eggs. *J. Embryol. Exp. Morphol.* 43: 221–227.

New, D. A. T. 1959. Adhesive properties and expansion of the chick blastoderm. *J. Embryol. Exp. Morphol.* 7: 146–164.

Nguyen, V. H., B. Schmid, J. Trout, S. A. Connors, M. Ekker and M. C. Mullins. 1998. Ventral and lateral regions of the zebrafish gastrula, including the neural crest progenitors, are established by a *bmp2b/swirl* pathway of genes. *Dev. Biol.* 199: 93–110.

Nonaka, S. and 7 others. 1998. Randomization of left-right asymmetry due to loss of nodal cilia generating leftward flow of extraembryonic fluid in mice lacking KIF3B motor protein. *Cell* 95: 829–837.

Nonaka, S., H. Shiratori, Y. Saijoh and H. Hamada. 2002. Determination of left-right patterning in the mouse by artificial nodal flow. *Nature* 418: 96–99.

Oppenheimer, J. M. 1936. Transplantation experiments on developing teleosts (*Fundulus* and *Perca*). *J. Exp. Zool.* 72: 409–437.

O'Sullivan, C. M., S. L. Rancourt, S. Y. Liu and D. E. Rancourt. 2001. A novel murine tryptase involved in blastocyst hatching and outgrowth. *Reproduction* 122: 61–71.

Pearson, H. 2002. Your destiny, from day one. *Nature* 418: 14–15.

Pedersen, R. A., K. Wu and H. Batakier. 1986. Origin of the inner cell mass in mouse embryos: Cell lineage analysis by microinjection. *Dev. Biol.* 117: 581–595.

Pedra, M. E., J. M. Icardo, M. Albajar, J. C. Rodriguez-Rey and M. A. Ros. 1998. Pitx2 participates in the late phase of the pathway controlling left-right asymmetry. *Cell* 94: 319–324.

Perea-Gomez, A., M. Rhinn and S.-L. Ang. 2001 Role of the anterior visceral endoderm in restricting posterior signals in the mouse embryo. *Int. J. Dev. Biol.* 45: 311–320.

Perona, R. M. and P. M. Wassarman. 1986. Mouse blastocysts hatch in vitro by using a trypsin-like proteinase associated with cells of mural trophectoderm. *Dev. Biol.* 114: 42–52.

Peyrieras, N., F. Hyafil, D. Louvard, H. L. Ploegh, and F. Jacob. 1983. Uvomorulin: a non-integral membrane protein of early mouse embryos. *Proc. Natl. Acad. Sci. USA* 80: 6274–6277.

Piko, L. and K. B. Clegg. 1982. Quantitative changes in total RNA, total poly(A), and ribosomes in early mouse embryos. *Dev. Biol.* 89: 362–378.

Piotrowska, K. and M. Zernicka-Goetz. 2001. Role for sperm in spatial patterning of the early mouse embryo. *Nature* 409: 517–521.

Plusa, B., K. Piotowska, and M. Zernicka-Goetz. 2002. Sperm entry position provides a surface marker for the first cleavage plane of the mouse zygote. *genesis* 32: 193–198.

Prather, R. S. 1989. Nuclear transfer in mammals and amphibia. *In* H. Schatten and G. Schatten (eds.), *The Molecular Biology of Fertilization*. Academic Press, New York, pp. 323–340.

Psychoyos, D. and C. D. Stern. 1996. Fates and migratory routes of primitive streak cells in the chick embryo. *Development* 122: 1523–1534.

Reiter, J. F., Y. Kikuchi and D. Y. R. Stanier. 2001. Multiple roles for Gata5 in zebrafish endoderm formation. *Development* 128: 125–135.

Renfree, M. B. 1982. Implantation and placentation. *In* C. R. Austin and R. V. Short (eds.), *Embryonic and Fetal Development*. Cambridge University Press, Cambridge, pp. 26–69.

Rijli, F. M., M. Mark, S. Lakkaraju, A. Dierich, P. Dollé and P. Chambon. 1993. A homeotic transformation is generated in the rostral branchial region of the head by disruption of *Hoxa-2*, which acts as a selector gene. *Cell* 75: 1333–1349.

Rodriguez-Esteban, C., J. Capdevila, A. N. Economides, J. Pascual, Á. Ortiz and J. C. Izpisúa-Belmonte. 1999. The novel Cer-like protein Caronte mediates the establishment of embryonic left-right asymmetry. *Nature* 401: 243–251.

Rosenquist, G. C. 1966. A radioautographic study of labeled grafts in the chick blastoderm: Development from primitive-streak stages to stage 12. *Carnegie Inst. Wash. Contrib. Embryol.* 38: 31–110.

Rosenquist, G. C. 1972. Endoderm movements in the chick embryo between the short streak and head process stages. *J. Exp. Zool.* 180: 95–104.

Rugh, R. 1967. *The Mouse*. Burgess, Minneapolis.

Russ, A. P. and 11 others. 2000. Eomesodermin is required for mouse trophoblast development and mesoderm formation. *Nature* 404: 95–99.

Ryan, A. K. and 14 others. 1998. Pitx2 determines left-right asymmetry of internal organs in vertebrates. *Nature* 394: 545–551.

Sakai, Y., C. Meno, H. Fujii, J. Nishino, H. Shiratori, Y. Saijoh, J. Rossant and H. Hamada. 2001. The retinoic acid-inactivating enzyme CYP26 is essential for establishing an uneven distribution of retinoic acid along the anterioposterior axis within the mouse embryo. *Genes Dev.* 15: 213–225.

Sampath, K. and 8 others. 1998. Induction of the zebrafish ventral brain and floorplate requires cyclops/nodal signalling. *Nature* 375: 185–189.

Schier, A. F. 2001. Axis formation and patterning in zebrafish. *Curr. Opin. Genet. Dev.* 11: 393–404.

Schier, A. F. and W. S. Talbot. 1998. The zebrafish organizer. *Curr. Opin. Genet. Dev.* 8: 464–471.

Schier, A. F. and W. S. Talbot. 2001. Nodal signaling and the zebrafish organizer. *Int. J. Dev. Biol.* 45: 289–297.

Schier, A. F., S. C. Neuhauss, K. A. Held, W. S. Talbot and W. Driever. 1997 The *one eyed pinhead* gene functions in mesoderm and endoderm formation in zebrafish and interacts with *no tail*. *Development* 124: 327–342.

Schlesinger, A. B. 1958. The structural significance of the avian yolk in embryogenesis. *J. Exp. Zool.* 138: 223–258.

Schmitz, B. and J. A. Campos-Ortega. 1994. Dorso-ventral polarity of the zebrafish embryo is distinguishable prior to the onset of gastrulation. *Wilhelm Roux Arch. Dev. Biol.* 203: 374–380.

Schneider, S., H. Steinbeisser, R. M. Warga and P. Hausen. 1996. β-catenin translocation into nuclei demarcates the dorsalizing centers in frog and fish embryos. *Mech. Dev.* 57: 191–198.

Schoenwolf, G. C. 1991. Cell movements in the epiblast during gastrulation and neurulation in avian embryos. *In* R. Keller, W. H. Clark, Jr. and F. Griffin (eds.), *Gastrulation: Movements, Patterns, and Molecules.* Plenum, New York, pp. 1–28.

Schoenwolf, G. C., V. Garcia-Martinez and M. S. Diaz. 1992. Mesoderm movement and fate during amphibian gastrulation and neurulation. *Dev. Dynam.* 193: 235–248.

Schulte-Merker, S., K. J. Lee, A. P. McMahon and M. Hammerschmidt. 1997. The zebrafish organizer requires *chordino*. *Nature* 387: 862–863.

Scott, M. 1992. Vertebrate homeobox gene nomenclature. *Cell* 71: 551–553.

Seidel, F. 1952. Die Entwicklungspotenzen einer isolierten Blastomere des Zweizellenstadiums im Säugetierei. *Naturwissenschaften* 39: 355–356.

Shinya, M., M. Furutani-Seiki, A. Kuroiwa and H. Takeda. 1999. Mosaic analysis with *oep* mutant reveals a repressive interaction between floor-plate and non-floor-plate mutant cells in the zebrafish neural tube. *Dev. Growth Diff.* 41: 135–142.

Shook, D. R., C. Majer and R. Keller. 2002. Urodeles remove mesoderm from the superficial layer by subduction through a bilateral primitive streak. *Dev. Biol.* 248: 220–239.

Skromne, I. and C. D. Stern. 2002. A hierarchy of gene expression accompanying induction of the primitive streak by Vg1 in the chick embryo. *Mech. Dev.* 114: 115–118.

Smith, J. L. and G. C. Schoenwolf. 1998. Getting organized: New insights into the organizer of higher vertebrates. *Curr. Top. Dev. Biol.* 40: 79–110.

Solnica-Krezel, L. and W. Driever. 1994. Microtubule arrays of the zebrafish yolk cell: Organization and function during epiboly. *Development* 120: 2443–2455.

Solnica-Krezel, L. and W. Driever. 2001. The role of the homeodomain protein Bozozok in zebrafish axis formation. *Int. J. Dev. Biol.* 45: 299–310.

Solursh, M. and G. M. Morriss. 1977. Glycosaminoglycan synthesis in rat embryos during the formation of the primary mesenchyme and neural folds. *Dev. Biol.* 57: 75–86.

Solursh, M. and J. P. Revel. 1978. A scanning electron microscope study of cell shape and cell appendages in the primitive streak region of the rat and chick embryo. *Differentiation* 11: 185–190.

Spemann, H. and H. Mangold. 1924. Induction of embryonic primordia by implantation of organizers from a different species. *In* B. H. Willier and J. M. Oppenheimer (eds.), *Foundations of Experimental Embryology.* Hafner, New York, pp. 144–184.

Spratt, N. T., Jr. 1946. Formation of the primitive streak in the explanted chick blastoderm marked with carbon particles. *J. Exp. Zool.* 103: 259–304.

Spratt, N. T., Jr. 1947. Regression and shortening of the primitive streak in the explanted chick blastoderm. *J. Exp. Zool.* 104: 69–100.

Spratt, N. T., Jr. 1963. Role of the substratum, supracellular continuity, and differential growth in morphogenetic cell movements. *Dev. Biol.* 7: 51–63.

Spratt, N. T., Jr. and H. Haas. 1960. Integrative mechanisms in development of early chick blastoderm. I. Regulated potentiality of separate parts. *J. Exp. Zool.* 145: 97–138.

Stern, C. D. and D. R. Canning. 1988. Gastrulation in birds: A model system for the study of cellular morphogenesis. *Experientia* 44: 651–657.

Stern, C. D., G. W. Ireland, S. E. Herrick, E. Gherardi, J. Gray, M. Perryman and M. Stoker. 1990. Epithelial scatter factor and development of the chick embryonic axis. *Development* 110: 1271–1284.

Storey, K. G., A. Goriely, C. M. Sargent, J. M. Brown, H. D. Burns, H. M. Abud and J. K. Heath. 1998. Early posterior neural tissue is induced by FGF in the chick embryo. *Development* 125: 473–484.

Strahle, U. and S. Jesuthasan. 1993. Ultraviolet irradiation impairs epiboly in zebrafish embryos: Evidence for a microtubule-dependent mechanism of epiboly. *Development* 119: 451–453.

Streit, A. and C. D. Stern. 1999. More to neural induction than inhibition of BMPs. *In* S. Mooney (ed.), *Cell Lineage and Fate Determination.* Academic Press, New York, pp. 437–449.

Streit, A., A. J. Berliner, C. Papanayotou, A. Sirulnik and C. D. Stern. 2000. Initiation of neural induction by FGF signaling before gastrulation. *Nature* 406: 74–78.

Strickland, S., E. Reich and M. I. Sherman. 1976. Plasminogen activator in early embryogenesis: Enzyme production by trophoblast and parietal endoderm. *Cell* 9: 231–240.

Sulik, K., D. B. Dehart, J. L. Carson, T. Vrablic, K. Gesteland and G. C. Schoenwolf. 1994. Morphogenesis of the murine node and notochordal plate. *Dev. Dynam.* 201: 260–278.

Supp, D. M., D. P. Witte, S. S. Potter and M. Brueckner. 1997. Mutation of an axonal dynein affects left-right asymmetry in *inversus viscerum* mice. *Nature* 389: 963–966.

Tanaka, S., T. Kunath, A.-K. Hadjantonakis, A. Nagy and J. Rossant. 1998. Promotion of trophoblast stem cell proliferation by FGF4. *Science* 282: 2072–2075.

Thomas, P. Q. and R. S. P. Beddington. 1996. Anterior primitive endoderm may be responsible for patterning the anterior neural plate in the mouse embryo. *Curr. Biol.* 6: 1487–1496.

Trinkaus, J. P. 1984. Mechanisms of *Fundulus* epiboly: A current view. *Am. Zool.* 24: 673–688.

Trinkaus, J. P. 1992. The midblastula transition, the YSL transition, and the onset of gastrulation in *Fundulus*. *Development* [Suppl.] 1992: 75–80.

Trinkaus, J. P. 1993. The yolk syncitial layer of *Fundulus*: Its origin and history and its significance for early embryogenesis. *J. Exp. Zool.* 265: 258–284.

Trinkaus, J. P. 1996. Ingression during early gastrulation of *Fundulus*. *Dev. Biol.* 177: 356–370.

Tuchmann-Duplessis, H., G. David and P. Haegel. 1972. *Illustrated Human Embryology*, Vol. 1. Springer-Verlag, New York.

Vakaet, L. 1984. The initiation of gastrula ingression in the chick blastoderm. *Am. Zool.* 24: 555–562.

Waddington, C. H. 1932. Experiments in the development of chick and duck embryos cultivated in vitro. *Phil. Trans. R. Soc. Lond.* [B] 13: 221.

Waddington, C. H. 1933. Induction by the primitive streak and its derivatives in the chick. *J. Exp. Zool.* 10: 38–46.

Warga, R. M. and C. B. Kimmel. 1990. Cell movements during epiboly and gastrulation in zebrafish. *Development* 108: 569–580.

Wiley, L. M. 1984. Cavitation in the mouse preimplantation embryo: Na/K ATPase and the origin of nascent blastocoel fluid. *Dev. Biol.* 105: 330–342.

Wilkinson, D. G., S. Bhatt, M. Cook, E. Boncinelli and R. Kruflauf. 1989. Segmental expression of Hox-2 homeobox-containing genes in the developing mouse hindbrain. *Nature* 341: 405–409.

Wolpert, L., R. Beddington, J. Brockes, T. Jessell, P. Lawrence and E. Meyerowitz. 1998. *Principles of Development.* Current Biology Ltd., London.

Yamanaka, Y. and 7 others. 1998. A novel homeobox gene, *dharma*, can induce the organizer in a non-cell-autonomous manner. *Genes Dev.* 12: 2345–2353.

Yamashita, S., C. Miyagi, A. Carmany-Rapey, T. Shimizu, R. Fuji, A. F. Schier and T. Hirano. 2002. Stat3 controls cell movements during zebrafish gastrulation. *Dev. Cell* 2: 363–375.

Yamazaki, K. and Y. Kato. 1989. Sites of zona pellucida shedding by mouse embryo other than mural trophectoderm. *J. Exp. Zool.* 249: 347–349.

Yokoyama, T., N . G. Copeland, N. A. Jenkins, C. A. Montgomery, F. F. B. Elder and P. A. Overbeek. 1993. Reversal of left-right symmetry: A situs inversus mutation. *Science* 260: 679–682.

Zhong, T. P., S. Childs, J. P. Leu and M. C. Fishman. 2001. Gridlock signalling pathway fashions the first embryonic artery. *Nature* 414: 216–220.

PART III Later Embryonic Development

12 The emergence of the ectoderm:
   The central nervous system and the epidermis

13 Neural crest cells and axonal specificity

14 Paraxial and intermediate mesoderm

15 Lateral plate mesoderm and endoderm

16 Development of the tetrapod limb

17 Sex determination

18 Metamorphosis, regeneration, and aging

19 The saga of the germ line

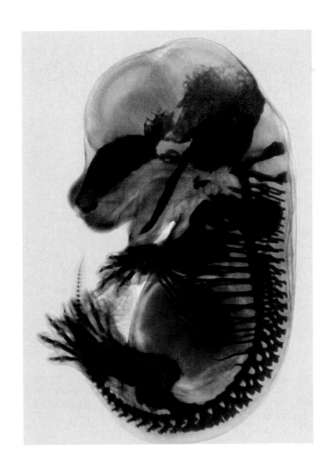

The cartilaginous skeleton of a 17-day mouse embryo. Meckel's cartilage is seen prominently in the lower jaw, extending from the middle ear to the distal tip of the mandible. Later, the proximal portion of this cartilage will ossify into the malleus and stapes bones of the middle ear. (Figure 11:45 courtesy of Dr. G. Morriss-Kay).

chapter 12 The emergence of the ectoderm: The central nervous system and the epidermis

"What is perhaps the most intriguing question of all is whether the brain is powerful enough to solve the problem of its own creation." So Gregor Eichele (1992) ended a review of research on mammalian brain development. The construction of an organ that perceives, thinks, loves, hates, remembers, changes, fools itself, and coordinates our conscious and unconscious bodily processes is undoubtedly the most challenging of all developmental enigmas. A combination of genetic, cellular, and organismal approaches is giving us a preliminary understanding of how the basic anatomy of the brain becomes ordered.

The fates of the vertebrate ectoderm are shown in Figure 12.1. In the past two chapters, we have seen how the ectoderm is instructed to form the vertebrate nervous system and epidermis. A portion of the dorsal ectoderm is specified to become neural ectoderm, and its cells become distinguishable by their columnar appearance. This region of the embryo is called the **neural plate**. The process by which this tissue forms a **neural tube**, the rudiment of the central nervous system, is called **neurulation**, and an embryo undergoing such changes is called a **neurula** (Figure 12.2). The neural tube will form the brain anteriorly and the spinal cord posteriorly. This chapter will look at the processes by which the neural tube and the epidermis arise and acquire their distinctive patterns.

Establishing the Neural Cells

In previous chapters, we have seen that neural cells become specified through their interactions with other cells. There are at least four stages through which the pluripotent cells of the epiblast or blastula become neural precursor cells, or **neuroblasts** (Wilson and Edlund 2001):

- **Competence**, wherein cells can become neuroblasts if they are exposed to the appropriate combination of signals.
- **Specification**, wherein cells have received the appropriate signals to become neuroblasts, but progression along the neural differentiation pathway can still be repressed by other signals.
- **Commitment** (**determination**), wherein neuroblasts have entered the neural differentiation pathway and will become neurons even in the presence of neural inhibitory signals.
- **Differentiation**, wherein the neuroblasts leave the mitotic cycle and express those genes characteristic of neurons.

As we have seen, the *specification* of the neural plate is primarily to BMP antagonists such as Noggin, chordin, and follistatin. However, the *competence* of the pluripotent cells to respond to inducing signals may be due to fibroblast growth factors. In the chick, the epiblast cells are not able to respond to BMP antagonists unless they have been previously exposed to some other organizer signal for 5 hours. This other signal is probably an FGF, since (1) FGF8 is made by the primitive streak precursor cells; (2) FGF8 can induce the expression of early (but not later) markers of neural precursor development; and (3) the inhibition of FGF signaling prevents neural

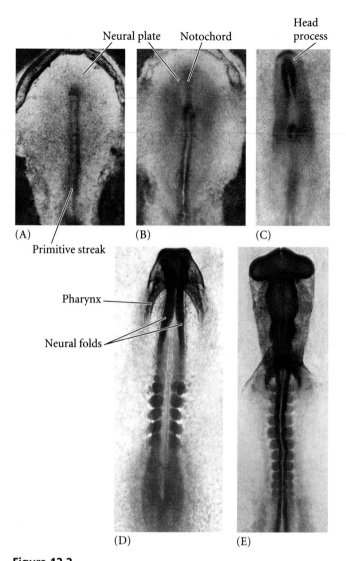

Figure 12.2
Neurulation in a chick embryo (dorsal view). (A) Flat neural plate. (B) Flat neural plate with underlying notochord (head process). (C) Neural groove. (D) Incipient neural tube. (E) Neural tube, showing the three brain regions and the spinal cord. (F) 24-hour chick embryo, as in (D). The cephalic (head) region has undergone neurulation, while the caudal (tail) regions are still undergoing gastrulation. (A–E, photographs courtesy of G. C. Schoenwolf; F after Patten 1971.)

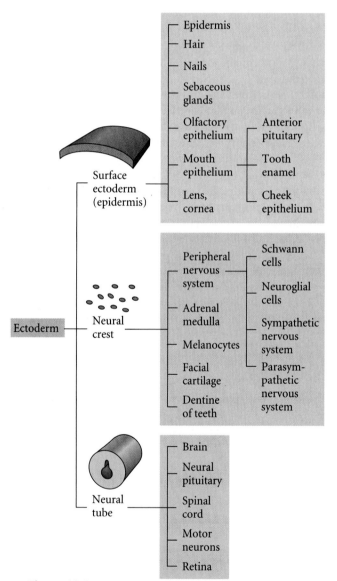

Figure 12.1
Major derivatives of the ectoderm germ layer. The ectoderm is divided into three major domains: the surface ectoderm (primarily epidermis), the neural crest (peripheral neurons, pigment, facial cartilage), and the neural tube (brain and spinal cord).

induction from occurring, even when Hensen's node is producing its BMP antagonists (Streit et al. 2000). Interestingly, these FGFs can be inhibited by Wnt signaling in the *lateral* epiblast cells (Wilson et al. 2001). This inhibition creates a boundary between the neurally competent epiblast cells responding to FGFs and the neurally incompetent epiblast cells synthesizing and responding to Wnts. The commitment of cells to a neuronal fate may be accomplished through the insulin-like growth factor pathway, but the evidence for this so far comes only from *Xenopus* (Pera et al. 2001).

(F)

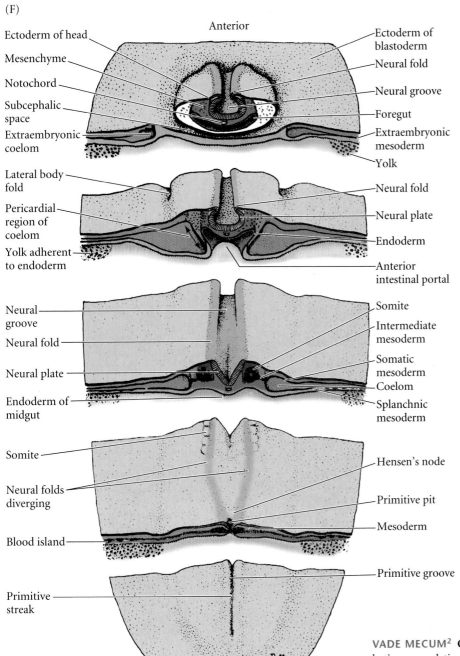

Anterior

Ectoderm of head
Mesenchyme
Notochord
Subcephalic space
Extraembryonic coelom

Ectoderm of blastoderm
Neural fold
Neural groove
Foregut
Extraembryonic mesoderm
Yolk

Lateral body fold
Pericardial region of coelom
Yolk adherent to endoderm

Neural fold
Neural plate
Endoderm
Anterior intestinal portal

Neural groove
Neural fold
Neural plate
Endoderm of midgut

Somite
Intermediate mesoderm
Somatic mesoderm
Coelom
Splanchnic mesoderm

Somite
Neural folds diverging
Blood island

Hensen's node
Primitive pit
Mesoderm

Primitive streak

Primitive groove

Patten

Posterior

Formation of the Neural Tube

There are two major ways of converting the neural plate into a neural tube. In **primary neurulation**, the cells surrounding the neural plate direct the neural plate cells to proliferate, invaginate, and pinch off from the surface to form a hollow tube. In **secondary neurulation**, the neural tube arises from the coalescence of mesenchyme cells to form a solid cord that subsequently hollows out (cavitates) to create a hollow tube. In general, the anterior portion of the neural tube is made by primary

neurulation, while the posterior portion of the neural tube is made by secondary neurulation. The complete neural tube forms by joining these two separately formed tubes together.

In birds, all the neural tube anterior to the twenty-eighth somite pair (i.e., everything anterior to the hindlimbs) is made by primary neurulation (Pasteels 1937; Catala et al. 1996). In mammals, secondary neurulation begins at the level of the sacral vertebrae (Schoenwolf 1984; Nievelstein et al. 1993). In amphibians such as *Xenopus*, only the tail neural tube is derived from secondary neurulation (Gont et al. 1993). and this pattern also occurs in fish (whose neural tubes had formerly been thought to be formed solely by secondary neurulation) (H. Sive and L. A. Hardaker, personal communication).

Primary neurulation

The events of primary neurulation in the chick and the frog are illustrated in Figures 12.2D,E and 12.3. This process divides the original ectoderm into three sets of cells: (1) the internally positioned neural tube, which will form the brain and spinal cord, (2) the externally positioned epidermis of the skin, and (3) the neural crest cells (see Figure 12.1). The **neural crest** cells form in the region that connects the neural tube and epidermis, but then migrate elsewhere; they will generate the peripheral neurons and glia, the pigment cells of the skin, and several other cell types.

VADE MECUM[2] **Chick neurulation.** By 33 hours of incubation, neurulation in the chick embryo is well underway. Both whole mounts and a complete set of serial cross sections through a 33-hour chick embryo are included in this segment so that you can see this amazing event. The serial sections can be displayed either as a continuum in movie format or individually, along with labels and color-coding that designates germ layers. **[Click on Chick-Mid]**

The process of primary neurulation appears to be similar in amphibians (Figure 12.4), reptiles, birds, and mammals (Gallera 1971). Shortly after the neural plate has formed, its edges thicken and move upward to form the **neural folds**, while a U-shaped **neural groove** appears in the center of the plate, dividing the future right and left sides of the embryo (see Figures

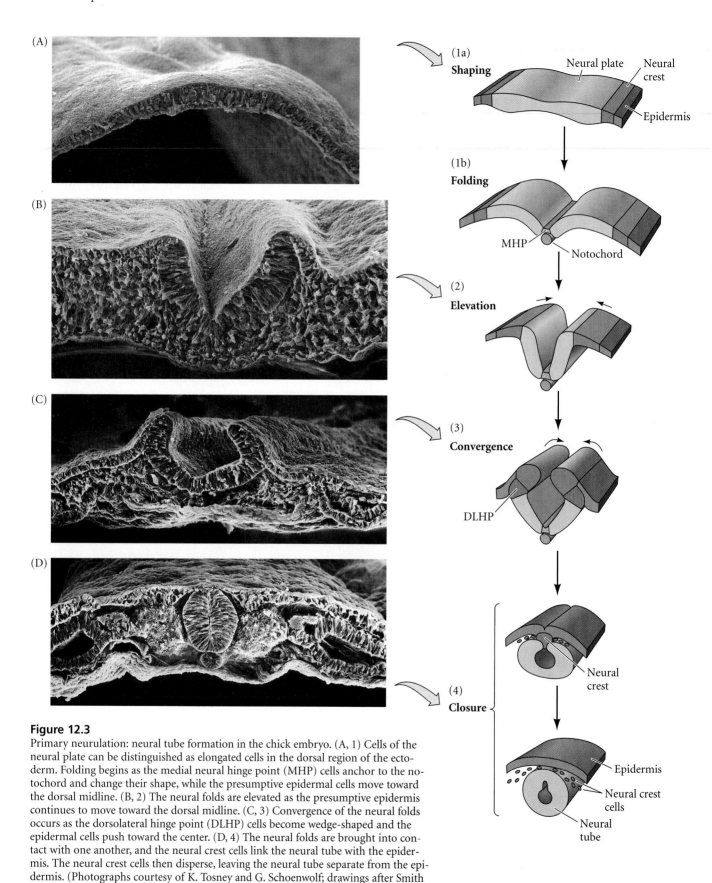

Figure 12.3

Primary neurulation: neural tube formation in the chick embryo. (A, 1) Cells of the neural plate can be distinguished as elongated cells in the dorsal region of the ectoderm. Folding begins as the medial neural hinge point (MHP) cells anchor to the notochord and change their shape, while the presumptive epidermal cells move toward the dorsal midline. (B, 2) The neural folds are elevated as the presumptive epidermis continues to move toward the dorsal midline. (C, 3) Convergence of the neural folds occurs as the dorsolateral hinge point (DLHP) cells become wedge-shaped and the epidermal cells push toward the center. (D, 4) The neural folds are brought into contact with one another, and the neural crest cells link the neural tube with the epidermis. The neural crest cells then disperse, leaving the neural tube separate from the epidermis. (Photographs courtesy of K. Tosney and G. Schoenwolf; drawings after Smith and Schoenwolf 1997.)

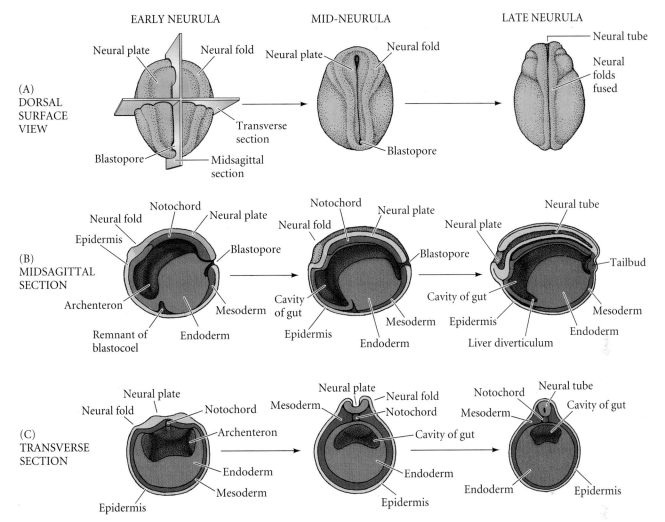

Figure 12.4
Three views of neurulation in an amphibian embryo, showing early (left), middle (center), and late (right) neurulae in each case. (A) Looking down on the dorsal surface of the whole embryo. (B) Sagittal section through the medial plane of the embryo. (C) Transverse section through the center of the embryo. (After Balinsky 1975.)

12.2C and 12.3). The neural folds migrate toward the midline of the embryo, eventually fusing to form the neural tube beneath the overlying ectoderm. The cells at the dorsalmost portion of the neural tube become the neural crest cells.

Primary neurulation can be divided into four distinct but spatially and temporally overlapping stages: (1) *formation* of the neural plate; (2) *shaping* of the neural plate; (3) *bending* of the neural plate to form the neural groove; and (4) *closure* of the neural groove to form the neural tube (Smith and Schoenwolf 1997; Colas and Schoenwolf 2001; see Figure 12.2).

FORMATION AND SHAPING OF THE NEURAL PLATE. The process of neurulation begins when the underlying dorsal mesoderm (and pharyngeal endoderm in the head region) signals the ectodermal cells above it to elongate into columnar neural plate cells (Smith and Schoenwolf 1989; Keller et al. 1992). Their elongated shape distinguishes the cells of the prospective neural plate from the flatter pre-epidermal cells surrounding them. As much as 50% of the ectoderm is included in the neural plate. The neural plate is shaped by the intrinsic movements of the epidermal and neural plate regions. The neural plate lengthens along the anterior-posterior axis, narrowing itself so that subsequent bending will form a tube (instead of a spherical capsule).

In both amphibians (see Figure 12.4) and amniotes, the neural plate lengthens and narrows by convergent extension, intercalating several layers of cells into a few layers. In addition, divisions of the neural plate cells are preferentially in the **rostral-caudal** (beak-tail; anterior-posterior) direction (Jacobson and Sater 1988; Schoenwolf and Alvarez 1989; Sausedo et al. 1997; see Figures 12.2 and 12.3). These events will occur even if the tissues involved are separated. If the neural plate is isolated, its cells converge and extend to make a thinner plate, but fail to roll up into a neural tube. However, if the "border region" containing both presumptive epidermis and neural plate tissue is isolated, it will form small neural folds in culture (Jacobson and Moury 1995; Moury and Schoenwolf 1995).

> **WEBSITE 12.1 Formation of the floor plate cells.** One of the major controversies in developmental neurobiology concerns the origin of the cells that form the ventral floor of the neural tube. It is possible that these cells are derived directly from the notochord and do not arise from the surface ectoderm.

BENDING OF THE NEURAL PLATE. The bending of the neural plate involves the formation of hinge regions where the neural plate contacts surrounding tissues. In birds and mammals, the cells at the midline of the neural plate are called the **medial hinge point** (**MHP**) **cells.** They are derived from the portion of the neural plate just anterior to Hensen's node and from the anterior midline of Hensen's node (Schoenwolf 1991a,b; Catala et al. 1996). The MHP cells become anchored to the notochord beneath them and form a hinge, which forms a furrow at the dorsal midline. The notochord induces the MHP cells to decrease their height and to become wedge-shaped (van Straaten et al. 1988; Smith and Schoenwolf 1989). The cells lateral to the MHP do not undergo such a change (Figure 12.3B, C). Shortly thereafter, two other hinge regions form furrows near the connection of the neural plate with the remainder of the ectoderm. These regions are called the **dorsolateral hinge points** (**DLHPs**), and they are anchored to the surface ectoderm of the neural folds. The DLHP cells, too, increase their height and become wedge-shaped.

Cell wedging is intimately linked to changes in cell shape. In the DLHPs, microtubules and microfilaments are both involved in these changes. Colchicine, an inhibitor of microtubule polymerization, inhibits the elongation of these cells, while cytochalasin B, an inhibitor of microfilament formation, prevents the apical constriction of these cells, thereby inhibiting wedge formation (Burnside 1973; Karfunkel 1972; Nagele and Lee 1987). As the neuroepithelial cells enlarge (by means of the microtubules), the microfilaments and their associated regulatory proteins accumulate at the apical ends of the cells (Zolessi and Arruti 2001). Constriction of these microfilaments allows the change of cell shape to occur. After the initial furrowing of the neural plate, the plate bends around these hinge regions. Each hinge acts as a pivot that directs the rotation of the cells around it (Smith and Schoenwolf 1991).

Meanwhile, extrinsic forces are also at work. The surface ectoderm of the chick embryo pushes toward the midline of the embryo, providing another motive force for the bending of the neural plate (see Figure 12.3C; Alvarez and Schoenwolf 1992; Lawson et al. 2001). This movement of the presumptive epidermis and the anchoring of the neural plate to the underlying mesoderm may also be important for ensuring that the neural tube invaginates *into* the embryo and not outward. If small pieces of neural plate are isolated from the rest of the embryo (including the mesoderm), they tend to roll inside

out (Schoenwolf 1991a). The pushing of the presumptive epidermis toward the center and the furrowing of the neural tube creates the neural folds.

CLOSURE OF THE NEURAL TUBE. The neural tube closes as the paired neural folds are brought together at the dorsal midline. The folds adhere to each other, and the cells from the two folds merge. In some species, the cells at this junction form the neural crest cells. In birds, the neural crest cells do not migrate from the dorsal region until after the neural tube has closed at that site. In mammals, however, the cranial neural crest cells (which form facial and neck structures; see Chapter 13) migrate while the neural folds are still being elevated (i.e., prior to neural tube closure), whereas in the spinal cord region, the neural crest cells do not migrate until closure has occurred (Nichols 1981; Erickson and Weston 1983).

The closure of the neural tube does not occur simultaneously throughout the ectoderm. This is best seen in those vertebrates (such as birds and mammals) whose body axis is elongated prior to neurulation. In amniotes, induction in the head starts before induction in the trunk, so in the 24-hour chick embryo, neurulation in the **cephalic** (head) region is well advanced, while the **caudal** (tail) region of the embryo is still undergoing gastrulation (see Figure 12.2F). The two open ends of the neural tube are called the **anterior neuropore** and the **posterior neuropore**.

Unlike neurulation in chicks, in which neural tube closure is initiated at the level of the future midbrain and "zips up" in both directions, neural tube closure in mammals is initiated at several places along the anterior-posterior axis (Golden and Chernoff 1993; Van Allen et al. 1993; Nakatsu et al. 2000). Different **neural tube defects** are caused when various parts of the neural tube fail to close (Figure 12.5). Failure of the human *posterior* neural tube regions to close at day 27 (or the rupture of the posterior neuropore shortly thereafter) results in a condition called **spina bifida**, the severity of which depends on how much of the spinal cord remains exposed. Failure to close the *anterior* neural tube regions results in a lethal condition, **anencephaly**, in which the forebrain remains in contact with the amnionic fluid and subsequently degenerates. Fetal forebrain development ceases, and the vault of the skull fails to form. The failure of the entire neural tube to close over the entire body axis is called **craniorachischisis**. Collectively, neural tube defects are not rare in humans, as they are seen in about 1 in every 500 live births. Neural tube closure defects can often be detected during pregnancy by various physical and chemical tests.

Human neural tube closure requires a complex interplay between genetic and environmental factors (Botto et al. 1999). Certain genes, such as *Pax3*, *sonic hedgehog*, and *openbrain*, are essential for the formation of the mammalian neural tube; but dietary factors, such as cholesterol and folic acid, also appear to be critical. It has been estimated that over 50% of human neural tube defects could be prevented by a pregnant woman's

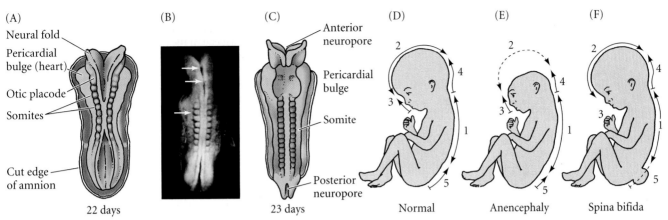

Figure 12.5

Neurulation in the human embryo. (A) Dorsal view of a 22-day (8-somite) human embryo initiating neurulation. Both anterior and posterior neuropores are open to the amniotic fluid. (B) A ten-somite human embryo showing regions of neural tube closure (arrows). (C) Dorsal view of a neurulating human embryo with only its neuropores open. (D) Regions of neural tube closure postulated by genetic evidence (superimposed on newborn body). (E) Anencephaly is caused by the failure of neural tube closure in region 2. (F) Spina bifida is caused by the failure of region 5 to fuse (or of the posterior neuropore to close). (B from Nakatsu et al., photograph courtesy of K. Shiota; D–F after Van Allen et al. 1993.)

taking supplemental folic acid (vitamin B_{12}), and the U.S. Public Health Service recommends that all women of childbearing age take 0.4 mg of folate daily to reduce the risk of neural tube defects during pregnancy (Milunsky et al. 1989; Czeizel and Dudas 1992; Centers for Disease Control 1992). The fetuses most at risk are those with genetic defects in their folate production—which may constitute about 10% of the population (Whitehead et al. 1995; DeMarco et al. 2000). While the mechanism by which folate aids neural tube closing is not understood, recent studies (Saitsu et al. 2002) have demonstrated a folate-binding protein on the most dorsal regions of the mouse neural tube immediately prior to fusion. But, folate deficiency appears to be only one risk factor for neural tube defects. Mothers in low socioeconomic groups appear to have a higher incidence of babies with neural tube defects, even when vitamin use is taken into account (Little and Elwood 1992a;

Wasserman et al. 1998). Moreover, there appears to be a seasonal variation in the incidence of neural tube defects that remains to be explained (Little and Elwood 1992b).

The neural tube eventually forms a closed cylinder that separates from the surface ectoderm. This separation is thought to be mediated by the expression of different cell adhesion molecules. Although the cells that will become the neural tube originally express E-cadherin, they stop producing this protein as the neural tube forms, and instead synthesize N-cadherin and N-CAM (Figure 12.6). As a result, the surface ectoderm and neural tube tissues no longer adhere to each other. If the surface ectoderm is experimentally made to express N-cadherin (by injecting N-cadherin mRNA into one cell of a 2-cell *Xenopus* embryo), the separation of the neural tube from the presumptive epidermis is dramatically impeded (Detrick et al. 1990; Fujimori et al. 1990).

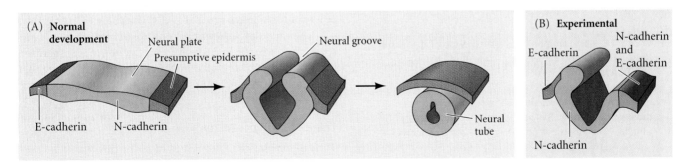

Figure 12.6

Expression of N- and E-cadherin adhesion proteins during neurulation in *Xenopus*. (A) Normal development. In the neural plate stage, N-cadherin is seen in the neural plate, while E-cadherin is seen on the presumptive epidermis. Eventually, the N-cadherin-bearing neural cells separate from the E-cadherin-containing epidermal cells. (The neural crest cells express neither N- nor E-cadherin, and they disperse.) (B) No separation of the neural tube occurs when one side of the frog embryo is injected with N-cadherin mRNA, so that N-cadherin is expressed in the epidermal cells as well as in the presumptive neural tube.

(A)

Medullary cord

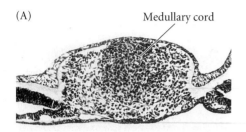

(B)

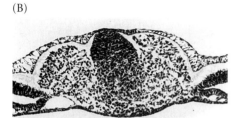

(C)

Neural tube

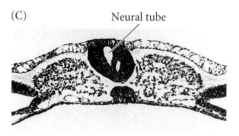

(D)

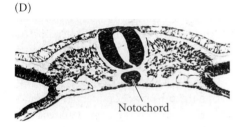

Notochord

Figure 12.7
Secondary neurulation in the caudal region of a 25-somite chick embryo. (A) The medullary cord forming at the most caudal end of the chick tailbud. (B) The medullary cord at a slightly more anterior position in the tailbud. (C) The neural tube is cavitating and the notochord forming. (D) The lumens coalesce to form the central canal of the neural tube. (From Catala et al. 1995; photographs courtesy of N. M. Le Douarin.)

WEBSITE 12.2 Neural tube closure. The closing of the neural tube is a complex event that can be influenced by both genes and environment. The interactions between genetic and environmental factors are now being untangled.

WEBSITE 12.3 Homologous specification of the neural tissue. The insect nervous system develops in a manner very different from that of the vertebrate nervous system. However, the genetic instructions for forming the central neural system and specifying its regions appear to be homologous. In *Drosophila*, neural fate often depends on the number of cell divisions preceding neuronal differentiation.

Secondary neurulation

Secondary neurulation involves the condensation of mesenchymal cells to form a medullary cord beneath the surface ectoderm, and the subsequent hollowing out of this cord into a neural tube (Figure 12.7). Knowledge of the mechanisms of

secondary neurulation may be important in medicine, given the prevalence of human posterior spinal cord malformations.

In frogs and chicks, secondary neurulation usually occurs in the lumbar (abdominal) and tail vertebrae. In both cases, it can be seen as a continuation of gastrulation. In the frog, instead of involuting into the embryo, the cells of the dorsal blastopore lip keep growing ventrally (Figure 12.8A, B). The growing region at the tip of the lip is called the **chordoneural hinge** (Pasteels 1937), and it contains precursors of both the posteriormost portion of the neural plate and the posterior portion of the notochord. The growth of this region converts the roughly spherical gastrula, 1.2 mm in diameter, into a linear tadpole some 9 mm long. The tip of the tail is the direct descendant of the dorsal blastopore lip, and the cells lining the blastopore form the **neurenteric canal**. The proximal part of the neurenteric canal fuses with the anus, while the distal portion becomes the **ependymal canal** (i.e., the lumen of the neural tube) (Figure 12.8C; Gont et al. 1993).

Differentiation of the Neural Tube

The differentiation of the neural tube into the various regions of the central nervous system occurs simultaneously in three different ways. On the gross anatomical level, the neural tube and its lumen bulge and constrict to form the chambers of the brain and the spinal cord. At the tissue level, the cell populations within the wall of the neural tube rearrange themselves to form the different functional regions of the brain and the spinal cord. Finally, on the cellular level, the neuroepithelial cells themselves differentiate into the numerous types of nerve cells (**neurons**) and supportive cells (**glia**) present in the body.

The early development of most vertebrate brains is similar, but because the human brain may be the most organized piece of matter in the solar system and is arguably the most interesting organ in the animal kingdom, we will concentrate on the development that is supposed to make *Homo* sapient.

The anterior-posterior axis

The early mammalian neural tube is a straight structure. However, even before the posterior portion of the tube has formed, the most anterior portion of the tube is undergoing drastic changes. In this region, the neural tube balloons into the three primary vesicles (Figure 12.9): forebrain (**prosencephalon**), midbrain (**mesencephalon**), and hindbrain (**rhombencephalon**). By the time the posterior end of the neural tube

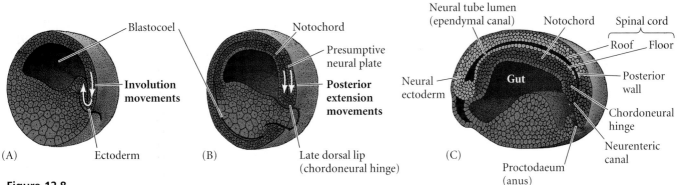

Figure 12.8
Movements of cells during secondary neurulation in *Xenopus*. (A) Involution of the mesoderm at the mid-gastrula stage. (B) Movements of the dorsal blastopore lip at the late gastrula/early neurula stage. Involution has ceased, and both the ectoderm and the mesoderm of the late blastopore lip move posteriorly. (C) Early tadpole stage, wherein the cells lining the blastopore form the neurenteric canal, part of which becomes the lumen of the secondary neural tube. (From Gont et al. 1993.)

closes, secondary bulges—the **optic vesicles**—have extended laterally from each side of the developing forebrain.

The prosencephalon becomes subdivided into the anterior **telencephalon** and the more caudal **diencephalon**. The telencephalon will eventually form the **cerebral hemispheres**, and the diencephalon will form the **thalamic** and **hypothalamic** brain regions, which receive neural input from the retina. Indeed, the retina itself is a derivative of the diencephalon.

The mesencephalon does not become subdivided, and its lumen eventually becomes the **cerebral aqueduct**. The rhombencephalon becomes subdivided into a posterior **myelencephalon** and a more anterior **metencephalon**. The myelencephalon eventually becomes the **medulla oblongata**, whose neurons generate the nerve centers ("nuclei") responsible for pain relay to the head and neck, auditory connections, tongue movement, and balance control, as well as respiratory, gastrointestinal, and cardiovascular movements. The metencephalon gives rise to the **cerebellum**, the part of the brain responsible for coordinating movements, posture, and balance.

The rhombencephalon develops a segmental pattern that specifies the places where certain nerves originate. Periodic swellings called **rhombomeres** divide the rhombencephalon into smaller compartments. The rhombomeres represent separate developmental "territories" in that the cells within each

Figure 12.9
Early human brain development. The three primary brain vesicles are subdivided as development continues. At the right is a list of the adult derivatives formed by the walls and cavities of the brain. (After Moore and Persaud 1993.)

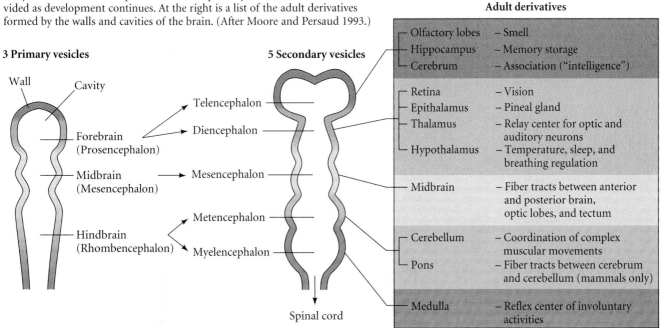

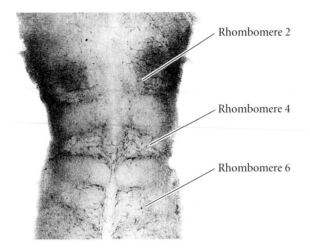

Figure 12.10
A 2-day embryonic chick hindbrain, splayed to show the lateral walls. Neurons were visualized with an antibody staining neurofilament proteins. Rhombomeres 2, 4, and 6 are distinguished by the high density of axons at this early developmental stage. (From Lumsden and Keynes 1989; photograph courtesy of A. Keynes.)

rhombomere can mix freely within it, but not with cells from adjacent rhombomeres (Guthrie and Lumsden 1991). Moreover, each rhombomere has a different developmental fate. The neural crest cells above each rhombomere will form **ganglia**—clusters of neuronal cell bodies whose axons form a nerve. The generation of the cranial nerves from the rhombomeres has been most extensively studied in the chick, in which the first neurons appear in the even-numbered rhom-

bomeres, r2, r4, and r6 (Figure 12.10; Lumsden and Keynes 1989). Neurons originating from r2 ganglia form the fifth (trigeminal) cranial nerve; those from r4 form the seventh (facial) and eighth (vestibuloacoustic) cranial nerves; and the ninth (glossopharyngeal) cranial nerve exits from r6.

The ballooning of the early embryonic brain is remarkable in its rate, in its extent, and in its being primarily the result of an increase in cavity size, not tissue growth. In the chick embryo, brain volume expands 30-fold between days 3 and 5 of development. This rapid expansion is thought to be caused by positive fluid pressure exerted against the walls of the neural tube by the fluid within it. It might be expected that this fluid pressure would be dissipated by the spinal cord, but this does not appear to happen. Rather, as the neural folds close in the region between the presumptive brain and the presumptive spinal cord, the surrounding dorsal tissues push in to constrict the neural tube at the base of the brain (Figure 12.11; Schoenwolf and Desmond 1984; Desmond and Schoenwolf 1986; Desmond and Field 1992). This occlusion (which also occurs in the human embryo) effectively separates the presumptive brain region from the future spinal cord (Desmond 1982). If the fluid pressure in the anterior portion of such an occluded neural tube is experimentally removed, the chick brain enlarges at a much slower rate and contains many fewer cells than are found in normal embryos. The occluded region of the neural tube reopens after the initial rapid enlargement of the brain ventricles.

The anterior-posterior patterning of the nervous system is controlled by a series of genes that include the Hox gene complexes. These genes were discussed more fully in Chapter 11 and will be brought up again in Chapter 23.

WEBSITE 12.4 Mapping the mes-encephalon. The chick cerebellum has a dual origin, with most of it coming from the metencephalon, but some portions coming from the mesencephalon. Genetic abnormalities can be traced to specific brain regions by transplantations between wild-type and mutant brain regions.

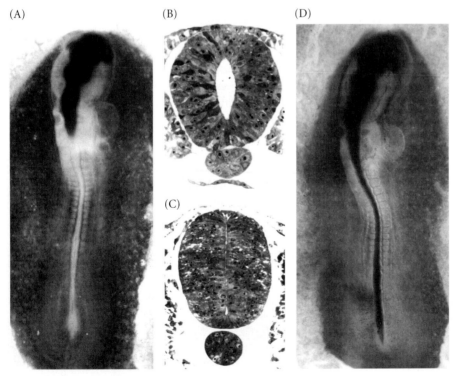

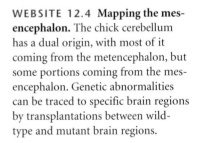

Figure 12.11
Occlusion of the neural tube allows expansion of the future brain region. (A) Dye injected into the anterior portion of a 3-day chick neural tube fills the brain region, but does not pass into the spinal region. (B, C) Sections of the chick neural tube at the base of the brain (B) before occlusion and (C) during occlusion. (D) Reopening of the occlusion after initial brain enlargement allows dye to pass from the brain region into the spinal cord region. (Photographs courtesy of M. Desmond.)

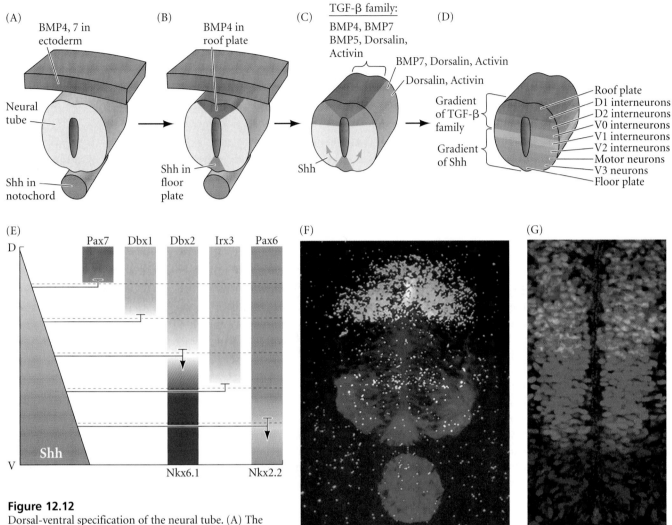

Figure 12.12
Dorsal-ventral specification of the neural tube. (A) The newly formed neural tube is influenced by two signaling centers. The roof of the neural tube is exposed to BMP4 and BMP7 from the epidermis, and the floor of the neural tube is exposed to Sonic hedgehog protein from the notochord. (B) Secondary signaling centers are established within the neural tube. BMP4 is expressed and secreted from the roof plate cells; Sonic hedgehog is expressed and secreted from the floor plate cells. (C) BMP4 establishes a nested cascade of TGF-β factors, spreading ventrally into the neural tube from the roof plate. Sonic hedgehog diffuses dorsally as a gradient from the floor plate cells. (D) The neurons of the spinal cord are given their identities by their exposure to these gradients of paracrine factors. The amounts and types of paracrine factors present cause different transcription factors to be activated in the nuclei of these cells, depending on their position in the neural tube. (E) Sonic hedgehog is confined to the ventral region of the neural tube by the TGF-β factors, and the gradient of Sonic hedgehog specifies the ventral neural tube by activating and inhibiting the synthesis of particular transcription factors. (F) Chick neural tube, showing areas of Sonic hedgehog (green) and the expression domain of the TGF-β-family protein dorsalin (blue). Motor neurons induced by a particular concentration of Sonic hedgehog are stained orange/yellow. (G) In situ hybridization for three transcription factors: Pax7 (red, characteristic of the dorsal neural tube cells), Pax6 (green), and Nkx6.1 (blue). Where Nkx6.1 and Pax6 overlap (yellow), the motor neurons become specified. (E courtesy of T. M. Jessell; F after Jessell 2000; G courtesy of J. Briscoe.)

WEBSITE 12.5 **Specifying the brain boundaries.** The Pax transcription factors and the paracrine factor FGF8 are critical in establishing the boundaries of the forebrain, midbrain, and hindbrain.

The dorsal-ventral axis

The neural tube is polarized along its dorsal-ventral axis. In the spinal cord, for instance, the *dorsal* region is the place where the spinal neurons receive input from sensory neurons, while the *ventral* region is where the motor neurons reside. In the middle are numerous interneurons that relay information between them. The dorsal-ventral polarity of the neural tube is induced by signals coming from its immediate environment. The ventral pattern is imposed by the notochord, while the dorsal pattern is induced by the epidermis (Figure 12.12).

Specification of the dorsal-ventral axis of the neural tube is initiated by two major paracrine factors. The first is Sonic hedgehog protein, originating from the notochord. The second set of factors are TGF-β proteins, originating in the dor-

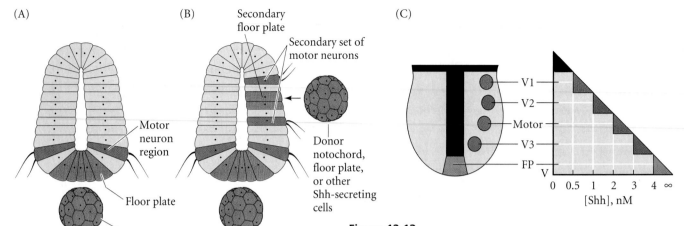

Figure 12.13

Cascade of inductions initiated by the notochord in the ventral neural tube. (A) Two cell types in the newly formed neural tube. Those closest to the notochord become the floor plate neurons; motor neurons emerge on the ventrolateral sides. (B) If a second notochord, floor plate, or any other Sonic hedgehog-secreting cell is placed adjacent to the neural tube, it induces a second set of floor plate neurons, as well as two other sets of motor neurons. (C) Relationship between Sonic hedgehog concentrations, the generation of particular neuronal types in vitro, and distance from the notochord. (A, B after Placzek et al. 1990; C after Briscoe et al. 1999.)

sal ectoderm. In both cases, these paracrine factors induce a second signaling center within the neural tube itself. Sonic hedgehog is secreted from the notochord and induces the medial hinge cells to become the **floor plate** of the neural tube. These floor plate cells also secrete Sonic hedgehog, and the protein from the floor plate cells forms a gradient that is highest at the most ventral portion of the neural tube (Roelink et al. 1995; Briscoe et al. 1999).

The dorsal fates of the neural tube are established by proteins of the TGF-β superfamily, especially bone morphogenetic proteins 4 and 7, dorsalin, and activin (Liem et al. 1995, 1997, 2000). Initially, BMP4 and BMP7 are found in the epidermis. Just as the notochord establishes a secondary signaling center—the floor plate cells—on the ventral side of the neural tube, the epidermis establishes a secondary signaling center by inducing BMP4 expression in the **roof plate** cells of the neural tube. The BMP4 protein from the roof plate induces a cascade of TGF-β proteins in adjacent cells (Figure 12.12C). Different sets of cells are thus exposed to different concentrations of TGF-β proteins at different times (the most dorsal being exposed to more factors at higher concentrations and at earlier times).

The paracrine factors interact to instruct the synthesis of different transcription factors along the dorsal-ventral axis of the neural tube. For instance, those cells adjacent to the floor plate that receive high concentrations of Sonic hedgehog (and hardly any TGF-β signal) synthesize the Nkx6.1 and Nkx2.2 transcription factors and become the ventral (V3) neurons. The cells dorsal to these, exposed to slightly less Sonic hedgehog and slightly more TGF-β factors, produce Nkx6.1 and Pax6 transcription factors. These cells become the motor neurons (Figure 12.13). The next two groups of cells, receiving progressively less Sonic hedgehog, become the V2 and V1 interneurons (Lee and Pfaff 2001; Muhr et al. 2001).

The importance of Sonic hedgehog in inducing and patterning the ventral portion of the neural tube can be shown experimentally. If notochord fragments are taken from one embryo and transplanted to the lateral side of a host neural tube, the host neural tube will form another set of floor plate cells at its sides (Figure 12.13B). These floor plate cells, once induced, induce the formation of motor neurons on either side of them. The same results can be obtained if the notochord fragments are replaced by pellets of cultured cells secreting Sonic hedgehog (Echelard et al. 1993; Roelink et al. 1994). Moreover, if a piece of notochord is removed from an embryo, the neural tube adjacent to the deleted region will have no floor plate cells (Placzek et al. 1990; Yamada et al. 1991, 1993). The importance of the TGF-β superfamily factors in patterning the dorsal portion of the neural tube was demonstrated by zebrafish mutants. Those mutants deficient in certain BMPs lacked dorsal and intermediate types of neurons (Nguyen et al. 2000).

WEBSITE 12.6 **Constructing the pituitary gland.** The developmental genetics of the pituitary have shown that (as in the trunk neural tube) paracrine signals elicit the expression of overlapping sets of transcription factors. These factors regulate the different cell types of the gland responsible for producing growth hormone, prolactin, gonadotropins, and other hormones.

Tissue Architecture of the Central Nervous System

The neurons of the brain are organized into layers (**cortices**) and clusters (**nuclei**), each having different functions and

(A)

Stage of cell cycle

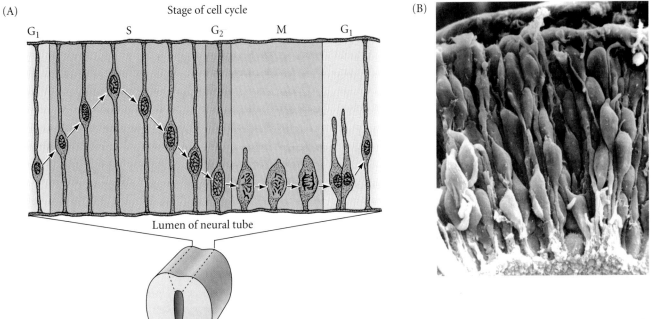

(B)

Figure 12.14
(A) Schematic section of a chick embryo neural tube, showing the position of the nucleus in a neuroepithelial cell as a function of the cell cycle. Mitotic cells are found near the inner surface of the neural tube, adjacent to the lumen. (B) Scanning electron micrograph of a newly formed chick neural tube, showing cells at different stages of their cell cycles. (A after Sauer 1935; B courtesy of K. Tosney.)

connections. The original neural tube is composed of a **germinal neuroepithelium**—a layer of rapidly dividing neural stem cells that is one cell layer thick. Sauer (1935) and others have shown that all the cells of the germinal epithelium are continuous from the luminal surface of the neural tube to the outside surface, but that the nuclei of these cells are at different heights, thereby giving the superficial impression that the neural tube has numerous cell layers. The nuclei move within their cells as they go through the cell cycle. DNA synthesis (S phase) occurs while the nucleus is at the outside edge of the neural tube, and the nucleus migrates luminally as the cell cycle proceeds (Figure 12.14). Mitosis occurs on the luminal side of the cell layer. If mammalian neural tube cells are labeled with radioactive thymidine during early development, 100% of them will incorporate this base into their DNA (Fujita 1964). Shortly thereafter, however, certain cells stop incorporating these DNA precursors, thereby indicating that they are no longer participating in DNA synthesis and mitosis. These cells then migrate and differentiate into neuronal and glial cells outside the neural tube (Fujita 1966; Jacobson 1968).

If dividing cells in the germinal neuroepithelium are labeled with radioactive thymidine at a single point in their development, and their progeny are found in the outer cortex in the adult brain, then those neurons must have migrated to their cortical positions from the germinal neuroepithelium. A neuroepithelial stem cell divides "vertically" instead of "horizontally." Thus, the daughter cell adjacent to the lumen remains connected to the luminal surface (and usually remains a stem cell), while the other daughter cell migrates and differentiates (Chenn and McConnell 1995; Hollyday 2001). The time of this vertical division is the last time the latter cell will divide, and is called that neuron's **birthday**. Different types of neurons and glial cells have birthdays at different times. Labeling at different times during development shows that the cells with the earliest birthdays migrate the shortest distances. Cells with later birthdays migrate through these layers to form the more superficial regions of the cortex. Subsequent differentiation depends on the positions these neurons occupy once outside the germinal neuroepithelium (Letourneau 1977; Jacobson 1991).

Spinal cord and medulla organization

As the cells adjacent to the lumen continue to divide, the migrating cells form a second layer around the original neural tube. This layer becomes progressively thicker as more cells are added to it from the germinal neuroepithelium. This new layer is called the **mantle** (or **intermediate**) **zone**, and the germinal epithelium is now called the **ventricular zone** (and, later, the **ependyma**) (Figure 12.15). The mantle zone cells differentiate into both neurons and glia. The neurons make connections among themselves and send forth axons away from the lumen, thereby creating a cell-poor **marginal zone**. Eventually, glial cells cover many of the axons in the marginal zone in myelin sheaths, giving them a whitish appearance. Hence, the mantle zone, containing the neuronal cell bodies, is often referred to as the **gray matter**; the axonal marginal layer is often called the **white matter**.

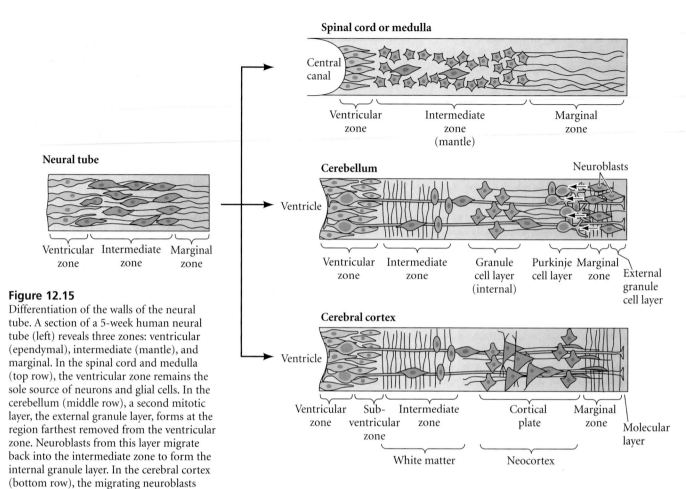

Figure 12.15
Differentiation of the walls of the neural tube. A section of a 5-week human neural tube (left) reveals three zones: ventricular (ependymal), intermediate (mantle), and marginal. In the spinal cord and medulla (top row), the ventricular zone remains the sole source of neurons and glial cells. In the cerebellum (middle row), a second mitotic layer, the external granule layer, forms at the region farthest removed from the ventricular zone. Neuroblasts from this layer migrate back into the intermediate zone to form the internal granule layer. In the cerebral cortex (bottom row), the migrating neuroblasts and glioblasts form a cortical plate containing six layers. (After Jacobson 1991.)

In the spinal cord and medulla, this basic three-zone pattern of ventricular (ependymal), mantle, and marginal layers is retained throughout development. The gray matter (mantle) gradually becomes a butterfly-shaped structure surrounded by white matter; and both become encased in connective tissue. As the neural tube matures, a longitudinal groove—the **sulcus limitans**—divides it into dorsal and ventral halves. The dorsal portion receives input from sensory neurons, whereas the ventral portion is involved in effecting various motor functions (Figure 12.16).

Cerebellar organization

In the brain, cell migration, differential neuronal proliferation, and selective cell death produce modifications of the three-zone pattern (see Figure 12.15). In the cerebellum, some neuronal precursors enter the marginal zone to form clusters of neurons called nuclei. Each nucleus works as a functional unit, serving as a relay station between the outer layers of the cerebellum and other parts of the brain. Other neuronal precursors migrate away from the germinal epithelium. These cerebellar neuroblasts migrate to the outer surface of the de-

veloping cerebellum and form a new germinal zone, the **external granule layer**, near the outer boundary of the neural tube.

At the outer boundary of the external granule layer (which is one to two cells thick), neuroblasts proliferate and come into contact with cells that are secreting BMP factors. The BMPs specify the postmitotic products of these neuroblasts to become the **granule neurons** (Alder et al. 1999). These granule neurons migrate back toward the ventricular (ependymal) zone, where they produce a region called the **internal granule layer**. Meanwhile, the original ventricular zone of the cerebellum generates a wide variety of neurons and glial cells, including the distinctive and large **Purkinje neurons**, the major cell type of the cerebellum. Purkinje neurons secrete Sonic hedgehog, which sustains the division of granule neuron precursors in the external granule layer (Wallace 1999). Each Purkinje neuron has an enormous **dendritic arbor** that spreads like a tree above a bulblike cell body. A typical Purkinje neuron may form as many as 100,000 connections (synapses) with other neurons, more than any other type of neuron studied. Each Purkinje neuron also emits a slender axon, which connects to neurons in the deep cerebellar nuclei.

Purkinje neurons are critical in the electrical pathway of the cerebellum. All electric impulses eventually regulate the

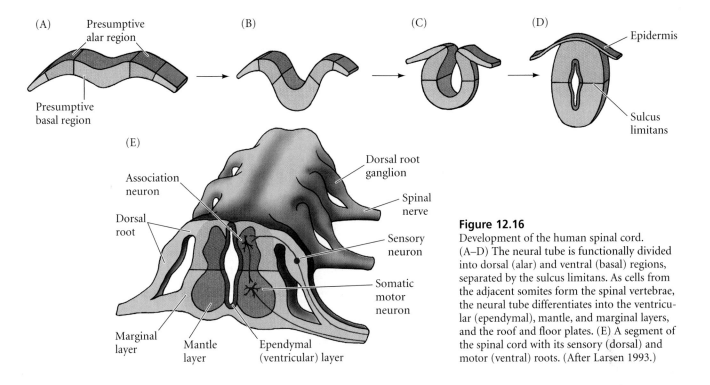

(A) Presumptive alar region
Presumptive basal region

(B)

(C)

(D) Epidermis
Sulcus limitans

(E)
Association neuron
Dorsal root
Dorsal root ganglion
Spinal nerve
Sensory neuron
Somatic motor neuron
Marginal layer
Mantle layer
Ependymal (ventricular) layer

Figure 12.16
Development of the human spinal cord. (A–D) The neural tube is functionally divided into dorsal (alar) and ventral (basal) regions, separated by the sulcus limitans. As cells from the adjacent somites form the spinal vertebrae, the neural tube differentiates into the ventricular (ependymal), mantle, and marginal layers, and the roof and floor plates. (E) A segment of the spinal cord with its sensory (dorsal) and motor (ventral) roots. (After Larsen 1993.)

activity of these neurons, which are the only output neurons of the cerebellar cortex. For this to happen, the proper cells must differentiate at the appropriate place and time. How is this accomplished? The development of spatial organization among neurons is critical for the proper functioning of the cerebellum.

One mechanism thought to be important for positioning young neurons within the developing mammalian brain is **glial guidance** (Rakic 1972; Hatten 1990). Throughout the cortex, neurons are seen to ride a "glial monorail" to their respective destinations. In the cerebellum, the granule cell precursors travel on the long processes of the **Bergmann glia** (Figure 12.17; Rakic and Sidman 1973; Rakic 1975). This neuronal-glial interaction is a complex and fascinating series of events, involving reciprocal recognition between glia and neuroblasts (Hatten 1990; Komuro and Rakic 1992). The neuron maintains its adhesion to the glial cell through a number of proteins, one of them an adhesion protein called **astrotactin**. If the astrotactin on a neuron is masked by antibodies to that protein, the neuron will fail to adhere to the glial processes (Edmondson et al. 1988; Fishell and Hatten 1991). Mice deficient in astrotactin have slow neuronal migration rates, abnormal Purkinje cell development, and problems coordinating their balance (Adams et al. 2002).

Much insight into the mechanisms of spatial ordering in the brain has come from the analysis of neurological mutations in mice. Over 30 mutations are known to affect the arrangement of cerebellar neurons. Many of these cerebellar mutants have been found because the phenotype of such mutants—namely, the inability to keep balance while walking—

can be easily recognized. For obvious reasons, these mutations are given names such as *weaver, reeler, staggerer,* and *waltzer.*

WEBSITE 12.7 Cerebellar mutations of the mouse. The mouse mutations affecting cerebellar function have given us remarkable insights into the ways in which the cerebellum is constructed. The *reeler* mutation, in particular, has been extremely important in our knowledge of how cerebellar neurons migrate.

Cerebral organization

The three-zone arrangement of the neural tube is also modified in the cerebrum. The cerebrum is organized in two distinct ways. First, like the cerebellum, it is organized vertically into layers that interact with one another (see Figure 12.15). Certain neuroblasts from the mantle zone migrate on glial processes through the white matter to generate a second zone of neurons at the outer surface of the brain. This new layer of gray matter will become the **neocortex.** The neocortex eventually stratifies into six layers of neuronal cell bodies (Figure 18); the adult forms of these layers are not completed until the middle of childhood. Each layer of the neocortex differs from the others in its functional properties, the types of neurons found there, and the sets of connections that they make. For instance, neurons in layer 4 receive their major input from the thalamus (a region that forms from the diencephalon), while neurons in layer 6 send their major output back to the thalamus.

Second, the cerebral cortex is organized horizontally into over 40 regions that regulate anatomically and functionally distinct processes. For instance, neurons in layer 6 of the visual

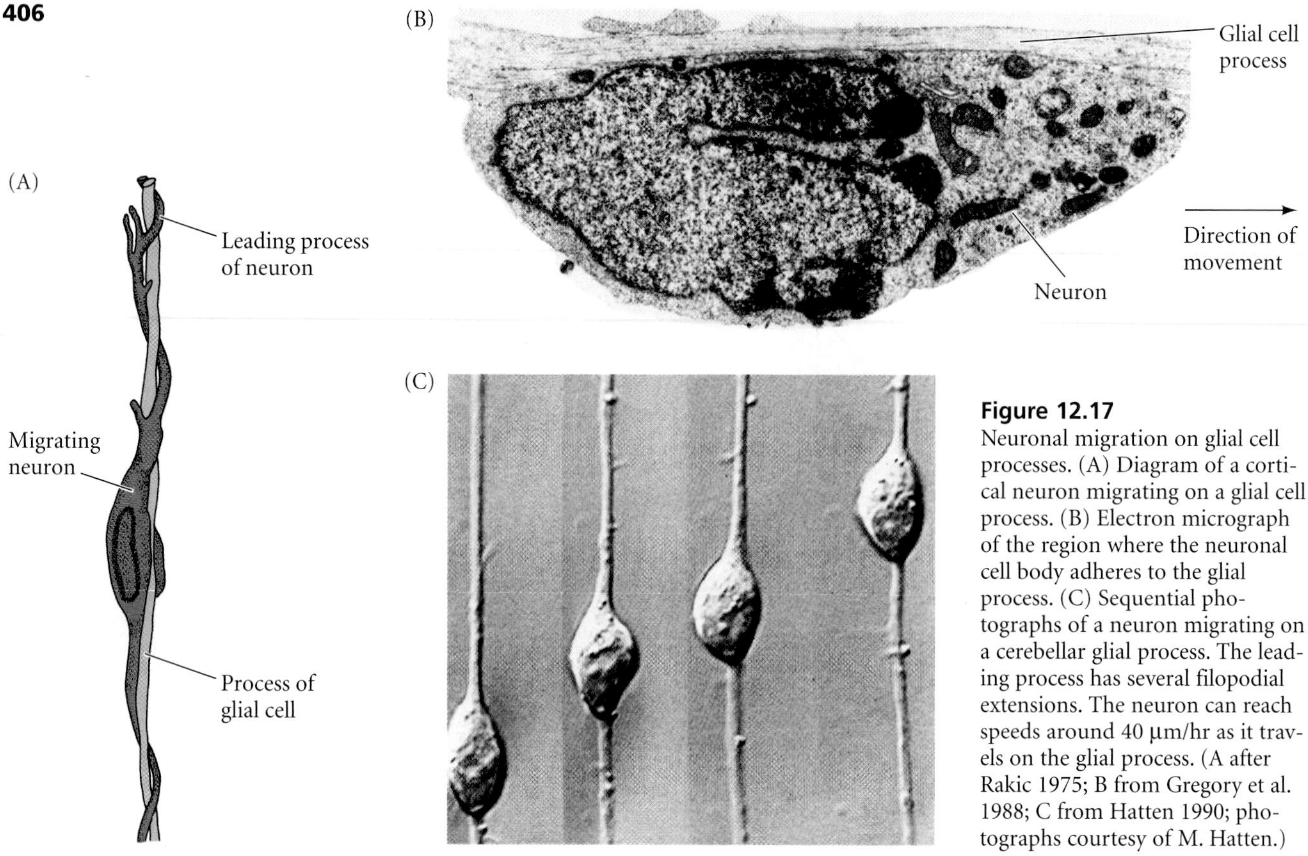

(A)

Leading process
of neuron

Migrating
neuron

Process of
glial cell

(B)

Glial cell
process

Neuron

Direction of
movement

(C)

Figure 12.17
Neuronal migration on glial cell
processes. (A) Diagram of a cortical neuron migrating on a glial cell
process. (B) Electron micrograph
of the region where the neuronal
cell body adheres to the glial
process. (C) Sequential photographs of a neuron migrating on
a cerebellar glial process. The leading process has several filopodial
extensions. The neuron can reach
speeds around 40 μm/hr as it travels on the glial process. (A after
Rakic 1975; B from Gregory et al.
1988; C from Hatten 1990; photographs courtesy of M. Hatten.)

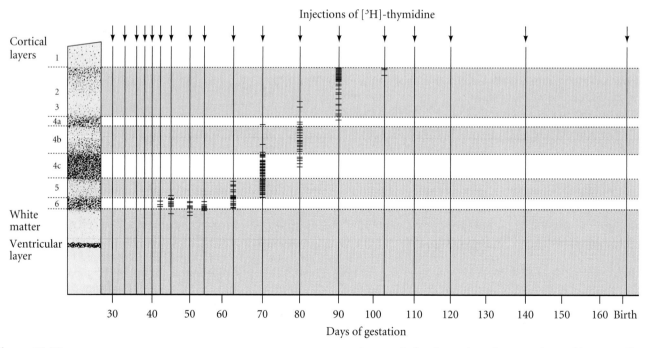

Injections of [³H]-thymidine

Cortical
layers

1
2
3
4a
4b
4c
5
6

White
matter

Ventricular
layer

30   40   50   60   70   80   90   100   110   120   130   140   150   160   Birth

Days of gestation

Figure 12.18
"Inside-out" gradient of cerebral cortex formation in the rhesus
monkey. The "birthdays" of cortical neurons were determined by
injecting pregnant animals intravenously with [³H]-thymidine at
certain times of gestation. Those fetal cells undergoing the S phase
of their final cell division cycle at the time of the injection were
heavily labeled. When labeled neurons were "born," they migrated
to various cortical regions, where they were detected by autoradiography of microscopic brain slices. Cells labeled at different times
were found to have migrated to different cortical regions. The figure
represents the positions of those neurons in the layers of the visual
cortex. The youngest neurons are found at the periphery of the developing cortex. Full gestation in rhesus monkeys is 165 days. (After
Rakic 1974.)

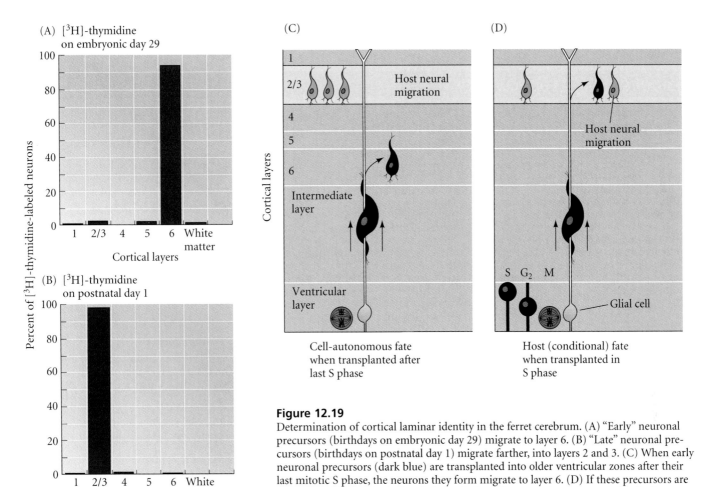

(A) [³H]-thymidine on embryonic day 29

y-axis: Percent of [³H]-thymidine-labeled neurons

x-axis: Cortical layers — 1, 2/3, 4, 5, 6, White matter

(B) [³H]-thymidine on postnatal day 1

x-axis: Cortical layers — 1, 2/3, 4, 5, 6, White matter

(C)

Cortical layers: 1, 2/3, 4, 5, 6, Intermediate layer, Ventricular layer

Host neural migration

Cell-autonomous fate when transplanted after last S phase

(D)

Host neural migration

S   G₂   M

Glial cell

Host (conditional) fate when transplanted in S phase

Figure 12.19
Determination of cortical laminar identity in the ferret cerebrum. (A) "Early" neuronal precursors (birthdays on embryonic day 29) migrate to layer 6. (B) "Late" neuronal precursors (birthdays on postnatal day 1) migrate farther, into layers 2 and 3. (C) When early neuronal precursors (dark blue) are transplanted into older ventricular zones after their last mitotic S phase, the neurons they form migrate to layer 6. (D) If these precursors are transplanted before or during their last S phase, however, they migrate (with the host neurons) to layer 2. (After McConnell and Kaznowski 1991.)

cortex project axons to the *lateral* geniculate nucleus of the thalamus, which is involved in vision (see Chapter 13), while layer 6 neurons of the auditory cortex (located more anteriorly than the visual cortex) project axons to the *medial* geniculate nucleus of the thalamus, which functions in hearing.

Neither the vertical nor the horizontal organization of the cerebral cortex is clonally specified. Rather, the developing cortex forms from the mixing of cells derived from numerous stem cells. After their final mitosis, most of the neuroblasts generated in the ventricular zone migrate outward along glial processes to form the **cortical plate** at the outer surface of the brain. As in the rest of the brain, those neurons with the earliest birthdays form the layer closest to the ventricle. Subsquent neurons travel greater distances to form the more superficial layers of the cortex. This process forms an "inside-out" gradient of development (Figure 12.18; Rakic 1974). A single stem cell in the ventricular layer can give rise to neurons (and glial cells) in any of the cortical layers (Walsh and Cepko 1988). But how do the cells know which layer to enter?

McConnell and Kaznowski (1991) have shown that the determination of laminar identity (i.e., which layer a cell migrates to) is made during the final cell division. Newly gener-

ated neuronal precursors transplanted after this last division from young brains (where they would form layer 6) into older brains whose migratory neurons are forming layer 2 are committed to their fate, and migrate only to layer 6. However, if these cells are transplanted prior to their final division (during mid-S phase), they are uncommitted, and can migrate to layer 2 (Figure 12.19).

The fates of neuronal precursors from older brains are more fixed. While the neuronal precursor cells formed early in development have the potential to become any neuron (at layers 2 or 6, for instance), later precursor cells give rise only to upper-level (layer 2) neurons (Frantz and McConnell 1996). Once the cells arrive at their final destination, it is thought that they produce particular adhesion molecules that organize them together as brain nuclei (Matsunami and Takeichi 1995). We still do not know the nature of the information given to the cell as it becomes committed.

WEBSITE 12.8 Constructing the cerebral cortex. Three genes have recently been shown to be necessary for the proper lamination of the mammalian brain. They appear to be important for cortical neural migration, and when mutated in humans can produce profound mental retardation.

The Unique Development of the Human Brain

The development of the human neocortex is strikingly plastic and an almost constant "work in progress." Whatever distinguishes humans from other primates must reside in the unique features of human development, especially in the development of the brain. At least five phenomena have been identified that distinguish the development of the human brain from that of other species, including other primates:

1. The retention of the fetal neuronal growth rate after birth.
2. The migration of cells from the prosencephalon to the diencephalon.
3. The activity of transcription.
4. The specific form of the *FOXP2* gene.
5. The continuation of brain maturation into adulthood.

Fetal neuronal growth rate after birth

If there is one important developmental trait that distinguishes humans from the rest of the animal kingdom, it is our retention of the fetal neural growth rate. Like human brains, apes' brains have a high growth rate before birth. After birth, however, this rate slows greatly in the apes; in contrast, humans have rapid brain growth for about two years after birth (Figure 12.20A; Martin 1990). During early postnatal development, we add approximately 250,000 neurons per minute (Purves and Lichtman 1985). The ratio of brain weight to body weight at birth is also similar for great apes and humans. By the time humans are adults, however, the ratio for humans is literally "off the chart" at 3.5 that of apes (Figure 12.20B; Bogin 1997).

At the cellular level, we find that no fewer than 30,000 synapses per second per square centimeter of cortex are formed during the first few years of human life. It is thought that this increase in neuron numbers may (1) generate new modules (addressable sites) that can acquire new functions, (2) store new memories for use in thinking and forecasting, and (3) enable learning by interconnecting among themselves and with prenatally generated neurons (Rose 1998; Barinaga 2003). It is in this early postnatal stage that intervention can raise IQ (reviewed in Wickelgren 1999). This stage also sees much of the maturation of the neural circuitry as determined by axon diameter and myelination.

The prodigious rate of continued neuron production has important consequences. Portmann (1941), Gould (1977), and Montagu (1962) have each made the claim that we are essentially extrauterine fetuses for the first year of life. Our actual gestation is 21 months if you follow the charts of ape maturity. In other words, humans are at 21 months what other apes are at birth. Our "premature" birth" is an evolutionary compromise between maternal pelvic width and the infant's head circumference and lung maturity. The mechanism for the retention of the fetal neural growth rate has been called *hypermorphosis*—the phyletic extension of development beyond its ancestral state (Vrba 1996).

WEBSITE 12.9 Neuronal growth and the invention of childhood. An interesting hypothesis claims that the caloric requirements of this brain growth necessitated a new stage of the human life cycle—childhood—wherein the child is actively fed by adults.

Migration of cells into the dorsal thalamus

Most neurons remain within the region of the brain in which they were formed. However, in the human brain (but not in the brains of mice or monkeys), certain neurons from the telencephalon migrate into the diencephalon. Here they enter the thalamus, an area involved in memory and problem solving that has grown especially large in the human brain (Figure 12.21; Letinic and Rakic 2001). The neurons from the telencephalon appear to contribute to those thalamic regions that distribute information to the cerebral cortex of the frontal lobes. The developing human thalamus chemotactically attracts the telencephalon neurons, but this chemotaxis is not seen in mice. The growth of the thalamus as a "relay station" that coordinates cortical functions may be one of the features distinguishing human brains from those of our animal cousins. The ability of this region to attract newly grown cortical neurons may help explain the unique status of the human brainhuman uniqueness.

High transcriptional activity

In 1975, Mary-Claire King and Allan Wilson showed that the proteins of the human and chimp brain were, within experimental limits, identical. But while the nearly 99% identity of human, chimpanzee, and bonobo genomes would suggest that all three be classified as species within the same genus (Goodman 1999), the morphological and behavioral differences are enormous. A. C. Wilson (quoted in Gibbons 1998) suggested that the difference between the species might reside in the

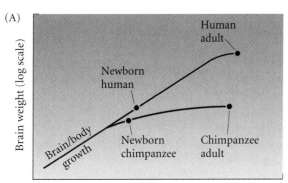

(A)

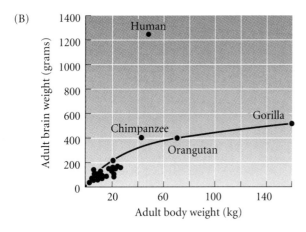

(B)

Figure 12.20
Retention of fetal neural growth rate in humans. (A) Whereas other primates, such as chimpanzees, complete their proliferaztion of neurons around the time of birth, the neurons of human newborns continue to proliferate at the same rate as the fetal brain neurons. (B) The human brain/body ratio (encephalization index) is about 3.5 times that of apes. (After Bogin 1997.)

amount of proteins made from their genes, and now there is evidence for this hypothesis. Using microarrays, Enard and colleagues (2002a) found that the quantity and types of genes expressed in humans and chimp livers and blood were indeed extremely similar. However, human *brains* produced over five times as much mRNA as the chimp brains. While this finding cannot rule out the existence of human-specific brain messages,* it does provide evidence that quantitative changes in gene expression may underlie some of our abilities.

The FOXP2 gene as critical for speech and language

Spoken language is a uniquely human trait and is presumed to be the prerequisite for the formation of cultures. Such language entails the fine-scale control of the larynx (voice box) and the mouth. Individuals who are heterozygous for mutations at the *FOXP2* locus have severe problems with language articulation and with forming sentences (Vargha-Khadem et al. 1995; Lai et al. 2001). This observation has provided genetic anthropologists with an interesting gene to study. Enard and colleagues (2002b) have shown that, although this gene is extremely well conserved throughout most of mammalian evolution, it has a unique form in humans. Indeed, the human *FOXP2* gene has accumulated two amino acid-changing mutations since our

---

*Humans lack a functional gene for cytidine monophospho-*N*-acetylneuraminic acid (CMP-NeuAc) hydroxylase due to a deletion of a 92-base pair exon (Chou et al. 1998; Irie et al. 1998). The result is that one of the most abundant cell surface carbohydrate moieties in most animals, *N*-glycolyl-neuraminic acid, is replaced in humans by *N*-acetylneuraminic acid. In neural development, neuraminic acids play roles in axon adhesion, but no known functional differences have yet been found between the two types of neuraminic acids with respect to species-specific development. This genetic difference is one of only a few that have been found between humans and other primates, although more will probably be found as microarrays become more widely used.

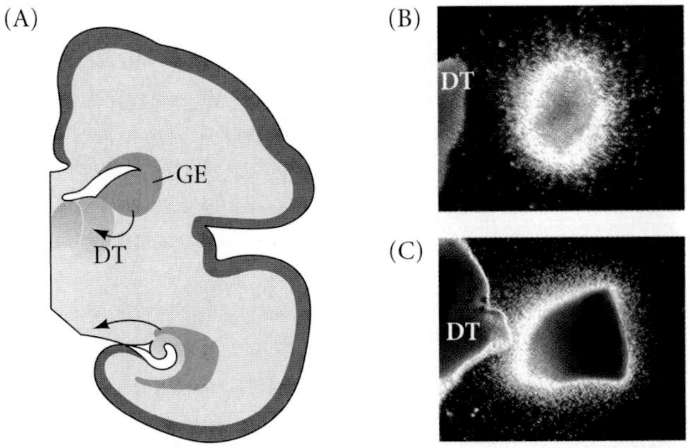

divergence from the common ancestor of humans and chimpanzees. One of these changes created a new phosphorylation site. *Foxp2* in the mouse is expressed in the developing brain, but its major site of expression is the lung (Shu et al. 2001). It is not known whether the amino acid changes in this putative transcription factor gave it a new function in the nervous system or if these changes were involved in the formation of language and culture. What is known is this protein is presently critical in our forming the orofacial movements and grammar characteristic of language.

Teenage brains: Wired and unchained

Until recently, most scientists thought that after the initial growth of neurons during fetal development and early childhood, there were no more periods of rapid neural proliferation. However, magnetic resonance imaging (MRI) studies have shown that the brain keeps developing until around puberty and that not all areas of the brain mature simultaneously (Giedd et al. 1999; Sowell et al. 1999). Moreover, soon after puberty, neuronal growth ceases and pruning occurs. This time of pruning correlates to the time when language acquisition becomes difficult. There is also a wave of myelin production ("white matter" from the glial cells that ensheathes neuronal axons) at this time. Myelination is critical for the proper functioning of the neural areas, and the greatest differences between brains in early puberty and those

Figure 12.21
Migration of cells from the telencephalon into the dorsal thalamus of the diencephalon. (A) Route of migration, showing migration path from the ganglionic eminence (GE) of the telencephalon to the dorsal thalamus (DT) of the diencephalon. (B) When pieces of the mouse GE are placed near mouse DT, the GE neurons move out symmetrically and do not respond to the presence of the DT. (C) When human GE is placed near human DT, the human GE neurons preferentially migrate toward the DT. (From Letinic and Rakic 2001.)

in early adulthood involve the frontal cortex (Sowell et al. 1999).

These differences in brain development may explain the extreme responses teens have to certain stimuli, as well as their ability to learn certain tasks. When functional MRI studies scanned subjects' brains when emotional pictures were shown on a computer screen, young teens activated the amygdala, a brain center than mediates fear and strong emotions. When older teens were shown the same pictures, most of their brain activity became centered in the frontal lobe, an area involved in more reasoned perceptions (Baird et al. 1999). The teenage brain is a very complicated and dynamic entity, and (as any parent knows) one that is not easily understood. But if one survives these years, the resulting brain is capable of making reasoned decisions, even in the onslaught of emotional situations.

Adult neural stem cells

Until recently, it was generally believed that once the nervous system was mature, no new neurons were "born." The neurons we formed *in utero* and during the first few years of life were all we could ever expect to have. However, the good news from recent studies is that the adult mammalian brain is capable of producing new neurons, and environmental stimulation can increase the number of these new neurons. In these experiments, researchers injected adult mice, rats, or marmosets with bromodeoxyuridine (BrdU), a nucleoside that resembles thymidine. BrdU will be incorporated into a cell's

DNA only if the cell is undergoing DNA replication; thus, any cell labeled with BrdU must have been undergoing mitosis during the time when it was exposed to BrdU. This labeling technique showed that thousands of new neurons were being made each day in adult mammals. Moreover, these new brain cells integrated with other cells of the brain, had normal neuronal morphology, and exhibited action potentials (Figure 12.22B; van Praag et al. 2002).

Injecting humans with BrdU is usually unethical, since large doses of BrdU are often lethal. However, in certain cancer patients, the progress of chemotherapy is monitored by

(A)

(B)

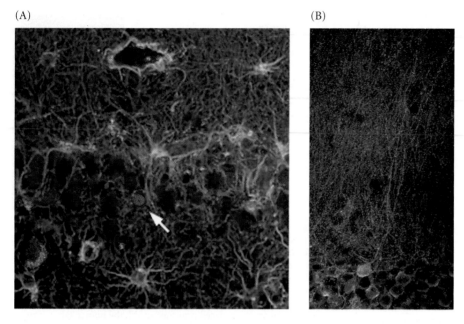

Figure 12.22
Evidence of adult neural stem cells. (A) Newly generated neuron in the adult human brain. This cell is located in the dentate gyrus of the hippocampus. The green staining, which indicates newly divided cells, is from a fluorescent antibody against BrdU (a thymidine analogue that is taken up only during S phase). The red fluorescence is from an antibody that stains only neural cells. Glial cells are stained purple. The overlap of green and red (arrow) shows a cell that is a newly formed neuron. (B) Newly generated adult mouse neurons have a normal morphology and receive synaptic inputs. The green cell is a newly formed neuron. The red spots are synaptophysin, a protein found on the dendrites at the synapses of axons from other neurons. (A from Erikksson et al. 1998, photograph courtesy of F. H. Gage; B form van Praag et al. 2002.)

transfusing the patient with a small amount of BrdU. Gage and colleagues (Erikkson et al. 1998) took postmortem samples from the brains of five such patients who had died between 16 and 781 days after the BrdU infusion. In all five subjects, they saw new neurons in the granular cell layer of the hippocampal dentate gyrus (a part of the brain where memories may be formed). The BrdU-labeled cells also stained for neuron-specific markers (Figure 12.22). Thus, although the rate of new neuron formation in adulthood may be relatively small, the human brain is not an anatomical *fait accompli* at birth, or even after childhood.

The existence of neural stem cells in adults is now well established for the olfactory epithelium (Kato et al. 2001), and it seems likely that the hippocampus can also form new neurons (Kempermann et al. 1997a,b; van Praag et al. 1999; Kornack and Rakic 1999). It appears that the stem cells producing these neurons are located in the ependyma (the former ventricular zone in which the embryonic neural stem cells once resided) or in the subventricular zone adjacent to it (Doetsch et al. 1999; Johansson et al. 1999; Cassidy and Frisén 2001).

These adult neural stem cells represent only about 0.3% of the ventricle wall cell population, but they can be distinguished from the other cells by their particular cell surface proteins (Rietze et al. 2001). These stem cells may play a role in replacing dead neurons.* During pregnancy, prolactin stimulates the production of neuronal progenitor cells in the subventricular zone of the adult mouse forebrain. These progenitors migrate to produce olfactory neurons that may be important for maternal behaviors of rearing offspring (Shingo et al. 2003). The existence of such stem cells in the cortex is more controversial. Some investigators (Gould et al. 1999a,b; Magavi et al. 2000) claim to have found them; other scientists (see Rakic 2002) question the existence of these cortical neural stem cells.

WEBSITE 12.10 **Parkinson disease.** Transplantation of human fetal neural stem cells has been used to alleviate the symptoms of Parkinson disease. While this therapy has worked in some people, it has exacerbated the symptoms in other patients. Other types of therapies, based on embryonic stem cells and paracrine factors, are also being tested.

Differentiation of Neurons

The human brain consists of over 10^{11} neurons associated with over 10^{12} glial cells. Those cells that remain integral components of the neural tube lining become **ependymal cells.** These cells give rise to the precursors of neurons and glial cells. As we have seen above, it is thought that the differentiation of these precursor cells is largely determined by the environment that they enter (Rakic and Goldman 1982) and that, at least in some cases, a given ependymal cell can give rise to both neurons and glia (Turner and Cepko 1987).

The brain contains a wide variety of neuronal and glial types (as is evident from a comparison of the relatively small granule cell with the enormous Purkinje neuron). The fine, branching extensions of the neuron that are used to pick up electric impulses from other cells are called **dendrites** (Figure 12.23). Some neurons develop only a few dendrites, whereas other cells (such as the Purkinje neurons) develop extensive dendritic trees. Very few dendrites can be found on cortical neurons at birth, but one of the amazing events of the first year of human life is the increase in the number of these re-

---

*The use of cultured neuronal stem cells to regenerate or repair parts of the brain will be considered in Chapter 21.

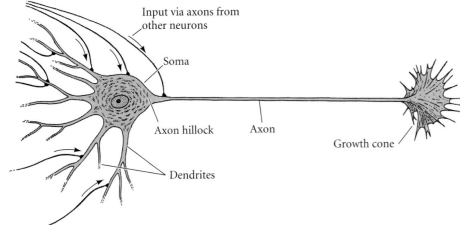

RECEPTOR

Figure 12.23
Diagram of a motor neuron. Electric impulses are received by the dendrites, and the stimulated neuron transmits impulses through its axon (which may be 2–3 feet long) to its target tissue. The axon is the cellular process through which the neuron sends its signals. The growth cone of the axon is both a loco-motor and a sensory apparatus. It actively explores the environment and picks up directional cues as to where to go. Eventually it will form a synapse with its target tissue.

ceptive processes. During this year, each cortical neuron develops enough dendritic surface to accommodate as many as 100,000 synapses with other neurons. The average cortical neuron connects with 10,000 other neural cells. This pattern of synapses enables the human cortex to function as the center for learning, reasoning, and memory, to develop the capacity for symbolic expression, and to produce voluntary responses to interpreted stimuli.

Another important feature of a developing neuron is its **axon** (sometimes called a **neurite**). Whereas dendrites are often numerous and do not extend far from the neuronal cell body, or **soma**, axons may extend for several feet. The pain receptors on your big toe, for example, must transmit their messages all the way to your spinal cord. One of the fundamental concepts of neurobiology is that the axon is a continuous extension of the nerve cell body. At the turn of the twentieth century, there were many competing theories of axon formation. Theodor Schwann, one of the founders of the cell theory, believed that numerous neural cells linked themselves together in a chain to form an axon. Viktor Hensen, the discov-

erer of the embryonic node, thought that the axon formed around preexisting cytoplasmic threads between the cells. Wilhelm His (1886) and Santiago Ramón y Cajal (1890) postulated that the axon was an outgrowth (albeit an extremely large one) of the neuronal soma. In 1907, Ross Harrison demonstrated the validity of the outgrowth theory in an elegant experiment that founded both the science of developmental neurobiology and the technique of tissue culture. Harrison isolated a portion of the neural tube from a 3-mm frog tadpole. (At this stage, shortly after the closure of the neural tube, there is no visible differentiation of axons.) He placed this neuroblast-containing tissue in a drop of frog lymph on a coverslip and inverted the coverslip over a depression slide so he could watch what was happening within this "hanging drop." What Harrison saw was the emergence of axons as outgrowths from the neuroblasts, elongating at about 56 μm/hr.

Nerve outgrowth is led by the tip of the axon, called the **growth cone** (Figure 12.24). The growth cone does not proceed in a straight line, but rather "feels" its way along the substrate. The growth cone moves by the elongation and contraction of pointed filopodia called **microspikes**. These microspikes contain microfilaments, which are oriented parallel to the long axis of the axon. (This mechanism is similar to

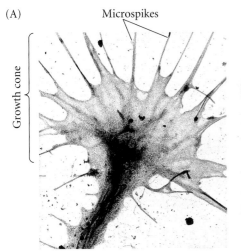

Figure 12.24
Axon growth cones. (A) Actin microspikes in an axon growth cone, seen by transmission electron microscopy. (B) Growth cone of the hawkmoth *Manduca sexta* during axon extension and pathfinding. The actin in the filopodia is stained green with fluorescent phalloidin, while the microtubules are stained red with a fluorescent antibody to tubulin. (A from Letourneau 1979; B courtesy of R. B. Levin and R. Luedemanan.)

the one seen in the filopodial microfilaments of secondary mesenchyme cells in echinoderms; see Chapter 8.) Treating neurons with cytochalasin B destroys the actin microspikes, inhibiting their further advance (Yamada et al. 1971; Forscher and Smith 1988). Within the axon itself, structural support is provided by microtubules, and the axon will retract if the neuron is placed in a solution of colchicine. Thus, the developing neuron retains the same mechanisms that we have already noted in the dorsolateral hinge points of the neural tube—namely, elongation by microtubules and apical shape changes by microfilaments.

As in most migrating cells, the exploratory microspikes of the growth cone attach to the substrate and exert a force that pulls the rest of the cell forward. Axons will not grow if the growth cone fails to advance (Lamoureux et al. 1989). In addition to their structural role in axonal migration, the microspikes also have a sensory function. Fanning out in front of the growth cone, each microspike samples the microenvironment and sends signals back to the soma (Davenport et al. 1993). As we will see in Chapter 13, the microspikes are the fundamental organelles involved in neuronal pathfinding.

Neurons transmit electric impulses from one region of the body to another. These impulses usually go from the dendrites into the soma, where they are focused into the axon. To prevent dispersion of the electric signal and to facilitate its conduction to the target cell, that part of the axon is insulated at intervals by glial cells. Within the central nervous system, axons are insulated at intervals by processes that originate from a type of glial cell called an **oligodendrocyte**. The oligodendrocyte wraps itself around the developing axon, then produces a specialized cell membrane called a **myelin sheath**. In the peripheral nervous system, a glial cell type called the **Schwann cell** accomplishes this myelination (Figure 12.25). The myelin sheath is essential for proper neural function, and demyelination of nerve fibers is associated with convulsions, paralysis, and several debilitating or lethal afflictions (such as multiple sclerosis).

In the *trembler* mouse mutant, the Schwann cells are unable to produce a particular protein component of the myelin sheath, so that myelination is deficient in the peripheral nervous system, but normal in the central nervous system. Conversely, in another mouse mutant, called *jimpy*, the central nervous system is deficient in myelin, while the peripheral nerves are unaffected (Sidman et al. 1964; Henry and Sidman 1988).

The axon must also be specialized for secreting a specific neurotransmitter across the small gap (synaptic cleft) that separates the axon of a cell from the surface of its target cell (the soma, dendrites, or axon of a receiving neuron or a receptor site on a peripheral organ). Some neurons develop the ability to synthesize and secrete acetylcholine, while other neurons develop the enzymatic pathways for making and secreting epinephrine, norepinephrine, octopamine, serotonin, γ-aminobutyric acid (GABA), dopamine, or some other neurotransmitter. Each neuron must activate those genes responsible for making the enzymes that can synthesize its neurotransmitter. Thus, neuronal development involves both structural and molecular differentiation.*

---

*The regeneration of neurons and their axons will be discussed in Chapter 18. The glial cells are probably very important in permitting or preventing axon regeneration.

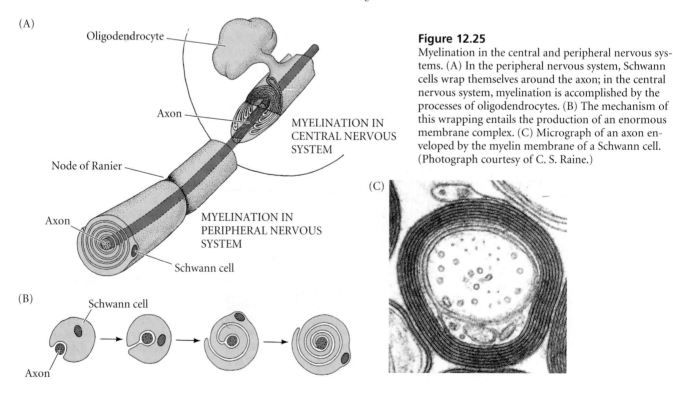

Figure 12.25
Myelination in the central and peripheral nervous systems. (A) In the peripheral nervous system, Schwann cells wrap themselves around the axon; in the central nervous system, myelination is accomplished by the processes of oligodendrocytes. (B) The mechanism of this wrapping entails the production of an enormous membrane complex. (C) Micrograph of an axon enveloped by the myelin membrane of a Schwann cell. (Photograph courtesy of C. S. Raine.)

Development of the Vertebrate Eye

An individual gains knowledge of its environment through its sensory organs. The major sensory organs of the head develop from interactions of the neural tube with a series of epidermal thickenings called the **cranial ectodermal placodes**. The most anterior of these are the two **olfactory placodes** that form the ganglia for the olfactory nerves, which are responsible for the sense of smell. The **auditory placodes** similarly invaginate to form the inner ear labyrinth, whose neurons form the acoustic ganglion, which enables us to hear. In this section, we will focus on the eye.

The dynamics of optic development

The induction of the eye was discussed in Chapter 6, and will only be summarized here (Figure 12.26). At gastrulation, the involuting endoderm and mesoderm interact with the adjacent prospective head ectoderm to give the head ectoderm a lens-forming bias (Saha et al. 1989). But not all parts of the head ectoderm eventually form lenses, and the lens must have a precise spatial relationship with the retina. The activation of the head ectoderm's latent lens-forming ability and the positioning of the lens in relation to the retina is accomplished by the **optic vesicle**. The optic vesicle extends from the diencephalon, and where it meets the head ectoderm, it induces the formation of a **lens placode**, which then invaginates to form the lens. The optic vesicle itself becomes the **optic cup**, whose two layers differentiate in different ways. The cells of the outer layer produce melanin pigment (being one of the few tissues other than the neural crest cells that can form this pigment) and ultimately become the **pigmented retina**. The cells of the inner layer proliferate rapidly and generate a variety of glia, ganglion cells, interneurons, and light-sensitive photoreceptor neurons. Collectively, these cells constitute the **neural retina**. The retinal ganglion cells are neurons whose axons send electric impulses to the brain. Their axons meet at the base of the eye and travel down the optic stalk, which is then called the **optic nerve**.

But how is it that a specific region of neural ectoderm is informed that it will become the optic vesicle? It appears that a group of transcription factors—*Six3*, *Pax6*, and *Rx1*—are expressed together in the most anterior tip of the neural plate. This single domain will later split into the bilateral regions that form the optic vesicles. Again, we see the similarities between the *Drosophila* and the vertebrate nervous system, for these three proteins are also necessary for the formation of the *Drosophila* eye. As discussed in Chapters 5 and 6, the Pax6 protein appears to be especially important in the development of the lens and retina. Indeed, it appears to be a common denominator for photoreceptive cells in all phyla. If the mouse *Pax6* gene is inserted into the *Drosophila* genome and activated randomly, *Drosophila* eyes form in those cells where the mouse Pax6 is being expressed (see Chapter 23; Halder et al. 1995). While *Pax6* is also expressed in the murine forebrain, hindbrain, and nasal placodes, the eyes seem to be most sensitive to its absence. In humans and mice, *Pax6* heterozygotes have small eyes, while homozygotic mice and humans (and *Drosophila*) lack eyes altogether (Jordan et al. 1992; Glaser et al. 1994; Quiring et al. 1994).

The separation of the single eye field into two bilateral fields depends upon the secretion of Sonic hedgehog. If the *sonic hedgehog* gene is mutated, or if the processing of this protein is inhibited, the single median eye field will not split. The result is **cyclopia**—a single eye in the center of the face (and usually below the nose) (Figure 12.27; see also Figure 6.26; Chiang et al. 1996; Kelley et al. 1996; Roessler et al.

(A) 4-mm embryo   (B) 4.5-mm embryo

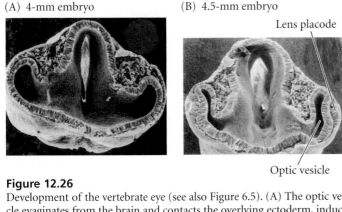

Lens placode

Optic vesicle

(C) 5-mm embryo   (D) 7-mm embryo

Lens vesicle   Retina   Lens

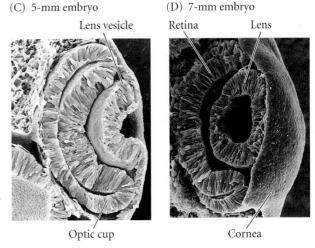

Optic cup   Cornea

Figure 12.26
Development of the vertebrate eye (see also Figure 6.5). (A) The optic vesicle evaginates from the brain and contacts the overlying ectoderm, inducing a lens placode. (B, C) The overlying ectoderm differentiates into lens cells as the optic vesicle folds in on itself, and the lens placode becomes the lens vesicle. (C) The optic vesicle becomes the neural and pigmented retina as the lens is internalized. (D) The lens vesicle induces the overlying ectoderm to become the cornea (A–C from Hilfer and Yang 1980, photographs courtesy of S. R. Hilfer; D, photograph courtesy of K. Tosney.)

(A) Wild-type

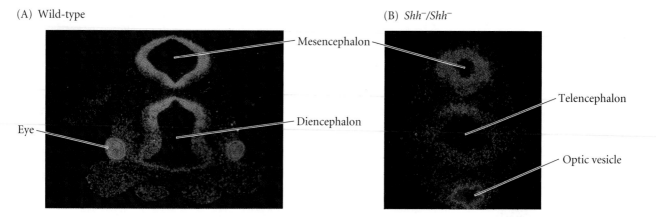

(B) *Shh⁻/Shh⁻*

Mesencephalon

Telencephalon

Diencephalon

Eye

Optic vesicle

(C)

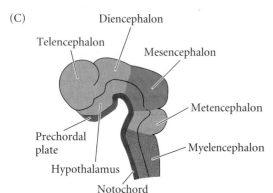

Diencephalon

Telencephalon

Mesencephalon

Metencephalon

Prechordal plate

Myelencephalon

Hypothalamus

Notochord

Figure 12.27

Sonic hedgehog separates the eye field into two bilateral fields. (A) A 12.5-day, wild-type mouse embryo and (B) a 12.5-day embryo lacking Sonic hedgehog. The expression of the *Otx2*, seen in red, highlights certain brain regions. In the mutant, no midline forms, and there is a single, continuous optic vesicle in the ventral region. The nose will form above it. (C) Drawing showing the location of the prechordal plate in the 12-day mouse embryo. (Photographs courtesy of P. A. Beachy and C. Chiang.)

1996). It is thought that Sonic hedgehog protein from the prechordal plate suppresses *Pax6* expression in the center of the embryo, dividing the field in two. The phenomenon of *human cyclopia* also involves Sonic hedgehog and will be discussed in Chapter 22.

Neural retina differentiation

Like the cerebral and cerebellar cortices, the neural retina develops into a layered array of different neuronal types (Figure 12.28). These layers include the light- and color-sensitive photoreceptor (rod and cone) cells, the cell bodies of the ganglion cells, and the bipolar interneurons that transmit electric stimuli from the rods and cones to the ganglion cells. In addition, the retina contains numerous Müller glial cells that maintain its integrity, as well as amacrine neurons (which lack large axons) and horizontal neurons that transmit electric impulses in the plane of the retina.

In the early stages of retinal development, cell division in the germinal layer and the migration and differential death of the resulting cells form the striated, laminar pattern of the neural retina. The formation of this highly structured tissue is one of the most intensely studied problems of developmental neurobiology. It has been shown (Turner and Cepko 1987) that a single neuroblast precursor cell from the retinal germinal layer can give rise to at least three types of neurons or to two types of neurons and a glial cell. This analysis was per-

formed using an ingenious technique to label the cells generated by one particular neuroblast precursor. Newborn rats (whose retinas are still developing) were injected in the back of their eyes with a virus that could integrate into their DNA. This virus contained a β-galactosidase gene (not normally present in the rat retina) that would be expressed only in the infected cells. A month after the rats' eyes were infected, the retinas were removed and stained for the presence of β-galactosidase. Only the progeny of the infected cells were stained blue. Figure 12.29 shows one of the strips of cells derived from an infected precursor cell. The stain can be seen in five rods, a bipolar neuron, and a Müller glial cell.

Lens and cornea differentiation

As it develops into a lens, the lens placode rounds up and contacts the new overlying ectoderm. The lens vesicle then induces the ectoderm to form the transparent cornea. Here, physical parameters play an important role in the development of the eye. Intraocular fluid pressure is necessary for the correct curvature of the cornea so that light can be focused on the retina. The importance of this pressure can be demonstrated experimentally: the cornea will not develop its characteristic curve if a small glass tube is inserted through the wall of a developing chick eye to drain away the intraocular fluid (Coulombre 1956, 1965). Intraocular pressure is sustained by a ring of scleral bones (probably derived from the neural crest), which acts as an inelastic restraint.

The differentiation of the lens tissue into a transparent membrane capable of directing light onto the retina involves changes in cell structure and shape as well as the synthesis of transparent, lens-specific proteins called crystallins (Figure

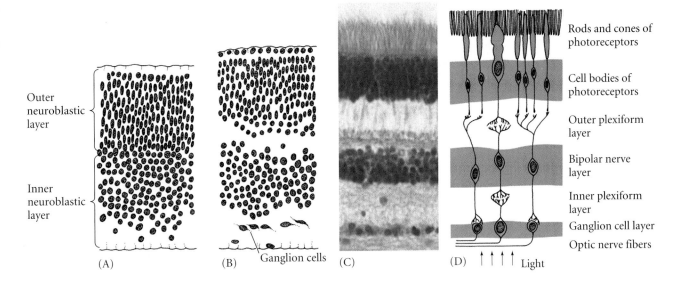

Outer neuroblastic layer

Inner neuroblastic layer

(A)   (B)   Ganglion cells   (C)   (D) ↑↑↑↑ Light

Rods and cones of photoreceptors

Cell bodies of photoreceptors

Outer plexiform layer

Bipolar nerve layer

Inner plexiform layer

Ganglion cell layer

Optic nerve fibers

Figure 12.28
Development of the human retina. Retinal neurons sort out into functional layers during development. (A, B) Initial separation of neuroblasts within the retina. (C) The three layers of neurons in the adult retina and the synaptic layers between them. (D) A functional depiction of the major neuronal pathway in the retina. Light traverses the layers until it is received by the photoreceptors. The axons of the photoreceptors synapse with bipolar neurons, which transmit electric signals to the ganglion cells. The axons of the ganglion cells join to form the optic nerve, which enters the brain. (A and B after Mann 1964; C, photograph courtesy of G. Grunwald.)

12.30). The cells at the inner portion of the lens vesicle elongate and, under the influence of the neural retina, become the lens fibers (Piatigorsky 1981). As these fibers continue to grow, they synthesize crystallins, which eventually fill up the cell and cause the extrusion of the nucleus. The crystallin-synthesizing fibers eventually fill the space between the two layers of the lens vesicle. The anterior cells of the lens vesicle constitute a germinal epithelium, which keeps dividing. These di-

viding cells move toward the equator of the vesicle, and as they pass through the equatorial region, they, too, begin to elongate (Figure 12.30D). Thus, the lens contains three regions: an anterior zone of dividing epithelial cells, an equatorial zone of cellular elongation, and a posterior and central zone of crystallin-containing fiber cells. This arrangement persists throughout the lifetime of the animal as fibers are continuously being laid down. In the adult chicken, the process of differentiation from an epithelial cell to a lens fiber takes 2 years (Papaconstantinou 1967).

Directly in front of the lens is a pigmented, muscular tissue called the iris. The iris muscles control the size of the pupil (and give an individual his or her characteristic eye color). Unlike the other muscles of the body (which are derived from the mesoderm), part of the iris is derived from the ectodermal layer. Specifically, this region of the iris develops from a portion of the optic cup that is continuous with the neural retina, but does not make photoreceptors.

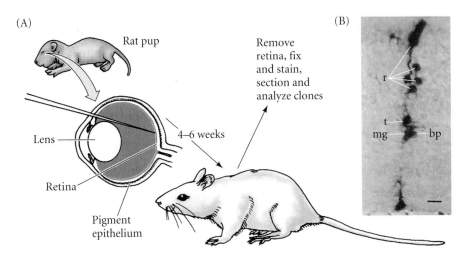

(A) Rat pup

Lens

Retina

Pigment epithelium

Remove retina, fix and stain, section and analyze clones

4–6 weeks

(B)

r

t

mg   bp

Figure 12.29
Determination of the lineage of a neuroblast in the rat retina. (A) Technique whereby a virus containing a functional β-galactosidase gene is injected into the back of the eye of a newborn rat to infect some of the retinal precursor cells. After a month to 6 weeks, the eye is removed and the retina is stained for the presence of β-galactosidase. (B) Stained cells forming a strip across the neural retina, including five rods (r), a bipolar neuron (bp), a rod terminal (t), and a Müller glial cell (mg). The identities of these cells were confirmed by Nomarski-phase contrast microscopy. (Scale bar, 20 μm.) (From Turner and Cepko 1987; photograph courtesy of D. Turner.)

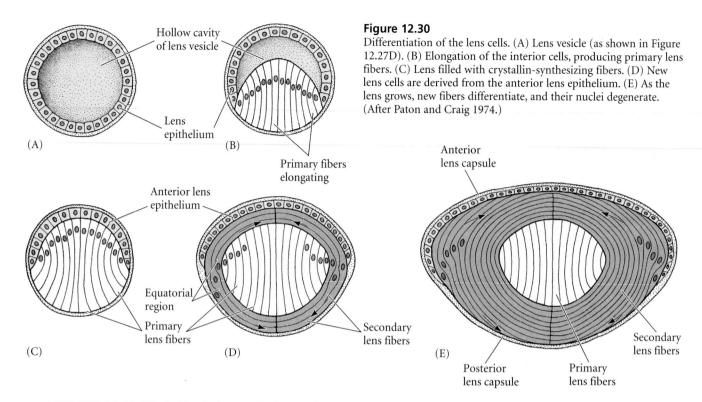

Figure 12.30
Differentiation of the lens cells. (A) Lens vesicle (as shown in Figure 12.27D). (B) Elongation of the interior cells, producing primary lens fibers. (C) Lens filled with crystallin-synthesizing fibers. (D) New lens cells are derived from the anterior lens epithelium. (E) As the lens grows, new fibers differentiate, and their nuclei degenerate. (After Paton and Craig 1974.)

WEBSITE 12.11 Why babies don't see well. The retinal photoreceptors are not fully developed at birth. As the child grows gets older, the density of photoreceptors increases, allowing far better discrimination and nearly 350 times the light-absorbing capacity that is present at birth.

The Epidermis and the Origin of Cutaneous Structures

The origin of epidermal cells

The cells covering the embryo after neurulation form the presumptive epidermis. Originally, this tissue is one cell layer thick, but in most vertebrates it shortly becomes a two-layered structure. The outer layer gives rise to the **periderm**, a temporary covering that is shed once the inner layer differentiates to form a true epidermis. The inner layer, called the **basal layer** (or **stratum germinativum**), is a germinal epithelium that gives rise to all the cells of the epidermis (Figure 12.31). The basal layer divides to give rise to another, outer population of cells that constitutes the **spinous layer**. These two epidermal layers together are referred to as the **Malpighian layer**. The cells of the Malpighian layer divide to produce the **granular layer** of the epidermis, so called because its cells are characterized by granules of the protein keratin. Unlike the cells remaining in the Malpighian layer, the cells of the granular layer do not divide, but begin to differentiate into epidermal skin cells, the **keratinocytes**. The keratin granules become more prominent as the keratinocytes of the granular layer age and migrate outward to form the **cornified layer** (stratum

corneum). These cells become dead, flattened sacs of keratin protein, and their nuclei are pushed to one edge of the cell.

The depth of the cornified layer varies from site to site, but it is usually 10 to 30 cells thick. Shortly after birth, the outer cells of the cornified layer are shed and are replaced by new cells coming up from the granular layer. Throughout life, the dead keratinocytes of the cornified layer are shed (humans lose about 1.5 grams of these cells each day*) and are replaced by new cells, the source of which is the mitotic cells of the Malpighian layer. Pigment cells (melanocytes) from the neural crest also reside in the Malpighian layer, where they transfer their pigment sacs (melanosomes) to the developing keratinocytes.

The epidermal stem cells of the Malpighian layer are bound to the basal lamina by their integrin proteins. However, as these cells become committed to differentiate, they downregulate their integrins and eventually lose them as they migrate into the spinous layer (Jones and Watt 1993).

Several growth factors stimulate the development of the epidermis. One of these is **transforming growth factor-α (TGF-α)**. TGF-α is made by the basal cells and stimulates their own division. When a growth factor is made by the same cell that receives it, that factor is called an **autocrine growth factor**. Such factors must be carefully regulated because if their levels are elevated, more cells are rapidly produced. In adult human

---

*Most of this skin becomes "house dust" on furniture and floors. If you doubt this, burn some dust; it will smell like singed skin.

Figure 12.31
Diagram of the layers of the human epidermis. The basal cells are mitotically active, whereas the fully keratinized cells characteristic of external skin are dead and are continually shed. The keratinocytes obtain their pigment through the transfer of melanosomes from the processes of melanocytes that reside in the basal layer. (After Montagna and Parakkal 1974.)

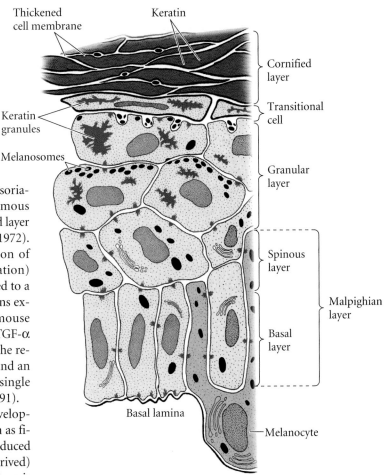

skin, a cell born in the Malpighian layer takes roughly 8 weeks to reach the cornified layer, and remains there for about 2 weeks. In individuals with psoriasis, a disease characterized by the exfoliation of enormous amounts of epidermal cells, a cell's time in the cornified layer is only 2 days (Weinstein and van Scott 1965; Halprin 1972). This condition has been linked to the overexpression of TGF-α (which occurs secondarily to an inflammation) (Elder et al. 1989). Similarly, if the TGF-α gene is linked to a promoter for keratin 14 (one of the major skin proteins expressed in the basal cells) and inserted into the mouse pronucleus, the resulting transgenic mice activate the TGF-α gene in their skin cells and cannot down-regulate it. The result is a mouse with scaly skin, stunted hair growth, and an enormous surplus of keratinized epidermis over its single layer of basal cells (Figure 12.32B; Vassar and Fuchs 1991).

Another growth factor needed for epidermal development is **keratinocyte growth factor** (KGF; also known as fibroblast growth factor 7), a paracrine factor that is produced by the fibroblasts of the underlying (mesodermally derived) dermis. KGF is received by the basal cells of the epidermis

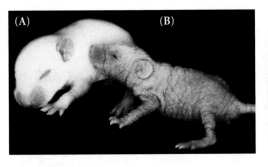

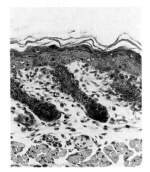

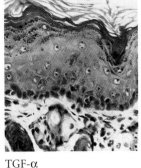

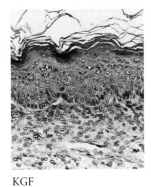

Wild-type   TGF-α   KGF

Figure 12.32
Growth factors and epidermal proliferation. (A) A wild-type mouse pup. (B) A littermate of (A) that is expressing high levels of TGF-α in its keratinocytes. It has scaly skin and very little hair. Below each mouse is a cross section through its skin. The mouse overexpressing TGF-α has extensive layers of keratinized epithelia, which it sheds. (C) A transgenic mouse expressing low levels of KGF in its keratinocytes. Note the sparsity of hair around the legs, eyes, and nose. The skin section shows an absence of hair follicles and an increased number of basal epidermal cells. (A, B from Vassar and Fuchs 1991; C from Guo et al. 1993, photographs courtesy of E. Fuchs.)

and is thought to regulate their proliferation. If the gene encoding KGF is fused with the keratin 14 promoter, the KGF becomes autocrine in the resulting transgenic mice (Figure 12.32C). These mice have a thickened epidermis, baggy skin, far too many basal cells, and no hair follicles, not even whisker follicles (Guo et al. 1993). In these mice, the basal cells are "forced" into the epidermal pathway of differentiation. The alternative pathway for basal cells leads to the generation of hair follicles.

Cutaneous appendages

The epidermis and dermis also interact at specific sites to create the sweat glands and the **cutaneous appendages**: hairs, scales, or feathers (depending on the species). In mammals, the first indication that a hair follicle primordium, or **hair germ**, will form at a particular place is an aggregation of cells in the basal layer of the epidermis. This aggregation is directed by the underlying dermal fibroblast cells and occurs at different times and different places in the embryo. It is probable that the dermal signals cause the stabilization of β-catenin in the ectoderm (Gat et al. 1998). The basal epidermal cells elongate, divide, and sink into the dermis. The dermal fibroblasts respond to this ingression of epidermal cells by forming a small node (the **dermal papilla**) beneath the hair germ. The dermal papilla then pushes up on the basal stem cells and stimulates them to divide more rapidly. The basal cells respond by producing postmitotic cells that will differentiate into the keratinized hair shaft (see Hardy 1992; Miller et al. 1993). Melanoblasts, which were present among the basal epidermal cells as they ingressed, differentiate into melanocytes and transfer their pigment to the hair shaft (Figure 12.33).

As this is occurring, two epithelial swellings begin to grow on the side of the hair germ. The cells of the lower swelling may retain a population of stem cells that will regenerate the hair shaft periodically when it is shed (Pinkus and Mehregan 1981; Cotsarelis et al. 1990). These cells also include a population of melanocyte stem cells which continue to produce melanocytes for most of the mammal's lifetime

(Nishimura et al. 2002). The cells of the upper swelling form the **sebaceous glands**, which produce an oily secretion, **sebum**. In many mammals, including humans, the sebum mixes with the shed peridermal cells to form the whitish vernix caseosa, which covers the fetus at birth. Thus, just as there is a pluripotent neural stem cell whose offspring become neural and glial cells, there appears to be a pluripotent epidermal stem cell whose progeny can become epidermis, sebaceous gland, or hair shaft.

The first hairs in the human embryo are of a thin, closely spaced type called **lanugo**. This type of hair is usually shed before birth and is replaced (at least in part, by new follicles) by the short and silky **vellus**. Vellus remains on many parts of the human body usually considered hairless, such as the forehead and eyelids. In other areas of the body, vellus gives way to "terminal" hair. During a person's life, some of the follicles that produced vellus can later form terminal hair and still later revert to vellus production. The armpits, for instance, have follicles that produce vellus until adolescence, then begin producing terminal shafts. Conversely, in normal masculine pattern baldness, the scalp follicles revert to producing unpigmented and very fine vellus hair (Montagna and Parakkal 1974).

The formation of hair (and feather) follicles requires a series of reciprocal inductive interactions between the dermal mesenchyme and the ectodermal epithelium. This dialogue appears to be initiated by a signal originating from the dermis that instructs the ectodermal cells to thicken and aggregate to form a placode (Hardy 1992; Millar 2002). The placode then

Figure 12.33
Development of the hair follicles in fetal human skin. (A) Basal epidermal cells become columnar and bulge slightly into the dermis. (B) Epidermal cells continue to proliferate, and dermal mesenchyme cells collect at the base of the primary hair germ to form a dermal papilla. (C) Hair shaft differentiation begins in the elongated hair germ. (D) The keratinized hair shaft extends from the hair follicle, the secondary bud forms the sebaceous gland, and beneath it is a region that may contain the hair stem cells for the next cycle of hair production. (E) Photograph of elongated hair germ. (After Hardy 1992 and Miller et al. 1993; photograph courtesy of W. Montagna.)

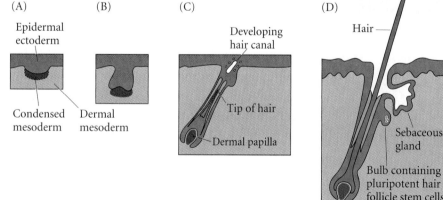
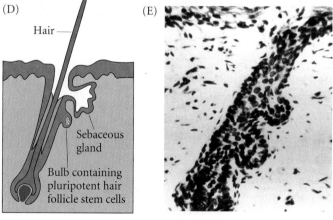

tells the dermal cells to condense, and the condensed dermal cells instruct the epidermis to make a hair shaft.

Developmental genetics of hair formation

The search for the molecular mechanisms of hair initiation and growth is one of the long-standing research programs in developmental biology. Recent advances have been made through the fusion of human genetics, transgenic mice, and new gene insertion techniques.

The discovery of how certain hairlessness phenotypes arise in humans led investigators to several important steps in the molecular pathway of hair formation. A syndrome called **X-linked anhidrotic ectodermal dysplasia** involves abnormalities of hair, sweat glands, mammary glands, and teeth. In each case, the epidermal placodes that normally produce these structures fail to form, suggesting that the products of this X-linked gene are involved in the developmental process that forms these placodes. The proteins encoded by this gene were identified by mRNAs that specifically hybridized to the region of the X chromosome whose mutations caused the syndrome. These proteins, called **ectodysplasins**, are synthesized in the embryonic ectoderm (Bayés et al. 1998; Mikkola et al. 1999; Montonen et al. 1998). The major ectodysplasin protein is shed from the cell surface by proteolytic cleavage, and it binds to its receptor on adjacent cells that are also in the ectoderm (Koppinen et al. 2001).

But what activates the expression of ectodysplasin in the ectoderm? A role for Wnt singlaing was strongly suggested by the work of Andl and his colleagues (2002). To ascertain whether Wnt signaling is critical for the initiation of follicle development, they inserted into mouse embryos a fused transgene consisting of the *dickkopf* gene attached to the promoter of the *keratin14* gene. Keratin14 is usually expressed in the basal layer of the epidermis, as we saw above. Dickkopf is a potent and soluble inhibitor of Wnt signaling (see Chapter 10) and can diffuse from the basal cells into the ectoderm above and into the dermis below.

Embryonic mice expressing Dickkopf protein in their skin had a complete absence of hair follicles, whiskers, teeth, and mammary glands (Figure 12.34). Indeed, the Wnt signal

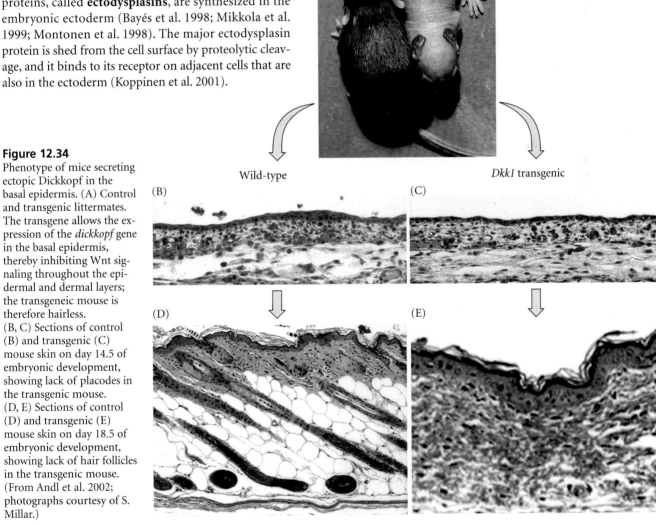

Figure 12.34
Phenotype of mice secreting ectopic Dickkopf in the basal epidermis. (A) Control and transgenic littermates. The transgene allows the expression of the *dickkopf* gene in the basal epidermis, thereby inhibiting Wnt signaling throughout the epidermal and dermal layers; the transgeneic mouse is therefore hairless. (B, C) Sections of control (B) and transgenic (C) mouse skin on day 14.5 of embryonic development, showing lack of placodes in the transgenic mouse. (D, E) Sections of control (D) and transgenic (E) mouse skin on day 18.5 of embryonic development, showing lack of hair follicles in the transgenic mouse. (From Andl et al. 2002; photographs courtesy of S. Millar.)

Wild-type

Dkk1 transgenic

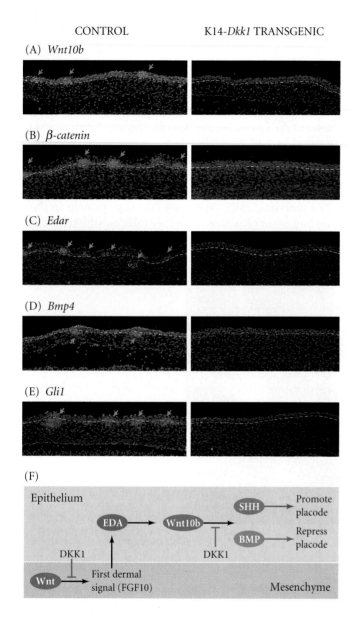

CONTROL   K14-*Dkk1* TRANSGENIC

(A) *Wnt10b*

(B) *β-catenin*

(C) *Edar*

(D) *Bmp4*

(E) *Gli1*

(F)

Epithelium

EDA → Wnt10b → SHH → Promote placode

BMP → Repress placode

DKK1   DKK1

Wnt ⊣ First dermal signal (FGF10)   Mesenchyme

Figure 12.35
Gene expression in control and transgenic Dickkopf-expressing mice. In situ hybridization for several proteins was performed around embryonic day 15 and is seen in red (arrows). (A) Wnt10b; (B) β-catenin; (C) ectodysplasin receptor; (D) BMP4; (E) Sonic hedgehog-inducible Gli1 (indicating the presences of Sonic hedgehog). (F) Outline of pathway regulating hair follicle development in mice. (After Andl et al. 2002; photographs courtesy of S. Millar.)

By in situ hybridization for various proteins in mutant mice, a hypothetical pathway has been outlined for the early development of the hair shaft (Figure 12.35F). The actions of a Wnt-family signal are required in the dermis to generate the first dermal message. The signal is probably not Wnt itself, however, but may be FGF10 (Huelsken et al. 2001; Tao et al. 2002). The dermal signal activates ectodysplasin synthesis in the ectoderm. This protein helps initiate placode formation and also activates another Wnt signal that comes from the placodes. This Wnt signal regulates the expression of Sonic hedgehog and BMPs. Sonic hedgehog activates dermal condensation and continues hair follicle development. The BMPs induce Dickkopf in the mesenchyme surrounding the hair follicle and repress hair follicle development in these neighboring cells.

WEBSITE 12.12 **Normal variation in human hair production.** The human hair has a complex life cycle. Moreover, some hairs (such as those of our eyelashes) grow short while other hairs (such as those of our scalp) grow long. The pattern of hair size and thickness (or lack thereof) is determined by paracrine and endocrine factors.

WEBSITE 12.13 **Mutations of human hair production.** In addition to normal variation, there are also inherited mutations that interfere with normal hair development. Some people are born without the ability to grow hair, while others develop hair over their entire bodies. These genetic conditions give us insights into the mechanisms of normal hair growth.

appears to be produced very early, since its absence prevents placode production and inhibits the expression of all the other genes involved in the dialogue that produces the hair follicle (Figure 12.35).

Snapshot Summary: The Central Nervous System and Epidermis

1. The neural tube forms from the shaping and folding of the neural plate. In primary neurulation, the surface ectoderm folds into a tube that separates from the surface. In secondary neurulation, the ectoderm forms a cord and then forms a cavity within it.

2. Primary neurulation is regulated by both intrinsic and extrinsic forces. Intrinsic wedging occurs within cells of the hinge regions to bend the neural plate. Extrinsic forces include the migration of the surface ectoderm towards the center of the embryo.

3. Neural tube closure is also the result of a mixture of extrinsic and intrinsic forces. In humans, various diseases can result if the neural tube fails to close.

4. The neural crest cells arise at the borders of the neural tube and surface ectoderm. They become located between the neural tube and surface ectoderm, and they migrate away from this region to become peripheral neural, glial, and pigment cells.

5. There is a gradient of maturity in many embryos (especially those of amniotes) so that the anterior develops earlier than the posterior.

6. The brain forms three primary vesicles: prosencephalon (forebrain), mesencephalon (midbrain), and rhombencephalon (hindbrain). The prosencephalon and rhombencephalon will become subdivided.

7. The brain expands through fluid secretion putting positive pressure on the vesicles.

8. The dorsal-ventral patterning of the neural tube is accomplished by proteins of the TGF-β superfamily secreted from the surface ectoderm and the roof plate of the neural tube, and by Sonic hedgehog protein secreted from the notochord and floor plate cells. Gradients of these proteins trigger the synthesis of particular transcription factors that specify the neuroepithelium.

9. The neurons of the brain are organized into cortices (layers) and nuclei (clusters).

10. New neurons are formed by the division of neural stem cells in the wall of the neural tube (called the ventricular zone). The resulting neural precursors, or neuroblasts, can migrate away from the ventricular zone and form a new layer, called the mantle zone (gray matter). Neurons forming later have to migrate through the existing layers. This process forms the cortical layers.

11. In the cerebellum, migrating neurons form a second germinal zone, called the external granule layer. Other neurons migrate out of the ventricular zone on the processes of glial cells.

12. The cerebral cortex in humans, called the neocortex, has six layers. Cell fates are often fixed as they undergo their last division. Neurons derived from the same stem cell may end up in different functional regions of the brain.

13. Human brains appear to differ from those of other primates by their retention of the fetal neuronal growth rate during early childhood, the migration of cells from the telencephalon to the diencephalon, the amount of transcriptional activity, the presence of certain *FoxP2* alleles, and by a spurt of neuronal growth and myelination that occurs during puberty.

14. Neural stem cells have been observed in the adult human brain. We now believe that humans can continue making neurons throughout life, although at nowhere near the fetal rate.

15. Dendrites receive signals from other neurons, while axons transmit signals to other neurons. The gap between cells where signals are transferred from one neuron to another (through the release of neurotransmitters) is called a synapse.

16. Axons grow from the nerve cell body, or soma. They are led by the growth cone.

17. The retina forms from an optic vesicle that extends from the brain. Pax6 plays a major role in eye formation, and the downregulation of Pax6 by Sonic hedgehog in the center of the brain splits the eye-forming region of the brain in half. If Sonic hedeghog is not expressed there, a single medial eye results.

18. The photoreceptor cells of the retina gather light and transmit an electric impulse through interneurons to the retinal ganglion cells. The axons of the retinal ganglion cells form the optic nerve.

19. The lens and cornea form from the surface ectoderm. Both must become transparent.

20. The basal layer of the surface ectoderm becomes the germinal layer of the skin. These basal cells divide to produce epidermal cells (keratinocytes).

21. Paracrine factors such as TGF-α and KGF (FGF7)are important in normal skin development.

22. Cutaneous appendages—hair, feathers, and scales—are formed by epithelial-mesenchymal interactions between the epidermis and the dermal mesoderm. The Wnt signaling pathway appears to play a critical role in this process.

Literature Cited

Adams, N. C., T. Tomoda, M. Cooper, G. Dietz and M. E. Hatten. 2002. Mice that lack astrotactin have slowed neuronal migration. *Development* 129: 965–972.

Alder, J., K. J. Lee, T. M. Jessell and M. E. Hatten. 1999. Generation of cerebellar granule neurons in vivo by transplantation of BMP-treated neural progenitor cells. *Nature Neurosci.* 2: 535–540.

Alvarez, I. S. and G. C. Schoenwolf. 1992. Expansion of surface epithelium provides the major extrinsic force for bending of the neural plate. *J. Exp. Zool.* 261: 340–348.

Andl, T., S. T. Reddy, T. Gaddapara and S. E. Millar. 2002. WNT signals are required for the initiation of hair follicle development. *Dev. Cell* 2: 643–653.

Baird, A. A. and 7 others. 1999. Functional magnetic resonance imaging of facial affect recognition in children and adolescents. *J. Am. Acad. Child Adolesc. Psychiatr.* 38: 195–199.

Balinsky, B. I. 1975. *Introduction to Embryology*, 4th Ed. Saunders, Philadelphia.

Barinaga, M. 2003. Newborn neurons search for meaning. *Science* 299: 32–34.

Bayés, M., A. J. Hartung, S. Ezer, J. Pispa, I. Thesleff, A. K. Srivastava and J. Kere. 1998. The anhidrotic ectodermal dysplasia gene (*EDA*) undergoes alternative splicing and encodes ectodysplasin-A with deletion mutations in collagenous repeats. *Hum. Mol. Genet.* 7: 1661–1669.

Bogin, B. 1997. Evolutionary hypotheses for human childhood. *Yrbk. Phys. Anthropol.* 40: 63–89.

Botto, L. D., C. A. Moore, M. J. Khoury and J. D. Erickson. 1999. Neural-tube defects. *New Engl. J. Med.* 341: 1509–1519.

Briscoe, J., L. Sussel, D. Hartigan-O'Connor, T. M. Jessell, J. L. R. Rubenstein and J. Ericson. 1999. Homeobox gene *Nkx2.2* and specification of neuronal identity by graded Sonic hedgehog signaling. *Nature* 398: 622–627.

Burnside, B. 1973. Microtubules and microfilaments in amphibian neurulation. *Am. Zool.* 13: 989–1006.

Cassidy, R. and J. Frisén. 2001. Stem cells on the brain. *Nature* 412: 690–691.

Catala, M., M.-A. Teillet and N. M. Le Douarin. 1995. Organization and development of the tail bud analyzed with the quail-chick chimaera system. *Mech. Dev.* 51: 51–65.

Catala, M., M.-A. Teillet, E. M. De Robertis and N. M. Le Douarin. 1996. A spinal cord fate map in the avian embryo: While regressing, Hensen's node lays down the notochord and floor plate thus joining the spinal cord lateral walls. *Development* 122: 2599–2610.

Centers for Disease Control. 1992. Recommendations for the use of folic acid to reduce the number of cases of spina bifida and other neural tube defects. *Morb. Mortal. Wkly. Rep.* 41: 1–7.

Chenn, A. and S. K. McConnell. 1995. Cleavage orientation and the asymmetric inheritance of Notch1 immunoreactivity in mammalian neurogenesis. *Cell* 82: 631–641.

Chiang, C., Y. Litingung, E. Lee, K. K. Young, J. E. Corden, H. Westphal and P. A. Beachy. 1996. Cyclopia and defective axial patterning in mice lacking *sonic hedgehog* gene function. *Nature* 383: 407–413.

Chou, H. H. and 9 others. 1998. A mutation in human CMP-sialic acid hydroxylase occurred after the *Homo-Pan* divergence. *Proc. Natl. Acad. Sci. USA* 95: 11751–11756.

Colas, J.-F. and G. C. Schoenwolf. 2001. Towards a cellular and molecular understanding of neurulation. *Dev. Dynam.* 221: 117–145.

Cotsarelis, G., T.-T. Sun and R. M. Lavker. 1990. Label-retaining cells reside in the bulge area of pilosebaceous unit: Implications for follicular stem cells, hair cycle and skin carcinogenesis. *Cell* 61: 1329–1337.

Coulombre, A. J. 1956. The role of intraocular pressure in the development of the chick eye. I. Control of eye size. *J. Exp. Zool.* 133: 211–225.

Coulombre, A. J. 1965. The eye. *In* R. DeHaan and H. Ursprung (eds.), *Organogenesis*. Holt, Rinehart & Winston, New York, pp. 217–251.

Czeizel, A. and I. Dudas. 1992. Prevention of first occurrence of neural tube defects by periconceptional vitamin supplementation. *N. Engl. J. Med.* 327: 1832–1835.

Davenport, R. W., P. Dou, V. Rehder and S. B. Kater. 1993. A sensory role for neuronal growth cone filopodia. *Nature* 361: 721–724.

De Marco, P. and 8 others. 2000. Folate pathway gene alterations in patients with neural tube defects. *Am. J. Med. Genet.* 95: 216–223.

Desmond, M. E. 1982. A description of the occlusion of the lumen of the spinal cord in early human embryos. *Anat. Rec.* 204: 89–93.

Desmond, M. E. and M. C. Field. 1992. Evaluation of neural fold fusion and coincident initiation of spinal cord occlusion in the chick embryo. *J. Comp. Neurol.* 319: 246–260.

Desmond, M. E. and G. C. Schoenwolf. 1986. Evaluation of the roles of intrinsic and extrinsic factors in occlusion of the spinal neurocoel during rapid brain enlargement in the chick embryo. *J. Embryol. Exp. Morphol.* 97: 25–46.

Detrick, R. J., D. Dickey and C. R. Kintner. 1990. The effects of N-cadherin misexpression on morphogenesis in *Xenopus* embryos. *Neuron* 4: 493–506.

Doetsch, F., I. Caillé, D. A. Lim, J. M. García-Verdugo and A. Alvarez–Buylla. 1999. Subventricular zone astrocytes are neural stem cells in the adult mammalian brain. *Cell* 97: 703–716.

Echelard, Y., D. J. Epstein, B. St.-Jacques, L. Shen, J. Mohler, J. A. McMahon and A. McMahon. 1993. Sonic hedgehog, a member of a family of putative signaling molecules, is implicated in the regulation of CNS polarity. *Cell* 75: 1417–1430.

Edmondson, J. C., R. K. H. Liem, J. C. Kuster and M. E. Hatten. 1988. Astrotactin: A novel neuronal cell surface antigen that mediates neuronal-astroglial interactions in cerebellar microcultures. *J. Cell Biol.* 106: 505–517.

Eichele, G. 1992. Budding thoughts. *The Sciences* (Jan.), 30–36.

Elder, J. T. and 8 others. 1989. Overexpression of transforming growth factor in psoriatic epidermis. *Science* 243: 811–814.

Enard, W. and 12 others. 2002a. Intra- and interspecific variation in primate gene expression patterns. *Science* 296: 340–343.

Enard, W. and 7 others. 2002b. Molecular evolution of *FOXP2*, a gene involved in speech and language. *Nature* 418: 869–872.

Erickson, C. A. and J. A. Weston. 1983. An SEM analysis of neural crest migration in the mouse. *J. Embryol. Exp. Morphol.* 74: 97–118.

Eriksson, P. S., E. Perfiliea, T. Björn-Erikksson, A.-M. Alborn, C. Nordberg, D. A. Peterson and F. H. Gage. 1998. Neurogenesis in the adult human hippocampus. *Nature Med.* 4: 1313–1317.

Fishell, G. and M. E. Hatten. 1991. Astrotactin provides a receptor system for glia-guided neuronal migration. *Development* 113: 755–765.

Forscher, P. and S. J. Smith. 1988. Actions of cytochalasins on the organization of actin filaments and microtubules in a neural growth cone. *J. Cell Biol.* 107: 1505–1516.

Frantz, G. D. and S. K. McConnell. 1996. Restriction of late cerebral cortical progenitors to an upper-layer fate. *Neuron* 17: 55–61.

Fujimori, T., S. Miyatani and M. Takeichi. 1990. Ectopic expression of N-cadherin perturbs histogenesis in *Xenopus* embryos. *Development* 110: 97–104.

Fujita, S. 1964. Analysis of neuron differentiation in the central nervous system by tritiated thymidine autoradiography. *J. Comp. Neurol.* 122: 311–328.

Fujita, S. 1966. Application of light and electron microscopy to the study of the cytogenesis of the forebrain. *In* R. Hassler and H. Stephen (eds.), *Evolution of the Forebrain*. Plenum, New York, pp. 180–196.

Gallera, J. 1971. Primary induction in birds. *Adv. Morphogenet.* 9: 149–180.

Gat, U., R. Das Gupta, L. Degenstein and E. Fuchs. 1998. De-novo hair follicle morphogenesis and hair tumors in mice expressing a truncated β-catenin in skin. *Cell* 95: 605–614.

Giedd, J. N. and 8 others. 1999. Brain development during childhood and adolescence: A longitudinal MRI study. *Nature Neurosci.* 2: 861–863.

Glaser, T., L. Jepeal, J. G. Edwards, S. R. Young, J. Favor and R. L. Maas. 1994. *PAX6* gene dosage effect in a family with congenital cataracts, aniridia, anophthalmia and central nervous system defects. *Nature Genet.* 7: 463–471.

Golden, J. A. and G. F. Chernoff. 1993. Intermittent pattern of neural tube closure in two strains of mice. *Teratology* 47: 73–80.

Gont, L. K., H. Steinbeisser, B. Blumberg and E. M. De Robertis. 1993. Tail formation as a continuation of gastrulation: The multiple cell populations of the *Xenopus* tailbud derive from the late blastopore lip. *Development* 119: 991–1004.

Goodman, M. 1999. The genomic record of humankind's evolutionary roots. *Am. J. Hum. Genet.* 64: 31–39.

Gould, E., A. Beylin, P. Tanapat, A. Reeves and T. J. Shors. 1999a. Learning enhances adult neurogenesis in the hippocampal formation. *Nature Neurosci.* 2: 260–265.

Gould, E., A. J. Reeves, M. S. A. Graziano and C. Gross. 1999b. Neurogenesis in the adult cortex of adult primates. *Science* 286: 548–552.

Gould, S. J. 1977. *Ontogeny and Phylogeny.* Harvard University Press, Cambridge, MA.

Gregory, W. A., J. C. Edmondson, M. E. Hatten and C. A. Mason. 1988. Cytology and neural-glial apposition of migrating cerebellar granule cells in vitro. *J. Neurosci.* 8: 1728–1738.

Guo, L., Q.-C. Yu and E. Fuchs. 1993. Targeting expression of keratinocyte growth factor to keratinocytes elicits striking changes in epithelial differentiation in transgenic mice. *EMBO J.* 12: 973–986.

Guthrie, S. and A. Lumsden. 1991. Formation and regeneration of rhombomere boundaries in the developing chick hindbrain. *Development* 112: 221–229.

Halder, G., P. Callaerts and W. J. Gehring. 1995. Induction of ectopic eyes by targeted expression of the *eyeless* gene in Drosophila. *Science* 267: 1788–1792.

Halprin, K. M. 1972. Epidermal "turnover time": A reexamination. *J. Invest. Dermatol.* 86: 14–19.

Hardy, M. H. 1992. The secret life of the hair follicle. *Trends Genet.* 8: 55–61.

Harrison, R. G. 1907. Observations on the living developing nerve fiber. *Anat. Rec.* 1: 116–118.

Hatten, M. E. 1990. Riding the glial monorail: A common mechanism for glial-guided neuronal migration in different regions of the mammalian brain. *Trends Neurosci.* 13: 179–184.

Henry, E. W. and R. L. Sidman. 1988. Long lives for homozygous *trembler* mutant mice despite virtual absence of peripheral nerve myelin. *Science* 241: 344–346.

Hilfer, S. R. and J.-J. W. Yang. 1980. Accumulation of CPC-precipitable material at apical cell surfaces during formation of the optic cup. *Anat. Rec.* 197: 423–433.

His, W. 1886. Zur Geschichte des mensch-lichen Rückenmarks und der Nervenwurzeln. *Ges. Wiss.* 13, S. 477.

Hollyday, M. 2001. Neurogenesis in the vertebrate neural tube. *Int. J. Dev. Neurosci.* 19: 161–173.

Huelsken, J., R. Vogel., B. Erdmann, G. Cotsarelis and W. Birchmeier. 2001. β-catenin controls hair follicle morphogenesis and stem cell differentiation in the skin. *Cell* 105: 533–545.

Irie, A., S. Koyama, Y. Kozutsumi, T. Kawasaki and A. Suzuki. 1998. The molecular basis for the absence of N-glycolylneuraminic acid in humans. *J. Biol. Chem.* 273: 15866–15871.

Jacobson, A. G. and J. G. Moury. 1995. Tissue boundaries and cell behavior during neurulation. *Dev. Biol.* 171: 98–110.

Jacobson, A. G. and A. K. Sater. 1988. Features of embryonic induction. *Development* 104: 341–359.

Jacobson, M. 1968. Cessation of DNA synthesis in retinal ganglion cells correlated with the time of specification of their central connections. *Dev. Biol.* 17: 219–232.

Jacobson, M. 1991. *Developmental Neurobiology*, 2nd Ed. Plenum, New York.

Jessell, T. M. 2000. Neuronal specification in the spinal cord: Inductive signals and transcriptional codes. *Nature Rev. Genet.* 1: 20–29.

Johansson, C. B., S. Momma, D. L. Clarke, M. Risling, U. Lendahl and J. Frisén. 1999. Identification of a neural stem cell in the adult mammalian central nervous system. *Cell* 96: 25–34.

Jones, P. H. and F. M. Watt. 1993. Separation of human epidermal stem cells from transit amplifying cells on the basis of differences in integrin function and expression. *Cell* 73: 713–724.

Jordan, T. and 7 others. 1992. The human *PAX6* gene is mutated in two patients with aniridia. *Nature Genet.* 1: 328–332.

Karfunkel, P. 1972. The activity of microtubules and microfilaments in neurulation in the chick. *J. Exp. Zool.* 181: 289–302.

Kato, T., K. Yokouchi, N. Fukushima, K. Kawagishi, Z. Li and T. Moriizumi. 2001. Continual replacement of newly-generated olfactory neurons in adult rats. *Neurosci. Lett.* 307: 17–20.

Keller, R., J. Shih, A. K. Sater and C. Moreno. 1992. Planar induction of convergence and extension of the neural plate by the organizer of *Xenopus*. *Dev. Dynam.* 193: 218–234.

Kelley, R. I. and 7 others. 1996. Holoprosencephaly in RSH/Smith-Lemli-Opitz syndrome: Does abnormal cholesterol metabolism affect the function of Sonic hedgehog? *Am. J. Med. Genet.* 66: 478–484.

Kempermann, G., H. G. Kuhn and F. H. Gage. 1997a. Genetic influence on neurogenesis in the dentate gyrus of adult mice. *Proc. Natl. Acad. Sci. USA* 94: 10409–10414.

Kempermann, G., H. G. Kuhn and F. H. Gage. 1997b. More hippocampal neurons in adult mice living in an enriched environment. *Nature* 386: 493–495.

King, M. C. and A. C. Wilson. 1975. Evolution at two levels in humans and chimpanzees. *Science* 188: 107–116.

Komuro, H. and P. Rakic. 1992. Selective role of N-type calcium channels in neuronal migration. *Science* 157: 806–809.

Koppinen, P., J. Pispa, J. Laurikkala, I. Thesleff and M. L. Mikkola. 2001. Signaling and subcellular localization of the TNF receptor Edar. *Exp. Cell Res.* 269: 180–192.

Kornack, D. R. and P. Rakic. 1999. Continuation of neurogenesis in the hippocampus of the adult macaque monkey. *Proc. Natl. Acad. Sci. USA* 96: 5768–5773.

Lai, C. S., S. E. Fisher, J. A. Hurst, F. Vargha-Khadem and A. P. Monaco. 2001. A *forkhead*-domain gene is mutated in a severe speech and language disorder. *Nature* 413: 519–523.

Lamoureux, P., R. E. Buxbaum and S. R. Heidemann. 1989. Direct evidence that growth cones pull. *Nature* 340: 159–162.

Larsen, W. J. 1993. *Human Embryology.* Churchill Livingstone, New York.

Lawson, A., H. Anderson and G. C. Schoenwolf. 2001. Cellular mechanisms of neural fold formation and morphogenesis in the chick embryo. *Anat. Rec.* 262: 153–168.

Lee, S.-K. and S. L. Pfaff. 2001. Transcriptional networks regulating neuronal identity in the developing spinal cord. *Nature Neurosci. Suppl.* 4: 1183–1191.

Letinic, K. and P. Rakic. 2001. Telencephalic origin of human thalamic GABAergic neurons. *Nature Neurosci.* 4: 931–936.

Letourneau, P. C. 1977. Regulation of neuronal morphogenesis by cell-substratum adhesion. *Soc. Neurosci. Symp.* 2: 67–81.

Letourneau, P. C. 1979. Cell substratum adhesion of neurite growth cones and its role in neurite elongation. *Exp. Cell Res.* 124: 127–138.

Liem, K. F., Jr., G. Tremmi, H. Roelink and T. M. Jessell. 1995. Dorsal differentiation of neural plate cells induced by BMP-mediated signals from epidermal ectoderm. *Cell* 82: 969–979.

Liem, K. F., Jr., G. Tremmi and T. M. Jessell. 1997. A role for the roof plate and its resident TGF-β-related proteins in neuronal patterning in the dorsal spinal cord. *Cell* 91: 127–138.

Liem, K. F., Jr, T. M. Jessell and J. Briscoe. 2000. Regulation of the neural patterning activity of Sonic hedgehog by secreted BMP inhibitors expressed by notochord and somites. *Development* 127: 4855–4866.

Little, J. and J. M. Elwood. 1992a. Ethnic origin and migration. *In* J. M. Elwood, J. Little and J. H. Elwood (eds.), *Epidemiology and Control of Neural Tube Defects.* Oxford University Press, Oxford, pp. 146–167.

Little, J. and J. M. Elwood. 1992b. Seasonal variation. *In* J. M. Elwood, J. Little and J. H. Elwood (eds.), *Epidemiology and Control of Neural Tube Defects.* Oxford University Press, Oxford, pp. 195–246.

Lumsden, A. and R. Keynes. 1989. Segmental patterns of neuronal development in the chick hindbrain. *Nature* 337: 424–428.

Magavi, S. S., B. R. Leavitt and J. D. Macklis. 2000. Induction of neurogenesis in the neocortex of adult mice. *Nature* 405: 951–955.

Mann, I. 1964. *The Development of the Human Eye.* Grune and Stratton, New York.

Manzanares, M. and R. Krumlauf. 2000. Raising the roof. *Nature* 403: 720–721.

Martin, R. D. 1990. *Primate Origins and Evolution: A Phylogenetic Reconstruction.* Princeton University Press, Princeton.

Matsunami, H. and M. Takeichi. 1995. Fetal brain subdivisions defined by T- and E- cadherins expressions: Evidence for the role of cadherin activity in region-specific, cell-cell adhesion. *Dev. Biol.* 172: 466–478.

McConnell, S. K. and C. E. Kaznowski. 1991. Cell cycle dependence of laminar determination in developing cerebral cortex. *Science* 254: 282–285.

Mikkola, M. L., J. Pispa, M. Pekkanen, L. Paulin, P. Nieminen, J. Kere and I. Thesleff. 1999. Ectodysplasin, a protein required for epithelial morphogenesis, is a novel TNF homologue and promotes cell-matrix adhesion. *Mech. Dev.* 88: 133–146.

Millar, S. E. 2002. Molecular mechanisms regulating hair follicle development. *J. Invest. Dermatol.* 118: 216–225.

Miller, S. J., R. M. Lavker and T.-T. Sun. 1993. Keratinocyte stem cells of corneal, skin, and hair follicle. *Semin. Dev. Biol.* 4: 217–240.

Milunsky, A., H. Jick, S. S. Jick, C. L. Bruell, D. S. Maclaughlen, K. J. Rothman and W. Willett. 1989. Multivitamin folic acid supplementation in early pregnancy reduces the prevalence of neural tube defects. *J. Am. Med. Assoc.* 262: 2847–2852.

Montagna, W. and P. F. Parakkal. 1974. The piliary apparatus. *In* W. Montagna (ed.), *The Structure and Formation of Skin.* Academic Press, New York, pp. 172–258.

Montagu, M. F. A. 1962. Time, morphology, and neoteny in the evolution of man. In M. F. A. Montagu (ed.) *Culture and Evolution of Man.* Oxford University Press, New York.

Montonen, O. and 10 others. 1998. The gene defective in anhidrotic ectodermal dysplasia is expressed in the developing epithelium, neuroectoderm, thymus, and bone. *J. Histochem. Cytochem.* 46: 281–289.

Moore, K. L. and T. V. N. Persaud. 1993. *Before We Are Born: Essentials of Embryology and Birth Defects.* W. B. Saunders, Philadelphia.

Morowitz, H. J. and J. S. Trefil. 1992. *The Facts of Life: Science and the Abortion Controversy.* Oxford University Press, New York.

Moury, J. D. and G. C. Schoenwolf. 1995. Cooperative model of epithelial shaping and bending during avian neurulation: Autonomous movements of the neural plate, autonomous movements of the epidermis, and interactions in the neural plate/epidermis transition zone. *Dev. Dynam.* 204: 323–337.

Muhr, J., E. Andersson, M. Persson, T. M. Jessell and J. Ericson. 2001. Groucho-mediated transcriptional repression establishes progenitor cell pattern and neuronal fate in the ventral spinal cord. *Cell* 104: 861–873.

Nagele, R. G. and H. Y. Lee. 1987. Studies in the mechanism of neurulation in the chick: Morphometric analysis of the relationship between regional variations in cell shape and sites of motive force generation. *J. Exp. Biol.* 24: 197–205.

Nakatsu, T., C. Uwabe and K. Shiota. 2000. Neural tube closure in humans initiates at multiple sites: Evidence from human embryos and implications for the pathogenesis of neural tube defects. *Anat. Embryol.* 201: 455–466.

Nguyen, V. H., J. Trout, S. A. Connors, P. Andermann, E. Weinberg and M. C. Mullins. 2000. Dorsal and intermediate neuronal cell types of the spinal cord are established by a BMP signaling pathway. *Development* 127: 1209–1220.

Nichols, D. H. 1981. Neural crest formation in the head of the mouse embryo as observed using a new histological technique. *J. Embryol. Exp. Morphol.* 64: 105–120.

Nievelstein, R. A. J., N. G. Hartwig, C. Vermeij-Keers and J. Valk. 1993. Embryonic development of the mammalian caudal neural tube. *Teratology* 48: 21–31.

Nishimura, E. K. and 9 others. 2002. Dominant role of the niche in melanocyte stem-cell fate determination. *Nature* 416: 854–860.

Papaconstantinou, J. 1967. Molecular aspects of lens cell differentiation. *Science* 156: 338–346.

Pasteels, J. 1937. Etudes sur la gastrulation des vertébrés méroblastiques. III. Oiseaux. IV. Conclusions générales. *Arch. Biol.* 48: 381–488.

Paton, D. and J. A. Craig. 1974. Cataracts: Development, diagnosis, and management. *CIBA Clin. Symp.* 26(3): 2–32.

Patten, B. M. 1971. *Early Embryology of the Chick*, 5th Ed. McGraw-Hill, New York.

Pera, E. M., O. Wessely, S. Y. Li and E. M. De Robertis. 2001 Neural and head induction by insulin-like growth factor signals. *Dev. Cell* 1: 655–665.

Piatigorsky, J. 1981. Lens differentiation in vertebrates: A review of cellular and molecular features. *Differentiation* 19: 134–153.

Pinkus, H. and A. H. H. Mehregan. 1981. *A Guide to Dermohistopathology.* Appleton Century Crofts, New York.

Placzek, M., M. Tessier-Lavigne, T. Yamada, T. Jessell and J. Dodd. 1990. Mesodermal control of neural cell identity: Floor plate induction by the notochord. *Science* 250: 985–988.

Portmann, A. 1941. Die Tragzeiten der Primaten und die Dauer der Schwangerschaft beim Menschen: Ein Problem der vergleichen Biologie. *Rev. Suisse Zool.* 48: 511–518.

Purves, D. and J. W. Lichtman. 1985. *Principles of Neural Development.* Sinauer Associates, Sunderland, MA.

Quiring, R., U. Walldorf, U. Kloter and W. J. Gehring. 1994. Homology of the *eyeless* gene of *Drosophila* to the *Small eye* gene in mice and *Aniridia* in humans. *Science* 265: 785–789.

Rakic, P. 1972. Mode of cell migration to superficial layers of fetal monkey neocortex. *J. Comp. Neurol.* 145: 61–84.

Rakic, P. 1974. Neurons in rhesus visual cortex: Systematic relation between time of origin and eventual disposition. *Science* 183: 425–427.

Rakic, P. 1975. Cell migration and neuronal ectopias in the brain. *In* D. Bergsma (ed.), *Morphogenesis and Malformations of Face and Brain.* Birth Defects Original Article Series, vol. 11, no. 7. Alan R. Liss, New York, pp. 95–129.

Rakic, P. 2002. Neurogenesis in adult primate neocortex: An evaluation of the evidence. *Nat. Rev. Neurosci.* 3: 65–71.

Rakic, P. and P. S. Goldman. 1982. Development and modifiability of the cerebral cortex. *Neurosci. Rev.* 20: 429–612.

Rakic, P. and R. L. Sidman. 1973. Organization of cerebellar cortex secondary to deficit of granule cells in *weaver* mutant mice. *J. Comp. Neurol.* 152: 133–162.

Ramón y Cajal, S. 1890. Sur l'origene et les ramifications des fibres neuveuses de la moelle embryonnaire. *Anat. Anz.* 5: 111–119.

Rietze, R. L., H. Valcanis, G. F. Brooker, T. Thomas, A. K. Voss and P. F. Bartlett. 2001. Purification of a pluripotent neural stem cell from the adult mouse brain. *Nature* 412: 736–739.

Roelink, H. and 10 others. 1994. Floor plate and motoneuron induction by *vhh-1*, a vertebrate homolog of *hedgehog* expressed by the notochord. *Cell* 76: 761–775.

Roelink, H., J. A. Porter, C. Chiang, Y. Tanabe, D. T. Chang, P. A. Beachy and T. M. Jessell. 1995. Floor plate and motor neuron induction by different concentrations of amino-terminal cleavage product of Sonic hedgehog autoproteolysis. *Cell* 81: 445–455.

Roessler, E. and 7 others. 1996. Mutations in the human *sonic hedgehog* gene cause holoprosencephaly. *Nature Genet.* 14: 357–360.

Rose, S. 1998. *Lifelines: Biology beyond Determinism.* Oxford University Press, Oxford.

Saha, M., C. L. Spann and R. M. Grainger. 1989. Embryonic lens induction: More than meets the optic vesicle. *Cell Diff. Dev.* 28: 153–172.

Saitsu, H., M. Ishibashi, H. Nakano and K. Shiota. 2002. Spatial and temporal expression of folate-binding protein 1 (Fbp1) is closely associated with anterior neural tube closure in mice. *Dev. Dynam.* 226: 112–117.

Sauer, F. C. 1935. Mitosis in the neural tube. *J. Comp. Neurol.* 62: 377–405.

Sausedo, R. A., J. Smith and G. C. Schoenwolf. 1997. Role of oriented cell division in shaping and bending of the neural plate. *J. Comp. Neurol.* 381: 473–488.

Schoenwolf, G. C. 1984. Histological and ultrastructural studies of secondary neurulation in mouse embryos. *Am. J. Anat.* 169: 361–374.

Schoenwolf, G. C. 1991a. Cell movements driving neurulation in avian embryos. *Development* 2 [Suppl.]: 157–168.

Schoenwolf, G. C. 1991b. Cell movements in the epiblast during gastrulation and neurulation in avian embryos. *In* R. Keller et al. (eds.), *Gastrulation.* Plenum, New York, pp. 1–28.

Schoenwolf, G. C. and I. S. Alvarez. 1989. Roles of neuroepithelial cell rearrangement and division in shaping of the avian neural plate. *Development* 106: 427–439.

Schoenwolf, G. C. and M. E. Desmond. 1984. Descriptive studies of the occlusion and reopening of the spinal canal of the early chick embryo. *Anat. Rec.* 209: 251–263.

Shu, W., H. Yang, L. Zhang, M. M. Lu and E. E. Morrisey. 2001. Characterization of a new subfamily of *winged-helix/forkhead* (*Fox*) genes that are expressed in the lung and act as transcriptional repressors. *J. Biol. Chem.* 276: 27488–27497.

Shingo, T. and 7 others. 2003. Pregnancy-stimulated neurogenesis in the adult female forebrain mediated by prolactin. *Science* 299: 117–120.

Sidman, R. L., M. M. Dickie and S. H. Appel. 1964. Mutant mice (*quaking* and *jimpy*) with deficient myelination in the central nervous system. *Science* 144: 309–312.

Smith, J. L. and G. C. Schoenwolf. 1989. Notochordal induction of cell wedging in the chick neural plate and its role in neural tube formation. *J. Exp. Zool.* 250: 49–62.

Smith, J. L. and G. C. Schoenwolf. 1991. Further evidence of extrinsic forces in bending of the neural plate. *J. Comp. Neurol.* 307: 225–236.

Smith, J. L. and G. C. Schoenwolf. 1997. Neurulation: Coming to closure. *Trends Neurosci.* 11: 510–517.

Sowell, E. R., P. M. Thompson, C. J. Holmes, T. L. Jernigan and A. W. Toga. 1999. In vivo evidence for post-adolescent brain maturation in frontal and striatal regions. *Nature Neurosci.* 2: 859–861.

Streit, A. and C. D. Stern. 1999. Establishment and maintenance of the border of the neural plate in the chick: Involvement of FGF and BMP activity. *Mech. Dev.* 82:51–66.

Streit, A., A. J. Berliner, C. Papanayoutou, A. Sirulnik and C. D. Stern. 2000. Initiation of neural induction by FGF signaling before gastrulation. *Nature* 406: 74–78.

Tao, H., Y. Yoshimoto, H. Yoshioka, T. Nohno, S. Noji and H. Ohuchi. 2002. FGF10 is a mes-

enchymally derived stimulator for epidermal development in the chick embryonic skin. *Mech. Dev.* 116: 39–49.

Turner, D. L. and C. L. Cepko. 1987. A common progenitor for neurons and glia persists in rat retina late in development. *Nature* 328: 131–136.

Van Allen, M. I. and 15 others. 1993. Evidence for multi-site closure of the neural tube in humans. *Am. J. Med. Genet.* 47: 723–743.

van Praag, H., G. Kempermann and F. H. Gage. 1999. Running increases cell proliferation and neurogenesis in the adult mouse dentate gyrus. *Nature Neurosci.* 2: 266–270.

van Praag, H., A. F. Schinder, B. R. Christie, N. Toni, T. D. Palmer and F. H. Gage. 2002. Functional neurogenesis in the adult hippocampus. *Nature* 415: 1030–1034.

van Straaten, H. W. M., J. W. M. Hekking, E. J. L. M. Wiertz-Hoessels, F. Thors and J. Drukker. 1988. Effect of the notochord on the differentiation of a floor plate area in the neural tube of the chick embryo. *Anat. Embryol.* 177: 317–324.

Vargha-Khadem, F., K. Watkins, K. Alcock, P. Fletcher and R. Passingham. 1995. Praxic and nonverbal cognitive deficits in a large family with a genetically transmitted speech and language disorder. *Proc. Natl. Acad. Sci. USA* 92: 930–933.

Vassar, R. and E. Fuchs. 1991. Transgenic mice provide new insights into the role of TGF-α during epidermal development and differentiation. *Genes Dev.* 5: 714–727.

Vrba, E. 1996. Climate, heterochrony, and human evolution. *J. Anthropol. Res.* 52: 1–28.

Wallace, V. A. 1999. Purkinje cell-derived Sonic hedgehog regulates granule neuron precursor cell proliferation in the developing mouse cerebellum. *Curr. Biol.* 9: 445–448.

Walsh, C. and C. L. Cepko. 1992. Widespread dispersion of neuronal clones across functional regions of the cerebral cortex. *Science* 255: 434–440.

Walsh, C. and C. L. Cepko. 1998. Clonally related cortical cells show several migration patterns. *Science* 241: 1342–1345.

Wasserman, C. R., G. M. Shaw, S. Selvin, J. B. Gould and S. L. Syme. 1998. Socioeconomic status, neighborhood social conditions, and neural tube defects. *Am. J. Pub. Health* 88: 1674–1680.

Weinstein, G. D. and E. J. van Scott. 1965. Turnover times of normal and psoriatic epidermis. *J. Invest. Dermatol.* 45: 257–262.

Whitehead, A. S. and 8 others. 1995. A genetic defect in 5,10-methylenetetrafolate reductase in neural tube defects. *Q. J. Med.* 88: 763–766.

Wickelgren, I. 1999. Nurture helps mold able minds. *Science* 283: 1832–1834.

Wilson, A. C. 1998. Quoted in A. Gibbons, Which of our genes makes us human? *Science* 281: 1432–1434.

Wilson, S. I. and T. Edlund. 2001. Neural induction: Toward a unifying mechanism. *Nature Neurosci.* 4: 1161–1168.

Wilson, S. I., A. Rydström, T. Trimborn, K. Willert, R. Nusse, T. M. Jessell and T. Edlund. 2001. The status of Wnt signaling regulates neural and epidermal fates in the chick embryo. *Nature* 411; 325–330.

Yamada, K. M., B. S. Spooner and N. K. Wessells. 1971. Ultrastructure and function of growth cones and axons of cultured nerve cells. *J. Cell Biol.* 49: 614–635.

Yamada, T., M. Placzek, H. Tanaka, J. Dodd and T. M. Jessell. 1991. Control of cell pattern in the developing nervous system: Polarizing activity of floor plate and notochord. *Cell* 64: 635–647.

Yamada, T., S. L. Pfaff, T. Edlund and T. M. Jessell. 1993. Control of cell pattern in the neural tube: Motor neuron induction by diffusible factors from notochord and floor plate. *Cell* 73: 673–686.

Zolessi, F. R. and C. Arruti. 2001. Apical accumulation of MARCKS in neural plate cells during neurulation in the chick embryo. *BMC Dev. Biol.* 1: 7.

13 *Neural crest cells and axonal specificity*

IN THIS CHAPTER WE CONTINUE our discussion of ectodermal development, focusing here on neural crest cells and axonal guidance. Neural crest cells and axonal growth cones (the motile tips of axons) share the property of having to migrate far from their source of origin to specific places in the embryo. They both have to recognize cues to begin this migration, and they both have to respond to signals that guide them along specific routes to their final destination. Recent research has discovered that many of the signals recognized by neural crest cells and by axonal growth cones are the same.

THE NEURAL CREST

Although derived from the ectoderm, the neural crest has sometimes been called the fourth germ layer because of its importance. It has even been said, somewhat hyperbolically, that "the only interesting thing about vertebrates is the neural crest" (quoted in Thorogood 1989). The neural crest cells migrate extensively to generate a prodigious number of differentiated cell types. These cell types include (1) the neurons and glial cells of the sensory, sympathetic, and parasympathetic nervous systems, (2) the epinephrine-producing (medulla) cells of the adrenal gland, (3) the pigment-containing cells of the epidermis, and (4) many of the skeletal and connective tissue components of the head. The fate of the neural crest cells depends, to a large degree, on where they migrate to and settle. Table 13.1 is a summary of some of the cell types derived from the neural crest.

Specification and Regionalization of the Neural Crest

Neural crest cells originate at the dorsalmost region of the neural tube. Transplantation experiments wherein a quail neural plate is grafted into chick non-neural ectoderm have shown that juxtaposing these tissues induces the formation of neural crest cells, and that both the prospective neural plate and the prospective epidermis contribute to the neural crest (Selleck and Bronner-Fraser 1995; see also Mancilla and Mayor 1996).

The boundary between neural plate and epidermal ectoderm is characterized by high levels of BMPs. Where these high levels of BMPs meet high levels of Wnt6 in the presumptive epidermis, the neural crest cells form. These two factors—BMPs and Wnt6—appear to cause the expression of particular transcription factors, including

Slug and **FoxD3**, in those cells that are to become the neural crest (Figure 13.1; Kos et al. 2001; Sasai et al. 2001; Garcia-Castro et al. 2002). FoxD3 appears to be critical for the specification of ectodermal cells as neural crest cells. When FoxD3 is inhibited from functioning, neural crest differentiation is inhibited. Conversely, when FoxD3 is expressed ectopically (by electroporating FoxD3 into neural plate cells), those neural plate cells begin to express proteins characteristic of the neural crest. The Slug protein appears to be necessary for neural crest cells to leave the epithelium and migrate.

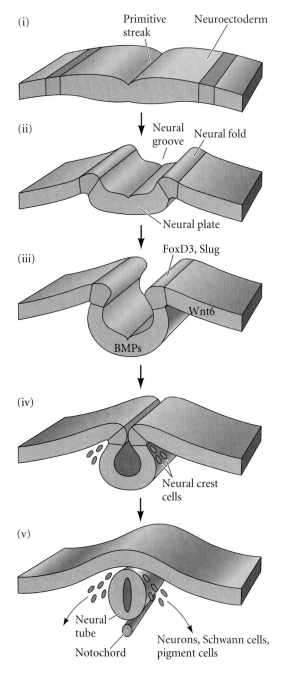

TABLE 13.1 Some derivatives of the neural crest

Derivative	Cell type or structure derived
Peripheral nervous system (PNS)	Neurons, including sensory ganglia, sympathetic and parasympathetic ganglia, and plexuses Neuroglial cells Schwann cells
Endocrine and paraendocrine derivatives	Adrenal medulla Calcitonin-secreting cells Carotid body type I cells
Pigment cells	Epidermal pigment cells
Facial cartilage and bone	Facial and anterior ventral skull cartilage and bones
Connective tissue	Corneal endothelium and stroma Tooth papillae Dermis, smooth muscle, and adipose tissue of skin of head and neck Connective tissue of salivary, lachrymal, thymus, thyroid, and pituitary glands Connective tissue and smooth muscle in arteries of aortic arch origin

Source: After Jacobson 1991, based on multiple sources.

WEBSITE 13.1 Avian neurulation. A companion to Figure 13.1, this website displays a thoughtful animation of the formation of the chick neural crest.

The neural crest can be divided into four main (but overlapping) domains, each with characteristic derivatives and functions (Figure 13.2):

- The **cranial** (**cephalic**) **neural crest** cells migrate dorsolaterally to produce the craniofacial mesenchyme that differentiates into the cartilage, bone, cranial neurons, glia, and connective tissues of the face. These cells enter the pharyngeal arches and pouches to give rise to thymic cells, the odontoblasts of the tooth primordia, and the bones of the middle ear and jaw.*
- The **trunk neural crest** cells take one of two major pathways. The early migratory pathway takes trunk neural crest cells ventrolaterally through the anterior half of each sclerotome. (**Sclerotomes** are blocks of mesodermal cells, derived from somites, that will differentiate into the vertebral cartilage of the spine.) Those trunk neural crest

---

*The **pharyngeal** (**branchial**) **arches** (see Figure 1.3) are outpocketings of the head and neck region into which neural crest cells migrate. The **pharyngeal pouches** form between these arches and become the thyroid, parathyroid, and thymus.

Figure 13.1
Schematic representation of neural crest formation in an amniote (chick) embryo, shown in cross section. Neural crest cells form at the junction between the Wnt6-expressing epidermal ectoderm and the BMPs produced by the presumptive neural ectoderm. (After Trainor and Krumlauf 2002.)

Figure labels (i–v):
(i) Primitive streak, Neuroectoderm
(ii) Neural groove, Neural fold, Neural plate
(iii) FoxD3, Slug, Wnt6, BMPs
(iv) Neural crest cells
(v) Neural tube, Notochord, Neurons, Schwann cells, pigment cells

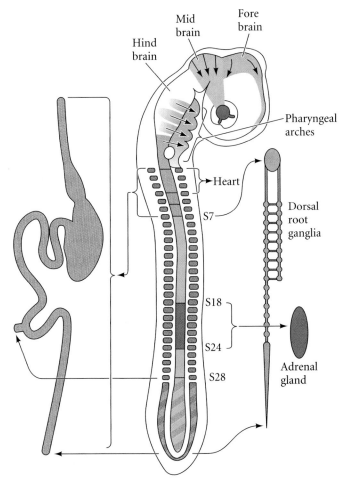

Figure 13.2
Regions of the neural crest. The cranial neural crest migrates into the pharyngeal arches and the face to form the bones and cartilage of the face and neck. It also produces the cranial nerves. The vagal neural crest (near somites 1–7) and the sacral neural crest (posterior to somite 28) form the parasympathetic nerves of the gut. The cardiac neural crest cells arise near somites 1–3; they are critical in making the division beeween the aorta and the pulmonary artery. Neural crest cells of the trunk (about somite 6 through the tail) make sympathetic neurons and pigment cells (melanocytes), and a subset of these (at the level of somites 18–24) form the medulla portion of the adrenal gland. (After Le Douarin 1982.)

cells that remain in the sclerotomes form the **dorsal root ganglia** containing the sensory neurons. Those cells that continue more ventrally form the sympathetic ganglia, the adrenal medulla, and the nerve clusters surrounding the aorta. Later, the trunk neural crest cells that become pigment-synthesizing **melanocytes** migrate dorsolaterally into the ectoderm and move through the skin toward the ventral midline of the belly.

- The **vagal** and **sacral neural crest** cells generate the **parasympathetic (enteric) ganglia** of the gut (Le Douarin and Teillet 1973; Pomeranz et al. 1991). The vagal (neck) neural crest lies opposite chick somites 1–7,

while the sacral neural crest lies posterior to somite 28. Failure of neural crest cell migration from these regions to the colon results in the absence of enteric ganglia and thus to the absence of peristaltic movement in the bowels.

- The **cardiac neural crest** is located between the cranial and trunk neural crests. In chick embryos, this neural crest region extends from the first to the third somites, overlapping the anterior portion of the vagal neural crest (Kirby 1987; Kirby and Waldo 1990). The cardiac neural crest cells can develop into melanocytes, neurons, cartilage, and connective tissue (of the third, fourth, and sixth pharyngeal arches). In addition, this region of the neural crest produces the entire muscular-connective tissue wall of the large arteries as they arise from the heart, as well as contributing to the septum that separates the pulmonary circulation from the aorta (Le Lièvre and Le Douarin 1975).

The Trunk Neural Crest

Migration pathways of trunk neural crest cells

The trunk neural crest is a transient structure, as its cells disperse soon after the neural tube closes. There are two major pathways taken by the migrating trunk neural crest cells (Figure 13.3A). Those cells that migrate along the **dorsolateral pathway** become melanocytes, the melanin-forming pigment cells. They travel between the epidermis and the dermis, entering the ectoderm through minute holes in the basal lamina (which they may make). Here they colonize the skin and hair follicles (Mayer 1973; Erickson et al. 1992). This pathway was demonstrated in a series of classic experiments by Mary Rawles (1948), who transplanted the neural tube and crest from a pigmented strain of chickens into the neural tube of an albino chick embryo (see Figure 1.11).

Fate mapping of the neural crest cells has shown that trunk neural crest cells can also take a **ventral pathway**. These cells become sensory (dorsal root) and sympathetic neurons, adrenomedullary cells, and Schwann cells (Weston 1963; Le Douarin and Teillet 1974). In birds and mammals (but not fish and frogs),* these cells migrate ventrally through the anterior, but not the posterior, section of the sclerotomes (Figure 13.3B–D; Rickmann et al. 1985; Bronner-Fraser 1986; Loring and Erickson 1987; Teillet et al. 1987). By transplanting quail neural tubes into chick embryos, Teillet and co-workers were able to mark neural crest cells both genetically and immunologically. The antibody marker recognized and labeled neural crest cells of both species; the genetic marker enabled the investigators to distinguish between quail and chick cells. These

---

*In fish, the neural tube forms by secondary neurulation. The neural crest cells emerge from a thickening of the ectoderm in the dorsal region of the neural keel. In the migration of fish neural crest cells, the sclerotome is not important; rather, the myotome appears to guide the migration of the crest cells ventrally (Morin-Kensicki and Eisen 1997).

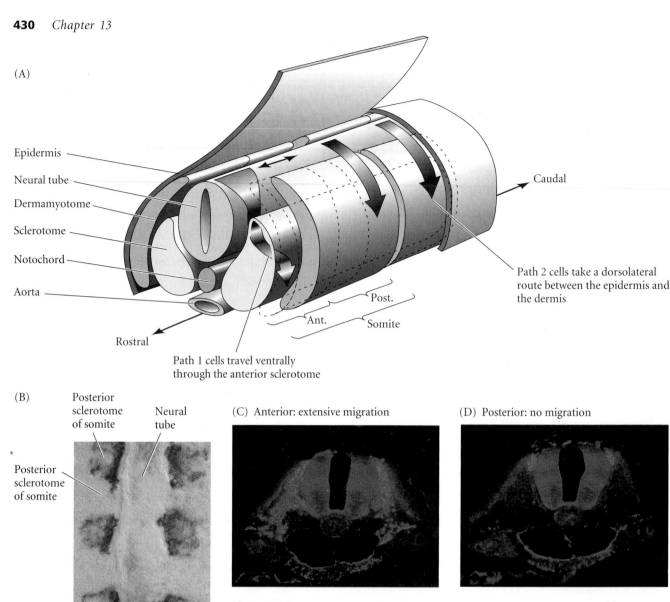

(A)

Epidermis

Neural tube

Dermamyotome

Sclerotome

Notochord

Aorta

Caudal

Path 2 cells take a dorsolateral route between the epidermis and the dermis

Post.

Ant.

Somite

Rostral

Path 1 cells travel ventrally through the anterior sclerotome

(B) Posterior sclerotome of somite   Neural tube

Posterior sclerotome of somite

(C) Anterior: extensive migration

(D) Posterior: no migration

Figure 13.3

Neural crest cell migration in the trunk of the chick embryo. (A) Schematic diagram of trunk neural crest cell migration. Cells taking path 1 (the ventral pathway) travel ventrally through the anterior of the sclerotome (that portion of the somite that generates vertebral cartilage). Those cells initially opposite the posterior portion of a sclerotome migrate along the neural tube until they come to an opposite anterior region. These cells contribute to the sympathetic and parasympathetic ganglia as well as to the adrenomedullary cells and dorsal root ganglia. Other trunk neural crest cells enter path 2 (the dorsolateral pathway) somewhat later. These cells travel along a dorsolateral route beneath the ectoderm, and become pigment-producing melanocytes. (Migration pathways are shown on only one side of the embryo.) (B) These fluorescence photomicrographs of longitudinal sections of a 2-day chick embryo are stained red with antibody to HNK-1, which selectively recognizes neural crest cells. Extensive staining is seen in the anterior, but not in the posterior, half of each sclerotome. (C, D) Cross sections through these areas, showing (C) extensive migration through the anterior portion of the sclerotome, but (D) no migration through the posterior portion. Here, the antibodies to HNK-1 are stained green. (B from Wang and Anderson 1997; C–D from Bronner-Fraser 1986; photographs courtesy of the authors.)

studies showed that neural crest cells initially located opposite the posterior region of a somite migrate anteriorly or posteriorly along the neural tube and then enter the anterior region of their own or an adjacent somite. These cells join with the neural crest cells that were initially opposite the anterior portion of the somite, and they form the same structures. Thus, each dorsal root ganglion is composed of three neural crest cell populations: one from the neural crest opposite the anterior portion of the somite and one from each of the neural crest regions opposite the posterior portions of the adjacent somites.

The mechanisms of trunk neural crest migration

EMIGRATION FROM THE NEURAL TUBE. Any analysis of migration (be it of birds, butterflies, or neural crest cells) has to ask four questions:

1. How is migration initiated?
2. How do the migratory agents know the route to travel?
3. What signals indicate that the destination has been reached and migration should end?
4. When does the migratory agent become competent to respond to these signals?

Neural crest cells initiate their migration from the neural folds through interactions of the neural plate with the presumptive epidermis. The neural crest cells, originally part of an epithelium, undergo an epithelial-to-mesenchymal transformation, wherein they lose their attachments to other cells and migrate away from the epithelial sheet. These changes can be mimicked by culturing neural plate cells with bone morphogenetic proteins 4 and 7, two proteins that are known to be secreted by the presumptive epidermis; see Chapter 12). In the presence of Wnt and FGF proteins, BMP4 and BMP7 induce the expression of the Slug protein and the RhoB protein in the cells destined to become neural crest (Figure 13.4; Nieto et al. 1994; Mancilla and Mayor 1996; Liu and Jessell 1998; LaBonne and Bronner-Fraser 1998). If either Slug or Rho is inactivated or inhibited from forming, the neural crest cells fail to emigrate from the neural tube.

In order for cells to leave the neural crest, there must be both pushes and pulls. The RhoB protein may be involved in establishing the cytoskeletal conditions that promote migration (Hall 1998). Upon activation, RhoB promotes actin polymerization into microfilaments and the attachment of these microfilaments to the cell membrane. However, the cells cannot leave the neural tube as long as they are tightly connected

to one another. One of the functions of the Slug protein is to activate the factors that dissociate the tight junctions binding the cells together (Savagne et al. 1997). Another factor in the initiation of neural crest cell migration is the loss of the N-cadherin that had linked the cells together. Originally found on the surface of the neural crest cells, this cell adhesion protein is downregulated at the time of cell migration. Migrating trunk neural crest cells have no N-cadherin on their surfaces, but they begin to express it again as they aggregate to form the dorsal root and sympathetic ganglia (Takeichi 1988; Akitaya and Bronner-Fraser 1992).

RECOGNITION OF SURROUNDING EXTRACELLULAR MATRICES. The route taken by migrating trunk neural crest cells is controlled by cues from the extracellular matrices they encounter (Newgreen and Gooday 1985; Newgreen et al. 1986). But what are the extracellular matrix molecules that enable or forbid migration? One set of proteins *promotes* migration. These proteins include fibronectin, laminin, tenascin, various collagen molecules, and proteoglycans, and they are seen throughout the matrix encountered by the neural crest cells. Upon migration, the chick trunk neural crest cells begin to express the α4β1 integrin protein, which binds to several of these extracellular matrix proteins. The expression of this integrin molecule is needed for the locomotion and survival of the newly freed neural crest cells. Without it, the cells leave the neural tube, but become disoriented and often undergo apoptosis (Testaz and Duband 2001). **Thrombospondin**, another extracellular matrix molecule, is found in the anterior, but not in the posterior, portion of the sclerotome (Tucker et al. 1999). Thrombospondin promotes neural crest cell adhesion and migration, and it may cooperate with fibronectin and laminin to promote neural crest cell migration through the anterior portion of the somite.

Equally important for the patterning of neural crest cell movement are those proteins that *impede* migration. The main proteins involved in the restriction of neural crest cell migration are the **ephrin proteins**. These proteins are expressed in the posterior section of each sclerotome, and wherever they are, neural crest cells do not go (Figure 13.5A). If neural crest cells are plated on a culture dish that contains alternate stripes of immobilized cell membrane proteins with or without ephrins, the cells will leave the ephrin-containing regions and move along the stripes that lack ephrin (Figure 13.5B; Krull et al. 1997; Wang and Anderson 1997). The neural crest cells recognize the ephrin proteins through their transmembrane **Eph receptors**. Thus, the neural crest cells

Figure 13.4
All migrating neural crest cells are stained red by antibody to HNK-1, as in Figure 13.3. The RhoB protein (green stain) is expressed in cells as they leave the neural crest. Cells expressing both HNK-1 and RhoB appear yellow. (After Liu and Jessell 1998; photograph courtesy of T. M. Jessell.)

(A)

ephrin | Neural crest cells

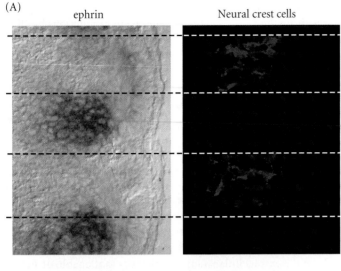

(B)   − + − + − + − + − + − + − + − +

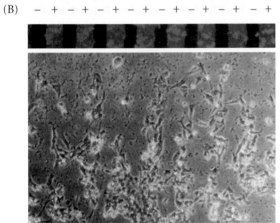

(C)

← Anterior

Motor axons

Posterior →

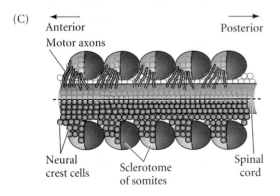

Neural crest cells | Sclerotome of somites | Spinal cord

Figure 13.5
Segmental restriction of neural crest cells and motor neurons by the ephrin proteins of the sclerotome. (A) Negative correlation between regions of ephrin in the sclerotome (dark blue stain, left) and the presence of neural crest cells (green HNK-1 stain, right). (B) When neural crest cells are plated on fibronectin-containing matrices with alternating stripes of ephrin, they bind to those regions lacking ephrin. (C) Composite scheme showing the migration of spinal cord neural crest cells and motor neurons through the ephrin-deficient anterior regions of the sclerotomes. (For clarity, the neural crest cells and motor neurons are each depicted on only one side of the spinal cord.) (A and B from Krull et al. 1997; C after O'Leary and Wilkinson 1999.)

contain an Eph receptor in their cell membranes, whereas the posterior portions of the sclerotomes contain ephrin proteins in their membranes. Binding to the ephrins activates the tyrosine kinase domains of the Eph receptors in the neural crest cells, and these kinases probably phosphorylate proteins that interfere with the actin cytoskeleton, which is critical for cell migration. In addition to ephrins, other proteins in the posterior portion of each sclerotome also contribute to the inhospitable nature of these regions (Krull et al. 1995). This patterning of neural crest cell migration generates the overall segmental character of the peripheral nervous system, reflected in the positioning of the dorsal root ganglia and other neural crest-derived structures.

Stem cell factor (mentioned in Chapter 6) is critical in allowing the continued proliferation of those neural crest cells that enter the skin, and it may also serve as an anti-apoptosis factor and a chemotactic factor. If stem cell factor is experimentally secreted from tissues, such as the cheek epithelium or the footpads, that do not usually synthesize this protein (and do not usually have melanocytes), neural crest cells will enter those regions and become melanocytes (Kunisada et al. 1998). Therefore, the migration of neural crest cells appears to be regulated both by the extracellular matrix and by soluble factors secreted at their potential destinations. As we will see later in this chapter, the movement of axonal growth cones can also be regulated by similar (and sometimes identical) cues.

WEBSITE 13.2 **The specificity of the extracellular matrix.** The importance of the extracellular matrix for neural crest cell migration was first shown in a series of creative experiments using mutant salamanders.

WEBSITE 13.3 **Mouse neural crest cell mutants.** Some of the most important insights into neural crest cell development and migration have come from studies of mutant mice. These mice can be recognized by their altered pigmentation, resulting from abnormalities of neural crest cell proliferation, migration, or differentiation.

Trunk neural crest cell differentiation

THE PLURIPOTENCY OF TRUNK NEURAL CREST CELLS. One of the most exciting features of neural crest cells is their pluripotency. A single neural crest cell can differentiate into any of several different cell types, depending on its location within the embryo. For example, the parasympathetic neurons formed by the vagal neural crest cells (adjacent to somites 1 and 7) produce acetylcholine as their neurotransmitter; they are therefore *cholinergic* neurons. These neurons will line the gut. The sympathetic neurons formed by the trunk (thoracic) neural crest cells produce norepinephrine; they are *adrenergic* neurons. But when chick vagal and thoracic neural crests are reciprocally transplanted, the former thoracic crest produces the cholinergic neurons of the parasympathetic ganglia, and the former vagal crest forms adrenergic neurons in the sympathetic ganglia (Le Douarin et al. 1975). Kahn and co-workers (1980) found that premigratory neural crest cells from both the thoracic and the vagal regions contain enzymes for synthesizing both acetylcholine and norepinephrine. Thus, the differentiation of a neural crest cell depends on its eventual location and not on its place of origin.

The pluripotency of some neural crest cells is such that even regions of the neural crest that never produce nerves in normal embryos can be made to do so under certain conditions. Cranial neural crest cells from the midbrain region normally migrate into the eye and interact with the pigmented retina to become scleral cartilage cells (Noden 1978). However, if this region of the neural crest is transplanted into the trunk region, it can form sensory ganglion neurons, adrenomedullary cells, glia, and Schwann cells (Schweizer et al. 1983).

The research we have just described studied the potential of populations of neural crest cells. It is still uncertain whether most of the individual cells that leave the neural crest are pluripotent or whether most are already restricted to certain fates. Bronner-Fraser and Fraser (1988, 1989) provided evidence that some, if not most, individual neural crest cells are pluripotent as they leave the crest. They injected fluorescent dextran molecules into individual avian neural crest cells while the cells were still above the neural tube, and then looked to see what types of cells their descendants became after migration. The progeny of a single neural crest cell could become sensory neurons, melanocytes, adrenomedullary cells, and glia (Figure 13.6).

However, D. J. Anderson's laboratory (Lo et al. 1997; Ma et al. 1998; Perez et al. 1999) found evidence that some populations of neural crest cells are committed very soon after leaving the neural tube, and that they have transcription factors that constrain which cell types they can produce. Similarly, Henion and Weston (1997) found that the initial neural crest population was a heterogeneous mixture of precursors, almost half of which generated clones containing a single cell type. The cells committed to becoming melanocytes were present as soon as cells began migrating, but these cells dispersed from the neural tube only after most of the neurogenic precursors had already done so.

As mentioned earlier, both BMPs and Wnts are involved in specifying neural crest cells. Jin and colleagues (2001) presented evidence that different ratios of these factors can specify different populations of neural crest cells. Early-migrating cells of the presumptive neural and glial lineages secrete Wnt inhibitors, and experimental placement of neural crest cells in Wnt proteins expands the number of melanocytes at the expense of neural and glial cells. Conversely, BMP is downregulated when the melanoblast lineages depart the neural tube. For the melanocyte lineage, ephrin can play a positive role in cell migration, rather than have the negative effect it exhibits for ventrally moving cells. The ephrin expressed along the dorsolateral migration pathway blocks the migration of early-migrating neural crest cells, but stimulates the migration of late-migrating cells. Indeed, Eph signaling appears to be critical for promoting this migration, since disruption of Eph signaling in late-migrating neural crest cells prevents their dorsolateral migration (Santiago and Erickson 2002).

> **WEBSITE 13.4 Committed neural crest cells.** Not all neural crest cells are pluripotent. The neural crest cells leaving the dorsal neural tube are a population of heterogeneous cells, some of which have a restricted potency.

FINAL DIFFERENTIATION OF THE TRUNK NEURAL CREST CELLS. The final differentiation of neural crest cells is determined in large part by the environment to which they migrate. It does not involve the selective death of those cells already committed to secreting a type of neurotransmitter other than the one called for (Coulombe and Bronner-Fraser 1987). Heart cells, for example, secrete a protein, leukemia inhibition factor (LIF), that can convert adrenergic sympathetic neurons into

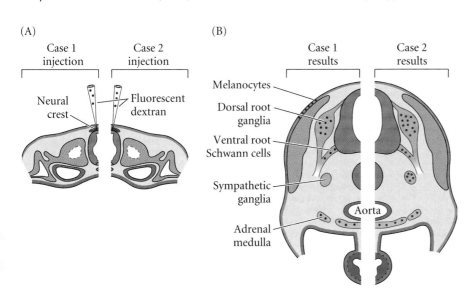

(A)

Case 1 injection   Case 2 injection

Neural crest   Fluorescent dextran

(B)

Case 1 results   Case 2 results

Melanocytes
Dorsal root ganglia
Ventral root Schwann cells
Sympathetic ganglia
Aorta
Adrenal medulla

Figure 13.6
Pluripotency of trunk neural crest cells. (A) A single neural crest cell is injected with highly fluorescent dextran shortly before migration of the neural crest cells is initiated. The progeny of this cell will each receive some of these fluorescent molecules. (B) Two days later, neural crest-derived tissues contain dextran-labeled cells descended from the injected precursor. The figure summarizes data from two different experiments (case 1 and case 2). (After Lumsden 1988a.)

cholinergic neurons without affecting their survival or growth (Chun and Patterson 1977; Fukada 1980; Yamamori et al. 1989). Similarly, bone morphogenetic protein 2 (BMP2), a protein secreted by the heart, lungs, and dorsal aorta, influences rat neural crest cells to differentiate into cholinergic neurons. Such neurons form the sympathetic ganglia in the region of these organs (Shah et al. 1996). While BMP2 may induce neural crest cells to become neurons, glial growth factor (GGF; neuregulin) suppresses neuronal differentiation and directs development toward glial fates (Shah et al. 1994). Another paracrine factor, endothelin-3, appears to stimulate neural crest cells to become melanocytes in the skin and adrenergic neurons in the gut (Baynash et al. 1994; Lahav et al. 1996). To distinguish between these two fates, the neural crest cells entering the skin also encounter Wnt proteins, which inhibit neural development and promote melanocyte differentiation (Dorsky et al. 1998). Similarly, the chick trunk neural crest cells that migrate into the region destined to become the adrenal medulla can differentiate in two directions. The presence of certain paracrine factors induces these cells to become sympathetic neurons (Varley et al. 1995), while those cells that also encounter glucocorticoids such as those made by the cortical cells of the adrenal gland differentiate into adrenomedullary cells (Figure 13.7; Anderson and Axel 1986; Vogel and Weston 1990). Thus, the fate of a neural crest cell can be directed by the tissue environment in which it settles.

The Cranial Neural Crest

The head, comprising the face and the skull, is the most anatomically sophisticated portion of the vertebrate body. It is the evolutionary novelty that separates the vertebrates from the other deuterostomes—echinoderms, tunicates, and lancelets (Northcutt and Gans 1983; Wilkie and Morriss-Kay 2001). The head is largely the product of the cranial neural crest, and the evolution of jaws, teeth, and facial cartilage occurs through changes in the placement of these cells (see Chapter 23). Brian Hall (2000) has gone so far as to suggest that we recognize the neural crest as a fourth germ layer and classify vertebrates as "cristata"—those animals with a neural crest.

Cranial neural crest cells have a repertoire of fates different from those of the trunk neural crest cells. While both types of neural crest cells can form melanocytes, neurons, and glia, only the cells of the cranial neural crest are able to produce cartilage and bone (Table 13.2). Moreover, if transplanted into the trunk region, the cranial neural crest participates in forming trunk cartilage that normally does not arise from neural crest components. This ability to form bone may have been a primitive property of the neural crest and may have been critical for forming the bony armor found in several extinct fish species (Smith and Hall 1993). In other words, the trunk crest has apparently lost the ability to form bone, rather than the cranial crest acquiring this ability. McGonnell and Graham (2002) have shown that bone-forming capacity may still be latent in the trunk neural crest, since if cultured in certain hormones and vitamins, the trunk cells become capable of forming bone and cartilage when placed into the head region.

Figure 13.7
Final differentiation of a trunk neural crest cell committed to becoming either a chromaffin (adrenomedullary) cell or a sympathetic neuron. Glucocorticoids appear to act at two places in this pathway to determine chromaffin cells. First, they inhibit the actions of those factors that promote neuronal differentiation, and second, they induce those enzymes characteristic of chromaffin cells. Those cells exposed sequentially to basic fibroblast growth factor (FGF2) and nerve growth factor (NGF) differentiate into sympathetic neurons.

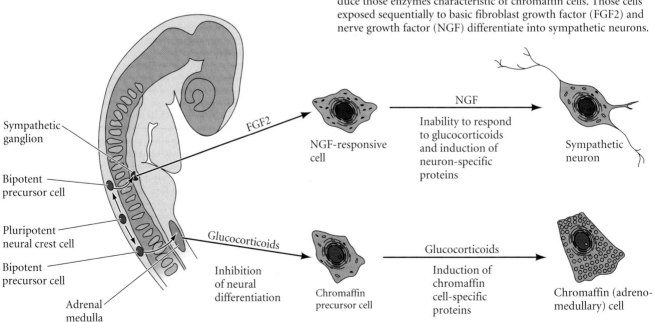

TABLE 13.2 Some derivatives of the pharyngeal arches

Pharyngeal arch	Skeletal elements (neural crest plus mesoderm)	Arches, arteries (mesoderm)	Muscles (mesoderm)	Cranial nerves (neural tube)
1	Incus and malleus (from neural crest); mandible, maxilla, and temporal bone regions (from crest dermal mesenchyme)	Maxillary branch of the carotid artery (to the ear, nose, and jaw)	Jaw muscles; floor of mouth; muscles of the ear and soft palate	Maxillary and mandibular divisions of trigeminal nerve (V)
2	Stapes bone of the middle ear; styloid process of temporal bone; part of hyoid bone of neck (all from neural crest cartilage)	Arteries to the ear region: cortico-tympanic artery (adult); stapedial artery (embryo)	Muscles of facial expression; jaw and upper neck muscles	Facial nerve (VII)
3	Lower rim and greater horns of hyoid bone (from neural crest)	Common carotid artery; root of internal carotid	Stylopharyngeus (to elevate the pharynx)	Glossopharyngeal nerve (IX)
4	Laryngeal cartilages (from lateral plate mesoderm)	Arch of aorta; right subclavian artery; original spouts of pulmonary arteries	Constrictors of pharynx and vocal cords	Superior laryngeal branch of vagus nerve (X)
6	Laryngeal cartilages (from lateral plate mesoderm)	Ductus arteriosus; roots of definitive pulmonary arteries	Intrinsic muscles of larynx	Recurrent laryngeal branch of vagus nerve (X)

Source: Based on Larsen 1992.

As mentioned in Chapter 12, the hindbrain is segmented along the anterior-posterior axis into compartments called rhombomeres. The cranial neural crest cells migrate ventrally from those regions anterior to rhombomere 6, taking one of three major pathways (Figures 1.3 and 13.8):

1. Neural crest cells from rhombomeres 1 and 2 migrate to the first pharyngeal arch (the mandibular arch), forming the jawbones as well as the incus and malleus bones of the ear. These neural crest cells are also pulled by the expanding epidermis to form the **frontonasal process**. The neural crest cells of the frontonasal process generate the bones of the face (Le Douarin and Kalcheim 1999).

2. Neural crest cells from rhombomere 4 populate the second pharyngeal arch, forming the hyoid cartilage of the neck as well as the stapes bone of the middle ear.

3. Neural crest cells from rhombomere 6 migrate into the third and fourth pharyngeal arches and pouches to form the thymus, parathyroid, and thyroid glands. If the neural crest is removed from those regions including rhom-

bomere 6, these structures fail to form (Bockman and Kirby 1984). Some of these cells migrate caudally to the clavicle (collarbone) where they settle at the sites that will be used for the attachment of certain neck muscles (McGonnell et al. 2001).

Neural crest cells from rhombomeres 3 and 5 do not migrate through the mesoderm surrounding them, but enter into the migrating streams of neural crest cells on either side of them. Those that do not enter these streams of cells will die (Graham et al. 1993, 1994; Sechrist et al. 1993). Observations of labeled neural crest cells from chick hindbrain, wherein individually marked cells were followed by cameras focusing through a Teflon membrane window in the egg, found that the migrating cells were "kept in line" by interactions with their environment (Kulkesa and Fraser 2000). In frog embryos, there is evidence that the separate streams are kept apart by ephrins. Blocking the activity of the Eph receptors causes cells from the different streams to mix together (Smith et al. 1997; Helbling et al. 1998).

(A)

Neural folds

Migrating neural crest cells

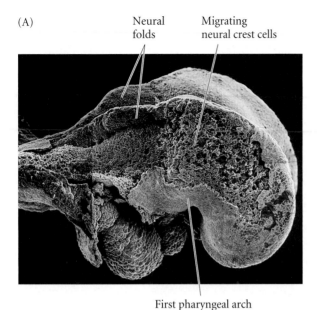

First pharyngeal arch

(B)

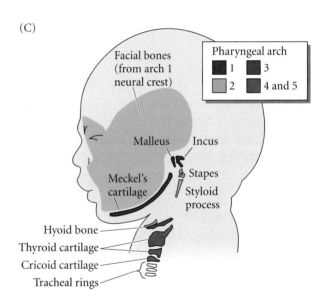

Figure 13.8
Cranial neural crest cell migration in the mammalian head. (A) Scanning electron micrograph of a rat embryo with part of the lateral ectoderm removed from the surface. Neural crest cells migrate prior to neural tube fusion in mammals. Such migration can be seen over the midbrain, and the column of neural crest cells migrating into the future first pharyngeal arch is evident. (B) The influence of mesoderm and ectoderm on the axial identity of cranial neural crest cells and the role of Hoxa-2 in regulating second arch morphogenesis. Neural crest cells (small circles) from rhombomere 2 do not express Hox genes as they enter into pharyngeal arch 1. Hoxa-2 is expressed in cranial crest cells migrating into pharyngeal arch 2, but not into pharyngeal arch 1, and it plays a major role in imposing second arch identity on second arch structures. The diagram at the right indicates the emerging pathways by which Hoxa-2 interacts with ectodermally secreted FGF8 and BMPs to regulate arch morphogenesis in the two arches. (C) Structures formed in the human face by the mesenchymal cells of the neural crest. The cartilaginous elements of the pharyngeal arches are indicated by colors, and the darker pink region indicates the facial skeleton produced by anterior regions of the cranial neural crest. (A from Tan and Morriss-Kay 1985, courtesy of S.-S. Tan; B after Trainor and Krumlauf 2001; C after Carlson 1999.)

Cranial neural crest cell migration and specification: The first wave

There appear to be two "waves" of cranial neural crest cell emigration. The first wave consists of a pluripotent population of neural crest-derived mesenchymal cells that migrate to form the cartilage and bones of the head and neck, as well as the stromal tissue of pharyngeal organs such as the thyroid and thymus. When *individual* mouse or zebrafish cranial crest cells are transplanted from one region of the hindbrain to another, they take on the characteristic Hox gene expression pattern of their host (new) region and lose their original gene expression pattern. However, if *groups* of cranial neural crest cells are so

transplanted, they tend to keep their original identity (Trainor and Krumlauf 2000; Schilling 2001).

There are reciprocal interactions whereby the neural crest cells and the pharyngeal arch regions mutually specify each other. First, the pharyngeal regions dictate the differentiation of the neural crest cells by modulating their Hox genes (Trainor and Krumlauf 2000; Schilling 2001). This determines their fates. Next, the neural crest cells instruct gene expression in the surrounding tissues. This influences the growth patterns of the arch—whether the bird's face will have a narrow beak or a duck-like bill (Noden 1991; Schneider and Helms 2003).

The Hox gene patterns of the neural crest cells appear to be specified by the mesoderm around them. The combinations of Hox genes expressed by the mesoderm in the various regions of cranial crest cell migration specify the fates of the crest cells (Trainor and Krumlauf 2001). The Hox expression pattern of the neural crest cells is critical to their fate in the head. When *Hoxa-2* is knocked out from mouse embryos, the neural crest cells of the second pharyngeal arch generate Meckel's cartilage, which is characteristic of the first pharyngeal arch (Gendron-Maguire et al. 1993; Rijli et al. 1993). Conversely, if *Hoxa-2* is expressed ectopically in the first arch, the neural crest cells of the first arch form second-arch cartilage (Grammatopolous et al. 2000; Pasqualetti et al. 2000). Thus, *Hoxa-2* appears to be essential for the proper development of the second-arch derivatives (Figure 13.8B).

Intramembranous ossification

The cranial neural crest cells form bones through **intramembranous ossification.** In the skull, neural crest-derived mesenchymal cells proliferate and condense into compact nodules. Some of these cells develop into capillaries; others change their shape to become **osteoblasts**, committed bone precursor cells (Figure 13.9). The osteoblasts secrete a collagen-proteoglycan **osteoid matrix** that is able to bind calcium. Those osteoblasts that become embedded in the calcified matrix become **osteocytes**, or bone cells. As calcification proceeds, bony spicules radiate out from the region where ossification began. Furthermore, the entire region of calcified spicules becomes surrounded by compact mesenchymal cells that form the **periosteum** (a membrane that surrounds the bone). The cells on the inner surface of the periosteum also become osteoblasts and deposit matrix parallel to the existing spicules. In this manner, many layers of bone are formed.

The mechanism of intramembranous ossification involves bone morphogenetic proteins and the activation of a transcription factor called **CBFA1**. Bone morphogenetic proteins (probably BMP2, BMP4, and BMP7) from the head epi-dermis are thought to instruct the neural crest-derived mesenchymal cells to become bone cells directly by causing them to express CBFA1 (Hall 1988; Ducy et al. 1997). CBFA1 appears to activate the genes for osteocalcin, osteopontin, and other bone-specific extracellular matrix proteins. This conclusion was confirmed by gene targeting experiments in which the mouse *Cbfa1* gene was knocked out (Komori et al. 1997; Otto et al. 1997). Mice homozygous for this deletion died shortly after birth without taking a breath, and their skeletons completely lacked bone. (Figure 13.9B,C; the mutants had only the cartilaginous skeletal model for the mesodermally produced bone that will be discussed in the next chapter). Mice that were heterozygous for the *Cbfa1* deletion showed skeletal defects similar to those of a human syndrome called cleidocranial dysplasia (CCD). In this syndrome, the skull sutures fail to close, growth is stunted, and the clavicle (collarbone) is often absent or deformed.[*] When DNA from patients with CCD was analyzed, each patient had either deletions or point mutations in the *CBFA1* gene. Therefore, it appears that cleidocranial dysplasia is caused by heterozygosity of the *CBFA1* gene (Mundlos et al. 1997).

---

[*]CCD may have been responsible for the phenotype of Thersites, the Greek soldier described in the *Iliad* as having "both shoulders humped together, curving over his caved-in chest, and bobbing above them his skull warped to a point…" (Dickman 1997).

Figure 13.9

Intramembranous ossification. (A) Mesenchyme cells condense and change shape to produce osteoblasts, which deposit osteoid matrix. Osteoblasts then become arrayed along the calcified region of the matrix. Osteoblasts that are embedded within the calcified matrix become osteocytes. (B, C) Targeting of the *Cbfa1* gene in mice prevents bone formation. Newborn mice were stained with alcian blue (for cartilage) and alizarin red (for bone). (B) Wild-type mouse. (C) The homozygous *Cbfa1* mutant littermate shows normal cartilage development, but an absence of ossification throughout the entire body. (From Otto et al. 1997; photographs courtesy of *Cell* and MIT Press.)

(A)

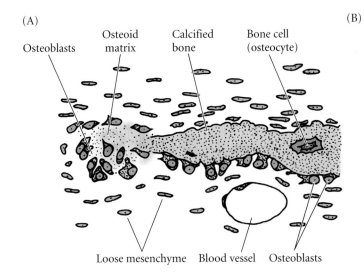

Osteoblasts · Osteoid matrix · Calcified bone · Bone cell (osteocyte)

Loose mesenchyme · Blood vessel · Osteoblasts

(B)

(C)

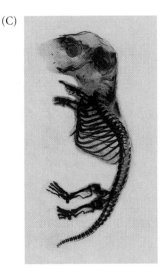

The vertebrate skull (cranium) is composed of the neurocranium (skull vault and base) and the viscerocranium (the jaws and other pharyngeal arch derivatives). While the neural crest origin of the viscerocranium has been well documented, the contributions of cranial neural crest cells to the skull vault have been more controversial. Some researchers have reported that the skull is entirely derived from the neural crest (Couly et al. 1993), whereas others (Le Lièvre 1978; Noden 1978, 1983) have reported that the skull bones are derived from both the neural crest and from the head mesoderm. In 2002, Jiang and colleagues did much to resolve this controversy by using a nonsurgical procedure that specifically marked cranial neural crest cells. They constructed transgenic mice that expressed β-galactosidase only in their cranial neural crest cells.* These cells would turn deep blue when stained for β-galactosidase. When the embryonic mice were stained, the cells

---

*These experiments were done using the Cre-Lox technique (see Chapter 5). The mice were heterozygous for both (1) a β-galactosidase allele that could be expressed only when Cre-recombinase was activated in that cell, and (2) a Cre-recombinase allele fused to a *Wnt1* promoter. Thus, the β-galactosidase gene was activated (blue stain) only in those cells expressing Wnt1—a protein that is activated in the cranial neural crest and in certain brain cells.

forming most of the skull—the nasal, frontal, alisphenoid, and squamosal bones—turned blue; the parietal bone did not (Figure 13.10). It appears, then, that while most of the skull is derived from the cranial neural crest, at least one major bone, the parietal bone, is made from the head mesoderm.

Innervation of the placodes: The second wave of cranial neural crest migration

In addition to the cranial neural crest cells, the anterior borders between the epidermal and neural ectoderm also form the **cranial placodes**, local and transient thickenings of the ectoderm in the head and neck. Cranial placodes are critical components of the sensory nervous system in the head, and they give rise to the sensory apparatus of the nose, ears, and taste receptors, as well as to the lens of the eye (Figure 13.11A,B; see Baker and Bronner-Fraser 2001). The cranial neural crest and cranial placodes may have originated from the same cell population during early vertebrate evolution (Northcutt and Gans 1983; Baker and Bronner-Fraser 1997). The placodes are induced by the neighboring tissue. For instance, the otic placode, which develops into the sensory cells of the inner ear, is induced in the region of ectoderm where the presumptive neural plate meets the presumptive epidermis. Here, FGF19 from the underlying mesoderm is received by both the presumptive otic vesicle and the adjacent neural plate. The neural plate is induced to express and secrete both Wnt8c and more FGF19. The FGF19 and Wnt8c act synergistically to induce the formation of the otic placode (Figure 13.11C; Ladher et al. 2000).

Figure 13.10
Cranial neural crest cells in embryonic mice, stained for β-galactosidase expression. (A–C) Cranial neural crest migration from day 6 to day 9.5, shown by the dark blue staining neural crest cells. (D) Dorsal view of day 17.5 embryonic mice showing lack of staining in the parietal bone. (From Jiang et al. 2002; photographs courtesy of G. Morriss-Kay.)

(A)
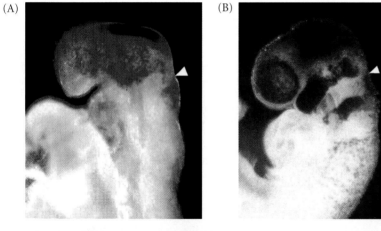

(B)

(C)
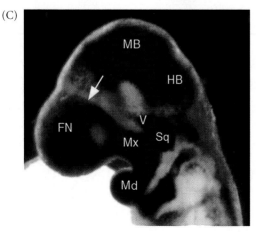

(D)
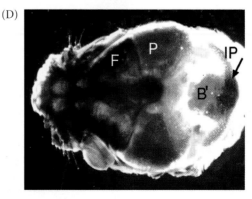

FN, frontal nasal process
MB, midbrain
HB, hindbrain
Mx, maxilla
Md, mandible
Sq, squamosal bone
F, frontal bone
P, parietal
B, mineralized bone (non-neural crest)
IP, interparietal bone
V, trigeminal ganglion

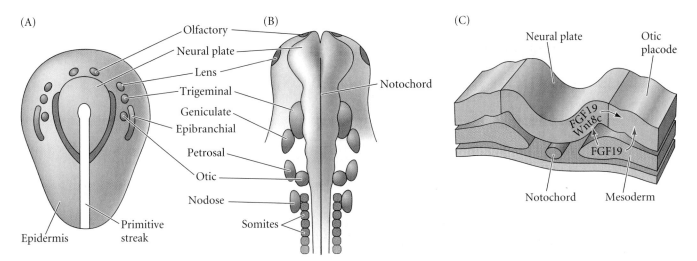

Figure 13.11
Fate map of the cranial placodes in the developing chick embryo. (A) Neural plate stage. (B) 8-somite stage. (C) Induction of the otic (inner ear) placode. The underlying mesoderm secretes FGF19, which is received by both the prospective otic placode and the adjacent neural plate. The neural plate is instructed by the FGF19 to secrete Wnt8c and more FGF19. These two paracrine factors work synergistically to induce *Pax2* and other genes that allow the cells to produce the otic placode and become sensory cells. (After Ladher et al. 2000.)

The epibranchial placodes form dorsally to where the pharyngeal pouches contact the epidermis. These structures split to form the geniculate, petrosal, and nodose placodes which give rise to the sensory neurons of the facial, glossopharyngeal, and vagal nerves, respectively. But how do these neurons, made in these placodes, find their way into the hindbrain? The cranial neural crest cells in the second "wave" of migration do not travel ventrally to enter the pharyngeal arches. Rather, they migrate dorsally to form glial cells (Weston and Butler 1966; Baker et al. 1997). These glia form the tracks that guide neurons from the epibranchial placodes to the hindbrain (Begbie and Graham 2001). The connections made by these neurons enable taste and other pharyngeal sensations to be appreciated. If these glial pathways are not made, the sensory neurons from the placodes will not find their way into the hindbrain. Therefore, the glial cells made by the second wave of cranial neural crest migration are critical in organizing the innervation of the hindbrain.

WEBSITE 13.5 Human facial development syndromes. Several human syndromes cause abnormalities of cranial neural crest cell migration. A region on chromosome 22 appears to be especially important in regulating normal facial and pharyngeal development.

WEBSITE 13.6 Kallmann syndrome. Some infertile men have no sense of smell. The relationship between sense of smell and male fertility was elusive until the gene for Kallmann syndrome was identified. The gene produces a protein that is necessary for the proper migration of both olfactory axons and hormone-secreting neurons from the olfactory placode.

Sidelights & Speculations

Tooth Development

During the morphogenesis of any organ, numerous dialogues take place between the interacting tissues. In epithelial-mesenchymal interactions, the mesenchyme influences the epithelium; the epithelial tissue, once changed by the mesenchyme, can secrete factors that change the mesenchyme. Such interactions continue until an organ forms, with organ-specific mesenchyme cells and organ-specific epithelia. Some of the most extensively studied epithelial-mesenchymal interactions are those that form the mammalian tooth. Here, the neural crest-derived mesenchyme cells become dentin-secreting **odontoblasts**, while the jaw epithelium differentiates into enamel-secreting **ameloblasts**. This process can be seen using the Cre-lox technique for transgenic mice mentioned earlier, wherein cranial neural crest cells stain blue (Figure 13.12. Chai et al. 2001).

Tooth development begins when the mandibular (jaw) epithelium causes neural crest-derived **ectomesenchyme** (i.e., mesenchyme produced from the ectoderm) to aggregate at specific sites. The polarity of the mandibular epithelium is determined by interactions between BMP4, which is located distally, and FGF8, which is located proximally (closest to the skull). Those teeth formed in the FGF8 regions will become molars, while those teeth that devel-

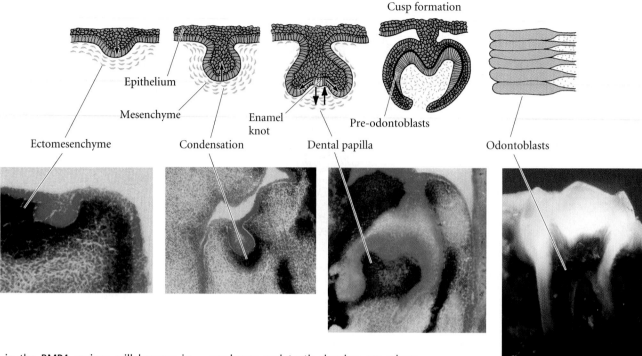

Cusp formation

Epithelium

Mesenchyme

Enamel knot

Pre-odontoblasts

Dental papilla

Ectomesenchyme

Condensation

Odontoblasts

op in the BMP4 regions will become incisors (Tucker et al. 1998). Soon afterward, the expression pattern of BMP4 and FGF8 changes, and the sites of the tooth primordia are determined by interactions between these same molecules in the epithelium. FGF8 induces Pax9 expression in the underlying ectomesenchyme, while BMP4 inhibits Pax9 expression. Pax9 is a transcription factor whose expression in the ectomesenchyme is critical for the initiation of tooth morphogenesis, and in Pax9-deficient mice, tooth development ceases early. The only places where ectomesenchyme

condenses and teeth develop are where FGF8 is present and BMPs are absent (Vainio et al. 1993; Neubüser et al. 1997). Thus, spaces develop between the teeth.

At this time, the epithelium possesses the potential to generate tooth structures out of several types of mesenchyme cells (Mina and Kollar 1987; Lumsden 1988b). However, this tooth-forming potential soon becomes transferred to the ectomesenchyme that has aggregated beneath it. These ectomesenchymal cells form the **dental papilla** and are now able to induce tooth morphogenesis in other epithelia (Kollar and Baird 1970). At

Figure 13.12
Changes in the cranial neural crest cell population as teeth are formed. Originally, the neural crest ectomesenchyme induces and maintains an ectodermal placode. The placode grows and causes the mesenchyme cells to condense beneath it. The mesenchyme cells push into the ectodermal bulge to create the dental papilla. These cells become the odontoblasts of the teeth and become part of the dentin-forming pulp. The figures below show neural crest cells stained blue; the ectodem has been counterstained pink. (After Thesleff and Sahlberg 1996; photographs from Chai et al. 2000, courtesy of Y. Chai.)

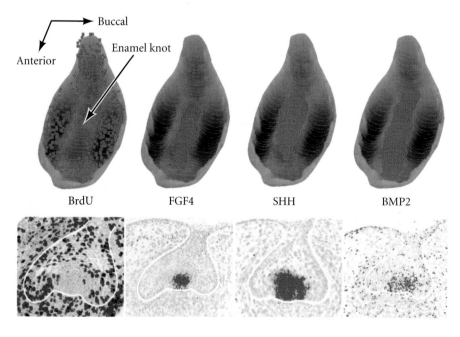

Anterior

Buccal

Enamel knot

BrdU

FGF4

SHH

BMP2

Figure 13.13
Concentration of paracrine growth and differentiation factors in the region where morphogenesis and differentiation are occurring in the 14-day embryonic mouse lower molar. The boundary of the tooth epithelium is shown in white. The paracrine factors are being secreted by the enamel knot, a mass of nondividing epithelial cells. (The concentration of BrdU shows that the cells of the enamel knot are not replicating DNA.) Above each in situ hybridization photograph is a serial reconstruction of the area of gene expression. (From Jernvall 1995; photographs courtesy of A. Vaahtokari, J. Jernvall, and I. Thesleff.)

this stage, the jaw epithelium has lost its ability to instruct tooth formation in other mesenchymes. Thus, the "odontogenic potential" has shifted from the epithelium to the mesenchyme. This shift in the odontogenic potential coincides with a shift in the synthesis of BMP4 from the epithelium to the ectomesenchyme.

As the dental mesenchyme cells condense, they are induced to synthesize the membrane protein syndecan and the extracellular matrix protein tenascin. These proteins (which can bind each other) appear at the time the epithelium induces mesenchymal aggregation, and Thesleff and her colleagues (1990) have proposed that these two molecules may interact to bring about this condensation. Moreover, after the ectomesenchyme has aggregated, it begins to secrete BMP4 as well as other growth and differentiation factors (FGF3, BMP3, HGF, and activin) (Wilkinson et al. 1989; Thesleff and Sahlberg 1996). These proteins from the ectomesenchyme induce a critical structure in the epithelium. This structure is called the **enamel knot**, and it functions as the major signaling center for tooth development (Jernvall et al. 1994). This group of cells appears as a nondividing population in the center of the growing cusps. In situ hybridization has demonstrated that the enamel knot is a source of Sonic hedgehog, FGF4, BMP7, BMP4, and BMP2 secretion (Figure 13.13; Koyama et al. 1996; Vaahtokari et al. 1996a). As a nondividing cell population secreting growth factors capable of being received by both the epithelium and the ectomesenchyme, the enamel knot is thought to direct the cusp morphogenesis of the tooth and to be critical in directing the evolutionary changes of tooth structure in mammals (Jernvall 1995).

As the dental mesenchyme cells begin to differentiate into odontoblasts, tenascin expression is induced at much higher levels and at the same sites as alkaline phosphatase expression. Both of these proteins have been associated with bone and cartilage differentiation, and they may promote the mineralization of the extracellular matrix (Mackie et al. 1987). Finally, as the odontoblast phenotype emerges, osteonectin and type I collagen are secreted as components of the extracellular matrix. The enamel knot disappears through apoptosis, responding to its own BMP4 (Vaahtokari et al. 1996b; Jernvall et al. 1998). By this step-like process, the cranial neural crest cells of the jaw are transformed into the dentin-secreting odontoblasts.

The Cardiac Neural Crest

The heart originally forms in the neck region, directly beneath the pharyngeal arches, so it should not be surprising that it acquires cells from the neural crest. However, the contributions of the neural crest to the heart have only recently been appreciated. The caudal region of the cranial neural crest is sometimes called the cardiac neural crest, since its cells (and only these particular neural crest cells) generate the endothelium of the aortic arch arteries and the septum between the aorta and the pulmonary artery (Figure 13.14; Kirby 1989; Kuratani and Kirby 1991; Waldo et al. 1998).

In mice, cardiac neural crest cells are peculiar in that they express the transcription factor Pax3. Mutations of *Pax3* result in fewer cardiac neural crest cells. This condition results in persistent truncus arteriosus (the failure of the aorta and pulmonary artery to separate), as well as defects in the thymus, thyroid, and parathyroid glands (Conway et al. 1997, 2000). Congenital heart defects in humans and mice often occur with defects in the parathyroid, thyroid, or thymus glands. It

Figure 13.14
The truncoconal septa that separate the truncus arteriosus into the pulmonary artery and aorta form from the cells of the cardiac neural crest. (A) Human cardiac neural crest cells migrate to pharyngeal arches 4 and 6 during the fifth week of gestation and enter the truncus arteriosus to generate the septa. (B) Quail cardiac crest cells were transplanted into the analogous region of a chick embryo, and the embryos were allowed to develop. The quail cardiac neural crest cells can be recognized by a quail-specific antibody, which stains them darkly. In the heart, these cells can be seen separating the truncus arteriosus (right) into the pulmonary artery and the aorta (left). (A after Kirby and Waldo 1990; B from Waldo et al. 1998, photographs courtesy of K. Waldo and M. L. Kirby.)

(A)

Migrating neural crest cells

Neural tube

Truncus arteriosus

Bulbus cordis

Descending aorta

Vitelline artery

Umbilical artery

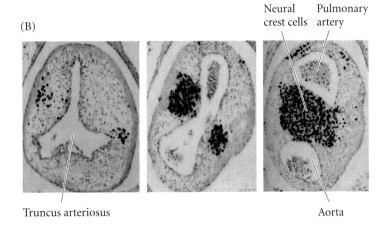

(B)

Neural crest cells   Pulmonary artery

Truncus arteriosus   Aorta

would not be surprising if all of these problems were linked to defects in the migration of cells from the neural crest.

> WEBSITE 13.7 **Communication between migrating neural crest cells.** Recent research has shown that neural crest cells might cooperate with one another as they migrate. There may be subtle communication between these cells through their gap junctional complexes, and this communication may be important for heart development.

NEURONAL SPECIFICATION AND AXONAL SPECIFICITY

Not only do neuronal precursor cells and neural crest cells migrate to their place of function, but so do the axons extending from the cell bodies of neurons. Unlike most cells, whose parts all stay in the same place, the neuron is able to produce axons that may extend for meters. As we saw in Chapter 12, the axon has its own locomotory apparatus, which resides in the growth cone. The growth cone has been called "a neural crest cell on a leash" because, like neural crest cells, it migrates and senses the environment. Moreover, it can respond to the same types of signals that migrating cells can sense. The cues for axonal migration, moreover, may be even more specific than those used to guide certain cell types to particular areas. Each of the 10^{11} neurons in the human brain has the potential to interact specifically with thousands of other neurons, and a large neuron (such as a Purkinje cell or motor neuron) can receive input from more than 10^5 other neurons (Figure 13.15; Gershon et al. 1985). Understanding the generation of this stunningly ordered complexity is one of the greatest challenges to modern science.

Goodman and Doe (1993) list eight stages of neurogenesis:

1. The induction and patterning of a neuron-forming (neurogenic) region
2. The birth and migration of neurons and glia
3. The specification of cell fates
4. The guidance of axonal growth cones to specific targets
5. The formation of synaptic connections
6. The binding of trophic factors for survival and differentiation
7. The competitive rearrangement of functional synapses
8. Continued synaptic plasticity during the organism's lifetime

The first two of these processes were the topics of the previous chapter. Here, we continue our investigation of the processes of neural development.

> WEBSITE 13.8 **The evolution of developmental neurobiology.** Santiago Ramón y Cajal, Viktor Hamburger, and Rita Levi-Montalcini helped bring order to the study of neural development by identifying some of the important questions that still occupy us today.

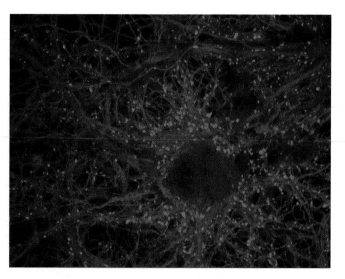

Figure 13.15
Connections of axons to a cultured rat hippocampal neuron. The neuron has been stained red with fluorescent antibodies to tubulin. The neuron appears to be outlined by the synaptic protein synapsin (stained green), which is present in the terminals of axons that contact it. (Photograph courtesy of R. Fitzsimmons and PerkinElmer Life Sciences.)

The Generation of Neuronal Diversity

Neurons are specified in a hierarchical manner. The first decision is whether a given cell is to become a neuron or epidermis. If the cell is to become a neuron, the next decision is what type of neuron it will be: whether it is to become a motor neuron, a sensory neuron, a commissural neuron, or some other type. After this fate is determined, still another decision gives the neuron a specific target. To illustrate this process of progressive specification, we will focus on the motor neurons of vertebrates.

Vertebrates form a dorsal neural tube by blocking a BMP signal, and the specification of neural (as opposed to glial or epidermal) fate is accomplished through the Notch-Delta pathway (see Chapter 6). The specification of the *type* of neuron appears to be controlled by the position of the neuronal precursor within the neural tube and by its birthday. As described in Chapter 12, neurons at the ventrolateral margin of the vertebrate neural tube become the motor neurons, while interneurons are derived from cells in the dorsal region of the tube. Since the grafting of floor plate or notochord cells (which secrete Sonic hedgehog protein) to lateral areas of the neural tube can re-specify dorsolateral cells as motor neurons, the specification of neuron type is probably a function of the cell's position relative to the floor plate. Ericson and colleagues (1996) have shown that two periods of Sonic hedgehog (Shh) signaling are needed to specify the motor neurons: an early period wherein the cells of the ventrolateral margin are instructed to become ventral neurons, and a later period (which includes the S phase of its last cell division) that instructs a ventral neuron to become a motor neuron (rather than an interneuron). The first decision is probably regulated by the secretion of Shh from the notochord, while the second is more likely regulated by Shh from the floor plate cells. Sonic hedgehog appears to specify motor neurons by inducing certain transcription factors at different concentrations (Ericson et al. 1992; Tanabe et al. 1998; see Figure12.13).

The next decision involves target specificity. If a cell is to become a neuron and, specifically, a motor neuron, will that motor neuron be one that innervates the thigh, the forelimb, or the tongue? The anterior-posterior specification of the neural tube is regulated primarily by the Hox genes from the hindbrain through the spinal cord and by specific head genes (such as *Otx*) in the brain. Within a region of the body, motor neuron specificity is regulated by the cell's age when it last divides. As discussed in Chapter 12, a neuron's birthday determines which layer of the cortex it will enter. As the younger cells migrate to the periphery, they must pass through neurons that differentiated earlier in development. As younger motor neurons migrate through the region of older motor neurons in the intermediate zone, they express new transcription factors as a result of a retinoic acid (or other retinoid) signal secreted by the early-born motor neurons (Sockanathan and Jessell 1998). These transcription factors are encoded by the **Lim genes** and are structurally related to those encoded by the Hox genes.

As a result of their differing birthdays and migration patterns, motor neurons form three major groupings. The cell bodies of the motor neurons projecting to a single muscle are clustered in a longitudinal column of the spinal cord called a "pool." This clustering is performed by different cadherins that become expressed on these different populations of cells (Landmesser 1978; Hollyday 1980; Price et al. 2002). The pools are grouped together into three larger columns according to their targets. In the thorax (Figure 13.16, top) motor neurons in the column of Terni (CT) project ventrally into the sympathetic ganglia. Motor neurons in the lateral motor column (LMC) project into the muscles. In the limb areas (Figure 13.16, bottom), the limb muscles are innervated by the LMC axons, with lateral neurons entering the dorsal musculature, while the more medial neurons innervate ventral limb musculature (Tosney et al. 1995). This arrangement of motor neurons is constant throughout the vertebrates.

The targets of these motor neurons are specified before their axons extend into the periphery. This was shown by Lance-Jones and Landmesser (1980), who reversed segments of the chick spinal cord so that the motor neurons were placed in new locations. The axons went to their original targets, not to the ones expected from their new positions (Figure 13.17). The molecular basis for this target specificity resides in the members of the LIM family of proteins that are induced during neuronal migration (see Figure 13.16; Tsushida et al. 1994;

Sharma et al. 2000). For instance, all motor neurons express Islet-1 and (slightly later) Islet-2. If no other LIM protein is expressed, the neurons project to the ventral body wall muscles. However, if Lim-3 is also synthesized, the motor neurons project dorsally to the axial muscles. If each of the motor neuron types is made to express Lim-3, they will all innervate the axial muscles (or at least attempt to; those axons that arrive late and find the sites filled will project to alternative targets.) Thus, each group of neurons is characterized by a particular constellation of LIM transcription factors.

WEBSITE 13.9 Horseradish peroxidase staining. Many of the fundamental discoveries of axon specificity used a technique wherein the plant enzyme horseradish peroxidase was injected into nerves.

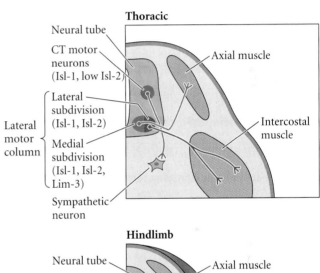

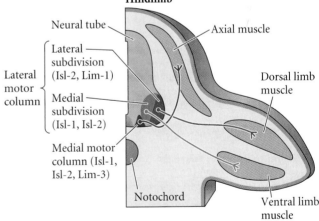

Figure 13.16
Motor neuron organization and LIM specification in the spinal cord. Neurons in the different columns express specific sets of LIM genes, and neurons within each column make similar pathfinding decisions. Motor neurons in the column of Terni (CT) project ventrally to the sympathetic ganglia. Neurons of the medial motor column project to the axial muscles, and neurons of the lateral motor column send axons to the limb musculature. Where these columns are subdivided, medial subdivisions project to ventral positions and lateral subdivisions send axons to dorsal regions of the target tissues. (After Tsushida et al. 1994; Tosney et al. 1995.)

Pattern Generation in the Nervous System

The functioning of the vertebrate brain depends not only on the differentiation and positioning of the neurons, but also on the specific connections these cells make among themselves and their peripheral targets. In some manner, nerves from a sensory organ such as the eye must connect to specific neurons in the brain that can interpret visual stimuli, and axons from the central nervous system must cross large expanses of tissue before innervating their target tissue. How does the neuronal axon "know" how to traverse numerous potential target cells to make its specific connection?

Ross G. Harrison (see Figure 4.3) first suggested that the specificity of axonal growth is due to pioneer nerve fibers, which go ahead of other axons and serve as guides for them* (Harrison 1910). This observation simplified, but did not solve, the problem of how neurons form appropriate patterns of interconnection. Harrison also noted, however, that axons must grow on a solid substrate, and he speculated that differences among embryonic surfaces might allow axons to travel in certain specified directions. The final connections would occur by complementary interactions on the target cell surface:

> *That it must be a sort of a surface reaction between each kind of nerve fiber and the particular structure to be in-*

---

*The growth cones of pioneer neurons migrate to their target tissue while embryonic distances are still short and the intervening embryonic tissue is still relatively uncomplicated. Later in development, other neurons bind to pioneer neurons and thereby enter the target tissue. Klose and Bentley (1989) have shown that in some cases, the pioneer neurons die after the other neurons reach their destination. Yet if the pioneer neurons are prevented from differentiating, the other axons do not reach their target tissue.

nervated seems clear from the fact that sensory and motor fibers, though running close together in the same bundle, nevertheless form proper peripheral connections, the one with the epidermis and the other with the muscle.... The foregoing facts suggest that there may be a certain analogy here with the union of egg and sperm cell.

Research on the specificity of neuronal connections has focused on two major systems: motor neurons, whose axons travel from the spinal cord to a specific muscle, and the optic system, wherein axons originating in the retina find their way back into the brain. In both cases, the specificity of axonal connections is seen to unfold in three steps (Goodman and Shatz 1993):

1. **Pathway selection**, wherein the axons travel along a route that leads them to a particular region of the embryo

2. **Target selection**, wherein the axons, once they reach the correct area, recognize and bind to a set of cells with which they may form stable connections

3. **Address selection**, wherein the initial patterns are refined such that each axon binds to a small subset (sometimes only one) of its possible targets

The first two processes are independent of neuronal activity. The third process involves interactions between several active neurons and converts the overlapping projections into a fine-tuned pattern of connections.

It has been known since the 1930s that the axons of motor neurons can find their appropriate muscles even if the neural activity of the axons is blocked. Twitty (who had been Harrison's student) and his colleagues found that embryos of the newt *Taricha torosa* secreted a toxin, tetrodotoxin, that

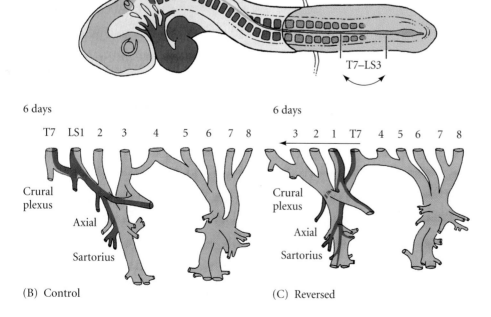

(A) 2–5 days

6 days

T7 LS1 2 3 4 5 6 7 8

Crural plexus

Axial

Sartorius

(B) Control

6 days

3 2 1 T7 4 5 6 7 8

Crural plexus

Axial

Sartorius

(C) Reversed

T7–LS3

Figure 13.17
Compensation for small dislocations of axonal initiation position in the chick embryo. (A) A length of spinal cord comprising segments T7–LS3 (seventh thoracic to third lumbosacral segments) is reversed in a 2.5-day embryo. (B) Normal pattern of axon projection to the muscles at 6 days. (C) Projection of axons from the reversed segment. The ectopically placed neurons eventually found their proper neural pathways and innervated the appropriate muscles. (From Lance-Jones and Landmesser 1980.)

blocked neural transmission in other species. By grafting pieces of *T. torosa* embryos onto embryos of other salamander species, they were able to paralyze the host embryos for days while development occurred. Normal neuronal connections were made, even though no neuronal activity could occur. At about the time the tadpoles were ready to feed, the toxin wore off, and the young salamanders swam and fed normally (Twitty and Johnson 1934; Twitty 1937). More recent experiments using zebrafish mutants with nonfunctional neurotransmitter receptors similarly demonstrated that motor neurons can establish their normal patterns of innervation in the absence of neuronal activity (Westerfield et al. 1990).

But the question remains: How are the axons instructed where to go?

Cell adhesion and contact guidance by attractive and permissive molecules

The initial pathway an axonal growth cone follows is determined by the environment the growth cone experiences. The extracellular environment can provide substrates upon which to migrate, and these substrates can provide navigational information to the growth cone. Some of the substrates the growth cone encounters will permit it to adhere to them, and thus promote axon migration. Other substrates will cause the growth cone to retract, and will not allow its axon to grow in that direction. Some of these substrates can give extremely specific cues, recognized by only a small set of neuronal growth cones, while other cues are recognized by large sets of neurons. The migrational cues of the substrate can come from anatomical structures, the extracellular matrix, or from adjacent cell surfaces.

Extracellular matrix cues often offer preferred adhesion substrates for migration. Growth cones prefer to migrate on surfaces that are more adhesive than their surroundings, and a track of adhesive molecules can direct them to their targets. The presence of adhesive extracellular matrix molecules delineates microscopic roads through the embryo on which axons travel (Akers et al. 1981; Gundersen 1987). Many of these roads appear to be paved with the glycoprotein **laminin** (Figure 13.18). Letourneau and co-workers (1988) have shown that the axons of certain spinal neurons travel through the neuroepithelium over a transient laminin-coated surface that precisely marks their path. Similarly, there is a very good correlation between the elongation of retinal axons and the presence of laminin on the neuroepithelial and glial cells in the embryonic mouse brain (Cohen et al. 1987; Liesi and Silver 1988). Punctate laminin deposits are seen on the glial cell surfaces along the pathway leading from the retinal ganglion cells to the optic tectum in the brain, whereas adjacent areas (where the retinal axons fail to grow) lack these laminin deposits. After the retinal axons have reached the tectum, the glial cells differentiate and lose their laminin. At this point, the retinal ganglion neurons that have formed the optic nerve lose their integrin receptor for laminin.

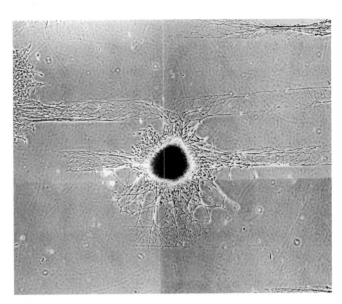

Figure 13.18
Outgrowth of sensory neurons placed on a patterned substrate consisting of parallel stripes of laminin applied to a background of type IV collagen. (From Gundersen 1987; photograph courtesy of R. W. Gundersen.)

Guidance by specific growth cone repulsion

In addition to specific adhesion, there is also the possibility of specific repulsion by extracellular substrates. Just as neural crest cells are inhibited from migrating across the posterior portion of a sclerotome, axons from the dorsal root ganglia and motor neurons pass only through the anterior portion of each sclerotome and avoid migrating through the posterior portion (Figure 13.19; also see Figure 13.5). Davies and colleagues (1990) showed that membranes isolated from the posterior portion of a somite cause the collapse of growth cones of these neurons (Figure 13.19B,C). Recent studies (Wang and Anderson 1997; Krull et al. 1999) have shown that these growth cones contain Eph receptors that are responsive to the ephrin proteins in the posterior sclerotome cells. Thus the same signals that pattern neural crest cell migration also pattern the spinal neuronal outgrowths.

In addition to the ephrins, the **semaphorin proteins** also guide growth cones by selective repulsion. These membrane proteins are seen throughout the animal kingdom and are responsible for steering many axons to their targets. They are especially important in making "turns" when an axon does not grow in a straight line, but must change direction. Semaphorin 1 is a transmembrane protein that is expressed in a band of epithelial cells in the developing insect limb. This protein appears to inhibit the growth cones of the Ti1 sensory neurons from moving forward, causing them to turn (Figure 13.20; Kolodkin et al. 1992, 1993). In *Drosophila*, semaphorin 2 is secreted by a single large thoracic muscle. In this way, the

(A)

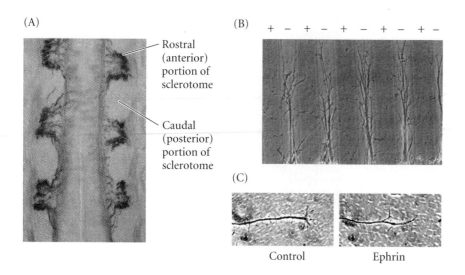

Rostral (anterior) portion of sclerotome

Caudal (posterior) portion of sclerotome

(B)

+ − + − + − + − + −

(C)

Control   Ephrin

Figure 13.19
Repulsion of dorsal root ganglion growth cones. (A) Motor axons migrating through the rostral (anterior), but not the caudal (posterior), compartments of each sclerotome. (B) In vitro assay, wherein ephrin stripes were placed on a background surface of laminin. Motor axons grew only where the ephrin was absent. (C) Inhibition of growth cones by ephrin after 10 minutes of incubation. The left-hand photograph shows a control axon subjected to a similar (but not inhibitory) compound; the right axon was exposed to an ephrin found in the posterior sclerotome. (From Wang and Anderson 1997; photographs courtesy of the authors.)

thoracic muscle prevents itself from being innervated by inappropriate axons (Matthes et al. 1995). Semaphorin 3, found in mammals and birds, is also known as collapsin (Luo et al. 1993). This secreted protein was found to collapse the growth cones of axons originating in the dorsal root ganglia. There are several types of neurons in the dorsal root ganglia whose axons enter the dorsal spinal cord. Most of these axons are prevented from traveling farther and entering the ventral spinal cord. However, a subset of these axons do travel ventrally through the other neural cells (Figure 13.21). These particular axons are not inhibited by semaphorin 3, while those of the other neurons are (Messersmith et al. 1995). This finding suggests that semaphorin/collapsin patterns sensory projections from the dorsal root ganglia by selectively repelling certain axons so that they terminate dorsally. A similar scheme is used in the brain, wherein semaphorin made in one part of the brain is used to prevent the entry of neurons coming from another region of the brain (Marín et al. 2001).

As any psychologist knows, there is often a thin line between attraction and repulsion. At the base of both phenomena is some sort of recognition event. This is also the case with neurons. When the receptor protein ephrin A7 recognizes ephrin A5 in the mouse neural tube, the result is attraction rather than repulsion. The interaction between these two proteins is critical for the closure of the neural tube. The EphA7 and EphA5 proteins cause *adhesion* of the neural plate cells, and deletion of either one results in a condition resembling anencephaly. The change from repulsion to adhesion is caused by alternative RNA processing. By using a different splice site, the mouse neural plate produces EphA7 that lacks the tyrosine kinase domain that transmits the repulsive signal (Holmberg et al. 2000). The result is that the cells recognize each other through these proteins, and no repulsion occurs. In a different way, semaphorins can also be chemoattractants. Semaphorin 3A is a classic chemorepellant for the axons coming from pyramidal neurons in the mammalian cortex. However, it is a chemoattractant for the *dendrites* of the same cells. In this way,

a target can "reach out" to the dendrites of these cells without attracting their axons as well (Polleux et al. 2000).

> **WEBSITE 13.10 The pathways of motor neurons.** To innervate the limb musculature, a motor axon extends over hundreds of cells in a complex and changing environment. Recent research has discovered several paths and several barriers that help guide these axons to their appropriate destinations.

Guidance by diffusible molecules

NETRINS AND THEIR RECEPTORS. The idea that chemotactic cues guide axons in the developing nervous system was first

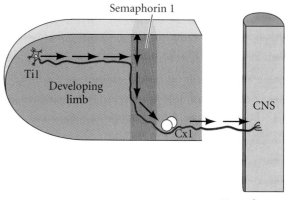

Semaphorin 1

Ti1

Developing limb

CNS

Cx1

Ventral nerve cord

Figure 13.20
The action of semaphorin 1 in the developing grasshopper limb. The axon of sensory neuron Ti1 projects toward the central nervous system. (The dark arrows represent sequential steps en route.) When it reaches a band of semaphorin 1-expressing epithelial cells, it reorients its growth cone and extends ventrally along the distal boundary of the semaphorin 1-expressing cells. When its filopodia connect to the Cx1 pair of cells, it crosses the boundary and projects into the CNS. When semaphorin 1 is blocked by antibodies, the growth cone searches randomly for the Cx1 cells. (After Kolodkin et al. 1993.)

(A)

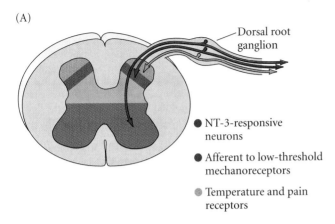

Dorsal root ganglion

● NT-3-responsive neurons

● Afferent to low-threshold mechanoreceptors

● Temperature and pain receptors

Figure 13.21
Semaphorin 3 as a selective inhibitor of axonal projections into the ventral spinal cord. (A) Trajectory of axons in relation to semaphorin 3 expression in the spinal cord of a 14-day embryonic rat. Neurotrophin 3-responsive neurons can travel to the ventral region of the spinal cord, but the afferent axons for the mechanoreceptors and for temperature and pain receptor neurons terminate dorsally. (B) Transgenic chick fibroblast cells that secrete semaphorin 3 inhibit the outgrowth of mechanoreceptor axons. These axons are growing in medium treated with NGF, which stimulates their growth, but are still inhibited from growing toward the source of semaphorin 3. (C) Neurons that are responsive to NT-3 for growth are not inhibited from extending toward the source of semaphorin 3 when grown with NT3. (A after Marx 1995; B and C from Messersmith et al. 1995, photographs courtesy of A. Kolodkin.)

(B)

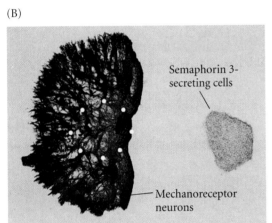

Semaphorin 3-secreting cells

Mechanoreceptor neurons

(C)

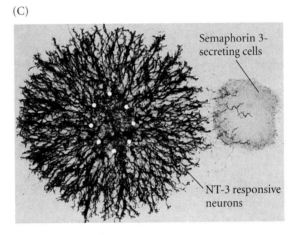

Semaphorin 3-secreting cells

NT-3 responsive neurons

proposed by Santiago Ramón y Cajal (1892). He suggested that the commissural neurons of the spinal cord might be told by diffusible factors to send axons from their dorsal positions to the ventral floor plate. Commissural neurons are interneurons that cross the ventral midline to coordinate right and left motor activities. Thus, they somehow must migrate to (and through) the ventral midline. The axons of these neurons begin growing ventrally down the side of the neural tube. However, about two-thirds of the way down, their direction changes, and they project through the ventrolateral (motor) neuron area of the neural tube toward the floor plate cells (Figure 13.22).

In 1994, Serafini and colleagues developed an assay that would allow them to screen for a diffusible molecule that might be guiding the commisurall neurons. When dorsal spinal cord explants from chick embryos were plated onto collagen gels, the presence of floor plate cells near them promoted the outgrowth of commissural axons. Serafini and his co-workers took fractions of embryonic chick brain homogenate and tested them to see if any of the proteins therein mimicked this activity. This process resulted in the identification of two proteins, **netrin-1** and **netrin-2**. Netrin-1 is made by and secreted from the floor plate cells, whereas netrin-2 is synthesized in the lower region of the spinal cord, but not in

the floor plate (Figure 13.22B). The chemotactic effects of these netrins were demonstrated by transforming chick fibroblast cells (which usually do not make these proteins) into netrin-producing cells using a vector containing an active netrin gene (Kennedy et al. 1994). Aggregates of these netrin-secreting fibroblast cells elicited commissural axon outgrowth from rat dorsal spinal cord explants, while control cells that were given the vector without the active netrin gene did not elicit such activity (Figure 13.23). It is possible that the commissural neurons first encounter a gradient of netrin-2, which brings them into the domain of the steeper netrin-1 gradient (see Figure 13.22A).

In these experiments, both netrins became associated with the extracellular matrix.* This association can play important roles. In vertebrates, netrin-1 may serve as both an attractive

---

*The binding of a soluble factor to the extracellular matrix makes for an interesting ambiguity between chemotaxis and migrating on preferred substrates (haptotaxis). Nature doesn't necessarily conform to the categories we create. There is also some confusion between the terms *neurotropic* and *neurotrophic*. Neuro*tropic* (Latin, *tropicus*, a turning movement) means that something attracts the neuron. Neuro*trophic* (Greek, *trophikos*, nursing) refers to a factor's ability to keep the neuron alive, usually by supplying growth factors. Since many agents have both properties, they are alternatively called "neurotropins" and "neurotrophins." In the recent literature, "neurotrophin" appears to be more widely used.

(A)

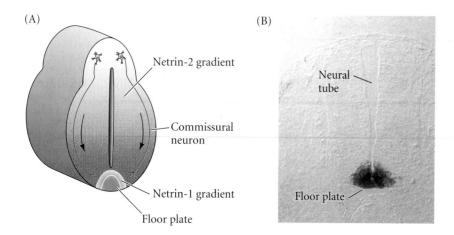

Netrin-2 gradient

Commissural neuron

Netrin-1 gradient

Floor plate

(B)

Neural tube

Floor plate

Figure 13.22

Trajectory of the commissural axons in the rat spinal cord. (A) Schematic drawing of a model wherein commissural neurons first experience a gradient of netrin-2 and then a steeper gradient of netrin-1. The commissural axons are thus chemotactically guided ventrally down the lateral margin of the spinal cord toward the floor plate. Upon reaching the floor plate, they change direction owing to contact guidance from the floor plate cells. (B) Autoradiographic localization of netrin-1 mRNA by in situ hybridization of antisense RNA to the hindbrain of a young rat embryo. The netrin-1 mRNA (dark area) is concentrated in the floor plate neurons. (B from Kennedy et al. 1994; photograph courtesy of M. Tessier-Lavigne.)

and a repulsive signal. The growth cones of *Xenopus* retinal neurons, for example, are attracted to netrin-1 and are guided to the head of the optic nerve by this diffusible factor. Once there, however, the combination of netrin-1 and laminin prevents the axons from departing from the optic nerve. It appears that the laminin of the extracellular matrix surrounding the optic nerve converts the netrin from being an attractive molecule to being a repulsive one (Höpker et al. 1999).

The netrins have numerous regions of homology with UNC-6, a protein implicated in directing the circumferential migration of axons around the body of *Caenorhabditis elegans*. In the wild-type nematode, UNC-6 induces axons from certain centrally located sensory neurons to move ventrally, and it induces some ventrally placed motor neurons to extend axons dorsally (Figure 13.24). In loss-of-function mutations of *unc-6*, neither of these axonal movements occurs (Hedgecock et al. 1990; Ishii et al. 1992; Hamelin et al. 1993). Mutations of the *unc-40* gene disrupt ventral (but not the dorsal) axonal migration, while mutations of the *unc-5* gene prevent

only the dorsal migration. Genetic and biochemical evidence suggests that UNC-5 and UNC-40 are portions of the UNC-6 receptor complex, and that UNC-5 can convert an attraction (mediated by UNC-40 alone) into a repulsion (Leonardo et al. 1997; Hong et al. 1999).

If UNC-6 is attractive to some neurons and repulsive to others, one might expect that this dual role might also be as-

Figure 13.23

Transformed chick fibroblast (COS) cells secreting netrins elicit axon outgrowth of commissural neurons from 11-day embryonic rat dorsal spinal cord explants. (A) Outgrowth of commissural axons is seen when a dorsal spinal cord explant (upper tissue) encounters a floor plate explant. (B) No outgrowth is seen when a dorsal spinal cord explant is exposed to aggregated chick fibroblast cells that have been transfected with the cloning vector alone (no netrin gene). (C, D) Commissural axon outgrowth in response to aggregated chick fibroblast cells that express the gene for (C) netrin-1 and (D) netrin-2. The identity of the outgrowths was confirmed by immunohistology, which showed commissural-specific antigens on the axons. Scale bar, 100 μm. (From Kennedy et al. 1994; photographs courtesy of M. Tessier-Lavigne.)

(A)

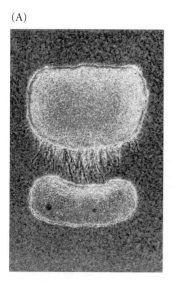

(B)

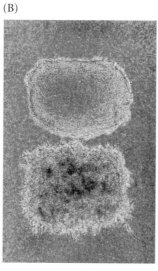

(C)

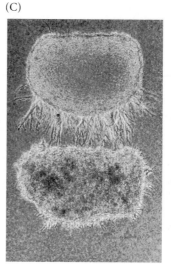

(D)

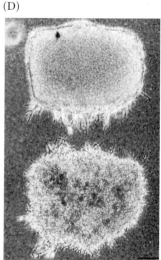

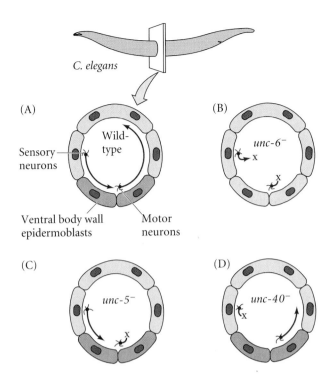

(A) Wild-type

Sensory neurons

Ventral body wall epidermoblasts

Motor neurons

(B) *unc-6⁻*

(C) *unc-5⁻*

(D) *unc-40⁻*

C. elegans

Figure 13.24

UNC expression and function in axonal guidance. (A) In the body of the wild-type *C. elegans* embryo, sensory neurons project ventrally and motor neurons project dorsally. The ventral body wall epidermoblasts expressing UNC-6 are darkly shaded. (B) In the *unc-6* mutant embryo, neither of these migrations occurs. (C) The *unc-5* loss-of-function mutation affects only the dorsal movements of the motor neurons. (D) The *unc-40* loss-of-function mutation affects only the ventral migration of the sensory growth cones. (After Goodman 1994.)

disease called rostral cerebellar malformation (Ackerman et al. 1997; Leonardo et al. 1997).

> **WEBSITE 13.11 Genetic control of neuroblast migration in *C. elegans*.** The homeotic gene *mab-5* controls the direction in which certain neurons migrate in the nematode. The expression of this gene can alter which way a neuron travels.

SLIT PROTEINS. Diffusible proteins can also provide guidance by repulsion. One important chemorepulsive molecule is the **Slit** protein. In *Drosophila*, the Slit protein is secreted by the neural cells in the midline, and it acts to prevent most neurons from crossing the midline from either side. The growth cones of *Drosophila* neurons express the **Roundabout (Robo)** protein, which is the receptor for the Slit protein. In this way, most *Drosophila* neurons are prevented from migrating across the midline. However, the commissural neurons that traverse the embryo from side to side find a way to avoid this repulsion. They accomplish this task by downregulating the Robo protein as they approach the midline. Once they have traveled across the middle of the embryo, they re-express this protein

cribed to netrins. Colamarino and Tessier-Lavigne (1995) have shown this to be the case by looking at the trajectory of the trochlear (fourth cranial) nerve. On their way to innervate an eye muscle, the axons of the trochlear nerve originate near the floor plate of the brain stem and migrate dorsally away from the floor plate region. This pathway is maintained when the brain stem region is explanted into collagen gel. The dorsal outgrowth of the trochlear neurons can be prevented by placing floor plate cells or transgenic, netrin-1-secreting chick fibroblast cells within 450 μm of the dorsal portion of the explant. This dorsal outgrowth is not prevented by dorsal explants of the neural tube or by chick fibroblast cells that do not contain the active netrin-1 gene (Figure 13.25). Therefore, netrins and UNC-6 appear to be chemotactic to some neurons and chemorepulsive to others.

There is some reciprocity in science, and just as research on vertebrate netrin genes led to the discovery of their *C. elegans* homologues, research on the nematode *unc-5* gene led to the discovery of the gene encoding the human netrin receptor. This turns out to be a gene whose mutation in mice causes a

Figure 13.25

Netrins inhibit the outgrowth of trochlear axons from explants of the mouse brain stem (the lower tissues). Trochlear axons, stained for trochlear axon-specific antigen, emerge dorsally and are not inhibited by (A) a dorsal spinal cord explant or by (B) unaltered chick fibroblast cells. They are inhibited by (C) transgenic fibroblast cells secreting netrin-1 and by (D) an explant from the netrin-secreting ventral floor plate of the spinal cord. (After Colamarino and Tessier-Lavigne 1995; photographs courtesy of M. Tessier-Lavigne.)

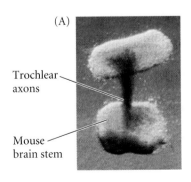

(A)

Trochlear axons

Mouse brain stem

(B)

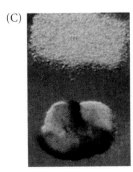

(C)

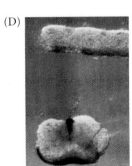

(D)

and become sensitive again to the midline inhibitory actions of Slit (Figure 13.26; Brose et al. 1999; Kidd et al. 1999).

Slit also functions in the vertebrate nervous system to guide neuronal growth cones. There are three Slit proteins in vertebrates, and they are used to turn away motor neurons and olfactory (smell) neurons (Brose et al. 1999; Li et al. 1999). This repulsive activity may be extremely important for the guidance of the olfactory neurons, since they are guided to specific regions of the telencephalon by repulsive cues emanating from the midline of the forebrain (Pini 1993).

WEBSITE 13.12 The early evidence for chemotaxis. Before molecular techniques, investigators using transplantation experiments and ingenuity found evidence that chemotactic molecules were being released by target tissues.

Target selection

Once a neuron reaches a group of cells wherein lie its potential targets, it is responsive to various proteins that are produced by the target cells. In addition to the proteins already mentioned, some target cells produce a set of chemotactic factors collectively called **neurotrophins**. These proteins include nerve growth factor (NGF), brain-derived neurotrophic factor (BDNF), neurotrophin-3 (NT-3), and NT-4/5. These proteins are released from potential target tissues and work at short ranges as either chemotactic factors or chemorepulsive factors (Paves and Saarma 1997). Each neurotrophin can promote the growth of some axons to its source while inhibiting other axons. For instance, sensory neurons from the rat dorsal root ganglia are attracted to sources of NT-3, but are inhibited from growing by BDNF (Figure 13.27).

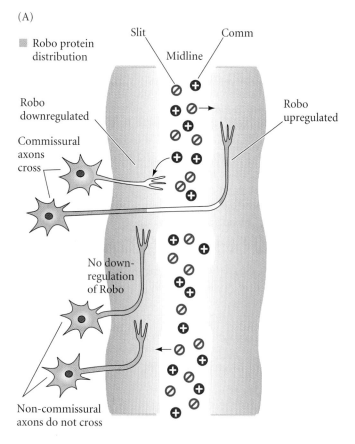

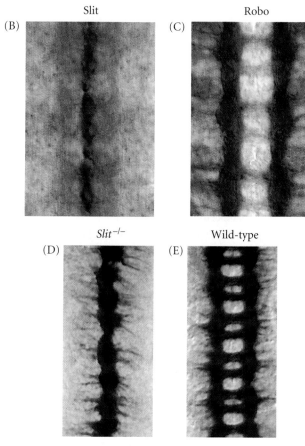

Figure 13.26
Chemotactic factors direct commissural axons to cross the midline of the *Drosophila* embryo while keeping other axons on one side of the midline. (A) The midline secretes Comm protein, which stimulates commissural axons, and Slit protein, which inhibits all axons that express Robo protein (the Slit receptor). When they reach the midline, commissural axons have little or no Robo protein. Thus, stimulated by Comm, these axons can cross the midline. Once across the midline, they re-express Robo and therefore cannot return. Non-commissural neurons express Robo continuously and therefore are inhibited from crossing the midline. (B) Expression of the Slit protein (as shown by antibody staining) in the midline neurons of *Drosophila*. (C) Expression of the Robo protein (as shown by antibody staining) along the neurons of the longitudinal tracts of the CNS axon scaffold. (D) Staining of the CNS axon scaffold with antibodies to all CNS neurons in a Slit loss-of-function mutant shows axons entering but failing to leave the midline (instead of running alongside it). (E) Wild-type CNS axon scaffold shows the ladder-like arrangement of neurons crossing the midline. (A after Thomas 1998; B–E from Kidd et al. 1999, photographs courtesy of C. S. Goodman.)

Although genetic and biochemical techniques enable us to look at the effect of one type of molecule at a time, we must remember that any growth cone is sensing a wide range of chemotactic and chemorepulsive molecules, both in solution and on the substratum upon which it migrates. Growth cones do not rely on a single type of molecule to recognize their target. Rather, they integrate the simultaneously presented attractive and repulsive cues and select their targets based on the combined input of these multiple cues (Winberg et al. 1998).

WEBSITE 13.13   The neurotrophin receptors. Neurotrophins can bind to high-affinity receptors or to low-affinity receptors, and the pattern of binding can determine whether the signal is stimulatory or inhibitory.

Forming the synapse: Activity-dependent development

When an axon contacts its target (usually either a muscle or another neuron), it forms a specialized junction called a **synapse**. Neurotransmitters from the axon terminal are released at these synapses to depolarize or hyperpolarize the membrane of the target cell across the synaptic cleft.

The construction of the synapse involves several steps (Figure 13.28; Burden 1998). When motor neurons in the spinal cord extend axons to muscles, growth cones that contact newly formed muscle cells migrate over their surfaces. When a growth cone first adheres to the cell membrane of a muscle fiber, no specializations can be seen in either membrane. However, the axon terminal soon begins to accumulate neurotransmitter-containing synaptic vesicles, the membranes of both cells thicken at the region of contact, and the synaptic cleft between the cells fills with an extracellular matrix that includes a specific form of laminin. This muscle-derived laminin specifically binds the growth cones of motor neurons and may act as a "stop signal" for axonal growth (Martin et al. 1995; Noakes et al. 1995). In at least some neuron-to-neuron synapses, the synapse is stabilized by N-cadherin. The activity of the synapse releases N-cadherin from storage vesicles in the growth cone (Tanaka et al. 2000).

In muscles, after the first contact is made, growth cones from other axons converge at the site to form additional synapses. During development, all mammalian muscles studied are innervated by at least two axons. However, this polyneuronal innervation is transient. During early postnatal life, all but one of these axon branches are retracted. This "address selection" is based on competition between the axons (Purves and Lichtman 1980; Thompson 1983; Colman et al. 1997). When one of the motor neurons is active, it suppresses the synapses of the other neurons, possibly through a nitric oxide-dependent mechanism (Dan and Poo 1992; Wang et al. 1995). Eventually, the less active synapses are eliminated. The remaining axon terminal expands and is ensheathed by a Schwann cell.

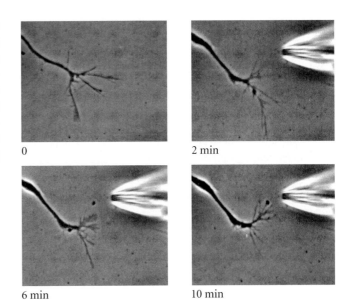

0   2 min

6 min   10 min

Figure 13.27
Embryonic axon from a rat dorsal root ganglion turning in response to a source of NT-3. The photographs document the turning over a 10-minute period. The same growth cone was insensitive to other neurotrophins. (After Paves and Saarma 1997; photographs courtesy of M. Saarma.)

Differential survival after innervation: Neurotrophic factors

One of the most puzzling phenomena in the development of the nervous system is neuronal cell death. In many parts of the vertebrate central and peripheral nervous systems, over half the neurons die during the normal course of development (see Chapter 6, especially Figure 6.28). Moreover, there do not seem to be great similarities in apoptosis patterns across species. For instance, in the cat retina, about 80% of the retinal ganglion cells die, while in the chick retina, this figure is only 40%. In the retinas of fish and amphibians, no retinal ganglion cells appear to die (Patterson 1992).

The apoptotic death of a neuron is not caused by any obvious defect in the neuron itself. Indeed, these neurons have differentiated and have successfully extended axons to their targets. Rather, it appears that the target tissue regulates the number of axons innervating it by limiting the supply of a neurotrophin. In addition to their roles as chemotrophic factors (see above), neurotrophins have been shown to regulate the survival of different subsets of neurons (Figure 13.29). NGF, for example, is necessary for the survival of sympathetic and sensory neurons. Treating mouse embryos with anti-NGF antibodies reduces the number of trigeminal sympathetic and dorsal root ganglion neurons to 20% of their control numbers (Levi-Montalcini and Booker 1960; Pearson et al. 1983). Furthermore, removal of these neurons' target tissues causes the death of the neurons that would have innervated them, and

Figure 13.28
Differentiation of a motor neuron synapse with a muscle. Parts (E) and (G) are depicted at a lower magnification than the others to give an overview of the region where axon meets muscle. (A) A growth cone approaches a developing muscle cell. (B) The axon stops and forms an unspecialized contact on the muscle surface. Agrin, a protein released by the neuron, causes acetylcholine (ACh) receptors to cluster near the axon. (C) Neurotransmitter vesicles enter the axon terminal, and an extracellular matrix connects the axon terminal to the muscle cell as the synapse widens. This matrix contains a nerve-specific laminin. (D) Other axons converge on the same synaptic site. (E) Overview of muscle innervation by several axons (seen in mammals at birth). (F) All axons but one are eliminated. The remaining axon can branch to form a complex junction with the muscle. Each axon terminal is sheathed by a Schwann cell process, and folds form in the muscle cell membrane. (G) Overview of the muscle innervation several weeks after birth. (After Hall and Sanes 1993; Purves 1994; Hall 1995.)

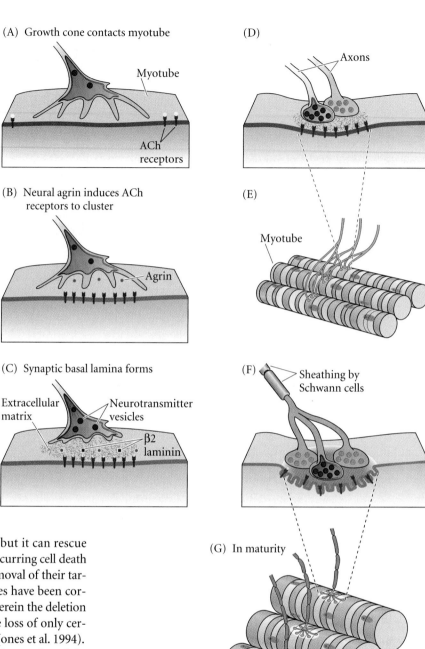

(A) Growth cone contacts myotube

Myotube

ACh receptors

(B) Neural agrin induces ACh receptors to cluster

Agrin

(C) Synaptic basal lamina forms

Extracellular matrix

Neurotransmitter vesicles

β2 laminin

(D)

Axons

(E)

Myotube

(F)

Sheathing by Schwann cells

(G) In maturity

there is a good correlation between the amount of NGF secreted and the survival of the neurons that innervate these tissues (Korsching and Thoenen 1983; Harper and Davies 1990). BDNF does not affect sympathetic or sensory neurons, but it can rescue fetal motor neurons in vivo from normally occurring cell death and from induced cell death following the removal of their target tissue. The results of these in vitro studies have been corroborated by gene knockout experiments, wherein the deletion of particular neurotrophic factors causes the loss of only certain subsets of neurons (Crowley et al. 1994; Jones et al. 1994).

Neurotrophic factors are continuously produced in adults, and loss of these factors may produce debilitating diseases. BDNF is required for the survival of a particular subset of neurons in the striatum (a region of the brain involved in modulating the intensity of coordinated muscle activity such as movement, balance and walking) and enables these neurons to differentiate and synthesize the receptor for dopamine. BDNF in this region of the brain is upregulated by huntingtin, a protein that is mutated in **Huntington disease**. Patients with Huntington disease have decreased production of BDNF, which leads to the death of striatal neurons (Guillin et al. 2001; Zuccato et al. 2001). The result is a series of cognitive abnormalities, involuntary muscle movements, and eventually death. Another neurotrophin, glial cell line-derived neurotrophic factor (GDNF), enhances the survival of an-

other group of neurons: the midbrain dopaminergic neurons whose destruction characterizes **Parkinson disease** (Lin et al. 1993). These neurons send axons to the striatum (whose ability to respond to their dopamine signals is dependent on BDNF). GDNF can prevent the death of these neurons in adult brains (see Lindsay 1995) and is being considered as a therapy for this disease (Zurn et al. 2001).

Figure 13.29
Effects of NGF (top row) and BDNF (bottom) on axonal outgrowths from (A) sympathetic ganglia, (B) dorsal root ganglia, and (C) nodose (taste perception) ganglia. While both NGF and BDNF had a mild stimulatory effect on dorsal root ganglion axonal outgrowth, the sympathetic ganglia responded dramatically to NGF and hardly at all to BDNF, while the converse was true of the nodose ganglia. (From Ibáñez et al. 1991.)

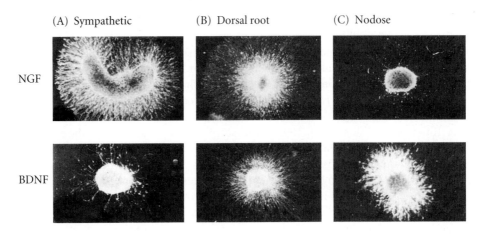

(A) Sympathetic (B) Dorsal root (C) Nodose

NGF

BDNF

The actual survival of any given neuron in the embryo may depend on a combination of agents. Schmidt and Kater (1993) have shown that neurotrophic factors, depolarization (activation), and interactions with the substrate all combine synergistically to determine whether a neuron survives. For instance, the survival of chick ciliary ganglion neurons in culture was promoted by FGF, laminin, or depolarization. However, FGF did not promote survival when laminin was absent, and the combined effects of laminin, FGF, and depolarization were greater than the summed effects of each of them (Figure 13.30). The neurotrophic factors and the other environmental agents appear to function by suppressing an apoptotic "suicide program" that would be constitutively expressed unless repressed by these factors (Raff et al. 1993). The survival of retinal ganglion cells in culture is dependent on neurotrophic factors, but these cells can respond to these factors only if they have been depolarized (Meyer-Franke et al. 1995). Moreover, since neuronal activity stimulates the production of neurotrophins by the active nerves, it is likely that neurons receiving a signal produce more neurotrophins (Thoenen 1995). These factors could have an effect on nearby synapses that are active (i.e., capable of responding to the neurotrophins), thereby stabilizing a set of active synapses to the exclusion of inactive ones.

Paths to glory:
Migration of the retinal ganglion axons

Nearly all the mechanisms for neuronal specification and axon specificity mentioned in this chapter can be seen in the ways in which individual retinal neurons send axons to the appropriate areas of the brain. Even when retinal neurons are transplanted far away from the eye, they are able to find these areas (Harris 1986). The ability of the brain to guide the axons of translocated neurons to their appropriate target sites implies that the guidance cues are not distributed solely along the normal pathway, but exist throughout the embryonic brain. Guiding an axon from a nerve cell body to its destination across the embryo is a complex process, and several different types of cues may be used simultaneously to ensure

that the correct connections are established. Although we are looking here at non-mammalian vertebrates, the processes we describe apply to mammals as well.

GROWTH OF THE RETINAL GANGLION AXON TO THE OPTIC NERVE. The first steps in getting the retinal ganglion axons to their specific regions of the optic tectum take place within the retina (Figure 13.31A). As the retinal ganglion cells differentiate, their position in the inner margin of the retina is determined by cadherin molecules (N-cadherin as well as retina-specific R-cadherin) on their cell membranes (Matsunaga et al. 1988; Inuzuka et al. 1991). Their axons grow along the inner surface of the retina toward the optic disc, the head of the optic nerve (Figure 13.31B). The mature human optic nerve will contain over a million retinal ganglion axons.

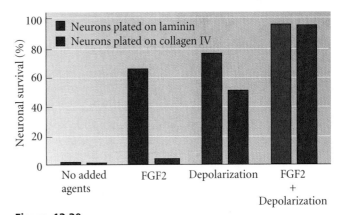

Figure 13.30
Interactions between the effects of substrate, depolarization, and the neurotrophin basic FGF (FGF2) in the survival of ciliary ganglion neurons. Neurons were plated either on laminin (a survival-enhancing substrate) or collagen IV (which does not enhance neuron survival) and observed after 24 hours of culture in the presence or absence of depolarization or FGF2. When cells were depolarized and grown in the presence of FGF2, it did not matter on what substrate they grew. However, when FGF2 was present without depolarization, the substrate made a large difference. (From Schmidt and Kater 1993.)

The adhesion and growth of the retinal ganglion axons along the inner surface of the retina may be governed by its laminin-containing basal lamina. However, simple adhesion to laminin cannot explain the directionality of this growth. N-CAM appears to be especially important here, since the directional migration of the retinal ganglion growth cones depends on the N-CAM-expressing glial endfeet at the inner retinal surface (Stier and Schlosshauer 1995). The secretion of netrin-1 by the cells of the optic disc (where the axons are assembled to form the optic nerve) probably plays a role in this migration as well. Mice lacking netrin-1 genes (or the genes for the netrin receptor found in the retinal ganglion axons) have poorly formed optic nerves, as many of the axons fail to leave the eye and grow randomly around the disc (Deiner et al. 1997). Condroitin sulfate proteoglycan, a repulsive factor for retinal neurons, may provide pushes toward the disc (Hynes and Lander 1992).

Upon their arrival at the optic nerve, the axons fasciculate with axons that are already present there. N-CAM is critical to this fasciculation, and antibodies against N-CAM (or removal of its polysialic acid component) cause the axons to enter the optic nerve in a disorderly fashion, which in turn causes them to emerge at the wrong positions in the tectum (Figure 13.31C; Thanos et al. 1984; Yin et al. 1995).

GROWTH OF THE RETINAL GANGLION AXON THROUGH THE OPTIC CHIASM. When the axons enter the optic nerve, they grow on glial cells toward the midbrain. In non-mammalian vertebrates,

the axons will go to a portion of the brain called the optic tectum. (Mammalian axons go to the lateral geniculate nuclei; this pathway will be discussed further in Chapter 22.) In vitro studies suggest that numerous cell adhesion molecules—N-CAM, cadherins, and integrins—play roles in orienting the axon toward the optic tectum (Neugebauer et al. 1988).

Upon entering the brain, mammalian retinal ganglion axons reach the optic chiasm, where they have to "decide" if they are to continue straight or if they are to turn 90° and enter the other side of the brain (Figure 13.31D). It appears that those axons that are not destined to cross to the other side of the brain are repulsed from doing so when they enter the optic chiasm (Godement et al. 1990). The basis of this repulsion appears to be the synthesis of ephrin on the neurons in the chiasm (Cheng et al. 1995; Marcus et al. 2000). Two guidance molecules, the L1 adhesion molecule and laminin, appear to promote the crossing of the chiasm.

On their way to the optic tectum, the axons of non-mammalian vertebrates travel on a pathway (the optic tract) over glial cells whose surfaces are coated with laminin (Figure 13.31E). Very few areas of the brain contain laminin, and the laminin in this pathway exists only when the optic nerve fibers are growing on it (Cohen et al. 1987).

TARGET SELECTION. When the axons come to the end of the laminin-lined optic tract (Figure 13.31F), they spread out and find their specific targets in the optic tectum. Studies on frogs and fish (in which retinal neurons from each eye project to the opposite side of the brain) have indicated that each retinal ganglion axon sends its impulse to one specific site (a cell or small group of cells) within the optic tectum (Sperry 1951). As shown in Figure 13.32, there are two optic tecta in the frog brain. The axons from the right eye form synapses with the left optic tectum, while those from the left eye form synapses in the right optic tectum. The growth of axons in the *Xenopus* optic tract appears to be mediated by fibroblast growth factors secreted by the cells lining the tract. The retinal ganglion axons express FGF receptors in their growth cones. However, as the axons reach the tectum, the amount of FGF rapidly diminishes, perhaps slowing down the axons and allowing them to find their targets (McFarlane et al. 1995).

The map of retinal connections to the frog optic tectum (the retinotectal projection) was detailed by Marcus Jacobson (1967). Jacobson created this map by shining a narrow beam of light on a small, limited region of the retina

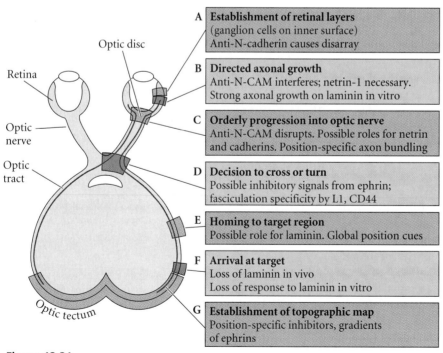

Figure 13.31
Multiple guidance cues direct the movement of retinal ganglion axons to the optic tectum. (After Hynes and Lander 1992.)

A | **Establishment of retinal layers** (ganglion cells on inner surface) Anti-N-cadherin causes disarray

B | **Directed axonal growth** Anti-N-CAM interferes; netrin-1 necessary. Strong axonal growth on laminin in vitro

C | **Orderly progression into optic nerve** Anti-N-CAM disrupts. Possible roles for netrin and cadherins. Position-specific axon bundling

D | **Decision to cross or turn** Possible inhibitory signals from ephrin; fasciculation specificity by L1, CD44

E | **Homing to target region** Possible role for laminin. Global position cues

F | **Arrival at target** Loss of laminin in vivo Loss of response to laminin in vitro

G | **Establishment of topographic map** Position-specific inhibitors, gradients of ephrins

and noting, by means of a recording electrode in the tectum, which tectal cells were being stimulated. The retinotectal projection of *Xenopus laevis* is shown in Figure 13.32. Light illuminating the ventral part of the retina stimulates cells on the lateral surface of the tectum. Similarly, light focused on the temporal (posterior) part of the retina stimulates cells in the caudal portion of the tectum. These studies demonstrated a point-for-point correspondence between the cells of the retina and the cells of the tectum. When a group of retinal cells is activated, a very small and specific group of tectal cells is stimulated. Furthermore, the points form a continuum; in other words, adjacent points on the retina project onto adjacent points on the tectum. This arrangement enables the frog to see an unbroken image. This intricate specificity caused Sperry (1965) to put forward the chemoaffinity hypothesis:

> *The complicated nerve fiber circuits of the brain grow, assemble, and organize themselves through the use of intricate chemical codes under genetic control. Early in development, the nerve cells, numbering in the millions, acquire and retain thereafter, individual identification tags, chemical in nature, by which they can be distinguished and recognized from one another.*

Current theories do not propose a point-for-point specificity between each axon and the neuron that it contacts. Rather, evidence now demonstrates that gradients of adhesivity (especially those involving repulsion) play a role in defining the territories that the axons enter, and that activity-driven competition between these neurons determines the final connection of each axon.*

ADHESIVE SPECIFICITIES IN DIFFERENT REGIONS OF THE OPTIC TECTUM.   There is good evidence that retinal ganglion cells can distinguish between regions of the optic tectum. Cells taken from the ventral half of the chick neural retina preferentially adhere to dorsal (medial) halves of the tectum, and vice versa (Roth and Marchase 1976; Gottlieb et al. 1976; Halfter et al. 1981).

Retinal ganglion cells are specified along the dorsal-ventral axis by a gradient of transcription factors. Dorsal retinal cells are characterized by high concentrations of the Tbx5 transcription factor, while ventral retinal cells have high levels of Pax2. These transcription factors are induced by paracrine factors (BMP4 and retinoic acid, respectively) from nearby tissues (Koshiba-Takeuchi et al. 2000). Misexpression of Tbx in the early chick retina results in marked abnormalities of the retinotectal projection. Therefore, the retinal ganglial cells are specified according to this location.

One gradient that has been identified functionally is a gradient of repulsion that is highest in the posterior tectum and weakest in the anterior tectum. Bonhoeffer and colleagues (Walter et al. 1987; Baier and Bonhoeffer 1992) prepared a "carpet" of tectal membranes having alternating stripes derived from the posterior and the anterior tecta. They then let cells from the nasal (anterior) or temporal (posterior) regions of the retina extend axons into this carpet. The ganglion cells from the nasal portion of the retina extended axons equally well on both the anterior and posterior tectal membranes. The neurons from the temporal side of the retina, however, extended axons only on the anterior tectal membranes (Figure 13.33). When the growth cone of a temporal retinal ganglion axon contacted a posterior tectal cell membrane, the filopodia of the growth cone withdrew, and the growth cone collapsed and retracted (Cox et al. 1990).

---

*In recent years, researchers have discovered dozens of mutations in zebrafish that affect the migration of the retinal ganglion axons to the tectum or the specificity of the retinotectal connections. These mutants are only now being analyzed, but they promise to provide major insights into the mechanisms by which sensory input enters the brain. The December 1996 (Volume 123) issue of *Development* contains several articles mapping the genes involved in the migration of the axon from the retina to the optic cortex. Over 30 mutant genes have been found that affect either the ability of zebrafish retinal ganglion axons to find the optic tectum or the ability of the axons to find their appropriate connections within the tectum (Karlstrom et al. 1997).

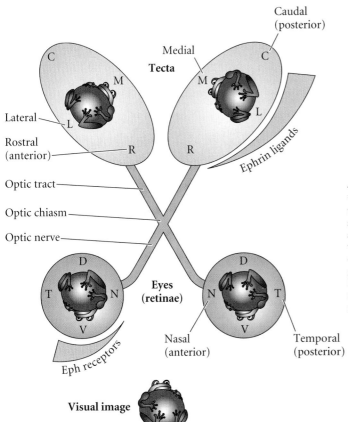

Figure 13.32
Map of the normal retinotectal projection in the adult *Xenopus*. The right eye innervates the left tectum, and the left eye innervates the right tectum. The dorsal portion of the retina (D) innervates the lateral (L) regions of the tectum. The nasal (anterior) region of the retina projects to the caudal (C) region of the tectum. (After Holt 2002; courtesy of C. Holt.)

Tectal membranes
Anterior
Posterior
Anterior
Posterior
Anterior
Posterior
Anterior

Figure 13.33
Differential repulsion of temporal retinal ganglion axons on tectal membranes. Alternating stripes of anterior and posterior tectal membranes were absorbed onto filter paper. When axons from temporal (posterior) retinal ganglion cells were grown on such alternating carpets, they preferentially extended axons on the anterior tectal membranes. (From Walter et al. 1987.)

The basis for this specificity appears to be gradients of ephrin proteins and their receptors. In the optic tectum, ephrin proteins (especially ephrins A2 and A5) are found in gradients that are highest in the posterior (caudal) tectum and decline anteriorly (rostrally) (Figure 13.34A). Moreover, cloned ephrin proteins have the ability to repulse axons (Figure 13.34B), and ectopically expressed ephrin will prohibit axons from the temporal (but not from the nasal) regions of the retina from projecting to where it is expressed (Drescher et al. 1995; Nakamoto et al. 1996). The complementary Eph receptors have been found on chick retinal ganglion cells, and they are expressed in a temporal-to-nasal gradient along the retinal ganglion axons (Cheng et al. 1995). There appear to be several Eph receptors and ligands in the tectum and retina, and they may play push-and-pull roles in guiding the temporal retinal ganglion axons to the anterior tectum and allowing the nasal retinal ganglion axons to project to the posterior portion of the tectum.

Activity-dependent synapse formation also appears to be involved in the final stages of retinal projection to the brain. In frog, bird, and rodent embryos treated with tetrodotoxin, axons will grow normally to their respective territories and will make synapses with the tectal neurons. However, the retinotec-

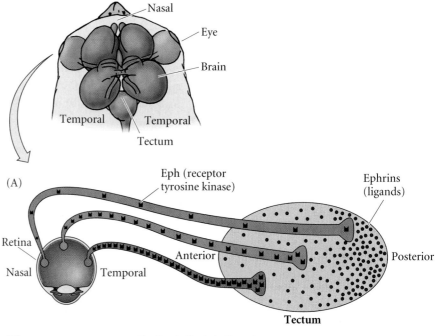

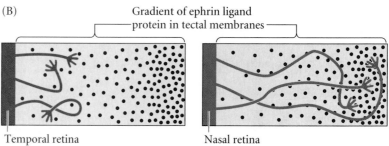

Figure 13.34
Differential retinotectal adhesion is guided by gradients of Eph receptors and their ligands. (A) Representation of the dual gradients of Eph receptor tyrosine kinase in the retina and its ligands (ephrin A2 and ephrin A5) in the tectum. (B) Experiment showing that temporal, but not nasal, retinal ganglion axons respond to a gradient of ephrin ligand in tectal membranes by turning away or slowing down. (After Barinaga 1995.)

tal map is a coarse one, lacking fine resolution. Just as in the final specification of motor neuron synapses, neural activity is needed for the point-for-point retinal projection onto the tectal neurons (Harris 1984; Fawcett and O'Leary 1985; Kobayashi et al. 1990). The fine-tuning of the retinotectal map involves the NMDA receptor, a protein on the tectal neurons. When the NMDA receptor is inhibited, the fine-scale resolution is not obtained (Debski et al. 1990). It appears that NMDA may be coordinating the interaction between nitric oxide (NO) and BDNF (Wu et al. 1994; Ernst et al. 1999; Cogen and Cohen-Cory 2000). Nitric oxide is involved in the elimination of mistargeted retinal axons, while BDNF may stabilize retinal axon connections. It appears that that NO induces growth cone collapse and retraction of developing retinal axons, whereas BDNF protects growth cones and axons from the effects of NO (Ernst et al. 2000). Exposure to both BDNF and NO, but not either factor alone, stabilized growth cones and axons. Activity-dependent synapse formation is extremely important during the development of the mammalian visual system, and this process will be detailed in Chapter 22.

The Development of Behaviors: Constancy and Plasticity

One of the most fascinating aspects of developmental neurobiology is the correlation of certain neuronal connections with certain behaviors. There are two remarkable aspects of this phenomenon. First, there are cases in which complex behavioral patterns are inherently present in the "circuitry" of the brain at birth. The heartbeat of a 19-day chick embryo quickens when it hears a distress call, and no other call will evoke this response (Gottlieb 1965). Furthermore, a newly hatched chick will immediately seek shelter if presented with the shadow of a hawk. An actual hawk is not needed; the shadow cast by a paper silhouette will suffice, and the shadow of no other bird will cause this response (Tinbergen 1951). There appear, then, to be certain neuronal connections that lead to "hard-wired" behaviors in vertebrates.

Equally remarkable are those instances in which the nervous system is so plastic that new experiences can modify the original set of neuronal connections, causing the creation of new neurons or the formation of new synapses between existing neurons. We will discuss neural plasticity at greater length in Chapter 22, but suffice it to say here that the brain does not stop developing at birth. The Nobel Prize-winning research of Hubel and Wiesel (1962, 1963) demonstrated that there is competition between the retinal neurons of each eye for targets in the cortex, and that their connections must be strengthened by experience. As mentioned in the preceding chapter, new experiences lead to the generation of new neurons in adult birds and mammals. Thus, the nervous system continues to develop in adult life, and the pattern of neuronal connections is a product of both inherited patterning and patterning produced by experience.

As one investigator (Purves 1994) concluded his analysis of brain development:

> Although a vast majority of this construction must arise from developmental programs laid down during the evolution of each species, neural activity can modulate and instruct this process, thus storing the wealth of idiosyncratic information that each of us acquires through individual experience and practice.

Snapshot Summary: Neural Crest Cells and Axonal Specificity

1. The neural crest is a transitory structure. Its cells migrate to become numerous different cell types.

2. Trunk neural crest cells can migrate dorsolaterally into the ectoderm, where they become melanocytes. They can also migrate ventrally, to become sympathetic and parasympathetic neurons and adrenomedullary cells.

3. The cranial neural crest cells enter the pharyngeal arches to become the cartilage of the jaw and the bones of the middle ear. They also form the bones of the frontonasal process, the papillae of the teeth, and the cranial nerves.

4. The cardiac neural crest enters the heart and forms the separation between the pulmonary artery and aorta.

5. The formation of the neural crest depends on interactions between the prospective epidermis and the neural plate.

Paracrine factors from these regions induce the formation of transcription factors that enable neural crest cells to emigrate.

6. The path a neural crest cell takes depends on the extracellular matrix it meets.

7. Trunk neural crest cells will migrate through the anterior portion of each sclerotome, but not through the posterior portion of a sclerotome. Ephrin proteins expressed in the posterior portion of each sclerotome appear to prevent neural crest cell migration.

8. Some neural crest cells appear to be capable of forming a large repertoire of cell types. Other neural crest cells may be committed to a fate even before migrating. The final destination of the neural crest cell can sometimes change its specification.

9. The fates of the cranial neural crest cells are to a great extent controlled by the Hox genes. They can acquire their Hox gene expression pattern through interaction with neighboring cells.

10. Teeth develop through an elaborate dialogue between the neural crest-derived mesenchyme and the jaw epithelium. The mesenchyme becomes the odontoblasts, while the epithelium generates the ameloblasts.

11. The major signaling center of the tooth is the enamel knot. It secretes several paracrine factors that regulate cell proliferation and differentiation in both the mesencyme and epithelium.

12. Motor neurons are specified according to their position in the neural tube. The LIM family of transcription factors plays an important role in this specification.

13. The targets of motor neurons are specified before their axons extend into the periphery.

14. The growth cone is the locomotor organelle of the neuron, and it senses environmental cues.

15. Axons can find their targets without neuronal activity.

16. Some proteins are generally permissive to neuron adhesion and provide substrates on which axons can migrate. Other substances prohibit migration.

17. Some growth cones recognize molecules that are present in very specific areas and therefore are guided by these molecules to their respective targets.

18. Some neurons are "kept in line" by repulsive molecules. If they wander off the path to their target, these molecules send them back. Some molecules, such as the semaphorins, are selectively repulsive to a particular set of neurons.

19. Some neurons sense gradients of a protein and are brought to their target by following these gradients. The netrins may work in this fashion.

20. Target selection can be brought about by neurotrophins, proteins that are made by the target tissue that stimulate the particular set of axons that can innervate it. In some cases, the target makes only enough of these factors to support a single axon.

21. Address selection is activity-dependent. An active neuron can suppress synapse formation by other neurons on the same target.

22. Retinal ganglial cells in frogs and chicks send axons that bind to specific regions of the optic tectum. This process is mediated by numerous interactions, and target selection appears to be mediated through ephrins.

23. Some behaviors appear to be innate ("hard-wired"), while others are learned. Experience can strengthen certain neural connections.

Literature Cited

Ackerman, S. L., L. P. Kozak, S. A. Przyborski, L. A. Rund, B. B. Boyer and B. B. Knowles. 1997. The *rostral malformation* gene encodes an UNC-5 protein. *Nature* 386: 838–842.

Akers, R. M., D. F. Mosher and J. E. Lilien. 1981. Promotion of retinal neurite outgrowth by substratum-bound fibronectin. *Dev. Biol.* 86: 179–188.

Akitaya, T. and M. Bronner-Fraser. 1992. Expression of cell adhesion molecules during initiation and cessation of neural crest cell migration. *Dev. Dynam.* 194: 12–20.

Anderson, D. J. and R. Axel. 1986. A bipotential neuroendocrine precursor whose choice of cell fate is determined by NGF and glucocorticoids. *Cell* 47: 1079–1090.

Baier, H. and F. Bonhoeffer. 1992. Axon guidance by gradients of a target-derived component. *Science* 255: 472–475.

Baker, C. V. and M. Bronner-Fraser M. 1997. The origins of the neural crest. II: An evolutionary perspective. *Mech. Dev.* 69: 13–29.

Baker, C. V. and M. Bronner-Fraser M. 2001. Vertebrate cranial placodes. I. Embryonic induction. *Dev. Biol.* 232: 1–61.

Baker, C. V., M. Bronner-Fraser, N. M. Le Douarin and M. A. Teillet. 1997. Early- and late-migrating cranial neural crest cell populations have equivalent developmental potential in vivo. *Development* 124: 3077–3087.

Barinaga, M. 1995. Receptors find work as guides. *Science* 269: 1668–1670.

Baynash, A. G., K. Hosoda, A. Giaid, J. A. Richardson, N. Emoto, R. E. Hammer and M. Yanagisawa. 1994. Interaction of endothelin-3 with endothelin-B receptor is essential for development of epidermal melanocytes and enteric neurons. *Cell* 79: 1277–1285.

Begbie, J. and A. Graham. 2001. Integration between the epibranchial placodes and the hindbrain. *Science* 294: 595–598.

Bockman, D. E. and M. L. Kirby. 1984. Dependence of thymus development on derivatives of the neural crest. *Science* 223: 498–500.

Bronner-Fraser, M. 1986. Analysis of the early stages of trunk neural crest migration in avian embryos using monoclonal antibody HNK-1. *Dev. Biol.* 115: 44–55.

Bronner-Fraser, M. and S. E. Fraser. 1988. Cell lineage analysis reveals multipotency of some avian neural crest cells. *Nature* 335: 161–164.

Bronner-Fraser, M. and S. E. Fraser. 1989. Developmental potential of avian trunk neural crest cells in situ. *Neuron* 3: 755–766.

Brose, K. and 7 others. 1999. Slit proteins bind robo receptors and have an evolutionary conserved role in axon guidance. *Cell* 96: 795–806.

Burden, S. J. 1998. The formation of neuromuscular synapses. *Genes Dev.* 12: 133–148.

Carlson, B. M. 1999. *Human Embryology and Developmental Biology.*, 2nd Ed. Mosby. St. Louis.

Chai, Y. and 8 others. 2000. Fate of the mammalian cranial neural crest during tooth and mandibular morphogenesis. *Development* 127: 1671–1679.

Cheng, H.-J., M. Nakamoto, A. D. Bergemann and J. G. Flanagan. 1995. Complementary gradients in expression and binding of ELF-1 and Mek4 in development of the topographic retinotectal projection map. *Cell* 82: 371–381.

Chun, L. L. Y. and P. H. Patterson. 1977. Role of nerve growth factor in development of rat sympathetic neurons in vitro: Survival, growth, and differentiation of catecholamine production. *J. Cell Biol.* 75: 694–704.

Cogen, J. and S. Cohen-Cory. 2000. Nitric oxide modulates retinal ganglion cell axon arbor remodeling in vivo. *J. Neurobiol.* 22: 120–133.

Cohen, J., J. F. Burne, C. McKinlay and J. Winter. 1987. The role of laminin and the laminin/fibronectin receptor complex in the outgrowth of retinal ganglial cell axons. *Dev. Biol.* 122: 407–418.

Colamarino, S. A. and M. Tessier-Lavigne. 1995. The axonal chemoattractant netrin-1 is also a chemorepellent for trochlear motor axons. *Cell* 81: 621–629.

Colman, H., J., Nabekura and J. W. Lichtman. 1997. Alterations in synaptic strength precede axon withdrawal. *Science* 275: 356–361.

Conway, S. J., D. J. Henderson and A. J. Copp. 1997. Pax3 is required for cardiac neural crest migration in the mouse: Evidence from the *Splotch* (Sp[2H]) mutant. *Development* 124: 505–514.

Conway, S. J., J. Bundy, J. Chen, E. Dickman, R. Rogers and B. M. Will. 2000. Decreased neural crest stem cell expansion is responsible for the conotruncal heart defects within the *Splotch* (Sp[2H])/*Pax3* mouse mutant. *Cardiovasc. Res.* 47: 314–328.

Coulombe, J. N. and M. Bronner-Fraser. 1987. Cholinergic neurones acquire adrenergic neurotransmitters when transplanted into an embryo. *Nature* 324: 569–572.

Couly, G. F., P. M. Coltey and N. Le Douarin. 1993. The triple origin of the skull in higher vertebrates: A study in quail-chick chimeras. *Development* 117: 409–429.

Cox, E. C., B. Müller and F. Bonhoeffer. 1990. Axonal guidance in chick visual system: Posterior tectal membranes induce collapse of growth cones from temporal retina. *Neuron* 2: 31–37.

Crowley, C. and 10 others. 1994. Mice lacking nerve growth factor display perinatal loss of sensory and sympathetic neurons yet develop basal forebrain cholinergic neurons. *Cell* 76: 1001–1011.

Dan, Y. and M.-M. Poo. 1992. Hebbian depression of isolated neuromuscular synapses in vitro. *Science* 256: 1570–1573.

Davies, J. A., G. W. M. Cook, C. D. Stern and R. J. Keynes. 1990. Isolation from chick somites of a glycoprotein fraction that causes collapse of dorsal root ganglion growth cones. *Neuron* 2: 11–20.

Debski, E. A., H. T. Cline and M. Constantine-Paton. 1990. Activity-dependent tuning and the NMDA receptor. *J. Neurosci.* 21: 18–32.

Deiner, M. S., T. E. Kennedy, A. Fazeli, T. Serafini, M. Tessier-Lavigne and D. W. Sretavan. 1997. Netrin-1 and DCC mediate axon guidance locally at the optic disk: Loss of function leads to optic nerve hypoplasia. *Neuron* 19: 575–589.

Dickman, S. 1997. No bones about a genetic switch for bone development. *Science* 276: 1502.

Dorsky, R. I., R. T. Moon and D. W. Raible. 1998. Control of neural crest cell fate by the Wnt signalling pathway. *Nature* 396: 370–372.

Drescher, U., C. Kremoser, C. Handwerker, J. Löschinger, M. Noda and F. Bonhoeffer. 1995. In vitro guidance of retinal ganglion cell axons by RAGS, a 25 kDa protein related to ligands for Eph receptor tyrosine kinases. *Cell* 82: 359–370.

Ducy, P., R. Zhang, V. Geoffroy, A. L. Ridall and G. Karsenty. 1997. Osf2/Cba1: A transcriptional activator of osteoblast differentiation. *Cell* 89: 747–754.

Erickson, C. A., T. D. Duong and K. W. Tosney. 1992. Descriptive and experimental analysis of the dispersion of neural crest cells along the dorsolateral pathway and their entry into ectoderm in the chick embryo. *Dev. Biol.* 151: 251–272.

Ericson, J., S. Thor, T. Edlund, T. J. Jessell and T. Yamada. 1992. Early stages of motor neuron differentiation revealed by expression of homeobox gene *islet-1*. *Science* 256: 1555–1560.

Ericson, J., S. Morton, A. Kawakami, H. Roelink and T. M. Jessell. 1996. Two critical periods of Sonic hedgehog signaling required for the specification of motor neuron identity. *Cell* 87: 661–673.

Ernst, A. F., H. H. Wu, E. E. el-Fakahany and S. C. McLoon. 1999. NMDA receptor-mediated refinement of a transient retinal projection during development requires nitric oxide. *J. Neurosci.* 19: 229–235.

Ernst, A. F., G. Gallo, P. C. Letourneau and S. C. McLoon. 2000. Stabilization of growing retinal axons by the combined signaling of nitric oxide and brain-derived neurotrophic factor. *J. Neurosci.* 20: 1458–1469.

Fawcett, J. W. and D. D. M. O'Leary. 1985. The role of electrical activity in the formation of topographic maps in the nervous system. *Trends Neurosci.* 8: 201–206.

Fukada, K. 1980. Hormonal control of neurotransmitter choice in sympathetic neuron cultures. *Nature* 287: 553–555.

Garcia-Castro, M. I., C. Marcelle and M. Bronner-Fraser. 2002. Ectodermal Wnt function as a neural crest inducer. *Science* 297: 848–851.

Gendron-Maguire, M., M. Mallo, M. Zhang and T. Gridley. 1993. *Hoxa2* mutant mice exhibit homeotic transformation of skeletal elements derived from cranial neural crest. *Cell* 75: 1317–1331.

Gershon, M. D., J. H. Schwartz and E. R. Kandel. 1985. Morphology of chemical synapses and pattern of interconnections. *In* E. R. Kandel and J. H. Schwartz (eds.), *Principles of Neural Science*, 2nd Ed. Elsevier, New York, pp. 132–147.

Godement, P., J. Salaun and C. A. Mason. 1990. Retinal axon pathfinding in the optic chiasm: Divergence of crossed and uncrossed fibers. *Neuron* 5: 173–186.

Goodman, C. S. 1994. The likeness of being: Phylogenetically conserved molecular mechanisms of growth cone guidance. *Cell* 78: 353–356.

Goodman, C. S. and C. Q. Doe. 1993. Embryonic development of the *Drosophila* central nervous system. *In* M. Bate and A. Martinez Arias, *The Development of* Drosophila melanogaster. Cold Spring Harbor Press, Cold Spring Harbor, NY, pp. 1131–1206.

Goodman, C. S. and C. J. Shatz. 1993. Developmental mechanisms that generate precise patterns of neuronal connectivity. *Neuron* 10 [Suppl.]: 77–98.

Gottlieb, D. I., K. Rock and L. Glaser. 1976. A gradient of adhesive specificity in developing avian retina. *Proc. Natl. Acad. Sci. USA* 73: 410–414.

Gottlieb, G. 1965. Prenatal auditory sensitivity in chickens and ducks. *Science* 147: 1596–1598.

Graham, A., I. Heyman and A. Lumsden. 1993. Even-numbered rhombomeres control the apoptotic elimination of neural crest cells from odd-numbered rhombomeres of the chick hindbrain. *Development* 119: 233–245.

Graham, A., P. Francis-West, P. Brickell and A. Lumsden. 1994. The signalling molecule BMP-4 mediates apoptosis in the rhombencephalic neural crest. *Nature* 372: 684–686.

Grammatopolous, G. A., E. Bell, L. Toole, A. Lumsden and A. S. Tucker. 2000. Homeotic transformation of branchial arch identity after *Hoxa2* overexpression. *Development* 127: 5367–5378.

Guillin, O., J. Diaz, P. Carroll, N. Griffon, J. C. Schwartz and P. Sokoloff. 2001. BDNF controls dopamine D_3 receptor expression and triggers behavioural sensitization. *Nature* 411: 86–89.

Gundersen, R. W. 1987. Response of sensory neurites and growth cones to patterned substrata of laminin and fibronectin in vitro. *Dev. Biol.* 121: 423–431.

Halfter, W., M. Claviez and U. Schwarz. 1981. Preferential adhesion of tectal membranes to anterior embryonic chick retina neurites. *Nature* 292: 67–70.

Hall, A. 1998. The Rho GTPases and the actin cytoskeleton. *Science* 279: 509–514.

Hall, B. K. 1988. The embryonic development of bone. *Am. Sci.* 76: 174–181.

Hall, B. K. 2000. The neural crest as a fourth germ layer and vertebrates as quadroblastic not triploblastic. *Evol. Dev.* 2: 3–5.

Hall, Z. W. 1995. Laminin α2 (S-laminin): A new player at the synapse. *Science* 269: 362–363.

Hall, Z. W. and J. R. Sanes. 1993. Synaptic structure and development: The neuromuscular junction. *Neuron* 10 [Suppl.]: 99–121.

Hamelin, M., Y. Zhou, M.-W. Su, I. M. Scott and J. G. Culotti. 1993. Expression of the *unc-5* guidance receptor in the touch neurons of *C. elegans* steers their axons dorsally. *Nature* 364: 327–330.

Harper, S. and A. M. Davies. 1990. NGF mRNA expression in developing cutaneous epithelium related to innervation density. *Development* 110: 515–519.

Harris, W. A. 1984. Axonal pathfinding in the absence of normal pathways and impulse activity. *J. Neurosci.* 4: 1153–1162.

Harris, W. A. 1986. Homing behavior of axons in the embryonic vertebrate brain. *Nature* 320: 266–269.

Harrison, R. G. 1910. The outgrowth of the nerve fiber as a mode of protoplasmic movement. *J. Exp. Zool.* 9: 787–848.

Hedgecock, E. M., J. G. Culotti and D. H. Hall. 1990. The *unc-5, unc-6,* and *unc-40* genes guide circumferential migrations of pioneer axons and mesodermal cells on the epidermis in *C. elegans. Neuron* 2: 61–85.

Helbling, P. M., C. T. Tran and A. W. Brändli. 1998. Requirement for EphA receptor signaling in the segregation of *Xenopus* third and fourth arch neural crest cells. *Mech. Dev.* 78: 63–79.

Henion, P. D. and J. A. Weston. 1997. Timing and pattern of cell fate restrictions in the neural crest lineage. *Development* 124: 4351–4359.

Hollyday, M. 1980. Organization of motor pools in the chick lumbar lateral motor column. *J. Comp. Neurol.* 194: 143–170.

Holmberg, J., D. L. Clarke and J. Frisén. 2000. Regulation of repulsion versus adhesion by different splice forms of an Eph receptor. *Nature* 408: 203–206.

Hong, K., L. Hinck, M. Nishiyama, M. M. Poo, M. Tessier-Lavigne and E. Stein. 1999. A ligand-gated association between cytoplasmic domains of UNC5 and DCC family receptors converts netrin-induced growth cone attraction to repulsion. *Cell* 97: 927–941.

Holt, C. 2002. Retinal-tectal projection. *Macmillan Encyclopedia of Life Sciences.* Macmillan, London.

Höpker, V. H., D. Shewan, M. Tessier-Lavigne, M.-M. Poo and C. Holt. 1999. Growth-cone attraction to netrin-1 is converted to repulsion by laminin-1. *Nature* 401: 69–73.

Hubel, D. H. and T. N. Wiesel. 1962. Receptive fields, binocular interaction and functional architecture in the cat's visual cortex. *J. Physiol.* 160: 106–154.

Hubel, D. H. and T. N. Wiesel. 1963. Receptive fields of cells in striate cortex of very young, visually inexperienced kittens. *J. Neurophysiol.* 26: 944–1002.

Hynes, R. O. and A. D. Lander. 1992. Contact and adhesive specificities in the associations, migrations, and targeting of cells and axons. *Cell* 68: 303–322.

Ibáñez, C. F., T. Ebendal and H. Persson. 1991. Chimeric molecules with multiple neurotrophic activities reveal structural elements determining the specificities of NGF and BDNF. *EMBO J.* 10: 2105–2110.

Inuzuka, H., S. Miyatani and M. Takeichi. 1991. R-cadherin: A novel Ca^{2+}-dependent cell-cell adhesion molecule expressed in the retina. *Neuron* 7: 69–79.

Ishii, N., W. G. Wadsworth, B. D. Stern, J. G. Culotti and E. M. Hedgecock. 1992. UNC-6, a laminin-related protein, guides pioneer axon migrations in *C. elegans. Neuron* 9: 873–881.

Jacobson, M. 1967. Retinal ganglion cells: Specification of central connections in larval *Xenopus laevis. Science* 155: 1106–1108.

Jernvall, J. 1995. Mammalian molar cusp patterns: Developmental mechanisms of diversity. *Acta Zool. Fennica* 198: 1–61.

Jernvall, J., P. Kettunen, I. Karavanova, L. B. Martin and I. Theseleff. 1994. Evidence for the role of the enamel knot as a control center in mammalian tooth cusp formation: Non-dividing cells express growth stimulating *Fgf-4* gene. *Int. J. Dev. Biol.* 38: 463–469.

Jernvall, J., T. Aberg, P. Kettunen, S. Keränen and I. Thesleff. 1998. The life history of an embryonic signaling center: BMP4 induces p21 and is associated with apoptosis in the mouse tooth enamel knot. *Development* 125: 161–169.

Jiang, X., S. Iseki, R. E. Maxson, H. M. Sucov and G. Morriss-Kay. 2002. Tissue origins and interactions in the mammalian skull vault. *Dev. Biol.* 241: 106–116.

Jin, E. J., C. A. Erickson, S. Takada and L. W. Burrus. 2001. Wnt and BMP signaling govern lineage segregation of melanocytes in the avian embryo. *Dev. Biol.* 233: 22–37.

Jones, K. R., I. Farinas, C. Backus and L. F. Reichart. 1994. Targeted disruption of the BDNF gene perturbs brain and sensory neuron development, but not motor neuron development. *Cell* 76: 989–999.

Kahn, C. R., J. T. Coyle and A. M. Cohen. 1980. Head and trunk neural crest in vitro: Autonomic neuron differentiation. *Dev. Biol.* 77: 340–348.

Karlstrom, R. O., T. Trowe and F. Bonhoeffer. 1997. Genetic analysis of axon guidance and mapping in the zebrafish. *Trends Neurosci.* 20: 3–8.

Kennedy, T. E., T. Serafini, J. R. de la Torre and M. Tessier-Lavigne. 1994. Netrins are diffusible chemotropic factors for commissural axons in the embryonic spinal cord. *Cell* 78: 425–435.

Kidd, T., K. S. Bland and C. S. Goodman. 1999. Slit is the midline repellent for the Robo receptor in *Drosophila. Cell* 96: 785–794.

Kirby, M. L. 1987. Cardiac morphogenesis: Recent research advances. *Pediatr. Res.* 21: 219–224.

Kirby, M. L. 1989. Plasticity and predetermination of mesencephalic and trunk neural crest

transplanted into the region of the cardiac neural crest. *Dev. Biol.* 134: 401–412.

Kirby, M. L. and K. L. Waldo. 1990. Role of neural crest in congenital heart disease. *Circulation* 82: 332–340.

Klose, M. and D. Bentley. 1989. Transient pioneer neurons are essential for forming an embryonic peripheral nerve. *Science* 245: 982–984.

Kobayashi, T., H. Nakamura and M. Yasuda. 1990. Disturbance of refinement of retinotectal projection in chick embryos by tetrodotoxin and grayanotoxin. *Dev. Brain Res.* 57: 29–35.

Kollar, E. J. and G. Baird. 1970. Tissue interaction in developing mouse tooth germs. II. The inductive role of the dental papilla. *J. Embryol. Exp. Morphol.* 24: 173–186.

Kolodkin, A. L., D. J. Matthes, T. P. O'Connor, N. H. Patel, D. Bentley and C. S. Goodman. 1992. Fasciclin IV: Sequence, expression, and function during growth cone guidance in the grasshopper embryo. *Neuron* 9: 831–845.

Kolodkin, A. L., D. J. Matthes and C. S. Goodman. 1993. The semaphorin genes encode a family of transmembrane and secreted growth cone guidance molecules. *Cell* 75: 1389–1399.

Komori, T. and 14 others. 1997. Targeted disruption of *Cba1* results in a complete lack of bone formation owing to maturational arrest of osteoblasts. *Cell* 89: 755–764.

Korsching, S. and H. Thoenen. 1983. Nerve growth factor in sympathetic ganglia and corresponding target organs of the rat: Correlation with density of sympathetic innervation. *Proc. Natl. Acad. Sci. USA* 80: 3513–3516.

Kos, R., M. V. Reedy, R. L. Johnson and C. A. Erickson. 2001. The winged-helix transcription factor FoxD3 is important for establishing the neural crest lineage and repressing melanogenesis in avian embryos. *Development* 128: 1467–1479.

Koshiba-Takeuchi, K. and 10 others. 2000. Tbx5 and retinotectal projection. *Science* 287: 134–137.

Koyama, E. and 10 others. 1996. Polarizing activity, Sonic hedgehog, and tooth development in embryonic and postnatal mouse. *Dev. Dynam.* 206: 59–72.

Krull, C. E., A. Collazo, S. E. Fraser and M. Bronner-Fraser. 1995. Segmental migration of trunk neural crest: Time lapse analysis reveals a role for PNA binding molecules. *Development* 121: 3733–3743.

Krull, C. and and 7 others. 1997. Interactions between Eph-related receptors and ligands confer rostrocaudal pattern to trunk neural crest migration. *Curr. Biol.* 7: 571–580.

Krull, C. E., J. Eberhart, R. McLennan, S. A. Koblar, D. P. Cerretti and E. B. Pasquale. 1999. Segmental patterning of the peripheral nervous system. *Dev. Biol.* 210: 203.

Kulesa, P. M. and S. E. Fraser. 2000. In ovo time-lapse analysis of chick hindbrain neural crest cell migration shows interactions during migration to the branchial arches. *Development* 127: 1161–1172.

Kunisada, T. and 8 others. 1998. Transgene expression of steel factor in the basal layer of epidermis promotes survival, proliferation, differentiation, and migration of melanocyte precursors. *Development* 125: 2915–2923.

Kuratani, S. C. and M. L. Kirby. 1991. Initial migration and distribution of the avian cardiac neural crest in the avian embryo: An introduction to the concept of the circumpharyngeal crest. *Am. J. Anat.* 191: 215–227.

LaBonne, C. and M. Bronner-Fraser. 1998. Neural crest induction in *Xenopus*: Evidence for a two-signal model. *Development* 125: 2403–2414.

Ladher, R. K., K. U. Anakwe, A. L. Gurney, G. C. Schoenwolf and P. H. Francis-West. 2000. Identification of synergistic signals initiating inner ear development. *Science* 290: 1965–1967.

Lahav, R., C. Ziller, E. Dupin and N. M. Le Douarin. 1996. Endothelin 3 promotes neural crest cell proliferation and mediates a vast increase in melanocyte number in culture. *Proc. Natl. Acad. Sci. USA* 93: 3892–3897.

Lance-Jones, C. and L. Landmesser. 1980. Motor neuron projection patterns in chick hindlimb following partial reversals of the spinal cord. *J. Physiol.* 302: 581–602.

Landmesser, L. 1978. The development of motor projection patterns in the chick hindlimb. *J. Physiol.* 284: 391–414.

Le Douarin, N. M. 1982. *The Neural Crest.* Cambridge University Press, New York.

Le Douarin, N. M. and C. Kalcheim. 1999. *The Neural Crest*, 2nd Ed. Cambridge University Press, Cambridge.

Le Douarin, N. M. and M.-A. Teillet. 1973. The migration of neural crest cells to the wall of the digestive tract in avian embryo. *J. Embryol. Exp. Morphol.* 30: 31–48.

Le Douarin, N. M. and M.-A. Teillet. 1974. Experimental analysis of the migration and differentiation of neuroblasts of the autonomic nervous system and of neuroectodermal mesenchyme derivatives, using a biological cell marking technique. *Dev. Biol.* 41: 162–184.

Le Douarin, N. M., D. Renaud, M.-A. Teillet and G. H. Le Douarin. 1975. Cholinergic differentiation of presumptive adrenergic neuroblasts in interspecific chimeras after heterotopic transplantation. *Proc. Natl. Acad. Sci. USA* 72: 728–732.

Le Lièvre, C. S. 1978. Participation of neural crest-derived cells in the genesis of the skull in birds. *J. Embryol. Exp. Morphol.* 47: 17–37.

Le Lièvre, C. S. and N. M. Le Douarin. 1975. Mesenchymal derivatives of the neural crest: Analysis of chimaeric quail and chick embryos. *J. Embryol. Exp. Morphol.* 34: 125–154.

Leonardo, E. D., L. Hinck, M. Masu, K. Keino-Masu, S. L. Ackerman and M. Tessier-Lavigne. 1997. Vertebrate homologues of *C. elegans* UNC-5 are candidate netrin receptors. *Nature* 386: 833–838.

Letourneau, P., A. M. Madsen, S. M. Palm and L. T. Furcht. 1988. Immunoreactivity for laminin in the developing ventral longitudinal pathway of the brain. *Dev. Biol.* 125: 135–144.

Levi-Montalcini, R. and B. Booker. 1960. Destruction of the sympathetic ganglia in mammals by an antiserum to the nerve growth factor protein. *Proc. Natl. Acad. Sci. USA* 46: 384–390.

Li, H.-S. and 12 others. 1999. Vertebrate slit, a secreted ligand for transmembrane protein Roundabout, is a repellent for olfactory bulb axons. *Cell* 96: 807–818.

Liem, K. F., Jr., G. Tremml and T. M. Jessell. 1997. A role for the roof plate and its resident TGFβ-related proteins in neuronal patterning of the dorsal spinal cord. *Cell* 91: 127–138.

Liesi, P. and J. Silver. 1988. Is astrocyte laminin involved in axon guidance in the mammalian CNS? *Dev. Biol.* 130: 774–785.

Lin, L.-F. H., D. H. Doherty, J. D. Lile, S. Bektesh and F. Collins. 1993. GDNF: A glial cell-line derived neurotrophic factor for midbrain dopaminergic neurons. *Science* 260: 1130–1132.

Lindsay, R. M. 1995. Neuron saving schemes. *Nature* 373: 289–290.

Liu, J.-P. and T. M. Jessell. 1998. A role for rhoB in the delamination of neural crest cells from the dorsal neural tube. *Development* 125: 5055–5067.

Lo, L., L. Sommer and D. J. Anderson. 1997. MASH1 maintains competence for BMP2-induced neuronal differentiation in post-migratory neural crest cells. *Curr. Biol.* 7: 440–450.

Loring, J. F. and C. A. Erickson. 1987. Neural crest cell migratory pathways in the trunk of the chick embryo. *Dev. Biol.* 121: 220–236.

Lumsden, A. G. S. 1988a. Multipotent cells in the avian neural crest. *Trends Neurosci.* 12: 81–83.

Lumsden, A. G. S. 1988b. Spatial organization of the epithelium and the role of neural crest cells in the initiation of the mammalian tooth germ. *Development* 103 [Suppl.]: 155–169.

Luo, Y., D. Raibile and J. A. Raper. 1993. Collapsin: A protein in brain that induces the collapse and paralysis of neuronal growth cones. *Cell* 75: 217–227.

Ma, Q., Z. Chen, I. del Barco Barrantes, J. L. de la Pompa and D. J. Anderson. 1998. Neurogenin-1 is essential for the determination of neuronal precursors for proximal cranial sensory ganglia. *Neuron* 20: 469–482.

Mackie, E. J., I. Thesleff and R. Chiquet-Ehrismann. 1987. Tenascin is associated with chondrogenic and osteogenic differentiation in vivo and promotes chondrogenesis in vivo. *J. Cell Biol.* 105: 2569–2579.

Mancilla, A. and R. Mayor. 1996. Neural crest formation in *Xenopus laevis*: Mechanisms of Xslug induction. *Dev. Biol.* 177: 580–589.

Marcus, R. C., G. A. Matthews, N. W. Gale, G. D. Yancopoulos and C. A. Mason. 2000. Axon guidance in the mouse optic chiasm: Retinal neurite inhibition by ephrin "A"-expressing hypothalamic cells in vitro. *Dev. Biol.* 221: 132–147.

Marín, O., A. Yaron, A. Bagri, M. Tessier-Lavigne and J. L. R. Rubenstein. 2001. Sorting of striatal and cortical interneurons regulated by semaphorin-neuropilin interactions. *Science* 293: 872–875.

Martin, P. T., A. J. Ettinger and J. R. Sanes. 1995. Synaptic localization domain in the synaptic cleft protein laminin β2 (s-laminin). *Science* 269: 413–416.

Marx, J. 1995. Helping neurons find their way. *Science* 268: 971–973.

Matsunaga, M., K. Hatta and M. Takeichi. 1988. Role of N-cadherin cell adhesion molecules in the histogenesis of neural retina. *Neuron* 1: 289–295.

Matthes, D. J., H. Sink, A. L. Kolodkin and C. S. Goodman. 1995. Semaphorin II can function as a selective inhibitor of specific synaptic arborizations. *Cell* 81: 631–639.

Mayer, T. C. 1973. The migratory pathway of neural crest cells into the skin of mouse embryos. *Dev. Biol.* 34: 39–46.

Mayor, R., N. Guerrero and C. Martinez. 1997. Role of FGF and Noggin in neural crest cell induction. *Dev. Biol.* 189: 1–12.

McFarlane, S., L. McNeill and C. E. Holt. 1995. FGF signaling and target recognition in the developing *Xenopus* visual system. *Neuron* 15: 1017–1028.

McGinnis, W. and R. Krumlauf. 1992. Homeobox genes and axial patterning. *Cell* 68; 283–302.

McGonnell, I. M. and A. Graham. 2002. Trunk neural crest has skeletogenic potential. *Curr. Biol.* 12: 767–771.

McGonnell, I. M., I. J. McKay and A. Graham. 2001. A population of caudally migrating cranial neural crest cells: Functional and evolutionary implications. *Development* 236: 354–363.

Messersmith, E. K., E. D. Leonardo, C. J. Shatz, M. Tessier-Lavigne, C. S. Goodman and A. Kolodkin. 1995. Semaphorin III can function as a selective chemorepellent to pattern sensory projections in the spinal cord. *Neuron* 14: 949–959.

Meyer-Franke, A., M. R. Kaplan, F. W. Pfrieger and B. A. Barres. 1995. Characterization of the signaling interactions that promote the survival and growth of developing retinal ganglion cells in culture. *Neuron* 15: 805–819.

Mina, M. and E. J. Kollar. 1987. The induction of odontogenesis in non-dental mesenchyme combined with early murine mandibular arch epithelium. *Arch. Oral Biol.* 32: 123–127.

Morin-Kensicki, E. M. and J. S. Eisen. 1997. Sclerotome development and peripheral nervous system segmentation in embryonic zebrafish. *Development* 124: 159–167.

Mundlos, S. and 13 others. 1997. Mutations involving the transcription factor CBFA1 cause cleiocranial dysplasia. *Cell* 89: 773–779.

Nakamoto, M. and 7 others. 1996. Topographically specific effects of ELF-1 on retinal axon guidance in vitro and retinal axon mapping in vivo. *Cell* 86: 755–766.

Neubüser, A., H. Peters, R. Ballig and G. R. Martin. 1997. Antagonistic interactions between FGF and BMP signaling pathways: A mechanism for positioning the sites of tooth formation. *Cell* 90: 247–255.

Neugebauer, K. M., K. J. Tomaselli, J. Lilien and L. F. Reichardt. 1988. N-cadherin, N-CAM, and integrins promote retinal neurite outgrowth on astrocytes in vitro. *J. Cell Biol.* 107: 1177–1187.

Newgreen, D. F. and D. Gooday. 1985. Control of onset of migration of neural crest cells in avian embryos: Role of Ca^{++}-dependent cell adhesions. *Cell Tissue Res.* 239: 329–336.

Newgreen, D. F., M. Scheel and V. Kaster. 1986. Morphogenesis of sclerotome and neural crest cells in avian embryos: In vivo and in vitro studies on the role of notochordal extracellular material. *Cell Tissue Res.* 244: 299–313.

Nieto, M. A., M. G. Sargent, D. G. Wilkinson and J. Cooke. 1994. Control of cell behavior during vertebrate development by *slug*, a zinc finger gene. *Science* 264: 835–839.

Noakes, P. G., M. Gautam, J. Mudd, J. R. Sanes and J. P. Merlie. 1995. Aberrant differentiation of neuromuscular junctions in mice lacking s-laminin/laminin β2. *Nature* 374: 258–262.

Noden, D. M. 1978. The control of avian cephalic neural crest cytodifferentiation. I. Skeletal and connective tissue. *Dev. Biol.* 69: 296–312.

Noden, D. M. 1983. The role of the neural crest in patterning of avian cranial skeletal, connective, and muscle tissues. *Dev. Biol.* 96: 144–165.

Noden, D. M. 1991. Cell movements and control of patterned tissue assembly during craniofacial development. *J. Craniofac. Genet. Dev. Biol.* 11: 192–213.

Northcutt, R. G. and C. Gans. 1983. The genesis of neural crest and epidermal placodes: A reinterpretation of vertebrate origins. *Q. Rev. Biol.* 58: 1–28.

Oakley, R. A., C. J. Lasky, C. A. Erickson and K. W. Tosney. 1994. Glycoconjugates mark a transient barrier to neural crest migration in the chicken embryo. *Development* 120: 103–114.

O'Leary, D. D. M. and D. G. Wilkinson. 1999. Eph receptors and ephrins in neural development. *Curr. Opin. Neurobiol.* 9: 65–73.

Otto, F. and 11 others. 1997. Cba1, a candidate gene for cleidocranial dysplasia syndrome, is essential for osteoblast differentiation and bone development. *Cell* 89: 765–771.

Pasqualetti, M., M. Ori, I. Nardi and F. M. Rijli. 2000. Ectopic *Hoxa2* induction after neural crest migration results in homeosis of jaw elements in *Xenopus*. *Development* 127: 5367–5378.

Patterson, P. H. 1992. Neuron-target interactions. *In* Z. Hall (ed.), *An Introduction to Molecular Neurobiology*. Sinauer Associates, Sunderland, MA, pp. 428–459.

Paves, H. and M. Saarma. 1997. Neurotrophins as in vitro growth cone guidance molecules for embryonic sensory neurons. *Cell Tissue Res.* 290: 285–297.

Pearson, J., E. M. Johnson, Jr. and L. Brandeis. 1983. Effects of antibodies to nerve growth factor on intrauterine development of derivatives of cranial neural crest and placode in the guinea pig. *Dev. Biol.* 96: 32–36.

Perez, S. E., S. Rebelo and D. J. Anderson. 1999. Early specification of sensory neuron fate revealed by expression and function of neurogenins in the chick embryo. *Development* 126: 1715–1728.

Pini, A. 1993. Chemorepulsion of axons in the developing mammalian central nervous system. *Science* 261: 95–98.

Polleux, F., T. Morrow and A. Ghosh. 2000. Semaphorin 3A is a chemoattractant for cortical apical dendrites. *Nature* 404: 567–573.

Pomeranz, H. D., T. P. Rothman and M. D. Gershon. 1991. Colonization of the post-umbilical bowel by cells derived from the sacral neural crest: Direct tracing of cell migration using an intercalating probe and a replication-deficient retrovirus. *Development* 111: 647–655.

Price, S. R., N. V. D. M. Garcia, B. Ranscht and T. M. Jessell. 2002. Regulation of motor neuron pool sorting by differential expression of type II cadherins. *Cell* 109: 205–216.

Purves, D. 1994. *Neural Activity and the Growth of the Brain*. Cambridge University Press, New York.

Purves, D. and J. W. Lichtman. 1980. Elimination of synapses in the developing nervous system. *Science* 210: 153–157.

Raff, M. C., B. A. Barres, J. F. Burne, H. S. Coles, Y. Ishizaki and M. D. Jacobson. 1993. Programmed cell death and the control of cell survival: Lessons from the nervous system. *Science* 262: 695–700.

Ramón y Cajal, S. 1892. Le Rètine de Vertèbrès. *La Cellule* 9: 119–258.

Rawles, M. E. 1948. Origin of melanophores and their role in development of color patterns in vertebrates. *Physiol. Rev.* 28: 383–408.

Rickmann, M., J. W. Fawcett and R. J. Keynes. 1985. The migration of neural crest cells and the growth of motor neurons through the rostral half of the chick somite. *J. Embryol. Exp. Morphol.* 90: 437–455.

Rijli, F. M., M. Mark, S. Lakkaraju, A. Dierich, P. Dollé and P. Chambon. 1993. A homeotic transformation is generated in the rostral branchial region of the head by a disruption of *Hoxa-2*, which acts as a selector gene. *Cell* 75: 1333–1349.

Roth, S. and R. B. Marchase. 1976. An in vitro assay for retinotectal specificity. *In* S. H. Barondes (ed.), *Neuronal Recognition*. Plenum, New York, pp. 227–248.

Santiago, A. and C. A. Erickson. 2002. Ephrin-B ligands play a dual role in the control of neural crest cell migration. *Development* 129: 3621–3632.

Sasai, N., K. Mizuseki and Y. Sasai. 2001. Requirement of FoxD3-class signaling for neural crest determination in *Xenopus*. *Development* 128: 2525–2536.

Savagne, R. P., K. M. Yamada and J. P. Thiery. 1997. The zinc finger protein slug causes desmosome dissociation, an initial and necessary step for growth factor-induced epithelial-mesenchymal transition. *J. Cell Biol.* 137: 1403–1419.

Schilling, T. 2001. Plasticity of zebrafish Hox expression in the hindbrain and cranial neural crest. *Dev. Biol.* 231: 201–216.

Schmidt, M. and S. B. Kater. 1993. Fibroblast growth factors, depolarization, and substrate interact in a combinatorial way to promote neuronal survival. *Dev. Biol.* 158: 228–237.

Schneider, R. A. and J. A. Helms. 2003. The cellular and molecular origins of beak morphology. *Science* 299: 565–568.

Schweizer, G., C. Ayer-Le Lièvre and N. M. Le Douarin. 1983. Restrictions in developmental capacities in the dorsal root ganglia during the course of development. *Cell Diff.* 13: 191–200.

Sechrist, J., G. N. Serbedzija, T. Scherson, S. E. Fraser and M. Bronner-Fraser. 1993. Segmental migration of the hindbrain neural crest does not arise from its segmental origin. *Development* 118: 691–703.

Selleck, M. A. and M. Bronner-Fraser. 1995. Origins of the avian neural crest: The role of neural plate-epidermal interactions. *Development* 121: 525–538.

Serafini, T., T. E. Kennedy, M. J. Galko, C. Mirayan, T. M. Jessell and M. Tessier-Lavigne. 1994. The netrins define a family of axon outgrowth-promoting proteins homologous to *C. elegans* UNC-6. *Cell* 78: 409–424.

Shah, N. M., M. A. Marchionni, I. Isaacs, P. Stroobant and D. J. Anderson. 1994. Glial growth factor restricts mammalian neural crest stem cells to a glial fate. *Cell* 77: 349–360.

Shah, N. M., A. K. Groves and D. J. Anderson. 1996. Alternative neural crest cell fates are instructively promoted by TGF-β superfamily members. *Cell* 85: 331–343.

Sharma, K., A. E. Leonard, K. Lettieri and S. L. Pfaff. 2000. Genetic and epigenetic mechanisms contribute to motor neuron pathfinding. *Nature* 406: 515–519.

Smith, A., V. Robinson, K. Patel and D. G. Wilkinson. 1997. The EphA4 and EphB1 receptor tyrosine kinases and ephrin-B2 ligand regulate targeted migration of branchial neural crest cells. *Curr. Biol.* 7: 561–570.

Smith, M. M. and B. K. Hall. 1993. A developmental model for evolution of the vertebrate exoskeleton and teeth: The role of cranial and trunk neural crest. *Evol. Biol.* 27: 387–448.

Sockanathan, S. and T. M. Jessell. 1998. Motor neuron-derived retinoid signal specifies subtype identity of spinal motor neurons. *Cell* 94: 503–514.

Sperry, R. W. 1951. Mechanisms of neural maturation. *In* S. S. Stevens (ed.), *Handbook of Experimental Psychology*. Wiley, New York, pp. 236–280.

Sperry, R. W. 1965. Embryogenesis of behavioral nerve nets. *In* R. L. DeHaan and H. Ursprung (eds.), *Organogenesis*. Holt, Rinehart & Winston, New York, pp. 161–186.

Stern, C. D., K. B. Artinger and M. Bronner-Fraser. 1991. Tissue interactions affecting the migration and differentiation of neural crest cells in the chick embryo. *Development* 113: 207–216.

Stier, H. and B. Schlosshauer. 1995. Axonal guidance in the chicken retina. *Development* 121: 1443–1454.

Takeichi, M. 1988. The cadherins: Cell-cell adhesion molecules controlling animal morphogenesis. *Development* 102: 639–656.

Tan, S.-S. and G. Morriss-Kay. 1985. The development and distribution of the cranial neural crest in the rat embryo. *Cell Tissue Res.* 240: 403–416.

Tanabe, Y., C. William and T. M. Jessell. 1998. Specification of motor neuron identity by the MNR2 homeodomain protein. *Cell* 95: 67–80.

Tanaka, H. and 8 others. 2000. Molecular modification of N-cadherin in response to synaptic activity. *Neuron* 25; 93–107.

Teillet, M.-A., C. Kalcheim and N. M. Le Douarin. 1987. Formation of the dorsal root ganglia in the avian embryo: Segmental origin and migratory behavior of neural crest progenitor cells. *Dev. Biol.* 120: 329–347.

Testaz, S. and J.-L. Duband. 2001. Central role of the α4β1 integrin in the coordination of avain truncal neural crest cell adhesion, migration, and survival. *Dev. Dynam.* 222: 127–140.

Thanos, S., F. Bonhoeffer and U. Rutishauser. 1984. Fiber-fiber interaction and tectal cues influence the development of the chicken retinotectal projection. *Proc. Natl. Acad. Sci. USA* 81: 1906–1910.

Thesleff, I. and C. Sahlberg. 1996. Growth factors as inductive signals regulating tooth morphogenesis. *Semin. Cell Dev. Biol.* 7: 185–193.

Thesleff, I., A. Vaahtokari and S. Vainio. 1990. Molecular changes during determination and differentiation of the dental mesenchyme cell lineage. *J. Biol. Bucalle* 18: 179–188.

Thoenen, H. 1995. Neurotrophins and neuronal plasticity. *Science* 270: 593–598.

Thomas, J. B. 1998. Axon guidance: Crossing the midline. *Curr. Biol.* 8: R102–R104.

Thompson, W. J. 1983. Synapse elimination in neonatal rat muscle is sensitive to pattern of muscle use. *Nature* 302: 614–616.

Thorogood, P. 1989. Review of developmental and evolutionary aspects of the neural crest. *Trends Neurosci.* 12: 38–39.

Tinbergen, N. 1951. *The Study of Instinct.* Clarendon Press, Oxford.

Tosney, K. W. and R. A. Oakley. 1990. The perinotochordal mesenchyme acts as a barrier to axon advance in the chick embryo: Implications for a general mechanism of axonal guidance. *Exp. Neurol.* 109: 75–89.

Tosney, K. W., K. B. Hotary and C. Lance-Jones. 1995. Specificity of motoneurons. *BioEssays* 17: 379–382.

Trainor, P. and R. Krumlauf. 2000. Plasticity in mouse neural crest cells reveals a new patterning role for cranial mesoderm. *Nature Cell Biol.* 2: 96–102.

Trainor, P. and R. Krumlauf. 2001. Hox genes, neural crest cells, and branchial arch patterning. *Curr. Opin. Cell Biol.* 13: 698–705.

Trainor, P. and R. Krumlauf. 2002. Riding the crest of the Wnt signaling wave. *Science* 297: 781–783.

Tsushida, T., M. Ensini, S. B. Morton, M. Baldassare, T. Edlund, T. M. Jessell and S. L. Pfaff. 1994. Topographic organization of embryonic motor neurons defined by expression of LIM homeobox genes. *Cell* 79: 957–970.

Tucker, A. S., K. L. Matthews and P. T. Sharpe. 1998. Transformation of tooth type induced by inhibition of BMP signaling. *Science* 282: 1136–1138.

Tucker, R. P., C. Hagios, R. Chiquet-Ehrismann, J. Lawler, R. J. Hall and C. A. Erickson. 1999. Thrombospondin-1 and neural crest migration. *Dev. Dynam.* 214: 312–322.

Twitty, V. C. 1937. Experiments on the phenomenon of paralysis produced by a toxin occurring in *Triturus* embryos. *J. Exp. Zool.* 76: 67–104.

Twitty, V. C. and H. H. Johnson. 1934. Motor inhibition in *Amblystoma* produced by *Triturus* transplants. *Science* 80: 78–79.

Vaahtokari, A., T. Aberg, J. Jernvall, S. Keränen and I. Thesleff. 1996a. The enamel knot as a signalling center in the developing mouse tooth. *Mech. Dev.* 54: 39–43.

Vaahtokari, A., T. Aberg and I. Thesleff. 1996b. Apoptosis in the developing tooth: Association with an embryonic signaling center and suppression by EGF and FGF-4. *Development* 122: 121–129.

Vainio, S., I. Karavanova, A. Jowett and I. Thesleff. 1993. Identification of BMP-4 as a signal mediating secondary induction between epithelial and mesenchymal tissues during early tooth development. *Cell* 75: 45–58.

Varley, J. E., R. G. Wehby, D. C. Rueger and G. D. Maxwell. 1995. Number of adrenergic and islet-1 immunoreactive cells is increased in avian trunk neural crest cultures in the presence of human recombinant osteogenic protein-1. *Dev. Dynam.* 203: 434–447.

Vogel, K. S. and J. A. Weston. 1990. The sympathoadrenal lineage in avian embryos. II. Effects of glucocorticoids on cultured neural crest cells. *Dev. Biol.* 139: 13–23.

Waldo, K., S. Miyagawa-Tomita, D. Kumiski and M. L. Kirby. 1998. Cardiac neural crest cells provide new insight into septation of the cardiac outflow tract: Aortic sac to ventricular septal closure. *Dev. Biol.* 196: 129–144.

Walter, J., S. Henke-Fahle and F. Bonhoeffer. 1987. Avoidance of posterior tectal membranes by temporal retinal axons. *Development* 101: 909–913.

Wang, H. U. and D. J. Anderson. 1997. Eph family transmembrane ligands can mediate repulsive guidance of trunk neural crest migration and motor axon outgrowth. *Neuron* 18: 383–396.

Wang, T., Z. Xie and B. Lu. 1995. Nitric oxide mediates activity-dependent synaptic suppression at developing neuromuscular synapses. *Nature* 374: 262–266.

Westerfield, M., D. W. Liu, C. B. Kimmel and C. Walker. 1990. Pathfinding and synapse formation in a zebrafish mutant lacking functional acetylcholine receptors. *Neuron* 4: 867–874.

Weston, J. A. 1963. A radiographic analysis of the migration and localization of trunk neural crest cells in the chick. *Dev. Biol.* 6: 274–310.

Weston, J. and S. L. Butler. 1966. Temporal factors affecting localization of neural crest cells in the chicken embryo. *Dev. Biol.* 14: 246–266.

Wilkie, A. O. M. and G. Morriss-Kay. 2001. Genetics of craniofacial development and malformation. *Nature Rev. Genet.* 2: 458–468.

Wilkinson, D. G., S. Bhatt and A. P. McMahon. 1989. Expression of the FGF-related proto-oncogene *int-2* suggests multiple roles in fetal development. *Development* 105: 131–136.

Winberg, M. L., K. J. Mitchell and C. S. Goodman. 1998. Genetic analysis of the mechanisms controlling target selection: Complementary and combinatorial functions of netrins, semaphorins, and IgCAMs. *Cell* 93: 581–591.

Wu, H. H., C. V. Williams and S. C. McLoon. 1994. Involvement of nitric oxide in the elimination of a transient retinotectal projection in development. *Science* 265: 1593–1596.

Yamamori, T., K. Fukada, R. Aebersold, S. Korsching, M.-J. Fann and P. H. Patterson. 1989. The cholinergic neuronal differentiation factor from heart cells is identical to leukemia inhibitory factor. *Science* 246: 1412–1416.

Yin, X., M. Watanabe and U. Rutishauser. 1995. Effects of polysialic acid on the behavior of retinal ganglion cell axons during growth into the optic tract and tectum. *Development* 121: 3439–3446.

Zuccato, C. and 12 others. 2001. Loss of huntingtin-mediated BDNF gene transcription in Huntington's disease. *Science* 239: 493–498.

Zurn, A. D., H. R. Widmur and P. Aebischer. 2001. Sustained delivery of GDNF: Towards a treatment for Parkinson's disease. *Brain Res. Rev.* 36: 222–229.

14 *Paraxial and intermediate mesoderm*

IN CHAPTERS 12 AND 13, we followed the various tissues formed by the vertebrate ectoderm. In this chapter and the next, we will follow the development of the mesodermal and endodermal germ layers. We will see that the endoderm forms the lining of the digestive and respiratory tubes, with their associated organs. The mesoderm generates all the organs between the ectodermal wall and the endodermal tissues.

The **trunk mesoderm** of a neurula-stage embryo can be subdivided into four regions (Figure 14.1). These subdivisions are thought to be specified along the mediolateral (center-to-side) axis by increasing amounts of BMPs (Pourquié et al. 1996; Tonegawa et al. 1997). The more lateral mesoderm of the chick embryo expresses higher levels of BMP4 than the midline areas, and one can change the identity of the mesodermal tissue by altering BMP expression.

1. The first region of trunk mesoderm is the **chordamesoderm**. This tissue forms the notochord, a transient organ whose major functions include inducing the formation of the neural tube and establishing the anterior-posterior body axis. The formation of the notochord on the future dorsal side of the embryo was discussed in Chapters 10 and 11.

2. The second region is the **paraxial mesoderm**, or **somitic dorsal mesoderm**. The term *dorsal* refers to the fact that the tissues developing from this region will be located in the back of the embryo, along the spine. The cells in this region will form **somites**, blocks of mesodermal cells on both sides of the neural tube, which will produce many of the connective tissues of the back (bone, muscle, cartilage, and dermis).

3. The **intermediate mesoderm** forms the urogenital system.

4. Farthest away from the notochord, the **lateral plate mesoderm** gives rise to the heart, blood vessels, and blood cells of the circulatory system, as well as to the lining of the body cavities and to all the mesodermal components of the limbs except the muscles. It also forms a series of extraembryonic membranes that are important for transporting nutrients to the embryo.

Anterior to the trunk mesoderm is a fifth mesodermal region, the **prechordal plate mesoderm**. This region provides the head mesenchyme that forms much of the connective tissues and musculature of the face.

PARAXIAL MESODERM: THE SOMITES AND THEIR DERIVATIVES

One of the major tasks of gastrulation is to create a mesodermal layer between the endoderm and the ectoderm. As shown in Figure 14.2, the formation of mesodermal and endodermal tissues is not subsequent to neural tube formation, but occurs synchronously. The notochord extends beneath the neural tube from the base of the head into the tail. On either side of the neural tube lie thick bands of mesodermal cells. These bands of paraxial mesoderm are referred to either as the **segmental plate** or the **unsegmented mesoderm**. As the primitive streak regresses and the neural folds begin to gather at the center of the embryo, the paraxial mesoderm separates into blocks of cells called somites. The paraxial mesoderm appears to be specified by the antagonism of BMP signaling by the Noggin protein. Noggin is usually synthesized by the early segmental plate mesoderm, and if Noggin-expressing cells are placed into the presumptive lateral plate mesoderm, the lateral plate tissue will be re-specified into somite-forming paraxial mesoderm (Figure 14.3; Tonegawa and Takahashi 1998).

Although somites are transient structures, they are extremely important in organizing the segmental pattern of vertebrate embryos. As we saw in the preceding chapter, the somites determine the migration paths of neural crest cells and spinal nerve axons. The somites give rise to the cells that form the vertebrae and ribs, the dermis of the dorsal skin, the skeletal muscles of the back, and the skeletal muscles of the body wall and limbs.

VADE MECUM[2] **Mesoderm in the vertebrate embryo.** The organization of the mesoderm in the neurula stage is similar for all vertebrates. You can see this organization by viewing serial sections of the chick embryo. **[Click on Chick-Mid]**

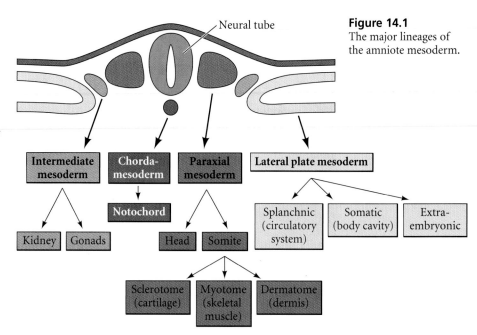

Figure 14.1
The major lineages of the amniote mesoderm.

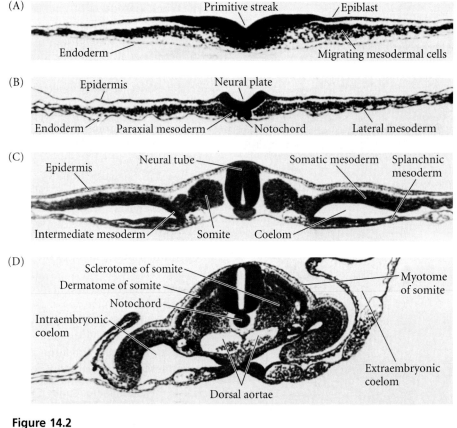

Figure 14.2
Gastrulation and neurulation in the chick embryo, focusing on the mesodermal component. (A) Primitive streak region, showing migrating mesodermal and endodermal precursors. (B) Formation of the notochord and paraxial mesoderm. (C, D) Differentiation of the somites, coelom, and the two aortae (which will eventually fuse). A–C, 24-hour embryos; D, 48-hour embryo.

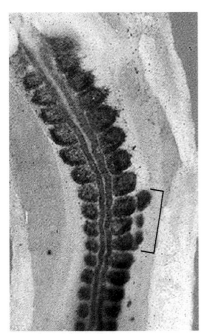

Figure 14.3
Specification of somites. Placing Noggin-secreting cells into a prospective region of chick lateral plate mesoderm will respecify that mesoderm into somite-forming paraxial mesoderm. Induced somites (bracketed) were detected by in situ hybridization with *Pax3*. (After Tonegawa and Takahashi 1998; photograph courtesy of Y. Takahashi.)

The Formation of Somites

The periodicity of somite formation

The important components of somitogenesis (somite formation) are periodicity, epithelialization, specification, and differentiation. The first somites appear in the anterior portion of the trunk, and new somites "bud off" from the rostral end of the paraxial mesoderm at regular intervals (Figure 14.4). Somite formation begins as paraxial mesoderm cells become organized into whorls of cells called **somitomeres**. The somitomeres become compacted and bound together by an epithelium and eventually separate from the presomitic paraxial mesoderm to form individual somites. Because individual embryos can develop at slightly different rates (as when chick embryos are incubated at slightly different temperatures), the number of somites present is usually the best indicator of how far development has proceeded. The total number of somites formed is characteristic of a species (50 in chicks, 65 in mice, and as many as 500 in some snakes).

Although we do not completely understand the mechanisms controlling the periodicity of somite formation, one of the key agents in this process is the Notch signaling pathway. When a small group of cells from a region constituting a presumptive somite boundary is transplanted into a region of unsegmented mesoderm that would not ordinarily be part of the

boundary area, a new boundary is created. The transplanted boundary cells instruct the cells anterior to them to epithelialize and separate. Moreover, non-boundary cells can acquire this boundary-forming ability if either an activated Notch protein or Lunatic fringe, a protein that activates Notch, is electroporated into those cells (Figure 14.5A–C; Sato et al. 2002).

These boundary cells establish *where* the border forms, but this information does not tell us *when* the border forms. It has been proposed (Dale et al. 2003) that one of the proteins activated by the Notch protein is also able to inhibit Notch, thus establishing a negative feedback loop. When this inhibitor is degraded, Notch would become active again. Such a cycle would create a "clock" whereby Notch would be turned on and off by a protein it itself induces. These off-and-on oscillations could provide the molecular basis for the periodicity of somite segmentation.

Mutations affecting Notch signaling have been shown to be responsible for aberrant vertebral formation in mice and humans. In humans, individuals with spondylocostal dysplasia have numerous vertebral and rib defects that have been linked to mutations of the *Delta-like3* gene, which encodes a ligand for Notch. Mice with knockouts of this gene have a phenotype similar to the human syndrome (Figure 14.5D,E; Bulman et al. 2000; Dunwoodie et al. 2002).

Somitogenesis has long been observed to occur simultaneously with the regression of the primitive streak. Recent research (Dubrulle et al. 2001; Sawada et al. 2001) suggests that

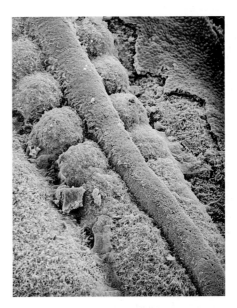

Figure 14.4
Neural tube and somites seen by scanning electron microscopy. When the surface ectoderm is peeled away, well-formed somites are revealed, as well as paraxial mesoderm (bottom right) that has not yet separated into distinct somites. A rounding of the paraxial mesoderm into a somitomere can be seen at the lower left, and neural crest cells can be seen migrating ventrally from the roof of the neural tube. (Photograph courtesy of K. W. Tosney.)

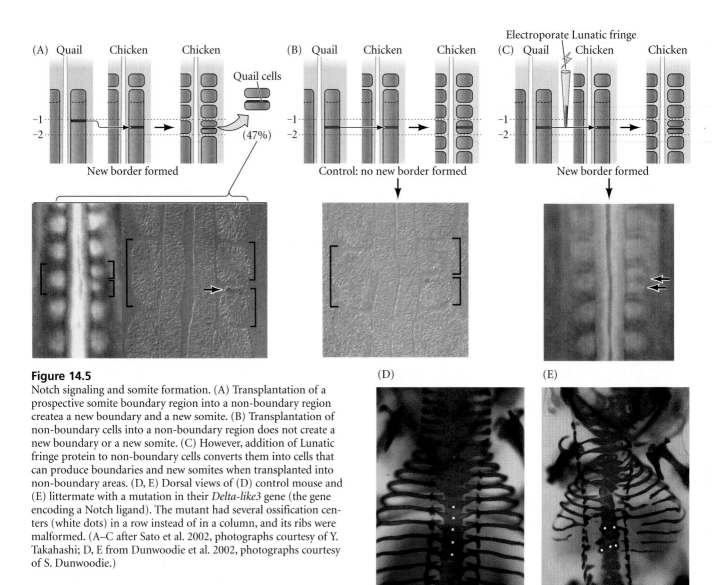

Figure 14.5
Notch signaling and somite formation. (A) Transplantation of a prospective somite boundary region into a non-boundary region createa a new boundary and a new somite. (B) Transplantation of non-boundary cells into a non-boundary region does not create a new boundary or a new somite. (C) However, addition of Lunatic fringe protein to non-boundary cells converts them into cells that can produce boundaries and new somites when transplanted into non-boundary areas. (D, E) Dorsal views of (D) control mouse and (E) littermate with a mutation in their *Delta-like3* gene (the gene encoding a Notch ligand). The mutant had several ossification centers (white dots) in a row instead of in a column, and its ribs were malformed. (A–C after Sato et al. 2002, photographs courtesy of Y. Takahashi; D, E from Dunwoodie et al. 2002, photographs courtesy of S. Dunwoodie.)

this is not merely a coincidence, and that Hensen's node regulates the timing of somitogenesis by regulating Lunatic fringe expression. As cells pass through Hensen's node, they secrete FGF8. This paracrine factor can prevent cells from expressing Lunatic fringe. As the node regresses, the FGF8-secreting cells of the node move more posteriorly, allowing the paraxial mesoderm cells that had been in the domain of FGF8 influence to become anterior to that zone. These paraxial mesoderm cells are now capable of becoming somites.

The separation of somites from the unsegmented mesoderm

If Notch signaling determines the placement of somite formation, then Notch must control a cascade of gene expression that ultimately separates the tissues. These genes are expressed in a cyclic fashion and function as an autonomous segmentation "clock" (Palmeirim et al. 1997; Jouve et al. 2000, 2002; see also

Chapter 5). One of these genes is *Hairy1*, a homologue of the *Drosophila* segmentation gene, which encodes a transcription factor.

Like other genes regulated by the Notch signaling pathway, the *Hairy1* gene is expressed in a dynamic manner. First, it is expressed in the caudal portion of each somite, and it persists in these caudal regions for at least 15 hours. Second, it is expressed in the presomitic segmental plate in a cyclic, wavelike manner, cresting every 90 minutes. Its expression is detected first in the caudalmost region of the presomitic mesoderm; this region of expression moves anteriorly as each somite forms. Eventually, like a wave leaving shells on a beach, its expression recedes caudally, leaving a thin band of expression at its most anterior reach. The caudalmost region of this anterior expression band correlates with the posterior termi-

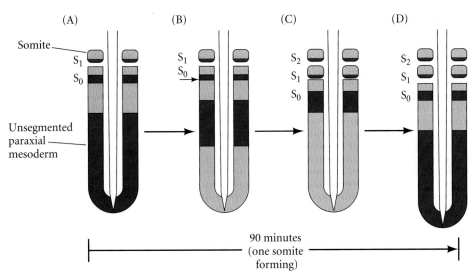

Figure 14.6
Somite formation correlates with the wavelike expression of the *Hairy1* gene in the chick. (A) Schematic representation of the posterior portion of a chick embryo somite. S_1 has just budded off the presomitic mesoderm. The expression of the *hairy1* gene (purple) is seen in the caudal half of this somite, as well as in the posterior portion of the presomitic mesoderm and in a thin band that will form the caudal half of the next somite (S_0). (B) A caudal fissure (small arrow) begins to separate the new somite from the presomitic mesoderm. The posterior region of *hairy1* expression extends anteriorly. (C) The newly formed somite is now referred to as S_1; it retains the expression of *hairy1* in its caudal half, as the posterior domain of *hairy1* expression moves farther anteriorly and shortens. The former S_1 somite, now called S_2, undergoes differentiation. (D) The formation of somite S_1 is completed, and the anterior region of what had been the posterior *hairy1* expression pattern is now the anterior expression pattern. It will become the caudal domain of the next somite. The entire process takes 90 minutes.

nus of the next somite to be formed. This process can be seen in Figure 14.6.

The *Hairy1* gene encodes a transcription factor, but it is not known what its targets are. Possible targets (directly or indirectly) are the genes for ephrin and its receptor. We saw in Chapter 13 that the Eph tyrosine kinase receptor proteins and their ephrin ligands are able to cause cell-cell repulsion between the posterior somite and migrating neural crest cells. Ephrin and Eph also may be critical for separating the somites. In the zebrafish and chick, the boundary between the most recently separated somite and the presomitic mesoderm forms between ephrin-B2 in the posterior of the somite and EphA4 in the most anterior portion of the presomitic mesoderm (Figure 14.7; Durbin et al. 1998). As somites form, this pattern of gene expression is reiterated caudally. Interfering with this signaling (by injecting embryos with RNA encoding dominant negative Ephs) leads to abnormal somite boundary formation. Eph signaling is thought to mediate cell shape changes, and such changes could be responsible for the separation of the presomitic mesoderm at the ephrin-B2/EphA4 border.

Although ephrin-Eph signaling may be responsible for converting the prepattern established by the Hairy1 protein in the presomitic mesoderm into actual somites, the mechanism by which this is done is not known. The somite boundaries do not follow the EphA4 pattern exactly, and it is possible that EphA4 function may be important after the somite has begun to pull apart from the segmental plate (Kulesa and Fraser 2002).

The epithelialization of the somite

Several studies in the chick have shown that the conversion of each somite from mesenchymal tissue into an epithelial block occurs even before the somite splits off. As seen in Figure 14.4, the cells of the somitomere are randomly organized as a mesenchymal mass. The synthesis of two extracellular matrix proteins, fibronectin and N-cadherin, links them into tissue blocks that will form tight junctions and generate their own basal laminae (Figure 14.8A,B; Ostrovsky et al. 1984; Lash and Yamada 1986; Hatta et al. 1987). These extracellular matrix proteins, in turn, may be regulated by the expression of the *Paraxis* gene. This gene encodes a transcription factor that is also expressed at the rostral (anterior) end of the unsegmented mesoderm of mouse embryos, and which is seen in precisely that region that will form the somite (Figure 14.8C). In *Paraxis*-deficient mice, somites segregate from the segmental plate, but they fail to epithelialize (Burgess et al. 1995; Barnes et al. 1997; Tajbakhsh and Spörle 1998).

Another important protein involved in this mesenchymal-epithelial transition is a small GTPase, Rac1. Rac1 can be activated by integrin when the integrin binds to fibronectin, and activated Rac1 can trigger dramatic changes in the actin cytoskeleton (Kjoller and Hall 1999). Such cytoskeletal changes are seen in the somite cells as they makes the transition from the mesenchymal to the epithelial state. Moreover, the electroporation of mutant Rac1 (but not other small GTPases) into the chick segmental plate prevents the somites from becoming epithelial (Takahashi et al. 2002).

(A)

Somites:
Anterior
Posterior

EphA4

ephrin-B2

Unsegmented
paraxial mesoderm

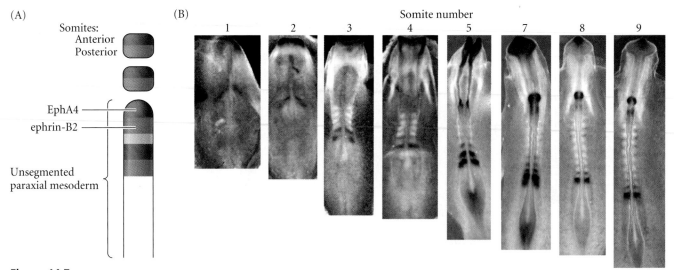

(B)   Somite number
1   2   3   4   5   7   8   9

Figure 14.7

Ephrin and its receptor constitute a possible cut site for somite formation. (A) Expression pattern of the receptor tyrosine kinase EphA4 (blue) and its ligand, ephrin-B2 (red) as somites develop. The somite boundary forms at the junction between the region of ephrin expression on the posterior of the last somite formed and the region of Eph expression on the anterior of the next somite to form. In the presomitic mesoderm, the pattern is created anew as each somite buds off. The posteriormost region of the next somite to form does not express ephrin until that somite is ready to separate. (B) In situ hybridization showing EphA4 (dark blue) expression as new somites are formed in the chick embryo. (A after Durbin et al. 1998; B, photograph courtesy of J. Kastner.)

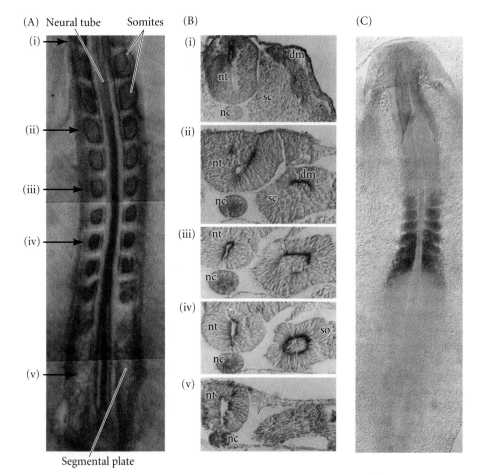

(A) Neural tube   Somites
(i)
(ii)
(iii)
(iv)
(v)

Segmental plate

(B)
(i)   dm
      nt
      nc   sc
(ii)
      nt
            dm
      nc   sc
(iii) nt
      nc
(iv) nt
            so
      nc
(v)   nt
      nc

(C)

Figure 14.8

Epithelialization and de-epithelialization in the somites of the 6-somite chick embryo. (A, B) Distribution of N-cadherin in the 2-day chick embryo during somitogenesis. (A) Whole-mount immunohistochemisty using an N-cadherin-specific antibody. (B) Sections of whole-mount immunostained embryo taken at the indicated positions along the embryonic axis indicated in (A). (i) A differentiated somite with little N-cadherin staining. (ii) De-epithelized somites and the emergence of sclerotome (iii) Early de-epithelization of somite with some loss of N-cadherin. (iv) Epithelialized somite showing N-cadherin lining the somitocoel. (v) Unsegmented mesoderm of the segmental plate showing minimal N-cadherin. N-Cadherin is detected in the neural tube throughout somitogenesis, with some down-regulation upon somite differentiation. dm, dermomyotome; nc, notochord; nt, neural tube; sc, sclerotome; so, somite; sp, segmental plate. In (C), in situ hybridization reveals the expression of Paraxis mRNA (red). Paraxis is seen both in the somites and in the region of the unsegmented mesoderm that will give rise to the next somite. (C from Barnes et al. 1997; photographs courtesy of R. Tuan.)

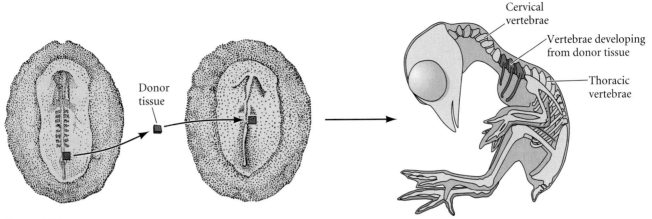

Cervical vertebrae

Vertebrae developing from donor tissue

Thoracic vertebrae

Donor tissue

Figure 14.9
The segmental plate mesoderm is determined by its position along the anterior-posterior axis before somitogenesis. When segmental plate mesoderm that would ordinarily form thoracic somites is transplanted into a region in a younger embryo (caudal to the first somite) that would ordinarily give rise to cervical (neck) somites, the grafted mesoderm differentiates according to its original position and forms ribs in the neck. (After Kieny et al. 1972.)

Specification of the somite along the anterior-posterior axis

Although all the somites look identical, they will form different structures at different positions along the anterior-posterior axis. For instance, the ribs are derived from somites. The somites that form the cervical vertebrae of the neck and the lumbar vertebrae of the abdomen are not capable of forming ribs; ribs are generated only by the somites forming the thoracic vertebrae. Moreover, the specification of the thoracic vertebrae occurs very early in development. If one isolates the region of chick segmental plate that will give rise to a thoracic somite and transplants this mesoderm into the cervical (neck) region of a younger embryo, the host embryo will develop ribs in its neck—but those ribs will form only on the side where the thoracic mesoderm has been transplanted (Figure 14.9; Kieny et al. 1972; Nowicki and Burke 2000).

As discussed in Chapter 11 (see Figure 11.46), the somites are specified according to the Hox genes they express. Mice that are homozygous for a loss-of-function mutation of *Hoxc-8* will convert a lumbar vertebra into an extra thoracic vertebra, complete with ribs (see Figure 11.44). The Hox genes are activated concomitantly with somite formation, and the embryo appears to "count somites" in setting the expression boundaries of the Hox genes. If FGF8 levels are manipulated to create extra (albeit smaller) somites, the Hox gene expression pattern will be activated earlier. The appropriate Hox gene expression will be activated in the appropriately numbered somite, even if it is in a different position along the anterior-posterior axis. Moreover, when mutations affect the autonomous segmentation clock, they also affect the activation of the appropriate Hox genes

(Dubrulle et al. 2001; Zakany et al. 2001). Once established, each somite retains its pattern of Hox gene expression, even if transplanted into another region of the embryo (Nowicki and Burke 2000). The regulation of the Hox genes by the segmentation clock should allow coordination between the formation and the specification of the new segments.

The derivatives of the somite

Somites form (1) the cartilage of the vertebrae and ribs, (2) the muscles of the rib cage, limbs, abdominal wall, back, and tongue, and (3) the dermis of the dorsal skin. In contrast to the early commitment of the mesoderm along the anterior-posterior axis, the commitment of the cells within a somite to their respective fates occurs relatively late, after the somite has already formed. When the somite is first separated from the presomitic mesoderm, any of its cells can become any of the somite-derived structures. However, as the somite matures, its various regions become committed to forming only certain cell types. The ventral-medial cells of the somite (those cells located farthest from the back but closest to the neural tube) undergo mitosis, lose their round epithelial characteristics, and become mesenchymal cells again. The portion of the somite that gives rise to these cells is called the **sclerotome**, and these mesenchymal cells ultimately become the cartilage cells (chondrocytes) of the vertebrae and part (if not all) of each rib (Figure 14.10A,B; see also Figure 14.2).

Fate mapping with chick-quail chimeras (Ordahl and Le Douarin 1992; Brand-Saberi et al. 1996; Kato and Aoyama 1998) has revealed that the remaining epithelial portion of the somite is arranged into three regions (Figure 14.10C,D). The cells in the two lateral portions of the epithelium (those regions closest to and farthest from the neural tube) constitute the muscle-forming region, or **myotome**. The cells of the myotome divide to produce a lower layer of muscle precursor cells, the **myoblasts**. The resulting double-layered structure is called the **dermamyotome**. Those myoblasts formed from the region closest to the neural tube form the **epaxial muscles** (the deep muscles of the back), while those myoblasts formed

(A) 2-day embryo

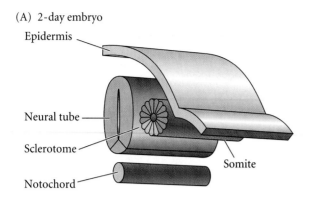

Epidermis

Neural tube

Sclerotome

Notochord

Somite

Figure 14.10
Diagram of a transverse section through the trunk of a chick embryo on days 2–4. (A) In the 2-day somite, the sclerotome cells can be distinguished from the rest of the somite. (B) On day 3, the sclerotome cells lose their adhesion to one another and migrate toward the neural tube. (C) On day 4, the remaining cells divide. The medial cells form an epaxial myotome beneath the dermamyotome, while the lateral cells form a hypaxial myotome. (D) A layer of muscle cell precursors (the myotome) forms beneath the epithelial dermamyotome. (A, B after Langman 1981; C, D after Ordahl 1993.)

(B) 3-day embryo

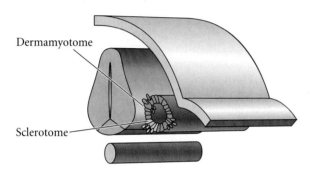

Dermamyotome

Sclerotome

(C) 4-day embryo

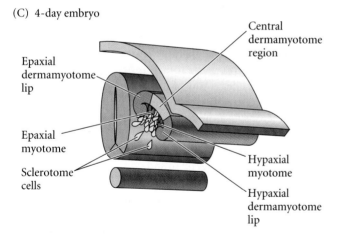

Central dermamyotome region

Epaxial dermamyotome lip

Epaxial myotome

Sclerotome cells

Hypaxial myotome

Hypaxial dermamyotome lip

(D) Late 4-day embryo

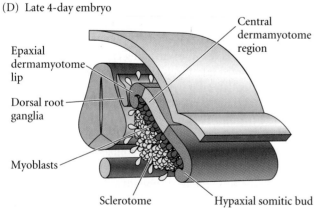

Central dermamyotome region

Epaxial dermamyotome lip

Dorsal root ganglia

Myoblasts

Sclerotome

Hypaxial somitic bud

in the region farthest from the neural tube produce the **hypaxial muscles** of the body wall, limbs, and tongue (see Figure 14.11; see also Christ and Ordahl 1995; Venters et al. 1999). The central region of the dorsal layer of the dermamyotome has classically been called the **dermatome**, and it generates the mesenchymal connective tissue of the back skin: the **dermis**. (The dermis of other areas of the body forms from other mesenchymal cells, not from the somites.) The dermatome is not physically distinct from the lateral myotome regions until the presumptive muscle cells migrate and differentiate.

Determination of the sclerotome and dermatome

Like the proverbial piece of real estate, the destiny of a somitic region depends on three things: location, location, and location.

The specification of the somite is accomplished by the interaction of several tissues. The ventral-medial portion of the somite is induced to become the sclerotome by paracrine factors, especially Sonic hedgehog, secreted from the notochord and the neural tube floor plate (Fan and Tessier-Lavigne 1994; Johnson et al. 1994). If portions of the notochord (or another source of Sonic hedgehog) are transplanted next to other regions of the somite, those regions, too, will become sclerotome cells. Sclerotome cells express a new transcription factor, Pax1, that is required for their differentiation into cartilage and whose presence is necessary for the formation of the vertebrae (Figure 14.11; Smith and Tuan 1996). They also express I-mf, an inhibitor of the myogenic bHLH family of transcription factors that initiate muscle formation (Chen et al. 1996). It appears that the notochord induces its surrounding mesenchyme cells to secrete epimorphin, and this epimorphin attracts sclerotome cells to the region around the notochord and neural tube (Oka et al. 2002).

The dermatome differentiates in response to two factors secreted by the neural tube: neurotrophin 3 (NT-3) and Wnt1. Antibodies that block the activities of NT-3 prevent the conversion of the epithelial dermatome into the loose dermal mesenchyme that migrates beneath the epidermis (Brill et al. 1995). In birds, this somite-derived dermis is responsible for feather induction (see Figure 6.7), and the dorsal region of the neural tube is critical for specifying this dermis. Removing or rotating the neural tube prevents this dermis from forming (Takahashi et al. 1992; Olivera-Martinez 2002).

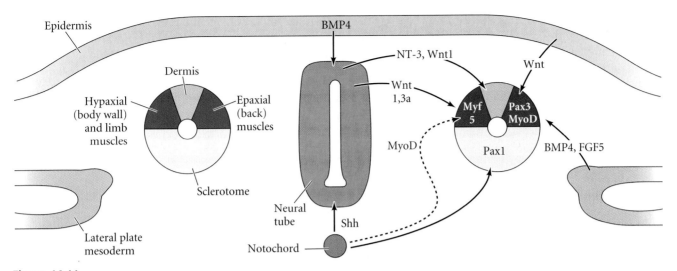

Figure 14.11
Model of major postulated interactions in the patterning of the somite. A combination of Wnts (probably Wnt1 and Wnt3a) is induced by BMP4 in the dorsal neural tube. These Wnt proteins, in combination with low concentrations of Sonic hedgehog from the notochord and floor plate, induce the epaxial myotome, which synthesizes the myogenic transcription factor Myf5. High concentrations of Sonic hedgehog induce Pax1 expression in those cells fated to become the sclerotome. Certain concentrations of neurotrophin-3 (NT-3) from the dorsal neural tube appear to specify the dermatome, while Wnt proteins from the epidermis, in conjunction with BMP4 and FGF5 from the lateral plate mesoderm, are thought to induce the hypaxial myotome. (After Cossu et al. 1996b.)

Determination of the myotome

In similar ways, the myotome is induced by at least two distinct signals. Studies involving transplantation and knockout mice indicate that the epaxial myoblasts coming from the medial portion of the somite are induced by factors from the neural tube, probably Wnt1 and Wnt3a from the dorsal region and low levels of Sonic hedgehog from the ventral region (Münsterberg et al. 1995; Stern et al. 1995; Ikeya and Takada 1998). The hypaxial myoblasts coming from the lateral edge of the somite are probably specified by a combination of Wnt proteins from the epidermis and bone morphogenetic protein 4 (BMP4) from the lateral plate mesoderm (Cossu et al. 1996a; Pourquié et al. 1996; Dietrich et al. 1998). These factors (see Figure 14.11) cause the myoblasts to migrate away from the dorsal region and delay their differentiation until they are in a more ventral position.

In addition to these positive signals, there are inhibitory signals that prevent a signal from affecting an inappropriate group of cells. For example, Sonic hedgehog not only activates sclerotome and myotome development, but also inhibits the BMP4 signal from the lateral plate mesoderm from extending medially and ventrally (thus preventing the conversion of sclerotome into muscle) (Watanabe et al. 1998). Similarly, Noggin is produced by the most medial portion of the dermamyotome and prevents BMP4 from giving these cells the migratory characteristics of hypaxial muscle (Marcelle et al. 1997).

And what happens to the notochord, that central mesodermal structure? After it has provided the axial integrity of the early embryo and has induced the formation of the dorsal neural tube, most of it degenerates by apoptosis. This apoptosis is probably signaled by mechanical forces (Aszódi et al. 1998). Wherever the sclerotome cells have formed a vertebral body, the notochordal cells die. However, in between the vertebrae, the notochordal cells form part of the intervertebral discs, the nuclei pulposi. These are the discs that "slip" in certain back injuries.

> **WEBSITE 14.1 Calling the competence of the somite into question.** When the *tbx6* gene was knocked out from mice, the resulting embryos had three neural tubes in the posterior of their bodies. Without the *tbx6* gene, the somitic tissue responded to the notochord and epidermal signals as if it were neural ectoderm.

> **WEBSITE 14.2 Cranial paraxial mesoderm.** Most of the head musculature does not come from somites. Rather, it comes from the cranial paraxial (prechordal plate) mesoderm. These cells originate adjacent to the sides of the brain, and they migrate to their respective destinations.

Myogenesis: The Development of Muscle

Specification and differentiation by the myogenic bHLH proteins

As we have seen, muscle cells come from two cell lineages in the somite. In both instances, paracrine factors instruct the myotome cells to become muscles by inducing them to synthesize the **MyoD** protein (see Figure 14.11; Maroto et al.

1997; Tajbakhsh et al. 1997; Pownall et al. 2002). The way this happens differs slightly between the hypaxial and epaxial lineages and between different vertebrate classes. In the lateral portion of the mouse dermamyotome, which forms the hypaxial muscles, factors from the surrounding environment induce the Pax3 transcription factor. In the absence of other inhibitory transcription factors (such as those found in the sclerotome cells), Pax3 then activates the *myoD* gene. In the medial region of the dermamyotome, which forms the epaxial muscles, MyoD is induced by the Myf5 protein. MyoD and Myf5 belong to a family of transcription factors called the **myogenic bHLH** (basic helix-loop-helix) **proteins** (sometimes also referred to as the **MRF**s, myogenic regulatory factors). The proteins of this family all bind to similar sites on the DNA and activate muscle-specific genes. For instance, the MyoD protein appears to directly activate the muscle-specific creatine phosphokinase gene by binding to the DNA immediately upstream from it (Lassar et al. 1989), and there are two MyoD-binding sites on the DNA adjacent to the genes encoding a subunit of the chicken muscle acetylcholine receptor (Piette et al. 1990). MyoD also directly activates its own gene. Therefore, once the *myoD* gene is activated, its protein product binds to the DNA immediately upstream of the *myoD* gene and keeps this gene active.

While Pax3 is found in several other cell types, the myogenic bHLH proteins are specific for muscle cells. Any cell making a myogenic bHLH transcription factor such as MyoD or Myf5 is committed to becoming a muscle cell. Transfection of genes encoding any of these myogenic proteins into a wide range of cultured cells converts those cells into muscles (Thayer et al. 1989; Weintraub et al. 1989).

WEBSITE 14.3 Myogenic bHLH proteins and their regulators. Because the MyoD protein and its relatives are so powerful they can turn nearly any cell into a muscle cell, the synthesis of this protein has to be inhibited at numerous steps. Numerous inhibitors of MyoD family gene expression and protein function have been found.

Muscle cell fusion

The myotome cells producing the myogenic bHLH proteins are the myoblasts—committed muscle cell precursors. Experiments with chimeric mice and cultured myoblasts showed that these cells align together and fuse to form the multinucleated **myotubes** characteristic of muscle tissue. Thus, the multinucleated myotube cells are the product of several myoblasts joining together and dissolving the cell membranes between them (Konigsberg 1963; Mintz and Baker 1967).

Muscle cell fusion begins when the myoblasts leave the cell cycle. As long as particular growth factors (particularly fibroblast growth factors) are present, the myoblasts will proliferate without differentiating. When these factors are depleted, the myoblasts stop dividing, secrete fibronectin onto their extracel-

lular matrix, and bind to it through α5β1 integrin, their major fibronectin receptor (Menko and Boettiger 1987; Boettiger et al. 1995). If this adhesion is experimentally blocked, no further muscle development ensues, so it appears that the signal from the integrin-fibronectin attachment is critical for instructing the myoblasts to differentiate into muscle cells (Figure 14.12).

The second step is the alignment of the myoblasts into chains. This step is mediated by cell membrane glycoproteins, including several cadherins and CAMs (Knudsen 1985: Knudsen et al. 1990). Recognition and alignment between cells takes place only if the cells are myoblasts. Fusion can occur even between chick and rat myoblasts in culture (Yaffe and Feldman 1965); the identity of the species is not critical.

The third step is the cell fusion event itself. As in most membrane fusions, calcium ions are critical, and fusion can be activated by calcium ionophores, such as A23187, that carry calcium ions across cell membranes (Shainberg et al. 1969; David et al. 1981). Fusion appears to be mediated by a set of metalloproteinases called **meltrins**. These proteins were discovered during a search for myoblast proteins that would be homologous to fertilin, a protein implicated in sperm-egg membrane fusion. Yagami-Hiromasa and colleagues (1995) found that one of these meltrins (meltrin-α) is expressed in myoblasts at about the same time that fusion begins, and that antisense RNA to the meltrin-α message inhibited fusion when added to myoblasts.

As the myoblasts become capable of fusing, another myogenic bHLH protein, **myogenin**, becomes active. While MyoD and Myf5 are active in the lineage specification of muscle cells, myogenin appears to mediate their differentiation (Bergstrom and Tapscott 2001). Myogenin binds to the regulatory region of several muscle-specific genes and activates their expression.

WEBSITE 14.4 Muscle formation. Two websites focus on muscle development. (1) The terms *hypaxial* and *epaxial* refer to the adult positions of the muscles. Recent studies suggest the use of a different nomenclature more in keeping with the embryonic derivation of the musculature. (2) Research on chimeric mice has shown that skeletal muscle becomes multinucleate by the fusion of cells, while heart muscle becomes multinucleate by nuclear divisions within a cell.

Osteogenesis: The Development of Bones

Some of the most obvious structures derived from the paraxial mesoderm are bones. We can only begin to outline the mechanisms of bone formation here; students wishing further details are invited to consult histology textbooks that devote entire chapters to this topic.

There are three distinct lineages that generate the skeleton. The somites generate the axial skeleton, the lateral plate mesoderm generates the limb skeleton, and the cranial neural

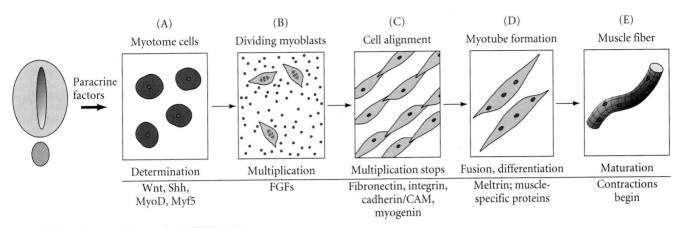

(A)	(B)	(C)	(D)	(E)
Myotome cells	Dividing myoblasts	Cell alignment	Myotube formation	Muscle fiber
Determination	Multiplication	Multiplication stops	Fusion, differentiation	Maturation
Wnt, Shh, MyoD, Myf5	FGFs	Fibronectin, integrin, cadherin/CAM, myogenin	Meltrin; muscle-specific proteins	Contractions begin

(F)

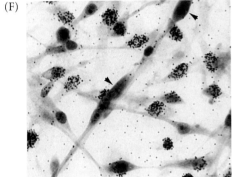

Figure 14.12

Conversion of myoblasts into muscles in culture. (A) Determination of myotome cells by paracrine factors. (B) Committed myoblasts divide in the presence of growth factors (primarily FGFs), but show no obvious muscle-specific proteins. (C–D) When the growth factors are used up, the myoblasts cease dividing, align, and fuse into myotubes. (E) The myotubes become organized into muscle fibers that spontaneously contract. (F) Autoradiograph showing DNA synthesis in myoblasts and the exit of fusing cells from the cell cycle. Phospholipase C can "freeze" myoblasts after they have aligned with other myoblasts, but before their membranes fuse. Cultured myoblasts were treated with phospholipase C and then exposed to radioactive thymidine. Unattached myoblasts continued to divide and thus incorporated the radioactive thymidine into their DNA. Aligned (but not yet fused) cells (arrows) did not incorporate the label. (A–E after Wolpert 1998; F from Nameroff and Munar 1976, photograph courtesy of M. Nameroff.)

crest gives rise to the branchial arch and craniofacial bones and cartilage.* There are two major modes of bone formation, or **osteogenesis**, and both involve the transformation of a preexisting mesenchymal tissue into bone tissue. The direct conversion of mesenchymal tissue into bone is called **intramembranous** (or **dermal**) **ossification**, and this was discussed in the preceding chapter. In other cases, the mesenchymal cells differentiate into cartilage, and this cartilage is later replaced by bone. The process by which a cartilage intermediate is formed and replaced by bone cells is called **endochondral ossification**.

Endochondral ossification

Endochondral ossification involves the formation of **cartilage** tissue from aggregated mesenchymal cells, and the subsequent replacement of cartilage tissue by bone (Horton 1990). This is the type of bone formation characteristic of the vertebrae, ribs, and limbs. The vertebrae and ribs form from the somites, while the limb bones (discussed in Chapter 16) form from the lateral plate mesoderm. An individual vertebra is not formed directly from a single somite (Aoyama and Asamoto 2000; Morin-Kensicki et al. 2002). Rather, a particular vertebra is formed from the posterior half of one somite and the anterior

half of the next somite. (In this manner, vertebrate segmentation resembles the parasegment-to-segment transition in insect development.)

The process of endochondral ossification can be divided into five stages (Figure 14.13). First, the mesenchymal cells commit to becoming cartilage cells. This commitment is caused by paracrine factors that induce the nearby mesodermal cells to express two transcription factors, Pax1 and **Scleraxis**. These transcription factors are thought to activate cartilage-specific genes (Cserjesi et al. 1995; Sosic et al. 1997). Thus, Scleraxis is expressed in the mesenchyme from the sclerotome, in the facial mesenchyme that forms cartilaginous precursors to bone, and in the limb mesenchyme (Figure 14.14).

During the second phase of endochondral ossification, the committed mesenchyme cells condense into compact nodules and differentiate into **chondrocytes**, the cartilage cells. N-cadherin appears to be important in the initiation of these condensations, and N-CAM seems to be critical for maintaining them (Oberlender and Tuan 1994; Hall and Miyake 1995). In humans, the *SOX9* gene, which encodes a DNA-binding protein, is expressed in the precartilaginous condensations. Mutations of the *SOX9* gene cause campomelic dysplasia, a rare disorder of skeletal development that results in deformities of most of the bones of the body. Most affected babies die from respiratory failure due to poorly formed tracheal and rib cartilage (Wright et al. 1995).

---

*Craniofacial cartilage development was discussed in Chapter 13 and will be revisited in Chapter 22; the development of the limbs will be detailed in Chapter 16.

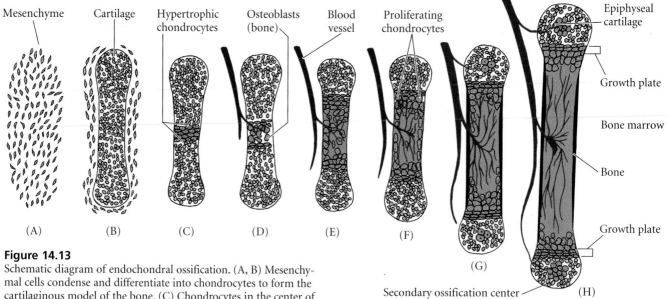

Figure 14.13
Schematic diagram of endochondral ossification. (A, B) Mesenchymal cells condense and differentiate into chondrocytes to form the cartilaginous model of the bone. (C) Chondrocytes in the center of the shaft undergo hypertrophy and apoptosis while they change and mineralize their extracellular matrix. Their deaths allow blood vessels to enter. (D, E) Blood vessels bring in osteoblasts, which bind to the degenerating cartilaginous matrix and deposit bone matrix. (F–H) Bone formation and growth consist of ordered arrays of proliferating, hypertrophic, and mineralizing chondrocytes. Secondary ossification centers also form as blood vessels enter near the tips of the bone. (After Horton 1990.)

During the third phase of endochondral ossification, the chondrocytes proliferate rapidly to form the cartilage model for the bone. As they divide, the chondrocytes secrete a cartilage-specific extracellular matrix. In the fourth phase, the chondrocytes stop dividing and increase their volume dramatically, becoming **hypertrophic chondrocytes**. These large chondrocytes alter the matrix they produce (by adding collagen X and more fibronectin) to enable it to become mineralized (calcified) by calcium carbonate. They also secrete the angiogenesis factor, VEGF, which can transform mesodermal mesenchyme cells into blood vessels (Gerber et al. 1999; Haigh et al. 2000; see Chapter 15). A number of events lead to the hypertrophy and mineralization (calcification) of the chondrocytes, including an initial switch from aerobic to anaerobic respiration, which alters their cell metabolism and mitochondrial energy potential (Shapiro et al. 1982). Hypertrophic chondrocytes secrete numerous small membrane-bound vesicles into the extracellular matrix. These vesicles contain enzymes that are active in the generation of calcium and phosphate ions and initiate the mineralization process within the cartilaginous matrix (Wu et al. 1997). The hypertrophic chondrocytes, their metabolism and mitochondrial membranes altered, then die by apoptosis (Hatori et al. 1995; Rajpurohit et al. 1999).

In the fifth phase, the blood vessels induced by VEGF invade the cartilage model. As the hypertrophic chondrocytes

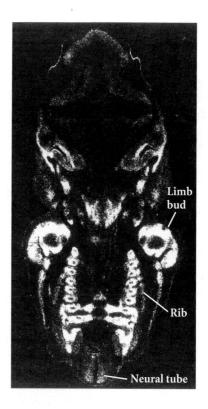

Figure 14.14
Localization of the *scleraxis* message (light areas) at the sites of chondrocyte formation in a 11.5-day mouse embryo. The *scleraxis* transcripts are seen in the condensing cartilage of the nose and face and in the precursors of the limbs and ribs. (After Cserjesi et al. 1995, photograph courtesy of E. Olson.)

Figure 14.15
Skeletal mineralization in 19-day chick embryos that developed (A) in shell-less culture and (B) inside an egg during normal incubation. The embryos were fixed and stained with alizarin red to show the calcified bone matrix. (From Tuan and Lynch 1983; photographs courtesy of R. Tuan.)

die, the cells that surround the cartilage model differentiate into **osteoblasts**. These cells express the **Cbfa1** (also called Runx2) transcription factor, which is necessary for the development of both intramembranous and endochondral bone (see Figure 13.9). The replacement of chondrocytes by bone cells is dependent on the mineralization of the extracellular matrix. This remodeling releases VEGF, and more blood vessels are made around the dying cartilage. These blood vessels bring in both osteoblasts and **chondroclasts** (which eat the debris of the apoptotic chondrocytes). If the blood vessels are inhibited from forming, bone development is significantly delayed (Yin et al. 2002; see Karsenty and Wagner 2002). The ostoblasts begin forming bone matrix on the partially degraded matrix and construct a **bone collar** around the dying cartilage cells (Bruder and Caplan 1989; Hatori et al. 1995; St. Jacques et al. 1999). Eventually, all the cartilage is replaced by bone. Thus, the cartilage tissue serves as a model for the bone that follows. The skeletal components of the vertebral column, pelvis, limbs, and portions of the skull are first formed of cartilage and later become bone.

The importance of the mineralized extracellular matrix for bone differentiation is clearly illustrated in the developing skeleton of the chick embryo, which utilizes the calcium carbonate of the eggshell as its calcium source. During development, the circulatory system of the chick embryo translocates about 120 mg of calcium from the shell to the skeleton (Tuan 1987). When chick embryos are removed from their shells at day 3 and grown in shell-less cultures (in plastic wrap) for the duration of their development, much of the cartilaginous skeleton fails to mature into bony tissue (Figure 14.15; Tuan and Lynch 1983).

WEBSITE 14.5 Paracrine factors, their receptors, and human bone growth. Mutations in the genes encoding paracrine factors and their receptors cause numerous skeletal anomalies in humans and mice. The FGF and Hedgehog pathways are especially important.

Osteoclasts

New bone material is added peripherally from the internal surface of the **periosteum,** a fibrous sheath containing connective tissue and capillaries that covers the developing bone. At the same time, there is a hollowing out of the internal region of the bone to form the bone marrow cavity. This destruction of bone tissue is carried out by **osteoclasts**, multinucleated cells that enter the bone through the blood vessels (Kahn and Simmons 1975; Manolagas and Jilka 1995). Osteoclasts are derived from the same precursors as macrophage blood cells, and they dissolve both the inorganic and the protein portions of the bone matrix (Ash et al. 1980; Blair et al. 1986). Each osteoclast extends numerous cellular processes into the matrix and pumps hydrogen ions out onto the surrounding material, thereby acidifying and solubilizing it* (Figure 14.16; Baron et al. 1985, 1986). The blood vessels also import the blood-forming cells that will reside in the marrow for the duration of the organism's life. The number and activity of osteoclasts must be tightly regulated. If there are too many active osteoclasts, too much bone will be dissolved, and **osteoporosis** will result. Conversely, if not enough osteoblasts are produced, the bones are not hollowed out for the marrow, and **osteopetrosis** results (Tondravi et al. 1997; Duong and Rodan 2001).

WEBSITE 14.6 Osteoclast differentiation. Hormones regulate the production of osteoclasts, and the hormonal changes of aging may cause osteoporosis by increasing the number of osteoclasts. The conversion of a macrophage stem cell into a osteoclast is regulated by osteoprotegerin and its ligand. It is thought that signals from osteoblasts instruct the progenitor cell to become an osteoclast.

INTERMEDIATE MESODERM: THE UROGENITAL SYSTEM

The intermediate mesoderm generates the urogenital system—the kidneys, the gonads, and their respective duct systems. Saving the gonads for our discussion of sex determination in Chapter 17, we will concentrate here on the development of the mammalian kidney.

---

*Given the physiology of the osteoclast, we can now appreciate H. L. Mencken's (1919) prescient intuition: "Life is a struggle, not against sin, not against the Money Power, not against malicious animal magnetism, but against hydrogen ions."

(A)

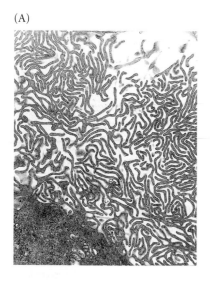

(B)

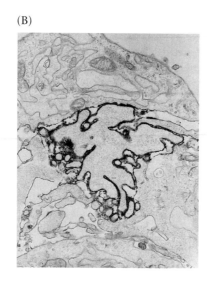

(C)

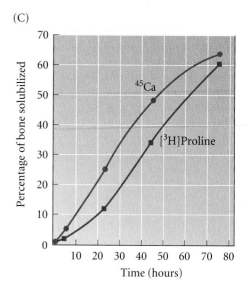

Figure 14.16

Osteoclast activity on the bone matrix. (A) Electron micrograph of the ruffled membrane of a chick osteoclast cultured on reconstituted bone matrix. (B) Section of ruffled membrane stained for the presence of an ATPase capable of transporting hydrogen ions from the cell. The ATPase is restricted to the membrane of the cell process. (C) Solubilization of inorganic and collagenous matrix components (as measured by the release of [^{45}Ca] and [^{3}H] proline, respectively) by 10,000 osteoclasts incubated on labeled bone fragments. (A and C from Blair et al. 1986; B from Baron et al. 1986. Photographs courtesy of the authors.)

these two proteins, the presumptive intermediate mesoderm undergoes apoptosis. It is likely, then, that the paraxial mesoderm sends some as-yet-unknown signal to induce *Pax2* and *Pax8* expression in the intermediate mesoderm, and these two transcription factors begin the production of the kidney.

The Specification of the Intermediate Mesoderm

The intermediate mesoderm of the chick embryo acquires its ability to form kidneys through its interactions with the paraxial mesoderm. Mauch and her colleagues (2000) showed that signals from the paraxial mesoderm induced pronephros formation in the intermediate mesoderm of the chick embryo. They cut developing embryos such that the intermediate mesoderm could not contact the paraxial mesoderm on one side of the body. That side of the body (where contact with the paraxial mesoderm was abolished) did not form kidneys, but the undisturbed side was able to form kidneys (Figure 14.17). The paraxial mesoderm appears to be both necessary and sufficient for the induction of kidney-forming ability in the intermediate mesoderm, since co-culturing lateral plate mesoderm with paraxial mesoderm caused pronephric tubules to form in the lateral plate mesoderm, and no other cell type would accomplish this.

Bouchard and colleagues (2002) have shown that mouse intermediate mesoderm lacking the genes encoding both the Pax2 and Pax8 transcription factors (but neither separately) are unable to form any kidney structures. Moreover, without

(A)　　　　　　　　　　　　　(B)

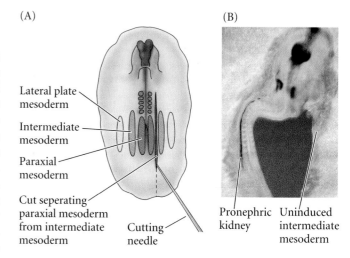

Lateral plate mesoderm

Intermediate mesoderm

Paraxial mesoderm

Cut seperating paraxial mesoderm from intermediate mesoderm

Cutting needle

Pronephric kidney　Uninduced intermediate mesoderm

Figure 14.17

Signals from the paraxial mesoderm induce pronephros formation in the intermediate mesoderm of the chick embryo. (A) The paraxial mesoderm was surgically separated from the intermediate mesoderm on the right side of the body. (B) As a result, a pronephric kidney (Pax2-staining tubules) developed only on the left side. (After Mauch et al. 2000; photograph courtesy of T. J. Mauch and G. C. Schoenwolf.)

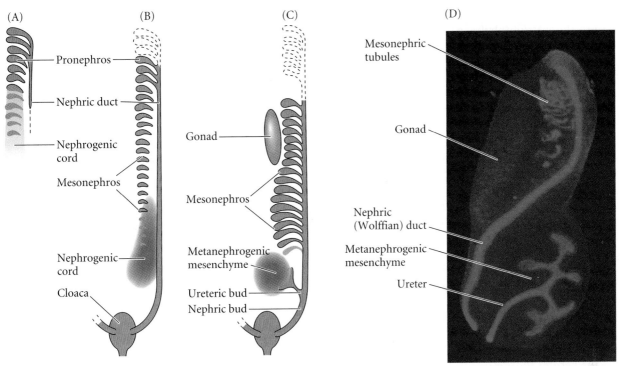

Figure 14.18
General scheme of development in the vertebrate kidney. (A) The original tubules, constituting the pronephros, are induced from the nephrogenic mesenchyme by the pronephric duct as it migrates caudally. (B) As the pronephros degenerates, the mesonephric tubules form. (C) The final mammalian kidney, the metanephros, is induced by the ureteric bud, which branches from the nephric duct. (D) The intermediate mesoderm of a 13-day mouse embryo showing the initiation of the metanephric kidney (bottom) while the mesonephros is still apparent. The duct tissue is stained with a fluorescent antibody to a cytokeratin found in the pronephric duct and its derivatives. (A–C after Saxén 1987; D, photograph courtesy of S. Vainio.)

Progression of Kidney Types

The importance of the kidney cannot be overestimated. As Homer Smith noted (1953), "our kidneys constitute the major foundation of our philosophical freedom. Only because they work the way they do has it become possible for us to have bone, muscles, glands, and brains." While this statement may smack of hyperbole, the human kidney is an incredibly intricate organ. Its functional unit, the **nephron**, contains over 10,000 cells and at least 12 different cell types, with each cell type located in a particular place in relation to the others along the length of the nephron.

The development of the mammalian kidney progresses through three major stages. The first two stages are transient; only the third and last persists as a functional kidney. Early in development (day 22 in humans; day 8 in mice), the **pronephric duct** arises in the intermediate mesoderm just ventral to the anterior somites. The cells of this duct migrate caudally, and the anterior region of the duct induces the adjacent mesenchyme to form the tubules of the initial kidney, the

pronephros (Figure 14.18A). While the pronephric tubules form functioning kidneys in fish and in amphibian larvae, they are not thought to be active in amniotes. In mammals, the pronephric tubules and the anterior portion of the pronephric duct degenerate, but the more caudal portions of the pronephric duct persist and serve as the central component of the excretory system throughout its development (Toivonen 1945; Saxén 1987). This remaining duct is often referred to as the **nephric** or **Wolffian duct**.

As the pronephric tubules degenerate, the middle portion of the nephric duct induces a new set of kidney tubules in the adjacent mesenchyme. This set of tubules constitutes the **mesonephros**, or mesonephric kidney (Figure 14.18B; Sainio and Raatikainen-Ahokas 1999). In some mammalian species, the mesonephros functions briefly in urine filtration, but in mice and rats, it does not function as a working kidney. In humans, about 30 mesonephric tubules form, beginning around day 25. As more tubules are induced caudally, the anterior mesonephric tubules begin to regress through apoptosis (although in mice, the anterior tubules remain while the posterior ones regress: Figure 14.18C, D). While it still remains unknown whether the human mesonephros actually functions to filter blood and make urine, the mesonephros provides important developmental functions during its brief existence. First, as we will see in Chapter 15, it is the source of hematopoietic stem cells necessary for blood cell development (Medvinsky and Dzierzak 1996; Wintour et al. 1996). Second, in male mammals, some of the mesonephric tubules persist to become the sperm-carrying tubes (the vas deferens and efferent ducts) of the testes (see Chapter 17).

The permanent kidney of amniotes, the **metanephros**, is generated by some of the same components as the earlier, transient kidney types (Figure 14.18C). It is thought to originate through a complex set of interactions between epithelial and mesenchymal components of the intermediate mesoderm. In the first steps, the **metanephrogenic mesenchyme** forms in posteriorly located regions of the intermediate mesoderm, and it induces the formation of a branch from each of the paired nephric ducts. These epithelial branches are called the **ureteric buds**. These buds eventually separate from the nephric duct to become the ureters that take the urine to the bladder. When the ureteric buds emerge from the nephric duct, they enter the metanephrogenic mesenchyme. The

ureteric buds induce this mesenchymal tissue to condense around them and differentiate into the nephrons of the mammalian kidney. As this mesenchyme differentiates, it tells the uretereric bud to branch and grow.

Reciprocal Interactions of Developing Kidney Tissues

The two intermediate mesodermal tissues—the ureteric bud and the metanephrogenic mesenchyme—interact and reciprocally induce each other to form the kidney (Figure 14.19). The metanephrogenic mesenchyme causes the ureteric bud to elongate and branch. The tips of these branches induce the loose mesenchyme cells to form epithelial aggregates. Each aggregated nodule of about 20 cells will proliferate and differentiate into the intricate structure of a renal nephron. Each nodule first elongates into a "comma" shape, then forms a characteristic S-shaped tube. Soon afterward, the cells of this epithelial structure begin to differentiate into regionally specific cell types, including the capsule cells, the **podocytes**, and the distal and proximal tubule cells.* While this is happening, the epithelializing nodules break down the basal lamina of the ureteric bud ducts and fuse with them. This fusion creates a connection between the ureteric bud and the newly formed

*The intricate coordination of nephron development with the blood capillaries the nephrons filter is accomplished by the secretion of VEGF from the podocytes. VEGF, as we will see in the next chapter, is a powerful inducer of blood vessels, and it causes endothelial cells from the dorsal aorta to form the capillary loops of the glomerular filtration apparatus (Aitkenhead et al. 1998; Klanke et al. 1998).

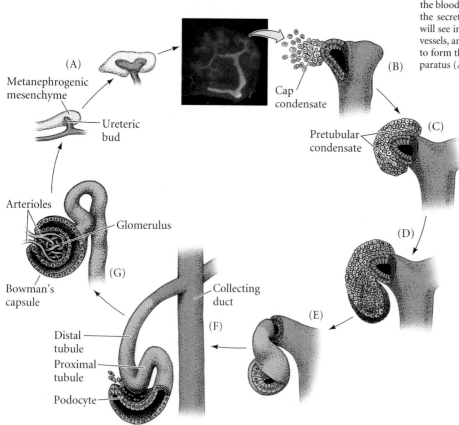

(A) Metanephrogenic mesenchyme

Ureteric bud

Cap condensate

Pretubular condensate

(B)

(C)

(D)

(E)

(F)

(G)

Arterioles

Glomerulus

Bowman's capsule

Collecting duct

Distal tubule

Proximal tubule

Podocyte

Figure 14.19
Reciprocal induction in the development of the mammalian kidney. (A) As the ureteric bud enters the metanephrogenic mesenchyme, the mesenchyme induces the bud to branch. (B–G) At the tips of the branches, the epithelium induces the mesenchyme to aggregate and cavitate to form the renal tubules and glomeruli (where the blood from the arteriole is filtered). When the mesenchyme has condensed into an epithelium, it digests the basal lamina of the ureteric bud cells that induced it and connects to the ureteric bud epithelium. A portion of the aggregated mesenchyme (the pretubular condensate) becomes the nephron (renal tubules and Bowman's capsule), while the ureteric bud becomes the collecting duct for the urine. (After Saxén 1987 and Sariola 2002.)

tubule, allowing material to pass from one into the other (Bard et al. 2001). These tubules derived from the mesenchyme form the secretory nephrons of the functioning kidney, and the branched ureteric bud gives rise to the renal collecting ducts and to the ureter, which drains the urine from the kidney.

Clifford Grobstein (1955, 1956) documented this reciprocal induction in vitro. He separated the ureteric bud from the metanephrogenic mesenchyme and cultured them either individually or together. In the absence of mesenchyme, the ureteric bud does not branch. In the absence of the ureteric bud, the mesenchyme soon dies. When they are placed together, however, the ureteric bud grows and branches, and nephrons form throughout the mesenchyme (Figure 14.20).

The mechanisms of reciprocal induction

The induction of the metanephros can be viewed as a dialogue between the ureteric bud and the metanephrogenic mesenchyme. As the dialogue continues, both tissues are altered. While there are several simultaneous dialogues between different groups of kidney cells (see Kuure 2000 and Bard 2002 for full reports on these conversations), there appear to be at least eight critical sets of signals operating in the reciprocal induction of the metanephros.

STEP 1: FORMATION OF THE METANEPHROGENIC MESENCHYME: HOX-11 AND WT1 Only the metanephrogenic mesenchyme has the competence to respond to the ureteric bud to form kidney tubules, and if induced by other tissues (such as embryonic salivary gland or neural tube tissue), this mesenchyme will respond by forming kidney tubules and no other structures (Saxén 1970; Sariola et al. 1982). Thus, the metanephrogenic mesenchyme cannot become any tissue other than nephrons.

Two sets of transcription factors may specify the metanephrogenic mesenchyme. The first set consists of the **Hoxa-11**, **Hoxc-11**, and **Hoxd-11** proteins. When the paralogous genes encoding these proteins are knocked out in mouse embryos, the differentiation of the metanephrogenic mesenchyme is arrested, and it cannot induce the ureteric bud to form (Patterson et al. 2001; Wellik et al. 2002). The second set of transcription factors includes **WT1**. Without WT1, the metanephrogenic mesenchyme lacks the competence to respond to ureteric bud inducers and these cells remain uninduced and die (Kreidberg et al. 1993). In situ hybridization shows that WT1 is normally first expressed in the intermediate mesoderm prior to kidney formation and is then expressed in the developing kidney, gonad, and mesothelium (Pritchard-Jones et al. 1990; van Heyningen et al. 1990; Armstrong et al. 1992). Although the metanephrogenic mesenchyme appears homogeneous, it may contain both mesodermally derived tissue and some cells of neural crest origin (Le Douarin and Tiellet 1974; Sariola et al. 1989; Sainio et al. 1994).

STEP 2: THE METANEPHROGENIC MESENCHYME SECRETES GDNF TO INDUCE AND DIRECT THE URETERIC BUD. The second signal in kidney development is a set of diffusible molecules that cause the ureteric buds to grow out from each of the two nephric ducts. Recent research has shown that **glial cell line-derived neurotrophic factor** (**GDNF**) is a critical component of this signal.* GDNF is synthesized in the metanephrogenic mesenchyme, and mice whose *gdnf* genes were knocked out died soon after birth from renal agenesis (lack of kidneys) (Moore et al. 1996; Pichel et al. 1996; Sánchez et al. 1996). The GDNF receptors (the Ret tyrosine kinase receptor and the GFRα co-receptor) are synthesized in the nephric ducts and later become concentrated in the growing ureteric buds (Figure 14.21; Schuchardt et al. 1996). Mice lacking the GDNF receptors die of renal agenesis. It is thought that GDNF may become expressed through a complex pathway initiated by the Pax2 and Hox-11 transcription factors (Xu et al. 1999; Wellik et al. 2002).

In another mouse mutation, *Danforth short-tail*, the ureteric bud is initiated but does not enter the metanephrogenic mesenchyme (Gluecksohn-Schoenheimer 1943). Here, too, the kidney does not form. This failure of the ureteric bud to grow has been correlated with the absence of Wnt11 expression in the tips of the ureteric bud. Wnt11 expression is maintained by proteoglycans made by the mesonephrogenic mes-

---

* This is the same compound that we saw in Chapter 13 as critical for the induction of dopaminergic neurons in the mammalian brain. We will meet GDNF again when we discuss sperm cell production in Chapter 19. GDNF is one busy protein.

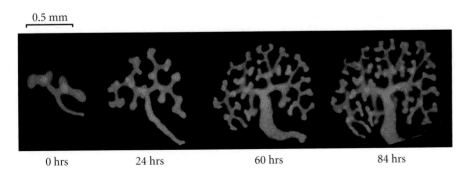

0.5 mm

0 hrs   24 hrs   60 hrs   84 hrs

Figure 14.20
Kidney induction observed in vitro. (A) A kidney rudiment from an 11.5-day mouse embryo was placed into culture. This transgenic mouse had a *GFP* gene fused to a *Hoxb-7* promoter, so it expressed green fluorescent protein in the Wolffian (nephric) duct and in the ureteric buds. Since GFP can be photographed in living tissues, the kidney could be followed as it developed. (After Srinivas et al. 1999; photographs courtesy of F. Costantini.)

(A)　　　　　　　　(B)　　　　　　　　(C)

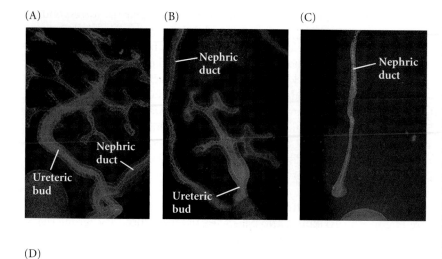

Figure 14.21
Ureteric bud growth is dependent on GDNF and its receptors. (A) The ureteric bud from a 11.5-day wild-type mouse embryonic kidney cultured for 72 hours has a characteristic branching pattern. (B) In embryonic mice heterozygous for a mutation of the gene encoding GDNF, the size of the ureteric bud and the number and length of its branches are reduced. (C) In mouse embryos missing both copies of the *gdnf* gene, the ureteric bud does not form. (Scale bars = 100 μm.) (D) The receptors for GDNF are concentrated in the posterior portion of the nephric duct. GDNF secreted by the metanephrogenic mesenchyme stimulates the growth of the ureteric bud from this duct. At later stages, the GDNF receptor is found exclusively at the tips of the ureteric buds. (A–C from Pichel et al. 1996, photographs courtesy of J. G. Pichel and H. Sariola; D after Schuchardt et al. 1996.)

(D)

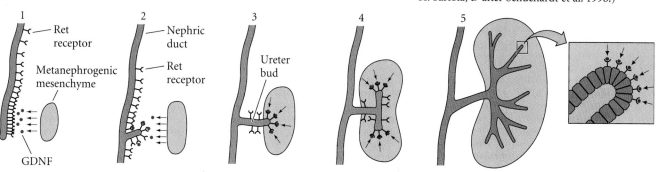

enchyme. It appears that once the ureteric bud has entered the mesenchyme, these mesenchymal proteoglycans stimulate its continued growth by maintaining its expression and secretion of Wnt11 (Davies et al. 1995; Kispert et al. 1996).

STEP 3: THE URETERIC BUD SECRETES FGF2 AND BMP7 TO PREVENT MESENCHYMAL APOPTOSIS. The third signal in kidney development is sent from the ureteric bud to the metanephrogenic mesenchyme, and it alters the fate of the mesenchyme cells. If left uninduced by the ureteric bud, the mesenchyme cells undergo apoptosis (Grobstein 1955; Koseki et al. 1992). However, if induced by the ureteric bud, the mesenchyme cells are rescued from the precipice of death and are converted into proliferating stem cells (Bard and Ross 1991; Bard et al. 1996). The factors secreted from the ureteric bud include fibroblast growth factor 2 (FGF2) and bone morphogenetic protein 7 (BMP7). FGF2 has three modes of action in that it inhibits apoptosis, promotes the condensation of mesenchyme cells, and maintains the synthesis of WT1 (Perantoni et al. 1995). BMP7 has similar effects, and in the absence of BMP7, the mesenchyme of the kidney undergoes apoptosis (see Figure 4.22; Dudley et al. 1995; Luo et al. 1995).

STEP 4: LIF AND WNT6 FROM THE URETERIC BUD INDUCE MESENCHYME CELLS TO AGGREGATE. The ureteric bud also causes dramatic changes in the behavior of the metanephrogenic mes-

enchyme cells, converting them into an epithelium. The newly induced mesenchyme synthesizes E-cadherin, which causes the mesenchyme cells to clump together. These aggregated nodes of mesenchyme now synthesize an epithelial basal lamina containing type IV collagen and laminin. At the same time, the mesenchyme cells synthesize receptors for laminin, allowing the aggregated cells to become an epithelium (Ekblom et al. 1994; Müller et al. 1997). The cytoskeleton also changes from one characteristic of mesenchyme cells to one typical of epithelial cells (Ekblom et al. 1983; Lehtonen et al. 1985).

The transition from mesenchymal to epithelial organization may be mediated by several molecules, including the expression of Pax2 in the newly induced mesenchyme cells. When antisense RNA to *Pax2* prevents the translation of the *Pax2* mRNA that is transcribed as a response to induction, the mesenchyme cells of cultured kidney rudiments fail to condense (Rothenpieler and Dressler 1993). Thus, Pax2 may play several roles during kidney formation.

In addition, FGF2 is needed to induce the aggregation of the mesenchyme cells, but it is not capable of turning these aggregates into epithelial cells (Karavanova et al. 1996). Leukemia inhibitory factor (LIF) is able to convert mesenchymal aggregates into kidney tubule epithelium—but only if they have been exposed to FGF2 (Figure 14.22; Barasch et al. 1999). The ureteric bud secretes FGFs and LIF, and the mesenchyme has receptors for these proteins.

(A) (B)

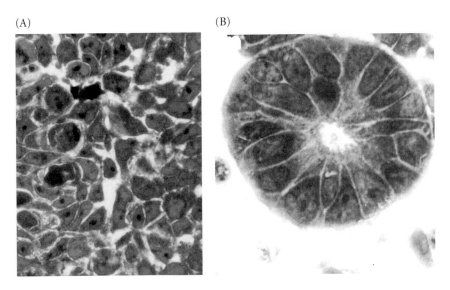

Figure 14.22
LIF induces kidney tubule formation. (A) Metanephrogenic mesenchyme treated with FGF2 or TGF-β will form clumps but will not form epithelia. (B) Tubular epithelium is induced when FGF-treated mesenchyme is exposed to ureteric bud secretions or LIF. (Photographs courtesy of J. Barasch.)

And finally, the mesenchyme has receptors for Wnt6, another protein made at the tip of the ureteric bud. Wnt6 appears to promote the condensation of mesenchyme in an FGF-independent way (Itäranta et al. 2002).

STEP 5: WNT4 CONVERTS AGGREGATED MESENCHYME CELLS INTO A NEPHRON. Once induced, and after it has started to condense, the mesenchyme begins to secrete Wnt4, which acts in an autocrine fashion to complete the transition from mesenchymal mass to epithelium (Stark et al. 1994; Kispert et al. 1998). Wnt4 expression is found in the condensing mesenchyme cells, in the resulting S-shaped tubules, and in the region where the newly epithelialized cells fuse with the ureteric bud tips. In mice lacking the Wnt4 gene, the mesenchyme becomes condensed but does not form epithelia. Therefore, the ureteric bud induces the changes in the metanephrogenic mesenchyme by secreting FGFs, LIF, and Wnt6; but these changes are mediated by the effects of the mesenchyme's secretion of Wnt4 on itself.

One molecule that may be involved in the transition from aggregated mesenchyme to nephrons is the Lim-1 homeodomain transcription factor (Karavanov et al. 1998). This protein is found in the mesenchyme cells after they have condensed around the ureteric bud, and its expression persists in the developing nephron (Figure 14.23). Two other proteins that may be critical for the conversion of the aggregated cells into a nephron are polycystins 1 and 2. These proteins are the products of the genes whose loss-of-function alleles give rise to human polycystic kidney disease. Mice deficient in these genes have abnormal, swollen nephrons (Ward et al. 1996; van Adelsberg et al. 1997).

STEP 6: SIGNALS FROM THE MESENCHYME INDUCE THE BRANCHING OF THE URETERIC BUD. Recent evidence has implicated several paracrine factors in the branching of the ureteric bud, and these factors probably work as pushes and pulls. Some factors may preserve the extracellular matrix surrounding the epithelium, thereby preventing branching from taking place. Conversely, other factors may cause the digestion of this extracellular matrix, permitting branching to occur.

The first candidate for regulating ureteric bud branching is GDNF (Sainio et al. 1997). GDNF from the mesenchyme not only induces the initial ureteric bud from the nephric duct but it can also induce secondary buds from the ureteric bud once the bud has entered the mesenchyme (Figure 14.24). The second candidate molecule is **transforming growth factor β1** (**TGF-β1**). When exogenous TGF-β1 is added to cultured kidneys, it prevents the epithelium from branching (Figure 14.25B; Ritvos et al. 1995). TGF-β1 is known to promote the synthesis of extracellular matrix proteins and to inhibit the metalloproteinases that can digest these matrices (Penttinen et al. 1988; Nakamura et al. 1990). Thus, it is possible that TGF-β1 stabilizes branches once they form.

A third molecule that may be important in epithelial branching is **BMP4** (Miyazaki et al. 2000). BMP4 is found in the mesenchymal cells surrounding the nephric duct, and

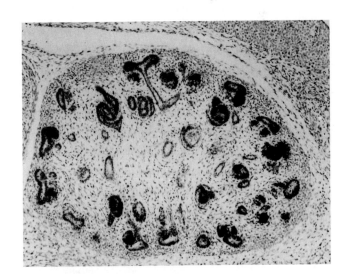

Figure 14.23
Lim-1 expression (dark stain) in a 19-day embryonic mouse kidney. In situ hybridization shows high levels of expression in the newly epithelialized comma-shaped and S-shaped bodies that will become nephrons. (From Karavanov et al. 1998; photograph courtesy of A. A. Karavanov.)

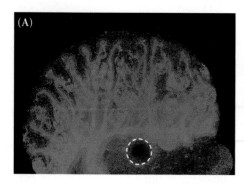 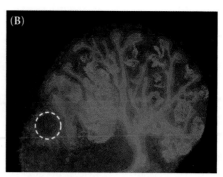

Figure 14.24

The effect of GDNF on the branching of the ureteric epithelium. The ureteric bud and its branches are stained orange (with antibodies to cytokeratin 18), while the nephrons are stained green (with antibodies to nephron brush border antigens). (A) 13-day embryonic mouse kidney cultured 2 days with a control bead (circle) has a normal branching pattern. (B) A similar kidney cultured 2 days with a GDNF-soaked bead shows a distorted pattern, as new branches are induced in the vicinity of the bead. (From Sainio et al. 1997; photographs courtesy of K. Sainio.)

BMP4 receptors are found in the epithelial tissue of the duct. The BMPs act to antagonize branching signals, and thus BMP4 restricts the branching of the duct to the appropriate sites. When the BMP4 signaling cascade is activated ectopically in embryonic mouse kidney rudiments, it severely distorts the normal branching pattern (Figure 14.25C; Ritvos et al. 1995).

The fourth molecule involved in branching is **collagen XVIII**. It is part of the extracellular matrix induced by the mesenchyme, and it may provide the specificity for the branching pattern (Lin et al. 2001). Collagen XVIII is found on the branches of the kidney epithelium, but not at the tips; in the developing lung, the reciprocal pattern is seen. This pattern is generated in part by GDNF, which downregulates collagen XVIII expression by the tips of the ureteric bud branches. When ureteric duct epithelium is incubated in lung mesenchyme, the collagen XVIII expression pattern seen is typical for that of the lung, and the branching pattern resembles that of the lung epithelium (Figure 14.25D).

STEPS 7 AND 8. DIFFERENTIATION OF THE NEPHRON AND GROWTH OF THE URETERIC BUD. The interactions we have described to this point create a **cap condensate** of metanephrogenic mesenchyme cells that covers the tips of the ureteric bud branches. A few cells (perhaps around five) of this aggregate, situated at the lateral edges of the cap, begin to proliferate rapidly and form the **pretubular condensate**. This pretubular condensate will give rise to the secretory nephron of the kidney. The other cells of the cap condensate regulate the subsequent branching of the ureteric bud so that more branches can be formed (Bard et al. 2001; Sariola 2002)

The transcription factor Foxb-2 is synthesized in the cap cells. When Foxb-2 is knocked out in mouse embryos, the resulting kidney lacks a branched ureteric tree (it branches only

Figure 14.25

Signaling molecules and the branching of the ureteric epithelium. (A) An 11-day embryonic mouse kidney cultured for 4 days in control medium has a normal branching pattern. (B) An 11-day mouse kidney cultured in TGF-β1 shows no branching until reaching the periphery of the mesenchyme, and the branches formed are elongated. (C) An 11-day mouse kidney cultured in activin (which activates the same receptor as BMP4) shows a marked distortion of branching. (D) Epithelial branching of kidney cells grown in lung mesenchyme takes on an appearance similar to that of lung epithelium. (A–C from Ritvos et al. 1995; D from Lin et al. 2001. Photographs courtesy of Y. Lin and S. Vainio.)

three or four times instead of the normal seven or eight, resulting in an 8- to 16-fold reduction in the number of branches), and the aggregates do not differentiate into nephrons (Hatini et al. 1996). The cap cells are also able to convert vitamin A to retinoic acid and use this compound to retain the expression of Ret (one of the GDNF receptors) in the ureteric bud (Batourina et al. 2002). The cap cells also have been found to secrete FGF7, a growth factor whose receptor is found on the ureteric bud. FGF7 is critical for maintaining ureteric epithelial growth and ensuring an appropriate number of nephrons in the kidney (Qiao et al. 1999).

In the developing kidney, we see an epitome of the reciprocal interactions needed to form an organ. We also see that we have only begun to understand how organs form.

Snapshot Summary: Paraxial and Intermediate Mesoderm

1. The paraxial mesoderm forms blocks of tissue called somites. Somites give rise to three major divisions: the sclerotome, the myotome, and the dermatome.

2. Somites are formed from the segmental plate (unsegmented mesoderm) by the interactions of several proteins. The Notch pathway is extremely important in this process, and Eph receptor systems may be involved in the separation of the somites from the unsegmented paraxial mesoderm. N-cadherin, fibronectin, and Rac1 appear to be important in causing these cells to become epithelial.

3. The dermatome of the somite forms the back dermis. The sclerotome of the somite forms the vertebral cartilage. In thoracic vertebrae, the sclerotome cells also form the proximal portions of the ribs.

4. The epaxial myotome forms the back musculature. The hypaxial myotome forms the muscles of the body wall, limb, diaphragm, and tongue.

5. The somite regions are specified by paracrine factors secreted by neighboring tissues. The sclerotome is specified to a large degree by Sonic hedgehog, which is secreted by the notochord and floor plate cells. The dermatome is specified by neurotrophin-3, secreted by the roof plate cells of the neural tube.

6. The two myotome regions are specified by different factors. The epaxial myotome is specified by Wnt proteins from the dorsal neural tube. The hypaxial myotome is specified by BMP4 (and perhaps other proteins) secreted by the lateral plate mesoderm. In both instances, myogenic bHLH transcription factors are induced in the hypaxial and epaxial myoblasts—the cells that will become muscles.

7. To form muscles, the myoblasts stop dividing, align themselves into myotubes, and fuse.

8. The major lineages that form the skeleton are the somites (axial skeleton), lateral plate mesoderm (appendages), and neural crest and head mesoderm (skull and face).

9. There are two major types of ossification. In intramembranous ossification, which occurs primarily in the skull and facial bones, mesenchyme is converted directly into bone. In endochondral ossification, mesenchyme cells become cartilage. These cartilagenous models are later replaced by bone cells.

10. The replacement of cartilage by bone during endochondral ossification depends upon the mineralization of the cartilage matrix.

11. Osteoclasts continually remodel bone throughout a person's lifetime. The hollowing out of bone for the bone marrow is accomplished by osteoclasts.

12. The intermediate mesoderm is specified through interactions with the paraxial mesoderm. It generates the kidneys and gonads.

13. The metanephric kidney of mammals is formed by the reciprocal interactions of the metanephrogenic mesenchyme and a branch of the nephric duct called the ureteric bud.

14. The metanephrogenic mesenchyme becomes competent to form nephrons by expressing WT1. It becomes able to secrete GDNF through the action of Hox11 proteins and Pax2. GDNF is secreted by the mesoderm and induces the formation of the ureteric bud.

15. The ureteric bud secretes FGF2 and BMP7 to prevent apoptosis in the metanephrogenic mesenchyme. Without these factors, this kidney-forming mesenchyme dies. FGF2 also makes the mesenchyme competent to respond to LIF.

16. The ureteric bud secretes LIF and WNT6, and these proteins induce the competent metanephrogenic mesenchyme to form epithelial tubules. As they form these tubules, the cells secrete Wnt4, which promotes and maintains their epithelialization.

17. The condensing mesenchyme secretes paracrine factors that mediate the branching of the ureteric bud. These factors include BMP4 and TGF-β2. The branching also depends upon the extracellular matrix of the epithelium.

Literature Cited

Aitkenhead, M., B. Christ, A. Eichmann, M. Feucht, D. J. Wilson and J. Wilting. 1998. Paracrine and autocrine regulation of vascular endothelial growth factor during tissue differentiation in the quail. *Dev. Dynam.* 212: 1–13.

Aoyama, H. and K. Asamoto. 2000. The developmental fate of the rostral/caudal half of a somite for vertebra and rib formation: experimental confirmation of the resegmentation theory using chick-quail chimeras. *Mech. Dev.* 99: 71–82.

Armstrong, J. F., K. Pritchard-Jones, W. A. Bickmore, N. D. Hastie and J. B. L. Bard. 1992. The expression of the Wilms' tumor gene, WT-1, in the developing mammalian embryo. *Mech. Dev.* 40: 85–97.

Ash, P. J., J. F. Loutit and K. M. S. Townsend. 1980. Osteoclasts derived from haematopoietic stem cells. *Nature* 283: 669–670.

Aszódi, A., D. Chan, E. Hunziker, J. F. Bateman and R. Fassler. 1998. Collagen II is essential for the removal of the notochord and the formation of intervertebral discs. *J. Cell Biol.* 143: 1399–1412.

Barasch, J. and 10 others. 1999. Mesenchymal to epithelial conversion in rat metanephros is induced by LIF. *Cell* 99: 377–386.

Bard, J. B. L. and A. S. A. Ross. 1991. LIF, the ES cell inhibition factor, reversibly blocks nephrogenesis in cultured mouse kidney rudiments. *Development* 113: 193–198.

Bard, J. B. L., J. A. Davies, I. Karavanova, E. Lehtonen, H. Sariola and S. Vainio. 1996. Kidney development: the inductive interactions. *Semin. Cell Dev. Biol.* 7: 195–202.

Bard, J. B. L., A. Gordon, L. Sharp and W. Sellers. 2001. The early stages of nephron formation in the developing mouse kidney. *J. Anat.* 199: 385–392.

Bard, J. B. L. 2002. Growth and death in the developing mammalian kidney: Signals, receptors, and conversations. *BioEssays* 24: 72–82.

Barnes, G. L., C. W. Hsu, B. D. Mariani and R. S. Tuan. 1997. Cloning and characterization of chicken paraxis: A regulator of paraxial mesoderm development and somite formation. *Dev. Biol.* 189: 95–111.

Baron, R., L. Neff, D. Louvard and P. J. Courtoy. 1985. Cell mediated extracellular acidification and bone resorption: Evidence for a low pH in resorbing lacuna and localization of a 100-kD lysosomal membrane protein at the osteoclast ruffled border. *J. Cell Biol.* 101: 2210–2222.

Baron, R., L. Neff, C. Roy, A. Boisvert and M. Caplan. 1986. Evidence for a high and specific concentration of (Na+,K+) ATPase in the plasma membrane of the osteoclast. *Cell* 46: 311–320.

Batourina, E. and 8 others. 2002. Distal ureter morphogenesis depends on epithelial cell remodeling mediated by vitamin A and Ret. *Nature Genet.* 32: 109–115.

Bergstrom, D. A. and S. J. Tapscott. 2001. Molecular distinction between specification and differentiation in the myogenic basic helix-loop-helix transcription factor family. *Mol. Cell. Biol.* 21: 2404–2412.

Bishop-Calame, S. 1966. Étude experimentale de l'organogenese du systéme urogénital de l'embryon de poulet. *Arch. Anat. Microsc. Morphol. Exp.* 55: 215–309.

Blair, H. C., A. J. Kahn, E. C. Crouch, J. J. Jeffrey and S. L. Teitelbaum. 1986. Isolated osteoclasts resorb the organic and inorganic components of bone. *J. Cell Biol.* 102: 1164–1172.

Bloom, W. and D. W. Fawcett. 1975. *Textbook of Histology*, 10th Ed. Saunders, Philadelphia.

Boettiger, D., M. Enomoto-Iwamoto, H. Y. Yoon, U. Hofer, A. S. Menko and R. Chiquet-Ehrismann. 1995. Regulation of integrin a5b1 affinity during myogenic differentiation. *Dev. Biol.* 169: 261–272.

Bouchard, M., A. Souabni, M. Mandler, A. Neubuser and M. Busslinger. 2002. Nephric lineage specification by Pax2 and Pax8. *Genes Dev.* 16: 2958–2970.

Brand-Saberi, B., J. Wilting, C., Ebensperger and B. Christ. 1996. The formation of somite compartments in the avian embryo. *Int. J. Dev. Biol.* 40: 411–420.

Brill, G., N. Kahane, C. Carmeli, D. von Schack, Y.-A. Barde and C. Kalcheim. 1995. Epithelial-mesenchymal conversion of dermatome progenitors requires neural tube-derived signals: Characterization of the role of neurotrophin-3. *Development* 121: 2583–2594.

Bruder, S. P. and A. I. Caplan. 1989. Cellular and molecular events during embryonic bone development. *Connect. Tiss. Res.* 20: 65–71.

Bulman, M. P. and 9 others. Mutations in the human delta homologue, DLL3, cause axial skeletal defects in spondylocostal dysostosis. *Nature Genet.* 24: 438–441.

Burgess, R., P. Cserjesi, K. L. Ligon and E. N. Olson. 1995. Paraxis: A basic helix-loop-helix protein expressed in paraxial mesoderm and developing somites. *Dev. Biol.* 168: 296–306.

Burgess, R., Rawls., D. Brown, A. Bradley and E. N. Olson. 1996. Requirement of the paraxis gene for somite formation and musculoskeletal patterning. *Nature* 384: 570–573.

Chen, C.-M., N. Kraut, M. Groudine and H. Weintraub. 1996. I-mf, a novel myogenic repressor, interacts with members of the MyoD family. *Cell* 86: 731–741.

Christ, B. and C. P. Ohrdahl. 1995. Early stages of chick somite development. *Anat. Embryol.* 191: 381–396.

Cossu, G., R. Kelly, S. Tajbakhsh, S. Di Donna, E. Vivarelli and M. Buckingham. 1996a. Activation of different myogenic pathways: myf-5 is induced by the neural tube and MyoD by the dor-

sal ectoderm in mouse paraxial mesoderm. *Development* 122: 429–437.

Cossu, G., S. Tajbakhsh and M. Buckingham. 1996b. How is myogenesis initiated in the embryo? *Trends Genet.* 12: 218–223.

Cserjesi, P. and 7 others. 1995. A basic helix-loop-helix protein that prefigures skeletal formation during mouse embryogenesis. *Development* 121: 1099–1110.

Dale, J. K., M. Maroto, M.-L. Dequeant, M. Malaper, M. McGrew and O. Pourquie. 2003. Periodic Notch inhibition by Lunatic Fringe underlies the chick segmentation clock. *Nature* 421: 275–278.

David, J. D., W. M. See and C. A. Higginbotham. 1981. Fusion of chick embryo skeletal myoblasts: Role of calcium influx preceding membrane union. *Dev. Biol.* 82: 297–307.

Davies, J. A., M. Lyon, J. Gallagher and D. R. Garrod. 1995. Sulphated proteoglycan is required for collecting duct growth and branching but not nephron formation during kidney development. *Development* 121: 1507–1517

Dickman, S. 1997. No bones about a genetic switch for bone development. *Science* 276: 1502.

Dietrich, S., F. R. Schubert, C. Healy, P. T., Sharpe and A. Lumsden. 1998. Specification of hypaxial musculature. *Development* 125: 2235–2249.

Dubrulle, J., M. J. McGrew and O. Pourquié. 2001. FGF signaling controls somite boundary position and regulates segmentation clock control of spatiotemporal Hox gene activation. *Cell* 106: 219–232.

Ducy, P., R. Zhang, V. Geoffroy, A. L., Ridall and G. Karsenty. 1997. Osf2/Cba1: A transcriptional activator of osteoblast differentiation. *Cell* 89: 747–754.

Dudley, A. T., K. M. Lyons and E. J. Robertson. 1995. A requirement for bone morphogenesis protein-7 during development of the mammalian kidney and eye. *Genes Dev.* 9: 2795–2807.

Dunwoodie, S. L., M. Clements, D. B. Sparrow, X. Sa, R. A. Conlon and R. S. P. Beddington. 2002. Axial skeletal defects caused by mutation in the spondylocostal dysplasia/pudgy gene *Dll3* are associated with disruption of the segmentation clock within the presomitic mesoderm. *Development* 129: 1795–1806.

Duong, L. T. and G. A. Rodan. 2001. Regulation of osteoclast formation and function. *Rev. Endocr. Metab. Disorders.* 2: 95–104.

Durbin, L. and 8 others. 1998. Eph signaling is required for segmentation and differentiation of the somites. *Genes Dev.* 12: 3096–3109.

Ekblom, P., I. Thesleff, L. Saxén, A. Miettinen and R. Timpl. 1983. Transferrin as a fetal growth factor: Acquisition of responsiveness related to embryonic induction. *Proc. Natl. Acad. Sci. USA* 80: 2651–2655.

Ekblom, P. and 8 others. 1994. Role of mesenchymal nidogen for epithelial morphogenesis in vitro. *Development* 120: 2003–2014.

Fan, C. M. and M. Tessier-Lavigne. 1994. Patterning of mammalian somites by surface ectoderm and notochord: Evidence for sclerotome induction by a hedgehog homolog. *Cell* 79: 1175–1186.

Gerber, H. P., T. H. Vu,. A. M. Ryan, J. Kowalski and Z. Werb. 1999. VEGF couples hypertrophic cartilage remodeling, ossification, and angiogenesis during endochondral bone formation. *Nature Med.* 5: 623–628.

Gluecksohn-Schoenheimer, S. 1943. The morphological manifestations of a dominant mutation in mice affecting tail and urogenital system. *Genetics* 28: 341–348.

Grobstein, C. 1955. Induction interaction in the development of the mouse metanephros. *J. Exp. Zool.* 130: 319–340.

Grobstein, C. 1956. Trans-filter induction of tubules in mouse metanephrogenic mesenchyme. *Exp. Cell Res.* 10: 424–440.

Haigh, J. J., H. P. Gerber, N. Ferrara and E. F. Wagner. 2000. Conditional inactivation of VEGF-A in areas of collagen 2a1 expression results in embryonic lethality in the heterozygous state. *Development* 127: 1445–1453.

Hall, B. K. 1988. The embryonic development of bone. *Am. Sci.* 76: 174–181.

Hall, B. K. and T. Miyake. 1995. Divide, accumulate, differentiate: Cell condensations in skeletal development revisited. *Int. J. Dev. Biol.* 39: 881–893.

Hatini, V., S. O. Huh, D. Herzlinger, V. C. Soares and E. Lai. 1996. Essential role of stromal morphogenesis in kidney morphogenesis revealed by targeted disruption of winged helix transcription factor, BF-2. *Genes Dev.* 10: 1467–1478.

Hatori, M., K. J. Klatte, C. C., Teixeira and I. M. Shapiro. 1995. End labeling studies of fragmented DNA in avian growth plate: Evidence for apoptosis in terminally differentiated chondrocytes. *J. Bone Miner. Res.* 10: 1960–1968.

Hatta, K., S. Takagi, H. Fujisawa and M. Takeichi. 1987. Spatial and temporal expression of N-cadherin cell adhesion molecule correlated with morphogenetic processes of chicken embryo. *Dev. Biol.* 120: 218–227.

Horton, W. A. 1990. The biology of bone growth. *Growth Genet. Horm.* 6(2): 1–3.

Huang, R., Q. Zhi, C. Schmidt, J. Wilting, B. Brand-Saberi and B. Christ. 2000. Sclerotomal origin of the ribs. *Development* 127: 527–532.

Ikeya, M. and S. Takada. 1998. Wnt signaling from the dorsal neural tube is required for the formation of the medial dermomyotome. *Development* 125: 4969–4976.

Itäranta, P., Y. Lin, J. Peräsaari, G. Roël, O. Destrée and S. Vainio. 2002. Wnt-6 is expressed in the ureter bud and induces kidney tubule development in vitro. *Genesis* 32: 259–268.

Johnson, R. L., E. Laufer, R. D. Riddle and C. Tabin. 1994. Ectopic expression of Sonic hedgehog alters dorsal-ventral patterning of somites. *Cell* 79: 1165–1173.

Jouve, C., I. Palmeirim, D. Henrique, J. Beckers, A. Gossler, D. Ish-Horowicz, and O. Pourquié. 2000. Notch signalling is required for cyclic expression of the hairy-like gene HES1 in the presomitic mesoderm. *Development* 127: 1421–1429.

Jouve, C., T. Imura and O. Pourquié. 2002. Onset of segmentation clock in the chick embryo: evidence for oscillations in the somite precursors in the primitive streak. *Development* 129: 1107–1117.

Kahn, A. J. and D. J. Simmons. 1975. Investigation of cell lineage in bone using a chimaera of chick and quail embryonic tissue. *Nature* 258: 325–327.

Karavanov, A. A., I. Karavanova, A., Perantoni and I. B. Dawid. 1998. Expression pattern of the *rat Lim-1* homeobox gene suggests a dual role during kidney development. *Int. J. Dev. Biol.* 42: 61–66.

Karavanova, I. D., L. F. Dove, J. H. Resau and A. O. Perantoni. 1996. Conditioned medium from a rat ureteric bud cell line in combination with bFGF induces complete differentiation of isolated metanephric mesenchyme. *Development* 122: 4159–4167.

Karsenti, G. and E. F. Wagner. 2002. Reaching a genetic and molecular understanding of skeletal development. *Dev. Cell* 2: 389–406.

Kato, N. and H. Aoyama. 1998. Dermamyomal origin of the ribs as revealed by extirpation and transplantation experiments in chick and quail embryos. *Development* 125: 3437–3443.

Kieny, M., A. Mauger and P. Segel. 1972. Early regionalization of somitic mesoderm as studied by the development of the axial skeleton of the chick embryo. *Dev. Biol.* 28: 142–161.

Kispert, A., S. Vainio, L. Shen, D. R. Rowitch and A. P. McMahon. 1996. Proteoglycans are required for maintenance of Wnt-11 expression in the ureter tips. *Development* 122: 3627–3637.

Kispert, A., S. Vainio and A. P. McMahon. 1998. Wnt-4 is a mesenchymal signal for epithelial transformation of metanephric mesenchyme in the developing kidney. *Development* 125: 4225–4234.

Kjoller, L. and A. Hall. 1999. Signaling to Rho GTPases. *Exp. Cell Res.* 253: 166–179.

Klanke, B., M. Simon, W. Rockl, H. A. Weich, H. Stolte and H. J. Grone. 2002. Effects of vascular endothelial growth factor (VEGF)/vascular permeability factor (VPF) on haemodynamics and permselectivity of the isolated perfused rat kidney. *Nephrol. Dial. Transplant.* 13: 875–885.

Knudsen, K. A. 1985. The calcium-dependent myoblast adhesion that precedes cell fusion is mediated by glycoproteins. *J. Cell Biol.* 101: 891–897.

Knudsen, K. A., S. A. McElwee and L. Myers. 1990. A role for the neural cell adhesion molecule, N-CAM, in myoblast interaction during myogenesis. *Dev. Biol.* 138: 159–168.

Komori, T. and 14 others. 1997. Targeted disruption of *Cba1* results in a complete lack of bone formation owing to maturational arrest of osteoblasts. *Cell* 89: 755–764.

Konigsberg, I. R. 1963. Clonal analysis of myogenesis. *Science* 140: 1273–1284.

Koseki, C., D. Herzlinger and Q. Al-Auqati. 1992. Apoptosis in metanephric development. *J. Cell Biol.* 119: 1327–1333.

Kreidberg, J. A., H. Sariola, J. M. Loring, M. Maeda, J. Pelletier, D. Housman and R. Jaenisch. 1993. WT-1 is required for early kidney development. *Cell* 74: 679–691.

Kulesa, P. M. and S. E. Fraser. 2002. Cell dynamics during somite boundary formation revealed by time-lapse analysis. *Science* 298: 991–995.

Kuure, S., R. Vuolteenaho and S. Vainio. 2000. Kidney morphogenesis: Cellular and molecular regulation. *Mech. Dev.* 92: 31–45.

Langman, J. 1981. *Medical Embryology*, 4th Ed. Williams & Wilkins, Baltimore.

Lash, J. W. and K. M. Yamada. 1986. The adhesion recognition signal of fibronectin: A possible trigger mechanism for compaction during somitogenesis. In R. Bellairs, D. H. Ede and J. W. Lash (eds.), *Somites in Developing Embryos*. Plenum, New York, pp. 201–208.

Lassar, A. B., J. N. Buskin, D. Lockshon, R. L. Davis, S. Apone, S. D. Hauschka and H. Weintraub. 1989. MyoD is a sequence-specific DNA binding protein requiring a region of myc homology to bind to the muscle creatine kinase enhancer. *Cell* 58: 823–831.

Le Douarin, N. and M.-A. Tiellet. 1974. Experimental analysis of the migration and differentiation of neuroblasts of the autonomic nervous system and of neuroectodermal derivatives, using a biological cell marking technique. *Dev. Biol.* 41: 162–184.

Lehtonen, E., I. Virtanen and L. Saxén. 1985. Reorganization of the intermediate cytoskeleton in induced mesenchyme cells is independent of tubule morphogenesis. *Dev. Biol.* 108: 481–490.

Lin, Y. and 8 others. 2001. Induced repatterning of type XVIII collagen expression in ureter bud from kidney to lung: association with sonic hedgehog and ectopic surfactant protein C. *Development* 128: 1573–1585.

Luo, G., C. Hofmann, A. L. J. J. Bronckers, M. Sohocki, A. Bradley and G. Karsenty. 1995. BMP-7 is an inducer of nephrogenesis and is also required for eye development and skeletal patterning. *Genes Dev.* 9: 2808–2820.

Maeshima A, Y. Nojima, and I. Kojima. 2001. The role of the activin-follistatin system in the developmental and regeneration processes of the kidney. *Cytokine Growth Factor Rev.* 12: 289–298.

Manolagas, S. and R. L. Jilka. 1995. Bone marrow, cytokines, and bone remodeling. *N. Engl. J. Med.* 332: 305–310.

Marcelle, C., M. R. Stark and M. Bronner-Fraser. 1997. Coordinate actions of BMPs, Wnts, Shh, and Noggin mediate patterning of the dorsal somite. *Development* 124: 3955–3963.

Maroto, M., R. Reshef, A. E. Münsterberg, S. Koester, M. Goulding and A. B. Lassar. 1997. Ectopic Pax-3 activates MyoD and Myf-5 expression in embryonic mesoderm and neural tissue. *Cell* 89: 139–148.

Mauch, T. J., G. Yang, M. Wright, D. Smith and G. C. Schoenwolf. 2000. Signals from trunk paraxial mesoderm induce pronephros formation in chick intermediate mesoderm. *Dev. Biol.* 220: 62–75.

Medvinsky, A. and E. Dzierzak. 1996. Definitive hematopoiesis is autonomously initiated by the AGM region. *Cell* 86: 897–906.

Mencken, H. L. 1919. Exeunt omnes. *Smart Set* 60: 138–145.

Menko, A. S. and D. Boettiger. 1987. Occupation of the extracellular matrix integrin is a control point for myogenic differentiation. *Cell* 51: 51–57.

Mintz, B. and W. W. Baker. 1967. Normal mammalian muscle differentiation and gene control of isocitrate dehydrogenase synthesis. *Proc. Natl. Acad. Sci. USA* 58: 592–598.

Miyazaki, Y., K. Oshima, A. Fogo, B. L. Hogan, and I. Ichikawa. 2000. Bone morphogenetic protein 4 regulates the budding site and elongation of the mouse ureter. *J. Clin. Invest.* 105: 863–873.

Moore, M. W. and 9 others. 1996. Renal and neuronal abnormalities in mice lacking GDNF. *Nature* 382: 76–79.

Morin-Kenmsicki, E. M., E. Melancon and J. S. Eisen. 2002. Segmental relationshiop between somites and vertebral column in zebrafish. *Development* 129: 3851–3860.

Müller, U., D. Wang, S. Denda, J. Menesis, R. Pedersen and L. Reichardt. 1997. Integrin alpha-8/beta-1 is critically important for epithelial-mesenchymal interactions during kidney morphogenesis. *Cell* 88: 603–613.

Mundlos, S. and 13 others. 1997. Mutations involving the transcription factor CBFA1 cause cleiocranial dysplasia. *Cell* 89: 773–779.

Münsterberg, A. E., J. Kitajewski, D. A. Bumcroft, A. P. McMahon and A. B. Lassar. 1995. Combinatorial signaling by sonic hedgehog and Wnt family members induce myogenic bHLH gene expression in the somite. *Genes Dev.* 9: 2911–2922.

Nakamura, T., S. Okuda, D. Miller, E. Ruoslahti and W. Border. 1990. Transforming factor-β (TGF-β) regulates production of extracellular matrix (ECM) components by glomerular epithelial cells. *Kidney Int.* 37: 221.

Nameroff, M. and E. Munar. 1976. Inhibition of cellular differentiation by phospholipase C. II. Separation of fusion and recognition among myogenic cells. *Dev. Biol.* 49: 288–293.

Nowicki, J. L. and A. C. Burke. 2000. Hox genes and morphological identity: Axial versus lateral patterning in the vertebrate mesoderm. *Development* 127: 4265–4275.

Oka, Y. Y. Sato, Y. Hirai, H. Tsuda and Y. Takahashi. 2002. Epimorphin promotes cartilage condensation/sorting during vertebral skeletogenesis. Abstract in *Dev. Bio.* 247: 476–477.

Oberlender, S. A. and R. S. Tuan. 1994. Expression and functional involvement of N-cadherin

in embryonic limb chondrogenesis. *Development* 120: 177–187.

Olivera-Martinez, I., S. Missier, S. Fraboulet, J. Thelu and D. Dhouailly. 2002. Differential regulation of the chick dorsal thoracic dermal progenitors from medial dermamyotome. *Development.* 129: 4763–4772.

Ordahl, C. P. 1993. Myogenic lineages within the developing somite. *In* M. Bernfield (ed.), *Molecular Basis of Morphogenesis.* Wiley-Liss, New York, pp. 165–170.

Ordahl, C. P. and N. Le Douarin. 1992. Two myogenic lineages within the developing somite. *Development* 114: 339–353.

Ordahl, C. P. and B. A. Williams. 1998. Knowing chops from chuck: Roasting MyoD redundancy. *BioEssays* 20: 357–362.

Ostrovsky, D., C. M. Cheney, A. W. Seitz and J. W. Lash. 1984. Fibronectin distribution during somitogenesis in the chick embryo. *Cell Diff.* 13: 217–223.

Otto, F. and 11 others. 1997. Cba1, a candidate gene for cleidocranial dysplasia syndrome, is essential for osteoblast differentiation and bone development. *Cell* 89: 765–771.

Palmeirim, I., D. Henrique, D. Ish-Horowicz and O. Pourquié. 1997. Avian hairy gene expression identifies a molecular clock linked to vertebrate segmentation and somitogenesis. *Cell* 91: 639–648.

Park, W.-J., G. Bellus and E. W. Jabs. 1995. Mutations in fibroblast growth factor receptors: Phenotypic consequences during eukaryotic development. *Am. J. Hum. Genet.* 57: 748–754.

Patterson, L. T., M. Pembaur and S. S. Potter. 2001. Hoxa11 and Hoxd11 regulate branching morphogenesis of the ureteric bud in the developing kidney. *Development* 128: 2153–2161.

Penttinen, R. P., S. Kobayashi and P. Bornstein. 1988. Transforming growth factor-β increases mRNA for matrix proteins in the presence and in the absence of the changes in mRNA stability. *Proc. Natl. Acad. Sci. USA* 85: 1105–1108.

Perantoni, A. O., L. F. Dove and I. Karavanova. 1995. Basic fibroblast growth factor can mediate the early inductive events in renal development. *Proc. Natl. Acad. Sci USA* 92: 4696–4700.

Pichel, J. G. and 11 others. 1996. Defects in enteric innervation and kidney development in mice lacking GDNF. *Nature* 382: 73–76.

Piette, J., J.-L. Bessereau, M. Huchet and J.-P. Changeaux. 1990. Two adjacent MyoD1-binding sites regulate expression of the acetylcholine receptor α-subunit gene. *Nature* 345:353–355.

Pourquié, O. and P. P. L. Tam. 2001. A nomenclature for prospective somites and phases of cyclic gene expression in the presomitic mesoderm. *Dev. Cell* 1: 619–620.

Pourquié, O. and 9 others. 1996. Lateral and axial signals involved in somite patterning: A role for BMP4. *Cell* 84: 461–471.

Pownall, M. E., M. K. Gustafsson and C. P. Emerson, Jr. 2002. Myogenic regulatory factors

and the specification of muscle progenitors in vertebrate embryos. *Annu. Rev. Cell Dev. Biol.* 18: 747–783.

Pritchard-Jones, K. and 11 others. 1990. The candidate Wilms' tumour gene is involved in genitourinary development. *Nature* 346: 194–197.

Qiao, J., R. Uzzo, T. Obara-Ishihara, L. Degenstein, E. Fuchs and D. Herzlinger. 1999. FGF-7 modulates ureteric bud growth and nephron number in the developing kidney. *Development* 126: 547–554.

Rajpurohit, R., K. Mansfield, K. Ohyama, D. Ewert and I. M. Shapiro. 1999. Chondrocyte death is linked to development of a mitochondrial membrane permeability transition in the growth plate. *J. Cell. Physiol.* 179: 287–296.

Ritvos, O., T. Tuuri, M. Erämaa, K. Sainio, K. Hilden, L. Saxén and S. F. Gilbert. 1995. Activin disrupts epithelial branching morphogenesis in developing murine kidney, pancreas, and salivary gland. *Mech. Dev.* 50: 229–245.

Rothenpieler, U. W. and G. R. Dressler. 1993. Pax-2 is required for mesenchyme-to-epithelium conversion during kidney development. *Development* 119: 711–720.

Shapiro, I.M, E. E. Golub, S. Kakuta, J. Hazelgrove, J. Havery, B. Chance and P. Frasca. 1982. Initiation of endochondral calcification is related to changes in the redox state of hypertrophic chondrocytes. *Science.* 217: 950–952.

Sainio, K. and A. Raatikainen-Ahokas. 1999. Mesonephric kidney: A stem cell factory. *Int. J. Dev. Biol.* 43: 435–439.

Sainio, K., D. Nonclercq, M. Saarma, J. Palgi, L. Saxén and H. Sariola. 1994. Neuronal characteristics of embryonic renal stroma. *Int. J. Dev. Biol.* 38: 77–84.

Sainio, K. and 10 others. 1997. Glial cell derived neurotrophic factor is required for bud initiation from ureteric epithelium. *Development* 124: 4077–4087.

St.-Jacques, B., M. Hammerschmidt and A. P. McMahon. 1999. Indian hedgehog signaling regulates proliferation and differentiation of chondrocytes and is essential for bone formation. *Genes Dev.* 13: 2072–2086.

Sánchez, M. P., I. Silos-Santiago, J. Frisén, B. He, S. A. Lira and M. Barbacid. 1996. Renal agenesis and the absence of enteric neurons in mice lacking GDNF. *Nature* 382: 70–73.

Santos, O. F. P., E. J. G. Baras, X.-M. Yang, K. Matsumoto, T. Nakamura, M. Park and S. K. Nigam. 1994. Involvement of hepatocyte growth factor in kidney development. *Dev. Biol.* 163: 525–529.

Sariola, H. 2002. Nephron induction revisited: From caps to condensates. *Curr. Opin. Nephrol. Hypertens.* 11: 17–21.

Sariola, H., P. Ekblom and L. Saxén. 1982. Restricted developmental options of the metanephric mesenchyme. In M. Burger and R. Weber (eds.), *Embryonic Development, Part B: Cellular Aspects.* Alan R. Liss, New York, pp. 425–431.

Sariola, H., K. Holm-Sainio and S. Henke-Fahle. 1989. The effect of neuronal cells on kidney differentiation. *Int. J. Dev. Biol.* 33: 149–155.

Sato, Y., K. Yasuda and Y. Takahashi. 2002. Morphological boundary forms by a novel inductive event mediated by Lunatic-Fringe and Notch during somatic segmentation. *Development* 129: 3633–3644.

Sawada, A., M. Shinya, Y.-J. Jiang, A. Kawakami, A. Kuroiwa and H. Takeda. 2001. Fgf/MAPK signalling is a crucial positional cue in somite boundary formation. *Development* 128: 4873–4880.

Saxén, L. 1970. Failure to demonstrate tubule induction in heterologous mesenchyme. *Dev. Biol.* 23: 511–523.

Saxén, L. 1987. *Organogenesis of the Kidney.* Cambridge University Press, Cambridge.

Saxén, L. and H. Sariola. 1987. Early organogenesis of the kidney. *Pediatr. Nephrol.* 1: 385–392.

Schuchardt, A., V. D-Agati, V. Pachnis and F. Constantini. 1996. Renal agenesis and hypodysplasia in *ret-k⁻* mutant mice result from defects in ureteric bud development. *Development* 122: 1919–1929.

Shainberg, A., G. Yagil and D. Yaffe. 1969. Control of myogenesis in vitro by Ca^{2+} concentration in nutritional medium. *Exp. Cell Res.* 58: 163–167.

Shapiro, I., K. DeBolt, V. Funanage, S. Smith and R. Tuan. 1992. Developmental regulation of creatine kinase activity in cells of the epiphyseal growth plate. *J. Bone Miner. Res.* 7: 493–500.

Smith, C. A. and R. S. Tuan. 1996. Functional involvement of Pax-1 in somite development: Somite dysmorphogenesis in chick embryos treated with Pax-1 paired-box antisense oligonucleotide. *Teratology* 52: 333–345.

Smith, H. 1953. *From Fish to Philosopher.* Little, Brown, Boston.

Sosic, D., B. Brand-Saberi, C. Schmidt, B. Christ and E. Olson. 1997. Regulation of paraxis expression and somite formation by ectoderm- and neural tube-derived signals. *Dev. Biol.* 185: 229–243.

Srinivas, S., M. R. Goldberg, T. Watanabe, V. D'Agati, Q. al-Awqati, and F. Costantini. 1999. Expression of green fluorescent protein in the ureteric bud of transgenic mice: a new tool for the analysis of ureteric bud morphogenesis. *Dev. Genet.* 24: 241 - 251.

Stark, K., S. Vainio, G. Vassileva and A. P. McMahon. 1994. Epithelial transformation of metanephric mesenchyme in the developing kidney regulated by Wnt-4. *Nature* 372: 679–683.

Stern, H. M., A. M. C. Brown and S. D. Hauschka. 1995. Myogenesis in paraxial mesoderm: Preferential induction by dorsal neural tube and by cells expressing *Wnt-1. Development* 121: 3675–3686.

Tajbakhsh, S. and R. Spörle. 1998. Somite development: Constructing the vertebrate body. *Cell* 92: 9–16.

Tajbakhsh, S., D. Rocancourt, G. Cossu and M. Buckingham. 1997. Redefining the genetic hierarchies controlling skeletal myogenesis: Pax-3 and Myf-5 act upstream of MyoD. *Cell* 89: 127–138.

Takahashi, Y., A.-H. Monsoro-Burg, M. Bontoux and N. M. Le Douarin. 1992. A role fo the Quox8 in the establishment of the dorsoventral pattern during vertebrate development. *Proc. Natl. Acad. Sci. USA* 89: 10237–10241.

Takahashi, Y., S. Kuroda, K. Kaibuchi, K. Yasuda and Y. Nakaya. 2002. A segmentation-specific role of small GTPases Rac 1 in mesenchymal-epithelial transition. *Dev. Biol.* 247: 52.

Thayer, M. J., S. J. Tapscott, R. L. Davis, W. E. Wright, A. B. Lassar and H. Weintraub. 1989. Positive autoregulation of the myogenic determination gene *MyoD1. Cell* 58: 241–248.

Toivonen, S. 1945. Uber die Entwicklung der Vor- und Uriniere beim Kaninchen. *Ann. Acad. Sci. Fenn. Ser. A.* 8: 1–27.

Tondravi, M. M., S. R. McKercher, K. Anderson, J. M. Erdmann, M. Quiroz, R. Maki and S. L. Teitelbaum. 1997. Osteopetrosis in mice lacking haematopoietic transcription factor PU.1. *Nature* 386: 81–84.

Tonegawa, A. and Y. Takahashi. 1998. Somitogenesis controlled by Noggin. *Dev. Biol.* 202: 172–182.

Tonegawa, A., N. Funayama, N. Ueno and Y. Takahashi. 1997. Mesodermal subdivision along the mediolateral axis in chicken controlled by different concentrations of BMP-4. *Development* 124: 1975–1984.

Torres, M., E. Gómez-Pardo, G. R. Dressler and P. Gruss. 1995. Pax-2 controls multiple steps of urogenital development. *Development* 121: 4057–4065.

Tuan, R. 1987. Mechanisms and regulation of calcium transport by the chick embryonic chorio-allantoic membrane. *J. Exp. Zool.* [Suppl.] 1: 1–13.

Tuan, R. S. and M. H. Lynch. 1983. Effect of experimentally induced calcium deficiency on the developmental expression of collagen types in chick embryonic skeleton. *Dev. Biol.* 100: 374–386.

Vainio, S., E. Lehtonen, M. Jalkanen, M. Bernfield and L. Saxén. 1989. Epithelial-mesenchymal interactions regulate the stage-specific expression of a cell surface proteoglycan, syndecan, in the developing kidney. *Dev. Biol.* 134: 382–391.

Vainio, S., M. Jalkanen, M. Bernfield and L. Saxén. 1992. Transient expression of syndecan in mesenchymal cell aggregates of the embryonic kidney. *Dev. Biol.* 152: 221–232.

van Adelsberg, J., S. Chamberlain and V. D'Agati. 1997. Polycystin expression is temporally and spatially regulated during renal development. *Am. J. Physiol.* 272: F602–F609.

van Heyningen, V. and 11 others. 1990. Role for Wilms tumor gene in genital development? *Proc. Natl. Acad. Sci. USA* 87: 5383–5386.

Venters, S. J., S. Thorsteindóttir and M. J. Duxton. 1999. Early development of the myotome in the mouse. *Dev. Dynam.* 216: 219–232.

Ward, C. J. and 8 others. 1996. Polycystin, the polycystic kidney disease protein, is expressed by epithelial cells in fetal, adult, and polycystic kidney. *Proc. Nat. Acad. Sci. USA* 93: 1524–1528.

Watanabe, Y., D. Duprez, A.-H. Monsoro-Burq, C. Vincent and N. Le Douarin. 1998. Two domains in vertebral development: Antagonistic regulation by SHH and BMP4 proteins. *Development* 125: 2631–2639.

Wellik, D. M., P. J. Hawkes and M. R. Capecchi. 2002. *Hox11* paralogous genes are essential for metanephric kidney induction. *Genes Dev.* 16: 1423–1432.

Weintraub, H., S. J. Tapscott, R. L. Davis, M. J. Thayer, M. A. Adam, A. B. Lassar and D. Miller. 1989. Activation of muscle-specific genes in pigment, nerve, fat, liver, and fibroblast cell lines by forced expression of *MyoD. Proc. Natl. Acad. Sci. USA* 86:5434–5438.

Wilkie, A. O. M., G. M. Morriss-Kay, E. Y. Jones and J. K. Heath. 1995. Functions of fibroblast growth factors and their receptors. *Curr. Biol.* 5: 500–507.

Wintour, E. M., A. Butkus, L. Earnest and S. Pompolo. 1996. The erythropoietin gene is expressed strongly in the mammalian mesonephric kidney. *Blood* 88: 3349–3353.

Wolpert, L. 1998. *Principles of Development.* Current Biology Press, London.

Woolf, A. S. and 8 others. 1995. Roles of hepatocyte growth factor/scatter factor and the Met receptor in the early development of the metanephros. *J. Cell Biol.* 128: 171–184.

Wright, E. and 8 others. 1995. The *Sry*-related gene *Sox9* is expressed during chondrogenesis in mouse embryos. *Nature Genet.* 9: 15–20.

Wu, L. N., B. R. Genge, D. G. Dunkelberger, R. Z. LeGeros, B. Concannon and R. E. Wuthier. 1997. Physiochemical characterization of the nucleational core of matrix vesicles. *J. Biol. Chem.* 272: 4404–4411.

Xu, P. X., J. Adams, H. Peters, M. C. Brown, S. Heaney and R. Maas. 1999. *Eya1*-deficient mice lack ears and kidneys and show abnormal apoptosis of organ primordia. *Nature Genet.* 23: 113–117.

Yaffe, D. and M. Feldman. 1965. The formation of hybrid multinucleated muscle fibres from myoblasts of different genetic origin. *Dev. Biol.* 11: 300–317.

Yagami-Hiromasa, T., T. Sato, T. Kurisaki, K. Kamijo, Y.-I. Nabeshima and A. Fujisawa-Sehara. 1995. A metalloprotease-disintegrin participating in myoblast fusion. *Nature* 377: 652–656.

Yin, M., E. Koyama, M. Zasloff and M. Pacifici. 2002. Antiangiogenic treatment delays chondrocyte maturation and bone formation during limb morphogenesis. *J. Bone Miner. Res.* 17: 56–65.

Zakany, J., M. Kmita, P. Alarcon, J.-L. de la Pompa and D. Duboule. 2001. Localized and transient transcription of Hox genes suggests a link between patterning and the segmentation clock. *Cell* 106: 207–217.

15 *Lateral plate mesoderm and endoderm*

IN THE CHAOS OF THE ENGLISH CIVIL WARS, William Harvey, physician to the King and discoverer of the blastoderm, was comforted by viewing the heart as the undisputed ruler of the body, through whose divinely ordained powers the lawful growth of the organism was assured. Later embryologists looked at the heart as more of a servant than a ruler, the chamberlain of the household who assured that nutrients reached the apically located brain and the peripherally located muscles. In either metaphor, the heart, circulation, and the digestive system were seen as being absolutely critical for development. As Harvey (1651) persuasively argued, the chick embryo must form its own blood without any help from the hen, and this blood is crucial in embryonic growth. How this happened was a mystery. "What artificer," he wrote, could create blood "when there is yet no liver in being?" The nutrition provided by the egg was also paramount to Harvey. His conclusion about the nutritive value of the yolk and albumen was that "The egge is, as it were, an exposed womb; wherein there is a substance concluded as the Representative and Substitute, or Vicar of the breasts."

This chapter will outline the mechanisms by which the circulatory system, the respiratory system, and the digestive system emerge in the amniote embryo. As we will see, the lateral plate mesoderm and the endoderm interact to create both the circulatory and the digestive organs.*

LATERAL PLATE MESODERM

On the peripheral sides of the two bands of intermediate mesoderm reside the **lateral plate mesoderm** (see Figures 14.1 and 14.2). Each plate splits horizontally into two layers. The dorsal layer is the **somatic (parietal) mesoderm**, which underlies the ectoderm and, together with the ectoderm, forms the **somatopleure**. The ventral layer is the **splanchnic (visceral) mesoderm**, which overlies the endoderm and, together with the endoderm, forms the **splanchnopleure** (Figure 15.1). The space between these two layers becomes the body cavity—the **coelom**—which stretches from the future neck region to the posterior of the body. During later development, the right-side and left-side coeloms fuse, and folds of tissue extend from the somatic

---

*Some scientists think that the mesoderm and endoderm were originally a single germ layer, the "mesendoderm," that accomplished all these functions. Recall from Chapter 8 that what in vertebrates is a broad territory of embryonic cells is actually the progeny of a single "mesentoblast" in many invertebrates. The signals that regulate the mesentoblast and the entire mesodermal and endodermal territory in vertebrates may be very similar (Maduro et al. 2001; Rodaway and Patient 2001).

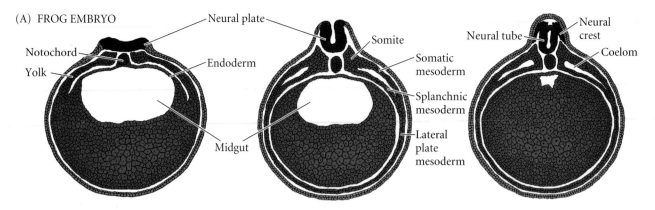

(A) FROG EMBRYO

Neural plate

Notochord
Yolk
Endoderm
Midgut

Somite
Somatic mesoderm
Splanchnic mesoderm
Lateral plate mesoderm

Neural tube
Neural crest
Coelom

(B) CHICK EMBRYO

Cuts for removal of embryo

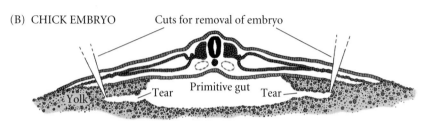

Yolk
Tear
Primitive gut
Tear

Figure 15.1
Comparison of mesodermal development in frog and chick embryos. (A) Neurula-stage frog embryos, showing progressive development of the mesoderm and coelom. (B) Transverse section of a chick embryo. (C) When the chick embryo is separated from its enormous yolk mass, it resembles the amphibian neurula at a similar stage. (A after Rugh 1951; B and C after Patten 1951.)

(C) CHICK "MADE INTO" FROG

Neural tube
Somite
Coelom
Primitive gut
Yolk

CHICK EMBRYO
(removed from yolk, edges pulled together)

FROG EMBRYO

an intricate system of blood vessels, the circulatory system provides nourishment to the developing vertebrate embryo. The circulatory system is the first functional unit in the developing embryo, and the heart is the first functional organ. The vertebrate heart arises from two regions of splanchnic mesoderm—one on each side of the body—that interact with adjacent tissue to become specified for heart development.

VADE MECUM[2] **Early heart development.** The vertebrate heart begins to function early in its development. You can see this in movies of the living chick embryo at early stages when the heart is little more than a looped tube.
[Click on Chick-Late]

Specification of heart tissue and fusion of heart rudiments

In amniote vertebrates, the embryo is a flattened disc, and the lateral plate mesoderm does not completely encircle the yolk sac. The presumptive heart cells originate in the early primitive streak, just posterior to Hensen's node and extending about halfway down its length. These cells migrate through the streak and form two groups of mesodermal cells lateral to (and at the same level as) Hensen's node (Figure 15.2A; Colas et al. 2000; Redkar et al. 2001). These groups of cells are called the **cardiogenic mesoderm**. The cells forming the atrial and

mesoderm, dividing the coelom into separate cavities. In mammals, the coelom is subdivided into the **pleural**, **pericardial**, and **peritoneal** cavities, enveloping the thorax, heart, and abdomen, respectively. The mechanism for creating the linings of these body cavities from the lateral plate mesoderm has changed little throughout vertebrate evolution, and the development of the chick mesoderm can be compared with similar stages of frog embryos (Figure 15.1).

> **WEBSITE 15.1 Coelom formation.** Coelom formation is readily visualized by animations. The animation presented here shows the expansion of the mesoderm during chick development.

The Heart

The circulatory system is one of the great achievements of the lateral plate mesoderm. Consisting of a heart, blood cells, and

(A)

(B)

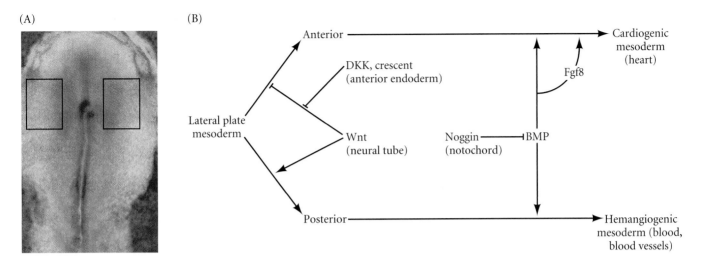

(C)

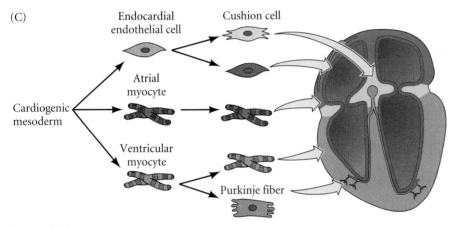

Figure 15.2
Heart-forming cells of the chick embryo. (A) Two symmetrical regions of cardiogenic (heart-forming) mesoderm form in the gastrulating chick embryo, determined by following individually labeled DiI-ltreated cells and by observing the expression patterns of particular genes such as smooth muscle α-actin. (B) Outline of the inductive interactions involving Wnt and BMP pathways, enabling the generation of heart and blood lineages form the lateral plate mesoderm. (C) The cardiogenic mesoderm contains the precursors of the three cell types of the endocardium and myocardium. The endocardium provides the endothelial lining of the heart as well as the cushion cells that form the valves. The atrial myocytes of the myocardium and the ventricular myocytes (some of which become the conducting Purkinje fibers) are also generated by the cardiogenic mesoderm. The neural crest cells and endothelium of the cardiac vessels are specified separately at different locations. (A after Redkar et al. 2001; C after Mikawa 1999.)

(Schultheiss et al. 1995; Nascone and Mercola 1995). It appears that Wnt proteins from the neural tube, especially Wnt3a and Wnt8, *inhibit* heart formation and *promote* blood formation. The anterior endoderm, however, produces Wnt inhibitors such as Cerberus, Dickkopf and Crescent, which prevent the Wnt proteins from binding to their receptors. BMPs from the endoderm promote both heart and blood development, and they can be blocked by Noggin and chordin coming from the notochord. Endodermal BMPs also induce FGF8 synthesis in the endoderm directly beneath the cardiogenic mesoderm, and FGF8 appears to be critical for the expression of cardiac proteins (Alsan and Schultheiss 2002). In this way, cardiac precursor cells are specified in the places where BMPs (lateral mesoderm and endoderm) and Wnt antagonists (anterior endoderm) coincide (Figure 15.2B; Marvin et al. 2001; Schneider and Mercola 2001; Tzahor and Lassar 2001).

ventricular musculature, the cushion cells of the valves, the Purkinje conducting fibers, and the endothelial lining of the heart are all generated from these two clusters (Mikawa 1999).

SPECIFICATION OF THE CARDIAC PRECURSOR CELLS. The specification of the cardiogenic mesoderm cells is induced by the endoderm adjacent to the heart through the BMP and FGF signaling pathways. The heart does not form if the anterior endoderm is removed. Moreover, isolated mesoderm from this region will form heart muscle when combined with anterior endoderm, but not if combined with posterior endoderm

MIGRATION OF THE CARDIAC PRECURSOR CELLS. When the chick embryo is only 18–20 hours old, the presumptive heart cells move anteriorly between the ectoderm and endoderm toward the middle of the embryo, remaining in close contact with the endodermal surface (Linask and Lash 1986). When these cells reach the lateral walls of the anterior gut tube, migration ceases. The directionality of this migration appears to be provided by the foregut endoderm. If the cardiac region endoderm is rotated with respect to the rest of the embryo, migration of the cardiogenic mesoderm cells is reversed. It is thought that the endodermal component responsible for this movement is an anterior-to-posterior concentration gradient of fibronectin. Antibodies against fibronectin stop the migration, while antibodies against other extracellular matrix components do not (Linask and Lash 1988a,b).

In the chick, the foregut is formed by the inward folding of the splanchnic mesoderm (Figure 15.3A, B). This movement brings the two cardiac tubes together. The bilateral origin of the heart can be demonstrated by surgically preventing the merger of the lateral plate mesoderm (Gräper 1907; De-Haan 1959). This manipulation results in a condition called **cardia bifida**, in which two separate hearts form, one on each side of the body (Figure 15.4A). The two endocardia lie witin the common tube for a short time, but these also will fuse.

In the zebrafish, the heart precursor cells migrate actively from the lateral edges toward the midline. Several mutations affecting endoderm differentiation disrupt this process, indicating that, as in the chick, the endoderm is critical for cardiac precursor migration. One particularly interesting zebrafish mutation is called *miles apart*. Its phenotype resembles the cardia bifida seen in experimentally manipulated chick embryos. The *miles apart* gene encodes a receptor for a cell surface sphyngolipid molecule, and it is expressed in the endoderm on either side of the midline (Figure 15.4B, C; Kupperman et al. 2000).

INITIAL CELL DIFFERENTIATION. The BMP pathway (in the absence of the Wnt signals) is critical in inducing the synthesis of the **Nkx2-5** transcription factor in the migrating cardiogenic mesoderm* (Figure 15.5; Komuro and Izumo 1993; Lints et al. 1993; Sugi and Lough 1994; Schultheiss et al. 1995; Andrée et al. 1998). Nkx2-5 is critical in instructing the mesoderm to become heart tissue, and it activates the synthesis of other transcription factors (especially members of the GATA and MEF2 families). Working together, these transcription factors activate the expression of genes encoding cardiac muscle-specific proteins (such as cardiac actin, atrial natriuretic factor, and the α-myosin heavy chains). Specification of the heart cells occurs gradually, in an anterior-to-posterior manner across the heart fields, such that the future ventricular cells become specified before the future atrial cells do (Markwald et al. 1998; Srivastava and Olson 2000).

Cell differentiation occurs independently in the two heart-forming primordia that are migrating toward each other. As they migrate, the ventral mesoderm cells begin to express N-cadherin on their apices, sort out from the dorsal mesoderm

*The Nkx2-5 homeodomain transcription factor is homologous to the Tinman transcription factor that is active in specifying the heart tube of *Drosophila*. Moreover, neither Tinman nor Nkx2-5 alone is sufficient to complete heart development in their respective organisms. Mice lacking Nkx2-5 start forming their heart tubes, but the tube fails to thicken or to loop (Lyons et al. 1995). Humans with a mutation in one of their *NKX2-5* genes have congenital heart malformations (Schott et al. 1998).

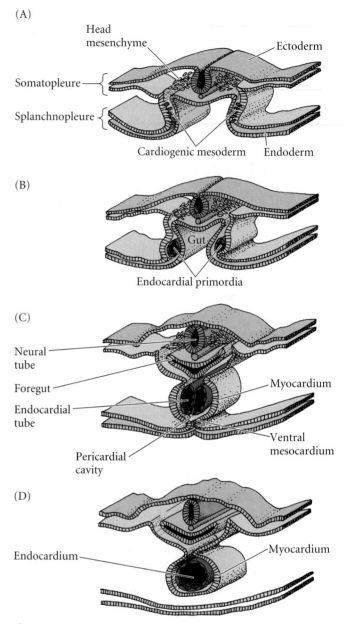

Figure 15.3
Formation of the chick heart from the splanchnic lateral plate mesoderm. The endocardium forms the inner lining of the heart, the myocardium forms the heart muscles, and the epicardium will eventually cover the heart. Transverse sections through the heart-forming region of the chick embryo are shown at (A) 25 hours, (B) 26 hours, (C) 28 hours, and (D) 29 hours. (After Carlson 1981.)

cells, and join together to form an epithelium. This joining will lead to the formation of the **pericardial cavity**, the sac in which the heart is formed (Linask 1992). A small population of these cells then downregulates N-cadherin and delaminates from the epithelium to form the **endocardium**, the lining of the heart that is continuous with the blood vessels (Figure

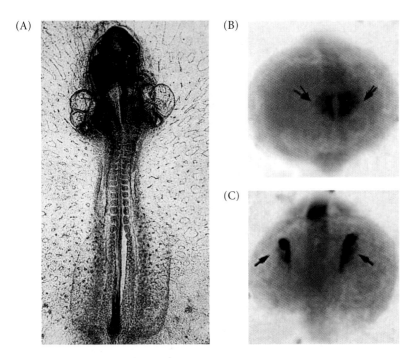

(A)

(B)

(C)

Figure 15.4
Migration of heart primordia. (A) Cardia bifida in a chick embryo, caused by surgically cutting the ventral midline, thereby preventing the two heart primordia from fusing. (B) Wild-type and (C) *miles apart* mutant, stained with probes for the cardiac myosin light chain. There is a lack of migration in the *miles apart* mutant. (A, photograph courtesy of R. L. DeHaan; B, C after Kupperman et al. 2000, courtesy of Y. R. Didier.)

15.3A, B).* The epithelial cells form the **myocardium** (Manasek 1968; Linask and Lash 1993; Linask et al. 1997), which will give rise to the heart muscles that will pump for the lifetime of the organism. The endocardial cells produce many of the heart valves, secrete the proteins that regulate myocardial growth, and regulate the placement of nervous tissue in the heart.

FUSION AND INITIAL HEARTBEATS. The fusion of the two heart primordia occurs at about 29 hours in chick development and at 3 weeks in human gestation (see Figure 15.3C,D). The myocardia unite to form a single tube. The two endocardia lie within this common tube for a short while, but eventually they also fuse. The unfused posterior portions of the endocardium become the openings of the **vitelline veins** into the heart (Figures 15.5 and 15.6). These veins will carry nutrients from the yolk sac into the **sinus venosus**, the posterior region where the two major veins fuse. The blood then passes through a valvelike flap into the atrial region of the heart. Contractions of the **truncus arteriosus** speed the blood into the aorta.

Fusion may initiate (or at least be simultaneous with) a second round of differentiation events. The mRNAs for certain cardiac muscle-specific proteins may have been transcribed earlier; but they remain untranslated until the fusion of the two heart rudiments into one tube (Colas et al. 2000).

---

*The endocardial cell population is distinct from the myocardial cell population even before gastrulation. Cell lineage studies using retroviral markers show that clones of myocardial cells during the pre-streak stages have no endocardial cells, and that no clone of endocardial cells has any myocardial cells. Thus, the cardiogenic mesoderm already has two committed populations of cells (Cohen-Gould and Mikawa 1996).

Pulsations of the heart begin while the paired primordia are still fusing. Heart muscle cells develop their own inherent ability to contract, and isolated heart cells from 7-day rat or chick embryos will continue to beat in petri dishes (Harary and Farley 1963; DeHaan 1967; Imanaka-Yoshida et al. 1998). Pulsations of the chick heart begin at about 33 hours of development (while the paired primordia are fusing).

The pulsations are made possible by the appearance of the sodium-calcium exchange pump in the muscle cell membrane; inhibiting this channel's function prevents the heartbeat from starting (Wakimoto et al. 2000; Linask et al. 2001). Eventually, the rhythmicity of the heartbeat becomes coordinated by the sinus venosus. The electric impulses generated here initiate waves of muscle contraction through the tubular heart. In this way, the heart can pump blood even before its intricate system of valves has been completed. Studies of mutations of cardiac cell calcium channels have implicated these channels in the pacemaker function (Rottbauer et al. 2001; Zhang et al. 2002).

LOOPING AND FORMATION OF HEART CHAMBERS. In 3-day chick embryos and 5-week human embryos, the heart is a two-chambered tube, with one atrium and one ventricle (Figure 15.6). In the chick embryo, the unaided eye can see the remarkable cycle of blood entering the lower chamber and being pumped out through the aorta. Looping of the heart converts the original anterior-posterior polarity of the heart tube into the right-left polarity seen in the adult organism. When this looping is completed, the portion of the heart tube destined to become the right ventricle lies anterior to the portion that will become the left ventricle.

Heart looping is dependent on the left-right patterning proteins (Nodal, Lefty-2) discussed in Chapter 11. Within the heart primordium, the Nkx2-5 protein regulates the Hand1 and Hand2 transcription factors. Although the Hand proteins appear to be synthesized throughout the early heart tube, Hand1 becomes restricted to the future left ventricle, and Hand2 to the right, as looping commences. Without these proteins, looping fails to occur normally, as the ventricles fail to

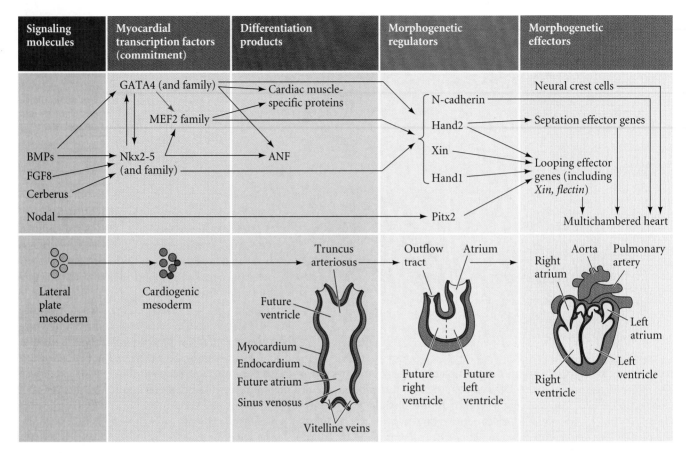

Signaling molecules	Myocardial transcription factors (commitment)	Differentiation products	Morphogenetic regulators	Morphogenetic effectors

Figure 15.5
Cascade of heart development. A correlation can be made between the morphological stages and the transcription factors present in the nucleus of the heart precursor cells. The cardioblasts are the committed heart precursor cells containing Nkx2-5 and GATA family proteins. These proteins convert the cardioblast into cardiomyocytes (heart muscle cells), which make cardiac muscle-specific proteins. These cardiomyocytes join together to form the cardiac tube. Under the influence of the Hand proteins, Xin, and Pitx2, the heart loops and the formation of the chambers begins.

form properly* (Srivastava et al. 1995; Biben and Harvey 1997). The Pitx2 transcription factor, activated solely in the left side of the lateral plate mesoderm (and eventually in the left side of the heart tube), is also critical for proper heart looping. Pitx2 appears to regulate the temporal expression of proteins such as the extracellular matrix protein **flectin,** which regulates the physical tension of the heart tissues on the different sides (Tsuda et al. 1996; Linask et al. 2002). Transcription factors Nkx2-5 and MEF2C also activate the *Xin* gene, whose protein

product, Xin (Chinese for "heart"), may mediate the cytoskeletal changes essential for heart looping (Wang et al. 1999).

WEBSITE 15.2 **Transcription factors and heart formation.** The formation of the heart and blood cells is presided over by several transcription factors whose combinations enable the different parts of the heart to form and allow different types of blood cells to develop.

The separation of atrium from ventricle is effected by several transcription factors that become restricted to either the anterior or the posterior portion of the heart tube (see Figure 15.6; Bao et al. 1999; Bruneau et al. 1999; Wang et al. 1999). This separation is accomplished when cells from the myocardium produce a factor (probably transforming growth factor β3) that causes cells from the adjacent endocardium to detach and enter the hyaluronate-rich "cardiac jelly" between the two layers (Markwald et al. 1977; Potts et al. 1991). In humans, these cells form an **endocardial cushion** that divides the tube into right and left atrioventricular channels (Figure 15.7). Meanwhile, the primitive atrium is partitioned by two **septa** that grow ventrally toward the endocardial cushion. The septa, however, have openings in them, so blood can still cross from one side into the other. This crossing of blood is needed

---

*Zebrafish, with only ventricle, have only one type of Hand protein. When the gene encoding this protein is mutated, the entire ventricular portion of the heart fails to form (Srivastava and Olson, 2000). Moreover, nongenetic agents are also critical in normal zebrafish heart formation. In the absence of high-pressure blood flow, the heart looping, chamber formation and valve development is impaired (Hove et al., 2003).

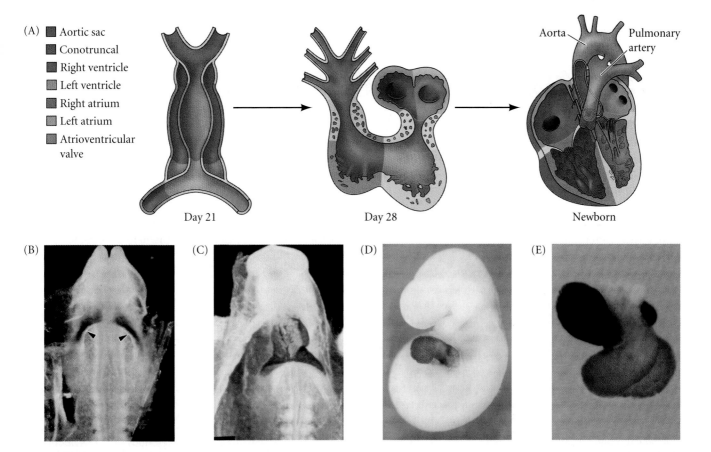

Figure 15.6

Cardiac looping and chamber formation. (A) Schematic diagram of cardiac morphogenesis in humans. On day 21, the heart is a single-chambered tube. The specification of the regions of the tube is shown in different colors. By day 28, cardiac looping has occurred, placing the presumptive atria anterior to the presumptive ventricles. (B, C) *Xin* expression in the fusion of left and right heart primordia. The cells fated to form the myocardium are shown by staining for the *Xin* message, whose protein product is essential for the looping of the heart tube. (B) Stage 9 chick neurula, in which expression of *Xin* (purple) is seen in the two symmetrical heart-forming fields (arrows). (C) Stage 10 chick embryo, showing the fusion of the two heart-forming regions prior to looping. (D, E) Specification of the atrium and ventricles occurs even before heart looping. The atrium and ventricles of the mouse embryo have separate types of myosin proteins, which allows them to be differentially stained. In these photographs, the atrial myosin is stained blue, while the ventricular myosin is stained orange. (D) In the tubular heart (prior to looping), the two myosins (and their respective stains) overlap at the atrioventricular channel joining the future regions of the heart. (E) After looping, the dark blue stain is seen in the definitive atria and inflow tract, while the orange stain is seen in the ventricles. The unstained region above the ventricles is the truncus arteriosus, which becomes separated into the aorta and pulmonary arteries. (A after Srivastava and Olson 2000; B, C from Wang et al. 1999, photographs courtesy of J. J.-C. Lin; E, F after Xavier-Neto et al. 1999, photographs courtesy of N. Rosenthal.)

Figure 15.7

Formation of the chambers of the heart. (A) Diagrammatic cross section of the human heart at 4.5 weeks. The atrial and ventricular septa are growing toward the endocardial cushion. (B) Cross section of the human heart during the third month of gestation. Blood can cross from the right side of the heart to the left side through the openings in the primary and secondary atrial septa. (After Larsen 1993.)

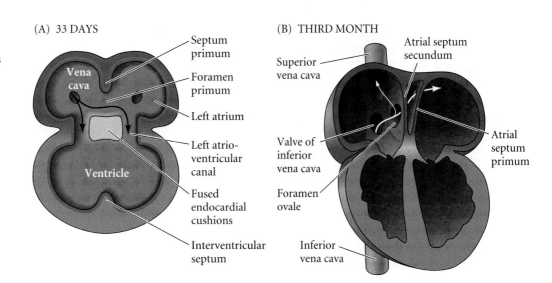

Redirecting Blood Flow in the Newborn Mammal

Although the developing mammalian fetus shares with the adult the need to get oxygen and nutrients to its tissues, the physiology of the fetus differs dramatically from that of the adult. Chief among the differences is the fetus's lack of functional lungs and intestines. All of its oxygen and nutrients must come from the placenta. This observation raises two questions. First, how does the fetus obtain oxygen from maternal blood? And second, how is blood circulation redirected to the lungs once the umbilical cord is cut and breathing becomes necessary?

Human Embryonic Circulation

The human embryonic circulatory system is a modification of that used in other amniotes, such as birds and reptiles. The circulatory system to and from the chick embryo and yolk sac is shown in Figure 15.8. Blood pumped through the dorsal aorta passes over the aortic arches and down into the embryo. Some of this blood leaves the embryo through the vitelline arteries and enters the yolk sac. Nutrients and oxygen are

absorbed from the yolk, and the blood returns through the vitelline veins to reenter the heart through the sinus venosus.

Mammalian embryos obtain food and oxygen from the placenta. Thus, although the embryo has vessels analogous to the vitelline veins, the main supply of food and oxygen comes from the umbilical vein, which unites the embryo with the placenta (Figure 15.9; see Figure 11.36 and front cover). This vein, which brings oxygenated and food-laden blood into the embryo, is derived from what would be the right vitelline vein in birds. The umbilical artery, carrying wastes to the placenta, is derived from what would have become the allantoic artery of the chick. It extends from the caudal portion of the aorta and proceeds along the allantois and then out to the placenta.

Fetal Hemoglobin

The solution to the fetus's problem of getting oxygen from its mother's blood involves the development of a specialized fetal hemoglobin. The hemoglobin in fetal red blood cells differs slightly from that in

adult corpuscles. Two of the four peptides—the alpha (α) chains—that make up fetal and adult hemoglobin chains are identical, but adult hemoglobin has two beta (β) chains, while the fetus has two gamma (γ) chains (Figure 15.10). The β-chains bind the natural regulator diphosphoglycerate, which assists in the unloading of oxygen. The γ-chain isoforms do not bind diphosphoglycerate as well, and therefore have a higher affinity for oxygen. In the low-oxygen environment of the placenta, oxygen is released from adult hemoglobin. In this same environment, fetal hemoglobin does not release oxygen, but binds it. This small difference in oxygen affinity mediates the transfer of oxygen from the mother to the fetus. Within the fetus, the myoglobin of the fetal muscles has an even higher affinity for oxygen, so oxygen molecules pass from fetal hemoglobin to the fetal muscles. Fetal hemoglobin is not deleterious to the newborn, and in humans, the replacement of fetal hemoglobin-containing blood cells with adult hemoglobin-containing blood cells is not complete until about 6 months after birth. (The molecular basis for this switch in globins was discussed in Chapter 5.)

From Fetal to Newborn Circulation

Once the fetus is no longer obtaining its oxygen from its mother, how does it restructure its circulation to get oxygen from its own lungs? During fetal development, an opening—the **ductus arteriosus**—di-

Figure 15.8
Circulatory system of the early avian embryo. (A) Construction of the vasculature in a 7-somite quail embryo, stained with a fluorescent antibody that recognizes endothelial cells. Blood islands, which will form the blood vessels and blood cells, can be seen at the edges. (B) Circulatory system of a 44-hour chick embryo. This view shows arteries in color; the veins are stippled. The sinus terminalis is the outer limit of the circulatory system and the site of blood cell generation. (See also Figure 1.2.) (A, photographic montage from Pardanaud et al. 1987, courtesy of F. Dieterlen-Lièvre; B after Carlson 1981.)

(A)

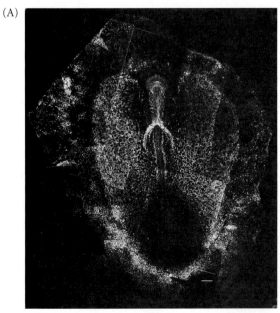

(B)

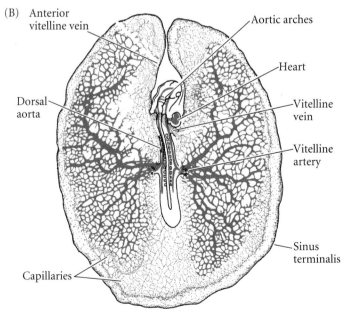

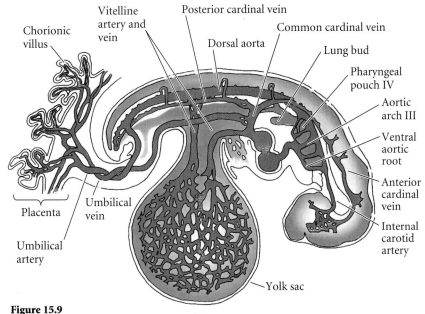

Figure 15.9
Circulatory system of a 4-week human embryo. Although at this stage all the major blood vessels are paired left and right, only the right vessels are shown. Arteries are shown in red, veins in blue. (After Carlson 1981.)

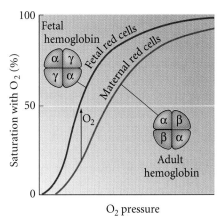

Figure 15.10
Adult and fetal hemoglobin molecules differ in their globin subunits. The fetal γ-chain binds diphosphoglycerate less avidly than does the adult β-chain. Consequently, fetal hemoglobin can bind oxygen more efficiently than can adult hemoglobin. In the placenta, there is a net flow (arrow) of oxygen from the mother's blood (which gives up oxygen to the tissues at lower oxygen pressures) to the fetal blood (which at the same pressure is still taking oxygen up).

verts blood from the pulmonary artery into the aorta (and thus to the placenta). Because blood does not return from the pulmonary vein in the fetus, the developing mammal has to have some other way of getting blood into the left ventricle to be pumped. This is accomplished by the **foramen ovale**, an opening in the septum separating the right and left atria. Blood can enter the right atrium, pass through the foramen into the left atrium, and then enter the left ventricle (Figure 15.11). When the first breath is drawn, blood pressure in the left side of the heart increases. This pressure closes the septa over the foramen ovale, thereby separating the pulmonary and systemic circulations. Moreover, the decrease in prostaglandins experienced by the newborn cause the muscles surrounding the ductus arteriosus to close that opening as well (Nguyen et al. 1997).* Thus, when breathing begins, the respiratory circulation is shunted from the placenta to the lungs.

*In some infants, the septa fail to close, and the foramen ovale is left open. Usually the opening is so small that such children have no physical symptoms, and the foramen eventually closes. If it does not close completely, however, and the secondary septum fails to form, the atrial septal opening may cause enlargement of the right side of the heart, which can lead to heart failure in early adulthood.

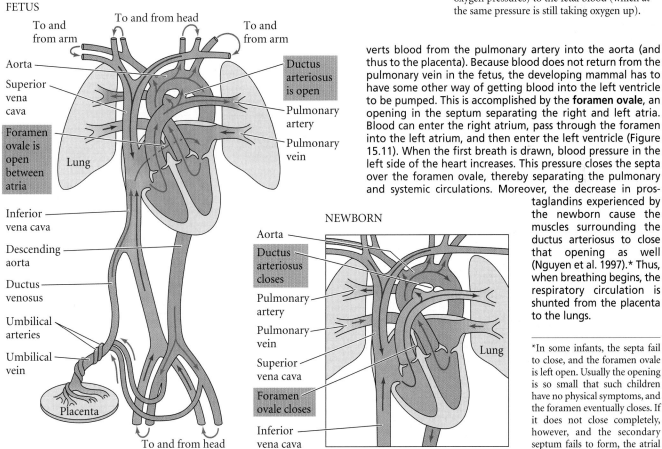

Figure 15.11
Redirection of human blood flow at birth. The expansion of air into the lungs causes pressure changes that redirect the flow of blood in the newborn infant. The ductus arteriosus squeezes shut, breaking off the connection between the aorta and the pulmonary artery, and the foramen ovale, a passageway between the left and right atria, also closes. In this way, pulmonary circulation is separated from systemic circulation.

for the survival of the fetus before circulation to functional lungs has been established. Upon the first breath, however, the septal openings close, and the left and right circulatory loops are established (see Sidelights & Speculations). With the formation of the septa (which usually occurs in the seventh week of human development), the heart is a four-chambered structure with the pulmonary artery connected to the right ventricle and the aorta connected to the left ventricle.

Formation of Blood Vessels

Although the heart is the first functional organ of the body, it does not even begin to pump until the vascular system of the embryo has established its first circulatory loops. Rather than sprouting from the heart, the blood vessels form independently, linking up to the heart soon afterward. Everyone's circulatory system is different, since the genome cannot encode the intricate series of connections between the arteries and veins. Indeed, chance plays a major role in establishing the microanatomy of the circulatory system. However, all circulatory systems in a given species look very much alike, because the development of the circulatory system is severely constrained by physiological, physical, and evolutionary parameters.

Constraints on the construction of blood vessels

The first constraint on vascular development is *physiological.* Unlike new machines, which do not need to function until they have left the assembly line, new organisms have to function even as they develop. The embryonic cells must obtain nourishment before there is an intestine, use oxygen before there are lungs, and excrete wastes before there are kidneys. All these functions are mediated through the embryonic circulatory system. Therefore, the circulatory physiology of the developing embryo must differ from that of the adult organism. Food is absorbed not through the intestine, but from either the yolk or the placenta, and respiration is conducted not

through the gills or lungs, but through the chorionic or allantoic membranes. The major embryonic blood vessels must be constructed to serve these extraembryonic structures.

The second constraint is *evolutionary.* The mammalian embryo extends blood vessels to the yolk sac even though there is no yolk inside (see Figure 15.9). Moreover, the blood leaving the heart via the truncus arteriosus passes through vessels that loop over the foregut to reach the dorsal aorta. Six pairs of these aortic arches loop over the pharynx (Figure 15.12). In primitive fish, these arches persist and enable the gills to oxygenate the blood. In adult birds and mammals, in which lungs oxygenate the blood, such a system makes little sense—but all six pairs of aortic arches are formed in mammalian and avian embryos before the system eventually becomes simplified into a single aortic arch. Thus, even though our physiology does not require such a structure, our embryonic condition reflects our evolutionary history.

The third set of constraints is *physical.* According to the laws of fluid movement, the most effective transport of fluids is performed by large tubes. As the radius of a blood vessel gets smaller, resistance to flow increases as r^{-4} (Poiseuille's law). A blood vessel that is half as wide as another has a resistance to flow that is 16 times greater. However, diffusion of nutrients can take place only when blood flows slowly and has access to cell membranes. So here is a paradox: the constraints of diffusion mandate that vessels be small, while the laws of hydraulics mandate that vessels be large. Living organisms

Figure 15.12
The aortic arches of the human embryo. (A) Originally, the truncus arteriosus pumps blood into the aorta, which branches on either side of the foregut. The six aortic arches take blood from the truncus arteriosus and allow it to flow into the dorsal aorta. (B) As development proceeds, arches begin to disintegrate or become modified (the dotted lines indicate degenerating structures). (C) Eventually, the remaining arches are modified and the adult arterial system is formed. (After Langman 1981.)

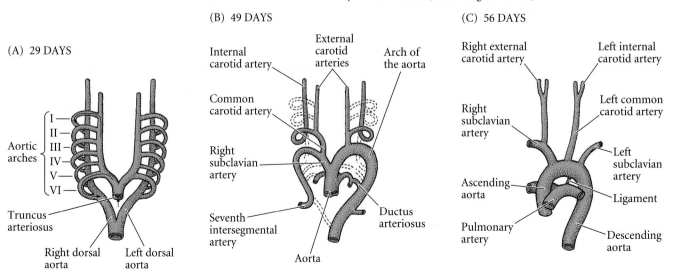

Figure 15.13
Origin and fate of the hemangioblast cells. Hemangioblasts are derived from mesodermal cells that are exposed to relatively high concentrations of certain bone morphogenetic proteins during early development. They give rise to angioblasts—the precursors of the vascular endothelial cells—and to the pluripotential hematopoietic stem cells that generate the blood cells and lymphocytes. (After Liao and Zon 1999.)

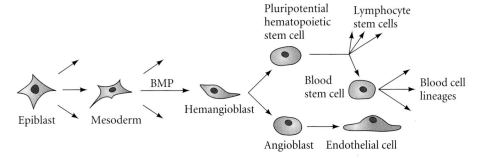

have solved this paradox by evolving circulatory systems with a hierarchy of vessel sizes (LaBarbera 1990). This hierarchy is formed very early in development (and is already well established in the 3-day chick embryo). In dogs, blood in the large vessels (aorta and vena cava) flows over 100 times faster than it does in the capillaries. With a system of large vessels specialized for transport and small vessels specialized for diffusion (where the blood spends most of its time), nutrients and oxygen can reach the individual cells of the growing organism.

But this is not the entire story. If fluid under constant pressure moves directly from a large-diameter tube into a small-diameter tube (as in a hose nozzle), the fluid velocity increases. The evolutionary solution to this problem was the emergence of many smaller vessels branching out from a larger one, making the collective cross-sectional area of all the smaller vessels greater than that of the larger vessel. Circulatory systems show a relationship (known as Murray's law) in which the cube of the radius of the parent vessel approximates the sum of the cubes of the radii of the smaller vessels. Computer models of blood vessel formation must take into account not only gene expression patterns but also the fluid dynamics of blood flow, if they are to show the branching and anastomosing of the arteries and veins (Gödde and Kurz 2001). The construction of any circulatory system negotiates among all of these physical, physiological, and evolutionary constraints.

Vasculogenesis: The initial formation of blood vessels

THE SITES OF VASCULOGENESIS. Blood vessel formation is intimately connected to blood cell formation. Indeed, blood vessels and blood cells are believed to share a common precursor, the **hemangioblast*** (Figure 15.13). Not only do blood vessels and blood cells share common sites of origin, but mutations of certain transcription factors in mice and zebrafish will delete both blood cells and blood vessels. In addition, the ear-

liest blood cells and the earliest blood vessel cells share many of the same rare proteins on their cell surfaces (Wood et al. 1997; Choi et al. 1998; Liao and Zon 1999).

Blood vessels are constructed by two processes, vasculogenesis and angiogenesis (Figure 15.14). During **vasculogenesis**, blood vessels are created de novo from the lateral plate mesoderm. In the first phase of vasculogenesis, groups of splanchnic mesoderm cells are specified to become hemangioblasts, the precursors of both the blood cells and the blood vessels† (Shalaby et al. 1997). These cells condense into aggregations that are often called **blood islands**. The inner cells of these blood islands become **hematopoietic stem cells** (**HSCs**, the stem cells that generate all the blood cell types), while the outer cells become **angioblasts**, the progenitor cells of the blood vessels. In the second phase of vasculogenesis, the angioblasts multiply and differentiate into **endothelial** cells, which form the lining of the blood vessels. In the third phase, the endothelial cells form tubes and connect to form the **primary capillary plexus**, a network of capillaries. The next stage of blood vessel formation is **angiogenesis**, this primary network will be remodeled and pruned into a distinct capillary bed, arteries, and veins (Risau 1997; Hanahan 1997).

In amniotes, formation of the primary vascular networks occurs in two distinct and independent regions. First, **extraembryonic vasculogenesis** occurs in the blood islands of the yolk sac. Second, **intraembryonic vasculogenesis** occurs within the embryo itself. The aggregation of hemangioblasts in the yolk sac is a critical step in amniote development, for the blood islands that line the yolk sac produce the veins that bring nutrients to the embryo and transport gases to and from the sites of respiratory exchange (Figure 15.15). In birds, these vessels are called the **vitelline veins**; in mammals, they are the **omphalomesenteric** (**umbilical**) veins. In the chick, blood islands are first seen in the area opaca, when the primitive streak is at its fullest extent (Pardanaud et al. 1987). They form cords of hemangioblasts, which soon become hollow.

---

*The prefix *hem-* (or *hemato-*) refers to blood (as in hemoglobin). Similarly, the prefix *angio-* refers to blood vessels. The suffix *-blast* denotes a rapidly dividing cell, usually a stem cell. The suffix *-poesis* refers to generation or formation, and is also the root of the word *poetry*. The adjectival form of *-poesis* is *-poietic*. So *hematopoietic* stem cells are those cells that generate the different types of blood cells. The Latin suffix *-genesis* (as in angiogenesis) means the same as the Greek *-poiesis*.

†Again, the endoderm plays a major role in lateral plate mesoderm specification. Here, the visceral endoderm interacts with the yolk sac mesoderm to induce the blood islands. The endoderm is probably secreting Indian hedgehog, a paracrine factor that activates BMP4 expression in the mesoderm. BMP4 expression feeds back on the mesoderm itself, causing it to form hemangioblasts (Baron 2001).

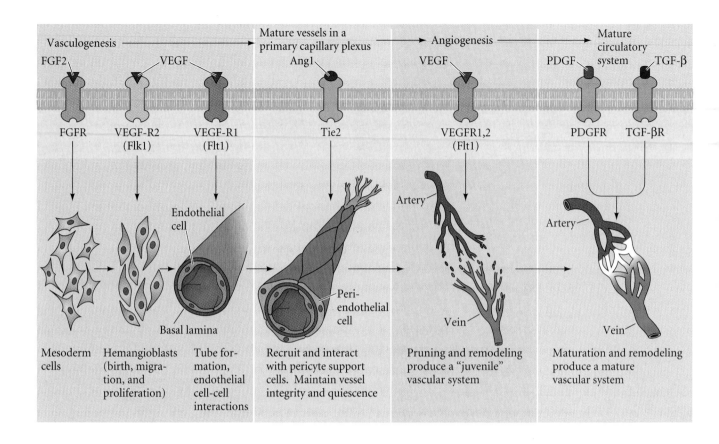

Vasculogenesis → Mature vessels in a primary capillary plexus → Angiogenesis → Mature circulatory system

FGF2 | VEGF | | Ang1 | VEGF | PDGF | TGF-β

FGFR | VEGF-R2 (Flk1) | VEGF-R1 (Flt1) | Tie2 | VEGFR1,2 (Flt1) | PDGFR | TGF-βR

Endothelial cell
Basal lamina
Peri-endothelial cell
Artery
Vein
Artery
Vein

Mesoderm cells | Hemangioblasts (birth, migration, and proliferation) | Tube formation, endothelial cell-cell interactions | Recruit and interact with pericyte support cells. Maintain vessel integrity and quiescence | Pruning and remodeling produce a "juvenile" vascular system | Maturation and remodeling produce a mature vascular system

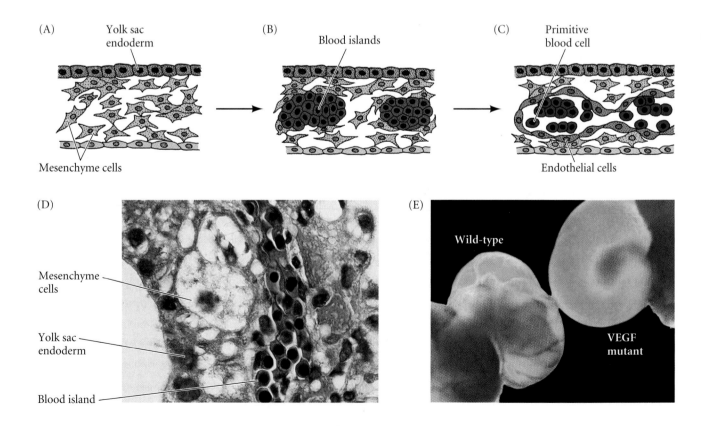

(A) Yolk sac endoderm
Mesenchyme cells

(B) Blood islands

(C) Primitive blood cell
Endothelial cells

(D)
Mesenchyme cells
Yolk sac endoderm
Blood island

(E)
Wild-type
VEGF mutant

◀ **Figure 15.14**
Vasculogenesis and angiogenesis. Vasculogenesis involves the formation of blood islands and the construction of capillary networks
from them. Angiogenesis involves the formation of new blood vessels by remodeling and building on older ones. Angiogenesis finishes the circulatory connections begun by vasculogenesis. The major
paracrine factors involved in each step are shown at the top of the
diagram, and their receptors (on the vessel-forming cells) are shown
beneath them. (After Hanahan 1997; Risau 1997.)

The outer cells of the blood islands become the flat endothelial cells lining the vessels. The central cells differentiate into
the embryonic blood cells. As the blood islands grow, they
eventually merge to form the capillary network draining into
the two vitelline veins, which bring food and blood cells to the
newly formed heart

The intraembryonic vascular networks usually arise from
single angioblast progenitor cells in the mesoderm surrounding a developing organ. These cells do not appear to be associated with blood cell formation, as is seen in the extraembryonic blood islands (Noden 1989; Risau 1995; Pardanaud et al.
1989). It is important to realize that these intraembryonic capillary networks arise within or around the organ itself and are
not extensions of larger vessels. Indeed, in some cases, the developing organ produces paracrine factors that induce blood
vessels to form only in its own mesenchyme (Auerbach et al.
1985; LeCouter et al. 2001).

In fish and amphibians, there is no yolk sac; hence there is
no extraembryonic vasculogenesis. In *Xenopus*, angioblasts
arise from the mesoderm and coalesce to form a vascular
plexus. Here, again, an interaction between the lateral plate
mesoderm and the endoderm is seen. The endoderm is not
required for the formation of the angioblasts. However, without the endoderm, angioblasts cannot form endothelial tubes
(Vokes and Krieg 2002).

GROWTH FACTORS AND VASCULOGENESIS. Three growth factors may be responsible for initiating vasculogenesis (see Figure 15.14). One of these, **basic fibroblast growth factor**
(**FGF2**), is required for the generation of hemangioblasts from
the splanchnic mesoderm. When cells from quail blastodiscs

◀ **Figure 15.15**
Vasculogenesis. Blood vessel formation is first seen in the wall of the
yolk sac, where (A) undifferentiated mesenchyme cells cluster to
form (B) blood islands. (C) The centers of these clusters form the
blood cells, and the outsides of the clusters develop into blood vessel endothelial cells. (D) Photograph of a human blood island in the
mesoderm surrounding the yolk sac. (The photograph is from a
tubal pregnancy—an embryo that had to be removed because it had
implanted in an oviduct rather than in the uterus.) (E) Yolk sacs of
a wild-type mouse and a littermate heterozygous for a loss-of-function mutation of VEGF. The mutant embryo lacks blood vessels in
its yolk sac and dies. (A–C after Langman 1981; D from Katayama
and Kayano 1999; E from Ferrara and Alitalo 1999. Photographs
courtesy of the authors.)

are dissociated in culture, they do not form blood islands or
endothelial cells. However, when these cells are cultured in
FGF2, blood islands emerge and form endothelial cells
(Flamme and Risau 1992). FGF2 is synthesized in the chick
embryonic chorioallantoic membrane and is responsible for
the vascularization of this tissue (Ribatti et al. 1995).

The second protein involved in vasculogenesis is **vascular
endothelial growth factor** (**VEGF**). VEGF appears to enable
the differentiation of the angioblasts and their multiplication
to form endothelial tubes. VEGF is secreted by the mesenchymal cells near the blood islands, and the hemangioblasts and
angioblasts have receptors for VEGF* (Millauer et al. 1993). If
mouse embryos lack the genes encoding either VEGF or the
major receptor for VEGF (the Flk1 receptor tyrosine kinase),
yolk sac blood islands fail to appear, and vasculogenesis fails
to take place (Figure 15.15E; Ferrara et al. 1996). Mice lacking
genes for the second receptor for VEGF (the Flt1 receptor tyrosine kinase) have blood islands and differentiated endothelial cells, but these cells are not organized into blood vessels
(Fong et al. 1995; Shalaby et al. 1995). As we saw in Chapter
14, VEGE is also important in forming blood vessels to the developing bone and kidney.

A third set of protein, the **angiopoietins**, mediates the interaction between the endothelial cells and the **pericytes**—
smooth muscle-like cells they recruit to cover them. Mutations of either angiopoietins or their receptor, Tie2, lead to
malformed blood vessels that are deficient in the smooth
muscles that usually surround them (Davis et al. 1996; Suri et
al. 1996; Vikkula et al. 1996; Moyon 2001).

Angiogenesis: Sprouting of blood vessels and remodeling of vascular beds

After an initial phase of vasculogenesis, angiogenesis begins. By
this process, the primary capillary networks are remodeled and
veins and arteries are made (see Figure 15.14). The extracellular matrix is extremely important in regulating angiogenesis.
First, VEGF acting alone on the newly formed capillaries
causes a loosening of cell-cell contacts and a degradation of the
extracellular matrix at certain points. The exposed endothelial
cells proliferate and sprout from these regions, eventually
forming a new vessel. New vessels can also be formed in the
primary capillary bed by splitting an existing vessel in two. The
loosening of cell-cell contacts may also allow the fusion of cap-

---

*VEGF needs to be regulated very carefully in adults, and recent studies indi-
cate that it can be affected by diet. The consumption of green tea has been associated with lower incidences of human cancer and the inhibition of tumor
cell growth in laboratory animals. Cao and Cao (1999) have shown that green
tea, and one of its components, epigallocatechin-3-gallate (EGCG), prevents
angiogenesis by inhibiting VEGF. Moreoverin mice that drank green tea instead of water (at levels similar to humans after drinking 2–3 cups of tea), the
ability of VEGF to stimulate new blood vessel formation was reduced by more
than 50%. The drinking of moderate amounts of red wine has been correlated with reduced coronary disease. Red wine has been shown to reduce VEGF
production in adults, and it appears to do so by inhibiting endothelin-1, a
compound that induces VEGF and which is crucial for the formation of ath-
erosclerotic plaques (Corder et al. 2001; Spinella et al. 2002).

illaries to form wider vessels—the arteries and veins. Eventually, the mature capillary network forms and is stabilized by TGF-β (which strengthens the extracellular matrix) and **platelet-derived growth factor** (**PDGF**), which is necessary for the recruitment of the pericyte cells that contribute to the mechanical flexibility of the capillary wall) (Lindahl et al. 1997).

It has recently been shown that the blood vessels contain powerful inhibitors of metalloproteases (which digest extracellular matrices), and that angiogenesis can only occur when these inhibitors are downregulated (Oh et al. 2001). One extremely important regulator of angiogenesis may be a particular form of collagen, **collagen XVIII**, the same collagen that is involved in epithelial branch formation (see Chapter 14). This collagen not only stabilizes the capillaries structurally, but also appears to defend the blood vessels chemically against external metalloproteases (Ergun et al. 2001). When collagen XVIII is cleaved by metalloproteases, its C-terminal 184 amino acids form a protein called **endostatin**. Endostatin has functions very different from those of collagen. It prevents angiogenesis by inhibiting cyclin expression and by interfering with the binding of VEGF to its receptor (Zatterstrom et al. 2000; Hanai et al. 2002; Kim et al. 2002). When new blood vessels are to be made, endostatin production is downregulated (Wu et al. 2001). Moreover, endostatin may prevent tumors from growing and spreading. Tumors are "successful" only when they are able to direct blood vessels into them. Therefore, tumors secrete angiogenesis factors. The ability to inhibit such factors may have important medical applications as a way of preventing tumor growth and metastasis (Folkman et al. 1971; Fidler and Ellis 1994; see Chapter 21).

WEBSITE 15.3 Angiogenesis in diabetes and tumor formation. Angiogenesis is a critical part of tumor formation and the effects of diabetes. Some newly discovered proteins, such as angiostatin, endostatin, and squalamine that can inhibit angiogenesis may provide cures for cancers.

Mechanisms of Arterial and Venous Differentiation

A key to our understanding of the mechanism by which veins and arteries form was the discovery that the primary capillary plexus in mice actually contains two types of endothelial cells. The precursors of the arteries contain ephrin-B2 in their cell membranes, and the precursors of the veins contain one of the receptors for this molecule, EphB4 tyrosine kinase, in their cell membranes (Figure 15.16; Wang et al. 1998). If ephrin-B2 is knocked out in mice, vasculogenesis occurs, but angiogenesis does not. It is thought that during angiogenesis EphB4 interacts with its ligand, ephrin-B2, in two ways. First, at the borders of the venous and arterial capillaries, it ensures that arterial capillaries connect only to venous ones. Second, in non-border areas, it ensures that the fusion of capillaries to make larger vessels occurs only between the same type of vessel.

In the zebrafish, the separation of arterial cells and venous cells occurs very early in development. The angioblasts develop in the posterior part of the lateral mesoderm, and they migrate to the midline of the embryo, where they coalesce to form the aorta (artery) and the cardinal vein beneath it (Figure 15.17). Zhong and colleagues (2001) followed individual angioblasts and found that, contrary to expectations, all the progeny of a single angioblast formed either veins or arteries, never both. In other words, each angioblast was already specified as to whether it would form aorta or cardinal vein. This specifica-

Figure 15.16
Model of the roles of ephrin and Eph receptors during angiogenesis. (A) Primary capillary plexus produced by vasculogenesis. The arterial and venous endothelial cells have sorted themselves out by the presence of ephrin-B2 or EphB4 in their respective cell membranes. (B) A maturing vascular network wherein the ephrin-Eph interaction mediates the joining of small branches (future capillaries) and may prevent fusion laterally. (After Yancopopoulos et al. 1998.)

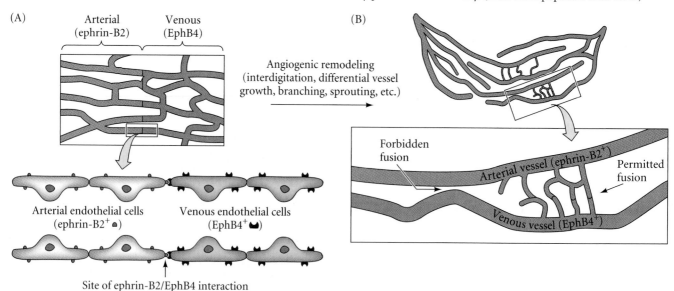

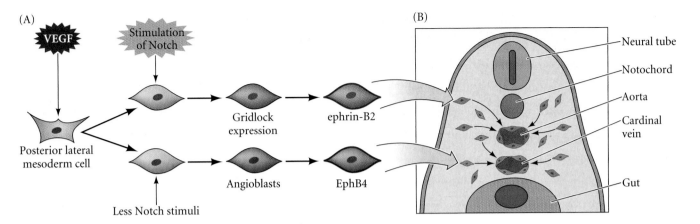

Figure 15.17
Blood vessel specification in the zebrafish embryo. (A) Angioblasts experiencing activation of Notch upregulate the Gridlock transcription factor. These cells express ephrin-B2 and become aorta cells. Those angioblasts experiencing significantly less Notch activation do not express Gridlock, and they become EphB4-expressing cells of the cardinal vein. (B) Once committed to forming veins or arteries, the cells migrate toward the midline of the embryo and contribute to forming the aorta or cardinal vein.

tion appears to be controlled by the Notch signaling pathway*
(Lawson et al. 2001, 2002). Repression of Notch signaling resulted in the loss of ephrin-B2-expressing arteries and their replacement by veins. Conversely, activation of Notch signaling suppressed venous development, causing more arterial cells to form. Activation of the Notch proteins in the membranes of the presumptive angioblasts causes the activation of the transcription factor **Gridlock**. Gridlock in turn activates the expression of Ephrin-B2 and other arterial markers, while those angioblasts with low amounts of Gridlock became EphB4-expressing vein cells.

What, then, controls Notch expression? Studies in zebrafish and mice have revealed that VEGF functions earlier than originally thought, and that it is critical in inducing the differentiation of arterial vessels. Reduction of VEGF activity in zebrafish prevents Notch activation and the subsequent development of arteries, while injection of *VEGF* mRNA into the cells of the posterior cardinal vein induces the expression of (arterial) ephrin-B2 in these cells (Lawson et al. 2002). This also explains the close association of arteries with the peripheral nerves in embryonic skin. Sensory neurons and their glia both express VEGF, and this VEGF is sufficient to induce arteries in nearby endothelial cells (Martin and Lewis 1989; Mukouyama et al. 2002). Weinstein and Lawson (2003) speculate that vascular beds are formed in a two-step process. First, new arteries form in response to VEGF. Second, these arteries then induce neighboring angioblasts (possibly through the ephrin/Eph interactions) to form the venous vessels that will provide the return for

the arterial blood (Figure 15.18A). This speculation fits well with the detailed observations of chick vascular development done (and exquisitely drawn) by Popoff (1894) and Isida (1956). The researchers found that the vitelline arteries appear first within the capillary network and that these capillaries appeared to induce veins on either side of them (Figure 15.18B–D).

Several organs make their own angiogenesis factors. Placental cells are especially adept at creating new blood vessels, since they are responsible for bringing the fetal and maternal blood supplies into close apposition. These cells secrete several angiogenesis factors, including **leptin**, a hormone involved in appetite suppression in the adult. However, leptin can also act locally to induce angiogenesis and cause endothelial cells to organize into tubes (Antczak et al. 1997; Sierra-Honigmann et al. 1998). The kidney vasculature is mainly derived from the sprouting of endothelial cells from the dorsal aorta during the intial steps of nephrogenesis. The developing nephrons secrete VEGF, thus allowing the blood vessels to enter the developing kidney and form the capillary loops of the glomerular apparatus (Kitamoto et al. 2002).

> **WEBSITE 15.4 Angioblast migration in the chick.** After their initial formation in the yolk sac mesoderm, angioblast stem cells are seen in the somites and splanchnic mesoderm. These stem cells populate different regions of the embryo. The lymphatic vessels develop from specialized endothelial cells in preexisting blood vessels.

The Development of Blood Cells

The stem cell concept

Many adult tissues are formed from cells that cannot be replaced. Most neurons and bones, for instance, cannot be replaced if they are damaged or lost. There are several populations of cells, however, that are constantly dying and being replaced. Each day, we lose and replace about 1.5 grams of skin cells and about 10^{11} red blood cells. The skin cells are

---

*The coordinated use of Notch and Eph signaling pathways is also used to regulate the production of neuroblasts and somites.

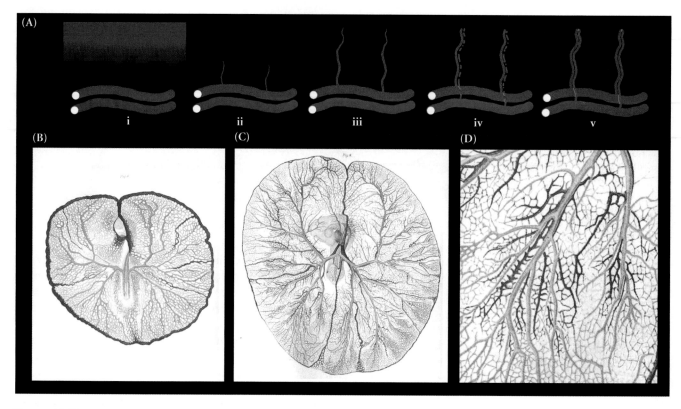

Figure 15.18

Model for blood vessel formation in the chick blastoderm. (A) In response to VEGF (green gradient), endothelial cells are induced to become arteries (red), and these arteries induce veins (blue) to form adjacent to them. New arterial vessels sprout from the arteries and then induce venous vessels adjacent to themselves. (B) In the chick embryo, a complex branched venous network emerges in the vascular region, with venous drainage at the periphery, via the marginal vein. (C) At later stages, collateral veins emerge adjacent to the arteries. (D) Higher magnification of the boxed region of (C). (After Weinstein and Lawson 2003; B–D modified from Popoff 1894, courtesy of N. D. Weinstein.)

sloughed off, and the red blood cells are killed in the spleen. Their replacements come from populations of stem cells.

As mentioned in earlier chapters, a **stem cell** is a cell that is capable of extensive proliferation, creating more stem cells (self-renewal) as well as more differentiated cellular progeny (Figure 15.19). Adult stem cells are, in effect, populations of embryonic cells within an adult organism, continuously producing both more stem cells as well as cells that can undergo further development and differentiation (Potten and Loeffler 1990). Our blood cells, intestinal crypt cells, epidermal cells, and (in males) spermatocytes are populations in a steady-state equilibrium in which cell production balances cell loss (Hay 1966). In most cases, stem cells can produce either more stem cells or more differentiated cells when body equilibrium is stressed by injury or by environmental factors. (This is seen by the production of enormous numbers of red blood cells when the body suffers from anoxia.)

The critical stem cell in hematopoiesis is the pluripotential hematopoietic stem cell. Often just referred to as the hematopoietic stem cell (HSC), this cell type can generate all the blood and lymph cells of the body. As we will see, the pluripotential HSC generates a series of intermediate stem cells whose potency is restricted to certain lineages.

Sites of hematopoiesis

Vertebrate blood development occurs in two phases: a transient **embryonic ("primitive") phase** of hematopoiesis and a subsequent **definitive ("adult") phase**. These phases differ in

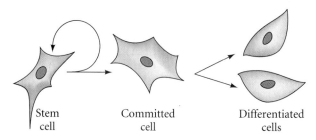

Figure 15.19

The concept of stem cells. A stem cell divides to produce another stem cell and a committed cell. The committed cell can divide again, but its progeny have the ability to form only a restricted set of differentiated cell types. A stem cell can also produce two stem cells or two committed cells. If a significant number of stem cells divide to produce two committed cells, the longevity of the stem cell lineage becomes limited.

their sites of blood cell production, in the timing of hematopoiesis, in the morphology of the cells produced, and even in the type of globin genes active in the red blood cells. The embryonic phase of hematopoiesis is probably utilized to provide the embryo with its initial blood cells and with its capillary network to the yolk. The definitive phase of hematopoiesis is used to generate more cell types and to provide the stem cells that will last for the lifetime of the individual.

Embryonic hematopoiesis is associated with the blood islands in the ventral mesoderm near the yolk sac. In chick embryos, the first blood cells are seen in those blood islands that form in the posterior marginal zone near the site of hypoblast initiation (Wilt 1974; Azar and Eyal-Giladi 1979). In *Xenopus*, the ventral mesoderm forms a large blood island that is the first site of hematopoiesis. BMPs are crucial in inducing the blood forming cells in all vertebrates studied. Ectopic BMP2 and BMP4 can induce blood and blood vessel formation in *Xenopus*, and interference with BMP signaling prevents blood formation (Maeno et al. 1994; Hemmati-Brivanlou and Thomsen 1995). In the zebrafish, the *swirl* mutation, which prevents BMP2 signaling, also abolishes ventral mesoderm and blood cell production (Mullins et al. 1996). As mentioned above, BMP4 is critical in the formation of the blood islands in the mammalian extraembryonic mesoderm.

This embryonic hematopoietic cell population, however, is thought to be transitory. The hematopoietic stem cells that last the lifetime of the organism are derived from the mesodermal area surrounding the aorta. This was shown by a series of elegant experiments performed by Dieterlen-Lièvre, who grafted the blastoderm of chickens onto the yolk of Japanese quail (Figure 15.20). Chick cells are readily distinguishable from quail cells because the quail cell nucleus stains much more darkly (owing to its dense nucleoli), thus providing a permanent marker for distinguishing the two cell types (see Figure 1.10). Using these "yolk sac chimeras," Dieterlen-Lièvre and Martin (1981) showed that the yolk sac stem cells do not contribute cells to the adult animal. Instead, the definitive stem cells are formed within nodes of mesoderm that line the mesentery and the major blood vessels. In the 4-day chick embryo, the aortic wall appears to be the most important source of new blood cells, and it has been found to contain numerous hematopoietic stem cells (Cormier and Dieterlen-Lièvre 1988).

Similarly, studies in fish, mammals, and frogs indicate that the definitive hematopoietic stem cells are formed near the aorta in a domain called the **aorta-gonad-mesonephros (AGM)** region. The first blood cells in the mouse embryo appear in the mesoderm around the yolk sac, but by day 11, pluripotential hematopoietic stem cells can be found in the AGM (Kubai and Auerbach 1983; Godlin et al. 1993; Medvinsky et al. 1993). These hematopoietic stem cells later colonize the fetal liver, and around the time of birth, stem cells from the liver populate the bone marrow, which then becomes the major site of blood formation throughout adult life. One of the proteins that distinguish the primitive HSCs from the definitive (adult) HSCs is **Hoxb-4**. The expression of Hoxb-4 is critical for the self-renewal of adult HSCs, and expression of this gene in yolk sac HSCs allows them to repopulate adult bone marrow, something that primitive HSCs cannot otherwise do (Antonchuk et al. 2002; Kyba et al. 2002).

One of the big questions about blood development concerns whether the AGM is the site of new HSC production or whether it serves as a reservoir for HSCs made elsewhere

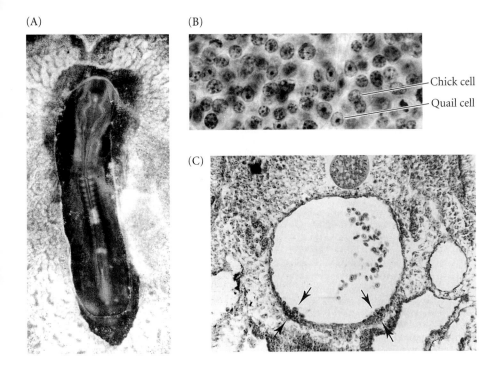

(A)

(B)

Chick cell
Quail cell

(C)

Figure 15.20
Blood cell mapping using chick-quail chimeras. (A) Photograph of a "yolk sac chimera," created by transplanting the blastoderm of a quail onto the yolk sac of a chick. (B) Photograph of chick and quail cells in the thymus of a chimeric animal, showing the difference in nuclear staining. The lymphoid cells are all chick, whereas the structural cells of the thymus are of quail origin. (C) Section through the aorta of a 3-day chick embryo, showing the cells (arrows) that give rise to the hematopoietic stem cells. If cells from this region are taken from quail embryos and placed in chick embryos, the chick embryos will have quail blood. (From Martin et al. 1978, and Dieterlen-Lièvre and Martin 1981; photographs courtesy of F. Dieterlen-Lièvre.)

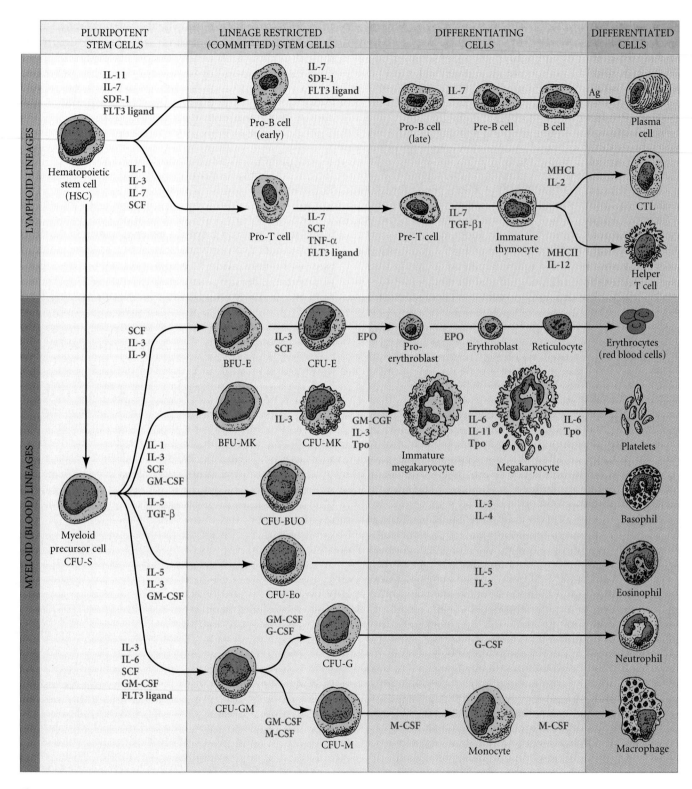

Figure 15.21

A model for the origin of mammalian blood and lymphoid cells. (Other models are consistent with the data, and this one summarizes features from several models.) Factors affecting each step of differentiation are shown in red. Ag, antigen; EPO, erythropoietin; G-CSF, granulocyte colony-stimulating factor; GM-CSF, granulocyte-macrophage colony-stimulating factor; IL, interleukin; LIF, leukemia inhibiting factor; M-CSF, macrophage colony-stimulating factor; MHCI, major histocampatability complex type I protein; MHCII, major histocompatability complex type II protein; SCF, stem cell factor; SDF-1, stromal-derived factor-1; TNF, tumor necrosis factor; Tpo, thrombopoietin. (After Nakauchi and Gachelin 1993; R&D Systems 1997.)

(such as in the yolk sac mesoderm). In the chick, there appears to be little if any contribution to adult circulation from the yolk blood islands, while in *Xenopus*, about 20% of the blood cells appear to be embryonic in origin. But in mammals, we do not yet know whether embryonic HSCs contribute to adult blood cells. While yolk sac HSCs cannot directly repopulate adult bone marrow, they appear to be able to repopulate the livers or marrow of newborn mice (Yoder et al. 1997) and may be induced to become definitive HSCs if they enter the AGM (Matsuoka et al. 2001). Thus, the AGM may be both a site for new HSC production as well as a site that can "update" embryonic HSCs to their new, adult, environment (Orkin and Zorn 2002).

Committed stem cells and their fates

The pluripotential hematopoietic stem cell is a remarkable cell, in that it is the common precursor of red blood cells (erythrocytes), white blood cells (granulocytes, neutrophils, and platelets), and lymphocytes. When transplanted into inbred, irradiated mice (who are genetically identical to the donor cells and whose own stem cells have been eliminated by radiation), pluripotential HSCs can repopulate the mouse with all the blood and lymphoid cell types. It is estimated that only about 1 in every 10,000 blood cells is a pluripotential HSC (Berardi et al. 1995). The descendants of these cells are shown in Figure 15.21.

The pluripotential HSC (sometimes called the CFU-M,L) appears to be dependent on the transcription factor SCL. Mice lacking this protein die from the absence of all blood and lymphocyte lineages. SCL may specify the ventral mesoderm to a blood cell fate, or it may enable the formation or maintenance of the HSC cells (Porcher et al. 1996; Robb et al. 1996).

The HSC cells give rise to **lineage-restricted stem cells** that produce either blood cells (the CFU-S cells) or the several lymphocytic stem cell types. The CFU-S are also pluripotent because their progeny, too, can differentiate into numerous cell types. The immediate progeny of the CFU-S, however, are *committed* lineage-restricted stem cells. Each can produce only *one* type of cell in addition to renewing itself. For instance, the **erythroid precursor cell** (BFU-E) is a committed stem cell that can form only red blood cells. Its immediate progeny (CFU-E) is capable of responding to the hormone **erythropoietin** to produce the first recognizable differentiated member of the erythrocyte lineage, the **proerythroblast**, a red blood cell precursor. Erythropoietin is a glycoprotein that rapidly induces the synthesis of the mRNA for globin (Krantz and Goldwasser 1965). It is produced predominantly in the kidney, and its synthesis is responsive to environmental conditions. If the level of blood oxygen falls, erythropoietin production is increased, leading to the production of more red blood cells. As the proerythroblast matures, it becomes an **erythroblast**, synthesizing enormous amounts of hemoglobin. Eventually, the mammalian erythroblast expels its nucleus, becoming a **reticulocyte**. Reticulocytes can no longer synthesize

globin mRNA, but they can still translate existing messages into globin. The final stage of differentiation is the **erythrocyte**, or mature red blood cell. In this cell, no division, RNA synthesis, or protein synthesis takes place. The DNA of the erythrocyte condenses and makes no further messages. Amphibians, fish, and birds retain the functionless nucleus; mammals extrude it from the cell.* The cell leaves the bone marrow and enters the circulation, where it delivers oxygen to the body tissues. Similarly, other lineage-restricted stem cells give rise to platelets, to granulocytes (neutrophils, basophils, and eosinophils), and to macrophages.

Hematopoietic inductive microenvironments

Different paracrine factors are important in causing hematopoietic stem cells to differentiate along particular pathways (see Figure 15.21). The paracrine factors involved in blood cell and lymphocyte formation are called **cytokines**. Cytokines can be made by several cell types, but they are collected and concentrated by the extracellular matrix of the stromal (mesenchymal) cells at the sites of hematopoiesis (Hunt et al. 1987; Whitlock et al. 1987). For instance, granulocyte-macrophage colony-stimulating factor (GM-CSF) and the multilineage growth factor interleukin 3 (IL-3) both bind to the heparan sulfate glycosaminoglycan of the bone marrow stroma (Gordon et al. 1987; Roberts et al. 1988). The extracellular matrix is then able to present these factors to the stem cells in concentrations high enough to bind to their receptors. In another example, the stem cell factor and VEGF responsible for initiating the proliferation of HSCs and vessel endothelia are usually tethered to the extracellular matrix in an unusable state. Signals that instruct the proliferation of HSCs and blood vessels often release metalloprotease 9, which then liberates both VEGF and stem cell factor from the extracellular matrix (Heissig et al. 2002).

The developmental path taken by the descendant of a pluripotential stem cell depends on which growth factors it meets, and is therefore determined by the stromal cells. Wolf and Trentin (1968) demonstrated that short-range interactions between stromal cells and stem cells determine the developmental fates of the stem cells' progeny. These investigators placed plugs of bone marrow in a spleen and then injected stem cells into it. Those CFU-S cells that came to reside in the spleen formed colonies that were predominantly erythroid, whereas those that came to reside in the marrow formed colonies that were predominantly granulocytic. In fact, colonies that straddled the borders of the two tissue

---

*In 1846, the young Joseph Leidy (then a struggling coroner, later the most famous biologist in America) was the first to use a microscope to solve a murder mystery. A man accused of killing a Philadelphia farmer had blood on his clothes and hatchet. The suspect claimed the blood was from chickens he had been slaughtering. Using his microscope, Leidy found no nuclei in these erythrocytes. Moreover, he found that if he let chick erythrocytes remain outside the body for hours, they did not lose their nuclei. Thus, he concluded that the blood stains could not have been chicken blood. The suspect subsequently confessed (Warren 1998).

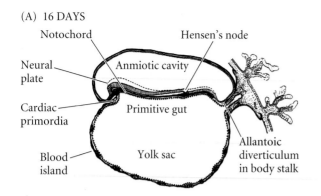

(A) 16 DAYS

Notochord
Hensen's node
Neural plate
Anmiotic cavity
Cardiac primordia
Primitive gut
Blood island
Yolk sac
Allantoic diverticulum in body stalk

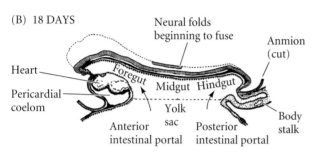

(B) 18 DAYS

Neural folds beginning to fuse
Anmion (cut)
Heart
Foregut   Midgut   Hindgut
Pericardial coelom
Yolk sac
Body stalk
Anterior intestinal portal
Posterior intestinal portal

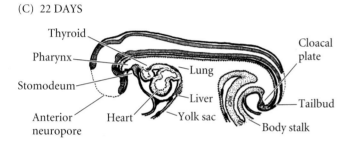

(C) 22 DAYS

Thyroid
Cloacal plate
Pharynx
Lung
Stomodeum
Liver
Anterior neuropore
Heart
Yolk sac
Tailbud
Body stalk

(D) 28 DAYS

Stomach
Lung bud
Pancreas
Thyroid
Stomodeum (now open)
Dorsal aorta
Notochord
Rathke's pouch
Allantois
Heart
Infundibulum
Brain
Yolk stalk
Proctodeum
Liver
Amnion (cut)

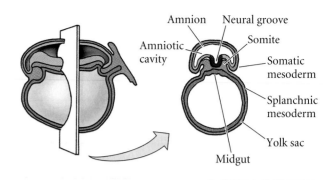

Amnion   Neural groove
Amniotic cavity
Somite
Somatic mesoderm
Splanchnic mesoderm
Yolk sac
Midgut

Figure 15.22
Formation of the human digestive system, depicted at about (A) 16 days, (B) 18 days, (C) 22 days, and (D) 28 days. (After Crelin 1961.)

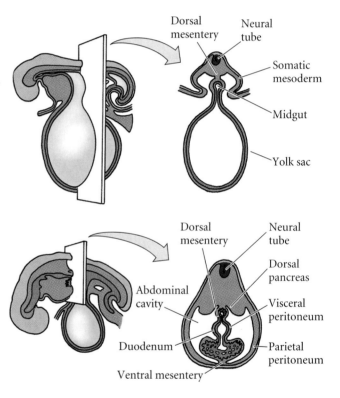

Dorsal mesentery   Neural tube
Somatic mesoderm
Midgut
Yolk sac

Dorsal mesentery   Neural tube
Dorsal pancreas
Abdominal cavity
Visceral peritoneum
Duodenum
Parietal peritoneum
Ventral mesentery

types were predominantly erythroid in the spleen and granu-locytic in the marrow. Such regions of determination are re-ferred to as **hematopoietic inductive microenvironments (HIMs)**.

ENDODERM

The endoderm has two major functions. The first function is to induce the formation of several mesodermal organs. As we have seen in this and earlier chapters, the endoderm is critical for instructing the formation of the notochord, the heart, the blood vessels, and even of the mesodermal germ layer. The second function of the embryonic endoderm is to construct the linings of two tubes within the vertebrate body. The first tube, extending throughout the length of the body, is the **di-gestive tube**. Buds from this tube form the liver, gallbladder, and pancreas. The second tube, the **respiratory tube**, forms as an outgrowth of the digestive tube, and it eventually bifurcates into two lungs. The region of the digestive tube anterior to the point where the respiratory tube branches off is called the **pharynx**. Epithelial outpockets of the pharynx give rise to the tonsils and the thyroid, thymus, and parathyroid glands, and eventually to the respiratory tube itself.

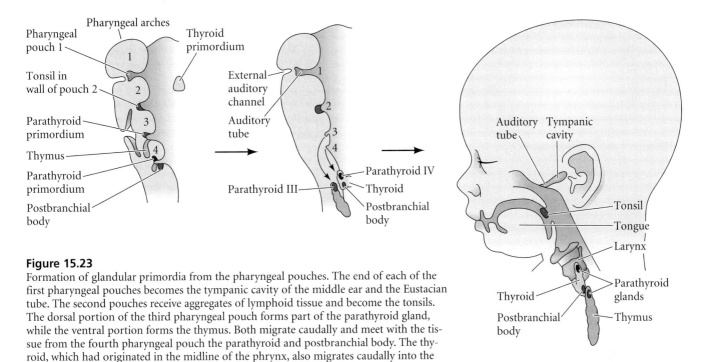

Figure 15.23
Formation of glandular primordia from the pharyngeal pouches. The end of each of the first pharyngeal pouches becomes the tympanic cavity of the middle ear and the Eustacian tube. The second pouches receive aggregates of lymphoid tissue and become the tonsils. The dorsal portion of the third pharyngeal pouch forms part of the parathyroid gland, while the ventral portion forms the thymus. Both migrate caudally and meet with the tissue from the fourth pharyngeal pouch the parathyroid and postbranchial body. The thyroid, which had originated in the midline of the phrynx, also migrates caudally into the neck region. (After Carlson 1999.)

The respiratory and digestive tubes are both derived from the primitive gut (Figure 15.22). As the endoderm pinches in toward the center of the embryo, the foregut and hindgut regions are formed. At first, the oral end of the gut is blocked by a region of ectoderm called the **oral plate**, or **stomodeum**. Eventually (at about 22 days in human embryos), the stomodeum breaks, thereby creating the oral opening of the digestive tube. The opening itself is lined by ectodermal cells. This arrangement creates an interesting situation, because the oral plate ectoderm is in contact with the brain ectoderm, which has curved around toward the ventral portion of the embryo. These two ectodermal regions interact with each other. The roof of the oral region forms **Rathke's pouch** and becomes the *glandular* part of the pituitary gland. The neural tissue on the floor of the diencephalon gives rise to the **infundibulum**, which becomes the *neural* portion of the pituitary. Thus, the pituitary gland has a dual origin, which is reflected in its adult functions.

The Pharynx

The anterior endodermal portion of the digestive and respiratory tubes begins in the pharynx. Here, the mammalian embryo produces four pairs of **pharyngeal pouches** (Figure 15.23). Between these pouches are the **pharyngeal arches**. The first pair of pharyngeal pouches becomes the auditory cavities of the middle ear and the associated Eustachian tubes. The second pair of pouches gives rise to the walls of the tonsils. The thymus is derived from the third pair of pharyngeal pouches; it

will direct the differentiation of T lymphocytes during later stages of development. One pair of parathyroid glands is also derived from the third pair of pharyngeal pouches, and the other pair is derived from the fourth. In addition to these paired pouches, a small, central diverticulum is formed between the second pharyngeal pouches on the floor of the pharynx. This pocket of endoderm and mesenchyme will bud off from the pharynx and migrate down the neck to become the thyroid gland. The respiratory tube sprouts from the pharyngeal floor, between the fourth pair of pharyngeal pouches, to form the lungs, as we will see below.

The Digestive Tube and Its Derivatives

Posterior to the pharynx, the digestive tube constricts to form the esophagus, which is followed in sequence by the stomach, small intestine, and large intestine. The endodermal cells generate only the lining of the digestive tube and its glands; mesenchyme cells from the splanchnic portion of the lateral plate mesoderm will surround the tube to provide the muscles for peristalsis.

As Figure 15.23 shows, the stomach develops as a dilated region of the gut close to the pharynx. More caudally, the intestines develop, and the connection between the intestine and yolk sac is eventually severed. At the caudal end of the intestine, a depression forms where the endoderm meets the overlying ectoderm. Here, a thin **cloacal membrane** separates the two tissues. It eventually ruptures, forming the opening that will become the anus.

(A)

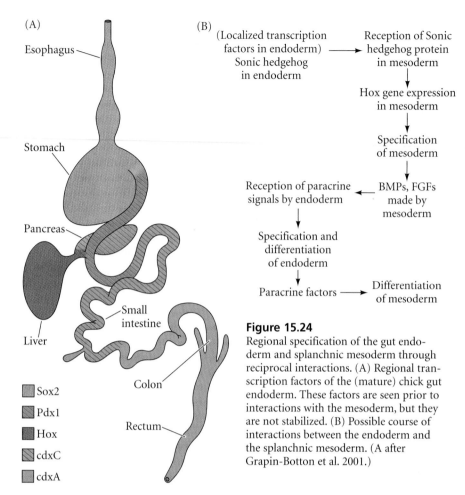

Esophagus

Stomach

Pancreas

Liver

Small intestine

Colon

Rectum

- ▢ Sox2
- ▨ Pdx1
- ▨ Hox
- ▨ cdxC
- ▢ cdxA

(B)

(Localized transcription factors in endoderm)
Sonic hedgehog in endoderm
→ Reception of Sonic hedgehog protein in mesoderm
↓
Hox gene expression in mesoderm
↓
Specification of mesoderm
↓
Reception of paracrine signals by endoderm ← BMPs, FGFs made by mesoderm
↓
Specification and differentiation of endoderm
↓
Paracrine factors → Differentiation of mesoderm

Figure 15.24
Regional specification of the gut endoderm and splanchnic mesoderm through reciprocal interactions. (A) Regional transcription factors of the (mature) chick gut endoderm. These factors are seen prior to interactions with the mesoderm, but they are not stabilized. (B) Possible course of interactions between the endoderm and the splanchnic mesoderm. (A after Grapin-Botton et al. 2001.)

Specification of the gut tissue

As the endodermal tubes form in tetrapods, the endodermal epithelium is able to respond differently to different regionally specific mesodermal mesenchymes. These responses enable the digestive tube and respiratory tube to develop different structures in different regions. Thus, as the digestive tube meets different mesenchymes, it differentiates into esophagus, stomach, small intestine, and colon (Okada 1960; Gumpel-Pinot et al. 1978; Fukumachi and Takayama 1980; Kedinger et al. 1990). The endoderm and the splanchnic lateral plate mesoderm undergo a complicated set of interactions. However, recent research into chick and frog gut formation is leading to a consensus on the steps that regionalize the digestive tube (Ishii et al. 1997; Horb and Slack 2001; Matsushita et al. 2002). Even to zebrafish gut, which is formed from assembling individual organ primordia (rather than arising from a single gut tube) shares the same molecular programs for generating the different gut tissues (Wallace and Pack 2003).

The gut appears to be regionally specified at a very early stage. Indeed, in the chick, the endoderm appears to be regionally specified even before it forms a tube. At this time, the endoderm expresses a set of transcription factors that are regionally specific. For instance, in the 1.5-day chick embryo, *CdxA*, a ho-

mologue of the *Drosophila Caudal* gene, is expressed in the region that will be the intestine, while *cSox2* is expressed in the precursors of the stomach and esophagus (Matsushita et al. 2002). Expression of these and other transcription factors will be retained throughout the development of the gut tube (Figure 15.24).

This regional specification of the gut tube is still labile, however, and the boundaries between the regions are uncertain. The stabilization of these boundaries results from interaction with the mesoderm. As the gut tube begins to form at the anterior and posterior ends, it induces the splanchnic mesoderm to become regionally specific. Roberts and colleagues (1988, 1995) have implicated Sonic hedgehog protein (Shh) in this specification. Early in development, Shh expression is limited to the posterior endoderm of the chick hindgut and the pharynx. As the tubes extend toward the center of the embryo, the domains of Shh expression increase, eventually extending throughout the gut endoderm. Shh is secreted in different concentrations at different sites, and its target appears to be the mesoderm surrounding the gut tube. The secretion of Shh by the hindgut endoderm induces a nested pattern of "posterior" Hox gene expression in the mesoderm. As in the vertebrae (see Chapter 11), the anterior borders of Hox gene expression delineate the morphological boundaries of the regions that will form the cloaca, large intestine, cecum, mid-cecum (at the midgut/hindgut border), and the posterior portion of the midgut (Roberts et al. 1995; Yokouchi et al. 1995). When Hox-expressing viruses cause the misexpression of these Hox genes in the mesoderm, the mesodermal cells alter the differentiation of the adjacent endoderm (Roberts et al. 1998). The Hox genes are thought to specify the mesoderm so that it can interact with the endodermal tube and specify its regions. Once the boundaries of the transcription factors are established, differentiation can begin.

Liver, pancreas, and gallbladder

The endoderm also forms the lining of three accessory organs that develop immediately caudal to the stomach. The **hepatic diverticulum** is a tube of endoderm that extends out from the foregut into the surrounding mesenchyme. The mesenchyme induces this endoderm to proliferate, to branch, and to form the glandular epithelium of the liver. A portion of the hepatic diverticulum (that region closest to the digestive tube) continues to function as the drainage duct of the liver, and a branch from this duct produces the gallbladder (Figure 15.25).

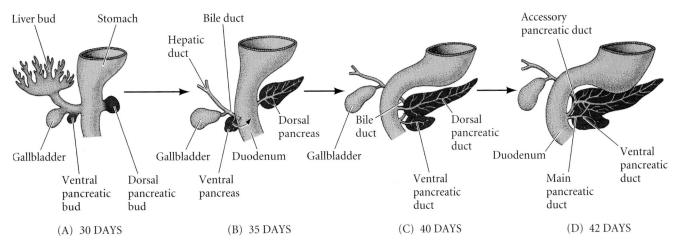

(A) 30 DAYS (B) 35 DAYS (C) 40 DAYS (D) 42 DAYS

The pancreas develops from the fusion of distinct dorsal and ventral diverticula. Both of these primordia arise from the endoderm immediately caudal to the stomach, and as they grow, they come closer together and eventually fuse. In humans, only the ventral duct survives to carry digestive enzymes into the intestine. In other species (such as the dog), both the dorsal and ventral ducts empty into the intestine.

Figure 15.25
Pancreatic development in humans. (A) At 30 days, the ventral pancreatic bud is close to the liver primordium. (B) By 35 days, it begins migrating posteriorly, and (C) comes into contact with the dorsal pancreatic bud during the sixth week of development. (D) In most individuals, the dorsal pancreatic bud loses its duct into the duodenum; however, in about 10% of the population, the dual duct system persists. (After Langman 1981.)

Sidelights & Speculations

Blood and Guts: The Specification of Liver and Pancreas

There is an intimate relationship between the splanchnic lateral plate mesoderm and the foregut endoderm. Just as the foregut endoderm is critical in specifying the cardiogenic mesoderm, the blood vessel endothelial cells induce the endodermal tube to produce the liver primordium and the pancreatic rudiments.

Liver Formation

The expression of liver-specific genes (such as the genes for α-fetoprotein and albumin) can occur anywhere in the gut tube, if that tube is exposed to cardiogenic mesoderm. However, this induction can occur only if the notochord is removed. If the notochord is placed by the portion of the endoderm normally induced by the cardiogenic mesoderm to become liver, the endoderm will not form liver (hepatic) tissue. Therefore, the developing heart appears to induce the liver to form, while the presence of the notochord inhibits liver formation (Figure 15.26). This induction is probably due to FGFs secreted by the developing heart cells (Le Douarin 1975; Gualdi et al. 1996; Jung et al. 1999).

However, Matsumoto and colleagues (2001) found that the heart is not the only

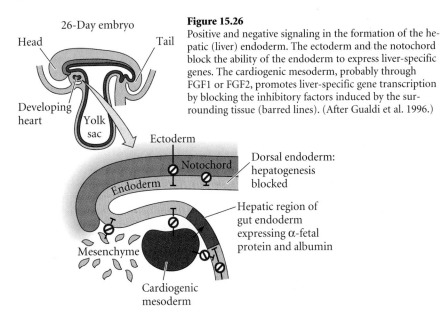

Figure 15.26
Positive and negative signaling in the formation of the hepatic (liver) endoderm. The ectoderm and the notochord block the ability of the endoderm to express liver-specific genes. The cardiogenic mesoderm, probably through FGF1 or FGF2, promotes liver-specific gene transcription by blocking the inhibitory factors induced by the surrounding tissue (barred lines). (After Gualdi et al. 1996.)

mesodermal derivative needed to form the liver. Blood vessel endothelial cells are also critical. If endothelial cells are not present in the area around the hepatic region of the

gut tube, the liver buds fail to form. This induction occurs even before the endothelial cells have formed tubes, so it does not have anything to do with getting nutrients or oxygen into this region. Thus, the blood vessel endothelial cells have a developmental function in addition to their circulatory roles: they induce the formation of the liver bud.

Pancreas Formation

The formation of the pancreas may be the flip side of liver formation. While the heart cells promote and the notochord prevents liver formation, the notochord may actively promote pancreas formation, and the heart may block the pancreas from forming. It has been hypothesized (Gannon and Wright 1999) that Pdx1 expression in the endoderm endows a particular region of the digestive tube (including future portions of the stomach, duodenum, liver, and pancreas rudiments) with the ability to become either pancreas or liver. One set of conditions (presence of heart, absence of notochord) induces the liver, while another set of conditions (presence of notochord, absence of heart) causes the pancreas to form.

The notochord activates pancreas development by repressing expression of the *Sonic hedgehog* gene in the endoderm (Apelqvist et al. 1997; Hebrok et al. 1998). (This was a surprising finding, since we saw in Chapter 13 that the notochord is a source of Sonic hedgehog protein and an inducer of further *shh* gene expression in ectodermal tissues.) Sonic hedgehog is expressed throughout the gut endoderm, except in the region that will form the pancreas. The notochord in this region of the embryo secretes FGF2 and activin, which are able to downregulate *shh* expression in the endoderm. If *shh* is experimentally expressed in this region, the tissue reverts to being intestinal (Jonnson et al. 1994; Ahlgren et al. 1996; Offield et al. 1996).

The lack of Shh in this region of the gut seems to enable it to respond to signals coming from the blood vessel endothelium. Indeed, pancreatic development is initiated at precisely those three locations where the foregut endoderm contacts the endothelium of the major blood vessels. It is at these points—where the endodermal tube meets the aorta and the vitelline veins—that the transcription factor Pdx1 is expressed (Figure 15.27A, B; Lammert et al. 2001). If the blood vessels are removed from this area, the Pdx1 expression regions fail to form, and the pancreatic endoderm fails to bud; if more blood vessels form in this area, more of the endodermal tube becomes pancreatic tissues.

The association of the pancreatic tissues with blood vessels is critical in the formation of the insulin-secreting cells of the pancreas. The *pdx1* gene appears to act in concert with another gene, *ngn3*, to form the endocrine cells of the pancreas, the islets of Langerhans. The exocrine cells (which produce digestive enzymes such as chymotrypsin) and the endocrine cells appear to have the same progenitor (Fishman and Melton 2002). The islet cells secrete VEGF to attract blood vessels, and these vessels surround the developing islet (Figure 15.27D).

The Pdx1 transcription factor is exceptionally important in pancreatic development. Pdx1 elicits budding from the gut epithelium, represses the expression of genes that are characteristic of other regions of the digestive tube, maintains the

Figure 15.27
Induction of *pdx1* gene expression in the gut epithelium. (A) In the chick embryo, *pdx1* (purple) is expressed in the gut tube is induced by contact with the aorta and vitelline veins. The regions of *pdx1* expression create the dorsal and ventral anlagen of the pancreas. (B) In situ hybridization of *pdx1* mRNA in a section through the region of contact between the blood vessels and the gut tube. The regions of *pdx1* expression show as deep blue. (C) In the mouse embryo, only the right vitelline vein survives, and it contacts the gut endothelium. Pdx expression is seen only on this side, and only one ventral pancreatic bud emerges. (D) Blood vessels (stained red with antibodies to PECAM1) direct islets (stained green with antibodies to insulin) to differentiate. The nuclei are stained blue with DAPI. (E) Lineage of pancreatic cells. All pancreatic cells express *pdx1*. The NGN3-expressing cells form the endocrine lineages. (A–C after Lammert et al. 2001; D after Grapin-Botton et al. 2001; photographs courtesy of E. Lammert and D. A. Melton.)

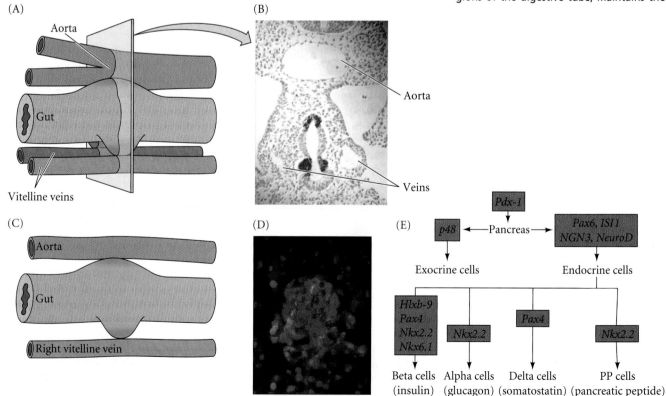

repression of Sonic hedgehog, initiates (but does not complete) islet cell differentiation, and is necessary (but not sufficient) for insulin gene expression (Grapin-Botton et al. 2001). Using an inducible Cre-Lox system to mark the progeny of cells expressing either NGN3 or Pdx1, Gu and colleagues (2002) demonstrated that the NGN3-expressing cells are specifically islet cell progenitors, whereas the cells expressing Pdx1 give rise to all three types of pancreatic tissue (exocrine, endocrine, and duct cells) (Figure 15.27E). Moreover, Horb and colleagues (2003) have shown that Pdx1 can respecify developing liver tissue into pancreas. When *Xenopus* tadpoles were given a *pdx1* gene attached to a promoter active in liver cells, Pdx1 protein was made in the liver, and the liver was converted into a pancreas containing both exocrine and endocrine cells.

The Respiratory Tube

The lungs are a derivative of the digestive tube, even though they serve no role in digestion. In the center of the pharyngeal floor, between the fourth pair of pharyngeal pouches, the **laryngotracheal groove** extends ventrally (Figure 15.28). This groove then bifurcates into the two branches that form the paired bronchi and lungs. The laryngotracheal endoderm becomes the lining of the trachea, the two bronchi, and the air sacs (alveoli) of the lungs. Sometimes this separation is not complete, and a baby is born with a connection between the two tubes. This digestive and respiratory condition is called a **tracheal-esophageal fistula**, and it is surgically repaired so the baby can breathe and swallow properly.

The production of the laryngotracheal groove is correlated with the presence of the transcription factor Tbx4 in the visceral endoderm. Tbx4 appears to induce the formation of the outgrowth and differentiation of the respiratory tube, and inhibiting Tbx4 results in the failure of lung bud formation (Sakiyama et al. 2003).

The lungs are among the last of the mammalian organs to fully differentiate. The lungs must be able to draw in oxygen at the newborn's first breath. To accomplish this, the alveolar cells secrete a surfactant into the fluid bathing the lungs. This surfactant, consisting of phospholipids such as sphingomyelin and lecithin, is secreted very late in gestation, and it usually reaches physiologically useful levels at about week 34 of human gestation. The surfactant enables the alveolar cells to touch one another without sticking together. Thus, infants born prematurely often have difficulty breathing and have to be placed on respirators until their surfactant-producing cells mature.

As in the digestive tube, the regional specificity of the mesenchyme determines the differentiation of the developing respiratory tube. In the developing mammal, the respiratory epithelium responds in two distinct fashions. In the region of the neck, it grows straight, forming the trachea. After entering the thorax, it branches, forming the two bronchi and then the lungs. The respiratory epithelium of an embryonic mouse can be isolated soon after it has split into two bronchi, and the two sides can be treated differently. Figure 15.29 shows the result of such an experiment. The right bronchial epithelium was allowed to retain its lung mesenchyme, whereas the left bronchus was surrounded with tracheal mesenchyme (Wessells 1970). The right bronchus proliferated and branched under the influence of the lung mes-

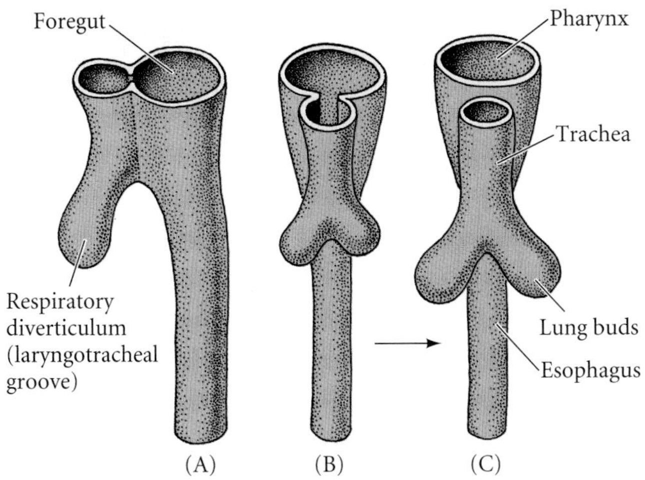

Figure 15.28
Partitioning of the foregut into the esophagus and respiratory diverticulum during the third and fourth weeks of human gestation. (A, B) Lateral and ventral views, end of week 3. (C) Ventral view, week 4. (After Langman 1981.)

Labels for Figure 15.28: Foregut, Pharynx, Trachea, Respiratory diverticulum (laryngotracheal groove), Lung buds, Esophagus, (A) (B) (C)

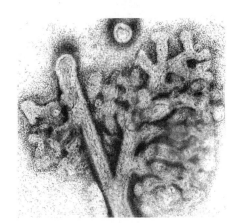

Figure 15.29
Ability of presumptive lung epithelium to differentiate with respect to the source of the inducing mesenchyme. After embryonic mouse lung epithelium had branched into two bronchi, the entire rudiment was excised and cultured. The right bronchus was left untouched, while the tip of the left bronchus was covered with tracheal mesenchyme. The tip of the right bronchus has formed the branches characteristic of the lung, whereas hardly any branching has occurred in the tip of the left bronchus. (From Wessells 1970; photograph courtesy of N. Wessells.)

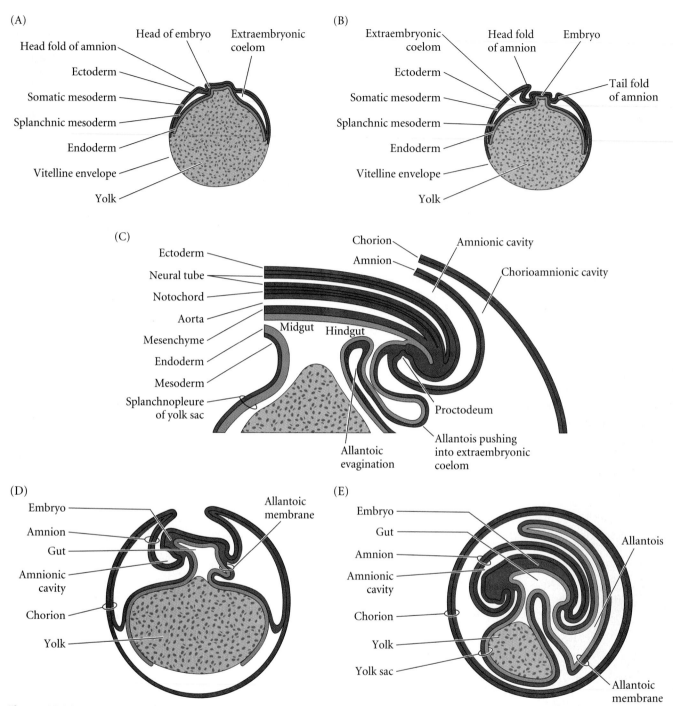

Figure 15.30
Schematic drawings of the extraembryonic membranes of the chick. The embryo is cut longitudinally, and the albumen and shell coatings are not shown. (A) A 2-day embryo. (B) A 3-day embryo.

(C) Detailed schematic diagram of the caudal (hind) region of the chick embryo, showing the formation of the allantois. (D) A 5-day embryo. (E) A 9-day embryo. (After Carlson 1981.)

enchyme, whereas the left bronchus continued to grow in an un-branched manner. Moreover, the differentiation of the respiratory epithelia into trachea cells or lung cells depends on the mesenchyme it encounters (Shannon et al. 1998).

WEBSITE 15.5 Induction of the lung. The induction of the lung involves interplay between FGFs and Shh. However, it appears to be different from the induction of either the pancreas or the liver.

The Extraembryonic Membranes

In reptiles, birds, and mammals, embryonic development has taken a new evolutionary direction—the amniote egg (see Figure 2.22). Reptiles evolved a mechanism for laying eggs on dry land; this adaptation freed them to explore niches that were not close to water. To cope with the challenges of terrestrial development, the embryo produces four sets of extraembryonic membranes to mediate between it and the environment (see Chapter 11). Even though most mammals evolved placentas that replaced shells, the basic pattern of extraembryonic membranes remains the same.

In developing reptiles, birds, and mammals, there initially is no distinction between embryonic and extraembryonic domains. However, as the body of the embryo takes shape, the epithelia at the border between the embryo and the extraembryonic domain divide unequally to create body folds that isolate the embryo from the yolk and delineate which areas are to be embryonic and which extraembryonic (Miller et al. 1994, 1999). These membranous folds are formed by the extension of ectodermal and endodermal epithelium underlain with lateral plate mesoderm. The combination of ectoderm and mesoderm, often referred to as the somatopleure (see Figure 15.3A), forms the amnion and chorion; the combination of endoderm and mesoderm—the splanchnopleure—forms the yolk sac and allantois. The endodermal or ectodermal tissue supplies functioning epithelial cells, and the mesoderm generates the essential blood supply to and from the epithelium. The formation of these folds can be followed in Figure 15.30.

The amnion and chorion

The first problem of a land-dwelling egg faces is desiccation. Embryonic cells would quickly dry out if they were not in an aqueous environment. This environment is supplied by the **amnion**. The cells of this membrane secrete **amnionic fluid**; thus, embryogenesis still occurs in water. This evolutionary adaptation is so significant and characteristic that reptiles, birds, and mammals are grouped together as the amniote vertebrates, or **amniotes**.

The second problem of a terrestrial egg is gas exchange. This exchange is provided for by the **chorion**, the outermost extraembryonic membrane. In birds and reptiles, this membrane adheres to the shell, allowing the exchange of gases between the egg and the environment. In mammals, as we have seen, the chorion has developed into the **placenta**, which has evolved endocrine, immune, and nutritive functions in addition to those of respiration.

The allantois and yolk sac

The third problem for a terrestrial egg is waste disposal. The **allantois** stores urinary wastes and also helps mediate gas exchange. In reptiles and birds, the allantois becomes a large sac, as there is no other way to keep the toxic by-products of metabolism away from the developing embryo. In some amniote species, such as chickens, the mesodermal layer of the allantoic membrane reaches and fuses with the mesodermal layer of the chorion to create the **chorioallantoic membrane**. This extremely vascular envelope is crucial for chick development and is responsible for transporting calcium from the eggshell into the embryo for bone production (Tuan 1987). In mammals, the size of the allantois depends on how well nitrogenous wastes can be removed by the chorionic placenta. In humans (in which nitrogenous wastes can be efficiently removed through the maternal circulation), the allantois is a vestigial sac. In pigs, however, the allantois is a large and important organ.

And finally, the land-dwelling egg must solve the problem of nutrition. The **yolk sac** is the first extraembryonic membrane to be formed, as it mediates nutrition in developing birds and reptiles. It is derived from splanchnopleural cells that grow over the yolk to enclose it. The yolk sac is connected to the midgut by an open tube, the **yolk duct**, so that the walls of the yolk sac and the walls of the gut are continuous. The blood vessels within the mesoderm of the splanchnopleure transport nutrients from the yolk into the body, for yolk is not taken directly into the body through the yolk duct. Rather, endodermal cells digest the protein in the yolk into soluble amino acids that can then be passed on to the blood vessels within the yolk sac. Other nutrients, including vitamins, ions, and fatty acids, are stored in the yolk sac and transported by the yolk sac blood vessels into the embryonic circulation. In these ways, the four extraembryonic membranes enable the amniote embryo to develop on land.

Snapshot Summary: Lateral Plate Mesoderm and Endoderm

1. The lateral plate mesoderm splits into two layers. The dorsal layer is the somatic (parietal) mesoderm, which underlies the ectoderm and forms the somatopleure. The ventral layer is the splanchnic (visceral) mesoderm, which overlies the endoderm and forms the splanchnopleure.

2. The space between the two layers of lateral plate mesoderm forms the body cavity, or coelom.

3. The heart arises from splanchnic mesoderm on both sides of the body. This region of cells is called the cardiogenic mesoderm. The cardiogenic mesoderm is specified by BMPs in the absence of Wnt signals.

4. The Nkx2-5 and GATA transcription factors are important in committing the cardiogenic mesoderm to become

heart cells. These cardiac precursor cells migrate from the sides to the midline of the embryo, in the neck region.

5. The cardiogenic mesoderm forms the endocardium (which is continuous with the blood vessels) and the myocardium (the muscular component of the heart).

6. The endocardial tubes form separately and then fuse. The looping of the heart transforms the original anterior-posterior polarity of the heart tube into a right-left polarity.

7. In mammals, fetal circulation differs dramatically from adult circulation. When the infant takes its first breath, changes in air pressure close the foramen ovale through which blood had passed from the right to the left atrium. At that time, the lungs, rather than the placenta, become the source of oxygen.

8. Blood vessel formation is constrained by physiological, evolutionary, and physical parameters. The subdividing of a large vessel into numerous smaller ones allows rapid transport of the blood to regions of gas and nutrient diffusion.

9. Blood vessels are constructed by two processes, vasculogenesis and angiogenesis. Vasculogenesis involves the condensing of splanchnic mesoderm cells to form blood islands. The outer cells of these islands become endothelial (blood vessel) cells. Angiogenesis involves the remodeling of existing blood vessels.

10. Numerous paracrine factors are essential in blood vessel formation. FGF2 is needed for specifying the angioblasts. VEGF is essential for the differentiation of the angioblasts. Angiopoietins allow the smooth muscle cells (and smooth muscle-like pericytes) to cover the vessels. Ephrin ligands and Eph receptor tyrosine kinases are critical for capillary bed formation.

11. The pluripotential hematopoietic stem cell generates other pluripotential stem cells, as well as lineage-restricted stem cells. It gives rise to both blood cells and lymphocytes.

12. In mammals, embryonic blood stem cells are provided by the blood islands near the yolk. The definitive adult blood stem cells come from the aorta-gonad-mesonephros region within the embryo.

13. The CFU-S is a blood stem cell that can generate the more committed stem cells for the different blood lineages. Hematopoietic inductive microenvironments determine the blood cell differentiation.

14. The endoderm constructs the digestive tube and the respiratory tube.

15. Four pairs of pharyngeal pouches become the endodermal lining of the Eustacian tubes, the tonsils, the thymus, and the parathyroid glands. The thyroid also forms in this region of endoderm.

16. The gut tissue forms by reciprocal interactions between the endoderm and the mesoderm. Sonic hedgehog from the endoderm appears to play a role in inducing a nested pattern of Hox gene expression in the mesoderm surrounding the gut. The regionalized mesoderm then instructs the endodermal tube to become the different organs of the digestive tract.

17. The endoderm helps specify the splanchnic mesoderm; the splanchnic mesoderm, especially the heart and the blood vessels, helps specify the endoderm.

18. The pancreas forms in a region of endoderm that lacks Sonic hedgehog expression. The Pdx1 transcription factor is expressed in this region.

19. The endocrine and exocrine cells of the pancreas have a common origin. The NGN3 transcription factor probably decides endocrine fate.

20. The respiratory tube is derived as an outpocketing of the digestive tube. The regional specificity of the mesenchyme it meets determines whether the tube remains straight (as in the trachea) or branches (as in the alveoli).

21. The chorion and amnion are made by the somatopleure. In birds and reptiles, the chorion abuts the shell and allows for gas exchange. The amnion in birds, reptiles, and mammals bathes the embryo in amnionic fluid.

22. The yolk sac and allantois are derived from the splanchnopleure. The yolk sac (in birds and reptiles) allows yolk nutrients to pass into the blood. The allantois collects nitrogenous wastes.

Literature Cited

Ahlgren, U., J. Jonnson and H. Edlund. 1996. The morphogenesis of the pancreatic mesenchyme is uncoupled from that of the pancreatic epithelium in IPF/PDX1-deficient mice. *Development* 122: 1409–1416.

Alsan, B. H. and T. M. Schultheiss. 2002. Regulation of avian cardiogenesis by Fgf8 signaling. *Development* 129: 1935–1943.

Andrée, B., D. Duprez, B. Vorbusch, H.-H. Arnold and T. Brand. 1998. BMP-2 induces ectopic expression of cardiac lineage markers and interferes with somite formation in chicken embryos. *Mech. Dev.* 70: 119–131.

Antczak, A. M., J. van Blerkom and A. Clark. 1997. A novel mechanism of VEGF, leptin, and TGFâ2 sequestration in human ovarian follicle cells. *Hum. Reprod.* 12: 2226–2234.

Antonchuk, J., G. Savageau and R. K. Humphries. 2002. HoxB4-induced expansion of adult hematopoietic stem cells ex vivo. *Cell* 109: 39–45.

Apelqvist, A., U. Ahlgren and H. Edlund. 1997. Sonic hedgehog directs specialised mesoderm differentiation in the intestine and pancreas. *Curr. Biol.* 7: 801–804.

Auerbach, R., L. Alby, L. Morrissey, M. Tu and J. Joseph. 1985. Expression of organ-specific antigens on capillary endothelial cells. *Microvasc. Res.* 29: 401–411.

Azar, Y. and H. Eyal-Giladi. 1979. Marginal zone cells—the primitive streak-inducing component of the primary hypoblast in the chick. *J. Embryol. Exp. Morphol.* 52: 79–88.

Bao, Z.-Z., B. G. Bruneau, J. G. Seidman, C. E. Seidman and C. L. Cepko. 1999. Regulation of chamber-specific gene expression in the developing heart by Irx4. *Science* 283: 1161–1164.

Baron, M. H. 2001. Induction of embryonic hematopoietic and endothelial stem/precursor cells by hedgehog-mediated signals. *Differentiation* 68: 175–185.

Berardi, A. C., A. Wang, J. D. Levine, P. Lopez and D. T. Scadden. 1995. Functional characterization of human hematopoietic stem cells. *Science* 267: 104–108.

Bergers, G., K. Javaherian, K. M. Lo, J. Folkman and D. Hanahan. 1999. Effects of angiogenesis inhibitors on multistage carcinogenesis in mice. *Science* 284: 808–812.

Biben, C. and R. P. Harvey. 1997. Homeodomain factor Nkx2-5 controls left-right asymmetric expression of bHLH gene *eHand* during murine heart development. *Genes Dev.* 11: 1357–1369.

Bruneau, B. G., M. Logan, N. Davis, T. Levi, C. J. Tabin, J. G. Seidman and C. E. Seidman. 1999. Chamber-specific cardiac expression of Tbx5 and heart defects in Holt-Oram syndrome. *Dev. Biol.* 211: 100–108.

Cao, Y. and R. Cao. 1999. Angiogenesis inhibited by drinking tea. *Nature* 398: 381.

Carlson, B. M. 1981. *Patten's Foundations of Embryology.* McGraw-Hill, New York.

Choi, H., M. Kennedy, A. Kazarov, J. C. Padaimitriou and G. Keller. 1998. A common precursor for hematopoietic and endothelial cells. *Development* 125: 725–732.

Cohen-Gould, L. and T. Mikawa. 1996. The fate diversity of mesodermal cells within the heart field during chicken early embryogenesis. *Dev. Biol.* 177: 265–273.

Colas, J. F., A. Lawson and G. C. Schoenwolf. 2000. Evidence that translation of smooth muscle alpha-actin mRNA is delayed in the chick promyocardium until fusion of the bilateral heart-forming regions. *Dev. Dynam.* 218: 316–330.

Corder, R., J. A. Douthwaite, D. M. Lees, N. Q. Khan, A. C. Viseu Dos Santos, E. G. Wood and M. J. Carrier. 2001. Endothelin-1 synthesis reduced by red wine. *Nature* 414: 863–864.

Cormier, F. and F. Dieterlen-Lièvre. 1988. The wall of the chick aorta harbours M-CFC, G-CFC, GM-CFC and BFU-E. *Development* 102: 279–285.

Crelin, E. S. 1961. Development of the gastrointestinal tract. *Clin. Symp.* 13: 68–82.

Davis, S. and 10 others. 1996. Isolation of angiopoietin-1, a ligand for the TIE2 receptor, by secretion trap expression cloning. *Cell* 87: 1161–1169.

DeHaan, R. L. 1959. Cardia bifida and the development of pacemaker function in the early chicken heart. *Dev. Biol.* 1: 586–602.

DeHaan, R. L. 1967. Regulation of spontaneous activity and growth of embryonic chick heart cells in tissue culture. *Dev. Biol.* 16: 216–249.

Dieterlen-Lièvre, F. and C. Martin. 1981. Diffuse intraembryonic hemopoiesis in normal and chimeric avian development. *Dev. Biol.* 88: 180–191.

Ergun, S. and 14 others. 2001. Endostatin inhibits angiogenesis by stabilization of newly formed endothelial tubes. *Angiogenesis* 4: 193–206.

Ferrara, N. and K. Alitalo. 1999. Clinical applications of angiogenic growth factors and their inhibitors. *Nature Med.* 5: 1359–1364.

Ferrara, N. and 8 others. 1996. Heterozygous embryonic lethality induced by targeted inactivation of the VEGF gene. *Nature* 380: 439–442.

Fidler, I. J. and L. M. Ellis. 1994. The implications of angiogenesis for the biology and therapy of cancer metastasis. *Cell* 70: 185–188.

Fishman, M. P. and D. A. Melton. 2002. Pancreatic lineage analysis using a retroviral vector in embryonic mice demonstrates a common progenitor for endocrine and exocrine cells. *Int. J. Dev. Biol.* 46: 201–207.

Flamme, I. and W. Risau. 1992. Induction of vasculogenesis and hematogenesis in vitro. *Development* 116: 435–439.

Folkman, J., E. Merler, C. Abernathy and G. Williams. 1971. Isolation of a tumor factor responsible for angiogenesis. *J. Exp. Med.* 133: 275–288.

Fong, G.-H., J. Rossant, M. Gertenstein and M. L. Breitman. 1995. Role of the Flt-1 receptor tyrosine kinase in regulating the assembly of vascular endothelium. *Nature* 376: 66–70.

Fukumachi, H. and S. Takayama. 1980. Epithelial-mesenchymal interaction in differentiation of duodenal epithelium of fetal rats in organ culture. *Experientia* 36: 335–336.

Gannon, M. and C. Wright. 1999. Endodermal patterning and organogenesis. In S. A. Moody (ed.), *Cell Lineage and Fate Determination.* Academic Press, San Diego. pp. 583–615.

Gödde, R. and H. Kurz. 2001. Structural and biophysical simulation of angiogenesis and vascular remodeling. *Dev. Dynam.* 220: 387–401.

Godlin, I. E., J. A. Garcia-Porrero, A. Coutinho, F. Dieterlen-Lièvre and M. A. R. Marcos. 1993.

Para-aortic splanchnopleura from early mouse embryos contain B1a cell progenitors. *Nature* 364: 67–70.

Gordon, M. Y., G. P. Riley, S. M. Watt and M. F. Greaves. 1987. Compartmentalization of a haematopoietic growth factor (GM-CSF) by glycosaminoglycans in the bone marrow microenvironment. *Nature* 326: 403–405.

Gräper, L. 1907. Untersuchungen über die Herzbildung der Vögel. *Wilhelm Roux Arch. Entwicklungsmech. Org.* 24: 375–410.

Grapin-Botton, A., A. R. Majithia and D. A. Melton. 2001. Key events of pancreas formation are triggered in gut endoderm by ectopic expression of pancreatic regulatory genes. *Genes Dev.* 15: 444–454.

Gu, G., J. Dubauskaite and D. A. Melton. 2002. Direct evidence for the pancreatic lineage: NGN3+ cells are islet progenitors and are distinct from duct progenitors. *Development* 129: 2447–2457.

Gualdi, R., P. Bossard, M. Zheng, Y. Hamada, J. R. Coleman and K. S. Zaret. 1996. Hepatic specification of the gut endoderm in vitro: Cell signaling and transcriptional control. *Genes Dev.* 10: 1670–1682.

Gumpel-Pinot, M., S. Yasugi and T. Mizuno. 1978. Différenciation d'épithéliums endodermiques associaés au mésoderme splanchnique. *Comp. Rend. Acad. Sci.* (Paris) 286: 117–120.

Hanahan, D. 1997. Signaling vascular morphogenesis and maintenance. *Science* 277: 48–50.

Hanai, J.-I. and 8 others. 2002. Endostatin causes G_1 arrest of endothelial cells through inhibition of cyclin D1. *J. Biol. Chem.* 277: 16464 –16469.

Harary, I. and B. Farley. 1963. In vitro studies on single beating rat heart cells. II. Intercellular communication. *Exp. Cell Res.* 29: 466–474.

Harvey, W. 1651. *Exercitationes de Generatione Animalium.* Jansson, Amsterdam. Quoted in J. Needham, *A History of Embryology.* Abelard-Schuman, New York. pp. 133–153.

Hay, E. 1966. *Regeneration.* Holt, Rinehart & Winston, New York.

Hebrok, M., S. Kim and D. A. Melton. 1998. Notochord repression of endodermal sonic hedgehog permits pancreas development. *Genes Dev.* 12: 1705–1713.

Heissig, B. and 11 others. 2002. Recruitment of stem and progenitor cells from bone marrow niche requires MMP-9 mediated release of Kit-ligand. *Cell* 109: 625–637.

Hemmati-Brivanlou, A. and G. H. Thomsen. 1995. Ventral mesodermal patterning in *Xenopus* embryos: Expression patterns and activities of BMP2 and BMP4. *Dev. Genet.* 17: 78–89.

Horb, M. E. and J. M. Slack. 2001. Endoderm specification and differentiation in *Xenopus* embryos. *Dev. Biol.* 236: 330–343.

Hunt, P., D. Robertson, D. Weiss, D. Rennick, F. Lee and O. N. Witte. 1987. A single bone marrow stromal cell type supports the in vitro growth of early lymphoid and myeloid cells. *Cell* 48: 997–1007.

Imanaka-Yoshida, K., K. A. Knudsen and K. K. Linask. 1998. N-cadherin is required for the differentiation and initial myofibrillogenesis of chick cardiomyocytes. *Cell Motil. Cytoskeleton* 39: 52–62.

Ishii, Y., K. Fukuda, H. Saiga, S. Matsushita and S. Yasugi. 1997. Early specification of intestinal epithelium in the chicken embryo: A study on the localization and regulation of CdxA expression. *Dev. Growth Diff.* 39: 643–653.

Isida, Z. 1956. Disvolvigo de kaleralaj vejnoj sur la ovoflavsako de kokojo. (Japanese with Esperanto summary.) *Kaibogaku Zasshi* 31: 334–348.

Jonnson, J., L. Carlsson, T. Edlund and H. Edlund. 1994. Insulin-promote-factor 1 is required for pancreas development in mice. *Nature* 371: 606–608.

Jung, J., M. Goldfarb and K. S. Zaret. 1999. Initiation of mammalian liver development from endoderm by fibroblast growth factors. *Science* 284: 1998–2003.

Katayama, I. and H. Kayano. 1999. Yolk sac with blood island. *New Engl. J. Med.* 340: 617.

Kedinger, M., P. M. Simon-Assman, F. Bouziges, C. Arnold, E. Alexandre and K. Haffen. 1990. Smooth muscle actin expression during rat gut development and induction in fetal skin fibroblastic cells associated with intestinal embryonic epithelium. *Differentiation* 43: 87–97.

Kim, Y. M. and 7 others. 2002. Endostatin blocks VEGF-mediated signaling via direct interaction with KDR/Flk-1. *J. Biol. Chem.* 277: 27872–27879.

Kitamoto, Y., H. Tokunaga, K. Miyamoto and K. Tomita. 2002. VEGF is an essential molecule for glomerular structuring. *Nephrol. Dial. Transplant.* 17 [Suppl. 9]: 25–27.

Komuro, I. and S. Izumo. 1993. Csx: A murine homeobox-containing gene specifically expressed in the developing heart. *Proc. Natl. Acad. Sci. USA* 90: 8145–8149.

Krantz, S. B. and E. Goldwasser. 1965. On the mechanism of erythropoietin induced differentiation. II. The effect on RNA synthesis. *Biochim. Biophys. Acta* 103: 325–332.

Kubai, L. and R. Auerbach. 1983. A new source of embryonic lymphocytes in the mouse. *Nature* 301: 154–156.

Kuo, C. J. and 10 others. 2001. Comparative evaluation of the antitumor activity of antiangiogenic proteins delivered by gene transfer. *Proc. Natl. Acad. Sci. USA* 98: 4605–4610.

Kupperman, E., S. An, N. Osborne, S. Waldron and D. Y. Stainier. 2000. A sphingosine-1-phosphate receptor regulates cell migration during vertebrate heart development. *Nature* 406: 192–195.

Kyba, M., R. C. R. Perlingeiro and G. Q. Daley. 2002. HoxB4 confers definitive lymphoid-myeloid engraftment potential on embryonic stem cell and yolk sac hematopoietic progenitors. *Cell* 109: 29–37.

LaBarbera, M. 1990. Principles of design of fluid transport systems in zoology. *Science* 249: 992–1000.

Lammert, E., O. Cleaver and D. A. Melton. 2001. Induction of pancreatic differentiation by signals from blood vessels. *Science* 294: 564–567.

Langman, J. 1981. *Medical Embryology*, 4th Ed. Williams & Wilkins, Baltimore.

Larsen, W. J. 1993. *Human Embryology*. Churchill-Livingstone, New York.

Lawson, N. D., N. Scheer, V. N. Pham, C. H. Kim, A. B. Chitnis, J. A. Campos-Ortega and B. M. Weinstein. 2001. Notch signaling is required for arterial-venous differentiation during embryonic vascular development. *Development* 128: 3675–3683.

Lawson, N. D., A. M. Vogel and B. M. Weinstein. 2002. Sonic hedgehog and vascular endothelial growth factor act upstream of the Notch pathway during arterial endothelial differentiation. *Dev. Cell* 3: 127–136.

LeCouter, J. and 13 others. 2001. Identification of an angiogenic mitogen selective for endocrine gland endothelium. *Nature* 412: 877–884.

Le Douarin, N. 1975. An experimental analysis of liver development. *Med. Biol.* 53: 427–455.|

Liao, E. C. and L. I. Zon. 1999. Conservation of themes in vertebrate blood development. *In* S. A. Moody (ed.), *Cell Lineage and Fate Determination*. Academic Press, San Diego, pp. 569–582.

Linask, K. K. 1992. N-cadherin localization in early heart development and polar expression of Na⁺,K⁺-ATPase, and integrin during pericardial coelom formation and epithelialization of the differentiating myocardium. *Dev. Biol.* 151: 213–224.

Linask, K. K. and J. W. Lash. 1986. Precardiac cell migration: Fibronectin localization at mesoderm-endoderm interface during directional movement. *Dev. Biol.* 114: 87–101.

Linask, K. K. and J. W. Lash. 1988a. A role for fibronectin in the migration of avian precardiac cells. I. Dose-dependent effects of fibronectin antibody. *Dev. Biol.* 129: 315–323.

Linask, K. K. and J. W. Lash. 1988b. A role for fibronectin in the migration of avian precardiac cells. II. Rotation of the heart-forming region during different stages and their effects. *Dev. Biol.* 129: 324–329.

Linask, K. K. and J. W. Lash. 1993. Early heart development: Dynamics of endocardial cell sorting suggests a common origin with cardiomyocytes. *Dev. Dynam.* 195: 62–66.

Linask, K. K., K. A. Knudsen and Y.-H. Gui. 1997. N-cadherin-catenin interaction: Necessary component of cardiac cell compartmental-ization during early vertebrate development. *Dev. Biol.* 185: 148–164.

Linask, K. K., M. D. Han, M. Artman and C. A. Ludwig. 2001. Sodium-calcium exchanger (NCX-1) and calcium modulation: NCX protein expression patterns and regulation of early heart development. *Dev. Dynam.* 221: 249–264.

Linask, K. K., X. Yu, Y. Chen and M. D. Han. 2002. Directionality of heart looping: Effects of *Pitx2c* misexpression on flectin asymmetry and midline structures. *Dev. Biol.* 246: 407–417.

Lindahl, P., B. R. Johansson, P. Levéen and C. Betscholtz. 1997. Pericyte loss and microaneurysm formation in PDGF-B-deficient mice. *Science* 277: 242–245.

Lints, T. J., L. M. Parsons, L. Hartley, I. Lyons and R. P. Harvey. 1993. Nkx2.5: A novel murine homeobox gene expressed in early heart progenitor cells and their myogenic descendants. *Development* 119: 419–431.

Lyons, I., L. M. Parsons, L. Hartley, R. Li, J. E. Andrews, L. Robb and R. P. Harvey. 1995. Myogenic and morphogenetic defects in the heart tubes of murine embryos lacking the homeobox gene *Nkx2.5*. *Genes Dev.* 9: 1654–1666.

Maduro, M. F., M. D. Meneghini, B. Bowerman, G. Broitman-Maduro and J. H. Rothman. 2001. Restriction of mesendoderm to a single blastomere by the combined action of SKN-1 and a GSK-3β homolog is mediated by MED-1 and -2 in *C. elegans*. *Mol. Cell* 7: 475–485.

Maeno, M., R. C. Ong, A. Suzuki, N. Ueno and H. F. Kung. 1994. A truncated BMP4 receptor alters the fate of ventral mesoderm to dorsal mesoderm: Roles of animal pole tissue in the development of ventral mesoderm. *Proc. Natl. Acad. Sci. USA* 91: 10260–10264.

Manasek, F. J. 1968. Embryonic development of the heart: A light and electron microscopic study of myocardial development in the early chick embryo. *J. Morphol.* 125: 329–366.

Markwald, R. R., T. P. Fitzharris and J. J. Manasek. 1977. Structural development of endocardial cushions. *Am. J. Anat.* 148: 85–120.

Markwald, R. R., T. Trusk and R. Moreno-Rodriguez. 1998. Formation and septation of the tubular heart: Integrating the dynamics of morphology with emerging molecular concepts. *In* M. de la Cruz and R. R. Markwald (ed.), *Living Morphogenesis of the Heart*. Birkhauser Press, Boston.

Martin, C., D. Beaupain and F. Dieterlen-Lièvre. 1978. Developmental relationships between vitelline and intraembryonic haemopoiesis studied in avian yolk sac chimeras. *Cell Diff.* 7: 115–130.

Martin, P. and J. Lewis. 1989. Origins of the neurovascular bundle: Interactions between developing nerves and blood vessels in embryonic chick skin. *Int. J. Dev. Biol.* 33: 379–387.

Marvin, M. J., G. Di Rocco, A. Gardiner, S. M. Bush and A. B. Lassar. 2001. Inhibition of *Wnt* activity induces heart formation from posterior mesoderm. *Genes Dev.* 15: 316–327.

Matsumoto, K., H. Yoshitomi, J. Rossant and K. S. Zaret. 2001. Liver organogenesis promoted by endothelial cells prior to vascular function. *Science* 294: 559–563.

Matsuoka, S. and 11 others. 2001 Generation of definitive hematopoietic stem cells from murine early yolk sac and paraaortic splanchnopleures by aorta-gonad-mesonephros region-derived stromal cells. *Blood* 98: 6–12.

Matsushita, S., Y. Ishii, P. J. Scotting, A. Kuroiwa and S. Yasugi. 2002. Pre-gut endoderm of chick embryos is regionalized by 1.5 days of development. *Dev. Dynam.* 223: 33–47.

Medvinsky, A. L., N. L. Samoylina, A. M. Müller and E. A. Dzierzak. 1993. An early pre-liver intraembryonic source of CFU-S in the developing mouse. *Nature* 364: 64–67.

Mikawa, T. 1999. Determination of heart cell lineages. *In* S. A. Moody (ed.), *Cell Lineage and Fate Determination*. Academic Press, San Diego. pp. 451–462.

Millauer, B., S. Wizigmann-Voos, H. Schnürch, R. Martinez, N. P. H. Müller, W. Risau and A. Ullrich. 1993. High-affinity VEGF binding and developmental expression suggest flk-1 as a major regulator of vasculogenesis and angiogenesis. *Cell* 72: 835–846.

Miller, S. A., K. L. Bresee, C. L. Michaelson and D. A. Tyrell. 1994. Domains of differential cell proliferation and formation of amnion folds in chick embryo ectoderm. *Anat. Rec.* 238: 225–236.

Miller, S. A. and 10 others. 1999. Domains of differential cell proliferation suggest hinged folding in avian gut endoderm. *Dev. Dynam.* 216: 398–410.

Moyon, D., L. Pardanaud, L. Yuan, C. Breant and A. Eichmann. 2001. Selective expression of angiopoietin 1 and 2 in mesenchymal cells surrounding veins and arteries of the avian embryo. *Mech. Dev.* 106: 133–136.

Mukouyama, Y. S., D. Shin, S. Britsch, M. Taniguchi and D. J. Anderson. 2002. Sensory nerves determine the pattern of arterial differentiation and blood vessel branching in the skin. *Cell* 109: 693–705.

Mullins, M. C. and 12 others. 1996. Genes establishing dorsoventral pattern formation in the zebrafish embryo: The ventral specifying genes. *Development* 123: 81–93.

Nakauchi, H. and G. Gachelin. 1993. Les cellules souches. *La Recherche* 254: 537–541.

Nascone, N. and M. Mercola. 1995. An inductive role for the endoderm in *Xenopus* cardiogenesis. *Development* 121: 515–523.

Nguyen, M.-T. and 7 others. 1997. The prostaglandin receptor EP$_4$ triggers remodelling of the cardiovascular system at birth. *Nature* 390: 78–81.

Noden, D. 1989. Embryonic origins and assembly of blood vessels. *Am. Rev. Respir. Dis.* 140: 1097–1103.

Offield, M. F. and 7 others. 1996. PDX-1 is required for pancreatic outgrowth and differenti-ation of the rostral duodenum. *Development* 122: 983–995.

Oh, J. and 16 others. 2001. The membrane-anchored MMP inhibitor RECK is a key regulator of extracellular matrix integrity and angiogenesis. *Cell* 107: 789–800.

Okada, T. S. 1960. Epithelio-mesenchymal interactions in the regional differentiation of the digestive tract in the amphibian embryo. *Roux Arch. Entwick. Mech.* 152: 1–21.

Orkin, S. H. and L. I. Zorn. 2002. Hematopoiesis and stem cells: Plasticity versus developmental heterogeneity. *Nature Immunol.* 3: 323–328.

Pardanaud, L., C. Altmann, P. Kitos, F. Dieterlen-Lièvre and C. Buck. 1987. Vasculogenesis in the early quail blastodisc as studied with a monoclonal antibody recognizing endothelial cells. *Development* 100: 339–349.

Pardanaud, L., F. Yassine and F. Dieterlen-Lièvre. 1989. Relationship between vasculogenesis, angiogenesis, and hemopoiesis during avian ontogeny. *Development* 105: 473–485.

Patten, B. M. 1951. *Early Embryology of the Chick*, 4th Ed. McGraw-Hill, New York.

Popoff, D. 1894. *Dottersack-gafässe des Huhnes*. Kreidl's Verlag, Wiesbaden.

Porcher, C., W. Swat, K. Rockwell, Y. Fujiwara, F. W. Alt and S. H. Orkin. 1996. The T cell leukemia oncoprotein SCL/tal-1 is essential for development of all hematopoietic lineages. *Cell* 86: 47–57.

Potten, C. S. and M. Loeffler. 1990. Stem cells: Attributes, spirals, pitfalls, and uncertainties. Lessons for and from the crypt. *Development* 110: 1001–1020.

Potts, J. D., J. M. Dagle, J. A. Walder, D. L. Weeks and R. B. Runyon. 1991. Epithelial-mesenchymal transformation of embryonic cardiac endothelial cells is inhibited by a modified antisense oligodeoxynucleotide to transforming growth factor β3. *Proc. Natl. Acad. Sci. USA* 88: 1516–1520.

R&D Systems. 1997. Human ELISAs for the cytokines of hematopoiesis. *Science* 278 (5340): Back cover.

Redkar, A., M. Montgomery and J. Litvin. 2001. Fate map of the early avian cardiac progenitor cells. *Development* 128: 2269–2279.

Ribatti, D., C. Urbinati, B. Nico, M. Rusnati, L. Roncali and M. Presta. 1995. Endogenous basic fibroblast growth factor is implicated in the vascularization of the chick embryo chorioallantoic membrane. *Dev. Biol.* 170: 39–49.

Risau, W. 1995. Differentiation of the endothelium. *FASEB J.* 9: 926–933.

Risau, W. 1997. Mechanisms of angiogenesis. *Nature* 386: 671–674.

Robb, L., N. J. Elwood, A. G. Elefanty, F. Köntgen, R. Li, L. D. Barnett and C. G. Begley. 1996. The *scl* gene product is required for the generation of all hematopoietic lineages in the adult mouse. *EMBO J.* 15: 4123–4129.

Roberts, D. J., R. L. Johnson, A. C. Burke, C. E. Nelson, B. A. Morgan and C. Tabin. 1995. Sonic hedgehog is an endodermal signal inducing Bmp-4 and Hox genes during induction and regionalization of the chick hindgut. *Development* 121: 3163–3174.

Roberts, D. J., D. M. Smith, D. J. Goff and C. J. Tabin. 1998. Epithelial-mesenchymal signaling during the regionalization of the chick gut. *Development* 125: 2791–2801.

Roberts, R., J. Gallagher, E. Spooncer, T. D. Allen, F. Bloomfield and T. M. Dexter. 1988. Heparan sulphate-bound growth factors: A mechanism for stromal cell mediated haemopoiesis. *Nature* 332: 376–378.

Rodaway, A. and R. Patient. 2001. Mesendoderm: An ancient germ layer? *Cell* 105: 169–172.

Rottbauer, W., K. Baker, Z. G. Wo, M. A. Mohideen, H. F. Cantiello and M. C. Fishman. 2001. Growth and function of the embryonic heart depend upon the cardiac-specific L-type calcium channel α1 subunit. *Dev. Cell* 1: 265–275.

Rugh, R. 1951. *The Frog: Its Reproduction and Development*. Blakiston, Philadelphia.

Sakiyama, J.-I., A. Yamagishi and A. Kuroiwa. 2003. Tbx4-Fgf10 system controls lung bud formation during chicken embryonic development. *Development* 130: 1225–1234.

Schneider, V. A. and M. Mercola. 2001. Wnt antagonism initiates cardiogenesis in *Xenopus laevis*. *Genes Dev.* 15: 304–315.

Schott, J.-J. and 8 others. 1998. Congenital heart disease caused by mutations in the transcription factor *NKX2-5*. *Science* 281: 108–111.

Schultheiss, T. M., S. Xydas and A. B. Lassar. 1995. Induction of avian cardiac myogenesis by anterior endoderm. *Development* 121: 4203–4214.

Shalaby, F., J. Rossant, T. P. Yamaguchi, M. Gertenstein, X.-F. Wu, M. L. Breitman and A. C. Schuh. 1995. Failure of blood-island formation and vasculogenesis in *flk-1*-deficient mice. *Nature* 376: 62–66

Shalaby, F. and 7 others. 1997. A requirement for Flk1 in primitive and definitive hematopoiesis and vasculogenesis. *Cell* 89: 981–990.

Shannon, J. M., L. D. Nielsen, S. A. Gebb and S. H. Randell. 1998. Mesenchyme specifies epithelial differentiation in reciprocal recombinants of embryonic lung and trachea. *Dev. Dynam.* 212: 482–494.

Sierra-Honigmann, M. R. and 10 others. 1998. Biological action of leptin as an angiogenic factor. *Science* 281: 1683–1686.

Spinella, F., L. Rosano, V. DiCastro, P. G. Natali and A. Bagnato. 2002. Endothelin-1 induces vascular endothelial growth factor by increasing hypoxia-inducible factor1α in ovarian carcinoma cells. *J. Biol. Chem.* 277: 27850–27855.

Srivastava, D. and E. N. Olson. 2000. A genetic blueprint for cardiac development. *Nature* 407: 221–232.

Srivastava, D., P. Cserjesi and E. N. Olson. 1995. A subclass of bHLH proteins required for cardiac morphogenesis. *Science* 270: 1995–1999.

Sugi, Y. and J. Lough. 1994. Anterior endoderm is a specific effector of terminal cardiac myocyte differentiation of cells from the embryonic heart forming region. *Dev. Dynam.* 200: 155–162.

Suri, C. and 7 others. 1996. Requisite role of angiopoietin-1, a ligand for the TIE2 receptor, during embryonic angiogenesis. *Cell* 87: 1171–1180.|

Tsuda, T., N. Philp, M. H. Zile and K. K. Linask. 1996. Left-right asymmetric localization of flectin in the extracellular matrix during heart looping. *Dev. Biol.* 173: 39–50.

Tuan, R. 1987. Mechanisms and regulation of calcium transport by the chick embryonic chorioallantoic membrane. *J. Exp. Zool.* [Suppl.] 1: 1–13.

Tzahor, E. and A. B. Lassar. 2001. *Wnt* signals from the neural tube block ectopic cardiogenesis. *Genes Dev.* 15: 255–260.

Vikkula, M. and 11 others. 1996. Vascular dysmorphogenesis caused by an activating mutation in the receptor tyrosine kinase TIE2. *Cell* 87: 1181–1190.

Vokes, S. A. and P. A. Krieg. 2002. Endoderm is required for vascular endothelial tube formation, but not for angioblast specification. *Development* 129: 775–785.

Wakimoto, K. and 19 others. 2000. Targeted disruption of Na⁺/Ca²⁺ exchanger gene leads to cardiomyocyte apoptosis and defects in heartbeat. *J. Biol. Chem.* 275: 36991–36998.

Wallace, K. N. and M. Pack. 2003. Unique and conserved aspects of gut development in zebrafish. *Dev. Biol.* 255: 12–29.

Wang, D.-Z. and 8 others. 1999. Requirement of a novel gene, *Xin*, in cardiac morphogenesis. *Development* 126: 1281–1294.

Wang, H. U., Z.-F. Chen and D. J. Anderson. 1998. Molecular distinction and angiogenic interaction between embryonic arteries and veins revealed by ephrin-B2 and its receptor Eph-B4. *Cell* 93: 741–753.

Warren, L. 1998. *Joseph Leidy: The Last Man Who Knew Everything.* Yale University Press, New Haven.

Weinstein, B. M. and N. D. Lawson. 2003. Arteries, veins, Notch, and VEGF. *Cold Spring Harb. Symp. Quant. Biol.* 67. In press.

Wessells, N. K. 1970. Mammalian lung development: Interactions in formulation and morphogenesis of tracheal buds. *J. Exp. Zool.* 175: 455–466.

Whitlock, C. A., G. F. Tidmarsh, C. Muller-Sieburg and I. L. Weissman. 1987. Bone marrow stromal cell lines with lymphopoietic activity express high levels of a pre-B neoplasia-associated molecule. *Cell* 48: 1009–1021.

Wilt, F. H. 1974. The beginnings of erythropoiesis in the yolk sac of the chick embryo. *Ann. NY Acad. Sci.* 241: 99–112.

Wolf, N. S. and J. J. Trentin. 1968. Hemopoietic colony studies. V. Effect of hemopoietic organ stroma on differentiation of pluripotent stem cells. *J. Exp. Med.* 127: 205–214.

Wood, H. B., G. May, L. Healy, T. Enver and G. M. Morriss-Kay. 1997. CD34 expression patterns during early mouse development are related to modes of blood vessel formation and reveal additional sites of hematopoiesis. *Blood* 90: 2300–2311.

Wu, P. and 7 others. 2001. Hypoxia downregulates endostatin production by human microvascular endothelial cells and pericytes. *Biochem. Biophys. Res. Comm.* 288: 1149–1154.

Xavier-Neto, J. and 7 others. 1999. A retinoic acid-inducible transgenic marker of sino-atrial development in the mouse heart. *Development* 126: 2667–2687.

Yancopopoulos, G. D., M. Klagsbrun and J. Folkman. 1998. Vasculogenesis, angiogenesis, and growth factors: Ephrins enter the fray at the border. *Cell* 93: 661–664.

Yoder, M. C., K. Hiatt and P. Mukherjee. 1997. In vivo repopulating hematopoietic stem cells are present in the murine yolk sac at day 9.0 postcoitus. *Proc. Natl. Acad. Sci. USA* 94: 6776–6780.

Yokouchi, Y., J. Sakiyama and A. Kuroiwa. 1995. Coordinate expression of *Abd-B* subfamily genes of the HoxA cluster in developing digestive tract of the chick embryo. *Dev. Biol.* 169: 76–89.

Zatterstrom, U. K., U. Felbor, N. Fukai and B. R. Olsen. 2000. Collagen XVIII/endostatin structure and functional role in angiogenesis. *Cell Struct. Funct.* 25: 97–101.

Zhang, Z., Y. Xu, H. Song, J. Rodriguez, D. Tuteja, Y. Namkung, H. S. Shin and N. Chiamvimonvat. 2002. Functional roles of Ca(v)1.3 (alpha(1D)) calcium channel in sinoatrial nodes: Insight gained using gene-targeted null mutant mice. *Circ. Res.* 90: 981–987.

Zhong, T. P., S. Childs, J. P. Lau and M. C. Fishman. 2001. Gridlock signaling pathway fashions the first embryonic artery. *Nature* 414: 216–219.

16 *Development of the tetrapod limb*

CONSIDER YOUR LIMB. First, consider its polarity. It has fingers or toes at one end, a humerus or femur at the other. In no instance among you or your friends will you find someone with fingers in the middle of their hand. Consider also the differences between your hands and your feet. The differences are subtle, difficult to put into words, but very obvious. If your fingers were replaced by toes, you would know it. But then consider how *similar* the bones of your feet are to the bones of your hand; it's easy to see that they share a common pattern. The bones of any vertebrate limb, be it arm or leg, wing or flipper, consist of a proximal **stylopod** (humerus/femur) adjacent to the body wall; a **zeugopod** (radius-ulna/tibia-fibula) in the middle region; and a distal **autopod** (carpals-fingers/tarsals-toes) (Figure 16.1). Last, consider the growth of your limbs. Each of your feet is very similar in size to your other foot. Each of your hands is remarkably similar in size to your other hand. So after about 20 years of growth, each of your feet turns out, independently, to be the same length.

These commonplace phenomena present fascinating questions to the developmental biologist. How can growth be so remarkably regulated? How is it that we have four limbs and not six or eight? How is it that the fingers form at one end of the limb and nowhere else? How is it that the little finger develops at one edge of the limb and the thumb at the other? How does the forelimb grow differently than the hindlimb?

All of these questions are really about pattern formation. **Pattern formation** is the set of processes by which embryonic cells form ordered spatial arrangements of differentiated tissues. The ability to carry out the processes of pattern formation is one of the most dramatic properties of developing organisms, and one that has provoked a sense of awe in scientists and laypeople alike. It is one thing to differentiate the chondrocytes and osteocytes that synthesize the cartilage and bone matrices, respectively; it is another thing to produce those cells in a temporal-spatial orientation that generates a functional bone. It is still another thing to make that bone a humerus and not a pelvis or a femur. The ability of limb cells to sense their relative positions and to differentiate with regard to those positions has been the subject of intense debate and experimentation.

The vertebrate limb is an extremely complex organ with an asymmetrical arrangement of parts in all three dimensions. The first dimension is the **proximal** (close)-**distal** (far) **axis**. The bones of the limb are formed by endochondral ossification. They are initially cartilaginous, but eventually, most of the cartilage is replaced by bone. Somehow the limb cells develop differently at early stages of development (when they make

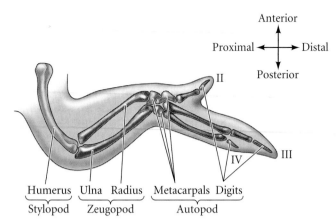

Figure 16.1
Skeletal pattern of the chick wing. According to convention, the digits are numbered II, III, and IV. (Digits I and V are not found in chick wings.)

the stylopod) then at later stages (when they make the autopod). The second axis is the **anterior-posterior axis**. Our little fingers, for instance, mark the *posterior* side, while our thumbs are at the *anterior* side. In humans, it is obvious that each hand develops as a mirror image of the other. One can imagine other arrangements to exist—such as the thumb developing on the left side of both hands—but these patterns do not occur. The third axis is the **dorsal-ventral** axis. The palm (ventral) is readily distinguishable from the knuckles (dorsal). In some manner, the complex three-dimensional pattern of the forelimb is routinely produced.

The fundamental problem of morphogenesis—how specific structures arise in particular places—is exemplified in limb development. Moreover, since the limbs, unlike the heart or brain, are not essential for embryonic or fetal life, one can experimentally remove or transplant parts of the developing limb, or create limb-specific mutants without interfering with the vital processes of the organism. Such experiments have shown that the basic "morphogenetic rules" for forming a limb appear to be the same in all tetrapods. Grafted pieces of reptile or mammalian limb buds can direct the formation of chick limbs, and regions of frog limb buds can direct the patterning of salamander limbs (Fallon and Crosby 1977; Sessions et al. 1989; see Hinchliffe 1991). Moreover, as will be detailed in Chapter 18, *regenerating* salamander limbs appear to follow the same rules as developing limbs (Muneoka and Bryant 1982). But what are these morphogenetic rules?

The positional information needed to construct a limb has to function in a three-dimensional coordinate system.* During the past decade, particular proteins have been identified that play a role in the formation of each of these limb axes. The proximal-distal (shoulder-finger; hip-toe) axis appears to

be regulated by the fibroblast growth factor (FGF) family of proteins. The anterior-posterior (thumb-pinky) axis seems to be regulated by the Sonic hedgehog protein, and the dorsal-ventral (knuckle-palm) axis is regulated, at least in part, by Wnt7a. The interactions of these proteins mutually support one another and determine the differentiation of cell types.

> **WEBSITE 16.1 The mathematical modeling of limb development.** The specification of the limb axes and the patterns in which limb outgrowth might occur were predicted mathematically before the actual molecular interactions were discovered.

Formation of the Limb Bud

Specification of the limb fields: Hox genes and retinoic acid

Limbs do not form just anywhere along the body axis. Rather, there are discrete positions where limbs are generated. The mesodermal cells that give rise to a vertebrate limb can be identified by (1) removing certain groups of cells and observing that a limb does not develop in their absence (Detwiler 1918; Harrison 1918), (2) transplanting groups of cells to a new location and observing that they form a limb in this new place (Hertwig 1925), and (3) marking groups of cells with dyes or radioactive precursors and observing that their descendants partake in limb development (Rosenquist 1971). Figure 16.2 shows the prospective forelimb area in the tailbud stage of the salamander *Ambystoma maculatum*. The center of this disc normally gives rise to the limb itself. Adjacent to it are the cells that will form the peribrachial flank tissue and the shoulder girdle. However, if all these cells are extirpated from the embryo, a limb will still form, albeit somewhat later, from an additional ring of cells that surrounds this area (and which

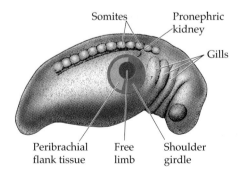

Figure 16.2
Prospective forelimb field of the salamander *Ambystoma maculatum*. The central area contains cells destined to form the limb per se (the free limb). The cells surrounding the free limb give rise to the peribrachial flank tissue and the shoulder girdle. The ring of cells outside these regions usually is not included the limb, but can form a limb if the more central tissues are extirpated. (After Stocum and Fallon 1983.)

---

*Actually, it is a four-dimensional system, in which time is the fourth axis. Developmental biologists get used to seeing nature in four dimensions.

would not normally form a limb). If this last ring of cells is included in the extirpated tissue, no limb will develop. This larger region, representing all the cells in the area capable of forming a limb, is called the **limb field**.

When it first forms, the limb field has the ability to regulate for lost or added parts. In the tailbud stage of *Ambystoma*, any half of the limb disc is able to generate an entire limb when grafted to a new site (Harrison 1918). This potential can also be shown by splitting the limb disc vertically into two or more segments and placing thin barriers between the segments to prevent their reunion. When this is done, each segment develops into a full limb. The regulative ability of the limb bud has recently been highlighted by a remarkable experiment of nature. In numerous ponds in the United States, multilegged frogs and salamanders have been found (Figure 16.3). The presence of these extra appendages has been linked to the infestation of the larval abdomen by parasitic trematode worms. The cysts of these worms apparently split the limb buds in several places while the tadpoles were forming their limb buds (Sessions and Ruth 1990; Sessions et al. 1999). Thus, like an early sea urchin embryo, the limb field represents a "harmonious equipotential system" wherein a cell can be instructed to form any part of the limb.

Interestingly, in all land vertebrates, there are only four limb buds per embryo, and they are always opposite each other with respect to the midline. Although the limbs of different vertebrates differ with respect to the somite level at which they arise, their position is constant with respect to the level of Hox gene expression along the anterior-posterior axis

(see Chapter 11). For instance, in fish (in which the pectoral and pelvic fins correspond to the anterior and posterior limbs, respectively), amphibians, birds, and mammals, the forelimb buds are found at the most anterior expression region of *Hoxc-6*, the position of the first thoracic vertebra* (Oliver et al. 1988; Molven et al. 1990; Burke et al. 1995). The lateral plate mesoderm in the limb field is also special in that it induces myoblasts to migrate out from the somites and enter the limb bud to become the limb musculature. No other region of the lateral plate mesoderm can do that (Hayashi and Ozawa 1995).

Retinoic acid appears to be critical for the initiation of limb bud outgrowth, since blocking the synthesis of retinoic acid with certain drugs prevents limb bud initiation (Stratford et al. 1996). Bryant and Gardiner (1992) suggested that a gradient of retinoic acid along the anterior-posterior axis might activate certain homeotic genes in particular cells and thereby specify them to be included in the limb field. The source of this retinoic acid is probably Hensen's node (Hogan et al. 1992). The specification of limb fields by retinoic acid-activated Hox genes might explain a bizarre observation made by (Mohanty-Hejmadi et al. 1992; Maden 1993): When the *tails* of tadpoles were amputated and the stumps exposed to retinoic acid during the first days of regeneration, the tadpoles regenerated several *legs* from the tail stump (Figure 16.4). It appears that the retinoic acid caused a homeotic transformation in the regenerating tail by respecifying the tail tissue as a limb-forming pelvic region (Müller et al. 1996).

Induction of the early limb bud: Wnt proteins and fibroblast growth factors

Limb development begins when mesenchyme cells proliferate from the somatic layer of the limb field lateral plate mesoderm (limb *skeletal* precursors) and from the somites (limb *muscle* precursors; Figure 16.5) These cells accumulate under the ectodermal tissue to create a circular bulge called a **limb bud**. Recent studies on the earliest stages of limb formation have shown that the signal for limb bud formation comes from the lateral plate mesoderm cells that will become the prospective limb mesenchyme. These cells secrete the paracrine factor **FGF10**. FGF10 is capable of initiating the limb-forming interactions between the ectoderm and the mesoderm. If beads containing FGF10 are placed ectopically beneath the flank ectoderm, extra limbs emerge (Figure 16.6) (Ohuchi et al. 1997; Sekine et al. 1999). FGF10 is originally produced throughout the lateral plate mesoderm, but immediately prior to limb formation, it becomes restricted to the regions of the lateral plate mesoderm where the limbs will form. This restriction in placement appears to be due to the actions of Wnt proteins (Wnt2b in the chick forelimb region;

Figure 16.3
The regulative ability of the limb field as demonstrated by an experiment of nature. This multilimbed Pacific tree frog, *Hyla regilla*, is the result of infestation of the tadpole-stage developing limb buds by trematode cysts. In this adult frog's skeleton, the cartilage is stained blue; the bones are stained red. (Photograph courtesy of S. Sessions.)

---

*Interestingly, Hox gene expression in at least some snakes (such as *Python*) creates a pattern in which each somite is specified to become a thoracic (ribbed) vertebra. The patterns of Hox gene expression associated with limb-forming regions are not seen (Cohn and Tickle 1999; see Chapter 23).

(A)

(B)

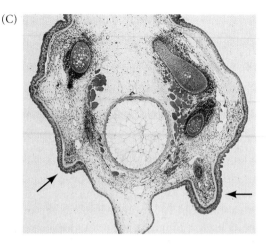

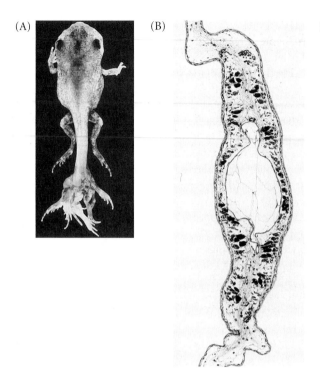

Figure 16.4

Retinoic acid causes legs to regenerate from an amputated tadpole tail. (A) The tail stump of a balloon frog tadpole treated with retinoic acid after amputation will form limbs at the amputation site. (B) Normal tail regeneration in a *Rana temporaria* tadpole 4 weeks after amputation. A small neural tube can be seen above a large notochord, and the muscles are arranged in packets. No cartilage or bone is present. (C) A retinoic acid-treated tadpole tail makes limb buds (arrows) as well as pelvic cartilage and bone. The cartilaginous rudiment of the femur can be seen in the right limb bud. (A from Mohanty-Hejmadi et al. 1992, courtesy of P. Mohanty-Hejmadi; B and C from Müller et al. 1996, courtesy of G. Müller.)

Wnt8c in the chick hindlimb region), which stabilize FGF10 expression at these sites (Figure 16.7; Ohuchi et al. 1997; Kawakami et al. 2001).

Specification of forelimb or hindlimb: *Tbx4* and *Tbx5*

The limb buds must be specified as either forelimb or hindlimb. How are these two limb bud types distinguished? In 1996, Gibson-Brown and colleagues made a tantalizing correlation: The gene encoding the **Tbx5** transcription fac-

Figure 16.5

Limb bud formation. (A) Proliferation of mesodermal cells from the somatic region of the lateral plate mesoderm causes the limb bud in the amphibian embryo to bulge outward. These cells generate the skeletal elements of the limb. Contributions of cells from the myotome provide the limb's musculature. (B) Entry of myotome cells (purple) into the limb bud. This computer reconstruction was made from sections of an in situ hybridization to the *myf5* mRNA found in developing muscle cells. If you can cross your eyes (or try focusing "past" the page, looking through it to your toes), the three-dimensionality of the stereogram will become apparent. (B, photograph courtesy of J. Streicher and G. Müller.)

(A)

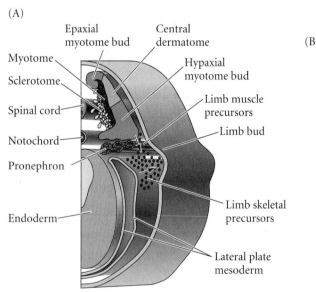

Epaxial myotome bud

Central dermatome

Myotome

Hypaxial myotome bud

Sclerotome

Spinal cord

Limb muscle precursors

Limb bud

Notochord

Pronephron

Endoderm

Limb skeletal precursors

Lateral plate mesoderm

(B)

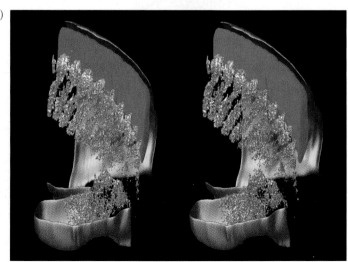

Figure 16.6
FGF10 expression and action in the developing chick limb. (A) FGF10 becomes expressed in the lateral plate mesoderm in precisely those positions (arrows) where limbs normally form. (B) When transgenic cells that secrete FGF10 are placed in the flanks of a chick embryo, the FGF10 can cause the formation of an ectopic limb (arrow). (From Ohuchi et al. 1997; courtesy of S. Noji.)

tor is transcribed in mouse forelimbs, while the gene encoding the closely related transcription factor **Tbx4** is expressed in hindlimbs.* Could these two transcription factors be involved in directing forelimb versus hindlimb specificity? The loss-of-function data were equivocal: humans heterozygous for the *TBX5* gene have Holt-Oram syndrome, characterized by abnormalities of the heart and upper limbs (Basson et al. 1996; Li et al. 1996). The legs are not affected, but neither are the arms transformed into a pair of legs.

In 1998 and 1999, however, several laboratories (Ohuchi et al. 1998; Logan et al. 1998; Takeuchi et al. 1999; Rodriguez-Esteban et al. 1999, among others) provided gain-of-function evidence

---

*Tbx stands for T-box. The *T* (*Brachyury*) gene and its relatives have a sequence that encodes this specific DNA-binding domain.

Figure 16.7
Molecular model for the initiation of the limb bud in the chick between 48 and 54 hours of gestation. (A) FGF10 expression is first seen throughout the lateral plate mesoderm. It becomes stabilized by Wnt8c to the area where hindlimbs form. (B) FGF10 synthesis is stabilized by the actions of Wnt2b to the region where the forelimbs will form. (C) FGF10 from these two regions of lateral plate mesoderm induces FGF8 in the apical ectodermal ridge (AER). This induction is accomplished through a pathway involving Wnt3a in the responding ectoderm. FGF8 secreted from the AER induces the continued mesodermal expression of FGF10. The roles of FGF8 in the intermediate mesoderm (IM) in inducing or maintaining Wnt expression are uncertain. SM, somatic mesoderm; SE, surface ectoderm. (After Kawakami et al. 2001.)

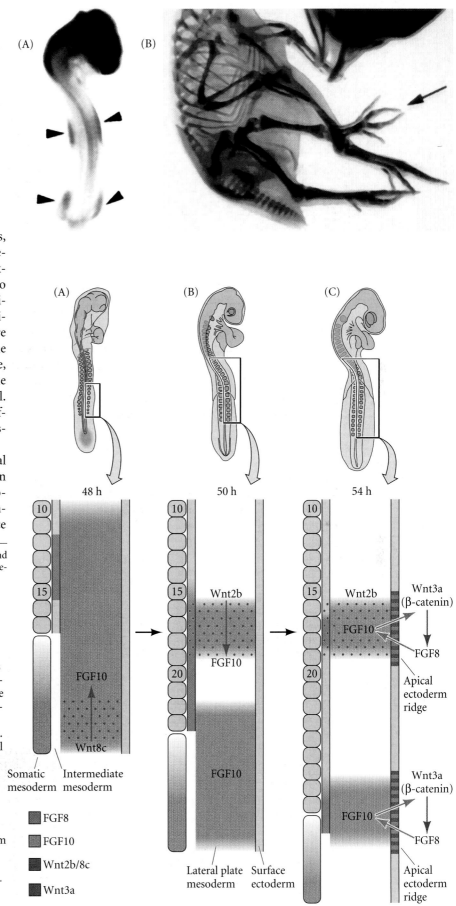

Stage 14/15
(early day 3)

(A) Normal

Forelimb
bud

→ Wing

Hindlimb
bud

→ Leg

(B) FGF induced

FGF

→ Wing

→ Chimera

→ Leg

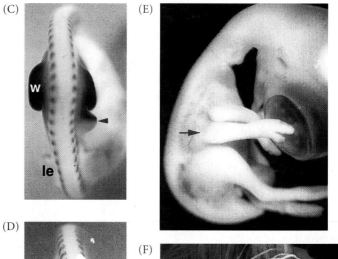

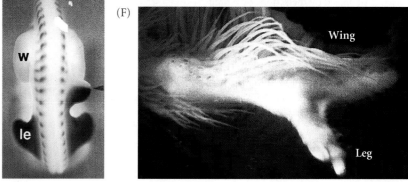

Figure 16.8

Specification of limb type by Tbx4 and Tbx5. (A) In situ hybridizations show that during normal chick development, Tbx5 (blue) is found in the anterior lateral plate mesoderm, while Tbx4 (red) is found in the posterior lateral plate mesoderm. Tbx5-containing limb buds produce wings, while Tbx4-containing limb buds generate legs. (B) If a new limb bud is induced with an FGF-secreting bead, the type of limb formed depends on which Tbx gene is expressed in the limb bud. If placed between the regions of *Tbx4* and *Tbx5* expression, the bead will induce the expression of *Tbx4* posteriorly and *Tbx5* anteriorly. The resulting limb bud will also express *Tbx5* anteriorly and *Tbx4* posteriorly and will generate a chimeric limb. (C) Expression of *Tbx5* in the forelimb (w, wing) buds and in the anterior portion of a limb bud induced by an FGF-secreting bead. (The somite level can be determined by staining for *Mrf4* mRNA, which is localized to the myotomes.) (D) Expression of *Tbx4* in the hindlimb (le, leg) buds and in the posterior portion of an FGF-induced limb bud. (E, F) A chimeric limb induced by an FGF bead contains anterior wing structures and posterior leg structures. (F) is at a later developmental stage, after feathers form. (A, B after Ohuchi et al. 1998, Ohuchi and Noji 1999; C–F, photographs courtesy of S. Noji.)

that Tbx4 and Tbx5 specify hindlimbs and forelimbs, respectively. First, if FGF-secreting beads were used to induce an ectopic limb between the chick hindlimb and forelimb buds, the type of limb produced was determined by the Tbx protein expressed. Those buds induced by placing FGF beads close to the hindlimb (opposite somite 25) expressed *Tbx4* and became hindlimbs. Those buds induced close to the forelimb (opposite somite 17) expressed *Tbx5* and developed as forelimbs (wings). Those buds induced in the center of the flank tissue expressed *Tbx5* in the anterior portion of the limb and *Tbx4* in the posterior portion of the limb. These limbs developed as chimeric structures, with the anterior resembling a forelimb and the posterior resembling a hindlimb (Figure 16.8). Moreover, when a chick embryo was made to express *Tbx4* throughout the flank tissue (by infecting the tissue with a virus that expressed *Tbx4*), limbs induced in the anterior region of the flank often became legs instead of wings. Thus, Tbx4 and Tbx5 appear to be critical in instructing the limbs to become hindlimbs and forelimbs, respectively.

However, the Tbx genes are not the complete story of limb specification. Transplantation experiments have shown that limb specification occurs earlier than the first appearance of Tbx4 or Tbx5 protein, and that there must be other genes that activate the *Tbx4* and *Tbx5* genes in their respective positions (Saito et al. 2002). Research is still going on to discover the mechanisms by which the forelimb becomes distinct from the hindlimb.

WEBSITE 16.2 Specifying forelimbs and hindlimbs.
While Tbx4 and Tbx5 are central to limb type specification, we still need to know how these two transcription factors become expressed in their respective limb buds, and what they do to make the limbs different.

Generating the Proximal-Distal Axis of the Limb

The apical ectodermal ridge

When mesenchyme cells enter the limb field, they secrete FGF10 that induce the overlying ectoderm to form a structure called the **apical ectodermal ridge** (**AER**) (Figure 16.9; Saunders 1948; Kieny 1960; Saunders and Reuss 1974). This ridge runs along the distal margin of the limb bud and will become a major signaling center for the developing limb. Its roles include (1) maintaining the mesenchyme beneath it in a plastic, proliferating phase that enables the linear (proximal-distal) growth of the limb; (2) maintaining the expression of those molecules that generate the anterior-posterior (thumb-pinky) axis; and (3) interacting with the proteins specifying the anterior-posterior and dorsal-ventral axes so that each cell is given instructions on how to differentiate.

The proximal-distal growth and differentiation of the limb bud is made possible by a series of interactions between

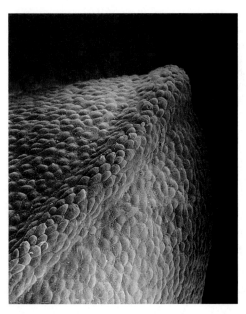

Figure 16.9
Scanning electron micrograph of an early chick forelimb bud, with its apical ectodermal ridge in the foreground. (Photograph courtesy of K. W. Tosney.)

the AER and the limb bud mesenchyme directly (200 μm) beneath it. This distal mesenchyme is often called the **progress zone** (**PZ**) mesenchyme, since its proliferative activity extends the limb bud (Harrison 1918; Saunders 1948). These interactions were demonstrated by the results of several experiments on chick embryos (Figure 16.10):

1. If the AER is removed at any time during limb development, further development of distal limb skeletal elements ceases.
2. If an extra AER is grafted onto an existing limb bud, supernumerary structures are formed, usually toward the distal end of the limb.
3. If leg mesenchyme is placed directly beneath the wing AER, distal hindlimb structures (toes) develop at the end of the limb. (However, if this mesenchyme is placed farther from the AER, the hindlimb (leg) mesenchyme becomes integrated into wing structures.)
4. If limb mesenchyme is replaced by nonlimb mesenchyme beneath the AER, the AER regresses and limb development ceases.

Thus, although the mesenchyme cells induce and sustain the AER and determine the type of limb to be formed, the AER is responsible for the sustained outgrowth and development of the limb (Zwilling 1955; Saunders et al. 1957; Saunders 1972; Krabbenhoft and Fallon 1989). The AER keeps the mesenchyme cells directly beneath it in a state of mitotic proliferation and prevents them from forming cartilage. Hurle and coworkers (1989) found that if they cut away a small portion of

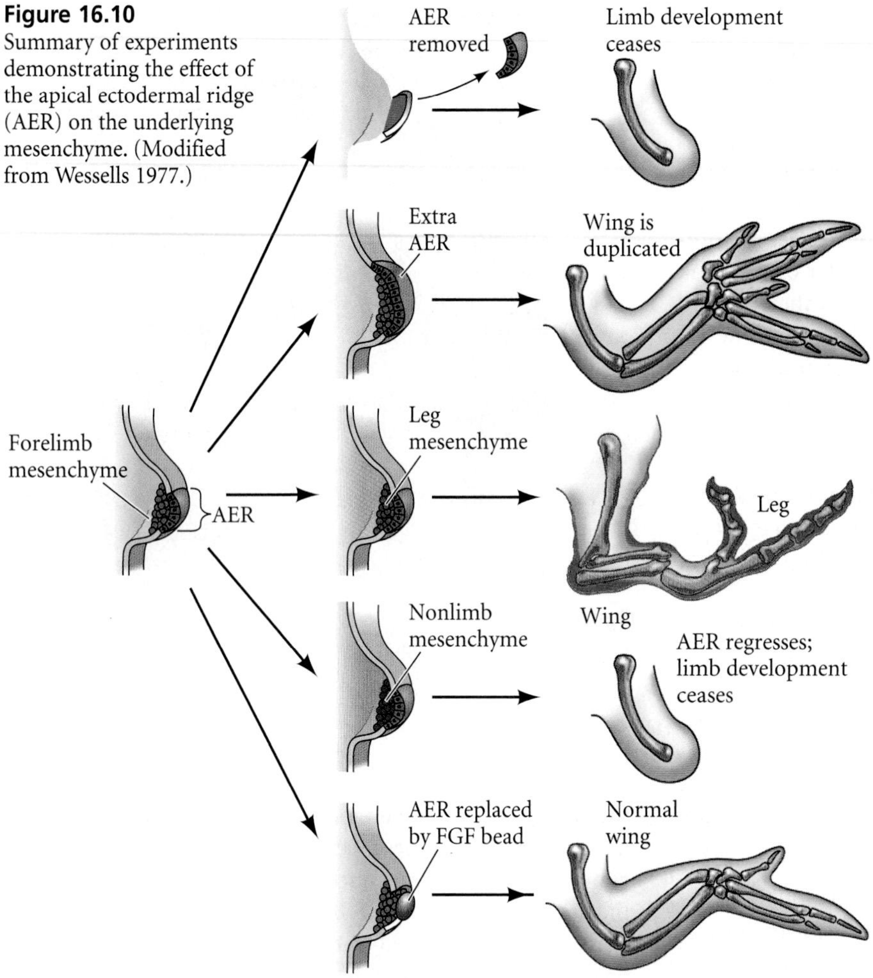

Figure 16.10
Summary of experiments demonstrating the effect of the apical ectodermal ridge (AER) on the underlying mesenchyme. (Modified from Wessells 1977.)

Forelimb mesenchyme

AER

AER removed — Limb development ceases

Extra AER — Wing is duplicated

Leg mesenchyme — Leg

Nonlimb mesenchyme — Wing — AER regresses; limb development ceases

AER replaced by FGF bead — Normal wing

sides of the embryo. The boundary where dorsal ectoderm and ventral ectoderm meet is critical to the placement of the AER. In mutants in which the limb bud is dorsalized and there is no dorsal-ventral junction (as in the chick mutant *limbless*), the AER fails to form and limb development ceases (Carrington and Fallon 1988; Laufer et al. 1997; Rodriguez-Esteban et al. 1997; Tanaka et al. 1997).

FGF10 produced by the underlying mesenchyme induces the formation of the AER, and at the same time induces the AER to synthesize and secrete FGF8 (Figure 16.11). FGF8 acts to maintain the mitotic state of the mesenchyme cells beneath it, causing these cells to keep expressing FGF10. Thus, a positive feedback loop is established: FGF10 in the mesenchyme induces FGF8 in the AER, and FGF8 in the AER maintains FGF10 expression in the mesenchyme. Each FGF activates the synthesis of the other (see Figure 16.7; Mahmood et al. 1995; Crossley et al. 1996; Vogel et al. 1996; Ohuchi et al. 1997; Kawakami et al. 2001). The continued expression of these FGFs maintains mitosis in the mesenchyme beneath the AER.

the AER in a region that would normally fall between the digits of the chick leg, an extra digit emerged at that position.*

FGFs in the induction and maintenance of the AER

FGFs are critical for the induction, maintenance, and function of the apical ectodermal ridge. The principal protein inducing the AER is FGF10 (Xu et al. 1998; Yonei-Tamura et al. 1999). FGF10 is capable of inducing the AER in the competent ectoderm between the dorsal and ventral

(A) (B) (C)

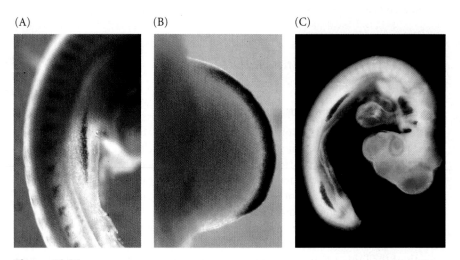

Figure 16.11
FGF8 in the apical ectodermal ridge. (A) In situ hybridization showing expression of FGF8 message in the ectoderm as the limb bud begins to form. (B) Expression of FGF8 RNA in the AER, the source of mitotic signals to the underlying mesoderm. (C) In the normal 3-day chick embryo, FGF8 is expressed in the apical ectodermal ridge of both the forelimb and hindlimb buds. It is also expressed in several other places in the embryo, including the pharyngeal arches. (A and B, photographs courtesy of J. C. Izpisúa-Belmonte; C, photograph courtesy of A. López-Martínez and J. F. Fallon.)

*When referring to the hand, one has an orderly set of names to specify each digit (*digitus pollicis, d. indicis, d. medius, d. annularis*, and d. *minimus*, respectively, from thumb to little finger). No such nomenclature exists for the pedal digits, but the plan proposed by Phillips (1991) has much merit. The pedal digits, from *hallux* to small toe, would be named *porcellus fori, p. domi, p. carnivorus, p. non voratus*, and *p. plorans domi*, respectively.

As shown in Figure 16.7, the *intermediate mesoderm* may stabilize the Wnt proteins that initiate the FGF cascade. While this has not been demonstrated on the molecular level, there is evidence from genetics and from embryological experiments suggesting that the Wolffian (nephric) duct may be important in limb induction. First, removing the Wolffian duct results in the absence of limb development (Muchmore 1957; Stephens and McNulty 1981; Geduspan and Solursh 1992; Smith et al. 1996). Second, studies of tissue explants from amphibian embryos suggest a link between limb formation and the presence of nephric tubules (Muchmore 1957; Lash and Saxén 1972). Fourth, there are several genetic and teratogenic syndromes in which both the kidney and limb fail to form.

WEBSITE 16.3 **Induction of the AER.** The induction of the AER is a complex event involving the interaction between the dorsal and ventral compartments of the ectoderm. The Notch pathway may be critical in this process. Misexpression of the genes in this pathway can cause absence or duplication of limbs.

Specifying the limb mesoderm: Determining the proximal-distal polarity of the limb

In 1948, John Saunders made a simple and profound observation: if the AER is removed from an early-stage wing bud, only a humerus forms. If the AER is removed slightly later, humerus, radius, and ulna form (Figure 16.12; Saunders 1948; Iten 1982; Rowe and Fallon 1982). Explaining how this happens, however, has not been easy. First it had to be determined whether the positional information for proximal-distal polarity resides in the AER or in the progress zone mesenchyme. Through a series of reciprocal transplantations, this specificity was found to reside in the mesenchyme. If the AER *had* provided the positional information—somehow instructing the undifferentiated mesoderm beneath it as to what structures to make—then older AERs combined with younger mesoderm should have produced limbs with deletions in the middle, while younger AERs combined with older mesoderm should have produced duplications of structures. This was not found to be the case; rather, normal limbs form in both experiments (Rubin and Saunders 1972). But when the entire progress zone (including both the mesoderm and the AER) from an early embryo is placed on the limb bud of a later-stage embryo, new proximal structures are produced beyond those already present. Conversely, when old progress zones are added to young limb buds, distal structures develop immediately, so that digits are seen to emerge from the humerus without an intervening ulna and radius (Figure 16.13; Summerbell and Lewis 1975).

But how does the mesenchyme specify the proximal-distal axis? Two major models have been proposed to account for this phenomenon. One emphasizes the dimension of *time*, while the other emphasizes the dimension of *space*.

The **progress zone model** postulates that each mesoderm cell is specified by the amount of time it spends dividing in the progress zone. The longer a cell spends in the progress zone, the more mitoses it achieves, and the more distal its specification becomes. Since the progress zone remains the same size, cells must exit the PZ constantly as the limb grows; once they leave, they can form cartilage. Thus, the first cells leaving the progress zone become the stylopod, while the last cells to leave it become the autopod (Figure 16.14A; Summerbell 1974; Wolpert et al. 1979). According to this model, removing the AER means that the progress zone cells can no longer divide and be further specified, so only the proximal structures form.

Although the progress zone model has been the prevailing means of explaining limb polarity, it does not

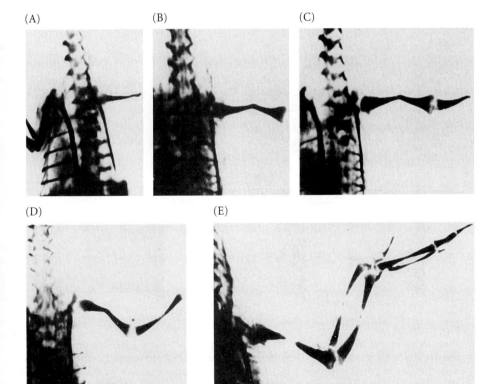

(A)   (B)   (C)

(D)   (E)

Figure 16.12
Dorsal views of chick skeletal patterns after removal of the entire AER from the right wing bud of chick embryos at various stages. The last photo (E) is of a normal wing skeleton. (From Iten 1982; photographs courtesy of L. Iten.)

(A)

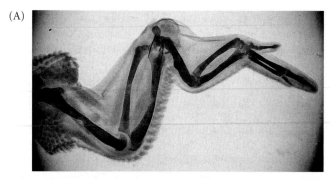

(B)

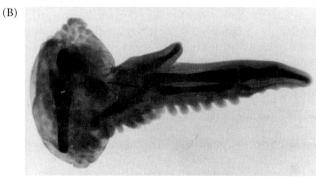

Figure 16.13
Control of proximal-distal specification by the progress zone mesenchyme. (A) Extra set of ulna and radius formed when an early wing-bud progress zone was transplanted to a late wing bud that had already formed ulna and radius. (B) Lack of intermediate structures seen when a late wing-bud progress zone was transplanted to an early wing bud. Hinges indicate the locations of the grafts. (From Summerbell and Lewis 1975; photographs courtesy of D. Summerbell.)

did not specify cell position. Moreover, cell division is seen throughout the limb bud, not solely in the progress zone.

Gail Martin and Cliff Tabin have proposed an **early allocation and progenitor expansion model** (Figure 16.14B) as an alternative to the progress zone model. Here, the cells of the entire early limb bud are already specified; subsequent cell divisions simply expand these cell populations. The effects of AER removal are explained by the observation of Rowe et al. (1982) that when the AER is taken off the limb bud, there is apoptosis for approximately 200 μm. If one removes the AER of an early limb bud, before the mesenchymal cell populations have expanded, the cells specified to be zygopod and autopod die, and only the cells already specified to be stylopod remain. In later embryonic stages, the limb bud grows and the 200 μm region of apoptosis merely deletes the autopod. Research is presently underway to distinguish the contributions of these mechanisms to A–P axis specification in the limb.

Hox genes, meis genes, and the specification of the proximal-distal axis

The products of the Hox genes have already played a role in specifying the place where the limbs will form. Now they will play a second role in specifying whether a particular mesenchymal cell will become stylopod, zeugopod, or autopod.

The 5′ (*AbdB*-like) portions (paralogues 9–13) of the Hoxa and Hoxd complexes appear to be active in the forelimb buds of mice. Based on the expression patterns of these genes, and on naturally occurring and gene knockout mutations, Davis and colleagues (1995) proposed a model wherein these Hox genes specify the identity of a limb region (Figure 16.15).

explain recent data from experiments in which the progress zone mesenchyme was prevented from dividing in mice lacking both *fgf8* and *fgf4* (both of which are expressed in the AER). In the limbs of these mice, the proximal elements were altogether missing, while the distal elements were present and often normal. This result could not be explained by a model wherein the specification of distal elements depended upon the number of mitoses in the progress zone (Sun et al. 2002). Moreover, when individual PZ cells were marked with a dye, the progeny of each cell were limited to a particular segment of the limb. The tip of the early limb bud provided cells only to the autopod, for instance (Dudley and Tabin 2002). This finding suggested that specification was early and that the time spent in the progress zone

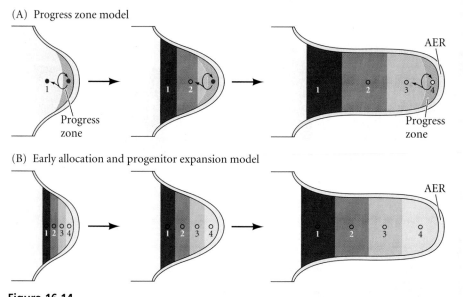

Figure 16.14
Two models for the mesodermal specification of the proximal-distal axis of the limb. (A) Progress zone model, wherein the length of time a mesodermal cell remains in the progress zone specifies its position. (B) Early allocation and progenitor expansion model, wherein the territories of the limb bud are established very early and the cells grow in each of these areas.

For instance, when they knocked out all four loci for the paralogous genes *Hoxa-11* and *Hoxd-11*, the resulting mice lacked the ulna and radius of their forelimbs (Figure 16.15A,B,). Similarly, knocking out all four *Hoxa-13* and *Hoxd-13* loci resulted in loss of the autopod (Fromental-Ramain et al. 1996). Humans homozygous for a *HOXD-13* mutation show abnormalities of the hands and feet wherein the digits fuse (Figure 16.15C), and human patients with homozygous mutant alleles of *HOXA-13* also have deformities of their autopods (Muragaki et al. 1996; Mortlock and Innis 1997). In both mice and humans, the autopod (the most distal portion of the limb) is affected by the loss of function of the most 5′ Hox genes.

The mechanism by which Hox genes could specify the proximal-distal axis is not yet understood, but one clue comes from the analysis of chicken *Hoxa-13*. Ectopic expression of this gene (which is usually expressed in the distal ends of developing chick limbs) appears to make the cells expressing it "stickier." This property might cause the cartilaginous nodules to condense in specific ways (Yokouchi et al. 1995; Newman 1996).

As the limb grows outward, the pattern of Hox gene expression changes. When the stylopod is forming, *Hoxd-9* and *Hoxd-10* are expressed in the most distal mesenchyme (Figure 16.16; Nelson et al. 1996). Here, they are joined by the products of the *meis* genes, which probably interact with the products of the Hox genes to specify the cells as proximal (Mercader et al. 1999, 2000; Capdevila et al. 1999). When the

zeugopod bones are forming, the pattern shifts remarkably, displaying a nested sequence of Hoxd gene expression. The posterior region expresses all the Hoxd genes from *Hoxd-9* to *Hoxd-13*, while only *Hoxd-9* is expressed anteriorly. In the third phase of limb development, when the autopod is forming, there is a further redeployment of Hox gene products. *Hoxd-9* is no longer expressed. Rather, *Hoxa-13* is expressed in the anterior tip of the limb bud and in a band marking the boundary of the autopod. *Hoxd-13* products join those of *Hoxa-13* in the anterior region of the limb bud, while *Hoxa-12*, *Hoxa-11*, and *Hoxd-10–12* are expressed throughout the posterior two-thirds of the limb bud. Kmita and colleagues (2002) have shown that the autopod-specific pattern of Hox gene expression is controlled by a particular enhancer region that interacts with each of the genes. The closer the Hox gene is to this enhancer, the more efficiently it is activated and the more product it makes. It is possible that the location of Hox genes

Figure 16.15
Deletion of limb bone elements by the deletion of paralogous Hox genes. (A) Wild-type mouse forelimb. (B) Forelimb of mouse made doubly mutant such that it lacked functional *Hoxa-11* and *Hoxd-11* genes. The ulna and radius are absent. (C) Human synpolydactyly (many fingers joined together) syndrome resulting from homozygosity for a mutation at the *HOXD-13* loci. The human syndrome also includes malformations of the urogenital system, which also expresses *HOXD-13*. (D) Hypothesis that the 5′ paralogues of Hox genes specify particular regions of the forelimb. (A, B, and D after Davis et al. 1995, photographs courtesy of M. Capecchi; C from Muragaki et al. 1996, photograph courtesy of B. Olsen.)

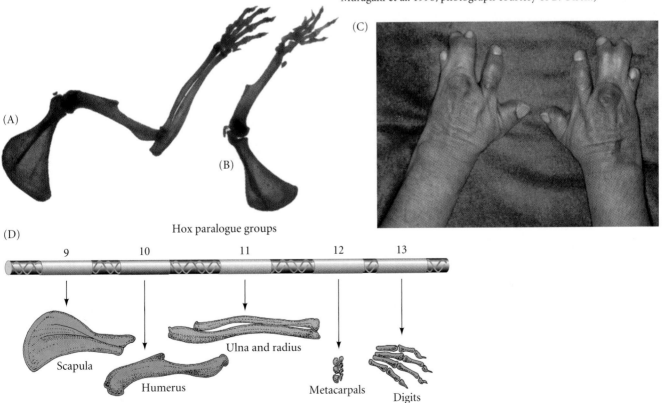

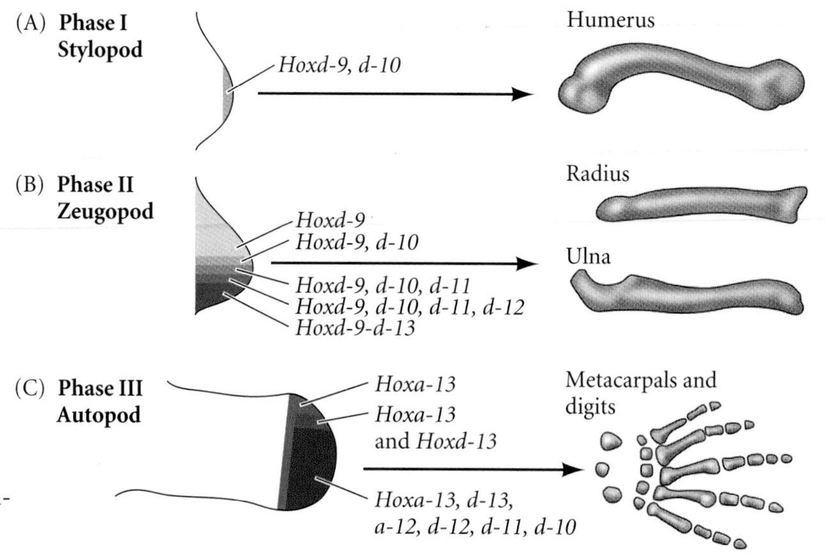

Figure 16.16
Hox gene expression changes during the formation of the tetrapod limb. (A) During the formation of the stylopod (phase I), *Hoxd-9* and *Hoxd-10* are expressed in the newly formed limb bud. (B) During zeugopod formation (phase II), there is a nested expression of Hoxd genes such that *Hoxd-9* through *Hoxd-13* are expressed in the posterior of the limb bud, while only *Hoxd-9* is expressed both anteriorly and posteriorly. (C) Inversion of Hox gene expression during autopod formation (phase III). *Hoxd-13* and *Hoxa-13* are expressed anteriorly and posteriorly, while *Hoxd-10* through *Hoxd-12* and *Hoxa-12* are expressed posteriorly. (After Shubin et al. 1997.)

(A) **Phase I Stylopod** — *Hoxd-9, d-10* → Humerus

(B) **Phase II Zeugopod** — *Hoxd-9* / *Hoxd-9, d-10* / *Hoxd-9, d-10, d-11* / *Hoxd-9, d-10, d-11, d-12* / *Hoxd-9-d-13* → Radius, Ulna

(C) **Phase III Autopod** — *Hoxa-13* / *Hoxa-13* and *Hoxd-13* / *Hoxa-13, d-13, a-12, d-12, d-11, d-10* → Metacarpals and digits

in the Hoxd cluster prevents the digits from forming anywhere but at the distal end of the limb.

> **WEBSITE 16.4 Hox genes and limb evolution.** The last stage of limb development—the formation of the autopod—is characterized by new expression patterns of Hox genes. This repatterning is not seen in fish fins, giving rise to the speculation that the limb evolved from the fin by changing Hox gene expression patterns.

> **WEBSITE 16.5 Dinosaurs and the origin of birds.** Whether birds are descendants of certain dinosaurs is a controversial subject that concerns hip structure, and whether the autopod of the dinosaur is made of the same digits as the bird autopod.

Specification of the Anterior-Posterior Limb Axis

The zone of polarizing activity

The specification of the anterior-posterior axis of the limb is the earliest change from the pluripotent condition. In the chick, this axis is specified shortly before a limb bud is recognizable. Hamburger (1938) showed that as early as the 16-somite stage, prospective wing mesoderm transplanted to the flank area develops into a limb with the anterior-posterior and dorsal-ventral polarities of the donor graft, not those of the host tissue.

Several experiments (Saunders and Gasseling 1968; Tickle et al. 1975) suggest that the anterior-posterior axis is specified by a small block of mesodermal tissue near the posterior junction of the young limb bud and the body wall. This region of the mesoderm has been called the **zone of polarizing activity** (**ZPA**). When this tissue is taken from a young limb bud and transplanted to a position on the anterior side of another limb bud (Figure 16.17), the number of digits of the resulting wing is doubled. Moreover, the structures of the extra set of digits are mirror images of the normally produced structures. The polarity has been maintained, but the information is now coming from both an anterior and a posterior direction.

Figure 16.17
When a ZPA is grafted to anterior limb bud mesoderm, duplicated digits emerge as a mirror image of the normal digits. (From Honig and Summerbell 1985; photograph courtesy of D. Summerbell.)

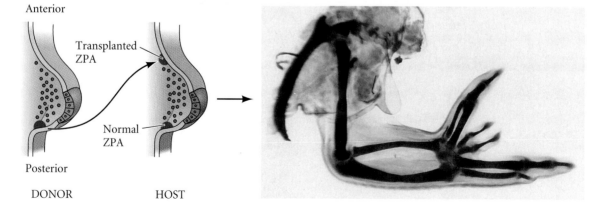

Anterior

Transplanted ZPA

Normal ZPA

Posterior

DONOR   HOST

Figure 16.18
Sonic hedgehog protein is expressed in the ZPA. In situ hybridization showing the sites of Sonic hedgehog expression (arrows) in the posterior mesoderm of the chick limb buds. These are precisely the regions that transplantation experiments defined as the ZPA. (Photograph courtesy of R. D. Riddle.)

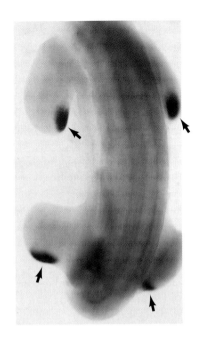

Specification of the ZPA

Two new questions emerged from this discovery. First, how does *sonic hedgehog* become expressed only in the posterior region of the limb bud? And second, what does it do once it is expressed? The *shh* gene appears to be activated by an FGF protein coming from the newly formed apical ectodermal ridge (Figure 16.20). FGF8 is secreted from the AER, and is capable of activating *shh* (Heikinheimo et al. 1994; Crossley and Martin 1995). But why doesn't FGF8 activate all the mes-

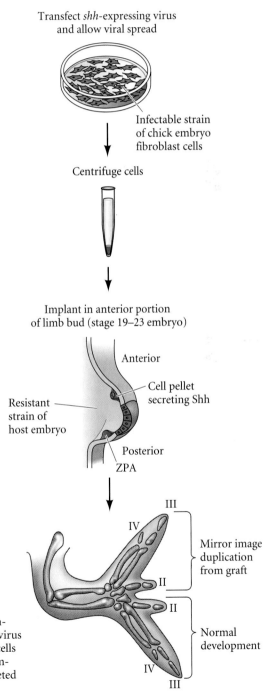

Transfect *shh*-expressing virus and allow viral spread

Infectable strain of chick embryo fibroblast cells

Centrifuge cells

Implant in anterior portion of limb bud (stage 19–23 embryo)

Anterior

Cell pellet secreting Shh

Resistant strain of host embryo

Posterior
ZPA

III
IV

Mirror image duplication from graft

II

II

Normal development

IV

III

Sonic hedgehog defines the ZPA

The search for the molecule(s) conferring this polarizing activity on the ZPA became one of the most intensive quests in developmental biology. In 1993, Riddle and colleagues showed by in situ hybridization that *sonic hedgehog* (*shh*), a vertebrate homologue of the *Drosophila hedgehog* gene, was expressed specifically in that region of the limb bud known to be the ZPA (Figure 16.18).

As evidence that this association between the ZPA and *sonic hedgehog* was more than just a correlation, Riddle and co-workers (1993) demonstrated that the secretion of Sonic hedgehog protein is sufficient for polarizing activity. They transfected embryonic chick fibroblasts (which normally would never synthesize the Shh protein) with a viral vector containing the *shh* gene (Figure 16.19). The gene became expressed and translated in these fibroblasts, which were then inserted under the anterior ectoderm of an early chick limb bud. Mirror-image digit duplications like those induced by ZPA transplants were the result. More recently, beads containing Sonic hedgehog protein were shown to cause the same duplications (López-Martínez et al. 1995). Thus, Sonic hedgehog appears to be the active agent of the ZPA.

Figure 16.19
Assay for polarizing activity of Sonic hedgehog. The *sonic hedgehog* gene was inserted adjacent to an active promoter of a chicken virus, and the recombinant virus was placed into cultured chick embryo fibroblast cells. The virally transfected cells were pelleted and implanted in the anterior margin of a limb bud of a chick embryo. The resulting limbs produced mirror-image digits, showing that the secreted protein had polarizing activity. (After Riddle et al. 1993.)

Figure 16.20

Feedback between the AER and the ZPA in the forelimb bud. Stage 16 (about 54 hours gestation) represents the same stage as (C) in Figure 16.7. The other two stages are about 5 hours apart. At stage 17, the newly induced AER secretes FGF8 into the underlying mesenchyme. The mesenchyme expressing *Hoxb-8* and *dHAND* is induced to express *shh*, thereby forming the ZPA in the posterior region of the fore-limb bud. At stage 18, Sonic hedgehog protein maintains FGF expression in the AER, while the FGFs from the AER maintain *shh* gene expression.

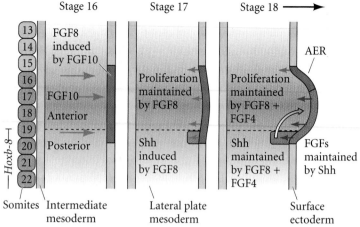

enchyme cells beneath the AER? The answer may re-side in the differential competence of certain mes-enchyme cells to respond to the FGF signal.

There are two transcription factors that exist in the posterior region of the limb bud that are not expressed in the anterior portion, and these factors may enable only the *posterior* mesoderm cells to express the *shh* gene. First, Charité and colleagues (1994) suggested that the **Hoxb-8** protein may be critical in providing this restricted competence. The *Hoxb-8* gene is normally expressed in the posterior half of the mouse forelimb bud. However, when Hoxb-8 protein is ectopically ex-pressed throughout the forelimb bud, both the anterior and posterior mesoderm express Sonic hedgehog protein, creating two ZPAs and mirror-image forelimb duplications. The second transcription factor in the posterior region is **dHAND**. The *dHAND* gene is expressed in the posterior mesoderm of the limb bud immediately before *shh* is expressed. In mice lacking *dHAND*, *shh* is not expressed in the limb buds and the limbs are severely truncated. Conversely, ectopic expression of dHAND protein in the anterior region of the mouse limb bud generates a second ZPA and duplications of the digits (Charité et al. 2000).

Specifying digit identity through cell interactions initiated by Sonic hedgehog

When the Sonic hedgehog protein was first shown to define the ZPA, it was thought to act as a morphogen. In other words, it was thought to diffuse from the ZPA where it was being synthe-sized and to form a concentration gradient from the posterior to the anterior of the limb bud. However, recent research has pro-vided evidence that Sonic hedgehog (or its active amino-termi-nal region) does not diffuse outside the ZPA (Yang et al. 1997). It is now thought that Sonic hedgehog works in at least two other ways.

First, Sonic hedgehog initiates and sustains a gradient of proteins such as BMP2 and BMP7 (Laufer et al. 1994; Kawa-kami et al. 1996; Drossopoulou et al. 2000), and these proteins act to specify the digits. Digit identity is not specified directly in each digit primordium. Rather, the identity of each digit is de-termined by the *interdigital* mesoderm. In other words, the identity of each digit is specified by the webbing between the digits—that region of mesenchyme that will shortly undergo apoptosis to free each digit. The interdigital tissue specifies the identity of the digit forming anteriorly to it (toward the thumb or big toe). Thus, when Dahn and Fallon (2000) removed the webbing between the cartilaginous condensations forming chick hindlimb digits 2 and 3, the second digit was changed into a copy of digit 1. Similarly, when the webbing on the other side of digit 3 was removed, the third digit formed a copy of digit 2 (Figure 16.21A–C). Moreover, the positional value of the webbing could be altered by changing the BMP levels in it. When beads containing BMP antagonists such as Noggin were placed in the webbing between digits 3 and 4, digit 3 was ante-riorly transformed into digit 2 (Figure 16.21 D–F).

The second way Sonic hedgehog works is to alter the ratio of repressor Gli3 to activator Gli3. As you may recall from Chapter 6, one of the mechanisms of Hedgehog signal-ing is to prevent the proteolysis that makes the repressor forms of the Gli/Ci set of transcription factors. When this proteolysis is prevented, the repressor forms of the Gli/Ci transcription factors are converted into the activator forms. Recent experiments (Litingtung et al. 2002; te Welscher et al. 2002) have shown that mice homozygous for loss-of-function mutations of *Gli3* (or *Gli3* and *Shh*) form 6–11 digits per limb, none of which has a normal identity (Figure 16.20G). Mutations of *shh* alone, however, generate truncated limbs with few, if any, digits. Thus, Sonic hedgehog appears to work through Gli3 to specify normal digit identities and to impose the pentadactyl (five-digit) constraint upon the limb. There is a 165-fold variation in the ratio of activator to repressor Gli3 across the mouse limb; this ratio is highest in the posterior. It appears that the repressor form of Gli3 is made throughout the mesoderm, but that Sonic hedgehog creates a gradient such that very little of the activator is made in the anterior portion of the limb bud.

However it works, Sonic hedgehog (directly or indirectly) regulates the expression of the 5′ Hoxd genes. The transition from phase I to phase II Hox expression patterns (see Figure 16.16) is coincident with *shh* expression in the ZPA. Moreover, transplantation of either the ZPA or other Sonic hedgehog-secreting cells to the anterior margin of the limb bud at this

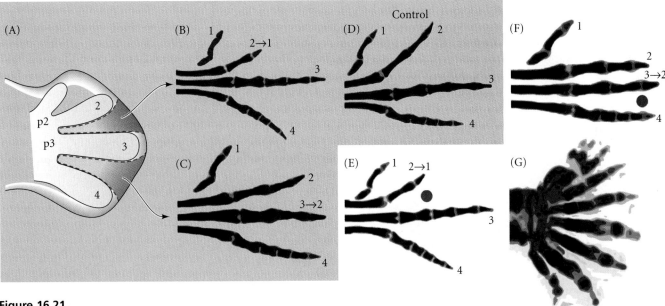

Figure 16.21
Regulation of digit identity by BMP concentrations in the interdigital space anterior to the digit and by Gli3. (A) Scheme for removal of interdigit (ID) regions. The results are shown in (B) and (C), respectively. (B) Removal of interdigital region 2 between digit primordia 2 (p2) and 3 (p3) causes digit 2 to change to the structure of digit 1. (C) Removing interdigital region 3 (between digit primordia 3 and 4) causes digit 3 to form the structures of digit 2. (D) Control digits and their interdigital spaces. (E, F) The same transformations as in (B) and (C) can be obtained by adding beads containing the BMP inhibitor Noggin to the interdigital regions. (E) When a Noggin-containing bead (green dot) is placed in interdigital region 2, the second digit is transformed into a copy of digit 1. (F) When the Noggin bead is placed in interdigital region 3, the third digit is transformed into a copy of digit 2. (G) Forelimb of a mouse homozygous for deletions of both *Gli3* and *shh* is characterized by extra digits of no specific type. (After Dahn and Fallon 2000; Litingtung et al. 2002; photographs courtesy of R. D. Dahn and J. F. Fallon.)

stage leads to the formation of mirror-image patterns of Hoxd gene expression and results in mirror-image digit patterns (Izpisúa-Belmonte et al. 1991; Nohno et al. 1991; Riddle et al. 1993).

Generation of the Dorsal-Ventral Axis

The third axis of the limb distinguishes the dorsal half of the limb (knuckles, nails) from the ventral half (pads, soles). In 1974, MacCabe and co-workers demonstrated that the dorsal-ventral polarity of the limb bud is determined by the ectoderm encasing it. If the ectoderm is rotated 180° with respect to the limb bud mesenchyme, the dorsal-ventral axis is partially reversed; the distal elements (digits) are "upside down." This suggested that the late specification of the dorsal-ventral axis of the limb is regulated by its ectodermal component.

One molecule that appears to be particularly important in specifying dorsal-ventral polarity is **Wnt7a**. The *Wnt7a* gene is expressed in the dorsal (but not the ventral) ectoderm of chick and mouse limb buds (Dealy 1993; Parr et al. 1993). In 1995, Parr and McMahon genetically deleted Wnt7a from mouse embryos. The resulting embryos had ventral footpads

on both surfaces of their paws, showing that Wnt7a is needed for the dorsal patterning of the limb (Figure 16.22).

Wnt7a is the first of the dorsal-ventral axis genes expressed in limb development. It induces activation of the *Lmx1* gene in the dorsal mesenchyme. This gene encodes a transcription factor that appears to be essential for specifying dorsal cell fates in the limb. If Lmx1 protein is expressed in the ventral mesenchyme cells, those cells develop a dorsal phenotype (Riddle et al. 1995; Vogel et al. 1995; Altabef and Tickle 2002). Mutants of *Lmx1* in humans and mice also show its importance for specifying dorsal limb fates. Knockouts of this gene in mice produce a syndrome in which the dorsal limb phenotype is lacking, and loss-of-function mutations in humans produce the nail-patella syndrome, a condition in which the dorsal sides of the limbs have been ventralized (i.e., no nails, no kneecaps) (Chen et al. 1998; Dreyer et al. 1998).

Coordinating the Three Axes

The three axes of the tetrapod limb are all interrelated and coordinated. Some of the principal interactions among the mechanisms specifying the axes are shown in Figure 16.23. In-

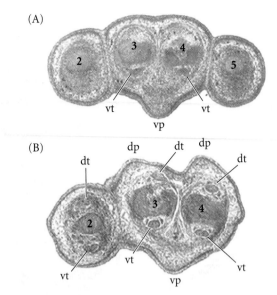

Figure 16.22
Dorsal-to-ventral transformations of limb regions in mice deficient for both *Wnt7a* genes. (A) Histological section (stained with hemotoxylin and eosin) of wild-type 15.5-day embryonic mouse forelimb paw. The ventral tendons and ventral footpads are readily seen. (B) Same section through a mutant embryo deficient in *Wnt7a*. Tendons and footpads are duplicated on what would normally be the dorsal surface of the paw. dt, dorsal tendons; dp, dorsal footpad; vp, ventral footpad; vt, ventral tendon. Numbers indicate digit identity. (From Parr and McMahon 1995; photographs courtesy of the authors.)

deed, the molecules that define one of these axes are often used to maintain another axis. For instance, Sonic hedgehog in the ZPA activates the expression of the *fgf4* gene in the posterior region of the AER (see Figure 16.20). Expression of FGF4 protein is important in recruiting mesenchyme cells into the progress zone, and it is also partially responsible (along with FGF8) in maintaining the expression of Sonic hedgehog in the ZPA (Li and Muneoka 1999). The Sonic hedgehog of the ZPA

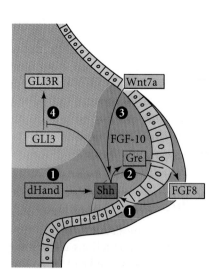

then sustains the AER FGFs by inducing Gremlin, a BMP inhibitor that prevents BMPs in the mesoderm from inhibiting the FGFs of the AER (Zúñiga et al. 1999). Therefore, the AER and the ZPA mutually support each other through the positive loop of Sonic hedgehog and FGFs (Todt and Fallon 1987; Laufer et al. 1994; Niswander et al. 1994).

The *Wnt7a*-deficient mice described above lacked not only dorsal limb structures but also posterior digits, suggesting that Wnt7a is also needed for the anterior-posterior axis. Yang and Niswander (1995) made a similar set of observations in chick embryos. These investigators removed the dorsal ectoderm from developing limbs and found that such an operation resulted in the loss of posterior skeletal elements from the limbs. The reason that these limbs lacked posterior digits was that *shh* expression was greatly reduced. Viral-induced expression of *Wnt7a* was able to replace the dorsal ectoderm and restore *shh* expression and posterior phenotypes. These findings showed that the synthesis of Sonic hedgehog is stimulated by the combination of FGF4 and Wnt7a proteins.

This coordination of activities in critical in regulating limb growth. (Imagine what would happen if the limb stopped growing along one axis but continued to grow along the others.) It appears that BMPs are responsible for shutting down the signaling from the AER and for simultaneously inhibiting theWnt7a signal along the dorsal-ventral axis (Pizette et al. 2001). The BMP signal eliminates growth and patterning along all three axes. When exogenous BMP is applied to the AER, the elongated epithelium of the AER reverts to a cuboidal epithelium and ceases to produce FGFs; and when BMP is inhibited by Noggin, the AER continues to persist days after it would normally have regressed (Gañan et al. 1998; Pizette and Niswander 1999). The mechanism by which the BMP signal is regulated is not yet fully understood, but it appears to involve the BMP antagonist protein Gremlin. Gremlin works against BMP to preserve the AER and the Wnt7a pathway (Merino et al. 1999).

Cell Death and the Formation of Digits and Joints

Sculpting the autopod

Cell death plays a role in sculpting the limb. Indeed, it is essential if our joints are to form and if our fingers are to become separate (Zaleske 1985). The death (or lack of death) of

Figure 16.23
Some of the molecular interactions by which limb bud formation and growth are initiated and maintained. Some of the major loops include (1) the establishment of the ZPA (Sonic hedgehog) by the AER (FGF8) and dHAND; (2) the induction of FGFs (in the AER) by Sonic hedgehog (in the ZPA) through Shh induction of Gremlin, which blocks the BMPs that could inhibit FGFs; (3) the maintenance of Sonic hedgehog by Wnt7a (in the dorsal ectoderm); (4) blocking by Sonic hedgehog the cleavage of Gli3 into its repressor form, creating a gradient of Gli3 activator and Gli3 repressor. (After te Welscher 2002.)

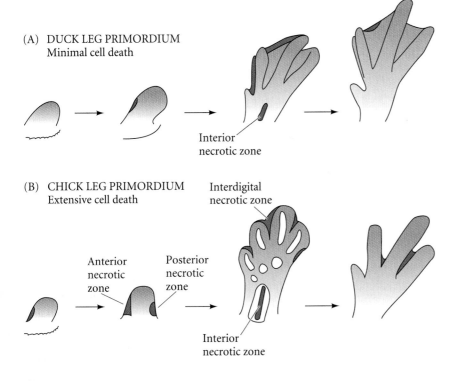

(A) DUCK LEG PRIMORDIUM
Minimal cell death

Interior
necrotic zone

(B) CHICK LEG PRIMORDIUM
Extensive cell death

Interdigital
necrotic zone

Anterior
necrotic
zone

Posterior
necrotic
zone

Interior
necrotic zone

Figure 16.24
Patterns of cell death in leg primordia of (A) duck and (B) chick embryos. Shading indicates areas of cell death. In the duck, the regions of cell death are very small, whereas there are extensive regions of cell death in the interdigital tissue of the chick leg. (After Saunders and Fallon 1966.)

apoptosis, and the death of the interdigital tissue is associated with the fragmentation of their DNA (Mori et al. 1995). The signal for apoptosis in the autopod is probably provided by the BMP proteins. BMP2, BMP4, and BMP7 are each expressed in the interdigital mesenchyme, and blocking BMP signaling (by infecting progress zone cells with retroviruses carrying dominant negative BMP receptors) prevents interdigital apoptosis (Zou and Niswander 1996; Yokouchi et al. 1996). Since these BMPs are expressed throughout the progress zone mesenchyme, it is thought that cell death would be the "default" state unless there were active suppression of the BMPs. This suppression may come from the Noggin protein, which is made in the developing cartilage of the digits and in the perichondrial cells surrounding it (Capdevila and Johnson 1998; Merino et al. 1998). If Noggin is expressed throughout the limb bud, no apoptosis is seen.

The regulation of BMPs is critical in creating the webbed feet of ducks (Merino et al. 1999). The interdigital regions of duck feet exhibit the same pattern of BMP expression as the webbing of chick feet. However, whereas the interdigital regions of the chick feet appear to undergo BMP-mediated apoptosis, developing duck feet synthesize the Gremlin protein and block this regional cell death (see Figure 23.22). Moreover, the webbing of chick feet can be preserved if Gremlin-soaked beads are placed in the interdigital regions (Figure 16.25).

Forming the joints

The function first ascribed to BMPs was the formation, not the destruction, of bone and cartilage tissue. In the developing limb, BMPs induce the mesenchymal cells either to undergo apoptosis or to become cartilage-producing chondrocytes—depending on the stage of development. The same BMPs can induce death or differentiation, depending on the age of the target cell. This "context dependency" of signal action is a critical concept in developmental biology. It is also critical for the formation of joints. Macias and colleagues (1997) have shown that during early limb bud stages (before cartilage condensation), beads secreting BMP2 or BMP7 cause apoptosis. Two days later, the same beads cause the limb bud cells to form cartilage.

In the normally developing limb, BMPs use both of these properties to form joints. BMP7 is made in the perichondrial cells surrounding the condensing chondrocytes and promotes

specific cells in the vertebrate limb is genetically programmed and has been selected for during evolution. One such case involves the webbing or nonwebbing of feet. The difference between a chicken's foot and that of a duck is the presence or absence of cell death between the digits (Figure 16.24A,B). Saunders and co-workers (1962; Saunders and Fallon 1966) have shown that after a certain stage, chick cells between the digit cartilage are destined to die, and will do so even if transplanted to another region of the embryo or placed in culture. Before that time, however, transplantation to a duck limb will save them. Between the time when the cell's death is determined and when death actually takes place, levels of DNA, RNA, and protein synthesis in the cell decrease dramatically (Pollak and Fallon 1976).

In addition to the **interdigital necrotic zone**, three other regions of the limb are "sculpted" by cell death. The ulna and radius are separated from each other by an **interior necrotic zone**, and two other regions, the **anterior** and **posterior necrotic zones**, further shape the end of the limb (Figure 16.25B; Saunders and Fallon 1966). Although these zones are referred to as "necrotic," this term is a holdover from the days when no distinction was made between necrotic cell death and apoptotic cell death (see Chapter 6). These cells die by

(A) 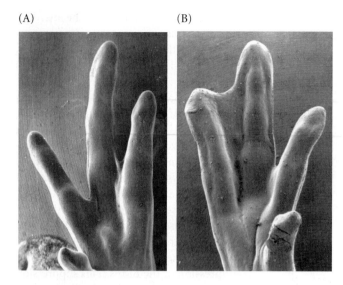 (B)

Figure 16.25
Inhibition of cell death by inhibiting BMPs. (A) Control chick limbs show extensive apoptosis in the space between the digits, leading to the absence of webbing. (B) However, when beads soaked with Gremlin protein are placed in the interdigital mesoderm, the webbing persists and generates a duck-like pattern. (After Merino et al. 1999; photographs courtesy of E. Hurle.)

(A)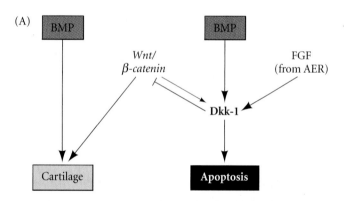

(B) 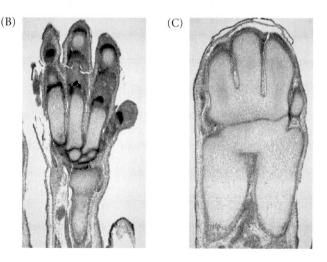 (C)

cartilage formation (Figure 16.26A; Macias et al. 1997). Two other BMP proteins, BMP2 and GDF5, are expressed at the regions between the bones, where joints will form (Figure 16.28B; Macias et al. 1997; Brunet et al. 1998). Mouse mutations have suggested that the function of these proteins in joint formation is critical. Mutations of *Gdf5* produce brachypodia, a condition characterized by a lack of limb joints (Storm and Kingsley 1999). In mice homozygous for loss-of-function alleles of *noggin*, no joints form, either. It appears that the BMP7 in these *noggin*-defective embryos is able to recruit nearly all the surrounding mesenchyme into the digits (Figure 16.26C,D; Brunet et al. 1998).

Wnt proteins and blood vessels also appear to be critical in joint formation. The conversion of mesenchyme cells into nodules of cartilage-forming tissue establishes where the bone boundaries are. The mesenchyme will not form such nodules in the presence of blood vessels, and one of the first indications of cartilage formation is the regression of blood vessels in the region wherein the nodule will form (Yin and Pacifici 2001). Wnt proteins can then initiate the changes in cell adhesion molecules (such as N-cadherin) that cause the mesenchyme cells to initiate chondrogenesis. The placement of these molecules may determine the number, shape, and size of these condensations (Hartmann and Tabin 2001; Tufan and Tuan 2001).

Continued Limb Growth: Epiphyseal Plates

If all of our cartilage were turned into bone before birth, we could not grow any larger, and our bones would be only as large as the original cartilaginous model. However, as the ossification front nears the ends of the cartilage model, the chondrocytes near the ossification front proliferate prior to undergoing hypertrophy, pushing out the cartilaginous ends of the bone. In the long bones of many mammals (including humans), endochondral ossification spreads outward in both directions from the center of the bone (see Figure 14.13). These cartilaginous areas at the ends of the long bones are called **epiphyseal growth plates**. These plates contain three regions: a region of chondrocyte proliferation, a region of mature chondrocytes, and a re-

Figure 16.26
Possible involvement of BMPs in stabilizing cartilage and apoptosis. (A) Model for the dual role of BMP signals in limb mesodermal cells. BMP can be received in the presence of FGFs (to produce apoptosis) or Wnts (to induce bone). When FGFs from the AER are present, Dickkopf (DKK-1) is activated. This protein mediates apoptosis and at the same time inhibits Wnt from aiding in skeleton formation. (B, C) The effects of Noggin. (B) 16.5-day autopod from a wild-type mouse, showing GDF5 expression (dark blue) at the joints. (C) 16.5-day *noggin*-deficient mutant mouse autopod, showing neither joints nor GDF5 expression. Presumably, in the absence of Noggin, BMP7 was able to convert nearly all the mesenchyme into cartilage. (A after Gratevald and Rüther 2002; B, C from Brunet et al. 1998, photographs courtesy of A. P. McMahon.)

(A) (B) (C)

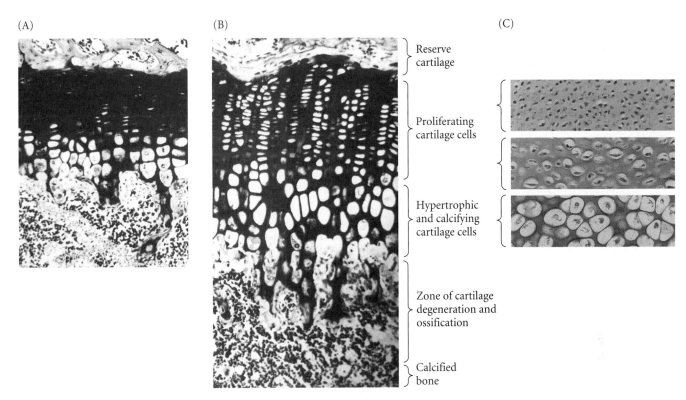

Reserve cartilage

Proliferating cartilage cells

Hypertrophic and calcifying cartilage cells

Zone of cartilage degeneration and ossification

Calcified bone

gion of hypertrophic chondrocytes (Figure 16.27; Chen et al. 1995). As the inner cartilage hypertrophies and the ossification front extends farther outward, the remaining cartilage in the epiphyseal growth plate proliferates. As long as the epiphyseal growth plates are able to produce chondrocytes, the bone continues to grow.

Figure 16.27
Proliferation of cells in the epiphyseal growth plate in response to growth hormone. (A) Epiphyseal growth plate in a young rat that was made growth hormone-deficient by removal of its pituitary. (B) Same region in the rat after injection of growth hormone. (C) Stained cartilage from specific regions of the epiphyseal growth plate. (A, B, from Bloom and Fawcett 1975, photographs by I. Gersh; C from Chen et al. 1995, photograph courtesy of P. Goetinck.)

Sidelights & Speculations

Control of Cartilage Maturation at the Growth Plate

Recent discoveries of human and murine mutations resulting in abnormal skeletal development have provided remarkable insights into how the differentiation, proliferation, and patterning of chondrocytes are regulated.

Fibroblast Growth Factor Receptors

The proliferation of the epiphyseal growth plate cells and facial cartilage can be halted by the presence of fibroblast growth factors (Deng et al. 1996; Webster and Donoghue 1996). These factors appear to instruct the cartilage precursors to differentiate rather than to divide. In humans, mutations of the receptors for fibroblast growth factors can cause these receptors to become activated prematurely. Such mutations give rise to the major types of human dwarfism.

Achondroplasia is a dominant condition caused by mutations in the transmembrane region of fibroblast growth factor receptor 3 (FGFR3). Roughly 95% of achondroplastic dwarfs have the same mutation of FGFR3, a base pair substitution that converts glycine to arginine at position 380 in the transmembrane region of the protein. In addition, mutations in the extracellular portion of the FGFR3 protein or in the tyrosine kinase intracellular domain may result in thanatophoric dysplasia, a lethal form of dwarfism that resembles homozygous achondroplasia (see Figure 6.23; Bellus et al. 1995; Tavormina et al. 1995). Mutations in FGFR1 can cause Pfeiffer syndrome, characterized by limb defects and premature fusion of the cranial sutures (craniosynostosis), which results in abnormal skull and facial shape. Different mutations in FGFR2 can give rise to various abnormalities of the limbs and face (Park et al. 1995; Wilkie et al. 1995).

Insulin-Like Growth Factors

The epiphyseal growth plate cells are very responsive to hormones, and their proliferation is stimulated by growth hormone and insulin-like growth factors. Nilsson and colleagues (1986) showed that growth hormone stimulates the production of insulin-like growth factor I (IGF-I) in the epiphyseal chondrocytes, and that these chondrocytes respond to it by proliferating. When they added growth hormone to the tibial growth plates of young mice who could not manufacture their own growth hormone (because their pituitaries had been removed), it stimulated the formation of IGF-I in the chondrocytes of the proliferative zone (see Figure 14.16). The combination of growth hormone and IGF-I appears to provide an extremely strong mitotic signal. It appears that IGF-I is essential for the normal growth spurt at puberty. The pygmies of the Ituri Forest of Zaire have normal growth hormone and IGF-I levels until puberty. However, at puberty, their IGF-I levels fall to about one-third of that of other adolescents (Merimee et al. 1987).

Estrogen Receptors

The pubertal growth spurt and the subsequent cessation of growth are induced by sex hormones (Kaplan and Grumbach 1990). At the end of puberty, high levels of estrogen or testosterone cause the remaining epiphyseal plate cartilage to undergo hypertrophy. These cartilage cells grow, die and are replaced by bone. Without any further cartilage formation, growth of these bones ceases, a process known as **growth plate closure.**

In conditions of precocious puberty, there is an initial growth spurt (making the individual taller than his or her peers), followed by the cessation of epiphyseal cell division (allowing that person's peers to catch up and surpass his or her height). In males, it was not thought that estrogen played any role in these events. However, in 1994, Smith and colleagues published the case history of a man whose growth was still linear despite his having undergone normal puberty. His epiphyseal plates had not matured, and at 28 years of age he still had proliferating chondrocytes. His "bone age"—the amount of ephiphyseal cartilage he retained—was roughly half his chronological age. This man was found to lack functional estrogen receptors. At present, at least three human males have been reported who either cannot make estrogens or who lack the estrogen receptor. All three are close to 7 feet tall and are still growing (Sharpe 1997). These cases show that estrogen plays a role in epiphyseal maturation in males as well as in females.

Thyroid hormone and parathyroid-related hormone are also important in regulating the maturation and hypertrophy program of the epiphyseal growth plate (Ballock and Reddi 1994). Thus, children with hypothyroidism are prone to developing growth plate disorders.

Extracellular Matrix Proteins

The extracellular matrix of the cartilage is also critical for the proper differentiation and organization of growth plate chondrocytes. Mutations that affect type XI collagen or the sulfation of cartilage proteoglycans can cause severe skeletal abnormalities. Mice with deficiencies of type XI collagen die at birth with abnormalities of limb, mandible, rib, and tracheal cartilage (Li et al. 1995). Failure to add sulfate groups to cartilage proteoglycans causes diastrophic dysplasia, a human dwarfism characterized by severe curvature of the spine, clubfoot, and deformed earlobes (Hästbacka et al. 1994).

While many of the "executives" of the limb bud formation have been identified, we still remain largely ignorant about how the orders of these paracrine factors and transcription factors are carried out. Niswander (2002) writes,

There is a very large gap in our understanding of how the activity of Shh, BMP, FGF, and Wnt genes influences, for example, where the cartilaginous condensation will form, how the elements are sculpted, how the

number of phalangeal elements are specified, and where the tendon/muscle will insert.

Limb development is an exciting meeting place for developmental biology, evolutionary biology, and medicine. Within the next decade, we can expect to know the bases for numerous congenital diseases of limb formation, and perhaps we will understand how limbs are modified into flippers, wings, hands, and legs.

Snapshot Summary: The Tetrapod Limb

1. The positions where limbs emerge from the body axis depend upon Hox gene expression.

2. The specification of a limb bud as hindlimb or forelimb is determined by Tbx4 and Tbx5 expression.

3. The proximal-distal axis of the developing limb is determined by the induction of the ectoderm at the dorsal-ventral boundary by an FGF (probably FGF10) from the mesenchyme. This induction forms the apical ectodermal ridge (AER). The AER secretes FGF8, which keeps the underlying mesenchyme proliferative and undifferentiated. This region of mesenchyme is called the progress zone.

4. As the limb grows outward, the stylopod forms first, then the zeugopod, and the autopod is formed last. Each phase of limb development is characterized by a specific pattern of Hox gene expression. The evolution of the autopod involved a reversal of Hox gene expression that distinguishes fish fins from tetrapod limbs.

5. The anterior-posterior axis is defined by the expression of Sonic hedgehog in the posterior mesoderm of the limb bud. This region is called the zone of polarizing activity (ZPA). If ZPA or Sonic hedgehog-secreting cells or beads are placed in the anterior margin of a limb bud, they establish a second, mirror-image pattern of Hox gene expression and a corresponding mirror-image duplication of the digits.

6. The ZPA is established by the interaction of FGFs from the AER with mesenchyme made competent to express Sonic hedgehog by its expression of dHAND and particular Hox genes. Sonic hedgehog acts, probably in an indirect manner, probably through the Gli factors, to change the expression of the Hox genes in the limb bud.

7. The identity of each digit is specified by BMP activity in the interdigital region posterior to it, and probably by the ratio of activator and repressor forms of Gli3.

8. The dorsal-ventral axis is formed in part by the expression of Wnt7a in the dorsal portion of the limb ectoderm. Wnt7a also maintains the expression level of Sonic hedgehog in the ZPA and of FGF4 in the posterior AER. FGF4 and Sonic hedgehog reciprocally maintain each other's expression.

9. Cell death in the limb is necessary for the formation of digits and joints. It is mediated by BMPs. Differences between the unwebbed chicken foot and the webbed duck foot can be explained by differences in the expression of Gremlin, a protein that antagonizes BMPs.

10. The BMPs are invoved both in inducing apoptosis and in differentiating the mesenchymal cells into cartilage. The effects of BMPs can be regulated by the Noggin and Gremlin proteins.

11. The ends of the long bones of humans and other mammals contain cartilagenous regions called epiphyseal growth plates. The cartilage in these regions proliferates so that the bone grows larger. Eventually, the cartilage is replaced by bone and growth stops.

Literature Cited

Altabef, M. and C. Tickle. 2002. Initiation of dorso-ventral axis during chick limb development. *Mech. Dev.* 116: 19–27.

Ballock, R. T. and A. H. Reddi. 1994. Thyroxine is the serum factor that regulates morphogenesis of columnar cartilage from isolated chondrocytes in chemically defined medium. *J. Cell Biol.* 126: 1311–1318.

Basson, C. T. and 13 others. 1996. Mutations in human *TBX5* cause limb and cardiac malformation in Holt-Oram syndrome. *Nature Genet.* 15: 30–35.

Bellus, G. A. and 8 others. 1995. A recurrent mutation in the tyrosine kinase domain of fibroblast growth factor receptor 3 causes hypochondroplasia. *Nature Genet.* 10: 357–359.

Brunet, L. J., J. A. McMahon, A. P. McMahon and R. M. Harland. 1998. Noggin, cartilage morphogenesis and joint formation in the mammalian skeleton. *Science* 280: 1455–1457.

Bryant, S. V. and D. M. Gardiner. 1992. Retinoic acid, local cell-cell interactions, and pattern formation in vertebrate limbs. *Dev. Biol.* 152: 1–25.

Burke, A. C., C. E. Nelson, B. A. Morgan and C. Tabin. 1995. Hox genes and the evolution of vertebrate axial morphology. *Development* 121: 333–346.

Capdevila, J. and R. L. Johnson. 1998. Endogenous and ectopic expression of *noggin* suggests a conserved mechanism for regulation of BMP function during limb and somite patterning. *Dev. Biol.* 197: 205–217.

Capdevila, J., T. Tsukui, C. Rodriguez Esteban, V. Zappavigna and J. C. Izpisúa Belmonte. 1999. Control of vertebrate limb outgrowth by the proximal factor Meis2 and distal antagonism of BMPs by Gremlin. *Mol. Cell* 4: 839–849.

Carrington, J. L. and J. F. Fallon. 1988. Initial limb budding is independent of apical ectodermal ridge activity: Evidence from a limbless mutant. *Development* 104: 361–367.

Charité, J., W. de Graaff, S. Shen and J. Duchamps. 1994. Ectopic expression of *Hoxb-8* causes duplications of the ZPA in the forelimb and homeotic transformation of axial structures. *Cell* 78: 559–601.

Charité, J., D. G. McFadden and E. N. Olsen. 2000. The bHLH transcription factor dHAND controls Sonic hedgehog expression and establishment of the zone of polarizing activity during limb development. *Development* 127: 2461–2470.

Chen, H. and 8 others. 1998. Limb and kidney defects in *Lmx1b* mutant mice suggest an involvement of *LMX1B* in human nail patella syndrome. *Nature Genet.* 19: 51–55.

Chen, Q., D. M. Johnson, D. R. Haudenschild and P. F. Goetinck. 1995. Progression and recapitulation of the chondrocyte differentiation program: Cartilage matrix protein is a marker for cartilage maturation. *Dev. Biol.* 172: 293–306.

Coates, M. I. 1994. The origin of vertebrate limbs. *Development* [Suppl.]: 169–180.

Cohn, M. J. and C. Tickle. 1999. Developmental basis of limblessness and axial patterning in snakes. *Nature* 399: 474–479.

Crossley, P. H. and G. R. Martin. 1995. The mouse *Fgf8* gene encodes a family of polypeptides and is expressed in regions that direct outgrowth and patterning in the developing embryo. *Development* 121: 439–451.

Crossley, P. H., G. Monowada, C. A. MacArthur and G. R. Martin. 1996. Roles for FGF8 in the induction, initiation, and maintenance of chick development of the tetrapod limb. *Cell* 84: 127–136.

Dahn, R. D. and J. F. Fallon. 2000. Interdigital regulation of digit identity and homeotic transformation by modulating BMP signaling. *Science* 289: 438–441.

Davis, A. P., D. P. Witte, H. M. Hsieh-Li, S. Potter and M. R. Capecchi. 1995. Absence of radius and ulna in mice lacking *hoxa-11* and *hoxd-11*. *Nature* 375: 791–795.

Dealy, C. N., A. Roth, D. Ferrari, A. M. C. Brown and R. A. Kosher. 1993. *Wnt-5a* and *Wnt-7a* are expressed in the developing chick limb bud in a manner that suggests roles in pattern formation along the proximodistal and dorsoventral axes. *Mech. Dev.* 43: 175–186.

Deng, C., A. Wynshaw-Boris, F. Zhou, A. Kuo and P. Leder. 1996. Fibroblast growth factor receptor-3 is a negative regulator of bone growth. *Cell* 84: 911–921.

Detwiler, S. R. 1918. Experiments on the development of the shoulder girdle and the anterior limb of *Amblystoma punctatum*. *J. Exp. Zool.* 25: 499–538.

Dreyer, S. D. and 7 others. 1998. Mutations in *LMX1B* cause abnormal skeletal patterning and renal dysplasia in nail patella syndrome. *Nature Genet.* 19: 47–50.

Drossopoulou, G., K. E. Lewis, J. J. Sanz-Ezguerro, N. Nikbaklt, A. P. McMahon, C. Hofman and C. Tickle. 2000. A model for anterior-posterior patterning of the vertebrate limb based on sequential long-and short-range Shh signalling and BMP signalling. *Development* 127: 1337–1348.

Dudley, A. and C. J. Tabin. 2002. A re-examination of the progress zone model of proximodistal patterning of the developing limb. *Nature*. In press.

Fallon, J. F. and G. M. Crosby. 1977. Polarizing zone activity in limb buds of amniotes. *In* D. A. Ede, J. R. Hinchliffe and M. Balls (eds.), *Vertebrate Limb and Somite Morphogenesis*. Cambridge University Press, Cambridge, pp. 55–59.

Fromental-Ramain, C., X. Warot, N. Messadecq, M. LeMeur, P. Dollé and P. Chambon. 1996. *Hoxa-13* and *Hoxd-13* play a crucial role in the patterning of the limb autopod. *Development* 122: 2997–3011.

Gañan, Y., D. Macias, R. D. Basco, R. Merino and J. M. Hurle. 1998. Morphological diversity of the avian foot is related with the pattern of *msx* gene expression in the developing autopod. *Dev. Biol.* 196: 33–41.

Geduspan, J. S. and M. Solursh. 1992. A growth-promoting influence from the mesonephros during limb outgrowth. *Dev. Biol.* 151: 212–250.

Gerard, M., D. Duboule and J. C. Zajkany. 1993. Cooperation of regulatory elements in the activation of the *Hoxd11* gene. *Compt. R. Acad. Sci.* III 316: 985–994.

Gibson-Brown, J. J., S. I. Agulnik, D. L. Chapman, M. Alexiou, N. Garvey, L. M. Silver and V. E. Papaioannou. 1996. Evidence of a role for T-box genes in the evolution of limb morphogenesis and the specification of forelimb/hindlimb identity. *Mech. Dev.* 56: 93–101.

Hamburger, V. 1938. Morphogenetic and axial self-differentiation of transplanted limb primordia of 2-day chick embryos. *J. Exp. Zool.* 77: 379–400.

Harrison, R. G. 1918. Experiments on the development of the forelimb of *Ambystoma*, a self-differentiating equipotential system. *J. Exp. Zool.* 25: 413–461.

Hartmann, C. and C. J. Tabin. 2001. Wnt-14 plays a pivotal role in inducing synovial joint formation in the developing appendicular skeleton. *Cell* 104: 341–351.

Hästbacka, J., A. de la Chapelle and M. M. Mahtani. 1994. The diastrophic dysplasia gene encodes a novel sulfate transporter: Positional cloning by fine-structure linkage disequilibrium mapping. *Cell* 78: 1073–1087.

Hayashi, K. and E. Ozawa. 1995. Myogenic cell migration from somites is induced by tissue contact with medial region of the presumptive limb mesoderm in chick embryos. *Development* 121: 661–669.

Heikinheimo, M., A. Lawshe, G. M. Shackleford, D. B. Wilson and C. A. MacArthur. 1994. Fgf-8 expression in the postgastrula mouse suggests roles in the development of the face, limbs, and central nervous system. *Mech. Dev.* 48: 129–138.

Hertwig, O. 1925. Haploidkernige Transplante als Organisatoran diploidkeniger Extremitaten be Triton. *Anat. Anz.* [Suppl.] 60: 112–118.

Hinchliffe, J. R. 1991. Developmental approaches to the problem of transformation of limb structure in evolution. *In* J. R. Hinchliffe (ed.), *Developmental Patterning of the Vertebrate Limb.* Plenum, New York, pp. 313–323.

Hinchliffe, J. R. 1994. Evolutionary biology of the tetrapod limb. *Development* [Suppl.] 163–168.

Hogan, B. L., M. Thaller, C. Thaller and G. Eichele. 1992. Evidence that Hensen's node is a source of retinoic acid synthesis. *Nature* 359: 237–241.

Honig, L. S. and D. Summerbell. 1985. Maps of strength of positional signaling activity in the developing chick wing bud. *J. Embryol. Exp. Morphol.* 87: 163–174.

Hurle, J. M., Y. Gañan and D. Macias. 1989. Experimental analysis of the in vivo chondrogenic potential of the interdigital mesenchyme of the chick limb bud subjected to local ectodermal removal. *Dev. Biol.* 132: 368–374.

Iten, L. E. 1982. Pattern specification and pattern regulation in the embryonic chick limb bud. *Am. Zool.* 22: 117–129.

Izpisúa-Belmonte, J.-C., C. Tickle, P. Dollé, L. Wolpert and D. Duboule. 1991. Expression of the homeobox *Hox-4* genes and the specification of position in chick wing development. *Nature* 350: 585–589.

Kaplan, S. L. and M. M. Grumbach. 1990. Pathophysiology and treatment of sexual precocity. *J. Clin. Endocrinol. Metab.* 71: 785–789.

Kawakami, Y. and 9 others. 1996. BMP signaling during bone pattern determination in the developing limb. *Development* 122: 3557–3566.

Kawakami, Y., J. Capdevila, D. Büscher, T. Itoh, C. Rodriguez Esteban and J. C. Izpisúa Belmonte. 2001. WNT signals control FGF-dependent limb initiation and AER induction in the chick embryo. *Cell* 104: 891–900.

Kieny, M. 1960. Rôle inducteur du mésoderm dans la différenciation précoce du bourgeon de membre chez l'embryon de poulet. *J. Embryol. Exp. Morphol.* 8: 457–467.

Kmita, M. N. Fraudeau, Y. Hérault and D. Duboule. 2002. Serial deletions and duplications suggest a mechanism for the colinearity of *Hoxd* genes in limbs. *Nature* 420: 145–150.

Krabbenhoft, K. M. and J. F. Fallon. 1989. The formation of leg or wing specific structures by leg bud cells grafted to the wing bud is influenced by proximity to the apical ridge. *Dev. Biol.* 131: 373–382.

Lash, J. W. and L. Saxén. 1972. Human teratogenesis: In vitro studies on thalidomide-inhibited chondrogenesis. *Dev. Biol.* 28: 61–70.

Laufer, E., C. E. Nelson, R. L. Johnson, B. A. Morgan and C. Tabin. 1994. Sonic hedgehog and Fgf-4 act through a signaling cascade and feedback loop to integrate growth and patterning of the developing limb bud. *Cell* 79: 993–1003.

Laufer, E. and 7 others. 1997. The *Radical fringe* expression boundary in the limb bud ectoderm regulates AER formation. *Nature* 386: 366–367.

Li, Q. Y. and 16 others. 1996. Holt-Oram syndrome is caused by mutations in *TBX5*, a member of the Brachyury (T) gene family. *Nature Genet.* 15: 21–29.

Li, S. and K. Muneoka. 1999. Cell migration and chick limb development: Chemotactic action of FGF-4 and the AER. *Dev. Biol.* 211: 335–347.

Li, Y. and 16 others. 1995. A fibrillar collagen gene, *Col11a1*, is essential for skeletal morphogenesis. *Cell* 80: 423–430.

Litingtung, Y., R. D. Dahn, Y. Li, J. F. Fallon and C. Chiang. 2002. Shh and Gli3 are dispensable for limb skeleton formation but regulate digit number and identity. *Nature* 418: 979–983.

Logan, M., H.-H. Simon and C. Tabin. 1998. Differential regulation of T-box and homeobox transcription factors suggests roles in controlling chick limb-type identity. *Development* 125: 2825–2835.

López-Martínez, A. and 7 others. 1995. Limb-patterning activity and restricted posterior localization of the amino-terminal product of sonic hedgehog cleavage. *Curr. Biol.* 5: 791–796.

MacCabe, J. A., J. Errick and J. W. Saunders, Jr. 1974. Ectodermal control of dorso-ventral axis in leg bud of chick embryo. *Dev. Biol.* 39: 69–82.

Macias, D., Y. Gañon, T. K. Sampath, M. E. Piedra, M. A. Ros and J. M. Hurle. 1997. Role of BMP2 and OP-1 (BMP7) in programmed cell death and skeletogenesis during chick limb development. *Development* 124: 1109–1117.

Maden, M. 1993. The homeotic transformation of tails into limbs in *Rana temporaria* by retinoids. *Dev. Biol.* 159: 379–391.

Mahmood, R. and 9 others. 1995. A role for FGF-8 in the initiation and maintenance of vertebrate limb outgrowth. *Curr. Biol.* 5: 797–806.

Mercader, N., E. Leonardo, N. Azpiazu, A. Serrano, G. Morata, C. Martinez and M. Torres. 1999. Conserved regulation of proximodistal limb axis development by *Meis1/Hth*. *Nature* 402: 425–429.

Mercader, N., E. Leonardo, M. E. Piedra, C. Martinez, M. A. Ros and M. Torres. 2000. Opposing RA and FGF signals control proximodistal vertebrate limb development through regulation of *Meis* genes. *Development* 127: 3961–3970.

Merimee, T. J., J. Zapf, B. Hewlett and L. L. Cavalli-Sforza. 1987. Insulin-like growth factors in pygmies: The role of puberty in determining final stature. *N. Engl. J. Med.* 316: 906–911.

Merino, R., Y. Gañan, D. Macias, A. N. Economides, K. T. Sampath and J. M. Hurle. 1998. Morphogenesis of digits in the avian limb is controlled by FGFs, TGFβs, and noggin through BMP signaling. *Dev. Biol.* 200: 35–45.

Merino, R, J. Rodriguez-Leon, D. Macias, Y. Gañan, A. N. Economides and J. M. Hurle. 1999. The BMP antagonist Gremlin regulates outgrowth, chondrogenesis and programmed cell death in the developing limb. *Development* 126: 5515–5522.

Mohanty-Hejmadi, P., S. K. Dutta and P. Mahapatra. 1992. Limbs generated at the site of tail amputation in marbled balloon frog after vitamin A treatment. *Nature* 355: 352–353.

Molven, A., C. V. E. Wright, R. Bremiller, E. M. De Robertis and C. B. Kimmel. 1990. Expression of a homeobox gene product in normal and mutant zebrafish embryos: Evolution of the tetrapod body plan. *Development* 109: 279–288.

Morgan, B. A. and C. Tabin. 1994. Hox genes and the evolution of vertebrate axial morphology. *Development* [Suppl.]: 181–186.

Mori, C., N. Nakamura, S. Kimura, H. Irie, T. Takigawa and K. Shiota. 1995. Programmed cell death in the interdigital tissue of the fetal mouse limb is apoptosis with DNA fragmentation. *Anat. Rec.* 242: 103–110.

Mortlock, D. P. and J. W. Innis. 1997. Mutation of *HOXA13* in hand-foot-genital syndrome. *Nature Genet.* 15: 179–181.

Muchmore, W. B. 1957. Differentiation of the trunk mesoderm in *Ambystoma maculatum*. *J. Exp. Zool.* 134: 293–314.

Müller, G., J. Streicher and R. Müller. 1996. Homeotic duplicate of the pelvic body segment in regenerating tadpole tails induced by retinoic acid. *Dev. Genes Evol.* 206: 344–348.

Muneoka, K. and S. V. Bryant. 1982. Evidence that patterning mechanisms in developing and regenerating limbs are the same. *Nature* 298: 369–371.

Muragaki, Y., S. Mundlos, J. Upton and B. Olsen. 1996. Altered growth and branching patterns in synpolydactyly caused by mutations in *HOXD13*. *Science* 272: 548–551.

Nelson, C. E. and 9 others. 1996. Analysis of Hox gene expression in the chick limb bud. *Development* 122: 1449–1466.

Newman, S. A. 1996. Sticky fingers: Hox genes and cell adhesion in vertebrate development of the tetrapod limb. *BioEssays* 18: 171–174.

Nilsson, A., J. Isgaard, A. Lindahl, A. Dahlström, A. Skottner and O. G. P. Isaksson. 1986. Regulation by growth hormone of number of chondrocytes containing IGF-I in rat growth plate. *Science* 233: 571–574.

Niswander, L. 2002. Interplay between the molecular signals that control vertebrate limb development. *Int. J. Dev. Biol.* 46: 877–881.

Niswander, L., S. Jeffrey, G. R. Martin and C. Tickle. 1994. A positive feedback loop coordinates growth and patterning in the vertebrate limb. *Nature* 371: 609–612.

Nohno, T. and 7 others. 1991. Involvement of the *Chox-4* chicken homeobox genes in determination of anteroposterior axial polarity during development of the tetrapod limb. *Cell* 64: 1197–1205.

Ohuchi, H. and S. Noji. 1999. Fibroblast growth factor-induced additional limbs in the study of initiation of limb formation, limb identity, myogenesis, and innervation. *Cell Tissue Res.* 296: 45–56.

Ohuchi, H. and 11 others. 1997. The mesenchymal factor, FGF10, initiates and maintains the outgrowth of the chick limb bud through interaction with FGF8, and apical ectodermal factor. *Development* 124: 2235–2244.

Ohuchi, H. and 7 others. 1998. Correlation of wing-leg identity in ectopic FGF-induced chimeric limbs with the differential expression of chick *Tbx5* and *Tbx4*. *Development* 125: 51–60.

Oliver, G., C. V. E. Wright, J. Hardwicke and E. M. De Robertis. 1988. A gradient of homeodomain protein in developing forelimbs of *Xenopus* and mouse embryos. *Cell* 55: 1017–1024.

Owen, R. 1849. *On the Nature of Limbs*. J. Van Voor, London.

Parr, B. A. and A. P. McMahon. 1995. Dorsalizing signal wnt-7a required for normal polarity of D-V and A-P axes of the mouse limb. *Nature* 374: 350–353.

Parr, B. A., M. J. Shea, G. Vassileva and A. P. McMahon. 1993. Mouse *Wnt* genes exhibit discrete domains of expression in early embryonic CNS and limb buds. *Development* 119: 247–261.

Phillips, J. 1991. Higgledy, piggledy. *N. Engl. J. Med.* 324: 497.

Pizette, S. and L. Niswander. 1999. BMPs negatively regulate structure and function of the limb apical ectodermal ridge. *Development* 126: 883–894.

Pizette S., C. Abate-Shen and L. Niswander. 2001. BMP controls proximodistal outgrowth, via induction of the apical ectodermal ridge, and dorsoventral patterning in the vertebrate limb. *Development* 128: 4463–4474.

Pollak, R. D. and J. F. Fallon. 1976. Autoradiographic analysis of macromolecular synthesis in prospectively necrotic cells of the chick limb bud. II. Nucleic acids. *Exp. Cell Res.* 100: 15–22.

Riddle, R. D., R. L. Johnson, E. Laufer and C. Tabin. 1993. Sonic hedgehog mediates the polarizing activity of the ZPA. *Cell* 75: 1401–1416.

Riddle, R. D., M. Ensini, C. Nelson, T. Tsuchida, T. M. Jessell and C. Tabin. 1995. Induction of the LIM homeobox gene *Lmx1* by *WNT7a* establishes dorsoventral pattern in the vertebrate limb. *Cell* 83: 631–640.

Rodriguez-Esteban, C., J. W. R. Schwabe, J. De La Peña, B. Foys, B. Eshelman and J. C. Izpisúa-Belmonte. 1997. Radical fringe positions the apical ectodermal ridge at the dorsoventral boundary of the vertebrate limb. *Nature* 386: 360–366.

Rodriguez-Esteban, C., T. Tsukui, S. Yonei, J. Magallon, K. Tamura and J. C. Izpisúa-Belmonte. 1999. T-box genes *Tbx4* and *Tbx5* regulate limb outgrowth and identity. *Nature* 398: 814–818.

Rosenquist, G. C. 1971. The origin and movement of the limb-bud epithelium and mesenchyme in the chick embryo as determined by radioautographic mapping. *J. Embryol. Exp. Morphol.* 25: 85–96.

Rowe, D. A. and J. F. Fallon. 1982. The effect of removing posterior apical ectodermal ridge of the chick wing and leg on pattern formation. *J. Embryol. Exp. Morphol.* 65 [Suppl.]: 309–325.

Rowe, D. A., J. M. Cairnes and J. F. Fallon. 1982. Spatial and temporal patterns of cell death in limb bud mesoderm after apical ectodermal ridge removal. *Dev. Biol.* 93: 83–91.

Rubin, L. and J. W. Saunders, Jr. 1972. Ectodermal-mesodermal interactions in the growth of limbs in the chick embryo: Constancy and temporal limits of the ectodermal induction. *Dev. Biol.* 28: 94–112.

Saito, D., S. Yonei-Tamura, K. Kano, H. Ide and K. Tamura. 2002. Specification and determination of limb identity: Evidence for inhibitory regulation of *Tbx* gene expression. *Development* 129: 211–220.

Saunders, J. W., Jr. 1948. The proximal-distal sequence of origin of the parts of the chick wing and the role of the ectoderm. *J. Exp. Zool.* 108: 363–404.

Saunders, J. W., Jr. 1972. Developmental control of three-dimensional polarity in the avian limb. *Ann. NY Acad. Sci. USA* 193: 29–42.

Saunders, J. W., Jr. and J. F. Fallon. 1966. Cell death in morphogenesis. In M. Locke (ed.), *Major Problems of Developmental Biology*. Academic Press, New York, pp. 289–314.

Saunders, J. W., Jr. and M. T. Gasseling. 1968. Ectodermal-mesodermal interactions in the origin of limb symmetry. In R. Fleischmajer and R. E. Billingham (eds.), *Epithelial-Mesenchymal Interactions*. Williams & Wilkins, Baltimore, pp. 78–97.

Saunders, J. W., Jr. and C. Reuss. 1974. Inductive and axial properties of prospective wing-bud mesoderm in the chick embryo. *Dev. Biol.* 38: 41–50.

Saunders, J. W., Jr., J. M. Cairns and M. T. Gasseling. 1957. The role of the apical ridge of ectoderm in the differentiation of the morphological structure and inductive specificity of limb parts of the chick. *J. Morphol.* 101: 57–88.

Saunders, J. W., Jr., M. T. Gasseling and L. C. Saunders. 1962. Cellular death in morphogenesis of the avian wing. *Dev. Biol.* 5: 147–178.

Sekine, K. and 10 others. 1999. Fgf10 is essential for limb and lung formation. *Nature Genet.* 21: 138–141.

Sessions, S. and S. B. Ruth. 1990. Explanation for naturally occurring supernumerary limbs in amphibians. *J. Exp. Zool.* 254: 38–47.

Sessions, S. K., D. M. Gardiner and S. V. Bryant. 1989. Compatible limb patterning mechanisms in urodeles and anurans. *Dev. Biol.* 131: 294–301.

Sessions, S. K., R. A. Franssen and V. C. Horner. 1999. Morphological clues from multilegged frogs: Are retinoids to blame? *Science* 284: 800–802.

Sharpe, R. M. 1997. Do males rely on female hormones? *Nature* 390: 447–448.

Shubin, N. H. and P. Alberch. 1986. A morphogenetic approach to the origin and basic organization of the tetrapod limb. *Evol. Biol.* 20: 319–387.

Shubin, N., C. Tabin and S. Carroll. 1997. Fossils, genes, and the evolution of animal limbs. *Nature* 388: 639–648.

Smith, D. M., R. D. Torres and T. D. Stephens. 1996. Mesonephros has a role in limb development and is related to thalidomide embryopathy. *Teratology* 54: 126–134.

Smith, E. P. and 8 others. 1994. Estrogen resistance caused by a mutation in the estrogen-receptor gene in a man. *N. Engl. J. Med.* 331: 1056–1061.

Sordino, P., F. van der Hoeven and D. Duboule. 1995. Hox gene expression in teleost fins and the origin of the vertebrate digits. *Nature* 375: 678–681.

Stephens, T. D. and T. R. McNulty. 1981. Evidence for a metameric pattern in the development of the chick humerus. *J. Embryol. Exp. Morphol.* 61: 191–205

Stocum, D. L. and J. F. Fallon. 1982. Control of pattern formation in urodele limb ontogeny: A review and hypothesis. *J. Embryol. Exp. Morphol.* 69: 7–36.

Storm, E. E. and D. M. Kingsley. 1999. GDF5 coordinates bone and joint formation during digit development. *Dev. Biol.* 209: 11–27.

Stratford, T., C. Horton and M. Maden. 1996. Retinoic acid is required for the initiation of outgrowth in the chick limb bud. *Curr Biol.* 6: 1124–1133.

Summerbell, D. 1974. A quantitative analysis of the effect of excision of the AER from the chick limb bud. *J. Embryol. Exp. Morphol.* 32: 651–660.

Summerbell, D. and J. H. Lewis. 1975. Time, place and positional value in the chick limb bud. *J. Embryol. Exp. Morphol.* 33: 621–643.

Sun, X., F. V. Mariani and G. M. Martin. 2002. Functions of FGF signaling from the apical ectodermal ridge in limb development. *Nature.* In press.

Takeuchi, J. K. and 8 others. 1999. *Tbx5* and *Tbx4* genes determine the wing/leg identity of limb buds. *Nature* 398: 810–814.

Tanaka, M., K. Tamura, S. Noji, T. Nohno and H. Ide. 1997. Induction of additional limb at the dorsal-ventral boundary of a chick embryo. *Dev. Biol.* 182: 191–203.

Tavormina, P. L. and 9 others. 1995. Thanatophoric dysplasia (types I and II) caused by distinct mutations in fibroblast growth factor receptor 3. *Nature Genet.* 9: 321–328.

te Welscher, P., A. Zuniga, S. Kuijper, T. Drenth, H. J. Goedemans, F. Meijlink and R. Zeller. 2002. Progression of vertebrate limb development through SHH-mediated counteraction of GLI3. *Science* 298: 827–830.

Tickle, C., D. Summerbell and L. Wolpert. 1975. Positional signaling and specification of digits in chick limb morphogenesis. *Nature* 254: 199–202.

Todt, W. L. and J. F. Fallon. 1987. Posterior apical ectodermal ridge removal in chick wing bud triggers a series of events resulting in defective anterior pattern. *Development* 101: 505–515.

Tufan, A. C. and R. S. Tuan. 2001. Wnt regulation of limb mesenchymal chondrogenesis is accompanied by altered N-cadherin-related functions. *FASEB J.* 15: 1436–1438.

van der Hoeven, F., J. Zakay and D. Duboule. 1996. Gene transpositions in the HoxD complex reveal a hierarchy of regulatory controls. *Cell* 85: 1025–1035.

Vogel, A., C. Rodriguez, W. Warnken and J.-C. Izpisúa-Belmonte. 1995. Dorsal cell fate specified by chick *Lmx1* during vertebrate development of the tetrapod limb. *Nature* 378: 716–720.

Vogel, A., C. Rodriguez and J.-C. Izpisúa-Belmonte. 1996. Involvement of FGF-8 in initiation, outgrowth, and patterning of the vertebrate limb. *Development* 122: 1737–1750.

Webster, M. K. and D. J. Donoghue. 1996. Constitutive activation of fibroblast growth factor receptor 3 by the transmembrane domain point mutation found in achondroplasia. *EMBO J.* 15: 520–527.

Wessells, N. K. 1977. *Tissue Interaction and Development.* Benjamin Cummings, Menlo Park, CA.

Wolpert, L., C. Tickle and M. Sampford. 1979. The effect of cell killing by x-irradiation on pattern formation in the chick limb. *J. Embryol. Exp. Morphol.* 50: 175–193.

Xu, X. L. and 7 others. 1998. Fibroblast growth factor receptor 2 (FGFR2)-mediated reciprocal regulatory loop between FGF8 and FGF10 is essential for limb induction. *Development* 125: 753–765.

Yang, Y. Z. and L. Niswander. 1995. Interaction between signaling molecules Wnt7a and Shh during vertebrate limb development: Dorsal signals regulate anteroposterior patterning. *Cell* 80: 939–947.

Yang, Y. Z. and 10 others. 1997. Relationship between dose, distance, and time in Sonic hedgehog-mediated regulation of anteroposterior polarity in the chick limb. *Development* 124: 4393–4404.

Yin, M. and M. Pacifici. 2001. Vascular regression is required for mesenchymal condensation and chondrogenesis in the developing limb. *Dev. Dynam.* 222: 522–533.

Yokouchi, Y., S. Nakazato, M. Yamamoto, Y. Goto, T. Kameda, H. Iba and A. Kuroiwa. 1995. Misexpression of *Hoxa-13* induces cartilage homeotic transformation and changes in adhesiveness in chick limb buds. *Genes Dev.* 9: 2509–2522.

Yokouchi, Y., J. Sakiyama, T. Kameda, H. Iba, A. Suzuki, N. Ueno and A. Kuroiwa. 1996. BMP-2/-4 mediate programmed cell death in chicken limb buds. *Development* 122: 3725–3734.

Yonei-Tamura, S., T. Endo, H. Yajima, H. Ohuichi, H. Ida and K. Tamura. 1999. FGF7 and FGF10 directly induce the apical ectodermal ridge in chick embryos. *Dev. Biol.* 211: 133–143.

Zaleske, D. J. 1985. Development of the upper limb. *Hand Clin.* 1985(3): 383–390.

Zou, H. and L. Niswander. 1996. Requirement for BMP signaling in interdigital apoptosis and scale formation. *Science* 272: 738–741.

Zúñiga, A., A. P. Haramis, A. P. McMahon and R. Zeller. 1999. Signal relay by BMP antagonism controls the SHH/FGF4 feedback loop in vertebrate limb buds. *Nature* 401: 598–602.

Zwilling, E. 1955. Ectoderm-mesoderm relationship in the development of the chick embryo limb bud. *J. Exp. Zool.* 128: 423–441.

chapter *17* *Sex determination*

Sexual reproduction is ... the masterpiece of nature.

ERASMUS DARWIN (1791)

It is quaint to notice that the number of speculations connected with the nature of sex have well-nigh doubled since Drelincourt, in the eighteenth century, brought together two hundred and sixty-two "groundless hypotheses," and since Blumenbach caustically remarked that nothing was more certain than that Drelincourt's own theory formed the two hundred and sixty-third.

J. A. THOMSON (1926)

HOW AN INDIVIDUAL'S SEX IS DETERMINED has been one of the great questions of embryology since antiquity. Aristotle, who collected and dissected embryos, claimed that sex was determined by the heat of the male partner during intercourse (Aristotle, ca. 335 B.C.E.). The more heated the passion, the greater the probability of male offspring; he thus counseled elderly men to conceive in the summer if they wished to have male heirs.

Aristotle promulgated a very straightforward hypothesis of sex determination: women were men whose development was arrested too early. The female was "a mutilated male" whose development had stopped because the coldness of the mother's womb overcame the heat of the father's semen. Women were therefore colder and more passive than men, and female sex organs had not matured to the point at which they could provide active seeds. This view was accepted both by the Christian Church and by the Graeco-Roman physician Galen* (whose anatomy texts were the standard for over a thousand years). Around the year 190 C.E., Galen wrote:

> *Just as mankind is the most perfect of all animals, so within mankind, the man is more perfect than the woman, and the reason for this perfection is his excess heat, for heat is Nature's primary instrument ... the woman is less perfect than the man in respect to the generative parts. For the parts were formed within her when she was still a fetus, but could not because of the defect in heat emerge and project on the outside.*

The view that women were but poorly developed men and that their genitalia were like men's, only turned inside out, remained very popular for over a thousand years. As late as 1543, Andreas Vesalius, the Paduan anatomist who overturned much of Galen's anatomy (and who risked censure by the church for arguing that men and women have the same number of ribs), held this view. The illustrations from his two major works, *De Humani Corporis Fabrica* and *Tabulae Sex*, show that he saw the female genitalia as internal representations of the male genitalia (Figure 17.1). Nevertheless, Vesalius' books sparked a revolution in anatomy, and by the end of the 1500s, anatomists had dismissed Galen's representation of female anatomy. During the 1600s and 1700s, females were seen as producing eggs that could transmit parental traits, and the physiology of the sex organs began to be studied. Still, there was no

---

*A Greek by birth, Galen first achieved fame as a physician to gladiators, from whose corpses he undoubtedly learned much of his anatomy. He traveled to Rome, where he became personal physician to the Emperor Marcus Aurelius and to his son and successor, Commodus.

(A)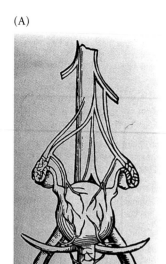

(B)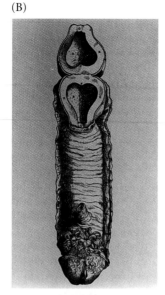

Figure 17.1
Vesalius' representations (1538, 1543) of the female reproductive organs. (A) Rendering of Galen's conception of the female reproductive tract, from the vagina to the uterus. (B) Rendering of the female reproductive system. (Reprinted in Schiebinger 1989.)

consensus about how the sexes became determined (see Horowitz 1976; Tuana 1988; Schiebinger 1989).

> WEBSITE 17.1 **Social critique of sex determination research.** In numerous cultures, women have been seen as the "default state" of men. Historians and biologists have shown that until recently such biases characterized the scientific study of human sex determination.

Until the twentieth century, the environment—temperature and nutrition, in particular—was believed to be important in determining sex. In 1890, Geddes and Thomson summarized all available data on sex determination and came to the conclusion that the "constitution, age, nutrition, and environment of the parents must be especially considered" in any such analysis. They argued that factors favoring the storage of energy and nutrients predisposed one to have female offspring, whereas factors favoring the utilization of energy and nutrients influenced one to have male offspring.

This environmental view of sex determination remained the only major scientific theory until the rediscovery of Mendel's work in 1900 and the rediscovery of the sex chromosome by McClung in 1902. It was not until 1905, however, that the correlation (in insects) of the female sex with XX sex chromosomes and the male sex with XY or XO chromosomes was established (Stevens 1905; Wilson 1905). This finding suggested strongly that a specific nuclear component was responsible for directing the development of the sexual phenotype. Thus, evidence accumulated that sex determination occurs by nuclear inheritance rather than by environmental happenstance.

Today we know that both environmental and internal mechanisms of sex determination can operate in different species. We will first discuss the chromosomal mechanisms of sex determination and then consider the ways in which the environment regulates the sexual phenotype.

CHROMOSOMAL SEX DETERMINATION IN MAMMALS

There are many ways chromosomes can determine the sex of an embryo. In mammals, the presence of either a second X chromosome or a Y chromosome determines whether the embryo is to be female (XX) or male (XY). In birds, the situation is reversed (Smith and Sinclair 2001): the male has the two similar sex chromosomes (ZZ), while the female has the unmatched pair (ZW). In flies, the Y chromosome plays no role in sex determination, but the ratio of X chromosomes to autosomes (the non-sex chromosomes) determines the sexual phenotype. In other insects (especially hymenopterans such as bees, wasps, and ants), fertilized, diploid eggs develop into females, while the unfertilized, haploid eggs become male (Beukeboom 1995). This chapter will discuss only two of the many chromosomal modes of sex determination: sex determination in placental mammals and sex determination in *Drosophila*.

Primary and Secondary Sex Determination in Mammals

Primary sex determination is the determination of the *gonads*. In mammals, primary sex determination is strictly chromosomal and is not usually influenced by the environment. In most cases, the female is XX and the male is XY. Every individual must have at least one X chromosome. Since the female is XX, each of her eggs has a single X chromosome. The male, being XY, can generate two types of sperm: half will bear a single X chromosome, half a Y. If the egg receives another X chromosome from the sperm, the resulting individual is XX, forms ovaries, and is female; if the egg receives a Y chromosome from the sperm, the individual is XY, forms testes, and is male.

The Y chromosome carries a gene that encodes a **testis-determining factor**. This factor organizes the gonad into a testis rather than an ovary. Unlike the situation in *Drosophila* (discussed later), the mammalian Y chromosome is a crucial factor for determining sex in mammals. A person with five X chromosomes and one Y chromosome (XXXXXY) would be male. Furthermore, an individual with a single X chromosome and no second X or Y (i.e., XO) develops as a female and begins making ovaries (although the ovarian follicles cannot be maintained; for a complete ovary, a second X chromosome is needed).

The formation of ovaries and of testes are both active, gene-directed processes. There is no "default state" in mammalian primary sex determination. Moreover, as we shall see, both diverge from a common precursor, the bipotential gonad.

Secondary sex determination affects the phenotype outside the gonads. This includes the male or female ductal systems and external genitalia. A male mammal has a penis, seminal vesicles, and prostate gland. A female mammal has a vagina, cervix, uterus, oviducts, and mammary glands. In many species, each sex has a sex-specific body size, vocal cartilage, and musculature. These secondary sex characteristics are usually determined by hormones secreted from the gonads. In the absence of gonads, however, the female phenotype is generated. When Jost (1953) removed fetal rabbit gonads before they had differentiated, the resulting rabbits had a female phenotype, regardless of whether they were XX or XY. They each had oviducts, a uterus, and a vagina, and each lacked a penis and male accessory structures.

The general scheme of mammalian sex determination is shown in Figure 17.2. If the Y chromosome is absent, the gonadal primordia develop into ovaries. The ovaries produce **estrogen**, a hormone that enables the development of the **Müllerian duct** into the uterus, oviducts, and upper end of the vagina (Fisher et al. 1998; Couse et al. 1999; Couse and Korach 2001). If the Y chromosome is present, testes form and secrete two major hormones. The first hormone—**anti-Müllerian duct hormone** (AMH; also referred to as Müllerian-inhibiting substance, MIS)—destroys the Müllerian duct. The second hormone—**testosterone**—masculinizes the fetus,

stimulating the formation of the penis, male ductal system, scrotum, and other portions of the male anatomy, as well as inhibiting the development of the breast primordia. Thus, the body has the female phenotype unless it is changed by the two hormones secreted by the fetal testes. We will now take a more detailed look at these events.

The Developing Gonads

The gonads embody a unique embryological situation. All other organ rudiments can normally differentiate into only one type of organ. A lung rudiment can become only a lung, and a liver rudiment can develop only into a liver. The gonadal rudiment, however, has two options when it differentiates: it can develop into either an ovary or a testis. The path of differentiation taken by this rudiment determines the future sexual development of the organism. But before this decision is made, the mammalian gonad first develops through a **bipotential (indifferent) stage**, during which time it has neither female nor male characteristics.

In humans, the gonadal rudiments appear in the intermediate mesoderm during week 4 and remain sexually indifferent until week 7. The gonadal rudiments are paired regions of the intermediate mesoderm; they form adjacent to the developing kidneys. The ventral portions of the gonadal rudiments are composed of the genital ridge epithelium. During the indifferent stage, the genital ridge epithelium proliferates and extends

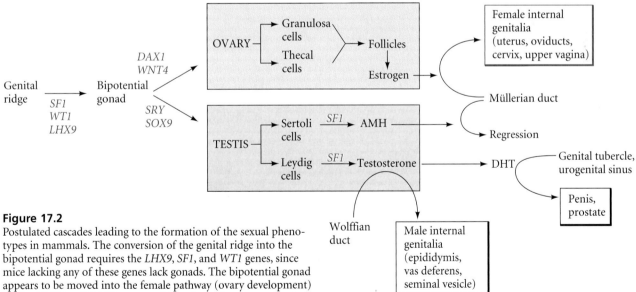

Figure 17.2
Postulated cascades leading to the formation of the sexual phenotypes in mammals. The conversion of the genital ridge into the bipotential gonad requires the *LHX9*, *SF1*, and *WT1* genes, since mice lacking any of these genes lack gonads. The bipotential gonad appears to be moved into the female pathway (ovary development) by the *WNT4* and *DAX1* genes and into the male pathway (testis development) by the *SRY* gene (on the Y chromosome) in conjunction with autosomal genes such as *SOX9*. The ovary makes thecal cells and granulosa cells, which together are capable of synthesizing estrogen. Under the influence of estrogen (first from the mother, then from the fetal gonads), the Müllerian duct differentiates into the female reproductive tract, and the offspring develops the secondary sex characteristics of a female. The testis makes two major hor-

mones. The first, anti-Müllerian duct hormone (AMH), causes the Müllerian duct to regress. The second, testosterone, causes the differentiation of the Wolffian duct into the male internal genitalia. In the urogenital region, testosterone is converted into dihydrotestosterone (DHT), and this hormone causes the morphogenesis of the penis and prostate gland. (After Marx 1995 and Birk et al. 2000.)

into the loose connective mesenchymal tissue above it (Figure 17.3A,B). These epithelial layers form the **sex cords**. The germ cells migrate into the gonad during week 6, and are sur-rounded by the sex cords. In both XY and XX gonads, the sex cords remain connected to the surface epithelium.

If the fetus is XY, the sex cords continue to proliferate through the eighth week, extending deeply into the connective tissue. These cords form a network of internal (medullary) sex cords and, at the most distal end, the thinner **rete testis** (Figure 17.3C, D). Eventually, the sex cords—now called **testis cords**—lose

INDIFFERENT GONADS

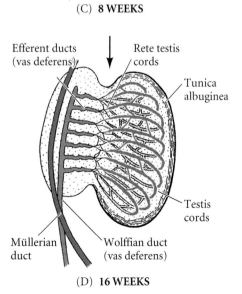

TESTIS DEVELOPMENT

Figure 17.3
Differentiation of human gonads shown in transverse section. (A) Genital ridge of a 4-week embryo. (B) Genital ridge of a 6-week indifferent gonad showing primitive sex cords. (C) Testis development in the eighth week. The sex cords lose contact with the cortical epithelium and develop the rete testis. (D) By the sixteenth week of development, the testis cords are continuous with the rete testis and connect with the Wolffian duct. (E) Ovary development in an 8-week human embryo, as the primitive sex cords degenerate. (F) The 20-week human ovary does not connect to the Wolffian duct, and new cortical sex cords surround the germ cells that have migrated into the genital ridge. (After Langman 1981.)

OVARIAN DEVELOPMENT

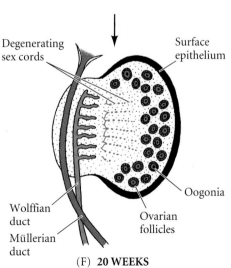

contact with the surface epithelium and become separated from it by a thick extracellular matrix, the **tunica albuginea**. Thus, when the germ cells enter the male gonads, they will develop in these testis cords, inside the organ. An inhibitor of meiosis appears to be made by the male cord cells and the germ cells will not initiate meiosis until puberty (McLaren and Southee 1997).

When the presumptive germ cells are inhibited from entering meiosis, they start developing as male germ cells, and they secrete a factor (probably a prostaglandin) that causes the testis cord cells to differentiate into **Sertoli cells** (Adams and McLaren 2002). The Sertoli cells of the fetal seminiferous tubules secrete anti-Müllerian hormone and will later support the development of sperm. At puberty, however, the cords hollow out to form the **seminiferous tubules** and the germ cells migrate to the periphery of these tubules, where they begin to differentiate into sperm (see Figure 19.19). In the mature seminiferous tubule, the sperm are transported from the inside of the testis through the rete testis, which joins the **efferent ducts**. These efferent tubules are the remnants of the mesonephric kidney, and they link the testis to the **Wolffian duct**, which used to be the collecting tube of the mesonephric kidney* (see Chapter 15). In males, the Wolffian duct differentiates to become the **epididymis** (adjacent to the testis) and the **vas deferens**, the tube through which the sperm pass into the urethra and out of the body. Meanwhile, during fetal development, the interstitial mesenchyme cells of the testes differentiate into **Leydig cells**, which make testosterone.

In females, the germ cells will reside near the outer surface of the gonad. Unlike the sex cords in males, which continue their proliferation, the initial sex cords of XX gonads degenerate. However, the epithelium soon produces a new set of sex cords, which do not penetrate deeply into the mesenchyme, but stay near the outer surface (cortex) of the organ. Thus, they are called **cortical sex cords**. These cords are split into clusters, with each cluster surrounding a germ cell (Figure 17.3E, F). The germ cells will become the ova, and the surrounding cortical sex cords will differentiate into the **granulosa cells**. The mesenchyme cells of the ovary differentiate into the **thecal cells**. Together, the thecal and granulosa cells will form the **follicles** that envelop the germ cells and secrete steroid hormones. Each follicle will contain a single germ cell, which will enter meiosis at this time. These germ cells are required for the gonadal cells to complete their differentiation into ovarian tissue[†] (McLaren 1991). In females, the Müllerian

duct remains intact, and it differentiates into the oviducts, uterus, cervix, and upper vagina. The Wolffian duct, deprived of testosterone, degenerates. A summary of the development of mammalian reproductive systems is shown in Figure 17.4.

The Mechanisms of Mammalian Primary Sex Determination

Several genes have been found whose function is necessary for normal sexual differentiation. Since the phenotype of mutations in sex-determining genes is often sterility, clinical studies have been used to identify those genes that are active in determining whether humans become male or female. Experimental manipulations to confirm the functions of these genes can be done in mice.

SRY: The Y chromosome sex determinant

In humans, the major gene for testis determination resides on the short arm of the Y chromosome. Individuals who are born with the short but not the long arm of the Y chromosome are male, while individuals born with the long arm of the Y chromosome but not the short arm are female. By analyzing the DNA of rare XX men and XY women, the position of the testis-determining gene was narrowed down to a 35,000-base-pair region of the Y chromosome located near the tip of the short arm. In this region, Sinclair and colleagues (1990) found a male-specific DNA sequence that could encode a peptide of 223 amino acids. This peptide is probably a transcription factor, since it contains a DNA-binding domain called the **HMG** (high-mobility group) **box**. This domain is found in several transcription factors and nonhistone chromatin proteins, and it induces bending in the region of DNA to which it binds (Giese et al. 1992). This gene is called *SRY* (sex-determining *region* of the *Y* chromosome), and there is extensive evidence that it is indeed the gene that encodes the human testis-determining factor. *SRY* is found in normal XY males and in the rare XX males, and it is absent from normal XX females and from many XY females. Another group of XY females was found to have point or frameshift mutations in the *SRY* gene; these mutations prevented the SRY protein from binding to or bending DNA (Pontiggia et al. 1994; Werner et al. 1995).

If the *SRY* gene actually does encode the major testis-determining factor, one would expect that it would act in the genital ridge immediately before or during testis differentiation. This prediction has been met in studies of the homologous gene found in mice. The mouse gene (*Sry*) also correlates with the presence of testes; it is present in XX males and absent in XY females (Gubbay et al. 1990; Koopman et al. 1990). The *Sry* gene is expressed in the somatic cells of the bipotential mouse gonad immediately before or during its differentiation into a testis; its expression then disappears (Hacker et al. 1995).

The most impressive evidence for *Sry* being the gene for testis-determining factor comes from transgenic mice. If *Sry* in-

---

*As discussed in Chapter 14, the mesonephric kidney is one of the three kidney types seen during mammalian development, but it does not function as a kidney in most mammals.

[†]There is a reciprocal relationship between the germ cells and the gonadal somatic cells. The germ cells are originally bipotential and can become either sperm or eggs. Once in the male or female sex cords, however, they are told either to remain in meiosis and become eggs or to cease meiosis and become sperm. These germ cells then further the differentiation of the gonad in either the male or female direction. The developing gonadal cells, moreover, then guide the differentiation of the germ cells. Thus, if XY germ cells are placed in XX gonads, they give rise to oocytes (Burgoyne et al. 1988).

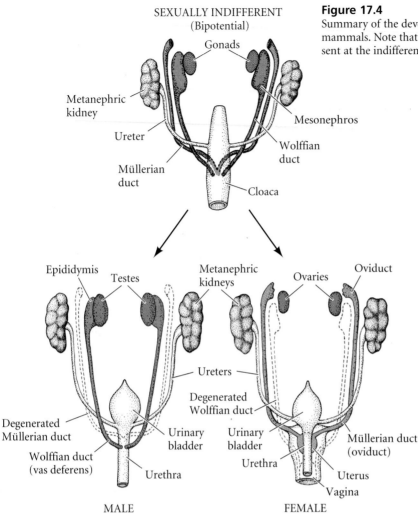

SEXUALLY INDIFFERENT
(Bipotential)

Gonads

Metanephric
kidney

Mesonephros

Ureter

Wolffian
duct

Müllerian
duct

Cloaca

Epididymis   Testes

Metanephric
kidneys

Ovaries   Oviduct

Ureters

Degenerated
Wolffian duct

Degenerated
Müllerian duct

Urinary
bladder

Urinary
bladder

Müllerian duct
(oviduct)

Wolffian duct
(vas deferens)

Urinary
bladder

Urethra

Urethra

Uterus

Urethra

Vagina

MALE   FEMALE

Figure 17.4
Summary of the development of the gonads and their ducts in mammals. Note that both the Wolffian and Müllerian ducts are present at the indifferent gonad stage.

GONADS		
Gonadal type	Testis	Ovary
Sex cords	Medullary (internal)	Cortical (external)
DUCTS		
Remaining duct for germ cells	Wolffian	Müllerian
Duct differentiation	Vas deferens, epididymis, seminal vesicle	Oviduct, uterus, cervix, upper portion of vagina

because the presence of two X chromosomes prevents sperm formation in XXY mice and men, and the transgenic mice lacked the rest of the Y chromosome, which contains genes needed for spermatogenesis.) Therefore, there are good reasons to think that *Sry/SRY* is the major gene on the Y chromosome for testis determination in mammals.

Sry/SRY is necessary but not sufficient for the development of the mammalian testis. Studies on mice (Eicher and Washburn 1983; Washburn and Eicher 1989; Eicher et al. 1996) have shown that the *Sry* gene of some strains of mice failed to produce testes when placed in a different strain of mouse. There has to be some interaction between Sry and some other factor(s). However, the cofactors and targets of the *Sry* gene are not yet known. It had been thought that Sry protein would bind to DNA, and that it would bend the DNA to bring distantly bound proteins of the transcription apparatus into close contact, enabling these proteins to interact and initiate transcription (Pontiggia et al. 1994; Werner et al. 1995). But no such proteins have been found. Recent evidence suggests that Sry might be acting in splicosomes at the level of RNA processing. Ohe and colleagues (2002) showed that human SRY protein colocalizes with splicosomes and that it functions in processing pre-mRNA (Figure 17.6). Such a speckled localization pattern (characteristic of splicing factors) has been seen in Sertoli cells (Poulat et al. 1995).

duces testis formation, then inserting *Sry* DNA into the genome of a normal XX mouse zygote should cause that XX mouse to form testes. Koopman and colleagues (1991) took the 14-kilobase region of DNA that includes the *Sry* gene (and presumably its regulatory elements) and microinjected this sequence into the pronuclei of newly fertilized mouse zygotes. In several instances, XX embryos injected with this sequence developed testes, male accessory organs, and penises (Figure 17.5). (Functional sperm were not formed—but they were not expected to, be-

WEBSITE 17.2   **Finding the male-determining genes.**
The mapping of the testis-determining factor to the SRY region took scientists more than 50 years to accomplish. Moreover, other testis-forming genes have been found on the autosomes

SOX9: An autosomal testis-determining gene

SRY may have more than one mode of action in converting the bipotential gonads into testes. It had been assumed for the

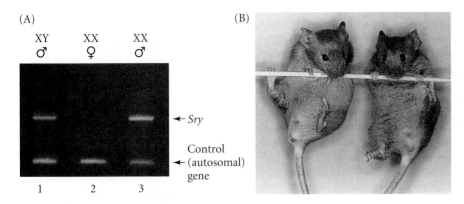

Figure 17.5
An XX mouse transgenic for *Sry* is male. (A) Polymerase chain reaction followed by electrophoresis shows the presence of the *Sry* gene in normal XY males and in a transgenic XX/*Sry* mouse. The gene is absent in a female XX littermate. (B) The external genitalia of the transgenic mouse are male (right) and are essentially the same as those in an XY male (left). (From Koopman et al. 1991; photographs courtesy of the authors.)

past decade that SRY worked directly in the genital ridge to convert the epithelium into male-specific Sertoli cells. However, no genes have been found whose expression is activated by the binding of Sry to their promoters, enhancers, or to the splicing regions of their transcripts. The most promising candidate for this role is another HMG-box protein, **SOX9**. *SOX9* is an autosomal gene that can also induce testis formation. XX humans who have an extra copy of *SOX9* develop as males, even if they have no *SRY* gene, and XX mice made transgenic

for *Sox9* develop testes (Figure 17.7; Huang et al. 1999; Vidal et al. 2001). Individuals having only one functional copy of this gene have a syndrome called campomelic dysplasia, a disease involving numerous skeletal and organ systems. About 75% of XY patients with this syndrome develop as phenotypic females or hermaphrodites (Foster et al. 1994; Wagner et al. 1994; Mansour et al. 1995). Therefore, it appears that *SOX9* can replace *SRY* in testis formation. This is not altogether surprising: While *Sry* is found specifically in mammals, *Sox9* is found throughout the vertebrates. *Sox9* may be the older and more central sex determination gene, and in mammals it may be activated by its relative, *Sry*. Thus, *Sry* may act merely as a "switch" to activate *Sox9*, and the Sox9 protein may initiate the conserved evolutionary pathway to testis formation (Pask and Graves 1999).

The Sox9 protein may act as both a splicing factor and a transcriptional regulator. Sox9 migrates into the nucleus at the time of sex determination. Here, it binds to a promoter site on the gene for the anti-Müllerian hormone, providing a critical link in the pathway toward a male phenotype (Arango et al. 1999; de Santa Barabara et al 2000). SOX9 also appears to be involved in RNA splicing and can replace missing splicing factors in experimental splicing assays (Ohe et al. 2002). The search is on for the genes that Sry or Sox9 might regulate (Koopman 2001).

Some major questions concerning male sex determination remain to be answered, including: (1) What activates the *Sry* gene? (2) What does the Sry protein activate? (3) How might this activation work? (4) Is the *Sox9* gene activated by Sry? and (5) What genes does Sox9 activate? As we will see, sex determination in *Drosophila* works largely via RNA processing; so if Sry and Sox9 act as RNA splicing factors, the mammalian and *Drosophila* schemes of sex determination may have more in common than originally thought.

Fibroblast growth factor 9

One of the genes regulated by Sry or Sox9 may encode a fibroblast growth factor. Recent studies (Capel et al. 1999) have suggested that Sry (or Sox9) works via an indirect mechanism: Sry in the genital ridge cells induces those cells to secrete a chemotactic factor that permits the migration of mesonephric cells into the XY gonad. These mesonephric cells then induce the gonadal epithelium to become Sertoli cells with male-specific gene expression patterns. The researchers found that when they cultured XX gonads with either XX or XY mesonephros, the mesonephric cells did not enter the gonads. However, when they cultured XX or XY mesonephrons with

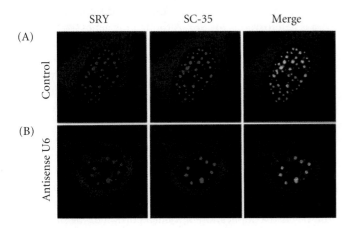

Figure 17.6
Co-localization of SRY and RNA processing factor SC-35. (A) Antibodies to SRY (red) and to RNA processing factor SC-35 (green) were incubated on male germ cell tumor cells that expressed SRY. The overlap between the green and red antibodies (yellow) demonstrates the co-localization of these proteins. (B) When the cells were injected with antisense oligonucleotides directed at a spliceosome small nuclear RNA, the spliceosomes aggregate together into larger clusters. The SRY protein co-localized with these new aggregations. (From Ohe et al. 2002.)

(A)   (B)   (C)   (D)

XY   XY   XX   XX
Wild-type   *Sox9* transgenic   Wild-type   *Sox9* transgenic

Newborn
gonadal
morphology

E11.5
Sox9

E16.5
AMH

Adult
gonad
histology

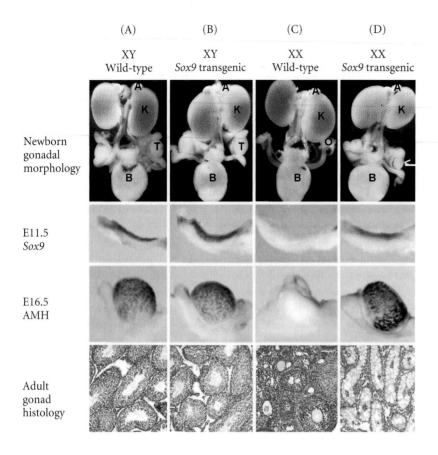

Figure 17.7
Ability of *Sox9* to generate testes. (A) A wild-type XY mouse embryo expresses *Sox9* in the genital ridge at 11.5 days, anti-Müllerian duct hormone (AMH) in the embryonic gonad Sertoli cells at 16.5 days, and eventually forms descended testes (T) with seminiferous tubules. (B) An XY embryo with the Sox9 transgene (a control for the effects of the transgene) also shows *Sox9* expression, AMH expression, and descended testes with seminiferous tubules. (C) The wild-type XX embryo shows neither *Sox9* expression nor AMH. It constructs ovaries with mature follicle cells. (D) An XX embryo with the *Sox9* transgene expresses the *Sox9* gene and has AMH in its 16.5-day Sertoli cells. It has descended testes, but the seminiferous tubules lack sperm (due to the presence of two X chromosomes in the Sertoli cells). K, kidneys; A, adrenal glands; B, bladder; T, testis; O, ovary. (From Vidal et al. 2001; photographs courtesy of A. Schedl.)

XY gonads, or with gonads from XX mice containing the *Sry* transgene, the mesonephric cells did enter the gonads (Figure 17.8A,B). There were strict correlations between the presence of Sry in the gonadal cells, mesonephric cell migration, and the formation of testis cords. Tilmann and Capel (1999) showed that mesonephric cells are critical for testis cord formation and that the migrating mesonephric cells can induce XX gonadal cells to form testis cords. It appears, then, that Sry may function, at least in part, indirectly to create testes by inducing mesonephric cell migration into the gonad.

A critical testis factor inducing the migration of mesonephric cells into the gonad is fibroblast growth factor 9 (Colvin et al. 2001). When the *Fgf9* gene is knocked out in mice, the homozygous mutants are almost all female. Analyses of these mice indicate that Fgf9 is important in stimulating the Leydig (testosterone-producing) cells to divide, promoting the differentiation of Sertoli cells and bringing the mesonephric cells into the gonad. Incubating XX gonads in Fgf9 allowed then to attract mesonephric cells (Figure 17.8).

SF1: The link between SRY and the male developmental pathways

Another protein that may be directly or indirectly activated by Sry is the transcription factor **Sf1** (steroidogenic factor 1). Sf1 is necessary to make the bipotential gonad, but while Sf1 lev-

els decline in the genital ridge of XX mouse embryos, the *Sf1* gene stays on in the developing testis. Sf1 appears to be active in masculinizing both the Leydig cells and the Sertoli cells. In the Sertoli cells, Sf1, working in collaboration with Sox9, is needed to elevate the levels of AMH transcription (Shen et al. 1994; Arango et al. 1999). In the Leydig cells, Sf1 activates the genes encoding the enzymes that make testosterone. In humans, the importance of SF1 for testis development and AMH regulation is demonstrated by an XY patient who is heterozygous for *SF1*. Although the genes for SRY and SOX9 are normal, this individual has malformed fibrous gonads and retains fully developed Müllerian duct structures (Achermann et al. 1999). It is thought that SRY (directly or indirectly) maintains *SF1* expression, and that the SF1 protein is then involved in the activation of AMH in the Sertoli cells and testosterone in the Leydig cells.

DAX1: A potential testis-suppressing gene on the X chromosome

In 1980, Bernstein and her colleagues described two sisters who were genetically XY. Their Y chromosomes were normal, but they had a duplication of a small portion of the short arm of the X chromosome. When similar cases were found, it was concluded that if there were two copies of this region on the active X chromosome, the *SRY* signal would be inhibited (Figure 17.9). Bardoni and her colleagues (1994) proposed that this region contains a gene encoding a protein that competes with the *SRY* factor and that is important in directing the development of the ovary. In XY embryos, this gene would be suppressed, but having two active copies of the gene would

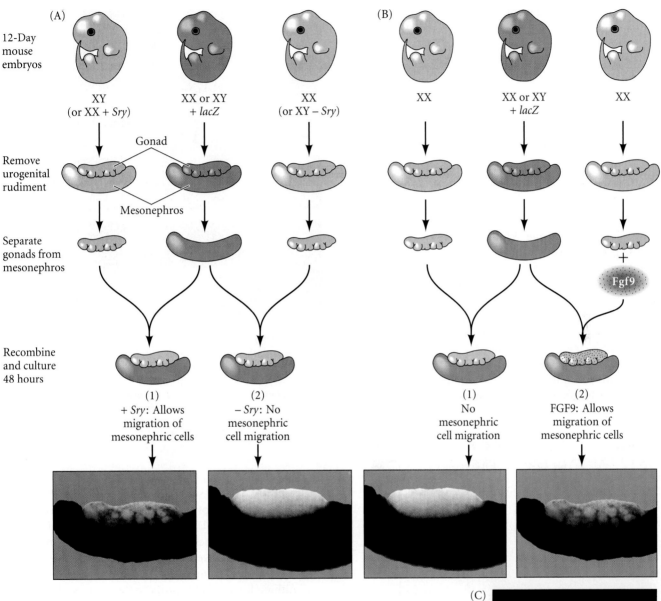

Figure 17.8

Migration of the mesonephric cells into *Sry*⁺ gonadal rudiments. In the experiment diagrammed, urogenital ridges (containing both the mesonephric kidneys and bipotential gonadal rudiments) were collected from 12-day embryonic mice. Some of the mice were marked with a β-galactosidase transgene (*lacZ*) that is active in every cell. Thus every cell of these mice turned blue when stained for β-galactosidase. The gonad and mesonephros were separated and recombined, using gonadal tissue from unlabeled mice and mesonephros from labeled mice. (A) Migration of mesonephric cells into the gonad was seen (1) when the gonadal cells were XY or (2) when they were XX with an *Sry* transgene. No migration of mesonephric tissue into the gonad was seen when the gonad contained either XX cells or XY cells in which the Y chromosome had a deletion in the *Sry* gene. The sex chromosomes of the mesonephros did not affect the migration. (B) Gonadal rudiments for XX mice could induce mesonephric cell migration if these rudiments had been incubated with FGF9. (C) Intimate relation between the mesonephric ducts and the developing gonad in the 16-day male mouse embryo. The duct tissue has been stained for cytokeratin-8. WD, Wolffian duct. (A, B after Capel et al. 1999, photographs courtesy of B. Capel; C from Sariola and Saarma 1999, photograph courtesy of H. Sariola.)

override this suppression. This gene, *DAX1*, has been cloned and shown to encode a member of the nuclear hormone receptor family (Muscatelli et al. 1994; Zanaria 1994). *Dax1* is initially expressed in the genital ridges of both male and female mouse embryos, and it is seen in male mice shortly after *Sry* expression. Indeed, in XY mice, *Sry* and *Dax1* are expressed in the same cells. Eventually, however, *Dax1* is expressed solely in the XX gonadal rudiment (Figure 17.10). *Dax1* appears to antagonize the function of *Sry* and *Sox9*, and it downregulates *Sf1* expression (Nachtigal et al. 1998; Swain et al. 1998).

WNT4: A potential ovary-determining gene on an autosome

The *Dax1* gene appears to be activated by the product of an ovary-determining gene, **Wnt4**. *Wnt4* is expressed in the mouse genital ridge while it is still in its bipotential stage. *Wnt4* expression then becomes undetectable in XY gonads (which become testes), whereas it is maintained in XX gonads as they begin to form ovaries. In transgenic XX mice that lack the *Wnt4* gene, the ovary fails to form properly, and its cells express testis-specific markers, including AMH- and testosterone-producing enzymes (Vainio et al. 1999). One possible target for Wnt4 is the gene encoding TAF$_{II}$105 (Freiman et al. 2002). This subunit of the TATA-binding protein for RNA polymerase binding is seen only in ovarian follicle cells. Fe-

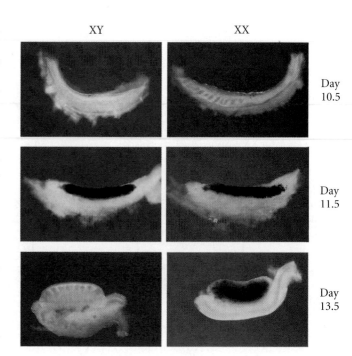

Figure 17.10

Dax1 expression in the mouse genital ridge. A *lacZ* gene was fused to the regulatory region of *Dax1* and the resulting transgene inserted into XY (left) and XX (right) mouse embryos. In both instances, *Dax1* was expressed at day 11.5 in the genital ridges. By day 13.5, however, *Dax1* expression persisted in the XX gonadal rudiments, but not in those of the XY embryos. (After Swain et al. 1998; photographs courtesy of R. Lovell-Badge.)

male mice lacking this subunit have no ovaries. In XY humans having a duplication of the *WNT4* region, DAX1 is overproduced and the gonads develop into ovaries (Jordan et al. 2001). Sry may form testes by repressing *Wnt4* expression in the genital ridge, as well as by promoting *Sf1*. One possible model is shown in Figure 17.11.

Although remarkable progress has been made in recent years, we still do not know what the testis- or ovary-determining genes are doing, and the problem of primary sex determination remains (as it has since prehistory) one of the great unsolved problems of biology.

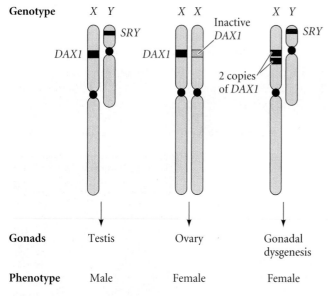

Figure 17.9

Phenotypic sex reversal in humans having two copies of the *DAX1* locus. *DAX1* (on the X chromosome) plus *SRY* (on the Y chromosome) produces testes. *DAX1* without *SRY* (since the other *DAX1* locus is on the inactive X chromosome) produces ovaries. Two active copies of *DAX1* (on the active X chromosome) plus SRY (on the Y chromosome) lead to a poorly formed gonad. Since the gonad makes neither AMH nor testosterone, the phenotype is female. (After Genetics Review Group 1995.)

Secondary Sex Determination: Hormonal Regulation of the Sexual Phenotype

Primary sex determination involves the formation of either an ovary or a testis from the bipotential gonad. This process, however, does not give the complete sexual phenotype. Secondary sex determination in mammals involves the development of the female and male phenotypes in response to hormones secreted by the ovaries and testes. Both female and male secondary sex determination have two major temporal

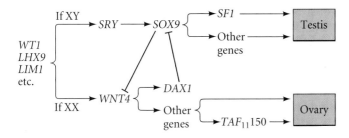

Figure 17.11
Possible mechanism for primary sex determination in mammals. While we do not know the specific interactions involved, this model attempts to organize the data into a coherent sequence. Other models are possible. In this model, *SRY* and *WNT4* are both activated in the gonad rudiment. If no SRY protein is present, Wnt4 activates ovary-forming genes, as well as the testis-suppressing gene *DAX1*. If SRY is present, it activates SOX9 which activates testes-forming genes such as SF1 and AMH, as well as suppressing WNT4.

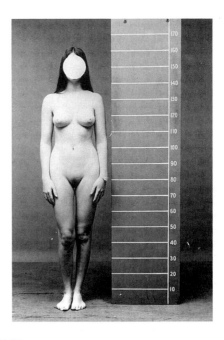

Figure 17.12
An XY individual with androgen insensitivity syndrome. Despite the XY karyotype and the presence of testes, such individuals develop female secondary sex characteristics. Internally, however, these women lack the Müllerian duct derivatives and have undescended testes. (Photograph courtesy of C. B. Hammond.)

phases. The first phase occurs within the embryo during organogenesis; the second occurs at puberty.

As mentioned earlier, if the bipotential gonads are removed from an embryonic mammal, the female phenotype is realized: the Müllerian ducts develop while the Wolffian ducts degenerate. This pattern is also seen in certain humans who are born without functional gonads. Individuals whose cells have only one X chromosome (and no Y chromosome) originally develop ovaries, but these ovaries atrophy before birth, and the germ cells die before puberty. However, under the influence of estrogen, derived first from the ovary but then from the mother and placenta, these infants are born with a female genital tract (Langman and Wilson 1982).

The formation of the male phenotype involves the secretion of two testicular hormones. The first of these hormones is AMH, the hormone made by the Sertoli cells that causes the degeneration of the Müllerian duct. The second is the steroid hormone testosterone, which is secreted from the fetal Leydig cells. This hormone causes the Wolffian duct to differentiate into the epididymis, vas deferens, and seminal vesicles, and it causes the urogenital swellings to develop into the scrotum and penis.

The existence of these two independent systems of masculinization is demonstrated by people with **androgen insensitivity syndrome**. These XY individuals have the *SRY* gene, and thus have testes that make testosterone and AMH. However, they lack the testosterone receptor protein, and therefore cannot respond to the testosterone made by their testes (Meyer et al. 1975). Because they are able to respond to estrogen made in their adrenal glands, they develop the female phenotype (Figure 17.12). However, despite their distinctly female appearance, these individuals do have testes, and even though they cannot respond to testosterone, they produce and respond to AMH. Thus, their Müllerian ducts degenerate.

These people develop as normal but sterile women,[*] lacking a uterus and oviducts and having testes in the abdomen.

Testosterone and dihydrotestosterone

Although testosterone is one of the two primary masculinizing hormones, there is evidence that it might not be the active masculinizing hormone in certain tissues. Testosterone appears

---

[*]Androgen insensitivity syndrome is one of several conditions called **pseudohermaphroditism**. In pseudohermaphrodites, there is only one type of gonad, but the secondary sex characteristics differ from what would be expected from the gonadal sex. In humans, male pseudohermaphroditism (wherein the gonadal sex is male and the secondary sex characteristics are female) can be caused by mutations in the androgen (testosterone) receptor or by mutations affecting testosterone synthesis (Geissler et al. 1994). Female pseudohermaphroditism (in which the gonadal sex is female but the person is outwardly male) can be caused by the overproduction of androgens in the ovary or adrenal gland. The most common cause of this condition is congenital adrenal hyperplasia, in which there is a genetic deficiency of an enzyme that metabolizes cortisol steroids in the adrenal gland. In the absence of this enzyme, testosterone-like steroids accumulate and can bind to the androgen receptor to masculinize the fetus (Migeon and Wisniewski 2001; Merke et al. 2002). Thus, pseudohermaphroditism is due to abnormalities of *secondary* sex determination.

"True" hermaphrodites contain both male and female gonadal tissue. Thus, in mammals, they result from abnormalities of *primary* sex determination. True hermaphroditism can occur, for instance, when a Y chromosome is translocated to an X chromosome. In those tissues where the translocated Y is on the active X chromosome, the Y chromosome will be active and the *SRY* gene will be transcribed; in those cells where the Y chromosome is on the inactive X chromosome, the Y chromosome will be inactive (Berkovitz et al. 1992; Margarit et al. 2000).

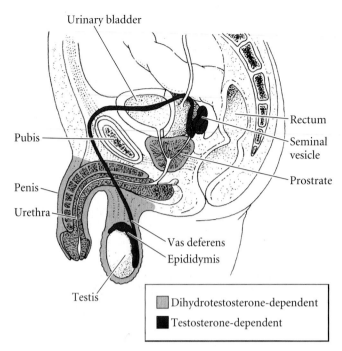

Figure 17.13
Testosterone- and dihydrotestosterone-dependent regions of the human male genital system. (After Imperato-McGinley et al. 1974.)

to be responsible for promoting the formation of the male reproductive structures (the epididymis, seminal vesicles, and vas deferens) that develop from the Wolffian duct primordium. However, it does not directly masculinize the male urethra, prostate, penis, or scrotum. These latter functions are controlled by **5α-dihydrotestosterone** (**DHT**; Figure 17.13). Siiteri and Wilson (1974) showed that testosterone is converted to 5α-dihydrotestosterone in the urogenital sinus and swellings, but not in the Wolffian duct. 5α-dihydrotestosterone appears to be a more potent hormone than testosterone.

The importance of 5α-dihydrotestosterone was demonstrated by Imperato-McGinley and her colleagues (1974). They found a small community in the Dominican Republic in which several inhabitants have a genetic deficiency of the enzyme 5α-ketosteroid reductase 2, the enzyme that converts testosterone to DHT. These individuals lack a functional gene for this enzyme (Andersson et al. 1991; Thigpen et al. 1992). Although XY children with this syndrome have functioning testes, they have a blind vaginal pouch and an enlarged clitoris. They appear to be girls and are raised as such. Their internal anatomy, however, is male: they have testes, Wolffian duct development, and Müllerian duct degeneration. Thus, it appears that the formation of the external genitalia is under the control of dihydrotestosterone, whereas Wolffian duct differentiation is controlled by testosterone itself. Interestingly, when the testes of these children produce more testosterone at puberty, the external genitalia are able to respond to the higher levels of the hormone, and they differentiate. The penis

enlarges, the scrotum descends, and the person originally thought to be a girl is shown to be a young man.

> **WEBSITE 17.3 Dihydrotestosterone in adult men.** The drug finasteride, which inhibits the conversion of testosterone to dihydrotestosterone, is being used to treat prostate growth and male pattern baldness.

> **WEBSITE 17.4 Insulin-like hormone 3.** In addition to testosterone, the Leydig cells secrete another hormone, insulin-like hormone 3 (Insl3). This hormone is required for the descent of the gonads into the scrotum. Males lacking this hormone are infertile because the testes do not descend. In females, lack of this hormone deregulates the menstrual cycle.

Anti-Müllerian duct hormone

Anti-Müllerian duct hormone (AMH), a member of the TGF-β family of growth and differentiation factors, is secreted from the Sertoli cells and causes the degeneration of the Müllerian duct (Tran et al. 1977; Cate et al. 1986). When fragments of fetal testes or isolated Sertoli cells are placed adjacent to cultured tissue segments containing portions of the Wolffian and Müllerian ducts, the Müllerian duct atrophies even though no change occurs in the Wolffian duct (Figure 17.14). AMH is thought to bind to the mesenchyme cells surrounding the

(A)

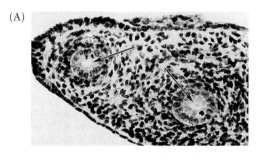

(B)

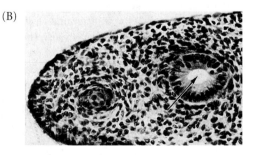

Figure 17.14
Assay for AMH activity in the anterior segment of a 14.5-day fetal rat reproductive tract. (A) At the start of the experiment, both the Müllerian duct (arrow at left) and Wolffian duct (arrow at right) are open. (B) After 3 days in culture with AMH-secreting tissue, the Wolffian duct (arrow) is open, but the Müllerian duct has degenerated and closed. (Photograph courtesy of N. Josso.)

Müllerian duct and to cause these cells to secrete a paracrine factor that induces apoptosis in the Müllerian duct epithelium (Trelstad et al. 1982; Roberts et al. 1999).

> **WEBSITE 17.5 Roles of AMH.** AMH may have other roles in sex determination besides causing the breakdown of the Müllerian ducts. It may cause sex reversal in some mammals, and may become useful as an anti-tumor drug.

Estrogen

Estrogen is needed for complete development of both the Müllerian and the Wolffian ducts. In females, estrogen secreted from the fetal ovaries appears sufficient to induce the differentiation of the Müllerian duct into its various components: the uterus, oviducts, and cervix. The extreme sensitivity of the Müllerian duct to estrogenic compounds is demonstrated by the teratogenic effects of **diethylstilbesterol (DES)**, a powerful synthetic estrogen that can cause infertility by changing the patterning of the Müllerian duct (see Chapter 21). In mice, DES can cause the oviduct epithelium to take on the appearance of the uterus, and the uterine epithelium to resemble that of the cervix (Ma et al. 1998).

Estrogen is needed for the fertility of both males and females. In female mice with knockouts of the estrogen receptors, the germ cells die in the adult, and the granulosa cells that had enveloped them start developing like Sertoli cells (Couse et al. 1999). In male mice with knockouts of the estrogen receptors, few sperm are made. One of the functions of the efferent duct (vas efferens) cells is to absorb about 90% of the water from the lumen of the rete testis. This absorption of water, which is regulated by estrogen, concentrates the sperm, giving them a longer life span and providing more sperm per ejaculate. If estrogen or its receptor is absent in mice, this water is not absorbed and the mouse is sterile (Hess et al. 1997). While blood concentrations of estrogen are higher in females than in males, the concentration of estrogen in the rete testis is even higher than that in female blood.

> **WEBSITE 17.6 The mechanisms of breast development.** Breast tissue has a sexually dimorphic mode of development. Testosterone inhibits breast development, while estrogen promotes it. Most breast development is accomplished after birth, and different hormones act during puberty and pregnancy to cause breast enlargement and differentiation.

Sidelights *&* Speculations

Sex Determination and Behaviors

The Organization/Activation Hypothesis

Does prenatal (or neonatal) exposure to particular steroid hormones impose permanent sex-specific changes on the central nervous system? Such sex-specific neural changes have been shown in regions of the brain that regulate "involuntary" sexual physiology. The cyclic secretion of luteinizing hormone by the pituitary in an adult female, for example, is dependent on the lack of testosterone during the first week of the animal's life. The luteinizing hormone secretion of female rats can be made noncyclic by giving them testosterone 4 days after birth. Conversely, the luteinizing hormone secretion of males can be made cyclical by removing their testes within a day of birth (Barraclough and Gorski 1962). It is thought that sex hormones may act during the fetal or neonatal stage of a mammal's life to organize the nervous system in a sex-specific manner, and that during adult life, the same hormones may have transitory, activational effects. This idea is called the **organization/activation hypothesis**.

Interestingly, the hormone chiefly responsible for determining the male neural pattern is **estradiol**, a type of estrogen.* Testosterone in fetal or neonatal blood can be converted into estradiol by the enzyme **P450 aromatase**. This conversion occurs in the hypothalamus and limbic system—two areas of the brain known to regulate hormone secretion and reproductive behavior (Reddy et al. 1974; McEwen et al. 1977). Thus, testosterone exerts its effects on the nervous system by being converted into estradiol in the brain. But the fetal environment is rich in estrogens from the gonads and placenta. What stops these estrogens from masculinizing the nervous system of a female fetus? Fetal estrogen (in both males and females) is bound by α-**fetoprotein**. This protein is made in the fetal liver and becomes a major component of the fetal blood and cerebrospinal fluid. It will bind and inactivate estrogen, but not testosterone.

Attempts to extend the organization/ activation hypothesis to "voluntary" sexual

---

*The terms *estrogen* and *estradiol* are often used interchangeably. However, estrogen refers to a class of steroid hormones responsible (among other functions) for establishing and maintaining specific female characteristics. Estradiol is one of these hormones, and in most mammals (including humans) it is the most potent of the estrogens.

---

behaviors are more controversial because there is no truly sex-specific behavior that distinguishes the two sexes of many mammals, and because hormonal treatment has multiple effects on the developing mammal. For instance, injecting testosterone into a week-old female rat will increase pelvic thrusting behavior and diminish lordosis—a receptive posture that stimulates mounting behavior in the male—when she reaches adulthood (Phoenix et al. 1959; Kandel et al. 1995). These behavioral changes can be ascribed to testosterone-mediated changes in the central nervous system, but they could also be due to hormonal effects on other tissues. Testosterone enables the growth of the muscles that allow pelvic thrusting. And since testosterone causes females to grow larger and causes their vaginal orifices to close, one cannot conclude that the lack of lordosis is due solely to testosterone-mediated changes in the neural circuitry (Harris and Levine 1965; De Jonge et al. 1988; Moore et al. 1992; Fausto-Sterling 1995). In addition, the effects of sex steroids on the brain are very complicated, and the steroids may be metabolized differently in different regions of the brain. Male mice lacking the testosterone receptor still retain a male-specific preoptic morphology in the brain, and male mice lacking the aromatase enzyme are capable of breeding (Breedlove 1992; Fisher et al. 1998).

These studies show that there is more to sex-specific morphology and behavior than steroid hormones. Despite best-selling books that pretend to know the answers, we have much more to learn regarding the relationship between development, steroids, and behavior. Moreover, extrapolating from rats to humans is a very risky business, as no sex-specific behavior has yet been identified in humans, and what is "masculine" in one culture may be considered "feminine" in another (see Jacklin 1981; Bleier 1984; Fausto-Sterling 1992). As one review (Kandel et al. 1995) concludes:

> There is ample evidence that the neural organization of reproductive behaviors, while importantly influenced by hormonal events during a critical prenatal period, does not exert an immutable influence over adult sexual behavior or even over an individual's sexual orientation. Within the life of an individual, religious, social, or psychological motives can prompt biologically similar persons to diverge widely in their sexual activities.

Male Homosexuality

Certain behaviors are often said to be part of the "complete" male or female phenotype. The brain of a mature man is said to be formed such that it causes him to desire mating with a mature woman, and the brain of a mature woman causes her to desire to mate with a mature man. However, as important as desires are in our lives, they cannot be detected by in situ hybridization or isolated by monoclonal antibodies. We do not yet know whether sexual desires are primarily instilled in us by our social education or are fundamentally "hard-wired" into our brains during our intrauterine development by genes or hormones or by other means.

In 1991, Simon LeVay proposed that part of the anterior hypothalamus of homosexual men has the anatomical form typical of women rather than of heterosexual men. The hypothalamus is thought to be the source of our sexual urges, and rats have a sexually dimorphic area in the anterior hypothalamus that appears to regulate their sexual behavior. Thus, this study—which was based on observations of human hypothalami removed and dissected post mortem—generated a great deal of publicity and discussion. Its major results are shown in Figure 17.15. The interstitial nuclei (neuron clusters) of the anterior hypothalamus (INAH) were divided into four regions. Three of them showed no signs of sexual dimorphism. However, one of them,

INAH3, showed a statistically significant difference in volume between males and females. The study found the male INAH3 to be, on average, more than twice as large as the female INAH3. Moreover, LeVay's data suggested that the INAH3 of homosexual men was similar in volume to that of women—less than half the size of INAH3 in heterosexual men. This finding, LeVay claimed, "suggests that sexual orientation has a biological substrate."

There have been several criticisms of LeVay's interpretation of these data. First, the conclusions are from populations, not individuals; one could argue that there is a statistical range and that men and women have the same general range. (Indeed, one of the INAH3 from a "homosexual" male was larger than all but one of those from the 16 "heterosexual" males in the study.) Second, the "heterosexual men" were not necessarily heterosexual, nor were the "homosexual men" necessarily homosexual: the brains came from corpses of persons whose sexual preferences were not definitively known. Homosexuality has many forms, and is probably not a single phenotype. The brains of the presumed "homosexual men" were taken from patients who died of AIDS, which brings up a third issue: AIDS affects the brain, and its effects on the hypothalamic neurons are not known.

Fourth, because the study was done on the brains of dead subjects, one cannot infer cause and effect. Such data show only correlations, not causation. It is as likely that behaviors can affect regional neuronal density as it is that regional neuronal density can affect behaviors. If one interprets the data as indicating that the INAH3 of male homosexuals is smaller than that of male heterosexuals, one still does not know whether that is a cause of homosexuality or a result of it. Indeed, Breedlove (1997) has shown that the density and size of certain neurons in rat spinal ganglia depend on the frequency of sexual intercourse. In this case, the behavior was affecting the neurons. Fifth, even if a difference in INAH3 does exist, there is no evidence that the difference has anything to do with sexuality. Sixth, these studies do not indicate when such differences (if they exist) emerge. The question of whether differences among the heterosexual male, female, and homosexual male INAH3 are established during embryonic development, shortly after birth, during the first few years of life, during adolescence, or at some other time was not addressed.

In 1993, a correlation was made between a particular DNA sequence on the X chromosome and a particular subgroup of male homosexuals: homosexual men who had a homosexual brother. Out of 40 pairs of homosexual brothers wherein one brother had inherited a particular region of the X chromosome from his mother, the

other brother had also inherited this region in 33 cases (Hamer et al. 1993). One would have expected both brothers to have done so in only 20 cases, on average. Again, this is only a statistical concordance, and one that could be coincidental. Moreover, the control (the incidence of the same marker in the "nonhomosexual" males of these families) was not reported, and the statistical bias of the observations has been called into question, especially since other laboratories have not been able to repeat the result (Risch et al. 1993; Marshall 1995). More recent studies (Hu et al. 1995; Rice et al. 1999) found little or no increase in the incidence of this DNA sequence when homosexual men were compared with their nonhomosexual brothers. Hu and colleagues concluded that this

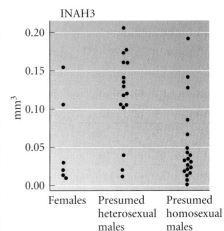

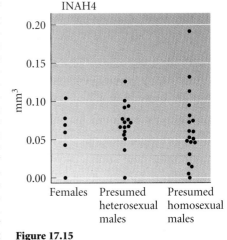

Figure 17.15
A portion of the data suggesting a biological basis for homosexuality. INAH4 and INAH3 are two groups of hypothalamic neurons. INAH4 shows no sexual dimorphism in volume, while INAH3 shows a statistically significant clustering, although the range is similar. The INAH3 from autopsies of men presumed to be homosexual cluster toward the female distribution. (After LeVay 1991.)

sequence is "neither necessary nor sufficient for a homosexual orientation." Thus, despite the reports of these studies in the public media, no "gay gene" has been found.

Genes encode RNAs and proteins, not behaviors. While genes may bias behavioral outcomes, we have no evidence for their "controlling" them. Observations of patients with schizophrenia, or of people whose personalities change radically after a religious conversion or a traumatic experience, indicate that a single genotype can support a wide range of personalities. This is certainly a problem with any definition of a "homosexual phenotype," since people can alternate between homosexual and heterosexual behavior, and the definition of what constitutes homosexual behavior differs between cultures (see Carroll and Wolpe 1996). Thus, whether the direction of sexual desires is indicated or biased by genes within the nucleus, by sex hormones during fetal development, or by experiences after birth is still an open question.

WEBSITE 17.7 **Sex and the central nervous system.** There is ample evidence that estrogens and testosterone can cause changes in the central nervous system. Birds appear especially susceptible to hormonally induced changes in their behaviors.

CHROMOSOMAL SEX DETERMINATION IN DROSOPHILA

The Sexual Development Pathway

Although both mammals and fruit flies produce XX females and XY males, their chromosomes achieve these ends using very different means. In mammals, the Y chromosome plays a pivotal role in determining the male sex. Thus, XO mammals are females, with ovaries, a uterus, and oviducts (but usually very few, if any, ova). In *Drosophila*, sex determination is achieved by a balance of female determinants on the X chromosome and male determinants on the autosomes. Normally, flies have either one or two X chromosomes and two sets of autosomes. If there is but one X chromosome in a diploid cell (1X:2A), the fly is male. If there are two X chromosomes in a diploid cell (2X:2A), the fly is female (Bridges 1921, 1925). In flies, the Y chromosome is not involved in determining sex. Rather, it contains genes that are active in forming sperm in adults. Thus, XO *Drosophila* are sterile males. Table 17.1 shows the different X-to-autosome ratios possible for *Drosophila* and the resulting sex.

In *Drosophila*, and in insects in general, one can observe **gynandromorphs**—animals in which certain regions of the body are male and other regions are female (Figure 17.16). This can happen when an X chromosome is lost from one embryonic nucleus. The cells descended from that cell, instead of being XX (female), are XO (male). Because there are no sex hormones in insects that integrate the phenotype for the entire body, each cell makes its own sexual "decision." The XO cells display male characteristics, whereas the XX cells display female traits. This situation provides a beautiful example of the association between insect X chromosomes and sex.

Any theory of *Drosophila* sex determination must explain how the X-to-autosome (X:A) ratio is "read" and how this information is transmitted to the genes controlling the male or female phenotypes (Cline 1993). As we will see, the X chromosome and the autosomes produce transcription factors that activate or repress the *Sex-lethal* gene, respectively. If there are two X chromosomes, the transcription factors that are encoded by those chromosomes prevail against the transcriptional repressors encoded by the autosomes. The *Sex-lethal* gene is thereby activated. The product of this gene then acts as a RNA processing factor that permits the synthesis of certain

Figure 17.16

Gynandromorphs. (A) Gynandromorph of *D. melanogaster* in which the left side is female (XX) and the right side is male (XO). The male side has lost an X chromosome bearing the wild-type alleles of eye color and wing shape, thereby allowing the expression of the recessive alleles *eosin eye* and *miniature wing* on the remaining X chromosome. (B) Photograph of a gynandromorphic Io moth, divided bilaterally into a rose-brown female half and a smaller, yellow male half. (A from Morgan and Bridges 1919, drawn by Edith Wallace. B; photograph by T. R. Manley, courtesy of *The Journal of Heredity*.)

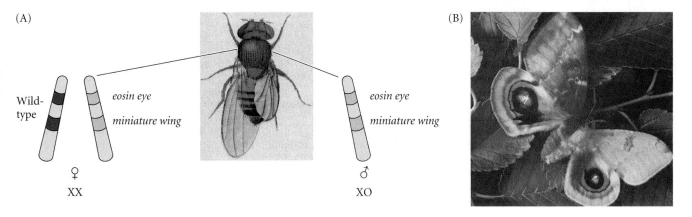

(A)

Wild-type

eosin eye

miniature wing

♀

XX

eosin eye

miniature wing

♂

XO

(B)

proteins. These proteins in turn act as RNA processing factors to modify the nuclear RNA (pre-mRNA) of the *doublesex* gene. Depending on how the *doublesex* RNA is spliced, two types of Doublesex proteins can be formed. Female Doublesex protein is formed in those flies where the two X chromosomes have activated the *Sex-lethal* gene; male Doublesex protein is formed in those flies where autosomes have prevented the activation of the *Sex-lethal* gene. The Doublesex proteins are involved in producing the sexual phenotype of the fly.

Mechanisms of Sex Determination

Research in the past two decades has revolutionized our view of *Drosophila* sex determination. Several genes with roles in sex determination have been found. Loss-of-function mutations in most of these genes—*Sex-lethal* (*Sxl*), *transformer* (*tra*), and *transformer-2* (*tra2*)—transform XX individuals into males. Such mutations have no effect on sex determination in XY males. Homozygosity of the *intersex* (*ix*) gene causes XX flies to develop an intersex phenotype having portions of male and female tissue in the same organ. The *doublesex* (*dsx*) gene is important for the sexual differentiation of both sexes. If *dsx* is absent, both XX and XY flies turn into intersexes (Baker and Ridge 1980; Belote et al. 1985a). The positioning of these genes in a developmental pathway is based on (1) the interpretation of genetic crosses resulting in flies bearing two or more mutations of these genes and (2) the determination of what happens when there is a complete absence of the products of one of these genes. Such studies have generated the model of the regulatory cascade seen in Figure 17.17.

The Sex-lethal *gene as the pivot for sex determination*

INTERPRETING THE X:A RATIO. The first phase of *Drosophila* sex determination is reading of the X:A ratio. What elements on the X chromosome are "counted," and how is this information used? It has been shown that high values of the X:A ratio are responsible for activating the feminizing switch gene **Sex-lethal** (**Sxl**). In XY cells, *Sxl* remains inactive during the early stages of development (Cline 1983; Salz et al. 1987). However, in XX *Drosophila*, *Sxl* is activated during the first 2 hours after fertilization, and this gene transcribes a particular embryonic type of *Sxl* mRNA that is found for only about 2 hours more (Salz et al. 1989). Once activated, the *Sxl* gene remains active because its protein product is able to bind to and activate its own promoter (Bell et al. 1991).

The female-specific activation of *Sxl* is thought to be stimulated by "**numerator proteins**" encoded by the X chromosome. These proteins are the X part (i.e., the numerator) of the X:A ratio. The numerator proteins bind to the "early" promoter of the *Sxl* gene to promote its transcription shortly after fertilization. Cline (1988) has demonstrated that these proteins include Sisterless-a and Sisterless-b. The "**denomina-**

TABLE 17.1 Ratios of X chromosomes to autosomes in different sexual phenotypes in *Drosophila melanogaster*

X chromosomes	Autosome sets	(A)X:A ratio	Sex
3	2	1.50	Metafemale
4	3	1.33	Metafemale
4	4	1.00	Normal female
3	3	1.00	Normal female
2	2	1.00	Normal female
2	3	0.66	Intersex
1	2	0.50	Normal male
1	3	0.33	Metamale

Source: After Strickberger 1968.

tor proteins" are autosomally encoded proteins such as Deadpan and Extramacrochaetae. These proteins block the binding or activity of the numerator proteins (Van Doren et al. 1991; Younger-Shepherd et al. 1992). The denominator proteins may actually be able to form inactive heterodimers with the numerator proteins (Figure 17.18). It appears, then, that the X:A ratio is measured by competition between X-encoded activators and autosomally encoded repressors of the early promoter of the *Sxl* gene.

MAINTENANCE OF *SXL* FUNCTION. Shortly after the initial *Sxl* transcription has taken place, a second, "late" promoter on the *Sex-lethal* gene is activated, and the gene is now transcribed in both males and females. However, analysis of cDNA from *Sxl* mRNA shows that the *Sxl* mRNA of males differs from the *Sxl* mRNA of females (Bell et al. 1988). This difference is the result of differential RNA processing. Moreover, the Sxl protein appears to bind to its own mRNA precursor to splice it in the female manner. Since males do not have any available Sxl protein when the late promoter is activated, their new *Sxl* transcripts are processed in the male manner (Keyes et al. 1992). The male *Sxl* mRNA is nonfunctional. While the female-specific *Sxl* message encodes a protein of 354 amino acids, the male-specific *Sxl* transcript contains a translation termination codon (UGA) after amino acid 48. The differential RNA processing that includes this termination codon in the male-specific mRNA is shown in Figures 17.18B and 17.19. In males, the nuclear transcript is spliced in a manner that yields eight exons, and the termination codon is within exon 3. In females, RNA processing yields only seven exons, and the male-specific exon 3 is spliced out as part of a large intron. Thus, the female-specific mRNA lacks the termination codon. The protein made by the female-specific *Sxl* transcript contains regions that are important for binding to RNA. There appear to be two major RNA targets to which the female-specific *Sxl* transcript binds. One of these is the pre-mRNA of *Sxl* itself; the other target is the pre-mRNA of *transformer*—the next gene on the pathway (Bell et al. 1988; Nagoshi et al. 1988).

The transformer *genes*

The *Sxl* gene regulates somatic sex determination by controlling the processing of the *transformer* (*tra*) gene transcript. The *tra* pre-mRNA is alternatively spliced to create a female-specific mRNA as well as a nonspecific mRNA that is found in both females and males. Like the male *Sxl* message, the nonspecific *tra* mRNA contains a termination codon early in the

message, making the protein nonfunctional (Boggs et al. 1987). In *tra*, the second exon of the nonspecific mRNA contains the termination codon. This exon is not utilized in the female-specific message (see Figure 17.19).

How is it that females make a different transcript than males? In females, the female-specific Sxl protein activates a 3′ splice site in the *transformer* pre-mRNA, causing it to be processed in a way that splices out the second exon. To do this, the Sxl protein blocks the binding of splicing factor U2AF to the nonspecific splice site of the *tra* message by specifically binding to the polypyrimidine tract adjacent to it (Figure 17.20; Handa et al. 1999). This causes U2AF to bind to the lower-affinity (female-specific) 3′ splice site and generate a female-specific mRNA (Valcárcel et al. 1993). The female-specific Tra product acts in concert with the product of the *transformer-2* (*tra2*) gene to help generate the female phenotype.

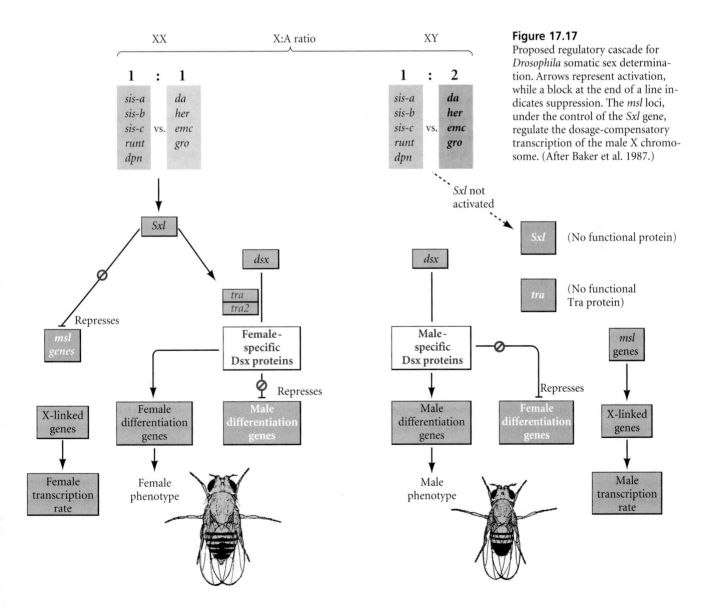

Figure 17.17
Proposed regulatory cascade for *Drosophila* somatic sex determination. Arrows represent activation, while a block at the end of a line indicates suppression. The *msl* loci, under the control of the *Sxl* gene, regulate the dosage-compensatory transcription of the male X chromosome. (After Baker et al. 1987.)

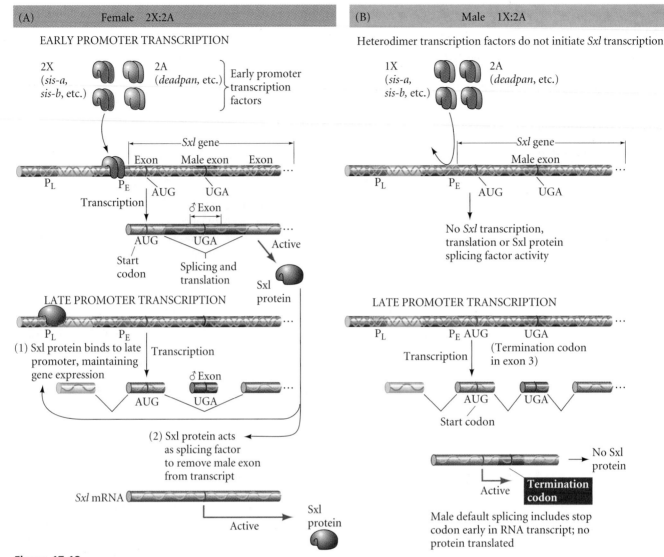

Figure 17.18

Differential activation of the *Sxl* gene in females and males. (A) In wild-type *Drosophila* with two X chromosomes and two sets of autosomes (2X:2A), the "numerator" proteins encoded on the X chromosomes (Sis-a, Sis-b, etc.) are not all bound by inhibitory "denominator" proteins derived from genes on the autosomes (such as *deadpan*). The numerator proteins activate the early promoter (P_E) of the *Sxl* gene. Eventually, in both males and females, constitutive transcription of *Sxl* eventually starts from the late promoter (P_L). If Sxl protein is already available (i.e., from early transcription), the

Sxl pre-mRNA is spliced to form the functional female-specific message. (B) In wild-type *Drosophila* with one X chromosome and two sets of autosomes (1X:2A), the numerator proteins are bound by the denominator proteins and cannot activate the early promoter. When the *Sxl* gene is transcribed from the late promoter, RNA splicing does not exclude the male-specific exon in the mRNA. The resulting message encodes a truncated and nonfunctional peptide, since the male-specific exon contains a translation termination codon. (After Keyes et al. 1992.)

Doublesex: *The switch gene of sex determination*

The *doublesex* (*dsx*) gene is active in both males and females, but its primary transcript is processed in a sex-specific manner (Baker et al. 1987). This alternative RNA processing is the result of the action of the *tra* and *tra2* gene products on the *dsx* gene (see Figure 5.30). If the Tra2 and female-specific Tra proteins are both present, the *dsx* transcript is processed in a female-specific manner (Ryner and Baker 1991). The female

splicing pattern produces a female-specific protein that activates female-specific genes (such as those of the yolk proteins) and inhibits male development. If no functional Tra is produced, a male-specific transcript of *dsx* is made. The male transcript encodes an active protein that inhibits female traits and promotes male traits.

In XX flies, the female Doublesex protein (Dsx^F)acts by combining with the product of the *intersex* gene to make a transcription factor that is responsible for female-specific

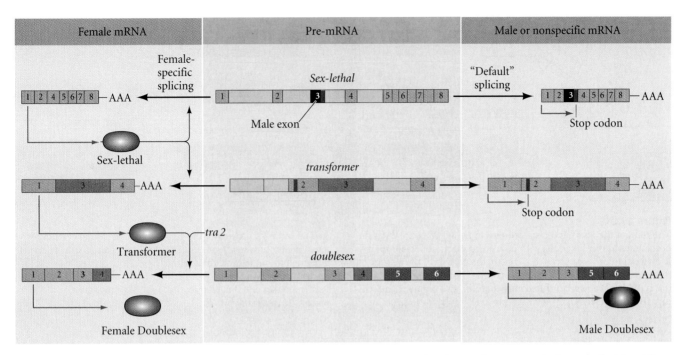

Figure 17.19
The pattern of sex-specific RNA splicing in three major *Drosophila* sex-determining genes. The pre-mRNAs are located in the center of the diagram and are identical in both male and female nuclei. In each case, the female-specific transcript is shown at the left, while the default transcript (whether male or nonspecific) is shown to the right. Exons are numbered, and the positions of the termination codons and poly(A) sites are marked. (After Baker 1989.)

traits. This Doublesex complex enhances the growth of the genital disc to make the female sex organs, represses the FGF genes responsible for making male accessory organs, activates the genes responsible for making yolk proteins, promotes the growth of the sperm storage duct, and modifies *bric-a-brac*

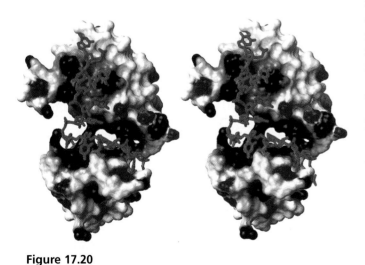

Figure 17.20
Stereogram showing binding of *tra* pre-mRNA by the cleft of the Sxl protein. The bound 12-nucleotide RNA (GUUGUUUUUUUU) is shown in yellow. The strongly positive regions are shown in blue, while the scattered negative regions are in red. It is worth crossing your eyes to get the three-dimensional effect. (From Handa et al. 1999; stereogram courtesy of S. Yokoyama.)

(*bab*) gene expression to give the female-specific pigmentation profile. In contrast, the male Doublesex protein (Dsx^M) acts directly as a transcription factor, and it directs the expression of male-specific traits. It causes the male region of the genital disc to grow at the expense of the female disc regions, it activates the FGF genes to produce the paragonia (a portion of the male genitalia), converts certain cuticular structures into claspers, and modifies the *bric-a-brac* gene to produce the male pigmentation pattern (Figure 17.21; Christiansen et al. 2002).

The functions of the Doublesex proteins can be seen in the formation of the *Drosophila* genitalia. Male and female genitalia in *Drosophila* are derived from separate cell populations of the larval **genital disc**. Dsx^M inhibits *wingless*, while Dsx^F inhibits the *Decapentaplegic* (*Dpp*) gene; these interactions determine the growth of the different regions of the genital disc (Keisman and Baker 2001; Keisman et al 2001; Figure 17.22). In XY flies, the *male* primordium (derived from the ninth abdominal segment) grows and differentiates into the testes, while the female primordium (derived from the eighth abdominal segment) is stunted and becomes a plate (tergite) on the back of the eighth abdominal segment. This is due to the inhibition of *wingless* in the eighth ("female") abdominal segment. In XX flies, the *female* primordium of the genital disc grows and differentiates into ovaries, while the male portion of that disc becomes paraovaria, the female accessory organs. This is due to the inhibition of *dpp* in the ninth ("male")

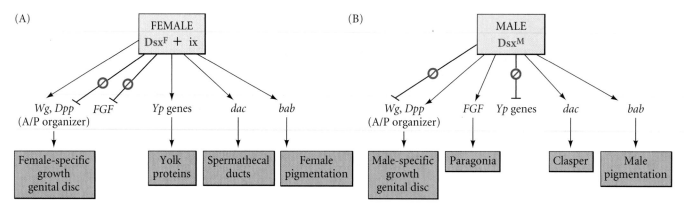

Figure 17.21
The roles of the Dsx^M and Dsx^F proteins in *Drosophila* sexual development. (A) Dsx^F functions with Intersex to promote female-specific expression of those genes that control the growth of the genital disc, the synthesis of yolk proteins, the formation of spermathecal ducts (which keep sperm stored after mating), and pigment patterning. (B) Conversely, Dsx^M acts as a transcription factor to promote the male-specific growth of the genital disc, the formation of male genitalia, the conversion of cuticle into claspers, and the male-specific pigmentation pattern. In addition, Dsx^F represses certain genes involved in specifying male-specific traits (such as the paragonia), and Dsx^M represses certain genes involved synthesizing female-specific proteins such as yolk protein. (After Christiansen et al. 2002.)

abdominal segment. If the *dsx* gene is absent (and thus neither transcript is made), both the male and the female primordia develop, and intersexual genitalia are produced.

According to this model, the result of the sex determination cascade comes down to what type of mRNA is processed from the *dsx* transcript. If the X:A ratio is 1, then *Sxl* makes a female-specific splicing factor that causes the *tra* gene transcript to be spliced in a female-specific manner. This female-specific protein interacts with the *tra2* splicing factor to cause the *doublesex* pre-mRNA to be spliced in a female-specific manner. If the *doublesex* transcript is not acted on in this way, it will be processed in a "default" manner to make the male-specific message.

Our discussion of sexual dimorphism in *Drosophila* has so far been limited to nonbehavioral aspects of development.

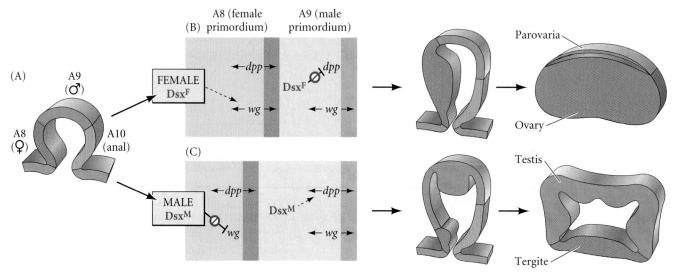

Figure 17.22
Action of Doublesex proteins to form the gonads of *Drosophila*. (A) The genital disc is derived from abdominal segments A8, A9, and A10. A8 forms the female portion of the disc, A9 forms the male portion of the disc, and A10 forms the anal structures (which will not be discussed here). (B) The female-specific Dsx protein inhibits *dpp* in the ninth ("male") abdominal segment, while enhancing *wingless* expression in the eighth ("female") abdominal segment. This causes the growth and development of the eighth abdominal segment into an ovary and the differentiation of the ninth abdominal segment into the parovarian accessory organs. (C) Conversely, the male-specific Doublesex protein inhibits *wingless* expression in the eighth abdominal segment, while enhancing *dpp* function in the ninth abdominal segment. This interaction allows the ninth segment to grow and develop into testes, while the eighth segment differentiates into a tergite. (After Keisman et al. 2001.)

There are sex-specific behaviors in *Drosophila* courtship that appear to be regulated by the products of the *fruitless* gene. As with *Doublesex* pre-mRNA, the Tra and Tra2 proteins splice *fruitless* pre-mRNA into a female-specific message; the default splicing pattern is male. Although no female behaviors appear to be influenced by *fruitless*, the courtship behavior of males appears to require the male-specific Fruitless protein at each stage (Ryner et al. 1996; Baker et al. 2001).

WEBSITE 17.9 Conservation of sex-determining genes. While the pathways of sex determination appear to differ between humans and fruit flies, the discovery of a human gene similar to *doublesex* and the discovery of FGF signaling for the recruitment of mesodermal cells into the testes of both humans and flies suggest common themes in the two pathways.

WEBSITE 17.10 Hermaphrodites. In *C. elegans* and many other invertebrates, hermaphroditism is the general rule. These animals are born with both ovaries and testes. In some fish, sequential hermaphroditism is seen, with an individual fish being female in some seasons and male in others. Among humans, hermaphrodites are rare and usually sterile.

ENVIRONMENTAL SEX DETERMINATION

Temperature-Dependent Sex Determination in Reptiles

While the sex of most snakes and most lizards is determined by sex chromosomes at the time of fertilization, the sex of most turtles and all species of crocodilians is determined by the environment after fertilization. In these reptiles, the temperature of the eggs during a certain period of development is the deciding factor in determining sex, and small changes in temperature can cause dramatic changes in the sex ratio (Bull 1980; Crews 2003). Often, eggs incubated at low temperatures (22–27°C) produce one sex, whereas eggs incubated at higher temperatures (30°C and above) produce the other. There is only a small range of temperatures that permits both males and females to hatch from the same brood of eggs. Figure 17.23 shows the abrupt temperature-induced change in sex ratios for the red-eared slider turtle. If a brood of eggs is incubated at a temperature below 28°C, all the turtles hatching from the eggs will be male. Above 31°C, every egg gives rise to a female. At temperatures in between, the brood will give rise to individuals of both sexes. Variations on this theme also exist. The eggs of the snapping turtle *Macroclemys*, for instance, become female at either cool (22°C or lower) or hot (28°C or above) temperatures. Between these extremes, males predominate.

One of the best-studied reptiles is the European pond turtle, *Emys obicularis*. In laboratory studies, incubating *Emys* eggs

at temperatures above 30°C produces all females, while temperatures below 25°C produce all-male broods. The threshold temperature (at which the sex ratio is even) is 28.5°C (Pieau et al. 1994). The developmental period during which sex determination occurs can be discovered by incubating eggs at the male-producing temperature for a certain amount of time and then shifting them to an incubator at the female-producing temperature (and vice versa). In *Emys*, the last third of development appears to be the most critical for sex determination, and it is thought that turtles cannot reverse their sex after this period.

Aromatase and the production of estrogens

The pathways to maleness and femaleness in reptiles are just being delineated. Unlike the situation in mammals, primary sex determination in reptiles (and birds) is hormone-dependent, and estrogen is essential for ovarian development. In reptiles, estrogen can override temperature and induce ovarian differentiation even at masculinizing temperatures. Similarly, injecting eggs with inhibitors of estrogen synthesis produces male offspring, even if the eggs are incubated at temperatures that usually produce females (Dorizzi et al. 1994; Rhen and Lang 1994). Moreover, the sensitive time for the effects of estrogens and their inhibitors coincides with the time when sex determination usually occurs (Bull et al. 1988; Gutzke and Chymiy 1988).

It appears that the enzyme **aromatase**, which converts testosterone into estrogen, is critical in temperature-dependent sex determination. The estrogen synthesis inhibitors used in the experiments mentioned above worked by blocking the aromatase enzyme, showing that experimentally low aromatase levels yield male offspring. This correlation appears to

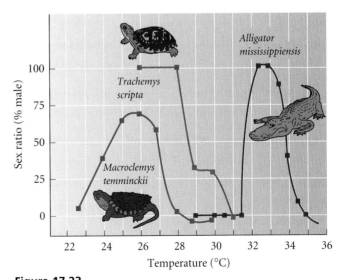

Figure 17.23
Temperature-dependent sex determination in three reptile species: the American alligator (*Alligator mississippiensis*), the red-eared slider turtle (*Trachemys scripta elegans*), and the alligator snapping turtle (*Macroclemys temminckii*). (After Crain and Guillette 1998.)

hold under natural conditions as well. The aromatase activity of *Emys* is very low at the male-promoting temperature of 25°C. At the female-promoting temperature of 30°C, aromatase activity increases dramatically during the critical period for sex determination (Desvages et al. 1993; Pieau et al. 1994). Temperature-dependent aromatase activity is also seen in diamondback terrapins, and its inhibition masculinizes their gonads* (Jeyasuria et al. 1994).

In these species, sex appears truly to reside in the brain. Aromatase is produced at two major sites in the embryo: the gonads and the brain. However, the gonadal aromatase level appears to be the same in both sexes. It's the *brain* aromatase activity that correlates with temperature and female sex. When embryos of *Trachemys* (red-eared slider turtle) or *Lepidochelys* (an olive Ridley turtle) were raised at high temperatures (to produce females), their brains had significantly higher aromatase activity than the brains of their siblings incubated at lower, male-producing, temperatures (Salame-Mendez et al. 1998; Willingham et al. 2000). This increase in brain estrogen production may induce neuroendocrine changes that result in gonadal steroid formation and ovary determination. The mechanisms by which this occurs remain largely unexplored.

Sex reversal, aromatase, and conservation biology

The evolutionary advantages and disadvantages of temperature-dependent sex determination are discussed in Chapter 22. Recent studies (Bergeron et al. 1994, 1999) have shown that polychlorinated biphenyl compounds (PCBs), a class of widespread pollutants introduced into the environment by humans, can act like estrogens. PCBs can reverse the sex of turtles raised at "male" temperatures. This knowledge may have important consequences in environmental conservation efforts to protect endangered species such as turtles and amphibians, in which hormones can effect changes in primary sex determination. Indeed, some reptile conservation biologists advocate using hormonal treatments to elevate the percentage of females in endangered species (www.reptileconservation.org).

In addition to deaths due to ozone depletion and fungal infections (see Chapter 3), amphibians may be at risk from herbicides that promote or destroy estrogens. One such case involves the development of hermaphroditic and demasculinized frogs after exposure to extremely low doses of the popular weed killer atrazine (Figure 17.24; Hayes et al. 2002a). Atrazine is the most widely used herbicide in the world; the United States alone uses 60 million pounds of it annually. Hayes and colleagues found that exposing tadpoles to concentrations as low as 0.1 part per billion produced gonadal and other sexual anomalies in male frogs. Many of the male frogs developing at doses 0.1 part per billion and higher had ovaries in addition to

testes. At concentrations of 1 part per billion atrazine, the vocal sacs of the male frogs (which the male must have in order to signal for potential mates) failed to develop properly.

Atrazine is thought to induce aromatase, the enzyme capable of converting testosterone into estrogen (Kniewald et al. 1995; Crain et al.1997). The testosterone levels of adult male frogs were reduced nearly 90% (to control female levels) by the exposure of the frogs to 25 parts per billion atrazine for 46 days. These are ecologically very relevant doses. The allowable amount of atrazine in our drinking water is 3 parts per billion, and atrazine levels can be as high as 224 parts per billion in streams in the midwestern United States (Battaglin et al. 2000; Barbash et al. 2001).

Given the amount of atrazine in the water and the sensitivity of frogs to this compound, this herbicide could be devastating to wild populations. Hayes and his colleagues collected leopard frogs and water at eight sites across the middle of the United States (Hayes et al. 2002b, 2003). They sent the water samples to two separate laboratories for the determination of atrazine, and they coded the frog specimens so that the technicians dissecting the gonads did not know from which site the animals came. The results showed that all but one site contained atrazine, and this was the only site from which the frogs had no gonadal abnormalities. At concentrations as low as 0.1 part per billion, leopard frogs displayed testicular dysgenesis (stunted growth) or conversion to ovaries. In many examples, oocytes were found in the testes (Figure 17.24).Concern over atrazine's apparent ability to disrupt sex hormones in both wildlife and humans has resulted in bans on the use of this herbicide by France, Germany, Italy, Norway, Sweden, and Switzerland (Dalton 2002).

Location-Dependent Sex Determination in Bonellia *and* Crepidula

As mentioned in Chapter 3, the sex of the echiuroid worm *Bonellia viridis* depends on where a larva settles. If a *Bonellia* larva lands on the ocean floor, it develops into a 10-cm-long female. If the larva is attracted to a female's proboscis, it travels along the tube until it enters the female's body. Therein it differentiates into a minute (1–3 mm long) male that is essentially a sperm-producing symbiont of the female (see Figure 3.1).

Another species in which sex determination is affected by the location of the organism is the slipper snail *Crepidula fornicata*. In this species, individuals pile up on top of one another to form a mound.Young individuals are always male. This phase is followed by the degeneration of the male reproductive system and a period of lability. The next phase can be either male or female, depending on the animal's position in the mound. If the snail is attached to a female, it will become male. If such a snail is removed from its attachment, it will become female. Similarly, the presence of large numbers of males will cause some of the males to become females. How-

---

*One remarkable finding is that the injection of an aromatase inhibitor into the eggs of an all-female parthenogenetic species of lizards causes the formation of males (Wibbels and Crews 1994).

(A) 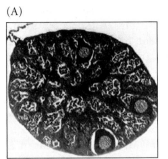 (B)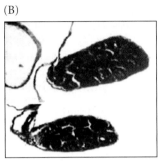

(C)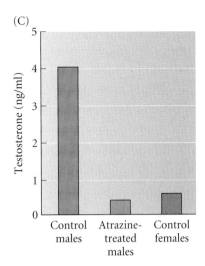

Figure 17.24
Demasculinization of frogs by low amounts of atrazine. (A) Testis of a frog from a natural site having 0.5 parts per billion (ppb) atrazine. The testis contains three lobules that are developing both sperm and an oocyte. (B) Two testes of a frog from a natural site containing 0.8 ppb atrazine. These organs show severe the testicular dysgenesis that characterized 28% of the frogs at that site. (C) Effect of a 46-day exposure to 25 ppb atrazine on plasma testosterone levels in sexually mature male *Xenopus*. Levels in control males were some tenfold higher than in control females; atrazine-treated males had plasma testosterone levels at or below those of the control females. (A, B after Hayes et al., 2003, photographs courtesy of T. Hayes; C after Hayes et al. 2002a.)

ever, once an individual becomes female, it will not revert to being male (Coe 1936). More examples of context-dependent sex determination will be studied in Chapter 22.

Nature has provided many variations on her masterpiece. In some species, including most mammals and insects, sex is determined by chromosomes; in other species, sex is a matter of environmental conditions. We are finally beginning to understand the mechanisms by which this masterpiece is created.

Snapshot Summary: Sex Determination

1. In mammals, primary sex determination (the determination of gonadal sex) is a function of the sex chromosomes. XX individuals are females, XY individuals are males.

2. The Y chromosome plays a key role in male sex determination. XY and XX mammals both have a bipotential gonad that makes the primary sex cords. In XY animals, these cords continue to be formed within the gonad, and eventually differentiate into the Sertoli cells of the testes. The interstitial mesenchyme becomes the Leydig cells.

3. In XX individuals, the internal sex cords degenerate, and a second set of cortical sex cords emerges. These remain on the periphery of the gonad. Germ cells enter the sex cords, but will not be released from the gonad until puberty. The epithelium of the sex cords becomes the granulosa cells; the mesenchyme becomes the thecal cells.

4. In humans, the *SRY* gene is the testis-determining factor on the Y chromosome. It synthesizes a nucleic acid-binding protein that may function as either a transcription factor as as an RNA splicing factor. It is thought to activate SOX9.

5. The *SOX9* gene can also initiate testes formation. It may also function as both a transcription and a splicing factor. It binds to the gene encoding anti-Müllerian duct hormone (AMH), and it may also be responsible for activating *Fgf9*.

6. Wnt4 and Dax1 are involved in ovary formation. Part of their function appears to be inhibiting the testis-forming pathway. The genes generating ovarian tissues are not as well characterized as those inducing testis formation.

7. Secondary sex determination in mammals involves the hormones produced by the developing gonads. In females, the Müllerian duct differentiates into the oviducts, uterus, cervix, and upper portion of the vagina. In male mammals, the Müllerian duct is destroyed by the AMH produced by the Sertoli cells, while the testosterone produced by the Leydig cells enables the Wolffian duct to differentiate into the vas deferens and seminal vesicle. In female mammals, the Wolffian duct degenerates because of the lack of testosterone, while estrogen permits the differentiation of the Müllerian duct.

8. The conversion of testosterone to dihydrotestosterone in the genital rudiment and prostate gland precursor enables the differentiation of the penis, scrotum, and prostate gland.

9. Individuals with mutations of these hormones or their receptors may have a distinction between their primary and secondary sex characteristics.

10. In *Drosophila*, sex is determined by the ratio of X chromosomes to autosomes, and the Y chromosome does not play a role in sex determination. There are no sex hormones, so each cell makes a sex determination decision.

11. The *Drosophila Sex-lethal* gene is activated in females (by proteins encoded on the X chromosomes) and is repressed in males (by factors encoded on the autosomes). Sxl protein acts as an RNA splicing factor to splice an inhibitory exon from the *transformer* transcript. Therefore, female flies have an active Tra protein, while males do not.

12. The Tra protein also acts as an RNA splicing factor to splice exons of the *doublesex* transcript. The *dsx* gene is transcribed in both XX and XY cells, but its pre-mRNA is processed to form different mRNAs, depending on whether Tra protein is present. The proteins translated from both *dsx* messages are active, and they activate or inhibit transcription of a set of genes involved in producing the sexually dimorphic traits of the fly.

13. In turtles and alligators, sex is often determined by the temperature during the time of gonad determination. Since estrogen is necessary for ovary development, it is possible that differing levels of aromatase (the enzyme that can convert testosterone into estrogen) distinguish male from female patterns of gonadal differentiation.

14. Aromatase may be activated by environmental compounds, causing demasculinization of the male gonads in those animals where primary sex determination can be effected by hormones.

15. In some species, such as *Bonellia* and *Crepidula*, sex is determined by the position of the individual with regard to other individuals of the same species.

Literature Cited

Achermann, J. C., M. Ito, P. Hindmarsh and J. Jameson. 1999. A mutation in the gene encoding steroidogenic factor-1 causes XY sex reversal and adrenal failure in humans. *Nature Genet.* 22: 125–126.

Adams, I. R. and A. McLaren. 2002. Sexually dimorphic development of mouse primordial germ cells: Switching from oogenesis to spermatogenesis. *Development* 129: 1155–1164.

Andersson, S., D. M. Berman, E. P. Jenkins and D. W. Russell. 1991. Deletion of steroid 5α-reductase 2 gene in male pseudohermaphroditism. *Nature* 354: 159–161.

Arango, N. A., R. Lovell-Badge and R. E. Behringer. 1999. Targeted mutagenesis of the endogenous mouse *Mis* gene promoter: In vivo definition of genetic pathways of vertebrate sexual development. *Cell* 99: 409–419.

Aristotle. ca. 335 B.C.E. The generation of animals. Translated by A. Platt. *In* J. Barnes (ed.), *The Complete Works of Aristotle*, Vol. 8. Princeton University Press, Princeton, NJ. 1984.

Baker, B. S. 1989. Sex in flies: The splice of life. *Nature* 340: 521–524.

Baker, B. S. and K. A. Ridge. 1980. Sex and the single cell. I. On the action of major loci affecting sex determination in *Drosophila melanogaster*. *Genetics* 94: 383–423.

Baker, B. S., R. N. Nagoshi and K. C. Burtis. 1987. Molecular genetic aspects of sex determination in *Drosophila*. *BioEssays* 6: 66–70.

Baker, B. S., B. J. Taylor and J. C. Hall. 2001. Are complex behaviors specified by directed regulatory genes? Reasoning from fruit flies. *Cell* 105: 13–24.

Barbash, J. E., G. P. Thelin, D. W. Kolpin and R. J. Gilliom. 2001. Major herbicides in ground water: Results from the National Water-Quality Assessment. *J. Environ. Qual.* 30: 831–845.

Bardoni, B. and 11 others. 1994. A dosage sensitive locus at chromosome Xp21 is involved in male to female sex reversal. *Nature Genet.* 7: 497–501.

Barraclough, C. A. and R. A. Gorski. 1962. Studies on mating behavior in the androgen-sterilized female rat in relation to the hypothalamic regulation of sexual behavior. *J. Endocrinol.* 25: 175–182.

Battaglin, W. A., E. T. Furlong, M. R. Burkhardt and C. J. Peter. 2000. Occurrence of sulfonylurea, sulfonamide, imidazolinone, and other herbicides in rivers, reservoirs, and ground water in the midwestern United States 1998. *Sci. Total Environ.* 248:123–133.

Bell, L. R., E. M. Maine, P. Schedl and T. W. Cline. 1988. *Sex-lethal*, a *Drosophila* sex determination switch gene, exhibits sex-specific RNA splicing and sequence similarity to RNA-binding proteins. *Cell* 55: 1037–1046.

Bell, L. R., J. I. Horabin, P. Schedl, and T. W. Cline. 1991. Positive autoregulation of *sex-lethal* by alternative splicing maintains the female determined state in *Drosophila*. *Cell* 65: 229–239.

Belote, J. M., M. B. McKeown, D. J. Andrew, T. N. Scott, M. F. Wolfner and B. S. Baker. 1985a. Control of sexual differentiation in *Drosophila melanogaster*. *Cold Spring Harb. Symp. Quant. Biol.* 50: 605–614.

Bergeron, J. M., D. Crews and J. A. MacLachlan. 1994. PCBs as environmental estrogens: Turtle sex determination as a biomarker of environmental contamination. *Environ. Health Perspect.* 102: 780–781.

Bergeron, J. M., E. Willingham, C. T. Osborn III, T. Rhen and D. Crews. 1999. Developmental synergism of steroidal estrogens in sex determination. *Environ. Health Perspect.* 107: 93–97.

Berkovitz, G. D. and 7 others. 1992. The role of the sex-determining region of the Y chromosome (SRY) in the etiology of 46,XX true hermaphrodites. *Hum. Genet.* 88: 411–416.

Bernstein, R., T. Jenkins, B. Dawson, J. Wagner, G. Devald, G. C. Koo and S. S. Wachtel. 1980. Female phenotype and multiple abnormalities in sibs with a Y chromosome and partial X-chromosome duplication: H-Y antigen and Xg blood group findings. *J. Med. Genet.* 17: 291–300.

Beukeboom, L. W. 1995. Sex determination in Hymenoptera: A need for genetic and molecular studies. *Bioessays* 17: 813–817.

Birk, O. S. and 11 others. 2000. The LIM homeobox gene *LHX9* is essential for mouse gonad formation. *Nature* 403: 909–913.

Bleier, R. 1984. *Science and Gender*. Pergamon, New York.

Boggs, R. T., P. Gregor, S. Idriss, J. M. Belote and M. McKeown. 1987. Regulation of sexual differentiation in *D. melanogaster* via alternative splicing of RNA from the *transformer* gene. *Cell* 50: 739–747.

Breedlove, S. M. 1992. Sexual dimorphism in the vertebrate nervous system. *J. Neurosci.* 12: 4133–4142.

Breedlove, S. M. 1997. Sex on the brain. *Nature* 389: 801.

Bridges, C. B. 1921. Triploid intersexes in *Drosophila melanogaster*. *Science* 54: 252–254.

Bridges, C. B. 1925. Sex in relation to chromosomes and genes. *Am. Nat.* 59: 127–137.

Bull, J. J. 1980. Sex determination in reptiles. *Q. Rev. Biol.* 55: 3–21.

Bull, J. J., W. H. N. Gutzke and D. Crews. 1988. Sex reversal by estradiol in three reptilian orders. *Gen. Comp. Endocrinol.* 70: 425–428.

Burgoyne, P. S., M. Buehr and A. McLaren. 1988. XY follicle cells in ovaries of XX-XY female mouse chimaeras. *Development* 104: 683–688.

Capel, B., K. H. Albrecht, L. L. Washburn and E. M. Eicher. 1999. Migration of mesonephric cells into the mammalian gonad depends on Sry. *Mech. Dev.* 84: 127–131.

Carroll, J. and P. R. Wolpe. 1996. *Sexuality and Gender in Society*. HarperCollins, New York.

Cate, R. L. and 18 others. 1986. Isolation of the bovine and human genes for Müllerian inhibiting substance and expression of the gene in animal cells. *Cell* 45: 685–698.

Christiansen, A. E., E. L. Keisman, S. M. Ahmad and B. S. Baker. 2002. Sex comes in from the cold: The integration of sex and pattern. *Trends Genet.* 18: 510–516.

Cline, T. W. 1983. The interaction between *daughterless* and *Sex-lethal* in triploids: A novel sex-transforming maternal effect linking sex determination and dosage compensation in *Drosophila melanogaster*. *Dev. Biol.* 95: 260–274.

Cline, T. W. 1988. Evidence that *sisterless-a* and *sisterless-b* are two of several discrete "numerator elements" of the X/A sex determination signal in *Drosophila* that switch *Sxl* between two alternative stable expression states. *Genetics* 119: 829–862.

Cline, T. W. 1993. The *Drosophila* sex determination signal: How do flies count to two? *Trends Genet.* 9: 385–390.

Coe, W. R. 1936. Sexual phases in *Crepidula*. *J. Exp. Zool.* 72: 455–477.

Colvin, J., R. P. Green, J. Schmal, B. Capel and D. M. Ornitz. 2001. Male-to-female sex reversal in mice lacking fibroblast growth factor 9. *Cell* 104: 875–889.

Couse, J. F. and K. S. Korach. 2001. Contrasting phenotypes in reproductive tissues of female estrogen receptor null mice. *Ann. N.Y. Acad. Sci.* 948: 1–8.

Couse, J. F., S. C. Hewitt, D. O. Bunch, M. Sar, V. R. Walker, B. J. Davis and K. S. Korach. 1999.

Postnatal sex reversal of the ovaries in mice lacking estrogen receptors alpha and beta. *Science* 286: 2328–2331.

Crain, D. A. and L. J. Guillette Jr. 1998. Reptiles as models of contaminant-induced endocrine disruption. *Anim. Reprod. Sci.* 53: 77–86.

Crain, D. A., L. J. Guillette, Jr., A. A. Rooney and D. B. Pickford. 1997. Alterations in steroidogenesis in alligators (*Alligator mississippiensis*) exposed naturally and experimentally to environmental contaminants. *Environ. Health Perspect.* 105: 528–533.

Crews, D. 2003. Sex determination: where environment and genetics meet. *Evol. Dev.* 5: 50–55.

Dalton, R. 2002. Frogs put in the gender blender by America's favorite herbicide. *Nature* 416: 665–666.

De Jonge, F. H., J.-W. Muntjewerff, A. L. Louwerse and N. E. Van de Poll. 1988. Sexual behavior and sexual orientation of the female rat after hormonal treatment during various stages of development. *Horm. Behav.* 22: 100–115.

De Santa Barbara, P., B. Moniot, F. Poulat and P. Berta. 2000. Expression and subcellular localization of SF-1, SOX9, WT1, and AMH proteins during early human testicular development. *Dev. Dynam.* 217: 293–298.

Desvages, G., M. Girondot and C. Pieau. 1993. Sensitive stages for the effects of temperature on gonadal aromatase activity in embryos of the marine turtle *Dermochelys coriacea*. *Gen. Comp. Endocrinol.* 92: 54–61.

Dorizzi, M., G. Richard-Mercier, G. Desvages and C. Pieau. 1994. Masculinization of gonads by aromatase inhibitors in a turtle with temperature-dependent sex determination. *Differentiation* 58: 1–8.

Eicher, E. M. and L. L. Washburn. 1983. Inherited sex reversal in mice: Identification of a new sex-determining gene. *J. Exp. Zool.* 228: 297–304.

Eicher, E. M. and L. L. Washburn. 1986. Genetic control of primary sex determination in mice. *Annu. Rev. Genet.* 20: 327–360.

Eicher, E. M. and 8 others. 1996. Sex-determining genes on mouse autosomes identified by linkage analysis of C57BL/6J-Y-POS sex reversal. *Nature Genet.* 14: 206–209.

Fausto-Sterling, A. 1992. *Myths of Gender*. Basic Books, New York.

Fausto-Sterling, A. 1995. Animal models for the development of human sexuality: A critical evaluation. *J. Homosexuality* 28: 217–236.

Fisher, C. R., K. H. Graves, A. F. Parlow and E. R. Simpson. 1998. Characterization of mice deficient in aromatase (ArKO) because of a targeted disruption of the *cyp19* gene. *Proc. Natl. Acad. Sci. USA* 95: 6965–6970.

Foster, J. W. and 11 others. 1994. Campomelic dysplasia and autosomal sex reversal caused by mutations in an SRY-related gene. *Nature* 372: 525–530.

Freiman, R. N., S. R. Albright, S. Zheng, W. C. Sha, R. E. Hammer, and R. Tjian. 2002. Re-

quirement of tissue-specific TBP-associated factor TAF$_{11}$105 in ovarian development. *Science* 293: 2084–2087.

Galen, C. ca. 200 C.E. *On the Usefulness of the Parts of the Body*. Translated by M. May. Cornell University Press, Ithaca, NY. 1968.

Geddes, P. and J. A. Thomson. 1890. *The Evolution of Sex*. Walter Scott, London.

Geissler, W. M. and 9 others. 1994. Male pseudohermaphroditism caused by mutations of testicular 17β-hydroxysteroid dehydrogenase 3. *Nature Genet.* 7: 34–39.

Genetics Review Group. 1995. One for a boy, two for a girl? *Curr. Biol.* 5: 37–39.

Giese, K., J. Cox and R. Grosschedl. 1992. The HMG domain of lymphoid enhancer factor 1 bends DNA and facilitates the assembly of functional nucleoprotein structures. *Cell* 69: 185–195.

Gubbay, J. and 8 others. 1990. A gene mapping to the sex-determining region of the mouse Y chromosome is a member of a novel family of embryonically expressed genes. *Nature* 346: 245–250.

Gutzke, W. H. N. and D. B. Chymiy. 1988. Sensitive periods during embryology for hormonally induced sex determination in turtles. *Gen. Comp. Endocrinol.* 71: 265–267.

Hacker, A., B. Capel, P. Goodfellow and R. Lovell-Badge. 1995. Expression of *Sry*, the mouse sex determining gene. *Development* 121: 1603–1614.

Hamer, D. H., S. Hu, V. L. Magnuson, N. Hu and A. M. L. Pattatucci. 1993. A linkage between DNA markers on the X chromosome and male sexual orientation. *Science* 261: 321–327.

Handa, N. and 7 others. 1999. Structural basis for recognition of the *tra* mRNA precursor by the Sex-lethal protein. *Nature* 398: 579–585.

Harris, G. W. and S. Levine. 1965. Sexual physiology of the brain and its experimental control. *J. Physiol.* 181: 379–400.

Hayes, T. B., A. Collins, M. Lee, M. Mendoza, N. Noriega, A. Stuart and A. Vonk. 2002a. Hermaphroditic, demasculinized frogs after exposure to the herbicide atrazine at low ecologically relevant doses. *Proc. Natl. Acad. Sci. USA* 99: 5476–5480.

Hayes, T., K. Haston, M. Tsui, A. Hoang, C. Haeffele and A. Vonk. 2002b. Herbicides: Feminization of male frogs in the wild. *Nature* 419: 895–896.

Hayes, T., K. Haston, M. Tsui, A. Hoang, C. Haeffele and A. Vonk. 2003. Atrazine-induced hermaphroditism at 0.1 ppb in American leopard frogs (*Rana pipiens*): laboratory and field evidence. *Envir. Health Perspec.* in press (D01: 10.1289/chp. 5932).

Hess, R. A., D. Bunick, K.-H. Lee, J. Bahr, J. A. Taylor, K. S. Korach and D. B. Lubahn. 1997. A role for oestrogen in the male reproductive system. *Nature* 390: 509–511.

Horowitz, M. C. 1976. Aristotle and women. *J. Hist. Biol.* 9: 183–213.

Hu, S. and 7 others. 1995. Linkage between sexual orientation and chromosome Xq28 in males but not in females. *Nature Genet.* 11: 248–256.

Huang, B., S. Wang, Y. Ning, A. N. Lamb and J. Bartley. 1999. Autosomal XX sex reversal caused by duplication of SOX9. *Am. J. Med. Genet.* 87: 349–353.

Imperato-McGinley, J., L. Guerrero, T. Gautier and R. E. Peterson. 1974. Steroid 5α-reductase deficiency in man: An inherited form of male pseudohermaphroditism. *Science* 186: 1213–1215.

Jacklin, D. 1981. Methodological issues in the study of sex-related differences. *Dev. Rev.* 1: 266–273.

Jeyasuria, P., W. M. Roosenburg and A. R. Place. 1994. Role of P-450 aromatase in sex determination of the diamondback terrapin, *Malaclemys terrapin*. *J. Exp. Zool.* 270: 95–111.

Jordan, B. K. and 9 others. 2001 Up-regulation of *WNT-4* signaling and dosage-sensitive sex reversal in humans. *Am. J. Hum. Genet.* 68: 1102–1109.

Jost, A. 1953. Problems of fetal endocrinology: The gonadal and hypophyseal hormones. *Recent Prog. Horm. Res.* 8: 379–418.

Kandel, E. R. and J. H. Schwartz. 1985. *Principles of Neural Science.* Elsevier, New York.

Kandel, E. R., J. H. Schwartz and T. M. Jessell. 1995. *Essentials of Neural Science and Behavior.* Appleton and Lange, Norwalk, CT.

Keisman, E. L. and B. S. Baker. 2001. The *Drosophila* sex determination hierarchy modulates *wingless* and *decapentaplegic* signaling to employ *dachshund* sex-specifically in the genital imaginal disc. *Development* 128: 1643–1656.

Keisman, E. L., A. E. Christiansen and B. S. Baker. 2001. The sex determination gene *doublesex* regulates the A/P organizer to direct sex-specific patterns of growth in the *Drosophila* genital imaginal disc. *Dev. Cell* 1: 215–225.

Keyes, L. N., T. W. Cline and P. Schedl. 1992. The primary sex determination signal of *Drosophila* acts at the level of transcription. *Cell* 68: 933–943.

Kniewald, J., V. Osredecki, T. Gojmerac, V. Zechner and Z. Kniewald. 1995. Effect of s-triazine compounds on testosterone metabolism in the rat prostate. *J. Appl. Toxicol.* 15: 215–218.

Koopman, P. 2001. The genetics and biology of vertebrate sex determination. *Cell* 105: 843–847.

Koopman, P., A. Münsterberg, B. Capel, N. Vivian and A. Lovell-Badge. 1990. Expression of a candidate sex-determining gene during mouse testis differentiation. *Nature* 348: 450–452.

Koopman, P., J. Gubbay, N. Vivian, P. Goodfellow and R. Lovell-Badge. 1991. Male development of chromosomally female mice transgenic for *Sry*. *Nature* 351: 117–121.

Langman, J. 1981. *Medical Embryology*, 4th Ed. Williams & Wilkins, Baltimore.

Langman, J. and D. B. Wilson. 1982. Embryology and congenital malformations of the female genital tract. *In* A. Blaustein (ed.), *Pathology of the Female Genital Tract*, 2nd Ed. Springer-Verlag, New York, pp. 1–20.

LeVay, S. 1991. A difference in hypothalamic structure between heterosexual and homosexual men. *Science* 253: 1034–1037.

Ma, L., G. V. Benson, H. Lim, S. K. Dey and R. L. Maas. 1998. *Abdominal B* (*AbdB*) Hoxa genes: Regulation in adult uterus by estrogen and progesterone and repression in Müllerian duct by synthetic estrogen diethylstilbesterol (DES). *Dev. Biol.* 197: 141–154.

Mansour, S., C. M. Hall, M. E. Pembrey and I. D. Young. 1995. A clinical and genetic study of campomelic dysplasia. *J. Med. Genet.* 32: 415–420.

Margarit, E., M. D. Coll, R. Olivo, D. Gómez, A. Soler and F. Ballesta. 2000. *SRY* gene transferred to the long arm of the X chromosome in a Y-positive XX true hermaphrodite. *Am. J. Med. Genet.* 90: 25–28.

Marshall, E. 1995. NIH's "gay gene" study questioned. *Science* 268: 1841.

Marx, J. 1995. Mammalian sex determination: Snaring the genes that divide sexes for mammals. *Science* 269: 1824–1825.

McClung, C. E. 1902. The accessory chromosome sex determinant? *Biol. Bull.* 3: 72–77.

McEwen, B. S., I. Leiberburg, C. Chaptal and L. C. Krey. 1977. Aromatization: Important for sexual differentiation of the neonatal rat brain. *Horm. Behav.* 9: 249–263.

McLaren, A. 1991. Development of the mammalian gonad: The fate of the supporting cell lineage. *Bioessays* 13: 151–156.

McLaren, A. and D. Southee. 1997. Entry of mouse embryonic germ cells into meiosis. *Dev. Biol.* 187: 107–113.

Merke, D. P., S. R. Bornstein, N. A. Avila and G. P. Chrousos. 2002. Future directions in the study and management of congenital adrenal hyperplasia due to 21-hydroxylase deficiency. *Ann. Intern. Med.* 136: 320–334.

Meyer, W. J., B. R. Migeon and C. J. Migeon. 1975. Locus on human X chromosome for dihydrotestosterone receptor and androgen insensitivity. *Proc. Natl. Acad. Sci. USA* 72: 1469–1472.

Migeon, C. J. and A. B. Wisniewski. 2000. Human sex differentiation: From transcription factors to gender. *Horm. Res.* 53: 111–119.

Moore, C. L. 1990. Comparative development of vertebrate sexual behavior: Levels, cascades, and webs. *In* D. A. Dewsbury (ed.), *Contemporary Issues in Comparative Psychology*. Sinauer Associates, Sunderland, MA, pp. 278–299.

Moore, C. L., H. Dou and J. M. Juraska. 1992. Maternal stimulation affects the number of motor neurons in a sexually dimorphic nucleus of the lumbar spinal cord. *Brain Res.* 572: 52–56.

Morgan, T. H. and C. B. Bridges. 1919. The origin of gynandromorphs. *In Contributions to the Study of* Drosophila. Publication no. 278. Carnegie Institution of Washington, Washington, DC, pp. 1–122.

Muscatelli, F. and 14 others. 1994. Mutations in the *DAX-1* gene give rise to both X-linked adrenal hypoplasia congenita and hypogonadotropic hypogonadism. *Nature* 372: 672–634.

Nachtigal, M. W., Y. Hirokawa, D. L. Enyeart-van Houten, J. N. Flanagan, G. D. Hammer and H. A. Ingraham. 1998. Wilms' tumor 1 and Dax-1 modulate the orphan nuclear receptor SF-1 in sex-specific gene expression. *Cell* 93: 445–454.

Nagoshi, R. N., M. McKeown, K. C. Burtis, J. M. Belote and B. S. Baker. 1988. The control of alternative splicing at genes regulating sexual differentiation in *D. melanogaster*. *Cell* 53: 229–236.

Ohe, K. E. Lalli, and P. Sassone-Corsi. 2002. A direct role of SRY and SOX proteins in pre-mRNA splicing. *Proc. Nat. Acad. Sci. USA* 99: 1146–1151.

Pask, A. and J. A. Graves. 1999. Sex chromosomes and sex-determining genes: Insights from marsupials and monotremes. *Cell. Mol. Life Sci.* 55: 864–875.

Phoenix, C. H., R. W. Goy, A. A. Gerall and W. C. Young. 1959. Organizing action of prenatally administered testosterone proprionate on the tissues mediating mating behavior in the female guinea pig. *Endocrinology* 65: 369–382.

Pieau, C., N. Girondot, G. Richard-Mercier, M. Desvages, P. Dorizzi and P. Zaborski. 1994. Temperature sensitivity of sexual differentiation of gonads in the European pond turtle. *J. Exp. Zool.* 270: 86–93.

Pontiggia, A., R. Rimini, P. N. Goodfellow, R. Lovell-Badge and M. E. Bianchi. 1994. Sex-reversing mutations affect the architecture of Sry/DNA complexes. *EMBO J.* 13: 6115–6124.

Poulat, F. and 7 others. 1995. Nuclear localization of the testis-determining gene product SRY. *J. Cell Biol.* 128: 737–748.

Reddy, V. R., F. Naftolin and K. J. Ryan. 1974. Conversion of androstenedione to estrone by neural tissues from fetal and neonatal rats. *Endocrinology* 94: 117–121.

Rhen, T. and J. W. Lang. 1994. Temperature-dependent sex determination in the snapping turtle: Manipulation of the embryonic sex steroid environment. *Gen. Comp. Endocrinol.* 96: 243–254.

Rice, G., C. Anderson, N. Risch and G. Ebers. 1999. Male homosexuality: Absence of linkage to microsatellite markers at Xq28. *Science* 284: 665–667.

Risch, N., E. Squires-Wheeler and B. J. B. Keats. 1993. Male sexual orientation and genetic evidence. *Science* 262: 2063–2065.

Roberts, L. M., Y. Hirokawa, M. W. Nachtigal and H. A. Ingraham. 1999. Paracrine-mediated apoptosis in reproductive tract development. *Development* 208: 110–122.

Ryner, L. C. and B. S. Baker. 1991. Regulation of *doublesex* pre-mRNA processing occurs by 3′ splice site activation. *Genes Dev.* 5: 2071–2085.

Ryner, L. C. and 8 others. 1996. Control of male sexual behavior and sexual orientation in *Drosophila* by the *fruitless* gene. *Cell* 87: 1079–1089.

Salame-Mendez, A. J. Herrera-Munoz, N. Morena-Mendoza and H. Merchant-Larios. 1998. Response of diencephalon but not the gonad to female-promoting temperature with elevated estradiol levels in the sea turtle *Lepidochelys olivacea*. *J. Exp. Zool.* 280: 304–313.

Salz, H. K., T. W. Cline and P. Schedl. 1987. Functional changes associated with structural alterations induced by mobilization of a P element inserted into the *Sex-lethal* gene of *Drosophila*. *Genetics* 117: 221–231.

Salz, H. K., E. M. Maine, L. N. Keyes, M. E. Samuels, T. W. Cline and P. Schedl. 1989. The *Drosophila* female-specific sex-determination gene, *Sex-lethal*, has stage-, tissue-, and sex-specific RNAs suggesting multiple modes of regulation. *Genes Dev.* 3: 708–719.

Sariola, H. and M. Saarma. 1999. GDNF and its receptors in the regulation of ureter branching. *Int. J. Dev. Biol.* 43: 413–418.

Schiebinger, L. 1989. *The Mind Has No Sex?* Harvard University Press, Cambridge, MA.

Shen, W.-H., C. C. D. Moore, Y. Ikeda, K. L. Parker and H. A. Ingraham. 1994. Nuclear receptor steroidogenic factor 1 regulates the Müllerian-inhibiting substance gene: A link to the sex determination cascade. *Cell* 77: 651–661.

Siiteri, P. K. and J. D. Wilson. 1974. Testosterone formation and metabolism during male sexual differentiation in the human embryo. *J. Clin. Endocrinol. Metab.* 38: 113–125.

Sinclair, A. H. and 9 others. 1990. A gene from the human sex-determining region encodes a protein with homology to a conserved DNA-binding motif. *Nature* 346: 240–244.

Smith, C. A. and A. H. Sinclair. 2001. Sex determination in the chicken embryo. *J. Exp. Zool.* 290: 691–699.

Stevens, N. M. 1905. Studies in spermatogenesis with especial reference to the "accessory chromosome." Publication no. 36. Carnegie Institution of Washington, Washington, DC.

Swain, A., V. Narvaez, P. Burgoyne, G. Camerino and R. Lovell-Badge. 1998. Dax1 antagonizes Sry action in mammalian sex determination. *Nature* 391: 761–767.

Thigpen, A. E., D. L. Davis, J. Imperato-McGinley and D. W Russell. 1992. The molecular basis of steroid 5α-reductase deficiency in a large Dominican kindred. *N. Engl. J. Med.* 327: 1216–1219.

Tilmann, C. and B. Capel. 1999. Mesonephric cell migration induces testis cord formation and Sertoli cell differentiation in the mammalian gonad. *Development* 126: 2883–2890.

Tran, D., N. Meusy-Dessolle and N. Josso. 1977. Anti-Müllerian hormone is a functional marker of foetal Sertoli cells. *Nature* 269: 411–412.

Trelstad, R. L., A. Hayashi, K. Hayashi and P. K. Donahoe. 1982. The epithelial-mesenchymal interface of the male rat Müllerian duct: Loss of basement membrane integrity and ductal regression. *Dev. Biol.* 92: 27–40.

Tuana, N. 1988. The weaker seed. *Hypatia* 3: 35–59.

Vainio, S., M. Heikkilä, A. Kispert, A. Chin and A. P. McMahon. 1999. Female development in mammals is regulated by Wnt4 signalling. *Nature* 397: 405–409.

Valcarcel, J., R. Singh, P. D. Zamore and M. R. Green. 1993. The protein Sex-lethal antagonizes the splicing factor U2AF to regulate alternative splicing of transformer pre-mRNA. *Nature* 362: 171–175.

Van Doren, H. M. Ellis and J. W. Posakony. 1991. The *Drosophila* extramacrochaetae protein antagonizes sequence-specific DNA binding by daughterless/achaete-scute protein complexes. *Development* 113: 245–255.

Vidal, V. P. I., M.-C. Chaboissier, D. G. de Rooij and A. Schedl. 2001. Sox9 induces testis development in XX transgenic mice. *Nature Genet.* 28: 216–217.

Wagner, T. and 13 others. 1994. Autosomal sex reversal and campomelic dysplasia are caused by mutations in and around the *SRY*-related gene *SOX9*. *Cell* 79: 1111–1120.

Washburn. L. L. and E. M. Eicher. 1989. Normal testis determination in the mouse depends on genetic interaction of a locus on chromosome 17 and the Y chromosome. *Genetics* 123: 173–179.

Werner, M. H., J. R. Huth, A. M. Groneborn and G. M. Clore. 1995. Molecular basis of human 46X,Y sex reversal revealed from the three-dimensional solution structure of the human SRY–DNA complex. *Cell* 81: 705–714.

Wibbels, T. and, D. Crews. 1994. Putative aromatase inhibitor induces male sex determination in a female unisexual lizard and in a turtle with temperature-dependent sex determination. *J. Endocrinol.* 141: 295–299.

Willingham, E., R. Baldwin, J. K. Skipper and D. Crews. 2000. Aromatase activity during embryogenesis in the brain and adrenal-kidney-gonad of the red-eared slider turtle, a species with temperature-dependent sex determination. *Gen. Comp. Endocr.* 119: 202–207.

Wilson, E. B. 1905. The chromosomes in relation to the determination of sex in insects. *Science* 22: 500–502.

Younger-Shepherd, S., H. Vaessin, E. Bier, L. Y. Jan and Y. N. Jan. 1992. *deadpan*, an essential pan-neural gene encoding an HLH protein, acts as a denominator in *Drosophila* sex determination. *Cell* 70: 911–922.

Zanaria, E. and 13 others. 1994. An unusual member of the nuclear hormone receptor superfamily responsible for X-linked adrenal hypoplasia congenita. *Nature* 372: 635–641.

chapter 18 Metamorphosis, regeneration, and aging

The old order changeth, yielding place to the new.
ALFRED LORD TENNYSON (1886)

The earth-bound early stages built enormous digestive tracts and hauled them around on caterpillar treads. Later in the life-history these assets could be liquidated and reinvested in the construction of an entirely new organism—a flying-machine devoted to sex.
CARROLL M. WILLIAMS (1958)

I'd give my right arm to know the secret of regeneration.
OSCAR E. SCHOTTÉ (1950)

DEVELOPMENT NEVER CEASES. Throughout life, we continuously generate new blood cells, lymphocytes, keratinocytes, and digestive tract epithelium from stem cells. In addition to these continuous daily changes, there are instances in which development after the embryo stages is obvious—sometimes even startling. One of these instances is metamorphosis, the transition from a larval stage to an adult stage. In many species that undergo metamorphosis, a large proportion of the animal's structure changes, and the larva and the adult are unrecognizable as being the same individual (see Figure 2.4). Another startling type of development in the adult is regeneration, the creation of a new organ after the original one has been removed. Some adult salamanders, for instance, can regrow limbs after these appendages have been amputated.

The third category of developmental change in adult organisms is a more controversial area. It encompasses those alterations of form associated with aging. Some scientists believe that the processes of degeneration are not properly part of the study of developmental biology. Other investigators point to genetically determined, species-specific patterns of aging and claim that gerontology, the science of aging, studies an important part of the life cycle and is therefore rightly included in developmental biology. Whatever their relationship to normative development, metamorphosis, regeneration, and aging are poised to be critical topics for the biology of the twenty-first century.

Metamorphosis: The Hormonal Reactivation of Development

In most species of animals, embryonic development leads to a larval stage with characteristics very different from those of the adult organism. Very often, larval forms are specialized for some function, such as growth or dispersal. The pluteus larva of the sea urchin, for instance, can travel on ocean currents, whereas the adult urchin leads a sedentary existence. The caterpillar larvae of butterflies and moths are specialized for feeding, whereas their adult forms are specialized for flight and reproduction, often lacking the mouthparts necessary for eating. The division of functions between larva and adult is often remarkably distinct (Wald 1981). *Cecropia* moths, for example, hatch from eggs and develop as wingless juveniles (caterpillars) for several months. All this development enables them to spend a day or so as fully developed winged insects, and to mate (quickly) before they die. The adults never eat, and in

TABLE 18.1 Summary of some metamorphic changes in anurans

System	Larva	Adult
Locomotory	Aquatic; tail fins	Terrestrial; tailless tetrapod
Respiratory	Gills, skin, lungs; larval hemoglobins	Skin, lungs; adult hemoglobins
Circulatory	Aortic arches; aorta; anterior, posterior, and common jugular veins	Carotid arch; systemic arch; cardinal veins
Nutritional	Herbivorous: long spiral gut; intestinal symbionts; small mouth, horny jaws, labial teeth	Carnivorous: short gut; proteases; large mouth with long tongue
Nervous	Lack of nictitating membrane; porphyropsin, lateral line system, Mauthner's neurons	Development of ocular muscles, nictitating membrane, rhodopsin; loss of lateral line system, degeneration of Mauthner's neurons; tympanic membrane
Excretory	Largely ammonia, some urea (ammonotelic)	Largely urea; high activity of enzymes of ornithine-urea cycle (ureotelic)
Integumental	Thin, bilayered epidermis with thin dermis; no mucous glands or granular glands	Stratified squamous epidermis with adult keratins; well-developed dermis contains mucous glands and granular glands secreting antimicrobial peptides

Source: Data from Turner and Bagnara 1976 and Reilly et al. 1994.

fact have no mouthparts during this brief reproductive phase of the life cycle. As might be expected, the juvenile and adult forms often live in different environments.

During metamorphosis, developmental processes are reactivated by specific hormones, and the entire organism changes morphologically, physiologically, and behaviorally to prepare itself for its new mode of existence. These changes are not solely ones of form. In amphibian tadpoles, for example, metamorphosis involves the maturation of liver enzymes, hemoglobin, and eye pigments, as well as the remodeling of the nervous and digestive systems. Thus, metamorphosis is often a time of dramatic developmental change affecting nearly the entire organism.

Amphibian Metamorphosis

Morphological changes associated with metamorphosis

Amphibians are so named for their ability to undergo metamorphosis—their appellation coming from the Greek *amphi* ("double") and *bios* ("life"). Amphibian metamorphosis is generally associated with morphological changes that prepare an aquatic organism for a primarily terrestrial existence. In **urodeles** (salamanders), these changes include the resorption of the tail fin, the destruction of the external gills, and a change in skin structure. In **anurans** (frogs and toads), the metamorphic changes are more dramatic, with almost every organ subject to modification (Table 18.1; see also Figure 2.4). Regressive changes include the loss of the tadpole's horny teeth and internal gills, as well as the destruction of the tail. At the same time, constructive processes such as limb develop-

ment and dermoid gland morphogenesis are also evident. The means of locomotion changes as the hindlimbs and forelimbs develop and the paddle tail regresses. The tadpole's cartilaginous skull is replaced by the predominantly bony skull of the frog. The horny teeth used for tearing pond plants disappear as the mouth and jaw take a new shape and the tongue muscle develops. Meanwhile, the long intestine characteristic of herbivores is remodeled to suit the more carnivorous diet of the adult frog. The gills regress and the gill arches degenerate. The lungs enlarge, and muscles and cartilage develop for pumping air in and out of the lungs. The sensory apparatus changes, too, as the lateral line system of the tadpole degenerates, and the eye and ear undergo further differentiation (see Fritzsch et al. 1988). The middle ear develops, as does the tympanic membrane characteristic of frog and toad outer ears. In the eye, both nictitating membranes and eyelids emerge.

When an animal changes its habitat and mode of nutrition, one would expect its jaw, digestive tube, and nervous system to undergo dramatic changes, and they certainly do. One readily observed consequence of anuran metamorphosis is the movement of the eyes* forward from their originally lateral position (Figure 18.1). The lateral eyes of the tadpole are typical of preyed-upon herbivores, whereas the frontally located eyes of the frog befit its more predatory lifestyle. To catch its prey, the frog needs to see in three dimensions. That is, it has to acquire a binocular field of vision wherein input from both eyes converges in the brain (see Chapter 13). In the tadpole, the right eye innervates the left side of the brain, and vice

---

*One of the most spectacular movements of eyes during metamorphosis occurs in flatfish such as flounder. Originally, the eyes are on opposite sides of the face. However, during metamorphosis, one of the eyes migrates across the head to meet the eye on the other side (see Hashimoto et al. 2002). This allows the fish to dwell on the bottom, looking upward.

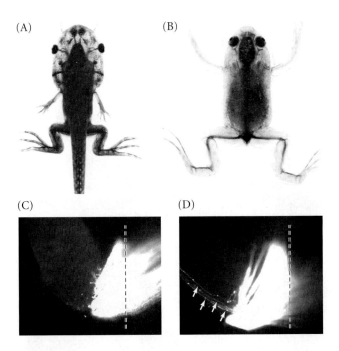

Figure 18.1
Eye migration and associated neuronal changes during metamorphosis of the *Xenopus laevis* tadpole. (A) The eyes of the tadpole are laterally placed, so there is relatively little binocular field of vision. (B) The eyes migrate dorsally and rostrally during metamorphosis, creating a large binocular field for the adult frog. (C, D). Retinal projections of metamorphosing tadpole. The dye DiI was placed on a cut stump of the optic nerve to label the retinal projection. (C) In early and middle stages of metamorphosis, axons project across the midline (dashed line) from one side of the brain to the other. (D) In late metamorphosis, ephrin-B is produced in the optic chiasm as certain neurons (arrows) are formed that project ipsilaterally. (A, B from Hoskins and Grobstein 1984; photographs courtesy of P. Grobstein; C, D from Nakagawa et al. 2000, photographs courtesy of C. E. Holt.)

newly formed adult muscle (Alley and Barnes 1983). Still other neurons, such as those innervating the tongue (a newly formed muscle not present in the larva), have lain dormant during the tadpole stage and first form synapses during metamorphosis (Grobstein 1987). Thus, the anuran nervous system undergoes enormous restructuring during metamorphosis. Some neurons die, others are born, and others change their specificity.

The shape of the skull is also significantly changed, and practically every structural component of the head is remodeled (Figure 18.2; Trueb and Hanken 1992; Berry et al. 1998). The most obvious change is that new bone is being made. The tadpole skull is primarily cartilage; the adult skull is primarily bone. Another outstanding change is the formation of the lower jaw. Here, Meckel's cartilage elongates to nearly double its original length, and dermal bone forms around it. While Meckel's cartilage is growing, the gills and branchial arch cartilage (that had been used for respiration in water) degenerate. Other cartilage, such as the ceratohyal cartilage (which will anchor the tongue), is extensively remodeled. Thus, as in the nervous system, some skeletal elements proliferate, some die, and some are remolded.

The *Xenopus* intestine changes dramatically, from a large structure used for a tadpole's herbivorous diet into a shorter structure more characteristic of carnivores. The formation and differentiation of this new intestinal epithelium is probably triggered by the transcription of the *bmp4* and *sonic hedgehog* genes. The *shh* gene is activated by thyroid hormones during metamorphosis (see below) in the intestine and stomach (Stolow and Shi 1995; Ishizuya-Oka et al. 2001). Therefore, the regional remodeling of the organs formed during metamorphosis may be generated by the reappearance of some of the same paracrine factors that modeled those organs in the embryo.

Biochemical changes associated with metamorphosis

In addition to the obvious morphological changes, important biochemical transformations occur during metamorphosis. In tadpoles, as in freshwater fish, the major retinal photopigment is porphyropsin. During metamorphosis, the pigment changes to rhodopsin, the characteristic photopigment of terrestrial and marine vertebrates (Wald 1945, 1981; Smith-Gill and Carver 1981; Hanken and Hall 1988). Tadpole hemoglobin is changed into an adult hemoglobin that binds oxygen more slowly and releases it more rapidly (McCutcheon 1936; Riggs 1951). The liver enzymes change also, reflecting the change in habitat. Tadpoles, like most freshwater fish, are **ammonotelic**—that is, they excrete ammonia. Like most terrestrial vertebrates, many adult frogs (such as the genus *Rana*, although not the more aquatic *Xenopus*) are **ureotelic**; they excrete urea, which requires less water than ammonia excretion. During metamorphosis, the liver begins to synthesize the urea cycle enzymes necessary to create urea from carbon dioxide

versa; there are no ipsilateral (same-side) projections of the retinal neurons. During metamorphosis, however, these additional ipsilateral pathways emerge, enabling input from both eyes to reach the same area of the brain (Currie and Cowan 1974; Hoskins and Grobstein 1985a).

In *Xenopus*, these new neuronal pathways result not from the remodeling of existing neurons, but from the formation of new neurons that differentiate in response to thyroid hormones (Hoskins and Grobstein 1985a,b). The ability of these axons to project ipsilaterally results from the induction of ephrin-B in the optic chiasm by the thyroid hormones (Nakagawa et al. 2000). Ephrin-B is also found in the optic chiasm of mammals (which have ipsilateral projections throughout life) but not in the chiasm of fish and birds (which have only contralateral projections). As shown in Chapter 13, ephrins can repel certain neurons, causing them to project in one direction rather than in another.

Some larval neurons, such as certain motor neurons in the tadpole jaw, switch their allegiances from larval muscle to

Figure 18.2
Changes in the *Xenopus* head during metamorphosis. Wholemounts were stained with Alcian blue to stain cartilage and alizarin red to stain bone. (A) Prior to metamorphosis, the branchial arch cartilages (open arrowheads) are prominent, Meckel's cartilage (arrows) are at the tip of the head, and the ceratohyal cartilages (arrowheads) are relatively wide and anteriorly placed. (B–D) As metamorphosis ensues, the branchial arch cartilage disappears, Meckel's cartilage elongates, the mandible (lower jawbone) forms around Meckel's cartilage, and the ceratohyal cartilage narrows and becomes more posteriorly located. (After Berry et al. 1998, photographs courtesy of D .D. Brown.)

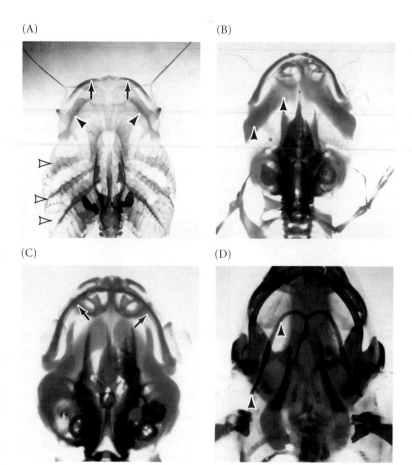

and ammonia (Figure 18.3; Cohen 1970; Atkinson et al. 1996).

Hormonal control of amphibian metamorphosis

The control of metamorphosis by **thyroid hormones** was demonstrated by Gudernatsch (1912), who discovered that tadpoles metamorphosed prematurely when fed powdered horse thyroid gland. In a complementary study, Allen (1916) found that when he removed or destroyed the thyroid rudiment of early tadpoles (thus performing a thyroidectomy), the larvae never metamorphosed, but grew into giant tadpoles. Subsequent studies (Saxén

Figure 18.3
Development of the urea cycle during anuran metamorphosis. (A) Major features of the urea cycle, by which nitrogenous wastes are detoxified and excreted with minimal water loss. (B) The emergence of urea cycle enzyme activities correlates with metamorphic changes in the frog *Rana catesbeiana*. (After Cohen 1970.)

(A)

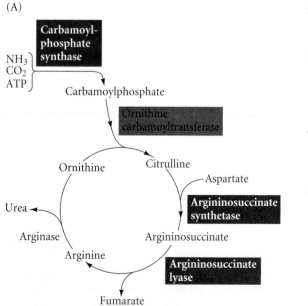

(B)

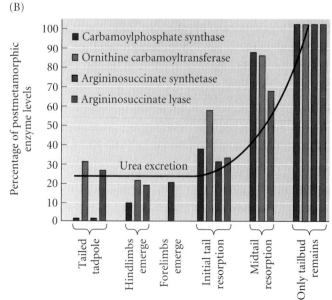

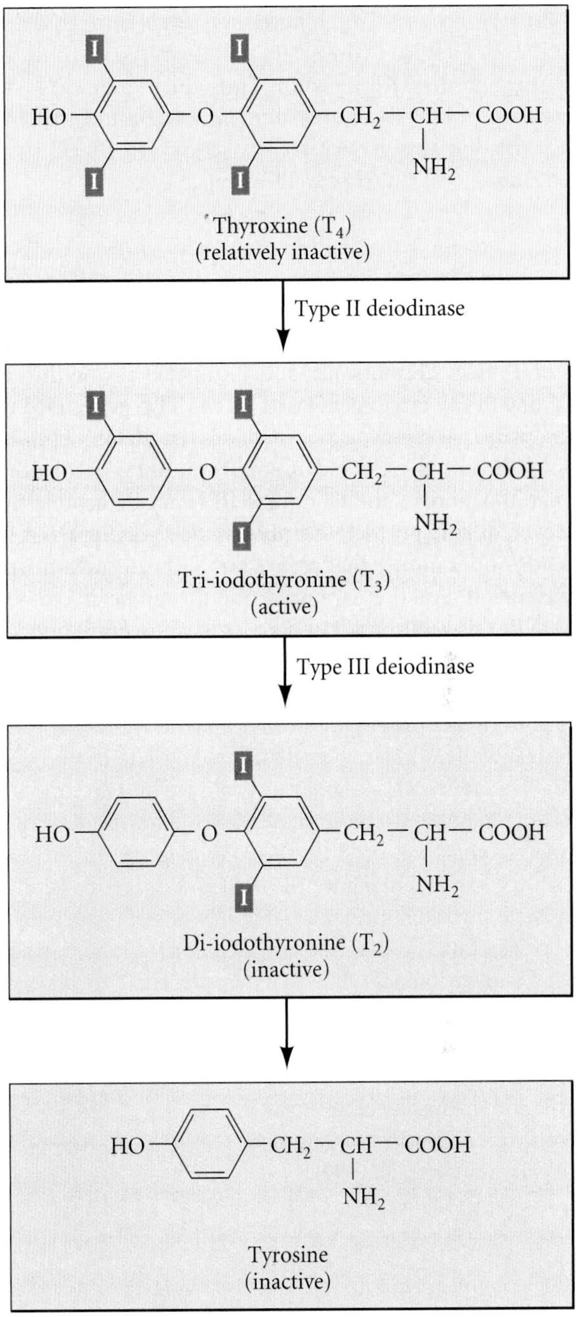

et al. 1957; Kollros 1961; Hanken and Hall 1988) showed that the sequential steps of anuran metamophosis are regulated by increasing amounts of thyroid hormone. Some events (such as the development of limbs) occur early, when the concentration of thyroid hormones is low, while other events (such as the regression of the tail and the remodeling of the intestine) occur later, after the thyroid hormones have reached higher concentrations. These observations gave rise to a **threshold model** wherein the different events of metamorphosis are triggered by different concentrations of thyroid hormones. Although this model remains useful, molecular studies have shown that amphibian metamorphosis is in fact more complex.

The metamorphic changes of frog development are brought about by (1) the secretion of the hormone **thyroxine (T$_4$)** into the blood by the thyroid gland; (2) the conversion of T$_4$ into a more active hormone, **tri-iodothyronine (T$_3$)** by the target tissues; and (3) the degradation of T$_3$ in the target tissues (Figure 18.4). T$_3$ binds to the nuclear **thyroid hormone receptors (TRs)** with much higher affinity than does T$_4$, and causes them to become transcriptional activators of gene expression. Thus, the levels of T$_3$ and TRs in the target tissues are essential for producing the metamorphic response in each tissue (Kistler et al. 1977; Robinson et al. 1977; Becker et al. 1997).

The concentration of T$_3$ in each tissue is regulated by the concentration of T$_4$ in the blood and by two critical intracellular enzymes that remove iodines from T$_4$ and T$_3$. **Type II deiodinase** removes an iodine atom from outer ring of the precursor (T$_4$) to convert it into the more active hormone T$_3$. **Type III deiodinase** removes an iodine atom from the inner ring of T$_3$ to convert it into an inactive compound that will eventually be metabolized to tyrosine (Becker et al. 1997). Tadpoles that are genetically modified to overexpress type III deiodinase in their target tissues never complete metamorphosis (Huang et al. 1999).

There are two types of thyroid hormone receptors as well. In *Xenopus*, **thyroid hormone receptor α (TRα)** is widely distributed throughout all tissues and is present even before the organism has a thyroid gland. **Thyroid hormone receptor β (TRβ)**, however, is the product of a gene that is directly activated by thyroid hormones. TRβ levels are very low before the advent of metamorphosis, and as the levels of thyroid hormone increase during metamorphosis, so do the intracellular levels of TRβ (Yaoita and Brown 1990; Eliceiri and Brown 1994). The TRs do not work alone, however, but form dimers with the retinoid receptor, RXR. Theses dimers bind thyroid hormones and can effect transcription (Mangelsdorf and Evans 1995; Wong and Shi 1995; Wolffe and Shi 1999). The T$_3$-TR-RXR complex is thought to bind to the chromatin of its target genes and activate those genes by a process involving histone acetylation (Sachs et al. 2001).

Metamorphosis is often divided into stages. During the first stage, **premetamorphosis**, the thyroid gland has begun to mature and is secreting low levels of T$_4$ (and very low levels of T$_3$). The initiation of thyroid secretion of T$_4$ may be brought

Figure 18.4
Metabolism of thyroxine (T$_4$) and tri-iodothyronine (T$_3$). Thyroxine serves as a prohormone. It is converted in the peripheral tissues to the active hormone T$_3$ by deiodinase II. T$_3$ can be inactivated by deiodinase III, which converts T$_3$ into di-iodothyronine and then to tyrosine.

about by corticotropin releasing hormone (CRH, which in mammals initiates the stress response). CRH may act directly on the frog pituitary, instructing it to release thyroid stimulating hormone (TSH), or it may act generally to make the body cells responsive to low amounts of T$_3$ (Denver, 1993, 2003).

The limb rudiments, which have high levels of both deiodinase II and TRα, can convert T_4 into T_3 and use it immediately through the TRα receptor. Thus, during the early stage of metamorphosis, the limb rudiments are able to receive thyroid hormone and use it to start leg growth (Becker et al. 1997; Huang et al. 2001; Schreiber et al. 2001).

As the thyroid matures, it secretes more thyroid hormones. This stage is called **prometamorphosis**. However, many of the major changes of metamorphosis (such as tail resorption, gill resorption, and intestinal remodeling) have to wait until **metamorphic climax**. At that time, the concentration of T_4 rises dramatically, and TRβ peaks inside the cells. Since one of the target genes of T_3 is the TRβ gene, TRβ may be the principal receptor that mediates the metamorphic climax. In the tail, there is only a small amount of TRα during prematamorphosis, and deiodinase II is not detectable then. However, during prometamorphosis, the rising levels of thyroid hormones induce higher levels of TRβ. At metamorphic climax, deiodinase II is expressed, and the tail begins to be resorbed. In this way, the tail undergoes its absorption only *after* the legs are functional (otherwise, the poor amphibian would have no means of locomotion). The wisdom of the frog is simple: never get rid of your tail before your legs are working.

Meanwhile, some tissues do not seem to be responsive to thyroid hormones. These tissues are protected by high concentrations of deiodinase III. For instance, thyroid hormones instruct the *ventral* retina to express ephrin-B and to generate the ipsilateral neurons shown in Figure 18.1D. The *dorsal* retina is not responsive to thyroid hormones, and it does not generate neurons. The dorsal retina appears to be insulated from thyroid hormones by expressing deiodinase III, which degrades the T_3 produced by deiodinase II. If deiodinase III is activated in the ventral retina, neurons will not proliferate during metamorphosis, and the ipsilateral axons will not be formed (Marsh-Armstrong et al. 1999; Kawahara et al. 1999).

Another organ that undergoes changes during a frog's metamorphosis is its brain, and one of the brain's functions is to downregulate metamorphosis once metamorphic climax has been reached. Thyroid hormones eventually produce negative feedback, shutting down the pituitary cells that instruct the thyroid to secrete them (Saxén et al. 1957; Kollros 1961; White and Nicoll 1981). Huang and colleagues (2001) have recently shown that at the climax of metamorphosis, deiodinase II expression is seen in those cells of the anterior pituitary that secrete thyrotropin, the hormone that activates thyroid hormone expression. The resulting T_3 suppresses the transcription of the thyrotropin gene, thereby initiating the negative feedback loop so that less thyroid hormone is made.

Regionally specific developmental programs

By regulating the amount of T_3 and TRs in their cells, the different regions of the body can respond at different times to the thyroid hormones. The type of response (proliferation, apoptosis, migration) is determined by other factors already present in the different tissues. The same stimulus causes some tissues to degenerate while causing others to develop and differentiate, as exemplified by the process of tail degeneration.

The degeneration of tail structures is relatively rapid, as the bony skeleton does not extend to the tail, which is supported only by the notochord (Wassersug 1989). The regression of the tail is brought about by apoptosis. After cell death occurs, macrophages collect in the tail region, digesting the debris with their own enzymes, especially collagenases and metalloproteinases. The result is that the tail becomes a large sac of proteolytic enzymes (Figure 18.5; Kaltenbach et al. 1979; Oofusa and Yoshizato 1991; Patterson et al. 1995).

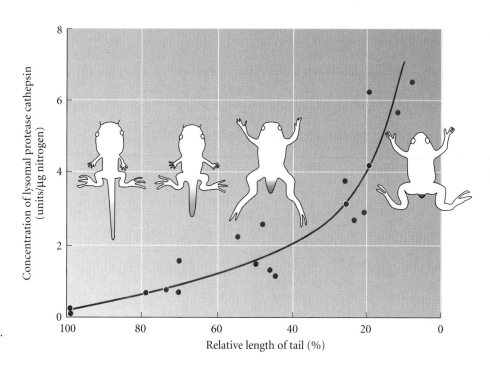

Figure 18.5
Increase in lysosomal protease activity during tail regression in *Xenopus laevis*. The lysosomal enzymes are thought to be responsible for digesting the tail cells. (After Karp and Berrill 1981.)

The tail epidermis acts differently than the head or trunk epidermis. During metamorphic climax, the larval skin is instructed to undergo apoptosis. The tadpole head and body are able to generate a new epidermis from epithelial stem cells. The tail epidermis, however, lacks these stem cells and fails to generate new skin (Suzuki et al. 2002).

Organ-specific responses to thyroid hormones have been dramatically demonstrated by transplanting a tail tip to the trunk region or by placing an eye cup in the tail (Schwind 1933; Geigy 1941). The tail tip placed in the trunk is not protected from degeneration, but the eye retains its integrity despite the fact that it lies within the degenerating tail (Figure 18.6). Thus, the degeneration of the tail represents an organ-specific programmed cell death response. Only specific tissues die when the signal is given. Such programmed cell deaths are important in molding the body. Interestingly, the degeneration of the human tail during week 4 of gestatiom resembles the regression of the tadpole tail (Fallon and Simandl 1978).

VADE MECUM[2] **Amphibian metamorphosis and frog calls.** For photographs of amphibian metamorphosis (and for the sounds of the adult frogs), check out the metamorphosis and frog call sections of the CD-ROM. **[Click on Amphibian]**

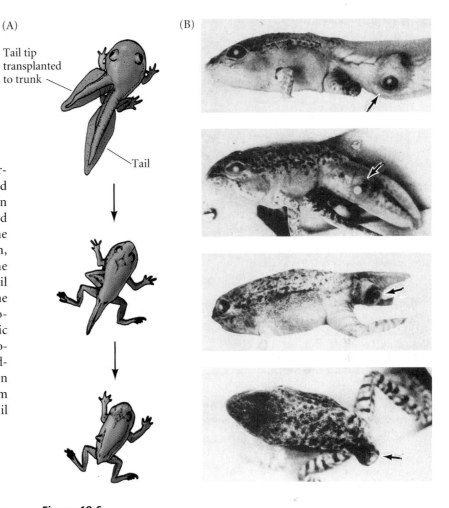

Figure 18.6
Regional specificity during frog metamorphosis. (A) Tail tips regress even when transplanted to the trunk. (B) Eye cups remain intact even when transplanted into the regressing tail. (After Schwind 1933.)

Sidelights & Speculations

Variations on the Theme of Amphibian Metamorphosis

Heterochrony

Most species of animals develop through a larval stage. However, some species have modified their life cycles by either greatly extending or shortening their larval period. The phenomenon wherein animals change the relative time of appearance and rate of development of characters present in their ancestors is called **heterochrony**. Here we will discuss three extreme types of heterochrony:

1. **Neoteny** refers to the retention of the juvenile form owing to the retardation of body development relative to that of the germ cells and gonads, which achieve maturity at the normal time.

2. **Progenesis** also involves the retention of the juvenile form, but in this case, the gonads and germ cells develop at a faster rate than normal and become sexually mature while the rest of the body is still in a juvenile phase.

3. In **direct development**, the embryo abandons the stages of larval development entirely and proceeds to construct a small adult.

Neoteny

In certain salamanders, sexual maturity occurs in what is usually considered a larval stage. The reproductive system and germ cells mature, but the rest of the body retains its juvenile form throughout life. In most such species, metamorphosis fails to occur and sexual maturity takes place in a "larval" body. The Mexican axolotl, *Ambystoma mexicanum*, does not undergo metamorphosis in nature because its pituitary gland does not release a thyrotropin (thyroid-stimulating hormone) to activate T_3 synthesis in its thyroid glands (Prahlad and DeLanney 1965; Norris et al. 1973; Taurog et al. 1974).

Figure 18.7
Metamorphosis in *Ambystoma*. (A) Normal adult *Ambystoma*, with prominent gills and broad tail. (B) Metamorphosed *Ambystoma* not seen in natural populations. This individual was grown in water supplemented with thyroxine. Its gills have regressed and its skin has changed significantly. (Photographs courtesy of K. Crawford.)

When investigators gave *A. mexicanum* either thyroid hormones or thyrotropin, they found that the salamander metamorphosed into an adult form not seen in nature (Figure 18.7; Huxley 1920). Other species, such as *A. tigrinum*, metamorphose only in response to cues from the environment. In part of its range, *A. tigrinum* is a neotenic salamander, paddling its way through the cold ponds of the Rocky Mountains. Its gonads and germ cells mature and the salamander mates successfully while the rest of the body retains its larval form. However, in the warmer regions of its range, the larval form is transitory, leading to the land-dwelling adult tiger salamander. Individuals from neotenic mountain populations of *A. tigrinum* can be induced to undergo metamorphosis simply by placing them in warm water. It thus appears that, in this species, some part of the salamander's metamorphic process will not function at low temperatures. Some salamanders are permanently neotenic, even in the laboratory. Whereas T_3 is able to produce the long-lost adult form of *A. mexicanum*, the neotenic species of *Necturus* and *Siren* remain unresponsive to thyroid hormones (Frieden 1981); their neoteny is permanent.

The molecular bases of these heterochronies are unknown. De Beer (1940) and Gould (1977) have speculated that neoteny is a major factor in the evolution of more complex taxa. By retarding the development of somatic tissues, neoteny may give natural selection a flexible substrate. According to Gould (1977, p. 283), neoteny may "provide an escape from specialization. Animals can relinquish their highly specialized adult forms, return to the lability of youth, and prepare themselves for new evolutionary directions."

Progenesis

In progenesis, gonadal maturation is accelerated while the rest of the body develops normally to a certain stage. Progenesis has enabled some salamander species to find new ecological niches. *Bolitoglossa occidentalis* is a tropical salamander that, unlike other members of its genus, lives in trees. This salamander's webbed feet and small body size suit it for arboreal existence, the webbed feet producing suction for climbing and the small body making such traction efficient. Alberch and Alberch (1981) have shown that *B. occidentalis* resembles juveniles of the related species *B. subpalmata* and *B. rostrata* (whose young are small, with digits that have not yet grown past their webbing; see Figure 23.18). *B. occidentalis* reaches sexual maturity at a much smaller size than its relatives, and this appears to have given it a phenotype that made tree-dwelling possible.

Direct Development

While some animals have extended their larval period of life, others have "accelerated" their development by abandoning their "normal" larval forms. This latter phenomenon, called **direct development**, is typified by frog species that lack tadpoles and by sea urchins that lack pluteus larvae. Elinson and his colleagues (del Pino and Elinson 1983; Elinson 1987) have studied *Eleutherodactylus coqui*, a small frog that is one of the most abundant vertebrates on the island of Puerto Rico. Unlike the eggs of *Rana* and *Xenopus*, the eggs of *E. coqui* are fertilized while they are still within the female's body. Each egg is about 3.5 mm in diameter (roughly 20 times the volume of a *Xenopus* egg). After the eggs are laid, the male gently sits on the developing embryos, protecting them from predators and desiccation (Taigen et al. 1984).

Early *E. coqui* development is like that of most frogs. Cleavage is holoblastic, gastrulation is initiated at a subequatorial position, and the neural folds become elevated from the surface. However, shortly after the neural tube closes, limb buds appear on the surface (Figure 18.8A,B). This early emergence of limb buds is the first indication that development is direct and will not pass through a limbless tadpole stage. Moreover, the development of *E. coqui* is modified such that the modeling of most of its features—including its limbs—does not depend on thyroid hormones. The thyroid does develop, however, and thyroid hormones appear to be critical for the eventual resorption of the tail (which is used as a respiratory, not a locomotor, organ) and of the primitive kidney (Lynn and Peadon 1955). It appears that the thyroid-dependent phase has been pushed back into embryonic growth (Hanken et al. 1992; Callery et al. 2001). What emerges from the egg jelly 3 weeks after fertilization is not a tadpole, but a little frog (Figure 18.8C). Such direct-developing frogs do not need ponds for their larval stages and can therefore colonize habitats that are inaccessible to other frogs. Direct development also occurs in other phyla, in which it is also correlated with a large egg. It seems that if nutrition can be provided in the egg, the life cycle need not have a food-gathering larval stage.

Tadpole-Rearing Behaviors

Most temperate-zone frogs do not invest time or energy in providing for their tadpoles. However, among tropical frogs, there are numerous species in which adult frogs take painstaking care of their tadpoles. An example is the poison arrow frog, *Dendrobates*, found in the rain forests of Central America. Most of the time, these highly toxic frogs live in the leaf litter of the forest floor. After the eggs are laid in a damp leaf, a parent (sometimes the male, sometimes the female) stands guard over the eggs. If the ground gets too dry, the frog will urinate on the eggs to keep them

(A)

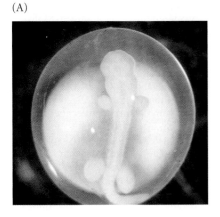

(B)

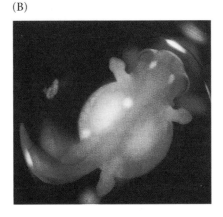

(C)

Figure 18.8
Direct development of the frog *Eleutherodactylus coqui*. (A) Limb buds are seen as the embryo develops on the yolk. (B) As the yolk is used up, the limb buds are easily seen. (C) Three weeks after fertilization, tiny froglets hatch. They are seen here in a petri dish and on a Canadian dime. (Photographs courtesy of R. P. Elinson.)

moist. When the eggs mature into tadpoles, the guarding parent allows them to wriggle onto its back (Figure 18.9A). The frog then climbs into the canopy until it finds a bromeliad with a small pool of water in its leaf base. Here it deposits one of its tadpoles, then goes back for another, and so on, until the brood has been placed in numerous small pools. Then, each day, the female returns to these pools and deposits a small number of unfertilized eggs into them, replenishing the dwindling food supply for the tadpoles until they finish metamorphosis (Mitchell 1988; van Wijngaarden and Bolanos 1992; Brust 1993). It is not known how the female frog remembers—or is informed about—where the tadpoles have been deposited.

Marsupial frogs carry their developing eggs in depressions in their skin, and they often brood their tadpoles in their mouths. When the tadpoles undergo metamorphosis, the frogs spit out their progeny. Even more impressive, the gastric-brooding frogs of Australia, *Rheobatrachus silus* and *R. vitellinus*, eat their eggs. The eggs develop into larvae, and the larvae undergo metamorphosis in the mother's stomach. About 8 weeks after being swallowed alive, about two dozen small frogs emerge from the female's mouth (Figure 18.9B; Corben et al. 1974; Tyler 1983). What stops the *Rheobatrachus* eggs from being digested or excreted? It appears that the eggs secrete prostaglandins that stop acid secretion and prevent peristaltic contractions in the stomach (Tyler et al. 1983). During this time, the stomach is fundamentally a uterus, and the frog does not eat. After the oral birth, the parent's stomach morphology and function return to normal. Unfortunately, both of these remarkable frog species are now feared extinct. No member of either species has been seen since the mid-1980s.

(A)

(B)

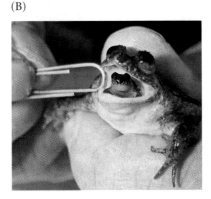

Figure 18.9
Parental care of tadpoles. (A) Tadpoles of the poison arrow frog *Dendrobates* are carried on their parent's back to small pools of water in the Peruvian rain forest canopy. (B) This female *Rheobatrachus* of Australia brooded over a dozen tadpoles in her stomach. They emerged after completing metamorphosis. The last time anyone saw a *Rheobatrachus* frog alive was in 1985. (A, photograph and M. Fogden/© M. and P. Fogden/Corbis; B, photograph courtesy of M. Tyler.)

Metamorphosis in Insects

Types of insect metamorphosis

Whereas amphibian metamorphosis is characterized by the remodeling of existing tissues, insect metamorphosis primarily involves the destruction of larval tissues and their replacement by an entirely different population of cells. Insects grow by molting—shedding the cuticle—and forming a new cuticle as their size increases. There are three major patterns of insect development. A few insects, such as springtails and mayflies, have no larval stage and undergo direct development. These are called the **ametabolous** insects (Figure 18.10A). These insects have a **pronymph** stage immediately after hatching, bearing the structures that have enabled it to get out of the

(A) AMETABOLOUS DEVELOPMENT (B) HEMIMETABOLOUS DEVELOPMENT (C) HOLOMETABOLOUS DEVELOPMENT

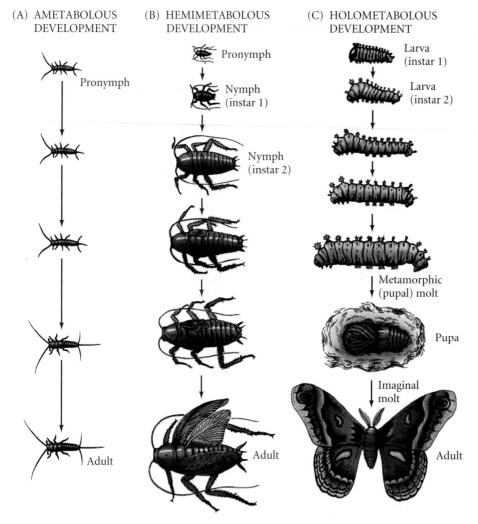

Figure 18.10
Modes of insect development. Molts are represented as arrows. (A) Ametabolous (direct) development in a silverfish. After a brief pronymph stage, the insect looks like a small adult. (B) Hemimetabolous (gradual) metamorphosis in a cockroach. After a very brief pronymph phase, the insect becomes a nymph. After each molt, the next nymphal instar looks more like an adult, gradually growing wings and genital organs. (C) Holometabolous (complete) metamorphosis in a moth. After hatching as a larva, the insect undergoes successive larval molts until a metamorphic molt causes it to enter the pupal stage. Then an imaginal molt turns it into an adult.

larval instars grow in a stepwise fashion, each instar being larger than the previous one. Finally, there is a dramatic and sudden transformation between the larval and adult stages. After the final instar, the larva undergoes a **metamorphic molt** to become a **pupa**. The pupa does not feed, and its energy must come from those foods it ingested as a larva. During pupation, the adult structures are formed and replace the larval structures. Eventually, an **imaginal molt** enables the adult (**imago**) to shed the pupal case and emerge. While the larva is said to *hatch* from an egg, the adult is said to *eclose* from the pupa.

Imaginal discs

In holometabolous insects, the transformation from juvenile into adult occurs within the pupal cuticle. Most of the old body of the larva is systematically destroyed by programmed cell death, while new adult organs develop from undifferentiated nests of **imaginal cells**. Thus, within any larva, there are two distinct populations of cells: the larval cells, which are used for the functions of the juvenile insect, and thousands of imaginal cells, which lie within the larva in clusters, awaiting the signal to differentiate.

There are three main types of imaginal cells:

1. The **imaginal discs**, whose cells will form the cuticular structures of the adult, including the wings, legs, antennae, eyes, head, thorax, and genitalia (Figure 18.11).
2. The **histoblast nests** are clusters of imaginal cells that will form the adult abdomen.

egg. But after this transitory stage, the insect begins to look like a small adult; after each molt, it is larger but unchanged in form (Truman and Riddiford 1999). Other insects, notably grasshoppers and bugs, undergo a gradual, **hemimetabolous** metamorphosis (Figure 18.10B). After spending a very brief period of time as a pronymph (whose cuticle is often shed as the insect hatches), the insect looks like an immature adult. This immature stage is called a **nymph**. The rudiments of the wings, genital organs, and other adult structures are present, and these structures become more mature with each molt. At the last molt, the emerging insect is a winged and sexually mature adult (**imago**).

In the **holometabolous** insects such as flies, beetles, moths, and butterflies, there is no pronymph stage (Figure 18.10C). The juvenile form that hatches from the egg is called a **larva**. The larva (a caterpillar, grub, or maggot) undergoes a series of molts as it becomes larger. The stages between these larval molts are called **instars**. The number of molts before becoming an adult is characteristic of a species, although environmental factors can increase or decrease the number. The

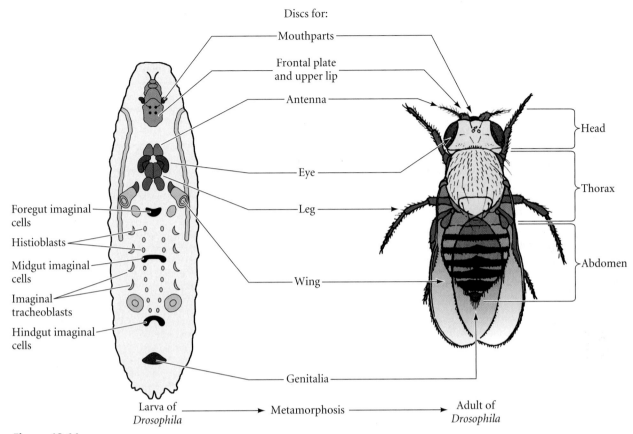

Discs for:

Mouthparts

Frontal plate and upper lip

Antenna

Eye

Leg

Wing

Genitalia

Foregut imaginal cells

Histioblasts

Midgut imaginal cells

Imaginal tracheoblasts

Hindgut imaginal cells

Larva of *Drosophila* → Metamorphosis → Adult of *Drosophila*

Head

Thorax

Abdomen

Figure 18.11
The locations and developmental fates of the imaginal discs and imaginal tissues in the third instar larva of *Drosophila melanogaster*. (After Kalm et al. 1995.)

3. Clusters of imaginal cells within each organ, which will proliferate to form the adult organ as the larval organ degenerates.

The imaginal discs can be seen in the newly hatched larva as local thickenings of the epidermis. Whereas most of the larval cells have a very limited mitotic capacity, imaginal discs divide rapidly at specific characteristic times. As the cells proliferate, they form a tubular epithelium that folds in upon itself in a compact spiral (Figure 18.12). The largest disc, that of the wing, contains some 60,000 cells, whereas the leg and haltere discs contain about 10,000 cells each (Fristrom 1972). At metamorphosis, these cells proliferate, differentiate, and elongate (Figure 18.12B).

The fate map and elongation sequence of the *Drosophila* leg disc are shown in Figure 18.13. At the end of the third instar, just before pupation, the leg disc is an epithelial sac connected by a thin stalk to the larval epidermis. On one side of the sac, the epithelium is coiled into a series of concentric folds "reminiscent of a Danish pastry" (Kalm et al. 1995). As pupation begins, the cells at the center of the disc telescope out to become the most distal portions of the leg—the claws

(A)

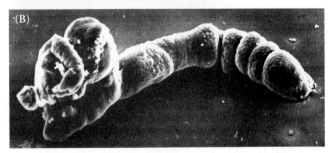

(B)

Figure 18.12
Imaginal disc elongation. Scanning electron micrograph of *Drosophila* third-instar leg disc (A) before and (B) after elongation. (From Fristrom et al. 1977; photograph courtesy of D. Fristrom.)

Embryo (specification)

Figure 18.13
Sequence of leg imaginal disc development in *Drosophila*. Specification of the disc type occurs within the embryo. The proliferation of the disc cells and the specification as to the type of leg cell each will produce is accomplished in the larval stages. The elongation of the disc takes place in the early pupal ("prepupa") stage, and the differentiation of the leg tissues occurs while the insect is a pupa. T_1, basitarsus; T_{2-5}, tarsal segments 2–5. (After Fristrom and Fristrom 1975; Kalm et al. 1995.)

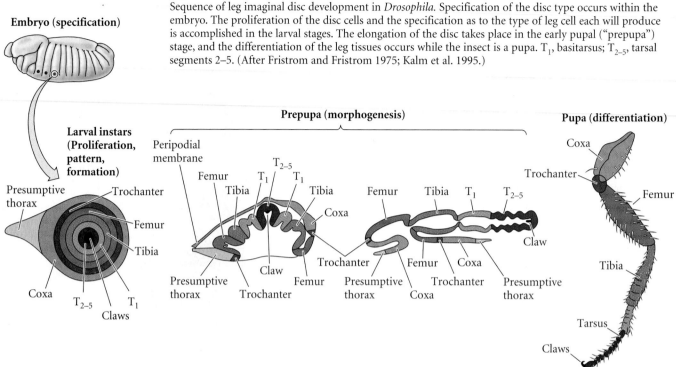

and the tarsus. The outer cells become the proximal structures—the coxa and the adjoining epidermis (Schubiger 1968). After differentiating, the cells of the appendages and epidermis secrete a cuticle appropriate for the specific region. Although the disc is composed primarily of epidermal cells, a small number of **adepithelial cells** migrate into the disc early in development. During the pupal stage, these cells give rise to the muscles and nerves that serve the structure.

SPECIFICATION AND PROLIFERATION. The specification of the general cell fates (that the disc is to be a leg disc and not a wing disc, for instance) occurs in the embryo. The more specific cell fates are specified in the larval stages, as the cells proliferate (Kalm et al. 1995). The type of leg structure generated is determined by the interactions between several genes in the imaginal disc. Figure 18.14 shows the expression of three genes involved in determining the proximal-distal axis of the fly leg. In the third-instar leg disc, the center of the disc secretes the highest concentration of two morphogens, Wingless (Wg) and Decapentaplegic (Dpp). High concentrations of these paracrine factors cause the expression of the *Distal-less* gene. Moderate concentrations cause the expression of the *dachshund* gene, and lower concentrations cause the expression of the *homothorax* gene. Those cells expressing *Distal-less* telescope out to become the most distal structures of the leg—the claw and distal tarsal segments. Those expressing *homothorax* become the most proximal structure, the coxa. Cells expressing *dachshund* become the femur and proximal tibia. Areas of overlap produce the trochanter and distal tibia (Abu-Shaar and Mann 1998).

These regions of gene expression are stabilized by inhibitory interactions between the protein products of these genes and of the neighboring genes. In this manner, the gradient of Wg and Dpp proteins is converted into discrete domains of gene expression that specify the different regions of the *Drosophila* leg.

EVERSION AND DIFFERENTIATION. The mature leg disc in the third instar of *Drosophila* does not look anything like the adult structure. It is determined, but not yet differentiated. Its differentiation requires a hormonal signal. This signal is a set of pulses of the hormone **20-hydroxyecdysone (20E)**. The first pulse, given in the late larval stages, initiates the formation of the pupa, arrests cell division in the disc, and initiates the cell shape changes that drive the eversion of the leg. Studies by Condic and her colleagues in the Fristrom laboratory have demonstrated that the elongation of imaginal discs occurs without cell division and is due primarily to cell shape changes within the disc epithelium (Condic et al. 1990). Using fluorescently labeled phalloidin to stain the peripheral microfilaments of leg disc cells, they showed that the cells of early third-instar discs are tightly arranged along the proximal-distal axis. Then, when the hormonal signal to differentiate is given, the cells change their shape, and the leg is everted, the central cells of the disc becoming the most distal (claw) cells of the limb. The leg structures will differentiate within the pupa, so that by the time the adult fly ecloses, they are fully formed and functional.

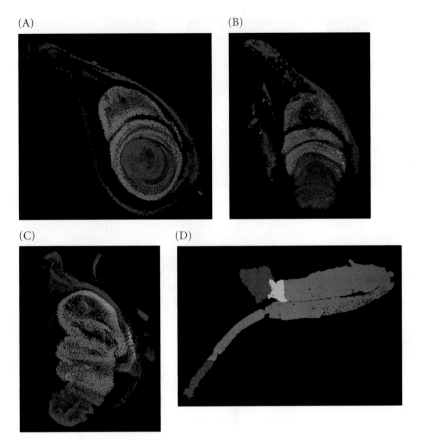

(A)   (B)

(C)   (D)

Figure 18.14

The fates of the imaginal disc cells are directed by transcription factors found in different regions. (A–D) Expression of transcription factor genes in the *Drosophila* leg disc. At the periphery, the *homothorax* gene (purple) establishes the boundary for the coxa. The expression of the *dachshund* gene (green) locates the femur and proximal tibia. The most distal structures, the claw and lower tarsal segments, arise from the expression domain of *Distal-less* (red) in the center of the imaginal disc. The overlap of *dachshund* and *Distal-less* appears yellow and specifies the distal tibia and upper tarsal segments. (A–C) Gene expression at successively later stages of pupal development. (D) Localization of expression domains of the genes onto a leg immediately prior to eclosion. The areas where there is overlap between expression domains are shown in yellow, aqua, and orange. (From Abu-Shaar and Mann 1998; photographs courtesy of R. S. Mann.)

Determination of the wing imaginal discs

The wing imaginal discs of *Drosophila* are distinguished from the fly's other imaginal discs by the expression of the *vestigial* gene (Kim et al. 1996). When this gene is expressed in any other imaginal disc, wing tissue emerges.

The axes of the wing are specified by gene expression patterns that divide the embryo into discrete, interacting compartments (Figure 18.15; Meinhart 1980; Caruso et al. 1993; Tabata et al. 1995). In the first instar, expression of the *en-*

Figure 18.15

Compartmentalization and anterior-posterior patterning in the wing disc. (A) In the first-instar larva, the anterior-posterior axis has been formed and is manifested by the expression of the *engrailed* gene in the posterior compartment. The Engrailed protein functions as a transcription factor to activate the *hedgehog* gene. Hedgehog acts as a short-range paracrine factor to activate the *decapentaplegic* gene in the anterior cells adjacent to the posterior compartment. The Decapentaplegic protein (Dpp) acts as a long-range paracrine factor spreading in both directions from its source. (B) Dpp forms a concentration gradient, wherein high concentrations of Dpp near the source activate the *spalt* and *omb* genes. Lower concentrations (near the periphery) activate *omb* without activating *spalt*.

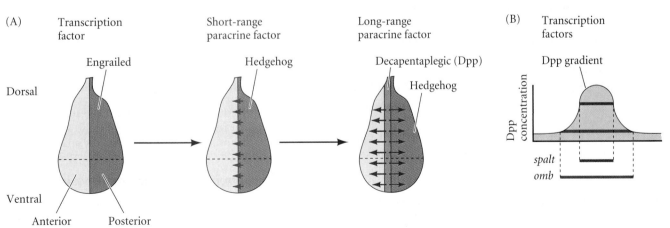

(A)   Transcription factor   Short-range paracrine factor   Long-range paracrine factor   (B)   Transcription factors

Engrailed   Hedgehog   Decapentaplegic (Dpp)   Dpp gradient

Dorsal   Hedgehog

Ventral

Anterior compartment   Posterior compartment

Dpp concentration

spalt
omb

Figure 18.16
Determining the dorsal-ventral axis. (A) The prospective ventral surface of the wing is stained by antibodies to the Vestigial protein (green), while the prospective dorsal surface is stained by antibodies to the Apterous protein (red). The region of yellow shows where the two proteins overlap. (B) The Wingless protein (purple) synthesized at this juncture organizes the wing disc along the dorsal-ventral axis. The expression of the Vestigial protein (green) is seen in those cells close to those expressing Wingless. (C) Anterior-posterior and dorsal-ventral boundaries. Apterous protein (red) is seen in the dorsal compartment; Vestigial protein (green) is seen in the ventral compartment. The yellow area is the overlap. Cubitus interruptus (blue) (induced by Dpp expression) marks the anterior compartment as distinct from the posterior one. (D) The dorsal and ventral portions of the wing disc telescope out to form the two-layered wing. Gene expression patterns are indicated on the double-layered wing. (Photographs courtesy of S. Carroll and S. Paddock.)

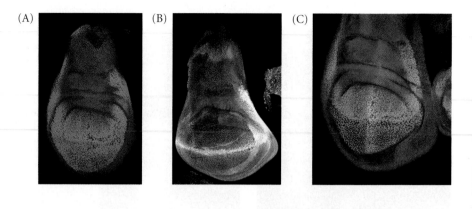

(A)   (B)   (C)

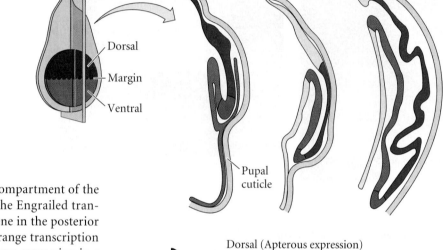

(D)
Anterior   Posterior

Dorsal
Margin
Ventral

Pupal cuticle

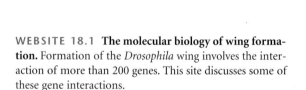

Dorsal (Apterous expression)

Ventral (Vestigial expression)

Margin (Wingless expression)

grailed gene distinguishes the posterior compartment of the wing from the anterior compartment. The Engrailed transcription factor activates the *hedgehog* gene in the posterior compartment, Hedgehog acts as a short-range transcription factor to turn on *decapentaplegic* (*dpp*) gene expression in a narrow range of cells in the anterior region of the wing disc. Dpp acts as a long-range diffusible paracrine factor, setting up a concentration gradient. High concentrations of Dpp activate both the *spalt* and *oculomotor blind* (*omb*) genes, whereas at low concentrations only *omb* is activated. These genes express transcription factors that activate the proteins that form the appropriate regions of the wing.

The dorsal-ventral axis of the wing is formed at the second instar stage by expression of the *apterous* gene in the prospective dorsal cells of the wing disc (Blair 1993; Diaz-Benjumea and Cohen 1993). Here, the upper layer of the wing is distinguished from the lower layer of the wing blade (see Bryant 1970; Garcia-Bellido et al. 1973). The *vestigial* gene remains "on" in the ventral portion of the wing disc (Figure 18.16). At the boundary between the dorsal and ventral compartments, the Apterous and Vestigial transcription factors interact to activate the *wingless* gene. Neumann and Cohen (1996) showed that Wingless protein acts as a growth factor to promote the cell proliferation that extends the wing. Wingless also helps establish the proximal-distal axis of the wing: high levels of Wingless activate the *Distal-less* gene, which specifies the most distal regions of the wing (Zecca et al. 1996; Neumann and Cohen 1996).

WEBSITE 18.1 **The molecular biology of wing formation.** Formation of the *Drosophila* wing involves the interaction of more than 200 genes. This site discusses some of these gene interactions.

WEBSITE 18.2 **Homologous specification.** If a group of cells in one imaginal disc are mutated such that they give rise to a structure characteristic of another imaginal disc (for instance, cells from a leg disc giving rise to antennal structures), the regional specification of those structures will be in accordance with their position in the original disc.

Hormonal control of insect metamorphosis

Although the detailed mechanisms of insect metamorphosis differ among species, the general pattern of hormonal action is very similar. Like amphibian metamorphosis, the metamorphosis of insects is regulated by systemic hormonal signals, which are controlled by neurohormones from the brain (for reviews, see Gilbert and Goodman 1981; Riddiford 1996). Insect molting and metamorphosis are controlled by two effector hormones: the steroid **20-hydroxyecdysone** (**20E**) and the lipid **juvenile hormone** (**JH**) (Figure 18.17). 20-hydroxyecdysone initiates and coordinates each molt and regulates the changes in gene expression that occur during metamorphosis. Juvenile hormone prevents the ecdysone-induced changes in gene expression that are necessary for metamorphosis. Thus, its presence during a molt ensures that the result of that molt is another larval instar, not a pupa or an adult.

The molting process is initiated in the brain, where neurosecretory cells release **prothoracicotropic hormone** (**PTTH**)

in response to neural, hormonal, or environmental signals. PTTH is a peptide hormone with a molecular weight of approximately 40,000, and it stimulates the production of ecdysone by the **prothoracic gland**. This ecdysone is modified in peripheral tissues to become the active molting hormone

Figure 18.17
Regulation of insect metamorphosis. (A) Structures of juvenile hormone, ecdysone, and the active molting hormone 20-hydroxyecdysone. (B) General pathway of insect metamorphosis. Ecdysone and juvenile hormone together cause molts that form another larval instar. When there is a lower concentration of juvenile hormone, the ecdysone-induced molt produces a pupa. When ecdysone acts in the absence of juvenile hormone, the imaginal discs differentiate, and the molt gives rise to an adult. (After Gilbert and Goodman 1981.)

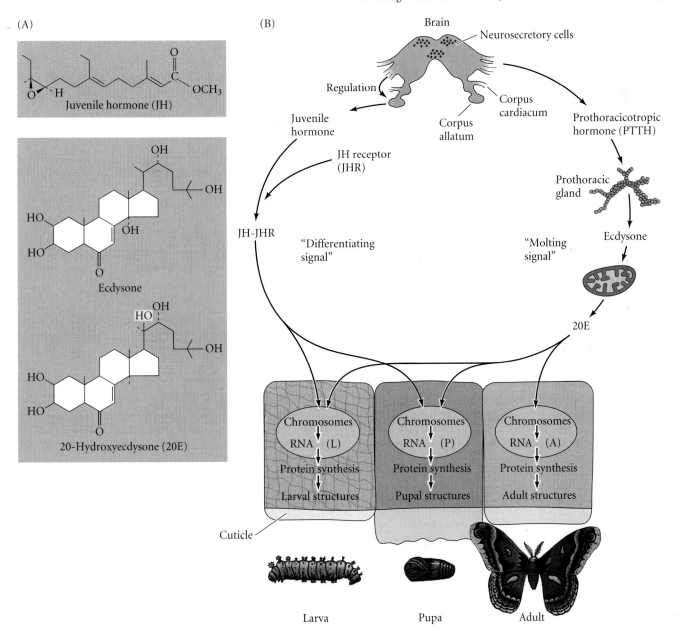

20E. Each molt is initiated by one or more pulses of 20E. For a larval molt, the first pulse produces a small rise in the 20E concentration in the larval hemolymph (blood) and elicits a change in cellular commitment in the epidermis. A second, larger pulse of 20E initiates the differentiation events associated with molting. These pulses of 20E commit and stimulate the epidermal cells to synthesize enzymes that digest and recycle the components of the cuticle.

Juvenile hormone is secreted by the **corpora allata**. The secretory cells of the corpora allata are active during larval molts but are inactive during the metamorphic molt. As long as JH is present, the 20E-stimulated molts result in a new larval instar. In the last larval instar, however, the medial nerve from the brain to the corpora allata inhibits these glands from producing JH, and there is a simultaneous increase in the body's ability to degrade existing JH (Safranek and Williams 1989). Both these mechanisms cause JH levels to drop below a critical threshold value. This triggers the release of PTTH from the brain (Nijhout and Williams 1974; Rountree and Bollenbacher 1986). PTTH, in turn, stimulates the prothoracic glands to secrete a small amount of ecdysone. The resulting 20E, in the absence of high levels of JH, commits the epidermal cells to pupal development. Larva-specific mRNAs are not replaced, and new mRNAs are synthesized whose protein products inhibit the transcription of the larval messages.

After the second ecdysone pulse, new pupa-specific gene products are synthesized (Riddiford 1982), and the subsequent molt shifts the organism from larva to pupa. It appears, then, that the first ecdysone pulse during the last larval instar triggers the processes that inactivate the larva-specific genes and prepare the pupa-specific genes to be transcribed. The second ecdysone pulse transcribes the pupa-specific genes and initiates the molt (Nijhout 1994). At the imaginal molt, when ecdysone acts in the absence of juvenile hormone, the imaginal discs differentiate, and the molt gives rise to an adult.

WEBSITE 18.3 **Insect metamorphosis.** Four websites discuss (1) the experiments of Wigglesworth and others who identified the hormones of metamorphosis and the glands producing them; (2) the variations that *Drosophila* and other insects play on the general theme of metamorphosis; (3) the remodeling of the insect nervous system during metamorphosis, and (4) a microarray analysis of *Drosophila* metamorphosis wherein several thousand genes were simultaneously screened.

The molecular biology of 20-hydroxyecdysone activity

ECDYSONE RECEPTORS. 20-hydroxyecdysone cannot bind to DNA by itself. Like amphibian thyroid hormones, 20E first binds to nuclear receptors. These proteins, called ecdysone receptors (EcRs), are almost identical in structure to the thyroid hormone receptors of amphibians. An EcR protein forms an active molecule by pairing with an Ultraspiracle (Usp) pro-

tein, the homologue of the amphibian RXR that helps form the active thyroid hormone receptor (Koelle et al. 1991; Yao et al. 1992; Thomas et al. 1993). In the absence of the hormone-bound EcR, the Usp protein binds to the ecdysone-responsive genes and inhibits their transcription.* This inhibition is converted into activation when the ecdysone receptor binds to the Usp (Schubiger and Truman 2000).

Although there is only one gene for EcR, the EcR mRNA transcript can be spliced in at least three different ways to form three distinct proteins. All three EcR proteins have the same domains for 20E and DNA binding, but they differ in their amino-terminal domains. The type of EcR in a cell may inform the cell how to act when it receives a hormonal signal (Talbot et al. 1993; Truman et al. 1994). All cells appear to have some EcRs of each type, but the strictly larval tissues and neurons that die when exposed to 20E are characterized by their abundance of the EcR-B1 isoform of the ecdysone receptor. Imaginal discs and differentiating neurons, on the other hand, show a preponderance of the EcR-A isoform. It is therefore possible that the different receptors activate different sets of genes when they bind 20E.

BINDING OF 20-HYDROXYECDYSONE TO DNA. During molting and metamorphosis, certain regions of the polytene chromosomes of *Drosophila* puff out in the cells of certain organs at certain times (see Figure 4.12; Clever 1966; Ashburner 1972; Ashburner and Berondes 1978). These chromosome puffs represent areas where DNA is being actively transcribed. Moreover, these organ-specific patterns of chromosome puffing can be reproduced by culturing larval tissue and adding hormones to the medium or by adding 20E to an early-stage larva. When 20E is added to larval salivary glands, certain puffs are produced and others regress (Figure 18.18). The puffing is mediated by the binding of 20E at specific places on the chromosomes; fluorescent antibodies against 20E find this hormone localized to the regions of the genome that are sensitive to it (Gronemeyer and Pongs 1980).

20E-regulated chromosome puffs occurring during the late stages of the third-instar *Drosophilia* larva (as it prepares to form the pupa) can be divided into three categories: "early" puffs that 20E induces rapidly; "intermolt" puffs that 20E causes to regress; and "late" puffs that are first seen several hours after 20E stimulation. For example, in the larval salivary gland, about six puffs emerge within a few minutes of hydroxyecdysone treatment. No new protein has to be made in order for these early puffs to be induced. A much larger set of puffs is induced later in development, and these late puffs do need protein synthesis to become transcribed. Ashburner (1974, 1990) hypothesized that the "early puff" genes make a protein product that is essential for the activation of the "late puff"

---

*The Ultraspiricle protein may be the receptor for juvenile hormone (JHR; see Figure 18.17), thereby suggesting mechanisms whereby JH can block 20E at the level of transcription (Jones et al. 2001).

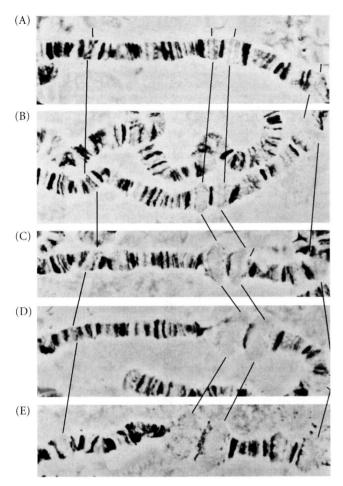

Figure 18.18
20E-induced puffs in cultured salivary gland cells of *D. melanogaster*. The chromosome region here is the same as in Figure 4.12. (A) Uninduced control. (B–E) 20E-stimulated chromosomes at (B) 25 minutes, (C) 1 hour, (D) 2 hours, and (E) 4 hours. (Photographs courtesy of M. Ashburner.)

VADE MECUM[2] **Chromosome squash.** How to do a chromosome squash using the *Drosophila* larval salivary gland. **[Click on Fruit Fly]**

genes and that, moreover, this early regulatory protein itself turns off the transcription of the early genes[†] (Figure 18.19). These insights have been confirmed by molecular analyses.

The early puffs include the genes for two important transcription factors, the Broad-Complex and E74B. Like the ecdysone receptor gene, the **Broad-Complex (BR-C)** gene can generate several different transcription factor proteins through differentially initiated and spliced messages. It appears that the variants of the ecdysone receptor may induce

the synthesis of particular variants of the BR-C protein. Organs such as the larval salivary gland that are destined for death during metamorphosis express the Z1 isoform; imaginal discs destined for differentiation express the Z2 isoform; and the central nervous system (which undergoes marked remodeling during metamorphosis) expresses all isoforms, with Z3 predominating (Emery et al. 1994; Crossgrove et al. 1996).

In addition to the restricted activities of the different isoforms of BR-C, there appear to be common processes that all of the isoforms accomplish. Restifo and Wilson (1998) provided evidence that these common functions are prevented by juvenile hormone. Deletions of the BR-C gene lead to faulty muscle development, retention of larval structures that would normally degenerate, abnormal nervous system morphology, and eventually the death of the larva (Restifo and White 1991). This syndrome is very similar to that induced by adding excess JH to *Cecropia* silkworm larvae (Riddiford 1972) or by adding juvenile hormone analogues to *Drosophila* larvae. Thus, it appears that juvenile hormone prevents ecdysone-inducible gene expression by interfering with the BR-C proteins. The BR-C and E74 proteins are themselves transcription factors, and their targets are being identified. These proteins can bind to the promoter of a late puff gene and directly regulate its transcription (Urness and Thummel 1995; Crossgrove et al. 1996).

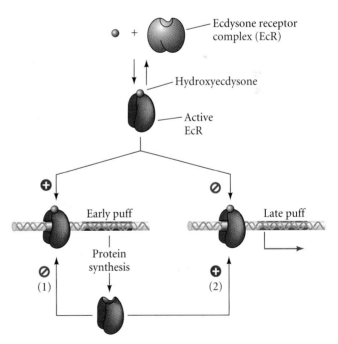

Figure 18.19
The Ashburner model of hydroxyecdysone regulation of transcription. 20E binds to its receptor, and the resulting compound binds to an early puff gene and a late puff gene. The early puff gene is activated, and its protein product (1) represses the transcription of its own gene and (2) activates the late puff gene, perhaps by displacing the ecdysone receptor. (After Richards 1992.)

[†]Since its discovery in 1954, when Butenandt and Karlson isolated 25 mg of ecdysone from 500 kg of silkworm moth pupae, 20-hydroxyecdysone has gone under several names, including ecdysterone, β-ecdysone, and crustecdysone. The observation that 20E controlled the transcriptional units of chromosomes was an extremely important and exciting discovery. This was our first real glimpse of gene regulation in eukaryotic organisms. At the time when this discovery was made, the only examples of transcriptional gene regulation were in bacteria.

THE DEATH OF LARVAL TISSUES. Larval cell death in *Drosophila* is driven by the stage- and tissue-specific induction of two "death" genes, *reaper* and *head involution defective* (*hid*). These two genes are coordinately induced by 20E. Genetic studies on the larval salivary gland have shown that both BR-C and E74 are required for the gland's cell death at metmorphosis. 20E induces BR-C and E74, and it helps transcribe the *reaper* gene. The BR-C protein is required to transcribe both the *hid* and *reaper* genes, and E74 is required for maximum levels of *hid* expression (Jiang et al. 2000).

COORDINATION OF RECEPTOR AND LIGAND. Like the story of amphibian metamorphosis, the stories of insect metamorphosis involve complex interactions between ligands and receptors. The "target tissues" are not merely passive recipients of the hormonal signal. Rather, they become responsive to hormones only at particular times. For example, when there is an ecdysone pulse at the middle of the fourth instar of the tobacco hornworm moth *Manduca*, the epidermis is able to respond to it, because it is expressing the ecdysone receptors. The wing discs, however, do not respond, as they do not synthesize their ecdysone receptors until the prepupal stage. Thus, the wing discs are unaffected by ecdysone until the prepual stage, at which time they synthesize the ecdysone receptors. Then they begin to grow and differentiate (Nijhout 1999).

> WEBSITE 18.4 **Precocenes and synthetic JH.** Given the voracity of insect larvae, it's amazing that any plant exists. However, many plants get revenge on their predators by making compounds that alter their metamorphoses and prevent the animals from developing or reproducing.

Regeneration

Regeneration—the reactivation of development in postembryonic life to restore missing tissues—is so "unhuman" that it has been a source of fascination to humans since the beginnings of biological science. It is difficult to behold the phenomenon of limb regeneration in newts or starfish without wondering why we cannot grow back our own arms and legs. What gives salamanders this ability we so sorely lack? Experimental biology was born in the efforts of eighteenth-century naturalists to document regeneration and to answer this question. The regeneration experiments of Tremblay (hydra), Réaumur (crustaceans), and Spallanzani (salamanders) set the standard for experimental research and for the intelligent discussion of one's data (see Dinsmore 1991). Réaumur, for instance, argued that crayfish had the ability to regenerate their limbs because their limbs broke easily at the joints.* Human limbs, he wrote, were not so vulnerable, so Nature provided us

not with regenerable limbs, but with "a beautiful opportunity to admire her foresight." Tremblay's advice to researchers who would enter this new field is pertinent even today. He tells us to go directly to nature and to avoid the prejudices that our education has given us.[†] Moreover, "one should not become disheartened by want of success, but should try anew whatever has failed. It is even good to repeat successful experiments a number of times. All that is possible to see is not discovered, and often cannot be discovered, the first time."

More than two centuries later, we are beginning to find answers to the great problem of regeneration, and we may soon be able to alter the human body so as to permit our own limbs, nerves, and organs to regenerate. This would mean that severed limbs could be restored, that diseased organs could be removed and regrown, and that nerve cells altered by age, disease, or trauma could once again function normally. To bring these treatments to humanity, we first have to understand how regeneration occurs in those species that have this ability. Our new knowledge of the roles of paracrine factors in organ formation, and our ability to clone the genes that produce those factors, have propelled what Susan Bryant (1999) has called "a regeneration renaissance." Since "renaissance" literally means "rebirth," and since regeneration can be seen as a return to the embryonic state, the term is apt in many ways.

There are three major ways by which regeneration can occur. The first mechanism involves the dedifferentiation of adult structures to form an undifferentiated mass of cells that then become respecified. This type of regeneration is called **epimorphosis** and is characteristic of regenerating limbs. The second mechanism is called **morphallaxis**. Here, regeneration occurs through the repatterning of existing tissues, and there is little new growth. Such regeneration is seen in hydra. A third, intermediate, type of regeneration can be thought of as **compensatory regeneration**. Here, the cells divide, but maintain their differentiated functions. They produce cells similar to themselves and do not form a mass of undifferentiated tissue. This type of regeneration is characteristic of the mammalian liver. We discussed the regeneration of flatworms (Chapter 3) earlier in the book. Here we will concentrate on salamander limb, hydra, and mammalian liver regeneration.

Epimorphic Regeneration of Salamander Limbs

When an adult salamander limb is amputated, the remaining cells are able to reconstruct a complete limb, with all its differentiated cells arranged in the proper order. In other words, the

---

*The relationship between adaptation and regeneration was confirmed in the phenomenon of the Wolffian regeneration of the salamander lens (see Website 4.4). The new attempts to coax human bone and neural tissue to regenerate are discussed in Chapter 21.

[†]The tradition of leaving one's books and classrooms and going directly to nature is very strong in developmental biology. There is a sign in Woods Hole Marine Biological Laboratory, the scene of some of the most important embryological research in America. The sign, attributed to Louis Agassiz reads, "Study Nature, Not Books." It hangs at the entrance to the library.

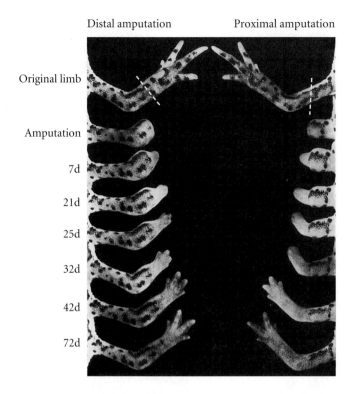

Distal amputation   Proximal amputation

Original limb

Amputation

7d

21d

25d

32d

42d

72d

Figure 18.20
Regeneration of a salamander forelimb. The amputation shown on the left was made below the elbow; the amputation shown on the right cut through the humerus. In both instances, the correct positional information is respecified and a normal limb is regenerated. (From Goss 1969; photograph courtesy of R. J. Goss.)

new cells construct only the missing structures and no more. For example, when a wrist is amputated, the salamander forms a new wrist and not a new elbow (Figure 18.20). In some way, the salamander limb "knows" where the proximal-distal axis has been severed and is able to regenerate from that point on. Salamanders accomplish this feat by cell dedifferentiation, proliferation, and respecification (see Brockes and Kumar 2002; Gardiner et al. 2002). This type of regeneration is called epimorphosis.

Formation of the apical ectodermal cap and regeneration blastema

Upon limb amputation, a plasma clot forms, and within 6 to 12 hours, epidermal cells from the remaining stump migrate to cover the wound surface, forming the **wound epidermis**. This single-layered structure is required for the regeneration of the limb, and it proliferates to form the **apical ectodermal cap**. Thus, in contrast to wound healing in mammals, no scar forms, and the dermis does not move with the epidermis to cover the site of amputation. The nerves innervating the limb degenerate for a short distance proximal to the plane of amputation (see Chernoff and Stocum 1995).

During the next 4 days, the cells beneath the developing cap undergo a dramatic dedifferentiation: bone cells, cartilage cells, fibroblasts, myocytes, and neural cells lose their differentiated characteristics and become detached from one another. Genes that are expressed in differentiated tissues (such as the *mrf4* and *myf5* genes expressed in muscle cells) are downregulated, while there is a dramatic increase in the expression of

genes such as *msx1*, that are associated with the proliferating progress zone mesenchyme of the embryonic limb (Simon et al. 1995). The formerly well-structured limb region at the cut edge of the stump thus forms a proliferating mass of indistinguishable, dedifferentiated cells just beneath the apical ectodermal cap. This dedifferentiated cell mass is called the **regeneration blastema**. These cells will continue to proliferate, and will eventually redifferentiate to form the new structures of the limb (Figure 18.21; Butler 1935).

The creation of the regeneration blastema depends on the formation of single, mononucleated cells. It is probable that the macrophages released into the wound site secrete metalloproteinases that digest the extracellular matrices holding epithelial cells together (Yang et al. 1999). But many of these cells are differentiated and have left the cell cycle. How do they regain the ability to divide? Microscopy and tracer dye studies have shown that when multinucleated myotubes (whose nuclei have been removed from the cell cycle: see Chapter 14) are introduced into a blastema, they give rise to labeled mononucleated cells that proliferate and can differentiate into many tissues of the regenerated limb (Hay 1959; Lo et al. 1993). It appears that myotube nuclei are forced to re-enter the cell cycle by a serum factor created by thrombin, the same protease that is involved in forming clots. Thrombin is released when the amputation is made, and when serum is exposed to thrombin, it forms a factor capable of inducing cultured newt myotubes to enter the cell cycle. Mouse myotubes, however, do not respond to this protein, although they will dedifferentiate in response to an extract of regenerating newt blastemas. This difference in responsiveness may relate directly to the difference in regenerative ability between salamanders and mammals* (Tanaka et al. 1999; McGann et al. 2001). The generation of a regeneration blastema may also depend upon the maintenance of ionic currents driven through the stump; if the electric field is suppressed, the regeneration blastema fails to form (Altizer et al. 2002).

---

*This thrombin-produced factor has not yet been isolated. But it is not the only difference between urodele and mammalian limbs. Another difference is that salamanders retain Hox gene expression in their appendages even when they are adults. It should be noted that, although most of the regeneration blastema comes from the dedifferentiation of the tissue at the edge of the stump, another source of cells is also possible. While multinucleated skeletal muscle tissue will dedifferentiate and supply uncommitted mononucleated cells to the blastema, the limb also contains muscle satellite cells—mononucleated cells committed to the muscle lineage—that may be used during regeneration. Thus, the regenerated limb's muscles may come both from the blastema and from these reserve cells (Dedkov et al. 2001; Zammit et al. 2002).

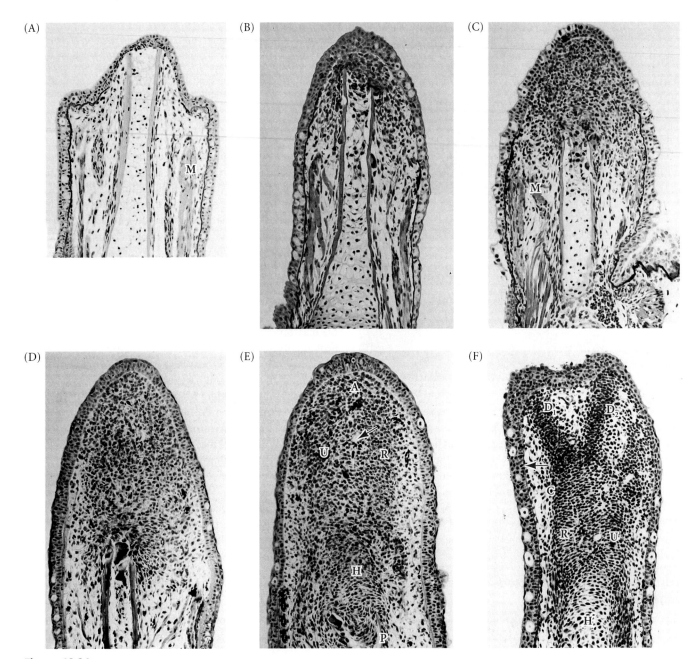

Figure 18.21

Regeneration in the larval forelimb of the spotted salamander *Ambystoma maculatum*. (A) Longitudinal section of the upper arm, 2 days after amputation. The skin and muscle have retracted from the tip of the humerus. (B) At 5 days after amputation, a thin accumulation of blastema cells is seen beneath a thickened epidermis. (C) At 7 days, a large population of mitotically active blastema cells lies distal to the humerus. (D) At 8 days, the blastema elongates by mitotic activity; much dedifferetiation has occured. (E) At 9 days, early redifferentiation can be seen. Chondrogenesis has begun in the proximal part of the regenerating humerus, H. The letter A marks the apical mesenchyme of the blastema, and U and R are the precartilaginous condensations that will form the ulna and radius, respectively. P represents the stump where the amputation was made. (F) At 10 days after amputation, the precartilaginous condensations for the carpal bones (ankle, C), and the first two digits (D₁, D₂) can also be seen. (From Stocum 1979; photographs courtesy of D. L. Stocum.)

Proliferation of the blastema cells: The requirement for nerves

The proliferation of the regeneration blastema depends on the presence of nerves. Singer (1954) demonstrated that a minimum number of nerve fibers must be present for regeneration to take place. It is thought that the neurons release factors that increase the proliferation of the blastema cells (Singer and Caston 1972; Mescher and Tassava 1975). One of these factors, **glial growth factor**, appears to be needed for maintaining a high rate of proliferation in the blastema cells (Brockes and Kinter 1986; Wang et al. 2000). **Fibroblast growth factor 2** (FGF2), another factor found in the axons innervating the amphibian limb. also appears to be critical to this process. FGF2 may play more than one role in regeneration. First, it may serve as an angiogenesis factor, since the regenerating tissues need and develop a blood supply very shortly after amputation (Rageh et al. 2002). Second, it may act to promote mitosis and patterning of the regenerating limb. Mescher and Gospodarowicz (1979) were able to restore mitosis to denervated blastemas by infusing FGFs into them, and Mullen and coworkers (1996) were able to obtain regeneration of denervated blastemas by implanting FGF2-containing beads in them. The FGF2 restored *Distal-less* gene expression to the ectoderm, probably making it responsive to some factor in the mesenchyme.

Proliferation of the blastema cells: The requirement for FGF10

The mesenchymal factor is probably FGF10. FGF10 is expressed in the mesenchyme of normal embryonic limb buds, and this same factor makes a critical difference in amphibian limb regeneration. In *Xenopus*, for instance, premetamorphic tadpoles are able to regenerate their hindlimbs, but later-stage tadpoles and adults are not. By means of reciprocal transplanation experiments, Yokoyama and colleagues (2000) found that the mesenchyme, not the ectoderm, was critical in regulating regeneration. A hindlimb would regenerate if it had the mesenchyme of a premetamorphic hindlimb and the ectoderm of an older hindlimb; it would not regenerate if it had

premetamorphic ectoderm and older mesenchyme. The later hindlimb mesenchyme did not synthesize FGF10. When Yokoyama and colleagues (2001) added beads of FGF10 to older hindlimbs, these limbs regenerated (Figure 18.22). Moreover, the FGF10 beads induced *fgf8* expression in the ectoderm overlying the bead.

Thus, the regeneration of the amphibian limb may be very similar to its original development. In both cases, there is feedback between FGF10 produced in the mesenchyme and FGF8 produced in the overlying ectoderm (AER in amniote limb development, apical ectodermal cap in amphibian regeneration). The initial condition is the expression of FGFs to make the ectoderm competent to express its own FGF. This function appears to be carried out by the lateral plate mesoderm during development and by neurons during regeneration.

Pattern formation in the regeneration blastema

The regeneration blastema resembles the progress zone mesenchyme of the developing limb in many ways. The dorsal-ventral and anterior-posterior axes between the stump and the regenerating tissue are conserved, and cellular and molecular studies have confirmed that the patterning mechanisms of developing and regenerating limbs are very similar. By transplanting regenerating limb blastemas onto developing limb buds, Muneoka and Bryant (1982) showed that the blastema cells could respond to limb bud signals and contribute to the developing limb. At the molecular level, just as Sonic hedgehog is seen in the posterior region of the developing limb progress zone mesenchyme, it is seen in the early posterior regeneration blastema (Imokawa and Yoshizato 1997; Torok et al. 1999). The initial pattern of Hox gene expression in regenerating limbs is not the same as that in developing limbs. However, the nested pattern of Hoxa and Hoxd gene expression characteristic of limb development is established as the limb regenerates (Torok et al. 1998).

(A)

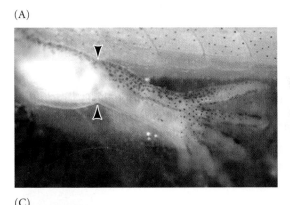

(B)

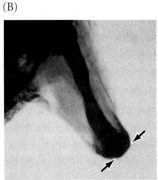

(C)

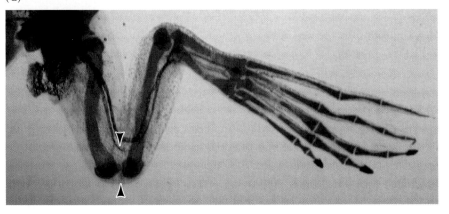

Figure 18.22
Restoration of regenerative ability of the *Xenopus* hindlimb by FGF10. (A) A late tadpole limb is amputated at knee level. The arrowheads show the site of amputation. (B) If beads lacking FGF10 are placed on the amputation site, there is no regeneration. Arrows mark the amputation site, and the limb cartilage is stained with alcian blue. (C) Beads containing FGF10 placed on the amputation site result in the regeneration of complete distal structures (some of which are clawed). Arrowheads mark the amputation site. (After Yokoyama et al. 2001; photographs courtesy of Dr. K. Tamura.)

(A)

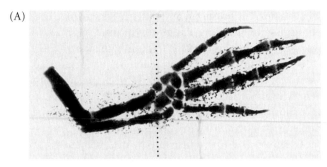

(B)

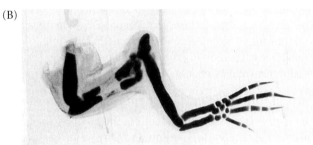

Figure 18.23
Effects of vitamin A (a retinoid) on regenerating salamander limbs. (A) Normal regenerated *Ambystoma mexicanum* limb (9×) with humerus, paired radius and ulna, carpals, and digits. The dotted line shows the plane of amputation through the carpal area. (B) Regeneration after amputation at the same location (5×), but after the regenerating animal had been placed in retinol palmitate (vitamin A) for 15 days. A new humerus, ulna, radius, carpal set, and digit set have emerged. (From Maden et al. 1982; photographs courtesy of M. Maden.)

This Hox gene expression may be controlled by retinoic acid. If regenerating animals are treated with sufficient concentrations of retinoic acid (or other retinoids), their regenerated limbs will have duplications along the proximal-distal axis (Figure 18.23; Niazi and Saxena 1978; Maden 1982). This response is dose-dependent and at maximal dosage can result in a complete new limb regenerating (starting at the most proximal bone), regardless of the original level of amputation. Dosages higher than this result in the inhibition of regeneration. It appears that the retinoic acid causes the cells to be respecified to a more proximal position (Figure 18.24; Crawford and Stocum 1988b; Pecorino et al. 1996). Moreover, the cells of the blastema can undergo **transdifferentiation** (Okada 1995; Tsonis et al. 1995)—cells that had been differentiated muscles can become cartilage tissue.

Retinoic acid is synthesized in the wound epidermis of the regenerating limb and forms a gradient along the proximal-distal axis of the blastema (Brockes 1992; Scadding and Maden 1994; Viviano et al. 1995). This gradient of retinoic acid is thought to activate Hoxa genes differentially across the blastema, resulting in the specification of pattern in the regenerating limb. Gardiner and colleagues (1995) have shown that the expression pattern of certain Hoxa genes in the distal cells of the regeneration blastema is changed by exogenous retinoic acid into an expression pattern characteristic of more proximal cells. It is probable that during normal regeneration, the wound epidermis/apical ectodermal cap secretes retinoic acid, which activates the genes needed for cell proliferation, downregulates the genes that are specific for differentiated cells, and activates a set of Hox genes that tells the cells where they are in the limb and how much they need to grow. The mecha-

(A)

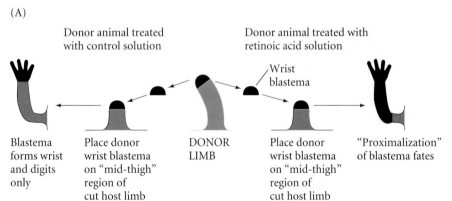

Donor animal treated with control solution — Donor animal treated with retinoic acid solution — Wrist blastema — Blastema forms wrist and digits only — Place donor wrist blastema on "mid-thigh" region of cut host limb — DONOR LIMB — Place donor wrist blastema on "mid-thigh" region of cut host limb — "Proximalization" of blastema fates

(B)

Figure 18.24
Proximalization of blastema respecification by retinoic acid. (A) When a wrist blastema from a recently cut axolotl forelimb is placed on a host hindlimb cut at the mid-thigh level, it will generate only the wrist. The host (whose own leg was removed) will fill the gap and regenerate the limb up to the wrist. However, if the donor animal is treated with retinoic acid, the wrist blastema will regenerate a complete limb and, when grafted, will fail to cause the host to fill the gap. (B) A wrist blastema from a darkly pigmented axolotl was treated with retinoic acid and placed on the amputated mid-thigh region of a golden axolotl. The treated blastema regenerated a complete limb. (After Crawford and Stocum 1998a,b; photograph courtesy of K. Crawford.)

nism by which the Hox genes do this is not known, but changes in cell-cell adhesion and other surface qualities of the cells have been observed (Nardi and Stocum 1983; Stocum and Crawford 1987; Bryant and Gardiner 1992).

Thus, in salamander limb regeneration, adult cells can go "back to the future," returning to an "embryonic" condition to begin the formation of the limb anew.

WEBSITE 18.5 **The polar coordinate and boundary models.** The phenomena of epimorphic regeneration can be seen formally as events that reestablish continuity among tissues that the amputation has severed. The polar coordinate and boundary models attempt to explain the numerous phenomena of limb regeneration.

WEBSITE 18.6 **Regeneration in annelid worms.** An easy laboratory exercise can discover the rules by which worms regenerate their segments. This website details some of those experiments.

Morphallactic Regeneration in Hydra

*Hydra**** is a genus of freshwater cnidarians. Most hydra are about 0.5 cm long. A hydra has a tubular body, with a "head" at its distal end and a "foot" at its proximal end. The "foot," or **basal disc**, enables the animal to stick to rocks or the undersides of pond plants. The hydra "head" consists of a conical **hypostome** region (containing the mouth) and a ring of tentacles (that catch food) beneath it. Hydra have only two epithelial cell layers, the ectoderm and endoderm, and they lack a true mesoderm. They can reproduce sexually, but do so only under adverse conditions, such as crowding or when the weather gets cold. They usually multiply by budding off a new individual (Martin 1997; Figure 18.25A). The buds form about two-thirds of the way down the body axis.

The body of the hydra is not as "stable" as are those of most organisms. In humans and flies, for instance, a skin cell in the trunk is not expected to migrate along the anterior-posterior axis of the body to be eventually sloughed off from the face or foot; but that is what precisely what happens in *Hydra*. The cells of the body column are constantly in mitosis, and the cells are eventually displaced to the extremities of the column, from which they are shed (Figure 18.25B; Campbell 1967a,b). Thus, each cell gets to play several roles, depending on how old it is, and the signals specifying cell fate must be active all the time.

If the hydra body column is cut into several pieces, each piece will regenerate a head at its original apical end and a foot at its original basal end. No cell division is required for this to happen, and the result is a small hydra. Since each cell retains its plasticity, each piece can re-form a smaller organism. This type of regeneration is known as morphallaxis.

The head activation gradient

The experiments described above show that every portion of the hydra body column along the apical-basal axis is potentially able to form both a head and a foot. However, the polarity of the hydra is coordinated by a series of morphogenetic gradients that permit the head to form only at one place and the basal disc to form only at another. Evidence for such gradients in hydras was first obtained from grafting experiments begun by Ethel Browne in the early 1900s. When hypostome tissue from one hydra is transplanted into the middle of another hydra, it forms a new apical-basal axis, with the hypostome extending outward (Figure 18.26A). When a basal disc is grafted to the middle of a host hydra, a new axis also forms, but with the opposite polarity, extending a basal disc (Figure 18.26B). When tissues from both ends are transplanted simul-

(A)

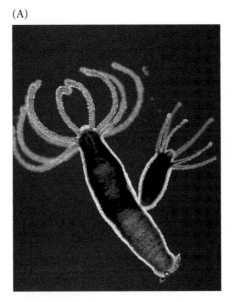

(B)

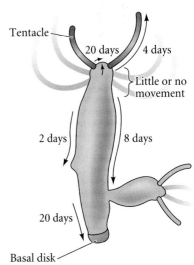

Figure 18.25
Budding in *Hydra*. (A) A new individual buds from the right side of an adult. (B) Cell movements in *Hydra* were traced by following the migration of labeled tissues. The arrows indicate the starting and leaving positions of the labeled cells. The bracketed regions are where no net cell movement took place. Cell division takes place throughout the body column except at the tentacles and foot (shaded). (A, photograph © Biophoto/Photo Researchers Inc.; B after Steele 2002.)

*The Hydra is another character from Greek mythology. Whenever one of this serpent's many heads was chopped off, it regenerated two new ones. Hercules finally defeated the Hydra by cauterizing the stumps of its heads with fire. Hercules had a longstanding interest in regeneration, for he was also the hero who finally freed the bound Prometheus, thus stopping his daily hepatectomies.

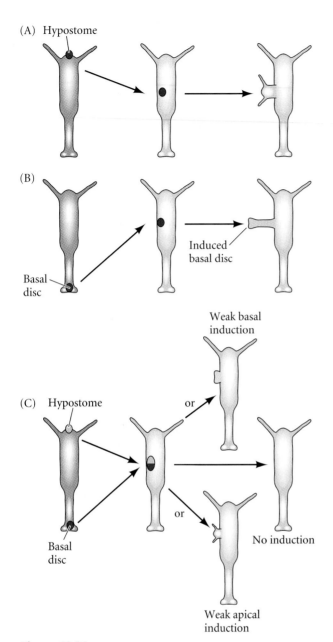

(A) Hypostome

(B)

Basal disc

Induced basal disc

(C) Hypostome

Basal disc

or

or

Weak basal induction

No induction

Weak apical induction

Figure 18.26
Grafting experiments demonstrating different morphogenetic capabilities in different regions of the *Hydra* apical-basal axis. (A) Hypostome tissue grafted onto a host trunk induces a secondary axis with an extended hypostome. (B) Basal disc tissue grafted onto a host trunk induces a secondary axis with an extended basal disc. (C) If hypostome and basal disc tissues are transplanted together, only weak inductions, if any, are seen. (After Newman 1974.)

taneously into the middle of a host, no new axis is formed, or the new axis has little polarity (Figure 18.26C; Browne 1909; Newman 1974). These experiments have been interpreted to indicate the existence of a **head activation gradient** (highest at the hypostome) and a **foot activation gradient** (highest at the basal disc).

The head activation gradient can be measured by implanting rings of tissue from various levels of a donor hydra into a particular region of the host trunk (Wilby and Webster 1970; Herlands and Bode 1974; MacWilliams 1983b). The higher the level of head activator in the donor tissue, the greater the percentage of implants that will induce the formation of new heads. The head activation factor is concentrated in the head and to decrease linearly toward the basal disc.

The head inhibition gradient

If the tissue of the hydra body column is capable of forming a head, why doesn't it do so? In 1926, Rand and colleagues showed that the normal regeneration of the hypostome is inhibited when an intact hypostome is grafted adjacent to the amputation site. Moreover, if a graft of subhypostomal tissue (from the region just below the hypostome, where there is a relatively high concentration of head activator) is placed in the same region of a host hydra, no secondary axis forms (Figure 18.27A). However, if one grafts subhypostomal tissue to a decapitated host hydra (Figure 18.27B), a second axis does form. The host head appears to make an inhibitor that prevents the grafted tissue from forming a head and secondary axis. A gradient of this inhibitor appears to extend from the head down the body column, and can be measured by grafting subhypostomal tissue into various regions along the trunks of host hydras. This tissue will not produce a head when implanted into the apical area of an intact host hydra, but it will form a head if placed lower on the host (Figure 18.27A,C). Thus, there is a gradient of head inhibitor as well as head activator (Wilby and Webster 1970; MacWilliams 1983a).

The hypostome as an "organizer"

Ethel Browne (1909; Lenhoff 1991) noted that the hypostome acted as an "organizer" of the hydra. This notion has been confirmed by Braun and Bode (2002), who demonstrated that (1) when transplanted, the hypostome can induce host tissue to form a second body axis, (2) the hypostome produces both the head activation and head inhibition signals, (3) the hypostome is the only "self-differentiating" region of the hydra, and (4) the head inhibition signal is actually a signal to inhibit the formation of new organizing centers.

By inserting small pieces of hypostome tissue into a host hydra that had been labeled with India ink (colloidal carbon), Braun and Bode found that the hypostome induced a new body axis and that almost all of the resulting head tissue came from *host* tissue, not from the differentiation of the donor tissue. In contrast, when tissues from other regions (such as the subhypostomal region) were grafted into a host trunk, the head and apical trunk of the new hydra were made from the grafted *donor* tissue (Figure 18.28). In other words, only the hypostome region could alter the fates of the trunk cells, causing them to become head cells. Braun and Bode also found that the signal did not have to emanate from a permanent graft. Even transient

(A) Intact host: No secondary axis induced

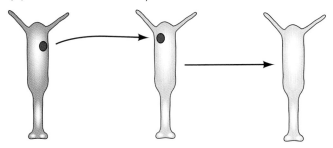

Figure 18.27

Grafting experiments providing evidence for a head inhibition gradient. (A) Subhypostomal tissue does not generate a new head when placed close to an existing host head. (B) Subhypostomal tissue generates a head if the existing host head is removed. A head also forms at the site where the host's head was amputated. (C) Subhypostomal tissue generates a new head when placed far away from an existing host head. (After Webster 1966.)

(B) Host's head removed: Secondary axis induced

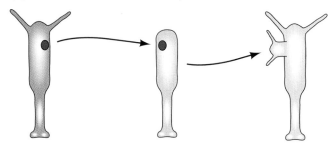

contact with the hypostome region was sufficient to induce a new axis from a host hydra. In these cases, *all* the tissue of the new axis came from the host. The head inhibitor appears to represses the effect of the inducing signal from the donor hypostome, and it normally functions to prevent other portions of the hydra from having such organizing abilities.

At least two genes are known to be active in the hypostome organizer area, and their expression there bodes well for the existence of an evolutionarily conserved set of signals that function as organizers throughout the animal kingdom. First, a *Hydra* Wnt protein is seen in the apical end of the early bud, and it defines the hypostome region as the bud elongates (Hobmayer et al. 2000; Figure 18.29). Second, the expression of another vertebrate organizer molecule, Goosecoid, is restricted to the *Hydra* hypostome region. Moreover, when the hypostome is brought into contact with the trunk of an adult hydra, it induces the expression of *Brachyury* (even though *Hydra* lacks mesoderm), just as vertebrate organizers do (Braun et al. 1999; Braun and Bode 2002).

(C) Intact host: Graft away from head region induces secondary axis

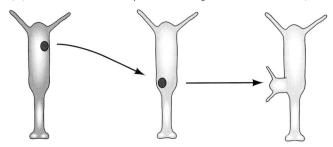

WEBSITE 18.7 Ethel Browne and the organizer. Spemann and Mangold brought the notion of the Organizer into embryology, and Spemann's laboratory helped make this idea a unifying notion in embryology. However, it has been argued that this idea had its origins in the experiments of Ethel Browne on *Hydra*.

(A)

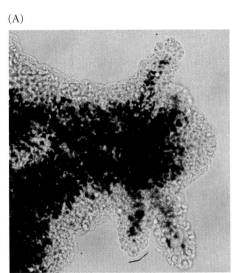

(B)

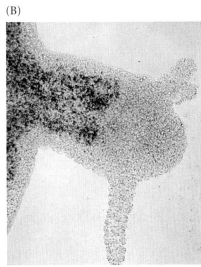

Figure 18.28

Formation of secondary axes following the transplantation of head regions into the trunk of a hydra. The endoderm of the host hydra had been stained with India ink. (A) When hypostome tissue is grafted onto a host trunk, it induces the trunk tissue to become tentacles and head. (B) When subhypostomal tissue is placed on the host trunk, it self-differentiates into a head and upper trunk. (From Braun and Bode 2002; photographs courtesy of H. R. Bode.)

(A)   (B)   (C)   (D)

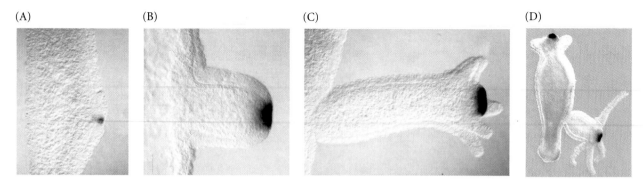

Figure 18.29
Expression of *Hydra Wnt* gene during budding. (A) Early bud. (B) Mid-stage bud. (C) Bud with early tenacles. (D) Adult with late bud. (From Hobmayer et al. 2000; photographs courtesy of T. W. Holstein.)

The basal disc activation and inhibition gradients

The basal disc has properties suggesting that it is the source of a foot inhibition gradient and a foot activation gradient (MacWilliams et al. 1970; Hicklin and Wolpert 1973; Schmidt and Schaller 1976; Meinhardt 1993; Grens et al. 1999). The inhibition gradients for the head and the foot may be important in determining where and when a bud can form. In young adult hydras, the gradients of head and foot inhibitors appear to block bud formation. However, as the hydra grows, the sources of these labile substances grow farther apart, creating a region of tissue about two-thirds down the trunk where levels of both inhibitors are minimal. This region is where the bud forms (Figure 18.30; Shostak 1974; Bode and Bode 1984; Schiliro et al. 1999).

Certain mutants of *Hydra* have defects in their ability to form buds, and these defects can be explained by alterations of the inhibition gradients. The *L4* mutant of *Hydra magnipapillata*, for instance, forms buds very slowly, and only after reaching a size about twice as long as wild-type individuals.

The amount of head inhibitor in these mutants was found to be much greater than in wild-type *Hydra* (Takano and Sugiyama 1983).

Several small peptides have been found that activate foot formation, but we do not know the mechanisms by which they arise and function (see Harafuji et al. 2001). However, the specification of cells as they migrate from the basal region through the body column may be mediated by a gradient of tyrosine kinase. The product of the *shin guard* gene is a tyrosine kinase that extends in a gradient from the ectoderm just above the basal disc through the lower region of the trunk. Buds appear to form where this gradient fades (Figure 18.30B). The *shin guard* gene appears to be activated through the product of the *manacle* gene, a putative transcription factor that is expressed earlier in the basal disc ectoderm.

The inhibition and activation gradients also inform the hydra "which end is up" and specify positional values along the apical-basal axis. When the head is removed, the head inhibitor no longer is made, and this causes the head activator to induce a new head. The region with the most head activator (i.e., the cells directly beneath the amputation site) will form the head. Once the head is made, it makes the head inhibitor, and the equilibrium is restored.

(A)

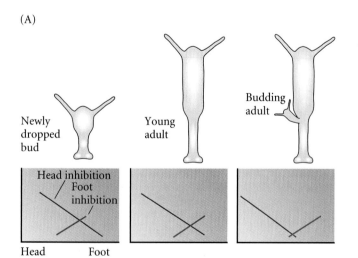

(B)

Figure 18.30
Bud location as a function of head and foot inhibition gradients. (A) Head inhibition (blue line) and foot inhibition (red line) gradients in newly dropped buds, young adults, and budding adults. (B) Expression of the Shin guard protein in a graded fashion in a budding hydra (A after Bode and Bode 1984; B after Bridge et al. 2000.)

Compensatory Regeneration in the Mammalian Liver

The liver's ability to regenerate seems to have been known since ancient times. According to Greek mythology, Prometheus's punishment (for giving fire to humans) was to be chained to a rock and to have an eagle eat a portion of his liver each day. His liver recovered from this **partial hepatectomy** each night, thereby providing a continuous food supply for the eagle and eternal punishment for Prometheus. Today the standard assay for liver regeneration is to remove (after anesthesia) specific lobes of the liver (i.e., a partial hepatectomy), leaving the others intact. The removed lobe does not grow back, but the remaining lobes enlarge to compensate for the loss of the missing liver tissue (Higgins and Anderson 1931). The amount of liver regenerated is equivalent to the amount of liver removed.

The liver regenerates by the proliferation of the existing tissues. Surprisingly, the regenerating liver cells do not fully dedifferentiate when they reenter the cell cycle. No regeneration blastema is formed. Rather, the five types of liver cells—hepatocytes, duct cells, fat-storing (Ito) cells, endothelial cells, and Kupffer macrophages—each begin dividing to produce more of themselves (Figure 18.31). Each type of cell retains its cellular identity, and the liver retains its ability to synthesize the liver-specific enzymes necessary for glucose regulation, toxin degradation, bile synthesis, albumin production, and other hepatic functions (Michalopoulos and DeFrances 1997).

As in the regenerating salamander limb, there is a return to some embryonic conditions in the regenerating liver. Fetal transcription factors and products are made, as are the cyclins that control cell division. But the return to the embryonic state is not as complete as in the amphibian limb. Although other paracrine and endocrine factors are necessary for liver regeneration, one of the most important proteins for returning liver cells to the cell cycle is **hepatocyte growth factor** (**HGF**). This protein, also known as **scatter factor** (see Chapter 11), induces many of the embryonic proteins. Within an hour after partial hepatectomy, the blood level of HGF has risen 20-fold (Lindroos et al. 1991). However, hepatocytes that are still connected to one another in an epithelium cannot respond to HGF. The trauma of partial hepatectomy may activate metalloproteinases that digest the extracellular matrix and permit the hepatocytes to separate and proliferate. These enzymes also may cleave HGF to its active form (Mars et al. 1995). The mechanisms by which these factors interact and by which the liver is told to stop regenerating after reaching the appropriate size remain to be discovered.

Aging: The Biology of Senescence

Entropy always wins. Each multicellular organism, using energy from the sun, is able to develop and maintain its identity for only so long. Then deterioration prevails over synthesis,

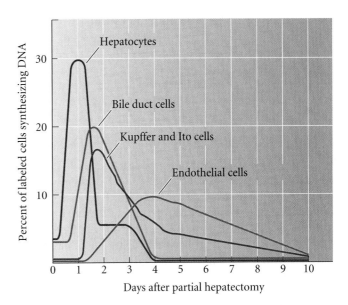

Figure 18.31
Kinetics of DNA synthesis in the five major cell types of the mammalian liver after partial hepatectomy. It is possible that, since the hepatocytes respond fastest, they are secreting paracrine factors that induce DNA replication in the other cells. (After Michalopoulos and DeFrances 1997.)

and the organism ages. **Aging** can be defined as the time-related deterioration of the physiological functions necessary for survival and fertility. The characteristics of aging—as distinguished from diseases of aging (such as cancer and heart disease)—affect all the individuals of a species. There are two major topics in research on aging. The first concerns how long an organism lives; the second concerns the physiological deterioration, or **senescence**, that characterizes old age. These topics are often seen as being interrelated.

Many evolutionary biologists (Medawar 1952; Kirkwood 1977) have denied that aging is part of the genetic repertoire of an animal. Rather, they consider senescence to be the default state occurring after the animal has fulfilled the requirements of natural selection. After its offspring are born and raised, the animal can die. Indeed, in many organisms, from moths to salmon, this is exactly what happens: as soon as the eggs are fertilized and laid, the adults die. However, as we will see below, recent studies have indicated that there are genetic components that regulate the rate of aging, and that altering the activity or expression of these genes can alter the life span of an individual. Indeed, recent evidence (summarized in Kenyon 2001) has shown that some of these genetic components might be evolutionarily conserved. Flies, worms, and mammals appear to use the same set of genes to promote survival and longevity.

Maximum life span and life expectancy

The maximum **life span** is a characteristic of a species; it is the maximum number of years a member of a species has been known to survive. The maximum human life span is estimated to be 121 years (Arking 1998). The life spans of tortoises and lake trout are both unknown, but are estimated to be more than 150 years.* The maximum life span of a domestic dog is about 20 years, and that of a laboratory mouse is 4.5 years. If a *Drosophila* fruit fly survives to eclose (in the wild, over 90% die as larvae), it has a maximum life span of 3 months.

However, a person cannot expect to live 121 years, and most mice in the wild do not live to celebrate their first birthday. **Life expectancy**, the length of time an individual of a given species can expect to live, is not characteristic of species, but of populations. It is usually defined as the age at which half the population still survives. A baby born in England in the 1780s could expect to live to be 35 years old. In Massachusetts during that same time, the life expectancy was 28 years. This was the normal range of human life expectancy for most of the human race in most times. Even today, in some areas of the world (Cambodia, Togo, Afghanistan, and several other countries) life expectancy is less than 40 years. In the United States, a person born in 1986 can expect to live around 70 years if male and around 80 years if female.[†]

Given that in most times and places, humans did not live much past 40 years, our awareness of human aging is relatively new. A 65-year-old person was rare in colonial America, but is commonplace today. Some survival curves for female *Homo sapiens* in the United States are plotted in Figure 18.32. In 1900, 50% of American women were dead by age 58. In 1980, 50% of American women were dead by age 81. Thus, the phenomena of senescence and the diseases of aging are much more common today than they were a century ago. In 1900, people did not have the "luxury" of dying from heart attacks or cancers, because these diseases are most likely to occur in people over the age of 50 years. Rather, people died (as they are still dying in many parts of the world) from infectious diseases and parasites (Arking 1998). Similarly, until recently, relatively few people exhibited the more general human senescent phenotype: graying hair, sagging and wrinkling skin, stiff joints, osteoporosis (loss of bone calcium), loss of muscle fibers and muscular strength, memory loss, eyesight deterioration, and the slowing of sexual responsiveness. As the melancholy Jacques notes in Shakespeare's *As You Like It*, those who did survive to senescence left the world "*sans* teeth, *sans* eyes, *sans* taste, *sans* everything."

---

*Some turtles may not only live long, but they might also escape senescence. Miller (2001) has shown that even the reproductive abilities of three-toed box turtles are unimpaired, and that a 60-year-old female turtle lays as many eggs annually as she ever did.

[†]You can see why the funding of Social Security is problematic in the United States. When it was created in 1935, the average working citizen died before age 65. Thus, he (and it usually was a he) was not expected to get back what he had paid into the system. Similarly, marriage "until death do us part" was an easier feat when death occurred in the third or fourth decade of life. The death rate of young women due to infections associated with childbirth was high throughout the world before antibiotics.

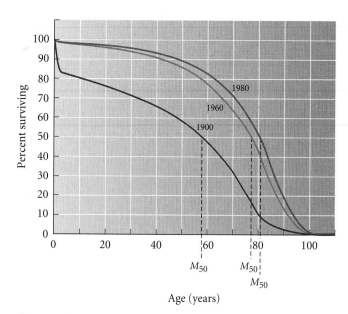

Figure 18.32
Survival curves for U.S. females in 1900, 1960, and 1980. M_{50} represents the age at which 50% of the individuals of each population survived. (After Arking 1998.)

Causes of aging

The general senescent phenotype is characteristic of each species. But what causes it? This question can be asked at many levels. Here we will be looking primarily at the cellular level of organization. While there is not yet a consensus on what causes aging (even at the cellular level), a theory is emerging that includes oxidative stress, hormones, and DNA damage.

GENERAL WEAR-AND-TEAR AND GENETIC INSTABILITY. "Wear-and-tear" theories of aging are among the oldest hypotheses proposed to account for the human senescent phenotype (Weismann 1891; Szilard 1959). As one gets older, small traumas to the body and its genome build up. At the molecular level, point mutations increase in number, and the efficiencies of the enzymes encoded by our genes decrease. If mutations occur in a part of the protein synthetic apparatus, the cell produces a large percentage of faulty proteins (Orgel 1963). If mutations were to arise in the DNA-synthesizing enzymes, the overall rate of mutation in the organism would be expected to increase markedly; Murray and Holliday (1981) have documented such faulty DNA polymerases in senescent cells.

Likewise, DNA repair may be important in preventing senescence. Certain premature aging syndromes in humans appear to be caused by mutations in such DNA repair enzymes (Sun et al. 1998; Shen and Loeb 2001). Individuals of species whose cells have more efficient DNA repair enzymes live longer (Figure 18.33; Hart and Setlow 1974).

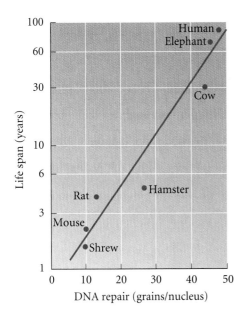

Figure 18.33
Correlation between life span and the ability of fibroblasts from various mammalian species to repair DNA. Repair capacity is represented in autoradiography by the number of grains from radioactive thymidine per cell nucleus. Note that the *y* axis (life span) is a logarithmic scale. (After Hart and Setlow 1974.)

OXIDATIVE DAMAGE. One major theory sees metabolism as the cause of aging. According to this theory, aging is a result of normal metabolism, which produces, as by-products, **reactive oxygen species (ROS)**. These ROS can oxidize and damage cell membranes, proteins, and nucleic acids. Some 2–3% of the oxygen atoms taken up by the mitochondria are reduced insufficiently and form the ROS set: superoxide ions, hydroxyl radicals, and hydrogen peroxide. Evidence that ROS are critical in the aging process includes the observation that *Drosophila* that overexpress enzymes that destroy ROS (catalase, which degrades peroxide, and superoxide dismutase) live 30–40% longer than do control flies (Orr and Sohal 1994; Parkes et al. 1998; Sunand Tower 1999).

Moreover, flies with mutations in the *methuselah* gene (named after the Biblical fellow said to have lived 969 years) live 35% longer than wild-type flies. The *methusaleh* mutants have enhanced resistance to paraquat, a poison that works by generating ROS within cells (Lin et al. 1998). These findings not only suggest that aging is under genetic control, but also provide evidence for the role of ROS in the aging process. In *C. elegans*, too, individuals with mutations that result in either the degradation of ROS or the prevention of ROS formation live much longer than wild-type nematodes (Larsen 1993; Vanfleteren and De Vreese 1996; Feng et al. 2001).

The evidence for ROS involvement in mammalian aging is not so clear. Mutations in mice that result in the lack of certain ROS-degrading enzymes do not cause premature aging

(Ho et al. 1997; Melov et al. 1998). However, there may be more genetic redundancy in mammals than in invertebrates, and other genes may be upregulated to produce related ROS-degrading enzymes. The oxidative damage hypothesis does not necessarily contradict the "wear-and-tear" hypothesis. Rather, it provides a specific mechanism that could be providing an important stress that accumulates as the organism matures.

MITOCHONDRIAL GENOME DAMAGE. The mutation rate in mitochondria is 10–20 times that of the nuclear DNA mutation rate (Johnson et al. 1999). It is thought that mutations in mitochondria could (1) lead to defects in energy production, (2) lead to the production of ROS by faulty electron transport, and/or (3) induce apoptosis. Age-dependent declines in mitochondrial function are seen in many animals, including humans (Boffoli et al. 1994). A recent report (Michikawa et al. 1999) shows that there are "hot spots" for age-related mutations in the mitochondrial genome, and that mitochondria with these mutations have a higher replication frequency than wild-type mitochondria. Thus, the mutants are able to outcompete the wild-type mitochondria and eventually dominate the cell and its progeny. Moreover, the mutations may not only allow more ROS to be made, but also may make the mitochondrial DNA more susceptible to ROS-mediated damage.

MUTATIONS CAUSING PREMATURE AGING SYNDROMES. Several genes have been shown to affect aging. In humans, Hutchinson-Gilford progeria syndrome causes children to age rapidly and to die (usually of heart failure) as early as 12 years of age (Figure 18.34). It is caused by a dominant mutant gene, and its symptoms include thin skin with age spots, resorbed bone mass, hair loss, and arteriosclerosis—all characteristics of the human senescent phenotype. A similar syndrome is caused by mutations of the *klotho* gene in mice (Kuro-o et al. 1997). The functions of the products of these genes are not known, but they are thought to be involved in suppressing the aging phenotypes. These proteins may be extremely important in determining the timing of senescence.*

Genetically programmed aging

One of the criticisms of the idea of genetic "programs" for aging asks how evolution could have selected for them. Once the organism has passed maturity and raised its offspring, it is "an excrescence" on the tree of life (Rostand 1962); natural selection presumably cannot act on traits that affect an organism

---

*There is a popular proposal that the shortening of telomeres—repeated DNA sequences at the ends of chromosomes—is responsible for senescence. Telomere shortening has been connected to a decreased ability of cells to divide. However, no correlation between telomere length and the life span of an animal (humans have much shorter telomeres than mice) has been found, nor is there a correlation between human telomere length and a person's age (Cristofalo et al. 1998; Rudolph et al. 1999; Karlseder et al. 2002). It has been suggested that telomere-dependent inhibition of cell division might serve primarily as a defense against cancer rather than as an "aging clock."

Figure 18.34
Children with progeria. Although not yet 8 years old, the child on the right has a phenotype similar to that of an aged person. The hair loss, fat distribution, and transparency of the skin are characteristic of the normal human aging pattern seen in elderly adults. (Photograph © Associated Press.)

only *after* it has reproduced. But "How can evolution select for a way to degenerate?" may be the wrong question. Evolution probably can't select for such traits. The right question may be, "How can evolution select for phenotypes that can postpone reproduction or sexual maturity?" There is often a trade-off between reproduction and maintenance, and in many species reproduction and senescence are closely linked.

Recent studies of *C. elegans* and *Drosophila* suggest that there *is* a conserved genetic pathway that regulates aging, and that it can be selected for during evolution. This pathway may be responsible for putting these organisms into a type of suspended animation during periods of adverse environmental conditions. Here, maintenance prevails over reproduction. This state is called **diapause**, and numerous organisms exhibit such a condition. A newly hatched *C. elegans* larva proceeds through four instars, after which it can become an adult; or, if the nematodes are overcrowded or if there is insufficient food, the larva can enter a nonfeeding, metabolically dormant **dauer larva** stage. This dauer larva is the diapause condition for *C. elegans*. It does not feed and it is resistant to oxidative stress. The nematode can remain in this diapause state for up to 6 months, rather than becoming an adult that lives only a few weeks. In this dauer state, the nematode has increased resistance to ROS, and when it comes out of this stage, it will live as long as if it had never been a dauer larva. If some of the genes in the pathway leading to dauer larva formation are mutated, adult development is allowed, but resistance to ROS is still in place. The resulting adults live two to four times as

long as wild-type adults (Figure 18.35; Friedman and Johnson 1988).

The pathway that regulates both dauer larva formation and longevity has been identified as the insulin signaling pathway (Guarente and Kenyon 2000; Gerisch et al. 2001). Favorable environments signal the activation of the insulin receptor homologue DAF2, and this receptor stimulates the onset of adulthood. Poor environments fail to activate the DAF2 receptor, and dauer formation ensues. While severe loss-of-function alleles in this pathway cause the formation of dauer larvae in any environment, weak mutations in the insulin signaling pathway enable the animals to reach adulthood and live longer than wild-type animals.

Downregulation of this insulin signaling pathway also has several other functions. First, it appears to influence metabolism, decreasing mitochondrial electron transport. When the DAF2 receptor is not active, organisms have decreased sensitivity to ROS (Feng et al. 2001; Scott et al. 2002). Second, it increases the production of enzymes that prevent oxidative damage, as well as DNA repair enzymes (Honda and Honda 1999; Tran et al. 2002). Third, it decreases fertility (Gems et al. 1998). From a classical point of view, the insulin pathway may mediate a trade-off between reproduction and survival/maintenance. Interestingly, many of the adverse effects of reduced insulin signaling (e.g., reduced fertility and metabolism) are not apparent in individuals with the weak *daf2* mutations that are also long-lived. This finding suggests that longevity may be uncoupled from infertility.

Another longevity signal that may act independently of the insulin signaling pathway, originates in the gonad. When the germ line cells are removed from *C. elegans*, the animals live longer. It is thought that the germ line stem cells produce a substance that blocks the effects of a longevity-inducing steroid hormone (Hsin and Kenyon 1999; Gerisch et al. 2001; Arantes-Oliviera 2002)

The insulin signaling pathway also appears to regulate life span in *Drosophila*, and flies with weak loss-of-function mutations of the insulin receptor gene live nearly 85% longer than wild-type flies (Clancy et al. 2001; Tatar et al. 2001). These long-lived mutants are sterile, and their metabolism resembles that of flies that are in diapause (Kenyon 2001). It is possible that this system also operates in mammals, but the mammalian insulin and insulin-like growth factor pathways are so integrated with embryonic development and adult metabolism that mutations often have numerous and deleterious affects (such as diabetes or Donahue syndrome). However, there is some evidence that the insulin signaling pathway does affect life span in mammals. First, mice with loss-of-function mutations of the insulin signaling pathway live longer than their wild-type littermates (see Partridge and Gems 2002). Second, dog breeds with low levels of insulin-like growth factor 1 (IGF1) live longer than those breeds with higher levels of this factor. Third, mice lacking one copy of their IGF1 receptor gene live about 25% longer

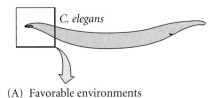

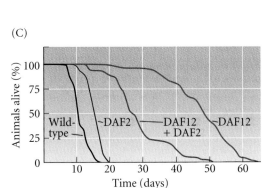

(A) Favorable environments

(B) Unfavorable environments

(C)

Figure 18.35

Proposed mechanism for extending the life span of *C. elegans* through the insulin signaling pathway. (A) Wild-type nematodes in a favorable environment make an insulin-like protein that binds to its receptor (DAF2) to initiate a pathway that interacts with a TGF-β-mediated signal to activate enzymes (DAF9) that synthesize a steroid hormone. This steroid is thought to be a ligand for the DAF12 protein. When DAF12 binds this hormone, it can activate the genes responsible for reproductive growth. (B) When food is not available, the insulin signal is not given, the steroid hormone is not made, and the DAF12 protein acts as an inhibitor to the genes it would have activated had it bound the hormone; in this situation, DAF12 activates the genes involved in dauer larva formation. (C) A graph showing the increase in longevity of *daf2* mutants. While DAF12 (steroid-binding transcription factor) mutants live slightly less long than wild-type, the DAF2 (insulin receptor) mutants live twice as long as wild-type. Worms with both sets of mutations live four times as long as wild-type worms. (After Gerisch et al. 2001.)

malian longevity (again, at the expense of fertility)—may extend life by reducing the levels of circulating insulin and IGF1 (Kenyon 2001; Roth et al. 2002; Holzenberger et al. 2003).

But within the genetic framework, chance plays a significant role. Even in *C. elegans*, aging is not totally a gene-regulated phenomenon. Within clonally identical nematodes fed an identical diet, some organisms live longer than others, and different organs deteriorate more rapidly in different individuals (Herndon et al. 2002).

As human life expectancy increases due to advances in our ability to prevent and cure disease, we are still left with a general aging syndrome that is characteristic of our species. Unless attention is paid to this general aging syndrome, we risk ending up like Tithonios, the miserable wretch of Greek mythology to whom the gods awarded eternal life, but not eternal youth. However, our new knowledge of regeneration is being put to use by medicine, and we may soon be able to ameliorate some of the symptoms of aging. This interaction of developmental biology and medicine may have far-reaching consequences, and it will be discussed further in Chapter 21.

than wild-type mice (and have higher ROS resistance but otherwise normal physiology and fertility). And finally, caloric restriction—one of the few known ways of increasing mam-

Snapshot Summary: Metamorphosis, Regeneration, and Aging

1. Amphibian metamorphosis includes both morphological and biochemical changes. Some structures are remodeled, some are replaced, and some new structures are formed.

2. The hormone responsible for amphibian metamorphosis is triiodothyronine (T_3). The synthesis of T_3 from thyroxine and the degradation of T_3 by deiodinases can regulate metamorphosis in different tissues. T_3 binds to thyroid hormone receptors and acts predominantly at the transcriptional level.

3. Many changes during amphibian metamorphosis are regionally specific. The tail muscles degenerate; the trunk muscles persist. An eye will persist even if transplanted into a degenerating tail.

4. Heterochrony involves changes in the relative rates of development of different parts of the animal. In neoteny, the larval form is retained while the gonads and germ cells mature at their normal rate. In progenesis, the gonads and germ cells mature rapidly while the rest of the body matures normally. In both instances, the animal can mate while retaining its larval or juvenile form.

5. In animals with direct development, the larval stage has been lost. Some frogs, for instance, form limbs while in the egg.

6. Ametabolous insects undergo direct development. Hemimetabolous insects pass through nymph stages wherein the immature organism is usually a smaller version of the adult.

7. In holometabolous insects, there is a dramatic metamorphosis from larva to pupa to sexually mature adult. In the stages between larval molts, the larva is called an instar. After the last instar, the larva undergoes a metamorphic molt to become a pupa. The pupa undergoes an imaginal molt to become an adult.

8. During the pupal stage, the imaginal discs and histoblasts grow and differentiate to produce the structures of the adult body.

9. The anterior-posterior, dorsal-ventral, and proximal-distal axes are sequentially specified by interactions between different compartments in the imaginal discs. The disc "telescopes out" during development, its central regions becoming distal.

10. Molting is caused by the hormone 20-hydroxyecdysone (20E). In the presence of high levels of juvenile hormone, the molt gives rise to another larval instar. In low concentrations of juvenile hormone, the molt produces a pupa; if no juvenile hormone is present, the molt is an imaginal molt.

11. The ecdysone receptor gene produces a nuclear RNA that can form at least three different proteins. The types of ecdysone receptors in a cell may influence the response of that cell to 20E. The ecdysone receptors bind to DNA to activate or repress transcription.

12. There are three major types of regeneration. In epimorphosis (such as regenerating limbs), tissues dedifferentiate into a regeneration blastema, divide, and redifferentiate into the new structure. In morphallaxis (characteristic of hydra), there is a repatterning of existing tissue with little or no growth. In compensatory regeneration (such as in the liver), cells divide but retain their differentiated state.

13. In the regenerating salamander limb, the epidermis forms an apical ectodermal cap. The cells beneath it dedifferentiate to form a blastema. The differentiated cells lose their adhesions and re-enter the cell cycle. FGF10 and factors from neurons appear to be critical in permitting regeneration to occur.

14. Salamander limb regeneration appears to use the same pattern formation system as the developing limb.

15. In *Hydra*, there appears to be a head activation gradient, a head inhibition gradient, a foot activation gradient, and a foot inhibition gradient. Budding occurs where these gradients are minimal.

16. The hypostome region of *Hydra* appears to be an organizer region that secretes paracrine factors to alter the fates of surrounding tissue.

17. The maximum life span of a species is the longest time an individual of that species has been observed to survive. Life expectancy is the age at which approximately 50 percent of the members of a given population still survive.

18. Aging is the time-related deterioration of the physiological functions necessary for survival and reproduction. The phenotypic changes of senescence (which affect all members of a species) are not to be confused with diseases of senescence, such as cancer and heart disease (which affect some individuals but not others).

19. Reactive oxygen species (ROS) can damage cell membranes, inactivate proteins, and mutate DNA. Mutations that alter the ability to make or degrade ROS can change the life span.

20. Mitochondria may be a target for proteins that regulate aging.

21. An insulin signaling pathway, involving a receptor for insulin and insulin-like proteins, may be an important component of genetically limited life spans.

Literature Cited

Abu-Shaar, M. and R. S. Mann. 1998. Generation of multiple antagonistic domains along the proximodistal axis during *Drosophila* leg development. *Development* 125: 3821–3830.

Alberch, P. and J. Alberch. 1981. Heterochronic mechanisms of morphological diversification and evolutionary change in the neotropical salamander *Bolitoglossa occidentalis* (Amphibia: Plethodontidae). *J. Morphol.* 167: 249–264.

Allen, B. M. 1916. Extirpation experiments in *Rana pipiens* larva. *Science* 44: 755–757.

Alley, K. E. and M. D. Barnes. 1983. Birthdates of trigeminal motor neurons and metamorphic re-organization of the jaw myoneural system in frogs. *J. Comp. Neurol.* 218: 395–405.

Altizer, A. M., S. G. Stewart, B. K. Albertson and R. B. Borgens. 2002. Skin flaps inhibit both the current of injury at the amputation surface and regeneration of that limb in newts. *J. Exp. Zool.* 293: 467–477.

Arantes-Oliviera, N., J. Apfeld, A. Dillin and C. Kenyon. 2002. Regulation of life-span by germ-line stem cells in *Caenorhabditis elegans. Science* 285: 502–505.

Arking, R. 1998. *The Biology of Aging*, 2nd Ed. Sinauer Associates, Sunderland, MA.

Ashburner, M. 1972. Patterns of puffing activity in the salivary glands of *Drosophila*. VI. Induction by ecdysone in salivary glands of *D. melanogaster* cultured in vitro. *Chromosoma* 38: 255–281.

Ashburner, M. 1974. Sequential gene activation by ecdysone in polytene chromosomes of *Drosophila melanogaster*. II. Effects of inhibitors of protein synthesis. *Dev. Biol.* 39: 141–157.

Ashburner, M. 1990. Puffs, genes, and hormones revisited. *Cell* 61: 1–3.

Ashburner, M. and H. D. Berondes. 1978. Puffing of polytene chromosomes. *In The Genetics and Biology of* Drosophila, Vol. 2B. Academic Press, New York, pp. 316–395.

Atkinson, B. G., C. Helbing and Y. Chen. 1996. Reprogramming of genes expressed in amphibian liver during metamorphosis. *In* L. I. Gilbert, B. G. Atkinson and J. R. Tata (eds.), *Metamorphosis: Postembryonic Reprogramming of Gene Expression in Amphibian and Insect Cells*. Academic Press, San Diego, pp. 539–566.

Becker, K. B., K. C. Stephens, J. C. Davey, M. J. Schneider and V. A. Galton. 1997. The type 2 and type 3 iodothyronine deiodinases play important roles in coordinating development in *Rana catesbeiana* tadpoles. *Endocrinology* 138: 2989–2997.

Berry, D. L., C. S. Rose, B. F. Remo and D. D. Brown. 1998. The expression pattern of thyroid hormone response genes in remodeling tadpole tissues defines distinct growth and resorption gene expression patterns. *Dev. Biol.* 203: 24–35.

Blair, S. S. 1993. Mechanisms of compartment formation: Evidence that non-proliferating cells do not play a role in defining the D/V lineage restriction in the developing wing of *Drosophila. Development* 119: 339–351.

Bode, P. M. and H. R. Bode. 1984. Patterning in *Hydra*. *In* G. M. Malacinski and S. V. Bryant (eds.), *Pattern Formation*. Macmillan, New York, pp. 213–241.

Boffoli, D., S. C. Sacco, R. Vergari, G. Solarino, G. Santacroce and S. Papa. 1994. Decline with age of the respiratory chain activity in human skeletal muscle. *Biochim. Biophys. Acta* 1226: 73–86.

Braun, M. and H. R. Bode. 2002. Characterization of the head organizer in hydra. *Development* 129: 875–884.

Braun, M., S. Sokol and H. R. Bode. 1999. Cngsc, a homologue of *goosecoid*, participates in the patterning of the head and is expressed in the organizer region of *Hydra. Development* 126: 5245–5254.

Brenner, S. 1996. Francisco Crick in Paradiso. *Curr. Biol.* 6: 1202.

Bridge, D. M., N. A. Stover and R. E. Steele. 2000. Expression of a novel receptor tyrosine kinase gene and a *paired*-like homeobox gene provides evidence of differences in patterning at the oral and aboral ends of hydra. *Dev. Biol.* 220: 253–262.

Brockes, J. P. 1992. Introduction of a retinoid reporter gene into the urodele limb blastema. *Proc. Natl. Acad. Sci. USA* 89: 11386–11390.

Brockes, J. P. and C. R. Kinter. 1986. Glial growth factor and nerve-dependent proliferation in the regeneration blastema of urodele amphibians. *Cell* 45: 301–306.

Brockes, J. P. and A. Kumar. 2002. Plasticity and reprogramming of differentiated cells in amphibian regeneration. *Nature Rev. Mol. Cell Biol.* 3: 566–574.

Browne, E. N. 1909. The production of small hydranths in hydra by insertion of small grafts. *J. Exp. Zool.* 7: 1–23.

Brust, D. G. 1993. Maternal brood care by *Dendrobates pumilio*: A frog that feeds its young. *J. Herpetol.* 26: 102–105.

Bryant, P. J. 1970. Cell lineage relationships in the imaginal wing disc of *Drosophila melanogaster. Dev. Biol.* 22: 389–411.

Bryant, S. 1999. A regeneration renaissance. *Semin. Cell Dev. Biol.* 10: 313.

Bryant, S. V. and D. M. Gardiner. 1992. Retinoic acid, local cell-cell interactions, and pattern formation in vertebrate limbs. *Dev. Biol.* 152: 1–25.

Butenandt, A. and P. Karlson. 1954. Über die Isolierung eines Metamorphosen-Hormons der Insekten in kristallisierter Form. *Z. Naturforsch.* B 9: 389–391.

Butler, E. G. 1935. Studies on limb regeneration in X-rayed *Ambystoma* larvae. *Anat. Rec.* 62: 295–307.

Callery. E. M., H. Fang and R. P. Elinson. 2001. Frogs without polliwogs: Evolution of anuran direct development. *BioEssays* 23: 233–241.

Campbell, R. D. 1967a. Tissue dynamics of steady state growth in *Hydra littoralis*. I. Patterns of cell division. *Dev. Biol.* 13: 487–502.

Campbell, R. D. 1967b. Tissue dynamics of steady state growth in *Hydra littoralis*. II. Patterns of tissue movement. *J. Morphol.* 1: 19–28.

Causo, J. P., M. Bate and A. Martinez-Arias. 1993. A *wingless*-dependent polar coordinate system in *Drosophila* imaginal discs. *Science* 259: 484–489.

Chernoff, E. A. G. and D. Stocum. 1995. Developmental aspects of spinal cord and limb regeneration. *Dev. Growth Diff.* 37: 133–147.

Clancy, D. J., D. Gems, L. G. Harshman, S. Oldham, H. Stocker, E. Hafen, S. J. Leevers and L.Partridge. 2001. Extension of life-span by loss of CHICO, a *Drosophila* insulin receptor substrate protein. *Science* 292: 104–106.

Clever, U. 1966. Induction and repression of a puff in *Chironomus tentans. Dev. Biol.* 14: 421–438.

Cohen, P. P. 1970. Biochemical differentiation during amphibian metamorphosis. *Science* 168: 533–543.

Condic, M. L., D. Fristrom and J. W. Fristrom. 1990. Apical cell shape changes during *Drosophila* imaginal leg disc elongation: A novel morphogenetic mechanism. *Development* 111: 23–33.

Corben, C. J., M. J. Ingram and M. J. Tyler. 1974. Gastric brooding: Unique form of parental care in an Australian frog. *Science* 186: 946–947.

Crawford, K. and D. L. Stocum. 1988a. Retinoic acid coordinately proximalizes regenerate pattern and blastema differential affinity in axolotl limbs. *Development* 102: 687–698.

Crawford, K. and D. L. Stocum. 1988b. Retinoic acid proximalizes level-specific properties responsible for intercalary regeneration in axolotl limbs. *Development* 104: 703–712.

Cristofalo, V. J., R. G. Allen, R. J. Pignolo, B. G. Martin and J. C. Beck. 1998. Relationship between donor age and relicative life span of human cells in culture: A reevaluation. *Proc. Natl. Acad. Sci. USA* 95: 10614–10619.

Crossgrove, K., C. A. Bayer, J. W. Fristrom and G. M. Guild. 1996. The *Drosophila* Broad Complex early gene directly regulates late gene transcription during the ecdysone-induced puffing cascade. *Dev. Biol.* 180: 745–758.

Currie, J. and W. M. Cowan. 1974. Evidence for the late development of the uncrossed retinothalamic projections in the frog *Rana pipiens. Brain Res.* 71: 133–139.

De Beer, G. 1940. *Embryos and Ancestors*. Clarendon Press, Oxford.

Dedkov, E. I., Y. Y. Kostrominova, A. B. Borisov and B. M. Carlson. 2001. Reparative myogenesis in long-term denervated skeletal muscle of adult rats results in a reduction of the satellite cell population. *Anat. Rec.* 263: 139–154.

del Pino, E. M. and R. P. Elinson. 1983. A novel development pattern for frogs: Gastrulation produces an embryonic disk. *Nature* 306: 589–591.

Denver, R. 1993. Acceleration of anuran amphibian metamorphosis by corticotropin-releasing factor-like peptides. *Gen. Comp. Endocrinol.* 91: 38–51.

Denver, R. 2003. The vertebrate neuroendocrine stress system and its role in orchestrating life history transitions. Society for Integrative and Comparative Biology, Annual Meeting. http://www.sicb.org/meetings/2003/schedule/abstractdetails.php3?id=39.

Diaz-Benjumea, F. J. and S. M. Cohen. 1993. Interaction between dorsal and ventral cells in the imaginal disc directs wing development in *Drosophila. Cell* 75: 741–752.

Dinsmore, C. E. (ed.) 1991. *A History of Regeneration Research: Milestones in the Evolution of a Science*. Cambridge University Press, New York.

Eliceiri, B. P. and D. D. Brown. 1994. Quantitation of endogenous thyroid hormone receptors alpha and beta during embryogenesis and metamorphosis in *Xenopus laevis. J. Biol. Chem.* 269: 24459–24465.

Elinson, R. P. 1987. Change in developmental patterns: Embryos of amphibians with large eggs. *In* R. A. Raff and E. C. Raff (eds.), *Development as an Evolutionary Process.* Alan R. Liss, New York, pp. 1–21.

Emery, I. F., V. Bedian and G. M. Guild. 1994. Differential expression of Broad-Complex transcription factors may forecast tissue-specific developmental fates during *Drosophila* metamorphosis. *Development* 120: 3275–3287.

Fallon, J. F. and B. K. Simandl. 1978. Evidence of a role for cell death in the disappearance of the embryonic human tail. *Am. J. Anat.* 152: 111–130.

Feng, J., F. Bussièe and S. Hekimi. 2001. Mitochondrial electron transport is a key determinant of life span in *Caenorhabditis elegans. Dev. Cell* 1: 633–644.

rieden, E. 1981. The dual role of thyroid hormones in vertebrate development and calorigenesis. *In* L. I. Gilbert and E. Frieden (eds.), *Metamorphosis: A Problem in Developmental Biology.* Plenum, New York, pp. 545–564.

Friedman, D. B. and T. E. Johnson. 1988. A mutation in the *age-1* gene in *Caenorhabditis elegans* lengthens life and reduces hermaphrodite fertility. *Genetics* 118: 75–86.

Fristrom, D. and J. W. Fristrom. 1975. The mechanisms of evagination of imaginal disks of *Drosophila melanogaster.* I. General considerations. *Dev. Biol.* 43: 1–23.

Fristrom, J. W. 1972. The biochemistry of imaginal disc development. *In* H. Ursprung and R. Nothiger (eds.), *The Biology of Imaginal Discs.* Springer-Verlag, Berlin, pp. 109–154.

Fristrom, J. W., D. Fristrom, E. Fekete and A. H. Kuniyuki. 1977. The mechanism of evagination of imaginal discs of *Drosophila melanogaster. Am. Zool.* 17: 671–684.

Fritzsch, B., U. Wahnschaffe and U. Bartsch. 1988. Metamorphic changes in the octavo-lateralis system of amphibians. *In* B. Fritzsch, M. Ryan, W. Wilczynski, T. Hetherington and W. Walkowiak (eds.), *The Evolution of the Amphibian Auditory System.* Wiley, Chichester, pp. 561–586.

Garcia-Bellido, A., P. Ripoll and G. Morata. 1973. Developmental compartmentalization of the wing disc of *Drosophila. Nature New Biol.* 245: 251–253.

Gardiner, D. M., B. Blumberg, Y. Konine and S. V. Bryant. 1995. Regulation of HoxA expression in developing and regenerating axolotl limbs. *Development* 121: 1731–1741.

Gardiner, D. M., T. Endo and S. Bryant. 2002. The molecular basis of amphibian limb regeneration: Integrating the old with the new. *Semin. Cell Dev. Biol.* 13: 345–352.

Geigy, R. 1941. Die metamorphose als Folge gewebsspezifischer determination. *Rev. Suisse Zool.* 48: 483–494.

Gems, D. and 7 others. 1998. Two pleiotropic classes of *daf-2* mutation affect larval arrest, adult behavior, reproduction and longevity in *Caenorhabditis elegans. Genetics* 150: 129–155.

Gerisch, B., C. Weitzel, C. Kober-Eisermann, V. Rottiers and A. Antebi. 2001. A hormonal signaling pathway influencing *C. elegans* metabolism, reproductive development, and life span. *Dev. Cell* 1: 841–851.

Gilbert, L. I. and W. Goodman. 1981. Chemistry, metabolism, and transport of hormones controlling insect metamorphosis. *In* L. I. Gilbert and E. Frieden (eds.), *Metamorphosis: A Problem in Developmental Biology.* Plenum, New York, pp. 139–176.

Goss, R. J. 1969. *Principles of Regeneration.* Academic Press, New York.

Gould, S. J. 1977. *Ontogeny and Phylogeny.* Harvard University Press, Cambridge, MA.

Grens, A., H. Shimizu, S. A. Hoffmeister, H. R. Bode and T. Fijisawa. 1999. The novel signal peptides, pedibin and Hym-346, lower positional value thereby enhancing foot formation in hydra. *Development* 126: 517–524.

Grobstein, P. 1987. On beyond neuronal specificity: Problems in going from cells to networks and from networks to behavior. *In* P. Shinkman (ed.), *Advances in Neural and Behavioral Development,* Vol. 3. Ablex, Norwood, NJ, pp. 1–58.

Gronemeyer, H. and O. Pongs. 1980. Localization of ecdysterone on polytene chromosomes of *Drosophila melanogaster. Proc. Natl. Acad. Sci. USA* 77: 2108–2112.

Guarante, L. and C. Kenyon. 2000. Genetic pathways that regulate ageing in model organisms. *Nature* 408: 255–262.

Gudernatsch, J. F. 1912. Feeding experiments on tadpoles. I. The influence of specific organs given as food on growth and differentiation: A contribution to the knowledge of organs with internal secretion. *Wilhelm Roux Arch. Entwicklungsmech. Org.* 35: 457–483.

Hanken, J. and B. K. Hall. 1988. Skull development during anuran metamorphosis. II. Role of thyroid hormones in osteogenesis. *Anat. Embryol.* 178: 219–227.

Hanken, J., M. W. Klymkowsky, C. H. Summers, D. W. Seufert and N. Ingebrigtsen. 1992. Cranial ontogeny in the direct-developing frog, *Eleutherodactylus coqui* (Anura: Leptodactylidae), analyzed using whole-mount immunohistochemistry. *J. Morphol.* 211: 95–118.

Harafuji, N., T. Takahashi, M. Hatta, H. Tezuka, F. Morishita, O. Matsushima and T. Fujisawa. 2001. Enhancement of foot formation in *Hydra* by a novel epitheliopeptide, Hym-323. *Development* 126: 437–446.

Hart, R. and R. B. Setlow. 1974. Correlation between deoxyribonucleic acid excision repair and life-span in a number of mammalian species. *Proc. Natl. Acad. Sci. USA* 71: 2169–2173.

Hashimoto, H. and 9 others. 2002. Isolation and characterization of a Japanese flounder clonal

line, *reversed,* which exhibits reversal of metamorphic left-right asymmetry. *Mech. Dev.* 111: 17–24.

Hay, E. D. 1959. Electron microscopic observations of muscle dedifferentiation in regenerating *Ambystoma* limbs. *Dev. Biol.* 1: 555–585.

Herlands, R. and H. Bode. 1974. The influence of tissue polarity on nematocyte migration in *Hydra attenuata. Dev. Biol.* 40: 323–339.

Herndon, L. A. and 8 others. 2002. Stochastic and genetic factors influence tissue-specific decline in ageing *C. elegans. Nature* 419: 808–814.

Hicklin, J. and L. Wolpert. 1973. Positional information and pattern regulation in hydra: Formation of the foot end. *J. Embryol. Exp. Morphol.* 30: 727–740.

Higgins, G. M. and R. M. Anderson. 1931. Experimental pathology of the liver. I. Restoration of the liver of the white rat following partial surgical removal. *Arch. Pathol.* 12: 186–202.

Ho, Y. S., J. L. Magnenat, R. T. Bronson, J. Cao, M. Gargano, M. Sugawara and C. D. Funk. 1997. Mice deficient in cellular glutathione peroxidase develop normally and show no increased sensitivity to hyperoxia. *J. Biol. Chem.* 272: 16644–16651.

Hobmayer, B., F. Rentzsch, K. Kuhn, C. M. Happel, C. C. von Laue, P. Snyder, U. Rothbäcker and T. W. Holstein. 2000. WNT signalling molecules act in axis formation in the diploblastic metazoan *Hydra. Nature* 407: 186–189.

Holzenberger, M. and 7 others. 2003. IGF-1 receptor regulates lifespan and resistance to oxidative stress in mice. *Nature* 421: 182–186.

Honda, Y. and S. Honda. 1999. The *daf-2* gene network for longevity regulates oxidative stress resistance and Mn-superoxide dismutase gene expression in *Caenorhabditis elegans. FASEB J.* 13: 1385–1393.

Hoskins, S. G. and P. Grobstein. 1984. Thyroxine induces the ipsilateral retinothalamic projection in *Xenopus laevis. Nature* 307: 730–733.

Hoskins, S. G. and P. Grobstein. 1985a. Development of the ipsilateral retinothalamic projection in the frog *Xenopus laevis.* II. Ingrowth of optic nerve fibers and production of ipsilaterally projecting cells. *J. Neurosci.* 5: 920–929.

Hoskins, S. G. and P. Grobstein. 1985b. Development of the ipsilateral retinothalamic projection in the frog *Xenopus laevis.* III. The role of thyroxine. *J. Neurosci.* 5: 930–940.

Hsin, H. and C. Kenyon. 1999. Signals from the reproductive system regulate the lifespan of *C. elegans. Nature* 399: 362–365.

Huang, H., N. Marsh-Armstrong and D. D. Brown. 1999. Metamorphosis is inhibited in transgenic *Xenopus laevis* tadpoles that overexpress type III deiodinase. *Proc. Natl. Acad. Sci. USA* 96: 962–967.

Huang, H., L. Cai, B. F. Remo and D. D. Brown. 2001. Timing of metamorphosis and the onset of the negative feedback loop between the thyroid gland and the pituitary is controlled by type II iodothyronine deiodinase in *Xenopus laevis. Proc. Natl. Acad. Sci. USA* 98: 7348–7353.

Huxley, J. 1920. Metamorphosis of axolotl caused by thyroid feeding. *Nature* 104: 436.

Imokawa, Y. and K. Yoshizato. 1997. Expression of *sonic hedgehog* gene in regenerating newt limb blastema recapitulates that in developing limb buds. *Proc. Natl. Acad. Sci. USA* 94: 9159–9164.

Ishizuya-Oka, A., S. Ueda, T. Amano, K. Shimizu, K. Suzuki, N. Ueno and K. Yoshizato. 2001. Thyroid-hormone-dependent and fibroblast-specific expression of BMP-4 correlates with adult epithelial development during amphibian intestinal remodeling. *Cell Tissue Res.* 303: 187–195.

Jiang, C., A. F. Lamblin, H. Steller and C. S. Thummel. 2000. A steroid-triggered transcriptional hierarchy controls salivary gland cell death during *Drosophila* metamorphosis. *Mol. Cell* 5: 445–455.

Johnson, F. B., D. A. Sinclair and L. Guarente. 1999. Molecular biology of aging. *Cell* 96: 291–302.

Jones, G., M. Wozniak, Y. Chu, S. Dhar and D. Jones. 2001. Juvenile hormone-III-dependent conformational changes of the nuclear receptor Ultraspiracle. *Insect Biochem. Mol. Biol.* 32: 33–39.

Kalm, L. von, D. Fristrom and J. Fristrom. 1995. The making of a fly leg: A model for epithelial morphogenesis. *BioEssays* 17: 693–702.

Kaltenbach, J. C., A. E. Fry and V. K. Leius. 1979. Histochemical patterns in the tadpole tail during normal and thyroxine-induced metamorphosis. II. Succinic dehydrogenase, Mg- and Ca-adenosine triphosphatases, thiamine pyrophosphatase, and 5′-nucleotidase. *Gen. Comp. Endocrinol.* 38: 111–126.

Karlseder, J., A. Smorgorzewska and T. de Lange. 2002. Senescence induced by altered telomere state, not telomere loss. *Science* 295: 2446–2449.

Karp, G. and N. J. Berrill. 1981. *Development.* McGraw-Hill, New York.

Kawahara, A., Y. Gohda and A. Hikosaka. 1999. Role of type III iodothyronine 5-deiodinase gene expression in temporal regulation of *Xenopus* metamorphosis. *Dev. Growth Diff.* 41: 365–373.

Kenyon, C. 2001. A conserved regulatory system for aging. *Cell* 105: 165–168.

Kim, J., K. D. Irvine and S. B. Carroll. 1995. Cell recognition, signal induction, and symmetrical gene activation at the dorsal-ventral boundary of the developing *Drosophila* wing. *Cell* 82: 795–802.

Kim, J., A. Sebring, J. J. Esch, M. E. Kraus, K. Vorwerk, J. Magee and S. B. Carroll. 1996. Integration of positional information and identity by *Drosophila vestigial* gene. *Nature* 382: 133–138.

Kirkwood, T. B. L. 1977. The evolution of aging. *Nature* 270: 301–304.

Kistler, A., K. Yoshizato and E. Frieden. 1977. Preferential binding of tri-substituted thyronine analogs by bullfrog tadpole tail fin cytosol. *Endocrinology* 100: 134–137.

Koelle, M. R., W. S. Talbot, W. A. Segraves, M. T. Bender, P. Cherbas and D. S. Hogness. 1991. The *Drosophila* EcR gene encodes an ecdysone receptor, a new member of the steroid receptor superfamily. *Cell* 67: 59–77.

Kollros, J. J. 1961. Mechanisms of amphibian metamorphosis: Hormones. *Am. Zool.* 1: 107–114.

Kuro-o, M. and 15 others. 1997. Mutation of the mouse *klotho* gene leads to a syndrome resembling ageing. *Nature* 390: 45–51.

Larsen, P. L. 1993. Aging and resistance to oxidative damage in *C. elegans. Proc. Natl. Acad. Sci. USA* 90: 8905–8909.

Lenhoff, H. M. 1991. Ethel Browne Harvey, Hans Spemann, and the discovery of the organizer phenomenon. *Biol. Bull.* 181: 72–80.

Lin, Y.-J., L. Seroude and S. Benzer. 1998. Extended life-span and stress resistance in the *Drosophila* mutant *methuselah. Science* 282: 943–946.

Lindroos, P. M., R. Zarnegar and G. K. Michalopoulos. 1991. Hepatocyte growth factor (hepatopoietin A) rapidly increases in plasma before DNA synthesis and liver regeneration stimulated by partial hepatectomy and carbon tetrachloride administration. *Hepatology* 13: 743–750.

Lo, D. C., F. Allen and J. P. Brockes. 1993. Reversal of muscle differentiation during urodele limb regeneration. *Proc. Natl. Acad. Sci. USA* 90: 7230–7234.

Lynn, W. G. and A. M. Peadon. 1955. The role of the thyroid gland in direct development of the anuran *Eleutherodactylus martinicenis. Growth* 19: 263–286.

MacWilliams, H. K. 1983a. Hydra transplantation phenomena and the mechanism of hydra head regeneration. I. Properties of head inhibition. *Dev. Biol.* 96: 217–238.

MacWilliams, H. K. 1983b. Hydra transplantation phenomena and the mechanism of hydra head regeneration. II. Properties of head activation. *Dev. Biol.* 96: 239–272.

MacWilliams, H. K., F. C. Kafatos and W. H. Bossert. 1970. The feedback inhibition of basal disk regeneration in *Hydra* has a continuously variable intensity. *Dev. Biol.* 23: 380–398.

Maden, M. 1982. Vitamin A and pattern formation in the regenerating limb. *Nature* 295: 672–675.

Mangelsdorf, D. J and R. M. Evans. 1995. The RXR heterodimers and orphan receptors. *Cell* 83: 841–850.

Mars, W. M., M. L. Liu, R. P. Kitson, R. H. Goldfarb, M. K. Gabauer and G. K. Michalopoulos.

1995. Immediate early detection of urokinase receptor after partial hepatectomy and its implications for initiation of liver regeneration. *Hepatology* 21: 1695–1701.

Marsh-Armstrong, N., H. Huang, B. F. Remo, T. T. Liu and D. D. Brown. 1999. Asymmetric growth and development of the *Xenopus laevis* retina during metamorphosis is controlled by type III deiodinase. *Neuron* 24: 871–878.

Martin, V. J. 1997. Cnidrians: The jellyfish and hydra. *In* S. F. Gilbert and A. M. Raunio (eds.) *Embryology: Constructing the Organism.* Sinauer Associates, Sunderland, MA, pp. 57–86.

McCutcheon, F. H. 1936. Hemoglobin function during the life history of the bullfrog. *J. Cell. Comp. Physiol.* 8: 63–81.

McGann, C. J., S. J. Odelberg and M. T. Keating. 2001. Mammalian myotube dedifferentiation induced by newt regeneration extract. *Proc. Natl. Acad. Sci. USA* 98: 13699–13704.

Medawar, P. B. 1952. *An Unsolved Problem in Biology.* H. K. Lewis, London.

Meinhardt, H. 1980. Cooperation of compartments for the generation of positional information. *Z. Naturforsch.* C 35: 1086–1091.

Meinhardt, H. 1993. A model for pattern formation of hypostome, tentacles, and foot in hydra: How to form structures close to each other, how to form them at a distance. *Dev. Biol.* 157: 321–333.

Melov, S., J. A. Schnieder, B. J. Day, D. Hinerfeld, S. S. Coskunra, J. D. Crapo and D. C. Wallace. 1998. A novel neurological phenotype in mice lacking mitochondrial superoxide dismutase. *Nature Genet.* 18: 59–63.

Mescher, A. L. and D. Gospodarowicz. 1979. Mitogenic effects of a growth factor derived from myelin on denervated regenerates of newt forelimbs. *J. Exp. Zool.* 207: 497–510.

Mescher, A. L. and R. A. Tassava. 1975. Denervation effects on DNA replication and mitosis during the initiation of limb regeneration in adult newts. *Dev. Biol.* 44: 187–197.

Michalopoulos, G. K. and M. C. DeFrances. 1997. Liver regeneration. *Science* 276: 60–66.

Michikawa, Y., F. Mazzucchelli, N. Bresolin, G. Scarlato and G. Attardi. 1999. Aging-dependent large accumulation of point mutations in the human mtDNA control region for replication. *Science* 286: 774–779.

Miller, J. K. 2001. Escaping senescence: Demographic data from the three-toed box turtle (*Terrapene carolina triungis*). *Exp. Gerontol.* 36: 829–832.

Mitchell, A. W. 1988. *The Enchanted Canopy.* Macmillan, New York.

Mullen, L. M., S. V. Bryant, M. A. Torok, B. Blumberg and D. M. Gardiner. 1996. Nerve dependency of regeneration: The role of Distal-less and FGF signaling in amphibian limb regeneration. *Development* 122: 3487–3497.

Muneoka, K. and S. V. Bryant. 1982. Evidence that patterning mechanisms in developing and regenerating limbs are the same. *Nature* 298: 369–371.

Murray, V. and R. Holliday. 1981. Increased error frequency of DNA polymerases from senescent fibroblasts. *J. Mol. Biol.* 146: 55–76.

Nakagawa, S., C. Brennan, K. G. Johnson, D. Shewan, W. A. Harris and C. E. Holt. 2000. Ephrin-B regulates the ipsilateral routing of retinal axons at the optic chiasm. *Neuron* 25: 599–610.

Nardi, J. B. and D. L. Stocum. 1983. Surface properties of regenerating limb cell: Evidence for gradation along the proximodistal axis. *Differentiation* 25: 27–31.

Neumann, C. J. and S. M. Cohen. 1996. Distinct mitogenic and cell fate specification functions of *wingless* in different regions of the wing. *Development* 122: 1781–1789.

Neumann, C. J. and S. M. Cohen. 1997. Long-range action of Wingless organizes the dorsal-ventral axis of the *Drosophila* wing. *Development* 121: 589–599.

Newman, S. A. 1974. The interaction of the organizing regions of hydra and its possible relation to the role of the cut end of regeneration. *J. Embryol. Exp. Morphol.* 31: 541–555.

Niazi, I. A. and S. Saxena. 1978. Abnormal hindlimb regeneration in tadpoles of the toad *Bufo andersonii* exposed to excess vitamin A. *Folia Biol.* (Krakow) 26: 3–8.

Nijhout, H. F. 1994. *Insect Hormones.* Princeton University Press, Princeton, NJ.

Nijhout, H. F. 1999. Control mechanisms of polyphenic development in insects. *BioScience* 49: 181–192.

Nijhout, H. F. and C. M. Williams. 1974. Control of moulting and metamorphosis in the tobacco hornworm, *Manduca sexta*: Cessation of juvenile hormone secretion as a trigger for pupation. *J. Exp. Biol.* 61: 493–501.

Norris, D. O., R. E. Jones and B. B. Criley. 1973. Pituitary prolactin levels in larval, neotenic, and metamorphosed salamanders (*Ambystoma tigrinum*). *Gen. Comp. Endocrinol.* 20: 437–442.

Okada, T. S. 1991. *Transdifferentiation: Flexibility to Cell Differentiation.* Oxford University Press, Oxford.

Oofusa, K. and K. Yoshizato. 1991. Biochemical and immunological characterization of collagenase in tissues of metamorphosing bullfrog tadpoles. *Dev. Growth Diff.* 33: 329–339.

Orgel, L. E. 1963. The maintenence of the accuracy of protein synthesis and its relevance to aging. *Proc. Natl. Acad. Sci. USA* 49: 517–521.

Orr, W. S. and R. S. Sohal. 1994. Extension of lifespan by overexpression of superoxide dismutase and catalase in *Drosophila melanogaster*. *Science* 263: 1128–1130.

Parkes, T. L., A. J. Elia, D. Dickinson, A. J. Hilliker, J. P. Phillips and G. L. Boulianne. 1998. Extension of *Drosophila* lifespan by overexpression of human SOD1 in motoneurons. *Nature Genet.* 19: 171–174.

Partridge, L. and D. Gems. 2002. Mechanisms of ageing: Public or private? *Nature Rev. Genet.* 3: 165–175.

Patterson, D., W. P. Hayes and Y. B. Shi. 1995. Transcriptional activation of the metalloproteinase gene *stromelysin-3* coincides with thyroid hormone-induced cell death during frog metamorphosis. *Dev. Biol.* 167: 252–262.

Pecorino, L. T., A. Entwistle and J. P. Brockes. 1996. Activation of a single retinoic acid receptor isoform mediates proximo-distal respecification. *Curr. Biol.* 6: 563–569.

Pfeiffer, S. and J. P. Vincent. 1999. Signalling at a distance: Transport of Wingless in the epidermis of *Drosophila*. *Semin. Cell Dev. Biol.* 10: 303–309.

Prahlad, K. V. and L. E. DeLanney. 1965. A study of induced metamorphosis in the axolotl. *J. Exp. Zool.* 160: 137–146.

Rageh, M. A., L. Mendenhall, E. E. Moussad, S. E. Abbey, A. L. Mescher and R. A. Tassava. 2002. Vasculature in pre-blastema and nerve-dependent blastema stages of regenerating forelimbs of the adult newt, *Notophthalmus viridiscens*. *J. Exp. Zool.* 292: 255–266.

Rand, H. W., J. F. Board and D. E. Minnich. 1926. Localization of formative agencies in *Hydra*. *Proc. Natl. Acad. Sci. USA* 12: 565–570.

Restifo, L. L. and K. White. 1991. Mutations in a steroid hormone-regulated gene disrupt the metamorphosis of the central nervous system in *Drosophila*. *Dev. Biol.* 148: 174–194.

Restifo, L. L. and T. G. Wilson. 1998. A juvenile hormone agonist reveals distinct developmental pathways mediated by ecdysone-inducible Broad Complex transcription factors. *Dev. Genet.* 22: 141–159.

Richards, G. 1992. Switching partners? *Curr. Biol.* 2: 657–658.

Riddiford, L. M. 1972. Juvenile hormone in relation to the larval-pupal transformation of the *Cecropia* silkworm. *Biol. Bull.* 142: 310–325.

Riddiford, L. M. 1982. Changes in translatable mRNAs during the larval-pupal transformation of the epidermis of the tobacco hornworm. *Dev. Biol.* 92: 330–342.

Riddiford, L. M. 1996. Molecular aspects of juvenile hormone action in insect metamorphosis. *In* L. I. Gilbert, J. R. Tata and B. G. Atkinson (eds.), *Metamorphosis: Postembryonic Reprogramming of Gene Expression in Amphibian and Insect Cells.* Academic Press, San Diego, pp. 223–251.

Riggs, A. F. 1951. The metamorphosis of hemoglobin in the bullfrog. *J. Gen. Physiol.* 35: 23–40.

Robinson, H., S. Chaffee and V. A. Galton. 1977. Sensitivity of *Xenopus laevis* tadpole tail tissue to the action of thyroid hormones. *Gen. Comp. Endocrinol.* 32: 179–186.

Rostand, J. 1962. *The Substance of Man.* Doubleday, Garden City, NJ. p. 9.

Roth, G. S. and 7 others. 2002. Biomarkers of caloric restriction may predict longevity in humans. *Science* 297: 811.

Rountree, D. B. and W. E. Bollenbacher. 1986. The release of the prothoracicotropic hormone in the tobacco hornworm, *Manduca sexta*, is controlled intrinsically by juvenile hormone. *J. Exp. Biol.* 120: 41–58.

Rudolph, K. L., S. Chang, H.-W. Lee, M. Blasco, G. J. Gottlieb, C. Greider and R. A. DePinho. 1999. Longevity, stress response, and cancer in aging telomerase-deficient mice. *Cell* 96: 701–712.

Sachs, L. M., T. Amano, N. Rouse and Y.-B. Shi. 2001. Involvement of histone deacetylase at two distinct steps in gene regulation during intestinal development in *Xenopus laevis*. *Dev. Dynam.* 222: 280–291.

Safranek, L. and C. M. Williams. 1989. Inactivation of the corpora allata in the final instar of the tobacco hornworm, *Manduca sexta*, requires integrity of certain neural pathways from the brain. *Biol. Bull.* 177: 396–400.

Saxén, L., E. Saxén, S. Toivonen and K. Salimäki. 1957. The anterior pituitary and the thyroid function during normal and abnormal development of the frog. *Ann. Zool. Soc. Fennicae Vanamo* 18 (4): 1–44.

Scadding, S. R. and M. Maden. 1994. Retinoic acid gradients during limb regeneration. *Dev. Biol.* 162: 608–617.

Schiliro, D. M., B. J. Forman and L. C. Javois. 1999. Interactions between the foot and bud patterning systems in *Hydra vulgaris*. *Dev. Biol.* 209: 399–408.

Schmidt, T. and H. C. Schaller. 1976. Evidence for a foot-inhibiting substance in hydra. *Cell Diff.* 5: 151–159.

Schreiber, A. M., B. Das, H. Huang, N. Marsh-Armstrong and D. D. Brown. 2001. Diverse developmental programs of *Xenopus laevis* metamorphosis are inhibited by a dominant negative thyroid hormone receptor. *Proc. Natl. Acad. Sci. USA* 98: 10739–10744.

Schubiger, G. 1968. Anlageplan, Determinationszustand, und Transdeterminationsleistungen der männlichen Vorderbeinschiebe von *Drosophila melanogaster*. *Wilhelm Roux Arch. Entwicklungsmech. Org.* 160: 9–40.

Schubiger, M. and J. W. Truman. 2000. The RXR ortholog USP suppresses early metamorphic processes in *Drosophila* in the absence of ecdysteroids. *Development* 127: 1151–1159.

Schwind, J. L. 1933. Tissue specificity at the time of metamorphosis in frog larvae. *J. Exp. Zool.* 66: 1–14.

Scott, B. A., M. S. Avidan and C. M. Crowder. 2002. Regulation of hypoxic death in *C. elegans* by the insulin/IGF receptor homolog DAF-2. *Science* 296: 2388–2391.

Shen, J.-C. and L. A. Loeb. 2001.Unwinding the molecular basis of the Werner syndrome. *Mech. Ageing Dev.* 122: 921–944.

Shostak, S. 1974. Bipolar inhibitory gradients' influence on the budding region of *Hydra viridis. Am. Zool.* 14: 619–632.

Simon, H. G., C. Nelson, D. Goff, E. Laufer, B. A. Morgan and C. Tabin. 1995. The differential expression of myogenic regulatory genes and *msx-1* during dedifferentiation and redifferentiation of regenerating amphibian limbs. *Dev. Dynam.* 202: 1–12.

Singer, M. 1954. Induction of regeneration of the forelimb of the postmetamorphic frog by augmentation of the nerve supply. *J. Exp. Zool.* 126: 419–472.

Singer, M. and J. D. Caston. 1972. Neurotrophic dependence of macromolecular synthesis in the early limb regenerate of the newt, *Triturus. J. Embryol. Exp. Morphol.* 28: 1–11.

Smith-Gill, S. J. and V. Carver. 1981. Biochemical characterization of organ differentiation and maturation. *In* L. I. Gilbert and E. Frieden (eds.), *Metamorphosis: A Problem in Developmental Biology.* Plenum, New York, pp. 491–544.

Steele, R. 2002. Developmental signaling in *Hydra:* What does it take to build a "simple" animal? *Dev. Biol.* 248: 199–219.

Stocum, D. L. 1979. Stages of forelimb regeneration in *Ambystoma maculatum. J. Exp. Zool.* 209: 395–416.

Stocum, D. L. and K. Crawford. 1987. Use of retinoids to analyse the cellular basis of memory in regenerating amphibian limbs. *Biochem. Cell Biol.* 65: 750–761.

Stolow, M. A. and Y. B. Shi. 1995. *Xenopus* sonic hedgehog as a potential morphogen during embryogenesis and thyroid hormone dependent metamorphosis. *Nucleic Acids Res.* 23: 2555–2562.

Sun, H., J. K. Jarow, I. D. Hickson and N. Mazels. 1998. The Blooms syndrome helicase unwinds G4 DNA. *J. Biol. Chem.* 273: 27587–27592.

Sun, J. and J. Tower. 1999. FLP recombinase-mediated induction of Cu/Zn-superoxide dismutase transgene expression can extend the life span of adult *Drosophila melanogaster* flies. *Mol. Cell Biol.* 19: 216–228.

Suzuki, K., R. Utoh, K. Kotani, M. Obara and K. Yoshizato. 2002. Lineage of anuran epidermal basal cells and their differentiation potential in relation to metamorphic skin remodeling. *Dev. Growth Diff.* 44: 225–238.

Szilard, L. 1959. On the nature of the aging process. *Proc. Natl. Acad. Sci. USA* 45: 30–45.

Tabata, T., E. Schwartz, E. Gustavson, Z. Ali and T. B. Kornberg. 1995. Creating a *Drosophila* wing de novo, the role of *engrailed*, and the compartment border hypothesis. *Development* 121: 3359–3369.

Taigen, T. L., F. H. Plough and M. M. Stewart. 1984. Water balance of terrestrial anuran (*Eleutherodactylus coqui*) eggs: Importance of paternal care. *Ecology* 65: 248–255.

Takano, J. and T. Sugiyama. 1983. Genetic analysis of developmental mechanisms in hydra. VIII. Head activation and head inhibition potentials of a slow-budding strain (L4). *J. Embryol. Exp. Morphol.* 78: 141–168.

Talbot, W. S., E. A. Swyryd and D. S. Hogness. 1993. *Drosophila* tissues with different metamorphic responses to ecdysone express different ecdysone receptor isoforms. *Cell* 73: 1323–1337.

Tanaka, E. M., D. Drechsel and J. P. Brockes. 1999. Thrombin regulates S-phase re-entry by cultured newt myotubes. *Curr. Biol.* 9: 792–799.

Tatar, M., A. Kopelman, D. Epstein, M. P. Tu, C. M. Yin and R. S. Garofalo. 2001. A mutant *Drosophila* insulin receptor homolog that extends life-span and impairs neuroendocrine function. *Science* 292: 107–210.

Taurog, A., C. Oliver, R. L. Porter, J. C. McKenzie and J. M. McKenzie. 1974. The role of TRH in the neoteny of the Mexican axolotl (*Ambystoma mexicanum*). *Gen. Comp. Endocrinol.* 24: 267–279.

Thomas, H. E., H. G. Stunnenberg and A. F. Stewart. 1993. Heterodimerization of the *Drosophila* ecdysone receptor with retinoid X receptor and *ultraspiracle. Nature* 362: 471–475.

Torok, M. A., D. M. Gardiner, N. H. Shubin and S. V. Bryant. 1998. Expression of HoxD genes in developing and regenerating axolotl limbs. *Dev. Biol.* 200: 225–233.

Torok, M. A., D. M. Gardiner, J.-C. Izpisúa-Belmonte and S. V. Bryant. 1999. *Sonic hedgehog (Shh)* expression in developing and regenerating axolotl limbs. *J. Exp. Zool.* 284: 197–206.

Tran, H. and 7 others. 2002. DNA repair pathways stimulated by the forkhead transcription factor FOXO3a through the Gadd45 protein. *Science* 296: 530–534.

Trueb, L. and J. Hanken. 1992. Skeletal development in *Xenopus laevis* (Anura: Pipidae). *J. Morphol.* 214: 1–41.

Truman, J. W. and L. M. Riddiford. 1999. The origins of insect metamorphosis. *Nature* 401: 447–452.

Truman, J. W., W. S. Talbot, S. E. Fahrbach and D. S. Hogness. 1994. Ecdysone receptor expression in the CNS correlates with stage-specific responses to ecdysteroids during *Drosophila* and *Manduca* development. *Development* 120: 219–234.

Tsonis, P. A., C. H. Washabaugh and K. Del Rio-Tsonis. 1995. Transdifferentiation as a basis for amphibian limb regeneration. *Semin. Cell Biol.* 6: 127–135.

Tyler, M. J. 1983. *The Gastric Brooding Frog.* Croom Helm, London.

Tyler, M. J., D. J. Shearman, R. Franco, P. O'Brien, R. F. Seamark and R. Kelly. 1983. Inhibition of gastric acid secretion in the gastric brooding frog, *Rheobatrachus silus. Science* 220: 607–610.

Urness, L. D. and C. S. Thummel. 1995. Molecular analysis of a steroid-induced regulatory hierarchy: The *Drosophila* E74A protein directly regulates L71-6 transcription. *EMBO J.* 14: 6239–6246.

Vanfleteren, J. R. and A. De Vreese. 1996. Rate of aerobic metabolism and superoxide production rate potential in the nematode *Caenorhabditis elegans. J. Exp. Zool.* 274: 93–100.

van Wijngaarden, R. and F. Bolanos. 1992. Parental care in *Dendrobates granuliferus* (Anura, Dendrobatidae) with a description of the tadpole. *J. Herpetol.* 26: 102–105.

Viviano, C. M., C. E. Horton, M. Maden and J. P. Brockes. 1995. Synthesis and release of 9-*cis*-retinoic acid by the urodele wound epidermis. *Development* 121: 3753–3762.

Wald, G. 1945. The chemical evolution of vision. *Harvey Lect.* 41: 117–160.

Wald, G. 1981. Metamorphosis: An overview. *In* L. I. Gilbert and E. Frieden (eds.), *Metamorphosis: A Problem in Developmental Biology.* Plenum, New York, pp. 1–39.

Wang, L., M. A. Marchionni and R. A. Tassava. 2000. Cloning and neuronal expression of a type III newt neuregulin and rescue of denervated, nerve-dependent newt limb blastemas by rhGGF2. *J. Neurobiol.* 43: 150–158.

Wassersug, R. J. 1989. Locomotion in amphibian larvae (or why aren't tadpoles built like fish). *Am. Zool.* 29: 65–84.

Webster, G. 1966. Studies on pattern regulation in hydra. 3. Dynamic aspects of factors controlling hypostome formation. *J. Embryol. Exp. Morphol.* 16: 123–141.

Weismann, A. 1891. The duration of life. *In* E. B. Poulton, S. Schonland and A. E. Shipley (eds.), *Essays on Heredity and Kindred Subjects,* 2nd Ed. Oxford University Press, Oxford, pp. 163–256.

White, B. H. and C. S. Nicoll. 1981. Hormonal control of amphibian development. *In* L. I. Gilbert and E. Frieden (eds.), *Metamorphosis: A Problem in Developmental Biology.* Plenum, New York, pp. 363–396.

Wilby, O. K. and G. Webster. 1970. Experimental studies on axial polarity in hydra. *J. Embryol. Exp. Morphol.* 24: 595–613.

Wolffe, A. P. and Y.-B. Shi. 1999. A hypothesis for the transcriptional control of amphibian metamorphosis by the thyroid hormone receptor. *Am. Zool.* 39: 807–817.

Wong, J. M. and Y. B. Shi. 1995. Coordinated regulation and transcriptional activation of *Xenopus* thyroid hormone and retinoid-X receptors. *J. Biol. Chem.* 270: 18479–18483.

Yang, E. V., D. M. Gardiner, M. R Carlson, C. A. Nugas and S. V. Bryant. 1999. Expression of *Mmp-9* and related matrix metalloproteinase genes during axolotl limb regeneration. *Dev. Dynam.* 216: 2–9.

Yao, T.-P., W. A. Segraves, A. E. Oro, M. Mc Keown and R. M. Evans. 1992. *Drosophila ultraspiracles* modulates ecdysone receptor function via heterodimer formation. *Cell* 71: 63–72.

Yaoita, Y. and D. D. Brown. 1990. A correlation of thyroid hormone receptor gene expression with amphibian metamorphosis. *Genes Dev.* 4: 1917–1924.

Yokoyama, H., S. Yonei-Tamura, T. Endo, J.-C. Izpisúa-Belmonte, K. Tamura and H. Ide. 2000. Mesenchyme with *fgf-10* expression is responsible for regenerative capacity in *Xenopus* limb buds. *Dev. Biol.* 219: 18–29.

Yokoyama, H., H. Ide and K. Tamura. 2001. FGF-10 stimulates limb regeneration ability in *Xenopus laevis. Dev. Biol.* 233: 72–79.

Zammit, P. S. and 7 others. 2002. Kinetics of myoblast proliferation show that resident satellite cells are competent to fully regenerate skeletal muscle fibers. *Exp. Cell Res.* 281: 39–49.

Zecca, M., K. Basler and G. Struhl. 1996. Direct and long-range action of a Wingless morphogen gradient. *Cell* 87: 833–844.

19 *The saga of the germ line*

WE BEGAN OUR ANALYSIS of animal development by discussing fertilization, and we will finish our studies of individual development by investigating **gametogenesis**, the processes by which the sperm and the egg are formed. Germ cells provide the continuity of life between generations, and the mitotic ancestors of our own germ cells once resided in the gonads of reptiles, amphibians, fish, and invertebrates.

In many animals, such as insects, roundworms, and vertebrates, there is a clear and early separation of germ cells from somatic cell types. In several other animal phyla (and throughout the entire plant kingdom), this division is not as well established. In these species (which include cnidarians, flatworms, and tunicates), somatic cells can readily become germ cells even in adult organisms. The zooids, buds, and polyps of many invertebrate phyla testify to the ability of somatic cells to give rise to new individuals (Liu and Berrill 1948; Buss 1987).

In those organisms in which there is an established germ line that separates from the somatic cells early in development, the germ cells do not arise within the gonad itself. Rather, their precursors—the **primordial germ cells** (**PGCs**)—arise elsewhere and migrate into the developing gonads. The first step in gametogenesis, then, involves forming the PGCs and getting them into the genital ridge as the gonad is forming. Our discussion of gametogenesis will include:

1. The formation of the germ plasm and the determination of the PGCs
2. The migration of the PGCs into the developing gonads
3. The process of meiosis and the modifications of meiosis for forming sperm and eggs
4. The differentiation of the sperm and egg
5. The hormonal control of gamete maturation and ovulation

Germ Plasm and the Determination of the Primordial Germ Cells

All sexually reproducing organisms arise from the fusion of gametes—sperm and eggs. All gametes arise from the primordial germ cells. In many instances (including frogs, nematodes, and flies), the primordial germ cells are specified autonomously by cytoplasmic determinants in the egg that are then parceled out to specific cells during cleavage. In other instances (such as salamanders and mammals), the germ cells are specified by interactions among neighboring cells. In those species wherein the determination of the primordial germ cells is brought about by the autonomous lo-

(A)

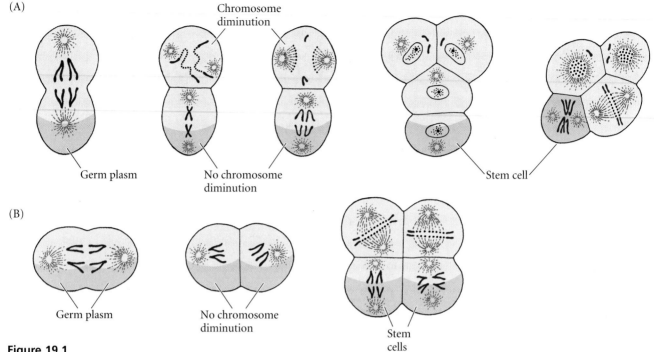

Chromosome
diminution

Germ plasm

No chromosome
diminution

Stem cell

(B)

Germ plasm

No chromosome
diminution

Stem
cells

Figure 19.1
Distribution of germ plasm during cleavage of (A) normal and (B) centrifuged zygotes of *Parascaris*. (A) The germ plasm is normally conserved in the most vegetal blastomere, as shown by the lack of chromosomal diminution in that particular cell. Thus, at the 4-cell stage, the embryo has one stem cell for its gametes. (B) When the first cleavage is displaced 90 degrees by centrifugation, both resulting cells have vegetal germ plasm, and neither cell undergoes chromosome diminution. After the second cleavage, these two cells give rise to germinal stem cells. (After Waddington 1966.)

calization of specific proteins and mRNAs, these cytoplasmic components are collectively referred to as the **germ plasm**.

Germ cell determination in nematodes

Theodor Boveri (1862–1915; see Figure 4.2) was the first person to observe an organism's chromosomes throughout its development. In so doing, he discovered a fascinating feature in the development of the roundworm *Parascaris aequorum* (formerly *Ascaris megalocephala*). This nematode has only two chromosomes per haploid cell, allowing for detailed observations of the individual chromosomes. The cleavage plane of the first embryonic division is unusual in that it is equatorial, separating the animal half from the vegetal half of the zygote (Figure 19.1A). More bizarre, however, is the behavior of the chromosomes in the subsequent division of these first two blastomeres. The ends of the chromosomes in the animal blastomere fragment into dozens of pieces just before this cell divides. This phenomenon is called **chromosome diminution**, because only a portion of the original chromosome survives. Numerous genes are lost when the chromosomes fragment, and these genes are not included in the newly formed nuclei (Tobler et al. 1972; Müller et al. 1996).

Meanwhile, in the vegetal blastomere, the chromosomes remain normal. During second cleavage, the animal cell splits meridionally while the vegetal cell again divides equatorially. Both vegetally derived cells have normal chromosomes. However, the chromosomes of the more animally located of these two vegetal blastomeres fragment before the third cleavage. Thus, at the 4-cell stage, only one cell—the most vegetal—contains a full set of genes. At successive cleavages, nuclei with diminished chromosomes are given off from this vegetalmost line until the 16-cell stage, when there are only two cells with undiminished chromosomes. One of these two blastomeres gives rise to the germ cells; the other eventually undergoes chromosome diminution and forms more somatic cells. The chromosomes are kept intact only in those cells destined to form the germ line. If this were not the case, the genetic information would degenerate from one generation to the next. The cells that have undergone chromosome diminution generate the somatic cells.

Boveri has been called the last of the great "observers" of embryology and the first of the great experimenters. Not content with observing the retention of the full chromosome complement by the germ cell precursors, he set out to test whether a specific region of cytoplasm protects the nuclei within it from diminution. If so, any nucleus happening to reside in this region should remain undiminished. Boveri (1910) tested this hypothesis by centrifuging *Parascaris* eggs shortly before their first cleavage. This treatment shifted the orientation of the mitotic spindle. When the spindle forms perpendicular to its normal orientation, both resulting blastomeres should contain some of the vegetal cytoplasm (see

Figure 19.2
The pole plasm of *Drosophila*. (A) Electron micrograph of polar granules from particulate fraction of *Drosophila* pole cells. (B) Scanning electron micrograph of a *Drosophila* embryo just prior to completion of cleavage. The pole cells can be seen at the right of this picture. (Photographs courtesy of A. P. Mahowald.)

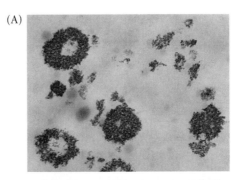

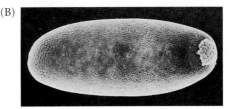

Figure 19.1B). Indeed, Boveri found that after the first division, neither nucleus underwent chromosomal diminution. However, the next division was equatorial along the animal-vegetal axis. Here the resulting animal blastomeres both underwent diminution, whereas the two vegetal cells did not. Boveri concluded that the vegetal cytoplasm contains a factor (or factors) that protects nuclei from chromosomal diminution and determines germ cells.

In the nematode *C. elegans*, the germ line precursor cell is the P4 blastomere. The **P-granules** enter this cell, and they appear critical for instructing it to become the germ line precursor (see Figure 8.43). The components of the P-granules include several transcriptional inhibitors and RNA-binding proteins, including homologues of the *Drosophila* Vasa and Nanos proteins, whose functions we will discuss below (Kawasaki et al. 1998; Seydoux and Strome 1999; Subramanian and Seydoux 1999).

> **WEBSITE 19.1 Mechanisms of chromosome diminution.** The somatic cells do not lose DNA randomly. Rather, specific regions of DNA are lost during chromosome diminution.

Germ cell determination in insects

In *Drosophila*, PGCs form as a group of cells (**pole cells**) at the posterior pole of the cellularizing blastoderm. These nuclei migrate into the posterior region at the ninth nuclear division, and they become surrounded by the **pole plasm**, a complex collection of mitochondria, fibrils, and **polar granules** (Figure 19.2; Mahowald 1971a,b; Schubiger and Wood 1977).

If the pole cell nuclei are prevented from reaching the pole plasm, no germ cells will be made (Mahowald et al. 1979).

Nature has provided confirmation of the importance of both the pole plasm and its polar granules. One of the components of the pole plasm is the mRNA of the **germ cell-less** (**gcl**) gene. This gene was discovered by Jongens and his colleagues (1992) when they mutated *Drosophila* and screened for females who did not have "grandoffspring." They assumed that if a female did not place functional pole plasm in her eggs, she could still have offspring, but those offspring would be sterile (since they would lack germ cells). The wild-type *gcl* gene is transcribed in the nurse cells of the fly's ovary, and its mRNA is transported into the egg. Once inside the egg, it is transported to the posteriormost portion and resides within what will become the pole plasm (Figure 19.3A). This message is translated into protein during the early stages of cleavage (Figure 19.3C). The *gcl*-encoded protein appears to enter the nucleus, and it is essential for pole cell production. Flies with mutations of this gene lack germ cells (Figure 19.3B,D).

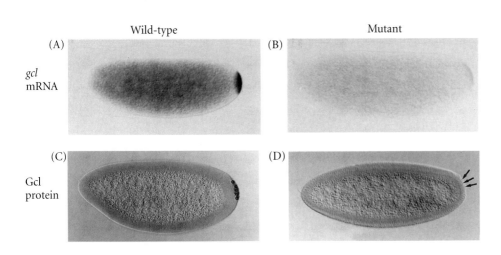

Wild-type	Mutant
(A) *gcl* mRNA	(B)
(C) Gcl protein	(D)

Figure 19.3
Localization of *germ cell-less* gene products in the posterior of the egg and embryo. (A, B) The *gcl* mRNA can be seen in the posterior pole of early-cleavage embryos produced by wild-type females (A), but not in embryos produced by *gcl*-deficient mutant females (B). (C, D) The protein encoded by the *gcl* gene can be detected in the germ cells at the cellular blastoderm stage of embryos produced by wild-type females (C), but not in embryos from mutant females (D). (From Jongens et al. 1992; photographs courtesy of T. A. Jongens.)

VADE MECUM[2] **Germ cells in the *Drosophila* embryo.**
This segment follows the primordial germ cells of the living *Drosophila* embryo from their formation as pole cells through gastrulation as they move from the posterior end of the embryo into the region of the developing gonad.
[Click on Fruit Fly]

A second set of pole plasm components are the posterior determinants mentioned in Chapter 9. **Oskar** appears to be the critical protein of this group, since expression of *oskar* mRNA in ectopic sites will cause the nuclei in those areas to form germ cells. The genes that restrict Oskar to the posterior pole are also necessary for germ cell formation (Ephrussi and Lehmann 1992; Newmark et al. 1997; Riechmann et al. 2002). Moreover, Oskar appears to be the limiting step of germ cell formation, since adding more *oskar* message to the oocyte causes the formation of more germ cells (Ephrussi and Lehmann 1992). Oskar functions by causing the localization of the proteins and RNAs necessary for germ cell formation. One of these RNAs is the *nanos* message, whose product is essential for posterior segment formation. **Nanos** is also essential for germ cell formation. Pole cells lacking Nanos do not migrate into the gonads and fail to become gametes. Nanos appears to be important in preventing mitosis and transcription during germ cell development (Kobayashi et al. 1996; Deshpande et al. 1999). Another one of these RNAs encodes **Vasa**, an RNA-binding protein. The mRNAs for this protein are seen in the germ plasm of many species.

A third germ plasm component, **mitochondrial ribosomal RNA (mtrRNA)**, was a big surprise. Kobayashi and Okada (1989) showed that the injection of mtrRNA into embryos formed from ultraviolet-irradiated eggs restores the ability of these embryos to form pole cells. Moreover, in normal fly eggs, the small and large mtrRNAs are located outside the mitochondria solely in the pole plasm of cleavage-stage embryos, where they appear as components of the polar granules (Kobayashi et al. 1993; Amikura et al. 1996; Kashikawa et al. 1999). While mtrRNA is involved in directing the formation of the pole cells, it does not enter them. The Tudor protein localizes to the germ plasm, and it appears to be critical for the export of these mtrRNAs from the mitochondria and into the cytoplasm (Amikura et al. 2001).

A fourth component of *Drosophila* pole plasm (and one that becomes localized in the polar granules) is a nontranslatable RNA called **polar granule component (PGC)**. While its exact function remains unknown, the pole cells of transgenic female flies making antisense RNA against PGC fail to migrate to the gonads (Nakamura et al. 1996).

WEBSITE 19.2 **The insect germ plasm.** The insect germinal cytoplasm was discovered as early as 1911, when Hegner found that removing the posterior pole cytoplasm of beetle eggs caused sterility in the resulting adults.

Germ cell determination in amphibians

Cytoplasmic localization of germ cell determinants has also been observed in vertebrate embryos. Bounoure (1934) showed that the vegetal region of fertilized frog eggs contains material with staining properties similar to those of *Drosophila* pole plasm (Figure 19.4). He was able to trace this cortical cytoplasm into the few cells in the presumptive endoderm that would normally migrate into the genital ridge. By transplanting genetically marked cells from one embryo into another of a differently marked strain, Blackler (1962) showed that these cells are the primordial germ cell precursors. The germ plasm of amphibians consists of germinal granules and a matrix around them. It contains many of the RNAs and proteins (including the large and small mitochondrial ribosomal RNAs) as the pole plasm of *Drosophila*, and they appear to repress transcription and translation (Kloc et al. 2002).

The early movements of the germ plasm in amphibians have been analyzed in detail by Savage and Danilchik (1993), who labeled the germ plasm with a fluorescent dye. They found that the germ plasm of unfertilized eggs consists of tiny "islands" that appear to be tethered to the yolk mass near the vegetal cortex. These germ plasm islands move with the vegetal yolk mass during the cortical rotation just after fertilization. After the rotation, the islands are released from the yolk mass and begin fusing together and migrating to the vegetal pole. Their aggregation depends on microtubules, and their movement to the vegetal pole depends on a kinesin-like protein that may act as the motor for germ plasm movement

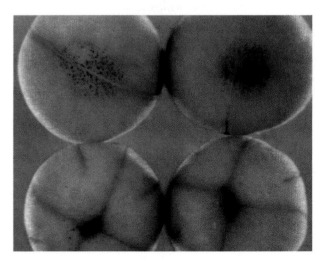

Figure 19.4
Germ plasm at the vegetal pole of frog embryos. In situ hybridization to the mRNA for *Xcat2* (the *Xenopus* homologue of *Nanos*) localizes the message in the vegetal cortex of first-cleavage (upper) and fourth-cleavage (lower) embryos. (After Kloc et al. 1998; photograph courtesy of L. Etkin.)

(Robb et al. 1996; Quaas and Wylie 2002). Savage and Danilchik (1993) found that UV light prevents vegetal surface contractions and inhibits the migration of germ plasm to the vegetal pole. Furthermore, the *Xenopus* homologues of *nanos* and *vasa* messages are specifically localized to the vegetal region (Forristall et al. 1995; Ikenishi et al. 1996; Zhou and King 1996). So, like the *Drosophila* pole plasm, the cytoplasm from the vegetal region of the *Xenopus* zygote contains the determinants for germ cell formation. Moreover, several of the components are the same in the two species.

The inert genome hypothesis

The components of the germ plasm have not all been catalogued. Indeed, in the birds and mammals, such a list has hardly even been started. Moreover, we still do not know the functions of the proteins (such as Vasa and Nanos) and nontranslated RNAs found in the germ plasm. One hypothesis (Nieuwkoop and Sutasurya 1981; Wylie 1999) is that the components of the germ plasm inhibit both transcription and translation, thereby preventing the cells containing it from differentiating into anything else. According this hypothesis, the cells become germ cells because they are forbidden to become any other type of cell. This suppression of transcription is seen in the germ cells of several species, including flies, frogs, and nematodes (Figure 19.5). Many of the proteins in the germ plasm (such as Gcl) act by inhibiting either transcription or translation (Leatherman et al. 2002).

Germ Cell Migration

Germ cell migration in amphibians

The germ plasm of anuran amphibians (frogs and toads) collects around the vegetal pole in the zygote. During cleavage, this material is brought upward through the yolky cytoplasm. Periodic contractions of the vegetal cell surface appear to push it along the cleavage furrows of the newly formed blastomeres. Germ plasm eventually becomes associated with the endodermal cells lining the floor of the blastocoel (Figure 19.4; see also 19.6A–E; Bounoure 1934; Ressom and Dixon 1988; Kloc et al. 1993). The PGCs become concentrated in the posterior region of the larval gut, and as the abdominal cavity forms, they migrate along the dorsal side of the gut, first along the dorsal mesentery (which connects the gut to the region where the mesodermal organs are forming) and then along the abdominal wall and into the genital ridges. They migrate up this tissue until they reach the developing gonads (Figure 19.6F). *Xenopus* PGCs move by extruding a single filopodium and then streaming their yolky cytoplasm into that filopodium while retracting their "tail." Contact guidance in this migration seems likely, as both the PGCs and the extracellular matrix over which they migrate are oriented in the direction of

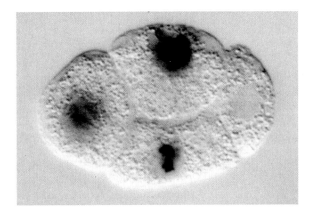

Figure 19.5
Inhibition of transcription in germ cell precursors of *C. elegans*. The photograph shows in situ hybridization to β-galactosidase mRNA expressed under control of the *pes-10* promoter. The *pes-10* gene is one of the earliest genes expressed in *C. elegans*. The P-blastomere that gives rise to the germ cells does not transcribe the gene. (From Seydoux and Fire 1994; photograph courtesy of G. Seydoux.)

migration (Wylie et al. 1979). Furthermore, PGC adhesion and migration can be inhibited if the mesentery is treated with antibodies against *Xenopus* fibronectin (Heasman et al. 1981). Thus, the pathway for germ cell migration in these frogs appears to be composed of an oriented fibronectin-containing extracellular matrix. The fibrils over which the PGCs travel lose this polarity soon after migration has ended.* As they migrate, *Xenopus* PGCs divide about three times, and approximately 30 PGCs colonize the gonads (Whitington and Dixon 1975; Wylie and Heasman 1993). These cells will divide to form the germ cells.

The primordial germ cells of urodele amphibians (salamanders) have an apparently different origin, which has been traced by reciprocal transplantation experiments to the regions of the mesoderm that involute through the ventrolateral lips of the blastopore. Moreover, there does not seem to be any particular localized "germ plasm" in salamander eggs. Rather, the interaction of the dorsal endoderm cells and animal hemisphere cells creates the conditions needed to form germ cells in the areas that involute through the ventrolateral lips (Sutasurya and Nieuwkoop 1974; Wakahara 1996). So in salamanders, the PGCs are formed by induction within the mesodermal region and presumably follow a different path into the gonads.

---

*This statement does not necessarily hold true for all anurans. In the frog *Rana pipiens*, the germ cells follow a similar route, but may be passive travelers along the mesentery rather than actively motile cells (Subtelny and Penkala 1984). The migration of fish PGCs follows a similar route, and there may be species differences as to whether the PGCs are active or passive travelers (Braat et al. 2000).

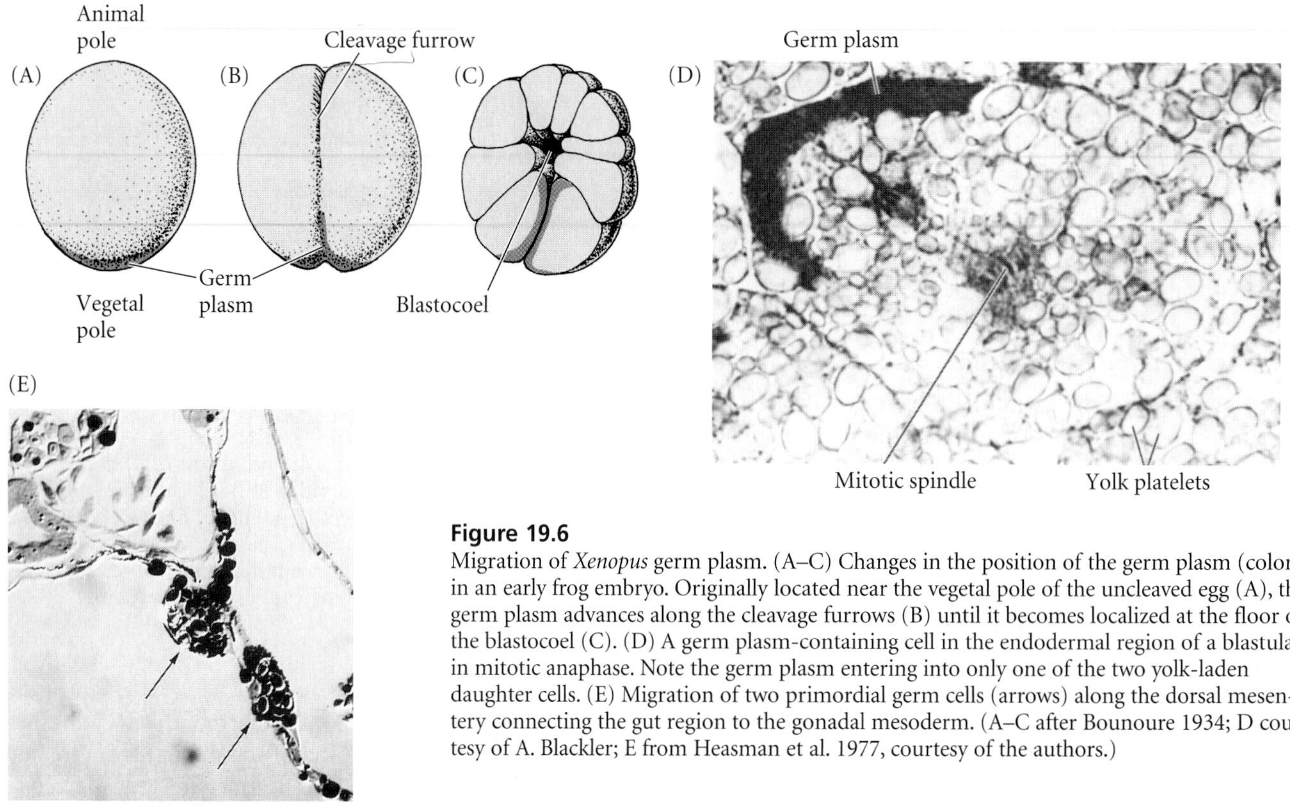

Figure 19.6

Migration of *Xenopus* germ plasm. (A–C) Changes in the position of the germ plasm (color) in an early frog embryo. Originally located near the vegetal pole of the uncleaved egg (A), the germ plasm advances along the cleavage furrows (B) until it becomes localized at the floor of the blastocoel (C). (D) A germ plasm-containing cell in the endodermal region of a blastula in mitotic anaphase. Note the germ plasm entering into only one of the two yolk-laden daughter cells. (E) Migration of two primordial germ cells (arrows) along the dorsal mesentery connecting the gut region to the gonadal mesoderm. (A–C after Bounoure 1934; D courtesy of A. Blackler; E from Heasman et al. 1977, courtesy of the authors.)

Germ cell formation and migration in mammals

There is no obvious germ plasm in mammals, and mammalian germ cells are not morphologically distinct during early development. Rather, germ cells are induced in the embryo. In mice, the germ cells form at the posterior region of the epiblast, at the junction of the extraembryonic ectoderm, epiblast, primitive streak, and allantois (Figure 19.7A). At day 6.5 of embryonic development, BMP4 and BMP8b from the extraembryonic ectoderm give certain cells in this area the ability to produce germ cells (Lawson et al. 1999; Ying et al. 2000). The cluster of cells capable of generating PGCs express *fragilis*, a gene encoding a particular transmembrane protein. However, these *fragilis*-expressing cells can form both PGCs and some somatic cells. In the center of this cluster of cells is a small group of cells that also expresses *stella*.* These cells are restricted to the germ cell fate (Saitou et al. 2002).

Based on differential staining of fixed tissue, it had long been thought that the mouse germ cell precursors migrated from the epiblast into the extraembryonic mesoderm and then back again into the embryo by way of the allantois (see

---

*It is not known what the Stella protein does. The sequence of this protein gives few clues other than that it contains some highly basic regions (and therefore might be able to associate with nucleic acids) and that it has a splicing factor motif in its carboxy terminus. It may therefore be involved in chromosome organization or RNA processing.

Chiquoine 1954; Mintz 1957; and earlier editions of this textbook). However, the ability to label mouse primordial germ cells with green fluorescent protein and to watch these living cells migrate has caused a reevaluation of the germ cell migration pathway in mammals (Anderson et al. 2000; Molyneaux et al. 2001). First, it appears that mammalian PGCs migrate directly into the endoderm from the posterior region of the primitive streak. (The cells that enter the allantois are believed to die.) These *stella*-expressing cells find themselves in the hindgut. Although they move actively, they cannot get out of the gut until about embryonic day 9. At that time, the PGCs exit the gut but do not yet migrate toward the genital ridges. By the following day, however, PGCs are seen migrating into the genital ridges (Figure 19.7E). By embryonic day 11.5, the PGCs enter the developing gonads. During this trek, they have proliferated from an initial population of 10–100 cells to the 2500–5000 PGCs present in the gonads by day 12 (Figure 19.8).

Like the PGCs of *Xenopus*, mammalian PGCs appear to be closely associated with the cells over which they migrate, and they move by extending filopodia over the underlying cell surfaces. These cells are also capable of penetrating cell monolayers and migrating through cell sheets (Stott and Wylie 1986). The mechanism by which the PGCs know the route of this journey is still unknown. Fibronectin is likely to be an im-

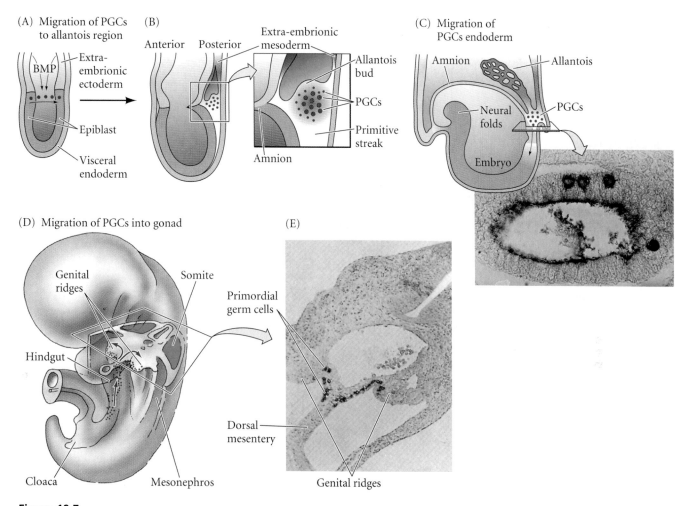

(A) Migration of PGCs to allantois region

BMP — Extra-embrionic ectoderm

Epiblast

Visceral endoderm

(B) Anterior Posterior Extra-embrionic mesoderm

Allantois bud

PGCs

Primitive streak

Amnion

(C) Migration of PGCs endoderm

Amnion — Allantois

Neural folds — PGCs

Embryo

(D) Migration of PGCs into gonad

Genital ridges Somite

Hindgut

Cloaca Mesonephros

(E)

Primordial germ cells

Dorsal mesentery

Genital ridges

Figure 19.7

Pathway for the migration of mammalian primordial germ cells. (A) In the mouse embryo, BMP signals (blue) from the extraembryonic ectoderm induce neighboring epiblast cells (purple circles) to become precursors of PGCs and extraembryonic mesoderm. During gastrulation (arrow), these cells come to reside in the posterior epiblast. (B) On embryonic day 7, these cells emerge from the posterior primitive streak. These cells express *fragilis* (green), and in the center of this cluster are PGC precursors that also express *stella* (dark green and red). (C) On day 8, these cells migrate into the definitive endoderm of the embryo. The inset shows four large PGCs in the hindgut of a mouse embryo stain positively for high levels of alkaline phosphatase. (D) The PGCs migrate through the gut and, dorsally, into the genital ridges. (E) Such alkaline phosphatase-staining cells can be seen entering the genital ridges around embryonic day 11. (A–C after Hogan 2002; C photograph from Heath 1978; E from Mintz 1957; photographs courtesy of the authors.)

(A) Female

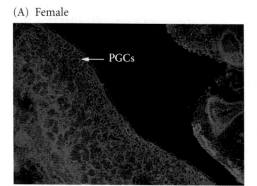

PGCs

(B) Male

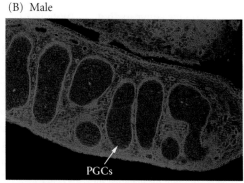

PGCs

Figure 19.8

Gonads from (A) female and (B) male mouse embryos stained at embryonic day 13.5 with antibodies to E-cadherin (red) and laminin (green). The PGCs of both sexes express high levels of E-cadherin and are arranged in cortical clusters in the ovary and in internal cords in the testes. (From Bendel-Stenzel et al. 2000; photographs courtesy of M. Bendel-Stenzel.)

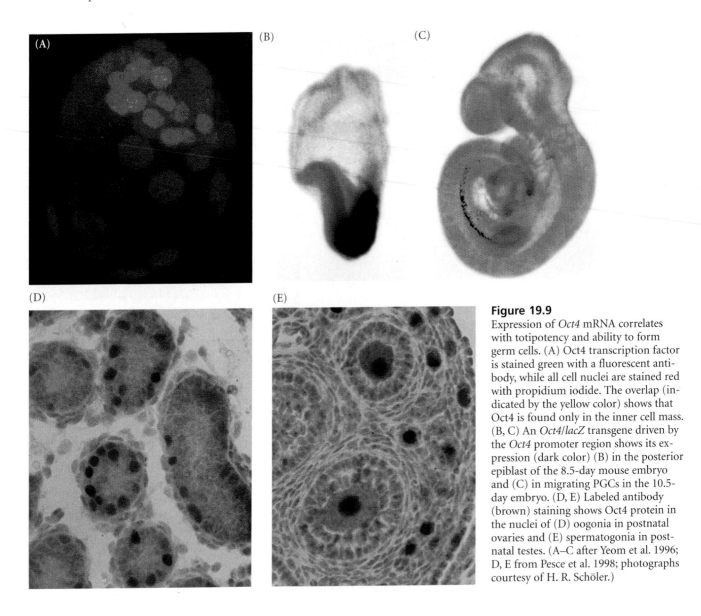

(A)

(B)

(C)

(D)

(E)

Figure 19.9

Expression of *Oct4* mRNA correlates with totipotency and ability to form germ cells. (A) Oct4 transcription factor is stained green with a fluorescent antibody, while all cell nuclei are stained red with propidium iodide. The overlap (indicated by the yellow color) shows that Oct4 is found only in the inner cell mass. (B, C) An *Oct4/lacZ* transgene driven by the *Oct4* promoter region shows its expression (dark color) (B) in the posterior epiblast of the 8.5-day mouse embryo and (C) in migrating PGCs in the 10.5-day embryo. (D, E) Labeled antibody (brown) staining shows Oct4 protein in the nuclei of (D) oogonia in postnatal ovaries and (E) spermatogonia in postnatal testes. (A–C after Yeom et al. 1996; D, E from Pesce et al. 1998; photographs courtesy of H. R. Schöler.)

portant substrate for PGC migration (ffrench-Constant et al. 1991), and germ cells that lack the integrin receptor for such extracellular matrix proteins cannot migrate to the gonads (Anderson et al. 1999). Directionality may be provided by a gradient of soluble protein.* In vitro evidence suggests that the genital ridges of 10.5-day mouse embryos secrete a diffusible TGF-β1-like protein that is capable of attracting mouse PGCs (Godin et al. 1990; Godin and Wylie 1991). Whether the genital ridge is able to provide such cues in vivo remains to be tested.

Although no germ plasm has been found in mammals, the retention of totipotency has been correlated with the expression of a nuclear transcription factor, Oct4. This factor is expressed in all of the early-cleavage blastomere nuclei, but its expression becomes restricted to the inner cell mass. During gastrulation, it becomes expressed solely in those posterior epiblast cells thought to give rise to the primordial germ cells. After that, this protein is seen only in the primordial germ cells, and later in oocytes (Figure 19.9; Yeom et al. 1996; Pesce et al. 1998). (Oct4 is not seen in the developing sperm after the germ cells reach the testes and become committed to sperm production.)

The proliferation of the PGCs appears to be promoted by stem cell factor, the same growth factor needed for the proliferation of neural crest-derived melanoblasts and hematopoietic stem cells (see Chapter 6). Stem cell factor is produced by the cells lining the migration pathway and remains bound to their cell membranes. It appears that the presentation of this protein on cell membranes is important for its activity. Mice

---

*In zebrafish, the primordial germ cells appear to be following a gradient of the SDF-1 protein, which is known to be a chemoattractant for mammalian lymphocytes (Doitsidou et al. 2002).

homozygous for mutations in the genes for either stem cell factor or its receptor (c-Kit) are deficient in germ cells (as well as melanocytes and blood cells) (see Dolci et al. 1991; Matsui et al. 1991). The addition of stem cell factor to PGCs taken from 11-day mouse embryos stimulates their proliferation for about 24 hours and appears to prevent programmed cell death that would otherwise occur (Godin et al. 1991; Pesce et al. 1993).

Sidelights *&* Speculations

EG Cells, ES Cells, and Teratocarcinomas

Embryonic germ (EG) cells

Stem cell factor increases the proliferation of migrating mouse primordial germ cells in culture, and this proliferation can be further increased by adding another growth factor, leukemia inhibition factor (LIF). However, the life span of these PGCs is short, and the cells soon die. But if an additional mitotic regulator—basic fibroblast growth factor (FGF2)—is added, a remarkable change takes place. The cells continue to proliferate, producing pluripotent embryonic stem cells with characteristics resembling those of the inner cell mass (Matsui et al. 1992; Resnick et al. 1992; Rohwedel et al. 1996). These PGC-derived cells are called **embryonic germ (EG) cells**, and they have the potential to differentiate into all the cell types of the body.

In 1998, John Gearhart's laboratory (Shamblott et al. 1998) cultured human EG cells. These cells were able to generate differentiated cells from all three primary germ layers, and they are presumably totipotent. Such cells could be used medically to create neural or hematopoietic stem cells, which might be used to regenerate damaged neural or blood tissues (see Chapter 4). EG cells are often considered embryonic stem (ES) cells, and the distinction of their origin is ignored.

Embryonic stem (ES) cells

Embryonic stem (ES) cells were described in Chapter 4; these cells are derived from the inner cell mass. ES cells and EG cells can be transfected with recombinant genes and inserted into blastocysts to create transgenic mice. Such a mammalian germ cell or stem cell contains within it all the information needed for subsequent development.

What would happen if such a cell became malignant? In one type of tumor, the germ cells become embryonic stem cells, like the FGF2-treated PGCs in the experiment above. This type of tumor is called a **teratocarcinoma**. Whether spontaneous or experimentally produced, a teratocarcinoma contains an undifferentiated stem cell

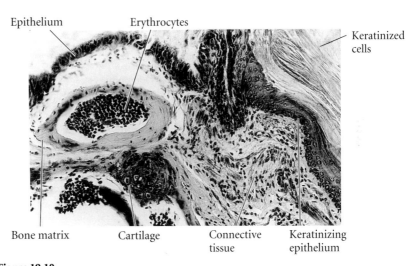

Epithelium   Erythrocytes   Keratinized cells

Bone matrix   Cartilage   Connective tissue   Keratinizing epithelium

Figure 19.10
Photomicrograph of a section through a teratocarcinoma, showing numerous differentiated cell types. (From Gardner 1982; photograph by C. Graham, courtesy of R. L. Gardner.)

population that has biochemical and developmental properties remarkably similar to those of the inner cell mass (Graham 1977). Moreover, these stem cells not only divide, but can also differentiate into a wide variety of tissues, including gut and respiratory epithelia, muscle, nerve, cartilage, and bone (Figure 19.10). Once differentiated, these cells no longer divide, and are therefore no longer malignant. Such tumors can give rise to most of the tissue types in the body. Thus, the teratocarcinoma stem cells mimic early mammalian development, but the tumor they form is characterized by random, haphazard development.

In 1981, Stewart and Mintz formed a mouse from cells derived in part from a teratocarcinoma stem cell. Stem cells that had arisen in a teratocarcinoma of an agouti (yellow-tipped) strain of mice were cultured for several cell generations and were seen to maintain the characteristic chromosome complement of the parental mouse. Individual stem cells descended from the tumor were injected into the blastocysts of

black-furred mice. The blastocysts were then transferred to the uterus of a foster mother, and live mice were born. Some of these mice had coats of two colors, indicating that the tumor cell had integrated itself into the embryo. This, in itself, is a remarkable demonstration that the tissue context is critical for the phenotype of a cell—a malignant cell was made nonmalignant.

But the story does not end here. When these chimeric mice were mated to mice carrying alleles recessive to those of the original tumor cell, the alleles of the tumor cell were expressed in many of the offspring. This means that the originally malignant tumor cell had produced many, if not all, types of normal somatic cells, and had even produced normal, functional germ cells! When such mice (being heterozygous for tumor cell genes) were mated with each other, the resultant litter contained mice that were homozygous for a large number of genes from the tumor cell (Figure 19.11).

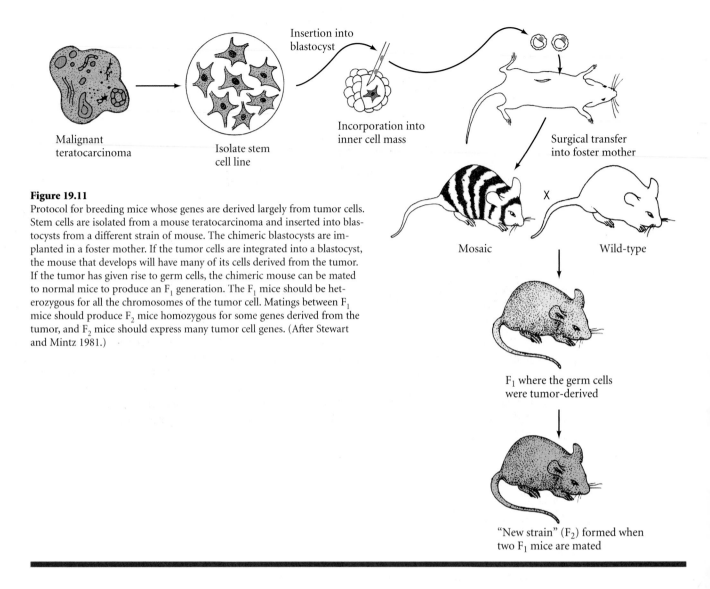

Figure 19.11
Protocol for breeding mice whose genes are derived largely from tumor cells. Stem cells are isolated from a mouse teratocarcinoma and inserted into blastocysts from a different strain of mouse. The chimeric blastocysts are implanted in a foster mother. If the tumor cells are integrated into a blastocyst, the mouse that develops will have many of its cells derived from the tumor. If the tumor has given rise to germ cells, the chimeric mouse can be mated to normal mice to produce an F_1 generation. The F_1 mice should be heterozygous for all the chromosomes of the tumor cell. Matings between F_1 mice should produce F_2 mice homozygous for some genes derived from the tumor, and F_2 mice should express many tumor cell genes. (After Stewart and Mintz 1981.)

Germ cell migration in birds and reptiles

In birds and reptiles, the primordial germ cells are derived from epiblast cells that migrate from the central region of the area pellucida to a crescent-shaped zone in the hypoblast at the anterior border of the area pellucida (Figure 19.12; Eyal-Giladi et al. 1981; Ginsburg and Eyal-Giladi 1987). This extraembryonic region is called the **germinal crescent**, and the PGCs multiply there.

Unlike those of amphibians and mammals, the PGCs of birds and reptiles migrate to the gonads primarily by means of the bloodstream (Figure 19.13). When blood vessels form in the germinal crescent, the PGCs enter those vessels and are carried by the circulation to the region where the hindgut is forming. Here they leave the circulation, become associated with the mesentery, and migrate into the genital ridges (Swift 1914; Mayer 1964; Kuwana 1993; Tsunekawa et al. 2000).

The PGCs of the germinal crescent appear to enter the blood vessels by **diapedesis**, a type of amoeboid movement common to lymphocytes and macrophages that enables cells to squeeze between the endothelial cells of small blood vessels. In some as yet undiscovered way, the PGCs are instructed to exit the blood vessels and enter the gonads (Pasteels 1953; Dubois 1969). Evidence for chemotaxis comes from studies (Kuwana et al. 1986) in which circulating chick PGCs were isolated from the blood and cultured between gonadal rudiments and other embryonic tissues. The PGCs migrated specifically into the gonadal rudiments during a 3-hour incubation.

Germ cell migration in Drosophila

During *Drosophila* embryogenesis, the primordial germ cells move from the posterior pole to the gonads in a manner similar to that of mammalian germ cells (Figure 19.14). The first step in this migration is a passive one, wherein the 30–40 pole cells are displaced into the posterior midgut by the movements of gastrulation. In the second step, the gut endoderm

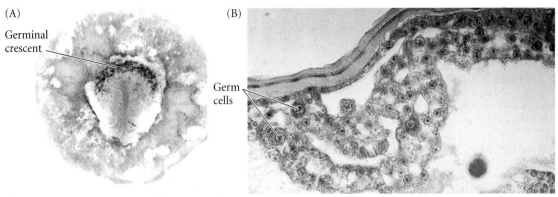

Figure 19.12
The germinal crescent of the chick embryo. (A) Germ cells of a definitive primitive streak-stage (stage 4, roughly 18-hour) chick embryo, stained (purple) for the chick Vasa homologue protein. The stained cells are confined to the germinal crescent. (B) Higher magnification of the stage 4 germinal crescent region, showing germ cells (stained brown) in the thickened epiblast. (Anterior is to the right.) (From Tsunekawa et al. 2000; photographs courtesy of N. Tsunekawa.)

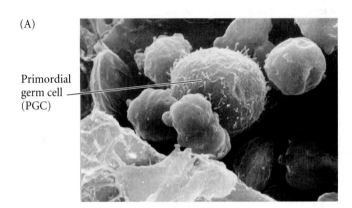

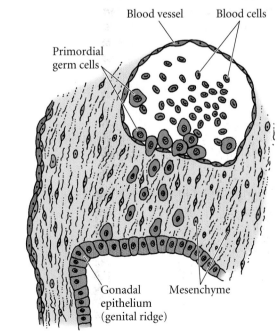

Figure 19.13
Migration of primordial germ cells in the chick embryo. (A) Scanning electron micrograph of a chick PGC in a capillary of a gastrulating embryo. The PGC can be identified by its large size and by the microvilli on its surface. (B) Diagram of transverse section near the prospective gonadal region of a chick embryo. Several PGCs within a blood vessel cluster next to the gonadal epithelium. One PGC is crossing through the blood vessel endothelium, and another PGC is already located adjacent to the gonadal epithelium. (C) Having passed through the endothelium of the dorsal aorta, chick germ cells (arrowheads) migrate toward the genital ridges of the embryo. (A from Kuwana 1993, courtesy of T. Kuwana; B after Romanoff 1960; C from Tsunekawa et al. 2000, courtesy of N. Tsunekawa.)

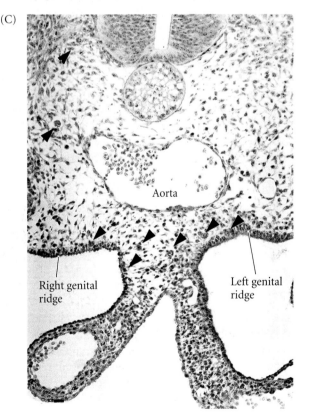

Figure 19.14
Migration of germ cells in the *Drosophila* embryo. The left column shows the germ plasm as stained by antibodies to Vasa, a protein component of the germ plasm (D has been counterstained with antibodies to Engrailed protein to show the segmentation, and E and F are dorsal views.) The right column diagrams the movements of the germ cells. (A) The germ cells originate from the pole plasm at the posterior end of the egg. (B) Passive movements carry the PGCs into the posterior midgut. (C) The PGCs move through the endoderm and into the caudal visceral mesoderm by diapedesis. The *wunen* (*wun*) gene product expressed in the endoderm expels the PGCs, while the product of the *columbus* (*clb*) gene expressed in the caudal mesoderm attracts them. (D–F) The movements of the mesoderm bring the PGCs into the region of the tenth through twelfth segments, where the mesoderm coalesces around them to form the gonads. (Photographs from Warrior et al. 1994, courtesy of R. Warrior; diagrams after Howard 1998.)

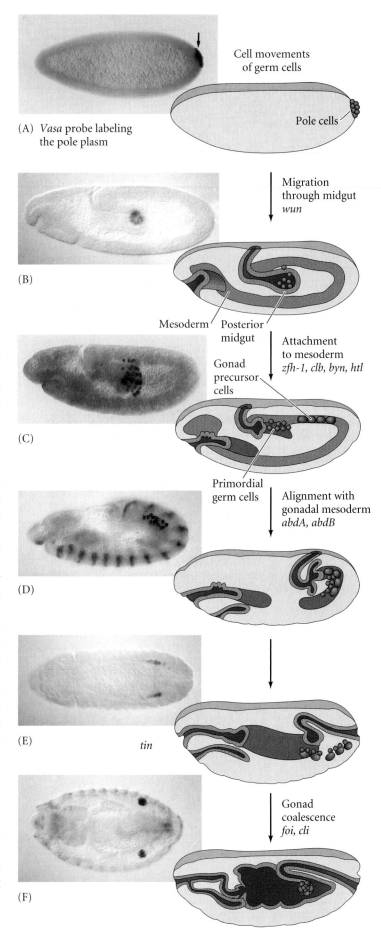

(A) *Vasa* probe labeling the pole plasm

Cell movements of germ cells

Pole cells

(B)

Migration through midgut
wun

Mesoderm Posterior midgut

Gonad precursor cells

(C)

Attachment to mesoderm
zfh-1, clb, byn, htl

Primordial germ cells

(D)

Alignment with gonadal mesoderm
abdA, abdB

(E) *tin*

(F)

Gonad coalescence
foi, cli

triggers active diapedesis in the PGCs, which travel through the blind end of the posterior midgut, migrating into the visceral mesoderm. In the third step, the PGCs split into two groups, each of which will become associated with a developing gonad primordium.

In the fourth step, the PGCs migrate to the gonads, which are derived from the lateral mesoderm of parasegments 10–12 (Warrior 1994; Jaglarz and Howard 1995; Broihier et al. 1998). This step involves both attraction and repulsion. The product of the *wunen* gene appears to be responsible for directing the migration of the PGCs from the endoderm into the mesoderm. This protein is expressed in the endoderm immediately before PGC migration, and it repels the PGCs. In loss-of-function mutants of this gene, the PGCs wander randomly (Zhang et al. 1997). Two proteins appear to be critical for the attracting the *Drosophila* PGCs to the gonads. One is the product of the *columbus* gene, the other is Hedgehog (Moore et al. 1998; Van Doren et al. 1998; Deshpande et al. 2001). These proteins are made in the mesodermal cells of the gonads. In loss-of-function mutants of either gene, the PGCs wander randomly from the endoderm, and if either gene is expressed in other tissues (such as the nerve cord), those tissues will attract the PGCs. In the last step, the gonad coalesces around the germ cells, allowing the germ cells to divide and mature into gametes.

Meiosis

Once in the gonad, the primordial germ cells continue to divide mitotically, producing millions of potential gamete precursors. The germ cells of both male and female gonads are then faced with the necessity of reducing their chromosomes from the diploid to the haploid condition. In the haploid condition, each chromosome is represented by only one copy, whereas diploid cells have two copies of each chromosome. To accomplish this reduction, the germ cells undergo **meiosis**. Meiosis differs from mitosis in that (1) meiotic cells undergo two cell divisions without an intervening period of DNA replication, and (2) homologous chromosomes (each consisting of two sister chromatids joined at a kinetochore) pair together and recombine genetic material. Meiosis is therefore at the center of sexual reproduction. As Villeneuve and Hillers (2001) conclude, "the very essence of sex is meiotic recombination." Yet for all its centrality in genetics, development, and evolution, we know surprisingly little about meiosis.

After the germ cell's last mitotic division, a period of DNA synthesis occurs, so that the cell initiating meiosis doubles the amount of DNA in its nucleus. In this state, each chromosome consists of two sister **chromatids** attached at a common kinetochore (centromere). (In other words, the diploid nucleus contains four copies of each chromosome.) Meiosis, shown in Figure 2.11, entails two cell divisions. In the first division, homologous chromosomes (e.g., the chromosome 3 pair in the diploid cell) come together and are then separated into different cells. Hence, the first meiotic division splits two homologous chromosomes between two daughter cells such that each cell has only one copy of each chromosome. But each of the chromosomes has already replicated (i.e., each has two chromatids). The second meiotic division then separates the two sister chromatids from each other. Consequently, each of the four cells produced by meiosis has a single (haploid) copy of each chromosome.

The first meiotic division begins with a long prophase, which is subdivided into five stages. During the **leptotene** (Greek, "thin thread") stage, the chromatin of the chromatids is stretched out very thinly, and it is not possible to identify individual chromosomes. DNA replication has already occurred, however, and each chromosome consists of two parallel chromatids. At the **zygotene** (Greek, "yoked threads")

stage, homologous chromosomes pair side by side. This pairing is called **synapsis**, and it is characteristic of meiosis. Such pairing does not occur during mitotic divisions. Although the mechanism whereby each chromosome recognizes its homologue is not known, synapsis seems to require the presence of the nuclear membrane and the formation of a proteinaceous ribbon called the **synaptonemal complex**. This complex is a ladder-like structure with a central element and two lateral bars (von Wettstein 1984; Schmekel and Daneholt 1995). The chromatin becomes associated with the two lateral bars, and the chromosomes are thus joined together (Figure 19.15B). Examinations of meiotic cell nuclei with the electron microscope (Moses 1968; Moens 1969) suggest that paired chromosomes are bound to the nuclear membrane, and Comings (1968) has suggested that the nuclear envelope helps bring together the homologous chromosomes. The configuration formed by the four chromatids and the synaptonemal complex is referred to as a **tetrad** or a **bivalent**.

During the next stage of meiotic prophase, the chromatids thicken and shorten. This stage has therefore been called the **pachytene** (Greek, "thick thread") stage. Individual chromatids can now be distinguished under the light microscope, and crossing-over may occur. **Crossing over** represents an exchange of genetic material whereby genes from one chromatid are exchanged with homologous genes from another. Crossing over may continue into the next stage, the **diplotene** (Greek, "double threads") stage. Here, the synaptonemal complex breaks down, and the two homologous

Figure 19.15
The synaptonemal complex and recombination. (A) Homologous chromosomes held together at the first meiotic prophase in a *Neottiella* oocyte. (B) Interpretive diagram of the synaptonemal complex structure. (C) Chiasmata in diplotene bivalent chromosomes of salamander oocytes. Kinetochores are visible as darkly stained circles; arrows point to the two chiasmata. (A from von Wettstein 1971, courtesy of D. von Wettstein; B after Schmekel and Daneholt 1995; C, photograph courtesy of J. Kezer)

(A)

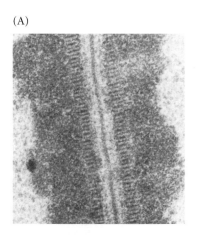

(B)

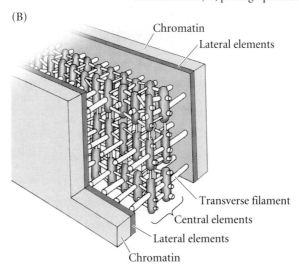

Chromatin
Lateral elements
Transverse filament
Central elements
Lateral elements
Chromatin

(C)

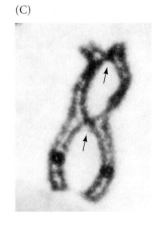

chromosomes start to separate. Usually, however, they remain attached at various points called **chiasmata**, which are thought to represent regions where crossing over is occurring (Figure 19.15C). The diplotene stage is characterized by a high level of gene transcription. In some species, the chromosomes of both male and female germ cells take on the "lampbrush" appearance characteristic of chromosomes that are actively making RNA (see below). During the next stage, **diakinesis** (Greek, "moving apart"), the kinetochores move away from each other, and the chromosomes remain joined only at the tips of the chromatids. This last stage of meiotic prophase ends with the breakdown of the nuclear membrane and the migration of the chromosomes to the **metaphase plate**.

During anaphase I, the homologous chromosomes are separated from each other in an independent fashion. This stage leads to telophase I, during which two daughter cells are formed, each cell containing one partner of each homologous chromosome pair. After a brief **interkinesis**, the second division of meiosis takes place. During this division, the kinetochore of each chromosome divides during anaphase so that each of the new cells gets one of the two chromatids, the final result being the creation of four haploid cells. Note that meiosis has also reassorted the chromosomes into new groupings. First, each of the four haploid cells has a different assortment of chromosomes. In humans, in which there are 23 different chromosome pairs, there can be 2^{23} (nearly 10 million) differ-

ent types of haploid cells formed from the genome of a single person. In addition, the crossing-over that occurs during the pachytene and diplotene stages of prophase I further increases genetic diversity and makes the number of different gametes incalculable.

If meiosis is the core of sexual reproduction, and sexual reproduction predominates among the eukaryotic kingdoms, then meiosis must be one of the central unifying themes of life. The study of meiosis is being facilitated by the observation that nearly all the genes and proteins used in yeast meiosis also function in mammalian meiosis. This observation has allowed the identification of a "core meiotic recombination complex" that is used by plants, fungi, and animals. This meiotic recombination complex is built on the backbone of the **cohesin complex**, the protein glue that keeps sister chromatids together during *mitotic* prophase and metaphase. In meiotic cells, this complex recruits another set of proteins that are involved in promoting pairing between the homologous chromosomes and allowing recombination to occur (Pelttari et al. 2001; Villeneuve and Hillers 2001). These recombination-inducing proteins are involved in making and repairing double-stranded DNA breaks.

Although the relationship between the synaptonemal complex and the cohesin-recruited recombination complex is not clear, it appears in mammals that the synaptonemal complex stabilizes the associations initiated by the recombination complex, giving a morphological scaffolding to the tenuous protein connections (Pelttari et al. 2001). If the synaptonemal complex fails to form, the germ cells arrest at the pachytene stage, and their chromosomes fragment (Figure 19.16). If the murine synaptonemal complex forms but lacks certain proteins, chiasmata formation fails, and the germ cells are often aneuploid, having multiple copies of one or more chromosomes (Tay and Richter 2001; Yuan et al. 2002).

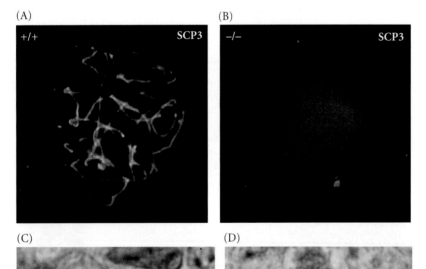

(A) +/+ SCP3 (B) −/− SCP3

(C) (D)

Figure 19.16

Importance of the synaptonemal complex. The CPEB protein is a constituent of the synaptonemal complex, and when this gene is knocked out in mice, the synaptonemal complex fails to form. (A, B) Staining for synaptonemal complex proteins in wild-type (A) and CPEB-deficient (B) mice. The synaptonemal complex is stained green, the DNA is stained blue. The synaptonemal complex is absent in the mutant nuclei, and their DNA is not organized into discrete chromosomes. (C, D) In preparations stained with hematoxylin-eosin, the DNA can seen to be in the pachytene stage in the wild-type cells (C), and fragmented in the mutant cells (D). (After Tay and Richter 2001; photographs courtesy of J. D. Richter.)

The events of meiosis appear to be coordinated through cytoplasmic connections between the dividing cells. Whereas the daughter cells formed by mitosis routinely separate from each other, the products of the meiotic cell divisions remain coupled to each other by **cytoplasmic bridges**. These bridges are seen during the formation of sperm and eggs throughout the animal kingdom (Pepling and Spradling 1998).

WEBSITE 19.3 Proteins involved in meiosis. The phenomenon of homologous pairing and crossing-over is being analyzed in several organisms and may involve DNA repair enzymes.

WEBSITE 19.4 Human meiosis. Nondisjunction, the failure of chromosomes to sort properly during meiosis, is not uncommon in humans. Its frequency increases with maternal age.

Sidelights & Speculations

Big Decisions: Mitosis or Meiosis? Sperm or Egg?

In many species, the germ cells migrating into the gonad are bipotential and can differentiate into either sperm or eggs, depending on their gonadal environment. When the ovaries of salamanders are experimentally transformed into testes, the resident germ cells cease their oogenic differentiation and begin developing as sperm (Burns 1930; Humphrey 1931). Similarly, in the housefly and mouse, the gonad is able to direct the differentiation of the germ cells (McLaren 1983; Inoue and Hiroyoshi 1986). Thus, in most organisms, the sex of the gonad and of its germ cells is the same.

But what about hermaphroditic animals, in which the change from sperm production to egg production is a naturally occurring physiological event? How is the same animal capable of producing sperm during one part of its life and oocytes during another part? Using *Caenorhabditis elegans*, Kimble and her colleagues identified two "decisions" that presumptive germ cells have to make. The first is whether to enter meiosis or to remain a mitotically dividing stem cell. The second is whether to become an egg or a sperm.

Recent evidence shows that these decisions are intimately linked. The mitotic/meiotic decision is controlled by a single nondividing cell at the end of each gonad, the **distal tip cell**. The germ cell precursors near this cell divide mitotically, forming the pool of germ cells; but as these cells get farther away from the distal tip cell, they enter meiosis. If the distal tip cell is destroyed by a focused laser beam, all the germ cells enter meiosis, and if the distal tip cell is placed in a different location in the gonad, germ line stem cells are generated near its new position (Figure 19.17; Kimble 1981; Kimble and White 1981). The distal tip cell extends long filaments that touch the distal germ cells. The extensions contain in their cell membranes the Lag-2 protein, a *C. elegans* homologue of Delta

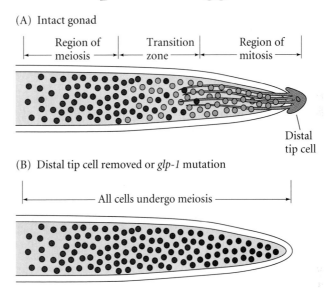

(A) Intact gonad

Region of meiosis | Transition zone | Region of mitosis

Distal tip cell

(B) Distal tip cell removed or *glp-1* mutation

All cells undergo meiosis

Figure 19.17
Regulation of the mitosis-or-meiosis decision by the distal tip cell of the *C. elegans* ovotestis. (A) Intact gonad early in development with regions of mitosis (light-colored cells) and meiosis. The membrane of the distal tip cell's extensions contains the *C. elegans* homologue of the Delta protein, while the PGCs contain the *C. elegans* homologue of Notch. (B) Gonad after laser ablation of the distal tip cell or mutation of the *glp-1* gene. All germ cells enter meiosis.

(Henderson et al. 1994; Tax et al. 1994; Hall et al. 1999). The Lag-2 protein maintains the germ cells in mitosis and inhibits their meiotic differentiation.

Austin and Kimble (1987) isolated a mutation that mimics the phenotype obtained when the distal tip cell is removed. It is not surprising that this mutation involves the gene encoding GLP-1, the *C. elegans* homologue of Notch—the receptor for Delta. All the germ cell precursors of nematodes homozygous for the recessive mutation of *glp-1* initiate meiosis, leaving no mitotic population. Instead of the 1500 germ cells usually found in the fourth larval stage of hermaphroditic development, these mutants produce only 5 to 8 sperm cells. When

genetic chimeras are made in which wild-type germ cell precursors are found within a mutant larva, the wild-type cells are able to respond to the distal tip cells and undergo mitosis. However, when mutant germ cell precursors are found within wild-type larvae, they all enter meiosis. Thus, the *glp-1* gene appears to be responsible for enabling the germ cells to respond to the distal tip cell's signal.*

---

*The *glp-1* gene appears to be involved in a number of inductive interactions in *C. elegans*. You will no doubt recall that GLP-1 protein is also needed by the AB blastomere for it to receive inductive signals from the EMS blastomere to form pharyngeal muscles (see Chapter 8).

After the germ cells begin their meiotic divisions, they still must become either sperm or ova. Generally, in each hermaphrodite gonad (called an ovotestis), the most proximal germ cells produce sperm, while the most distal (near the tip) become eggs (Hirsh et al. 1976). This means that the germ cells entering meiosis early become sperm, while those entering meiosis later become eggs. The genetics of this switch are currently being analyzed. The laboratories of Hodgkin (1985) and Kimble (Kimble et al. 1986) have isolated several genes needed for germ cell pathway selection, but the switch appears to involve the activity or inactivity of *fem-3* mRNA. Figure 19.18 presents a scheme for how these genes might function. During early development, the *fem* genes, especially *fem-3*, are critical for the specification of sperm cells. Loss-of-function mutations of these genes convert XX nematodes into females (i.e., spermless hermaphrodites). As long as the FEM proteins are made in the germ cells, sperm are produced. FEM protein is thought to activate the *fog* genes (whose loss-of-function mutations cause the feminization of the germ line and eliminate spermatogenesis). The *fog* gene products activate the genes involved in transforming the germ cell into sperm and also inhibit those genes that would otherwise direct the germ cells to initiate oogenesis.

Oogenesis can begin only when *fem* activity is suppressed. This suppression appears to act at the level of RNA translation. The 3′ untranslated region (3′ UTR) of the *fem-3* mRNA contains a sequence that binds a repressor protein during normal development. If this region is mutated such that the repressor cannot bind, the *fem-3* mRNA remains translatable, and oogenesis never occurs. The result is a hermaphrodite body that produces only sperm (Ahringer and Kimble 1991; Ahringer et al. 1992). The *trans*-acting repressor of the *fem-3* mes-

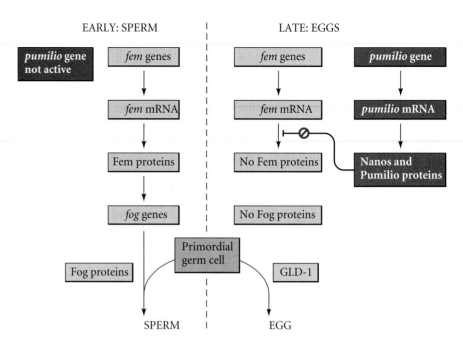

Figure 19.18
Model of sex determination switch in the germ line of *C. elegans* hermaphrodites. Sex determination in somatic tissues shows a hierarchy of negative regulation. In the early larva, Pumilio is not synthesized, and the *fem* mRNA is able to be translated. The FEM proteins activate the *fog* genes, whose proteins cause the germ cells to undergo spermatogenesis. Later in development, *pumilio* is activated, and combines with Nanos to make a repressor of *fem* translation. Without FEM, the GLD-1 protein can function to make certain the germ cell undergoes oogenesis.

sage is a combination of the Nanos and Pumilio proteins (the same combination that represses *hunchback* message translation in *Drosophila*). The up-regulation of Pumilio expression may be critical in regulating the germ line switch from spermatogenesis to oogenesis, since Nanos is made constitutively. Nanos appears to be necessary in *C. elegans* (as it is in *Drosophila*) for the survival of all germ line cells (Kraemer et al. 1999).

WEBSITE 19.5 **Germ line sex determination in *C. elegans*.** The establishment of whether a germ cell is to become a sperm or an egg involves multiple levels of inhibition. Translational regulation is seen in several of these steps.

Spermatogenesis

Forming the haploid spermatid

While the reductive divisions of meiosis are conserved in every eukaryotic kingdom of life, the regulation of meiosis in mammals differs dramatically between males and females. The differences between **oogenesis**, the production of eggs, and **spermatogenesis**, the production of sperm, are outlined in Table 19.1.

Spermatogenesis is the production of sperm from the primordial germ cells. Once the vertebrate PGCs arrive at the genital ridge of a male embryo, they become incorporated into the sex cords. They remain there until maturity, at which

time the sex cords hollow out to form the seminiferous tubules, and the epithelium of the tubules differentiates into the Sertoli cells. The initiation of spermatogenesis during puberty is probably regulated by the synthesis of BMP8b by the spermatogenic germ cells, the **spermatogonia**. When BMP8b reaches a critical concentration, the germ cells begin to differentiate. The differentiating cells produce high levels of BMP8b, which can then further stimulate their differentiation. Mice lacking BMP8b do not initiate spermatogenesis at puberty (Zhao et al. 1996).

The spermatogenic germ cells are bound to the Sertoli cells by N-cadherin molecules on both cell surfaces and by galactosyltransferase molecules on the spermatogenic cells

TABLE 19.1 Sexual dimorphism in mammalian meioses

Female oogenesis	Male spermatogenesis
Meiosis initiated once in a finite population of cells	Meiosis initiated continuously in a mitotically dividing stem cell population
One gamete produced per meiosis	Four gametes produced per meiosis
Completion of meiosis delayed for months or years	Meiosis completed in days or weeks
Meiosis arrested at first meiotic prophase and reinitiated in a smaller population of cells	Meiosis and differentiation proceed continuously without cell cycle arrest
Differentiation of gamete occurs while diploid, in first meiotic prophase	Differentiation of gamete occurs while haploid, after meiosis ends
All chromosomes exhibit equivalent transcription and recombination during meiotic prophase	Sex chromosomes excluded from recombination and transcription during first meiotic prophase

Source: Handel and Eppig 1998.

that bind a carbohydrate receptor on the Sertoli cells (Newton et al. 1993; Pratt et al. 1993). The Sertoli cells nourish and protect the developing sperm cells, and spermatogenesis—the developmental pathway from germ cell to mature sperm—occurs in the recesses between the Sertoli cells (Figure 19.19). The processes by which the PGCs generate sperm have been studied in detail in several organisms, but we will focus here on spermatogenesis in mammals.

After reaching the gonad, the PGCs divide to form **type A₁ spermatogonia**. These cells are smaller than the PGCs and are characterized by an ovoid nucleus that contains chromatin associated with the nuclear membrane. Type A_1 spermatogonia are found adjacent to the outer basement membrane of the sex cords. They are stem cells, and at maturity, they are thought to divide to make another type A_1 spermatogonium as well as a second, paler type of cell, the type A_2 spermatogonium. Thus, each type A_1 spermatogonium is a stem cell capable of regenerating itself as well as producing a new cell type. The A_2 spermatogonia divide to produce the A_3 spermatogonia, which then beget the type A_4 spermatogonia. It is possible that each of the type A spermatogonia are stem cells, capable of self-renewal.

The A_4 spermatogonium has three options: it can form another A_4 spermatogonium (self-renewal); it can undergo cell death (apoptosis); or it can differentiate into the first committed stem cell type, the **intermediate spermatogonium**. Intermediate spermatogonia are committed to becoming spermatozoa, and they divide mitotically once to form **type B spermatogonia**. These cells are the precursors of the spermatocytes and are the last cells of the line that undergo mitosis. They divide once to generate the **primary spermatocytes**—the cells that enter meiosis.

The transition between spermatogonia and spermatocytes appears to be mediated by glial cell line-derived neurotrophic

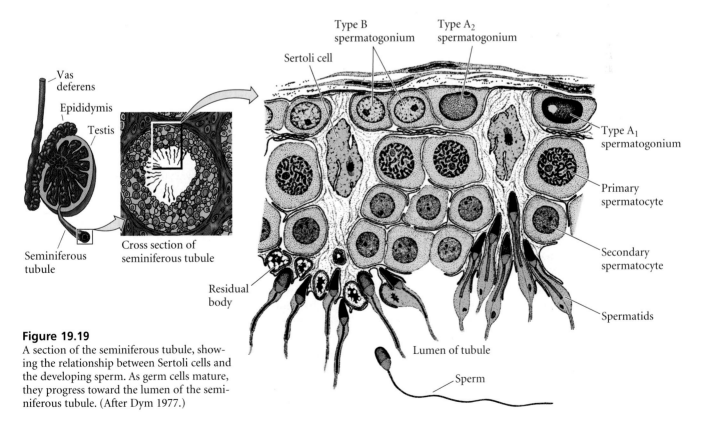

Figure 19.19
A section of the seminiferous tubule, showing the relationship between Sertoli cells and the developing sperm. As germ cells mature, they progress toward the lumen of the seminiferous tubule. (After Dym 1977.)

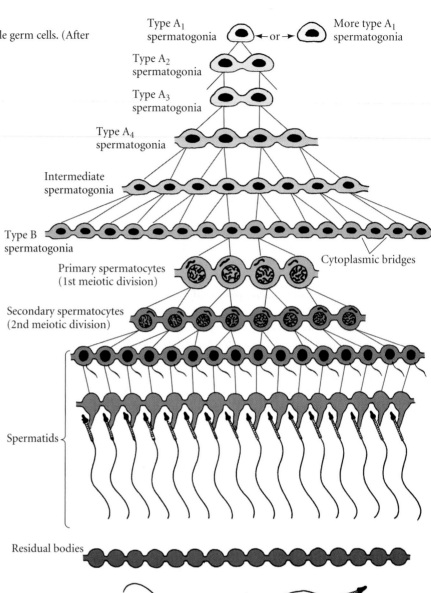

Figure 19.20
The formation of syncytial clones of human male germ cells. (After Bloom and Fawcett 1975.)

Type A_1 spermatogonia ← or → More type A_1 spermatogonia

Type A_2 spermatogonia

Type A_3 spermatogonia

Type A_4 spermatogonia

Intermediate spermatogonia

Type B spermatogonia

Cytoplasmic bridges

Primary spermatocytes (1st meiotic division)

Secondary spermatocytes (2nd meiotic division)

Spermatids

Residual bodies

Sperm cells

factor (GDNF), which is secreted by the Sertoli cells. GDNF levels determine whether the dividing spermatogonia remain spermatogonia or enter the pathway to become spematocytes. Low levels of GDNF favor the differentiation of the spermatogonia, while high levels favor self-renewal of the stem cells (Meng et al. 2000). Since GDNF is upregulated by follicle-stimulating hormone (FSH), GDNF may serve as a link between the Sertoli cells and the endocrine system, and it provides a mechanism for FSH to instruct the testes to produce more sperm (Tadokoro et al. 2002).

Looking at Figure 19.20, we find that during the spermatogonial divisions, cytokinesis is not complete. Rather, the cells form a syncytium whereby each cell communicates with the others via cytoplasmic bridges about 1 μm in diameter (Dym and Fawcett 1971). The successive divisions produce clones of interconnected cells, and because ions and molecules readily pass through these cytoplasmic bridges, each cohort matures synchronously. During this time, the spermatocyte nucleus often transcribes genes whose products will be used later to form the axoneme and acrosome.

Each primary spermatocyte undergoes the first meiotic division to yield a pair of **secondary spermatocytes**, which complete the second division of meiosis. The haploid cells thus formed are called **spermatids**, and they are still connected to one another through their cytoplasmic bridges. The spermatids that are connected in this manner have haploid nuclei, but are functionally diploid, since a gene product made in one cell can readily diffuse into the cytoplasm of its neighbors (Braun et al. 1989).

During the divisions from type A_1 spermatogonium to spermatid, the cells move farther and farther away from the basement membrane of the seminiferous tubule and closer to its lumen (see Figure 19.19). Thus, each type of cell can be found in a particular layer of the tubule. The spermatids are located at the border of the lumen, and here they lose their cytoplasmic connections and differentiate into spermatozoa. In humans, the progression from spermatogonial stem cell to mature spermatozoa takes 65 days (Dym 1994).

WEBSITE 19.6 Gonial syncytia: Bridges to the future. The products of meiotic divisions are connected by cytoplasmic bridges. The functions of these connections may differ between those cells producing sperm and those producing eggs.

Spermiogenesis: The differentiation of the sperm

The mammalian haploid spermatid is a round, unflagellated cell that looks nothing like the mature vertebrate sperm. The next step in sperm maturation, then, is **spermiogenesis** (or **spermateliosis**), the differentiation of the sperm cell. For fertilization to occur, the sperm has to meet and bind with an egg, and spermiogenesis prepares the sperm for these functions of motility and interaction. The process of mammalian sperm differentiation was shown in Figure 7.2. The first step is the construction of the acrosomal vesicle from the Golgi apparatus. The acrosome forms a cap that covers the sperm nucleus. As the acrosomal cap is formed, the nucleus rotates so that the cap will be facing the basement membrane of the seminiferous tubule. This rotation is necessary because the flagellum, which is beginning to form from the centriole on the other side of the nucleus, will extend into the lumen. During the last stage of spermiogenesis, the nucleus flattens and condenses, the remaining cytoplasm (the residual body or "cytoplasmic droplet") is jettisoned, and the mitochondria form a ring around the base of the flagellum.

One of the major changes in the nucleus is the replacement of the histones by protamines. Transcription of the genes for protamines is seen in the early haploid spermatids, although translation is delayed for several days (Peschon et al. 1987). Protamines are relatively small proteins that are over 60% arginine. During spermiogenesis, the nucleosomes dissociate, and the histones of the haploid nucleus are eventually replaced by protamines. This causes the complete shutdown of transcription in the nucleus and facilitates its assuming an almost crystalline structure. The resulting sperm then enter the lumen of the tubule.

In the mouse, development from stem cell to spermatozoon takes 34.5 days. The spermatogonial stages last 8 days, meiosis lasts 13 days, and spermiogenesis takes up another 13.5 days. In humans, sperm development takes nearly twice as long. Because type A_1 spermatogonia are stem cells, spermatogenesis can occur continuously. Each day, some 100 million sperm are made in each human testicle, and each ejaculation releases 200 million sperm. Unused sperm are either resorbed or passed out of the body in urine. During his lifetime, a human male can produce 10^{12} to 10^{13} sperm (Reijo et al. 1995).

WEBSITE 19.7 Gene expression during spermatogenesis. Transcription occurs both from the diploid spermatocyte nucleus and from the haploid spermatid nucleus. Posttranscriptional control is also important in regulating sperm gene expression.

WEBSITE 19.8 The Nebenkern. Sperm mitochondria are often highly modified to fit the streamlined cell. The mitochondria of flies fuse together to form a structure called the Nebenkern; this fusion is controlled by the *fuzzy onions* gene.

Oogenesis

Oogenic meiosis

Oogenesis—the differentiation of the ovum—differs from spermatogenesis in several ways. Whereas the gamete formed by spermatogenesis is essentially a motile nucleus, the gamete formed by oogenesis contains all the materials needed to initi-

ate and maintain metabolism and development. Therefore, in addition to forming a haploid nucleus, oogenesis also builds up a store of cytoplasmic enzymes, mRNAs, organelles, and metabolic substrates. While the sperm becomes differentiated for motility, the egg develops a remarkably complex cytoplasm.

The mechanisms of oogenesis vary among species more than those of spermatogenesis. This variation should not be surprising, since patterns of reproduction vary so greatly among species. In some species, such as sea urchins and frogs, the female routinely produces hundreds or thousands of eggs at a time, whereas in other species, such as humans and most mammals, only a few eggs are produced during the lifetime of an individual. In those species that produce thousands of ova, the germ cells, called **oogonia**, are self-renewing stem cells that endure for the lifetime of the organism. In those species that produce fewer eggs, the oogonia divide to form a limited number of egg precursor cells. In the human embryo, the thousand or so oogonia divide rapidly from the second to the seventh month of gestation to form roughly 7 million germ cells (Figure 19.21). After the seventh month of embryonic development, however, the number of germ cells drops precipitously. Most oogonia die during this period, while the remaining oogonia enter the first meiotic division (Pinkerton et al. 1961). These latter cells, called **primary oocytes**, progress through the first meiotic prophase until the diplotene stage, at which point they are maintained until the female matures. With the onset of puberty, groups of oocytes periodically resume meiosis. Thus, in the human female, the first part of meiosis begins in the embryo, and the signal to resume meiosis is not given until roughly 12 years later. In fact, some oocytes are maintained in meiotic

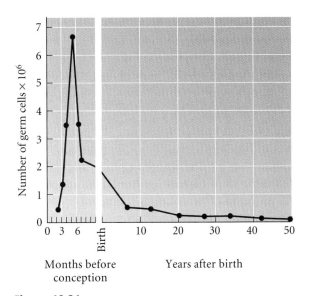

Figure 19.21
Changes in the number of germ cells in the human ovary over the life span. (After Baker 1970.)

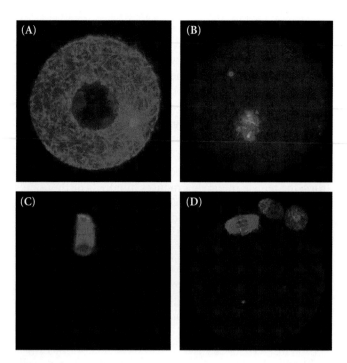

Figure 19.22
Meiosis in the mouse oocyte. The tubulin of the microtubules is stained green; the DNA is stained blue. (A) Mouse oocyte in meiotic prophase. The large haploid nucleus (the germinal vesicle) is still intact. (B) The nuclear envelope of the germinal vesicle breaks down as metaphase begins. (C) Meiotic anaphase I, wherein the spindle migrates to the periphery of the egg and releases a small polar body. (D) Meiotic metaphase II wherein the second polar body is given off (the first polar body has also divided). (From De Vos 2002; photographs courtesy of L. De Vos.)

prophase for nearly 50 years. As Figure 19.21 indicates, primary oocytes continue to die even after birth. Of the millions of primary oocytes present at birth, only about 400 mature during a woman's lifetime.

Oogenic meiosis also differs from spermatogenic meiosis in its placement of the metaphase plate. When the primary oocyte divides, its nucleus, called the **germinal vesicle**, breaks down, and the metaphase spindle migrates to the periphery of the cell. At telophase, one of the two daughter cells contains hardly any cytoplasm, whereas the other cell retains nearly the entire volume of cellular constituents (Figure 19.22). The smaller cell is called the **first polar body**, and the larger cell is referred to as the **secondary oocyte**. During the second division of meiosis, a similar unequal cytokinesis takes place. Most of the cytoplasm is retained by the mature egg (ovum), and a second polar body receives little more than a haploid nucleus. Thus, oogenic meiosis conserves the volume of oocyte cytoplasm in a single cell rather than splitting it equally among four progeny.

In a few species of animals, meiosis is greatly modified such that the resulting gamete is diploid and need not be fer-

tilized to develop. Such animals are said to be **parthenogenetic** (Greek, "virgin birth"). In the fly *Drosophila mangabeirai*, one of the polar bodies acts as a sperm and "fertilizes" the oocyte after the second meiotic division. In other insects (such as *Moraba virgo*) and in the lizard *Cnemidophorus uniparens*, the oogonia double their chromosome number before meiosis, so that the halving of the chromosomes restores the diploid number. The germ cells of the grasshopper *Pycnoscelus surinamensis* dispense with meiosis altogether, forming diploid ova by two mitotic divisions (Swanson et al. 1981). All of these species consist entirely of females. In other species, haploid parthenogenesis is widely used not only as a means of reproduction, but also as a mechanism of sex determination. In the Hymenoptera (bees, wasps, and ants), unfertilized eggs develop into males, whereas fertilized eggs, being diploid, develop into females. The haploid males are able to produce sperm by abandoning the first meiotic division, thereby forming two sperm cells through second meiosis.

Maturation of the oocyte in amphibians

The egg is responsible for initiating and directing development, and in some species (as we saw above), fertilization is not even necessary. The accumulated material in the oocyte cytoplasm includes energy sources and energy-producing organelles (the yolk and mitochondria); the enzymes and precursors for DNA, RNA, and protein syntheses; stored messenger RNAs; structural proteins; and morphogenetic regulatory factors that control early embryogenesis. A partial catalogue of the materials stored in the oocyte cytoplasm is shown in Table 19.2, while a partial list of stored mRNAs is shown in Table 5.2. Most of this accumulation takes place during meiotic prophase I, which is often subdivided into two phases, **previtellogenesis** (Greek, "before yolk formation") and **vitellogenesis**.

The eggs of fish and amphibians are derived from an oogonial stem cell population that can generate a new cohort of oocytes each year. In the frog *Rana pipiens*, oogenesis takes

TABLE 19.2 Cellular components stored in the mature oocyte of *Xenopus laevis*

Component	Approximate excess over amount in larval cells
Mitochondria	100,000
RNA polymerases	60,000–100,000
DNA polymerases	100,000
Ribosomes	200,000
tRNA	10,000
Histones	15,000
Deoxyribonucleoside triphosphates	2,500

Source: After Laskey 1979.

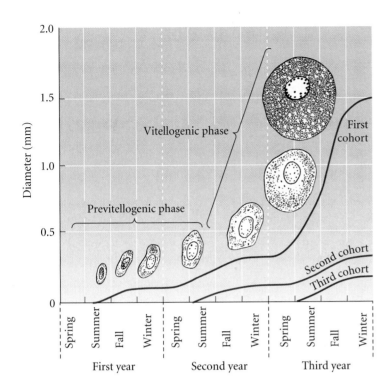

Figure 19.23
Growth of oocytes in the frog. During the first 3 years of life, three cohorts of oocytes are produced. The drawings follow the growth of the first-generation oocytes. (After Grant 1953.)

nears an end, the oocyte cytoplasm becomes stratified. The cortical granules, mitochondria, and pigment granules are located at the periphery of the oocyte, within the actin-rich cortex. Within the inner cytoplasm, distinct gradients emerge. While the yolk platelets become more heavily concentrated at the vegetal pole of the oocyte, the glycogen granules, ribosomes, lipid vesicles, and endoplasmic reticulum are found toward the animal pole. Even specific mRNAs stored in the cytoplasm become localized to certain regions of the oocyte.

WEBSITE 19.9 Hormonal control of yolk production. Vitellogenesis in amphibians is mediated primarily by estrogen. Estrogen instructs the liver to express and secrete vitellogenin, and this protein is absorbed from the blood by the young oocyte.

WEBSITE 19.10 Transporting the Vg1 mRNA. The Vera protein specifically binds to the 3′ UTR of the *Vg1* message. Vera may link *Vg1* mRNA to a set of endoplasmic reticulum vesicles that are translocated to the vegetal cortex.

WEBSITE 19.11 Establishment of egg polarity. In several species, the developing oocyte is a flagellated cell whose flagellum marks the future animal pole of the egg. This flagellum is lost during oogenesis.

3 years. During the first 2 years, the oocyte increases its size very gradually. During the third year, however, the rapid accumulation of yolk in the oocyte causes the egg to swell to its characteristically large size (Figure 19.23). Eggs mature in yearly batches, with the first cohort maturing shortly after metamorphosis; the next group matures a year later.

Vitellogenesis occurs when the oocyte reaches the diplotene stage of meiotic prophase. Yolk is not a single substance, but a mixture of materials used for embryonic nutrition. The major yolk component in frog eggs is a 470-kDa protein called **vitellogenin**. It is not made in the frog oocyte (as are the major yolk proteins of organisms such as annelids and crayfish), but is synthesized in the liver and carried by the bloodstream to the ovary (Flickinger and Rounds 1956; Danilchik and Gerhart 1987).

As the yolk is being deposited, the organelles also become arranged asymmetrically. The cortical granules begin to form from the Golgi apparatus; they are originally scattered randomly through the oocyte cytoplasm, but later migrate to the periphery of the cell. The mitochondria replicate at this time, dividing to form a "mitochondrial cloud." These millions of mitochondria will be apportioned to the different blastomeres during cleavage. (In *Xenopus*, new mitochondria will not be formed until after gastrulation is initiated.) As vitellogenesis

Completion of amphibian meiosis: Progesterone and fertilization

Amphibian primary oocytes can remain in the diplotene stage of meiotic prophase for years. This state resembles the G_2 phase of the mitotic cell division cycle (see Chapter 8). Resumption of meiosis in the amphibian oocyte requires progesterone. This hormone is secreted by the follicle cells in response to gonadotropic hormones secreted by the pituitary gland. Within 6 hours of progesterone stimulation, **germinal vesicle breakdown (GVBD)** occurs, the microvilli retract, the nucleoli disintegrate, and the chromosomes contract and migrate to the animal pole to begin division. Soon afterward, the first meiotic division occurs, and the mature ovum is released from the ovary by a process called **ovulation**. The ovulated egg is in second meiotic metaphase when it is released (Figure 19.24).

How does progesterone enable the egg to break its dormancy and resume meiosis? To understand the mechanisms by which this activation is accomplished, it is necessary to briefly review the model for early blastomere division (see Chapter 8). Entry into the mitotic (M) phase of the cell cycle (in both meiosis and mitosis) is regulated by **mitosis-promoting factor**, or **MPF** (originally called "maturation-promoting factor" after its meiotic function). MPF contains two subunits, **cyclin B** and the **p34** protein. The p34 protein is a cyclin-dependent-kinase—its activity is dependent upon the presence

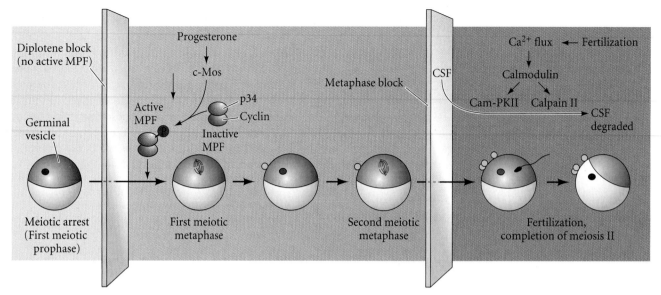

Figure 19.24
Schematic representation of *Xenopus* oocyte maturation, showing the regulation of meiotic cell division by progesterone and fertilization. Oocyte maturation is arrested at the diplotene stage of first meiotic prophase by the lack of active MPF. Progesterone activates the production of the c-mos protein. This protein initiates a cascade of phosphorylation that eventually phosphorylates the p34 subunit of MPF, allowing the MPF to become active. The MPF drives the cell cycle through the first meiotic division, but further division is blocked by CSF, a compound containing c-mos and cdk2. Upon fertilization, calcium ions released into the cytoplasm are bound by calmodulin and are used to activate two enzymes, calmodulin-dependent protein kinase II and calpain II, which inactivate and degrade CSF. Second meiosis is completed, and the two haploid pronuclei can fuse. At this time, cyclin B is resynthesized, allowing the first cell cycle of cleavage to begin.

of cyclin. Since all the components of MPF are present in the amphibian oocyte, it is generally thought that progesterone somehow converts a pre-MPF complex into active MPF.

The mediator of the progesterone signal is the **c-mos** protein. Progesterone reinitiates meiosis by causing the egg to polyadenylate the maternal *c-mos* mRNA that has been stored in its cytoplasm (Sagata et al. 1988; Sheets et al. 1995; Mendez et al. 2000). This message is translated into a 39-kDa phosphoprotein. This c-mos protein is detectable only during oocyte maturation and is destroyed quickly upon fertilization. Yet during its brief lifetime, it plays a major role in releasing the egg from its dormancy. The c-mos protein activates a phosphorylation cascade that phosphorylates and activates the p34 subunit of MPF (Ferrell and Machleder 1998; Ferrell 1999). The active MPF allows the germinal vesicle to break down and the chromosomes to divide. If the translation of *c-mos* is inhibited by injecting *c-mos* antisense mRNA into the oocyte, germinal vesicle breakdown and the resumption of oocyte maturation do not occur.

However, oocyte maturation then encounter a second block. MPF can take the chromosomes only through the first meiotic division and the prophase of the second meiotic division. The oocyte is arrested again in the metaphase of the second meiotic division. This metaphase block is caused by the combined actions of c-mos and another protein, cyclin-dependent kinase 2 (cdk2; Gabrielli et al. 1993). These two proteins are subunits that together form **cytostatic factor** (**CSF**), which is found in mature frog eggs, and which can block cell cycles in metaphase (Matsui 1974). It is thought that CSF prevents the degradation of cyclin.

The metaphase block is broken by fertilization. Evidence suggests that the calcium ion flux attending fertilization enables the calcium-binding protein **calmodulin** to become active. Calmodulin, in turn, can activate two enzymes that inactivate CSF: calmodulin-dependent protein kinase II, which inactivates the cdk2 kinase, and calpain II, a calcium-dependent protease that degrades c-mos (Watanabe et al. 1989; Lorca et al. 1993). This action promotes cell division in two ways. First, without CSF, cyclin can be degraded, and the meiotic division can be completed. Second, calcium-dependent protein kinase II also allows the centrosome to duplicate, thus forming the poles of the meiotic spindle (Matsumoto and Maller 2002). The coordination of fertilization and meiosis appears to be intimately coordinated through the release and binding of calcium ions.

Gene transcription in oocytes

In most animals (insects being a major exception), the growing oocyte is active in transcribing genes whose products are necessary for cell metabolism, for oocyte-specific processes, or for early development before the zygote-derived nuclei begin to function. In mice, for instance, the growing diplotene oocyte is actively transcribing the genes for zona pellucida

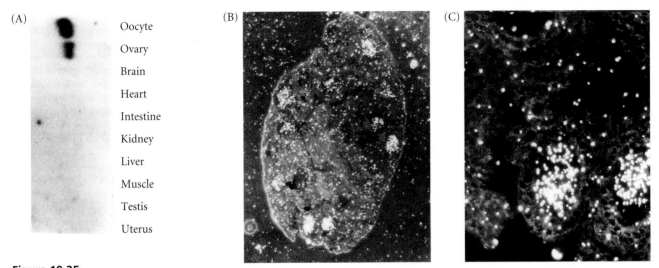

Figure 19.25

Expression of the *ZP3* gene in the developing mouse oocyte. (A) Northern blot of *ZP3* mRNA accumulation in the tissues of a 13-day mouse embryo. A radioactive probe to the ZP3 message found it expressed only in the ovary, and specifically in the oocytes. (B) When the luciferinase reporter gene is placed onto the *ZP3* promoter and inserted into the mouse genome, luciferinase message is seen only in the developing oocytes of the ovary. (C) Higher magnification of a section of (B), showing two of the ovarian follicles containing maturing oocytes. (A from Roller et al. 1989; B and C from Lira et al. 1990; photographs courtesy of P. Wassarman.)

proteins ZP1, ZP2, and ZP3. Moreover, these genes are transcribed only in the oocyte and not in any other cell type (Figure 19.25; Roller et al. 1989; Lira et al. 1990; Epifano et al. 1995).

The amphibian oocyte has certain periods of very active RNA synthesis. During the diplotene stage, certain chromosomes stretch out large loops of DNA, causing the chromosome to resemble a lampbrush (a handy instrument for cleaning test tubes in the days before microfuges). These **lampbrush chromosomes** (Figure 19.26A) can be revealed as sites of RNA synthesis by in situ hybridization. Oocyte chromosomes can

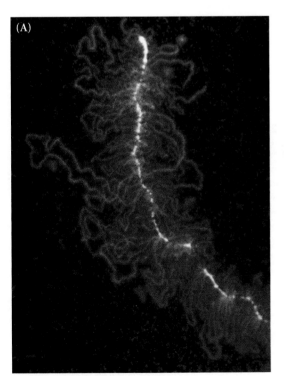

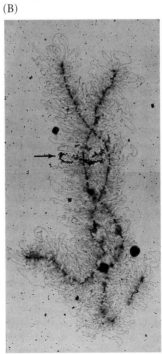

Figure 19.26

In amphibian oocytes, lampbrush chromosomes are active in the diplotene germinal vesicle during first meiotic prophase. (A) Lampbrush chromosome of the salamander *Notophthalmus viridescens*. Extended DNA (white) loops out and is transcribed into RNA (red). (B) Localization (arrow) of histone gene transcription on a lampbrush chromosome of the newt *Triturus cristatus*. The genes have been visualized by in situ hybridization and autoradiography. (A courtesy of M. B. Roth and J. Gall; B from Old et al. 1977, courtesy of H. G. Callan.)

be incubated with a radioactive RNA probe and autoradiography used to visualize the precise locations where genes are being transcribed. Figure 19.26B shows diplotene chromosome I of the newt *Triturus cristatus* after incubation with radioactive histone mRNA. It is obvious that a histone gene (or set of histone genes) is being transcribed on one of the loops of this lampbrush chromosome. Electron micrographs of gene transcripts from lampbrush chromosomes also enable one to see chains of mRNA coming off each gene as it is transcribed (see Hill and MacGregor 1980).

In addition to mRNA synthesis, ribosomal RNA and transfer RNA are also transcribed during oogenesis. Figure 19.27A shows the pattern of rRNA and tRNA synthesis during *Xenopus* oogenesis. Transcription appears to begin in early (stage I, 25–40 μm) oocytes, during the diplotene stage of meiosis. At this time, all the rRNAs and tRNAs needed for protein synthesis until the mid-blastula stage are made, and all the maternal mRNAs needed for early development are transcribed. This stage lasts for months in *Xenopus*. The rate of ribosomal RNA production is prodigious. The *Xenopus* oocyte genome has over 1800 genes encoding 18S and 28S rRNA (the two large RNAs that form the ribosomes), and these genes are selectively amplified such that there are over 500,000 genes making ribosomal RNA in the oocyte (Figure 19.27B; Brown and Dawid 1968). When the mature (stage VI) oocyte reaches a certain size, its chromosomes condense, and the rRNA genes are no longer transcribed. This "mature oocyte" condition can also last for months. Upon hormonal stimulation, the oocyte completes its first meiotic division and is ovulated. The mRNAs stored by the oocyte now join with the ribosomes to initiate protein synthesis. Within hours, the second meiotic division has begun, and the secondary oocyte has been fertilized. The embryo's genes do not begin active transcription until the mid-blastula transition (Newport and Kirschner 1982).

As we have seen in Chapter 5, the oocytes of several species make two classes of mRNAs—those for immediate use in the oocyte and those that are stored for use during early development. In sea urchins, the translation of stored oocyte messages (maternal mRNAs) is initiated by fertilization, while in frogs, the signal for such translation is initiated by progesterone as the egg is about to be ovulated. One of the results of the MPF activity induced by progesterone may be the phosphorylation of proteins on the 3′ UTR of stored oocyte mRNAs. The phosphorylation of these factors is associated with the lengthening of the poly(A) tails of the stored messages and their subsequent translation (Paris et al. 1991).

WEBSITE 19.12 **Synthesizing oocyte ribosomes.** Ribosomes are almost a "differentiated product" of the oocyte, and the *Xenopus* oocyte contains 20,000 times as many ribosomes as somatic cells do. Gene repetition and gene amplification are both used to transcribe these enormous amounts of rRNA.

(A)

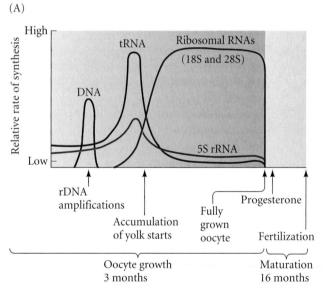

Figure 19.27
Ribosomal RNA production in *Xenopus* oocytes. (A) Relative rates of DNA, tRNA, and rRNA synthesis in amphibian oogenesis during the last 3 months before ovulation. (B) Transcription of the large RNA precursor of the 28S, 18S, and 5.8S ribosomal RNAs. These units are tandemly linked together, with some 450 per haploid genome. (A after Gurdon 1976; B courtesy of O. L. Miller, Jr.)

(B)

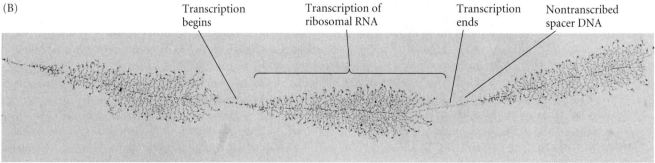

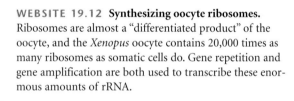

Meroistic oogenesis in insects

There are several types of oogenesis in insects, but most studies have focused on those insects, such as *Drosophila* and moths, that undergo **meroistic oogenesis**, in which cytoplasmic connections remain between the cells produced by the oogonium. In *Drosophila*, each oogonium divides four times to produce a clone of 16 cells connected to one another by **ring canals**. The production of these interconnected cells, called **cystocytes**, involves a highly ordered array of cell divisions (Figure 19.28). Only those two cells having four interconnections are capable of developing into oocytes, and of those two, only one becomes the egg. The other begins meiosis but does not complete it. Thus, only one of the 16 cystocytes can become an ovum. All the other cells become **nurse cells**. As it turns out, the cell destined to become the oocyte is that cell residing at the most posterior tip of the egg chamber, or **ovariole**, that encloses the 16-cell clone. However, since the nurse cells are connected to the oocyte by cytoplasmic bridges, the entire complex can be seen as one egg-producing unit.

The oocytes of meroistic insects do not pass through a transcriptionally active stage, nor do they have lampbrush chromosomes. Rather, RNA synthesis is largely confined to the nurse cells, and the RNA made by those cells is actively transported into the oocyte cytoplasm (see Figure 9.13A). Oogenesis takes place in only 12 days, so the nurse cells are very metabolically active during this time. They are aided in their transcriptional efficiency by becoming polytene. Instead of having two copies of each chromosome, they replicate their chromosomes until they have produced 512 copies. The 15 nurse cells pass ribosomal and messenger RNAs as well as

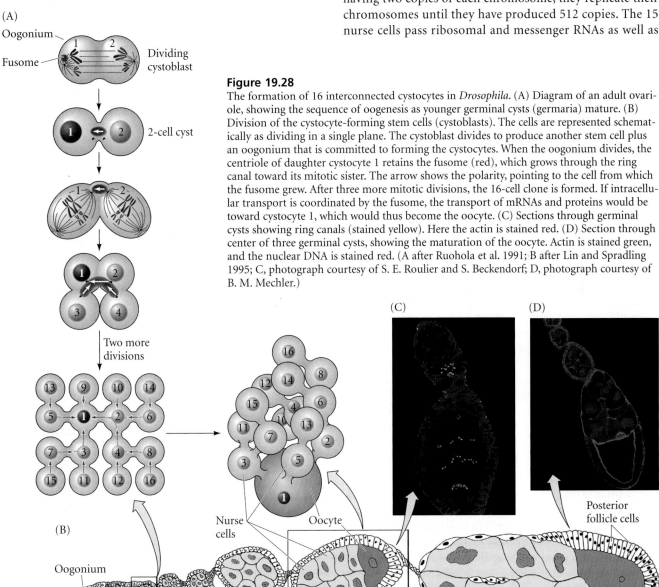

Figure 19.28
The formation of 16 interconnected cystocytes in *Drosophila*. (A) Diagram of an adult ovariole, showing the sequence of oogenesis as younger germinal cysts (germaria) mature. (B) Division of the cystocyte-forming stem cells (cystoblasts). The cells are represented schematically as dividing in a single plane. The cystoblast divides to produce another stem cell plus an oogonium that is committed to forming the cystocytes. When the oogonium divides, the centriole of daughter cystocyte 1 retains the fusome (red), which grows through the ring canal toward its mitotic sister. The arrow shows the polarity, pointing to the cell from which the fusome grew. After three more mitotic divisions, the 16-cell clone is formed. If intracellular transport is coordinated by the fusome, the transport of mRNAs and proteins would be toward cystocyte 1, which would thus become the oocyte. (C) Sections through germinal cysts showing ring canals (stained yellow). Here the actin is stained red. (D) Section through center of three germinal cysts, showing the maturation of the oocyte. Actin is stained green, and the nuclear DNA is stained red. (A after Ruohola et al. 1991; B after Lin and Spradling 1995; C, photograph courtesy of S. E. Roulier and S. Beckendorf; D, photograph courtesy of B. M. Mechler.)

proteins into the oocyte cytoplasm, and entire ribosomes may be transported as well. The mRNAs do not associate with polysomes, which suggests that they are not immediately active in protein synthesis (Paglia et al. 1976; Telfer et al. 1981).

The meroistic ovary confronts us with some interesting problems. If all the cystocytes are connected so that proteins and RNAs shuttle freely among them, why should they have different developmental fates? Why should one cell become the oocyte while the others become "RNA-synthesizing factories," sending mRNAs, ribosomes, and even centrioles into the oocyte? Why is the flow of protein and RNA in one direction only?

As the cystocytes divide, a large, spectrin-rich structure called the **fusome** forms and spans the ring canals between the cells (Figure 19.28D). It is constructed asymmetrically, as it always grows from the spindle pole that remains in one of the cells after the first division (Lin and Spradling 1995; de Cuevas and Spradling 1998). The cell that retains the greater part of the fusome during the first division becomes the oocyte. It is not yet known if the fusome contains oogenic determinants, or if it directs the traffic of materials into this particular cell.

Once the patterns of transport are established, the cytoskeleton becomes actively involved in transporting mRNAs from the nurse cells into the oocyte cytoplasm (Cooley and Theurkauf 1994). An array of microtubules that extends through the ring canals is critical for oocyte determination. In the nurse cells, the Exuperantia protein binds *bicoid* message to the microtubules and transports it to the anterior of the oocyte (Cha et al. 2001; see Chapter 8). If the microtubular array is disrupted (either chemically or by mutations such as *bicaudal-D* or *egalitarian*), the nurse cells gene products are transmitted in all directions, and all 16 cells differentiate into nurse cells (Gutzeit 1986; Theurkauf et al. 1992, 1993; Spradling 1993). The Bicaudal-D and Egalitarian proteins are probably core components of a dynein motor system that transports mRNAs and proteins throughout the oocyte (Bullock and Ish-Horowicz 2001). It is possible that some compounds transported from the nurse cells into the oocyte become associated with transport proteins, such as dynein and kinesin, which would enable them to travel along the tracks of microtubules extending through the ring canals (Theurkauf et al. 1992; Sun and Wyman 1993). The *oskar* message, for instance, is linked to kinesin through the Barentsz protein, and kinesin can transport the *oskar* message to the posterior of the oocyte (van Eeden et al. 2001).

Actin may become important for maintaining the polarity of transport during later stages of oogenesis. Mutations that prevent actin microfilaments from lining the ring canals prevent the transport of mRNAs from the nurse cells to the oocyte, and disruption of the actin microfilaments randomizes the distribution of mRNA (Cooley et al. 1992; Watson et al. 1993). Thus, the cytoskeleton controls the movement of organelles and RNAs between nurse cells and oocyte such that developmental cues are exchanged only in the appropriate direction.

WEBSITE 19.13 *Drosophila* spermatogenesis. *Drosophila* sperm are derived from stem cells that maintain a long-term capacity to divide. The renewal of stem cells is specified by the STAT pathway, which responds to a specific support cell signal.

Maturation of the mammalian oocyte

Ovulation in mammals follows one of two patterns, depending on the species. One type of ovulation is stimulated by the act of copulation. Physical stimulation of the cervix triggers the release of gonadotropins from the pituitary. These gonadotropins signal the egg to resume meiosis and initiate the events that will expel it from the ovary. This mechanism ensures that most copulations will result in fertilized ova, and animals that utilize this method of ovulation—such as rabbits and minks—have a reputation for procreative success.

Most mammals, however, have a periodic ovulation pattern, in which the female ovulates only at specific times of the year. This ovulatory period is called **estrus** (or its English equivalent, "heat"). In these animals, environmental cues, most notably the amount and type of light during the day, stimulate the hypothalamus to release gonadotropin-releasing hormone. This factor stimulates the pituitary to release the gonadotropins—follicle-stimulating hormone (FSH) and luteinizing hormone (LH)—that cause the ovarian follicle cells to proliferate and secrete estrogen. The estrogen enters certain neurons and evokes the pattern of mating behavior characteristic of the species. The gonadotropins also stimulate follicular growth and initiate ovulation. Thus, mating behavior and ovulation occur close together.

Humans have a variation on the theme of periodic ovulation. Although human females have cyclical ovulation (averaging about once every 29.5 days) and no definitive yearly estrus, most of human reproductive physiology is shared with other primates. The characteristic primate periodicity in maturing and releasing ova is called the **menstrual cycle** because it entails the periodic shedding of blood and endothelial tissue from the uterus at monthly intervals.* The menstrual cycle represents the integration of three very different cycles:

---

*The periodic shedding of the uterine lining is a controversial topic. Some scientists speculate that menstruation is an active process, with adaptive significance in evolution. Profet (1993) proposed that menstruation is a crucial immunological adaptation, protecting the uterus against infections contracted from semen or other environmental agents. Strassmann (1996) suggested that the cyclicity of the endometrium is an energy-saving adaptation that is important in times of poor nutrition. Vaginal bleeding would be a side effect of this adaptive process. Finn (1998) claimed that menstruation has no adaptive value and is necessitated by the immunological crises that are a consequence of bringing two genetically dissimilar organisms together in the uterus. Martin (1992) pointed out that it might even be wrong to think of there being a single function of menstruation, and that its roles might change during a woman's life cycle.

1. The ovarian cycle, the function of which is to mature and release an oocyte.
2. The uterine cycle, the function of which is to provide the appropriate environment for the developing blastocyst.
3. The cervical cycle, the function of which is to allow sperm to enter the female reproductive tract only at the appropriate time.

These three functions are integrated through the hormones of the pituitary, hypothalamus, and ovary.

VADE MECUM[2] Oogenesis in mammals. The development of the mammalian ovum and its remarkable growth during the primary oocyte stage are the subject of photographs and QuickTime movies of histological sections through a mammalian ovary.
[Click on Gametogenesis]

The majority of the oocytes within the adult human ovary are maintained in the diplotene stage of the first meiotic prophase (often referred to as the **dictyate state**). Each oocyte is enveloped by a primordial follicle consisting of a single layer of epithelial granulosa cells and a less organized layer of mesenchymal thecal cells (Figure 19.29). Periodically, a group of primordial follicles enters a stage of follicular growth. During this time, the oocyte undergoes a 500-fold increase in volume (corresponding to an increase in oocyte diameter from 10 μm in a primordial follicle to 80 μm in a fully developed follicle).

Concomitant with oocyte growth is an increase in the number of granulosa cells, which form concentric layers around the oocyte. This proliferation of granulosa cells is mediated by a paracrine factor, GDF9, a member of the TGF-β family (Dong et al. 1996). Throughout this growth period, the oocyte remains in the dictyate stage. The fully grown follicle thus contains a large oocyte surrounded by several layers of granulosa cells. The innermost of these cells will stay with the ovulated egg, forming the **cumulus**, which surrounds the egg in the oviduct. In addition, during the growth of the follicle, an **antrum** (cavity) forms, which becomes filled with a complex mixture of proteins, hormones, and other molecules. Just as the maturing oocyte synthesizes paracrine factors that allow the follicle cells to proliferate, the follicle cells secrete growth and differentiation factors (TGF-β2, VEGF, leptin, FGF2) that allow the oocyte to grow and bring blood vessels into the follicular region (Antczak et al. 1997).

At any given time, a small group of follicles is maturing. However, after progressing to a certain stage, most oocytes and their follicles die. To survive, the follicle must be exposed to a wave of gonadotropic hormone release, "catch the wave" at the right time, and ride it until it peaks. Thus, for oocyte maturation to occur, the follicle needs to be at a certain stage of development when the waves of gonadotropin arise.

The first day of vaginal bleeding is considered to be day 1 of the menstrual cycle (Figure 19.30). This bleeding represents the sloughing off of endometrial tissue and blood vessels that

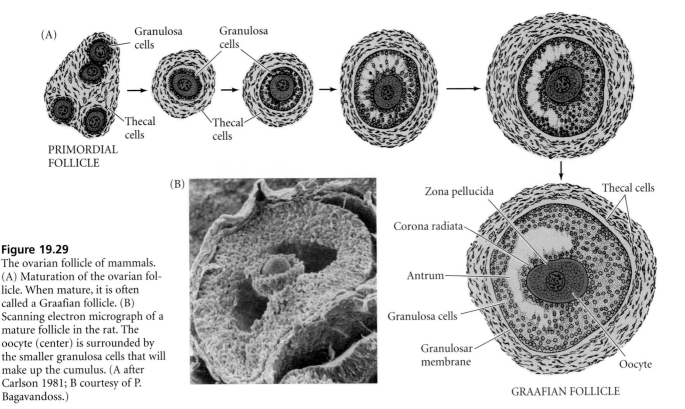

Figure 19.29
The ovarian follicle of mammals. (A) Maturation of the ovarian follicle. When mature, it is often called a Graafian follicle. (B) Scanning electron micrograph of a mature follicle in the rat. The oocyte (center) is surrounded by the smaller granulosa cells that will make up the cumulus. (A after Carlson 1981; B courtesy of P. Bagavandoss.)

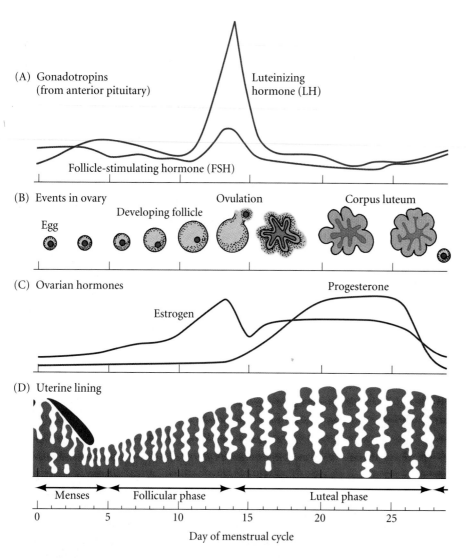

(A) Gonadotropins (from anterior pituitary)

Luteinizing hormone (LH)

Follicle-stimulating hormone (FSH)

(B) Events in ovary

Egg

Developing follicle

Ovulation

Corpus luteum

(C) Ovarian hormones

Estrogen

Progesterone

(D) Uterine lining

Menses

Follicular phase

Luteal phase

Day of menstrual cycle

Figure 19.30
The human menstrual cycle. The coordination of (B) ovarian and (D) uterine cycles is controlled by (A) the pituitary and (C) the ovarian hormones. During the follicular phase, the egg matures within the follicle, and the uterine lining is prepared to receive a blastocyst. The mature egg is released around day 14. If a blastocyst does not implant in the uterus, the uterine wall begins to break down, leading to menstruation.

would have aided the implantation of the blastocyst. In the first part of the cycle (called the **proliferative** or **follicular phase**), the pituitary starts secreting increasingly large amounts of FSH. Any maturing follicles that have reached a certain stage of development respond to this hormone with further growth and cellular proliferation. FSH also induces the formation of LH receptors on the granulosa cells. Shortly after this period of initial follicle growth, the pituitary begins secreting LH. In response to LH, the dictyate meiotic block is broken. The nuclear membranes of competent oocytes break down, and the chromosomes assemble to undergo the first meiotic division. One set of chromosomes is kept inside the oocyte, and the other ends up in the small polar body. Both are encased by the zona pellucida, which has been synthesized by the growing oocyte. It is at this stage that the egg will be ovulated.

The two gonadotropins, acting together, cause the follicle cells to produce increasing amounts of estrogen, which has at least five major activities in regulating the further progression of the menstrual cycle:

1. It causes the uterine endometrium to begin its proliferation and to become enriched with blood vessels.
2. It causes the cervical mucus to thin, thereby permitting sperm to enter the inner portions of the reproductive tract.
3. It causes an increase in the number of FSH receptors on the granulosa cells of the mature follicles while causing the pituitary to lower its FSH production. It also stimulates the granulosa cells to secrete the peptide hormone inhibin, which also suppresses pituitary FSH secretion.
4. At low concentrations, it inhibits LH production, but at high concentrations, it stimulates it.
5. At very high concentrations and over long durations, estrogen interacts with the hypothalamus, causing it to secrete gonadotropin-releasing hormone.

As estrogen levels increase as a result of follicular production, FSH levels decline. LH levels, however, continue to rise as more estrogen is secreted. As estrogen continues to be made (days 7–10), the granulosa cells continue to grow.

Starting on day 10, estrogen secretion rises sharply. This rise is followed at midcycle by an enormous surge of LH and a smaller burst of FSH. Experiments with female monkeys have shown that exposure of the hypothalamus to greater than 200 pg of estrogen per milliliter of blood for more than 50 hours results in hypothalamic secretion of gonadotropin-releasing hormone. This factor causes the subsequent release of FSH and LH from the pituitary. Within 10–12 hours after the gonadotropin peak, the egg is ovulated (Figure 19.31; Garcia et al. 1981).

Although the detailed mechanism of ovulation is not yet known, the physical expulsion of the mature oocyte from the follicle appears to be the result of an LH-induced increase in collagenase, plasminogen activator, and prostaglandins within the follicle (Lemaire et al. 1973). Prostaglandins may cause localized contractions in the smooth muscles of the ovary and may also increase the flow of water from the ovarian capillaries, increasing fluid pressure in the antrum (Diaz-Infante et al. 1974; Koos and Clark 1982). If ovarian prostaglandin synthesis is inhibited, ovulation does not take place. In addition, collagenase and the plasminogen activator protease loosen and digest the extracellular matrix of the follicle (Beers et al. 1975; Downs and Longo 1983). The mRNA for plasminogen activator has been dormant in the oocyte cytoplasm. LH causes this message to be polyadenylated and translated into this powerful protease (Huarte et al. 1987). The result of LH, then, is increased follicular pressure coupled with the degradation of the follicle wall. A hole is digested through which the ovum can burst.

Following ovulation, the **luteal phase** of the menstrual cycle begins. The remaining cells of the ruptured follicle, under the continued influence of LH, become the **corpus luteum**. (They are able to respond to this LH because the surge in FSH stimulates them to develop even more LH receptors.) The corpus luteum secretes some estrogen, but its predominant secretion is **progesterone**. This steroid hormone circulates to the uterus, where it completes the job of preparing the uterine tissue for blastocyst implantation, stimulating the growth of the uterine wall and its blood vessels. Blocking the progesterone receptor with the synthetic steroid **mifepristone** (RU486) stops the uterine wall from thickening and prevents the implantation of a blastocyst* (Couzinet et al. 1986; Greb et al. 1999).

Progesterone also inhibits the production of FSH, thereby preventing the maturation of any more follicles and ova. (For this reason, a combination of estrogen and progesterone has

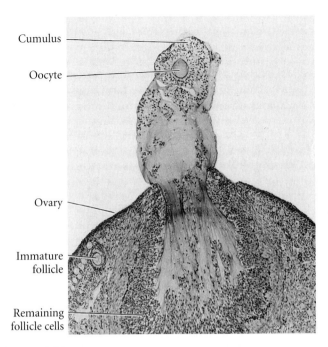

Cumulus

Oocyte

Ovary

Immature follicle

Remaining follicle cells

Figure 19.31
Ovulation in the rabbit. The ovary of a living, anesthetized rabbit was exposed and observed. When the follicle started to ovulate, the ovary was removed, fixed, and stained. (Photograph courtesy of R. J. Blandau.)

been used in birth control pills. The growth and maturation of new ova are prevented as long as FSH is inhibited.)

If the ovum is not fertilized, the corpus luteum degenerates, progesterone secretion ceases, and the uterine wall is sloughed off. With the decline in serum progesterone levels, the pituitary secretes FSH again, and the cycle is repeated. However, if fertilization occurs, the trophoblast secretes a new hormone, **luteotropin**, which causes the corpus luteum to remain active and serum progesterone levels to remain high. Thus, the menstrual cycle enables the periodic maturation and ovulation of human eggs and allows the uterus to periodically develop into an organ capable of nurturing a developing organism for 9 months.

WEBSITE 19.14 **The reinitiation of mammalian meiosis.** The hormone-mediated disruption of communication between the oocyte and its surrounding follicle cells may critical in the resumption of meiosis in female mammals.

We are now back where we began: the stage is set for fertilization to take place. The egg and the sperm will both die if they do not meet. As F. R. Lillie recognized in 1919, "The elements that unite are single cells, each on the point of death; but by their union a rejuvenated individual is formed, which constitutes a link in the eternal process of Life."

---

*This is why RU486 is used for postconception birth control. RU486 is thought to compete for the progesterone receptor inside the nucleus. RU486 can bind to the progesterone site in the receptor, and the receptor-RU486 complex appears to form heterodimers with the normal progesterone-carrying progesterone receptor. When this RU486-progesterone complex binds to progesterone-responsive enhancer elements on the DNA, transcription from these sites is inhibited (Vegeto et al. 1992; Spitz and Bardin 1993).

Snapshot Summary: The Germ Line

1. The precursors of the gametes—the sperm and eggs—are the primordial germ cells. They form outside the gonads and migrate into the gonads during development.

2. In many species, a distinctive germ plasm exists. It often contains the Oskar, Vasa, and Nanos proteins or the mRNAs encoding them.

3. In *Drosophila*, the germ plasm becomes localized in the posterior of the embryo and forms pole cells, the precursors of the gametes. In frogs, the germ plasm originates in the vegetal portion of the oocyte.

4. In amphibians, the germ cells migrate on fibronectin matrices from the posterior larval gut to the gonads. In mammals, a similar migration is seen, and fibronectin pathways may also be used. Stem cell factor is critical in this migration, and the germ cells proliferate as they travel.

5. In birds, the germ plasm is first seen in the germinal crescent. The germ cells migrate through the blood, then leave the blood vessels and migrate into the genital ridges.

6. Germ cell migration in *Drosophila* occurs in several steps involving passive translocation, repulsion from the endoderm, and attraction to the gonads.

7. Before meiosis, the DNA is replicated and the resulting sister chromatids remain bound at the kinetochore. Homologous chromosomes are connected through the synaptonemal complex.

8. The first division of meiosis separates the homologous chromosomes. The second division of meiosis splits the kinetochore and separates the chromatids.

9. The meiosis/mitosis decision in nematodes is regulated by a Delta protein homologue in the membrane of the distal tip cell. The decision for a germ cell to become either a sperm or an egg is regulated at the level of translation of the *fem-3* message.

10. Spermatogenic meiosis in mammals is characterized by the production of four gametes per meiosis and by the absence of meiotic arrest. Oogenic meiosis is characterized by the production of one gamete per meiosis and by an arrest at first meiotic prophase to allow the egg to grow.

11. In some species, meiosis is modified such that a diploid egg is formed. These species can produce a new generation parthenogenetically, without fertilization.

12. The egg not only synthesizes numerous compounds, but also absorbs material produced by other cells. Moreover, it localizes many proteins and messages to specific regions of the cytoplasm, often tethering them to the cytoskeleton.

13. The *Xenopus* oocyte transcribes actively from lampbrush chromosomes during the first meiotic prophase.

14. In *Drosophila*, nurse cells make mRNAs that enter the developing oocyte. Which of the cells derived from the primordial germ cell becomes the oocyte and which become nurse cells is determined by the fusome and the pattern of divisions.

15. In mammals, the hormones of the menstrual cycle integrate the ovarian cycle, the uterine cycle, and the cervical cycle. This integration allows the uterus to be ready to receive an embryo shortly after ovulation occurs.

Literature Cited

Ahringer, J. and J. Kimble. 1991. Control of the sperm-oocyte switch in *Caenorhabditis elegans* hermaphrodites by the *fem-3* 3′-untranslated region. *Nature* 349: 346–348.

Ahringer, J., T. A. Rosenquist, D. N. Lawson and J. Kimble. 1992. The *C. elegans* sex determining gene, *fem-3*, is regulated post-transcriptionally. *EMBO J.* 11: 2303–2310.

Amikura, R., S. Kobayashi, H. Saito and M. Okada. 1996. Changes in subcellular localization of mtrRNA outside mitochondria in oogenesis and early embryogenesis of *Drosophila melanogaster*. *Dev. Growth Diff.* 38: 489–498.

Amikura, R., K. Hanyu, M. Kashikawa and S. Kobayashi. 2001. Tudor protein is essential for the localization of mitochondrial RNA in polar granules of *Drosophila* embryos. *Mech. Dev.* 107: 97–104.

Anderson, R. and 8 others. 1999. Mouse primordial germ cells lacking β1 integrins enter the germline but fail to migrate normally to the gonads. *Development* 126: 1655–1664.

Anderson, R., T. K. Copeland, H. Schöler, J. Heasman and C. Wylie. 2000. The onset of germ cell migration in the mouse embryo. *Mech. Dev.* 91: 61–68.

Antczak, M., J. van Blerkom and A. Clark. 1997. A novel mechanism of vascular endothelial growth factor, leptin, and transforming growth factor-β2 sequestration in a population of human ovarian follicle cells. *Hum. Reprod.* 12: 2226–2234.

Austin, J. and J. Kimble. 1987. GLP-1 is required in the germ line for regulation of the decision between mitosis and meiosis in *C. elegans*. *Cell* 51: 589–599.

Baker, T. G. 1970. Primordial germ cells. *In* C. R. Austin and R. V. Short (eds.), *Reproduction in Mammals, Vol. 1: Germ Cells and Fertilization.* Cambridge University Press, Cambridge, pp. 1–13.

Beers, W. H., S. Strickland and E. Reich. 1975. Ovarian plasminogen activator: Relationship to ovulation and hormonal regulation. *Cell* 6: 387–394.

Bendel-Stenzel, M. R., M. Gomperts, R. Anderson, J. Heasman and C. Wylie. 2000. The role of cadherins during primordial germ cell migration and early gonad formation in the mouse. *Mech. Dev.* 91: 143–152.

Blackler, A. W. 1962. Transfer of primordial germ cells between two subspecies of *Xenopus laevis*. *J. Embryol. Exp. Morphol.* 10: 641–651.

Bloom, W. and D. W. Fawcett. 1975. *Textbook of Histology*, 10th Ed. Saunders, Philadelphia.

Bounoure, L. 1934. Recherches sur lignée germinale chez la grenouille rousse aux premiers stades au développement. *Ann. Sci. Zool.* Ser. 17, 10: 67–248.

Boveri, T. 1910. Über die Teilung centrifugierter Eier von *Ascaris megalocephala*. *Wilhelm Roux Arch. Entwicklungsmech. Org.* 30: 101–125.

Braat, A. K., J. E. Speksnijder and D. Zivkovic. 2000. Germ line development in fishes. *Int. J. Dev. Biol.* 43: 745–760.

Braun, R. E., R. R. Behringer, J. J. Peschon, R. L. Brinster and R. D. Palmiter. 1989. Genetically haploid spermatids are phenotypically diploid. *Nature* 337: 373–376.

Broihier, H. T., L. A. Moore, M. Van Doren, S. Newman and R. Lehmann. 1998. *zfh-1* is required for germ cell migration and gonadal mesoderm development in *Drosophila*. *Development* 125: 655–666.

Brown, D. D. and I. B. Dawid. 1968. Specific gene amplification in oocytes. *Science* 160: 272–280.

Bullock, S. L. and D. Ish-Horowicz. 2001. Conserved signals and machinery for RNA transport in *Drosophila* oogenesis and embryogenesis. *Nature* 414: 611–616.

Burns, R. K., Jr. 1930. The process of sex-transformation in parabiotic *Ambystoma*. I. Transformation from female to male. *J. Exp. Zool.* 55: 123–169.

Buss, L. W. 1987. *The Evolution of Individuality*. Princeton University Press, Princeton, NJ.

Carlson, B. M. 1981. *Patten's Foundations of Embryology*. McGraw-Hill, New York.

Cha, B. J., B. S. Koppetsch and W. E. Theurkauf. 2001. In vivo analysis of *Drosophila bicoid* mRNA localization reveals a novel microtubule-dependent axis specification pathway. *Cell* 106: 35–46.

Chiquoine, A. D. 1954. The identification, origin, and migration of the primordial germ cells in the mouse embryo. *Anat. Rec.* 118: 135–146.

Comings, D. E. 1968. The rationale for an ordered arrangement of chromatin in the interphase nucleus. *Am. J. Hum. Genet.* 20: 440–460.

Cooley, L. and W. E. Theurkauf. 1994. Cytoskeletal functions during *Drosophila* oogenesis. *Science* 266: 590–596.

Cooley, L., E. Verheyen and K. Ayers. 1992. *chickadee* encodes a profilin required for intercellular cytoplasm transport during *Drosophila* oogenesis. *Cell* 69: 173–184.

Couzinet, B., N. Le Strat, A. Ulmann, E. E. Baulieu and G. Schaison. 1986. Termination of early pregnancy by the progesterone antagonist RU486 (Mifepristone). *N. Engl. J. Med.* 315: 1565–1570.

Danilchik, M. V. and J. C. Gerhart. 1987. Differentiation of the animal-vegetal axis in *Xenopus laevis* oocytes: Polarized intracellular translocation of platelets establishes the yolk gradient. *Dev. Biol.* 122: 101–112.

de Cuevas, M. and A. C. Spradling. 1998. Morphogenesis of the *Drosophila* fusome and its implications for oocyte specification. *Development* 125: 2781–2789.

Deshpande, G., G. Calhoun, J. Yanowitz and P. D. Schedl. 1999. Novel functions of *nanos* in downregulating mitosis and transcription during the development of the *Drosophila* germline. *Cell* 99: 271–281.

Deshpande, G., L. Swanhart, P. Chiang and P. Schedl. 2001. Hedgehog signaling in germ cell migration. *Cell* 106: 759–769.

De Vos, L. 2002. "Méiose." http://www.ulb.ac.be/sciences/biodic/ImCellulles3.html.

Diaz-Infante, A., K. H. Wright and E. E. Wallach. 1974. Effects of indomethacin and PGF2a on ovulation and ovarian contraction in the rabbit. *Prostaglandins* 5: 567–581.

Doitsidou, M. and 8 others. 2002. Guidance of primordial germ cell migration by the chemokine SDF-1. *Cell* 111: 647–659.

Dolci, S. and 8 others. 1991. Requirement for mast cell growth factor for primordial germ cell survival in culture. *Nature* 352: 809–811.

Dong, J., D. F. Albertini, K. Nishimori, T. R. Kumar, N. Lu and M. Matzuk. 1996. Growth differentiation factor-9 is required during early ovarian folliculogenesis. *Nature* 383: 531–534.

Downs, S. and F. J. Longo. 1983. Prostaglandins and preovulatory follicular maturation in mice. *J. Exp. Zool.* 228: 99–108.

Dubois, R. 1969. Le mécanisme d'entrée des cellules germinales primordiales dans le réseau vasculaire, chez l'embryon de poulet. *J. Embryol. Exp. Morphol.* 21: 255–270.

Dumont, J. N. 1978. Oogenesis in *Xenopus laevis*. VI. Route of injected tracer transport in follicle and developing oocyte. *J. Exp. Zool.* 204: 193–200.

Dym, M. 1977. The male reproductive system. *In* L. Weiss and R. O. Greep (eds.), *Histology*, 4th Ed. McGraw-Hill, New York, pp. 979–1038.

Dym, M. 1994. Spermatogonial cells of the testis. *Proc. Nat. Acad. Sci. USA* 91: 11287–11289.

Dym, M. and D. W. Fawcett. 1971. Further observations on the number of spermatogonia, spermatocytes, and spermatids connected by intercellular bridges in the mammalian testis. *Biol. Reprod.* 4: 195–215.

Ephrussi, A. and R. Lehmann. 1992. Induction of germ cell formation by *oskar*. *Nature* 358: 387–392.

Epifano, O., L.-F. Liang, M. Familari, M. C. Moos, Jr. and J. Dean. 1995. Coordinate expression of the three zona pellucida genes during mouse oogenesis. *Development* 121: 1947–1956.

Eyal-Giladi, H., M. Ginsburg and A. Farbarou. 1981. Avian primordial germ cells are of epiblastic origin. *J. Embryol. Exp. Morphol.* 65: 139–147.

Ferrell, J. E., Jr. 1999. *Xenopus* oocyte maturation: New lessons from a good egg. *BioEssays* 21: 833–842.

Ferrell, J. E., Jr. and E. M. Machleder. 1998. The biochemical basis of an all-or-none cell fate switch in *Xenopus* oocytes. *Science* 280: 895–899.

ffrench-Constant, C., A. Hollingsworth, J. Heasman and C. Wylie. 1991. Response to fibronectin of mouse primordial germ cells before, during, and after migration. *Development* 113: 1365–1373.

Finn, C. A. 1998. Menstruation: A nonadaptive consequence of uterine evolution. *Q. Rev. Biol.* 73: 163–173.

Flickinger, R. A. and D. E. Rounds. 1956. The maternal synthesis of egg yolk proteins as demonstrated by isotopic and serological means. *Biochem. Biophys. Acta* 22: 38–72.

Forristall, C., M. Pondel, L. Chen and M. L. King. 1995. Patterns of localization and cytoskeletal association of two vegetally localized RNAs, *Vg1* and *Xcat-2*. *Development* 121: 201–208.

Gabrielli, B., L. M. Roy and J. L. Maller. 1993. Requirement for cdk2 in cytostatic factor-mediated metaphase II arrest. *Science* 259: 1766–1769.

Garcia, J. E., G. S. Jones and G. L. Wright. 1981. Prediction of the time of ovulation. *Fertil. Steril.* 36: 308–315.

Gardner, R. L. 1982. Manipulation of development. *In* C. R. Austin and R. V. Short (eds.), *Embryonic and Fetal Development*. Cambridge University Press, Cambridge, pp. 159–180.

Ginsburg, M. and H. Eyal-Giladi. 1987. Primordial germ cells of the young chick blastoderm originate from the central zone of the area pellucida irrespective of the embryo-forming process. *Development* 101: 209–219.

Godin, I. and C. C. Wylie. 1991. TGFβ1 inhibits proliferation and has a chemotactic effect on mouse primordial germ cells in culture. *Development* 113: 1451–1457.

Godin, I., C. Wylie and J. Heasman. 1990. Genital ridges exert long-range effects on primordial germ cell numbers and direction of migration in culture. *Development* 108: 357–363.

Godin, I., R. Deed, J. Cooke, K. Zsebo, M. Dexter and C. C. Wylie. 1991. Effects of the *steel* gene product on mouse primordial germ cells in culture. *Nature* 352:807–809.

Graham, C. E. 1977. Teratocarcinoma cells and normal mouse embryogenesis. *In* M. I. Sherman (ed.), *Concepts of Mammalian Embryogenesis*. M.I.T. Press, Cambridge, MA, pp. 315–394.

Grant, P. 1953. Phosphate metabolism during oogenesis in *Rana temporaria*. *J. Exp. Zool.* 124: 513–543.

Greb, R. R., L. Kiesel, A. K. Selbmann, M. Wehrmann, G. D. Hodgen, A. L. Goodman and D. Wallwiener. 1999. Disparate actions of mifepristone (RU 486) on glands and stroma in the primate endometrium. *Hum. Reprod.* 14: 198–206.

Gurdon, J. B. 1976. *The Control of Gene Expression in Animal Development*. Harvard University Press, Cambridge, MA.

Gutzeit, H. O. 1986. The role of microfilaments in cytoplasmic streaming in *Drosophila* follicles. *J. Cell Sci.* 80: 159–169.

Hall, D. H. and 7 others. 1999. Ultrastructural features of the adult hermaphrodite gonad of *Caenorhabditis elegans:* Relationship between the germ line and soma. *Dev. Biol.* 212: 101–123.

Heasman, J., T. Mohun and C. C. Wylie. 1977. Studies on the locomotion of primordial germ cells from *Xenopus laevis* in vitro. *J. Embryol. Exp. Morphol.* 42: 149–162.

Heasman, J., R. D. Hynes, A. P. Swan, V. Thomas and C. C. Wylie. 1981. Primordial germ cells of *Xenopus* embryos: The role of fibronectin in their adhesion during migration. *Cell* 27: 437–447.

Heath, J. K. 1978. Mammalian primordial germ cells. *Dev. Mammals* 3: 272–298.

Henderson, S. T., D. Gao, E. J. Lambie and J. Kimble. 1994. *Lag-2* may encode a signaling ligand of Glp-1 and Lin-12 receptors of *C. elegans. Development* 120: 2913–2924.

Hill, R. S. and H. C. MacGregor. 1980. The development of lampbrush chromosome-type transcription in the early diplotene oocytes of *Xenopus laevis:* An electron microscope analysis. *J. Cell Sci.* 44: 87–101.

Hirsh, D., D. Oppenheim and M. Klass. 1976. Development of the reproductive system of *Caenorhabditis elegans. Dev. Biol.* 49: 200–219.

Hodgkin, J., T. Doniach and M. Shen. 1985. The sex determination pathway in the nematode *Caenorhabditis elegans:* Variations on a theme. *Cold Spring Harb. Symp. Quant. Biol.* 50: 585–593.

Hogan, B. 2002. Decisions, decisions! *Nature* 418: 282–283.

Howard, K. 1998. Organogenesis: *Drosophila* goes gonadal. *Curr. Biol.* 8: R415–R417.

Huarte, J., D. Belin, A. Vassalli, S. Strickland and J.-D. Vassalli. 1987. Meiotic maturation of mouse oocytes triggers the translation and polyadenylation of dormant tissue-type plasminogen activator mRNA. *Genes Dev.* 1: 1201–1211.

Humphrey, R. R. 1931. Studies of sex reversal in *Ambystoma*. III. Transformation of the ovary of *A. tigrinum* into a functional testis through the influence of a testis resident in the same animal. *J. Exp. Zool.* 58: 333–365.

Ikenishi, K., T. S. Tanaka and T. Komiya. 1996. Spatio-temporal distribution of the protein of the *Xenopus vasa* homologue (*Xenopus vasa-like gene-1, XVLG1*) in embryos. *Dev. Growth Diff.* 38: 527–535.

Inoue, H. and T. Hiroyoshi. 1986. A maternal-effect sex-transformation mutant of the housefly, *Musca domestica* L. *Genetics* 112: 469–481.

Jaglarz, M. K. and K. R. Howard. 1995. The active migration of *Drosophila* primordial germ cells. *Development* 121: 3495–3503.

Jongens, T. A., B. Hay, L. Y. Jan and Y. N. Jan. 1992. The *germ cell-less* gene product: A posteriorly localized component necessary for germ cell development in *Drosophila. Cell* 70: 569–584.

Kammerman, S. and J. Ross. 1975. Increase in numbers of gonadotropin receptors on granulosa cells during follicle maturation. *J. Clin. Endocrinol.* 41: 546–550.

Kashikawa, M., R. Amikura, A. Nakamura and S. Kobashi. 1999. Mitochondrial small ribosomal RNA is present on polar granules in early cleavage embryos of *Drosophila melanogaster. Dev. Growth Diff.* 41: 495–502.

Kawasaki, I., Y.-H. Shim, J. Kirschner, J. Kaminker, W. B. Wood and S. Strome. 1998. PGL-1, a predicted RNA-binding component of germ granules, is essential for fertility in *C. elegans. Cell* 94: 635–645.

Kimble, J. E. 1981. Strategies for control of pattern formation in *Caenorhabditis elegans. Phil. Trans. R. Soc. Lond.* [B] 295: 539–551.

Kimble, J. E. and J. G. White. 1981. Control of germ cell development in *Caenorhabditis elegans. Dev. Biol.* 81: 208–219.

Kimble, J., M. K. Barton, T. B. Schedl, T. A. Rosenquist and J. Austin. 1986. Controls of postembryonic germ line development in *Caenorhabditis elegans. In* J. Gall (ed.), *Gametogenesis and the Early Embryo.* Alan R. Liss, New York, pp. 97–110.

Kloc, M. and L. Etkin. 1995. Two distinct pathways for the localization of RNAs at the vegetal cortex in *Xenopus* oocytes. *Development* 121: 287–297.

Kloc, M., G. Spohr and L. Etkin. 1993. Translocation of repetitive RNA sequences with the germ plasm in *Xenopus* oocytes. *Science* 262: 1712–1714.

Kloc, M., C. Larabell and L. Etkin. 1996. Elaboration of the messenger transport organizer pathway for localization of RNA to the vegetal cortex of *Xenopus* oocytes. *Dev. Biol.* 180: 119–130.

Kloc, M., C. Larabell, A. P. Y. Chan and L. D. Etkin. 1998. Contributions of METRO pathway localized molecules to the organization of the germ cell lineage. *Mech. Dev.* 75: 81–93.

Kloc, M. and 7 others. 2002. Three-dimensional ultrastructural analysis of RNA distribution within germinal granules of *Xenopus. Dev. Biol.* 241: 79–93.

Kobayashi, S. and M. Okada. 1989. Restoration of pole-cell forming ability to UV-irradiated *Drosophila* embryos by injection of mitochondrial lrRNA. *Development* 107: 733–742.

Kobayashi, S., R. Amikura and M. Okada. 1993. Presence of mitochondrial large ribosomal RNA outside mitochondria in germ plasm of *Drosophila melanogaster. Science* 260: 1521–1524.

Kobayashi, S., M. Yamada, M. Asaoka and T. Kitamura. 1996. Essential role of the posterior morphogen nanos for germline development in *Drosophila. Nature* 380: 708–711.

Koos, R. D. and M. R. Clark. 1982. Production of 6-keto-prostaglandin F_{1a} by rat granulosa cells in vitro. *Endocrinology* 111: 1513–1518.

Kraemer, B., S. Crittenden, M. Gallegos, G. Moulder, R. Barstead, J. Kimble and M. Wickens. 1999. NANOS-3 and FBF proteins physically interact to control the sperm-oocyte switch in *Caenorhabditis elegans. Curr. Biol.* 9: 1009–1018.

Kuwana, T. 1993. Migration of avian primordial germ cells toward the gonadal anlage. *Dev. Growth Diff.* 35: 237–243.

Kuwana, T., H. Maeda-Suga and T. Fujimoto. 1986. Attraction of chick primordial germ cells by gonadal anlage in vitro. *Anat. Rec.* 215: 403–406.

Lawson, K. A. and 7 others. 1999. Bmp4 is required for the generation of primordial germ cells in the mouse embryo. *Genes Dev.* 13: 424–436.

Leatherman, J. L., L. Levin, J. Boero and T. A. Jongens. 2002. Germ cell-less acts to repress transcription during the establishment of the Drosophila germ cell lineage. *Curr. Biol.* 12: 1681–1685.

Lemaire, W. J., N. S. T. Yang, H. H. Behram and J. M. Marsh. 1973. Preovulatory changes in concentration of prostaglandin in rabbit Graafian follicles. *Prostaglandins* 3: 367–376.

Lillie, F. R. 1919. *Problems of Fertilization.* University of Chicago Press, Chicago.

Lin, H. and A. C. Spradling. 1995. Fusome asymmetry and oocyte determination in *Drosophila. Dev. Genet.* 16: 6–12.

Lira, A. A., R. A. Kinloch, S. Mortillo and P. A. Wassarman. 1990. An upstream region of the mouse *ZP3* gene directs expression of firefly luciferinase specifically to growing oocytes in transgenic mice. *Proc. Natl. Acad. Sci. USA* 87: 7215–7219.

Liu, C. K. and N. J. Berrill. 1948. Gonophore formation and germ cell origin in *Tubularia. J. Morphol.* 83: 39–60.

Lorca, T., F. H. Cruzalegui, D. Fesquet, J.-C. Cavadore, J. Méry, A. Means and M. Dorée. 1993. Calmodulin-dependent protein kinase II mediates inactivation of MPF and CSF upon fertilization of *Xenopus* eggs. *Nature* 366: 270–273.

Mahowald, A. P. 1971a. Polar granules of *Drosophila*. III. The continuity of polar granules during the life cycle of *Drosophila. J. Exp. Zool.* 176: 329–343.

Mahowald, A. P. 1971b. Polar granules of *Drosophila*. IV. Cytochemical studies showing loss of RNA from polar granules during early stages of embryogenesis. *J. Exp. Zool.* 176: 329–343.

Mahowald, A. P., J. H. Caulton and W. J. Gehring. 1979. Ultrastructural studies of oocytes and embryos derived from female flies carrying the *grandchildless* mutation in *Drosophila subobscura. Dev. Biol.* 69: 118–132.

Martin, E. 1992. *The Woman in the Body: A Cultural Analysis of Reproduction.* Beacon Press, Boston.

Matsui, Y. 1974. A cytostatic factor in amphibian: Its extraction and partial characterization. *J. Exp. Zool.* 187: 141–147.

Matsui, Y., D. Toksoz, S. Nishikawa, S.-I. Nishikawa, D. Williams, K. Zsebo and B. L. M. Hogan. 1991. Effect of Steel factor and leukemia inhibitory factor on murine primordial germ cells in culture. *Nature* 353: 750–752.

Matsui, Y., K. Zsebo and B. L. M. Hogan. 1992. Derivation of pluripotential embryonic stem cells from murine primordial germ cells in culture. *Cell* 70: 841–847.

Matsumoto, Y. and J. L. Maller. 2002. Calcium, calmodulin and CaMKII requirement for initiation of centrosome duplication in *Xenopus* egg extracts. *Science* 295: 499–502.

Maurice, J. 1991. Improvements seen for RU-486 abortions. *Science* 254: 198–200.

Mayer, D. B. 1964. The migration of primordial germ cells in the chick embryo. *Dev. Biol.* 10: 154–190.

McLaren, A. 1983. Does the chromosomal sex of a mouse cell affect its development? *Symp. Brit. Soc. Dev. Biol.* 7: 225–227.

Mendez, R., L. E. Hake, T. Andresson, L. E. Littlepage, J. V. Ruderman and J. D. Richter. 2000. Phosphorylation of CPE binding factor by Eg2 regulates translation of *c-mos* mRNA. *Nature* 404: 302–307.

Meng, X. and 13 others. 2000. Regulation of cell fate decision of undifferentiated spermatognia by GDNF. *Science* 287: 1489–1493.

Mintz, B. 1957. Embryological development of primordial germ cells in the mouse: Influence of a new mutation. *J. Embryol. Exp. Morphol.* 5: 396–403.

Moens, P. B. 1969. The fine structure of meiotic chromosome polarization and pairing in *Locusta migratoria. Chromosoma* 28: 1–25.

Molyneaux, K. A., J. Stallock, K. Schaible and C. Wylie. 2001. Time-lapse analysis of living mouse germ cell migration. *Dev. Biol.* 240: 488–498.

Moore, L. A., H. T. Broihier, M. Van Doren, L. B. Lunsford and R. Lehmann. 1998. Identification of genes controlling germ cell migration and embryonic gonad formation in *Drosophila. Development* 125: 667–678.

Moses, M. J. 1968. Synaptonemal complex. *Annu. Rev. Genet.* 2: 363–412.

Mowry, K. L. and D. A. Melton. 1992. Vegetal messenger RNA localization directed by a 340-nt RNA sequence element in *Xenopus* oocytes. *Science* 255: 991–994.

Müller, F., V. Bernard and H. Tobler. 1996. Chromatin diminution in nematodes. *BioEssays* 18: 133–138.

Nakamura, A., R. Amikura, M. Mukai, S. Kobayashi and P. F. Lasko. 1996. Requirement for a noncoding RNA in *Drosophila* polar granules for germ cell establishment. *Science* 274: 2075–2079.

Newmark, P. A., S. E. Mohr, L. Gong and R. E. Boswell. 1997. *mago nashi* mediates the posterior follicle cell-to-oocyte signal to organize axis formation in *Drosophila. Development* 124: 3194–3207.

Newport, J. and M. Kirschner. 1982. A major developmental transition in early *Xenopus* embryos: II. Control of the onset of transcription. *Cell* 30: 687–696

Newton, S. C., O. W. Blaschuk and C. F. Millette. 1993. N-cadherin mediates Sertoli cell-spermatogenic cell adhesion. *Dev. Dynam.* 197: 1–13.

Nieuwkoop, P. and L. Sutasurya. 1981. *Primordial Germ Cells of Chordates.* Cambridge University Press, Cambridge, pp. 186–187.

Old, R. W., H. G. Callan and K. W. Gross. 1977. Localization of histone gene transcripts in newt lampbrush chromosomes by in situ hybridization. *J. Cell Sci.* 27: 57–80.

Paglia, L. M., J. Berry and W. H. Kastern. 1976. Messenger RNA synthesis, transport, and storage in silkmoth ovarian follicles. *Dev. Biol.* 51: 173–181.

Palka, J. 1989. The pill of choice? *Science* 245: 1319–1323.

Paris, J., K. Swenson, H. Piwnice-Worms and J. D. Richter. 1991. Maturation-specific polyadenylation: In vitro activation by $p34^{cdc2}$ and phosphorylation of a 58-kD CPE-binding protein. *Genes Dev.* 5: 1697–1708.

Pasteels, J. 1953. Contributions à l'étude du developpement des reptiles. I. Origine et migration des gonocytes chez deux Lacertiens. *Arch. Biol.* 64: 227–245.

Pelttari, J. and 10 others. 2001. A meiotic chromosomal core consisting of cohesion complex proteins recruits DNA recombination proteins and promotes synapsis in the absence of an axial element in mammalian meiotic cells. *Mol. Cell. Biol.* 21: 5667–5677.

Pepling, M. E. and A. C. Spradling. 1998. Female mouse germ cells form synchronously dividing cysts. *Development* 125: 3323–3328.

Pesce, M., M. G. Farrace, M. Piacentini, S. Dolci and M. De Felici. 1993. Stem cell factor and leukemia inhibitory factor promote primordial germ cell survival by suppressing programmed cell death (apoptosis). *Development* 118: 1089–1094.

Pesce, M., X. Wang, D. J. Wolgemuth and H. Schöler. 1998. Differential expression of the Oct-4 transcription factor during mouse germ cell differentiation. *Mech. Dev.* 71: 89–98.

Peschon, J. J., R. R. Behringer, R. L. Brinster and R. D. Palmiter. 1987. Spermatid-specific expression of protamine-1 in transgenic mice. *Proc. Natl. Acad. Sci. USA* 84: 5316–5319.

Pinkerton, J. H. M., D. G. McKay, E. C. Adams and A. T. Hertig. 1961. Development of the human ovary: A study using histochemical techniques. *Obstet. Gynecol.* 18: 152–181.

Pratt, S. A., N. F. Scully and B. D. Shur. 1993. Cell surface β-1,4-galactosyltransferase on primary spermatocytes facilitates their initial adhesion to Sertoli cells in vitro. *Biol. Reprod.* 49: 470–482.

Profet, M. 1993. Menstruation as a defense against pathogens transported by sperm. *Q. Rev. Biol.* 68: 335–385.

Quaas, J. and C. Wylie. 2002. Surface contraction waves (SCWs) in the *Xenopus* egg are required for the localization of the germ plasm and are dependent upon maternal stores of the kinesin-like protein Xklp1. *Dev. Biol.* 243: 272–280.

Reijo, R. and 12 others. 1995. Diverse spermatogenic defects in humans caused by Y chromosome deletions encompassing a novel RNA-binding protein gene. *Nature Genet.* 10: 383–393.

Resnick, J. L., L. S. Bixler, L. Cheng and P. J. Donovan. 1992. Long-term proliferation of mouse primordial germ cells in culture. *Nature* 359: 550–551.

Ressom, R. E. and K. E. Dixon. 1988. Relocation and reorganization of germ plasm in *Xenopus* embryos after fertilization. *Development* 103: 507–518.

Riechmann, V., G. J. Gutierrez, P. Filardo, A. R. Nebreda and A. Ephrussi. 2002. Par-1 regulates stability of the posterior determinant Oskar by phosphorylation. *Nature Cell Biol.* 4: 337–342.

Rivier, C., J. Rivier and W. Vale. 1986. Inhibin-mediated feedback control of follicle-stimulating hormone secretion in the female rat. *Science* 234: 205–208.

Robb, D. L., J. Heasman, J. Raats and C. Wylie. 1996. A kinesin-like protein is required for germ plasm aggregation in *Xenopus. Cell* 87: 823–831.

Rohwedel, J., U. Sehlmeyer, J. Shan, A. Meister and A. M. Wobus. 1996. Primordial germ cell-derived mouse embryonic germ (EG) cells in vitro resemble undifferentiated stem cells with respect to differentiation capacity and cell cycle distribution. *Cell Biol. Int.* 20: 579–587.

Roller, R. J., R. A. Kinloch, B. Y. Hiraoka, S. S.-L. Li and P. M. Wassarman. 1989. Gene expression during mammalian oogenesis and early embryogenesis: Quantification of three messenger RNAs abundant in full grown mouse oocytes. *Development* 106: 251–261.

Romanoff, A. L. 1960. *The Avian Embryo.* Macmillan, New York.

Ruohola, H., K. A. Bremer, D. Baker, J. R. Swedlow, L. Y. Jan and Y. N. Jan. 1991. Role of neurogenic genes in establishment of follicle cell fate and oocyte polarity during oogenesis in *Drosophila. Cell* 66: 433–449.

Sagata, N., M. Oskarsson, T. Copeland, J. Brumbaugh and G. F. Vande Woude. 1988. Function of *c-mos* proto-oncogene product in meiotic maturation in *Xenopus* oocytes. *Nature* 335: 519–525.

Saitou, M., S. C. Barton and A. Surani 2002. A molecular programme for the specification of germ cell fate in mice. *Nature* 418: 293–300.

Savage, R. M. and M. V. Danilchik. 1993. Dynamics of germ plasm localization and its inhibition by ultraviolet irradiation in early cleavage *Xenopus* eggs. *Dev. Biol.* 157: 371–382.

Schmekel, K. and B. Daneholt. 1995. The central region of the synaptonemal complex revealed in three dimensions. *Trends Cell Biol.* 5: 239–242.

Schubiger, G. and W. J. Wood. 1977. Determination during early embryogenesis in *Drosophila melanogaster. Am. Zool.* 17: 565–576.

Seydoux, G. and A. Fire. 1994. Soma-germline asymmetry in the distributions of embryonic RNAs in *Caenorhabditis elegans. Development* 120: 2823–2834.

Seydoux, G. and S. Strome. 1999. Launching the germline in *Caenorhabditis elegans:* Regulation of gene expression in early germ cells. *Development* 126: 3275–3283.

Shamblott, M. J. and 8 others. 1998. Derivation of pluripotent stem cells from cultured human primordial germ cells. *Proc. Natl. Acad. Sci. USA* 95: 13726–13731.

Sheets, M. D., M. Wu and M. Wickens. 1995. Polyadenylation of *c-mos* mRNA as a control point in *Xenopus* meiotic maturation. *Nature* 374: 511–516.

Spitz, I. M. and C. W. Bardin. 1993. Mifepristone (RU486): A modulator of progestin and glucocorticoid action. *N. Engl. J. Med.* 329: 404–412.

Spradling, A. C. 1993. Germline cysts: Communes that work. *Cell* 72: 649–651.

Stewart, T. A. and B. Mintz. 1981. Successful generations of mice produced from an established culture line of euploid teratocarcinoma cells. *Proc. Natl. Acad. Sci. USA* 78: 6314–6318.

Stott, D. and C. C. Wylie. 1986. Invasive behaviour of mouse primordial germ cells in vitro. *J. Cell Sci.* 86: 133–144.

Strassmann, B. I. 1996. The evolution of the endometrial cycles and menstruation. *Q. Rev. Biol.* 71: 181–220.

Subramanian, K. and G. Seydoux. 1999. *nos-1* and *nos-2,* two genes related to *Drosophila nanos,* regulate primordial germ cells development and survival in *Caenorhabditis elegans. Development* 126: 4861–4871.

Subtelny, S. and J. E. Penkala. 1984. Experimental evidence for a morphogenetic role in the emergence of primordial germ cells from the endoderm of *Rana pipiens. Differentiation* 26: 211–219.

Sun, Y.-A. and R. J. Wyman. 1993. Reevaluation of electrophoresis in the *Drosophila* egg chamber. *Dev. Biol.* 155: 206–215.

Sutasurya, L. A. and P. D. Nieuwkoop. 1974. The induction of primordial germ cells in the urodeles. *Wilhelm Roux Arch. Entwicklungsmech. Org.* 175: 199–220.

Swanson, C. P., T. Merz and W. J. Young. 1981. *Cytogenetics: The Chromosome in Division, Inheritance and Evolution.* Prentice-Hall, Englewood Cliffs, NJ.

Swift, C. H. 1914. Origin and early history of the primordial germ-cells in the chick. *Am. J. Anat.* 15: 483–516.

Tadokoro, Y., K. Yomogida, H. Ohta, A. Tohda and Y. Nishimune. 2002. Homeostatic regulation of germinal stem cell proliferation by the GDNF/FSH pathway. *Mech. Dev.* 113: 29–39.

Tax, F. E., J. J. Yeargers and J. H. Thomas. 1994. Sequence of *C. elegans* Lag-2 reveals a cell-signaling domain shared with Delta and Serrate of *Drosophila. Nature* 368: 150–154.

Tay, J. and J. D. Richter. 2001. Germ cell differentiation and synaptonemal complex formation are disrupted in CPEB knockout mice. *Dev. Cell* 1: 201–213.

Telfer, W. H., R. I. Woodruff and E. Huebner. 1981. Electrical polarity and cellular differentiation in meroistic ovaries. *Am. Zool.* 21: 675–686.

Theurkauf, W. E., S. Smiley, M. L. Wong and B. M. Alberts. 1992. Reorganization of the cytoskeleton during *Drosophila* oogenesis: Implications for axis specification and intercellular transport. *Development* 115: 923–936.

Theurkauf, W. E., B. M. Alberts, Y. N. Jan and T. A. Jongens. 1993. A control code for microtubules in the differentiation of *Drosophila* oocytes. *Development* 118: 1169–1180.

Tobler, H., K. D. Smith and H. Ursprung. 1972. Molecular aspects of chromatin elimination in *Ascaris lumbricoides. Dev. Biol.* 27: 190–203.

Tsunekawa, N., M. Naito, T. Nidhida and T. Noce. 2000. Isolation of chicken *vasa* homolog gene and tracing the origin of primordial germ cells. *Development* 127: 2741–2750.

Van Doren, M., H. T. Broihier, L. A. Moore and R. Lehmann. 1998. HMG-CoA reductase guides migrating primordial germ cells. *Nature* 396: 466–469.

Van Eeden, F. J. M., I. M. Palacios, M. Petronczki, M. J. D. Weston and D. St. Johnston. 2001. Barentsz is essential for the posterior localization of *oskar* mRNA and co-localizes with it to the posterior pole. *J. Cell Biol.* 154: 511–523.

Vegeto, E., G. F. Allan, W. T. Schrader, M.-J. Tsai, D. P. McDonnell and B. W. O'Malley. 1992. The mechanism of RU486 antagonism is dependent on the conformation of the carboxy-terminal tail of the human progesterone receptor. *Cell* 69: 703–713.

Villeneuve, A. and K. J. Hillers. 2001. Whence meiosis? *Cell* 106: 647–650.

von Wettstein, D. 1971. The synaptonemal complex and four-strand crossing over. *Proc. Natl. Acad. Sci. USA* 68: 851–855.

von Wettstein, D. 1984. The synaptonemal complex and genetic segregation. *In* C. W. Evans and H. G. Dickinson (eds.), *Controlling Events in Meiosis.* Cambridge University Press, Cambridge, pp. 195–231.

Waddington, C. H. 1966. *Principles of Development and Differentiation.* Macmillan, New York.

Wakahara, M. 1996. Primordial germ cell development: Is the urodele pattern closer to mam-

mals than to anurans? *Int. J. Dev. Biol.* 40: 653–659.

Warrior, R. 1994. Primordial germ cell migration and the assembly of the *Drosophila* embryonic gonad. *Dev. Biol.* 166: 180–194.

Watanabe, N., G. F. Vande Woude, Y. Ikawa and N. Sagata. 1989. Specific proteolysis of the *c-mos* proto-oncogene product by calpain on fertilization of *Xenopus* eggs. *Nature* 342: 505–517.

Watson, C. A., I. Sauman and S. J. Berry. 1993. Actin is a major structural and functional element of the egg cortex of giant silkmoths during oogenesis. *Dev. Biol.* 155: 315–323.

Whitington, P. M. and K. E. Dixon. 1975. Quantitative studies of germ plasm and germ cells during early embryogenesis of *Xenopus laevis. J. Embryol. Exp. Morphol.* 33: 57–74.

Woodruff, T. K., J. D'Agostino, N. B. Schwartz and K. E. Mayo. 1988. Dynamic changes in inhibin messenger RNAs in rat ovarian follicles during the reproductive cycle. *Science* 239: 1296–1299.

Wylie, C. 1999. Germ cells. *Cell* 96: 165–174.

Wylie, C. C. and J. Heasman. 1993. Migration, proliferation, and potency of primordial germ cells. *Semin. Dev. Biol.* 4: 161–170.

Wylie, C. C., J. Heasman, A. P. Swan and B. H. Anderton. 1979. Evidence for substrate guidance of primordial germ cells. *Exp. Cell Res.* 121: 315–324.

Yeom, Y. I. and 7 others. 1996. Germline regulatory element of *Oct-4* specific for the totipotent cycle of embryonal cells. *Development* 122: 881–894.

Ying, Y., X. M. Liu, A. Marble, K. A. Lawson and G. Q. Zhao. 2000. Requirement of Bmp8b for the generation of primordial germ cells in the mouse. *Mol. Endocrinol.* 14: 1053–1063.

Yisraeli, J. K., S. Sokol and D. A. Melton. 1990. A two-step model for the localization of a maternal mRNA in *Xenopus* oocytes: Involvement of microtubules and microfilaments in translocation and anchoring of Vg1 mRNA. *Development* 108: 289–298.

Yuan, L., M.-R. Hoja, J. Wilbertz, K. Nordqvist and C. Höög. 2002. Female germ cell aneuploidy and embryo death in mice lacking the meiosis-specific protein SCP3. *Science* 296: 1115–1118.

Zhang, N., J. Zhang, K. J. Purcell, Y. Cheng and K. Howard. 1997. The *Drosophila* protein Wunen repels migratory germ cells. *Nature* 385: 64–67.

Zhao, G.-Q., K. Deng, P. A. Labosky, L. Liaw and B. L. M. Hogan. 1996. The gene encoding bone morphogenetic protein 8B is required for the initiation and maintenance of spermatogenesis in the mouse. *Genes Dev.* 10: 1657–1669.

Zhou, Y. and M. L. King. 1996. Localization of *Xcat-2* RNA, a putative germ plasm component, to the mitochondrial cloud in *Xenopus* stage I oocytes. *Development* 122: 2947–2953.

PART IV Further Ramifications of Developmental Biology

20 An overview of plant development
21 Medical implications of developmental biology
22 Environmental regulation of animal development
23 Developmental mechanisms of evolutionary change

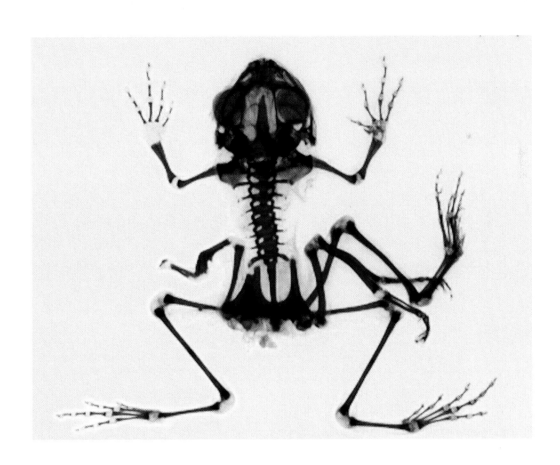

Environmental factors can alter development. When the larvae of trematode parasites enter the hindlimb fields of developing tadpoles, they split these fields. The cells within the separated fields regulate to form two or more limbs coming from the same site. The inability to reject these parasites may involve exposure of the tadpole to pollutants or ultraviolet radiation. (Figure 16.3 courtesy of Dr. S. Sessions).

chapter 20 An overview of plant development

THE FUNDAMENTAL QUESTIONS in developmental biology are similar for plants and animals. Their developmental strategies, which have evolved over millions of years, have many elements in common; however, some of the challenges and solutions found in plants are sufficiently unique to warrant a separate discussion in this chapter.

The term *plant* loosely encompasses many organisms, from algae to flowering plants (angiosperms). While this chapter focuses primarily on the flowering plants, their developmental strategies are best understood in an evolutionary context. Recent phylogenetic studies show a common lineage for all green plants, distinct from the red and brown plants (Figure 20.1). Land plants have their origins in the freshwater green algae (specifically the charophytes), and the transition to land correlates with the evolution of an increasingly protected embryo. Mosses, ferns, gymnosperms (conifers, cycads, and ginkgos), and angiosperms (flowering plants) all develop from protected embryos. Two examples of embryo protection are the seed coat that first appeared in the gymnosperms and the fruit that characterizes the angiosperms. As we have seen, embryo protection is also a theme in animal development. What are the fundamental differences between animals and land plants?

1. **Plant cells do not migrate.** Plant cells are trapped within rigid cellulose walls that generally prevent cell and tissue migration. Plants, like most metazoan animals, develop three basic tissue systems (dermal, ground, and vascular), but do not rely on gastrulation to establish this layered system of tissues. Plant development is highly regulated by the environment, a strategy that is adaptive for a stationary organism.

2. **Plants have sporic meiosis rather than gametic meiosis.** That is, spores, not gametes, are produced by meiosis in plants. Gametes are produced by mitotic divisions following meiosis.

3. **The life cycle of land plants (as well as many other plants) includes both diploid and haploid multicellular stages.** This type of life cycle is referred to as **alternation of generations** and results in two different multicellular body plans over the lifecycle of an individual. The evolutionary trend has been toward a reduction in the size of the haploid generation.

4. **Germ cells are not set aside early in development.** While this is also the case in several animal phyla, it is the case for all plants.

5. **Plants undergo extended morphogenesis.** Clusters of actively dividing cells called **meristems**, which are similar to stem cells in animals, persist long after maturity. Meristems allow for reiterative development and the formation of new structures throughout the life of the plant.

(A)

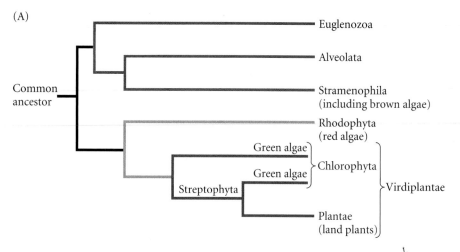

(B)

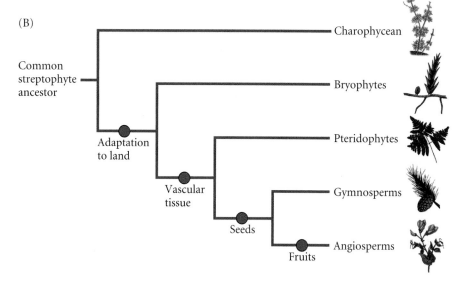

Figure 20.1
Evolutionary origins of plants. (A) Land plants and green algae are more closely related to each other than they are to red and brown algae. The green algae are not monophyletic, and only one branch gave rise to the land plants. The Streptophyta is a monophyletic lineage that includes one line of green algae, along with the multicellular lineages that were once considered the kingdom Plantae. A new kingdom, Virdiplantae, has been proposed that includes all green algae and the land plants. (B) Adaptation to land resulted in the evolution of vascular tissue, seeds, and fruits.

6. **Plants have tremendous developmental plasticity.** Although cloning in animals illustrates their plasticity, plants depend heavily on this developmental strategy. For example, if a shoot is grazed by herbivores, meristems in the leaf often grow out to replace the lost part. (This strategy has similarities to the regeneration seen in some animals.) Whole plants can even be regenerated from some single cells. In addition, a plant's form (including branching, height, and relative amounts of vegetative and reproductive structures) is greatly influenced by environmental factors such as light and temperature, and a wide range of morphologies can result from the same genotype. This amazing level of plasticity may help compensate for the plant's lack of mobility.

7. **Plants are more tolerant of aneuploidy and polyploidy than animals.** Aneuploidy and polyploidy tend to be developmentally harmful to animals. When plants are aneuploid or polyploid, the consequences can be adaptive. Many flowers found in the florist shop and the wheat used for bread flour are examples of successful polyploids.

Huge amounts of "hitchhiking" DNA, often in the form of retrotransposons, can also account for enlarged genomes in both plants and animals. For example, half of the maize (corn) genome appears to be made up of foreign DNA (SanMiguel et al. 1996). Comparisons of the completed genomes of *Arabidopsis* (a model plant system), rice, humans, and mice should provide insight into the possible developmental significance (including similarities or differences) of the vast amounts of noncoding DNA in both plants and animals (*Arabidopsis* Genome Initiative 2000; Genome International Sequencing Consortium 2001; Goff et al. 2002; Mouse Genome Sequencing Consortium 2002; Yu et al 2002).

8. **Developmental mechanisms evolved independently in plants and animals.** Genome-level comparisons indicate that there is minimal homology between genes and proteins used to establish body plans in plants and animals (Meyerowitz 2002). Although Hox and MADS box genes were present in the last common ancestor of plants and animals, the MADS box family controls many major developmental regulatory processes in plants, but not in animals. Hox-like genes, however, are important regulators of both plant and animal development.

Despite the major differences between plants and animals, developmental genetic studies are revealing some commonalities in the logic of pattern formation in the two groups, along with evolutionarily distinct solutions to the problem of creating three-dimensional form from a single cell.

Plant Life Cycles

Plants have both multicellular haploid and multicellular diploid stages in their life cycles. Embryonic development is seen only in the diploid generation. The embryo, however, is produced by the fusion of gametes, which are formed only by the haploid generation. Understanding the relationship between the two generations is important in the study of plant development.

In plants, gametes develop in the multicellular haploid **gametophyte** (Greek *phyton*, "plant"). Fertilization gives rise to a multicellular diploid **sporophyte**, which produces haploid spores via meiosis. This type of life cycle is called a **haplodiplontic** life cycle (Figure 20.2). It differs from the **diplontic** life cycle of most animals, in which only the gametes are in the haploid state. In haplodiplontic life cycles, gametes are not the direct result of a meiotic division. Diploid sporophyte cells undergo meiosis to produce haploid **spores**. Each spore then goes through mitotic divisions to yield a multicellular, haploid gametophyte. Mitotic divisions within the gametophyte are required to produce the gametes. The diploid sporophyte results from the fusion of two gametes. Among the Plantae, the gametophytes and sporophytes of a species have distinct morphologies (in some algae they look alike). How a single genome can be used to create two unique morphologies is an intriguing puzzle.

All plants alternate generations. There is an evolutionary trend from sporophytes that are nutritionally dependent on autotrophic (self-feeding) gametophytes to the opposite—gametophytes that are dependent on autotrophic sporophytes. This trend is exemplified by comparing the life cycles of a moss, a fern, and an angiosperm (see Figures 20.3–20.5).

(Gymnosperm life cycles bear many similarities to those of angiosperms; the distinctions will be explored in the context of angiosperm development.)

The "leafy" moss you walk on in the woods is the gametophyte generation of that plant (Figure 20.3). Mosses are **heterosporous**, which means they make two distinct types of spores; these develop into male and female gametophytes. Male gametophytes develop reproductive structures called **antheridia** (singular, antheridium) that produce sperm by mitosis. Female gametophytes develop **archegonia** (singular, archegonium) that produce eggs by mitosis. Sperm travel to a neighboring plant via a water droplet, are chemically attracted to the entrance of the archegonium, and fertilization results.[*] The embryonic sporophyte develops within the archegonium, and the mature sporophyte stays attached to the gametophyte. The sporophyte is not photosynthetic. Thus both the embryo and the mature sporophyte are nourished by the gametophyte. Meiosis within the capsule of the sporophyte yields haploid spores that are released and eventually germinate to form a male or female gametophyte.

Ferns follow a pattern of development similar to that of mosses, although most (but not all) ferns are **homosporous**. That is, the sporophyte produces only one type of spore within a structure called the **sporangium** (Figure 20.4). A single gametophyte can produce both male and female sex organs. The greatest contrast between the mosses and the ferns is that both the gametophyte and the sporophyte of the fern photosynthesize and are thus autotrophic; the shift to a dominant sporophyte generation is taking place.[†]

At first glance, angiosperms may appear to have a diplontic life cycle because the gametophyte generation has been reduced to just a few cells (Figure 20.5). However, mitotic division still follows meiosis in the sporophyte, resulting in a multicellular gametophyte, which produces eggs or sperm. All of this takes place in the organ that characterizes the angiosperms: the flower. Male and female gametophytes have distinct morphologies (i.e., angiosperms are heterosporous), but the gametes they produce no longer rely on water for fertilization. Rather, wind or members of the animal kingdom deliver the male gametophyte—**pollen**—to the female gametophyte. Another evolutionary innovation found in the gymnosperms and angiosperms is the production of a seed coat, which adds an extra layer of protection around the embryo. A further protective layer, the fruit, is unique to the angiosperms and aids in the dispersal of the enclosed embryos by wind or animals.

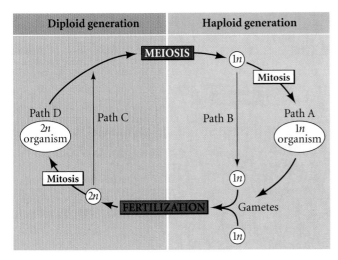

Figure 20.2
Plants have haplodiplontic life cycles that involve mitotic divisions (resulting in multicellularity) in both the haploid and diploid generations (paths A and D). Most animals are diplontic and undergo mitosis only in the diploid generation (paths B and D). Multicellular organisms with haplontic life cycles follow paths A and C.

---

[*]Have you ever wondered why there are no moss trees? Aside from the fact that the gametophytes of mosses (and other plants) do not have the necessary structural support and transport systems to attain tree height, it would be very difficult for a sperm to swim up a tree!

[†]It *is* possible to have tree ferns, for two reasons. First, the gametophyte develops on the ground, where water can facilitate fertilization. Second, unlike mosses, the fern sporophyte has vascular tissue, which provides the support and transport system necessary to achieve substantial height.

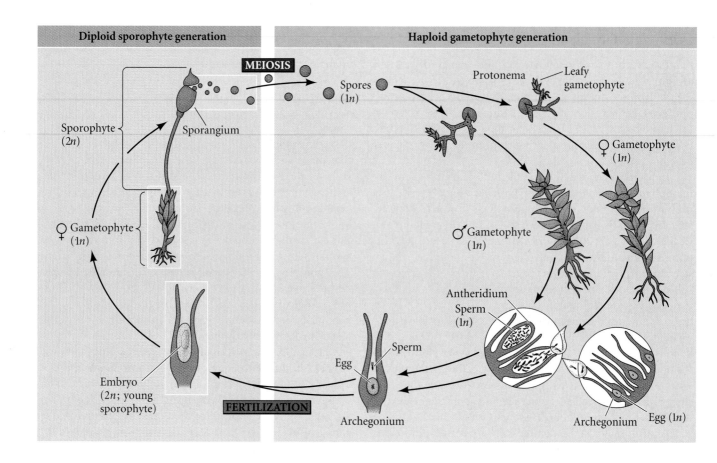

Diploid sporophyte generation

Sporophyte (2n)

Sporangium

MEIOSIS

Spores (1n)

Protonema — Leafy gametophyte

♀ Gametophyte (1n)

Haploid gametophyte generation

♂ Gametophyte (1n)

Antheridium
Sperm (1n)

Sperm

Egg

Archegonium

Egg (1n)

Archegonium

Embryo (2n; young sporophyte)

FERTILIZATION

♀ Gametophyte (1n)

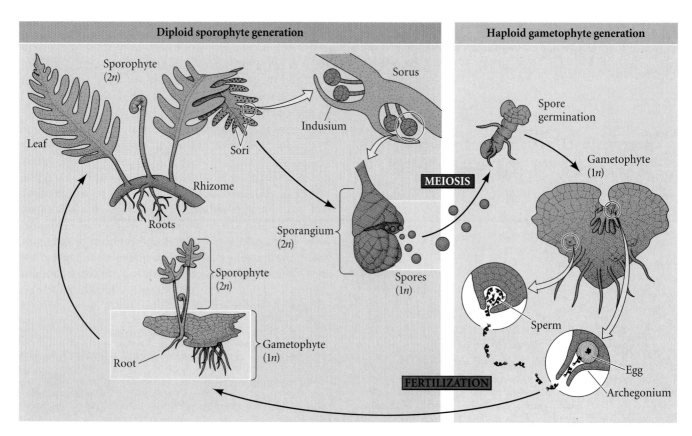

Diploid sporophyte generation

Sporophyte (2n)

Leaf

Sori

Rhizome

Roots

Sorus

Indusium

Sporangium (2n)

Spores (1n)

MEIOSIS

Haploid gametophyte generation

Spore germination

Gametophyte (1n)

Sperm

Egg

Archegonium

FERTILIZATION

Sporophyte (2n)

Root

Gametophyte (1n)

◀ **Figure 20.3**

Life cycle of a moss (genus *Polytrichum*). The sporophyte generation is dependent on the photosynthetic gametophyte for nutrition. Cells within the sporangium of the sporophyte undergo meiosis to produce male and female spores, respectively. These spores divide mitotically to produce multicellular male and female gametophytes. Differentiation of the growing tip of the gametophyte produces antheridia in males and archegonia in females. The sperm and eggs are produced mitotically in the antheridia and archegonia, respectively. Sperm are carried to the archegonia in water droplets. After fertilization, the sporophyte generation develops in the archegonium and remains attached to the female gametophyte.

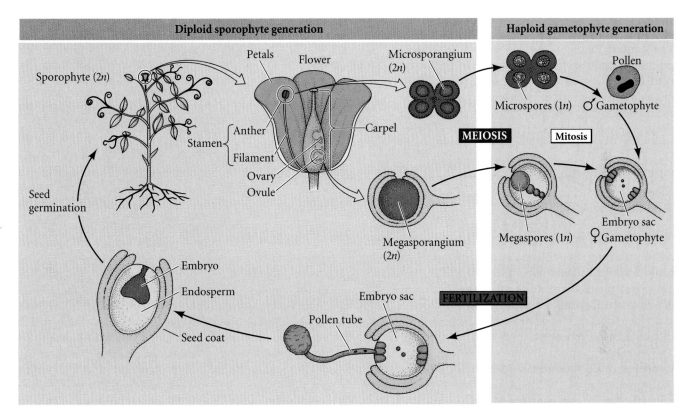

▲ **Figure 20.5**

Life cycle of an angiosperm, represented here by a pea plant (genus *Pisum*). The sporophyte is the dominant generation, but multicellular male and female gametophytes are produced within the flowers of the sporophyte. Cells of the microsporangium within the anther undergo meiosis to produce microspores. Subsequent mitotic divisions are limited, but the end result is a multicellular pollen grain. The megasporangium is protected by two layers of integuments and the ovary wall. Within the megasporangium, meiosis yields four megaspores—three small and one large. Only the large megaspore survives to produce the embryo sac. Fertilization occurs when the pollen germinates and the pollen tube grows toward the embryo sac. The sporophyte generation may be maintained in a dormant state, protected by the seed coat.

◀ **Figure 20.4**

Life cycle of a fern (genus *Polypodium*). The sporophyte generation is photosynthetic and is independent of the gametophyte. The sporangia are protected by a layer of cells called the indusium. This entire structure is called a sorus. Meiosis within the sporangia yields a haploid spore. Each spore divides mitotically to produce a heart-shaped gametophyte, which differentiates both archegonia and antheridia on one individual. The gametophyte is photosynthetic and independent, although it is smaller than the sporophyte. Fertilization takes place when water is available for sperm to swim to the archegonia and fertilize the eggs. The sporophyte has vascular tissue and roots; the gametophyte does not.

The remainder of this chapter provides a detailed exploration of angiosperm development from fertilization to senescence. Keep in mind that the basic haplodiplontic life cycle seen in the mosses and ferns is also found in the angiosperms, continuing the trend toward increased nourishment and protection of the embryo.

Gamete Production in Angiosperms

Like those of mosses and ferns, angiosperm gametes are produced by the gametophyte generation. Angiosperm gametophytes are associated with flowers. The gametes they produce join to form the sporophyte. The study of embryonic development in plants is therefore the study of early sporophyte development. In angiosperms, the sporophyte is what is commonly seen as the plant body. The shoot meristem of the sporophyte produces a series of vegetative structures. At a certain point in development, internal and external signals trigger a shift from vegetative to reproductive (flower-producing) development (see reviews by McDaniel et al. 1992, Levy and Dean 1998, and Simpson et al. 1999).

Once the meristem becomes floral, it initiates the development of floral parts sequentially in whorls of organs modified from leaves (Figure 20.6). The first and second whorls become **sepals** and **petals**, respectively; these organs are sterile. The pollen-producing **stamens** are initiated in the third whorl of the flower. The **carpel** in the fourth whorl contains the female gametophyte. The stamens contain four groups of cells, called the **microsporangia** (pollen sacs), within an **anther**. The microsporangia undergo meiosis to produce **microspores**. Unlike most ferns, angiosperms are heterosporous, so the prefix *micro* is used to identify the spores that mitotically yield the male gametophytes—pollen grains. The inner wall of the pollen sac, the **tapetum**, provides nourishment for the developing pollen.

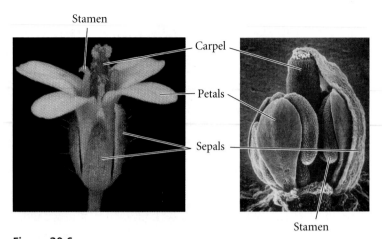

Figure 20.6
An *Arabidopsis* flower, illustrating the major organs (petals, sepals, stamen, and carpel) of the typical angiosperm flower. The micrograph on the right shows the dissected flower. Tremendous morphological variation is possible in all four organs; this variation appears to be related to reproductive strategies. (Photographs courtesy of J. Bowman.)

Pollen

The pollen grain is an extremely simple multicellular structure. The outer wall of the pollen grain, the **exine**, is composed of resistant material provided by both the tapetum (sporophyte generation) and the microspore (gametophyte generation). The inner wall, the **intine**, is produced by the microspore. A mature pollen grain consists of two cells, one within the other (Figure 20.7). The **tube cell** contains a **generative cell** within it. The generative cell divides to produce two sperm. The tube cell nucleus guides pollen germination and the growth of the pollen tube after the pollen lands on the stigma of a female gametophyte. One of the two sperm will fuse with the egg cell to produce the next sporophyte generation. The second sperm will participate in the formation of the endosperm, a structure that provides nourishment for the embryo.

(A)

(B)

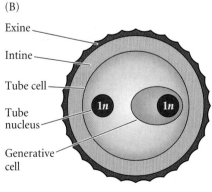

Figure 20.7
(A) Pollen grains have intricate surface patterns, as seen in this scanning electron micrograph of aster pollen. (B) A pollen grain consists of a cell within a cell. The generative cell will undergo division to produce two sperm cells. One will fertilize the egg, and the other will join with the polar nuclei, yielding the endosperm.

The ovary

The fourth whorl of organs within the flower forms the **carpel**, which gives rise to the female gametophyte (Figure 20.8). The carpel consists of the **stigma** (where the pollen lands), the **style**, and the **ovary**. Following fertilization, the ovary wall will develop into the **fruit**. This unique angiosperm structure provides further protection for the developing embryo and also enhances seed dispersal by frugivores (fruit-eating animals). Within the ovary are one or more **ovules** attached by a **placenta** to the ovary wall. Fully developed ovules are called **seeds**. The ovule has one or two outer layers of cells called the **integuments**. These enclose the **megasporangium**, which contains sporophyte cells that undergo meiosis to produce **megaspores** (see Figure 20.5). There is a small opening in the integuments, called the **micropyle**, through which the pollen tube will grow. The integuments—an innovation first appearing in the gymnosperms—develop into the **seed coat**, which protects the embryo by providing a waterproof physical barrier. When the mature embryo disperses from the parent plant, diploid sporophyte tissue accompanies the embryo in the form of the seed coat and the fruit.

Within the ovule, meiosis and unequal cytokinesis yield four megaspores. The largest of these megaspores undergoes three mitotic divisions to produce a 7-celled **embryo sac** with eight nuclei (Figure 20.9). One of these cells is the egg. The two **synergid cells** surrounding the egg may be evolutionary remnants of the archegonium (the female sex organ seen in mosses and ferns). The **central cell** contains two or more **polar nuclei**, which will fuse with the second sperm nucleus and develop into the polyploid endosperm. Three **antipodal cells** form at the opposite end of the embryo sac from the synergids and degenerate before or during embryonic development. These cells have no known function. Genetic analyses of female gametophyte development in maize and *Arabidopsis** are providing insight into the regulation of the specific steps in this process (Drews et al. 1998).

Pollination

Pollination refers to the landing and subsequent germination of the pollen on the stigma. Hence it involves an interaction between the gametophyte generation of the male (the pollen) and the sporophyte generation of the female (the stigmatic surface of the carpel). Pollination can occur within a single flower (self-fertilization), or pollen can land on a different flower on the same or a different plant. About 96% of flowering plant species produce male and female gametophytes on the same plant. However, about 25% of these produce two different types of flowers on the same plant, rather than **perfect flowers** containing both male and female gametophytes. **Staminate** flowers lack carpels, while **carpellate** flowers lack stamens. Maize plants, for example, have staminate (tassel) and

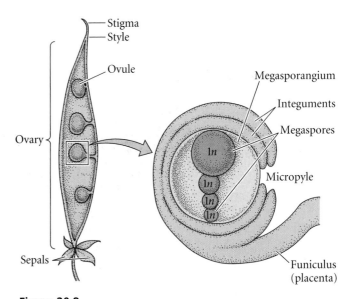

Figure 20.8
The carpel contains one or more ovules; these contains megasporangia protected by two layers of integument cells. The megasporangia divide meiotically to produce haploid megaspores, All of the carpel is diploid except for the megaspores, which divide mitotically to produce the embryo sac (the female gametophyte).

carpellate (ear) flowers on the same plant. Such species are considered to be **monoecious** (Greek *mono*, "one"; *oecos*, "house"). The remaining 4% of species (willows, for example) produce staminate and carpellate flowers on separate plants. These species are considered to be **dioecious** ("two houses"). Only a few plant species have true sex chromosomes. The terms "male" and "female" are most correctly applied only to the gametophyte generation of heterosporous plants, not to the sporophyte (Cruden and Lloyd 1995).

The arrival of a viable pollen grain on a receptive stigma does not guarantee fertilization. **Interspecific incompatibility**

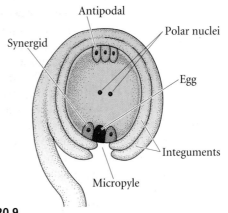

Figure 20.9
The embryo sac is the product of three mitotic divisions of the haploid megaspore. Two of the nuclei are contained within the central cell; the other six cells contain one haploid nucleus each.

---

*A small weed in the mustard family, *Arabidopsis* is used as a model system because of its very small genome.

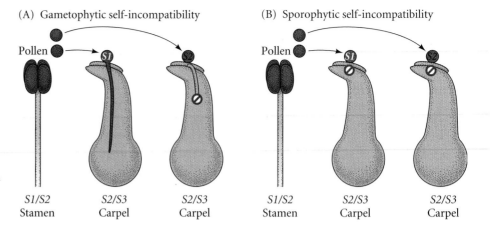

Figure 20.10
Self-incompatibility. *S1*, *S2*, and *S3* are different alleles of the self-incompatibility (*S*) locus. (A) Plants with gametophytic self-incompatibility reject pollen only when the genotype of the pollen matches one of the carpel's two alleles. (B) In sporophytic self-incompatibility, the genotype of the pollen parent, not just of the haploid pollen grain itself, can trigger an incompatibility response.

refers to the failure of pollen from one species to germinate and/or grow on the stigma of another species (for a review, see Taylor 1996). **Intraspecific incompatibility** is incompatibility that occurs within a species. **Self-incompatibility**—incompatibility between the pollen and the stigmas of the same individual—is an example of intraspecific incompatibility. Self-incompatibility blocks fertilization between two genetically similar gametes, increasing the probability of new gene combinations by promoting outcrossing (pollination by a different individual of the same species). Groups of closely related plants can contain a mix of self-compatible and self-incompatible species.

Several different systems for self-incompatibility have evolved (Figure 20.10). Recognition of self depends on the multiallelic self-incompatibility (*S*) locus (Dodds et al. 1996; Gaude and McCormick 1999; Nasrallah 2002). Gametophytic self-incompatibility occurs when the *S* allele of the pollen grain matches either of the *S* alleles of the stigma (remember that the stigma is part of the diploid sporophyte generation, which has two *S* alleles). In this case, the pollen tube begins developing, but stops before reaching the micropyle. Sporophytic self-incompatibility occurs when one of the two *S* alleles of the pollen-producing sporophyte (not the gametophyte) matches one of the *S* alleles of the stigma. Most likely, sporophyte contributions to the exine are responsible.

The *S* locus consists of several physically linked genes that regulate recognition and rejection of pollen. An *S* gene has been cloned that codes for an RNAse (called S RNAse), which in the gametophytically self-incompatible petunia pistil is sufficient to recognize and reject self-pollen (Lee et al. 1994). The pollen component recognized by S RNAse is most likely a different gene in the *S* locus, but it has not yet been positively identified in either gametophytically or sporophytically self-incompatible plants. In sporophytic self-incompatibility, a ligand on the pollen is thought to bind to a membrane-bound kinase receptor in the stigma, starting a signaling process that leads to pollen rejection. In *Brassica*, one of the proteins coded for by the *S* locus genes is a transmembrane serine-threonine kinase (SRK) that functions in the epidermis of the stigma and binds a cysteine-rich peptide (SCR) from the pollen (Figure 20.11; Kachroo et al. 2001).

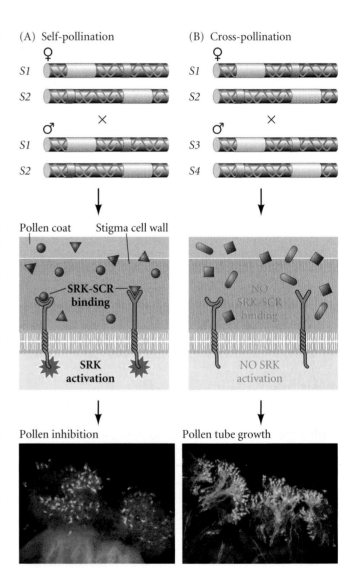

Figure 20.11
Receptor-ligand self-recognition is the key to self-incompatibility in *Brassica*. Allelic variability in both the *SRK* and *SRC* genes leads to a variety of possible combinations of ligand and receptor proteins. Unlike other self-recognition systems (including immunity and mating) self-incompatibility results from the binding of SRK and SRC proteins of self (from allelic *S* loci) rather than nonself. (After Nasrallah 2002; photographs courtesy of June Nasrallah.)

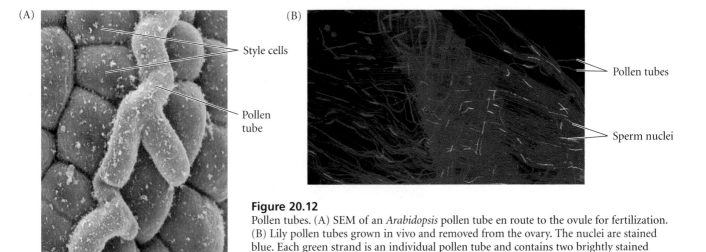

(A)

Style cells

Pollen tube

(B)

Pollen tubes

Sperm nuclei

Figure 20.12
Pollen tubes. (A) SEM of an *Arabidopsis* pollen tube en route to the ovule for fertilization. (B) Lily pollen tubes grown in vivo and removed from the ovary. The nuclei are stained blue. Each green strand is an individual pollen tube and contains two brightly stained sperm nuclei and a fainter tube cell nucleus. Note the huge number of pollen tubes, all "racing" to fertilize a single egg (and the polar nuclei). (Photographs courtesy of E. Lord.)

There are numerous examples of plant populations that have switched from self-incompatible to self-fertilizing systems. Changes in the *S* locus, specifically the *SKR* and *SCR* genes, could account for these evolutionary changes. The Nasrallahs (Nasrallah et al. 2002) created self-incompatible *Arabidopsis thaliana* plants (normally self-compatible) by introducing the *SKR* and *SCR* genes that encode self-recognizing proteins from *A. lyrata* (a self-incompatible species). This experiment demonstrates that *A. thaliana* still has all of the downstream components of the signal cascade that can lead to pollen degradation. The mechanism of pollen degradation is unclear, but appears to be highly specific.

If the pollen and the stigma are compatible, the pollen takes up water (hydrates) and the pollen tube emerges. The pollen tube grows down the style of the carpel toward the micropyle (Figure 20.12). The tube nucleus and the sperm cells are kept at the growing tip by bands of callose (a complex carbohydrate). This may possibly be an exception to the "plant cells do not move" rule, as the generative cell(s) appear to move ahead via adhesive molecules (Lord et al. 1996). Pollen tube growth is quite slow (up to a year) in gymnosperms, while in some angiosperms the tube can grow as rapidly as 1 cm per hour.

Calcium has long been known to play an essential role in pollen tube growth (Brewbaker and Kwack 1963). Calcium accumulates in the tip of the pollen tube, where open calcium channels are concentrated (Jaffe et al. 1975; Trewavas and Malho 1998). There is direct evidence that pollen tube growth in the field poppy is regulated by a slow-moving calcium wave controlled by the phosphoinositide signaling pathway (Figure 20.13; Franklin-Tong et al. 1996). Calcium influx occurs at both the tip of the pollen tube and on the shanks and altered calcium influx is observed when the pollen tube is self-incompatible with the style (Franklin-Tong et al. 2002). Cytoskeletal investigations show that organelle positioning during pollen

tube growth depends on interactions with cytoskeletal components. This must link to signaling, but the specifics are still unknown (Cai and Cresti 1999).

Genetic approaches have been useful in investigating how the growing pollen tube is guided toward unfertilized ovules. In *Arabidopsis*, the pollen tube appears to be guided by a long-

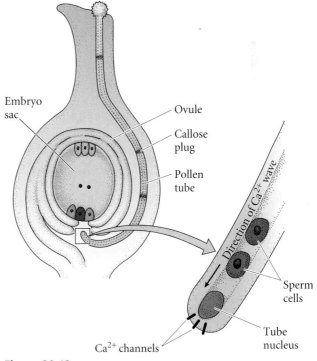

Embryo sac

Ovule

Callose plug

Pollen tube

Direction of Ca²⁺ wave

Sperm cells

Tube nucleus

Ca²⁺ channels

Figure 20.13
Calcium and pollen tube tip growth. After compatible pollen germinates, the pollen tube grows toward the micropyle. Calcium plays a key role in the growth of the tube. (After Franklin-Tong et al. 1996.)

distance signal from the ovule (Hulskamp et al. 1995; Wilhelmi and Preuss 1999). Analysis of pollen tube growth in ovule mutants of *Arabidopsis* indicates that the haploid embryo sac is particularly important in the long-range guidance of pollen tube growth. Mutants with defective sporophyte tissue in the ovule but a normal haploid embryo sac appear to stimulate normal pollen tube development.

While the evidence points primarily to the role of the gametophyte generation in pollen tube guidance, diploid cells may make some contribution. Two *Arabidopsis* genes, *POP2* and *POP3*, have been identified that specifically guide pollen tubes to the ovule with no other apparent effect on the plant (Wilhelmi and Preuss 1996, 1999). These genes function in both the pollen and the pistil, thus implicating the sporophyte generation in the guidance system.

The two synergid cells in the embryo sac (see Figure 20.15) may attract the pollen tube as the final step in pollen guidance. In *Torenia fournieri*, the embryo sac protrudes from the micropyle and can be cultured. In vitro, it can attract a pollen tube. Higashiyama and colleagues (2001) used a lasar beam to destroy individual cells in the embryo sac and then tested whether or not pollen tubes were still attracted to the embryo sac. When both synergids were destroyed, pollen tubes were not attracted to the embryo sac, but a single synergid was sufficient to guide pollen tubes.

Fertilization

The growing pollen tube enters the embryo sac through the micropyle and grows through one of the synergids. The two sperm cells are released, and a **double fertilization** event occurs (reviewed by Southworth 1996). One sperm cell fuses with the egg, producing the zygote that will develop into the sporophyte. The second sperm cell fuses with the bi- or multinucleate central cell, giving rise to the **endosperm**, which nourishes the developing embryo. This second event is not true fertilization in the sense of male and female gametes undergoing syngamy (fusion). That is, it does not result in a zygote, but in nutritionally supportive tissue. (When you eat popcorn, you are eating "popped" endosperm.) The other accessory cells in the embryo sac degenerate after fertilization.

The zygote of the angiosperm produces only a single embryo; the zygote of the gymnosperm, on the other hand, produces two or more embryos after cell division begins, by a process known as cleavage embryogenesis. Double fertilization, first identified a century ago, is generally restricted to the angiosperms, but it has also been found in the gymnosperms *Ephedra* and *Gnetum*, although no endosperm forms. Friedman (1998) has suggested that endosperm may have evolved from a second zygote "sacrificed" as a food supply in a gym-

Figure 20.14
Amborella trichopoda. This plant is more closely related to the first angiosperm than any other extant species. (Photograph courtesy of Sandra K. Floyd.)

nosperm with double fertilization. Investigations of the most closely related extant relative of the basal angiosperm, *Amborella*, should provide information on the evolutionary origin of the endosperm (Figure 20.14; Brown 1999). An analysis of embryo sacs of extant basal angiosperms has shown that it is equally probable that the first angiosperm had a 4-nuclei embryo sac as that it had an 8-nuclei embryo sac (Figure 20.15; Williams and Friedman 2002). The critical cell to consider is the central cell, which is fertilized by the second

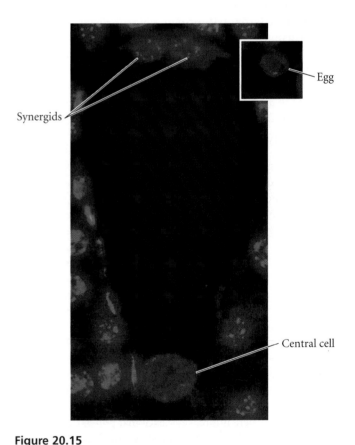

Figure 20.15
Ancestral angiosperms may have had 2*n* endosperms. The embryo sac of *Nuphar* has a single nucleus in its central cell which is fertilized to produce a 2*n* endosperm. DAPI staining was used to show that the DNA content was 1*n*, not 2*n*. Because this is a section of tissue, the egg cell is hidden behind the two synergids and is shown in the insert. (From Williams and Freidman 2002; photograph courtesy of William Freidman.)

sperm to create the endosperm. In 8-nuclei embryo sacs, there are seven cells. The central cell contains two nuclei and, when fertilized, produces a triploid endosperm. In *Nuphar*, a basal angiosperm, the embryo sac consists of four nuclei and the central cell has a single nucleus that, when fertilized, develops into a 2*n* endosperm. The 2*n* endosperm provides convincing evidence that the 4-celled embryo sac in *Nuphar* does not result from the degradation of four nuclei. If other cells had degraded, a 3*n* endosperm would be predicted.

Fertilization is not an absolute prerequisite for angiosperm embryonic development (Mogie 1992). Embryos can form within embryo sacs from haploid eggs and from cells that did not divide meiotically. This phenomenon is called **apomixis** (Greek, "without mixing") and results in viable seeds. The viability of the resulting haploid sporophytes indicates that ploidy alone does not account for the morphological distinctions between the gametophyte and the sporophyte. Embryos can also develop from cultured sporophytic tissue. These embryos develop with no associated endosperm, and they lack a seed coat.

Embryonic Development

Experimental studies

The angiosperm zygote is embedded within the ovule and ovary and thus is not readily accessible for experimental manipulation. The following approaches, however, can yield information on the formation of the plant embryo:

- **Histological studies** of embryos at different stages show how carefully regulated cell division results in the construction of an organism, even without the ability to move cells and tissues to shape the embryo.
- **Culture experiments** using embryos isolated from ovules and embryos developing de novo from cultured sporophytic tissue provide information on the interactions between the embryo and surrounding sporophytic and endosperm tissue.
- **In vitro fertilization experiments** provide information on gamete interactions.
- **Biochemical analyses** of embryos at different stages of development provide information on such things as the stage-specific gene products necessary for patterning and establishing food reserves.
- **Genetic and molecular analyses of developmental mutants** characterized using the above approaches have greatly enhanced our understanding of embryonic development.
- **Clonal analysis** involves marking individual cells and following their fate in development (see Poethig 1987 for details on the methodology). For example, seeds heterozygous for a pigmentation gene may be irradiated so that a certain cell loses the ability to produce pigment. Its descendants will form a colorless sector that can be identified and related to the overall body pattern.

Embryogenesis

In plants, the term **embryogenesis** covers development from the time of fertilization until dormancy occurs. The basic body plan of the sporophyte is established during embryogenesis; however, this plan is reiterated and elaborated after dormancy is broken. Embryogenesis presents several major challenges:

1. To establish the basic body plan. Radial patterning produces three tissue systems, and axial patterning establishes the apical-basal (shoot-root) axis.
2. To set aside meristematic tissue for postembryonic elaboration of the body structure (leaves, roots, flowers, etc.).
3. To establish an accessible food reserve for the germinating embryo until it becomes autotrophic.

Embryogenesis is similar in all angiosperms in terms of the establishment of the basic body plan (see Figure 20.18; Steeves and Sussex 1989). There are differences in pattern elaboration, however, including differences in the precision of cell division patterns, the extent of endosperm development, cotyledon development, and the extent of shoot meristem development (Esau 1977; Johri et al. 1992).

Polarity is established in the first cell division following fertilization. The establishment of polarity has been investigated using brown algae as a model system (Belanger and Quatrano 2000). The zygotes of these plants are independent of other tissues and amenable to manipulation. The initial cell division results in one smaller cell, which will form the rhizoid (root homologue) and anchor the rest of the plant, and one larger cell, which gives rise to the thallus (the main body of the sporophyte). The point of sperm entry fixes the position of the rhizoid end of the apical-basal axis. This axis is perpendicular to the plane of the first cell division. F-actin accumulates at the rhizoid pole (Kropf et al. 1999). However, light or gravity can override this fixing of the axis and establish a new position for cell division (Figure 20.16; Alessa and Kropf 1999).

Once the apical-basal axis is established, secretory vesicles are targeted to the rhizoid pole of the zygote (Figure 20.17). These vesicles contain material for rhizoid outgrowth, with a cell wall of distinct macromolecular composition. Targeted secretion may also help orient the first plane of cell division. Maintenance of rhizoid versus thallus fate early in development depends on information in the cell walls (Brownlee and Berger 1995). Cell wall information also appears to be important in angiosperms (reviewed in Scheres and Benfey 1999).

The basic body plan of the angiosperm laid down during embryogenesis also begins with an asymmetrical cell division,* giving rise to a **terminal cell** and a **basal cell** (Figure 20.18). The terminal cell gives rise to the **embryo proper**. The basal cell forms closest to the micropyle and gives rise to the **suspensor**. The **hypophysis** is found at the interface between

---

*Asymmetrical cell division is also important in later angiosperm development, including the formation of guard cells of leaf stomata and of different cell types in the ground and vascular tissue systems.

Figure 20.16
Axis formation in the brown alga *Pelvetia compressa*. (A) An F-actin patch (orange) is first formed at the point of sperm entry (the blue spot marks the sperm pronucleus). (B) Later, light was shone in the direction of the arrow. The sperm-induced axis was overridden, and an F-actin patch formed on the dark side, where the rhizoid will later form. (Photographs courtesy of W. Hables.)

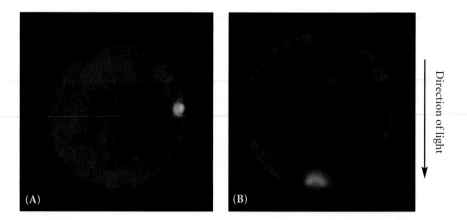

(A) (B)

Direction of light

the suspensor and the embryo proper. In some species it gives rise to a portion of the root cells. (The suspensor cells divide to form a filamentous or spherical organ that degenerates later in embryogenesis.) In both gymnosperms and angiosperms, the suspensor orients the absorptive surface of the embryo toward its food source; in angiosperms, it also appears to serve as a nutrient conduit for the developing embryo. Culturing isolated embryos of scarlet runner beans (*Phaseolus coccineus*) with and without the suspensor has demonstrated the need for a suspensor through the heart stage in dicots (Figure 20.19; Yeung and Sussex 1979). Embryos cultured with a suspensor are twice as likely to survive as embryos cultured without an attached suspensor at this stage. The suspensor may be a source of hormones. In scarlet runner beans, younger embryos without a suspensor can survive in culture if they are supplemented with the growth hormone gibberellic acid (Cionini et al. 1976).

As the establishment of apical-basal polarity is one of the key achievements of embryogenesis, it is useful to consider how the suspensor and embryo proper develop unique morphologies. As early as the 4-cell stage in scarlet runner bean development, there is distinct transcription in apical and basal cells (Weterings et al. 2001). Genes isolated from a scarlet runner bean genomic library have been shown to be selectively expressed in suspensor or embryo cells. When these genes were introduced into transgenic tobacco plants, the same patterns of expression, delineating the suspensor and embryo, were observed using in situ hybridization. In *Arabidopsis*, *MERISTEM LAYER 1* (*AtML1*) is expressed in the apical

Figure 20.17
Asymmetrical cell division in brown algae. Time course from 8 to 25 hours after fertilization, showing algal cells stained with a vital membrane dye to visualize secretory vesicles, which appear first, and the cell plate, which begins to appear about halfway through this sequence. (Photographs courtesy of K. Belanger.)

8 hours after fertilization

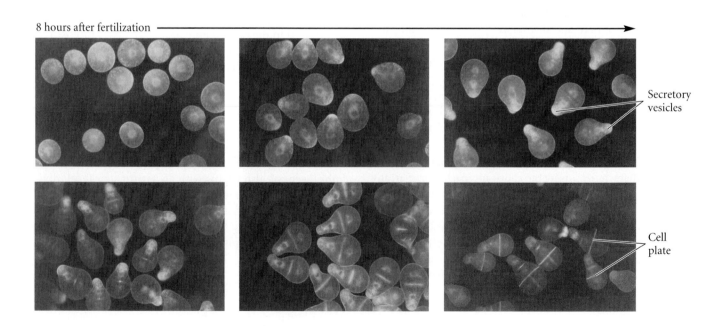

Secretory vesicles

Cell plate

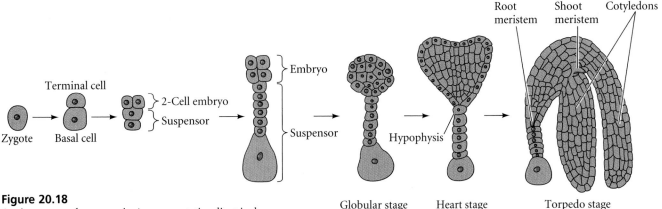

Figure 20.18
Angiosperm embryogenesis. A representative dicot is shown; a monocot would develop only a single cotyledon. While there are basic patterns of embryogenesis in angiosperms, there is tremendous morphological variation among species.

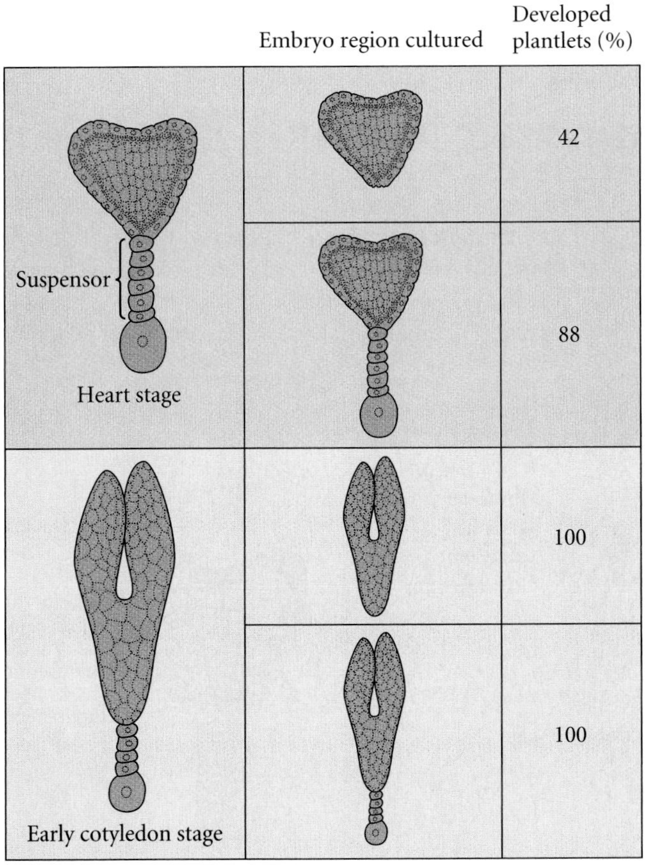

Embryo region cultured		Developed plantlets (%)
Heart stage (with Suspensor)		42
		88
Early cotyledon stage		100
		100

Figure 20.19
Role of the suspensor in dicot embryogenesis. Culturing scarlet runner bean embryos with and without their suspensors has demonstrated that the suspensor is essential at the heart stage, but not later. (After Yeung and Sussex 1979.)

daughter cell, but not the basal daughter cell at the two cell stage (Lu et al. 1996).

The study of mutant embryos in maize and *Arabidopsis* has been particularly helpful in sorting out the different developmental pathways of embryos and suspensors. Investigations of suspensor mutants (*sus1*, *sus2*, and *raspberry1*) of *Arabidopsis* have provided genetic evidence that the suspensor has the capacity to develop embryo-like structures (Figure 20.20; Schwartz et al. 1994; Yadegari et al. 1994). In these mutants, abnormalities in the embryo proper appear prior to suspensor abnormalities.* Earlier experiments in which the embryo proper was removed also demonstrated that suspensors could develop like embryos (Haccius 1963). A signal from the embryo proper to the suspensor may be important in maintaining suspensor identity and blocking the development of the suspensor as an embryo. Molecular analyses of these and other genes are providing insight into the mechanisms of communication between the suspensor and the embryo proper.

The *SUS1* gene has been renamed *DCL1* (*DICER-LIKE 1*) because its predicted protein sequence is structurally like that of *Dicer* in *Drosophila melanogaster* and *DCR-1* in *C. elegans* (Schauer et al. 2002). These proteins are part of an RNA-processing enzyme that may control the translation of developmentally important mRNAs. This exciting discovery should lead to a better understanding of the regulation of development beyond the level of transcriptional control.

Intriguingly, *DCL1* has several alleles that were originally assumed to be completely different genes regulating very different developmental processes. *DCL1* alleles include *SIN1* alleles. These mutants affect ovule development (discussed in the next paragraph) and the transition from vegetative to reproductive development. The *carpel factory* (*caf-1*) allele of *DCL1* causes indeterminancy in floral meristems, leading to extra

---

*Another intriguing characteristic of these mutants is that cell differentiation occurs in the absence of morphogenesis. Thus, cell differentiation and morphogenesis can be uncoupled in plant development.

(A)　　　　　　　　　　(B)　　　　　　　　　　(C)

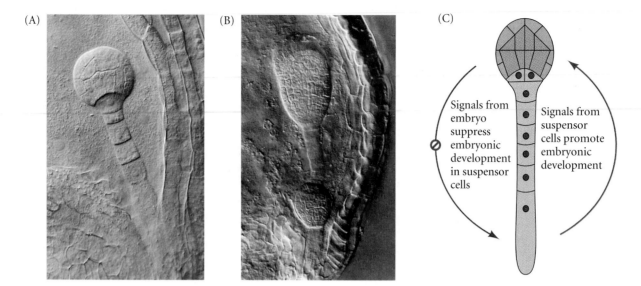

Figure 20.20
The *SUS* gene (a *DCL1* allele) suppresses embryonic development in the suspensor. (A) Wild-type embryo and suspensor. (B) *sus* mutant with suspensor developing like an embryo (arrow). (C) Model showing how the embryo proper suppresses embryonic development in the suspensor and the suspensor provides feedback information to the embryo. (Photographs courtesy of D. Meinke.)

whorls of carpels. Based on *Drosophila* Dicer protein function, *DCL1* may be involved in cleaving small, noncoding RNAs into even smaller, 21- to 25-nucleotide, single-stranded RNA products that could cleave mRNAs and affect translation. Many questions about the role of small RNAs as possible developmental signals are arising from the work on *DCL1* alleles.

Maternal effect genes play a key role in establishing embryonic pattern in animals (see Chapter 9). The role of extrazygotic genes in plant embryogenesis is less clear, and the question is complicated by at least three potential sources of influence: sporophytic tissue, gametophytic tissue, and the polyploid endosperm. All of these tissues are in close association with the egg/zygote (Ray 1998). Endosperm development could also be affected by maternal genes. Sporophytic and gametophytic maternal effect genes have been identified in *Arabidopsis*, and it is probable that the endosperm genome influences the zygote as well. The first maternal effect gene identified, *SHORT INTEGUMENTS 1* (*SIN1*), must be expressed in the sporophyte for normal embryonic development (Ray et al. 1996). Two transcription factors (FBP7 and FBP11) are needed in the petunia sporophyte for normal endosperm development (Columbo et al. 1997). A female gametophytic maternal effect gene, *MEDEA* (after Euripides' Medea, who killed her own children), has protein domains similar to those of a *Drosophila* maternal effect gene (Grossniklaus et al. 1998). Curiously, *MEDEA* is in the Polycomb gene group, whose products alter chromatin (directly or indirectly) and

affect transcription. *MEDEA* affects an imprinted gene (see Chapter 5) that is expressed by the female gametophyte and by maternally inherited alleles in the zygote, but not by paternally inherited alleles (Vielle-Calzada et al. 1999). How significant maternal effect genes are in establishing the sporophyte body plan is still an unanswered question.

Radial and axial patterns develop as cell division and differentiation continue (Figure 20.21; see also Bowman 1994 for detailed light micrographs of *Arabidopsis* embryogenesis). The cells of the embryo proper divide in transverse and longitudinal planes to form a **globular stage** embryo with several tiers of cells. Superficially, this stage bears some resemblance to cleavage in animals, but the nuclear to cytoplasmic ratio does not necessarily increase. The emerging shape of the embryo depends on

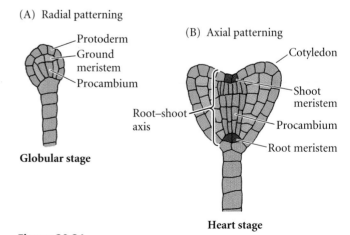

Figure 20.21
Radial and axial patterning. (A) Radial patterning in angiosperms begins in the globular stage and results in the establishment of three tissue systems. (B) The axial pattern (shoot-root axis) is established by the heart stage.

regulation of the planes of cell division and expansion, since the cells are not able to move and reshape the embryo. Cell division planes in the outer layer of cells become restricted, and this layer, called the **protoderm**, becomes distinct. Radial patterning emerges at the globular stage as the three tissue systems (dermal, ground, and vascular) of the plant are initiated. The **dermal tissue** (epidermis) will form from the protoderm and contribute to the outer protective layers of the plant. **Ground tissue** (cortex and pith) forms from the ground meristem, which lies beneath the protoderm. The **procambium**, which forms at the core of the embryo will give rise to the **vascular tissue** (xylem and phloem), which will function in support and transport. The differentiation of each tissue system is at least partially independent. For example, in the *keule* mutant of *Arabidopsis*, the dermal system is defective while the inner tissue systems develop normally (Mayer et al. 1991).

The globular shape of the embryo is lost as **cotyledons** ("first leaves") begin to form. Dicots have two cotyledons, which give the embryo a heart-shaped appearance as they form. The axial body plan is evident by this **heart stage** of development. Hormones (specifically, auxins) may mediate the transition from radial to bilateral symmetry (Liu et al. 1993). In monocots, such as maize, only a single cotyledon emerges.

In many plants, the cotyledons aid in nourishing the plant by becoming photosynthetic after germination (although those of some species never emerge from the ground). In some cases—peas, for example—the food reserve in the endosperm is used up before germination, and the cotyledons serve as the nutrient source for the germinating seedling.* Even in the presence of a persistent endosperm (as in maize), the cotyledons store food reserves such as starch, lipids, and proteins. In many monocots, the cotyledon grows into a large organ pressed against the endosperm and aids in nutrient transfer to the seedling. Upright cotyledons can give the embryo a torpedo shape. In some plants, the cotyledons grow sufficiently long that they must bend to fit within the confines of the seed coat. The embryo then looks like a walking stick. By this point, the suspensor is degenerating.

The *Arabidopsis LEAFY COTYLEDON 1* gene was first identified by a mutant with leaflike cotyledons (Meinke 1994). *LEC1* is necessary to maintain the suspensor early in development, to specify cotyledon identity, to initiate maturation, and to prevent early germination of the seed. It belongs to the *LEAFY COTYLEDON 1-LIKE* gene class, which is unique among the embryogenesis genes in acting throughout the course of embryo development (Harada 2001; Kwong et al. 2003).

The **shoot apical meristem** and **root apical meristem** are clusters of stem cells that will persist in the postembryonic plant and give rise to most of the sporophyte body (see Jurgens 2001 for a review of apical-basal pattern formation). The root meri-

stem is partially derived from the hypophysis in some species. All other parts of the sporophyte body are derived from the embryo proper. Genetic evidence indicates that the formation of the shoot and root meristems is regulated independently. From an evolutionary perspective, this is not surprising. One of the major adaptations to land involved the evolution of a root system (the nonvascular plants, including the mosses, did not develop root systems). This independence is demonstrated by the *dek23* maize mutant and the *shootmeristemless* (*STM*) mutant of *Arabidopsis*, both of which form a root meristem but fail to initiate a shoot meristem (Clark and Sheridan 1986; Barton and Poethig 1993). The *STM* gene, which has a homeodomain, is expressed in the late globular stage, in cells that will form the shoot meristem. Genes that specifically affect the development of the root axis during embryogenesis have also been identified. Mutations of the *HOBBIT* gene in *Arabidopsis* (Willemson et al. 1998), for example, affect the hypophysis derivatives and eliminate root meristem function.

While it is clear that root and shoot developmental programs are different, what triggers root or shoot development is less tractable. The *TOPLESS* gene in *Arabidopsis* may provide some clues. A single mutant allele has been identified that converts a shoot into a root, but how the wild-type gene functions is still a puzzle (Long et al. 2002).

The shoot apical meristem will initiate leaves after germination and, ultimately, the transition to reproductive development. In *Arabidopsis*, the cotyledons are produced from general embryonic tissue, not from the shoot meristem (Barton and Poethig 1993). In many angiosperms, a few leaves are initiated during embryogenesis. In the case of *Arabidopsis*, clonal analysis points to the presence of leaves in the mature embryo, even though they are not morphologically well developed (Irish and Sussex 1992). Clonal analysis has demonstrated that the cotyledons and the first two true leaves of cotton are derived from embryonic tissue rather than an organized meristem (Christianson 1986).

Clonal analysis experiments provide information on cell fates, but do not necessarily indicate whether or not cells are determined for a particular fate. Cells, tissues, and organs are shown to be determined when they have the same fate in situ, in isolation, and at a new position in the organism (see McDaniel et al. 1992 for more information on developmental states in plants). Clonal analysis has demonstrated that cells that divide in the wrong plane and "move" to a different tissue layer often differentiate according to their new position. Position, rather than clonal origin, appears to be the critical factor in embryo pattern formation, suggesting some type of cell-cell communication (Laux and Jurgens 1994). Microsurgery experiments on somatic carrot embryos demonstrate that isolated pieces of embryo can often replace the missing complement of parts (Schiavone and Racusen 1990; Scheres and Heidstra 1999). A cotyledon removed from the shoot apex will be replaced. Isolated embryonic shoots can regenerate a new root; isolated root tissue regenerates cotyledons, but is

---

*Mendel's famous wrinkled-seed mutant (the *rugosus* or *r* allele) has a defect in a starch branching enzyme that affects starch, lipid, and protein biosynthesis in the seed and leads to defective cotyledons (Bhattacharyya et al. 1990).

less likely to regenerate the shoot axis. Although most embry-onic cells are pluripotent and can generate organs such as cotyledons and leaves, only meristems retain this capacity in the postembryonic plant body.

Dormancy

From the earliest stages of embryogenesis, there is a high level of zygotic gene expression. As the embryo reaches maturity, there is a shift from constructing the basic body plan to creating a food reserve by accumulating storage carbohydrates, proteins, and lipids. Genes coding for seed storage proteins were among the first to be characterized by plant molecular biologists because of the high levels of specific storage protein mRNAs that are present at different times in embryonic development. The high level of metabolic activity in the developing embryo is fueled by continuous input from the parent plant into the ovule. Eventually metabolism slows, and the connection of the seed to the ovary is severed by the degeneration of the adjacent supporting sporophyte cells. The seed desiccates (loses water), and the integuments harden to form a tough seed coat. The seed has entered **dormancy**, officially ending embryogenesis. The embryo can persist in a dormant state for weeks or years, a fact that affords tremendous survival value. There have even been examples of seeds found stored in ancient archaeological sites that germinated after thousands of years of dormancy!

Maturation leading to dormancy is the result of a precisely regulated program. The *viviparous* mutation in maize, for example, produces genetic lesions that block dormancy (Figure 20.22; Steeves and Sussex 1989). The apical meristems of *viviparous* mutants behave like those of ferns, with no pause before producing postembryonic structures. The embryo continues to develop, and seedlings emerge from the kernels on the ear attached to the parent plant. Recently, a group of plant genes have been identified that belong to the Polycomb group, which regulates early development in mammals, nematodes, and insects (Preuss 1999). These genes encode chromatin silencing factors, which may play an important role in seed formation.

Plant hormones are critical in dormancy, and linking them to genetic mechanisms is an active area of research. The hormone **abscisic acid** is important in maintaining dormancy in many species. **Gibberellins**, another class of hormones, are important in breaking dormancy.

Germination

The postembryonic phase of plant development begins with **germination**. Some dormant seeds require a period of **after-ripening** during which low-level metabolic activities continue to prepare the embryo for germination. Highly evolved interactions between the seed and its environment increase the odds that the germinating seedling will survive to produce an-

Figure 20.22
The *viviparous* mutant of maize. Each kernel on this maize ear contains an embryo. These embryos do not go through a dormant phase, but begin germinating while still on the ear. (From McCarty et al. 1989; photograph © American Society of Plant Biologists and reprinted with permission.)

other generation. Temperature, water, light, and oxygen are all key in determining the success of germination. **Stratification** is the requirement for chilling (5°C) to break dormancy in some seeds. In temperate climates, this adaptation ensures germination only after the winter months have passed. In addition, many seeds have maximum germination rates at moderate temperatures of 25°–30°C and often will not germinate at extreme temperatures. Seeds such as lettuce require light (specifically, the red wavelengths) for germination; these seeds will not germinate so far below ground that they use up their food reserves before photosynthesis is possible.

Desiccated seeds may be only 5–20% water. **Imbibition** is the process by which the seed rehydrates, soaking up large volumes of water and swelling to many times its original size. The **radicle** (primary embryonic root) emerges from the seed first to enhance water uptake; it is protected by a root cap produced by the root apical meristem. Water is essential for metabolic activity, but so is oxygen. A seed sitting in a glass of water will not survive. Some species have such hard protective seed coats that they must be **scarified** (scratched or etched)

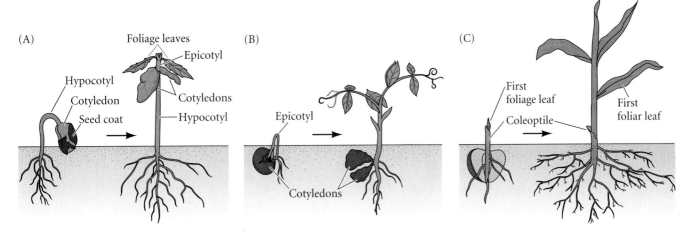

(A) Foliage leaves / Epicotyl / Hypocotyl / Cotyledon / Seed coat / Cotyledons / Hypocotyl

(B) Epicotyl / Cotyledons

(C) First foliage leaf / Coleoptile / First foliar leaf

Figure 20.23
Meristems have delicate cells that are susceptible to damage during germination. A variety of strategies have evolved for protecting the shoot meristem during germination. (A) Some cotyledons protect the meristem as the shoot emerges from the soil. (B) Bent epicotyls (and sometimes hypocotyls) push through the soil. (C) Monocots utilize the coleoptile, a leaflike structure that sheathes the young shoot tip.

before water and oxygen can cross the barrier. Scarification can occur by the seed being exposed to the weather and other natural elements over time, or by its exposure to acid as the seed passes through the gut of a frugivore (fruit eater). The frugivore thus prepares the seed for germination, as well as dispersing it to a site where germination can take place.

During germination, the plant draws on the nutrient reserves in the endosperm or cotyledons. Interactions between the embryo and endosperm in monocots use gibberellin as a signal to trigger the breakdown of starch into sugar. As the shoot reaches the surface, the differentiation of chloroplasts is triggered by light. Seedlings that germinate in the dark have long, spindly stems and do not produce chlorophyll. This environmental response allows plants to use their limited resources to reach the soil surface, where photosynthesis will be productive.

The delicate shoot tip must be protected as the shoot pushes through the soil. Three strategies for protecting the shoot tip have evolved (Figure 20.23):

1. Cotyledons protect the shoot tip.
2. The epicotyl (the stem above the cotyledons) bends so that stem tissue, rather than the shoot tip, pushes through the soil.
3. In monocots, a special leaflike structure, the coleoptile, forms a protective sheath around the shoot tip.

Vegetative Growth

When the shoot emerges from the soil, most of the sporophyte body plan remains to be elaborated. Figure 20.24 shows the basic parts of the mature sporophyte plant, which will emerge from meristems.

Meristems

As has been mentioned, meristems are clusters of cells that allow the basic body pattern established during embryogenesis to be reiterated and extended after germination. Meristem-

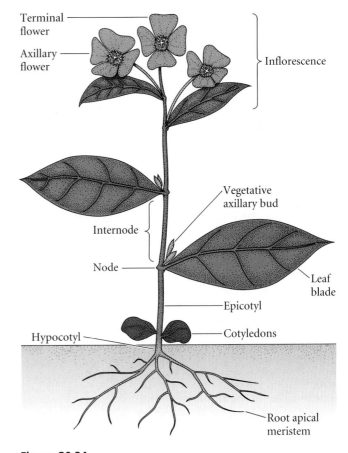

Terminal flower / Axillary flower / Inflorescence / Vegetative axillary bud / Internode / Node / Leaf blade / Epicotyl / Hypocotyl / Cotyledons / Root apical meristem

Figure 20.24
Morphology of a generalized angiosperm sporophyte.

Figure 20.25
Shoot and root apical meristems. Both shoots and roots develop from apical meristems, undifferentiated cells clustered at their tips. In roots, a root cap is also produced, which protects the meristem as it grows through the soil. The lateral organs of the shoot (leaves and axillary branches) have a superficial origin in shoot apical meristems. Lateral roots are derived from pericycle cells deep within the root.

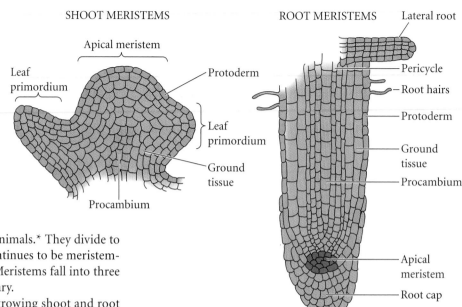

atic cells are similar to stem cells in animals.* They divide to give rise to one daughter cell that continues to be meristematic and another that differentiates. Meristems fall into three categories: apical, lateral, and intercalary.

Apical meristems occur at the growing shoot and root tips (Figure 20.25). Root apical meristems produce the root cap, which consists of lubricated cells that are sloughed off as the meristem is pushed through the soil by cell division and elongation in more proximal cells. The root apical meristem also gives rise to daughter cells that produce the three tissue systems of the root. New root apical meristems are initiated from tissue within the core of the root and emerge through the ground tissue and dermal tissue. Root meristems can also be derived secondarily from the stem of the plant; in the case of maize, this is the major source of root mass.

The shoot apical meristem produces stems, leaves, and reproductive structures. In addition to the shoot apical meristem initiated during embryogenesis, axillary shoot apical meristems (axillary buds; see Figure 20.24) derived from the original one form in the axils (the angles between leaf and stem). Unlike new root meristems, these arise from the surface layers of the meristem.

Angiosperm apical meristems are composed of up to three layers of cells (labeled L1, L2, and L3) on the plant surface (Figure 20.26). One way of investigating the contributions of different layers to plant structure is by constructing chimeras. Plant chimeras are composed of layers having distinct

genotypes with discernible markers. When L2, for example, has a different genotype than L1 or L3, all pollen will have the L2 genotype, indicating that pollen is derived from L2. Chimeras have also been used to demonstrate classical induction in plants, in which, as in animal development, one layer influences the developmental pathway of an adjacent layer.

The size of the shoot apical meristem is precisely controlled by intercellular signals, most likely between layers of the

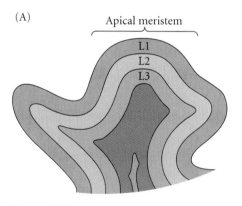

Figure 20.26
Organization of the shoot apical meristem. (A) Angiosperm meristems have two or three outer layers of cells that are histologically distinct (here labeled L1, L2, and L3). While cells in certain layers tend to have certain fates, they are not necessarily committed to those fates. If a cell is shifted to a new layer, it generally develops like the other cells in that layer. (B) The fates of the cell layers can be seen in a chimeric tobacco plant. One portion of the meristem contains three layers of wild-type cells, while the other portion has an L2 that lacks chlorophyll. This section of the meristem has given rise to the variegated leaves. In wild-type plants, the L1 layer always lacks chlorophyll (except in guard cells), but in this plant the L2, too, is genetically unable to produce chlorophyll; the L3 remains green. The L3 does not contribute to the outer edges of leaves, which is why they appear white in this plant. (Photograph courtesy of M. Marcotrigiano.)

*The similarities between plant meristem cells and animal stem cells may extend to the molecular level, indicating that stem cells existed before plants and animals pursued separate phylogenetic pathways. Homology has been found between genes required for plant meristems to persist and genes expressed in *Drosophila* germ line stem cells (Cox et al. 1998).

meristem (reviewed by Doerner 1999). Mutations in the *Arabidopsis CLAVATA* (*CLV*) genes, for example, lead to increased meristem size and the production of extra organs. *CLV1*, *CLV2*, and *CLV3* all limit the number of undifferentiated stem cells in both vegetative and floral meristems. The CLV1 protein is a serine/threonine kinase that, along with the receptor-like transmembrane CLV2 protein, forms a receptor for the CLV3 protein, which is localized in the extracellular space between cell layers of the meristem (Figure 20.27; Clark et al. 1997; Jeong et al. 1999; Rojo et al. 2002). *POLTERGEIST* (*POL*) and *WUSCHEL* (*WUS*) are redundant in function and appear to keep genes in an undifferentiated state (Yu et al. 2000). It is possible that *POL* expression is a target in CLV1 signal transduction. *STM*, like *POL* and *WUS*, plays an important role in maintaining an undifferentiated population of meristematic cells and *STM* may positively regulate *WUS* (Clark and Schiefelbein 1997). The interactions of these gene products balance the rate of cell division (which enlarges the meristem) and the rate of cell differentiation in the periphery of the meristem (which decreases meristem size) (Meyerowitz 1997).

Lateral meristems are cylindrical meristems found in shoots and roots that result in secondary growth (an increase in stem and root girth by the production of vascular tissues). Monocot stems do not have lateral meristems, but often have **intercalary meristems** inserted in the stems between mature tissues. The popping sound you can hear in a cornfield on a summer night is actually caused by the rapid increase in stem length due to intercalary meristems.

Root development

Radial and axial patterning in roots begins during embryogenesis and continues throughout development as the primary root grows and lateral roots emerge from the pericycle cells deep within the root. Clonal analyses and laser ablation experiments eliminating single cells have demonstrated that root cells are plastic and that position is the primary determinant of fate in early root development. Analyses of root radial organization mutants have revealed genes with layer-specific activity (Scheres et al. 1995; Scheres and Heidstra 1999). We will illustrate these findings by looking at two *Arabidopsis* genes that regulate ground tissue fate.

Wild-type *Arabidopsis* has two layers of root ground tissue. The outer layer becomes the cortex; the inner layer becomes the endodermis, which forms a tube around the vascular tissue core. The *SCARECROW* (*SCR*) and *SHORT-ROOT* (*SHR*) genes have mutant phenotypes with one, instead of two, layers of root ground tissue (Benfey et al. 1993). The *SCR* gene is necessary for an asymmetrical cell division in the initial layer of cells, yielding a smaller endodermal cell and a larger cortex cell (Figure 20.28). The *scr* mutant expresses markers for both cortex and endodermal cells, indicating that differentiation progresses in the absence of cell division (Di Laurenzio et al. 1996). *SHR* is responsible for endodermal cell specification. Cells in the *shr* mutant do not develop endodermal features.

Axial patterning in roots may be morphogen-dependent, paralleling some aspects of animal development. A variety of experiments have established that the distribution of the plant hormone auxin organizes the axial pattern. A peak in auxin concentration at the root tip must be perceived for normal axial patterning (Sabatini et al. 1999).

As discussed earlier, distinct genes specifying root and shoot meristem formation have been identified; however, root and shoot development may share common groups of genes that regulate cell fate and patterning (Benfey 1999). This appears to be the case for the *SCR* and *SHR* genes. In the shoot, these genes are necessary for the normal gravitropic response, which is dependent on normal endodermis formation (a defect in mutants of both genes; see Figure 20.28C). It is important to keep in mind that there are a number of steps between establishment of the basic pattern and elaboration of that pattern into anatomical and morphological structure. Uncovering the underlying control mechanisms is likely to be the most productive strategy in understanding how roots and shoots develop.

Shoot development

The unique aboveground architectures of different plant species have their origins in shoot meristems. Shoot architecture is affected by the amount of axillary bud outgrowth. Branching patterns are regulated by the shoot tip—a phenomenon called apical dominance—and plant hormones appear to be the factors responsible. Auxin is produced by young leaves and transported toward the base of the leaf. It can suppress the outgrowth of axillary buds. Grazing and flowering often release buds from apical dominance, at which time branching occurs. **Cytokinins** can also release buds from apical domi-

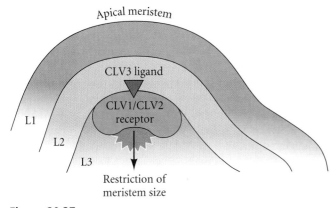

Figure 20.27
CLAVATA genes are necessary for limiting the pool of undifferentiated meristem cells. CLV3 protein is expressed in L1 and L2. It moves locally to bind to the CLV1/CLV2 receptor in L3 cells. This binding triggers a signal transduction cascade that regulates the number of undifferentiated cells in the meristem and balances the roles of *STM* and *WUS* in keeping cells undifferentiated and dividing.

(A)

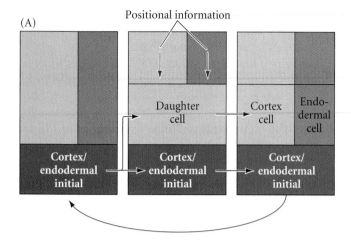

Positional information

Daughter cell

Cortex cell

Endo-dermal cell

Cortex/ endodermal initial

Cortex/ endodermal initial

Cortex/ endodermal initial

(B) Root

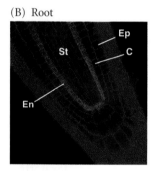

(C) Shoot

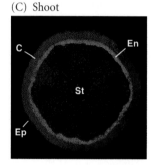

(D) Wild-type

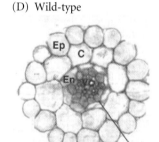

(E) *scr* mutant

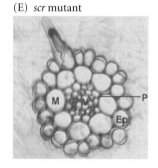

Figure 20.28
SCR and *SHR* regulate endodermal differentiation in root radial development. (A) Diagram of normal cell division yielding cortical and endodermal cells. *SCR* regulates this asymmetrical cell division. (B, C) *SCR* expression in root and shoot. The *SCR* promoter is linked to the gene for GFP (green fluorescent protein). (D–F) Cross sections of primary roots of (D) wild-type *Arabidopsis*, (E) *scr* mutant, and (F) *shr* mutant. Ep, epidermis; C, cortex; En, endodermis; M, mutant layer; P, pericycle; V, vascular tissue; St, stele. (A after Scheres and Heidstra 1999; B–F, photographs courtesy of P. Benfey.)

(F) *shr* mutant

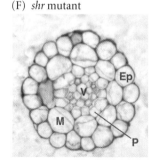

nance. Axillary buds can initiate their own axillary buds, so branching patterns can get quite complex. Branching patterns can be regulated by environmental signals. For example, an expansive canopy in an open area maximizes light capture, and asymmetrical tree crowns form when two trees grow very close to each other. In addition to its environmental plasticity, shoot architecture is genetically regulated. In several species, genes have now been identified that regulate branching patterns.

Leaf primordia (clusters of cells that will form leaves) are initiated at the periphery of the shoot meristem (see Figure 20.25). The union of a leaf and the stem is called a **node**, and stem tissue between nodes is called an **internode** (see Figure 20.24). In a simplistic sense, the mature sporophyte is created by stacking node/internode units together. **Phyllotaxy**, the positioning of leaves on the stem, involves communication among existing and newly forming leaf primordia. Leaves may be arranged in various patterns, including a spiral, 180-degree alternation of single leaves, pairs, and whorls of three or more leaves at a node (Jean and Barabé 1998). Experimentation has revealed a number of mechanisms for maintaining geometrically regular spacing of leaves on a plant, including chemical and physical interactions of new leaf primordia with the shoot apex and with existing primordia (Steeves and Sussex 1989).

It is not clear how a specific pattern of phyllotaxy gets started. Descriptive mathematical models can replicate the observed patterns, but reveal nothing about the mechanism. Biophysical models (for example, of the effects of stress/strain on deposition of cell wall material, which affects cell division and elongation) attempt to bridge this gap. Developmental genetics approaches are promising, but few phyllotactic mutants have been identified. One candidate is the *terminal ear* mutant in maize, which has irregular phyllotaxy. The wild-type gene is expressed in a horseshoe-shaped region, with a gap where the leaf will be initiated (Veit et al. 1998). The plane of the horseshoe is perpendicular to the axis of the stem.

Leaf development

Leaf development includes commitment to become a leaf, establishment of leaf axes, and morphogenesis, giving rise to a tremendous diversity of leaf shapes. Culture experiments have assessed when leaf primordia become determined for leaf development. Research on ferns and angiosperms indicates that the youngest visible leaf primordia are not determined to make a leaf; rather, these young primordia can develop as shoots in culture (Steeves 1966; Smith 1984). The programming for leaf development occurs later. The radial symmetry of the leaf primordium becomes dorsal-ventral, or flattened, in all leaves. Two other axes, the proximal-distal and lateral, are also established. The unique shapes of leaves result from regulation of cell division and cell expansion as the leaf blade develops. There are some cases in which selective cell death (apoptosis) is involved in the shaping of a leaf, but differential cell growth appears to be a more common mechanism (Gifford and Foster 1989).

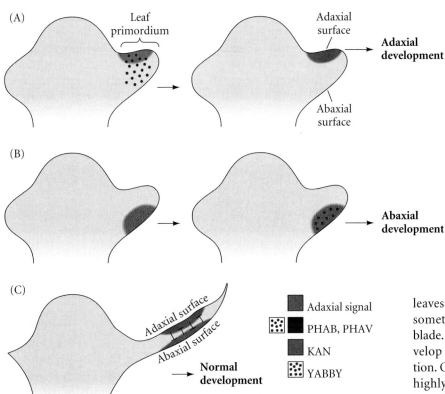

Figure 20.29
Patterning of adaxial and abaxial leaf surfaces of *Arabidopsis*. (A) *PHAB* and *PHAV* RNAs are uniformly distributed throughout the leaf primordium. A signal on the adaxial side of a leaf primordium activates PHAB and PHAV proteins, leading to the transcription of the genes needed for adaxial development. (B) Transient expression of *KAN* on the abaxial side of the leaf leads to the expression of *YABBY* genes, which in turn lead to the transcription of other abaxial genes. (C) PHAB and PHAV block *KAN* adaxial activity, and KAN, in turn, blocks abaxial activity of *PHAB* and *PHAV*. Expression of *PHAB*, *PHAV*, and *KAN* genes is transient and occurs early in leaf development.

THE ESTABLISHMENT OF DORSAL-VENTRAL PATTERNING IN LEAVES.

Plant biologists refer to the surface of the leaf that is closest to the stem as the adaxial side and the more distant surface as the abaxial side. As the leaf primordia form, the *Arabidopsis* genes *PHABULOSA* (*PHAB*) and *PHAVOLUTA* (*PHAV*) are initially transcribed uniformly in the primordia. The PHAB and PHAV proteins are postulated to be receptors for an adaxial signal that leads to the accumulation of PHAB and PHAV on the adaxial leaf surface (Figure 20.29A; McConnell et al. 2001). The sequences of the *PHAB* and *PHAV* genes make it likely that they are activated by a lipid ligand, and that activation is followed by the development of adaxial cells. *KANADI* (*KAN*) initiates abaxial cell differentiation in *Arabidopsis* (Kerstetter et al. 2001). In situ hybridization shows that *KAN* is transiently expressed on the abaxial side of cotyledons, leaves, and initiating floral organ primordia (Figure 20.29B). *KAN* contains a GARP domain found in transcription factors. Abaxial fate also appears to be specified by members of the *YABBY* gene family (Siegfried et al. 1999). The KAN protein turns on the *YABBY* genes. *KAN* gene expression is restricted to the abaxial side by PHAB and PHAV, while KAN protein restricts *PHAB* and *PHAV* gene expression to the adaxial side (Figure 20.29C).

THE GENETICS OF LEAF MORPHOGENESIS

Leaves fall into two categories, simple and compound (Figure 20.30; see review by Sinha 1999). There is much variety in simple leaf shape, from smooth-edged leaves to deeply lobed oak leaves. Compound leaves are composed of individual leaflets (and sometimes tendrils) rather than a single leaf blade. Whether simple and compound leaves develop by the same mechanism is an open question. One perspective is that compound leaves are highly lobed simple leaves. An alternative perspective is that compound leaves are modified shoots. The ancestral state for seed plants is believed to be compound, but for angiosperms it is simple. Compound leaves have arisen multiple times in the angiosperms, and it is not clear if these events are reversions to the ancestral state.

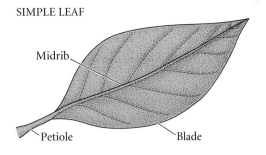

SIMPLE LEAF

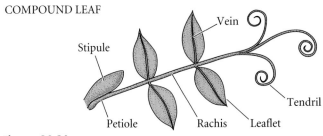

COMPOUND LEAF

Figure 20.30
Simple and compound leaves.

Developmental genetic approaches are being applied to leaf morphogenesis. The Class I *KNOX* genes are homeobox genes that include *STM* and the *KNOTTED 1* (*KN1*) gene in maize. Gain-of-function mutations of *KN1* cause meristem-like bumps to form on maize leaves. In wild-type plants, this gene is expressed in meristems. *KNOX* genes stimulate meristem initiation and growth. While some *YABBY* genes specify abaxial leaf identity, others function by downregulating *KNOX* genes, thereby restricting where meristems form (Kumaran et al. 2002). If too much *KN1* is present, *YABBY* is insufficient to control *KN1* expression.

When *KN1*, or its tomato homologue *LeT6*, has its promoter replaced with a promoter from cauliflower mosaic virus and is inserted into the genome of tomato, the gene is expressed at high levels throughout the plant, and the leaves become "supercompound" (Figure 20.31; Hareven et al. 1996; Janssen et al. 1998). Simple leaves become more lobed (but not compound) in response to the overexpression of *KN1*, consistent with the hypothesis that compound leaves may be an extreme case of lobing in simple leaves (Jackson 1996). The role of *KN1* in shoot meristem and leaf development, however, is consistent with the hypothesis that compound leaves are modified shoots. Looking at patterns of *KN1* expression over a broad range of angiosperm taxa, Bharathan and colleagues (2002) demonstrated that *KN1* expression is associated with compound leaf primordia, but is not present in the developing primordia of simple leaves (Figure 20.32). These data support the conclusion that compound leaves maintain shootlike activity, at least for some time. How *KN1* works in meristems and in leaf primordia is an area of active research. In tomato leaves, *KN1* causes a decrease in the expression of a gene needed for the production of the plant hormone gibberellic acid (Hay et al. 2002).

A second gene, *LEAFY*, which is essential for the transition from vegetative to reproductive development, also appears to play a role in compound leaf development. It was identified in *Arabidopsis* and snapdragon (in which it is called *FLORICAULA*) and has homologues in other angiosperms. The pea homologue (*UNIFOLIATA*) has a mutant phenotype in which compound leaves are reduced to simple leaves (Hofer and Ellis 1998). This finding is also indicative of a regulatory relationship between shoots and compound leaves. In an intriguing evolutionary twist, *UNI* appears to have taken on the role of *KN1* in the leaf development of the pea (Bharathan et al. 2002). In this highly derived legume, *UNI* appears to have been co-opted from its role in flowering.

In some compound leaves, developmental decisions about leaf versus tendril formation are also made. Mutations of two leaf-shape genes can individually and in sum dramatically alter the morphology of the compound pea leaf (Figure 20.33). The *acacia* mutant (*tl*) converts tendrils to leaflets; *afila* (*af*) converts leaflet to tendrils (Marx 1987). The *af tl* double mutant has a complex architecture and resembles a parsley leaf.

At the microscopic level, the patterning of stomata (openings for gas and water exchange) and trichomes (hairs) across the leaf is also being investigated. In monocots, the stomata form in parallel files, while in dicots the distribution appears more random. In both cases, the patterns appear to maximize the evenness of stomata distribution. Genetic analysis is providing insight into the mechanisms regulating this distribution. A common gene group appears to be working in

Figure 20.31
Overexpression of Class 1 *KNOX* genes in tomato plants. The photograph shows the single leaves of (A) a wild-type plant, (B) a *mouse ears* mutant, with increased leaf complexity, and (C) a transgenic plant that uses a viral promoter to overexpress the tomato homologue (*LeT6*) of the *KN1* gene from maize. (Photographs courtesy of N. Sinha.)

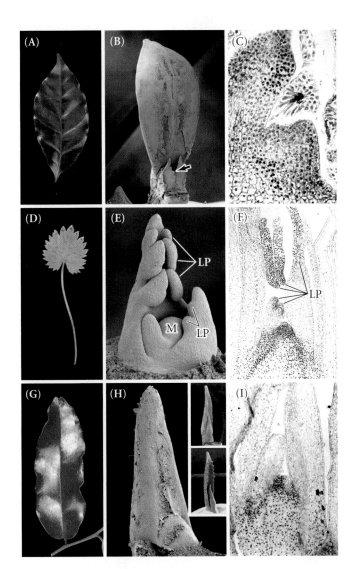

Figure 20.32

KNOX expression is known to occur in leaf primordia in species making complex leaves. Here we show that species with simple leaves can also express *KNOX* in leaf primordia that start out complex but undergo a secondary alteration to become simple. (A) Coffee simple leaf. (B) SEM of primordia (arrow) showing early complexity. (C) Expression of *KNOX* genes in these early leaves. (D) Anise simple leaf. (E) SEM showing early complexity. (F) *KNOX* expression in these primordia. (G) *Amborella*, showing a simple leaf that is simple throughout development (H) and has no *KNOX* expression (I). LP, leaf primordium; M, meristem. (From Bharathan et al. 2002; photographs courtesy of Neelima Sinha.)

both shoots and roots, affecting the distribution pattern of both trichomes and root hairs (Benfey 1999).

The Vegetative-to-Reproductive Transition

Unlike some animal systems, in which the germ line is set aside during early embryogenesis, the germ line in plants is established only after the transition from vegetative to reproductive development—that is, after flowering. The vegetative and reproductive structures of the shoot are all derived from the shoot meristem formed during embryogenesis. Clonal analysis indicates that no cells in the shoot meristem of the embryo are set aside to be used solely in the creation of reproductive structures (McDaniel and Poethig 1988). In maize, irradiating seeds causes changes in the pigmentation of some cells. These seeds give rise to plants that have visually distinguishable sectors descended from the mutant cells. Such sectors may extend from the vegetative portion of the plant into the reproductive regions (Figure 20.34), indicating that maize embryos do not have distinct reproductive compartments.

Maximal reproductive success depends on the timing of flowering—and on balancing the number of seeds produced with the resources allocated to individual seeds. As in animals, different strategies work best for different organisms in different environments. There is a great diversity of flowering patterns among the over 300,000 angiosperm species, yet there appears to be an underlying evolutionary conservation of flowering genes and patterns of flowering regulation.

A simplistic explanation of the flowering process is that a signal from the leaves moves to the shoot apex and induces flowering. In

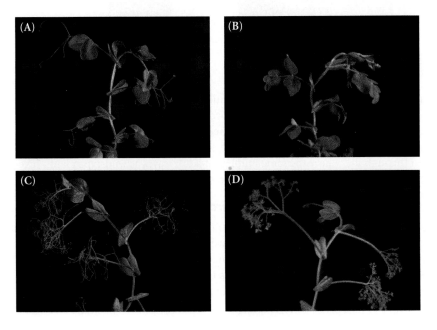

Figure 20.33

Leaf morphology mutants in peas. (A) Wild-type pea plant. (B) The *tl* mutant, in which tendrils are converted to leaflets. (C) The *af* mutant, in which leaflets are converted to tendrils. (D) An *af tl* double mutant, which has a "parsley leaf" phenotype.

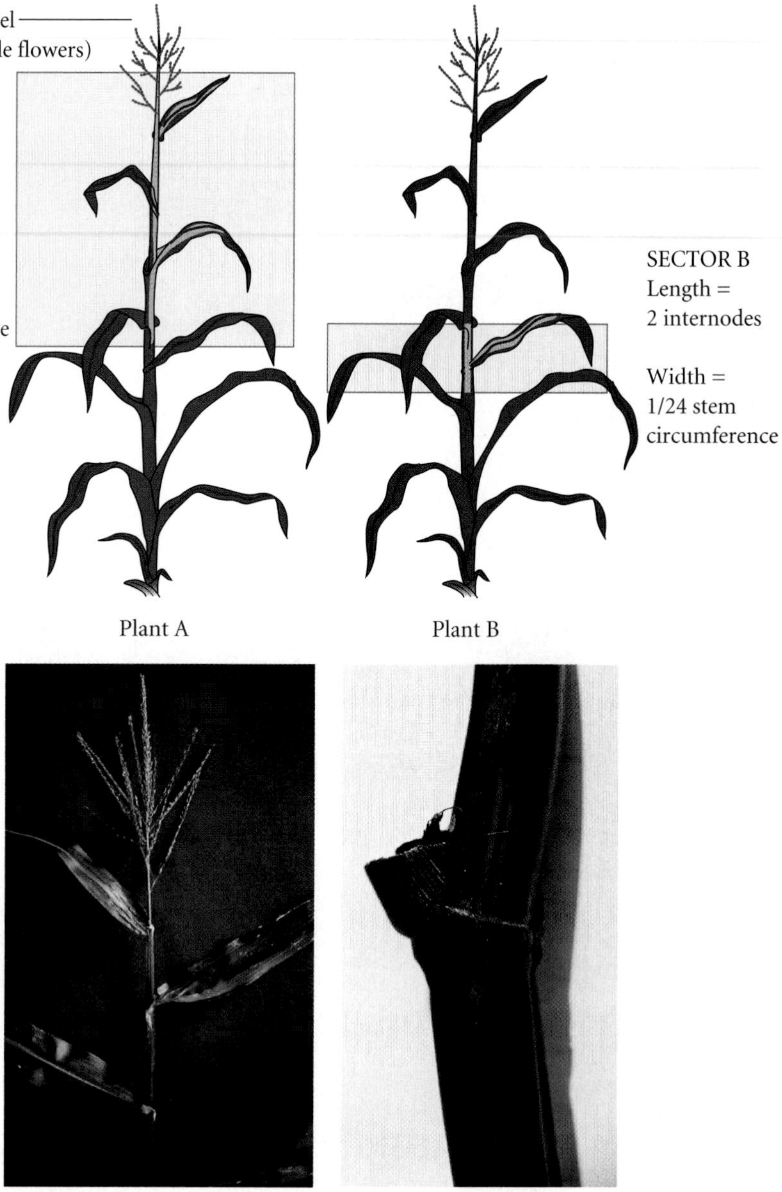

Figure 20.34
Clonal analysis can be used to construct
a fate map of a shoot apical meristem in
maize. Seeds that are heterozygous for
certain pigment genes (anthocyanins)
are irradiated so that the dominant allele
is lost in a few cells (a chance occurrence). All cells derived from the somatic
mutant will be visually distinct from the
nonmutant cells. Plants A and B have
mutant sectors that reveal the fate of
cells in the shoot meristem of the seed.
The mutant sector in A includes both
vegetative and reproductive (tassel) internodes. Thus there is no distinct developmental compartment that forms the tassel. The
mutant sector in A is longer and wider than the mutant sector in B. This indicates that more cells were
set aside to contribute to the lower than to the upper
internodes in the shoot meristem in the seed. The actual number of cells can be calculated by taking the
reciprocal of the fraction of the stem circumference
the sector occupies. Sector A contributes to 1/8 of the
circumference of the stem; thus 8 cells were fated to
contribute to these internodes in the seed meristem.
Sector B is only 1/24 of the stem circumference; thus
24 cells were fated to contribute to these internodes.
In this example, only cells derived from the L1 are
being analyzed. It is also important to consider the
possible contributions of the L2 and L3 cell layers of
the shoot meristem. (Data from McDaniel and Poethig 1988; photographs courtesy of C. McDaniel.)

Tassel (male flowers)

SECTOR A
Length = 4 internodes
Width = 1/8 stem circumference

SECTOR B
Length = 2 internodes
Width = 1/24 stem circumference

Plant A   Plant B

some species, this flowering signal is a response
to environmental conditions. The developmental pathways leading to flowering are regulated
at numerous control points in different plant
organs (roots, cotyledons, leaves, and shoot
apices) in various species, resulting in a diversity of flowering times and reproductive architectures (Figure 20.35). The nature of the flowering signal, however, remains unknown. There
is evidence that RNA with developmental functions can move through the phloem, and the roles of very
small pieces of RNA in signaling developmental mechanisms
has recently been recognized. Whether these will play a role in
flowering or not is an open question (Schauer et al. 2002).

Some plants, especially woody perennials, go through a
juvenile phase, during which the plant cannot produce reproductive structures even if all the appropriate environmental
signals are present (Lawson and Poethig 1995). The transition
from the juvenile to the adult stage may require the acquisition
of competence by the leaves or meristem to respond to an internal or external signal (McDaniel et al. 1992; Singer et al.
1992; Huala and Sussex 1993). Even in herbaceous plants, there
is a phase change from juvenile to adult vegetative growth. For
example, the *early phase change* (*epc*) gene in maize is required

to maintain the juvenile state (Vega et al. 2002). Mutant *epc*
plants flower early because they have fewer juvenile leaves,
which are distinguished from adult leaves by the presence of
wax and the absence of hairs. These *epc* mutants, however,
have the same number of adult leaves as wild-type plants. Juvenile traits are not expressed in the absence of *epc*.

Control of the reproductive transition

Grafting and organ culture experiments, mutant analyses, and
molecular analyses give us a framework for describing the reproductive transition. Grafting experiments have identified
the sources of signals that promote or inhibit flowering and
have provided information on the developmental acquisition
of meristem competence to respond to these signals (Figure

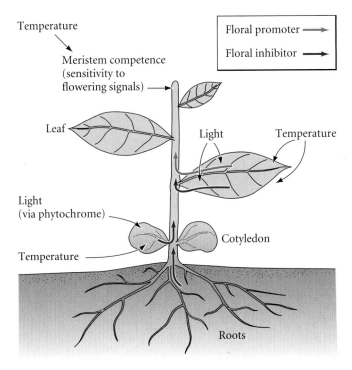

Temperature

Meristem competence
(sensitivity to
flowering signals)

Leaf

Light

Temperature

Light
(via phytochrome)

Cotyledon

Temperature

Roots

Figure 20.35
Regulation of the vegetative-to-reproductive transition. Internal and external factors regulate whether a meristem produces vegetative or reproductive structures. Not all of the regulatory mechanisms shown are used in all species, and some species flower independently of external environmental signals. Signals promoting or inhibiting flowering can move from the roots, cotyledons, or leaves to the shoot apex, where meristem competence determines whether or not the plant will respond to the signals. Leaves may also need to develop competence to respond to environmental signals before they can produce floral promoters.

pass on photoperiodic signals, however. The **phytochrome** pigments transduce these signals from the external environment. The structure of phytochromes is modified by red and far-red light, and these changes can initiate a cascade of events leading to the production of either floral promoter or floral inhibitor (Deng and Quail 1999). Leaves, cotyledons, and roots have been identified as sources of floral inhibitors in some species (McDaniel et al. 1992; Reid et al. 1996). A critical balance between inhibitor and promoter is needed for the reproductive transition.

20.36; Lang et al. 1977; Singer et al. 1992; McDaniel et al. 1996; Reid et al. 1996). Analyses of mutants and molecular characterization of genes are yielding information on the mechanics of these signal-response mechanisms (Levy and Dean 1998; Hempel et al. 2000).

Leaves produce a graft-transmissible substance that induces flowering. In some species, this signal is produced only under specific **photoperiods** (day lengths), while other species are day-neutral and will flower under any photoperiod (Zeevaart 1984). Not all leaves may be competent to perceive or

Figure 20.36
The interaction of competence and signal strength. *Nicotiana tabacum* (*Nt*) is a day-neutral plant that flowers when the meristem gains competence to respond to internal signals. *N. silvestris* (*Ns*) is a long-day plant that flowers when the floral signal(s) reach a critical level. These grafting experiments illustrate that a young *Nt* shoot is less competent to respond to the *Nt* flowering signal than an older *Nt* shoot. Young *Nt* shoots respond quickly to the flowering signal from flowering *Ns* plants, but flower later when grafted to young *Ns* plants with a lower level of signal. The scion is the shoot that is grafted on to the stock; the stock is the rooted portion of the plant from which the shoot has been excised. (After Singer et al. 1992.)

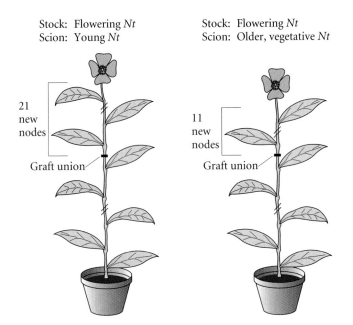

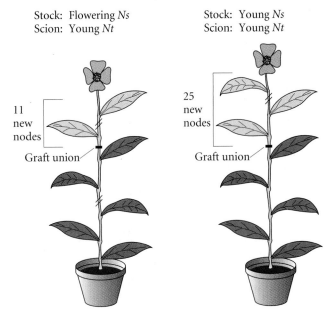

Stock: Flowering *Nt*
Scion: Young *Nt*

21 new nodes

Graft union

Stock: Flowering *Nt*
Scion: Older, vegetative *Nt*

11 new nodes

Graft union

Stock: Flowering *Ns*
Scion: Young *Nt*

11 new nodes

Graft union

Stock: Young *Ns*
Scion: Young *Nt*

25 new nodes

Graft union

In some species, meristems change in their competence to respond to flowering signals during development (Singer et al. 1992). **Vernalization**, a period of chilling, can enhance the competence of shoots and leaves to perceive or produce a flowering signal. The reproductive transition depends on both meristem competence and signal strength (see Figure 20.36). Shoot tip culture experiments in several species (including tobacco, sunflower, and peas) have demonstrated that determination for reproductive function can occur before reproductive morphogenesis (reviewed in McDaniel et al. 1992). That is, isolated shoot tips that are determined for reproductive development but are morphologically vegetative will produce the same number of nodes before flowering in situ and in culture (Figure 20.37).

The "black box" between environmental signals and the production of a flower is vanishing rapidly, especially in the model plant *Arabidopsis* (Figure 20.38). The signaling pathways from light via different phytochromes to key flowering genes are being elucidated. *CONSTANS* (*CO*) responds to daylength, promoting flowering under long-day conditions. Some of the genes that *CO* activates are also activated by transcription factors in other, non-light-dependent flowering pathways. Molecular explanations are revealing redundant pathways that ensure that flowering will occur. In *Arabidopsis*, three separable pathways leading to flowering are being elucidated. Light-dependent, vernalization-dependent, and autonomous pathways that regulate the floral transition have been genetically dissected. The autonomous pathway involves the gibberellins (plant hormones). The availability of the *Arabidopsis* genome sequence now makes it possible to search on a broader scale for classes of genes that regulate critical events such as the time of flowering (Ratcliffe and Riechmann 2002). Another promising approach is the search for quantitative trait loci (QTL) that control responses to environmental and hormonal factors. In *Arabidopsis*, two lines, including a wild accession that had natural variation in light and hormone response, were used to identify new genes (Borevitz et al. 2002). Once QTL are mapped, the genomic map makes it more likely that function can be associated with a specific gene.

Inflorescence meristems

The ancestral angiosperm is believed to have formed a terminal flower directly from the terminal shoot apex (Stebbins 1974). In modern angiosperms, a variety of flowering patterns exist in which the terminal shoot apex is indeterminate, but axillary buds produce flowers. This observation introduces an intermediate step into the reproductive process: the transition of a vegetative meristem to an **inflorescence meristem**, which initiates axillary meristems that can produce floral organs, but does not directly produce

Figure 20.37
Determination for reproductive development. In this experiment, buds three nodes from the base of pea plants were treated when the plants had three expanded leaves. These buds were determined for reproductive development, and produced the same number of new nodes (A) when they were allowed to grow in situ by removing the terminal shoot; (B) when they were removed from the plant and cultured; and (C) when only the bud meristem was cultured. If these buds had not been committed to reproductive development, they would have developed like vegetative buds and produced many more nodes before flowering.

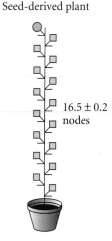

Seed-derived plant

16.5 ± 0.2 nodes

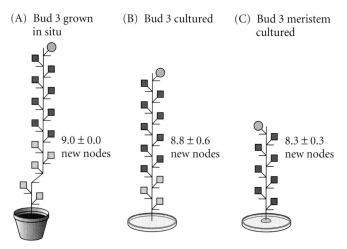

(A) Bud 3 grown in situ — 9.0 ± 0.0 new nodes

(B) Bud 3 cultured — 8.8 ± 0.6 new nodes

(C) Bud 3 meristem cultured — 8.3 ± 0.3 new nodes

floral parts itself. The inflorescence is the reproductive backbone (stem) that displays the flowers (see Figure 20.24). The inflorescence meristem probably arises through the action of a gene that suppresses terminal flower formation. The *CEN-*

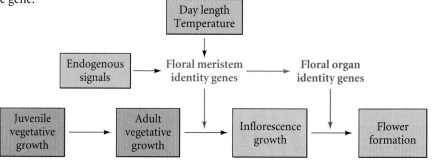

Figure 20.38
Different phases of shoot growth are regulated by environmental and endogenous signals. Adult meristems can express floral meristem identity genes in response to day length, temperature, and internal signals, including gibberellins. These genes turn on additional genes which leads to the expression of floral organ identity genes, essential for the production of sepals, petals, stamens, and carpels.

TRORADIALUS (*CEN*) gene in snapdragons suppresses terminal flower formation (Bradley et al. 1996). It suppresses expression of *FLORICAULA* (*FLO*), which specifies floral meristem identity. Curiously, the expression of *FLO* is necessary for *CEN* to be turned on.

The *Arabidopsis* homologue of *CEN* (*TERMINAL FLOWER 1* or *TFL1*) is expressed during the vegetative phase of development as well, and has the additional function of delaying the commitment to inflorescence development (Bradley et al. 1997). Overexpression of *TFL1* in transgenic *Arabidopsis* extends the time before a terminal flower forms (Ratcliffe et al. 1998). *TFL1* must delay the reproductive transition. Garden peas branch one more time than snapdragons do before forming a flower. That is, the axillary meristem does not directly produce a flower, but acts as an inflorescence meristem that initiates floral meristems. Two genes, *DET* and *VEG1*, are responsible for this more complex inflorescence, and only when both are nonfunctional is a terminal flower formed (Figure 20.39; Singer et al. 1996).

Floral meristem identity genes

The next step in the reproductive process is the specification of floral meristems—those meristems that will actually produce flowers (Weigel 1995). In *Arabidopsis*, *LEAFY* (*LFY*), *APETALA 1* (*AP1*), and *CAULIFLOWER* (*CAL*) are **floral meristem identity genes** (Figure 20.40). *LFY* is the homologue of *FLO* in snapdragons, and its upregulation during development is key to the transition to reproductive development (Blázquez et al. 1997; Blázquez and Weigel 2000). The LFY transcription factor can actually move between cell layers in the meristem before activating other flowering genes (Sessions et al. 2000). Analysis of the promoter of *LFY* reveals that there are separate sites for activation by the daylength pathway and the autonomous pathway. The autonomous pathway works through the binding of gibberellin to the *LFY* promoter.

Expression of floral meristem identity genes is necessary for the transition from an inflorescence meristem to a floral meristem. Strong *lfy* mutants tend to form leafy shoots in the axils where flowers form in wild-type plants; they are unable to make the transition to floral development. If *LFY* is overexpressed, flowering occurs early. For example, when aspen was transformed with a *LFY* gene that was expressed throughout the plant, the time to flowering was dramatically shortened from years to months (Weigel and Nilsson 1995). *AP1* and *CAL* are closely related and redundant genes. The *cal* mutant looks like the wild-type plant, but *ap1 cal* double mutants produce inflorescences

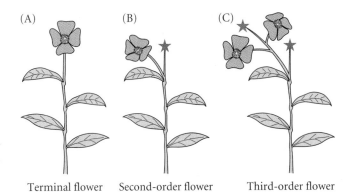

Terminal flower   Second-order flower   Third-order flower

Figure 20.39
Regulation of inflorescence branching architecture. (A) In the simplest angiosperms, a terminal flower forms directly from the terminal shoot apex. (B) *Arabidopsis* and snapdragons produce flowers on axillary branches (second-order flowers). (C) In peas, the axillary branches grow out and initiate flowers (third-order flowers), but do not directly produce flowers. Recent evidence suggests that more complex branching patterns in inflorescences are the result of suppressing the expression of flowering genes in meristems. The stars indicate meristems where floral gene expression is suppressed. In *Arabidopsis* and snapdragons (A), a single gene suppresses flowering in the first-order meristem. In peas (B), two genes are necessary to suppress flowering in the first- and second-order meristems.

Figure 20.40
Floral meristem identity mutants. (A) Wild-type *Arabidopsis*. (B) *leafy* mutant. (C) *apetala1* mutant. (D) *leafy apetala1* double mutant. (Photographs courtesy of J. Bowman.)

Figure 20.41
Arabidopsis double mutant of *ap1* and *cal*. Since *cal* alone gives a wild-type phenotype, the double mutant demonstrates the redundancy of these two genes in the flowering pathway. (Photograph courtesy of J. Bowman.)

that look like cauliflower heads (Figure 20.41). More recently another gene, *FRUITFUL* (*FUL*), with close sequence similarity to *AP1* and *CAL* has been characterized. The FUL gene family is closely related to the AP1 gene family in angiosperms and most likely arose as a result of gene duplication and divergence. It is also partially redundant to *AP1* and *CAL* (Ferrandiz et al. 2000). *CAL*, however, arose only within the *Brassica* through an *AP1* gene duplication event, and appears to have accompanied the domestication of cauliflower and broccoli (Purugganan et al. 2000).

Organ identity genes: The ABC model

Floral meristem identity genes initiate a cascade of gene expression that turns on region-specifying (cadastral) genes, which further specify pattern by initiating transcription of floral **organ identity genes** (Weigel 1995). *SUPERMAN* (*SUP*) is an example of a cadastral gene in *Arabidopsis* that plays a role in specifying boundaries for organ identity gene expression. Three classes (A, B, and C) of organ identity genes are necessary to specify the four whorls of floral organs (Figures 20.42 and 20.43; Coen and Meyerowitz 1991). They are homeotic genes (but not Hox genes; rather, most are members of the **MADS box** gene family, which had its origins before the divergence of animals and plants). In *Arabidopsis*, these genes include *AP2*, *AGAMOUS* (*AG*), *AP3*, and *PISTILLATA* (*PI*). Class A genes (*AP2*) alone specify sepal development. Class A genes and class B genes (*AP3* and *PI*) together specify

(A)
Wild-type

(B)
ap2 (A⁻)

(C)
ap3 and pi (B⁻)

(D)
ag (C⁻)

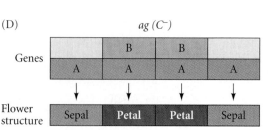

Figure 20.42
Wild-type and mutant phenotypes of the *Arabidopsis* class A (*ap2*), class B (*ap3*, *pi*), and class C (*ag*) floral organ identity genes. (Model proposed by Coen and Meyerowitz 1991; photographs courtesy of J. Bowman.)

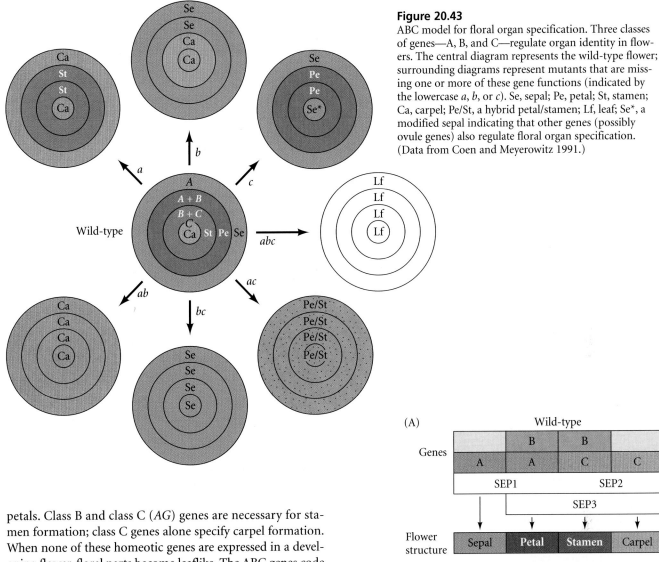

Figure 20.43
ABC model for floral organ specification. Three classes of genes—A, B, and C—regulate organ identity in flowers. The central diagram represents the wild-type flower; surrounding diagrams represent mutants that are missing one or more of these gene functions (indicated by the lowercase *a*, *b*, or *c*). Se, sepal; Pe, petal; St, stamen; Ca, carpel; Pe/St, a hybrid petal/stamen; Lf, leaf; Se*, a modified sepal indicating that other genes (possibly ovule genes) also regulate floral organ specification. (Data from Coen and Meyerowitz 1991.)

petals. Class B and class C (*AG*) genes are necessary for stamen formation; class C genes alone specify carpel formation. When none of these homeotic genes are expressed in a developing flower, floral parts become leaflike. The ABC genes code for transcription factors that initiate a cascade of events leading to the actual production of floral parts.

Although the ABC model is compelling, it is not sufficient to support the hypothesis that flowers evolved from leaves. Overexpression of the ABC genes in leaves does not produce petals or other flower parts. About a decade after the ABC model was proposed, a fourth class of floral organ identity genes, *SEPALLATA* (*SEP*), was identified (Figure 20.44; see review by Jack 2001). These MADS box genes can convert a leaf into a petal when ectopically expressed in the leaf. In the absence of *SEP* function, flowers become whorls of sepals. SEP transcription factors form dimers with ABC or SEP transcription factors to initiate floral development.

In addition to the ABC and SEP genes, class D genes are now being investigated that specifically regulate ovule development. The ovule evolved long before the other angiosperm floral parts and, while its development is coordinated with that of the carpel, one would expect more ancient, independent pathways to exist.

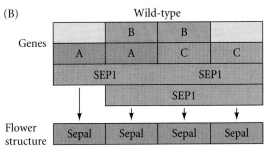

Figure 20.44
The *SEPALLATA* (*SEP*) genes are a fourth class of floral organ identity genes. (A) *Arabidopsis sep1⁺ sep2⁺ sep3⁺* plants produce whorls of sepals rather than the other three floral organs. (B) Ectopic expression of all three *SEP* genes in leaves causes the leaves to convert to petals, as shown in (C).

Transcription of floral organ identity genes is actually the beginning rather than the end of flower development. One of the amazing attributes of angiosperms is the tremendous diversity of flower phenotypes, many of which have evolved to attract specific pollinators. Adaxial/abaxial asymmetry in flowers is one contributing factor to diverse floral morphologies. Phylogenetic evidence indicates that this trait arose independently many times, as well as being lost many times. The cloning of the *CYCLOIDEA* (*CYC*) gene in snapdragon has led to extensive discussion about the molecular mechanisms involved in the evolution of asymmetry (Donoghue et al. 1998; Cubas et al. 2001). In snapdragon, *CYC* expression is observed on the adaxial side of the floral primordium, early in development. In the *cyc* mutants, flowers have a more radial symmetry. Did *CYC* evolve independently numerous times, or are there many ways to make an asymmetric flower? Putative *CYC* orthologues have been identified in other species, and one plausible explanation for the multiple origins of asymmetry is that this gene was recruited for the same function multiple times. The combination of developmental and phylogenetic approaches to the study of patterning in plants promises to provide insight into the origins of the morphological novelties that are so extensive among the angiosperms.

Senescence

Flowering and senescence (a developmental program leading to death) are closely linked in many angiosperms. In some species, individual flower petals senesce following pollination. Orchids, which stay fresh for long periods of time if they are not pollinated, are a good example. Fruit ripening (and ultimately overripening) is an example of organ senescence. Whole-plant senescence leads to the death of the entire sporophyte generation. **Monocarpic** plants flower once and then senesce. **Polycarpic** plants such as the bristlecone pine can live for thousands of years (4,900 years is the current record) and flower repeatedly.

In polycarpic plants, death is by accident; in monocarpic plants, it appears to be genetically programmed. Flowers and fruits play a key role in the process, and their removal can sometimes delay senescence. In some legumes, senescence can be delayed by removing the developing seed—in other words, the embryo may trigger senescence in the parent plant. During flowering and fruit development, nutrients are reallocated from other parts of the plant to support the development of the next generation. The reproductive structures become a nutrient sink, and this can lead to whole-plant senescence.

Snapshot Summary: Plant Development

1. Land plants have their origins in the charophytes, a lineage of freshwater green algae.

2. Plants are characterized by alternation of generations; that is, their life cycle includes both diploid and haploid multicellular generations.

3. A multicellular diploid sporophyte produces haploid spores via meiosis. The spores divide mitotically to produce a haploid gametophyte. Mitotic divisions within the gametophyte produce the gametes. The diploid sporophyte results from the fusion of two gametes.

4. Land plants have evolved mechanisms to protect embryos. Angiosperm embryos develop deeply embedded in parent tissue. The parent tissue provides nutrients but only minimal patterning information. This evolutionary theme of an increasingly protected embryo is shared by both plants and animals.

5. In angiosperms, the male gamete (pollen) arrives at the style of the female gametophyte and effects fertilization through the pollen tube. Two sperm cells move through the pollen tube; one joins with the ovum to form the zygote, and the other is involved in the formation of the endosperm.

6. Early embryogenesis is generally characterized by the establishment of the shoot-root axis and by radial patterning yielding three tissue systems. Pattern emerges by the regulation of planes of cell division and the directions of cell expansion, since plant cells do not move during development.

7. As the angiosperm embryo matures, a food reserve is established. Only the rudiments of the basic body plan are established by the time embryogenesis ceases and the seed enters dormancy.

8. Pattern is elaborated during postembryonic development, when meristems construct the reiterative structures of the plant.

9. Unlike animals, the plant germ line is not set aside early in development. Coordination of signaling among leaves, roots, and shoot meristems regulates the transition to the reproductive state.

10. Floral meristem identity genes and organ identity genes enable the angiosperm flower to display a tremendous amount of morphological diversity

11. Reproduction may be followed by genetically programmed senescence of the parent plant.

Literature Cited

Alessa, L. and D. L. Kropf. 1999. F-actin marks the rhizoid pole in living *Pelvetia compressa* zygotes. *Development* 126: 201–209.

Arabidopsis Genome Initiative. 2000. Analysis of the genome sequence of the flowering plant *Arabidopsis thaliana*. *Nature* 408: 796–815.

Barton, M. K. and R. S. Poethig. 1993. Formation of the shoot apical meristem in *Arabidopsis thaliana*: An analysis of development in the wild-type and in the *shoot meristemless* mutant. *Development* 119: 823–831.

Belanger, K. D. and R. S. Quatrano. 2000. Polarity: The role of localized secretion. *Curr. Opin. Plant Biol.* 3: 67–72.

Benfey, P. N. 1999. Is the shoot a root with a view? *Curr. Opin. Plant Biol.* 2: 39–43.

Benfey, P. N., P. J. Linstead, K. Roberts, J. W. Schiefelbein, M.-T. Hauser and R. A. Aeschbacher. 1993. Root development in *Arabidopsis*: Four mutants with dramatically altered root morphogenesis. *Development* 119: 53–70.

Bharathan, G., T. E. Goliber, C. Moore, S. Kessler, T. Pham and N. R. Sinha. 2002. Homologies in leaf form inferred from *KNOX1* gene expression during development. *Science* 296: 1858–1860.

Bhattacharyya, M. K., A. M. Smith, T. H. N. Ellis, C. Hedley and C. Martin. 1990. The wrinkled-seed character of pea described by Mendel is caused by a transposon-like insertion in a gene encoding starch-branching enzyme. *Cell* 60: 115–122.

Blázquez, M. A. and D. Weigel. 2000. Integration of floral inductive signals in *Arabidopsis*. *Nature* 404: 889–892.

Blázquez, M. A., L. N. Soowai, I. Lee and D. Weigel. 1997. *LEAFY* expression and flower initiation in *Arabidopsis*. *Development* 124: 3835–3844.

Borevitz, J. O. and 9 others. 2002. Quantitative trait loci controlling light and hormone response in two accessions of *Arabidopsis thaliana*. *Genetics* 160:683–696

Bowman, J. 1994. *Arabidopsis: An Atlas of Morphology and Development*. Springer-Verlag, New York.

Bradley, D., R. Carpenter, L. Copsey, C. Vincent, S. Rothstein and E. Coen. 1996. Control of inflorescence architecture in *Antirrhinum*. *Nature* 379: 791–797.

Bradley, D., O. Ratcliffe, C. Vincent, R. Carpenter and E. Coen. 1997. Inflorescence commitment and architecture in *Arabidopsis*. *Science* 275: 80–83.

Brewbaker, J. L. and B. H. Kwack. 1963. The essential role of calcium ions in pollen germination and pollen tube growth. *Am. J. Bot.* 50: 859–863.

Brown, K. S. 1999. Deep Green rewrites evolutionary history of plants. *Science* 285: 990–991.

Brownlee, C. and F. Berger. 1995. Extracellular matrix and pattern in plant embryos: On the lookout for developmental information. *Trends Genet.* 11: 344–348.

Cai, G. and M. Cresti. 1999. Rethinking cytoskeleton in plant reproduction: Towards a biotechnological future? *Sex. Plant Reprod.* 12: 67–70.

Christianson, M. L. 1986. Fate map of the organizing shoot apex in *Gossypium*. *Am. J. Bot.* 73: 947–958.

Cionini, P. G., A. Bennici, A. Alpi and F. D'Amato. 1976. Suspensor, gibberellin, and in vitro development of *Phaseolus coccineus* embryos. *Planta* 131: 115–117.

Clark, J. K. and W. F. Sheridan. 1986. Developmental profiles of the maize embryo-lethal mutants *dek22* and *dek23*. *J. Hered.* 77: 83–92.

Clark, S. E. and Schiefelbein. 1997. Expanding insights into the role of cell proliferation in plant development. *Trends Cell Biol.* 7: 454–458

Clark, S.E., R. W. Williams and E. M. Meyerowitz 1997. The *CLAVATA1* gene encodes a putative receptor kinase that controls shoot and floral meristem size in *Arabidopsis*. *Cell* 89: 575–585.

Coen, E. S. and E. M. Meyerowitz. 1991. The war of the whorls: Genetic interactions controlling flower development. *Nature* 353: 31–37.

Columbo, L., J. Franken, A. P. Van der Krol, P. E. Wittich, H. J. Dons and G. C. Angenent. 1997. Downregulation of ovule specific MADS box genes from petunia results in maternally controlled defects in seed development. *Plant Cell* 9: 703–715.

Cox, D. N., A. Chao, J. Baker, L. Chang, D. Qiao and H. Lin. 1998. A novel class of evolutionarily conserved genes defined by piwi are essential for stem cell self-renewal. *Genes Dev.* 12: 3715–3727.

Cruden, R. W. and R. M. Lloyd. 1995. Embryophytes have equivalent sexual phenotypes and breeding systems: Why not a common terminology to describe them? *Am. J. Bot.* 82: 816–825.

Cubas, P., E. Coen and J. M. Martinex-Zapater. 2001. Ancient asymmetries in the evolution of flowers. *Curr. Biol.* 11: 1050–1052.

Deng, X. W. and P. H. Quail. 1999. Signalling in light-controlled development. *Semin. Cell Dev. Biol.* 10: 121–129.

Di Laurenzio, L. and 8 others. 1996. The *SCARECROW* gene regulates an asymmetric cell division that is essential for generating the radial organization of the *Arabidopsis* root. *Cell* 86: 423–433.

Dodds, P. N., A. E. Clarke and E. Newbigin. 1996. A molecular perspective on pollination in flowering plants. *Cell* 85: 141–144.

Doerner, P. 1999. Shoot meristems: Intercellular signals keep the balance. *Curr. Biol.* 9: R377–R380.

Donoghue, M. J., R. H. Ree and D. A. Baum. 1998. Phylogeny and the evolution of flower symmetry in the Asteridae. *Trends Plant Sci.* 3: 311–317

Drews, G. N., D. Lee and C. A. Christensen. 1998. Genetic analysis of female gametophyte development and function. *Plant Cell* 10: 5–17.

Esau, K. 1977. *Anatomy of Seed Plants*, 2nd Ed. John Wiley and Sons, New York.

Ferrandiz, C., Q. Gu, R. Martienssen and M. F. Yanofsky. 2000. Redundant regulation of meristem identity and plant architecture by *FRUITFUL*, *APETALA1* and *CAULIFLOWER*. *Development* 127: 725–734.

Franklin-Tong, V. E., B. K. Drobak, A. C. Allan, P. A. C. Watkins and A. J. Trewavas. 1996. Growth of pollen tubes in *Papaver rhoeas* is regulated by a slow-moving calcium wave propagated by inositol 1,4,5-trisphosphate. *Plant Cell* 8: 1305–1321.

Franklin-Tong, V. E., T. L. Holdaway-Clarke, K. R. Straatman, J. G. Kunkel and P. K. Hepler. 2002. Involvement of extracellular calcium influx in the self-incompatibility response of *Papaver rhoeas*. *Plant J.* 29: 333–345.

Friedman, W. E. 1998. The evolution of double fertilization and endosperm: An "historical" perspective. *Sex. Plant Reprod.* 11: 6–16.

Gaude, T. and S. McCormick. 1999. Signaling in pollen-pistil interactions. *Semin. Cell Dev. Biol.* 10: 139–147.

Genome International Sequencing Consortium. 2001. Initial sequencing and analysis of the human genome. *Nature* 409: 860–921

Gifford, E. M. and A. S. Foster. 1989. *Morphology and Evolution of Vascular Plants*. W. H. Freeman, New York.

Goff, S. A. and 53 others. 2002. A draft sequence of the rice genome (*Oryza sativa* L. ssp. *japonica*). *Science* 296: 92–100.

Grossniklaus, U., J. Vielle-Calzada, M. A. Hoeppner and W. B. Gagliano. 1998. Maternal control of embryogenesis by *MEDEA*, a polycomb group gene in *Arabidopsis*. *Science* 280: 446–450.

Haccius, B. 1963. Restitution in acidity-damaged plant embryos: Regeneration or regulation? *Phytomorphology* 13: 107–115.

Harada, J. J. 2001. Role of *Arabidopsis LEAFY COTYLEDON* genes in seed development. *J. Plant Physiol.* 158: 405–409.

Hareven, D., T. Gutfinger, A Pornis, Y. Eshed and E. Lifschitz. 1996. The making of a compound leaf: Genetic manipulation of leaf architecture in tomato. *Cell* 84: 735–744.

Hay, A., H. Kaur, A. Phillips, P. Hedden, S. Hake and M. Tsiantis. 2002. The gibberellin pathway mediates *KNOTTED-1* type homeobox function in plants with different body plans. *Curr. Biol.* 12: 1557–1565.

Hempel, F. D., D. R. Welch and L. J. Feldman. 2000. Floral induction and determination: Where is flowering controlled? *Trends Plant Sci.* 5: 17–21.

Higashiyama, T., S. Yabe, N. Sasaki, Y. Nishimura, S. Miyagishima, H. Kyroiwa and T. Kuroiwa. 2001. Pollen tube attraction by the synergid cell. *Science* 293: 1480–1483.

Hofer, J. M. I. and T. H. N. Ellis. 1998. The genetic control of patterning in pea leaves. *Trends Plant Sci.* 3: 434–444.

Huala, E. and I. M. Sussex. 1993. Determination and cell interactions in reproductive meristems. *Plant Cell* 5: 1157–1165.

Hulskamp, M., K. Schneitz and R. E. Pruitt. 1995. Genetic evidence for a long-range activity that directs pollen tube guidance in *Arabidopsis. Plant Cell* 7: 57–64.

Irish, V. F. and I. M. Sussex. 1992. A fate map of the *Arabidopsis* embryonic shoot apical meristem. *Development* 115: 745–754.

Jack, T. 2001. Relearning our ABCs: New twists on an old model. *Trends Plant Sci.* 6: 310–316.

Jackson, D. 1996. Plant morphogenesis: Designing leaves. *Curr. Biol.* 6: 917–919.

Jaffe, L. A., M. H. Weisenseel and L. F. Jaffe. 1975. Calcium accumulation within the growing tips of pollen tubes. *J. Cell Biol.* 67: 488–492.

Janssen, B.-J., L. Lund and N. Sinha. 1998. Overexpression of a homeobox gene *LeT6* reveals indeterminate features in the tomato compound leaf. *Plant Physiol.* 117: 771–786.

Jean, R. V. and D. Barabé. 1998. *Symmetry in Plants.* World Scientific Publishing, River Edge, NJ.

Jeong, S., A. E. Trotochaud and S. E. Clark. 1999. The *Arabidopsis CLAVATA2* gene encodes a receptor-like protein required for the stability of the *CLAVATA1* receptor-like kinase. *Plant Cell* 11: 1925–1933.

Johri, B. M., K. B. Ambegaokar and P. S. Srivastava. 1992. *Comparative Embryology of Angiosperms.* Springer-Verlag, New York.

Jurgens, G. 2001. Apical-basal pattern formation in *Arabidopsis* embryogenesis. *EMBO J.* 20: 3609–3616.

Kachroo, A., C. R. Schopfer, M. E. Nasrallah and J. B. Nasrallah. 2001 Allele-specific receptor-ligand interactions in *Brassica* self-incompatibility. *Science* 293: 1824–1826.

Kerstetter, R., K. Bollman, R. A. Taylor, K. Bomblies and R. S. Poethig. 2001. *KANADI* regulates organ polarity in *Arabidopsis. Nature* 411: 704–709.

Kropf, D. L., S. R. Bisgrove and W. E. Hable. 1999. Establishing a growth axis in fucoid algae. *Trends Plant Sci.* 4: 490–494.

Kumaran, M. K., J. L. Bowman and V. Sundaresan. 2002. *YABBY* polarity genes mediate the repression of *KNOX* homeobox genes in *Arabidopsis. Plant Cell* 14: 2761–2770.

Kwong, R. W., A. Q. Bui, H. Lee, L. W. Kwong, R. L. Fischer, R. B. Goldberg and J. J. Harada. 2003. *LEAFY COTYLEDON1-LIKE* defines a class of regulators essential for embryo development. *Plant Cell* 15: 5–18

Lang, A., M. K. Chailakhyan and I. A. Frolova. 1977. Promotion and inhibition of flower formation in a day-neutral plant in grafts with short-day and long-day plants. *Proc. Natl. Acad. Sci. USA* 74: 2412–2416.

Laux, T. and G. Jurgens. 1994. Establishing the body plan of the *Arabidopsis* embryo. *Acta Bot. Neer.* 43: 247–260.

Lawson, E. J. R. and R. S. Poethig. 1995. Shoot development in plants: Time for a change. *Trends Genet.* 11: 263–268.

Lee, H.-S., S. Huang and T.-H. Kao. 1994. S proteins control rejection of incompatible pollen in *Petunia inflata. Nature* 367: 560–563.

Levy, Y. Y. and C. Dean. 1998. The transition to flowering. *Plant Cell* 10: 1973–1989.

Liu, C.-M., Z.-H. Xu and N.-H. Chua. 1993. Auxin polar transport is essential for the establishment of bilateral symmetry during early plant embryogenesis. *Plant Cell* 5: 621–630.

Long, J. A., S. Woody, S. Poethig, E. M. Meyerowitz and M. K. Barton. 2002. Transformation of shoots into roots in *Arabidopsis* embryos mutant at the *TOPLESS* locus. *Development* 129: 2797–2806.

Lord, E. M., L. L. Walling and G. Y. Jauh. 1996. Cell adhesion in plants and its role in pollination. *In* M. Smallwood, J. P. Knox and D. J. Bowles (eds.), *Membranes: Specialized Functions in Plants.* Bios Scientific Publishers, Oxford.

Lu, P., R. Porat, J. A. Nadeau and S. D. O'Neill. 1996. Identification of a meristem L1 layer-specific gene in *Arabidopsis* that is expressed during embryonic pattern formation and defines a new class of homeobox genes. *Plant Cell* 8: 2155–2168.

Marx, G. A. 1987. A suite of mutants that modify pattern formation in pea leaves. *Plant Mol. Biol. Reprod.* 5: 311–335.

Mayer, U., R. A. Torres Ruiz, T. Berleth, S. Misera and G. Jurgens. 1991. Mutations affecting body organization in the *Arabidopsis* embryo. *Nature* 353: 402–406.

McCarty, D. R., C. B. Carson, P. S. Stinard and D. S. Robertson. 1989. Molecular analysis of *viviparous-1*: An abscisic acid-insensitive mutant of maize. *Plant Cell* 1: 523–532.

McConnell, J. R., J. Emery, Y. Eshed, N. Bao, J. Bowman and M. K. Barton. 2001. Role of PHABULOSA and PHAVOLUTA in determining radial patterning in shoots. *Nature* 411: 709–713

McDaniel, C. N. and R. S. Poethig. 1988. Cell-lineage patterns in the shoot apical meristem of the germinating maize embryo. *Planta* 175: 13–22.

McDaniel, C. N., S. R. Singer and S. M. E. Smith. 1992. Developmental states associated with the floral transition. *Dev. Biol.* 153: 59–69.

McDaniel, C. N., L. K. Hartnett and K. A. Sangrey. 1996. Regulation of node number in day-neutral *Nicotiana tabacum:* A factor in plant size. *Plant J.* 9: 55–61.

Meinke, D. W., L. H. Franzmann, T. C. Nickle and E. C. Yeung. 1994. *LEAFY COTYLEDON* mutants of *Arabidopsis, Plant Cell* 6: 1049–1064.

Meyerowitz, E. M. 1997. Genetic control of cell division patterns in developing plants. *Cell* 88: 299–308.

Meyerowitz, E. M. 2002. Plants compared to animals: The broadest comparative study of development. *Science* 295: 1482–1485.

Mogie, M. 1992. *The Evolution of Asexual Reproduction in Plants.* Chapman and Hall, New York.

Mouse Genome Sequencing Consortium. 2002. Initial sequencing and comparative analysis of the mouse genome. *Nature* 420: 520–562.

Nasrallah, J. B. 2002. Recognition and rejection of self in plant reproduction. *Science* 296: 305–308.

Nasrallah, M. E., P. Liu and J. B. Nasrallah. 2002. Generation of self-incompatible *Arabidopsis thaliana* by transfer of two *S* locus genes from *A. lyrata. Science* 297: 247–249

Poethig, R. S. 1987. Clonal analysis of cell lineage patterns in plant development. *Am. J. Bot.* 74: 581–594.

Preuss, D. 1999. Chromatin silencing and *Arabidopsis* development: A role for polycomb proteins. *Plant Cell* 11: 765–768.

Purugganan, M. D., A. L. Boyles and J. I. Suddith. 2000. Variation and selection at the *CAULIFLOWER* floral homeotic gene accompanying the evolution of domesticated *Brassica oleracea. Genetics* 155: 855–862.

Ratcliffe, O. J. and J. L. Riechmann. 2002. *Arabidopsis* transcription factors and the regulation of flowering time: A genomic perspective. *Curr. Issues Mol. Biol.* 4: 77–91.

Ratcliffe, O. J., I. Amaya, C. A. Vincent, R. Rothstein, R. Carpenter, E. S. Coen and D. J. Bradley. 1998. A common mechanism controls the life cycle and architecture of plants. *Development* 125: 1609–1615.

Ray, A. 1998. New paradigms in plant embryogenesis: Maternal control comes in different flavors. *Trends Plant Sci.* 3: 325–327.

Ray, S., G. T. Golden and A. Ray. 1996. Maternal effects of the *short integument* mutation on embryo development in *Arabidopsis. Dev. Biol.* 180: 365–369.

Reid, J. B., I. C. Murfet, S. R. Singer, J. L. Weller and S. A. Taylor. 1996. Physiological genetics of flowering in *Pisum. Semin. Cell Dev. Biol.* 7: 455–463.

Rojo, E., V. K. Sharma, V. Koveleva, N. V. Raikhel and J. C. Fletcher. 2002. CLV3 is localized to the extracellular space, where it activates the *CLAVATA* stem cell signaling pathway. *Plant Cell* 14: 969–977.

Sabatini, S. and 10 others. 1999. An auxin-dependent distal organizer of pattern and polarity in the *Arabidopsis* root. *Cell* 99: 463–472.

SanMiguel, P. and 9 others. 1996. Nested retrotransposons in the intergenic regions of the maize genome. *Science* 274: 765–768.

Schauer, S. E., S. E. Jacobsen, D. W. Meinke and A. Ray. 2002. *DICER-LIKE1*: Blind men and elephants in *Arabidopsis* development. *Trends Plant Sci.* 17: 487–491.

Scheres, B. and P. N. Benfey. 1999. Asymmetric cell division in plants. *Annu. Rev. Plant Physiol. Plant Mol. Biol.* 50: 505–537.

Scheres, B. and R. Heidstra. 1999. Digging out roots: Pattern formation, cell division, and morphogenesis in plants. *Curr. Top. Dev. Biol.* 45: 207–247.

Scheres, B., L. Di Laurenzio, V. Willemsen, M.-T. Hauser, K. Janmaat, P. Weisbeek and P. N. Benfey. 1995. Mutations affecting the radial organisation of the *Arabidopsis* root display specific defects throughout the embryonic axis. *Development* 121: 53–62.

Schiavone, F. M. and R. H. Racusen. 1990. Microsurgery reveals regional capabilities for pattern reestablishment in somatic carrot embryos. *Dev. Biol.* 141: 211–219.

Schwartz, B. W., E. C. Yeung and D. W. Meinke. 1994. Disruption of morphogenesis and transformation of the suspensor in abnormal suspensor mutants of *Arabidopsis*. *Development* 120: 3235–3245.

Sessions, A., M. F. Yanfosky and D. Weigel. 2000. Cell-cell signaling and movement by the floral transcription factors LEAFY and APETALA1. *Science* 289: 779–781.

Siegfried, K. R., Y. Eshed, S. F. Baum, D. Otsuga, G. N. Drews and J. L. Bowman. 1999. Members of the *YABBY* gene family specify abaxial cell fate in *Arabidopsis*. *Development* 126: 4117–4128.

Simpson, G. G., A. R. Gendall and C. Dean. 1999. When to switch to flowering. *Annu. Rev. Cell Dev. Biol.* 15: 519–550.

Singer, S. R., C. H. Hannon and S. C. Huber. 1992. Acquisition of competence for floral determination in shoot apices of *Nicotiana*. *Planta* 188: 546–550.

Singer, S. R., S. L. Maki, J. Sollinger, H. Mullen, J. Fick and A. McCall. 1996. Suppression of terminal flower formation in pea: Evolutionary implications. *Dev. Biol.* 175: 377.

Sinha, N. 1999. Leaf development in angiosperms. *Annu. Rev. Plant Physiol. Plant Mol. Biol.* 50: 419–446.

Smith, R. H. 1984. Developmental potential of excised primordial and expanding leaves of *Coleus blumei* Benth. *Am. J. Bot.* 71: 114–120.

Southworth, D. 1996. Gametes and fertilization in flowering plants. *Curr. Top. Dev. Biol.* 34: 259–279.

Stebbins, L. 1974. *Flowering Plants: Evolution Above the Species Level*. Belknap Press, Cambridge, MA.

Steeves, T. A. 1966. On the determination of leaf primordia in ferns. *In* E. G. Cutter (ed.), *Trends in Plant Morphogenesis*. Longman, London, pp. 200–219.

Steeves, T. A. and I. M. Sussex. 1989. *Patterns in Plant Development*, 2nd Ed. Cambridge University Press, New York.

Taylor, C. B. 1996. More arresting developments: *S* RNases and interspecific incompatibility. *Plant Cell* 8: 939–941.

Trewavas, A. J. and R. Malho. 1998. Ca^{2+} signalling in plant cells: The big network. *Curr. Opin. Plant Biol.* 1: 428–433.

Vega, S. H., M. Sauer, J. A. J. Orkwiszewski and R. S. Poethig. 2002. The *early phase change* gene in maize. *Plant Cell* 14: 133–147.

Veit, B., S. P. Briggs, R. J. Schmidt, M. F. Yanofsky and S. Hake. 1998. Regulation of leaf initiation by the *terminal ear 1* gene of maize. *Nature* 393: 166–168.

Vielle-Calzada, J., J. Thomas, C. Spillane, A. Coluccio, M. A. Hoeppner and U. Grossniklaus. 1999. Maintenance of genomic imprinting at the *Arabidopsis MEDEA* locus requires zygotic DDM1 activity. *Genes Dev.* 13: 2971–2982.

Weigel, D. 1995. The genetics of flower development: From floral induction to ovule morphogenesis. *Annu. Rev. Genet.* 29: 19–39.

Weigel, D. and O. Nilsson. 1995. A developmental switch sufficient for flower initiation in diverse plants. *Nature* 377: 495–500.

Weterings, K., N. R. Apuya, Y. Bi, R. L. Fischer, J. J. Harada and R. B. Goldberg. 2001. Regional localization of suspensor mRNAs during early embryo development. *Plant Cell* 13: 2409–2425.

Wilhelmi, L. K. and D. Preuss. 1996. Self-sterility in *Arabidopsis* due to defective pollen tube guidance. *Science* 274: 1535–1537.

Wilhelmi, L. K. and D. Preuss. 1999. The mating game: Pollination and fertilization in flowering plants. *Curr. Opin. Plant Biol.* 2: 18–22.

Willemsen, V., H. Wolkenfelt, G. de Vrieze, P. Weisbeek and B. Scheres. 1998. The *HOBBIT* gene is required for the formation of the root meristem in the *Arabidopsis* embryo. *Development* 125: 521–531.

Williams, J. H. and W. E. Friedman. 2002. Identification of diploid endosperm in an early angiosperm lineage. *Nature* 415: 522–526.

Yadegari, R. and 8 others. 1994. Cell differentiation and morphogenesis are uncoupled in *Arabidopsis raspberry* embryos. *Plant Cell* 6: 1713–1729.

Yeung, E. C. and I. M. Sussex. 1979. Embryogeny of *Phaseolus coccineus*: The suspensor and the growth of the embryo-proper in vitro. *Z. Pflanzenphysiol.* 91: 423–433.

Yu, J. and 99 others. 2002. A draft sequence of the rice genome (*Oryza sativa* L. ssp. *indica*). *Science* 296: 79–92.

Zeevaart, J. A. D. 1984. Photoperiodic induction, the floral stimulus, and floral-promoting substances. *In* D. Vince-Prue, B. Thomas and K. E. Cockshull (eds.), *Light and the Flowering Process*. Academic Press, Orlando, pp. 137–142.

21 Medical implications of developmental biology

ONE OF THE NEWEST AREAS OF DEVELOPMENTAL BIOLOGY involves the medical uses of our newly acquired knowledge of mammalian development. This knowledge has numerous medical applications, including:

- The regulation of fertility.
- The identification of genetic defects that affect development.
- The identification of teratogenic compounds that affect development.
- The identification of factors in the maternal environment that may influence the health of the fetus.
- The realization that cancers can be disruptions of developmental regulation and might be cured through developmental processes.
- Attempts to detect developmental diseases (including cancers) and cure them through cloning, stem cell therapy, and genetic engineering.
- The attempt to cure traumatic and degenerative disease through regeneration.*

When one discusses the medical aspects of any science, one must take social issues, as well as purely scientific ones, into account. Discussions concerning how genetic diseases, birth defects, cancers, or infertility might be controlled also raise issues concerning social, political, and economic motives and opportunities.

INFERTILITY

Diagnosing Infertility

Infertility—the inability to achieve or sustain pregnancy—is not a disease in the usual sense of the word. It is not a symptom or condition that prevents the physical well-being of the infertile individuals or couples. However, since the desire to have children can be exceptionally strong for biological and social reasons (indeed, empires have fallen due to infertility), it is certainly an important condition in our society, and it is usually managed in the context of clinical medicine.

There is no precise definition of infertility. Rather, infertility is the absence of pregnancy after an appropriate duration of attempting conception by regular inter-

---

*Much of the material in this chapter is of a medical nature, and while it seeks to provide general information about these topics, this chapter is not intended to provide medical advice for specific persons or disorders. This chapter also does not propose to condense all of medical embryology into a single unit. Rather, it attempts to summarize particular principles that will be useful in studying the normal or abnormal development of any human organ. For more complete medical embryology texts, see Dudek and Fix (1998), Moore et al. (1998), Carlson (1999), Sadler and Langman (2000), and Larsen et al. (2001).

course. This duration differs between and within cultures. Population conception curves reveal that 80% of couples will have achieved a conception by 12 months and 90% by 18 months. The remaining 10% may be labeled as having infertility, or "subfertility," but in the following 6 months about half of this "infertile" group achieves pregnancy.

Infertility can be caused by failure to ovulate a mature oocyte, by few or defective sperm, by physical blockage of the male or female ducts, or by incompatibilities between the sperm and the milieu of the egg or the reproductive tract (McVeigh and Barlow 2000). While there are numerous treatments for women that can lead to the maturation and ovulation of oocytes, there are relatively few treatments for men who are not making sufficient sperm. In women, exogenous gonadotropins or anti-estrogenic drugs (clomiphene or tamoxifen), can be used to stimulate the ovaries. In men, sperm may be concentrated and injected either into the oocyte or into the reproductive tract near the oocyte.

In Vitro Fertilization (IVF)

In vitro fertilization (IVF) is an infertility treatment in which eggs and sperm are retrieved from the male and female partners and placed together in a petri dish for fertilization. After the eggs have begun dividing, they are transferred into the female partner's uterus, where implantation and embryonic development occur as in a normal pregnancy.

IVF was developed in the early 1970s to treat infertility caused by blocked or damaged fallopian tubes. The first IVF baby, Louise Brown, was born in England in 1978. Since then, the number of IVF procedures performed each year has increased, and the success rate has improved significantly. The success rates compare favorably to natural pregnancy rates in any given month when the woman is under age 40 and there are no sperm problems (see Trounson and Gardner 2000).

Because IVF was the first **assisted reproductive technology** (**ART**) developed and widely publicized, many people mistakenly believe that IVF is the only treatment option for infertile couples. Actually, less than 3 percent of all patients who seek treatment for infertility receive IVF. Most infertile couples respond well to less complicated treatment options, such as hormonal therapies and artificial insemination. IVF remains the most commonly used of the ART procedures.

The IVF procedure

The IVF procedure has four basic steps, outlined below and in Figure 21.1.

- **Step 1: Ovarian stimulation and monitoring.** Having several mature eggs available for IVF increases the possibility that at least one will result in a pregnancy. Typically, gonadotropins or anti-estrogens are used to "hyperstimulate" the ovaries to produce several mature oocytes.
- **Step 2: Egg retrieval.** Once the follicle has matured (but not yet ruptured), the physician attempts to retrieve as many eggs as possible. Although they are usually called eggs, the female gametes about to be ovulated are actually metaphase II oocytes (see Chapter 19). The physician retrieves the oocytes by guiding an aspiration pipette to each mature follicle and sucks up the oocyte. Once the oocytes are recovered, those that are mature and healthy are transferred to a sterile container to await fertilization in the laboratory.
- **Step 3: Fertilization.** A semen sample is collected from the male partner approximately 2 hours before the fe-

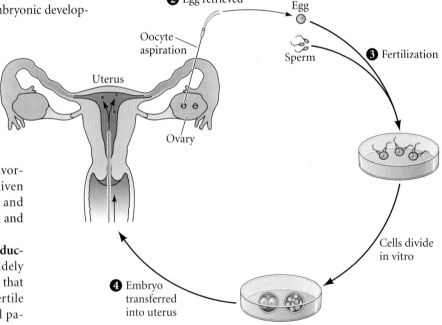

Figure 21.1

In vitro fertilization. (A) The IVF process can be divided into four basic steps: (1) ovarian stimulation, (2) egg retrieval; (3) fertilization and (4) transfer of the embryo into the uterus. (B) Assisted hatching, whereby a hole is poked in the zona pellucida, is a procedure to help ensure the embryo implants in the uterus. (B, photograph courtesy of The Institute for Reproductive Medicine and Science of St. Barnabas, Livingston, NJ.)

male partner's oocytes are retrieved. These sperm are then processed (a procedure called sperm washing) using various techniques. Sperm washing capacitates the sperm and selects only the healthiest and most active sperm in the sample. The selected sperm are then placed in a petri dish with the oocytes, and the gametes are incubated at body temperature. If fertilization is successful, the eggs will begin to divide, and the resulting embryos will shortly be ready to be transferred into the uterus

• **Step 4: Embryo transfer.** Embryo transfer is not a complicated procedure and can be performed without anesthesia. The embryos are placed in a tubular instrument called a catheter. The physician then inserts the catheter through the female partner's vagina and cervix in order to place the embryos directly into the uterus. Normal implantation and maturation of at least one embryo is required to achieve pregnancy.

In cases in which fertilization has been achieved in vitro, but after a number of cycles, implantation into the uterus fails, the physician may suggest "assisted hatching," in which a small hole is lysed in the zona pellucida prior to inserting the embryo into the uterus (Figure 21.1B). This procedure ensures that the embryo will be able to hatch from the zona pellucida in time to adhere to the uterus.

Variations on IVF

In addition to the sperm-meets-egg-in-a-dish way of performing IVF, other techniques have become available. **Intracytoplasmic sperm injection (ICSI)** was developed to treat couples who had a low probability of achieving fertilization due to the male partner's extremely low numbers of normal viable sperm (a condition called **oligospermia**). In ICSI, a single sperm is injected into the cytoplasm of an egg (Figure 21.2). Male infertility may be caused by genetic factors leading to poor sperm

production or by blockages or abnormalities of the ejaculatory ducts. Men who have had a severe injury to their reproductive organs, a vasectomy, or chemotherapy or radiation for testicular cancers may have few sperm in the ejaculate. ICSI, when combined with in vitro fertilization, allows these couples a more favorable probability of achieving conception.

Gamete intrafallopian transfer, or **GIFT,** was developed in 1984 as another variation on in vitro fertilization. Here, sperm are injected into the oviduct at the time when the oocytes are ovulated. This technique is often used by couples with unexplained infertility in which the female partner has at least one open fallopian tube. GIFT is also recommended for couples whose infertility is due to cervical or immunological factors that prevent the sperm from reaching the oocyte in the oviduct. The main difference between IVF and GIFT is that GIFT fertilization occurs naturally within the female partner's body, instead of in the laboratory as with IVF.

Zygote intrafallopian transfer, or **ZIFT**, is another variation on IVF. ZIFT is also called **tubal embryo transfer**. As in IVF, fertilization takes place outside the body in a petri dish. The resulting zygotes are then transferred into the female partner's fallopian tube. The location where fertilization takes place and the physician's ability to observe and confirm fertilization are the main differences between ZIFT and GIFT. With GIFT, the actual fertilization cannot be observed, because the eggs and sperm are united inside the female partner's fallopian tube.

Success rates and complications of IVF

The rate of delivery of live babies per oocyte retrieval depends on the age of the female partner. Some recent statistics suggest that approximately 31 couples out of every 100 who try one retrieval with IVF are likely to achieve pregnancy and delivery. Compared with the one in four (or 25%) probability of achieving conception each cycle for normal healthy couples using unprotected intercourse, IVF offers improved chances of conception to some infertile couples. The success rate drops to 25.5%, however, for women 35–37 years of age, and to 17.1% for women 38–40. After 40 years of age, the success rate is less than 5% (CDC 2002a). This decline is most likely due to the declining quality of eggs as women advance in age.

Although the IVF procedure is quite successful in achieving pregnancy, it does carry the risk of multiple births. It is well known that the transfer of more embryos increases the chance of pregnancy, but it also increases the risk of multiple births. The rate of multiple births depends upon the age of the woman and the number of embryos transferred. When 3 embryos were transferred, the multiple-birth rate was 46% for women aged 20–29. The rate was only 39% for women aged 40–44 when 7 or more embryos were transferred. The risk of multiple births is a serious concern because multiple-birth infants are predisposed to many health problems, including malformations, infant death, premature delivery, and low birth weight (Lipshultz and Adamson 1999; Schieve et al. 1999; Bhattacharya and Templeton 2000; Gleicher et al. 2000). Babies born prematurely and

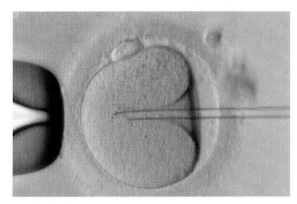

Figure 21.2
Intracytoplasmic sperm injection. In this technique, a single sperm is injected into the cytoplasm of the egg. The suction pipette holding the egg is on the left. The injection pipette (with the sperm) is on the right. (Photograph courtesy of The Institute for Reproductive Medicine and Science of St. Barnabas, Livingston, NJ.)

Sidelights & *Speculations*

Ethical Issues and Assisted Reproductive Technology

ART began as a way of allowing infertile couples to become pregnant. This technique has been successful, in that the rates of deliveries are just about equal to those achieved by normal fertilization. However, this technology has raised several ethical (and legal) concerns (see Purdy 2001).

• **Whom does it assist?** Couples may pay $10,000 for a cycle of ICSI in the United States. Is ART only for the wealthy? If a woman with infertility knows that she could possibly have a genetically related child if she were wealthy, does this frustrate more women than it helps? If curing infertility is our goal, then should our focus be on high-tech medicine or on public health efforts to eliminate sexually transmitted diseases (one of the leading causes of infertility)?

• **Is there a "right" to have a genetically related child?** Are procedures designed to allow 50-year-old women to have babies the best use of our medical resources?

• **Why is there a "need" to have genetically related children?** Is this desire "biological" or is it being manufactured by the advertising done by fertility clinics competing with one another in the present market?

• **What is the status of a frozen embryo?** Is throwing away the extra embryos produced by IVF equivalent to abortion? Who has the right to keep the embryos if the couple should divorce? (And is the biological father obligated to make child support payments if the embryo is implanted and comes to term?)

• **Are current ART procedures safe for mothers and offspring?** While the link between hormones and reproductive cancers has been known for years, it is not known whether women undergoing extensive cycles of ART are at risk for cancers (see Pappert 2000.) High incidences of congenital anomalies among children born through ART have been reported, but these may be due to the fact that the parents had problems themselves (see Wennerholm et al. 2000).

• **Should infertility clinics be regulated?** In contrast to Great Britain, where there are strict laws regulating what infertility clinics can do and how they must report their results, infertility clinics in the United States are not under federal or state regulations. It is often difficult to compare success records or health records between clinics.

---

at low birth weight are at risk for cerebral palsy and chronic respiratory problems. In addition, mothers who carry multiple infants are also at risk for many health conditions and complications (e.g., high blood pressure, diabetes), and the costs for multiple pregnancies are also greatly increased.

GENETIC ERRORS OF HUMAN DEVELOPMENT

If you think it is amazing that any one of us survives to be born, you are correct. It is estimated that one-half to two-thirds of all human conceptions do not develop successfully to term (Figure 21.3). Many of these embryos express their abnormality so early that they fail to implant in the uterus. Others implant, but fail to establish a successful pregnancy. Thus, most embryos are spontaneously aborted, often before the woman even knows she is pregnant (Boué et al. 1985). Edmonds and co-workers (1982), using a sensitive immunological test that can detect the presence of human chorionic gonadotropin (hCG) as early as 8 or 9 days after fertilization, monitored 112 pregnancies in normal women. Of these hCG-determined pregnancies, 67 failed to be maintained.

Defects in the lungs or limbs, however, are not deleterious to the fetus (which does not depend on those organs while inside the mother), but can threaten

life once the baby is born. Over 5% of all human babies born have recognizable malformations, some of them mild, some very severe (McKeown 1976).

Congenital ("present at birth") abnormalities and losses of embryos and fetuses prior to birth have both intrinsic and extrinsic causes. Those abnormalities that are caused by genetic events (mutations, aneuploidies, translocations) are

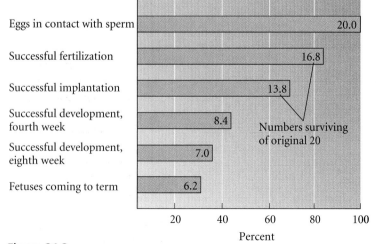

Figure 21.3
The fate of 20 hypothetical human eggs in the United States and western Europe. Under normal conditions, only 6.2, or fewer, of the original 20 eggs would be expected to develop successfully to term. (After Volpe 1987.)

(A)

(B)

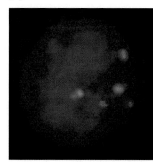

Figure 21.4
Down syndrome. (A) Down syndrome, caused by a third copy of chromosome 21, is characterized by a particular facial pattern and by mental retardation, the absence of a nasal bone, and often heart and gastrointestinal defects. (B) The procedure shown here tests for chromosome number using fluorescently labeled probes that bind to DNA on chromosomes 21 (pink) and 13 (blue). This person has Down syndrome (trisomy 21), but the normal two copies of chromosome 13. (A, photograph ©Laura Dwight/Corbis; B, photograph courtesy of Vysis, Inc.)

called **malformations** (Opitz et al. 1987). For instance, aniridia (absence of the iris), caused by a mutation of the *PAX6* gene, is a malformation. **Trisomy 21** causes several malformations (most of them minor, such as facial muscle changes, but some of them major, such as heart abnormalities, gut abnormalities, and cognitive problems), collectively called **Down syndrome** (Figure 21.4). Certain genes on chromosome 21 are thought to encode transcription factors, and the extra copy of chromosome 21 probably causes an overproduction of these regulatory proteins. Such overproduction would cause the misregulation of genes necessary for heart, muscle, and nerve formation (Nishigaki et al. 2002).

Most early embryonic and fetal demise is probably due to chromosomal abnormalities that interfere with normal developmental processes. As many as 50% of human ova have chromosomal abnormalities that would result in non-viable fetuses (Opitz 1987). Even an extra copy of the extremely small chromosome 21 causes the misregulation of numerous developmental functions. Indeed, people with Down syndrome are some of the only people with autosomal trisomies to survive beyond the first weeks of infancy (and often reach adulthood). People with trisomies 13 and 18 (Patau syndrome and Edward syndrome, respectively) can sometimes live for years with proper medical care.

Until recently, the molecular study of human genetics focused almost exclusively on inborn errors in metabolic and structural proteins. Thus, diseases of enzymes, collagens, and globins predominated. But these proteins are the final products of differentiated cells. Errors in proteins involved with development—transcription factors, paracrine factors, and elements of signal transduction pathways—were little understood. Diseases resulting from such errors were often catego-

rized as "malformations" or "congenital anomalies," and their causes remained unknown. In the past decade, however, we have made great gains in our knowledge of human development, and we now know that many of these malformations are caused by mutations of genes encoding transcription factors and signal transduction proteins (Epstein et al. 2003).

Identifying the Genes for Human Developmental Anomalies

One cannot experiment on human embryos, nor can one selectively breed humans to express a particular mutant phenotype. How, then, can we find the genes that are involved in normal human development and whose mutations cause malformations? Two techniques have revolutionized human embryology within the past decade.

The first technique is **positional gene cloning** (Collins 1992; Scambler 1997). Here, linkage (pedigree) analysis highlights a region of the genome where a particular mutant gene is thought to reside. By cloning overlapping sequences from that region to make a DNA map, researchers can hope to find a region of DNA that differs between people who have the malformation and people who lack it. Then, by sequencing that region of the genome, the gene can be located.

As an example, we will look at the discovery of the human *Aniridia* gene, whose absence causes defects in normal eye development. People heterozygous for this gene have little or no iris in their eyes. Mice have a similar phenotype, and mice homozygous for this condition are stillborn and have neither eyes nor nose. Using DNA from people with aniridia, Ton and his colleagues (1991) found a region of chromosome 11 that was present in unaffected individuals, but completely or partially absent in individuals with aniridia. Moreover, this region of DNA contained a region that could be a gene (i.e., it had a promoter site as well as sequences connoting introns and exons). To determine whether this region was a gene that was active in eye development, they did a Northern blot and

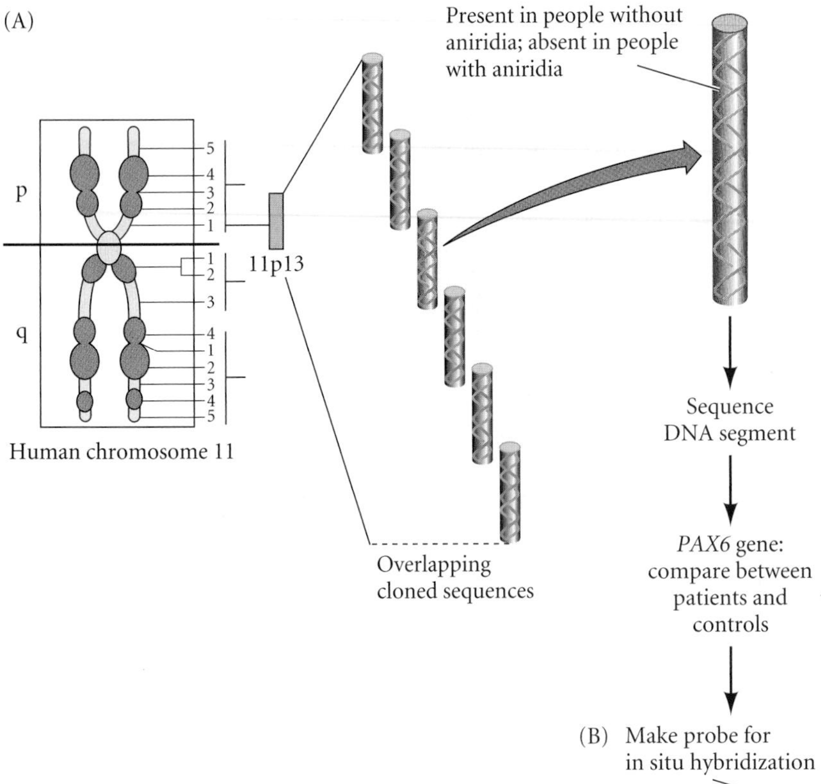

(A)

Human chromosome 11

11p13

Present in people without
aniridia; absent in people
with aniridia

Overlapping
cloned sequences

Sequence
DNA segment

PAX6 gene:
compare between
patients and
controls

(B) Make probe for
in situ hybridization

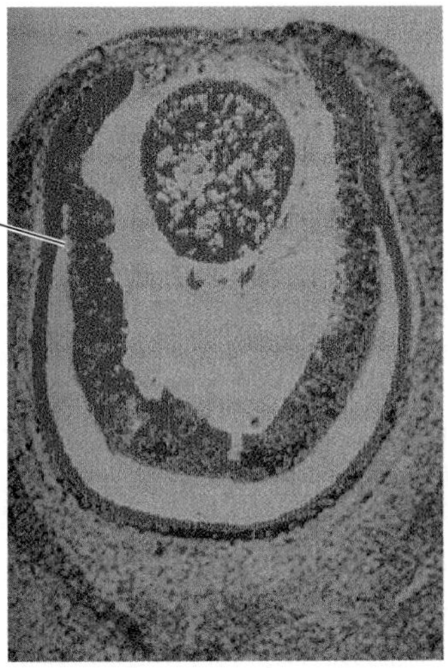

Figure 21.5
Positional cloning of the human *PAX6* gene.
(A) Principles of positional cloning. A gene is determined, by pedigree analysis and somatic cell genetics, to be located in a certain region of the
genome—in this case, band 13 of the short arm of
chromosome 11. Overlapping cloned sequences
that span this region are compared between people who have the condition of interest and people
who do not. In this case, certain people with
aniridia lacked a particular region of DNA. This
region was sequenced and found to contain the
human homologue of the *Pax6* gene known in
mice. Comparisons of the sequences found in
controls with those found in people with aniridia
indicated that the people with aniridia either
lacked a copy of this gene or had mutations in one
of their copies. (B) In situ hybridization of an antisense RNA probe with a human fetal eye shows
PAX6 expression (yellow) in the retina, presumptive iris, and surface ectoderm. (From Ton et al.
1991; photograph courtesy of G. F. Saunders.)

in situ hybridizations, using this region of DNA as a probe. As
Figure 4.13F shows, this probe found complementary mRNA
primarily in tissues from the brain and eye regions of the
body. Thus this fragment of DNA contained a gene that was
indeed expressed in the brain and eye. Sequencing this fragment showed that it was the human *PAX6* gene: the human
homologue of the *Pax6* gene already known in mice. This
finding was confirmed through in situ hybridization (Figure
21).

The positional gene cloning approach is currently being
used to map those genes responsible for the congenital anomalies seen in Down syndrome. Down syndrome is the most
frequent genetic disorder in humans, and as mentioned
above, it is usually caused by a complete extra copy of chromosome 21. However, there are rare cases of Down syndrome
wherein only a small region of chromosome 21 is present in
three copies. This region has been called the **Down syndrome
critical region** (DSCR). One gene that has been found in this
region is the *DSCR1* gene. It is predominantly expressed in the
brain, heart, and skeletal muscle. Recent studies (Fuentes et al.
2000; Ermak et al. 2002) have shown that it is overexpressed
in people with Down syndrome and that it encodes a protein
that inhibits calcineurin, a major calcium regulatory protein.*

The second technique is **candidate gene mapping**. This
approach is similar to positional gene cloning, but here, one

starts with a correlation between the genetic mapping of a particular syndrome and a gene associated with a similar phenotype in other species. As an example, we will take another condition that produces small eyes: Waardenburg syndrome type 2.
This autosomal dominant condition is characterized by deafness, heterochromatic (multicolored) irises, and a white forelock (Figure 21.6A). By analyzing the pedigrees of several families whose members had this condition, Hughes and her
colleagues (1994) showed that it is caused by a mutation in a
gene on the short arm of chromosome 3, between bands 12.3
and 14.4. A strikingly similar condition is found in mice, in
which it is called microphthalmia. Mutations of the *microphthalmia* gene (*Mitf*) cause a dominant syndrome involving
deafness, a white patch of fur, and eye abnormalities (Figure
21.6B). Could Waardenburg syndrome type 2 be caused by mutations in the human equivalent of the mouse *Mitf* gene? The

---

*Interestingly, overproduction of DSCR1 in adults has been implicated in
Alzheimer disease (Ermak et al. 2001).

(A)

(B)

Figure 21.6
Microphthalmia syndrome in humans and mice. (A) Human patients with Waardenburg syndrome type 2. (B) Mice with *microphthalmia* mutation. The relationship between these two syndromes was ascertained by showing that patients with Waardenburg syndrome type 2 had mutations in a gene that is homologous with the mouse *Mitf* gene. (A photograph from Partington 1959; B photograph courtesy of D. Fischer.)

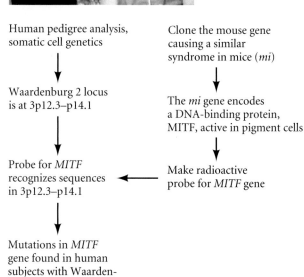

Human pedigree analysis, somatic cell genetics

↓

Waardenburg 2 locus is at 3p12.3–p14.1

↓

Probe for *MITF* recognizes sequences in 3p12.3–p14.1

↓

Mutations in *MITF* gene found in human subjects with Waardenburg syndrome 2 and not in people without this syndrome

Clone the mouse gene causing a similar syndrome in mice (*mi*)

↓

The *mi* gene encodes a DNA-binding protein, MITF, active in pigment cells

↓

Make radioactive probe for *MITF* gene

gene for holoprosencephaly. By searching for the sonic hedgehog gene, it was determined that several (but not all) families with holoprosencephaly have mutations in this gene (Nanni et al. 1999; Odent et al. 1999).

WEBSITE 21.1 Human embryology and genetics. These websites connect you to tutorials in human development as well as to the Online Mendelian Inheritance in Man (OMIM), which details all human genetic conditions.

The Nature of Human Syndromes

As we have seen, human infants are sometimes born with malformations, which range from life-threatening to relatively benign. Often these malformations are linked into syndromes (see Chapter 1) and therefore show pleiotropy.

Pleiotropy

The production of several effects by one gene or pair of genes is called **pleiotropy** (Figure 21.7). For instance, Waardenburg syndrome type 2, as mentioned above, involves iris defects, pigmentation abnormalities, deafness, and inability to produce the normal number of mast cells (a type of blood cell). The skin pigment, the iris of the eye, the inner ear tissue, and the mast cells of the blood are not related to one another in such a way that the absence of one would produce the absence of the others. Rather, all four parts of the body independently use the MITF protein as a transcription factor. This type of pleiotropy has been called **mosaic pleiotropy** because the affected organ systems are separately affected by the abnormal gene function (Grüneberg 1938; Hadorn 1955).

While the eye pigment, body pigment, and mast cell manifestations of Waardenburg syndrome type 2 are separate

mouse *Mitf* gene was cloned, and was found (by sequencing and in situ hybridization) to encode a DNA-binding protein that is expressed in the pigment cells of the eyes, ears, and hair follicles of embryonic mice (Hemesath et al. 1994; Hodgkinson et al. 1994; Nakayama et al. 1998). The human version of *Mitf* thus became a candidate gene for this syndrome.

The mouse *Mitf* gene was then used to make a probe to look for similar genes in the human genome. The probe found a human homologue of the mouse *microphthalmia* gene, and this sequence (*MITF*) mapped to the exact same region of chromosome 3 as Waardenburg syndrome type 2 (Tachibana et al. 1994; Tassabehji et al. 1994). When people with Waardenburg syndrome were studied, it was found that they each had mutations of the *MITF* gene. Thus, Waardenburg syndrome type 2 was correlated with mutations in a DNA-binding protein encoded by the human *MITF* locus.

Recently, candidate gene mapping has been supplemented with data from developmental biology. For instance, the phenotype of knockouts of the mouse *sonic hedgehog* gene resembles that of humans with holoprosencephaly (cyclopia; see Figure 6.26). This finding made *sonic hedgehog* a candidate

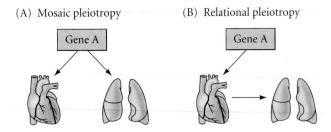

(A) Mosaic pleiotropy (B) Relational pleiotropy

Figure 21.7
Mosaic and relational pleiotropy. (A) In mosaic pleiotropy, a gene is independently expressed in several tissues. Each tissue needs the gene product and develops abnormally in its absence. (B) In relational pleiotropy, a gene product is needed by only one particular tissue. However, a second tissue needs a signal from the first tissue in order to develop properly. If the first tissue develops abnormally, the signal is not given, so the second tissue develops abnormally as well.

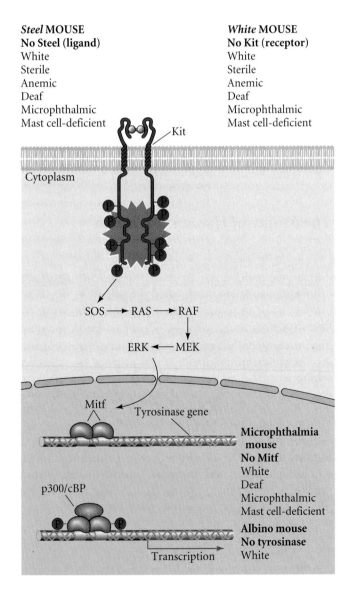

Figure 21.8
Phenotypes of mice with mutations along the pigment synthesis pathway. The *Steel* and *White* mice have mutations in the genes encoding stem cell factor (the ligand) and Kit protein (the receptor), respectively. These proteins activate the *Mitf* transcription factor through the pathway shown in Figure 6.15. *Mitf* mutations give rise to the *microphthalmia* phenotype, which contains a subset of those anomalies seen upstream in the pathway. The albino mutation is farther down the pathway and contains a subset of those conditions found in the *microphthalmia* mutants.

events, other parts of the syndrome are not. For instance, the failure of *MITF* expression in the pigmented retina prevents this structure from fully differentiating. This, in turn, causes a malformation of the choroid fissure of the eye, resulting in the drainage of vitreous humor. Without this fluid, the eye fails to enlarge (hence the name *microphthalmia*, which means "small eye"). This type of pleiotropy, in which several developing tissues are affected by the mutation even though they do not express the mutated gene, is called **relational pleiotropy** (see Grüneberg 1938; Hadorn 1955).

Genetic heterogeneity

Another important feature of syndromes is that mutations in different genes can produce the same phenotype. If the genes are part of the same signal transduction pathway, a mutation in any of them can give a similar result. The production of similar phenotypes by mutations in different genes is called **genetic heterogeneity**. Cyclopia, for example, can be produced by mutations in the *sonic hedgehog* gene or by mutations in the genes controlling cholesterol synthesis. Since cholesterol binds to the Sonic hedgehog protein to activate it as a paracrine factor (see Chapter 6), mutations that interfere with the synthesis of cholesterol also prevent the activation of the Sonic hedgehog protein. Therefore, different kinds of mutations can produce a similar suite of abnormalities. Similarly, as we saw in Chapter 6, murine mutations in the stem cell factor (*Steel*) gene produce a syndrome resembling that produced by mutations in the gene for its receptor, the Kit protein (*White*). Since mutations of either of these genes prevent *Mitf* from being activated, they produce a phenotype similar to that of the *Mitf*-deficient mouse (Figure 21.8; see also Figure 1.15. But here, there is a difference. Since the Kit and stem cell factor proteins are also used by migrating germ cells and blood cell precursors (which do not use Mitf), mice with mutations of the genes encoding Kit and stem cell factor also have fewer gametes and blood cells. Albinism, which is produced by a loss-of-function mutation of the tyrosinase gene, also gives a white phenotype, but it does not preclude mast cell function or fertility.

Phenotypic variability

Not only can different mutations produce the same phenotype, but the same mutation can produce a different phenotype in different individuals (Wolf 1995, 1997; Nijhout and Paulsen

1997). This phenomenon, called **phenotypic variability**, is a result of the integration of genes into complex networks. Genes are not autonomous entities. Rather, they interact with other genes and gene products to make pathways and networks.

Bellus and colleagues (1996) analyzed the phenotypes derived from the same mutation in the *FGFR3* gene in 10 independent human families. These phenotypes ranged from relatively mild anomalies to potentially lethal malformations. Similarly, Freire-Maia (1975) reported that within one family, the homozygous state of a mutant gene affecting limb development caused phenotypes ranging from severe phocomelia (lack of limb development shortly after the most proximal bone, resulting in a "flipper-like stump") to a mild abnormality of the thumb. The severity of a mutant gene's effect often depends on the *other* genes in the pathway.

Phenotypic variability in women can also be caused by statistical variation in X chromosome inactivation. One recent example is the case of two identical twins, each of whom carried one normal allele for an X-linked clotting factor and one mutant allele for the same factor. Due to random X chromosome inactivation, one twin inactivated a large percentage of the X chromosome containing the normal allele, and she had severe hemophilia (the inability to clot one's blood after injury). The other twin, who inactivated a lower percentage of the X chromosome with the normal allele, had a milder case (Tiberio 1994; Valleix et al. 2002).

WEBSITE 21.2 Holoprosencephaly. Holoprosencephaly can be caused by several genes in the Hedgehog pathway as well as by environmental factors. This syndrome has an enormous phenotypic variability as well as genetic heterogeneity.

Mechanisms of dominance

Whether a syndrome is dominant or recessive can now be explained in many cases at a molecular level. First, it must be recognized that there are many syndromes that are referred to as "dominant" only because the homozygous condition is lethal to the embryo and the fetus is never born; therefore, the homozygous condition is not seen. Second, there are at least three ways of achieving a "dominant" phenotype.

The first mechanism of dominance is **haploinsufficiency**. This merely means that one wild-type copy of the gene (the heterozygous condition) is not enough to produce the amount of gene product required for normal development. For example, individuals with Waardenburg syndrome type 2 have roughly half the wild-type amount of MITF. This is not enough for full pigment cell proliferation, mast cell differentiation, or inner ear development. Thus, an aberrant phenotype results when only one of the two copies of this gene is absent or nonfunctional.

The second mechanism of dominance is **gain-of-function mutations**. Thus, thanatophoric dysplasia (as well as milder forms of dwarfism such as achondroplasia) results from mutations in the *FGFR3* gene that causes the mutant FGF receptor 3 to be constitutively active (instead of being activated only by

FGFs; see Chapter 6). This activity is enough to cause the premature differentiation and death of cartilage in the growth plates of the long bone, leading to the shortening of these bones.

The third mechanism of dominance is a **dominant negative allele**. When the active form of a protein is made up of multiple subunits, all of the subunits may have to be wild-type in order for the protein to function. In such cases, a mutation in just one allele—the dominant negative allele—can make the protein nonfunctional. A dominant negative allele isn't merely nonfunctional; it is deleterious. A dominant negative allele is the cause of **Marfan syndrome**, a disorder of the extracellular matrix. Marfan syndrome results in joint and connective tissue anomalies, not all of which are necessarily disadvantageous. Increased height, disproportionately long limbs and digits, and mild to moderate joint laxity are characteristic of this syndrome. However, patients with Marfan syndrome may also experience vertebral column deformities, myopia, loose or dislocated lenses, and (most importantly) aortic problems that may lead to aneurysm (tearing of the aorta) later in life. The mutation is in the gene encoding fibrillin, a secreted glycoprotein that forms multimeric microfibrils in elastic connective tissue. The presence of even small amounts of mutant fibrillin prohibits the association of wild-type fibrillin into microfibrils. El-dadah and colleagues (1995) have shown that when a mutant human gene for fibrillin is transfected into fibroblast cells that already contain two wild-type genes, the incorporation of fibrillin into the extracellular matrix is inhibited.

We are beginning to understand the molecular bases for many of the inherited malformation syndromes in humans. This understanding constitutes a critically important synthesis of developmental biology, medical genetics,* and pediatric medicine.

Gene Expression and Human Disease

Candidate gene mapping and positional gene cloning techniques have brought together medical embryology and medical genetics. This fusion has enabled scientists and physicians to understand normal human development as well as the causes of many malformations. As would be expected, alterations in gene expression can occur at the levels of transcription, RNA processing, translation, and posttranslational modification.

Inborn errors in transcriptional regulation

The best-studied group of genes are those genes involved in regulating transcription. These genes include genes encoding transcription factors and the components of signal transduction cascades. Many of these genes are mentioned in Table

---

*Although there is significant overlap between the disciplines, *human genetics* is the study of human variability, while *medical genetics* is the study of abnormal human variability. *Clinical genetics* is the medical specialty that cares for individuals with abnormal variability, and *genetic counseling* is that profession that supports in all respects the roles of the clinical geneticists and the needs of the patients and their families (J. M. Opitz, personal communication.)

21.1, and several of them (e.g., *PAX6, MITF*) have already been discussed.

Inborn errors of nuclear RNA processing

It is estimated that at least 35% of human genes produce RNAs that can be alternatively spliced (Croft et al. 2000). Therefore, even though the human genome may contain only 35,000–80,000 genes, its **proteome** (encompassing all the proteins encoded by the genome) is probably far more complex.

MUTATIONS IN SPLICE SITES   One gene that has a remarkable pattern of splicing is *Dscam* (see Chapter 5). The gene's abbreviation stands for "*D*own *s*yndrome *c*ell *a*dhesion *m*olecule," and it is located in that region of chromosome 21 that is associated with the neurological symptoms of Down syndrome. Mammalian *Dscam* is expressed abundantly in the nervous system during development, especially in axonal and dendritic processes within the cerebellum, hippocampus, and olfactory bulb. The Dscam protein is most likely involved in cell-cell interactions during axonal-dendritic development and in the maintenance of functional neuronal networks. There appear to be several isoforms of this protein, and they are expressed in different subsets of neurons (Yamakawa et al. 1998; Barlow et al. 2001).

Mutations in the splice sites of genes can prevent certain protein isoforms from arising. It is estimated that 15% of all point mutations that result in human genetic disease create such splice site abnormalities (Krawczak et al. 1992; Cooper and Mattox 1997) One such disease is congenital adrenal hyperplasia. This syndrome is often caused by the absence of a functional gene for 21-steroid hydroxylase (*CYP21*). In the absence of this enzyme, steroids that would normally be used to make cortisol are shunted into making more testosterone. The result is that girls with this condition are phenotypically masculinized. In many instances, people with this condition have a point mutation in the splice site for the second intron of the *CYP21* gene. This mutation prevents the intron from being skipped and makes an ineffective enzyme. Mutations that create new splice sites or prevent the use of existing ones also result in nonfunctional proteins (Hutchinson et al. 2001).

MUTATIONS IN SPLICING FACTORS   There are some proteins and small nuclear RNAs that are used throughout the body to effect differential pre-mRNA splicing, and there are some proteins that appear to regulate differential splicing in a manner characteristic of a particular cell type. If the genes encoding these cell type-specific splicing factors are mutated, one could expect several cell type-specific isoforms to be aberrant. This appears to be the case in one of the leading causes of hereditary infant mortality, spinal muscular atrophy. Here, mutations in the gene encoding the SMN (survival of motor neurons) protein prevent the maintenance of motor neurons. This protein is involved in splicing nRNAs in this subset of neurons (Pellizzoni et al. 1999).

Inborn errors of translation

Many genetic diseases are due to mutations that create translation termination codons. For instance, androgen insensitivity syndrome (see Figure 17.12) can be caused by a guanine-to-adenine transition at nucleotide 2682 of the gene encoding the androgen receptor. This mutation changes the coding of codon 717 (in the middle of the message) from tryptophan to a translation termination codon (Sai et al. 1990). This truncated receptor thereby lacks most of its androgen-binding domain.

Other mutations can alter the longevity of an mRNA, which can greatly affect the number of protein molecules synthesized from it. For example, ***Hemoglobin Constant Spring*** is a naturally occurring mutation wherein the translation termination codon of the α-globin gene has been mutated to an amino acid codon, and the translation continues for 31 more codons (Wang et al. 1995). This readthrough results in destabilization of the α-globin mRNA, causing a reduction of greater than 95% in α-globin gene expres-

TABLE 21.1 Some genes encoding human transcription factors and phenotypes resulting from their mutation

Gene	Mutation phenotype
Androgen receptor	Androgen insensitivity syndrome (Ch. 20)
AZF1	Azoospermia
CBFA1	Cleidocranial dysplasia (Ch. 15)
CSX	Heart defects
EMX2	Schizencephaly (Ch. 23)
Estrogen receptor	Growth regulation problems, sterility (Ch. 15)
Forkhead-like 15	Thyroid agenesis, cleft palate
GLI3	Grieg syndrome (Ch. 16)
HOXA-13	Hand-foot-genital syndrome (Ch. 16)
HOXD-13	Polysyndactyly (Ch. 16)
LMX1B	Nail-patella syndrome (Ch. 16)
MITF	Waardenburg syndrome type 2 (Chs. 5, 21)
PAX2	Renal-coloboma syndrome (Ch. 15)
PAX3	Waardenburg syndrome type 1 (Ch. 13)
PAX6	Aniridia (Chs. 5, 21)
PTX2	Reiger syndrome (Ch. 11)
PITX3	Congenital cataracts
POU3F4	Deafness and dystonia
SOX9	Campomelic dysplasia, male sex reversal (Chs. 15, 20)
SRY	Male sex reversal (Ch. 20)
TBX3	Schinzel syndrome (ulna-mammary syndrome)
TBX5	Holt-Oram syndrome (Ch. 16)
TCOF	Treacher-Collins syndrome
TWIST	Seathre-Chotzen syndrome
WT1	Urogenital anomalies (Ch. 15)

(A)

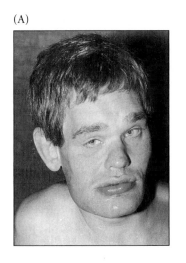

(B)

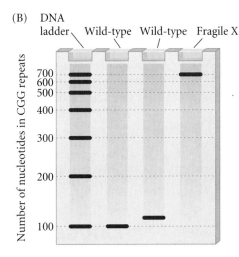

Figure 21.9
Fragile X syndrome. (A) A 26-year-old man with fragile X syndrome. Affected individuals often have a large head with relatively large forehead, ears, and jaws. (B) The number of CGG repeats is critical in determining whether the *FMR1* gene is transcribed in the nervous system. This number can be determined by the polymerase chain reaction. PCR usually reveals 6–50 CGG repeats in the *FMR1* gene. However, in people with fragile X syndrome, over 200 repeats are seen. The product of this gene appears to be necessary for the translation of certain neuronal messages; without it, these messages are not brought to the ribosomes. (A, photograph from De Boulle et al. 1993.)

sion from the affected locus. The resultant clinical disease is one of the types of **α-thalassemias**.

In many instances, certain proteins are used to stabilize particular messenger RNAs or to bring them to the ribosomes. The most prevalent form of inherited mental retardation, **fragile X syndrome**, may result from a mutation in a gene whose product is critical for the translation of certain nerve-specific messages. Fragile X syndrome is usually caused by the expansion and hypermethylation of CGG repeats in the 5′ untranslated region of the *FMR1* gene. Whereas most individuals have *FMR1* alleles containing from 6 to 50 CGG repeats, people with fragile X syndrome have alleles with over 200 such repeats (Figure 21.9). This CGG expansion blocks the transcription of this gene. The *FMR1* gene encodes an RNA-binding protein that appears to be critical for the *translation* of certain messages. Normally, nearly 85% of the FMR1 protein molecule is associated with translating polysomes. FMR1 proteins with mutations in the RNA-binding domains are not observed with cytoplasmic polysomes, and such mutations produce severe forms of fragile X syndrome (Feng et al. 1997a,b). Recent studies (Brown et al. 2001; Darnell et al. 2001) have shown that a particular subset of mouse brain messenger RNA requires the FMR1 protein for proper translation. Most of these messages are involved with synapse function or neuronal development. It is probable that FMR1 binds specific mRNAs, either regulating their translation or targeting them to the dendrite where they may await the signal for translation.

Preimplantation Genetics

Amniocentesis, chorionic villus sampling, and preimplantation genetics

One of the consequences of in vitro fertilization and the ability to detect genetic mutations early in development is a new area of medicine called **preimplantation genetics**. Preimplantation genetics seeks to test for genetic disease *before* the embryo enters the uterus.

Many genetic diseases can be diagnosed before a baby is born. This **prenatal diagnosis** can be done by chorionic villus sampling at 8–10 weeks of gestation or by amniocentesis around the fourth or fifth month of pregnancy. **Chorionic villus sampling** involves taking a sample of the placenta, while **amniocentesis** involves taking a sample of the amnionic fluid. In both cases, fetal cells from the sample can be grown and then analyzed for the presence or absence of certain chromosomes, genes, or enzymes. However useful these procedures have been in detecting genetic disease, they have brought with them many ethical concerns. First, if the fetus is found to have a genetic disease, the only means of prevention is aborting the pregnancy. Second, the waiting time between knowledge of being pregnant and the results from amniocentesis or chorionic villus sampling can create a "tentative pregnancy." Many couples do not announce their pregnancy during this stressful period for fear that it might have to be terminated (Rothman et al. 1995).

One way around this problem is to screen the embryonic cells before the embryo is even implanted in the womb. Preimplantation genetics is performed on embryos generated by in vitro fertilization. While the embryos are still in the petri dish, at the 6– to 8-cell stage, a small hole is made in the zona pellucida, and two blastomeres are removed from the embryo (Figure 21.10). Since the mammalian egg undergoes regulative development, their removal does not endanger the embryo. The isolated blastomeres can then be tested immediately. The polymerase chain reaction technique allows the presence or absence of certain genes to be determined, and **fluorescent in situ hybridization (FISH)** can be used to determine whether the normal numbers and types of chromosomes are present (Kanavakis and Traeger-Synodinos 2002; Miny et al. 2002). Results are often available within 2 days. The presumptive wild-type embryos can be implanted into the uterus, while the presumptive mutant embryos are discarded. For many couples, it is easier to consider implanting

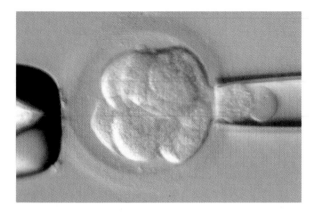

Figure 21.10
Preimplantation genetics is performed on one or two blastomeres (seen here in the pipette) taken from an early blastocyst. The polymerase chain reaction is then used to determine whether certain genes in these cells are present, absent, or mutant. (Photograph courtesy of The Institute for Reproductive Medicine and Science of St. Barnabas, Livingston, NJ.)

only those embryos that are most likely to be healthy than aborting those fetuses that are most likely to produce malformed or nonviable children.

> WEBSITE 21.3 **Bioethics: What is "normal"?** The ability to diagnose genetic disease prenatally has allowed the selective termination of pregnancy if the fetus is abnormal. However, while certain diseases produce stillbirths and babies who die shortly after delivery, other diseases are not as life-threatening. This creates moral dilemmas as to whether to continue the pregnancy.

Sperm separation and sex selection

Preimplantation genetics, while circumventing the ethical problems of amniocentesis and chorionic villus sampling, raises ethical issues of its own. The same procedures that allow preimplantation genetics also enable the physician to know the sex of the embryo. Sometimes parents wish to have this information; sometimes they do not. However, knowing the sex of an embryo before implantation raises the possibility that parents could decide the sex of their offspring by having only embryos of the desired sex implanted. Different countries and even different hospitals have different policies as to whether to permit preimplantation diagnosis solely for the purpose of sex determination.

Another way to accomplish sex selection is through sperm selection. The X chromosome is substantially larger than the Y chromosome; therefore, human sperm cells containing an X chromosome contain nearly 3% more total DNA than sperm cells containing a Y chromosome. This DNA difference can be measured, and the X- and Y-bearing sperm cells separated based on their size/mass ratio, using a flow cytometer. The separated sperm can then be used for artificial insemination (or in vitro fertilization). Recent studies have shown that this technique is

about 90% reliable for sorting X-bearing sperm, and about 78% reliable for sorting Y-bearing sperm (Stern et al. 2002).

Sex selection is seen as a way of preventing X-linked diseases, but it has been used primarily as a method of family balancing. Opponents of sex selection point to its possible use to prevent the birth of girls in cultures where women are not as highly valued as men.

> WEBSITE 21.4 **Bioethics: Sex selection.** Sex selection can be used as a form of social engineering. Is it moral to select the sex of your child? Under some circumstances, is it moral not to do so?

TERATOGENESIS: ENVIRONMENTAL ASSAULTS ON HUMAN DEVELOPMENT

In addition to genetic mutations that can affect development, we now know that numerous environmental factors can disrupt development. The summer of 1962 brought two portentious discoveries. The first was the disclosure by Rachel Carson (1962) that the pesticide DDT was destroying bird eggs and was preventing reproduction in several species (see Chapter 22). The second (Lenz 1962) was the discovery that thalidomide, a sedative drug used to manage pregnancies, could cause limb and ear abnormalities in the fetus (see Chapter 1). These two discoveries showed that the embryo was vulnerable to environmental agents. This was underscored in 1964, when an epidemic of rubella (German measles) spread across America. Adults showed relatively mild symptoms when infected by this virus, but over 20,000 fetuses infected by rubella became either blind, deaf, or both. Many of these infants were also born with heart defects or mental retardation (CDC 2002b).

Abnormalities caused by exogenous agents are called **disruptions**. The agents responsible for these disruptions are called **teratogens**.* Most teratogens produce their effects only during certain critical periods of development. Human development is usually divided into two periods, the **embryonic period** (to the end of week 8) and the **fetal period** (the time in utero thereafter). It is during the embryonic period that most of the organ systems form; the fetal period is generally one of growth and modeling. The period of maximum susceptibility to teratogens is between weeks 3 and 8, since that is when most organs are forming. The nervous system, however, is constantly forming, and remains susceptible throughout de-

---

*In some cases, the same condition can be either a disruption (caused by an exogenous agent) or a malformation (caused by a genetic mutation). Chondrodysplasia punctata is a congenital defect of bone and cartilage, characterized by abnormal bone mineralization, underdevelopment of nasal cartilage, and shortened fingers. It is caused by a defective gene on the X chromosome. An identical phenotype is produced by exposure of fetuses to the anticoagulant (blood-thinning) compound warfarin. It appears that the defective gene is normally responsible for producing an arylsulfatase protein (CDPX2) necessary for cartilage growth. The warfarin compound inhibits this same enzyme (Franco et al. 1995).

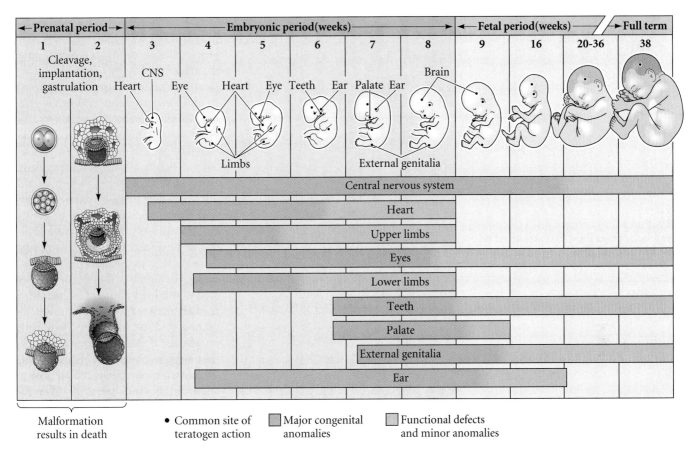

Figure 21.11
Periods (weeks of gestation) and degrees of sensitivity of embryonic organs to teratogens. (After Moore and Persaud 1993.)

velopment. Before week 3, the embryo is not usually susceptible to teratogens. During this time, a substance either damages most or all of the cells of an embryo, resulting in its death, or only a few cells, allowing the embryo to recover. Figure 21.11 indicates the times at which various organs are most susceptible to teratogens.

> WEBSITE 21.5 **Thalidomide as a teratogen.** The drug thalidomide caused thousands of babies to be born with malformed arms and legs, and it provided the first major evidence that drugs could induce congenital anomalies. The mechanism of its action is still hotly debated.

Teratogenic Agents

Different agents are teratogenic in different organisms. A partial list of agents that are teratogenic in humans is given in Table 21.2. The largest class of teratogens includes drugs and chemicals, but viruses, radiation, hyperthermia, and metabolic conditions in the mother can also act as teratogens.

Some chemicals that are naturally found in the environment can cause birth defects. Even in the pristine alpine meadows of the Rocky Mountains, teratogens are found. Here grows the skunk cabbage *Veratrum californicum*, upon which sheep sometimes feed. If pregnant ewes eat this plant, their fetuses tend to develop severe neurological disruptions, including cyclopia, the fusion of two eyes in the center of the face (see Figure 6.26). Two products made by this plant, **jervine** and **cyclopamine**, inhibit cholesterol synthesis in the fetus and prevent Sonic hedgehog from functioning. Indeed, this condition resembles human genetic syndromes (discussed above and in Chapter 12) in which the genes for either Sonic hedgehog or cholesterol-synthesizing enzymes are mutated (Beachy et al. 1997; Opitz et al. 2002). The affected organism dies shortly after birth (as a result of severe brain defects, including the lack of a pituitary gland).

Quinine and alcohol, two other substances derived from plants, can also cause developmental disruptions. Quinine ingested by a pregnant mother can cause deafness, and alcohol can cause physical and mental retardation in the infant, as we will see below. Nicotine and caffeine have not been proved to cause congenital anomalies, but women who are heavy smokers (20 cigarettes a day or more) are likely to have infants that are smaller than those born to women who do not smoke. Smoking also significantly lowers the number, quality, and motility of

TABLE 21.2 Some agents thought to cause disruptions in human fetal development[a]

DRUGS AND CHEMICALS	IONIZING RADIATION (X-RAYS)
Alcohol	
Aminoglycosides (Gentamycin)	**HYPERTHERMIA**
Aminopterin	**INFECTIOUS MICROORGANISMS**
Antithyroid agents (PTU)	Coxsackie virus
Bromine	Cytomegalovirus
Cortisone	Herpes simplex
Diethylstilbesterol (DES)	Parvovirus
Diphenylhydantoin	Rubella (German measles)
Heroin	*Toxoplasma gondii* (toxoplasmosis)
Lead	*Treponema pallidum* (syphilis)
Methylmercury	**METABOLIC CONDITIONS IN THE MOTHER**
Penicillamine	Autoimmune disease (including Rh incompatibility)
Retinoic acid (Isotretinoin, Accutane)	Diabetes
Streptomycin	Dietary deficiencies, malnutrition
Tetracycline	Phenylketonuria
Thalidomide	
Trimethadione	
Valproic acid	
Warfarin	

Source: Adapted from Opitz 1991.

[a]This list includes known and possible teratogenic agents and is not exhaustive.

sperm in the semen of males who smoke at least 4 cigarettes a day (Kulikauskas et al. 1985; Mak et al. 2000; Shi et al. 2001).

In addition, hundreds of new artificial compounds come into general use each year in our industrial society. Pesticides and organic mercury compounds have caused neurological and behavioral abnormalities in infants whose mothers have ingested them during pregnancy. Moreover, drugs that are used to control diseases in adults may have deleterious effects on fetuses. Valproic acid, for example, is an anticonvulsant drug used to control epilepsy. It is known to be teratogenic in humans, as it can cause major and minor spinal defects. Barnes and colleagues (1996) have shown that valproic acid decreases the level of *Pax1* transcription in chick somites. This decrease causes malformation of the somites and corresponding malformations of the vertebrae and ribs.

Retinoic acid as a teratogen

In some instances, even a compound that is involved in normal metabolism can have deleterious effects if present in large amounts at particular times. Retinoic acid is important in forming the anterior-posterior axis of the mammalian embryo as well as in forming the limbs (see Chapters 11 and 16). In normal development, retinoic acid is secreted from discrete cells and works in a small area. However, if retinoic acid is present in large amounts, cells that normally would not receive high concentrations of this molecule are exposed to it and will respond to it.

13-*cis*-retinoic acid has been useful in treating severe cystic acne and has been available for that purpose (under the trade name Accutane®) since 1982. Because the deleterious effects of administering large amounts of vitamin A or its analogues to pregnant animals have been known since the 1950s (Cohlan 1953; Giroud and Martinet 1959; Kochhar et al. 1984), the drug carries a label warning that it should not be used by pregnant women. However, about 160,000 women of childbearing age (15 to 45 years) have taken this drug since it was introduced, and some of them have used it during pregnancy. Lammer and his co-workers (1985) studied a group of women who inadvertently exposed themselves to retinoic acid and who elected to remain pregnant. Of their 59 fetuses, 26 were born without any noticeable anomalies, 12 aborted spontaneously, and 21 were born with obvious anomalies. The affected infants had a characteristic pattern of anomalies, including absent or defective ears, absent or small jaws, cleft palate, aortic arch abnormalities, thymic deficiencies, and abnormalities of the central nervous system.*

This pattern of multiple congenital anomalies is similar to that seen in rat and mouse embryos whose pregnant mothers have been given these drugs. Goulding and Pratt (1986) placed 8-day mouse embryos in a solution containing 13-*cis*-retinoic acid at a very low concentration (2×10^{-6} *M*). Even at this concentration, approximately one-third of the embryos developed a very specific pattern of anomalies, including dramatic reduction in the size of the first and second pharyngeal arches (see Figure 11.45). In normal mice, the first arch eventually forms the maxilla and mandible of the jaw and two ossicles of the middle ear, while the second arch forms the third ossicle of the middle ear as well as other facial bones.

The basis for this developmental disruption appears to reside in the drug's ability to alter the expression of the Hox genes and thereby respecify portions of the anterior-posterior axis and inhibit neural crest cell migration from the cranial region of the neural tube (Moroni et al. 1994; Studer et al. 1994). Radioactively labeled retinoic acid binds to the cranial

---

*Retinoic acid is a critical public health concern because there is significant overlap between the population using acne medicine and the population of women of childbearing age, and because it is estimated that half of the pregnancies in America are unplanned (Nulman et al. 1997). Vitamin A is itself teratogenic in megadose amounts. Rothman and colleagues (1995) found that pregnant women who took more than 10,000 international units of vitamin A per day (in the form of vitamin supplements) had a 2% chance of having a baby born with disruptions similar to those produced by retinoic acid.

neural crest cells and arrests both their proliferation and their migration (Johnston et al. 1985; Goulding and Pratt 1986). This binding seems to be specific to the cranial neural crest cells, and the teratogenic effect of the drug is confined to a specific developmental period (days 8–10 in mice; days 20–35 in humans). Animal models of retinoic acid teratogenesis have been extremely successful in elucidating the mechanisms of teratogenesis at the cellular level (see Chapter 11).

WEBSITE 21.6 **Mechanisms of retinoic acid teratogenesis.** Within the cell, numerous retinoid-binding proteins interact to influence the ability of retinoic acid to transcribe particular genes.

Alcohol as a teratogen

In terms of the frequency of its effects and its cost to society, the most devastating teratogen is undoubtedly ethanol. In 1968, Lemoine and colleagues noticed a syndrome of birth defects in the children of alcoholic mothers. This **fetal alcohol syndrome** (**FAS**) was confirmed by Jones and Smith (1973). Babies with FAS were characterized by small head size, an indistinct philtrum (the pair of ridges that runs between the nose and mouth above the center of the upper lip), a narrow upper lip, and a low nose bridge. The brain of such a child may be dramatically smaller than normal and often shows defects in neuronal and glial migration (Figure 21.12; Clarren 1986). There is also prominent abnormal cell death in the frontonasal process and the cranial nerve ganglia (Sulik et al. 1988). Fetal alcohol syndrome is the third most prevalent type of mental retardation (behind fragile X syndrome and Down syndrome) and affects 1 out of every 500–750 children born in the United States (Abel and Sokol 1987).

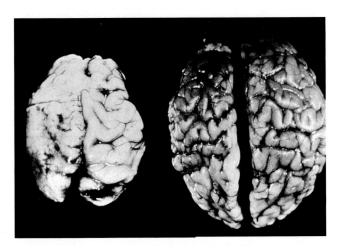

Figure 21.12
Comparison of a brain from an infant with fetal alcohol syndrome (FAS, left) with a brain from a normal infant of the same age (right). The brain from the infant with FAS is smaller, and the pattern of convolutions is obscured by glial cells that have migrated over the top of the brain. (Photographs courtesy of S. Clarren.)

Children with fetal alcohol syndrome are developmentally and mentally retarded, with a mean IQ of about 68 (Streissguth and LaDue 1987). FAS patients with a mean chronological age of 16.5 years were found to have the functional vocabulary of 6.5-year-olds and to have the mathematical abilities of fourth graders. Most adults and adolescents with FAS cannot handle money or their own lives, and they have difficulty learning from past experiences. Moreover, in many cases of FAS, the behavioral abnormalities exist without any gross physical changes in head size or reductions in IQ (J. M. Opitz, personal communication).

There is great variation in the ability of mothers and fetuses to metabolize ethanol, and it is thought that 30–40% of the children born to alcoholic mothers who drink during pregnancy will have FAS. It is also thought that lower amounts of ethanol ingestion by the mother can lead to fetal alcohol effect, a condition that is less severe than FAS, but which lowers the functional and intellectual abilities of the affected person.*

A mouse model system has been used to explain the effects of ethanol on the face and nervous system. When mice are exposed to ethanol at the time of gastrulation, it induces the same range of developmental defects as in humans. As early as 12 hours after the mother ingests alcohol, abnormalities of development are observed. The midline facial structures fail to form, allowing the abnormally close proximity of the medial processes of the face. Forebrain anomalies are also seen, and the most severely affected fetuses lack a forebrain entirely (Sulik et al. 1988).

Studies on these mice suggest that ethanol may induce its teratogenic effects by more than one mechanism. First, anatomical evidence suggests that neural crest cell migration is severely impaired. Instead of migrating and dividing, ethanol-treated neural crest cells prematurely initiate their differentiation into facial cartilage (Hoffman and Kulyk 1999). Second, ethanol can cause the apoptosis of neurons. One way it can do this is by generating superoxide radicals that can oxidize cell membranes and lead to cytolysis (Figure 21.13A–C; Davis et al. 1990; Kotch et al. 1995). Ethanol-induced apoptosis can delete millions of neurons from the developing forebrain, frontonasal (facial) process, and cranial nerve ganglia. Third, alcohol may directly interfere with the ability of the cell adhesion molecule L1 to hold cells together. Ramanathan and colleagues (1996) have shown that ethanol can block the adhesive function of the L1 protein in vitro at levels as low as 7 mM, a concentration of ethanol produced in the blood or brain with a single drink (Figure 21.13D). Moreover, mutations in the human *L1* gene cause a syndrome of mental retardation and malformations similar to that seen in severe cases of fetal alcohol syndrome.

---

*For a remarkable account of raising a child with fetal alcohol syndrome, as well as an analysis of FAS in Native American culture in the United States, read Michael Dorris's *The Broken Cord* (1989), in which the personal and sociological effects of FAS are well integrated with the scientific and economic data.

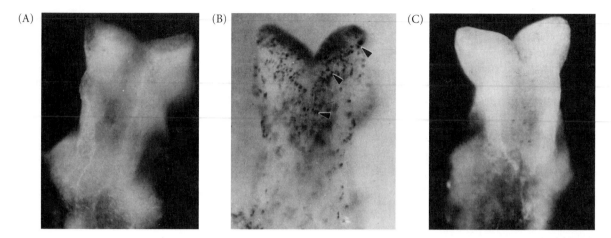

Figure 21.13
Possible mechanisms producing fetal alcohol syndrome (FAS).
(A–C) Cell death caused by ethanol-induced superoxide radicals.
Staining with Nile blue sulfate shows areas of cell death. (A) Control
9-day mouse embryo head region. (B) Head region of ethanol-treated embryo, showing areas of cell death. (C) Head region of embryo
treated with both ethanol and superoxide dismutase, an inhibitor of
superoxide radicals. The enzyme prevents the alcohol-induced cell
death. (D) The inhibition of L1-mediated cell adhesion by ethanol.
(A–C from Kotch et al. 1995, photographs courtesy of K. Sulik; D
after Ramanathan et al. 1996.)

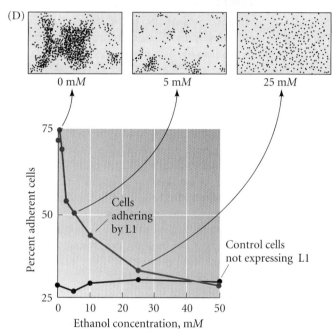

In the chick embryo, a fourth mechanism has been found
by which ethanol can disrupt development (Ahlgren et al.
2002). Ethanol injected into the amnion of a chick embryo results in the apoptosis of cranial neural crest cells and in failure
to form the frontonasal process. This apoptosis was correlated
with a loss of *sonic hedgehog* gene expression in the pharyngeal arches. Moreover, if Sonic hedgehog-secreting cells were
placed in the head mesenchyme at this time, they prevented
the ethanol-induced apoptosis of the cranial neural crest cells.
Together, these studies show that ethanol is exceptionally dangerous to the developing vertebrate because it is so widespread, because it can get through the placenta, and because it
can operate in several ways.

> WEBSITE 21.7 **Our knowledge of alcohol's teratogenic-
> ity.** A hundred years ago, alcohol was considered dangerous
> to the fetus. Fifty years ago, it was considered harmless, and
> today it is considered very dangerous. Sociological studies
> look at how these assessments were made.

Endocrine disruptors

Endocrine disruptors (sometimes called hormone mimics or
environmental signal modulators) are exogenous chemicals
that interfere with the normal function of hormones. Endocrine disruptors can disrupt hormonal function in several
ways:

- They can mimic the effects of a natural hormone by
  binding to that hormone's receptors. DES (diethylstilbestrol) is one such example, as we will see below.
- They can block the binding of a hormone to its receptor,
  or they can block the synthesis of the hormone. Finasteride, a drug used to prevent male pattern baldness and
  enlargement of the prostate gland, is an anti-androgen,
  since it blocks the synthesis of dihydrotestosterone. Dihydrotestosterone (as we saw in Chapter 17) is necessary
  for the development of the male external genitalia,
  prostate differentiation, and the descent of the testes; so
  women are warned not to handle this drug if they are
  pregnant.
- They can interfere with the transport of a hormone or its
  elimination from the body. **Polychlorinated biphenyls**
  (**PCBs**; see below) act in this manner, as we will see in
  the next chapter.

This chapter will discuss the known and possible effects of endocrine disruptors on human development. The next chapter, on the environmental regulation of development, will provide evidence that environmental endocrine disruptors are causing disruptions in the development of wild animals. Some endocrine disruptors are the products of natural plant biochemistry. Others, such as stabilizers that are used in the manufacture of polystyrene and other plastics, are not.

One of the most potent environmental estrogens has been **diethylstilbestrol**, a drug that was prescribed to approximately 5 million pregnant women from the 1940s through the 1960s to prevent miscarriage and premature labor (see Palmund 1996; Bell 1986). Although research from the mid-1950s showed that DES had no beneficial effects on pregnancy, it continued to be prescribed until the early 1970s, when it was shown that women exposed to this drug in utero had an increased chance of having cervical cancers and morphological abnormalities of the reproductive tract. Male offspring who had been exposed to DES in utero often had abnormal genitalia (see Robboy et al. 1982; Mittendorf 1995).

The female reproductive tract abnormalities resulting from prenatal DES exposure involve changes in the cell types along the Müllerian duct. In particular, there is a loss of the boundary between the oviduct and the uterus due to the lack of the uterotubal junction. Moreover, the distal Müllerian ducts often fail to come together to form a single cervical canal. Symptoms similar to the human DES syndrome occur in mice exposed to DES in utero.

Ma and colleagues (1998) showed that the effects of DES on the female mouse reproductive tract could be explained as the result of altered *Hoxa-10* expression in the Müllerian duct. They showed that estrogen and progesterone are able to regulate the 5′ genes of the Hoxa cluster (*Hoxa-9, Hoxa-10, Hoxa-11,* and *Hoxa-13*). Normally, these genes are regulated in a nested fashion throughout the Müllerian duct. *Hoxa-9* message is detected throughout the uterus and continues about halfway through the presumptive oviduct. *Hoxa-10* expression exhibits a sharp anterior boundary at the junction between the future uterus and the future oviduct. *Hoxa-11* has strong expression in the anterior regions where *Hoxa-10* is expressed, but weakens in the posterior regions. *Hoxa-13* expression is seen only in the cervix (Figure 21.14).

To determine whether DES changed Hox gene expression patterns, DES was injected under the skin of pregnant mice, and the fetuses were allowed to develop almost to birth. When the fetuses from the DES-injected mothers were compared with fetuses from mothers that had not received DES, it was seen that DES almost completely repressed the expression of *Hoxa-10* in the Müllerian duct (Figure 21.15). This repression was most pronounced in the stroma (mesenchyme) of the duct, the place where experimental embryologists had localized the effect of DES (Boutin and Cunha 1997). The case for DES acting through repression of *Hoxa-10* is strengthened by the phenotype of the *Hoxa-10* knockout mouse (Benson et al.

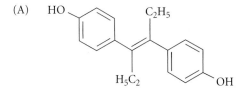

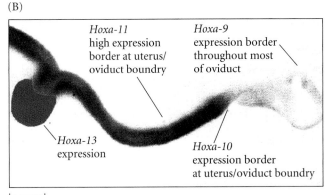

Figure 21.14
(A) The chemical structure of diethylstilbestrol (DES). (B) 5′ Hoxa gene expression in the reproductive system of a normal 16.5-day embryonic female mouse. A whole-mount in situ hybridization of the *Hoxa-13* probe is shown here. *Hoxa-9* expression extends from the cervix through the uterine anlage to about halfway up the Müllerian duct. *Hoxa-10* expression has a sharp anterior border at the transition between the presumptive uterus and the oviduct. *Hoxa-11* has the same anterior border as *Hoxa-10*, but its expression diminishes closer to the cervix. The expression of *Hoxa-13* is found only in the cervix and upper vagina. (After Ma et al. 1998.)

1996; Ma et al. 1998), in which there is a transformation of the proximal quarter of the uterus into oviduct tissue, and there are abnormalities of the uterotubal junction.

The link between Hox gene expression and uterine morphology is the Wnt proteins. Wnt proteins are associated with cell proliferation and protection against apoptosis, and the reproductive tracts of DES-exposed female mice also resemble those of *Wnt7a* knockouts. Miller and colleagues (1998) have shown that the Hox genes and the Wnt genes communicate to keep each other activated. Moreover, the Hox and Wnt proteins are involved in the specification and morphogenesis of the reproductive tissues (Figure 21.16). DES, however, acting through the estrogen receptor, represses the *Wnt7a* gene. This repression prevents the maintenance of the Hox gene expression pattern, and it prevents the activation of another Wnt gene, *Wnt5a*, which encodes a protein necessary for cell proliferation.

(A)

(B)

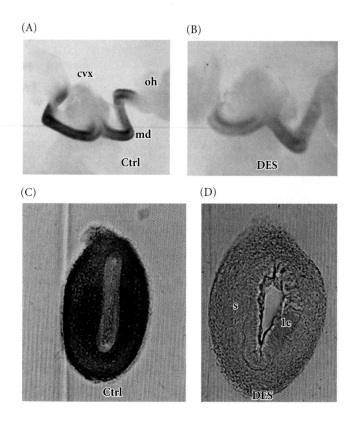

Figure 21.15

In situ hybridization of a *Hoxa-10* probe shows that DES exposure represses *Hoxa-10*. (A) Normal 16.5-day embryonic female mice show *Hoxa-10* expression from the boundary of the cervix through the uterus primordium and through most of the oviduct. (B) In mice exposed prenatally to DES, this expression was severely repressed. (C) In control female mice at 5 days after birth (when the reproductive tissues are still forming), a section through the uterus shows abundant expression of the *Hoxa-10* gene in the uterine stroma. (D) In female mice that were given high doses of DES at 5 days after birth, *Hoxa-10* gene expression in the mesenchyme is almost completely suppressed. cvx, cervix; md, Müllerian duct; o, ovary; le, luminal epithelium; s, stroma. (After Ma et al. 1998.)

(C)

(D)

Other teratogenic agents

Over 50,000 artificial chemicals are currently used in the United States, and about 200 to 500 new compounds are being made each year (Johnson 1980). Although teratogenic compounds have always been with us, the risks increase as more and more untested compounds enter our environment. Most industrial chemicals have not been screened for their teratogenic effects. Standard screening protocols are expen-

sive, long, and subject to interspecies differences in metabolism. There is still no consensus on how to test a substance's teratogenicity for human embryos.

HEAVY METALS. Heavy metals, such as zinc, lead, and mercury, are powerful teratogens. Industrial pollution has resulted in high concentrations of heavy metals in the environment in many places. In the former Soviet Union, the unregulated "industrial production at all costs" approach has left a legacy of soaring birth defect rates. In some regions of Kazakhstan, heavy metals are found in high concentrations in drinking water, vegetables, and the air. In these places, nearly half the people tested have extensive chromosome breakage. In some areas, the incidence of birth defects has doubled since 1980 (Edwards 1994). Lead and mercury can cause damage to the developing nervous system. The polluting of Minamata Bay, Japan, with mercury in 1956 produced brain deficiencies both by transmission of the mercury across the placenta and by its transmission through mother's milk. It appears that mercury is selectively absorbed by regions of the cerebral cortex (Eto 2000; Kondo 2000; Eto et al. 2001).

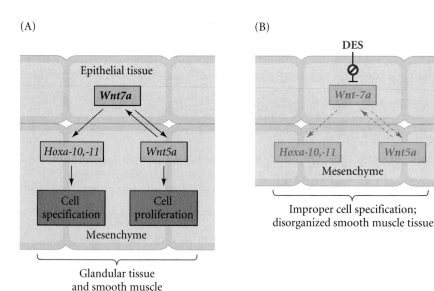

Figure 21.16

Misregulation of Müllerian duct morphogenesis by DES. (A) During normal morphogenesis, the *Hoxa-10* and *Hoxa-11* genes in the mesenchyme are activated and maintained by Wnt7a from the epithelium. Wnt7a also induces Wnt5a in the mesenchyme, and Wnt5a both maintains Wnt7a expression and causes mesenchymal cell proliferation. Together, these factors specify and order the morphogenesis of the uterus. (B) DES, acting through the estrogen receptor, blocks Wnt7a expression. The proper activation of the Hox genes and *Wnt5a* in the mesenchyme does not occur, leading to a radically altered morphology. (After Kitajewsky and Sassoon 2000.)

The DES Story: Is It Happening Again?

The DES tragedy is a complex story of public policy, medicine, and development (Palmund 1996; Bell 1986). However, according to Frederick vom Saal, Ana Soto, and others, the public and the pharmaceutical industry have learned little from the DES tragedy, and we may be experiencing the same endocrine disruption today, only on a much larger scale. These researchers claim that some of the major constituents of plastics are estrogenic compounds, that they are present in doses large enough to have profound effects on sexual development and behavior, and that the plastics industry is fighting against public awareness of this issue.

Nonylphenol

Estrogenic compounds are in the food we eat and in the plastic wrapping that surrounds them. The discovery of the estrogenic effect of plastic stabilizers was made in a particularly alarming way. Investigators at Tufts University Medical School had been studying estrogen-responsive tumor cells, which require estrogen in order to proliferate. Their studies were going well until 1987, when the experiments suddenly went awry. Their control cells began to show high growth rates suggesting stimulation comparable to that of the estrogen-treated cells. Thus, it was as if someone had contaminated the medium by adding estrogen to it. What was the source of contamination? After spending 4 months testing all the components of their experimental system, the researchers discovered that the source of estrogen was the plastic containers that held their water and serum. The company that made the containers refused to describe its new process for stabilizing the polystyrene plastic, so the scientists had to discover it themselves.

The culprit turned out to be *p*-**nonylphenol**, a compound that is also used to harden the plastic of the pipes that bring us water and to stabilize the polystyrene plastics that hold water, milk, orange juice, and other common liquid food products (Soto et al. 1991; Colburn et al. 1996). This compound is also the degradation product of detergents, household cleaners, and contraceptive creams. Nonyphenol has been shown to alter reproductive physiology in female mice and to disrupt sperm function. It is also correlated with developmental anomalies in wildlife (Fairchild et al. 1999; Hill et al. 2002; Kim et al. 2002; Adeoya-Osiguwa et al. 2003).

Bisphenol A

Bisphenol A was actually synthesized as an estrogenic compound in the 1930s. In the early years of hormone research, the actual steroid hormones were very difficult to isolate, so chemists manufactured synthetic analogues that would accomplish the same tasks. Bisphenol A was synthesized by Dodds, who demonstrated it to be estrogenic in 1936. (Two years later, Dodds synthesized diethylstilbestrol.) Later, polymer chemists realized that bisphenol A could be used in plastic production, and today it is one of the top 50 chemicals in production. Four corporations in the United States make almost 2 billion pounds of it each year for use in the resin lining in most cans, the polycarbonate plastic in baby bottles and children's toys, and dental sealant. Its modified form, tetrabromo-bisphenol A, is the major flame retardant on the world's fabrics.

However, vom Saal (2000) asserts that this chemical is not fixed in plastic forever. He asserts that if you place water in an old polycarbonate rat cage, you can measure up to 100 mg per liter of bisphenol A in the water. That is a biologically active amount—a concentration that will reverse the sex of a frog placed in that solution. Similarly, when products containing bisphenol A are thrown into a landfill, bisphenol A leaches out of them.

vom Saal and his colleagues assert that their experiments demonstrate that environmentally relevant concentrations of bisphenol A (and other estrogenic compounds) can cause disruptions in the morphology of the fetal sex organs, low sperm counts, and behavioral changes when these fetuses become adults (vom Saal et al. 1998; Palanza et al. 2002). Vom Saal and his colleagues have implicated bisphenol A (and related estrogenic compounds such as dioxins) in a suite of trends, including the lowering of human sperm counts, the increase in prostate enlargement, and the decrease in the age of female sexual maturation.

The age at which girls begin to express adult female sexual characteristics has declined over the past hundred years (probably due to nutrition), but new research suggests that puberty is starting extremely early (Herman-Giddens et al. 1997). Work in vom Saal's laboratory (Howdeshell et al. 1999) showed that when female mice are exposed in utero to low doses of bisphenol A, those mice undergo sexual maturity faster than unexposed mice.

In 1992, Carlsen and colleagues reported a large average decline in sperm counts from studies around the world. This finding has recently been confirmed by Swan and colleagues (2000), who also come to the conclusion that, on average, human sperm density and quality have declined significantly over the past 5 decades.

These studies also implicate bisphenol A (and other estrogenic compounds) in causing the increase in prostate enlargement seen in men. In the United States, 65% of men at the age of retirement have enlarged prostate glands, and 40,000 men die of prostate cancer each year. As an assay, researchers used the mouse prostate, a gland whose size is sensitive to estrogens. vom Saal and colleagues (1998) found that when they gave pregnant mice 2 parts per billion bisphenol A—that is, 2 nanograms per gram of body weight—for the 7 days at the end of pregnancy (equivalent to the period when human reproductive organs are developing), the treated animals showed an increase in prostate size of about 30%. Indeed, the whole area normally stimulated by dihydrotestosterone was enlarged. The prostate size was increased and the sperm count of the mice lowered. Adult male mice exposed to small amounts of bisphenol A have enlarged prostates, and bisphenol A increases the rate of mitosis in human prostate cells (Figure 21.17; Gupta 2000; Ramos et al. 2001; Wetherill 2002). Female mice exposed to bisphenol A in utero had alterations in the organization of their breast tissue and ovaries and altered estrous cyclicity as adults (Markey et al. 2003).

But is there any evidence that bisphenol A reaches the fetus in concentrations that matter? Unfortunately, recent studies (Ikezuki et al. 2002; Schöenfelder et al. 2002) have shown that bisphenol A in the human placenta is neither eliminated nor metabolized into an inactive compound; rather, it accumulates to concentrations that can alter development in laboratory animals.

The plastics industry counters that vom Saal's evidence is meager and that his experiments cannot be repeated (Cagen et al. 1999; see Lamb 2002). However, a review of that industry's own study (claiming that mice exposed in utero to bisphenol A do not have enlarged prostates or low sperm counts) points out that the positive control of the industry-sponsored research did not produce the expected effects.

Figure 21.17

Bisphenol A reduces androgen receptor and prostatic acid phosphatase expression in the ventral prostate of rats exposed to it in utero. (A) Androgen receptor expression in periductal stromal cells of ventral rat prostates. Two bisphenol A-treated groups (whose mothers were given 25 µg/kg/day or 250 µg/kg/day bisphenol A (BPA) from embryonic day 8 to birth) had smaller percentages of androgen receptor-containing stromal cells than did controls rats. (B) Prostatic acid phosphatase expression in epithelial cells in the ventral prostates of rats . In the control animals, many more glandular epithelial cells were positive for prostatic acid phosphatase than in the bisphenol A-treated groups. Means with different letters differ significantly ($P < 0.01$). (After Ramos et al. 2001.)

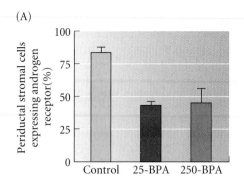

(A)

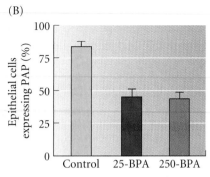

(B)

Pesticides

Bell and colleagues (2001) used California state records and epidemiological research to show that deaths due to congenital anomalies increased significantly in those areas sprayed with pesticides. The greatest risk of fetal death due to congenital anomalies was seen when the pesticides were sprayed during the third to eighth week of the pregnancy. Those children exposed to pesticides in utero were found to have less mature motor skills than those not exposed to pesticides (Guillette et al. 1998).

One of the most important teratogenic pesticides is DDT, which is discussed at length in the following chapter. In humans, DDT has been linked to pre-term births and immature babies (Longnecker et al. 2001).

Genistein

Another estrogenic chemical that is starting to cause concern is **genistein**. Genistein is the major estrogenic compound found in soy and soy products (such as tofu). For people eating a normal diet, there does not appear much to worry about (indeed, this compound may protect against certain cancers). However, researchers at the National Institute of Environmental Health Science (Newbold et al. 2001) are worried that infants who are born to vegan or vegetarian mothers and who are given solely soy-based formulae to drink may develop abnormalities of the reproductive system and thyroid glands. Their concerns are based on experiments showing that mice given doses of genistein equivalent to those such babies receive showed a significant incidence of uterine cancers. Similarly, mice born of mothers given injections of genistein had thymuses that weighed only 20% as much as the weight of controls (Yellayi et al. 2002).

Genetic-Environmental Interactions

The observation that a substance may be teratogenic in one strain of mice, but not in another strongly suggests that the effect of an endogenous substance on development has a genetic component (see Chapter 22). Recent evidence suggests that different alleles in the human population can influence whether a substance is benign or dangerous to the fetus. For example, among the general population, there is only a slight risk that heavy smoking by a mother will cause facial malformations in her fetus. However, if the fetus has a particular allele (A_2) of the gene for the growth factor TGF-β, tobacco smoke absorbed through the placenta can raise the risk of cleft lip and palate tenfold (Shaw et al. 1996). Similarly, low birth weight in infants born to smoking mothers is associated with particular polymorphisms of two genes that are involved in metabolizing cyclic hydrocarbons (X. Wang et al. 2002). Different alleles encoding the enzyme alcohol dehydrogenase-2 result in differing abilities to degrade ethanol. Whether heavy maternal alcohol consumption leads to fetal alcohol syndrome or fetal alcohol effect may depend on the types of alcohol dehydrogenase and aldehyde dehydrogenase isozymes in the mother and fetus (McCarver-May 1996; Eriksson et al. 2001). Thus, whether or not a compound is "teratogenic" depends on many things, including the genes of the individuals exposed to it.

WEBSITE 21.8 **Our stolen future.** This website monitors the environmental effects of endocrine disruptors. It is a political and consumer action site as well as a scientific clearinghouse for endocrine disruption. Run by the authors of the book *Our Stolen Future*, it also provides links to the websites of people who disagree with them.

PATHOGENS AS TERATOGENIC AGENTS. Another class of teratogens includes viruses and other pathogens. Gregg (1941) first documented the fact that women who had rubella (German measles) during the first third of their pregnancy had a 1 in 6 chance of giving birth to an infant with eye cataracts, heart malformations, or deafness. This was the first evidence that the mother could not fully protect the fetus from the outside environment. The earlier the rubella infection occurred during the pregnancy, the greater the risk that the embryo would be malformed. The first 5 weeks of development appear to be the most critical, because that is when the heart, eyes, and ears are being formed. The rubella epidemic of 1963–1965 probably resulted in over 10,000 fetal deaths and 20,000 infants with birth defects in the United States (CDC, 2002b). Two other viruses, cytomegalovirus and the herpes simplex virus, are also teratogenic. Cytomegalovirus infection of early embryos is nearly always fatal, but infection of later embryos can lead to blindness, deafness, cerebral palsy, and mental retardation.

Bacteria and protists are rarely teratogenic, but two of them can damage human embryos. *Toxoplasma gondii*, a protist carried by rabbits and cats (and their feces), can cross the placenta and cause brain and eye defects in the fetus. *Treponema pallidum*, the bacterium that causes syphilis, can kill early fetuses and produce congenital deafness and facial damage in older ones.

IONIZING RADIATION. Ionizing radiation can break chromosomes and alter DNA structure. For this reason, pregnant women are told to avoid unnecessary X-rays, even though there is no evidence for congenital anomalies resulting from diagnostic radiation (Holmes 1979). Heat from high fevers is also a possible teratogen.

While we know the causes of certain malformations, most congenital abnormalities cannot yet be explained. For instance, congenital cardiac anomalies occur in about 0.5–1% of all live births. Genetic causes are responsible for about 8% of these heart abnormalities, and about 2% can be explained by known teratogens. That leaves 90% of them unexplained (O'Rahilly and Müller 1992). We still have a great deal of research to do.

> **WEBSITE 21.9 Other teratogenic agents.** Certain behavior-modifying drugs are teratogenic, while others appear not to be. Conflicting studies debate whether caffeine, cannabis, and cocaine are teratogenic.

> **WEBSITE 21.10 Maternal effects on adults.** Recent epidemiological data suggest that nutritional stress in a pregnant mother may predispose her offspring to certain diseases when they are adults.

DEVELOPMENTAL BIOLOGY AND THE FUTURE OF MEDICINE

Developmental Cancer Therapies

Cancer as a disease of altered development

Many cancers result from an accumulation of mutations that puts certain cells back into the cell cycle and makes them unresponsive to their cellular environments (see Lengauer et al. 1998; Sandberg and Chen 2002). This genetic approach to malignancy has explained the formation of numerous tumor types, but it is not the whole story. As Folkman and colleagues (2000) recently noted, the genetic approach to cancer therapy must be complemented by a developmental approach that sees that "epigenetic, cell-cell, and extracellular interactions are also pivotal in tumor progression." Indeed, many tumor cells have normal genomes, and whether or not they are malignant depends upon their environment.

The most remarkable of these cases is the **teratocarcinoma**, which is a tumor of germ cells or stem cells (see Chapter 19; Illmensee and Mintz 1976; Stewart and Mintz 1981). Teratocarcinomas are malignant growths of cells that resemble the inner cell mass of the mammalian blastocyst, and they can kill the organism. However, if a teratocarcinoma cell is placed on the inner cell mass of a mouse blastocyst, it will integrate into the blastocyst, lose its malignancy, and divide normally. Its cellular progeny can become part of numerous embryonic organs. Should its progeny form part of the germ line, sperm or egg cells can be formed from the tumor cell and transmit its genome to the next generation (see Figure 19.11)!

The surroundings of the cell are critical in determining malignancy. Cell division is a "normal" function of cells, and in many cases, tissue interactions are required to prevent cells from dividing. Thus, there are tumors that arise through defects in tissue architecture (Sonnenschein and Soto 1999, 2000; Bissel et al. 2002). Recent studies have shown that some tumors can be caused by altering the structure of the tissue, and that these tumors can actually be suppressed by restoring an appropriate tissue environment (Coleman et al. 1997; Weaver et al. 1997; Sternlicht et al. 1999). Many carcinogens (cancer-forming agents) do not function by making mutations. Asbestos, for instance, causes cancers, and probably does so by disturbing tissues, not genes.

Differentiation therapy

Both carcinogenesis and teratogenesis are diseases of tissue organization and intercellular communication. Recent studies have shown that some defects in signaling pathways can cause cancers, while other defects in the same pathways can cause malformations. By understanding the developmental events involved, one can design better therapies for these diseases. Medulloblastomas, the most common malignant brain tumors in children, appear to be due to abnormalities of the Hedgehog signaling pathway. In these brain tissues, the Smoothened protein is constitutively activated, due to either mutations in Patched (the normal negative regulator of Smoothened) or faulty methylation of the *Patched* gene. Berman and colleagues (2002) showed that medulloblastomas in mice and humans can be halted by cyclopamine, a teratogen that functions by *blocking* the Hedgehog pathway. Similarly, the teratogen valproic acid has been shown to reduce several tumors. The same properties that make valproic acid a teratogen may also make it useful for destroying tumor tissue (Blaheta et al. 2002).

Knowledge of developmental pathways can also explain the resistance of some tumors to conventional chemotherapy. Melanomas for instance, are tumors of the pigment cells. Recent studies (McGill et al. 2002) have demonstrated that MITF, in addition to activating tyrosinase and other melanin-forming genes (see above), also activates the anti-apoptosis gene *BLC2*. That is probably why melanomas are so resistant to treatment.

Another area of cancer therapy in which developmental biology has played an important role has been "**differentiation therapy**" for acute promyelocytic leukemia (APL). In 1978, Pierce and colleagues noted that cancer cells were in many ways reversions to embryonic cells, and hypothesized that cancer cells should revert to normalcy if they were made to differentiate. That same year, Sachs (1978) discovered that certain leukemias could be controlled by making their cells differentiate rather than proliferate. One of these leukemias is APL, which is caused by a somatic recombination between chromosomes 15 and 17 in the precursor cells that give rise to neutrophils (see Figure 15.21). The fusion of these chromosomes creates a chimeric transcription factor, one of whose parts is retinoic acid receptor-α. The expression of this chimeric transcription factor causes the cell to become malignant (Miller et al. 1992; Grignani et al. 1998). Normally, retinoic acid receptor-α is involved in the differentiation of myeloid precursor cells into neutrophils, but the chimeric transcription factor appears to prevent this from occurring. Treatment of APL patients with all-*trans*-retinoic acid causes remission of the APL in more than 90% of cases, since the additional retinoic acid is able to effect the differentiation of the leukemic cells into normal neutrophils (Hansen et al. 2000; Fontana and Rishi 2002).

The concept of differentiation therapy is now being extended to other types of cancers (see Altucci and Gronemeyer 2001). It appears that tumor cells can become "addicted" to the high expressions of transcriptional regulators that spur their rapid cell division. Jain and colleagues (2002) have found that a brief inactivation of growth regulators such as Myc tells the tumor cells to either differentiate or die (Figure 21.18). An important question, then, is how to bring about such brief inactivation of growth regulators. The answer may be provided by RNA interference technology. Xia and colleagues (2002) demonstrated the efficacy of a viral-mediated delivery mechanism that caused the specific silencing of targeted genes through the expression of **small interfering RNA (siRNA)**. Adenovirus vectors were used to transfect tumor cells with genes encoding a small antisense transcript targeted to a specific region of the gene of interest. As a result, the protein encoded by that gene was not translated.

Angiogenesis inhibition

Another area in which knowledge of development could contribute to cancer therapies is the inhibition of angiogenesis. Judah Folkman (1974) has estimated that as many as 350 billion mitoses occur in the human body every day. With each cell division comes the chance that the resulting cells will be malignant. Yet very few tumors actually do develop in any individual. Folkman has suggested that cells capable of forming tumors develop at a certain frequency, but that most are never able to form observable tumors. The reason is that a solid tumor, like any other rapidly dividing tissue, needs oxygen and nutrients to survive. Without a blood supply, potential tumors either die or remain dormant. Such "microtumors" remain as a stable cell population wherein dying cells are replaced by new cells.

The critical point at which a node of cancerous cells becomes a rapidly growing tumor occurs when it becomes vascularized. A microtumor can expand to 16,000 times its original volume in the 2 weeks after vascularization. Without a blood supply, however, no growth is seen (Folkman 1974; Ausprunk and Folkman 1977). To accomplish this vascularization, the microtumor secretes substances called **tumor angiogenesis factors.** These often include the same factors that produce blood vessel growth in the embryo—VEGF, FGF2, placenta-like growth factor, and others. The tumor angiogenesis factors stimulate mitosis in endothelial cells and direct the cell differentiation into blood vessels in the direction of the tumor.

Tumor angiogenesis can be demonstrated by implanting a piece of tumor tissue within the layers of a rabbit or mouse cornea. The cornea itself is not vascularized, but it is surrounded by a vascular border, or limbus. The tumor tissue induces blood vessels to form and grow toward the tumor (Figure 21.19; Muthukkaruppan and Auerbach 1979). Once the blood vessels enter the tumor, the tumor cells undergo explosive growth, eventually bursting the eye. Other adult solid tissues do not induce blood vessels to form. It might therefore be possible to block tumor development by blocking angiogenesis. Folkman and his colleagues have pioneered attempts to find and use angiogenesis inhibitors. Table 21.3 lists several natural and artificial angiogenesis inhibitors. These compounds act to prevent the endothelial cells from responding to the angiogenetic signal of the tumor.* Interestingly, thalidomide, the teratogen responsible for birth defects in the 1960s,

---

*The consumption of green tea has been associated with lower incidences of human cancer and the inhibition of tumor cell growth in laboratory animals. Cao and Cao (1999) have shown that green tea, as well as one of its components, epigallocatechin-3-gallate (EGCG), prevents FGF2- and VEGF-induced angiogenesis.

(A)

(B)

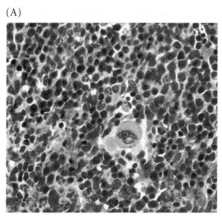 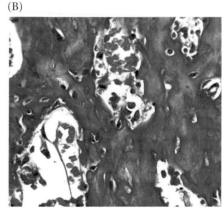

Figure 21.18
Differentiation of osteogenic sarcomas in mice due to inactivation of the Myc transcription factor. (A) Osteogenic sarcomas injected into mice remained undifferentiated and metastasized. (B) When transcription of the Myc transcription factor (which promotes cell division) was briefly blocked, the tumor cells differentiated into mature bone-like cells. (After Jain et al. 2002.)

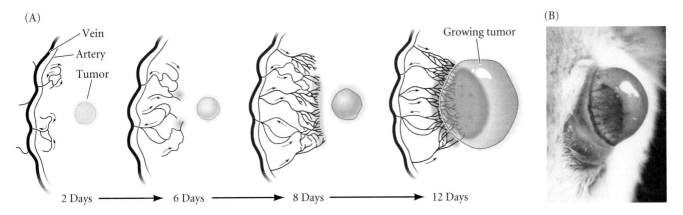

Figure 21.19
New blood vessel growth to the site of a mammary tumor transplanted into the cornea of an albino mouse. (A) Sequence of events leading to the vascularization of the tumor on days 2, 6, 8, and 12. The veins and arteries in the limbus surrounding the cornea both supply blood vessels to the tumor. (B) Photograph of living cornea of an albino mouse, with new blood vessels from the limbus approaching the tumor graft. (From Muthukkaruppan and Auerbach 1979; photograph courtesy of R. Auerbach.)

is on this list. Thalidomide has been found to be a potent anti-angiogenesis factor that can reduce the growth of cancers in rats and mice (D'Amato et al. 1994; Dredge et al. 2002). One of the advantages of these compounds is that the tumor cells are unlikely to evolve resistance to them, since the tumor cell

itself is not the target of these agents (Boehm et al. 1997; Kerbel and Folkman 2002).

Hanahan and Folkman (1996) have hypothesized that tumor angiogenesis is mediated by a change in the balance between angiogenesis factors and angiogenesis inhibitors. Observations of human tumor progression and gene knockouts in mice suggest that tumor angiogenesis may be mediated either by a decrease in the production of angiogenesis inhibitors or an increase in the production of angiogenesis factors. The signal for new VEGF synthesis in tumor cells might even be the hypoxia the cells experience in the center of a premalignant cell mass (Shweiki et al. 1992).

Cancer is not the only disease characterized by ectopic angiogenesis. The destruction of the retina in people with severe diabetes and in premature babies is probably due to hypoxia-induced VEGF, which stimulates new blood vessel formation (see Barinaga 1995).

Cancer and congenital malformations are opposite sides of the same coin. Both involve disruptions of normal development. Moreover, as we have seen, agents that have been known to cause congenital malformations—thalidomide, retinoic acid, and cyclopamine—can be used as drugs to prevent cancers. Just as they disrupt normal development, they can also disrupt the caricature of development caused by the tumor cells.

Gene Therapy

The technologies of gene therapy may give us the ability to genetically modify our bodies (and the bodies of our children) within the next decade. The ability to insert new genes into a fertilized egg and the ability to produce human

TABLE 21.3 Some natural and artificial angiogenesis inhibitors

Compound	Mechanism of action
Angiostatin	Binds to ATP synthase, angiomotin, and annexin II on endothelial cells; inhibits endothelial cell proliferation and migration
Bevacizumab (Avastin)	Recombinant humanized monoclonal antibody against vascular endothelial growth factor (VEGF)
Arresten	Believed to bind integrin $\alpha_5\beta_3$; inhibits endothelial cell proliferation, migration, tube formation, and neovascularization
Canstatin	Believed to bind integrin $\alpha_5\beta_3$; inhibits endothelial cell proliferation, migration, and tube formation
Combretastatin	Targets microtubules; induces reorganization of the actin cytoskeleton and early membrane blebbing in human endothelial cells
Endostatin	Believed to target integrin $\alpha_5\beta_3$; inhibits endothelial cell proliferation and migration. May induce apoptosis of proliferating endothelial cells
NM-3	Inhibits VEGF; has been shown to selectively inhibit endothelial cell proliferation, sprouting, and tube formation in vitro.
Thalidomide	Inhibits FGF2- and VEGF-mediated angiogenesis
Tumstatin	Binds integrin $\alpha_5\beta_3$; inhibits endothelial cell proliferation and neovascularization
2-Methoxyestradiol	Inhibits microtubule formation in proliferating endothelial cells, resulting in endothelial apoptosis
Vitaxin	Humanized monoclonal antibody against integrin $\alpha_5\beta_3$

Source: After Kerbel and Folkman 2002.

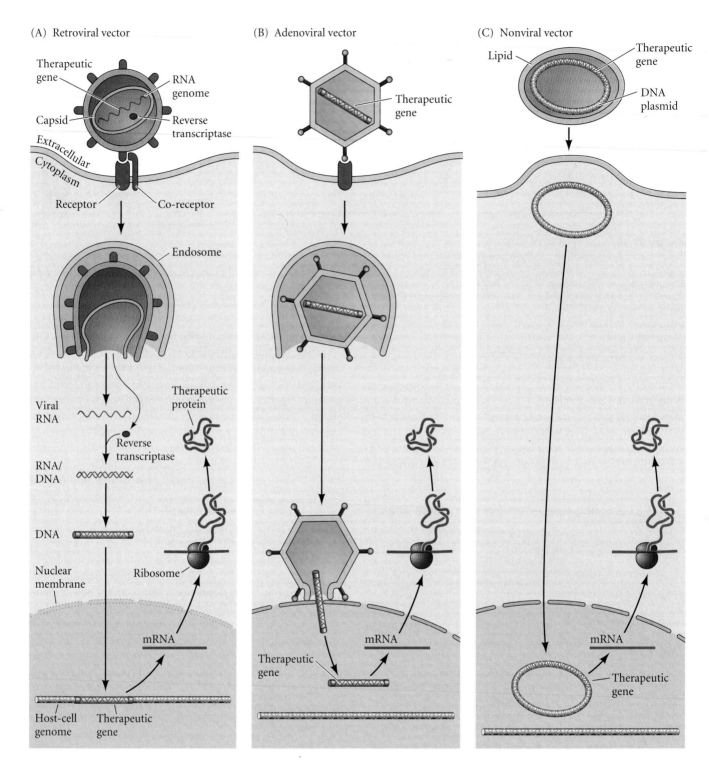

Figure 21.20

Vectors used in gene therapy. (A) When a gene (green) is placed in a retroviral vector, the virus is taken into the cell and releases its RNA into the cell. The reverse transcriptase of the virus copies a double-stranded DNA from the RNA, and this DNA is integrated into the host genome. The inserted gene can then be transcribed. (B) The adenovirus genome is already a double-stranded DNA. This vector uncoats itself on the host cell's nuclear envelope. The genes it carries can then be expressed without being incorporating into the host genome. (C) DNA inserted into a lipid membrane must be inserted in the host cell artificially. However, once in the host cell, the DNA can enter the host nucleus and be transcribed. (After Eck 1999.)

embryonic stem cells, however, have raised concerns that the human genome could be manipulated. Some people look forward to the time when genetic diseases might be treated by gene therapy and eliminated from subsequent generations as well. Other people (and in some cases, the same people) worry that the ability to manipulate genes will result in misguided attempts to enhance human physical or mental abilities.

Somatic cell gene therapy

Somatic cell gene therapy raises the hope of curing a person's genetic disease by inserting a wild-type gene that would be activated at the appropriate times and places. This type of gene therapy is currently being tested in over 600 laboratories (Roberts 2002). In such experiments, a committed stem cell (such as a blood stem cell or liver stem cell) is cultured, given the new gene, and reinserted into the body (Anderson 1998; Gage 1998; Ye et al. 1999). The new gene is usually packaged in a viral or lipid vector that enables it to enter the cell and be transcribed (Figure 21.20). This technique was used in the small interfering RNA study mentioned above.

There have been stunning successes in this field, as well as some prominent failures. In 1990, W. French Anderson and his colleagues (Blaese et al. 1995; Onodera et al. 1998) treated several patients suffering from immunodeficiency due to a mutation in their adenosine deaminase (ADA) gene. Their own T cells (which lack the ADA enzyme) were transfected with a wild-type ADA gene packaged in a retrovirus. These patients became immunocompetent, although they still have to return for treatment. Somatic cell gene therapy has also been used successfully to treat genetic liver disease and hemophilia (Roth et al. 2001). However, in 1999, an adult patient undergoing gene therapy for a kidney-specific enzyme defect died 4 days after the treatment. The cause of his death remains a mystery, but it is possible that the adenovirus vector stimulated an allergic reaction that shut down the patient's lungs. Similarly, a promising study targeting another inherited immune deficiency was halted when one of the patients developed leukemia-like symptoms (Check 2002).

Somatic cell gene therapy is akin to standard medical treatment wherein the *individual* is treated, so the ethical issues it raises are akin to those involved in human tests of any new drug or surgical procedure in market-driven economies. Some of the public concerns focus on the possibility of genetic enhancements to individuals. There is a very fine line between treatment and enhancement. The same treatments that could cure genetic diseases and cancers could also be used to alter the phenotypes of "normal" individuals. Insertion of certain genes into muscles, for instance, could give recipients a stronger musculature or make them less likely to experience muscle fatigue. If genes were found that could make people live longer or retain their intelligence or memories longer, should they be given to those who could afford such treatments? Are baldness, short stature, and mild obesity diseases, or just conditions that our society has defined as subop-

timal? The answers to these questions may mean a great deal to the commercial success of genetic engineering.

Germ line gene therapy

In contrast to somatic cell gene therapy, which seeks to cure individuals, **germ line gene therapy** seeks to eliminate "bad" genes both from the individual *and* from that person's descendants. Germ line gene therapy can be accomplished in two ways: (1) by modifying a germ cell or fertilized egg such that the new genome is in each cell of the individual's body and is therefore transferred to next generation through the germ cells, or (2) by modifying embryonic stem cells and making a body that contains a high percentage of cells derived from these genetically altered blastomeres.

Genes have been routinely added, subtracted, and replaced in mice for the past decade (see Chapter 5). There is little doubt that such procedures could be done in humans should we desire to do so. Already, a transgenic rhesus monkey has been born that carries the gene for green fluorescent protein (GFP) in each cell of its body (Figure 21.21; Chan et al. 2001). A retroviral vector carrying the gene for GFP was injected into the space between the oocyte and the zona pellucida. The retrovirus inserted itself into the egg DNA, and

Figure 21.21
ANDi (backwards for "inserted DNA"), a transgenic rhesus monkey. The gene for green fluorescent protein is present in each of his nuclei.

the monkey that developed from the egg had GFP genes (from a jellyfish) in all of its cells.

Germ line gene therapy raises many important social and ethical issues. In a symposium on Engineering the Human Genome, James D. Watson was explicit about using gene therapy techniques for human enhancement: "I mean, if we could make better human beings by knowing how to add genes, why shouldn't we do it?" (Stock and Campbell 1998). Critics and commentators (Haraway 1997; Silver 1997; Rifkin 1998; Fukuyama 2002) claim that if this technology were successful, it would result in a large gap between the haves and the have-nots. Those who could afford gene therapy could have "prettier," healthier, perhaps even smarter children. Those who could not afford it would be left out of this bright future. Thus, genetic engineering of the germ line could exacerbate social stratification, reify economic differences into biological ones, and create a genetic underclass.* Moreover, what we think of today as being a "good" genetic trait might not turn out to be so good in a future environment.

There are also scientific issues that might mediate against successful germ line gene therapy. One of these concerns is the possibility that the introduction of a new gene might knock out a previously functioning gene. Viral insertion mechanisms may prefer to splice viral vectors into genes rather than into noncoding regions of DNA (Schröder et al. 2002). The *inversion of embryonic turning* gene (Chapter 11) was discovered when Yokoyama and colleagues (1993) inadvertently knocked out this gene by inserting a transgene into a mouse embryo.

> **WEBSITE 21.11 Bioethics: Genetic enhancement.** The prospects of germ line gene therapy have elicited numerous ethical concerns. This website includes a chapter dedicated to the ongoing debate in this area.

Stem Cells and Therapeutic Cloning

Cloning

The cloning of mammals was described in Chapter 4. The cloning of large mammals has applications in agriculture, and the ability to clone transgenic animals producing large quantities of human proteins has important applications in medicine (see Di Berardino 2001). However, as mentioned in Chapter 4, mammalian clones are not usually healthy animals. There are many reasons for the abnormal development of cloned embryos, including the faulty activation of imprinted genes, the failure of histone modifications, and methylation deficiencies (Wilmut et al. 2002). Microarray technology (Humphreys et al. 2002) has documented that cloned mice have abnormal patterns of gene expression in many of their

organs. One of the most important genes for establishing pluripotency, the *Oct4* gene, is aberrantly expressed in the blastocysts of most mouse clones made by somatic cell nuclear transfer, and the pattern of X chromosome inactivation is aberrant in cow clones (Boiani et al. 2002; Xue et al. 2002).

Cloning does not appear to be a good reproductive technology for humans. As Hans Schöler (quoted in Glausiusz 2002) has noted, "To obtain one normal organism, you're paving the way with a lot of dead or malformed fetuses." As mentioned in Chapter 4, Ian Wilmut, the person in charge of the group that cloned Dolly the sheep, has similar negative views concerning human reproductive cloning. However, some cloning technology may find important applications in "therapeutic cloning" done to generate stem cells, rather than live births.

> **WEBSITE 21.12 Bioethics: The ethics of human cloning.** When, if ever, should reproductive cloning of humans be allowed?

Embryonic stem cells and therapeutic cloning

Embryonic stem cells (ES cells) are pluripotent, can be cultured indefinitely in an undifferentiated state, and retain their developmental potential after prolonged culture (see Chapter 4). Currently, human ES cells can be obtained by two major techniques (Figure 21.22A,B). First, they can be derived from the inner cell masses of human embryos, such as those left over from in vitro fertilization (Thomson et al. 1998; see above). Second, they can be generated from germ cells derived from spontaneously aborted fetuses (Gearhart 1998). In both cases, the embryonic stem cells are pluripotent, since they are able to differentiate in culture to form more restricted stem cell types. Other methods for obtaining stem cells are in the experimental stages (Figure 21.22C,D).

The importance of pluripotent stem cells in medicine is potentially enormous. The hope is that human ES cells can be used to produce new neurons for people with degenerative brain disorders (such Alzheimer or Parkinson disease) or spinal cord injuries, to produce a new pancreas for people with diabetes, or to produce new blood cells for people with anemias. People with deteriorating hearts might be able to have the damaged tissue replaced with new heart cells, and those suffering from immune deficiencies might be able to replenish their failing immune systems. Such therapies have already worked in mice. Murine ES cells have been cultured under conditions causing them to form insulin-secreting cells, muscle stem cells, glial stem cells, and neural stem cells (Figure 21.23A, B; Brüstle et al. 1999; McDonald et al. 1999). Dopaminergic neurons derived from ES cells have been shown to significantly reduce the symptoms of Parkinson disease in rodents (Bjorklund et al. 2002; Kim et al. 2002).

Human embryonic stem cells differ in some ways from their murine counterparts in their growth requirements. In

---

*If this sounds like science fiction, remember that scientists have successfully cloned several species of mammals—something that *was* science fiction until 1997.

Figure 21.22
Four major ways of obtaining human pluripotent stem cells. Methods A and B have been documented to work; methods C and D remain experimental. (A) Cells from the inner cell mass of a blastocyst are cultured and become pluripotent embryonic stem cells. (B) The primordial germ cells a fetus are harvested and become pluripotent embryonic stem cells. (C) Adult stem cells are obtained and grown in a manner that allows them to become pluripotential. (D) "Therapeutic cloning," wherein the nucleus of a somatic cell is transferred into an enucleated oocyte. The oocyte is activated and gives rise to a blastocyst, whose inner cell mass is harvested and cultured to become pluripotent embryonic stem cells. (After NIH 2000.)

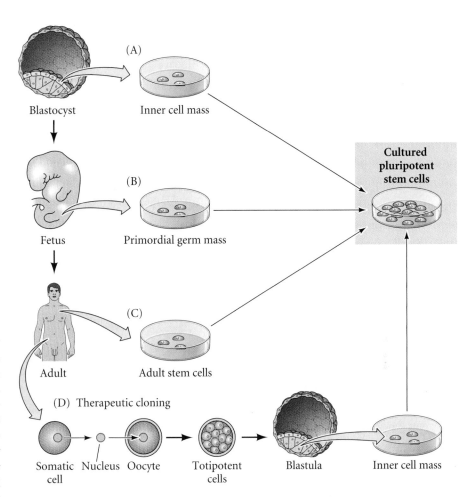

most ways, however, they are very similar, and have a similar, if not identical, pluripotency. Like mouse ES cells, human ES cells can be directed down specific developmental paths. For example, Kaufman and his colleagues (2001) directed human ES cells to become blood-forming stem cells by placing them on mouse bone marrow or endothelial cells. These ES-derived hematopoietic stem cells could further differentiate into numerous types of blood cells (Figure 21.23C).

One big difference between human and mouse experimentation is that mice can be inbred and made genetically identical, but humans cannot. Moreover, as human ES cells differentiate, they express significant amounts of the major histocompatability proteins that can cause immune rejection (Drukker et al. 2002). To get around the problem of host rejection, human ES cells could be modified, or somatic cell nuclear transfer could be used to ensure that the stem cells are genetically identical to the person who would be receiving their progeny. This brings us to another potential way of obtaining embryonic stem cells: **therapeutic cloning** (see Figure 21.22D). In this technique, a nucleus from the patient is inserted into an enucleated oocyte (as in cloning). The resulting embryo is grown in vitro until it has developed an inner cell mass. Cells from the inner cell mass are then cultured to generate stem cells that are genetically identical to the patient. In therapeutic cloning (sometimes called "somatic nuclear transplantation"), the result is the production of stem cells, not the production of a new child. While most medical and biological organizations do not advocate reproductive cloning, most favor therapeutic cloning. This distinction has led to debates that encompass major social issues.

WEBSITE 21.13 Bioethics: When does human life begin? Scientists have provided several answers to this question, depending on whether one privileges the genome, the individual, or the nervous system as the sine qua non of human life.

WEBSITE 21.14 Bioethics: Stem cell science ethics and politics. Different nations entertain different ideas about whether it is moral or expedient to study human embryonic stem cells. These sites look at the current debates about whether adult stem cells are pluripotent and how stem cell therapy should or should not be utilized.

Adult stem cells

As mentioned earlier in this book, numerous organs contain multipotent stem cells, even in the adult. Multipotent stem cells can give rise to a limited set of adult tissue types. There are several reasons why they are not as easy to use as pluripotent embryonic stem cells. First, they appear to have a relatively low rate of cell division and do not proliferate readily. Second, they are difficult to isolate, and are often fewer than one out of every thousand cells in an organ. However, when researchers have been able to isolate and culture such cells,

(A)

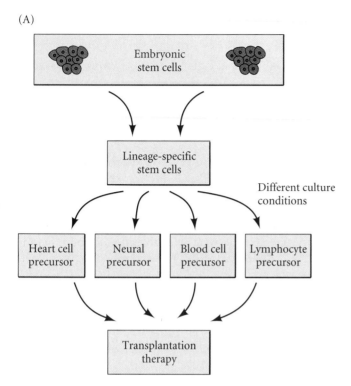

(B)

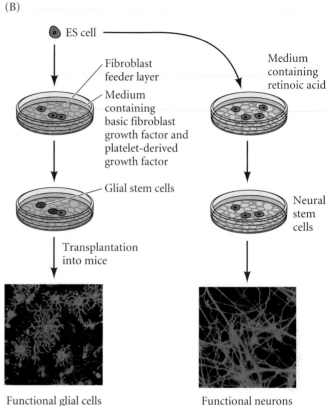

Functional glial cells   Functional neurons

Figure 21.23
Embryonic stem cell therapeutics. (A) Human embryonic stem cells (ES cells) can differentiate into lineage-specific stem cells, which can then be transplanted into a host. (B) The differentiation of mouse ES cells into lineage-restricted (neuronal and glial) stem cells can be accomplished by altering the media in which the ES cells grow. (C) Blood cells developing from human embryonic stem cells cultured on mouse bone marrow. (A after Gearhart 1998; B, photographs from Brüstle et al. 1999 and Wickelgren 1999, courtesy of O. Brüstle and J. W. McDonald; C, photograph courtesy of University of Wisconsin.)

(C)

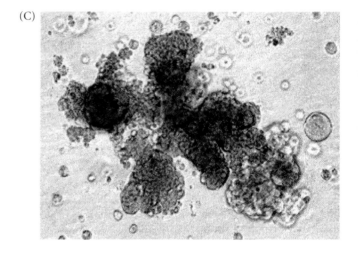

they have proved to be very useful (see Figure 21.22C). Carvey and colleagues (2001) have shown that when neural stem cells from the mesencephalon of adult rats are cultured in a mixture of paracrine factors, including interleukin-1 and interleukin-11, they differentiate into dopaminergic neurons that can cure the rat version of Parkinson disease. Moreover, investigators are also finding ways to selectively enhance the proliferation of particular stem cells in culture (Wu et al. 2002).

Whether multipotent stem cells can become pluripotent stem cells is controversial. It appears that most (if not all) adult multipotent stem cells are restricted to forming only a few cell types (Wagers et al. 2002). When hematopoietic stem cells were marked with green fluorescent protein (GFP) and placed in mice, their labeled descendants were found throughout the blood of these animals, but not in any other tissue.

However, some scientists claim that under some culture conditions, multipotent stem cells are able to generate progeny that can produce many different cell types. Jiang and col-

leagues (2002a) isolated a population of stem cells from the bone marrow of mice that, when injected into adult mice, gave rise to blood cells (as expected), but also to neural cells and endodermal cells (Figure 21.24). LaBarge and Blau (2002) have shown that these cells can also become muscle cells. When mouse bone marrow was labeled with green fluorescent protein and transplanted into recipient mice, GFP was subsequently found in muscle stem cells and later (after trauma to the muscles), in mature myocytes. These **mesenchymal stem**

(A) (B)

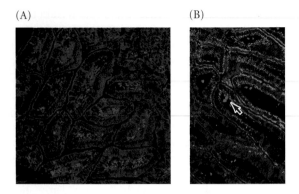

Figure 21.24
Colonization of adult organs by bone marrow mesenchymal stem cells. Intestines of (A) control mouse and (B) a mouse inoculated with mesenchymal stem cells. Green indicates the presence of the β-galactosidase protein used to mark the mesenchymal stem cells from the bone marrow. The arrow points to a few such cells that have contributed to the adult intestine. (After Jiang et al. 2002a; photograph courtesy of C. M. Verfaillie.)

cure genetic diseases that affect specific organs or tissues. For example, mice deficient in the *Rag2* gene are unable to recombine their DNA to make antibodies. Rideout and his colleagues (2002) took tail tip cells from a *Rag2*-deficient mouse and implanted the nuclei from those cells into enucleated mouse oocytes (Figure 21.25). The renucleated oocytes were activated and gave rise to blastocysts, from which *Rag2*-deficient ES cells were cultivated. One of the mutant *Rag2* genes in the ES cells was then converted into a wild-type gene by homologous recombination. The "repaired" ES cells were grown, labeled, and cultured to produce hematopoietic stem cells. These hematopoietic stem cells were injected back into the mouse from which the nuclei were originally taken, and they repopulated the mouse's immune system. Mature antibody-producing cells were detectable within a month of the transplantation. Transgenic human embryonic stem cells have recently been produced (Zwacka and Thompson 2003.)

cells had other unexpected abilities: they were able to integrate into the inner cell mass of a blastocyst, and they expressed *Oct4*—just like ES cells. This type of stem cell has also been found among human bone marrow stem cells (Jiang et al. 2002b; Korbling et al. 2002). When biopsies were taken from individuals who had received hematopoietic stem cell transplants, hepatocytes and skin cells were found to be derived from the donor cells. These cells may represent an extremely small population of pluripotent stem cells that persist into adult life, or they might be normal multipotent organ-specific stem cells that regain pluripotency under the culture conditions. Other investigators have criticized these findings, saying that the stem cells do not contribute any new tissue. Rather, the adult stem cells fuse with existing cells, thereby importing their dye to an already existing muscle or liver cell (Terada et al. 2002; Ying et al. 2002). This important question of whether multipotent stem cells can become pluripotent remains open. The research into embryonic and adult stem cells has produced a continuously surprising and unexpected set of observations.

Transgenic stem cells

The combination of somatic cell nuclear transfer and gene therapy can be used to

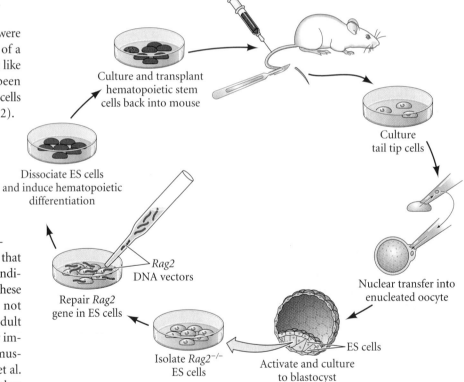

Figure 21.25
Repair of *Rag2* deficiency by therapeutic cloning. Tail tip cells from a *Rag2*-deficient (*Rag2⁻/⁻*) mouse are cultured, and the nuclei of these somatic cells are transplanted into activated enucleated oocytes. The oocytes develop to the blastocyst stage, and the *Rag2*-deficient inner cell mass cells are isolated and grown as stem cells. When these stem cells are incubated with wild-type *Rag2* DNA, homologous recombination allows some of these cells to replace the mutant allele with a wild-type allele. The repaired ES cells are aggregated, marked with a GFP gene, and cultured in a manner that directs them to differentiate into hematopoietic and lymphocytic stem cells. When these cells are injected into the mouse, they can restore its immune system by producing *Rag2*-positive cells that can make antibodies.

Regeneration therapy

Humans do not regenerate organs. Children can regenerate their fingertips, but even this ability is lost in adults. The ability to regenerate damaged human organs would constitute a medical revolution. Researchers are attempting to find ways of activating developmental programs that were used during organogenesis in the adult. One avenue of this research program is the above-mentioned search for adult stem cells that are still relatively undifferentiated, but which can form particular cell types if given the appropriate environment. The second avenue is a search for the environments that will allow such cells to initiate cell and organ formation. We will look at two of these medical efforts.

BONE REGENERATION. While fractured bones can heal, bone cells in adults usually do not regrow to bridge wide gaps. The finding that the same paracrine and endocrine factors involved in endochondral ossification are also involved in fracture repair (Vortkamp et al. 1998) raises the possibility that new bone could grow if the proper paracrine factors and extracellular environment were provided. The problem is how to deliver these compounds to a particular location over a long period of time.

One solution to the problem of delivery was devised by Bonadio and his colleagues (1999), who developed a collagen gel containing plasmids carrying the human parathyroid hormone gene. The plasmid-impregnated gel was placed in the gap between the ends of a broken dog tibia or femur. As cells migrated into the collagen matrix, they incorporated the plasmid and made parathyroid hormone. A dose-dependent increase in new bone formation was seen in about a month (Figure 21.26). This type of treatment has the potential to help people with large bone fractures as well as those with osteoporosis.

Another approach is to find the right mixture of paracrine factors to recruit stem cells and produce normal bone. Peng and colleagues (2002) genetically modified muscle stem cells to secrete BMP4 or VEGF. These cells were placed in gel matrix discs, which were implanted in wounds made in mouse skulls. The researchers found that certain ratios of BMP4 and VEGF were able to heal the wounds by making new bone. Similarly, BMP2 has been used to heal large fractures of primate mandibles and rabbit femurs (Li et al. 2002; Marukawa et al. 2002).

NEURONAL REGENERATION. While the central nervous system is characterized by its ability to change and make new connections, it has very little regenerative capacity. The motor neurons of the peripheral nervous system, however, have significant regenerative powers, even in adult mammals. The regeneration of motor neurons involves the regrowth of a severed axon, not the replacement of a missing or diseased cell body. If the cell body of a motor neuron is destroyed, it cannot be replaced.

(A) Treated fracture

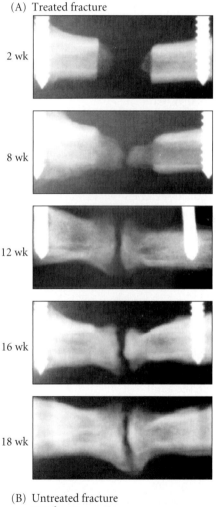

2 wk

8 wk

12 wk

16 wk

18 wk

(C) Whole bone
53 wk

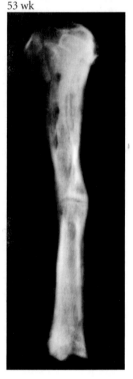

(B) Untreated fracture
24 wk

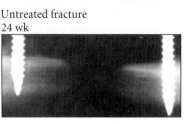

Figure 21.26
Bone formation from collagen matrix containing plasmids bearing the human parathyroid hormone. (A) A 1.6 cm gap was made in a dog femur and stabilized with screws. Plasmid-containing gel was placed on the edges of the break. Radiographs of the area at 2, 8, 12, 16, and 18 weeks after the surgery show the formation of bone bridging the gap at 18 weeks. (B) Control fracture (no plasmid in the gel) at 24 weeks. (C) Whole bone a year after surgery, showing repaired region. (After Bonadio et al. 1999; photographs courtesy of J. Bonadio.)

The myelin sheath that covers the axon of a motor neuron is necessary for its regeneration. This sheath is made by the Schwann cells, a type of glial cell in the peripheral nervous system (see Chapter 12). When an axon is severed, the Schwann cells divide to form a pathway along which the axon can regrow from the proximal stump. This proliferation of the Schwann cells is critical for directing the regenerating axon to the original Schwann cell basement membrane. If the regrowing axon can find that basement membrane, it can be guided to its target and restore the original connection. In turn, the regenerating neuron secretes mitogens that allow the Schwann cells to divide. Some of these mitogens are specific to the developing or regenerating nervous system (Livesey et al. 1997).

The neurons of the central nervous system cannot regenerate their axons under normal conditions. Thus, spinal cord injuries can cause permanent paralysis. As mentioned in Chapter 12, one strategy to get around this block is to find ways of enlarging the population of adult neural stem cells and to direct their development in ways that circumvent the lesions caused by disease or trauma. The neural stem cells found in adult mammals may be very similar to embryonic neural stem cells and may respond to the same growth factors (Johe et al. 1996; Johansson et al. 1999).

Another strategy for CNS neural regeneration is to create environments that encourage axonal growth. Unlike the Schwann cells of the peripheral nervous system, the myelinating glial cells of the central nervous system, the oligodendrocytes, produce substances that inhibit axon regeneration (Schwab and Caroni 1988). Schwann cells transplanted from the peripheral nervous system into a CNS lesion are able to encourage the growth of CNS axons to their targets (Keirstead et al. 1999; Weidner et al. 1999). Three substances that inhibit axonal outgrowth have been isolated from oligodendrocyte myelin: myelin-associated glycoprotein, Nogo-1. and oligodendrocyte-myelin glycoprotein (Mukhopadyay et al. 1994; Chen et al. 2000; GrandPré et al 2000; K. Wang et al. 2002). Antibodies against Nogo-1 and the blocking of the Nogo receptor allow partial axon regeneration after spinal cord injury (Schnell and Schwab 1990; GrandPré et al. 2002). Moreover, at sites of CNS injury, a glial scar forms, containing chondroitin sulfate proteoglycans in its extracellular matrix. These compounds inhibit axonal outgrowth as well. When Bradbury and colleagues (2002) delivered chondroitinase to lesioned spinal cords of adult rats, the axons were able to grow past the lesions. Research into CNS axon regeneration may become one of the most important contributions of developmental biology to medicine.

Snapshot Summary: Medical Implications of Developmental Biology

1. In vitro fertilization involves retrieving oocytes from the ovary and mixing them with sperm. The dividing eggs are then inserted into the uterus.

2. Intracytoplasmic sperm injection (ICSI) involves injecting a sperm directly into the oocyte cytoplasm.

3. In gamete intrafallopian transfer (GIFT), sperm are injected into the oviduct when the oocytes are ovulated.

4. Most human fertilized eggs do not successfully come to term. Chromosomal abnormalities are the major cause of early fetal death. Congenital anomalies can be due to gene mutations, chromosomal aneuploidy, or teratogens, or to a combination of genetic predisposition and environmental effects.

5. Positional gene cloning starts with pedigree analysis and makes a DNA map of the region differing between people having and lacking a particular phenotype.

6. Candidate gene mapping starts with a correlation between a genetic region and a gene whose mutation causes a similar phenotype in another species.

7. Pleiotropy occurs when several different effects are produced by a single gene. In mosaic pleiotropy, each effect is caused independently by the expression of the same gene in different tissues. In relational pleiotropy, abnormal gene expression in one tissue influences other tissues, even though those other tissues do not express that gene.

8. Genetic heterogeneity occurs when mutations in more than one gene can produce the same phenotype.

9. Phenotypic variability arises when the same gene can produce different defects (or differing severities of the same defect) in different individuals.

10. Dominant inheritance can be caused by haploinsufficiency, in which expression of a single copy of a gene is not sufficient to produce the wild-type phenotype; by gain-of-function mutations, in which a pathway is activated independently of its normal initiator; or by dominant negative alleles, mutant alleles encoding a subunit of a protein whose dysfunction makes the entire protein nonfunctional.

11. Inborn errors of gene expression can cause abnormalities at the levels of transcription, RNA processing, translation, and posttranslational modification.

12. Fragile X syndrome, the leading cause of mental retardation in humans, is caused by an overabundance of CGG repeats in the *FMR1* gene. These repeats prevent the *FMR1* gene from being transcribed. The protein encoded by *FMR1* appears to target certain neuronal genes to the ribosome for translation.

13. Preimplantation genetics involves testing for genetic abnormalities in early embryos in vitro, and implanting only those embryos that may develop normally. Selection for sex is also possible using preimplantation genetics.

14. Teratogenic agents include certain chemicals such as ethanol, retinoic acid, and valproic acid, as well as heavy metals, certain pathogens, and ionizing radiation.

15. Endocrine disruptors can bind to or block hormone receptors or block the synthesis, transport, or excretion of hormones.

16. DES is a powerful endocrine disruptor. Presently, bisphenol A and other PCBs are being considered as possible agents of precocious puberty in women and low sperm counts in men.

17. Environmental estrogens can cause reproductive system anomalies by suppressing Hox gene expression and Wnt pathways.

18. Cancer can be seen as a disease of altered development. Some tumors revert to nonmalignancy when placed in environments that fail to support rapid cell division.

19. Differentiation therapy makes tumor cells nonmalignant by exposing them to factors that cause them to stop dividing and to differentiate.

20. Disrupting tumor-induced angiogenesis may become an important means of stopping tumor progression.

21. Somatic cell gene therapy aims to cure an individual's genetic disease by adding genes to some of that person's somatic cells.

22. Germ line gene therapy aims to cure the genetic disease of a person and that person's descendants by repairing the gene in the germ cells.

23. Therapeutic cloning involves inserting the nucleus of a somatic cell from an individual into an enucleated activated oocyte, growing the embryo to blastocyst stage, and making stem cells from the inner cell mass of that embryo. The pluripotent stem cells would not be rejected by that individual.

24. Adult stem cells may provide alternative ways of growing differentiated cells to repair damaged tissue. Mesenchymal stem cells from the bone marrow appear to be able to form many cell types. Most adult stem cells have a limited developmental repertoire and are difficult to culture.

25. By altering conditions to resemble those in the embryo and by providing surfaces on which adult stem cells might grow, some adult stem cells might be "tricked" into regenerating bones and neurons.

Literature Cited

Abel, E. L. and R. J. Sokol. 1987. Incidence of fetal alcohol syndrome and economic impact of FAS-related anomalies. *Drug Alcohol Depend.* 19: 51–70.

Adeoya-Osiguwa, S.A., S. Markoulaki, V. Pocock, S. R. Milligan and L. R. Fraser. 2003. 17b Estradiol and environmental estrogens significantly affect mammalian sperm function. *Hum. Reprod.* 18: 100–107.

Ahlgren, S. C., V. Thakur and M. Bronner-Fraser. 2002. Sonic hedgehog rescues cranial neural crest from cell death induced by ethanol. *Proc. Natl. Acad. Sci. USA* 99: 10476–10481.

Altucci, L. and H. Gronemeyer. 2001. The promise of retinoids to fight against cancer. *Nature Rev. Cancer* 1: 181–193.

Anderson, H. F. 1998. Human gene therapy. *Nature* 392: 25–30.

Assady, S., G. Maor, M. Amit, J. Itskovitz-Eldor, K. L. Skorecki and M. Tzukerman. 2001. Insulin production by human embryonic stem cells. *Diabetes* 50: 1691–1697.

Ausprunk, D. H. and J. Folkman. 1977. Migration and proliferation of endothelial cells in preformed and newly formed blood vessels during tumor angiogenesis. *Microvasc. Res.* 14: 53–65.

Barinaga, M. 1995. Shedding light on blindness. *Science* 267: 452–453.

Barlow, G. M., B. Micales, G. E. Lyons and J. R. Korenberg. 2001. Down syndrome cell adhesion molecule is conserved in mouse and highly expressed in the adult mouse brain. *Cytogenet. Cell Genet.* 94: 155–162.

Barnes, G. L. Jr., B. D. Mariani and R. S. Tuan.1996.Valproic acid-induced somite teratogenesis in the chick embryo: Relationship with *Pax-1* gene expression. *Teratology* 54: 93–102.

Beachy, P. A. and 7 others. 1997. Multiple roles of cholesterol in Hedgehog protein biogenesis and signaling. *Cold Spring Harb. Symp. Quant. Biol.* 62: 191–204.

Bell, E. M., I. Hertz-Picciotto and J. J. Beaumont. 2001. A case-control study of pesticides and fetal death due to congenital anomalies. *Epidemiology* 12:148–156.

Bell, S. E. 1986. A new model of medical technology development: A case study of DES. *Sociol. Health Care* 4: 1–32.

Bellus, G. A., K. Gaudenz, E. H. Zackai, L. A. Clarke, J. Szabo, C. A. Francomano and M. Muenke. 1996. Identical mutations in three fibroblast growth factor receptor genes in autosomal dominant craniosynostosis syndromes. *Nature Genet.* 14: 174–176.

Benson, G. V., H. Lim, B. C. Paria, I. Satokata, S. K. Dey and R. Maas. 1996. Mechanisms of female infertility in *Hoxa-10* mutant mice: Uterine homeosis versus loss of maternal *Hoxa-10* expression. *Development* 122: 2687–2696.

Berman, D. M. and 10 others. 2002. Medulloblastoma growth inhibition by Hedgehog pathway blockade. *Science* 297: 1559–1561.

Bhattacharya, S. and A. Templeton. 2000. In treating infertility, are multiple pregnancies unavoidable? *N. Engl. J. Med.* 343: 58–60.

Bissell, M. J., D. C. Radisky, A. Rizki, V. M. Weaver and O. W. Petersen. 2002. The organizing principle: Microenvironmental influences in the normal and malignant breast. *Differentiation* 70: 537–546.

Bjorklund, L. M. and 9 others. 2002. Embryonic stem cells develop into functional dopaminergic neurons after transplantation in a Parkinson rat model. *Proc. Natl. Acad. Sci. USA* 99: 2344–2349.

Blaese, R. M. and 9 others. 1995. T lymphocyte-directed gene therapy for ADA-SCID: Initial trial results after 4 years. *Science* 270: 475–480.

Blaheta, R. A., H. Nau, M. Michaelis and J. Cinatl, Jr. 2002. Valproate and valproate analogues: Potent tools to fight against cancer. *Curr. Med. Chem.* 9: 1417–1433.

Boehm, T., J. Folkman, T. Browder and M. S. O'Reilly. 1997. Antiangiogenic therapy of experimental cancer does not induce acquired drug resistance. *Nature* 390: 404–407.

Boiani, M., S. Eckardt, H. R. Scholer and K. J. McLaughlin. 2002. Oct4 distribution and level in mouse clones: Consequences for pluripotency. *Genes Dev.* 16: 1209–1219.

Bonadio, J., E. Smiley, P. Patil and S. Goldstein. 1999. Localized, direct plasmid gene delivery in vivo: Prolonged therapy results in reproducible tissue regeneration. *Nature Med.* 5: 753–759.

Boué, A., J. Boué and A. Gropp. 1985. Cytogenetics of pregnancy wastage. *Adv. Hum. Genet.* 14: 1–57.

Boutin, E. L. and G. R. Cunha. 1997. Estrogen-induced proliferation and cornification are uncoupled in sinus vaginal epithelium associated with uterine stroma. *Differentiation* 62: 171–178.

Bradbury, E. J. and 7 others. 2002. Chondroitinase ABC promotes functional recovery after spinal cord injury. *Nature* 416: 636–640.

Brown, V. and 10 others. 2001. Microarray identification of FMRP-associated brain mRNAs and altered mRNA translational profiles in fragile-X syndrome. *Cell* 107:477–487.

Brüstle, O. and 7 others. 1999. Embryonic stem cell-derived glial precursors: A source of myelinating transplants. *Science* 285: 754–756.

Cagen, S. Z. and 10 others. 1999. Normal reproductive organ development in CF-1 mice following prenatal exposure to bisphenol A. *Toxicol. Sci.* 50: 36–44.

Cao, Y. and R. Cao. 1999. Angiogenesis inhibited by drinking tea. *Nature* 398: 381.

Carlsen, E., A. Giwercman, N. Keiding and N. Skakkebæk. 1992. Evidence for decreasing quality of semen during past 50 years. *Brit. Med J.* 305: 609–613.

Carlson, B. M. 1999. *Human Embryology and Developmental Biology*, 2nd Ed. Mosby, New York.

Carson, R. 1962. *Silent Spring.* Houghton Mifflin, New York.

Carvey, P. M., Z. D. Ling, C. E. Sortwell, M. R. Pitzer, S. O. McGuire, A. Storch and T. J. Collier. 2001. A clonal line of mesencephalic progenitor cells converted to dopamine neurons by hematopoietic cytokines: A source of cells for transplantation in Parkinson's disease. *Exp. Neurol.* 171: 98–108.

CDC (Centers for Disease Control). 2002a. ART 1997: National Summary. http://www2.cdc.gov/nccdphp/drh/art97/nation97.asp.

CDC (Centers for Disease Control). 2002b. Rubella near elimination in the United States. Press release at http://www.cdc.gov/od/oc/media/pressrel/r020429.htm

Chan, A. W. S., K. Y. Chong, C. Martinovich, C. Simerly and G. Schatten. 2001. Transgenic monkeys produced by retroviral gene transfer into mature oocytes. *Science* 291: 309–312.

Check, E. 2002. A tragic setback. *Nature* 420: 116–118.

Chen, M. S. and 7 others. 2000. Nogo-A is a myelin-associated neurite outgrowth inhibitor and an antigen for monoclonal antibody IN-1. *Nature* 403: 434–439.

Clarren, S. K. 1986. Neuropathology in the fetal alcohol syndrome. *In* J. R. West (ed.), *Alcohol and Brain Development.* Oxford University Press, New York.

Cohlan, S. Q. 1953. Excessive intake of vitamin A as a cause of congenital anomalies in the rat. *Science* 117: 535–537.

Colburn, T., D. Dumanoski and J. P. Myers. 1996. *Our Stolen Future.* Dutton, New York.

Coleman, H., J. Nabekura and J. W. Lichtman. 1997. Alterations in synaptic strength preceding axon withdrawal. *Science* 275: 356–361

Coleman, W. B., A. E. Wennerberg, G. J. Smith and J. W. Grisham. 1993. Regulation of the differentiation of diploid and some aneuploid rat liver epithelial (stemlike) cells by the hepatic microenvironment. *Am. J. Pathol.* 142: 1373–1382.

Collins, F. S. 1992. Positional cloning: Let's not call it reverse anymore. *Nature Genet.* 1: 3–6.

Cooper, T. A. and W. Mattox. 1997. The regulation of splice-site selection, and its role in human disease. *Am. J. Hum. Genet.* 61: 259–266.

Creech Kraft, J. 1992. Pharmacokinetics, placental transfer, and teratogencity of 13-*cis* retinoic acid, its isomer and metabolites. *In* G. M. Morriss-Kay (ed.), *Retinoids in Normal Development and Teratogenesis.* Oxford University Press, Oxford, pp. 267–280.

Croft, L., S. Schandorff, F. Clark, K. Burrage, P. Arctander and J. S. Mattick. 2000. ISIS, the intron information system, reveals the high frequency of alternative splicing in the human genome. *Nature Genet.* 24: 340–341.

D'Amato, R. J., M. S. Loughnan, E. Flynn and J. Folkman.1994. Thalidomide is an inhibitor of angiogenesis. *Proc. Natl. Acad. Sci. USA* 91: 4082–4085.

Darnell, J. C., K. B. Jensen, P. Jin, V. Brown, S. T. Warren and R. B. Darnell. 2001. Fragile X mental retardation protein targets G quartet mRNAs important for neuronal function. *Cell* 107: 489–499.

Davis, W. L., L. A. Crawford, O. Cooper, G. R. Farmer, D. Thomas and B. L. Freeman. 1990. Ethanol induces the generation of reactive free radicals by neural crest cells in culture. *J. Craniofac. Genet. Dev. Biol.* 10: 277–293.

De Boulle, K. and 9 others. 1993. A point mutation in the *FMR-1* gene associated with fragile X mental retardation. *Nat. Genet.* 3: 31–35.

Di Berardino, M. A. 2001. Animal cloning: The route to new genomics in agriculture and medicine. *Differentiation* 68: 67–83.

Dorris, M. 1989. *The Broken Cord.* Harper and Row, New York.

Dredge, K. and 7 others. 2002. Novel thalidomide analogues display anti-angiogenic activity independently of immunomodulatory effects. *Brit. J. Cancer* 87: 1166–1172.

Drukker, M. and 8 others. 2002. Characterization of the expression of MHC proteins in human embryonic stem cells. *Proc. Natl. Acad. Sci. USA* 99: 9864–9869.

Dudek, R. W. and J. T. Fix. 1998. *Embryology.* Lippincott/Williams and Wilkins, Baltimore.

Eck, S. L. 1999. The prospects for gene therapy. *Hosp. Pract.* 34: 67–75.

Edmonds, D. K., K. S. Lindsay, J. F. Miller, E. Williamson and P. J. Wood. 1982. Early embryonic mortality in women. *Fertil. Steril.* 38: 447–453.

Edwards, M. 1994. Pollution in the former Soviet Union: Lethal legacy. *Natl. Geog.* 186(2): 70–115.

Eldadah, Z. A., T. Brenn, H. Furthmayr and H. C. Dietz. 1995. Expression of a mutant human fibrillin allele upon a normal human or murine genetic background recapitulates a Marfan cellular phenotype. *J. Clin. Invest.* 95: 874–880.

Epstein, C. J., R. P. Erickson and A. Wynshaw-Boris (eds.). 2003. General principles of differentiation and morphogenesis. In *The Molecular Basis of Inborn Errors of Development.* Oxford University Press, New York. In press.

Eriksson, C. J. and 21 others. 2001. Functional relevance of human *adh* polymorphism. *Alcohol Clin. Exp. Res.* 25 (Suppl.): 157S–163S.

Ermak, G., T. E. Morgan and K. J. Davies. 2001. Chronic overexpression of the calcineurin inhibitory gene *DSCR1* (Adapt78) is associated with Alzheimer's disease. *J. Biol. Chem.* 276: 38787–38794.

Ermak, G., C. D. Harris and K. J. Davies. 2002. The DSCR1 (Adapt78) isoform 1 protein calcipressin 1 inhibits calcineurin and protects against acute calcium-mediated stress damage, including transient oxidative stress. *FASEB J.* 16: 814–824.

Eto, K. 2000. Minamata disease. *Neuropathology* 20 (Suppl.): S14–S19.

Eto, K. and 7 others. 2001. Methylmercury poisoning in common marmosets: A study of selective vulnerability within the cerebral cortex. *Toxicol. Pathol.* 29: 565–573.

Fairchild, W. L., E. O. Swansburg, J. T. Arsenault and S. B. Brown. 1999. Does an association between pesticide use and subsequent declines in catch of Atlantic salmon (*Salmo salar*) represent a case of endocrine disruption? Environ. Health Perspect. 107: 349–358.

Feng, Y., D. Absher, D. E. Eberhart, V. Brown, H. E. Malter and S. T. Warren. 1997a. FMRP associates with polyribosomes as an mRNP, and the I304N mutation of severe fragile X syndrome abolishes this association. *Mol. Cell* 1: 109–118.

Feng, Y., C. A. Gutekunst, D. E. Eberhart, H. Yi, S. T. Warren and S. M. Hersch. 1997b. Fragile X mental retardation protein: Nucleocytoplasmic shuttling and association with somatodendritic ribosomes. *J. Neurosci.* 17: 1539–1547.

Folkman, J. 1974. Tumor angiogenesis. *Adv. Cancer Res.* 19: 331–358.

Folkman, J. P., P. Hahnfeldt and L. Hlatky. 2000. Cancer: Looking outside the genome. *Nature Rev. Cell Biol.* 1: 76–79.

Fontana, J. A. and A. K. Rishi. 2002. Classical and novel retinoids: Their targets in cancer therapy. *Leukemia* 16: 463–472.

Franco, B. and 12 others. 1995. A cluster of sulfatase genes on Xp22.3: Mutations in chondrodysplasia punctata (CDPX) and implications for warfarin embryopathy. *Cell* 81: 15–21.

Freire-Maia, N. 1975. A heterozygote expression of a "recessive" gene. *Hum. Hered.* 25: 302–304.

Fuentes, J. J., L. Genesca, T. J. Kingsbury, K. W. Cunningham, M. Perez-Riba, X. Estivill and S. de la Luna. 2000. *DSCR1*, overexpressed in Down syndrome, is an inhibitor of calcineurin-mediated signaling pathways. *Hum. Mol. Genet.* 9: 1681–1690.

Fukuyama, F. 2002. *Our Post-Human Future: Consequences of the Biotechnology Revolution.* Farrar, Straus and Giroux, NY.

Gage, F. H. 1998. Cell therapy. *Nature* 392: 18–24.

Gearhart, J. 1998. New potential for human embryonic stem cells. *Science* 282: 1061–1062.

Giroud, A. and M. Martinet. 1959. Teratogenese pur hypervitaminose A chez le rat, la souris, le cobaye, et le lapin. *Arch. Fr. Pediatr.* 16: 971–980.

Glausiusz, J. 2002. The woes of the clones. *Discover* 23(8): 11.

Gleicher, N., D. M. Oleske, I. Tur-Kaspa, A. Vidali and V. Karande. 2000. Reducing the risk of high-order multiple pregnancy after ovarian stimulation with gonadotropins. *N. Engl. J. Med.* 343: 2–7.

Goulding, E. H. and R. M. Pratt. 1986. Isotretinoin teratogenicity in mouse whole embryo culture. *J. Craniofac. Genet. Dev. Biol.* 6: 99–112.

GrandPré, T., F. Nakamura, T. Vartanian and S. M. Strittmatter. 2000. Identification of the Nogo inhibitor of axon regeneration as a reticulon protein. *Nature* 403: 439–442.

GrandPré, T., S. Li and S. M. Strittmatter. 2002. Nogo-66 receptor antagonist peptide promotes axonal regeneration. *Nature* 417: 547–551.

Gregg, N. M. 1941. Congenital cataract following German measles in the mother. *Trans. Opthalmol. Soc. Aust.* 3: 35.

Grignani, F. and 14 others. 1998. Fusion proteins of the retinoic acid receptor-alpha recruit histone deacetylase in promyelocytic leukaemia. *Nature* 391: 815–818.

Grüneberg, H. 1938. An analysis of the "pleiotropic" effects of a new lethal mutation in the rat (*Mus norwegicus*). *Proc. R. Soc. Lond. B* 125: 123–144.

Guillette, E. A., M. M. Meza, M. G. Aquilar, A. D. Soto and I. E. Garcia. 1998. An anthropological approach to the evaluation of preschool children exposed to pesticides in Mexico. *Environ. Health Perspect.* 106: 347–353.

Gupta, C. 2000. Reproductive malformation of the male offspring following maternal exposure to estrogenic chemicals. *Proc. Soc. Exp. Biol. Med.* 224: 61–68.

Hadorn, E. 1955. *Letalfaktoren in ihrer Bedeutung für Erbpathologie und Genphysiologie der Entwicklung.* Thieme Verlag, Stuttgart.

Hanahan, D. and J. Folkman. 1996. Patterns and emerging mechanisms of the angiogenic switch during tumorigenesis. *Cell* 86: 353–364.

Hansen, L. A., C. C. Sigman, F. Andreola, S. A. Ross, G. J. Kelloff and L. M. De Luca. 2000. Retinoids in chemoprevention and differentiation therapy. *Carcinogenesis* 21: 1271–1279.

Haraway, D. 1997. *Modest-Witness@Second-Millenium.* Routledge, New York.

Hemesath, T. J. and 9 others. 1994. Microphthalmia, a critical factor in melanocyte development, defines a discrete transcription factor family. *Genes Dev.* 8: 2770–2780.

Herman-Giddens, M. E., E. J. Slora, R. C. Wasserman, C. J. Bourdony, M. V. Bhapkar, G. G. Koch and C. M. Hasemeir. 1997. Secondary sexual characteristics and menses in young girls seen in office practice: A study from the pediatric research in office settings network. *Pediatrics* 99: 505–512.

Hill, M., C. Stabile, L. K. Steffen and A. Hill. 2002. Toxic effects of endocrine disrupters on freshwater sponges: common developmental abnormalities. Environ. Pollut. 117: 295–300.

Hodgkinson, C. A., K. J. Moore, A. Nakayama, E. Steingrimsson, N. G. Copeland, N. A. Jenkins and H. Arnheiter. 1994. Mutations at the mouse *microphthalmia* locus are associated with defects in a gene encoding a novel basic helix-loop-helix-zipper protein. *Cell* 74: 395–404.

Hoffman, L. M. and W. M. Kulyk. 1999. Alcohol promotes in vitro chondrogenesis in embryonic facial mesenchyme. *Int. J. Dev. Biol.* 43: 167–74.

Holmes, L. B. 1979. Radiation. *In* V. C. Vaughan, R. J. McKay and R. D. Behrman (eds.), *Nelson Textbook of Pediatrics*, 11th Ed. Saunders, Philadelphia.

Howdeshell, K., A. K. Hotchkiss, K. A. Thayer, J. G. Vandenbergh and F. S. vom Saal. 1999. Plastic bisphenol A speeds growth and puberty. *Nature* 401: 762–764.

Hughes, A. E., V. E. Newton, X. Z. Liu and A. P. Read. 1994. A gene for Waardenburg Syndrome type II maps close to the human homologue of the *microphthalmia* gene at chromosome 3 p12–p14.1. *Nature Genet.* 7: 509–512.

Humphreys, D. and 8 others. 2002. Abnormal gene expression in cloned mice derived from embryonic stem cell and cumulus cell nuclei. *Proc. Natl. Acad. Sci. USA* 99: 12889–12894.

Hutchinson, S., B. P. Wordsworth and P. A. Handford. 2001. Marfan syndrome caused by a mutation in *FBN1* that gives rise to cryptic splicing and a 33-nucleotide insertion in the coding sequence. *Hum. Genet.* 109: 416–420.

Ikezuki, Y., O. Tsutsumi, Y. Tahai, Y. Kamzi and Y. Taketa. 2002. Determination of bisphenol A concentrations in human biological fluids reveals significant prenatal exposure. *Human Reprod.* 17: 2839–2841.

Illmensee, K. and B. Mintz. 1976. Totipotency and normal differentiation of single teratocarcinoma cells cloned by injection into blastocysts. *Proc. Natl. Acad. Sci. USA* 73: 549–553.

Jain, M. and 8 others. 2002. Sustained loss of neoplastic phenotype by brief inactivation of MYC. *Science* 297: 102–104.

Jiang, Y. and 15 others. 2002a. Pluripotency of mesenchymal stem cells derived from adult marrow. *Nature* 418: 41–49.

Jiang, Y., B. Vaessen, T. Lenvik, M. Blackstad, M. Reyes and C. M. Verfaillie. 2002b. Multipotent progenitor cells can be isolated from postnatal murine bone marrow, muscle, and brain. *Exp. Hematol.* 30: 896–904.

Johansson, C. B., S. Mamma, D. L. Clarke, M. Rislig, U. Lendahl and J. Frisén. 1999. Identification of a neural stem cell in the adult mammalian central nervous system. *Cell* 96: 25–34.

Johe, K. K., T. G. Hazel, T. Muller, M. M. Dugich-Djordjevic and R. D. G. McKay. 1996. Single factors direct the differentiation of stem cells from the fetal and adult central nervous system. *Genes Dev.* 10: 917–927.

Johnson, E. M. 1980. Screening for teratogenic potential: Are we asking the proper questions? *Teratology* 21: 259.

Johnston, M. C., K. K. Sulik, W. S. Webster and B. L. Jarvis. 1985. Isotretinoin embryopathy in a mouse model: Cranial neural crest involvement. *Teratology* 31: 26A.

Jones, K. L. and D. W. Smith. 1973. Recognition of the fetal alcohol syndrome. *Lancet* 2: 999–1001.

Kanavakis, E. and J. Traeger-Synodinos. 2002. Preimplantation genetic diagnosis in clinical practice. *J. Med. Genet.* 39: 6–11.

Kaufman, D. S., E. T. Hanson, R. L. Lewis, R. Auerbach and J. A. Thomson. 2001. Hematopoietic colony-forming cells derived from human embryonic stem cells. *Proc. Natl. Acad. Sci. USA* 98: 10716–10721.

Keirstead, H. S., S. V. Morgan, M. J. Wilby and J. W. Fawcett. 1999. Enhanced axonal regeneration following combined demyelination plus Schwann cell transplantation therapy in the injured adult spinal cord. *Exp. Neurol.* 159: 225–236.

Kerbel, R. and J. Folkman. 2002. Clinical translation of angiogenesis inhibitors. *Nature Rev. Cancer* 2: 727–739.

Kim, J. H. and 10 others. 2002. Dopamine neurons derived from embryonic stem cells function in an animal model of Parkinson's disease. *Nature* 418: 50–56.

Kitajewsky, J. and D. Sassoon. 2000. The emergence of molecular gynecology: Homeobox and

Wnt genes in the female reproductive tract. *BioEssays* 22: 902–910.

Kim, H. S. and 8 others. 2002. Comparative estrogenic effects of p-nonylphenol by 3-day uterotrophic assay and female pubertal onset assay. Reprod. Toxicol. 16: 259–268.

Kochhar, D. M., J. D. Penner and C. Tellone. I. 1984. Comparative teratogenic activities of two retinoids: Effects on palate and limb development. *Teratogen. Carcinogen. Mutagen.* 4: 377–387.

Kondo, K. 2000. Congenital Minamata disease: Warnings from Japan's experience. *J. Child Neurol.* 15: 458–464.

Korbling, M. and 7 others. 2002. Hepatocytes and epithelial cells of donor origin in recipients of peripheral-blood stem cells. *N. Engl. J. Med.* 346: 738–746.

Kotch, L. E., S.-Y. Chen and K. K. Sulik. 1995. Ethanol-induced teratogenesis: Free radical damage as a possible mechanism. *Teratology* 52: 128–136.

Krawczak, M., J. Reiss and D. N. Cooper. 1992. The mutational spectrum of single base-pair substitutions in messenger RNA splice junctions of human genes: Causes and consequences. *Hum. Genet.* 90: 41–54.

Kulikauskas, V., A, B, Blaustein and R. J. Ablin. 1985. Cigarette smoking and its possible effects on sperm. *Fertil. Steril.* 44: 526–528.

LaBarge, M. A. and H. M. Blau. 2002. Biological progression from adult bone marrow to mononucleate muscle stem cell to multinucleate muscle fiber in response to injury. *Cell* 111: 589–601.

Lamb, J. 2002. Why you should ignore the baby bottle scare. http://www.quackwatch.org/04 ConsumerEducation/babybottle.html.

Lammer, E. J. and 11 others. 1985. Retinoic acid embryopathy. *N. Engl. J. Med.* 313: 837–841.

Larsen, W. J. 2001. *Human Embryology*, 3rd Ed. Churchill-Livingstone, New York.

Lemoine, E. M., J. P. Harousseau, J. P. Borteyru and J. C. Menuet. 1968. Les enfants de parents alcooliques: Anomalies observées. *Oest. Med.* 21: 476–482.

Lengauer, C., K. W. Kinzler and B. Vogelstein. 1998. Genetic instabilities in human cancers. *Nature* 396: 643–649.

Lenz, W. 1962. Thalidomide and congenital abnormalities. *Lancet* 1: 45. (First reported at a 1961 symposium.)

Li, G. and 7 others. 2002. Bone consolidation is enhanced by rhBMP-2 in a rabbit model of distraction osteogenesis. *J. Orthop. Res.* 20: 779–788.

Linney, E. 1992. Retinoic acid receptors: Transcription factors modulating gene expression, development, and differentiation. *Curr. Top. Dev. Biol.* 27: 309–350.

Lipshultz, L. and D. Adamson. 1999. Multiple-birth risk associated with in vitro fertilization: Revised guidelines. *J. Am. Med. Assoc.* 282: 1813–1814.

Livesey, F. J., J. A. O'Brien, M. Li, A. G. Smith, L. J. Murphy and S. P. Hunt. 1997. A Schwann cell mitogen accompanying regeneration of motor neurons. *Nature* 390: 614–618.

Longnecker, M. P., M. A. Klebanoff, H. Zhou and J. W. Brock. 2001. Association between maternal serum concentration of DDT and metabolite DDE and pre-term and small-for-gestational-age babies at birth. *Lancet* 358: 110–114.

Ma, L., G. V. Benson, H. Lim, S. K. Dey and R. Maas. 1998. *Abdominal B (AbdB)* Hoxa genes: Regulation in adult uterus by estrogen and progesterone and repression in Müllerian duct by the synthetic estrogen diethylstilbestrol (DES). *Dev. Biol.* 197: 141–154.

Mak, V., K. Jarvi, M. Buckspan, M. Freeman, S. Hechter and A. Zini. 2000. Smoking is associated with the retention of cytoplasm by human spermatozoa. *Urology* 56: 463–466.

Markey, C. M., M. A. Coombs, C. Sonnenschein and A. M. Soto. 2003. Mammalian development in a changing environment: Exposure to endocrine disruptors reveals the developmental plasticity of steroid hormone target organs. *Evo. Dev.* 5: 67–75.

Marukawa, E., L. Asahina, M. Oda, L. Seto, M. Alam and S. Enomoto. 2002. Functional reconstruction of the non-human primate mandible using recombinant human bone morphogenetic protein-2. *Int. J. Oral Maxillofac. Surg.* 31: 287–295.

McCarver-May, D. G. 1996. Genetic differences in alcohol dehydrogenase and fetal alcohol effects. Abstracts of the Ninth International Congress of Human Genetics. *Brazil J. Genet.* 19: 73.

McDonald, J. W. and 7 others. 1999. Transplanted embryonic stem cells survive, differentiate, and promote recovery in injured rat spinal cord. *Nature Med.* 5: 1410–1412.

McGill, G. G. and 14 others. 2002. *Bcl2* regulation by the melanocyte master regulator Mitf modulates lineage survival and melanoma cell viability. *Cell* 109: 707–718.

McKeown, T. 1976. Human malformations: An introduction. *Br. Med. Bull.* 32: 1–3.

McKinney, J. D., J. Fawkes, S. Jordan, K. Chae, S. Oatley, R. E. Coleman and W. Briner. 1985. 2,3,7,8-Tetrachlorodibenzo-p-dioxin (TCDD) as a potent and persistent thyroxine agonist: A mechanistic model for toxicity based on molecular reactivity. *Environ. Health Perspect.* 61: 41–53.

McVeigh, E. and D. Barlow. 2000. Infertility: Causes, investigation, and management. *Encyclopedia of Life Sciences*. http://www.els.net.

Miller, C., K. Degenhardt and D. A. Sassoon. 1998. Fetal exposure to DES results in de-regulation of *Wnt7a* during uterine morphogenesis. *Nature Genet.* 20: 228–230.

Miller, W. H., Jr. and 7 others. 1992. Reverse transcription polymerase chain reaction for the rearranged retinoic acid receptor alpha clarifies diagnosis and detects minimal residual disease in acute promyelocytic leukemia. *Proc. Natl. Acad. Sci. USA* 89: 2694–2698.

Miny, P., S. Tercanli and W. Holzgreve. 2002. Developments in laboratory techniques for prenatal diagnosis. *Curr. Opin. Obstet. Gynecol.* 14: 161–168.

Mittendorf, R. 1995. Teratogen update: Carcinogenesis and teratogenesis associated with exposure to diethylstilbestrol (DES) in utero. *Teratology* 51: 435–445.

Moore, K. L. and T. N. N. Persaud. 1993. *Before We Are Born: Essentials of Embryology and Birth Defects*. W. B. Saunders, Philadelphia.

Moore, K. L., T. V. N. Persaud and W. Schmitt. 1998. *The Developing Human: Clinically Oriented Embryology*. W. B. Saunders, Philadelphia.

Moroni, M. C., M. A. Vigano and F. Mavilio. 1994. Regulation of human *Hoxd-4* gene by retinoids. *Mech. Dev.* 44: 139–154.

Morse, D. C., E. K. Wehler, W. Wesseling, J. H. Koeman and A. Brouwer. 1996. Alterations in rat brain thyroid hormone status following pre- and postnatal exposure to polychlorinated biphenyls (Aroclor 1254). *Toxicol. Appl. Pharmacol.* 136: 269–279.

Mukhopadyay, G., P. Doherty, F. Walsh, P. R. Crocker and M. Filbin. 1994. A novel role for myelin-associated glycoproteins as an inhibitor of axonal regeneration. *Neuron* 13: 757–767.

Muthukkaruppan, V. R. and R. Auerbach. 1979. Angiogenesis in the mouse cornea. *Science* 205: 1416–1418.

Nakayama, A., M. T. Nguyen, C. C. Chen, K. Opdecamp, C. A. Hodgkinson and H. Arnheiter. 1998. Mutations in *microphthalmia*, the mouse homolog of the human deafness gene *MITF*, affect neuroepithelial and neural crest-derived melanocytes differently. *Mech Dev.* 70: 155–166.

Nanni, L. and 16 others.1999. The mutational spectrum of the *Sonic hedgehog* gene in holoprosencephaly: *SHH* mutations cause a significant proportion of autosomal dominant holoprosencephaly. *Hum. Mol. Genet.* 8: 2479–2488.

Newbold, R. R., E. P. Banks, B. Bullock and W. N. Jefferson. 2001. Uterine adenocarcinoma in mice treated neonatally with genistein. *Cancer Res.* 61: 4325–4328.

NIH (National Institutes of Health). 2000. *Stem Cells: A Primer*. http://www.nih.gov/news/stem-cells/primer.htm.

Nijhout, H. F. and S. M. Paulsen. 1997. Developmental models and polygenic characters. *Am. Nat.* 149: 394–405.

Nishigaki, R. and 8 others. 2002. An extra human chromosome 21 reduces *mlc-2a* expression in chimeric mice and Down syndrome. *Biochem. Biophys. Res. Commun.* 295: 112–118.

Nulman, I. and 8 others. 1997. Neurodevelopment of children exposed in utero to antidepressant drugs. *N. Engl. J. Med.* 336: 258–262.

Odent, S. and 10 others. 1999. Expression of the *Sonic hedgehog (Shh)* gene during early human development and phenotypic expression of new mutations causing holoprosencephaly. *Hum. Mol. Genet.* 8: 1683–1689.

Onodera, M. and 13 others. 1998. Successful peripheral T-lymphocyte-directed gene transfer for a patient with severe combined immune deficiency caused by adenosine deaminase deficiency. *Blood* 91: 30–36.

Opitz, J. 1987. Prenatal and perinatal death: The future of developmental pathology. The Farber Lecture. *Pediatr. Pathol.* 7:363–394.

Opitz, J. et al. 1987 Aneuploidy in mouse and man. In F. Vogel and K. Sperling (eds.), *Human Genetics: Proceedings of the 7th International Congress, Berlin 1986.* Springer-Verlag, Berlin, pp 382–385.

Opitz, J. M., E. Gilbert-Barness, J. Ackerman and A. Lowichik. 2002. Cholesterol and development: The RSH ("Smith-Lemli-Opitz") syndrome and related conditions. *Pediatr. Pathol. Mol. Med.* 21: 153– 181.

O'Rahilly, R. and F. Müller. 1992. *Human Embryology and Teratology.* Wiley-Liss, New York.

OSF (Our Stolen Future website). 2002. http://www.ourstolenfuture.org/Myths/vomsaal.htm.

Palanza, P., K. L. Howdeshell, S. Parmigiani and F. S. vom Saal. 2002. Exposure to a low dose of bisphenol-A during fetal life or in adulthood alters maternal behavior in mice. *Environ. Health Perspect.* (Suppl.) 110(3): 415–422.

Palmund, I. 1996. Exposure to a xenoestrogen before birth: The diethylstilbestrol experience. *J. Psychosom. Obstet. Gynecol.* 17: 71–84.

Pappert, A. 2000. What price pregnancy? *Ms.*, June/July.

Partington, M. W. 1959. Mother and daughter show the white forelock of Waardenburg's syndrome. *Arch. Dis. Childhood* 34: 154–157.

Pellizzoni, L., B. Charroux and G. Dreyfuss. 1999. *SMN* mutants of spinal muscular atrophy patients are defective in binding to snRNP proteins. *Proc. Natl. Acad. Sci. USA* 96: 11167–11172.

Peng, H., V. Wright, A. Usas, B. Gearhart, H. C. Shen, J. Cummins and J. Huard. 2002. Synergistic enhancement of bone formation and healing by stem cell-expressed VEGF and bone morphogenetic protein-4. *J. Clin. Invest.* 110: 751–759.

Pierce, G. B., R. Shikes and L. M. Fink. 1978. *Cancer: A Problem of Developmental Biology.* Prentice-Hall, Englewood Cliffs, NJ.

Pöpperl, H. and M. S. Featherstone. 1993. Identification of retinoic acid response element upstream from the mouse *Hox-4.2* gene. *Mol. Cell. Biol.* 13: 257–265.

Purdy, L. 2001. Bioethics of new assisted reproduction. Encyclopedia of Life Sciences. http://www.els.net.html.

Ramanathan, R., M. F. Wilkemeyer, B. Mittel, G. Perides and M. E. Charness. 1996. Alcohol inhibits cell-cell adhesion mediated by human L1. *J. Cell Biol.* 133: 381–390.

Ramos, J. G., J. Varayoud, C. Sonnenschein, A. M. Soto, M. Munoz De Toro and E. H. Luque. 2001. Prenatal exposure to low doses of bisphe-

nol A alters the periductal stroma and glandular cell function in the rat ventral prostate. *Biol. Reprod.* 65: 1271–1277.

Rideout, W. M. III, K. Hochedlinger, M. Kyba, G. Q. Daley and R. Jaenisch. 2002. Correction of a genetic defect by nuclear transplantation and combined cell and gene therapy. *Cell* 109: 17–27.

Rifkin, J. 1998. *The Biotech Century: Harnessing the Gene and Remaking the World.* Tarcher/Putnam, New York.

Robboy, S. J., R. H. Young and A. L. Herbst,. 1982. Female genital tract changes related to prenatal diethylstilbestrol exposure. In A. Blaustein (ed.), *Pathology of the Female Genital Tract*, 2nd Ed. Springer-Verlag, New York, pp. 99–118.

Roberts, J. P. 2002. Psst! Gene therapy research lives. *The Scientist*, June 24, 12–14.

Roth, D. A., N. E. Tawa, Jr., J. M. O'Brien, D. A. Treco, and R. F Selden. 2001. Nonviral transfer of the gene encoding coagulation factor VIII in patients with severe hemophilia A. *N. Engl. J. Med.* 344: 1735–1742.

Rothman, K. J., L. L. Moore, M. R. Singer, U.-S. D. T. Nguyen, S. Mannino and A. Milunsky. 1995. Teratogenicity of high vitamin A intake. *N. Engl. J. Med.* 333: 1369–1373.

Ruberte, E., P. Dollé, A. Krust, A. Zalent, G. Morriss-Kay and P. Chambon. 1990. Specific spatial and temporal distribution of retinoic acid receptor transcripts during mouse embryogenesis. *Development* 108: 213–222.

Ruberte, E. and 7 others. 1991. Retinoic acid receptors in the embryo. *Semin. Dev. Biol.* 2: 153–159.

Sachs, L. 1978. Control of normal cell differentiation and the phenotypic reversion of malignancy in myeloid leukemias. *Nature* 274: 535–539.

Sadler, T. W. and J. Langman, J. 2000. *Langman's Medical Embryology*, 8th Ed. Lippincott/ Williams and Wilkins, Baltimore.

Sai, T. and 11 others. 1997. An exonic point mutation of the androgen receptor gene in a family with complete androgen insensitivity. *Am. J. Hum. Genet.* 46: 1095–1110.

Sai, T. et al. 1990. An exonic point mutation of the androgen receptor gene in a family with complete androgen insensitivity. *Am. J. Hum. Genet.* 46:1095–1100.

Sandberg, A. A. and Z. Chen. 2002. Cytogenetics and molecular genetics of human cancer. *Am. J. Med. Genet.* 115: 111–112.

Scambler, P. J. 1997. Positional cloning and analysis of loci implicated in human birth defects. In P. Thorogood (ed.), *Embryos, Genes, and Birth Defects*. John Wiley, New York, pp. 33–47.

Schieve, L. A., H. B. Peterson, S. F. Meikle, G. Jeng, I. Danel, N. M. Burnett and L. S. Wilcox. 1999. Live-birth rates and multiple-birth risk using in vitro fertilization. *J. Am. Med. Assoc.* 282: 1832–1838.

Schnell, L. and M. E. Schwab. 1990. Axonal regeneration in the rat spinal cord produced by an antibody against myelin-associated neurite growth inhibitor. *Nature* 343: 269–272.

Schönfelder, G., W. Wittfoht, H. Hopp, C. E. Talsness, M. Paul and I. Chahoud. 2002. Parent bisphenol A accumulation in the human maternal-fetal-placental unit. *Environ. Health Perspect.* 110: A703–A707.

Schröder, A. R., P. Shinn, H. Chen, C. Berry, J. R. Ecker and F. Bushman. 2002. HIV-1 integration in the human genome favors active genes and local hotspots. *Cell* 110: 521–529.

Schwab, M. E. and P. Caroni. 1988. Oligodendrocyte and CNS myelin are non-permissive substrates for neuron growth and fibroblast spreading in vitro. *J. Neurosci.* 8: 2381–2393.

Shaw, G. M., C. R. Wasserman, E. J. Lammer, C. D. O'Malley, J. C. Murray, A. M. Basart and M. M. Tolarova. 1996. Orofacial clefts, parental cigarette smoking, and transforming growth factor-α gene variants. *Am. J. Hum. Genet.* 58: 551–561.

Shi, Q., E. Ko, L. Barclay, T. Hoang, A. Rademaker and R. Martin. 2001. Cigarette smoking and aneuploidy in human sperm. *Mol. Reprod. Dev.* 59: 417–421.

Shweiki, D., A. Itin, D. Soffer and E. Keshet. 1992. Vascular endothelial growth factor induced by hypoxia may mediate hypoxia initiated angiogenesis. *Nature* 359: 843–845.

Silver, L. M. 1997. *Remaking Eden: Cloning and Beyond in the Brave New World.* Avon, New York.

Sonnenschein, C. and A. M. Soto. 1999. *The Society of Cells: Cancer and the Control of Cell Proliferation.* Bios Publishers, Oxford.

Sonnenschein, C. and A. M. Soto. 2000. Somatic mutation theory of carcinogenesis: Why it should be dropped and replaced. *Mol. Carconogen.* 29: 205–211.

Soto, A., H. Justicia, J. Wray and C. Sonnenschein. 1991. *p*-nonylphenol: An estrogenic xenobiotic released from "modified" polystyrene. *Environ. Health Perspect.* 92: 167–173.

Stern, H., S. Wiley, R. Matken, D. Karabinus and K. Blauer. 2002. Microsort™ babies: 1994–2002 preliminary postnatal follow-up results. *Fertil. Steril.* 78(S1): S133.

Sternlicht, M. D. and 8 others. 1999. The stromal proteinase MMP3/stromelysin-1 promotes mammary carcinogenesis. *Cell* 98: 137–46.

Stewart, T. A. and B. Mintz. 1981. Successive generations of mice produced from an established culture line of euploid teratocarcinoma cells. *Proc. Natl. Acad. Sci. USA* 78: 6314–6318.

Stock, G. and J. Campbell (ed). Engineering the Human Genome: Summary Report. Science, Technology, and Society Program, UCLA. http://www.ess.ucla.edu:80/huge/report.html July 11 1998.

Streissguth, A. P. and R. A. LaDue. 1987. Fetal alcohol: Teratogenic causes of developmental disabilities. *In* S. R. Schroeder (ed.), *Toxic Substances and Mental Retardation.* American

Association of Mental Deficiency, Washington, DC, pp. 1–32.

Studer, M., H. Pöpperl, H. Marshall, A. Kuroiwa and R. Krumlauf. 1994. Role of a conserved retinoic acid response element in rhombomere restriction of *Hoxb-1*. *Science* 265: 1728–1732.

Sulik, K. K., C. S. Cook and W. S. Webster. 1988. Teratogens and craniofacial malformations: Relationships to cell death. *Development* 103 (Suppl.): 213–231.

Swan, S. H., E. P. Elkin and L. Fenster. 2000. The question of declining sperm density revisited: An analysis of 101 studies published 1934–1996. *Environ. Health Perspect.* 108: 961–966.

Tachibana, M and 9 others. 1994. Cloning of *MITF*, the human homolog of the mouse *microphthalmia* gene, and the assignment to human chromosome 3, region p14.1–p12.4. *Hum. Mol. Genet.* 3: 553–557.

Tassabehji, M., V. E. Newton and A. P. Read. 1994. Waardenburg syndrome type 2 caused by mutations in the human *microphthalmia* (*MITF*) gene. *Nature Genet.* 8: 251–255.

Terada N. and 9 others. 2002. Bone marrow cells adopt the phenotype of other cells by spontaneous cell fusion. *Nature* 416: 542–545.

Thomson, J. A., J. Itskovitz-Eldor, S. S. Shapiro, M. A. Waknitz, J. J. Swiegiel, V. S. Marshall and J. M. Jones. 1998. Embryonic stem cell lines derived from human blastocysts. *Science* 282: 1145–1147.

Tiberio, G. 1994. MZ female twins discordant for X-linked diseases: A review. *Acta Genet. Med. Gemellol.* 43: 207–214.

Ton, C. T. and 14 others. 1991. Positional cloning and characterization of a paired-box and homeobox-containing gene from the aniridia region. *Cell* 67: 1059–1074.

Trounson, A. O. and D. K. Gardner (eds.). 2000. *Handbook of In Vitro Fertilization*, 2nd Ed. CRC Press, Boca Raton, FL.

Valleix, S., C. Vinciguerra, J. M. Lavergne, M. Leuer, M. Delpech and C. Negrier. 2002. Skewed X-chromosome inactivation in monochorionic diamniotic twin sisters results in severe and mild hemophilia A. *Blood* 100: 3034–3036.

Volpe, E. P. 1987. Developmental biology and human concerns. *Am. Zool.* 27: 697–714.

vom Saal, F. S. 2000. Very low doses of bisphenol A and other estrogenic chemicals alters development in mice. Endocrine Disruptors and Pharmaceutical Active Compounds in Drinking Water Workshop. http://www.cheec.uiowa.edu /conferences/edc_2000/vom_saal.html.

vom Saal, F. S., P. S. Cooke, D. L. Buchanan, P. Palanza, K. A. Thayer, S. C. Nagel, S. Parmigiani and W. F. Welshons. 1998. A physiologically based approach to the study of bisphenol A and other estrogenic chemicals on the size of reproductive organs, daily sperm production, and behavior. *Toxicol. Ind. Health* 14: 239–260.

Vortkamp, A., S. Pathi, G. M. Peretti, E. M. Caruso, D. J. Zaleske and C. J. Tabin. 1998. Recapitulation of signals regulating embryonic bone formation during postnatal growth and in fracture repair. *Mech. Dev.* 71: 65–76.

Wagers, A. J., R. I. Sherwood, J. L. Christensen and I. L. Weissman. 2002. Little evidence for developmental plasticity of adult hematopoietic stem cells. *Science* 297: 2256–2259.

Wang, K., C. V. Koprivica, J. A. Kim, R. Sivasankaran, Y. Guo, R. L. Neve and Z. He. 2002. Oligodendrocyte-myelin glycoprotein is a Nogo receptor ligand that inhibits neurite outgrowth. *Nature* 417: 941–944.

Wang, X., M. Kiledjian, I. M. Weiss and S. A. Liebhaber. 1995. Detection and characterization of a 3′ untranslated region ribonucleoprotein complex associated with human á-globin mRNA stability. *Mol. Cell. Biol.* 15: 1769–1777.

Wang, X. and 9 others. Maternal cigarette smoking, metabolic gene polymorphism and infant birth weight. *J. Am. Med. Assoc.* 287: 195–202.

Weaver, V. M., O. W. Petersen, F. Wang, C. A. Larabell, P. Briand, C. Damsky and M. J. Bissell.1997. Reversion of the malignant phenotype of human breast cells in three-dimensional culture and in vivo by integrin-blocking antibodies. *J. Cell Biol.* 137: 231– 245.

Weidner, N., A. Blesch, R. J. Grill and M. H. Tuszynski. 1999. Nerve growth factor-hypersecreting Schwann cell grafts augment and guide spinal cord axonal growth and remyelinate central nervous system axons in a phenotypically appropriate manner that correlates with expression of L1. *J. Comp. Neurol.* 413: 495–506.

Wetherill, Y. B., C. E. Petse, K. R. Monk, A. Puga and K. E. Knudson. 2002. Xenoestrogen bisphenol-A induces inappropriate androgen receptor activation and mitogenesis in prostatic adenocarcinoma cells. *Mol. Cancer Ther.* 1: 515–524.

Wennerholm, U. B., C. Bergh, L. Hamberger, K. Lundin, L. Nilsson, M. Wikland and B. Kallen. 2000. Incidence of congenital malformations in children born after ICSI. *Hum. Reprod.* 15: 944–948.

Wickelgren, I. 1999. Stem cells. Rat spinal cord function partially restored. *Science* 286: 1826–1827.

Wilmut, I. and 7 others. 2002. Somatic cell nuclear transfer. *Nature* 419: 583–586.

Wolf, U. 1995. The genetic contribution to the phenotype. *Hum. Genet.* 95: 127–148.

Wolf, U. 1997. Identical mutations and phenotypic variation. *Hum. Genet.* 100: 305–321.

Wu, P., Y. I. Tarasenko, Y. Gu, L. Y. Huang, R. E. Coggeshall and Y. Yu. 2002. Region-specific generation of cholinergic neurons from fetal human neural stem cells grafted in adult rat. *Nature Neurosci.* 5: 1271–1278.

Xia, H., Q. Mao, H. L. Paulson and B. L. Davidson. 2002. siRNA-mediated gene silencing in vitro and in vivo. *Nature Biotech.* 20: 1006–1010.

Xue, F. and 9 others. 2002. Aberrant patterns of X chromosome inactivation in bovine clones. *Nature Genet.* 31: 216–220.

Yamakawa, K., Y. K. Huot, M. A. Haendelt, R. Hubert, X. N. Chen, G. E. Lyons and J. R. Korenberg. 1998. *DSCAM*: A novel member of the immunoglobulin superfamily maps in a Down syndrome region and is involved in the development of the nervous system. *Hum. Mol. Genet.* 7: 227–237.

Ye, X. and 8 others. 1999. Regulated delivery of therapeutic proteins after in vivo somatic cell gene transfer. *Science* 283: 88–91.

Yellayi, S. and 9 others. 2002. A phytoestrogen, genestein, induces thymic and immune changes: a human health concern? *Proc. Natl. Acad. Sci. USA* 99: 7616–7621.

Ying, Q. L., J. Nichols, E. P. Evans and A. G. Smith. 2002. Changing potency by spontaneous fusion. *Nature* 416: 545–548.

Yokoyama, T., N. G. Copeland, N. A. Jenkins, C. A. Montgomery, F. F. Elder and P. A. Overbeek.1993. Reversal of left-right asymmetry: A situs inversus mutation. *Science* 260: 679–82.

Yu, V. C. and 9 others. 1991. RXRb: A coregulator that enhances binding of retinoic acid, thyroid hormone, and vitamin D receptors to their cognate response elements. *Cell* 67: 1251–1266.

Zwacka, T. P. and J. A. Thompson. 2003. Homologous recombination in human embryonic stem cells. *Nature Biotech.* 21: doi10. 1038/nbt788.

chapter 22 Environmental regulation of animal development

We may now turn to consider adaptations towards the external environment; and firstly the direct adaptations ... in which an animal, during its development, becomes modified by external factors in such a way as to increase its efficiency in dealing with them.

C. H. WADDINGTON (1957)

Why should our bodies end at our skin, or include at best other beings encapsulated by skin?

DONNA HARAWAY (1991)

I T HAD LONG BEEN THOUGHT that the environment played only minor roles in development. Nearly all developmental phenomena were believed to be regulated by genes, and those organisms whose development was significantly controlled by the environment were considered interesting oddities. However, recent studies have shown that the environmental context plays significant roles in the development of almost all species, and that animal and plant genomes have evolved to respond to environmental conditions (Table 22.1). Moreover, symbiotic associations, wherein the genes of one organism are regulated by the products of another organism, appear to be the rule rather than the exception.

One of the reasons developmental biologists have largely ignored the environment is that one criterion for selecting which animals to study has been their ability to develop regularly in the laboratory. Given adequate nutrition and temperature, these "model systems"—*C. elegans*, fruit flies, sea urchins, *Xenopus*, chicks, and laboratory mice—develop independently of their environment (Bolker 1995). These animals can give one the erroneous impression that everything needed to form the embryo is within the fertilized egg. Today, with new concerns about the loss of organismal diversity and the effects of environmental pollutants, there is renewed interest in the regulation of development by the environment (see van der Weele 1999; Gilbert 2001).

There are numerous examples (and *Homo sapiens* provides some of the best) wherein the environment plays a critical role in determining the organism's phenotype. In Chapters 3 and 17, we encountered environmentally regulated sex determination and morphologies. In these and many other cases, the environment can elicit different phenotypes from the same nuclear genotype. The genetic ability to respond to such environmental factors has to be inherited, of course, but in these cases it is the environment that directs the formation of the particular phenotype.

In this chapter, we will discuss how organisms use environmental cues in the course of their normal development, as well as how exogenous compounds found in the environment can divert development from its usual path and cause congenital abnormalities.

The Environment as Part of Normal Development

If the environment contains predictable components (such as gravity) or predictable changes (such as seasons), these elements can become part of the development of the organism. Any developing animal can expect to encounter a 1G gravitational field, bacteria, and fungi as part of its environment. Many species have used these as agents for normal development.

TABLE 22.1 Some topics covered by ecological developmental biology

CONTEXT-DEPENDENT NORMAL DEVELOPMENT

A. Morphological polyphenisms

    1. Nutrition-dependent (*Nemoria*, hymenoptera castes, sea urchin larvae)

    2. Temperature-dependent (*Arachnia, Bicyclus*)

    3. Density-dependent (locusts)

    4. Stress-dependent (*Scaphiopus*)

B. Sex determination polyphenisms

    1. Location-dependent (*Bonellia, Crepidula*)

    2. Temperature-dependent (*Menidia*, turtles)

    3. Social-dependent (wrasses, gobys)

C. Predator-induced polyphenisms

    1. Adaptive predator-avoidance morphologies (*Daphnia, Hyla*)

    2. Adaptive immunological responses (*Gallus, Homo*)

    3. Adaptive reproductive allocations (ant colonies)

D. Stress-induced bone formation

    1. Prenatal (fibular crest in birds)

    2. Postnatal (patella in mammals; lower jaw in humans?)

E. Environmentally responsive neural systems

    1. Experience-mediated visual synapses (*Felix*, monkeys)

    2. Cortical remodeling (phantom limbs; learning)

CONTEXT-DEPENDENT LIFE CYCLE PROGRESSION

A. Larval settlement

    1. Substrate-induced metamorphosis (bivalves, gastropods)

    2. Prey-induced metamorphosis (gastropods, chitons)

    3. Temperature/photoperiod-dependent metamorphosis

B. Diapause

    1. Overwintering in insects

    2. Delayed implantation in mammals

C. Sexual/asexual progression

    1. Temperature/photoperiod-induced (aphids, *Megoura*)

    2. Temperature/colony-induced (*Volvox*)

D. Symbioses/parasitism

    1. Blood meals (*Rhodnius, Aedes*)

    2. Commensalism (*Euprymna/Vibrio*; eggs/algae; *Paleon/Alteromonas*; Mammalian gut microbiota)

    3. Parasites (*Wollbachia* in wood lice)

E. Developmental plant-insect interactions

ADAPTATIONS OF EMBRYOS AND LARVAE TO ENVIRONMENTS

A. Egg protection

    1. Sunscreens against radiation (*Rana*, sea urchins)

    2. Plant-derived protection (*Utetheisa*)

B. Larval protection

    1. Plant-derived protection (*Danaus*; tortoise beetles)

TERATOGENESIS

A. Chemical teratogens

    1. Natural compounds (retinoids, alcohol, lead)

    2. Synthetic compounds (thalidomine, warfarin)

    3. Hormone mimics (diethylstilbesterol, PCBs)

B. Infectious agents

    1. Viruses (*Coxsackie, Herpes, Rubella*)

    2. Bacteria (*Toxoplasma, Treponema*)

C. Maternal conditions

    1. Malnutrition

    2. Diabetes

    3. Autommunity

Source: After Gilbert 2001.

Note: This list should not be thought to be inclusive. For example, the list is limited to animals; plant developmental plasticity and many plant-animal interactions have not been included here.

Gravity and pressure

In Chapters 10 and 11 we saw that gravity is critical for frog and chick axis formation. Moreover, there are several bones whose formation is dependent on stresses occasioned by the movement of the embryo. Such stresses are known to be responsible for the formation of the human patella (kneecap) after birth. In vertebrates, cartilage cell differentiation and cartilage matrix production depend on a mechano-sensitive interaction among a number of genes and gene products. One of the most important of these genes is *Sox9*, which is upregulated by compressive force (Takahashi et al. 1998). The Sox9 protein activates numerous bone-forming genes (see Chapter 14). Tension forces also activate bone morphogenesis proteins and align chondrocytes (Bard 1990; Sato et al. 1999; Ikegame et al. 2001). Recent studies (Q. Wu et al. 2001) implicate *indian hedgehog* as a key signaling molecule that is stimulated by stress and which activates the bone morphogenetic proteins.

In the chick, several bones do not form if embryonic movement in the egg is suppressed. One of these bones is the **fibular crest**, which connects the tibia to the fibula and allows the force of the iliofibularis muscle to pull directly from the femur to the tibia. This direct connection is thought to be important in the evolution of birds, and the fibular crest is a universal feature of the bird hindlimb (Müller and Steicher 1989). When the bird is prevented from moving within its egg, this bone fails to develop (Figure 22.1; K. C. Wu et al. 2001; Müller 2003). Therefore, even in the formation of important features such as bones, the environment can play a critical role.

Developmental symbiosis

In some cases, the development of one individual is brought about by the presence of organisms of a different species. In some organisms, this relationship has become **symbiotic**

(A)

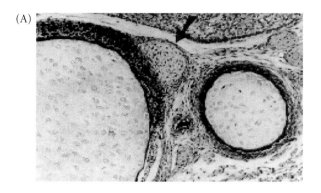

(B) (C)

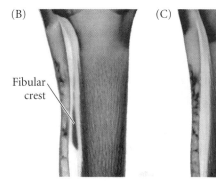

Fibular crest

Figure 22.1

Activity-induced formation of the fibular crest. The fibular crest (syndesmosis tibiofibularis) is formed when the movement of the embryo in the egg puts stress on the tibia. (A) Transverse section through the 10-day embryonic chick limb, showing the condensation (arrow) that will become the fibular crest. (B) 13-day chick embryo showing fibular crest forming between the tibia and fibular bones. (C) Absence of fibular crest in the connective tissue of a 13-day embryo whose movement was inhibited. The blue dye stains cartilage, while the red dye stains the bone elements. (After Müller 2003; Photographs courtesy of G. Müller.)

bacteria bind to a ciliated epithelium that extends into this cavity. This epithelium only binds *V. fisheri*, allowing other bacteria to pass through. The bacteria induce the apoptotic death of these epithelial cells, their replacement by a nonciliated epithelium, and the differentiation of the surrounding epithelial cells into storage sacs for the bacteria (Figure 22.2; McFall-Ngai and Ruby 1991; Montgomery and McFall-Ngai 1995).

Symbioses between egg masses and photosynthetic algae are critical for the development of several species. Clutches of amphibian and snail eggs, for example, are packed together in

(Sapp 1994): the symbionts become so tightly integrated into the host organism that the host cannot develop without them. Recent evidence (McFall-Ngai 2002) indicates that developmental symbioses are extremely common and may constitute the "rule" rather than the exceptional case.

One of the best-studied examples of developmental symbiosis is that between the squid *Euprymna scolopes* and the luminescent bacterium *Vibrio fischeri*. The adult *Euprymna* is equipped with a light organ composed of sacs containing these luminous bacteria. The juvenile squid, however, does not contain these light-emitting symbionts, nor does it have a structure to house them. Rather, the squid acquires the bacteria from the seawater pumped through its mantle cavity. The

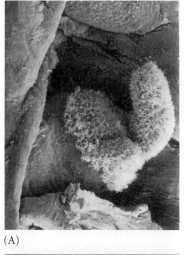

(A)

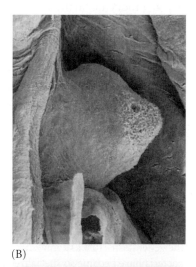

(B)

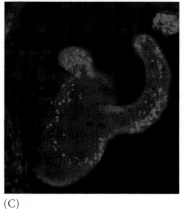

(C)

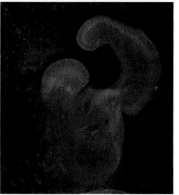

(D)

Figure 22.2

Symbiosis in the squid *Euprymna*. (A, B) Scanning electron micrographs of a light organ primordium of a 3-day-old juvenile squid *E. scolopes*. (A) Light organ of an uninfected juvenile. (B) Light organ of a juvenile infected with the symbiotic *V. fischeri* bacteria. Regression of the epithelium is obvious. (C) Bacteria-induced apoptosis is shown by acridine orange staining at 12 hours after infection of the juvenile squid with the bacteria. The bright green areas indicate regions of cell death. (D) Light organ of a squid grown in the absence of *V. fischeri*. No areas of apoptosis are seen. (From Montgomery and McFall-Ngai 1995; photographs courtesy of M. McFall-Ngai.)

(A)

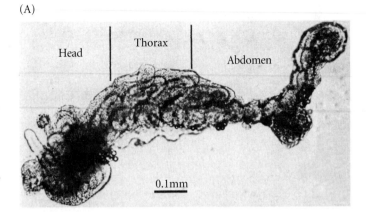

(B)

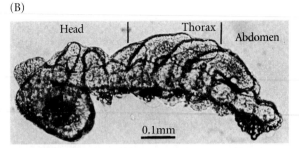

Figure 22.3
Microbial symbionts are necessary for gut formation in the leafhopper *Euscelis incisus*. (A) Control embryo, with symbionts, has normal gut formation. (B) Abnormal, gut-deficient embryo formed after antibiotics have eliminated most of the symbiotic bacteria from the egg. (From Schwemmler 1974; photographs courtesy of W. Schwemmler.)

tight masses. The supply of oxygen limits the rate of development, so embryos on the inside of the cluster develop more slowly than those near the surface (Strathmann and Strathmann 1995). While there is a steep gradient of oxygen from the outside of the cluster to deep within it, the embryos seem to get around this problem by coating themselves with a thin film of photosynthetic algae. In clutches of amphibian and snail eggs, photosynthesis from this algal "fouling" enables net oxygen production in the light, while respiration exceeds photosynthesis in the dark (Bachmann et al. 1986; Pinder and Friet 1994; Cohen and Strathmann 1996). Thus, the algae "rescue" the eggs by their photosynthesis.

Symbioses between eggs and bacteria can protect the eggs from fungal pathogens. Lobster and shrimp eggs, for instance, are prone to fungal infection. (As anyone who owns an aquarium knows, uneaten fish food soon become surrounded by a halo of filamentous fungi.) The chorions of these crustacean eggs actually attract bacteria that produce fungicidal compounds (Gil-Turnes et al. 1989).

An even tighter link between morphogenesis and symbiosis is exemplified by the parasitic wasp *Asobara tabida* and the leafhopper *Euscelis incisus*. In these insects, symbiotic bacteria are found within the egg cytoplasm and are transferred through the generations, just like mitochondria. In the leafhopper, these bacteria have become so specialized that they can multiply only inside the leafhopper's cytoplasm, and the host has become so dependent on the bacteria that it cannot complete embryogenesis without them. In fact, the bacterial symbionts appear to be essential for the formation of the embryonic gut. If the bacteria are surgically or metabolically removed from the eggs by feeding antibiotics to larvae or adults, the symbiont-free oocytes develop into embryos that lack an abdomen (Figure 22.3; Sander 1968; Schwemmler 1974, 1989). In *Asobara*, the bacteria enable the wasp to complete yolk production and egg maturation (Dedeinde et al. 2001). If the symbionts are removed (by adding antibiotics), eggs are not produced.

Even mammals maintain developmental symbioses with bacteria. Bacteria actually regulate some of our intestinal genes. The polymerase chain reaction technique is able to identify bacterial species that cannot be cultured, and microarray analyses can show changes in the expression of a large population of genes. These techniques have revealed a remarkable complexity in our "selves." Human bacterial symbionts have particular geographic distributions within us. The 400 or so bacterial species of the human colon are stratified into specific regions along the length and diameter of the gut tube, where they can attain densities of 10^{11} cells per milliliter (Savage 1977; Hooper et al. 1998). We never lack these microbial components; we pick them up from the reproductive tract of our mother as soon as the amnion bursts. We have co-evolved to share our spaces with them, and we have even co-developed such that our cells are primed for their docking, and their cells are primed to induce gene expression from our nuclei (Bry et al. 1996).

Bacteria-induced expression of mammalian genes was first demonstrated in the mouse gut. Umesaki (1984) noticed that a particular fucosyl transferase enzyme characteristic of mouse intestinal villi was induced by bacteria, and more recent studies (Hooper et al. 1998) have shown that the intestines of germ-free mice can initiate, but not complete, their differentiation. For complete development, the microbial symbionts of the gut are needed. Microarray analysis of mouse intestinal cells (Figure 22.4A; Hooper et al. 2001) has shown that normally occurring gut bacteria can upregulate the transcription of several mouse genes, including those encoding colipase, which is important in nutrient absorption; angiogenin-3, which helps form blood vessels; and sprr2a, a small, proline-rich protein that is thought to fortify matrices that line the intestine. Stappenbeck and colleagues (2002) have demonstrated that in the absence of particular intestinal microbes, the capillaries of the small intestinal villi fail to develop their complete vascular networks (Figure 22.4B, C). Intestinal microbes also appear to be critical for the maturation of the mouse gut-asso-

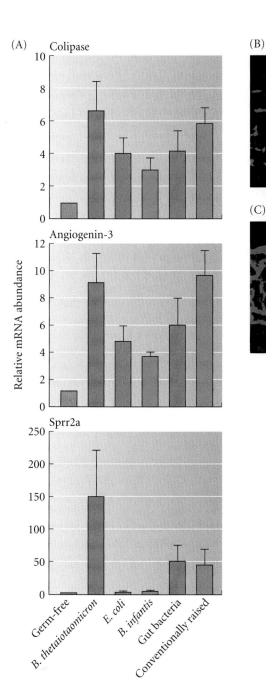

(A)

Colipase

Angiogenin-3

Sprr2a

Relative mRNA abundance

Germ-free
B. thetaiotaomicron
E. coli
B. infantis
Gut bacteria
Conventionally raised

(B)

(C)

Figure 22.4

Specificity of host genome responses to different bacteria. (A) Mice raised in "germ-free" environments were either left alone or inoculated with one or more types of bacteria. After 10 days, their intestinal mRNAs were isolated and tested on microarrays. Mice grown in "germ-free" conditions had very little expression of the genes encoding colipase, angiogenin-3, or sprr2a. Several different bacteria—*Bacteroides thetaiotaomicron*, *E. coli*, *Bifidobacterium infantis*, and an assortment of gut bacteria harvested from conventionally raised mice—induced the genes for colipase and angionenin-3. *B. thetaiotaomicron* appeared to be totally responsible for the 205-fold increase in *sprr2a* expression over that of germ-free animals. (B) Confocal microscope section of an intestinal villus capillary bed in a mouse raised for 6 weeks in germ-free conditions. The capillaries are stained green. (C) Capillary network of an intestinal villus of a mouse raised for 6 weeks in germ free conditions, then inoculated with conventional gut microbes 10 days before examination. The capillary network has fully developed. (A after Hooper et al. 2001; B, C after Stappenbeck et al. 2002; photograph courtesy of J. L. Gordon.)

Larval settlement

Environmental cues are critical to metamorphosis in many species, and some of the best-studied examples are the settlement cues used by marine larvae. A free-swimming marine larva often needs to settle near a source of food or on a firm substrate on which it can metamorphose. Thus, if prey or substrates give off soluble molecules, these molecules can be used by the larvae as cues to settle and begin metamorphosis. Among the molluscs, there are often very specific cues for settlement (Table 22.2; Hadfield 1977). In some cases, the prey supply the cues, while in other cases the substrate gives off molecules used by the larvae to initiate settlement. These cues may not be constant, but they need to be part of the environment if further development is to occur* (Pechenik et al. 1998).

The larvae of the red abalone, *Haliotis rufescens*, settle only when they physically contact coralline red algae. A brief contact is all that is required for competent larvae to stop swimming and begin metamorphosis. The chemical agent responsible for this change has not yet been isolated, but a receptor that recognizes an algal peptide induces metamorphosis in compe-

ciated lymphoid tissue (Cebra 1999). In short, mammals have coevolved with bacteria to the point that our bodily phenotypes do not fully develop without them.

WEBSITE 22.1 Developmental symbioses and parasitism. Some embryos acquire protection and nutrients by forming symbiotic associations with other organisms. The mechanisms by which these associations form are now being elucidated. In other situations, one species uses material from another to support its development. Blood-sucking mosquitoes are examples of such parasites.

---

*The importance of substrates for larval settlement and metamorphosis was first demonstrated in 1880, when William Keith Brooks, an embryologist at Johns Hopkins University, was asked to help the ailing oyster industry of Chesapeake Bay. For decades, oysters had been dredged from the bay, and there had always been a new crop to take their place. But recently, each year brought fewer oysters. What was responsible for the decline? Experimenting with larval oysters, Brooks discovered that the American oyster (unlike its better-studied European cousin) needed a hard substrate on which to metamorphose. For years, oystermen had thrown the shells back into the sea, but with the advent of suburban sidewalks, the oystermen were selling the shells to the cement factories. Brooks's solution: throw the shells back into the bay. The oyster population responded, and the Baltimore wharves still sell their descendants.

TABLE 22.2 Specific settlement substrates of snail larvae

Species	Substrate
Nassarius obsoletus	Mud from adult habitat
Philippia radiata	*Porites lobata* (a cnidarian)
Adalaria proxima	*Electra pilosa* (a bryozoan)
Doridella obscura	*Electra crustulenta* (a bryozoan)
Phestilla sibogae	*Porites compressa* (a cnidarian)
Rostanga pulchra	*Ophlitaspongia pennata* (a sponge)
Trinchesia aurantia	*Tubularia indivisa* (a cnidarian)
Elysia chlorotica	Primary film of microorganisms from adult habitat
Haminoea solitaria	Primary film of microorganisms from adult habitat
Aplysia californica	*Laurencia pacifica* (a red alga)
Aplysia juliana	*Ulva* spp. (green algae)
Aplysia parvula	*Chondrococcus hornemanni* (a red alga)
Stylocheilus longicauda	*Lyngbya majuscula* (a cyanobacterium)
Onchidoris bilamellata	Living barnacles

Source: Hadfield 1977.

tent larvae. Larvae that are not competent to begin metamorphosis do not appear to have this receptor. The receptor is thought to be linked to a G protein similar to those found in vertebrates, and the activation of this G protein may be necessary for inducing larval settlement and metamorphosis (Morse et al. 1984; Baxter and Morse 1992; Degnan and Morse 1995).

Sex in its season

Several species of aphids have a fascinating life cycle wherein an egg hatched in the spring gives rise to several generations of parthenogenetically (asexually) reproducing females. During the autumn, however, a particular type of female is produced whose eggs can give rise to both males and sexual females. These sexual forms mate, and their eggs are able to

survive the winter. When the overwintering eggs hatch, each one gives rise to an asexual female.

Some of the mysteries of this type of development were solved in 1909 by Thomas Hunt Morgan (before he started working on fruit flies). Morgan analyzed the chromosomes of the hickory aphid through several generations (Figure 22.5). He found that the diploid number of female aphids is 12. In parthenogenetically reproducing females, only one polar body is extruded from the developing ovum during oogenesis, so the diploid number of 12 is retained in the egg. This type of egg develops without being fertilized. In the females that give rise to eggs that become male or female, a modification of oogenesis occurs. In the female-producing eggs, 6 chromosome pairs enter the sole polar body; the diploid number of 12 is thereby retained. In the male-producing eggs, however, an extra chromosome pair enters the polar body. The male diploid number is thus 10. The resulting males and females are sexual and produce gametes by complete meiotic divisions. The females produce oocytes with a haploid set of 6 chromosomes. The males, however, divide their 10 chromosomes to produce some sperm with a haploid number of 4 and other sperm with a haploid number of 6. The sperm with 4 chromosomes degenerate. The sperm with 6 chromosomes fertilize the eggs to restore the diploid chromosome number of 12. These eggs overwinter, and when they hatch in the spring, parthenogenetic females emerge.

Morgan solved one riddle, but the riddle of how autumn weather regulates whether the female reproduces sexually or parthenogenetically remains unsolved. Similarly, we do not know what regulates whether a diploid oocyte gives rise to male- or female-producing eggs. Moreover, the same environmental factors are used differently by other aphid species (Hardie 1981; Hardie and Lees 1985). But it is not known how environmental changes become transformed into titers of JH,

Figure 22.5
Chromosomal changes during the life cycle of the hickory aphid. Autumn weather induces the production of males and females, which mate to produce the overwintering eggs.

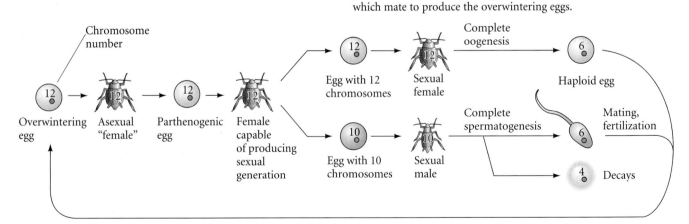

or how the autumn weather (or perhaps declining hours of sunlight) causes the differential movement of chromosomes into the polar body.

Diapause: Suspended development

Many species of insects and mammals have evolved a developmental strategy called diapause to survive periodically harsh conditions. **Diapause** is a suspension of development that can occur at the embryonic, larval, pupal, or adult stage, depending on the species (see Chapter 18). Diapause is not a physiological response brought about by harsh conditions. Rather, it is induced by stimuli (such as changes in the duration of daylight) that *presage* a change in the environment—cues beginning before the severe conditions actually arise. Diapause is especially important for temperate-zone insects, enabling them to survive the winter. The overwintering eggs of the hickory aphid (mentioned above) provide an example of this strategy. The development in the egg is suspended over the winter, so the larvae do not hatch when food is unavailable. In this case, diapause occurs during early development. The silkworm moth *Bombyx mori* similarly overwinters as an embryo, entering diapause just before segmentation. The gypsy moth *Lymantia dispar* initiates its diapause as a larva, and needs an extended period of cold weather to end diapause, which is why this pest is not found in the southern regions of Europe or the United States.

Over 100 mammalian species undergo diapause. The two most common mammalian strategies are delayed fertilization (the sperm are stored for later use) and delayed implantation (the blastocyst remains unimplanted within the uterus, and the rate of cell divisions diminishes or vanishes). Some species have *seasonal* diapause, so that embryos conceived in autumn will be born in spring rather than winter; in other species, diapause is induced by the presence of a newborn who is still getting milk. In the tammar kangaroo, *Macropus eugenii*, diapause can be induced by suckling-induced prolactin release, but it can also be induced by prolactin synthesized in response to changes in day length. In both cases, progesterone seems to be the signal that restores implantation and embryonic growth. Different groups of mammals use different hormones to induce or break diapause, but the result is the same: diapause lengthens the gestation period, allowing mating to occur and young to be born at times and seasons appropriate to the habitat of that species (Renfree and Shaw 2000).

WEBSITE 22.2 **Complex environmental effects on development.** The life cycles of certain insects are controlled by several environmental cues whose intersection provides a delicate timing mechanism.

WEBSITE 22.3 **Mechanisms of diapause.** Light and temperature are critical for the induction and maintenance of diapause. Different species use different signals for this event.

Phenotypic Plasticity: Control of Development by Environmental Conditions

Polyphenism and reaction norms

In most developmental interactions, the genome provides the specific instructions, while the environment is permissive. Dogs will generate dogs and cats will beget cats, even if they live in the same house. However, in most species, there are instances in development wherein the environment plays the instructive role and the genome is merely permissive. The ability of an individual to express one phenotype under one set of circumstances and another phenotype under another set is called **phenotypic plasticity.***

There are two main types of phenotypic plasticity: polyphenism and reaction norms. A **polyphenism** refers to discontinuous ("either/or") phenotypes elicited by the environment. Migratory locusts, for instance, exist in two mutually exclusive forms: a short-winged, uniformly colored solitary phase and a long-winged, brightly colored gregarious morph. Cues in the environment (mainly population density) determine which morphology a young locust will develop (Figure 22.6; see Pener 1991). Similarly, the nymphs of planthoppers can develop in two ways, depending on their environment. High population densities and the presence of certain plant communities lead to the production of migratory insects, in which the third thoracic segment produces a large hindwing. Low population densities and other food plants lead to the development of flightless planthoppers, in which the third thoracic segment develops into a haltere-like vestigial wing (Raatikainen 1967; Denno et al. 1985). The seasonal coat color changes in arctic animals are another example of polyphenism.†

In other cases, the genome encodes a *range* of potential phenotypes, and the environment selects the phenotype that is usually the most adaptive. For instance, constant and intense labor can make our muscles grow larger; but there is a genetically defined limit to how much muscular hypertrophy is possible. Similarly, the microhabitat of a young salamander can cause its color to change (again, within genetically defined limits). Such a continuous range of phenotypes that can be expressed by a single genotype across a range of environmental conditions is called a **reaction norm** (Woltereck 1909; Schmalhausen 1949; Stearns et al. 1991). The reaction norm is

---

*The ability of environmental cues to induce phenotypic change should be considered "tertiary induction." Primary induction involves the establishment of a single field within the embryo (such that one egg gives rise to just one embryo). Secondary induction refers to those cascades of inductive events within the embryo by which the organs are formed. Tertiary induction is the induction of developmental changes by factors in the environment.

†Although seasonal polyphenism is usually considered adaptive, there are times when it does not increase the fitness of the organism. For instance, the photoperiod can cause a hare's color to change from brown to white, but if it doesn't snow, the hare will be conspicuous against the dark background.

(A)

Stationary
morph

Migratory
morph

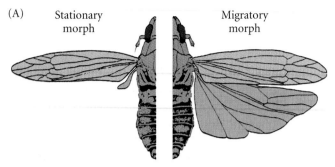

(B)

(C)

Figure 22.6
Density-induced polyphenism in planthoppers and grasshoppers. (A) Composite diagram showing the short-winged (left) and long-winged (right) forms of the planthopper *Prokelisia marginata*. The long-winged migratory morph is an excellent flier; the short-winged morph is flightless. (B, C) Density-induced changes in the "plague locust" *Schistocerca gregaria*. (B) Low-density morph, showing green pigmentation and miniature wings. (C) High-density morph, showing new pigmentation and wing and leg development. (A after Denno et al. 1985; B, C from Tawfik et al. 1999, photographs courtesy of S. Tanaka.)

thus a property of the genome and can also be selected. Different genotypes are expected to differ in the direction and amount of plasticity that they are able to express (Gotthard and Nylin 1995; Via et al. 1995).

Seasonal polyphenism in butterflies

Two dramatic examples of polyphenism were described in Chapter 3. The two phenotypes of the butterfly *Araschnia levana* are so different that Linnaeus classified them as two different species (see Figure 3.3), and the phenotype of the moth *Nemoria arizonaria* depends on its diet (see Figure 3.4). This type of polyphenism is not uncommon among insects. Throughout much of the Northern Hemisphere, one can see polyphenism in butterflies of the family Pieridae (the cabbage whites) between individuals that eclose during the long days of summer and those that eclose at the beginning of the season, in the short, cooler days of spring. The hindwing pigments of the short-day forms are darker than those of the long-day butterflies. This pigmentation has a functional advantage during the cool months of spring: darker pigments absorb sunlight more efficiently than light ones, raising the body temperature more rapidly (Figure 22.7; Shapiro 1968 1978; Watt 1968 1969; Hoffmann 1973; see also Nijhout 1991).

In tropical parts of the world, there is often a hot wet season and a cool dry season. In Africa, the Malawian butterfly *Bicyclus anynana* has a polyphenism that is adaptive to seasonal changes. It occurs in two phenotypes,

or **morphs**. The dry (cool) season morph is a mottled brown and gray butterfly that survives by hiding in dead leaves on the forest floor. In contrast, the wet (hot) season morph routinely flies. Unlike the cool-season form, it has prominent ventral eyespots that deflect attacks from predatory birds and lizards (Figure 22.8).

The factor determining the seasonal pigmentation of this butterfly is the temperature during pupation. Low temperatures produce the dry-season morph; high temperatures produce the wet-season morph (Brakefield and Reitsma 1991). The mechanism by which temperature regulates the *Bicyclus* phenotype is becoming known. In the late larval stages, the transcription of the *distal-less* gene in the wing imaginal discs

Figure 22.7
Polyphenic variation in *Pontia* (Pieridae) butterflies. The top row shows summer morphs: *P. protodice* female (left) and male (center); *P. occidentalis* male (right). The bottom row shows spring morphs: *P. protodice* female (left) and male (center); *P. occidentalis* male (right). (Photograph courtesy of T. Valente.)

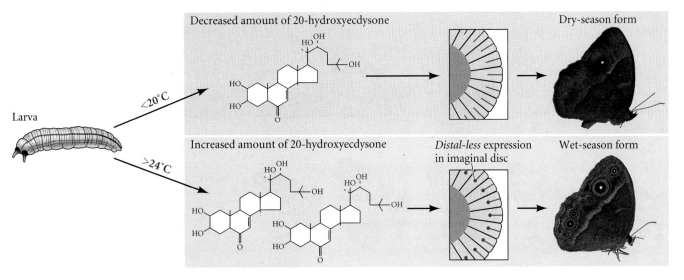

Figure 22.8
Phenotypic plasticity in *Bicyclus anynana* is regulated by temperature. High temperature (either in the wild or in controlled laboratory conditions) allows the accumulation of 20-hydroxyecdysone (20E), a hormone that is able to sustain *Distal-less* expression in the pupal imaginal disc. The region of *Distal-less* expression becomes the focus of each eyespot. In cooler weather, 20-hydroxyecdysone is not formed, *Distal-less* expression in the imaginal disc begins but is not sustained, and eyespots fail to form. (Photographs courtesy of S. Carroll and P. Brakefield.)

fourth and fifth instars experience a photoperiod (hours of daylight) that is longer or shorter than a particular critical day length. Below this critical day length, ecdysone levels are low and the butterfly that ecloses from its pupa has the orange wings characteristic of spring butterflies. Above that critical point, ecdysone is made and the summer pigmentation forms.

is restricted to a set of cells that will become the signaling center of each eyespot. In the early pupa, the higher temperatures elevate the formation of 20-hydroxyecdysone (20E; see Chapter18). This hormone (in a manner not yet described) sustains and expands the expression of *distal-less* in those regions of the wing imaginal disc. (Brakefield et al. 1996; Koch et al. 1996). Distal-less protein is believed to be the activating signal that determines the size of the spot (Figure 22.9). In the dry season, the cold temperature prevents the accumulation of 20E in the pupa, and the foci of Distal-less signaling are not sustained. In the absence of the Distal-less signal, the eyespots do not form.

The importance of hormones for mediating environmental signals controlling wing phenotypes has been documented in the *Araschnia* butterfly mentioned in Chapter 3. *Araschnia* develops alternative phenotypes depending on whether the

(A)　　　　　　　　(B)

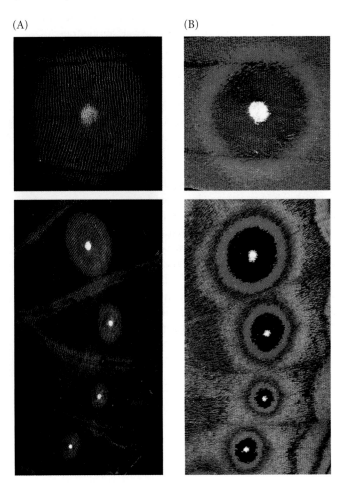

Figure 22.9
Distal-less and Spalt proteins define concentric areas corresponding to the rings of adult eyespots. (A) The transcription factors Distal-less (purple), Spalt (green), and their overlap (white) define the areas of the expanding pupal wing imaginal disc that correspond to the colored scales of the adult. (B) Portion of ventral hindwing of an adult *Bicyclus*, showing four of the seven eyespots and the large ventral eyespot of the forewing. (After Brunetti et al., 2001; photographs courtesy of S. Paddock and S. B. Carroll.)

Normal
summer form

Normal
spring form

Figure 22.10
Hormonal regulation mediates the environmentally controlled pigmentation of *Araschnia*. In the wild, different generations experience significantly different photoperiods. In the short photoperiod (below the critical day length), there is no pulse of 20E during early pupation, and the spring form of the butterfly is generated. When these spring butterflies mate, the larvae experience a long photoperiod and generate the summer pigmentation. In the laboratory, injections of 20E at different times during pupation can induce both phenotypes, as well as intermediate phenotypes not seen in the wild. (From Nijhout 2003; photograph courtesy of H. F. Nijhout.)

The summer form can be induced in diapause (spring) pupae by injecting 20E into the pupae. Moreover, by altering the timing of such injections, one can generate a series of intermediate forms not seen in the wild (Figure 22.10; Koch and Bückmann 1987; Nijhout 2003).

> **WEBSITE 22.4 Polyphenisms in butterflies.** Environmental cues appear to be sensed by the nervous system and transmitted through the body by juvenile hormone or ecdysone.

Nutritional polyphenism

Not all polyphenisms are controlled by the seasons. In bees, the size of the female larva at its metamorphic molt determines whether the individual is to be a worker or a queen. A larva fed nutrient-rich "royal jelly" retains the activity of her corpora allata during her last instar stage. The juvenile hormone secreted by these organs delays pupation, allowing the resulting bee to emerge larger and (in some species) more specialized in her anatomy (Brian 1974 1980; Plowright and Pendrel 1977). The JH level in larvae destined to become queens is 25 times that in larvae destined to become workers, and applying JH to worker larvae can transform them into queens as well (Wirtz 1973; Rachinsky and Hartfelder 1990). Using DNA macroarray technology, Evans and Wheeler (1999, 2001) identified five genes that were active in young larvae and in those larvae destined to become workers, but which were downregulated in those larvae destined to become queens. These genes were involved in controlling metabolic rates and responsiveness to hormones.

Similarly, ant colonies are predominantly female, and the females can be extremely polymorphic. The reproductive females ("gynes" or "queens") have functional ovaries and wings; the workers do not. These striking differences in anatomy and physiology are regulated through juvenile hormone (Passera 1985; Wheeler 1991). The influence of the environment on hormone levels and gene expression in ants was analyzed by Abouheif and Wray (2002), who found that nutrition-induced JH levels regulated wing formation.

Caste determination in the ant *Pheidole morrisi* has three main stages. First, males are distinguished from females at fertilization. Males are derived from unfertilized eggs. The haploid larvae so formed give rise to winged and reproductively active male adult. Females are derived from fertilized eggs, but the environment elicits one of three morphologically distinct female forms. The second switch point occurs during embryogenesis. If a diploid embryo experiences relatively high temperatures and long day length, JH level in the resulting larva will be high and the larva will form a reproductively active, winged adult. If the embryo does not experience this seasonal elevation in JH, it becomes a worker. The third switch point involves larval nutrition. If the worker larvae are fed well, they experience a rise in JH and will become major workers (soldiers); if they are not so well fed, they will not experience a surge of JH and will become minor workers (Figure 22.11A–D.)

Juvenile hormone level appears to be critical for the formation of wings from the imaginal discs. In the queen, both the forewing and the hindwing disc undergo normal development, expressing the same genes as *Drosophila* wing discs (Figure 22.11E–J; see Chapter 18). However, in the wing imaginal discs of workers, some of these genes remain unexpressed. The hindwing imaginal discs of soldiers and minor workers do not express *engrailed* (necessary for anterior-posterior compartmentation), *wingless* (necessary for growth and specification of the wing margin), or *spalt* (which helps convert the anterior-posterior information into the adult wing). Thus, the hindwing fails to form in both sets of workers. In

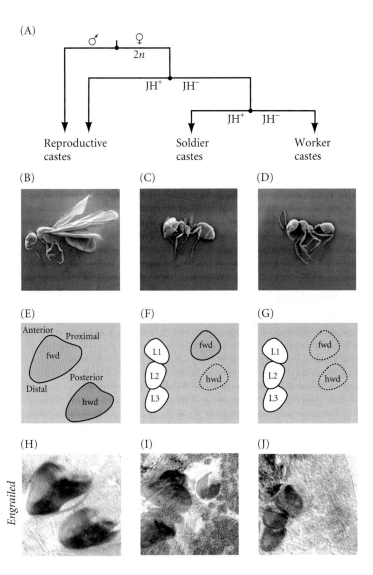

Figure 22.11

Polyphenism and gene expression differences in the ant *Pheidole morrisi*. (A) Caste determination at three switchpoints during development. At the point of sex determination, fertilized eggs become female and unfertilized eggs become male. The second switch point is environmentally regulated. Those female embryos experiencing higher temperature and longer photoperiod experience higher levels of juvenile hormone (JH) and become winged, reproductively active queens (B). Embryos not experiencing these conditions become workers. The third switch involves larval nutrition. Those larvae that are well fed become major workers (soldiers) with a vestigial forewing (C), and larvae that are not well fed become completely wingless minor workers (D). (E–G) Locations of the forewing discs (fwd), hindwing discs (hwd), and leg discs (L1-3) in each caste. The discs showing some gene expression are indicated in white. (H–J) In situ hybridization showing expression of the Engrailed protein in these imaginal discs. The wing discs in the queen larvae expressed all these genes. The soldier larvae expressed none of these genes in the hindwing disc, and misexpressed *spalt* (no medial expression zone) in the forewing disc. The minor worker larvae expressed none of these genes in either hindwing or forewing discs. (After Abouheif and Wray 2002; photographs courtesy of E. Abouheif.)

major workers (soldiers), the forewing imaginal discs express all the genes except *spalt*. The result is a malformed and nonfunctional forewing. The forewing imaginal discs of the minor workers do not express *wingless*, *engrailed*, or *spalt*, resulting in the complete absence of wings. Abouheif and Wray (2002) also found that although the end result—winged queens and wingless workers—is the same in numerous ant species, the actual genes downregulated in the workers differed from species to species.

WEBSITE 22.5 Nutritional polyphenism in the dung beetle. The shape of the male dung beetle is controlled by the amount of dung it eats as a larva. This maternal provision of food determines the level of juvenile hormone, and JH effects the growth of imaginal discs. The growth of imaginal discs regulates the adult's shape, and the shape of the beetle determines its mating strategy and just about everything else important in its life.

Environment-dependent sex determination

As we saw in Chapter 17, there are many species in which the environment determines whether an individual is to be male or female. The temperature dependence of sex determination in fish and reptiles has provided the best studied cases. Figure 17.23 displays some of the patterns of temperature-dependent sex determination in reptiles. This type of environmental sex determination has advantages and disadvantages. One advantage is that it probably gives the species the benefits of sexual reproduction without tying the species to a 1:1 sex ratio. In crocodiles, in which temperature extremes produce females while moderate temperatures produce males, the sex ratio may be as great as 10 females to each male (Woodward and Murray 1993). The major disadvantage of temperature-dependent sex determination may be its narrowing of the temperature limits within which a species can exist. This means that thermal pollution (either locally or due to global warming) could conceivably eliminate a species in a given area (Janzen and Paukstis 1991). Ferguson and Joanen (1982) speculate that dinosaurs may have had temperature-dependent sex determination and that their sudden demise may have been caused by a slight change in temperature that created conditions wherein only males or only females hatched from their eggs.

Charnov and Bull (1977) argued that environmental sex determination would be adaptive in those habitats characterized by patchiness—a habitat having some regions where it is more advantageous to be male and other regions where it is more advantageous to be female. Conover and Heins (1987) provided evidence for this hypothesis. In certain fish species, females benefit from being larger, since larger size translates into higher fecundity. If you are a female Atlantic silverside (*Menidia menidia*), it is advantageous to be born early in the breeding season, which allows you a longer feeding season and thus allows you to grow larger. In the males, size is of no importance. Conover and Heins showed that in the southern range of *Menidia*, females are indeed born early in the breeding season. Temperature appears to play a major role in this pattern. However, in the northern reaches of its range, the species shows no environmental sex determination. Rather, a 1:1 sex ratio is generated at all temperatures (Figure 22.12). The researchers speculated that the more northern populations have a very short feeding season, so there is no advantage for females in being born earlier. Thus, this fish has environmental sex determination in those regions where it is adaptive and genotypic sex determination in those regions where it is not.

Temperature isn't the only environmental factor that can affect sex determination in fish. Many fish can change their sex based on social interactions (Godwin et al. 2002). The sex of the blue-headed wrasse, a Panamanian reef fish, depends on the other fish it encounters. If a wrasse larva reaches a reef where a male lives with many females, it develops into a female. When the male dies, one of the females (usually the largest) becomes a male. Within a day, its ovaries shrink and

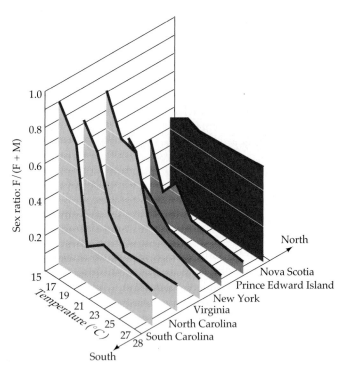

Figure 22.12
Relationship between temperature during the period of sex determination and sex ratio [F:(F + M)] in *Menidia menidia*. In fish collected from the northernmost portion of its range (Nova Scotia), temperature had little effect on sex determination. Among fish collected at more southerly locations (especially from Virginia through South Carolina), however, temperature had a large effect. (After Conover and Heins 1987.)

its testes grow. If the same wrasse larva had reached a reef that had no males or that had territory undefended by a male, it would have developed into a male (Warner 1984). These changes appear to be mediated by neuropeptides in the hypothalamus of these fish (Godwin et al. 2000, 2003).

The marine goby *Trimma okinawae* is one of the few fish that can change its sex more than once and in either direction. A female goby can become male if the male of the group dies. However, if a larger male enters the group, such males revert to being female. Grober and Sunobe (1996) induced females to become males, males to become females, and females to become males and then females again—merely by changing their companions. A goby can change its sex in about 4 days, and these changes appear to be mediated by the same hypothalamic neuropeptides that are correlated with wrasse sex determination.

Polyphenisms for alternative environmental conditions

Most studies of adaptation concern the roles adult structures play in enabling the individual to survive in otherwise precarious or hostile environments. However, the developing animal, too, has to survive in its habitat, and its development must adapt to the conditions of its existence.

The spadefoot toad, *Scaphiopus couchii*, has a remarkable strategy for coping with a very harsh environment. These toads are called out from hibernation by the thunder that accompanies the first spring storm in the Sonoran desert. (Unfortunately, motorcycles produce the same sounds, causing the toads to come out of hibernation only to die in the scorching Arizona sun.) The toads breed in temporary ponds formed by the rain, and the embryos develop quickly into larvae. After the larvae metamorphose, the young toads return to the desert, burrowing into the sand until the next year's storms bring them out.

Desert ponds are ephemeral pools that can either dry up quickly or persist, depending on the initial depth and the frequency of the rainfall. One might envision two alternative scenarios confronting a tadpole in such a pond: either (1) the pond persists until you have time to metamorphose and you live, or (2) the pond dries up before your metamorphosis is complete, and you die. *S. couchii* (and several other amphibians), however, have evolved a third alternative. The timing of their metamorphosis is controlled by the pond. If the pond persists at a viable level, development continues at its normal rate, and the algae-eating tadpoles develop into juvenile toads. However, if the pond is drying out and getting smaller, some of the tadpoles embark on an alternative developmental pathway. They develop a wider mouth and powerful jaw muscles, which enables them to eat (among other things) other *Scaphiopus* tadpoles (Figure 22.13). These carnivorous tadpoles metamorphose quickly, albeit into a smaller version of the juvenile spadefoot toad. But they survive while other *Scaphiopus* tadpoles perish from desiccation (Newman 1989, 1992).

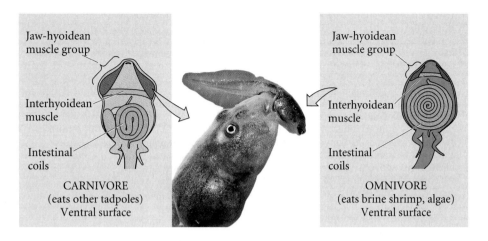

Figure 22.13
Polyphenism in the tadpoles of the spadefoot toad, *Scaphiopus couchii.* The typical morph is an omnivore, usually eating insects and algae. When ponds are drying out quickly, however, a carnivorous (cannibalistic) morph forms. It develops a wider mouth, larger jaw muscles, and an intestine modified for a carnivorous diet. The center photograph shows a cannibalistic tadpole eating a smaller pondmate. (Photograph © Thomas Wiewandt; drawings courtesy of R. Ruibel.)

Sidelights & Speculations

Genetic Assimilation

There is a trade-off for the carnivorous spadefoot toad. Although cannibalistic tadpoles survive the drought, their survival rate after metamorphosis is lower than that of tadpoles that develop more slowly. If there is no significant trade-off for a morph induced by hostile environmental conditions, we might expect it to become the predominant form of the species.

C. H. Waddington and I. I. Schmalhausen independently made this prediction to explain how some species have evolved rapidly in particular directions (see Gilbert 1994). Both scientists were impressed by the calluses of the ostrich. Most mammalian skin has the ability to form calluses on areas that are abraded by the ground or some other surface.* The skin cells respond to friction by proliferating. While such examples of environmentally induced callus formation are widespread, the ostrich is born with calluses where it will touch the ground (Figure 22.14).

*Until the late twentieth century, writers could be recognized by the calluses on their fingers. Thus, in *The Red Headed League,* Sherlock Holmes correctly surmised that the red-headed man had been hired as a scrivener.

Figure 22.14
Ventral side of an ostrich; arrows mark the calluses. (After Waddington 1942).

Waddington and Schmalhausen hypothesized that since the skin cells are already competent to be induced by friction, they could be induced by other things as well. As ostriches evolved, a mutation (or a particular combination of alleles) appeared that enabled the skin cells to respond to some substance within the embryo. Waddington (1942) wrote:

> *Presumably its skin, like that of other animals, would react directly to external pressure and rubbing by becoming thicker. ... This capacity to react must itself be dependent upon genes. ...It may then not be too difficult for a gene mutation to occur which will modify some other area in the embryo in such a way that it takes over the function of external pressure, interacting with the skin so as to 'pull the trigger' and set off the development of callosities.*

By this transfer of induction from an external inducer to an internal inducer, a trait that had been induced by the environment became part of the genome of the organism and could be selected. Waddington called this phenomenon "genetic assimilation," while Schmalhausen (1949) called it "stabilizing selection." Both scientists had used orthodox embryology and orthodox genetics to explain phenomena that had been considered cases of Lamarckian "inheritance of acquired characteristics."

A shift from environmental stimulus to genetic stimulus might explain sex determination in *Menidia* and caste determination in ants. Similarly, the preexisting developmental plasticity of arm length in feeding echinoderm larvae may have bridged the transition from pluteus (feeding) larvae to nonfeeding larvae that lack ciliated arms. This change in the allocation of resources between larval and juvenile structures parallels that seen in the food reserves stored in the egg. Thus, the changes already present as adaptations to external food resources may have become genetically fixed in those species whose larvae do not need to hunt for food (Strathmann et al. 1992).

If genetic assimilation is the genetic fixation of one of two or more phenotypes that had been adaptively expressed, then butterflies would be a good place to look for further examples. Brakefield and colleagues (1996) showed that they could genetically fix the different morphs of the adaptive polyphenism of *Bicyclus*, and Shapiro (1976) showed that the short-day (cold-weather) adaptive phenotype of several butterflies is the same as the single genetically produced phenotype of related species or subspecies living at higher altitudes or latitudes. Thus, the genetic assimilation of morphs originally produced through developmental plasticity may result in the origin of new species (West-Eberhard 1989).

Genetic assimilation may play an important role in providing a bias for evolutionary change. If an organism inherits a reaction norm, the developmental pathways leading to a particular phenotype are already in place, and all that evolution need do is supply a constant initiator of those pathways. In the next chapter, we will discuss some of the molecular evidence for genetic assimilation.

WEBSITE 22.6 Genetic assimilation and phenocopies. Genetic assimilation has been documented in the laboratory, where the ability to react to environmental stimuli has been transferred to embryonic inducers.

The signal for accelerated metamorphosis appears to be the change in water volume. In the laboratory, *Scaphiopus* tadpoles are able to sense the removal of water from aquaria, and their acceleration of metamorphosis depends on the rate at which the water is removed. The stress-induced corticotropin-releasing hormone signaling system appears to modulate this effect (Denver et al. 1998; Denver 1999). As in nearly all cases of polyphenism, the developmental changes are mediated through the endocrine system. Sensory organs send a neural signal to regulate hormone relase. The hormones then can alter gene expression in a coordinated and relatively rapid fashion.

Predator-induced polyphenisms

Imagine an animal who is often confronted by a particular predator. One could then imagine an individual who could recognize soluble molecules secreted by that predator and use those molecules to activate the development of structures that would make this individual less palatable to the predator.

This ability to modulate development in the presence of predators is called **predator-induced defense**, or **predator-induced polyphenism**.

To demonstrate predator-induced polyphenism, one has to show that the phenotypic modification is caused by the presence of the predator and that the phenotypic modification increases the fitness of its bearers when the predator is present (Adler and Harvell 1990; Tollrian and Harvell 1999). For instance, several rotifer species will alter their morphology when they develop in pond water in which their predators were cultured (Figure 22.15; Dodson 1989; Adler and Harvell 1990). The predatory rotifer *Asplanchna* releases into its water

Figure 22.15
Predator-induced defenses. Typical (upper row) and predator-induced (lower row) morphs of various organisms are shown. The numbers beneath each column represent the percentage of organisms surviving predation when both induced and uninduced individuals were presented with predators (in various assays). (Data from Adler and Harvell 1990 and references cited therein.)

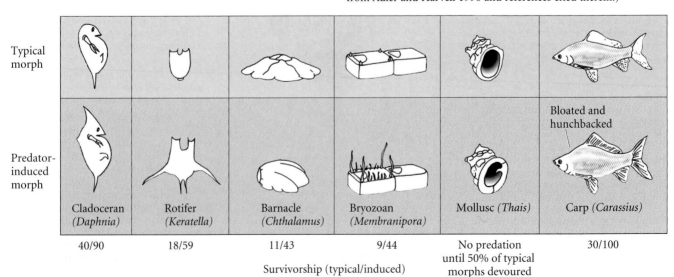

	Cladoceran (*Daphnia*)	Rotifer (*Keratella*)	Barnacle (*Chthalamus*)	Bryozoan (*Membranipora*)	Mollusc (*Thais*)	Carp (*Carassius*)
Typical morph						
Predator-induced morph						Bloated and hunchbacked
	40/90	18/59	11/43	9/44	No predation until 50% of typical morphs devoured	30/100

Survivorship (typical/induced)

a soluble compound that induces the eggs of a prey rotifer species, *Keratella slacki*, to develop into individuals with slightly larger bodies and anterior spines 130% longer than they otherwise would be, making the prey more difficult to eat. The snail *Thais lamellosa* develops a thickened shell and a "tooth" in its aperture when exposed to the effluent of the crab species that preys on it. In a mixed snail population, crabs will not attack the thicker snails until more than half of the normal-morph snails are devoured (Palmer 1985). Figure 22.15 shows the typical and predator-induced morphs for several species. In each case, the induced morph is more successful at surviving the predator, and soluble filtrate from water surrounding the predator is able to induce the changes. Chemicals that are released by a predator and can induce defenses in its prey are called **kairomones**.

The predator-induced polyphenism of the parthenogenetic water flea *Daphnia* is beneficial not only to itself, but also to its offspring. When *Daphnia cucullata* encounter the predatory larvae of the fly *Chaeoborus*, their "helmets" grow to twice their normal size (Figure 22.16). This increase lessens the chances that *Daphnia* will be eaten by the fly larvae. This same helmet induction occurs if the *Daphnia* are exposed to extracts of water in which the fly larvae had been swimming. Agrawal and colleagues (1999) have shown that the offspring of such an induced *Daphnia* are born with this same altered head morphology. It is possible that the *Chaeoborus* kairomone regulates gene expression both in the adult and in the developing embryo. We still do not know how *Daphnia* evolved the ability to make receptors that bind the kairomone or to utilize the kairomone to generate an adaptive morphological change.

Colonial invertebrates display some spectacular developmental responses to predators. These sessile animals cannot escape predators by moving away, and predators often treat them like plants, eating modules without killing off the entire colony. *Membranipora membranacea* is a widely distributed bryozoan often seen on kelp. It is grazed upon by certain nudibranch molluscs that suck in the modules at the periphery of the colony. When exposed to such predation, the modules near the nudibranch develop spines. These spines interfere with the predator's ability to establish the suction needed to feed. Spines can also be induced within 3 days by treating a colony with chemical extracts from the predator (Harvell 1986, 1999.)

Predator-induced polyphenism is not limited to invertebrates. The carp *Carassius carassius* is able to respond to the presence of a predatory pike only if the pike has already eaten other fish. The carp grows into a pot-bellied, hunchbacked morph that will not fit into the pike's jaws. However, as in most predator-induced defenses, there is a trade-off (otherwise one would expect the induced morph to become the normal phenotype). In this case, the induced morphol-ogy puts a drag on swimming efficiency, and the fatter fish cannot swim as well (Brönmark and Pettersson 1994).

Predator-induced polyphenisms are abundant among amphibians. Tadpoles found in ponds or reared in the presence of other species may differ significantly from tadpoles reared by themselves in aquaria. For instance, newly hatched wood frog (*Rana sylvetica*) tadpoles reared in tanks containing the predatory larval dragonfly *Anax* (confined in mesh cages so that they cannot eat the tadpoles) grow smaller than those reared in similar tanks without predators. Moreover, their tail musculature deepens, allowing faster turning and swimming speeds (van Buskirk and Relyea 1998). The addition of more predators to the tank causes a continuously deeper tail fin and tail musculature, and in fact what initially appeared to be a polyphenism may be a reaction norm that can assess the number (and type) of predators.

McCollum and Van Buskirk (1996) have shown that in the presence of its predators, the tail fin of the tadpole of the gray tree frog *Hyla chrysoscelis* grows larger and becomes bright red. This phenotype allows the tadpole to swim away faster and to deflect predator strikes toward the tail region. The trade-off is that noninduced tadpoles grow more slowly and survive better in predator-free environments.

WEBSITE 22.7 **Inducible caste determination in ant colonies.** In some species of ants, the loss of soldier ants creates conditions that induce more workers to become soldiers.

(A) (B) (C)

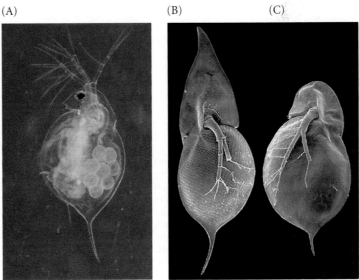

Figure 22.16
Predator-induced polyphenism in *Daphnia*. (A) *Daphnia* is an all-female species, producing eggs (visible within the adult organism) parthenogenetically. (B, C) Scanning electron micrographs showing predator-induced (B) and normal (C) morphs of the same clone. (A, photograph courtesy of R. Tollrian; B, C, photographs courtesy of A. A. Agrawal.)

Mammalian Immunity as a Predator-Induced Response

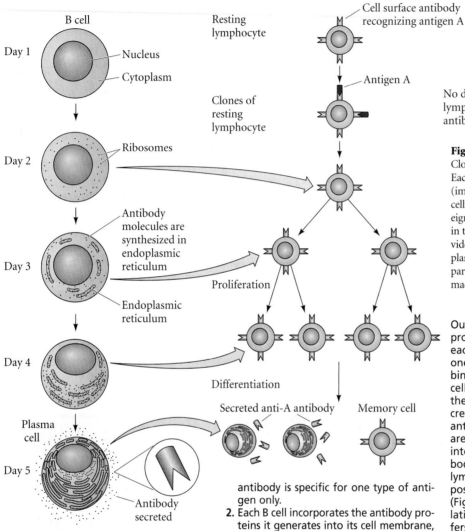

Figure 22.17
Clonal selection model of antibody formation. Each B cell makes a particular type of antibody (immunoglobulin) protein and displays it on its cell surface. When an antigen (a molecule foreign to the body) binds to the antibody proteins in the B cell membrane, the B cell is able to divide and differentiate into an antibody-secreting plasma cell. The plasma cell secretes only the particular type of antibody that was originally made by the B cell.

Out of the 10 million types of antibody proteins the cell can possibly synthesize, each B cell makes only one type. That is, one B cell may be making antibodies that bind to poliovirus, while a neighboring B cell might be making antibodies to diphtheria toxin. B cells are continually being created and destroyed. However, when an antigen binds to a set of B cells, these cells are stimulated to divide and differentiate into plasma cells (that secrete the antibody) and memory cells (that populate lymph nodes and respond rapidly when exposed to the same antigen later in life) (Figure 22.17). Thus, each person's constellation of plasma cells and memory cells differs depending on which antigens he or she has encountered. Identical twins have different populations of B cell descendants in their spleens and lymph nodes. This predator-induced defense induced in our immune systems is transgenerational. The passage of antibodies through the placenta and the transfer of antibodies through milk give the newborn protection as well.

WEBSITE 22.8 **The induction of immune cells.** The details of the mammalian immune response involve inductive interactions between B cells, T cells, macrophages, and antigens. The HIV kills the T cells responsible for allowing B cells to differentiate.

If predator-induced polyphenism is an adaptive response to potential threats, then the mammalian immune system is its highest achievement. Our main predators are microbes, and our immune system can recognize and destroy nearly all of them. When we are exposed to a foreign molecule (called an **antigen**), we manufacture **antibodies** and secrete them into our blood serum (see Chapter 4). These antibodies bind to the antigen, inactivating or eliminating it. The basis for the immune response is summarized in the five major postulates of the clonal selection hypothesis (Burnett 1959):

1. Each **B lymphocyte (B cell)** can make one, and only one, type of antibody. This antibody is specific for one type of antigen only.
2. Each B cell incorporates the antibody proteins it generates into its cell membrane, with the specificity-bearing side outward.
3. Antigens are presented (usually by macrophages) to the antibodies on the B cell membranes.
4. Only those B cells that bind the antigen can differentiate into antibody-secreting **plasma cells**. Plasma cells divide repeatedly, produce an extensive rough endoplasmic reticulum, and synthesize enormous amounts of antibody molecules that are secreted into the blood.
5. The specificity of the antibody made by a plasma cell is exactly the same as that which was on the surface of the B cell from which it differentiated.

The type of antibody molecule on the surface of a B cell is determined by chance.

Learning: An Environmentally Adaptive Nervous System

In Chapter 13 we saw that neuronal activity can be a critical factor in deciding which synapses are retained by the adult organism. Here we will extend that discussion to highlight those remarkable instances in which new experiences modify the original set of neuronal connections, causing the creation of new neurons, or the formation of new synapses between existing neurons.

The formation of new neurons

Since neurons, once formed, do not divide, the "birthday" of a neuron can be identified by treating the organism with radioactive thymidine. Normally, very little radioactive thymidine is taken up into the DNA of a neuron that has already been formed. However, if a new neuron differentiates by cell division during the treatment, it will incorporate radioactive thymidine into its DNA.

Such new neurons are seen to be generated when male songbirds first learn their songs. Juvenile zebra finches memorize a model song and then learn the pattern of muscle contractions necessary to sing a particular phrase. In this learning and repetition process, new neurons are generated in the hyperstriatum of the finch's brain. Many of these new neurons send axons to the archistriatum, which is responsible for controlling the vocal musculature (Nordeen and Nordeen 1988). These changes are not seen in males who are too old to learn the song, nor are they seen in juvenile females (who do not sing these phrases). In white-crowned sparrows, where song is regulated by photoperiod and hormones, the exposure of adult males to long hours of light and to testosterone induces over 50,000 new neurons in their vocal centers (Tramontin et al. 2000). The neural circuitry of these birds' brains shows seasonal plasticity. Testosterone is believed to increase the level of brain-derived neurotropic factor (BDNF) in the song-producing vocal centers. If female birds are given BDNF, they also produce more neurons there (Rasika et al. 1999).

The cerebral cortices of young rats reared in stimulating environments are packed with more neurons, synapses, and dendrites than are found in rats reared in isolation (Turner and Greenough 1983). Even the adult brain continues to develop in response to new experiences. When adult canaries learn new songs, they generate new neurons whose axons project from one vocal region of the brain to another (Alvarez-Buylla et al. 1990). Studies on adult rats and mice indicate that environmental stimulation can increase the number of new neurons in the dentate gyrus (Kempermann et al. 1997a,b; Gould et al. 1999; van Praag et al. 1999). Similarly, when adult rats learn to keep their balance on dowels, their cerebellar Purkinje neurons develop new synapses (Black et al. 1990). Thus, the pattern of neuronal connections is a product of inherited patterning and patterning produced by experiences. This interplay between innate and experiential development

has been detailed most dramatically in studies on mammalian vision.

Experiential changes in mammalian visual pathways

Some of the most interesting research on mammalian neuronal patterning concerns the effects of sensory deprivation on the developing visual system in kittens and monkeys. The paths by which electric impulses pass from the retina to the brain in mammals are shown in Figure 22.18. Axons from the retinal ganglion cells form the two optic nerves, which meet at the optic chiasm. As in *Xenopus* tadpoles, some axons go to the opposite (contralateral) side of the brain, but, unlike most other vertebrates, mammalian retinal ganglion cells also send inputs into the same (ipsilateral) side of the brain (see Chapter 13). These axons end at the two lateral geniculate nuclei. Here the input from each eye is kept separate, with the uppermost and anterior layers receiving the axons from the contralateral eye, and the middle of the layers receiving input from the ipsilateral eye. The situation becomes even more complex as neurons from the lateral geniculate nuclei connect with the neurons of the visual cortex. Over 80 percent of the neural cells in the visual cortex receive input from both eyes. The result is binocular vision and depth perception.

Another remarkable finding is that the retinocortical projection pattern is the same for both eyes. If a certain cortical neuron is stimulated by light flashing across a region of the left eye 5 degrees above and 1 degree to the left of the fovea,* it will also be stimulated by a light flashing across a region of the right eye 5 degrees above and 1 degree to the left of the fovea. Moreover, the response evoked in the cortical neuron when both eyes are stimulated is greater than the response when either retina is stimulated alone.

Hubel, Wiesel, and their co-workers (see Hubel 1967) demonstrated that the development of the nervous system depends to some degree on the experience of the individual during a critical period of development. In other words, not all neuronal development is encoded in the genome; some is the result of learning. Experience appears to strengthen or stabilize some neuronal connections that are already present at birth and to weaken or eliminate others. These conclusions come from studies of partial sensory deprivation. Hubel and Wiesel (1962, 1963) sewed shut the right eyelids of newborn kittens and left them closed for 3 months. After this time, they unsewed the right eyelids. The cortical neurons of these kittens could not be stimulated by shining light into the right eye. Almost all the inputs into the visual cortex came from the left eye only. The behavior of the kittens revealed the inadequacy of their right eyes—when the left eyes of these kittens were covered, they became functionally blind. Because the lateral geniculate neurons appeared to be stimulated by input

---

*The fovea is a depression in the center of the retina where only cones are present (rods and blood vessels are absent). Here it serves as a convenient landmark.

(A)

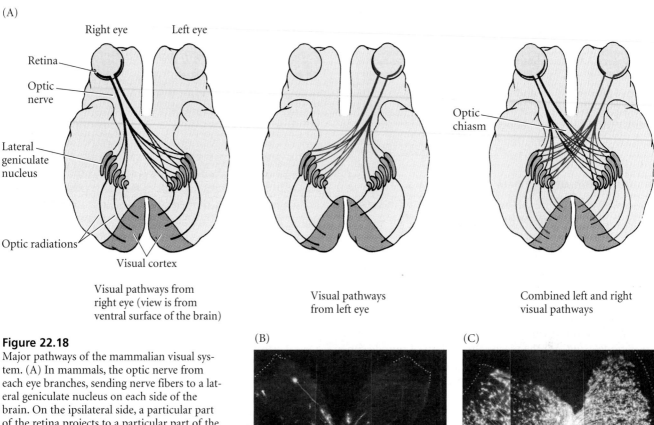

Visual pathways from right eye (view is from ventral surface of the brain)

Visual pathways from left eye

Combined left and right visual pathways

Figure 22.18
Major pathways of the mammalian visual system. (A) In mammals, the optic nerve from each eye branches, sending nerve fibers to a lateral geniculate nucleus on each side of the brain. On the ipsilateral side, a particular part of the retina projects to a particular part of the lateral geniculate nucleus. On the contralateral side, the lateral geniculate nucleus receives input from all parts of the retina. Neurons from each lateral geniculate nucleus innervate the visual cortex on the same side. (B, C) Isolated (and filleted) retinas from a 16-day mouse embryo show (B) ipsilateral and (C) contralateral projections from the retinal ganglion cells. The fluorescent carbocyanine dye DiI was inserted behind the optic chiasm and was allowed to enter the retinal axons. The dye diffuses along the axons, tracing them as to their origin. Ipsilateral projections mostly come from a single part of the retina (in this case, the ventrotemporal region). Contralateral projections to the same site come from all over the retina. (B, C from Colello and Guillery 1990; photographs courtesy of the authors.)

(B)

(C)

from both right and left eyes, the physiological defect appeared to be in the connections between the lateral geniculate nuclei and the visual cortex. Similar phenomena have been observed in rhesus monkeys, where the defect has been correlated with a lack of protein synthesis in the lateral geniculate neurons innervated by the covered eye (Kennedy et al. 1981).

Although it would be tempting to conclude that the blindness resulting from these experiments was due to a failure to form the proper visual connections, this is not the case. Rather, when a kitten or monkey is born, axons from lateral geniculate neurons receiving input from each eye overlap ex-

tensively in the visual cortex (Hubel and Wiesel 1963; Crair et al. 1998). However, when one eye is covered early in the animal's life, its connections in the visual cortex are taken over by those of the other eye (Figure 22.19). The axons compete for connections, and experience plays a role in strengthening and stabilizing the connections that are made. Thus, when *both* eyes of a kitten are sewn shut for 3 months, most cortical neurons can still be stimulated by appropriate illumination of one eye or the other. The critical time in kitten development for this validation of neuronal connections begins between 4 and 6 weeks after birth. Monocular deprivation up to the fourth week produces little or no physiological deficit, but through the sixth week, it produces all the characteristic neuronal changes. If a kitten has had normal visual experience for the first 3 months, any subsequent monocular deprivation (even for a year or more) has no effect. At that point, the synapses have been stabilized.

(A)

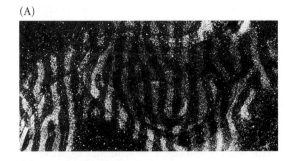

(B)

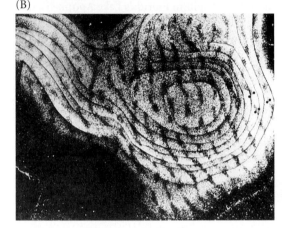

Figure 22.19
(A, B) Dark-field autoradiographs of monkey striate (visual) cortex 2 weeks after one eye was injected with [³H]proline in the vitreous humor. Each retinal neuron takes up the radioactive label and transfers it to the cells with which it forms synapses. (A) Normal labeling pattern. The white stripes indicate that roughly half the columns took up the label, while the other half did not—a pattern reflecting that half the cells were innervated by the labeled eye and half by the unlabeled eye. (B) Labeling pattern when the unlabeled eye was sutured shut for 18 months. Axonal projections from the normal (labeled) eye have taken over the regions that would normally have been innervated by the sutured eye. (C, D) Drawings of axons from the lateral geniculate nuclei of kittens in which one eye was occluded for 33 days. The terminal branching of axons receiving input from the occluded eye (C) was far less extensive than that of axons receiving input from the nonoccluded eye (D). (A and B from Wiesel 1982, photograph courtesy of T. Wiesel; C and D after Antonini and Stryker 1993.)

(C)   (D)

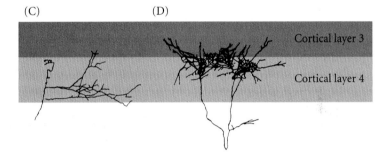

Cortical layer 3

Cortical layer 4

Two principles, then, can be seen in the patterning of the mammalian visual system. First, the neuronal connections involved in vision are present even before the animal sees. Second, experience plays an important role in determining whether or not certain connections persist.* Just as experience refines the original neuromuscular connections, experience plays a role in refining and improving the visual connections. It is possible, too, that adult functions such as learning and memory arise from the establishment and/or strengthening of different synapses by experience. As Purves and Lichtman (1985) remark,

> *The interaction of individual animals and their world continues to shape the nervous system throughout life in ways that could never have been programmed. Modification of the nervous system by experience is thus the last and most subtle developmental strategy.*

WEBSITE 22.9 **The phantom limb phenomenon.** Individuals who have a limb amputated sometimes feel pain in the absent appendage. This phenomenon appears to be caused by a reorganization of the human cerebral cortex following the amputation.

---

*Studies (Colman et al. 1997) have shown that differences in neurotransmitter release result in changes in synaptic adhesivity and cause the withdrawal of the axon providing the weaker stimulation. Studies in mice (Huang et al. 1999; Katz 1999) suggest that brain-derived neurotropic factor (BDNF) is crucial during the critical period.

Endocrine Disruptors

If the developing organism is sensitive to environmental factors, it also becomes vulnerable to agents in the environment that can disrupt normal development. Compounds that can disrupt normal development are called **teratogens**. Some of these compounds were mentioned in Chapters 1 and 21. One class of teratogens will be discussed here: the set of compounds called **endocrine disruptors**, which are exogenous chemicals that interfere with the normal function of hormones.

Endocrine disruptors can alter hormonal function in many ways. As mentioned in Chapter 21, they can mimic the effects of natural hormones (as in DES); they can block the synthesis of a hormone or block the binding of a hormone to its receptor (as in finasteride); or they can interfere with the transport or elimination of a hormone (as in PCBs).

Developmental toxicology and endocrine disruption are relatively new fields of research. While traditional toxicology has pursued the environmental causes of death, cancer, and genetic damage, endocrine disruptor research and the field of developmental toxicology focus on the roles environmental chemicals may play in altering development by disrupting normal endocrine function during both pre- and postnatal development (Bigsby et al. 1999).

WEBSITE 22.10 **Environmental endocrine disruptors.** The Wingspread Consensus Statement of 1991 began a move by scientists to influence government policy concerning potential endocrine disruptors. This site looks at that statement and at some of the policies presently being implemented.

Environmental estrogens

There are probably no greater controversies in the field of toxicology than whether chemical pollutants are responsible for congenital malformations in wild animals, the decline of sperm counts in men, and breast cancer in women. One major focus of this research is whether pesticide pollutants can function as estrogens in adult and developing mammals. Estrogen is more than just a "sex hormone." In both sexes, estrogens are used to regulate muscle and skeletal growth, maintain bone density, develop the organs of the immune system, and help maintain the nervous system. Thus, any chemical that activates or suppresses estrogen receptors can potentially affect several reproductive and nonreproductive organs.

There are several sources of environmental estrogens. The first source consists of naturally occurring estrogenic compounds. Besides the estrogens that mammalian embryos acquire from the maternal blood, estrogenic compounds are also found in certain plants (such as soybeans). These plant-derived estrogens are referred to as **phytoestrogens**. The second source of estrogenic compounds is drugs that are specifically designed to work as estrogens. These drugs include the synthetic estrogens used in birth control pills as well as drugs such as diethylstilbesterol (DES), which caused birth defects in humans when administered to pregnant women from 1948–1971(Hill 1997; Palanza et al. 2001; see Chapter 21). The third and fourth categories are pesticides and industrial compounds. These are seen as becoming increasingly significant sources of environmental estrogens, since Americans use some 2 billion pounds of pesticides each year, and some pesticide residues stay in the food chain for decades. Moreover, industrial chemicals such as polychlorinated biphenyls (PCBs, discussed in Chapter 21 and below) and bisphenol A (BPA, found in plastic-coated food cans and microwave pizzas) are ubiquitous by-products of our technological culture.

In 1962, Rachel Carson, a fisheries biologist, published *Silent Spring*, which became one of the most influential books of the twentieth century. Carson warned that pesticides were destroying wildlife, that DDT in particular appeared to be destroying shorebird populations, and that pesticides were becoming a staple of the American diet. For this, she was reviled by the agricultural chemicals industry and called a fanatic, a Communist, and worse. However, subsequent research bore out her claims, and when peregrine falcons and bald eagles were found to be endangered because of DDT-induced fragility of their eggshells (Cooke 1973), the use of this pesticide was banned in the United States.

Recent evidence has shown that DDT (dichloro-diphenyl-trichloroethane) and its chief metabolic by-product, DDE (which lacks one of the chlorine atoms), can act as estrogenic compounds, either by mimicking estrogen or by inhibiting androgen effectiveness (Davis et al. 1993; Kelce et al. 1995). Although banned in the United States in 1972, DDT has an environmental half-life of about 15 years. This means it may take 100 years or more for some concentrations of DDT

in the soil to get below active levels. DDE is a more potent estrogen than DDT and is able to inhibit androgen-responsive transcription at doses comparable to those found in contaminated soil in the United States and other countries. DDT and DDE have been linked to such environmental problems as the decrease in alligator populations in Florida, the feminization of fish in Lake Superior, the rise in breast cancers, and the worldwide decline of human sperm counts (Carlsen et al. 1992; Keiding and Skakkebaek 1993; Stone 1994; Swan et al. 1997). Guillette and co-workers (1994; Matter et al. 1998) have linked a pollutant spill in Florida's Lake Apopka (a discharge including DDT, DDE, and numerous other polychlorinated biphenyls) to a 90% decline in the birth rate of alligators and reduced penis size in the young males.

> **WEBSITE 22.11 Rachel Carson and the ban on DDT.** Even before the age of molecular biology, Rachel Carson pointed out that DDT was having a disastrous effect on bird populations. DDT caused egg shells to thin, and birds would often crush their eggs when sitting on them. Her book, *Silent Spring*, caused the political movement leading to the banning of DDT in the United States.

Dioxin, a by-product of the chemical processes used to make pesticides and paper products, has been linked to reproductive anomalies in male rats. The male offspring of females rats exposed to this planar, lipophilic molecule when pregnant have reduced sperm counts, smaller testes, and fewer male-specific sexual behaviors. Fish embryos seem particularly susceptible to dioxin and related compounds, and it has been speculated that the amount of these compounds in the Great Lakes during the 1940s was so high that none of the lake trout hatched there during that time survived (Figure 22.20; Hornung et al. 1996; Zabel and Peterson 1996; Johnson et al. 1998).

Polychlorinated biphenyls (**PCBs**) were widely used as refrigerants before they were banned in the 1970s, when they were shown to cause cancer in rats. However, they still remain circulating through the food chain (in both water and sediments), and they have been blamed for the widespread decline in the reproductive capacities of otters, seals, mink, and fish. An environmental estrogen, some PCBs resemble diethylstilbesterol in shape, and they may affect the estrogen receptor as DES does, perhaps by binding to another site on the estrogen receptor.

A similar molecule, methoxychlor, is found in many pesticides. Pickford and colleagues (1999) found that methoxychlor blocked progesterone-induced oocyte maturation in *Xenopus* at concentrations similar to those found in the environment. Such blockage would severely inhibit the fertility of the frogs, and it may be a component of the worldwide decline in amphibian populations.

Some scientists say that these claims are exaggerated. Tests on mice had shown that litter size, sperm concentration, and development were not affected by environmentally relevant concentrations of environmental estrogens. However, recent work by Spearow and colleagues (1999) has shown a remark-

(A)

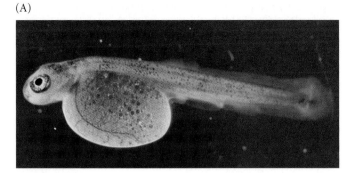

(B)

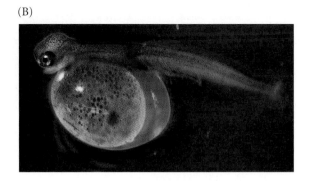

Figure 22.20
Lake trout 4 weeks after hatching. (A) Normal larva with its golden yellow yolk sac. (B) Dioxin-exposed larva exhibiting a blue yolk sac. The yolk sac has swelled with water and has numerous sites of hemorrhage. Such fish often have reduced growth, as well as heart and facial anomalies. (Photograph courtesy of R. E. Peterson.)

Another factor involved in establishing safety limits is the interaction of environmental estrogens. Silva and her colleagues (2002) have shown that if cells are exposed to a *mixture* of environmental estrogens, each of which is at a concentration that induces an estrogen-responsive gene very weakly, this mixture can induce a response that is much more than the additive responses of the individual components (Figure 22.22).

able genetic difference in sensitivity to estrogen among different strains of mice. The strain that had been used for testing environmental estrogens, the CD-1 strain, is at least 16 times more resistant to endocrine disruption than the most sensitive strains, such as B6. When estrogen-containing pellets were implanted beneath the skin of young male CD-1 mice, very little happened. However, when the same pellets were placed beneath the skin of B6 mice, their testes shrank, and the number of sperm seen in the seminiferous tubules dropped dramatically (Figure 22.21). This widespread range of sensitivities has important consequences for determining safety limits for humans.

Figure 22.21
Effects of estrogen implants on different strains of mice. The graph shows the percentage of seminiferous tubules containing elongated spermatozoa. (The mean ± standard error is for an average of six individuals). The photographs show cross sections of the testicles and are all at the same magnification. 40 µg of estradiol did not affect spermatogenesis in the CD-1 strain, but as little as 2.5 µg of estradiol almost completely abolished spermatogenesis in the B6 strain. (After Spearow et al. 1999; photographs courtesy of J. L. Spearow.)

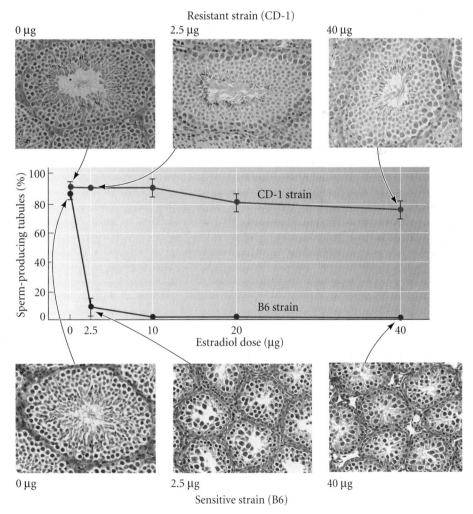

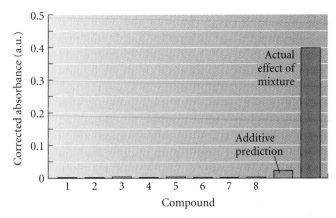

Figure 22.22
Ability of a mixture of estrogenic compounds at low concentrations to activate an estrogen-responsive gene. Plasmids containing an estrogen response element attached to a *lacZ* gene were added to cells, and each of eight environmental estrogens was added at concentrations that only weakly induced the expression of the *lacZ* gene. However, when a mixture of these eight compounds was added, the response was far greater than the additive values of the separate responses. (After Silva et al. 2002.)

Environmental thyroid hormone disruptors

In addition to being environmental estrogens, the structure of some PCBs resembles that of thyroid hormones (Figure 22.23), and exposure to them alters serum thyroid hormone levels in humans. Hydroxylated PCB have high affinities for the thyroid hormone serum transport protein transthyretin and can block thyroxine from binding to this protein. This leads to the elevated excretion of thyroid hormones. Thyroid hormones are critical for the growth of the cochlea of the inner ear, and rats whose mothers were exposed to PCBs had poorly developed cochleas and hearing defects (Goldey and Crofton, cited in Stone 1995; Cheek et al. 1999).

Chains of causation

Whether in law or in science, establishing chains of causation is a demanding and necessary task. In developmental toxicology, numerous endpoints must be checked, and many different levels of causation must be established (Crain and Guillette 1998; McNabb et al. 1999). For instance, researchers might ask whether the pollutant spill in Lake Apopka was responsible for the decline in the alligator population there. To establish this, they might first ask how the specific chemicals in the spill could contribute to physiological anomalies in alligators, and what would be the consequences of these anomalies. Table 22.3 shows the postulated chain of causation.

After observing that number of alligators in Lake Apopka had declined, *population level* observations revealed that there was decrease in the number of alligators being born (the birth rate). At the *organism level*, researchers found unusually high levels of estrogens in the female alligators and unusually low levels of testosterone in the males. On the *tissue and organ level*, the researchers observed elevated production of estrogens from the juvenile testes, malformations of the testes and penis, and changes in enzyme activity in the female gonads. On the *cellular level*, ovarian abnormalities were correlated with unusually elevated estrogen levels. These cellular changes could be explained at the *molecular level* by the fact that many of the chemicals in the pollutant spill bind to alligator estrogen and progesterone receptors, and that these chemicals are able to circumvent the cell's usual defenses against the overproduction of steroid hormones (Crain et al. 1998). Thus, the pollutant chemicals can be linked to a reproductive anomaly that could explain the decreased birth rate among alligators in the lake.

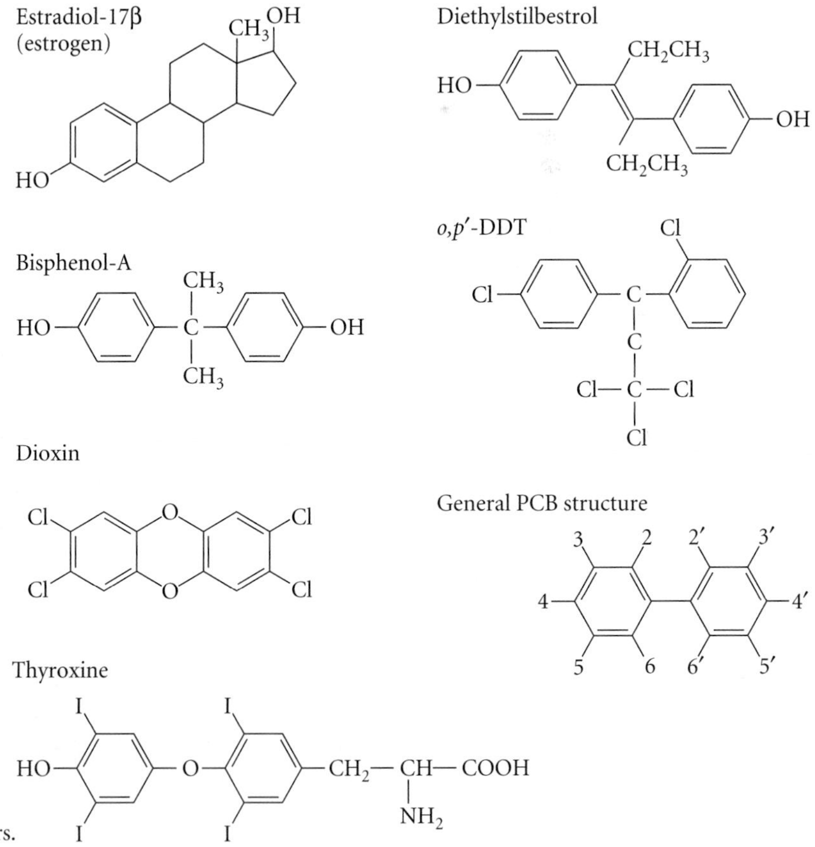

Figure 22.23
Structures of some hormones and endocrine disruptors.

TABLE 22.3 Chain of causation linking contaminant spill in Lake Apopka to endocrine disruption in juvenile alligators

Level	Evidence
Population	The juvenile alligator population in Lake Apopka has decreased.
Organism	Juvenile Apopka females have elevated circulating levels of estradiol-17β.
	Juvenile Apopka males have depressed circulating concentrations of testosterone.
Tissue/organ	Juvenile Apopka females have altered gonad aromatase activity.
	Juvenile Apopka males have poorly organized seminiferous tubules.
	Juvenile Apopka males have reduced penis size.
	Testes from juvenile Apopka males have elevated estradiol (estrogen) production.
Cellular	Juvenile Apopka females have polyovular follicles that are characteristic of estrogen excess.
Molecular	Many contaminants bind the alligator estrogen receptors and progesterone receptor.
	Many of these contaminants do not bind to the alligator cytosol proteins that blockade excess hormones.

Source: After Crain and Guillette 1998.

Developmental Biology Meets the Real World

In a recent review of insect development, Fred Nijhout (1999) concluded,

> *A single genotype can produce many phenotypes, depending on many contingencies encountered during development. That is, the phenotype is an outcome of a complex series of developmental processes that are influenced by environmental factors as well as by genes.*

Development usually occurs in a rich environmental milieu, and most animals are sensitive to environmental cues. As we have seen in this chapter the environment may determine sexual phenotype, may induce remarkable structural and chemical adaptations according to the season, may induce specific morphological changes that allow an individual to escape predation, and can induce caste determination in insects. The environment can also alter the structure of our neurons and the specificity of our immunocompetent cells. Unfortunately, the environment can also be the source of chemicals that disrupt normal developmental processes.

This concept that the environment is critical in phenotype production has many implications. First, the developmental plasticity of the nervous system assures that each person is an individual. Our brain adds experience to endowment. Fears that cloning could produce "thousands of Hitlers," for example, are unfounded. Not only have no genes been identified for bigotry, demagoguery, oratory skill, or political canniness, but one would have to reconstruct Hitler's personal, social, and political milieus to even come close to replicating the dictator's personality. Wolpe (1997) has pointed out that thinking that a genetically identical clone of Hitler would become a bigoted dictator is buying into the same genetic essentialism that made Hitler so evil. Similarly,

Gould (1997) pointed out that even Eng and Chang Bunker, the well-publicized conjoined twins who most likely shared all their genes and certainly shared their environment, became very different people. One was cheerful and abstained from all liquor. The other was morose and alcoholic (which was a problem, since they shared the same liver). The plasticity of the nervous system enables us to be individuals and "allows us to escape the tyranny of our genes" (Childs 1999). As mentioned in Chapter 4, even sheep cloned from the same embryo (and therefore having the same genome) can be very different from one another. We can therefore give a definite answer to the question posed by Wolpert (p. 51) in 1994:

> *Will the egg be computable? That is, given a total description of the fertilized egg—the total DNA sequence and the location of all proteins and RNA—could one predict how the embryo will develop?*

The answer has to be, "No." One cannot reduce phenotype completely to the inherited genes. Experience must be added to endowment.

The idea that one's phenotype is controlled in part by other species (as in developmental symbioses and predator-induced polyphenisms) certainly links us to our environment. Our "self" is partially constructed by "others," and co-development may be as important a concept as coevolution. We must learn to look for developmental signals coming not only from within but also from outside the organism. Our methods might have to expand. If animals in the wild develop differently than they would in our laboratories, we may have to go outdoors to study developmental phenomena—certainly not the standard operating procedure of past research. Relyea and Mills (2001) found that, while certain pesticides were harmless to tadpoles in the laboratory, the same concentrations of pesticide were deadly when the tadpoles underwent predator-induced responses.

Deformed Frogs

Throughout the United States and southern Canada, there is a dramatic increase in the number of deformed frogs and salamanders in what seem to be pristine woodland ponds. In some local ponds, it is estimated that 60% of certain species of amphibians have visible malformations (Figure 22.24A; Ouellet et al. 1997; NARCAM 2002). These deformities include extra or missing limbs, missing or misplaced eyes, deformed jaws, and malformed hearts and guts. In recent years, three of the hypotheses that have been put forward to explain the phenomenon appear to be gaining acceptance.

The first hypothesis proposes a combination of causes. The proximate cause of limb deformities may be the infection of the larval limb buds by trematode parasites (Stopper et al. 2002; see Figure 16.3). Under most conditions, trematode larvae can be destroyed by the tadpole's immune system. In some ponds, however, it appears that tadpoles have acquired an immune deficiency syndrome. Kiesecker (2002) has shown that frogs in habitats contaminated by certain pesticides are less able to resist parasite infection. He found that neither pesticides alone nor trematodes alone were sufficient to cause limb deformities in wild populations of frogs, but that the combination of the two factors resulted in significant proportions of the frog population showing limb deformities. At least in some species, trematode infestation has to be coupled with pollutants in order for the limb anomalies to be observed. The pollutants may be suppressing the tadpole's immune resistance to the trematode larvae.

But some lakes containing high proportions of malformed frogs do not appear to be infested with trematodes. A second hypothesis proposes that ultraviolet radiation can cause these malformations. Ankley and colleagues (2002) have shown that whereas UV irradiation can cause the death of amphibian embryos (see Chapter 3), those individuals that survive have high proportions of limb deformities. The spectrum of limb abnormalities seen in frogs that develop from UV-exposed embryos and tadpoles is somewhat different than that seen in natural populations (Meteyer et al. 2000), but this explanation may account for some of limb abnormalities.

There are other (nonlimb) malformations that do not seem to be explainable by this route, however, and a third possibility is that retinoid-like pesticides are causing physiological disruptions. The spectrum of abnormalities seen in natural populations of deformed frogs resembles some of the malformations caused by exposing tadpoles to retinoic acid (Crawford and Vincenti 1998; Gardiner and Hoppe 1999); and Ndayibagira and colleagues (2002) have shown that retinoic acid can induce all the limb malformations seen in wild populations of some species of deformed frogs. Moreover, Grün and colleagues (2002), in the same laboratory, have purified an endocrine-disrupting retinoid from the water of a lake that had a high incidence of malformed amphibians.

If retinoids are in the water, how did they get there? Certain herbicides, such as atrazine (Hayes et al. 2002, 2003; Tavera-Mendoza et al. 2002; see Figure 17.24) are thought to be responsible for gonadal malformations in many frog species, and some scientists speculate that certain pesticides sprayed for mosquito and tick control might be responsible for the limb and eye deformities (Hilleman 1996; Ouellet et al. 1997). One possibility for the activation of retinoic acid pathways involves the insecticidal compound methoprene. Methoprene is a juvenile hormone mimic that inhibits mosquito pupae from metamorphosing into adults. Because vertebrates do not have juvenile hormone, it was assumed that this pesticide would not harm fish, amphibians, or humans. Indeed, this has been found to be the case: methoprene, itself, does not have teratogenic properties. However, on exposure to sunlight, methoprene breaks down into products that have significant teratogenic activity in frogs (Figures 22.24B,C). These compounds have a structure similar to that of retinoic acid and will bind to the retinoid receptor (Har-

(A)

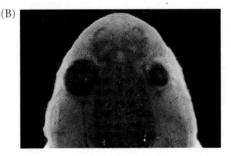

(B)

(C)

Methoprene

Sunlight

Water

Methoprenoic acid (MA)

(D)

Retinoic acid

Figure 22.24

Teratogenesis in frogs. (A) Wild green frog (*Rana clamitans*) with an eye deformity, collected in New Hampshire in 1999 by K. Babbitt. (B) *Xenopus* tadpole with eye deformities caused by incubating newly fertilized eggs in water containing methoprenic acid, a by-product of methoprene. (C) One of several pathways by which methoprene can decay into teratogenic compounds such as methoprenic acid. (D) An isomer of retinoic acid showing the structural similarities to methoprenic acid. (A, photograph courtesy of K. Babbitt and K. Reed; B, photograph courtesy of J. Bantle; C, D after La Claire et al. 1998.)

mon et al. 1995; La Claire et al. 1998). When *Xenopus* eggs are incubated in water containing these compounds, the tadpoles are often malformed, and show a spectrum of deformities similar to those seen in the wild (La Claire et al. 1998).

WEBSITE 22.12 Deformed frogs and salamanders. The original observations of malformed frogs were made by public school science classes in Minnesota. Since then, considerable efforts have made to find the causes

for both the recent decline of amphibian populations and the developmental anomalies being discovered in these animals. There are several websites for the recording and analysis of amphibian malformations.

Indeed, recent events may necessitate the enlistment of developmental biology in the service of conservation biology, a field where it has not been active. The pesticide studies of Hayes, Relyea and Mills, LeClaire, and others indicate that anthropogenic chemicals may be depleting natural populations by killing embryos or fetuses, or by rendering adults unable to mate. The studies of Morreale and colleagues (1982) show that the practices of conservation biology must incorporate an understanding of the way development works in the wild.

The study of environmental regulation of development also opens up a whole new world to developmental biologists. Ecologists know numerous stories of developmental plasticity—animals obtaining different jaw morphologies with dif-

ferent diets (Corruccini and Beecher 1982; Stearns 1989; Hegrenes 2001), animals resorbing organs under certain conditions (Piersma and Gill 1998), and frogs that can change their development within minutes of a predator's presence (Warkentin 1995, 2000). The proximate causation of these incredible developmental changes has not been studied.

By returning to the "real world," developmental biology is becoming the science that integrates ecology, evolution, genetics, cell biology, and physiology. In 1973, evolutionary biologist Leigh Van Valen claimed that evolution can be defined as "the control of development by ecology." We have reached a point where we can now begin to study the mechanisms by which this happens.

Snapshot Summary: The Environmental Regulation of Development

1. Development is sometimes cued to normal circumstances that the organism can expect to find in its environment. The larvae of many marine species do not begin metamorphosis until they find a suitable substrate. In other instances, symbiotic relationships between two or more species are necessary for the complete development of one or more of the species.

2. Developmental plasticity makes it possible for environmental circumstances to elicit different phenotypes from the same genotype.

3. Some species exhibit polyphenisms, in which distinctly different phenotypes are evoked by different environmental cues. Many species have a broad reaction norm, wherein the genotype can respond in a graded way to environmental conditions.

4. Seasonal cues such as photoperiod, temperature, or type of food can alter development in ways that make the organ-

ism more fit under the conditions it encounters. Changes in temperature also are responsible for determining sex in several organisms, including many reptiles and fish.

5. Predator-induced polyphenisms have evolved such that prey species can respond morphologically to the presence of a specific predator. In some instances, this induced adaptation can be transmitted to the progeny of the prey.

6. The differentiation of immunocompetent cells and the formation of synapses in the visual system are examples of experience influencing the phenotype.

7. Numerous compounds present in the environment may act as hormone mimics or antagonists. These compounds may disrupt normal development by interfering with the endocrine system.

8. Genetic differences can predispose individuals to being affected differently by teratogens.

Literature Cited

Abouheif, E. and G. A. Wray. 2002. Evolution of the gene network underlying wing polyphenism in ants. *Science* 297: 249–252.

Adler, F. R. and C. D. Harvell. 1990. Inducible defenses, phenotypic variability, and biotic environments. *Trends Ecol. Evol.* 5: 407–410.

Agrawal, A. A., C. Laforsch and R. Tollrian. 1999. Transgenerational induction of defenses in animals and plants. *Nature* 401: 60–63.

Alvarez-Buylla, A., J. R. Kirn and F. Nottebohm. 1990. Birth of projection neurons in adult avian brain may be related to perceptual or motor learning. *Science* 249: 1444–1446.

Ankley, G. T., S. A. Diamond, J. E. Tietge, G. W. Holcombe, K. M. Jensen, D. L. Defoe and R. Peterson. 2002. Assessment of risk of solar ultraviolet radiation to amphibians. I. Dose-dependent induction of hindlimb malformations in the northern leopard frog (*Rana pipiens*). *Environ. Sci. Technol.* 36: 2853–2858.

Antonini, A. and M. P. Stryker. 1993. Rapid remodeling of axonal arbors in the visual cortex. *Science* 260: 1818–1821.

Bachmann, M. D., R. G. Carlton, J. M. Burkholder and R. G. Wetzel. 1986. Symbiosis between salamander eggs and green algae: Microelectrode measurements inside eggs demonstrate effects of photosynthesis on oxygen concentrations. *Can. Zool.* 64: 1586–1588.

Bard, J. B. L. 1990. Traction and the formation of mesenchymal condensations in vivo. *BioEssays* 12: 389–395.

Baxter, G. T. and D. E. Morse. 1992. Cilia from abalone larvae contain a receptor-dependent G-protein transduction system similar to that in mammals. *Biol. Bull.* 183: 147–154.

Beck, S. D. 1980. *Insect Photoperiodism*, 2nd Ed. Academic Press, New York.

Bigsby, R. and 8 others. 1999. Evaluating the effects of endocrine disrupters on endocrine function during development. *Environ. Health Perspect.* [Suppl.] 107: 613–618.

Black, J. E., K. R. Issacs, B. J. Anderson, A. A. Alcantara and W. T. Greenough. 1990. Learning causes synaptogenesis, whereas motor activity causes angiogenesis, in cerebellar cortex of adult rats. *Proc. Natl. Acad. Sci. USA* 87: 5568–5572.

Blaustein, A. and L. K. Beldin. 2003. Amphibian defense against UV-B radiation. *Evol. Dev.* In press.

Bolker, J. A. 1995. Model systems in developmental biology. *BioEssays* 17: 451–455.

Borovsk, D., D. A. Carlson, P. R. Griffin, J. Shabanowitz and D. F. Hunt. 1990. Mosquito oostatic factor: A novel decapeptide modulating trypsin-like enzyme biosynthesis in the midgut. *FASEB J.* 4: 3015–3020.

Brakefield, P. M. and N. Reitsma. 1991. Phenotypic plasticity, seasonal climate, and the population biology of *Bicyclus* butterflies (Satyridae) in Malawi. *Ecol. Entomol.* 16: 291–303.

Brakefield, P. M. and 7 others. 1996. Development, plasticity, and evolution of butterfly eyespot patterns. *Nature* 384: 236–242.

Brian, M. V. 1974. Caste differentiation in *Myrmica rubra*: The role of hormones. *J. Insect Physiol.* 20: 1351–1365.

Brian, M. V. 1980. Social control over sex and caste in bees, wasps and ants. *Biol. Rev.* 55: 379–415.

Brönmark, C. and L. Pettersson. 1994. Chemical cues from piscivores induce a change in morphology in crucian carp. *Oikos* 70: 396–402.

Brunetti, C. R., J. E. Selegue, A. Monteiro, V. French, P. M. Brakefield and S. B. Carroll. 2001. The generation and diversification of butterfly eyespot color patterns. *Curr. Biol.* 11: 1578–1585.

Bry, L., P. G. Falk and J. L. Gordon. 1996. Genetic engineering of carbohydrate biosynthetic pathways in transgenic mice demonstrates cell cycle-associated regulation of glycoconjugate production in small intestinal epithelial cells. *Proc. Natl. Acad. Sci. USA* 93: 1161–1166.

Burnett, F. M. 1959. *The Clonal Selection Theory of Immunity*. Vanderbilt University Press, Nashville.

Carlsen, E., A. Giwercman, N. Keiding and N. E. Skakkebaek. 1992. Evidence for decreasing quality of semen during past 50 years. *Brit. Med. J.* 305: 609–613.

Carson, R. 1962. *Silent Spring*. Houghton-Mifflin, New York.

Cebra, J. J. 1999. Influences of microbiota on intestinal immune system development. *Am. J. Clin. Nutr.* 69 (Supplement). 1046S–1051S.

Charnov, E. L. and J. J. Bull. 1977. When is sex environmentally determined? *Nature* 266: 828–830.

Cheek, A. O., K. Kow, J. Chen and J. A. McLachlan. 1999. Potential mechanisms of thyroid disruption in humans: Interaction of organochlorine compounds with thyroid receptor, transthyretin, and thyroid-binding globulin. *Environ. Health Perspect.* 107: 273–278.

Childs, B. 1999. *Genetic Medicine: A Logic of Disease*. John Hopkins University Press.

Cohen, C. S. and R. R. Strathmann. 1996. Embryos at the edge of tolerance: Effects of environment and structure of egg masses on supply of oxygen to embryos. *Biol. Bull.* 190: 8–15.

Colello, R. J. and R. W. Guillery. 1990. The early development of retinal ganglion cells with uncrossed axons in the mouse: Retinal position and axon course. *Development* 108: 515–523.

Colman, H., J. Nabekura and J. W. Lichtman. 1997. Alterations in synaptic strength preceding axon withdrawal. *Science* 275: 356–361.

Conover, D. O. and S. W. Heins. 1987. Adaptive variation in environmental and genetic sex determination in a fish. *Nature* 326: 496–498.

Cooke, A. S. 1973. Shell thinning in avian eggs by environmental pollutants. *Environ. Pollut.* 4: 85–152.

Corruccini, R. S. and C. L. Beecher. 1982. Occlusal variation related to soft diet in a nonhuman primate. *Science* 218: 74–76.

Crain, D. A. and L. L. Guillette, Jr. 1998. Reptiles as models of contaminant–induced endocrine disruption. *Anim. Reprod. Sci.* 53: 77–86.

Crain, D. A., N. Noriega, P. M. Vonier, S. F. Arnold, J. A. McLachlan and L. J. Guillette Jr. 1998. Cellular bioavailability of natural hormones and environmental contaminants as a function of serum and cytosolic binding factors. *Toxicol. Ind. Health* 14: 261–273.

Crair, M. C., D. C. Gillespie and M. P. Stryker. 1998. The role of visual experience in the development of columns in the cat visual cortex. *Science* 279: 566–570.

Crawford, K. and D. M. Vincenti. 1998. Retinoic acid and thyroid hormone may function through similar and competitive pathways in regenerating axolotls. *J. Exp. Zool.* 282: 724–738.

Davis, D. L., H. L. Bradlow, M. Wolff, T. Woodruff, D. G. Hoel and H. Anton-Culver. 1993. Xenoestrogens as preventable causes of breast cancer. *Environ. Health Perspect.* 101: 372–377.

Dedeinde, F., F. Vavre, F. Fleury, B. Loppin, M. E. Hochberg and Boulétreau. 2001. Removing symbiotic *Wolbachia* bacteria specifically inhibits oogenesis in a parasitic wasp. *Proc. Natl. Acad. Sci. USA* 98: 6247–6252.

Degnan, B. M. and D. E. Morse. 1995. Developmental and morphogenetic gene regulation in *Haliotis rufescens* larvae at metamorphosis. *Am. Zool.* 35: 391–398.

Denno, R. F., L. W. Douglass and D. Jacobs. 1985. Crowding and host plant nutrition: Environmental determinants of wing form in *Prokelisia marginata*. *Ecology* 66: 1588–1596.

Denver, R. J. 1999. Evolution of the corticotropin-releasing hormone signaling system and its role in stress-induced phenotypic plasticity. *Neuropeptides: Structure and Function in Biology and Behavior. Ann. N.Y. Acad. Sci.* 897: 46–53.

Denver, R. J., N. Mirhadi and M. Phillips. 1998. Adaptive plasticity in amphibian metamorphosis: Response of *Scaphiopus hammondii* tadpoles to habitat desiccation. *Ecology* 79: 1859–1872.

Dodson, S. 1989. Predator-induced reaction norms. *BioScience* 39: 447–452.

Evans, J. D. and D. E. Wheeler. 1999. Differential gene expression between developing queens and workers in the honey bee, *Apis mellifera*. *Proc. Natl. Acad. Sci. U.S.A.* 96: 5575–5580.

Evans, J. D. and D. E. Wheeler. 2001. Expression profiles during honeybee caste determination. *Genom. Biol.* 2: 1001.1–1001.6.

Fallon, A. M., H. H. Hagedorn, G. R. Wyatt and H. Laufer. 1974. Activation of vitellogenin synthesis in the mosquito *Aedes aegypti* by ecdysone. *J. Insect Physiol.* 26: 829–1823.

Ferguson, M. W. J. and T. Joanen. 1982. Temperature of egg incubation determines sex in *Alligator mississippiensis*. *Nature* 296: 850–853.

Gardiner, D. M. and D. M. Hoppe. 1999. Environmentally induced limb malformations in mink frogs (*Rana septentrionalis*). *J. Exp. Zool.* 284: 207–216.

Gilbert, S. F. 1994. Dobzhansky, Waddington, and Schmalhausen: Embryology and the Modern Synthesis. In M. B. Adams (ed.), *The Evolution of Theodosius Dobzhansky: Essays on His Life and Thought in Russia and America*. Princeton University Press, Princeton, pp. 143–154.

Gilbert, S. F. 2001. Ecological developmental biology: Developmental biology meets the real world. *Dev. Biol.* 233: 1–12.

Gil-Turnes, M. S., M. E. Hay and W. Fenical. 1989. Symbiotic marine bacteria chemically defend crustacean embryos from a pathogenic fungus. *Science* 246: 116–118.

Godwin, J., R. Sawby, R. R. Warner, D. Crews and M. S. Grober. 2000. Hypothalamic arginine vasotocin mRNA abundance variation across sexes and with sex change in a coral reef fish. *Brain Behav. Evol.* 55: 77–84.

Godwin, J., J. A. Luckenbach and R. J. Borski. 2003. Ecology meets endocrinology: Environmental sex determination in fishes. *Evol. Dev.* 5: 40–49.

Gotthard, K. and S. Nylin. 1995. Adaptive plasticity and plasticity as an adaptation: A selective review of plasticity in animal morphology and life history. *Oikos* 74: 3–17.

Gould, E., A. Beylin, P. Tanapat, A. Reeves and T. J. Shors. 1999. Learning enhances adult neurogenesis in the hippocampal formation. *Nature Neurosci.* 2: 260–265.

Gould, S. J. 1997. Individuality. *Sciences* 37: 14–16.

Grober, M. S. and T. Sunobe. 1996. Serial adult sex change involves rapid and reversible changes in forebrain neurochemistry. *Neuroreport* 7: 2945–2949.

Grün, F., A. Ndayibagira, D. Gardiner, D. Hoppe and B. Blumberg. 2002. HPLC purification of an endocrine disrupting retinoid activity from a Minnesota lake with a high incidence of malformed amphibians. Gordon Conference on Environmental Endocrine Disruptors, p. 1. Quoted with permission.

Guillette, L. J., T. S. Gross, G. R. Masson, J. M. Matter, H. F. Percival and A. R. Woodward. 1994. Developmental abnormalities of the gonad and abnormal sex hormone concentrations in juvenile alligators from contaminated and control lakes in Florida. *Environ. Health Perspect.* 102: 680–688.

Hadfield, M. G. 1977. Metamorphosis in marine molluscan larvae: An analysis of stimulus and response. In R.-S. Chia and M. E. Rice (eds.), *Settlement and Metamorphosis of Marine Invertebrate Larvae.* Elsevier, New York, pp. 165–175.

Hagedorn, H. H. 1983. The role of ecdysteroids in the adult insect. In G. Downer and H. Laufer (eds.), *Endocrinology of Insects.* Alan R. Liss, New York, pp. 241–304.

Hardie, J. 1981. Juvenile hormone and photoperiodically controlled polymorphism in *Aphis fabae*: Postnatal effects on presumptive gynoparae. *J. Insect Physiol.* 27: 347–355.

Hardie, J. and A. D. Lees. 1985. Endocrine control of polymorphism and polyphenism. In G. A. Kerkut and L. I. Gilbert (eds.), *Comprehensive Insect Physiology, Biochemistry, and Pharmacology,* vol. 8. Pergamon Press, Oxford, pp. 441–490.

Harmon, M. A., M. F. Boehm, R. A. Heyman and D. J. Mangelsdorf. 1995. Activation of mammalian retinoid-X receptors by the insect growth regulator methoprene. *Proc. Natl. Acad. Sci. USA* 92: 6157–6160.

Hart, M. W. and R. R. Strathmann. 1994. Functional consequences of phenotypic plasticity in echinoid larvae. *Biol. Bull.* 186: 291–299.

Harvell, C. W. 1986. The ecology and evolution of inducible defences in a marine bryozoan: Cues, costs, and consequences. *Am. Nat.* 128: 810–823.

Harvell, C. W. 1999. Complex biotic environments: Coloniality and the hereditary variation for inducible defences. In R. Tollrian and C. D. Harvell (eds.), The *Ecology and Evolution of Inducible Defenses.* Princeton University Press, Princeton, NJ, pp. 231–244.

Hayes, T. B., A. Collins, M. Lee, M. Mendoza, N. Noriega, A. A. Stuart and A. Vonk. 2002. Hermaphroditic, demasculinized frogs after exposure to the herbicide atrazine at low ecologically relevant doses. *Proc. Natl. Acad. Sci. USA* 99: 5476–5480.

Hayes, T. B., K. Haston, M. Tsui, A. Hoag, C. Haeffele and A. Vonk. 2003. Atrazine-induced hermaphroditism at 0,1 ppb in Americal leopard frogs (*Rana pipiens*): Laboratory and field evidence. *Environ. Health Res.* Doi: 10.1289/ehp.59232. In press.

Hegrenes, S. 2001. Diet-induced phenotypic plasticity of feeding morphology in the orange-spotted sunfish, *Lepomis humilis. Ecol. Freshw. Fish.* 10: 35–42.

Hill, M. 1997. *Understanding Environmental Pollution.* Cambridge University Press, Cambridge.

Hilleman, B. 1996. Frog deformities pose a mystery. *Chem. Engnr. News* 74: 24.

Hoffmann, R. J. 1973. Environmental control of seasonal variation in the butterfly *Colias eurytheme.* I. Adaptive aspects of a photoperiodic response. *Evolution* 27: 387–397.

Hooper, L. V., L. Bry, P. G. Falk and J. I. Gordon. 1998. Host-microbial symbiosis in the mammalian intestine: Exploring an internal ecosystem. *BioEssays* 20: 336–343.

Hooper, L. V., M. H. Wong, A. Thelin, L. Hansson, P. G. Falk and J. I. Gordon. 2001. Molecular analysis of commensal host-microbial relationships in the intestine. *Science* 291: 881–884.

Hornung, M. W., E. W. Zabel and R. E. Peterson. 1996. Toxic equivalency factors of polybrominated dibenzo-*p*-dioxin, dibenzofuran, biphenyl, and polyhalogenated diphenyl ether congeners based on rainbow trout early life stage mortality. *Toxicol. Appl. Pharmacol.* 140: 227–234.

Huang, Z and 7 others. 1999. BDNF regulates the maturation of inhibition and the critical period of plasticity in mouse visual cortex. *Cell* 98: 739–755.

Hubel, D. H. 1967. Effects of distortion of sensory input on the visual system of kittens. *Physiologist* 10: 17–45.

Hubel, D. H. and T. N. Wiesel. 1962. Receptive fields, binocular interaction and functional architecture in the cat's visual cortex. *J. Physiol.* 160: 106–154.

Hubel, D. H. and T. N. Wiesel. 1963. Receptive fields of cells in striate cortex of very young, visually inexperienced kittens. *J. Neurophysiol.* 26: 944–1002.

Ikegame, M., O. Ishibashi, T. Yoshizawa, J. Shimomura, T. Komori, H. Ozawa, and H. Kawashima. 2001. Tensile stress induces bone morphogenetic protein 4 in preosteoblastic and fibroblastic cells, which later differentiate into osteoblasts leading to osteogenesis in the mouse calvariae in organ culture. *J. Bone Miner. Res.* 16: 24–32.

Janzen, F. J. and G. L. Paukstis. 1991. Environmental sex determination in reptiles: Ecology, evolution, and experimental design. *Q. Rev. Biol.* 66: 149–179.

Johnson, R. D. and 10 others. 1998. Toxicity of 2,3,7,8-tetrachlorodibenzo-*p*-dioxin to early life stage brook trout (*Salvelinus fontinalis*) following parental dietary exposure. *Environ. Toxicol. Chem.* 17: 2408–2421.

Katz, L. C. 1999. What's critical for the critical period in the visual cortex? *Cell* 99: 673–676.

Keiding, N and N. E. Skakkebaek. 1993. Are estrogens involved in falling sperm counts and disorders of the male reproductive tract? *Lancet* 341: 1392–1395.

Kelce, W. R., C. R. Stone, S. C. Laws, L. E. Gray, J. A. Kemppainen and E. M. Wilson. 1995. Persistent DDT metabolite p,p′-DDE is a potent androgen receptor antagonist. *Nature* 375: 581–585.

Kempermann, G., H. G. Kuhn and F. H. Gage. 1997a. More hippocampal neurons in adult mice living in an enriched environment. *Nature* 386: 493–495.

Kempermann, G., H. G. Kuhn and F. H. Gage.1997b. Genetic influence on neurogenesis in the dentate gyrus of adult mice. *Proc. Natl. Acad. Sci. USA* 94: 10409–10414.

Kennedy, C., S. Suda, C. B. Smith, M. Miyaoka, M. Ito and L. Sokoloff. 1981. Changes in protein synthesis underlying functional plasticity in immature monkey visual system. *Proc. Natl. Acad. Sci. USA* 78: 3950–3953.

Kiesecker, J. M. 2002. Synergism between trematode infection and pesticide exposure: A link to amphibian limb deformities in nature? *Proc. Natl. Acad. Sci. USA* 99: 9900–9904.

Koch, P. B. and D. Bückmann, 1987. Hormonal control of seasonal morphs by the timing of ecdysteroid release in *Arachnia levana* L. (Nymphalidae: Lepidoptera). *J. Insect Physiol.* 33: 823–929.

Koch, P. B., P. M. Brakefield and F. Kesbeke. 1996. Ecdysteroids control eyespot size and wing color pattern in the polyphenic butterfly *Bicyclus anynana* (Lepidoptera: Satyridae). *J. Insect Physiol.* 42: 223–230.

La Claire, J. J., J. A. Bantle and J. Dumont. 1998. Photoproducts and metabolites of a common insect growth regulator produce developmental deformities in *Xenopus. Environ. Sci. Technol.* 32: 1453–1461.

Matter, J. M. and 8 others. 1998. Effects of endocrine-disrupting contaminants in reptiles: Alligators. In R. J. Kendall, R. L. Dickerson, J. P. Geisy and W. A. Suk (eds.), *Principles and Processes for Evaluating Endocrine Disruptions in Wildlife.* SETAC Press, Pensacola, FL, pp. 267–289.

McCollum, S. A. and J. Van Buskirk. 1996. Costs and benefits of a predator induced polyphenism in the gray treefrog *Hyla chrysoscelis*. *Evolution* 50: 583–593.

McFall-Ngai, M. J. 2002. Unseen forces: The influence of bacteria on animal development. *Dev. Biol.* 242: 1–14.

McFall-Ngai, M. J. and E. G. Ruby. 1991. Symbiont recognition and subsequent morphogenesis as early events in an animal-bacterial mutualism. *Science* 254: 1491–1494.

McNabb, A. and 7 others. 1999. Basic physiology. In R. T. Di Giulio and D. E. Tillitt (eds.), *Reproductive and Developmental Effects of Contaminants in Oviparous Vertebrates*. SETAC Press, Pensacola, FL, pp. 113–223.

Meteyer, C. U. and 9 others. 2000. Hind limb malformations in free-living northern leopard frogs (*Rana pipiens*) from Maine, Minnesota, and Vermont suggest multiple etiologies. *Teratology* 62: 151–171.

Montgomery, M. K. and M. J. McFall-Ngai. 1995. The inductive role of bacterial symbionts in the morphogenesis of a squid light organ. *Am. Zool.* 35: 372–380.

Morgan, T. H. 1909. Sex determination and parthenogenesis in phylloxerans and aphids. *Science* 29: 234–237.

Morreale, S. J., G. J. Ruiz, J. R. Spotila and E. A. Standora. 1982. Temperature-dependent sex determination: Current practices threaten conservation of sea turtles. *Science* 216: 1245–1247.

Morse, A. N. C., C. A. Froyd and D. E. Morse. 1984. Molecules from cyanobacteria and red algae that induce larval settlement and metamorphosis in the mollusc *Haliotis rufescens*. *Mar. Biol.* 81: 293–298.

Müller, G. B. 2003. Embryonic motility: Environmental influences on evolutionary innovation. *Evo. Dev.* 5: 56–60.

Müller, G. B. and J. Steicher. 1989. Ontogeny of the syndesmosis tibiofibularis and the evolution of the bird hindlimb: A caenogenetic feature triggers phenotypic novelty. *Anat. Embryol.* 179: 327–339.

NARCAM. 2002. North American Reporting Center for Amphibian Malformations. Northern Prairie Wildlife Research Center, Jamestown, ND. http://www.nprwc.usgs.gov/narcam.

Ndayibagira, A., F. Grün, B. Blumberg and D. M. Gardiner. 2002. Retinoids induce all the limb malformations observed in wild populations of deformed frogs. Gordon Conference on Environmental Endocrine Disruptors. P. 1 Quoted with permission.

Newman, R. A. 1989. Developmental plasticity of *Scaphiopus couchii* tadpoles in an unpredictable environment. *Ecology* 70: 1775–1787.

Newman, R. A. 1992. Adaptive plasticity in amphibian metamorphosis. *BioScience* 42: 671–678.

Nijhout, H. F. 1991. *The Development and Evolution of Butterfly Wing Patterns*. Smithsonian Institution Press, Washington, D.C.

Nijhout, H. F. 1994. *Insect Hormones*. Princeton University Press, Princeton, NJ.

Nijhout, H. F. 1999. Control mechanisms of polyphenic development in insects. *BioScience* 49: 181–192.

Nijhout, H. F. 2003. Development and evolution of adaptive polyphenisms. *Evo. Dev.* 5: 9–18.

Nordeen, K. W. and E. J. Nordeen. 1988. Projection neurons within a vocal pathway are born during song learning in zebra finches. *Nature* 334: 149–151.

Ouellet, M., J. Bonin, J. Rodriguez, J. L. DesGanges and S. Lair. 1997. Hindlimb deformities (ectromelia, ectrodactyly) in free-living anurans from agricultural habitats. *J. Wildlife Dis.* 33: 95–104.

Palanza, P., S. Parmigiani and F. S. vom Saal. 2001. Effects of prenatal exposure to low doses of diethylstilbestrol, *o'p'*-DDT, and methochlor on postnatal growth and neurobehavioral development in male and female mice. *Horm. Behav.* 40: 252–265.

Palmer, A. R. 1985. Adaptive value of shell variation in *Thais lamellosa*: Effect of thick shells on vulnerability to and preference by crabs. *Veliger* 27: 349–356.

Passera, L. 1985. Soldier determination in ants of the genus *Pheidole*. In J. A. L. Watson, B. M Okot-Kotber and C. Noirot (eds.), *Caste Determination in Social Insects*. Pergamon, Oxford, pp. 331–346.

Pechenik, J. A., D. E. Wendt and J. N. Jarrett. 1998. Metamorphosis is not a new beginning. *BioScience* 48: 901–910.

Pener, M. P. 1991. Locust phase polymorphism and its endocrine relations. *Adv. Insect Physiol.* 3: 1–79.

Pfennig, D. W. 1992a. Polyphenism in spadefoot toad tadpoles as a locally adjusted evolutionarily stable strategy. *Evolution* 46: 1408–1420.

Pfennig, D. W. 1992b. Proximate and functional causes of polyphenism in an anuran tadpole. *Funct. Ecol.* 6: 167–174.

Pickford, D. B. and I. D. Morris. 1999. Effects of endocrine-disrupting contaminants on amphibian oogenesis: Methoxychlor inhibits progesterone-induced maturation of *Xenopus laevis* oocytes in vitro. *Environ. Health Perspect.* 107: 285–292.

Piersma, T. and R. E. Gill, Jr. 1998. Guts don't fly: Small digestive organs in obese Bar-tailed Godwits. *Auk* 115: 196–203.

Pinder, A. W. and S. C. Friet. 1994. Oxygen transport in egg masses of the amphibians *Rana sylvatica* and *Ambystoma maculatum*: Convection, diffusion, and oxygen production by algae. *J. Exp. Biol.* 197: 17–30.

Plowright, R. C. and B. A. Pendrel. 1977. Larval growth in bumble-bees. *Can. Entomol.* 109: 967–973.

Purves, D. and J. W. Lichtman. 1985. *Principles of Neural Development*. Sinauer Associates, Sunderland, MA.

Raatikainen, M. 1967. Bionomics, enemies, and population dynamics of *Javesella pellucida* (F.) (Homoptera, Delphaidae). *Annales Agric. Fenniae* 6: 1–49.

Rachinsky, A. and K. Hartfelder. 1990. Corpora allata activity, a prime regulating element for caste-specific juvenile hormone titre in honey bee larvae (*Apis mellifera carnica*). *J. Insect Physiol.* 36: 329–349.

Rasika, S., A. Alvarez-Buylla and F. Nottebohm. 1999. BDNF mediates the effects of testosterone on the survival of new neurons in the adult brain. *Neuron* 22: 53–62.

Relyea, R. A. and N. Mills. 2001. Predator-induced stress makes the pesticide carbaryl more deadly to grey treefrog tadpoles (*Hyla versicolor*). *Proc. Natl. Acad. Sci. USA* 2491–2496.

Renfree, M. B. and G. Shaw. 2000. Diapause. *Annu. Rev. Physiol.* 62: 353–375.

Sander, K. 1968. Entwicklungsphysiologische Untersuchungen am embryonalen Mycetom von *Euscelis plebejus* F. (Homoptera, Ciciadina). I. *Dev. Biol.* 17: 16–38.

Sapp, J. 1994. *Evolution by Association: A History of Symbiosis*. Oxford University Press, New York.

Sato, M., T. Ochi, T. Nakase, S. Hirota, Y. Kitamura, S. Nomura and N. Yasui. 1999. Mechanical tension-stress induces expression of bone morphogenetic protein (BMP)-2 and BMP-4, but not BMP-6, and GDF-5 mRNA, during distraction osteogenesis. *J. Bone Miner. Res.* 14: 1084–1095.

Savage, D. C. 1977. Microbial ecology of the gastrointestinal tract. *Annu. Rev. Microbiol.* 31: 107–133.

Schmalhausen, I. I. 1949. *Factors of Evolution: The Theory of Stabilizing Selection*. University of Chicago Press, Chicago.

Schwemmler, W. 1974. Endosymbionts: Factors of egg patterning. *J. Insect Physiol.* 20: 1467–1474.

Schwemmler, W. 1989. Insect symbiosis as a model system for egg cell differentiation. In W. Schwemmler and G. Gassner (eds.), *Insect Endosymbiosis*. CRC Press, Boca Raton, FL, pp. 37–53.

Shapiro, A. M. 1968. Photoperiodic induction of vernal phenotype in *Pieris protodice* Boisduval and Le Conta (Lepidoptera: Pieridae). *Wasmann J. Biol.* 26: 137–149.

Shapiro, A. M. 1976. Seasonal polyphenism. *Evol. Biol.* 9: 259–333.

Shapiro, A. M. 1978. The evolutionary significance of redundancy and variability in phenotypic induction mechanisms of pierid butterflies (Lepidoptera). *Psyche* 85: 275–283.

Silva, E., N. Rajapakse and A. Kortenkamp. 2002. Something from "nothing": Eight weak estrogenic chemicals combined at concentration below NOECs produce significant mixture effects. *Environ. Sci. Technol.* 36: 1751–1756.

Spearow, J. L., P. Doemeny, R. Sera, R. Leffler and M. Barkley. 1999. Genetic variation in susceptibility to endocrine disruption by estrogen in mice. *Science* 285: 1259–1261.

Stappenbeck, T. S., L. V. Hooper and J. I. Gordon. 2002. Developmental regulation of intestinal angiogenesis by indigenous microbes via Paneth cells. *Proc. Natl. Acad. Sci. USA* 99: 15451–15455.

Stearns, S. C. 1989. The evolutionary significance of phenotypic plasticity: Phenotypic sources of variation among organisms can be described by developmental switches and reaction norms. *BioScience* 30: 436–446.

Stearns, S. C., G. de Jong and R. A. Newman. 1991. The effects of phenotypic plasticity on genetic correlations. *Trends Ecol. Evol.* 6: 122–126.

Stone, R. 1994. Environmental estrogens stir debate. *Science* 265: 308–310.

Stone, R. 1995. Environmental toxicants under scrutiny at Baltimore meeting. *Science* 267: 1770–1771.

Stopper, G. F., L. Hecker, R. A. Franssen and S. K. Sessions. 2002. How trematodes cause limb deformities in amphibians. *J. Exp. Zool.* 294: 252–263.

Strathmann, R. R. and M. F. Strathmann. 1995. Oxygen supply and limits on aggregation of embryos. *J. Mar. Biol. Assoc. UK* 75: 413–428.

Strathmann, R. R., L. Fenaux and M. F. Strathmann. 1992. Heterochronic developmental plasticity in larval sea urchins and its implication for evolution of nonfeeding larvae. *Evolution* 46: 972–986.

Swan, S. H., E. P. Elkin and L. Fenster. 1997. Have sperm densities declined? A reanalysis of global trend data. *Environ. Health Perspect.* 105:1228–1232.

Takahashi, I. and 7 others. 1998. Compressive force promotes *sox9*, type II collagen and aggrecan and inhibits IL-1β expression resulting in chondrogenesis in mouse embryonic limb bud mesenchymal cells. *J. Cell. Sci.* 111: 2067–2076.

Tavera-Mendoza, L., S. Ruby, P. Brousseau, M. Fournier, D. Cyr and D. Marcogliese. 2002. Response of the amphibian tadpole *Xenopus laevis* to atrazine during sexual differentiation of the ovary. *Environ. Toxicol. Chem.* 21: 1264–1267.

Tawfik, A. I. and 9 others. 1999. Identification of the gregarization-associated dark-pigmentotropin in locusts through an albino mutant. *Proc. Natl. Acad. Sci USA* 96: 7083–7087.

Tevini, M. 1993. *UV-B Radiation and Ozone Depletion: Effects on Humans, Animals, Plants, Mi-*

croorganisms and Materials. Lewis Publishers, Boca Raton, FL.

Tollrian, R. and C. D. Harvell. 1999. *The Ecology and Evolution of Inducible Defenses.* Princeton University Press, Princeton, NJ.

Tramontin, A. D., V. N. Hartman and E. A. Brenowitz. 2000. Breeding conditions induce rapid and sequential growth in adult avian song control circuits: A model for seasonal plasticity in the brain. *J. Neurosci.* 20: 854–861.

Turner, A. M. and W. T. Greenough. 1983. Synapses per neuron and synaptic dimensions in occipital cortex of rats reared in complex, social, or isolation housing. *Acta Stereologica* 2 [Suppl. 1]: 239–244.

Umesaki, Y. 1984. Immunohistochemical and biochemical demonstration of the change in glycolipid composition of the intestinal epithelial cell surface in mice in relation to epithelial cell differentiation and bacterial association. *J. Histochem. Cytochem.* 32: 299–304.

Van Buskirk, J. and R. A. Relyea. 1998. Natural selection for phenotypic plasticity: Predator-induced morphological responses in tadpoles. *Biol. J. Linn. Soc.* 65: 301–328.

van der Weele, C. 1999. *Images of Development: Environmental Causes in Ontogeny.* SUNY Press, Albany, NY.

van Praag, H., G. Kempermann and F. H. Gage. 1999. Running increases cell proliferation and neurogenesis in the adult mouse dentate gyrus. *Nature Neurosci.* 2: 266–270.

Van Valen, L. 1973. Festschrift. *Science* 180: 488.

Via, S., R. Gomulkiewicz, G. De Jong, S. M. Scheiner, C. D. Schlichting and P. H. Van Tienderen. 1995. Adaptive phenotypic plasticity: Consensus and controversy. *Trends Ecol. Evol.* 10: 212–217.

Waddington, C. H. 1942. Canalization of development and the inheritance of acquired characteristics. *Nature* 150: 563–565.

Warkentin, K. M. 1995. Adaptive plasticity in hatching age: A response to predation risk trade-offs. *Proc. Natl. Acad. Sci. USA* 92: 3507–3510.

Warkentin, K. M. 2000. Wasp predation and wasp-induced hatching of red-eyed treefrog eggs. *Anim. Behav.* 60: 503–510.

Warner, R. R. 1984. Mating behavior and hermaphroditism in coral reef fishes. *Am. Sci.* 72: 382–389.

Watt, W. B. 1968. Adaptive significance of pigment polymorphism in *Colias* butterflies. I. Variation of melanin in relation to thermoregulation. *Evolution* 22: 437–458.

Watt, W. B. 1969. Adaptive significance of pigment polymorphism in *Colias* butterflies. II. Thermoregulation and periodically controlled melanin production in *Colias eurytheme. Proc. Natl. Acad. Sci. USA* 63: 767–774.

West-Eberhard, M. J. 1989. Phenotypic plasticity and the origins of diversity. *Annu. Rev. Ecol. Syst.* 20: 249–278.

Wheeler, D. 1991. The developmental basis of worker caste polymorphism in ants. *Am. Nat.* 138: 1218–1238.

Wiesel, T. N.1982. Postnatal development of the visual cortex and the influence of environment. *Nature* 299: 583–591.

Wirtz, P. 1973. Differentiation in the honeybee larva. *Meded. Landb. Hogesch. Wagningen* 73–75, 1–66.

Wolpe, P. R. 1997. If I am only my genes, what am I? *Kennedy Inst. Ethics J.* 7: 213–230.

Wolpert, L. 1994. Do we understand development? *Science* 266: 571–572.

Woltereck, R. 1909. Weitere experimentelle Untersuchungen über Artveränderung, speziell über das Wesen quantitativer Artunderscheide bei Daphniden. *Versuch. Deutsch. Zool. Ges.* 1909: 110–172.

Woodward, D. E. and J. D. Murray. 1993. On the effect of temperature-dependent sex determination on sex ratio and survivorship in crocodilians. *Proc. R. Soc. Lond.* [B] 252: 149–155.

Wu, K. C., J. Streicher, M. L. Lee, B. I. Hall and G. B. Müller. 2001. Role of motility in embryonic development: I. Embryo movements and amnion contractions in the chick and the influence of illumination. *J. Exp. Zool.* 291: 186–194.

Wu, Q., Y. Zhang and Q. Chen. 2001. *Indian hedgehog* is an essential component of mechanotransduction complex to stimulate chondrocyte proliferation. *J. Biol. Chem.* 276: 35290–35296.

Zabel, E. W. and R. E. Peterson. 1996. TCDD-like activity of 2,3,6,7-tetrachloroxanthene in rainbow trout early life stages and in a rainbow trout gonadal cell line (RTG-2). *Environ. Toxicol. Chem.* 15: 2305–2309.

chapter 23 · Developmental mechanisms of evolutionary change

WHEN WILHELM ROUX ANNOUNCED the creation of experimental embryology in 1894, he broke many of the ties that linked embryology to evolutionary biology. However, he promised that embryology would someday return to evolutionary biology, bringing with it new knowledge of how animals were generated and how evolutionary changes might occur. He stated that "an ontogenetic and a phylogenetic developmental mechanics are to be perfected." Roux thought that research into the developmental mechanics of individual embryos (the ontogenetic branch) would proceed faster than the phylogenetic (evolutionary) branch, but he predicted that "in consequence of the intimate causal connections between the two, many of the conclusions drawn from the investigation of individual development [would] throw light on the phylogenetic processes." A century later, we are now at the point of fulfilling Roux's prophecy. Developmental biology is returning to evolutionary biology, forging a new discipline, **evolutionary developmental biology**, sometimes called "evo-devo." This return is producing a new model of evolution that integrates both developmental genetics and population genetics to explain the diversity of life on earth.

The fundamental principle of this new evolutionary synthesis is that evolution is caused by heritable changes in the development of organisms. This view can be traced back to Darwin, and it is compatible with and complementary to the view of evolution based on population genetics that evolution is caused by changes in gene frequency between generations. The merging of the developmental genetic approach to evolution with the population genetic approach is creating a more complete evolutionary biology that is beginning to explain the origin of both species and higher taxa (Raff 1996; Gerhart and Kirschner 1997; Hall 1999; Carroll et al. 2001; Wilkins 2002).

"Unity of Type" and "Conditions of Existence"

Charles Darwin's synthesis

In the nineteenth century, debates over the origin of species pitted two ways of viewing nature against each other. One view, championed by Georges Cuvier and Charles Bell, focused on the *differences* among species that allowed each species to adapt to its environment. Thus, the hand of the human, the flipper of the seal, and the wings of birds and bats were seen as marvelous contrivances, each fashioned by the Creator, to allow these animals to adapt to their "conditions of existence." The other view, championed by Étienne Geoffroy Saint-Hilaire and Richard Owen, was that "unity of type" (the *similarities* among organisms, which Owen called "homologies") was critical. The human hand, the seal's flipper, and the wings of bats and birds were all modifications of the same basic plan (see Figure 1.13). In discovering that plan, one could find the form upon which the Creator designed these animals. The adaptations were secondary.

Darwin acknowledged his debt to these earlier debates when he wrote in 1859, "It is generally acknowledged that all organic beings have been formed on two great laws—Unity of Type, and Conditions of Existence." Darwin went on to explain that his theory would explain unity of type by descent from a common ancestor. The changes creating the marvelous adaptations to the conditions of existence would be explained by natural selection. Darwin called this concept **descent with modification.** As mentioned in Chapter 1, Darwin noted that the homologies between the embryonic and larval structures of different phyla provided excellent evidence for descent with modification. He also argued that adaptations that depart from the "type" and allow an organism to survive in its particular environment develop late in the embryo. Thus, Darwin recognized two ways of looking at descent with modification. One could emphasize the *common descent* by pointing out embryonic homologies between two or more groups of animals, or one could emphasize the *modifications* by showing how development was altered to produce structures that enabled animals to adapt to particular conditions.

WEBSITE 23.1 **Lillie and Wilson.** In the late 1800s, two eminent embryologists came forth with proposals relating embryology to molluscan evolution. Wilson stressed embryological homologies as showing common descent; Lillie stressed embryological adaptations as showing natural selection. Both approaches are still operating today.

WEBSITE 23.2 **Haeckel's biogenetic law.** In the early 1900s, a fusion of evolution and embryology was wrongly interpreted to support a linear (as opposed to a branched) model of evolution. The interpretation of Ernst Haeckel was that every organism evolved by the terminal addition of a new stage to the end of the last "highest" organism. Thus, he saw the entire animal kingdom as representing truncated steps of human development.

"Life's splendid drama"

Until the present decade, "many invertebrate biologists saw the reconstruction of relationships among the phyla as an insoluble dilemma. … Indeed, as late as 1990, a comprehensive summary concluded that the relationships between most of the higher animal groups were entirely unresolved" (Erwin et al. 1997). However, in the 1990s, a broad consensus on the general form of a phylogenetic tree of life began to emerge among paleontologists, molecular biologists, and developmental geneticists (see Winnepenninckx et al. 1998; Adoutte

et al. 1999; Erwin 1999). This consensus (one representation of which is shown in Figure 23.1A) came about from (1) improved methods of using DNA for phylogenetic analysis, taking into account its variation within groups of animals; (2) new data on conserved regulatory gene sequences such as the Hox genes, which are usually stable within phyla but can diverge between phyla; (3) morphological evidence for the related nature of some structures that had once been thought to be distinct; and (4) computer programs that can sort out enormous amounts of data, without privileging any particular set of relationships over others. The results, which surprised many scientists, can be summarized as follows:

- The animal kingdom can be divided into Porifera (sponges), Cnidaria and Ctenophora (jellyfish and comb jellies), and the Bilateria. The Porifera lack any coherent epithelium or any symmetry. The Cnidaria and Ctenophora are diploblastic (with two epithelial layers, lacking mesoderm) and have radial symmetry. The Bilateria are triploblastic (with true endoderm, mesoderm, and ectoderm), and have bilateral symmetry.

- The Bilateria can be divided into two groups, the deuterostomes and the protostomes. The protostomes are those invertebrates that form their mouth from the blastopore opening; the deuterostomes form the mouth secondarily, and the blastopore marks the site of the anus. Deuterostomes include the echinoderms and the chordates. (Although the adults display radial symmetry, echinoderm larvae are originally bilateral.)

- The protostomes can be divided into two groups, the **Ecdysozoa** (animals whose bodies are covered by an exoskeleton, which therefore molt as they grow) and the **Lophotrochozoa** (animals that have most or all of their soft tissues in contact with the environment and which generally use cilia in feeding or locomotion). Each of these two groups is **monophyletic**, meaning that the phyla in each of them all share a common ancestor.

- Nematodes and flatworms (which lack true coeloms) had formerly been considered basal groups, perhaps ancestral to both the protostomes and the deuterostomes. However, the new studies have shown that the nematodes belong to the Ecdysozoa and the flatworms to the Lophotrochozoa.

WEBSITE 23.3 **The emergence of embryos.** How did individual cells come to sacrifice their individual potentials and generate embryos? How did gastrulation evolve? How did the deuterostome gut emerge from the protostome gut? The answers may involve predation and the inability to divide and be ciliated at the same time.

WEBSITE 23.4 **How taxonomic groups are classified.** The advent of cladistics has put some order into the various ways of classifying animals. This does not mean, however, that there is unanimous agreement on the results.

(A)

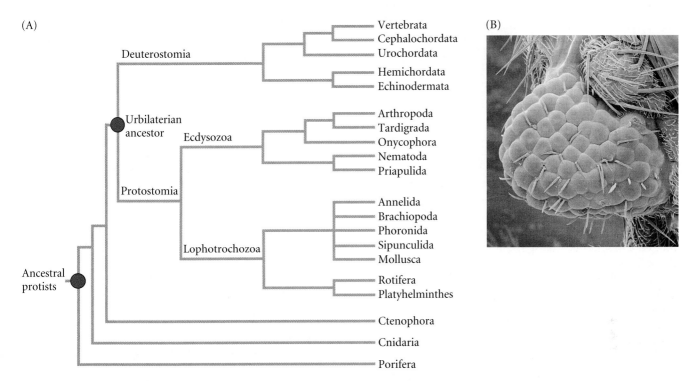

(B)

Figure 23.1
Relationships among phyla. (A) A current phylogeny and (B) evidence of the evolutionary conservation of a regulatory gene. The *Pax6* gene for eye development is an example of a gene ancestral to both protostomes and deuterostomes. The micrograph shows ommatidia emerging in the leg of a fruit fly (a protostome) in which mouse (deuterostome) *Pax6* cDNA was expressed in the leg disc. (A based on Adoutte et al. 2000; B from Halder et al. 1995, photograph courtesy of W. J. Gehring and G. Halder.)

The search for the Urbilaterian ancestor

It is doubtful that we will find a fossilized representative of the ancestral phylum that gave rise to both the deuterostomes and the protostomes. This hypothetical animal is sometimes called the **Urbilaterian ancestor** or the **PDA** (protostome-deuterostome ancestor). Since such an animal probably had neither a bony endoskeleton (a deuterostome trait) nor a hard exoskeleton (characteristic of ecdysozoans), it would not have fossilized well. However, we can undertake what Sean Carroll (quoted in DiSilvestro 1997) has called "paleontology without fossils." The logic of this approach is to find homologous genes that are performing the same functions in both a deuterostome (usually a chick or a mouse) and a protostome (generally an arthropod such as *Drosophila*). Many such genes have been found (Table 23.1), and their similarities of structure and function in protostomes and deuterostomes make it likely that these genes emerged in an animal that is ancestral to both groups.

The Pax6 protein, for example, plays a role in forming eyes in both vertebrates and invertebrates (see Chapters 4 and 5).

Ectopic expression of Pax6 results in extra eyes in both *Drosophila* and *Xenopus*—representatives of the protostomes and deuterostomes, respectively (Chow et al. 1999; see Figure 5.15 and Carroll et al. 2001, pp. 114–117). Moreover, the ectopic expression of a deuterostome (mouse) *Pax6* gene in a fly larva induces ectopic fly eyes (Figure 23.1B), and the ectopic expression of the *Drosophila Pax6* gene in *Xenopus* ectoderm induces eye development in the frog tadpole (Halder et al. 1995; Onuma et al. 2002). Therefore, it is a safe assumption that the same *Pax6* gene is involved in eye production in both deuterostomes and protostomes. Moreover, at least three other genes—*sine oculis, eyes absent,* and *dachshund*—are also used to form eyes in both *Drosophila* and vertebrates (Jean et al. 1998; Relaix and Buckingham 1999). Since it is extremely unlikely that deuterostomes and protostomes would have evolved *Pax6* (and other genes) independently—and used them independently for the same function—it is very likely that the PDA possessed a *Pax6* gene and used it for generating eyes.

Another gene shared by deuterostomes and protostomes is the homeobox-containing gene *tinman*. The Tinman protein is expressed in the *Drosophila* splanchnic mesoderm, eventually residing in the region of the cardiac mesoderm. Loss-of-function mutants of *tinman* lack a heart (hence its name, after the Wizard of Oz character) (Bodmer 1993). In mice, the homologous gene is called *Nkx2-5*, and it, too, is originally expressed in the splanchnic mesoderm and then continues to be expressed in those cells that form the heart tubes (see Chapter 15; Manak and Scott 1994). Thus, although the heart of vertebrates and the heart of insects have hardly

TABLE 23.1 Developmental regulatory genes conserved between protostomes and deuterostomes

Gene	Function	Distribution
achaete-scute group	Cell fate specification	Cnidarians, *Drosophila*, vertebrates
Bcl2/Drob-1/ced9	Programmed cell death	*Drosophila*, nematodes, vertebrates
Caudal	Posterior differentiation	*Drosophila*, vertebrates
delta/Xdelta-1	Primary neurogenesis	*Drosophila*, *Xenopus*
Distal-less/DLX	Appendage formation (proximal-distal axis)	Numerous phyla of protostomes and deuterostomes
Dorsal/NFκB	Immune response	*Drosophila*, vertebrates
forkhead/Fox	Terminal differentiation	*Drosophila*, vertebrates
Fringe/radical fringe	Formation of limb margin (apical ectodermal ridge in vertebrates)	*Drosophila*, chick
Hac-1/Apaf/ced 4	Programmed cell death	*Drosophila*, nematodes, vertebrates
Hox complex	Anterior-posterior patterning	Widespread among metazoans
lin-12/Notch	Cell fate specification	*C. elegans*, *Drosophila*, vertebrates
Otx-1, Otx-2/Otd, Emx-1, Emx-2/ems	Anterior patterning, cephalization	*Drosophila*, vertebrates
Pax6/eyeless; Eyes absent/eya	Anterior CNS/eye regulation	*Drosophila*, vertebrates
Polycomb group	Controls Hox expression/ cell differentiation	*Drosophila*, vertebrates
Netrins, Split proteins, and their receptors	Axon guidance	*Drosophila*, vertebrates
RAS	Signal transduction	*Drosophila*, vertebrates
sine occulus/Six3	Anterior CNS/eye pattern formation	*Drosophila*, vertebrates
sog/chordin, dpp/BMP4	Dorsal-ventral patterning, neurogenesis	*Drosophila*, *Xenopus*
tinman/Nkx 2-5	Heart/blood vascular system	*Drosophila*, mouse
vnd, msh	Neural tube patterning	*Drosophila*, vertebrates

Source: After Erwin 1999.

anything in common except their ability to pump fluids, they both appear to be predicated on the expression of the same gene, *Nkx2-5/tinman*. Therefore, it is probable that the PDA had a circulatory system with a pump based on the expression of the *Nkx2-5/tinman* gene.

Another set of genes shared by deuterostomes and protostomes are those for the transcription factors involved in head formation (Finkelstein and Boncinelli 1994; Hirth and Reichert 1999). In *Drosophila*, the brain is composed of three segments, called **neuromeres**. These neuromeres are specified by three transcription factors. The genes encoding these factors are *tailless* (*tll*) and *orthodenticle* (*otd*), which are expressed predominantly in the anteriormost neuromere, and *empty spiracles* (*ems*), which is expressed in the posterior two neuromeres (Monaghan et al. 1995; Hirth et al. 1998). Loss-of-function mutations of *otd* eliminate the anteriormost neuromere of the developing *Drosophila* embryo, and loss-of-function mutations of *ems* eliminate the second and third neuromeres (Hirth et al. 1995). In frogs and mice, the homologues of these genes (*Otx-1, Otx-2, Emx-1, Emx-2*) are also expressed in the brain (Simeone et al. 1992), although the exact patterns of transcription are not identical (Figure 23.2). The *Otx-2* gene has been experimentally knocked out (Acampora et al. 1995;

Matsuo et al. 1995; Ang et al. 1996), and the resulting mice have neural and mesodermal head deficiencies anterior to rhombomere 3. In humans, mutations of *EMX2* lead to a rare condition known as schizencephaly, in which there are clefts ripping through the entire cerebral cortex (Brunelli et al. 1996). Even though the *Drosophila otd* and *ems* genes are specified by the Bicoid and Hunchback gradients and the mammalian *Otx* and *Emx* genes are induced by the anterior dorsal mesoderm and endoderm, it appears that the same genes are used for determining the anterior brain regions.

It is therefore likely that the ancestor of all bilaterian organisms had sensory organs based on Pax6, a heart based on *tinman*, and a head based on *otd*, *ems*, and *tll*. It also had something else: an anterior-posterior polarity based on the expression of Hox genes. The analysis of Hox genes has given us critical clues as to how morphological changes could occur through alterations of development.

Hox Genes: Descent with Modification

As mentioned throughout this book, the expression of Hox genes provides the basis for anterior-posterior axis specification throughout the animal kingdom. This means that the enor-

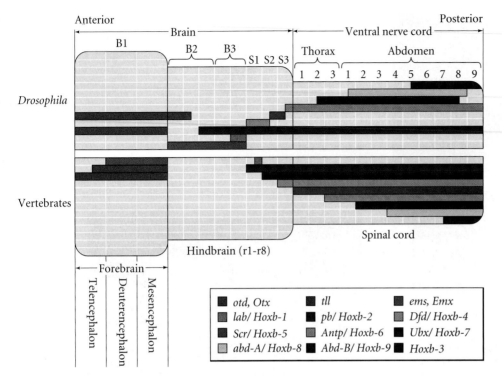

Figure 23.2
Expression of regulatory transcription factors in *Drosophila* and in vertebrates along the anterior-posterior axis. The *Drosophila* genes *ems*, *tll*, and *otd* are expressed in the anterior regions of the brain, as are the homologous genes of vertebrates. The Hox complex genes are expressed in *Drosophila* and in vertebrates in similar patterns in the hindbrain and spinal cord. (After Hirth and Reichert 1999.)

mous variation of morphological form in the animal kingdom is underlain by a common set of instructions. Indeed, Hox genes provide one of the most remarkable pieces of evidence for deep homologies among all the animals of the world. As mentioned in Chapter 11, not only are the Hox genes of different phyla homologous, but they are in the same order on their respective chromosomes (see Figure 11.42). The expression patterns are also remarkably similar between the Hox genes: the genes at the 3′ end are expressed anteriorly, while those at the 5′ end are expressed more posteriorly* (see Figure 23.2).

As if this evidence of homology were not enough, Malicki and colleagues (1992) demonstrated that the human *HOXB4* gene could mimic the function of its *Drosophila* homologue, *Deformed*, when introduced into *Dfd*-deficient *Drosophila* embryos. Slack and his colleagues (1993) postulated that the Hox gene expression pattern defines the development of all animals and that the pattern of Hox gene expression is constant for all phyla.†

If the underlying Hox gene expression pattern is uniform, how did the differences among the phyla emerge? There appear to be at least five ways in which alterations in Hox pro-

tein binding or expression might lead to evolutionary changes (Gellon and McGinnis 1998; Hughes and Kaufman 2002):

- Changes in the Hox protein-responsive elements of downstream genes
- Changes in a Hox gene that give its protein new properties
- Changes in Hox gene transcription patterns within a region of the body
- Changes in Hox gene transcription patterns between regions of the body
- Changes in the number of Hox genes

CHANGES IN HOX-RESPONSIVE ELEMENTS OF DOWNSTREAM GENES: WHY A BUTTERFLY HAS FOUR WINGS BUT A FLY HAS ONLY TWO. One of the most obvious differences between a fruit fly and a butterfly is that the fly has two halteres where the butterfly has a pair of hindwings (Figure 23.3). The expression pattern of the Hox genes, however, does not differ between a butterfly larva and a fly larva. In both cases, the *Ultrabithorax* (*Ubx*) gene is expressed throughout the imaginal discs of the third thoracic segment (from which the hindwing and haltere are derived; see Figure 9.30). What distinguishes a haltere from a hindwing is the response of the target genes. The Ubx protein downregulates several genes in the *Drosophila* imaginal discs. Many of these same genes are not regulated by Ubx in butterflies. Moreover, some other genes regulated by Ubx in *Drosophila* are regulated differently in butterflies (Carroll et al. 1995; Weatherbee et al. 1999). Thus, the wing differences between dipterans (two-winged insects

---

*The conservation of Hox genes and their colinearity demands an explanation. One recent proposal (Kmita 2000, 2002) contends that the Hox genes "compete" for a remote enhancer that recognizes the Hox genes in a polar fashion. This enhancer most efficiently activates the Hox genes at the 5′ end. If the positions of the Hox genes are changed by recombination or deletion, then different genes become activated in different regions of the body, and the morphology changes.

†The following discussion will use Hox genes primarily from the arthropods and the chordate phyla. Other examples are possible, but since this book has provided background in these groups, we will use them as our examples.

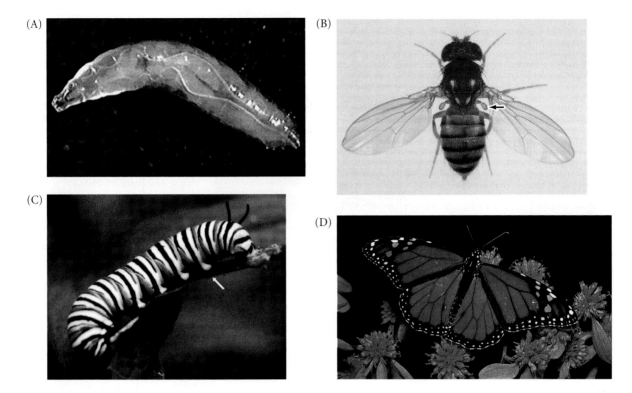

Figure 23.3
Differences in larval and adult morphology due to Hox gene differences. (A, B) Larva and adult of *Drosophila*, a dipteran. An arrow points to one of the halteres of the adult. The larva lacks prolegs; its anterior end is at the left. (C, D) Larva and adult of *Danaus plexippus*, the monarch butterfly, a lepidopteran. The anterior of the caterpillar is at the left, and a proleg is indicated by an arrow. The adult has hindwings rather than halteres. The regulation of Hox genes determines the presence of prolegs, and the targets of Hox genes determine whether the third thoracic segment is to generate halteres or hindwings. (A courtesy of M. Tyler; B courtesy of E. B. Lewis; C courtesy of G. Savage; D by Bill Beatty/Visuals Unlimited.)

such as flies) and lepidopterans (butterflies and moths) can be attributed to the different ways in which potential target genes in the imaginal discs respond to the Ubx protein.

CHANGES IN HOX GENE TRANSCRIPTION PATTERNS WITHIN A BODY REGION: WHY CATERPILLARS HAVE ABDOMINAL PROLEGS BUT MAGGOTS DO NOT. Among arthropods, differences in limb morphology may be caused by Hox gene expression differences. Insects have six legs as adults, with a pair arising from each of the three thoracic segments. In *Drosophila*, *the Distal-less* (*Dll*) *gene* is critical for providing the proximal-distal axis of the appendages (see Figure 18.15). *Distal-less* expression occurs in the cephalic and thoracic appendage-forming discs, but it is excluded in the abdomen by the abdA and Ubx homeodomain proteins. Thus, the appendages grow into legs and wings in the thorax and into jaws in the head. No limbs ever develop in the abdomen.

Butterfly and moth larvae, however, are characterized by rudimentary abdominal legs called **prolegs** (see Figure 23.3C). Panganiban and her colleagues (1994) cloned the *Distal-less* homologue from the buckeye (*Precis*) butterfly and mapped its expression during development. During the early portion of *Precis* embryogenesis, the *Dll* expression pattern is the same as it is in *Drosophila*. During gastrulation, *Dll* expression is seen first in the head regions and in the thoracic regions that will give rise to the leg imaginal discs. However, as development proceeds, the *Dll* gene of *Precis* becomes expressed in the third through sixth abdominal segments (Figure 23.4). Whereas *Dll* expression is seen in both the proximal ring and the "socks" of the true thoracic legs, the expression of *Dll* in the abdomen is restricted to the proximal ring. Thus, the lepidopteran prolegs appear to be homologous to the proximal portion of the thoracic legs.*

The presence of larval prolegs and *Dll* expression in the *Precis* abdominal segments suggests that *Dll* is regulated differently in dipterans and lepidopterans. Two possibilities come to the fore: first, that the *Dll* genes of *Precis* are not repressed by the abdA and Ubx homeodomain proteins; and second, that the expression of these repressing genes is somehow abrogated in the abdominal regions of *Precis*. Warren and co-workers (1994) showed that *Drosophila* and *Precis* embryos

*The expression of *Dll* in the maxilla and labial segments in both *Drosophila* and *Precis* is interesting because it is consistent with paleontological evidence (Kukalova-Peck 1992) that, although these jaw structures originated from limb primordia, distal limb elements have been lost from all arthropod jaws.

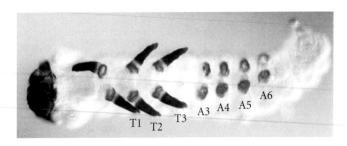

Figure 23.4
Distal-less gene expression in the larva of the buckeye butterfly, *Precis*. By 40 percent of the way through embryonic development, *Dll* expression in *Precis* has diverged significantly from that of *Drosophila* in that Dll expression is also seen in abdominal segments 3–6. (From Panganiban et al. 1994; photograph courtesy of the authors.)

have the same initial pattern of *abdA* and *Ubx* gene expression. However, about 20 percent of the way through *Precis* embryogenesis, the expression of *abdA* and *Ubx* is downregulated in small patches of segments A3–A6, the abdominal segments that give rise to the prolegs (Figure 23.5). Shortly thereafter, the *Dll* genes are expressed in those "holes." It is not known what molecules are used to downregulate *abdA* and *Ubx* gene expression in the regions of *Dll* expression. The

Polycomb group genes are the best suspects, since they can repress both genes in *Drosophila*.

CHANGES IN THE STRUCTURE AND PROPERTIES OF THE PROTEINS ENCODED BY HOX GENES: WHY INSECTS HAVE SIX LEGS WHILE CENTIPEDES HAVE MANY MORE. Insects have just six legs, but most other arthropod groups (think of spiders, millipedes, centipedes, and shrimp) have many more. How is it that the insects came to form legs only in their three thoracic segments? The answer seems to reside in the above-mentioned relationship between Ultrabithorax protein and the *Distal-less* gene.

Throughout most families of the arthropod phylum, Ubx protein does not inhibit the *Distal-less* gene. However, in the insect lineage, a mutation occurred in the *Ubx* gene wherein the original 3′ end of the protein-coding region was replaced by a group of nucleotides encoding a stretch of about 10 alanine residues (Figure 23.6; Galant and Carroll 2002; Ronshaugen et al. 2002). This polyalanine region functions as a repressor of *Distal-less* transcription. When a shrimp *Ubx* gene is experimentally modified to encode this poly(A) region, it, too, represses the *Distal-less* gene. The ability of insect Ubx to inhibit *Distal-less* thus appears to be the result of a gain-of-function mutation that characterizes the insect lineage.

CHANGES IN HOX GENE EXPRESSION BETWEEN BODY SEGMENTS: HOW SHRIMP DIFFER FROM LOBSTERS. There are substantial differences in Hox gene expression patterns between insects and crustaceans, and there are also significant differences in Hox gene expression patterns among the different crustacean groups. Crustaceans are characterized by a pre-gnathal head (similar to the insect acron), gnathal (jawed) head segments, six thoracic segments, genital segments, abdominal seg-

(A)

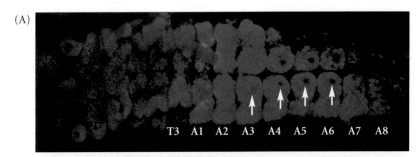

(B)

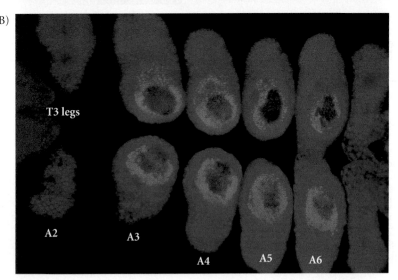

Figure 23.5
"Holes" in the expression of *abdA* and *Ubx* in the abdomen of the larval butterfly *Precis*. The abdA and Ubx proteins are stained green. The Distal-less protein is stained red, and areas of overlap appear yellow. (A) In the early caterpillar, the thoracic limbs (in segments T1–T3) and jaws are seen to express the Distal-less protein. Some abdominal segments (A3–A6) begin to have holes (indicated by arrows) in the expression domains of *abdA* and *Ubx*. (B) In a later-stage caterpillar, these holes have become regions of *Distal-less* expression. (From Warren et al. 1994; photographs courtesy of B. Warren, S. Paddock, and S. Carroll.)

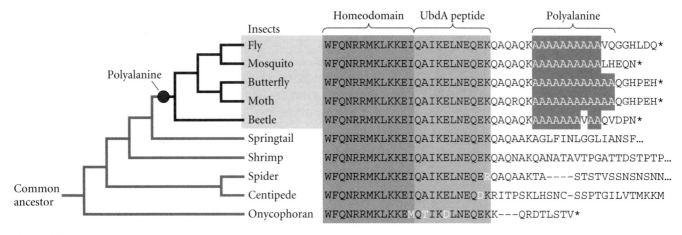

Figure 23.6
Changes in the Ultrabithorax protein associated with the insect clade in the evolution of arthropods. Among the arthropods, only the insects have a Ubx protein that is able to repress *Distal-less* gene expression and thereby inhibit abdominal legs. This ability to repress *Distal-less* is due to a mutation that is seen only in the insect *Ubx* gene. (After Galant and Carroll 2002; Ronshaugen et al. 2002.)

ments, and a telson (Figure 23.7). Each of the thoracic segments expresses *Antp, Ubx,* and *abdA,* and these genes appear to be interchangeable in the crustaceans. Indeed, the thoracic segments all look alike, and there is no specialization among them. In the arthropod lineage that gave rise to the insects, however, each of these genes took on different (but sometimes overlapping) functions (Averof and Akam 1995).

But within the crustacean lineages, there are interesting variations on this theme. Averof and Patel (1997) have shown that if a thoracic segment does not express both *Ubx* and *abdA,* the anterior locomotor limb becomes a feeding ap-

pendage called a **maxilliped**. Thus, brine shrimp such as *Artemia* have a uniform expression of *Ubx* and *abdA* in their thoracic segments, and they lack maxillipeds. Lobsters such as *Homarus* lack *Ubx* and *abdA* expression in their first and second thoracic segments, and these segments have paired maxillipeds (Figure 23.8). The fossil record suggests that the earliest crustaceans lacked maxillipeds and had uniform thoracic

Figure 23.7
Hox gene expression and morphological change in crustaceans. At the bottom of the figure, the domains of Hox gene expression specifying the various structures of a brine shrimp (crustacean) and a grasshopper (insect) are coded in color. Whereas the Hox gene expression domains segregate in the insect thorax and abdomen, they coincide in the crustacean thorax. Above them is a hypothetical model for the divergence of insect and crustacean body plans from a common ancestor. The *Antennapedia, Ultrabithorax,* and *abdominal A* genes are very similar and are thought to have emerged by gene duplication from a single gene in a distant ancestor of the arthropods. *Abdominal B* is expressed in the segments destined to become genitalia. Paleontological evidence suggests that the ancestral arthropod had several identical thoracic segments, similar to those of extant crustaceans. In both cases, *Antennapedia, Ultrabithorax,* and *abdominal A* expression is seen throughout the thorax. The crustaceans later evolved tail segments. Insects, however, diversified their thoracic segments and used these Hox genes to specify the segments differently. The ancestors of the insects and crustaceans are hypothetical reconstructions based on fossils of the middle Cambrian. (After Averof and Akam 1995 and Manton 1977.)

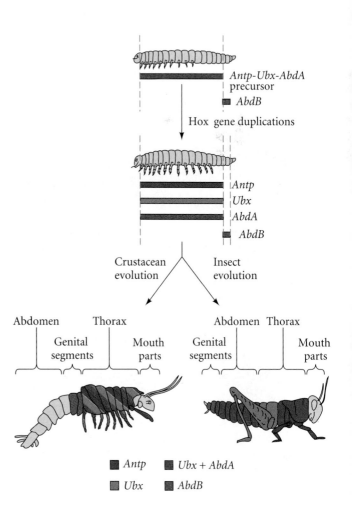

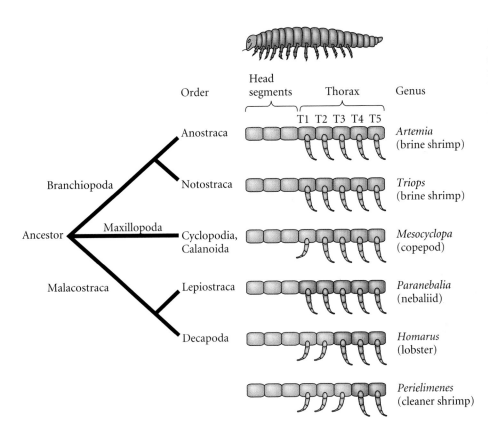

Figure 23.8
Schematic representation of the expression of *Ubx* and *abdA* (green) in the thoracic segments of different types of crustaceans. The generation of maxillipeds occurs in those thoracic segments that do not express either one of these two homeodomain proteins. (After Averof and Patel 1997.)

segments. This would mean that the presence of maxillipeds is a derived characteristic that evolved in several crustacean lineages.

CHANGES IN HOX GENE EXPRESSION BETWEEN BODY SEGMENTS: HOW THE SNAKE LOST ITS LIMBS.

As described in Chapter 11, the expression pattern of Hox genes in vertebrates determines the type of vertebral structure formed. Thoracic vertebrae, for instance, have ribs, while cervical (neck) vertebrae and lumbar vertebrae do not. The type of vertebra produced is specified by the Hox genes expressed in the somite.

One of the most radical alterations of the vertebrate body plan is seen in snakes. Snakes evolved from lizards, and they appear to have lost their legs in a two-step process. Both paleontological and embryological evidence support the view that snakes first lost their forelimbs and later lost their hindlimbs (Caldwell and Lee 1997; Graham and McGonnell 1999). Fossil snakes with hindlimbs, but no forelimbs, have been found. Moreover, while the most derived snakes (such as vipers) are completely limbless, the more primitive snakes (such as boas and pythons) have pelvic girdles and rudimentary femurs.

The missing forelimbs can be explained by the Hox expression pattern in the anterior portion of the snake. In most vertebrates, the forelimb forms just anterior to the most anterior expression domain of *Hoxc-6* (Gaunt 1994; Burke et al. 1995; Gaunt 2000). Caudal to that point, *Hoxc-6*, in combination with *Hoxc-8*, helps specify vertebrae to be thoracic. During early python development, *Hoxc-6* is not expressed in the absence of *Hoxc-8*, so the forelimbs do not form. Rather, the combination of *Hoxc-6* and *Hoxc-8* is expressed for most of the length of the organism, telling the vertebrae to form ribs throughout most of the body (Figure 23.9; Cohn and Tickle 1999).

The loss of hindlimbs has apparently occurred by a different mechanism. Hindlimb buds do begin to form in some snakes, such as pythons, but they do not produce anything more than a femur. This appears to be due to the lack of *sonic hedgehog* expression by the limb bud mesenchyme. Sonic hedgehog is needed both for the polarity of the limb and for the maintenance of the apical ectodermal ridge (AER). Python hindlimb buds lack the AER, and the phenotype of the python hindlimb resembles that of mouse embryos with loss-of-function mutations of *sonic hedgehog* (Chiang et al. 1996).

HOX GENES AND ATAVISMS.

As we have seen, alteration of Hox gene expression can change one type of vertebra into another. In some instances, the mutation of a Hox gene can produce "atavistic" conditions, wherein the organism resembles an earlier evolutionary stage. Deletion or misregulation of the *Hoxa-2* genes in mice, for instance, results in a partial transformation of the second pharyngeal arch into a copy of the first pharyngeal arch. The mutant fetuses lack the stapes and styloid bones formed from the second arch, but have extra malleus, incus, tympanic, and squamosal bones. They also possess a rodlike cartilage that has no counterpart in normal

(A)

(B)

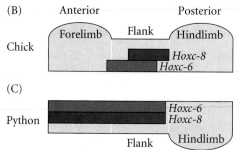

Chick

(C)

Python

Figure 23.9
Loss of limbs in snakes. (A)
Skeleton of an embryo of the
garter snake, *Thamnophis*,
stained with alcian blue.
Ribbed vertebrae are seen from
the head to the tail. (B, C) Hox
expression patterns in the chick
(B) and the python (C). (Photograph courtesy of A. C.
Burke; B, C after Cohn and
Tickle 1999.)

mice, but looks like the pterygoquadrate cartilage thought to have been present in therapsids, the reptilian group that gave rise to the mammals (Figure 23.10; Rijli et al. 1993; Lohnes et al. 1994; Mark et al. 1995). Thus, major evolutionary changes can be correlated with the alteration of Hox gene expression in different parts of the embryo.

CHANGES IN HOX GENE NUMBER: HOW TO GET MORE CELL TYPES The number of Hox genes may play a role in permitting the evolution of complex structures. All invertebrates have a single Hox complex per haploid genome. In the simplest invertebrates—such as sponges—there appears to be only one or two Hox genes in this complex (Degnan et al. 1995; Schierwater and Kuhn 1998). More complex invertebrates, such as insects, have numerous Hox genes in this complex. Comparing the Hox genes of chordates, arthropods, and molluscs suggests that there was a common set of seven Hox genes in the Urbilaterian ancestor of the protostomes and deuterostomes. Indeed, in invertebrate deuterostomes (echinoderms and amphioxus, an invertebrate chordate), there is only one Hox complex, which looks very much like that of the insects (Figure 23.11; Holland and Garcia-Fernández 1996).

By the time the earliest vertebrates (agnathan fishes) evolved, there were at least four Hox complexes. The transition from amphioxus to early fish is believed to be one of the major leaps in complexity during evolution (Amores et al. 1998; Holland 1998; see below). This transition involved the evolution of the head, the neural crest, new cell types (such as osteoblasts and odontoblasts), the brain, and the spinal cord. As we saw in Chapter 11, the regionalization of the brain and

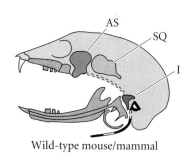

Wild-type mouse/mammal

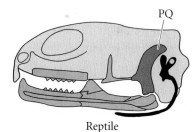

Reptile

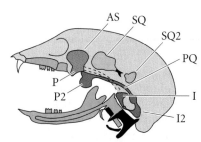

Hoxa-2 mutant mouse

Figure 23.10
Representation of skeletal elements derived from the first pharyngeal arch (gray) and the second pharyngeal arch (black) in normal and *Hoxa-2* mutant mouse, and in reptiles. AS, alisphenoid; I, incus; I2, duplicated incus; P and P2, normal and duplicated pteroid cartilage; PQ, pterygoquadrate cartilage; SQ, squamosal; SQ2, duplicated squamosal. (After Mark et al. 1995.)

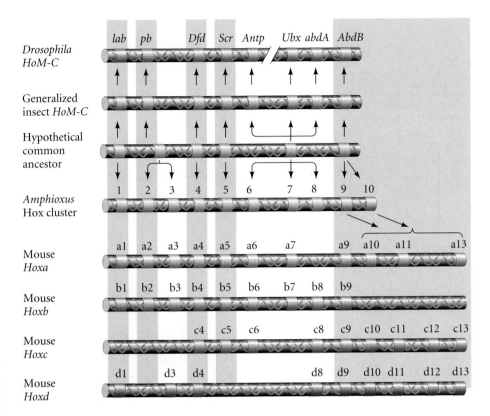

Figure 23.11
Postulated ancestry of the Hox genes from a hypothetical ancestor of both deuterostomes and prototomes. Amphioxus has only one Hox cluster, similar to that of insects. Vertebrates have four Hox clusters, none of which is complete. (After Holland and Garcia-Fernández 1996.)

spinal cord is dependent upon Hox genes, and the regional specification of the somitic segments depends on the paralogous members of the different Hox clusters. Holland (1998) speculates that the generation of new structures was allowed by the fourfold duplication of the Hox gene complex as well as by duplications of the *Distal-less* gene, which allowed them to serve new functions.

WEBSITE 23.5 Gene expression data and macroevolutionary debates. Three of the major debates in evolution are (1) whether the birds are descended from dinosaurs, (2) how the tetrapod limb evolved from the fish fin, and (3) how turtles are related to other reptiles. Scientists are attempting to integrate paleontological data and gene expression data to resolve these questions.

Sidelights & Speculations

How the Chordates Got a Head

The vertebrate head is derived largely from the neural crest. Thus, Hall (2000) considers vertebrates to be not merely triploblastic animals, but *quadroblastic*, with the neural crest constituting a fourth germ layer, and Holland and Chen (2001) propose calling vertebrates and their fossilized precursors "cristozoa," the crest-animals. But how did the neural crest arise?

Once a pattern of expression has been determined for the genes in a given structure, one can attempt to figure out how that structure evolved by looking at the patterns of gene expression in related animals that lack the structure. Only vertebrates have a neural crest, and the cranial neural crest is responsible for forming the bones and cartilage of the face and much of the skull (see Chapter 14). While we do not know how neural crest cells arose, it appears that they evolved from cells at the neural plate/epidermal boundaries of an ancestral protochordate. Both the urochordate tunicates and the cephalochordate amphioxus express in their dorsal midline ectoderm many of the same genes expressed in vertebrate neural crest cells

(Holland and Holland 2001; Wada 2001).

Amphioxus is an invertebrate chordate that has a notochord, somites, and a hollow neural tube. It lacks a brain and facial structures, and, most importantly, it lacks neural crest cells. In amphioxus, the neural plate/epidermal border contains cells expressing several of the same genes expressed in vertebrate neural crest cells—*BMP2, Pax3/7, Msx, Dll,* and *Snail*. However, the protochordate cells expressing these genes do not migrate, nor do they differentiate into a wide range of tissues (Figure 23.12A, B; Holland and Holland 2001).

(A)

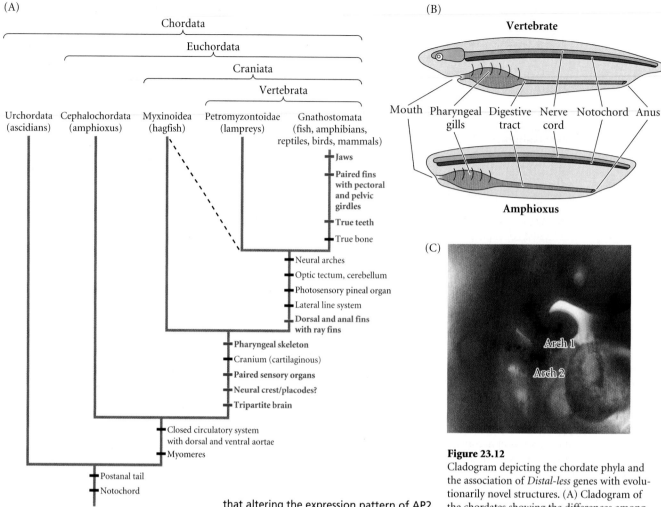

(B)

Vertebrate

Mouth Pharyngeal Digestive Nerve Notochord Anus
gills   tract   cord

Amphioxus

(C)

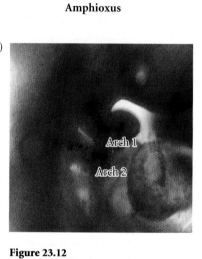

Figure 23.12
Cladogram depicting the chordate phyla and the association of *Distal-less* genes with evolutionarily novel structures. (A) Cladogram of the chordates showing the differences among chordate groups on the right. The characters associated with *Distal-less* expression are printed in red. The position of hagfishes is controversial and is therefore represented as a dashed line. (B) Schematic diagram of amphioxus and vertebrate body plans. (C) Expression in the mouse embryo of a *lacZ* reporter transgene attached to an amphioxus Hox enhancer. The Hox enhancer causes expression of the transgene in the neural crest cells of the first and second pharyngeal arches—those arches that will form the jaws. (A after Neidert et al. 2001; B after Finnerty 2000; C after Manzanares et al. 2000.)

In addition, many of the *cis*-regulatory elements needed for the positioning of neural crest cells in the vertebrate body are also present in amphioxus. Manzanares and colleagues (2000) added regulatory regions of amphioxus Hox genes onto a reporter *lacZ* gene and observed the expression of these chimeric genes in mouse and chick embryos. Remarkably, some of the amphioxus Hox regulatory sequences caused the expression of the reporter gene in the neural crest cells and neural placodes of mice and chicks (Figure 23.12C).

Two transcription factors, AP2 and Distal-less, may have been critical in the formation of neural crest cells. Vertebrate AP2 is expressed in non-neural, cranial neural crest, and neural tube cells, even in jawless fish. It appears to be essential for *Hoxa-2* expression in the neural crest cells. However, AP2 expression is confined to the non-neural ectoderm of amphioxus. Meulemans and Bronner-Fraser (2002) speculate

that altering the expression pattern of AP2 was critical in establishing the neural crest cell lineage.

Holland and colleagues (1996) have suggested that the origin of the neural crest involves the duplication and divergence of the *Distal-less* genes. *Distal-less* is found throughout the animal kingdom, and it is expressed in those tissues that stick out from the body axis, notably limbs and antennae (Panganiban et al. 1997). But in vertebrates, *Distal-less* has acquired new functions. Like *Drosophila*, amphioxus has only one copy of the *Distal-less* gene per haploid genome and, also as in *Drosophila*, this gene is expressed in the epidermis and central nervous system. Vertebrates, however, have five or six closely related copies of *Distal-less*, all of which probably originated from a single ancestral gene that resembles the one in amphioxus (Price 1993; Boncinelli 1994; Neidert et al. 2001). These *Distal-less* homologues have found new functions. Some are expressed in the mesoderm, a place where *Distal-less* is not expressed in amphioxus. At least three of the vertebrate *Distal-less* genes function in the patterning of neural crest cells, and delet-

ing these genes results in the absence or malformation of the pharyngeal arches, face, jaws, teeth, and vestibular apparatus (Qiu et al. 1997; DePew et al. 1999). Although it remains to be proved, it is possible that a new type of *Distal-less* gene could have played a major role in allowing migratory ectodermal cells of amphioxus to evolve into neural crest cells.

Homologous Pathways of Development

One of the most exciting findings of the past decade has been the discovery not only of homologous regulatory genes, but also of homologous signal transduction pathways (Zuckerkandl 1994; Gilbert 1996; Gilbert et al. 1996), many of which have been mentioned earlier in this book. In different organisms, these pathways are composed of homologous proteins arranged in a homologous manner. In this respect, the homology is similar to that of a human forearm and a seal flipper. Both the parts—the proteins—and the structures they make up—the pathways—are homologous.

Homologous signal transduction pathways form the infrastructure of development. However, the targets of these pathways may differ among organisms. For example, the Dorsal-Cactus pathway used by *Drosophila* to specify dorsal-ventral polarity is also used by the mammalian immune system to activate inflammatory proteins (see Figure 9.38). This does not mean that the *Drosophila* blastoderm is homologous to the human macrophage. It merely means that there is a very ancient pathway that predates the deuterostome-protostome split, and that this pathway can be used in different systems. The pathways are homologous; the organs they form are not.

Signal transduction pathways undergo descent with modification just as physical structures do. This phenomenon is readily seen in the Wnt pathway that we have discussed throughout the book. Figure 23.13 shows how the Wnt pathway is used in several different organisms. The pathways are homologous, but not identical. They are thought to have originated in a common ancestral pathway that predates the deuterostome-protostome split. Indeed, Wnt proteins are important in the axis-forming ability of cnidarians (see Figure 18.35), organisms that are ancestral to the split between protostomes and deuterostomes.

Earlier in this chapter, we discussed genes such as *Pax6* and *tinman* that appear to have had their developmental functions before the protostome-deuterostome split. We have also discussed homologous pathways that may or may not be used in similar structures. However, there appear to be some pathways that are used to form the same structure in all animals. When homologous pathways made of homologous parts are used for the same function in both protostomes and deuterostomes, they are said to have "deep homology" (Shubin et al. 1997).

Instructions for forming the central nervous system

One example of deep homology has already been discussed in earlier chapters. First, as seen in Chapter 10, the chordin/

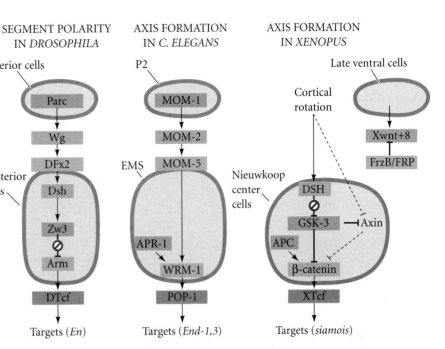

Figure 23.13
Three modifications of the Wnt pathway. Each pathway proceeds vertically. The cells secreting the Wingless protein are labeled (anterior cells in the *Drosophila* parasegment, the P2 cell in C. elegans). The responding cells are shown beneath them (posterior cells in the *Drosophila* parasegment, the EMS cell of *C. elegans*, and the Nieuwkoop center cells of *Xenopus*. In *Xenopus*, the Disheveled (DSH) protein is thought to be brought to the Nieuwkoop center by the cortical rotation of the cytoplasm. The proteins on each level are homologous to one another. (After Cadigan and Nusse 1997.)

BMP4 pathway demonstrates that in both vertebrates and invertebrates, chordin/Short-gastrulation (Sog) inhibits the lateralizing effects of BMP4/Decapentaplegic (Dpp), thereby allowing the ectoderm protected by chordin/Sog to become the neurogenic ectoderm. These reactions are so similar that *Drosophila* Dpp protein can induce ventral fates in *Xenopus* and can substitute for the Sog protein (Holley et al. 1995).

In addition to this central inhibitory reaction, there are other reactions that add to the deep homology of the instructions for forming the protostome and deuterostome neural tube. For instance, the spread of Dpp in *Drosophila* is aided by Tolloid, a metalloprotease that degrades Sog. The gradient of Dpp concentration from dorsal to ventral is created by the opposing actions of Tolloid (increasing Dpp) and Sog (decreasing it) (Figure 23.14; Marqués et al. 1997). In *Xenopus* and zebrafish, the homologues of Tolloid (Xolloid and BMP1, respectively) have the same function: they degrade chordin. The gradient of BMP4 from ventral to dorsal is established by the antagonistic interactions of Xolloid or BMP1 (increasing BMP4) and chordin (decreasing BMP4) (Blader et al. 1997; Piccolo et al. 1997). Even the regulators of BMP stability are conserved between these two groups of animals and function

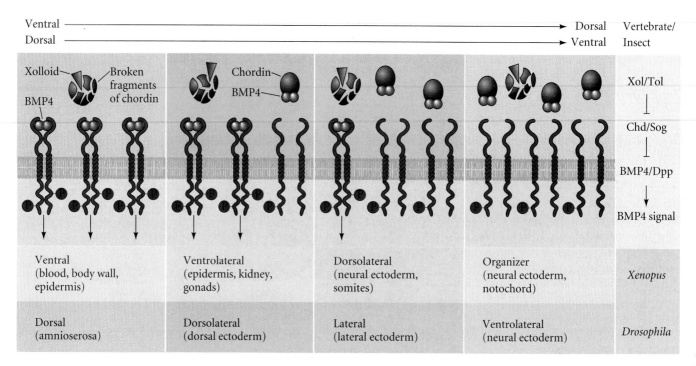

Figure 23.14
Homologous pathways specifying neural ectoderm in protostomes (*Drosophila*) and deuterostomes (*Xenopus*). Both pathways involve a source of chordin/Sog (the organizer in *Xenopus*, the presumptive neural ectoderm in *Drosophila*) and a source of BMP4/Dpp (the ventral mesoderm in frogs, the presumptive amnioserosa in flies). In both instances, the gradient is shaped by a constant supply of Tolloid/Xolloid, which degrades Sog/chordin. In both cases, the neural ectoderm forms where BMP4 signaling is prevented. (After Dale and Wardle 1999.)

in the same way (Larrain et al. 2001). Thus, it appears that nature may have figured out how to make a nervous system only once. The protostome and deuterostome nervous systems, despite their obvious differences, seem to be formed by the same set of instructions.

Instructions for appendage formation

It is also possible that nature provides only one set of instructions for forming appendages (Shubin et al. 1997). Nothing could be a better example of analogy than vertebrate and insect legs. Fly legs and vertebrate limbs have little in common except their function. They have no structural similarities. Insect legs are made of chitin and have no inner skeleton. They are formed by the telescoping out of ectodermal imaginal discs (see Chapter 18). Vertebrate limbs, on the other hand (no pun intended), have no chitin, but possess a bony endoskeleton; they are created by the interaction of ectoderm and mesoderm (see Chapter 16). However, the genetic instructions to form these two distinctly different types of limbs are extremely similar.

As we have seen, Sonic hedgehog protein is usually expressed in the posterior part of the vertebrate limb bud. If it is expressed in the anterior part of the bud, mirror-image duplications arise (see Figure 16.20; Riddle et al. 1993). In *Drosophila*, Hedgehog protein is expressed in the posterior portion of the wing or leg imaginal disc. If it is expressed anteriorly, mirror-image duplications of the wing will form (Figure 23.15; Basler and Struhl 1993; Ingham 1994). Furthermore, certain genes regulated by the Hedgehog proteins have been conserved as well (Marigo et al. 1996). Thus, the anterior-posterior axis appears to be specified in the same way in vertebrate and in insect appendages. The dorsal-ventral axis appears to be similarly specified. The ventral limb compartments of both insects and vertebrates appear to be specified by the expression of the *engrailed* gene (Davis et al. 1991; Loomis et al. 1996), while the dorsal compartment is defined by *apterous* (in insects) or its relative *Lmx1* (in vertebrates) (Figure 23.16).

The formation of the proximal-distal axis also proceeds similarly in vertebrate and invertebrate appendages. In insects, the wing margin forms at the border between the dorsal cells and the ventral cells. The dorsal cells express Apterous protein, and this activates the expression of Fringe. The interaction of the Fringe protein with the ventral cells leads to the growth of the wing blade outward from the body wall. Unexpectedly, a very similar cascade induces the outgrowth of the vertebrate limb. The apical ectodermal ridge (AER) forms at the junction of the dorsal cells with the ventral cells. The dorsal cells express Radical fringe, a vertebrate homologue of the Fringe protein. This protein is critical in forming the AER. It is interesting that the ways in which these proteins become lo-

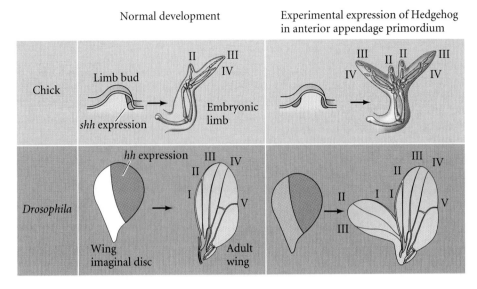

Normal development

Experimental expression of Hedgehog in anterior appendage primordium

Figure 23.15

Homology of signaling pathways in the formation of the anterior-posterior axes in *Drosophila* and chick appendages. A chick limb bud expresses Sonic hedgehog protein in its posterior region. If Sonic hedgehog is also expressed in an anterior region, the limb develops a mirror-image duplication of the anterior-posterior axis. A *Drosophila* wing disc expresses Hedgehog in its posterior compartment. If Hedgehog is also expressed in the anterior compartment, the wing develops a mirror-image duplication of the anterior-posterior axis. (After Ingham 1994.)

calized differ enormously between these phyla. Radical fringe, for instance, is induced by fibroblast growth factors, and it induces the AER to produce more fibroblast growth factors for the outgrowth of the limb. Fibroblast growth factors play no role in insect limb development. Yet the instructions for limb outgrowth and polarity appear to be essentially the same in insects and vertebrates. Nature may have evolved the mechanism to form an appendage only once, in the PDA, and both arthropods and vertebrates use modifications that process to this day.*

---

*In addition to these similarities, there are also unambiguous differences between insect and vertebrate appendage formation. For instance, insects use neither FGFs nor Hox genes to pattern the proximal-distal axis (see Wilkins 2002 for a full discussion of this issue). One interesting possibility is that both types of limb arose from modifications of those programs used to form some other distal outgrowth in the PDA, such as genitalia (Knoll and Carroll 1999).

Figure 23.16

Deep homology of the limbs. The same set of proteins is used to establish the polarity of limbs in both deuterostomes (chick) and protostomes (*Drosophila*). The top panels represent chick limb buds with the dorsal region on top and the apical ectodermal ridge facing the viewer. The bottom panels represent the *Drosophila* wing disc with the dorsal region upward and its anterior side to the left. (A) Proximal-distal axes are specified by the Distal-less protein in the most distal region of the limb bud or disk. This protein is expressed at the junction where the Fringe-containing dorsal cells meet the ventral cells. (B) Dorsal-ventral patterning is specified by the expression of a LIM protein, either Apterous (*Drosophila*) or Lmx1 (chick), in the dorsal portion of the disk or bud. A Wnt protein (Wingless in *Drosophila*, Wnt7A in the chick) induces this expression. (C) Anterior-posterior patterning is accomplished by the expression of Hedgehog in the posterior of the disk or bud. Hedgehog, in turn, activates a BMP that can relay a signal to other cells.

(A) Proximal-distal patterning: Distal-less in most distal region

(B) Dorsal-ventral patterning: Lim protein in dorsal region specified by Wnt protein

(C) Anterior-posterior patterning: Hedgehog in posterior induces BMP to signal

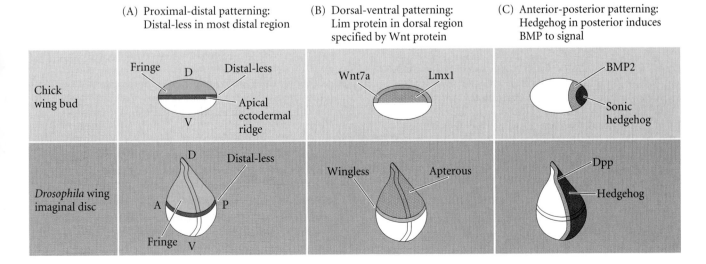

Modularity: The Prerequisite for Evolution through Development

How can the development of an embryo change when development is so finely tuned and complex? How can such change occur without destroying the entire organism? It was once thought that the only way to promote evolution was to add a step to the end of embryonic development, but we now know that even early stages can be altered to produce evolutionary novelties. The reason why changes in development can occur is that the embryo, like the adult organism, is composed of **modules** (Riedl 1978; Bonner 1988; Bolker 2000).

Organisms are constructed of units that are coherent within themselves and yet part of a larger unit. Thus, cells are parts of tissues, which are parts of organs, which are parts of systems, and so on. Such a hierarchically nested system has been called a level-interactive modular array (Dyke 1988). Development occurs through a series of discrete and interacting modules (Gilbert et al. 1996; Raff 1996; Wagner 1996; von Dassow and Munro 1999). Developmental modules include morphogenetic fields (for example, those described for the limb or eye), signal transduction pathways (such as those mentioned above), imaginal discs, cell lineages (such as the inner cell mass or trophoblast), insect parasegments, and vertebrate organ rudiments. Even the regions that form the enhancers of genes are modular. Thus, if a particular gene loses or gains a modular enhancer element, the organism containing that particular allele will express that gene in different places or at different times than those organisms retaining the original allele. This can cause different morphologies to develop (Sucena and Stern 2000). The modularity of enhancer elements allows the particular sets of genes to be activated together and permits a particular gene to become expressed in several discrete places. Modular units allow certain parts of the body to change without interfering with the functions of other parts.

The fundamental principle of modularity allows three processes to alter development: dissociation, duplication and divergence, and co-option (Raff 1996). Since modules are found on all levels, from molecular to organismal, it is not surprising that one sees these processes operating at all levels of development.

> **WEBSITE 23.6 Modularity as a principle of evolution.** Complex structures are created by the assortment of preexisting modules. It is foolish to consider a protein as a collection of atoms. It is an ordered assembly of amino acids that have already formed from atoms. Modularity allows evolution to occur by forming components that can be individually modified.

> **WEBSITE 23.7 Correlated progression.** In many cases, modules must co-evolve. The upper and lower jaws, for instance, have to fit together properly. If one changes, so must the other. If the sperm-binding proteins on the egg change,

then so must the egg-binding proteins on the sperm. This site looks at correlated changes during evolution.

> **WEBSITE 23.8 Correlated progression in domestic animals.** Domestication appears to be selection for neotenic conditions. In selecting for behavioral plasticity, changes in skull shape and pigment patterns are also produced. This phenomenon can also be seen in current attempts to domesticate wild wolves and foxes.

Dissociation: Heterochrony and allometry

The modularity of development allows changes that are either spatial or temporal. By means of mutation or environmental perturbation, one part of the embryo can change without the other parts changing.

Heterochrony is a shift in the relative timing of two developmental processes from one generation to the next. In other words, one module can change its time of expression relative to the other modules of the embryo. This temporal modularity of development can be appreciated when we look at early amniote embryos and see the development of the eyes relative to other regions of the body (Figure 23.17). The eyes of birds and lizards initiate development earlier than the eyes of mammals and are therefore proportionately much larger than mammalian eyes when they reach the pharyngula stage (Jeffery et al. 2002). We have already come across this concept in our discussion of neoteny and progenesis in relation to metamorphosis (see Chapter 18). In salamanders, heterochronies in which the larval stage is retained are caused by genetic changes in the ability to induce or respond to the hormones initiating metamorphosis. Other heterochronic phenotypes, however, are caused by the heterochronic expression of certain genes. For instance, the direct development of some sea urchins involves the early activation of adult genes and the suppression of larval gene expression (Raff and Wray 1989). Heterochrony can also give larval characteristics to an adult organism, as in the small size and webbed feet of arboreal salamanders (Figure 23.18) or the fetal growth rate of human newborn brain tissue (see Chapter 12).

> **WEBSITE 23.9 Heterochrony in evolution.** Heterochrony is an important means of dissociating the development of one portion of the body from another. It has been seen to play critical roles in the evolution of direct-developing sea urchins, arboreal salamanders, and furless apes.

Another consequence of modularity is **allometry**. Allometry occurs when different parts of an organism grow at different *rates* (Huxley and Teissier,1936; Gayon 2000; see Chapter 1). Allometry can be very important in forming variant body plans within a phylum. Such differential changes in growth rate can involve altering a target cell's sensitivity to growth fac-

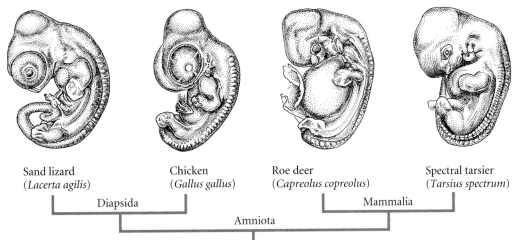

Sand lizard
(*Lacerta agilis*)

Chicken
(*Gallus gallus*)

Roe deer
(*Capreolus copreolus*)

Spectral tarsier
(*Tarsius spectrum*)

Figure 23.17
Allometry and modularity in the vertebrate eye. In the diapsid amniotes (reptiles and birds), eye development starts relatively early,
and by the pharyngula stage, the embryo is characterized by large
eyes. In mammals, however, eye development is initiated later, and
the eye is much smaller relative to the body. (After Jeffery et al.
2002; drawings by Nancy Haver.)

tors or altering the amounts of growth factors produced.
Again, the vertebrate limb can provide a useful illustration.
Local differences among chondrocytes cause the central toe of
the horse to grow at a rate 1.4 times that of the lateral toes
(Wolpert 1983). This means that as the horse grew larger during evolution, this regional difference caused the five-toed
horse to become a one-toed horse. A particularly dramatic example of allometry in evolution comes from skull development. In the very young (4- to 5-mm) whale embryo, the nose
is in the usual mammalian position. However, the enormous
growth of the maxilla and premaxilla (upper jaw) pushes over
the frontal bone and forces the nose to the top of the skull
(Figure 23.19). This new position of the nose (blowhole) allows the whale to have a large and highly specialized jaw apparatus and to breathe while parallel to the water's surface (Slijper 1962).

Allometry can also generate evolutionary novelty by
small, incremental changes that eventually cross some developmental threshold (sometimes called a bifurcation point). A
change in quantity eventually becomes a change in quality
when such a threshold is crossed. It has been postulated that
this type of mechanism produced the external fur-lined cheek
pouches of pocket gophers and kangaroo rats that live in

deserts. These external pouches differ from internal ones in
that they are fur-lined and have no internal connection to the
mouth. They allow these animals to store seeds without running the risk of desiccation. Brylski and Hall (1988) have dissected the heads of pocket gopher and kangaroo rat embryos
and have looked at the way the external cheek pouch is constructed. When data from these animals were compared with
data from animals that form internal cheek pouches (such as
hamsters), the investigators found that the pouches are
formed in very similar manners. In both cases, the pouches
are formed within the embryonic cheeks by outpocketings of
the cheek (buccal) epithelium into the facial mesenchyme. In
animals with internal cheek pouches, these evaginations stay
within the cheek. However, in animals that form external
pouches, the elongation of the snout draws up the outpocketings into the region of the lip. As the lip epithelium rolls out
of the oral cavity, so do the outpockets. What had been inter-

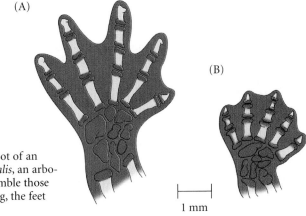

(A)

(B)

1 mm

Figure 23.18
Heterochrony in *Bolitoglossa* can create a tree-climbing salamander. (A) Foot of an
adult *B. rostratus*, a terrestrial salamander. (B) Foot of an adult *B. occidentalis*, an arboreal salamander. The feet, skull, and body size of *B. occidentalis* adults resemble those
of juvenile *B. rostratus*. Since the digits have not expanded past the webbing, the feet
can produce suction to climb trees. (After Alberch and Alberch 1981.)

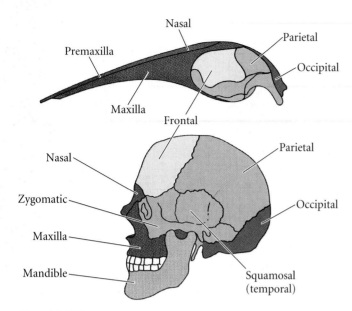

Figure 23.19
Allometric growth in the whale head. An adult human skull is shown for comparison. The whale's upper jaw (maxilla) has pushed forward, causing the nose to move to the top of the skull. The mandible is not shown. (The premaxilla is present in the early human fetus, but it fuses with the maxilla by the end of the third month of gestation. The human premaxilla was discovered by Goethe, among others, in 1786.) (After Slijper 1962.)

nal becomes external. The fur lining is probably derived from the external pouches' coming into contact with dermal mesenchyme, which can induce hair to form in epithelia (see Chapter 12).

The transition from internal to external pouch is one of threshold. The placement of the evaginations—anteriorly or posteriorly—determines whether the pouch is internal or external. There is no "transitional stage" displaying two openings, one internal and one external.* One could envision this externalization occurring by a chance mutation or concatenation of alleles that shifted the outpocketings to a slightly more anterior location. Such a trait would be selected for in desert

---

*The lack of such transitional forms is often cited by creationists as evidence against evolution. For instance, in the transition from reptiles to mammals, three of the bones of the reptilian jaw became the incus and malleus, leaving only one bone (the dentary) in the lower jaw (see Chapter 1 and below). Gish (1973), a creationist, says that this is an impossible situation, since no fossil has been discovered showing two or three jaw bones and two or three ear ossicles. Such an animal, he claims, would have dragged its jaw on the ground. However, such a specific transitional form (and there are over a dozen documented transitional forms between reptilian and mammalian skulls) need never have existed. Hopson (1966) has shown on embryological grounds how the bones of the jaw could have divided and been used for different functions, and Romer (1970) has found reptilian fossils wherein the new jaw articulation was already functional while the older bones were becoming useless. There are several species of therapsid reptiles that had two jaw articulations, with the stapes brought into close proximity with the upper portion of the quadrate bone (which would become the incus).

environments, where dessication is a constant risk. As Van Valen reflected in 1976, evolution can be defined as "the control of development by ecology."

Duplication and divergence

Modularity also allows duplication and divergence. The duplication part of this process allows the formation of redundant structures, and divergence allows these structures to assume new roles. One of the copies can maintain the original role while the others are free to mutate and diverge functionally. This process can happen at numerous levels. The Hox genes, TGF-β family genes, MyoD family genes, *Distal-less* genes, and globin genes each probably started as a single gene that was duplicated several times. After every duplication, mutations caused divergence that gave the members of each family new functions. Thus, Morange (2001, p. 33) concludes, "The most important difference between the genome of a fruit fly and that of a human is therefore not that the human has new genes but that where the fly only has one gene, our species has multigene families." At the tissue level, one sees duplication and divergence in the somites that give rise to the cervical, thoracic, and lumbar vertebrae.

Co-option

No one structure is destined for a single particular purpose. A pencil can be used for writing, but it can also be used as a toothpick, a dagger, a hole-punch, or a drumstick. On the molecular level, the gene *engrailed* is used for segmentation in the *Drosophila* embryo, is used later to specify its neurons, and is used in the larval stages to provide an anterior-posterior axis to imaginal discs. Similarly, a protein that functions as an enolase or alcohol dehydrogenase enzyme in the liver can function as a structural crystallin protein in the lens (Piatigorsky and Wistow 1991). In other words, preexisting units can be **co-opted** (recruited; redeployed) for new functions. Sometimes, whole signaling pathways are co-opted from one system to another. For instance, the pathway by which Hedgehog protein induces Engrailed protein to pattern and extend the insect wing is later used within the wing blades to make the eyespots of butterflies and moths. Distal-less, another protein used to extend the wing imaginal disc, is later used to form the center of such eyespots (Figure 23.20).

Co-option can also be seen on the morphological level. Wings have evolved three times during vertebrate evolution, and in each case, different forearm structures were modified for an entirely new function: a structure originally used for walking has been recruited into a structure suitable for flying. A famous case of co-option is the use of embryonic jaw parts in the creation of the mammalian middle ear, as described in Chapter 1 (see Figure 1.14; Gould 1990). First, the gill arches of jawless fishes became the jaws of their descendants; then, millions of years later, the upper elements of the reptilian jawbone became the malleus and incus (hammer and stirrup) bones of the mammalian middle ear (Figure 23.21).

Figure 23.20
Co-option of signal transduction pathways in the formation of butterfly eyespots. (A) The interaction between Hedgehog (green) and Engrailed (blue) is first seen in the insect blastoderm, and is then modified during the creation of a new structure, the wing. In the wing disc, Hedgehog in the posterior compartment induces Engrailed expression in the cells immediately anterior to the anterior/posterior boundary. Within the butterfly wing, the same pathway is used to create the eyespot. In the posterior of the wing, Hedgehog is locally downregulated so that re-expression can induce Engrailed. (B, C) In *Bicyclus anynana* (see Figures 22.8 and 22.9), the signal to produce the eyespot also activates several other genes that were used for wing formation. The white central portion of the eyespot expresses Engrailed, Spalt, and Distal-less. The cells that will become the black region surrounding the center express Spalt and Distal-less. The peripheral ring of cells that will generate the orange pigment expresses only Engrailed. (After Keys et al. 1999 and Brunetti et al. 2001; photograph courtesy of S. B. Carroll and S. Paddock.)

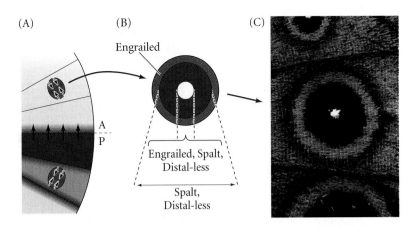

Generating Evolutionary Novelty

In 1940, Richard Goldschmidt wrote that the accumulation of small genetic changes was not sufficient to generate evolutionarily novel structures such as the neural crest, teeth, the turtle shell, feathers, or cnidocysts. He claimed that such evolution could occur only through inheritable changes in the genes that regulated development. For years, those genes remained unknown. In 1977, the idea that change within regulatory genes was critical to evolution was extended by François Jacob, the Nobel Prize winner who helped establish the operon model of gene regulation. First, Jacob said that evolution worked with what it had: it acted by combining existing parts in new ways rather than by creating new parts. Second, he predicted that such "tinkering" occurred in the genes that constructed the embryo, not in the genes that functioned in adults (Jacob 1977).

HOW DUCKS GOT THEIR WEBBED FEET. In the past decade, scientists from around the world have confirmed this developmental model of the generation of novel structures. Some of this research has been noted in previous chapters. For example, in Chapter 16, the difference between the webbed duck foot and the clawed chicken foot was ascribed to the expression of *gremlin* in the duck's interdigital space. The Gremlin protein is an inhibitor of BMPs, and BMP signaling is thought to be responsible for the apoptosis in the interdigital region (Figure 23.22; Merino et al. 1999).

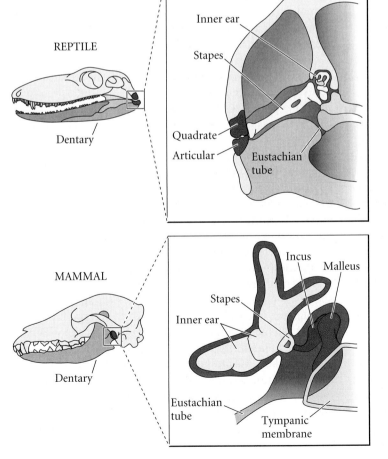

Figure 23.21
Evolution of the mammalian middle ear bones from the reptilian jaw. The quadrate and articular bones of reptiles were part of the lower jaw. Sound could be transmitted from these bones via the large stapes. When the dentary bone grew and took over the jaw functions of these two bones, the articular bone became the malleus and the quadrate bone became the incus. (After Romer 1949.)

Figure 23.22
Regulation of chick limb apoptosis by BMPs. Autopods of chick feet (left column) and duck feet (right column) are shown at similar stages. The in situ hybridizations show that while BMPs are expressed in both the chick and duck hindlimbs webbing, the duck limb shows expression of Gremlin protein (arrows) in the webbing as well. Gremlin is an inhibitor of BMPs. The pattern of cell death (shown by neutral red dye accumulation) becomes distinctly different in the two species. (Photographs courtesy of J. Hurle and E. Laufer.)

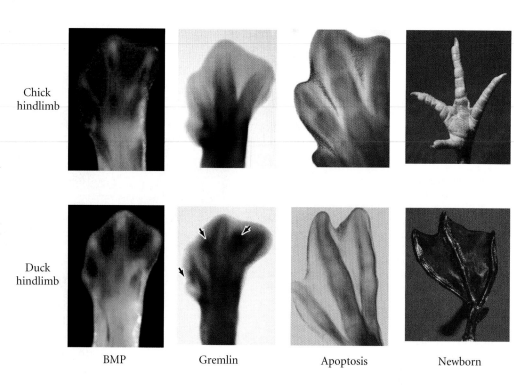

Chick hindlimb

Duck hindlimb

BMP   Gremlin   Apoptosis   Newborn

HOW BIRDS GOT THEIR FEATHERS. The avian feather is an excellent example of an evolutionary novelty. Feathers evolved only among the reptiles and are currently found only in the birds, which descended from that group. Although it had long been thought that feathers emerged as a modification of reptilian scales (see Maderson 1972; Maderson and Alibardi 2000; Prum et al. 1999), the mechanism that produces feathers has remained elusive. Harris and his colleagues (2002) have provided a developmental mechanism for feather evolution, showing that the feather most likely evolved from the archosaurian (dinosaur/bird ancestor) scale through an alteration of the expression pattern of the Sonic hedgehog and BMP2 proteins.

Both the scale and the feather start off the same way, with BMP2 and Shh expressed in separate domains. However, in the feather, both expression domains shift to the distal region of the appendage. Subsequently, this feather-specific pattern becomes repeated serially around the proximal-distal axis. The interaction between BMP2 and Shh then causes each of these regions to form its own axis—the barbs of the feather (Figure 23.23). Moreover, when this serially repeated pattern of Shh or BMP was experimentally modified, the feather pattern was modified in a predictable manner (Harris et al. 2002; Yu et al. 2002).

HOW MAMMALS CHANGED THEIR MOLARS. Stephen J. Gould (1989) joked that paleontologists believe mammalian evolution occurs when two teeth mate to produce slightly altered descendant teeth. Since enamel is far more durable than ordinary bone, teeth often remain after all the other bones have decayed. Indeed, tooth morphology has been critical to mammalian ecology and classification. Changes in the cusp pattern of molars is seen to be especially important in allowing the radiation of mammals into new ecological niches. What mechanism allows the mammalian molars to change their form so rapidly?

Jukka Jernvall and colleagues (Jernvall et al. 2000; Salazar-Ciudad and Jernvall 2002) pioneered a computer-based approach to phenotype production using Geographic Information Systems (GIS)—the same technology ecologists use to map the vegetation of hillsides. They have used this technology to map gene expression patterns in incipient tooth buds (literally turning a mountain into a molar). Their studies have shown that gene expression patterns forecast the exact location of the tooth cusps. They have also shown that the differences between the molars of mice and voles are predicated on differences in gene expression patterns (Figure 23.24).

Next, Salazar-Ciudad and Jernvall proposed a mathematical reaction-diffusion model (see Chapter 1) that would explain the differences in gene expression between these two species. Small changes in a gene network, again working through the interactions of the BMPs and Shh proteins, were crucial. Shh and FGFs (produced by the enamel knot signaling centers) inhibit BMP production, while BMP production stimulates both the production of more BMPs and the synthesis of its inhibitors, the FGFs and Shh proteins. This would create regions of activators (BMPs) that blocked epithelial proliferation, and regions of inhibitors (FGFs, Shh) that blocked BMP synthesis and independently stimulated mesenchymal proliferation. The result is a pattern of gene activity

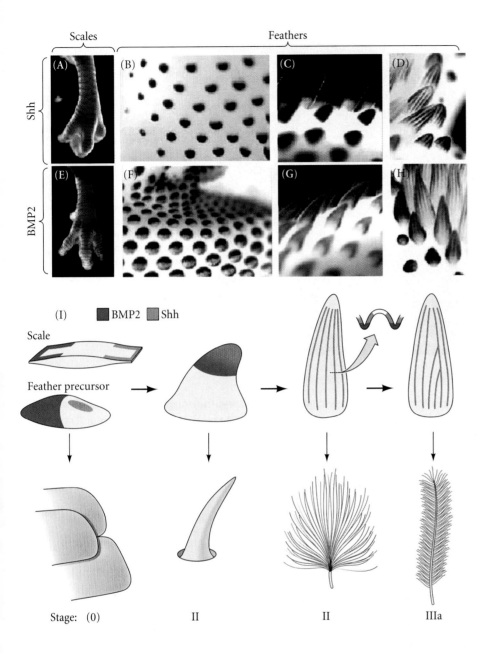

Scales · Feathers

Shh · BMP2

(A) (B) (C) (D)
(E) (F) (G) (H)

(I) ■ BMP2 ■ Shh

Scale

Feather precursor

Stage: (0)   II   II   IIIa

Figure 23.23
Expression patterns of BMP2 and Shh in feathers and scales. (A–H) Whole mount in situ hybridizations show the placement of Shh and BMP2 in developing chick feathers and scales. (A, E) Shh and BMP2 are seen at opposite ends of the developing scale. In feather buds, the pattern is initially the same as in the scales (B, F) but both genes become expressed at the distal tip of the developing feather (C, G) and then form bands along the feather rudiment (D, H). (I) Model for the evolution of the feather by changes in the pattern of BMP2 and Shh expression. Stage 0 shows the Shh and BMP2 expression in the scale (above) and feather bud (below). Stage 1 represents a tubular feather as evolved from an archosaurian scale. The Shh and BMP2 expression patterns are postulated to be at the tip. Stage 2 represents the emergence of a branched feather evolved by further changing the expression patterns of BMP2 and Shh to form rows along the proximal-distal axis. In stage 3a, changes in feather morphology evolved by altering the pattern to produce a central rachis. (After Harris et al. 2002.)

that changes as the shape of the tooth changes, and vice-versa (Figure 23.25).

Using this model, the large differences between mouse and vole molars can be generated by small changes in the binding constants and diffusion rates of the BMP and Shh proteins. A small increase in the diffusion rate of BMP4 and a stronger binding constant of its inhibitor is sufficient to change the vole pattern of tooth growth into that of the mouse. Thus, large morphological changes can result from very small changes in initial conditions. Another conclusion is that all the cells can start off with the same basic set of instructions; the specific instructions emerge as the cells interact. This mathematical model also predicts that some types of teeth are much more likely to evolve in certain ways and not

in others, and that certain shapes are more likely to evolve than others. These predictions conform to what paleontologists have concluded about mammalian evolution.*

HOW FISH GOT THEIR JAWS. When the vertebrate face first evolved, it did not have jaws. These earliest vertebrates were similar to today's **cyclostomes**—the lampreys and hagfish. They have a completely cartilagenous skeleton and a skull that does not differ greatly from the vertebrae. In gnathostomes

---

*One of the critical differences between the Jernvall and Salazar-Ciudad model and other reaction-diffusion models is that it allows the parameters to change while the tooth is developing. This assumption gives a great amount of flexibility to the developing system, allowing it to change at relatively rapid rates during evolution.

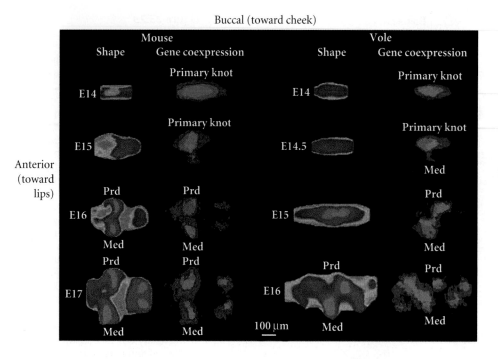

Buccal (toward cheek)

Figure 23.24

GIS analysis of gene activity in the formation of the first set of cusps in mouse and vole molars. For both species, GIS mapping of molar shape is shown on the left and the expression of *fgf4* and *shh* (two genes expressed from the enamel knots) are shown at the right. The gene expression pattern on embryonic day 15 predicts the formation of the new cusp seen on day 16; the gene expression pattern on day 16 predicts the formation of cusps in those areas on day 17. Similarly, in the vole molar, whose cusps are diagonal to one another, gene expression predicts cusp formation. (After Jernvall et al. 2000; photograph courtesy of J. Jernvall.)

(jawed vertebrates), cranial neural crest cells enter the pharyngeal arches, and those neural crest cells migrating into the first pharyngeal arch form the mandible (lower jaw) and the maxillary process (roof of the mouth). A migration of neural crest cells takes place in cyclostomes, but the first pharyngeal arch does not form Meckel's cartilage or the mandible derived from it. Instead, the neural crest cells form a round, jawless mouth. It appears that two major events had to occur in order for this group of first pharyngeal arch neural crest cells to become a jaw: they had to have a permissive environment, and they had to receive a set of new instructions.

The *permissive environment* came via the removal of a barrier. The cyclostomes form a naso-hypophyseal plate from which the nasal epithelium and the hypophysis (pituitary) both develop. This epithelial plate forms a barrier to neural

crest cell migration, and the only way for the neural crest cells to migrate forward is to go beneath it. These neural crest cells then form the upper lip of the cyclostome mouth. In gnathostomes, the naso-hypophyseal plate has separated into the nasal placode and Rathke's pouch, leaving a space between these structures through which the neural crest cells could mi-

Figure 23.25

Basic model for cusp development in mice and voles. (A) Experimentally derived gene network wherein BMPs activate their own production as well as the production of their inhibitors, Shh and FGFs. The FGFs and Shh stimulate cell proliferation; the BMPs inhibit it. (B) Predicted and observed results from this model. The model can generate the final and intermediate forms of molar development in mice and voles, and the difference between mouse and vole molars can be reproduced by slight alterations in the rate of BMP diffusion and binding to inhibitors. (After Salazar-Ciudad and Jernvall 2002; photographs courtesy of J. Jernvall.)

(A)
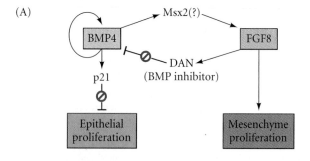

(B)
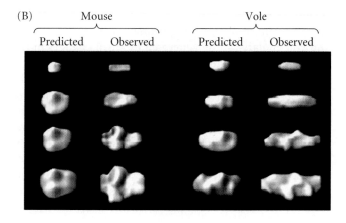

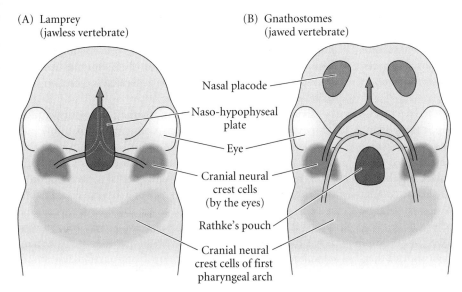

(A) Lamprey
(jawless vertebrate)

(B) Gnathostomes
(jawed vertebrate)

Nasal placode

Naso-hypophyseal plate

Eye

Cranial neural crest cells (by the eyes)

Rathke's pouch

Cranial neural crest cells of first pharyngeal arch

Figure 23.26
Neural crest cell migration patterns in lampreys and gnathostomes. (A) In the lamprey, the upper lip develops from cranial neural crest cells that migrate rostrally underneath the naso-hypophyseal plate (which will form both the nasal placode and the pituitary). (B) In gnathostomes, the naso-hyophyseal tissue has already split into Rathke's pouch (the pituitary rudiment) and the nasal (olfactory) placode. The cranial neural crest cells migrate rostrally between these two structures to form part of the lower jaw cartilage. The upper jaw forms from those cranial neural crest cells that migrate rostrally and fuse anterior to Rathke's pouch. (After Kuratani et al. 2001.)

grate. The cells that migrate into this region form the mandible (Figure 23.26; Kuratani et al. 2001). Thus, the difference allowing the formation of the jaw may be a difference in the timing of the separation between the rudiments of the nasal placodes and the rudiments of the pituitary. If the separation is early, jaws are possible. If the separation is late, a barrier exists that prevents neural crest cell migration into the region that would form the mandible. Such a shift in timing may be due to a slight change in the timing of a particular tissue interaction (Shigetani et al. 2002).

The *new set of instructions* come from a shift in Hox gene expression. In lampreys, there is Hox gene expression in the first pharyngeal arch (Cohn 2002); in gnathostomes, there is no such expression. Moreover, if Hox genes are ectopically expressed in the first pharyngeal arches of fish, frog, or chick embryos, jaw development is severely inhibited (Alexandre et al. 1996; Pasqualetti et al. 2000; Grammatopoulos et al. 2000). (Indeed, the structures look reminiscent of the ancestral hyoid arches). Thus, Hox genes appear to prevent the neural crest cells of the first pharyngeal arch from forming jaws. When Hox genes are inhibited, a new set of genetic instructions is made available.

Developmental Constraints

There are only about three dozen animal phyla, which represent the major body plans of the animal kingdom. One can easily envision other types of body plans and imagine animals that do not exist (science fiction writers do it all the time). Why aren't more body plans found among the animals? To answer this, we have to consider the constraints that development imposes on evolution.

The number and forms of possible phenotypes that can be created are limited by the interactions that are possible among molecules.* These molecular interactions also allow

change to occur in certain directions more easily than in others. Collectively, the restraints on phenotype production are called **developmental constraints**. These constraints on evolution fall into three major categories: physical, morphogenetic, and phyletic.

> **WEBSITE 23.10 Why are there no new animal phyla?** It appears that the three dozen or so known phyla were all created by 500 million years ago. It may be the case that no new phylum has emerged since the late Cambrian. What is the evidence for the early formation of the phyla, and are any body plans left unused?

Physical constraints

There are immutable physical constraints on the construction of an organism. The laws of diffusion, hydraulics, and physical support allow only certain mechanisms of development to occur. One cannot have a vertebrate on wheeled appendages (of the sort that Dorothy saw in Oz) because blood cannot circulate to a rotating organ; this entire possibility of evolution has been closed off. Similarly, structural parameters and fluid dynamics forbid the existence of 5-foot-tall mosquitoes.

The elasticity and tensile strength of tissues is also a physical constraint. In the sperm of *Drosophila*, for example, the type of tubulin that can be used in the axoneme is constrained by the need for certain physical properties in the exceptionally long flagellum (Nielsen and Raff 2002). The six cell behaviors used in morphogenesis (cell division, growth, shape change,

---

*G. W. Leibniz, probably the philosopher who most influenced Darwin, noted that existence must be limited not only to the possible but to the *compossible*—that is, whereas numerous things can come into existence, only those that are mutually compatible will actually exist (see Lovejoy 1964). So although many developmental changes are possible, only those that can integrate into the rest of the organism (or which can cause a compensatory change in the rest of the organism) will be seen.

migration, death, and matrix secretion) are each limited by physical parameters, and thereby provide limits on what structures animals can form. Interactions between different sets of tissues involve coordinating the behaviors of cell sheets, rods, and tubes in a limited number of ways (Larsen 1992).

Morphogenetic constraints

Rules for morphogenetic construction also limit the phenotypes that are possible (Oster et al. 1988). Bateson (1894) and Alberch (1989) noted that when organisms depart from their normal development, they do so in only a limited number of ways. Some of the best examples of these types of constraints come from the analysis of limb formation in vertebrates. Holder (1983) pointed out that although there have been many modifications of the vertebrate limb over 300 million years, some modifications (such as a middle digit shorter than its surrounding digits) are not found. Moreover, analyses of natural populations suggest that there is a relatively small number of ways in which limb changes can occur (Wake and Larson 1987). If a longer limb is favorable in a given environment, the humerus may become elongated, but one never sees two smaller humeri joined together in tandem, although one could imagine the selective advantages that such an arrangement might have. This observation suggest a limb construction scheme that follows certain rules.

The rules governing the architecture of the limb may be the rules of the reaction-diffusion model (Newman and Frisch 1979; Miura and Shiota 2000a,b). Oster and colleagues (1988) found that the reaction-diffusion model can explain the known morphologies of the limb and could explain why other morphologies are forbidden. The reaction-diffusion equations predicted the observed succession of bone development from stylopod (humerus/femur) to zeugopod (ulna-radius/tibia-fibula) to autopod (hand/foot). If limb morphology is indeed determined by the reaction-diffusion mechanism, then spatial features that cannot be generated by reaction-diffusion kinetics will not occur.

Evidence supporting this mathematical model comes from experimental manipulations, comparative anatomy, and cell biology. When an axolotl limb bud is treated with the anti-mitotic drug colchicine, the number of cells in the limb bud is reduced. In these experimental limbs, there is not only a reduction in the number of digits, but a loss of certain digits in a certain order, as predicted by the mathematical model and by the "forbidden" morphologies. Moreover, these losses of specific digits produce limbs very similar to those of certain salamanders whose limbs develop from particularly small limb buds (Figure 23.27; Alberch and Gale 1983, 1985).

Phyletic constraints

Phyletic constraints on the evolution of new structures are historical restrictions based on the genetics of an organism's

Figure 23.27
Relationship between cell number and number of digits in salamanders. (A) The hindlimb of an axolotl (*Ambystoma mexicanum*) with its five symmetrical digits. (B, C) Digits on the axolotl hindlimb after the hindlimb bud was incubated in colchicine to reduce cell number. (D, E) Hindlimbs of two wild salamanders, each having a smaller limb bud than most other salamanders: (D) *Hemidactylium scutatum* and (E) *Proteus anguinus*. The parallels between the experimental variation and the natural variation can be seen; the common denominator is reduced cell numbers in the limb buds. (After Oster et al. 1988.)

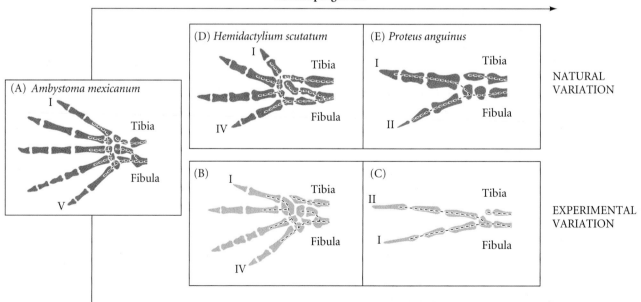

development (Gould and Lewontin 1979). Once a structure comes to be generated by inductive interactions, it is difficult to start over again. The notochord, for example, which is functional in adult protochordates such as amphioxus (Berrill 1987), is considered vestigial in adult vertebrates. Yet it is transiently necessary in vertebrate embryos, where it specifies the neural tube. Similarly, Waddington (1938) noted that, although the pronephric kidney of the chick embryo is considered vestigial (since it has no ability to concentrate urine), it is the source of the ureteric bud that induces the formation of a functional kidney during chick development (see Chapter 14).

One fascinating example of a phyletic constraint is the lack of variation among marsupial limbs. While eutherian limbs show a dramatic range of diversity (claws, wings, paddles, flippers, hands), marsupial limbs are pretty much the same in all marsupial species. Sears (2002) has documented that the necessity for the marsupial fetus to crawl into its mother's pouch has constrained limb development such that the limb musculature and cartilage develop very early into a structure that can grasp and crawl. Any other variation has effectively been eliminated.

As genes acquire new functions during evolution, they may become involved in more than one module, making change difficult. Galis (Galis 1999; Galis and Metz 2001) speculates that mammals have only seven cervical vertebrae (while birds may have dozens) because the Hox genes that specify these vertebrae have become linked to cell proliferation in mammals. Thus, changes in Hox gene expression that might facilitate evolutionary changes in the skeleton might inadvertently misregulate cell proliferation and lead to cancers. She supports this speculation with epidemiological evidence showing that changes in skeletal morphology correlate with childhood cancer. Similarly, Galis and colleagues (Galis et al. 2002) provide evidence that the reason the segment polarity gene network is conserved in all types of insects is that these genes play roles in several different pathways. Such pleiotropy constrains the possibilities for alternative mechanisms, since it makes change difficult.

Until recently, it was thought that the earliest stages of development would be the hardest to change, because altering them would either destroy the embryo or generate a radically new phenotype. But recent work (and the reappraisal of older work; see Raff et al. 1991) has shown that certain alterations can be made to early cleavage without upsetting the final form. Evolutionary modifications of cytoplasmic determinants in mollusc embryos can give rise to new types of larvae that still metamorphose into molluscs, and changes in sea urchin cytoplasmic determinants can generate sea urchins that develop directly (with no larval stage), but still become sea urchins.

The earliest stages of development, then, appear to be extremely plastic. The later stages are very different from species to species, as the different phenotypes of mice, zebrafish, snakes, and newts amply demonstrate. There is something in the middle of development, however, that appears to be invariant. Sander (1983) and Raff (1994) have argued that the formation of new body plans (*Baupläne*) is inhibited by the global consequences of induction during the **phylotypic stage**—the stage that typifies a phylum. For instance, the late neurula, also known as the **pharyngula**, is the phylotypic stage that appears to be critical for all vertebrates (see Figures 23.17 and 1.5; Slack et al. 1993). And in fact, while all the vertebrates arrive at the pharyngula stage, they do so by very different means. Birds, reptiles, and fish arrive there after meroblastic cleavages of different sorts; amphibians get there by way of radial holoblastic cleavage; and mammals reach the same stage after constructing a blastocyst, chorion, and amnion (see Chapter 11).

Before the vertebrate pharyngula stage, there are few inductive events, and most of them are on global scales (involving axis specification). In these early stages of development, there is a great deal of regulation, so small changes in morphogen distributions or the position of cleavage planes can be accommodated (Figure 23.28; Henry et al. 1989). After the pharyngula stage, there are a great many inductive events, but almost all of them occur within discrete modules. The lens induces the cornea, but if it fails to do so, only the eye is affected. But during the phylotypic pharyngula stage, the modules interact. Failure to have the heart in a certain place can affect the induction of eyes (see Figure 6.4). Failure to induce the mesoderm in a certain region leads to malformations of the kidneys, limbs, and tail. By searching the literature on congenital anomalies, Galis and Metz (2001) have documented that the pharyngula is much more vulnerable than any other stage. Moreover, based on patterns of multiple organ anomalies within the same person, they concluded that these mal-

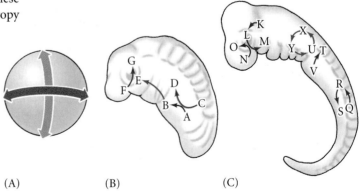

(A)   (B)   (C)

Figure 23.28
Mechanism for the bottleneck at the pharyngula stage of vertebrate development. (A) In the cleaving embryo, global interactions exist, but there are very few of them (mainly to specify the axes of the organism). (B) At the neurula to pharyngula stages, there are many global inductive interactions. (C) After the pharyngula stage, there are even more inductive interactions, but they are primarily local in effect, confined to their own modules. (After Raff 1994.)

formations are due to the interactivity of the modules at this stage. Thus, this phylotypic stage that typifies the vertebrate phylum appears to constrain its evolution. Once an organism becomes a vertebrate, it is difficult (if not impossible) to evolve into anything else.

> **WEBSITE 23.11 Changing embryonic traits through natural selection.** Just as changes in embryos can produce new phenotypes, so natural selection on adults can favor certain types of embryos that produce favorable adult phenotypes. This selection for adult traits may explain the con-

servation of some patterns of development as well as deviations from the norm.

> **WEBSITE 23.12 Alternative mechanisms for evolutionary developmental biology.** Evolution is accomplished through heritable changes in development. In this textbook, these heritable changes are assumed to be those that alter gene expression patterns. However, other models have been proposed in which there is horizontal transmission of genetic information between phyla, or in which there is inheritance of cytoplasmic properties.

Sidelights & Speculations

Canalization and the Release of Developmental Constraints

Morphogenetic constraints are one way of preventing certain variations. It is not the only way, however. Surprisingly, many newly occurring mutations, even in vital genes, do not affect development. Rather, development appears to be buffered so that slight abnormalities of genotype or slight perturbations in the environment do not lead to the formation of abnormal phenotypes (Waddington 1942; Siegal and Bergman 2002). This phenomenon, called **canalization**, serves as an additional constraint on the evolution of new phenotypes.

It is difficult for a mutation to actually affect development (Nijhout and Paulsen 1997). It is the rare mutation that is 100 percent penetrant. Canalization allows mutations to accrue in the genotype without being expressed in the phenotype (and therefore without being immediately accessible to natural selection). Thus, in the short term, it limits the variability of the phenotype by promoting cryptic genetic variation. However, in the long term, canalization can act as a capacitor for phenotypic change, since it allows mutant alleles to accumulate in a genome without being expressed. Such cryptic genetic variability can be made manifest by a change in the environmental conditions, and can then be selected.

Canalization can be the result of genetic redundancy. One of the major recent discoveries of developmental biology has been the stability of the phenotype even after the deletion of major developmentally important genes (Wilkins 1997; Morange 2001). In many instances, the loss of function of a particular gene is compensated for by the activation of another gene, sometimes not even of the same structural family as the one deleted. In other instances, there is already another protein in the cell whose activities are partially re-

dundant to those of the protein encoded by the deleted gene (Erickson 1993; Wilkins 1997). Nowak and colleagues (1997) have provided mathematical models to explain how such redundancy can be selected for and how redundancy can be made evolutionarily stable.

Canalization can also result from the buffering capacity of heat shock proteins. It has long been known that stress, in the form of environmental factors such as temperature, can overpower the buffering systems of development and alter the phenotype. The altered phenotype then becomes subject to natural selection, and if selected, will eventually appear without the stress that originally induced it. Waddington called this phenomenon genetic assimilation (see Chapter 22). For instance, when Waddington subjected *Drosophila* larvae of a certain strain to high temperatures, they lost their wing crossveins. After a few generations of repeated heat shock, the crossveinless phenotype continued to be expressed even without the heat shock treatment. While Waddington's results look like a case of "inheritance of acquired characteristics," there is no evidence for that view. Certainly, the crossveinless phenotype was not an adaptive response to heat. Nor did the heat shock cause the mutations that led to it. Rather, the heat shock overcame the buffering systems, allowing preexisting mutations to give rise to the crossveinless phenotype.

In 1998, Suzanne Rutherford and Susan Lindquist showed that a major agent responsible for this buffering was the "heat shock protein" **Hsp90**. Hsp90 binds to a set of signal transduction molecules that are inherently unstable. When it binds to them, it stabilizes their tertiary structure so that they can respond to upstream signaling molecules. Heat shock, however, causes other proteins in the cell to become unsta-

ble, and Hsp90 is diverted from its normal function (of stabilizing the signal transduction proteins) to the more general function of stabilizing any of the cell's now partially denatured peptides (Jakob et al. 1995; Nathan et al. 1997). Since Hsp90 was known to be involved with inherently unstable proteins and could be diverted by stress, the researchers suspected that it might be involved in buffering developmental pathways against environmental contingencies.

Evidence for the role of Hsp90 as a developmental buffer first came from mutations of *Hsp83*, the gene for Hsp90. Homozygous mutations of Hsp83 are lethal in *Drosophila*. Heterozygous mutations increase the proportion of developmental abnormalities in the population into which they are introduced. In populations of *Drosophila* heterozygous for mutant *Hsp83*, deformed eyes, bristle duplications, and abnormalities of legs and wings appeared (Figure 23.29). When different mutant alleles of *Hsp83* were brought together in the same flies, both the incidence and severity of the abnormalities increased. Abnormalities were also seen when a specific inhibitor of Hsp90 (geldanamycin) was added to the food of wild-type flies, and the types of defects seen differed between different stocks of flies. The abnormalities observed did not show simple Mendelian inheritance, but were the outcome of the interactions of several gene products. Selective breeding of the flies with the abnormalities led over a few generations to populations in which 80–90% of the progeny had the mutant phenotype. Moreover, these mutant progeny did not all have the *Hsp83* mutation. In other words, once the mutation in *Hsp83* had allowed cryptic mutations to be expressed, selective matings could retain the abnormal phenotype even in the absence of abnormal Hsp90.

(A)

(B)

(C)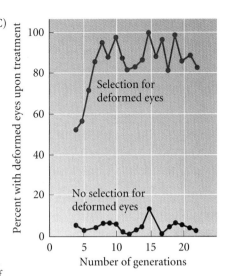

Figure 23.29
Hsp90 buffers development. (A, B) Developmental abnormalities in *Drosophila* associated with mutations in the *Hsp83* gene include (A) deformed eyes and (B) thickened wing veins. (C) The deformed eye trait seen in (A) was selected by breeding only those individuals expressing the trait. This abnormality was not observed in the original stock, but it can be seen in a high proportion of the descendants of individuals are mated to heterozygous *Hsp83* flies. The strong response to selection showed that even though the population was small, it contained a large amount of hidden genetic variation. (After Rutherford and Lindquist 1998; photographs courtesy of the authors.)

Thus, Hsp90 is probably a major component of the buffering system that enables the canalization of development. It provides a way to resist phenotype fluctuations that would otherwise result form slight mutations or slight environmental changes. Hsp90 might also be responsible for allowing mutations to accumulate, but keeping them from being expressed until the environment changes. No individual mutation would change the phenotype, but mating would allow these mutations to be "collected" by members of the population. An environmental change (anything that might stress the cells) might thereby release the hidden phenotypic possibilities of the population. In other words, transient decreases in Hsp90 (resulting from its aiding stress-damaged proteins) would uncover preexisting genetic variations that would produce morphological variations. Most of these morphological variations would probably be deleterious, but some might be selected for in the new environment. Such release of hidden morphological variation may be responsible for the many examples of rapid speciation found in the fossil record.

A New Evolutionary Synthesis

In 1922, Walter Garstang declared that **ontogeny** (an individual's development) does not recapitulate **phylogeny** (evolutionary history); rather, it creates phylogeny. Evolution is generated by heritable changes in development. "The first bird," said Garstang, "was hatched from a reptile's egg." Thus, when we say that the contemporary one-toed horse evolved from a five-toed ancestor, we are saying that heritable changes occurred in the differentiation of the limb mesoderm into chondrocytes during embryogenesis in the horse lineage. Evolution, said Richard Goldschmidt (1940), is the result of heritable changes in development, and this is as true for whether a fly has two or three bristles on its back as for whether an appendage is to become a fin or a limb.

This view of evolution as the result of hereditary changes affecting development was largely lost during the 1940s, when the Modern Synthesis of population genetics and evolutionary biology formed a new framework for research in evolutionary biology. The Modern Synthesis has been one of the greatest intellectual achievements of biology. By merging the traditions of Darwin and Mendel, evolution within a species could be explained: Diversity within a population arose from the random production of mutations, and the environment acted to select the most fit phenotypes. Those individuals most capable of reproducing would transmit the genes that gave them their advantage. Such genes included, for example, those encoding enzymes with higher rates of synthesis and globins with greater oxygen-carrying capacity. It was assumed that the same kinds of changes (genetic or chromosomal mutations) that caused evolution within a species also caused the evolution of new species. There would need to be an accumulation of these mutations, and a mechanism of reproductive isolation to enable them to accumulate in new combinations, if a new phenotype was to be produced.

Not only could the Modern Synthesis explain evolution within a species remarkably well, it also explained medically relevant questions such as why certain alleles that seem deleterious (the hemoglobin gene variant that can result in sickle cell anemia, for example) might be selected for in certain populations. The population genetic approach to evolution was summed up by one of its foremost practitioners and theorists, Theodosius Dobzhansky, when he declared, "Evolution is a change in the genetic composition of populations. The study of the mechanisms of evolution falls within the province of population genetics" (Dobzhansky 1951).

When the Modern Synthesis was formulated, developmental biology and developmental genetics were not established sciences. As the Synthesis became strengthened by systematics, plant biology, and paleontology, there was little that developmental biology could add. The questions raised in evolutionary biology could be answered without recourse to developmental biology, and the developmental approach became excluded from the Modern Synthesis (Hamburger 1980; Gottlieb 1992; Dietrich 1995; Gilbert et al. 1996). It was thought that population genetics could explain evolution, so morphology and development were seen to play little role in modern evolutionary theory (Adams 1991). In other words, **macroevolution** (the large morphological changes seen between species, classes, and phyla) could be explained by the mechanisms of **microevolution**, which is evolution within a species and can be described as "differential adaptive values of genotypes or deviations from random mating or both these factors acting together" (Torrey and Feduccia 1979).

However, the population genetic model of evolution contained some major assumptions that have now been called into question.

- **Gradualism**. The supposition that all evolutionary changes occur gradually was debated by Darwin and his friends. Thomas Huxley, for instance, accepted evolution, but he felt that Darwin had burdened his theory with an unnecessary assumption of gradualism. A century later, Eldredge and Gould (1972), Stanley (1979), and others postulated **punctuated equilibrium** as an alternative to the gradualism that characterized the Modern Synthesis. According to this theory, species were characterized by their morphological stability. Evolutionary changes tended to be rapid, not gradual. At the same time, molecular studies (King and Wilson 1975) showed that 99% of the DNA of humans and chimpanzees was identical, demonstrating that a small change in DNA could cause large and important morphological changes. New findings in paleontology and molecular biology prompted scientists to consider seriously the view that mutations in regulatory genes can create large changes in morphology in a relatively short time.

- **Extrapolation of microevolution to macroevolution**. The idea that accumulations of small mutations result in changes leading to higher taxa has also been criticized. Goldschmidt (1940) began his book *The Material Basis of Evolution* by asking the population genetic evolutionary biologists to try to explain the evolution of hair in mammals, feathers in birds, and the poison apparatus of snakes. Interestingly, both Goldschmidt and Waddington saw homeotic mutations as the kind of genetic change that could change one structure into another and possibly create new structures or new combinations of struc-

tures. These mutations would not be in the structural genes, but in the regulatory genes.

- **Lack of genetic similarity in disparate organisms**. We have come a long way from when Ernst Mayr (1966) could state, concerning macroevolution: "Much that has been learned about gene physiology makes it evident that the search for homologous genes is quite futile except in very close relatives." Indeed, when one considers the Hox genes, the signal transduction pathways, and the families of paracrine factors, adhesion molecules, and transcription factors, the opposite has been seen to be the case. Adult organisms may have dissimilar structures, but the genes instructing the formation of these structures are extremely similar.

The population genetic model was formulated to explain natural selection. It is based on genetic differences in adult organisms competing for reproductive advantage. The developmental genetic model has been formulated to account for phylogeny—evolution above the species level. It is based on the similarities among regulatory genes that are active in embryos and larvae. We are still approaching evolution in the two ways that Darwin recognized. Both views involve descent with modification, and one can emphasize either the similarities or the differences between taxa. Thus, when confronted with the question of how the arthropod body plan arose, Hughes and Kaufman (2003) begin their study, "To answer this question by invoking natural selection is correct—but insufficient. The fangs of a centipede…and the claws of a lobster accord these organisms a fitness advantage. However, the crux of the mystery is this: From what developmental genetic changes did these novelties arise in the first place?"

There is emerging now a rapprochement between the population genetic accounts of evolution and the developmental genetic accounts of evolution. The population genetic approach has focused on variation within populations, while the developmental genetic approach has focused on variation between populations (Amundson 2001; Gilbert 2000). Similarly, population geneticists have been looking primarily at genes in adults competing for reproductive success, while developmental geneticists have been looking at genes involved in forming embryonic and larval organs. These differences are becoming blurred as both population geneticists and developmental geneticists begin looking at the regulatory genes that control development (see Arthur 1997; Macdonald and Goldstein 1999; Zeng et al. 1999). The two approaches complement each other: While the population genetic approach focuses on the survival of the fittest, the developmental genetic approach to evolution is more concerned with the *arrival* of the fittest.

Evolutionary developmental biology is already beginning to provide answers to classic evolutionary genetic questions. Evolutionary biologists have long studied problems such as mimicry and industrial melanism. Now, the genes involved in these processes are being identified so that the mechanisms of under-

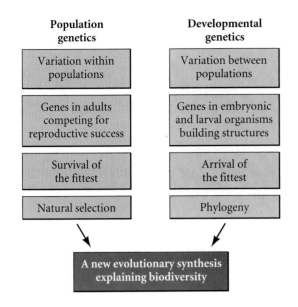

Population genetics	Developmental genetics
Variation within populations	Variation between populations
Genes in adults competing for reproductive success	Genes in embryonic and larval organisms building structures
Survival of the fittest	Arrival of the fittest
Natural selection	Phylogeny

A new evolutionary synthesis explaining biodiversity

Figure 23.30
A newly emerging evolutionary synthesis. The classic approach to evolution has been that of population genetics. This approache emphasized variations within a species that enabled certain adult individuals to reproduce more frequently; thus, it could explain natural selection. The developmental approach looks at variation between populations, and it emphasizes the regulatory genes responsible for organ formation. It is better able to explain evolutionary novelty and constraint. Together, these two approaches constitute a more complete genetic approach to evolution.

lying phenomena can be explained (Koch et al. 1998; Brakefield 1998). To explain evolution, both the population genetic and the developmental genetic accounts are required (Figure 23.30).

Leaving developmental biology out of evolutionary biology has left evolutionary biology open to attacks by creationists. According to Behe (1996), population genetics cannot explain the origin of structures such as the eye, so Darwinism must be false.* How could such a complicated structure have emerged by a collection of chance mutations? If a mutation caused a change in the lens, how could it be compensated for by changes in the retina? Mutations would serve only to destroy complex organs, not create them. However, once one adds development to the evolutionary synthesis, one can see how the eye can develop through induction, and that the concepts of modularity and correlated progression can readily ex-

*Behe (1996) makes this point explicitly, using the example of the eye. Although he attempts to disprove the theory of evolution by using the eye as an example, he never once mentions the studies on Pax6. Rather, Behe mentions theories from the 1980s (based solely on population genetics) and puts them forth as contemporary science.

plain such a phenomenon (Waddington 1940; Gehring 1998). Moreover, when one sees that the formation of eyes in all known phyla is based on the same signal transduction pathway, using the *Pax6* gene, it is not difficult to see descent with modification forming the various types of eyes. This was much more difficult before the similarity of eye instructions had been discovered. Indeed, one study based in population genetics claimed that photoreceptors of eyes had arisen independently more than 40 times during the history of the animal kingdom (Salvini-Plawen and Mayr 1977).

WEBSITE **23.13** **"Scientific Creationism" and "Intelligent Design."** Modern creationists often deny that genes play a role in development, because this would enable evolution to occur. These websites look at "scientific creationism" and "intelligent design" and includes a lecture on how evolutionary developmental biology provides some of the best evidence against creationism.

In his review of evolution in 1953, J. B. S. Haldane expressed his thoughts about evolution with the following developmental analogy: "The current instar of the evolutionary theory may be defined by such books as those of Huxley, Simpson, Dobzhansky, Mayr, and Stebbins [the founders of the Modern Synthesis]. We are certainly not ready for a new moult, but signs of new organs are perhaps visible." This recognition of developmental ideas "points forward to a broader synthesis in the future." We have finally broken through the old pupal integument, and a new, broader, developmentally inclusive evolutionary synthesis is taking wing.

Snapshot Summary: Evolutionary Developmental Biology

1. Evolution is caused by the inheritance of changes in development. Modifications of embryonic or larval development can create new phenotypes that can then be selected.

2. Darwin's concept of "descent with modification" explained both homologies and adaptations. The similarities of structure are due to common ancestry (homology), while the modifications were due to natural selection (adaptation to the environmental circumstances).

3. The Urbilaterian ancestor can be extrapolated by looking at the developmental genes that are common to protostomes and deuterostomes and that perform similar functions. These include the Hox genes that specify body segments; the *tinman* gene that regulates heart development; the *Pax6* gene that specifies those regions able to form eyes; and the genes that instruct head and tail formation.

4. Changes in the sequence of Hox genes can give them new properties that may have significant developmental effects. The constraint on insect anatomy of having only six legs is one of them.

5. Changes in the *targets* of Hox genes can alter the effects of those genes. In response to the Ubx protein, for example, the third thoracic segment specifies halteres in flies but hindwings in butterflies.

6. Changes of Hox gene expression *within* a region can alter the structures formed by that region. For instance, changes in the expression of Ubx and abdA in insects regulate the production of prolegs in the abdominal segments of the larvae.

7. Changes in Hox gene expression *between* body regions can alter the structures formed by each region. In crustaceans, different Hox expression patterns enable the body to have or to lack maxillipeds on its thoracic segments.

8. Changes in Hox gene expression are correlated with the limbless phenotypes of snakes.

9. Changes in Hox gene *number* may allow Hox genes to take on new functions. Large changes in the numbers of Hox genes correlate with major transitions in evolution.

10. Duplications of genes may also enable these genes to become expressed in new places. The formation of new cell types may result from duplicated genes whose regulation has diverged.

11. In addition to structures and genes being homologous, signal transduction pathways can be homologous. In these cases, homologous proteins are organized in homologous ways. These pathways can be used for different developmental processes both in different organisms and within the same organism.

12. The modularity of development allows parts of the embryo to change without affecting other parts.

13. The dissociation of one module from another is shown by heterochrony (a shift in the timing of the development of one region with respect to another) and by allometry (a shift in the growth rates of different parts of the organism relative to one another).

14. Allometry can create new structures (such as the pocket gopher cheek pouch) by crossing a threshold.

15. Duplication and divergence are important mechanisms of evolution. On the genetic level, the Hox genes and many other gene families started as single genes that were duplicated. The divergent members of such a gene family can assume different functions.

16. Changes in gene expression appear to account for the loss of limbs in snakes, the retention of webbing in ducks' feet, the emergence of feathers, and the evolution of differently shaped molars.

17. Co-option (recruitment) of existing genes and pathways for new functions is a fundamental mechanism for creating new phenotypes. One such case is the use of the limb development signaling pathway to form eyespots in butterfly wings.

18. Developmental constraints prevent certain phenotypes from occurring. Such constraints may be physical (no rotating limbs), morphogenetic (no middle digit smaller than its neighbors), or phyletic (no neural tube without a notochord).

19. The Hsp90 protein enables cells to accumulate mutations without expressing them. When the organism is stressed during development, these "cryptic mutations" can emerge in new phenotypes.

20. The merging of the population genetic model with the developmental genetic model of evolution is creating a new evolutionary synthesis that can account for macroevolutionary as well as microevolutionary phenomena.

Literature Cited

Acampora, D., S. Mazan, Y. Lallemand, V. Avantaggiato, M. Maury, A. Simeone and P. Brulet. 1995. Forebrain and midbrain regions are deleted in *Otx2*−/− mutants due to a defective anterior neuroectoderm specification during gastrulation. *Development* 121: 3279–3290.

Adams, M. 1991. Through the looking glass: The evolution of Soviet Darwinism. *In* L. Warren and H. Koprowski (eds.), *New Perspectives in Evolution.* Liss/Wiley, New York, pp. 37–63.

Adoutte, A., G. Balavoine, N. Lartillot and R. de Rosa. 1999. Animal evolution: The end of the intermediate taxa? *Trends Genet.* 15: 104–108.

Adoutte, A., G. Balavoine, N. Lartillot, O. Lespinet, B. Prud'homme and R. de Rosa. 2000. The new animal phylogeny: Reliability and implications. *Proc. Natl. Acad. Sci. USA* 97: 4453–4456.

Alberch, P. 1989. The logic of monsters: Evidence for internal constraints in development and evolution. *Geobios* (Lyon) mémoires spécial 12: 21–57.

Alberch, P. and J. Alberch. 1981. Heterochronic mechanisms of morphological diversification and evolutionary change in neotropical salamander *Bolitoglossa occidentalis* (Amphibia: Plethodontidae). *J. Morphol.* 167: 249–264.

Alberch, P. and E. Gale. 1983. Size dependency during the development of the amphibian foot: Colchicine induced digital loss and reduction. *J. Embryol. Exp. Morphol.* 76: 177–197.

Alberch, P. and E. Gale. 1985. A developmental analysis of an evolutionary trend: Digit reduction in amphibians. *Evolution* 39: 8–23.

Alexandre, D., J. D. Clarke, E. Oxtoby, Y. L. Yan, T. Jowett and N. Holder. 1996. Ectopic expression of *Hoxa-1* in the zebrafish alters the fate of the mandibular arch neural crest and phenocopies a retinoic acid-induced phenotype. *Development* 122: 735–746.

Amores, A. and 12 others. 1998. Zebrafish Hox clusters and vertebrate genome evolution. *Science* 282: 1711–1714.

Amundson, R. 2001. Adaptation and development: On the lack of common ground. *In* E. Sober and S. Orzack (eds.), *Adaptation and Optimality*. Cambridge University Press, Cambridge, pp. 303–334.

Ang, S. L., O. Jin, M. Rhinn, N. Daigle, L. Stevenson and J. Rossant. 1996. A targeted mouse *otx2* mutation leads to severe defects in gastrulation and formation of axial mesoderm and to deletion of rostral brain. *Development* 122: 243–252.

Arthur, W. 1997. *The Origin of Animal Body Plans: A Study in Evolutionary Developmental Biology*. Cambridge University Press, New York.

Averof, M. and M. Akam. 1995. Hox genes and the diversification of insect and crustacean body plans. *Nature* 376: 420–423.

Averof, M. and N. H. Patel. 1997. Crustacean appendage evolution associated with changes in Hox gene expression. *Nature* 388: 682–686.

Basler, K. and W. Struhl. G. 1993. Compartment boundaries and the control of *Drosophila* limb pattern by hedgehog protein. *Nature* 368: 208–214.

Bateson, W. 1894. *Materials for the Study of Variation*. Cambridge University Press, Cambridge.

Behe, M. J. 1996. *Darwin's Black Box: The Biochemical Challenge to Evolution*. Simon and Schuster, New York.

Berrill, N. J. 1987. Early chordate evolution. I. *Amphioxus*, the riddle of the sands. *Int. J. Invert. Reprod. Dev.* 11: 1–27.

Blader, P., S. Rastegar, N. Fischer and U. Strähle. 1997. Cleavage of the BMP4 antagonist chordin by zebrafish Tolloid. *Science* 278: 1937–1940.

Bodmer, R. 1993. The gene *tinman* is required for specification of the heart and visceral muscles in *Drosophila*. *Development* 118: 719–729.

Bolker, J. A. 2000. Modularity in development and why it matters to evo-devo. *Am. Zool.* 40: 770–776.

Boncinelli, E. 1994. Early CNS development: *Distal-less* related genes and forebrain development. *Curr. Opin. Neurobiol.* 4: 29–36.

Bonner, J. T. 1988. *The Evolution of Complexity*. Princeton University Press, Princeton, NJ.

Brakefield, P. M. 1998. The evolution-development interface and advances with the eyespot patterns of *Bicyclus* butterflies. *Heredity* 80: 265–272.

Brunelli, S., A. Faiella, V. Capra, V. Nigro, A. Simeone, A. Cama and E. Boncinelli. 1996. Germline mutations in the homeobox gene *EMX2* in patients with severe schizencephaly. *Nature Genet.* 12: 94–96.

Brunetti, C. R., J. E. Selegue, A. Monteiro, V. French, P. M. Brakefield and S. B. Carroll. 2001. The generation and diversification of butterfly eyespot color pattern. *Curr. Biol.* 11: 1578–1585.

Brylski, P. and B. K. Hall. 1988. Ontogeny of a macroevolutionary phenotype: The external cheek pouches of geomyoid rodents. *Evolution* 42: 391–395.

Burke, A. C., A. C. Nelson, B. A. Morgan and C. Tabin. 1995. Hox genes and the evolution of vertebrate axial morphology. *Development* 121: 333–346.

Cadigan, K. M. and R. Nusse. 1997. Wnt signaling: A common theme in animal development. *Genes Dev.* 11: 3286–3305.

Caldwell, M. W. and M. S. Y. Lee. 1997 A snake with legs from the marine Cretaceous of the Middle East. *Nature* 386: 705–709.

Carroll, S. B., S. D. Weatherbee and J. A. Langeland. 1995. Homeotic genes and the regulation and evolution of insect wing number. *Nature* 375: 58–61.

Carroll, S. B., J. K. Grenier and S. D. Weatherbee. 2001. *From DNA to Diversity: Molecular Genetics and the Evolution of Animal Design*. Blackwell, Oxford.

Chiang, C., Y. Litingtung, E. Lee, K. E. Young, J. L. Cordoen, H. Westphal and P. A. Beachy. 1996. Cyclopia and axial patterning in mice lacking *sonic hedgehog* gene function. *Nature* 383: 407–413.

Chow, R. L., C. R. Altmann, R. A. Lang and A. Hemmati-Brivanlou. 1999. Pax6 induces ectopic eyes in a vertebrate. *Development* 126: 4213–4222.

Cohn, M. J. 2002. Lamprey Hox genes and the origin of jaws. *Nature* 416: 386–387.

Cohn, M. J. and C. Tickle. 1999. Developmental basis of limblessness and axial patterning in snakes. *Nature* 399: 474–479.

Dale, L. and F. C. Wardle. 1999. A gradient of BMP activity specifies dorsal-ventral fates in early *Xenopus* embryos. *Semin. Cell Dev. Biol.* 10: 319–326.

Darwin, C. 1859. *The Origin of Species*. John Murray, London.

Davis, C. A., D. P. Holmyard, K. J. Millen and A. L. Joyner. 1991. Examining pattern formation in mouse, chicken, and frog embryos with an En-specific antiserum. *Development* 111: 287–298.

Degnan, B. M., S. M. Degnan, A. Giusti and D. E. Morse. 1995. A *Hox/hom* homeobox gene in sponges. *Gene* 155: 175–177.

DePew, M. J., J. K. Liu, J. E. Long, R. Presley, J. J. Meneses, R. A. Pedersen and J. L. R. Rubenstein. 1999. Dlx5 regulates regional development of the branchial arches and sensory capsules. *Development* 126: 3831–3846.

Dietrich, M. 1995. Richard Goldschmidt's "heresies" and the evolutionary synthesis. *J. Hist. Biol.* 28: 431–461.

DiSilvestro, R. L. 1997. Out on a limb. *BioScience* 47: 729–731.

Dobzhansky, Th. 1951. *Genetics and the Origin of Species*, 3rd Ed. Columbia University Press, New York.

Dyke, C. 1988. *The Evolutionary Dynamics of Complex Systems*. Oxford University Press, New York.

Eldredge, N. and S. J. Gould. 1972. Punctuated equilibria: An alternative to phyletic gradualism. *In* T. J. M. Schopf (ed.), *Models of Paleobiology*. Freeman, Cooper & Co., San Francisco, pp. 82–115.

Erickson, H. P. 1993. Gene knockouts of *c-src*, *TGFβ-1*, and tenascin suggest superfluous nonfunctional expression of proteins. *J. Cell Biol.* 120: 1079–1081.

Erwin, D. H. 1999. The origin of bodyplans. *Am. Zool.* 39: 617–629.

Erwin, D. H., J. Valentine and D. Jablonski. 1997. The origin of animal body plans. *Am. Sci.* 85: 126–137.

Finkelstein, R. and E. Boncinelli. 1994. From fly head to mammalian forebrain: The story of OTD and OTX. *Trends Genet.* 10: 310–315.

Finnerty, J. R. 2000. Head start. *Nature* 408: 778–780.

Galant, R. and S. B. Carroll. 2002. Evolution of a transcription repression domain in an insect Hox protein. *Nature* 415: 910–913.

Galis, F. 1999. Why do almost all mammals have seven cervical vertebrae? Developmental constraints, *Hox* genes, and cancer. *J. Exp. Zool./Mol. Dev. Evol.* 285: 19–26.

Galis, F. and J. A. Metz. 2001. Testing the vulnerability of the phylotypic stage: On modularity and evolutionary conservation. *J. Exp. Zool./Mol. Dev. Evol.* 29: 195–204.

Galis, F., T. J. M. van Dooren and J. A. Metz. 2002. Conservation of the segmented germband stage: Robustness or pleiotropy? *Trends Genet.* 18: 504–509.

Garstang, W. 1922. The theory of recapitulation: A critical restatement of the biogenetic law. *Zool. J. Linn. Soc.* 35: 81–101.

Gaunt, S. J. 1994. Conservation in the Hox code during morphological evolution. *Int. J. Dev. Biol.* 38: 549–552.

Gaunt, S. J. 2000. Evolutionary shifts of vertebrates structures and Hox expression up and down the axial series of segments: A consideration of possible mechanisms. *Int. J. Dev. Biol.* 44: 109–117.

Gayon, J. 2000. History of the concept of allometry. *Am. Zool.* 40: 7480–758.

Gehring, W. J. 1998. *Master Control Genes in Development and Evolution: The Homeobox Story*. Yale University Press, New Haven.

Gellen, G. and W. McGinnis. 1998. Shaping animal body plans in development and evolution by modulation of Hox expression patterns. *BioEssays* 20: 116–125.

Gerhart, J. and M. Kirschner. 1997. *Cells, Embryos, and Evolution*. Blackwell Science, Oxford.

Gilbert, S. F. 1996. Cellular dialogues in organogenesis. *In* M. E. Martini-Neri, G. Neri and J. M. Opitz (eds.), *Gene Regulation and Fetal Development: Proceedings of the Third International Workshop on Fetal Genetic Pathology 1993.* Wiley-Liss, New York, pp. 1–12.

Gilbert, S. F. 1998. Conceptual breakthroughs in developmental biology. *J. Biosci.* 23: 169–176.

Gilbert, S. F. 2000. Genes classical and genes developmental: The different uses of genes in evolutionary synthesis. *In* P. J. Beurton, R. Falk and H. J. Rheinberger (eds.), *The Concept of the Gene in Development and Evolution.* Cambridge University Press, Cambridge, pp. 178–192.

Gilbert, S. F., J. M. Opitz and R. A. Raff. 1996. Resynthesizing evolutionary and developmental biology. *Dev. Biol.* 173: 357–372.

Gish, D. T. 1973. *Evolution? The Fossils Say No!* Creation-Life Publishers, San Diego.

Goldschmidt, R. B. 1940. *The Material Basis of Evolution.* Yale University Press, New Haven.

Gottlieb, G. 1992. *Individual Development and Evolution: The Genesis of Novel Behavior.* Oxford University Press, New York.

Gould, S. J. 1989. *Wonderful Life.* W. W. Norton Co., New York.

Gould, S. J. 1990. An earful of jaw. *Nat. Hist.* 1990(3): 12–23.

Gould, S. J. and R. C. Lewontin. 1979. The spandrels of San Marcos and the Panglossian paradigm: A critique of the adaptationist programme. *Proc. R. Soc. Lond. B* 205: 581–598.

Graham, A. and I. McGonnell. 1999. Developmental evolution: This side of paradise. *Curr. Biol.* 9: R630–R632.

Grammatopoulos, G. A., E. Bell, L. Toole, A. Lumsden and A. S. Tucker. 2000. Homeotic transformation of branchial arch identity after *Hoxa2* overexpression. *Development* 127: 5355–5365.

Greer, J. M, J. Puet, K. R. Thomas and M. Capecchi. 2000. Maintenance of functional equivalence during paralogous Hox gene evolution. *Nature* 403: 661–664.

Haldane, J. B. S. 1953. Foreword. *In* R. Brown and J. F. Danielli (eds.), *Evolution: Society of Experimental Biology Symposium 7.* Cambridge University Press, Cambridge, pp. ix–xix.

Halder, G., P. Callaerts and W. J. Gehring. 1995. Induction of ectopic eyes by targeted expression of the *eyeless* gene in *Drosophila. Science* 267: 1788–1792.

Hall, B. K. 1999. *Evolutionary Developmental Biology*, 2nd Ed. Kluwer, Boston.

Hall, B. K. 2000. The neural crest as a fourth germ layer and vertebrates as quadroblastic, not triploblastic. *Evol. Dev.* 2: 3–5.

Hamburger, V. 1980. Embryology and the Modern Synthesis in evolutionary theory. *In* E. Mayr and W. Provine (eds.), *The Evolutionary Synthesis: Perspectives on the Unification of Biology.*

Cambridge University Press, New York, pp. 97–112.

Harris, M. P., J. F. Fallon and R. O. Prum. 2002. Shh-Bmp2 signaling module and the evolutionary origin and diversification of feathers. *J. Exp. Zool.* 294: 160–176.

Henry, J. J., S. Amemiya, G. A. Wray and R. A. Raff. 1989. Early inductive interactions are involved in restricting cell fates of mesomeres in sea urchin embryos. *Dev. Biol.* 136: 140–153.

Hirth, F. and H. Reichert. 1999. Conserved genetic programs in insect and mammalian brain development. *BioEssays* 21: 677–684.

Hirth, F., S. Therianos, T. Loop, W. J. Gehring, H. Reichart and K. Furukubo-Tokunaga. 1995. Developmental defects in brain segmentation caused by mutations of the homeobox genes *orthodenticle* and *empty spiracles* in *Drosophila. Neuron* 15: 769–778.

Hirth, F., B. Hartmann and H. Reichert. 1998. Homeotic gene action in embryonic brain development of *Drosophila. Development* 125: 1579–1589.

Holder, N. 1983. Developmental constraints and the evolution of vertebrate limb patterns. *J. Theor. Biol.* 104: 451–471.

Holland, L. Z. and N. D. Holland. 2001. Evolution of neural crest and placodes: Amphioxus as a model for the ancestral vertebrate? *J. Anat.* 199: 85–98.

Holland, N. D. and J. Chen. 2001. Origin and early evolution of the vertebrates: New insights from advances in molecular biology, anatomy, and paleobiology. *BioEssays* 23: 142–151.

Holland, N. D. and L. Z. Holland. 1999. Amphioxus and the utility of molecular genetic data for hypothesizing body part homologies between distantly related animals. *Am. Zool.* 39: 630–640.

Holland, N. D., G. Panganiban, E. L. Henyey and L. Z. Holland. 1996. Sequence and developmental expression of *AmphiDll*, an amphioxus *Distal-less* gene transcribed in the ectoderm, epidermis, and nervous system: Insights into evolution of craniate forebrain and neural crest. *Development* 122: 2911–2920.

Holland, P. W. H. 1998. Major transitions in animal evolution: A developmental genetic perspective. *Am. Zool.* 38: 829–842.

Holland, P. W. H. and J. Garcia-Fernández. 1996. Hox genes and chordate evolution. *Dev. Biol.* 173: 382–395.

Holley, S., P. D. Jackson, Y. Sasai, B. Lu, E. M. De Robertis, F. M. Hoffmann and E. L. Ferguson. 1995. A conserved system for dorsal-ventral patterning in insects and vertebrates involving sog and chordin. *Nature* 376: 249–23.

Hopson, J. A. 1966. The origin of the mammalian middle ear. *Am. Zool.* 6: 437–450.

Hughes, C. GG. and T. C. Kaufman. 2002. Hox genes and the evolution of the arthropod body plan. *Evo. Dev.* 4: 459–499.

Huxley, J. and G. Teissier. 1936. Terminology of relative growth. *Nature* 137: 780–781.

Ingham, P. W. 1994. Hedgehog points the way. *Curr. Biol.* 4: 345–350.

Jacob, F. 1977. Evolution and tinkering. *Science* 196: 1161–1166.

Jakob, U., H. Lilie, I. Meyer and J. Buchner. 1995. Transient interaction of Hsp90 with early unfolding intermediates of citrate synthase: Implications for heat shock in vivo. *J. Biol. Chem.* 270: 7288–7294.

Jean, D., K. Ewan and P. Gruss. 1998. Molecular regulators involved in vertebrate eye development. *Mech. Dev.* 76: 3–18.

Jeffery, J. E., O. R. P. Beninda-Emonds, M. I. Coates and M. K. Richardson. 2002. Analyzing evolutionary patterns in amniote embryonic development. *Evol. Dev.* 4: 292–302.

Jernvall, J., S. V. Keranen and I. Thesleff. 2000. Evolutionary modification of development in mammalian teeth: Quantifying gene expression patterns and topography. *Proc. Natl. Acad. Sci. USA* 97: 14444–14448.

Keys, D. N. and 8 others. 1999. Recruitment of a *hedgehog* regulatory circuit in butterfly eyespot evolution. *Science* 283: 532–534.

King, M. C. and A. C. Wilson. 1975. Evolution at two levels in humans and chimpanzees. *Science* 188: 107–116.

Kmita, M. F., F. van der Hoeven, J. Zákány, R. Krumlauf and D. Duboula. 2000. Mechanisms of Hox gene colinearity: Transposition of the anterior *Hoxb-1* gene into the posterior HoxD complex. *Genes Dev.* 14: 198–211.

Kmita, M. F., N. Frandeau, Y. Hérault and D. Duboule. 2002. Serial deletions and duplications suggest a mechanism for the collinearity of *Hoxd* genes in limbs. *Nature* 420: 145–150.

Knoll, A. H. and S. B. Carroll. 1999. Early animal development: Emerging views from comparative biology and geology. *Science* 256: 622–627.

Koch, P. B., D. N. Keys, T. Rocheleau, K. Aronstein, M. Blackburn, S. B. Carroll and R. H. ffrench-Constant. 1998. Regulation of dopa decarboxylase expression during color pattern formation in wild-type and melanic tiger swallowtail butterflies. *Development* 125: 2303–2313.

Kukalova-Peck, J. 1992. The Uniramia do not exist: The ground plan of the Pterygota as revealed by Permian Diahanopterodea from Russia (Insecta: Paleodictyopteroidea). *Can. J. Zool.* 70: 236–255.

Kuratani, S., Y. Nobusada, N. Horigome and Y. Shigetani. 2001. Embryology of the lamprey and evolution of the vertebrate jaw: Insights from molecular and developmental perspectives. *Philos. Trans. Roy. Soc. Lond. B* 356: 1615–1632.

Larrain, J., M. Oelgeschlager, N. I. Ketpura, B. Reversade, L. Zakin and E. M. De Robertis. 2001. Proteolytic cleavage of Chordin as a switch for the dual activities of Twisted gastrulation in BMP signaling. *Development* 128: 4439–4447.

Larsen, E. W. 1992. Tissue strategies as developmental constraints: Implications for animal evolution. *Trends Ecol. Evol.* 7: 414–417.

Lohnes, D. and 7 others. 1994. Function of retinoic acid receptors in development. I. Craniofacial and skeletal abnormalities in RAR double mutants. *Development* 120: 2723–2743.

Loomis, C. A., E. Harris, J. Michaud, J. Wurst, W. Hanks and A. L. Joyner. 1996. The mouse *Engrailed-1* gene and ventral limb patterning. *Nature* 382: 360–363.

Lovejoy, A. O. 1964. *The Great Chain of Being.* Harvard University Press, Cambridge, MA.

Macdonald, S. J. and D. B. Goldstein. 1999. A quantitative genetic analysis of male sexual traits distinguishing the sibling species *Drosophila simulans* and *D. sechellia. Genetics* 153: 1683–1699.

Maderson, P. F. A. 1972. On how an archosaurian scale might have given rise to an avian feather. *Am. Nat.* 106: 424–428.

Maderson, P. F. A. and L. Alibardi. 2000. The development of the sauropsid integument: A contribution to the problem of the origin and evolution of feathers. *Am. Zool.* 40: 513–529.

Malicki, J., L. C. Cianetti, C. Peschle and W. McGinnis. 1992. Human *HOX4B* regulatory element provides head-specific expression in *Drosophila* embryos. *Nature* 358: 345–347.

Manak, J. R. and M. P. Scott. 1994. A class act: Conservation of homeodomain protein functions. *Development* [Suppl.] 61–71.

Manton, S. M. 1977. *The Arthropoda: Habits, Functional Morphology, and Evolution.* Clarendon Press, Oxford.

Manzanares, M., H. Wada, N. Itasaki, P. A. Trainor, R. Krumlauf and P. W. H. Holland. 2000. Conservation and elaboration of Hox gene regulation during evolution of the vertebrate head. *Nature* 408: 854–857.

Marigo, V., R. L. Johnson, A. Vortkamp and C. J. Tabin. 1996. Sonic hedgehog differentially regulates expression of GLI and GLI3 during development of the tetrapod limb. *Development* 180: 273–283.

Mark, M., F. M. Rijli and P. Chambon. 1995. Alteration of Hox gene expression in the branchial region of the head causes homeotic transformations, hindbrain segmentation defects, and atavistic changes. *Semin. Cell Dev. Biol.* 6: 275–284.

Marqués, G., M. Musacchio, M. J. Shimell, K. Wunnenburg-Stapleton, K. W. Y. Cho and M. B. O'Connor. 1997. The Dpp activity gradient in the early *Drosophila* embryo is established through the opposing actions of the Sog and Tld proteins. *Cell* 91: 417–425.

Matsuo, I., S. Kuratani, C. Kimura, N. Takeda and S. Aizawa. 1995. Mouse *Otx2* functions in the formation and patterning of the rostral head. *Genes Dev.* 9: 2646–2658.

Mayr, E. 1966. *Animal Species and Evolution.* Harvard University Press, Cambridge, MA.

Merino, R., J. Rodríguez-Leon, D. Macias, Y. Ganan, A. N. Economides and J. M. Hurle. 1999. The BMP antagonist Gremlin regulates outgrowth, chondrogenesis and programmed cell death in the developing limb. *Development* 126: 5515–5522.

Meulemans, D. and M. Bronner-Fraser. 2002. Amphioxus and lamprey *AP-2* genes: Implications for neural crest evolution and migration patterns. *Development* 129: 4953–4962.

Miura, T. and K. Shiota. 2000a. TGF-β2 acts as an "activator" molecule in reaction-diffusion model and is involved in cell sorting phenomenon in mouse limb micromass culture. *Dev. Dynam.* 217: 241–249.

Miura, T. and K. Shiota. 2000b. Extracellular matrix environment influence chondrogenic pattern formation in limb bud micromass culture: Experimental verification of theoretical models. *Anat. Rec.* 258: 100–107.

Monaghan, P., E. Grau, D. Bock and G. Schulz. 1995. The mouse homolog of the orphan nuclear receptor tailless is expressed in the developing brain. *Development* 121: 839–851.

Morange, M. 2001. *The Misunderstood Gene.* Harvard University Press, Cambridge, MA.

Nathan, D. F., M. H. Vos and S. Lindquist. 1997. In vivo functions of the *Saccharomycetes cerevisiae* Hsp90 chaperone. *Proc. Natl. Acad. Sci. USA* 94: 12949–12956.

Neidert, A. H., V. Virupannavar, G. W. Hooker and J. A. Langeland. 2001. Lamprey *Dlx* genes and early vertebrate evolution. *Proc. Natl. Acad. Sci. USA* 98: 1665–1670.

Newman, S. A. and H. L. Frisch. 1979. Dynamics of skeletal pattern formation in the developing chick limb. *Science* 205: 662–668.

Nielsen, M. G. and E. C. Raff. 2002. The best of all worlds or the best possible world? Developmental constraint in the evolution of β-tubulin and the sperm tail axoneme. *Evol. Dev.* 4: 303–315.

Nijhout, H. F. and S. M. Paulsen. 1997. Developmental models and polygenic characters. *Am. Nat.* 149: 394–405.

Nowak, M. A., M. C. Boerlijst, J. Cooke and J. M. Smith. 1997. Evolution of genetic redundancy. *Nature* 388: 167–171.

Onuma, Y., S. Takahashi, M. Asashima, S. Kurata and W. J. Gehring. 2002. Conservation of *Pax6* function and upstream activation by Notch signaling in eye development of frogs and flies. *Proc. Natl. Acad. Sci. USA* 99: 2020–2025.

Oster, G. F., N. Shubin, J. D. Murray and P. Alberch. 1988. Evolution and morphogenetic rules: The shape of the vertebrate limb in ontogeny and phylogeny. *Evolution* 42: 862–884.

Panganiban, G., L. Nagy and S. B. Carroll. 1994. The role of the *Distal-less* gene in the development and evolution of insect limbs. *Curr. Biol.* 4: 671–675.

Panganiban, G. and 13 others. 1997. The origin and evolution of animal appendages. *Proc. Natl. Acad. Sci. USA* 94: 5162–5166.

Pasqualetti, M., M. Ori, I. Nardi and F. M. Rijli. 2000. Ectopic *Hoxa2* induction after neural crest migration results in homeosis of jaw elements in *Xenopus. Development* 127: 5367–5378.

Piatigorsky, J. and G. Wistow. 1991. The recruitment of crystallins: New functions precede gene duplication. *Science* 252: 1078–1079.

Piccolo, S., E. Agius, B. Lu, S. Goodman, L. Dale and E. M. De Robertis. 1997. Cleavage of chordin by the Xolloid metalloprotease suggests a role for proteolytic processing in the regulation of Spemann organizer activity. *Cell* 91: 407–416.

Price, M. 1993. Members of the *Dlx* and *Nkx2* gene families are regionally expressed in the developing forebrain. *J. Neurobiol.* 24: 1385–1399.

Prum, R. O. 1999. Development and evolutionary origin of feathers. *J. Exp. Zool.* 285: 291–306.

Qiu, M. S. and 8 others. 1997. Role of the Dlx homeobox genes in proximodistal patterning of the branchial arches: Mutations of *Dlx-1, Dlx-2,* and *Dlx-1* and *-2* alter morphogenesis of proximal skeletal and soft tissue structures derived from the first and second arches. *Dev. Biol.* 185: 165–184.

Raff, R. A. 1994. Developmental mechanisms in the evolution of animal form: Origins and evolvability of body plans. *In* S. Bengston (ed.), *Early Life on Earth.* Columbia University Press, New York, pp. 489–500.

Raff, R. A. 1996. *The Shape of Life: Genes, Development, and the Evolution of Animal Form.* University of Chicago Press, Chicago.

Raff, R. A. and G. A. Wray. 1989. Heterochrony: Developmental mechanisms and evolutionary results. *J. Evol. Biol.* 2: 409–434.

Raff, R. A., G. A. Wray and J. J. Henry. 1991. Implications of radical evolutionary changes in early development for concepts of developmental constraint. *In* L. Warren and H. Korpowski (eds.), *New Perspectives in Evolution.* Liss/Wiley, New York, pp. 189–207.

Relaix, F. and M. Buckingham. 1999. From insect eye to vertebrate muscle: Redeployment of a regulatory network. *Genes Dev.* 13: 3171–3178.

Riddle, R. D., R. L. Johnson, E. Laufer and C. Tabin. 1993. Sonic hedgehog mediates the polarizing activity of the ZPA. *Cell* 75: 1401–1416.

Riedl, R. 1978. *Order in Living Systems: A Systems Analysis of Evolution.* John Wiley and Sons, New York.

Rijli, F. M., M. Mark, S. Lakkaraju, A. Dierich, P. Dollé and P. Chambon. 1993. A homeotic transformation is generated in the rostral branchial region of the head by the disruption of *Hoxa-2,* which acts as a selector gene. *Cell* 75: 1333–1349.

Romer, A. S. 1949. *The Vertebrate Body.* Saunders, Philadelphia.

Romer, A. S. 1970. The Chanares (Argentina) Triassic reptile fauna VI. A chiniquodontid cynodont with an incipient squamosal-dentary jaw articulation. *Breviora* 344: 1–18.

Ronshaugen, M., N. McGinnis and W. McGinnis. 2002. Hox protein mutation and macroevolution of the insect body plan. *Nature* 415: 914–917.

Roux, W. 1894. The problems, methods and scope of developmental mechanics. In *Biological lectures of the Marine Biology Laboratory, Woods Hole*. Ginn, Boston, pp. 149–190.

Rutherford, S. L. and S. Lindquist. 1998. Hsp90 as a capacitor for morphological evolution. *Nature* 396: 336–342.

Salazar-Ciudad, I. and J. Jernvall. 2002. A gene network model accounting for development and evolution of mammalian teeth. *Proc. Natl. Acad. Sci. USA* 99: 8116–8120.

Salvini-Plawen, L. V. and E. Mayr. 1977. Evolution of photoreceptors and eyes. *Evol. Biol.* 10: 207–263.

Sander, K. 1983. The evolution of patterning mechanisms: Gleanings from insect embryogenesis and spermatogenesis. *In* B. C. Goodwin, N. Holder, and C. C. Wylie(eds.), *Development and Evolution*. Cambridge University Press, Cambridge, pp. 137–159.

Schierwater, B. and K. Kuhn. 1998. Homology of Hox genes and the zootype concept in early metazoan evolution. *Mol. Phylogenet. Evol.* 9: 375–381.

Sears, K. E. 2002. Reproductive mode and its effect on ontogeny and evolution in therian mammals: Have marsupials been constrained? *J. Vert. Paleontol.* 22 (3) 105A.

Shigetani, Y., F. Sugahara, Y. Kawakami, Y. Murakami, S. Hirano and S. Kuratani.2002. Heterotopic shift of epithelial-mesenchymal interactions in vertebrate jaw evolution. *Science* 296: 1316–1319.

Shubin, N., C. Tabin and S. Carroll. 1997. Fossils, genes, and the evolution of animal limbs. *Nature* 388: 639–648.

Siegal, M. L. and A. Bergman. 2002. Waddington's canalization revisited: Developmental stability and evolution. *Proc. Natl. Acad. Sci. USA* 99: 10528–10532.

Simeone, A., M. Gulisano, D. Acampora, A. Stornaiuolo, M. Rambaldi and E. Boncinelli. 1992. Two vertebrate homeobox genes related to *Drosophila empty spiracles* are expressed in the embryonic cerebral cortex. *EMBO J.* 11: 2541–2550.

Slack, J. M. W., P. W. H. Holland and C. F. Graham. 1993. The zootype and the phylotypic stage. *Nature* 361: 490–492.

Slijper, E. J. 1962. *Whales.* A. J. Pomerans (transl.). Basic Books, New York.

Stanley, S. M. 1979. *Macroevolution: Pattern and Process.* W. H. Freeman, San Francisco.

Sucena, E. and D. Stern. 2000. Divergence of larval morphology between *Drosophila sechellia* and its sibling species caused by *cis*-regulatory evolution of *ovo/shaven-baby*. *Proc. Natl. Acad. Sci. USA* 97: 4530–4534.

Torrey, T. W. and A. Feduccia. 1979. *Morphogenesis of the Vertebrates.* Wiley, New York.

Van Valen, L. M. 1976. Energy and evolution. *Evol. Theor.* 1: 179–229.

Von Dassow, G. and E. Munro. 1999. Modularity in animal development and evolution: Elements of a conceptual framework for evo-devo. *J. Exp. Zool./Mol. Dev. Evol.* 285: 307–325.

Wada, H. 2001. Origin and evolution of the neural crest: A hypothetical reconstruction of its evolutionary history. *Dev. Growth Diff.* 43: 509–520.

Waddington, C. H. 1938. The morphogenetic function of a vestigial organ in the chick. *J. Exp. Biol.* 15: 371–376.

Waddington, C. H. 1940. *Organisers and Genes.* Cambridge University Press, Cambridge.

Waddington, C. H. 1942. Canalization of development and the inheritence of acquired characters. *Nature* 150: 563.

Wagner, G. P. 1996. Homologues, natural kinds, and the evolution of modularity. *Am. Zool.* 36: 36–43.

Wake, D. B. and A. Larson. 1987. A multidimensional analysis of an evolving lineage. *Science* 238: 42–48.

Warren, R. W., L. Nagy, J. Selegue, J. Gates and S. Carroll. 1994. Evolution of homeotic gene regulation and function in flies and butterflies. *Nature* 372: 458–461.

Weatherbee, S. D., H. F. Nijhout, L. W. Grunert, G. Halder, R. Galant, J. Selegue and S. Carroll. 1999. *Ultrabithorax* functions in butterfly wings and the evolution of insect wing patterns. *Curr. Biol.* 9: 109–115.

Wilkins, A. S. 1997. Canalization: A molecular genetic perspective. *BioEssays* 19: 257–262.

Wilkins, A. S. 2002. *The Evolution of Developmental Pathways.* Sinauer Associates, Sunderland, MA.

Winnepenninckx, B. M. H., Y. van de Peer and T. Backeljau. 1998. Metazoan relationships on the basis of 18S rRNA sequences: A few years later. *Am. Zool.* 38: 888–906.

Wolpert, L. 1983. Constancy and change in the development and evolution of pattern. *In* B. C. Goodwin, N. Holder and C. C. Wylie (eds.), *Development and Evolution*. Cambridge University Press, Cambridge, pp. 47–57.

Yu, M., P. Wu, R. B. Widelitz and C. M. Chuong. 2002. The morphogenesis of feathers. *Nature* 420: 308–312.

Zeng, Z. B., C. H. Kao and C. J. Basten. 1999. Estimating the genetic architecture of quantitative traits. *Genet. Res.* 74: 279–289.

Zuckerkandl, E. 1994. Molecular pathways to parallel evolution. I. Gene nexuses and their morphological correlates. *J. Mol. Evol.* 39: 661–678.

Sources for chapter-opening quotations

Chapter 1
Cezanne, P. Quoted in J. Gasquet, 1921. *Cezanne.* Paris, pp. 79–80.

Oppenheimer, J. M. 1955. Analysis of development: Problems, concepts, and their history. In *Analysis of Development*, B. H. Willier, P. A. Weiss and V. Hamburger (eds.). Saunders, Philadelphia, pp. 1–24.

Chapter 2
Bonner, J. T. 1965. *Size and Cycle: An Essay on the Structure of Biology.* Princeton University Press, Princeton, NJ, p. 3.

"The Circle of Life." Music by Elton John. Lyrics by Tim Rice. Copyright 1994 Wonderland Music Company, Inc. All rights reserved. Used by permission.

Chapter 3
Bard, J. 1997. Explaining development. *BioEssays* 20: 598–599.

Morgan, T. H. 1898. "Some Problems of Regeneration." Biological Lectures Delivered at the Marine Biology Laboratory, Woods Hole, 1898, p. 207.

Chapter 4
Brenner, S. 1979. Quoted in H. F. Judson, 1979. *The Eighth Day of Creation.* Simon and Schuster, New York, p. 205.

Ozick, C. 1989. *Metaphor and Memory.* Alfred A. Knopf, New York, p. 111.

Chapter 5
Claude, A. 1974. The coming of age of the cell. Nobel lecture, reprinted in *Science* 189: 433–435.

Waddington, C. H. 1956. *Principles of Embryology.* Macmillan, New York, p. 5.

Chapter 6
Butler, O. 1998. *Parable of the Talents.* Warner Books, New York, p. 3.

Harrison, R. G. 1933. Some difficulties of the determination problem. *Am. Nat.* 67: 306–321.

Chapter 7
Darwin, C. 1871. *The Descent of Man.* Murray, London, p. 893.

Whitman, W. 1855. "Song of Myself." In *Leaves of Grass and Selected Prose.* S. Bradley (ed.), 1949. Holt, Rinehart & Winston, New York, p. 25

Chapter 8
Just, E. E. 1939. *The Biology of the Cell Surface.* Blakiston, Philadelphia, p. 288.

Wolpert, L. 1986. *From Egg to Embryo: Determinative Events in Early Development.* Cambridge University Press, Cambridge, p. 1.

Chapter 9
Kohler, R. E. 1994. *Lords of the Fly:* Drosophila *Genetics and the Experimental Life.* University of Chicago Press, Chicago, p. 33.

Schultz, J. 1935. Aspects of the relation between genes and development in *Drosophila.* Am. Nat. 69: 30–54.

Chapter 10
Rostand, J. 1960. *Carnets d'un Biolgiste.* Librairie Stock, Paris.

Spemann, H. 1943. *Forschung und Leben.* Quoted in T. J. Horder, J. A. Witkowski and C. C. Wylie, 1986. *A History of Embryology.* Cambridge University Press, Cambridge, p. 219.

Stern, C. 1936. Genetics and ontogeny. *Am. Nat.* 70: 29-35.

Chapter 11
Doyle, A. C. 1891. "A Case of Identity." In *The Adventures of Sherlock Holmes.* Reprinted in *The Complete Sherlock Holmes Treasury,* 1976. Crown, New York, p. 31.

Holub, M. 1990. "From the Intimate Life of Nude Mice." In *The Dimension of the Present Moment.* Trans. D. Habova and D. Young. Faber and Faber, London, p. 38

Chapter 12
Levi-Montalcini, R. 1988. *In Praise of Imperfection.* Basic Books, New York, p. 90.

Thomas, L. 1979. "On Embryology." In *The Medusa and the Snail,* Viking Press, New York, p. 157.

Chapter 13
Ramon y Cajal, S. 1937. *Recollections of My Life.* Trans. E. H. Craigie and J. Cano. MIT Press, Cambridge, MA, pp. 36–37.

Whitehead, A. N. 1934. *Nature and Life.* Cambridge Univesity Press, Cambridge, p. 41.

Chapter 14
Coleridge, S. T. 1885. *Miscellenea.* Bohn, London, p. 301

Whitman, W. 1867. "Inscriptions." In *Leaves of Grass and Selected Prose.* S. Bradley (ed.), 1949. Holt, Rinehart & Winston, New York, p. 1.

Chapter 15
Goethe, J. W. von. 1805. *Faust,* Part I. Trans. R. Jarrell, 1976.

Harvey, W. 1628. *Exercitio Anatomica de Motu Cordis et Sanguinis Animalibus.* Reprinted in 1928, C. C. Thomas, Baltimore, p. A2.

Chapter 16
Darwin, C. 1859. *On the Origin of Species.* New American Library, New York, p. 403.

Fuentes, C. 1989. *Christopher Unborn.* Trans. A. MacAdam. Farrar, Straus, and Giroux, New York, p. 281.

Chapter 17
Darwin, E. 1791. Quoted in M. T. Ghiselin, 1974. *The Economy of Nature and the Evolution of Sex.* University of California Press, Berkeley, p. 49.

Thomson, J. A. 1926. *Heredity.* Putnam, New York, p. 477.

Chapter 18
Schotte, O. Quoted in R. J. Goss, 1991. The natural history (and mystery) of regeneration. In *A History of Regeneration Research,* C. E. Dinsmore (ed.). Cambridge Univesity Press, Cambridge, p. 12.

Tennyson, A. 1886. *Idylls of the King,* 1958 ed. Macmillan, London, p. 292.

Williams, C. M. 1959. Hormonal regulation of insect metamorphosis. In *The Chemical Basis of Development,* W. D. McElroy and B. Glass (eds.). Johns Hopkins University Press, Baltimore, p. 794.

Chapter 19
Eliot, T. S. 1942. "Little Gidding." In *Four Quartets.* Harcourt, Brace and Company, New York, 1943, p. 39. Copyright T. S. Eliot.

Hadorn, E. 1955. *Developmental Genetics and Lethal Factors.* London 1961, p. 105.

Chapter 20
Thompson, D. W. 1942. *On Growth and Form.* Cambridge University Press, Cambridge.

Chapter 21
Carson, R. 1962. Quoted in B. Watson, 2002. Sounding the alarm. *Smithsonian* 33(6): 115.

Gibson, W. 1999. Quoted in M. Peyser, "The Home of the Gay," *Newsweek* March 1, p. 50.

Chapter 22
Haraway, D. H. 1991. "A Cyborg Manifesto." In *Simians, Cyborgs, and Women: The Reinvention of Nature.* Routledge, New York. p, 178.

Waddington, C. H. 1957. *The Strategy of the Genes.* Allen & Unwin, London, pp. 154–155.

Chapter 23
Rostand, J. 1962. *The Substance of Man.* Doubleday, Garden City, New York, p. 12.

Rushdie, S. 1989. *The Satanic Verses.* Viking, New York, p. 8.

Author index

Entries in **boldface** type indicate bibliographic citations.

Abe, K., 124, **141**
Abel, E. L., 697, **714**
Abouheif, E., 730, 731, **745**
Abu-Shaar, M., 586, 587, **606**
Acampora, D., 754, **781**
Achermann, J. C., 554, **570**
Ackerman, S. L., 449, **458**
Adachi, A., 166, **179**
Adams, D., 42, **47**
Adams, J. M., 164, **175**
Adams, M., 778, **780**
Adams, N. C., 405, **421**
Adams, N. L., 54, **77**
Adamson, D. 685, **717**
Adeoya-Osiguwa, S. A., 701, **714**
Adler, F. R., 734, **745**
Adoutte, A., 752, 753, **780**
Afzelius, B. A., 186, **214**, 382, **384**
Agius, E., 311, 325, 326, **338**
Agrawal, A., 91, **105**, 735, **745**
Ahlgren, U., 514, **518**, 698, **714**
Ahringer, J., 628, **642**
Aitkenhead, M., 480, **486**
Akam, M., 286, 288, 289, **299**, 758, **781**
Akers, R. M., 445, **458**
Akhtar, A., 124, **138**
Akitaya, T., 431, **458**
Alberch, J., 276, 780, 582, **606**
Alberch, P. **545**, 582, **606**, 767, 774, **780**
Alberts, B. M., 264, 265, **300, 301**
Alder, J., 404, **421**
Alessa, L., 659, **679**
Alexandre, D., 773, **780**
Alexandrov, D. A.
Alfandari, D., 317, **338**
Alibardi, L., 770, **783**
Alitalo, K., 503, **519**
Allen, B. M., 578, **606**
Allen, G. E., 82, **105**
Alley, K. E., 577, **606**
Allman, J. A., 63, **77**
Alsan, B. H. 493, **518**
Altabef, M., 537, **543**
Altizer, A. M., 593, **606**
Altmann, C. R., 96, **105**
Altucci, L., 704, **714**

Alvarez, I. S., 362, **384**, 395, 396, **422, 424**
Alvarez-Buylla, A., 737, **745**
Alves, A.-P., 192, **214**
Ambros, V., 134, **140**
Amikura, R., 616, **642**
Amiri, A., 253, **259**, 294, **299**
Amores, A., 760, **781**
Amundson, R., 778, **781**
Ancel, P., 210, **214**
Andersen, F. G., 116, 117, **138**
Anderson, D. J., 120, **141**, 430, 434, 445, 446, 451, **458, 463**
Anderson, H. F., 707, **714**
Anderson, K. V., 272, 291, 295, **299**
Anderson, K., 17, **23**
Anderson, R., 601, **608**, 618, 620
Anderson, W. F., 707
Andersson, S., 558, **570**
Andl, T., 419, 420, **422**
Andrée, B., 494, **518**
Ang, S. L., 754, **781**
Angerer, L., 129, **259**
Angerer, R., 129
Angier, N., 117, **138**, 465
Anjard, C., 44, **47**
Ankley, G. T., 744, **745**
Antczak, A. M., 505, **518**
Antczak, M., 639, **642**
Antonini, A., **746**
Aoyama, H., 471, 475, **486, 487**
Apelqvist, A., 514, **519**
Appel, T. A., 331, **338**
Arabidopsis Genome Initiative, 650, **679**
Arango, N. A., 553, 554, **570**
Arantes-Olivera, N., 604, **607**
Aréchaga, J., 317, **339**
Arendt, D., 354, 355, **384**
Ariizumi, T., 64, **78**
Aristotle, 4, 5, 6, **23**, 62, 354, 547, **570**
Arking, R., 602, **607**
Armstrong, J. F., 481, **486**
Armstrong, N., 236, **259**
Armstrong, P. B., 73, **77**
Arney, K. L., 123, **138**
Arnheiter, H., 115, 158
Arnoult, C., 193, **214**
Arruti, C., 396, **425**
Artavanis-Tsakonis, S., 167, **175**
Arthur, W., 778, **781**

Asai, R., 20, 21, 22, **23**
Asamoto, K., 475, **486**
Asams, I. R., **570**
Asashima, M., 65, 66, **77**
Ash, P. J., 477, **486**
Ashburner, M., 590, 591, **607**
Ashworth, D., 88, **105**
Assady, S., **714**
Aszódi, A., 473, **486**
Atkinson, B. G., 578, **607**
Atkinson, J. W., 245, **259**
Atonchuk, J., 507, **519**
Auerbach, R., 150, 503, 507, **519**, **520**, 704, 705, **717**
Ausprunk, D. H., 704, **714**
Austin, C. R., 188, 193, 203, **214**
Austin, J., 627, **642**
Averof, M., 758, 759, **781**
Axel, R., 434, **458**
Aza-Blanc, P., 162, **175**
Azar, Y., 507, **519**
Azhar, M., 43, **47**

Babbitt, K., 744
Babcock, D. F., 191, **215**
Bachiller, D., 375, 376, **384**
Bachmann, M. D., 724, **746**
Bachvarova, R. F., 361, **384**,
Baehrecke, E. H., **175**
Baek, H. J., 112, **138**
Bagnara, J. T., 576, **611**
Bagavandoss, P., 639
Baier, H., **458**
Baird, A. A., 409, **422**
Baird, G., 440, **460**
Baker, B. S., 562, 563, 564, 565, 567, **570**
Baker, B., 131, **138**
Baker, C. V., 438, 439, **458**
Baker, T. G., 631, **642**
Baker, W. W., 264, 474, **488**
Balinsky, B. I., 246, **259**, 310, **338**, 358, **384**, 394, **422**
Balling, R., 379, **384**
Ballock, R. T., 542, **543**
Baltzer, F., 52, **77**, 83, **105**
Banerjee, U., 156, **175**
Bantle, J., 744
Bao, Z.-Z., 496, **519**
Barabé, D., 668, **680**
Barasch, J., 482, 482, **486**
Barbash, J. E., 568, **570**

Bard, J. B. L., 51, 145, **175**, 481, 482, 484, 486, 722, **746**
Bardin, C. W., 641, **646**
Bardoni, B., 456, **458**, 554, **570**
Barinaga, M., 165, **175**, 408, **422**, 705, **714**
Barker, D. D., 277, **299**
Barlow, D., 684, **717**
Barlow, D. P., 123, **138**
Barlow, G. M., 692, **714**
Barlow, P., 366, **384**
Barnes, G. L., 469, 470, 486, 696, **714**
Barnes, M. D., 577, **606**
Barnett, T., 92, **105**
Baron, M. H., 501, **519**
Baron, R., 477, 478, **486**
Barr, M. L., 124, 125, **138**
Barraclough, C. A., 559, **570**
Bartolomei, M. S., 123, **138**
Barton, M. K., 663, **679**
Barton, S. C., 212, **218**
Basler, K., 156, **175**, 764, **781**
Bassing, C. H., 91, **105**
Basson, C. T., 527, **543**
Bate, M., 266
Bates, W. R., 250, **259**
Bateson, W., **299**, 774, **781**
Batourina, E., 485, **486**
Battaglin, W. A., 568, **570**
Baumgartner, S., 278, **299**
Baxter, G. T., **746**
Bayés, M., 419, **422**
Baynash, A. G., 434, **458**
Beachy, P. A., 163, **175**, 414, 695, **714**
Beams, H. W.,306, **338**, 348, **384**
Beck, S. D. 25, **48, 746**
Beckendorf, S., 637, **642**
Becker, K. B., 579, 580, **607**
Beddington, R. S. P., 375, **384, 388**
Bedford, J. M., 209
Beermann, W., **105**
Beers, W. H., 641, **642**
Begbie, J., 439, **458**
Begon, M., 15, **23**
Behe, M. J., 779, **781**
Behrens, J., 161, **175**
Beimesche, S., 118, **138**
Belanger, K. D., 659, 660, **679**
Belding, L. K., **746**
Bell, A. C., 123, **138**

Bell, E. M., 702, **714**
Bell, C., 751,
Bell, L. R., 563, **570**
Bell, S. E., 699, 701, **714**
Bellairs, R., 354, 355, 356, **384**
Bellus, G. A., 690, **714**
Belote, J. M., 131, **138**, 562, **570**
Bendel-Stenzel, M. R., 619, **642**
Benfey, P. N., 659, 667, 668, 671, **679, 681**
Benirschke, K., 126, **138**
Benoff, S., 193, **214**
Benson, G. V., 379, **385**, 699, **714**
Bentley, D., 444, **460**
Bentley, J. K., 190, **214**
Bentley, N. J., 115, **138**
Ben-Yosef, D., 208, **214**
Berardi, A. C., 509, **519**
Berg, L. K., 234, **259**
Berger, F., 659, **679**
Berger, S., 31
Bergeron, J. M., 568, **570**,
Bergers, G., **519**
Bergman, A., 776, **784**
Bergman, K., 33, 35, **48**
Bergsten, S. E., 277, **299**
Bergstrom, D. A., 474, **486**
Berkovitz, G. D., 557, **570**
Berleth, T., 274, **299**
Berman, D. M., 703, **714**
Bernirschke, K., 60
Bernstein, R., 554, **570**
Berondes, H. D., 590, **607**
Berridge, M. J., 206, **214**
Berrill, N. J., 580, **609**, 613, **644**, 775, **781**
Berry, D. L., 577, 578, **607**
Berset, T., 166, **175**
Bertocchini, F., 356, **384**
Bertram, E. G., 124, **138**
Bestor, T. H., 123, **138**
Bestor, T. M., 209, **214**
Beukeboom, L. W., 548, **570**
Bevilacqua, A., 172, **175**
Bhanot, P., 285, **299**
Bharathan, G., 670, 671, **679**
Bhattacharyya, M. K., 663, **679**, 685, **714**
Bi, G.-Q., 192, **214**
Bianchi, D. W., 369, 371, **384**
Biben, C., 496, **519**
Biedler, L., 306
Bienz, M., 288, **301**
Bier, E., 296
Bigsby, R., 739, **746**
Bijtel, J. H., 334, **338**
Binns, W., 163, **177**
Birk, O. S., 549, **570**
Bisgrove, B. W., 236, **259**
Bissell, M. J., 169, 170, **175, 178**, 703, **714**
Bitgood, M. J., 151, **175, 176**

Björklund, L. M., 708, **714**
Black, J. E., 737, **746**
Blackler, A. W., 616, 618, **642**
Blackwell, T. K., 255, **259**
Blader, P., 346, **385**, 763, **781**
Blaese, R. M., 707, **714**
Blaheta, R. A., 703, **715**
Blair, H. C., 477, 478, **486**
Blair, S. S., 588, **607**
Blandau, R. J., 641
Blau, H. M., 710, **717**
Blaustein, A. R., 54, 55, 56, **77, 746**
Blázquez, M. A., 675, **679**
Bleier, R., 560, **570**
Bleil, J. D., 196, 197, 201, **214**
Blitz, I. L., 327, **339**
Bloom, W., 486, 630, **643**
Blow, J. J., 306, **340**
Blumberg, B., 336, **339**
Blumberg, D. D., 44, **49**
Blumenbach, J. F., 7
Bockman, D. E., 435, **458**
Bode, H. R., 598, 599, 600, **608**
Bode, P. M., 600
Bodmer, R., 753
Boehm, T., 705, **715**
Boettger, T., 356, 361, 362, **385**
Boettiger, D., 474, **486, 488**
Boffoli, D., **607**
Bogart, J. P., 203, **214**
Boggs, R. T., 563, **571**
Bogin, B., 408, **422**
Boiani, M., 708, **715**
Bokor, P., 285, **299**
Bolanos, F., 583, **611**
Bolker, J. A., 721, 766, **781, 746**
Bollenbacher, W. E., 590, **610**
Bonadio, J., 712, **715**
Boncinelli, E., 377, **385**, 754, 762, **781**
Bonder, E. M., 223, **259**
Bonhoeffer, F., **458**
Bonner, J. T., 25, 40, **48**, 766, **781**
Bonner-Fraser, M., 427, 431, 438, **458, 461, 462**
Bonnet, C., 6, **23**
Booker, B., 451, **461**
Borevitz, J. O., 674, **679**
Borland, R. M., 367, **385**
Borovsk, D., **746**
Borsani, G., 126, **138**
Botto, L. D. 396, **422**
Boucaut, J. -C., 71, **77**, 316, 317, **339**
Bouchard, M., 478, **486**
Boué, A., 686
Bounoure, L., 616, 617, 618, **643**
Boutin, E. L., 699, **715**
Bouwmeester, T., 331, **339**
Boveri, T., 81, 82, 83, 188, 199, 200, 203, **214**, 231, **259**, 614, **643**

Bowerman, B., 253, 255, 256, **259**
Bowman, J., 654, 662, 675, 676, **679**
Bowtell, D. D. L., 156, 176
Boycott, A. E., 240, **259**
Braat, A. K., 617, **643**
Bradbury, E. J., 713, **715**
Bradley, A, 152, **178**
Bradley, D., 674, 675, **679**
Bradley, P. L., 297, **299**
Brakefield, P. M., 728, 729, 934, 779, **781**
Brand-Saberi, B., 471, **486**
Brannon, M., 324, 325, **339**
Braun, M., 598, 599, **607**
Braun, R. E., 630, **643**
Breedlove, S. M., 560, **571**
Breitbart, R. A., 130, **138**
Brendza, R. P., 272, 276, **299**
Brennan, J., 375, **385**
Brenner, C. A., 368, **385**
Brenner, S., 59, 81, 165, 251, **259**
Brent, R., 290, **300**
Brewbaker, J. L., 657, **679**
Brian, M. V., 730, **746**
Bridge, D. M., 600, **607**
Bridges, C. B., 263, 561, **571**
Briggs, R., 85, **105, 106**
Brill, G., 472, **486**
Bringas, P., Jr., 149, **179**
Brinster, R. L., 100, **106**
Briscoe, J., 160, **176**, 401, 402, **422**
Brivanlou, A. H., 154
Brock, H. W., 288, **301**
Brockdorrf, N., 126, **138**
Brockes, J. P., 593, 594, 596, **607**
Broihier, H. T., **643**
Brönmark, C., 735, **746**
Bronner-Fraser, M., 429, 430, 431, 433, **458, 459, 783**
Brooks, W. K., 18, **23**, 725
Brose, K., 450, **458**
Brown, D. D., 108, 127, 126, **141**, 578, 579, **607, 611**, 636, **643**
Brown, K. S., 658, **679**
Brown, N. L., 155, 156, **176**
Brown, S., **299**
Brown, V., 692, **715**
Browne, E. N., 597, 598, **607, 609**
Brownlee, C., 659, **679**
Bruder, S. P., 477, **486**
Brugge, J. S., 171, **176**
Bruneau, B. G., 496, **519**
Brunelli, S., 754, **781**
Brunet, L. J., 540, **543**
Brunetti, C. R., 729, 769, **746, 781**
Brush, S., 83, **105**
Brust, D. G., 583, **607**
Brüstle, O., 708, 709, **715**
Bry, L., 724, **746**
Bryant, P. J., 588, **607**

Bryant, S. V., 524, 525, **543, 545**, 592, 595, 597, **607, 609**
Brylski, P., 767, **781**
Buchmann, D., 53, **78**
Buckingham, M., 753, **783**
Bull, J. J., 567, **571**, 725, **746**
Bullock, S. L., 638, **643**
Bulman, M. P., 467, **486**
Bunker, C., 743,
Bunker, E., 743
Buratowski, S., 111, 112, **138**
Burden, S. J., 451, **458**
Burdsal, C. A., 369, **385**
Burgess, R., 469, **486**
Burgoyne, P. S., 551, **571**
Burian, R., 84, **105**
Burke, A. C., 380, **385**, 471, 488, 525, **543**, 759, 760, **781**
Burke, R. D., 236, **259**
Burkholder, G. D., 92, **105**
Burnett, F. M., 736, **746**
Burns, R. K., Jr.627, **643**
Burnside, B., 396, **422**
Burtis, K. C., 132, **139**
Busa, W. B., 206, **214**
Buss, L. W., 613, **643**
Busslinger, M., 122, **138**
Butenandt, A., 591, **607**
Butler, E. G., 593, **607**
Butler, O., 143
Butler, S. L., 439, **463**

C. elegans, **259**
Cadigan, K. M., 161, **176**, 763, **781**
Cagen, S. Z., 701, **715**
Cai, G., 657, **679**
Caldwell, M. W., 759, **781**
Cales, C., 154, **176**
Callan, H. C., 635
Callery, E. M., 582, **607**
Calvin, H. I., 209, **214**
Cameron, R. A., 233, **259**
Campbell, J. 708, **718**
Campbell, R. D., 597, **607**
Campos-Ortega, J. A., 167, **177**, 267, 268, **299**, 349, **387**
Canning, D. R., 360, **388**
Cao, R., 503, **519**, 705, **715**
Cao, Y., 503, **519**
Cao, Z., 295, **299**, 705, **715**
Capdevila, J.,533, 539, 54
Capecchi, M. R., 99, **105**, 377, 378, 379, **385, 387**, 533
Capel, B., 553, 554, 555, **571, 573**
Caplan, A. I., 477, **486**
Capovilla, M., 281, **299**
Capuron, A., 320, **339**
Carlsen, E., 701, **715746**
Carlson, B. M., 7, **23**, 306, **339**, 436, **458**, 494, 498, 499, 511, 516, 639, **643**, 683, **715**
Caroni, P., 713, **718**

Carrington, J. L., 530, **543**
Carroll, D. J., 201, 206, 207, 208, **215**
Carroll, E. J., Jr., 187, **215**
Carroll, J., 561, **571**
Carroll, S. B., 297, **303**, 729, 751, 753, 755, 757, 758, 765, 769, **781**
Carroll, S., 156, 269, 288, 588
Carson, D. D., 76, **77**, 368, **385**
Carson, R., 683, 694, **715**
Carthew, R. W., 157, **176**, 740, **746**
Carver, V., 577, **611**
Carvey, P. M., 710, **715**
Casal, J., 285, **299**
Casanova, J., 277, 278, 286, **299**
Casares, F., 287, 288, **299**
Casey, B., 382, **386**
Cassidy, R., 410, **422**
Cassirer, E., 7, **23**
Castellano, L. E., 192, **215**
Castelli-Gair, J., 288, 289
Caston, J. D., 594, **611**
Catala, M., 393, 396, 398, **422**
Cate, R. L., 558, **571**
Cather, J. N., 240, 244, **259, 262**
Causo, J. P., 587, **607**
Cebra, J. J., 725
Cecconi, F., 165, **176**
Celotto, A. M., 130, **138**
Centers for Disease Control, 397, **422**, 685, 694, 702
Centerwall, W. R., 126, **138**
Cepko, C. L., 407, 410, 414, 415, **425**
Cezanne, P., 3
Cha, B. J., 272, 275, **299**, 638, 643
Chabry, L. M., 247, **259**
Chabry, L. M., 57, **77**
Chai, Y., 439, 440, **459**
Chaillet, J. R., 124, **138**
Chambers, E. L., 200, 202, 207, **217**
Chambers, E. L., **215**
Chan, S. K., 276, **299**, 707, **715**
Chandler, D. E., 187, **217**
Chandler, D. E., 203, **215**
Chang, M. C., 193, **215**
Chapman, D. B., 381, **386**
Charité, J., 536, **543**
Charnov, E. L., 731, **746**
Chasan, R., 294, **299**
Chauvet, S., 290, **300**
Check, E., 707, **715**
Cheek, A. O., 742, **746**
Chen, C.-M., 472, **486**
Chen, H., 537, **543**
Chen, J., 761, **782**
Chen, J.-L., 112, **138**
Chen, K. C., 156, **176**
Chen, M. S., 713, **715**
Chen, Q., 541, **543**

Chen, W. T., 169, **176**
Chen, Y. P., 336, **339**
Chen, Z., 703, **718**
Cheng, H.-J., 454, 456, **459**
Cheng, J., 197
Cheng, L. Y., **260**
Cheng, P. F., 336, **342,**
Chenn, A., 403, **422**
Chernoff, E. A. G., 593, **607**
Chernoff, G. F., 396, **422**
Cherr, G. N., 195, **215**, 234, 235, **259**
Chiang, C., 162, **176**, 413, 414, 759, **781**
Chien, C. T., 156, **176**
Childs, B., 743, **746**
Chiquoine, A. D., 618, **643**
Chisaka, O., 377, 379, **385**
Chitnis, A., 166, **176**
Cho, K. W. Y., 327, 336, **339**
Chong, J. A., 120, **138**
Chou, H. H., 409, **422**
Chou, R. L., 120, **138**
Chou, W. H., 157, **176**
Chou, W., **138**
Chow, R. L., 753, **781**
Christ, B., 472, **486**
Christian, J. L., 331, **339**
Christianson, A. E., 565, 566, **571**
Christianson, M. L., 663, **679**
Chun, J. J. M., 91, **105**
Chun, L. L. Y., 434, **459**
Chuong, C.-M., 147, **179**
Churchill, A., 6, 8, **23**
Chymiy, D. B., 567, **571**
Cionini, P. G., 660, **679**
Clack, J. A., 16, **23**
Clancy, D. J., 604, **607**
Clark, E. A., 171, **176**
Clark, J. K., 667, **679**
Clark, M. R., 641, **644**
Clarren, S. K., 697, **715**
Claude, A., 107
Clegg, K. B., 366, **387**
Clement, A. C., 240, 244, **259**
Clermont, Y., 185, **215**
Clever, U., 590, **607**
Cline, T. W., 131, **138**, 561, 562, **571**
Coates, M. I., **543**
Coe, W. R., 569, **571**
Coen, E. S., 676, 677, **679**
Cohen, C. S., 724, **746**
Cohen, J., 445, 454, 459
Cohen, M. M., Jr., 17, **23**
Cohen, P. P., 578, **607**
Cohen, S. M., 276, **300**, 588, **607**, **610**
Cohen-Cory, S., **459**
Cohen-Dayag, A., 194, **215**
Cohen-Gould, L., 495, **519**
Cohlan, S. Q., 696, **715**

Cohn, M. J., 525, **543**, 759, 760, **781**
Colamarino, S. A., 449, **459**
Colas, J. –F., 395, **422**, 492, 495, **519**
Colburn, T., 701, 71
Colello, R. J., 738, **746**
Coleman, H., 703, **715**
Colman, H., 451, **459**, 739, **746**, 739, **746**
Colman, W. B., **715**
Collas, P., 209, **217**
Collier, J. R., 245, **259**
Collignon, J., 382, **385**
Collins, F. D., 195
Collins, F. S., 687, **715**
Columbo, L., 662, **679**
Colvin, J., **571**
Colwin, A. L., 191, **215**
Colwin, L. H., 191, **215**
Comings, D. E., 625, **643**
Condic, M. L., 586, **607**
Condie, B. G., 378, **385**
Conklin, E. G., 10, 11, **23**, 58, 246, 247, **259**
Conlon, R. A., 379, **385**
Conner, S., 202, **215**
Conover, D. O., 731, 732, **746**
Constantini, F., 112, **141**, 481, **486**
Conway, S. J., 441, **459**
Cook, S. P., 191, **215**
Cooke, A. S., 740, **746**
Cooke, J., 311, **339**
Cooley, L., 638, **643**
Cooper, M. K., 163, **176**
Cooper, T. A., 692, **715**
Corben, C. J., 583, **607**
Corden, R., 503, **519**
Cormier, F., 507, **519**
Cormier, P., 209, **215**
Correia, L. M., 187, **215**
Corselli, J., 193, **215**
Cory, S., 164, **175**
Coschigano, K. T., 132, **138**
Cossu, G., 473, **486**
Cotsarelis, G., 418, **422**
Coulombe, J. N., 433, **459**
Coulombre, A. J., 414, **422**
Couly, G. F., 438, **459**
Couse, J. F., 549, 559, **571**
Coutinho, C., 46, **48**
Couzinet, B., 641, **643**
Cowan, W. M., 577, **607**
Cox, D. N., 666, **679**
Cox, E. C., 455, **459**
Craig, J. A., 416, **424**
Craig, M. M., 240, **259**
Crain, D. A., 567, 568, **571**, 742, 743, **746**
Crair, M. C., 738, **746**
Crampton, H. E., 240, 243, 244, **259**

Crawford, K., 582, 596, **607**, 744, **746**
Creech Kraft, J., **715**
Crelin, E. S., 510
Cresti, M., 657, **679**
Crews, D., 567, 568, **571**
Cristofalo, V. J., 603, **607**
Croft, L., 130, **138**, 692, **715**
Crosby, G. M., 524, **543**
Cross, J. C., 76, **77**
Cross, N. L., 193, 200, **215, 216**
Crossgrove, K., 591, **607**
Crossley, P. H., 150, **176**, 530, 535, **543**
Crowley, C., 452, **459**
Cruden, R. W., 655, **679**
Cruz, Y. P., 372, **385**
Cserjesi, P., 475, 476, **486**
Cubas, P., 678
Cubitt, A. B., 44, **48**
Cummins, J. M., 210, **215**
Cunha, G. R. 699, **715**
Currie, J., 577, **607**
Cuvier, G., 14, 751
Cvekl, A., 116, 117, 118, **138**, 145, 147, 467
Czeizel, A., 397, **422**

DAmato, R. J., 705, **715**
Darribère, T., 317, **339**
da Silva, A. M., 42, **48**
Dahanukar, A., 277, **299**
Dahn, R. D., 536, 537, **543**
Daily, D., 264
Dale, J. K., 467, 469, **486**
Dale, L., 322, 329, **339**, 764, **781**
Dallon, J. C., 41, **48**
Dalton, R., 568, **571**
Dan, J. C., 191, **215**
Dan, K., 227, 237, **259**
Dan, Y., 451, **459**
Daneholt, B., 625, **646**
Daniel, C. W., 153, **176**
Danilchik, M. V., 28, 29, 223, **259**, 315, **340,** 616, 617, 633, **643, 646**
Dan-Sohkawa, M., 227, **259**
Darnell, D. K., 13, **23**, 358, **385**
Darnell, J. C., 692, **715**
Darnell, J. E., 111, **141**, 154, 172, **176**
Darras, S., 249, **259**
Darwin, C., 9, 14, 22, **23**, 54, 174, 523, 547, 752, 773, 777, **781**
Davenport, R. W., 412, **422**
David, J. D., 474, **486**
Davidson, E. H., 56, 58, **77**, 227, 229, 230, 231,232, **259, 261**
Davidson, L. A., 317, **339**
Davidson, N., 92, **106**
Davies, A. M., 452, **460**
Davies, J. A., 445, 459482, **486**

Davis, A. P., 532, 533, **543**
Davis, B. K., 193, **215**
Davis, C. A., 764, **781**
Davis, D. L., 740, **746**
Davis, G. S., 74, **77**
Davis, I., 285, **303**
Davis, S., 503, **519**
Davis, W. L., 697, **715**
Dawid, I. B., 636, **643**
Dazy, A.-C., 32, **48**
de Beer, G. R., 63, 67, **78**, 582, **607**
de Cuevas, M., 638, 643
de Graaf, V., 9
De Jonge, F. H., 559, **571**
de la Chappelle, A., 374, **385**
De Marco, P., 397, **422**
De Robertis, E. D. P., 67, **77**, 186, **215**, 314
De Robertis, E. M., 317, 321, 327, 330, 331, 332, 336, **339**, 376
De Santa Barbara, P., 553, **571**
De Vreese, A., 603, **611**
Dealy, C. N., 537, **543**
Dean, C., 654, 673, **680**
Debski, E. A., **459**
DeChiara, T. M., 123, **138**
Decker, C. J., 132, **138**
Decker, G. L., 191
Dedeinde, F., 724, **746**
Dedkov, E. I., 593, **607**
DeFrances, M. C., 601, **609**
Degelmann, A., 277, **299**
Degnan, B. M., 726, **746**, 760, **781**
DeHaan, R. L., 494, 495, **519**
Deiner, M. S., 454, **459**
del Pino, E. M., 582, **607**
DeLanney, L. E., 581, **610**
DeLuca, S. M., 357, **385**
Deng, C., 161, **176**, 541, **543**
Deng, X. W., **679**
Denno, R. F., 727, 728
Denver, R., 579, **607**, 734
DePew, M. J., 762, **781**
Descartes, R., 6, 19
Deshpande, G., 616, 624, **643**
DeSimone, D. W., 169, 317, **341**
Desmond, M. E., 400, **422, 425**
Dessain, S., 287, **299**
Desvages, G., 568, **571**
Detrick, R. J., 75, **77**, 397, **422**
Detwiler, S. R., 524, **543**
De Vos, L., 632, **643**
Devreotes, P. N., 41, **49**
Di Gregorio, A., 249, **259**
Di Laurenzio, L., 667, **679**
Dias, M. S., 361, **385**
Diaz-Benjumea, F. J., 588, **607**
Diaz-Infante, A., 641, **643**
DiBerardino, M. A., 86, 708, **715**
Dickman, S., 437, **459**
Dickson, B., 157, **176**
Didier, Y. R., 495, **519**

Dienhart, M., 347, **385**
Dieterlen-Lièvre, F., 498, 507, **519**
Dietrich, M., 778, **781**
Dietrich, S., 473, **486**
DiNardo, S., 285, 286, **299**, 330
Dinsmore, C. E., 592, **607**
DiSilvestro, R. L., 753, **781**
Dixon, G. H., 111, **139**
Dixon, J. E., 333, **339**
Dixon, K. E., 617, **645, 646**
Dobzhansky, Th., 778, 779, **781**
Dodds, P. N., 656, **679**
Dodson, S., **746**
Doe, C. Q., 289, **300**, 442, **459**
Doerner, P., 667, **679**
Doetsch, F., 410, **422**
Doherty, P., **176**
Doitsidou, M., 620, **643**
Dokucu, M. E., 156, **176**
Dolci, S., 621, **643**
Domingos, P. M., 335, 336, **339**
Dong, J., 288, **299**, 639, **643**
Doniach, T., 333, **339**
Donoghue, D. J., 161, **179**, 541, **546**
Donoghue, M. J., **679**
Dorizzi, M., 567, **571**
Dorris, M., 697, **715**
Dorsky, R. I., **176**, 434, **459**
Dosch, R., 330, **339**
Downs, S., 641, **643**
Dredge, K., 705, **715**
Drescher, U., 456, **459**
Dressler, G. R., 482, **488**
Drews, G. N., 655, **679**
Dreyer, S. D., 537, **543**
Driesch, H., 61, 62, 63, **77**
Driever, W., 135, **138**, 274, 275, 276, 277, **300**, 346, 349, 350, 353, **385, 388**
Driscoll, D. J., 124, **138**
Drossopoulou, G., 536, **543**
Drukker, M., 709, **715**
Du, S. J., 347, 347, **385**
Duband, J. -L., 431, **463**
Dubois, R., 622, **643**
Dubrulle, J., 467, 471, **486**
Ducibella, T., 204, 207, **215**
Ducy, P., 437, 459486
Dudas, I., 397, **422**
Dudek, R. W., 683, **715**
Dudley, A., 100, 103, **105**, 482, 486, 532, **543**
Dufour, S., 168, **176**
Dulhunty, A. F., 172, **178**
Dumais, J., 31, **48**
Dumas, J. B., 9, 184, **218**
Dumont, J. N., **643**
Dunn, L., 85
Dunwoodie, S. L., 467, 468, **486**
Duong, L. T., 477, **486**

Dupé, V., 336, **339**
Durbin, L., 469, 470, **486**
Dutt, A., 76, **77**
Dyce, J. M., 367, **385**
Dyke, C., 766, **781**
Dym, M., 170, 629, 630, **643**
Dyson, S., 65, **78**
Dzierzak, E., 480, **488**

Early, A., 44, **48**
Echelard, Y., 402, **422**
Eck, S. L., 706, **715**
Edgar, B. A., 133, **138**, 223, **259**, 266, 282, **300**
Edlund, T., 391, **425**
Edmonds, D. K., 686, **715**
Edmondson, J. C., 405, **422**
Edwards, M., 700, **715**
Eichele, G., 391, **422**
Eicher, E. M., 552, **571, 573**
Eisen, A., 202, **215**
Eisen, J. S., 429, **461**
Eisenbach, M., 193, 194, **215**
Ekblom, P., 482, **486**
Eldadah, Z. A., 691, **715**
Elder, J. T., 417, **422**
Eldredge, N., 778, **781**
Eliceiri, B. P., 579, **607**
Elinson, R. P., 200, 210, 211, 212, 213, **217**, 312, **340**, 582, 583, **608**
Ellis, L. M., 504, **519**
Eliot, T. S., 613
Elwood, J. M., **423**
Emery, I. F., 591, **608**
Enard, W., 409, **422**
Endo, Y. G., 197, **215**
Engleka, M. J., 324, 325, **339**
Ephrussi, A., 616, **643**
Epifano, O., 635, **643**
Epstein, C. J., 687, **715**
Epstein, J., **139**
Ergun, S., 504, **519**
Erickson, C. A., 429, 432, **459, 461**, 462396, **422**
Erickson, H. P., 776, **781**
Ericson, J., 443, 45
Eriksson, C. J., 687, **715**
Erikksson, P. S., 410, **422**
Erives, A., 248, **259**
Ermak, G., 688, **715**
Ernest, S., **385**, 346
Ernst, A. F., 457, **459**
Erwin, D. H., 752, 754, **781**
Esau, K., **679**
Escalante, R., 43, **48**
Essner, J. J., 383, **385**
Etkin, L., **644**
Eto, K., 700, **715**
Ettensohn, C. A., 227, 229, 235, 236, 237, 238, **259, 260**

Evans, J. D., 730, **746**
Evans, J. P., 198, **215**
Evans, M. J., 374, **385**
Evans, R. M., 579, **609**
Evans, T., 222, 254, 256, **260**
Ewald, A., 309
Eyal-Giladi, H., 354, 355, 356, 360, 507, **519**, 622, **643**

Faber, M., 62, **78**
Fabian, B., 360, **385**
Fairchild, W. L., 701, **715**
Faist, A. M., 383, **385**
Faix, J., **48**
Falck, P., 8
Fallon, A. M., **746**
Fallon, J. F., 148, 164, **179**, 524, 529, 530, 532, 536, 537, 538, 539, **543, 545, 546**, 581, **608**
Fan, C. M., 151, **176**, 324, **339**, 472, **487**
Farach, M. C., 76, **78**
Farley, B., 495, **519**
Fässler, P. E., 317, 320, **339, 343**
Faure, S., 363, **385**
Fausto-Sterling, A., 559, **571**
Fawcett, D. W., 630, **643**
Fawcett, J. W., 457, **459**
Featherstone, M. S., **718**
Feduccia, A., 778, **784**
Fekany, K., 353, **385**
Feldman, M., 474, **489**
Fell, P. E., 46, **48**
Feng, J., 603, 604, **608**
Feng, Y., 692, **715**
Ferguson, E. L., 295, **300**
Ferguson, M. W. J., 52, **78**, 731, **746**
Ferguson-Smith, A. C., 124, **139**
Ferrandiz, C., 6767, **679**
Ferrandon, D., 135, **139**, 274
Ferrara, N., 503, **519**
Ferré-D'Amaré, A. R., 115, **139**
Ferrell, J. E., Jr., 634, **643**
Ferris, C. D., 206, **215**
ffrench-Constant, C., 620, **643**
Fidler, I. J., 504, **519**
Field, M. C., 400, **422**
Field, R., 88
Fink, R. D., 233, 234, **260**
Finkelstein, R., 276, **300**, 754, **781**
Finn, C. A., 638, **643**
Finnerty, J. R., 762, **781**
Fire, A., **105, 105**, 617, **646**
Fischer, D., 689
Fischer, J.-L., 56, **78**
Fishell, G., 405, **422**
Fisher, C. R., 549, **571**, 560, **571**
Fisher, S. J., 371, **385**
Fishman, M. P., 514, **519**
Fitzsimmons, R., 442
Fix, J. T., 683, **715**

Flaherty, D., 235
Flamme, I., 503, **519**
Flavell, R. D., 122, **141**
Fleck, L., 42, **48**
Fleischman, R. A., 17
Fleming, T. P., 366, 367, **385**
Flickinger, R. A., 633, **643**
Florman, H. M., 192, 195, 196, 197, **215**
Florschütz, P. A., 313, **341**
Floyd, S. K., 658
Fluck, R. A., 352, **385**
Foe, V. E., 264, 265, 295, **300**
Foerder, C. A., 201, **215**
Fol, H., 184, **215**
Foley, A. C., 356, **385**
Folkman, J., 504, **519**, 704, 705, **714, 716**
Folkman, J. P., 703, **716**
Foltz, K. R., 196, **215**
Fong, G.-H., 503, **519**
Fontana, J. A., 704, **716**
Forlani, S., 293, **300**
Forristall, C., 617, **643**
Forscher, P., 412, **422**
Foster, A. S., 668, **679**
Foster, J. W., 553, **571**
Foty, R. A., 74, **78**
Fowler, D., 21
Franco, B., 694, **716**
François, V., 296, **300**
Franklin, L. E., 191, **215**
Franklin-Tong, V. E., 657, **679**
Frantz, G. D., 407, **422**
Fraser, S. E., 12, **24**, 433, 435, 458 460469, **487**
Fraser, S. F., 309, 315
Freiman, R. N., 556, **571,**
Freeman, G., 240, **260**
Freire-Maia, N., 690, **716**
Frieden, E., 582, **608**
Friedman, W. E., 658, **679, 681**
Friet, S. C., 724, **748**
Frigerio, G., 274, **300**
Frisch, H. L., 774, **783**
Frisch, S. M., 169, **176**
Frischer, L. E., 286, **300**
Frisén, J., 410, **422**
Fristrom, D., 585, 586, **608**
Fristrom, J. W., 585, 586, **608**
Fritzsch, B., 576, **608**
Fromental-Ramain, C., 533, **543**
Fu, M.-F., 363, **385**
Fuchs, E., 417, **425**
Fuentes, C. 523, 688, **716**
Fujimori, T., 75, **78**, 397, **422**
Fujimoto, S., 91, **105**
Fujioka, M., 281, 282, **300**
Fujisawa, H., 227, **259**
Fujita, S., 403, **422**
Fujiwara, M., 144, 145, **176**
Fukada, K., 434, **459**

Fukui, A., 65, 66, **78**
Fukumachi, H., 512, **519**
Fukuzawa, M., 44, **48**
Fukuyama, F., 708, **716**
Fullilove, S. L., 265, **300**
Funayama, N., 332, **339**
Furriols, M., 277, **300**
Furlong, E. E. M., 296, **300**
Furuichi, T., S., 206, **215**
Furuta, Y., 145, **176**

Gabay, L., 277, 278, **300**
Gabrielli, B., 634, **643**
Gachelin, G., 508, **521**
Gage, F. H., 410, 707, **716**
Gagnon, M. L., 128, 129, **139**
Galant, R., 757, 758, **781**
Galantino-Homer, H. L., 193, **215**
Galen, C., 547, **571**
Galileo, D. S., 234, **260**
Galindo, R. L., 294, **300,**
Gall, J., 635
Galis, F., 775, **781**
Gallera, J., 393, **422**
Gamse, J., 333, **339**
Gañan, Y., 538, **543**
Gannon, M., 514, **519**
Gans, G. 434, **462**
Garbers, D. L., 190, 209, **215, 218**
Garcia, E., 32, **48**
Garcia, J. E., 641, **643**
Garcia-Bellido, A., 588, **608**
Garcia-Castro, M. I., 428, **459**
Garcia-Fernández, J., 760, 761, **782**
Gardiner, D. M., 202, **216**, 525, **543**, 593, 596, 597, **608**, 744
Gardiner, R. C., 373, **385**
Gardner, D. K., 684, **719**
Gardner, R. L., 100, **105**, 381, **385**, 621, **643**
Garey, J. R., 45
Garstang, W., 777, **781**
Gartler, S. M., 126, **139**
Gat, U., 418, **422**
Gaude, T., 656, **679**
Gaudet, J., 257, **260**
Gaul, U., 281, **300**
Gaunt, S. J., 380, **385**, 759, **781**
Gavis, E. R., 135, **139**, 276, 277, **299, 300**
Gawantka, V., 330, **339**
Gayon, J., 766, **781**
Gearhart, J., 621, **643**, 708, 710, **716**
Gedamu, L., 111, **139**
Geddes, P., 548, **571**
Geduspan, J. S., 531, **543**
Gehring, A. J., 282
Gehring, W. J., 119, 120, **140**, 753, 779, **781**
Geigy, R., 581, **608**
Geissler, W. M., 557, **571**

Gelbart, W. M., 297, **301**
Gellen, G., 755, **781**
Gems, D., 604, **608, 610**
Gendron-Maguire, M., 377, **385**, 437, **459**
Genetics Review Group, 556, **571**
Genome International Sequencing Consortium, **679**
Geoffroy Saint-Hilaire, E., 331, 752
George-Weinstein, M., 362, **385**
Gerard, M., **544**
Gerber, H. P., 476, **487**
Gerhart, J. C., 211,216, 222, 311, 322, **339, 342**, 633, **643**, 751, **781**
Gerisch, B., 604, 605, **608**
Gerisch, G., 42, **48**
Gersh, I., 541
Gershon, M. D., 442, **459**
Gibson-Brown, J. J., **544**
Giedd, J. N., 409, **422**
Giese, K., 551, **571**
Gifford, E. M., 668, **679**
Gilbert, L. I., 589, **608**
Gilbert, S. F., **47, 48**, 62, **78**, 82, 84, **105, 216**, 326, 331, **339**, 721, 722, 733, **746**, 761, 766, 778, **782**
Gilbert-Barness, E., 161, **176**
Gilkey, J. C., 202, **216**
Gill, R. E. Jr., 745, **748**
Gil-Turnes, M. S., 724, **746**
Gilula, N. B., 172, **178**
Gimlich, R. L., 312, 322, **339**
Ginsburg, G. T., 44, **48**
Ginsburg, M., 622, **643**
Giroud, A., 696, **716**
Gish, D. T., 768, **782**
Giudice, G., 71, **78**
Giusti, A. F., 194, 207, **216**
Glabe, C. G., 194, 195, 196, 201, **216**
Glaser, T., 116, **139**, 413, **422**
Glausiusz, J., 708, **716**
Glavic, A, 330, **340**
Gleicher, N., 685, **716**
Gleicher, N., 685, **716**
Glinka, A., 331, 332, **340**
Glueckson-Schoenheimer, S., 13, 84, 85, **105**, 481, **487**
Gödde, R., 501, **519**
Godement, P., 454, **459**
Godin, I., 620, 621, **643**
Godlin, I. E., 507, **519**
Godsave, S. F., 329, **340**
Godt, D., 76, **78**
Godwin, J., 732, **747**
Goethe, W., 491, 768
Goetinck, P., 541
Goff, S. A., 650, **679**
Golden, J. A., 396, **422**
Goldman, P. S., 410, **424**

Goldschmidt, R. B., 769, 777, 778, **782**
Goldstein, B., 253, 255, 256, **260**, 778, **783**
Goldwasser, E., 509, **520**
Gont, L. K., 326, **340**, 393, 398, 399, **422**
González-Crespo, S., 295, **300**
González-Martínez, M. T., 192, **216**
González-Reyes, A., 76, **78**, 272, 288, **300**
Gooday, D., 431, **462**
Goodenough, U. W., 34, **48**
Goodman, C. S., 442, 449, 450, **459**
Goodman, M., 408, **423**
Goodman, W., 589, **608**
Goodrich, E. S., 16, **23**
Goodwin, B., 18, **24**
Gordon, J. L., 725
Gordon, M. Y., 509, **519**
Gorski, R. A., 559, **570**
Gospodarowicz, D., 595, **609**
Goss, R. J., 593, **608**
Gosse, R. J, 64, **78**
Gossler, A., 100, **105**
Gotthard, K.728, **747**
Gottlieb, D. I., 455, **459**
Gottlieb, G., 457, **459**, 778, **782**
Gouilleux, F., 160, **176**
Gould, E., 6, 15, **23**, 410, **423**, 737, **747**
Gould, M., 200, **216**
Gould, S. J., 408, **423**, 582, **608**, 743, **747**, 778, 775, **782**
Goulding, E. H., 696, 697, **716**
Goulding, M. D., 381, **385**
Gould-Somero, M., 200, **216**
Goustin, A. S., 149, **176**
Govind, S., 296, **303**
Graff, J. M., **176**
Graham, A., 434, 435, 439, **458, 461**
Graham, C., 621, **643**, 759, **782**
Grainger, R., 113, 144, 146, 145, **176, 177**
Grammatopolous, G. A., 437, **459**, 773, **782**
Granato, M., 345, **386**
GrandPré, T., 713, **716**
Grant, P., 633, **643**
Gräper, L., 494, **519**
Grapin-Botton, A. 512, 514, 515, **519**
Graveley, B. R., 130, **138**
Graves, J. A., 553, **572**
Greb, R. R., 641, **643**
Green, D. R., 165, **176**
Green, G. R., 209, **216**
Green, J. B. A., 67, **78**
Greenough, W. T., 737, **749**
Greenspan, R. J., 167, **177**

Greenwald, I., 167, 168, **176**
Greer, J. M., 378, **386, 782**
Gregg, N. M., 702, **716**
Gregory, W. A., 406, **423**
Grens, A., 600, **608**
Grey, R. D., 202, **216**
Grignani, F., 704, **716**
Grindley, J. C., 98, **105**
Gritsman, K., 353, **386**
Grober, M. S., 732, **747**
Grobstein, C., 149, 150, **176**, 481, 482, 487, 577, **608**
Gronemeyer, H., 590, **608**, 704, **714**
Groner, B., 160, **176**
Gross, P. R., 208, **216**
Grosshans, H., 134, **139**
Grossniklaus, U., 276, **300**, 662, **679**
Groudine, M. 122, **139**
Groudine, M.
Gruenbaum, Y., 123, **139**
Grumbach, M. M., 542, **544**
Grün, F., 744, **747**
Gruneberg, H., 689, 690, **716**
Grunwald, G., 415
Grunz, H., 326, 329, **340**
Gruss, P., 379, **386**
Gu, G., 515, **519**
Gualdi, R., 513, **519**
Guarente, L., 604, **608**
Gubbay, J., 551, **571**
Gudernatsch, J. F., 578, **608**
Guerrier, P., 245, **260**
Guger, K. A., 323, **340**
Guilette, L. J., 740, **747**
Guilette, L. L. Jr., 567, **571**
Guillery, R. W., 738, **746**
Guillette, E. A., 702, **716**Jr., 743, **746**
Guillin, O., 452, **459**
Gulyas, B. J., 366, **386**
Gumbiner, B. M., 75, **78**, 323, **340**
Gumpel-Pinot, M., 512, **519**
Gundersen, R. W., 445, **459**
Guo, L., 104, **105**, 417, 418, **423**
Gupta, C., 701, **716**
Gurdon, J. B., 149, **176**
Gurdon, J. B., 65–67, **78, 79**, 87, **105**, 636, **643**
Gustafson, T., 227, 233, **260**
Guthrie, S., 400, **423**
Gutzeit, H. O., 638, **644**
Gutzke, W. H. N., 567, **571**
Guyette, W. A., 132, **139**
Gwatkin, R. B. L., 193, **216**

Haas, H., 360, **388**
Habener, J.-F., 117, **139**
Hables, W., 660
Haccius, B., 661, **679**
Hacker, A., 551, **571**

Hadfield, M. G., 725, 726, **747**
Hadley, M. A., 170, **176**
Hadorn, E., 613, 689, 690, **716**
Hafen, E., 156, **175, 176**, 281, **300**
Haffter, P., 346, 347, **386**
Hafner, M., 202, **216**
Hagedon, H. H., **747**
Hahn, H., 162, **176**
Haigh, J. J., 476, **487**
Hakamori, S., 169, **176**
Hakem, R., 166, **176**
Haldane, J. B. S., 779, **782**
Halder, G., 119, 120, **139**, 413, **423**, 753, **782**
Haley, S. A., 201, **216**
Halfter, W., 455, **459**
Hall, A., 431, 454, 469, **487**
Hall, B. K., 434, 437, **459, 462**, 475, **487**, 577, 579, **608**, 767, **781**, 761, **782**
Hall, D. H., 627, **644**
Hall, H. G., 236, **260**
Hall, Z. W., 452, **460**
Hallmann, A., 38, **48**
Halpern, M. E., 352, **386**
Halprin, K. M., 417, **423**
Hamaguchi, M. S., 209, 210, **216**
Hamburger, V., 62, **78**, 320, **340**, 534, **544**, 778, **782**
Hamburgh, M., 149, **177**
Hamelin, M., 448, **460**
Hamer, D. H., 560, **571**
Hämmerling, J., 31, **48**
Hammerschmidt, M., 329, **340**, 352, **386**
Hammes, A., 130, **139**
Hammond, C. B., 557
Hammond, S. M., 104, **105**
Hampsey, M., 112, **142**
Han, M., 256, **260**
Hanahan, D., 501, 503, **519**, 705, **716**
Hanai, J. -I., 504, **519**
Handa, N., 563, 565, **571**
Hanes, S. D., 289, 290, **300**
Hanken, J., 577, 579, 582, **608, 611**
Hanna, L. A., 367, **386**
Hansen, C. S., 329, **340**
Hansen, L. A., 704, **716**
Hara, K., 305, **340**
Harada, J. J., 663, **679**
Harafuji, N., 600, **608**
Harary, I., 495, **519**
Haraway, D. J., 62, **78**, 708, **716**, 721
Hardie, J., 726, **747**
Hardin, J. D., 223, 228, 236, 237, 238, **260**, 313, **340**
Harding, K. W., 279, 281, 282, 286, 288, **300, 301**
Hardman, P., 153, **177**
Hardy, D. M., 190, **216**
Hardy, M. H., 418, **423**

Hareven, D., 670, **679**
Harkey, M. A., 236, **260**
Harland, R. M., 328, **342**
Harmon, M. A., 744, **747**
Harper, S., 452, **460**
Harrington, A., 62, **78**
Harris, G. W., 559, **571**
Harris, M. P., 770, 771, **782**
Harris, W. A., 320, **342**, 453, 457, **460**
Harrison, R. G., 56, **78**, 83, **105**, 143, 149, **177**, 206, **216**, 326, 411, **423**, 444, **460**, 524, 525, 529, **544**
Hart, A. C., 156, **177**
Hart, M. W., **747**
Hart, R., 602, 603, **608**
Hartenstein, V., 167, **177**, 267, 268
Hartfelder, K., 739, **748**
Hartmann, C., 540, **544**
Hartsoeker, N., 183, **216**
Harvell, C. D., 734, **745, 749**
Harvell, C. W., 735, **747**
Harvey, R. P., 6, 9, **23**, 496, **519**
Harvey, W., 491, **519**
Harwood, J., 84, **105**
Hashimoto, C., 294, **300, 301**, 575, **608**
Hästbacka, J., 542, **544**
Hastie, N. D., 130, **139**
Hatini, V., 485, **487**
Hatori, M., 476, 477, **487**
Hatta, K., 75, **78**, 469, **487**
Hatten, M. E., 405, 406, **422, 423**
Haver, N., 767
Hawley, S. H., 331, **340**
Hay, A., **679**
Hay, E. D., 145, **177**, 506, **519**, 593, **608**
Hayashi, K., 525, **544**
Hayes, T. B., 568, 569, **571**, 744, **745, 747**
He, T.-C., 161, **177**
He, X., 323, **340**
Heard, E., **139**
Heasman, J., 28, 75, **78**, 103, **105**, 169, **177**, 307, 322, **340**, 617, 618, **644**, 64
Heath, J. K., 619, **644**
Heberlein, U., 155, **177**
Hebrok, M., 514, **519**
Hedgecock, E. M., 448, **460**
Heemskerk, J., 285, 286, **300, 301**, Hegrenes, S., **745, 747**
Heidstra, R., 663, 667, 668, **681**
Heikinheimo, M., 535, **544**
Heinecke, J. W., 205, **216**
Heins, S. W., 731, 732, **746**
Heissig, B., 509, **519**
Heitzler, P., 167, **177**
Helbling, P. M., 435, **460**
Helde, K. A., 349, **386**
Heldin, C.-H., 159, **177**

Hemesath, T. J., 115, **139**, 159, **177**, 689, **716**
Hemler, M. E., 198, **218**
Hemmati-Brivanlou, A., 327, 329, 330, **340, 343**, 507, 51
Hempel, F. D., **680**
Henderson, S. T., 627, **644**
Hengartner, M. O., 165, **177**
Henion, P. D., 433, **460**
Henry, J. J., 145, **177**, 233, 245, **260**, 412, **423**, 775, **782**
Hensen, V., 411
Herlands, R., 598, **608**
Herman-Giddens, M. E., 701, **716**
Herndon, L. A., 605, **608**
Hertwig, O., 54, 62, **78**, 184, 209, **216**
Hess, R. A., 559, **571**
Heuser, J., 203, **215**
Hicklin, J., 600, **608**
Higashiyama, T., 658, **680**
Higgins, G. M., 601, **608**
Hilfer, S. R., 413, **423**
Hill, M., 701, **716**, 740, **747**
Hill, R. E., 98
Hill, R. J., 157, **177**
Hill, R. S., 636, **644**
Hilleman, B., 744, **747**
Hillers, K. J., 625, 626, **646**
Hinchliffe, J. R., 524, **544**
Hiramoto, Y., 209, 210, **216**
Hirano, T., 354
Hird, S. N., **260**
Hirohashi, N., 192, 195, **216**
Hiroma, Y., 119
Hiroyoshi, T., 627, **644**
Hirsch, D., 628, **644**
Hirsch, M. S., 169, **177**
Hirsch, N., 305, **340**
Hirsh, D., 251, 257, **260**
Hirsh, I., **261**
Hirth, F., 754, 755, **782**
His, W., 411, **423**
Hitt, A. L., 169, **178**
Ho, R. K., 350, **386**
Ho, Y. S., 603, **608**
Hobmayer, B., 599, 600, **608**
Hoch, M., 281, **301**
Hodgkin, J. T., 628, **644**
Hodgkinson, C. A., 689, **716**
Hodgman, R., 133, **139**
Hodor, P. G., 234, **260**
Hofer, J. M. I., **680**
Hoffman, L. M., 697, **716**
Hoffmann, R. J., 728, **747**
Hogan, B. L. M., 145, 153, **176, 177**, 525, **544**, 619, **644**
Hogness, D. S., 289, **301**
Holder, N., 774, **782**
Holland, L. Z., 761, **782**
Holland, N. D., 760, 761, 762, **782**
Holland, P. W. H., 760, 761, **782**

Holley, S. A., 330, 331, **340**, 763, **782**
Holliday, R., 602, **610**
Hollyday, M., 443, **460**403, **423**
Holmberg, J., 446, **460**
Holmes, L. B., 703, **716**
Holmes, S., 67
Holowacz, T., 312, **340**
Holt, C., 455, **460**, 577
Holtfreter, H., 312, 313, 327, **340**
Holtfreter, J., 70, 71, 73, **79**
Holtzer, H., 145, **177**
Holy, J., 200, 210, **216**
Holzenberger, M., 605, **608**
Honda, S., 604, **608**
Honda, Y., 604, **608**
Hong, C. C., 294, **301**
Hong, K., 448, **460**
Honig, L. S., 534, **544**
Hooper, L. V., 724, 725, **747**
Höpker, V. H., 448, **460**
Hoppe, D. M., 744, **746**
Hoppe, P. E., 167, **177**
Hopson, J. A., 768, **782**
Horb, M. E., 512, 515, **519**
Horder, T., 84
Hornung, M. W., 740, **747**
Horowitz, D. S., 129, **139**
Horowitz, M. C., 548, **572**
Hörstadius, S., 64, **78**, 229, 230, **260**
Horton, W. A., 474, 476, **487**
Horvitz, H. R., 165, **179**, 251, **262**
Horwitz, A., **177**
Hoskins, S. G., 577, **608**
Hotchkiss, R. D., 121, **139**
Hotta, K., 249, **260**
Houliston, E., 212, **216**
Howard, E. W., **260**
Howard, K., 624, **644**
Howard, K. R., 624, **644**
Howdeshell, K., 701, **716**
Howe, A., 170, **177**
Hozumi, N., 90, **105**
Hsin, H., 604, **608**
Hsiung, F., 156, **177**
Hsu, Y.-R., 159, **177**
Hu, S., 560, **572**
Huala, E., 672, **680**
Huang, B., 553, **572**, 580, **608,**
Huang, G. Y., 171, **177**
Huang, H., 579, 580, **608**
Huang, Z., 739, **747**
Huarte, J., 641, **644**
Hubel, D. H., 457, **460**, 737, 738, **747**
Huelsken, J., 420, **423**
Hughes, A. E., **716**
Hughes, C. G., 755, 778, **782**
Hui, H., 119
Hulskamp, M., 658, **680**
Hummel, K. P., 381, **386**

Humphrey, R. R., 627, **644**
Humphreys, D., 88, **105**, 708, **716**
Humphreys, T., 128, **139**, 208, **216**
Hunt, P., 377, **386, 519**
Hunter, C. P., 104, **106**, 255, **260**
Hunter, R. H. F., 194, **216**
Hurle, E., 540
Hurle, J. M., **544**, 770
Hurt, E., 132, **141**
Hussain, M. A., 117, **139**
Hutchinson, S., 692, **716**
Hutter, H., 256, **260**
Huxley, J. S., 19, **23**, 63, 67, **78**, 582, **608**, 766, 779
Hyatt, B. A., 337, **340**
Hylander, B. L., 191, 192, 201, **218**
Hynes, R. O., 454, **460**

Ibáñez, C. F., 453, **460**
Ibrahim, H., 313, **340**
Ichtchenko, K., 130, **139**
Iemura, S.-I., 329, **340**
Ihle, J. N., 159, **177**
Ikegame, M., 722, **747**
Ikenishi, K., 617, **644**
Ikeya, M., 472, **487**
Ikezuki, Y., 701, **716**
Illmensee, K., 703, **716**
Imai, K. S., 249, **260**
Imanaka-Yoshida, K., 495, **520**
Imokawa, Y., 595, **609**
Imperato-McGinley, J., 558, **572**
Ingersoll, E. P., 227, **259**
Ingham, P. W., 285, 288, **301,** 764, 765, **782**
Ingram, V. M, 123, **138**
Innis, J. W., 533, **544**
Inoue, H., 627, **644**
Insall, R. H., 43, **48**
Inuzuka, H., 453, **460**
Irie, A., 409, **423**
Irish, V. F., 297, **301,** 663, **680**
Irvine, R. F., 206, **219**
Ishii, N., 448, **460**, 512, **520**
Ishizuya-Oka, A., 577, **609,**
Isida, Z., 505, **520**
Itäranta, P., 483, **487**
Iten, L. E., 531, **544**
Iwabuchi, M., 306, 307, **340**
Iwao, Y., 200, **216**
Izpisúa-Belmonte, J. C., 337, 365, 530, 537, **544**
Izumo, S., 494, **520**

Jack, T., 677, **680**
Jäckle, H., 280, 281, **300, 301, 302**
Jacklin, D., 560, **572**
Jackson, D., 670, **680**
Jacob, F., 769, **782**
Jacobs, P. A., 212, **216**
Jacobson, A. G., 144, 145, 146, **177**, 265, **300**, 395, 396, **423**

Jacobson, M., 165, **177**, 403, 404, **423**, 428, 454, **460**
Jaenisch, R., 88, **106**, 126
Jaffe, L. A., 200, 201, 204, 205, 209, **216**, 657, **680**
Jaffe, L. F., 44, **48**
Jaglarz, M. K., 624, **644**
Jain, M., 704, **716**
Jakob, U., 776, **782**
Janssen, B.-J., 670, **680**
Janzen, F. J., 731, **747**
Jares, P., 306, **340**
Jaynes, J. B., 281
Jean, D., 753, **782**
Jean, R. V., 668, **680**
Jeffery, J. E., 466, 767, **782**
Jeffery, W. R., 249, 250, 251, **259**, 260
Jelalian, K., **140**
Jenkins, N., 115
Jeong, S., 667, **680**
Jeppesen, P., 127, **139**
Jepson, K., 120, **139**
Jermyn, K. A., 39, **48**
Jernvall, J., 440, 441, **460**, 770, 771, 772, **782, 784**
Jessell, T. M. 431, 443, **461**, 401, **423**
Jesuthasan, S., 349, **388**
Jeyasuria, P., 568, **572**
Jiang, C., 592, **609**
Jiang, J., 295, 296, **301**
Jiang, X., 438, **460**
Jiang, Y., 710, 711, **716**
Jiang, Y.-H., **139**
Jilka, R. L., 477, **487**
Jin, E. J., 433, **460**
Joanen, T., 52, **78**, 731, **746**
Johansson, C. B., 410, **423**, 713, **716**
Johe, K. K., 713, **716**
Johnson, E. M., 700, **716**
Johnson, F. B., **609**
Johnson, H. H., 445, **463**
Johnson, J. E., 514, **520**
Johnson, J., 209, **216**
Johnson, R. D., 740, **747**
Johnson, R. L., 40, **48**, 151, 162, 163, **177**, 487, 539, **543**
Johnston, M. C., 697, **716**
Johri, B. M., 659, **680**
Jokiel, P. L., 54, **78**
Jones, C. M., 149, **177**, 329, **340**
Jones, G., 590, **609**
Jones, K. L., 697, **716**
Jones, K. R., 452, **460**
Jones, P. H., 416, **423**
Jones, P. L., 122, **139**
Jongens, T. A., 615, **644**
Jordan, B. K., **572**
Jordan, T., 413, **423**
Joseph, E. M., 308, **340**
Josso, N., 558

Jost, A., 549, **572**
Jouve, C., 468, **487**
Jung, J., 513, **520**
Jurand, A., 369, **386**
Jürgens, G., 276, **300**, 663, **680**
Jursnich, V. A., 132, **139**
Just, E. E., 70, 71, **78**, 83, 84, **106**, 190, 201, 203, **216**, 221, 290, **301**

Kachroo, A., 656, **680**
Kadokawa, Y., 74, 76, **78**
Kafri, T., 124, **139**
Kahn, A. J., 477, **487**
Kahn, C. R., 433, **460**
Kaji, K., 198, **216**
Kalcheim, C., 435, **461**
Kalderon, D., **301**
Kalimi, G. H., 172, **177**
Kallunki, P., 120, 121, **139**
Kalm, L. von, 585, 586, **609**
Kalt, M. R., 306, **340**
Kaltenbach, J. C., 580, **609**
Kalthoff, K., 270, **301**
Kamachi, Y., 117, **139**
Kamen, R., 132, **141**
Kammandel, B., 113, **139**
Kammerman, S., **644**
Kanavakis, E., 693, **716**
Kandel, E. R., 559, 560, **572**
Kandler-Singer, I., 270, **301**
Kane, D. A., 348, **386**
Kant, I, 7
Kaplan, S. L., 542, **544**
Karavanov, A. A., 483, **487**
Karavanova, I. D., 482, **487**
Karfunkel, P., 396, **423**
Karlseder, J., 603, **609**
Karlson, P., 591, **607**
Karlstrom, R. O., 455, **460**
Karp, G. C., 234, **260**, 580, **608**
Karr, T. L., 264, 265, 269, 282, 283, **301**
Karsenty, G., 477, **487**
Kashikawa, M., 616, **644**
Kastern, W. H., 133, **139**
Kastner, J., 470
Katayama, I., 503, **520**
Kater, S. B., 453, **462**
Kato, N., 88, **106**, 471, **487**
Kato, T., 410, **423**
Kato, Y., 368, **388**
Katz, L. C., 739, **747**
Katz, W. S., 157, 158, **177**
Kaufman, D. S., 709, **716**
Kaufman, M. H., 212, **216**, 374, **385**
Kaufman, T. C., 242, **261**, 275, 280, 286, 287, **301**, 755, 778, **782**
Kawahara, A., 117, 353, **386**, 580, **609**

Kawakami, Y., 526, 527, 530, 536, **544**

Kawasaki, I., 615, **644**

Kay, G. F., 127, **139**

Kay, R. R., 43, 44, **48**

Kayano, H., 503, **520**

Kaznowski, C. E., 407, **424**

Kedinger, M., 512, **520**

Keeler, R. F., 163, **177**

Keiding, N., 740, **747**

Keirstead, H. S., 713, **716**

Keisman, E. L., 565, 566, **572**

Keith, D. H., 127, **139**

Kelce, W. R., 740, **747**

Keller, E. F., 40, **48**, 84, **106**, 395, **423**

Keller, R. E., 308, 309, 313, 314, 315, 316, 326, **341**, **343**

Kelley, R. I., 163, **177**, 413, **423**

Kelly, G. M., 352, **386**

Kempermann, G., 410, **423**, 737

Kemphues, K. J., 104, **105**, 253, **260**

Kennedy, C., 738, **747**

Kennedy, T. E., 447, 448, **460**

Kenny, A. P., **260**

Kenwrick, S., 601, **608**

Kenyon, C., 158, **177**, 255, **260**, 604, 605, **608**

Kerbel, R., 705, **716**

Kerr, K., 274, **301**

Kerstetter, R., 669, **680**

Keshet, I., 122, **139**

Kessel, M., 306, **338**, 379, 380, **386**

Kessel, R. G., 348, **384**

Kessler, D. S., 324, 324, 325, **339**, **340**

Keyes, L. N., 562, 564, **572**

Keynes, R., 377, **386**, 400, **423**

Keys, D. N., 769, **782**

Kezer, J., 625

Khaner, O., 360, 361, **386**

Kidd, S., 294, 301.

Kidd, T., 450, **460**

Kiecker, C., 334, 335, **340**

Kieny, M., 471, 487, 529, **544**

Kiesecker, J. M., 55, **78**, 744, **747**

Kim, H. S., 701, **716**

Kim, G. J., 249, **260**

Kim, J., 587, **609**

Kim, J. H., 708, **716**

Kim, -S. K., 148, 158, **179**

Kim, S.-H., 313, 315, **341**

Kim, Y. M., 504, **520**

Kimberly, E. L., 236, **260**

Kimble, J., 157, **177**, 251, **260**, 627, 628, **642**, **644**

Kimelman, D., 323, 324, 325, **339**, **341**, **342**

Kimmel, A. R., 44, **48**

Kimmel, C. B., 345, 346, 348, 349, 350, 351, **386**, **388**

Kimura, C., 376, **386**

Kimura, Y., 208, **216**

King, M. C., 408, **423**, 778, **782**

King, T. J., 85, 86, **105**, **106**

Kingsley, D. M., 540, **546**

Kinsey, W. H., 207, **216**

Kinter, C. R., 594, **607**

Kintner, C. R., 76, **78**, 333, **339**

Kirby, C. M., 253, **260**, 429, **460**

Kirby, M. L., 441

Kirk, D. L., 33, 35–38, **48**

Kirkwood, T. B. L., 601, **609**

Kirschner, M., 636, 645751, **781**

Kirschner, W. W., 222, 223, 223, **261**, 266, **302**, 310, 311, **341**

Kishimoto, Y., 352, **386**

Kispert, A., 482, 483, **487**

Kistler, A., 579, **609**

Kitamoto, Y., 505, **520**

Kitajewsky, J., 700, **716**

Kiyama, J., 515, **520**

Kjoller, L., 435, 441, 458, **460**, 461469, **487**

Klanke, B., 480, **487**

Kleene, K. C., 128, **139**

Klein, C., 42, **48**

Klingler, M., 277, **301**

Kloc, M., 616, **644**

Kloppstech, K., 32, **48**

Klose, M., 444, **460**

Kmita, M. F., 533, **544**, 755, **782**

Knecht, D. A., 40, 42, **48**

Kniewald, 568, **572**

Knipple, D. C., 281, **301**

Knoll, A. H., 765, **782**

Knoll, J. H. M., 123, **139**

Knudsen, K., 169, **177**, 474, **487**

Ko, K. 515, **520**

Kobayashi, T., 457, **460**, 616, **644**

Koch, P. B., 53, **78**, 729, 730, **747**, 779, **782**

Kochav, S. M., 360, **386**

Kochert, G., **49**

Kochhar, D. M., 696, **717**

Koelle, M. R., 590, **609**

Koga, M., 158, **177**

Kohler, R., 263, **301**

Kollar, E. J., 440, **460**, **461**

Kollros, J. J., 579, 580, **609**

Kolm, P. J., 334, 336, **341**

Kolodkin, A. L., 445, 446, 447, **460**

Kölreuter, J. G., 6, **23**

Komori, T., 437, 460487

Komuro, H., 405, **423**

Komuro, I., 494, **520**

Kondo, K., 700, **717**

Kondo, S., 20, **23**

Konigsberg, I. R., 474, **487**

Konjin, T. M., 40, **48**

Koopman, P., 551, 552, 553, **572**

Koos, R. D., 641, **644**

Kopf, G. S., 190, 193, **219**

Kopf, G. S., **217**

Koppinen, 419, **423**

Korbling, M., 711, **717**

Koreach, K. S., 549, **571**

Korinek, V., 161, **177**

Kornack, D. R., 410, **423**

Kornberg, R. D., 107, 112, **139**, **140**, 282, 283, **301**

Korsching, S., 452, **460**

Kos, R., 428, **460**

Kosaki, K., 382, **386**

Koseki, C., 482, **487**

Koshiba-Takeuchi, K., 455, **460**

Koshida, S., 351, **386**

Kosman, D., 296, 297, **301**

Kotch, L. E., 697, 698, **717**

Kowalevsky, A., 14, **23**

Koyama, E., 441, **460**

Koziol-Dube, 166

Kozlowski, D. J., 12, **23**

Krabbenhoft, K. M., 529, **544**

Kraemer, B., 628, **644**

Krainer, A. R., 129, **139**

Kramer, K. L., 337, **340**

Krantz, S. B., 509, **520**

Kraut, R., 281, **301**

Krawczak, M., 692, **717**

Kreidberg, J. A., 481, **487**

Krieg, P. A., 503, **522**

Kropf, D. L., 659, 679, **680**

Krull, C. E., 431, 432, 460

Krumlauf, R., 377, 378, **387**, 428, 436, 437, **461**, **463**

Krutch, J, W., 37, **48**

Kubai, L., 507, **520**

Kuhn, K., 760, **784**

Kuida, K., 165, 166, **177**

Kukalova-Peck, J., 756, 757, **782**

Kulesa, P. M., 435, 460469, **487**

Kulikauskas, V., 696, 71

Kulyk, W. M., 697, **716**

Kumar, A., 593, **607**

Kumaran, M. K., **680**

Kunisada, T., 432, **461**

Kunkle, M., 209, **217**

Kuo, C. J., **520**

Kupperman, E., 494, 495, **520**

Kurata, T., 330, **341**

Kuratani, S. C., 441, **461**, 773, **782**

Kuretake, S., 207, **217**

Kuro-o, M., 603, **609**

Kurz, H., 501, **519**

Kuure, S., 481, **487**

Kvist, U., 209, **217**

Kwack, B. H., 657, **679**

Kwan, K.-M., 118, **139**

Kwong, R. W., 663, **680**

Kyba, 507, **520**

La Due, R. A., 697, **718**

LaBarbera, M., 501, **520**

LaBarge, M. A., 710, **717**

LaBonne, C., 431, **461**

Ladher, R. K., 438, **461**

Laemmli, U. K., 122

LaFleur, G. J., Jr., 201, **217**

Lagos-Quintana, M., 134, **139**

Lahav, R., 434, **461**

Lai, C. S., 409, **423**

Lamb, J. 701, **717**

Lambert, J. D., 226, 243, **260**, 244, 245

Lammer, E. J., 696, **717**

Lammert, E., 514, **520**

Lamoureux, P., 412, **423**

Lance-Jones, C., 443, 444, 46

Lander, A. D., 285, **301**, 454, **460**

Landmesser, L., 443, 444, **461**

Landstrom, U., 308, **341**

Lane, M. C., 236, 237, **260**, 272, **301**

Lang, A., 673, **680**

Lang, J. W., 567, **572**

Langeland, J., 345, 346, 349, 350, 351, **386**

Langman, J., 373, **386**, 472, **487**, 500, 503, 512, 513, 515, **520**, 550, 557, **572**, 683, **718**

Lappi, D. A., 150, **177**

Larabell, C. A., 323, **341**

Larrain, J., 764, **782**

Larsen, E. W., 774, **783**

Larsen, P. L., 603, **609**

Larsen, W. J., 370, 371, **386**, 405, **423**, 497, **520**, 683, **717**

Larson, A., 774, **784**

Lash, J. W., 169, **178**, 359, **386**, 469, 487, 493, 494, 495, **520**, 531, **544**

Lassar, A. B., 487, 493, **522**

Latinkic, B. V., **339**

Lau, N. C., 134, **139**

Laufer, E., 770

Laufer, L., 151, 530, 536, 538, **544**

Laurent, M. N., 324, **341**

Laux, T., 663, **680**

Law, R. D., 348, **386**

Lawn, R. M., 109, **140**

Lawrence, P. A., 64, **78**, 278, 279

Lawson, A., 356, 361, 362, **386**, 396, **423**

Lawson, E. J. R., 356, 672, **680**

Lawson, K. A., 369, **386**, 618, **644**

Lawson, N. D., 505, 506, **520**

Layton, W. M., Jr., 381, **386**

Lazebnik, Y. A., 165, **177**

Le Clair, J. J., 744, **745**, **747**

Le Douarin, N. M., 13, **23**, 358, **386**, 429, 433, 435, **463**, 471, 481, 487, 488, 513, **520**,

Le Lièvre, C. S., 429, 438, **461**

Le Mouellic, H., 378, 379, **386**

Le Naour, F., 198, **217**

Leatherman, J. L., 617, **644**

Leblond, C. P., 185

Lechleiter, J. D.
Lecourtois, M., **177**
Lee, C.-H., 75, **78**
Lee, H. Y., 396, **424**
Lee, H.-S., 656, **680**
Lee, J. E., 323, **341**
Lee, J. T., 127, 134, **140**
Lee, M. S. Y., 759, **781**
Lee, R. C., **140**
Lee, S. –K., **177,** 402, **423**
Lee, S.-J., 206, **217**
Lee, T. I., **140**
Lees, A. D., 726, **747**
Leeuwenhoek, A. van., 183, **217**
Leevers, S. J., 154, **178**
Lefebvre, R., 194, **217**
Lehmann, R., 135, **139,** 167, **178,** 276, 277, **300, 301, 303,** 616, **643**
Lehtonen, E., 482, **487**
Leibniz, G. W., 773
Leidy, J., 509
Lemaire, P., 324, **341**
Lemaire, W. J., 641, **644**
Lemaitre, B., 295, **301**
Lemoine, E. M., 697, **717**
LeMosy, E. K., 294, **301**
Lengauer, C., 703, **717**
Lengyel, J. A., 276
Lenhoff, H. M., 598, **609**
Lennarz, W. J., 191, 195, 194, 196, **216**
Lenoir, T., 7, **23**
Lenz, W., 17, **23,** 694, **717**
Leonardo, E. D., 448, 449, **461** 251, **260**
Lepage, T., 228, **260**
Leptin, M., 295, 296, **301**
Letinic, K., 408, 409, **423**
Letourneau, P. C., 403, 411, **423,** 445, **461**
Letterio, J. J., 153, **178**
Leung, C., 348, **386**
Leutert, T. R, 52, **78**
LeVay, S., 560, **572**
Levi-Montalcini, R., 391, 451, **461**
Levin, M., 363, 364, 365, **386**
Levin, R. B., 411
Levine, A. E., 191, **217**
Levine, M. S., 248, **259,** 269, 279, 281, 282, 288, 295, **300, 301**
Levken, A. C., 353, **386**
Levy, Y. Y., 654, 673, **680**
Lewis, E. B., 285, 286, 287, **301,** 756
Lewis, J., 505, **520**
Lewis, J. H., 531, 532, **546**
Lewis, P. M., 163, **178**
Lewontin, R. C., 775, **782**
Leyns, L., 331, 332, **341**
Leyton, L., 197, **217**

Li, E., **178**
Li, G., 712, **717**
Li, H. -S., 99, **106, 178,** 450, **461**
Li, Q.Y., 527, **544**
Li, S., 538, **544**
Li, W., **178**
Li, Y., 542, **544**
Liao, E. C., 501, **520**
Lichtman, J. W., 408, **424,** 451, **462,** 739, **748**
Lieberfarb, M. E., 133, **140**
Liem, K. F., Jr., 402, **423, 461**
Liesi, P., 445, **461**
Lillie, F. R., 7, **23,** 83, 84, **106,** 242, **260,** 641, **644**
Lin, C. R., 363, **386**
Lin, H., 637, 638, **644**
Lin, J. J.-C., 497
Lin, L.-F. H., 452, **461**
Lin, R., 256, **260**
Lin, Y., 194, **217,** 484, **487**
Lin, Y.-J., 603, **609**
Linask, K. K., 493, 494, 495, **520**
Linask, K. L., 169, **178,** 359, 495, 496, **520**
Lindahl, P., 504, **520**
Lindquist, S., 777, **784**
Lindroos, P. M., 601, **609**
Lindsay, R. M., 452, **461**
Linney, E., **717**
Lints, T. J., 494, **521**
Lipshultz, L., 685, **717**
Lira, A. A., 635, **644**
Litingtung, Y., 536, 537, **544**
Litt, M. D., 122, 123, **140**
Little, J., 397, **423**
Liu, C. K., 613, **644**
Liu, C.-M., 663, **680**
Liu, J.-K., 169, **178**
Liu, J.-P., 431, **461**
Livesey, F. J., 713, **717**
Lloyd, R. M., 655, **679**
Lo, C., 172, **177**
Lo, D. C., 593, **609**
Lo, L., 433, **461**
Loeb, J., 602, **610**
Loeffler, M., 506, **521**
Logan, C. Y., 228, 229, 231, 237, **261**
Logan, M., 363, 365, **387,** 527, **544**
Lohnes, D., 760, **783**
London, C, 333, **341**
Long, J. A., 663, **680**
Longnecker, M. P., 702, **717**
Longo, F. J., 198, 202, 204, 209, 210, **217,** 641, **643**
Loomis, C. A., 764, **783**
Loomis, W. F., 41–43, **48**
Lopez, L. C., 193, **217**
López-Martínez, A., 530, 535, **544**
Lorca, T., 634, **644**
Lord, E. M., 657, **680**

Loredo, G., 125
Loredo, R., 125
Loring, J. F., 429, **461**
Lough, J., 494, **521**
Lovejoy, A. O., 773, **783**
Lovell-Badge, R., 556
Løvtrup, S., 308, **341**
Lowe, L. A., 382, **387**
Lu, N., 127, **140**
Lu, P., 661, **680**
Lucchesi, J. C., 124, **140**
Luckett, W. P., 369, **387**
Luedemanan, R., 411, **423**
Luger , K., 108, **140**
Lumsden, A., 336, 339, 377, **386,** 400, **423**
Lumsden, A. G. S., 433, 440, **461**
Luna, E. J., 169, **178**
Lundelius, J. W., 240, **260**
Luo, G., 482, **487**
Luo, M. L., 132, **140**
Luo, Y., 446, **461**
Luttmer, S., 202, 204, **217**
Lynch, M. H., 477, **489**
Lynn, W. G., 582, **609**
Lyon, M. F., 124, **140**
Lyons, I., 494, **520**

Ma, C., 155, **178**
Ma, L., 559, **572,** 699, **717**
Ma, Q., 433, **461**
Ma, Q. F., 323, **341**
MacCabe, J. A., 537, **544**
Macdonald, P. M., 274, **301**
Macdonald, S. J., 778, **783**
MacDougall, C., 131, **140**
MacGregor, H. C., 636, **644**
Machleder, E. M., 634, **643**
Macias, D., 539, 540, **544**
Mackie, E. J., 441, **461**
MacWilliams, H. K., 598, 599, **609**
Maden, M., 525, **544,** 596, **609, 610**
Maderson, P. F. A., 770, **783**
Maduro, M. F., 255, **261,** 491, **520**
Maeno, M., 507, **520**
Maeshima, A., **487**
Magavi, S. S., 410, **423**
Mahmood, R., 530, **544**
Mahowald, A. P., 303, 615, **644**
Maimonides (Moshe ben Maimon), 143, **178**
Maître-Jan, A., 6, **23**
Mak, V., 696, 700, **717**
Malacinski, G. M., 307, **342**
Malebranche, N., 6
Malho, R., 657, **681**
Malicki, J., 755, **783**
Malinda, K. M., 236, **261**
Maller, J. L., 634, **645**
Malpighi, M., 6, 7, **23**
Manak, J. R., 753, **783**

Manasek, F. J., 495, **520**
Mancilla, A., 427, 430, **461**
Mandoli, D. F., 31, **48**
Manes, M. E., 210, **217**
Mangelsdorf, D. J., 579, **609**
Mango, S. E., 257, **261**
Mangold, H., 320, 322, **342,** 351, **388**
Mangold, O., 334, **341**
Maniatis, T., 112, 132, **140, 141**
Manley, N. R., 377, **387**
Manley, T. R., 561
Mann, I., 415, **423**
Mann, R. S., 288, 289, **299,** 301, 587, **606**
Manning, J. E., 124, **140**
Manolagas, S., 477, **487**
Mansour, S., 553, **572**
Manton, S. M., 758, **783**
Manzanares, M., **423,** 762, **783**
Mao, B., 332, **341**
Marahrens, Y., 127, **140**
Marcelle, C., 473, **487**
Marchase, R. B., 455, **462**
Marcus, N. H., 62, **78**
Marcus, R. C., 454, **461**
Margarit, E., 557, **572**
Marigo, V., 764, **783**
Marín, O., 446, **461**
Markert, C. L., 374, **387**
Markey, C. M., 701, **717**
Markwald, R. R., 494, 496, **520**
Maroto, M., 473, **487**
Marqués, G., 763, **783**
Marrari, Y., 212, **217**
Mars, W. M., 601, **609**
Marsden, M., 168, 317, **341**
Marshall, E., 560, **572**
Marsh-Armstrong, N., 580, **609**
Martin, C., 507, **520**
Martin, E., 638, **645**
Martin, G. R., 374, 383, **387,** 535, **543**
Martin, J. R., 277, **301**
Martin, P., 505, **520**
Martin, P. T., 451, **461**
Martin, R. D., 408, **424**
Martin, V. J., 597, **609**
Martindale, M. Q., 245, **260**
Martinet, M., 696, **716**
Martinez-Arias, A., 266, 278, 279, 288, **299, 301**
Martins, G. G., 237, 238, **261**
Martins-Green, M., 169, **178**
Marukawa, E., 712, **717**
Maruyama, Y. K., 231, **261**
Marvin, M. J., 493, **520**
Marx, G. A., 670, **680**
Marx, J., 447, **461,** 549, **572**
Matsui, Y., 621, 634, **645**
Matsumoto, K., 513, **520,** 634, **645**
Matsunaga, M., 453, **461**

Matsunami, H., 407, **424**

Matsuo, I., 754, **783**

Matsuoka, M., 91, **106**

Matsuoka, S., 509, **521**

Matsushita, S., 512, **521**

Matter, J. M., 740, **747**

Matthes, D. J., 446, **461**

Mattox, W., 692, **715**

Mauch, T. J., 478, **488**

Maurice, J., 645

Mavilio, F., 121, 122, **140**

Mayer, D. B., 622, **645**

Mayer, T. C., 429, **461**

Mayer, U., 663, **680**

Mayor, R., 431, **461**

Mayr, E., 778, 779, **783, 784**

Mayr, W. R., 374, **387**

Mazia, D., 133, **141**, 198, **218**

Mazmumdar, A., 265, **301**

Mazummdar, M., 265, **301**

McArthur, M., 122, **140**

McBride, W. G., 17, **23**

McCarty, D. R., 669, **680**

McCarver-May, D. G., 702, **717**

McClaren, A., 627, **645**

McClay, D. R., 228, 229, 230, 231, 233, 234, 236, 237, 238, **261**

McClendon, J. F., 63, **78**

McClung, C. E., 548, **572**

McCollum, S. A., 735

McConnell, S. K., 403, 407, **422, 424**

McCormick, F., 154, **178**

McCormick, S., 656, **679**

McCulloh, D. H., 200, 207, **217**

McCutcheon, F. H., 577, **609**

McDaniel, C. N., 654, 663, 671, 672, 673, 674

McDonald, J. W., 67, **78**, 708, 710, **717**

McEwen, B. S., 559, **572**

McEwen, D. G., 161, **178**

McFall-Ngai, M. J., 723, **748**

McFarlane, S., 454, **461**

McGann, C. J., 593, **609**

McGill, G. G., 165, **178** 703, **717**

McGinnis, W., 377, 378, **387, 461**, 755, **781**

McGonnell, I., M., 434, 435, **461**, 759, **782**

McGrath, J., 212, **217**

McKeon, J., 288, **301**

McKeown, T., 686, **717**

McKinnell, R. G., 86, **106**

McKinney, J. D., **717**

McLaren, A., 551, **570, 572**

McMahon, A. P., 151, 152, **175, 178**, 537, 538, 540, **545**

McNabb, A., **748**

McNulty, T. R., 531, **545**

McPherson, S. M., 206, **217**

McVeigh, E., 684, **717**

Mead, K. S., 54, **78**, 205, **217**

Meade, H. M., 89, **106**

Mechler, B. M., 637

Medawar, P. B., 601, **609**

Medvinsky, A. L., 480, 488, 507, **521**

Mehregan, A. H. H., 418, **424**

Meier, S., 145, **178**

Meinhardt, H., 20, 21, **23**, 587, **609**

Meinke, D. W., 662, **680**

Meizel, S., 192, **217**

Meldolesi, J., 130, **141**

Mello, C. C., 255, 256, 257, **261**

Melov, S., 603, **609**

Melton, D. A., 135, 149, **178**, 308, 327, 329, **340**, 514, **519**, 645

Mencken, H. L., 477, **488**

Mendel, G., 59, 81–83, **106**, 548, 777

Mendez, R., 133, 134, **140**, 634, **645**

Meng, X., 630, **645**

Menko, A. S., 474, **488**

Meno, C., 382, 383, **387**

Mercader, N., 533, **544**

Mercola, M., 493, **521**

Merimee, T. J., 542, **544**

Merino, R., 538, 539, 540, **544**, 769, **783**

Merke, D. P., 557, **572**

Mermoud, J. E., 127, **140**

Mescher, A. L., 594, 595, **609**

Messersmith, E. K., 446, 447, **461**

Meteyer, C. U., 744, **748**

Metz, C. B., 190, 194, **217**

Metz, J. A., 775, **781**

Meulemans, D., **783**

Meyer, W. J., 557, **572**

Meyer-Franke, A., 453, **461**

Meyerowitz, E. M., 650, 667, 676, 677, **679, 680**

Meyers, E. N., 383, **387**

Michalopoulos, G. K., 601, **609**

Michikawa, Y., 603, **609**

Migeon, B. R., 124, 126, 127, **138, 140**

Migeon, C. J., 557, **572**

Mikawa, T., 493, 495, **519, 521**

Mikkola, M. L., 419, **424**

Millar, S. E., 418, 419, 420, **424**

Millauer, B., 503, **521**

Miller, B. S., 189, **217**

Miller, C., 699, **717**

Miller, D. J., 202, **217**

Miller, J. K., 602, **609**

Miller, J. R., 236, **261**, 324, **341**

Miller, O. L., Jr., 636

Miller, R. L., 189, 190, **217**

Miller, S. A., 164, **178**, 516, **521**

Miller, S. J., 418, **424**

Mills, N., 743, **745, 748**

Miller, W. H., Jr,. 704, **717**

Milunsky, A., 397, **424**

Mina, M., 440, **461**

Minganti, A., 10, 11, **24**, 58, **79**, 247, **261**

Minshull, J., 222, **261**

Minsuk, S. B., 308, **341**

Mintz, B., 374, 474, 488, 618, 619, 621, 622, **645, 646**, 703, **716, 718**

Miny, P. 693, **717**

Mitchell, A. W., 583, **609**

Mitrani, E., 361, **387**

Mittendorf, R., 699, **717**

Miura, T., 774, **783**

Miyado, K., 198, **217**

Miyake, T., 475, **487**

Miyazaki, S.-I., 207, **217**

Miyazaki, Y., 483, **488**

Mizzen, C. A., 116, **140**

Moens, P. B., 625, **645**

Mogie, M., 659, **680**

Mohanty-Hejmadi, P., 525, 526, **544**

Mohler, J., 285, **301**

Mohri, T., 207, **217**

Molenaar, M., 324, **341**

Moller, C. C., 202, **217**

Molven, A., 525, **544**

Molyneaux, K. A., 618, **645**

Monaghan, P., 754, **783**

Monk, M., 124, **140**

Monroy, A., 73, **78**

Montagna, W., 417, 418, **424**

Montagu, M. F. A., 408, **424**

Montell, D. J., 293, **301**

Montgomery, A. M. P., 169, **178**

Montgomery, M. K., 723, **748**

Montonen, O., 419, **424**

Moon, R. T., 323, 325, 331, 333, **339, 341**

Moore, C. L., 559, **572**

Moore, K. L., 20, **23**, 399, **424**, 683, 695, **717**

Moore, L. A., 624, **645**

Moore, M. W., 481, **488**

Morange, M., 84, **106**, 768, 776, **783**

Morata, G., 288, **300**

Morgan, B. A., **544**

Morgan, R., 315, **341**

Morgan, T. H., 51, 63, **78**, 241, 161, 241, **261**, 561, **572**, 726, **748**

Mori, C., 539, **544**

Morin-Kensicki, E. M., 429, **461**, 475, **488**, 696, **717**

Morreale, S.J., **745, 748**

Check out Morril p. 40

Morrill, J. B., 234, 235, 237, 240, **260**

Morris, I. D., 740, **748**

Morriss, G. M., 370, **388**

Morriss-Kay, G., 380, **387**, 434, 438, **463**

Morse, D. C., **717**

Morse, D. E., 726, **746**

Mortlock, D. P., 533, **544**

Moscona, A. A., 71, 73, 74, **78**

Moses, K., 156, **177**

Moses, M. J., 625, **645**

Moury, J. D., 396, **423**

Moury, J. G., 396, **423**

Mouse Genome Sequencing Consortium, 650, **680**

Moustafa, L. A., 100, **106**

Mowry, K. L., **645**

Moy, G. W., 194, 195, **217, 219**

Moyon, D., 503, **521**

Mozingo, N. M., 187, **217**

Muchmore, W. B., 531, **544**

Muhr, J., 402, **424**

Mukhopadyay, G., 713, **717**

Mukouyama, Y., 505, **521**

Müller, F., 14, **23**, 614, **645**, 703, 718

Müller, G. B., 722, **748**

Müller, G., 525, 526, **545**

Müller, J., 14, 288, **301**

Müller, K., 42, **48**

Muller, L. M., 595, **609**

Müller, M., 289, **301**

Müller, U., 482, **488**

Mullins, M. C., 507, **521**

Mulnard, J. G., 367, **387**

Multigner, L., 186, **217**

Munar, E., 475, **488**

Mundlos, S., 437, **462**

Muneoka, K., 524, 538, **544, 545**, 595, **609**

Munro, E., 766, **784**

Münsterberg, A. E., 473, **488**

Muragaki, Y., 533, **545**

Murray, J. D., 40, **49**, 731, **749**

Murray, V., 602, **610**

Murre, C., 290

Muscatelli, F., 556, **572**

Muschler, J., 171, **178**

Muta, M., 116, **140**

Muthukkarapan, V. R., 150, **178**, 704, 705, **717**

Myers, L. C., 112, **140**

Myles, D. G., 194, 198, **218**

Nabel-Rosen, H., 130, 132, **140**

Nachtigal, M. W., 556, **572**

Nagel, M., **343**

Nagele, R. G., 396, **424**

Nagoshi, R. N., 131, **140**, 563, **572**

Nagy, L., 226, 243, 244, 245, **260**

Nakagawa, S., 577, **610**

Nakamoto, M., 456, **462**

Nakamura, A., 616, **645**, 689, **717**

Nakamura, O., 321, **341**

Nakamura, T., 483, **488**

Nakatani, Y., 249, **261**

Nakatsu, T., 396, **424**

Nakatsuji, N., 316, **341**

Nakauchi, H., 508, **521**
Nakayama, A., 115, **140**, 158, **178**
Nambu, J. R., 296, **301**
Nameroff, M., 475, **488**
Nan, X., 122, **140**
Nanjundiah, V., 40, 43, **48**
Nanni, L., 689, **717**
NARCAM, 744, **748**
Nardi, J. B., 597, **610**
Nascone, N., 493, **521**
Nasrallah, J. B., 656, **680**
Nasrallah, M. E., 657, **680**
Nathan, D. F., 776, **783**
National Institutes of Health, 68, **78**, 709, **717**
Ndaybagira, A., 744, **748**
Needham, J., 62, **78**
Neidert, A. H., 762, **783**
Nelson, C. E., 533, **545**
Neubüser, A., 440, **462**
Neugebauer, K. M., 454, **462**
Neumann, C. J., 588, **610**
Neumann, D., 25, **48**
Neuman-Silberberg, F. S., 293, **301**
New, D. A. T., 354, 359
Newbold, R. R., 702, **717**
Newgreen, D. F., 431, **462**
Newman, R. A., 732, **748**
Newman, S. A., 533, **545**, 598, **610**, 774, **783**
Newmark, P. A., 616, **645**
Newport, J., 636, **645**
Newport, J. W., 222, 223, **261**, 266, **302**, 310, 311, **341**
Newrock, K. M., 244, **261**
Newton, S. C., 629, **645**
Nguyen, M.-T., **521**
Nguyen, V. H., 352, **387**, 402, **424**
Niazi, I. A., 596, **610**
Nicholls, R. D., 123, **140**
Nichols, D. H., 396, **424**
Nicolis, G., 20
Nicoll, C. S., 580, **611**
Niehrs, C., 327, 334, 335, **340, 341**
Nielsen, M. G., 773, **783**
Niessing, D., 276, **302**
Nieto, M. A., 430, **462**
Nieuwkoop, P., 617, **645**
Nieuwkoop, P. D., 307, 313, 321, 325, **341**, 617, **646**
Nievelstein, R. A. J., 393, **424**
Nigg, E. A., 222, **261**
Nijhout, H. F., 62, **78**, 590, 592, **610**, 690, **717**, 728, 730, 743, **748**, 776, **783**
Nilsson, A., 541, **545**
Nilsson, O., 675, **681**
Nishida, H., 11, **23**
Nishida, H., 248, 249, 250, **259**, **261**
Nishigaki, R., 687, **717**
Nishimura, E. K., 418, **424**

Nishizuka, Y., 206, **217**
Niswander, L., 536, 538, 539, 542, **545, 546**
Noakes, P. G., 451, **462**
Noda, Y. D., 192, 198, 199, **219**
Noden, D., 503, **521**
Noden, D. M., 433, **462**
Nohno, T., 537, **545**
Nohno, T. W., 147, **178**
Noji, S., 527, 529, **545**
Noll, M., **299**
Nomura, M., **178**
Nonaka, S., 382, **387**
Nonaka, S., 602, **610**
Nordeen, E. J., 737, **748**
Nordeen, K. W., 737, **748**
Norris, D. O., 581, **610**
Norris, D. P., **140**
Northcutt, R. G., 434, 438, **462**
Northrop, J., 330, **341**
Nose, A., 74, **78**
Notenboom, R. G. E., 169, **178**
Nowack, E., 17, 18, **23**
Nowak, M. A., 776, **783**
Nowicki, J. L., 471, **488**
Nübler-Jung, K., 354, 355, **384**
Nulman, I., **717**
Nusse, R., 161, **176**, 763, **781**
Nüsslein-Volhard, C. 69, **78, 138**, 135, 269, 271, 274, 275, 276, 277, 279, 291, **299, 300, 301**, **302**, 345, **386**
Nyhart, L. K., 56, **78**
Nylin, S., 728, **747**

O'Farrell, P. H., 266
O'Kane, C. J., 119, **140**
O'Leary, D. D. M., 432, **462**
O'Rahilly, R., 703, **718**
O'Sullivan, C. M., 368, **387**
Oakley, R. A., **462, 463**
Oberlender, S. A., 76, **79**, 475, **488**
Odent, S. 689, **717**
Oelgeschläger, M., 329, **341**
Offield, M. F., 113, **140**, 514, **521**
Ogawa, K., 185, **217**
Ogino, H., 145, **178**
Ogryzko, V. V., 116, **140**
Oh, J., 504, **521**
Ohama, K., 212, **217**
Ohe, K. E., 553, **572**
Ohkawara, B., 153, **178**
Ohshima, Y., 158, **177**
Ohuchi, H., 525, 526, 527, 528, 530, **545**
Oka, Y., 472, **488**
Okada, M., 512, **521**, 616, **644**
Okazaki, K., 237, **259**
Old, R. W., 635, **645**
Oliphant, G., 193, **219**
Oliver, G., 525, **545**
Olivera-Martinez, I., 472, **488**
Oliveri, P., 230, 232, **261**

Olsen, B., 533
Olson, E., 476
Olson, F. N., 494, 496, 497, **521**
Olsson, L., 8
Onodera, M., 707, **717**
Onuma, Y., 753, **783**
Oofusa, K., 580, **610**
Opitz, J. M., 64, **79**, 161, **176**, 687, 696, 691, 695, 697, **718**
Oppenheimer, J. M., 3, 84, **106**, 350, 351, **387**
Ordahl, C. P., 471, 472, **488**
Orgel, L. E., 602, **610**
Orkin, S. H., 509, **521**
Orphanides, G., 136, **140**
Orr, N. H., 86
Orr, W. S., 603, **610**
Oster, G. F., 774, **783**
Othmer, H. G., 41, **48**
Otte, A. P., 333, **341**
Otting, G., 289, **302**
Otto, F., 437, **462, 488**
Ouellet, M., 744, **748**
Owen, R., **545**, 752
Oyama, M., 44, **49**
Ozawa, E., 525, **544**
Ozick, C., 81

Pääbo, S. O., 116
Pacifici, M., 540, **546**
Pack, M., 512, **522**
Paddock, S., 156, 269, 588, 729, 757, 769
Padgett, R. W., 153, **178**
Paglia, L. M., 638, **645**
Palanza, P., 701, **718**, 740, **748**
Palka, J., **645**
Palmeirim, I., 468, **488**
Palmer, A. R., 735, **748**
Palmund, I., 699, 701, **718**
Palumbi, S. R., 194, **217**
Pan, D., 33, **49**, 296, **302**
Pander, C., 7, 8, **23**
Panganiban, G., 756, 762, **783**
Pankratz, M. J., 282, **302**
Panning, B., 127, **140**
Panzer, S., 297, **302**
Papacostantinou, J., 415, **424**
Pappert, A., 686, **718**
Parakkal, P. F., 417, 418, **424**
Pardanaud, L., 498, 501, 503, **521**
Paris, J., 636, **645**
Park, W.-J., **488**
Parker, R., 132, **138**
Parkes, T. L., 603, **610**
Parr, B. A., 537, 538, **545**
Paroush, Z., 277, 278, **302**
Parr, B. A., 537, 538, **545**
Partington, M. W., 689, **718**
Partridge, L., 604, **610**
Parvis, F., 118, **140**
Pask, A., 553, **572**
Pasqualetti, M. 437, **462**, 773, **783**

Passera, L., 730, **748**
Pasteels, J., 393, 398, **424**, 622, **645**
Patapoutian, A., 113
Patel, N. H., 759, **781**
Patient, R., 491, **521**
Paton, D., 416, **424**
Patten, B. M., 9, **23**, 392, **424**, 492, **521**
Patterson, D., 580, **610**
Patterson, L. T., **488**
Patterson, P. H., 434, 451, **459, 462**
Paukstis, G. L., 731, **747**
Paulsen, S. M., 690, **717**, 776, **783**
Paves, H., 450, 451, **462**
Payne, J. E., 194, **219**
Peadon, A. M., 582, **609**
Pearson, H., 381, **387**
Pearson, J., 451, **462**
Pechenik, J. A., 725, **748**
Pecorino, L. T., 596, **610**
Pedersen, R. A., 367, 372, **385, 387**
Pedra, M. E., 383, **387**
Peifer, M., 161, **178**, 290, **302**
Pelliter, J., 626, **645**
Pellizzoni, L., 692, **718**
Peltari, J., 626, **645**
Pendrel, B. A., 730, **748**
Pener, M. P., 727, **748**
Peng, H., **712, 718**
Penkala, J. E., **646**
Pennisi, E., 116, **140**, 162, **178**
Penny, G. D., 127, **140**
Penttinen, R. P., 483, **488**
Pepling, M. E., 627, **645**
Pera, E. M., 336, **341**, 392, **424**
Peracchia, C., 172, **178**
Perantoni, A. O., 482, **488**
Perea-Gomez, A., 375, 376, **387**
Perez, S. E., 433, **462**
Peri, F., 293, **302**
Perona, R. M., 369, **387**
Perreault, S. D., 209, **217**
Perrimon, N., 276
Perry, A. C. F., 207, **217**
Persaud, T. V. N., 399, **424**, 695, **717**
Perucho, M., 99, **106**
Pesce, M., 620, 621, **645**
Peschon, J. J., 631, **645**
Peterson, R. E., 740, 741, **749**
Petters, R. M., 374, **387**
Pettersson, L., 735, **746**
Peyrieras, N., 366, **387**
Pfaff, S. L., 402, **423**
Phillips, C., 310, 333, **342**
Phillips, J., **545**
Phillips, K., 54, **79**
Phoenix, C. H., 559, **572**
Piatigorsky, J., 116, 117, **138**, 147, **176**, 415, **424**, 768, **783**
Piccolo, S., 329, 331, **341**, 763, **783**
Pichel, J. G., 481, 482, **488**

Pickford, D. B., 740, **748**
Pieau, C., 567, **572**
Pierce, G. B., 703, **718**
Pierce, S. B., 323, **342**
Pierschbacher, M. D., 169, **178**
Piersma, T., **745, 748**
Piette, J., 474, **488**
Pigliucci, M., 54, **79**
Pignoni, F., 277, **302**
Piko, L., 366, **387**
Pinder, A. W., 724, **748**
Pines, M., 253, **261**
Pini, A., 450, **462**
Pinkerton, J. H. M., 631, **645**
Pinkus, H., 418, **424**
Pinto-Correia, C., 6, **23, 217**
Piotrowska, K., 381, **387**
Pizette, S., 538, **545**
Placzek, M., 402, **424**
Plaza, S., 117, **140**, 288, **302**
Plowright, R. C., 730, **748**
Plusa, B., 381, **387**
Poccia, D., 209, **217**
Poccia, E. L., **216**
Poethig, R. S., 659, 663, 671, **679, 680**
Pollak, R. D., 539, **545**
Polleux, F., 446, **462**
Pomeranz, H. D., 429, **462**
Pommerville, J., 36, **49**
Pongs, O., 590, **608**
Pontiggia, A., 551, 552, **572**
Poo, M.-M., 451, **459, 718**
Popoff, D., 505, 506, **521**
Porcher, C., 509, **521**
Porter, D. C., 209, **217**, 285, **302**
Portmann, A., 408, **424**
Potten, C. S., 506, **521**
Potts, J. D., 496, **521**
Poulat, F., 552, **572**
Poulson, D. F., 167, **178**
Pourquié, O., 465, 469, 473, **488**
Powers, J. H., 38, **49**
Pownall, M. E., 474, **488**
Prahlad, K. V., 581, **610**
Prather, R. S., 89, **106**, 366, **387**
Pratt, R. M., 696, 697, **716**
Pratt, S. A., 629, **645**
Presaraari, J., 152
Preuss, D., 658, 664, **680, 681**
Prevost, J. L., 9, 184, **218**
Price, E. R., 116, **140**, 158, 159, **178, 302**
Price, J. V., 293
Price, M., 762, **783**
Price, S. R., 443, **462**
Priess, R. A., 254, 256, 257, **261**
Prigogine, I., 20, **23**
Primakoff, P., 194, 197, 198, **218**
Pritchard-Jones, K., 481, **488**
Profet, M., 638, **645**
Prum, R. O., 770, **783**

Prusinkiewicz, P., 21
Przibram, H., 18, **23**
Psychoyos, D., 357, **387**
Purcell, S. M., 308, **342**
Purdy, L., 686, **718**
Purugganan, M. D., 676, **680**
Purves, D., 408, **424**, 451, 452, 457, **462**, 739, **748**

Qiao, J., 485, **488**
Qiu, M. S., 762, **783**
Quaas, J., 617, **645**
Quail, P. H., **679**
Quatrano, R. S., 659, 660, **679**
Quiring, R., 413, **424**
Qureshi, S. T., 295, **302**

R&D Systems, 508, **521**
Raatikainen, M., 727, **748**
Raatikainen-Ahokas, A., 479, **488**
Rachinsky, A., 730, **748**
Racusen, R. H., 663, **681**
Raff, E. C., 773, **783**
Raff, M. C., 453, **462**
Raff, R. A., 233, 242, 244, **260**, 751, 766, 775, **783**
Rageh, M. A., 595, **610**
Raich, W. B., 257, **261**
Raine, C. S., 412
Raje, N., 17, **23**
Rajpurohit, R., 476, **488**
Rakic, P., 405, 406, 407, 408, 409, 410, **423, 424, 425**
Ralt, D., 194, **218**
Ramanathan, R., 697, 698, **718**
Ramarao, C. S., 190, **218**
Ramón y Cajal, S., 411, **424**, 427, 442, 447, **462**
Ramos, J. G., 701, 702, **719**
Rand, H. W., **610**
Ransick, A., 229, 230, **261**
Raper, K. B., 40, **49,**
Rappolee, D. A., 93, **106**
Rasika, S., 737, **748**
Raskolb, C., 290, **302**
Ratcliffe, O. J., **680**
Rathke, H., 8
Raunio, A. M., **47, 48**
Rawles, M. E., 13, **24**, 429, **462**
Ray, A., 662, **680**
Ray, K., 28, 29
Ray, R. P., 293, 297, **302**
Ray, S., 662, **680**
Rea, S., 122, **140**
Reach, M., 294, **302**
Ready, D. F., 156, **179**
Reaher, C., 368, 372
Reaume, A. G., 171, **178**
Rebagliati, M. R., 134, **140**
Recanzone, G., 320, **342**
Reddi, A. H., 542, **543**

Reddy, V. R., 559, **572**
Redkar, A., 492, 493, **521**
Reed. L., 744
Rees, B. B., 208, **218**
Reichert, C. B., 16, **24**, 754, 755, **782**
Reid, J. B., 673, **680**
Reijo, R., 631, **645**
Reilly, K. M., 149, **178**
Reinberg, D., 136, **140**
Reinitz, J., 276, 283, **302**
Reinke, R., 156, **178**
Reiter, J. F., 353, **387**
Reitsma, N., 726, **746**
Relaix, F., 753, **783**
Relyea, R. A., 735, 743, **745, 748, 749**
Ren, D., 194, **218**
Render, J., 240, 241, **261**
Renfree, M. B., 372, **387**, 727, **748**
Resnick, J. L., 621, **645**
Ressom, R. E., 617, **645**
Restifo, L. L., 591, **610**
Revel, J. P., 358, **388**
Reverberi, G., 10, 11, **24**, 58, **79, 261**
Reynolds, G. T., 202, **215**
Rhen, T., 567, **572**
Ribatti, D., 503, **521**
Rice, G., 560, **572**
Rice, T., 25
Richards, G., 591, **610**
Richardson, M. K., 10, **24**
Richter, J. D., 133, 134, **140**, 626, **646**
Rickmann, M., 429, **462**
Riddiford, L. M., 584, 589, 590, 591, **611**
Riddihough, G., 290, **302**
Riddle, R. D., 151, **178**, 535, 537, **545**, 764, **783**
Rideout, W. M. III, 123, **140**, 711, **718**
Ridge, K. A., 562, **570**
Riechmann, V., 616, **645**
Riedl, R., 760, 766, **783**
Rietze, R. L., 410, **424**
Rifkin, J., 708, **718**
Riggs, A. F., 577, **610**
Rijli, F. M., 377, **387**, 437, **462, 783**
Ris, H., 13, **24**
Risau, W., 501, 503, **519, 521**
Risch, N., 560, **572**
Rishi, A. K., 704, **716**
Ritvos, O., 153, **178**, 483, 484, **488**
Rivera-Pomar, R., 276, 280, 281, **302**
Rivier, C., **645**
Robb, D. L., 617, **645**
Robb, L., 509, 52
Robboy, S. J., 699, **718**
Roberts, D. J., 509, 512, **521**

Roberts, D. S., 164, **178**
Roberts, J. P. 707, **718**
Roberts, L. M., 559, **572**
Roberts, R., 512, **521**
Robertson, A., 40, **49**
Robertson, E. J., 103, 376, **384**
Robins, D. M., 99, **106**
Robinson, H., 579, **610**
Rocheleau, C. E., 256, **261**
Rock, F. L., 295, **302**
Rodan, G. A., 477, **486**
Rodaway, A., 491, **521**
Roder, L., **302**
Rodriguez, I., 164, **178**
Rodriguez-Esteban, C., 364, 365, **387**, 527, 530, **545**
Roe, S., 6, **24**
Roegiers, F., 247, **261**
Roelink, H., 402, **424**
Roessler, E., 163, **178**, 413, **424**
Rohwedel, J.,621, 64
Rojo, E., 667, **680**
Roller, R. J., 635, **645**
Romanoff, A. L., 623, **645**
Romer, A. S., 768, 769, **783, 784**
Ronshaugen, M., 757, 758, **784**
Roopra, A., 120, **140**
Rose, S., 408, **424**
Rosenberg, U. B., 102, 103, **106**
Rosenquist, G. C., 12, **24**, 356, **387**, 524
Rosenthal, E., 133, **141**
Rosenthal, L., 130, **141**
Rosenthal, N., 497
Roskelley, C. D., 171, **178**
Ross, A. S. A., 482, **486**
Ross, J., **644**
Rostand, J., 305, 603, **610**, 751
Roth, D. A., 707, **718**
Roth, G. S., 605, **610**
Roth, M. B., 635
Roth, S., 272, 291, **302**, 455, **462**
Rothenpieler, U. W., 482, **487**
Rothman, K. J., 692, 696, **718**
Rottbauer, W., 495, **521**
Roulier, S. E., 637
Rounds, D. E., 633, **643**
Rountree, D. B., 590, **610**
Rousseau, F., 160, **178**
Roux, W., 51, 61–63, **79**, 117, **141**, 210, **218**, 751, **784**
Rowe, D. A., 531, 532, **545**
Rowning, B., 212
Ruberte, E., **718**
Rubin, G. M., 100, **106**, 157, 167, 168, **176**
Rubin, L., 531, **545**
Ruby, E. G., 723, **748**
Rudolph, K. L., 603, **610**
Ruffins, S. W., 236, **261**
Rugh, R, 26, **49**, 492, **521**
Ruibel, R., 733

Runft, L. L., 205, 208, **218**
Ruohola, H., 637, **645**
Ruoslahti, E., 169, 176178
Rushdie, S., 751
Rushlow, C. A., 269, 291, 297, **302**
Russ, A. P., 367, **387**
Ruth, S. B., 525, **545**
Rutherford, S. L., 767, 777, **784**
Ryan, A., 337, **342**, 363, **387**
Ryner, L. C., 564, 567, **573**

Saarma, M., 450, 451, **462,** 555, **573**
Sabatini, S., 667, **680**
Sachs, L. M., 579, **610**, 703, **718**
Sackerson, C., 281, **302**
Sadler, T. W., 683, **718**
Sado, T., 127, **141**
Safranek, L., 590, **610**
Sagata, N., 634, **645**
Saha, M. S., **178**, 413, **424**
Sahlberg, C., 440, **463**
Sai, T., 692, **718**
Saiki, R. K., **106**
Sainio, K., 172, **179**, 479, 481, 483, 484, **488**
Saito, Y., 131, **141**, 529, **545**
Saitsu, H., 397, **424**
Saka, Y., 315, 316
Sakai, Y., 379, **387**
Salame-Mendez, A. J., 568, **573,** 618, **645**
Salazar-Ciudad, I., 18, **24,** 770, 771, 772, **784**
Saling, P., 195, 197, **217**
Saling, P. M., **218**
Sallés, F. J., 133, **141**
Salvini-Plawen, L. V., 779, **784**
Salz, H. K., 562, **573**
Sampath, K., 353, **387**
Sánchez, M. P., 481, **488**
Sánchez-Herrero, E., 286, 287, **299, 302**
Sandberg, A. A., 703, **718**
Sander, K., 84, **106,** 270, **301, 302,** 317, 320, **339, 342,** 724, **748,** 775, **784**
Sander, M., 117, **141**
Sanes, J. R., 452, **459**
Saneyoshi, T., 330, **342**
Sanford, J. P., 124, **141**
Santiago, A., 433, **462**
Santos, L. L., 235, 237, **261**
Sapp, J., 723, **748**
Sardet, C., 202, 204, 206, **218**
Sargent, T. D., 329, **340**
Sariola, H., 480, 481, 482, 484, 488, 555, **573**
Sasai, N., 428, **462**
Sasai, Y., 328, 330, 331, 332, **339, 342**
Sassoon, D., 700, **716**
Sater, A. K., **177**, 395, **423**

Sato, S. M., 329, **342**
Sato, Y., 467, 468, **489**
Satoh, N., 249, **261**
Satou, Y., 249, **261**
Saudou, F., 165, **179**
Sauer, F., **141**
Sauer, F. C., 403, **424**
Saunders, C. M., 208, **218**
Saunders, G. F., 688
Saunders, J. W., Jr., 148, 164, **179,** 529, 531, 534, 539, **545**
Sausedo, R. A., 395, **424**
Savage, D. C., 724, **748**
Savage, G., 756
Savage, R., 333, **342**
Savage, R. M., 616, 617, **646**
Savagne, R. P., 431, **462**
Sawada, A., 467, **489**
Sawada, K., 247, 248, 249, **261**
Saxén, L., 64, **79**, 146, **179**, 319, 326, 327, 334, 335, **339, 342,** 479, 481, 488, 531, **544,** 578, 580, **610**
Saxena, S., 596, **610**
Scadding, S. R., 596, **610**
Scambler, P. J., 687, **718**
Schaap, P., 44, **49**
Schackmann, R. W., 192, **218**
Schatten, G., 185, 198, 199, 204, 209, 210, 211, **218,** 247, **261**
Schatten, H., 198, **218**
Schauer, S. E., 661, 672, **681**
Schedl, A., 554
Schejter, E. D., 265, **302**
Scheres, B., 659, 663, 667, 668, **681**
Schetzer, J. W., 266
Schiavone, F. M., 663, **681**
Schiebinger, L., 548, **573**
Schier, A. F., 282, **302,** 352, 353, **387**
Schierenberg, E., 257, **261, 262**
Schierwater, B., 760, **784**
Schieve, L. A., 685, **718**
Schiliro, D. M., 600, **610**
Schilling, T., 436, **462**
Schlesinger, A. B., 359, **387**
Schlichting, C. D., 54, **79**
Schlissel, M. S., 108, **141**
Schmalhausen, I. I., 54, **79,** 727, 733, **748**
Schmekel, K., 625, **646**
Schmidt, J., 330, **342**
Schmidt, M., 453, **462**
Schmidt, T., 600, **610**
Schmitz, B., 349, **387**
Schmucker, D., 130, **141**
Schnabel, R., 256, **260**
Schneider, S., 323, **340,** 352, 353, **386, 387**
Schneider, V. A., 493, **521**
Schnell, L., 713, **718**
Schneuwly, S., 286, **302**
Schnieke, A. E., 89, **106**

Schnorrer, F., 274, **302**
Schoenherr, C. J., 120, **141**
Schoenwolf, G. C., 315, **340,** 356, 357, 361, 362, 371, 383, **388,** 391, 392, 393, 394, 395, 396, 400, **422, 424, 425,** 478
Schöler, H., 708
Schönfelder, G., 701, **718**
Schott, J.-J., 494, **521**
Schotté, O. E., 148, **179,** 575
Schreiber, A. R., 708, 718M., 580, **610**
Schroeder, E. H., 166, **179**
Schroeder, T. E., 188, **218,** 223, 237, **261, 262**
Schubiger, G., 264, 586, 590, **610,** 615, **646**
Schuchardt, A., 481, 482, **489**
Schüler, H., 620
Schulte-Merker, S., 352, **388**
Schultheiss, T. M., 493, 494, **518, 521**
Schultz, J., 263
Schulz, C., 281, **302**
Schüpbach, T., 277, 293, 294, 297, **302**
Schürfeld, M., 313, 314, 316, **343**
Schwab, M. E., 713, **718**
Schwalm, F., 266, **302**
Schwartz, B. W., 661, **681**
Schwartz, M., 210, **218**
Schwarz, K., 91, **106**
Schweiger, H. G., 32, **48**
Schweisguth, F., 166, **177**
Schweizer, G., 433, **462**
Schwemmler, W., 724, **748**
Schwind, J. L., 581, **610**
Scott, B. A., 604, **610**
Scott, M., 377, **387**
Scott, M. P., 163, **177,** 770, **783**
Sears, K. E., 775, **784**
Sechrist, J., 435, **462**
Segal, L. A., 40, **48**
Seidel, F., 373, **388**
Seidel, F., 373, **388**
Sekine, K., 525, **545**
Selker, E. U., 122, **141**
Selleck, M. A., 427, **462**
Sen, J., 294, **302**
Sengal, P. 148
Serafini, T., 447, **462**
Serikawa, K. A., 32, **49**
Sessions, A., 675, **681**
Sessions, S. K., 524, 525, **545,** 648
Setlow, R. B., 602, 603, **608**
Seydoux, G., 167, **179,** 253, 255, **262,** 615, 617, **646**
Shaffer, B. M., 40, **49**
Shah, N. M., 434, **462**
Shaham, S., 165, **179**
Shainberg, A., 474, **489**
Shalaby, F., 501, 503, **521**
Shaller, H. C., 600, **610**
Shamblott, M. J., 621, **646**

Shannon, J. M., 516, **521**
Shapira, E., K., 333, **342**
Shapiro, A. M., 728, 734, **748**
Shapiro, B. M., 190, 192, 201, 205, **215, 218**
Shapiro, I. M., 476, 488, **489**
Sharma, K., 443, **462**
Sharp, D. H., 283, **302**
Sharpe, C. R., 333, **342**
Sharpe, R. M., 542, **545**
Shatkin, A. J., 111, **141**
Shatz, C. J., 444, **459**
Shaw, G., 132, **141,** 727, **748**
Shaw, G. M., 702, **718**
Sheardown, S. A., 126, 127, **141**
Shearer, J., 207, **218**
Sheets, M. D., 634, **646**
Sheiness, D., 111, **141**
Shelton, C. A., 294, **302**
Shen, J. –C., 602, **610**
Shen, S. S., 206, 207, 208, **217**
Shen, W.-H., 554, **573**
Sherwood, D. R., 230, **262**
Shi, D.-L., 316, **342**
Shi, Q., 696, **718**
Shi, X., 197, **218**
Shi, Y. B., 577, 579, **611**
Shiang, R., 160, **179**
Shick, J. M., 54, **77**
Shigetani, Y., 773, **784**
Shih, C., 154, **179**
Shimizu, K., 65, **79**
Shimomura, H., 191, **218**
Shin, S. H., 162, **179**
Shin, T., 88, **106**
Shingo, T., 410, **425**
Shinya, M., 352, 353, **388**
Shinyoji, C., 185, **218**
Shiota, K., 397, 774, **783**
Shitara, H., 210, **218**
Shook, D. R., 308, **342,** 356, **388**
Shostak, S., 600, **611**
Shu, W., 409, **425**
Shubin, N. H., 534, **545,** 762, 764
Shulman, J. M., 161, **179**
Shweiki, D., 705, **718**
Sidman, R. L., 405, 412, **424, 423, 425**
Siebeck, O., 54, **79**
Siegal, M. L., 776, **784**
Siegert, F., 41, **49**
Siegfried, E., 285, **303**
Siegfried, K. R., 669, **681**
Sierra-Honigmann, M. R., 505, **521**
Sies, H., 42, **49**
Signer, E. N., 88, **106**
Silva, E., 741, 742, **748**
Silver, J., 445, **461**
Silver, L. M., 708, **718**
Simandl, B. K., 581, **608**
Simeone, A., 754, **784**

Simerly, C., 200, 209, 211, **218**
Simmonds, A. J., 284, **302**
Simmons, D. J., 477, **487**
Simon, H. G., 593, **611**
Simon, J., 288, **303**
Simpson, G. G., 654, **681**
Simpson, P., 167, **177**
Simpson-Brose, M., 276, **303**
Simske, J. S., 158, **179**
Sinclair, A. H., 548, **573**, 551, **573**
Singer, M., 594, **611**
Singer, S. R., 672, 673, 674, 675, **681**
Sinha, N., 669, 670, 671, **681**
Sive, H. L., 333, 334, **339**, 336, **341**, **342**
Skakkebaek, N. E., 740, **747**
Skeath, J. B., 297, **303**
Skiba, F., 257, **262**
Skromne, I., I., 361, **388**
Slack, F. J., 134, **139**, 278
Slack, J. M. W., 56, **79**, **303**, 320, 322, 329, **339**, **340**, **342**, 512, **519**, 755, 775, **784**
Slavkin, H. C., 149, **179**
Slijper, E. J., 767, 768, **784**
Sluder, G., 209, **218**
Small, S., 282, **303**
Smibert, C. A., 134, **141**, 274, 277, **301**, **303**
Smith, A., 435, 446
Smith, C. A., 472, 489, 548, **573**
Smith, D. M., 531, **545**
Smith, D. W., 697, **716**
Smith, E. P., 542, **545**
Smith, E. R., 124, **141**
Smith, H., 479, **489**
Smith, J. C., 307, 315, 316, 320, **342**
Smith, J. L. 357, **388**, 394, 395, 396, **425**
Smith, L. D., 86, **106**
Smith, M. M., 434, **462**
Smith, P., 253, **262**
Smith, R. H., 668, **681**
Smith, S. J., 412, **422**
Smith, T. T., 193, **218**
Smith, W. C., 328, 329, **342**
Smith-Gill, S. J., 577, **611**
Snell, W. J., 33, **49**
Sockanathan, S., 443, **462**
Socolovsky, M., 165, **179**
Sohal, R. S., 603, **610**
Sokol, R. J., 697, **714**
Sokol, S. Y., 324, **339**
Solnica-Krezel, L., 349, 353, **388**
Solter, D., 212, **217**
Solursh, M., 234, **260**, 358, 370, **388**, 531, **543**
Sonnenschein, C., 703, **718**
Sopta, M., 111, **141**
Sordino, P., **545**

Sosic, D. B., 475, **489**
Sosnowski, B. A., 131, **141**
Soto, A., 701, **718**
Soto, A. M., 703, **718**
Southee, D., 551, **572**
Southworth, D., 658, **681**
Sowell, E. R., 409, **425**
Spearow, J. L., 740, 741, **748**
Speksnijder, J. E., 247, **262**
Spemann, H., 67, **79**, 83–86, **106**, 148, 149, **179**, 305, 317, 318, 319, 320, 321, 322, **338**, **342**, 351, **388**
Sperry, R. W., 454, 455, **462**
Spindler, K.-D., 25, **48**
Spinella, F., 503, **521**
Spitz, I. M., 641, **646**
Spofford, W. R., 334, **342**
Spörle, R., 469, **489**
Spradling, A. C., 100, **106**, 638, **643**, **644**, 645 **646**
Spratt, N. T., Jr., 357, 359, 360, **388**
Spritz, R. A., 16, **24**, 159, **179**
Srinivas, S., 481, **489**
Srinivasan, S., 296. **303**
Srivastava, D., 494, 496, 497, **521**
St. Jacques, 477, **488**
St. Johnston, D., 76, **78**, 271, 293, **303**
Stancheva, I., 331, **342**
Standart, N., 133, **141**
Stanley, S. M., 778, **784**
Stanojević, D., 282, 283, **303**
Stappenbeck, T. S., 724, 725, **749**
Stark, K., 483, **489**
Stearns, S. C., 54, **79**, 727, **749**
Stears, R. L., 194, **218**
Stebbins, L., 674, **681**
Stebbins-Boaz, B., 133, **141**
Steele, R., 597, **611**
Steeves, T. A., 659, 664, 668, **681**
Steicher, J., 722, **748**
Stein, D., 294, **299**
Steinberg, M. S., 72, 74, 75, **79**
Steingrímsson, E., 115, **141**
Steinhardt, R. A., 202, 204, 205, 208, **218**, **219**
Stephano, J. L., 200, **216**
Stephanson, E. C., 273, **303**
Stephens, S., 209, **218**
Stephens, T. D., 531, **545**
Stern, C. D., 305, 356, 357, 360, 361, 364, 365, **387**, **388**, **463**
Stern, D., 766, **784**
Stern, H., 694, **718**
Stern, H. M., 152, **179**, 473, **489**
Sternberg, P. W., 157, 158, **177**, **179**
Sternlicht, M. D., 703, **718**
Stevens, L. M., 277, **303**
Stevens, N. M., 82, 83, **106**, 548, **573**

Steward, R., 291, 294, 296, **303**
Stewart, T. A., 621, 622, **646**, **718**
Stier, H., 454, **463**
Stock, G., 708, **718**
Stocum, D. L., 524, **546**, 593, 594, 596, 597, **607**, **611**
Stokoe, D., 154, **179**
Stolow, M. A., 577, **611**
Stone, R., 740, 742, **749**
Stopper, G. F., 744, **749**
Storey, B. T., 193, 195, **215**, **218**
Storm, E. E., 540, **546**
Stott, D., 618, **646**
Strähle, U., 346, 349, **385**, **388**
Sträßer, K., 132, **141**
Strassmann, B. I., 638, **646**
Stratford, T., 525, **546**
Strathmann, M. F., 724, **749**
Strathmann, R. R., 724, 734, **746**, **747**, **749**
Strauss, E., 337, **342**
Streicher, J., 526
Streissguth, A. P., 697, **718**
Streit, A., 356, 362, **388**, 392, **425**
Streuli, C. H., 169, **179**
Strickberger, M. W., 32, 33, **49**,
Strickland, S., 368, **388**
Strome, S., 253, 254, **262**, 615, **646**
Struhl, G., 166, **179**, 276, 277, 281, 286, **299**, **301**, **303**, 764, **781**
Studer, M., 696, **718**
Stumpf, H., 64, **79**
Sturtevant, A., 263
Sturtevant, M. H., 240, **262**
Su, W.-C. S., 161, **179**
Suarez, S. S., 193, 194, **218**
Subramanian, K., 615, **646**
Subtelny, S., **646**
Sucena, D., 766, **784**
Sugi, Y., 494, **521**
Sugimoto, M., 125, **141**
Sugiyama, K., **262**, 600, **611**
Sulik, K., 369, 371, 375, 383, **388**, 697, 698, **719**
Sullivan, W., 264, 265, **303**
Sulston, J., 165
Sulston, J. E., 251, 253, **262**
Summerbell, D., 531, 532, 534, **544**, **546**
Summers, R. G., 191, 192, 198, 201, **218**, 227, **262**
Summerton, J., 102, **106**
Sumper, M., 38, **49**
Sun, Y. -A., 638, **646**
Sun, H., 602, **611**
Sund, X., 532, **546**
Sundin, O., 99
Sunobe, T., 732, **747**
Supp, D. M., 382, **388**
Surani, A. M., 124, **139**
Surani, M. A. H., 212, **218**
Suri, C., 503, **522**

Sussex, I. M., 659, 660, 661, 664, 668, 663, 672, **680**, **681**
Sutasurya, L. A., 617, **645**, **646**
Sutovsky, P., 185, **218**
Suzuki, A., 329, 330, **342**
Suzuki, F. 197
Suzuki, K., 581, **611**
Swain, A., 556, **573**
Swalla, B. J., 251, **260**
Swan, S. H., 701, **719**, 740, **749**
Swann, K., 206, **218**
Swanson, C. P., 632, **646**
Sweet, H. C., 229, 230, 240, **262**
Swenson, K. L., 222, **262**
Swift, C. H., 622, **646**
Sze, L. C., 222, **262**
Szilard, L., **611**

Tabata, T., 587, **611**
Tabin, C., 151, 152, **179**, 364, 365, 544
Tabin, C. J., 532, 540, **543**, **544**
Tachibana, I., 198, **218**
Tachibana, M., 689, **719**
Tacke, L., 329, **340**
Tadokoro, Y., 630, **646**
Tagaki, N., 124, 126, **141**
Taigen, T. L., 582, **611**
Tajbakhsh, S., 469, 474, **489**
Takada, S., 473, **487**
Takagi, N., 125, **141**
Takahashi, I., 722, **749**
Takahashi, Y., 466, 467, 468, 471, 472, **489**
Takano, J., 600, **611**
Takasaki, H., 321, **341**
Takayama, K.,135, **141**,
Takayama, S., 512, **519**
Takeuchi, J. K., 527, **546**
Takeichi, M., 74, 75, **78**, **79**, 407, **424**, 431, **463**
Talbot, P., 193, **215**
Talbot, W. S., 352, 353, **387**, 590, **611**
Tam, P. P. L., 469, **488**
Tamaru, H., 122, **141**
Tamkun, J. W., 169, **179**
Tamura, K., 595
Tan, P. B., 158, **179**
Tan, S.-S., 436, **463**
Tanabe, Y., **463**
Tanaka, E. M., 593, **611**
Tanaka, H., 451, **463**
Tanaka, M., 530, **546**
Tanaka, S., 367, **388**, 728
Tao, H., 420, **425**
Tapscott, S. J., 474, **486**
Tassava, R. A., 594, **609**
Tatar, M., 604, **611**
Taurog, A., 581, **611**
Tautz, D., 277, 281, **303**
Tavera,-Mendoza, L., 744, **749**

Tavormina, P. L., 541, **546**
Tawfik, A. I., 728, **749**
Tax, F. E., 627, **646**
Tay, J., 626, **646**
Taylor, C. B., 657, **681**
Teillet, M.-A., 13, 429, **461, 463**
Teissier, G., 766, **782**
Telfer, W. H., 638, **646**
Templeton, A., 685, **714**
Tennyson, A. L., 575
Tepass, U., 76, **78**
Terada, N., 711, **719**
Terasaki, M., 198, 202, 204, 206, 218, 448, 449, **459**
Tessier-Lavigne, M. S., 151, **176,** 472, **487**
Testaz, S., 431, **463**
Tevini, M., **749**
teWelscher, P., 536, 538, **546**
Thanos, S., 454, **463**
Thayer, M. J., 474, **489**
Thesleff, I., 440, 441, **463**
Theurkauf, W. E., 265, 638, **643, 646**
Thiery, J. –P., 317
Thigpen, A. E., 558, **573**
Thisse, B., 296, **303**
Thoenen, H., 452, 453, **460, 463**
Thoma, F., 108, **141**
Thomas, H. E., **611**
Thomas, J. B., 450, **463**
Thomas, J. D., 107, **139**
Thomas, J. O., 122, **140**
Thomas, L., 164, **179,** 391
Thomas, P. Q., 374, **388**
Thompson, C. R. L., 44, **49**
Thompson, D. W., 18, 19, **24,** 649
Thompson, J. V., 14
Thompson, J. N., 254, 256, **261**
Thompson, W. J., 451, **463**
Thomsen, G. H., 330, **340,** 507, **519**
Thomson, J. A., 547, 548, **571,** 708, **719**
Thomson, J. N., 254, 256, **261**
Thorogood, P.,16, **24,** 427,463
Thorpe, C. J., 256, **262**
Thummel, C. S., 591, **611**
Thut, C., 145, **179**
Tian, M., 132, **141**
Tiberio, G., 691, **719**
Tickle, C., 525, 534, 537, **543, 546,** 759, 760, **781**
Tiellet, M.-A., 481, **487**
Tilghman, S. M., 123, **138**
Tilmann, C., 554, **573**
Tilney, L. G., 186, 192, **218**
Timmins, W. N., 82
Tinbergen, N., 457, **463**
Ting-Berreth, S. A., 147, **179**
Tobler, H., 614, **646**
Todt, W. L., 538, **546**
Toivonen, S., 64, **79,** 319, 327, 334, 335, 336, **342,** 479, **489**

Tollrian, R., 734, 735, **749**
Tombes, R. M., 190, **219**
Tomchick, K. J., 41, **49**
Tomes, C. N., 197, **219**
Tomlinson, A., 156, **179**
Toms, D. A., 17, **24**
Tomlinson, A., 156, **179**
Ton, C. T., 94, **106,** 687, 688, **719**
Tondravi, M. M., 477, **489**
Tonegawa, S., 90, **105,** 465, 466, **489**
Töpfer-Petersen, E., 193, **219**
Torok, M. A., 595, **611**
Torres, M., **489**
Torrey, T. W., 778, **784**
Tosney, K., 394, 403, 413, 443, **463,** 467
Townes, P. L., 70–72, **79**
Toyota, M., 197
Traeger-Synodinos, J., 693, **716**
Trainor, P., 428, 436, 437, **463**
Tran, D. N., 558, **573**
Tran, H., 604, **611**
Treisman, R., 132, **141**
Trelsted, R. L., 169, 559, **573**
Trentin, J. J., 509, **522**
Trewavas, A. J., 657, **681**
Trinkaus, J. P., **262,** 349, 350, 351, **388**
Trounson, A. O., 684, **719**
Trudel, M., 112, **141**
Trueb, L., 577, **611**
Truman, J. W., 584, 590, **611**
Tsonis, P. A., 596, **611**
Tsuda, T., 496, **522**
Tsunekawa, N., 622, 623
Tsushida, T., 443, **463**
Tuan, R., 517, **522**
Tuan, R. S., 76, **79,** 470, 472, 475, 477, 488, **489,** 540, **546**
Tuana, N., 548, **573**
Tuchmann-Duplessis, H., 366, **388**
Tucker, A. S., 440, **463**
Tucker, R. P., 431, **463**
Tufan, A. C., 540, **546**
Tung, T. C., 248, **262**
Turing, A. M., 19–21, **24**
Tur-Kaspa, I., 194, **215**
Turner, A. M.,737, **749**
Turner, B. M., 127, **139**
Turner, D. L., 410, 414, 415, **425**
Turner, F. R., 264, 267, **303**
Turner, P. R., 206, **219**
Turner, R. S., Jr., 44, **49**
Twitty, V. C., 326, **342,** 445, **463**
Tyler, M. J., 583, **611,** 756
Tyler, M. S., 266, **303**
Tyson, J. J., 40, **49**
Tzahor, E., 493, **522**
————————————
Ueda, N., 96
Ueno, N., 97, **340**

Ullrich, B., 130, **141**
Ulrich, A. S., 198, **219**
Umesaki, Y., 724, **749**
Urness, L. D., 591, **611**
Uzzell, T. M., 203, **219**
————————————
Vaahtokari, A., 152, 164, **179,** 441, **463**
Vachon, G., 288, **303**
Vacquier, V. D., 189, 191, 192, 194, 195, 198, 201, 209, **216,** 236, **260,** 440, **463**
Vainio, S., 152, 479, 484, 556, **573**
Vakaet, L., 356, **388**
Valcárcel, J., 131, **141,** 563, **573**
Valente, T., 728
Valentine, S. A., 297, **303**
Valleiz, S., 691, **719**
Valles, J. M., Jr., 306, 307, **342**
van Adelsberg, J., 483, **489**
Van Allen, M. I., 396, 397, **425**
van Buskirk, C.
Van Buskirk, J., 735, **749**
van den Biggelaar, J. A. M., 244, **262**
van der Hoeven, F., **546**
van der Ploeg, L. H. T, **141,** 122
van der Weele, C., 53, **79,** 721, **749**
Van Doren, H. M., 562, **573**
Van Doren, M., 624, **646**
van Driel, R., 44, **49**
van Dyke, M. A., 290, **303**
Van Eeden, F., 293, **302,** 638, **646**
van Gogh, V., 649
van Heyningen, V., 481, **489**
van Praag, H., 409, 410, **425,** 737, **749**
van Scott, E. J., 417, **425**
Van Steirtinghem, A., 208, **219**
van Straaten, H. W. M., 396, **425**
Van Valen, L. M., **745, 749, 784**
van Wijngaarden, R., 583, **611**
Vanfleteren, J. R., 603, **611**
Vani, K., 285, **301**
Vargha-Khadem, F., 409, **425**
Varley, J. E., 434, 463
Vassar, R., 417, **425**
Vaux, D. L., 165, **179**
Vega, S. H., 672, **681**
Vegeto, E., 641, **646**
Veit, B., 668, **681**
Venkatesh, T., 155
Venters, S. J., 472, **489**
Verdonk, N. H., 244, **262**
Verfaillie, C. M., 711
Vesalius, A., 547, 548
Via, S.728, **749**
Vicente, J. J., 43, **48**
Vidal, V. P. I., 553, 554, **573**
Vielle-Calzada, J., 662, **681**
Vikkula, M., 503, **522**
Vilela-Silva, A. C., 192, **219**

Villa, A., 91, **106**
Villeneuve, A., 625, 626, **646**
Vincent, J. P., 210, **219,** 311, 322, **342**
Vincenti, D. M., 744, **746**
Vintenberger, P., 210
Visconti, P. E., 193, **219**
Vissing, J., 210, **218**
Viviano, C. M., 596, **611**
Vogel, A., 530, 537, **546**
Vogel, K. S., 434, **463**
Vogel-Höpker, A., 145, 150, 151, **179**
Vogt, W., 11, 12, **24**
Vokes, S. A., 503, **522**
Volpe, E. P., 686, **719**
vom Saal, F. S., 701, **719**
von Baer, K. E., 8, 9, 14, 15, 22, **24**
von Dassow, G., 766, **784**
von Haller, A., 6, 9
von Kolliker, A., **219**
von Wettstein, D., 625, **646**
Vonica, A., 229, **262**
Vortkamp, A., 712, **719**
Vrba, E., 408, **425**
————————————
Wada, H., **784**
Waddington, C. H., 84, 85, **106,** 107, 144, **179,** 356, 360, 361, 362, **388,** 614, **646,** 721, 733, **749,** 775, 776, 779, **784**
Wagenaar, E. B., 133, **141**
Wagers, A. J., 710, **719**
Wagner, E. F., 477, **487**
Wagner, G. P., 766, **784**
Wagner, T., 553, **573**
Wagner, T. E., 100, **106**
Wakahara, M., 617, **646**
Wakayama, T., 88, **106**
Wake, D. B., 774, **784**
Wakimoto, B. T., 286, **303,** 495, **522**
Wald, G., 575, 577, **611**
Waldo, K. L., 429, 441, **460, 463**
Walker, M. D., 112
Wallace, K. N., 512, **522**
Wallace, V. A., 404, **425**
Wallberg, A. E., 166, **179**
Wallenfang, M. R., 253, **262**
Wallingford, J. B., 313, 315, **342**
Walsh, C., 407, **425**
Walsh, F. S., 176
Walter, J., 455, 456, **463**
Wan, J. S., **106**
Wang, C., **303**
Wang, D.-Z., 496, 497, **522**
Wang, H. U., 430, 431, 445, 446, 451, **463,** 504, **522**
Wang, J., 42, **49**
Wang, K., 713, **719**
Wang, L., 253, **262,** 594, **611**
Wang, M. R., 44, **49**
Wang, N., **49**

Wang, S., 166, **179**, 331, **343**
Wang, T., **463**
Wang, X., 692, 702, **719**
Wang, Y., 16, **24**, 194, **219**
Ward, C. J., 483, **489**
Ward, C. R., 190, **219**
Ward, G. E., 190, 191, **219**
Wardle, F. C., 764, **781**
Warga, R. M., 349, **388**
Warkentin, K. M., **745, 749**
Warner, A. E., 172, 173, **179**
Warner, R. R., 732, **749**
Warren, L., 509, **522**
Warren, R. W., 756, 757, **784**
Warrior, R., **646**
Wartiovaara, J., 327, **342**
Wary, K. K., 171, **179**
Washburn, L. L., 552, **571, 573**
Wassarman, P., 635
Wassarman, P. M., 196, 197, 201, 202, **219**, 368, **387**
Wasserman, C. R., 397, **425**
Wasserman, S. A., 294
Wassersug, R. J., 580, **611**
Watanabe, N., 209, **219**, 634, **646**
Watanabe, Y., 473, **489**
Watchmaker, G., 62, 234
Watson, C. A., 638, **646**
Watson, J. W., 708
Watt, F. M., 416, **423**
Watt, W. B., 728, **749**
Weatherbee, S. D., 288, **303**, 755, **784**
Weaver, V. M., 703, **719**
Webster, G., 18, **24**, 598, 599, **611**
Webster, M. K., **179**, 541, **546**
Weidner, N., 713, **719**
Weigel, D., 675, 676, **679, 681**
Weijer, C. J., 41, **49**
Weinberg, E. S., 12
Weinberg, R. A., 154, **179**
Weinstein, B. M., 505, 506, **522**
Weinstein, G. D., 417, **425**
Weintraub, H., 108, 122, **139, 141**, 474, **489**
Weismann, A., 15, **24**, 52, 59, 61, 62, 63, **79**, 602, **611**
Weiss, P., 67, 68, **79**
Weiss, R. L., 34, **48**
Weller, D., 102, **106**
Wellik, D. M., 481, **489**
Welsh, J. H., 242, **262**
Wennerholm, U. B., 686, **719**
Wensink, P., 132, **138**
Wensink, P. C., 92
Werner, M. H., 551, **573**
Wessel, G. M., 201, **216**, 202, 234, **262**
Wessells, N. K., 146, **179**, 515, **522**, 530, **546**
Wessley, O., 333, **343**

West, G. B., 19, **24**
West-Eberhard, M., 734, **749**
Westerfield, M., 445, **463**
Weston, J., 13, 14, **24**, 396, **422**
Weston, J. A., 429, 434, 439, **460, 463**
Weterings, K., 675, **681**
Wetherill, Y. B., 701, **719**
Wetmur, J. G., 92, **106**
Whalen, A. M., 294, **303**
Wharton, R. P., 276, **303**
Wheeler, D., 730, **746, 749**
Whitaker, M., 204, 206, 208, **218, 219**
White, B. H., 580, **611**
White, J. C., 223, **262**, 627, **644**
White, K. P., 94, **106**
White, R. J., 322, **343**
Whitehead, A. N., 427, 397, **425**
Whiteley, A. M., 236, **260**
Whitington, P. M., 617, **646**
Whitlock, C. A., 509, **522**
Whittaker, J. R., 58, 59, **79**, 247, 248, **262**
Whittle, R., 288, **301**
Wianny, F., 104, **106**
Wibbels, T., 568, **573**
Wickelgren, I., 408, **425, 719**
Wickens, M., 135, **141**
Wieschaus, E., 265, 271, 277, 279, 290, **302**
Wiesel, T. N., 457, **460,** 737, 738, **749**
Wightman, B., 134, **141**
Wikramanayake, A. H., 229, **262**
Wilby, O. K., 598, **611**
Wilcox, A. J., 193, **219**
Wiley, L. M., 367, **388**
Wilhelmi, L. K., **681**
Wilkie, A. O. M., 434, **463, 489**
Wilkie, G. S., 285, **303**
Wilkins, A. S., 59, **79**, 751, 776, **784**
Wilkinson, D. G., **106**, 377, **388**, 432, 441, **462, 463**
Wilkinson, H. A., 167, **179**
Willard, H. F., 127, **138**
Willemsen, V. H., 663, **681**
Willer, B. H., 14
Williams, C. M., 575, 590, **610**
Williams, E. J., 171, **179**
Williams, J. G., 39, **48**
Williams, J. H., 658, **681**
Williams, M. E., 16, **24**
Williams, S. C., 113, **141**
Willingham, E., 568, **573**
Wilmut, I., 86, 88, 89, **106,** 708, 719
Wilson, A. C., 408, **423**, 778, **782**
Wilson, D. B., 557, **572**
Wilson, E. B., 31, **48**, 57, 60, **79**, 81, 82, **106**, 223, 239, 243, **262**, 548, **573**

Wilson, H. V., 44, **49**
Wilson, N. F., 33, **49**
Wilson, P., 314, 315, **343**
Wilson, P. A., 329, **343**
Wilson, S. I., 391, 392, **425**
Wilson, T., 132, **141**
Wilson, W. L., 193, **219**
Wilstow, G., 768, **783**
Wilt, F., 507, **522**
Winberg, M. L., 451, **463**
Winfree, A. T., 20, **24**
Winklbauer, R., 313, 314, 315, 316
Winkler, M. M., 208, **219**
Winnepenninckx, B. M. H., 752, **784**
Winsor, M. P., 14, **24**
Winter, C. G., 161, **180**
Wintour, E. M., **489**
Wirtz, P., 730, **749**
Wisniewski, A. B., 557, **572**
Witte, O. N., 159, **180**
Wold, B. J., 113, 128, **141**
Wolf, N. S., 509, **522**
Wolf, S. F., 127, **141**
Wolf, U., 690, **719**
Wolfe, S. L., 108, **141**
Wolff, C., 276, **303**
Wolff, K. F., 6, **24**
Wolffe, A. P., 579, **611**
Wolpe, P. R., 561, 743, **749**
Wolpert, L., 64, 65, 67, **79**, 221, 227, 233, **260**, 360, **388**, 475, 489, 531, **546**, 600, **608**, 743, 767, **784**
Wolsky, A., 64, **78**
Woltereck, R., 54, **79**, 727, **749**
Wong, E. F. S., 40, **49**
Wong, J. M., 579, **611**
Woo, K., 12, **24**
Wood, H. B., 501, **522**
Wood, W. B., 253, **262**
Wood, W. J., 615, **646**
Woodruff, T. K., **646**
Woodward, D. E., 731, **749**
Woolf, A. S., **489**
Woychik, N. A., 112, **142**
Wray, G. A., 228, 229, **262**, 730, 731, **745, 766, 783**
Wreden, C., 277, **303**
Wright, C., 514, **519**
Wright, E., 475, **489**
Wu, H. H., 457, **463**
Wu, K. C., **749**
Wu, L. H., 276, **303**
Wu, L. N., 476, **489**
Wu, M., 159, **180**
Wu, P., 504, **522**, 710, **719**
Wu, X., 288, **303**
Wylie, C., 617, **645, 646**
Wylie, C. C., 331, **343**, 617, 618, 620, **643, 646**

Wyman, R. J., 638, **646**

Xavier-Neto, J., 497, **522**
Xia, H., 704, **719**
Xu, P. X., 481, **489**
Xu, W., 116, **142**
Xu, X. L., 530, **546**
Xu, Z., 207, **219**
Xue, F., 708, **719**

Yadegari, R., 661, **681**
Yaffe, D., 474, **489**
Yagami-Hiromasa, T., 474, **489**
Yamada, K. M., 412, **425**, 469, **487**
Yamada, T., 151, **180**, 402, **425**
Yamagata, K., 197, **219**
Yamagishi, H., 91, **105**
Yamakawa, K., 131, **142**, 692, **719**
Yamamori, T., 434, **463**
Yamamoto, T. S., 330, **343**
Yamanaka, Y., 353, **388**
Yamashita, S., 353, 354, **388**
Yamazaki, K., 368, **388**
Yanagimachi, R., 189, 192, 198, 199, **219**
Yancopopoulos, G. D., 504, **522**
Yang, E. V., 593, **611**
Yang, J.-J. W., 310, **343**, 413, **423**
Yang, Y. Z., 536, 538, **546**
Yao, T.-P., 590, **611**
Yaoita, Y., 579, **611**
Yasuda, K., 145, **178**
Yasumoto, K. -I., 115, **142**
Ye, X., 707, **719**
Yellayi, S., 702, **719**
Yeo, -Y. C., 151
Yeom, Y. I., 620, **646**
Yeung, E. C., 660, 661, **681**
Yin, M., 477, **489**, 540, **546**
Yin, X., 454, **463**
Ying, Q. L., 711, **719**
Ying, Y., 618, **646**
Yisraeli, J. K., **646**
Yoder, M. C., 509, **522**
Yokouchi, Y., 512, **522**, 533, 539, **546**, 708, **719**
Yokoyama, S., 565
Yokoyama, T., 381, **388**, 595, **611**
Yonei-Tamura, S., 530, **546**
York, R. H., Jr., 54, **78**
Yoshida, H., 165, **180**
Yoshida, M., 189, **219**
Yoshizato, K., 580, 595, **609, 610**
Yost, C. M., 323, **343**
Yost, H. J., 337, **343**
Young, R. A., **140**
Younger-Shepherd, S., 562, **573, 719**
Yu, J., 650, **681**
Yu, M., 770, **784**
Yuan, L., 626, **646**
Yuh, C.-H., 231, 232, **262**

Zabel, E. W., 740, **749**

Zakany, J., 471, **489**

Zaleske, D. J., 530, **546**

Zammit, P. S., 593, **612,**

Zanaria, E., 556, **573**

Zangerl, R., 16, **24**

Zatterstrom, U. K., 504, **522**

Zecca, M., 288, **612**

Zeevaart, J. A. D., 673, **681**

Zeng, Z. B., 778, **784**

Zernicka-Goetz, M., 104, **106,** 381, **387**

Zeschingk, M., 124, **142**

Zhang, J., 308, **343**

Zhang, N., 624, **646**

Zhang, Z., 495, **522**

Zhao, G.-Q., 628, **646**

Zhao, K., 123, **142**

Zhong, T. P., 346, **388,** 504, **522**

Zhou, B., 297, **303**

Zhou, Y., 617, **646**

Zimmerman, L. B., 328, 329, **343**

Ziomek, C., 367

Zipursky, A. L., 156, **178**

Zolessi, F. R., 396, **425**

Zon, L. I., 501, **520**

Zorn, L. I., 509, **521**

Zou, H., 539, **546**

Zuccato, C., 452, **463**

Zuckerkandl, E., 763, **784**

Zúñiga, A., 538, **546**

Zurn, A. D., 452, **463**

Zwacka, T. P., 711, **719**

Zwilling, E. 529, 546

Zygar, C. A., 145, **180**

Subject index

In-text definitions of terms are indexed in **boldface** type.

Italic type indicates the information will be found in an illustration.

The designation "*n*" indicates the information is found in a footnote.

A23187 (calcium ionophore), 202, 205
AB cells, *C. elegans*, 253–257
Abalone, 725
Abaxial side, leaf, 669
ABC model, plant organ identity, 676–677
abdominal A (abdA) gene, 287, 757, *757–759*
Abdominal B (AbdB) gene, 287, 377
Abembryonic region, 380–381
Abortion, preimplantation genetics, 693
Abscisic acid, 664
Accutane, 696
Acetabularia, 31–32
Acetylation, of nucleosomes, 116
Acetylcholine, 412, 432
Acetylcholinesterase, 58, *59*
N-Acetylglucosamine, 201–202
Acetyltransferases, 124, 159, 166
achaete (ac) gene, 297
Achaete-scute gene complex, 297
Achondroplasia, 161, 542, 691
Acoustic ganglion, 413
Acron, 277, 278
Acrosomal process, 184, 192, 195
Acrosomal vesicle, 184, *185,* 631
Acrosome, 184, *185,* 630–631
Acrosome reaction, 191–192, 197
Actin, 33, 224
    cardiac protein, 494
    egg cortex, 188
    fertilization cone, 198
    transport to oocyte, 638
Actin microfilaments, 264–265, 315
α-Actinin, 169
Activation, gamete metabolism, 206–208
Activation gradients, hydra, 597–600
Activin, 65, 153
    avian right-left axis formation, 363
    controlled experiments, 67
    diffusibility, 149*n*
    follistatin inhibition, 329
    gradients in *Xenopus,* 65–66
    pancreas formation, 514
    tooth development, 441

Activity-dependent synapse formation, visual system, 457
Acute promyelocytic leukemia (APL), 703–704
Adaptor protein, 154
Adaxial side, leaf, 669
Address selection, axonal, **444**
Adepithelial cells, insects, 586
Adhesive glycoproteins, 189
ADMP protein, *326*
Adrenergic neurons, 432–433
Adrenomedullary cells, 434, *434*
Aequorin, 202
AER. *See* Apical ectodermal ridge
After-ripening, 664
AGAMOUS (AG) genes, 676
Aggregation
    Dictyostelium, 40, *41*
    histotypic, 71, *72, 73*
Aging, 523, **601**
    causes, 602–603
    genetically programmed, 603–605
AGM. *See* Aorta-gonad-mesonophros
Albinism, 690
Albumin, 360
Alcohol (Ethanol), 695, 697–698
Alkaline phosphatase, teeth, 441
Allantois, *46,* **47,** 367, **517**
Alligators
    environmental sex determination, 52, 567
    population decline, 740, 742, *743*
Allometry, 17–18, 766–768, *767, 768*
Alternation of generations, 649, 651–654, *651–653*
Alternative RNA splicing, *129,* **129–130**
Alveoli, 515
Alzheimer's disease, 165*n,* 166, 708
Amacrine neurons, 414
Amborella trichopoda, 658, *658*
Ambystoma, 308, *See also* Amphibians; Salamander
    A. gracile, 55
    A. jeffersonianum, 203*n*
    A. macrodactylum, 55
    A. maculatum, prospective forelimb, *524*
    A. mexicanum, 6, 581–582, *582, 596, 774*
    A. platineum, 203*n*
    A. tigrinum, 582
Ameloblasts, 439

Ametabolous metamorphosis, 583
AMH. *See* Anti-Müllerian duct hormone
γ-Aminobutyric acid (GABA), axons, 412
Ammonium, differentiation in *Dictyostelium,* 44
Ammonotelic excretion, 577
Amniocentesis, 693
Amnion, 47, 367, *371, 373,* **517**
Amnionic fluid, 4, 369, **517**
Amnionic cavity, 369
Amnionic ectoderm, *370*
Amnioserosa, 297
Amniote egg, 46–47, **517**
Amniotes, membranes, *516,* **517**
Amphibian axis formation
    anterior-posterior axis, 325–326, 334–336
    dorsal-ventral axis, 311–312, 321, 323–324
    induction, 320–321, 321*n,* 334–337
    mechanisms of, 321–325
    organizer, 325–333
    progressive determination, 317–319
Amphibian metamorphosis, 30
    biochemical changes, 577–578
    hormonal control, 578–580
    morphological changes, 576–577
    regional specificity, 580–581, *581*
    types and behavior, 581–582
Amphibians
    cleavage, 221, 305–307, *306*
    cloning studies, 85–87
    deformities, 744–745
    egg cytoplasm, 210–213, 310–312, *632*
    egg polarity, 308
    gastrulation, *27, 29,* 307–317, *314*
    gene transcription, 634–636
    germ cells, 616–617
    herbicides and sex determination, 568
    interspecific induction studies, 148–149
    jaw homology, 13–14
    limb regeneration, 525, *526,* 595
    meiosis, 633–634
    mid-blastula transition, 223, 310–311
    mRNA selective translation, 133
    neurulation, 392–393, *395, 397, 398, 399*
    oogenesis, 632–634, *632*
    predator-induced polyphenisms, 735
    ultraviolet radiation, 54–56
Amphioxus, 12, 45, 761–762
Ampulla, sperm, 193, *194*

Anal region, avian, 358
Analogous structures, 13
Anaphase I/II, meiosis, 626
Anatomical approaches, developmental biology, 3
Anax (Dragonfly), 735
Anchor cell, 157–158
Androgen insensitivity syndrome, 692
Androgen sensitivity syndrome, 557
Anencephaly, 396, *397*
Aneuploidy, plants, 650
Angelfish, 19
Angelman syndrome, 123
Angioblasts, 501, 503, 505
Angiogenesis, 150, 503–505, *503*
   bone development, 476–477
   inhibition and tumors, 704–705, *705*
Angiopoietins, 503
Angiosperms. *See also* Plant development
   life cycle, 651, *653*
Animal cap, 315
Animal hemisphere, 26
   cell fate map, 308
Animal models, 15
Animal phyla, *45, 753, 773*
Animal pole, 224, 313, 316
Animal posterior cytoplasm, 247, 250
Animal-vegetal axis
   egg cytoplasm reorganization, 210, *211, 213*
   mammalian egg, *381*
Aniridia, 687–688
Aniridia gene, 687
Annelids, *45,* 46, 597
Antennapedia (antp) **gene, 285**–288, 290, *758*
Antennapedia protein, 286, 288, 289
Anterior necrotic zone, 539
Anterior neuropore, 396
Anterior organizing center, 273–276, 278
Anterior-posterior axis, 226
   amphibian, 325–326
   birds, 360–362
   C. elegans, 253
   Cartesion coordinate model, 297
   Drosophila, 135, 264, 268, *270–271, 273*
   fish, 353–354
   homeodomain, 289
   homeotic selector genes, 285–290
   homologies, *765*
   Hox genes, 377, 379–381, *381, 755*
   limb development, 524, 534–537
   mammalian, 375–380
   maternal effect genes, 269–270
   maturity gradient, 358
   mRNAs, 271–273
   neural tube differentiation, 398–400
   patterning, *274,* 376, 379–380, *381*
   primitive streak defines, 356

   retinoic acid disruption, 686
   sea urchin, 231
   segmentation genes, 278–285
   somitogenesis, 471
   specification, 334–336
   tunicates, 247, 250
Anterior visceral endoderm (AVE), 375–376
Anther, 654
Antheridia, 651
Anti-Müllerian duct hormone (AMH), 549, 551, 554, 557–559
Antibodies, 736
Antigen, 736
Antipodal cells, 655
Antisense RNA, 97, 101–104, 127, 134
Antrum, 639
Ants, 730–731, *731,* 734, 735
Anurans. *See also* Amphibians; Frogs; Toads
   metamorphosis, 30, **576**–583
Anus, 511
   deuterostomes vs protostomes, 46
   sea urchin, 236, 239
Aorta, 441
Aorta-gonad-mesonophros region, 507–509
Aortic arches, 500, *500*
AP-1 protein, 173
AP2 genes, 676
AP2 protein, vertebrate head, 762
AP3 genes, 676
Apaf-1 gene, 165–166
Apaf-1 protein, 164, *165*
APC protein, 161
APETALA 1 (APL) genes, 675
Aphids, seasonality, 726–727
Apical-basal axis, hydra, 597, *598*
Apical-basal polarity, plants, 659–661
Apical dominance, 667–668
Apical ectodermal cap, 593–594
Apical ectodermal ridge (AER), 529
   axis coordination, 538
   evolutionary comparisons, 759, 764–765
   progress zone, 529–532, *531, 536*
   proximal-distal axis, 529–532
Apical meristems, 666–667
Apical surfaces, 312
APL. *See* Acute promyelocytic leukemia
Apomixis, 659
Apoptosis, 164–166
   amphibian metamorphosis, 580–581
   erythropoietin, 164
   of hypertrophic chondrocytes, 476–477
   insect metamorphosis, 592
   integrins, 169–170
   interdigital tissue, 538–539, *540, 541, 770*
   mesonephros, 480
   neuronal, 451–453
apterous gene, 588, 764
APX-1 protein, 256–257

Arabidopsis
   embryogenesis, 660–663
   gametophyte development, 655
   pollen tube, 657–658, *657*
   reproductive transition, 675–676
   vegetative growth, 667, 669, 670
Araschnia levana, 53, 728, *730*
Arbacia punctulata, 190–191, *208, See also* Sea urchin fertilization
Archegonia, 651
Archenteron
   amphibian, 309–310, 313, 315, 334
   invagination, 236–239
   sea urchin, **236**
Archeocytes, 44
Area opaca, 354, *355*
Area pellucida, 354, *355*
Aristotle, 3–4, 547
Arkadia protein, 375
armadillo, *60*
armadillo (arm) gene, *282*
Aromatase, 559–560, **567**–568
ART. *See* Assisted reproductive technology
Arteries, 441
Arthropods, 16, *45*
Articular bone, 13
Ascaris, 184
Ascidians, *45,* 246
Asexual reproduction, 36–37, 39
Asobara tabida (Wasp), 724
Asplanchna (Rotifer), 734–735
Assisted reproductive technology (ART), 684, 686
Asters, sperm, 209, *210*
Astrotactin, 405
Asymmetrical cell division, 659, *660*
Asymmetry
   avian right-left axis, 354
   evolution in plants, 678
   human, *382*
   mammalian right-left, 381–383
Atavisms, 759–760
Atlantic silverside fish, 731–732
Atlas, 378
Atonal transcription family, 156
Atrazine, sex determination, 568, *569*
Atrial natriuretic factor, 494
Atrium, heart formation, 496, *497*
Auditory cortex, 407
Auditory placodes, 413
Autocrine growth factor, 416
Autocrine regulation, 149*n*
Autocrine signaling, differentiation, 173
Autonomous development, 319
Autonomous segmentation clock, 468, 471
Autonomous specification, 56–58
   C. elegans, 254–255
   sea urchin, 229–230
   tunicates, 247–249

Autopod, 523, 538–539
Autoradiography, 9–10, 93, *94*
Auxins, 663, 667
Avian embryo. *See* Birds
Axes, definition by primitive streak, 356
Axial patterning, plants, 659, *662*, 663, 667
Axial protocadherin, 313
Axillary meristem, 675
Axis formation, 226–227, 659, *See also*
    Anterior-posterior axis; Dorsal-
    ventral axis; Proximal-distal axis;
    Right-left axis
Axolotl, neoteny, 581–582
Axonal guidance, 431–432, *444*, 445,
    448–450, *451*, 455
Axonal specificity, 442–457
    address selection, 451
    cell adhesion, 445
    diffusible molecules, 446–450
    labeled pathways hypothesis, 446
    motor neurons, 442–443
    neural crest, 427–429, 432–433, 436–437
    repulsion/attraction, 445–450
    retinal ganglion axons, 453–457, 455n
    target selection, 443–**444**, 450–451,
        454–455
Axoneme, 184–186, *186*, 630
Axons. *See also* Motor neurons; Neurons
    activity-dependent synapse development,
        451
    fasciculation, 454
    formation and growth, **411**–412
    regeneration, 712–713

B lymphocytes (B cells), *57*, 90–91, **736**
B7 glycoprotein, 173
Bacteria, symbioses, 723–725
Baldness, 418
Barentsz protein, 638
Barnacles, *12*
Barr body, 124, *125*, 126–127
Basal cell, plants, **659**
Basal cell carcinomas, 162
Basal cell nevus syndrome, 162
Basal disc, hydra, **597**, 600
Basal lamina, 169, 233–236
Basal layer, 416, *417*
Basal surfaces, 312
Basal transcription factors, 111, **112**
Basic helix-loop-helix proteins (bHLH), 474
Bat wings, 13
Bateson, William, 286
Bcl-2 genes, **165**
BCL2 gene, 703
BclXL protein, *164*
BDNF. *See* Brain-derived neurotrophic factor
Beckwith-Wiedemann growth syndrome,
    123n
Bees, nutritional polyphenism, 730

Behavior
    and genes, 559–561
    sex-specific, 561, 567
Behavioral patterns, neuronal connections,
    457
Bell, Charles, 751
Bergmann glia, 405
Bf1 gene, *335*
BFU-E cell, 509
bHLH. *See* basic Helix-Loop-Helix
bicaudal-D gene, 638
Bicaudal-D protein, 638
bicoid (bcd) gene, 133, **272**, 274
***bicoid* mRNA**, 134, **272**, *274*, *275*, 638
Bicoid protein, 269
    anterior organizing center, 272–276, *275*
    cytoplasmic localization of, 134
    gap genes, 278, 279
    homeodomain, 289–290
    Hunchback protein gradient, 277
    syncytial specification, 69
    terminal gene region, 277
Bicyclus anynana, 728–729, 734, *769*
Bifurcation point, 767
Bilaminar germ disc, 369
Bilateral cleavage, 224–*225*
    holoblastic, 246
Bilateral symmetry, 46
Bilateria, 45, 46, 752
Bildungstrieb, 5–6
Bindin, 194–195, *196*, 198, 207–208
Bioethics, 694, 708
    assisted reproductive technology, 686
    cloning, 708
    gene therapy, 707–708
    human life, beginning, 709
    preimplantation genetic intervention,
        693–694
    stem cells, 709
Bipolar neurons, eye, 414, *415*
Bipotential stage, 549
Bird wings, 13
Birds
    axis formation, 356, 358–362, *364*, *365*
    cleavage, 224, 354–355
    dinosaur origin, 534
    egg, 187, 354, *355*
    epiblast, 358, 360
    extraembryonic membranes, 517
    gastrulation, 356–360
    germ cell migration, 622, *623*
    neuronal patterning, 737
    neurulation, *392*, *393*, *394*, 396, 398, 428
    sex determination, 548
Birth control, 641
Birth defects
    cardiac neural crest cells, 441–442
    causes of, 14–16

    frequency of, 14
    teratology, 3, 15–16
Birthday, neurons, 403
Bisphenol A (BPA), 701–702, *702*, 740
Bithorax gene complex, 286, 288, 290
Bivalents, 625
Black widow spider, 130n
Blastema, 593–597
Blastocoel, 27, *28*
    amphibian, 306–307
    avian, 356–357
    C. elegans, 257
    mammalian, 367, *370*
    mesodermal migration, 316–317
    sea urchin, **227–228**, 233–239
    tunicates, 247
Blastocyst, 367, *370*, 381
Blastoderm, 4
    avian, 355, *355*, 356, *357*
    cellular, 264–265
    fish, **348**, *348*–349
    syncytial, 264, *265*
Blastodisc, 347, 355, 360, *370*
Blastomeres, 25, 221
    amphibian, 306–307
    autonomous specification, 57–58, *59*
    biphasic cell cycle, 222
    C. elegans, 251–257
    cleavage furrow, 223–224
    conditional specification, 58–59, 61–63
    embryonic stem cells, 374
    fate map in snail, *241*
    mammalian, 366–367
    molluscs, 239–245
    sea urchin, *231*
    tunicates, 247–249
Blastopore, 27, **236**
    amphibian, 309–312
    C. elegans, 257
    deuterostomes, 239
    tunicate, 250
Blastula, 25, *28*
    amphibian, 306–307
    fish, *349*
    gastrulation, 226
    mollusc, 239–245
    sea urchin, 227–228, *231*, *233*
Blood cell formation
    from stem cells, 68, *508*
    osteogenesis, 477
Blood flow
    embryo to newborn, 498–499
    fluid dynamics, 500–501
Blood islands, 501, 503
Blood vessels
    arterial/venous differentiation, 504–505
    bone development, 476–477
    constraints on, 500–501
    extraembryonic membranes, 370–372

formation of, 501–505, *506*
germ cell migration, 622, *623*
joint formation, 540
liver induction, 513–514
Blue-headed wrasse, 732
Blumenbach, Johann Friedrich, 5
BMP. *See* Bone morphogenetic protein
BMP antagonists, 392
bmp gene, 353
bmp4 gene, 330
amphibian metamorphosis, 577
bmp7 gene, 100–101, *102, 103*
Bolitoglossa, 582, *767*
Bombyx mori, 727
Bon protein, fish, 353
Bone
cranial neural crest, 434
regeneration, 712
Bone collar, 477
Bone development. *See* Osteogenesis
Bone growth, 160–161
abnormalities, 541–542
Bone marrow, hematopoietic stem cells, 507, 509
Bone matrix, 477, *478*
Bone morphogenetic protein (BMP), 153
anterior-posterior limb polarity, 536, *537*
avian axis formation, 363, *364*
avian neural induction, 362
bone ossification, 437
cerebellar cell development, 404
hair follicle development, 420
heart development, 493–494, *493*
interdigital tissue, 539–540, *541*
joint formation, 539–540
neural crest cell migration, 431, 433
neural crest, cranial, 439–440
neural crest formation, 427
neural tube dorsal patterning, *401*, 402
node, *376*
organizer inhibitor proteins, 327–330
Smad signal transduction pathway, 159
tooth evolution, 770–771
Bone morphogenetic protein 1 (BMP1), 153
Bone morphogenetic protein 2 (BMP2)
apoptosis of interdigital tissues, 539
bone regeneration, 712
digit specification, 536, *537*
feathers, 770, *771*
joint formation, 539–540
organizer diffusible proteins, 328, 330
tooth development, 441
trunk neural crest cell differentiation, 434
Bone morphogenetic protein 2B (BMP2B), fish, 352
Bone morphogenetic protein 3 (BMP3), tooth development, 441
Bone morphogenetic protein 4 (BMP4), 145
apoptosis, 164, 539

bone regeneration, 712
decapanetaplegic protein, 153
diffusibility, 149*n*
epidermis induction, **329**–330
homologues, 763
interdigital tissues, 539
kidney development, 483–484
mesodermal identity, 465
neural crest cell migration, 431
neural tube dorsal patterning, *401*, 402
organizer diffusible proteins, 328, 329
retinal ganglion axons, 455
somitic cell fates, 473
tooth development, 439, 441
vasculogenesis, 501*n*
Bone morphogenetic protein 7 (BMP7)
apoptosis of interdigital tissues, 539
digit specification, 536
joint formation, 539–540
kidney development, 482
knockout mice, 100–101
neural crest cell migration, 431
neural tube dorsal patterning, *401*, 402
tooth development, 441
Bone morphogenetic protein 8b (BMP8b), spermatogenesis, 628–631
Bonellia, 52, 568
Bonnet, Charles, 4
Borozok protein, 353
boss gene. *See bride of sevenless* gene
Boss protein, 156
Bottle cells, 36, *37*, **309**–310, 312–313
Boundary models, 597
Boveri, Theodor, 81–82, *83*, 199, 203, 614–615
Bowman's capsule, *480*
Brachet's cleft, 316
Brachiopoda, *45*
Brachypodia, 540
Brachyury gene, 65–67, 84, 248, 370
Brachyury protein, 599
Brain-derived neurotrophic factor (BDNF)
axonal specificity, 450, 452, *453*, 457
neuronal patterning, 737, 739*n*
Brain development, 398–402
anterior-posterior axis, 398–400
apoptosis, *165*
cell numbers, 410
cerebral organizaton, 405–407
Chick, 400
human, 408–409
neuronal differentiation, 410–412
tissue architecture, 402–410
tunicates, 247
Brain maturation, human, 409
Branchial arches, 6
Branching patterns, shoot development, 668
Brassica, 656, *656*, 676
Brenner, Sydney, 59*n*, 165, 251

bride of sevenless (boss) gene, **156**, *157*
Bridges, Calvin, 263
Briggs, Robert, 85
Bristlecone pine, 678
Broad-Complex (BR-C) transcription factor, 591–592
Bromodeoxyuridine (BrdU), 409–410
Bronchi, 515
Brood pouch, 242
Brooks, William Keith, 725*n*
Brown algae, 659, *660*
Browne, Ethel, 597–599
Bryozoans, 46, 735
Buccinum undatum, 243
Budding, hydras, 597, *597*–600
Bufo boreas, 54–56
Bullfrogs, 26
Butterflies, 52–53
genetic assimilation, 734
Hox genes, 755–757, *756, 757*
polyphenism, 728–730, *729, 730*
wings, 13
buttonhead (btd) gene, 276, *282*
bZip transcription factor, 255

C-cadherin, 75
C. elegans. See *Caenorhabditis elegans*
c-fos gene, 173
c-Fos protein, 173
c-Jun protein, 173
c-mos mRNA, 634
c-mos protein, 634
c-myc gene, 170
C segment, immunoglobulin genes, 90–91
cactus gene, 291
Cactus protein, *292*, 294–297
Cadastral genes, 676
Cadherins, 74–76, *See also* E-cadherin; N-cadherin
EP, 75, 307
epithelial (E), 74–75, 397
FGF receptors, 171
homophilic binding, 75
muscle cell formation, 474
neural (N), **75**–76
other adhesion systems, 76
placental (P), **74**–76
protocadherins, 75
retinal ganglion axons, 454
Caenorhabditis elegans
aging, 602–604, *605*
apoptosis, 164–165
axis formation, 253–254
axonal guidance, 449
cell lineage chart, *252*
cell number, 251, 253, 257
developmental studies, 251, 257
gastrulation, 257
gene silencing, 104

germ cell determination, 614–615, 627–628

mRNA translation, 134, *135*

rotational holoblastic cleavage, 251, 253

structure, *252*

vulval induction, 157–158, 167

Caffeine, teratogenesis, 695

Calcineurin, 688

Calcium

channels, 197

cortical granule reaction, 202, *204*

differentiation in *Dictyostelium,* 44

egg activation, 204–208

endochondral ossification, 476–477

ionophores, 202, 205

pollen tube growth, 657, *657*

Calcium flux, 313–315

Calcium ions, 634

Callose, 657

Calmodulin, 634

Calmodulin-dependent protein kinase II, 634

Calpain II, 634

Calponin protein, 315

cAMP, 40, *41,* 44

Camptomelic dysplasia, 475, 553

CAMs, muscle cell formation, 474

Canalization, 776

Cancer

acute promyelocytic leukemia (APL), 703–704

APC protein, 161

colon, 161

gene enhancers, 120

and green tea, 503*n*

medulloblastoma, 703

melanomas, 703

osteogenic sarcoma, *704*

teratocarcinoma, 621, *622,* 703

Cancer therapies, 703–705

Candidate gene mapping, 688–689

5′ Cap, mRNA translation, 133

Cap cells, 484–485

Cap condensate, 484

Cap sequence, *110,* 111

Capacitation, 186, 192–193

Capillary networks, 501, 503–504

Capsule cells, 480

Carassius carassius (Carp), 735

Cardia bifida, *494,* **494,** *495*

Cardiac development. *See* Heart

Cardiac muscle-specific proteins, 494

Cardiac neural crest, 428, 441–442

Cardiogenic mesoderm, 492–493, *493*

caronte gene, 363, *364*

Caronte protein, 363, *364*

Carp, 735

Carpel, 654, *655,* **655**

carpel factory (caf-1) allele, 661

Carpellate flowers, 655

Carson, Rachel, 694, 740

Cartesian coordinate model, gene expression, 297

Cartilage

bone and joint formation, 475–477, 540

epiphyseal growth plate maturation, 541–542

facial, 434

gravity and pressure, 722

somite origins, 471–472

Cascades, of inductions, 67

Cascades frog, 56

Casein

integrins and prolactin, 170

mRNA and prolactin, 132

STAT pathway and prolactin, 160

caspace-9 gene, 166

Caspase-3 protease, 165

Caspase-9 protease, *164,* **165–166**

Catenin(s), 74, *75*

β-Catenin

avian axis formation, 361

dorsal signal, 312

fish, 352–353

gradient in amphibians, 334

Nieuwkoop center, **323–325,** *326*

sea urchin endoderm, 230, *231, 232*

Wnt pathway, 161, 229

Caterpillar larva, 575

Cats

cloning, 88, *89*

neuronal patterning, 737–738

X chromosome inactivation, *125*

caudal (cad) gene, 272

Caudal gradient, amphibians, 334

caudal mRNA, **272,** *274,* 276

Caudal protein, 269, *273, 276, 278,* 281

homeodomain, 289

posterior fates, *276*

Caudal region, neurulation, 396

CAULIFLOWER (CAL) gene, 675, 676

Causation, toxicology, 742, *743*

Caveolin, *171*

Cavitation, 367

Cbfa1 protein, 477

CBP protein, *163*

CD9 protein, 198

CD28 protein, *173*

CD95 protein, *164,* 166

cdc2 protein, 306

CdxA gene, 512

Cecropia, 575, 591

Cecum, 512

ced-3 **gene, 165**

CED-3 protein, 165

ced-4 **gene, 165**

CED-4 protein, 165

ced-9 **gene, 165**

CED-9 protein, 165

Cell adhesion

activation of FGF receptors, *171*

alcohol damage, 697

cadherins, 74–76

differential cell affinity, 70–74

neural crest, 431

thermodynamic model, 71–74

Cell adhesion molecules

amphibians, 307, 313

axonal migration, 445

Dictyostelium, 40–43

neural tube closure, 397

retinal ganglion axon growth, 454–457

Cell affinity, differential, 70–74

Cell-cell communication

cell death pathways, 164–165

cell surface receptors, 154–163, 169–171

cross-talk, 172–173

extracellular matrix, 168–171

induction and competence, 143–149

juxtacrine signaling, 166–172

paracrine factors, 149–153

signal transduction pathways, 154–163, 171–172

Cell-cycle

blastomeres, 222

cloning, 87

neuroepithelium cell, *403*

somatic, *222*

Cell death. *See also* Apoptosis

Drosophila metamorphosis, 592

Volvox, 37–38

Cell determination. *See also* Determination

C. elegans, 251–257

and chance, 167

molluscs, 242–245

plants, 663

sea urchins, 228–233

Cell differentiation

DNA methylation, 122

heart, 494–495

maintenance of, 173–174

mollusc blastomeres, 242–245

Cell division

cytoskeletal mechanisms, 223–224

DNA methylation, 123

germinal neuroepithelium, 403, *403*

karyokinesis, 223–224

mammalian embryo, 364

mid-blastula transition, 265–266

Cell fates, 226, 229–230, *231, See also* Fate maps

cerebral development, 407

hydras, 597

somites, 471–473

trunk neural crest, 433–434

Cell lineages, 8, *252,* 465, *466*

Cell membrane. *See* Plasma membrane

Cell migration. *See also* Gastrulation
central nervous system, 403–405
cytoplasmic control, 226
fibronectin, 168–169, 316–317
germ cells, 550–551, 617–624
heart formation, 493–495
human dorsal thalamus, 408
neural crest cells, 11, *12*, 396, 429–432, *429n, 430, 431*, 435–439, *436*
organogenesis, 26
Cell numbers
brain, 410
C. elegans, 251, 253, 257
during cleavage, 222–223
Cell-specific transcription factors, 112
Cell specification, 56–59, *See also* Axonal specificity
by *Drosophila* genes, *286*
heart tissue, 492–495
sea urchin, 229–230, *231, 232*, 233
Cell-substrate adhesion, 168–171
Cell surface receptors, 154–163, 169–171
Hedgehog pathway, 161–163
JAK-STAT pathway, 159–161
Notch protein, 166
paracrine factors, 154
RTK pathway, 154–159
Smad pathway, 159
Wnt pathway, 161
Cell-to-cell induction, RTK pathway, 155–158
Cell types
germ layers, *3*
neural crest, 427–428, 434, *435*
Cell wall information, plants, 659
Cell wedging, 396
Cellular blastoderm, 264–265, *265*
Cellular ecosystem, 67–68
Cellular slime mold. See *Dictyostelium*
Central cell, 655, 658
Central nervous system
brain development, 398–402
cerebral organization, 405–407
deep homology, 763–764, *764*
myelination, *412*
neural regeneration, 712–713
neural tube, 393–402
neuronal differentiation, 410–412
optic development, 413–414
Pax6 transcription factor, 116–117
tissue architecture, 402–410
Centriole, sperm, 209, 212, 247, 253, 311
Centrolecithal eggs, 224–*225*
CENTRORADIALIS (CEN) gene, 675
Centrosome, 209
Cephalic furrow, 266, *267*
Cephalic neural crest, 428
Cephalic region, neurulation, **396**
Cephalochordata, *45*
Ceratohyal cartilage, 577, 578

cerberus mRNA, 331
Cerberus protein, 331–332
cardiac specification, *493*
chick, 360
mammalian axis formation, 375
organizer, *326*
Cerebellum, 399, 404–405
Cerebral aqueduct, 399
Cerebral cortex, 405–407
Cerebral hemispheres, 399
Cerebrum, 395–407
CFU-E cell, 509
CFU-S cells, 509
Chabry, Laurent, 57–58
Chaeoborus, 735
Chains of causation, 742, *743*
Chance, cell determination and, 167
Chang and Eng (conjoined twins), 743
Chemoaffinity hypothesis, 455
Chemotaxis
axonal migration, 446–451
Dictyostelium, 40, *41*
egg-sperm, **189–191**, 194
growth cones, 451
neural crest cell migration, 431–432
Chiasmata, 626
Chick. *See also* Birds
amniote egg, *46*
anatomical depictions of development, *5*
asymmetry pathway, 382
axis formation, 356, *364, 365*
axonal guidance, *444, 448*, 455
blood cell formation, *507*
bone development and pressure, *723*
brain development, 400
cadherins and limb formation, 76
circulatory system, *498, 506*
cleavage, 354–355
development, 4, 354, 360
endochondral ossification, *477*
endoderm specification, 512
epithelial-mesenchymal interactions, 147–148
extracellular matrices, *169*
fate mapping, 11, *12*, 358
Fibroblast growth factors (FGFs), *151*
gastrulation, 356–360, *357, 358, 359*
germ cell migration, 622, *623*
β-Globin genes, *122*, 123
heart formation, 492–495, *497*
hindbrain, *400*
histotypic aggregation, *73*
Hox genes expression, 380
lens development, 415
limb apoptosis, *540*
limb development, 527–528, 531, *532, 539, 540*
mesodermal development, *492*
neural crest, 11, *12*, 428, *430*, 441

neuroepithelium cell cycle, *403*
neurulation, *392, 393, 394*, 396, 398
notochord, 6, *7*
Pax6 gene, *99*
somitogenesis, 467–473, *467–472*
sonic hedgehog gene, *152*
wing skeleton, *524*
Chimeras, 100–101, 374
blood cell formation, *507*
fate mapping, **10–11**
gene targeting, 100–101, *102, 103*
germ-line, *294*
human, 374
interspecific induction, 149
plant, 666, *666*
somite differentiation, 471
tumor-derived, 621, *622*
Chimeric mice, 100–101, *102*
Chimeric proteins, 289
Chlamydomonas, 33, *34*, 38
Chloroplasts, germination, 665
Cholesterol, 162–163, 690, 695
Cholinergic neurons, 432–434
Chondroclasts, 477
Chondrocytes, *57*, **475–477**
apoptosis inhibition, 169–170
bone growth and modeling, 475, 540–541
JAK-STAT pathway and FGFR3, **160–161**
joint formation, 540
somite origins, 471
Chondrodysplasia punctata, 694*n*
Chondroitin sulfate proteoglycan (CSPG), 236, *237*, 454, 713
Chordamesoderm, 247, 310, 465
avian, **357**
fish, 351–352
Chordates, 12, 46, 761–762
chordin gene, *97*, 328, 362, 376
chordin mRNA, 328, *329*
Chordin protein, 326, 328–329, 333, 392
cardiac specification, *493*
Hensen's node, 362
homologues, 330–331, 763
mammalian axis formation, 375
chordino gene, 352–353
Chordino protein, 352
Chordoneural hinge, 398
Chorioallantoic membrane, 46, 517
Chorion, 517
amniote egg, 46, **47**
mammals, 366–367, 369, 371–372, *373*
Chorionic villi, *370, 371*, 372
Chorionic villus sampling, 693
Chromaffin cells, *434*
Chromatids, 275–276
Chromatin, 107–108
DNA methylation, 122–123
during fertilization, 209
mid-blastula transition, 266, 311

Chromosomal sex determination
  Drosophila, 131–132, 561–567
  mammalian, 548–561
Chromosomal theory of inheritance, 82–83
Chromosome elimination, invertebrates, 126
Chromosome puffs, 590–591, *591*
Chromosomes
  diminution, 614–615
  dosage compensation, 124–126
  lampbrush, 627, *635*, **635**–636
  meiosis, 624–628
  polytene, 92, 637
Ci protein. *See* Cubitus interruptus protein
Cilia, blastula, 228
Ciliary cells, 382–383
Ciliary ganglion, *453*
Circulatory system, 491–492
  blood cell formation, 505–510
  blood vessel formation, 500–505
  extraembryonic membranes, 372, *372*
  fetal to newborn blood flow, 498–499
  heart development, 492–500
Cis-regulatory elements, 112*n*, 762
CiVegTR transcription factor, 249
Cladistics, 752
Clams, cleavage, 242
Class D genes, 677
Class I *KNOX* genes, 670
CLAVATA (CLV) gene, 667
Cleavage, 25, 221–226
  amphibian, 305–307
  Aristotle's observations, 3–4
  avian, 354–355
  C. elegans, 251–257
  cell division rates, 221–222
  cell fates, 226
  Drosophila, 264–266
  evolutionary adaptation in snails, 242
  fish, 347–349
  frogs, 27, *28*, 306
  human, 368
  mammalian, 363–368
  mid-blastula transition, **223**
  mitotic mechanisms, 223–224
  mollusc, 239–245
  sea urchin, 227–233
  transition to, 222–223
  tunicates, 246–250
  types of, 224, *225*
Cleavage embryogenesis, plants, 658
Cleavage furrow, 223–224, 305–306
Cleavage and polyadenylation specificity factor (CPSF), 133
Clinical genetics, 691*n*
Cloaca, 512
Cloacal membrane, 511
Clomiphene, 684
Clonal analysis experiments, plants, 663–664
Clonal selection, 736

Cloning, 743
  amphibian, 85–87
  bioethics, 708
  DNA methylation, 122–123
  genes, 93*n*
  mammals, 87–90
  nuclear potency, 85–87
  therapeutic, 708–711
Clunio marinus (Fly), 25
Clypeaster japonicus, gamete fusion, *210*
Cnemidophorus uniparens, 632
Cnidaria, 45, 46, 752
CNS. *See* Central nervous system
Co-development, 743
Co-option, evolutionary, **768**
Coat color, X chromosome inactivation, *125*, 126
Coelom, 46, 228–229, **491**–492
Cohesin complex, 626
Coiling, snails, 240, *241*
Colchicine, *224*, 396
Coleoptile, 665
Collagen, 168, 368
  trunk neural crest, 431
Collagen I, tooth development, 441
Collagen IV, basal lamina, 169
Collagen X, endochondral ossification, 476
Collagen XI, extracellular matrix, 542
Collagen XVIII
  angiogenesis, **504**
  kidney development, **484**
Collagenase, 368, 641
Collapsin, axonal migration, 446
Colon cancer, 161
Colonial invertebrates, 735
columbus gene, 624
Column of Terni (CT), 443
Comb jellies, 45, 46
Comm protein, *450*
Commissural neurons, 447–449, *450*
Commitment, 56–58, *See also* Specification
  amphibian gastrulation, 319, *319*
  neural crest, 433
  neuroblasts, **391**
  neurons, 407
  progressive, 43
  stem cells, 68–69, 509
Committed stem cells, 68–69
Compaction, 366–368
Comparative embryology, 3–5
  cell migration, 11–12
  fate mapping, 8–11
  gene expression patterns, 380, *381*
  primary germ lines and early organs, 6
  von Baer's principles, 7
Compensatory regeneration, 592, 601
Competence, 143–144
  neuroblasts, **391**–392
Competence factor, 144

Complementary DNA (cDNA), 93, *94*, 96–97
Composite embryo, 374, *374*
Compound eye, *155*
Compound leaves, 669–670
Computer modeling, pattern formation, 18, *19, 20*
Concentration gradients, 63
  activin gradients in *Xenopus*, 65–66
  French flag analogy, 64–65
  LIN-3 protein, 158
  regeneration, 64
Conditional development, 319
Conditional specification, 58–68
  C. elegans, 255–257
  morphogen gradients, 63–66
  morphogenetic fields, 67–68
  sea urchin, 229–230
  tunicates, 247–249
"Conditions of Existence," 751–752
Cone cells, 414
Congenital abnormalities, 686–687
Congenital adrenal hyperplasia, 557*n*, 692
Conjoined twins, 374, 743
Conjugation, 32–33
Conklin, Edwin G., 8
Connexins, 171–172
Conservation biology, 568, 745
Constant regions, 90–91
Contact guidance, germ cell migration, 617
Contractile ring, 223–224
Constraints, on evolution, 773–776
Convergent extension, 237, 313–315, 395
  avian, 356
  fish, 353
  zebrafish, *351*
Cornea, 145–147, 414–415
Cornified layer, 416, *417*
Corona radiata, 188
Corpora allata, 590
Corpus luteum, 641
Correlative evidence, 42
Cortex, egg, 188
Cortex, root, 667, *668*
Cortex protein, 133
Cortical granule reaction, 201–202, *204*
Cortical granule serine protease, 201
Cortical granules, *188, 189*
Cortical plate, 407
Cortical rotation
  amphibian gastrulation, 311–312, 316
  axis formation, 324
  Nieuwkoop center, 321–322
Cortical sex cords, 551
Cortices, 402–403
Cos2 protein, *163*
Cotyledons, 663
Cows, cloning, 88
CPEB Protein, 133, *626*

CPSF. *See* Cleavage and polyadenylation specificity factor
Cranial ectodermal placodes, 413
Cranial nerves, 400
Cranial neural crest, 428, 434–442
    cell migration, 435–437, *436*
    cell types, 434, *435*
    retinoic acid disruption, 696–697
    second wave migration, 438–439
    skull, 438
    tooth development, 439–441
Cranial paraxial mesoderm, 473
Cranial placodes, 438–439
Craniofacial development
    cranial neural crest, 434–442
    embryonic homology, 13–14
Craniorachischisis, 396
Craniosynostosis, 541
Cre-lox technique, **118,** *119*
Creatine phosphokinase gene, 474
Creationism, 779
Crepidula, 246, 568–569
Crescent protein, cardiac specification, *493*
Cristozoa, 761
Cross-talk, cell communication, **172–173**
Crustaceans, Hox gene expression, 757–759, *758, 759*
Crypto protein, axis formation, 376
Crystallin gene, 113, 116–117, *117*
Crystallins, 414–415, *416*
CSF. *See* Cytostatic factor
CSL protein family, 166
Ctenophores, 45, 46, 752
Cubitus interruptus (Ci) protein, *156,* 161–163
cubitus interruptusD (CiD) gene, *282*
Cumulus, 187, *189,* **639**
Cutaneous appendages, 418–420
Cutaneous structures, induction, 148
Cuvier, Georges, 751
Cyclic adenosine 3′5′-monophosphate. *See* cAMP
Cyclic AMP. *See* cAMP
Cyclin, 209
Cyclin B, 222–223, 306, **634**
Cyclin-dependent kinase, 222–223, 306, **634**
cyclinD1 gene, 170
CYCLOIDEA (CYC) gene, 678
Cyclopamine, 163*n,* **695**
Cyclopia, 162–163, **413–414,** 689, **690**
Cyclops protein, fish, 353
Cyclostomes, 771
CyIIIa genes, 128, *129*
CYP21 gene, 692
Cystein-rich peptide (SCR), 656
Cytochalasin B, 224, 396, 412
Cytochalasin D, 253
Cytokines, hematopoiesis, **509**
Cytokinesis, 223–224

Cytokinins, plants, 667–668
Cytomegalovirus, 702
Cytoplasm
    developmental morphogenesis, 31–32
    egg, 133, 186–187, 631–633, *632*
    egg reorganization, 210–213
    morphogenetic determinants, 242–245
    mRNA expression, 134–135
    sperm, 184
    syncytial specification, 69
Cytoplasmic bridges, 626–627, 630
Cytosine, methylation, 121
Cytoskeleton. *See also* Microfilaments
    integrins, 169
    mitosis, 223–224
    transport to oocyte, 638
Cytostatic factor (CSF), 634
Cytotrophoblast, 149*n,* **370**

da Vinci, Leonardo, 372*n*
dachshund gene, 586, *587*
DAF proteins, 604, *605*
daf2 mutations, 604, *605*
DAG. *See* Diacylglycerol
D'Alton, Eduard, 5, 6
Danaus plexippus, 756
Danforth short-tailed mouse, 481
Danio rerio, 345
Darwin, Charles, 12, 13, 751–752, 778
Dasypus novemcinctus, 60
Dauer larva stage, 604
DAX1 gene, *549,* **554–556**
DCL1 (DICER-LIKE 1), 661, 662
DDE, 740
DDT (Dichloro-diphenyl-trichloroethane), 694, 740
Deadpan protein, 562
Death, 36–37
decapentaplegic (dpp) gene, 290, *296, 297, 330, 587,* 588
Decapentaplegic (Dpp) protein, 153, 155, 297
    homologues, 763
    insect metamorphosis, 586, *587,* 588
Decidua, 369
Decondensation, sperm nucleus, 209, 210
Dedifferentiation, 592–593
Deep cells, 348, 354, 358
Deep homology, 439
Defect experiments, 61
Definitive phase, hematopoiesis, **506–509**
deformed (dfd) gene, 286, 377
Deiodinase II/III, 579–580
dek23 maize mutant, 663
Delamination, 226
Delta-like 3 gene, 467, *468*
Delta protein, 166, *168,* 257
Demasculinization of frogs, 568, *568*

Demecolcine, 253
Dendrites, 410–411
Dendritic arbor, 404
Dendrobates (Poison arrow frog), 582, *583*
"Denominator proteins", 562, *564*
Dental papilla, *439,* **440**
Dentalium, 243
Depolarization, neuronal survival, 453
Dermal mesenchyme, **147**
Dermal ossification, 475
Dermal papilla, 418
Dermal tissue, plants, **663**
Dermamyotome, 471, *472,* 473–474
Dermatome, 472
Dermis, 147, **471–472**
DES. *See* Diethylstilbesterol
Descartes, René, 4, *17*
The Descent of Man (Darwin), 12
Descent with modification, 752
desert hedgehog (dhh) gene, 151
DET gene, 675
Determination, 56, 167, *See also* Cell Determination; Sex Determination
    amphibian axes, 317–319
    dorsal lip tissue, 320–321
    fish blastoderm, 348–349
    germ cells, 613–617
    imaginal discs, 587–588, *587*
    neuroblasts, **391**
    segmentation in *Drosophila,* 278–285
    sperm centriole, 311
Deuterostomes, 45, 46–47, 239, 752
Development, 2
    behaviors and neuronal connections, 457
    C. elegans studies, 251, 257
    mathematical modeling, 16–20
Developmental biology, 1–2
    anatomical approaches, 3
    environmental studies, 51–56
    new evolutionary synthesis, 777–779
    real world, 743–745
Developmental buffering, 776–777
Developmental cancer therapies, 703–705
Developmental constraints, 773–776
"Developmental force," 5
Developmental genetics, 107
    amphibian studies, 305
    determining gene function, 98–104
    differential gene expression, 92–93
    Drosophila studies, 263
    early attempts at, 84–85
    genomic alterations, 90–91
    genomic equivalence, 85–89, 317–318
    hair formation, 419–420
    leaf morphogenesis, 670
    RNA localization techniques, 93–98
Developmental northern blotting, 93, *94*

Developmental pathways. *See* Signal trans-
 duction pathways
Developmental patterns
 animal embryogenesis, 25–26
 Dictyostelium, 39–44
 frog, 26–30, *27–29*
 metazoan, 44–47
 unicellular protists, 30–33
 Volvox, 36–38
Developmental plasticity, plants, 650
Developmental toxicology, 739, 742
Dextral coiling, 240, *241*
dHAND gene, 536
dHAND protein, 536
DHT. *See* 5a-Dihydrotestosterone
Diabetes, 708
Diacylglycerol (DAG), 206
Diakinesis, 626
Diamond back terrapin, 568
Diapause, 604, **727**
Diapedesis, 613, *623,* 624
Diastrophic dysplasia, 542
Dicer protein, 104, 135
dickkopf gene, fish, 353
Dickkopf protein, *326,* **331–332,** 419–420,
 493
Dictyate state, 639–640
Dictyostelium
 aggregation, 40, *41*
 cell adhesion, 40–43
 differentiation, 43–44
 life cycle, 39–40, *41*
Diencephalon, 399
Diethylstilbesterol (DES), 559, 698–699, 740
Differential adhesion hypothesis, 71–72
Differential gene expression, 92–93
Differential RNA processing, 129–130
Differentiation, 2, 56, 173–174
Differentiation-inducing factor (DIF), 34
Differentiation therapy, 703–704
Digestive system, 491
Digestive tube, 510–515
Digits, specification, 536, *537*
Digoxigenin, 98, *99*
5a-Dihydrotestosterone (DHT), 558
diI, 228
Dimers, TGF-β peptides, 153
Dinosaurs, 534, 731
Dioecious species, **655**
Dioxin, 740, *741*
Diploblastic animals, 6, *45,* **46**
Diplontic life cycle, **651**
Diplotene stage, 625, 633, 639
Direct development, 581, **582**
Discoidal cleavage, 224–*225,* **348**
 meroblastic, *348,* 354, *355*
Disheveled protein, 161, 324
Dispermic eggs, *200*
Disruptions, 15

Dissociation, evolution, 766–768
Distal-less (Dll) gene, 288–289, 586, 588, 595
 evolutionary duplication, 768
 expression patterns, 756–757, *757, 758*
 polyphenism, 728–729, *729*
 vertebrate head, 762, *762*
Distal-less protein, 289
Distal structures, avian embryo, **356**
Distal tip cell, 627–628
Dizygotic twins, 373
DNA (deoxyribonucleic acid)
 (cytosine-5)-methyltransferase, 123
 in apoptosis, 165*n*
 chromatin structure, 107–108
 homeobox, 286*n,* 289–290
 in mid-blastula transition, 311
 muscle-specific genes, 474
 polymerase chain reaction (PCR), 93–95
 polytene chromosomes, 92
 synthesis during egg activation, 208
 transcription factors, 114–117
 transfer techniques, 99–100
 ultraviolet radiation, 54
DNA-binding, homeodomain proteins,
 289–290, *290*
DNA-binding domain, 114–116
DNA methylation, 121–123, 127
DNA microarrays, 96
DNA regulatory elements
 cis- and *trans-,* 112*n*
 enhancers, 112–114
 promoters, 111–112, *112*
 silencers, 120–121
 techniques for studying, 118–120
 transcription factors, 114–121
DNA replication, amphibian, 306, *307*
DNA-RNA hybridization, 92, 93–94
DNase protection assay, 118
Dobzhansky, Theodosius, 777
Dolly, cloning of sheep, 87–88
Dominance, genetic anomalies, 691
Dominant negative allele, 691
Dominant negative protein, 313*n*
Dopamine, axons, 412
Dopaminergic neurons, 452
Dormancy, plants, **664**
Dorris, Michael, 697*n*
Dorsal blastopore lip, 27, *29*
 avian equivalent, 361
 blastopore, 312
 chick equivalent, 356
 chordin mRNA, 328
 fish equivalent, 350–351
 frog blastopore, **310**
 the organizer, 320
Dorsal ectoderm, neural tissue, 332
dorsal gene, 291
Dorsal mesoderm
 amphibian, *325,* 326

avian neural induction, 362
 frogs, 27
 the organizer, 328
 specification in fish, 353
Dorsal protein, 291
 gradient effects, 295–297
 translocation, 291–293
Dorsal root, spinal cord, *405*
Dorsal root ganglia, 429–432, 446
Dorsal side, fish embryo, 349–350
Dorsal-ventral axis, 226–227
 amphibian, 311–312, 321, 323–324
 avian, 356–357, 360
 C. elegans, 253–254
 Cartesion coordinate model, 297
 cytoplasmic movements, 210
 Drosophila, 264, 290–297, *292*
 fish, 351–353
 homology, 764, *765*
 insect wing, 588, *588*
 leaf morphogenesis, 669
 limb development, **524,** 537, *538*
 mammalian, 380–381
 mollusc, 245
 neural tube differentiation, 401–402
 primitive streak, 356
 retinal ganglion axonal growth, 455
 sea urchin, 231–232
 tunicates, 250
Dorsolateral hinge points (DLHPs), 396
Dorsolateral pathway, 429–431
Dosage compensation, 124–126
Double fertilization, 658
Double stranded RNA (dsRNA), 103–104
***doublesex (dsx)* gene,** 131–132, 562,
 564–567, *565*
Doublesex protein, 562, *566*
Down syndrome, *687,* **687,** 688
Down Syndrome Critical Region (DSCR),
 688
Dpp mRNA, snails, *243*
Dragonfly, 735
Dri gene, sea urchin, *232*
Driesch, Hans, 61–63
Drosophila. See also Insect metamorphosis
 aging, 603–604
 anterior-posterior polarity, 135, 268–290,
 269
 axonal migration, 449, *450*
 BMP4/chordin homologies, 330–331
 cleavage, 221, 264–266
 D. mangabeirai, 632
 developmental genetics, 263
 dorsal-ventral polarity, 290–297, *292*
 dosage compensation, 124
 eye, *155,* 413
 gastrulation, 266–268, 295–296
 gene transfer with P elements, 99–100
 genetic studies, 251

germ cells, 615–616, 622–624, *624*
head formation, *275*, 285
heat shock protein, 776–777
hedgehog gene, 151*n*
Hom-C genes, 377, *378*
homeotic selector genes, 285–290
Hox genes, 755–757, *756*
Krüppel gene, 102, *103*
maternal effect genes, 269–270, *274*
metamorphosis, 266*n*, *585*, 586, *587*, *591*
methuselah mutants, 603
mid-blastula transition, 223, 265–266
morphogen gradients, 64
morphogenetic fields, 67
mouse gene homologues, 377
nervous system specification, 398
neuroblast determination, 167
oocyte mRNA, 132, 133, *133*
oogenesis, *637*
Pax6 gene expression, 119–120, *120*
photoreceptor induction, 155–157
polytene chromosomes, 92
RNA splicing, 130
segmentation genes, 278–285
semaphorin proteins, 445–446
sex determination, 131–132, 561–567
spermatogenesis, 638
syncytial specification, 69
Drugs, mammalian cloning, 89, *90*
Dscam gene, 130, 692
DSCR1 gene, 688
dsRNA. *See* Double stranded RNA
Duck, webbed feet, *540*, 769, *770*
Ductus arteriosus, 498–499
Dumas, J. B., 184
Dung beetles, polyphenism, 731
Duplication and divergence, 768
Dwarfism, 160–161, 541–542
Dynein protein, 185–186, 272, *273*
   asymmetry, 382
   transport to oocyte, 638

E-cadherin, 74–75, 397
   avian gastrula, 357
   mammalian, 366, 370
   mouse PGCs, *619*
E74B transcription factor, 591–592
Ear
   co-option in evolution, 768, 768*n*, *769*
   embryonic homology, 13–14
Early allocation and progenitor model, 532
early phase change (epc) gene, 672
easter (ea) gene, 294
Easter protein, *292*
Ecdysone. *See* 20-Hydroxyecdysone
Ecdysone receptors (EcRs), 590
Ecdysozoa, *45, 46,* **752**
Echinoderms, *45, 46, 366*
Echiuroid worms, 52, 568

Eclosion, 584, 728
Ecological developmental biology, *722*
Ecology, 745
EcR isoforms, 590
Ectoderm, 6, 26
   amphibian, 308, 310, 315–316, 327–330,
      332–333
   avian, 358–359
   conversion to neural tissue, 332, *333*
   differentiation, *333*
   diploblasts, 46
   Drosophila, 266
   eye development, 144–145, *147*
   fish, 351–352
   formation of, 226
   induction by the organizer, 327–328
   major derivatives of, *392*
   mammalian, 370
   neural crest, 427
   neurulation, 393, *394*, 395–397
   organizer, 320–321
   origin, 27
   Pax6 transcription factor, 116–117
   sea urchin, 228, 229
   selective affinity, 71
   tunicates, 247
Ectodysplasins, 419, *420*
Ectomesenchyme, 439–440
Ectopic pregnancy, 368
Edward syndrome, 687
δEF proteins, 117
Efferent ducts, 480, **551**
eFGF protein, 322
egalitarian gene, 638
Egalitarian protein, 638
EGF. *See* Epidermal growth factor
Egg, 26, 631, *See also* Oocyte; Ovum
   acrosome reaction, 191, 197
   amniote, 46–47
   animal-vegetal axis formation, 633
   anterior-posterior polarity, 268, *269*, 270
   bindin receptors, 194–195, *196*
   cytoplasm reorganization, 210–213,
      311–312
   cytoplasmic contents, 187
   discovery of mammalian, 6, 7*n*
   dispermic, *200*
   fusion with sperm, 197–203, 209
   genomic imprinting, 124
   hemispheres of, 27
   mammalian fertilization, 192–193
   metabolic activation, 203–209
   morphogenetic determinants, 248–249
   nuclear potency studies, 85–87
   numbers of, 631
   "Organizing centers," 270
   polarity in amphibians, 308
   species-specific recognition, 194–197
   sperm, 189–191, 194–197

   stages of maturation, *188*
   structure of, 186–189
   symbioses, 723–724
   syncytial specification, *69*
   ultraviolet radiation, 54–56
   in vitro fertilization, 684–685
   yolk distribution, 224
Egg jelly, 189, 191
EGTA, egg activation experiments, 205
Electrical impulse, neurons, 412–414
Electrophoresis, 93*n*, *94*, 118
Electroporation, 99
Eleutherodactylus coqui, 582, *583*
Elk-1 protein, 173
Embôitment, 4
Embryo
   composite, 374
   evolutionary emergence, 752
   specification of induction, 334
   uterine implantation, *366*, 368
   in vitro fertilization, 685
Embryo proper, plants, *659*–660, 662–663
Embryo sac, plants, *655*, **655**, 658–659
Embryogenesis
   animals, **25–26**
   plants, 659–664, *661*
Embryology, 2, *See also* Comparative em-
      bryology; Experimental embryolo-
      gy
   evolutionary, 12–14
   genetics, 81–85
   medical, 14–16
   reductionist methodology, 62*n*
   wholist organicism, 62*n*
Embryonic-abembryonic axis, mammalian
      egg, *381*
Embryonic epiblast, 369
Embryonic germ cells (EG), 621
Embryonic hemoglobin. *See* fetal hemoglobin
Embryonic period, gestation, **694**
Embryonic phase, hematopoiesis, **506–507**
Embryonic region, **380**–381
Embryonic shield, fish, **350,** 353
Embryonic stem (ES) cells, 100, *102,* 374,
      621, 708–709
empty spiracles (ems) gene, 276, *282,* 377
   evolutionary conservation, 754
EMS cell, *C. elegans,* 254–257
EMX2 gene, 754
Emys, 567
Enamel knot, 441
Encapsulation, 4
endo16 gene, *232*
Endocardial cushion, 496
Endocardium, 493, *494*–495, 495*n*
Endochondral ossification, 475–477
Endocrine disruptors, 698–699, 739–743
Endocrine factors, **149**
Endocrine hormone genes, 112

Endoderm, 6, 26
amphibian, 308, 310, 313, 315
anterior visceral, 345–376, *376*
avian, 357–358
avian precursors, 356
C. elegans, 255
digestive tube, 511–515
diploblasts, 46
Drosophila, 266
extraembryonic membranes, 517
fish, 351, 353
formation of, 226
functions, 510
mammalian, 369–370
origin, 27
pharyngeal, 357, 511
respiratory tube, 515–516
sea urchin, 228–230
selective affinity, 71
specification, *232,* 333, 512, *512*
tunicates, 247
Endodermis, root, 667, *668*
Endomesoderm, 330–332
Endometrium, 368, 371, 639–640
Endoplasmic reticulum, 202, *204, 206*
Endosperm, 654, 658
Endosperm development, **658**–659, 662
Endostatin, angiogenesis, **504**
Endothelial cells, blood vessels, **501**
Endothelins, 429, 434
Energids, 264
engrailed (en) gene, *282, 284,* 285, 587–588,
    730–731, 764
Engrailed protein, 285, *587, 588*
Enhancer trap, *119,* **119**
Enhancers, 112–114
cancers, 120
functions of, 114
methylation, 121–122
Pax6 gene, 113–114, *113,* 119–120
primary pair-rule genes, 282
regulation of, 123–124
techniques for identifying, **119**
transcription factors, 112, 114–117
Entelechy, 62
Enteric ganglia, 429
Enterocoelous body cavity, 46
Entwicklungsmechanik, 51
Enveloping layer (EVL), 348, 349
Environment
adaptation of embryos and larvae, 52–56
developmental symbiosis, 722–725
diapause, 727
endocrine disruptors, 739–743
genetic assimilation, 733–734
gravity and pressure, 722, *723*
larval settlement, 725–726, *726*
neuronal patterning and experience,
    737–739

phenotypic plasticity, 52–54, 727–736
predator-induced defences, 734–736, *734*
seasonality, 726–727
sex determination, 52, 567–569, 730–731
teratogens, 163*n,* 740
ultraviolet radiation, 54–56
Environmental developmental biology,
    51–56, *722,* 743–745
Environmental integration, 2
Eomesodermin, mammalian, **367**
EP-cadherin, 75, 307
Epaxial muscles, 472
Epaxial myoblasts, 473
Ependyma, 403, *404, 405,* 410, **410**
Ependymal canal, 398
Eph proteins, 446, 469, *470,* 504–505
Eph receptors, 469, *470*
axonal migration, **431**–433, 435, 445–446,
    456
EphA4 gene, *469*
Ephdra, 658
Ephrin proteins
axonal migration, 435, 445–446, 454
cell migration, **431**–**432,** *432,* 469, *470*
Ephrin-B proteins, *470,* 504–505, 577
Epi1 protein, 333
Epiblast
cell heterogeneity, 360
chick, 358, 360
fish, 350–351
mammalian, 369
signaling centers, 375, *376*
Epiboly, 226
amphibian, 309, *310,* 315–316
avian, 358–359
C. elegans, 257
fish blastoderm, 349–350, *351*
molluscs, 245, *246*
tunicates, 250, *251*
Epibranchial placodes, 439
Epicardium, *494*
Epicotyl, 665
Epidermal ectoderm, 427
Epidermal growth factor (EGF), 153, 293
Epidermal placodes, 419–420
Epidermis, 147
amphibian, 327–330
cell determination, 167
cutaneous appendages, 418–419
development, 418
Drosophila, 297
epithelial-mesenchymal interactions,
    147–148
layers, *417*
selective affinity, 70–71
somite origins, 472
Epididymis, 551
Epigenesis, 4–5
Epimorphin protein, 472

Epimorphosis, salamander limbs, **592**–597
Epinephrine, axons, 412
Epiphyseal growth plate, 540, 541–542, *541*
Epithelial cadherin, 74–75
Epithelial cells, 70, 147, 168–169
Epithelial-mesenchymal interactions,
    147–148
avian embryo, 356–357
tooth development, 439
Epithelialization
kidney development, 482–483
somitogenesis, 469–471, *470*
Equivalence group (VPCs), 157, 374
ERK kinase, 154, 159
Erythroblast, 509
Erythrocytes, *57,* **509**
Erythroid precursor cells, 123, **509**
Erythropoietin, 164, **509**
ES. *See* Embryonic stem cells
Escherichia coli, reporter genes, 112–113
Esophagus, 511
"Essential force," 4
Estradiol, 559
Estrogen
environmental, 740–741, *742*
environmental sex determination,
    567–568
estradiol, 559
estrus cycle, 640–641
frogs, 27
mammalian sex determination, **549,**
    559–561
ovulation, 641
Estrogen receptors, 542
Estrogenic compounds, 701–702
Estrus, 638
Ethanol. *See* Alcohol
Ethics. *See* Bioethics
Ets gene, *232*
Euchromatin, 124
Eudorina, 34, 35
Eukaryotes, gene structure, 107–111
Euprymna scolopes (Squid), 723
European map butterfly, 53, 728, *730*
Euscelis incisus, 724
Eustachian tubes, 511
even-skipped (eve) gene, *281,* 282, *282*
transcription in *Drosophila, 283*
Even-skipped protein, 285
Evidence, types of, 42, 67
Evolution, 2
and aging, 603–604
homeotic selector genes, 377
Evolutionary biology
Darwin's synthesis, 751–752
developmental constraints, 773–777
generating novelty, 769–773
homologous developmental pathways,
    763–765

homologous genes, 753–754, *754*
homologous processes, 330–331
homologous structures, 13–14
Hox genes, 754–762
modularity principle, 766–769
phylogenetic relationships, 752–753, *753*
Urbilatarian ancestor hypothesis, 753–754
Evolutionary developmental biology, 330–331, **751**
developmental genetic model, 777–779
generation of novelty, 769–773
Evolutionary embryology, 3, 12–14
homologous structures, 13–14
Evolutionary relationships, animal phyla, *45*
Exine, 654
Exocrine protein genes, 112
Exocytosis
acrosome reaction, 191, 192*n*
cortical granule reaction, 202, *203*
Exons, 108–110, *110*, 128, 130
Experimental embryology, 51
amphibian studies, 305
cell specification studies, 56–69
environmental studies, 51–56
External granule layer, 404
External YSL, fish, **348**
Extracellular envelope, 187, *188*
Extracellular matrix, 168–171, 368
angiogenesis, 503
axonal guidance, 431–432, 445
endochondral ossification, 476–477
epiphyseal growth plate, 542
integrins, 169–171
netrins, 447
primary mesenchyme ingression, 233–236
somitogenesis, 469
synapses, 451
Extradenticle (Exd) gene, 290
Extraembryonic endoderm, 369
Extraembryonic membranes, 370–372, *371*
amniotes, *516*, 517
signaling centers, 375
Extraembryonic vasculogenesis, 501
Extramacrochaetae protein, 562
Extrazygotic genes, plants, 662
exuperantia (exu) gene, *272*, 274, *275*, 276
Exuperantia protein, 272, 638
Eye
allometry and heterochrony, 766, *767*
development, 413–416
induction, 143–145, *147*, *150*
Notch pathway, 166
Pax6 gene, 93
photoreceptor differentiation, 155–157
Eye migration, metamorphosis, 576*n*, 577
eyeless gene, 288

F-actin, 659, *660*
Facial cartilage, 434–435

Facial development syndromes, 439
Facial nerve, placode origin, 439
Fasciculation, retinal ganglion axons, 454
Fast block to polyspermy, 199–**200**, *201*
Fat-storing (Ito) cells, 601
Fate maps, 8–**11**
Drosophila, *291*, 295
fish, 348–349
frog blastula, 308–309
germ layers, 307–308
mollusc, 243–244
mouse epiblast, 369
observing living embryos, 8
sea urchin, 228–233
snail, *241*, 244–245
tunicate, 246–247
Xenopus, 308–309
zebrafish, 351
FBP transcription factors, 662
Feathers, evolution of, 770, *771*
fem genes, 628
Female gametophytes, 655
Female phenotype, 549–552
Female pronucleus, 209
Female sex determination, *152*
Female-specific RNA, 131–132
Ferns, life cycle, 651, *652*
Fertilin, 474
Fertilization, 25, 183–184, 189
activation of egg, 203–209
activation of metabolism, 206–208
blocks to polyspermy, 198–202
frogs, 26
fusion of genetic material, 208–209
gamete fusion, 197–203
germ plasm theory, 59–61
human, *211*, 641
mammalian, 192–193, 195–197, 198, *199*, 201–202, 209–210
metaphase block, 634
plants, 658–659
reorganization of egg cytoplasm, 210–213
salamanders, 203*n*
sperm-egg recognition, 189–197
in vitro fertilization, 684–685
volvocaceans, 38
Fertilization cone, 198, *198*
Fertilization envelope, 201, *202, 203*
Fetal alcohol syndrome (FAS), 697, 697, *698*
Fetal hemoglobin, 122, 498, *499*
Fetal period, 694
α-**Fetoprotein, 559**
FGFs. *See* Fibroblast growth factor(s)
fgf4 gene, limb development, 538
fgf8 gene, 363, *364*, 369, 595
FGF8 signaling, 383
Fgf9 gene, 554
FGFR3 gene, 691
Fibrillin, 691

Fibroblast growth factors (FGFs), 150–**152**
avian neural induction, 362
binding, *151*, 152
heart development, 493, *493*
kidney development, 482–483, 485
limb development, 524, 530–532, 535, 538
mammalian, 369–370
multiple pathways, 173
neural plate competence, 392
neuron survival, 453
retinal ganglion axons, 454
trunk neural crest, 431
Wnt signaling, 336
Fibroblast growth factor 1 (FGF1), 150
Fibroblast growth factor 2 (FGF2), 150, 639
kidney development, 482, 482–483
neuronal survival, *453*
pancreas formation, 514
salamander regeneration, 595
trunk neural crest, *434*
Fibroblast growth factor 3 (FGF3), 441
Fibroblast growth factor 4 (FGF4)
limb development, 538
mammalian, 367
trunk neural crest, 441
Fibroblast growth factor 5 (FGF5), somitic cell fates, *473*
Fibroblast growth factor 7 (FGF7), 150
kidney development, 485
Fibroblast growth factor 8 (FGF8), 150, *151*, 363, *364*, 392, 471
anterior-posterior limb polarity, 535
cardiac specification, *493*
limb development, *527*, 530–531
mammalian right-left asymmetry, 383
somitogenesis, 468
tooth development, 439–440
Fibroblast growth factor 9 (FGF9), 554
Fibroblast growth factor 10 (FGF10), 420
limb development, **525**–526, *527*, 529–531
limb regeneration, 594–595, *595*
Fibroblast growth factor 19 (FGF19), placode induction, 438, *439*
Fibroblast growth factor receptors (FGFRs), 150, 160–161
epiphyseal growth plate, 541
Fibroblast growth factor receptor 2 (FGFR2), bone growth abnormalities, 541
Fibronectin, 146, 168–169, 368
avian gastrulation, 359
cardiac cell migration, 494
endochondral ossification, 476
frog embryo, *168*
germ cell migration, 617–620
mesenchyme epithelialization, 469
mesodermal migration, 316–317, *317*
muscle cell formation, 474
trunk neural crest, 431
Fibronectin receptors, 169–171

Fibropellins, 236
Fibular crest, 722, *723*
Fiddler crab, 17, *18*
Field of organization, 67
Filopodia, 233–239, 411–412
   avian gastrulation, 359
   secondary mesenchyme, 238
   target, 238
Finasteride, 698
First polar body, 632
FISH. *See* Fluorescent in situ hybridization
Fish
   axis formation, 351–354, *352*
   cleavage, 224, 347–349
   dioxin, 740, *741*
   environmental sex determination, 731–732
   gastrulation, 349–351, *350*
   gill homology, 13–14
   neural crest cell migration, 429*n*
   neurulation, 393
   retinal ganglion axons, 454–455
5′ Cap, mRNA translation, 133
Flagellum, 184–186, 631
Flatworms, *45*, 46, 64–65
Flk1 receptor tyrosine kinase, 503
floating head gene, zebrafish, 352
Floor plate, 402, 449
Floral meristem identity genes, 675–676
Floral organ identity genes, 676–678
Floral promoters/inhibitors, 673
FLORICAULA (FLO) gene, 670, 675
Flower structure, 654
Flowering signal, 672–675, *673*
Floxed technique, 118
Flt1 receptor tyrosine kinase, 503
Fluid dynamics, blood flow, 500–501
Fluorescent dyes, 10, 202
Fluorescent in situ hybridization (FISH), 693
FMR1 gene, 693
fog gene, 628
Fol, Herman, 184
Folate receptor genes, 123
Folic acid, 397
Follicle cells (ovarian), *57*
   Drosophila, 292, 293–294
   mammalian, **551**, 639–641, *639*
Follicle-stimulating hormone (FSH), 630, 638, 640–641
Follicles (hair), 418–419
Follicular phase, 640–641
Follistatin, *326*, 329–330, 392
Food reserves, seeds, 664
Foot activation/inhibition, hydra, **598**, 600
Foramen ovale, 499
Foregut, 236, 356
Forelimbs
   development, *524*, 525–529, *528*
   homologous, **13**
   snakes, 759

Founder cells, 253, *257*
Fovea, 737*n*
Foxb-2 protein, 484–485
Foxd3 protein, 367, 428
FOXP2 gene, brain development, 409
Fragile X syndrome, *693*, 693
fragilis gene, 618, *619*
Fraternal twins, 373
Frizzled receptor family, 161
Frogs. *See also* Amphibians; *Rana*; *Xenopus*
   asymmetry pathway, 382
   axis determination, 210
   axonal guidance, 448
   deformities, 744–745
   fibronectin, *168*
   gap junctions, 172
   herbicides and sex determination, 568, *569*
   interspecific induction studies, 149
   life cycle, 26–30, *27*
   mesodermal development, *492*
   metamorphosis, 576–583
   nuclear potency studies, 86–87
   parental care, 582–583
   reporter genes, 113
   retinal ganglion axons, 454–455
   ultraviolet radiation, 54–56
Frontonasal process, 435
Frugivores, seed germination, 665
Fruit, 655, 678
Fruit fly. See *Drosophila*
FRUITFUL (FUL) gene, 676
fruitless gene, 567
frzb mRNA, *332*
Frzb protein, 324, *326*, 331–332
fs(1)Nasrat (fs(1)N) gene, *272*
fs(1)polehole (fs(1)ph) gene, *272*
FSH. *See* Follicle-stimulating hormone
ftz gene. *See fushi tarazu* gene
Ftz protein. *See* Fushi tarazu protein
Fundulus, blastula, 349
Fura-2, 202
fused (fu) gene, *282*
Fused protein, *163*
fushi tarazu (ftz) gene, 278, *279*, *280*, *282*, *282*, *283*
Fushi tarazu (Ftz) protein, 282, 285, 289
Fusion
   gametes, 197–203, 208–210, *211*
   kidney development, 480–481
   muscle cell formation, 474
   plasma membrane, *190*, *191*, 197–198
   syngamy in plants, 658
   vascular networks, *504*
"Fusogenic" proteins, 198
Fusome, 638
fuzzy onions gene, 631
Fyn proteins, *171*

G protein, 154
GABA (γ-aminobutyric acid), 412
Gain-of-function evidence, 42
Gain-of-function mutations, 691
GAL4 transcription factor, 119–120
β-Galactosidase, **112**–113, *119*
   retinal development, 414, *415*
Galactosyltransferase protein, 197, 628
Galen, 547
Gall bladder, 512
Gallus (Chicken), 354
Gamete intrafallopian transfer (GIFT), 685
Gametes, 26, 613
   activation of metabolism, 206–208
   egg activation, 203–209
   egg structure, 186–189
   fusion of genetic material, 208–210
   fusion of sperm and egg, 197–203
   germ plasm theory, 60–61
   mammalian, 192–193
   plant, 649, 651, 654–658
   protist, 33
   sperm-egg recognition, 189–197
   sperm structure, 183–186
Gametogenesis, 26, 189, **613**
   germ cells, 613–624
   hermaphrodites, 627–628
   meiosis, 624–628
   oogenesis, 631–641
   spermatogenesis, 628–631
   tumor-derived cells, 621, *622*
Gametophyte, 651, 654
Gametophytic self-incompatibility, 656
Ganglia, 400
Ganglion cells, 414
Gap genes, 269, 276, **278**–281, *282*
   homeotic selector genes, 286–288
   terminal regions, 277
   transition to segmentation, 278–279
Gap junctions, 171–172
GARP domain, plants, 669
Gastric-brooding frogs, 583
Gastropods, 16
Gastrula, 26, *233*
Gastrulation, 26–27, **226**
   amphibian, *27*, *29*, 307–317, *314*
   avian, 356–360, *357*, *358*, *359*
   C. elegans, 257
   Drosophila, 266–268, 295–296
   mammalian, 368–374
   mesodermal components, 466
   mollusc, 245, *246*
   sea urchin, 233–239
   tunicates, 250–251
gastrulation defective (gd) gene, 294
Gastrulation defective protein, *292*
GATA transcription factors, *494*, *496*
GATA5 protein, fish, 353

GataE gene, sea urchin, *232*
"Gay gene," 561
Gdf5 gene, *540*
GDFs. *See* Growth and differentiation factors
GDNF. *See* Glial cell line-derived neurotrophic factor
GDNF receptors, 481, *482*, 485
Gel mobility shift assay, 118
Gene(s)
    cloning, 93*n*
    developmental function of, 98–104
    evolutionary conservation, 752–761, *754*
    identifying anomalies, 687–689
    for immunoglobulins, 90–91
    nomenclature, 15*n*
Gene activation, 230, 326–327
Gene expression, 92–93, *See also* Transcription; Translation
    avian embryo, 361, *362*
    BMP4 effect, 330
    Cartesion coordinate model, 297
    differential gene transcription, 107–121
    differential protein modification, 135–136
    differential RNA processing, 127–132
    DNA methylation, 121–123
    dosage compensation, 124–126
    extracellular matrix, 169–171
    hair follicle development, *420*
    homeotic selector genes, 285–289, *287*, *289*
    hox gene patterns, 380–381
    human disease, 691–693
    maintenance of, 173–174
    regulation, *136*
    spermatogenesis, 631
    in transgenic clones, 89, *90*
Gene knockouts, 100–101, *102, 103, See also* Knockout mice
Gene silencing, 103–104
Gene structure
    anatomy, 107–114
    enhancers, 111
    human β-globin gene, *109, 110*
    promoters, 111–112
Gene targeting, 100–101, *102, 103*
    Hox genes, 377–379
Gene theory, embryological origins, 81–85
Gene therapy, 705–708, *706,* 711
Gene transcription. *See* Transcription
The Generation of Animals (Aristotle), 3
Generative cell, 654
Genetic assimilation, 733–734, 776
Genetic counseling, 691*n*
Genetic essentialism, 743
Genetic heterogeneity, 690
Genetic marking, 10–11
Genetic screening, zebrafish, 345, *347*
Genetic variability, cryptic, 776–777

Genetics
    and aging, 602–605
    Drosophila studies, 251, 263
    embryological origins, 81–85
    human genetic anomalies, 686–694
    leaf morphogenesis, 669–671
    preimplantation, 693–694
Geniculate placodes, 439
Genital ridge epithelium, 549
Genome, 85
Genome sequencing, *C. elegans,* 251
Genomic equivalence, 85–89, 317–318
    lymphocyte exception, 90–91
Genomic imprinting, 123–124
Genotype, range of phenotypes, 53–54
Germ band, 266, *267*
germ cell-less (gcl) **gene,** *615*
Germ cells, 26
    avian, 357
    C. elegans lineage, 255
    determination, 551*n,* 613–617
    egg structure and development, 186–189
    germ plasm, 613–617
    migration, 550–551, 617–624
    plants, 649
    sperm maturation, *185*
    therapeutic cloning, 708–709
    volvocaceans, 35
    X chromosome reactivation, 126
Germ layers, 26
    amphibian fate map, 307–308
    avian, 356
    cell types of, *3*
    derivation of organs, 7
    early identification of, 6
    fish, 349–351
    formation, 226
    selective affinity, 70–71, *72, 73*
Germ line
    germ cell determination, 613–617
    germ cell migration, 617–624
    germ plasm, 613–617
    hermaphrodites, 627–628
    longevity, 604
    meiosis, 624–628
    oogenesis, 631–641
    plants, 671–672
    spermatogenesis, 628–631
    tumor-derived cells, 621
Germ line gene therapy, 707–708
Germ line switch, 628
Germ plasm, 614, 616–617
Germ plasm theory, 59–61
Germ ring, 350
Germinal crescent, 357, *622, 623*
Germinal epithelium, 415, 416
Germinal neuroepithelium, 403
Germinal vesicle, 632

Germinal vesicle breakdown (GVBD), 633–634
Germination, plants, 664–665
Gerontology, 575
Gestation
    human brain development, 408
    sensitivity to teratogens, *695*
GFP. *See* Green fluorescent protein
GFRα co-receptor, 481
GGF. *See* Glial growth factor
giant (gt) gene, 279–281, *282*
Giant protein, 281
Gibberellins, 660, 664, 665, 670, 675
GIFT. *See* Gamete intrafallopian transfer
Gill arches, 6, 13–14
Gli proteins, 162
GLI3 gene, 162
GLI3 protein, 162
Glia, cranial neural crest, 434
Glial cell line-derived neurotrophic factor (GDNF), 153, 452, **481**–483, *482, 484,* 629–630
    joint formation, 540, *541*
Glial cells, 295–297, **398,** 405, 412
    birthday, 403
Glial growth factor (GGF), trunk neural crest, 434
Glial guidance, 405, *406,* 407
Glial pathways, cranial neural crest migration, 439
Glioblasts, *404*
γ-Globin/ε-Globin, 122
Globin genes, 121–122, *122*
β-Globin genes
    regulation, *122, 123*
    structure, 108, *109, 110*
Glochidia larva, **242**
Glossopharyngeal nerve, 439
glp-1 gene, 256, 627
glp-1 mRNA, *C. elegans,* 133
GLP-1 protein, 256–257, 627
Glucagon gene, 116–117
Glucagon-secreting alpha cells, 515
Glucocorticoids, 434
Glueckson-Schoenheimer, Salome, 84, *85*
Glutathione, 209
Glycogen metabolism, 161
Glycogen synthase kinase 3 (GSK-3), 161, 323–324
Glycoproteins
    adhesion in *Dictyostelium,* 40, *41*
    adhesive, 189
    extracellular matrix, 168–169
    muscle cell formation, 474
Glycosyltransferase, 76
GM-CSF. *See* Granulocyte-macrophage colony-stimulating factor
Gnathostomes, 771–773
Gnetum, 658

Goldschmidt, Richard, 769, 777, 778
Gonadotropins, 638–640
Gonads
    development, 549–551, *550*, *552*
    primary sex determination, 548–551
Gonidia, 36, *38*
Gonium (Protist), 34, *35*
gooseberry (gsb) gene, *282*
goosecoid gene, 65, 66, 325, *326*, 353
goosecoid mRNA, 326–327
Goosecoid protein, 324, *326*, 327, 599
Gradient models, conditional cell specification, 64–66
Gradients. *See also* Protein gradients
    Drosophila embryo, 270–278, *274*, *275*, *276*
Gradualism, 778
Granular layer, 416, *417*
Granule neurons, 404
Granulocyte-macrophage colony-stimulating factor (GM-CSF), 509
Granulosa cells, 551, 639–640
Grasshoppers
    axonal migration, *446*
    parthenogenesis, 632
Grauzone protein, 133
Gravity, avian axis formation, 360, *361*
Gray crescent, 210, *211*, 308, **318–319**
Gray matter, 403, 404–405
Great Lakes, dioxin, 740
Green fluorescent protein (GFP), 113, 346, *346*, 707–708
Green frog, *744*
Green tea, and cancer, 503*n*
Gremlin protein
    limb development, 538–539
    webbed feet, *540*, 769
Grex, 39
Gridlock transcription factor, blood vessel specification, **505**
Grieg cephalopolysyndactyly, 162
Groucho protein, 277, *278*, 297
Ground tissue, plants, **663**, 667
Growth, 2, 16–17
Growth cones, 411–412
    axonal specificity, 445–446, 448
    chemotaxis, 451
Growth and differentiation factors (GDFs), 149, 639
Growth factors
    autocrine, 416
    axon formation, 411–412
    and cancer, 704
    epidermis, 416–418
    fibroblast, 392, 420
    keratinocyte (KGF), 417–418
    TGF-α, **416**–418
    TGF-β, *401*, 402
Growth plate closure, 542
GS box, *159*

GSK-3. *See* Glycogen synthase kinase 3
GTPase-activating protein (GAP), 154
Guanine nucleotide releasing factor (GNRP), 154
Guidance, axonal growth, 431–432, 445–450
Guidance molecules, 455–456
Guinea pigs, sperm binding, 197*n*
Gurdon, John, 86–87
gurken gene, **293**–294
Gurken protein, 272, *273*, 292
Gut, right-left axis, 337
Gut tissue, specification, 512, *512*
Gymnosperms, embryogenesis, 658
Gynandromorphs, 561

Haeckel's biogenetic law, 752
Hair development, 418–420, *418*
Hair germ, 418
Hairlessness, phenotypes, 419
hairy (h) gene, *156*, *282*
Hairy 1 protein, 469
Hairy1 gene, 468–469, *469*
Haldane, J. B. S., 779
Haliotis rufescens, 725
Haller, Albrecht von, 4
Halocynthia roretzi, 248
Halteres, 286, *288*
Hämmerling, J., 31
Hamsters
    acrosome reaction, *192*
    egg, *189*
    gametes, *199, 207*–208
Hand proteins, heart chambers, 495
Haplodiplontic life cycle, *651*, **651**
Haploid condition, 624–625
Haploinsufficiency, 691
Haptotaxis, 447*n*
Harrison, Ross Granville, 83, *84*, 411, 444
Hartsoeker, Nicolas, 183–184
Harvey, William, 4
Hatched blastula, sea urchin, **228**
Head
    cranial neural crest, 434–436, *436*
    sperm, 184
    vertebrates, 434–436, 761–762
Head activation gradient, 597–598
Head formation, 315, 330–332, 334–336, 375
head involution defective (hid) gene, 592
Head mesenchyme, 357, 465
Head mesoderm, avian, 356
Head process, avian, **357**, *359*
Heart development, 492–500
    cardiac neural crest cells, 441
    connexin-43, 172
    formation of chambers, 495–500
    fusion of primordia, 495
    looping, 363, *364, 365*, 381
    molluscs, 244
    right-left axis, 337

snail, 241
    specification of tissue, 492–495
Heart stage, 663
Heartbeat, initiation, 495
Heat shock protein (Hsp90), 776–777
Heavy chains, 90–91
Heavy metals, 700
hedgehog (hh) gene, 151, *282, 284, 285*, 346, *587, 588*
Hedgehog pathway, 161–163, 703
Hedgehog protein family, 151–152
    Drosophila, 283, 285, *286*, 624
    homology, 764
    photoreceptor induction, 155, *156*
Held-out-wings gene, 130, 132
Helix-loop-helix (bHLH) transcription factor, 115
Hemangioblasts, *501*, **501**
Hematopoiesis, sites, 506–509
Hematopoietic inductive micro-environments (HIMs), 509–510
Hematopoietic stem cells (HSCs), 480, **501**, **506–509**, 709
Hemichordata, 45
Hemidactylium scutatum, 774
Hemimetabolous metamorphosis, 584
Hemoglobin
    adult and fetal forms, 122
    amphibian metamorphosis, 577
    production steps, *110*
HemoglobinConstantSpring gene, **692**
Hemophilia, 691, 707
Hen's teeth, 149
Hensen, Viktor, 411
Hensen's node, 361, *362, 363*, 396
    avian, **356**–357
    limb specification, 525
    mammalian equivalent, 369*n*
    regression, 358
    right-left axis formation, 363, *364, 365*
    somitogenesis, 467–468
Heparan sulfate glycosaminoglycan, 509
Heparan sulfate receptors, 368
Hepatectomy, partial, 601
Hepatic diverticulum, 512
Hepatocyte growth factor (HGF), 153, 441, 601
Hepatocytes, *57*, 601
Herbicides, 568
Hermaphrodites, 557*n*, 558, 567, 627–628
Herpes simplex virus, 702
Hertwig, Oscar, 62, 184
Hesx-1 protein, 375
Heterochromatin, 124, 127
Heterochrony
    amphibian metamorphosis, **581**–582
    evolutionary biology, 766, *767*
Heterogamy, 38
Heterogeneous nuclear RNA (hnRNA), 128
Heterospory, 651

HGF. *See* Hepatocyte growth factor
Hindbrain, 357, 435, 439
Hindgut, snakes, 236
Hindlimbs, 526–529, 759
Hinge regions, 396
Hippocampus, 410
His, Wilhelm, 411
Histoblast nests, insect, **584**
Histone(s), 107–108
 deacetylation, **120**
 nucleosomes, 122–123
 sperm flagellum, 186
 X chromosome inactivation, 127, *127*
Histone acetyltransferase, 116, *159*, 166
Histone deacetylases, 122
Histotechniques, 11
Histotypic aggregation, 71, *72, 73*
HMG (high mobility group) box, 551, 553
HNF-3β protein, 257, *326*
Hnf4α transcription factor, 118
HNK-1, *431*
HOBBIT gene, 663
Holoblastic cleavage, 3, 224, *225*
 rotational, 251–253
Holometabolous metamorphosis, 584
Holoprosencephaly, 689, 691
Holt-Oram syndrome, 527
Holtfreter, Johannes, 312
Homeobox, 286*n*, **289–290**
Homeodomain, 286*n*, **289–290**
Homeotic complex (Hom-C), 286, 289, 290
Homeotic genes, plants, 676
Homeotic mutants, 286
Homeotic selector (Hom-C) genes, 279,
  285–290
 evolutionary conservation, 377
 initiating and maintaining gene expres-
  sion, 286–288, *287*
 maternal effect genes, 270
 realistor genes, 288–289
Homologous chromosomes, 625–626
Homologous genes, 752–761, *754*
Homologous processes, 330–331
 nuclear transport, *295*
Homologous structures, 13–14
Homophilic binding, 75
Homosexuality, 560–561
Homospory, 651
homothorax gene, 288, 586
Homunculi, 184
Hormones. *See also* Plant hormones
 amphibian metamorphosis, 578–580
 endocrine disruptors, 739–743
 insect metamorphosis, 588–590
 phenotypic plasticity, 729–730, *730, 731,*
  734
 sex determination, 556–559
 sexual behavior, 559–561
Horse, allometric evolution, 767

Horvitz, Bob, 165
Hotchkiss, R. D., 121
Hox genes, 377–379
 atavistic mutations, 759–760
 changes between body segments, 757–759
 changes in number, 760–761, *761*
 changes in target genes, 755–756
 changes within body segments, 756–757
 cranial neural crest, 436–437, *436*
 developmental constraints, 775
 dorsal axis, 377
 evolutionary conservation, 377, 754–762
 evolutionary duplication, 768
 experimental analysis of, 377–380
 gut tissue specification, 512
 limb development, 525, 532–534, *533, 534*
 limb regeneration, 595–597
 plants, 650
 somite specification, 471
 transcription patterns, *378*
 vertebrate jaw, 773
HOXA-HOXD series genes (human), 377
HOXA13 gene, 533
HOXD13 gene, limb development, 533
Hoxa-Hoxd series genes (mouse), 377
Hoxa series genes
 cranial neural crest, *436*, 437
 limb development, 532–534, *533, 534*
Hoxa-2 gene, 759, *760*
Hox-10 gene, DES, 699
Hoxb-8 gene, 536
Hoxc-6 series genes, 525, 759
Hoxc-8 gene, 471, 758
Hoxd series genes, limb development,
  532–534, *533, 534,* 536
Hox11 proteins, 481
Hoxb-4 protein, 507
Hoxb-8 protein, 536
HSCs. *See* Hematopoietic stem cells
Hsp83 gene, 776–777
Hsp90 (heat shock protein), 776–777
huckebein (hkb) gene, 277, *278,* 279, *282*
Human(s). *See also* Bioethics; Medicine
 aging and life expectancy, 602–605, *602*
 allometric growth, *18*
 brain development, 398–402, 408–409
 chimeric, 374
 cloning proteins, 89, *90*
 congenital deafness, 346
 cyclopia syndrome, 163
 digestive system development, *510*
 dispermic egg, *200*
 egg number, 631
 embryonic and extraembryonic tissues,
  370–372
 epidermis, *417*
 facial development syndromes, 439
 fertilization to implantation, *211,*
  366–368, *366*

 fetal to newborn blood flow, 498–499
 genomic imprinting, 123
 heart and blood formation, *497, 499, 500,*
  502
 kidneys, 479
 neural tube defects, 396, *397*
 pancreatic development, *513*
 retinal development, *415*
 secondary neurulation, 398
 tissue formation, *370*
 transcription factor genes, *692*
 twins, *373, 374,* 743
hunchback (hb) gene, 276, 279, 281, *282*
***hunchback* mRNA, 272,** 276, 277
Hunchback protein, 269, *273,* 276–277,
  280–282
Huntingtin protein, 452
Huntington disease, 452
Hutchinson-Gilford progeria syndrome, 603,
  604
Huxley, Thomas, 778
Hyalin protein, 189
Hyaline layer, 201, *203, 235,* 236
Hyaluronic acid, 368, 370
Hyaluronidase, 145
Hybrids, 100*n*
Hydatidiform mole, 212
Hydra(s), 46, 63, 597–600
20-Hydroxyecdysone (20E), 586, *589,*
  589–592, *591*
 DNA binding, 590, 591*n*
 seasonal polyphenism, 729, *729*
Hyla chrysoscelis, 735
Hyla regilla, 54, 56
 limb field, *525*
Hymenoptera, meiosis, 632
Hyoid cartilage, cranial neural crest, 435
Hyomandibular bone, 14
Hypaxial muscles, 472
Hypaxial myoblasts, 473
Hyperactivated sperm, 194
Hypertrophic chondrocytes, 476
Hypoblast, 257
 avian, *355,* **356,** 357, *358,* 360
 fish, **350–351**
 mammalian, **369,** *370,* 371, 380
Hypodermis, 167
Hypophysis, 659, 663
Hypostome, hydra, **597–599**
Hypothalamus, 399, 560, 638

I-mf protein, somite differentiation, 472
IκB protein, homology to Cactus protein,
  295
ICM. *See* Inner cell mass
ICSI. *See* Intracytoplasmic sperm injection
Identical twins, 373–374
Igf genes, 123
Ilyanassa, 240, 243, 244

Imaginal discs, 584–588
    determination, 587–588
    Drosophila, 266n, *585, 586, 587*
Imaginal molt, 584
Imaginal tissues, 584–588, *586, 587*
Imago, 584
Imbibition, 664
Immortal animals, 30
Immune deficiency, therapies, 707, 708
Immunity, as predator-induced defense, 736
Immunoglobulins, genes, **90–91**
Implantation, mammalian embryo, 368
In situ hybridization, 97–98, *99*
    whole-mount, 98, *99*
In vitro fertilization (IVF), 684–686, 708
Inborn errors in metabolism, 687
Incus, 13, 435
indian hedgehog (ihh) gene, 151
Indian hedgehog protein, 501n
Indifferent stage, 549
Individuality, 743
Induction, 6, 143–145
    amphibian axis formation, 317–319
    by Hensen's node, *362*
    cascades, 143–147
    cell-to-cell, 155–158
    epidermal, 327
    **epithelial-mesenchymal interactions,
        147–149**
    eye, 413
    genetic specificity, 148–149
    instructive and permissive interactions,
        145–146
    interspecific, 148–149
    juxtracrine signaling, 166–172
    mesoderm, 321–322
    neural crest, 427
    neural tube patterning, 402
    paracrine factors, 149–153
    primary, 320–321
    reciprocal, 480–485, *480, 481*
    regional specificity, 147–148
Industrial compounds, 740
Inert genome hypothesis, 617
Infertility, 683–686
Inflorescence meristems, 674–675
Infundibulum, 511
Ingression, 226
    avian embryo, 356–358
    fish gastrulation, 350
    primary mesenchyme, 233, *235*
Inheritance
    early debates on, 83–84
    genomic imprinting, 123–124
    germ plasm theory, 59–61
    preformationist viewpoint, 4n
Inhibition gradients, hydra, 598–600
Initiation factors. *See* Translation initiation
    factors

Initiator tRNA, 134
Inner cell mass (ICM), 367, 369
    dorsal-ventral axis, 380–381
    mammals, 76
    stem cells, 374
Inositol 1,4,5-trisphosphate (IP3), 206–207
Insect metamorphosis, 583–584
    hormonal control, 588–592, *589, 591*
    imaginal tissues, 584–588, *586, 587*
Insects. *See also* Butterflies; *Drosophila*
    axonal migration, *446, 450, 450*
    cleavage, 224
    germ cell determination, 615–616
    leg number, 757, *758*
    oogenesis, 637–638
    parthenogenesis, 632
Instars, 584
Instructive interaction, 145–146
Insulators, gene transcription, *122*, 123–124
Insulin gene, 116–118
Insulin-like growth factors (IGFs), 336
Insulin-like growth factor I (IGF-I),
        541–542, 604–605
Insulin-like growth factor II (Igf2), 123–124
Insulin secreting cells, 514–515
Insulin signaling pathway, longevity,
        604–605, *605*
integrated gene, 152
α4β1 Integrin, trunk neural crest, 431
α5β1 Integrin, muscle cell formation, 474
Integrins, 169–171
    mesodermal migration, 317
    muscle cell fusion, 474
    retinal ganglion axons, 454
    trunk neural crest, 431
Integuments, 655
Intercalary meristems, 667
Intercalation, 313–316, *316*
Interdigital mesoderm, 536
Interdigital necrotic zone, 539
Interior necrotic zone, 539
Interkinesis, meiosis, **626**
Interleukin 1 receptor, 295
Interleukin 2, 173
Interleukin-3 (IL-3), 509
Intermediate mesoderm, 465
    kidney formation, 477–485
    limb development, 531
    specification, 478
Intermediate spermatogonium, 629
Intermediate zone, 403, *404, 407*
Internal granule layer, 404
Internal YSL, 348
Interneurons, 401–402, 414, 447–449, *450*
Internode, plants, **668**
intersex (ix) gene, 564
Interspecific incompatibility, 655
Interstitial nuclei of the anterior hypothala-
        mus (INAH), 560

Intervertebral discs, 473
Intestinal development, 511
    C. elegans, 255, *256*
    molluscs, 244
    sea urchin, 236
    snail, 241
Intine, 654
Intracytoplasmic sperm injection (ICSI),
        208, **685**
Intraembryonic vasculogenesis, 501
Intramembranous ossification, *437, 437,*
        475
Intraocular pressure, cornea, 414
Intraspecific incompatibility, 656
Introns, 108–110, *109, 110*
Invagination, 226
    amphibian gastrulation, 309, 312–313
    Drosophila blastoderm, 264–265
    sea urchin archenteron, 236–239
    tunicates, 250, *251*
Inversion, 36, *37*
inversion of embryonic turning (inv) gene,
        381, 708
Involuting marginal zone (IMZ), 313, 315,
        316
Involution, 226
    amphibian gastrulation, 309, 312–315
    fish gastrulation, 350, *351*
    tunicates, 250, *251*
Ionizing radiation, 702–703
IP3. *See* Inositol 1,4,5-trisphosphate
Iris, 415
Islet proteins, 443
Islets of Langerhans, 514
Isogamy, 38
Isolation experiments, 61
Isolecithal eggs, 224, *225*
Isometric growth, 16
Ito (fat-storing) cells, 601
Ituri Forest pygmies, 542
IVF. *See* In vitro fertilization

———————————————

J segment, immunoglobulin genes, 90–91
Jacob, François, 769
Jagged protein, 166
JAK protein, 159–160
JAK-STAT pathway, 159–161, 164
Jaws, 438
    co-option in evolution, 768, 768n, 769
    cranial neural crest, 434, 435
    embryonic homology, 13–14
    evolution in vertebrates, 771–773, *773*
Jefferson salamander, 203n
Jellyfish, *45, 46*
Jervine, 163n, 695
jimpy mouse mutant, 412
Joints, formation, 539–540
Just, Ernest E., 83, *84,* 203, 290
Juvenile hormone (JH), 589–591, 590n
    nutritional polyphenism, 730, *731*

Juvenile phase, plants, **672**
Juxtacrine interactions, 149
    extracellular matrix, 168–171
    gap junctions, 171–172
    Notch pathway, **166**

Kairomones, 735
Kallmann syndrome, 439
KANADI (KAN) gene, 669
Kangaroo rats, cheek pouches, 767–768
Kangaroo, tammar, 727
Kant, Immanuel, 5
Kartagener triad syndrome, 185–186
Karyokinesis, 223–224
Keratella clacki, 735
Keratin 14, 416, *417,* 418–419
keratin 14 gene, 419
Keratinocyte growth factor (KGF), 417–418
Keratinocytes, 57, **416,** *417*
5α-Ketosteroid reductase, 558
keule mutant, plants, 663
Kidney(s)
    developmental stages, 479–480, *479*
    intermediate mesoderm, 477–478
    mesonephric kidney, 551*n*
    reciprocal induction, 480–485, *480, 481*
    snail larva, 241
    vertebrates, *479, 480*
    Wnt4 proteins, *152*
Kinesin protein, 272, 638
King, Thomas, 85
Kit gene, 15, 14, *15*
Kit protein, 15, **158, 159,** 690
klotho gene, 603
knirps (kni) gene, 279, 281, *282*
Knirps protein, 282
Knockout mice, 100–101, *102, 103,* 375–379
    conditional knockouts, 118, *119*
    floxed technique, 118
    kidney development, 481, 485
    Wnt7a gene, 537–538, *538*
KNOTTED 1 (KN1) gene, 670
KNOX genes, 670, *671*
Koller's sickle, 355, *356,* 361
Kolliker, A. von, 184
Kölreuter, Joseph, 4*n*
Krox20 gene, 335
Krüppel (Kr) gene, 102, *103,* 278–281, *282*
Krüppel protein, 161, 282
Kupffer macrophages, 601

L1 gene, 120, *121,* 697
L1/L2/L3 layers, apical meristems, 666
L1 protein, 120, 454
L4 mutant, hydra, 600
labial (lab) gene, 286, 377
Lactoferrin, 170
lacZ gene, 112, 118, 120, 282
LAG-2 protein, 167, 627

Lake Apopka, 740, 742, *743*
Laminar identity, cerebrum, *407*
Laminin, 146, **168–170,** 368, *619*
    axonal guidance, 445, 448
    neuron survival, 453
    retinal ganglion axon growth, 454
    synapse development, 451
    trunk neural crest, 431
Lampbrush chromosomes, 635
Lampsilis ventricosa, 242
Lanugo, 418
Large intestine, 511, 512
Larva, 26
    caterpillar, 575
    evolutionary embryology, 12–13
    insect, 584
    metamorphosis, 575
    nauplius, *12*
    pluteus, 62, *233, 234,* 575
Larval settlement, 725–726, *726*
Laryngotracheal groove, 515
Lateral geniculate nucleus, 407, 454, 737
Lateral meristems, 667
Lateral motor column (LMC), 443
Lateral plate mesoderm, 465, 491–492
    avian, 363, *365*
    blood cell formation, 505–510
    blood vessel formation, 500–505
    coelomic cavities, 492
    heart development, 492–510
    limb musculature, 525, *526*
    myogenesis, 473
    right-left asymmetry, 383
    tissue derivatives, *493*
Le Douarin, Nicole, 11
Leader sequence, *109,* **110**
Leaf, initiation, 663
Leaf development, 326, *671*
    dorsal-ventral pattern, 669
Leaf morphogenesis, genetics, 669–671
Leaf primordia, 668
LEAFY (LFY) gene, 670, 675
LEAFY COTYLEDON (LEC1) genes, 663
Leeuwenhoek, Anton van, 184
LEF DNA-binding protein, 161
Left-right axis. *See* Right-left axis
lefty-2 gene, 363, *364*
Lefty proteins, *364,* 383
Leibniz, G. W., 773*n*
Leidy, Joseph, 509*n*
Lens cells, *57*
    differentiation, 414–415, *416*
    eye development, 144–145, *147*
    paracrine induction, *150*
Lens placode, *413,* 414
Lens vesicle, 415, *416*
Leopard frog, 26, 86–87
leopard gene, 19, *20*
Lepidochelys, 568

Leptin, 505, 639
Leptotene stage, meiosis, **625**
LET-23 receptor tyrosine kinase, 157
lethal of scute (l'sc) gene, 297
Leukemia inhibition factor (LIF), 433, 482, *483*
Level-interactive modular array, 766
Leydig cells, *57,* **551,** 554
LH. *See* Luteinizing hormone
LHX9 gene, *549*
LIF. *See* Leukemia inhibition factor
Life cycles
    aphids, *726*
    Dictyostelium, 39–40
    frog, 26–30, *27*
    human, 30
    plants, 651–654
    Volvox, 36–38
Life expectancy, 602–605, *605*
Life span, 601–602
Ligands, 154
Light, germination, 664, 665
Light chains, immunoglobulins, 90–91
Lillie, Frank R., 5, 83, *84,* 195, 242, 641, 752
Lim-1 protein, 375, 481–483
Lim **genes, 443**
LIM proteins, 443
Limb development, 523–524
    AER, 529–530, *529*–531
    anterior-posterior axis, 534–537
    coordination among axes, 537, *538*
    deep homologies, 764–765, *765*
    developmental constraints, 774–775
    digit and joint formation, 538–540
    dorsal-ventral axis, 537, *538*
    epiphyseal growth plates, 540–542
    forelimb/hindlimb specification, 526–529
    Hedgehog pathway, 162
    limb bud, 523–529, **525,** *526, 527*
    proximal-distal axis, 529–534
Limb disc, salamander, 524–525
Limb field, 524–525
limbless Chick mutant, 530
lin genes, 134, *135,* 167
LIN proteins, 134, 157–158, 166, 167
Lineage-restricted stem cells, 509
Liver
    formation, 512–514
    hematopoietic stem cells, 507
    regeneration, 601
Liver disease, gene therapy, 707
Liver-specific genes, 118
Lmx1 gene, 537, 764
Location-dependent sex determination, 568–569
Loeb, Jacques, 195
Longevity, 602–605, *605*
Lophotrochozoa, *45,* **46,** 752
Loss-of-function evidence, 42

loxP gene, 118, *119*
Lunatic fringe (Lfng) gene, *469*
Lunatic fringe protein, 467–468
Lungs, 510–511, 515–516
Luteal phase, 641
Luteinizing hormone, 559, 638, 640–641
Luteotropin, 641
Lymphocytes
 apoptosis, *164*, 166
 differentiation, 173, 295
 immunoglobulin synthesis, 90–91
Lytechinus, 228, 234, 237, 238

Maarfan syndrome, 691
McClendon, J. F., 63
macho-1 mRNA, 248
Macho-1 transcription factor, **248**–249
Macroarray, RNA localization, **96**
Macroclemys, 567
Macroevolution, 761, 778
Macromeres
 amphibian, 306
 mollusc, 239–241, 244–245
 sea urchin, 227–228
Macropus eugenii, 727
Mad protein, 159
MADS box genes, 650, **676**, 677
L-Maf transcription factor, 145, 150, *151*
Mahler, Alma, 16*n*
Maimonides, 143*n*
Maize
 clonal fate map, 671, *672*
 embryogenesis, 661
 gametophyte development, 655
Male gametophytes, 654
Male infertility, 209*n*
Male mammalian sex determination,
  551–554, *556*, 557–559
Male pattern baldness, 418
Male phenotype, mammalian, 549–552, *558*
Male pronucleus, 209
Male specific RNA, 131–132
Malebranch, Nicolas, 4
Malformations, 14, 687, 691
Malleus bone, 14, 435
Malpighi, Marcello, 4, *5*
Malpighian layer, 416, *417*
Mammalian development
 aging, 602–605
 apoptosis pathways, 164–166
 axis formation, 375–380
 cleavage, 363–368
 cloning, 87–90
 composite, 374, *374*
 cortical granule reaction, 201–202
 cranial neural crest cells, *436*
 cutaneous appendages, 418–419
 developmental symbioses, 724–725, *725*
 difficulty of study, 363–364

evolution of molars, 770–771, *772*
extraembryonic membranes, 370–372,
  517
fetal to newborn circulation, 498–499
gametes and fertilization, 192–198,
  201–202, 207–210, *211*
gastrulation, 368–374
germ cell migration, 618–621, *619*
implantation in uterus, 76
infertility, human, 683–686
jaw homology, 13–14
modifications for development within another organism, 368–370
neurulation, 393, 393–398
nonequivalence of pronuclei, 212
oogenesis, 638–641
regeneration, 593*n*, 601
respiratory tube, 515
sperm capacitation, *193*
spermatogenesis, 628–631
stem cell descendents, *508*
transgenic animals, 89, *90*, 100–101, *102*,
  103
twinning, 373–374
uterine implantation, 366, *366*, 368
X chromosome inactivation, 124–126
Mammalian sex determination
 behavior and, 559–561
 gonad development, 549–556, *550*, *552*
 primary determination, 548–549, 551–556
 secondary determination, 549, 556–561
Mammalian visual system, 737–738, *739*
Mammary gland cells, 170–171
manacle gene, 600
Manduca (Moth), 592
Mangold, Hilde, 320, 320*n*
Mantises, 16
Mantle, snail, 241
Mantle zone, 403, *404*, *405*
Marginal zone, 309, 403, *404*, *405*
Marginal zone cells, 312–313, **354–356**
Marine goby, 732
mariner gene, 346
Marsupial frogs, 583
Marsupium, 242
Maskin protein, mRNA selective transla-
  tion, **133**
Maternal effect genes, 269–270, *272*
 Bicoid gradient, 273–276
 dorsal-ventral polarity, 291
 Drosophila, 272, *274*
 gap genes, 279–281
 molecular model, *269*, 270–273, *270*, *271*
 Nanos gradient, 276–277
 plants, 662
 terminal gene group, 277–278
Maternal effects on adults, 703
Mathematical modeling, 3, 16–20
Mating types, protist, 33
Mbx transcription factor, 117

Meckel's cartilage, 13, 14, 577, 578
MeCP2 protein, 122
med genes, 255
MED transcription factors, 255
MEDEA gene, 662
Medial geniculate nucleus, 407
Medial hinge point (MHP) cells, 396
Medial motor column, 443
Medial structures, avian embryo, 356
Mediator complex, 112
Medical embryology, 14–16
Medical genetics, 691*n*
Medicine
 cancer therapies, 703–705
 gene therapy, 705–708
 genetic anomalies, 686–694
 infertility, 683–686
 regeneration therapy, 712–713
 stem cells, 708–711
 teratogens, 694–703
 therapeutic cloning, 708–711
Medulla oblongata, 399, 403–404
Medullary cord, 398
Medulloblastoma, 703
MEF2 transcription factors, 494, 496
Megasporangium, 655
Megaspores, 655
Meiosis, *34–35,* **624–628**
 amphibian, 633–634
 oogenic, 631–632
 sexual reproduction, 33
 spermatogenic, 629–630
 sporic, 649
 timing in mammals, 551
Meiotic recombination complex, 626
meis genes, 533
MEK kinase, 154
Melanin, 413
 epidermal, 416
Melanoblasts, *158,* 418
Melanocytes, 11, *57,* 416, *417,* 418
 neural crest, 429, 432–433, 434
 RTK pathway, 158–159
Melanomas, 703
Melanosomes, 416, *417*
Meltrins, muscle cell fusion, **474**
Membrane potential
 fast block to polyspermy, 199–200, *201*
 resting potential, 200
 sodium ions, 199–200
 sperm capacitation, 193
Membranipora membranacea, 735
Memory cells, 736
Mencken, H. L., 477*n*
Mendel, 663*n*
Menidia menidia, 731–732, 734
Menstrual cycle, 638–641, *640*
Mercury, 700

MERISTEM LAYER 1 (AtML1) gene, 660–661
Meristems, 649, 665–667, *665*
    flowering, 674–676, *674–676*
Meroblastic cleavage, 3–4, 224, *225*, **347–348**
    avian, *355*
Meroistic oogenesis, 637–638
Mesectoderm, 296, 297
Mesencephalon, 398–400
Mesenchymal stem cells, 710–711, *711*
Mesenchyme, 70, 147
    archenteron invagination, 236–239
    epithelial-mesenchymal interactions, 147–148
    epithelialization during somitogenesis, 469–470
    head, 357, 465
    kidney formation, 479–480
    metanephrogenic, 480–485
    osteogenesis, 475
    primary, 228, 233–236
    progress zone, 529–531
    sea urchin gastrulation, 233–236
    secondary, 228, 237–239
    skeletogenic, **233**–236, *236*
    tooth development, 439–441
    tunicates, 247
Mesendoderm, 230, 491*n*
Mesentoblast, invertebrates, 241, **244**, 491*n*
Mesoderm, 6, 26
    amphibian, 308, 310, 313, 315, *315*–317
    avian, 356–358
    C. elegans gastrulation, 257
    dorsal, 326, 328, 362
    dorsalization, 328
    Drosophila, 266, 295–297
    endodermal interactions, 512
    extraembryonic, 371
    flatworms, 46
    formation of, 226
    frog and chick comparison, *492*
    frogs, 27
    induction by Nieuwkoop center, 321–322
    interdigital, 536
    intermediate, 465, 477–485
    lateral plate, 465
    major lineages of, 465, *466*
    mammalian, 369
    the organizer, 320–321, *326*
    origin, 27, 239
    paraxial, 351, 465
    prechordal plate, 310, 351, **465**
    Radiata, 46
    sea urchin, 229
    segmental plate, 466, 468–469, *471*
    selective affinity, 71
    somites, 27
    somitic dorsal, 465

    tunicates, 247
    unsegmented, 466, 468–469, *471*
Mesodermal mantle, 315
Mesodermal organs, 239
Mesodermal somites, avian, *357, 359*
Mesolecithal eggs, *225*
Mesomeres, sea urchin, **227–228**
Mesonephric cell migration, 553–554, *555*
Mesonephric kidney, 551*n*
Mesonephros, 479
Messenger RNA (mRNA)
    developmental morphogenesis, 31–32
    differential longevity, 132
    during egg activation, 208–209
    microarrays and macroarrays, 96–97
    oocytes, 132, *133*, 134–135, 187, 636, 638
    selective inhibition of, 132–134
    tissue specific, 92
Metalloproteinases, 483, 504, 509
Metamorphic climax, 580
Metamorphic molt, 584
Metamorphosis, 575–576
    amphibian, 30, 576–583, *582, 583*
    insects, 583–592
    morphological changes, 576–577
Metanephrogenic mesenchyme, 480–485
    apoptosis prevention, 482
    cell-line origins, 481
Metanephros, 480
    reciprocal induction, 480–485, *480, 481*
Metaphase block, 634
Metaphase plate, meiosis, **626**
Metaplasia, 87
Metazoans, 44–47
Metencephalon, 399
Methoprene, 744–745, *744*
Methoxychlor, 740
methuselah gene, 603
Methyl transferases, nucleosomes, 123–124
5-Methylcytosine, 121
Methylcytosine, 121
mex-3 gene, 255
MEX-3 protein, 255
Mice. *See also* Knockout mice
    alcohol damage, 697
    axis formation, 375–383
    basement membrane-directed gene expression, 170–171
    cerebellar mutations, 405
    cleavage, 363–367
    cloning, 88, 708
    connexins, 171–172
    Danforth short-tailed mutation, 481
    egg, *189*
    embryo escape from zone pellucida, 368
    epidermal mutations, 417–418
    estrogen sensitivity, 740–741, *741*
    eye development, 413, *413*–414, *414*
    fertilization, *190*

    gamete activation, 207–208
    gastrulation, 368–374
    genomic imprinting, 123
    germ cell migration, 618–621
    hair follicle mutations, 419–420
    heart formation, *497*
    Hox genes, 377–380, *381*
    joint formation, 540
    kidney development mutants, 481–483
    lens induction, *147, 150*
    microphthalmia, 115, 159
    neural crest mutants, 437, 441
    neurological mutations, 412
    neurulation, 393
    nonequivalence of pronuclei, 212
    Pax6 gene, 98
    piebaldism, 15
    reporter genes, 112–113
    sperm-binding protein, 195–197
    Sry gene, 551–552
    Steel mutation, 159
    transgenic, 100–101, *102*, 707–708
    White mutation, 159
    Wnt4 mutations, *152*
    X chromosome inactivation, 124
Micro-RNAs, 135
Microarrays, RNA localization, 96–97
Microevolution, 778
Microfilaments, 188, *223*, 224
    microspikes, 411–412
    transport to oocyte, 638
    trunk neural crest, 431
Microinjection, 99
Micromeres
    amphibian, 306
    mollusc, 239–241
    sea urchin, 227–230, *232*
    snail fate map, *241*
Microphthalmia, 688, *689*
microphthalmia (Mitf) gene, 159, 688–689
Microphthalmia transcription factor (MITF), 115–116, *115*, 158–159, 165, 689–690, 703
Microphthalmia transcription factor (*Mitf*) mRNA, *115*
Micropyle, 655
Microspikes, 411–412
Microsporangia, 654
Microspores, 654
Microtubules
    axons, 411–412
    egg cytoplasm reorganization, 212, *213*
    fish gastrulation, 349
    gamete fusion, 209, *210*
    mitosis and cell division, 223–224
    sperm flagellum, 184–186
    transport to oocyte, 638
Microvilli, 188, *198*, 201
Mid-blastula transition, 223

amphibian, 223, **310**–311
Drosophila, 265–266
fish, 348
gene transcription, 636
Midbrain, avian, 357
Middle ear, 14, 435
Midgut, 236
Mifepristone, 641
miles apart gene, mutation, **494**, *495*
Minamata Bay, 700
Mineralization, bone development, 476–477
Mississippi alligator, 52, *567*
MITF. *See* Microphthalmia transcription factor
Mitochondria
sperm flagellum, 186
transmission during fertilization, 209–210
Microarrays and macroarrays, RNA localization, 96–97
Mitochondrial damage, aging, 603
Mitochondrial ribosomal RNA (mtrRNA), 616
Mitosis, 221–224
Mitosis promoting factor (MPF), 222–223, 306, **633**–634
Mitotic spindle, 223
Model systems, 227
Modern synthesis, evolution, 777–778
Modified shoot hypothesis, 670
Modularity, evolutionary principle, 766
Modules, 766
Molecular biology
DNA-RNA hybridization, 92
northern blotting, 93, *94*
polymerase chain reaction, 93–94, *95*
situ hybridization, 97–98, *99*
Molluscs, 16, *45*, 46
autonomous specification, *57*
axis formation, 245
cell determination, 242–245
evolution of cleavage, 242
fate map, 244–245
gastrulation, 245, *246*
polar lobe, 242–245
spiral holoblastic cleavage, 239–245
Molt, insect metamorphosis, 584
MOM proteins, 256
Monkeys, 707, 737–738, *739*
Monocarpic plants, **678**
Monoecious, 655
Monophyletic groups, 752
Monospermy, 198
Monozygotic twins, 373–374
"Monster," *14n*
Moraba virgo, 632
Morgan, Thomas Hunt, 63–64, 81–83, 263, 726
Morphallaxis, hydras, **592**, 597–600
Morphogenesis, 2, 174
cell adhesion studies, 51–76

cellular processes, 70
Maimonides on, 143*n*
major questions, 70
protists, 31–32, 39–44
symbiosis, 724
Morphogenetic constraints, 774
Morphogenetic determinants, 56–57, 187
mollusc cytoplasm, 242–245
tunicate cytoplasm, 248–249
Morphogenetic fields, 67–68
Morphogenetic furrow, 155–156
Morphogenetic gradients, *Drosophila,* 270–278
Morphogens, 63
Drosophila head, *275*, 285
movement of, 285
Morpholino antisense oligomeres, 102–103
Morphs, 53, 728
Morula, 306, 366
Mosaic development, 56, *57,* 242–243, **319**
Mosaic pleiotropy, 689
Moss, life cycles, 651, *652*
Moths
environmental adaptation, 53, 728
larval stage, 26
Motility, of sperm, 184–186
Motor neurons, 401–402
activity-dependent synapse development, 451, *452*
axonal migration, 443
differential survival, 452
regeneration, 712–713
specification of, 442–443
spinal cord, *405*
Mouse. *See* Mice
Mouth
deuterostomes vs protostomes, 46
sea urchin, 239
snail, 241
MPF. *See* Mitosis promoting factor
mrf genes, 593
MRFs (Myogenic regulatory factors), 474
mRNA. *See* Messenger RNA
msx1 gene, 593
Msx1 protein, 332
Mucopolysaccharides, 189
Müller glial cells, 414, *415*
Müller, Johannes, 12
Müllerian duct, 549, 557–559
DES, 699, *700*
Müllerian inhibitory factor, 153
Multicellularity
Dictyostelium, 39–44, 43
Volvocaceans, 34–39
Multiple births, 373–374, 685–686, 743
Multipotent stem cells, **367,** 709–711
Murine development. *See* Mice
Murray's law, 501
Muscle(s). *See also* Myocardium; Myogenesis

cardiac muscle-specific proteins, 494
development of, 472–475
innervation, 451
snail, 241
somite origins, 471–472
Muscle precursors, limb, 525
Mutagenesis, zebrafish, 345
Mutations, aging syndromes, 603
Myc transcription factor, 704, *704*
Mycosporine amino acid pigments, 54
Myelencephalon, 399
Myelin-associate glycoprotein, 713
Myelin sheath, 412, 713
Myelination, *412*
Myf5 protein. *See* Myogenic transcription factor
Myoblasts, 471, 473–474, *475,* 525, *526*
Myocardium, *493, 494, ***495**
Myocytes, *57*
myoD gene, 474, 768
MyoD protein, 173, **473**–474
Myogenesis, 473–474, *475*
Myogenic bHLH proteins, 472, 473–**474**
Myogenic transcription factor (Myf5), *473,* 474
Myogenin, muscle cell formation, **474**
Myoplasm, tunicates, 247–249
α-Myosin heavy chains, 494
Myosin VIIA protein, 346
Myotome, 471, *472,* 473–474
Myotubes, 474, 593
Myxamoebae, 39

N-cadherin, 75–76, 397
endochondral ossification, 475
heart cell differentiation, 494–495
mesenchyme epithelialization, 469, *470*
retinal ganglion axons, 453
spermatogenetic germ cells, 628
synapse development, 451
trunk neural crest, 431
N-CAM
endochondral ossification, 475
retinal ganglion axons, 454
NADPH, egg activation, 205
Naegleria, 32
Nail-patella syndrome, 537
nanos (nos) gene, 133, 134, *272,* 276
Nanos gradient, maternal effect, 276–277
nanos mRNA, 134, *272, 273, 274,* 616–617
Nanos protein, 269, 272–273, **616**–617, 628
cytoplasmic localization, 135
posterior organizing center, 276–277
syncytial specification, 69
Nauplius larvae, *12*
Nautilus, 16, *17*
Nebenkern, 631
Necrosis, 164*n*
Necturus, 582

Negative controls, 67
Nematodes, *45, See also Caenorhabditis elegans*
  germ cells, 614–615
Nemoria arizonaria, 53, 728
Neocortex, 405
Neoteny
  amphibian metamorphosis, **581–582**
  heterochrony, 766
Neottiella, 625
Nephric duct, 479
Nephrons, 479–481, 484–485
Nerve growth factor (NGF), *434,* 450–452, *453*
Nervous system
  homeotic selector genes, 286, 288
  sex-specific patterns, 559
Nervous system pattern generation
  address selection, 451
  behavioral patterns, 457
  cell adhesion, 445
  differential neuronal cell death, 451, 453
  diffusible molecules, 446–450
  overview of, 444–445
  plasticity, 457
  retinal ganglion axons, 453–457
  specific repulsion, 446–450
  target selection, 443, **444,** 450–451
Netrin proteins, 446–449, *448*
Netrin receptor, 454
Neural cadherin. *See* N-cadherin
Neural cells, establishment of, 391–392
Neural crest, 11
  alcohol damage, 697
  cardiac, 441–442
  cell migration, 11, *12,* 396, 429–432, *430, 431,* 438–439
  cell types derived from, 427–428
  commitment, 433
  cranial, 434–442
  fish, 429*n*
  formation, *428*
  frogs, **27**
  functional domains, 428–429
  major derivatives of, *392*
  neurulation, 393, *394,* 395–396
  regions, *429*
  RTK pathway, 158
  specification, 427–429, 432–433, 436–437
  tooth development, 439–441, *440*
  trunk, 429–434
  vertebrate head, 761–762
  vertebrate jaws, 771–773, *773*
Neural development. *See also* Neural tube
  behavioral patterns, 457
  cell differentiation, 410–412
  CNS tissue architecture, 402–410
  Drosophila, 266

Hedgehog signal transduction pathway, 162
  homeodomain, 289
  induction, *327,* 328
  Notch pathway, 166–167
  pattern generation, 444–445
  specification of neurons, 442–443
  stages of, 442
  trunk neural crest, 432–434
  tunicates, 251
  vertebrate eye, 413–416
Neural differentiation, 296
Neural ectoderm, 295–296, *296,* 297
  specification in fish, 353
Neural folds, *29,* **393,** *394,* 395–396, 431
Neural genes, silencers, 120
Neural groove, 393, *394,* 395
Neural induction
  amphibian, 327–330, 332, 334–336, *335, 337*
  avian, *362, 363*
Neural keel, 351
Neural plate, 391, 431
  amphibian, 333
  folding and shaping, 393, *394,* 395–398
  neural crest formation, 427–428
Neural restrictive silencer element (NRSE), 120, *121*
Neural restrictive silencer factor (NRSF), 120
Neural retina, 413, *414*
Neural tube, **27,** *29,* **391**
  closure, 396–398
  defects, *396, 397*
  dermatome differentiation, 472, *473*
  emigration of trunk neural crest cells, 431
  fish, 352, 354
  formation of, 393–398
  major derivatives of, *392*
  neural tissue development, 402–410
  occlusion, *400*
  organizer in amphibians, 320–321
  Sonic hedgehog protein, 151
  tunicates, 247, 251
Neural tube differentiation, 398–402, *404*
  anterior-posterior axis, 398–400
  dorsal-ventral axis, 401–402
"Neuralizing" signaling, amphibian, 334
Neuregulin, trunk neural crest, 434
Neurenteric canal, 398
Neurexin genes, 130, 130*n*
Neurites, 411
Neuroblasts, 404–405
  determination of, *166,* **167**
  Drosophila, 297
  establishment of, 391–392
  retinal development, 414, *415*
Neurocranium, 438
NeuroD gene, *333*

NeuroD protein, 332, *333*
Neurogenin, 332, 333
Neuromeres, *Drosophila,* **754**
Neuronal growth rate, human brain, 408
Neuronal path finding, 412
Neuronal patterning, environmental, 737–738, *739*
Neurons, *57*
  alcohol damage, 697
  apoptosis, *164*
  birthday, 403, 442, 737
  central nervous system, 398, 401–403
  differential survival, 451–453, *453*
  differentiation, 410–412
  electrical impulses, 412–414
  glial guidance, 405, *406,* 407
  neural crest, 434
  regeneration, 594–595, 712–713
  retinal, 414
  specification of, 442–443
  stem cells, 409–410, *410,* 710–713
Neurotransmitters, axons, 412
Neurotrophin 3 (NT-3), 472, *473*
Neurotrophins, 153, 447*n,* **450–453,** *451*
Neurula, 27, *29,* **391**
Neurulation, 391–398, *399*
  avian gastrula, *357*
  bias toward, 333
  cadherins, 76
  mesodermal components, *466*
  primary, 391–398
  secondary, 393, 398, *399*
Newts. *See also* Salamanders
  axis determination, 317–319
  neuronal growth experiments, 444–445
  primary embryonic induction, 320–321
NF-κB protein, homology to Dorsal protein, 295
NGF. *See* Nerve growth factor
NGN3 protein, 515
Nicotiana, 673
Nicotine, teratogenesis, 695–696
Nieuwkoop center, *321,* **321–322**
  avian equivalent, 360
  fish, 352–353
  gene activation, 326–327
  induction, 322–325
Nieuwkoop, Pieter, 321
Nitric oxide (NO), 451, 457
NKx transcription factors, 402
Nkx2-5 transcription factor, heart development, **494–496**
NKX2-5 gene, 494*n,* 753–754
NMDA (*N*-methyl D-aspartate) receptor, 457
NO. *See* Nitric oxide
no tail gene, 352
Nocodazole, *224*
nodal gene, 337, 361, 363, *364, 365, 375*

Nodal protein, 153
  avian right-left axis formation, 363, *364*
  diffusibility, 149*n*
  Hensen's node, 360, 362
  mammalian axis formation, 375–376
  mammalian right-left asymmetry, 383
Nodal-related protein 3 (Xnr-3), 329
Nodal-related proteins
  fish, 353
  the organizer, 322, 325, *326*
Node, 369, 375
  BMP antagonists, *376*
  ciliary cells, 382–383
  equivalent of Hensen's node, 369*n*
  plants, **668**
Nodose placodes, 439
noggin gene, 328, 353, 376, 540, *541*
noggin mRNA, localization, 328, *328*
Noggin protein, 326, 332, 392
  cardiac specification, *493*
  Hensen's node, 362
  interdigital tissue death, 539
  limb development, 538–539, *541*
  mammalian axis formation, 375
  the organizer, 328
  somitic cell fates, 466, *467*, 473
Nogo-1, 713
Nondisjunction, humans, 627
Noninvoluting marginal zone (NIMZ), 315
Nonylphenol, 701
Norepinephrine, 412, 432
Northern blotting, 93, *94*
Notch-Delta pathway, 442
Notch genes, 166, *469*
Notch pathway
  blood vessel specification, 505
  sea urchin, 230
  somitogenesis, 467, *468*
Notch protein, 166, *168*
  C. elegans, 256–257
  somitogenesis, 467
Notochord, 6, 308, **310,** 315
  avian, 356, 357, *357*
  degeneration of, 473
  fish, 351–352, **351**–353
  frogs, 27
  intervertebral discs, 473
  mammalian, 369, 381
  murine, *371*
  neural tube ventral patterning, 402
  phlyletic constraint, 775
  sclerotome formation, 472
  sonic hedgehog protein, 151
  tunicates, 247, 251
Notophthalmus, 635
Nuclear equivalence. *See* Genomic equivalence
Nuclear hormone receptor family, DAX1, 556

Nuclear RNA (nRNA), 128–129, 132
Nuclear transport, *295*
Nuclei (neural), 402–404
Nuclei pulposi, 473
Nucleosomes, 107–**108**
  acetylation of, 116
  stabilization, 121
Nucleotide sequence, human β-globin gene, *109*
Nucleus
  control of developmental morphogenesis, 30–32
  egg, 187
  fusion during fertilization, 209–210, *211*
  potency studies, 85–87
  sperm, 184
  spermiogenesis, 631
Nudel protein, *292*, 294
"Numerator proteins", 562, *564*
Nuphar, 658, 659
Nurse cells, *274,* 276, **637**–638
Nüsslein-Volhard, Christiane, 271
Nutritional polyphenism, 730–731
Nymph, 584

Oct4 gene, 708, 711
Oct4 protein, 367, 620
Octopamine, axons, 412
Oculomotor blind (omb) gene, *587,* 588
odd-paired (opa) gene, *282*
odd-skipped (odd) gene, *282*
Odd-skipped protein, 285
Odontoblasts, 439
"Odontogenetic potential," 441
Olfactory epithelium, 410
Olfactory neurons, axonal guidance, 450
Olfactory placodes, 413
Oligodendrocyte-myelin glycoprotein, 713
Oligodendrocytes, 412, 713
Oligonucleotides, 93–94
  antisense, 248
Oligospermia, 685
Oliva porphyria, 19
Ommatidia, 155
Omphalomesenteric veins, 501
On the Generation of Living Creatures (Harvey), 4
On the Origin of Species (Darwin), 12
Oncogenes, *RAS* gene, 154
Ontogeny, 777
Oocyte, 186–**189,** *See also* Egg; Ovum
  amphibian, 632–633, *634*
  anterior-posterior axis, 270–272
  cadherins, *75*
  cloning studies, 85–86
  cytoplasmic contents, 187
  Dorsal protein translocation, *292*
  gene transcription, 634–636
  meiosis, 631–632

  microtubules, in *Drosophila,* 272
  mRNA in, *133,* 134–135, 635–636, 637–638
  mRNA stored, 132, *133*
  mRNA transported to, 637–638
  sperm chemotaxis, 189–191, 194
  in vitro fertilization, 684–685
  X chromosome reactivation, 126
Oocyte-Follicle messages, 293–294
Oogamy, 38
Oogenesis, *629,* **631**–641
  amphibian, 632–634
  C. elegans, 628
  gene transcription, 634–636
  insects, 637–638
  mammalian, 638–641
  meiosis, 631–632
  parthenogenetic, 632
Oogonia, 631
openbrain gene, 396
Opsin proteins, 157
Optic chiasm, retinal ganglion axons, 454
Optic cup, 145, *147,* **413**
Optic development, 413–416
Optic disc, retinal ganglion axons, 453
Optic nerve, 413, 453–454
Optic stalk, 413
Optic tectum, retinal ganglion axons growth, 453–457, *454*
Optic vesicles, 146, 399, *413,* 414
  lens induction, 144–145, *146*
Oral plate, 511
Organ identity genes, floral, **676**–678
Organization/activation hypothesis, 559
Organizer, 320–**321,** 325
  BMP inhibitors, 327–330
  chick, 361
  diffusible proteins, 327–332
  fish equivalent, 351, *352,* 352*n*
  functions of, 325–333
  hydra, 598–600, *599*
  induction, 322–326, *325*
  mammalian equivalent, 375
  model, *330*
  neural induction, 334
  Nieuwkoop center, 321–322
Organizer proteins, 326, *326,* 353
"Organizing centers," insect egg, 270
Organogenesis, 26
orthodenticle (otd) gene, 276, *282,* 377, 754
Orthopyxis caliculata, 190
oskar (osk) gene, *272*
oskar mRNA, *273,* 638
Oskar protein, 272, *273,* 277, **616**
Ossification
  endochondral, 475–**477**
  intramembranous, 475
Osteid matrix, 437
Osteoblasts, 57, 437, 477

Osteoclasts, **477**, *478*
Osteocytes, **437**
Osteogenesis, **474–477**, 722
Osteogenic sarcoma, *704*
Osteoid matrix, 477
Osteonectin, tooth development, 441
Osteopetrosis, **477**
Osteoporosis, **477**, 712
Ostrich, 733
Otic placode, induction, 438, *439*
Otx-2 gene, 315, 327, *335*, 754
   eye development, *414*
Otx-2 protein, *146*, 315, *326*, 375–376
Otx transcription factor, sea urchin, *232*
Our stolen future, 702
Outgrowth theory, axon formation, 411
Ovarian nurse cells, *Drosophila*, *274*, 276
Ovariole, **637**
Ovary
   development, 551, 554–556
   infertility treatment, 684
   plant, **655**
   primary sex determination, 548, 551,
      554–556
   secondary sex determination, 559
Oviduct, sperm translocation, 192–193
Oviparity, **3**
Ovotestis, *C. elegans*, 628
Ovoviviparity, **3**
Ovulation, 633, 638–641, *641*
Ovules, **655**
Ovum, **186–189**, 551, 631–632, *See also* Egg;
     Oocyte
   hermaphrodites, 627–628
Owen, Richard, 752
Oxidative damage, aging, 603
Oxygen reduction, egg activation, 205
Oysters, 725*n*

P-cadherin, **74**
P cells, *C. elegans*, 253–257
P elements, **99**–100
P-granules, **253**
p34 protein, **633**
p300/CBP Histone acetyltransferace protein,
    116, 159
p450 aromatase, **559**–560, 567–568
PABP. *See* Poly(A) tail binding protein
Pachytene stage, meiosis, **625**
Pacific tree frog, 54, *56*
   limb field, *525*
Pair-rule genes, **269**, **278**, 281–283
   homeotic selector genes, 286–288
   listed, *282*
paired (prd) gene, *282*
Paired protein, 285
pal-1 mRNA, 255
PAL-1 protein, **255**
Pallister-Hall syndrome, 162

Pancreas, 513–515
   transcription factors, 112, 116–117
Pancreas-specific genes, 118
Pancreatic islet cells, *57*
Pander, Christian, 6
Pandorina, 34, *35*
pangolin (pan) gene, *282*
par-1 gene, 104
PAR proteins, 253
Paracrine factors, **149–153**
   cell surface receptors, 154
   fibroblast growth factors, 150–151
   Hedgehog family, 151–152
   maintenance of differentiation, 173
   TGF-β superfamily, 153
   Wnt family, 152
Paracrine interaction, **149**
Paralogous group, **377**–379
Paramecia, 32
Parascaris aequorum, 614–615
Parasegments, **278**
Parasitic wasp, 724
Parasitism, developmental symbioses, 725
Parasympathetic ganglia, **429**
Parathyroid glands, 510–511
   cranial neural crest, 435
Parathyroid-related hormone, epiphysial
    growth plate, 542
Paraxial mesoderm, **465**
   amphibian, 313
   fish, **351**
   kidney induction, 478
   muscle formation, 473–474
   myogenesis, 473–474
   osteogenesis, 474–477
   somitogenesis, 466–472
Paraxial protocadherin, **313**, *314*
Paraxis gene, 469
Paraxis protein, 470, *470*
"Parazoans," 44
Parietal mesoderm, **491**
Parkinson disease, 410, **452**, 708
Parthenogenesis, **212**, **632**
Partial hepatectomy, **601**
Patau syndrome, 687
patched (ptc) gene, 162, *282*, 703
Patched protein, 161–163, 703
Patella (Mollusc), *57*
Pathway selection, axons, **444**
Pattern baldness, 418
Pattern formation, **523**
   anterior-posterior in *Drosophila*, *274*
   computer modeling, 18, *19*, *20*
   dorsal gradient, 294–297
   gene regulation in *Drosophila*, 298
   limb regeneration, 595–597
   Turing's reaction-diffusion model, 18–20
Patterning
   mammalian axis formation, 375
   in plants, 659, *662*, *663*, *667*, *668*

Patterning molecules, sea urchin cleavage,
    226
Pax1 protein
   cartilage formation, 472, *473*
   endochondral ossification, 475
Pax2 protein
   kidney induction, 478, 481–482
   retinal ganglion axons, 455
Pax3 gene, 396
Pax3 protein
   cardiac neural crest, 441
   myogenesis, 474
Pax6 gene, 93, 98, 688
   enhancers, 113–114, *113*, 119–120
   evolutionary conservation, 753–754, *753*
   eye development, 413
   mutations in rats, 144–145
   regulation of, 117
   targeted expression, 119–120, *120*
Pax6 protein, 93, 98, 402
   assays, 118
   eye development, 144, 413
Pax6 protein, transcription factor, **116–117**
Pax8 gene, 478
Pax8 protein, kidney induction, 478
Pax9 protein, trunk neural crest, 440
Pbx1 protein, 117
PCBs. *See* Polychlorinated biphenyls
PCR. *See* Polymerase chain reaction
PDA (protostome-deuterostome ancestor),
    753–754
PDGF. *See* Platelet-derived growth factor
Pdx proteins, 117, 118
pdx1 gene, 514, *514*
Pdx1 protein, 514, 515
Pea plant, life cycle, *653*
Pelle protein, *292*, 294, 295
Pelvetia compressa, 660
Penaeus (Shrimp), 12
Peregrine falcons, 740
Perfect flowers, **655**
Pericardial cavity, **492**, **494**
Pericycle cells, 667, *668*
Pericytes, vasculogenesis, **503**
Periderm, 348, 416
Periosteum, **437**, 477
Peripheral nervous system. *See also* Axon;
    Spinal cord
   myelination, *412*
Peritoneal cavity, **492**
Permissive interaction, **148**
Pesticides, 740, 743, 745
Petals, **654**
Petrosal placodes, 439
Pfeiffer syndrome, 541
PGCs. *See* Primordial germ cells
PH-20 protein, 197*n*
pH
   avian axis formation, 360
   egg activation, 208

PHA-4 transcription factor, 257
PHABULOSA (PHAB) gene, 669
Phantom midge, 735
Pharmaceutical companies, mammalian
    cloning, 89
Pharyngeal arches, 6, 428*n*, **511**, 759, 772
    derivatives, *435, 436, 436,* 438
Pharyngeal development, 254–255, 257
Pharyngeal ectoderm, 359
Pharyngeal endoderm, 326, 357
Pharyngeal endomesoderm, 313*n*, 330–332
Pharyngeal pouches, 428*n, 511,* **511**
Pharyngula, *775,* **775**
Pharynx, 510, 511
Phaseolus coccineus, 660, *661*
PHAVOLUTA (PHAV) gene, 669
Pheidole morrisi, 730, *731*
Phenotypes
    in cloning, 88–89
    developmental stability, 776
    individuality, 743
Phenotypic plasticity, 52–54, **727–736**
Phenotypic variability, 690–**691**
Phloem, 663
Phocomelia, 15, *16,* 691
**Phosphatidylinositol 4,5-bisphosphate
    (PIP2), 206**
Phosphoinositide signaling pathway, 657
Phospholipase C (PLC), 206–**208**
Phospholipids, gamete metabolism, 206
Photolyase, 54–55
Photoperiods, 673
Photoreceptor cells, 413–414, 416
Photoreceptors, 155–157
Phyletic constraints, 774–776
Phyllotaxy, 668
Phylogeny, 777
Phylotypic stage, 775
Phytochrome pigments, **673**
Phytoestrogens, 740
pie-1 gene, 255
PIE-1 protein, 255
Piebaldism, 14, *15*
Pigment cells, 416
    migration of precursors, 11
Pigmented retina, 413
Pigments, pattern formation, 18, *19*
PIP2. *See* Phosphatidylinositol 4,5-bisphos-
    phate
pipe gene, **294**
PISTILLATA (PI) gene, 676
Pisum, 653
Pituitary gland
    developmental genetics, 402
    dual origin, 511
    frogs, 26
    ovulation, 641
pitx2 gene, amphibian right-left axis, **337**
pitx2 gene
    avian, 363, *364, 365*

mammalian right-left asymmetry, 383
    human asymmetry, 363*n*
Pitx2 protein
    amphibian, **337**
    avian, 363, *364*
PKA protein, *163*
Placenta, 47*n*, **367,** 369
    angiogenesis factors, 505
    extraembryonic membranes, 371–372,
        517
    plant, 655
Placental cadherin, 75
Placodes, 413, 419–420, 438–439
Planar induction, ectodermal signaling, 333
Plant development
    animal development compared to,
        649–650
    dormancy, 664
    embryonic development, 659–664
    experimental studies, 659
    fertilization, 658–659
    gamete production, 654–658
    germination, 664–665
    life cycles, 651–654, *651–653*
    senescence, 678
    vegetative growth, 665–671
    vegetative-to-reproductive transition,
        671–678
Plant hormones
    abscisic acid, 664
    auxins, 663, 667
    gibberellins, 660, 664, 665, 675
Planthoppers, polyphenism, 727
Plants, phylogeny, 649, *650*
Plasma cells, 736
Plasma membrane
    egg, 187
    fast block to polyspermy, 199–200, *201*
    gamete fusion, 197–198
    slow block to polyspermy, 201–202
    sperm capacitation, 193
Plasminogen activator, 368, 641
Platelet-derived growth factor (PDGF), 504
Platyhelminthes, *45*
PLC. *See* Phospholipase C
Pleiotropy, 689–**690,** 775
Pleodorina, 34, *35*
Plethodon (Amphibian), *55*
Pleural cavity, 492
Pluripotency
    neural crest, 432–433, *433,* 436
    plants, 664
Pluripotent stem cells, 68, 708–711, *709*
    epidermal, 418
    hematopoietic, 506, 710–711
Pluripotentiality, mammalian, **367**
Pluteus larva, 62, *233, 234,* 575
Pmar1 **gene,** 230, *232*
PMZ. *See* Posterior marginal zone
Pocket gophers, 767–768

Podocytes, 480
Point-for-point specificity hypothesis, 455
Polar body, *189, 381,* 632
Polar coordinate models, 597
Polar granule component, 616
Polar granules, 615–616
Polar lobe, 242–**245**
Polar nuclei, 655
Polarity, morphogen gradients, 63–64
Pole cells, 615
    Drosophila, 264
Pole plasm, 615
Pollen, 651, *654,* **654**
Pollen tube, 657–658, *657*
Pollination, 655–658
Polly, transgenic sheep, 89
POLTERGEIST (POL) gene, 667
Poly(A) tail, *109, 110,* **110–111**
    mRNA selective translation, 134
    mRNA stability, 132
    Poly(A) tail binding protein (PABP), 134
Polyadenylation, 110
Polycarpic plants, **678**
Polychlorinated biphenyls (PCBs), 698, 740
    effect on sex determination, 568
    environmental effects, 740, 742
Polycomb genes, 757
Polycomb protein, 288
Polycomb gene group, plants, 662, 664
Polycystic disease, 483
Polycystins, 483
Polyinvagination islands, 356
Polymerase chain reaction (PCR), 93–**95**
Polyphenism, 727–736
Polyploidy, plants, 650
Polypodium, 652
Polyspermy, 198–199
    blocks to, 199–202
    cortical granule reaction, 202–203
Polytene chromosomes, 637
Polytrichium, 652
Pontia (Butterfly), *728*
pop-1 gene, 256
POP genes, plants, 658
Population genetics, evolutionary model,
    777–779
Porifera, 44–47, *45,* 752
Porphyropsin, amphibian metamorphosis,
    577
Positional gene cloning, 687–688
Positive controls, 67
Posterior marginal zone (PMZ), 360
Posterior necrotic zone, 539
Posterior neuropore, 396
Posterior-organizing center, 276–278
Posterior vegetal cytoplasm, 250
"Posteriorizing" signaling, amphibian, 334
Potassium ions, fast block to polyspermy,
    199–200
Prader-Willi syndrome, 123

Pre-messenger RNA (pre-mRNA), 128
Prechordal plate, 313
Prechordal plate mesoderm, 310, 465, 473
  avian, 357
  fish embryonic shield, 351
Precis, Hox gene expression, 756–757, *757*
Precursor cells, 68
Predator-induced defense, 734–736
Predator-induced polyphenism, 734–736
Preformation, 4–5
Pregnancy
  teratogens, 15–16
  tubal, 368
Preimplantation genetics, 693–694
Premetamorphosis, amphibian, 579
Prenatal diagnosis, 693
Presenilin-1 protease, 166
Presomitic mesoderm. *See* Unsegmented
  mesoderm
Pretubular condensate, 484
Previtellogenesis, 632
Prevost, J. L., 184
Priapulida, 45
Primary capillary plexus, 501, 503, 504
Primary embryonic induction, 320–321,
  321*n*
Primary hypoblast, avian, 356
Primary mesenchyme, 228, *232,* **233**–236
Primary neurulation, 391–398
Primary oocytes, 631
Primary pair-rule genes, listed, *282*
Primary sex determination, 548–549,
  551–556
Primary spermatocytes, 629–630, *630*
Primers, 93–94, *95*
Primitive endoderm, 369
Primitive groove, 356
Primitive gut, 236, 511
Primitive knot, 356
Primitive pit, 356
Primitive streak, 356–358, *357, 358, 359,*
  360, 361
  mammalian, 369–370, 375
Primordial germ cells (PGCs), 613
  determination, 613–617
  hermaphrodites, 627–628
  meiosis, 624
  migration, 617–624
  spermatogenesis, 628–629
Prism-stage larva, *233*
proboscipedia (pb) gene, 286, 377
Procambium, 663
Proerythroblast, 509
Progenesis, amphibian metamorphosis, 581,
  582
Progenitor cells, 68
Progeria syndrome, 603, *604*
Progesterone, 641
  amphibian oogenesis, 633–634
  frogs, 26

Progress zone, 529–532
Progress zone model, 531–532
Progressive commitment, *Dictyostelium, 43*
Progressive determination, amphibian axes,
  317–319
Prolactin, 132, 160, 171
Proliferative phase, 640
Prometamorphosis, 580
Prometheus, 601
Promoter regions, *Drosophila, 281*
Promoters, 108, **111–112,** 122
Pronephric duct, 480
Pronephros, 478, 480, 775
Pronucleus, 26
  genetic fusion, 209–210, *211*
  nonequivalence of male and female, 212
Pronymph stage, 583
Prophase I, meiosis, 625–626
Prosencephalon, 398
Prostaglandins, 641
Protamines, 631
Protein drugs, mammalian cloning, 89, *90*
Protein gradients
  control of gene expression, *275, 276,* 285
  dorsal patterning, 294–297
  germ cell migration, 620*n*
  the organizer, 325
Protein kinase A, 272, *273*
Protein kinase C, 206
Protein phosphorylation, sperm capacita-
  tion, 193
**Protein-protein interaction domain,
  114**–115
Protein synthesis
  during egg activation, 208
  frog oocyte, 636
Proteins. *See also specific proteins*
  differential modification, 135–136
  numbers of, 130, *131*
  oocyte cytoplasm, 187
Proteoglycans, 168
  kidney induction, 481–482
  retinal axon growth, 454
  trunk neural crest, 431
Proteome, 130, 692
Proteus anguinus, 774
Prothoracic gland, 589
**Prothoracicotropic hormone (PTTH),
  589**–590
Protists, differentiation, 32
Protocadherins, 75, *314*
Protoderm, 663
Protostomes, 45, 46, 752
Proximal-distal axis
  homology, 764, *765*
  insect wing, 588
  limb development, **523**–524, 529–534
Pseudohermaphroditism, 557*n*
Psoriasis, 417

Psuedoplasmodium, 39
Pterygotequadrate cartilage, 760
Puberty, growth spurt, 542
Pulmonary artery, 441
Pumilio protein, 272, 277
  germ line switch, 628
Punctuated equilibrium, 778
Pupa, 584
Purkinje neurons, 404–405
Pycnoscelus surinamensis, 632
Pygmies, 542
Python, 759

Quadrate bone, 13
Quail
  blood cell formation, *507*
  cell fate mapping, 11
Quinine, teratogenesis, 695

R-Cadherin, retinal ganglion axons, 453
Rabbit, ovulation, *641*
Radial cleavage, 224, *225*
Radial patterning, plants, 659, *662, 663,* 667,
  668
Radiata, *45,* 46
Radicle (primary embryonic root), **664**
Radioactive labeling, 9–10
Raf kinase, 154
Rag2 gene, 711, *711*
Ramón y Cajal, Santiago, 411, 447
Rana. See also Amphibians; Frogs; *Xenopus*
Rana cascadae, 55–56
Rana catesbiana, 26
Rana clamitans, 744
Rana pipiens, 210, *211*
  life cycle, 26, *27*
  nuclear potency studies, 86–87
  rate of cell division, *222*
Rana sylvetica, 735
Rana temporaria, 526
Rana variegatus, 55
Ras G protein, 154, 171
Rat
  axonal guidance, *448, 451*
  dioxin, 740
  eye development, 144–145
  neuronal patterning, 737
  RNA splicing, *130*
Rathke, Heinrich, 6
Rathke's pouch, 6, 511
Reaction-diffusion model
  limb morphology, 774
  tooth evolution, 770–771
  Turing's, **18**–20
Reaction norm, 54, 727–728
Reactive oxygen species (ROS), aging,
  603–605
Realistor genes, 288–289
reaper gene, 592

Réaumur, regeneration experiments, 592
Receptor gradients, 65
Receptor tyrosine kinase pathway (RTK),
  154–159
  cell-to-cell induction, 155–158
  crosstalk, 173
  diagram, *155*
  integrins, 171
Reciprocal induction, 145
  kidney development, 480–485, *480, 481*
Recognition
  protists, 33
  sperm-egg, 189–197
Recombinases, 91
Recombination, of immunoglobulin genes,
  91
Recombination experiments, 61
Red blood cells. *See* Erythrocytes
Reductionism, 62*n*
reeler mouse mutant, 405
regA gene, 36–37
Regeneration, 575, 592
  annelids, 597
  bone, 712
  flatworms, 63–64
  hydras, 597–600
  mammalian liver, 601
  neurons, 712–713
  retinoic acid, 525, *526*
  salamander limbs, 592–597
Regeneration blastema, 593–597
Regeneration therapy, 712–713
Regional specificity
  amphibian metamorphosis, 580–581, *581*
  of induction, 334–337, *334, 335*
Regulation, 59
  cell cycle during cleavage, 222–223
  Dictyostelium, **40**
  inflorescence, *675*
  production of chimeras, 374
Regulative development, 59, 319
Regulatory genes, developmental genetic
  model, 778
Relational pleiotropy, 689–**690**
Relay amplification cascade, 67
Renal collecting ducts, 481
Repair enzymes, for UV damage, 54–55
Repiratory tube, 510
Reporter genes, 112–**113,** 118–119
Repression, gene transcription, 121–122
Reproduction, 2, 32–**33**
Reproductive cloning, 709
Reproductive transition, plants, 671–678,
  673, 674
Reptiles
  cleavage, 224
  environmental sex determination, 52,
    567–568, 731
  extraembryonic membranes, 517

germ cell migration, 617
  jaw homology, 13–14
Repulsion/attraction, axonal specificity,
  445–446, 454–456
Resact, 190–**191**
Residual body, sperm, 631
Respiratory system, 491
Respiratory tube, 515–516
Resting membrane potential, 200
Ret protein, 481, 485
Rete testis, 550
Reticulocyte, 509
Retina
  development, 145, 399, 413, 414, *415*
  differential cell adhesion, *73*
  photoreceptor induction, 156
Retinal ganglion axons
  adhesive specificities, tectum, 455–456
  target selection, 454–455
  through the optic chiasm, 454
  to the optic nerve, 453–457, *454*
13-*cis*-Retinoic Acid, 696
Retinoic acid
  acute promyelocytic leukemia, 704
  anterior-posterior axis, 336
  limb field specification, 525, *526*
  limb regeneration, 596, *596*
  retinal ganglion axons, 455
  teratogenic effects, 379, *380*, 696–697,
    744–745
Retinoid receptor (RXR), 579
Retinotectal projections, 454–455, *455*
Retroviral vectors, 99
Reverse transcriptase (RT) enzyme, 93
Reverse transcriptase-polymerase chain re-
    action (RT-PCR), 93–**94,** *95*
Rheobatrachus (Frog), 583
RhoB protein, 431
Rhodopsin, 577
Rhombencephalon, 398–400
rhomboid gene, 296
Rhombomeres, 399–400, 435
Ribosomal RNA (rRNA), oocytes, 187, 636
Ribs, 471, 475
Ridley turtle, 568
Rieger syndrome, 363*n*
Right-left axis, 227
  amphibian, 337
  avian, 363, *364, 365*
  C. elegans, 254
  mammalian, 381–383
  primitive streak defines, 356
  sea urchin, 231
  tunicates, 250
Ring canals, 637
RNA (ribonucleic acid)
  antisense, **101**–**104**
  differential processing, 127–132
  localization techniques, 93–98

northern blotting, *94*
  see also Heterogeneous nuclear RNA;
    Messenger RNA; Nuclear RNA;
    Ribosomal RNA; Small nuclear
    RNA; Small regulatory RNA;
    Transfer RNA
RNA interference (RNAi), 103–**104,** 704
RNA polymerases, 111–112
RNA processing
  Drosophila sex determination, 562–567,
    564, 565
  inborn errors, 692–693
  plant embryogenesis, 661–662
RNA splicing, 128–132
  genetic errors, 692–693
Robo (Roundabout) protein, *450,* **450**
Rod cells, 414, *416*
Roof plate, 402
Root apical meristem, 663, *666, 666*
Root cap, 666
Root development, 667
ROS. *See* Reactive oxygen species
Rostral-caudal direction, neural plate cells,
  395
Rostral cerebellar malformation, 450
Rotation. *See* Cortical rotation
Rotational cleavage, 224, *225*
  holoblastic, 251, 253
  mammals, 363–368
Rotifers, *45,* 734–735
rough (ro) gene, *157*
Roundworm, 184
Roux, Wilhelm, 51, 61, 63, 117, 751
RT enzyme. *See* Reverse transcriptase en-
    zyme
RT-PCR. *See* Reverse transcriptase-poly-
    merase chain reaction
RTK. *See* Receptor tyrosine kinase pathway
RTK-kinase cascade, 277
RU486, 641*n*
Rubella, 694
Rules of evidence, 42, 67
runt (run) gene, *282*
Runt protein, 285
Runx2 protein, 477
Rx1 protein, 413

S gene, plant self-incompatibility, 656
Sacral neural crest, 429
Saint-Hilaire, Etienne Geoffroy, 331, 752
Salamanders. *See also Ambystoma;*
    Amphibians; Newts; *Triturus*
  allometry, *767*
  bottle cells, 312
  cleavage, 305, 308*n*
  deformities, 744
  dorsal marginal zone, *313*
  fertilization, 203*n*
  mesodermal migration, 316
  metamorphosis, 576–577

morphogenetic constraint, 774, *774*
pharyngeal arches, *6*
prospective forelimb, *524*
regeneration of limbs, 592–597
reproduction, 203*n*
specification of induction, 334
Salivary glands, *Drosophila*, 297, *297*
Sander, Klaus, 270
Scaphiopus couchii, 732–734, *733*
SCARECROS (SCR) gene, 667, *668*
Scarification (seeds), **664**
Scarlet runner beans, 660, *661*
Scatter factor, 357, 601
Schizocoelous body cavity, 46
Schultz, Jack, 263
Schwann cells, 412, 713
Schwann, Theodor, 411
SCL transcription factor, 509
Scleral bones, 414
Scleraxis mRNA, *476*
Scleraxis protein, 475
Sclerotomes, 428, *432*, *471*, 472, *472*
SCR. *See* Cystein-rich peptide
scute (sc) gene, 297
SDF1/SDF2. *See* Spore differentiation factors
Sea squirts. *See* Tunicates
Sea urchin fertilization, 184, *205*
 acrosome reaction, 191–192, *192*
 cortical granule reaction, 201–202, 202
 egg, *187*, *188*, *200*, 203–204, *207*, 208–209
 gamete fusion, 197–199
 genetic fusion, 209, *210*
 plasma membrane fusion, *190*, *191*
 prevention of polyspermy, 198–202
 species-specific recognition, 194–195
 sperm chemotaxis, 190–191
Sea urchins
 archenteron invagination, 236–239
 cleavage, *223*, *224*, 227–233
 development, *233*
 differential RNA processing, 128, *129*
 fate map, 228–233
 gastrulation, 233–239
 specification, 61–63, 228–233, *230*
 ultraviolet radiation, 54
Seasonality, 726–727
Sebaceous glands, 418
Sebum, 418
Second polar body, 631
Secondary hypoblast, 356
Secondary mesenchyme, 228, *230*, 237–239
Secondary neurulation, 393, 398, *399*
Secondary oocytes, 632
Secondary pair-rule genes, *282*
Secondary sex determination, 549, 556–561
Secondary spermatocytes, 630
Securin, 209
Seed coat, 655
Seeds, 655, 664

Segment polarity genes, 269–270, **279**, *282*, 283–285
Segmental plate mesoderm, 466, 468–469, *471*
Segmentation. *See also* Somitogenesis
 determination, 278–279
 Drosophila, 266–268, 278–285, *279*
 information theory, 278*n*
Segmentation genes, 278–279
 Drosophila, *282*
 gap genes, 269, 278–281
 pair-rule genes, 269, **278**, 281–283
 segment polarity genes, 270, 279, 283–285
Selective affinity, 70–71
"Self," 743
Self-incompatibility, *656*, *656*
Semaphorin proteins, 445–446, *447*
Seminiferous tubules, 551, *629*, 631
Senescence, 601–605
Senescence (Plants), 678
Sensory apparatus, 438
Sensory neurons, 401–402
Sensory organs, formation, 413
SEPALLATA (SEP) genes, *677*
Sepals, 654
Septa, heart formation, 496, 500
Serine proteases, 294
Serine-threonine kinase (SRK), 656, 667
Serotonin, axons, 412
Serrate protein, 166
Sertoli cells, 174, **551**, 553, 554, 628–630
Sev protein, 156
sevenless (sev) gene, **156**, *157*
Sevenless-in-Absentia protein, 156–157
Sex, 32–33
 Dictyostelium, 40
 Volvox, 37–38
sex combs reduced (scr) gene, 286, 297
Sex combs reduced (Scr) protein, 290
Sex cords, 550–551, 628
Sex determination
 chromosomal, 548–549
 Drosophila, 131–132, 561–567
 environmental, 52, 567–569, 730–731
 germ cells, 627–628
 historical perspective, 547–548
 location-dependent, 568–569
 mammalian, 548–561, *549*
Sex-determination genes, 567
Sex hormones, growth spurt, 542
Sex-lethal (Sxl) gene, 131, **561–563**, *564*, *565*
Sex selection, 694
Sexual behavior, 559–561
Sexual dimorphism, 52
Sexual inducer protein, 38
Sexual phenotypes, mammalian, 549–552, *558*
Sexual reproduction
 meiosis, 625

types of, 38
unicellular protists, 32–33
 Volvox, 36–38
SF1/Sf1 gene, *549*, 554, 556
SF2 protein, 129
Sheep
 cloning, 87–89
 transgenic, 89, *90*
Sheep cyclopia syndrome, 163
Shell gland, chick, 360
Sherlock Holmes, 67
shh gene. *See sonic hedgehog* gene
shin guard gene, 600
Shoot apical meristem, 663, 666–667, *666*
Shoot development, 667–668
Shoot tips, germination, 665
shootmeristemless (STM) mutant, 663, 667
short gastrulation (sog) gene, *296*, *296*
Short-gastrulation (SOG) protein, 330, 763
SHORT INTEGUMENTS 1 (SIN1) gene, 662
SHORT-ROOT (SHR) gene, 667, *668*
Shrimp, *12*
Siamese (conjoined) twins, , 374, 743
siamois gene, 324–325
Signal transduction pathways, 154
 and cancer, 703
 evolutionary co-option, 768, *769*
 Hedgehog pathway, 161–163
 homologies, 439
 JAK-STAT pathway, 159–161
 juxtracrine signaling, 166–172
 RTK pathway, 154–159
 Smad pathway, 159
 Wnt pathway, 161
Signaling centers, mammalian axis formation, 375–376
Silencers, 120–121, *121*
Silent Spring (Rachel Carson), 740
Silkworm moth, 26, 727
Silver salamander, 203*n*
SIN1 gene, 661
Sinistral coiling, 240, *241*
Sinus venosus, 495
Sipunculida, *45*
Siren, 582
siRNA. *See* Small interfering RNA
Sisterless proteins, 562
situs inversus viscerum (iv) gene, 381–382
Six3 protein, 413
Skeletal precursors, limb, 525
Skeletogenic mesenchyme, 233–236
Skeletogenic micromeres, 229, *232*
Skeleton, osteogenesis, 474–477
Skin. *See* Epidermis
skn-1 gene, 255, 257
SKN-1 protein, 254–255
Skull, 438
Slimb protein, *163*
Slipper snail, 568–569
Slit protein, axonal guidance, *450*, **450**

sloppy-paired (slp) gene, 282, 284, 285
Slow block to polyspermy, 205–206
Slug protein, 428, 431
Smad family of transcription factors, 159
Small interfering RNA (siRNA), 704
Small intestine, 511
Small nuclear RNA (snRNA), 129, 692
Small regulatory RNAs (sRNA), 135
Smaug protein, 277
SMN (survival of motor neurons) protein, 692
Smoothened protein, 161–163
snail (cSnR) gene
   avian, 363, *364*
   Drosophila, 295, 296, 297
   mammalian, 370
   tunicate, 248
   zebrafish, *351*
Snail protein
   Drosophila, 296, 297
   mammalian asymmetry, 383
Snails
   cleavage, 224, 239–245
   coiling, 240, *241*
   fate map, *241,* 244–245
   gastrulation, 245, *246*
   pattern formation, *19*
   polar lobe, 242–245
   predator-induced polyphenism, 735
   sex determination, 568–569
   spiral growth, 16
snake (snk) gene, 291, 294
Snake limbs, Hox genes, 759, *760*
Snake protein, *292*
Snapdragon, 670, 675, 678
Snapping turtle, 567
Sodium ions
   egg activation, 208
   fast block to polyspermy, 199–200
Sodium pumps, 367
Soma, 411
Somatic cells, 26
   cell cycle, *222*
   totipotency, 86–87
   volvocaceans, 35
Somatic cell gene therapy, 707
Somatic mesoderm, 491
Somatic motor neurons, spinal cord, *405*
Somatic nuclear transfer, 711
somatic regulator A gene, 36
Somatopleure, 491, 517
Somatostatin gene, 116–118
Somatostatin-secreting delta cells, 515
Somite formation. *See* Somitogenesis
Somites, 465–473, *See also* Somitogenesis
   determination of cell fates, 471–473, *473*
   frogs, 27

Somitic dorsal mesoderm, 465
Somitogenesis, 467–473, *467–473*
   differentiation, 471–473
   epithelialization, 469–471
   periodicity, 467–468
   specification, 467, 471
Somitomeres, 467
sonic hedgehog (shh) gene, 151, *152,* 396, 689, 690, 695
   alcohol damage, 698
   amphibian metamorphosis, 577
   anterior-posterior limb polarity, 535–537, 759
   dorsal-ventral axis, 362
   fish, *352*
   eye development, 413–414
   neural crest cells, 396
   pancreas formation, 514
   right-left axis, 363, *364, 365*
Sonic hedgehog protein, 151–152, 162–163, *326*
   assymetric pathway, 382
   avian axis formation, 363, *364*
   cerebellar development, 404
   enamel knot, 441
   expression in chick embryo, *148*
   eye development, 413–414
   feather evolution, 770, *771*
   gut tissue specification, 512
   hair follicle development, 420
   homology, 764
   limb development and regeneration, 524, 535–538, 595
   motor neuron specification, 443
   neural tube ventral patterning, 401–402
   sclerotome formation, 472
   tooth evolution, 770–771
Sox proteins, 117, 145, 553–554
cSox2 gene, 512
SOX9/Sox9 gene, 475, *549,* **552**–554
Sp1 transcription factor, 116
Spadefoot toad, 732–734, *733*
Spallanzani, Lazzaro, 184
   regeneration experiments, 592
spalt gene, 587, 588, 730–731
Spalt protein, seasonal polyphenism, *729*
Spätzle protein, *292,* 294
Specification, 56, *See also* Autonomous specification; Axonal specificity; Conditional specification; Syncytial specification
   anterior-posterior axis, 334–336
   C. elegans, 251–257
   dorsal-ventral axis, 322–323
   endoderm, 333
   fish, 353
   floral meristems, 675–676
   gut tissue, 512
   homeotic selector genes, 286–289
   induction, 334

insect metamorphosis, 586
left-right axis, 337
mammalian gastrulation, 369–370
neural plate, 392
neuroblasts, 391–392
somitogenesis, *467,* 471
transcription factors, 118
tunicates, 247–250
Spemann, Hans, 83, *84,* 317–321, 338
Sperm, 26
   acrosome reaction, 191, *192,* 197
   asters, 209, *210*
   capacitation, 192–193
   chemotaxis, 189–191, 194
   discovery of, 183–184
   egg activation, 206–208
   egg fusion, 197–203, 210, *211*
   egg reorganization, 311–312
   genomic imprinting, 124
   hermaphrodites, 627–628
   hyperactivation, 194
   mammalian fertilization, 192–193
   maturation of, 184–186
   numbers of, 631
   oviduct translocation, 192–193
   polyspermy blocks, 199–202
   species-specific recognition, 194–197
   spermatogenesis, 628–631
   structure of, 184–186
   in vitro fertilization, 684–685
   zona pellucida binding, 195–197
Sperm-activating peptide, 190–191
Sperm cells, plants, 654, 657, 658
Sperm centriole, 209, 212
   axis determination, 311
   C. elegans, 253
   tunicates, 247
Sperm entry, mammalian axis formation, 381
Sperm nucleus, decondensation, 209, *210*
Sperm separation, 694
Spermateliosis, 630
Spermatids, 630
Spermatocytes, 629–630
Spermatogenesis, 628–**631,** *630,* 638
Spermatogonia, 628–630, *629*
Spermatozoa, 183, 630
Spermiogenesis, 630–631
Spicule formation, 234, *235,* 236
Spina bifida, 396, *397*
Spinal cord, 401
   commissural neurons, 447, *450*
   development, 403–404, *405*
   dorsal-ventral polarity, 402
   injuries, 708, 713
   intervertebral discs, 473
Spinal muscular atrophy, 692
Spinous layer, 416, *417*
Spiral cleavage, 224, *225*

adaptation in clams, 242
  holoblastic, 239–245
Spiral growth, 16, *17*
Spiral holoblastic cleavage, 239
Spiralia, *45*
SpKrl gene, sea urchin, *232*
Splanchnic mesoderm, 491, *494,* 513
  digestive tube interactions, 511–512, *512*
Splanchnopleure, 491, 517
Spliceosome-RNA complex, 132
Spliceosomes, 129
Splicing isoforms, 130
Splicing, of nRNA, 128–130
Spondylocostal dysplasia, 467–468
Sponges, 44–47, *45*
Sporangium, 651
Spore differentiation factors (SDF1/SDF2),
    44, 620*n*
Sporophyte, 651, 654, *665*
Sporophytic self-incompatibility, 656
Squid, 723
Squint protein, 353
SRK. *See* Serine-threonine kinase
SRY/Sry gene, 549, 551–554, *555,* 556
Stabilizing selection, 733
staggerer mouse mutant, 405
Stamens, 654
Staminate flowers, 655
Stapes bone, 14, 435
STAT pathway, 173
STAT protein, 159–161
Stat3 protein, fish, 353, *354*
Stat5 transcription factor, 164–165
Staufen protein, 272
Steel gene, 159, 690
stella gene, 618, *619*
Stem cells
  adult, 409–410, 709–711
  bioethics, 709
  C. elegans, 253, 256–257
  commitment, 68–69, 509
  embryonic, *102,* 374, 621
  gene therapy, 707
  hematopoietic, 480, 505–510
  melanocyte, 418
  mesenchymal, 710–711, *711*
  multipotent, 709–711
  neural, 409–410, 710, 713
  plants, 663
  pluripotent, 68, 418, 506, 708–711, *709*
  therapeutic cloning, 708–711
  transgenic, 711
Stem cell factor, 153, 159, 432, 620–621, 690
Stereoblastula, mollusc, **239**
Steroid hormones, and sex, 559–561
21-Steroid hydroxylase (*CYP21*), 692
Stevens, Nettie, 82–83
Stigma, 655
Stomach, 511

Stomodeum, 511
Storage protein mRNAs, plants, 664
Stratification, 664
Stratum corneum, 416
Stratum germinativum, 416
Striatum, neuronal survival, 452
Stripe mRNA, 132
Stripe protein, 132
Stroke, apoptosis, 165*n*
Stromal cells, hematopoiesis, 509
Stromelysin, 368
Strongylocentrotus, 194, *See also* Sea urchin
  cell fate map, *229*
  gastrulation, *235*
  RNA splicing, 128, *129*
  ultraviolet radiation, *54*
Strypsin, 368
Sturtevant, Alfred, 263
Styela partita, 246, 247
  fate map, 8, *9*
  yellow crescent, *248*
Style, 655
Stylopod, 523
Subgerminal cavity, avian, *355, 355,* 360
Substrates
  axonal guidance, 445
  neuronal survival, *453*
Sulcus limitans, 404, *405*
Sulston, Jonathan, 165
Sunscreens, 54
Superfamilies, 153*n*
Superficial cleavage, 224, *225,* 264
SUPERMAN (SUP) genes, 676
Surface ectoderm, 397
  major derivatives of, *392*
  neurulation, 396
Surface tension, 74
Surfactant, lungs, 515
SUS1 gene, 661
Suspensor, 659–661, *661, 662*
swallow (swa) gene, *272,* 274–276
Swallow protein, 272
Sxl protein, *Drosophila* sex determination,
    131, 562–563
Symbiosis, 722–725
Sympathetic neurons, trunk neural crest,
    434
Synapses (neural), 411, **451–453,** 456–457,
    739
Synapsis (chromosomal), 625
Synaptic cleft, 412
Synaptonemal complex, 625–626
Syncytial blastoderm, 264, *265*
Syncytial cables, *235*
Syncytial clones, *630*
Syncytial specification, 58, 69
Syncytiotrophoblast, 370–371
Syncytium, 69
Syndecan, 441

Syndromes, 14
Synergid cells, 655, 658
Syngamy (fusion), plants, 658
Synpolydactyly, human, 533
Syphilis, 702

T cells (T lymphocytes), *57,* 90, 173
Tadpole (Amphibian)
  metamorphosis, 30, 576–583
  parental care, 582–583
  pesticides, 743–745
  structure, 27–30, *29*
Tadpoles (Tunicates), 251
$TAF_{11}250$, 116
tailless (tll) gene, 277, *278, 279, 282*
  evolutionary conservation, 754
Tailless protein, 281
Talin, 169
Tamoxifen, 684
Tapetum, 654
Taq polymerase, 95
Target selection, axons, 443–444, 450–451,
    454–455
Taricha torosa, 146, 444
TATA-binding protein (TBP), 111, *112*
  Wnt4 gene, 556
TATA box, *111,* **111**
Taxonomy, 752
TBP-associated factors (TAFs), 112, 112, *116*
Tbx genes, 528, *529*
Tbx protein, retinal ganglion axons, 455
Tbx4 protein, hindlimb specification,
    526–529, *528*
Tbx4 transcription factor, 515
TBX5 gene, 527
Tbx5 protein, forelimb specification,
    526–529, *528*
tbx6 gene, 473, 370
TCF DNA-binding protein, 161
Tcf3, dorsal axis formation, **324**
teashirt gene, 290
Teeth. *See* Tooth
Telencephalon, 399
Telolecithal egg, 224, *225*
  avian, 354
  fish, 347
Telomeres, theory of aging, 603*n*
Telophase I, meiosis, 626
Telson, 277, *278*
Temperature-dependent sex determination,
    567–568
Tenascin, 431, 441
Teratocarcinoma, 621, *622,* **703**
Teratogenesis, 694–703, *744*
Teratogens, 15–16, 163*n,* **694, 739**
  alcohol, 695, 697–698
  caffeine, 695
  cancer therapy, 705
  diethylstilbesterol (DES), 559, 698–699,
    699, 740

embryonic sensitivity, *695*
experiments using, 379, *380*
heavy metals, 700
ionizing radiation, 702
listed, *696*
pathogens, 702
quinine, 695
retinoic acid, 696–697
rubella, 684
thalidomide, 15–16, 694–695, 704–705
valproic acid, 696
Teratology, 3, 14–16
Terminal boundary region, 278
torso signaling, *278*
Terminal cell, 659
terminal ear maize mutant, 668
TERMINAL FLOWER 1 (TFL1) gene, 675
Terminal gene group, 277–278
Terminal hair, 418
Tertiary induction, 727*n*
Testes
development, 550–551
primary sex determination, 548, 550–551
secondary sex determination, 556–558
Testis cords, 550
Testis-determining factors, 548, 551–554
Testosterone, 549, 551, 557–559
Tetraclita, 12
Tetrads, 625
Tetrodotoxin, 444–445, 456–457
TFIIA-TFIIH, 111–112
TGF-β protein. *See* Transforming growth
factor-β superfamily
Thais lamellosa, 735
Thalamus, 399, 407
α-Thalessemias, *693*
Thalidomide, 15–16, 694–705
Thamnophis, 760
Thanatophoric dysplasia, 160, *161,* 541, 691
The Adventure of Black Peter (Sherlock
Holmes), 67
The Broken Cord (Michael Dorris), 697*n*
Thecal cells, 551
Therapeutic cloning, 708–711
Therapsids, 760
Thermococcus littoralis, 93
Thermodynamic model, cell adhesion,
71–74
Thermus aquaticus, 93, *95*
Thompson, J. V., 12
Thorax, *Drosophila,* 268
Thr gene, sea urchin, *232*
Threshold model, metamorphosis, **579**
Thrombin, limb regeneration, 593, 593*n*
Thrombospondin, trunk neural crest, 431
Thryoid hormone, environmental disrup-
tors, 742
Thymidine, 9–10
Thymus gland, 435, 510–511

Thyroid gland, 435, 510–511
Thyroid hormone receptors (TRs), 579–580
Thyroid hormone receptor α (TRα),
579–580
Thyroid hormone receptor β (TRβ),
579–580
Thyroid hormones
epiphysial growth plate, 542
metamorphosis, 30, **578**–580, *579*
Thyroxine (T4), 145
Thyroxine (T$_4$), 579
Ti1 protein, 446
Tie2 receptor, 503
tinman gene, 753–754
Tinman transcription factor, 494*n*
Tissue environment, and cancer, 703
Tissue systems, plants, 663
Tissues
formation in mammals, 369–370, *369*
histotypic aggregation, 71, *72, 73*
Tithonios, 605
Toads
accelerated metamorphosis, 732–734, *733*
ultraviolet radiation, 54–56
Tobacco hornworm moth, 133
Toivonen, Sulo, 334
toll gene, 294
Toll protein, 292, 294
tolloid gene, 296
Tolloid protein, homologues, 763
Tonsils, origin, 510–511
Tooth development, 434, 439–441, *440*
apoptosis, 164
cranial neural crest, 434, 439–441, *440*
Sonic hedgehog protein, 152
Tooth evolution, mammals, 770–771, *772*
TOPLESS gene, 663
Torenia fournieri, 658
torpedo gene, 293–294
Torpedo protein, *292*
torso (tor) gene, *272, 277,* 278, *278*
torso-like gene, 277
torso mRNA, 277
Torso protein, 277, 278
Torsolike protein, 277, *278*
Totipotency, 85–87
blastomeres, 374
Toxoplasma gondii, 702
Toxopneustes lividus, 184
Tra protein, 567
Trachea, 515
Tracheal-esophageal fistula, 515
Trachemys, 568
Trans-activating domain, 114, 116
Trans-regulatory elements, 112*n*
Transcription
DNA methylation, 121–122, *121–123*
dosage compensation, 124–126
Drosophila genes, *283, 284*

enhancers, 112, 114–117
genetic errors, 691–692, *692*
inhibition in germ cells, 617
initiation in blastula, 311
introns and exons, 108–110, *109, 110*
oocytes, 634–636
promoters, 111–112
signal pathways cross-talk, 173
silencers, 120, *121*
stabilization, 281, 282, 288
Transcription activity, human brain development, 408
Transcription factors, *111,* **111–112,**
114–117
basal, 111–112
cascades of, 117, 120
cell-specific, 112
classification, 154
combinatorial factors, 116–117, 120
enhancers, 112, 114–117
families, *114*
heart formation, 495–496, *496*
human, *692*
maintenance of differentiation, 173
organizer-specific, *326, 327*–330
techniques for studying, 118–120
Transcription initiation site, 108, *109, 110,*
111
Transcriptional regulation, 26*n,* 32
Transfection, 99
Transfer RNA (tRNA)
insect oocyte, 638
oocyte, 187, 636
transformer (tra) genes, 131, 562–563, *565*
transformer mRNa, 131
Transformer protein, 131
Transforming growth factor-α (TGF-α),
416–418
Transforming growth factor-β (TGF-β) superfamily, 154, 558, 639
avian axis formation, 361
capillary networks, 504
evolutionary duplication, 768
kidney development, 483
neural tube dorsal patterning, *401,* 402
the organizer, 325, *326*
Smad signal transduction pathway, 159
Transgenes, 89
Transgenic animals, 89, *90,* 708
chimeric mice, 100–101, *102, 103*
gene targeting experiments, 100–101, *102,*
103
gene therapy, 707–708, *707*
sheep, 89, *90*
TGF-α overexpression, 417
Transgenic cells, DNA insertion techniques,
99–100
Transgenic stem cells, 711
Translation
cytoplasmic localization, 135–136

differential mRNA longevity, 132
hunchback mRNA, *277*
inborn errors, 692–693
selective inhibition of, *134*
Translation initiation factors (eIF4E/eIF4G), 133, 134
Translation initiation site, 108
Translation termination codon, *109,* **110**
Translational regulation, 26*n*, 32
Translocation, sperm, 192–193, *194*
Transplantation experiments, 61
Transposable elements, 99
Tree frogs, 54, *56, 525,* 735
Trefoil-stage embryo, 244
Trematodes, deformed frogs, 744
Tremblay, regeneration experiments, 592
trembler mouse mutant, 412
Treponema pallidum (Bacterium), 702
Tri-iodothyronine (T$_3$), 579
Trimma okinawae, 732
Triploblastic organisms, 6
Trisomy 21, 687
Trithorax protein, 173, 288
Triturus, 55, 308, 320, 320*n*, 334*n*, 635, 636, *See also* Amphibians; Salamanders
tRNA. *See* Transfer RNA
Trochlear nerve, 449
Trochophore larva, 244, **244***n*
Trochus spiral cleavage, *239*
Trophectoderm, mammals, 366
Trophoblast cells, 368–369, 370, *370*
dorsal-ventral axis, 380–381
mammals, 76, **366**
node formation, 375
α-Tropomyosin proteins, *130*
TRs. *See* Thyroid hormone receptors
Truncoconal septa, *441*
Truncus arteriosus, 441, **495,** 500
Trunk, avian, 357
trunk (trk) gene, *272*
Trunk mesoderm, regions, **465,** *466*
Trunk neural crest, 428–429
bone, 434
cell migration, 429–432
differentiation, 432–434
Tsix gene, 127
Tubal embryo transfer, 685
Tubal pregnancy, 368
Tube cell, 654
Tube nucleus, 657
Tube protein, *292,* 294
Tubule cells, *57,* 480
Tubulin, 185, 223–224
tudor (tud) gene, *272*
Tumor angiogenesis factor, 704–705
Tumors
and angiogenesis, 504, **704–705**
embryonic stem cells, 621, *622*
Patched protein, 162

RAS gene mutations, 154
suppression, 161
Tunica albuginea, 551
Tunicates
autonomous specification, 57–58, *59*
axis formation, 248–250
bilateral holoblastic cleavage, 246
cell determination, 246–249
egg cytoplasm reorganization, 246–247
evolutionary classification, 12
fate map, 8, *9,* 246–247
gastrulation, 250–251
morphogenetic determinants, 248–249
Turing, Alan, 17–19
Turritopsis, 30
Turtles
life expectancy, 602*n*
temperature-dependent sex determination, 567–568
Twins, 373–374, 743, *See also* Multiple births
twist gene, 295, 297
Twist protein, *295,* 296
Type A/Type B spermatogenesis, 629–630
Type I-XVIII collagens. *See under* Collagen
Tyrosinase family proteins, 115
Tyrosine kinase, 150, *151,* 432, 600, *See also* Receptor tyrosine kinase pathway (RTK)

U1 sRNA, 129
U2AF protein, 129, 562
Uca pugnax, 17, *18*
Ultrabithorax (Ubx) gene, 286, *287,* 288, 756–757, *757–759*
Ultrabithorax (Ubx) protein, 288–289, 290
Hox gene regulation, 755–757, *757*
Ultraspiracle (Usp) protein, 590, 590*n*
Ultraviolet radiation, 54–55, 744
Umbilical artery, 498
Umbilical cord, 371
Umbilical vein, 498, **501**
UNC-6 protein, 448–449, *449*
unc genes, 448–449, *449*
Unicellular protists, 31–33
UNIFOLIATA (UNI) gene, 670
Unio, 242
Unionids, 242
"Unity of Type," 751–752
Unsegmented mesoderm, 466, 468–469, *471*
Untranslated regions (UTRs), *109,* **110,** 132, 133, 636
Upstream promoter elements, 112
Urbilaterian ancestor (PDA), 331, **753–754**
Urea, amphibian metamorphosis, 577, *578*
Ureteric buds, 480–484
branching, 483–484
GDNF induction, 481–482, *482*
mesenchyme apoptosis, 482
Ureters, 481
Urochordata, *45*

Urodeles, 576–577, *See also* Amphibians; Salamanders
Urogenital organs, Wnt4 proteins, *152*
Urogenital system, intermediate mesoderm, 477–485
Urotelic excretion, 577
Uterine precursor cells, 167
Uterus
embryo implantation, 76, *366,* 368
extraembryonic membranes, 371–372
ovulation, 641
UTR. *See* Untranslated regions

V segment, immunoglobulin genes, 90–91
Vagal nerve, placode origin, 439
Vagal neural crest, 429
valois (val) gene, *272*
Valproic acid, 696
Variable regions, 90–91
Vas deferens, 480, **551**
vasa (vas) gene, *272*
Vasa protein, 616–617
Vascular endothelial growth factor (VEGF), 639
blood vessel specification, 505
bone regeneration, 712
endochondral ossification, 476–477
nephron formation, 480*n*
vasculogenesis, **503**
Vascular tissue, plants, **663,** 667
Vascularization, tumors, 704
Vasculogenesis, 501–503, *502, See also* Angiogenesis
VEG1 gene, 675
vega 1 gene, 353
Vegetal cells
amphibian axis, *321, 322*
amphibian gastrulation, 311–312, 315
the organizer, 322
sea urchin, 230, *231*
tunicates, 250
Vegetal hemisphere, 26
amphibian egg, 305–306
cell fates, 308
Vegetal plate
invagination, 236, *237*
primary mesenchyme, 233
sea urchin, **228**
Vegetal pole, 224
Vegetal posterior cytoplasm, 247, 250
Vegetative growth, plants, 665–671
VEGF. *See* Vascular endothelial growth factor
VegT protein, 308, 311, 322, *326*
Veliger larva, *245*
Vellus, 418
Velum, snail, 241
Venom, 130*n*
Ventral furrow, 266, *267*
Ventral lip, frog blastopore, 310
Ventral neurons, 402

Ventral pathway, 429–431
Ventricles
brain development, 400
heart formation, 496, *497*
Ventricular zone, 403, *405, 407*
Vera protein, 633
Veratrum californicum, 163, 695
Vernix caseosa, 418
Vertebrae
Hox genes, 378, *379*, 380, *381*
intervertebral discs, 473
somite origins, 467, 471–472
Vertebrates, *45*
allometry, 766–768, *767, 768*
cadherins, 74–76
differences between embryos, *8*
DNA methylation, 121–123
environmental sex determination, 52, 567–568
eye development, 413–416
fate maps, *8*
gamete activation, 207–208
head, 434–436, 761–762
Hedgehog family proteins, 151
Hox gene complex, 759, 760–761, 775
jaw evolution, 771–773, *773*
jaw homology, 13–14
limb structure, 523–524
neural growth cone guidance, 450
specification of motor neurons, 442–443
von Baer's principles, 7
Vesalius, Andreas, 547
Vesicle exocytosis, 191, *192n*
Vesicles, brain development, 398, *399*
vestigial gene, 587, *588*
Vestigial protein, 588
Vg1 gene, avian, 361
Vg1 mRNA, *135*, 633
Vg1 protein, 149*n*, 153, **308**, 325, *326*, 337, 361
Vibrio fischeri (Bacterium), 723
Villi, chorionic, *370, 371*, 372
vis essentialis, 4
Visceral mesoderm, 491
Viscerocranium, 438
Vision
chemoaffinity hypothesis, 455
retinotectal projections, 454–455
Visual cortex, 405–407
mammalian neuronal patterning, 737–738
Vital dyes, 9, *10*
Vitalism, 62*n*
Vitamin A, 696*n*
Vitamin B12, 397
Vitelline envelope, 187, *188*, 201, *203*
avian gastrulation, 359
Vitelline veins, 495, **501**, 503
Vitellogenesis, 632–633
Vitellogenin, 26, **633**

Vitronectin, 169
Viviparity, 3
viviparous maize mutation, 664
Volvocaceans, 34–39
Volvox
asexual/sexual reproduction, 36–37
death, 36–37
differentiation, 34–37
life cycles, 36–38
von Baer, Karl Ernst, 6, 7, *7*
VPCs. *See* Vulval precursor cells
Vulva, *C. elegans,* 157–158, 167
Vulval precursor cells (VPCs), 157–158

Waardenburg syndrome type 2, 688–689, *689*, 691
Waddington, Conrad Hal, 84, *85*
waltzer mouse mutant, 405
Warfarin, 694*n*
Water flea, 735
Watson, James D., 708
Wear-and-tear theories of aging, 602
weaver mouse mutant, 405
Webbed feet, *540*, 769, *770*
Weismann, August, 59–62
Western toad, 54–55
Whale, allometric evolution, 767, *768*
Whey acidic protein, 170
White-crowned sparrows, 737
White gene, 159, 690
White matter, 403–404
Whittaker, J. R., 58
Wholist organicism, 62*n*
Wieschaus, Eric, 271
Wilmut, Ian, 87–89
Wilson, E. B., 81–82, 224, 752
Wing formation, *Drosophila*, 587–588, *588*
wingless (wg) gene, 152, *282, 284*, 285, 288, 566, 588, 730–731
Wingless protein (Wg), 283, *284*, 285, *286*, 586
Wings
analogous and homologous, 13
Drosophila, 286*n*, *287, 288*
evolutionary co-option, 768
wnt gene, 353
Wnt7a gene
DES, 699
limb development, 537–538
***Wnt4* gene,** *152*, 549, **556**
Wnt proteins, 152, 473
cardiac specification, *493*
effects of, 161
fish, 353–354
hydra regeneration, 599, *600*
kidney development, 481–483
limb development, 525–526, 531, 537, 540
trunk neural crest, 431, 433–434

Wnt1 protein
dermatome differentiation, 472, *473*
myotome determination, 473
Wnt2b protein, Chick forelimb development, 525
Wnt3a protein, 473
limb development, *527*
Wnt6 protein
kidney development, 483
neural crest, 427
Wnt7a protein, limb specification, 524
Wnt8 protein, 230, *232*, **352**, 353
Wnt8c protein
chick hindlimb development, 526
placode induction, 438, *439*
Wnt11 expression, kidney induction, 481–482
Wnt signal transduction pathway, 161
amphibian axis formation, *323*
anterior-posterior axis, 334–336, *335*
C. elegans, 256
dorsal-ventral axis, *323*
hair follicle development, 419–420
homologies, 439
neurulation, 392
organizer diffusible inhibitors, 330–332
Wolff, Kaspar Friedrich, 4
Wolffian duct, 479, **551**, 557–559
Wound epidermis, 593
WT1 gene, *549*
WT1 protein, 130, **481**
wunen gene, 624
WUSCHEL(WUS) gene, 667

X chromosome
Drosophila, 124, 561–562
mammalian sex determination, 548–549
X chromosome inactivation, 124–127, *125, 127*, 691
X-linked anhidrotic ectodermal dysplasia, 419
X-to-autosome ratio, 561–562, *562*
XANF1 protein, *326*
Xenopus. See also Amphibians; Frogs
activin gradients, 65–66
angioblast formation, 503
atrazine effects, *569*
cadherins, 75, *76*
cleavage, 221, 306–307
development, *28–29*, 305*n*
dorsal-ventral axis, 312
epidermal inducer, 329
eye migration, 577
fate map, 307–309
gap junctions, 171, *172*
germ cell migration, 617, *618*
head metamorphosis, *578*
lens induction, 143–144, *146*
limb regeneration, 595, *595*

mesodermal migration, 315, 316
methoxychlor, 740
Nieuwkoop center, *321*, 322
oocyte maturation, *634*
oocyte RNA, *134*, *135*
retinal ganglion axons, 455
right-left axis, 337
somatic cell totipotency, 86–87
tail regression, *580*
ultraviolet radiation, 54–56
Xenopus Brachyury (Xbra) gene, 322
Xenopus brain factor-2 (Xbf2), 332
Xenopus calponin gene, 315
Xenopus Nodal-related (*Xnr*) genes, *326*, 337
Xenopus Nodal-related (Xnr) proteins, 96, *96*, *326*
XFD1 protein, *326*
Xfrzb1 gene, *335*
Xin protein, heart, 496, *497*
Xist gene, 126–127
   X chromosome inactivation, 126–127
Xist RNA, X chromosome inactivation, 126–127
Xlim1 protein, 324, *326*
Xnot protein, *326*
Xnr proteins. See *Xenopus* Nodal-related proteins
Xvex genes, 333
Xwnt8 protein, 331, 334
Xylem, 663

Y chromosome
   Drosophila, 561
   mammalian sex determination, 548–549
   SRY gene, 551–552
YABBY gene family, 669–670
Yellow crescent, tunicates, **246**–249
Yolk, 26, *28*, 187
   amniote egg, 46
   amphibian egg, 305–306, 633
   avian egg, 354, *355*
   distribution, 224–225
   enclosure by ectoderm, 358–359
   fish egg, 348, *349*
Yolk cell, fish, **348**
Yolk duct, 517
Yolk platelets, 633
Yolk plug, 310
Yolk sac, 46, 369, **517**
   avian, 356
   Hnf4α transcription factor, 118
   mammalian, 367
Yolk syncytial layer (YSL), 348, 349
YSL. *See* Yolk syncytial layer

Z1/Z2/Z3 isoforms (BR-C protein), 591
Zebra finches, neuronal patterning, 737
Zebrafish. *See also* Fish
   axis formation, 351–354, *352*
   axis patterning compared to *Xenopus*, 353
   blood vessel differentiation, 504–505
   chordino mutant, 329
   cleavage, 347–349
   developmental studies, 345–347
   dorsal-ventral neural tube patterning, 402
   fate mapping, *10*
   gastrulation, 349–351, *351*
   miles apart mutation, 494
   pattern formation, 19, *20*
   retinal ganglion axon specificity, 455n
   SDF-1 in germ cell migration, 620n
Zebras, 19
zen gene, 297
zerknüllt gene, 296, 297
Zeugopod, 523
ZIFT. *See* Zygote intrafallopian transfer
Zinc finger transcription factor, 120
Zona pellucida, 187, *189*, 195–197, 201–202, **368,** 640
Zone of polarizing activity (ZPA), *534–536*, **534–539,** *538*
ZP1/ZP2/ZP3 proteins, 195–197, 201, 202, 635
ZP3 gene, *635*
Zygote, 2
   brown algae, 659
   mammalian cleavage, 366
   nucleus, 209
   plant, 658
   sea urchin fate map, *233*
   volvocaceans, 37–39
Zygote intrafallopian transfer (ZIFT), 685
Zygotene stage, meiosis, **625**

About the Book

Editor: Andrew D. Sinauer

Project Editor: Carol Wigg

Production Manager: Christopher Small

Multimedia and Ancillaries: Jason Dirks, Gayle Sullivan, Mara Silver

Electronic Bookbuilders: Jefferson Johnson, Janice Holabird

Illustration Program: J/B Woolsey Associates, Dragonfly Media Group

Book Design: Susan Brown Schmidler, Jefferson Johnson

Cover Design: Jefferson Johnson

Copy Editor: Norma Roche

Author Index: Chelsea D. Holabird

Subject Index: Acorn Indexing

Book and Cover Manufacture: Courier Companies, Inc.

5.11 The medical importance of X chromosome inactivation
5.12 Differential nRNA censoring
5.13 An inside-out gene
5.14 The mechanism of differential nRNA splicing
5.15 Mechanisms of mRNA translation and degradation
5.16 The discovery of stored mRNAs
5.17 Other examples of translational regulation of gene expression

6.1 Hen's teeth
6.2 FGF binding
6.3 Functions of the Hedgehog family
6.4 Wnts: An ancient family
6.5 Eye formation: A conserved pathway
6.6 FGFR mutations
6.7 The uses of apoptosis
6.8 Notch mutations
6.9 Connexin mutations
VM^2 *Drosophila* imaginal discs
VM^2 Cyclopia induced in zebrafish

7.1 Leeuwenhoek and images of homunculi
7.2 The origins of fertilization research
7.3 The egg and its environment
7.4 The Lillie-Loeb dispute over sperm-egg binding
7.5 Building the egg's extracellular matrix
7.6 Blocks to polyspermy
7.7 Sperm decondensation
VM^2 Gametogenesis
VM^2 Sea urchin fertilization
VM^2 E. E. Just
VM^2 Amphibian fertilization

8.1 Regulating the cell cycle
8.2 Sea urchin cell specification
8.3 Alfred Sturtevant and the genetics of snail coiling
8.4 Modifications of cell fate in spiralian eggs
8.5 The experimental analysis of tunicate cell specification
8.6 P-granule migration
8.7 The PAR proteins
8.8 Mechanisms of cytoplasmic localization in *C. elegans*
VM^2 Cleavage patterns
VM^2 Gastrulation movements
VM^2 Sea urchin development

9.1 *Drosophila* fertilization
9.2 The regulation of *Drosophila* cleavage
9.3 The early development of other insects
9.4 Evidence for gradients in insect development
9.5 Christiane Nüsslein-Volhard and the molecular approach to development
9.6 Mechanism of *bicoid* mRNA localization
9.7 Asymmetrical spread of morphogens
9.8 Getting a head in the fly
9.9 Homeotic genes and their protein products
VM^2 *Drosophila* development

10.1 Demonstrating tissue affinities
10.2 Migration of the mesodermal mantle
10.3 Embryology and individuality
10.4 Spemann, Mangold, and the organizer
10.5 Mesoderm induction
10.6 GBP
10.7 Early attempts to locate the organizer molecules
10.8 The specification of the endoderm
10.9 Planar induction
10.10 Regional specification
VM^2 Amphibian development

11.1 GFP zebrafish movies and photographs
11.2 Epiblast cell heterogeneity
11.3 Mechanisms of compaction and the formation of the inner cell mass
11.4 Human cleavage and compaction
11.5 The mechanisms of implantation
11.6 Placental functions
11.7 Nonidentical monozygotic twins
11.8 Conjoined twins
11.9 Why do mammals have only seven cervical vertebrae?
VM^2 Zebrafish development
VM^2 Chick development
VM^2 Retinoic acid as a teratogen

12.1 Formation of the floor plate cells
12.2 Neural tube closure
12.3 Homologous specification of the neural tissue
12.4 Mapping the mesencephalon
12.5 Specifying the brain boundaries
12.6 Constructing the pituitary gland
12.7 Cerebellar mutations of the mouse
12.8 Constructing the cerebral cortex
12.9 Neuronal growth and the invention of childhood
12.10 Parkinson disease
12.11 Why babies don't see well
12.12 Normal variation in human hair production
12.13 Mutations of human hair production
VM^2 Chick neurulation

13.1 Avian neurulation
13.2 The specificity of the extracellular matrix
13.3 Mouse neural crest cell mutants
13.4 Committed neural crest cells
13.5 Human facial development syndromes
13.6 Kallmann syndrome
13.7 Communication between migrating neural crest cells
13.8 The evolution of developmental neurobiology
13.9 Horseradish peroxidase staining
13.10 The pathways of motor neurons
13.11 Genetic control of neuroblast migration in *C. elegans*
13.12 The early evidence for chemotaxis
13.13 The neurotrophin receptors

14.1 Calling the competence of the somite into question
14.2 Cranial paraxial mesoderm